AF616325

High-Temperature Ordered Intermetallic Alloys V

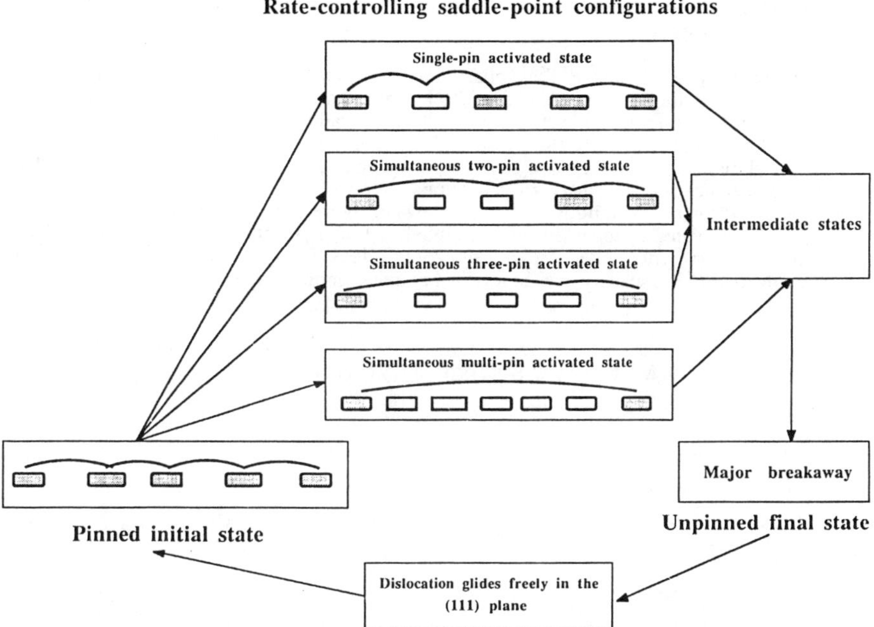

The viewgraph represents a flow chart describing the main features of a thermally activated "unpinning" mechanism for $[\bar{1}01]$ screw dislocations in the (111) plane of a Ni_3Al-type $L1_2$ compound via distinct alternate saddle-point configurations. The anomalous increase in the yield stress with increase in temperature observed in these compounds is caused largely due to a thermally activated "unpinning" mechanism that impedes the movement of screw dislocations by enabling short segments to cross-slip to the (010) plane at random positions. The figure also represents how the two thermally activated processes are coupled in steady state by an intermediate configuration in which the screw dislocation glides freely in the (111) plane after achieving major breakaway and before it becomes pinned again.

Figure courtesy of M. Khantha, J. Cserti and V. Vitek.

MATERIALS RESEARCH SOCIETY SYMPOSIUM PROCEEDINGS VOLUME 288

High-Temperature Ordered Intermetallic Alloys V

Symposium held November 30-December 3, 1992, Boston, Massachusetts, U.S.A.

EDITORS:

Ian Baker

Thayer School of Engineering
Dartmouth College
Hanover, New Hampshire, U.S.A.

Ram Darolia

GE Aircraft Engines
Cincinnati, Ohio, U.S.A.

J. Daniel Whittenberger

NASA-Lewis Research Center
Cleveland, Ohio, U.S.A.

Man H. Yoo

Oak Ridge National Laboratory
Oak Ridge, Tennessee, U.S.A.

MATERIALS RESEARCH SOCIETY
Pittsburgh, Pennsylvania

The figures on the title page and section dividers are representative of the sides and viewgraphs utilized during the presentations at the Fall 1992 symposium on high temperature ordered intermetallics.

Single article reprints from this publication are available through University Microfilms Inc., 300 North Zeeb Road, Ann Arbor, Michigan 48106

Materials Research Society Symposium Proceedings : ISSN: 0272-9172

CODEN: MRSPDH

Published by:

Materials Research Society
9800 McKnight Road
Pittsburgh, Pennsylvania 15237
Telephone (412) 367-3003
Fax (412) 367-4373

High-Temperature Ordered Intermetallic Alloys V

ISSN: 1067-9995

Manufactured in the United States of America

Contents

*Invited Paper

PART III: DEFECTS AND MICROSTRUCTURE

PART IV: DEFORMATION AND FRACTURE

PART V: MECHANICAL PROPERTIES

PART VI: PROCESSING

PART VIII: MULTIPHASE MATERIALS, COMPOSITES AND JOINING

Preface

These proceedings represent the written record of the High-Temperature Ordered Intermetallic Alloys V Symposium which was held in conjunction with the 1992 Fall Materials Research Society meeting in Boston, Massachusetts. This symposium, which was the fifth in the series originated by C.C. Koch, C.T. Liu and N.S. Stoloff in 1984, was very successful with 86 oral presentations over four days, and approximately 140 posters given during two lively evening sessions. Such a response, in view of the increasing number of conferences being held on intermetallics each year, reveals the continued high regard for this series of symposia. Of the 181 papers in this volume, approximately 31% are contributions from outside the U.S.A.; this compares with approximately 28% non-U.S.A. papers in the 1991 (Vol. 213) proceedings, 22% in 1989 (Vol. 133), 15% in 1987 (Vol. 81) and only 8% from the first symposium (Vol. 39).

Examination of the past four High-Temperature Ordered Intermetallic Alloys proceedings in combination with the present volume indicates an active and still growing interest in this field, as can be seen in the following figure.

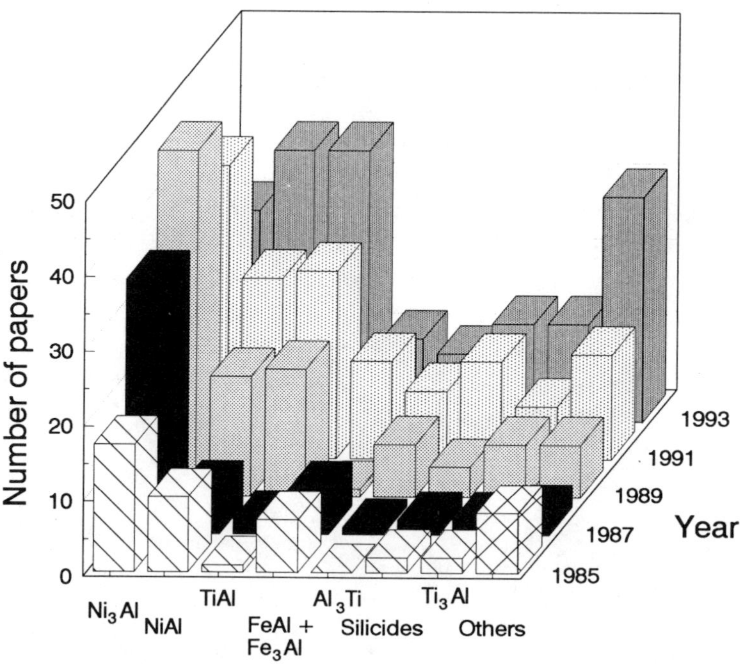

Major areas of work on intermetallics as indicated by number of papers published in the five MRS High-Temperature Ordered Intermetallic Alloys proceedings. (Courtesy of Randy Bowman and Michael Nathal of the NASA Lewis Research Center, Cleveland, Ohio.)

The figure indicates that the vast majority of work in the past ten years on intermetallics has involved nickel and titanium aluminides. Of these systems only the study of Ni_3Al appears to be declining; whereas effort on NiAl, TiAl, and to some extent Ti_3Al is expanding. While work is continuing on the iron aluminides, titanium trialuminides and silicides, the above figure does not indicate major thrusts or growth in these systems. Lastly, the factor of three increase in the number of papers on "other" intermetallics in this proceedings over the 1991 edition indicates a renewed effort to identify systems with higher melting points, higher strengths, lower densities, etc. than are offered by the current nickel/titanium aluminides. Clearly, intermetallic alloys represent a vibrant field of research which should continue into the future.

The following record of the 1992 symposium includes papers on the oxidation resistance of intermetallics as well as preliminary attempts to join these materials. Both of these areas were emphasized in the Call for Papers and represent our attempt to help promote engineering use of intermetallics. The traditional areas, including atomic structure and defects, alloy development, phase stability and mechanical properties, are well represented, and they reveal sophistication in these areas through refined modeling and/or significant numbers of experimental measurements.

Ian Baker
Ram Darolia
Dan Whittenberger
Man Yoo

January 1993

Acknowledgments

All papers appearing in this proceedings were reviewed on-site following MRS procedures. This process was facilitated by the following session chairs who acted as key readers for the manuscripts.

J.B. Darby, Jr.
E.M. Schulson
M.A. Crimp
D.M. Dimiduk
S.H. Whang
J.A. Horton, Jr.
Y. Mishima
D.G. Morris
Y. Umakoshi
A.H. Rosenstein
M.V. Nathal
J.J. Lewandowski
K. Sadananda
R.W. Cahn
R. Gibala
D.B. Miracle
C.G. McKamey
D.F. Lahrman
T.L. Lin
R.D. Field
K.S. Kumar
M.A. Morris
S.V. Raj
R.D. Noebe
H.L. Fraser
V.K. Vasudevan
J.S. Livingston
C.M. Kennifick

We gratefully acknowledge the financial support for the High-Temperature Ordered Intermetallic Alloys V symposium by the Oak Ridge National Laboratory, the U.S. Department of Energy, the Office of Naval Research, the General Electric Company and the NASA Lewis Research Center.

MATERIALS RESEARCH SOCIETY SYMPOSIUM PROCEEDINGS

Volume 258—Amorphous Silicon Technology—1992, M.J. Thompson, Y. Hamakawa, P.G. LeComber, A. Madan, E. Schiff, 1992, ISBN: 1-55899-153-0

Volume 259—Chemical Surface Preparation, Passivation and Cleaning for Semiconductor Growth and Processing, R.J. Nemanich, C.R. Helms, M. Hirose, G.W. Rubloff, 1992, ISBN: 1-55899-154-9

Volume 260—Advanced Metallization and Processing for Semiconductor Devices and Circuits II, A. Katz, Y.I. Nissim, S.P. Murarka, J.M.E. Harper, 1992, ISBN: 1-55899-155-7

Volume 261—Photo-Induced Space Charge Effects in Semiconductors: Electro-optics, Photoconductivity, and the Photorefractive Effect, D.D. Nolte, N.M. Haegel, K.W. Goossen, 1992, ISBN: 1-55899-156-5

Volume 262—Defect Engineering in Semiconductor Growth, Processing and Device Technology, S. Ashok, J. Chevallier, K. Sumino, E. Weber, 1992, ISBN: 1-55899-157-3

Volume 263—Mechanisms of Heteroepitaxial Growth, M.F. Chisholm, B.J. Garrison, R. Hull, L.J. Schowalter, 1992, ISBN: 1-55899-158-1

Volume 264—Electronic Packaging Materials Science VI, P.S. Ho, K.A. Jackson, C-Y. Li, G.F. Lipscomb, 1992, ISBN: 1-55899-159-X

Volume 265—Materials Reliability in Microelectronics II, C.V. Thompson, J.R. Lloyd, 1992, ISBN: 1-55899-160-3

Volume 266—Materials Interactions Relevant to Recycling of Wood-Based Materials, R.M. Rowell, T.L. Laufenberg, J.K. Rowell, 1992, ISBN: 1-55899-161-1

Volume 267—Materials Issues in Art and Archaeology III, J.R. Druzik, P.B. Vandiver, G.S. Wheeler, I. Freestone, 1992, ISBN: 1-55899-162-X

Volume 268—Materials Modification by Energetic Atoms and Ions, K.S. Grabowski, S.A. Barnett, S.M. Rossnagel, K. Wasa, 1992, ISBN: 1-55899-163-8

Volume 269—Microwave Processing of Materials III, R.L. Beatty, W.H. Sutton, M.F. Iskander, 1992, ISBN: 1-55899-164-6

Volume 270—Novel Forms of Carbon, C.L. Renschler, J. Pouch, D. Cox, 1992, ISBN: 1-55899-165-4

Volume 271—Better Ceramics Through Chemistry V, M.J. Hampden-Smith, W.G. Klemperer, C.J. Brinker, 1992, ISBN: 1-55899-166-2

Volume 272—Chemical Processes in Inorganic Materials: Metal and Semiconductor Clusters and Colloids, P.D. Persans, J.S. Bradley, R.R. Chianelli, G. Schmid, 1992, ISBN: 1-55899-167-0

Volume 273—Intermetallic Matrix Composites II, D. Miracle, J. Graves, D. Anton, 1992, ISBN: 1-55899-168-9

Volume 274—Submicron Multiphase Materials, R. Baney, L. Gilliom, S.-I. Hirano, H. Schmidt, 1992, ISBN: 1-55899-169-7

Volume 275—Layered Superconductors: Fabrication, Properties and Applications, D.T. Shaw, C.C. Tsuei, T.R. Schneider, Y. Shiohara, 1992, ISBN: 1-55899-170-0

Volume 276—Materials for Smart Devices and Micro-Electro-Mechanical Systems, A.P. Jardine, G.C. Johnson, A. Crowson, M. Allen, 1992, ISBN: 1-55899-171-9

Volume 277—Macromolecular Host-Guest Complexes: Optical, Optoelectronic, and Photorefractive Properties and Applications, S.A. Jenekhe, 1992, ISBN: 1-55899-172-7

Volume 278—Computational Methods in Materials Science, J.E. Mark, M.E. Glicksman, S.P. Marsh, 1992, ISBN: 1-55899-173-5

MATERIALS RESEARCH SOCIETY SYMPOSIUM PROCEEDINGS

Volume 279—Beam-Solid Interactions—Fundamentals and Applications, M.A. Nastasi, N. Herbots, L.R. Harriott, R.S. Averback, 1993, ISBN: 1-55899-174-3

Volume 280—Evolution of Surface and Thin Film Microstructure, H.A. Atwater, E. Chason, M. Grabow, M. Lagally, 1993, ISBN: 1-55899-175-1

Volume 281—Semiconductor Heterostructures for Photonic and Electronic Applications, D.C. Houghton, C.W. Tu, R.T. Tung, 1993, ISBN: 1-55899-176-X

Volume 282—Chemical Perspectives of Microelectronic Materials III, C.R. Abernathy, C.W. Bates, D.A. Bohling, W.S. Hobson, 1993, ISBN: 1-55899-177-8

Volume 283—Microcrystalline Semiconductors—Materials Science & Devices, Y. Aoyagi, L.T. Canham, P.M. Fauchet, I. Shimizu, C.C. Tsai, 1993, ISBN: 1-55899-178-6

Volume 284—Amorphous Insulating Thin Films, J. Kanicki, R.A.B. Devine, W.L. Warren, M. Matsumura, 1993, ISBN: 1-55899-179-4

Volume 285—Laser Ablation in Materials Processing—Fundamentals and Applications, B. Braren, J. Dubowski, D. Norton, 1993, ISBN: 1-55899-180-8

Volume 286—Nanophase and Nanocomposite Materials, S. Komarneni, J.C. Parker, G.J. Thomas, 1993, ISBN: 1-55899-181-6

Volume 287—Silicon Nitride Ceramics—Scientific and Technological Advances, I-W. Chen, P.F. Becher, M. Mitomo, G. Petzow, T-S. Yen, 1993, ISBN: 1-55899-182-4

Volume 288—High-Temperature Ordered Intermetallic Alloys V, I. Baker, J.D. Whittenberger, R. Darolia, M.H. Yoo, 1993, ISBN: 1-55899-183-2

Volume 289—Flow and Microstructure of Dense Suspensions, L.J. Struble, C.F. Zukoski, G. Maitland, 1993, ISBN: 1-55899-184-0

Volume 290—Dynamics in Small Confining Systems, J.M. Drake, D.D. Awschalom, J. Klafter, R. Kopelman, 1993, ISBN: 1-55899-185-9

Volume 291—Materials Theory and Modelling, P.D. Bristowe, J. Broughton, J.M. Newsam, 1993, ISBN: 1-55899-186-7

Volume 292—Biomolecular Materials, S.T. Case, J.H. Waite, C. Viney, 1993, ISBN: 1-55899-187-5

Volume 293—Solid State Ionics III, G-A. Nazri, J-M. Tarascon, M. Armand, 1993, ISBN: 1-55899-188-3

Volume 294—Scientific Basis for Nuclear Waste Management XVI, C.G. Interrante, R.T. Pabalan, 1993, ISBN: 1-55899-189-1

Volume 295—Atomic-Scale Imaging of Surfaces and Interfaces, D.K. Biegelson, D.S.Y. Tong, D.J. Smith, 1993, ISBN: 1-55899-190-5

Volume 296—Structure and Properties of Energetic Materials, R.W. Armstrong, J.J. Gilman, 1993, ISBN: 1-55899-191-3

Prior Materials Research Society Symposium Proceedings available by contacting Materials Research Society

PART I

Overviews

Dislocation Networks in NiAl

Rotation of Subgrains

A subgrain boundary formed by an array of non-parallel dislocations results in a rotation between subgrains about the axis, $\mathbf{b}_1 \times \mathbf{b}_2$.

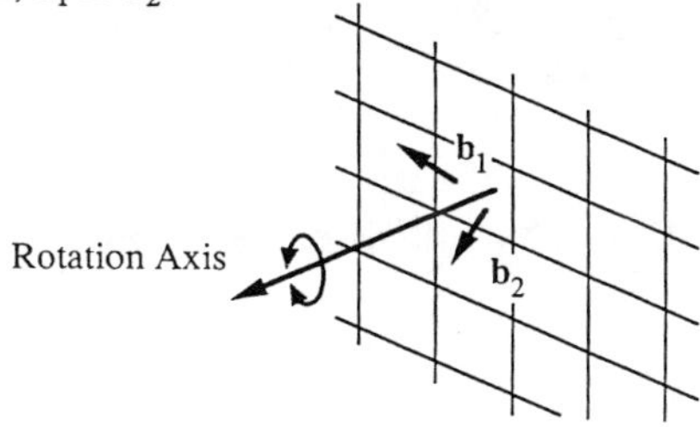

Rotation Axis	Dislocations in the Network $\mathbf{b}_1$	$\mathbf{b}_2$	Gliding Dislocations, $\mathbf{b}$
[100]*	[010]	[001]	[011]
[010]*	[001]	[100]	[101]
[001]	[100]	[010]	—

*Observed Rotations.

Rotation Axis Corresponds to Undeformed Diagonal of Cross-Section.

There is no resolved shear stress for slip of $\mathbf{b} = \langle 100 \rangle$ dislocations in NiAl single crystals loaded along the hard [001] axis. However, dislocation networks containing $\mathbf{b} = \langle 100 \rangle$ dislocations are common in hard oriented NiAl crystals deformed at high temperatures. These $\mathbf{b} = \langle 100 \rangle$ networks form low angle subgrain boundaries. The rotation between subgrains is about the axis defined by the cross product of the dislocations in the boundary. The axis of rotation between subgrains has been measured with Laue x-ray photography. In samples tested along [001], only [100] or [010] rotations are observed. The dislocations that must make up the networks producing these rotations are those which would result from the decomposition of the $\mathbf{b} = \langle 011 \rangle$ dislocations which experience of glide forces under [001] loading. The rotation of subgrains in hard oriented NiAl suggests that the observed $\mathbf{b} = \langle 100 \rangle$ dislocations are the result of the decomposition of $\mathbf{b} = \langle 011 \rangle$ dislocations.

Figure courtesy of K.R. Forbes, U. Glatzel, R. Darolia and W.D. Nix.

RECENT ADVANCES IN ORDERED INTERMETALLICS

C. T. LIU
Metals and Ceramics Division, Oak Ridge National Laboratory, Oak Ridge, TN 37831-6115

ABSTRACT

This paper briefly summarizes recent advances in intermetallic research and development. Ordered intermetallics based on aluminides and silicides possess attractive properties for structural applications at elevated temperatures in hostile environments; however, brittle fracture and poor fracture resistance limit their use as engineering materials in many cases. In recent years, considerable efforts have been devoted to the study of the brittle fracture behavior of intermetallic alloys; as a result, both intrinsic and extrinsic factors governing brittle fracture have been identified. Recent advances in first-principles calculations and atomistic simulations further help us in understanding atomic bonding, dislocation configuration, and alloying effects in intermetallics. The basic understanding has led to the development of nickel, iron, and titanium aluminide alloys with improved mechanical and metallurgical properties for structural use. Industrial interest in ductile intermetallic alloys is high, and several examples of industrial involvement are mentioned.

INTRODUCTION

Ordered intermetallics based on aluminides and silicides constitute a unique class of metallic materials for structural use at elevated temperatures in hostile environments. Their promising properties include excellent elevated-temperature strength, resistance to oxidation and corrosion, and relatively low density and high melting point [1-12]. However, most intermetallics exhibit brittle fracture and low ductility at ambient temperatures, and poor fracture resistance and limited fabricability restrict their use as engineering materials in many cases. The recent search for new high-temperature materials has stimulated a great deal of interest in development of ordered intermetallics for structural use. The progress made during the past 10 years has been recorded in several proceedings and books:

1. High-Temperature Ordered Intermetallic Alloys I, II, II, and IV, ed. Koch et al., 1985; Stoloff et al., 1987; Liu et al., 1989; and Johnson et al., 1991; respectively [1-4];
2. High Temperature Aluminides and Intermetallics, ed. Whang et al., 1990 and 1992 [5,6];
3. Intermetallic Compounds — Structure and Mechanical Properties, ed. Izumi, 1991 [7];
4. Ordered Intermetallics—Physical Metallurgy and Mechanical Behavior, ed. Liu et al., 1992 [8];
5. Microstructure/Property Relationships in Titanium Aluminides and Alloys, ed. Kim and Boyer, 1991 [9];
6. The Deformation Behavior of Intermetallic Superlattice Compounds, Yamaguchi and Umakoshi, 1990 [10];
7. Ordered Intermetallics, Liu et al., 1990 [11]; and
8. Ordered Alloys, ed. Stoloff, 1984 [12].

At present, there is world-wide interest in ordered intermetallics; as a result, many new results have been generated each year. Because of page limitation, a systematic review of the recent progress on ordered intermetallics is not feasible in this brief paper. Consequently, this paper will include only some highlights of the recent progress made in understanding brittle fracture behavior and improving the mechanical properties of intermetallic alloys at ambient and elevated temperatures. Readers are urged to go over the invited papers in this MRS proceedings [13], which provide comprehensive reviews of recent progress in different alloy systems and property areas. A number of intermetallic systems have been developed to the stage where they are ready for engineering use, and examples of industrial interest and potential use of these intermetallics are also briefly mentioned in this paper.

Mat. Res. Soc. Symp. Proc. Vol. 288. ©1993 Materials Research Society

IRON ALUMINIDES AND ENVIRONMENTAL EFFECTS

Iron aluminides based on Fe_3Al (D0$_3$) and FeAl (B2) have excellent oxidation and corrosion resistance because they are capable of forming protective oxide scales at elevated temperatures in hostile environments [14]. In addition, these aluminides offer low material cost, low density, and conservation of strategic elements (such as chromium). The major drawbacks of the aluminides are their poor ductility and fracture resistance at ambient temperatures and their poor strength and creep resistance at temperatures above 600°C. The aluminides were known to be brittle at room temperature for more than 40 years; however, the major cause of their brittleness was not identified until recently [15].

Recent studies have shown that Fe_3Al and FeAl aluminides are intrinsically quite ductile and that the poor ductility commonly observed in air tests is caused mainly by an extrinsic effect—environmental embrittlement [15,16]. The effect of test environment on tensile properties is shown in Fig. 1 for FeAl containing 36.5 at. % Al. The yield strength is insensitive to environment, and the ultimate tensile strength correlates with the tensile elongation, which depends strongly on test environment. The aluminide had ductilities of 2% in air, 6% in vacuum, and 17.6% in dry oxygen. The water-vapor test confirmed the low ductility found in the air tests, indicating that moisture in air is the embrittling agent. The increase in ductility from 2 to 18% is accompanied by a change in fracture mode from transgranular cleavage in air to mainly grain-boundary separation in dry oxygen. This observation suggests that cleavage planes in FeAl are more susceptible to embrittlement than are the grain boundaries.

Environmental embrittlement has been explained by the following chemical reaction:

$$2Al + 3H_2O \rightarrow Al_2O_3 + 6H\,. \tag{1}$$

The reaction of moisture in air with aluminum atoms at crack tips results in the generation of high-fugacity atomic hydrogen that rapidly penetrates into crack tips and causes severe embrittlement. The fact that the yield strength is insensitive to ductility and test environment is consistent with the mechanisms of hydrogen embrittlement observed in other ordered intermetallic alloys [17-25]. The highest ductility is generally obtained in dry oxygen environment (rather than in vacuum) because oxygen reacts with the aluminum to form aluminum oxide directly [26], thereby suppressing the moisture/aluminum reaction and the generation of atomic hydrogen in Eq. 1:

$$2xM + yO_2 \rightarrow 2M_xO_y\,. \tag{2}$$

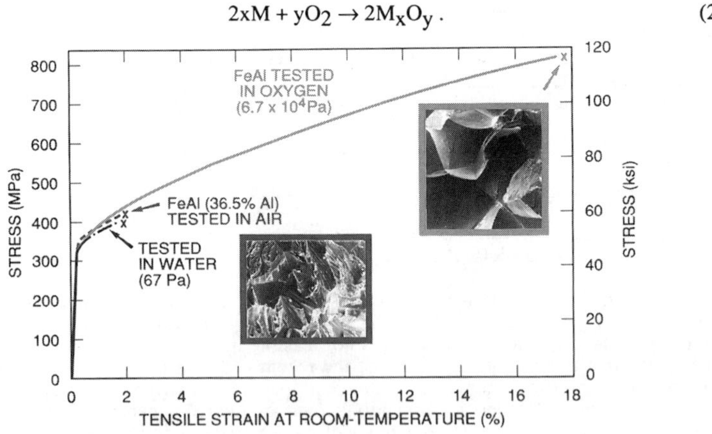

Fig. 1. Effect of test environment on the room-temperature ductility and fracture behavior of FeAl (36.5 at. % Al) [11].

It should be noted that the maximum degree of moisture-induced hydrogen embrittlement occurs around ambient temperatures [27]. At higher temperatures less hydrogen is concentrated at crack tips, and in situ protective oxide films can form more readily on specimen surfaces, while at lower temperatures the aluminum-moisture reaction is slowed, and the equilibrium moisture content in air also is lowered.

The environmental sensitivity of FeAl is markedly reduced when the aluminum concentration is higher than 38 at. % [28-30]. For Fe-43 at. % Al, the ductility is almost nil in air as well as in dry oxygen; all specimens fail intergranularly. The lack of an environmental effect is explained by the fact that grain boundaries in FeAl alloys with Al > 38% are intrinsically brittle. Therefore, environmental embrittlement and intrinsic grain-boundary brittleness must both be recognized in order to establish strategies for reducing overall brittleness. It has been demonstrated that the intrinsic grain-boundary brittleness in FeAl, as well as other intermetallics, can be alleviated by microalloying with boron [28,29], which tends to segregate to the boundaries and enhance their cohesive strength.

Similar air embrittlement has been observed in Fe_3Al alloys [16,31-34]. The moisture-induced hydrogen affects not only tensile properties, but also the fatigue and crack growth behaviors [33]. Under cyclic loading conditions, vacuum or oxygen environments raise the fatigue threshold and reduce crack growth rates at ambient temperatures (Fig. 2).

The understanding of the cause of brittleness in FeAl and Fe_3Al has led to new directions in the design of ductile iron-aluminide alloys. The schemes used to improve the ductility of the iron aluminides include [14,35,36]:

(1) formation of protective oxide scales on surfaces by alloying with chromium and/or preoxidizing in air;
(2) refinement of grain structure by thermomechanical treatment;
(3) refinement of grain structure by second-phase particles, such as formation of zirconium borides and carbide by alloying with Zr, B, and C;
(4) enhancement of grain-boundary cohesion by microalloying with boron; and
(5) reducing hydrogen solubility and diffusivity by alloying additions (possibly boron).

Figure 3 compares the tensile curve of binary Fe_3Al with ductile Fe_3Al (28 at. % Al) alloys developed by both thermomechanical treatment and alloying additions [14]. The ductile Fe_3Al alloys showed a high ductility of 16% when tested in air at room temperature. The strength of Fe_3Al and FeAl alloys at elevated temperatures can be improved by alloying with Mo, Nb, and Zr [14,36]. The recent development of ductile and strong Fe_3Al- and FeAl-base alloys is summarized in the paper by McKamey et al. [14] and the report by Liu et al. [36].

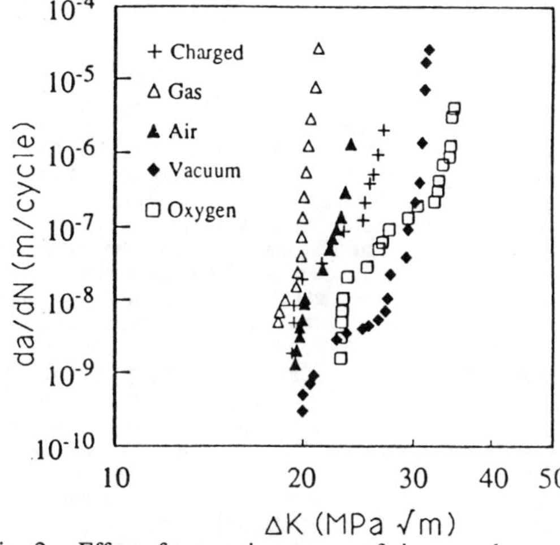

Fig. 2. Effect of test environment on fatigue-crack-growth rate (da/dN) of a $D0_3$-Fe_3Al alloy (FA-129:Fe-28Al-5.0Cr-0.5Nb-0.2C, at. %) tested at room temperature [33].

Ni_3Al AND GRAIN-BOUNDARY FRACTURE

Ni_3Al with the $L1_2$ structure is the most important strengthening constituent in nickel-base superalloys. This is because this aluminide has excellent strength, in addition to good oxidation resistance, at elevated temperatures. The aluminide is a

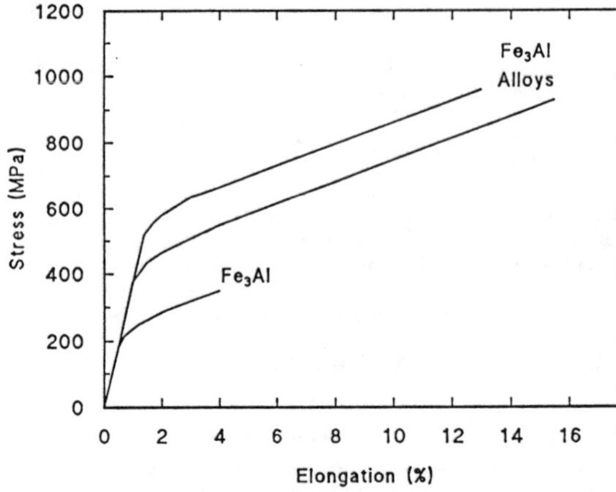

Fig. 3. Comparison of the stress-strain curve of binary Fe_3Al (28 at. % Al) with ductile Fe_3Al alloys developed by thermomechanical treatment and alloying additions [14]. All specimens were tested at room temperature in air.

model material to study the yield anomaly and brittle grain-boundary fracture. The detailed mechanism governing the positive temperature dependence of flow strength in Ni_3Al and other $L1_2$ ordered intermetallics is given in the review paper by Hirsch in this proceedings [13]. The previous work on brittle grain-boundary fracture and ductilizing effect of boron was given in the viewpoint set in *Scr. Metall.*, 1991 [37]. This paper will focus on some new evidence in connection with the brittle grain-boundary fracture in Ni_3Al and other $L1_2$ intermetallics discovered in the last couple of years.

The grain boundary in binary Ni_3Al was considered to be intrinsically brittle mainly because no appreciable impurities at grain boundaries had been detected by Auger analyses [38-40]. In 1991, Liu et al. [41,42] had made a first attempt to link brittle intergranular fracture with environmental embrittlement in binary Ni_3Si and Ni_3Al. Table 1 shows the tensile properties of recrystallized polycrystalline Ni_3Al alloys produced by repeated cold forging and 1000°C annealing. The aluminides containing 23.5 to 24% Al showed only 2.5 to 2.6% elongation in air but 7.2 to 8.2% in dry oxygen, an increase in ductility by a factor of ~3. These results clearly demonstrate that binary Ni_3Al alloys are prone to environmental embrittlement at room temperature. Thus, the extrinsic factor—environmental embrittlement—is a major cause of low ductility and brittle intergranular fracture in binary Ni_3Al [41].

The environmental embrittlement is apparently not the sole source of grain-boundary brittleness in Ni_3Al, because the elimination of the environmental effect by testing in dry oxygen does not lead to extensive ductility (e.g. > 30%) and suppression of intergranular fracture in Ni_3Al. Another factor governing the brittle intergranular fracture is the poor grain-boundary cohesion due to high ordering energy and the large difference in electronegativity and valence electron between nickel and aluminum atoms [43-47]. Atomistic simulation calculations suggest that the high ordering energy in Ni_3Al reduces atomic relaxation in the grain-boundary region and causes a formation of columnar "cavities," which serve as suitable sites for nucleation and growth of intergranular cracks [48-51]. This is illustrated in Fig. 4, where the calculated structure of the $\Sigma = 29$ (520)/[001] symmetrical tilt boundaries in weakly ordered Cu_3Au is compared with that in strongly ordered Ni_3Al. The formation of the cavities is visible along the grain boundary in Ni_3Al.

It is difficult to assess the relative importance of the two causes of brittle grain-boundary fracture in Ni_3Al. Takasugi et al. [20] first reported the environmental effect on ductility reduction in beryllium-doped Ni_3Al, but not in binary Ni_3Al (24% Al). The Ni_3Al alloy doped with 1 at. % Be showed a tensile ductility of 5% in vacuum but 1% in air. On the other hand, no environmental effect was detected in cast binary Ni_3Al, which exhibited very limited plastic deformation at room temperature. Manganese at a level of 15 at. % was added for enhancement of the grain-boundary cohesion in Ni_3Al [53]. Masahashi et al. [53] found that the tensile ductility of $Ni_3(Al,Mn)$ is highly susceptible to environmental embrittlement at room temperature at different strain rates. Note that the environment effect was thought previously to be associated with the ternary alloying additions added to Ni_3Al (rather than in connection with binary Ni_3Al) until the recent finding of a clear environmental effect in fabricated and recrystallized Ni_3Al and Ni_3Si [41,42].

Table I. Effect of test environment on room-temperature tensile properties of binary Ni_3Al (Ni-23.5% Al) and Zr-doped Ni_3Al (Ni-22.65%-0.26% Zr)

Test environment	Tensile ductility (%)	Yield strength (MPa)	Ultimate tensile strength (MPa)
	Ni-23.5% Al [41]		
Air	2.5	193	230
Oxygen	8.2	194	351
	Ni-22.65% Al-0.26% Zr [54]		
Water	8.7	322	528
Air	13.2	324	661
Oxygen	50.6	326	1451

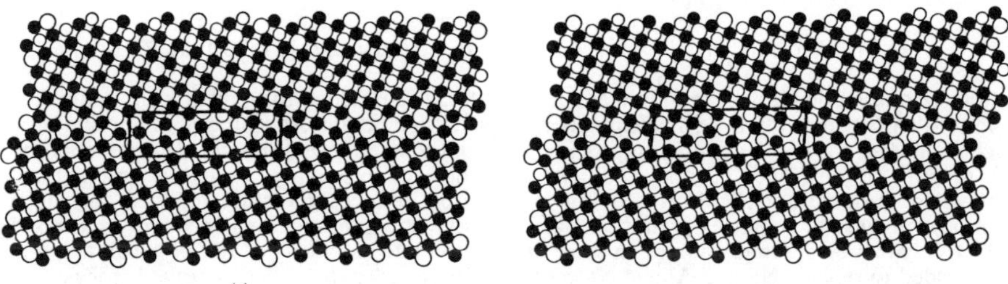

Fig. 4. Calculated structures of the Σ = 29 (520)/[001] symmetrical tilt boundaries in: *(a)* Cu_3Au and *(b)* Ni_3Al projected onto the (001) plane. In these figures atoms are depicted as shaded circles of two different sizes. The larger circles correspond to Au or Al and smaller circles to Cu or Ni, respectively. The shading distinguishes two different (002) layers in the [001] period [50].

More recently, George et al. [54] have studied the environmental effect in a carefully prepared Ni_3Al alloy with the composition of Ni-22.65% Al-0.26% Zr, where zirconium was added to enhance the grain-boundary properties. In this case, the polycrystalline material was prepared from recrystallization of cold-worked single crystals. As shown in Table 1, the polycrystalline specimens exhibited a room-temperature tensile ductility of 8.7 in water, 13.2% in air, and 50.6% in oxygen. These results demonstrate that, by adding zirconium to eliminate the intrinsic source of grain-boundary brittleness, the Ni_3Al alloy becomes very sensitive to test environment. The ductility increases from 8.7 to 50.6% by testing the alloy in a dry environment. Moisture-induced hydrogen environment has been observed in many other $L1_2$ intermetallics, including Ni_3Si [42], $Ni_3(Si,Ti)$ [55], Co_3Ti [52], and $(Fe,Co)_3V$ [56,57] alloys.

Boron has been found to be most effective in improving the tensile ductility of Ni_3Al (<25% Al) tested in air at room temperature [38,58]. In view of the recent finding of environmental embrittlement, the ductilizing effect of boron was reassessed by testing boron-doped Ni_3Al (24% Al) in various environments. As shown in Table 2 [59], all specimens, including the one tested in water, are ductile with tensile elongation more than 35% and fracture in a transgranular mode. These results indicate that boron-doped Ni_3Al is not sensitive to test environment at room temperature, consistent with the early data reported by Takasugi et al. [20]. Thus, boron is effective in eliminating environmental embrittlement in Ni_3Al. Since boron and hydrogen both occupy interstitial sites, it is reasonable to assume that the strong segregation of

Table II. Effect of test environment on room-temperature tensile properties of B-doped Ni_3Al(24% Al) [59]

Test environment	Elongation (%)	Strength (MPa)	
		Yield	Ultimate
Oxygen	42.8	289	1315
Air	39.3	280	1241
Water	36.8	288	1120

boron to Ni_3Al grain boundaries would block the diffusion of hydrogen along the boundaries and thus alleviate hydrogen embrittlement. Boron-free Ni_3Al (24% Al) showed a ductility of only 7.2% in dry oxygen while boron-doped Ni_3Al exhibited 50.6%. This comparison suggests that boron segregation also enhances the grain-boundary cohesion in Ni_3Al.

The Ni_3Al phase is capable of dissolving substantial alloying additions that strongly affect the mechanical and metallurgical properties of Ni_3Al. Recent alloy design efforts have led to the development of Ni_3Al-base alloys with the following composition range for structural use at elevated temperatures in hostile environments [60-62]:

Ni-14 to 18Al-6 to 9Cr-1 to 4Mo-0.01 to 1.5Zr/Hf-0.01 to 0.20B (at. %) (3)

In these aluminide alloys, chromium at a level of 6 to 9% is added for reducing environmental embrittlement in oxidizing environments at elevated temperatures. Zirconium and hafnium additions are most effective in improving the high-temperature strength via solid-solution hardening effects. Molybdenum additions are added for improving strength at ambient and elevated temperatures. Microalloying with boron reduces moisture-induced hydrogen embrittlement and enhances grain-boundary cohesive strength, resulting in sharply increased ductility at ambient temperatures. In some cases, moderate amounts (<20%) of cobalt and iron are added to replace Ni, and Al and Ni, respectively, in order to further improve hardness and corrosion resistance [63,64]. The alloys with optimum properties usually contain 5 to 15 vol % of the disordered γ phase, which has the beneficial effect of reducing environmental embrittlement in oxidizing atmospheres and improving creep properties at elevated temperatures.

Cast aluminide alloys usually possess a coarse grain structure which lowers the yield strength at ambient temperatures. The strength can be effectively increased by alloying with molybdenum via solid-solution hardening. The aluminide alloys prepared by investment casting possess good mechanical properties. The strength of a cast Ni_3Al alloy (IC-221M: Ni-15.9Al-8.0Cr-0.8Mo-1.0Zr-0.03B, at. %) is comparable to the superalloy IN-713C at room temperature but is much higher at higher temperatures (e.g., 1000°C). The high-cycle fatigue life of IC-221M is longer than that of IN-713C by more than two orders of magnitude at 650°C in air [65]. The cast aluminide alloys usually have a tensile elongation of 10 to 30% at room and elevated temperatures.

Limited effort has been devoted to the development of single crystals and directionally solidified Ni_3Al alloys. Han et al. [66] have recently developed a strong Ni_3Al-base alloy prepared by directional solidification (DS). The alloy (density = 7.91 g/cm^3) has a simple composition of Ni-16.3Al-8.2Mo-0.2B (at. %), with molybdenum as the major solid-solution strengthener. It has a yield strength of 990 MPa at 700°C, 600 MPa at 1000°C, and 520 MPa at 1050°C, all of which appear to be higher than those of existing nickel-base superalloys in single-crystal or DS forms. Table 3 shows that the Ni_3Al alloy is better than PWA 1422 in creep resistance at high temperatures, particularly above 1000°C. The excellent mechanical properties of the Ni_3Al alloy are attributed mainly to solid-solution hardening by molybdenum and second-phase strengthening by 15 to 20% γ phase (with a lattice misfit of 1.24% between γ' and γ phases). This result has demonstrated the possibility of the development of promising Ni_3Al-base alloys for structural applications above 1000°C. Additional development work is certainly needed for further optimizing the alloy composition and properties.

Table III. Comparison of creep properties of an advanced DS Ni_3Al alloy (Ni-16.3Al-8.24Mo-0.26B, at. %) with a strong commercial alloy PWA 1422 (Ni-11.23Al-11.89Cr-4.93Co-3.95Ta-1.76Ti-1.29W, at.%) (data from [66])

Creep condition		Rupture life (h)	
Temperature (°C)	Stress (MPa)	Ni_3Al alloy	PWA 1422
760	765	190	100
1040	137	170	100
1100	88	254	43

NiAl AND MINOR ALLOY ADDITIONS

Nickel aluminide containing more than about 41 at. % Al starts to form a single-phase ordered B2 structure based on the body-centered cubic (bcc) lattice. In terms of thermophysical properties, B2 NiAl offers more potential for high-temperature applications than $L1_2$ Ni_3Al [67-70]. It has a higher melting point (1638°C), a substantially lower density (5.86 g/cm^2), a higher Young's modulus (294 GPa), and a distinctly higher thermal conductivity (76 W/m-K) at ambient temperatures. In addition, NiAl has excellent oxidation resistance at high temperatures. In the 1950s and 1960s, NiAl alloys were employed as coating material for hot components in corrosive environments. The oxidation resistance of NiAl can be further improved by alloying with yttrium and other refractory elements such as zirconium and hafnium [71,72].

The structural use of NiAl suffers from two major drawbacks: poor fracture resistance at ambient temperatures and low strength and creep resistance at elevated temperatures. Single crystals of NiAl are quite ductile in compression, but both single-crystal and polycrystalline NiAl appear to be brittle in tension at ambient temperatures. The nickel aluminide exhibits mainly <100> slip, rather than <111> slip as commonly observed for bcc materials [73-75]. The insufficient deformation modes, poor cleavage resistance, and brittle grain-boundary fracture [76] are all considered to be the major causes of low tensile ductility in NiAl. The aluminide shows a sharp increase in ductility above 400°C and becomes very ductile above 500°C at conventional strain rates [77,78]. In general, hot fabrication of NiAl at elevated temperatures presents no major problems. The brittle grain-boundary fracture in polycrystalline NiAl can be readily suppressed by microalloying with boron; however, the suppression of intergranular fracture does not lead to increase in ductility as found by George and Liu recently [76].

Because of the excellent high-temperature capability of NiAl, considerable effort has been devoted to understanding brittle fracture and improving mechanical properties of NiAl during the past years. A striking result was reported recently by Darolia et al. [69,79], who observed that the tensile ductility of <110> single-crystal NiAl can be substantially increased by alloying with less than 1% of alloying additions. As shown in Fig. 5, the room-temperature tensile ductility of <110> NiAl (a soft orientation) is increased from 1% to as high as 6% by adding about 0.2 at. % Fe. The ductility decreases sharply when the iron content is more than 0.4%. A similar effect, but smaller in scale, is observed for molybdenum and gallium. The ductilizing effect of iron and molybdenum is not detected in polycrystalline NiAl [80], possibly because of strain incompatibility at grain boundaries in association with the deformation of NiAl polycrystals. However, molybdenum-modified NiAl with a wrought microstructure showed a room-temperature ductility higher than that of unalloyed NiAl [80].

Hack, Brzeski, and Darolia [81] have recently reported some interesting results that contribute to our basic understanding of deformation and fracture in single-crystal NiAl. Their study indicates that the fracture resistance of <110> crystals can be dramatically improved by controlled heat treatments. The NiAl crystals furnace-cooled from a homogenization treatment at 1300°C exhibited a tensile elongation of 1% and a fracture toughness of 2.4 MPa m$^{1/2}$ at room temperature. Surprisingly, the room-temperature ductility increased to 7% and the toughness went up to 16.7 MPa m$^{1/2}$ when NiAl crystals were reheated to 400°C followed by air cooling. The beneficial effect of the 400°C heat treatment disappeared as the specimens were cooled down slowly inside a furnace. Hack et al. [81] attribute the low ductility and poor toughness of

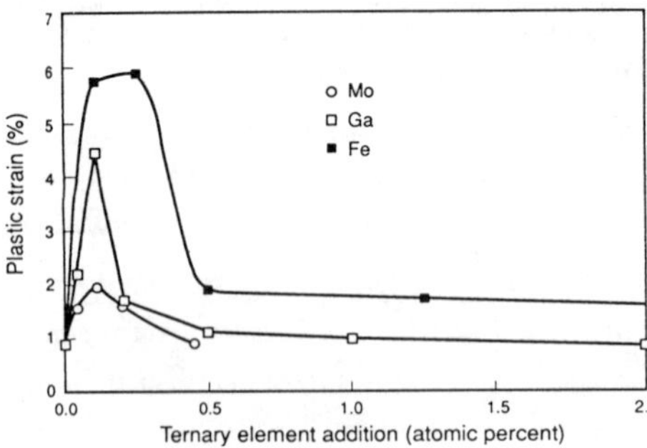

Fig. 5. Significant improvement in room-temperature tensile ductility of <110> NiAl by alloying with iron, gallium and molybdenum [69].

single-crystal NiAl to strain-age embrittlement involving pinning of mobile dislocation by interstitials such as carbon and oxygen, similar to strain-aging embrittlement in high-strength steels [82]. The interstitial content in these NiAl crystals is reported to be only about 100 wt ppm carbon and 50 wt ppm oxygen. The implication of this study is very significant; that is, NiAl single crystals basically possess reasonable intrinsic ductility and toughness at room temperature, and their poor fracture resistance is mainly caused by an extrinsic effect—strain-aging embrittlement by interstitial impurities [82]. Based on this embrittling mechanism, it is expected that the fracture resistance of NiAl crystals can be effectively improved by either reducing the interstitial content or scavenging interstitials by certain alloying additions. Certainly, further research should be directed to this area.

Because of strong affinity between nickel and aluminum atoms, NiAl can dissolve only limited amounts of solutes, with a solubility limit typically less than a couple of percents. A number of investigators have found that NiAl in single-crystal or polycrystal forms can be effectively hardened by adding <1% solutes. Recently, Noebe, Bowman, and Nathal [70] have compiled the solute-solution hardening data, which are shown as a function of solute radii in Fig. 6. The interstitial elements [76] boron and carbon and the substitutional elements [70,83] yttrium, zirconium, molybdenum, and lanthanum all have a hardening rate ($\Delta\sigma_y/\Delta c$) > 1500 MPa per solute atom percent. For instance, alloying with 1% Zr increases the yield strength (σ_y) of NiAl by 400 MPa (58 ksi). Note that, because of the extremely low solubility limit of certain elements, the hardening effect detected in many NiAl alloys actually comes from precipitation of fine second-phase particles, as shown by atom-probe imaging [84] and other microstructural analysis [70,80,83].

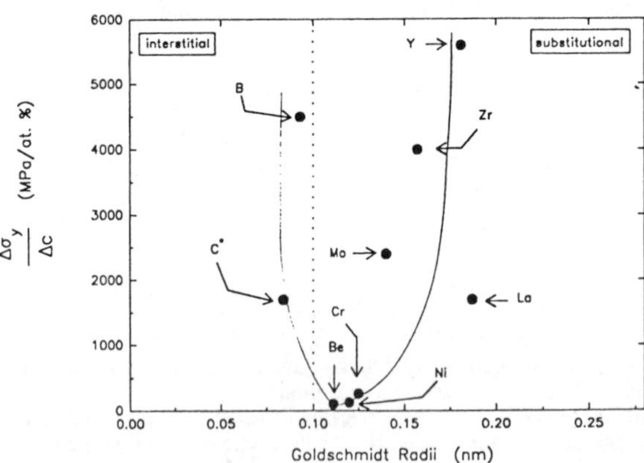

Fig. 6. The relationship between hardening rate and atom size for various alloying elements in NiAl [70].

At present, there appears to be no major difficulty in achieving high strength and creep resistance in NiAl alloys by controlling crystal orientation and alloying additions [80,85]. The remaining problem for structural use of NiAl alloys is their poor impact resistance at ambient and elevated temperatures. NiAl alloys typically show a fracture toughness of 4 to 6 MPa $m^{1/2}$ at room temperature [70,81,86]. They generally have extensive tensile ductilities above 600°C at conventional strain rates; however, their impact resistance remains poor at elevated temperatures. The poor fracture resistance is possibly related to the fact that all thermally activated deformation processes operating at elevated temperatures are suppressed by high strain-rate deformation [70,87,88], such as impact testing.

γ TITANIUM ALUMINIDES AND MICROSTRUCTURAL CONTROL

Titanium aluminides based on $D0_{22}$-$TiAl_3$, γ-TiAl, and α_2-Ti_3Al possess some unique properties for structural applications, particularly for aerospace and aeronautic industries [9,11,89]. This is due to their sufficiently low material density and high retention of strength and creep resistance at elevated temperatures. $TiAl_3$-base trialuminide alloys with the Ll_2 crystal structure are of current interest; however, these alloys remain brittle in tension at ambient temperatures [90,91]. The poor fracture resistance simply keeps these alloys outside the realm of engineering materials, and there is no indication of breakthroughs in this area at present. Ti_3Al-base alloys (such as Ti-24Al-11Nb and Ti-24Al-10Nb-3V-1 Mo) and composites, on the other hand, suffer mainly from structural instability and cracking problems when exposed to hostile environments under cyclic loads [9,92,93]. Currently, most efforts on titanium aluminides have been focused on γ-TiAl alloys which commonly contain up to 20% of the α_2-Ti_3Al phase. The γ-base alloys are superior to α_2-base alloys because of their higher elevated-temperature strength, better oxidation resistance, and relatively lower material density.

First-principles total-energy quantum mechanical calculations have been advanced to the stage that they are able to help us in understanding the fundamental deformation behavior, in addition to the prediction of phase stability, of order intermetallics [8,94]. Recently, Fu and Yoo [95-99] have calculated elastic constants, shear-fault energies, Griffith's cleavage strength, and point defects in TiAl and other intermetallics, using the full-potential linearized augmented plane-wave (FLAPW) method. Table 4 summarizes the calculated results of elastic constants and shear-fault energies for TiAl. Since no experimental data are reported at the present, the calculated elastic constants are the only set of data available for single-crystal TiAl. The calculated shear and bulk moduli of TiAl agree well with the experimental ones, indicating the accuracy of the calculations. The high shear anisotropy (C_{44}/C_{66}) reflects the formation of strong p-d bonds between titanium and aluminum atoms. The twin-boundary energy in TiAl is quite low as compared with other calculated shear-fault energies. This is consistent with the experimental observation of twinning deformation active in TiAl alloys tested at various temperatures [100-102]. Thus, the first-principles calculations are effective in predicting the fundamental mechanical behavior in TiAl as well as other ordered intermetallics.

Table IV. Calculated elastic constants (C_{ij}), bulk modulus (B), shear modulus (G), and shear fault energies for TiAl [99]

Elastic constants (10^{11} N/m^2)						Bulk (10^{11} N/m^2)	Shear (10^{11} N/m^2)
C_{11}	C_{12}	C_{13}	C_{33}	C_{44}	C_{66}	B	G
1.90	1.05	0.95	1.85	1.20	0.50	1.25	0.70

Shear fault energies (mJ/m^2)

$APB_{(010)}$	$APB_{(111)}$	SISF	SESF	TWIN
430	510	90	80	60

Yamaguchi and his associates [10,100,101,103] have conducted a systematic study of the mechanical properties of polysynthetically twinned (PST) TiAl crystals (Ti-49 at. % Al) produced in an optical-floating zone furnace. The crystals have a two-phase structure containing γ (the main phase) and α_2 (the minor phase) lamellae. The γ and α_2 lamellae have the orientation relationship of (111) γ // (0001)α_2 and [$1\bar{1}0$] γ // <1120> α_2, where (111) in the γ-phase is assigned to be parallel to the lamellar boundaries. Figure 7 shows a plot of the room-temperature yield strength as a function of the angle ϕ between the lamellar boundaries and tensile axis [100]. The orientation of two crystal groups (1) and (2) is indicated in the stereographic projection of tensile axis orientations shown in Fig. 7. The yield strength depends strongly on the angle ϕ, but the orientation dependence is not exactly symmetrical with respect to $\phi = 45°$. The variation of yield strength with ϕ can be explained by the two main deformation modes (i.e., the true twinning of the {111}<[112] type and slip on {111}<[110]), both of which are operational for the hard and easy modes of deformation.

The tensile elongation to fracture is also strongly dependent on the angle ϕ, as indicated in Fig. 8 [100]. A tensile elongation as large as 20%, which is far larger than other reported values for TiAl-base alloys, has been obtained for the soft orientation with $\phi = 31°$. Fracture occurs in a brittle manner, independent of the ductility. PST crystals failed mainly along various lamellar boundaries in the γ-phase, except for the $\phi = 0°$ crystal where fracture occurs across the lamellar boundaries. It is interesting to note that polycrystalline duplex structures produced from master ingots used for preparation of PST crystals exhibited an average elongation of 1 to 2%, which is in good agreement with the reported room-temperature ductilities for nearly stoichiometric TiAl.

Mechanical properties of TiAl alloys are sensitive to alloy composition and microstructure. A great amount of development work has been focused recently on improving the mechanical properties of TiAl alloys by alloying additions and microstructural control [9-11,100,102]. These efforts have led to the development of TiAl-base alloys with the following general compositions (at. %) for structural uses at elevated temperatures [104]:

$$\text{Ti-(46-49)Al-(0-3)(Cr, Mn, V) -(0.5-6)(Nb, W, or Mo)-(0-1)(Si, B, N, Ni, etc.)} \quad (4)$$

The TiAl alloys with 46 to 49% Al commonly contain the α_2 phase, the amount of which depends on aluminum concentration, heat treatment, and microstructural control. Cr, Mn, and V at levels up to 3% are added to enhance the ductility of TiAl through different proposed mechanisms including promotion of twinning, reduction in the c/a ratio, and lowering of aluminum level in the γ-phase, etc. Nb, W, and Mo are effective in improving the strength and oxidation resistance of TiAl alloys at elevated temperatures. Alloying with up to 1% of Si, B, N, and Ni generally enhances alloy castability and refines the grain structure. Silicon additions are also reported to be beneficial for oxidation resistance.

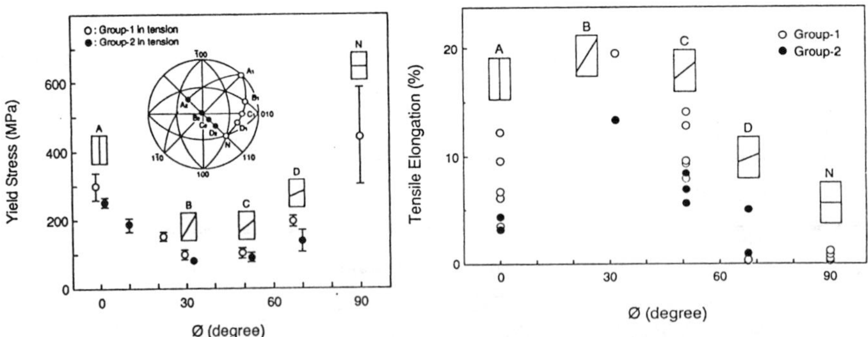

Fig. 7. Plot of the room-temperature yield strength of PST TiAl crystals tested in tension as a function of the angle Ø [100].

Fig. 8. Plot of the room-temperature tensile elongation of PST TiAl crystals as a function of the angle Ø [100].

Figure 9 schematically illustrates the relations between microstructure and tensile properties/fracture toughness at room temperature. Microstructures in TiAl alloys are expressed in terms of the relative volume fractions of equiaxed gamma (G), fully lamellar structure (L), and duplex structure of G and L (G + L). The grain size and microstructures can be controlled by thermomechanical treatment, with feature temperatures corresponding to various phase stabilities in the Ti-Al system. The general trend in Fig. 9 indicates that the G+L microstructure with a fine grain size gives the best tensile ductility (~4%) but lowest fracture toughness (~10 MPa $m^{1/2}$). On the other hand, the fully lamellar (L) structure with a coarse grain size gives almost the highest fracture toughness (~30 MPa $m^{1/2}$) but the lowest tensile ductility (<1%). Also, the TiAl alloys with a coarse lamellar structure generally exhibit excellent creep resistance at elevated temperatures.

The inverse relationship between tensile ductility and fracture toughness at room temperature provides a challenge to materials engineers for microstructural design of TiAl alloys. To date, Kim and his associates [102,104,105] have made significant progress in modifying the lamellar structure by innovative thermomechanical treatment or processing. The desired microstructures should be close to nearly fully lamellar structure with a finer grain size. Table 5 compares expected mechanical properties from designed microstructures with duplex and fully lamellar structures in TiAl alloys based on the compositions shown in Eq. (4). Limited results generated so far have demonstrated the feasibility of achieving optimum properties from designed microstructures.

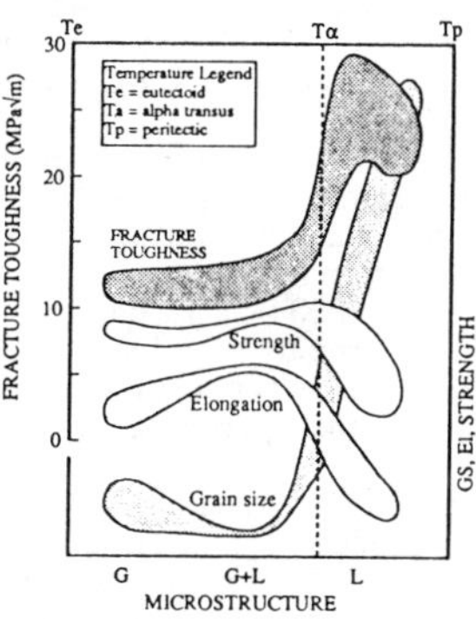

Fig. 9. Schematic relations among microstructure (G, G+L, L), grain size, fracture toughness and tensile properties of TiAl alloys [102].

INDUSTRIAL INTEREST AND POTENTIAL USE OF INTERMETALLIC ALLOYS

Recent research and development efforts on ordered intermetallics have resulted in substantial improvement in their mechanical and metallurgical properties. The intermetallic alloys with improved properties offer advantages over many conventional materials for structural applications at elevated temperatures in hostile environments. As compared with Ni-base superalloys, Ni_3Al alloys developed so far have better fatigue resistance, better oxidation resistance, relatively lower density, and higher

Table V. Effect of microstructure on mechanical properties of TiAl alloys with compositions listed in Eq. (4) [104]

Microstructure	YS (MPa)	UTS (MPa)	Plastic Elong. (%)	K_{IC} (MPa $m^{1/2}$)	Creep
Duplex[a]	350 to 550	400 to 650	1.5 to 3.5	10 to 15	Moderate
FL[b]	290 to 450	300 to 500	0.2 to 0.9	22 to 35	Excellent
Designed[c]	450 to 600	500 to 720	0.6 to 2.5[d]	20 to 30[d]	Very good[d]

[a]Duplex: A mixture of equiaxed gamma and lamellar structures.
[b]FL: Fully lamellar structure.
[c]Designed: Designed microstructure—nearly lamellar structure with a finer grain size.
[d]Expected values.

high-temperature strength under high strain-rate deformation [60-62,66]. Iron-aluminide alloys with excellent oxidation and corrosion resistance and low material density and cost are superior to ferritic steels and austenitic stainless steels for many structural applications [14]. NiAl alloys possess better oxidation resistance, lower material density, higher melting point, and better thermal conductivity than nickel-base superalloys [70]. In comparison with titanium-base alloys, γ titanium aluminide alloys offer the advantages of lower material density and better elevated-temperature strength and oxidation resistance [9,11,102].

The unique properties of many intermetallic alloys have drawn attention recently from industries for structural applications. At present, the successful development of intermetallic alloys is critically dependent on industrial participation in material processing, data base generation, and structural utilization of these alloys as hot components in engineering systems. A brief description of the current industrial interest and potential use of the intermetallic alloys is outlined below:

1. Diesel-engine turbocharger rotors: Cummins Engine Company, PCC Airfoils, and Oak Ridge National Laboratory (ORNL) are jointly pursuing the use of Ni_3Al alloys for turbocharger rotors in diesel-engine trucks [106]. The aluminide alloy IC-221M (Ni-15.9 Al-8.0 Cr-0.8 Mo-1.0 Zr-0.03 B, at. %) with a good castability has been selected for this application. Aluminide rotors have been successfully produced by investment casting at PCC (Fig. 10). This castable aluminide alloy is expected to replace IN-713C with the major benefits of improved fatigue life and potential for lower cost.
2. High-temperature dies and molds: The good high-temperature oxidation resistance, together with excellent strength at high strain rates, makes Ni_3Al alloys attractive as die material for isothermal forging and mold material for glass processing. At present, Ni_3Al alloys are being evaluated for these applications in several companies [106].
3. Furnace fixtures for heat treating auto parts: Because of its resistance to carburizing and oxidizing atmospheres, the Ni_3Al alloy IC-221M is currently being evaluated by General Motors (Saginaw Division) and ORNL for use as a fixture material for heat treatment of auto parts in high-temperature furnaces. The Saginaw Division has the world's largest heat-treatment facility with a heat-treatment capability of about 6000 tons per day. The HU alloy (Fe-40 Ni-20 Cr-2.5 Si-2.0 Mn-0.5 Mo-0.4 C, wt %) is currently used for furnace fixtures, which cracked badly after ~500 thermal cycles. Ni_3Al alloy furnace fixtures have been successfully made by sand casting, and the evaluation of their performance is in progress [107].
4. Rollers for steel slab heating furnaces: The high-temperature strength, together with good oxidation and corrosion resistance of Ni_3Al alloys, can produce significant savings in energy costs by not requiring water cooling and in material costs by extending the life four to six times over the current material in use [106].

Fig. 10. Ni_3Al-alloy turbocharger rotor made by investment casting [106].

5. Turbine blades for jet-engine applications: General Electric has a major effort on developing single-crystal NiAl alloy turbine blades for new-generation jet-engine applications [69,85,108]. The NiAl-base alloys are selected because of their good high-temperature capability, low material density, high stiffness, and excellent thermal conductivity. Figure 11 shows a high-pressure turbine blade machined from a single-crystal NiAl alloy ingot. The NiAl alloys developed so far have adequate high-temperature strength and creep resistance, and further improvement in fracture toughness and impact resistance at ambient and elevated temperatures is being pursued.
6. Lightweight auto parts: Toyota Motor Corporation has reported a dramatic reduction in grain size of cast TiAl alloy ingots by additions of ≥0.3 wt % N [109]. Toyota is now developing lightweight auto parts using grain-refined TiAl alloys.
7. Vanes, blades, and turbocharger rotors for engine applications: Several companies including Ishikawajima-Harima Heavy Industries (IHI) are currently developing and processing TiAl-base alloys for engine applications [109]. Various engine parts made from TiAl alloys are successfully fabricated by innovative casting methods. Figure 12 shows various turbocharger rotors, turbine blades, and vanes made from TiAl alloys containing Fe, V, and B and fabricated at IHI. The beneficial effects of these elements are mentioned in the paragraph describing the TiAl alloy compositions [see Eq. (4)].

BRIEF SUMMARY AND REMARKS

At present, there is world-wide interest in research and development of ordered intermetallic alloys for structural use at elevated temperatures in hostile environments. As a result, a great deal of knowledge has been gained in understanding physical metallurgy and mechanical behavior of intermetallic alloys based on nickel, iron, and titanium aluminides. The alloy design efforts have led to the development of useful intermetallic alloys based on Ni_3Al, NiAl, Fe_3Al, FeAl, Ti_3Al, and TiAl systems for structural applications. Industrial interest in these intermetallic alloys with improved mechanical and metallurgical properties is high, and several examples of industrial involvement in processing and utilization of these intermetallic alloys are mentioned. Currently, the knowledge and experience gained from the aluminide development have been extended to silicides and other intermetallic systems, including disilicides (e.g., $MoSi_2$) [110,111], A15 compounds (e.g., Nb_3Al) [112,113], Laves-phase (e.g., Cr_2Nb) [114,115] alloys, etc.

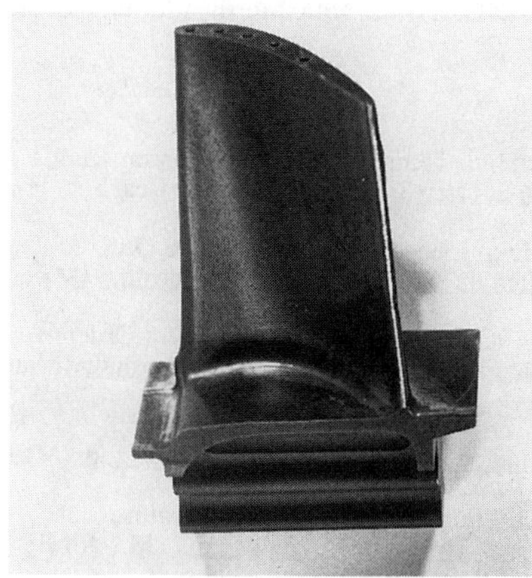

Fig. 11. A high-pressure turbine blade machined from a single-crystal NiAl ingot [108].

Ordered intermetallics have been employed or have the potential to be used in many other areas besides structural applications [11]. Molybdenum disilicide has been used commercially as electrical heating elements in high-temperature furnaces since 1956 [116]. NiTi alloys, referred to as Nitinol [117-119], are currently the major shape-memory material for systems control in the building, automobile, and automation industries. Considerable efforts are now being devoted to the development of new shape-memory alloys based on intermetallics (e.g., NiAl/Ni_3Al) for use at temperatures above ambient (>100°C) [120]. Many ordered intermetallics possess special attractive properties for magnetic, optical, and electronic applications. A prominent example is the use of $NdFe_{14}B_2$ as ahardmagnetic material for high-energy applications

Fig. 12. Cast turbocharger rotor, turbine blade and turbine vane made from TiAl alloys (doped with Fe, V, B) and fabricated at IHI [109].

[121,122]. Epitaxially grown $CoSi_2$ and $NiSi_2$ are used in novel devices like the metal base transistors and the permeable base transistors [123]. A-15 compounds such as Nb_3Sn and Nb_3Al are attractive as superconducting materials for industrial applications. It is our hope that the knowledge and experience gained from the development of structural intermetallic alloys will be applied to improve the performance of non-structural intermetallic materials in the coming years.

ACKNOWLEDGMENTS

The author is grateful to Drs. M. Yamaguchi, Y-W. Kim, R. Darolia, M. H. Yoo, C. L. Fu, and K. S. Kumar for unpublished materials and valuable discussions. Thanks are due to E. P. George and J. H. Schneibel for reviewing the manuscript and Millie Atchley, Patty Boyd, and Connie Dowker for manuscript preparation. The author also thanks K. Spence for editing. This research was sponsored by the Division of Materials Sciences, Assistant Secretary for Conservation and Renewable Energy/Office of Industrial Technologies/Advanced Industrial Concepts (AIC) Materials Program, and Fossil Energy AR&TD Materials Program, U.S. Department of Energy under contract DE-AC05-84OR21400 with Martin Marietta Energy Systems, Inc.

REFERENCES

1. C. C. Koch, C. T. Liu, and N. S. Stoloff, ed. "High Temperature Ordered Intermetallic Alloys," in Proceedings of Materials Research Society Symposium (Mater. Res. Soc. Proc. 39, Pittsburgh, PA, 1985).
2. N. S. Stoloff, C. C. Koch, C. T. Liu, and O. Izumi, ed. "High Temperature Ordered Intermetallic Alloys II," in Proceedings of Materials Research Society Symposium (Mater. Res. Soc. Proc. 81, Pittsburgh, PA, 1987).
3. C. T. Liu, A. I. Taub, N. S. Stoloff, and C. C. Koch, ed. "High Temperature Ordered Intermetallic Alloys III," in Proceedings of Materials Research Society Symposium (Mater. Res. Soc. Proc. 133, Pittsburgh, PA, 1989).
4. L. A. Johnson, D. P. Pope, and J. O. Stiegler, ed. "High Temperature Ordered Intermetallic Alloys IV," in Proceedings of Materials Research Society Symposium (Mater. Res. Soc. Proc. 213, Pittsburgh, PA, 1991).
5. S. H. Whang, C. T. Liu, D. P. Pope, and J. O. Stiegler, ed. "High Temperature Aluminides and Intermetallics," in Proceedings of TMS/ASM Symposium (TMS-AIME, Warrendale, PA, 1990).
6. S. H. Whang, C. T. Liu, D. P. Pope, and J. O. Stiegler, ed. "High-Temperature Aluminides Intermetallics," *Mater. Sci. Eng.* **A152/A153** (1992).
7. O. Izumi, ed. "Intermetallic Compounds - Structure and Mechanical Properties," in Proceedings of JIMIS-6 (Japan Institute of Metals, Tokyo, 1991).

8. C. T. Liu, R. W. Cahn, and G. Sauthoff, ed. "Ordered Intermetallics - Physical Metallurgy and Mechanical Behavior," NATO ASI Series, Vol. 213 (Kluwer Academic Publishers, Boston, MA, 1992).
9. Y. W. Kim and R. R. Boyer, ed. "Microstructure/Properties Relationships in Titanium Aluminides and Alloys" (TMS-AIME, Warrendale, PA, 1991).
10. M. Yamaguchi and Y. Umakoshi, "The Deformation Behavior of Intermetallic Superlattice Compounds," *Prog. Mater. Sci.* **34**(1), 1 (1990).
11. C. T. Liu, J. O. Stiegler, and F. H. Froes, "Ordered Intermetallics," in Metals Handbook, 10th ed., Vol. 2 (ASM, Materials Park, OH, 1990), pp. 913-42.
12. N. S. Stoloff, *Int. Met. Rev.* **29**(3), 123 (1984).
13. I. Baker, R. Darolia, J. D. Whittenberger, and M. H. Yoo, ed. "High Temperature Ordered Intermetallic Alloys V," in Proceedings of Materials Research Society Symposium (Mater. Res. Soc. Proc., Pittsburgh, PA, 1993).
14. C. G. McKamey, J. H. DeVan, P. F. Tortorelli, and V. K. Sikka, *J. Mater. Res.* **6**, 1779 (1991).
15. C. T. Liu, E. H. Lee, and C. G. McKamey, *Scr. Metall.* **23,** 875 (1989).
16. C. T. Liu, C. G. McKamey, and E. H. Lee, *Scr. Metall.* **24**, 385 (1990).
17. T. Takasugi and O. Izumi, *Acta Metall.* **34**, 607 (1986).
18. N. Masahashi, T. Takasugi, and O. Izumi, *Metall Trans. A* **19A**, 353 (1988).
19. O. Izumi and T. Takasugi, *J. Mater. Res.* **3**, 426 (1988).
20. T. Takasugi, N. Masahashi, and O. Izumi, *Scr. Metall.* **20**, 1317 (1986).
21. N. Masahashi, T. Takasugi, and O. Izumi, *Acta Metall.* **36**, 1823 (1988).
22. T. Takasugi and O. Izumi, *Scr. Metall.* **19**, 903 (1985).
23. A. K. Kuruvilla, S. Ashok, and N. S. Stoloff, in Proceedings of the Third International Congress on Hydrogen in Metals, Vol. 2, 1982), p. 629.
24. A. K. Kuruvilla and N. S. Stoloff, *Scr. Metall.* **19**, 83 (1985).
25. G. M. Camus, N. S. Stoloff, and D. J. Duquette, *Acta Metall.* **37**, 1497 (1989).
26. C. T. Liu and M. Takeyama, *Scr. Metall.* **24,** 1583-86 (1990).
27. C. T. Liu, C. L. Fu, E. P. George, and G. S. Painter, *ISIJ Int.* (October 1991).
28. C. T. Liu and E. P. George, *Scr. Metall.* **24**, 1285 (1990).
29. C. T. Liu and E. P. George, pp. 527-32 in ref. 4 (1991).
30. D. J. Gaydosh and M. V. Nathal, *Scr. Metall.* **24**, 1281 (1990).
31. R. J. Lynch, L. A. Heldt, and W. W. Milligan, *Scr. Metall.* **25**, 2147 (1991).
32. M. Shea, A. Castagna, and N. S. Stoloff, p. 607 in ref. 4 (1991).
33. A. Castagna and N. S. Stoloff, *Scr. Metall.* **26**, 673 (1992).
34. N. S. Stoloff, unpublished results, Rensselaer Polytechnic Institute (1992).
35. C. T. Liu, p. 321 in ref. 6 (1992).
36. C. T. Liu, V. K. Sikka, and C. G. McKamey, "Alloy Development of FeAl Aluminide Alloys for Structural Use in Corrosive Environments," ORNL Report, Martin Marietta Energy Systems, Inc., Oak Ridge Natl. Lab., Oak Ridge, TN, 1993 (unpublished).
37. C. T. Liu, *Scr. Metall.* **25**, 1231 (1991).
38. C. T. Liu, C. L. White, and J. A. Horton, *Acta Metall.* **33**, 213 (1985).
39. T. Takasugi, E. P. George, D. P. Pope, and O. Izumi, *Scr. Metall.* **19**, 551 (1985).
40. T. Ogura, S. Hanada, T. Masumoto, and O. Izumi, *Metall. Trans.* **16A**, 441 (1985).
41. C. T. Liu, *Scr. Metall.* **27**, 25 (1992).
42. C. T. Liu and W. C. Oliver, *Scr. Metall.* **25**, 1933 (1991).
43. T. Takasugi and O. Izumi, *Acta Metall.* **33**, 1247 (1985).
44. T. Takasugi, O. Izumi, and N. Masahashi, *Acta Metall.* **33**, 1259 (1985).
45. A. I. Taub, C. L. Briant, S. C. Huang, K. M. Chang, and M. R. Jackson, *Scr. Metall.* **20**, 129 (1986).
46. A. I. Taub and C. L. Briant, p. 343 in ref. 2 (1987).
47. A. I. Taub and C. L. Briant, *Acta Metall.* **35**, 1597 (1987).
48. V. Vitek and S. P. Chen, *Scr. Metall.* **25**, 1237 (1991).
49. V. Vitek, S. P. Chen, A. F. Voter, J. J. Kruisman, and J. Th. M. De Hosson, "Grain Boundary Chemistry and Intergranular Fracture," edited by G. S. Was and S. M. Bruemmer, *Mater. Sci. Forum* **46**, 237 (1989).
50. M. Yan, V. Vitek, and G. J. Ackland, p. 335-70 in ref. 8 (1992).
51. J. J. Kruisman, V. Vitek, and J. Th. M. De Hosson, *Acta Metall.* **36**, 2729 (1989).
52. T. Takasugi and O. Izumi, *Acta Metall.* **34**, 607 (1986).
53. N. Masahashi, T. Takasugi, and O. Izumi, *Metall. Trans.* **19A**, 353 (1988).

54. E. P. George, C. T. Liu and D. P. Pope, *Scr. Metall.* **27**, 365-70 (1992).
55. T. Takasugi, H. Suenaga, and O. Izumi, *J. Mater. Sci.* **26**, 1179 (1991).
56. C. Nishimura and C. T. Liu, *Scr. Metall.* **27**, 1307-11 (1992).
57. C. Nishimura and C. T. Liu, *Scr. Metall.* **25**, 791 (1991).
58. A. Aoki and O. Izumi, *Nippon Kinzoku Gakkaishi* **43**, 1190 (1979).
59. C. T. Liu, Martin Marietta Energy Systems, Inc., Oak Ridge Natl. Lab., Oak Ridge, TN, August 1992 (unpublished).
60. C. T. Liu and V. K. Sikka, *J. Met.* **38**, 19 (1986).
61. C. T. Liu, V. K. Sikka, J. A. Horton, and E. H. Lee, "Alloy Development and Mechanical Properties of Nickel Aluminide (Ni_3Al) Alloys," ORNL-6483, Martin Marietta Energy Systems, Inc., Oak Ridge Natl. Lab., Oak Ridge, TN, August 1988.
62. C. T. Liu, U.S. Patent No. 5,108,700 April 1992).
63. C. T. Liu, in Micon 86 (ASTM, Philadelphia, PA, 1988) p. 222.
64. A. I. Taub, K.-M. Chang, and C. T. Liu, *Scr. Metall.* **20**, 1613 (1986).
65. B. G. Gieseke and V. K. Sikka, Martin Marietta Energy Systems, Inc., Oak Ridge Natl. Lab., Oak Ridge, TN, August 1992 (unpublished).
66. Y. F. Han, S. H. Li, S. Ma, and Y. N. Tan, paper presented at First Pacific Rim International Conference on Advanced Materials and Processing, Hongzhou, China, June 23-27, 1992 (unpublished).
67. K. Vedula, V. Pathare, I. Aslamidis, and R. H. Titran, pp. 411-421 in Ref. 1.
68. J. L. Smialek, *Metall. Trans. A* **9A**, 309 (1978).
69. R. Darolia, *J. Met.* **43**(3), 44 (1991).
70. R. D. Noebe, R. R. Bowman, and M. V. Nathal, accepted for publication in *Int. Met. Rev.* (1993).
71. J. A. Nesbitt, E. J. Vinarcik, C. A. Barrett and J. Doychak, *Mater. Sci. Eng.* **A153**, 561-66 (1992).
72. C. A. Barrett, *Oxid. Met.* **30**, 361 (1988).
73. A. Ball and R. E. Smallman, *Acta Metall.* **14**, 1517 (1966).
74. N. J. Zaluzec and H. L. Fraser, *Scr. Metall.* **8**, 1049 (1974).
75. I. Baker and E. M. Schulson, *Metall. Trans. A* **15A**, 1129 (1984).
76. E. P. George and C. T. Liu, *J. Mater. Res.* **5**, 754 (1990).
77. K. H. Hahn and K. Vedula, *Scr. Metall.* **23**, 7 (1989).
78. E. M. Grala, in Mechanical Properties of Intermetallic Compounds, edited by J. H. Westbrook (Wiley, New York, 1960), p. 368.
79. R. Darolia, D. F. Lahrman, and R. D. Field, *Scr. Metall.* **26**, 1007 (1992).
80. C. T. Liu, J. A. Horton, E. H. Lee, and E. P. George, "Alloying Effects on Mechanical and Metallurgical properties of NiAl," Martin Marietta Energy Systems, Inc., Oak Ridge Natl. Lab., ORNL Report, Oak Ridge, TN, 1993 (unpublished).
81. J. E. Hack, J. M. Brzeski, and R. Darolia, *Scr. Metall.* **27**, 1259 (1992).
82. R. E. Reed-Hill in Physical Metallurgy Principles, 2nd ed. (Van Nostrand, New York, 1973).
83. R. R. Bowman, R. D. Noebe, S. V. Raj, and I. E. Locci, *Metall. Trans. A* **23A**, 1493 (1992).
84. R. Jayaram and M. K. Miller, *Surf. Sci.* **266**, 310 (1992).
85. R. Darolia, General Electricm Aircraft Engines, 1992 (private communication).
86. J. D. Rigney and J. J. Lewandoski, *Mater. Sci Eng. A***149**, 143-51 (1992).
87. R. D. Noebe, C. L. Cullers, R. R. Bowman, *J. Mater. Res.* **7**, 605 (1992).
88. D. F. Lahrman, R. D. Field, and R. Darolia, p. 603 in ref. 4 (1991).
89. H. Lipsitt, p. 351 in ref. 1 (1985).
90. K. S. Kumar and S. A. Brown, *Philos. Mag. A* **65**, 91 (1992).
91. E. P. George, J. A. Horton, W. D. Porter, and J. H. Schneibel, *J. Mater. Res.* **5**, 1639 (1990).
92. S. L. Draper, P. K. Brindley, and M. V. Nathal, *Metall. Trans. 23*A 2541 (1992).
93. D. B. Miracle, D. L. Anton, and J. A. Graves, in Intermetallic Matrix Composite II, Vol. 273 (MRS, Pittsburgh, PA, 1992).
94. G. M. Stocks and A. Gonis, "Alloy Phase Stability and Design," NATO ASI Series, Vol. 163 (Kluwer Academic Publishers, Boston, MA, 1992).
95. C. L. Fu and M. H. Yoo, *Philos. Mag. Lett.* 62, 159 (1990).
96. M. H. Yoo and C. L. Fu, *ISIJ Int.* **31**, 1049 (1991).
97. M. H. Yoo, C. L. Fu, and J. K. Lee, p. 545 in ref. 4 (1991).

98. C. L. Fu and M. H. Yoo, p. 155 in ref. 8 (1992).
99. M. H. Yoo and C. L. Fu, Martin Marietta Energy Systems, Inc., Oak Ridge Natl. Lab., Oak Ridge, TN, 1992 (private communication).
100. H. Inui, M. H. Oh, A. Nakamura, and M. Yamaguchi to be published in *Acta Metall.* (1993).
101. M. Yamaguchi and H. Inui, p. 217 in ref. 8 (1992).
102. Y-W. Kim, p. 777 in ref. 4 (1991), and *Acta Metall.* **40**, 1121 (1992).
103. T. Fujiwara, A. Nakamura, M. Hosomi, S. R. Nishitani, Y. Shirai, and M. Yamaguchi, *Philos. Mag. A* **61**, 591 (1990).
104. Y-W. Kim, Universal Energy Systems, Inc., 1992 (private communication).
105. K. S. Chan and Y-W. Kim, *Trans. Metall. A* **23A**, 1663 (1992).
106. V. K. Sikka, J. T. Mavity, and K. Anderson, *Mater. Sci. Eng.* **A-153**, 712 (1992).
107. V. K. Sikka, Oak Ridge Natl. Lab., Oak Ridge, TN, 1992 (private communication).
108. R. Darolia, D. F. Lahrman, R. D. Field, J. R. Dobbs, K. M. Chang, E. H. Goldman, and D. G. Konitzer, p. 679 in ref. 8 (1992).
109. M. Yamaguchi, Kyoto University, Kyoto, Japan, 1992 (private communication).
110. P. Meschter and D. S. Schwartz, *J. Met.* **41**, 52-55 (1989).
111. S. Maloy, A. H. Heuer, J. J. Lewandowski, and J. Petrovic, *J. Am. Ceram. Soc.* **74**, 2704 (1991).
112. T. Fujiwara, K. Yasuda, and H. Kodama, pp. 633-37 in ref. 7 (1991).
113. D. L. Anton and D. M. Shah, pp. 361-71 in ref. 3 (1989).
114. M. Takeyama and C. T. Liu, *Mater. Sci. Eng.* **A132**, 61 (1991).
115. J. D. Livingston, *Phys. Status. Solidi A* **131**, 415 (1992).
116. Kanthal Super Handbook, Kanthal Furnace Products, 1986.
117. W. J. Buehler and F. I. Wang, *Ocean Eng.* **1**, 105-20 (1986).
118. I. M. Schetky, *Sci. Am.* **241**, 74-82 (1979).
119. C. T. Liu, H. Kunsmann, K. Otsuka, and M. Wuttig, ed., "Shape-Memory Materials and Phenomena–Fundamental Aspects and Applications," in Proceedings of Materials Research Society Symposium (MRS, Pittsburgh, PA, 1992).
120. E. P. George, C. T. Liu, C. J. Sparks, Ming-Yuan Kao, J. A. Horton, H. Kunsmann, and T. King, pp. 121-28 in ref. 119 (1992).
121. A. Hutton, *J. Met.* **44(3)**, 11 (1992).
122. M. Sagawa, S. Hirosawa, H. Yamamoto, S. Fujimura, and Y. Matsura, *Jpn. J. Appl. Phys.* **26**, 785 (1987).
123. K. S. Kumar, "Silicides Technology and Applications," in Intermetallic Compounds: Principles and Practice, edited by J. H. Westbrook and R. L. Fleischer (John Wiley and Sons , New York, 1993).

BULK AND DEFECT PROPERTIES OF ORDERED INTERMETALLICS: A FIRST-PRINCIPLES TOTAL-ENERGY INVESTIGATION

C. L. FU, Y.-Y. YE,* AND M. H. YOO
Metals and Ceramics Division, Oak Ridge National Laboratory, P.O. Box 2008, Oak Ridge, TN 37831

ABSTRACT

First-principles quantum mechanical calculations based on local-density-functional theory have been used to investigate the fundamental factors that govern the deformation and fracture behavior of ordered intermetallic alloys. Unlike in Ni_3Al, the calculated elastic constants and shear fault energies indicate that anomalous yield strength behavior is not likely to occur in Ni_3Si. From the calculated Griffith strength and a phenomenological theory relating fracture toughness to ideal cleavage strength, Ni_3Si is predicted to be ductile with respect to cleavage fracture. For TiAl, we find the absence of structural vacancies due to the strong Ti-Al bonding and similar atomic radii for Ti and Al. For NiAl, the defect structure is found to be dominated by two types of defects - monovacancies on the Ni sites and substitutional antisite defects on the Al sites. For FeAl, on the other hand, we find a more complex defect structure, which is closely related to the importance of electronic structure effect in FeAl. More importantly, we predict the strong tendency for vacancy clustering in FeAl due to the large binding energy found for divacancies. Effects of thermomechanical history on microhardness are discussed in terms of the calculated results.

1. INTRODUCTION

During the past decade the theoretical tools for studying the properties of materials have been developed to the point that it is possible to understand many of their properties from first principles. This understanding has been used both in interpretive and predictive modes. It is now well established that the bonding behavior in intermetallics has various characteristics, which include metallic bonding, directional bonding, strongly hybridized states from p- and d-electron, and charge transfer effect. These electronic structure factors can manifest themselves in the atomic-level bonding interactions, which, in turn, determine the physical and mechanical properties of alloys as well as their response to intrinsic and extrinsic defects. Thus, there is a growing recognized need, in the field of intermetallics, for the type of atomistic details that first principles calculations can provide, and which have proved so useful in aiding the understanding of the interactions that govern alloy behavior.

The framework of modern first-principle calculations of the ground state properties of solids is based on the local density functional (LDF) theory [1]. With this theory the complicated many-body problem of interacting nuclei and electrons is mapped onto a set of decoupled single particle-like differential equations, i.e. effective one-particle Schrödinger equations that describe the interaction of a single electron with the assembly of nuclei and electrons through an 'effective' one-particle potential. This effective potential is given as the sum of the electrostatic Coulomb potential, related to the electron charge density by Poisson's equation, and the local exchange-correlation potential as obtained from many-body theory of homogeneous and interacting electron gas.

The advent of accurate LDF calculations for complex systems like surfaces, interfaces, and superlattice structures means that the comparison between experiment and theory is no longer limited to empirical parameter-dependent models. In addition, the first-principles calculations offer information on electronic structure from which a clearer understanding of experimental results can be made. We solve the LDF equations either by a full-potential linearized augmented plane wave (FLAPW) method [2] or by a mixed-basis pseudopotential method [3,4]. Both methods represent a major advance in applying LDF theory to crystalline solids in that the LDF equations are solved without any shape approximation to the potential or charge density and a high degree of variational freedom (and precision) can be obtained. The uniqueness of these features makes it possible to determine the energetics associated with shape deformation and lattice defects (point and extended defects) accurately. Since much more extensive discussions can be found in the literature, we shall only briefly outline the techniques we used.

In the FLAPW method the 'natural' form of the variational basis is adopted to describe electronic wave functions, i.e., plane waves in the interstitial region, a product of radial functions and spherical harmonics inside the spheres (around the nuclei), and in the vacuum region (for surface calculation) a product of numerical functions which depend only on the coordinates normal to the surface and two-dimensional plane waves. In the mixed-basis pseudopotential method, an energy independent basis set containing both plane waves and a Bloch sum of localized numerical orbitals is used to represent the electronic wave functions.

Three-dimensional translational symmetry is imposed for the bulk electronic structure calculation. The surface is modelled by a single (free standing) slab consisting of a few atomic layers (5-15) with two-dimensional translational symmetry parallel to the surface and vacuum below and above the slab. For the calculation of shear stacking fault [antiphase boundary (APB), superlattice intrinsic and extrinsic stacking faults (SISF and SESF), and twin boundary] energies, we use a supercell approach by introducing periodic stacking faults. As long as these periodic stacking faults do not interact with each other, we then have the good representation of the energetics of a planar fault.

The calculated elastic constants, various shear fault energies, defect self-energies, and cleavage energies were used in conjunction with continuum modelling to predict the mechanical behavior and to understand the underlying electronic mechanisms of observed properties.

2. $L1_2$-TYPE Ni-BASE INTERMETALLICS

a) Yield Strength Anomaly

One of the most unique aspects of mechanical behavior in intermetallics is the yield strength anomaly. The best studied case is Ni_3Al, in which the strength anomaly can be explained by the cross-slip-pinning (CSP) mechanism. The two driving forces for a $1/2[10\bar{1}]$ superpartial to cross slip from the primary (111) slip plane to (010) are the anisotropy of antiphase boundary (APB) energies [5] and the interaction torque between the two superpartials [6] due to the elastic shear anisotropy, $A=2C_{44}/(C_{11}-C_{12})$. According to the CSP mechanism, the driving force per unit length of a superpartial to cross slip is

$$F = \frac{\gamma_1}{\sqrt{3}}\left(\frac{3A}{A+2}\right) - \gamma_0 + \tau_0 b \tag{1}$$

where τ_0 is the resolved shear stress on the (010) cross-slip plane. γ_1 and γ_0 are the (111) and (010) APB energies, respectively. The energy criterion is then defined, for the case of [001] orientation of uniaxial loading where τ_0=0, as

$$\left(\frac{3A}{A+2}\right)\frac{\gamma_1}{\gamma_0} > \sqrt{3} \tag{2}$$

This criterion is satisfied for Ni_3Al, with the calculated results of γ_1 / γ_0 =1.25 and A=2.9 for Ni_3Al (c.f., Tables I and II) [7]. The calculated APB energies (140 mJ/m^2 and 175 mJ/m^2 for the (010) and (111) planes, respectively) and CSF energy (225 mJ/m^2) are very consistent with the experimental results [8].

The yield strength anomaly is observed also in Ni_3Si and its alloys [9-11]. One outstanding feature associated with this anomaly in $Ni_3(Si,Ti)$ is that the yield strength starts increasing from low temperatures almost linearly with temperature [11]. Furthermore, the peak temperature (where the maximum of τ_y or σ_y occurs) can be raised by Ti addition. These observations make Ni_3Si an ideal alloy for a critical examination of the physical mechanism for the yield strength anomaly, in particular, the driving forces associated with the CSP mechanism (e.g., elastic shear anisotropy and APB anisotropy).

In a crystal of cubic symmetry, there exist three independent elastic constants (C_{11}, C_{12}, and C_{44}). We determine these three elastic constants from the bulk modulus, and the distortion energies associated with orthorhombic and tetragonal distortions. The calculated results obtained by use of the FLAPW method are given in Table I as denoted by (a). The data denoted by (b) in Table I were obtained using the mixed-basis pseudopotential method. An alternative approach to determine the $C'=(C_{11}-C_{12})/2$ value is to calculate the distortion energy associated with the [110] shear on the $(1\bar{1}0)$ plane.

A comparison of these two sets of calculated elastic constants (c.f. Table I) shows a striking consistency between the two independent first-principles methods and confirms the numerical accuracy of our calculations. The large elastic moduli (E=3.43 and G=1.34 in units of Mbar) are directly related to the directional (and covalent) bond formation between Ni and Si. Both methods yield a similar elastic anisotropy factor which has a value $A \cong 2$.

Table I. Elastic Constants of Ni-Base Intermetallics of the $L1_2$ Type

Alloys		$(10^{11}N/m^2)$			
		C_{11}	C_{12}	C_{44}	A
Ni_3Al		2.35	1.45	1.32	2.93
Ni_3Si	(a)	3.76	2.00	1.67	1.91
	(b)	3.63	2.05	1.72	2.18
Ni_3Ti		2.90	1.94	1.28	2.67
$Ni_{78}Si_{11}Ti_{11}$*		2.55	1.64	1.29	2.83

(a) FLAPW (b) mixed-basis pseudopotential method
*Reference 12

No elastic constants have been reported for binary Ni_3Si. Recently, the elastic constants of $Ni_{78}Si_{13}Ti_9$ have been measured in the temperature range of 78-338 K [12]. The temperature dependence of the elastic constants of $Ni_3(Si,Ti)$ was found to be generally small. Apparently, the addition of 9 at. % of Ti at Si-sites in Ni-rich compound lowered the magnitudes of all three elastic constants from those of Ni_3Si, and raised the A value by about 40%. It is conceivable that the addition of Ti considerably reduces the directional bonding (and increases the metallic character) in Ni_3Si.

It is not possible at present to calculate the elastic constants of $Ni_3(Si,Ti)$ due to the local distortions associated with the solid solution effect of Ti in Ni_3Si. However, valuable information can be obtained from the study of elastic behavior of Ni_3Ti with the $L1_2$ (hypothetical) structure. The calculation has been carried out using the mixed-basis pseudopotential method. The calculated results are also listed in Table 1. The results show that the replacement of Si by Ti in Ni_3Si considerably reduces the elastic moduli and increases the A value—a trend consistent with experimental measurement.

For the calculation of planar shear fault energies, we use a supercell approach by introducing periodic stacking faults. The APBs are produced by introducing a $1/2[\bar{1}01]$ displacement on the (010) and (111) planes. In this investigation we use twelve atomic layers for the (010) type and six atomic layers for the (111) type in separating adjacent APBs. A SISF has a stacking sequence of ABCBCABC (the matrix has an ABC stacking sequence) on the close-packed (111) plane. Again, two independent first-principle total-energy methods (i.e., FLAPW and mixed basis pseudopotential) are used to calculate these fault energies. The atomic relaxations near the interfaces are taken into account. The results of our calculations are shown in Table II.

Table II. Calculated Shear Fault Energies of Ni-Base Intermetallics of the $L1_2$ Type

Alloy	APB (100)	APB (111)	SISF	CSF
Ni_3Al	140	175	40	225
Ni_3Si	707	625	460	710
Ni_3Ti ($L1_2$)	-160 (unstable)	550		
Ni_3(Si,Ti)	360			

For Ni_3Si, in general, we find rather high shear fault energies. Very consistent results for the fault energies are also obtained from our two independent theoretical approaches (c.f. Table II). The high APB energy on the (010) plane is surprising (with a high value of $\simeq$ 700 mJ/m^2), considering that the introduction of an (010) APB does not alter the nearest-neighbor coordination of atoms at the APB interface. Again, this suggests the strong directional Ni-Si bonding. For the (010) APB, we do not find any appreciable atomic relaxations from the bulk truncated atomic positions near the interface. For the (111) APB, the atomic relaxations are very limited to four atomic layers adjacent to the APB interface. The effect of lattice relaxations lowers the (111) APB energy by about 60 mJ/m^2. Again, this result is surprising in that it puts the (111) APB energy (= 620 mJ/m^2) lower than that of the (010) APB, even though the (111) plane is the close-packed plane and the introduction of a (111) APB does change the nearest-neighbor coordination of interface atoms. The results of our calculation cast serious doubt on the cross-slip pinning (CSP) mechanism in explaining strength anomaly simply on the basis of the anisotropy of APB energies. This strong Ni-Si bonding behavior is also manifested in the high SISF energy (c.f. Table II).

The equilibrium width of the (111) APB ribbon between a pair of screw superpartials is obtained by $d_1 = K_s b^2 / (2\pi\gamma_1)$, where $K_s = \sqrt{C_{44}C'}$, $b = a_o / \sqrt{2}$, and a_o is the lattice parameter. Using γ_1 = 620 mJ/m^2 for Ni_3Si and the calculated elastic constants, one finds $d_1 \cong$ 2 nm which would be too small to be resolved by TEM. The range of d_1 = 3.2-4.4 nm in $Ni_{78}Si_{11}Ti_{11}$ determined using the TEM weak beam analysis [11] suggests that the APB energy can be drastically reduced by the titanium addition to the range of γ_1 = 154-211 mJ/m^2.

For Ni_3Ti with the $L1_2$ (hypothetical) structure, we obtain an unstable (010) APB ($\simeq$−160 mJ/m^2) and a high (111) APB energy ($\simeq$ 550 mJ/m^2). These suggest that phase transformation occurs if most of the Si atoms are replaced by Ti in Ni_3Si. We have also determined the (010) APB energy of Ni_3(Si,Ti) (with every other Si-site replaced by Ti) and a value of 360 mJ/m^2 is obtained. On the other hand, there is no appreciable change in the (111) APB energy between Ni_3Si and Ni_3Ti (with the hypothetical $L1_2$ structure). Thus, our calculations indicate that the increased driving force for CSP in Ni_3(Si,Ti) stems from both increasing shear anisotropy factor and decreasing (010) APB energy. However, the existing models based on CSP mechanism appear to be unsatisfactory to explain the yield strength anomaly in the binary Ni_3Si.

It is worth noting that the elastic moduli of Ni_3Si are about 50% larger than those of Ni_3Al. Furthermore, there is a four-fold increase in the shear fault energies in going from Ni_3Al to Ni_3Si. From an electronic structure point of view, the difference of these two systems comes from an extra p-electron for Ni_3Si. One can understand the effect of this extra p-electron from a charge density difference contour plot between Ni_3Si and Ni_3Al on the (001) plane (c.f., Fig. 1). Clearly, the charge density difference plot shows the polarization of this extra p-electron directly along the Ni-Si direction (e.g., solid contour lines). This type of charge polarization gives rise to the directional/covalent Ni-Si bonding, which results in high elastic moduli and shear fault energies. We can also investigate the effect of this directional bonding on the ductility of Ni_3Si (in a binary, single crystal form) by considering Pugh's criterion and fracture strength.

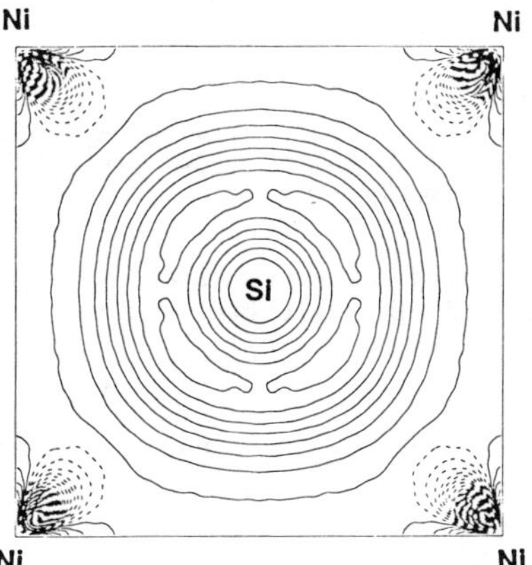

Fig. 1. Charge difference contour plot between Ni_3Si and Ni_3Al on the (001) plane which shows the polarization of extra p-electron directly along the Ni-Si direction (i.e., directional/covalent bonding).

b) Cleavage Energy and Strength

In Table III we list the ideal cleavage energies (G_c), Griffith strengths (k_{IG}), and Pugh's criterion for Ni_3Al and Ni_3Si. The ideal cleavage energy is defined as the total surface energy of the two cleaved surface planes. The Griffith strength (for a mode I crack in an elastic-perfectly brittle solid) is determined from the calculated G_c and elastic compliance constants. As an indicator of intrinsic ductility, we first consider Pugh's criterion (i.e., the quotient of bulk modulus to shear modulus). This simple criterion based on elastic moduli gives both Ni_3Al and Ni_3Si the same plasticity and ductility. Next, we consider the cleavage fracture. We find that there is an increase in the ideal cleavage energy in going from Ni_3Al to Ni_3Si (the increase for the (100) cleavage energy is more noticeable, since more Ni-Si bonds are broken on this plane). The same degree of increase (about 20% increase) is also reflected in the calculated Griffith strength (c.f., Table III). Therefore, from an ideal cleavage fracture point of view, Ni_3Si is predicted to be ductile with respect to cleavage fracture. In this respect, it is interesting to ask the question if a large k_{IG} value can lead to a high fracture toughness (K_{IC})? Unfortunately, there is no formal theory in fracture mechanics which can correctly take into account the large amount of plastic work on an atomistic level. Furthermore, it is very difficult to model the strong coupling of k_{IG} (or intrinsic bond strength) with the dislocation field near the crack tip. Nevertheless, one can get much insight into the correlation between k_{IG} and K_{IC} by considering a phenomenological theory developed by Gerberich et al. [13]. In their analysis, the experimentally measured fracture toughness values are plotted as a function of the calculated Griffith strengths by Yoo and Fu [14] for Ni-, Fe-, Ti-aluminides, and trialuminides. The result of their analysis suggests that the relationship of K_{IC} and k_{IG} can be well fitted by a relationship

$$K_{IC} = \frac{1}{\beta}\exp\left\{\frac{k_{IG}^{2}}{\alpha\sigma_{y}}\right\} \tag{3}$$

This equation has been used widely in the analysis of fracture toughness and hydrogen embrittlement of Fe-base alloys. In this equation α and β are assumed to be material independent constants, and σ_y is the yield strength which typically has a value of 200-300 MPa for aluminides. The important implication of Eq. (3) is that the macroscopic thresholds depend on the ideal cleavage energies through an exponentially dependent function. This finding is not entirely unexpected, since the stronger the 'atomic glue' at a crack tip, the more the plastic work is allowed around it. The most significant finding is that a 2-fold increase in the cleavage strength can lead to a 10^2 increase in the fracture toughness. Thus, the goal of alloy design is to find those alloys with high values of k_{IG} and ways of trapping hydrogen so as to minimize the decrease in k_{IG}. Based on our calculations, we suggest that Ni_3Si is a promising candidate for structural applications due to its high k_{IG} value.

Table III. The calculated ideal cleavage energies (G_c) and Griffith strength (k_{IG}) for Ni_3Al and Ni_3Si. Also listed is the ratio of bulk modulus to shear modulus (i.e., Pugh's criterion).

	B/G	(hkl)	G_c (J/m^2)	[uvw]	k_{IG} (MPa-m$^{1/2}$)
Ni3Al	2.0	111	4.6	$1\bar{1}0$	1.06
				$11\bar{2}$	1.09
		100	5.8	001	1.11
				011	1.17
Ni_3Si	2.0	111	5.1	$1\bar{1}0$	1.37
				$11\bar{2}$	1.38
		100	7.2	001	1.56
				011	1.59

3. TiAl OF THE $L1_0$ STRUCTURE

a) Bonding Charge Density

Recently, there has been a growing interest in measuring the bonding charge density by electron and x-ray diffraction techniques [15]. The bonding charge density is the response of electron distribution, referred to the atomic charge density, in the presence of crystal field. In Fig. 2 we present the bonding charge density of TiAl on the (001) and (100) planes. The solid (dashed) contour lines in Fig. 2 denote contours of increased (decreased) electronic density as atoms are brought together to form a crystal. Fig. 2(a) clearly shows the formation of directional d-bond between nearest-neighbor Ti atoms, as expected from the open d-shell of Ti atoms. The most remarkable feature in the bonding charge density, however, is the polarization of p-electrons at the Al sites pointing directly along the [001] direction. It is noticeable that the distribution of bonding charge on the Al site is highly non-spherical, i.e., depleting of charge on the (001) pure-Al plane accompanied with a build-up of charge in the [001] direction which can result in an enhancement of the Ti-Al bonding. The bonding charge density gives the indication for bond formation and the tendency for bond enhancement. However, it is important to be able to calculate the bond strength associated with the bonding charge density in order to gain a more in-depth understanding of the interplay between bonding and strength. We can further investigate the bonding behavior in TiAl from both the calculated elastic constants and the energetics associated with shear stacking faults.

The calculated elastic constants of TiAl are listed in Table IV [16]. A large C_{44} value in TiAl (which corresponds to the (001)[100] shear) is indicative of the strong cohesion between Ti and Al layers (as shown in the bonding charge density plot). It is this large Ti-Al bond bending force which hinders the mobility of ordinary- and super-dislocations. The calculated shear and Young's moduli (by Hill's average; c.f., Table IV) are in excellent agreement with experiments [17,18]. However, up to now, there has been no experimental measurement on these six independent elastic constants.

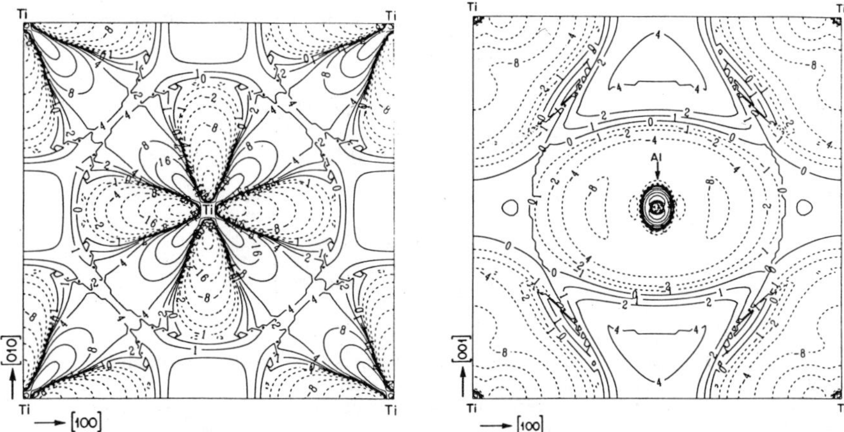

Fig. 2. The bonding charge density of TiAl on the (001) and (100) planes in units of 10^{-3} e/(a.u.)^3. The charge density difference plot (referred to overlapping atomic densities) shows the polarization of p-electrons on the Al sites.

Table IV. Calculated elastic constants, bulk and shear moduli (in units of 10^{11} N m^{-2}) of TiAl

C_{11}	C_{12}	C_{13}	C_{33}	C_{44}	C_{66}	B	G
1.90	1.05	0.90	1.85	1.20	0.50	1.25	0.70

We find high APB energies on the (100) and (111) planes (430 and 510 mJm^{-2}, respectively), and relatively low twinning, SISF and SESF energies on the (111) planes (60, 80, and 90 mJm^{-2}, respectively) [16]. The high APB energies are attributed to the large bond bending force as a result of the p-electron polarization on the Al sites. The introduction of an APB changes not only the character of directional bonding, but also strongly affects the layer (or superlattice) structure of TiAl. By contrast, the boundary interface created by twinning does not change the Ti-Al or Ti-Ti nearest-neighbor bonding (the twinning energy comes from the change of bond angles between second-nearest neighbors) and a lower boundary energy is expected. Locally, SISF and SESF can be viewed as 'microtwins'; therefore the values of the corresponding fault energies are very similar to that of twinning.

In TiAl, which has relatively low SISF and twin boundary energies as compared to its APB energies, the twin-slip conjugate relationship makes an important contribution to the strain compatibility for blunting of a Mode-I type crack. Recently, clear experimental evidence has been reported that deformation twinning in TiAl can lead to an appreciable increase in the fracture resistance [19].

b) Point Defects

The behavior of point defects is of great relevance for the study of solid solution hardening effect and the understanding of ordering behavior in intermetallics. In our approach to calculate point defect concentration in TiAl [20], we assume dilute point defect concentration (i.e., non-interacting defect) and consider only the configuration entropy for the entropy term. The defect concentrations are then determined from the minimization of grand potential, which leads to (for a binary AB compound)

$$n_A^v = \frac{1}{2}\left[\frac{e^{-(E_v^A+\mu_A)/k_BT}}{1+e^{-(E_v^A+\mu_A)/k_BT}}\right], \tag{4}$$

and

$$n_{AB} = \frac{1}{2}\left[\frac{e^{-(E_{AB}+(\mu_A-\mu_B))/k_BT}}{1+e^{-[E_{AB}+(\mu_A-\mu_B)]/k_BT}}\right] \tag{5}$$

for the defect concentrations of vacancies and antisite defects at site A in thermal equilibrium, respectively. In Eqs. (4) and (5), μ is the chemical potential, and $(E_v^A+\mu_A)$ and $(E_{AB}+(\mu_A-\mu_B))$ are the effective vacancy and antisite formation energies, respectively. The internal energies of defects (i.e., E_v^A and E_{AB}) are determined from first-principles calculation.

For stoichiometric TiAl the calculated defect concentration as a function of temperature is shown in an Arrhenius plot (c.f., Fig. 3). No structural vacancies are predicted from our calculation, which is consistent with the recent experimental result of positron annihilation by Shirai and Yamaguchi [21]. The vacancy formation energies at stoichiometry are obtained to be 1.95 eV and 2.46 eV on the Ti and Al sublattices, respectively. The high vacancy formation energy on the Ti sublattices, for example, gives a vacancy concentration of 10^{-6} - 10^{-7} at 1500K. Thus, the point defects in TiAl at stoichiometry are of Schottky defect type with antisite defects on both sublattices. The antisite defect concentration is predicted to be larger than 10^{-4} above 1000K (c.f., Fig. 3). The corresponding formation energy for antisite defect is 0.72 eV.

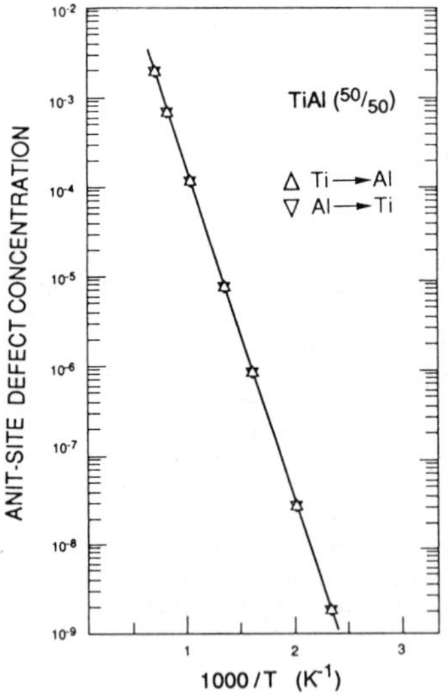

Fig. 3. Arrhenius plot for point defect concentration of TiAl at stoichiometry.

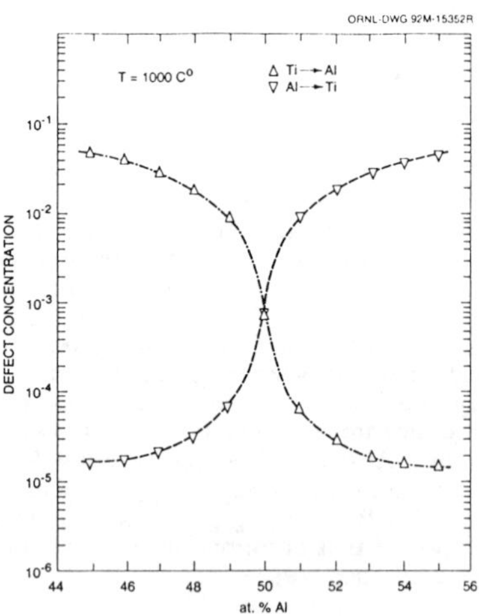

Fig. 4. The dependence of point defect structure of TiAl at 1300K on atomic percentage of Al.

As in other close packed intermetallic alloys (e.g., Ni_3Al), the dominant point defect types in off-stoichiometric TiAl are of antisite defects on both sublattices. Fig. 4 shows the dependence of the concentration of various point defect types on the atomic percentage of Al at high temperature (1000°C). In addition to dominant substitutional antisite defects, at this temperature, the vacancy concentration (mainly on the Ti sublattices) is about 10^{-7} for Al-rich TiAl and 10^{-9} for Ti-rich TiAl.

The effect of lattice distortion (i.e., local tetragonal distortion) due to point defects is found to be generally small in TiAl. For Ti-rich TiAl, there is a tendency to decrease the interatomic distance for substitutional defects with the first neighboring shell of Ti atoms which corresponds to a decrease of c/a ratio from 1.01 to 1.00. On the other hand, for Al-rich TiAl, the lattice relaxation of the neighboring Al shell is in the direction to increase the c/a ratio to 1.02. The direction of atomic relaxation occurs mainly in the [001] direction. It should be mentioned, however, that a decrease of c/a ratio for Ti-rich TiAl does not warrant a increase in ductility, since, chemically, the structure is still not cubic. The absence of structural vacancies in TiAl is due to two facts: (1) Ti and Al have similar atomic radii; and (2) the presence of strong TiAl bonding favors a close packed structure.

4. B2-TYPE ALUMINIDES: FeAl AND NiAl

FeAl and NiAl are representatives of two classes of B2-type aluminides which have contrasting mechanical behavior [22,23]. For example, the primary deformation mode of NiAl is <010> slip, whereas FeAl shows normal <111> slip behavior; the cleavage habit plane is reported to be (110) for NiAl, but a (100) plane is observed for FeAl. While the brittle fracture of FeAl is shown to be related to the environmental embrittlement [24,25], the brittleness of NiAl is often thought to be associated with intrinsic effect (such as a low cleavage energy and the lack of slip systems). Furthermore, the mechanical properties of FeAl (e.g., microhardness) are very sensitive to thermomechanical treatment, whereas NiAl appears to be not as sensitive as the case of FeAl [26].

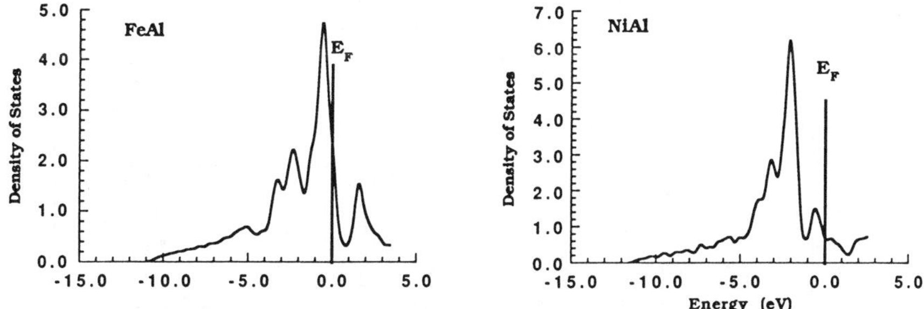

Fig. 5. Density-of-states of FeAl and NiAl. E_F is the position of Fermi level.

FeAl and NiAl are also dissimilar in the electronic structure. The density-of-states of B2 aluminides (c.f., Fig. 5) is distinguished by a 'pseudo gap' separating the bonding (between transition metals and aluminum) and non-bonding states. In FeAl, the electronic state has unfilled d-shell and the Fermi level lies in the Fe-Al bonding state region. Thus, the bonding strength (and mechanical behavior) is very sensitive to the intrinsic and extrinsic defects. By contrast, NiAl has a more closed d-shell (and a narrower band width) and the Ni-Al bonding region is well below the Fermi level. It is worth noting that the interatomic interaction in NiAl is very much dominated by the strong short-range nearest-neighbor interaction. It is this insufficient metallic character which makes NiAl brittle.

a) Point Defects in NiAl

To understand the ordering behavior and microhardness in NiAl and FeAl, we have to consider the point defect structure [27]. Among four kinds of point defects (i.e., substitutional antisites and structural vacancies on both sublattices), we only find the presence of two kinds of defects for NiAl at stoichiometry (c.f., Fig. 6), i.e., substitutional antisite defect on the Al sublattices and structural vacancy on the Ni sublattice. The vacancy concentration at high temperature (above 1000°C) is found to be 10^{-3} - 10^{-4}, which is in good agreement with experiments [28]. (Note that the theoretical prediction tends to be near the lower bound of experimental data, since thermal equilibrium is a built-in condition in the theoretical formalism). The effective vacancy formation energy at the Ni sites is 0.93 eV and for antisite defect at the Al sites is 0.97 eV. At stoichiometry, we find that the point defect configuration for NiAl is of triple-defect type (i.e., two vacancies accompanied with an antisite defect). Other kinds of defect configuration (e.g., Schottky defect) are found to be unstable due to higher defect formation energies (larger than 2 eV) associated with antisite defects at the Ni sites and vacancies at the Al sites.

In sharp contrast to the point defect structure in close packed lattices (e.g., TiAl), the point defect configuration for off-stoichiometric NiAl is drastically different for Ni-rich and Al-rich cases. In short, as shown in Fig. 7, we find: (1) substitutional antisite defects at the Al sublattices to favor the metallic bonding among Ni atoms for Ni-rich case; and (2) structural vacancies at the Ni sublattices so that nearest-neighbor Al-Al repulsion (due to size effect) can be avoided for Al-rich NiAl. At high temperatures (c.f., Fig. 7), there still exists a residual vacancy concentration at the Ni sites (about 10^{-5}) for Ni-rich NiAl; there is a two orders of magnitude difference between vacancy and antisite defect concentrations for Al-rich NiAl.

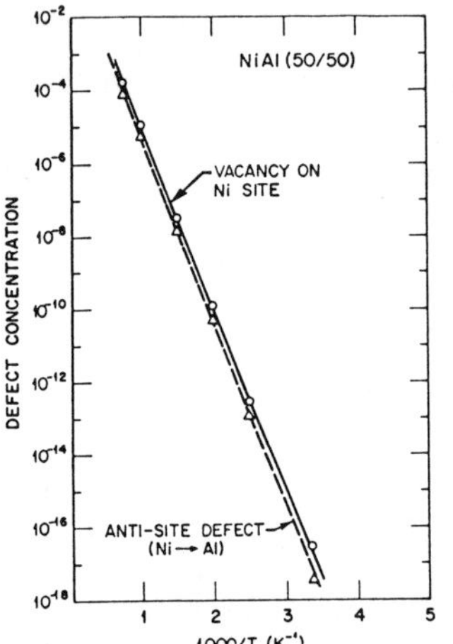

Fig. 6. Arrhenius plot for point defect concentration of NiAl at stoichiometry.

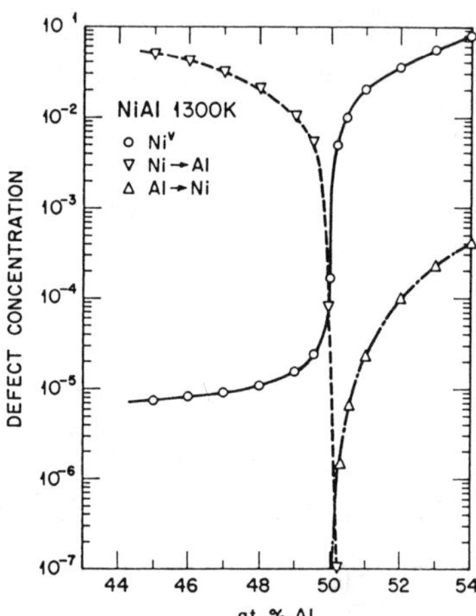

Fig. 7. The dependence of point defect configuration of NiAl on atomic percentage of Al at 1300K.

b) Point Defects in FeAl

Based on the non-interacting point defect model, similar to the case of NiAl, we find low vacancy formation energy at the Fe sites (0.97 eV) and low antisite defect formation energy at the Al sites (0.95 eV) for FeAl at stoichiometry. However, a basic difference between these two alloys is found to be the antisite formation energy at the transition metal sites. For FeAl, we find a relatively low antisite formation energy at the Fe sites with a value of 1.04 eV. Clearly, the electronic structure effect plays a far more important role in determining the point defect structure in the case of FeAl (other than the size effect). Indeed, as shown in Fig. 8 for the concentration of point defects at high temperatures, there is a competition between vacancies and antisite defects for Al-rich FeAl. On the other hand, for Fe-rich FeAl, the constitutional antisite defect at the Al sites is the dominant defect type. A vacancy concentration of 10^{-4} is obtained at stoichiometry for FeAl at 1300K. This value is too low by two orders of magnitude compared with the experimental data in the literature [26,28]. However, it should be noted that the calculation simulates a thermal equilibrium condition (which corresponds to sufficiently long anneal time) under the assumption that the defects are well separated (i.e., large vacancy clusters can be annealed out by forming fault dislocation loops) . Thus, in order to rationalize the difference between the present calculation and experiments, and to understand the strong dependence of hardness on heat treatment for FeAl, we have to examine the possibilities of defect clustering and complexes in FeAl and NiAl (which are not accounted for in the non-interacting defect model).

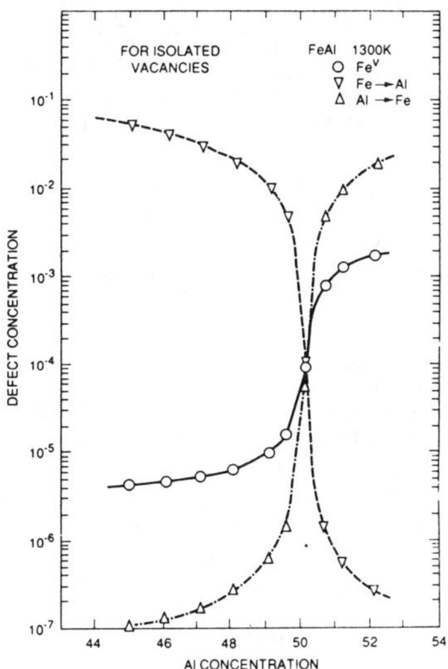

Fig. 8. The dependence of point defect configuration of FeAl on atomic percentage of Al at 1300K near stoichiometry (based on non-interacting defect model).

For FeAl, we predict a high divacancy binding energy with a value of 0.57 eV. (In this case, two vacancies are separated by the distance of a lattice constant). The implications of our result are: (1) there is a strong tendency for vacancy clustering; and (2) the vacancies can be annealed out into dislocations or grain boundaries [26]. Thus, it is not surprising that the microhardness of FeAl shows strong dependence on thermomechanical treatment [26]. The energetics of other types of defects (such as antisite defects) are found to be far less sensitive to the distance between defects than in the case of vacancies. The larger lattice distortion found for vacancies at the Fe (or Ni) sites also implies that the microhardness is larger for the case of the Al-rich side in B2-type aluminides. By contrast, for NiAl, we find that the interaction between vacancies tends to be weakly repulsive with a divacancy binding energy of -0.1 eV. Thus, a monovacancy configuration is the stable defect structure and its related mechanical properties are less sensitive to the thermal annealing time. Our calculation correctly explains (and predicts) the contrasting mechanical behavior between NiAl and FeAl.

ACKNOWLEDGEMENT

Research sponsored by the Division of Materials Science, Office of Basic Energy Sciences, U.S. Department of Energy under contract DE-AC05-84OR21400 with Martin Marietta Energy Systems, Inc.

REFERENCES

*Permanent address: Dept. of Physics, Wuhan University, Hubei, Wuhan, PRC; currently at Dept. of Materials Science and Engineering, University of Tennessee, Knoxville,

1. P. Hohenberg and W. Kohn, Phys. Rev. B 136, 864 (1964).
2. E. Wimmer, H, Krakauer, M. Weinert, and A. J. Freeman, Phys. Rev. B 24, 864 (1981).
3. S. G. Louie, K. M. Ho, and M. L. Cohen, Phys. Rev. B 19, 1774 (1979).
4. K. M. Ho, C. L. Fu, and B. N. Harmon, Phys. Rev. B 29, 1575 (1984).
5. P. A. Flinn, Trans, AIME 218, 145 (1960).
6. M. H. Yoo, Scripta Met. 20, 915 (1986).
7. The results presented here for Ni_3Al are better converged results as compared to those reported before (e.g., see C. L. Fu, International Symposium on Intermetallics Compounds - Structure and Mechanical Properties, edited by O. Izumi, (Sendai, The Japan Institute of Metals) ;. p. 387).
8. K. J. Hemker, B. Viguer, R. Schänblin, and M. J. Mills, in this proceedings.
9. R. Lowrie, Trans. AIME 194, 1093 (1952).
10. T. Suzuki, Y. Oya, and S. Ochiai, Metall. Trans. 15A, 173 (1984).
11. T. Takasugi and M. Yoshida, Phil. Mag. A 65, 613 (1992).
12. H. Yasuda, T. Takasugi, and M. Koiwa, Acta Met. 40, 381 (1992).
13. W. W. Gerberich, H. Huang, and P. G. Marsh, NASA Conf. on Advanced Earth-to-Orbit Propulsion Technology, Marshall Space Flight Center, Huntsville, Alabama, May 1992.
14. M. H. Yoo and C. L. Fu, Mater. Sci. and Eng. A153, 470 (1992).
15. For example, see A. G. Fox and M. A. Tabbernor, Acta Metall. 39, 669 (1991).
16. C. L. Fu and M. H. Yoo, Phil. Mag. Lett. 62, 159 (1990).
17. H. A. Lipsitt, D. Schechtman, and R. E. Schafrik, Metall. Trans. A6, 1991 (1975).
18. R. E. Schafrik, Metall. Trans A8, 1003 (1977).
19. H. Deve and A. G. Evans, Acta Metall. 39, 1171 (1991).
20. C. L. Fu and M. H. Yoo, Intermetallics, to be published.
21. Y. Shirai and M. Yamaguchi, Mater. Sci. and Eng. A152, 173 (1992).
22. C. L. Fu and M. H. Yoo, Acta Metall. 40, 703 (1992).
23. M. H. Yoo and C. L. Fu, Scripta Metall. 25, 2345 (1991).
24. C. T. Liu, E. H. Lee, and C. G. McKamey, Scripta Metall. 23, 875 (1989).
25. C. L. Fu and G. S. Painter, J. Mater. Res. 6, 719 (1991).
26. P. Nagpal and I. Baker, Metall. Trans. 21A, 2281 (1990).
27. C. L. Fu, Y. Y. Ye, and M. H. Yoo, to be published.
28. For example, see J. P. Neumann, Y. A. Chang, and C. M. Lee, Acta Metall. 24, 593 (1976); and references therein.

THE YIELD STRESS ANOMALY IN $L1_2$ ALLOYS

PETER B. HIRSCH
University of Oxford, Department of Materials, Parks Road, Oxford OX1 3PH, England.

ABSTRACT

The implications of the basic assumptions of the local pinning theories for the yield stress anomaly in $L1_2$ alloys are discussed. An alternative theory is presented in which the superpartials on (111) cross-slip on to (010) to form long locks lying partially on (111) and (010) or completely on (010) planes (Kear-Wilsdorf locks). The ends of the locks are joined by glissile superkinks. The yield stress is controlled by superkinks bypassing the screw dislocation locks, and the increase of the yield stress with increasing temperature is due to the decrease of the lengths of superkinks. The theory accounts satisfactorily for the mechanical properties including the small strain-rate dependence of the yield stress and is consistent with electron microscope observations.

INTRODUCTION

Several detailed models have been developed to explain the anomalous yield stress observed in some $L1_2$ ordered alloys in which the yield stress for slip on $(11\bar{1})$ increases with increasing temperature [1, 2, 3, 4, 5, 6, 7, 8]. In all these models, the $[10\bar{1}]$ dislocations on (111) are dissociated into two $1/2[10\bar{1}]$ superpartials bounding on APB and the strengthening results from the pinning of screw dislocations, which, owing to a lower APB energy on {001} than on {111} [9] and a torque between the superpartials due to elastic anisotropy [10] can reduce their energy by cross-slipping from the primary glide plane (111) to (010). The activation energy for cross-slip from (111) to (010) is assumed by all the models (except [1]) to be that derived by Paidar, Pope and Vitek (PPV) [2]. In this model short screw segments of the leading partial cross-slip on (010) by b/2, where b is the Burger's vector of the superpartial (Fig 1a). This process involves the constriction of the superpartial, which is initially dissociated on (111) into two Shockley partials bounding a complex stacking fault. A double jog of height b/2 is formed and the process is completed by redissociation on the $(1\bar{1}1)$ cross-slip plane (Fig. 1b). In all the theories the cross-slipped dislocations act as obstacles to further slip on (111); in broad terms the yield stress increases with increasing temperature because the number of cross-slip events, and therefore of obstacles, increases with increasing temperature. Furthermore, in all the theories the orientation dependence and tension/compression asymmetry of the yield stress are predicted correctly from the PPV activation energy.

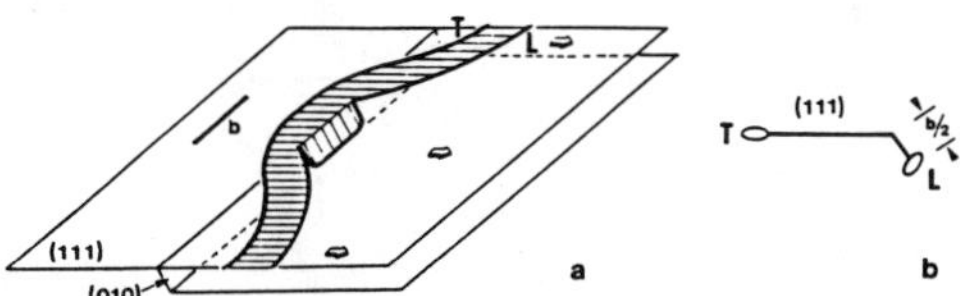

Fig. 1 Pinning configuration in the PPV model after cross-slip of b/2 on (010)

However, the models differ in respect of the evolution of the barriers after the first cross-slip step takes place, the nature of the unlocking process, and that of the the slip which is producing the strain. There are two types of model; in the first, following Takeuchi and Kuramoto [1], the screws are pinned locally, the screws between the pinning points advance until a critical angle in reached, whereupon breakaway occurs. In the steady state it is assumed that a new obstacle is formed wherever one is destroyed (Fig. 2). The yield stress τ at temperature T is given by [1,2].

$$\tau = A \exp - (H_p/3kT) \quad (1)$$

where H_p is the activation energy for locking, and A is a constant. In [1,2] the breakaway mechanism is athermal, and at the yield stress (steady state) the screws move with free flight velocity. This model does not explain the small strain-rate dependence of the yield stress [11,12]. Vitek and Sodani [3] therefore modified the model by assuming that unpinning occurs by a thermally activated process, in which the activation energy for unlocking at zero stress is an adjustable parameter and is taken to be ~3ev to fit the experimental data. An expression similar to (1) is obtained but without the factor 3 in the denominator. No mechanism is suggested for such a high activation energy, and the assumption that the cross-slipped segment does not expand significantly beyond the saddle point configuration is highly questionable [7] and is inconsistent with the TEM in situ observations of Molenat and Caillard [13] which show that long lengths of screw dislocations cross-slip "instantaneously" during deformation. The theory does not account for two different regimes in the dependence of activation volume on temperature, observed experimentally [11,14], which suggests two mechanisms of unpinning. A new version of PPV [4] has been published recently in which thermally activated break-away is again postulated, but breakaway now occurs by unpinning at one, two or three pinning points simultaneously depending on temperature, giving rise to three different regimes in the variation of activation volume with temperature. The activation energy for unpinning is again taken as ~3ev, but no mechanism is proposed. In all these four theories the strain is generated by moving screws. The yield stress increases with increasing temperature, because the distance l between pinning points (Fig. 2) decreases.

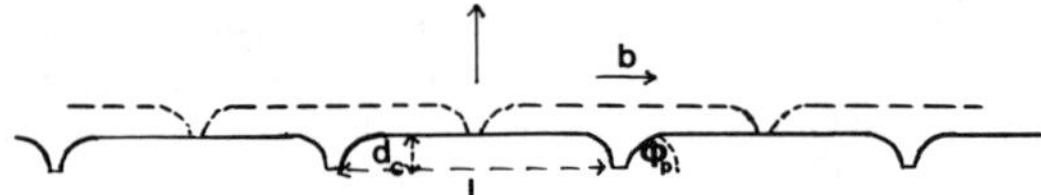

Fig. 2 Dynamic break-away model [1,2]

Mills et al [8] have used an elegant computer simulation method to study the evolution of the dislocation structure on the assumption that the pinning rate is controlled by H_p, local pinning occurs, and the pinned segment is unstable above a critical breaking angle $\Phi_o = 22°$, and stable for angles less than Φ_o. This critical angle is much larger than the breaking angles estimated in [6]. (See also section 2). Much correlated pinning is observed, and it appears that the strain is nucleated by a few large superkinks with edge character [15,16,17].

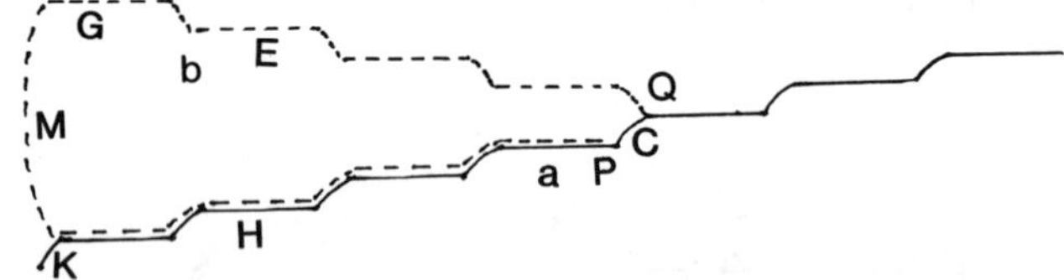

Fig. 3 a (full line) CHK original locked configuration; b (dashed lined) CEGKP after unpinning of C. Along CK dislocations annihilate or form dipoles and APB tubes. Note the "free" edge segment M.

In contrast to all these models, in the theory described in [5,6,7], the elementary cross-slipped segments extend rapidly laterally (beyond the saddle point) before the critical breaking angle is reached; further cross-slip along the same segment stabilises the long cross-slipped screws (see section 3), generating structures shown schematically in Fig. 3, where long cross-slipped screws, e.g., H are linked by glissile superkinks, e.g., C. There will be a statistical distribution of lengths of superkinks, and new slip will be nucleated when the stress is sufficient to move the longest kinks. Unpinning of C in Fig. 3a would generate the dashed structure, where dislocations in the screw orientation become locked again by cross-slip. Existing screws are effectively bypassed; they do not move again after being locked. In this model slip is essentially generated in bursts by unpinning superkinks and the generation of glissile edge segments. The athermal component of the yield stress is related to the height l of the critical superkink by [5,6,7,8,15,16,18]

$$\tau_g = p\ Gb/l \tag{2}$$

where p is a constant which depends on the mechanism of unpinning. The increase in yield stress with increasing temperature is due to the decrease in l. Thus while for the point pinning models the characteristic length controlling the yield stress is the length of glissile screws between pinning points, in the long screw-barrier models it is the length of the glissile edge segments (superkinks). The computer simulation model [8] falls into the second category, although the pinning is assumed to be localised; this arises from the strong correlation between pinning events leading essentially to long obstacles [8]. Which of these two types of model is a better approximation to reality? This question will be considered in section 2, and the detailed superkink model [5,6,7] will be discussed and compared with experiment in subsequent sections.

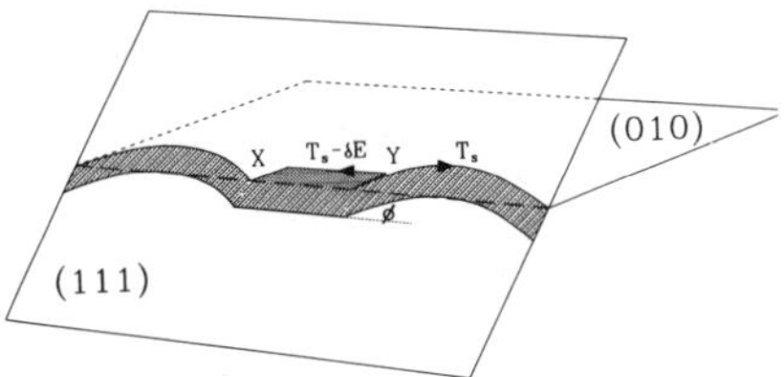

Fig. 4 Stability of cross-slipped segment

THE STABILITY OF THE CROSS-SLIPPED SEGMENT

Consider the stability of a cross-slipped segment XY in Fig. 4. We assume that the motion of the two superpartials is coupled; the effective forces acting on the jog at Y are (T_s - δE) to the left and $T_s\cos\Phi$ to the right, where T_s is the line tension of the whole dislocation and δE is the reduction in energy per unit length of the leading superpartial after cross-slip of b/2 on (010). The force per unit length on the leading partial for cross-slip on (010) is

$$F_{AC} = E\,\gamma_1 - \gamma_0 + \tau_c b \tag{3}$$

where E is an anisotropy factor due to Yoo [10], whose value for Ni_3Al is 1.075, γ_1, and γ_o are the APB energies on (111) and (010) respectively, τ_c is the component of the shear stress on (010) and b is the Burger's vector of the superpartial. The reduction in energy per unit length of the cross-slipped segment is $\delta E = F_{AC}\ b/2$. The jog Y is in mechanical equilibrium when [6]

$$T_s - \delta E = T_s\cos\Phi_o \tag{4}$$

or

$$\sin\Phi_o/2 = (\delta E/2T_s)^{1/2} \tag{5}$$

Using the Hirth and Lothe [19] expression for small bow-outs, for l/b ~ 100, $T_s \sim 1.35Gb^2$. As an example we use values of γ_1, γ_o for Ni_3Al quoted in [4], γ_1=120, γ_o=100mJm^{-2}. At 650°K, τ_p ~ 100MPa in the [1,2,20] orientation, $F_{AC} \sim 3.1 \times 10^{-2} Jm^{-2}$ and with G=65GPa, $\delta E/2T_s \sim 3.5\times10^{-4}$ and $\Phi_o \sim 2.14°$, about 1/10 of the value assumed in [8] in the computer simulations. Now for $\Phi < \Phi_o$ the cross-slipped segment will expand, while for $\Phi > \Phi_o$ it will contract. If we assume that the cross-slipped segment is formed when the dislocation is bowed out in a convex configuration (i.e. opposite to the curvature in Fig. 4), the two jogs will move apart over an angular range $-\Phi_o < \Phi < \Phi_o$, and when $\Phi > \Phi_o$ the double jog contracts. Fig. 5 a,b,c, shows how the structure develops; the length of the cross-slipped segment depends on relative velocity of the bowed out edge dislocation and jog A. The local pinning models implicitly assume that the jog velocity is slow, so that the critical value Φ_o is reached when the jogs have moved only just past the saddle point, whereas in the superkink model [5,6,7] the jogs move fast and long lengths of cross-slipped screws are produced before $\Phi = \Phi_o$. The latter is consistent with the in situ TEM observations which show that elemetary jogs are highly glissile [13], and leads naturally to the formation of Kear-Wilsdorf (KW) locks as observed in many TEM studies.

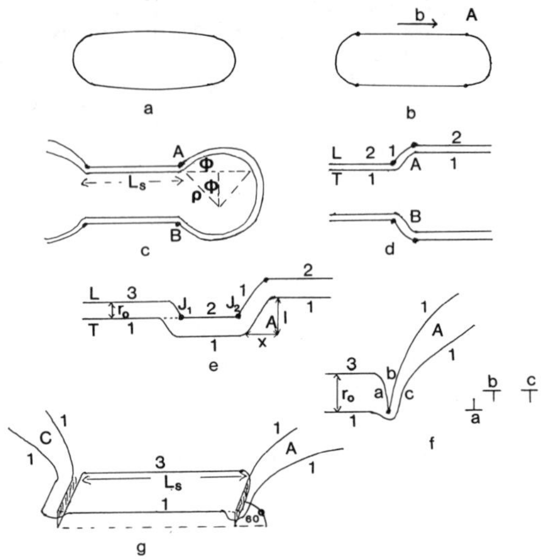

Fig. 5 abc: Formation of long cross-slipped screw segment;
d: bowed out dislocation has cross-slipped, forming superkink A;
ef: formation of dipole barrier by interaction of A with superkink terminating twice cross-slipped screw LT. Numbers refer to different levels of (111) planes.

If the motion of the jogs is controlled by an activation energy W_m (>kT) over a barrier, the nucleation rate J of double jogs is given by [19]

$$J = \frac{(\delta E - 2T_s \sin^2\Phi/2)}{kT} \nu_o \exp - (H_p/kT) \exp - (W_m/kT) \quad (6)$$

where H_p is the activation energy as derived in [2] or [4], ν_o is the Debye frequency. (If $W_m < kT$, expressions (6), (7) apply with the exponential equal to unity.) Fig. 6 illustrates the variation of the double jog energy ΔF with distance x between the jogs. The first point to note is that the experimentally determined activation energy includes W_m; secondly the jogs will still drift in the direction governed by $(\Phi - \Phi_o)$, but their velocity v_j is controlled by W_m, i.e. [19]

$$v_j = b^2\frac{(\delta E - 2T_s \sin^2\Phi/2)}{kT} \nu_o \exp - (W_m/KT) \quad (7)$$

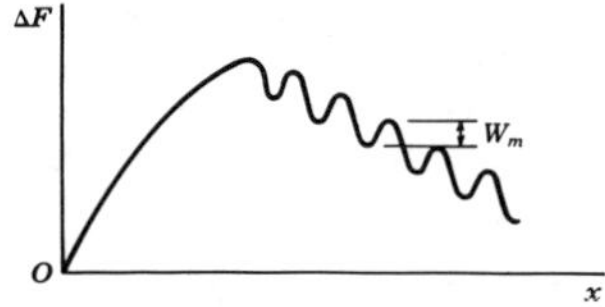

Fig. 6 Energy of double jog (ΔF) as a function of jog separation x [19].

It is interesting to note that in [4] the distance between pinning points along the screws between 600 and 650K is given as 120b, τ_p = 100MPa, and using the line tension approximation $\Phi \sim 7.9°$. This is considerably larger than Φ_o and the double jog will therefore collapse at a rate dependent on W_m. Although the value of W_m is not known, it must be considerably smaller than the total activation energy for pinning ($H_p + W_m$), i.e. W_m could be a few tenths of an ev at the most. The strain-rate is given in [4] as $\dot{\epsilon} = \dot{\epsilon}_o \exp - (H_u/kT)$ where H_u is the activation energy for unpinning, and $\dot{\epsilon}_o$ is taken as $10^{12}s^{-1}$ [4]. The same formulation can be used for the collapse mechanism discussed here, where H_u is replaced by W_m and $\dot{\epsilon}_o$ is multiplied by $nb(2T_s \sin^2\Phi/2 - \delta E)/kT$, where n = number of steps before collapse takes place. For the local pinning model $n \sim 1$, and for the numerical values used above the driving force term is ~ 1 at 600°K. For $W_m \sim 0.5$ev, which is implausibly large, we find $\dot{\epsilon} \sim 10^{+7}s^{-1}$ compared with $10^{-3}s^{-1}$ to $10^{-2}s^{-1}$ for the large H_u ($\sim$3ev) postulated in [4], i.e. the collapse by the migrating jog is a much faster process. We therefore suggest that the local pinning model with the numbers suggested in [4] is untenable. The fundamental problem with the local pinning theories is that the barries are too weak and unstable. If the pinning points are far apart so that $\Phi > \Phi_o$, the double jogs collapse; if they are close so that $\Phi < \Phi_o$, barriers extend and the local pinning model no longer applies. Both these processes are fast compared to unpinning by a mechanism with large activation energy (except for the special case when $\Phi \sim \Phi_o$ which is likely to be unstable). What is required is a mechanism to stabilise the jogs. This will be discussed in the following section. The computer simulation of Mills et al [8] is very powerful and seems to be the best way of studying the nature of the anomaly. However, at present it does not allow the double jogs to extend for $\Phi < \Phi_o$.

FORMATION OF STABLE BARRIERS

Suppose a second cross-slip step (by b/2 on (010)) is nucleated before the angle Φ in Fig. 5c reaches the critical value. Fig. 7 shows that the leading partial is now dissociated on the (111) plane (a distance b along (010) below the original (111) plane) and advances by a distance equal to the separation between the two superpartials. Fig. 5 d,e,f, shows the developing structure projected on to the (111) plane; both superpartials are shown, and in (d) the bowed out dislocation has cross-slipped again, forming the superkink A. This moves to the left and meets the part of the dislocation which has cross-slipped twice (e), the leading partial of which is at a different level. The result is the formation of a dipole ab in (f), consisting of a line of vacancies or interstitials [6,7]. This constitutes a strong lock and stabilises the screw barrier. Unless the superkink breaks free the screw will continue to cross-slip further forming KW type locks.

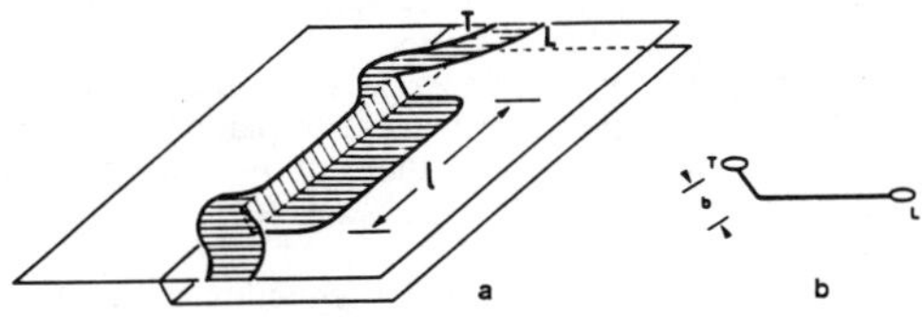

Fig. 7 Pinning configuration after cross-slip of b on (010).

Consideration of the forces acting on the superpartials, shows that at relatively low stresses, KW locks may be formed by continuing slip (in steps of b/2) on (010) with one or more intermediate steps on (111) planes, as shown in Fig. 8 for a centre triangle (top sequence) and [001] orientation (bottom sequence) respectively. With increasing stress the sequences shown in Fig. 9 become more frequent, where either (a) the dislocation cross-slips by b on (010) and then advances on (111) by a distance equal the separation of the partials, or (b) transforms into a glissile dislocation before cross-slipping again [6,7]. Repeated jumps on (111) such as those in Figs. 8 or 9a have been observed frequently during in situ TEM observations at room temperature [13]. At higher stresses the activation energies for cross-slip by steps w of several b may become comparable with or less than that for $w/b=1/2$, and KW locks may be formed in a few or even one jump. This case is discussed in [7], but detailed calculations remain to be done for large w/b. When KW locks are formed in single jumps without intermediate slip steps on (111), it is assumed that the long jogs (with edge component) are glissile during the expansion of the loop on (010), but that when they come to rest they transform into a pure edge relatively sessile configuration such as a Lomer-Cottrell lock. If the crystal is unloaded in the low temperature regime, those screws not in the KW configuration will continue to slip on (010) until a low energy configuration partly on (111) and (010) is reached. Internal stresses will tend to favour the formation of KW locks, which will be observed by post deformation TEM.

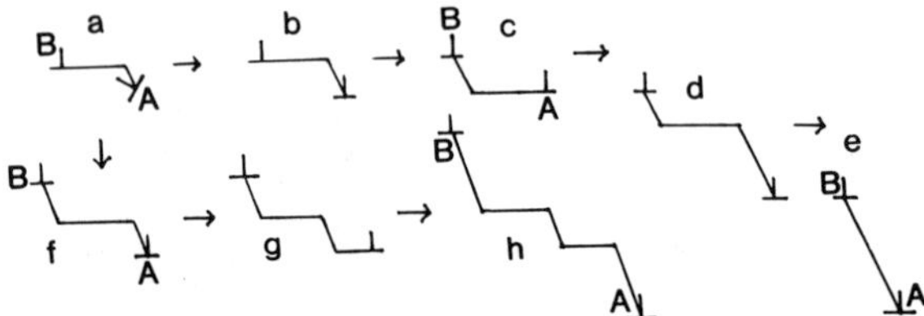

Fig. 8 Sequence of intermediate steps in formation of screw dislocation locks; abcde for centre of triangle, afgh for [001] orientation.

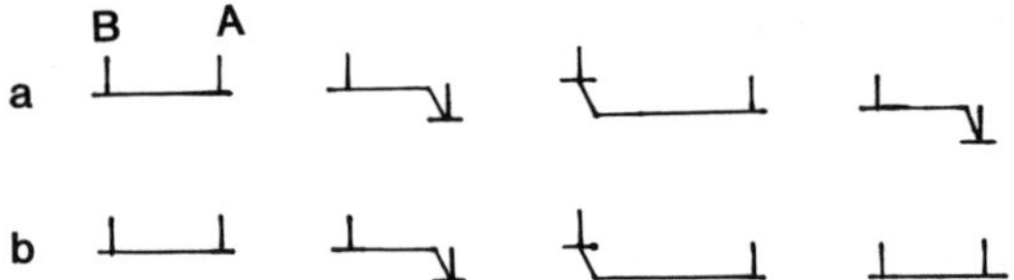

Fig. 9 a) At a certain stress the screw will tend to advance in steps of APB width on (111); b) above that stress a glissile configuration will tend to be formed [6,7].

UNLOCKING OF SUPERKINKS

At low temperatures it is envisaged that the rate controlling unlocking process is that to unlock the configuration in Fig. 5fg. The mechanism is shown in Fig. 10 [6,7]; at a critical angle Φ_o the superkink HI on (111) bypasses the lock. It moves to the left forming a short segment at Y which cross-slips from plane 1 to plane 3 annihilating the screw there of opposite sign (b). The second dislocation I then follows and generates a screw segment CD of opposite sign to that of (c); CD bows out (d) removing the APB. The final configuration in (e) is entirely glissile, leaving behind two dipoles (lines of vacancies or interstitials) BE and DF. The driving force is the reduction in energy by transforming the original screw superdislocation into a screw dipole. The bypass process will occur when $\Phi_o \sim 53°$, and this determines p in equation (2), which is the "athermal" component of the yield stress; p is estimated to be between 0.3 and 0.4 [6,7]. However there is an additional thermally activated step, i.e. cross-slip of dislocation H along XY in Fig. 10b. This will have a rather small activation energy. The stress for bypassing will therefore have a large "athermal" and a small temperature and strain-rate dependent component.

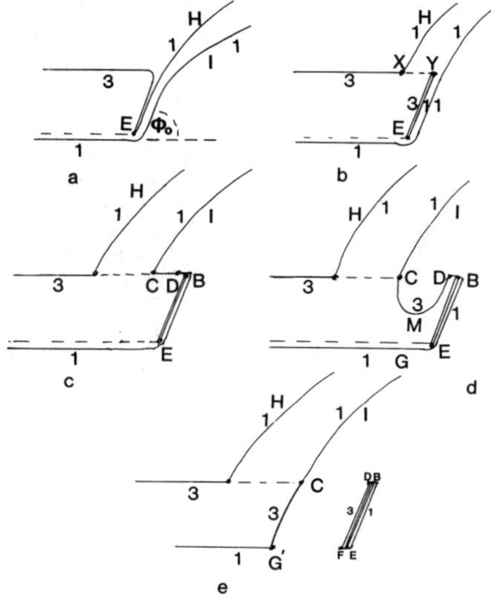

Fig. 10 Bypassing of locked structure by superkink; numbers indicate different levels of (111) planes.

It should be noted that after the dislocation has bypassed, the only debris remaining along CK in Fig. 3 behind consists of the lines of interstitials and vacancies shown in Fig. 10e. This mechanism also generates long edge dislocations, M, in Fig. 3, which may account for the partial reversibility of the yield stress [20,21] and the absence of the anomaly in the micro-strain region [11].

Many of the cross-slipped screws will continue to cross-slip to form complete or partial KW locks, while waiting for a superkink to be unpinned. KW locks can also be bypassed by mechanisms described in detail in [6,7]. Fig. 11 shows the simplest geometry, where a KW lock was formed in a single jump without an intermediate step on (111). Here leading dislocation H (11a) forms a kink leading to a constriction and the switching of partials (11b); the lock is then bypassed, the two dislocations L and M being replaced by a screw dipole (11c). The "athermal" stress is again given by equation (2), with values of p similar to those for the bypass mechanism in Fig. 10. When slip steps on (111) occur before the KW lock is completed the bypass mechanism leaves behind dipoles of screw superdislocations which can gradually annihilate forming APB tubes [22]. The debris left behind in Fig. 3 after the bypass is completed therefore consists of screw dipoles and APB tubes joined by edge dipoles. APB tubes are commonly observed in TEM studies of Ni_3X alloys (e.g. [18]). A full discussion of the bypass mechanism is given in [6,7].

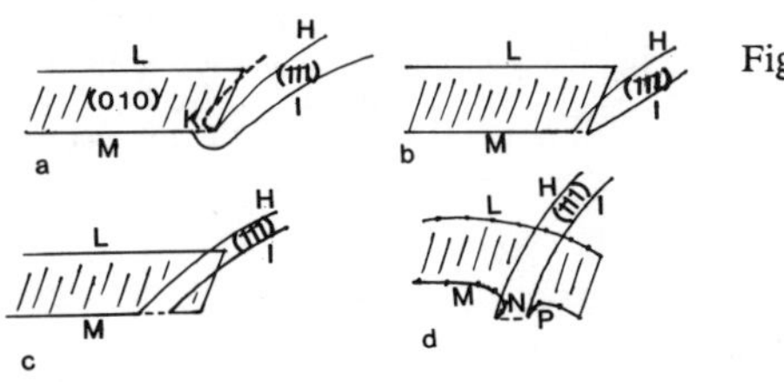

Fig. 11 Bypassing of KW locks formed in single jumps. a) H advances forming kink, leading to constriction and switching partials; b), c) by pass mechanism leaves screw dipole; d) bypassing bowed out KW lock; dots represent glissile jogs.

An important difference between this bypass mechanism and that described in Fig. 10 is that the former mechanism always occurs at the trailing partial of the locked dislocation and involves the switching of partials, while the latter occurs at the leading partial and does not involve switching. Switched partials are often observed by TEM; examples are shown in [7,18]. We expect the mechanism in Fig. 10 to be rate controlling at low temperatures, and that in Fig. 11 at high temperatures.

THE YIELD STRESS/TEMPERATURE/STRAIN RATE RELATION

To derive an expression for the yield stress a steady state configuration has been considered, where following the unpinning of a superkink (length l), one long screw barrier consisting of n locked screws of length L_s is bypassed and unzipped, and another is formed after the glissile dislocation advances by an average of $\sim nl$. The assumption is made that $L_s = C/\tau$, where C is a constant. The yield stress τ is then given by an expression similar to equation (1), where A is now slightly strain-rate dependent. With C obtained from room temperature TEM observations, and reasonable estimates made for other parameters, the pre-experimental factor is found to be ~2770MPa, in good agreement with data obtained by Dimiduk [21] for 8 binary and ternary alloys, interpolated for Ni_3Al. The orientation dependence of τ and the tension/compression asymmetry is the same as that predicted by PPV, since the activation energy is the same. The strain-rate ($\dot{\varepsilon}$) dependence at constant temperature can be obtained by differentiating (1). It is found that

$$\delta\tau/\tau \sim kT\,(\delta ln\dot{\varepsilon})/3pqGb^3\,(l'_c/b) \qquad (8)$$

where l'_c is the critical length for cross-slip to occur in the unlocking mechanism in Fig. 10, or for the constriction to occur in the mechanism in Fig. 11. $q \equiv v/ll'_c b$, where v is the activation volume, has been calculated for various values of p [6,7]. For $l'_c/b \sim 1$, $\delta\tau/\tau \sim 0.3\%$ at 300K for a change of strain-rate of ten times, which is of the same order as that observed experimentally [11]. (The estimate of $\delta\tau/\tau$ at constant temperature given in (8) is one third that given in [6,7], which is derived at constant structure). The reason for the very small strain rate dependence is the very large "athermal" stress component τ_g which has to be exceeded in the bypass process. The activation volumes are large. Bonneville et al. [14] have measured v as a function of T for Ni_3 (Al 1% T_a); their results are shown in Fig. 12 showing two regimes in which v decreases with increasing temperature. We identify these two regimes as corresponding to unlocking at the leading (Fig. 10) and trailing (Fig. 11) partials respectively. Good agreement is obtained by taking $l'_c/b = 1$ and 2.5 for the low and high temperature mechanisms respectively; the theoretical results are indicated by crosses and squares.

The superkink size distribution is expected to follow an exponential law [23,7]. Measurements in Ni_3Ga show that this is indeed the case, except that at low temperatures there is an excess of superkink sizes in the range equal to the superpartial separation [23]. This is expected from the type of motion described in [13] and Figs. 8 and 9a. It turns out that the values of l predicted from the experimentally determined yield stresses, and using equation (2), are close to the tail of the distribution of superkink lengths, two to three times larger than the average. This is consistent with the model, since unpinning will occur preferentially at the largest superkinks.

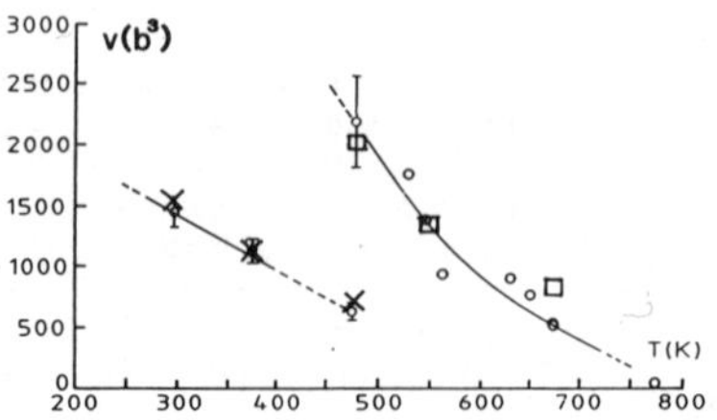

Fig. 12 Variation of activation volume with temperature for Ni_3 (Al 1% Ta) [14]. Crosses and squares are theoretical estimates.

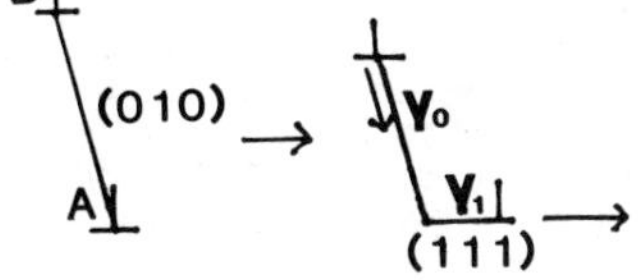

Fig. 13 Breakdown of KW lock.

THE PEAK STRESS

At sufficiently high stresses, the KW locks become unstable, as shown in Fig. 13 which is a projection normal to the screw direction. If A and B are in equilibrium on (010), then the force on A parallel to (111) due to elastic interaction from B in $\gamma_o \cos\alpha$. A can therefore glide off on $(1\bar{1}1)$ provided the stress exceeds the saturation stress

$$\tau_{sat} = (\gamma_1 - \gamma_o \cos\alpha)/b \qquad (9)$$

where $\alpha = 54°44'$. The rate of this transfer from (010) to (111) is determined by the rate of which B glides down (010), which is a thermally activated process. Equation (9) is an upper limit to the saturation stress, but is likely to be a good approximation for the [001] orientation, for which the component of applied shear stress τ_c on (010) is zero. For other orientations the saturation stress is likely to be less because of the work done by τ_c on B during the activated process, an effect which will increase with increasing τ_c. The saturation stress is therefore predicted to decrease as the orientation changes from [001] to $[\bar{1}11]$, in agreement with experiment [24,25,18]. Table I gives estimates of saturation stress from (9) using values of γ_o, γ_1 from weak beam experiments.

Table I: Estimates of saturation stress for $Ni_3(AlX)$ crystals with [001] orientation

Alloy	γ_1 mJm^{-2}	γ_o mJm^{-2}	τ_{sat} MPa	τ_{sat} MPa (exptal)
Ni_3Al [26]	111 ± 15	90 ± 5	236 ± 72	300 (slightly Ni rich) [29]
Ni_3Al 0.25 at % Hf [25]	150 ± 20	120 ± 20	324 ± 126	320 (10° from [001]) [25]
Ni_3Al 1 at % Hf [28]				370 (near [001]) [28]
Ni_3Al 1.5 at % Hf[25]	190 ± 20	170 ± 20	365 ± 126	410 (10° from [001]) [25]
Ni_3Al 1% Ta [27]	237 ± 32	200 ± 23	488 ± 180	400 [30]

It is clear that very accurate values of γ_1, γ_o are needed to predict the saturation stress with any certainty, and to test (9) TEM and mechanical property measurements should be carried out on the same crystals/composition. The agreement for all the alloys is within the experimental error. It should be noted that according to (9), it is the absolute value of γ_1 which is particularly important in determining τ_{sat}.

CONCLUSIONS

The local pinning models of the yield stress anomaly are unsatisfactory because the barriers are too weak and unstable. However, assuming the jogs bounding the screw element cross-slipped from (111) to (010) are glissile, extended barriers are formed whose ends are stabilised by further cross-slip steps. These screw barriers are connected by edge segments, the superkinks [15,16,17]. Slip is produced by unpinning of superkinks resulting in bypassing of the screw barriers. The mechanism involves a large "athermal" stress component, and a small temperature and strain-rate dependent component from a thermally activated process. The lengths of the superkinks are given by the distances travelled by screws before becoming locked. These lengths decrease with increasing temperature, resulting in the yield stress anomaly. The theory accounts for the orientation dependence and tension/compression asymmetry of the yield stress in terms of the PPV activation energy for cross-slip. It predicts a very small strain rate dependence, and two unlocking mechanisms are identified, in different temperature ranges, depending on whether bypassing takes place at the leading or trailing screw partial of the cross-slipped screw. In each of these regimes, the activation volume decreases with increasing stress in accord with experimental observations [14]. The bypass mechanisms lead to the formation of "free" edge dislocations, which can account for the behaviour in the microstrain region and for the partial reversibility of the flow stress. The main dislocation structures are predicted to be KW locks on (010). At low temperatures, these are formed by mechanisms which involve one or more intermediate slip steps on (111), depending on stress, the width of the steps being of the order of or less than the APB ribbon width of the superpartials dissociated on (111). The bypass mechanisms lead to the production of APB tubes and switched superpartials. TEM observations are consistent with all these predictions. The peak stress is identified with the stress at which KW locks become unstable and for crystals in the [001] orientation is predicted to depend in a simple manner on the APB energies on (111) and (010).

NOTE

After this paper was completed, an important paper by Mills and Chrzan [31] became available on a dynamical simulation of dislocation motion in $L1_2$ alloys. In this elegant study local pinning is assumed, but extended barriers roughly along the screw direction arise because of strong correlation of the pinning events. The overall geometry of the pinned structures, the generation of slip by unpinning at long superkinks resulting in unzipping locked screws and the superkink size distribution are broadly similar to the processes envisaged in the present model. [31] provides insight into the transient creep and a possible workhardening mechanism, neither of which are considered in the present model. An important difference in concept is that in [31] the small strain-rate dependence can be explained assuming an athermal unlocking process, whereas in the present theory it is assumed to be controlled by a thermal activated process, i.e. essentially by a temperature dependent critical bypass angle. This needs to be considered further, particularly the changes expected when the strain-rates are increased or decreased during the test; furthermore as yet [31] does not seem to explain the two different regimes observed in stress relaxation tests [14]. With regard to workhardening, it is not clear how the simulation explains the characteristic temperature and orientation dependence [18]. It is possible that a thermally activated softening process occurs in addition to the hardening indicated by the simulations.

REFERENCES

[1] S. Takeuchi and E. Kuramoto, *Acta Metall.*, **21**, 415 (1973)
[2] V. Paidar, D.P. Pope and V. Vitek, *Acta Metall.*, **32**, 435 (1984)
[3] V. Vitek and Y. Sodani, *Scripta Metall. Mater.*, **25**, 939 (1991)
[4] M. Khantha, J. Cserti and V. Vitek, *Scripta Metall. Mater.*, **27**, 481 (1992)
[5] P.B. Hirsch, *Scripta Metall. Mater.*, **25**, 1725 (1991)
[6] P.B. Hirsch, *Phil. Mag. A*, **65**, 569 (1992)
[7] P.B. Hirsch, *Progr. Mats. Sci.*, **36**, 63 (1992)
[8] M.J. Millls, D.C. Chrzan, K.J. Hemker and W.D. Nix, in **Modelling the Deformation in Crystalline Solids**, Eds. T. Lowe and R. Rollett, TMS Publ. 0423 (1991).
[9] P. Flinn, *Trans. TMS-AIME*, **218**, 145 (1960)
[10] M.H. Yoo, *Scripta Metall.*, **20**, 915 (1986)
[11] P.H. Thornton, R.G. Davies and T.L. Johnston, *Metall. Trans.*, **1**, 207 (1970)
[12] P.B. Hirsch, *J. de Phys. III*, **1**, 989 (1991).
[13] G. Molénat and D. Caillard, *Phil. Mag. A*, **64**, 1291 (1991).
[14] J. Bonneville, N. Baluc and J.L. Martin in **Int'l Symp. on Intermetallic Compounds - Structure and Mechanical Properties**, JIMIS-6, p323, edited by O. Izumi (The Japan Institute of Metals).

[15] M.J. Mills, N. Baluc and H.P. Karnthaler in **High Temperature Ordered Intermetallics III**, edited by C.C. Kock, C.T. Liu, N.S. Stoloff and A.I. Taub, p203 (MRS Proc., **133**, 1989).
[16] P. Veyssière in **High Temperature Ordered Intermetallics III**, edited by C.C. Kock, C.T. Liu, N.S. Stoloff and A.I. Taub, p175 (MRS Proc., **133**, 1989).
[17] Y.Q. Sun and P.M. Hazzledine, *Phil. Mag. A*, **58**, 603 (1988).
[18] Y.Q. Sun, D. Phil thesis, University of Oxford (1990).
[19] P.B. Hirsch and J. Lothe, Teory of Dislocations, 1st Ed. (McGraw-Hill, New-York 1968) p493.
[20] R.G. Davies and N.S. Stoloff, *Trans. TMS AIME*, **233**, 714 (1965).
[21] D.M. Dimiduk, D. Phil thesis, Carnegie-Mellon University (1989).
[22] C.T. Chou and P.B. Hirsch, *Phil. Mag.*, **44**, 1415 (1981).
[23] A. Couret, Y.Q. Sun and P.B. Hirsch, *Phil. Mag.* (1992), in press.
[24] Y. Umakoshi, D.P. Pope and V. Vitek, *Acta Metall.*, **32**, 449 (1984).
[25] C. Neveu, D. Phil thesis, University of Paris-Sud (1991).
[26] J. Douin, P. Veyssière and P. Beauchamp, *Phil. Mag.*, **A54**, 375 (1986).
[27] N. Baluc, R. Schaublin and K.J. Hemker, *Phil. Mag. Lett.*, **64**, 327 (1991).
[28] F.E. Heredia and D.P. Pope, *J. de Phys. III*, **1** (6), 1055 (1991).
[29] F.E. Heredia and D.P. Pope, *Acta. Metall. Mater.*, **39**, 2027 (1991).
[30] L.R. Curwick, D. Phil thesis, University of Minnesota (1972).
[31] M.J. Mills and D.C. Chrzan, *Acta. Metall. Mater.*, **40**, 3051 (1992).

HIGH TEMPERATURE DEFORMATION OF SINGLE CRYSTALS OF NiAl

Keith R. Forbes, Uwe Glatzel*, R. Darolia** and William D. Nix
Department of Materials Science and Engineering, Stanford University, Stanford, CA 94305.
* Institut für Metallforschung, BH18, Technische Universität Berlin, 1000 Berlin 12, FRG.
** Engineering Materials Technology Laboratories, GE Aircraft Engines,
1 Newman Way, Cincinnati, OH 45215.

Abstract

The high temperature deformation properties of single crystals of stoichiometric NiAl have been studied in tension creep and in constant strain rate compression at temperatures between 850 and 1200°C. Samples were tested in a "soft", [223], orientation and the "hard", [001], orientation. The samples exhibit a strong orientation dependence of the strength and show other revealing deformation characteristics. The activation energy for steady state flow in both hard and soft orientations is near that for lattice self diffusion. Soft oriented crystals reach steady state rapidly and develop little dislocation substructure. Deformation of these soft oriented crystals occurs by the glide of **b**=<001> dislocations.

The creep curves of hard oriented crystals show pronounced sigmoidal creep, suggesting that the dislocations move in a sluggish manner, multiply in the early stages of creep and, ultimately, lead to strain hardening. Hard oriented crystals also develop extensive dislocation substructure during creep. This dislocation substructure is composed of **b**=<100> dislocations, which have no resolved shear stress for glide. Evidence for $\{10\bar{1}\}$<101> glide in hard oriented crystals is presented and a model is developed by which the decomposition of gliding **b**=<101> dislocations can produce the observed **b**=<100> dislocation networks. The increased creep resistance of hard oriented crystals compared to soft oriented crystals is described in terms of the differences in dislocation mobility and substructure formation for deformation in these directions.

Introduction

The search for new materials for high temperature structural applications (such as in gas turbine engines) has created strong interest in the development of intermetallic alloys. The intermetallic alloy NiAl offers many advantages over current materials used for these purposes. The density of NiAl is 5.95 g/cm^3, about 30% lower than the density of nickel-based superalloys. NiAl also has excellent oxidation resistance and very high thermal conductivity. Strong bonding between nickel and aluminum results in a high melting temperature (1638°C) and the potential for good high temperature strength. The utilization of NiAl as a high temperature structural material is currently hindered by limited ductility at low temperatures and poor strength at high temperature. These limitations are being addressed by investigations into the alloying of NiAl[1, 2, 3]. As a result of these studies, the prospects for technical applications of NiAl-based alloys at high temperatures are promising.

NiAl has an ordered cubic B2 crystal structure based on the simple cubic structure with an Al atom at the body center of a Ni cube (Fig. 1). The lattice parameter is 2.887Å. The minimum lattice translation distance that preserves the atomic ordering is <100>, which is observed as the Burgers vector for easy slip in NiAl. Dislocations with the <100> Burgers vectors glide primarily on {001} planes, although {011}<100> slip systems have been observed for deformation in some orientations[4, 5, 6]. These slip systems produce only three independent slip systems, not enough to accommodate arbitrary plastic deformations. The limited number of slip systems results in a large anisotropy in the plastic deformation of NiAl. Single crystals oriented for deformation along <001> have an unusually high flow stress and are called "hard" oriented crystals and crystals oriented along non-<001> axes are called "soft" oriented crystals. Since loading along a <001> axis produces no macroscopic resolved shear stress on any slip system with **b**=<100>, plastic deformation in hard crystals must occur either by the activation of additional slip systems or by climb processes.

Previous creep studies of NiAl single crystals have confirmed the high plastic resistance of hard oriented crystals. No significant differences in the shapes of the creep curves for hard and

Figure 1. The B2 crystal structure of NiAl showing the <100> Burgers vectors.

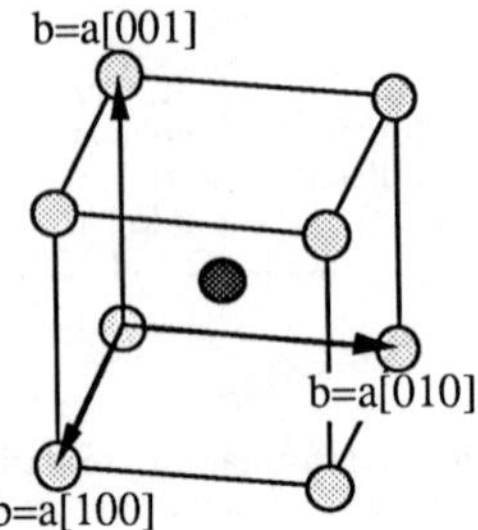

soft orientations have been reported. Between 750 and 1000°C, normal primary transients have been typically observed prior to extensive steady state creep[7, 8], although some inverted primaries have also been reported[7].

The controlling deformation processes in hard oriented crystals have not been clearly established in these studies. Slip trace analysis and transmission electron microscope (TEM) observations suggest that {101}<101> slip occurs in hard oriented crystals[8, 9]. However, the TEM evidence for glide of **b**=<101> dislocations is complicated by the observation of a large number of **b**=<100>dislocations. Dislocations with **b**=<101> are typically found in tangled networks with **b**=<100> dislocations, and it has been suggested that they do not contribute to deformation, but are only the result of dislocation reactions[10]. Evidence that climb of **b**=<100> dislocations contributes significantly to deformation of NiAl at high temperatures has been presented[11]. The stress and temperature dependence of deformation has also suggested more complex processes, such as the constriction of super dislocations[12].

Subgrains bounded by well-developed networks of **b**=<100> dislocations are observed in crystals deformed in the hard orientation[8, 13]. The interiors of these subgrains have a low dislocation density. Some interactions between **b**=<101> dislocations and **b**=<100> networks have been reported[13]. However, the formation of these networks and their contribution to deformation processes has not been explored.

In this paper we focus on the high temperature deformation processes in both hard and soft oriented NiAl crystals and on the role of dislocation network formation in these processes. We identify the principal factors that cause the large anisotropy of the high temperature strength of this intermetallic compound.

Sample Preparation

The [223] orientation was chosen as the soft orientation for study because it has a high resolved shear stress on the (110)[001] slip system and should promote glide on a single slip system. Hard and soft oriented crystals were tested in tension creep and in constant strain rate compression at temperatures between 850 and 1200°C. A comparison is made not only of the steady state deformation rates but also of the deformation transients for each orientation. After testing, the dislocation structures produced by high temperature deformation were studied with TEM.

Single crystals of stoichiometric NiAl were produced at the Engineering Materials Technology Laboratories of General Electric Aircraft Engines. These crystals, which were nominally 25mm x 38mm x 100mm, were grown in argon by the Bridgman method and then homogenized in an argon atmosphere at 1316°C for 50 hours. Each crystal was oriented using Laue back reflection and tension and compression specimens were electrode discharge machined (EDM) with the desired crystallographic orientations. Compression samples were cut 7.62mm long with a 3.81mm x 3.81mm cross section. Sheet tension samples were cut with a nominal gauge length of 20.3mm and a cross-section of 3.10mm x 3.10mm. All specimens were low stress ground to remove the recast layer resulting from the EDM cutting. Specimens were mechanically polished to 25 microns and then electropolished. The specimens were polished at -10°C and 25V for 5 minutes in a stirred electrolic solution of 10 vol.% perchloric acid in ethanol.

The compression specimens were deformed in an air furnace between two flat alumina platens. Extensometry was attached directly to these platens in order to measure strain at the sample without the complications of subtracting machine displacements. Tension specimens were held with TZM molybdenum grips which contacted the 2.38mm radius shoulders on the specimens. In order to protect the grips from oxidation, tension tests were conducted under a vacuum of about 10^{-6} torr. When the sample is loaded, some grip slippage occurs and the relative displacement of the grips does not give an accurate measurement of the strain in the sample. Therefore, extensometers were attached to pins passing through the sample in the grip region in order to measure the strain in the sample.

After testing, 3mm diameter cylinders were EDM cut from the samples. These cylinders were mechanically ground to discs 200μm thick and then electropolished to form thin foils for transmission electron microscopy (TEM) observations. Electropolishing was accomplished at -10°C and 25V in a solution of 10 vol.% perchloric acid in ethanol.

Steady State Creep Rates

The activation energy for creep of NiAl was measured using a temperature change test at constant compressive stress. The samples were first deformed at a constant strain rate until a steady state stress level was reached. The crosshead control was then switched to a computer-controlled routine that maintained a constant true stress on the sample. The temperature was then raised or lowered as the strain rate was monitored. As soon as the temperature stabilized at a new level, a new steady state strain rate was measured. The advantage of this method is that the activation energy can be measured over a wide temperature range using only one sample. Using this technique, the activation energy for creep of soft oriented crystals was found to be 330±30 kJ/mol and that for the hard oriented crystals was found to be 305±15 kJ/mol. Separate

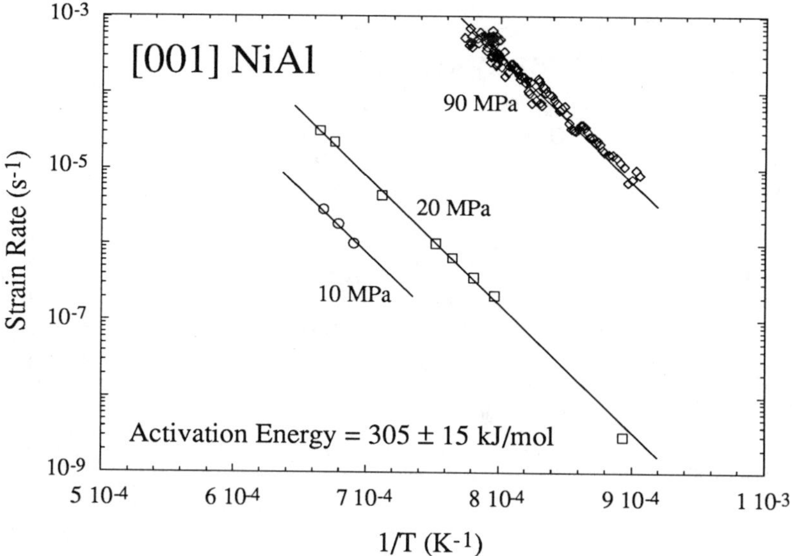

Figure 2. The activation energy of [001] NiAl calculated from tests at constant stress and varying temperature. The 90 MPa test is from one compression sample which was cooled at constant stress.

samples of hard oriented crystals were also tested in tension at constant stress at temperatures of 1175, 1200 and 1225°C. The activation energy results for tension and compression tests are shown in Fig. 2 and are found to be in good agreement. Previously reported values of the activation energy for creep of NiAl range between 260 and 340 kJ/mol[6, 14, 15]. Though typically constant at high temperatures, activation energies for hard oriented crystals have been reported to increase with temperature above 500°C[12]. We find the activation energy for deformation to be constant between 850 and 1200°C and to be independent of orientation. The measured creep activation energy is close to that for lattice diffusion of Ni in NiAl[16], 308±10 kJ/mol. Thus creep deformation of NiAl at these temperatures appears to be limited by diffusional processes.

The temperature dependence of the creep rate can be removed by normalizing the creep strain rate by the lattice diffusivity. The complete normalization can be derived from the Dorn equation:

$$\frac{\dot{\varepsilon} kT}{D \mu b} = A \left(\frac{\sigma}{\mu} \right)^n$$

where $\dot{\varepsilon}$ is the strain rate, D is the lattice diffusivity, μ is the shear modulus, b is the Burgers vector, kT has the usual meaning, and where the constant A and the stress exponent, n, are parameters which characterize the creep process. The shear modulus used for this normalization is C_{44}, the shear modulus corresponding to shear loading on {100} planes and in <001> directions[17].

The normalized creep rates for both hard and soft orientations are plotted in Fig. 3. The results confirm the strong plastic anisotropy of NiAl. The hard oriented crystals have steady state strain rates that are typically two orders of magnitude less than those of the soft oriented crystals. For both crystal orientations, the stress exponent is near 3 at low stresses and increases with increasing stress.

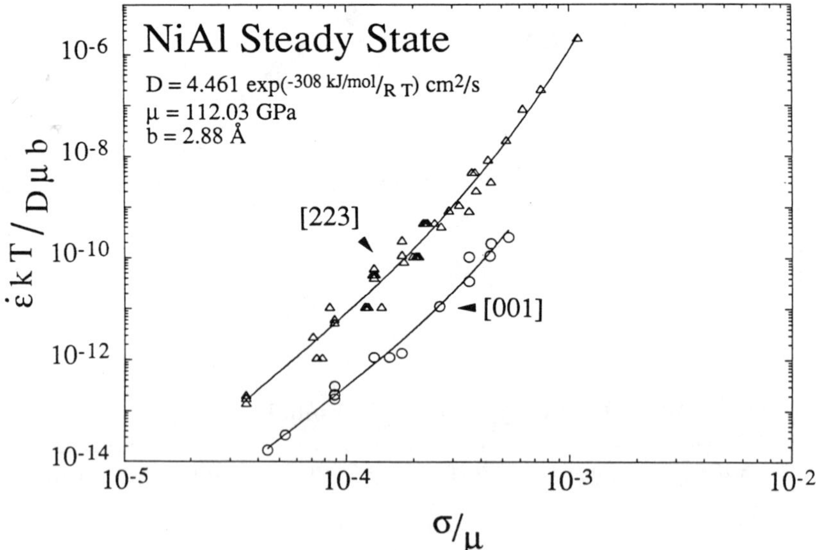

Figure 3. The normalized steady state creep rates for hard [001] oriented and soft [223] oriented NiAl single crystals.

Deformation in the Soft Orientation

Figure 4 shows a typical creep curve for a NiAl single crystal tested in the soft [223] orientation. The soft orientation shows no initial transient and virtually no strain hardening throughout the test. The strain rate varies little during the entire creep test. Compression tests also reveal no strain hardening. The lack of transients and the absence of strain hardening suggests that the dislocation structure changes very little during deformation. Most pure FCC metals exhibit a large amount of primary creep and strain hardening, which are characteristic of substructure controlled deformation. The creep curves of soft NiAl crystals are more similar to those of BCC alloys, which have small transients and where deformation is more strongly influenced by the mobility of dislocations.

Observations with TEM confirm that the dislocation density in [223] oriented samples does not increase significantly during deformation, remaining near 10^7 cm^{-2}. Dislocation tangles and networks are not typically formed except at high temperatures and low strain rates where they occur to a limited extent[18]. In general, **b**=<100> dislocations are free to glide completely through the crystal, unimpeded by dislocation substructure. The deformation curves and TEM observations suggest that the mobility of **b**=<100> dislocations is most important in controlling the deformation of soft oriented NiAl crystals.

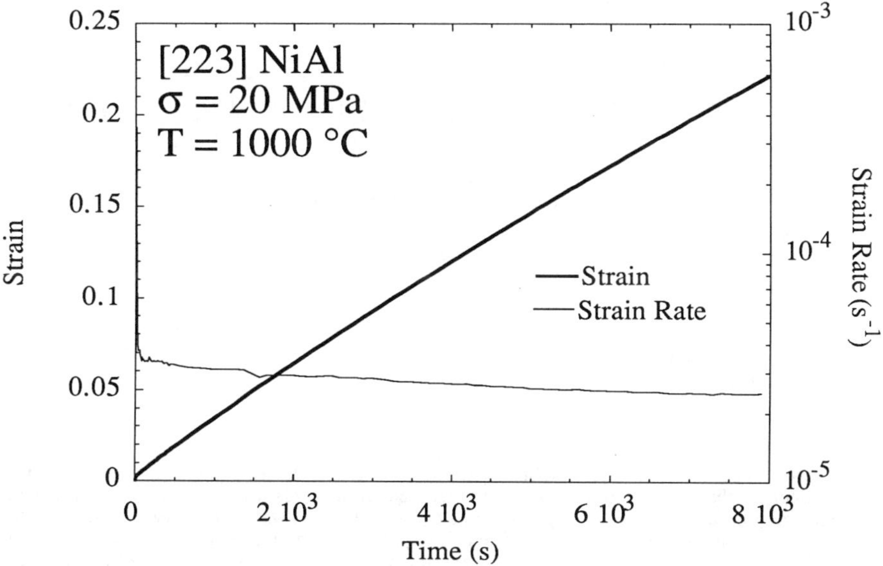

Figure 4. The response a of soft [223] oriented NiAl crystal in tension creep.

Deformation in the Hard Orientation

The shapes of the creep curves for hard oriented crystals differ significantly from the creep curves for soft oriented crystals, as shown in Fig. 5. Hard oriented crystals exhibit a pronounced sigmodial creep curve. The sigmodial transient suggests that deformation in hard oriented crystals occurs by the movement of dislocations with very low mobility. The shape of the creep curve can be explained using the Orowan equation and assuming that the dislocation velocity decreases with increasing dislocation density. In the Orowan equation, the strain rate, $\dot{\varepsilon}$, is the product of the

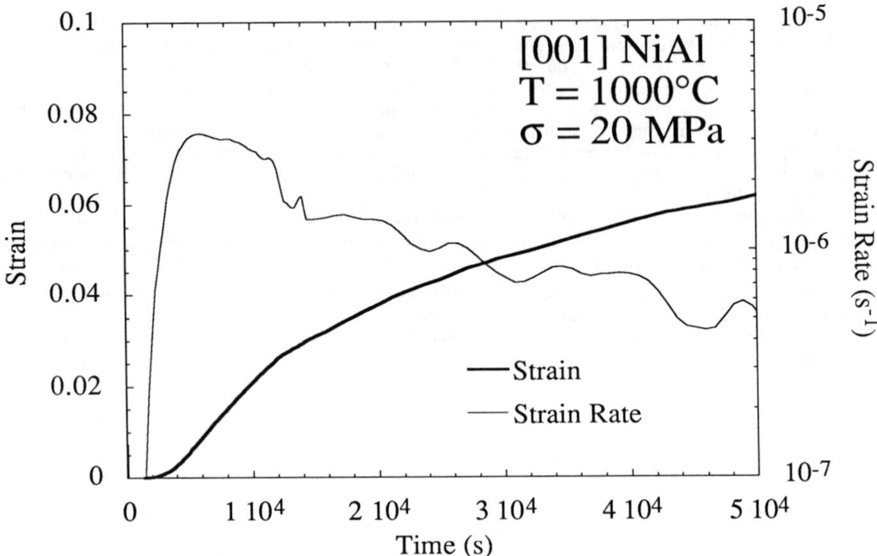

Figure 5. The response a of hard [001] oriented NiAl crystal in tension creep.

mobile dislocation density, ρ, the Burgers vector, b, and the velocity average, v, of the dislocations.

$$\dot{\varepsilon} = \rho \, b \, v$$

We expect the density of mobile dislocations in the as-grown crystals to be low, so that the product on the right hand side of the Orowan equation gives a low initial strain rate. During deformation, dislocations multiply and this causes the strain rate to increase. If the mobility of dislocations is low, multiplication occurs slowly, and the strain rate increases gradually. But as dislocations multiply they are more likely to interact and to decrease the overall mobility of the dislocations. As the average dislocation velocity drops, the strain rate first reaches a maximum and then begins to decrease. This kind of sigmodial creep is characteristic of materials with sluggish dislocation motion and low initial dislocation density. It does not depend on the details of the multiplication process or of the process which causes the dislocation mobility to decline[19].

The creep curve shown in Fig. 5 exhibits extensive strain hardening. The strain rate for this crystal decreased by an order of magnitude from its maximum, over a creep strain of about 8%. Strain hardening in hard oriented crystals persists even in creep tests at 1200°C, though it is not as extensive. In constant strain rate compression tests, strain hardening is also prevalent and steady state is reached only at high temperatures and low strain rates. These observations of strain hardening suggests that dislocation substructure must be forming in the course of deformation. Transient tests on hard oriented crystals also suggest that dislocation motion is limited by substructure formation[20].

Observations of Networks

Hard oriented crystals examined with TEM after tensile creep or compression deformation reveal extensive dislocation substructure. Dislocation networks of **b**=<100> dislocations, such as those shown in Fig. 6, are found throughout the deformed crystals. These networks have been shown to be in low energy configurations and are stable with respect to perturbations within the network and to dislocation reactions[21]. The **b**=<100> networks observed in NiAl form irregular low-angle subgrain boundaries with diameters between 20 and 200 μm[18]. The subgrain interiors have a relatively low dislocation density, near that found in undeformed crystals. Observations of samples deformed in the temperature range 1000°C to 1200°C reveal few **b**=<101> dislocations and many of these do not lie on their glide plane. Only a few isolated dislocations are observed; typically short segments of **b**=<101> dislocations are in tangled configurations with other **b**=<100> dislocations[18].

The networks in hard oriented crystals are found to be composed of only two types of **b**=<100> dislocations. In a sample deformed along the hard [001] direction, one of the dislocations in the networks is always **b**=[001], which is the only **b**=<100> dislocation that would experience a climb force for the given loading. The other dislocation is either **b**=[100] or **b**=[010] (which cannot be distinguished with reference only to the loading axis), but not both. If one network contains **b**=[100] dislocations, then no network is found in the crystal containing **b**=[010] dislocations. In that case, approximately half of the dislocations in the networks are **b**=[001] dislocations and the other half are **b**=[100] dislocations.

The **b**=<100> networks will produce a rotation between adjoining subgrains. The axis of rotation between subgrains separated by a network of intersecting dislocations is given by the cross product of their Burgers vectors[22]. Thus a network of **b**=[001] and **b**=[100] dislocations will produce a rotation between subgrains about the [010] axis. This rotation can be seen in Laue X-ray photographs. Figure 7(a) describes the orientation of the [001] hard oriented crystals that have been cut with {110} faces so that the [100] and [010] axes lie along the diagonals of the cross section. The Laue photograph of the undeformed crystal in Fig. 7(b) shows that the sample is oriented within 2° of the [001] axis. After being deformed in tension creep to 10% strain, the Laue spots are elongated in the [100] direction, indicating subgrain rotations about the [010] axis. Both compression and tension samples show subgrain rotations about only one of the transverse <100> axes, called the [010] axis in this discussion; rotations in one crystal typically occur about only one of the [100] and [010] axes and not about the [001] loading axis. The Laue photographs confirm that networks formed in crystals oriented for deformation along [001] are composed of only **b**=[001] and **b**=[100] dislocations.

Glide of b=<101> Dislocations

Deformation processes in hard oriented NiAl crystals must produce a large number of **b**=<100> dislocations that then form the observed dislocation networks. The networks always include **b**=[001] dislocations, which could multiply by climb under a [001] loading. However, the other dislocations in the networks have no climb forces at all and would not be expected to move and multiply by climb. Thus, it seems unlikely that the networks could form by climb processes alone.

An alternative process of multiplication of **b**=<100> dislocations is one involving the decomposition of **b**=<101> dislocations.

$$\mathbf{b}=[101] \Rightarrow \mathbf{b}=[100] + \mathbf{b}=[001]$$

Glide of **b**=<101> has been observed in hard oriented NiAl crystals at lower temperatures and there is no reason to think that this glide process would become inoperative at higher temperatures. Dislocation lines of **b**=<101> have been observed to leave **b**=<100> debris during glide and to interact with **b**=<100> networks. Since **b**=<100> have low line energies[4, 21], a decomposition would not involve a large energy change. In fact, the decomposition of an edge **b**=<101> would result in no line energy increase at all. Also recent high resolution TEM observations of edge **b**=<101> cores have revealed two core configurations, one of which is decomposed into **b**=<100> dislocations[23]. Although we have not observed the decomposition of

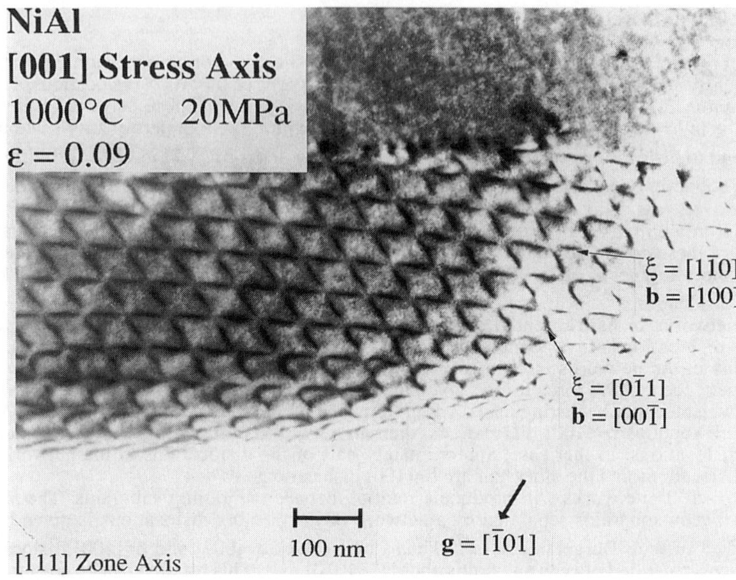

Figure 6. A TEM micrograph of the **b**=<100> network found in NiAl crystals after deformation along the hard [001] direction.

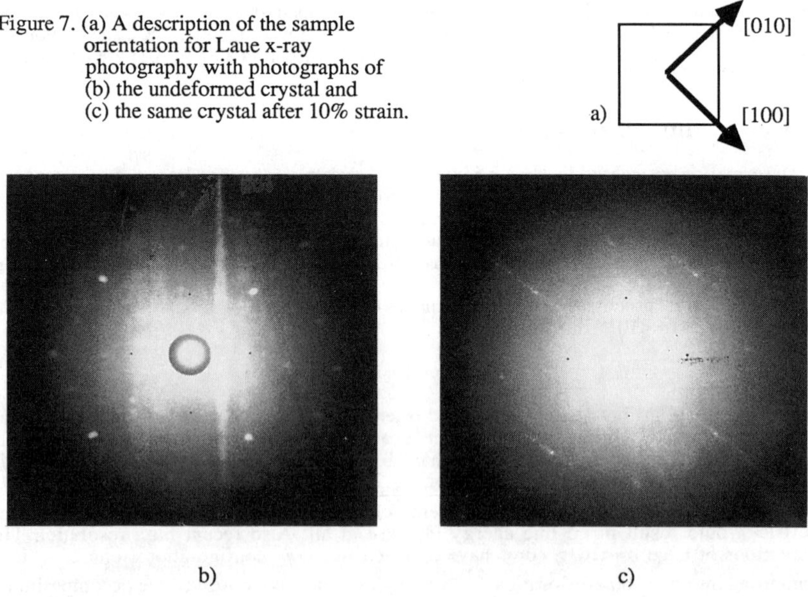

Figure 7. (a) A description of the sample orientation for Laue x-ray photography with photographs of (b) the undeformed crystal and (c) the same crystal after 10% strain.

b=<101> dislocations, current information suggests that such decomposition is possible. The observed networks could be formed by the glide and subsequent decomposition of **b**=<101> dislocations.

Although there are few **b**=<101> dislocations in hard oriented NiAl crystals deformed at high temperatures and no observable slip traces corresponding to $\{10\bar{1}\}$<101> glide, there is still evidence that this glide occurs. Normal loads along the [001] axis would produce high resolved shear stresses on four $\{10\bar{1}\}$<101> slip systems. However the activation of **b**=[101] dislocations would inhibit the glide of **b**=[011] dislocations by latent hardening effects. Latent hardening has been observed in single crystals with the rock-salt crystal structure, which also deforms on $\{10\bar{1}\}$<101> slip systems[24]. In a manner similar to these crystals, activation of the $(10\bar{1})[101]$ and $(101)[10\bar{1}]$ slip systems in NiAl should inhibit glide on the $(01\bar{1})[011]$ and $(011)[01\bar{1}]$ slip systems. Figure 8(a) shows that the exclusion of one set of slip systems results in an asymmetry in deformation and a change in the shape of the sample cross section. Under a [001] tensile load, glide on the $(10\bar{1})[101]$ and $(101)[10\bar{1}]$ slip systems would result in a shrinking of the cross section in the [100] direction but no change in the [010] direction. The cross section of the samples tested is illustrated in Fig. 8(b) where the [010] and [100] axes are along the diagonals of the undeformed cross section. Figure 8(c) shows the cross section of a sample after tensile loading to 15% strain. After testing in tension, the diagonal along the [100] direction has decreased by about 15% while the diagonal along the [010] direction remains unchanged. Hard crystals tested in compression show an increase along the [100] diagonal of an amount consistent with the axial strain in the sample.

It is important to note that the climb of **b**=<100> dislocations would not produce the change in the sample cross section which is observed. Because of the cubic symmetry of the cross section, long range diffusion of atoms, which is neccesary for climb, would not be expected to occur preferentially from any one of the surfaces. The change in the shape of the cross section after deformation is consistent with glide of **b**=[101] and **b**=$[10\bar{1}]$ dislocations only, but not with the climb of **b**=<100> dislocations.

Excluding the $(01\bar{1})[011]$ and $(011)[01\bar{1}]$ slip systems by latent hardening also affects the networks that would form by decomposition into **b**=<100> dislocations. Decomposition of **b**=[101] and **b**=$[10\bar{1}]$ dislocations would produce only **b**=[100] and **b**=[001] dislocations and no **b**=[010] dislocations. The networks that are observed in hard oriented NiAl crystals consist of exactly these dislocations, such that the rotation axis determined from Laue photographs coincides with the undeformed diagonal of the cross section. Glide on $(10\bar{1})[101]$ and $(101)[10\bar{1}]$ slip systems will not deform the [010] diagonal of the sample cross section, and may decompose into the **b**=[100] and **b**=[001] dislocations that produce networks, resulting in rotations about the

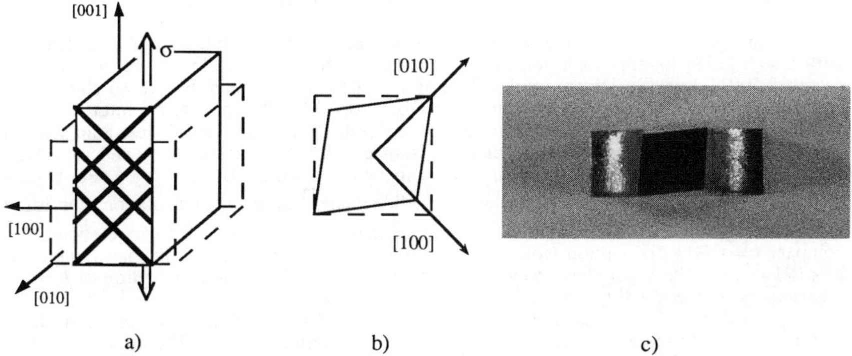

Figure 8. (a) Slip Selection by latent hardening produces a cross-section shape change. b) A description of the orientation of the sample cross section during tension creep with (b) a photograph of the sample after 15% strain showing the change in shape of the cross section.

same [010] diagonal. Thus, both the cross section shape changes and the rotation of subgrains suggest that glide of **b**=<101> occurs in hard oriented crystals in an asymmetric manner due to latent hardening.

A Model of Network Formation

Our observations suggest that deformation of hard oriented NiAl crystals occurs by the glide of **b**=<101> dislocations, which may decompose to form networks of **b**=<100> dislocations. The model illustrated in Fig. 9 is a possible sequence by which the glide and decomposition of **b**=<101> dislocations can produce such networks. Consider a section of a NiAl single crystal with a [001] tensile axis, as shown in Fig. 9(a). The top half of this section is allowed to deform by glide of the $(101)[10\bar{1}]$ slip system and in the bottom half, glide occurs on the $(10\bar{1})[101]$ system. These are the slip systems allowed by latent hardening in NiAl. A (111) plane separates the two sections so that as deformation occurs both **b**=$[10\bar{1}]$ and **b**=[101] dislocations are laid out on this plane, the plane of the dislocation network. The operation of $(101)[10\bar{1}]$ slip in the top section would lay out a right hand screw dislocation on the (111) plane; slip in the bottom section would lay out a dislocation of mixed character. The plan view of the (111) network plane in Fig. 9(b) clarifies the line directions of these dislocations and the resulting intersecting grid.

In Fig. 9(c) the **b**=<101> dislocations are allowed to decompose. This decomposition will increase the total energy of the network since the line energy of the **b**=<101> dislocations is less than that of the decomposed **b**=<100> dislocations. However the resulting **b**=<100> network in Fig. 9(f) has a significantly lower energy than the initial **b**=<101> network in Fig. 9(b). Thus, once an activation energy has been overcome and the dislocations have decomposed, reactions can take place at the nodes with parallel Burgers vectors (Fig. 9(c)). These reactions will cause the network to transform into the configuration shown in Fig. 9(d). A large array of dislocations that have been laid down by these glide processes and decomposed will result in the grid illustrated in Fig. 9(e). The line energy of these dislocations can be reduced by straightening the dislocation lines. In doing so the **b**=<100> dislocations will lie near <110> directions, which is known to be a low energy direction for these dislocations. The resulting network shown in Fig. 9(f) is exactly that found in TEM observations, as shown in Fig. 6.

Dislocation networks in deformed NiAl do not always lie on {111} planes. Networks seem to generally lie on {111} or {110} planes but other configurations are also observed. The model presented in Fig. 9 illustrates the formation of one network, but other networks on other planes could also be modeled in a similar manner. The model does show that gliding **b**=<101> dislocations can form networks that can lower their energy by decomposing into **b**=<100> networks.

Discussion: Increased Creep Resistance in Hard Oriented Crystals

Our analysis suggests that, at temperatures between 850 and 1200°C, NiAl single crystals with a soft [223] loading axis deform by the glide of **b**=<100> dislocations, and that hard oriented, [001], crystals deform by the glide of **b**=<101> dislocations. Figure 10 compares the creep curves of the soft and hard oriented crystals deformed at the same temperature and stress. The maximum strain rate of the hard oriented crystal is an order of magnitude less than that of the soft oriented crystal. This increased creep resistance will have both mobility and substructure components. The **b**=<100> dislocation core has been observed in HRTEM to be very compact, whereas the **b**=<101> core is extended out of its glide plane[23]. Atomistic calculations confirm that the low energy **b**=<101> core is extended in such a manner as to make glide very difficult[25]. The compact **b**=<100> dislocation would be expected to have greater mobility than the extended **b**=<101> core. The sigmodial creep transient is further confirmation that glide of **b**=<101> dislocations is indeed a sluggish process.

However, Fig. 10 indicates that the low mobility of **b**=<101> dislocations is not the only factor causing the increased creep resistance of hard oriented crystals. The strain rate of hard oriented crystals drops significantly from its maximum as the crystal strain hardens. This hardening is attributed to the formation of the dislocation substructure that is observed in TEM. The glide and decomposition of **b**=<101> dislocations form an extensive **b**=<100> substructure

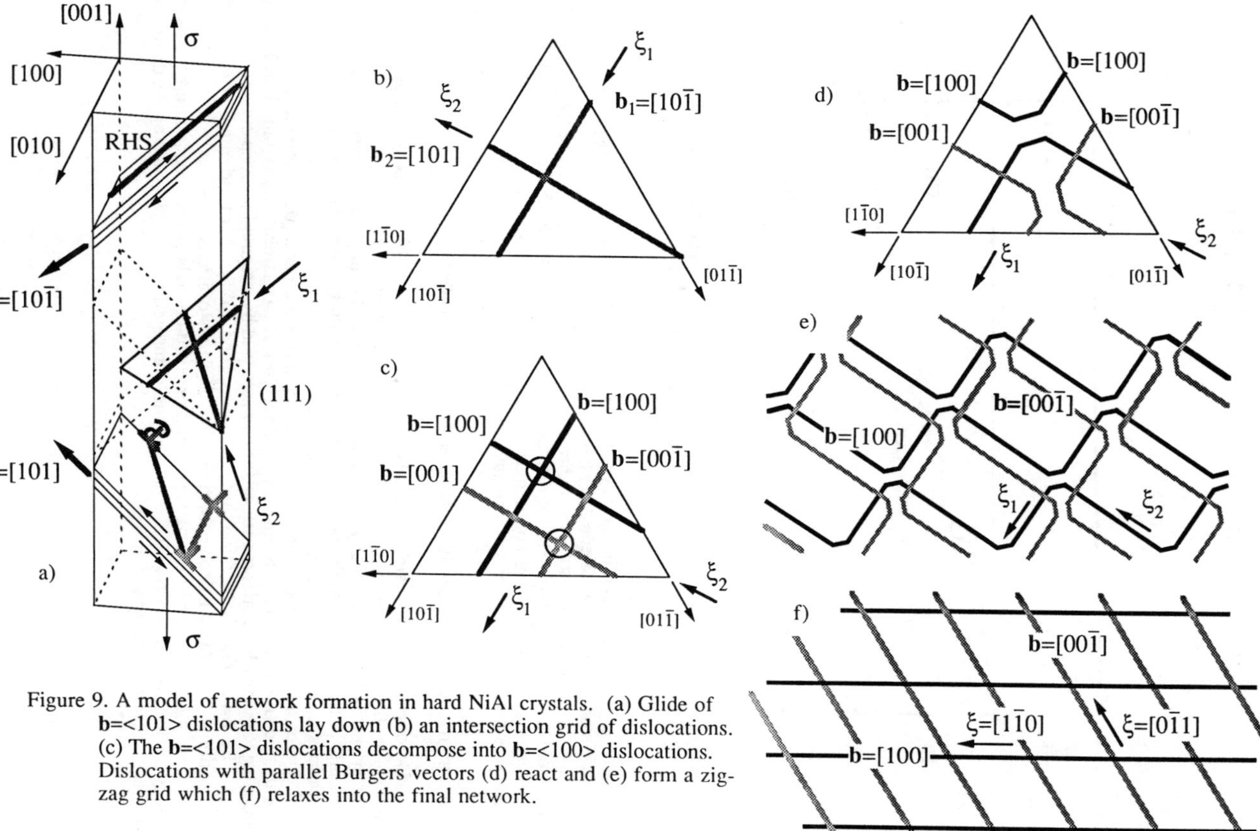

Figure 9. A model of network formation in hard NiAl crystals. (a) Glide of **b**=<101> dislocations lay down (b) an intersection grid of dislocations. (c) The **b**=<101> dislocations decompose into **b**=<100> dislocations. Dislocations with parallel Burgers vectors (d) react and (e) form a zig-zag grid which (f) relaxes into the final network.

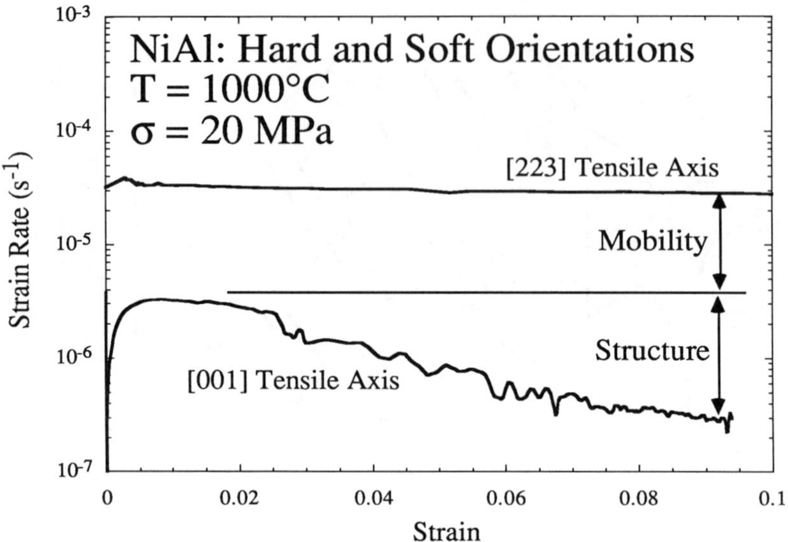

Figure 10. A comparison of the creep curves of NiAl crystals loaded in hard and soft directions at the same stress and temperature.

that restricts glide processes in the crystal. No significant dislocation substructure or strain hardening is observed in soft oriented crystals.

Deformation in hard and soft oriented crystals differs both in the mobility of the gliding dislocations and in the formation of the dislocation substructure. Although the exact contribution of mobility and substructure has not been determined, both contribute significantly to the high temperature creep resistance of hard oriented crystals.

Conclusions: Deformation in Soft and Hard orientations

1. Deformation of NiAl single crystals loaded along the soft [223] axis occurs by single slip and involves the glide movement of **b**=[001] dislocations. High temperature creep deformation is characterized by immediate steady state creep with no significant primary creep transient. Strain hardening is minimal in both tension creep and constant strain rate compression and few dislocations are observed in the deformed crystals.

2. Deformation of NiAl single crystals loaded along the hard, [001], orientation occurs by glide of two **b**=<101> dislocations; glide of the **b**=<0$\bar{1}$1> dislocations is inhibited by latent hardening. Creep of hard oriented crystals is characterized by a sigmodial creep curve.

3. The **b**=<100> dislocation networks observed in the hard oriented crystals after deformation result from the decomposition of **b**=<101> dislocations.

4. The high deformation resistance of hard oriented crystals at high temperatures is caused both by the low mobility of **b**=<101> dislocations, compared to **b**=<100>, and by the dislocation substructure formed during deformation.

Acknowledgments

The authors gratefully acknowledge the support of the Air Force Office of Scientific Research under AFOSR Grant No. F49620-92-J-0009. The support of Dr. Alan Rosenstein of AFOSR is very much appreciated. One of the authors (UG) also gratfully acknowledges the Alexander von Humboldt Foundation for their support with a scholarship in the Feodor-Lynen Program.

References

1. D. B. Miracle, S. Russell and C. C. Law in *High Temperature Ordered Intermetallic Alloys III*, edited by C. T. Liu, A. I. Taub, N. S. Stoloff and C. C. Koch (Mater. Res. Soc. Proc., **133**, Pittsburgh, PA, 1989) pp. 225-230.
2. R. Darolia, D. F. Lahrman, R. D. Field and A. J. Freeman in *High Temperature Ordered Intermetallic Alloys III*, edited by C. T. Liu, A. I. Taub, N. S. Stoloff and C. C. Koch (Mater. Res. Soc. Proc., **133**, Pittsburgh, PA, 1989) pp. 113-118.
3. R. Darolia, *Journal of Metals* **43**, 44 (1991).
4. A. Ball and R. E. Smallman, *Acta Metall.* **14**, 1517 (1966).
5. A. Ball and R. E. Smallman, *Acta Metall.* **14**, 1349 (1966).
6. R. R. Vandervoort, A. K. Mukerjee and J. E. Dorn, *Transactions of the ASM* **59**, 930 (1966).
7. P. R. Strutt and R. A. Dodd in *Structural Application and Physical Metalurgy*, edited by Kear, Sims and Stoloff (1970) pp. 475-503.
8. J. Bevk, R. A. Dodd and P. R. Strutt, *Metallurgical Transactions* **4**, 159 (1973).
9. R. D. Field, D. F. Lahrman and R. Darolia, *Acta metall. mater.* 2951 **39** 2951(1990).
10. N. J. Zaluzec and H. L. Fraser, *Scripta metallurgica* **8**, 1049 (1974).
11. R. D. Noebe, R. R. Bowman, C. L. Cullers and S. V. Raj in *High Temperature Ordered Intermetallic Alloys IV*, edited by L. A. Johnson, D. P. Pope and J. O. Stiegler (Mater. Res. Soc. Proc., **213**, Pittsburgh, PA, 1990), pp. 589-596.
12. R. T. Pascoe and C. W. A. Newey, *Metal Science Journal* **5**, 50 (1971).
13. P. R. Strutt, R. A. Dodd and G. M. Rowe, *Proc. 2nd Int. Conf. on Metals and Alloys*, (Am. Soc. Metals, **3**, Metals Park, 1970) p1057.
14. W. J. Yang and R. A. Dodd, *Metal Science Journal* **7**, 41 (1973).
15. J. D. Whittenberger, *Journal of Materials Science* **22**, 394 (1987).
16. G. F. Hancock and B. R. McDonnell, *Phys. Stat. Sol.* **4**, 143 (1971).
17. R. J. Wasileski, *Transactions of the Meturlurgical Society of AIME* **236**, 455 (1966).
18. U. Glatzel, K. R. Forbes and W. D. Nix, these Proceedings.
19. P. Haasen in *Dislocation Dynamics* edited by A. R. Rosenfield, G. T. Hahn, A. L. Bement and R. I. Jaffe (McGraw-Hill, New York, 1967) pp 701-722.
20. K. R. Forbes and W. D. Nix, these Proceedings..
21. U. Glatzel, K. R. Forbes and W. D. Nix, *accepted for publ. in Phil.Mag.*
22. D. Hull and D. J. Bacon, *Introduction to Dislocations* , 3rd ed. (Pergamon Press, Oxford, 1984) p. 185.
23. M. J. Mills and D. B. Miracle, *submitted to Acta metall. mater.*
24. T. H. Alden, *Trans. AIME* **230**, 649 (1964).
25. M. J. Mills, M. S. Daw and D. B. Miracle, these Proceedings.

FATIGUE OF INTERMETALLIC COMPOUNDS AND THEIR COMPOSITES

N.S. Stoloff, T.R. Smith and A. Castagna
Materials Engineering Department Rensselaer Polytechnic Institute,
Troy, NY 12180-3590.

ABSTRACT

Recent studies of fatigue behavior of intermetallic compounds are reviewed. Emphasis is upon fatigue damage leading to crack initiation, effects of environment on crack propagation and behavior of intermetallic matrix composites. Future research directions are suggested.

INTRODUCTION

Improvements in low temperature ductility and high temperature creep resistance remain as major goals for most research studies on intermetallic compounds. Nevertheless, use of intermetallics in aerospace or other structural applications will require knowledge of their behavior under cyclic loading conditions. It already has been established that several Ll_2 intermetallics, including Ni_3Al+B, display very good resistance to stress controlled fatigue (high cycle or crack growth) at room temperature[1]. Other intermetallics, such as Fe_3Al[2] and Ti_3Al[3], display very rapid crack growth, possibly due to high notch sensitivity or susceptibility to the environment (e.g. moisture or hydrogen[4]). Moreover, at elevated temperatures, interactions with the test environment (typically oxygen) or with creep components arising from the load cycle can adversely influence fatigue behavior[5,6]. With respect to strain controlled fatigue, detailed studies of crack initiation have been carried out on crystals of Ni_3Al+B[7-9] and NiAl[10]. Low cycle fatigue behavior of polycrystalline Ni_3Al[11] and NiAl[12,13] alloys also have been reported recently.

The purpose of this paper is to review recent progress in understanding of fatigue behavior, especially with respect to crack initiation and growth.

CRACK INITIATION

Most work on crack initiation has been carried out on Ni_3Al single crystals. A series of studies in our laboratory have provided evidence that crack initiation in Ni_3Al is preceded by an accumulation of polygonal loops, Fig. 1[14], dislocation dipoles and point defect clusters, as well as the formation of persistent slip bands (PSB) and intrusion/extrusion surface morphology[7-9]. The dipoles are generated primarily by nonconservative motion of jogged superdislocations and their break-up by pinch-off, see Fig. 2[14]. Dragging of jogs, impedance of dislocation motion by dipoles and dislocation interactions contribute to cyclic hardening.

The resemblance of the strain-controlled cyclic behavior of Ni_3Al to copper with respect to point defect formation and the development of intrusions and extrusions has

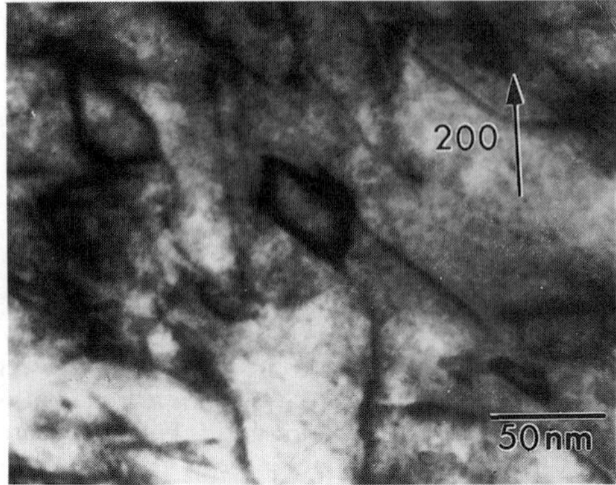

Fig. 1. Polygonal dislocation loops in fatigued Ni_3Al[14].

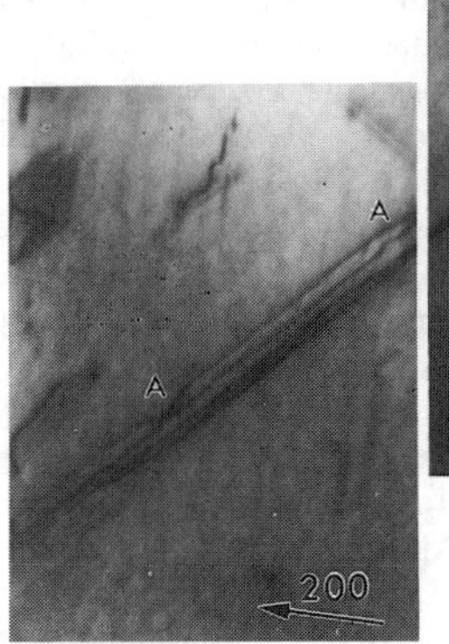

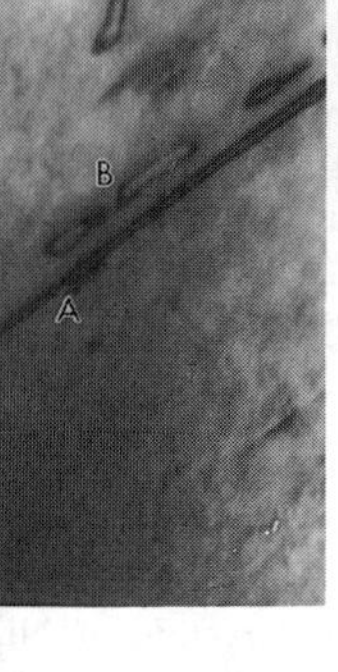

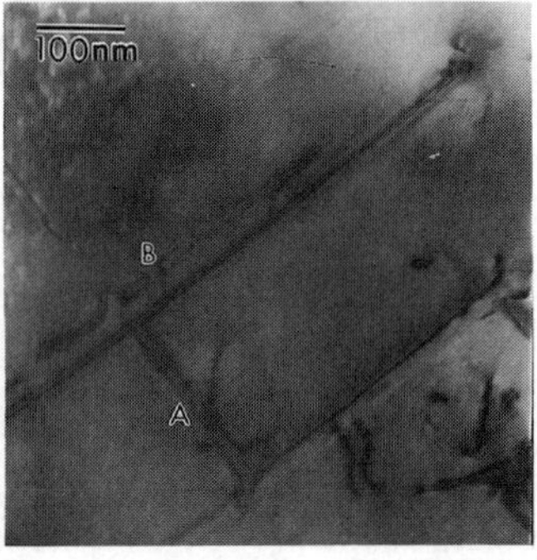

Fig. 2. Dislocation dipole breakup by pinch off in Ni_3Al[14].

been commented upon elsewhere[8]. The most apparent difference between Ni_3Al and copper crystals is the formation of a "ladder" structure in the latter only[15].

Although studies of crack initiation in other intermetallics are not as detailed, it is clear that extrusion-intrusion formation prior to the development of cracks is a general phenomenon. Such surface morphologies have been seen in Ll_2 Ni_3Fe (polycrystals)[16], and B2 FeCo-2%V polycrystals[17] and B2 NiAl (single crystals[10]). Fig. 3 shows an incipient crack at an intrusion-extrusion pair in an NiAl crystal[10]. Dislocation loops have been seen in the interior of the same crystal, Fig. 4[10]. Note the zig-zag appearance of the dislocations, which have been identified as of <001> type. This configuration probably arises from the large elastic anisotropy in NiAl. Loops have been observed also in monotonically deformed NiAl, so it is not possible as yet to determine their significance (if any) with respect to crack nucleation in NiAl.

The formation of vacancy clusters has been linked previously to crack initiation in Ni_3Al single crystals[8]. This model suggests that the condensation of point defect clusters at the PSB-matrix interface plays a role in the initiation of cracks. Thus the micrographs

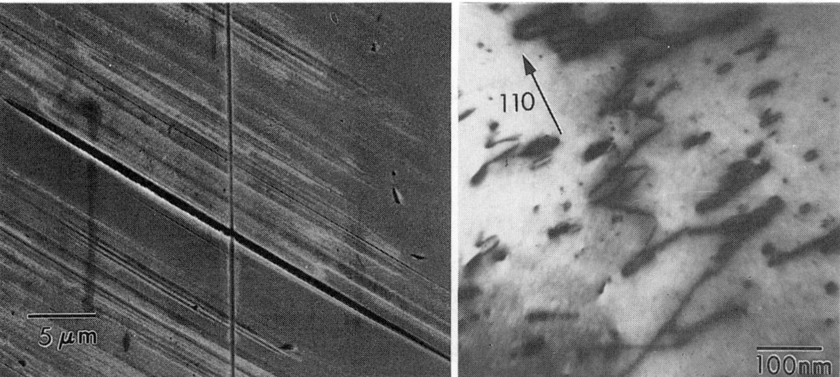

Fig. 3. Incipient crack at intrusion/ extrusion pair in NiAl[10].

Fig. 4. Loop formation in NiAl[10].

presented here provide additional evidence for the formation of defect clusters. Further, the formation of these clusters has been shown to be energetically favorable[9].

Studies of crack initiation in NiAl also have included cyclic hardening data. Single crystals tested at room temperature display hardening to at least 400 cycles at each of several strain amplitudes[10]. Hardening is continuous in polycrystals at room temperature, and a doubling of stress to maintain a plastic strain range of 0.002 was noted in just over 400 cycles[12]. Final fracture stress under cyclic conditions was at least 60% higher than the monotonic tensile strength[12]. At 700K, hardening occurs rapidly to about 30 cycles, and then very slowly to 3000 cycles[12]. At this temperature an extrusion-intrusion surface slip band morphology was noted. However, the only internal dislocation substructure reported in that work was a set of slip bands with a spacing similar to that of the surface bands. No clear evidence of a flow stress asymmetry between compression and tension has been found for NiAl, unlike Ni_3Al, which displays a marked, orientation-dependent asymmetry[8,19]. Also, unlike NiAl, cyclic hardening in Ni_3Al crystals saturates within the first 100 cycles[8].

CRACK PROPAGATION

Ni_3Al Alloys

We have carried out detailed studies of crack propagation in two aluminides, an Ni_3Al alloy[6] and Fe_3Al[2,4,20]. In both alloy systems environmental effects are very important: for Ni_3Al due to oxygen and for Fe_3Al due to moisture and or hydrogen. Crack growth rates for Ni_3Al alloys as well as other $L1_2$ alloys are lower than for nickel-base superalloys[1]. Also, crack growth rates increase with increasing temperature in spite of a rise in flow stress over the same temperature range. Crack tip embrittlement by oxygen seems to be a major factor, although creep processes also influence growth rates at temperatures above about 700°C[5]. Fig. 5 shows that crack growth rates, da/dN, increase with decreasing frequency in both vacuum and air for alloy IC 221 (Ni-9w%Al-8%Cr,1.8%Zr,0.02%B) at 800°C but that rates always are higher in air[5]. While increased

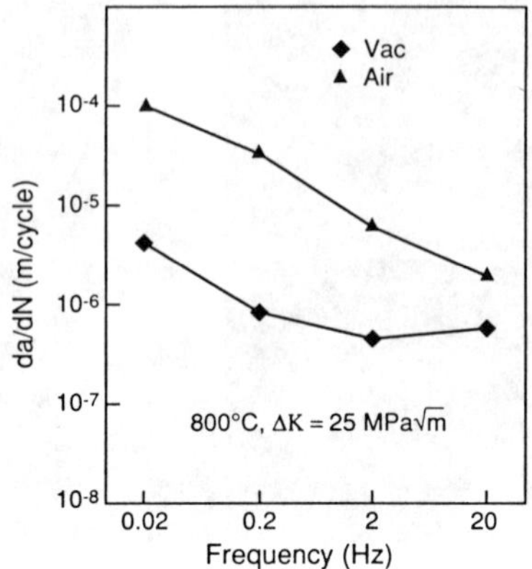

Fig. 5. Effect of frequency and environment on crack growth of Ni_3Al alloy IC 221[6].

growth rates at low frequency in vacuum are indicative of a creep effect, the higher growth rates in air undoubtedly reflect a contribution of environmental embrittlement. Intergranular fracture occurs in air at all test frequencies, while transgranular fracture is noted in vacuum.

Fe_3Al Alloys

Binary Fe_3Al alloys generally are intermediate in growth rates (in air) between superalloys and $L1_2$ intermetallics, e.g. Ni_3Al or $(Fe,Ni)_3V$, at low ΔK levels; at high ΔK, crack growth is more rapid in Fe_3Al[1].

Recent work has shown that hydrogen[20] or moisture[21] is very detrimental to fracture ductility of iron aluminides. Our own research has extended the study of environmental effects to cyclic loading. Crack growth rates for an Oak Ridge-developed Fe_3Al alloy, 28.6a%Al,4.8%Cr,0.5%Nb,0.21%C, bal Fe (FA-129), are shown in Fig. 6a) for the partially disordered B2 condition, and in Fig. 6b) for the fully ordered DO_3 condition, both at room temperature[22]. Note that the lowest crack growth rates are observed with oxygen or vacuum environments for both ordered conditions. Also apparent are considerably higher slopes of the crack growth curves for the DO_3 condition, Fig. 6b). Values for the slopes as well as apparent threshold and critical stress intensities for fracture for each environment and ordered condition may be found in Table I. In every respect the DO_3 condition is less resistant to cyclic crack growth and fracture, especially in each of the aggressive environments: air, hydrogen gas or after precharging. It is now generally considered that moisture in contact with iron aluminides breaks down by chemical reaction at the surface of the sample, thereby liberating hydrogen, which is the specific embrittling agent[22,23].

Chromium reportedly improves the ductility of Fe_3Al in monotonic tension tests in air[24], but offers no benefit under cyclic loading conditions. Growth rates are actually higher for FA-129 than for a binary alloy with similar Al content, both tested in the DO_3 condition[22]. This unexpected result may be due to the influence of chromium on the repeated formation and rupture of the oxide film on a freshly exposed crack surface.

The influence of temperature on crack growth rates of FA-129 is shown in Fig. 7[22]. Growth rates decrease between 25°C and 150°C in the B2 condition, Fig. 7a) as well

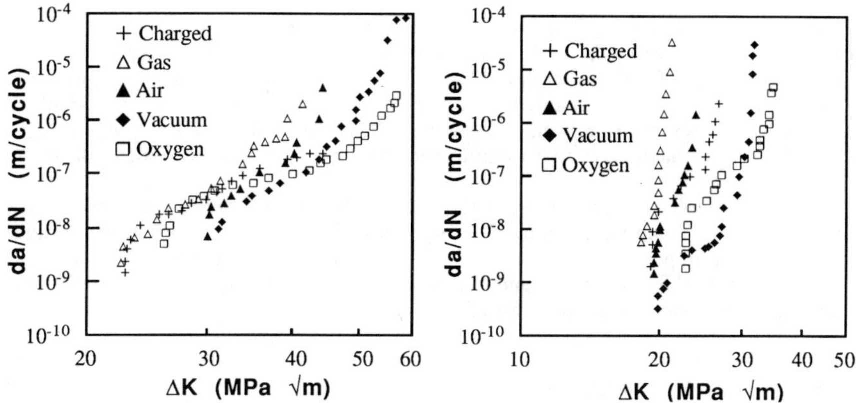

Fig 6. Crack growth of Fe_3Al alloy FA-129, 25°C[22].
a) B2 condition; b) DO_3 condition

Table I. Fatigue Crack Growth Data for FA129 at 25°C[22]

Condition	Fracture Surface	m	ΔK_{TH} (MPa√m)	ΔK_{TH} (MPa√m)
B2 Oxygen	Dimpled, few striations	2.9	26	61
B2 Vacuum	TG, many striation	5.1	31	59
B2 Air	TG, many striation	9.4	30	44
B2 Charged	TG, few striations	6.8	23	44
B2 Gas	TG, few striations	8.6	22	41
DO_3 Oxygen	Dimpled + Cleavage	8.8	23	35
DO_3 Vacuum	TG, few striations	5.2	20	32
DO_3 Air	Mixed, few striations	22.8	19	24
DO_3 Charged	Mixed, no striations	9.1	19	27
DO_3 Gas	Mixed, no striations	37.6	18	21

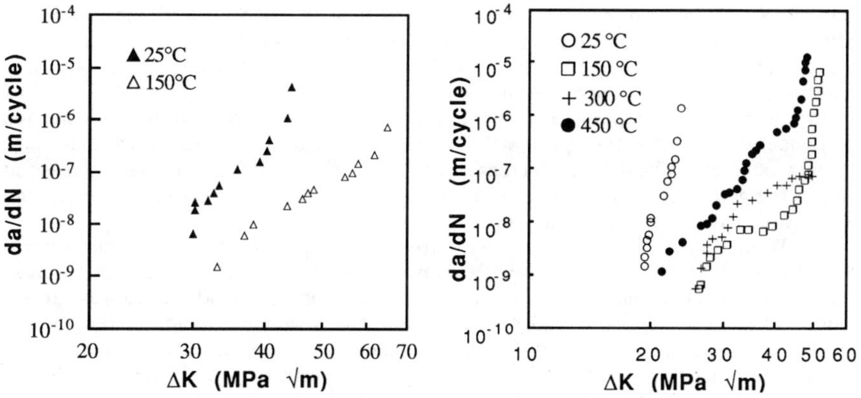

Fig. 7. Influence of temperature on crack growth of FA-129[22].
a) B2 condition; b) DO_3

as in the DO_3 condition, Fig. 7b). However, at higher temperatures growth rates rise again for the DO_3 condition. (Tests of the B2 condition are limited to low temperatures due to the occurrence of DO_3 ordering). Even at 450°C, crack growth rates are lower than at room temperature, presumably due to the fact that hydrogen diffuses too rapidly at high temperatures to cause embrittlement. Therefore, it appears that hydrogen embrittlement is maximized near room temperature, as is the case for structural steels.

Fractographic features in Fe_3Al alloys differ somewhat with environment and type of order[4,22]. At room temperature, fracture is usually transgranular, often with fatigue striations superimposed on cleavage facets. However, air tends to cause some intergranular cracking during fatigue. At elevated temperatures, in the DO_3 condition, cracking is more fibrous in appearance.

Ti_3Al and Ti_3Al Composites

Although crack growth phenomena in Ti_3Al alloys have been studied extensively, little had been published until recently. One of the most interesting aspects of crack growth behavior is the potent effect of microstructure, see Fig. 8[25]. Williams et al[3] have shown that crack growth rates of Ti-24-11 and Ti-25-10-3-1 (super α_2) at 25°C are lower than that for the commercial alloy Ti-6246 in the near threshold region, but above a ΔK of about 10MPa√m growth rates are more rapid in the intermetallics. Note that the super α_2, which is stronger than the Ti-24-11 alloy, has a much higher growth rate than either of the others. Davidson et al[27] report about the same threshold for super α_2, but show that the range of thresholds in α+ß alloys is higher, and crack growth rates are lower than for super α_2 at 20°C, see Fig. 9. As temperature increases to 650°C relative crack growth rates of Ti-24-11 change in a complex manner. However, at 650°C, the growth rate of Ti-24-11 is higher than that of either IMI 834 or the nickel-base superalloy IN 100. The influence of microstructure on fatigue life is as great or greater than compositional or temperature effects. For example, super α_2 that has been solutioned in the ß range has a higher growth rate at 650°C than the same alloy with a 20% equiaxed α, transformed ß microstructure[26]. There is a further improvement in fatigue crack growth resistance when the alloy is cycled in vacuum.

Further evidence for an environmental effect on crack growth of α_2 at both room and elevated temperature has been provided by Aswath and Suresh[28] and Balsone et al[29]. There is an effect of frequency on crack growth rates in laboratory air and in vacuum (10^{-5}torr) for Ti-24-11 with coarse Widmanstatten α_2 surrounded continuously by transformed ß (this is a microstructure with relatively high FCG resistance[28,29]). Crack growth is much more rapid in air than in vacuum. Similar observations have been made at temperatures in the range 650-800°C[28,29]. Balsone et al[29] have shown that crack growth rates of Ti-24Al-11Nb decrease between 25°C and 250°C, but then increase at higher temperatures. Crack growth rates in air decreased with increasing frequency, and were further decreased by testing in vacuum.

Wessels et al[30] also have reported strong microstructural effects on crack growth at 425°C of a Ti-24.5Al-8-Nb-2Mo-2Ta alloy. A ß-heat treated sample with a coarse aligned lath or colony microstructure produced by slow cooling provided the lowest growth rates and highest threshold (about 6 MPa√m). The lowest threshold, 3 MPa√m, and highest growth rate was exhibited by a salt quenched α_2+ß sample. Vekaturaman et al[31] report that microstructure has little effect on crack growth rates of Ti-16w%Al-10%Nb in air, but that testing in vacuum lowered the crack growth rate. Therefore, it may be concluded that hydrogen embrittles α_2 alloys at low temperatures, while oxygen is detrimental at elevated temperatures, similar to the behavior of Ni_3Al alloys. In general,

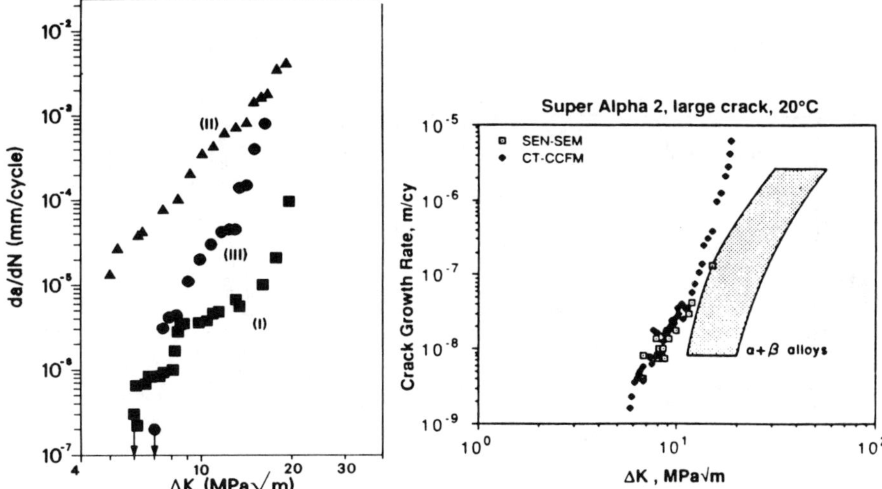

Fig. 8. Effect of microstructure on crack growth of Ti_3Al at 25°C[25]. Coarse Widmanstätten α_2:[I]; α_2 in Widmanstätten matrix:[II]; coarse equiaxed α_2:[III].

Fig. 9. Crack growth in Ti-25-10-3-1 (super α_2) compared to $\alpha+\beta$ alloys[27].

fatigue crack growth rates in α_2 increase with moist environments (probably due to the release of hydrogen), lowered test frequency, increased hold times and increased test temperature.

Perhaps the single factor most likely to influence the fatigue crack growth resistance of titanium aluminides is the use of fibrous reinforcements. Fig. 10 shows that although the crack growth rates of α_2 and γ are higher than for the P/M Ni base superalloy IN 100, the longitudinally oriented composite of α_2 (Ti-24Al-11Nb) with continuous SCS-6 SiC fibers is far more crack growth resistant than is IN 100[32]. Unfortunately, the transversely oriented α_2 composite displays the highest growth rate of all. (Note: the specimen geometry differed in that thin tubes of Ti-48Al-1V (γ) were utilized; also, the frequency of tests of the α_2 composite was 3.33 Hz, which would be expected to result in a somewhat lower growth rate than at 0.2 Hz). It is noteworthy that the fracture toughness of the longitudinally tested α_2/SCS-6 composite is about 70 MPa√m at 650°C, while IN 100 displays a K_{IC} of about 100 MPa√m. Transverse specimens of the α_2 composite as well as the γ alloy, on the other hand, display about the same level of K_{IC}, approximately 15 MPa√m.

TiAl

Data for TiAl(γ) alloys are even more limited than for Ti_3Al. Note in Fig. 10 that the growth rate of Ti-48Al-1V is comparable to that of super α_2[32]. Soboyejo et al[33] have reported the influence of microstructure and temperature on fatigue crack growth behavior of a P/M Ti-48%Al alloy tested in air. Crack growth at room temperature was slightly slower than for a Ti-6Al-4V alloy, see Fig. 11[33]. At 700°C, crack growth was even slower, probably due to oxide-induced crack closure.

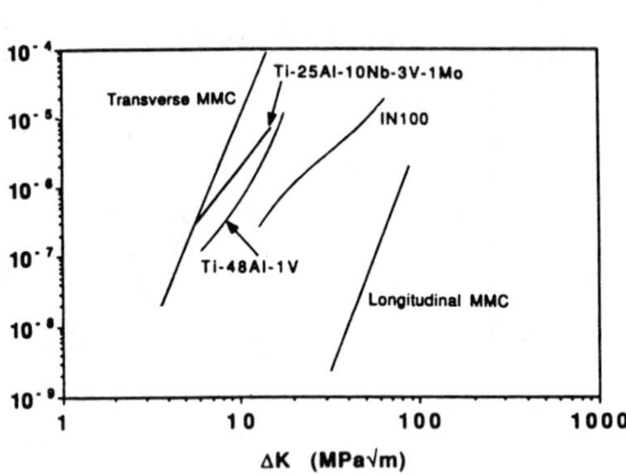

Fig. 10. Comparison of crack growth rates in α_2, γ and α_2/SCS-6 composites[32].

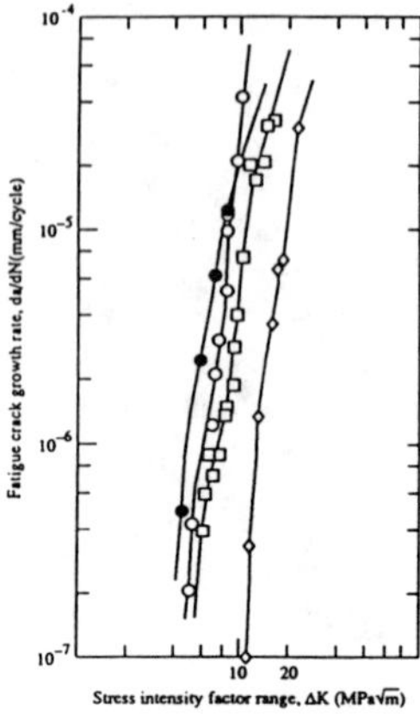

Fig. 11. Crack growth rates in Ti-48%Al-compared to Ti-6Al-4V[33]. □ Ti-48Al 298°K, R=0.1; ○ Ti-48Al 298°K, R=0.3; ● Ti-6Al-4V mill annealed forging, R=0.1; ◇ 973°K, R=0.1.

Ductile-Phase Toughened Composites

With the exception of the α_2-SCS-6 system, crack growth data have been reported only for composites reinforced with ductile phases, as listed in Table II. All composites listed have been tested only at room temperature in air. As in the case of monotonic tests, particles do not confer much advantage due to the lack of crack-stopping ability of low aspect ratio reinforcements. For example, cracks grow at very low stress intensities in $MoSi_2$-Nb composites, as shown in Fig. 12[34]. For this reason it has been suggested that increasing the aspect ratio, together with a reduction in particle-matrix interface strength, are necessary for improved crack growth resistance in these composites[34].

Research on Nb_3Al-Nb composites has shown that microstructures developed for superior toughness under monotonic conditions may not result in improved fatigue resistance[35]. Under cyclic loading conditions, the effects of crack bridging are decreased and fatigue cracks grow at low stress intensity levels, as in the case of the $MoSi_2$ system. In fact, cracks grow in Nb at much lower rates than in the Nb_3Al-Nb composites[35].

The most detailed study of crack growth in ductile-phase toughened composites has been carried out on γ-TiAl reinforced with 25 or 50μm TiNb particles[36,37]. As in the case of Nb_3Al-Nb composites, cracks can propagate subcritically at very low stress intensities, less than 6MPa√m, see Table III and Fig. 13, compared to a fracture toughness of 25MPa√m. However, the Paris slopes, m, are considerably reduced by TiNb[37]. A particularly striking observation was that the crack growth resistance of monolithic TiAl appears to be diminished by the particles, particularly at near-threshold levels, as shown in Fig. 13. However, crack growth occurs in pure TiNb at stress intensities below

3MPa√m with no evidence of a threshold[37]. The composite displays a clear threshold, similar to the behavior of metals. Closure measurements in the composite and in monolithic TiAl are similar, with maximum closure levels at about 40% of K_{max}. In pure TiNb, for which the threshold, if any, is lowest, crack closure is high, approaching levels of about 0.7 K_{max}[37].

Table II. Fatigue Crack Growth of Composite Systems

Matrix	Reinforcement	Form	Vol. Fract. %	Ref.
γ TiAl	TiNb	particle	5-20	36,37
$MoSi_2$	Nb	particle	20	34
NiAl-25v%B_4C	304SS	tube	40	38
Nb_3Al	Nb	fiber	20	35
Ti-24Al-11Nb	SCS-6	fiber	-----	32

Table III. Summary of Cyclic Crack-Growth Data in TiAl+TiNb Composites, TiAl and TiNb at R = 0.1

Material	ΔK_{TH}(MPa√m)	Slope m	Constant C
Monolithic γ-TiAl	5.8	29.4	$9.7x10^{-31}$
TiAl + 5% TiNb(50μm)	4.5	17.6	$5.3x10^{-18}$
TiAl + 10% TiNb(50μm)	5.6	14.1	$1.1x10^{-16}$
TiAl + 20% TiNb(50μm)	5.0	9.6	$2.0x10^{-13}$
TiAl + 20% TiNb(25μm)	5.3	9.7	$2.5x10^{-13}$
Monolithic ß-TiNb	2.9	3.9	$7.4x10^{-9}$

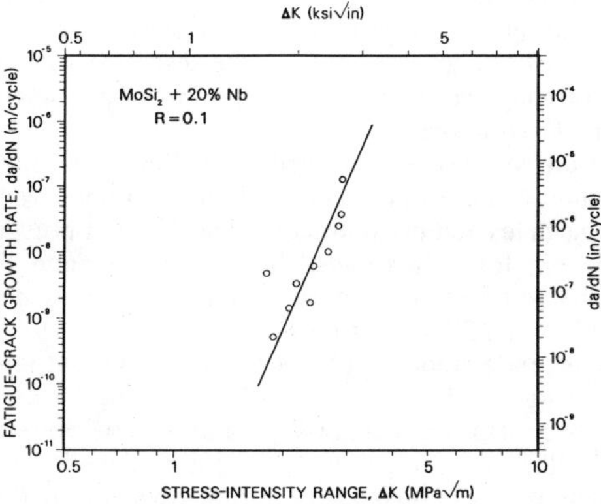

Fig. 12. Crack growth in $MoSi_2$/Nb(p) composites, room temperature, air[34].

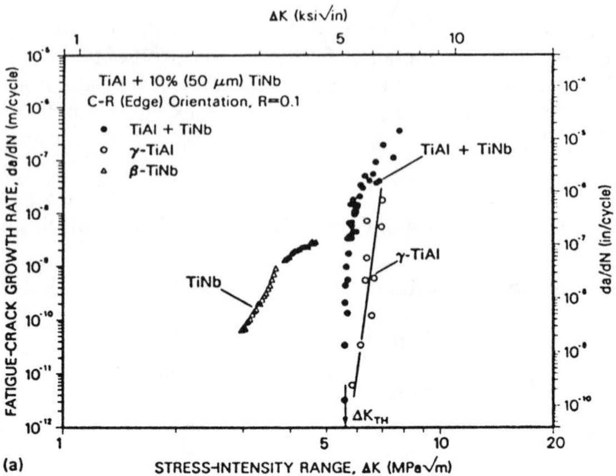

Fig. 13. Variation in cyclic behavior of TiAl+10v%TiNb (50μm thick) composite, compared to monolithic TiAl and TiNb, room temperature[37].

DISCUSSION

Studies of fatigue behavior of intermetallics continue to lag behind other mechanical property evaluations. Most of the published fatigue literature of the past five years pertains to Ni_3Al and Ti_3Al, with isolated reports of fatigue data for Fe_3Al and TiAl. It is clear that fatigue behavior of aluminides is very much affected by the external environment, just as is the case for other fracture-related properties. Hydrogen or moisture adversely affects the low temperature HCF and crack growth behavior of Ni_3Al, Fe_3Al and Ti_3Al alloys. At elevated temperatures oxygen is very detrimental to the fatigue resistance of Ni_3Al and Ti_3Al alloys. Chromium is highly beneficial in improving the tensile ductility of Ni_3Al in air[24]; under cyclic loading conditions some protection against oxygen penetration may be provided, but to some extent this is overcome by the much longer times necessary to carry out fatigue tests. During the latter, diffusion of oxygen to the crack tip may occur readily. Compositional and microstructural effects are particularly strong in Ti_3Al alloys[25,28-30].

The research on composites reviewed here shows that ductile phase toughening under monotonic conditions does not necessarily lead to improved fatigue crack growth resistance. Ductile particles and discontinuous fibers cannot prevent crack growth under cyclic conditions at very low ΔK values. Therefore, it seems likely that continuous, aligned composites are required for improved cyclic crack growth resistance, as in in the case of α_2 reinforced with SCS-6 SiC fibers[32].

One of the least understood aspects of cyclic behavior of intermetallic compounds is the sequence of events leading to crack initiation. Studies on polycrystals show that grain boundaries, pores, inclusions and/or slip bands can be preferred sites for initiation, depending upon the alloy system and processing technique[1]. Although not explicitly reviewed here, slip bands are preferred sites in $(Fe,Ni)_3V$, FeCo-V and wrought Ni_3Al+B at room temperature while grain boundaries are the dominant sites in unnotched Fe_3Al at room and elevated temperatures[2].

Single crystal studies on Ni_3Al[8] and NiAl[10] show clearly that persistent slip bands form readily and are sites for early crack initiation. Although a few other studies of fatigue in single crystals of Ni_3Al have been carried out[19,39], none have included detailed surface damage observations.

There is a need for work on single crystals of other ordered alloys to establish the relations among ordering energy, crystal orientation and crack initiation sites. Also, the role of cyclic hardening (or softening) in crack initiation remains unclear. In non-ordered metals, initiation is connected with the onset of saturation. However, in the few studies carried out to date on ordered alloys, wide differences in cyclic hardening behavior have been noted. For example, while a flow stress asymmetry has been noted in single crystals of Cu_3Au[18], Ni_3Al[8,19], and Ni_3Ge[26], the effect is very small in Cu_3Au and is not always in the same direction in the three systems. For Cu_3Au and Ni_3Ge the flow stress is higher in compression than in tension, while in Ni_3Al the reverse is true. Hardening saturates early in Ni_3Al crystals, but not in NiAl single crystals or polycrystals. Differences in ordering energy among the various intermetallics, as well as different orientations of crystals in each investigation, could be responsible for some of the discrepancies, but clearly much more work needs to be done in this field.

ACKNOWLEDGEMENTS

This research was supported by the National Science Foundation under Grant No. DMR-8911975 and by Department of Energy Fossil Energy Program under Oak Ridge Subcontract No. 19X-SF521C. Many discussions with Professor K. Rajan of Rensselaer have been of great benefit. The author is grateful to Dr. C. Liaw of Pratt and Whitney Aircraft, Division of United Technologies Corporation for supplying single crystals of Ni_3Al+B, to Drs. K.M. Chang and R. Darolia of General Electric Company for providing single crystals of NiAl and to Dr. V.K. Sikka of Oak Ridge National Laboratory for providing alloy FA-129.

REFERENCES

1. N.S. Stoloff, G.E. Fuchs, A.K. Kuruvilla and S.J. Choe in High Temperature Ordered Intermetallic Alloys II, Mat. Res. Soc. Symp. Proc. **81**, Pittsburgh, PA, 1987, pp. 247-261.
2. G.E. Fuchs and N.S. Stoloff, Acta Metall. **36**, 1381 (1988).
3. K.A. Williams, S.J. Balsone, M.A. Stucke and J.M. Larsen, presented at Aeromat 90, Long Beach, CA, 1990, pp. 21-24.
4. A. Castagna and N.S. Stoloff, Scripta Metall. **26**, 673 (1992).
5. G. Camus, D.J. Duquette, N.S. Stoloff, J. Mater. Res. **5**, 950, (1990).
6. W. Matuszyk, G. Camus, D.J. Duquette and N.S. Stoloff, Metall. Trans. A, **21A**, 2967 (1990).
7. L.M. Hsiung and N.S. Stoloff in High Temperature Ordered Intermetallic Alloys III, Mat. Res. Soc. Symp. Proc. **133**, Pittsburgh, PA, 1989, pp. 261-267.
8. L.M. Hsiung and N.S. Stoloff, Acta Metall. **38**, 1191 (1990).
9. L.M. Hsiung and N.S. Stoloff, Acta Metall. **40**, 2993 (1992).

10. T.R. Smith, C.G. Kallingal, K. Rajan and N.S. Stoloff, Scripta Metall. **27**, 1389 (1992).
11. D.E. Gordon and C.K. Unni, J. Mater. Res. (1992) in press.
12. C.L. Cullers and S.D. Antolovich in Superalloys 92, Proc. 7 Springs Conf. on Superalloys, TMS, Warrendale, PA, 1992, pp. 351-359.
13. R.D. Noebe and B.A. Lerch, Scripta Metall. **27**, 1161 (1992).
14. C.G. Kallingal, T.R. Smith, N.S. Stoloff and K. Rajan, Scripta Metall. **27**, 1407 (1992).
15. J.G. Antonopolous, L.M. Brown and A.T. Winter, Phil. Mag. **34**, 549 (1976).
16. S.J. Choe and N.S. Stoloff, Rensselaer Polytechnic Institute, Troy, NY, 1987 (unpublished).
17. N.S. Stoloff, S.J. Choe and K. Rajan, Scripta Metall. **26**, 331 (1992).
18. K.H. Chien and E.A. Starke, Jr., Acta Metall. **23**, 1173 (1975).
19. N.R. Bonda, D.P. Pope and C. Laird, Acta Metall. **35**, 2371 (1987).
20. A. Castagna, M. Shea and N.S. Stoloff in High Temperature Ordered Intermetallic Alloys IV, Mat. Res. Soc. Proc. **213**, Pittsburgh, PA, 1991, pp. 609-616.
21. C.T. Liu, C.G. McKamey and E.H. Lee, Scripta Metall. **24**, 385 (1990).
22. A. Castagna and N.S. Stoloff in High Temperature Ordered Intermetallic Alloys V, Mat. Res. Soc. Symp. Proc., 1992, in press.
23. L. Heldt, TMS Fall Meeting, Chicago, Illinois, Nov. 1992, unpublished.
24. C.G. McKamey, J.A. Horton and C.T. Liu, Scripta Metall. **22**, 1679 (1988).
25. P.B. Aswath and S. Suresh, Metall. Trans. **22A**, 817 (1991).
26. L.M. Hsiung, M.S. Thesis, New Mexico Institute of Technology, Socorro, New Mexico, (1986).
27. D.L. Davidson, J.B. Campbell and R.A. Page, Metall. Trans. A **22A**, 377 (1991).
28. P.B. Aswath and S. Suresh, Mat. Sci. and Eng. **A114**, L5 (1989).
29. S.J. Balsone, D.C. Maxwell, M. Khobaib and T. Nicholas in Fatigue 90, edited by H. Kitagawa and T. Tanaka (Materials and Components Eng. Publ. Ltd., **III**, Warley, UK (1990) pp. 1905-1910.
30. J.F. Wessels, B.J. Marquardt and D.D. Krueger, presented at TMS-AIME Symp. on Creep and Fracture of Titanium Aluminides, TMS, Indianapolis, IN, 1989.
31. S. Vekaturaman, AFWAL-TR-87-4103, Air Force Materials Lab, Wright Patterson Air Force Base, Ohio, (1987).
32. J.M. Larsen, K.A. Williams, S.J. Balsone and M.A. Stucke in High Temperature Aluminides and Intermetallics, edited by S.H. Whang et al, TMS-AIME, Warrendale, PA, 1990, pp. 521-556.
33. W.O. Soboyejo, P.B. Aswath and J.E. Deffeyes, Mat. Sci. and Eng. A. **A138**, 95, 1991.
34. K.T. Venkateswara Rao, W.O. Soboyejo and R.O. Ritchie, submitted to Metall. Trans. A, (1992).
35. L. Murugesh, K.T. Venkateswara Rao, L.C. DeJonghe and R.O. Ritchie in Intermetallic Composites II, Mat. Res. Soc. Proc. **273**, 1992, in press.
36. K.T. Venkateswara Rao, G.R. Odette and R.O. Ritchie, Acta Metall. **40**, 353 (1992).
37. K.T. Venkateswara Rao and R.O. Ritchie, submitted to Mat. Sci. and Eng. A. (1992).
38. Vincent C. Nardone, Metall. Trans. A. **23A**, 563 (1992).
39. J.E. Doherty, A.F. Giamei and B.H. Kear, Metall. Trans. A **6A**, 2195 (1975).

FRACTURE AND TOUGHNESS OF INTERMETALLICS

HORST VEHOFF
MPI für Eisenforschung, Max Planck Str. 1, D-4000 Düsseldorf, Germany

ABSTRACT

The difficulties to obtain valid fracture toughness values in brittle intermetallics are discussed. Different intermetallic alloys with the same specimen size were tested in four point bending to allow the direct comparison of the brittleness of different alloys. The influence of grain size, phase distribution, and temperature on the fracture toughness was measured. The fracture toughness of many intermetallic alloys was found to be extremely rate sensitive. This is discussed in view of dynamic models of the brittle/ductile transition developed recently. NiAl single crystals with <100>-specimen axis showed the largest toughness at room temperature and compared to other orientations the highest transition temperature. The apparent activation energy for the brittle/ductile transition depends on orientation. Evidence was given that in multiphase alloys the yielding of a second phase initiates interfacial fracture. For polycrystals it was shown that different processes can cause brittle/ductile transitions.

INTRODUCTION

New intermetallic materials should combine the advantages in material behavior of ceramics (high strength at high temperature) with a high fracture toughness and sufficient ductility at room temperature. Therefore in many countries the development of new and improved intermetallic alloys is currently under way as is documented in several conference proceedings on intermetallic alloys [1, 2].

The most challenging problem is still the brittleness at room temperature. The purpose of this overview is to review briefly our current understanding of fracture and problems involved to test small quantities of brittle materials. Different methods for precracking and notching are described together with a description of the method used to obtain fracture toughness values. Fracture toughness data for intermetallic single crystals and for multiphase alloys are presented. The influence of temperature, strain rate and texture on fracture toughness is discussed in view of recently developed models for fracture.

TOUGHNESS TESTING IN BRITTLE MATERIALS

Intermetallics, depending on crystal structure, microstructure or processing, show fracture toughness values which either rise sharply or smoothly with temperature. For toughness testing in the brittle/ductile transition range, standards known from metals, like precracking in fatigue and a sufficient specimen size according to the ASTM-standard must be used. Many high temperature intermetallics show brittle behavior at lower temperatures with toughness values below 5 $MPa\sqrt{m}$. In these cases, procedures known from the testing of ceramics must be applied. It is known, that in ceramics the notch radius, r, has only a minor influence on the toughness values, Munz et al [3] have found that K_{IC} is independent of r for radii below 66 μm. For intermetallics corresponding data are not measured till today. But the results presented in Fig. 4 show that in NiAl single crystals spark eroded and precracked specimens had the same low temperature toughness, therefore the low temperature toughness obtained in this way seems to be valid as long as no plasticity is involved.

For intermetallics, the large amount of material necessary to obtain valid K_{IC}-values for materials with a higher toughness is not available in most cases. Four point bending tests of notched bars, using standards applied successfully for ceramics, are a good compromise to obtain a valid low temperature toughness. The data given in this report are obtained in this way. The rig used is shown in Fig. 1. It is designed according to the work of Kleinlein [4]. The rig can move freely in three axes to avoid constraints produced by imperfect machining of specimens. The specimens with dimensions of 4 x 6 x 50 mm^3 were notched by spark erosion if not otherwise noted. The K_{IC}- values were evaluated according to the relationship

$$K_{IC} = F_{crit} \cdot \frac{3(l-e)}{2Bw^2} \cdot \sqrt{a} \cdot Y\left(\frac{a}{w}\right) \tag{1}$$

where the geometry function Y(a/w) is given by Gross and Srawley [5],

$$Y\left(\frac{a}{W}\right) = 1.99 - 2.47\left(\frac{a}{W}\right) + 12.97\left(\frac{a}{W}\right)^2 - 23.17\left(\frac{a}{W}\right)^3 + 24.8\left(\frac{a}{W}\right)^4 \tag{2}$$

l and e are the distances between the lower and higher support roller, respectively. The other constants have the usual meaning. A typical curve of K_{IC} vs. temperature for a NiAl single crystal with <100>{110}-orientation is given in Fig. 1. The first direction denotes the direction of notching and the second the orientation of the nominal fracture surface perpendicular to specimen axis (Fig. 2). Valid K_{IC}- values for the chosen geometry are obtained

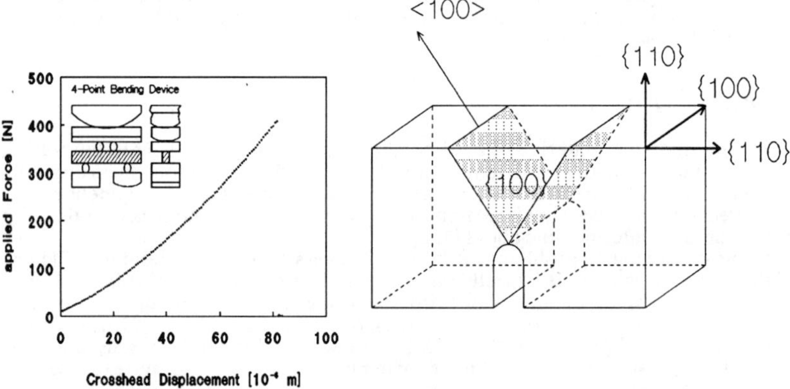

Fig. 1: Load vs. displacement curve and four point bending device

Fig. 2: 4-point-bending specimen with the orientation <100>{110}

$$a, W-a, B > 2.5\left(\frac{K_{IC}}{\sigma_y}\right)^2 \tag{3}$$

when the specimen dimensions are calculated from eq. 3 according to ASTM-Standard E399 [6].

All data given in this report are measured in the same way, therefore the relative strengths of the materials can be compared directly. The standard tests were done with a loading rate of $\dot{K} \approx 0.04 MPa\sqrt{m}/s$ which corresponds to a displacement rate of 0.2 μm/s.

However, most intermetallics tested show a brittle/ductile transition with rising temperature or decreasing strain rate. For a given displacement rate, the local loading rate at the crack tip is much lower in notched specimens than in precracked specimens. This can be most easily seen from the Inglis solution of an elliptical crack [7], for which the stress concentration at the tip and hence, the local loading rate, is directly proportional to the aspect ratio of the crack. Hence precracked specimens should be used to measure brittle/ductile transitions.

But, the controlled initiation of sharp precracks in small, brittle specimens turned out to be very difficult. Recently, we were successful in introducing sharp fatigue cracks into NiAl-single crystals. Compression-compression fatigue tests were conducted with notched specimens as originally proposed by Suresh [8]. The stress amplitude was slowly risen until a precrack formed. Fig. 3 shows a precrack obtained in this way [9]. In Fig. 4 the BDT of notched (70 μm notch radius) and precracked NiAl single crystals is compared. Notched and precracked crystals had the same fracture toughness as long as no plasticity was involved. But, due to precracking, the brittle/ductile transition was shifted by more than 150 K towards higher temperatures. For precracked specimens and a given loading rate initiation can be easier since the local stresses are higher, in addition the local loading rate is much higher compared to a notched specimen. Both effects would result in a shift of the BDT to higher temperatures. Which effect dominates depends on the loci of crack nucleation and must be examined in detail.

The cracks produced by this technique are not atomically sharp. Further efforts are necessary to develop a technique for making atomically sharp cracks which then allows a comparison between experiments and the computer simulations described below [10]. The brittle/ductile transitions presented in this work are obtained from notched specimens. Depending on the rate sensitivity of different materials, the real transitions will occur at higher temperatures. However, it is believed that the strain rate dependence of the BDT is ,at least qualitatively, reproduced correctly for notched specimens.

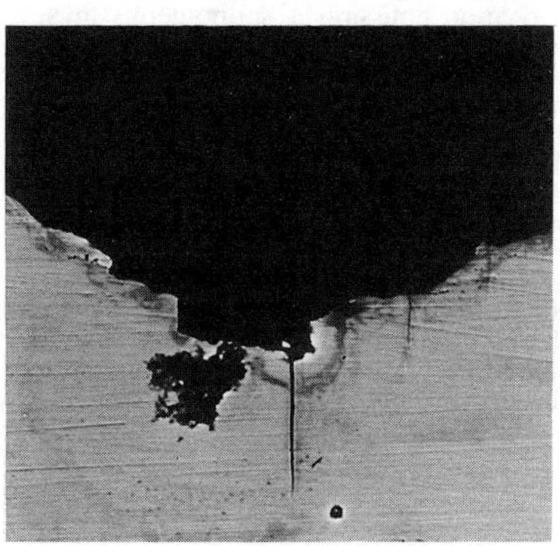

Fig. 3: Typical precrack obtained by compression fatigue in NiAl single crystals

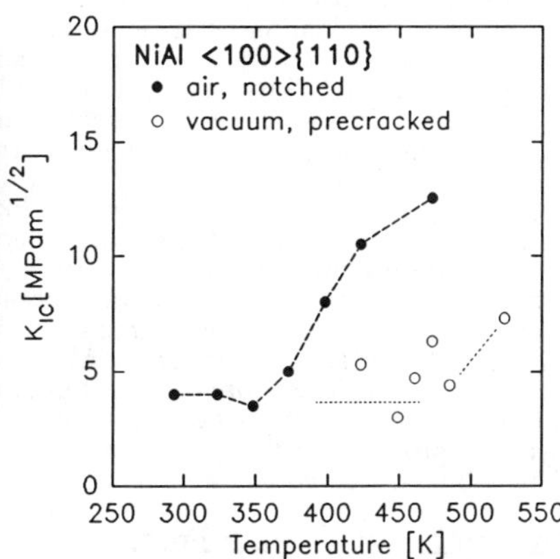

Fig. 4: K_{IC} as a function of temperature for notched and precracked NiAl single crystals, displacement rate 0.2 μm/s

MECHANISMS OF FRACTURE

In order to develop intermetallics with high toughness at low temperatures and good strength at higher temperatures the origin of the low temperature brittleness and the severe brittleness of most intermetallics subjected to impact loading must be understood. Two classes of materials must be distinguished, one class suffers embrittlement due to interfacial or intergranular fracture, the other class even shows brittle fracture as single crystals.

Intergranular fracture of intermetallics was reviewed recently by Schulson and Baker [11] and Takasugi and Izumi [12, 13] and will be mentioned only briefly here. The interest in intergranular fracture started with the work of Aoki and Izumi [14] who showed that Boron can drastically improve the toughness of Ni_3Al. Since then much experimental and theoretical work was focused on the question in which way composition, order and minor segregants like boron and carbon can alter the grain boundary strength [15, 16, 17]. Two origins are discussed for grain boundary weakness. In the first place, it is assumed that order can persist up to the boundary which then results in a more open grain boundary structure. The boundaries are inherently weak and are prone to segregation. On the second place, the passage of dislocations through ordered boundaries can be more difficult. This would result in higher stresses at the boundary for a given applied external stress and again facilitates fracture. For both arguments, boundaries in materials which are slightly of the stoichiometric composition should be stronger. This is in agreement with many findings, but the reason of grain boundary brittleness and the role of impurities is still an open question [11].

For single crystals, two theoretical approaches must be distinguished. The one approach is based on the Griffith model and addresses the question which materials are inherently brittle or ductile. The other approach is a kinetic one and asks, which loading rate can a material support at a given temperature without failing by cleavage. Surface energies and elastic constants can now be calculated by first principle calculations, hence the ideal fracture strength of a material can be predicted [18]. In the framework of the shielding concept, these predictions are also important for predicting the ductile vs. brittle behavior of metals [19].

The ductile vs. brittle behavior is not only controlled by the surface energy but also by the energy necessary to nucleate a dislocation at a crack tip. The nucleation is either spontaneous, for example in Cu, or the dislocations have to overcome an activation barrier. According to the Rice-Thomson approach [20] for the ductile vs. brittle behavior of metals, the critical combination of k_e at which dislocation emission is spontaneous must be compared with the critical load for cleavage, which is calculated by equating the elastic energy release rate G to 2 γ. Anderson and Rice [21], using a refined description of the Rice-Thomson model, have calculated estimates for the critical stress intensity factor k_e for several metals. They found that most fcc metals beside Ir should be intrinsically ductile, most bcc metals, however, intrinsically brittle. In addition, they found that the results of the calculation depended strongly on the loading condition and crystal orientation which indicates that ductile vs brittle behavior depends on the microstructure at the crack tip.

This interpretation of ductile vs. brittle crack response suggests that either dislocation emission or atomistic brittle cleavage occurs at a tip of a micro crack. However, intermetallic single crystals can be extremely brittle even when the yield stress is low as shown by Schneibel and Hazzledine [22] for Al_3Sc. Hence, even low yield stresses do not guarantee ductile behavior. Therefore the effect of temperature on the emission criteria and on the dislocation velocity must be considered. The brittle/ductile transition of a metal is a dynamic process which depends on loading rate, dislocation kinetics, and temperature. This will be discussed below.

Dynamic Models of Brittle Fracture

A crack will emit a dislocation when the local stress exceeds k_e. However, according to recent refinements of the BCS model [23], the emission of a dislocation is not sufficient for ductile fracture. A brittle crack can be shielded from the applied stress by other dislocations. For mode III, it can be shown analytically that the resulting stress at the crack tip depends on details of the dislocation distribution at the tip which yields [24]

$$k = \frac{3}{\pi}\sqrt{\frac{2}{\pi}}\sigma_f\sqrt{c}(\ln(4d/c) + 4/3) \quad (4)$$

where σ_f is the friction stress, c the length of the dislocation free zone and d-c the length of the pile up.

Gerberich and co-workers simulated the temperature dependence of the brittle/ductile transition by adapting eq. 4 for mode I, substituting σ_f with the macroscopic yield stress, σ_{ys}, and d by the plastic zone size. The dislocation free zone size, c, was obtained by empirically redistributing the dislocations at the crack tip to keep the local stress intensity at zero when the next dislocation is emitted. Details of the applied procedure are given in [25]. Despite to the severe approximations, this model sheds light on the possible role of the yield stress in the ductile/brittle transition. The model was applied to the fracture toughness of intermetallics [26]. Fig. 8 shows the results together with fracture toughness values obtained by Reuss [27] and Wunderlich, Kremser, Frommeyer [28] (open symbols). The k_{IG} values were calculated according to the equation

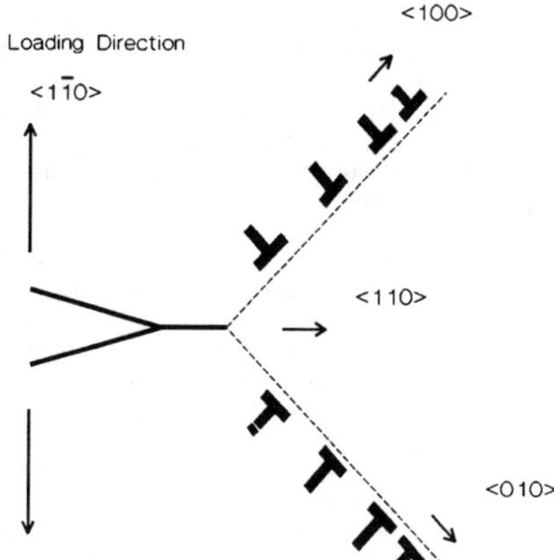

Fig. 5: Geometry for the simulation in Fig. 6

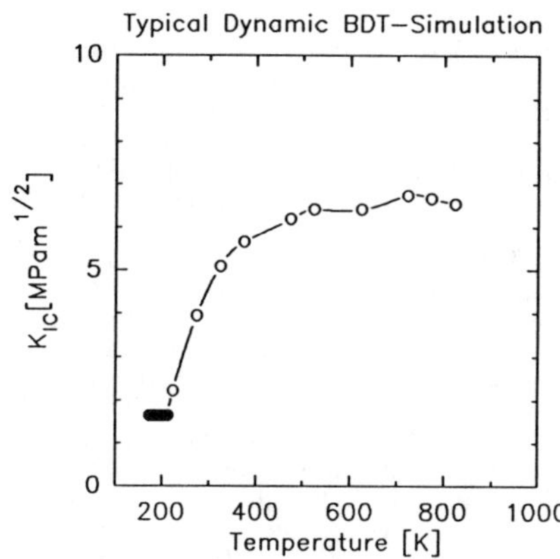

Fig.6: Typical dynamic simulation of a brittle/ductile transition in an artificial material with the velocity law of Si and the elastic properties of NiAl

$$K_{IC} = \frac{1}{\beta'} \exp\left\{ \frac{k_{IG}^2}{\alpha'' \sigma_{ys}} \right\} \tag{5}$$

with the fit constants α'' and β' taken from [26]. The yield stresses and fracture toughness values were taken from data obtained just below and above the BDT-temperature, and k_{IG} was calculated according to eq. 5. The data for the TiAl alloys agree reasonably with the proposed fit since the yield stresses were around 300 MPa. The soft and hard orientation of NiAl, however, were found to be far off the proposed curve since the yield stresses were much lower or, respectively higher than 300 MPa for which the solid line in Fig. 8 was plotted. The fit constants α'' and β' depend on details of the model, hence must be calculated for each material separately. Work is under way to examine how far this will improve the predictions.

In principle, this simulation calculates dislocation arrangements for an applied K giving the stress for the emission of the next dislocation which is compared with the Griffith stress. For metals, the assumption of local equilibrium seems to be reasonable since dislocation velocity equations for metals have a large stress exponent in contrast to semi-conductors. However, as will be discussed below, dynamic simulations of the dislocation emission from crack tips as conducted by Hirsch, Roberts and Samuel [29] and by Brede and Haasen [30] always yield distributions which depend on the rate law and are never in equilibrium before fracture.

The response of a crack tip to the external loading rate is a dynamic process. Possible dislocation sources along the crack tip emit dislocations, these dislocations shield the source until the emitted dislocation has propagated far enough to raise the stress at the source above the value for the emission of the next dislocation. During this process, the local k_I can rise above k_{IG}.

The brittle/ductile transition depends on the emission criterion, the dislocation velocity and the number and distribution of dislocation sources. The dynamic three-dimensional problem is not yet solved. But, based on experimental results obtained with Si single crystals, several one- and two-dimensional models were developed by Hirsch, Roberts and Samuel [29] and by Brede [31] which will be reviewed briefly below and discussed together with measurements of the brittle/ductile transition in intermetallic alloys. The Brede model was refined recently by Maeda [32] who incorporated crack blunting. This resulted in a steeper brittle/ductile transition without additional assumptions.

The main problem of the dynamic models compared to the equilibrium approximations is the lack of reliable velocity laws for almost all materials, except for Si. Therefore, till today, predictions based on dynamic simulations cannot be obtained for intermetallics. But simulations can be used to discuss the influence of different velocity laws, emission criteria and dislocation barriers on the brittle/ductile transition. An example is given in Fig. 6 with the geometry shown in Fig. 5 for a material having the geometry and elastic properties of NiAl in the hard orientation but the velocity law of Si.

This geometry allows a simple, one-dimensional treatment of the stress intensities in which dislocations on different slip planes do not interact. In the simulation a high loading rate (2m/s) was chosen in order to reduce computing time to test different rate laws and emission criteria, quickly. Only a single source of edge dislocations at the crack tip is assumed. To adapt this model to NiAl, the distance of the source to the crack tip was chosen to be about 0.3nm according to calculations of the core radius of dislocations in NiAl [33]. The dislocations are emitted into the material, when the local k exceeds the critical emission stress intensity, k_e. In agreement with experimental results, an inverse pile-up and a dislocation-free zone near the crack tip was observed.

The rate law, emission location and criterion were varied systematically to study their influence on the BDT. Different rate laws can be formulated for metals and semiconductors, depending on the kind of barriers which hinder dislocation motion. The rate law given below was chosen for the simulation since it is assumed that near the crack tip dislocation motion is only opposed by the Peierls stress. These simulations give a guide for necessary experiments to understand the BDT in different intermetallics. In contrast to the equilibrium calculations by Marsh et al [25] our calculation showed that varying the rate laws had the largest influence on the BDT. Different emission criteria only gave small changes in the BDT for a given rate law as long as emission is possible. In principle, two possibilities exist to vary the velocity law:

$$v = v_o \cdot \tau^m \cdot e^{\frac{-U}{kT}} \qquad (6)$$

Either the activation energy U or the stress exponent m can be varied. Increasing the activation energy or the stress exponent will increase the BDT accordingly. Taking a high stress exponent as expected for metals [34], in the ductile region the emitted dislocations reached a quasi-equilibrium position more quickly than with a lower exponent, which resulted in a larger dislocation free zone in front of the crack tip. This zone had to be filled with dislocations until cleavage occurred. From the simulation it seems, that a higher stress exponent also increases the plastic zone size and fracture toughness of the material because of the higher sensitivity of the outer dislocations to a newly emitted dislocation. Simulations which treat large amounts of dislocations are in progress in order to understand the behavior of intermetallics and the influence of boundaries and crack injection on the BDT [35].

In this context another approximation should be mentioned here which was put forward by Jokl et Al [36], the concept of injected cracks. They considered a crack which nucleates in front of a loaded notch for example at a phase boundary. This crack which is initially atomically sharp will see in principle a step loading. Under this condition, the crack can either blunt or propagate by cleavage, and cleavage is possible even in materials for which dislocations are extremely mobile. With this model they were able to describe the brittle/ductile transition in mild steels. The model should be equally successful to describe the brittle fracture of intermetallic alloys.

FRACTURE OF SINGLE CRYSTALS

Cleavage Fracture

Order and crystal structure can limit the number of possible slip systems in intermetallics below five. Hence brittle behavior is expected for polycrystals. For ductile fracture in single crystals, however, only two independent slip systems are necessary. Therefore the inherent ductility of an alloy can be only tested with single crystals. Materials which are brittle in tension and compression, only in tension or always ductile must be distinguished. Many $L1_2$ alloys are ductile as single crystals whereas B_2 alloys often are even brittle as single crystals but show limited ductility in compression at room temperature.

Most extensively examined were NiAl single crystals. They are of technical interest, and they can be easily grown by the Bridgman technique. The orientation dependence of K_{IC} was measured by Chang, Darolia and Lipsitt [37] and by Reuss and Vehoff [27]. Both groups determined the fracture toughness for samples tested along <110> to be 4-5 $MPa\sqrt{m}$, and 8 $MPa\sqrt{m}$ is obtained for samples tested along <001>. The cleavage plane for NiAl tested in <100> and <110>- direction is {110} for impact loading but less clearly defined at lower loading rates. For predeformed samples cleavage planes near <115> are reported [37]. Trace analysis of specimens with <100>{110}-orientation showed inclined planes which are locally a mixture of {100} and {110}-planes [9]. In general, the cleavage planes in NiAl, even at 77 K are less well defined than in carbon steels or refractory metals. Therefore the cleavage energies of different planes must be very similar. But calculations exist only for {110} and {100}-planes with a 30% lower energy for {110} [38].

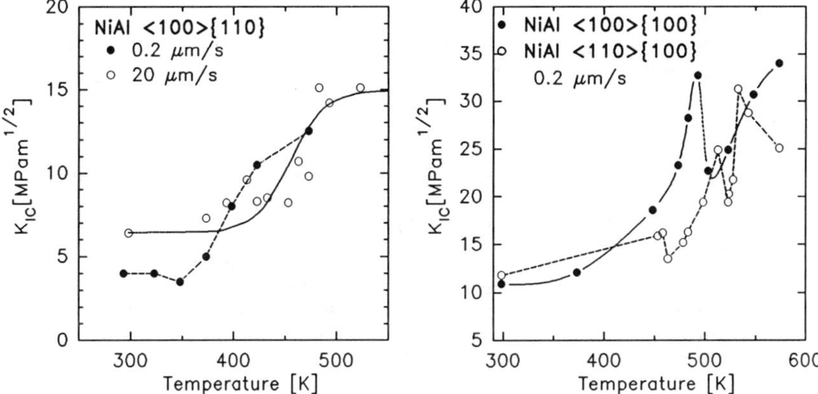

Fig. 7a,b: Fracture toughness of NiAl single crystals as a function of temperature and displacement rate

Figure 7 shows the temperature dependence of K_{IC} for NiAl single crystals with different orientations. Crystals with <hkl>{100} orientations had a factor of two higher K_{IC}-values and a more than 100 K higher BDT temperature. However, the transition temperature must be considered with care since recent data obtained on precracked specimens showed a large shift of the BDT towards higher temperatures (Fig. 4).

Ductile/Brittle Transition

In Fig. 9 the displacement rate for the onset of ductile fracture is plotted vs 1/T for a two phase alloy Ti_5Si_3/Ti_3Al (the plot was obtained from the data given in Fig. 10 [39]), for NiAl single crystals of the two different orientations shown in Fig. 7, and for FeSi single crystals [40]. For the two phase alloy Ti_5Si_3/Ti_3Al, the curve indicates the onset of interfacial fracture due to yielding of the Ti_3Al-phase. In all three systems the onset of yielding is extremely rate sensitive with an increasing activation energy with increasing structure complexity. Since no measurements of dislocation velocities are known for these alloys, these plots only indirectly support the view that the BDT in intermetallics can be described in similar ways as in Si.

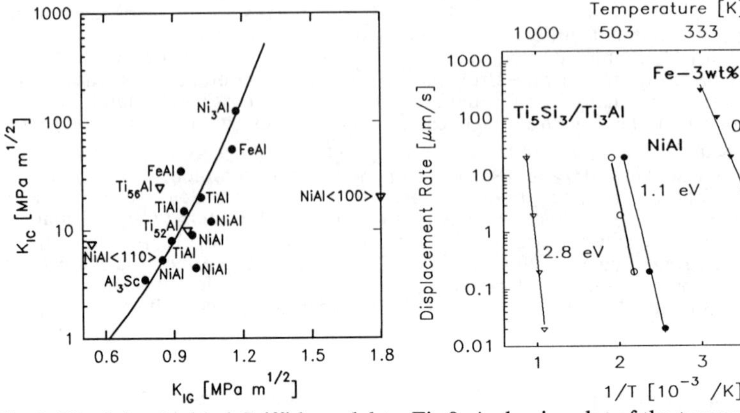

Fig. 8: Fit of the shielded Griffith model to aluminide fracture toughness data [26]

Fig.9: Arrhenius plot of the temperature and displacement rate dependence of the ductile/brittle transition

In combination with the strong shift in BDT for precracked specimens compared to notched specimens, it can be concluded, that at least for NiAl and FeSi the BDT is strongly influenced by the local loading rate at the crack tip in accordance with the predictions of a dynamic model. However, the friction stress produced by long range stresses of other dislocations changes with temperature and strain rate and according to eq. 5, this can explain the shift of K_{IC} with temperature and strain rate as well. Better experiments with ideal precracks and well defined orientations in combination with dislocation rate measurements are needed to decide if the BDT is controlled either by the local processes at the crack tip or by the operation of external sources partly responsible for the temperature and rate dependence of the yield stress which in turn alters the inverse dislocation pile ups at the crack tip.

FRACTURE IN INTERMETALLIC ALLOYS

In general, the nucleation and growth of intergranular cracks depend on the cohesive strength and on the stress state at the boundary, which is affected by dislocations which can cross or pile up at the boundary depending on the boundary orientation. The cohesive strength depends on the grain boundary structure, the degree of disorder and on the chemical composition. In addition, for an interfacial crack the stress state at the crack tip is altered by the different elastic properties of the adjacent phases or grains resulting in a multiaxial stress state even for uniaxial loading [41].

Decreasing the grain boundary size, even below 5 μm in NiAl, does not ductilize the alloy at room temperature [42]. Only for special bicrystal orientations additional slip systems were found to operate near the boundary at room temperature [43]. But obviously these systems cannot operate at enough boundaries to inhibit brittle fracture. Another concept to obtain more resistant alloys, is to combine a ductile phase with a more brittle phase. This was tried by Wunderlich, Machon and Sauthoff [44] for NiAl-NbNiAl. Fracture toughness data for this alloy are given in [45]. The eutectic composition of this alloy had the best mechanical data, but no improvement in ductility compared to NiAl was obtained. TEM examinations showed that microcracks had formed in the NiAlNb-phase during cooling. These cracks reflect at the phase boundary of NiAl for angles of incidence below 55°. At angles above 55°, the crack is stopped at the interphase by producing a plastic zone in the NiAl phase [44].

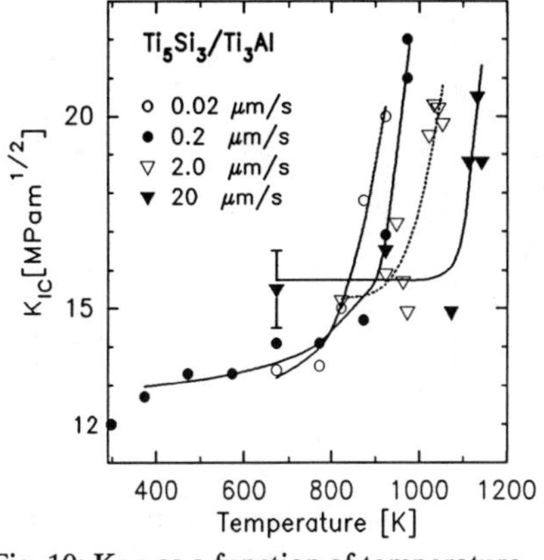

Fig. 10: K_{IC} as a function of temperature and strain rate for Ti_5Si_3/Ti_3Al

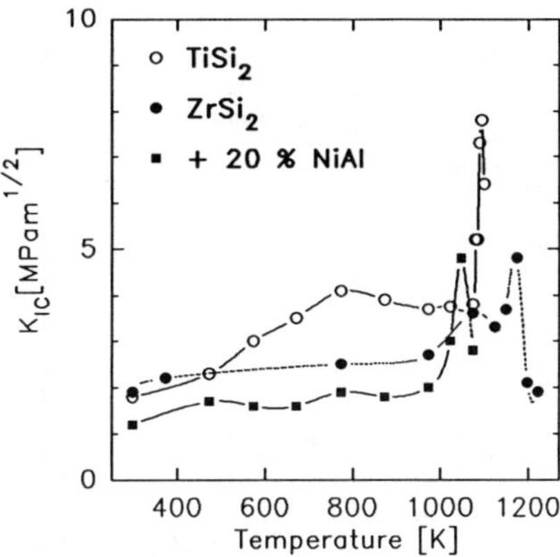

Fig. 11: K_{IC} as a function of temperature for different disilicides

Figure 10 shows the fracture toughness of eutectic Ti_5Si_3/Ti_3Al as a function of temperature. K_{IC} rises sharply for temperatures above 800 K. This rise depends strongly on the displacement rate. Below 800 K the crack propagates along cleavage planes whereas interfacial fracture was observed in the transition range [39]. In this temperature range the Ti_3Al phase starts to yield. The temperatures and displacement rates for this transition are plotted in Fig. 9 (open triangle). Fig. 11 shows the temperature dependence of the fracture toughness for disilicides. Both alloys show a steep ductile/brittle transition above 1000 K similar to Si. Alloying with NiAl decreases the fracture toughness and pronounced interfacial fracture was observed (filled squares). Hence in intermetallic alloys, the phase and grain boundaries must be strong and at least one phase must have enough ductility to inhibit crack propagation. In addition the size of the brittle phases must be small enough that cracks which nucleated in the brittle phase remain below the critical size for unstable fracture in the ductile phase.

ENVIRONMENTAL EFFECTS

Many intermetallics showed strong embrittlement due to hydrogen. Liu and co-workers [46] examined the effects of hydrogen on the brittleness of CoTi alloys. In tests with Co_3Ti single crystals they showed that hydrogen alters the glide processes at the crack tip due to the formation of stacking faults which were not observed in a vacuum environment. These faults hinder the emission and motion of dislocations which results in brittle fracture. Experiments with NiAl single crystals have shown that charging does not affect the fracture behavior of NiAl. However, in tests in hydrogen gas stable crack growth was observed in precracked specimens whereas in vacuum, the specimens failed catastrophically under otherwise identical conditions. Figure 12 shows the effect of gas phase charging on the fracture toughness of polycrystals. The specimens were heated in hydrogen gas at 1273 K and then water cooled. Hydrogen embrittles the grain boundaries up to temperatures of 800 K. Therefore hydrogen must be either strongly trapped at the boundaries and lower the cohesive strength in NiAl or interfacial cracks must form due to internal gas pressure at internal faults. Without hydrogen the same treatment does not change the fracture properties. A detailed examination of the role of hydrogen concentration and texture on the fracture toughness as well as outgasing experiments to test for irreversible internal damage due to hydrogen are under way [9].

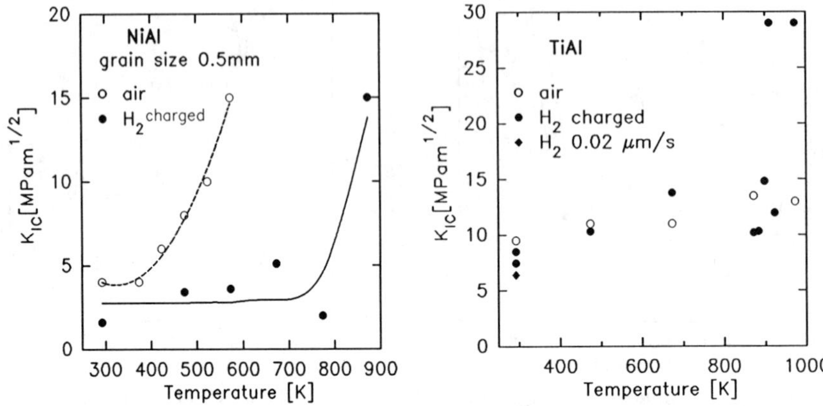

Fig. 12: K_{IC} as a function of temperature for charged and uncharged NiAl polycrystals

Fig. 13: K_{IC} as a function of temperature for charged and uncharged TiAl

The fracture toughness of TiAl/Ti_3Al alloys was observed to be independent of temperature. However, some alloys showed compared to air in vacuum a markedly higher fracture toughness for all temperatures [45]. Tests in water vapor, hydrogen, and oxygen revealed that the normal humidity embrittles the alloy by hydrogen. With increasing temperature the effect of hydrogen diminishes but the alloy starts to corrode by stress corrosion cracking due to local oxidation. Similar tests were done recently with the single phase alloy TiAl [9]. The results are given in Fig. 13. Neither hydrogen charging, nor tests in hydrogen gas showed a degradation of the material. Only after pronounced electrochemical charging a small embrittled zone could be detected in this alloy, possibly due to hydride formation but this could not be confirmed up to now. The data showed conclusively that single phase TiAl has a higher environmental resistance than the TiAl/Ti_3Al alloys, however, the overall toughness is lower.

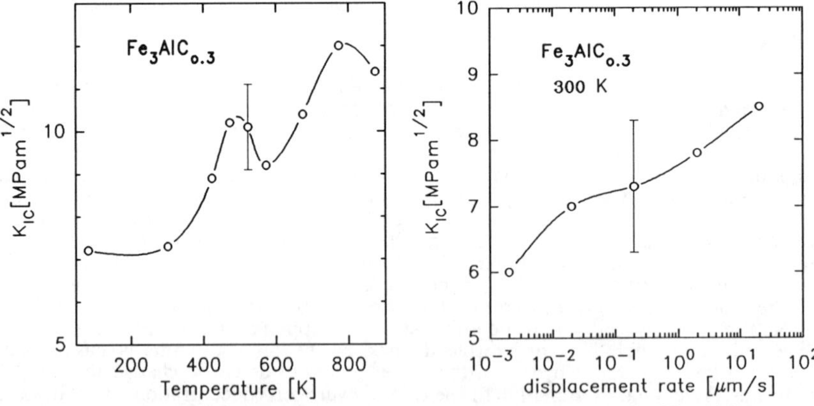

Fig. 14a,b: K_{IC} as a function of temperature and displacement rate for $Fe_3AlC_{0.3}$

Figure 14 shows the temperature dependence of the fracture toughness of the Fe_3AlC/Fe_3Al phase. Details of the alloy treatment and mechanical behavior are given by Jung and Sauthoff [47] The alloy showed stable intergranular crack growth for all temperatures tested. At room temperature, the toughness increased with increasing strain rate (Fig. 14b), a strong indication that the moisture embrittlement of Fe_3AlC/Fe_3Al is equally severe like in Fe_3Al [48]. Detailed examinations of the environmental influence on the fracture toughness of this alloy are under way.

SUMMARY

The influences of specimen size, notching and testing method on the fracture toughness of intermetallics are discussed. Data for the temperature and strain rate dependence of the fracture toughness of intermetallic alloys and single crystals were reviewed. Depending on the alloy system and on structure, different reasons for brittle/ductile transitions were identified, like the yielding of a second phase in multiphase alloys, the operation of additional slip systems with rising temperature in polycrystals, or the strain rate sensitivity of the yield stress in single crystals. Apparent activation energies for these transitions were measured and discussed in view of recent models of fracture. Evidence was given that the activation of additional slip systems at higher temperatures can reduce the toughness of NiAl single crystals by blocking the further emission of dislocations which then results in a second BDT. Examples are given for the effect of hydrogen on the brittle fracture of intermetallics. It was found, that hydrogen embrittles NiAl polycrystals severely but has only a minor influence of the brittleness of single crystals. Single phase TiAl showed only some embrittlement under severe charging conditions and behaved much better in hydrogen than the two phase alloy $TiAl/Ti_3Al$.

Further studies on the effects of micro alloying, phase distribution and interface structure on fracture are necessary to improve our understanding of the toughness of intermetallics. From the fundamental point of view, fracture experiments with well defined geometries are necessary to obtain toughness values in the brittle/ductile transition range which can be compared with theories. For dynamic modelling, the lack of experimental data for disclocation kinetics is distressing. Crack nucleation at interfaces and the role of interfaces in blocking the further emission of dislocations from crack tips, must be studied in detail in experiments and by computer simulations to obtain a deeper understanding of the influence of the microstructure on the brittle/ductile transition.

Acknowledgements The author likes to thank the group of G. Frommeyer for providing some of the titanium aluminides and silicides, Dave Pope for many helpful discussions regarding the brittle/ductile response of materials, Frau B. Schaff for the growth of single crystals, and W. Vogt and S. Reuss for conducting the experiments.

REFERENCES

1. High-Temperature Ordered Intermetallic Alloys, I, II, III and IV, Mat. Res. Soc., Pittsburgh (1985, 1987, 1989 and 1991)
2. Ordered Intermetallics - Physical Metallurgy and Mechanical Behaviour, eds. C.T. Liu, R.W. Cahn, G. Sauthoff, Kluwer Academic Publishers, Dordrecht (1992)
3. D. Munz, R.T. Bubsey, J.L. Shannon, J.Am.Cer.Soc 63, 300, (1980)
4. F.W. Kleinlein, Dissertation, Erlangen (1980)
5. B. Gross, J.E. Srawley NASA TN-D, 2603 (1965)
6. ASTM Annual Book of Standards, Part 10, American Society for Testing Materials, Philadelphia, E 399-618 (1981)
7. C.E. Inglis, Trans.Instn Nav. Archit., LV,1 ,219 (1913)
8. S. Suresh, J.R. Brockenbrough, Acta. metall. 36, 1455 (1988)
9. G. Bergmann, H. Vehoff: work in progress
10. P. Specht, M. Brede, work in progress
11. E.M. Schulson, I. Baker, Ordered Intermetallics - Physical Metallurgy and Mechanical behaviour, NATO ASI Series, Kluwer Academic Publ. Dordrecht, 371 (1992)
12. T. Takasugi, O. Izumi, Acta metall. 33, 1247 (1985)
13. T. Takasugi, O. Izumi, N. Masahashi, Acta metall. 33, 1259 (1985)

14. K. Aoki, O. Izumi, Trans. J. Japan Inst. of Met. 43, 1190 (1979)
15. E.P. George, C.T. Liu, J. Mater. Res. 5, 754 (1990)
16. G.J. Ackland, V. Vitek, MRS, Vol. 133, 105 (1989)
17. T. Takasugi, MRS, Vol. 213, 403 (1919)
18. M.H. Yoo, C.L. Fu, Scripta metall. mater. 25, 2345 (1991)
19. I.H. Lin, R. Thomson, Acta metall. 34, 187 (1986)
20. J.R. Rice, R. Thomson, Phil. Mag. 29, 73 (1974)
21. P.M. Anderson, J.R. Rice, Scripta metall., 20, 1467 (1986)
22. J.H. Schneibel, P.M. Hazzledine, MRS Vol. 213, 323 (1991)
23. S.M. Ohr, S.-J. Chang, J. Appl. Phys. 53, 5645 (1982)
24. R. Thomson, Script. metall. 20, 1473 (1986)
25. P. G. Marsh, W. Zielinski, H. Huang, W.W. Gerberich, Acta metall. mater. 40, 2883 (1992)
26. W.W. Gerberich, H. Huang, P.G. Marsh, NASA 1992, Conference on Advanced Earth-to-Orbit Propulsion Technology, Marshall Space Flight Center, Alabama, May 1992
27. S. Reuss, H. Vehoff, Proceedings of the 2nd European Conference on Advanced Materials and Processes, EUROMAT 2, The Institute of Materials, 313, (1992)
28. W. Wunderlich, T. Kremser, G. Frommeyer, Z. Metallkde. 11, 802 (1990)
29. P.B. Hirsch, S.G. Roberts, J. Samuels, Proc. R. Soc. Lond. A 421, 25 (1989)
30. M. Brede, P. Haasen, ICSMA 9, eds. D.G. Brandon, R. Chaim, A. Rosen, Haifa, Freund Publishing House, London, 813 (1991)
31. M. Brede, Acta metall. mater. 41, 211 (1993)
32. K. Maeda, Scripta metall. mater. 27, 805 (1992)
33. M. Yamaguchi, Y. Umakoshi, Script. metall. 9, 637 (1975)
34. R.T. Pascoe, C.W.A. Newey, Metal Sci. J. 2, 138 (1968)
35. P. Ochmann, H. Vehoff, work in progress
36. M.L. Jokl, V. Vitek, C.J. McMahon, P. Burgers, Acta metall. 37, 87 (1989)
37. K.-M Chang, R. Darolia, H.A. Lipsitt, MRS Vol. 213, 597 (1991)
38. M.H. Yoo, C.L. Fu, Scripta metall. mater. 25, 2345 (1991)
39. S. Reuss, Dissertation, Aachen (1991)
40. H. Vehoff, P. Neumann, Acta metall. 28, 265 (1980)
41. J.R. Rice, Transactions ASME, 55, 98 (1988)
42. R.D. Noebe, R.R. Bowman, C.L. Cullers, S.V. Raj, MRS, Vol. 213, 589 (1991)
43. D.B. Miracle, Acta metall. mater. 39, 1457 (1991)
44. W. Wunderlich, L. Machon, G. Sauthoff, Z. Metallkde. 83, 9 (1992)
45. H. Vehoff, Ordered Intermetallics - Physical Metallurgy and Mechanical Behaviour, NATO ASI Series, Kluwer Academic Publishers, Dordrecht, 299 (1992)
46. Y. Liu, T. Takasugi, O. Izumi, T. Yamada, Acta metall. 37, 507 (1989)
47. I. Jung, G. Sauthoff, Z. Metallkde. 80, 490 (1989)
48. C.T. Liu, Ordered Intermetallics - Physical Metallurgy and Mechanical Behaviour, NATO ASI Series, Kluwer Academic Publishers, 321 (1992)

SINGLE CRYSTAL PROCESSING OF INTERMETALLICS FOR STRUCTURAL APPLICATIONS

EDWARD H. GOLDMAN
GE Aircraft Engines, Cincinnati, OH 45215

ABSTRACT

A number of techniques are available for making metals, non-metals, and intermetallic materials into high-purity single crystals. The most common of these for producing large crystals involve solidification from the melt. The high melting temperatures of most intermetallics of interest for structural applications result in the expected problems of achieving the required high temperatures and temperature gradients while containing the molten material in a chemically, thermally and mechanically stable environment. Processes which have produced intermetallic single crystals, and the materials which have been crystallized, are reviewed. The largest known single crystals of a high temperature intermetallic have been produced in alloys based on NiAl using a modified Bridgman-type directional solidification process, an evolution of the process commonly used to create large jet engine turbine airfoils in Ni-base superalloys. Issues related to processing are described, and the resultant solidification structures are compared with those typical of superalloys. Finally, the prospects for the various processes, and the advances required to push them toward more practical applications, are addressed.

INTRODUCTION

The high melting temperature, low density, and good environmental resistance of certain intermetallic compounds make them attractive materials for high temperature (>2000F [1100C]) applications, such as those in the hot section of jet engines. The traditional barriers for use of intermetallics as structural materials have been low room temperature plasticity and low high temperature strength. Directional solidification is a tool that can help to overcome these barriers. In particular, high temperature mechanical properties such as creep resistance are generally grain boundary-limited; the absence of boundaries normal to the induced stress can significantly improve performance at temperatures approaching the melting point. This is the case for Ni-base superalloy turbine blades in jet engines, where equiaxed castings are replaced by the now-common columnar-grain (CG) and single crystal (SX) castings (Figure 1).

Figure 1. Turbine blades with equiaxed (left), columnar grain (middle), and single crystal (right) structures. Blades in front are hollow, with thin walls and serpentine cooling passages. (Photo courtesy of Howmet Corp.)

In addition, the absence of grain boundaries opens the door to alloying additions which may otherwise be impossible due to segregation at grain boundaries, or, as in the case of superalloys, may allow the removal of alloying elements which are no longer required for grain boundary enhancement and which reduce the melting point or otherwise compromise properties or processibility. In some intermetallic systems, the absence of grain boundaries may improve low temperature ductility and/or toughness.

Mat. Res. Soc. Symp. Proc. Vol. 288. ©1993 Materials Research Society

The first portion of this paper will describe methods that have been used to produce SX structures in high temperature intermetallics and their potential for producing crystals large enough for industrial applications. The crystal growth techniques are well documented elsewhere (for example, [1] and [2]) and will be discussed here only in as much detail as is required to elucidate strengths and weaknesses. Also, melting technology for intermetallics, a critical step prior to solidification processing, was recently reviewed [3], and will not be discussed in detail here.

The second portion of this paper describes recent progress in producing industrial-sized SX castings of alloys based on NiAl. NiAl is a very attractive material for the hot section of jet engines, owing to its 1) low density, high thermal conductivity, and high melting point compared to current superalloys (0.67x, 3 to 5x, and 1.25x respectively), 2) simple crystal structure (CsCl) and low brittle-to-ductile transition temperature compared to other intermetallics, and 3) excellent oxidation resistance. Recent efforts have also demonstrated high temperature rupture strengths exceeding those of SX superalloys. The characteristics, benefits, and state-of-the-art properties of SX NiAl alloys are summarized elsewhere [4].

CRYSTAL GROWTH OF HIGH TEMPERATURE INTERMETALLICS

Melt Growth

Since J. Czochralski first described a process for dipping a seed (a glass tube, held by a silk thread) into a molten pool and withdrawing a mass of metal (Figure 2) [5], many processes that use dipping and pulling from a melt have evolved, and have been dubbed "Czochralski" (Cz) processes. Processing has progressed from Czochralski's original 0.039-inch (1 mm) diameter "crystal threads" of Sn, Pb, and Zn to high-purity silicon crystals as large as 10 inches (25.4 cm) in diameter used in the electronics industry (Figure 3). A schematic of the process for producing silicon crystals is shown in Figure 4.

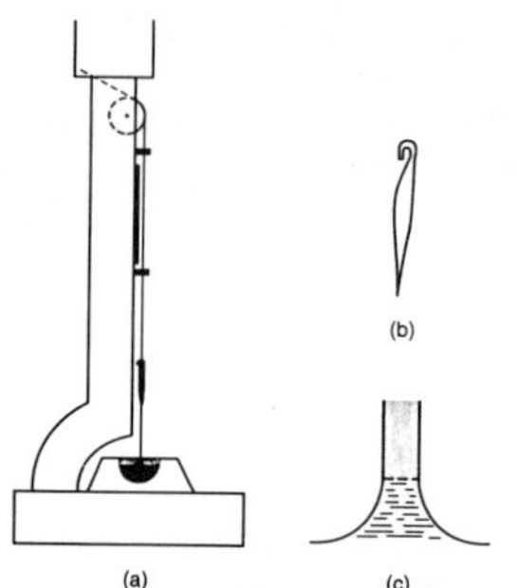

Figure 2. Schematic diagram of Czochralski crystal pulling set-up (after Czochralski [5]): a) overall, b) glass seed, c) enlargement of solidification zone.

Figure 3. Large silicon boules grown by the Cz process. (Photo courtesy of Ferrofluidics Corp.)

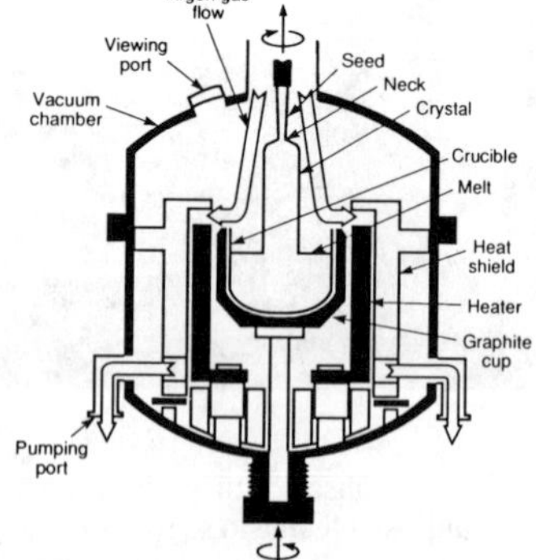

Figure 4. Schematic diagram of Cz apparatus used to make silicon crystals.

The greatest attribute of Cz melt growth for use with high temperature materials that have large liquidus-solidus ranges is its high thermal gradient compared to processes that require molds — heat extraction is not restricted by an insulating ceramic material. In addition, the process can produce very long crystals in a continuous manner, and as such, large volumes of material can be produced under steady-state conditions, e.g. the thermal gradient and the position and shape of the solidification front remain essentially constant throughout a run. The major drawback of this process is that it requires a container for the molten material. For this reason, few high temperature materials have been grown by what could be termed conventional Cz. NiAl is an exception, however, because of its tolerance for Al_2O_3 as a containment material. Single crystals of NiAl as large as 1.5 inches (38.1 mm) in diameter (Figure 5) have been grown at Ferrofluidics Corporation in a standard Cz unit whose hot zone was modified to accommodate the unique characteristics of NiAl [6].

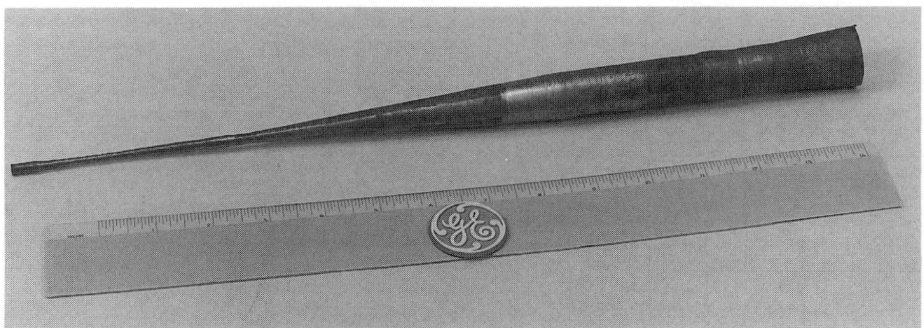

Figure 5. Large NiAl single crystal grown by the Cz process. Crystal is 1.5 inches (38.1 mm) in diameter at right; seed is at left.

Electromagnetic levitation of the melt (Figure 6) has been used with great success to circumvent containment problems for many intermetallic systems. Recent examples include TiAl and NiAl [7], VSi_2, $NbSi_2$, and $TaSi_2$ [8], Cr_3Si-, Nb_3Si-, and V_3Si eutectics [9], and $MoSi_2$ [10]. The amount of material is typically very small, however, and only small SX (~0.25 inch in diameter) have been produced. Large scale skull melt crystal growth has been proposed [11], and would be a logical next step.

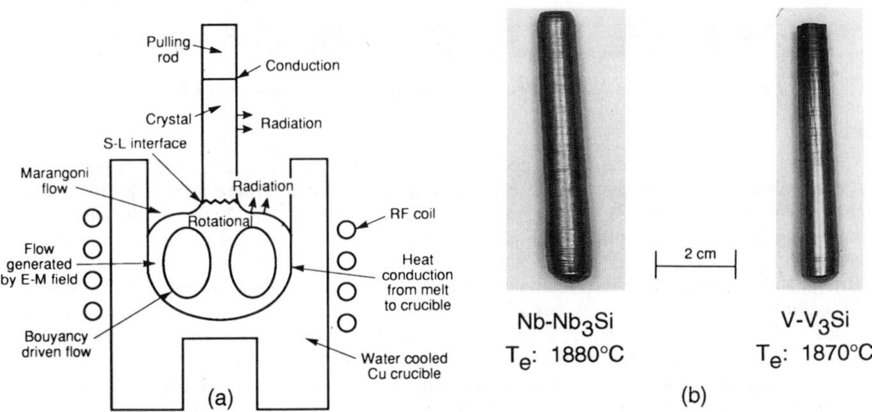

Figure 6. Cold crucible Cz process: a) schematic of the process and b) Nb-Nb_3Si and V-V_3Si eutectics directionally solidified therein. (Photo courtesy of B. Bewlay).

The growth of shaped crystals would be desirable for certain applications, but prospects using Cz are slim, particularly since rotation of the crystal is often required to effect uniform temperature distributions and/or to counteract liquid convection. Another melt growth process — Edge-Defined Film-Fed Growth (EFG) [12] — does have the capability to produce shaped cross sections, however. It uses a heated die whose top surface is shaped; the molten material wets the top of the die, and the solidifying material assumes the shape of the die edge (Figure 7). This process has been used to produce intricate shapes in SX Al_2O_3 and other high temperature materials. More recently, NiAl SX rods (0.3 inch [7.6 mm] in diameter), and sheets of NiAl-Mo eutectics were produced using a modified EFG process (Figure 8) [4]. Due to the high thermal gradients achieved in this process, an aligned eutectic structure could be maintained at higher withdrawal rates (compared to Bridgman processing) for the NiAl-Mo sheet; this resulted in the very fine fiber spacing shown in the figure. Again, this process is limited by the crucible and die materials.

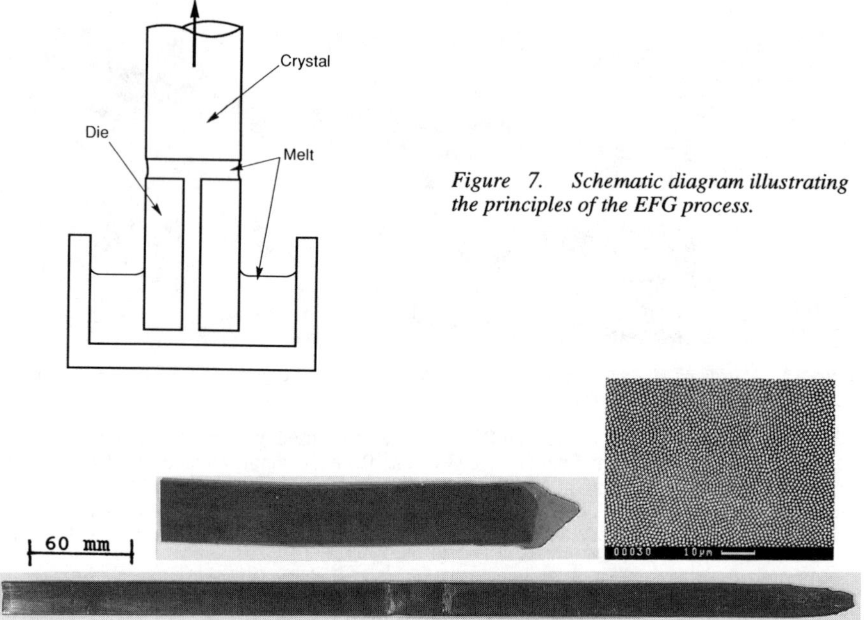

Figure 7. Schematic diagram illustrating the principles of the EFG process.

Figure 8. Large NiAl-Mo eutectic bars grown by the modified EFG process.

Bridgman

In the mid-1920's, P. W. Bridgman reported a solidification process that involved withdrawing from a furnace a mold (Figure 9) that actively choked off the growth of multiple grains [13]. His first experiments involved Sb, Bi, Tl, Cd, and Zn; crystals as large as 0.87 inch (22 mm) in diameter were produced. While others used similar methods — for example, Obreimov and Schubnikow used glass tubes with thin capillaries at one end [14] — processes that use grain selectors are typically known today as "Bridgman" processes. One incarnation is the workhorse for producing CG and SX castings in the turbine engine industry. Figure 10 shows a very large (>15 inches [38.1 cm] long) CG superalloy turbine blade for a land-based power generation unit; SX turbine blades of similar size have also been produced. The process is shown schematically in Figure 11, and is described in more detail later in this paper. Molds can be simple ceramic tubes or complex-shaped composite shells produced by the lost-wax process; the Bridgman process can be used to produce shapes from molten materials that are as intricate as the mold that contains them. Ceramic cores are also routinely incorporated to produce hollow castings. Of course, reaction of the molten material with the mold and core is limiting for many high temperature intermetallics.

Figure 9. Mold configuration used by Bridgman to produce large single crystals (after Bridgman [12]).

Figure 10. Large directionally solidified superalloy turbine blade casting for a land-based gas turbine, produced by the Bridgman process. (Photo courtesy of Howmet Corp.)

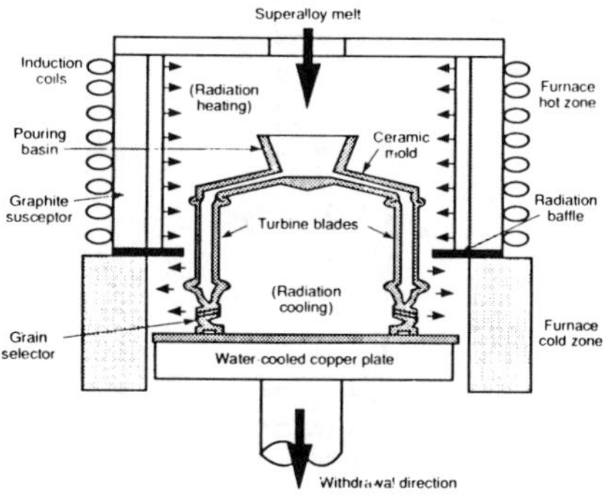

Figure 11. Schematic diagram of the Bridgman process used to produce single crystal turbine blades.

A great deal of work has been done on directional casting of NiAl and its alloys using Bridgman-type processes; these will be described later in this paper. A number of other intermetallics have been also made into SX, including Al_3Sc, Al_3Ti, Ni_3Ge, Ni_3Al, and Co-, Fe-, and RuAl [15-19].

Float Zone

Float zone (FZ) techniques have been the methods of choice for producing single crystals of very high temperature and highly reactive materials. First practiced in the 1950's [20], they use a polycrystalline bar of relatively pure material that is heated locally to create a small molten zone. The zone is moved along the length of the workpiece (by moving the heat source or the workpiece) at a rate that allows directional solidification to occur behind it (Figure 12). As in melt growth, heat extraction from the solidifying bar is not hindered by a mold material, resulting in very high thermal gradients. Electron beam, induction, laser, optical imaging, and other sources have been used to create the molten zone. Directional structures have been produced in a variety of high temperature intermetallics, including Ni_3Al, TiAl, and Al_3Ti using induction sources [21-24] and Ni_3Al, NiAl, TiAl and others using optical imaging sources [25, 26]. Figure 13 shows an optical FZ set-up; a NiAl crystal produced in this unit is shown in Figure 14. $MoSi_2$ has been attempted with limited success [27].

Since the molten zone is supported only by surface tension, its physical stability is a key challenge, however, particularly for alloys that melt over a range of temperatures; this greatly limits the cross section of material that can be crystallized. Cross sections of about 0.25 inch in diameter are typical for conventional FZ processes. The size limitation has been addressed through the use of electromagnetic fields to help support the molten zone [28, 29]. Crystals 1.0 inch in diameter are reported as typical; TiAl, Ni_3Al, and Nb-Al alloys, and others have been attempted, with varying degrees of single crystallinity. This appears to be the most promising form of FZ for producing larger and perhaps even shaped cross sections.

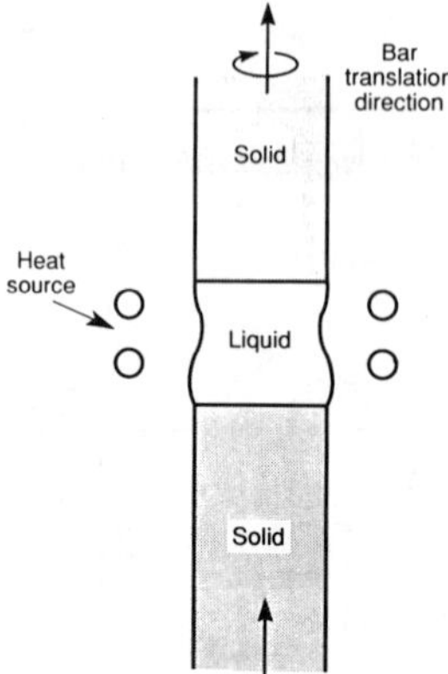

Figure 12. Schematic diagram illustrating the float zone concept.

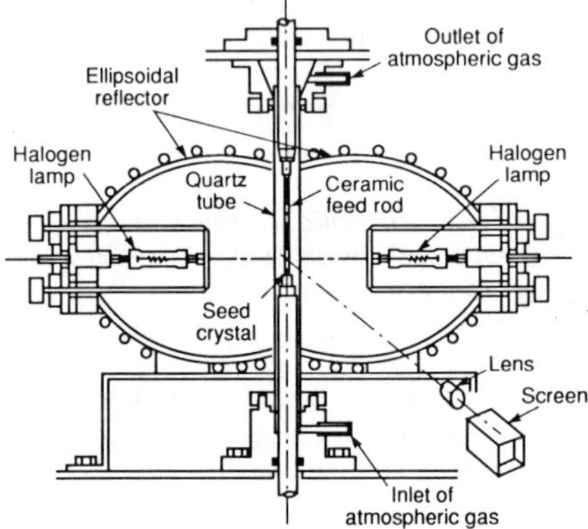

Figure 13. Schematic diagram of an optical imaging float zone furnace.

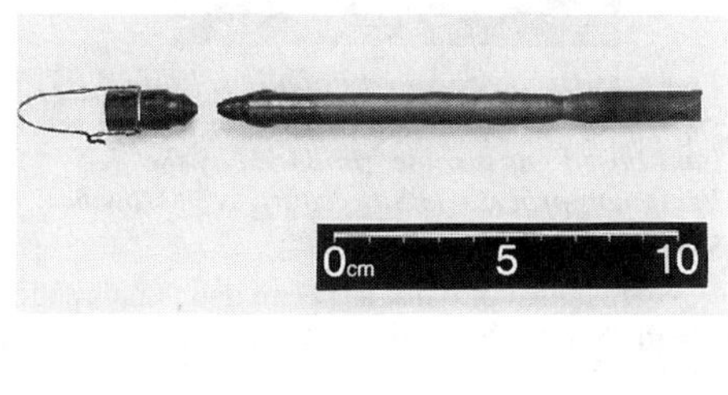

Figure 14. NiAl single crystal grown in an optical imaging float zone furnace.

Arc zone refining is a type of FZ process that uses melting of some, but not all, of the cross section of the sample; it relies on abnormal grain growth in the solid to create a single-grain structure [30]. It has been used successfully on some refractory metals, but thus far has been limited to materials with BCC structures that melt congruently.

Pedestal Growth

A somewhat forgotten, but potentially useful, technique for producing crystals without the use of containers is based on the Verneuil method, first used to produce single crystals of alumina [31], and more recently to produce other high temperature materials [27]. It involves the slow, continuous deposition of material onto the molten surface of a platform of already-solidified material. While this process successfully avoids containment issues, process control is quite difficult.

Other Ways to Make Single-Crystalline Bodies

In addition to processes that involve solidification from the melt, processes that involve solid and gas phase reactions may also be of importance, particularly where liquid-container interactions are a problem. In particular, directional recrystallization (DR), or similar processes that involve massive grain growth, have been used to produce single grain structures in many materials systems, including superalloys. In fact, turbine blade fabrication by bonding sheets of DR superalloys was seriously pursued [32]; at the time, the cost of such a process was deemed excessive compared to alternative processes, and work was stopped. However, such processing may make sense in situations where no easy alternatives exist.

BRIDGMAN FOR LARGE NiAl ALLOY SINGLE CRYSTAL SHAPES

While small specimens of NiAl-Mo and -Cr eutectics were produced using a Bridgman method as far back as the early 1970's [33], it was not until the mid-1980's that relatively large SX test bars of NiAl alloys were produced. This achievement was the springboard for serious development of alloys in this class.

The following sections describe some of the lab-scale efforts at GE Aircraft Engines and subsequent efforts to scale up to production-like sizes and quantities.

Molds

The high melting point and low RT plasticity of NiAl alloys present special problems for mold materials. High purity Al_2O_3 crucibles can be used when simple shapes are required, but more intricate geometries require use of the lost wax process. Typical shell molds for advanced superalloys contain Al_2O_3 particles held together largely by mullite (Al_2O_3-SiO_2). As shown in Figure 15, at the temperatures required to process NiAl alloys (> 3000F [1650C]), the constituents of the mold are very close to their use limits. Thus, SiO_2 levels in the mold are kept as low as possible when processing NiAl; not only does SiO_2 lower the softening temperature of the mold, it can also be chemically reduced through reaction with Al and/or other alloying elements in the molten alloy. However, the mold, which must be very strong at the casting temperature, must also be weak enough at low temperatures to avoid cracking of the low-plasticity alloy due to differential contraction during cooling. This is made difficult by the high degree of sintering that occurs at the high casting temperature. Typical mold removal methods must also be reevaluated in this context. A typical mold used to produce test bars is shown in Figure 16.

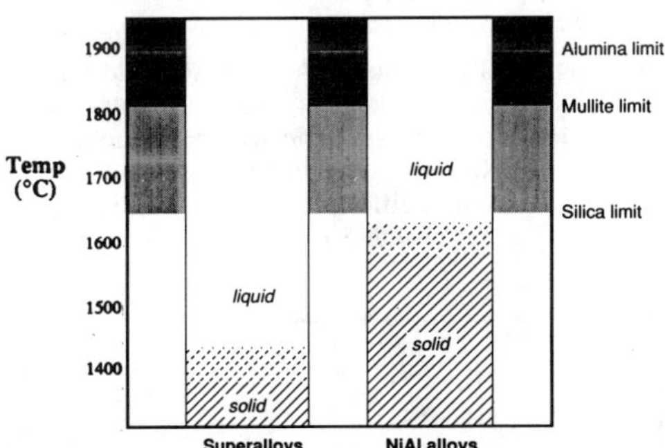

Figure 15. Temperature limits for investment casting mold and core materials.

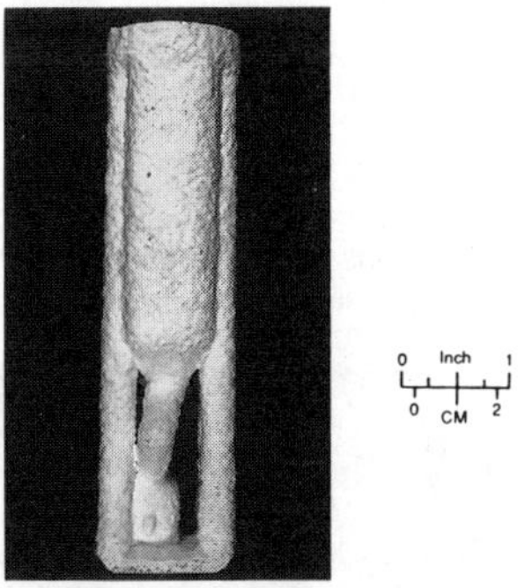

Figure 16. Alumina-silica shell mold, produced by the lost-wax method, used to grow NiAl single crystal bars in Bridgman furnaces.

Melting

Vacuum induction melting using Al_2O_3-SiO_2 crucibles has been successful in producing both lab- and pilot-scale melts of NiAl alloys. Heats as large as 700 pounds (318 kg) have been melted with good control of alloying elements and very low levels of undesirable elements.

Directional Solidification

NiAl SX bars as large as 1.5 x 1.5 x 4.0 inches (3.8 x 3.8 x 10.2 cm) have been produced in modified Bridgman units at GE Aircraft Engines; Figure 17 shows a typical bar. A major challenge in producing a SX structure in these alloys is to generate and maintain a sufficient thermal gradient at the solidification front. The first part of this challenge is to achieve sufficient superheat in the melt — NiAl alloys melt at approximately 3000F (1650C) with melting ranges between 0 and 250F (0 and 120C), depending on composition. The second part of the challenge comes from the very high thermal conductivity of these alloys (as high as 45 BTU/hr-ft-F [78 watt/m-K] at 1800F [982C]), a property whose enormous benefit in performance — improved cooling efficiencies and lower localized thermal stresses, resulting in improved airfoil thermal fatigue resistance — is perhaps matched by its drawback in processing — reduced ability to maintain large thermal gradients required to produce a SX structure. The position and shape of the solidification front, other key factors in producing single crystals, are also strong functions of conductivity. Figure 18 shows the effect of conductivity on the thermal gradient and on the position of the solidification front, as calculated in finite element computer simulations of the solidification process [34].

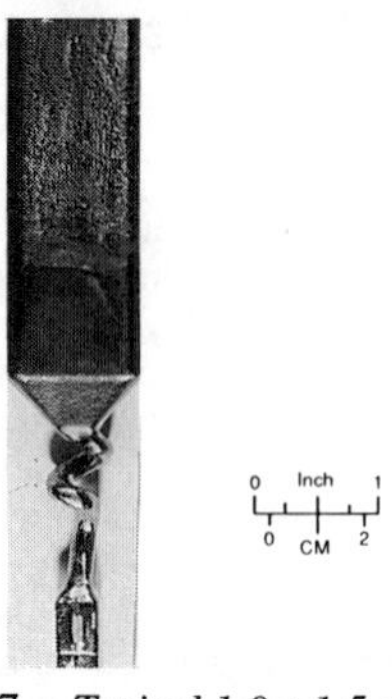

Figure 17. Typical 1.0 x 1.5 x 4.0 inch (2.5 x 3.8 x 10.2cm) single crystal NiAl bar produced by the Bridgman process.

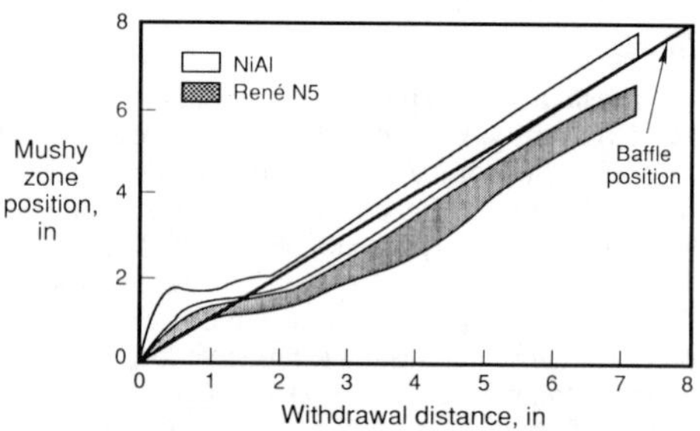

Figure 18. Calculations of the effects of alloy thermal conductivity on the position of the solidification front in bars produced by the Bridgman process.

In most NiAl alloys, solidification is dendritic with typically, but not exclusively, fourfold symmetry. As expected, dendrite morphology is a function of the casting parameters; Figure 19 shows typical microstructures. The dendrite arm spacings observed in NiAl alloys are typically larger than those in superalloys processed under similar conditions (Figure 20). In alloys containing large amounts of alloying additions that segregate interdendritically, this can result in difficulties in fully solutioning the segregated phase(s); this has not been a problem in NiAl alloys of interest.

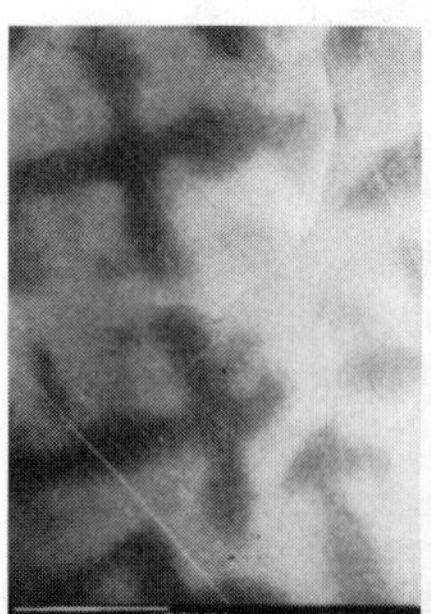

Figure 19. Scanning electron photomicrographs (backscattered electrons) of transverse sections through NiAl alloy bars that were directionally solidified at 0.5 in/hr (left) and 20 in/hr (right) (0.02 and 0.85 cm/sec, respectively).

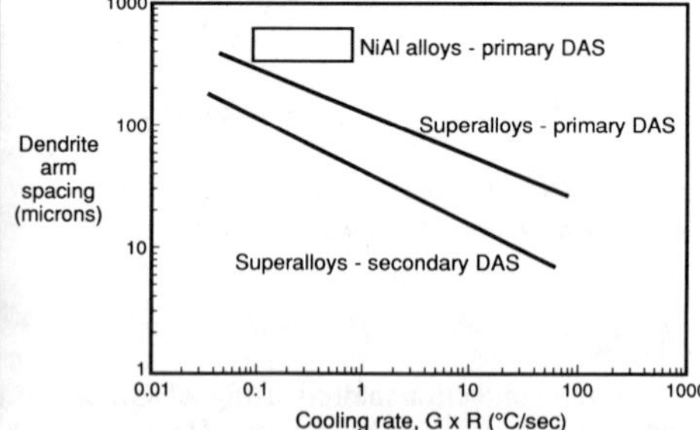

Figure 20. Relationship between cooling rate and dendrite arm spacing for single crystal superalloys and NiAl alloys.

Also, unlike superalloys, nearly all of the alloying elements of interest are of higher density than NiAl, but *depress* the melting point. Thus, the interdendritic regions are generally enriched with heavy alloying elements. Figure 21 shows segregation typical of alloys containing large amounts of transition elements]. The segregation of heavy elements to the interdendritic regions perhaps accounts for the absence of freckle defects — chains of small equiaxed grains commonly observed in SX superalloy castings — in NiAl alloy castings. These defects are thought to be caused by convective currents that are set up in the liquid between the dendrites by rejection of lower density elements into the interdendritic fluid; the equiaxed grains are nucleated on dendrite arms that are broken off by the passing currents.

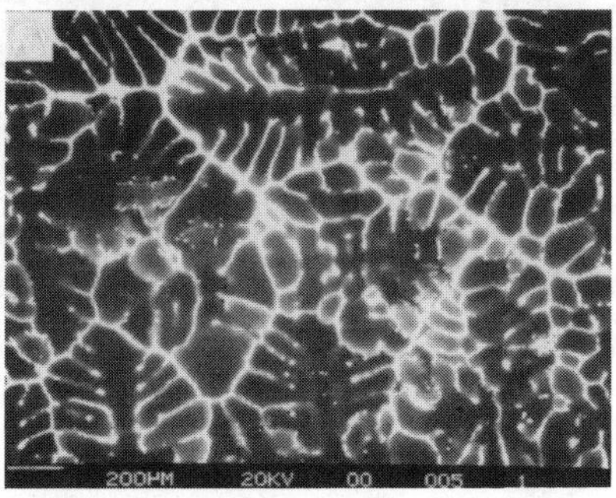

Figure 21. Scanning electron photomicrograph (backscattered electrons) of transverse section through a directionally solidified NiAl alloy bar containing large amounts of a transition element.

Jet Engine Components

Figure 22 shows SX turbine blade shapes produced in NiAl alloys. The turbine blade shown in Figure 22a was fabricated from a rectangular SX bar of a ternary NiAl alloy. It is typical of the blades used in turboshaft (helicopter) engines. The SX casting in Figure 22b is a solid high pressure turbine (HPT) blade casting similar in size to that used on engines in the 30,000 to 40,000 pound thrust range. Figure 23 shows larger HPT blade shapes, produced at PCC Airfoils, Inc. Some parts were overstocked to avoid cracking of the thin platform areas; crack prevention in thick-to-thin section transitions remains a challenge.

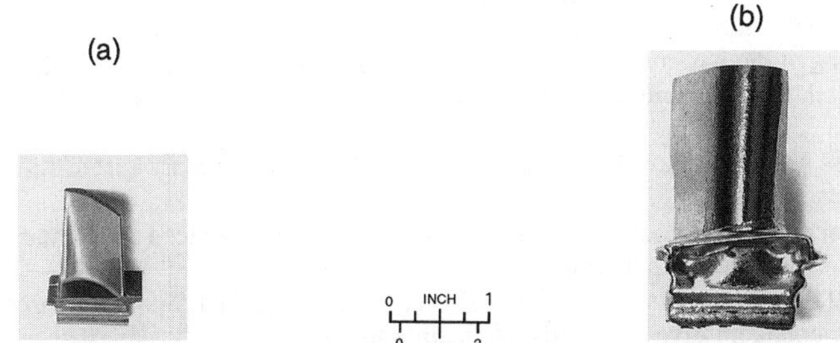

Figure 22. NiAl turbine blades: a) small turbine blade that was machined from a single crystal bar; b) solid, ner net shape, single crystal NiAl high pressure turbine blade casting.

Note that all of the blade castings shown are solid. While many low pressure turbine blades are indeed solid, HPT blades are typically hollow, with thin walls (as thin as 0.030 inch [0.8 mm]) and serpentine passages that guide cooling air through the blade. The cooling passages are produced using a ceramic core inside the shell mold. However, differential contraction of the NiAl around the core during cooling from the casting temperature generally results in cracking of the NiAl. While turbine blades and vanes can still be produced by casting solid SX pieces, machining the cooling passages, and bonding them together, it may be desirable to produce one-piece castings. Figure 24 shows a NiAl tube that

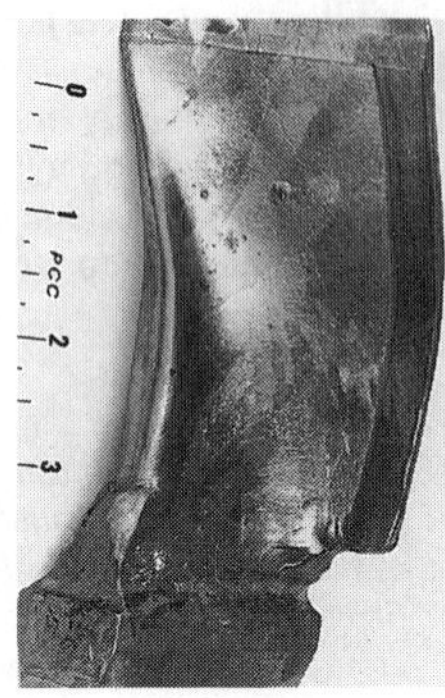

Figure 23. Large NiAl alloy high pressure turbine blade castings that were directionally solidified in a production-scale Bridgman furnace. (Scale in inches.)

Figure 24. Thin-wall NiAl tube that was directionally solidified around a core with no evidence of cracking. (Scale in inches.) (Photo courtesy of Howmet Corp.)

was directionally solidified around a core designed specifically for NiAl [36]. This is the first indication that cored casting may be possible with the NiAl alloys, in spite of low room temperature plasticity.

DIRECTIONS/NEEDS

For a wider range of directionally solidified intermetallics to be taken seriously as structural materials, the following are needed:

- Engineered containment materials. For NiAl alloys, evolutionary advances in alumina shell technology will be required as alloy development continues; revolutionary advances are required for cores, although some directions exist. For the more reactive intermetallics, it appears that containment for times longer than seconds is unlikely, particularly for complex shapes. However, the use of composite molds — non-reactive surface layer covering a backup structure — may have potential for certain intermetallics.
- Determination of the physical limits to producing larger cross-sections by levitation float zone and cold crucible melt growth techniques. Eventually, scale up processes to exploit maximum capabilities.
- Containerless process(es) for producing foil or sheet that could be consolidated and/or fabricated into structures.
- Fabrication schemes for producing 1) structures from sheet, as practiced for composites, and 2) complex geometries, such as hollow airfoils.
- Fresh look at directional recrystallization. This could be attractive for particular intermetallic systems that lend themselves to massive grain growth.

SUMMARY

Small single crystals of many high temperature intermetallics of current interest — Ni_3Al, NiAl, Cr_3Si, Al_3Ti, Ni_3Ge, TiAl, and others — have been made by Bridgman, float zone, and Czochralski methods. NiAl alloy single crystals as large as 1.5 inches thick have been produced by a modified Bridgman method and used to fabricate jet engine turbine components. The NiAl alloys solidify much like the superalloys; differences in casting parameters, dendrite morphology, and propensity for grain defects can be largely attributed to the high thermal conductivity and low density of the NiAl alloys.

Most high temperature intermetallics cannot be made into single crystals of significant size using current processes; container-dependent processes are limited by reactivity; containerless processes have limited size and geometry capabilities. Containerless processes must be pushed toward becoming more production-like — if larger cross sections and/or geometries that can be fabricated into useful shapes cannot be achieved, these potentially revolutionary materials will languish as tools for understanding physical properties and polycrystalline behavior, and as titillating curiosities.

ACKNOWLEDGEMENTS

The efforts to produce large NiAl single crystals by Bridgman processing were performed by J. A. Oti and K. O. Yu of PCC Airfoils, Inc. as part of USAF Contract #F33615-90-C 5938, Capt. R. H. Lilley and T. Broderick, Program Managers. G. McCabe, M. Agnello, P. D. Hines, J. Hopkins, C. Muncy, R. Smashey, and W. Walston have been instrumental in developing NiAl single crystal casting technology at GE Aircraft Engines.

REFERENCES

1. J.J. Gilman, editor, The Art and Science of Growing Crystals (John Wiley & Sons, Inc., New York, 1963).
2. R.F. Bunshah, editor, Techniques of Metals Research, Vol 1, Part 2 (Interscience Publishers, New York, 1968).
3. S. Sen and D.M. Stefanescu, JOM **43** (5), 30-34 (1991).
4. R. Darolia, D.F. Lahrman, R.D. Field, J.R. Dobbs, K.M. Chang, E.H. Goldman, D.G. Konitzer, in Ordered Intermetallics — Physical Metallurgy and Mechanical Behaviour, edited by C.T. Liu, et al (Kluwer Academic Publishers, Netherlands, 1992), p. 679.
5. J. Czochralski, Z. Phys. Chem. **92**, 219 (1916).
6. R. Corderman, E.H. Goldman, C. Chartier, unpublished research, 1990.
7. B.P. Bewlay (private communication).
8. U. Gottlieb, O. Laborde, O. Thomas, A. Rouault, J.P. Senateur and R. Madar, Appl. Surf. Sci. **53**, 247 (1991).
9. K.M. Chang, B.P. Bewlay, J.A.Sutliffe, M.R.Jackson, JOM **44** (6), 59-63 (1992).
10. O. Thomas, J.P.Senateur, R. Madar, O. Laboide, E. Rosencher, Solid State Commun. **55**, 629 (1985).
11. Jewett, D.N., US Patent No. 4,659,421 (1987).
12. H. E. LaBelle, Jr., J. Crystal Growth **50** (1980).
13. P.W. Bridgman, Proc. Am. Acad. Arts and Sciences **60** (6), 305 (1925).
14. I. Obreimov and L. Schubnikow, Z. Phys. **25**, 31 (1924).
15. J.H. Schneibel and P.M. Hazzledine in High-Temperature Ordered Intermetallic Alloys IV, edited by L.A. Johnson, et al (Mater. Res. Soc. Proc. **213**, Boston, MA, 1990) p.323.
16. Z.L. Wu, D.P. Pope, V. Vitek, ibid, p.487.
17. H.R.P. Inoue, C.M. Wayman, T. Saburi, ibid, p.521.
18. S. Miura, T. Hayashi, M. Takekawa, Y. Mishima, T. Suzuki, ibid, p.623.
19. K.-M. Chang, R.Darolia, H.A. Lipsitt, ibid, p.597.
20. Keck, P.H., M.J.E. Golay, Phys. Rev. **89**, 1297 (1953).
21. N. Fat-Halla, S. Bahi, T. Kawabata, O. Izumi, Mat. Sci, & Eng. **61**, 227 (1983).
22. E.P. George, W.D. Porter, H.M. Henson, W.C. Oliver, B.F. Oliver, J. Mater Res. **4**, 78 (1989).
23. T. Hirano, S.-S. Chung, Y. Mishima, T. Suzuki, in High-Temperature Ordered Intermetallic Alloys IV, edited by L.A. Johnson, et al (Mater. Res. Soc. Proc. **213**, Boston, MA, 1990) p. 635.
24. Savitskii, E.M., G.S. Burkhanov, I.M. Zalivin, Problem Prochnosti **11**, 111 (1972).
25. S. Chang and D.P. Pope (private communication).
26. H. Inui, A. Nakamura, M. Yamaguchi, in High-Temperature Ordered Intermetallic Alloys IV, edited by L.A. Johnson, D.P. Pope and J.O. Stiegler (Mater. Res. Soc. Proc. 213, Boston, MA, 1990) p. 569.
27. T.A. Lograsso, Mat. Sci & Eng. **A155**, 115 (1992).
28. B.F. Oliver, Trans. AIME **1**, 960 (1963).
29. R.D. Reviere, B.F. Oliver, D.D. Bruns, Mat. & Manuf. Proc. **4** (1), 103 (1989).
30. T.A. Lograsso, F.A. Schmidt, J. Crystal Growth **110**, 363 (1991).
31. Verneuil, A., Compt. Rend. **135**, 791 (1902).
32. Manufacturing Technology for Advanced Propulsion Materials, Final Report, F33615-85-C-5152 (1987).
33. J.L.Walter, H.E. Cline, Met. Trans. **4**, 33 (1973).

34. K.O. Yu, J.A. Oti and W. S. Walston, in High-Temperature Ordered Intermetallic Alloys V, (Mater. Res. Soc. Proc., Boston, MA, 1992), to be published.
35. R.D. Field, R. Darolia, D.F. Lahrman, Scripta Met. **23**, 1469 (1989).
36. P.R. Aimone, B. Kilinski and B. London, in Processing and Fabrication of Advanced Materials for High Temperature Applications, (TMS-AIME Proc., Chicago, IL, 1992), to be published.

PROCESSING AND HIGH TEMPERATURE DEFORMATION OF Nb_3Al

Y.MURAYAMA, T.KUMAGAI* AND S.HANADA

Institute for Materials Research, Tohoku University, Sendai 980, Japan
*Now at National Research Institute for Metals, Tokyo 153, Japan

ABSTRACT

Alloys based on the ordered intermetallic Nb_3Al were fabricated by powder metallurgy processing, infiltration processing and clad-chip extrusion processing. By controlling processing variables, single phase Nb_3Al and two phase Nb_3Al-Nb solid solution alloys were obtained with various microstructures. The single phase Nb_3Al deforms by {001}<100> slip above 1400K and stress-strain curves exhibit deformation softening after showing a high peak stress. Transmission electron microscopic observations revealed that dislocations are dissociated into two partial dislocations which bound a stacking fault and the dissociation width between the partial dislocations depends on aluminum content in Nb_3Al. The deformation softening was found to be caused by dynamic recrystallization. The two phase alloy shows similar stress-strain curves to the single phase alloys. However, sub-boundaries and dislocation networks were observed without recrystallized grains.

INTRODUCTION

Nb_3Al, with the A15 structure, is a candidate as a high temperature structure material for use above 1500K. It has a high melting temperature, 2333K, low specific gravity, $7.26Mg/m^3$, and high strength at elevated temperatures[1]. However, an arc-melted button with a near-stoichiometric composition is easily cracked during cooling after solidification because of severe brittleness[2]. Therefore, it is difficult to fabricate structural parts of the near-stoichiometric Nb_3Al by ingot metallurgy. Cracking in an arc-melted button is suppressed with decreasing Al content and sound castings can be produced in Nb-rich Nb_3Al alloys consisting of Nb_3Al and Nb solid solution(Nb_{ss})[2]. Microstructural observations have revealed that the solidification structure of these castings is very coarse and inhomogeneous, that is, the Nb_{ss} phase exists inhomogeneously at grain boundaries and within grains of Nb_3Al[2,3]. Moreover, this structure is affected by solidification conditions. Thus, sample preparation due to ingot metallurgy is not appropriate for experiments to investigate mechanical properties of Nb_3Al alloys at elevated temperatures as a function of composition in single phase Nb_3Al or in alloys containing Nb_{ss}. Consequently, the need to develop other process in which cracking can be suppressed and microstructure can be controlled is apparent.

Fabrication processes of Nb_3Al as a superconducting material have been extensively studied[4-12]. In this study three compound fabrication processes, i.e., the powder metallurgy process, the infiltration process and the clad-chip extrusion process, are investigated to obtain structure-controlled Nb_3Al alloys without micro- and macrocracks. Compressive properties of the obtained alloys at high temperatures are reported .

POWDER METALLURGY PROCESS

Nb/Al composites for superconducting Nb_3Al wires have been produced from elementary Nb and Al powders . However, in this process oxide formation during processing is significant. This paper demonstrates the processing of prealloyed powders and fabrication of bulk Nb_3Al suitable for compression tests.

According to a recent study on the determination of the Nb-Al binary phase diagram[13], single phase Nb_3Al at 1873K exists in the composition range from approximately 21 to 23 mol%Al. Considering the loss of aluminum due to volatilization during arc-melting and due to formation of alumina during ball-milling and HIPing, Nb_3Al alloy buttons with nominal compositions of 22~27mol%Al, which is slightly higher in Al content than the compositions of the Nb_3Al single phase range[13], were arc-melted in order to examine the high temperature deformation behavior of single phase Nb_3Al alloys. They were crushed and ground to powders under 200 mesh. The powders were compacted at room temperature in a uniaxial hydraulic pressing machine. The green compacts were canned in evacuated glass tubes at 1673K. The canned samples were HIPed at 1873K and 147MPa for 1.8ks. Compositions of the HIPed samples were chemically analysed. Alumina was insoluble in acid in chemical analysis. Fig.1 shows an optical micrograph of the HIPed Nb_3Al alloy with a chemical composition of Nb-23.2 mol%Al. One can see black, spherical particles of about 1 μm in diameter (phase A) and fine, gray grains of about 2 μm in diameter (phase N) in Nb_3Al. Microstructure and composition of these phases were examined with an analytical transmission electron microscope. Energy dispersive X-ray spectroscopy and electron diffraction revealed that the sample consisted of Nb_3Al, Nb_{ss} and alumina. The black particles and gray grains in Fig.1 corresponded to alumina particles and Nb_{ss} grains, respectively. It has been found from microscopic observations that the HIPed Nb_3Al alloys contain a small amount of Nb_{ss} or Nb_2Al depending on composition and alumina which is independent of composition. The Nb_3Al alloys in the chemically analysed composition range from 21 to 23 mol%Al consist of near-single phase Nb_3Al with a grain size of about 10 μm (a very small amount of Nb_{ss} was detected).

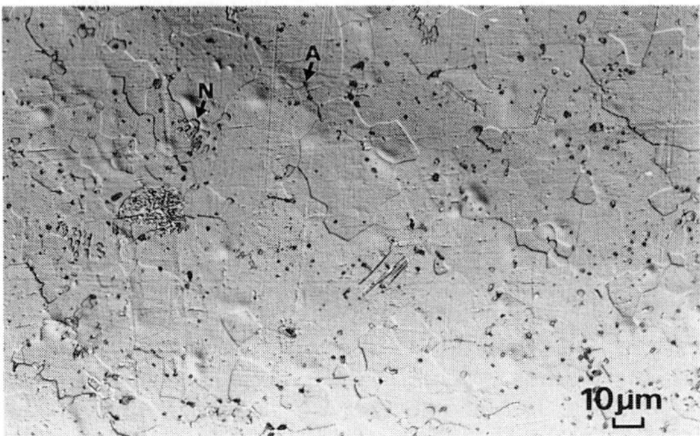

Fig.1 Optical micrograph of the near-single phase Nb_3Al fabricated by the powder metallurgy process.

Fig. 2 shows the composition dependence of yield stress in compression at 1473K and an initial strain rate of $1.67\times10^{-4}s^{-1}$, where chemically analysed compositions are used. Yield stresses are remarkably high compared with conventional superalloys or Ni_3Al based alloys, and increase with increasing Al content in the compositions from 21 to 23 mol%Al. Fig.3 shows deformation microstructure of the Nb-24.2 mol%Al compressed to a peak strain (see Fig.5) at 1473K and $1.67\times10^{-4}s^{-1}$. A lot of planar faults can be seen. Murayama et al.[14] have observed similar planar faults in deformed Nb-23.2 mol%Al. They found that the fault consists of two partial dislocations with the Burgers vector a/2<100> and a CSF with the displacement vector a/2<100>. Marieb et al.[3] have also observed planar faults in Nb_3Al deformed at 1473K. Their arc-melted alloy having the composition of Nb-18mol%Al consisted of Nb_3Al and Nb_{ss}. Since the sample was rapidly cooled on a water-cooled copper hearth after arc-melting, the composition of Nb_3Al in the alloy may be close to Nb-23mol%Al, referring to the Nb-Al phase diagram in [13]. If this is the case, their observation is consistent with the present result. On the other hand, no widely extended CSF is observed in the deformed Nb-21.5 mol% Al, as shown in Fig.4. Weak-beam observations of dislocations in Fig.4 revealed that each dislocation was of a<100> and also dissociated into two a/2<100> partial dislocations with CSF, although the dissociation width is narrow. No widely extended planar fault can be seen in deformation microstructure of Nb-21mol%Al fabricated with the infiltration process,

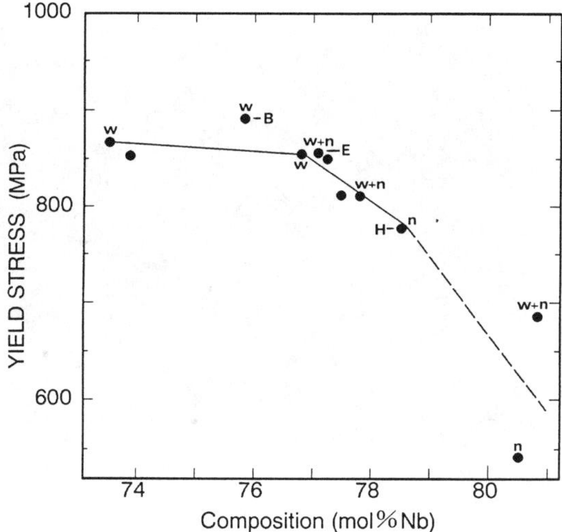

Fig.2 Composition dependence of yield stress for the Nb_3Al alloys fabricated by the powder metallurgy process.

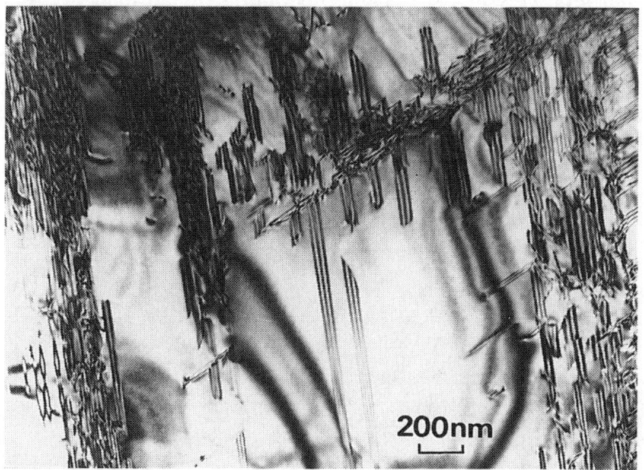

Fig.3 Transmission electron micrograph of Nb-24.2mol%Al deformed to a peak strain at 1473K and $1.67 \times 10^{-4} s^{-1}$.

which will be described in the following section. Moreover, Aindow et al.[15] have also observed that many of dislocations are present as pairs with a narrow separation in melt-spun Nb-18mol%Al ribbons. Thus, it is suggested that the dissociation of dislocations in Nb_3Al depends on composition. Accordingly, deformation microstructures in the Nb_3Al alloys with various compositions were observed and it was found that there were two types of dissociation, i.e., wide and narrow spacings between partial dislocations. These are indicated by w and n in Fig.2. It is evident that the dissociation width is wide in Nb_3Al with high Al content, while it is narrow in Nb_3Al with low Al content. In the Nb_3Al alloys with higher Al content than 23mol%, where the alloys are composed of two phases, Nb_3Al and Nb_2Al, yield stress does not increase significantly. It may be difficult to explain this composition dependence of yield

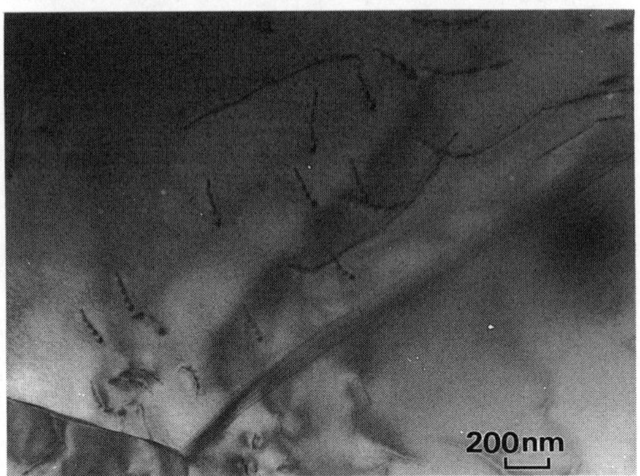

Fig.4 Transmission electron micrograph of Nb-21.5mol%Al deformed to a peak strain at 1473K and $1.67 \times 10^{-4}s^{-1}$.

stress, because no information is available on the temperature dependence of yield stress of a single phase Nb_2Al. Recently, a directionally solidified Nb_2Al- $NbAl_3$ alloy, which has microstructure consisting of long $NbAl_3$ fibers in Nb_2Al matrix, has been found to deform in a ductile manner under relatively low stress at temperatures higher than 1473K[16]. Therefore, it is unlikely that Nb_2Al grains in the Nb_3Al alloys act as hard particles during high temperature deformation in this experiment. In the Nb_3Al alloys with lower Al content than 21mol%, where the alloys are composed of two phases, Nb_3Al and Nb_{ss}, yield stress decreases significantly with decreasing Al content, which can be explained by the increase in volume fraction of Nb_{ss}.

Composition dependence of flow behavior at high strains may be related to the dissociation width. Fig.5 shows stress-strain curves of the Nb_3Al alloys with three typical

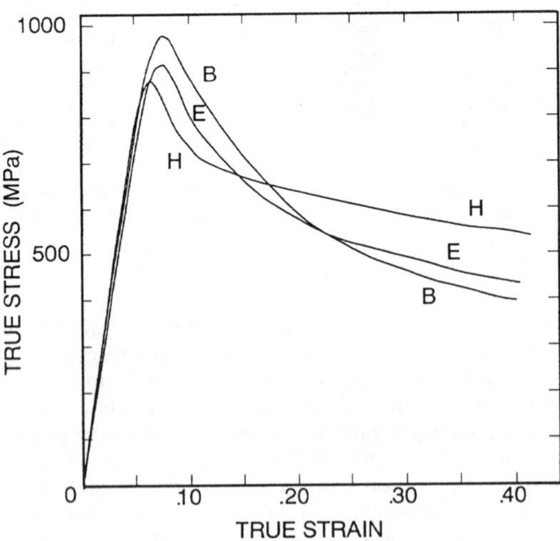

Fig.5 Stress-strain curves for the Nb_3Al alloys deformed at 1473K and $1.67 \times 10^{-4}s^{-1}$.

compositions (B: Nb-24.2 mol%Al, E : Nb-22.9mol%Al, H: Nb-21.5 mol%Al). The alloys B and E with high Al content have a high peak stress and exhibit remarkable softening. At high strains many fine grains less than 1 μm were found at initial grain boundaries. Stacking faults and partial dislocations could be observed in some of the fine grains. Therefore, it is considered that they were produced by dynamic recrystallization, which will be discussed elesewhere[17]. Similar stress-strain curves characterized by a peak stress and remarkable softening have been reported in some single phase intermetallics, such as $Fe_3(Si,Al)$[18] and Ni_3Al[19]. These intermetallics exhibited superplasticity accompanied with dynamic recrystallization. Therefore, superplasticity is expected to occur in Nb_3Al.

INFILTRATION PROCESS

This process has been used to improve superconducting properties of Nb_3Al wires[20]. Niobium powders with - 200+250 mesh(from 59 to 118 μm) were compacted by cold isostatic pressing(CIP) at various pressures to produce porous Nb rods with different densities. The rods were sintered for 600s at 2523K under a vacuum of 4×10^{-3}Pa to provide green compacts with desired porosities and then immersed in a molten Al bath maintained at 973K in a vacuum chamber. Immediately after the immersion Ar gas was charged into the chamber to a pressure of 0.2 MPa. The compacts were taken out of the Al bath after 180s and Nb/Al composites with various ratios in volume of Nb to Al could be fabricated depending upon the porosity of the compacts. Holding for more than 180s produced inevitably intermetallic phases at interfaces between Nb and Al, which made the composites undeformable to high reductions. The composites were put into a Nb tube sheath, extruded and rod-rolled . The rolled rods were bundled in a Nb tube sheath, extruded and rerolled to obtain fine, layered structure of Nb and Al. Repeatedly rod-rolled composites were heat treated in a vacuum. Near-single phase Nb_3Al and two phase Nb_3Al alloys containing Nb_{ss} could be produced by controlling CIP pressures and heat treatment conditions. A low CIP pressure leads to a highly porous compact which increase the amount of infiltrated Al and eventually leads to the formation of single phase Nb_3Al, as shown in Fig.6. Decreasing the amount of infiltrated Al resulted in a two phase

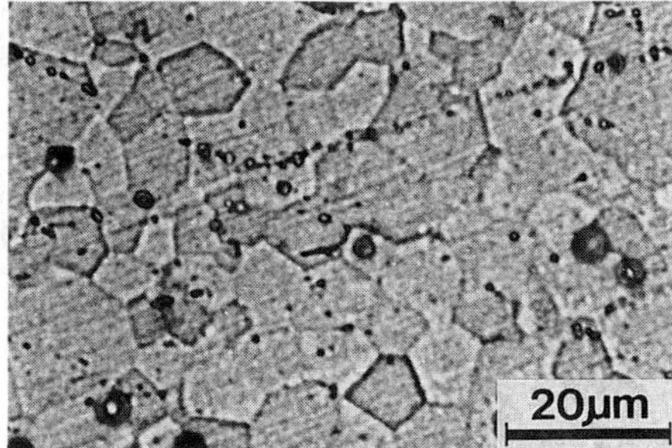

Fig.6 Optical micrograph of the near-single phase Nb_3Al fabricated by the infiltration process.

Nb_3Al alloy. A typical microstructure of an as-heat treated alloy is shown in Fig.7, where optical micrographs, Fig.7(a) and (b), are taken on sections perpendicular and parallel to the rolling direction, respectively. Dark areas, which are isolated in Fig.7(a) and elongated along the rolling direction in Fig.7(b), are Nb_{ss}. The white matrix is Nb_3Al with a grain size of 10 μm. Fig.8 shows a transmission electron micrograph of the two phase alloy in Fig.7. No dislocation was observed in Nb_3Al grains, whereas a Nb_{ss} grain contains a very high density of dislocations. It seems that these dislocations were not introduced by mishandling, but by a difference in coefficient of thermal expansion between two phases. Analytical electron microscopy indicated that Al content was 21±1mol% in Nb_3Al of Figs.6 and 7 and 10±1mol% in Nb_{ss} of Fig.7.

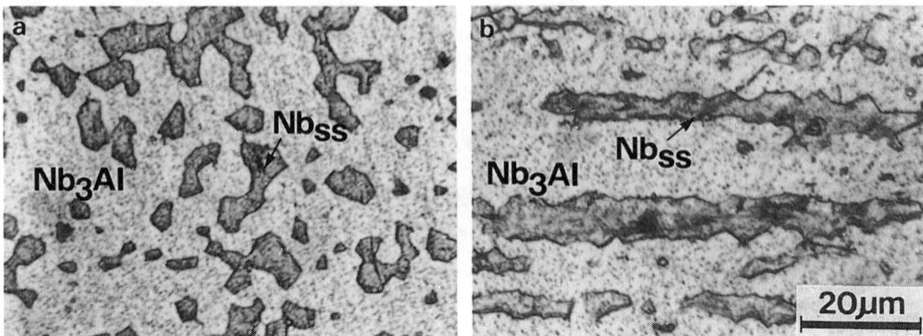

Fig.7 Optical micrographs of the two phase Nb_3Al alloy fabricated by the infiltration process. (a) a transverse section. (b) a longitudinal section.

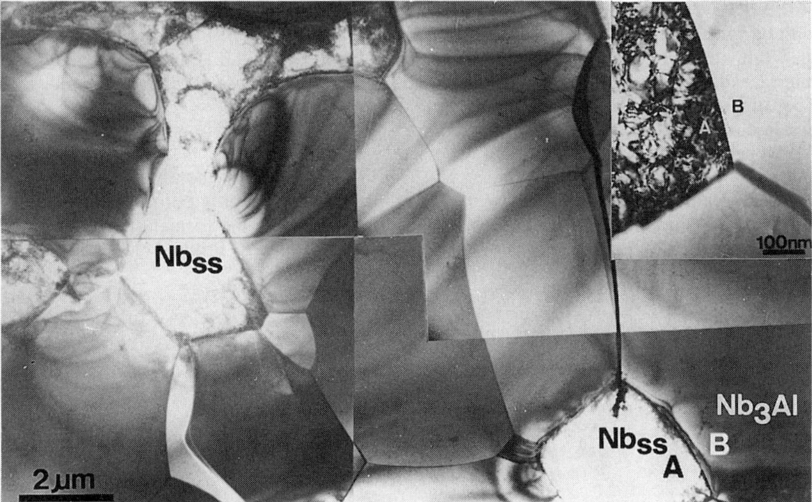

Fig.8 Transmission electron micrograph of the two phase alloy in Fig.7(a).

Yield stress of single phase Nb_3Al and two phase Nb_3Al alloys compressed in the extrusion direction at an initial strain rate of $1.8 \times 10^{-4} s^{-1}$ is shown in Fig. 9 as a function of volume content of Nb_{ss}. The figure demonstrates that the yield stress increases with decreasing temperature and volume content of Nb_{ss}. The plot of yield stress at the lowest temperature in each curve indicates the critical temperature for plastic deformation. Below this temperature the sample exhibits brittle fracture before yielding. That is, the ductile-brittle transition temperature (DBTT) is suggested to be around 1300K for the single phase Nb_3Al, while it is around 1000K for the two phase Nb_3Al alloy containing 31 vol% Nb_{ss}. The single phase Nb_3Al has high yield stresses, e.g., 1200MPa at 1373K, 800 MPa at 1473 K and 500MPa at 1573K, which are much higher than those of conventional superalloys. Yield stresses are lower in the two phase alloys than in the single phase alloy. At a lower strain rate yield stress decreased and DBTT was lowered[21].

Fig.10 shows a transmission electron micrograph of Nb_3Al-35vol% Nb_{ss} deformed to

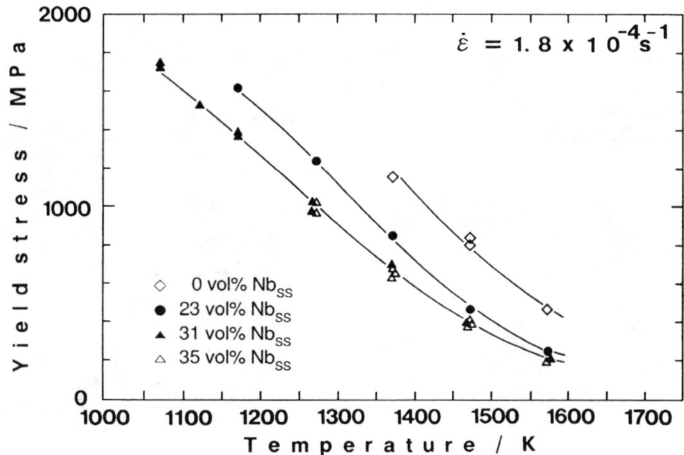

Fig.9 Temperature dependence of yield stress for the Nb_3Al alloys fabricated by the infiltration process.

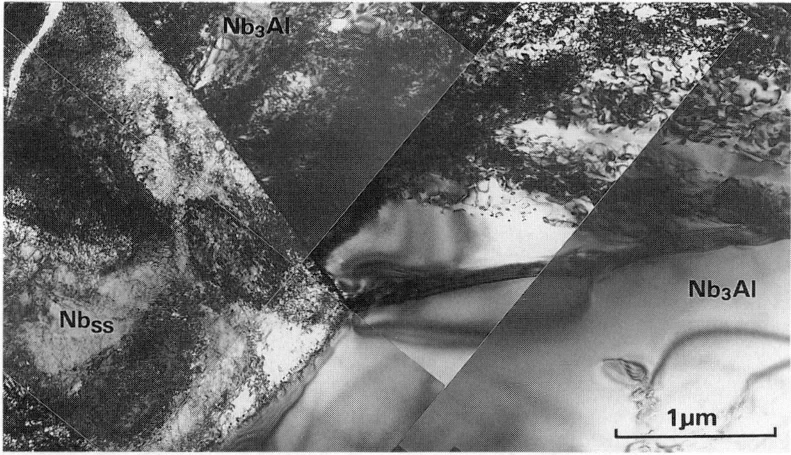

Fig.10 Deformation microstructure in a Nb_3Al grain adjacent to a Nb_{ss} grain in the Nb_3Al-35vol%Nb_{ss} deformed to 0.03 true strain at 1123K and $1.8\times10^{-4}s^{-1}$.

0.03 true strain at 1123K and $1.8\times10^{-4}s^{-1}$. A high density of dislocations can be observed predominantly in a Nb_3Al grain which is adjacent to a Nb_{ss} grain. Since at 1123K Nb_{ss} has low yield stress and Nb_3Al has very high yield stress, the high density of dislocations in the Nb_3Al grain may be explained by the enhancement of plasticity in Nb_3Al, which is induced by the existence of the adjacent Nb_{ss} grain. On the other hand, inhomogeneous deformation bands are frequently observed in the vicinity of a Nb_3Al-Nb_3Al grain boundary, as shown in Fig.11. Crack initiation at this boundary is considered to be most likely to occur by high stress concentration which develops there. Actually the sample was broken at low plastic strain. Fig. 12 shows a transmission electron micrograph of single phase Nb_3Al deformed to 0.03 true

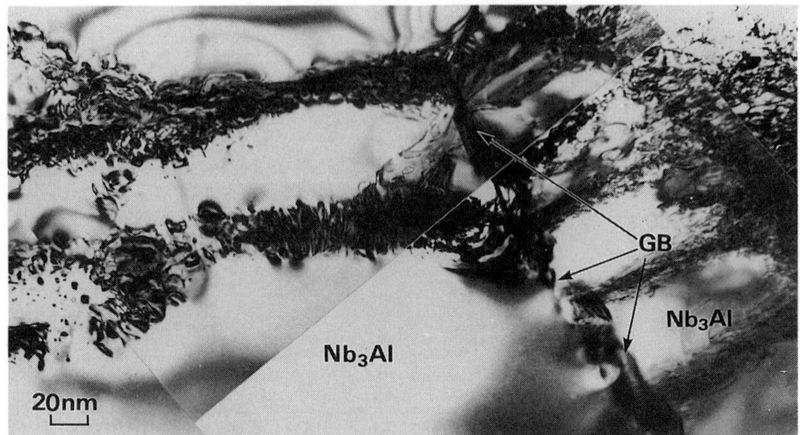

Fig.11 Deformation microstructure at a Nb_3Al - Nb_3Al boundary in the Nb_3Al-35vol%Nb_{ss} deformed to 0.03 true strain at 1123K and $1.8\times10^{-4}s^{-1}$.

Fig.12 Deformation microstructure in the near-single phase Nb_3Al deformed to 0.03 true strain at 1473K and $1.8\times10^{-3}s^{-1}$.

strain at 1473 K and $1.8\times10^{-3}s^{-1}$. In contrast to Fig.11, tangled dislocations are uniformly distributed in all the grains, indicating low stress concentration at grain boundaries.

Above 1474K, ductile deformation to high strains was possible both in the single phase Nb_3Al and the two phase Nb_3Al alloys. Fig.13 shows deformation microstructure of the two phase Nb_3Al alloy compressed to 0.3 true strain at 1473K and $1.8\times10^{-4}s^{-1}$. One can see

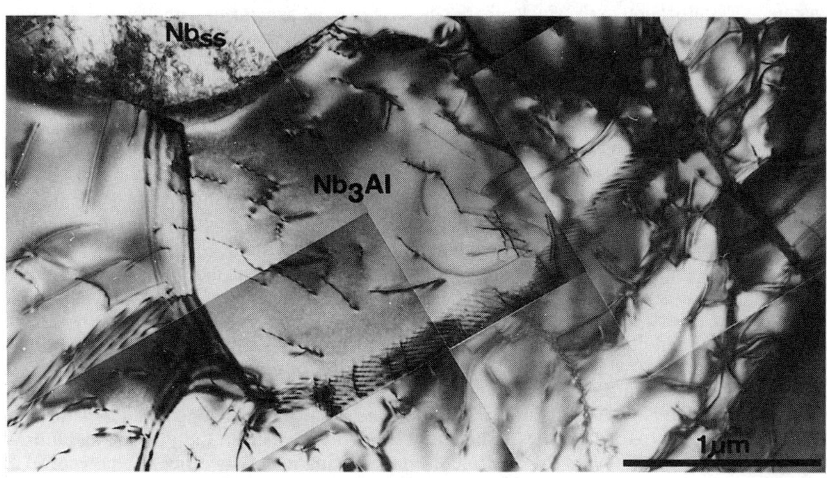

Fig.13 Deformation microstructure in the Nb_3Al-23vol%Nb_{ss} deformed to 0.3 true strain at 1473K and $3.3\times10^{-4}s^{-1}$.

relatively uniform microstructure consisting of many sub-boundaries and dislocation networks, which indicates that restoration processes can occur predominantly during deformation. On the contrary, inhomogeneous distribution of dislocations is produced in the single phase Nb_3Al, as shown in Fig.14. Furthermore, small grains with a grain size of about 1 μ m can be seen in regions with a high density of dislocations. Since these small grains contain some dislocations, it is suggested that they are dynamically recrystallized grains. This result on the occurrence of

Fig.14 Deformation microstructure in the near-single phase Nb_3Al deformed to 0.3 true strain at 1473K and $1.8\times10^{-4}s^{-1}$.

dynamic recrystallization in the single phase Nb_3Al is in good agreement with that in the sample prepared from prealloyed powders in the preceding section. As described above, Al content in Nb_3Al fabricated by the infiltration process is 21±1at% , even though the alloy is single or two phases. This is because widely extended CSF was not observed in deformation microstructures of the Nb_3Al alloys fabricated by the infiltration process.

CLAD-CHIP EXTRUSION PROCESS

As described above, near-single phase Nb_3Al is obtained with the powder metallurgy process using prealloyed powders and with the infiltration process using Nb powders. However, the Nb_3Al alloys contain small amounts of Nb_{ss} and alumina. Although Nb_{ss} was found to be dissolved by further annealing, it was difficult to suppress the formation of alumina in both the processes in spite of powder processing in Ar and degassing. Deformation behavior of Nb_3Al may be affected by alumina. We have tried to prepare Nb_3Al without any second phases with the clad-chip extrusion process[22], which was recently developed to fabricate superconducting Nb_3Al thin wires. The process consists of clad-rolling Nb and Al plates, cutting the clad-sheet into pieces, filling up the pieces into a container, extruding the billet, rolling the extruded bar, and diffusion annealing, and it has the following characteristics: (1) deformability of Nb/Al composites is better in this process than in the powder metallurgy process, since relatively thick Nb plates are used as starting materials instead of powders, (2) chemical composition after diffusional reaction can be easily controlled by changing the initial thickness of Nb and Al plates.

In fabrication of Nb_3Al, a 1mm thick Nb sheet and 0.14~0.16mm thick Al foil were used as starting materials. Three layered Al/Nb/Al sheets were clad by cold rolling at 70% reduction per pass to 0.2mm in total thickness. Chipped pieces were poured into a Cu-20% Zn tube. After evacuation the billet was extruded at a reduction ratio of 5 and followed by rod-rolling. After the sheath was removed, the rod-rolled Nb/Al composites were bundled in a double sheath consisting of inside sheath Nb and outside sheath Cu-Zn and re-extruded. Then the extruded rod was rod-rolled. Kirkendall voids were produced after heat treatments of the rods. HIPing eliminated these voids and produced single phase Nb_3Al, as shown in Fig.15. Thus, this process appears to be very suitable for the fundamental investigation of mechanical properties of single phase Nb_3Al . Compressive tests are now in progress.

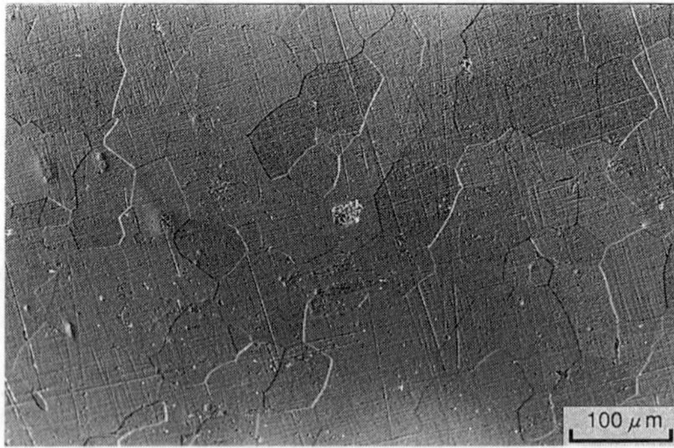

Fig.15 Optical micrograph of single phase Nb_3Al fabricated by the clad- chip extrusion process.

SUMMARY

Nb_3Al alloys with various compositions were fabricated by the powder metallurgy process, the infiltration process and the clad-chip extrusion process. Near-single phase Nb_3Al alloys with compositions between 21 and 23 mol%Al, which were fabricated by the powder metallurgy process, exhibited much higher yield strength above 1400K compared with conventional superalloys. On deformation of the alloys at 1470K widely extended CSFs bounded by two a/2<100> partial dislocations are introduced on {001} in the Nb-23mol%Al, while the dissociation width decreases remarkably in the Nb-21 mol%Al. In this composition range yield stress increases with increasing Al content.

Nb_{ss} with various volume fractions could be distributed in Nb_3Al with the infiltration process. DBTT is lowered in the two phase alloys, although yield strength decreases with increasing the volume fraction of Nb_{ss}. At high strains dynamic recrystallization occurs in near-single phase alloys, whereas dynamic recovery occurs in Nb_3Al of the two phase alloys.

Fully single phase Nb_3Al was produced with the clad-chip extrusion process.

ACKNOWLEDGMENTS

This work is partly supported by Grant-in-Aid for Scientific Research on Priority Area from the Ministry of Education, Science and Culture, Japan and in part by the research grant from R & D Institute of Metals and Composites for Future Industries. The authors would like to thank Dr. S.Saito and Mr. Y.Abe for technical assistance of the infiltration process and clad-chip extrusion process.

REFERENCES

1. Y.Umakoshi, Bulletin Japan Inst. Metals, 30, 72-79(1991).
2. T.Kumagai, Doctoral Thesis, Tohoku University, (1992).
3. T.N.Marieb, A.D.Kaiser, S.R.Nutt, D.L.Anton and D.M.Shah, in High-Temperature Ordered Intermetallic Alloys IV, edited by L.A.Johnson, D.P.Pope and J.O.Stiegler (Mater. Res. Soc. Proc. 213, Pittsburgh, PA, 1991) pp.329-336.
4. S.Foner and E.J.McNiff, Jr.: Phisica, 55, 534-539 (1971).
5. R.Akihama, R.J.Murphy and S.Foner: IEEE Trans.Magn., MAG-17, 274-277 (1981).
6. C.L.H.Thieme, H.Zhang, J.Otubo, S.Pourrahimi, B.B.Schwartz and S.Foner: IEEE Trans.Magn., MAG-19, 567-569(1983).
7. C.L.H. Thieme, S.Pourrahimi, B.B.Schwartz and S.Foner: Appl. Phys. Lett., 44 260-262(1984).
8. K.Watanabe, K.Noto and Y.Muto: IEEE Trans. Magn., MAG-23, 1428-1431(1987).
9. M.R.Picus, J.T.Holthis and M.Rosen: Filamentary A-15 Superconductors, Ed.by M.Suenaga and A.F.Clark (Plenum Press, New York, 1980) pp.331-353.
10. S.Ceresara, M.V.Ricci, N.Sacchetti and G.Sacerdoti: IEEE Trans. Magn., MAG-11, 263-265(1975).
11. D.Dew-Hughes: IEEE Trans. Magn., MAG-17, 561-564(1981).
12. K.Inoue, Y.Iijima and T.Takeuchi: Appl. Phys.Lett., 52, 1724-1725(1988).
13. R.Suyama and K.Hashimoto, in High-performance Materials for Severe Environments, edited by R & D Institute of Metals and Composites for Future Industries (Japan Industrial Technology Association, Tokyo, 1992) pp. 141-150.
14. Y.Murayama, S.Hanada, K.Obara and K.Hiraga, Phil. Mag. A, in press.
15. M.Aindow, J.Shyue, T.A.Gaspar and H.L.Fraser, Phil. Mag. Lett. 64, 59-65(1991).
16. T.Kumagai and S.Hanada, Mater. Sci. Eng. A152, 349-355(1992).
17. Y.Murayama, S.Hanada and K.Obara, Mater. Sci. Eng. A, in press.
18. S.Hanada, T.Sato, S.Watanabe and O.Izumi, J. Japan Inst. Metals, 45. 1293-1299(1981).
19. M.S.Kim, S.Hanada, S.Watanabe and O.Izumi, Mater.Trans., JIM, 30, 77-85(1989).

20. S.Saito, I.Yoshii, K.Ikeda and S.Hanada, Mater.Trans., JIM, 31, 501-503(1990).
21. T.Kumagai, S.Hanada and S.Saito, in Intermetallic Compounds-Structure and Mechanical Properties-, edited by O.Izumi, (JIMIS-6 Proc.,Sendai, 1991) pp.1039-1044.
22. S.Saito, S.Ikeda, K.Ikeda and S.Hanada, J. Japan Inst. Metals, 53, 458-463(1989). 54,737-740(1990). 55, 85-91(1991).

PREDICTION OF THE HIGH TEMPERATURE OXIDATIVE LIFE OF INTERMETALLICS

James A. Nesbitt and Carl E. Lowell
NASA Lewis Research Center, Cleveland, OH 44135

ABSTRACT

A method is presented to predict the oxidative life of intermetallics. The method is demonstrated by predicting the lifetimes of several aluminides undergoing cyclic oxidation at 1200°C. For NiAl and NiAl-Zr alloys, the lifetimes were predicted at several other temperatures as well. Using a critical surface recession failure criterion, it is shown that several aluminides (e.g., NiAl and FeAl) have long-term oxidation resistance (~10,000 hours) to approximately 1150°C. Aluminides containing reactive elements (e.g., NiAl-Zr alloys) have long-term oxidation resistance to approximately 1200°C. The method is also applied to predict the oxidative lifetime of $MoSi_2$ at temperatures of 1200°-1400°C which shows that the oxidation resistance of $MoSi_2$ is significantly better than that for the aluminides. For the same failure criterion, it is shown that $MoSi_2$ can exhibit long-term oxidation resistance (~10,000 hours) at temperatures in excess of 1400°C. Use of the method to predict the maximum use temperature is also demonstrated for NiAl and NiAl-Zr alloys.

INTRODUCTION

Several intermetallics are of considerable interest for high temperature structural applications because of their relatively high specific strength (strength/density). However, as with most materials for use at high temperatures, oxidation resistance is a concern since fast growing surface oxides result in unacceptably high rates of metal consumption. Oxidation protection is afforded by the formation of oxide scales which grow at an acceptably low rate such that oxidation does not limit the useful life of a component far short of the mechanical life (e.g., fatigue or creep life). Very few oxides grow sufficiently slowly as to be considered acceptable in protecting components for high temperature applications in aero gas turbine engines. The parabolic rate constant for an oxide, k_p, relates the weight gain during oxidation to time. Consequently, these rate constants give an indication of the growth rate of an oxide scale and can be used to rank the "protection" afforded by an oxide. An Arrhenius plot showing k_p values for several oxides of interest is shown in Fig 1. This figure demonstrates the superiority of Al_2O_3 and SiO_2 scales over other oxides. Thus, two intermetallics under current consideration, NiAl and $MoSi_2$, have excellent inherent oxidation resistance based on Al_2O_3 and SiO_2 scales and have been the foundation of coating schemes to protect other materials for several decades (e.g., aluminide coatings for superalloys and $MoSi_2$ coatings for refractory metals). In contrast, other intermetallics of interest, such as Ti-24Al-11Nb[*] alloys, have poor inherent oxidation resistance due to significant TiO_2 formation and will likely require

[*] All alloy compositions are given in atomic percent.

a protective coating for high temperature applications. Hence, current oxidation protection for aero gas turbine applications, whether by coatings or an inherent oxidation resistance of the material itself, has been based on the formation of either Al_2O_3 or SiO_2 scales.

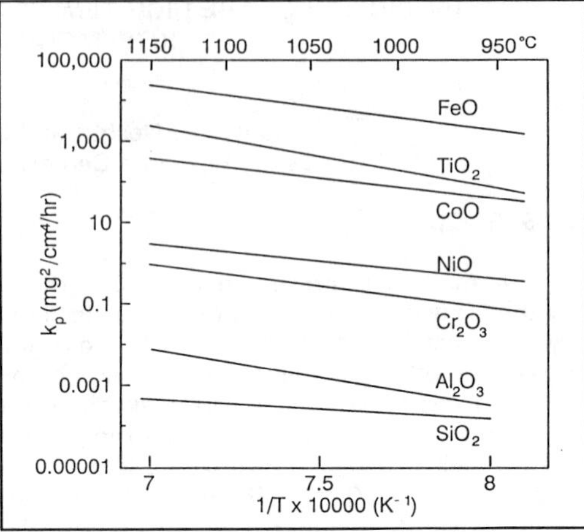

Figure 1 Arrhenius plot showing typical parabolic rate constants for several oxides [1-3].

Many recent studies of the oxidation of intermetallics have been attempts at forming protective Al_2O_3 or SiO_2 scales on new and promising intermetallics. Numerous studies have also examined an accelerated attack (pest) which occurs at low temperatures in certain intermetallics (e.g., $MoSi_2$ [4-6] and $NbAl_3$ [7]) resulting in even higher rates of metal consumption. In addition to initially establishing a protective oxide scale, several studies have also examined the long-term durability of these scales and in particular, the ability of the alloy to reform the protective oxide when the scale is damaged. The most common damage inflicted on the oxide scale occurs during thermal cycling where, due to the thermal expansion mismatch between the substrate and oxide, the scale can undergo cracking and even spalling within the scale or to the substrate. The main concern with this type of damage is the healing of the scale by the reformation and growth of the protective oxide. Alumina scales appear much more susceptible to this type of degradation, whereas SiO_2 scales seem more tolerant of the stresses imposed during thermal cycling. Consequently, several studies in the past have been concerned with the length of time at which an oxide scale, primarily Al_2O_3, continues to provide protection during repeated damage. This transition from protective to nonprotective scale can be used to define the "oxidative life." However, other criteria, such as loss of a certain weight of Al or Si from a sample [8], or a specific amount of surface recession can also be used as a failure criterion to define oxidative life [9].

The purpose of the present paper is to demonstrate a method of predicting the oxidative life of intermetallics. Since limited oxidation data exists on many intermetallics of relatively recent interest, the scope of this paper will be limited to those intermetallic alloys which readily form either Al_2O_3 or SiO_2 scales and for which oxidation testing for a reasonable period has been performed (i.e., testing beyond that necessary to demonstrate the initial formation of a protective oxide scale). This type of oxidation data exists for several aluminides, primarily NiAl, and to a lesser extent for FeAl, CoAl, and certain $TiAl_3$ alloys. In addition, sufficient oxidation testing has also been performed on $MoSi_2$ to allow life prediction for this silica former.

OXIDATION TESTING

Two types of oxidation testing are typically employed to evaluate the oxidation resistance of various alloys. Isothermal oxidation studies are performed by holding a sample at a single temperature for an extended period of time whereas in cyclic oxidation studies a sample is repeatedly raised to an elevated temperature for a certain time period, such as 1 hour, and then allowed to cool, typically to room temperature. Cyclic oxidation testing may be performed manually but is more easily accomplished with an automated furnace rig or burner rig which can cycle several samples at once. Automated furnace rigs, such as those used at the NASA Lewis Research Center [10], allow the number of cycles, the length of the heating cycle, and the length of the cool down period to be set. The samples are periodically removed and weighed according to a predefined schedule. Since most cyclic oxidation testing of intermetallics has been performed in furnaces, burner rig testing will not be discussed.

Isothermal Oxidation. Isothermal oxidation is generally performed by suspending a sample from a microbalance into a furnace at the desired temperature. The sample and balance are typically enclosed in a sealed system allowing a specific gas to gently flow past the sample (e.g., high purity O_2, or gas mixtures with a low oxygen partial pressure) [11,12]. At the elevated temperature, the Al or Si is selectively oxidized and the scale grows and thickens with time. The balance records the oxide growth as a weight gain as shown schematically in Fig 2a. For protective Al_2O_3 and SiO_2 scales, the specific weight (ΔW) increases parabolically with time (t), (i.e., $\Delta W = (k_p t)^{1/2}$ where typical values for k_p are shown in Fig 1). Plotting ΔW versus $t^{1/2}$ yields $k_p^{1/2}$ as the slope (Fig 2b). The time dependence of the oxidation rate, or growth rate, can be determined by taking the derivative of the ΔW versus time curve (Fig 2a) as shown in Fig 2c. The weight of material (Al or Si) consumed from the sample (W_m), and the rate of material consumption (dW_m/dt) can easily be determined from Figs 2a and 2c through multiplication by a simple proportionality constant (η) relating the weight of Al or Si to the weight of oxygen in the oxide (i.e., $W_m = \eta\ (k_p t)^{1/2}$ and $dW_m/dt = \frac{1}{2}\eta\ (k_p/t)^{1/2}$) where $\eta = 1.1242$ for Al_2O_3 and 0.878 for SiO_2. Hence, the amount and rate of Al or Si consumption during isothermal oxidation is proportional to $k_p^{1/2}$.

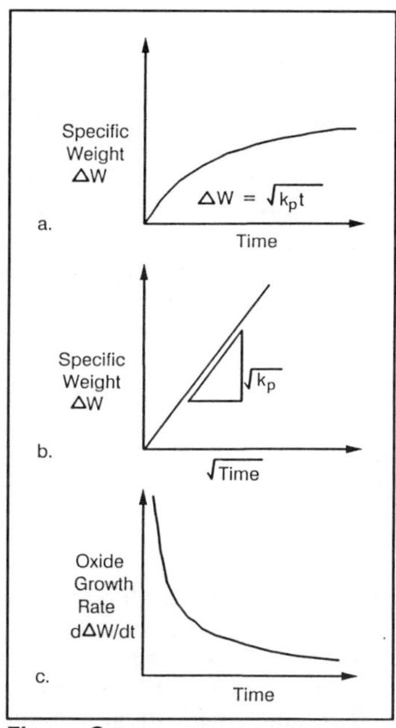

Figure 2 Schematic plots of the isothermal oxidation process, (a) ΔW vs time, (b) ΔW vs square root of time, (c) $d\Delta W/dt$ vs time.

Cyclic Oxidation. Several studies to date indicate that SiO_2 scales are not damaged by thermal cycling [13-15]. Thus, the following discussion of cyclic oxidation will only be applicable to alumina formers. On heating, Al is selectively oxidized to form an Al_2O_3 scale. However, on cooling, the scale tends to crack and spall due to the thermal expansion mismatch between the oxide and metal substrate. This spalling is generally random over the surface of the sample with some oxide loss occurring within the oxide scale while some losses occur to bare metal. Increased spalling occurs at the corners and edges due to geometric stress concentrations at these locations. The parameter measured during cyclic oxidation is the sample weight which, when combined with the initial sample weight and area, yields the specific weight change ΔW. This specific weight change, generally plotted versus the time or number of cycles, has a positive component due to the oxygen in the oxide which is retained on the surface and a negative component resulting from the loss of metal in the oxide which has spalled from the sample. The interplay of these two components results in what is commonly referred to as "paralinear type" weight change behavior represented by an initially positive weight change eventually followed by a linear weight loss. The positive portion of the curve may last for only a few cycles (poorly adherent scale or high temperature) or several hundred cycles (good scale adherence or low temperature). This type of behavior, showing the measured weight change (solid symbols) and the presumed weight gain due to oxide growth and weight loss due to oxide spallation on cooling for 5 cycles is shown schematically in Fig 3. The total weight of metal consumed W_m, the associated rate of metal consumption each cycle dW_m/dt, and the weight of oxide retained on the surface W_r, which includes the weight of both Al and O, is also shown.

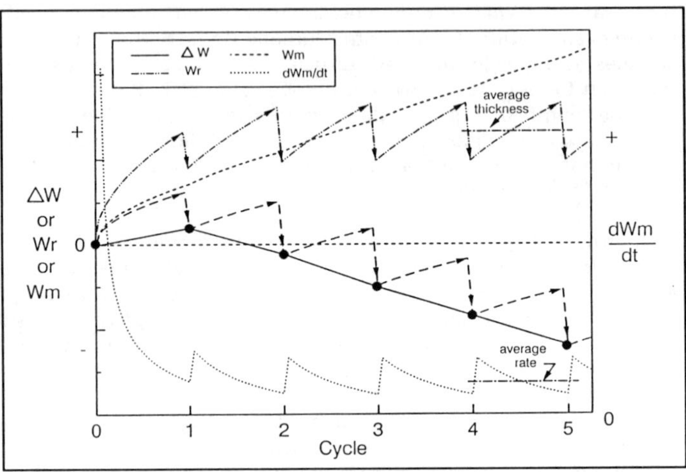

Figure 3. Schematic plot for the cyclic oxidation of alumina formers. Solid circles indicate measured weights, upward-angled arrows indicate weight gains due to oxide growth while down-pointing arrows indicate weight loss due to oxide spalling.

Two factors are known to strongly affect the cyclic oxidation behavior of NiAl alloys, namely oxidizing temperature and reactive element additions. As expected, increased temperatures result in more rapid oxide growth at the elevated temperature and greater oxide spalling on cooling with an overall result of more rapid weight losses. The effect of increased temperatures on the cyclic oxidation behavior of other aluminides would be expected to be similar. Measured weight changes for a Ni-47Al alloy cyclically oxidized at 1100°, 1150° and 1200°C are shown in Fig 4 where much greater weight losses at the higher temperatures are apparent. Small reactive element additions to NiAl alloys greatly improve the cyclic oxidation behavior by making the oxide scales much more adherent during thermal cycling. Figure 4 also shows the weight change for the 1200°C cyclic oxidation of the same alloy with the addition of 0.05Zr. The consequence of the Zr addition is that much less

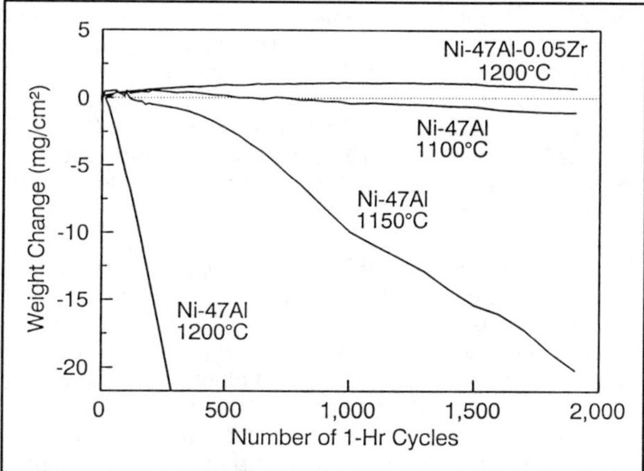

Figure 4 Effect of temperature and Zr additions on the cyclic oxidation behavior of a Ni-47Al alloy [8,16].

of the Al_2O_3 scale is lost during thermal cycling such that the weight change is still positive after nearly 2000 1-hr cycles. Although very small additions are beneficial, larger additions of 0.3% and above result in unacceptably large amounts of internal oxidation and catastrophic weight gains [17,18]. This beneficial effect of small reactive element additions is not unique to NiAl but was observed years ago and has been the subject of considerable research for over a decade, particularly with MCrAlY alloys (M = Ni,Co or Fe). A similar reactive element effect for other intermetallics would be expected although different optimum levels may be found for the different alloys [19,20]. Many mechanisms have been proposed to explain the increased adherence of the oxide scale and the effect continues to be observed and examined with new alloy compositions [21-23].

LIFE PREDICTION METHODOLOGY

Imperative in predicting the oxidative life of any alloy is the selection of a failure criterion. Three failure criteria for NiAl have been used in the past and are (1) a critical amount of surface recession, (2) a critical minimum Al concentration, (3) the breakdown of the protective scale resulting in the formation of $NiAl_2O_4$ [9,24]. Obviously, other failure criteria could be defined but each criterion may not be appropriate for different alloy systems. For instance, there is a large Al solubility in β NiAl alloys and a significant amount of Al can be removed (to below 40%) while maintaining the β phase, whereas for $MoSi_2$, which is a line compound with essentially no solubility range for Si, the removal of Si can quickly result in the formation of Mo_5Si_3. Hence, a suitable failure criterion must be selected when comparing the oxidative lifetimes for different alloys. Probably the most widely applicable failure criterion for various intermetallic alloys is that of a critical surface recession. All aluminides, as well as $MoSi_2$, will reduce in thickness as Al or Si is removed from a sample or component during oxidation, assuming no porosity formation in the sample. Therefore, a critical surface recession criterion will be used to predict the oxidative lifetimes for the various intermetallics where sufficient data exist.

Ideally, surface recession, ξ, can be assumed directly proportional to the loss of Al or Si during oxidation. Dividing W_m by the density, ρ, gives an approximation of the surface recession (i.e.,$\xi \approx W_m/\rho$). The rate of surface recession, $d\xi/dt$ is derived in a similar manner using dW_m/dt. Use of the density to estimate the surface recession assumes that the molar volume of the solute (Al or Si) is equal to that of the alloy. During isothermal oxidation, dW_m/dt can be determined directly using k_p such that $dW_m/dt = \frac{1}{2}\eta\,(k_p/t)^{1/2}$. During cyclic oxidation, determining dW_m/dt is more difficult. The rate may be determined directly by measuring the sample weight change and the amount of oxide which has spalled from the sample. This procedure has been performed for a high test temperature (1400°C) for short test periods (less than 200 1-hr cycles) [9] but at lower temperatures where the amount of oxide growth and spallation each cycle is much smaller, this direct measurement becomes impractical.

A more reasonable means of acquiring the rate of metal consumption is to predict it using an oxide spalling model. One such model which has evolved over the past decade is referred to as "COSP" [25]. In it's simplest form, this model tracks the oxide growth and amount of oxide spallation which occurs each cycle and, in so doing, predicts the sample weight change, ΔW, the weight of metal consumed, W_m, the weight of oxide retained on the sample surface, W_r, and the rate of metal consumption, dW_m/dt. The two basic premises of the model are that (1) the oxide grows parabolically with time during the high-temperature portion of each cycle {although this requirement can easily be modified to incorporate any oxide growth law}, and (2), that the amount of oxide which spalls on cooling is dependent on the current oxide thickness. The justification for the second premise is that, as expected intuitively from simple stress arguments, the thicker the oxide scale, the larger the fraction of oxide spall. Although the model assumes parabolic oxide growth, the amount which grows each cycle is determined by the current average oxide thickness. Hence, the amount of oxide which grows on the second cycle will be less than that grown on the first cycle since some oxide will be present on the surface at the start of the second cycle resulting in a lower growth rate. This effect can be seen in Fig 3 by comparing dW_m/dt during the first and second cycle. Although somewhat difficult to see, the rate during the third cycle is slightly less than that during the second. However, the values for dW_m/dt for subsequent cycles become more and more similar. The explanation for this similarity is that the oxide thickness, averaged over the duration of the cycle, reaches a "steady state" value which is constant from cycle to cycle, as indicated by W_r, the weight of oxide retained on the surface in the figure. The basis for this "steady state" thickness is the prediction that the oxide which grows and spalls each cycle become constant and equal. This predicted steady state oxide thickness and equal oxide growth and spallation each cycle produces the linear weight change behavior observed experimentally.

The COSP spalling model requires two parameters for each alloy for a given test condition, the parabolic rate constant k_p and a spalling parameter Q_o. While k_p can be measured independently with an isothermal oxidation test, Q_o is very difficult to measure independently for alumina formers because of the very small amounts of oxide which grow and spall each cycle. Hence, use of the model involves selecting Q_o to match the predicted ΔW values with the measured ΔW values for a given test. Then, the values for dW_m/dt predicted by the spalling model can be used to predict the rate of surface recession. The use of the COSP model is demonstrated in Fig 5 where the weight change for two Ni-50Al samples cyclically oxidized at 1200°C is shown. The weight change ΔW predicted by the COSP model after adjusting Q_o for agreement is shown as well as the predicted amount of metal consumption W_m.

A simpler means may be employed to estimate the rate of metal consumption during cyclic oxidation. Since the average amount of oxide on the surface becomes constant when the weight change curves become linear, the weight loss each cycle reflected in the weight change curve is simply the weight of metal in the oxide which forms and spalls from the surface each cycle. Hence, dW_m/dt can be estimated from the slope of the measured ΔW curves in the linear region (i.e., $dW_m/dt = -d\Delta W/dt$), as indicated in Fig 5. The simplifying assumption can then be made to extend this rate to the beginning of the test, as indicated as the approximated W_m shown in Fig 5. The error associated with this approximation amounts to the weight of metal in the oxide which remains on the sample surface. As shown, W_m predicted by COSP is always greater than that using the approximation with the difference becoming constant when ΔW becomes linear (i.e., when the steady state oxide thickness is reached). Hence, for the data shown in Fig 5, the difference is 6% after 500 cycles but only 3% after 1000 cycles.

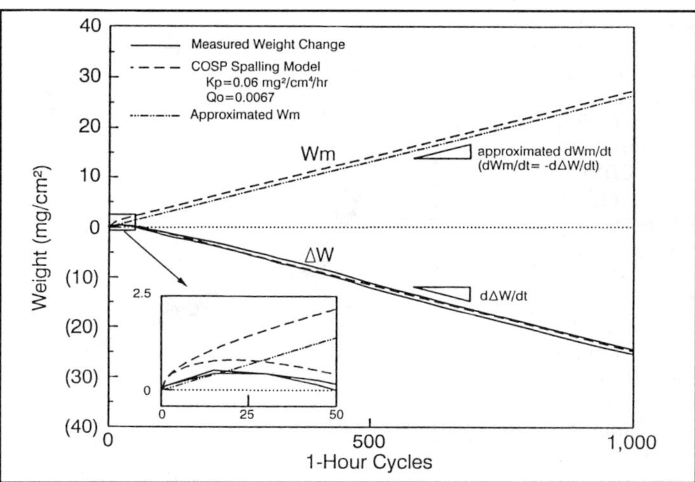

Figure 5. Measured weight change ΔW, for the 1200°C cyclic oxidation of two Ni-50Al alloys [24]. Predicted weight of metal consumed W_m and ΔW using the COSP model and the linear approximation of W_m.

OXIDATION DATA

Isothermal. Bands indicating reported values of several studies for the parabolic rate constants, k_p, for Al_2O_3 [11,12] and SiO_2 [2] are shown in Fig 6. As shown, two forms of Al_2O_3 form on NiAl, the higher-temperature α phase and the faster growing, lower-temperature θ phase. Limited data for α-Al_2O_3 on $TiAl_3$ [3] all fall within the band for Al_2O_3 on NiAl. Similarly, higher-temperature k_p values for Al_2O_3 on a Nb-68Al-7Cr-0.5Y alloy [16] also fall within this band although the 1000°C value falls within the θ-Al_2O_3 band.

At temperatures greater than approximately 1000°C, SiO_2 scales on $MoSi_2$ grow slower than Al_2O_3 scales on any of the aluminides. Although the high-temperature oxidation of $MoSi_2$ is excellent, the low-temperature oxidation (375°-500°C) has been plagued with a rapid and catastrophic mode of oxidation referred to as "pest." The pest phenomena has been observed in single and polycrystalline material and commonly results in voluminous sample distortions and/or a reduction of the test sample to oxide powder. Although there is not general agreement on the mechanism of pesting, it is apparent that the protective SiO_2 scale is never established on the surface of the samples. Over the past decade, the bulk of the oxidation studies of $MoSi_2$ have focussed on this pesting problem. Since in this paper the oxidative life of $MoSi_2$ will only be predicted at high temperatures (>1200°C), the problem of pest will not be considered although it continues to be an obstacle to the use of $MoSi_2$ in aero engines [4-6].

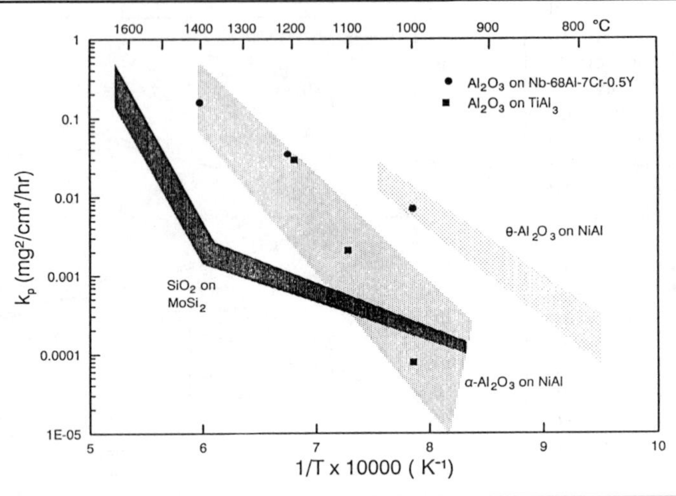

Figure 6 Ranges of k_p values for select Al_2O_3 formers and SiO_2 formation on $MoSi_2$ as a function of temperature.

Cyclic. Isothermal growth rates do not adequately reflect the rates of Al consumption during cyclic oxidation of the aluminides. As indicated above, alumina scales crack and spall during oxidation accompanied by thermal cycling producing "paralinear" weight change curves. The rates of Al consumption for the 1200-°C cyclic oxidation of several NiAl [16,24], NiAl-Zr [9,13,24], FeAl [18], CoAl [18] and $TiAl_3$-8Cr [26] alloys were estimated from the linear portion of the weight change curves, and are shown in Fig 7. At this temperature, the rates of Al consumption can be ranked from lowest to highest as NiAl-Zr : $TiAl_3$-8Cr : FeAl : NiAl : CoAl.

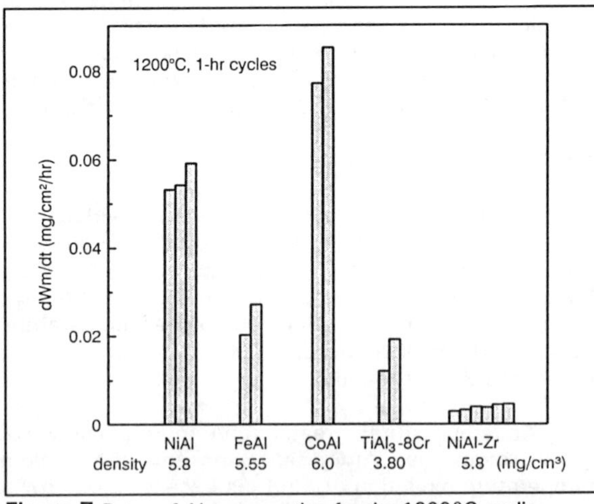

Figure 7 Rates of Al consumption for the 1200°C cyclic oxidation of several Al_2O_3 formers approximated from ΔW data.

Sufficient data were available to examine the temperature dependence of the rate of Al consumption for NiAl and NiAl-Zr alloys. The rates for these alloys, processed by a variety of methods and tested over a period of more than 5 years, are shown in Fig 8. It is significant that this rate data, taken from many sources and fabricated by various methods, and for the NiAl-

Zr data, representing alloy compositions from 40 to 50% Al, generally fall about a single line for the binary alloys and a single line for the NiAl-Zr alloys. Hence, it is suggested that this rate, estimated from measured weight change data, can be used as a useful statistic for comparing the cyclic oxidation behavior of various alloys and at various temperatures.

Cyclic oxidation testing of $MoSi_2$ at 1200° and 1500°C for 144 hours (55 minute cycle) has recently been performed, however no oxide spalling was observed [13]. The SiO_2 scale was glassy and amorphous which is typical of SiO_2 scales formed on $MoSi_2$ (see [4,14] for a more detailed discussion of the high-temperature oxidation of $MoSi_2$). Past work also indicates that the scale remains attached and adherent and continues to grow parabolically with time. Thus, the rate of Si consumption is simply that given above for the case of isothermal oxidation and is dependent on the time. This time dependence makes comparison to the rates for Al consumption difficult. However, average rates for 1,000 hours were determined by integrating the rate over this time span and are compared to those for the Ni aluminides in Fig 8. The average rate for Si consumption after 10,000 hours is even less, being a factor of 3 lower than those for 1000 hours. The data in Fig 8 clearly shows that the rates for Si consumption in $MoSi_2$ are considerably less than those for any of the Ni aluminides.

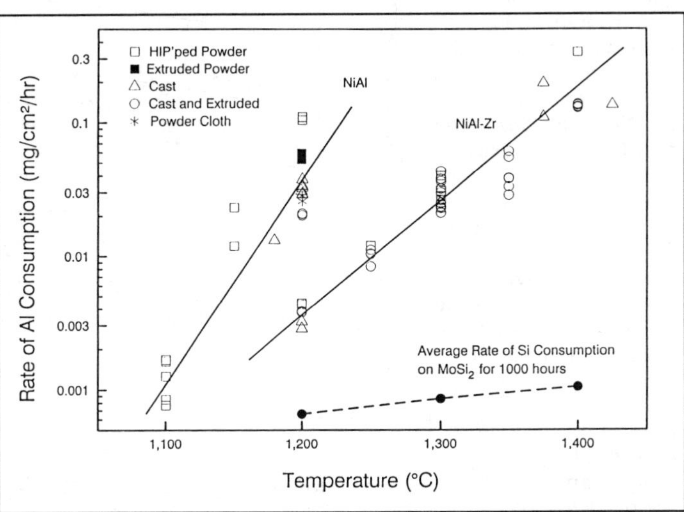

Figure 8 Approximate rates of Al consumption for NiAl and NiAl-Zr alloys as a function of temperature. Average rates for Si consumption on $MoSi_2$ for 1000 hour period is also shown.

LIFE PREDICTION

The rates of Al consumption for the cyclic oxidation of the aluminides and the k_p values for SiO_2 formation and growth on $MoSi_2$, coupled with the density, are sufficient to predict the oxidative life of these materials. The failure criterion used to predict the lifetime was based on the surface recession and was defined as a reduction of 10% of the original sample thickness (5% per side). The density of each of the aluminides is shown in Fig 7. The predicted lifetimes for the NiAl, NiAl-Zr, FeAl, CoAl and $TiAl_3$-8Cr alloys, using the rates for Al consumption shown in Fig 7, are shown in Fig 9. The predicted lifetimes for the NiAl and NiAl-Zr alloys at various temperatures, using the rates of temperature dependence of the rates shown in Fig 8, are shown in Fig 10. Using the k_p values shown in Fig 6 and a density of 6.28 mg/cm³ [27], the oxidative lifetime for $MoSi_2$ is also shown in Fig 10. As expected, the oxidative lifetimes for $MoSi_2$ greatly exceed those for the Ni aluminides. The four-fold increase in thickness shown in Figs 9 and 10 for the aluminides generally increase the life by a factor of four. However, the same thickness increase for $MoSi_2$ increases the life by a factor of 16.

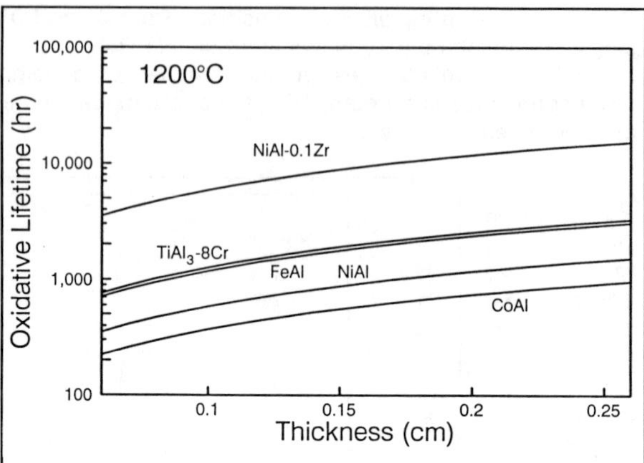

Figure 9 Predicted oxidative lifetimes as a function of initial thickness for several Al_2O_3 formers undergoing cyclic oxidation at 1200°C.

The prediction methodology can also be reversed allowing one to solve for the temperature for a given lifetime. For example, for an application involving thermal cycling of Ni aluminides with a desired lifetime of 10,000 hours, a temperature can be predicted at which the surface recession will be 10% of the original thickness after 10,000 hours. This temperature can then be interpreted as the maximum use temperature for that alloy for the desired lifetime and initial thickness. Maximum use temperatures for the Ni aluminides for a lifetime of 10,000 hours of cyclic operation is shown in Fig 11.

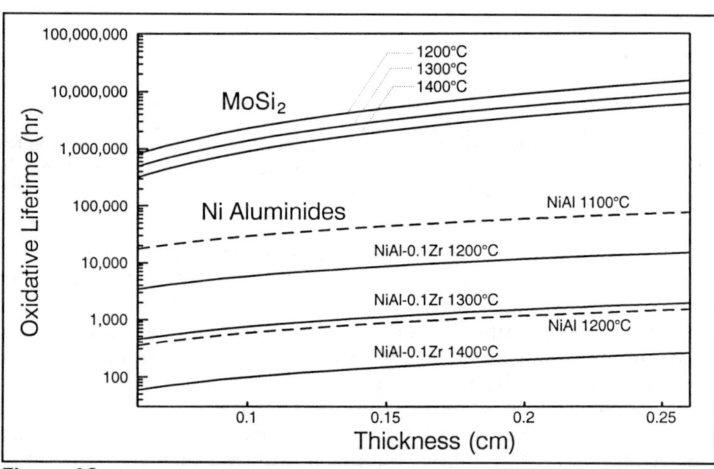

Figure 10 Predicted maximum use temperature for Ni aluminides as a function of initial thickness for a 10,000 lifetime.

It is important to note the potential shortcomings of long term oxidative life predictions. An oxidative life prediction capability is desirable in order to limit extensive oxidation testing. However, the time consuming nature of life testing makes verification of the predictions difficult or impractical. For instance, it is predicted

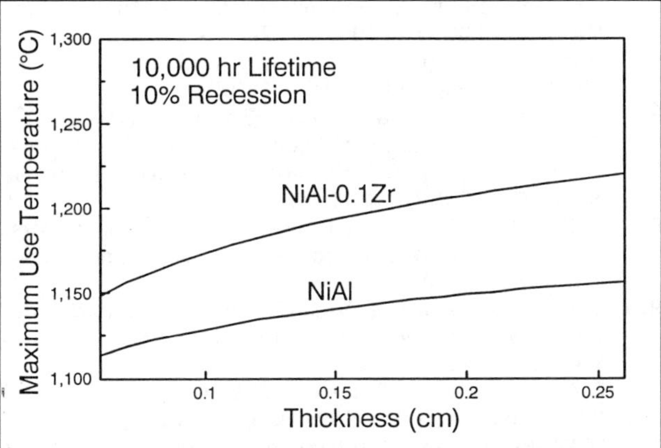

Figure 11 Predicted oxidative lifetimes as a function of initial thickness for $MoSi_2$ and Ni aluminides at various temperatures.

that the surface of a 0.1 cm thick sample of $MoSi_2$ is reduced by 10% after 300,000 hours, even at a temperature of 1400°C. The recession at reasonable experimental times (e.g., 1000 hrs) is too small to measure. Hence, verification of the predictions, even for some of the aluminides at the lower temperatures, is not experimentally practical. Similarly, it has been assumed in each of the above predictions that only the protective oxide scale forms and grows on the surface in a uniform manner. Nonplanar oxide growth on the surface can occur at defects resulting in pits which grow at a faster rate resulting in accelerated failure. In addition, the rate of Al or Si loss is greater at edges and corners of samples than from a large surface resulting in a shorter oxidative lifetime by whatever failure criterion is used. Finally, the extremely long oxidative lifetimes predicted for $MoSi_2$ are based on the observation that the SiO_2 scale did not spall during thermal cycling. However, at very long times or at very high temperatures, there is evidence of devitrification of the scale [4,14] which could lead to higher growth rates or to spalling. Hence, one must be aware of the assumptions stated or implied in making oxidative life predictions and also the inability to perform extensive verification testing.

CONCLUSIONS

Several aluminides (e.g., NiAl and FeAl) have long-term oxidation resistance (~10,000 hours) to approximately 1150°C. Aluminides containing reactive elements (e.g., NiAl-Zr alloys) have long-term oxidation resistance to approximately 1200°C.

The oxidation resistance of $MoSi_2$ is significantly better than that for the aluminides.

$MoSi_2$ can exhibit long-term oxidation resistance (~10,000 hours) at temperatures in excess of 1400°C.

REFERENCES

1. J.L. Smialek and G.H. Meier in *Superalloys II,* edited by C.T. Sims, N.S. Stoloff and W.C. Hagel (John Wiley and Sons, New York, NY, 1987) p. 293.
2. E. Fitzer and J. Schlichting in *High Temperature Corrosion,* edited by R. Rapp (NACE, Houston, TX, 1983) p. 604.

3. J.L. Smialek, M.A. Gedwill and D.L. Humphrey, in HITEMP Review 1990, NASA CP 10051 (1990) p. 17-1.
4. D.A. Berztiss, R.R. Cerchiara, E.A. Gulbransen, F.S. Pettit and G.H. Meier, Mater. Sci. Eng. A155, 165 (1992).
5. P.J. Meschter, Metall. Trans. 23A, 1763 (1992).
6. T.C. Chou and T.G. Nieh, submitted to J. Mater. Res.
7. J. Doychak, S.V. Raj, I.E. Locci and M.G. Hebsur, in HITEMP Review 1991, NASA CP 10082, (1991) p. 18-1.
8. J. Doychak, C.A. Barrett and J.L. Smialek, in *Corrosion & Particle Erosion at High Temperatures,* edited by V. Srinivasan and K. Vedula (TMS, Warrendale, PA, 1989) p. 487.
9. J.A. Nesbitt and E.J. Vinarcik, in *Damage and Oxidation Protection in High Temperature Composites,* edited by G.K. Haritos and O.O. Ochoa (ASME, New York, NY, 1991), p. 9.
10. C.A. Barrett and C.E. Lowell, JTEVA, 10, 273 (1982).
11. G.C. Rybicki and J.L. Smialek, Oxid. Met. 31, 275 (1989).
12. M.W. Brumm and H.J. Grabke, Corros. Sci. 33, 1677 (1992).
13. J. Cook, A. Khan, E. Lee and R. Mahapatra, Mater. Sci. Eng. A155, 183 (1992).
14. J. Schlichting, High Temp.-High Pressures 10, 241 (1978) [NASA TM 76529, 1981].
15. P.J. Meschter, Mater. Res. Soc. Symp. Proc., 213, 1027 (1991).
16. C.A. Barrett, Oxid. Met., 30, 361 (1988).
17. J.A. Nesbitt, C.A. Barrett, J. Doychak and E.J. Vinarcik, HITEMP Review 1990, NASA CP 10051 (1990) p. 18-1.
18. C.A. Barrett and R.H. Titran, NASA TM 105620, April 1992.
19. M.G. Hebsur, J.R. Stevens, J.L. Smialek, C.A. Barrett and D.S. Fox, in *Oxidation of High-Temperature Intermetallics,* edited by T. Grobstein and J. Doychak (TMS, Warrendale, PA, 1988) p. 171.
20. J.L. Smialek, J. Doychak and D.J. Gaydosh, in *Oxidation of High-Temperature Intermetallics,* edited by T. Grobstein and J. Doychak (TMS, Warrendale, PA, 1988) p. 83.
21. A.M. Huntz, Mater. Sci. Eng., 87, 251 (1987).
22. D.P. Whittle and J. Stringer, Phil. Trans. R. Soc. Lond. A295, 309 (1980).
23. J.L. Smialek, Metall. Trans., 22A, 739 (1991).
24. J.A. Nesbitt, E.J. Vinarcik, C.A. Barrett, and J. Doychak, Mater. Sci. Eng., A153, 561 (1992).
25. C.E. Lowell, C.A. Barrett, R.W. Palmer, J.V. Auping and H.B. Probst, Oxid. Met., 36, 81 (1991).
26. L.J. Parfitt, J.L. Smialek, J.P. Nic and D.E. Mikkola, Scripta Met. et Mat., 25, 727 (1991).
27. T.C. Chou and T.G. Nieh, Scripta Met. & Mat., 27, 19 (1992).

PROGRESS IN THE DEVELOPMENT AND UNDERSTANDING OF Ti_3Al BASED INTERMETALLIC MATERIALS

HARRY A. LIPSITT
Department of Mechanical and Materials Engineering,
Wright State University, Dayton, Ohio 45435

ABSTRACT

Recent results on the development of Ti_3Al base materials, including orthorhombic alloys, are reviewed. Included in this discussion are new data and ideas concerning alloy development, phase equilibria, processing for improved toughness and ductility, environmental degradation, oxidation and corrosion resistance, and coating development. Several new families of alloys have been developed in recent years and the microstructure/processing/property relationships in these alloys are reviewed. It is shown that errors in processing this class of alloy can lead to microstructural instabilities under service conditions. Finally, some immediate needs and some directions for further development are suggested.

INTRODUCTION

Over the period of the past two decades, military and industrial interest in Ti_3Al (α_2) base materials has generally lagged behind the interest in TiAl (γ) base materials. This has been because of the apparent "potential" for the broader application of the lighter, stiffer, more oxidation resistant γ base alloys. This attitude assumes that the ease of solving the various developmental problems is nearly equivalent in the two cases. However, the technical literature shows that this assumption is not valid. Perhaps this is because of the many similarities between the α_2 alloys and conventional titanium alloys. Nevertheless, the appearance of a (nearly) useful α_2 alloy occurred about 10 years before the same event in the γ base systems. However, when one attempts to understand alloy behavior, it becomes clear that the development of the alloys had significantly preceded the real understanding of their metallurgy. Today, studies to provide new understanding and studies to provide new alloy compositions proceed together. As a result, there is more information to review than can be digested in a single article. Fortunately, there have been several recent reviews of the literature of Ti_3Al [1-3] base materials, so it is not necessary for this review to begin at the beginning. Even more fortunately, Dr. Dipankar Banerjee, Assistant Director, Defence Metallurgical Research Laboratory, Hyderabad, India, has recently completed a most extensive review of our basic understanding of Ti_3Al base materials [4] and it seemed unnecessary to repeat what he had done so well. Therefore, using that article as a point of departure, this review will attempt to define the current state of understanding of the nearly useful materials that have been developed. Thus, the present review will concentrate on material properties and behavior in a service environment, on property goals and deficiencies, and on research necessary to validate the use of Ti_3Al base materials for aircraft turbine engine service.

ALLOY DEVELOPMENT

The α_2 base alloys which have been studied most are the Ti-24Al-11Nb composition reported by Blackburn, Ruckle, and Bevan, in 1978 [5] and the Ti-25Al-10Nb-3V-1Mo composition reported by Blackburn and Smith in 1982 [6]. (All compositions reported in this paper are in atomic percent.) Until recently these were the only compositions which had received widespread attention. In 1989, several new compositions were reported based on the results of two research efforts devoted to improving the toughness of α_2 base alloys. At the completion of an extensive program, Blackburn and Smith [7] reported the composition Ti-25Al-17Nb-1Mo to be one with an optimum property balance. This ingot alloy was isothermally beta forged and quenched from the forging press into an 815°C salt bath, held for 30 minutes and air cooled. The structure consisted of a relatively large prior beta grain size with relatively fine transformed α_2 platelets. The R.T. yield strength of this alloy was in the range 900-1000 MPa, the ultimate strength was in the range 1000-1100 MPa, R.T. elongation 1.0-2.3%, and the fracture toughness was 18.7 MPa$\sqrt{m}$ at R.T. and 48.4 MPa$\sqrt{m}$ at 315°C. Creep, fatigue, and impact data were also reported for this alloy. The creep and fatigue behavior were outstanding as compared to conventional titanium alloys [7].

At the completion of a parallel development program, Marquardt, et al [8], reported several high toughness compositions which are important not so much in their own right, but because they led to the development of a new composition, Ti-24.5Al-12.5Nb-1.5Mo which is a most interesting alloy [9,10]. This material was produced as an $\alpha_2 + \beta$ forged (820 kg) ingot product which was given a final forging reduction either in the β or in the $\alpha_2 + \beta$ region. The β forged material was given a direct salt quench to 815°C and held for 30 minutes. The $\alpha_2 + \beta$ forged material was solution treated at four temperatures to yield a variation in volume fraction of primary α_2 from 0-12%. The microstructure of these alloys consisted of a matrix of fine transformed α_2 platelets containing a uniform dispersion of primary α_2 globules. The $\alpha_2 + \beta$ treated materials gave R.T. tensile yield strengths in the range 800-1050 MPa, ultimate strengths in the range 970-1150 MPa, elongations of 1.2-3.1% and R.T. fracture toughness values (K_{Ic}) in the range 16.6-19.0 MPa$\sqrt{m}$. The creep resistance of these materials is good, but the steady state creep rate decreases and the percent primary creep increases as the volume fraction of primary α_2 increases. Other microstructural and processing studies on this alloy family have been reported [11-13].

Maurer, Saqib, and Lipsitt [12] studied the effects of aging following solution treating and quenching the alloys Ti-24.5Al-10.5Nb-1.5Mo and Ti-24.5Al-14.5Nb-1.5Mo. Both alloys were found to be unstable in this condition although the aging kinetics were much slower in the 14.5Nb alloy. Without a stabilization treatment, the 10.5Nb alloy developed a nodular orthorhombic ("O" phase) precipitate at short aging times at 650°C. At longer times (up to one month) and/or higher temperatures (up to 850°C) a coarse cellular precipitation occurred. An intermediate temperature stabilization (30 minutes at 760°C) immediately after solution treatment prevented both the nodular orthorhombic phase appearance and cellular coarsening. This behavior is similar to that previously reported for both the Ti-24Al-11Nb [5] and Ti-25Al-10Nb-3V-1Mo alloys [14]. This is an important observation since it means that the leaner Nb bearing alloys in this family of alloys cannot be quenched to room temperature following solution treatment and prior to aging. These materials must first be stabilized at an intermediate temperature to avoid storing excess energy in the structure which can lead to cellular coarsening during elevated temperature service. An example situation where this is important is if this alloy were to be welded and would quickly cool to room temperature.

Recently Bartz, Marquardt, and Lipsitt [13] determined the TTT curves for the alloys Ti-24.5Al-(10.5, 12.5, and 14.5Nb)-1.5Mo for periods up to one month between the temperatures of 648°C and 1010°C. They showed that the phase transformation sequence at higher temperatures is B2 → B2 + α_2. At temperatures lower than about 900°C a series of transformations occur including B2 → B2 + O and B2 + O → B2 + O + α_2. The effect of increasing Nb content is to slow the transformation kinetics and to raise the temperature limit of stability of the orthorhombic phase. However, it is important to note that for the Nb composition range of these alloys, the volume fraction of the orthorhombic phase is continuously changing at temperatures above 650°C and thus, the material properties will change continuously with time in this temperature range. Since the presence of the "O" phase contributes to some of the good properties of this alloy family, it is necessary to determine the effects of the above described time/temperature dependent microstructural changes on the mechanical properties of the material if the intended use temperature is above 650°C.

At this writing these alloys, Ti-24.5Al-(10.5, 12.5, and 14.5Nb)-1.5Mo, have not been fully characterized and further studies are in order. The fatigue and impact behavior of this alloy family have not been published. The environmental degradation of this family has also not been reported nor has it been shown that coatings are available to protect these materials in service.

Finally, the influence of oxygen content on the beta transus temperature of this alloy family should be determined. Recently, Szaruga, et al [15], showed that oxygen was a strong α_2 stabilizer in Ti-25Al-10Nb-3V-1Mo in the oxygen content range between 710-1000 wppm oxygen. They showed that the beta transus temperature of Ti-25Al-10Nb-3V-1Mo was a strong function of oxygen content (increasing ~17.2°C per 100 wppm oxygen to 1154°C) in this range. Above 1000 wppm oxygen the beta transus temperature was a very weak function (increasing ~3.5°C per 100 wppm oxygen) of the oxygen content. The temperature of the B2/bcc order-disorder transition was also a function of the oxygen content, being a maximum of 1154°C at 1000 wppm oxygen and falling at a rate of ~2.2°C per 100 wppm oxygen with oxygen increase to 1800 wppm oxygen. Thus, it might be expected that the beta transus temperature of every α_2 + B2/β alloy will be a strong function of the oxygen content of the alloy. Since one may wish to adjust the heat treatment schedule in many cases in relation to the beta transus temperature, it is necessary to understand the nature and magnitude of the effect of oxygen content on the beta transus temperature of every new α_2 base alloy family.

Si Containing α_2Base Alloys

Another new alloy family has recently been reported as a result of the work of Arrell, Flower, and Kerry [16] and of Kerry [17]. The first reference describes a series of alloys patterned after the Ti-24Al-11Nb alloy, but containing nominally only 20 at.% Al and various additions of silicon. The Si addition is shown to have a marked beneficial effect on both the ambient temperature ductility and the elevated temperature creep rupture life. The research was performed on forged 200g buttons. The authors have shown that the Si addition modifies the alloy partitioning between the α_2 and β phases by stabilizing the primary α_2 phase. In the α_2 + β solution treated and aged condition (1 hour/1050°C + 2 hours/625°C) the microstructures consisted of fairly equiaxed primary α_2 in a transformed β matrix. In this condition, the room temperature tensile properties of the alloys showed a maximum strength (1178 MPa) and elongation (4.2%) with a 0.5 at.% Si addition and a slightly reduced strength (1054 MPa) and ductility (2.8%) with a 0.9 at.% Si addition. However, the stress rupture life (250 MPa/625°C) is very much increased at the 0.9 at.% Si level (299 vs 71 hours). It is not clear why increased stress rupture life occurs with increasing amounts of primary α_2, but it could be due to strengthening the α_2 by retaining Si in solid solution or simply be due to increased material ductility.

The authors do not present elevated temperature tensile data, primary and secondary creep data, fracture toughness, or fatigue data and these are of vital importance. However, perhaps the most critically necessary data are those related to the stability of the microstructure during long term exposure at service temperatures. These data are especially important since the amount of Si employed in this alloy family (nearly 1 atomic percent) is very much higher than the known solubility of Si in alpha titanium [18] and it is possible that some microstructural changes will occur upon extended exposure at service temperatures. The effect of the Si addition on environmental degradation should be determined. In addition, these results need to be verified using larger ingots.

Kerry [17] used a constant Si level of 0.9 at.% and explored the behavior of a series of Ti-23Al-11(Nb + Mo)-0.9Si alloys containing 0-4 at.% Mo. As above, this work was performed on forged 200g ingots. The forged buttons were α_2 + β solution treated for one hour and subsequently aged for 1 hour at 800°C. The heat treated materials contained 10-20% primary α_2 in a matrix of α_2 laths surrounded by β. As the Mo content increased, the size of the α_2 laths decreased and the matrix became increasingly refined. The R.T. tensile strength of the alloys increased with increasing Mo content, being about 1200 MPa at 2% Mo, however, a maximum elongation of 4.0% was achieved at both 1 and 2 at.% Mo. The stress rupture life (250MPa/650°C) increased sharply (300 to 900 hrs) between 1 and 2 at.% Mo. Thus, the composition with the best balance of properties was Ti-23Al-9Nb-2Mo-0.9Si. This is an interesting result because it illustrates clearly that an appropriate property balance may be achieved by using a reduced Al content and a higher than usual Mo content. (This point has been recently verified in a series of alloys, Ti-(19-27)Al-10V. The results show that high strength with the highest ductility may be achieved at low Al contents [19].) This study confirms many previous results concerning the virtues of Mo as a strengthening/stiffening alloying addition for α_2 base alloys. The Mo addition also resulted in a ductility increase apparently due to the considerable microstructural refinement. This result needs to be further studied, preferably using ingots of significant size. In addition, data on the usual properties (tension, creep, fatigue, fracture toughness, and environmental degradation) as a function of temperature are needed. As above, perhaps the most critical data for alloys with an addition of a slowly diffusing β stabilizing species is the determination of the microstructural stability following long term exposure at service temperatures. Finally, since only by chance will the optimum Si level determined in a quaternary alloy system also be the optimum level in a quinary system, it would be important to employ a statistical approach to determine the optimum levels of Al, Mo, and Si, the critical elements in this alloy family.

Orthorhombic Base Alloys

Another relatively new alloy family, perhaps not α_2 based, but clearly related to those alloys, is the alloy family with compositions near the stable orthorhombic or "O" phase composition, nominally Ti-25Al-25Nb or Ti_2AlNb. The "O" phase has an ordered orthorhombic structure which is a slightly distorted DO_{19} or Ti_3Al structure [20]. It differs from the DO_{19} structure in that one of the three Ti sublattices in the DO_{19} structure is preferentially occupied by Nb atoms. Banerjee, Rowe, and Hall [21] have shown that the overall arrangement of "a" dislocations is very similar to that found in α_2 following deformation at room temperature. However, in addition, profuse "c" component slip is observed, in contrast to the situation found in α_2 base materials. This frequent observation of "c + a/2" slip indicates that the stress required to initiate slip on this system may be nearly the same as required for basal slip, resulting in easier, more homogeneous deformation in this material and greater tensile ductility. In addition, the alloys of this family have higher strength, specific strength, and fracture toughness than alloys based on Ti_3Al [22-24]. One of the alloys in this family, Ti-22Al-27Nb, shows a R.T. tensile yield strength of 1290 MPa, an ultimate strength of 1415 MPa, and 3.5% elongation with a fracture toughness (K_{Ic}) of 28 MPa$\sqrt{m}$. At 650°C the average tensile yield strength was 1120 MPa, the average ultimate strength was 1275 MPa, with 5.0% elongation. The ingot alloy was given an extrusion reduction of 12.5:1 at 1050°C and heat treated at 815°C for one hour. The microstructure contained Widmanstatten O + B2 phases.

These alloy development results are quite recent and much remains to be accomplished. A few creep data are available, but no fatigue data have been published. Oxidation, corrosion, environmental degradation, and coating information are also not available. Vanadium has been added to this alloy family with a benefit in ductility along with a loss in strength [24]. However, an optimum vanadium level has not been established, nor has the effect of the vanadium addition on other properties (including oxidation) been determined. Given the known beneficial effect of Mo additions in α_2 base alloys, the effect of this alloying element on "O" phase materials should also be explored. This alloy family is subject to the same surface interaction with oxygen as are all the α_2 base alloys and a similar embrittlement and cracking behavior is observed [25]. It is believed that these alloys contain stable phases, however, the longest heat treatment has been 600 hours at 900°C. The true stability of these phases should be established with exposures of at least 3000 hours at 900°C. The influence of oxygen on the phase boundary positions needs to be established. The beta transus temperature is a strong function of the oxygen content in these alloys [26] as it is in α_2 base materials and this is a critical parameter needed to adjust solution treatment temperatures to maintain a reasonably fine grain size. In addition, the effect of oxygen and nitrogen levels on the mechanical behavior of these alloys must become known. Oxygen and nitrogen are powerful strengthening and embrittling additions in α and α_2 base materials and are likely to have a similar effect in the "O" phase alloys.

A subset of this family of alloys are those materials which contain less Nb than the

Ti-22Al-27Nb composition discussed above. The prototypical alloy of this set is the composition Ti-22Al-23Nb. In the usual heat treated condition, and with oxygen contents above 1000 wppm, this alloy is a three phase material containing α_2, "O" (the ordered orthorhombic phase), and B2 (the ordered β phase). However, when this composition is produced with oxygen contents in the range 250-1000 wppm, the alloy is a two phase, O + B2, material [27]. This alloy is attractive partly because of the ease with which it may be rolled to foil (for foil/fiber/foil composites). Graves, Smith, and Rhodes have reported the properties and behavior of this alloy as a monolith [28] and Smith, Graves, and Rhodes have published data where this alloy is used as the matrix for a multi-ply SiC reinforced composite panel [29]. In the first named paper [28], a 36 kg ingot was forged and rolled through a series of reductions to produce 2 mm sheet and foil of two thicknesses, 89 and 127 μm, all with oxygen contents above 1300 wppm. The sheet and foil exhibited the presence of the expected three phases and evidence of banding that may have resulted from the necessity to use elemental Nb for the alloy preparation instead of a master alloy. Following a heat treatment to simulate composite consolidation the R.T. yield strength of the foil was in the range 860-1030 MPa, the ultimate strength was in the range 1070-1170 MPa, with elongations of 4.1-5.6%. In a companion paper [29], the foil was processed to produce unreinforced 4-ply "neat" panels and 35 volume % 4-ply unidirectional SiC (SCS-6) reinforced composite panels. Although many of the mechanical property data obtained in this study have not been published, several interesting points have emerged. In "neat" specimens exposed for 100 hours at temperatures in the range 315-815°C and tested at R.T. following exposure, the tensile strength does not seem to be remarkably affected by the exposure and remains in the 900-975 MPa range although the elongation does decrease to values in the range 4-5%. Another interesting datum is that no-load thermal cycling the composite in air for 500 cycles between 150-815°C does not result in the cracking or serious property degradation reported for Ti-24Al-11Nb matrix composites [30,31].

There are even fewer published data available for the Ti-22Al-23Nb family of alloys than for the Ti-22Al-27Nb alloys. Thus creep, fatigue, and fracture toughness data are needed immediately. As above, the effects of V and Mo on strength and ductility should be explored. There is an unexpected, apparent difference in the surface interaction effect of oxygen between the two alloys of this class; this should be clarified. Oxygen has been shown to be an α_2 stabilizer in Ti-22Al-23Nb (as it is in Ti-25Al-10Nb-3V-1Mo [15]); determination of its effect in the Ti-22Al-27Nb system has already been recommended.

Although many of the data needed to fully characterize the systems discussed above are not yet available, it is possible to gain an impression of the progress made by examining the specific yield strength of example materials from several of the categories discussed above. This information is shown in Figure 1 [24].

The attractive properties of these alloys have driven the need to understand the phase relationships and phase transformations, so that the effects of processing and heat treatment can be understood. A number of studies have been published which contributed significantly to our understanding of the phase equilibria in the Ti-Al-Nb system [32-35]. The 900°C isotherm [35] is helpful in understanding the microstructures developed in these alloys. This is shown in Figure 2. This diagram establishes that at 900°C the ordered orthorhombic phase has substantial stability lying between approximately 25-28 at.% Al for Nb contents from 15-30 at.%. The beta transus for these alloys is approximately 1100-1200°C. In addition, the diagram makes clear the relationships between the several classes of alloys which have been discussed above. The first alloy category includes alloys containing a total β stabilizer content of 10-12 at.%. These include the Ti-24Al-11Nb alloy [5], the improved toughness alloys such as Ti-24.5Al-12.5Nb-1.5Mo, reported by Marquardt, et al [9], and the Si bearing alloys reported by Arrell, et al [16], and by Kerry [17]. A second alloy category includes those with a total β stabilizer content of 14-18 at.%. Alloys such as Ti-25Al-10Nb-3V-1Mo [6] and Ti-25Al-17Nb-1Mo [7] fall into this category. The third alloy category includes all the materials with a total β stabilizer content in excess of 23 at.% which contain the O + B2 phases at equilibrium.

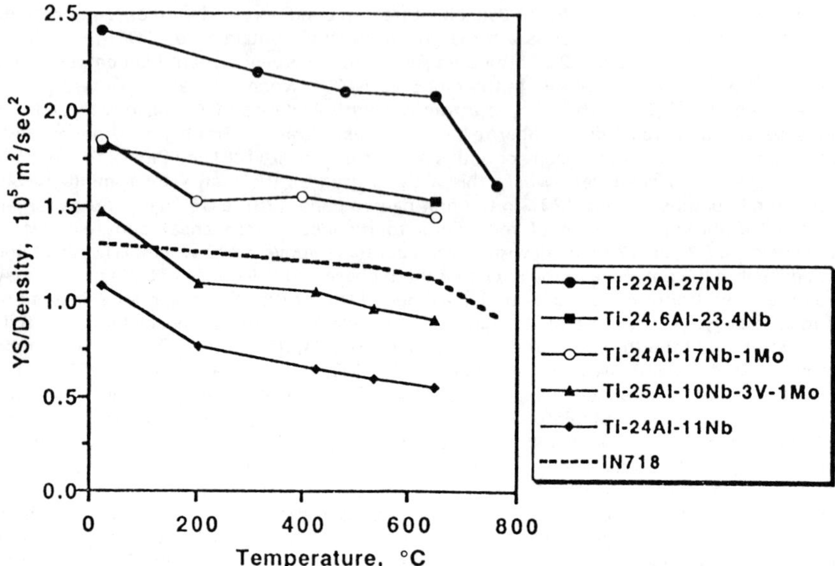

Figure 1. Specific yield strength of several Ti_3Al and Ti_2AlNb base alloys relative to IN 718 nickel base superalloy [24].

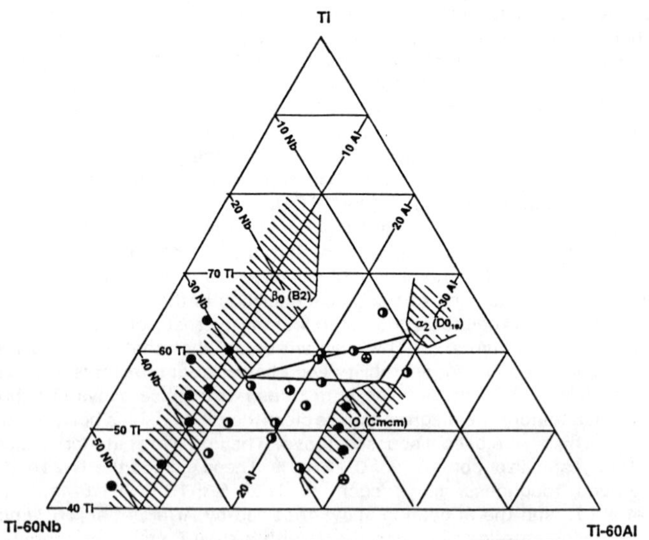

Figure 2. 900°C isothermal ternary section from the Ti-Al-Nb equilibrium diagram [35].

ENVIRONMENTAL INTERACTION AND DEGRADATION

Environmentally induced cracking during elevated temperature tensile or creep testing has been observed in most of the Ti_3Al base alloys studied to date [5,36,37]. Environmentally induced surface cracks initiate at multiple sites around the specimen circumference and grow together to form cracks that extend completely around the specimen and over the entire gauge length. These cracks propagate inward without regard to microstructure and attain a depth at least equal to the depth of an oxygen enriched layer on the specimen surface. This layer is severely hardened and embrittled by the oxygen penetration [36]. The presence of these cracks reduces tensile ductility and final specimen failure in creep often proceeds from these cracks. The cracking phenomenon does not occur in vacuum and specimen failure is intergranular. This cracking phenomenon is unacceptable in a high temperature structural material because it is responsible for reduced ductility in tension and provides the initiation sites for creep failure. Thus, a coating should be employed if these alloys are to be successful as turbine engine structural materials.

Residual fatigue life at room temperature is also affected by environmental exposure [38,39]. In a recently completed study [39], pre-exposures were performed in the temperature range from 538-760°C and the effect of this exposure was determined on a β processed and β solution treated Ti-25Al-10Nb-3V-1Mo material. Samples were tested following pre-exposures in hot air, hot salt, and molten sulfate salt environments. Air and molten salt pre-exposures at temperatures above 649°C resulted in a 100x decrease in the low cycle fatigue (LCF) life, see Figure 3, but had only a slight debit in the high cycle fatigue (HCF) strength. Hot salt effects were also severe; a 1 hour, no stress, exposure at 649°C causes a severe loss in LCF life. A creep pre-exposure does not result in loss of HCF life. The severe LCF debit following exposure in hot air and molten salt may be attributed to the poor oxidation resistance of this alloy as well as the tendency to form the brittle strain-intolerant α_2 stabilized outer case discussed above. The sensitivity to hot salt is the result of a pitting attack on the sample surface similar in nature to that observed on conventional $\alpha + \beta$ titanium alloys. Once again, it cannot be emphasized too strongly in the face of the data above; these materials should not be considered for long term turbine engine service without the development of thin adherent coatings which will protect the base alloys from the ravages of oxygen and hot salt attack.

Oxidation

The oxidation resistance of these materials is also of some concern. Recent published data [40] show that the α_2 base alloys have better oxidation resistance than conventional titanium alloys over a broad temperature range to more than 1000°C. However, it is important to consider the change in oxidation resistance that accompanies the development of the most recent alloys that exhibit a better balance of properties. In a recent publication, Schaeffer [41] presents static oxidation data for Ti-25Al-10Nb-3V-1MO and Ti-24.5Al-12.5Nb-1.5Mo and compares these data to equivalent data for the conventional alloy Ti-6Al-2Sn-4Zr-2Mo (wt.%)(Ti-6242). The report gives static isothermal oxidation data for these alloys between 593°C and 705°C, a sample of which is shown in Figure 4. It is clear that the Ti-24.5Al-12.5Nb-1.5Mo alloy oxidizes at significantly slower rate than do Ti-25Al-10Nb-3V-1Mo or Ti-6242. However, even more important than the difference in the rates of oxidation is the reason for that difference. The Ti-24.5Al-12.5Nb-1.5Mo alloy forms alumina (Al_2O_3) on the external surface which grows more slowly than rutile (TiO_2). Ancillary data show that an interstitial layer rich in C, N, and O forms on α_2 alloys oxidized in air, therefore, it is the oxidation behavior of this layer that must be considered. A SIMS profile on the surface layer of the Ti-24.5Al-12.5Nb-1.5Mo alloy clearly shows an Al enrichment of about 2.5 times compared to the Al level on the surface layer of the Ti-25Al-10Nb-3V-1Mo alloy. The oxidation rate of the Ti-24.5Al-12.5Nb-1.5Mo alloy is so much slower than for the Ti-25Al-10Nb-3V-1Mo alloy that the details of the microstucture may still be seen on the surface following 100 hours isothermal air oxidation at 649°C! This effect, i.e., the growth of alumina in air as the exterior oxidation layer on

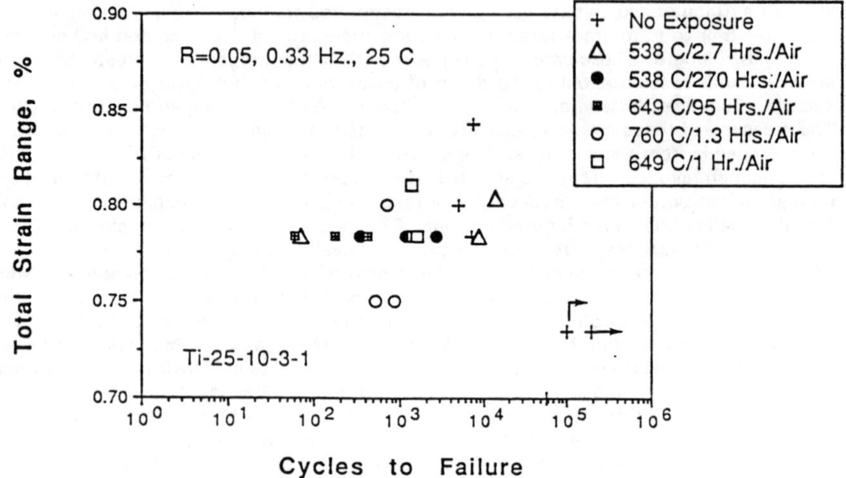

Figure 3. Effect of air pre-exposure on the 25°C LCF life of Ti-25Al-10Nb-3V-1Mo. The various pre-exposures result in an 100x LCF debit at room temperature [34].

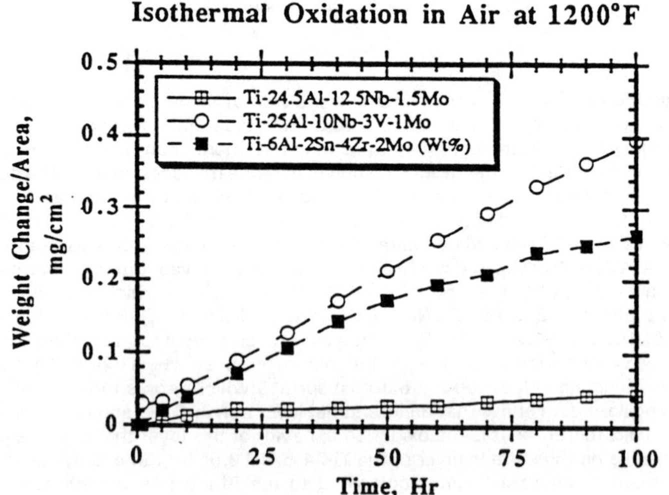

Figure 4. Isothermal air oxidation at 649°C of two recent α_2 base alloys compared to a conventional titanium alloy [41].

an α_2 base alloy, is strange and unexpected. It may also be a significant breakthrough! It is unexpected because alumina formation is not favored on binary Ti-Al alloys containing 25 at.% Al [42,43]. Because of its importance, this observation demands further exploration and verification. The effect should be explored on other heats of the same and similar alloys. It should be explored under various cyclic conditions, since these conditions are much more critical for turbine engine structural materials. A simple experiment to determine if this alloy can be used as a protective coating on other similar, but less oxidation resistant, alloys is in order. A study could be done to determine if ion implantation of the correct species, i.e., simulating this composition on the surface of another alloy, causes the same effect or causes damage.

Coatings

An early report [44] on the potential for development of coatings for titanium aluminides centered on coatings from within the Ti-Al system, assuming those would be the most easily compatible coating materials. Ternary element additions that were shown to result in alumina formation upon oxidation in air (at 1100-1400°C) included Re(<6.1 at%), Mo(6.5), Cr(7.5), Fe(<8.8), W(>9.4), and V(13.8). At these levels, and with an Al content greater than 43 at%, a critical concentration is reached at which a continuous alumina scale will form beneath the rapidly growing rutile scale that forms first. Thus, the base metal oxide becomes a transient oxide which ceases to grow when the continuous alumina scale forms under it. When the concentration of the third element was further increased beyond the minimum necessary values cited above, the amount of transient oxide formed was decreased until only a pure alumina scale was formed. At very high temperatures (1100-1400°C), the parabolic rate constants for oxidation of alloys that formed a single layer alumina scale were the same as found for NiAl. At a more modest temperature (800°C), the parabolic rate constants were several orders of magnitude higher. Although an inner scale of alumina forms at 800°C in all these cases, it does so only after a transient oxide scale of rutile forms initially. It is the formation of this outer, faster growing scale that increases the overall oxidation kinetics.

The approach suggested above is now being followed by McCarron, et al, [45]. A re-examination of the results above [40] leads quickly to the conclusion that Cr is the best third element to add to a Ti-Al alloy to cause the formation of an alumina scale. The results to date have shown that Ti-Al-Cr alloys with Cr contents from 6-35 at% and Al contents from 47-55 at.% form thin continuous alumina scales in air at temperatures between 760 and 1000°C. Oxidation studies have shown that the Ti-Al-Cr alloys have excellent cyclic oxidation resistance and that they exhibit only limited interdiffusion with the (gamma) alloy substrate with Cr being the inward diffusing species. However, the diffusion of Cr is slow, showing a rate that would yield about 25 μm penetration after 5000 hours at 1000°C. The coatings have been applied by a number of standard techniques with encouraging results.

These are important results, but the virtues of these coatings in service situations must quickly be assessed. Assuming that these results have shown that the formation of an alumina film on a titanium aluminide is possible, it should be determined immediately if the presence of this coating is deleterious to the mechanical properties of the aluminide alloys, and if so, how serious is the property debit. Next, it should quickly be determined how coated specimens behave under the action of hot air, hot salt, and molten salt [38,39]. In the present paper, it has been assumed that the application of α_2 base alloys would be at temperatures and in environments where a profusion of circumferential cracks would form and lead to a pronounced debit in ductility, creep, and fatigue strength. It is important that some coating compositions have been developed which should be able to resist the ravages of hot air, hot salt, and molten salt attack as well as the standard (NiAl base) coatings used on nickel base superalloys. However, these coatings will be of little value if the very presence of the coating seriously debits the valuable mechanical properties of the titanium aluminide base alloys or if the coating is so brittle that it cannot accomodate cyclic stresses and strains in service.

CONCLUSIONS

Recent data on alloy development, alloy behavior, environmental degradation, oxidation, and the development of protective coatings have been reviewed. It has been shown that the development efforts on α_2 base alloys are alive and well and that the current rate of progress is truly exciting. However, along with this excitement must come some sobering thoughts. Uppermost in our minds must remain the fact that these materials are being developed primarily for use in the hot gas stream of an aircraft turbine engine. The materials scientist must remember that only materials with a suitable balance of properties will be chosen for this demanding task. A large amount of reliable data are necessary before this determination may be made. A goal of this paper has been to try to indicate what new data are necessary and which data are needed to assure (or to prevent) the use of α_2 base alloys as turbine engine structural materials.

ACKNOWLEDGEMENTS

The author wishes to thank his many colleagues with whom he has over the years discussed the development of materials based on the titanium aluminides. The attitudes expressed in this paper have been formed over the entire two decades during which this difficult and elusive research and development goal has been pursued. Very special thanks are due to those colleagues who critiqued this paper and helped to give it coherence: Jim Chesnutt, Jeff Graves, Jim Hall, Blair London, Brian Marquardt, Cecil Rhodes, Grant Rowe, and Jon Schaeffer.

REFERENCES

1. H.A. Lipsitt, in High Temperature Ordered Intermetallic Alloys, edited by C.C. Koch, C.T. Liu, and N.S.Stoloff (Mater. Res. Soc. Proc. **39**, Pittsburgh, PA, 1985) pp. 351-364.

2. D.A. Koss, D. Banerjee, D.A. Lukasak, and A.K. Gogia, in High Temperature Aluminides and Intermetallics, edited by S.H. Whang, C.T. Liu, D.P. Pope, and J.O. Stiegler (TMS, Warrendale, PA, 1990) pp. 175-196; R.G. Rowe, ibid., pp.375-401;Y.W. Kim and F.H. Froes, ibid., pp. 465-492;J.M. Larsen, K.A. Williams, S.J. Balsone, and M.A. Stucke, ibid. pp. 521-556.

3. C.H. Ward, International Metallurgical Reviews, in press, 1992.

4. D. Banerjee, in Intermetallic Compounds: Principles and Practice, edited by J.H. Westbrook and R.L. Fleischer (John Wiley, Ltd., London, 1993), to be published.

5. M.J. Blackburn, D.L. Ruckle, and C.E. Bevan, AFML-TR-78-18 (1978).

6. M.J. Blackburn and M.P. Smith, AFML-TR-82-4086 (1982).

7. M.J. Blackburn and M.P. Smith, WRDC-TR-89-4095 (1989).

8. B.J. Marquardt, G.K. Scarr, J.C. Chesnutt, C.G. Rhodes, and H.L. Fraser, WRDC-TR-89-4133 (1989).

9. B.J. Marquardt and D.D. Kreuger, in the Proceedings, 7th World Conference on Titanium, edited by F.H. Froes and I. Caplan (TMS-AIME, Warrendale, PA, 1993), to be published.

10. J.C. Chesnutt, in Superalloys 92, edited by S.D. Antolovich, R.W. Strusrud, R.A. MacKay, D.L. Anton, T. Khan, R.D. Kissinger, and D.L. Klarstrom (TMS-AIME, Warrendale, Pa, 1992), pp.381-389.

11. George T. Gray, Sun Ig Hong, and B.J. Marquardt, in the Proceedings, 7th World Conference on Titanium, edited by F.H. Froes and I. Caplan (TMS-AIME, Warrendale, PA, 1993), to be published.

12. Norbert O. Maurer, Mohammad Saqib, and Harry A. Lipsitt, in the Proceedings, 7th World Conference on Titanium, edited by F.H. Froes and I. Caplan (TMS-AIME, Warrendale, PA, 1993), to be published.

13. A. Bartz, B. Marquardt, and H.A. Lipsitt, presented at the TMS Annual Meeting, Denver, Co., 22 February 1993 (unpublished).

14. C.H. Ward, J.C. Williams, and A.W. Thompson, Scripta Metallurgica et Materialia **24**, 617 (1990).

15. A. Szaruga, M. Saqib, R. Omlor, and H.A. Lipsitt, Scripta Metallurgica et Materialia **26**, 787 (1992).

16. D.J. Arrell, H.M. Flower, and S. Kerry, in the Proceedings, 7th World Conference on Titanium, edited by F.H. Froes and I. Caplan (TMS-AIME, Warrendale, PA, 1993), to be published.

17. S. Kerry, in the Proceedings, 7th World Conference on Titanium, edited by F.H. Froes and I. Caplan (TMS-AIME, Warrendale, PA, 1993), to be published.

18. J.L. Murray, in Binary Alloy Phase Diagrams, 2nd Edition, edited by T.B. Massalski, H. Okamoto, P.R. Subramanian, and L. Kacprzak (ASM International, Materials Park, OH, 1990), pp.3367-3371.

19. I. Ehrhart and A. Vassell, in the Proceedings, 7th World Conference on Titanium, edited by F.H. Froes and I. Caplan (TMS-AIME, Warrendale, PA, 1993), to be published.

20. D. Banerjee, Acta Met. et Mat. **36**, 871 (1988).

21. D. Banerjee, R.G. Rowe, and E.L. Hall in High Temperature Ordered Intermetallic Alloys - IV, edited by L.A. Johnson, D.P. Pope, and J.O. Stiegler (Mater. Res. Soc. Proc. **213**, Pittsburgh, PA, 1991) pp.285-290.

22. R.G. Rowe, in Microstructure/Property Relationships in Titanium Aluminides and Alloys, edited by Y-W. Kim and R.R. Boyer (TMS-AIME, Warrendale, PA, 1991) p. 387.

23. R.G. Rowe, D.G. Konitzer, A.P. Woodfield, and J.C. Chesnutt in High Temperature Ordered Intermetallic Alloys - IV, edited by L.A. Johnson, D.P. Pope, and J.O. Stiegler (Mater. Res. Soc. Proc. **213**, Pittsburgh, PA, 1991) pp. 703-708.

24. R.G. Rowe, in the Proceedings, 7th World Conference on Titanium, edited by F.H. Froes and I. Caplan (TMS-AIME, Warrendale, PA, 1993), to be published.

25. R.G. Rowe (private communication), 16 November 1992.

26. C.G. Rhodes (private communication), 17 November 1992.

27. J.A. Graves (private communication), 18 November 1992.

28. J.A. Graves, P.R. Smith, and C.G. Rhodes in Intermetallic Matrix Composites - II, edited by D.L. Anton, D.B. Miracle, and J.A. Graves, (Mater. Res. Soc. Proc. **273**, Pittsburgh, PA, 1992) pp.31-42.

29. P.R. Smith, J.A. Graves, and C.G. Rhodes in Intermetallic Matrix Composites - II, edited by D.L. Anton, D.B. Miracle, and J.A. Graves, (Mater. Res. Soc. Proc. **273**, Pittsburgh, PA, 1992) pp.43-52.

30. S.M. Russ, Metall. Trans. A **21**, 1592 (1990).

31. W.C. Revelos and P.R. Smith, Metall. Trans. A **23**, 587 (1992).

32. D. Banerjee, T.K. Nandy, A.K. Gogia, K. Muraleedharan, in the Proceedings, Sixth World Conference on Titanium, edited by P. Lacombe, R. Tricot, and G. Beranger (J. de Physique, Les Ulis, France, 1989) pp.1091-1096.

33. J.H. Perepezko, Y.A. Chang, L.E. Seitzman, J.C. Lin, N.R. Bonda, T.J. Jewett, and J.C. Mishurda, in High Temperature Aluminides and Intermetallics, edited by S.H. Whang, C.T. Liu, D.P. Pope, and J.O. Stiegler (TMS, Warrendale, PA, 1990) pp.19-47.

34. K. Muraleedharan, A.K. Gogia, T.K. Nandy, D. Banerjee, and S. Lele, Metall. Trans. **23A**, 401 (1992).

35. R.G. Rowe, D. Banerjee, K. Muraleedharan, M. Larsen, E.L. Hall, D.G. Konitzer, and A.P. Woodfield, in the Proceedings, 7th World Conference on Titanium, edited by F.H. Froes and I. Caplan (TMS-AIME, Warrendale, PA, 1993), to be published.

36. S.J. Balsone, in Oxidation of High Temperature Intermetallics, edited by T. Grobstein and J. Doychak (TMS-AIME, Warrendale, PA, 1988) pp. 219-234.

37. C.H. Ward and S.J. Balsone, in Microstructure/Property Relationships in Titanium Aluminides and Alloys, edited by Y-W. Kim and R.R Boyer (TMS-AIME, Warrendale, PA, 1991) pp. 373-386.

38. D.P. DeLuca, B.A. Cowles, F.K. Haake, and K.P. Holland, WRDC-TR-89-4136 (1990).

39. P.S. Godavarti, M.D. Lipshutz, J.G. Snow, and J.A. Hall, in the Proceedings, 7th World Conference on Titanium, edited by F.H. Froes and I. Caplan (TMS-AIME, Warrendale, PA, 1993), to be published.

40. G.H. Meier, D. Appalonia, R.A. Perkins, and K.T. Chiang, in Oxidation of High Temperature Intermetallics, edited by T. Grobstein and J. Doychak (TMS-AIME, Warrendale, PA, 1989) pp. 185-193; K.E. Wiedeman, S.N. Sankaran, R.K. Clark, and T.A. Wallace, ibid., pp. 195-206; G. Welsch and A.I. Kahveci, ibid., pp.207-218.

41. J.C. Schaeffer, General Electric Report R92-AEB-248, September 29, 1992, submitted to Scripta Met. et Mat., October 1992.

42. A. Rahmel and P.J. Spencer, J. Oxid. Metals, **35**, 53 (1991).

43. K.L. Luthra, General Electric-CRD Tech. Infor. Ser. 90CRD177, August 1990.

44. R.A. Perkins, K.T. Chiang, G.H. Meier, and R. Miller, in Oxidation of High Temperature Intermetallics, edited by T. Grobstein and J. Doychak (TMS-AIME, Warrendale, Pa, 1989) pp.157-169.

45. R.L. McCarron, J.C. Schaeffer, G.H. Meier, D. Bertztiss, R.A. Perkins, and J. Cullinan, in the Proceedings, 7th World Conference on Titanium, edited by F.H. Froes and I. Caplan (TMS-AIME, Warrendale, PA, 1993), to be published.

LATTICE DEFECTS AND PLASTIC DEFORMATION OF $CoSi_2$

M. YAMAGUCHI, Y. SHIRAI and H. INUI
Department of Metal Science and Technology, Kyoto University Kyoto 606, Japan

ABSTRACT

Current knowledge of the properties of vacancies and the slip behavior and mechanisms of $CoSi_2$ is reviewed based on the results of our recent studies on $CoSi_2$. The concentration of thermal vacancies in $CoSi_2$ is much higher than that in ordinary metals and alloys with melting points comparable with that of $CoSi_2$. Vacancy defects are easily retained in $CoSi_2$ even after air cooling from high temperatures. An annealing stage observed at around 310 K after electron irradiation is concluded to occur by the migration of vacancies to form secondary defects. Peculiar phenomena recently reported on $CoSi_2$ such as an anisotropy of electrical resistivity and the climbing of dislocations at room temperature can be understood on the basis of the current knowledge of defect properties. Slip in $CoSi_2$ occurs along <100> on {001} at low temperatures. The selection of {001}<100> as the primary slip system in $CoSi_2$ can be interpreted in terms of high covalency of Co-Si bonding. a<100> dislocations have a strong tendency to align along their edge orientation and are dissociated into two a / 2<100> partial dislocations separated by a stacking fault on {001}. {001}<100> slip is augmented by {111}<110> and {110}<110> slip at high temperatures. Thermal activation analysis of deformation indicates that while deformation at low temperatures is controlled by the Peierls mechanism, the greatly increased concentration of thermal vacancies influence the mobility of dislocations at high temperatures.

INTRODUCTION AND STATUS

$CoSi_2$ is one of the transition metal disilicides which have recently received considerable interest as materials for ohmic contacts and gate electrodes to Si in VLSI devices because of their low electrical resistivity and chemical stability. In particular, $CoSi_2$ crystallizes in a cubic (C1 type) structure resembling the structure of Si (A4 type) and is lattice matched (within 1.2%) with Si. Thus, single crystalline thin films of $CoSi_2$ can be formed epitaxially on a silicon wafer. $CoSi_2$ is attractive not only for such microelectronics applications but also as a high temperature structural material. Although the melting point of $CoSi_2$ is not high (Tm=1599 K) in comparison with the disilicides of transition metals of groups IV-VI, it has low density (4.95 g/cm^3) and excellent oxidation resistance [1]. Furthermore, $CoSi_2$ possibly provides a sufficient number of equivalent slip systems and, hence, some ductility even at low temperatures because of its cubic structure.

The first report of the deformation behavior of $CoSi_2$ appeared in 1968. Sauer and Freise [2] examined deformation modes of single crystal and coarse-grained polycrystalline cobalt silicides, $CoSi_2$, CoSi and Co_2Si using hardness indentations and concluded that the cubic $CoSi_2$ deformed primarily by slip on {001} planes at room temperature. More recently, there has been an increase in research activity on the mechanical properties of $CoSi_2$. Anton and Shah [1] conducted elevated temperature characterization on $CoSi_2$ and some other high temperature compounds, which includes ductile/brittle transition temperature (DBTT) determination, minimum creep rate analysis, elastic modulus, tensile strength and cyclic oxidation testing. The DBTT of $CoSi_2$ was reported to be in the range of 1273-1473 K. The creep resistance of $CoSi_2$ at 1273-1473 K was revealed to be inferior to that of other high-temperature intermetallic compounds because of its relatively low melting point. Takeuchi et al.[3] studied the strain rate and temperature dependences of yield stress of polycrystals of $CoSi_2$ and $(Co,Ni)Si_2$ and suggested that the deformation was controlled by the Peierls mechanism. Ito et al.[4] made a systematic study on the room temperature deformation of single crystals of $CoSi_2$ and clarified that the primary slip system at room temperature was {001}<100> and appreciable fracture strains of about 2-4% was obtained in single crystals with a non-zero Schmid factor for {001}<100> slip. Anongba and Steinemann [5] studied the deformation behavior of [$\bar{1}$13]-oriented single crystals of $CoSi_2$ and observed the occurrence of cube slip and a marked temperature dependence of flow stress. Suzuki and Takeuchi [6] examined dislocation structures in polycrystalline specimens of $CoSi_2$ deformed at 700 K and in those crashed at room temperature and reported that the active slip system at room temperature was {001}<100>, however, a / 2<110> dislocations were also observed in specimens deformed at 700 K. They estimated the energy of the stacking faults with a displacement vector of a / 2<100> on {001} in $CoSi_2$ to be 17±4 mJ/m^2 from the dissociation width of a<100> dislocations. Thus, recent publications have provided wide knowledge of the mechanical properties of $CoSi_2$. However, a deeper understanding of the deformation mechanisms of $CoSi_2$ requires more work, for example a detailed study on the operative slip modes, their critical shear stresses and corresponding dislocation structures in a wide temperature range and for various crystal orientations, is needed.

Systematic data are lacking not only on the dislocation and slip mechanisms but also on the lattice defects in general. Ditchek [7] and recently Hirano and Kaise [8] grew single crystals of some transition metal disilicides including $CoSi_2$ and measured their electrical resistivity. Since $CoSi_2$ is cubic, the resistivity should be isotropic. However, the resistivity of $CoSi_2$ was found to weakly depend on crystal orientation; the resistivity for the [111] direction is slightly higher than those for the [001] and [112] directions [8]. Such an anisotropy of electrical resistivity was reported by Ditchek [7] as well and was thought to be of an extrinsic origin. Furthermore, Hirano and Kaise [8] reported that the residual resistivity of $CoSi_2$ was high and the residual resistivity ratio was low in comparison with those of other transition metal disilicides. This might be due to the existence of growth faults, quenched-in vacancies or vacancy-related secondary defects. In fact, Suzuki and Takeuchi [6] reported that straight dislocations in specimens crashed at room temperature changed into helical ones during leaving the specimens at room temperature. They suggested that a large amount of excess vacancies were quenched in their specimens which were prepared from an as-arc melted crystal and such quenched-in vacancies caused the climbing of dislocations. We need more information on the lattice defects in $CoSi_2$, in particular, the formation and migration of vacancies. Since the properties of vacancies influence not only deformation at high temperatures but also the properties of $CoSi_2$ as a VLSI device material, in particular, if excess vacancies can be easily quenched-in and migrate at room temperature.

In this paper, the state of our current knowledge of the properties of vacancies and the slip behavior and mechanisms of $CoSi_2$ is reviewed based on the results of our recent experiments in the corresponding field. We have been studying the lattice defects and deformation of $CoSi_2$ not only because we are interested in $CoSi_2$ itself but also because we think we may gain a better insight into silicides and related compounds through the extensive basic studies of $CoSi_2$ since complete information for no individual silicide systems is available. The readers are also recommended to refer to the recent review by Shah et al. [9] for the silicides as structural materials.

VACANCIES AND THEIR CLUSTERS

Recently, the vacancies and vacancy clusters in TiAl compounds with non-stoichiometric compositions were investigated using positron lifetime spectroscopy [10], which can most sensitively detect and differentiate submicroscopic defects such as vacancies and related secondary defects [11]. The results of the study [10] successfully demonstrated that deviations from the stoichiometric composition in TiAl were compensated by anti-site atoms without forming constitutional vacancies.

More recently, we have applied positron lifetime spectroscopy to investigate the properties of vacancies and their clusters in $CoSi_2$ [12]. A $CoSi_2$ rod was produced by melting pure cobalt (99.9 %) and silicon (99.9999 %) in a plasma arc-furnace. The $CoSi_2$ rod was remelted to grow a single crystal. Specimens of 9 mm x 12 mm x 0.8 mm were cut from the single crystal. The specimens were fully annealed for 6 hours at 1273 K and slowly cooled to room temperature in evacuated quartz capsules and then chemically polished. Then, the following three experiments were made to introduce vacancies and their small clusters in the specimens:

(1) Water- and air-cooling quench from temperatures above 610 K.
(2) Irradiation with 10 KeV electrons below 250 K to a dose of 2.5 x 10^{18} e^-/cm^2.
(3) Irradiation with 2.0 MeV protons below 100 K to a dose of 2.2 x 10^{18} p/cm^2.

Figure 1 shows the obtained values of mean positron lifetime for the quenched specimens as a function of quenching temperature. The mean positron lifetime is seen to begin to increase at 760 K and approach a saturation value of 172 ps. Comparison of water-quenched and air-cooled specimens shows very similar behavior, i.e. little difference in positron mean lifetime. We may thus draw the following conclusions :

(1) The concentration of thermal vacancies in $CoSi_2$ exceeds the vacancy-detection limit by positron lifetime spectroscopy at about 700 K (0.44Tm), while in ordinary metals and

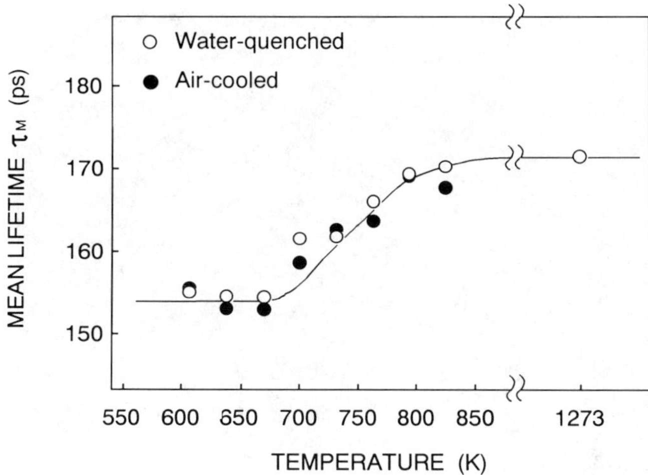

Fig. 1. Mean positron lifetime in quenched $CoSi_2$ as a function of quenching temperature [12].

alloys, the existence of thermal vacancies can be detected by positron lifetime spectroscopy only at temperatures higher than 0.6Tm. Above 800 K the concentration of thermal vacancies in $CoSi_2$ becomes so high that the specimens quenched from temperatures higher than 800 K contain enough quenched-in defects to trap practically all positrons injected into the specimens.

(2) Thermal vacancies are easily quenched-in even at the rate of cooling by air.

(3) Assuming that all thermal vacancies at each temperature are quenched-in, the vacancy formation enthalpy in $CoSi_2$, Hv can be estimated to be 1.1 ± 0.7 eV based on the result shown in Fig. 1.

These results are rather surprising, but are in agreement with the recent observation by Suzuki and Takeuchi [6] that helical dislocations are formed at room temperature. Their button ingots were prepared on a water-cooled copper hearth using an arc-melting furnace, and hence their ingots were considered to contain a considerable amount of excess-vacancies or their small clusters. The existence of such quenched-in defects may be a cause of electrical resistivity anomalies in $CoSi_2$ such as an anisotropy of resistivity and a high value of residual resistivity at 4.2 K [7, 8]. Constitutional vacancies and antisite atoms may cause such anomalies, however, no evidence of constitutional vacancies was found by positron annihilation [13].

Low temperature irradiation with 1 MeV electrons gives rise to the increase in positron lifetime in $CoSi_2$ from 154 ps for the fully annealed state to 180 ps. Then, the positron lifetime decreases from 180 ps to 172 ps upon the subsequent isochronal annealing at 310 K [12]. Judging from the electron energy and dose, it is quite reasonable to conclude that only Frenkel pairs are produced by the irradiation. Interstitial atoms themselves do not affect positron annihilation parameters. However, migrating interstitials recombine with vacancies and reduce the concentration of vacancies. Thus, interstitial migration is expected to result in a decrease in the relative intensity of the lifetime component corresponding to the vacancies. On the other hand, the migration of vacancies should cause a change in positron lifetime owing to the formation of vacancy clusters. Our observations correspond to the latter case [12]. We believe, the interstitial atoms are more mobile than vacancies as in pure metals, and immediately disappear at vacancies or other sinks during electron irradiation below 250 K. Thus, the lifetime value of 180 ps observed for as-electron irradiated specimens should correspond to the positron lifetime at vacancies in $CoSi_2$. Vacancies in the electron irradiated specimens begin to migrate and form secondary defects with a smaller lifetime value of 172 ps at 310 K [12]. This lifetime value exactly coincide with the saturation lifetime value above 800 K in Fig. 1. This indicates that the quenched-in defects in the water-quenched and air-cooled specimens in Fig. 1

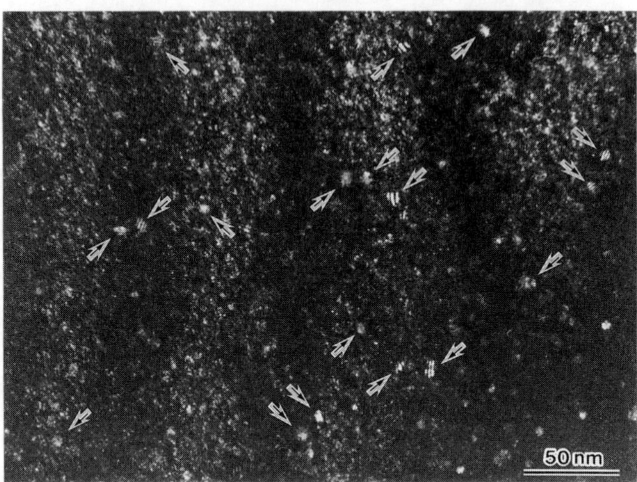

Fig. 2. Faulted-loop like defects in a specimen of $CoSi_2$ which was electron irradiated and subsequently annealed at 310 K.

are such secondary defects. Helical dislocations observed by Suzuki and Takeuchi [6] may be formed by vacancies gradually evaporated from such secondary defects at room temperature.

Figure 2 shows a microstructure from a specimen which was electron-irradiated and subsequently annealed at 310 K. Defects which look like faulted dislocation loops (marked by arrows) are observed. They might be secondary faults with the lifetime value of 172 ps. However, we can not yet rule out that they are formed during foil preparation by ion-thinning. Further investigation on the faulted loop-like defects is now in progress.

Low temperature irradiation with 2 MeV protons causes a more marked increase of positron lifetime than electron irradiation. The positron lifetime in the proton-irradiated $CoSi_2$ is 200 ps which is much longer than the lifetime at vacancies. This is believed to be due to the formation of uncollapsed small vacancy clusters, similarly to the case of proton-irradiated TiAl [10].

ELASTIC CONSTANTS

Only three independent elastic constants are necessary in order to describe the elastic properties of a cubic crystal. Recently, Tanaka et al.[14] have measured the three elastic constants of $CoSi_2$ and their temperature dependence in the range of 300-1173 K using a ultrasonic method. The values of elastic constants at room temperature are [14]:

$$C_{11}=225\ \text{GPa},\quad C_{12}=138\ \text{GPa}\quad \text{and}\quad C_{44}=84\ \text{GPa}.$$

The C_{11} and C_{12} values of $CoSi_2$ are comparable with those of the Ni_3Al based $L1_2$ compounds. However, $CoSi_2$ has a smaller value of C_{44} and hence a smaller anisotropic ratio $(2C_{44} / (C_{11}\text{-}C_{12}))$ than the Ni_3Al-based $L1_2$ compounds. All the three elastic constants decrease gradually with increasing temperature.

SLIP SYSTEMS

The operative slip systems in $CoSi_2$ single crystals have been systematically determined at room temperature by Ito et al.[4] and recently at temperatures in the range of 293-1273 K by the same authors. The primary slip system in $CoSi_2$ is {001}<100>. The primary slip systems of other compounds isostructural with $CoSi_2$ are reported to be {001}<110> for CaF_2 [15], BaF_2 [15] and SrF_2 [16], and {111}<110> for TiH_2 [17] and ZrH_2 [18]. The difference in the primary slip system among those compounds has been attributed to the difference in the

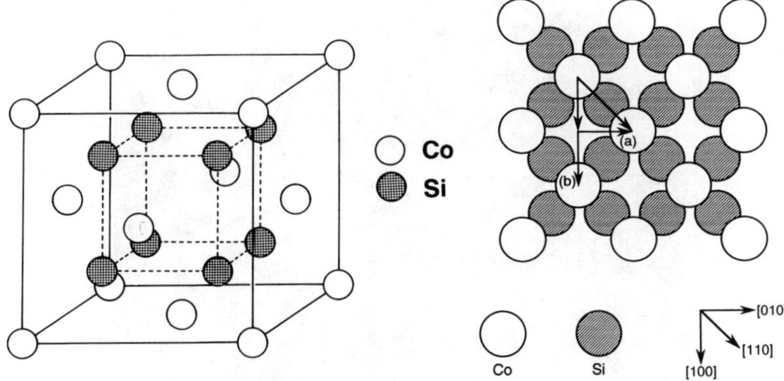

Fig. 3. Crystal structure of $CoSi_2$ and the atomic arrangement on (001) planes.

electrostatic nature of the dislocation core [15], i.e. the former compounds are ionic in bonding character while the latter are metallic. $CoSi_2$ has both a high covalency and metallicity. The occurrence of slip on {001}<100> in $CoSi_2$ has been interpreted in the light of the covalency and the dislocation dissociation modes in $CoSi_2$ [4]. Figure 3 shows the crystal structure of $CoSi_2$ and the atomic arrangement on (001) planes. In Fig. 3, the atomic layer composed of silicon atoms (hatched circles) is located above and below that composed of cobalt atoms (open circles). It is known that there is strong directional covalent bonding between cobalt and silicon atoms in $CoSi_2$ [19]. Sauer and Freise [2] discussed the selection of the primary slip system of $CoSi_2$ among {001}<110>, {110}<110> and {110}<111> in terms of the distortion of the strong Co-Si covalent bonding in the slip process and concluded that {001}<110> is preferred to the others since the distortion of the Co-Si bonding is minimum during {001}<110> slip if perfect dislocations (**b**=a / 2<110>) are dissociated into partials as follows (path (a) in Fig. 3)

$$a/2[110] \rightarrow a/2[100] + (SF) + a/2[010] \quad (1)$$

This results in no change in the self energy of the dislocation line if the square of the Burgers vectors is used as a first approximation, but there is a net increase in the total energy upon dissociation due to the contribution of the stacking fault. Sauer and Freise [2] concluded that in spite of this energy increase owing to the stacking fault, this dissociation is still favored since the path along <100> gives the minimum distortion of the Co-Si covalent bonding for each partial dislocation, as shown in Fig. 3. In contrast, if the primary slip system is {001}<100>, perfect dislocations (**b**=a<100>) can be dissociated into partials as follows (path (b) in Fig. 3)

$$a[100] \rightarrow a/2[100] + (SF) + a/2[100] \quad (2)$$

In this case slip also occurs along the [100] direction so that the distortion of the Co-Si covalent bonding is also minimized for each partial dislocation. However, in addition, the total energy of the dislocation is reduced upon dissociation, in spite of the energy increase owing to the presence of a stacking fault, since the self-energy of the two partial dislocations is one half that of the perfect dislocation. This is probably the main reason why the {001}<100> slip system is the primary system in $CoSi_2$.

Figure 4 shows a typical dislocation structure in a crystal deformed by slip on {001}<100>. The crystal was deformed in compression along [011] and the primary slip occurred on (001) along [010]. The thin foil was cut parallel to (001). Long and straight dislocations are seen to lie on (001). Their Burgers vector is [010] and hence it is seen that these dislocations have a strong tendency to align along the edge orientation. This tendency has been observed in the dislocation structures in the specimens deformed by slip on {001}<100> at room temperature. The implications of these observations have yet to be investigated. Figure 5 shows a weak-beam image of an edge dislocation in a (001) foil prepared from a [123]-oriented crystal deformed by slip on (001)[100] at 673 K. The edge dislocation is seen to be dissociated into two partial dislocations. Two partials are in contrast for **g**(reflection

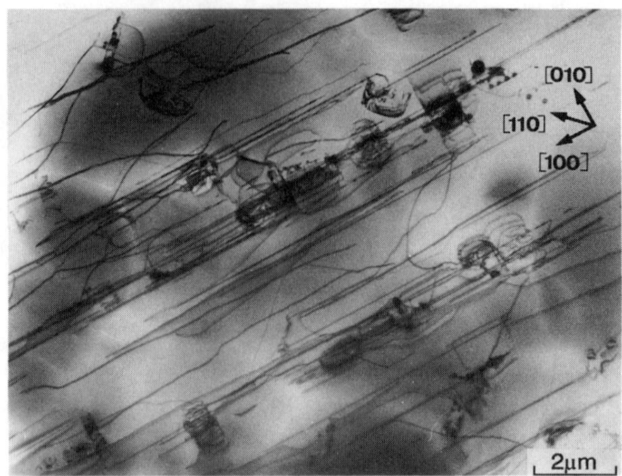

Fig. 4. Dislocation structure in a $CoSi_2$ single crystal deformed along [011] at 673 K.

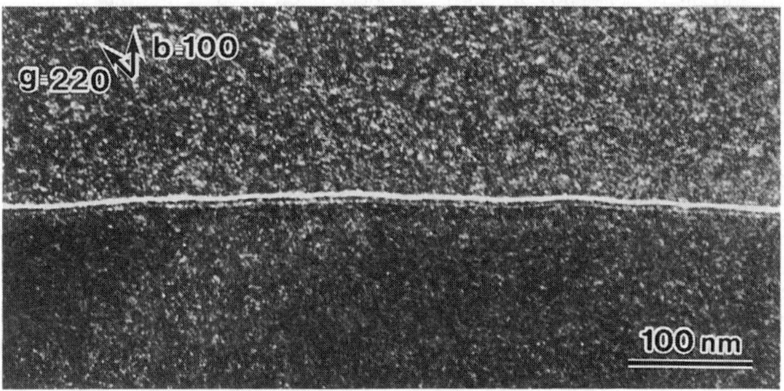

Fig. 5. Weak-beam image of a [100] dislocation dissociated into two a / 2[100] partials.

vector)=220, $2\bar{2}0$ and 200 but disappear simultaneously for **g**=020 and 022, and their separation is not changed when **g** is reversed. This combination leads to the total Burgers vector a[100] and the dissociation into two a / 2[100] partial dislocations (eq. (2)) separated by the stacking fault on (001). The separation width is about 6 nm. Using the anisotropic elasticity and the values of elastic constants at room temperature, the stacking fault energy is estimated to be about 170 mJ / m^2.

In crystals with orientations such as <123> and <135> in mid-field of the stereographic unit triangles, slip occurs always on {001}<100> in the temperature range of 293-1273 K. However, in crystals with orientations such as <001> and <110>, slip occurs along <110> on {001}, {111} and {110} planes at high temperatures. [001]-oriented crystals, in which plastic deformation occurs only at temperatures higher than 773 K, is deformed by slip along <110> on {111} below 973 K and {110} above 1073 K. [011]-oriented crystals exhibit a slip mode transition from {001}<100> to {111}<110> in the range of 973-1073 K. However, more work is needed to clarify the mechanisms of such slip mode transitions. Two possible dissociations have been proposed for a / 2[110] dislocations [2]. The first is according to the reaction of eq. (1). The second is according to the reaction

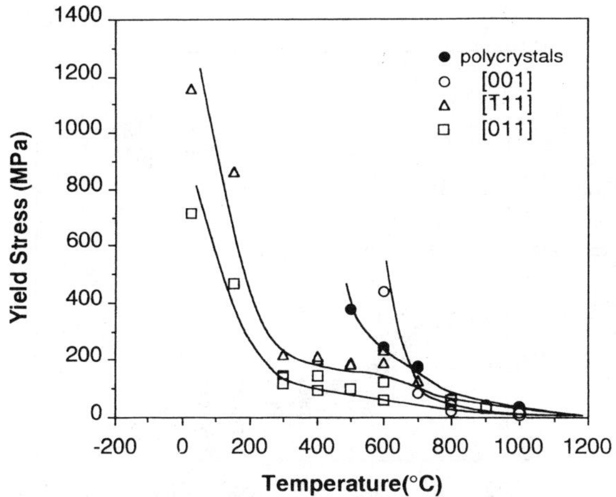

Fig. 6. Yield stress of $CoSi_2$ single crystals as a function of temperature.

$$a/2\,[110] \rightarrow a/4\,[111] + (SF) + a/4\,[11\bar{1}] \qquad (3)$$

which involves a stacking fault on $(1\bar{1}0)$. Slip on {001} occurs always along <100>, and hence the dissociation given by eq. (1) does not seem to occur in $CoSi_2$. The dissociation of eq. (3) may occur in $CoSi_2$, however, neither Suzuki and Takeuchi [6] nor the present authors have obtained any evidence to indicate the dissociation of a / 2<110> dislocations on {110} planes. The energy of the stacking fault on {110} involved in the reaction of eq. (3) is thought to be considerably high.

YIELD STRESS AND ITS TEMPERATURE DEPENDENCE

Yield stress for single crystals with orientations [001], [011] and $[\bar{1}11]$ is plotted as a function of temperature in Fig. 6 together with the results for polycrystalline specimens. Yield stress of single crystals of $CoSi_2$ is seen to strongly depend on both crystal orientation and temperature. Yield stress for [001] orientation is much higher than those for other orientations below 973 K. This indicates that the critical resolved shear stresses for the slip systems other than {001}<100> is considerably high in this temperature range. Polycrystals of $CoSi_2$ can be deformed only at temperatures higher than 773 K. The reasons for this are two-fold: firstly, for {001}<100> slip, which is the easiest slip mode in $CoSi_2$, there are no ways for choosing five independent slip systems; and secondly, the slip systems to augment the three independent {001}<100> have considerably higher critical resolved shear stresses than that for slip on {001}<100>. Takeuchi et al.[3] also reported a strong temperature dependence of yield stress of polycrystals of $CoSi_2$. They made yield stress measurements at temperatures higher than 700 K. The yield stresses reported by Takeuchi et al.[3] are lower than those shown in Fig. 6. There would be a difference, for example, in grain size between the specimens of the two research groups. Figure 7 shows the temperature dependence of the critical resolved shear stress for slip on {001}<100>.

In order to obtain information on the deformation mechanisms of $CoSi_2$, thermal activation analysis was conducted. The strain rate sensitivity of the flow stress ($\partial\sigma/\partial\ln\dot{\varepsilon}$) was measured by suddenly increasing the strain rate by a factor of ten in the temperature range of 873-1273 K for $[\bar{1}23]$-oriented single crystals, and the activation volume and activation energy for deformation were evaluated using the conventional method. The strain rate sensitivity decreases with increasing temperature. Both the activation volume and activation enthalpy decrease with increasing effective stress. The value of the activation enthalpy is in the range of 3.0-4.0 eV at the effective stress level smaller than 20 MPa (i.e., at temperatures above 973 K) and in the range of 0.3-0.4 eV at the effective stress level larger than 40 MPa (i.e., at temperatures below 773 K) for $[\bar{1}23]$-oriented crystals. The values of the activation volume are

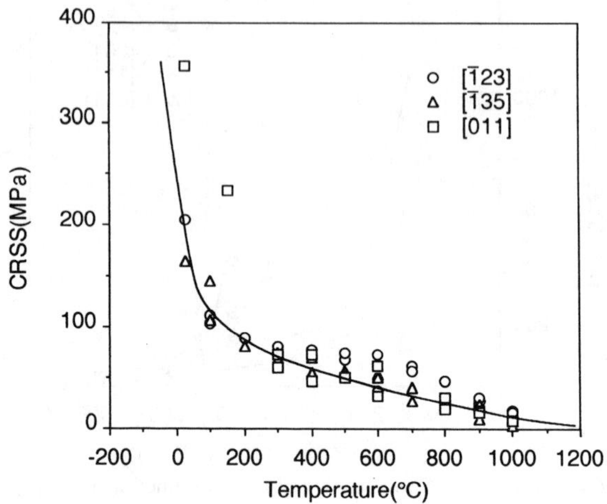

Fig. 7. Critical resolved shear stress for slip on {001}<100> in $CoSi_2$ single crystals.

in the range of 10-100 b^3 (**b**=a / 2<100>) in the temperature range of 373-873 K. This suggests that slip on {001}<100> is controlled by the Peierls mechanism. At temperature above 973 K, the activation volume rapidly increases with increasing temperature. In this temperature range, the concentration of thermal vacancies is greatly increased (Fig. 1) and they are thought to influence the mobility of dislocations. This might be one of the reasons for the rapidly increasing ductility of the compound at temperatures above 973 K.

CONCLUSIONS

In this paper, we have reviewed a wide cross-section of the studies on the properties of vacancies and the deformation behavior of $CoSi_2$. The primary slip system of $CoSi_2$ is {001}<100>, which fact can be interpreted in terms of a high covalency in Co-Si bonding. In view of slip and deformation behavior of $CoSi_2$, this compound is quite different from ordinary metals and alloys. However, considering the properties of vacancies, $CoSi_2$ rather resembles ordinary metals and alloys. The formation enthalpy of vacancy is about 1.1 eV, which is comparable with those of ordinary metals, and the concentration of thermal vacancies becomes high enough to be detected by positron lifetime measurements at 0.44 Tm, which is lower than the corresponding temperature of about 0.6 Tm in ordinary metals and alloys. There is evidence to indicate that excess vacancies can be easily quenched-in and they are mobile at around room temperature. Not only the mechanical properties but also the electrical properties of $CoSi_2$ are believed to be influenced by such properties of vacancies in the compound. Understanding the slip and deformation behavior and the properties of vacancies in $CoSi_2$ in a rational way may provide a new insight into not only $CoSi_2$ but also other transition metal disilicides.

ACKOWLEDGMENTS

This work was supported by Grant-in-Aid for Scientific Research on the Priority Area "Intermetallic Compounds as New High Temperature Structural Materials" from the Ministry of Education, Science and Culture, Japan and in part by the research grant from the R & D Institute of Materials and Composites for Future Industries and the NEDO International Joint Research Grant for the Intermetallics Research Team.

REFERENCES

1. D.L. Anton and D.M. Shah, in High-Temperature Ordered Intermetallic Alloys III, edited by C.T. Liu, A.I. Taub, N.S. Stoloff and C.C. Koch (Mater. Res. Soc. Proc. **133**, Pittsburgh, PA, 1989) p.361.
2. R.W. Sauer and E.J. Freise, in Anisotropy in Single-Crystal Refractory Compounds, Vol.1, edited by F.W. Vahldiek and S.A. Merson (Plenum, New York, 1968) p.459.
3. S. Takeuchi, T. Hashimoto and T. Shibuya, in Intermetallic Compounds - Structure and Mechanical Properties -, edited by O. Izumi (Japan Inst. Metals, 1991) p.645 ; J. Mater. Sci. **27**, 1380 (1992).
4. K. Ito, H. Inui, T. Hirano and M. Yamaguchi, Mater. Sci. Eng. A, **152**, 153 (1992).
5. P. Anongba and S. Steinemann, presented at 1991 Colloque Plastcité at Metz.
6. K. Suzuki and S. Takeuchi, to be published in Intermetallics, **1** (1992).
7. B.M. Ditchek, J. Cryst. Growth, **69**, 207 (1984).
8. T. Hirano and M. Kaise, J. Appl. Phys., **68**, 627 (1990).
9. D.M. Shah, D. Berczik, D.L. Anton and R. Hecht, Mater. Sci. Eng. A, **155**, 45 (1992).
10. Y. Shirai and M. Yamaguchi, Mater. Sci. Eng. A, **152**, 173 (1992).
11. See for example, J. Takamura, Y. Shirai, K. Furukawa and F. Nakamura, Mater. Sci. Forum, **15-18**, 809 (1987).
12. Y. Ito, Y. Shirai, Y. Yamada and M. Yamaguchi, in this proceedings.
13. A.G Balogh, L. Bottyan, G. Brauer, I. Dezsi and B. Molnar, J. Phys. F, **16**, 1725 (1986).
14. K. Tanaka, H. Numakura and M. Koiwa, private communication.
15. A.G. Evans and P.L. Pratt, Phil. Mag., **21**, 951 (1970).
16. T.S. Liu and C.H. Li, J. Appl. Phys., **35**, 3325 (1964).
17. P.E. Irving and C.J. Beevers, J. Mater. Sci., **7**, 23 (1972).
18. K.G.Barraclogh and C.J. Beevers, J. Mater. Sci., **4**, 518 (1969).
19. S. Geller and V.M. Wolontis, Acta Crystallogr., **8**, 83 (1955).

Prospects, Promises and Properties of Refractory Intermetallics

Donald L. Anton* and Dilip M. Shah**
*United Technologies Research Center, 410 Silver Lane, E. Hartford, CT 06108
**Pratt & Whitney, 400 Main St., E. Hartford, CT 06108

ABSTRACT

Extensive research activity, over the last 15 years, has been conducted on structure/property relationships and processing of intermetallic compounds for high temperature use. Progress has been made in improving a number of properties of these compounds; however, the demanding balance of properties required (high strength, good strength retention at temperatures exceeding 1000°C, low density, damage tolerance at ambient temperatures, good creep and stress rupture characteristics and environmental stability at high temperatures) are not likely to be achieved in a monolithic (single phase) compound.

Initial work on intermetallic matrix composites has proven to be quite promising. It has already been shown that compounds, brittle at room temperature, may be toughened by the inclusion of appropriate reinforcements, either strong or tough and ductile. Both artificial and natural or *in-situ* composite fabrication techniques have been used to manufacture these composite systems. The properties of two specific intermetallic matrix composite systems, $NiAl/Al_2O_3$ and Cr_2Nb/Nb are summarized to elucidate their strengths, weaknesses and potential. Candidate composite systems are also discussed along with the rationale behind their selection.

INTRODUCTION

Superalloys, used to the limit of their temperature capabilities in both the turbine and combustor of advanced jet aircraft, are being considered for replacement by refractory intermetallics such as NiAl and $MoSi_2$. Higher gas path operating temperatures will lead to greater thrust-to-weight ratios, higher fuel efficiencies and lower NOX emissions. Take off noise levels will also be limited through new high temperature nozzle designs. All of these problems, if solved, will lead to a new generation of turbo-fan engines which will meet the ever challenging needs of our industrialized society [1].

This paper will endeavor to briefly review the state-of-the-art of refractory intermetallic materials being considered for aerospace applications [2]. For our purposes, refractory intermetallics will be defined as those compounds with melting points in excess of 1600°C. In particular, the compounds NiAl, Cr_2Nb, $MoSi_2$, Mo_5Si_3, Cr_3Si and Nb_2Al, which hold the greatest potential at this time for use as structural materials, will be discussed. A number of shortcomings will become evident in all of these materials and both artificial and *in-situ* composite approaches put forth to solve these obstacles. This will be followed by a description of the fundamental aspects encountered in high temperature composites, with special attention paid to the particular problems encountered with intermetallic matrix composites.

MONOLITHIC PROPERTIES

Creep Strength

A significant amount of creep data on refractory intermetallics has been generated over the past eight years since publication of the first *High Temperature Ordered Intermetallic Alloys* proceedings published in 1985 [3]. A plot of density normalized stress for 1% creep in 300 hours, as a function of temperature is given in Fig. 1 [2] for a number of refractory intermetallics. Also given in this figure for comparison is data for PWA 1480, a state-of-the-art superalloy single crystal. The "target strength" for a typical candidate replacement material in turbine engines is also shown.

All of the compounds evaluated here lie in a band having strengths within one order of magnitude below PWA 1480 at 1100°C, but extending considerably beyond this temperature where the dotted line indicates that PWA 1480 loses its strength as the temperature approaches the γ' solvus. The compound $MoSi_2$ lies at the upper bound of this band and would appear to be an extension of the P&W 1480 curve through 1400°C. Co_2Nb defines the lower end of this band below 1200°C, while both Cr_2Nb and Cr_3Si lose their strength more rapidly with temperature above 1200°C. Nb_2Al had a very low creep activation energy, and was anomalous in that it maintained very high creep resistance with increasing temperature.

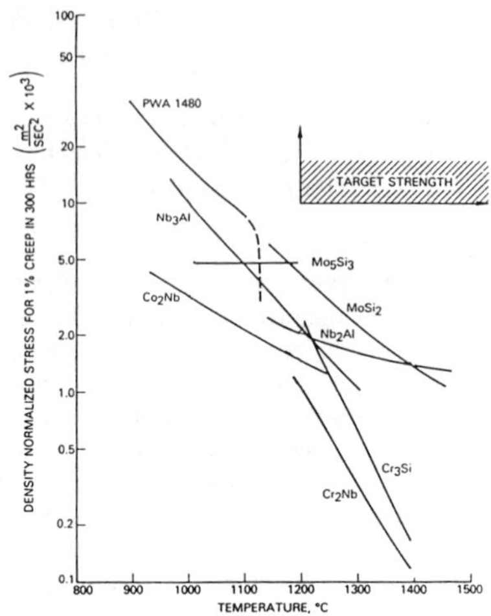

Figure 1 Summary of creep data for refractory intermetallics compared with a superalloy single crystal PWA 1480.

Tensile Strength and Fracture Toughness

Tensile test data as shown in Fig. 2a gives a plot of strain to failure vs. temperature which best describes the ductile brittle transition temperature, DBTT, of these materials. Here it is shown that DBTT's are generally confined to the range of 1000 to 1200°C. A couple of these compounds, Cr_3Si and Nb_2Al, did not display a DBTT below 1200°C, the limit of this test methodology. NiAl was anomalous in this respect in that its DBTT was in the range of 400-500°C. This relatively low DBTT will be shown to play a critical role in toughening strategies later in this paper.

Fig. 2b graphically depicts the ultimate strengths for these compounds, ranked in descending order of 1200°C strength. Co_2Nb was the strongest of these compounds by far,

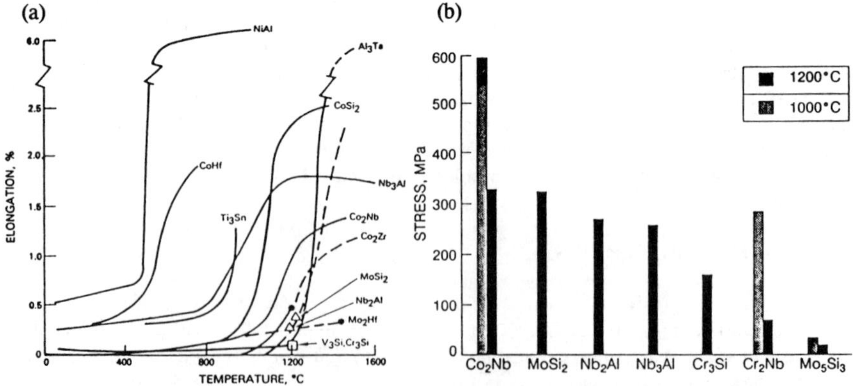

Figure 2 Mechanical test summary showing (a) DBTT and (b) ultimate tensile strengths for selected intermetallic compounds [2].

with $MoSi_2$ and Nb_2Al also giving good strengths below their DBTT's. Mo_5Si_3 had the lowest strength. In this last case, microscopic examination of transverse sections of the specimen showed a multitude of cracks emanating from the grain boundaries into their interiors. While the formation of these cracks may not affect the compressive properties of the material, such as in the creep test evaluation reported previously, tensile loading resulted in very low load failures.

Cyclic Oxidation

From the plot of weight change per unit area vs. number of cycles from 1150°C to room temperature, presented in Fig. 3, the outstanding oxidation resistance of $MoSi_2$ with minimal weight change is clearly demonstrated. This outstanding oxidation resistance is generally attributed to the high volatility of MoO_3 leaving a glassy SiO_2 layer at the surface. Of moderately less oxidation resistance is Cr_3Si where a limited amount of SiO_2 was detected on the surface along with an adherent Cr_2O_3 layer.

The other compounds with noncatastrophic oxidation, Co_2Nb and Cr_2Nb, are based exclusively on transition elements. In these cases, some degree of $Co_4Nb_2O_9$ and $CrNbO_4$ scale adherence was achieved, respectively, along with spallation of CoO and Cr_2O_3. As a realistic reference, the cyclic oxidation resistance of these two compounds, is comparable to that of conventional nickel base superalloys at this temperature.

The remaining compounds, Nb_3Al, Nb_2Al and Mo_5Si_3 for which data is not presented in Fig. 3 catastrophically oxidized within the first few cycles. The aluminum activity in both of the aluminides is clearly insufficient to form a protective scale. The remaining oxides were analyzed as $AlNb_{11}O_{29}$ and $AlNbO_4$. Mo_5Si_3 left α-crystobalite, tetragonal SiO_2, with loss expected through vaporization of MoO_3.

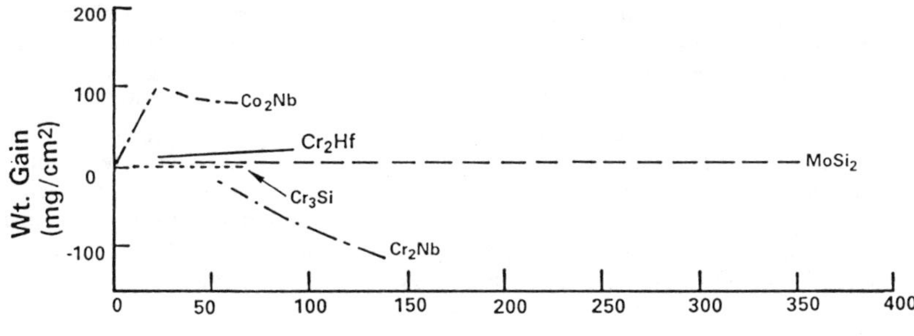

Figure 3 Cyclic oxidation results for noncatastrophically oxidizing compounds.

Composite Approach Necessary

These refractory intermetallics clearly show promising properties, especially when one considers that only binary compositions have been reviewed. As data for ever higher order alloys have been generated their potential becomes ever more promising. Creep resistance has been shown to be greatly enhanced through ternary alloy modification [2,3] while having negligible affect on oxidation resistance. The low intrinsic fracture toughness of these materials has not been changed through solid solution elemental additions. A composite approach must then be considered in an effort to increase fracture toughness and room temperature cleavage strength.

TOUGHNESS THROUGH COMPOSITES

Definition

Those intermetallic compounds closest to commercial utilization, Ti_3Al, TiAl as well as Ni_3Al as used in superalloys, are actually multi-component systems with highly refined microstructures consisting of a majority strong and some times brittle intermetallic phase in close association with a more ductile phase. This is reminiscent of pearlite, the eutectoid microstructure developed between cementite, Fe_3C, and ferrite, a microstructure common in many plain carbon steels and known for its toughness. Multiphase systems such as these can be considered composites since the two constituent phases usually have differing moduli, thermal expansion and ductility. This realization leads us to define the term *COMPOSITE*, which in the present context will not be straightforward. For our purposes, a composite will be defined as:

> *any multi-component structure for which the phases are manipulated in geometry or volume fraction so as to obtain a desired mix of mechanical properties.*

Under this broad definition, many practical engineering materials such as titanium and nickel alloys would be defined as composites. We must make this definition, because for intermetallic based materials to become useful, their properties, especially toughness, will need to be enhanced through the proper manipulation of either artificially manufactured composites in which particles, whiskers or fibers are added, or through natural or *in-situ* composites in which our knowledge of the phase diagrams can be put into practice to yield tough, strong microstructures.

Artificial Composites

As noted above, a composite can be obtained through either the traditional means of infiltrating a matrix around the reinforcement, an artificial composite, or the reinforcement can be introduced via either solidification or solid state precipitation, an *in-situ* composite. Artificial composites have the advantage of a wide latitude of component selection, limited to materials which are commercially available. Thus any matrix can theoretically be mixed or infiltrated into any reinforcement. Additionally, the reinforcement can be geometrically arranged into the most suitable arrangement, thus providing the maximum isotropic or anisotropic properties. This freedom comes with the high processing costs inherent when the reinforcement must be aligned and constrained during the infiltration process. Additional constraint comes from thermodynamic compatibility between the two phases at the consolidation temperatures, which can be significantly greater that the use temperature.

In-Situ Composites

Either solidification or precipitation can be used to naturally separate phases. This is a particularly cost effective means of composite fabrication since diffusion is used to align and separate the reinforcement. As long as the phase diagram is well understood, not an easy task in complex alloy systems, directional solidification, forging and heat treatment can be used separate and align the reinforcement into the desired geometry. By nature of their origin, these composites are also thermodynamically stable; however, their induced morphologies can be degraded by diffusional mechanisms. Naturally separated composites are limited in their constituents by the phase diagrams, which also limits reinforcement volume fraction, chemistry and morphology. Our understanding of the pertinent phase diagrams is critical if we are to exploit this group of composite materials.

Strong Phase Reinforcement

Artificial composite techniques are being studied on a wide scale in the ceramics industry where toughness has also been identified as the limiting engineering quantity. One of the most successful applications of ceramic composite technology is referred to as COMPGLAS*. This is a family of fiber reinforced glass and glass ceramic matrix composites with various combinations of different fibers and glass or glass-ceramic matrices. Their fabrication is similar to that of resin matrix composites, since the glass matrix can be readily deformed and flowed in its low viscosity state at elevated temperatures. The continuous fiber reinforced COMPGLAS* [5] results in material characterized by high strength, stiffness and with performance similar to resin matrix composites with temperature capabilities of up to 1350°C. Despite the brittle nature of the glass matrix, COMPGLAS* exhibits high *toughness*, and fractures upon loading in a non-catastrophic manner, which is essential for gas turbine applications. A number of important factors have been utilized in these systems which results in this balance of properties. In short they are (i) the thermodynamic compatibility of the reinforcements, typically SiC or C with the glass oxide matrices, (ii) the low thermal expansion mismatch between both matrix and reinforcement and finally (iii) the coating of non-reactive carbon on the SiC fibers results in a very weak matrix/fiber bond allowing long lengths of the reinforcement to bear load during crack growth. These three major considerations will be discussed in more detail in the following section with specific attention paid to intermetallic systems.

An alternate method of utilizing a strong reinforcement is to approach our matrix as a metal, which will need to be strengthened at high temperatures. This assumes that a reasonable degree of ductility occurs at low temperatures, but that strengthening and rigidity are required at elevated temperatures. This compositing approach has been used to strengthen NiAl which was shown previously to have a low DBTT. It is critical in this application that the reinforcement maintain good strength at high temperatures and that load be efficiently transferred from the matrix to the reinforcement. Efficient load transfer requires either a chemical or mechanical bond between the matrix and reinforcement. The most widely utilized metal matrix composite, Aluminum/SiC, is typical of this alloying approach, although not a high temperature system. Here, chemical stability allows for liquid metal infiltration while the significant thermal expansion mismatch induces thermal elastic clamping of the aluminum around the reinforcement to effect load transfer.

Ductile Phase Reinforcements

While most artificial composites have incorporated strong reinforcements, we must not be limited to this line of thought. The introduction in the late 1960's of rubber spheres into a brittle polystyrene matrix and ABS resins [8] has led to the development of a whole field of high impact resistant polymers. Energy absorbing or ductile phase composites can also be fabricated through artificial as well as *in-situ* means. The *Super Alpha-2* titanium aluminides are representative of the *in-situ* composite approach where a more ductile phase, β-titanium, is incorporated into the matrix, Ti_3Al, through alloy and process modification. These composites have the advantage of natural separation of the reinforcement from the matrix, thereby resulting in uniform microstructures with minimal occurrence of debilitating defects such as damaged reinforcements. Excellent matrix infiltration and superb reinforcements alignment also results.

Composite Requirements

Having identified two classes of intermetallics, strong and ductile phase reinforced, as well as two processing approaches, artificial and *in-situ*, we need to approach the selection of the required combination in a systematic fashion. To do this, we must first understand all of the restraints which will be required. Materials in gas turbines must exist in a highly stressed, high temperature, oxidative and thermally cyclic environment. Candidate materials must then have high tensile and creep strength, be resistant to oxidation attack, thermally stable over long time periods and maintain their integrity through repeated thermal excursions from ambient to operating conditions.

Thermodynamic Stability

Composite thermodynamic compatibility, gained either through reinforcement coating or occurring naturally between the two phases of the composite, will be required. This is especially true for the refractory intermetallics which are expected to operate at temperatures above 1200°C. At these very high temperatures, chemical interdiffusion between incompatible materials will not result in the stability and endurance that will be required for confident application of these materials. It has been our experience that thermodynamic calculations, while leading to general trends in materials compatibilities, are unreliable in predicting actual reactions. This is due to the limited amount of reliable thermodynamic data available at the high processing and operating temperatures required. Thus, empirical results need to be generated at both processing and utilization temperatures to assess interface strength and compatibility.

Thermal Expansion Compatibility

The cyclic thermal environment imposes restraints on the difference in thermal expansion the matrix and reinforcement can have. In actuality, this difference can not be more than a few per cent. Under severe and frequent thermal excursions, the matrix and reinforcement will expand and contract at different rates with the resultant strain given as $\epsilon_T = \Delta\alpha \cdot \Delta T$. As a very rough estimation, for 1200°C thermal cycles, and an allowable cyclic strain of 0.1%, the maximum thermal expansion mismatch, $\Delta\alpha$, allowable is 8.3×10^{-7}°C^{-1}. To put this in perspective, Fig. 3 gives the thermal expansion for candidate matrices and reinforcements. On this chart, our allowable $\Delta\alpha$ does not give one much latitude in reinforcement selection. In the case of strong fiber reinforced composites, it should also be remembered that it is preferred to have the coefficient of thermal expansion, CTE, of the matrix lower than the reinforcement. Under these conditions the matrix will be driven into compression upon cooling, thus closing any cracks which may have developed.

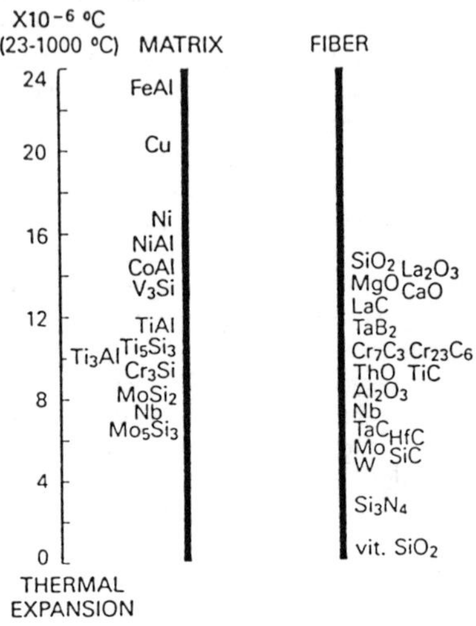

Figure 4 Thermal expansion comparison of various intermetallics and reinforcements.

SPECIFIC INTERMETALLIC COMPOSITES

Artificial Composites

Typical of the artificially reinforced intermetallic matrix composites is NiAl reinforced with alumina. This is a thermodynamically stable system where the reinforcement is readily available in polycrystalline yarn and monofilament single crystal.

Tensile test results obtained for alumina fiber reinforced NiAl [7] show a great enhancement of tensile strength with the incorporation of aligned, NiAl/AFP and chopped NiAl/CFP FP* fibers, as can be seen in Fig. 5. In a -10μm grain size, the matrix material is plastic at ambient temperatures [7]. A composite of this type would be *metal matrix* in character, having a room temperature ductile matrix. With the introduction of 35% FP alumina fiber, the composite became a brittle system with tensile strength increased signifi-

(a)

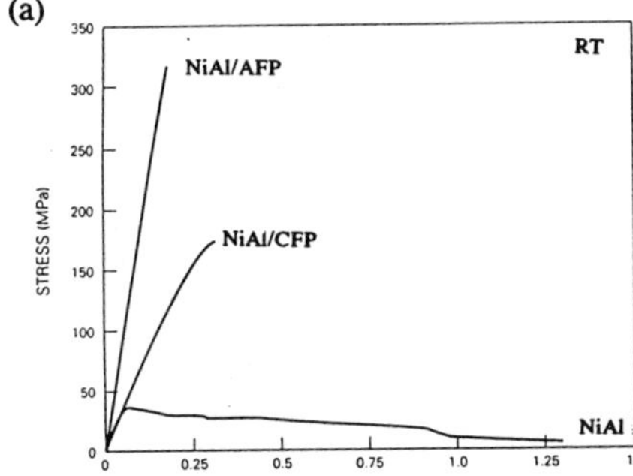

(b)

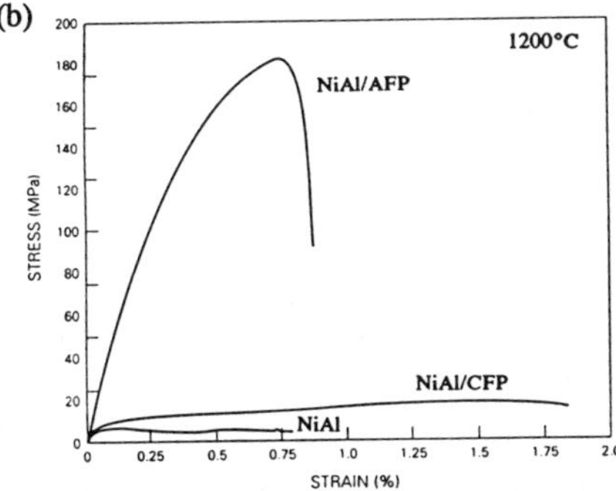

Figure 5 Comparison of bend test stress strain data for NiAl and both chopped and aligned FP alumina composites at (a) 25, and (b) 1200°C.

cantly, as seen in Fig. 5. The strengthening increment of the aligned FP composite is quite striking at 1200°C. Where both the monolithic and chopped FP composite show very poor strength, the aligned FP material has substantial strength. Since the matrix is very plastic at these high temperatures, significant plasticity is also observed in the aligned FP composite.

The creep strength of both NiAl/AFP and NiAl/CFP composites was substantially increased through incorporation of the reinforcement. Monolithic NiAl data obtained at 1027°C can be used for comparison [8]. Here, minimum creep rate values of $4x10^{-7}$, 10^{-6} and 10^{-3} sec.$^{-1}$ were obtained at 8, 14 and 70 MPa respectively. Creep testing of FP reinforced NiAl revealed a one to two order of magnitude decrease in minimum creep rate for the chopped FP composites and a three order of magnitude strengthening for the aligned FP composites [7]. At 1200°C and 69 MPa (10 ksi) loading, the minimum creep rate for NiAl/CFP was $4.20x10^{-6}$sec.$^{-1}$ while at 1200 °C and 13.8 MPa (2 ksi) loading of NiAl/CFP, a minimum creep rate of $6.2x10^{-8}$sec.$^{-1}$ resulted.

These advantages come at the cost of enhanced brittle behavior at all temperatures leading to reduced ductilities and lowered fracture toughness. The fiber pull-out mechanism was not achieved in this composite system, and it is expected that the large thermal expansion mismatch between the matrix and fiber induce a clamping stress around the reinforcements.

The other refractory intermetallic coming under ever increasing development is $MoSi_2$. This compound is quite brittle under ambient conditions, and thus needs to be toughened utilizing the *ceramic matrix* approach. Complete chemical stability has been demonstrated with SiC reinforcements [10]. The incorporation of 20% SiC whiskers has resulted in a two order of magnitude decrease in the minimum creep rate of unalloyed $MoSi_2$ [11]. While both ultimate [12] and creep strength have been shown to be greatly enhanced through reinforcement with SiC whiskers, a substantial increase of the fracture toughness has yet to demonstrated. This is most likely due to the great thermal expansion mismatch between the matrix at approximately 15 ppm °C^{-1} and SiC at approximately 5 ppm °C^{-1}.

Two other $MoSi_2$ based composites which are of interest utilize either refractory metal wires [13,14] or alumina as the reinforcement. The former maintains a reasonably good thermal expansion mismatch while being highly reactive. Composites of this type will require fiber coating to maintain high temperature stability. The alumina reinforced $MoSi_2$ system has an ideally matched thermal expansion, and while chemical compatibility is not total, a thin layer of mullite forms at the fiber-matrix interface, and the reaction stops [15].

Natural Composites

Bend test results are presented in Fig. 6 for Cr_2Nb/Nb with 30% Nb. While room temperature failure strain was not measured for Cr_2Nb, the failure strain and bend strength were determined for the *in-situ* composite. In spite of a K_{IC} of 3.5 MPa√m for the *in-situ*, the fractograph presented in Fig. 7 shows plastic stretching of ductile niobium particles. It can be noted that the Nb particles debond from the matrix readily, which allows for enhanced plastic deformation in each particle due to minimal particle constraint.

A comparison of minimum creep rate versus stress is given in Fig. 8 where it is shown that at both 1200 and 1400°C greatly enhanced creep resistance is obtained for the ductile particle reinforced intermetallic. At 1400°C, a drop in the minimum creep rate of two orders of magnitude is evident. The mechanism behind the observed creep strengthening may be related to specimen grain size and not the Nb particles directly. The monolithic intermetallic could only be processed via a powder route, resulting in a 40μm grain size, while the two phase material was easily castable with resultant millimeter size grains.

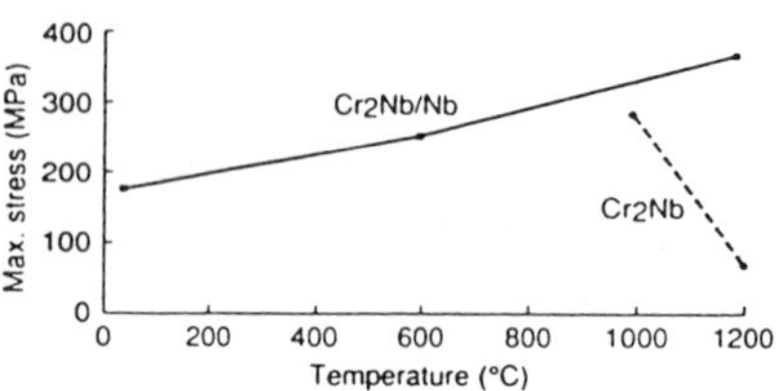

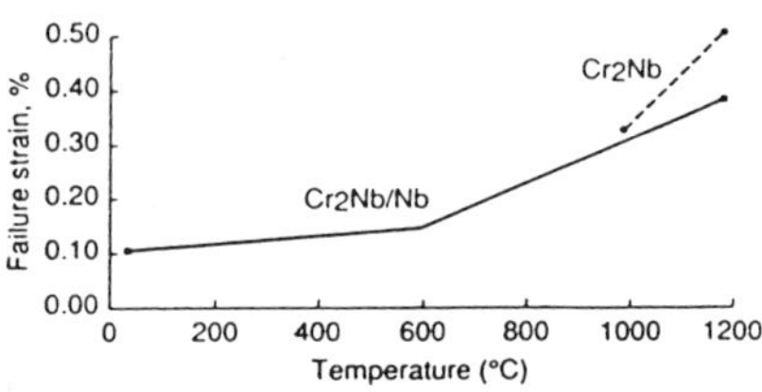

Figure 6 Comparison of monolithic and Nb reinforced Cr_2Nb in (a) maximum stress and (b) failure strain.

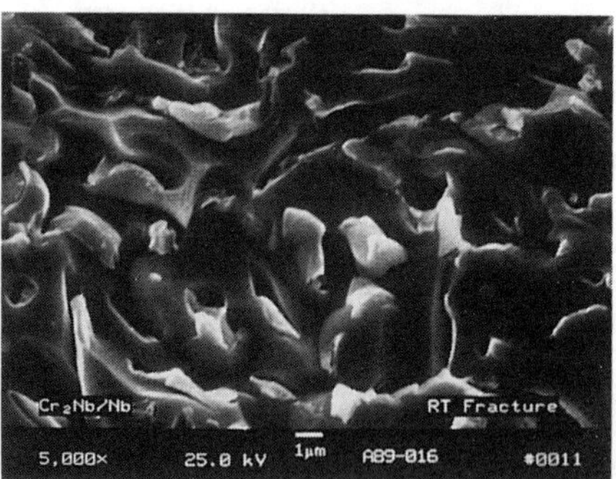

Figure 7 Fractograph of Cr_2Nb/Nb showing plastic tearing of ductile niobium particles.

The Search for Other *In-Situ* Composites

The search for other effective refractory intermetallic ductile particle strengthened systems is in its infancy. One of the first such systems to be developed was an NiAl matrix reinforced with either α Mo or Cr precipitates [16-18]. In these systems, the precipitate morphology was controlled through directional solidification. Recent work [19-21] has shown a doubling of the fracture toughness of these materials, with Mo additions being by far most effective.

Other *in-situ* composites studied include Nb_5Si_3/Nb [22], CoAl/Co [23], Cr_3Si/Cr [24], Nb_3Al/Nb [25] and Ti_5Si_3 /Nb[26], to name just a few. With the understanding gained from these studies, and parallel theoretical work aimed at defining the optimum ductile phase characteristics [27-29], it has become evident that microstructural control through processing is of primary importance.

A number of points have become self evident in evaluating various *in-situ* composite systems. The first is that it would be prudent to confine our search to the low density refractory metal (i.e. Nb, Mo and Cr) aluminides and silicides. This gives us our best opportunity of developing a low density, oxidation resistant material. Within the confines of these five elements, systems must be chosen which are in thermodynamic stability with the terminal solid solution, most likely Nb or Mo, since they have the highest potential for ductility. By constructing the possible quaternary phase diagrams, it is a relatively simple exercise to define which in-situ composite systems can be obtained. A simple construct of the Cr-Mo-Nb-Si quatern-ary is given in Fig. 9 where the known ternaries are used. The minds eye can connect the obvious phase field continuations such as $(Nb,Mo)_2Al$ - $(Nb,Cr)_2Al$ or $(Nb,Cr)_3Al$ - $(Mo,Cr)_3Al$ to help predict possible ternary compositions. This example of the quaternary phase diagrams graphically depicts the wide latitude the alloy designer will have in obtaining solid solution strengthened compounds.

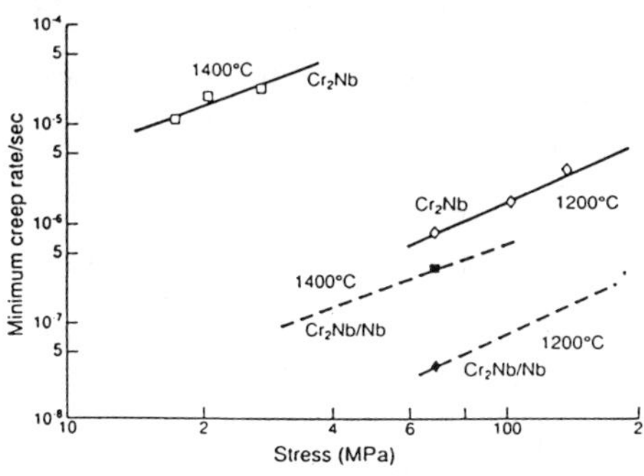

Figure 8 Comparison of minimum creep rates for monolithic and Nb reinforced Cr_2Nb.

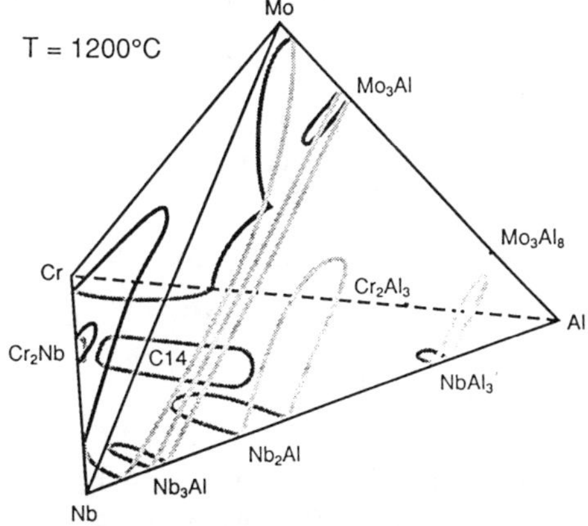

Figure 9 Cr-Mo-Nb-Al quaternary phase diagram showing extent of known ternary phases.

CONCLUSIONS

A number of refractory intermetallics, with melting points above 1600°C have been shown to be strong at temperatures well above their DBTT as well as oxidation resistant. Their generally brittle character has limited their development in spite of a protracted effort in identifying toughening mechanisms. Two composite approaches, artificial and natural or *in-situ*, have been identified which can be used to introduce either strong or ductile reinforcements into these room temperature brittle materials. Since their use temperature will be high, and the temperature will vary greatly during operation, the resulting composites must be thermodynamically stable as well as matched in thermal expansion coefficient. Examples were given of properties for strong fiber reinforced NiAl composite fabricated artificially as well as in-situ fabricated ductile Nb phase reinforced Cr_2Nb. The former, while thermodynamically stable, maintained a large thermal expansion mismatch which precluded fiber pull-out. Thus a strong composite with minimally enhanced fracture toughness was produced. The latter composite system resulted in greatly enhanced strength at both room and elevated temperatures with minimal toughness enhancement. A methodology is suggested to determine other intermetallic matrix composite systems which may be considered for future development.

ACKNOWLEDGEMENTS

This work is the result of a continuous learning process funded in part through WRDC/MLLM contracts F33615-90-C-5957, F33615-88-C-5405 and F33615-87-C-5214. The insightful discussions with Drs. D. Miracle, D. Dimiduk, and Mr. S. Mazdiyasni and Mr. T. Broderick were invaluable, especially in clarifying salient points during the conduct of this research.

REFERENCES

1. B.L. Koff, *The Next 50 Years in Jet Propulsion*, Presented at AIAA Meeting, Dayton, OH, August 23, 1989.
2. D.L. Anton and D.M. Shah, Intermetallic Compounds -Structure and Mechanical Properties-, O. Izumi ed., (The Japan Institute of Metals, Sedai, Japan, 1991), pp.379-86.
3. High Temperature Ordered Intermetallic Alloys, C.C. Koch et. al. eds., (MRS, Pittsburgh, PA, 1985).
4. D.L. Anton and D.M. Shah, "Development Potential of Advanced Intermetallic Materials", WRDC Final Report, Contract No. F33615-87-C-5214, 1990.
5. K.M. Prewo, J.J. Brennan and G.K. Layden, Ceram. Bul., 65, 305 (1986).
6. C.B. Bucknall, Brit. Plastics, 40, 118 (1967).
7. D.L. Anton and D.M. Shah, Intermetallic Matrix Composites II, D.B. Miracle et.al. eds. (MRS, Pitsburgh, PA, 1992), pp.157-64.
8. Stephens, J.R., High Temperature Ordered Intermetallic Alloys, C.C. Koch et. al. eds., (MRS, Pittsburgh, PA, 1985), p. 381.
9. R.D. Noebe, R.R. Bowman and J.I. Eldridge, Intermetallic Matrix Composites, eds. D.L. Anton et. al. (MRS, Pitsburgh, PA, 1992), pp.323-32.
10. F.D. Gac and J.J. Petrovich, J. Amer. Ceram. Soc., **68**, c200 (1985).
11. K. Sadenanda, C.R. Feng, H. Jones and J.R. Petrovich, High Temperature Structural Silicides, Eds. A.K. Vasudevan et. al., Elsevir Science Publishers, Amsterdam, (1992).
12. D.H. Carter, *SiC Whisker Reinforced $MoSi_2$*, Los Alamos Nat. Labs. Report LA-11411-T, (1988).
13. R.G. Castro, R.W. Smith, A.D. Rollett and P.W. Stanek, op. cit. 9, 101.
14. M.J. Maloney and R.J. Hecht, op. cit. 9, p.19.
15. D.M. Shah, D.L. Anton and C.W. Muson, op. cit. 7, p. 333.
16. J.L. Walter and H.E. Cline, Met. Trans., **1**, 1221(1970).
17. H.E. Cline and J.W. Walter, Met. Trans., **1**, 2907 (1970).
18. H.E. Cline, J.W. Walter, E. Lifshin and R.R. Russell, Met. Trans., **2**, 189 (1971).
19. P.R. Subramanian, M.G. Mendiratta, D.B. Miracle and D.M. Dimiduk, op. cit. 7, p. 147.
20. Keh-Min Chang, op. cit. 5, p. 191.
21. F.E. Heredia and J.J. Valencia, op. cit. 5, p. 197.
22. M.G, Mendiratta, J.J. Lewandowski and D.M. Dimiduk, Met. Trans. A, **22**, 1573 (1991).
23. M.A. Przystupa and T.H. Courtney, Met. Trans. A, **13,** 873 (1982).
24. J.W. Newkirk and J.A. Sago, op. cit. 7, p. 183.
25. D.L. Anton and D.M. Shah, op. cit. 7, p. 45.
26. G. Frommeyer, R. Rozenkranz and C. Ludecke, Z. Metallkde., 81, 30 (1990).
27. L.S. Sigle, P.A. Mataga, B.J. Dalgleish, R.M. McMeeking and A.G. Evans, Acta Metall., **36**, 945 (1988).
28. M. Bannister, H. Shercliff, G. Bao, F. Zok and M.F. Ashby, Acta Metall., **40**, 1531 (1992).
29. K.S. Ravichandran, Scripta Metall., **26**, 1389 (1992).

PART II

Alloy Structure and Phase Stability

SCHEMATICS OF CGBS

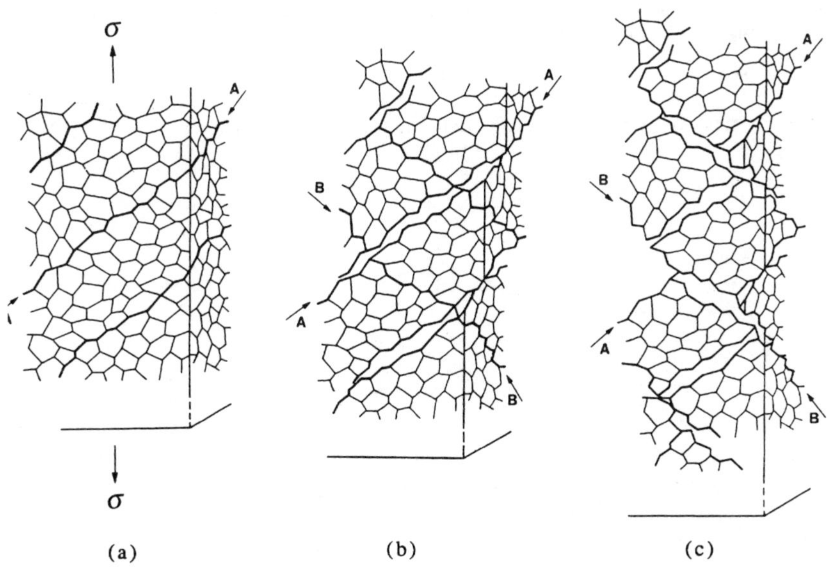

Schematics of superplastic deformation of a tensile specimen by means of cooperative grain boundary sliding (GBS) along two systems of grain boundary surfaces, A and B (arrowed): (a) undeformed polycrystal; (b) after sliding along system A; (c) after sliding along system A and then B. In addition to the illustration of a possible way of strain accumulation by CBGS, the schematics can also be used to interpret some microstructural features observed in superplastically deformed materials including intermetallics: (1) dynamic grain growth; (2) void formation; (3) phase stringer formation and apparent change in phase volume fraction in case of two phase materials; (4) breaking-up of lath-like particles.

Figure courtesy of H.S. Yang, M.G. Zelin and A.K. Mukherjee.

A FIRST-PRINCIPLES STUDY OF THE PHASE STABILITY OF FCC-AND HCP-BASED Ti-Al ALLOYS

Mark Asta[1], Mark van Schilfgaarde[2] and Didier de Fontaine[1]
[1]Department of Materials Science and Mineral Engineering, University of California at Berkeley, Berkeley, CA 94720, and Materials Sciences Division, Lawrence Berkeley Laboratory, Berkeley CA 94720
[2]SRI International, Menlo Park, CA 94025

ABSTRACT

In this paper we present results of a first-principles study of phase stability and structural and thermodynamic properties of fcc- and hcp-based Ti-Al alloys. In particular, the full-potential linear muffin tin orbital method has been used to determine heats of formation and other zero-temperature properties of 9 fcc and 7 hcp ordered superstructures as well as fcc and hcp Ti and Al. From these results a set of effective cluster interactions are determined which are used in a cluster variation method calculation of the solid-state portion of the composition-temperature phase diagram for fcc- and hcp-based alloys.

INTRODUCTION

Despite the many attractive properties of Ti-Al compounds, such as low density and high elevated-temperature strength, these alloys have found only limited use as high-temperature structural materials due primarily to their brittleness at room temperature [1]. Substantial research has been devoted to improving the ductility of Ti-Al alloys without appreciably altering other desirable properties. In particular, through ternary additions, and by using novel processing techniques, alloy designers have worked to stabilize compounds with high (cubic) symmetry and to develop materials with multiphase microstructures [1]; somewhat improved ductilities have been achieved. Clearly an understanding of the relative stability and the properties of equilibrium and metastable phases is needed to guide research in the Ti-Al system.

In the present paper we summarize the results of a first-principles study of structural and thermodynamic properties and the composition-temperature (c-T) phase diagram of fcc- and hcp-based Ti-Al alloys. In this type of study where the atomic numbers of the elements are the only information input, calculated phase diagrams cannot be expected to agree completely with experiment regarding absolute values of the temperatures of equilibrium reactions. However, *ab-initio* calculations are valuable since alloy properties for stable and *metastable* phases can be calculated *as a function of composition and temperature*, the relative stability of competing structures can be reliably predicted, and a fundamental understanding of these results can be obtained in terms of the electronic structure and bonding mechanisms.

COMPUTATIONAL APPROACH

The computational approach taken in this study is described in detail in reference 2 and will only be briefly summarized here. The total energy of elemental Ti and Al and of 9 fcc and 7 hcp ordered alloy compounds were calculated as a function of the atomic volume using the full-potential [3] linear muffin tin orbitals [4] method (FLMTO). It has been shown previously [2] that structural relaxations are large for Al-rich alloys with low symmetry structures and that the proper $TiAl_3$ ground state can only be predicted from first-principles calculations if the effect of tetragonal distortion on the total energy of the $D0_{22}$ structure is taken into account [5,6]. Therefore, at a given volume all structural degrees of freedom were relaxed for each compound considered in our calculations. The total energy versus volume relation was fit to a third-order polynomial from which we determined equilibrium volumes (Ω), bulk moduli (B) and heats of formation (ΔH). The heat of formation is defined as the difference between the energy of a compound at its equilibrium volume and the concentration-weighted average of the energies of hcp Ti and fcc Al (the pressure-volume contribution to ΔH is negligible in this system).

For the purposes of studying phase stability, the results of the FLMTO total energy calculations were used to obtain volume-dependent effective cluster interactions (ECI's) [7] for fcc- and hcp-based Ti-Al alloys as described in detail in reference 2. These ECI's formed the input into cluster variation method (CVM) [8] calculations from which free energies were determined and used to construct equilibrium phase boundaries. Specifically, the configurational entropy of fcc- and hcp-based phases was treated using the tetrahedron-octahedron [9,10] approximation of the CVM.

RESULTS

In figures 1 (a)-(d) the FLMTO calculated heats of formation, atomic volumes and bulk moduli of elemental Ti and Al as well as of 7 hcp and 9 fcc superstructure alloy compounds are displayed. Dashed and dotted lines in figures 1 (b)-(d) correspond to results for random mixtures of Ti and Al on the sites of an hcp structure and fcc lattice, respectively. Results for these random alloys were obtained from ECI's, which in turn were determined from the values of ΔH, Ω and B for the ordered alloy compounds [2]. Descriptions of the structures of the alloy compounds considered in this study can be found in references 2 and 11.

Calculated heats of formation are plotted versus the atomic concentration of Al in figure 1(a). Out of all of the structures considered, only hcp Ti, fcc Al, $D0_{19}$ (denoted as H1 in figure 1) Ti_3Al, $L1_0$ (F4) TiAl and $D0_{22}$ (F1) $TiAl_3$ are found to be stable in agreement with experimental findings [12]. The $L1_2$ (F2) structure is found to be energetically competitive at 25 and 75 at. % Al where it is unstable by only 0.7 and 1.7 mRy/atom, respectively. It should be emphasized that although most of the compounds considered in the FLMTO calculations will not be observed in the phase diagram, their heats of formation are needed in order to obtain the ECI's which are used to determine thermodynamic properties; furthermore, all structures must be fully relaxed with respect to all structural degrees of freedom in the total energy calculations in order to describe correctly phase stability in this system.

In figure 1 (b) the heats of formation of the energetically stable compounds are again plotted against the concentration of Al, and the results of calorimetry experiments [13] are also included for comparison (experimental data are plotted as solid symbols). The agreement between the calculated and experimental results is excellent. The dashed and dotted lines in figure 1 (b) show that the disordered hcp alloy is predicted to be stable for concentrations of Al less than 50 at. % only, in agreement with the experimental phase diagram [12].

The stable ordered compounds (H1, F4, F1) and the random alloys are shown in figure 1 (c) to have calculated molar volumes which deviate strongly from values given by Vegard's law. The minimum value of the molar volume is found for $TiAl_3$ compounds, in agreement with experimental measurements [14]. Calculated bulk moduli, figure 1 (d), for ordered and random alloys are all larger than for hcp Ti and fcc Al, indicating strong bonding between unlike atoms in this system. Also, the values of the moduli for all Ti-Al alloys shown are closer to those for pure Ti than for Al. In general, the values of Ω and B are found to have an interesting dependence on the concentration. Furthermore, by comparing dashed and dotted lines with open symbols in figure 1 (b)-(d), it is found that atomic ordering on a given lattice causes an increase in the magnitude of ΔH by as much as 30 %, as well as causing up to a 3 % decrease and 10 % increase in the values of Ω and B, respectively.

In figure 2 the solid-state portion of the phase diagram is shown for fcc and hcp Ti-Al alloys, as calculated by the CVM with ECI's determined from the results presented in figure 1. In constructing the phase diagram only the free energies of the phases with $D0_{22}$ (F1), $L1_2$ (F2), $L1_0$ (F4), $D0_{19}$ (H1), $B19$ (H3), fcc and hcp structures were considered. For compositions less than 50 at. % Al, the topology of the calculated phase diagram is in good agreement with experimental measurements [12] although the order-disorder temperature [12] of the $D0_{19}$ Ti_3Al phase is predicted to be 600 ^{o}C too high. A narrow region of stability is found for a phase with an $L12$ structure at elevated temperatures around 35 at. % Al.

At Al-rich compositions, we find the $L1_0$ and $D0_{22}$ phases to be strongly ordered to well above their experimentally observed melting points, and the phase field of the $TiAl_3$ ($D0_{22}$) phase

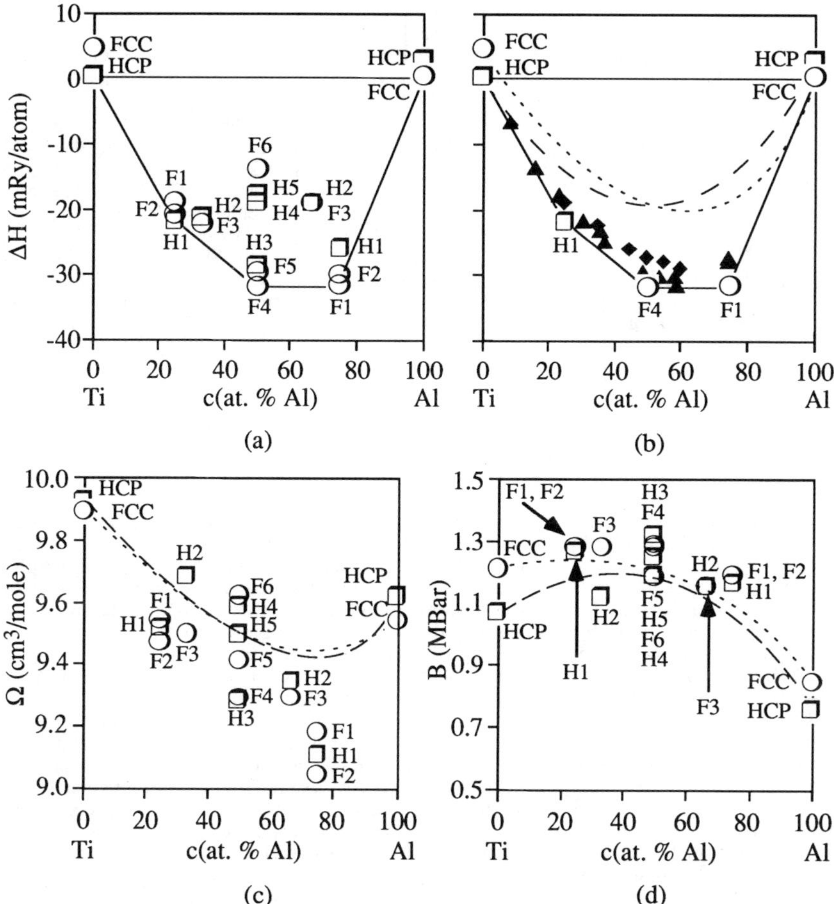

Figure 1: (a) Calculated heats of formation versus the atomic concentration of Al. The solid lines connect those structures which are energetically stable with respect to others at the same composition and with respect to phase separation to those at other concentrations. (b) Calculated ground state formation energies are replotted along with experimentally measured values [13] indicated by solid triangle and diamond symbols. Dashed and dotted lines are the heats of formation for random (disordered) hcp and fcc alloys, respectively. (c) Calculated molar volumes of ordered structures are plotted as open symbols. Dashed and dotted lines are again for hcp- and fcc-based random alloys. (d) Calculated bulk moduli for ordered and disordered structures. In (a)-(d) open squares and circles are for hcp- and fcc-based structures and the symbols have the following meaning: H1=$D0_{19}$, H2=Cmcm, H3=$B19$, H4=$P\bar{6}m2$, H5=Pmmn, F1=$D0_{22}$, F2=$L1_2$, F3=$MoPt_2$-type, F4=$L1_0$, F5=Phase "40" and F6=$L1_1$. The fcc (F1-F6) and hcp (H1-H5) superstructures are described in references 2 and 11, respectively.

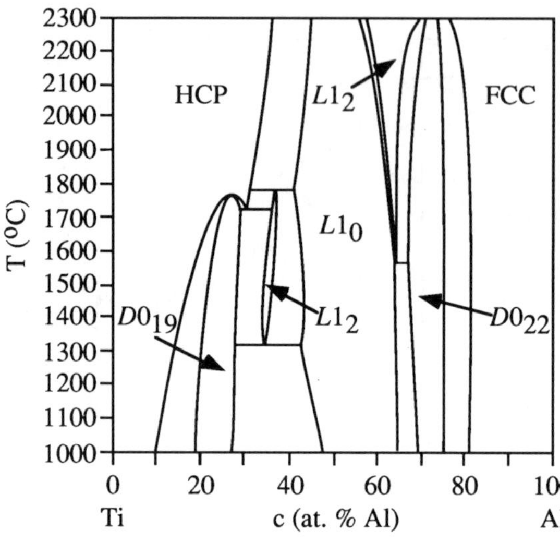

Figure 2: Calculated compostion-temperature phase diagram for fcc- and hcp-based Ti-Al alloys. Equilibrium phase boundaries were calculated using the CVM and ECI's determined from first-principles FLMTO total energy calculations. The stability regions for ordered phases are indicated by the Strukturbericht designation for their structures: $L1_2$, $L1_0$, $D0_{22}$ and $D0_{19}$. The labels HCP and FCC indicate regions of disordered solid solutions.

is predicted to be much narrower than that for TiAl ($L1_0$); these observations are also in agreement with the assessed phase diagram [12]. Another $L12$ phase is found to be stabilized between the $D0_{22}$ and $L1_0$ compounds as the temperature is raised; this feature of the phase diagram was missed in a previous calculation [2]. Experimentally, the portion of the phase diagram between 50 and 75 at. % Al is not well established; a series of complex fcc-based long period superstructures (LPS), not considered in our study, are known to be present [12,15,16] although precise phase boundaries for these LPS have not been determined.

A shortcoming of the calculated phase diagram is our prediction that solid solution of Ti in Al is stable over an extended composition range at low temperatures while experimentally it is found that Ti is soluble only up to a maximum of 2 at. % below the melting point of Al [12]. We have found that this discrepancy can be explained by the fact that several ordered superstructures with compositions between 75 and 100 at. % Al are predicted by our ECI's to be lower in energy than phase separation between fcc Al and the $D0_{22}$ $TiAl_3$ compound. The $D0_{22}$ structure at 75 at. % Al is known to undergo significant structural relaxations, and it is likely that local atomic relaxations in the Al-rich solid-solution give rise to a large elastic energy term which is not well described in our calculations. More FLMTO calculations would be needed to understand better this portion of the phase diagram.

DISCUSSION

The dependence of the alloy properties on the atomic configuration for compounds with fixed Al concentrations on a given lattice, as shown in figures 1 (a)-(d), is indicative of an interesting relationship between the state of atomic order and the electronic structure of Ti-Al alloys. In particular, for fcc-based $TiAl_3$ compounds the $D0_{22}$ and $L1_2$ structures give rise to nearly identical bulk moduli and to values of ΔH and Ω which are very similar when compared to those of the random fcc alloy. Hong et al. [17] calculated the electronic structures and total energies of $D0_{22}$ and $L1_2$ $TiAl_3$ compounds (the $D0_{19}$ structure was also considered) and it was found that the detailed nature of the charge distributions (bonding) and the features of the total electronic densities of states (dos) were relatively similar. By contrast, the dos of the random fcc alloy for $TiAl_3$, as calculated using the coherent potential approximation and the LMTO method [18], does not show the characteristic pronounced pseudo-gap near the Fermi level [17] which is found in the dos for the $D0_{22}$ and $L1_2$ structures [19]. For both the $D0_{22}$ and $L1_2$, one half of the nn bonds are between the majority species while the other half are between unlike atoms, no nn bonds are found between the minority component (Ti in the case of $TiAl_3$); these structures are distinct only when next-nearest-neighbor atomic arrangements are considered. Therefore, it appears that in order to significantly alter the electronic structure of the $TiAl_3$ compound, it is necessary to stabilize phases with more disordered arrangements of atoms in the nn shell than are found in both the $D0_{22}$ and $L1_2$ structures. This observation may be relevant to alloy designers seeking to synthesize ductile Al-rich Ti-Al intermetallic compounds since Fu [20] has argued that the brittleness of $L1_2$-based $TiAl_3$ alloys can be understood based on the electronic structure of these compounds.

From the results of the FLMTO total energy calculations shown in figure 1 (a), the $L1_2$ is found to be competitive with both the $D0_{22}$ and $D0_{19}$ structures. Furthermore, the calculated phase diagram, figure 2, shows that for the Ti_3Al ($TiAl_3$) compound, the tendency to form an $L1_2$ phase increases as the temperature is raised and the concentration of Al is increased (decreased). Although the calculated phase boundaries in figure 2 are sensitive to small inaccuracies in the structural energy differences as calculated by the FLMTO and represented by the ECI's, the tendency to stabilize the $L1_2$ with respect to the $D0_{22}$ and $D0_{19}$ structures as the concentration of Al approaches the equiatomic value is a robust result which agrees with experimental findings for Al-rich alloys.

In particular, the presence of the phase with an $L1_2$ structure in the calculated phase diagram for Al concentrations between 50 and 75 at. % Al is consistent with the fact that a variety of fcc-based LPS are observed experimentally in this composition range [15,16]. The $D0_{22}$ can be obtained from the $L1_2$ structure by periodically introducing <1/2 1/2 0> antiphase boundaries (APB's) on each {001} plane. Other LPS can be obtained by periodically arranging such APB's in different ways in an $L1_2$ structure [15]. From the calculated region of stability for the Al-rich $L1_2$ and $D0_{22}$ structures in figure 2, we can conclude that the APB energy is initially negative, increases and becomes positive as the concentration of Al is decreased. When the APB energy goes to zero, an infinite number of LPS (differing only by the number and arrangement of APB's) have an energy which is degenerate with the $D0_{22}$ and $L1_2$ structures; this degeneracy is lifted and specific LPS can be stabilized by long-ranged interactions (eighth nearest neighbor or greater in the case of the fcc lattice) [21]. Furthermore, it has been found that when Fe, Ni, Cu, Mn, or Cr is substituted for Al near $TiAl_3$, an $L1_2$ structure is stabilized [22], and the width of the phase field has been shown to decrease and vanish as the temperature is lowered for Fe and Ni substituted alloys [23]. Therefore, it can be concluded that raising the temperature and the concentration of *d*-electrons acts to stabilize the $L1_2$ with respect to the $D0_{22}$ structure for Al-rich alloys, in agreement with the calculated phase diagram.

The calculated region of stability for the Ti-rich $L1_2$ structure arises from the very small energy difference between $D0_{19}$ and $L1_2$ Ti_3Al. This energy difference is found to decrease with the addition of Al and the $L1_2$ structure becomes stable with respect to the $D0_{19}$ for $c \approx 35$ at. % Al; upon further additions of Al, the $L1_0$ structure becomes lower in energy. No experimental evidence for a Ti-rich $L1_2$-based phase appears to exist in the literature. Most experimental studies of ternary additions to Ti-rich alloys have been devoted to examining the effect of transition metals such as Nb which act to stabilize bcc-based phases [24]. It would, therefore, be interesting to add elements which enhance the formation of fcc phases to see if the $L1_2$ structure can be observed for Ti-rich Ti-Al alloys.

Although the general topology of the calculated phase diagram shown in figure 2 agrees well with the assessed experimental diagram [12], the theoretical value of the order-disorder temperature for the $D0_{19}$ Ti_3Al is about 600 ^{o}C too high. This discrepancy is probably at least partly due to the neglect of the contributions of vibrational and electronic excitations to the free energy in our calculations. Furthermore, local atomic relaxations in the disordered hcp and fcc phases are likely to be important in this system due to the large heats of formation of the compounds, implying strong Ti-Al bonding. Although the effect of *structural* relaxation on the alloy energies has been included in the determination of the ECI's, a proper treatment of *local* atomic relaxations in the disordered phase requires that the contribution to the energy of long-ranged interactions, which have been neglected in this study, be included [25]. Although the local atomic relaxations and vibrational and electronic excitations will affect the calculated temperatures of the transformations, we expect that the topology of the phase diagram is dominated by chemical effects, and will remain relatively unchanged.

ACKNOWLEDGEMENTS

The research at the University of California was supported by the Director, Office of Energy Research, Office of Basic Energy Sciences, Materials Sciences Division of the U.S. Department of Energy under Contract No. DE-AC03-76F00098 and by the Institute for Scientific Computing Research at the Lawrence Livermore National Laboratory, Livermore, CA. We would also like to thank Dr. Marcel Sluiter for numerous helpful comments and suggestions.

REFERENCES

1. R. W. Cahn, MRS Bulletin 16, 18 (1991); R. L. Fleischer, D. M. Dimiduk and H. A. Lipsitt, Ann. Rev. Mater. Sci. 19, 231 (1989).
2. M. Asta, D. de Fontaine, M. van Schilfgaarde, M. Sluiter and M. Methfessel, Phys. Rev. B 46, 5055 (1992).
3. M. Methfessel, Phys. Rev. B 38, 1537 (1988).
4. O.K. Andersen, O. Jepsen and D. Glötzel, in Highlights of Condensed Matter Theory, edited by F. Bassani et al. (North Holland, Amsterdam 1985).
5. D. M. Nicholson, G. M. Stocks, W. M. Temmerman, P. Sterne and D. G. Pettifor, in High Temperature Ordered Intermetallic Alloys III, edited by C. T. Liu, A. I. Taub, N. S. Stoloff and C. C. Koch (Mater. Res. Soc. Proc. 133, Pittsburgh, PA 1989) pp. 17-22.
6. Prabhakar P. Singh, M. Asta, D. de Fontaine and M. van Schilfgaarde, in Alloy Phase Stability and Design, edited by G.M. Stocks, D.P. Pope and A.G. Giamei (Mater. Res. Soc. Proc. 186, Pittsburgh, PA 1991) pp. 41-46.
7. M. Asta, C. Wolverton, D. de Fontaine and H. Dreyessé, Phys. Rev. B 44, 4907 (1991).
8. R. Kikuchi, Phys., Rev. 81, 988 (1951); J. M. Sanchez, F. Ducastelle and D. Gratias, Physica 128A, 334 (1984)..
9. J. M. Sanchez and D. de Fontaine, Phys. Rev. B 17, 2926 (1978).
10. D. Gratias, J.M. Sanchez and D. de Fontaine, Physica 113A, 315 (1982).
11. R. McCormack, M. Asta, D. de Fontaine and G. Ceder, submitted to Phys. Rev. B.
12. U.R. Kattner, J.-C. Lin and Y.A. Chang, Met. Trans. A 23A, 2081 (1992).
13. O. Kubaschewski and W. A. Dench, Acta Metall. 3, 339 (1955); O. Kubaschewski and G. Heymer, Trans. Faraday Soc. 56, 473 (1960).
14. P. Villars and L. D. Calvert, Pearson's Handbook of Crystallographic Data for Intermetallic Phases (American Society for Metals, Metals Park, OH, 1985).
15. A. Loiseau, G. Van Tendeloo, R. Portier and F. Ducastelle, J. Physique 46, 595 (1985).
16. J.C. Schuster and H. Ipser, Z. Metallkd. 81, 389 (1990).
17. T. Hong, T. J. Watson-Yang, A. J. Freeman, T. Oguchi and Jian-hua Xu, Phys. Rev. B 41, 12462 (1990).
18. Prabhakar P. Singh, D. de Fontaine and A. Gonis, Phys. Rev. B 44, 8578 (1991).
19. M. Asta and Prabhakar P. Singh, 1992 (unpublished).
20. C.L. Fu, J. Mater. Res. 5, 971 (1990).
21. D. de Fontaine and J. Kulik, Acta Metall. 33, 145 (1985).
22. A. Raman, K. Schubert, Z. Metallk. 56, 40 and 99 (1965); P. Virdis, U. Zwicker, Z. Metallk. 62, 46 (1971); H. Mabuchi, K. Hirakawa, H. Tsuda, Y. Nakayama, Scripta Met. 24, 505 (1990); H. Mabuchi, K. Hirukawa, Y. Nakayama, Scripta Met. 23, 1761 (1989)
23. S. Mazdiyashi, D.B. Miracle, D.M. Dimiduk, M.G. Mendiratta, and P.R. Subramanian, Scripta Met. 23, 327-331 (1989).
24. J.H. Perepezko, Y.A. Chang, L.E. Seitzman, J.C. Lin, N.R. Bonda, T.J. Jewett, and J.C. Mishurda, in High Temperature Aluminides and Intermetallics, edited by S.H. Whang. C.T. Liu, D.P. Pope, and J.O. Stiegler (Min., Met. and Met. Soc., 1990), and references therein.
25. D.B. Laks, L.G. Ferreira, S. Froyen and A. Zunger, submitted to Phys. Rev. B.

PHASE STABILITY OF $MoSi_2$ WITH Cr ADDITIONS

P. S. Frankwicz, J. H. Perepezko, and D. L. Anton*. Dept. of Materials Science and Engineering, University of Wisconsin-Madison, 1509 University Ave., Madison WI. 53706 USA; *United Technologies Research Center, East Hartford CT. 06108 USA.

ABSTRACT

The phase stability of $MoSi_2$ with Cr additions has been investigated in order to explore the issues of ternary solubility and structural stability of $MoSi_2$. The solidification microstructure of $MoSi_2$-rich alloys, along the $MoSi_2$-$CrSi_2$ ternary section, displays a two phase mixture of primary $MoSi_2$ ($C11_b$) and intercellular ternary $CrSi_2$ (C40). The development of the phase equilbria between the $C11_b$ and C40 disilicides, as observed in this system, is characteristic of a broad class of intersilicide reactions involving $MoSi_2$. The issues of chemical reactivity and structural stability of $MoSi_2$ composite designs underscores the importance of phase equilibria investigations. The solubility of Cr in annealed $MoSi_2$ was observed to be on the order of 3 atomic percent. Past studies demonstrated that Ti and Ta have limited solution in $MoSi_2$; the minor solubility of Cr in $MoSi_2$ corroborates the trend of limited solubility of transition metals in the $MoSi_2$ ($C11_b$) structure. The relatively small changes in the lattice parameters of $MoSi_2$ with Cr additions point to an inability of the $C11_b$ disilicide structure to accommodate the lattice perturbation resulting from solute atoms. The observations of this investigation suggest that the phase stability of $MoSi_2$ is primarily controlled by geometrical factors.

INTRODUCTION

Current investigations of high temperature $MoSi_2$ based intermetallic matrix composites are directed toward the development of multiphase designs with enhanced mechanical properties while maintaining the superior oxidation resistance of $MoSi_2$ [1]. The viability of $MoSi_2$-based composites during high temperature service will depend on the critical understanding of interfacial reactions and intersilicide compound formation. To achieve this goal, the ternary phase equilibria and miscibility relationships, which govern the structure and chemical reactivity of $MoSi_2$ in a multiphase environment, must be established. The phase reaction sequence of $MoSi_2$ with several transition metals (*i.e.* TM = Nb, Cr, Ti and Ta) involves the formation of ternary silicide complexes $(Mo,TM)Si_2$ and $(Mo,TM)_5Si_3$ [2]. The ternary alloying of $MoSi_2$ with transition metal additions may add some degree of solid solution strengthening and effect microstructural toughing *via* methods for second phase precipitation within the $MoSi_2$ matrix. In the present study, the Mo-Cr-Si system has been chosen to investigate the solubility of Cr in $MoSi_2$ and the associated ternary phase equilibria in the vicinity of the $MoSi_2$-$CrSi_2$ section.

The Mo-Cr Si ternary system has been investigated by Nowotny [3] and Ageev [4] and an isothermal section at 1300°C is shown in figure 1. Nowotny *et al.* has reported, based on the analysis of annealed (1300°C 20h) powder compacts, $MoSi_2$ and $CrSi_2$ display mutual solubility and are separated by a two phases field ($C11_b$ + C40) along the $MoSi_2$-$CrSi_2$ section. The unit cell volume of $MoSi_2$ was observed to decrease from 0.080 nm^3 to 0.077 nm^3 for $MoSi_2$ with 13 atomic percent Cr. The reciprocal solubility limits for annealed (1300°C 20h) powder compacts along the $MoSi_2$-$CrSi_2$ ternary section were observed to be 16 atomic percent Cr in tetragonal $MoSi_2$ and 10 atomic percent Mo in hexagonal $CrSi_2$ structure [3]. Ageev *et al.* have reported an extensive solubility of 15 atomic percent Cr in $MoSi_2$ at 1300°C and the existence of a two phase field between $MoSi_2$ and $(Cr,Mo)_5Si_3$

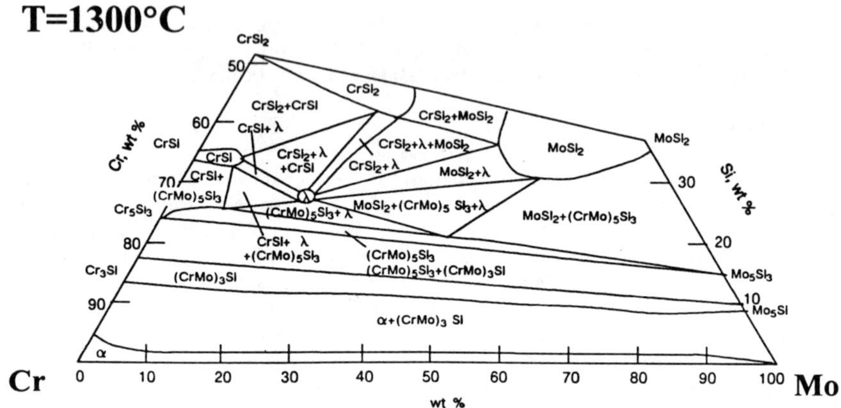

Figure 1: A partial isothermal section of the Mo-Cr-Si section at 1300°C according to Ageev [4]. The structures of the phases $MoSi_2$ and $CrSi_2$ are $C11_b$ and C40, respectively.

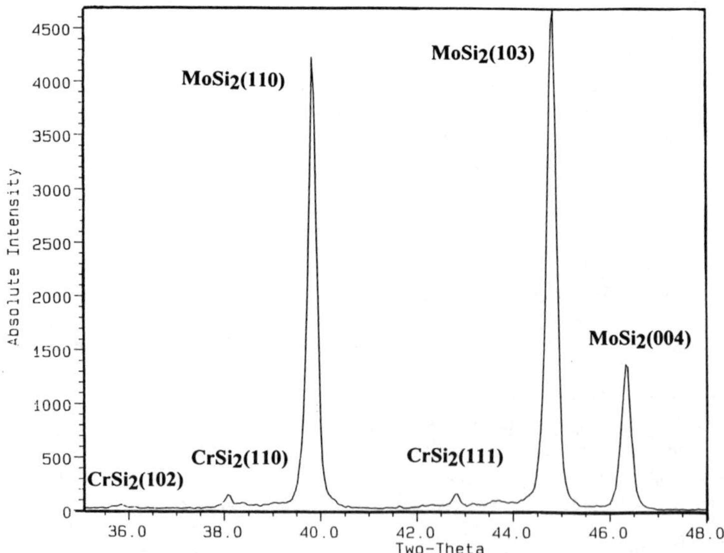

Figure 2: Powder x-ray diffraction pattern of 05Cr-28Mo-67Si alloy which has been annealed 100h at 1600°C.

silicide complexes. The liquid to solid phase equilibria in the Mo-Cr-Si ternary system is not well understood. In investigations of the mechanical properties of $(Mo_{0.97}Cr_{0.03})Si_2$ single crystals, the improvement of yield stress during compression testing (1300°C< T <1500°C) compared to $MoSi_2$ has been observed [5]. These findings highlight the possibility of the modification of mechanical properties of $MoSi_2$ through ternary alloying .

EXPERIMENTAL PROCEDURES

The Cr-Mo-Si alloys were prepared using non-consumable electrode arc melting. Research grade Si (99.9999%), Cr (99.99%) and Mo (99.95%) were used as starting materials. Stoichiometric mixtures were arc melted in Ti-gettered high purity Ar; the complete ingots were remelted 3 to 5 times to promote homogenization. The alloys were annealed in a $MoSi_2$ furnace using Ti-gettered high purity Ar processing atmospheres. All alloy compositions are reported in atomic percent.

Powder X-ray diffraction (XRD) analysis with Cu Kα radiation has been used for crystalline phase identification and determination of lattice parameter constants. The space group nomenclatures for the $C11_b$ and C40 *Strukturbericht* designations are I4/mmm and $P6_222$, respectively. The microstructures of mechanically polished wafers were examined with optical microscopy and scanning electron microscopy (SEM) in secondary and backscattered electron imaging modes. Chemical composition analysis was obtained using energy dispersive spectroscopy (EDS) with standards in the SEM.

RESULTS AND DISCUSSION

The solidification microstructures, compositions and crystal structures of ternary alloys in the vicinity of $MoSi_2$ and $CrSi_2$ have been examined in order to assess the phase selection and equilibria of ternary disilicides. The x-ray powder diffraction patterns from arc melt processed and high temperature annealed $MoSi_2$-rich alloys (figure 2) display a two phase x-ray diffraction pattern of $MoSi_2$ ($C11_b$) + $CrSi_2$ (C40). A summary of lattice parameters for the $C11_b$ and C40 crystalline phases is presented in Table I. The solidification morphology of arc melt processed 05Cr-28Mo-67Si displays a two phase mixture of ($MoSi_2$ + $CrSi_2$). Scanning electron micrographs of arc melt processed 05Cr-28Mo-67Si (figure 3a) exhibit light contrast dendrites of $MoSi_2$ surrounded by an interdendritic ternary $CrSi_2$. The arc melt processed solidification morphology of $MoSi_2$ displays conformal interfaces indicative of prior liquid/solid interfaces and would suggest the $MoSi_2$ dendrites have formed as a primary solid. The level of Cr solubility in $MoSi_2$ was observed to be 3 atomic percent after thermal annealing.

The $CrSi_2$ terminus side features the formation of a hexagonal ternary disilicide with relatively extensive Mo solution in comparison to the reciprocal solubility behavior exhibited in $MoSi_2$-rich alloys. The chemical analysis of annealed $CrSi_2$-rich alloys (1000°C 400h and 1375°C 98h) would indicate that a continuous series of solid solutions exists between binary $CrSi_2$ and the ternary composition of 17Cr-16Mo-67Si. Therefore, binary $CrSi_2$ exhibits relatively large solubility for Mo. The composition analysis of the solidification microstructures of 25Cr-07Mo-68Si and 29Cr-03Mo-68Si alloys (figure 3b) point to a complex solidification pathway which involves the formation of a primary solid $(Cr_x,Mo_{0.33-x})Si_2$, where $0.33 \gtrsim x \gtrsim 0.16$, with cellular morphology surrounded by intercellular Si and ($CrSi_2$ + Si) eutectic structure. The solubility of Mo in the annealed $CrSi_2$ ternary phase was observed to be on the order of 16 atomic percent Mo.

This investigation of the response of $MoSi_2$ to ternary additions of Cr corroborates the trend of limited solubility of transition metal additions in the $MoSi_2$ ($C11_b$) crystalline phase. The phase equilibria and lattice parameter trends observed in the $TiSi_2$-$MoSi_2$ and $CrSi_2$-

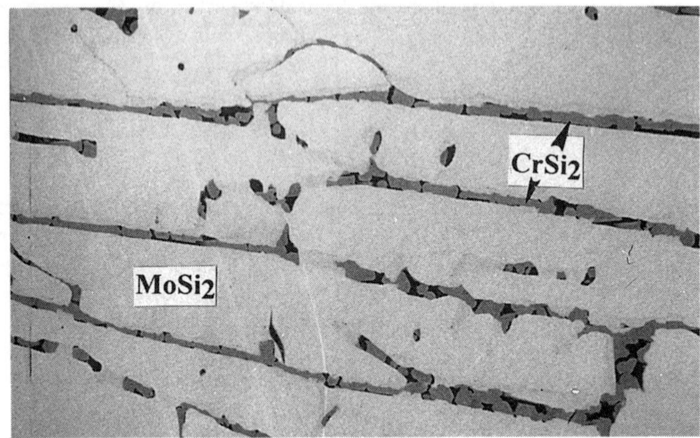

20μm

Figure 3a: Backscattered electron micrograph of as-cast 05Cr-28Mo-67Si. The solidification microstructure displays $MoSi_2$ ($C11_b$) dendrites (light contrast), interdendritic $CrSi_2$ (C40) and Si (dark pores).

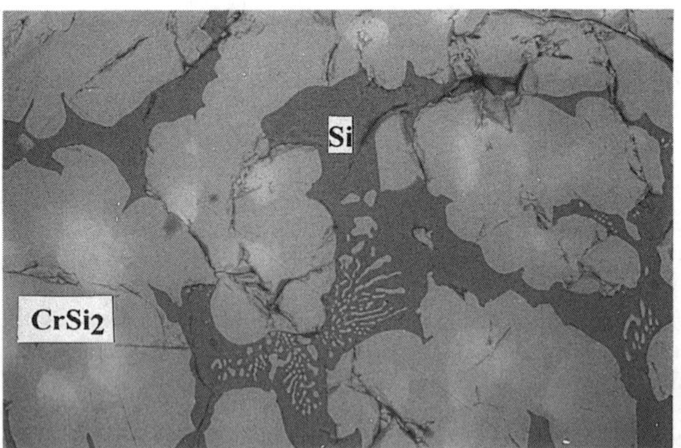

100μm

Figure 3b): Backscattered electron micrograph of 25Cr-07Mo-68Si after cooling at 20°C min^{-1} from the liquid. The solidification microstructures displays ternary $CrSi_2$ cells (light contrast), intercellular Si and areas of ($CrSi_2$ + Si) eutectic morphology.

TABLE I [1]
Lattice Parameter Summary [2]

	a_o(nm)	c_o(nm)	c_o/a_o
$C11_b$-Phase			
$MoSi_2$*	0.3207	0.7855	2.449
30Mo-03Cr-67Si ●	0.3191	0.7816	2.449
32Mo-01Ti-67Si■	0.3205	0.7856	2.451
WSi_2∅	0.3214	0.7833	2.437
28W -05Cr-67Si▲	0.3198	0.7804	2.440
32W -01Ti-67Si▲	0.3204	0.7833	2.445
C40-Phase			
$CrSi_2$∅	0.4429	0.6372	1.439
15Mo-20Cr-65Si●	0.4457	0.6459	1.449

UW

Heat treatments of samples:
* annealed 1600°C 100h; ■ annealed 1400°C 100h;
● annealed 1375°C 98h; ▲ annealed 1350°C 125h; ∅ as cast.

[1] WSi_2-based alloys included from a study of this investigation.
[2] lattice parameter determination error did not exceed (± 0.0002 nm).

TABLE II
Transition Metal Solubility in $MoSi_2$

Solute	Solubility Level	Goldschmidt Radius (CN=10)	Reference
Cr	3.0 at. %	0.127nm	[this study]
W	(complete)[1]	0.139nm	[3]
Mo		0.139nm	
Ti	1.5 at. %	0.144nm	[6]
Ta	1.0 at. %	0.145nm	[7]

UW

[1] The WSi_2-$MoSi_2$ ternary section has been reported to be a series of solid solutions [3].

$MoSi_2$ ternary systems would suggest a simple substitution defect mechanism of transition metal solute on the Mo sublattice of the $C11_b$ structure. Table II summarizes the observed levels of transition metal solution in $MoSi_2$ and correlates these solubility trends with interpolated solute Goldschmidt radii for a coordination number of ten. This coordination number reflects the Mo site nearest neighbor environment of the {110} $C11_b$ structure. The atomic size correlation would suggest that a transition metal solute that possesses a smaller Goldschmidt radius than Mo (R_{cn10} = 0.139nm) will have an greater solubility in $MoSi_2$ compared to those solute atoms with a radius equal to or larger than Mo. A preliminary investigation of Al additions to $MoSi_2$ would suggest that alloying additions which may perturb the Si sublattice of the $C11_b$ structure are subject to similar solute size effects as the Mo-sublattice. The importance of the solute size factor is further substantiated by the similar solubility and lattice parameter trends exhibited by the homotypic disilicide $WiSi_2$ ($C11_b$). In a parallel study, the solubility of Ti and Cr in annealed WSi_2 ($C11_b$) were observed to be 1 at.% and 5 at.%, respectively.

These observations of solubility trends with transition metal additions to $MoSi_2$ have served to raise fundamental issues regarding the phase stability of the $C11_b$ crystalline prototype. The $C11_b$ crystalline prototype displays a pronounced bimodal population distribution of the lattice parameter ratio (c_0/a_0) centered about the values of 2.45 and 3.50. The mode at $c_0/a_0 \cong 3.50$ is primarily composed of noble metal / transition metal pairs such as $CuZr_2$ and $AuHf_2$. The refractory disilicides $MoSi_2$, WSi_2 and $ReSi_2$ reside at lower ratio (c_0/a_0) value of 2.45. The relatively minor changes in the a_0 and c_0 lattice parameters of $MoSi_2$ and WSi_2 with small Cr and Ti additions (Table I) point to an inability of the $C11_b$ disilicide structure to accommodate the lattice perturbation resulting from different size solute atoms. Carlsson and Meschter have calculated (using augmented-spherical-wave method) a total-energy difference, $\Delta E = E(C11_b) - E(C40)$, for $MoSi_2$ to be on the order of $\Delta E = -0.1$ kJ/g atom [8]. A phase stability criterion of $MoSi_2$ based solely on electronic effects would be tenuous in view of the relatively small $MoSi_2$ structural-energy difference. The phase stability of $MoSi_2$ ($C11_b$) structure appears to be closely related to maintaining the lattice parameter ratio (c_0/a_0) at the characteristic value of 2.45, which optimizes the geometry of the {110} planes , along the [001] and [331] directions, that are involved in the body centered cubic stacking sequence. The underlying geometrical factors of the phase stability of the $C11_b$ disilicide structure would predict a limited potential for the substitutional ternary alloying of $MoSi_2$ with transition metals.

CONCLUSIONS

The experimental results corroborate the trend of limited solubility of transition metals (group IVB-VIB) in the $MoSi_2$ ($C11_b$) structure. The solubility of Cr in $MoSi_2$ was observed to be limited to 3 atomic percent. The solidification morphology of $MoSi_2$ indicates that it has formed as a primary solid from the liquid. The lattice parameter ratio (c_0/a_0) at the value of 2.45 apparently represents a characteristic phase stability criterion for $MoSi_2$ ($C11_b$), which is unaffected by the ternary alloying of $MoSi_2$ with Ti and Cr. The current investigation suggests the phase stability of $MoSi_2$ is primarily controlled by geometrical factors.

PSF and JHP acknowledge the support of DARPA/ARO (DAAL 03-90-G-0183) for this research.

REFERENCES:

1. High Temperature Structural Silicides, edited by A. K. Vasudevan and J. J. Petrovic (Elseviers, Amsterdam, 1992).
2. J. J. Petrovic, R. E. Honnell and A. K. Vasudevan, in Intermetallic Matrix Composites, edited by D L. Anton, P. L. Martin, D. B. Miracle and R. McMeeking (Materials Research Society, Symp. Proc. Vol. 194, Pittsburgh, 1990) pp. 123-130.
3. H. Nowotny, R. Kieffer and H. Schachner, Mh. Chem. 83, 1243 (1952).
4. Diagrammy Sostoianiia Metallicheskikh Sistem: Vol. 10, edited by N. V. Ageev (Viniti Press, Moscow, 1964) p.182.
5. Y. Umakoshi, T. Hirano, T. Sakagami and T. Yamane, in High Temperature Aluminides and Intermetallics, edited by S. H. Whang, C.T. Liu, D. P. Pope and J. O. Stiegler (The Minerals, Metals and Materials Society, Warrendale, PA, 1990) pp. 111-129.
6. P. S. Frankwicz and J. H. Perepezko, in High Temperature Ordered Intermetallics Alloys IV, edited by L. A. Johnson, D. P. Pope and J. O. Stiegler (Materials Research Society, Symp. Proc. Vol. 213, Pittsburgh, 1991) pp. 169-174.
7. W. J. Boettinger, J. H. Perepezko and P. S. Frankwicz, Mat. Sci. and Engin. A155, 33 (1992).
8. A. E. Carlsson and P. J. Meschter, J. Mater. Res. 7, 1512 (1991).

SOLID SOLUTION HARDENING OF INTERMETALLIC COMPOUNDS

ROBERT L. FLEISCHER
Rensselaer Polytechnic Institute, Troy, NY 12180-3590 (formerly General Electric Research & Development Center, Schenectady, NY 12301)

ABSTRACT

Solid solution hardening is sought and observed in two B2(cP2) CsCl and two C15(cF24) Cu_2Mg type structures. New data are presented for the high-temperature compounds AlCo, AlRu, and Cr_2Zr; prior data of Livingston on Cu_2Mg are analyzed. From lattice parameters and specific gravities, cell occupancy numbers have been measured and used to infer likely defect types and concentrations. The effects of constitutional defects in binary compounds correlate with traditional substitutional solution hardening from elastic interactions due to size differences and modulus differences. The hardening from ternary solutes is similar in magnitude, but the defect structures are complicated and not yet adequately understood.

INTRODUCTION

Hardness and strength can be attained in pure metals by solid-solution hardening. Although the primary mechanisms are thought to be known in dilute metal alloys, the mechanisms in intermetallic compounds are not established for most structure types. The mere existence of solution hardening is not widely demonstrated. Prior work however, includes data for the B2 compound AgMg [1], the $L1_2$ structure Ni_3Al [2], and preliminary results for the C15 cubic Laves phase Cu_2Mg [3,4]. The new work that is summarized here is on the B2 compounds AlCo [5] and AlRu [6], and a C15 phase, Cr_2Zr [7].

We wish to improve understanding of solid-solution hardening in intermetallics (in contrast to dilute alloys) by testing whether the same elastic interactions between solute atoms and dislocations dominate, but with the effective solute concentration increased in intermetallics by the substitutions or vacancies that are created by deviations from stoichiometry. To make this test, we wish first to identify the sites of constitutional solutes in off-stoichiometric intermetallics, and then to quantify the elastic interactions with dislocations of these solutes and of ternary additions. Data needed are the changes in lattice parameter, density, and shear modulus that accompany alloying and off-stoichiometry. Knowledge of the density and lattice parameter indicate whether the excess of one constituent is accommodated by substitution in the other sublattice or by creation of vacancies.

A special feature and complication, of intemetallics is that off-stoichiometry leads to constitutional defects--substitutions or vacancies that are created by deviations from stoichiometry. Thus, in order to understand hardening, we need to identify the sites of constitutional solutes in off-stoichiometric intermetallics and to quantify the elastic interactions with dislocations both of these solutes and of ternary additions.

Hardening by substitutional solute is generally of what is called the gradual hardening type, usually due to two mechanisms: Atoms that alter the lattice parameter b have effectively a different size from the matrix atoms [8] and produce hardening that increases with the size misfit ε_b(=d(lnb/dc), where c is the concentration of solute). Atoms that alter the shear modulus G are regarded as tiny inclusions of different elastic moduli [9], which produce hardening that increases with ε_G (=d(lnG)/dc). For quantitative completeness it is necessay to include both interactions and appropriately sum them to predict hardening and infer the relative strengths of the two effects [10].

DEFECT STRUCTURES AND INTERACTION PARAMETERS

Numerous data on binary and ternary compounds have been measured for lattice parameters, elastic moduli, and specific gravities for AlCo and AlCo plus Mn, Re, or Ti; AlRu and AlRu plus Co, Fe, or Ti; and Cr_2Zr and Cr_2Zr plus Mo or Hf[5-7]. They allow interaction parameters ε_b and ε_b to be found, and cell occupancy numbers N_c to be calculated.

In some cases the variations of the lattice parameter b alone implies the nature of defects. For example, Fig. 1 shows b for Cr_2Zr as a function of off-stoichiometry or of substitution of Hf for Zr or Mo for Cr in stoichiometric Cr_2Zr. The effect of Mo is indistinguishable from that of excess Zr, and neither excess Cr nor substituted Hf alter b measurably. These effects (or lack of effects) were taken as evidence of simple substitutions in a lattice that is controlled by the size of atoms in the Cr sub-lattice [7].

In binary AlCo and AlRu the cell occupancy behaviors are informative (Figs. 2 and 3) [5,6]. As in other B2 alummides, Al-rich alloys tend to compensate for the excess by creating vacancies in the other sublattice [11]. The decrease in N_c for Al-rich AlCo fits precisely Co-vacancy predictions (Fig. 2). For Co-rich alloys most excess Co is inferred to reside in the Al sublattice, but vacancies are also produced--possibly quenched in during cooling after annealing at 1350°C. The AlRu case (Fig. 3) is less clear. Vacancies also occur for Al-rich compositions, but apparently fewer than one per excess Ru atom. Single-phase Ru-rich AlRu alloys do not exist.

Ternary alloy behavior is complex and diverse. As Figs. 4 and 5 show, N_c can be constant or can either decrease or increase. The simplest cases are for substitutions in originally stoichiometric alloys (the solid dots in the two figures). Thus substitutions of Mn, Re, and Ti in AlCo and of Co in AlRu produce vacancies, while Fe and Ti in AlRu leave N_c essentially unchanged. Without a clear understanding of the defects in ternary alloys, quantitative correlation with hardening effects is elusive.

SOLUTION HARDENING

Hardening by substitutional solutes is observed in AlCo[5], AlRu. [6] Cu_2Mg[3,4], and Cr_2Zr [7], as shown in Figs. 6-10 and Table 1.

B2(cP2)CsCl-type Alloys: Since most intermetallics are brittle at room temperature, elevated-temperatures are used in testing hardening in such materials. Fig. 6 shows hardening by constitutional defects in binary AlCo at 750°C, and Fig. 7 displays weakening at 950°C. The latter is expected at high

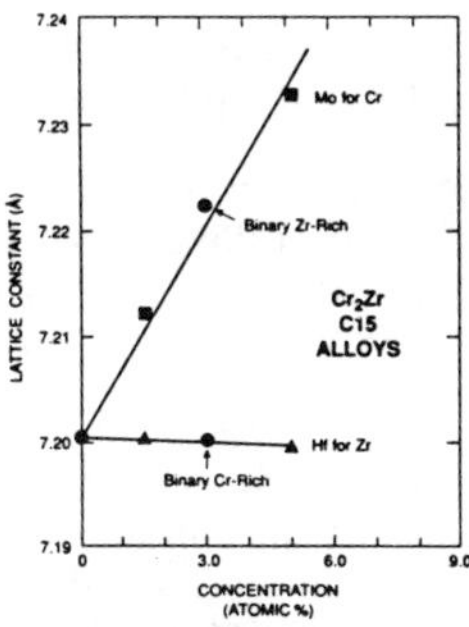

Fig. 1. Lattice parameter as a function of composition for Cr_2Zr alloys [8].

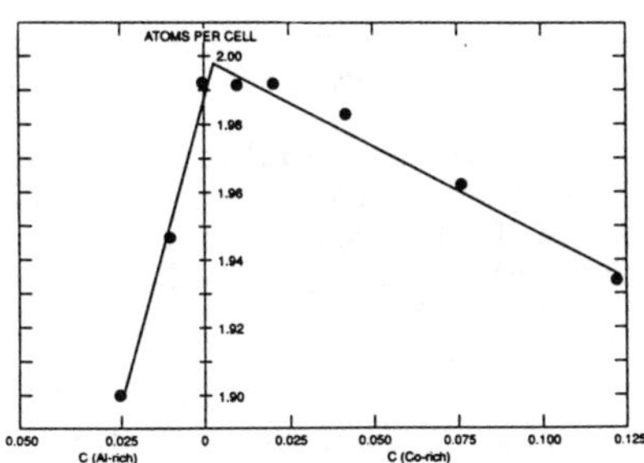

Fig. 2. Cell occupancy for binary Al-Co B2 alloys calculated from ρ and b as a function of composition. Each excess Al atom produces a vacant lattice site on the Co sub-lattice. Excess Co leads to increasing but lesser vacancy concentrations [6].

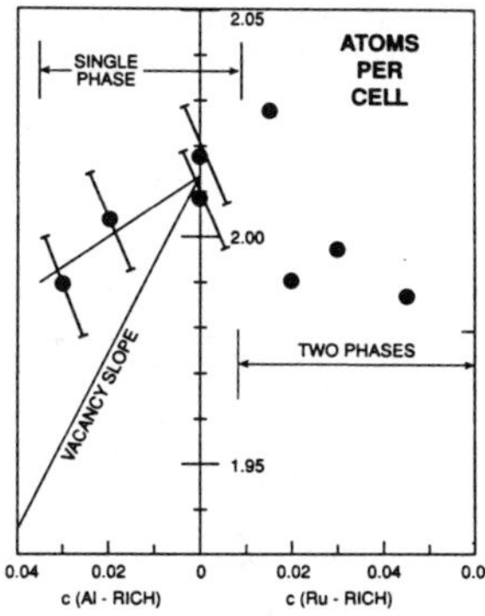

Fig. 3. Cell occupancy for binary Al-Ru B2 alloys calculated from ρ and b as a function of composition. Excess Al atoms produce vacancies, but at less than a one-for-one basis. Uncertainties arise from the specific gravities. Data in the two-phase region are formal results only [7].

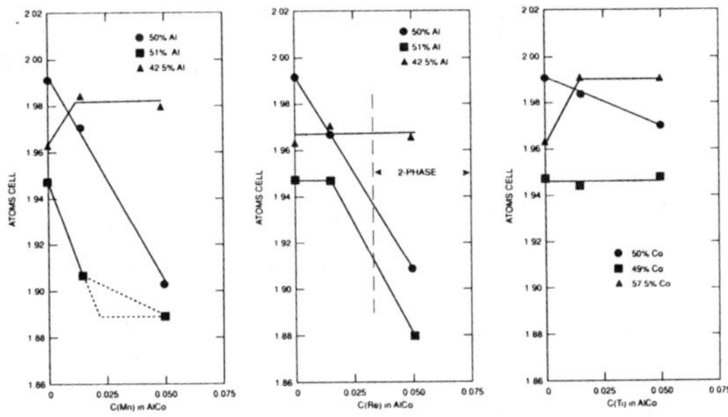

Fig. 4. Cell occupancy calculated for ternary Al-Co-(Mn, Re, Ti) alloys. For substitutions in the AlCo stoichiometric alloy, ternary elements produce vacant lattice sites. Data for 5% Re alloys are formal results only, since the structures are not single phase [6].

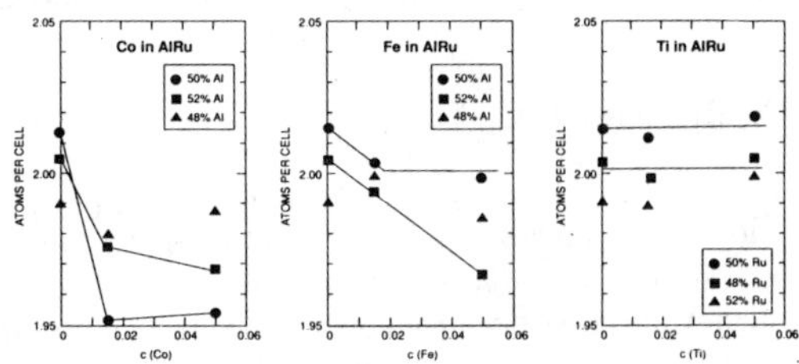

Fig. 5. Cell occupancy calculated for ternary Al-Ru-(Co, Fe, Ti) alloys. Data for the filled triangles are formal results only, since the structures are not single phase [7].

temperature from enhanced diffusion of defects produced by off-stoichiometry. This effect was seen and so interpreted for B2 AgMg by Westbrook [12].

AlRu is unusual in being plastic (in compression) at room temperature for many compositions. Fig. 8 shows the hardening inferred at 23°C, normalized to a strain of 7%. Because work hardening is great, such a normalization is needed to show effects clearly [6]. Except for Cu in 50% Al alloys, hardening is evident. Similiarly, Fig. 9 shows hardening by ternary solute in AlCo.

C15(cF24)Cu_2Mg-type alloys: In Cr_2Zr the maximum temperature available for compression testing was 950°C. It is not high enough to achieve plastic flow in all of the alloys (most clearly SSC-6 in Table I.) Microhardness provides a crude supplement to the compression data, as recorded in Table I. Amid the scatter it appears that hardening occurs from constitutional solute (alloys SSC-2 and SSC-3) and from ternary solute, Mo and Hf (alloys SSC-4 through SSC-7).

Data from Livingston el al. [3,4], replotted in Fig. 10[7], also show definite hardening effects from several ternary additions. The inset shows that the hardening does not correlate well with solely size effects, but are simply accounted for if valence differences are included. Remember that in copper, valence-difference effects were attributed to the changes in modulus they produced [10]. Thus the hardening in Cu_2Mg is consistent with elastic interactions that are due to the superposition of modulus and size differences.

Table I

Fracture Stress And Microhardness At 950°c

Sample	Stress (MPa)	Vickers Hardness (10 MPa)
SSC-1 $Cr_{67}Zr_{33}$	266	254
SSC-2 $Cr_{64}Zr_{36}$	276(a)	186
SSC-3 $Cr_{69}Zr_{31}$	314	400
SSC-4 $Cr_{65.5}Mo_{1.5}Zr_{33}$	232	313
SSC-5 $Cr_{62}Mo_{5}Zr_{33}$	280	277
SSC-6 $Cr_{67}Hf_{1.5}Zr_{31.5}$	106	416
SSC-7 $Cr_{67}Hf_{5}Zr_{28}$	277	479

(a) Yield stress; fractured at 225 MPa at 750°C.

COMPARISON OF HARDENING TO THAT OF DILUTE METAL ALLOYS

How do the stress increases compare to experience with substitutional alloys? An approximate relation [13] for the flow stress increment from solution hardening is $\sigma=Gc^{1/2}\varepsilon_s^{3/2}/\alpha$ where $\varepsilon_s = |\varepsilon_G - 3\varepsilon_b|$, and in copper $\alpha \approx 700$. The hardening results can most sensibly be compared to defect concentrations for binary alloys, where cell occupancies are best understood. The ratio of hardening on either side of stoichiometry in AlCo is in the corect proportions, and the absolute values fit the equation with $\alpha=240$. That ratio is not available in AlRu (since Ru-rich B2's do not exist), but the values for Al-rich compositions fit $\alpha \approx 150$. The data for Cr_2Zr scatter too much for quantitative evaluation, but are of roughly

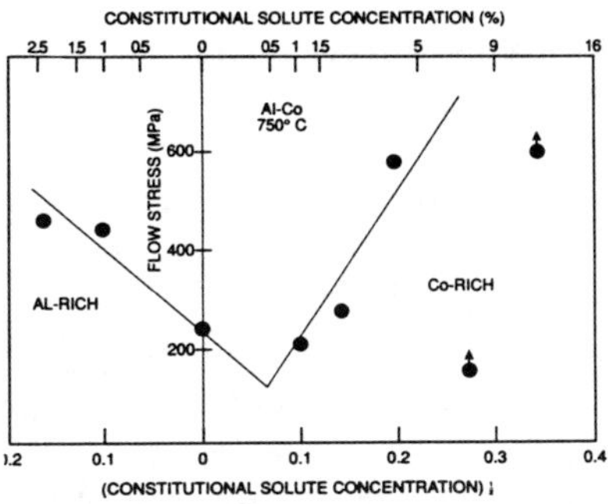

Fig. 6. Flow stress versus composition for binary Al-Co B2 alloys at 750°C. Results for brittle compositions are given as lower limits [6].

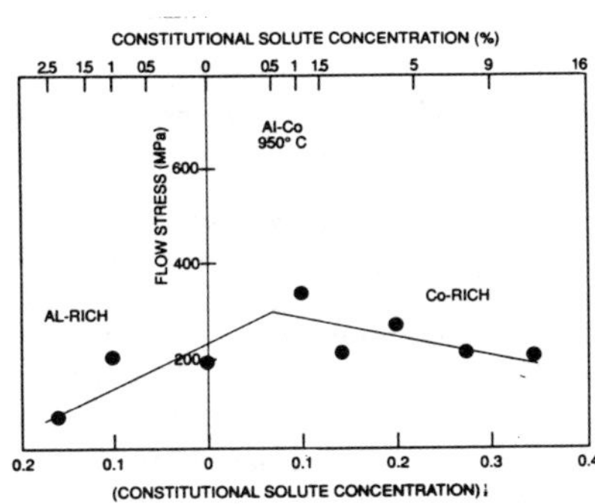

Fig. 7. Flow stress versus composition for binary Al-Co B2 alloys at 950°C [6].

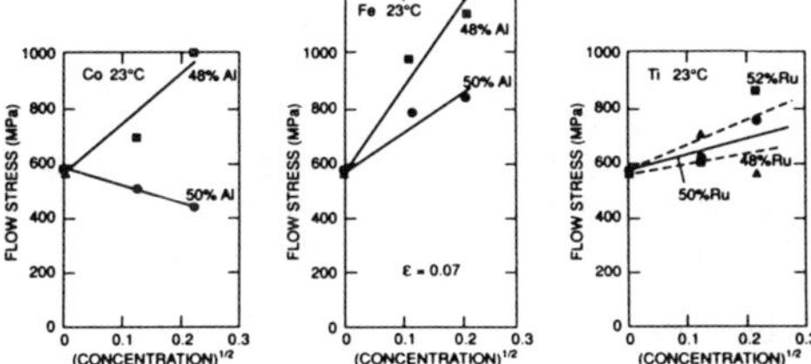

Fig. 8. Flow stress versus solute concentration at 23°C for AlRu alloys, adjusted to a strain of 0.07 [7].

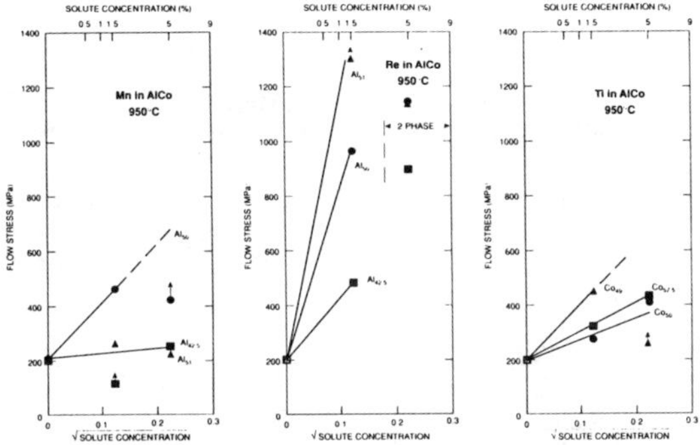

Fig. 9. Effects of Mn, Re, and Ti on the flow stress of $Al_{51}Co_{49}$, AlCo, and $Al_{42.5}Co_{57.5}$ at 950°C. Stresses for brittle failure are shown as lower limits [6].

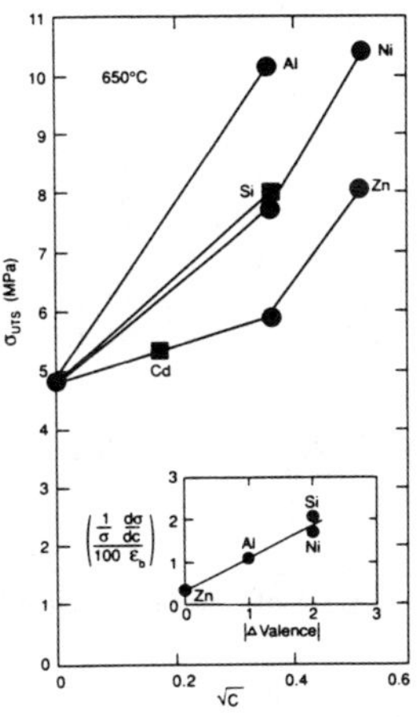

Fig. 10. Maximum stress in compression of Cu_2Mg and alloys at 650°C (from Livingston [3,4]. The inset shows that hardening per unit lattice parameter change varies, but correlates with the valence difference from Mg. The hardening rates are average initial slopes from the main figure [8].

similar magnitudes. For Cu_2Mg, ε_G has not been measured, so that comparison with predictions of the stress increment are not possible.

For ternary elements in AlCo the average hardening rates are similar ($\alpha \approx 160$), but correlations with the relative values of ε_s are garbled. Possible complications with ternary substitutions are legion, since often (but not always) they are accompanied by vacancies whose sublattice locations are unknown, as is whether they are paired or clustered, with or without the solute that led to their production. In AlRu the solutes produce more rapid hardening ($\alpha \approx 20$), but in consistent relative amounts for the measured ε_b and ε_G's. This low value of α can be interepreted as a hint that at least some of the defects are paired in such a manner as to produce stress dipoles - which interact strongly with dislocations. In contrast, the α values of 150 or 250 in comparison with 700 in Cu are thought to be due to the larger slip vectors that exist in the intermetallics, since they have larger unit cells.

CONCLUSIONS

Defects have been partially characterized in several intermetallic sytems and solid solution hardening observed in each. Although many gaps in our understanding remain, the hardening is consistent in magnitude with that expected from conventional elastic interactions between dislocations and atomic defects. The self-consistency of effects is good in binary alloys, but much remains to be done toward identifying the defects in ternary solution-hardening systems before thorough quantitative tests can be done.

ACKNOWLEDGEMENT

The beginnings of this work were supported by the Office of Naval Reserch under contract N00014-86-0353, work unit 431N002; it was completed with the support of the General Electric Research and Development Center.

References

1. D.L Wood and J.H. Westbrook, Trans, AIME 224, 1024 (1992).
2. Y. Mishima, S. Ochiai, H. Hamao, M. Yodogawa, and T. Suzuki, Japan. Inst. Met. 27, 648 (1986).
3. J.D. Livingston, E.L. Hall, and E.F. Koch, MRS Conf. Proc. 133, 243 (1989).
4. J.D. Livingston, "Deformation and Defects in Laves-Phase Intermetallic Compounds," Gen. Elect. Report 88CRD001 to the U.S. Dept. of Energy, Grant DE-FG02-87ER45296.A00(1987)
5. R.L. Fleischer, J. Mat. Res., 7 (Jan. 1993).
6. R.L. Fleischer, Acta Met. Mat., 41 (1993).
7. R.L. Fleischer, Scripta Met. Mat. 27, 799 (1992).
8. N.F. Mott and F.R.N. Nabarro, Proc. Phys. Soc. 52, 86 (1940).
9. C. Crussard, Metaux and Corrosion 25, 203 (1950).
10. R.L. Fleischer, Acta Met. 11, 203 (1963).
11. Y.A. Chang and J.P. Neumann, Prog. Solid State Chem. 14, 221 (1982).
12. J.H. Westbrook, J. Electrochem. Soc. 104, 369 (1957).
13. R.L. Fleischer, Chapter 3, pp. 93-162 in The Strengthening of Metals, D. Peckner, ed., Reinhold Press, New York (1964).

ENERGY OF PLANAR FAULTS AS A FUNCTION OF COMPOSITION IN BINARY AND TERNARY TiAl ALLOYS

C. Woodward, J. M. MacLaren* and D. M. Dimiduk**
UES, Inc., 4401 Dayton-Xenia Rd., Dayton, OH 45432
*Department of Physics, Tulane University, New Orleans, LA 70118
**Wright Laboratory, WL/MLLM, Wright Patterson AFB, OH 45433-6533

ABSTRACT

Establishing the chemical dependence of thermally activated processes which govern plasticity in intermetallic alloys requires that the dislocation dissociation reactions be determined as a function of composition. A major parameter governing such reactions is the relative fault stability as a function of composition. Here the results of first principles electronic structure calculations, using the layer Korringa-Kohn-Rostoker method, are reported for planar faults in γ TiAl at various compositions. The influence of dilute substitutional impurities on the fault energies is treated using the coherent potential approximation. The variation of fault energies as a function of binary composition (Ti_xAl_{1-x} where $52 \leq x \leq 49$) and the addition of transition metals (Cr, Mn and Nb at 2% concentration) are presented. The influence of this chemical dependence on the stability of <101] super-dislocations is discussed, along with expected trends in the flow stress behavior.

INTRODUCTION

TiAl is a promising candidate for aerospace applications, however practical applications of this material are limited by its poor strength and low ductility at room temperature.[1] While TiAl undergoes a brittle to ductile transition at 700 °C, the mechanism which mediates this transition is not well understood. Like some other high temperature aluminides the flow stress of TiAl increases with temperature, peaking at ~900 K.[2] In the related $L1_2$ alloys this flow behavior is thought to be controlled by the formation of dislocation glide barriers, whose creation is enhanced with increasing temperature.[3] Several dislocation barriers have been proposed for TiAl and the role of these dislocation configurations in determining flow behavior is subject to speculation.[4,5] Understanding and influencing the defects that control flow and fracture is critical in finding the practical limitations of this intermetallic alloy.

Early TiAl alloy development produced several materials with a ductility of 2%(Ti-52Al and Ti-52Al-1V),[6] and recently alloy ductility and strength have been improved by the addition of small concentrations of various ternary solutes such as Cr, Mn and V.[7,8,9] The dependence of material properties on composition has been rationalized by a number of heuristic arguments: changes in electronic structure, reduced covalent bonding (covalency), enhanced metallic bonding, changes in unit cell volume, reduced tetragonality, site occupancy, a decrease in the magnitude and anisotropy of the Peierls stress, twinning, a reduction of the stacking fault energies and an optimum presence of the two phases.[10] At the present time we have some empirical correlation but not a fundamental understanding as to how, or why, these changes in composition improve the materials properties of these alloys. For TiAl the most ductile materials are known to have significant 1/2<110] dislocations and twinning activity, while <101] dislocations are less active. The mobility and core structure of ordinary (1/2<110]) and superdislocations in TiAl is dictated in part by dislocation core stability. Variations in fault energies with composition must influence the core structure and modify activation energies for cross slip which depend on the constriction of partial dislocations. By similar mechanisms fault energies influence dislocation multiplication, storage and recovery processes. In this paper we investigate how chemistry influences the relative stability of possible glissile and sessile configurations of the <101] superdislocations in TiAl.

A number of sessile dislocation structures have been proposed by B. A. Greenberg et al. in order to explain the yield stress versus temperature peak in TiAl.[5] These barriers are formed by the splitting of <101] superdislocations into nonplanar configurations, some of which are depicted in figure 1. They include 'roof' barriers, Kear-Wilsdorf locks, and 'C' barriers, 'C' barriers will not be considered in this discussion. Greenberg studied the stability and activation energy to form these dislocations using isotropic elasticity theory and by assuming the traditional hierarchy of planar fault energies. Where available, experimentally determined, fault energies were also used. In this study we revisit these elasticity calculations and, using fault energies determined from electronic structure methods, determine the influence of composition on the relative stability of these dislocation core structures. Changes in the fault energies can produce a reordering of the hierarchy of energies for the various dislocation core structures. We also have evaluated additional configurations that are similar to the dislocations shown in figure 1, but with the trailing partials (δC and Bδ) reversed to form an SESF or complex extrinsic stacking fault (CESF). This was motivated by the recent high resolution TEM observations of Hemker et al., who have observed the modified (KW1) core structure in TiAl.[11]

FIGURE 1 Possible <101] superdislocation core structures.

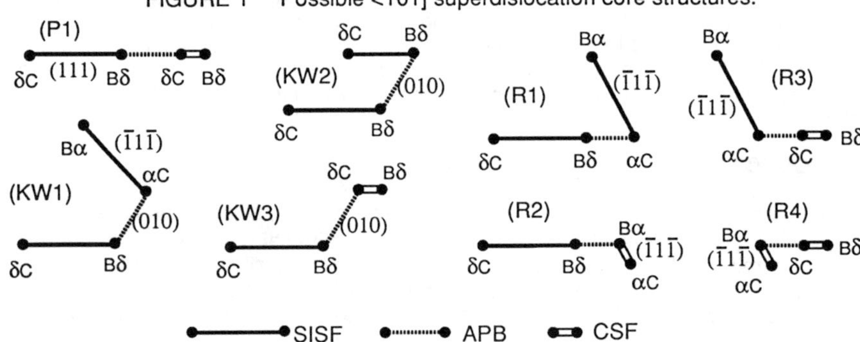

THE LKKR-CPA METHOD

The layered Korringa Kohn Rostoker method (LKKR) is a self-consistent electronic structure algorithm based on multiple scattering theory (MST) which allows for the simulation of planar defects in an otherwise perfect crystal.[12] Combining the LKKR method with the coherent potential approximation (CPA) allows for the simulation of substitutionally disordered alloys, where the composition can be continuously varied over some range of interest. Within the CPA an effective periodic medium is constructed which has the property that the electron scatters in the coherent potential as it would on average in the disordered medium.[13] The total energy for the disordered system is given by the concentration average of contributions from each component species and follows the philosophy adopted by Johnson et al.[14]. The interstitial charge density in the intermetallic alloys is not as constant as that for simple close packed metals and we have found that the errors inherent in the muffin tin approximation to the coulomb energy are significant in TiAl. These errors can be partially corrected using the atomic-sphere approximation where integrals over the Wigner Sietz cell are approximated by integrals over the Wigner Sietz sphere. The total energy is then computed directly as a sum of kinetic, coulomb and exchange-correlation contributions. A more detailed discussion of this method and its application to $Cu_{1-x}Zn_x$ $(0.0 < x < 0.3)$ can be found in the literature.[15] For the calculations reported here equal radii were used for the different atoms, the partial wave basis was expanded up to $\ell=3$, 24 energy points, 18 special k points and 23 plane waves were used to calculate the electronic structure. This was sufficient to converge the defect energies in both ordered and disordered alloys to within

$10mJ/m^2$, an acceptable error for determining trends in the faults energies as a function of composition.

RESULTS

γ TiAl (a_0=3.989 Å, c/a_0=1.02) has an ordered FCC crystal structure, composed of alternating (001) planes of Ti and Al atoms. We have previously calculated the electronic structure of stoichiometric TiAl, determining the energies of various planar faults and the bonding near those faults.[16] In this paper we investigate the effect of composition on three planar faults, the superlattice intrinsic stacking fault (SISF), the (111) antiphase boundary (APB), and the complex stacking fault (CSF). The self-consistent electronic structure was calculated for binary and ternary alloys for Al concentrations ranging from 48 to 51 atomic percent, and three ternary additions, Mn, Cr and Nb, at the level of 2 atomic percent. Site occupancy for the binary and ternary compositions were taken from available experimental results, though these experimental findings remain controversial.[8,17,18] Table 1 shows the ratio of site occupancy for the ternary additions, here the α and β sites refer to the Ti and Al sites respectively. For Al rich alloys Al was substituted on the α sites and in Ti rich alloys Ti was substituted on β sites, no vacancies were allowed in these calculations.

Table 1.
Ratio of site occupancy for ternary elements

Element	Cr			Mn			Nb		
%Al	>50	50	<50	>50	50	<50	>50	50	<50
α/β	2/1	1/1	0/1	1/0	1/1	1/2	1/0	1/0	2/1

The results for these alloy calculations are shown in Figure 2. For all compositions the (111)APB energy has the largest energy, consistent with our previous calculations on stoichiometric TiAl.[16] Little WB-TEM data on the equilibrium spreading of partial dislocations is available for this range of composition. However, G. Hug and I. Phan have reported SISF and CSF energies for Ti-48.5Al-3Cr of 110-120 mJ/m^2 and 170-180 mJ/m^2, respectively, and SISF and (111)APB energies for Ti-47.2Al-1.3Mn of 63 mJ/m^2 and 199 mJ/m^2, respectively.[19] The SISF and CSF energies are in reasonable agreement with the LKKR-CPA results. Near neighbor violation analysis (bond counting) arguments suggest that the dependence of the APB energy with anti-site concentration, c, should be proportional to (1-2c) for the binary alloy.[16] The (111)APB energy peaks at the stoichiometric composition for the binary alloy, consistent with these arguments. Of the three ternary additions, Mn has the largest effect on the planar fault energies at these compositions. At 48 atomic percent Al, which is within or near the mixed α_2/γ phase region for this alloy, the APB, CSF and SISF energies are reduced by ~40%, 40% and 70%, respectively, from the stoichiometric fault energies. Several studies have shown that for stoichiometric TiAl the twin energy is approximately one half the SISF energy.[16,20] Given a low activation energy barrier for twinning, the reduced SISF energy implies that twinning should become more active in off stoichiometric Mn doped TiAl. This is consistent with experimental observations.[21] Details of the electronic structure which make Mn effective in reducing the planar fault energies will be discussed elsewhere.

The hierarchy of these fault energies is inconsistent with the traditional ordering of (111) planar faults.[22] However, we have shown that for an $L1_0$ alloy with sufficient directional bonding the APB energy should be approximately the sum of the SISF and CSF energies.[16] These arguments are based on near neighbor violation analysis using a three body atomistic model. Previous studies of the stability of <101] superdislocations in TiAl assumed the CSF to have the largest energy.[5] We have re-evaluated the relative stability of these various nonplanar configurations of the superdislocation using the

FIGURE 2. (111)APB, CSF and SISF energies as a function of composition.

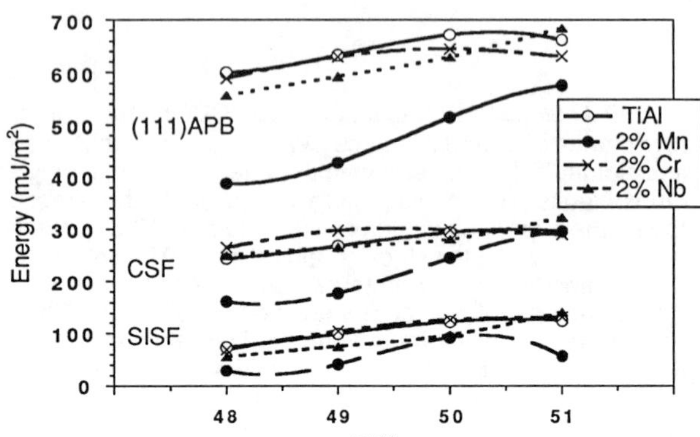

calculated fault energies for $Ti_xAl_{1-x}Mn_2$. The interaction energy of partial dislocations that do not share the same glide plane is not well defined within anisotropic elasticity theory.[23] Instead, as in the studies by Greenberg, we have used the Voigt average of the elastic constants[20] within isotropic elasticity to estimate the relative energy of these dislocation core structures. The dislocation configurations we considered are shown in figure 1. The relatively large (111)APB energy suggests the possibility that no stable planar <101] screw dislocation exists at low temperature. For stoichiometric TiAl the (111)APB energy (670 mJ/m^2) is large compared to the cube plane APB energy (350 mJ/m^2) and it is likely that the planar configuration, P1, will transform to include out-of-plane components.[16] This cross slip may take one of two forms, an extension of the APB onto a cube or octahedral plane, or a volume of distorted lattice extending onto adjacent (111) planes. In the former case we find that the ratio of octahedral to cube APB energies to be ~1.9. This APB anisotropy is large enough to allow for thermally assisted cross slip provided that the mechanism of cross-slipping is not substantially different from the $L1_2$ alloys.[3] For now we assume that the APB and CSF energies are within a range of values that are consistent with the formation of the barriers shown in figure 1.

The dislocation energy of the various barriers, relative to the dissociated planar <101](111) screw dislocation ($E_{sessile}$ - $E_{glissile}$), is shown figure 3a. The fault energies from Ti-xAl-2Mn and an (010)APB energy of 250 mJ/m^2 were used to generate these plots. At this (010)APB energy the Kear-Wilsdorf (KW) barrier with an SISF on two octahedral planes, KW1, has a lower energy than 'roof' barrier, R1. The energy of the alternative KW locks, KW2 and KW3, can be reduced to the energy of the 'roof' barrier, R1, by lowering the (010)APB energy to the values shown in figure 3b. Given the sharp reduction in (111)APB energy for Ti rich Mn doped TiAl a (010)APB energy of 100 mJ/m^2 may be possible in the composition range we have considered, though this has not been specifically determined. However, if the (010)APB energy is reduced then the energy of the lock KW1 would also be reduced and it would continue to be the lowest energy configuration we have considered.

We have also evaluated the elastic energy of the 'roof' barrier, R1, and the KW lock ,KW1, with the trailing SISF (bounded by δC and $B\delta$) replaced with a CESF, and SESF. Here we assume that the SESF and CESF energies can be approximated by the energies of the SISF and CSF respectively, consistent with previous electronic structure results for the SESF and SISF energies in stoichiometric TiAl. Again we find that the modified KW lock (KW1) has the lowest energy of this family of <101] super-dislocations. The low energy of this core structure is consistent with the reported observation of this substructural defect by high resolution electron microscopy.[11].

FIGURE 3a. The elastic energy of various <101](111) screw superdislocation core structures relative to the dissociated planar dislocation, P1. Labels correspond to configurations shown in figure 1.
FIGURE 3b. The (010)APB energy necessary to lower the energy of the KW2 and KW3 barriers to the energy of the 'roof' barrier, R1.

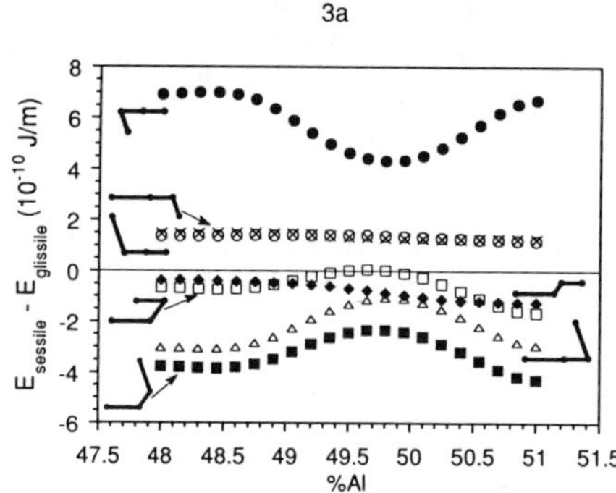

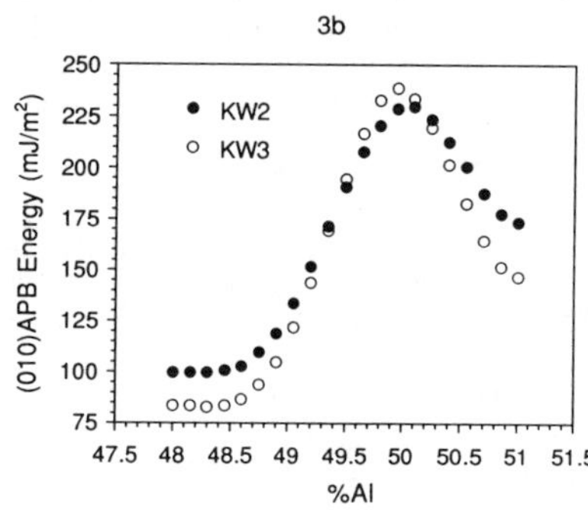

The hierarchy of dislocation core energies found in this study is consistent with the previous work of Greenberg et al. The relative energy of the dislocations is insensitive to the magnitude of the CSF energy. Also, we observe little shuffling of the core energies as a function of composition. We expect that the activation energy to create or destroy any of these barriers is also dependent on the fault energies. While we will not derive the activation energies in this paper, we can comment on the trends expected in activation energy as a function of the calculated fault energies.

The activation energy of the various nonplanar faults shown in figure 1 is dependent on the constriction energy of the partials bounding the leading CSF and the driving force for cross slip. This constriction energy decreases with increasing Al concentration, see figure 2. Similarly, the height of the R1 and KW1 barriers decrease as the SISF energy and Al content increase, up to 50 atomic percent. Therefore, the activation energy to annihilate these barriers by constricting the partials bounding the SISF (cross slip) also decreases. This implies that a higher temperature is needed in Ti-48Al-2Mn relative to Ti-50Al to form and annihilate these superdislocation barriers. These barriers are extremely stable against annihilation by cross slip, and other mechanisms may be active in the destruction of these barriers. If we assume that the cross slip process is limited by the constriction energy, and that the positive yield stress is driven by cross slip events then the peak yield stress of Ti-48Al-2Mn will be shifted to higher temperatures and the slope of the yield stress will be reduced relative to Ti-50Al. Alternatively, Ti-51Al-2Mn has a similar CSF energy and a smaller SISF energy than Ti-50Al. In this case the slope of the flow stress vs. temperature for Ti-51Al-2Mn is expected to be the same as Ti-50Al for low temperatures with the yield stress peak coming at a higher temperature. Finally, if the (111)APB energy is large enough it may spread the planar <101] dislocation core onto more than one (111) plane. Atomistic simulations of the <101] superdislocation indicate that the lower bound to the APB and CSF energies for this to occur are 770 mJ/m^2 and 580 mJ/m^2 respectively.[24]

SUMMARY

We have calculated the energy of planar faults associated with <101] superdislocations, for near stoichiometric γ TiAl. Of the three ternary additions considered in this study (Mn, Cr and Nb), Mn had the strongest effect, reducing the SISF, CSF and (111)APB energies by more the 40% from the stoichiometric values. The influence of these variations in fault energy, as a function of chemistry, on the stability of <101] dislocation core structures was evaluated using isotropic elasticity theory. The lowest energy core structure was found to be the KW lock, (KW1), for an (010)APB energy below 250 mJ/m^2. Trends in the fault energies suggest that doping Ti rich TiAl with Mn will increase the activation energy for forming both 'roof' and KW barriers, while simultaneously increasing the energy needed to destroy these barriers. This may result in a shift of the peak temperature of yield stress to higher temperatures and a decrease in the work hardening rate. Similarly, Mn doped Al rich TiAl should have a higher T_{max} than Ti-50Al. Finally, the R1 and KW1 barriers are stable relative to the other dislocations configurations and the destruction of these barriers is difficult.

ACKNOWLEDGEMENTS

The authors gratefully acknowledge numerous useful discussions with Drs. S. Rao, T.A. Partharasarthy and P.M. Hazzledine. This work was sponsored by the U.S. Air Force under the contract #F33615-91-C-5663.

REFERENCES

[1] H.A. Lipsitt, D. Shechtman, R.E. Schafrik, Metall. Trans. **6A**, 1991 (1975).
[2] T. Kawabata, T. Kanai, and O. Izumi, Acta Metall. **33**, 1355 (1985).
[3] V. Paidar, D.V. Pope and V. Vitek, Acta Metall. **32**, 435 (1984).
[4] G. Hug, A. Loiseau and P. Veyssiere, Phil. Mag. **A57**, 499 (1988)
[5] B.A. Greenberg, O.V. Antonova, V.N. Indenbaum, L.E. Karkina, A.B. Notkin, M.V. Ponomarev and L.V. Smirnov, Acta Metall. mater. **39**, 233 (1991).
[6] M.J. Blackburn and M.P. Smith, U.S.A.F. Tech. Report No. AFML-TR-79-4056, 1979.
[7] S.C. Huang and E.L. Hall in High-Temperature Ordered Intermetallic Alloys III, edited by C.T. Liu, A.I. Taub, N.S. Stoloff, and C.C. Koch, (Mater. Res. Soc. Proc. **133**, Pittsburgh, PA 1989) pp. 329-334.
[8] T. Tsujimoto and K. Hashimoto in High-Temperature Ordered Intermetallic Alloys III, edited by C.T. Liu, A.I. Taub, N.S. Stoloff, and C.C. Koch, (Mater. Res. Soc. Proc. **133**, Pittsburgh, PA 1989) pp. 391-396.
[9] T. Kawabata, T. Tamura and O. Izumi in High-Temperature Ordered Intermetallic Alloys III, edited by C.T. Liu, A.I. Taub, N.S. Stoloff, and C.C. Koch, (Mater. Res. Soc. Proc. **133**, Pittsburgh, PA 1989) pp. 391-396.
[10] For a review of this work see Y.W. Kim and D.M. Dimiduk, JOM, **43**, 40 (1991).
[11] K.J. Hemker, B. Viguier and M.J. Mills, private communications.
[12] J.M. MacLaren, S. Crampin, and D.D. Vvedensky, Phys. Rev. **B40**, 12164 (1989).
[13] see for example, J.S. Faulkner in Progress in Materials Science, edited by J.W. Christian, P. Hassen and T.B. Massalski, (Pergamon, New York, 1973), p. 385.
[14] D.D. Johnson, D.M. Nicholson, F.J. Pinsky, B.L. Gyorffy and G.M. Stocks, Phys. Rev. **B41**, 9701 (1990).
[15] J.M MacLaren, A. Gonis and G. Schadler, Phys. Rev. **B45**, 14392 (1992).
[16] C. Woodward, J.M. MacLaren and S. Rao, J. Mater. Res. **7**, 1735 (1992).
[17] H. Doi, K. Hashimoto, K. Kasahara and T. Tsujimoto, Mat. Trans, JIM, **31,** 975 (1990).
[18] D. Shindo, A. Chiba, K. Hiraga and S. Hanada in Proceedings of the International Symposium on Intermetallic Compounds, (Sendai, Japan, 1991), p. 87.
[19] G. Hug and I. Phan, private communication.
[20] C.L. Fu and M.H. Yoo, Philos. Mag. Lett. **62**, 159 (1990).
[21] G. Hug and P. Veyssiere in International Symp. on Electronic Microscopy an Plasticity and Fracture Research of Materials, (Dresden, October, 1989).
[22] M.J. Marcinkowski, in Electron Microscopy and Strength of Crystals, edited by G. Thomas and J. Washburn (Interscience Publishers, 1963), p. 431.
[23] J. W. Steeds in Anisotropic Elasticity Theory of Dislocations, (Clarendon Press, Oxford, 1973) Pg. 31.
[24] C. Woodward and S. I. Rao, to be published.

THE IMPORTANCE OF MICROSTRUCTURAL INSTABILITY IN DETERMINING THE MECHANICAL BEHAVIOUR OF CUBIC TITANIUM TRIALUMINIDES

D.G. MORRIS, S. GUNTHER and R. LERF
Institute of Structural Metallurgy, Avenue de Bellevaux 51, University of Neuchâtel, NEUCHATEL 2000, SWITZERLAND.

ABSTRACT

Cubic trialuminides deform by the movement of <110> dislocations which are clearly dissociated as APB superdislocations at high temperatures, with debate about whether these are dissociated as APB or SISF superdislocations at low temperatures. These materials are characterized by the following mechanical behaviour: (i) strength variable with ternary element addition or titanium content; (ii) mild strength anomaly at high temperature, sometimes; (iii) serrations in stress-strain curve at intermediate temperatures; (iv) low tensile ductility ($\approx$0) at room temperature, increasing at higher temperatures, with an intermediate temperature minimum.

These properties are explained by the fine microstructure and its variation locally and with temperature: (a) a tetragonal component of order in ternary alloys in addition to the basic $L1_2$ order, varying with ternary element and content; (b) precipitation (sometimes on dislocations) during high temperature testing - accounting for the strength anomaly; (c) extra-ordinarily rapid solute collection at dislocations and APB's - accounting for strain aging and minimum ductility phenomena; (d) strain relaxation at crack tips restricted to single slip planes by the structural modification produced by shear - making a major contribution to brittleness.

All these processes are particularly acute in the titanium trialuminides because of the structural instabilities involved, and the fast kinetics of atom rearrangement; thus low temperatures during testing or during cooling after prior heat treatment play a major role. Similar effects may be of importance in other intermetallics such as FeAl, TiAl.

RESULTS AND DISCUSSION

Strength of Cubic Al_3Ti Alloys

The yield or flow strength of cubic Al_3Ti materials has been reported on numerous occasions (e.g. 1-5). The most important characteristics of strength variations are: - (i) the observation of a weak stress anomaly at test temperatures of about 500-600°C (1,3,4,5) - sometimes observed as a strength peak, other times as a stress plateau, other times not clearly detected; (ii) a significant flow stress increase on testing at low temperatures (77K and below) (2,4); (iii) significant variations from one alloy system to another and from one observer to another (1,3,4,5,6). These results have received various interpretations;: - (i) operation of cube cross slip processes leading to Kear-Wilsdorf locks causing the stress anomaly (5); (ii) superdislocation dissociation into a pair of superShockley partials separated by a SISF fault - the cores of these dislocations are known to be non-planar and lead to significant low temperature strengthening, as in the case of the Pt_3Al and Co_3Ti intermetallics (7-9); (iii) differing bond strengths according to choice of ternary element (6) or depending on the Ti content (10) leading to strength variations between different alloys.

The strength considered is usually the flow strength (at 0.2% plastic strain), typically measured in compression. Where bend (11,5,12) or tension (2) tests have been performed, failure occurs shortly after the initiation of plastic flow at a stress similar to that in compression. Failure in compression is geometrically restrained and occurs by shattering after large nominal compressive strains (>10%) as barrelling starts and local tensile stresses are induced.

Fig. 1 shows the results of compression tests performed at various temperatures at a strain rate of about 2 x 10^{-4} /s on a variety of trialuminide alloys of $L1_2$ structure. The precise compositions of these alloys are given in Table I. Several characteristic features may be noted and correlated with the alloy composition, in particular with the titanium content. At low temperatures the yield or flow stress increases as the temperature falls - this increase is similar in extent for all the alloys tested, can best be correlated with elastic moduli changes, and is not the same as the very rapid stress increase measured by Wu et al (4). At temperatures above about 350°C a given alloy may show a stabilisation in strength (alloy Cr I) or a stress anomaly (alloys Fe I and Cr II) - these effects are not dependent on the choice of ternary element but on the Ti content, which must be 26.9% or above here for such a stabilisation or anomaly to be observed. The stress level at the intermediate temperatures (0 - 300°C) increases steadily with Ti content. This influence of Ti content on yield stress is clearly seen in Fig. 2. In particular, for Ti contents above about 27% the stress anomaly is seen as higher yield stress at 700°C than at 500°C. These strength variations may be analysed in terms of the phases present, their distribution, and dislocation microstructure.

TABLE I

Composition of alloys analysed by mechanical testing (atomic %)

Alloy Designation	Aluminium Content	Titanium Content	Ternary Addition
CrI	65.0	26.9	8.1 Cr
CrII	64.5	27.5	8.0 Cr
FeI	64.0	28.0	8.0 Fe
FeIII	66.5	25.5	8.0 Fe
Mn	66.7	25.7	7.6 Mn

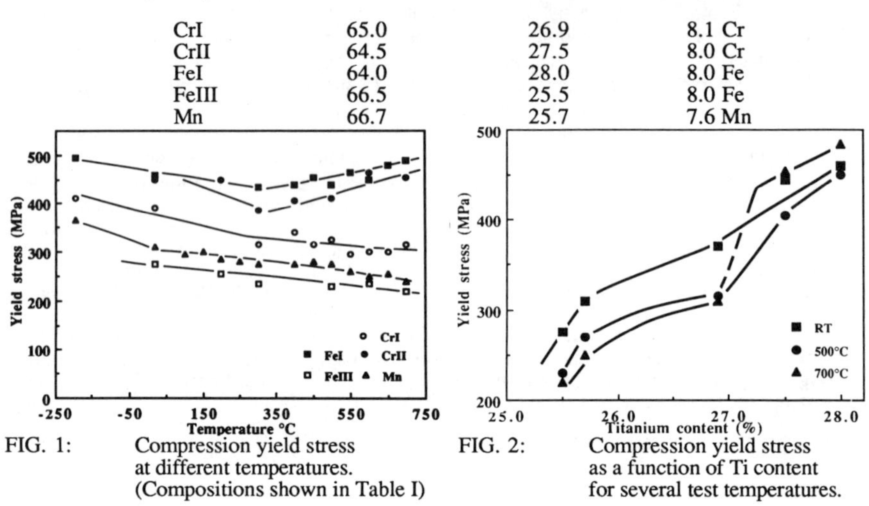

FIG. 1: Compression yield stress at different temperatures. (Compositions shown in Table I)

FIG. 2: Compression yield stress as a function of Ti content for several test temperatures.

Low temperature strengthening

The low temperature strengthening observed in Fig. 1 will be examined in terms of the flow stress measured at 77K divided by the flow stress at 300 K. For the materials examined here this term takes values of 1.05 - 1.2. A question of interest is whether this low temperature strengthening can be ascribed to elastic modulus changes or whether other strengthening mechanisms are involved. Similar mild strength increases at low temperatures have been reported by others (1,2,5), with on some occasions the strength increasing by much larger amounts, for example by 1.5 (2,4). The extent of stress increase can not be correlated with either the stress level at room temperature nor with the ternary addition - for example a small stress increase (1,5) and a large stress increase (4) has been reported for alloys containing Fe, and small stress increases at low temperature have been observed both for alloys rather strong (1,5) and rather weak (2) at room temperature. A strength increase of about 1.1 may be expected based on elastic moduli changes between room temperature and 77K (5) and this may be the sole cause of the low temperature strengthening observed here. On the other hand, the precise nature of the dislocation cores and their dissociation in the ordered lattice may determine the strength variation. In the case of intermetallics such as Ni_3Al where superdislocation dissociation by the formation of an APB fault occurs, the formation of strong Kear-Wilsdorf locks at high temperature can lead to strengthening; such locks are not formed at low temperatures and such low temperature strengthening will not occur. A greater strengthening increase at low temperatures is found in materials such as Pt_3Al where superdislocations are dissociated to produce an SISF fault, leaving non-planar dislocation configurations which are sessile at low temperature (7-9). It is therefore of relevance to establish the dislocation core nature in the cubic Al_3Ti alloys.

There has been considerable discrepancy regarding dislocation dissociation in the cubic Al_3Ti alloys, with some reports of dissociation to 2 x 1/2 <110> + APB (5,13,14) and other times reports of dissociation to 2 x 1/3 <211> + SISF (15,16). While it may be possible that different dissociation schemes may be followed by alloys of nominally the same, but actually slightly differing compositions, it has also been claimed recently that the SISF dissociation analyses have been incorrectly performed (17), and only the APB dissociation scheme has actually been confirmed. The present work confirms the dissociation of dislocations in a cubic Al_3Ti alloy deformed at room temperature leading to an APB fault. Similar dislocation morphologies and dissociations are seen after deforming at 77K where dislocations are still APB dissociated and appear slightly straighter, oriented along crystallographically close-packed atom directions. This suggests that dislocation core or Peierls mechanisms may play a role in affecting dislocation mobility at such low temperatures but this is still a minor effect.

A likely cause of the somewhat larger-than-expected low temperature strengthening seen in the literature (2,4) and perhaps in Fig. 1 is hardening by either solute or precipitate, based on the order-solute associated with tetragonal ordering of solute, as discussed elsewhere (18), or on fine precipitation of Al_2Ti needles in the case of the stronger Ti-rich materials. Such solute effects are known to lead to significant strengthening at very low temperatures.

High temperature strengthening anomaly

Strengthening at high temperatures has previously been explained by the formation of Kear-Wilsdorf locks following cube cross slip processes (5), similar to the case of Ni_3Al. This deduction has been supported by the observation of many superdislocations on cube planes after high temperature deformation. Two recent observations point to another source of high temperature strengthening, however, and correlate with the influence of Ti content on the absolute strength value shown in Fig. 1 as well as the extent of the strength anomaly.

After high temperature deformation, many superdislocations are seen to be mobile on cube planes for all the cubic trialuminides examined, including those such as the Al-Ti-Mn alloy, Fig. 3, which showed no stress anomaly at high temperatures (the Mn alloy in Fig. 1). The presence of superdislocations on cube planes, and extensive cross slip between octahedral planes and between octahedral and cube planes was previously noted for a Al-Ti-Fe alloy (5) and taken to imply the formation of strong Kear-Wilsdorf barriers in this alloy which showed a stress rise at high temperatures (the FeI alloy in Fig. 1). Under such circumstances, cross slip between octahedral and cubic planes, as well as glide on both types of planes, characterises the easy plastic deformation occurring in these alloys when deformed at high temperature and is not a cause of strengthening. The strengthening shown by some alloys during high temperature deformation can instead be associated with fine precipitation of Al_2Ti phase. Fig. 4 presents a 28% Ti alloy which showed no precipitation before testing but shows fine precipitation after deformation at 700°C. The "anomalous" strengthening seen can more correctly be described as an age-hardening effect occurring during the 15-30 mins heating time of the mechanical test.

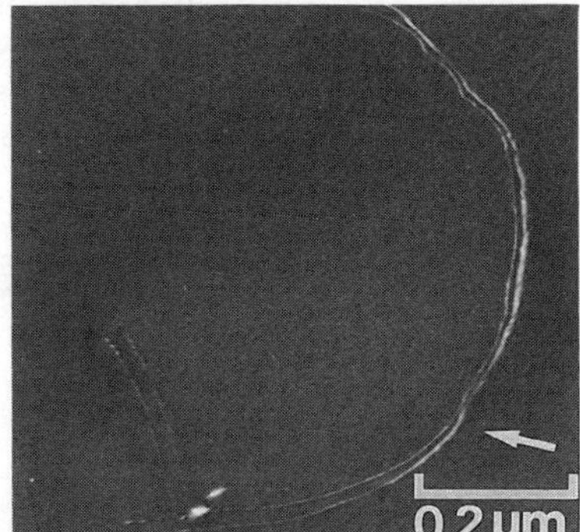

FIG. 3: Dislocations in Al-Ti-Mn deformed 2% at 700°C. The superdislocation lies on the cube plane at the screw part and on an octahedral plane for the edge part. Foil (110) **g** vector 002.

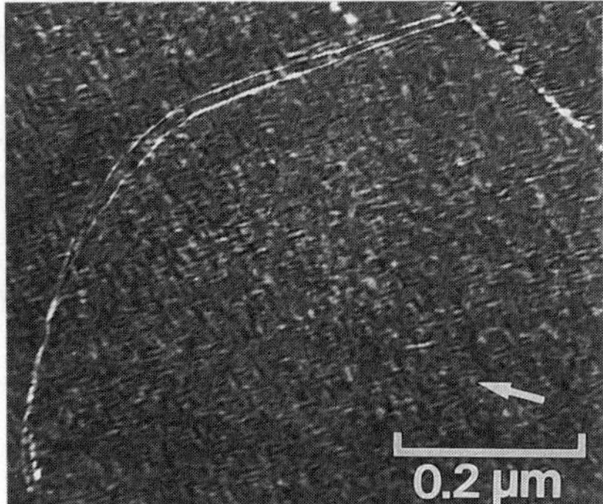

FIG. 4: Precipitation of Al_2Ti in cubic Al-Ti-Fe alloy. This material has 28% Ti and shows precipitation after deformation at 700°C. Foil (011), **g** vector $\bar{1}\bar{1}1$.

Solute and Precipitate Effects on Mechanical Behaviour

In addition to leading to an overall increase in strength at ambiant temperatures and to the strength increase at high temperatures, concurrent solute movement during deformation leads to strain instabilities such as the Portevin-Le Chatelier effect (5,19), seen in Fig. 5. Such serrations are not visible when testing at room temperature, are visible at intermediate temperature, and disappear again at very high temperatures - corresponding to insufficient atomic movement, sufficient movement to block a dislocation, and excessively high atomic mobility. For a given alloy and temperature the stress amplitude of an instability increases with the strain, and the instability will be visible over a strain range, ε_0 to ε_1, values which both increase at lower temperatures. Alloys richer in Ti, such as the Al-Ti-Cr II and Al-Ti-Fe I alloys, showed more intense instabilities over a wider temperature range. The Al-Ti-Mn and Al-Ti-Cr I alloys, of Ti content intermediate between the high and low values, showed intermediate levels of serrations on their stress-strain curves. It is also of interest to note that a Al-Ti-V alloy of DO_{22} crystal structure showed no observable instabilities. The reason for this is not clear and several possibilities may be suggested. On testing at low temperatures, namely 400°C and below, the dislocations are not dissociated, and any dislocation locking effects associated with the APB (see below) cannot operate. A second possible reason for the absence of dislocation pinning is that the alloy composition lies within a well-defined single phase DO_{22} region in the ternary diagram (as suggested by Yamaguchi et al (20)) and is less susceptible to compositional instabilities than the near-line compound Al_3Ti-based alloys.

Equivalent stress-strain instabilities can also be induced by deliberate strain ageing experiments, as illustrated in Fig. 6. These instabilities appear readily, in time periods of 1-60 mins, on ageing at temperatures in the range 150-350°C, at the lower and of the range where dynamic serrations are seen, and corresponding to only about 0.25 - 0.35 of the melting temperature of this alloy. As summarized in Table II, the intensity of the static strain ageing effect correlates with that of the dynamic serrations, and also with the extent of high temperature anomalous strengthening. It is clear that all three effects are related to the same cause - namely high Ti content and a tendency to Ti segregation and precipitation of phases such as Al_2Ti.

Parallel studies on superdislocation arrangements have shown that the separation of the partial dislocations of an APB-dissociated superdislocation increases dramatically with increase of temperature (5,14). This has been noted on studying samples tested at different temperatures as well as during in situ heating studies (21,22). Such studies show that changes occur over time periods of minutes (hours) at temperatures in the range 300-500°C (200-300°C). Similar changes in dislocation arrangements have also been seen for a wide range of intermetallics, and this effect appears to be general (23,24). The kinetics of relaxation observed here are clearly the same as these documented for static strain ageing, and have been interpreted as an order relaxation at the APB, similar that proposed initially by Brown (25). As shown later by Popov et al (26,27) the extent of relaxation depends also on the precise alloy composition, in particular how far off the stoïchiometric A_3B composition is the alloy under consideration. Fig. 7 shows the variation of

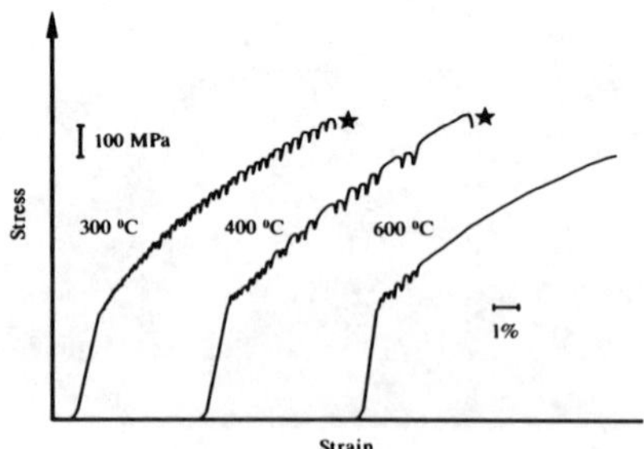

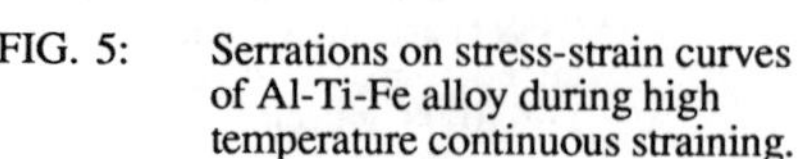

FIG. 5: Serrations on stress-strain curves of Al-Ti-Fe alloy during high temperature continuous straining.

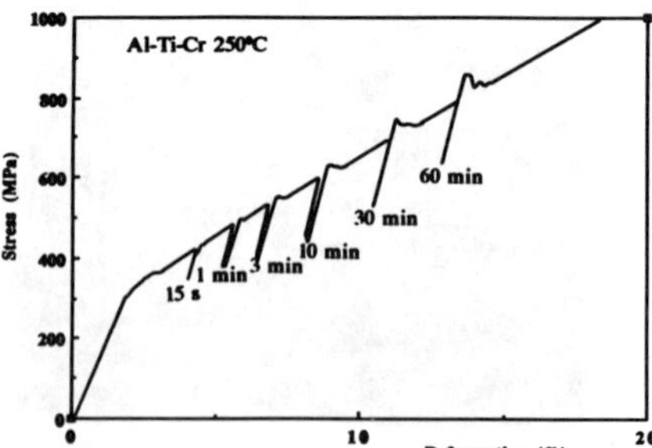

FIG. 6: Discontinuous yielding induced by strain ageing a Al-Ti-Cr alloy.

TABLE II
COMPARISON OF ALLOYS - DISLOCATION RELAXATION - SERRATIONS

ALLOY (%Ti)	Superdislocation partial separation (nm) RT	500°C	700°C	Anomalous stress increase (MPa) 500°C	700°C	Serration amplitude (MPa) 500°C	Static aging peak (MPa) 200°C 1h
Al-Ti-Fe I (28.0%)	3	7	23	50	115	32	78
Al-Ti-Cr II (27.5%)	2	7	23	60	160	30	>60
Al-Ti-Fe III (25.5%)	3	5	5.5	≈0	0-30	10	≈10
Al-Ti-V (25%Ti+V)	0	0	4.5	≈0	≈0	0	≈0-5
Al-Ti-Mn (25.7%)	4.5	7	25	≈0	0-30	20	18

superdislocation dissociation distance on annealing for several of the trialuminide alloys where it can be seen that the extent of APB relaxation (by how much the partial separation increases) depends on the alloy, and is greatest for those alloys significantly hyperstoïchiometric in Ti (>25%). Detailed examination shows incipient precipitation at the dislocation core, Fig. 8, even though there was no sign of precipitation in the bulk crystal. The atomic plane arrangement seen is the same as that of Al_2Ti precipitates. Such incipient precipitation at the dislocation core will obviously reduce the elastic energy locally and hence the increased separation of the partials of the superdislocation implies that the APB energy must be very significantly reduced.

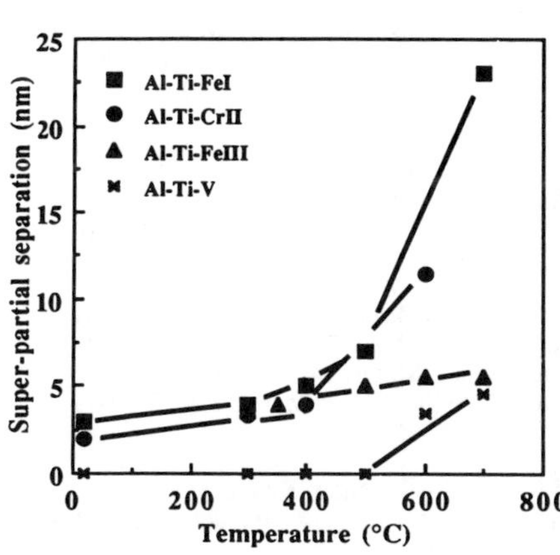

FIG. 7: Variation in separation of the partial dislocations of superdislocations in trialuminides examined during in situ annealing. All materials have the $L1_2$ crystal structure with the exception of the Al-Ti-V (DO_{22} structure).

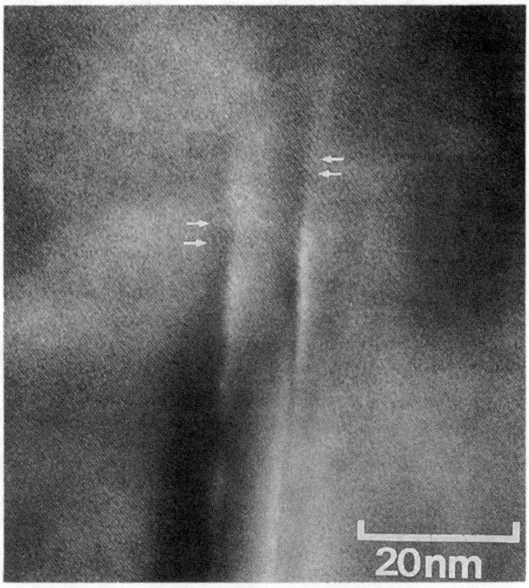

FIG. 8: High resolution electron micrograph showing a superdislocation in Al-Ti-Cr (CrI) deformed 1% at 500°C. Foil orientation (001). At the dislocation cores a change in lattice spacing corresponds to fine Al_2Ti precipitate formation.

CONCLUSIONS

A conclusion of this analysis is that slight excess of Ti beyond the stoïchiometric $(Al,X)_3Ti$ composition - or beyond the limit of the single phase $L1_2$ region, which must be very close to

25.0%Ti for all the alloys here - leads to very rapid atom arrangements at dislocation cores and at APB's which will significantly affect dislocation mobility leading to high temperature strengthening and intermediate temperature unstable locking. The rapid fcrmation of incipient precipitation probably also affects room temperature and low temperature properties. Kinetics of precipitation are extremely rapid because of the small extent and scale of atomic movement required, which happens also at or close to dislocation cores. Considering by way of example the composition change Al_3Ti to Al_2Ti, it is clear that the removal of one Al atom only over a distance of one lattice parameter can already create a unit cell nucleus of the new phase. Considering either in situ relaxation kinetics (21,22) or the quantitive strain ageing data we can deduce diffusion coefficients of the order of 10^{-20} m^2/s at 200-300°C. This is much higher than estimates of bulk diffusion kinetics (10^{-31} - 10^{-27} m^2/s) for an fcc metal at $T/Tm \approx 0.3 - 0.35$, but similar to crude estimates of pipe diffusion kinetics (10^{-21} - 10^{-16} m^2/s) on the same material (these estimates are based on a similar pre-exponential factor of about $10^{-4} m^2/s$ for both bulk and pipe diffusion and an activation energy for pipe diffusion about 0.5-0.6 that of bulk diffusion). Such rapid diffusion kinetics suggest finally that significant ageing effects on dislocations will occur during material preparation and even during material storage at ambiant temperature before testing, and the effects of such fine, localised dislocation pinning may be very common.

It is the presence of fine ordered regions of tetragonal order in the $L1_2$ lattice(20) or fine Al_2Ti precipitates that produce strength increases at low temperatures. Strength increases at high temperatures are correlated directly with the appearance of fine precipitates during testing. It would be of interest to examine alloys containing smaller amounts of Ti which could be softer, show different dislocation mobility, and be less flaw or crack sensitive than prior materials.

REFERENCES

1. K.S. Kumar and J.R. Pickens, Scripta Metall. **22**, 1015 (1988).
2. K.S. Kumar and S.A. Brown, Acta Metall. Mater. **40**, 1923 (1992).
3. S. Zhang, J.P. Nic, W.W. Milligan and D.E. Mikkola, Scripta Metall. Mater. **24**, 1441 (1990).
4. Z.L. Wu, D.P. Pope and V. Vitek, Scripta Metall. Mater. **24**, 2187 (1990).
5. R. Lerf and D.G. Morris, Acta Metall. Mater. **39**, 2419 (1991).
6. J.P. Nic, S. Zhang and D.E. Mikkola, Scripta Metall. **24**, 1099 (1990).
7. G. Tichy, V. Vitek and D.P. Pope, Phil. Mag. **A53**, 467 (1986).
8. G. Tichy, V. Vitek and D.P. Pope, Phil. Mag. **A53**, 485 (1986).
9. T. Takasugi, S. Hirakawa, O. Izumi, S. Ono and S. Watanabe, Acta Metall. **35**, 2015 (1987).
10. M.B. Winnicka and R.A. Varin, Scripta Metall. Mater. **25**, 2297 (1991).
11. D.G. Morris, R. Lerf and G. Hollrigl, EUROMAT 91, Proc. 2nd European Conf. on Advanced Structural Materials, Vol. 2, The Institute of Metals, London, p. 398 (1992).
12. S.Zhang, J.P. Nic and D.E. Mikkola, Scripta Metall. Mater. **24**, 57 (1990).
13. E.P. George, J.A. Horton, W.D. Porter and J.H. Schneibel, J. Mater. Res. **5**, 1639 (1990).
14. D.G. Morris, J. Mater. Res. **7**, 303 (1992).
15. H. Gengxiang, C. Shipu, W. Xiaohua and C. Xiaofu, J. Mater. Res. **6**, 957 (1991).
16. H. Inui, D.E. Luzzi, W.D. Porter, D.P. Pope, V. Vitek and M. Yamaguchi, Phil. Mag. **A65**, 245 (1992).
17. P. Veyssiere and D.G. Morris, Phil. Mag., in press.
18. D.G. Morris and S. Gunther, Acta Metall. et Mater., **40**, 3065 (1992).
19. L. Potez, G. Lapasset and L.P. Kubin, Scripta Metall. Mater. **26**, 841 (1992).
20. M. Yamaguchi, Y. Umakoshi and T. Yamane, Mater. Res. Soc. Symp. Proc., Vol. 81, Materials Research Society, Pittsburgh, p. 275 (1987).
21. D.G. Morris, Scripta Metall. Mater. **25**, 713 (1991).
22. D.G. Morris, Phil. Mag. **A65**, 389 (1992).
23. A. Korner and G. Schoeck, Phil. Mag. **A61**, 909 (1991).
24. D.G. Morris, Scripta Metall. Mater. **26**, 733 (1992).
25. N. Brown, Phil. Mag. **4**, 693 (1959).
26. L.E. Popov, E.V. Kozlov and N.S. Golosov, Phys. Stat. Sol. **13**, 569 (1966).
27. N.S. Golosov, L. Ya. Pudan and L.E. Popov, Phys. Stat. Sol. **11**, 123 (1972).

THE TEMPERATURE DEPENDENCE OF GRAIN BOUNDARY SEGREGATION IN B-DOPED Ni_3Al BICRYSTALS

URSULA OTTERBEIN, SIEGFRIED HOFMANN AND MANFRED RÜHLE

Max-Planck-Institut für Metallforschung, Institut für Werkstoffwissenschaft, Seestr. 92, D-7000 Stuttgart 1, Germany.

ABSTRACT

Equilibrium grain boundary segregation at symmetrical <110> tilt grain boundaries of Ni_3Al bicrystals containing 0.17 at% B and traces of S has been studied by Auger electron spectroscopy and compared directly to surface segregation by applying a special specimen geometry. From the experimentally determined interfacial concentrations after annealing at different temperatures, segregation free energies for B and S at both types of interfaces have been determined using the Langmuir-McLean equation. The effect of B and S segregation on the grain boundary cohesive energy has been calculated.

INTRODUCTION

Intermetallics with $L1_2$ structure such as Ni_3Al possess scientifically interesting features at high temperatures like anomalous yield strength-temperature behaviour and large Youngs modulus. Ni_3Al serves as model material for understanding $L1_2$ properties.

Polycrystalline Ni_3Al possesses very poor ductility due to intrinsic brittleness of its grain boundaries. However, small additions of B enhance the ductility by up to 50% in Ni rich samples[1]. B segregates to grain boundaries in both Ni-rich and Al-rich samples[2]. Ni grain boundary enrichement has been also experimentally observed and is independent of B doping[3,4].The reason for the ductilizing effect of B is not yet clear. Some authors[3,5] think that B segregation enhances the cohesive strength of the grain boundaries, whereas others[6,7] assume that disordering near the grain boundaries eases transmission of slip across the grain boundaries.

In this study the segregation behavior of B and S in Ni_3Al was examined. Because previous experiments[3] on polycrystalline samples showed large differences in segregation levels from boundary to boundary due to anisotropy[8], only bicrystals with well defined orientations were studied. The change in cohesive energy with B was determined from the differences in impurity concentrations at grain boundaries and free surfaces. This requires the use of a special structure of the bicrystalline interfaces, which enables analysis of adjacent free surface and grain boundary regions under identical conditions.

CONNECTION BETWEEN SEGREGATION TO INTERFACES AND COHESIVE ENERGY

For adiabatic separation the ideal work of fracture per unit area of the grain boundary is equal to the cohesive energy γ_{coh} which is given by the difference of energies of the grain boundary γ_{GB} and the two free surfaces $2\gamma_{FS}$ ($\gamma_{coh} = 2\gamma_{FS} - \gamma_{GB}$ in the case of symmetrical boundaries)[9]:

The interfacial energies γ_Φ (index Φ denotes GB or FS) are changed by the segregation of impurities by $\Delta\gamma_\Phi \approx N_\Phi G_\Phi$[10], where N_Φ is the number of impurity atoms per unit area and G_Φ their Gibbs free energy at the interface. G_Φ should be independent of N_Φ. For $N_{GB} = 2N_{FS}$ and following Rice et al.[11] the change of γ_{coh} by impurity segregation results in

$$\Delta\gamma_{coh} \approx N_{GB}(G_{FS}-G_{GB}) \quad (1)$$

N_{GB} can be estimated as $c_{GB}/\langle a_V\rangle^2$ for substitutional solutes and $c_{GB}/((1-c_{GB})\langle a_V\rangle^2)$ for interstitials, where $\langle a_V\rangle$ is the mean distance between intrinsic bulk atoms.

Equilibrium segregation is described by the Langmuir-McLean equation[12]

$$\ln\left(\frac{\frac{c_\Phi}{c_{\Phi max} - c_\Phi}}{\frac{c_V}{c_{Vmax} - c_V}}\right) = \frac{-\Delta G_{\Phi V}}{RT} \qquad (2)$$

where $\Delta G_{\Phi V} = G_\Phi - G_V$ is the Gibbs free energy of interfacial segregation, c_Φ, $c_{\Phi max}$ and c_V, c_{Vmax} the atomic concentrations and their maximum possible values at the interface and in the bulk, respectively. Measuring c_{GB}, c_{FS} and c_V allows the determination of $G_{GB} - G_V$ and $G_{FS} - G_V$ resulting in values for $\Delta\gamma_{coh}(T)$ according to equation (1).

EXPERIMENTAL

Sample preparation

Ni_3Al bicrystals with <011>{211} and <011>{311} symmetrical tilt boundaries and {211} and {311} surfaces, respectively, were prepared by UHV diffusion bonding of $Ni_{76}Al_{24}$ single crystals, doped with 0.169±0.015 at% B and containing traces of S ($1.06\cdot10^{-3}$ at%). Before diffusion bonding one of every pair of single crystals was photolithographically structured[13] and etched (fig. 1). This geometry enables direct comparison of grain boundary and surface segregation under identical conditions.

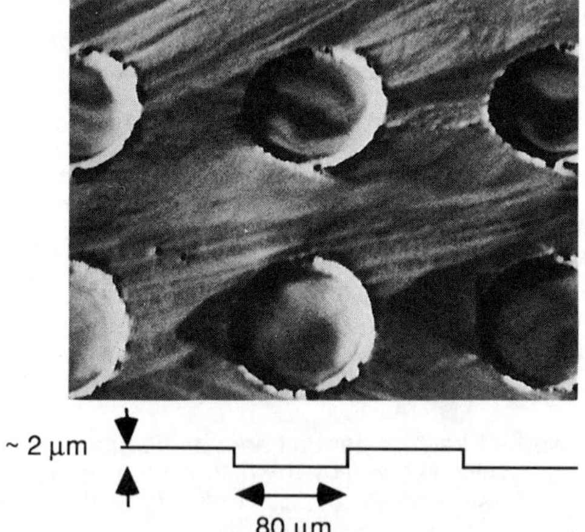

Fig. 1: Structured bicrystalline interface after fracture

Before diffusion bonding the single crystal surfaces were sputter cleaned by Ar ions and the purity was controlled by AES. The bonding itself took place at 1150 °C for 1 h, the base pressure was $4\cdot10^{-10}$ mbar, the pressure at 1150 °C was $6\cdot10^{-9}$ mbar. The scatter of the relative orientation of the single crystals after bonding was within ±2°.

Bicrystalline samples of dimensions 8 x 3 x 1 mm^3 with the boundary perpendicular to the sample axis were preannealed at 1000° C for 24h in $Ar+20\%H_2$ atmosphere and water quenched. To establish equilibrium segregation the samples were again annealed at 500 °C for 96 h, at 600 °C for 24 h or at 800 °C for 30 min and water quenched. The segregation levels of B for one grain boundary orientation at the three temperatures were measured for different annealing times, to find out when thermal equilibrium had been reached.

Grain boundary fracture was facilitated by notching the samples and by hydrogen embrittlement.

AES Measurements

The bicrystals were broken at room temperature in the fracture stage of a PHI 600 scanning Auger multiprobe spectrometer at a system pressure lower than 10^{-9} mbar. At every fracture surface Auger spectra of 5 to 10 different regions of about 50x50 μm^2 area were recorded at GBs and free surfaces using 10keV primary electron energy and 1 μA primary current. From the survey spectra (fig. 2) the Auger peak-to-peak heights (APPH´s) of Ni (61eV), S (151 eV), B (179 eV), C (272 eV) and O (510 eV) were determined.

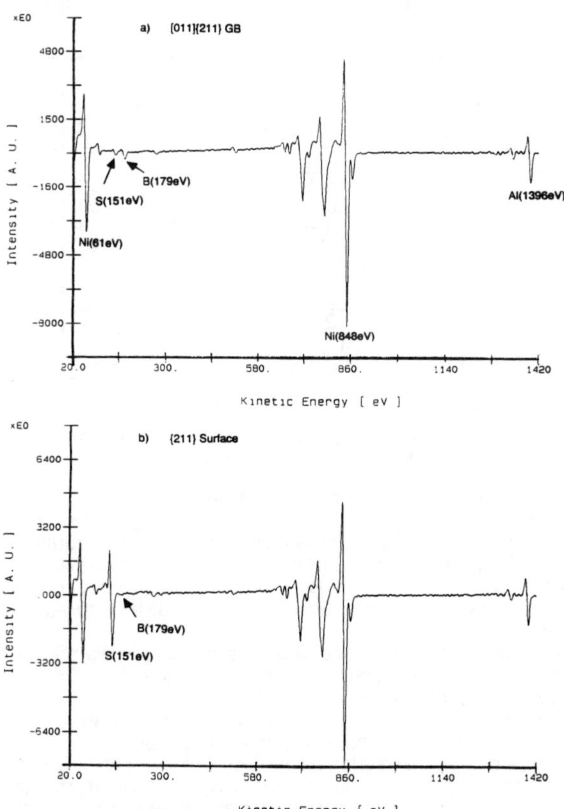

Fig. 2: Derivative Auger spectra of a) the <011>{211} GB and b) the {211} surface of a B-doped $Ni_{76}Al_{24}$ bicrystal with sulfur impurity.

Relating the APPH's of B and S to that of Ni gave some scatter in the data even by evaluating the APPH's of only one sample at different locations. It was found, that even at a pressure of $6 \cdot 10^{-10}$ mbar during fracture and analysis, O and C from residual gas were slowly covering the fracture surface. From the covering profiles of two samples, the covering coefficients for each element were determined. The atomic concentrations were then calculated considering these covering coefficients, and also inelastic mean free paths[14] and measured intensity standards for each element. In the multiplex spectral region of 47-70 eV the Ni (61eV) and the Al (68eV) overlapped peaks were least squares fitted to standards of Ni and Al to determine the relative concentrations.

RESULTS AND DISCUSSION

Using the determined atomic concentrations and supposing c_{GBmax} and c_{Vmax} to be 0.5 for B as for interstitials in fcc crystals, equation (2) was solved for the different temperatures to get segregation Gibbs free energies $\Delta G_{\Phi V}$. The change of cohesive energy $\Delta\gamma_{coh}$ by B and S segregation was calculated by equation (1). The results are listed in table 1. In fig. 3 the temperature dependence of B segregation at GBs is shown.

orientation	type	T [°C]	c_Φ(B) [at%]	$-\Delta G_{\Phi V}$ [kJ/mol]	$\Delta\gamma_{coh}$ [mJ/m^2]	c_Φ(S) [at%]	$-\Delta G_{\Phi V}$ [kJ/mol]	$\Delta\gamma_{coh}$ [mJ/m^2]
<110>{211}	GB	500	4.5 ± 0.8	22 ± 1	22	1.8 ± 0.9	48 ± 2	-14
		600	4.4 ± 0.7	24 ± 1	26	1.6 ± 1.0	53 ± 3	-16
		800	3.7 ± 0.9	28 ± 2	16	2.4 ± 1.0	69 ± 3	-17
<110>{311}	GB	500	2.6 ± 0.6	18 ± 1	11	0.2 ± 0.1	35 ± 3	- 4
		600	2.7 ± 0.5	20 ± 1	6	1.1 ± 0.8	51 ± 4	-12
		800	2.5 ± 0.8	23 ± 2	5	1.4 ± 0.8	64 ± 4	-14
{211}	S	500	1.6 ± 0.6	15 ± 2		10.2 ± 3	60 ± 1	
		600	1.4 ± 0.5	16 ± 2		11.6 ± 3	68 ± 2	
		800	1.9 ± 0.7	22 ± 3		7.8 ± 3	80 ± 2	
{311}	S	500	1.1 ± 0.3	12 ± 1		13.0 ± 2	61 ± 1	
		600	1.7 ± 0.5	17 ± 2		10.2 ± 3	67 ± 2	
		800	1.5 ± 0.5	20 ± 2		7.9 ± 1	80 ± 1	

Table 1: Equilibrium segregation levels of B and S at different interfaces of Ni_3Al, determined Gibbs free energies of segregation at GBs and free surfaces $\Delta G_{\Phi V}$ and the change in GB cohesive energy by B and S $\Delta\gamma_{coh}$.

Fig. 3 and table 1 clearly show that more B segregates to the {211} than to the {311} grain boundary. This is expected, because in anology to fcc lattices[15] the {311} boundary in $L1_2$ should be of lower GB energy than the {211} boundary. At free surfaces the level of B segregation is much smaller and is just above the detection limit of AES (≈0.5 at% for B). In contrast to B, S segregates strongly to free surfaces and weakly to GBs. The concentrations of B at GBs and S at free surfaces decrease with increasing temperature.

The values of the segregation Gibbs free energy $\Delta G_{\Phi V}$ seem to be small in comparison to the theoretically determined value of -183 kJ/mol for B segregation to <001> {210} symmetric tilt boundaries[16].

$\Delta G_{\Phi V}$ may contain terms of site competition and mutual interaction of the different segregants[17]. Because of the relatively low segregation levels and in view of the assumption that B occupies interstitial octahedral sites whereas S occupies substitutional sites[16], site competition is of minor importance. However, mutual interaction of the segregants in this multicomponent system cannot be excluded and deserves further study.

The values of $\Delta\gamma_{coh}$ show that the increase in strength of the {211} boundary is greater than that of the more stable {311} boundary. The absolute values are smaller than 1 % of the average GB cohesive energy γ_{coh} of 2998 mJ/m^2 determined theoretically by Chen et al.[16]. This is very small for effectively altering the fracture properties of Ni_3Al.

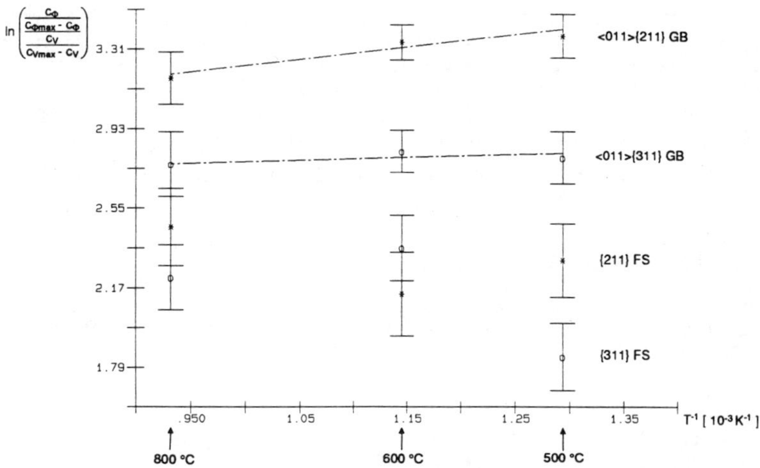

Fig. 3: B segregation in dependence of equilibrium temperature at <011>{211} and <011>{311} GBs and {211} and {311} free surfaces of $Ni_{76}Al_{24}$ bicrystals.

The concentrations of Al and Ni are listed in table 2. The Al concentration at surfaces is smaller than that at GBs and that at GBs is smaller than the bulk concentration. In contrast, the Ni concentration is similar at GBs and surfaces (see fig. 4) suggesting that S occupies Al sites at interfaces.

From the relative atomic concentrations of Al and Ni the secondary effect of B, namely attracting excess Ni to the GBs, is qualitatively supported by this study (see table 2). A more detailed quantification with additional samples is in progress.

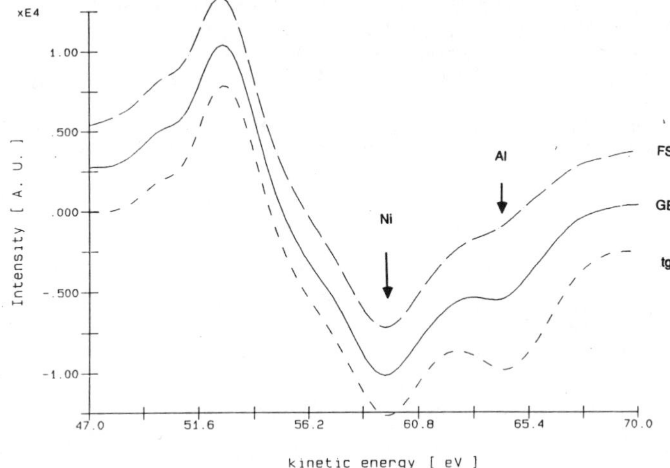

Fig. 4: Low energy Auger spectrum of a {211} surface, of a <011>{211} GB in a bicrystal of B-doped $Ni_{76}Al_{24}$ and, for comparison, a transgranular (tg) fracture surface of a polycrystal with the same composition.

	<011>{211}	{211}	<011>{311}	{311}
c_{Ni} [at%]	75±2	74±2	73±2	72±2
c_{Al} [at%]	20±2	13±2	23±2	16±2
$\frac{c_{Al}}{c_{Al}+c_{Ni}}$ [%]	21±2	15±2	24±2	18±2

Table 2: Atomic concentrations of Al and Ni at GBs and surfaces in B-doped $Ni_{76}Al_{24}$ annealed at 600 °C for 24 h.

ACKNOWLEDGEMENTS

The authors like to thank Dr. Chuang for the single crystal, D. Korn for the UHV diffusion bonding, Dr. P. Lejček for valuable discussions and the KSB-Stiftung for financial support.

[1]C. T. Liu, C. L. White and J.A.Horton, Acta metall. **33**, 213 (1985)
[2]C. L. Briant and A. I. Taub, Acta metall. **36**, 2761 (1988)
[3]E. P. George, C. T. Liu and R. A. Padgett, Scripta Metall. **23**, 979 (1989)
[4]J. E. Krzanowski, Scripta Metall. **23**, 1219 (1989)
[5]M. J. Mills, Scripta Metall. **23**, 2061 (1989)
[6]I. Baker and E. M. Schulson, Scripta Metall. **23**, 1883 (1989)
[7]H. Kung, D. R. Rasmussen and S. L. Sass, MSC Report #7035
[8]S. Hofmann, in Surface Segregation Phenomena, edited by P. A. Dowben & A. Miller (CRC Press, Boca Raton, 1990), p. 107
[9] C. L. White, Defect and Diffusion Forum **59**, 47 (1988)
[10] J. R. Smith and T. V. Cianciolo, Surf. Sci. **210**, L229 (1989)
[11]J. R. Rice and J.-S. Wang, Mat. Sci. & Eng. **A107**, 23 (1989)
[12]D. McLean, Grain Boundaries in Metals (Clarendon Press, Oxford 1957)
[13]T. Muschik, S. Hofmann and W. Gust, Scripta Metall. **22**, 349 (1988)
[14]M. P. Seah and W. Dench, Surf. Interf. Anal. **1**, 2 (1979)
[15]G. C. Hasson and C. Goux, Scripta Metall. **5**, 889 (1971)
[16]S. P. Chen, A. F. Voter, R. C. Albers, A. M. Boring and P. J. Hay, Scripta Metall. **23**, 217 (1989)
[17]M. Guttmann and D. McLean, in Interfacial Segregation, edited by W. C. Johnson and J. M. Blakely (American Society for Metals, Ohio, 1977), p. 261-348

SEGREGATION TO AND PHASE TRANSITION AT Σ5 (310)/[001] TILT GRAIN BOUNDARIES IN $Ni_{3-x}Al_{1+x}$ ALLOYS

R. Najafabadi and D. J. Srolovitz
University of Michigan, Department of Materials Science & Eng., Ann Arbor, MI 48109

ABSTRACT

The free energy simulation method and EAM type potentials are employed to study segregation to the Σ5 (310)/[001] tilt grain boundary and its free energy in ordered $Ni_{3-x}Al_{1+x}$ alloys. For 300K≤T≤900K, there is weak Al and Ni segregation to the grain boundary in the stoichiometric and Ni-rich (76.6 at. %) alloys, respectively. In the Al-rich (73.5 at. %) alloy, however, there is strong Al segregation. Al segregation induces the formation of a thin, ordered AlNi layer at the boundary. The width of this layer grows as the temperature is decreased to approximately 6 Å at 300K. The grain boundary segregation decreases the grain boundary free energy by less than 10% in the stoichiometric and Ni-rich alloys, and by 30% in the Al-rich alloys, respectively. These results suggest a new model to explain the observed brittleness of grain boundaries in polycrystalline Ni_3Al. In this model, the brittleness of Ni_3Al alloys is attributed to the formation of an intrinsically brittle, thin AlNi layer at grain boundaries. This brittle layer can be removed by increasing the grain boundary Ni content - either by changing the alloy composition or due to cosegregation with B.

INTRODUCTION

Ordered intermetallic Ni_3Al alloys have received considerable attention due to their high temperature mechanical properties. The focus of these studies [1-15] has been to understand the intergranular fracture behavior of the polycrystalline Ni_3Al at low temperatures. Experimentally, it has been demonstrated that the stoichiometric or Ni-rich alloys can be ductilized by doping with boron. The presence of boron in these alloys enhances the degree of Ni segregation to the grain boundaries [4]. It has been suggested that the B induced ductilization of Ni_3Al is related to this nickel enrichment at the grain boundaries. Two models [16,17] have been proposed to explain this presumed association. In one model [e.g. 16], an increase in the grain boundary cohesive energy is attributed to the nickel enrichment at grain boundaries. The other model [17] suggests that a compositionally disordered thin layer of up to 20 nm in thickness forms at the grain boundaries upon Ni segregation. It is speculated that the formation of this disordered phase at the grain boundary facilitates the motion of dislocations across the grains and prevents the initiation and growth of cracks along the grain boundary [18].

In our previous atomistic simulations of [001] twist grain boundaries in ordered Ni_3Al [14], we showed that the nickel enrichment at the grain boundaries in Ni-rich alloys was very weak and it did not significantly changes the grain boundary cohesive energy. It was also shown, that the nickel segregation to the boundaries did not induce the formation of a compositionally disordered phase, in agreement with other atomistic calculations [13]. Furthermore, in a recent field ion microscope/atom probe study [18], the formation of such a phase at boundaries has been questioned.

To address the generality of our previous results [14] (which were conducted on the highly symmetric, high angle Σ5 and Σ13 twist grain boundaries), the present investigation focuses on the effect of temperature and bulk composition on the chemical, structural, and thermodynamic properties of Σ5 (310)/[001] tilt grain boundaries in ordered Ni_3Al. Although this boundary is still a "special" boundary in that it has similar translational periodicity in the plane of the boundary as that of the twist boundary, it is of lower symmetry and therefore could be viewed as more representative of a "general" grain boundary than the Σ5 twist boundary. In the following section, we briefly describe the free energy simulation method. Its application to the Σ5 (310)/[001] tilt grain boundary (36.9 misorientation angle) in $Ni_{3-x}Al_{1+x}$ (x=-0.016, 0, and +0.015) are presented in the Results section.

METHOD

The present simulations are based upon the minimization of the free energy of the system with respect to atomic positions and the composition profile. The approximate free energy functional consists of a description of atomic vibrations, the Local Harmonic (LH) model, a Bragg-Williams description of the configurational entropy and an Embedded Atom Method type [8] description of the atomic interactions.

The local harmonic (LH) model has been applied with considerable success to both perfect and defected single component solids [19,20]. In this model, the classical vibrational contribution to the free energy for a single component system is given by

$$A_v = k_B T \sum_{i=1}^{N} \sum_{\beta=1}^{3} \ln\left[\frac{h\omega_{i\beta}}{2\pi k_B T}\right] \tag{1}$$

where k_B is Boltzmann's constant, h is Planck's constant, and w_{i1}, w_{i2}, and w_{i3} are the three vibrational eigenfrequencies of atom i. These frequencies may be determined in terms of the local dynamical matrix of each atom $D_{iab} = (\partial^2 E/\partial x_{ia}\partial x_{ib})$, where the x_{ia} and x_{ib} correspond to atomic displacements of atom i in the a and b directions, respectively.

In order to study binary alloys within the LH model frame work, each atom is replaced with an "effective atom" of mass $m_i = x_a(i)m_a + x_b(i)m_b$. Here, $x_a(i)$ is the probability that atomic site i is occupied by an atom of type a and correspondingly $x_b(i) = 1 - x_a(i)$ is the probability that the same atomic site is occupied by an atom of type b. The vibrational contribution to the free energy is determined from the appropriately averaged local dynamical matrix (see reference [21] for details). Within the Bragg-Williams approximation, the configurational entropy is written as

$$S_c = -k_B \sum_{i=1}^{N} \{x_a(i) \ln[x_a(i)] + x_b(i) \ln[x_b(i)]\} \tag{2}$$

In the present simulations, we employ the reduced Grand Canonical ensemble, where the total number of atoms remains fixed but the relative amounts of each atomic species varies. The appropriate thermodynamic potential for this type of ensemble is the reduced Grand potential:

$$\Omega = A + \Delta\mu \sum_{i=1}^{N} x_a(i) = E + A_v - TS_c + \Delta\mu \sum_{i=1}^{N} x_a(i) \tag{3}$$

where A is the Helmholtz free energy, E is the static energy (obtained from the interatomic potential), and Dm is the difference in chemical potential between atoms of type a and b. Given Dm, the equilibrium concentration at each site may be determined by minimizing W with respect to those concentrations. The grain boundary free energy (g_{gb}), in this ensemble, is defined as the difference in reduced Grand potential energies between the system containing a grain boundary and the defect-free system at the same chemical potential difference and temperature.

The simulation cell used here has been described in detail elsewhere [20]. Briefly, the grain boundary is embedded in a perfect crystal at the desired temperature and composition. In the plane of the boundary, a periodic border condition is enforced. Normal to the boundary plane the simulation cell is bounded by two infinite, mobile but rigid blocks of structurally perfect crystals. The extent of the grain boundary region between the two blocks of perfect crystal is increased during the course of the simulation to a size such that the free energy and other properties of the grain boundary remain unchanged within the preset convergence criteria.

In order to separate the effects of segregation from that due merely to atomic structural relaxation, we have performed two types of simulations for each temperature and bulk composition. In the first, the composition of each atomic site remains fixed at a value corresponding to the sublattice composition in the perfect crystal (i.e., no segregation) and the free energy is minimized with respect to the atom positions. In the second, the composition of each site is allowed to vary along with the atomic positions to reach the minimum free energy.

RESULTS

The free energy minimization method, described above, was used to investigate the compositional changes in the vicinity of the $\Sigma 5$ (310)/[001] tilt grain boundary in $Ni_{3-x}Al_{1+x}$ ordered alloys for three different bulk compositions, namely 73.5, 75.0, and 76.6 at. % Ni. These compositions are within the stability range of Ni_3Al phase as predicted by the free energy minimization model using the present interatomic potentials [8]. The experimental stability range for Ni_3Al is approximately 74-76 at. %. However, we note that with these interatomic potentials, we find that the Ni_3Al phase field extends significantly further toward the Ni-rich side of the phase diagram than is experimentally observed. The effect of temperature on grain boundary segregation was determined by performing simulations at 300, 600, and 900K for these three bulk compositions.

The starting structure for all our simulations was determined by minimizing the total energy of an unrelaxed $\Sigma 5$(310)/[001] grain boundary in the stoichiometric Ni_3Al with respect to the atom positions and the relative position of the two grains. In this minimization, concentration at each site was either one (i.e., pure Ni) or zero (pure Al). In Figure 1, we show the unrelaxed and relaxed structure of the boundary. In addition to 0.112a grain boundary expansion (a is the Ni_3Al lattice constant at zero temperature) in the z direction, the upper grain has shifted relative to the lower one by 0.50a and 0.17a in the x and y direction, respectively.

For the $\Sigma 5$ grain boundary in the Ni-rich alloy, the excess nickel concentration profiles of the (310) atomic planes near the boundary are shown in Figure 2a for T=300 and 900K. The excess nickel concentration is defined as the difference in the averaged nickel concentration in the (310) plane and a corresponding (310) plane in the bulk. The geometrical grain boundary plane was originally between planes 0 and 1. At T=300K, plane #3, which was initially an ordered Al-Ni plane, is enriched in nickel by about 47%. The Ni concentrations on the other (310) planes near the boundary are nearly the same as those in the bulk, except for the Ni-concentrations on planes #6 and #-6 which each increased by about 7%. When the temperature is increased from 300K to 900K, the nickel enrichment on plane #3 decreased to approximately 11% while the other (310) planes remain at essentially the same concentrations as in the bulk.

The concentration profiles at the stoichiometric bulk composition for 300 and 900K are shown in Figure 2b. Except for plane #3, almost no Ni or Al segregation to the other (310) planes near the boundary is observed. Plane #3, which was initially almost pure Ni, has become an ordered, plane with average Ni compositions of 51 and 30% at 300 and 900K, respectively. These results clearly show that there is a strong Al segregation to the grain boundary in this alloy. However, this segregation is limited to a single (310) plane.

The calculated excess concentration profile in the Al-rich alloy are shown in Figure 3 at 300 and 900K. At T=900K, planes #-4 and #-2 which were initially mixed ordered planes have been enriched in Al by 48% and 18%, respectively. Plane #1, which was initially an almost pure Ni plane, has become enriched in Al by approximately 16% at the same temperature. When the temperature is decreased to 300K, planes #-6, #-4, and #-2 become almost pure Al, while the Ni concentration on plane #1 is reduced to nearly 50% and it becomes ordered. The net segregation of Al to the grain boundary changes from 0.82 to 2.00 (310) Al monolayers when the temperature decreases from 900 to 300K.

The results for the Al-rich alloy composition suggest that a new phase is being formed at the grain boundary as the temperature is decreased from 900 to 300K. Figure 3 shows that the region immediately below the boundary is a stacking sequence of alternating Al and Ni (310) planes. This region extends up to seven (310) planes at 300K. The three (310) planes immediately above the grain boundary are ordered with nickel compositions of about 50%. Together, on both sides of the boundary there is a region of ordered AlNi phase which includes ten (310) planes. In Figure 4, we show the site concentrations in the plane perpendicular to the tilt axis at 300K. The grayness of the circles represents the nickel concentration at each site. The region delineated by dashed lines shows the new, ordered phase that formed upon Al segregation to the boundary.

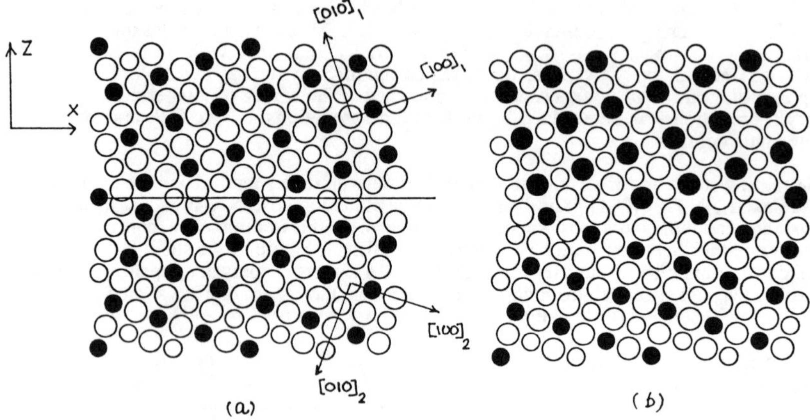

Figure 1. The $\Sigma5$ (310)/[001] tilt grain boundary structure. Open and filled circles are Ni and Al atoms, respectively. Different circle sizes represent different (002) planes. (a) the unrelaxed and (b) relaxed structure.

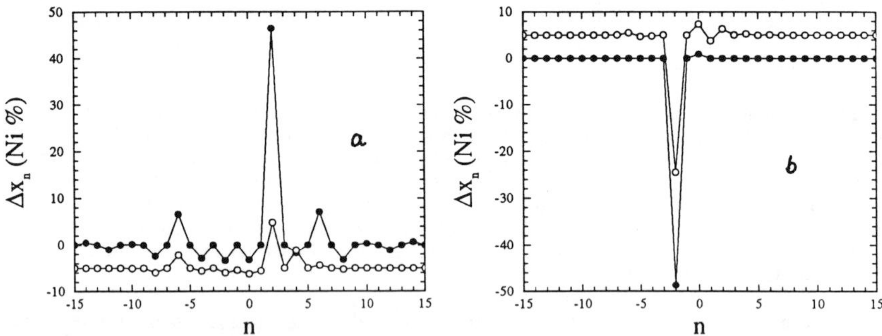

Figure 2. Excess nickel composition profiles of (310) planes plotted as a function of plane number from the interface for the $\Sigma5$ grain boundary. Δx_n is the difference in the Ni concentration on the nth (310) plane and its corresponding plane in the bulk. Open and filled symbols are for 900 and 300 K, respectively. (T=900 shifted by 5%). (a) the Ni-rich (b) the stoichiometric bulk composition

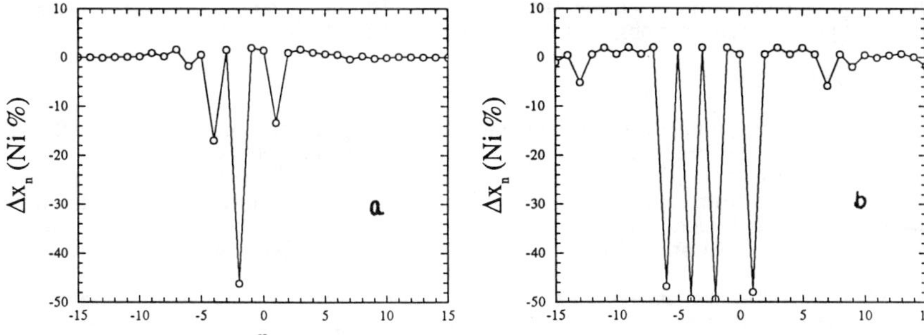

Figure 3. Excess nickel composition profiles of (310) planes plotted as a function of plane number from the interface for the $\Sigma 5$ grain boundary at the Al-rich bulk composition. (a) T=900 and (b) T=300K.

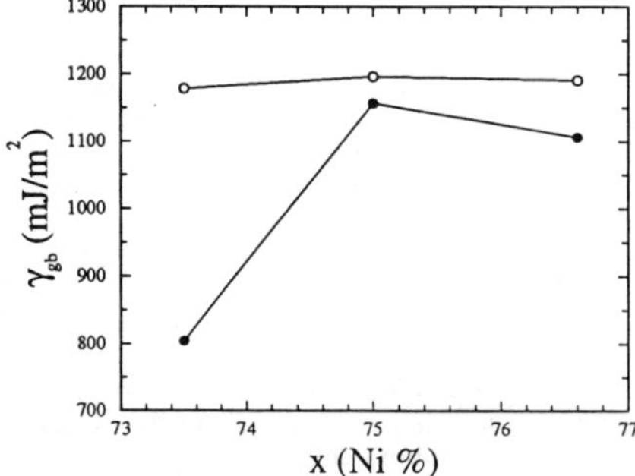

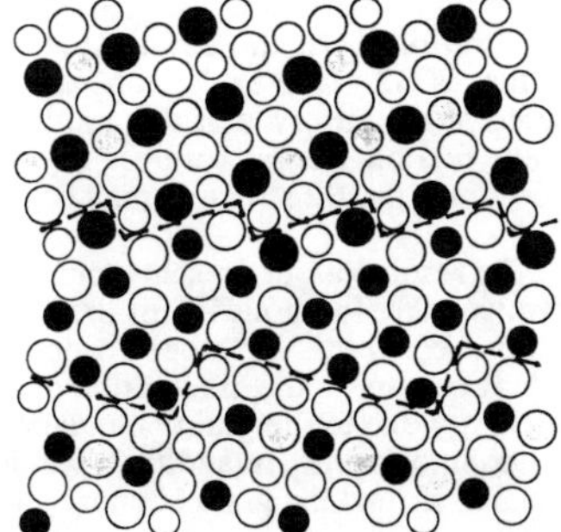

Figure 4. Atomic site concentration on two (002) planes of the grain boundary for Al-rich bulk composition at 300 K. The darker the circle, the higher the Al concentration at the site. The delineated region shows the presence of an ordered AlNi phase at the boundary.

Figure 5. Grain boundary free energy at T=300K as a function of Ni bulk composition. Solid and dashed lines represent the calculations with and without segregation, respectively.

The effect of segregation on the grain boundary free energy at 300K is shown in Fig. 5 as a function of alloy composition. The dashed line is the grain boundary free energy in the absence of segregation and the solid line is the free energy of the system in which segregation is allowed to occur. For the Al-rich alloy, segregation is accompanied by a pronounced decrease in the free energy (about 30%). On the other hand, the free energy decreases by 9 and 5% in the Ni-rich and stoichiometric alloys when segregation occurs. The same trend in the drops in the free energies upon segregation is observed at 900K but with a smaller overall magnitude.

DISCUSSION

The free energy minimization method was used to study the properties of $\Sigma 5$ (310)/[001] tilt grain boundary in ordered $Ni_{3-x}Al_{1+x}$ alloys. For the Ni-rich bulk composition ($300K \leq T \leq 900K$), strong Ni segregation was observed at a single (310) plane. The net segregation is, however, rather limited. No disordering in the grain boundary region in the temperature range studied here was observed. This is consistent with our previous simulation results [14] and Monte Carlo calculations [13] for the [001] twist boundaries in the same alloys.

While strong Al segregation to a single (310) plane in stoichiometric Ni_3Al is observed, the net Al segregation is rather small. This is in contrast to our previous twist boundary study [14], in which neither Al nor Ni were found to segregate. This is also contrary to the predicted Ni segregation to this $\Sigma 5$ tilt boundary, based upon zero-temperature calculations [8]. This may be attributable to different translation states (i.e., shifts in the plane of the boundary) of the grain boundaries in the two studies.

In the Al-rich alloy (73.5% Ni), the composition profile (Figure 3) shows significant Al segregation to the boundary region. The magnitude of this segregation decreases with increasing temperature. This is in agreement with our previous, twist boundary results [14]. Figure 3 and 4 show that the formation of an ordered AlNi phase at the boundary accompanies this Al segregation. The width of this phase grows from four to ten (310) planes as the temperature is reduced from 900 to 300K. The extent of this phase, normal to the boundary plane, is different on the two sides of the grain boundary. On one side of the boundary, the (310) planes alternate between nearly pure Ni and Al (310), while on the other side, each (310) plane is ordered (i.e., contains both Ni and Al sublattices). This new ordering in the region adjacent to the grain boundary also introduces two new interfaces: i.e., between the ordered grain boundary phase and the bulk Ni_3Al phase). This new phase is a distorted $L1_0$ structure of equal Ni and Al concentrations. At the equiatomic AlNi alloy composition, the equilibrium perfect crystal structure is B2. Due to the coherency constraints imposed by the surrounding $L1_2$ phase, the ordered phase observed at the grain boundary cannot be B2 structure, but instead takes on the closely related $L1_0$ structure. Perhaps, if the coherency constraints are relaxed by misfit dislocations, the B2 structure may be observed instead of the $L1_0$ structure.

The results of our present atomistic study for the $\Sigma 5$ (310)/[001] tilt grain boundary and our previous study (without B) for the twist grain boundaries [14] do not support either of the two models [16,17] that have been proposed to explain the ductilization of polycrystalline Ni-rich Ni_3Al. Contrary to the assumption made in the first model [16], the reduction in grain boundary free energy due to the grain boundary segregation is less for Ni-rich alloys than it is for Al-rich alloy compositions. In addition, we do not observe the formation of a thin layer of disordered fcc nickel around the grain boundary, as is assumed in the second model [17].

We do, however, observe the formation of a new phase at the grain boundary in the Al-rich alloy. One may speculate that the brittle fracture behavior of Ni_3Al polycrystalline solids is not due to the intrinsic brittleness of grain boundaries in ordered intermetallic alloys of the $L1_2$ structure. Perhaps, the formation of a brittle AlNi ordered phase at grain boundaries is responsible for the intergranular fracture behavior of this alloy at Al-rich compositions. In this light, the effect of alloying Ni_3Al with boron may be to remove this $L1_0$ phase from the grain boundaries. This may be associated with cosegregation of Ni along with the B, since we have shown that increasing the Ni content of the boundary destabilizes the $L1_0$ boundary phase. If, indeed, the brittleness associated with sub-stoichiometric Ni_3Al is associated with the formation of a brittle $L1_0$ AlNi phase at the boundary, then Ni-rich grain boundaries (whether due to an increase in the bulk Ni content or due to B addition) should be less prone to promote brittle fracture.

Acknowledgment - This research was supported by the Division of Materials Science of the Office of Basic Energy Sciences of the U. S. Department of Energy, Grant #FG02-88ER-45367.

References

1. D. D. Sieloff, S. S. Brenner, and H. Ming Jian, MRS Symp. Proc. 133, 213 (1989).
2. D. D. Sieloff, S. S. Brenner, and M.G. Burke, MRS Symp. Proc. 83, 87 (1987).
3. I. Baker, E. M. Schulson, J. R. Michael, and S. J. Pennycook, Phil. Mag. B 62, 659 (1990).
4. I. Baker, E. M. Schulson, and J. R. Michael, Phil. Mag. B 57, 379 (1988).
5. E. P. George, C. T. Liu, and R. A. Padgett, Scripta Metall. 23, 979 (1989).
6. I. Baker and E. M. Schulson, Scripta Metall. 23, 1883 (1989).
7. H. Kung, D. R. Rasmussen, and S. L. Sass, submitted to Acta Metall. Mater.
8. S. P. Chen, D. J. Srolovitz, and A. F. Voter, J. Mat. Res. 4, 62 (1989).
9. S. P. Chen, A. F. Voter, R. C. Albers, A. M. Boring, and P. J. Hay, J. Mat. Res. 5, 955 (1990).
10. G. J. Ackland and V. Vitek, MRS Symp. Proc. 133, 106 (1989).
11. J. J. Kruisman, V. Vitek, and J. Th. M. De Hosson, Acta Metall. 36, 2729 (1988).
12. S. M. Foiles and M. S. Daw, J. Mat. Res. 2, 5 (1987).
13. S. M. Foiles, paper presented at TMS Fall Meeting, Detroit (1990).
14. R. Najafabadi, H. Y. Wang, D. J. Srolovitz, and R. LeSar, MRS Symp. Proc. 213, 51 (1990).
15. M. Yan, V. Vitek, and G. J. Ackland , preprint.
16. A. I. Taub and C. L. Briant, Acta Metall. 35, 427 (1987).
17. E. M. Schulson, I. Baker, and H. J. Frost, MRS Symp. Proc. 83, 1195 (1987).
18. S. S. Brenner and H. Ming-Jian, Scipta Met. 24, 671 (190).
19. R. LeSar, R. Najafabadi, and D. J. Srolovitz, Phys. Rev. Lett. 63, 624 (1989).
20. R. Najafabadi, D. J. Srolovitz, and R. LeSar, J. Mater. Res. 5, 2663 (1990).
21. R. Najafabadi, H. Y. Wang, D. J. Srolovitz, and R. LeSar, Acta Metall. Mater., in press.
22. A. F. Voter and S. P. Chen, Mat. Res. Soc. Symp. Proc. 82, 175 (1987).

THE EFFECT OF TEMPERATURE ON THE STRUCTURE OF GRAIN BOUNDARIES IN Ni_3Al WITH AND WITHOUT BORON

DONGLIANG LIN (T. L. LIN), JIAN SUN, DA CHEN AND MIN LU
Department of Materials Science and Engineering, Shanghai Jiao Tong University, Shanghai 200030, P. R. China

ABSTRACT

Monte Carlo computer simulations with embedded atom method potentials are used to study the structure and energy of symmetric tilt grain boundaries in Ni_3Al at high temperatures and compared with those at absolute zero, which is conducted with static relaxation method. The effect of stoichiometry and boron addition on the structure and energy of grain boundary are also investigated. The simulation results show that there exists more compositional disorder at grain boundaries than in the bulk in non-stoichiometric Ni_3Al with and without boron. At absolute zero the grain boundary consists of periodically distributed structural units which are distorted with the increasing temperature. The effect of temperature on the structure and energy of grain boundaries in Ni_3Al with and without boron was discussed.

INTRODUCTION

The main reason for studies of grain boundaries in intermetallic compound Ni_3Al is that the grain boundaries are intrinsically brittle, and the segregation of boron to grain boundaries dramatically improves the ductility in polycrystalline Ni_3Al. Recently, there has been some atomic simulation studies of the structure and property of grain boundaries in intermetallic Ni_3Al [1-4]. It has been shown that the intrinsic brittleness of grain boundaries in ordered compound is related to the preference for chemical order [1], and the presence of boron at grain boundaries increases the grain boundary cohesion, in particular at Ni-rich grain boundaries [2]. These results suggest that boron doping at grain boundaries should decrease the tendency for brittle fracture in polycrystalline Ni_3Al. However, most of these calculations have been conducted at zero temperature. The temperature effects are particularly important because compositional disordering and spatial disordering at grain boundaries are expected for high temperatures. The Monte Carlo and Free Energy simulations have been conducted to calculate these questions [3,4].

In the present study, Monte Carlo (MC) simulation with embedded atom method (EAM) potentials are used to study the structure and energy of symmetric tilt grain boundaries (GBs) [101] $\Sigma3$ / (121) in Ni_3Al at 500K and 1200 K, and compared with those at zero temperature, which is conducted with energy gradient method. The effect of stoichiometry and boron addition on the structure and energy of GBs are also considered in this study. Finally, the effect of temperature on the structure and energy of the GBs was discussed.

COMPUTATIONAL PROCEDURES

Monte Carlo simulations were performed empolying EAM potentails on bi-crystal slab, which used the N. Metropolis *et al.* algorithm [5] based on a stochastic process which, by moving atoms one at a time, generates a Boltzman-weighted chain of phase configurations of a given N-atom system. We adopted the canonical ensemble where the total numbers of atoms, bulk composition, volume and temperature were held constant. At each step of the calculation chain, two features of changes were empolyed in calculation separately. They are (a) moving an atom from its geometric position; (b) changing the position of some antisite defects. The first change allows for relaxing atomic configuration. The atomic relaxations are a result of thermal vibrations of atoms and structural changes which occur at or near grain boundaries. The second change is for studying the equilibrium position of antisite defects in ordered intermetallics, which deduce the segregation of Ni or Al to grain boundaries in Ni_3Al with various Al contents. The magnitude of the maximum atomic displacement was given as the value with which the ratio of successful changes to total attempts is about half or more. Three dimensional periodic boundary condition was empolyed in the calculation.

When a new configuration is established at each step, the total energy of the system is computed by EAM. The EAM potentials is a semiempirical technique for computing the energy of an arbitary arrangement of atoms. In this method, the energy is modeled as having two contributions: the energy to embed an atom into the local electron density provided by the remainder of the atoms and an electrostatic interaction represented by pair interaction. In this work the EAM potentials of Ni-Al-B systems described by S.P.Chen *et al.* [2] is used to calculate the energy of the system after each micro-MC step.

The decision whether or not a particular change is retained, is based on the computation of the relative probablity (Δp) of two configuration. If $\Delta p > 1$, the new configuration is always retained, while if $\Delta p \leqslant 1$, the new configuration is retained with the probability Δp, i.e., when $\Delta p > \xi$, the new configuration is also retained. Here ξ is a random number between 0 and 1. Repeating the procedure a large number of times results in a set of atomic configurations corresponding to thermal equilibrium. Whether or not a a equilibrium state is achieved is judged by monitoring the evolution of total energy of system. Figure 1 shows the evolution of the total energy during the Monte Carlo simulation, which demonstrates that 10^6 micro-MC steps is sufficient to obtain an equilibrium state. The statical errors can be calculated using the following expression:

$$\langle(\Delta E)^2\rangle = \frac{1}{M(M-1)} \sum_{i=N+1}^{N+M} (E_i^2 - \langle E \rangle^2) \quad (1)$$

where N is the number of micro-MC steps for reaching at the equilibrium state, and M is the number of micro-MC steps for being analyzed. In this case, $\langle(\Delta E)^2\rangle$ equals 8.35 $(eV)^2$.

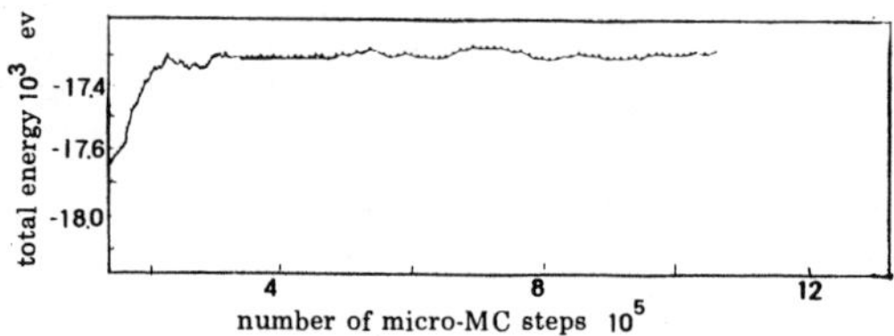

Fig.1 The evaluation of the total energy with the number of configurational chains.

The symmetric tilt grain boundary, whose tilt axis is in the [101] direction, and θ is 70.53^0 corresponding to $\Sigma=3$ and grain boundary plane (121) was constructed by coincident-site lattice model. Boron is inserted into the interstitial sites at the grain boundary. Both rigid body translation for two direction vertical to grain boundary plane and energy gradient method was used to calculate the structure and energy of grain boundaries with and without boron at zero temperature. Since the $L1_2$ structure consists of four interpentrating simple cubic lattices, there are three types of boundaries corresponding to the stoichiometric, Ni-rich and Al-rich composition in the grain boundaries which have been denoted as 50/100, 100/100 and 50/50 in ref. [6]. In the current calculation, the 50/100 type grain boundaries were chosen to study their properties as a function of temperature and bulk composition. The calculated structure at zero temperature is used as the starting configuration for Monte Carlo simulation at high temperatures.

RESULTS AND DISCUSSION

The Structures of Grain Boundaries

The Structural unit model [2,7] has been successful in describing the structures and energies of grain boundaries. In our calculation, the grain boundaries in Ni_3Al with and without boron at zero temperature consist of the periodically distributed structural units, as indicated in Figure 2. When boron is inserted into the interstitial site at grain boundaries, the structural units change, but still keep periodical. Figure 3 and Figure 4 show the grain boundary structure with and without boron at 500K and 1200K after Monte Carlo simulation.

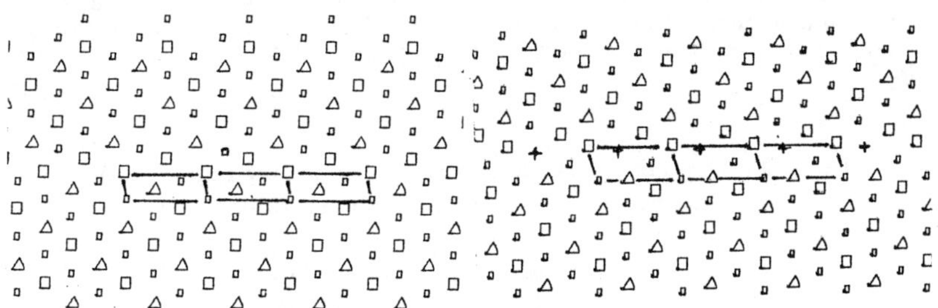

Fig.2 The structures of grain boundaries with and without boron at zero temperature. □ Ni △ Al + B

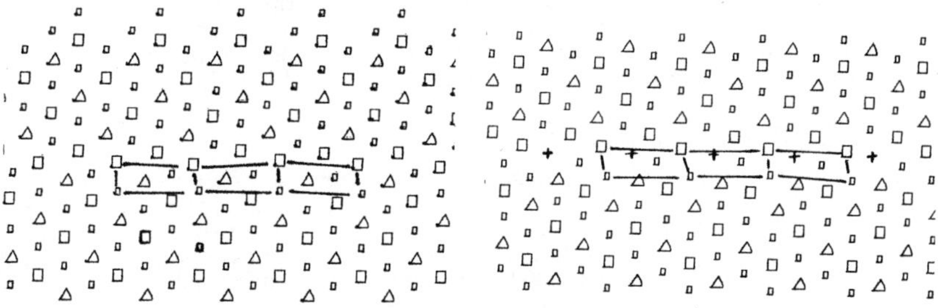

Fig.3 The structures of grain boundaries with and without boron at 500 K after MC simulation.

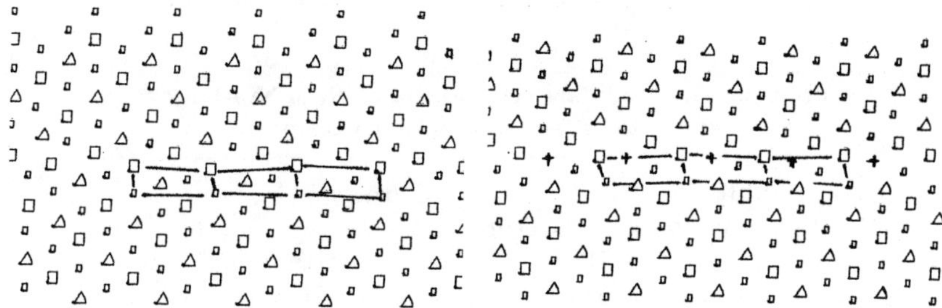

Fig.4 The structures of grain boundaries with and without boron at 1200 K after MC simulation.

The atomic structure of grain boundaries can also be described by the structural unit model. However, during the relaxation, the originally perfect shape of the structural units is distorted randomly at high temperature. The distortion in the bulk crystal is very weaker than that in the grain boundary region. The random distortion increases with the increasing temperature and with boron dopants at grain boundaries. The distortion at grain boundaries and their adjacency can be described by the parameter D:

$$D = \left(\sum_{i=1}^{n} (\mathbf{r}_i^t - \mathbf{r}_i^0)^2 \right)^{1/2} \qquad (2)$$

where n is the number of calculated atoms. $\mathbf{r}_i^t$ is the displacement of an atom at high temperature after Monte Carlo relaxation. $\mathbf{r}_i^0$ is the displacement of an atom at zero temperature after static relaxation. The parameter D in the grain boundary region with and without boron at different temperature is shown in Table I.

Table I. The parameter D(Å) at grain boundary region with and without boron at high temperatures

	500K	1200K
GB	2.548	3.654
GB (B)	3.378	4.378

The value of the parameter D is higher at 1200K than that at 500K, and at the same temperatures, the value of D is higher at boron doped grain boundary than that at grain boundary without boron. The structure distortion at high temperature can be seen from the change of the parameter D.

Segregation and Local Disorder at GBs

In this calculation, some Ni or Al antisite defects is distributed randomly in the crystal initially when the composition of Ni_3Al deviate the stoichiometry. After the Monte Carlo relaxation, one can see that the equilibrium position of the antisite defects in the grain boundary, which results in the segregation of Ni or Al to grain boundaries in hypostoichiometric or hyperstoichiometric Ni_3Al. These results were also obtained by S. M. Foils' calculation [3], which suggests that the grain boundary acts as a sink for the Ni in Ni-rich alloys and for the Al in Al-rich alloys. The local composition of the grain boundary can be described in term of excess of Ni on the grain boundary. The grain boundary excess of Ni as a function of bulk composition in Ni_3Al with and without boron at different temperatures are shown in Table II.

Table II. The grain boundary excess of Ni in Ni_3Al alloys with and without boron at different temperatures

Bulk	Ni excess /Å²			
Composition	500K		1200K	
	GB(B)	GB	GB(B)	GB
Ni-24at%Al	0.023	0.020	0.015	0.014
Ni-25at%Al	0.003	0.002	0.001	0.002
Ni-26at%Al	-0.018	-0.019	-0.013	-0.013

It is shown that there is Ni segregation, almost no segregation, and Al segregation to the grain boundaries with and without boron corresponding to the Ni-rich, stoichiometric and Al-rich composition of the bulk. As is expected, the segregation of Ni or Al to the grain boundary decreases with increasing temperature. The presence of boron at grain boundaries does not induced any compositional disorder at grain boundaries. These results are consistent with the previous calculation [3] and have been observed experimentally in Ni_3Al alloys [8,9]. Microdiffration and EDS-microanalysis of grain boundaries in Ni_3Al alloys show that there is a local compositional disorder in the vicinity of grain boundaries, which is related to the Ni enrichment or depletion in grain boundaries in hypostoichiometri or hyperstoichiometric Ni_3Al and the degree of order in stoichiometric Ni_3Al still keeps higher in grain boundaries because there is no Ni or Al enrichment in grain boundaries [9]. Because boron only ductilies the hypostoichiometric Ni_3Al effectively, the ductilizing effect of boron on Ni_3Al alloys can not be explained by the formation of the local disorder in the grain boundary region. Our calculation may support the conclusion that the presence of boron enhance the cohensive strength of grain boundaries suggested by S.P.Chen and A.F.Voter based on static relaxations [2].

The Energies of Grain Boundaries

The energies of the grain boundaries with and without boron at different temperatures relaxed by static relaxation and Monte Carlo simulation are also calculated. Table III shows that the grain boundary energies increase with the increasing temperature. This is consistant with the change of the grain boundary structure. When the temperature increases, the grain boundary atoms are of higher mobility under thermal activation and the grain boundary energy increases either . The reason that the energies of the grain boundaries with boron dopants are higher than those without boron seems that the reference state was not considered in calculation.

Table III The energies of the grain boundaries at different temperatures (mJ/m^2)

Temperature	0 K	500 K	1200 K
GB	895	1800	3154
GB (B)	1263	2154	3501

The intrinsic brittleness of grain boundaries in Ni_3Al at room temperature has been considered to relate to the preference for chemical order [1]. This phenomenon also connects with the low mobility of atoms at grain boundaries at room temperature. When the temperature increases, the tendency of brittleness of grain boundaries becomes lower, because the grain boundary energy and the grain boundary distortion increase, resulting in the high mobility of atoms at grain boundaries.

SUMMARY

The static relaxation and Monte Carlo simulation empolying EAM potentials have been used to study [101] $\Sigma3$ / (121) symmetric tilt grain boundaries in Ni_3Al with and without boron at absolute zero and high temperatures.

(1) At absolute zero, the grain boundary structures consist of the periodically distributed structural units, however, they become distorted randomly with the increasing temperature and boron segregation. The random high distortion at grain boundaries reflects the spatial disorder tendency of atoms at grain

boundaries to some extent at the high temperature.

(2) When the composition of Ni_3Al deviates from the stoichiometry, there is the Ni and Al segregations to the hypostoichiometric and hyperstoichiometric grain boundaries with and without boron, respectively, which results in the chemical disorder in the vicinity of grain boundaries. These segregations decrease with the increasing temperature. However, there is no Ni or Al segregation in stoichiometric Ni_3Al. The local disorder at grain boundaries is not affected by the presence of boron at grain boundaries, therefore, the presence of such a local disorder region at grain boundaries is not a main factor to control the ductility of Ni_3Al.

(3) The change of grain boundary energy is consistent with that of grain boundary structure with the increasing temperature. These changes reflect the mobility of atoms at grain boundaries. The low brittle tendency of grain boundaries in Ni_3Al at the higher temperature can be attributed to the high mobility of atoms at grain boundaries. Furthermore, the benefits of boron to the grain boundary brittleness in Ni_3Al may also be explained by the high mobility of atoms at the grain boundary with boron dopants.

REFERENCE

1. V. Vitek and S.P. Chen, Scripta Metall. *et* Mater. **25**, 1237 (1991).
2. S.P. Chen, A.F. Voter, R.C. Alkers, A.M. Boring and P.J. Hay, J. Mater. Res. **5**, 955 (1990).
3. R. Najafabadi, H.Y. Wang, D.J. Srolovitz and R. Lesar in High Temperature Ordered Intermetallics edited by L.A. Johnson, D.P. Pope and J.O. Stiegler, (Mat. Res. Soc. Symp. **213**, Pittsburge, PA, 1991), p. 51.
4. S.M. Foils in High Temperature Ordered Intermetallics edited by N.S. Stoloff, C.C. Koch, C.T. Liu and O. Izumi (Mat. Res. Soc. Symp. **81**, Pittsburge, PA, 1987), p. 51.
5. N. Metropolis, A.W. Rosenbluth, M.C. Rosenbluth, A.H. Teller and E. Feller, J. Chem. Phys. **21,** 1087 (1953).
6. S.P. Chen, D.J. Srolovitz and A.F. Voter, J. Mater. Res. **4**, 62 (1989).
7. G.J. Wang and V. Vitek, Acta Metall. **34**, 1941 (1986).
8. James E. Krzanowski, Scripta Metall. **22**, 1219 (1988).
9. J. Sun and T. L. Lin, paper presented at MRS Fall Meeting, Boston (1992).

DEFORMATION MECHANISMS IN A LAVES PHASE

Yaping Liu, Samuel M. Allen and James D. Livingston

Department of Materials Science and Engineering
Massachusetts Institute of Technology
Cambridge, MA 02139, USA

ABSTRACT

The stress-induced phase transformation between C36 and C15 structures in Fe_2Zr is studied by electron microscopy. The nucleus of the transformation is believed to be some pre-existing C15 layers in C36 particles. Microstructural evidence for three mechanisms of growth of a new phase were found: Fault accumulation and rearrangement, moving of individual partial dislocations between two phases, and the migration of microscopic ledges composed of a series of Shockley partials between C36 and C15. Plastic deformation by slip on non-basal planes of C36 caused by indentation is studied.

I. INTRODUCTION

Stress-induced phase transformation has been found to be one of the martensitic transformation mechanisms in steels, many alloys and even semiconductors. However, there has been comparatively little work on this mechanism in Laves phases, probably due to the difficulty involved in the shear deformation in the phases of complex structure. Y. Ohba and N. Sakuma[1] observed a gum elastic deformation in $MgCu_2$ Laves phase when it was rapidly cooled from the melt, which was considered to be due to the stress-induced transformation, at a condition of very low stress and room temperature. Our previous work[2] used X-ray diffraction and electron microscopy to show that such a transformation could occur in a two-phase Fe-Zr alloy during uniaxial compression in room temperature. The present paper provides further results and discusses observations related to the nucleation and growth mechanism of the transformation in the Fe-10 at% Zr alloy.

Laves phases have three structure types: Cubic C15 ($MgCu_2$), hexagonal C14 ($MgZn_2$) or dihexagonal C36 ($MgNi_2$). The structure type appearing in a Laves phase is mainly determined by the electronic factor. Many studies suggested favorable ranges in terms of valence electron-to-atom ratios for the formation of Laves phases[3]. In Fe-Zr system, Kai et al.[4] found from their results of X-ray and magnetic measurements that two single-phase regions of dihexagonal ($MgNi_2$-type) and of cubic ($MgCu_2$) Laves phase are located in two separate composition ranges from 27.3 to 31.4 at% Zr and from 32.8 to 34.0 at% Zr, respectively. Because of the eutectic point at 8.8 at% Zr, The 10 at% Zr alloy used in the present study contains substantial amounts of both eutectic and pro-eutectic Fe_2Zr Laves phases.

II. EXPERIMENTAL

The Fe-10 at% Zr alloy used in the study was prepared by arc melting. Samples for compression testing were then cut from the arc-cast ingots using an electric-discharge machine. The size of the samples was 5×5×5 mm. Samples were encapsulated in a vacuum of 10^{-6} Torr and annealed at 1190°C for 48 hours. Compression experiments were performed at room temperature with a crosshead speed of 2.5×10^{-3} cm/min. Indentations were performed at room temperature in a DM 400 Microhardness Tester, and 10 × 10 arrays of 50 g Vickers indentations were made on the central part of alloy disks which are 3 mm in diameter and 300 μm in thickness.

TEM specimens approximately 0.4 mm thick were cut from the undeformed and compressed samples using a diamond saw, ground to a thickness of 0.1 mm, and electropolished in a solution of 10% perchloric acid-90% methanol. Ion milling was also used in some samples to get addional thin area for high resolution microscopy. The indentation samples were polished and dimpled to 30 μm followed by atom-mill thinning to electron transparency. A JEOL 200CX and an Akashi EM-002B high-resolution transmission electron microscope operating at an accelerating voltage of 200 kV were used in the microstructure analysis. A Vg HB5 scanning transmission electron microscope with a minimum probe size of 0.5 nm was used in the compositional micro-analysis. The thin sections cut from undeformed and compressed samples were also used in X-ray diffraction experiments, which were performed on an RIGAKU RU300 diffractometer with Cu-K_α radiation and a rotating anode. In order to minimize the influence of any possible texture produced by the compression, X-ray specimens were cut at random orientations from lateral faces of the compressed samples.

III. RESULTS AND DISCUSSION

A typical nominal stress-strain curve is shown in Fig.1. Samples of the annealed alloy were deformed in compression at room temperature to strains of 46-48% without any macroscopic cracks or load drops. The 0.2% offset yield strength $\sigma_{0.2}$ of the alloy is about 730 MPa. Microstructural analysis by TEM revealed that the α-Fe matrix was heavily deformed, as it contained a high density of dislocation arrays, networks and tangles. On the other hand, SEM revealed many cracks in the pro-eutectic Laves particles. The chemical compositions of both the eutectic and pro-eutectic Fe_2Zr were determined to be in the range of 29-30 at% Zr, by STEM composition micro-analysis.

X-ray diffraction provided evidence that microstructure changes had occurred within the Laves phase. Comparison of the X-ray diffraction pattern of samples before and after compression in Fig.2 shows that the dominant structure of the Laves phase in undeformed samples is C36. However, after deformation, the intensity of peaks of C15 increased. Repeated experiments on many samples showed the same result: the volume fraction of C15 structure increased after compression. To study the detail and mechanism of the phase transformation, the microstructures of both compressed and undeformed alloys were observed using TEM. Transmission electron microscopy revealed extensive faulting in the C36 Laves phase before deformation. A high-resolution lattice image is shown in Fig.3, in which the stacking sequence of close-packed planes is indicated by the line superimposed on the figure. The typical dihexagonal C36 structure is a mixture composing a half hexagonal (h) and a half cubic (c) components, with a stacking sequence of "hchc". The existence of the faults changes the stacking sequence of some of the layers in the figure, as indicated by the arrows. Closer examination of the faults reveals that they can be divided into two types. In the first type two or more (but usually no more than 7-8) cubic type layers stack continuously, changing the stacking sequence into, say, "hcchc" as indicated by arrow A in the figure. This can be interpreted as an insertion of a "c" layer or a removal of a "h" layer. The second type is a continuous stacking of "h" layers, but

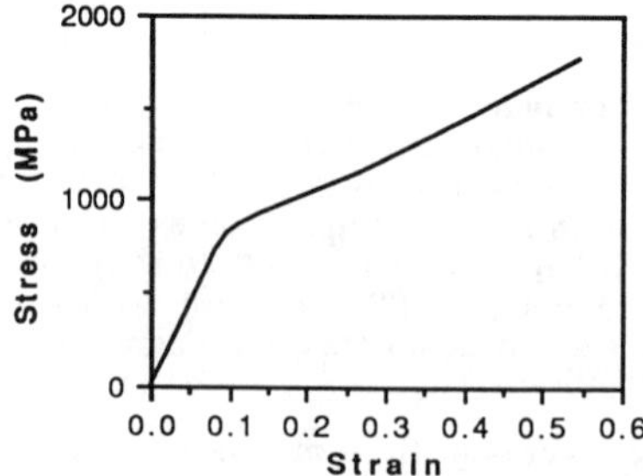

Fig.1 Stress-strain curve of room temperature compression test of Fe-10 at% Zr alloy.

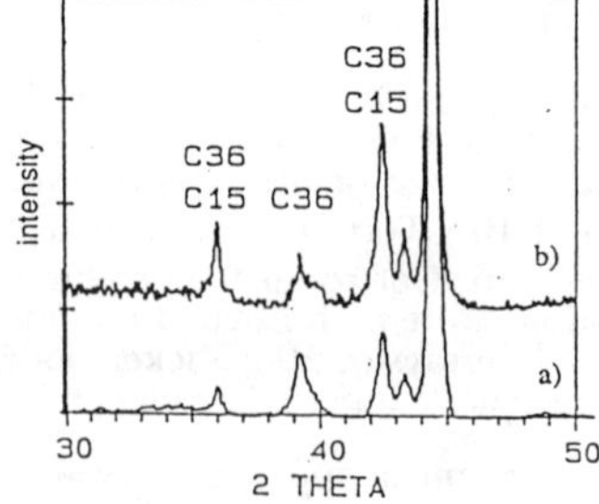

Fig.2 X-ray diffraction patterns of (a) undeformed and (b) compressed alloy.

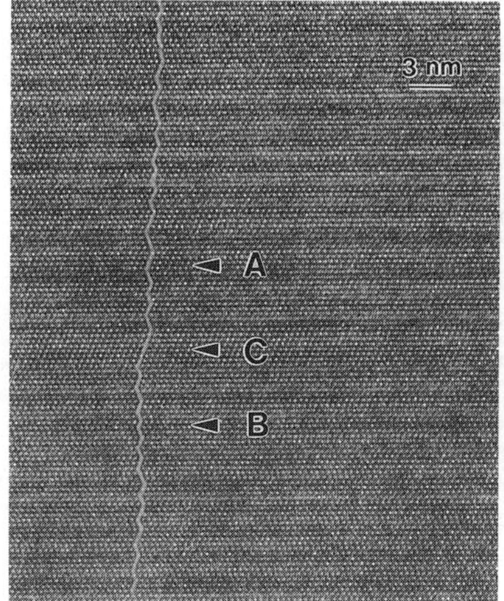

Fig.3 Stacking sequence of undeformed C36 planes showing the faults (A and B) and a nucleus of C15 structure (C).

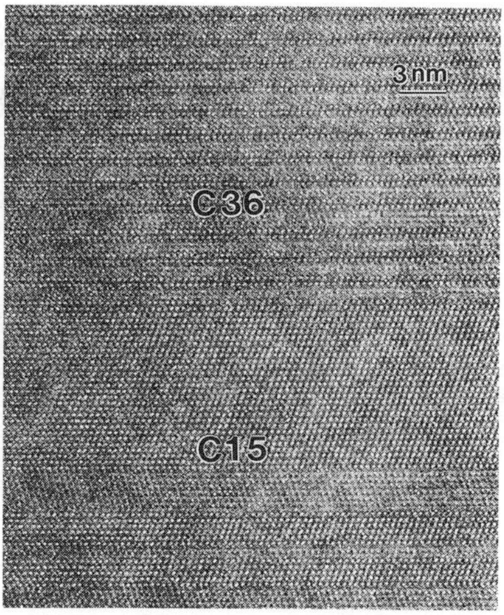

Fig.4 New phase C15 formed in a C36 particle caused by the compression.

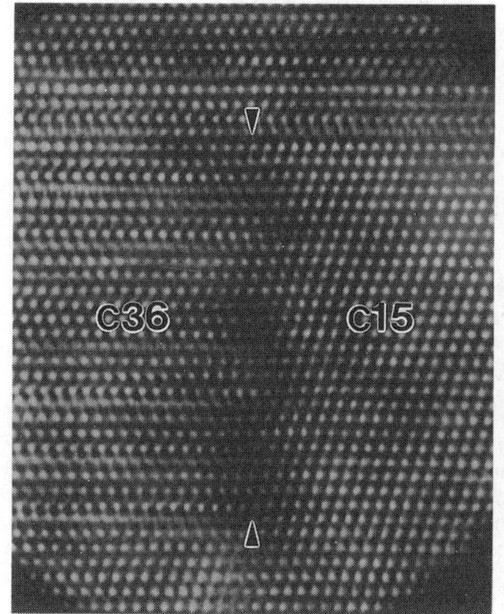

Fig.5 Computer enhanced picture of ledge interface between C15 and C36.

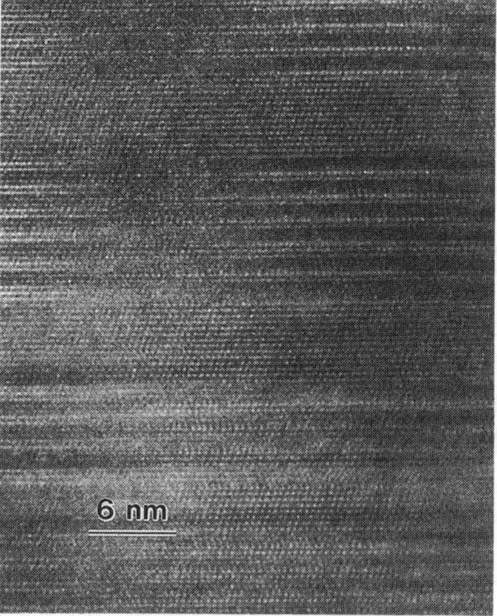

Fig.6 Growth mechanism: Gathering of faults and narrow C15 bands near a C15 nucleus.

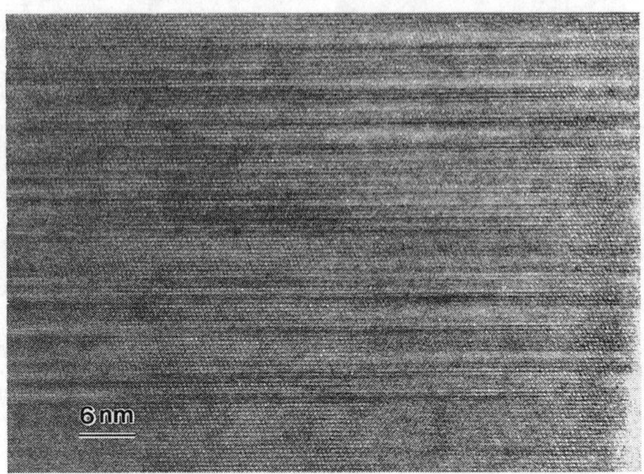

Fig 7 Growth mechanism: Moving of single partials between two phases. To the left , There are more and more C15; to the right, more and more C36.

usaully no more than two layers, as indicated by arrow B in the figure. This is a insertion of a "h" layer or a removal of a "c" layer. The first type is more common than the second one, causing the composition of the mixture to deviate from the ideal 50% hexagonal and 50% cubic components. Statistical calculation on many micrographs shows the mixture is: 54% cubic and 46% hexagonal stacking componants.

The high density of stacking faults implies a low stacking fault energy and hence easy nucleation and growth of a new phase. It should be noted that the continuous stacking of "c" layers, indicated by C in Fig.3, is frequently observed in the alloy. The pre-existing C15 layers should be the nuclei for the phase transformation. It is interesting that the layer numbers of the C15 nuclei in many particles are almost all the same (7-8 layers).

After compression, the new phase C15 was found in the Laves particles by electron microscope in a band morphology. An example is shown in Fig.4. Looking along the zone axis of $[2\overline{1}\,\overline{1}\,0]_{C36}/[011]_{C15}$ in the figure one can easily distinguish between the C15 area which has nearly pure "c" stacking, and the C36 area which has the stacking sequence "hchc".

The phase transformation between C36 and C15 is a shear transformation, which occurs by the glide of Shockley partial dislocations of Burgers vector b = a/6 <211> [5,6] and involves two partials every four layers[2]. By observing and analyzing numerous pictures of Laves particles containing both C15 and C36 structures, some mechanisms of growth of a new phase are suggested: Firstly, the transformation proceeds gradually from one narrow band to another adjacent band, by the motion of an interface with microscopic ledges composed of a series of Shockley partials between C36 and C15. This interface can be observed in a computer enhanced TEM picture in Fig.5, indicated by two arrows. The locations of the partial dislocations are also marked in the figure. This mechanism has been discussed in detail in [2].

Besides the migration of the ledge interface, the new phase can grow by a widening mechanism. This was frequently observed on dark-field images, which showed the contrast of high density faults and narrow C15 bands gathering near the C15 area. Faults may increase in density due to the stress concentration near the nucleus in the compression, and some of the faults may rearrange to a periodic sequence in which there are two faults every four layers. These rearranged layers thus become some narrow C15 bands, which have lower energy. An example is shown in Fig.6. These faults and narrow bands may extend into a large region of the new phase. In this mechanism, the growth direction is normal to the coherent interface between C36 and C15.

Fig.7 shows the third mechanism of growth of the new phase. In the figure, C36 and C15 are located at right and left ends along the "basal planes", respectively. They are seperated by a "transition area" which consists of a series stacking faults and partial dislocations. In the area close to the right, the stacking is more like C36; to the left, more like C15. In this case, the new phase can grow by means of the moving of single partial dislocations. Then the single C15 components in the transtion area grow normal to the basal plane, and the large C15 area grows parallel to the basal plane to the right to extend the new phase area.

Microhardness testing showed that the hardness of the alloy was 350 HV. The diagonal length and the depth of the indents were 16.9 μm and 2.4 μm, respectively. No crack was found at corners or other areas of the any of the indents, indicating a considerable toughness value of the two-phase alloy. Other than twinning and faults on C15 area and transformation in C36 area caused by compression[2], the indentation can produce other types of deformation. Low magnification TEM pictures revealed a large shear strain occuring on non-basal planes of the lamellar Laves phase. These shears are approximately normal to basal plane. The distance between two single shears is typically several hundred angstroms. This can be seen in Fig.8. High-resolution TEM pictures, such as Fig.9, show that the slips may not be restricted on a certain plane, and the shear distance may vary from 1 basal plane thickness to more than ten basal planes. These can be shown by the slip line on the left and that on the right in Fig.9, respectively. Comparison of the Lamellar structure at different areas shows that the shears usually occur within 20 μm from the indents, and can not be found at the areas further away. This confirms the shears are caused by the indentation. This type of deformation needs further study.

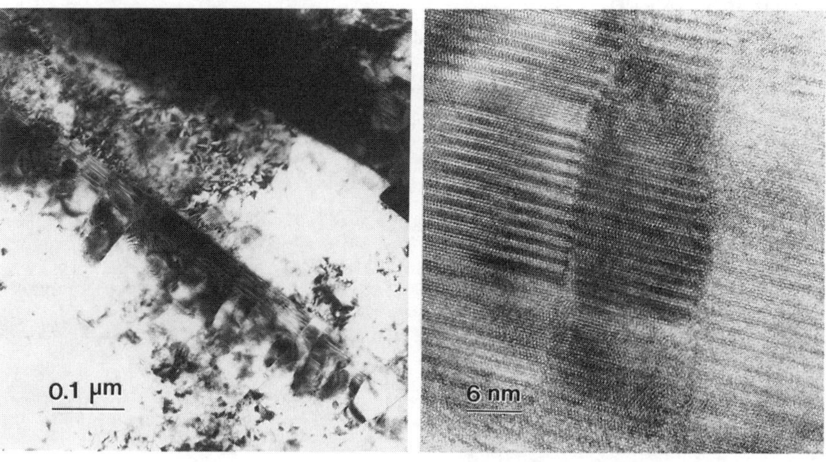

Fig.8 Slips on lamellar C36 caused by indentation. Slip planes are normal to basal plane.

Fig.9 High-resolution TEM picture of slips on C36 caused by indentation, showing the slip distance varying from one to more than ten basal panes.

IV. SUMMARY

Nucleus for the phase transformation from C36 to C15 was found to be the pre-existing C15 thin layers which may cause a stress concentration near the interface. Besides the microscopic ledge interface migration mechanism, the new phase can grow normal to the coherent interface by creating faults to widen the C15 band, or grow along basal plane by the moving of some individual dislocations to lengthen the new phase. Non-basal plane shear slip was found to be another form of deformation caused by indentation. The slip planes are aproximately normal to basal plane.

ACKNOWLEDGEMENT

Financial support from the Department of Energy, Division of Basic Energy Sciences, Grant No. DE-FG02-90ER45426, is greatly appreciated. We also acknowledge the helpful discussions with Kathy Chen and the providing of computer enhancement of TEM micrographs by Ernest L. Hall.

REFERENCES

[1] Y. Ohba and N. Sakuma, Acta Metall., Vol. 37, No. 9, 2377, (1989).

[2] Yaping Liu, James D. Livingston, and Samuel M. Allen, Met. Trans. A, Vol. 23A, No 12, (1992).

[3] J. H. Westbrook, Intermetallic Compounds, Robert E. Krieger Publishing Company, 1977.

[4] K. Kai, T. Nakamichi and M. Yamamoto, J. Phys. Soc. Japan, Vol. 25, 1192, (1968).

[5] C. W. Allen and K. C. Liao, Phys. Stat. Sol. (a) 673-681, 74, (1982).

[6] C. W. Allen, H. R. Kolar and J. C. H. Spence, Proceedings of The International Conference on Martensitic Transformations, The Japan Institute of Metals, Sendai, Japan, 1986, p. 186-191.

[7] G. B. Olson and M. Cohen, Ann. Rev. Mater. Sci., 11:1-30, (1981).

[8] J. D. Livingston, Phys. Stat. Sol.(a) 131, 415, (1992).

[9] J. D. Livingston and E. L. Hall, J. Mater. Res., Jan., Vol. 5, No. 1, 5-8, (1990).

SOME EFFECTS OF COMPOSITION AND MICROSTRUCTURE ON THE B2↔DO_3 ORDERED PHASE TRANSITION IN Fe_3Al ALLOYS

P.J. Maziasz, C.G. McKamey, O.B. Cavin, C.R. Hubbard, and T. Zacharia
Oak Ridge National Laboratory, P.O. Box 2008 Oak Ridge, TN 37831-6114

ABSTRACT

Binary Fe-28 at.% Al alloys (Fe_3Al-type) transform from B2→$D0_3$ ordered phases at 550°C upon cooling. More complex Fe-28Al alloys can have B2↔$D0_3$ transition temperatures that range from 530 to 670°C. The kinetics of the B2→$D0_3$ ordered phase transition also appear to be affected by alloy composition and by matrix dislocation microstructure. The binary and Nb-containing ternary alloys have rapid B2→$D0_3$ transition kinetics during cooling. However, the ternary or more complex alloys that contain Cr have relatively slower phase transition kinetics during similar cooling. Deformation of a complex Fe-28Al-5Cr alloy in the B2 phase regime creates a very high matrix dislocation concentration, which coincides with complete suppression of the B2→$D0_3$ phase transition during cooling.

INTRODUCTION

Iron-aluminide alloys near the Fe_3Al composition transform from an imperfectly ordered B2 phase to a more fully ordered $D0_3$ phase at about 550°C during cooling [1]. There is evidence that alloying additions to hyperstoichiometric Fe_3Al alloys (i.e. Fe-28 at.% Al) can affect the B2↔$D0_3$ phase transition temperature [1,2]. Ternary Fe-28Al alloys that contain 5-6 at.% Cr have been found to retain nearly 100% of the B2 phase after cooling to room temperature, whereas binary alloys consist mainly of the $D0_3$ ordered phase at room-temperature [1,3,4]. Generally, Fe_3Al-type alloys containing chromium have better room-temperature ductility and better environmental embrittlement resistance than similar alloys without chromium [5,6]. To date, the benefits of chromium additions have been attributed to better and more rapid oxide scale formation during cooling (environmental embrittlement resistance), and/or to solid-solution softening effects (higher ductility). However, more recent studies of room-temperature tensile ductility and fatigue-crack growth, and the environmental embrittlement effects of hydrogen, in complex Fe-28Al-5Cr alloys indicate that heat-treatments which produce the B2 phase coincide with better mechanical/environmental behavior [4,7]. A complex Fe-28Al-5Cr alloy has also been found to be weldable [8], and is currently the focus of industrial scale-up efforts for structural applications of Fe_3Al type aluminides [9,10]. The purpose of this study is to survey various heat-treated or creep-tested specimens of several different Fe_3Al type alloys for evidence that metallurgical variables can indeed affect the B2→$D0_3$ phase transition during cooling.

EXPERIMENTAL PROCEDURE

Small 500 g laboratory heats of the binary and ternary Fe_3Al alloys were arc-melted and drop-cast. Alloy compositions and designations are given in Table 1. After forging at 1000°C and hot-rolling at 800°C into sheet stock, these alloys were hot-rolled at 600-650°C to produce 0.76 mm thick sheet. A larger heat (165 kg) of a similar, but more complex Fe-28Al-5Cr alloy (FA-129) was cast commercially [9], but then similarly hot-rolled into 0.76 mm sheet.

Room-temperature and high-temperature X-ray diffraction (XRD and HTXRD, respectively) data were obtained using Scintag $2\theta/\theta$ and θ-θ diffractometers (Cu K_α X-rays). The HTXRD used a Buehler constant-current power supply, a precise furnace controller (<±1°C), a thermocouple at the specimen, and an ultra-high-purity flowing helium atmosphere. XRD/HTXRD measurements were made on sheet specimens of the various alloys which were chemically etched to remove surface oxides. The B2-$D0_3$ phase transition temperature was determined by recording the intensity of the d_{111} diffraction peak of the $D0_3$ phase during heating. Transition temperatures, at which d_{111} goes to zero, are accurate to within ±10°C. The transition temperature was confirmed by then cooling the specimen to observe the $D0_3$ -

d_{111} peak reappear. Because the grain size was large, the specimen goniometer was rocked to be certain that peak intensity was not lost due to crystallite misorientation.

Transmission electron microscopy (TEM) was used to observe the anti-phase domain and boundary (APD and APB, respectively) microstructures of the the DO_3 and B2 ordered phases, and to identify them using electron diffraction. Microstructural analysis was performed using JEOL 2000FX (200 kV) and Philips CM-30 (300 kV) TEMs. Specimens of most alloys (except FA-78) were examined after heat-treatments of 6.5 d at 500°C to establish the DO_3 ordered phase structure. Alloys pretreated to contain the DO_3 phase were also creep-rupture tested in the B2 phase regime (625-650°C/34-69 MPa). TEM specimens were taken from the gage-sections of these specimens to observe any effects of high-temperature deformation and cooling on the ordered-phase microstructure. Finally, a piece of as-hot-rolled 0.76 mm sheet of the FA-129 alloy was examined by TEM after gas tungsten-arc (GTA) welding to observe the effects of heating and resolidification on the ordered-phase microstructure.

RESULTS

HTXRD Data

Room-temperature XRD showed the binary Fe-28Al alloy (FA-61) to be mainly DO_3 phase (Fig. 1a), and HTXRD measurements showed the $DO_3 \leftrightarrow B2$ transition temperature to be 550°C (Table 2). The phase transformation was often reversible, with the signature DO_3 - d_{111} peak reappearing upon cooling below 550°C. If that peak reappeared within the few minutes required to record the HTXRD pattern, the transition kinetics were classed as reversible.

The ternary alloys had higher $DO_3 \leftrightarrow B2$ transition temperatures than the binary FA-61 alloy, while the more complex, large-heat alloy (FA-129) had a slightly lower transition temperature (Table 2). The Fe-28Al-1Nb (FA-79) and Fe-28Al-2Mo (FA-62) alloys had $DO_3 \leftrightarrow B2$ transition temperatures of 606 and 601 °C, respectively. The FA-79 alloy had rapidly reversible phase transition kinetics (Fig. 2), while the reverse transformation of the FA-62 alloy was more sluggish. The Fe-28Al-6Cr (FA-78) alloy was the only alloy found to be 100% B2 in as-received (as-hot-rolled plus heat-treatments at 700-850°C) condition via room-temperature XRD (Fig. 1b). HTXRD measurements showed no evidence for the DO_3 - d_{111} peak even after 5-6 h at 600-650°C. This specimen was then given a heat-treatment of 5 d at 500°C to produce the DO_3 phase. HTXRD heating measurements then showed a $DO_3 \leftrightarrow B2$ transition temperature of 670°C for FA-78. The phase transformation was not reversible after waiting for several hours during HTXRD measurements, suggesting very sluggish transition kinetics. Finally, the complex Fe-28Al-5Cr (FA-129) alloy was found to have a $DO_3 \leftrightarrow B2$ transition temperature of about 530°C with a reversible phase transition upon cooling (Table 2).

TEM Microstructural Analysis

All of the Fe_3Al alloys examined (except FA-78) after heat-treatments (4-6.5 d at 500°C) to produce the DO_3 showed the characteristic APB structure and d_{111} spots of the ordered DO_3 phase (Fig. 3). Details of heat-treatment/mechanical-test condition and a summary of the observed microstructure for each alloy are given in Table 3. APD sizes varied with alloy composition, with the binary Fe-28Al having the coarsest structure, and the FA-62, FA-79 and FA-129 alloys having finer APD structures (Table 3). All alloys in the DO_3-phase condition had low matrix dislocation contents ($<10^{12}$ m/m^3). Only the FA-79 alloy showed a much coarser B2-APB structure superimposed upon the finer DO_3-APB structure. The binary FA-61 alloy was also examined after a heat-treatment of 2 h at 700°C (air-cooled) to produce the B2 phase [11], but TEM analysis showed this material to still have a small amount (<25%) of very fine DO_3 APDs within an otherwise B2 matrix (Table 3, Fig. 4a). The as-hot-rolled (600-650°C) specimen of the complex Fe-28Al-5Cr alloy (FA-129) showed a very high concentration of 2-fold superdislocation networks ($>5 \times 10^{15}$ m/m^3) and a B2 matrix, with no evidence of the DO_3 phase (Fig. 5a and 5b).

To further investigate effects of B2-phase deformation on the $B2 \rightarrow DO_3$ ordered phase transition during cooling, specimens of the FA-61, FA-62 and FA-79 alloys that had been creep-rupture tested at 625-650°C after the DO_3 pretreatment were also examined. From previous studies, TEM examination indicates that the examined gage portion had remained in the furnace

Table 2 - HTXRD Measurements of $D0_3 \leftrightarrow B2$ Ordered Phase Transition Temperature in Fe_3Al Type Alloys

Alloy	$D0_3 \leftrightarrow B2$ Transition Temperature (°C)	Comments
FA-61	550	Reversible, rapid kinetics
FA-62	601	Reversible
FA-79	606	Reversible, rapid kinetics
FA-78	670	Sluggish kinetics, difficult to reverse on cooling
FA-129	530	Reversible

Table 1 - Alloy Compositions (at.%)

FA-61	Fe-28 Al
FA-79	Fe-28 Al-1 Nb
FA-62	Fe-28 Al-2 Mo
FA-78	Fe-28 Al-6 Cr
FA-129	Fe-28 Al-5 Cr-0.5 Nb-0.2 C

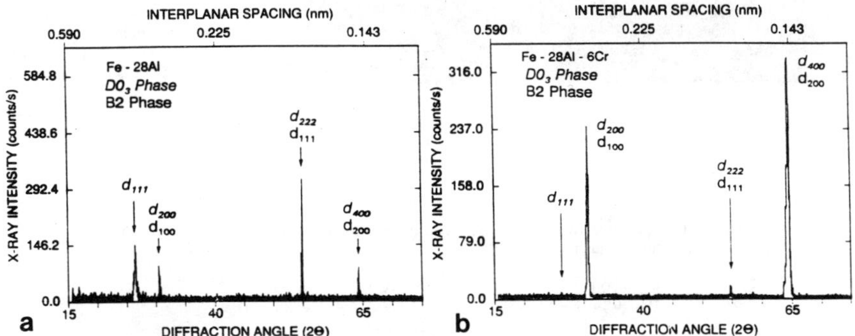

Fig. 1 - Room-temperature XRD of a) binary Fe-28Al (FA-61) alloy showing mainly the $D0_3$-ordered phase and b) ternary Fe-28Al-5Cr (FA-78) showing mainly the B2-phase.

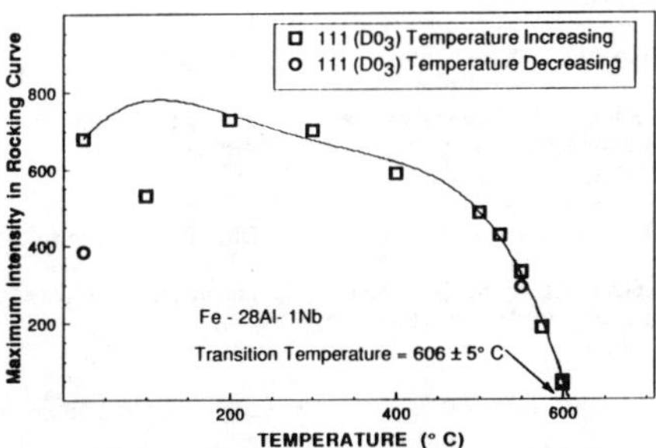

Fig. 2 - HTXRD intensity of the $D0_3$-d_{111} diffraction peak plotted as a function of specimen heating and cooling temperature for the ternary Fe-28Al-1Nb (FA-79) alloy, showing the B2-$D0_3$ phase transition temperature.

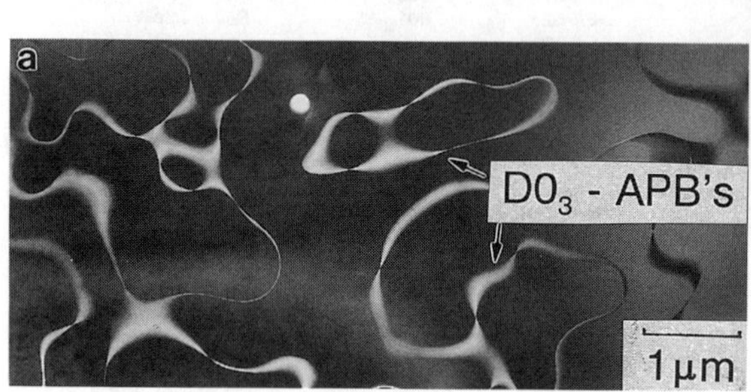

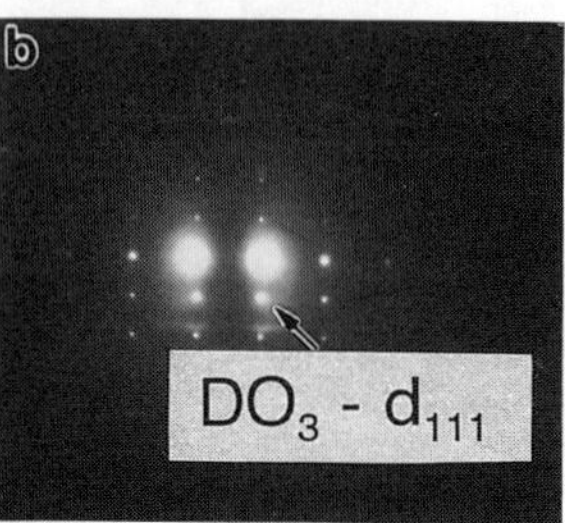

Fig. 3 - TEM analysis of binary Fe-28Al (FA-61) alloy showing a) $D0_3$-APB structure and b) electron diffraction of $D0_3$-d_{111} (arrow) on a Z=[112] zone, after a 6.5 d at 500°C heat-treatment.

Table 3 - Effects of Heat-Treatment, Processing or Mechanical Testing on the B2/$D0_3$ Ordered Phase Microstructure

Alloy	Condition	TEM Observations
FA-61	1h-850°C + 4d-500°C	$D0_3$ - 0.4 to 2 μm APD size.
	1h-750°C + 6.5d-500°C	$D0_3$ - 0.5 to 3 μm APD size.
	1h-700°C	Mixed B2-$D0_3$ structure, mainly B2 matrix with 11.5 to 45 nm $D0_3$ APD size.
	6.5d-500°C + creep-rupture at 625°C/34 MPa (t_r-447h, ϵ_t-131%)	Mainly $D0_3$ - 42 to 146 nm APD size.
FA-62	1h-900°C + 6.5d-500°C	$D0_3$ - 0.1 to 1 μm APD size.
	6.5d-500°C + creep-rupture at 650°C/69 MPa (t_r-91h, ϵ_t-49%)	Mainly B2, faintly detectable APDs (?), <10 nm in size.
FA-79	1h-1000°C + 6.5d-500°C	$D0_3$ - 0.1 to 1 μm APD size.
	6.5d-500°C + creep-rupture at 650°C/69 MPa (t_r-1932, ϵ_t-102%)	Mainly $D0_3$ - 42 to 146 nm APD size.
FA-129	1h-700°C + 4d-500°C	$D0_3$ - 0.13 to 0.36 μm APD size.
	hot-rolled at 600-650°C (base metal)	B2 - very dense dislocation network concentration.
	hot-rolled at 600-650°C, + GTA welding (HAZ)	Mainly B2 - recovered network dislocation structure, with some $D0_3$ in small, recrystallized grains.
	hot-rolled at 600-650°C, + GTA welding (fusion zone)	Mainly $D0_3$ - 0.1 to 1 μm APD size.

after rupture, so that these specimens were furnace-cooled (about 2 h to cool from 650 to 400°C when the furnace is turned off) rather than air-quenched [12]. The creep-tested Fe-28Al and Fe-28Al-1Nb alloys were both mainly DO_3 phase (Fig. 4b), with almost identical APD structures despite significant differences in creep parameters, deformation and remnant dislocation structure (Table 3). By comparison, the Fe-28Al-2Mo alloy showed a mainly B2-phase structure after creep-testing, with no clear DO_3-APB structure, but with faint contrast suggesting the existance of very fine DO_3-APDs (<10 nm in size) (Table 3).

Finally, as another test of the idea that the heavy dislocation structure induced by hot-rolling of the FA-129 alloy was somehow stabilizing the B2 phase during cooling, portions of a GTA-weld specimen of the same base-metal were also examined. For comparison with the as-hot-rolled base-metal portion, specimens were examined from the heat-affected-zone (HAZ) and the fusion zone (FZ) of the welded specimen. The HAZ showed mainly a recovered dislocation structure that was B2-phase. However, there were also some small, recrystallized areas (about 5-10 vol.%), with dislocation-free grains encompassing coarse NbC particles. These small recrystallized grains showed evidence of DO_3-ABPs. The FZ showed an even lower dislocation density, dissolution of many of the coarse NbC matrix particles and reprecipitation of NbC along grain boundaries, and some new, fine precipitates along dislocations in the matrix. The FZ specimen also appeared to be entirely DO_3-phase, with an APD structure somewhat coarser than the the DO_3-APD structure produced via isothermal heat-treatment (Table 3, Fig. 5c and 5d).

DISCUSSION and SUMMARY

The measurements of B2↔DO_3 transition temperatures using HTXRD in this work are consistent with measurements of that same parameter by others for the binary Fe-28Al and ternary Fe-28Al-2Mo alloys using different techniques [2]. Furthermore the B2↔DO_3 transition temperature for the binary agrees with that of published Fe-Al alloy phase diagrams [1]. Recent work [3,4] suggesting that chromium stabilizes retention of the B2 phase during cooling had not shown directly the significant increase in B2↔DO_3 transition temperature, or given direct evidence for sluggish transformation kinetics upon cooling suggested from this work. The low transition temperature for the complex FA-129 alloy is somewhat surprising, but could reflect effects of carbon or synergistic effects between the various alloying elements.

The relative microstructural behavior of the various alloys is consistent with the HTXRD measurements of B2↔DO_3 transition temperatures. The microstructural evidence for relative B2↔DO_3 phase transition kinetics during cooling is also consistent with the relative ranking of such kinetics among the various alloys for the HTXRD measurements. The TEM evidence suggest very rapid transition kinetics for the binary Fe-28Al alloy, particularly since some amount of DO_3 phase is present even after rapid cooling in air after the B2-phase heat-treatment. The lack of DO_3 phase in the Fe-28Al-2Mo from TEM analysis of the creep-tested specimens is also consistent with the more sluggish B2↔DO_3 phase reversibility measured via HTXRD relative to the other alloys.

The TEM evidence for the dislocation structure produced during high-temperature deformation contributing additional stability to the B2 phase and retarding the kinetics of the B2→DO_3 transformation during cooling in the FA-129 alloy appears to be new information. The relative microstructural behavior in the welded specimen, with some DO_3-phase occurring in recrystallized grains in the HAZ while mainly DO_3-phase is present in the resolidified FZ, is self-consistent. This data supports the concept that variations in processing/heat-treatment parameters can affect the B2↔DO_3 phase transition behavior. These results are significant because recent mechanical-properties studies show the relative benefits of the B2 compared to the DO_3 phase for improved room-temperature ductility and resistance to environmental embrittlement [4,7,9,10].

In summary, it appears that Mo and Cr additions raise the B2↔DO_3 transition temperature of Fe-28Al alloys, and retard the transformation kinetics during cooling. However, a complex Fe-28Al-5Cr alloy with additions of Nb and C has a lower transition temperature than the binary alloy. TEM data indicates that a heavy dislocation concentration due to deformation in the B2-phase regime completely suppresses the B2→DO_3 transition during cooling, so that a completely B2-phase structure is retained at room-temperature. To the degree that this dislocation

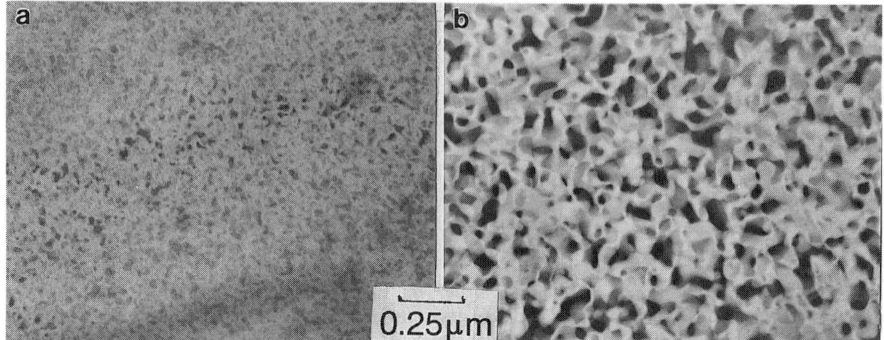

Fig. 4 - TEM images of $D0_3$-APD structure of binary Fe-28Al (FA-61) alloy after a) heat-treatment for 2h at 700°C (air-cooled) to establish the B2-phase, and b.) creep-testing at 625°C/34 MPa for 447 h (furnace-cooled).

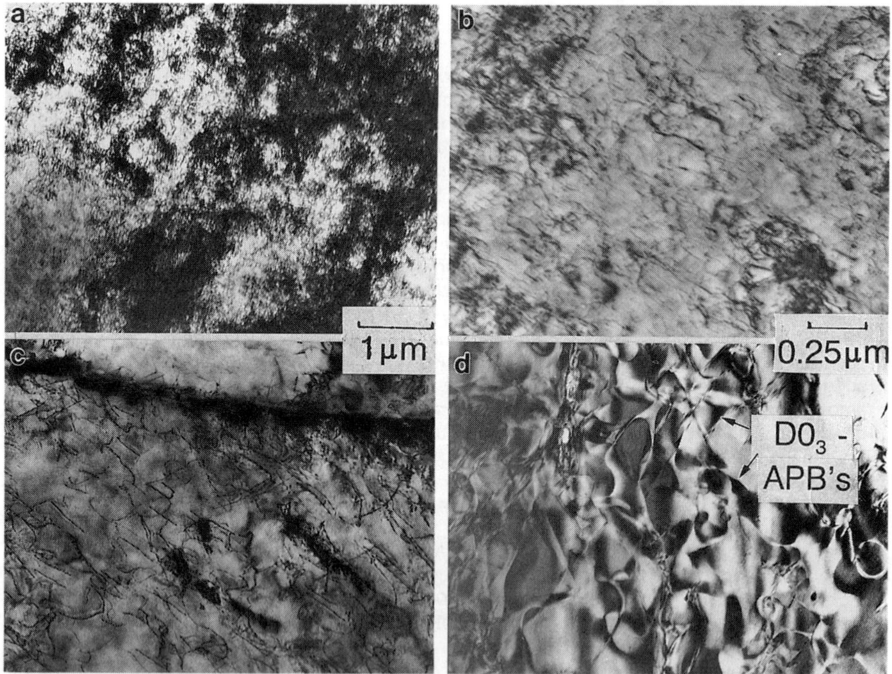

Fig. 5 - TEM analysis of complex Fe-28Al-5Cr (FA-129) alloy showing as-hot-rolled (600-650°C) base-metal at a) lower magnification and b) higher magnification with heavy dislocation networks and B2-phase matrix, and showing the fusion-zone after GTA welding at c) lower magnification with a lower dislocation content and d) higher magnification with $D0_3$-APD structure.

structure is recovered or removed, the B2→DO_3 transition kinetics become rapid enough that this phase transformation is no longer suppressed during cooling.

ACKNOWLEDGEMENTS

This research was sponsored by the U.S. Department of Energy, by the Assistant Secretary for Conservation and Renewable Energy, Office of Industrial Technologies, Advanced Industrial Concepts (AIC) Division, AIC Materials Program, and by the Office of Transportation Technologies as part of the High Temperature Materials Laboratory User Program, and by the Fossil Energy AR&TD Materials Program under contract DE-AC05-84OR21400 with Martin Marietta Energy Systems, Inc.

REFERENCES

1. C.G. McKamey, J.H. DeVan, P.F. Tortorelli, and V.K. Sikka, J. Mater. Res. 6, 1779 (1991).
2. C.G. McKamey and J.A. Horton, Metall. Trans. 20A, 751 (1989).
3. P.J. Maziasz, C.G. McKamey and C.R. Hubbard, Mat. Res. Soc. Symp. Proc. Vol. 186 (1991) p. 349.
4. Y. Huang, W. Yang, and G. Chen, elsewhere in this proceedings.
5. C.G. McKamey, J.A. Horton and C.T. Liu, Scripta Met. 22, 1679 (1988).
6. C.G. McKamey and C.T. Liu, Scripta Met. et Mat. 24, 2119 (1990).
7. A. Castagna and N.S. Stoloff, Scripta Met. et Mat. 26, 673 (1992).
8. S.A. David and T. Zacharia, to be published in Welding J., 1992.
9. V.K. Sikka, C.G. McKamey, C.R. Howell and R.H. Baldwin, Oak Ridge National Laboratory Report ORNL/TM-11796 (March, 1991).
10. V.K. Sikka, in Proc. First Internat. Conf. Heat-Resistant Materials, ASM-International, Materials Park, OH (1991) p. 141.
11. C.T. Liu, C.G. McKamey and E.H. Lee, Scripta Met. et Mat. 24, 385 (1990).
12. C.G. McKamey, P.J. Maziasz, and J.W. Jones, J. Mater. Res. 7, 2089 (1992).

ELECTRON MICROSCOPY OBSERVATIONS ON THE INFLUENCE OF BORON ADDITIONS ON STOICHIOMETRIC NiAl

T.-C. WU and S. L. SASS
Cornell University, Department of Materials Science and Engineering, Ithaca, NY 14853

ABSTRACT

The microstructure and local chemistry in single crystal and polycrystalline stoichiometric NiAl, with and without boron additions, were investigated using transmission and analytical electron microscopy. Plate-like precipitates were present in the boron-doped NiAl. Since these precipitates were not observed in the boron-free material, they must be due to the formation of borides. After irradiation with 400kV electrons, fine scale changes in the image contrast and the appearance of diffuse scattering in the diffraction pattern indicate that a phase transformation has been induced by the electron beam.

INTRODUCTION

The B2-structured intermetallic, NiAl, has received considerable attention because of its potential for high temperature structural applications[1]. However, lack of ductility at room temperature is still its major drawback and restricts its use for high performance applications. A great deal of effort has been directed to overcome the intrinsic brittleness of grain boundaries in polycrystalline NiAl by microalloying with boron [2]. Unlike the Ni_3Al system, even though boron strengthens the boundaries and suppresses intergranular fracture in NiAl alloys, its low temperature ductility is still not greatly improved [3]. One suggestion to explain the continuing poor ductility is that boron causes solid solution hardening in the NiAl and in this manner raises the yield stress to above the transgranular fracture stress. Recently, Jayaram and Miller [4] reported the presence of a low number density of $(Ti,V,Cr)B_2$ precipitates in stoichiometric boron-doped NiAl alloys examined by Atom Probe Field Ion Microscopy (APFIM). These particles are believed to cause the increase in the yield stress by precipitation hardening.

In this paper, the microstructures of polycrystalline and single crystal NiAl alloys were studied using conventional transmission electron microscope (CTEM) and high resolution transmission electron microscope (HRTEM). The local chemistry of precipitate particles observed in boron-doped NiAl was investigated by analytical electron microscope. During the course of this research microstructural changes induced in the NiAl by the high energy electron beam were observed, and these will also be described.

EXPERIMENTAL PROCEDURE

Polycrystalline alloys with composition of Ni-50 at%Al, both doped with 300 wppm boron and boron-free, were supplied by Oak Ridge National

Laboratory [2]. The alloys were arc-melted and homogenized as drop-casted ingots in vacuum at 1100 °C for 24 hrs, and then hot-extruded at 900 °C. NiAl single crystals with composition of Ni-50 at%Al, both doped with 500 wppm boron and boron-free, were supplied by GE Aircraft Engines. They were grown by the Bridgman method and homogenized at 1316 °C for 50 hrs in an Ar atomosphere [5]. TEM specimens were prepared by electropolishing in the Struers Tenupol using a 10% perchloric acid-methanol solution at room temperature. CTEM images were obtained using a JEOL 1200EX and HRTEM images were obtained using a JEOL 4000EX, operating at 200kV or 400kV. The chemistry of boron-doped specimens was examined by a Vacuum Generators HB 501 scanning transmission electron microscopy operating at 100kV. X-ray energy dispersive spectra were collected using a Link AN-10000 Series analyzer.

EXPERIMENTAL RESULTS

Polycrystalline NiAl

The image of a boron-doped NiAl alloy, taken at 200kV, is shown in Fig. 1(a); particles with lengths of 10-20 nm and width less than 10 nm are present. The high resolution image in Fig. 1(b) shows that the long sides and lattice fringes of the particle are parallel to the [100] direction. The selected area diffraction (SAD) pattern in Fig. 1(c) shows that the electron beam is incident along the [001]. Weak streaked reflections along [010] present in Fig. 1(c) are consistent with the presence of thin platelets. Extensive examination of boron-free specimens did not find the particles that were present in boron-doped alloys.

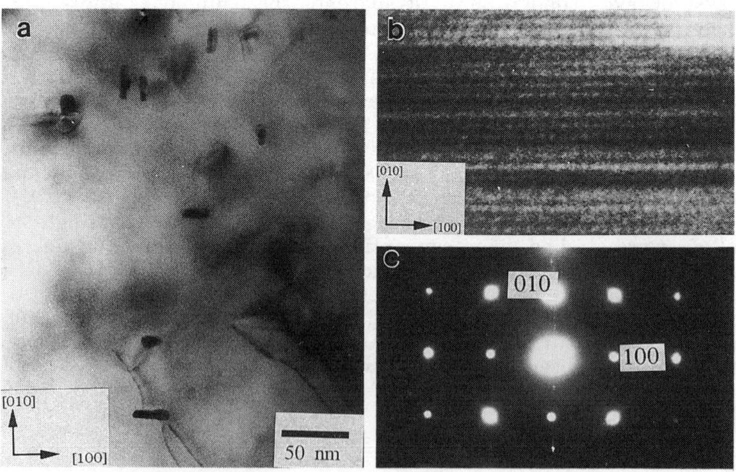

Fig. 1 Boron-doped polycrystalline NiAl (a) Bright-field image (b) HRTEM image taken at 200 kV (c) Diffraction pattern for (b).

The particle shown in Fig. 2(a) in boron-doped NiAl was examined by analytical electron microscopy. The corresponding X-ray spectra obtained by placing the probe both on the particle and in the nearby matrix are shown in Fig. 2(b). Peaks of Ti K_α, V K_α, Cr K_α and Mn K_α are found only from the particle while a peak of Fe K_α is found in both the particle and the matrix. Since the Fe Kα peak from the particle has higher intensity than from the matrix, the particle must be enriched in Fe, as well as Ti, V, Cr and Mn.

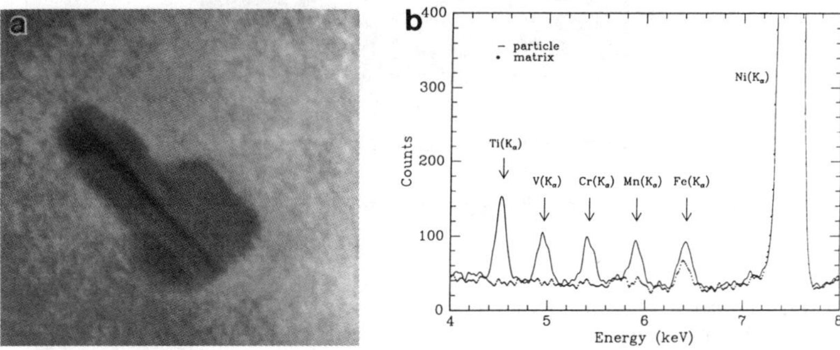

Fig. 2 (a) Bright-field image of the particle examined by analytical electron microscopy (b) X-ray spectra from the particle and the nearby matrix.

Single Crystal NiAl

A CTEM image and the corresponding SAD pattern of a boron-doped NiAl single crystal are shown in Fig. 3(a) and (b), respectively. Linear features parallel to <100> directions, coming from particles, are present in Fig. 3(a). One set of sharp fine lines of intensities is present in Fig. 3(b), with extra streaks along the <100> direction.

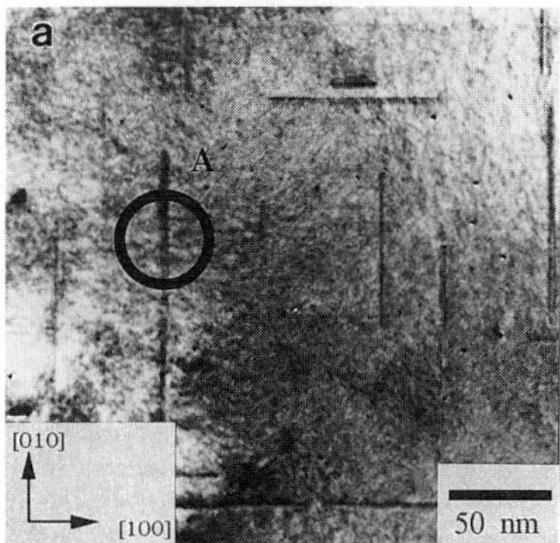

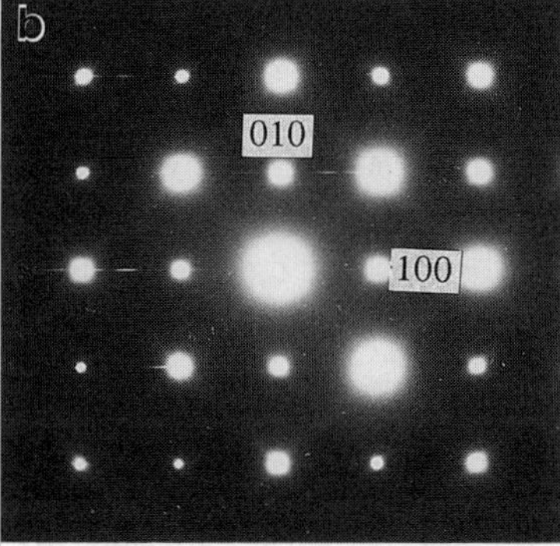

Fig. 3 Boron-doped single crystal NiAl (a) Bright-field image (b) Diffraction pattern from the area marked A in (a).

The lattice image, taken at 200kV, is shown in Fig. 4, and is consistent with the linear features in Fig. 3(a) being from plate-like particles. Extensive examination of large areas of boron-free and boron-doped specimens indicates that these plate-like particles are present only in boron-doped NiAl alloys.

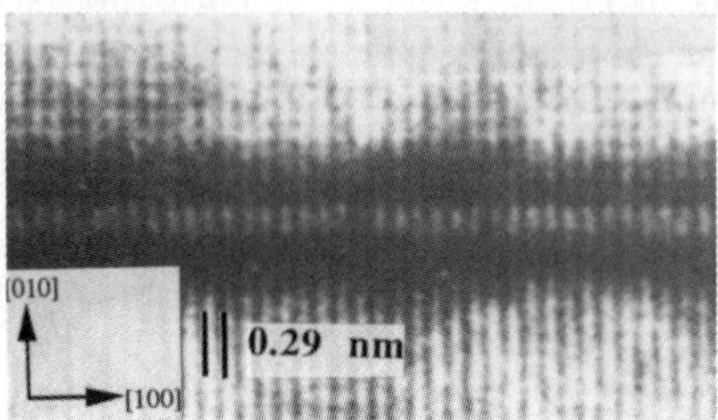

Fig. 4 HRTEM image from the precipitate in the boron-doped single crystal NiAl.

The particle shown in Fig. 5(a) in boron-doped NiAl was examined by analytical electron microscopy. The X-ray spectra obtained by placing the probe on the particle and in the nearby matrix are shown in Fig. 5(b). A small Fe K_{α} peak is observed from the particle.

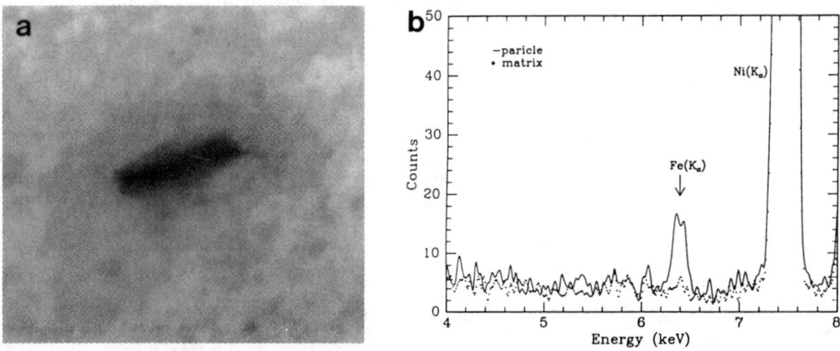

Fig. 5 (a) Bright-field image of the plate-like particle examined by analytical electron microscopy (b) X-ray spectra from the particle and the nearby matrix.

Influence of the Electron Beam

The image of a boron-free polycrystalline NiAl alloy, taken at 400kV, is shown in Fig. 6(a), with a HRTEM insert. The corresponding SAD pattern is in Fig. 6(b). Tweed structure is observed with fine striations parallel to the traces of {110} planes. These striations are found in polycrystalline and single crystal NiAl alloys with and without boron additions. This fine structure is very similar to previous observations made on Ni-37 at%Al alloys with the CsCl-structure, which is in the region of the phase diagram where pre-martensitic transformations were reported [6]. In the SAD pattern in Fig. 6(b), streaks along <110> directions are present with broad intensity maxima at 1/2<110>-type positions.

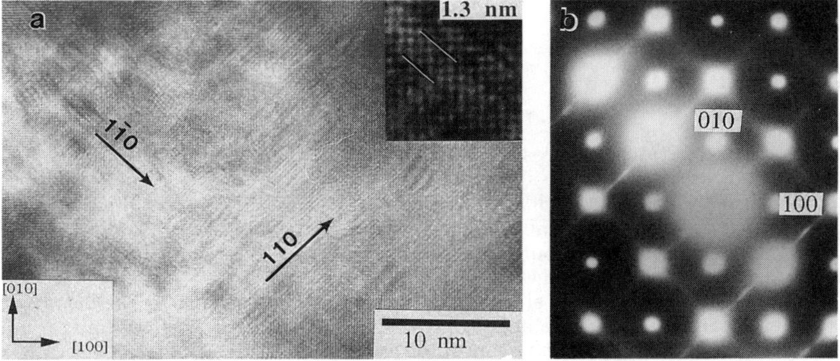

Fig. 6 (a) Bright-field image from boron-free polycrystalline NiAl, with HRTEM insert (b) Diffraction pattern for (a).

DISCUSSION

This study has shown that when boron is present in NiAl, particles are also present, while if the NiAl is boron-free, no particles are observed. In both boron-doped polycrystalline and single crystal alloys, the particle plates have sides parallel to <100> directions; in the single crystal their traces appear as long straight lines, while in the polycrystals, they are shorter and irregular. The difference may result from the single crystal being annealed at a much higher temperature and for a longer time than the polycrystalline alloy.

APFIM studies on polycrystalline NiAl alloys, doped with 100 wppm and 300 wppm boron, reported precipitates as large as ~20 nm in diameter [4]. The precipitates in Fig. 1(a) have lengths of 10-20 nm in agreement with the APFIM observations. Analytical electron microscopy results show that the observed particles in boron-doped polycrystalline NiAl contain Ti, Cr, V and Mn and possibly Fe, which is similar to the observation reported by APFIM [4], except for the presence of Fe and Mn. Particles present in boron-doped NiAl single crystals contain Fe. Thus, for the polycrystalline NiAl from Oak Ridge, the nominal composition of the precipitate is $(Ti,V,Cr,Mn,Fe)B_2$, and for the single crystalline NiAl from General Electric, the nominal composition is FeB_2.

All lattice images from polycrystalline and single crystal NiAl alloys, taken at 400kV, show a similar tweed structure which is not seen at 200kV. The electron diffraction pattern taken at 400kV also show extra diffraction effects, with diffuse lines and diffuse maxima which were not present at 200kV. It seems clear that irradiating with 400kV electrons induces a fine scale change of structure in the stoichiometric NiAl. The high energy electrons may be stimulating the formation of ω-like defects [7] that Georgopoulos and Cohen[8] have suggested to be present in off-stoichiometric NiAl alloys, based on their analysis of the diffuse X-ray scattering from these alloys. For the present study of the effect of boron on the microstructure of NiAl, the electron irradiation induced structural change is clearly an artifact. It may be worthwhile studying in its own right, however, as an electron irradiation induced phase transformation.

CONCLUSIONS

(1) Precipitates are present in NiAl single crystals and polycrystals doped with boron. Since these precipitates were not seen in boron-free NiAl alloys, it is likely that they are a boride compound with nominal composition of FeB_2 in the NiAl single crystals and $(Ti,V,Cr,Mn,Fe)B_2$ in the polycrystalline NiAl. These particles may contribute to the increase in yield stress that is observed when boron is added to NiAl.

(2) The high energy electron beam stimulates a phase transformation which is manifested by the appearance of tweed structure in the image and extra diffuse scattering in the electron diffraction patterns.

ACKNOWLEDGEMENTS

This research was supported by the Department of Energy under Grant No. DE-FG02-85ER45211. We are especially grateful to Dr. C. T. Liu of Oak Ridge National Laboratory and Dr. R. Darolia of General Electric Aircraft Engines for providing us with the polycrystalline and single crystal NiAl alloys, respectively. The use of the Electron Microscopy and X-ray Facilities of the Materials Science Center at Cornell University, which is supported by National Science Foundation, is gratefully acknowledged. The assistance of R. Keyse, M. Thomas and M. Rich is greatly appreciated.

REFERENCES

1. R. Darolia, JOM 3, 44 (1991)
2. E. P. George and C. T. Liu, J. Mater. Res. 5, 754 (1990)
3. E. P. George, C.T. Liu and J. J. Liao, MRS Symp. Proc. 186 (1990)
4. R. Jayaram and M. K. Miller, Surf. Sci. 266, 310 (1992); MRS Symp. Proc. 239, 445 (1992)
5. R. Darolia, D. Lahrman and R. Field, Scripta Met. 26, 1007 (1992)
6. L. E. Tanner, A. R. Pelton, G. VanTendeloo, D. Schryvers and M. E. Wall, Scripta Met. 24, 1731 (1990)
7. T. S. Kuan and S. L. Sass, Acta Met. 24, 1053 (1976)
8. P. Georgopoulos and J. B. Cohen, Acta Met. 29, 1535 (1981)

THE $\alpha \rightarrow \gamma$ TRANSFORMATION DURING CONTINUOUS COOLING IN Ti-48 At% Al ALLOYS

G. Ramanath* and Vijay K. Vasudevan*
*Department of Materials Science and Engineering, University of Cincinnati, OH 45221

ABSTRACT

The $\alpha \rightarrow \gamma$ transformation in a Ti-48 at% Al alloy was studied by a novel computer interfaced control-cum-data acquisition technique. *In situ*, real time, high speed measurements of temperature were made in the system engineered for this purpose. The samples were heated by controlled direct resistance heating and gas-jet quenched. Various cooling rates were achieved by control algorithms and controlling the pressure of the quenching medium. Cooling curves and thermal arrest data were used to determine the transformation temperatures and characteristics as a function of cooling rate. Subsequent microstructural analysis by optical microscopy was used to confirm the occurrence of the various transformation modes, namely, lamellar, Widmanstatten and massive. The transformation start and completion temperatures for the various reactions were determined for different cooling rates, and continuous cooling transformation diagrams were determined. The various results are presented and discussed in light of thermodynamic and phase diagram considerations.

INTRODUCTION

Alloys based on the intermetallic TiAl are presently receiving considerable attention as structural materials for elevated temperature aerospace applications owing to their unique combination of attractive properties such as light weight, high strength, high melting temperature, superior creep and oxidation resistance [1]. The most promising of the alloys developed so far with regard to room temperature ductility and high temperature strength are based on the Ti-48Al composition (compositions in the text are in atomic percent) with ternary or quaternary additions [2, 3]. The properties of these alloys are extremely sensitive to microstructure. While many aspects relating to the phase equilibria are well understood, the transformation mechanisms and microstructure development during continuous cooling are just beginning to be understood. Recent studies have shown that cooling rate has a major effect on the transformation of the α phase. Wang et al. [4, 5] have reported that at low cooling rates the lamellar morphology prevails, at intermediate cooling rates both a Widmanstatten and a feathery microstructure appear whereas at very high cooling rates a massive transformation dominates. Other studies have reported similar effects of cooling rate and confirmed the occurrence of these various microstructures [6, 7]. However, the actual temperatures at which the different reactions occur and their dependence on the cooling rate are not known. The present paper addresses these issues in light of preliminary results of an experimental study initiated to understand the temperature and cooling rate dependence of the $\alpha \rightarrow \gamma$ transformation.

EXPERIMENTAL

A novel computer controlled high-speed Temperature and Electrical Resistivity Measurement System (TERMS) was designed and constructed for the purpose of studying phase transformations in alloys by means of *in situ* measurements under controlled isothermal, continuous heating, continuous cooling conditions or a combination of these. A schematic diagram of the system is shown in Figure 1. The system broadly consists of the following 1) a heating assembly which includes a computer controlled DC power supply and a heating chamber in which the sample is heated by its electrical resistance, 2) a computer triggered quenching device designed to let in a jet of purified helium gas at a predetermined flow rate and pressure at the sample while shutting off the power, 3) fast-response thermocouples and resistance probes spotwelded to the sample, and 4) hardware and the software interfaces for closed loop temperature control, data acquisition and analysis. A detailed description of the TERMS is given elsewhere [8].

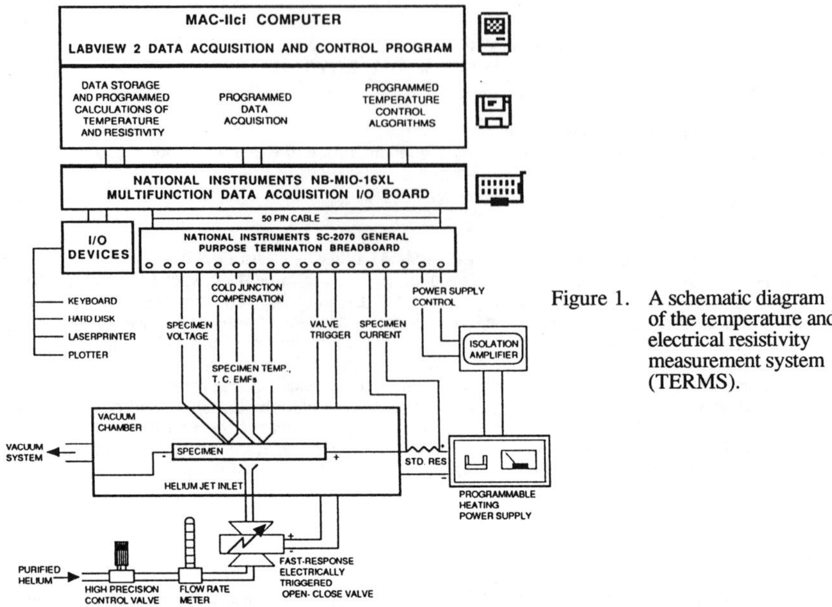

Figure 1. A schematic diagram of the temperature and electrical resistivity measurement system (TERMS).

Ti-48Al alloys with two kinds of starting microstructures namely, completely lamellar and duplex were used for the study. Samples of dimensions (20mm x 3.5mm x 0.5mm) were sliced and heated in the TERMS to the completely α region (α-transus 1380°C), namely 1400°C, solutionized there and quenched at different cooling rates ranging from about 150 °C/s to 850 °C/s and data was acquired at high speeds and analysed. The cooling rates were obtained by measuring the slope of the curve in the region before the first distinct change was observed. Any changes observed in the cooling curves such as variations in slope and thermal arrest were recorded and correlated with microstructural features observed by optical microscopy.

RESULTS

Previous studies have shown that the decomposition of the α phase is sensitive to the cooling rate[4-7]. Wang et al [4] have shown that at low cooling rates, the γ phase forms in a lamellar morphology, at intermediate cooling rates, a Widmanstatten morphology appears and at high cooling rates a massive morphology prevails. The results of the present study confirm these observations. The results also indicate that at very high cooling rates, the massive reaction is arrested by the $\alpha \rightarrow \alpha_2$ transformation. Figure 2 shows four representative cooling curves corresponding to 240 °C/s, 370 °C/s, 530 °C/s and 650 °C/s cooling rates. Optical micrographs of samples cooled at a wide range of cooling rates are shown in Figures 3 and 4. At low cooling rates less than about 210 °C/s, the sample transformed to the equilibrium lamellar structure(γ_L) shown in Figure 3a. At slightly higher cooling rates, the microstructure observed was a mixture of Widmanstatten-like structure(γ_w) and packets of dark etching regions identified to be massive (γ_m) near the grain boundaries in a background of lamellar gamma(γ_L) grains, Figure 3b. Another feature observed at these cooling rates is that in most cases, the γ_m was separated from the grain boundary by the γ_L structure. This sample was cooled at the rate of 230 °C/s. As shown in the cooling curve, Figure 2, corresponding to this microstructure, the first change in slope(γ_{Ls}) is associated with the formation of γ_L from α and the second change in slope is associated with the start of either the Widmanstatten or the massive modes of transformation.

As the cooling rate was gradually increased further, the "equilibrium" lamellar morphology disappeared, and along with γ_w, a "feathery" massive morphology consisting of fine needles was

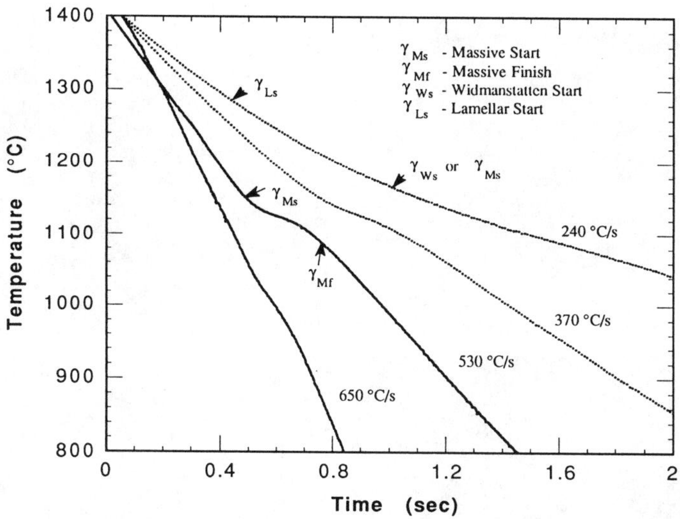

Figure 2. Cooling curves of Ti-48Al alloy samples cooled at various rates from 1400°C.

observed. The micrograph in Figure 3c shows a medley of this "feathery" microstructure and the Widmanstatten morphology with no trace of the equilibrium γ_L. The "feathery" microstructure has been reported by Wang et al [5]. While the nature of this morphology is not yet known, it may be speculated to be a morphology of γ intermediate to the Widmanstatten and massive morphologies. This microstructure was obtained at a cooling rate of 370 °C/s. The corresponding cooling curve in Figure 2 shows a distinct thermal arrest with no previous slope changes. The start and end of the arrest were interpreted to be the start and finish temperatures of the transformation(s). As the cooling rate was increased further, the γ_w structure and the feathery morphology disappeared altogether and the sample transformed completely to γ_m. Figure 3d shows the γ_m microstructure of a sample cooled a rate of 430 °C/s. At cooling rates higher than about 500 °C/s, the γ_m morphology persisted, but with a marked difference. In the midst of darkly etched γ_m phase packets of light etching regions of α_2 were observed. Figure 4a shows the micrograph of a sample cooled at 530 °C/s with small regions of α_2 scattered in a background of the massive phase. From this and the almost flat thermal arrest observed, Figure 2, it is clear that while the $\alpha \rightarrow \alpha_2$ transformation competes to arrest the massive transformation, the latter is still the dominant mode at this cooling rate. As the cooling rate was increased further, the volume fraction of the γ_m phase decreased and that of the α_2 increased. This is indicated in Figure 4b and 4c which show the microstructure of samples cooled at 560 °C/s and 650 °C/s, respectively.

DISCUSSION

In this study, the temperature dependence of the decomposition of the α phase under continuous cooling conditions in a binary Ti-48Al alloy by the different modes namely, lamellar, Widmanstatten and massive is documented. The start temperature for the different modes are reported for the first time. The preliminary results are summarized in the form of the tentative CCT diagram shown in Figure 5. The dashed lines indicate tentative plots of the curves. A more detailed and complete picture of the CCT diagram, including the finish temperatures for the reactions, will be presented in a forthcoming publication. For the purpose of clarity and simplicity, the cooling curves shown in this diagram are just schematic representations of the cooling rates and not the original curves with arrests and slope changes. The estimated positions of the α/γ and α/α_2 T_O temperatures and the eutectoid reaction temperature T_E are also shown. At cooling rates of approximately about 200 to 300 °C/s, the start temperature of the lamellar mode of transformation

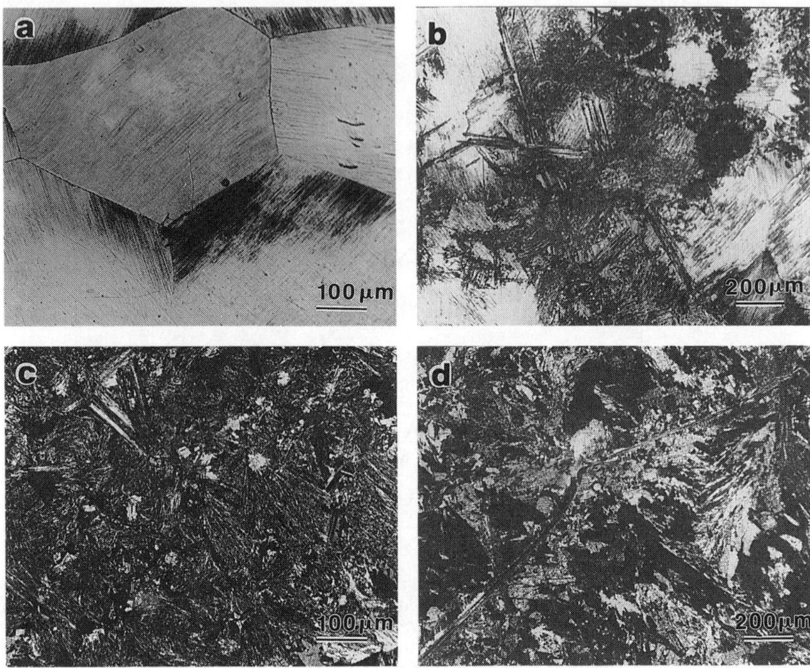

Figure 3. Optical micrographs of Ti-48Al alloy samples cooled at various rates from 1400°C. (a) 210°C/s, (b) 240°C/s, (c) 370°C/s, and (d) 430°C/s.

varies steeply between 1300 to 1250 °C with increasing cooling rate. Owing to this high sensitivity to cooling rate, it is inferred that this is the nose of α transforming to γ_L. At the cooling rates wherein the γ_w and the γ_m phases are observed, the reaction start temperature of one or both of these phases appeared to be fairly isothermal as shown. As all the microstructures in samples cooled at intermediate rates have both γ_w and γ_m morphologies, it is not yet clear if the start temperature corresponding to thermal arrests or changes in slope observed at these cooling rates is the result of one of the modes or a cumulative effect of both. If the latter is true, then it may be inferred that the Widmanstatten mode of transformation observed at the lower cooling rates might be an incipient form of the massive reaction that occurs at higher cooling rates. On the other hand, it is possible that the Widmanstatten mode might have a higher start temperature intermediate to the γ_{Ls} and the γ_{Ms}, in which case the second slope change would be an exclusive effect of the massive reaction. Electrical resistivity measurements are underway to investigate this issue. At cooling rates greater than about 400 °C/s, with increasing cooling rate, the start temperature of the massive reaction showed a perceptible depression.

At low cooling rates, γ nucleated at the grain boundaries starts to grow as lamellae at temperatures shown as γ_{Ls} in Figure 5 and proceeds to completion if the cooling curve meets the γ_{Lf} before it hits any other transformation curve. An interesting point to note is that at intermediate cooling rates, the massive phase formed is separated from the grain boundary by γ_L. At these cooling rates, it is possible that the nucleated γ grows initially in the lamellar mode till sufficient undercooling for either the Widmanstatten or the massive modes of transformation is reached. At this point, the untransformed α decomposes directly to γ by one of these modes. At even higher cooling rates, the lamellar mode is not observed probably because of its sluggish kinetics coupled

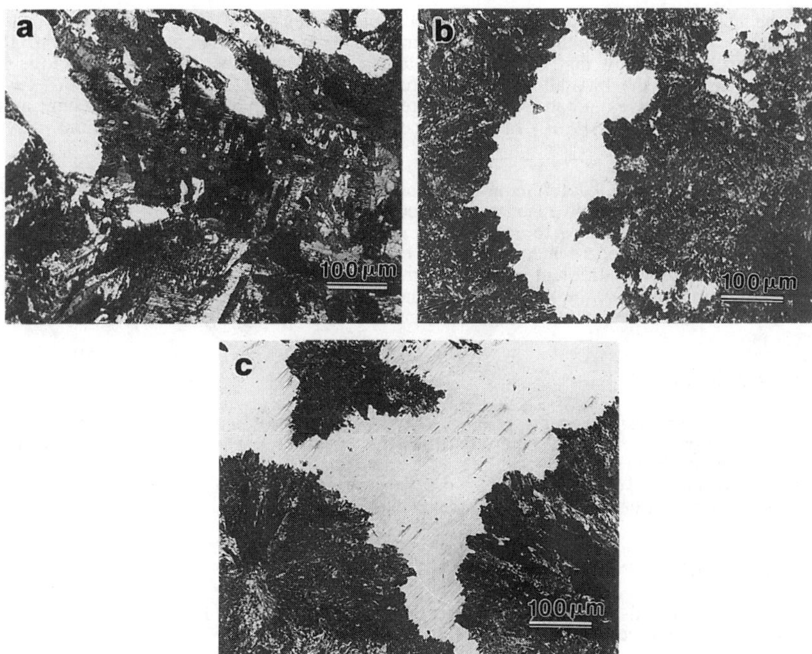

Figure 4. Optical micrographs of Ti-48Al alloy samples cooled at various rates from 1400°C. (a) 530°C/s, (b) 560°C/s, and (c) 650°C/s.

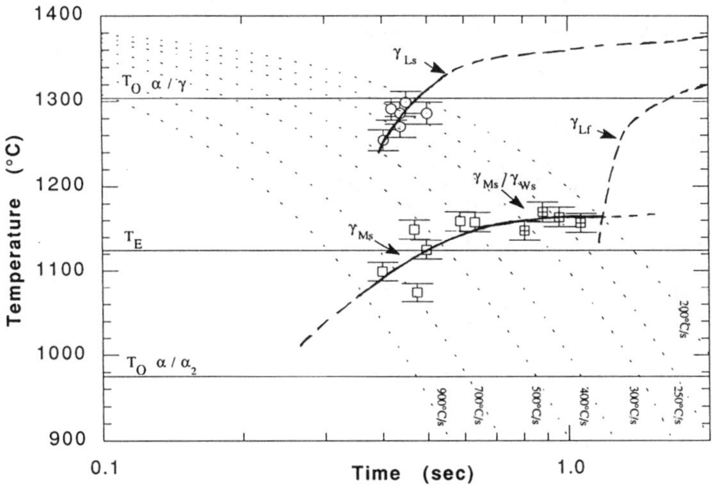

Figure 5. Tentative CCT diagram proposed for Ti-48Al alloy.

with the fact that minimal time is spent in the lamellar region of the CCT curve. Hence, the α decomposes completely by either the Widmanstatten or the massive modes of transformation. The present results also show that, for a wide range of cooling rates, the start temperature for the massive reaction in the 48Al alloy is usually around ~1150°C, i.e., in the equilibrium $\alpha+\gamma$ two-phase field. This gives support for the contention that the massive reaction occurs upon undercooling below the T_O [9, 10], rather than the metastable extension of the equilibrium γ solvus line [11].

Finally, it is clear that further investigation is required to establish the start temperatures of the Widmanstatten and massive modes as a function of both Al content, and over a wide range of cooling rates from very slow to very fast, to understand the temperature dependence of the decomposition of the α phase under continuous cooling. These aspects together with the kinetics of the decomposition in different modes, determination of growth rates of the massive phase by quantitative microscopy, the nature of the massive-parent interface, nucleation and growth mechanisms of the different morphologies of γ are the subject of ongoing research.

CONCLUSIONS

The decomposition of the α in a Ti-48Al alloy was studied using in situ high speed temperature measurements. The study confirmed the cooling rate dependence of transformations reported previously [4]. Results of the start temperatures of the massive, Widmanstatten and lamellar structure were documented for the first time. A tentative CCT diagram was generated and discussed in terms of observed features in the microstructure.

ACKNOWLEDGEMENTS

Support for this research from an Air Force Engineering Foundation Research Initiation Grant and from the Alcoa Foundation is deeply appreciated.

REFERENCES

1. H. A. Lipsitt, MRS. Symp. Proc., 39, 351 (1985).
2. M. J. Blackburn and M. P. Smith, United States Patent, No.4,294,615 (1981).
3. Y. W. Kim, MRS Symp. Proc., 213, 777 (1991).
4. P. Wang, G. B. Viswanathan and V. K. Vasudevan, Metall. Trans., 23A, 790 (1992).
5. P. Wang and V. K. Vasudevan, Scripta Metall. et Mater., 27, 89 (1992).
6. S. A. Jones and M. J. Kaufmann, Acta Metall. et Mater., in press (1992).
7. P. McQuay, D. M. Dimiduk, L. S. Semiatin, Scripta Metall. et Mater., 25, 1689 (1991)
8. G. Ramanath and V. K. Vasudevan, to be published.
9. T. B. Massalski, in: Phase Transformations, p.433, Metals Park, OH (1970).
10. J. H. Perepezko, Metall. Trans., 15A, 437, (1984).
11. M. Hillert, Metall. Trans., 15A, 411 (1984).

EFFECT OF COOLING RATE ON DECOMPOSITION OF THE α PHASE IN Ti-(43-50) At.% Al ALLOYS:

Ping Wang and Vijay K.Vasudevan
Dept. of Materials Science and Engineering, University of Cincinnati, Cincinnati, OH 45221

ABSTRACT

The effect of cooling rate on transformations of the disordered, hcp α phase in TiAl alloys containing 43 to 50 at.% Al is reported. Cast and homogenized samples of the alloys were solutionized in the completely α region and cooled at various rates, and the microstructures examined by various microscopical techniques. The results indicate a very strong effect of cooling rate on transformations, and depending on it the γ forms from the α in three morphologies. At low cooling rates, a lamellar morphology is observed, at intermediate rates both Widmanstatten and feathery morphologies appear, and at very high rates a massive transformation dominates. The crystallography of the various forms of γ, together with orientation relationships, compositions, and nature of the parent/product interfaces have been examined. Controlled heat treatments were also performed to further elucidate the cooling rate dependence of the lamellar and Widmanstatten structures. These various results are presented and the temperature and composition dependence of transformations are discussed in light of phase diagram considerations.

INTRODUCTION

It is well known that the properties of TiAl alloys are a strong function of microstructure. The most promising alloys, which are based on the Ti-48 Al composition (compositions in the text refer to at.%) with ternary or quarternary additions, are characterized by the two-phase Ti_3Al+TiAl (α_2+γ) lamellar microstructure (1-4). In alloys containing >~42Al, the γ phase in this structure is produced from the high temperature disordered hexagonal α phase upon cooling [4], and so it is important to understand the possible modes and mechanisms of decomposition of the latter phase. In several recent studies on Ti-(47-48)Al-based binary [5-7] and ternary/quarternary alloys [8,9], it has been shown that cooling rate has a major effect on the transformation of the α phase and depending on it the γ forms in three morphologies. At low cooling rates, the lamellar morphology prevails, at intermediate rates a Widmanstatten morphology appears and at very high cooling rates a massive transformation dominates. The purpose of this paper is to report on the composition dependence of transformations in TiAl alloys cooled at various rates from the fully α region.

EXPERIMENTAL

The alloy compositions studied were Ti-43, 46, 48 and 50 Al, same as in a previous study [5,6], where detailed alloy preparation methods, homogenization procedures, sample protection schemes, and chemistries before and after heat treatment have been presented. The actual Al contents in these were 43.19, 46.54, 47.86 and 49.40 Al , respectively. Cut and protected samples of the alloys were heated to the completely α region, namely, 1325, 1360, 1400°C and 1435-1450°C for the 43, 46, 48 and 50 Al alloys, respectively. After holding for various times to 1h, the samples were cooled at different rates, namely, furnace cool (FC), air cool (AC), oil quench (OQ), water quench (WQ) and iced brine quench (IBQ). The microstructures were observed by optical (OM) and scanning electron microscopy (SEM) with back scattered electron (BSE) imaging for atomic number contrast and microprobe analysis for compositions of phases. Thin foils of the samples were observed in a Philips CM20 transmission electron microscope (TEM) operated at 200 kV, using bright field (BF), dark field (DF) and selected area diffraction (SAD) modes.

RESULTS

The optical microstructures of the various alloys cooled at different rates are presented in Figures 1 through 4. The micrograph of the WQ Ti-43.19Al was featureless, and TEM examination showed that the structure was composed fully of α_2 domains. The AC sample, although mostly featureless, revealed a few lamellae, presumably of γ, primarily restricted to the grain boundaries. This result suggests that at this Al content even air cooling can partially suppress γ formation and that the γ forms directly from α; the latter is in agreement with other findings [7].

The optical microstructures of the Ti-46.54Al alloy cooled at various rates are shown in Figures 1a–d. The FC microstructure appeared fully lamellar. The AC microstructure, Figure 1a, appears similar to that of the Ti-47.20Al alloy reported earlier [5], with the exception that fewer regions of the Widmanstatten structure are present. This structure usually appears as plates in several orientations (Figure 1a). A few deeply etched regions can also be seen within grains in Figure 1a. The microstructures of the OQ, WQ and IBQ samples appear very similar to those of the Ti-47.20Al alloy [5] and consist of dark, deeply etched regions emerging from the grain boundaries in a featureless matrix. Previously [5,6], it has been established by TEM that the light featureless regions in Figures 1b–d were α_2, whereas the dark regions were massively formed γ, and the compositions of these were determined to be the same by SEM-EPMA. The massive γ has nucleated largely at the prior α grain boundaries and then spread rapidly into the grains in a spheroidal fashion, although occasionally, it is also found within grains, at some distance from the grain boundaries. Closer inspection of the latter sometimes reveals a fine striated structure emerging from the grain boundaries and terminating at locations where the massive γ commences. It is also interesting to note that the volume fraction of the massive γ appears to be dependent on cooling rate, being highest in the OQ sample and decreasing with further increase in cooling rate.

Figure 1. Optical microstructures of Ti-46.54Al alloy held 40 mins at 1360°C and cooled. (a) AC, (b) OQ, (c) WQ, and (d) IBQ.

Optical microstructures of the Ti-47.86 and Ti-49.40Al alloys cooled at various rates are shown in Figures 2a–d and 3a–d, respectively. Compared with the Ti-47.20 [5] and 46.54 Al (Figure 1a) alloys , the AC microstructures in these higher Al alloys (Figures 2a and 3a) appear with a larger volume fraction of the Widmanstatten structure. Also, some dark regions, presumably massive γ, are visible inside grains. The OQ sample of the Ti-47.86Al alloy, Figure 2b, shows a mixed microstructure consisting of lamellar, Widmanstatten and massive regions, thus indicating that this cooling rate is not sufficiently high to produce completely massive γ. Also visible is a feathery lamellar morphology, which appears to fan-out from various locations The OQ microstructure of the 49.40Al alloy, Figure 3a, seems to be mostly composed of the lamellar and Widman-

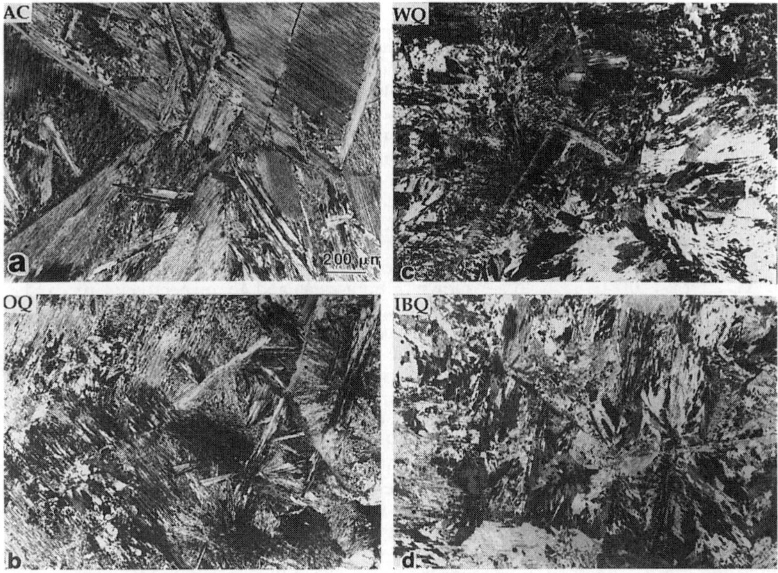

Figure 2. Optical microstructures of Ti-47.86Al alloy held 1h at 1400°C and cooled. (a) AC, (b) OQ, (c) WQ, and (d) IBQ.

Figure 3. Optical microstructures of Ti-49.40Al alloy held 1h at 1435°C and cooled. (a) AC, (b) OQ, (c) WQ, and (d) IBQ.

statten morphologies. These results indicate quite clearly the difficulty in suppressing these morphologies and the requirement of very high cooling rates to produce fully massive γ as alloy Al content increases. Four orientations of the Widmansatten/plate-like structure can be clearly seen in the AC and OQ micrographs of both the 47.86 and 49.40Al alloys, particularly in Figure 3b. The microstructures of the WQ (Figures 2c and 3c) and IBQ samples (Figures 2a and 3d) appear to have massively transformed to γ practically throughout the samples; some regions of retained α_2 can, however, be noted. The above microstructures, especially those of the 47.86Al alloy, appear very similar to those reported recently by McQuay et al in Ti-48Al-(2-3)Cr alloys [9].

In order to firmly establish that the Widmanstatten morphology appeared due to an increase in cooling rate and not due to some other effects, an extruded 47.86Al alloy was furnace and air cooled from a number of temperatures both above and below the α-transus (1380°C). The as-extruded microstructure was relatively fine-grained, fully lamellar. The microstructures following FC and AC after a 1400°C, 1h hold are shown in Figures 4a and 4b, respectively. As can be seen, the FC microstructure is large-grained fully lamellar, with little or no evidence of the Widmanstatten structure. In contrast, the AC microstructure shows not only the lamellar structure emanating from the grain boundaries, but also the Widmanstatten morphology in several different orientations within grains and traversing across over large distances. In BSE images of this sample, contrast differences were observable only in the lamellar structure present near grain boundaries, whereas images from within grains, where the Widmanstatten structure was present, showed no contrast variations, i.e., were uniformly dark. The microstructures following AC after holding for 1h at 1375 and 1365°C (both within the $\alpha+\gamma$ region) are shown in Figures 4c and 4d, respectively. Surprisingly, in contrast with the behavior following AC from above the α-transus (Figure 4b), fewer regions of the Widmanstatten structure are present in the 1375°C case and no evidence for this structure is seen in the 1365°C treatment. The microstructures in the latter show small amounts of fine, primary γ grains together with grains of the typical lamellar structure but of a much finer grain size than when cooled from the fully α region (Figure 4a). These microstructures were also essentially similar to those in samples FC from the lower temperatures. A TEM micrograph of the 1400°C, 1h AC sample is shown in Figure 5a. The large feature in the micrograph, which corresponds to the Widmanstatten structure, was established to completely γ by electron diffraction. The regions surrounding the large Widmanstatten plate in the micrograph correspond to the lamellar structure. However, no low index orientation relations between the lamellar and Widmanstatten structures were found by electron diffraction. Another example of two Widmanstatten plates/colonies is shown in Figure 5b. These colonies are entirely composed of γ as seen by the number of $<110>\gamma$ SAD patterns taken from the various regions indicated in Figure 5b. This type of microstructure and associated diffraction patterns resemble those seen in the massive γ structure in high Al alloys [6].

DISCUSSION

The results presented above and in previous reports [5,6] firmly establish that the decomposition of the α phase (and hence microstructure) depends strongly on cooling rate. In the present study, the effect of Al content in binary alloys has been examined. The results show that 1) the lamellar γ morphology prevails at low cooling rates over a wide range of Al contents from 43 to 50 Al, and 2) the Widmanstatten/plate-like/feathery and massive γ morphologies dominate at medium and very high cooling rates, respectively, only in alloys containing 46.54 to 50 Al; and that the volume fraction of both increase with increase in Al content in this range. Previously [6], an explanation for the composition dependence of the massive reaction was given based on phase diagram considerations. In that explanation, the T_0 α/γ line was taken to be the upper limit for the massive reaction and the T_0 α/α_2 line as the lower limit. The difference between these then represents the temperature range, ΔT, available for nucleation and growth of the massive γ for various alloy Al contents. Since the values of ΔT (hence available reaction time) increase rapidly with Al content, both the observation of the massive reaction only in alloys with $\geq$46.54Al, as well as the increase in the volume fraction of the massive product with increase in Al, could be accounted for [6].

To gain insight into the nucleation/growth mechanisms of the massive phase, it is important to ascertain the nature of the parent/massive interface. An example is shown in the the TEM micrograph in Figure 6, taken from the WQ 46.54Al alloy, where the α_2, massive γ and the interface between them can be seen. Although detailed studies have not been carried out, preliminary examination has indicated an absence of low index orientation relationships between the two phases. This, ofcourse, does not preclude the existence of orientation relationships during the

Figure 4. Optical microstructures of Ti-47.86Al alloy held 1h at various temperatures and cooled. (a) 1400°C, FC, (b) 1400°C, AC, (c) 1375°C, AC, and (d) 1365°C, AC.

nucleation stage, nor the existence of low energy interfaces in the absence of such a relationship [10]. Thus, coherent nucleation could still have occurred, suggesting that the early stages of nucleation must be studied. The massive-parent interface appears straight and reveals steps; the detailed structure and crystallography have not been determined yet. Stacking faults emerging from the interface into the α_2 can also be seen, but it is not yet clear if they are involved in the nucleation and growth of the massive γ, as in the case of the lamellar structure. Also, it is not known whether or not growth of the massive γ is accomplished by a spontaneous process involving random atom transfer across incoherent interfaces or by a ledge mechanism.

Another important feature concerns the mechanism of formation of the Widmanstatten γ structure which appears at intermediate cooling rates in the high Al alloys. At the same cooling rate, the volume fraction of this structure is observed to increase with an increase in alloy Al content. Furthermore, the ability of this structure to form appears to be sensitive to the temperature from which samples are cooled, i.e., it is more prevalent when samples are cooled from the completely α region than when cooled from the α+γ region. Reasons for this behavior are unclear at present. While morphologically the Widmanstatten and massive γ structures appear different, there is uncertainty in distinguishing between the two in terms of mechanisms and temperature dependence, since the composition of the Widmanstatten γ has not been ascertained. One possible way in which

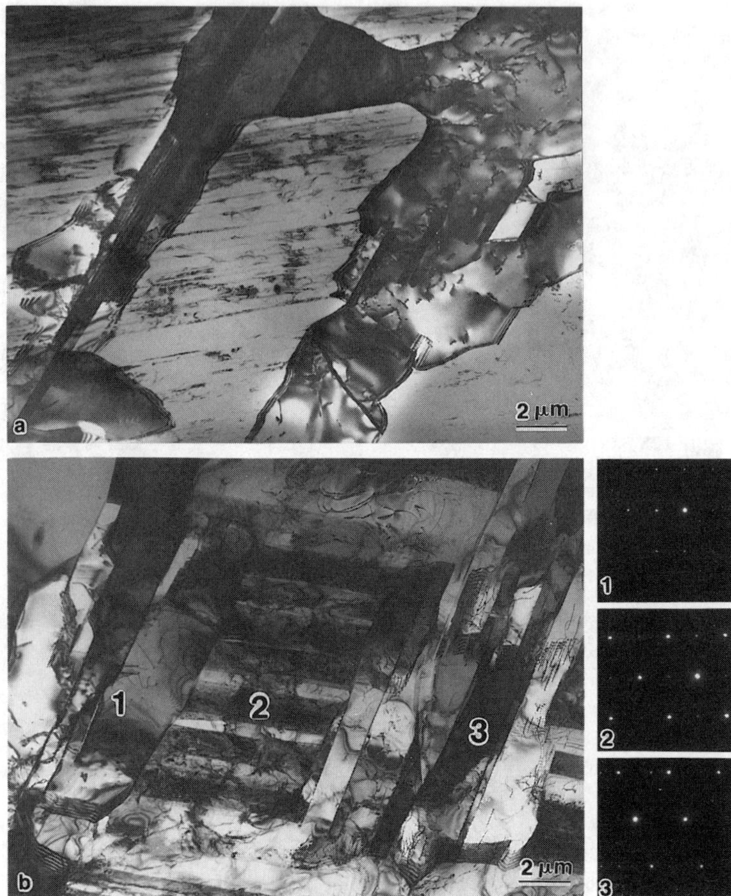

Figure 5. TEM micrographs showing the Widmanstatten γ structure in an extruded Ti-47.86Al alloy held at 1400°C for 1h and AC. (a) BF micrograph; (b) BF micrograph of two Widmanstatten colonies with $<110>\gamma$ SAD patterns from various indicated regions.

this structure forms is as follows. During heating the initial homogenized alloys to the completely α region, particularly those that initially have a large volume fraction of primary γ grains, the existing γ could transform to α on each of the four $\{111\}$ planes to produce different orientations of α platelets. On subsequent cooling at intermediate rates, each of these α platelets could transform to γ in a massive fashion, which then appears as the Widmanstatten structure in several orientations. This structure would be expected even under slow cooling conditions, which is not generally observed, thus lending limited support for this mechanism. A second possibility is that although the solutionizing temperature is expected to place the alloy in the fully α region, it may actually be in a two-phase $\beta+\alpha$ region because of the uncertainty in the location of the phase boundaries at these temperatures. Thus, during cooling, the b.c.c. β will transform to Widmanstatten α with six possible orientations because of the orientation relationship between β and α; the α on subsequent cooling transforms to the Widmanstatten γ structure in a massive fashion. Once again, this process would be expected to produce the Widmanstetten γ structure even under slow cooling conditions, which is not generally observed. Furthermore, it cannot explain the increase in the extent of the

Widmanstatten structure with an increase in cooling rate. Another reason why both of the above two mechanisms seem unlikely can be given as follows. Thus, if indeed the various orientations of Widmanstatten γ originate from differently oriented α platelets which have formed by one of the above two pathways, there is difficulty explaining why these α platelets on cooling would transform completely to γ rather than the typical $\alpha+\gamma$ lamellar structure as in the surrounding regions (which were also previously completely α and have experienced the same cooling rate). A third possibility is that as the sample is cooled at intermediate rates and its temperature drops, quenching stresses accumulate and produce differently oriented twins on $\{10\bar{1}2\}$type planes in the α. Since by this time the temperature may have dropped sufficiently below T_0 α/γ, the twinned α platelets transform in a massive manner to produce the Widmanstatten γ structure in different orientations. A fourth possibility is that as the cooling rate increases from slow to moderate, the lamellar γ morphology breaks down because of its inability to grow at speeds commensurate with the heat extraction rates. Under these conditions, it may be possible for the γ to form and grow by advance of an semi-coherent/incoherent interface. A final possibility is that as the samples are cooled at intermediate rates and the temperature continuously drops, the lamellar γ structure nucleates at the grain boundaries and grows into the grains and the non-consumed α within grains being sufficiently undercooled relative to T_0 α/γ transforms in a massive manner to produce regions that appear Widmanstatten. While this process can account for the observation of patchy massive γ regions within grains at intermediate cooling rates, it is difficult to state if it is also responsible for the Widmanstatten γ structure, because this structure on an optical scale appears in several orientations with relatively straight and apparently crystallographic sides. It is clear, however, that before the exact mechanism can be established, the crystallography/ composition of the Widmanstatten structure and its relationship with the surrounding lamellar regions need to be determined. Obviously, more detailed investigations are needed.

Figure 6. TEM micrograph showing α_2, massive γ and the interface between them in a Ti-46.54Al alloy WQ from 1360°C after holding for 40 minutes.

Finally, it is clear that before the temperature dependence of the Widmanstatten and massive reactions can be ascertained positively, experimental determination of the temperature of start of the reactions is required, which is considered elsewhere in these proceedings [11]. Determination of the crystallography, nature of the parent-product interfaces, orientation relationships, defect structures, and the nucleation and growth mechanisms of the different morphologies of γ are the subject of ongoing research.

CONCLUSIONS

In summary, the cooling rate dependence of the decomposition of the α phase and hence of microstructure in Ti-(43-50) Al alloys have been studied. In particular, evidence for the occurrence of Widmanstatten/plate-like/feathery and massive γ morphologies at intermediate and very high cooling rates, respectively, has been presented for alloys with 46.54 to 50 Al. The composition dependence of these morphologies has been rationalized in terms of phase diagram considerations. Preliminary examination of the parent/product interfaces and orientation relationships has been carried out and possible mechanisms of formation of the different morphologies have been provided.

ACKNOWLEDGEMENTS

Support for this research from a Air Force Engineering Foundation Research Initiation Grant, the Alcoa Foundation and the State of Ohio Edison Materials Technology Center is deeply appreciated. The authors are thankful to the Air Force Materials Laboratory, WPAFB, Dayton for use of the SEM facilities and to Mr. Cameron Begg for his help with the SEM/microprobe work.

REFERENCES

1. M. J. Blackburn and M. P. Smith, United States Patent, No. 4,294,615 (1981).
2. S. C. Huang and E. L. Hall, MRS Symp. Proc., 133, 373 (1989).
3. Y. W. Kim, MRS Symp. Proc., 213, 777 (1991); Acta Metall. Mater., 40, 1121 (1992).
4. D. S. Shih, S. C. Huang, G. K. Scarr, H. Jang and J. C. Chesnutt: in: Microstructure/ Property Relations in Titanium Aloys and Titanium Aluminides, Y. W. Kim and R. R. Boyer (eds.), p. 135, TMS-AIME, Warrendale, PA (1991).
5. P. Wang, G. B. Viswanathan and V. K. Vasudevan, Metall. Trans., 23A, 790 (1992).
6. P. Wang and V. K. Vasudevan, Scripta Metall. Mater., 27, 89 (1992).
7. S. A. Jones and M. J. Kaufmann, Acta Metall. et Mater., in press (1992).
8. P. McQuay, D. M. Dimiduk, and S. L. Semiatin, Scripta Metall. et Mater., 25, 1689 (1991).
9. P. McQuay, D. M. Dimiduk, H. A. Lipsitt and S. L. Semiatin, in: Proceedings of the Sixth World Titanium Conference, in press (1992).
10. J. C. Caretti and H. R. Bertorello, Acta Metall., 31, 325 (1983).
11. G. Ramanath and V. K. Vasudevan, paper in these proceedings (1992).

EFFECT OF ALLOYING ON PHYSICAL PROPERTIES OF NiAl

W.S. WALSTON AND R. DAROLIA
GE Aircraft Engines, Cincinnati, OH 45215

ABSTRACT

Several alloying additions were made to NiAl and the effect on physical properties was evaluated. The values obtained were compared to available data on NiAl and NiAl alloys. Negligible effects were observed on thermal expansion and dynamic modulus with the additions studied. The high thermal conductivity of NiAl can be advantageously utilized in many applications, and it was found that additions of Hf lowered thermal conductivity only slightly, while larger decreases were found with other additions. However for the NiAl alloys currently under consideration for applications, significant advantages in physical properties are still maintained over nickel-base superalloys.

INTRODUCTION

The unique physical properties of NiAl make it an attractive choice for high temperature structural applications. Significant improvements in strength have recently been attained by alloying additions, such that these alloys are being considered to replace nickel-base superalloys in the turbine section of aircraft jet engines. The impact of these alloying additions on the physical properties is important in determining if NiAl alloys still maintain significant advantages over superalloys. Of primary interest are density and thermal conductivity, although thermal expansion and modulus are important as well. There have been relatively few studies on the physical properties of NiAl and even fewer on the effect of alloying additions. This paper presents data on different levels of several additions, primarily Hf, Ti and Re.

EXPERIMENTAL PROCEDURES

The alloys in this study were single crystal castings made using a modified Bridgman technique. In general, these alloys contain 50 at.% Ni with alloying additions substituted for Al. Unless otherwise noted, specimens were EDM wirecut in the <001> orientation. All specimens received a 1315°C/50 hr homogenization heat treatment, which also served to solution the primary precipitates.

Density was determined by the water displacement method. Resonance measurements for calculating the dynamic Young's modulus were performed and no corrections made for thermal expansion. Thermal expansion measurements were performed on a dilatometer at heating rates of 3°C/min. Specific heat was measured using a differential scanning calorimeter, and thermal diffusivity was measured using the laser flash diffusivity method. Neither of these sets of data were corrected for thermal expansion, and thus the thermal conductivity was not corrected for thermal expansion, as is common practice. Thermal conductivity was calculated as a function of temperature from the specific heat, thermal diffusivity and room temperature density data.

RESULTS AND DISCUSSION

Density

Several investigators have measured the density of NiAl as a function of stoichiometry.[1,2,3] The room temperature density of stoichiometric NiAl is 5.90 g/cm^3. The effects of substituting Ni,[1,2] Hf and Ti for Al can be described by the following linear relationships in g/cm^3:

$$\rho_{NiAl} = 3.05 + 0.057(\text{at.\% Ni}) \quad \text{for 50-60 at.\% Ni} \quad (1)$$
$$\rho_{NiAl + Hf} = 5.90 + 0.15(\text{at.\% Hf}) \quad \text{for 0-3 at.\% Hf} \quad (2)$$
$$\rho_{NiAl + Ti} = 5.90 + 0.025(\text{at.\% Ti}) \quad \text{for 0-10 at.\% Ti} \quad (3)$$

Elastic modulus

For polycrystalline NiAl, Young's modulus has been measured as a function of stoichiometry for Ni concentrations from 45 at.% to 62.5 at.%.[1,4] There is some disagreement over the behavior from 45 at.% to 50 at.%, but it is believed that Young's modulus increases slightly with higher Ni contents. At 56 at.% Ni, Young's modulus is about 10% higher than at 50 at.% Ni. Between 56 at.% Ni and 60 at.% Ni, Young's modulus drops precipitously, which is believed due to premartensitic transformations that occur in this composition range.[4]

Based on the variation observed as a function of stoichiometry, it was thought that alloying could also affect Young's modulus. Dynamic Young's modulus data as a function of orientation are shown in Figure 1. This figure compares results by Wasilewski[5] on a Ni-50.6Al alloy, ultrasonic data on a stoichiometric NiAl crystal supplied by GEAE[6] and our data on stoichiometric NiAl. The data on GEAE single crystals is consistently about 10% lower than that of Wasilewski. The slopes of the two data sets as a function of temperature are equivalent, and the difference could stem from the fact that Wasilewski's values are relative to his initial room temperature value. There are also material processing and thermal history differences which could play a role.

The effect of alloying on dynamic Young's modulus is shown in Figure 2 for an alloy with 0.5 at.% Hf, while static Young's modulus was obtained for an alloy with 3.0 at.% Re from available tensile data. There was no effect of alloying with Hf or Re additions, however it is believed that larger or different alloying additions could produce significant changes in elastic behavior based on the results on stoichiometric NiAl.

Several investigators have measured the elastic constants, C_{ij}, for NiAl,[5,7,8] and two have evaluated the effect of stoichiometry with Ni contents ranging from 45 at.% to 63 at.%.[7,8] It was found that C_{11} decreased with increasing Ni contents, while C_{12} and C_{44} increased. There is no data on the effect of alloying on the elastic constants.

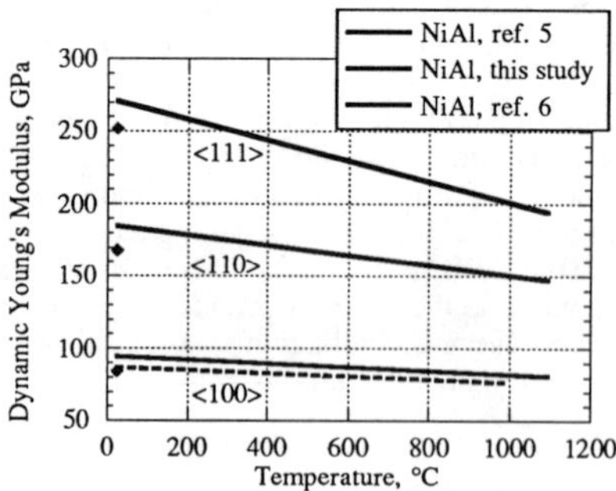

Figure 1. Young's modulus as a function of orientation.

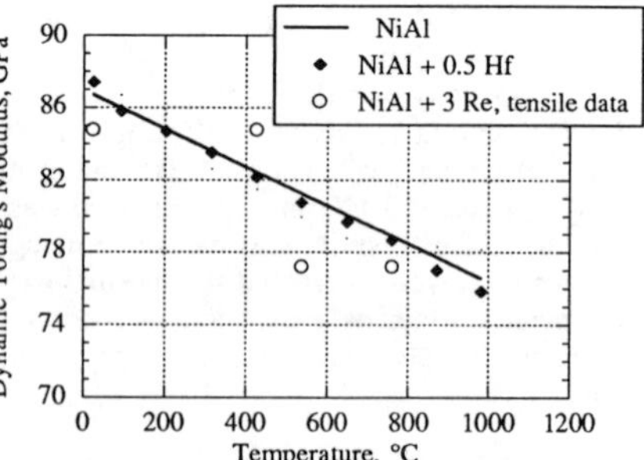

Figure 2. <001> Young's modulus of NiAl and NiAl alloys.

Thermal Expansion

The coefficient of thermal expansion for single crystal, stoichiometric NiAl has been measured in the <001> and <111> orientations. As expected, there is little difference between the two orientations due to cubic symmetry. Figure 3 compares these values for stoichiometric NiAl to others in the literature.[9-11] Obviously, there is quite a spread in the data. The effect of stoichiometry on thermal expansion has been studied by Clark and Whittenberger,[9] and no significant effect was observed for Al contents from 44 at.% to 52.7 at.%. Another study on Ni-55Al shows a small increase in thermal expansion compared to binary measurements.[10] Figure 3 also shows data on alloyed NiAl. The values of the coefficient of thermal expansion for alloyed NiAl are slightly higher than most for stoichiometric NiAl.

It is known that the content of vacancies can have a significant effect on thermal expansion.[7] However, while this could explain some of the scatter in the data due to differences in processing and thermal history, it cannot explain the observation that there was no effect of stoichiometry.[9] Clearly, more work is needed to understand the scatter and any possible effect of alloying.

Thermal Diffusivity

The thermal diffusivity of single crystal, stoichiometric NiAl has been measured in the <001> and <111> orientations. The thermal diffusivity in the <001> orientation is about 5% higher than the <111> orientation up to 500°C, at which point the data merge. This difference is not believed significant, and the data have been averaged in Figure 4. This figure shows that at low temperatures NiAl has an decreasing slope, while all alloys have a increasing slope. This is also true for measurements on other NiAl alloys,[12] but these differences currently are not understood. Figure 4 also shows that there are significant effects of alloying additions; all of which decrease thermal diffusivity. The largest decrease in thermal diffusivity occurs as a result of Ti additions, although Re also significantly reduces this property. Hf additions have little effect, and the 0.5 at.% Hf alloy is actually higher than NiAl. However this is likely due simply to scatter in the data.

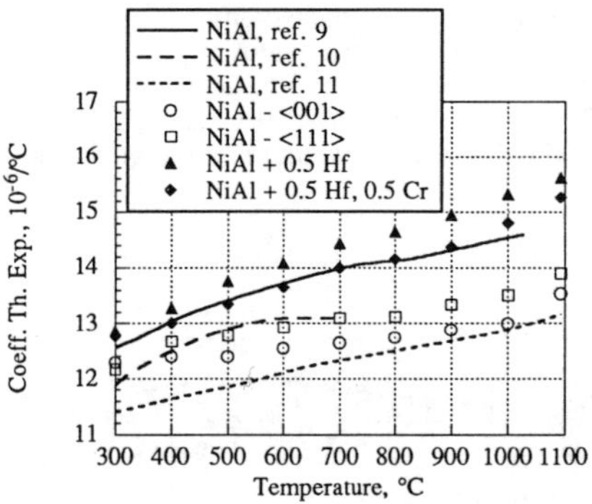

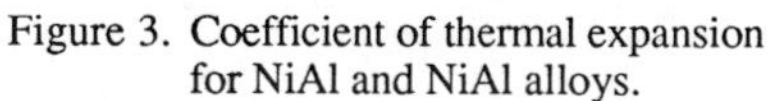
Figure 3. Coefficient of thermal expansion for NiAl and NiAl alloys.

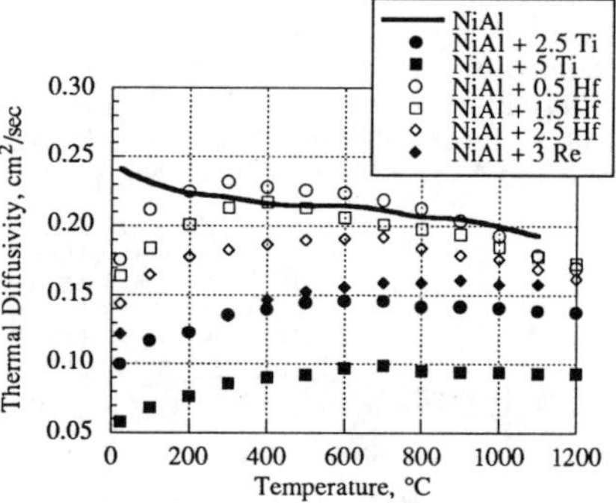

Figure 4. Thermal diffusivity of NiAl and NiAl alloys with Re, Ti and Hf.

Specific Heat

The specific heat of single crystal, stoichiometric NiAl has been measured in the <001> and <111> orientations, and no difference was observed. To the authors' knowledge, there is no other data on the specific heat of stoichiometric NiAl, however there have been studies of alloyed NiAl.[12,13] Figures 5 and 6 show the data of this study on alloyed NiAl compared to stoichiometric NiAl. This data is consistent with other available data[12,13] which shows a slight decrease in specific heat due to alloying. Figure 5 shows that 3.0 at.% Re causes about a 5% decrease in specific heat at all temperatures studied, while a smaller decrease is observed for up to 5.0 at.% Ti. In Figure 6, it can be seen that Hf additions cause up to a 10% decrease in specific heat.

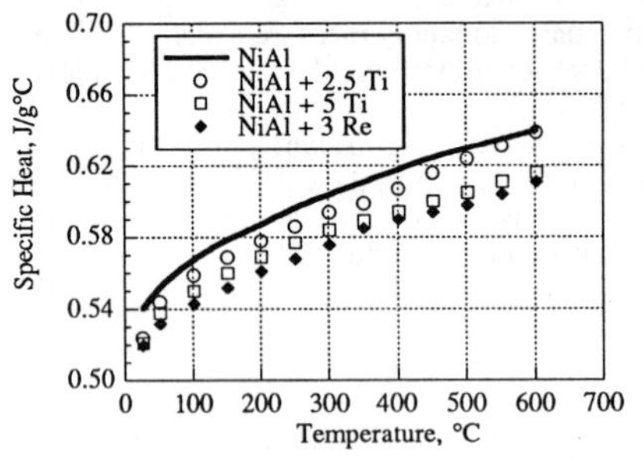

Figure 5. Specific heat of NiAl and NiAl alloys with Re and Ti additions.

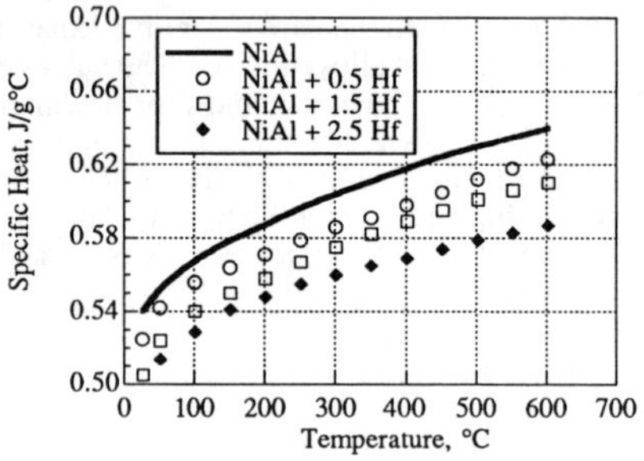

Figure 6. Specific heat of NiAl and NiAl alloys with Hf additions.

Thermal Conductivity

Thermal conductivity is a product of density, specific heat and thermal diffusivity, thus the alloying trends discussed previously hold true. There are other measurements of thermal conductivity for stoichiometric NiAl,[11,14] however these values are either much higher or much lower than those of this study and are not consistent with other measurements on alloyed NiAl.[12,13] Figures 7 and 8 mimic the behavior of the thermal diffusivity plot of Figure 4. The addition of 3 at.% Re decreases the property about 25% at 600°C. Similar to the thermal diffusivity measurements, the effect of Hf at 600°C ranged from none to a decrease of only 15% with a 2.5 at.% addition. As expected, Ti decreased the thermal conductivity most significantly with a 55% decrease at 600°C for 5 at.% Ti.

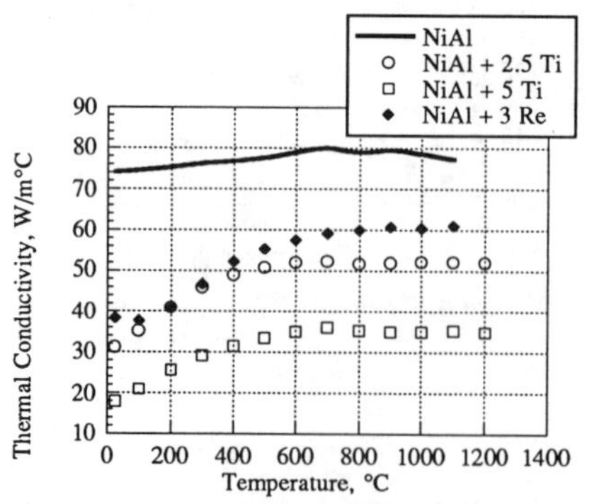

Figure 7. Thermal conductivity of NiAl and NiAl alloys with Re and Ti.

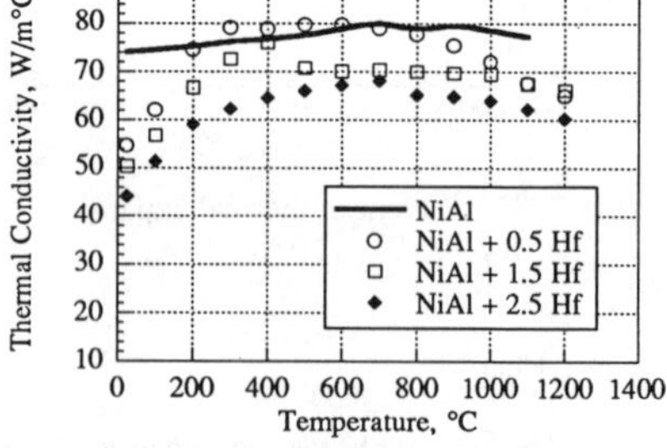

Figure 8. Thermal conductivity of NiAl and NiAl alloys with Hf.

Comparison to Superalloys

Among potential applications for NiAl, one of the most promising is that of turbine airfoils in aircraft jet engines. In such an application, comparisons with nickel-base superalloys are necessary to determine payoffs. Table I compares stoichiometric NiAl, NiAl alloys and an advanced superalloy, such as single crystal René N5. The NiAl alloy column represents alloys with small additions of generally less than 5 at.%. Most of the work has been done on additions which form β', so the data is skewed towards this class of alloys. References are provided for NiAl when not discussed previously in this paper. In comparing NiAl and NiAl alloys to superalloys, there are several physical property advantages to point out. The 25-30% lower density and 3x-5x higher thermal conductivity of NiAl alloys are especially important advantages for turbine airfoils. The coefficient of thermal expansion is similar to superalloys and is an important attribute of NiAl in many applications. There are also large differences in electrical resistivity which could be utilized in some applications.

One interesting property is matrix/precipitate mismatch which is slightly negative for γ/γ' superalloys and positive for β/β' NiAl alloys. It is well known that this negative mismatch for superalloys produces rafting of the precipitates normal to the stress axis during creep rupture tests. Interestingly, we have observed the positive mismatch in β/β' NiAl alloys results in rafting of the β' precipitates parallel to the stress axis. We are currently exploring the implications of this rafting behavior on creep strength.

Table I. Physical properties of NiAl, NiAl alloys and an advanced superalloy.

Property	units	Temp, °C	NiAl	NiAl Alloys[1]	Adv. Superalloy
Matrix Crystal Structure			B2	B2	FCC
Bonding			Covalent/Metallic	Covalent/Metallic	Metallic
Melting Point	°C		1682[2]	1610-1676[3]	1390
Matrix Lattice Parameter	Å	RT	2.887[4]	2.888-2.900[3,4]	3.580
Matrix/Ppt. Lattice Mismatch	%	RT	N/A	1 to 6 (β')[3,4]	0 to -0.5 (γ')
Density	g/cm^3	RT	5.9	up to 6.30	8.60
Young's Modulus, Polycrystal	GPa	RT	188	188	205
Young's Modulus - <001>	GPa	RT	88	88	130
Anisotropy Factor		RT	3.25	3.25[5]	2.72
Shear Modulus, Polycrystal	GPa	RT	71.5[6]	71.5[5]	74
Poisson's Ratio		RT	0.313[6]	0.313[5]	0.380
Thermal Expansion	10^{-6}/°C	600	13.2	13.7	13.5
Specific Heat	J/g°C	600	0.64	0.61-0.64	0.46
Thermal Diffusivity	cm^2/sec	600	0.22	0.10-0.22	0.033
Thermal Conductivity	W/m°C	600	76	35-76	15
Electrical Resistivity	μohm•cm	RT	8-10[7]	10-30[8]	120-140

[1] NiAl alloys containing primarily β' precipitates and less than 5 at.% alloying addition.
[2] GEAE DTA data show that the melting point of NiAl is about 40°C higher than literature data.[19]
[3] Reference 15.
[4] Reference 16.
[5] Estimated based on no effect on Young's modulus.
[6] Reference 17.
[7] Reference 18.
[8] Estimated based on stoichiometric effect in [7].

SUMMARY

Several physical properties have been measured as a function of alloying content for NiAl alloys. The following key points have been covered:

1. No effect of alloying on Young's modulus was observed, although other alloying levels or combinations could have some effect.
2. Large scatter exists for thermal expansion measurements, but there does not appear to be significant effects of stoichiometry or alloying.
3. No significant effects of single crystal orientation on thermophysical properties was observed.
4. Specific heat values were only slightly decreased by additions of Hf, Ti or Re.
5. Thermal diffusivity and thermal conductivity are sensitive to alloying, with Ti additions producing the largest decrease.
6. NiAl alloys maintain significant physical property advantages over superalloys, especially in density and thermal conductivity.

REFERENCES

1. M.R. Harmouche and A. Wolfenden, J. Test. Eval. **15**, 101 (1987).
2. P. Georgopoulous and J.B. Cohen, Scr. Met. **11**, 147 (1977).
3. A. Taylor and N.J. Doyle, J. Appl. Cryst. **5**, 201 (1972).
4. N. Rusovic and H. Warlimont, Phys. Stat. Sol. (a) **53**, 283 (1979).
5. R.J. Wasilewski, Trans. AIME **236**, 455 (1966).
6. J. Travisonno, John Carroll University, unpublished research, 1990.
7. N. Rusovic and H. Warlimont, Phys. Stat. Sol. (a) **44**, 609 (1977).
8. L. Zhou, P. Cornely, J. Trivisonno and D. Lahrman, in IEEE 1990 Ultrasonics Symposium Proceedings, (**3**, IEEE, New York, NY, 1990) pp. 1309.
9. R.W. Clark and J.D. Whittenberger, in Thermal Expansion 8, edited by T.A. Hahn, (Plenum Press, New York, N.Y., 1984) pp. 189.
10. E.G. Ivanov, Izvestiya Akademii Nauk SSSR, Metally **No. 2**, 168 (1986).
11. R.H. Singleton, A.V. Wallace and D.G. Miller, in Summary of the Eleventh Refractory Composites Working Group Meeting, (1966, AFML-TR-66-179) pp. 717.
12. R.E. Taylor, H. Groot and J. Larimore, "Thermophysical Properties of Aluminides", Thermophysical Properties Research Laboratory, TPRL 506, 1986.
13. M. Kaufman, GE Aircraft Engines, unpublished research, 1982.
14. M.G.X. Gigliotti and R.L. Fleischer, GE Corporate Research and Development, unpublished research, 1991.
15. GE Aircraft Engines, unpublished research, 1989-1992.
16. M. Takeyama, C.T. Liu and J. C.J. Sparks, in JIMIS 6, (Sandai, Japan, 1991) pp. 871.
17. C.A. Moose, M.S. Thesis, Pennsylvania State University, 1991.
18. H. Jacobi, B. Vassos and H.J. Engell, J. Phys. Chem. Solids **30**, 1261 (1969).
19. M.F. Singleton, J.L. Murray and P. Nash, in Binary Alloy Phase Diagrams, edited by T.B. Massalski, (American Society for Metals, Metals Park, OH, 1986) pp. 140.

ACKNOWLEDGMENTS

This work represents a several year effort at GE Aircraft Engines. Contributions have been made by many individuals, in particular, J. Dobbs, D.F. Lahrman and R.D. Field. Support for part of this work came from the Air Force Wright Laboratory and Naval Air Warfare Center Aircraft Division Trenton under contract F33615-90-C-2006.

THE STABILITY OF B2 COMPOUNDS IN Ti-MODIFIED Nb-Al ALLOYS

J. SHYUE, D-H. HOU, S.C. JOHNSON, M. AINDOW* AND H.L. FRASER
Department of Materials Science and Engineering
The Ohio State University, 2041 College Road, Columbus, OH 43210

* Now at School of Metallurgy and Materials, and the IRC in Materials for High Performance Application, The University of Birmingham, Elms Road, Birmingham B15 2TT, UK

ABSTRACT

It has been found that additions of Ti to Nb_3Al results in the formation of phases with the B2 crystal structure in as-cast samples. The microstructure and phase stability of two alloys, namely Nb-15Al-10Ti and Nb-15Al-40Ti (in at.%), have been studied. Heat-treatment of the Nb-15Al-10Ti alloy at 1100°C results in the precipitation of an A15 phase in the B2 matrix. A particular orientation relationship exists between the two phases. Heat-treatment above 1600°C results in the dissolution of the second phase. The ALCHEMI technique has been employed to assess qualitatively the distribution of atom types over the two sublattices of the B2 compound.

INTRODUCTION

In binary Nb-Al alloys with <25% Al there are two crystal structures which are predicted by the equilibrium phase diagram; the A2 (bcc) Nb solid solution and the A15 compound Nb_3Al. In rapidly solidified ribbons of binary Nb-18Al alloys, however, grains with the B2 crystal structure have been observed [1,2]. Prolonged heat-treatment of rapidly solidified Nb-18Al ribbons resulted in significant dislocation activity [2] suggesting that the presence of this new phase may give rise to improvements in ductility. If this were the case then it would be useful if a stable, as opposed to a metastable, B2 compound could be produced by, for example, the addition of ternary alloying elements. It has been reported [3-5] that additions of Nb to Ti_3Al result in the formation of the β-phase with the B2 crystal structure. In this B2 structure, it has been shown that Nb occupies the same sub-lattice as Ti [6], and since for binary Nb-Ti alloys there are continuous solid solutions from pure Nb to pure Ti, it seems reasonable to suggest that a stable or metastable B2 phase field may exist for a significant range of intermediate compositions between Nb_3Al and Ti_3Al. In the present study we have investigated this possibility by adding Ti to Nb_3Al-based alloys. Results obtained from two alloys are presented in this paper; Nb-15Al-10Ti (Alloy 1) and Nb-15Al-40Ti (Alloy 2). It will be shown that indeed the material adopts the B2 crystal structure and the nature of the domain structures and lattice occupancies are discussed. The deformation characteristics of these alloys are discussed in a seperate paper [7].

EXPERIMENTAL PROCEDURE

A binary (Nb-Al) ingot was first produced by arc-melting and then remelted with Ti to form the ternary alloys. Thermal homogenization was performed in a graphite furnace in an inert atmosphere. Wet chemical analyses were performed independently as references. Bulk compositional analyses were performed using quantitative electron probe microanalysis (EPMA) and an EDS system attached to a JEOL JSM 820. Quantitative x-ray microanalyses were also performed on TEM foils. The B2 order-disorder transition phenomena of single phase Alloy 1 and 2 were studied by holding specimens at desired temperatures followed by water quenching and/or furnace cooling. TEM specimens were produced by twin jet electropolishing using 10% H_2SO_4 in methanol at -40°C and 90mA. TEM work was carried on a JEOL 200CX for imaging and quantitative chemical analysis and a Philips CM20 for electron channeling microanalysis [8]. Both were operated at an accelerating voltage of 200kV. The order-disorder transition was determined by the presence of anti-phase domain boundaries and of 001 superlattice reflections in diffraction patterns.

RESULTS AND DISCUSSION

Microstructures of the Ti-modified Nb-Al alloys

General microstructrues

In the as-cast form, both alloys consist mainly of a phase with the B2 crystal structure. Dark field images formed with 001 superlattice reflections showed a very refined (unresolvable) microstructure in Alloy 1 while fine anti-phase domains were observed in Alloy 2. In the case of Alloy 1, heat-treatment at 1100°C results in the precipitation of a second phase. Figure 1(a) is a typical optical micrograph obtained from a specimen heat-treated at 1100°C for 50 hours. The second phase, which has the A15 crystal structure, adopts a plate-like morphology; this changes significantly after annealing at 1500°C as shown in Figure 1(b). It seems likely that this morphological change occurs by partial dissolution of the A15 phase which starts to form at approximately 1550°C in this alloy. Homogenization performed at 1600°C or higher resulted in single phase B2 material with a very large grain size. The matrix of heat-treated Alloy I is an Nb-rich solid solution having either B2 or *bcc* structure depending on its local composition . The volume fraction and grain size of the A15 phase increased significantly as the duration of heat treatment at 1100°C increased from 20hrs (≈20%) to 50hrs (≈60%). Specimens heat treated for 100 hours did not show any further change in the volume fraction and morphology of the A15 phase. Very fine needle-like precipitates were found in Alloy 2 after annealing at 800°C whilst a 1000°C heat-treatment resulted in total dissolution of the second phase. Preliminary EDS data suggests that this phase has composition as Ti_2NbAl.

A15/B2 orientation relationship

Both TEM and SEM images from heat-treated Alloy 1 show that the A15/B2 interfaces tend to be straight and faceted suggesting that there may be a specific orientation relationship between the two phases. An example of the faceted interphase boundary is shown in Figure 2. A typical A15/B2 orientation relationship has been determined to be $[100]_{A15}//[1\bar{2}1]_{B2}$ and $(0\bar{2}1)_{A15}//(10\bar{1})_{B2}$. It is worth noting that this relationship is probably not unique since boundaries have been observed across which $<100>_{A15}//<100>_{B2}$ which is not consistent with the deduced orientation relationship.

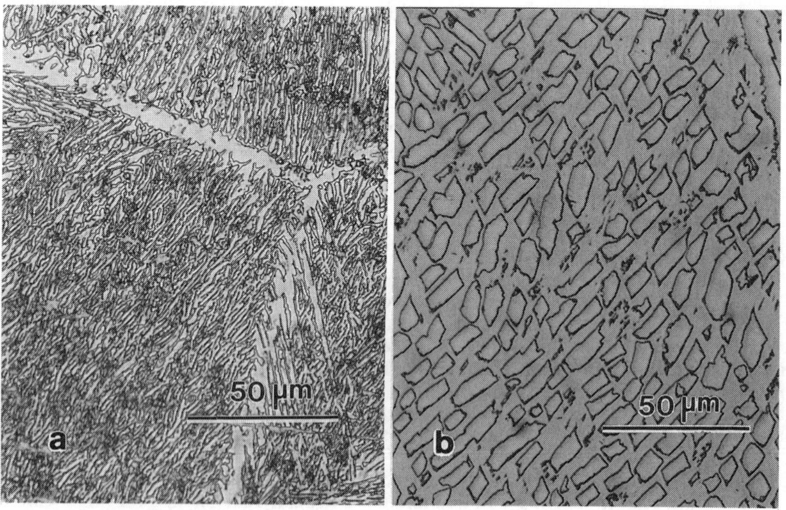

Figure 1. Optical micrograph of heat-treated Alloy 1 showing a) plate-like structure and the distribution of A15 phase, b) same as a) pluse 1500°C-5hrs, the A15 morphology changed remarkably.

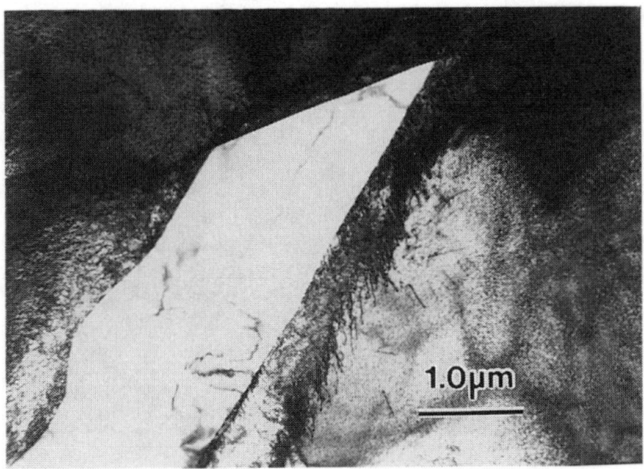

Figure 2. TEM image showing the typical A15/B2 interphase boundaries.

Compositional measurements

Both Alloy 1 and 2 have 60 ppm N and 400 ppm C and O. Wet chemical analysis also indicated appreciable (approximately 3 at.%) Al macro-segregation at the center (Al-rich) and edge of Alloy 1 ingot. Microsegregation of Al and Ti was noted from EDS-SEM results, where compositions of as-cast ingots were determined to be Nb-14Al-11Ti for Alloy 1 and Nb-17Al-38Ti for Alloy 2. The EPMA results from A15 phase and the matrix from heat-treated Alloy 1 samples are given as

Eleme nt	Heat Treated Alloy A15 Phase	matrix	As-Cast Alloy
Nb	73.4±1.3 at.%	79.4±1.7 at.%	76.5±2.3 at.%
Ti	9.0±1.0 at.%	9.1±1.4 at.%	10.4±0.7 at.%
Al	17.6±1.2 at.%	11.5±1.6 at.%	13.1±1.1 at.%

Although the Alloy 1 ingot was compositionally inhomogeneous on a macroscopic scale, the results collected at various locations by different analytical methods indicate consistently that the A15 phase is Al-rich relative to its parent phase. As noted, the matrix can exhibit either the B2 or *bcc* crystal structures. It is important to understand the role that composition plays in the stability of these phases. Diffraction patterns were recorded at the same locations as the x-ray spectra so that a correlation between composition and microstructure could be obtained. A typical set of EDS-TEM data acquired from a number of heat-treated (1100°C-50hrs) Alloy 1 foils are listed below. The results suggest that Al concentration is likely to be one of the factors which determine the B2 order-disorder transition. Figure 3 shows that the domains get much smaller and then dissappear at the A15/B2 interface, presumably due to localized Al depletion.

A15: Nb-9 at.%Ti-22 at.%Al
B2: Nb-12 at.%Ti-12 at.%Al
A2: Nb-12 at.%Ti-7.5 at.%Al

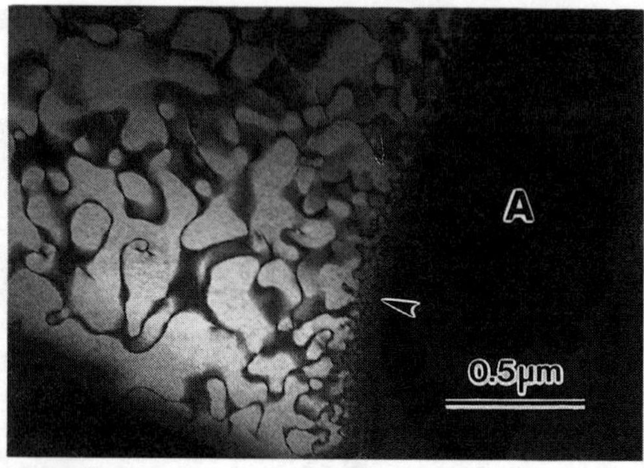

Figure 3. Dark field TEM micrograph obtained using an 001_{B2} type superlattice relection. The anti-phase domains in Alloy 1 (heat-treated at 1100°C).

Site occupancies in B2 structure

The B2 structure is usually exhibited by alloys with compositions AB. Since the composition of the B2 phase in the present study is rather rich in Nb, and contains three elements, it is interesting to determine the site occupancies of the three elements on the two sub-lattices. Two orientations were chosen for channeling-enhanced microanalysis; in one, {100} type superlattice reflections provide chemical information on {100} layer, and in the other {110} type reflections yield {110} plane chemistry. The {100} planes, consisting of only one type of site, are chemically layered while the {110} planes, containing both sites, are expected to be homogeneous and can be used as a reference. The intensity ratios acquired under channeling conditions are listed as the following

	Reflectin g Condition	Sense of Deviatio n	Al/Nb	Ti/Nb	Al/Ti
Alloy 1	100	+s	0.1282±0.0020	0.1836±0.0026	0.6987±0.0128
		- s	0.1202±0.0020	0.1817±0.0024	0.6615±0.0130
	110	+s	0.1300±0.0019	0.1853±0.0025	0.7001±0.0116
		- s	0.1293±0.0019	0.1874±0.0025	0.6901±0.0117
Alloy 2	100	+s	0.2505±0.0021	1.4160±0.0068	0.1769±0.0014
		- s	0.2331±0.0019	1.4924±0.0069	0.1562±0.0012
	110	+s	0.2380±0.0019	1.4650±0.0069	0.1624±0.0013
		- s	0.2322±0.0019	1.4552±0.0067	0.1596±0.0012

The 110 channeling data indicate that the Al/Nb, Ti/Nb and Al/Ti intensity ratios do not change on reversing the sign of the deviation. Thus, it is evident that the {110} planes are not chemically layered, as was anticipated for the B2 structure. The 100 channeling enhanced microanalysis data from Alloy 2 show that both Al/Nb, Ti/Nb and Al/Ti intensity ratios are significantly unequal. Both Al/Nb and Al/Ti are high when Ti/Nb is low and the opposite is true when the sense of deviation from the exact Bragg condition is reversed. The ALCHEMI data indicate that Ti atoms occupy different sites from the Al atoms preferentially in Alloy 2. In the case of Alloy 1, the Al/Nb intensity ratio changed noticeably but the differences in the Ti/Nb and the Al/Ti ratios were too small to be significant. This may suggest that Ti atoms replace Nb atoms randomly on both sites but we cannot ignore the possibility that there may be some contribution from adjacent domains. Further studies are being performed on single phase Alloy 1 with significantly larger anti-phase domains to test this hypothesis.

The order-disorder transition

In the as-cast foils from Alloy 1, dark-field TEM micrographs obtained using superlattice reflections did not reveal the presence of anti-phase domains, but showed very small bright spots. A fine distribution of anti-phase domains was observed in samples aged at 1600°C for 5 hours as shown in Figure 4(a); the size of these domains is ≈ 100-200Å. The domain size increased only slightly following further aging at 1500°C for 20 hours (Figure 4b), but increased markedly when heat-treated at 900°C for 2 hours. The interpretation of these results is that the order-disorder temperature lies below 1500°C. Thus, domain sizes following heat-treatment at 1600°C and 1500°C are similar since both samples would have been in the disordered state at the heat-treatment temperatures. The domains in the samples aged at 900°C are, however, much larger since they result from growth of the APB domains which are already present in the material. If the aging is performed at 700°C (Figure 4d), the domain size is still fairly small, presumably because of the sluggish kinetics afforded at this temperature. In the case of Alloy 2, the anti-phase domains were larger than those in Alloy 1 of comparable aging history. Figure 5 shows a series of dark-field micrographs from Alloy 2 of various aging temperatures. A marked change in domain size indicates that the order-disorder transition temperature of Alloy 2 lies between 1020°C and 1120°C.

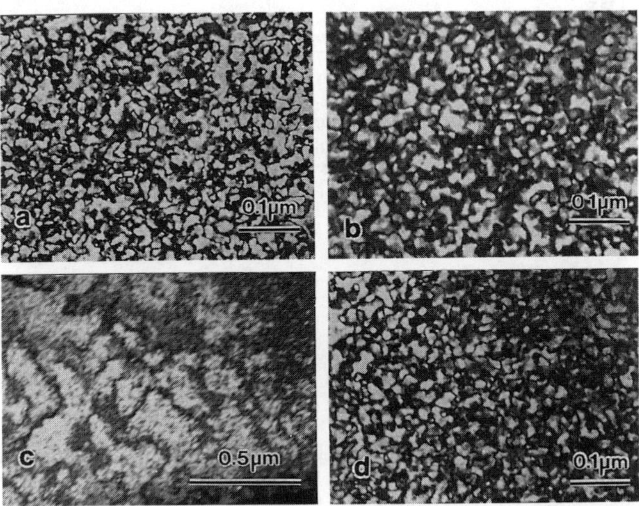

Figure 4. Anti-phase domain sizes in Alloy 1, imaged with $<001>_{B2}$ superlattice reflections, as a function heat treatment a) 1600°C-5hrs furnace cool, b) same as (a) plus 1500°C-5hrs furnace cool, c) same as (a) plus 900°C-2hrs, water quench, d) same as (a) plus 700°C-2hrs water quench.

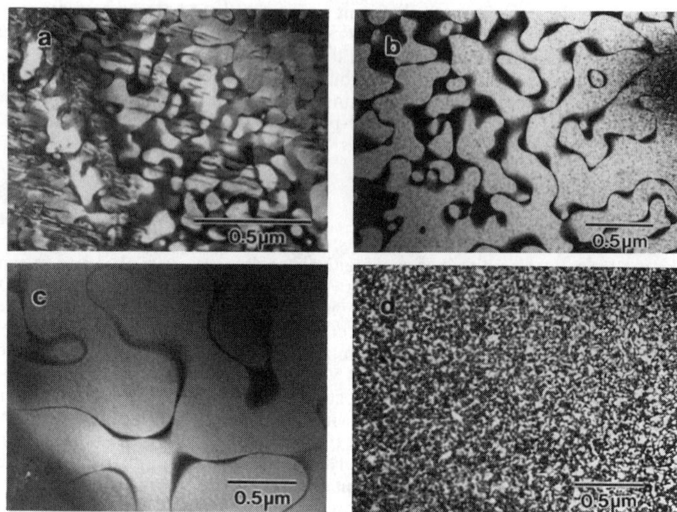

Figure 5. Anti-phase domain sizes in Alloy 2, imaged with 001_{B2} type superlattice reflections as a function heat treatment (a) as-cast Alloy 2, (b) 920°C-10min, water quench, (c) 1020°C-10min, water quench, (d)1120°C-10min, water quench.

SUMMARY

The Ti-modified Nb-Al alloys consist of single phase with B2 crystal structure in their as-cast form. When Alloy 1 was heat-treated at 1100°C a second phase having the A15 crystal structure formed preferentially along the grain boundaries with a plate-like morphology. One of the orientation relationships observed between the A15 and B2 phases is $[100]_{A15}//[1\bar{2}1]_{B2}$ with $(0\bar{2}1)_{A15}//(10\bar{1})_{B2}$. The lattice parameters of the A15 and as-cast B2 phases of Alloy 1 were deduced from powder x-ray spectra. A needle-like second phase, presumably Ti_2NbAl, was observed in 800°C heat-treated Alloy 2. Microchemical analyses indicate that the Al content influences the B2 order-disorder transition. The Ti atoms in Alloy 2 appeared to lie on a different sub-lattice from the Al atoms in the B2 crystal structure.

ACKNOWLEDGMENT

This work was supported by the Office of Naval Research, Dr. G. Yoder Program Manager.

REFERENCE

1. H. Kohmoto, J. Shyue, M. Aindow and H.L. Fraser, submitted to *Scripta Metall.* (1992).
2. J. Shyue, Ph. D. Thesis, The Ohio State University. (1992)
3. S. M. L. Sastry and H. A. Lipsitt, *Met. Trans. A*, **8A**, 1543 (1977).
4. L. A. Bendersky and W. J. Boettinger, *Mat. Res. Soc. Symp. Proc.*, **133**, 45 (1989).
5. T. J. Jewett, J. C. Lin, N. R. Bonda, L. E. Seitzman, K. C. Hsieh, Y. A. Chang, J. H. Perepezko, *Mat. Res. Soc. Symp. Proc.*, **133**, 69 (1989).
6. D.E. Laughlin, R. Sinclair and L.E. Tanner, *Scripta Met.*, **14**, 373 (1980).
7. J. Shyue, D-H. Hou, S.C. Johnson, M. Aindow and H.L. Fraser, these proceedings.
8. J.C.H. Spence and J. Tafto, *J. Microscopy*, **130**, pt2, 147 (1983).

PART III

Defects and Microstructure

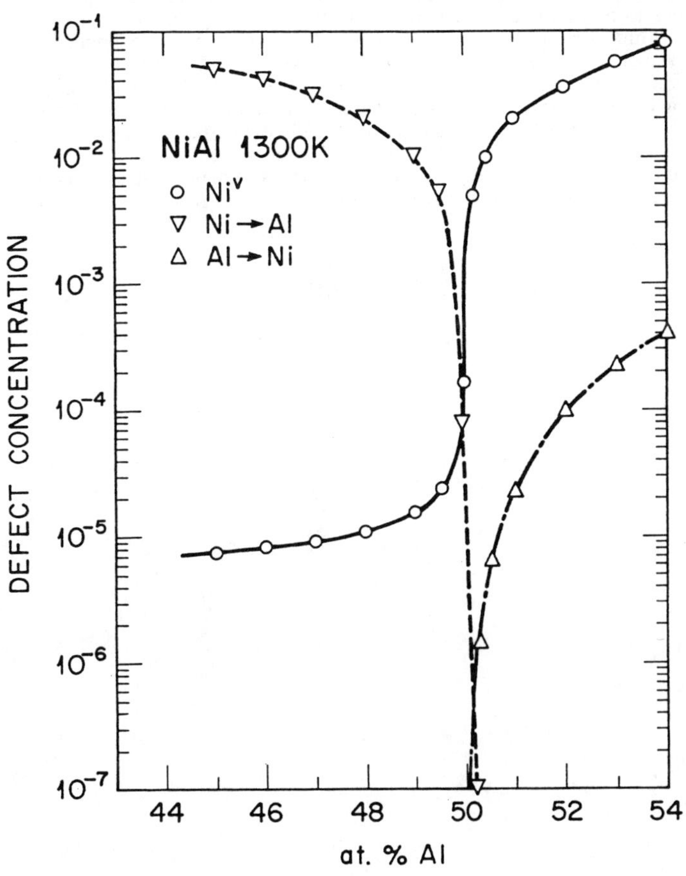

First principles calculations permit both the type and concentration of point defects in intermetallics to be estimated. The figure illustrates the expected behavior for NiAl at 1300 K as a function of aluminium content. Of the four possible point defects (two types of vacancies and two types of antisite defects) the calculations revealed that the concentration of vacancies on the Al sublattice were negligibly small and essentially independent of the Al level. The vacancy concentration on the Ni sublattice, on the other hand, is about 10−5 for the Ni-rich side even though there exists a high concentration of Ni atoms on Al sites. In the Al-rich side, the major point defects are vacancies at the Ni sublattice, and there is a two-orders of magnitude difference between the vacancies and antisite defect concentrations.

Figure courtesy of C.L. Fu, Y.-Y. Ye and M.H. Yoo.

STUDY OF POINT DEFECTS IN INTERMETALLIC COMPOUND Ni_3Al

DONGLIANG LIN (T.L. Lin) AND JIAN SUN
Department of Materials Science and Engineering, Shanghai JiaoTong University, Shanghai 200030, P. R. China

ABSTRACT

Formation energies of point defects in Ni_3Al were calculated by EAM and static relaxations. The equilibrium equation of point defects in $L1_2$ type of intermetallic compounds was established. The concentration of point defects at 1000 K as a function of bulk composition and the effect of temperature on them was calculated for Ni_3Al. The concentration of vacancies in Ni_3Al with various Al and boron contents was also investigated by means of positron annihilation to verify the calculated results.

INTRODUCTION

Aoki and Izumi[1] first discovered that microalloying with boron suppresses intergranular fracture and improve the ductility of Ni_3Al, which generated considerable interest in both grain boundary phenomena and the lattice defect in Ni_3Al. The results studied by Liu et al.[2] showed that an increase of aluminium from 24 at% to 25.2 at% results in a reduction of boron segregation to grain boundary by 45%. Since the grain boundary properties in Ni_3Al are sensitive to the boron level, the boundary remains brittle when the amount of boron segregation is insuffient. Thus, it is highly important to clarify the role of off-stoichiometry in the transport of boron to the grain boundaries. Mass transport usually occurs by atomic migration via point defects. The positron annihilation study of vacancies in Ni_3Al by Dasgupta et al.[3] suggested that there exists small concentration of constitutional Ni vacancies essentially independent of composition in the stoichiometric or hyperstoichiometric alloys. The constitutional vacancies combined with the boron dopant may limit the long range transport of boron to grain boundaries. However, the properties of constitutional point defects in Ni_3Al is not identified theoretically.

In the current study, The formation energies of point defects in Ni_3Al, such as antisite defects, vacancies, interstitials were calculated by EAM and static relaxations. The equilibrium equation of point defects in $L1_2$ type of intermetallic compounds was established and the effect of temperature on four types of point defects was studied. The concentration of vacancies in Ni_3Al with various Al and boron contents was also investigated by means of positron annihilation to verify the calculated results.

THEORETICAL CALCULATION

In this calculation, the embedded atom potentials reported in ref.[4] is used to calculate the formation energies of antisite defects, vacancies and interstitials with the energy gradient method. A cubic cell containing 1728 atoms with periodic boundary condition and constant pressure is adopted. The details of calculations was decribed elsewhere[5]. The calculated results are shown in table I.

Table I Formation energies of the relevent intrinsic point defects in Ni_3Al

Type of defect	ΔEf (ev)
Al Vacancy	1.92
Ni Vacancy	1.47
Al→Ni anti-site	0.56
Ni→Al anti-site	0.59
Al Octahedral	4.53
Ni Octahedral	3.78

The calculated results are very similar to those reported in ref.[6]. It can be seen that the formation energies of interstitials are much higher than those of vacancies and antisite defects, so the concentration of interstitials in Ni_3Al is negligible.

The equilibrium concentration of point defects can be determined in an ensemble with fixed temperature and bulk composition of Ni_3Al. Consider a system with N lattice sites that can be divided into two sublattice defined as A and B. Each lattice site will be occupied either by an atom appropriate to the sublattice, by a vacancy, or by an atom of the opposite type (an antisite defect). Assume this system in which the concentration (atomic percent) of A sublattice sites occupied by A atoms is X_{AA}, the concentration of B sublattice sites occupied by B atoms is X_{BB}, the concentration of vacancies on the A sublattice sites and B sublattice sites is X_A^v and X_B^v, the concentration of A sublattice sites occupied by B atoms (A antisite defects) is X_{AB} , and the concentration of B sublattice sites occupied by A atoms (B antisite defects) is X_{BA}. For A_3B compound, the concentration of atoms X_A can vary within the composition range, but crystal structure of the compound still keep unchanged. Therefore there exist two structure relations among the concentrations of various point defects.

$$(X_{AA}+X_{BA})/(X_{AA}+X_{BA}+X_{BB}+X_{AB})=X_A \tag{1}$$

$$(X_{AA}+X_A^v+X_{AB})/(X_{BB}+X_B^v+X_{BA})=3 \tag{2}$$

Combine Eq.(1) with Eq.(2) to obtain Eq.(3)

$$X_{AB}+(1-X_A)X_A^v-X_{BA}-X_AX_B^v+X_A-\frac{3}{4}=0 \tag{3}$$

The concentration of various point defects and bulk composition of the compound have been linked in Eq.(3). For noninteracting defects, the energy of the system is

$$E=E_0+X_A^vE_A^v+X_B^vE_B^v+X_{AB}E_{AB}+X_{BA}E_{BA} \tag{4}$$

where $E_0=(X_{AA}+X_{AB}+X_A^v)E_A+(X_{BB}+X_{BA}+X_B^v)E_B$, E_0 is the energy of the ideal lattice, E_A^v and E_B^v are the formation energies of vacancies on the A and B sublattice, and E_{AB} and E_{BA} the formation energies of A antisite defects and B antisite defects respectively, which are shown in table Ⅰ. The configurational entropy of the system is suggested by Foiles et al[7].

$$S=\frac{3}{4}[f(\frac{4}{3}X_{AB})+f(\frac{4}{3}X_A^v)]+\frac{1}{4}[f(4X_{BA})+f(4X_B^v)] \tag{5}$$

where f(x) is the ideal entropy function

$$f(x)=-K[x\ln(x)+(1-x)\ln(1-x)] \tag{6}$$

Only consider the vibration entropy introduced by two vacancies

$$S_f^v=X_A^vS_{f\text{-}A}^v+X_B^vS_{f\text{-}B}^v \tag{7}$$

The change of free energy of the system is

$$\Delta F=\Delta E-T(S+S_f^v) \tag{8}$$

Among four unknown concentration, X_A^v, X_B^v, X_{AB} and X_{BA}, only three are independent owing to the structure relation Eq.(3). In present calculation, X_A^v, X_B^v, and X_{AB} or X_A^v, X_B^v, and X_{BA} are chosen as the independent variables. The equilibrium state of the system at a given temperature and bulk composition corresponds to a minimum of the free energy.

$$\frac{\partial\Delta F}{\partial X_A^v}=0,\ \frac{\partial\Delta F}{\partial X_B^v}=0,\ \frac{\partial\Delta F}{\partial X_{AB}}=0\ (\text{ or }\frac{\partial\Delta F}{\partial X_{BA}}=0\ ). \tag{9}$$

The concentrations of defects can be expressed by

$$\begin{aligned}
X_A^v&=\frac{3}{4}A(\frac{4X_{BA}}{1-4X_{BA}})^{X_A-1}\exp[(-E_A^v+(X_A-1)E_{BA})/KT]\\
X_B^v&=\frac{1}{4}A(\frac{4X_{BA}}{1-4X_{BA}})^{X_A}\exp[(-E_B^v+X_AE_{BA})/KT]\\
X_{AB}&=\frac{3}{4}(\frac{4X_{BA}}{1-4X_{BA}})^{-1}\exp[(-E_{AB}-E_{BA})/KT]
\end{aligned} \tag{10}$$

$$X_{BA}=X_{AB}+(1-X_A)X_A^v-X_AX_B^v+X_A-\frac{3}{4}$$

when $X_A>3/4$

$$X_A^v=\frac{3}{4}A(\frac{4X_{AB}}{3-4X_{AB}})^{1-X_A}\exp[(-E_A^v+(1-X_A)E_{AB})/KT]$$
$$X_B^v=\frac{1}{4}A(\frac{4X_{AB}}{3-4X_{AB}})^{-X_A}\exp[(-E_B^v-X_AE_{AB})/KT] \quad (11)$$
$$X_{BA}=\frac{1}{4}(\frac{4X_{AB}}{3-4X_{AB}})^{-1}\exp[(-E_{AB}-E_{BA})/KT]$$
$$X_{AB}=X_{BA}+(X_A-1)X_A^v+X_AX_B^v-X_A+\frac{3}{4}$$

when $X_A<3/4$

where $A=\exp[(S_{f\text{-}A}^v)/K]=\exp[(S_{f\text{-}B}^v)/K]$. Taking the value A to be 4.5, the same as that in pure nickel[8]. The equations (9) and (10) can be solved by iterated method. The Eq. (9) and (10) show that for ordered alloy, the effective vacancy energy depends on the overall composition of the alloy, but the formation energy of antisite defects dose not. For hypostoichiometric Ni_3Al with 24 at% Al content, the effective formation energy of Ni vacancy and Al vacancy is 1.60 eV and 1.50 eV respectively. The nominal vacancy formation energies with the value 1.6±0.2 eV for this alloy determined by Wang et al. by means of positron annihilation[8].

The concentration of four types of point defects at 1000 K as a function of bulk composition has been calculated as shown in Fig.1. It is shown that the concentration of Ni antisite defects are much higher than that of the other three defects in hyperstoichiometric Ni_3Al, and the concentration of vacancies is also higher on Ni sublattice than on the Al sublattice in this alloy. The concentration of four types of defects changes significantly near the stoichiometric composition. In hypostoichiometric Ni_3Al, the concentration of Al antisite defects is the highest one, which is the same as the Ni antisite in hyperstoichiometric Ni_3Al and less vacancies are found. The concentration of two antisite defects is the lowest in stoichiometric Ni_3Al. However, the total concentration of two vacancies reaches the lowest in hypostoichiometric Ni_3Al with 75.9 at% Ni content, and increases with Al content from 24.1 at% to 28 at%.

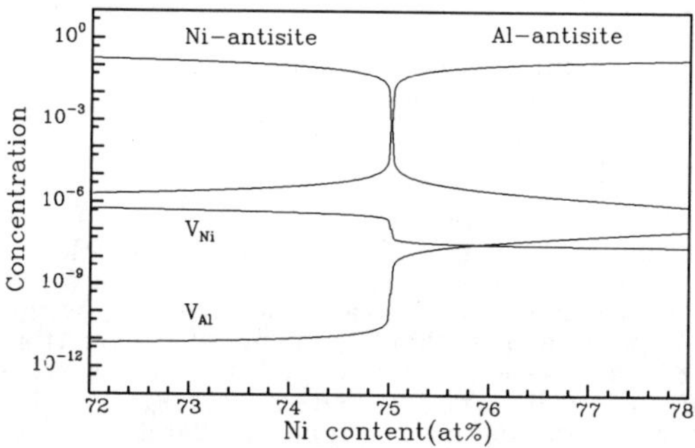

Fig. 1 The calculated concentration of four types point defects at 1000 K as a function of bulk compoistion.

In order to study the properties of the constitutional point defects in Ni_3Al alloys, the relations of the concentration of defects and temperature were calculated respectively for Ni_3Al alloys with three different bulk compositions($Ni_{76}Al_{24}$, $Ni_{75}Al_{25}$ and $Ni_{74}Al_{26}$). The calculated results are shown in Fig. 2a-2c. It can be seen in Fig.2a that except for Al antisite defects, the concentration of two types of vacancies and Ni antisite defects increase markedly when the temperature increases in $Ni_{76}Al_{24}$ alloy. As is well known, the thermal concentration of defects is susceptible to temperature. However, unlike other three defects, the concentration of Al antisite defects is almost independent on temperature. Actually it decreases very slightly with the temperature increasing. In $Ni_{76}Al_{24}$ the excess Ni atoms will occupy the Al sublattices in order to maintain the stability of the $L1_2$ structure of the alloy. Therefore the Al antisite defects are of the character of the constitutional point defects resulted from the deviation from stoichiometry, but other three defects are the thermal point defects introduced by the thermal motion of atoms in $Ni_{76}Al_{24}$. For $Ni_{74}Al_{26}$ alloy, it can be deduced that the Ni antisite defects are the constitutional defects, but two types of vacancies and Al antisite defects are the thermal defects as shown in Fig.2c. However, four types of defects in stoichiometric Ni_3Al belong to thermal defects as shown in Fig.2b. Our results show that the vacancies in Ni_3Al are simply thermal defects, not the constitutional defects.

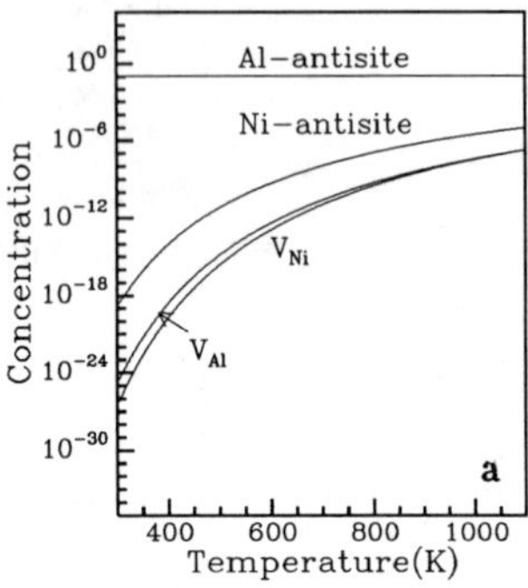

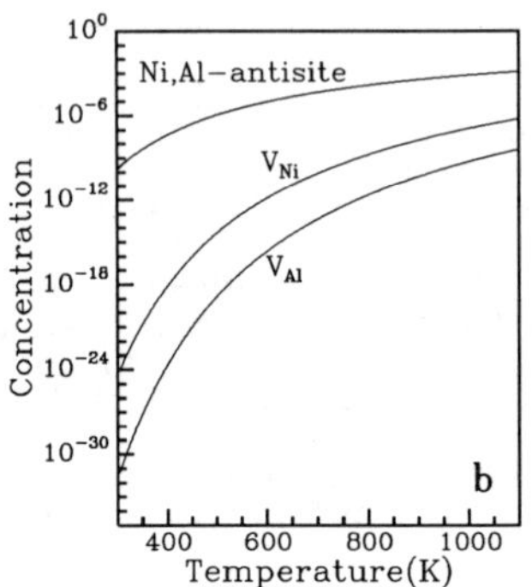

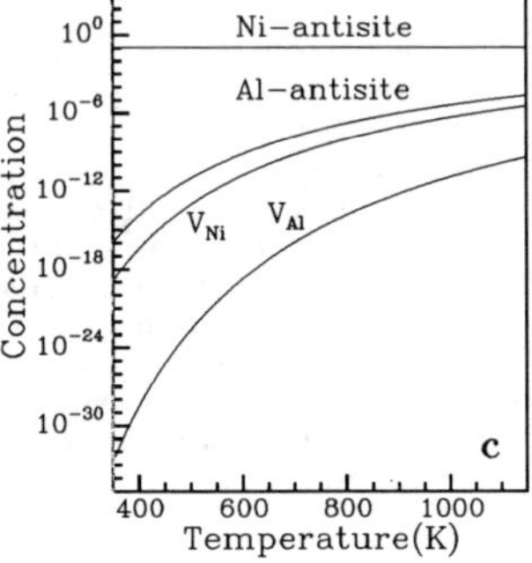

Fig. 2 The calculated concentration of four types of point defects vs temperature.

EXPERIMENTAL STUDY

The alloys used in this study prepared by directional solidification with various aluminium and boron contents. All of them annealed at 1423K for 50 hours followed by furnace cooling. Their compositions are (a)Ni-24 at%Al with and without 700ppm wt boron;(b)Ni-25 at%Al with and without 700ppm wt boron;(c)Ni-26 at%Al with and without 700ppm wt boron. Rectangular plate samples of $10\times10\times1$ mm were cut off normal to the growth direction, then machined and polished. All samples had very large single phase grains, approximately several mm in size, which rendered the grain boundary contribution negligible to the positron annihilation. The positron lifetime was measured by the conventional fast-fast coincidence type. The system resolution function was approximated by a gaussian having a FWHM of 245 ps. A ^{22}Na source of about 15 μci contained in an envelope made from a thin nickel foil was sandwiched in between two samples. More than 10^5 counts were accumulated in the life-time measureing for each samples. The lifetime data for each samples were analysed using POSITRONFIT computer program. These analyses were conducted using three-term fit without any restrictions.

The life-time spectrum for each example was represented with three lifetime components, attributable to the Bloch state, the vacancy trapped state and the positron source respectively. The bulk life-time of all six Ni_3Al samples is very closed to 116 ps found by Wang et al.. The trapped state life-time of samples is 230 ps or more in the present work, which is highger than 180 ps measured by Wang. The intensity of the source term of all samples is very small (approximately 1 %). According to the trapping model, the trapping rate for Ni_3Al alloys was calculated as a function of alloy composition shown in Fig.3. On the other hand, the trapping rate can be assumed to be $K = C_v \mu_v$, where C_v is the concentration of vacancy, and μ_v is the trapping factor. Figure 3 shows that the trapping rate or the concentration of two vacancies in Ni_3Al doped with or without boron increase with the increase of Al content from 24 at% to 26 at%. The trapping rates in Ni_3Al decrease when it is doped with boron. It suggests that a part of the vacancies are filled with boron atoms or atom clusters, and the effective concentration of vacancies is reduced. Because positrons are insensitive to the presence of antisite defects, the concentration of antisite defects in Ni_3Al can not be determined.

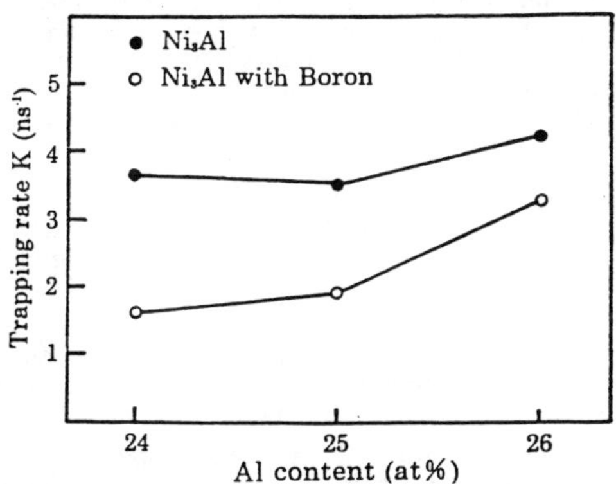

Fig. 3 The trapping rate in Ni_3Al alloys vs Al content.

DISSCUSION

Positron annihilation measurements performed on Ni_3Al with various Al and boron contents by Dasgupta[3] show that positron trapping was found in the stoichiometric and hyperstoichiometric Ni_3Al, but no positron trapping in pure hypostoichiometric Ni_3Al. The intensity of vacancy trapped state lifetime in pure Ni_3Al alloys in this study is slightly larger than that in Dasgupta's work. The reason is probably that the samples was not annealed perfectly. This reason can also be used to explain why the trapped state lifetime is longer than 180 ps determined by Wang et al. in quenched specimen of Ni_3Al. The defects may act as small vacancy clusters or vacancy boron complexes in the case of the alloys doped with boron. Anyway, our results still reflect the change tendency of vacancies in Ni_3Al as function of bulk composition. The intensity of trapped state lifetime in boron-doped Ni_3Al is lower than in pure Ni_3Al alloys, which contradict Dasgupta's results. It is easy to understand that this is due to the capture of vacancies by dopant[9]. The vacancy concentration in pure stoichiometric and hyperstoichiometric Ni_3Al estimated by Dasgupta would be 2 $\times 10^{-6}$. However, since the limitation of sensitivity of such measurements is approximately 10^{-7} at best for determination of vacancy concentration[10], the lower concentration of vacancies near or beyond the limitation can not be excluded. The results in both Dasgupta's and the present work show that the concentration of vacancies increases as the increase of Al content from 24 at% to 26 at% in Ni_3Al. which is in good agreement with the calculated results. However, it can not be deduced that constitutional vacancies are present in stoichiometric and hyperstoichiometric Ni_3Al alloys, but not in hypo-stoichiometric Ni_3Al. The calculated results apparently show that there only exist constitutional antisite defects, but not any types of constitutional vacancies in non-stoichiometric Ni_3Al. The calculated equilibrium concentration of vacancies is much lower than that determined by positron annihilation at room temperature, which indicates that it is not easy to reach the real equilibrium state for intermetallic compound because of difficult diffusion of atoms in compound.

On the other hand, the concentration of vacancies within overall composition of Ni_3Al is significantly lower than the doped-boron concentration considered here. Although vacancies interact with the boron dopant, which limits the long range tranport of boron to grain boundary, the difference of concentration of vacancies in Ni_3Al alloys could not be considered as a main reason in explaining the effect of stoichiomtry on the segregaton of boron. Computer simulation results studied by Chen et al.[11] using embedded atom potentials showed that boron segregates preferentially to the grain boundary, especially to the Ni-rich grain boundary, instead of the surface and interstital site. Futhermore, in Ni-rich grain boundary together with boron makes the grain boundary stronger than the bulk. Therefore, the segregation of boron on grain boundary mainly controlled by grain boundary structure and chemistry, not by the types and distributions of point defects in Ni_3Al alloys.

CONCLUSIONS

1. In hypostoichiometric Ni_3Al, the Al antisite defects are the constitutional point defects, and the Ni antisite defects are the constitutional defects in hyperstoichiometric Ni_3Al. Two types of vacancies simply belong to the thermal point defects. There do not exist any types of constitutional point defects in stoichiometric Ni_3Al.
2. The experimental results by positron annihilation show that the concentration of vacancies in Ni_3Al alloys increases with increasing Al content from 24 at% to 26 at% which is in good agreement with the calculated results.
3. Although the vacancies interact with the boron dopent, which limits the long range tranport of boron to grain boundary, the difference of vacancy concentration in Ni_3Al alloys could not be considered as a main reason in explaining the effect of stoichiometry on the segregaton of boron.

REFERENCE

1. K. Aoki and O. Izumi, Nippon Kinzoku Gakkaishi **43**, 1190 (1979).
2. C. T. Liu, C. L. White, and J. A. Horton, Acta Metall. **33**, 213 (1985).
3. A. Dasgupta, L.C. Smedskjaer, D.G. Legnini, and R.W. Siegel, Mater. Lett. **3**, 457 (1985).
4. A.F. Voter and S.P. Chen, in High Temperature Ordered Intermetallic Alloys, edited by C.C. Koch, C.T. Liu and N.S Stoloff (Mater. Res. Soc. **39,** Pittsburgh, PA, 1985) pp. 175.
5. T.L. Lin and Da Chen, Journal De Physique, collogue C-**1**, 227 (1990).
6. A. Caro, M. Victoria, and R.S. Averback, J. Mater. Res. **5**, 1409 (1990).
7. S.M. Foiles and M.S. Daw, J. Mater. Res. **2**, 5 (1987).
8. T.M. Wang, M. Shimotomai, and M. Doyama, J. Phys, F:Phys. **14**, 37 (1984).
9. P. Hantojarvi, J. Johanson, P. Morse, L. Pollanen, A. Vehanen, and J. Yli-Kauppila, in Point Defects and Defect Interactions in Metals, edited by J. Takamura, 1982, pp. 504.
10. P. Hantojarvi, in Positron in Solids, Springer-Verlag, 1979.
11. S.P. Chen, A.F. Voter, R.C. Albers, A.M. Boring and P.J. Hay, J. Mater. Res. **5**, 955 (1990).

HRTEM OBSERVATION AND EAM CALCULATION OF DISLOCATION CORES IN NiAl

M. J. MILLS*, M. S. DAW*, S. M. FOILES* and D.B. MIRACLE**
*Sandia National Laboratories, Livermore, CA 94550
**Wright Laboratory, Wright-Patterson AFB, OH 45433

ABSTRACT

High resolution transmission electron microscopy (HRTEM) has been used to observe the core structures of *a*<100> and *a*<110> edge dislocations in bicrystals of stoichiometric NiAl. The images indicate that the core of *a*<100> edge dislocations are compact. Two different core structures have been observed for *a*<110> edge dislocations. In one configuration, the *a*<110> dislocation is decomposed into two *a*<100> dislocations, while in the other it is climb-dissociated into partial dislocations. Both configurations are non-planar on the scale of about 1 nm. EAM calculations have been performed to determine the relative energies and stabilities of these configurations. The implications of these core configurations are briefly discussed with respect to the macroscopic flow behavior of NiAl.

INTRODUCTION

The deformation mechanisms and mechanical properties of NiAl (B2 or CsCl structure) have been studied extensively due to its potential for high temperature applications, and because of the attractiveness of NiAl as a model system for deformation in a high symmetry ordered cubic compound. A striking feature of deformation in NiAl single crystals is the large anisotropy of its mechanical properties. It has been well-established that deformation occurs principally by the motion of *a*<100> dislocations in polycrystals and single crystals loaded along non-<100> axes (defined as "soft" orientations) [1]. The yield strength for "soft" orientations is relatively low above room temperature (RT) and increases moderately below RT [2]. Coupled with this relatively weak temperature dependence is a mild strain rate dependence at lower temperatures [3]. When deformed along a <100> axis (defined as the "hard" orientation), there is no resolved shear stress on *a*<100> dislocations in these crystals. At lower temperatures, single crystals deformed in tension are brittle, and a brittle-to-ductile transition temperature (BDTT) occurs at about 400°C [3]. Yielding in the "hard" orientation at RT occurs at a stress approximately ten times larger than in the "soft" orientation [4]. Above the BDTT, *a*<110> dislocations begin to operate, and the yield strength decreases sharply with temperature and increases with strain rate [3].

These macroscopic observations suggest that the anisotropy in the mechanical properties of NiAl single crystals might be due to a difference in the mobility of *a*<100> and *a*<110> dislocations, which are operative in the "soft" and "hard" orientations, respectively. This investigation has sought to determine whether the apparent difference in *a*<100> and *a*<110> mobilities is due to the core structure of these dislocations. High resolution transmission electron microscopy (HRTEM) has been used recently to examine for the first time the core structure of dislocations in NiAl [5]. In this paper, these results will be compared with Embedded Atom Method (EAM) calculations which have been used to determine the stable, lowest energy core configurations. The cores of the a<100> and a<110> dislocations are found to be quite different, offering a fundamental explanation for the anisotropic mechanical properties of single crystals in the "hard" and "soft" orientations.

Experimental and Calculational Procedures

Oriented bicrystals of stoichiometric NiAl were deformed to produce a<100> and a<110> dislocations. The bicrystals consisted of two single crystals (denoted A and B) diffusion-bonded together. The compression axis and the grain boundary normal were parallel to the [001] direction and the [232] direction in crystals A and B, respectively. This bicrystal geometry was based on a previous investigation [6] which showed that a high density of $a[01\bar{1}]$ dislocations were produced in crystal A near the bicrystal boundary. Dislocations of the type a[010] and $a[00\bar{1}]$ were also present in crystal A, in low numbers near the boundary and with increasing density far (>3 mm) from the boundary. Thin foils were prepared by slicing crystal A normal to the [100] direction so that $a[01\bar{1}]$ and a<010> edge dislocations could be viewed "end-on". HRTEM images were obtained using a JEOL 4000EX operating at 400kV.

The EAM calculations [7] were performed utilizing potentials developed specifically for NiAl [8]. The static calculations were conducted by first establishing the atomic positions around the cores of the dislocations assuming anisotropic elasticity. The typical computational supercell was a cylinder with a 10 nm radius. The atoms in an outer ring of the supercell (1nm thick) were held fixed during the relaxations to maintain the long range elastic stress field around the dislocation. In order to compare the energies of core structures which might have different local compositions, the chemical potentials of Ni and Al were used, referenced to the stoichiometric NiAl (see [9] for an example of this procedure applied to Ni_3Al). In practice, this was accomplished by transforming the functions such that the energy for each Ni and Al atom in perfect crystal was equal to their respective chemical potentials, without altering the total energy or structure after minimization. Image simulations were conducted using the Electron Microscopy Software package [10]. The important microscope parameters used for the simulations were: spherical aberration coefficient (C_s) = 1.0 nm, spread of defocus (D) = 10 nm and semi-angular beam divergence (J_c) = 0.7 mrad. The defocus values for the observed images were determined by analysis of optical diffractograms and the crystal thicknesses were estimated by obtaining the best qualitative match between the observed and simulated image contrast in the perfect crystal near the disolcation cores.

RESULTS AND DISCUSSION

Shown in Figure 1a is an HRTEM image of the core of an a[010] dislocation. The lowest energy a[010] core calculated using the EAM is shown in Figure 1b and a simulated image based on this structure for the approximate conditions used to obtain the observed image is shown in Figure 1c. For these conditions, the image intensities correspond to the tunnels between atomic columns in perfect crystal. The same reference position is indicated by an arrow in these figures. Since this dislocation is in the edge orientation, a Burgers circuit around the core (not shown) yields a closure failure of a a[010], corresponding to the complete displacement vector. This core is characterized by the termination of two (020) fringes at the core. By sighting along the (002) fringes, it can be seen that they are distorted, but continuous across the dislocation core. Therefore, glide of this a[010] dislocation would occur on this plane. The compact nature of the

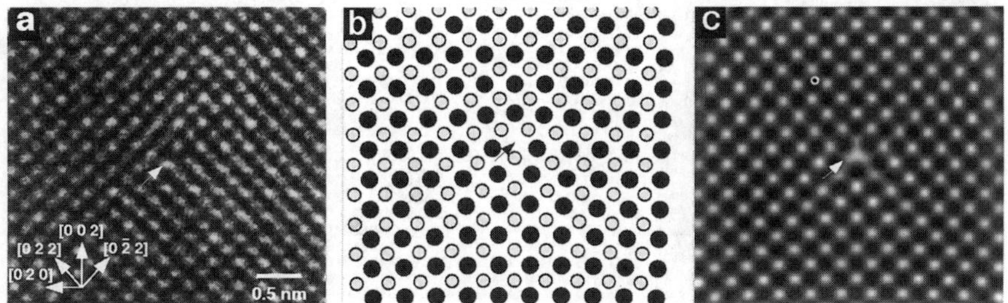

Figure 1: (a) HRTEM image of an $a[010]$ edge dislocation in NiAl. (b) Statically minimized EAM calculated structure for this dislocation. Small atom symbols are Al. (c) Simulated HRTEM image based on the calculated structure in (b) using a microscope defocus of 80 nm and a crystal thickness of 4.0 nm. The arrows indicate the same reference position in the three figures.

core is particularly evident by sighting along the $(02\bar{2})$ and (022) fringes. While significant bending of these lattice fringes is apparent at the core in both the observed and simulated images, no identifiable spreading or dissociation of the core is seen in any of the projections. The structure shown in Figure 1b has a triangular arrangement of Al atoms at the core. This core has a calculated energy of 1.69×10^{-9} J/m (for a radius of 3.5 nm), while a similar, stable core with Ni atoms at its center has an energy of 1.82×10^{-9} J/m.

Two distinctly different core images have been found for the $a[01\bar{1}]$ dislocation. An example of the first core type is shown in Figure 2a. This core image is characterized by two $(02\bar{2})$ fringes which terminate on different (022) planes, indicating that this dislocation does not have a singular core. In fact, analysis of the displacements at the core indicates that the $a[01\bar{1}]$ dislocation has *decomposed* into two perfect a<100> dislocations via the reaction:

$$a[01\bar{1}] = b_1 + b_2 = a[010] + a[00\bar{1}]. \quad (1)$$

The approximate locations of b_1 and b_2 are indicated in Figure 2a. Note that supplementary (022) fringes (see arrows) are also seen extending to the left at b_1, and to the right at b_2. The distance between b_1 and b_2 is only about 1.2 nm. However, since b_1 and b_2 do not lie in a common slip plane for this dislocation line direction, glide of this decomposed $a[01\bar{1}]$ core is not possible.

The result of an EAM calculation for a decomposed core is shown in Figure 2b, with a corresponding simulated image shown in Figure 2c. This calculated core structure is notably different from previously published calculations for the $a[01\bar{1}]$ core, which indicated that the core should be singular [11-12]. The calculation shown in Figure 2b was performed by establishing the initial (unrelaxed) atom positions assuming the decomposed configuration observed experimentally [5]. Similar relaxed structures can also be obtained by mimicking a climb process through the removal of atomic columns from above an initially singular core. The physical motivation behind these latter calculations is that under an applied stress in the "hard" orientation, the two a<010> dislocations will tend to separate if climb is operative. The energy of this decomposed core is 3.6 x 10^{-9} J/m, which is almost the same as that for the singular core (3.5 x 10^{-9} J/m), indicating that there is little energetic penalty for this decomposition. The difference in the clarity of the observed

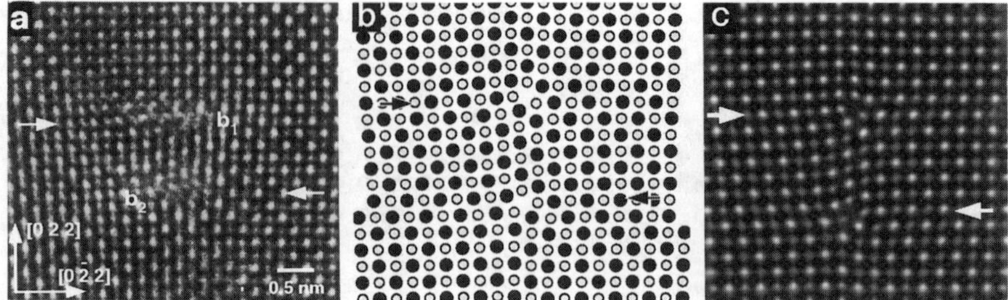

Figure 2: (a) HRTEM image of an $a[01\bar{1}]$ dislocation which is decomposed into an $a[010]$ (indicated as $\mathbf{b_1}$) and an $a[00\bar{1}]$ (indicated as $\mathbf{b_2}$) dislocation. (b) Statically minimized EAM calculated structure for this dislocation. (c) Simulated HRTEM image based on the calculated structure in (b) using a microscope defocus of 60 nm and a crystal thickness of 7.0 nm. The arrows indicate the extra (022) half-planes characteristic of this core.

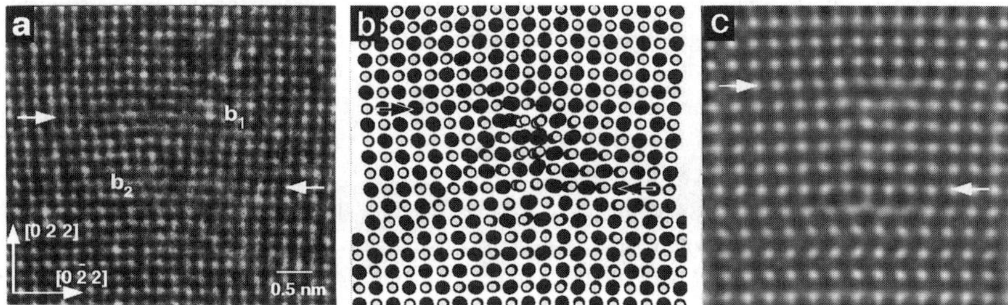

Figure 3: (a) HRTEM image of an $a[01\bar{1}]$ dislocation which is dissociated into two partial dislocations indicated as $\mathbf{b_1}$ and $\mathbf{b_2}$ dislocation. (b) Statically minimized EAM calculated structure for a dissociated dislocation in a 4.0 nm thick slab with free surfaces normal to the line direction. Note the misalignment of the atomic columns at the core, and the presence of an APB between the partials. (c) Simulated HRTEM image based on the calculated structure in (b) using a microscope defocus of 60 nm and a crystal thickness of 7.0 nm. The arrows indicate the continuous (022) half-planes characteristic of this core.

and simulated images near the cores of the individual a<010> dislocations is perhaps due to the presence of kinks along the dislocation lines.

The second core type for the $a[01\bar{1}]$ dislocation shown in Figure 3a, which is actually the most frequently observed configuration, is also characterized by two terminating $(02\bar{2})$ fringes. But, in contrast to the decomposed core of Figure 2a, the (022) fringes are now continuous (see arrows). Analysis of the images indicates that the $a[01\bar{1}]$ dislocation has *dissociated* into two partial dislocations which could occur via two possible reactions:

$$a[01\bar{1}] = a/2[\bar{1}1\bar{1}] + \text{fault} + a/2[11\bar{1}] \tag{2}$$

$$a[01\bar{1}] = a/2[01\bar{1}] + \text{fault} + a/2[01\bar{1}]\ . \tag{3}$$

In both reactions, a fault would be produced between the partial dislocation cores. In reaction (2) the fault is an antiphase boundary, while in reaction (3) a local orthorhombic structure is formed.

The EAM calculations have shed light on the relative stabilities and energies of these cores, as well as the effect of the thin TEM foil geometries. The calculations indicate that both of the dissociation schemes (reactions 2 and 3) yield core structures which are *unstable* for a periodic (infinitely long) dislocation line. Starting with either reaction (2) or (3), and assuming a dissociation distance of five (022) planes (about 1 nm) as indicated by the observations, both spontaneously transform into a decomposed configuration (reaction 1) during static minimization. This transformation occurs by a simple shear process which eliminates the fault between the partials. Under an applied stress in the "hard" orientation, a shear stress exists favoring this transformation. Consequently, the EAM calculations indicate that the decomposed core is the one relevant to bulk deformation behavior. This conclusion is consistent with estimates using anisotropic elastic calculations of the relative energies of the decomposed and dissociated configurations [5,13].

EAM calculations have also been performed in which free surfaces are created normal to the line direction, thereby simulating the geometry of the thin TEM foil. The relaxed "through-thickness" structure for one such calculation which used the dissociated reaction (2) as a starting structure is shown in Figure 3b. The corresponding simulated image for conditions similar to those used to obtain the observed image is shown in Figure 3c. It can be seen that the (022) fringes remain continuous across the core of both partials in the simulated image (see arrows), indicating that this dissociated configuration has been *stabilized* in the thin foil. As a result of these EAM calculations, we conclude that the observed dissociated cores are probably artifacts of the thin-foil geometry.

These HRTEM observations show that the core structures for the a<100> and a<110> dislocations are remarkably different. This difference may be a fundamental reason for the anisotropic mechanical behavior of NiAl single crystals. The compact nature of the *a*<100> core indicates that there should be only a small lattice resistance to its glide motion. Consequently, it is not surprising that these dislocations are responsible for deformation in "soft" orientations. In contrast, the decomposed $a[01\bar{1}]$ core is unable to move in a conservative manner for deformation in the "hard" orientation since there is no resolved shear stress on the individual *a*<010> dislocations. A possible mechanism for the motion of the decomposed core has recently been proposed [5] in which glide of the $a[01\bar{1}]$ dislocation is accomplished by a cooperative, short-range diffusion process in which one of the *a*<010> dislocations acts as a source and the other a sink for vacancies. By this mechanism, the overall $a[01\bar{1}]$ dislocation would appear to be moving viscously on its glide plane, while its mobility would be controlled by the short-range diffusion mechanism. Although this mechanism is attractive since it can potentially explain the strong temperature and strain rate dependence of the yield strength, portions of $a[01\bar{1}]$ dislocations with edge character will tend to completely decompose when climb is active since there is no attractive interaction between the individual *a*<010> dislocations based on both elasticity theory [13] and the atomistic calculations described above. The edge orientation may therefore be effectively sessile, and flow might then be accomplished by the kink-like motion of non-edge components of the $a[01\bar{1}]$ dislocation for which there is an attractive interaction based on elasticity theory.

CONCLUSIONS

These HRTEM observations of dislocation cores in NiAl have revealed a fundamental reason for the anisotropic mechanical properties in the "soft" and "hard" orientations. The *a*<100> dislocation cores are compact, with no apparent dissociation. Therefore, these dislocations should be relatively mobile on {100} slip planes, which is consistent with the characteristics of flow in the "soft" orientations. Two different *a*<110> core structures have been observed, but only one — the decomposed core — is believed to be relevant to "bulk" deformation based on the thin-foil and periodic EAM calculations. The decomposed core, which consists of two *a*<010> dislocations with orthogonal Burgers vectors, has no possibility for conservative glide motion in the "hard" orientation. This non-planar arrangement for the *a*<110> edge dislocations will severely impede their mobility and may be responsible for the high yield strengths, as well as the strong temperature and strain rate sensitivities above the BDTT in the "hard" orientation.

ACKNOWLEDGMENTS

The authors would like to thank W. D. Nix, K. R. Forbes and M. I. Baskes for a number of stimulating discussions. Special thanks are due to U. Glatzel, who kindly shared the results of his elasticity calculations. MJM, MSD and SMF would like to acknowledge the support of the U. S. Department of Energy, Office of Basic Energy Sciences under Contract No. DE-AC04-76DP00789. Partial support for MJM was also provided by a research grant from the Air Force Wright Laboratory.

REFERENCES

[1] M. H. Loretto and R. J. Wasilewski; *Phil. Mag.*, **23**, 1311 (1971).
[2] I. Baker and E. M. Schulson; *Metall. Trans.*, **15A**, 1129 (1984).
[3] R. D. Field, D. F. Lahrman, and R. Darolia; *Acta metall. mater.*, **39**, 2951 (1991).
[4] C. H. Lloyd and M. H. Loretto; *phys. stat. sol.*, **39**, 163 (1970).
[5] M. J. Mills and D. B. Miracle, *Acta metall. mater.*, **41**, 85 (1993).
[6] D. B. Miracle; *Acta metall. mater.*, **39**, 1457 (1991)
[7] M. S. Daw and M. I. Baskes, *Phys. Rev. B*, **29**, 6443 (1984).
[8] S. I. Rao, C. Woodward, and T. A. Parthasarathy, *MRS Proceedings*, **213**, 125 (1991).
[9] S. M. Foiles and M. S. Daw, *J. Mater. Res.*, **2**, 5 (1987).
[10] P. Stadelmann, *Ultramicroscopy*, **21**, 131 (1987).
[11] D. Farkas, R. Pasianot, E. J. Savino and D. B. Miracle, *MRS Proceedings*, **213**, 223 (1991).
[12] T. A. Parthasarathy, S. I. Rao and D. M. Dimiduk, *Phil. Mag. A*, in press (1992).
[13] U. Glatzel, Unpublished research.

PLANAR GROWTH FAULTS IN Nb_3Al

L.S. SMITH, T-T. CHENG AND M. AINDOW
IRC in Materials for High Performance Applications, and School of Metallurgy and Materials, The University of Birmingham, Edgbaston, Birmingham, B15 2TT, UK.

ABSTRACT

A transmission electron microscopy study of planar growth faults in the A15 intermetallic phase Nb_3Al is presented. These faults are not observed in plasma melted material but a high density of these features is observed in arc-melted material which is cooled more slowly. The faults are found most frequently on planes {100}, often changing from one plane to another along their length. It has not been possible to obtain a full diffraction contrast analysis from the faults but HREM images indicate that they are intrinsic with a displacement vector **R**=1/4<120>. Since this vector does not lie in the plane, they cannot form by glide processes alone and therefore probably arise when vacancies coalesce into sheets on {100}. Since this value of **R** is consistent with the removal of an Al-rich plane {004}, it is argued that the formation of these faults may help to stabilise the A15 phase to Nb-rich compositions.

INTRODUCTION

Several intermetallic compounds of the general formula A_3B possess the cubic β-W structure (A15). This class of compound is best known for their superconducting properties at liquid helium temperatures. Some of these compounds, notably Nb_3Al, Cr_3Si and V_3Si, are also of interest for aerospace applications since they have high melting points, low densities, small lattice parameters and are tolerant to significant deviations from stoichiometry. The exploitation of these properties has been prevented by the limited ductility of A15 compounds, but the reasons for this are not well established. The purpose of the present research program is to investigate the defect microstructures and deformation mechanisms in Nb_3Al based alloys.

In this paper we present a TEM study on the nature of extended stacking faults in as-cast specimens of Nb_3Al-based alloys. Theoretical studies have suggested that faults in A15 alloys could have displacements **R** equal to 1/2 <111> [1] and extended faults have been observed experimentally in Nb_3Al [2,3]. In recent work [4,5], it has been shown that two different types of fault can occur in Nb_3Al; narrow ribbons with **R** = 1/2<100> formed by the dissociation of perfect dislocations and large growth faults with **R** = 1/4<021>. In this paper we consider these latter faults in detail: TEM images from such faults in binary Nb-Al alloys with ≈25% Al are presented and their origins and effects on deformation behaviour are discussed.

EXPERIMENTAL PROCEDURE

Ingots with nominal compositions Nb-25at.% Al were prepared by two different routes, both using elemental starting materials. Small (≈50g) buttons were prepared by arc melting under vacuum - the ingots were remelted twice to promote homogeneity. Larger (≈ 500g) ingots were produced by plasma melting under a positive overpressure of Ar plus He. The ingots were remelted four times, flipping between each remelt, to give a homogeneous mixture. Specimens for TEM were obtained by trepanning discs 3mm in diameter from slices 150μm thick and then electropolishing to perforation with 10 vol.% sulphuric acid in methanol at ≈-25°C and 100mA. For HREM imaging, surface contamination layers were removed by Ar^+ ion beam milling briefly using a low accelerating voltage. The foils were examined in a Philips CM20ST transmission electron microscope operating at an accelerating voltage of 200kV.

RESULTS

Samples obtained from plasma melted Nb-25Al buttons consisted of two phases; the majority A15 phase and small spheroidal A2 particles. These latter particles are very niobium rich and their presence indicates that the peritectic reaction, α + Liquid $\leftrightarrow$ Nb_3Al, has not proceeded to completion. The only stacking faults present in the A15 grains were ribbons, ≈20nm wide, of shear faults with displacement vectors **R**, given by **R** = 1/2<100>. As discussed previously [4,6], these are produced by the dissociation of screw dislocations with Burgers vectors **b** = <100> into two partial dislocations on common planes {012}thus;

$$[100] \leftrightarrow 1/2\,[100] + \mathrm{CSF} + 1/2\,[100]$$

A region containing one such feature is presented in Figure 1 which is a weak beam dark field image obtained using **g** = 210. The outer segments are in screw orientations and appear to constrict at various points along their length as the dissociation plane changes from one {012} to another. The central segment is, however in edge orientation; it is dissociated similarly but in this case the reaction can only occur on (001).

Samples obtained from the arc-melted Nb-25Al buttons consisted only of the A15 phase. In this case two types of faults were observed, thin ribbons from dissociations as before and large planar faults extending in many cases over ≥1μm. A bright field TEM image obtained from a typical region with the beam direction **B** close to $[1\bar{2}0]$ is presented in Figure 2. Two sets of large planar faults can be observed at ≈90° to one another; stereographic analysis of the fault planes shows that these are (100) and (010). Many such analyses were performed and it was found that these defects lie predominantly on {001} with occasional segments on {012} and on {011}. These defects show rather complex behaviour; the fringe contrast often changes abruptly and peculiar asymmetric fringes are observed. They usually terminate at other large faults or with small segments on other planes at their ends and occasional "staircase faults" are observed whose habit plane changes from one {001} to another.

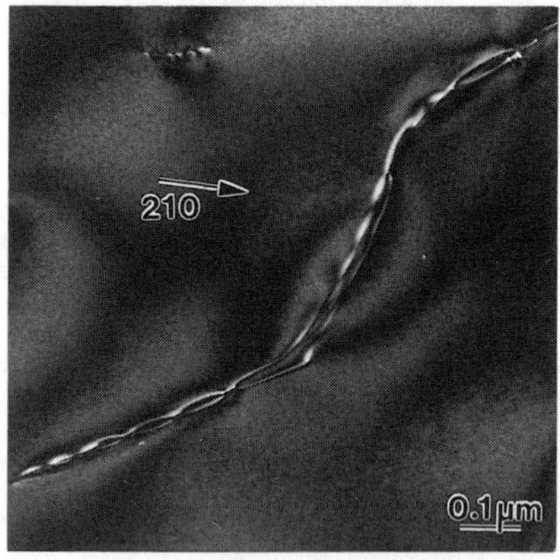

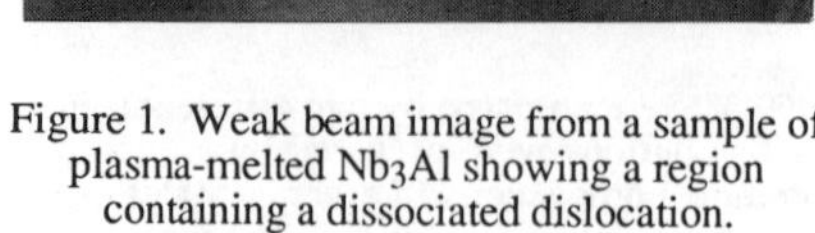

Figure 1. Weak beam image from a sample of plasma-melted Nb_3Al showing a region containing a dissociated dislocation.

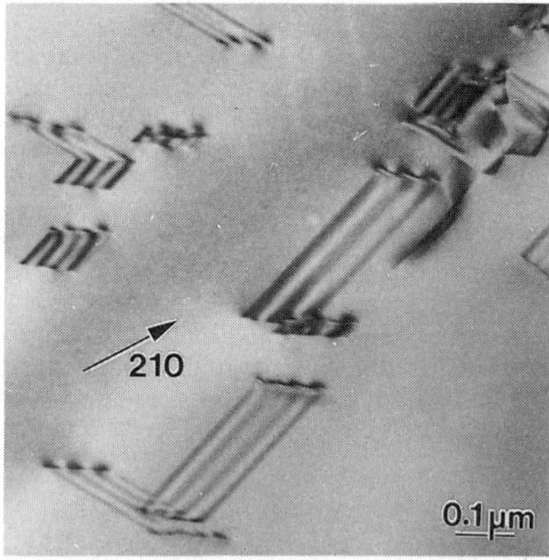

Figure 2. Bright field image from a sample of arc-melted Nb_3Al showing a region containing many planar faults

Diffraction contrast analysis of the displacement vector **R** at these faults is difficult since, of all the available diffraction conditions, **g**, only 200, 210 and 222 type give modest extinction distances at 200kV (or indeed 100kV) and give rise to sufficiently sharp dark field images. Thus many of the diffraction conditions which would be useful in such an analysis are not viable. Moreover, as shown in Figure 2, few of the faults terminate in a simple fashion at a single partial dislocation. Figures 3(a)-(e) are weak beam dark field images from a region where the surface normal is ≈9° away from [001], obtained with **g** = 210, $\bar{2}01$, 021, $0\bar{2}0$ and 200 respectively. The area contains several faults; most of these lie on (100) or (010) and terminate at other faults, but one large fault lies on (001) and terminates at a simple partial dislocation (marked X in Figure 3(a)). If the terminating partial dislocation has a Burgers vector **b**, we assume **R** = **b** = [h k l]. Both the dislocation and the fault are out of contrast for **g** = 200 (Figure 3(e)) so **g.R** = 0 and $h = 0$. For **g** = $0\bar{2}0$ the fault is out of contrast but the bounding dislocation is in contrast (Figure 3(d)) indicating that **g.R** = integer. Since both the dislocation and the fault are in contrast for **g** = 210 (Figure 3(a)), $k = 1/2$. Similarly, the dislocation and fault are in contrast for **g** = $\bar{2}01$ (Figure 3(b)), so $l \neq$ integer or 0. It was not possible to obtain the magnitude of l by this approach but it seems reasonable to expect $l < 1/2$ since the fringe contrast for **g** = 210 (Figure 3(a)) is significantly stronger than that for **g** = $\bar{2}01$ (Figure 3(b)).

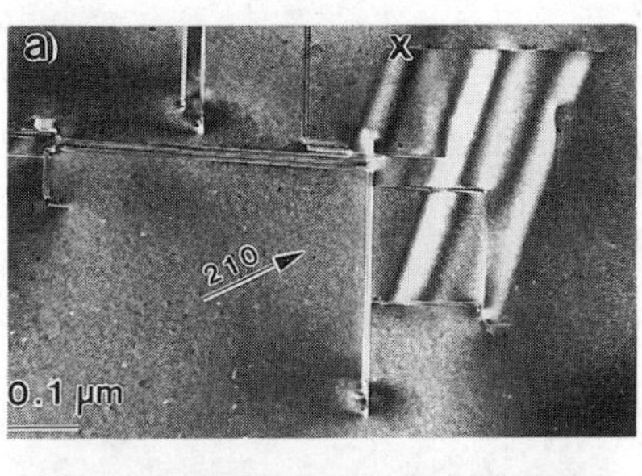

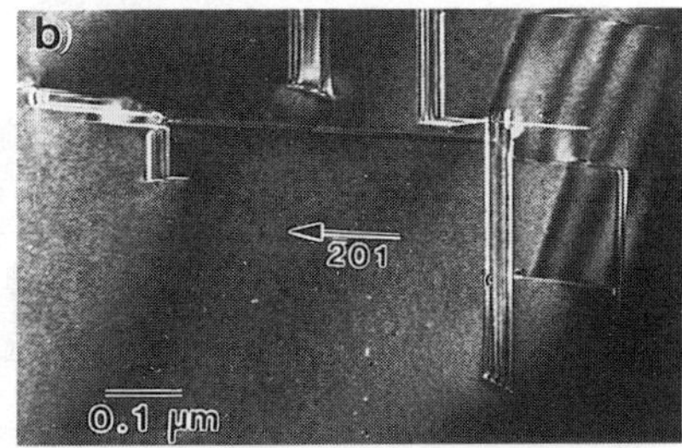

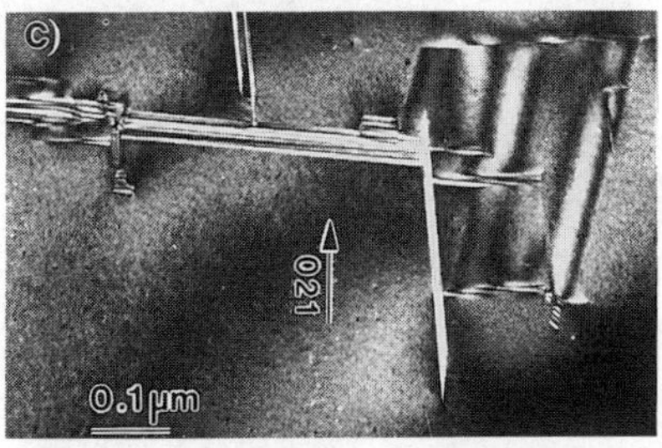

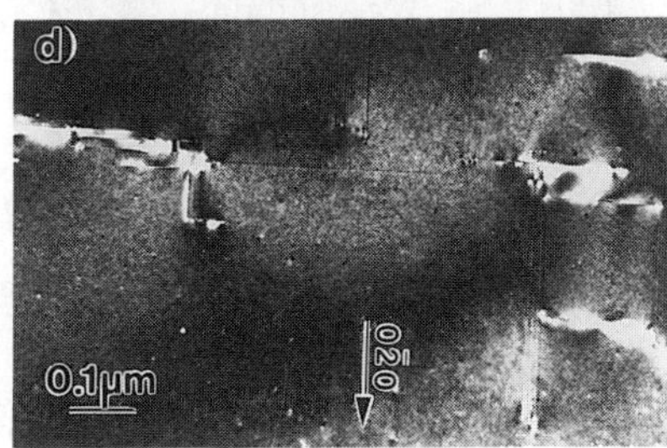

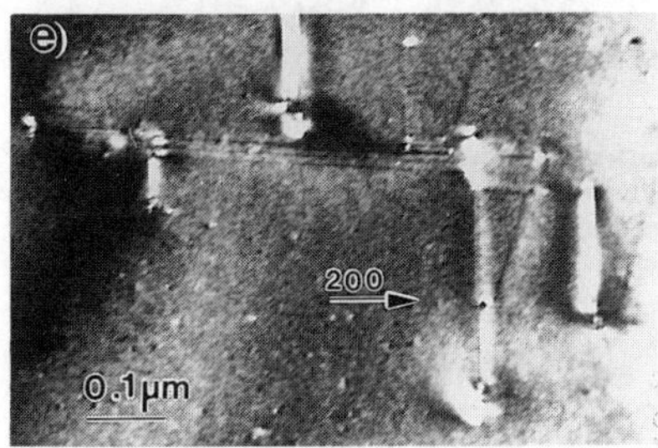

Figure 3. weak beam dark field images from a region containing several planar faults. The largest of these lies on (001).

High resolution lattice images obtained with the beam direction, **B**, parallel to <100> show these faults edge on (Figure 4). Two different displacements are observed and one example of each is shown in Figures 4(a) and (b) - we note that in Figure 4(a) the fault terminates in a simple partial dislocation at the position marked by the large white arrow. Both images were obtained with B // [100] and show a fault on (010). The projected displacements across the faults are 1/4 [010] and 1/4 [012]; thus for a displacement vector **R** = [*hkl*], the values are [*h* 1/4 0] and [*h* 1/4 1/2], respectively with the component in h unknown since it lies parallel to the beam direction. As these faults lie on equivalent planes it is reasonable to assume that the HREM images are simply different projections of equivalent displacements. Thus we deduce that the faults will have **R** = 1/4 <012>, which is consistent with the diffraction contrast data. Since this vector connects Al to Nb sites in the Nb_3Al structure, the fault will have APB character, however the Nb and Al sub-lattices are not equivalent so the defect is a complex stacking fault. In both cases (and for all other faults observed in such images) there is a component 1/4 <010> of R normal to the habit plane which corresponds to a contraction; i.e. these are intrinsic stacking faults formed by the removal of an {004}. We should emphasise that isolated simple faults such as those observed in Figures 4(a) and (b) are relatively rare, with more complex structures usually being observed. The most common of these is a double fault such as that shown in Figure 4(c); two parallel faults with identical displacements are lying on planes less than 1nm apart. It is not clear why such closely spaced faults should be formed but their presence could account for some of the more unusual fringe contrast observed in bright and dark field images.

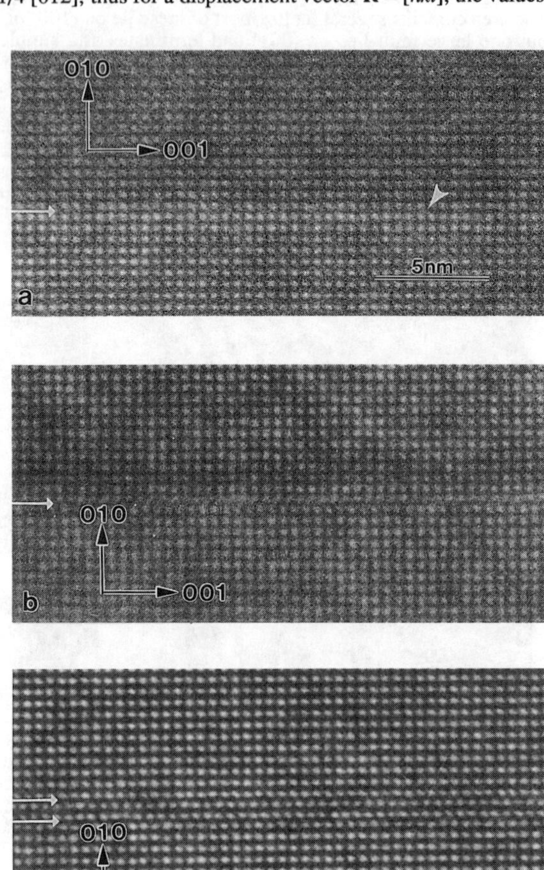

Figure 4. HREM of planar faults "edge on"; (a) and (b) show projected displacements of 1/4[010] and 1/4[012], respectively, (c) shows a double fault.

DISCUSSION

All of the faults have a displacement of 1/4<001> normal to the fault plane, and since there is no in-plane displacement which would give an equivalent arrangement, they must arise by the coalescence of point defects rather than by glide processes i.e. they are growth faults. Moreover, all of the faults observed experimentally appear to be intrinsic in character, thus the point defects involved must be vacancies. As the elastic constants in Nb_3Al are very high ($G \approx 36GPa$), vacancies probably coalesce on planes where the resultant fault will have low energy, in order to minimise the strain energy associated with a random configuration of vacancies. It is significant that faults were observed in arc melted material but not in plasma melted material with the same composition. The plasma melted sample had cooled from the melt quite rapidly as indicated by the incomplete peritectic reaction; this may have inhibited the migration of vacancies and thus prevented fault formation. The arc-melted sample had, however, been allowed to cool extremely slowly giving a completely transformed microstructure and a large number of planar faults. It is interesting to compare these observations with those of Marieb et al. [2]; they saw no faults in arc-melted Nb-18Al in the as cast condition. Following deformation at 1200°C, a significant number of large planar faults had formed on planes {001}. If indeed the faults observed in their experiments correspond to those observed here then they cannot have arisen by deformation in the way suggested but could instead be the result of diffusion at the deformation temperature.

The A15 crystal structure has a simple cubic lattice with a complex eight atom basis as shown in Figure 5(a). Following Reynaud and Lamine [7] we can describe the structure in terms of two different types of layer parallel to the cube faces (Figure 5(b)); those containing two Nb and one Al atom per unit cell intersected denoted ***a*** or ***b***, and those containing one Nb atom denoted α. These stack in an ****....a*** α ***b*** α ***a*** α ***b*** α ***a....*** sequence and thus the layers correspond to {004} planes in the structure. Clearly there are two different intrinsic faults which could be produced by the removal of an {004}. If an α layer was removed to produce the fault, an ***a*** layer would be adjacent to a ***b*** layer, but whilst this would give a displacement of 1/4<001> normal to the layers, there is no reason for a shear displacement to arise across the fault plane. If, however, one of the more closely packed ***a*** or ***b*** layers was removed to produce the fault, two α layers would be adjacent to one another. This would give little or no displacement since it is usually assumed that the Nb atoms are in contact along the chains parallel to <001>; a subsequent shear of the layers by 1/2<010> across the plane, however, would also give an appropriate displacement normal to the plane. The net displacement produced in this latter case would be 1/4 [021] as observed experimentally and thus these intrinsic faults probably correspond to missing ***a*** or ***b*** layers in the structure.

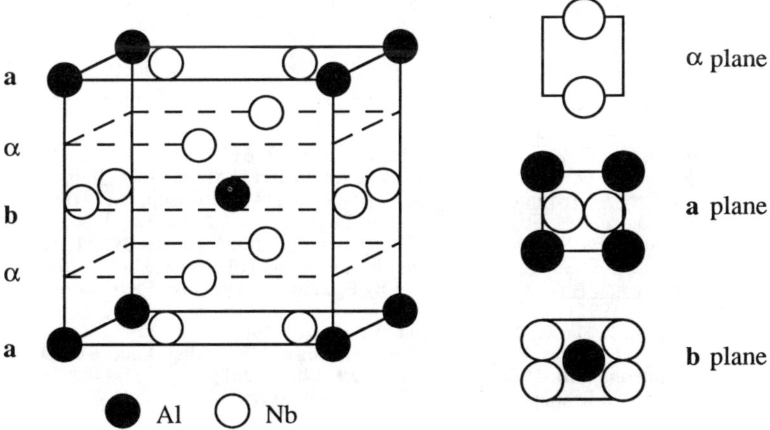

Figure 5. Arrangement of atoms in the {004} layers which stack to form the A15 crystal structure (after Reynaud and Lamine [7]).

It has been suggested by Sudareva et al. [3] that the presence of planar faults may play a role in "absorbing deviations from stoichiometry". None of the layers in the A15 structure have the stoichiometric composition and thus the removal of a layer to produce a stacking fault will cause a local change in composition. If the observed faults correspond to the removal of ***a*** or ***b*** layers which contain Nb and Al atoms in the ratio 2:1, the overall effect of a large number of such faults would be to give an increase in the Nb content of the crystal. We note that the A15 phase field in the binary Nb-Al system extends from 75% Nb to 82% Nb, i.e. the equilibrium phase is tolerant to deviations from stoichiometry for Nb-rich compositions but not for Al-rich ones. It is tempting to speculate that the ability to form planar faults may play a role in the stability of the Nb-rich compositions. Whilst such a link may seem tenuous, a survey of the literature indicates that A15 phases in which large planar faults are formed (e.g. Nb_3Sn [8]) have much wider phase fields than those in which no faults have been observed (e.g. V_3Si). If such a link were to exist then one would expect the fault density in well annealed material to be a function of niobium content and we are now performing experiments to investigate this. The mechanisms by which these defects form and their influence on the plastic deformation characteristics of Nb_3Al are also the subject of further study.

CONCLUSIONS

Plasma melted Nb_3Al cools rapidly giving a partially transformed microstructure with only narrow ribbons of fault produced by the dissociation of screw dislocations. In arc-melted material, which cools more slowly, large intrinsic planar faults with **R** = 1/4<021> are present, mainly on {001}. Since there is a component 1/4<001> perpendicular to the fault plane in each case they must be growth faults. It is proposed that these form as a result of vacancy coalescence and, since they correspond to the removal of Al-rich {004} planes, that they could stabilise the A15 phase in Nb-rich compositions.

ACKNOWLEDGEMENTS

This work has been carried out with financial support from SERC. The authors would like to thank Dr. J. Shyue, Prof. H.L. Fraser and Prof. M.H. Loretto for helpful discussions. Financial support to attend this meeting was provided by the Institute of Physics EMAG group (LSS + MA), C.R. Barber Trust Fund (LSS) and the Institute of Materials Andrew Carnegie Grant Scheme (LSS).

REFERENCES

[1] U. Essman and G. Zerweck, Phys. Stat. Sol. (b) **57**, 611 (1973)..

[2] T.N. Marieb, A.D. Kaiser, S.R. Nutt, D.L. Anton and D.M. Shah, in High Temperature Ordered Intermetallic Alloys IV edited by L. Johnson, D.P. Pope and J.O. Stiegler (Mater. Res. Soc. Proc. **213**, Pittsburgh, PA, 1991) pp. 329-336.

[3] S.V. Sudareva, N.N. Buynov and Y.P. Romanov, Phys. Met. Metall., **44**, 114 (1978).

[4] M. Aindow, T-T. Cheng, R. Beanland, J. Shyue, and H.L. Fraser, in Electron Microscopy and Analysis 1991 edited by F.J. Humphreys (Inst. Phys. Conf. Ser.,**119**, Bristol, UK, 1991) pp. 249-252.

[5] J. Shyue, Ph.D. Thesis, The Ohio State University (1992).

[6] M. Aindow, J. Shyue, T.A. Gaspar and H.L. Fraser, Phil. Mag. Lett., **64**, 59 (1991).

[7] F. Reynaud and A.B. Lamine, Acta Met., **29**, 1485 (1981).

[8] Y. Uzel and H. Diepers, Z. Phys., **258**, 126 (1973).

ANTIPHASE BOUNDARY TUBES IN ORDERED INTERMETALLIC COMPOUNDS

ALFONSO H.W. NGAN*, I.P. JONES AND R.E. SMALLMAN
School of Metallurgy and Materials, University of Birmingham, Birmingham B15 2TT, England.
* Now at Department of Materials, University of Oxford, Oxford OX1 3PH, England.

ABSTRACT

TEM observations suggest the presence of edge-type strain fields around antiphase boundary tubes in deformed $L1_2$ compounds such as Ni_3Al and Fe_3Ge. The present paper aims at discussing the origin of these strain fields. Some preliminary results concerning the thermal instability of the tubes will also be discussed.

INTRODUCTION

Antiphase boundary (APB) tubes are a type of defect specific to ordered structures. Geometrically speaking, APB tubes are polygonal cylinders whose faces are bounded by APB's. They were first proposed by Vidoz and Brown [1] to account for the generally high work-hardening rates in well-ordered intermetallic compounds. In their formation mechanism, an APB tube may result from the glide of an APB-dissociated superdislocation which has been jogged previously by a forest dislocation. Other formation mechanisms which involve the cross-slip of screw superdislocations were proposed subsequently ([2] & [4]). Experimentally, the first evidence of APB tubes was found in highly deformed Fe-35.5%Al by Crawford [3] using TEM methods. Chou and Hirsch [4] later formulated a contrast formation mechanism which considered only the step-wise and rigid nature of the crystal displacement across each of the APB's forming the tube. Because no atomic relaxation of the APB's was taken into account and also, the joining of APB's at the corners of the tubes does not require dislocations, no continuous strain field was then thought to be associated with the tubes. In Chou and Hirsch's mechanism, APB tubes can therefore be imageable via superlattice reflections only; when imaged under fundamental reflections, they should always be out of contrast since $\mathbf{g} \cdot \mathbf{R}_F$, where $\mathbf{g}$ is the diffraction vector and $\mathbf{R}_F$ the APB displacement, is an integer.

As Sun [5] has shown, however, APB tubes do in fact exhibit detectable contrast when imaged under certain fundamental reflections. Sun ascribed this supplementary contrast in fundamental reflections to the strain field caused by the surface tension effect of the APB's forming the tube. However, the contrast which he predicted was two orders of magnitude lower than that observed experimentally. The present paper therefore suggests a different interpretation of the strain fields of APB tubes with the aim of explaining quantitatively the tube contrast.

The present paper also reports some preliminary results concerning the thermal instability of APB tubes. Potential effects of the observed thermal instability upon work-hardening behaviour will be mentioned briefly.

EXPERIMENTAL

The observations described in the present paper were made on two $L1_2$ compounds: Ni_3Al and Fe_3Ge. The specimen preparation procedures and details regarding compression tests and TEM observations can be found in references [6] and [7].

THE STRAIN FIELD OF AN APB TUBE

Fig. 1 shows the microstructure of a Ni_3Al specimen after deformation at room temperature using a nominal strain rate of $1 \times 10^{-4}s^{-1}$ to a plastic strain of ~8.5%. In fig. 1(a), when imaged using the fundamental reflection 220, APB tubes are seen to appear as straight and short lines parallel to the dotted line which is along the $[1\bar{1}0]$ direction. In fig.1(b), the previously visible tubes go out of contrast when imaged using the $2\bar{2}0$ reflection which is parallel to the tube axis. As first suggested by Sun [5], this observed contrast is characteristic

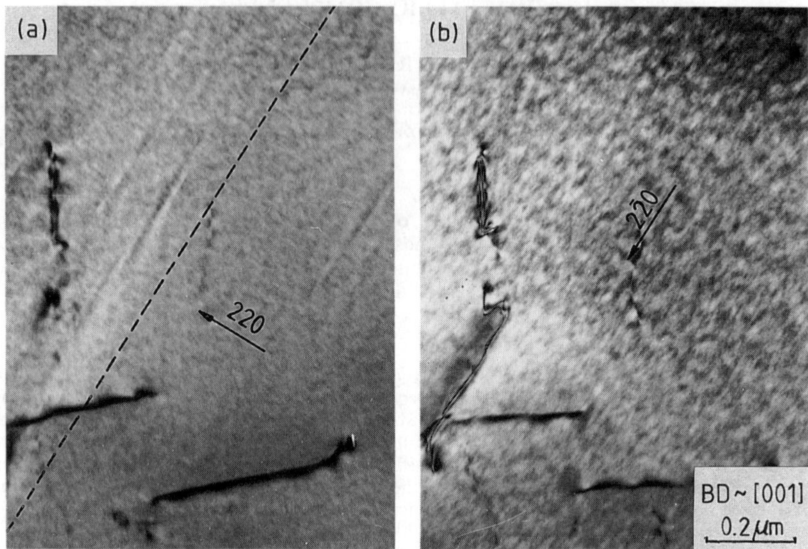

Fig. 1- Bright field transmission electron photomicrographs of APB tubes in room-temperature-deformed Ni_3Al.
(a) Tubes (parallel to dotted line) in contrast
(b) Tubes out of contrast

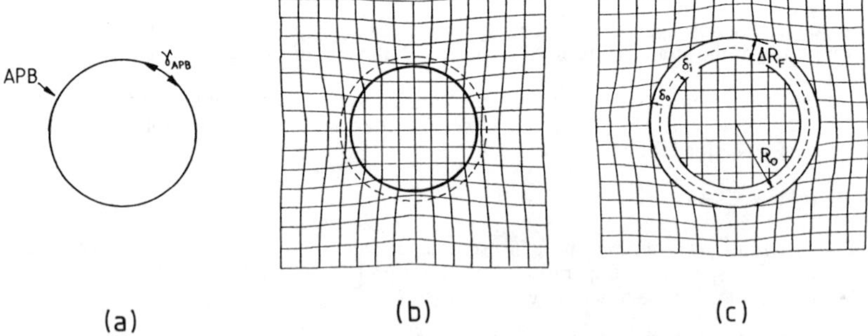

Fig. 2 - Strain field of a circular APB tube produced by APB relaxation
(a) Unrelaxed state
(b) Relaxation according to Sun [5]
(c) Relaxation due to extrinsic supplementary fault displacement

of an edge-type strain field around the tube. Sun's explanation of the strain-field is a "surface tension-like" effect of the APB forming the tube and is schematicised in fig. 2(a)&(b) for a hypothetical circular tube.

The present authors [8] have compared the experimental contrast as observed in deformed Fe_3Ge with that computed according to Sun's mechanism. As fig. 3 shows, however, Sun's contrast (which is too weak to see in fig. 3(b)) is much lower than that observed experimentally (fig. 3(a)) and therefore a new explanation is required for the unexpectedly large tube strain field.

The new explanation we suggest here involves the concept of a supplementary fault displacement [9]. Consider first a flat APB. Because of the wrong atomic environment, the lengths and angles of the atomic bonds across the APB are expected to be different from the situation when there had been no APB. As an approximation, this phenomenon can be modelled by incorporating a supplementary displacement $\Delta\mathbf{R_F}$ at the APB, so that apart from displacing relative to each other by the basic fault vector $\mathbf{R_F}$, the two half crystals on either side of the APB displace additionally by $\Delta\mathbf{R_F}$. This $\Delta\mathbf{R_F}$ concept has been found useful in explaining quantitatively the supplementary contrast of "thermal" APB's in well-annealed Fe_3Ge under fundamental reflections (fig. 4 and [10]). In the case of an APB tube, apart from the $\Delta\mathbf{R_F}$ at the APB which is imageable by fundamental reflections, a long-range strain field will also be set up. This is schematicised in fig. 2(c). As is the case in fig. 4, $\Delta\mathbf{R_F}$ is considered to be extrinsic and to decompose into two components: δ_i compressing the material inside the tube and δ_o expanding the material outside the tube. In more realistic modelling, the circular tube assumption can be waived and the fact that in an actual tube, the APB's would lie on specific crystallographic planes can also be taken into consideration. As an example, the contrast predicted for a polygonal-shaped tube is shown in fig. 3(b) (the curve marked "Tube"); in which case, the $\Delta\mathbf{R_F}$'s at various APB's have magnitudes ~5% of the lattice parameter ([8] & [10]). The agreement with experiment (fig. 3(a)) is extremely good.

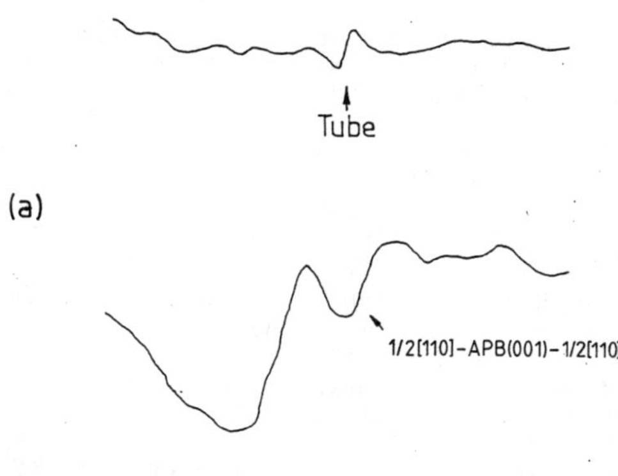

Fig. 3 - Tube image vs superdislocation image in Fe_3Ge
(a) Experimental
(b) Theoretical

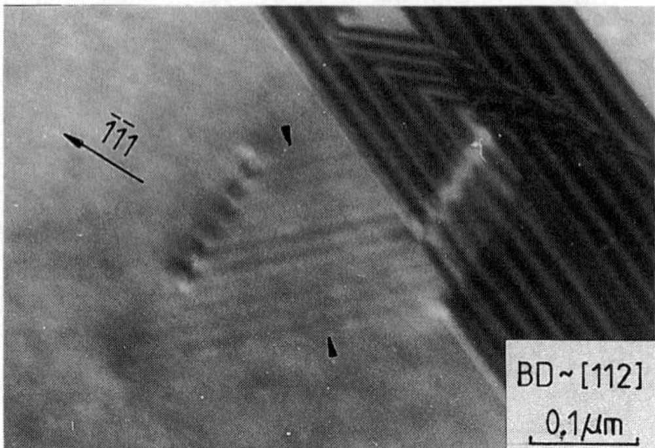

Fig. 4 - Bright field transmission electron photomicrograph showing supplementary contrast at thermal APB's in Fe_3Ge imaged using a fundamental reflection.

In the circular tube approximation shown in fig. 2(c), the strain field is edge-type; in other words, the elastic displacement component along the tube axis is zero. In a (true) polygonal-shaped tube, the $\Delta \mathbf{R_F}$ on various crystallographic APB's will in general not be strictly perpendicular to the tube axis, ie., the resultant strain field will not in general be edge-type. However, in the specific case of $L1_2$, the APB's on any tube are expected to lie either on {111} or {001} planes, and in both cases, simple crystallographic considerations suggest that $\Delta \mathbf{R_F}$ is strictly perpendicular to the tube axis [8]. Hence, APB tubes in $L1_2$ compounds are expected to have edge-type strain fields, as is observed experimentally (fig. 1).

THERMAL INSTABILITY OF APB TUBES

Another interesting phenomenon concerning APB tubes is their low thermal stability. Fig. 5(a) shows the microstructure of a freshly deformed Fe_3Ge specimen; APB tubes can clearly be seen as faint straight lines along the [110] direction. Fig. 5(b) shows the same area imaged under similar conditions, but the specimen has now been kept at room temperature for about 6 months. The *hitherto* present tubes have now disappeared. Observations on all Fe_3Ge and Ni_3Al specimens which have been deformed at low temperatures and afterwards kept at room temperature for a long period of time reveal the same phenomenon - the APB tubes gradually disappeared. Another evidence for the low thermal stability of the tubes is that all specimens which have been deformed at high temperatures, 573K or above, for Fe_3Ge using strain rates of $\sim 1 \times 10^{-4} s^{-1}$, contained no tubes at all. It seems that all the APB tubes produced during deformation annihilated almost instantly at such high temperatures. Similar low thermal stability has also been observed in Ni_3Ga by Sun ([5] and private communication).

The low thermal stability of the tubes is not at all surprising, taking into account the tubes' large self energies and the ease with which they annihilate. Consider first the self energy, which is also the driving force for annihilation. It comprises in principle two parts: the fault (APB) energy E_F and the strain energy E_S. For a circular tube with radius r, E_F per unit length of tube is given by $2\pi r\gamma$, where γ is the APB energy. The strain energy E_S per unit length is given analytically by

$$\frac{\pi E(1-\nu)(\Delta R_F)^2}{4(1+\nu)(1-2\nu)}$$

where E and ν are the Young's modulus and Poisson's ratio respectively. As Ngan [10] has shown, for tubes with reasonable sizes (say, $r \sim 2$nm), E_S is small compared with E_F ($E_S \sim$ a few % of E_F). The self energy is therefore dominated by E_F. Taking a model $\gamma \sim 200$ mJm^{-2}, r ~ 2nm, the tube self energy is thus $\sim 2 \times 10^{10}$eV/m, which is of the same order of magnitude as that for a dislocation.

The annihilation of a tube is however much easier than that of a dislocation. Unlike dislocations, the annihilation of a tube requires only very localised diffusion - the atoms inside

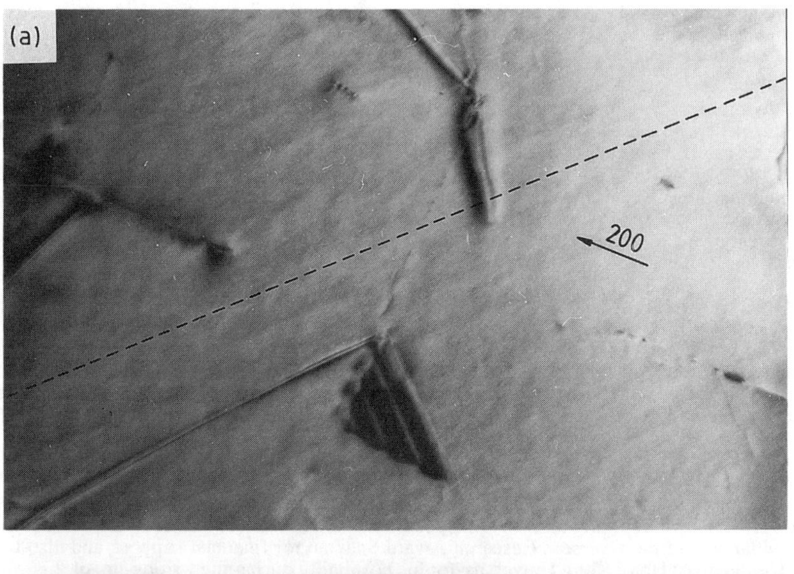

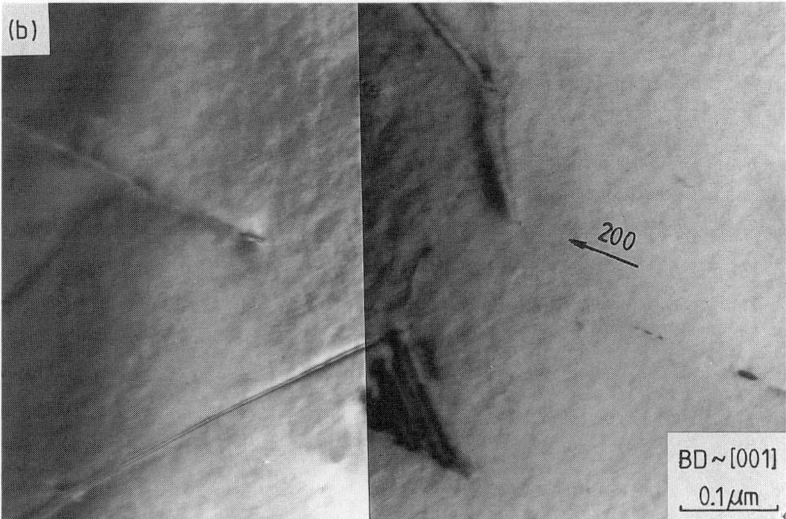

Fig. 5 - Bright field transmission electron photomicrographs of the thermal instability of APB tubes.

(a) Microstructure of a Fe_3Ge specimen freshly defomred at room temperature. APB tubes are present.

(b) Same area as in (a) imaged using similar diffraction condition about 6 months later. APB tubes have disappeared.

the tube merely have to migrate by the vector $\mathbf{R_F}$ along the tube, which is 1/2<110>, or one atom spacing, for $L1_2$. After annihilation, a vacancy and an interstitial cluster will be left at either end of the tube. However, as shown by Ngan [10], the energies of both clusters are expected to be small compared with the self energy of a tube with reasonable length (say a few μm's) and therefore the driving force for annihilation is little affected.

To summarise, while detailed modelling of the annihilation process is still waiting to be performed, with such a large driving force and an evidently small obstacle, the annihilation kinetics can be expected to be very fast, especially at high temperatures. As mentioned in the *Introduction*, APB tubes are currently thought to be closely connected to the high work-hardening rates in intermetallic compounds. However, all the existing theories in this regard assume that the tubes can stay after they have been formed. In the light of the present observation of low thermal stability, it would therefore seem necessary to re-consider the work-hardening effects of the tubes at high temperatures.

CONCLUSION

The atomic relaxation of the APB's forming an APB tube is expected to give rise to a long-range strain field which is imageable by certain fundamental reflections in the TEM. APB tubes were observed to have low thermal stability because of their large self energies and small barriers for annihilation.

ACKNOWLEDGEMENT

The authors thank Prof. J.F. Knott, FRS FEng for the provision of laboratory facilities and Dr. H. Yu for preparing the Ni_3Al specimens. A.H.W.N. is grateful to the Croucher Foundation and the Overseas Research Award Scheme for financial support, and also to Dr. B.J. Duggan of Hong Kong University for his hospitality during the writing-up of this work.

REFERENCES

1. A.E. Vidoz and L.M. Brown, Phil. Mag., **7**, 1167 (1962).
2. P.M. Hazzledine and P.B. Hirsch in High-Temperature Ordered Intermetallic Alloys II, edited by N.S. Stoloff, C.C. Koch, C.T. Liu and O. Izumi (Mater. Res. Soc. Proc. **81**, Pittsburgh, PA, 1987) pp.75-85.
3. R.C. Crawford, Phil. Mag. **33**, 529 (1976).
4. C.T. Chou and P.B. Hirsch, Proc. Roy. Soc. A, **387**, 91 (1983).
5. Y.Q. Sun, Phil. Mag. A, **65**, 287 (1992).
6. H. Yu, PhD Thesis, University of Birmingham, 1991.
7. A.H.W. Ngan, I.P. Jones and R.E. Smallman, Phil. Mag. A, **65**, 1003 (1992).
8. A.H.W. Ngan, I.P. Jones and R.E. Smallman, Phil. Mag. (in press)
9. E. Haque, M.H. Loretto and I.P. Jones, EMAG75 Proceedings, 429 (1975).
10. A.H.W. Ngan, PhD Thesis, University of Birmingham, 1992.

STUDIES OF THE PROPERTIES OF VACANCIES AND THEIR CLUSTERS IN $CoSi_2$ BY POSITRON LIFETIME SPECTROSCOPY

Y. ITO, Y. SHIRAI, Y. YAMADA and M. YAMAGUCHI
Department of Metal Science and Technology, Kyoto Univ., Kyoto, Japan

ABSTRACT

Properties of thermal vacancies in $CoSi_2$ and defects in electron irradiated and proton irradiated $CoSi_2$ have been studied by positron lifetime spectroscopy. It has been found that thermal vacancies in $CoSi_2$ are easily quenched in. The effective formation enthalpy of vacancies has been estimated to be about 1.1 eV. The recovery of electron irradiated $CoSi_2$ occurs in two stages; the first stage around 310 K attributed to the migration of vacancies forming secondary defects and the final stage between 670 and 700 K, where the secondary defects dissolve. Since the positron lifetime at the secondary defects is smaller than that at vacancies, the secondary defects must be collapsed vacancy clusters, such as faulted dislocation loops. In the recovery of proton irradiated $CoSi_2$, microvoids, which are annealed out above 820 K, as well as collapsed vacancy clusters are also formed. The three dimensional vacancy clustering is probably due to the existence of implanted hydrogen atoms.

INTRODUCTION

Transition metal silicides have recently received much attention in Si device applications [1]. Among them, $CoSi_2$ can be used as Shottkey barriers, ohmic contacts and gate electrodes because of its low electrical resistivity and high temperature stability. On the other hand, $CoSi_2$ is expected to be used as new high temperature structural materials because of its low density and excellent oxidation resistance [2-4]. However, properties of point defects in $CoSi_2$, which affect the electrical and mechanical properties of the material, have scarecely been studied.

We have studied vacancies and their clusters in $CoSi_2$ by positron lifetime spectroscopy. Positrons are sensitively trapped by vacancy-type defects and positron lifetime depends on defect species [5]. In this paper, we report the recovery of electron irradiated and proton irradiated $CoSi_2$ and estimate the effective formation enthalpy of vacancies in $CoSi_2$.

EXPERIMENTAL PROCEDURE

A $CoSi_2$ rod was produced by melting pure cobalt (99.9 %) and silicon (99.9999 %) in a plasma arc-furnace. The $CoSi_2$ rod was remelted in an ASGAL optical floating zone furnace to grow a single crystal. Specimens of 9 mm x 12 mm x 0.8 mm were cut from the single crystal. The specimens were fully annealed for 6 hours at 1273 K in evacuated quartz capsules and chemically polished.

Quenched-in defects were introduced by water- or air-cooling from high temperatures above 610 K. After each cooling positron lifetime measurements were performed at 100 K in a cryostat.

Electron irradiation was carried out using a Van de Graff type accelerator at the Department of Nuclear Engineering, Kyoto University. Specimens were mounted in a cryostat and irradiated with 1.0 MeV electrons below 250 K to a dose of 2.5 x 10^{18} e^-/cm^2. Proton irradiation was carried out with 2.0 MeV protons below 100 K to a dose of 2.2 x 10^{18} p/cm^2 using the Van de Graff type accelerator at the Department of Nuclear Engineering, Kyoto University. Electron or proton irradiated specimens were isochronally annealed for 15 min at steps of 30 K. After each annealing, positron lifetime measurements were performed at 100 K in a cryostat.

The positron lifetime spectrometer employed is a fast-slow coincidence system with an instrumental time resolution of 200 ps FWHM. A $^{22}NaCl$ positron source of about 20 μCi was sandwiched between two $CoSi_2$ specimens. Measured lifetime spectra were analyzed in terms of one or two lifetime components after source correction, using the computer programs, Resolution [6] and Positronfit Extended [7].

RESULTS

Quenching

The positron lifetime spectra obtained from specimens water- or air-cooled from high temperatures were analyzed in terms of one lifetime component to get mean positron lifetimes. Fig. 1 shows the obtained values of mean positron lifetime as a function of quenching temperature. It is seen that above 670 K the mean positron lifetime increases remarkably with temperature and approach to a saturation value of 172 ps, and that there is little difference between water-quenching and air-cooling. This means that the concentration of quenched-in defects become high enough to trap positrons effectively above 670 K and increases with temperature. Above 790 K, most positrons are trapped at defects where the lifetime is 172 ps. Even the rate of cooling by air seems to be fast enough to quench the defects in $CoSi_2$ specimens.

Electron Irradiation

Fig. 2 shows the change in mean positron lifetime on the isochronal annealing of electron irradiated specimens resulting from the one-component-analysis of the measured lifetime spectra. After isochronal annealing at 670 and 700 K, specimens were furnace-cooled to room temperature not to introduce additional quenched-in defects into specimens, because even air-cooling from temperatures above 670 K increases positron lifetime, as shown in Fig. 1. Mean positron lifetimes thus obtained are shown with solid circles and labelled as `Furnace cooled` in Fig. 2. The positron lifetime for specimens fully-annealed at 1273 K and furnace-cooled is 154 ps, which is shown with a broken line in Fig. 2. It is obvious that electron irradiation increases the positron lifetime in $CoSi_2$ because of irradiation induced defects, and that the recovery starts in the stage around 310 K and is completed in the stage between 670 and 700 K.

To investigate the recovery process in more detail, the two-component-analysis of measured lifetime spectra were made. However, it was found that there was little second component. This indicates that below 700 K most positrons are trapped by defects in specimens and measured lifetime spectra consist of the defects' component only. Therefore, the change in mean positron lifetime can be regarded as the change in the lifetime of the defects' component.

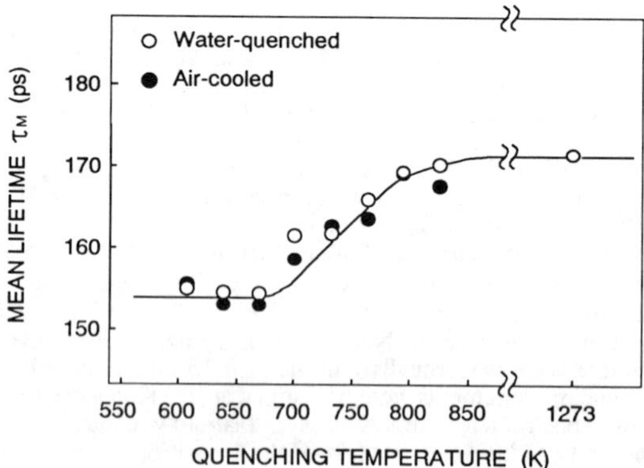

Fig. 1. The changes of mean positron lifetime for $CoSi_2$ water-quenched (open circles) or air-cooled (solid circles) from each temperature.

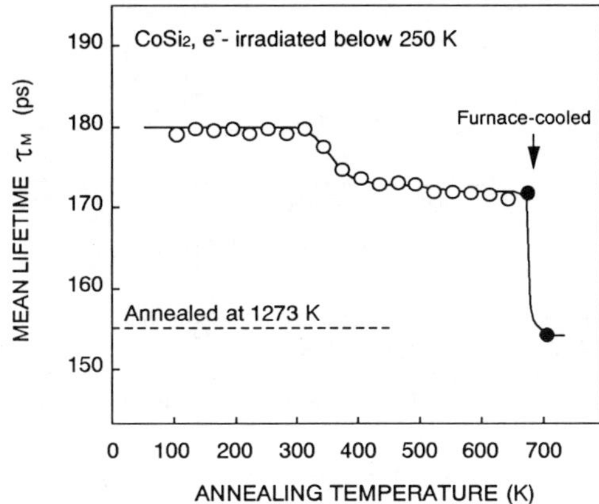

Fig. 2. The change in mean positron lifetime on the isochronal annealing of $CoSi_2$ irradiated below 250 K with 1 MeV electrons to a dose of 2.5 x 10^{18} e^-/cm^2. The lifetimes at 670 and 700 K shown with solid circles were obtained by furnace-cooling after isochronal annealing at these temperatures. Positron lifetime for fully annealed specimens is shown with a broken line.

Proton Irradiation

The depth of damaged layer in $CoSi_2$ corresponding to the range of 2.0 MeV protons is estimated to be 20 μm or less. On the other hand, positrons emitted from a ^{22}Na positron source penetrate to the depth of around 500 μm and thereby the measured lifetime spectra contain the lifetime component of the non-damaged substrate. The measured lifetime spectra of as-irradiated specimens were analyzed into two components with the lifetime for the substrate fixed to the value of 154 ps of fully-annealed specimens, and the relative intensity of the substrate component was found to be about 60 %. Assuming that the thickness of the damage layer does not change during isochronal annealing, the measured lifetime spectra were analyzed into two components, with the constant lifetime parameters for the substrate component.

The change in mean positron lifetime for the damaged layer on the isochronal annealing is shown in Fig. 3. Proton irradiation causes a more marked increase of positron lifetime than electron irradiation. The recovery starts around 200 K. Strange increase of mean lifetimes at 670 and 700 K is due to additional defects quenched in by air-cooling after isochronal annealing at these temperatures, and is consistent with our result shown in Fig. 1. Above 730 K, specimens were furnace-cooled to room temperature after each isochronal annealing for the reason mentioned in the section of electron irradiation. The recovery is not completed even at 820 K.

In order to investigate the recovery process in more detail, the measured lifetime spectra were reanalyzed to resolve the damaged layer component into two components of defects and matrix. Fig. 4 shows the changes in the lifetime at defects, τ_d , and the relative intensity, I_d , on the isochronal annealing. Here, I_d is renormalized so as to show the relative intensity in the damaged layer. The values of τ_d after proton irradiation are higher than those after electron irradiation. Between 670 and 730 K, a sharp decrease of I_d and a marked increases of τ_d are observed. This means that most of defects dissolve around 700 K but there remain defects at which the lifetime is about 400 ps, which are stable up to above 820 K .

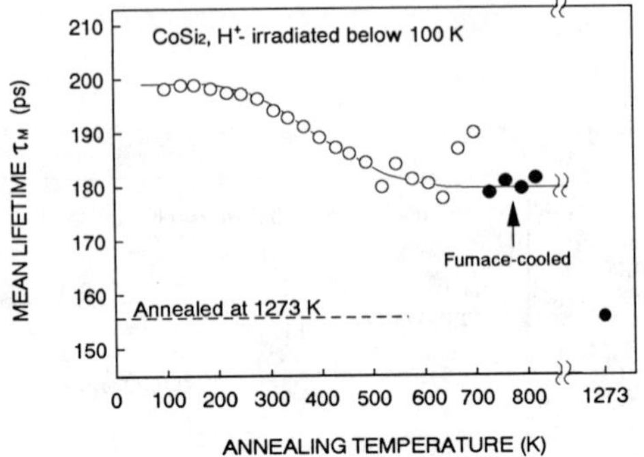

Fig. 3. The change in mean positron lifetime on the isochronal annealing of $CoSi_2$ irradiated below 100 K with 2 MeV protons to a dose of 2.2 x 10^{18} p /cm^2. The lifetimes above 730 K shown with solid circles were obtained by furnace-cooling after isochronal annealing at these temperatures. Positron lifetime for fully annealed specimens is shown with a broken line.

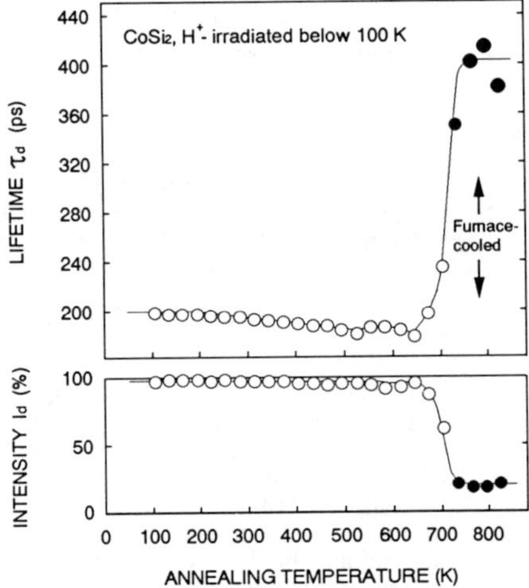

Fig. 4. The changes in the positron lifetime at defects in the damaged layer (τ_d) and the relative intensity of the defects' component (I_d) on the isochronal annealing of proton irradiated $CoSi_2$.

DISCUSSION

Vacancy Migration

Low temperature irradiation with 1 MeV electrons increases the positron lifetime in $CoSi_2$, as shown in Fig. 2. Judging from the electron energy and the dose, only Frenkel pairs are expected to be formed by the electron irradiation. The recovery of electron irradiated $CoSi_2$ proceeds in two stages, around 310 and 670 K. Which species, vacancies or interstitial atoms, migrate in the first stage? Since the migration of vacancies results in the formation of vacancy clusters, the migration of vacancies can be observed through the changes in the lifetime of the defects' component. On the other hand, the migration of interstitial atoms to which positrons are insensitive should be observed through the decrease in the intensity of the defects' component owing to the recombination of interstitial atoms with vacancies. In the present case, the mean positron lifetime which can be regarded as the lifetime of the defects' component noticeably decrease around 310 K. Therefore, the change in mean positron lifetime in the first stage is attributed to the migration of vacancies. We believe that in this stage the migration of vacancies at which the lifetime is 180 ps occurs, resulting in the formation of vacancy clusters at which the lifetime is 172 ps. Suzuki and Takeuchi have recently observed that straight dislocations in $CoSi_2$ pulverized at room temperature change into helical ones during leaving it at room temperatures. This suggests that excess vacancies are mobile at room temperature and causes the climbing of dislocations [4]. Our result agrees well with theirs. We think that self-interstitial atoms migrate and disappear at sinks during the electron irradiation below 250 K because interstitial atoms are much more mobile than vacancies in metals.

Vacancy Formation

$CoSi_2$ water- or air-cooled from temperatures above 670 K retains defects as shown in Fig. 1. The saturation value of mean positron lifetime in Fig. 1 is 172 ps, which is smaller than that at vacancies, 180 ps, and equal to that at secondary defects formed in electron irradiated $CoSi_2$ as shown above. This means that thermal vacancies are quenched in not as single vacancies but as vacancy clusters owing to the migration of vacancies during quenching. Little difference between water-quenching and air-cooling clearly indicates that most of thermal vacancies are easily quenched-in without disappearing at sinks.

Assuming that all thermal vacancies at each quenching temperature are quenched-in, we can estimate the vacancy formation enthalpy in $CoSi_2$, H_V, based on the result shown in Fig. 1. Thermal vacancies can be formed at both Co and Si sites and the vacancy formation enthalpy should depend on which of the two sites they are formed at. Thus, the value obtained below should be regarded as an averaged effective one.

The total concentration of thermal vacancies, C_V, may be approximated as

$$C_V = \exp(S_V / k_B) \exp(-H_V / k_B T) \tag{1}$$

where S_V is the vacancy formation entropy, and k_B is the Boltzmann's constant. The relation between the mean positron lifetime, τ_M, and the concentration of vacancy clusters, C_{Vn}, is given by

$$(\tau_M - \tau_b) / (\tau_{Vn} - \tau_M) = \tau_b \nu_{Vn} C_{Vn} \tag{2}$$

where τ_b is the positron lifetime in the bulk, 154 ps, τ_{Vn} is the lifetime at vacancy clusters, 172 ps, and ν_{Vn} is the specific trapping rate of them. Assuming that $\nu_{Vn} C_{Vn} \approx \nu_V C_V$, eqs. (1) and (2) give

$$\ln [(\tau_M - \tau_b) / (\tau_{Vn} - \tau_M)] = -H_V / k_B T + S_V / k_B + \ln(\tau_b \nu_V) \tag{3}$$

where ν_V is the specific trapping rate of vacancies. The Arrhenius plot based on eq. (3) gives that H_V is 1.1 ± 0.7 eV.

Secondary Defects

Vacancies in electron irradiated $CoSi_2$ migrate and form vacancy clusters in the first stage around 310 K. The vacancy clusters are stable up to the final stage around 700 K, as shown in Fig. 2. The lifetime at the vacancy clusters is 172 ps and smaller than that at vacancies. It is known that the positron lifetime at collapsed small vacancy clusters, faulted dislocation loops or stacking fault tetrahedra, is smaller than that at vacancies [6]. Therefore, we think that vacancies in $CoSi_2$ are apt to form collapsed vacancy clusters.

However, low temperature irradiation with 2 MeV protons causes the formation of uncollapsed small vacancy clusters besides the collapsed ones. As shown in Fig. 4, τ_d in as-irradiated specimens is about 200 ps which is much higher than the lifetime at vacancies, showing that the proton-irradiated specimens contain uncollapsed small vacancy clusters.

The recovery of proton irradiated $CoSi_2$ occurs in two stages around 200 and 700 K, as shown Figs. 3 and 4. The former stage starts at lower temperature than 310 K at which vacancies migrate. We can not conclude what occurs in the former stage at present, but divacancies or some other small vacancy clusters introduced by proton irradiation may migrate at lower temperature than vacancies as in fcc metals. In the latter stage around 700 K, I_d decreases and at the same time, τ_d increases remarkably up to 400 ps. This indicates that defects introduced by proton irradiation, vacancies and uncollapsed small vacancy clusters, grow not only into faulted-loops or collapsed vacancy clusters, which are only stable up to 700 K, but also into microvoids, which are stable above 820 K. Three dimensional vacancy clustering, which does not occur in electron irradiated specimens as shown in Fig. 2, may hence be due to an extrinsic factor, the coexistence of implanted hydrogen atoms. Recently, we observed similar three dimensional vacancy clustering in the presence of hydrogen atoms in proton irradiated TiAl [9]. We believe that the interaction between vacancies and hydrogen atoms stabilizes three dimensional clusters in $CoSi_2$.

SUMMARY

We have performed positron lifetime measurements to study defects in quenched, electron irradiated and proton irradiated $CoSi_2$ single crystals. The results obtained are summarized as follows.

Quenching of $CoSi_2$ specimens from high temperatures above 670 K increases positron lifetime. This shows that thermal vacancies in $CoSi_2$ are easily quenched-in and form secondary defects. This secondary defects are collapsed small vacancy clusters and stable up to 700 K. The effective formation enthalpy of vacancies in $CoSi_2$ has been estimated to be about 1.1 eV.

The recovery of electron irradiated $CoSi_2$ occurs in two stages. In the first stage around 310 K, vacancies introduced by electron irradiation migrate and form secondary defects. This secondary defects dissolve in the final stage between 670 and 700 K. This secondary defects are collapsed vascancy clusters, and probably faulted loops.

Proton irradiation introduces uncollapsed vacancy clusters and they grow not only into faulted-loops but also into microvoids which have not been observed in electron irradiated $CoSi_2$. It is thought that such three dimensional vacancy clustering occurs in the existence of hydrogen atoms. Microvoids are not annealed out below 820 K.

ACKNOWLEDGEMENTS

The authors wish to thank Professor N. Imanishi and Mr. K. Yoshida, Department of Nuclear Engineering, Kyoto University, for their kind assistance in electron and proton irradiation. The authors also thank Dr. S. Ito and Dr. M. Tozaki, Radioisotope Research Center, Kyoto University, for their support in preparing positron sources. This work was supported by Grant-in Aid from Ministry and education, Science and Culture for Scientific Research on the Priority Area Intermetallic Compounds.

REFERENCES

1 S. P. Murarka, Silicedes for VLSI Applications, (Academic, New York, 1983).
2 K. Ito, H. Inui, T. Hirano and M. Yamaguchi, Mater. Sci. & Eng. A152, 153 (1992).
3 S. Takeuchi and T. Hashimoto, J. Mat. Sci. 27, 1380 (1992).
4 K. Suzuki and S. Takeuchi, to be published.
5 P. Hautojärvi, Mat. Res. Sym. Proc., Vol. 82, 3 (1987).
6 P. Kirkegaard, M. Eldrup, O. E. Morgensen and N. Pedersen, Computer Phys. Comun., 23, 307 (1981)
7 P. Kirkegaard and M. Eldrup, Computer Phys. Comun., 3, 240 (1972); 7, 401 (1974).
8 J. Takamura, Y. Shirai, K. Furukawa and F. Nakamura, Mat. Sci. Forum, 15-18, 809 (1987).
9 Y. Shirai and M. Yamaguchi, Mater. Sci. & Eng. A152, 173 (1992).

MÖSSBAUER STUDY ON B2 INTERMETALLIC COMPOUND Fe-40Al AND ITS Mn OR Ti CONTAINING ALLOYS

DINGQIANG LI, PEIEN LI, DEFANG SUN AND DONGLIANG LIN (T. L. LIN)
Institute of Materials Science and Engineering, Shanghai Jiao Tong University, Shanghai 200030, P. R. China

ABSTRACT

The Mössbauer measurements at room temperature have been carried out on the B2 intermetallic compound Fe-40Al and its Mn or Ti containing alloys. The results show that all the Mössbauer spectra of Fe-40Al and its Mn or Ti containing alloys are approximately one-peak singlets. The Fe atoms in these alloys show no significant magnetic moment. The spectra of Mn or Ti containing alloys have been fitted with two Lorentzian lines by the method of least squares. By studying the isomer shifts and other parameters of these fitted lines, it have been shown that Mn atoms occupy both Fe and Al sublattices, Ti atoms occupy preferentially Fe sublattice, however, a few of Ti atoms occupy Al sublattice only as Ti concentration is over certain amount (~5-at%).

INTRODUCTION

B2 intermetallic compound FeAl offers an advantage for structural uses at elevated temperatures because of its excellent oxidation and corrosion resistance and relatively low density and cost [1]. However, a major drawback for B2 FeAl compound is its poor ductility and low fracture toughness at room temperatures. Recently, alloying additions is considered as an important way to improve its ductility. B2 FeAl compound has a wide range of composition [2], so it is very suitable for improving its properties by means of alloying additions [3]. Fe-40Al (40-at%Al) compound has not only higher strength but also better ductility, so it is chosen to be studied.

The sublattice occupances of ternary addition atoms usually play a key role in improving properties of alloys [4]. Mössbauer effect offers an effective measurement method to show the microscopic environments of atoms. S. Nasu *et al.* [5] studied the phases and defects of system Fe-Al alloys by means of the Mössbauer effect. It pointed out that the isomer shift of the ordered stoichiometric FeAl alloy was 0.28 ± 0.10mm/s. G. K. Wertheim *et al.* [6] showed that all the Mössbauer spectra of the B2 Fe-Al alloys were approximately one-peak singlets, no significant magnetic transition, but there were distorts in the singlet of $Fe_{1.1}Al_{0.9}$ (Fe-45Al). It confirmed that some Fe atoms occupied Al sublattice. These Fe atoms had magnetic moments. But the number of these Fe atoms was so few that the magnetic property was not significant. I. Vincze *et al.* [7] thought that the first and second iron neighbors of the impurity affect the Mö ssbauer spectra. The objective of this study is to determine the sublattice occupances of Mn or Ti addition atoms in the intermetallic compound Fe-40Al by means of the Mössbauer effect.

EXPERIMENTAL

Seven alloys (-at%): (1)Fe-40Al, (2)Fe-40Al-1Mn, (3)Fe-40Al-5Mn, (4)Fe-40Al-10 Mn, (5)Fe-40Al-1Ti, (6)Fe-40Al-5Ti, (7)Fe-40Al-10Ti were prepared with pure iron (99.9%), pure aluminum (99.99%), pure manganese (99.9%) or pure titanium (99.9%). They were arc melted in an argon atmosphere and annealed for 10h at 1000℃ for homogenizing, 5h at 700℃ for B2 ordering, 24h at 400℃ for relieving vacancies. The specimens in size of $\phi 15\times0.3$mm were cut from these alloy buttons, then ground into 50μm-thick wafers following the requirements from Ref. [8]. A Mössbauer spectrometer was utilized with a multichannel analyser and a ^{57}Co in Pd source. All measurements were carried out at room temperature.

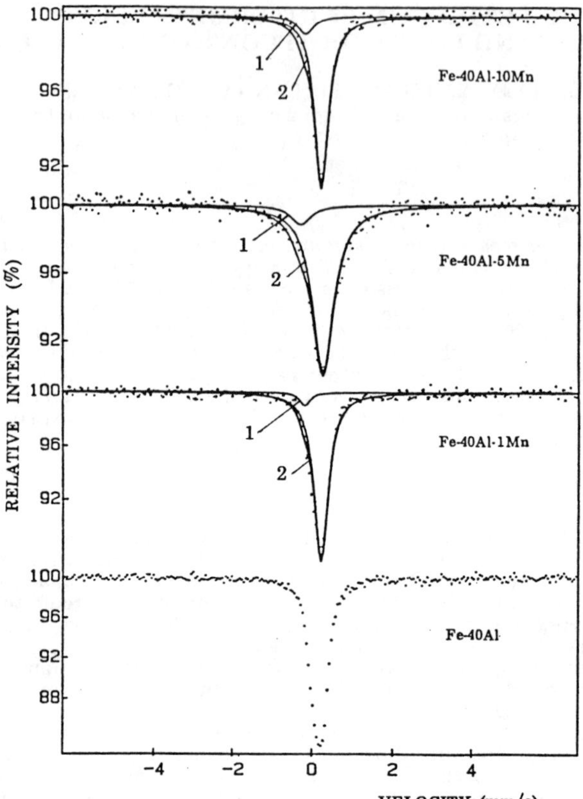

Figure 1. The Mössbauer spectra of Fe-40Al compound and its Mn containing alloys and their fitted lines

RESULTS

Figs.1 and 2 show the Mössbauer spectra of seven alloys. All spectra are approximately one-peak singlets and all their isomer shifts are positive. These Mössbauer spectra have been fitted with two Lorentzian lines by the method of least squares. The fomular is

$$y = A_1/\{1+[(x-P_1)/B_1]\} + A_2/\{1+[(x-P_2)/B_2]\} \quad (1)$$

Where A_1, A_2 are the amplitudes, P_1, P_2 are positions and B_1, B_2 are the breadths of half height of the first and second Lorentzian lines respectively. The fitted lines are shown in Figs.1 and 2 and the parameters of these lines are shown in Tab. I . The isomer shifts of the two Lorentzian lines (line 1, line 2) for each alloy are δ_1 and δ_2 from left to right ($\delta_1 < \delta_2$) and they are shown in Tab. II .

The two isomer shifts show no significant linear relation to the concentrations of ternary additions, but there are some relations between the isomer shifts and concentrations of ternary additions. In Mn containing alloys, the intensities of lines 1 and 2 decrease, δ_1 decreses and δ_2 increases as the Mn concentration increases from 1 to 5-at%. δ_1 increases and δ_2 decreases as the Mn concentration increases from 5 to 10-at%. In Ti containing alloys, the intensity of line 1

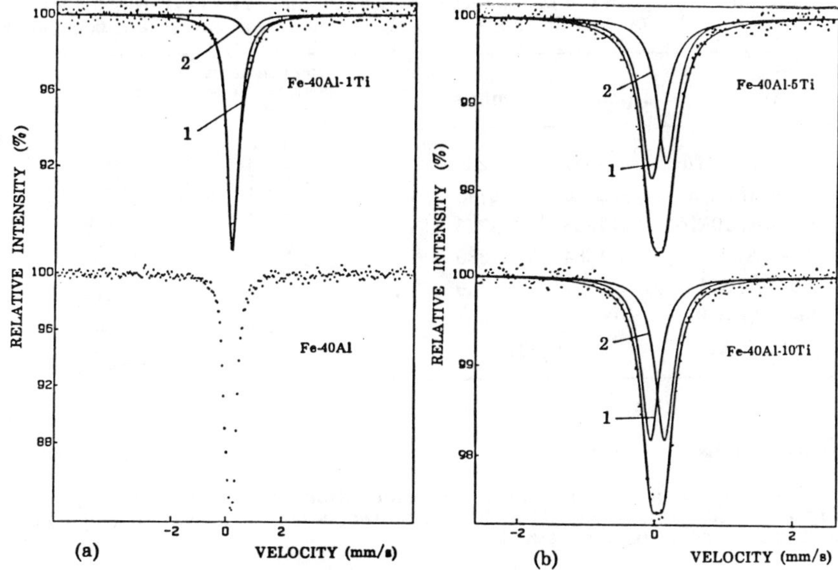

Figure 2. The Mössbauer spectra of Fe-40Al compound and its Ti containing alloys and their fitted lines (a) and (b) were measured using two different devices

decreases significantly and that of line 2 increases significantly as Ti concentration increases from 1 to 5-at%. Both δ_1 and δ_2 decrease obviously. As Ti concentration increases from 5 to 10-at%, the intensity of line 1 has no appreciable change and that of line 2 increases and δ_1 and δ_2 have a little change.

Table Ⅰ. The parameters of the Lorentzian lines fitted with the Mössbauer spectra of the Mn or Ti containing Fe-40Al alloys

Alloy	Fe-40Al-1Mn		Fe-40Al-5Mn		Fe-40Al-10Mn	
Line	1	2	1	2	1	2
A	-2954	-35751	-1351	-11739	-2024	-19530
B	6.28	8.31	11.83	12.44	9.10	8.01
P	121.00	129.00	119.00	130.00	121.00	129.00
I (%)	0.059	0.941	0.099	0.901	0.105	0.895
S (%)	0.0101	0.1222	0.0113	0.0986	0.0089	0.0854
Alloy	Fe-40Al-1Ti		Fe-40Al-5Ti		Fe-40Al-10Ti	
Line	1	2	1	2	1	2
A	-13923	-1215	-79957	-72405	-52084	-52105
B	10.03	12.81	19.51	16.85	15.69	15.68
P	128.50	139.00	124.50	134.50	124.69	134.46
I (%)	0.900	0.100	0.561	0.439	0.500	0.500
S (%)	0.1210	0.0106	0.0185	0.0167	0.0181	0.0181

Table II. The isomer shifts of the Lorentzian lines fitted with the Mössbauer spectra of Fe-40Al and its alloys

Alloy	δ_1(mm/s)	δ_2(mm/s)
Fe-40Al-1Mn	-0.1841	0.2239
Fe-40Al-5Mn	-0.2816	0.2745
Fe-40Al-10Mn	-0.1853	0.2246
Fe-40Al-1Ti	0.1984	0.7339
Fe-40Al-5Ti	-0.0621	0.1449
Fe-40Al-10Ti	-0.0582	0.1441
Fe-40Al	0.2237	

Table III. The probabilities of the most common near and next near neighbor arrangements for the Fe-40Al compound

Iron neighbors (near, next near)	Probability
0 , 6	0.1398
1 , 6	0.2796
2 , 6	0.2447
3 , 6	0.1223
4 , 6	0.0382
5 , 6	0.0076
8 , 0	0.0437
8 , 1	0.0655
8 , 2	0.0410
8 , 3	0.0137
8 , 4	0.0026
	0.999

DISCUSSION

The Mössbauer spectra can give the information on the charge density at the Fe nucleus [9], thus its localized changes on the atomic scale can be detected and analysed by means of this method.

The Fe-40Al Compound

G. K. Werthen and J. H. Wernick [6] gave a method in calculating the probability of the most common near and next near neighbor arrangements for $Fe_{1+x}Al_{1-x}$ compounds. For Fe-40Al, $x=0.2$, the neighbor arrangements are given in Tab.III.

The structure of B2 Fe-40Al compound is conveniently described as cubic Fe lattice with an interpenetrating cubic Al lattice, in which the Fe sublattice is entirely occupied by Fe atoms, but Al sublattice is randomly occupied by Al and Fe atoms in the ratio 0.8 to 0.2. All Fe atoms on the Al sublattice have 8 Fe near neighbors, and dominantly Al next near neighbors. Fe atoms on the Fe sublattice may have 0, 1, 2, ···, 8 Fe near neighbors, and all Fe next near neighbors. For such a small deviation from stoichiometry the number of Fe atom neighbors will be small, i.e. 0, 1, 2 or 3. As a result, such an ordered compound has some Fe atoms with all Fe near neighbors and others with both Al and Fe near neighbors and almost none with equal numbers of Fe and Al near neighbors.

The results show that the breadth of Mössbauer spectrum of Fe-40Al is larger than that of the stoichiometric FeAl compound in Ref. [6], the isomer shift of the Fe-40Al compound is 0.22 mm/s, lower than that of the stoichiometric FeAl compound in Ref. [5], which was 0.28 mm/s. There is a magnetic hyperfine interaction in the Fe-40Al compound, presumably because the 6 next near neighbors of Al atom are occupied by both Al and Fe atoms. The Fe atoms with one or more Fe atoms in the Al near neighbor shell produce the low energy satellites which may actually be the result of small hyperfine splitting. It is apparent that only a small fraction of Fe atoms (those in Al sublattice) has appreciable magnetic moment. The other Fe atoms have no magnetic moment, or too small to produce resolved splitting, so the spectrum of ordered Fe-40Al compound is still approximately a simple one-peak singlet and its isomer shift is lower than that of the stoichiometric FeAl compound.

The Mn Or Ti Containing Fe-40Al Alloys

The Mössbauer spectra of all Mn or Ti containing Fe-40Al alloys are approximately one-peak singlets, not ferromagnetic sextets at room temperature. The result shows that there is no significant magnetic hyperfine interaction and the Fe atoms have no significant magnetic moment. The electrical field of the Fe nucleus is asymmetric, and it is more obvious when alloy contains 5-at% Mn. It is probably related to that Mn is an antiferromagnet and its atoms occupy the near or next near neighbors, resulting in the change of the electrical field and quadrupole splitting. But the intensity of quadrupole splitting is too low to produce multi-peak singlets, and it can only broaden these spectra. This is not included in the spectra fitting.

The isomer shift reveals the effect of near, next near neighbors on the charge density at Fe nucleus. Its formular is

$$\delta = (4\pi/5)\, Z e^2 R^2 (\Delta R/R)\, S(Z)\, [\, |\psi(0)|_s^2 - |\psi(0)|_a^2 \,] \qquad (2)$$

Where δ is the isomer shift of a Mössbauer spectrum; Z is the nuclear charge number; R is the radius of ground state nucleus; ΔR is the difference of nucleus radius between excited and ground states, for ^{57}Fe, $\Delta R < 0$; S(Z) is the relative error for the charge density; $|\psi(0)|_s^2$, $|\psi(0)|_a^2$ are the charge densities of emitting source and absorber respectively.

Both Mn and Ti are 3d transition metal elements. The 3d electrons act to screen the Fe nucleus from the atomic 3s and 4s electrons which contribute to the charge density of s electrons at the nucleus [10]. When Mn or Ti atoms occupy the near neighbors of Fe to substitute Fe atoms, the charge density at the Fe nucleus will increase, so $\Delta|\psi(0)|^2 > 0$, δ decreases. When Al sublattice is occupied by Mn or Ti atom, the donation of their electrons to the conduction band is less than that of Al, the 3d electron charge density falls and the s electrons at the Fe nucleus rise, thus leading to the decrease of the overall isomer shifts.

The asymmetric profile of these Mössbauer spectra is an indication of the distribution in the charge density at the Fe nucleus surrounded with different combinations of neighboring atoms. Moreover, each of these singlets was best fitted with two lines reflecting the existence of two Fe atom sites, each with a different distribution of atomic surrounding and thus different isomer shifts δ_1 and δ_2.

1. The Mn Containing Alloys

The low isomer shift δ_1 corresponds to the charge density of Fe atoms surrounded with heavier Fe clustering while the high δ_2 corresponds to that of Fe atoms surrounded with Fe , Mn and Al atoms [9]. They are exhibited in Fig.1 and Tab. I . The intensity of the lower energy peak increases as the Mn concentration increases, displacing more Fe atoms into the near neighbors of Fe forming an Fe rich configuration. As the Mn concentration increases from 1 to 5-at%, δ_1 decreases but δ_2 increases. This originates from that most of Mn atoms preferentially substitute the Fe atom sites among the near neighbors of Fe atoms and very few of Mn atoms substitute the Al atom sites among the near neighbors of Fe atoms, which increases the charge density at the Fe nucleus and δ_1. In the neighbors of Fe atoms surrounded with Fe, Mn and Al atoms, Mn atoms mainly substitute Al atoms and partially substitute Fe atoms. Hence, the whole charge density decreases at the Fe nucleus and δ_2 increases.

δ_1 increases and δ_2 decreases as the Mn concentration increasess from 5 to 10-at%. This probably results from that, in the neighbors of Fe atoms surrounded with all Fe atoms, Fe atoms are substituted mainly by Al atoms. But in the neighbors of Fe atoms surrounded with Fe, Mn and Al atoms, Fe and Al atoms are substituted by Mn atoms. These result in the increase of charge density at the Fe nucleus surrounded with all Fe atoms and the decrease of charge density at Fe nucleus surrounded with Fe, Mn and Al atoms, thus increases δ_2 and decreases δ_1 respectively. The decrease in the intensity of Mössbauer spectrum corresponds to the decrease in the Fe concentration and the change of intensity of the fitted line corresponds to the Fe atoms surrounded with two different atoms clusters. On the whole, Mn atoms occupy not only Fe but also Al sublattice.

M. A. Kobeissi [9] pointed out that two different surrundings at Fe atoms showed the existence of two different magnetizations. This corresponds to the theory of coexistence of a spin-glass phase and a phase with long-range order [10].

2. The Ti Containing Alloys

Similar to Mn containing alloys, the Mössbauer spectra of Ti containing Fe-40Al alloys were fitted with Lorentzian lines 1 and 2. The low isomer shift δ_1 corresponds to the charge density of Fe atoms surrounded with heavier Fe atom clustering while the high δ_2 corresponds to the charge density of Fe atoms surrounded with Fe, Ti and Al atoms. As Ti concentration is 1-at%, the intensity of line 2 is much lower than that of line 1. As Ti concentration increases, the intensity of line 2 increases. But as Ti concentration is up to 5-at%, both δ_1 and δ_2 decrease. This shows that Ti atoms occupy Fe sublattice. As Ti concentration increases to 10-at%, Ti atoms occupy a few of Al and Fe sites in the neighbors of Fe atoms surrounded with Fe, Ti and Al atoms, this results in the change of δ_2. In the neighbors of Fe atoms surrounded with all Fe atoms, both Ti and Al substitute Fe atoms. As a whole, this decreases the charge density at the Fe nucleus, but slightly increases δ_2. These show that Ti atoms occupy preferentially Fe sublattice. However, a few of Ti atoms occupy Al sublattice only as Ti concentration is over certain amount (~5-at%).

CONCLUSIONS

1. At room temperature, the Mössbauer spectrum of B2 Fe-40Al compound is approximately a one-peak singlet, Fe atoms show no significant magnetic moment in this compound. Its isomer shift is lower than that of the stoichiometric FeAl compound. It originats from that more Fe atoms occupy the near neighbor sites of Fe atoms.

2. The Mössbauer spectra of B2 Fe-40Al alloys with Mn addition are approximately one-peak singlets at room temperature. Mn atoms occupy not only Al but also Fe sublattice.

3. The Mössbauer spectra of B2 Fe-40Al alloys with Ti addition are also approximately one-peak singlets at room temperature. Ti atoms occupy preferentially Fe sublattice, however, a few of Ti atoms occupy Al sublattice only as Ti concentration is over certain amount (~5-at%).

REFERENCES

1. C. T. Liu, C. G. McKamey and E. H. Lee, *Scripta* Metall. Mater. **24**, 385-390 (1990).
2. C. G. McKamey, J. H. Devan, P. F. Tortorelli and V. K. Sikka, J. Mater. Res. **6**, 1779-1805 (1991).
3. P. E. Li, D. Xu and T. L. Lin, Ordnance Mater. Sci. & Eng. (in Chinese) **15**(2), 10-16 (1992).
4. T. E. Granshaw, J. Phys. F: Met. Phys. **18**, 43-48 (1988).
5. S. Nasu, U. Gonser and R. S. Preston, J. Physique **41** C, 385-386 (1980).
6. G. K. Wertheim and J.H.Wernick, *Acta* Met. **15**, 297-302 (1967).
7. I. Vincze and I. A. Cambell, J. Phys. F: Metal. Phys. **3**, 647-663 (1973).
8. Y. F. Xia and Y. Chen, Mössbauer Spectroscopy and Its Applications (in Chinese), (Science Publishers, Beijing, P. R. China, 1987), p. 94.
9. M. A. Kobeissi, J. Phys.: Condens. Matter **3**, 4983-4998 (1991).
10. Marc. Gobay, Phys. Rew. Lett. **47**, 201-204 (1981).

JERKY MOTION OF DISLOCATIONS IN Ni_3Al AND Ni_3Ga. EXPERIMENTAL OBSERVATIONS AND MODELLING.

GUY MOLENAT*, DANIEL CAILLARD*, ALAIN COURET* AND VACLAV PAIDAR**
*CEMES-LOE/CNRS, B.P. 4347, 31055 Toulouse Cedex, France,
**Institute of Physics, Na Slovance 2, 180 40 Praha 8, Czechoslovakia.

ABSTRACT

Recent in situ and post mortem observations in Ni_3Al and Ni_3Ga alloys have yielded consistent results concerning the structure and the movement of screw superdislocations in {111} planes. The different stresses involved in the observed cross-slip processes have been computed, and the outline of a new model of stress anomaly has been formulated on the basis of these results.

INTRODUCTION

Octahedral glide in Ni_3Al has been studied extensively over the last ten years, because it is a model material for studying the dislocation behaviour and the mechanical properties of many intermetallic alloys and advanced alloys with high technological potentials. In particular, it exhibits an anomalous increase of the yield stress with temperature, the origin of which is still being debated.

Several models which have been summarized in comprehensive review articles (see for instance [1]) give possible explanations for the yield stress anomaly. These models are consistent with either some aspects of the mechanical properties or microscopic observations. It was however shown recently [2] that no model can account for all the macroscopic and microscopic characteristics observed.

It is thus the aim of this paper to discuss recent microscopic TEM observations and new observations, with an emphasis on the consistent results obtained using both post mortem observations and in situ deformations. A mechanism is proposed for the glide of screw dislocations, and the stress necessary to activate this mechanism has been calculated. The outline of a new model for the yield stress anomaly is described.

EXPERIMENTAL OBSERVATIONS AND INTERPRETATIONS

Experimental observations have been made in Ni_3Al and Ni_3Ga single crystals. The results obtained can be directly compared because these two materials have very similar mechanical properties and dislocation substructures. Post mortem observations have been made in Ni_3Ga crystals in collaboration with Oxford University, and in situ straining experiments were conducted in Ni_3Al crystal supplied by ONERA.

The results obtained can be divided into two temperature ranges corresponding to the lower and higher parts of the yield stress anomaly domain.

In the low temperature range (300 K-473 K), in situ observations have revealed a high density of rectilinear screw superdislocations dissociated into two superpartials separated by an APB ribbon which lies very often in a plane close to the primary octahedral glide plane. These dislocations are seen to move jerkily over distances which are often equal to the width of the APB ribbon, in such a way that the trailing superpartials take the exact place of the leading superpartials [3], as illustrated in fig. 1.

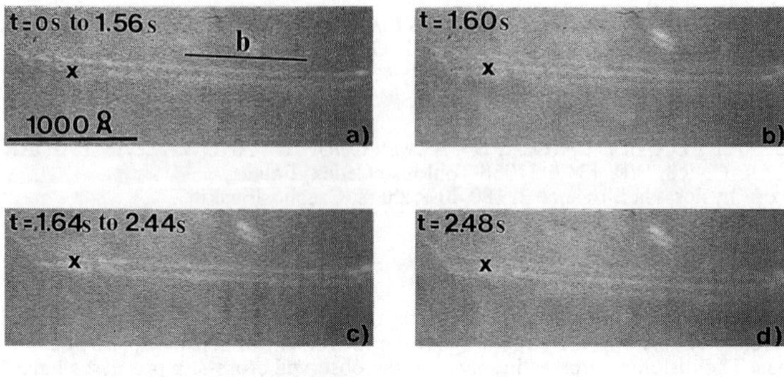

Figure 1 - Screw superdislocation dissociated in a plane close to the (111) glide plane, gliding by jumps over a distance equal to the width of the APB ribbon. In situ experiment at T=300K in Ni_3Al, b is the Burgers vector and X is a fixed point.
a. Locked position before the first jump; b. Jump with the starting and final positions on the same frame because of the remanence of the fluorescent screen; c. Locked position after the jump; d. Second jump with the starting and final positions; e. Locked position after the second jump.

Jerky movements of screws by nucleation and propagation of macrokink pairs are sometimes observed. These observations are however rather scarce because macrokinks are seen to be very mobile along screw dislocations and they disappear very easily at the surfaces.

The high frequency of jumps over a length scaling with the dissociation width of superdislocations is corroborated by a statistical analysis of the macrokink height measured post mortem [4] which exhibits an anomalous high density of macrokinks with height between 5 and 10 nm (fig. 2).

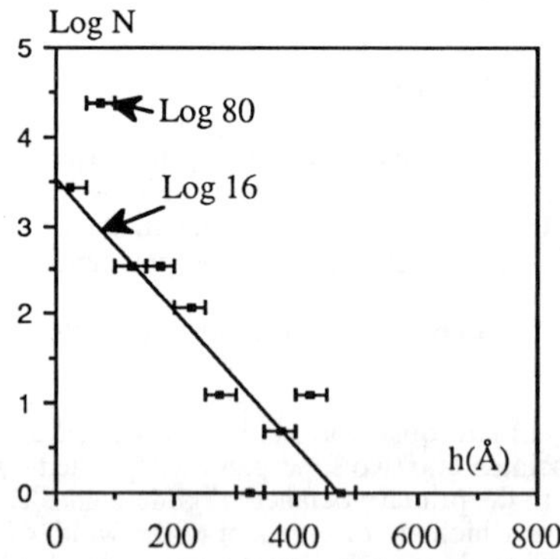

Figure 2 - Macrokink heights in Ni_3Ga deformed at 300K (post mortem observations) plotted on a semi-logarithm scale [4]. N is the number of macrokinks having a height h. Note a deviant point on a normal exponential decrease, for 5nm < h < 10nm. Total number of measurements : 155.

The high density of locked rectilinear screw dislocations with their APB lying in the octahedral glide plane, and the jerky movement of such dislocations over distances which often correspond to their dissociation width, have been interpreted by a short range cross-slip process of the leading superpartial into the cube plane, according to the schematic illustration in fig. 3.

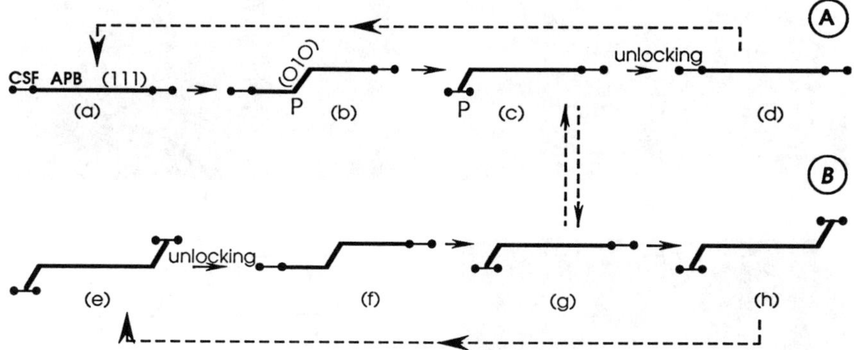

Figure 3 - Locking and unlocking of an extended configuration.
Cycle A : jumps over variable distances (P is a fixed point).
Cycle B : jumps over a length scaling with the dissociation width.
Dislocation movement occurs to the right.

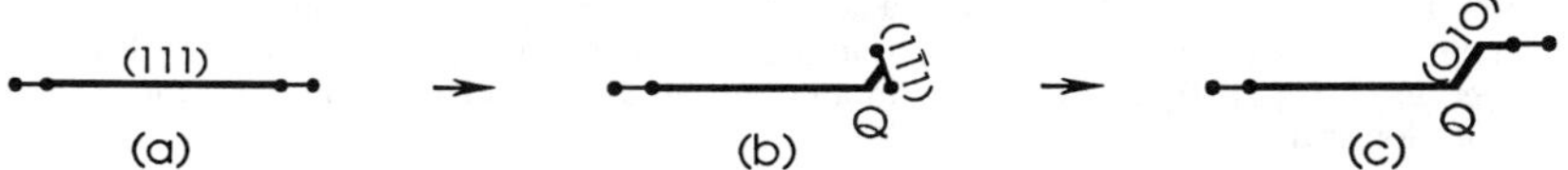

Figure 4 - Locking and unlocking of a contracted configuration.
Q is a fixed point. Dislocation movement occurs to the right.
Configuration c is followed by configurations fig. 3b and c.

Starting from a glissile configuration with both superpartials extended in the octahedral glide plane (fig. 3a), a short range cross-slip process of the leading superpartial (fig. 3b) immediately locks the trailing superpartial (fig. 3c). Here, we assume that the leading superpartial is redissociated in the primary octahedral plane. A redissociation in the cross-slip plane may lead to another situation shown in fig. 4, which however does not explain the observations.

The configuration in fig. 3c is sessile till the trailing superpartial cross-slips in turn, following the same path as the leading superpartial (fig. 3d). The glissile configuration is thus recovered and the dislocation can glide over a distance which depends only on its probability of cross-slipping again (cycle A). If, however, the leading superpartial cross-slips before the trailing superpartial unlocks (fig. 3g, h), a series of jumps over the dissociation width is obtained (Cycle B : fig. 3e-h). Recent post mortem observations [5] indicate that the amount of cross-slip in the cube plane at room temperature is small, of the order of one or a few interatomic distances. The set of experimental results are thus explained. It should be noted that this short-range cross-slip process into the cube plane is different from that proposed earlier by Paidar, Pope and Vitek [6] since it extends over the whole screw dislocations.

At high temperatures (473 K - peak in flow stress), in situ experiments have indicated that dislocations glide over long distances and multiply very rapidly (in less than 1/50 s) in slip bands till they lock themselves along the screw orientation. Such rectilinear screw dislocations are dissociated in a plane close to the cube cross-slip plane, i.e. in the so-called Kear-Wilsdorf configuration. It is however difficult to get detailed information about the initiation of such a glide process, because only very few locked dislocations are seen to unlock themselves.

Recent post mortem observations [5] have provided important additional information on the unlocking process. They have revealed a large density of Kear-Wilsdorf dislocations, many of them being strongly curved in the cube plane. They have also shown the first evidence of the unlocking of Kear-Wilsdorf dislocations by cross-slip onto the primary octahedral plane in a sample which has been cut in a plane normal to the primary octahedral plane. This cross-slip process can be observed in fig. 5.

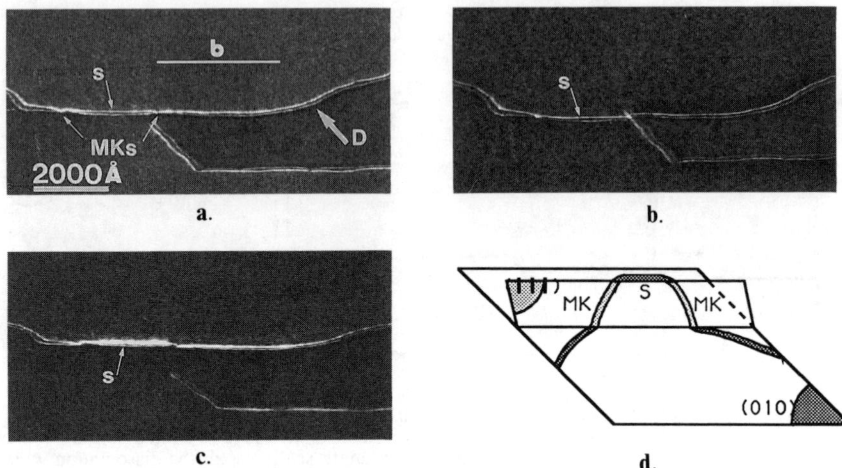

a. b. c. d.

Figure 5 - Cross-slip from the (010) plane onto the (111) primary plane [5]. Post mortem observation after deformation of Ni_3Ga at 673K. Weak beam condition, **g**//**b** = [$\bar{1}$01].
a. (111) plane nearly edge on (electron beam direction **B** // [1$\bar{3}$1]); b. (111) tilted in one direction (**B** // [0$\bar{1}$0]); c. (111) tilted in an opposite direction (**B** // [212]); d. corresponding dislocation structure.

The dislocation loop "D" is curved and dissociated in the cube plane. A screw segment "s", limited by macrokinks "MKs" is lined up with the other screw parts of the loop when the primary octahedral plane is seen edge on (fig. 5a). It is however shifted downwards in fig. 5b and upwards in fig. 5c, as the sample is tilted in opposite directions. The configuration of the dislocation is thus as illustrated schematically in fig. 5d, showing that the screw segment "s" has cross-slipped onto the octahedral plane and has dissociated again in the cube plane. Such a cross-slip process is thought to be at the origin of the unlocking of Kear-Wilsdorf dislocations.

DISCUSSION OF EXPERIMENTAL RESULTS

The good correlation found between results of in situ deformation and post mortem observations significantly enhances their reliability. In addition, post mortem observations of other authors have also shown results which are consistent with those described above. For instance, the observation of rectilinear screw dislocations dissociated in octahedral planes at and below room temperature by Bontemps [7] is consistent with our in situ observations between 300 K and 473 K. Observations of dislocations which are dissociated partly in the octahedral plane and partly in the cube cross-slip plane by Bontemps and Veyssière [8], Korner [9], Lin and Wen [10], in the low temperature range, also indicate that short range cross-slip is possible.

All these results indicate that different amounts of cross-slip onto the cube plane which extends over the whole screw dislocations exist at different temperatures (for a more detailed discussion see [5]). They also show that all the corresponding locked dislocations can unlock themselves by an homogeneous cross-slip process. This has been shown directly in the case of Kear-Wilsdorf dislocations in fig. 5. This conclusion also holds in the case of the short range cross-slips, since the jumps over distances equal to the dissociation width can be explained only by an homogeneous cross-slip mechanism (fig. 3).

This set of results can thus be used as a reliable basis for modeling the behaviour of dislocations during plastic deformation.

CALCULATION OF UNLOCKING STRESSES

The form of the <101> superlattice dislocation core, which is considered to be dissociated into two superpartials connected by a ribbon of APB, is affected essentielly by three different forces. The superpartials are attracted by the APB surface tension, repelled by the elastic interaction force between the dislocations with the same Burgers vector, and both of them are under the action of the Peach-Koehler force due to the applied stress. In general, the overall force acting on a dislocation does not lie in the (111) glide plane and may cause cross-slip of a screw superpartial into the (010) plane. On this plane, the APB energy is thought to have a minimum value, and the tangential interaction force is zero in the anisotropic elasticity. Moreover, with the exception of the [001] loading axis, the cross-slip of screw dislocations onto (010) is also facilitated by the applied stress.

In one of our recent papers [11], the driving force for cross-slip was calculated assuming that a superpartial, composed of two Shockley partials connected by a ribbon of complex stacking fault (CSF), is freely mobile on the same octahedral glide plane as the CSF. On the other hand, the superpartials with the CSF on a different plane from the glide plane are sessile. When the forces acting on a glissile superpartial on the (111) glide plane are balanced, nevertheless, the force component into the cube cross-slip plane may not be zero.

The total cross-slip driving forces were calculated [11] for different loading axis orientations, applied stress levels and cross-slip distances onto the (010) plane. It has been shown that in all the cases the driving force increases with the cross-slip distance. Two types of locked superlattice dislocations can be distinguished, namely, a sessile leading superpartial connected with a glissile trailing superpartial (contracted configuration, in fig. 4b) and a glissile leading superpartial connected with a sessile trailing partial (extended configuration, in fig.3c and 3g). Notice that the separation of superpartials in the locked configuration of fig. 3h is determined by the configuration preceding the immobilization of the leading superpartial (fig. 3g). Generally, the cross-slip driving force acting on the leading superpartial is higher than on the trailing one due to the contribution of the applied stress.

Let us discuss now the unlocking of superpartials as the controlling mechanism of dislocation motion. In the case of a contracted configuration, when the sessile leading superpartial is unlocked by its core transformation, which may take place during a cross-slip event, it will move a distance of the APB width provided the trailing superpartial remains glissile (fig. 4). In the case of the extended configuration (fig. 3), when the trailing superpartial is unlocked and the leading is glissile being dissociated on the (111) plane but different from the APB glide plane, the superlattice dislocation jumps forward again only the APB width (fig. 3e-h). The motion over a larger distance takes place only when the trailing superpartial is unlocked at the moment when the leading one is glissile on the APB glide plane (fig. 3a-d).

In order to compare the microscopic observations of dislocations with their macroscopic behaviour we have calculated in [11] for the case of the extended configuration the overall force acting on the trailing superpartial in the (010) plane, f_{cs}, as a function of the shear stress resolved on the (111) glide plane, τ_p, and as a function of the cross-slip distance, w. Here the inverse problem has been solved by a different presentation, i.e. τ_p has been plotted as a function of w, for different values of f_{cs} (fig. 6b). Since f_{cs} can be identified with the total cross-slip driving force necessary to move the trailing superpartial in the (010) plane, τ_p is the critical shear stress for unlocking the superdislocation. The calculated values of τ_p, normalized by the ratio γ_1/b ($t_p = \tau_p\, b/\gamma_1$) are shown in fig. 6b for five different magnitudes of the critical cross-slip driving force f_{cs}. With the exception of the APB energy on (111), γ_1, which was assumed to be of the same magnitude as the APB energy on (010), γ_0, the same hypotheses were used as in [11] : ratio of the Schmid factors on (010) and (111) $N=\sqrt{3}/2$, magnitude of the Burgers vector b=0.252nm, the elastic constants $c_{11} = 228$, $c_{12} = 150$, $c_{44} = 128$ GPa, giving the anisotropy factor A = 3.3 and the APB energy $\gamma_0 = \gamma_1 = 120$ mJm^{-2}.

In addition to the five curves described above, we have plotted (dashed line in fig. 6b) the critical stress at which the dislocation is likely to move distances larger than the APB width. The critical stress calculated in [11] is defined by the condition that the cross-slip driving force (f'_{cs}), on the glissile leading superpartial of the extended configuration, is equal to that required for unlocking the sessile trailing superpartial (f_{cs}) (fig. 6a). At a lower stress f'_{cs} is larger than f_{cs} and hence it will be more probable that the leading superpartial cross slips before the trailing one starts to move and so the motion to a larger distance could not happen.

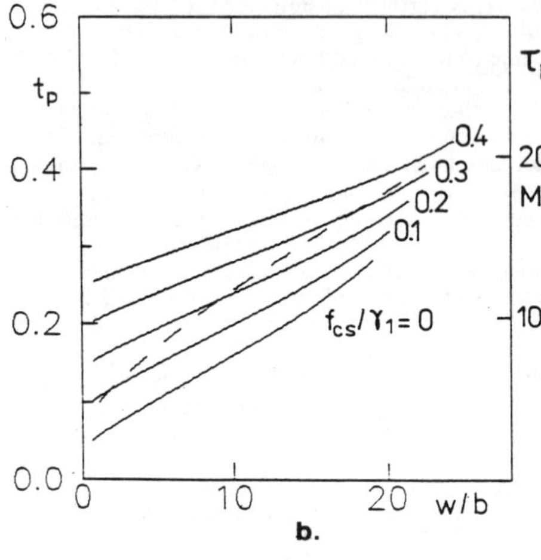

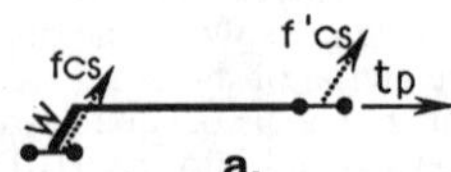

Figure 6 - a. Stress acting on the extended configuration. f_{cs} and f'_{cs} are the cross-slip driving forces acting respectively on the trailing and leading superpartials, w is the amplitude of the deviation in the (010) cross-slip plane, t_p is the normalized shear stress resolved in the (111) glide plane.

b. t_p and τ_p for dislocation unlocking as a function of w, for 5 values of f_{cs}/γ_1. The critical stress under which the dislocations can continue to move distances larger than over the APB width is depicted by the dashed line.

OUTLINE OF A NEW MODEL OF ANOMALY

On the basis of the experimental results, the yield stress at room temperature is proposed to be controlled by the mechanisms described in fig. 3 with a small value of w. On the other hand, it is reasonable to assume that the amount of cross-slip onto the cube plane, w, increases with increasing temperature (this hypothesis is supported by several observations).

Under such conditions, the curve in fig. 6b indicates that, whatever the exact process involved is (jumps over the dissociation width or not), the stress τ_p necessary to keep the dislocations moving has to increase with increasing temperature. Moreover, τ_p has the same order of magnitude as the CRSS for the octahedral glide. In spite of the fact that the anomalous temperature dependence of the flow stress has not been directly addressed in this paper, the obtained results describing the behaviour of dissociated screw dislocations may serve as a basis on which the theoretical models and experimental observations can be put together.

REFERENCES

1. D.P. Pope and S.S. Ezz, Int. Met. Reviews, **29**, 136 (1984).
2. D. Caillard, A. Couret and G. Molénat, Proc. of the Int. Conf. on Fundamental Aspects of Dislocation Interactions, Ascona, Switzerland, 1992, Mat. Sci. and Eng, in the press.
3. G. Molénat and D. Caillard, Phil. Mag., **A64**, 1291 (1991).
4. A. Couret, Y.Q. Sun and P.B. Hirsch, Phil. Mag. A, in the press.
5. G. Molénat, D. Caillard, Y. Sun and A. Couret, Proc. of the Int. Conf. on Fundamental Aspects of Dislocation Interactions, Ascona, Switzerland, 1992, Mat. Sci. and Eng, in the press.
6. V. Paidar, D.P. Pope and V. Vitek, Acta Met., **34**, 435 (1984).
7. C. Bontemps, Ph.D. Thesis, Université de Paris-Sud, France, 1991.
8. C. Bontemps and P. Veyssière, Phil. Mag., **A61**, 259 (1991).
9. A. Korner, Phil. Mag. Letter, **66**, 141 (1992).
10. D. Lin and M. Wen, Proc. of the Int. Conf. on Fundamental Aspects of Dislocation Interactions, Ascona, Switzerland, 1992, Mat. Sci. and Eng, in the press.
11. V. Paidar, G. Molénat and D. Caillard, Phil. Mag. A, in the press.

TEM STUDIES OF THE INFLUENCE OF THE ORIENTATION ON THE DISLOCATION STRUCTURES IN PLASTICALLY DEFORMED Ni_3Al SINGLE CRYSTALS

H. P. Karnthaler, C. Rentenberger and E. Mühlbacher
University of Vienna, Inst. für Festkörperphysik, Boltzmanngasse 5, A-1090 Vienna, Austria.

ABSTRACT

Ni_3Al shows like many other $L1_2$ alloys an anomalous increase of the yield stress up to a peak temperature of about 800°C. This increase is generally explained by a thermal activated cross slip process of superlattice screw dislocations from (111) to (010). To study the influence of the external stress on this process two orientations were selected: compression axis $[\bar{1}23]$ and [001]. They show a large difference of the Schmid factors for cube cross slip.

Single crystals of Ni_3Al were deformed at RT and 400°C, and TEM foils cut parallel to (111) and (010) were studied. At RT which is below the anomalous temperature regime the dislocation structures are in both cases similar to those of fcc structures without an order. At 400°C the TEM results show a distinct difference: dislocations bowing out on (010) are a typical feature for the $[\bar{1}23]$ axis; whereas in the case of the [001] axis dislocation reactions and long straight screws are predominant. They occur in the form of KW locks and contain frequently SISF converted KW locks. Therefore, despite of the different appearance, in both cases strong obstacles are contained in the observed dislocation structures. To overcome these strongly locked segments an Orowan type source model is proposed. When the stress is sufficiently high the dislocation can bow out between obstacles, move on and leave the locked segments behind, which are stored.

INTRODUCTION

Ni_3Al shows like many other $L1_2$ alloys an anomalous increase of the yield stress up to a certain peak temperature which is about 800°C in the case of binary Ni_3Al. In this temperature regime below the peak the present studies were carried out. The thermal activated cross slip of superlattice screw dislocations (Burgers vector $b=a\langle 110\rangle$) from their (111) glide plane onto the cube cross slip plane is a major ingredient of all the theories that try to explain the anomalous yield strength behaviour (e.g. [1], [2], [3]). On a macroscopic scale the experimental results show that from the observation of the slip lines there is no indication of the cube cross slip system being activated. The microscopical results (yielded by transmission electron microscopy, TEM) seem to be conflicting; some authors [4], [5] report the bowing out of dislocations on the cube cross slip planes whereas the results of Dimiduk [6] show the occurrence of straight screws (in a Kear Wilsdorf lock configuration). Since different orientations of the compression axis were used in these studies it is the purpose of this paper to carry out a systematical investigation. Therefore the dislocation structures observed after plastic deformation in crystals with single slip orientation ($[\bar{1}23]$ axis) are compared with those occurring in crystals with multiple slip orientation ([001] axis). The influence of the orientation on the dislocation structures is of further interest since a large difference in creep life time has been reported when a cube orientation of the compression axis is used instead of a single slip orientation [7].

EXPERIMENTAL PROCEDURES

The binary alloy Ni_3Al (nominal composition 77 at% Ni, 23 at% Al) was made from pure components. Single crystals were grown using a modified Bridgman technique. The crystals were carefully oriented by Laue x-ray methods and cut by spark erosion to make the compression test samples (3x3x7mm). For the compression axis two different orientations were selected: (i) $[\bar{1}23]$, as a single slip orientation and

(ii) [001], as a multiple slip orientation with cube axis.

The different orientations lead to different Schmid factors of the relevant octahedral glide systems and their cube cross slip systems which are listed in the table. (N. B.: In the case of the [001] axis the Schmid factors on all the cube systems are zero.)

	$[\bar{1}01](111)$	$[\bar{1}01](1\bar{1}1)$	$[\bar{1}01](010)$	$[\bar{1}10](111)$	$[101](\bar{1}11)$
$[\bar{1}23]$	0.47	0	0.40	0.35	0.35
[001]	0.41	0.41	0	0	0.41

Table: Schmid factors of the two different orientations used in this study

The plastic deformation of the specimens was carried out in a compression test using a strain rate between 1.3×10^{-4} s^{-1} and 2.2×10^{-4} s^{-1} up to a total plastic strain of about 4% (except the RT $[\bar{1}23]$ sample: 12%). The specimens of both orientations were tested at RT and at 400°C, respectively. RT is below and 400°C is in the middle of the temperature regime of the yield stress anomaly. The values of the critical resolved shear stress are 27±5 and 48±5 MPa at RT, 104±10 and 101±10 MPa at 400°C for the $[\bar{1}23]$ and [001] axis, respectively.

For the TEM investigation the $[\bar{1}23]$ samples were sliced parallel to the primary glide plane (111) and its cube cross slip plane (010), respectively. In the case of the [001] samples one of the four equivalent octahedral glide planes and its cube cross slip plane (containing the compression axis) were selected. Thin foils for TEM investigation were prepared by standard jet polishing methods and studied in a Philips CM30 ST operating at 150 and 300kV, respectively. To resolve both the superpartial and the Shockley partial dislocations, dark field images were taken using a weak Bragg reflection in g(4g) or g(5g) condition with a diffraction vector g=<220>.

EXPERIMENTAL RESULTS

Room temperature deformation

Fig. 1 shows the dislocation structures in the specimen with $[\bar{1}23]$ compression axis. In all the samples oriented for single slip dislocations of the primary octahedral slip system $[\bar{1}01](111)$ are predominant; they are arranged in the form of dipoles and bundles of dipoles with a character near edge orientation stabilized by a few secondary dislocations on inclined planes. The dislocation structure is rather inhomogeneous and the internal stresses bending the foil near the bundles indicate that some of the dislocations have no dipole partner. Some dislocations with SISF dissociation are observed, most of them lie along a close packed direction (mainly along **b**). Primary screw dislocations are very rare and KW locks are even less frequent.

Fig. 2 shows the dislocation structures in the specimen with [001] compression axis. Glide occurs on four octahedral planes simultaneously and in each of them two Burgers vectors ($\mathbf{b}_1$,$\mathbf{b}_2$) are activated. Therefore the three-dimensional dislocation arrangement is very different to the one in the specimen of Fig. 1; still the dislocation structure within the individual glide plane is rather similar. Dipoles and bundles of dipoles near edge and 60 degree character (of both **b**) form a very inhomogeneous structure; dislocations on inclined planes and therefore dislocation reactions are of course more frequent. Several dislocations with SISF faults are observed. Screw dislocations are very rare.

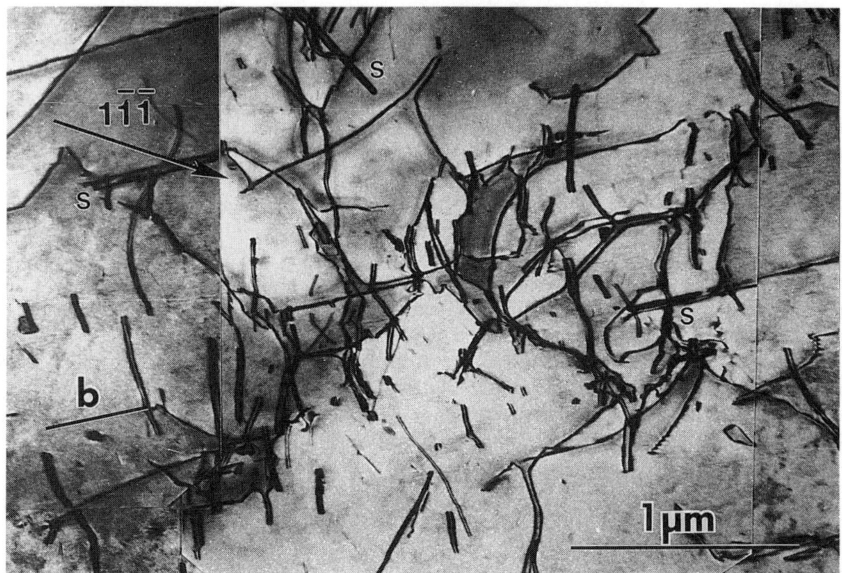

Figure 1: Ni_3Al, [$\bar{1}$23] compression axis, RT, FN=[111]. Inhomogeneous dislocation structure of dipoles and bundles of dipoles of primary **b**, screws are rare, some SISF (at S).

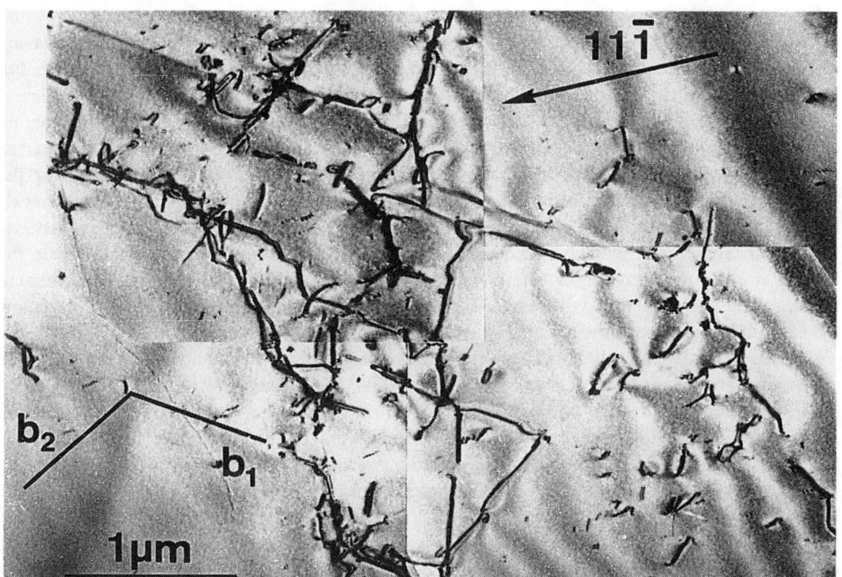

Figure 2: Ni_3Al, [001] compr. axis, RT, FN=[111]. Inhomogeneous dislocation structure, mainly dipoles of $\mathbf{b_1}$ and $\mathbf{b_2}$, dislocation reactions with dislocations on inclined planes.

Deformation at 400°C

Fig. 3 shows dislocations in a specimen with [$\bar{1}$23] axis. The structure is very homogeneous and the primary **b** is predominant; the most striking feature are the segments bowed out on the cube cross slip plane (010). (Some of them are marked with C.) Most of the primary dislocations contain a segment on (010), frequently half loops and sometimes complete loops are observed. The APB fault of the segments C lies on the cube plane. Segments C near screw are prevailing but some segments near edge character are observed. Straight screws (KW locks) are not frequent since most dislocations near screw are bowed out on (010).

Fig. 4 and 5 are taken from a specimen with deformation axis [001]. In Fig. 4 the dislocation structure is homogeneous and straight screws (of both **b**) are predominant. It should be pointed out that the half loops lie on (111) planes and not on cube planes (as in Fig. 3). Dislocation reactions and narrow loops of screws are frequent, some edge dipoles are observed. In Fig. 5 the partials of superdislocations are resolved. KW locks and SISF converted KW locks are a predominant feature whereas only a few dislocations with slight bow outs on their cube cross slip planes have been observed.

DISCUSSION

The room temperature experiments correspond to the temperature regime below the anomaly. The observed dislocation structures are similar to those of pure fcc metals [8]. The formation of a dipole structure indicates that cross slip must be frequent and it is assumed that it occurs via octahedral cross slip planes. (As expected tubes are connected to the edge dipoles.) The dislocation structures for both the single slip and the multiple slip orientation are basically similar (cf. Fig. 1 and 2). The numerous forest interactions in the case of Fig. 2 might account for the higher value of the flow stress.

In the temperature regime of the yield stress anomaly the observed dislocations are completely different from those at RT. For both orientations the structures are rather homogeneous since they repeat themselves within a small scale (about 1μm). This indicates that intrinsic properties of the dislocations (like the frequency of forming locked segments by cube cross slip) are more important than the interaction of dislocation groups and the internal stresses caused by them. The details of the dislocation structures look different for the different orientations of the deformation axis. In the single slip case segments C bowed out on (010) are predominant. Since no slip lines of glide on (010) are observed it is safe to assume that the C segments move short distances only. They lie on a non compact plane where the Peierls stresses are very high. The C segments would need a large thermal activation to multiply and therefore they will remain sessile. Only when the deformation speed is very low the C segments will be activated as it is the case during inverse creep [7].

In the crystals with a [001] compression axis straight screws are dominating the observed dislocation structure. They contain segments of KW locks and KW locks with pulled out SISF. (In a similar form the pulling out of SISF from screws was observed in polycrystalline samples and described as 'flagging'[9].) In addition several dislocation reactions are observed. It is assumed that the segments with SISF and those with reactions will act as strong obstacles and that they cannot be removed by a superkink source mechanism [10] as easily as KW locks.

Therefore, despite of the different appearance, in both cases ([$\bar{1}$23]and [001] axis) strong obstacles are contained in the observed dislocation structures. To overcome these strongly locked segments an Orowan type source model is proposed, which will be published

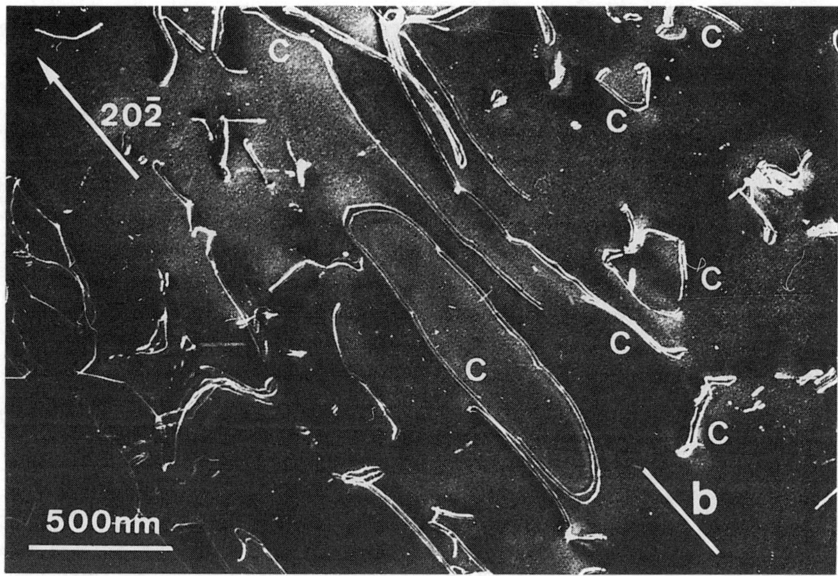

Figure 3: Ni_3Al, [$\bar{1}$23] compr. axis, 400°C, FN=[010]. Homogeneous dislocation structure of primary dislocations **b**, partially bowed out on the cube cross slip plane forming a loop-like structure on (010).

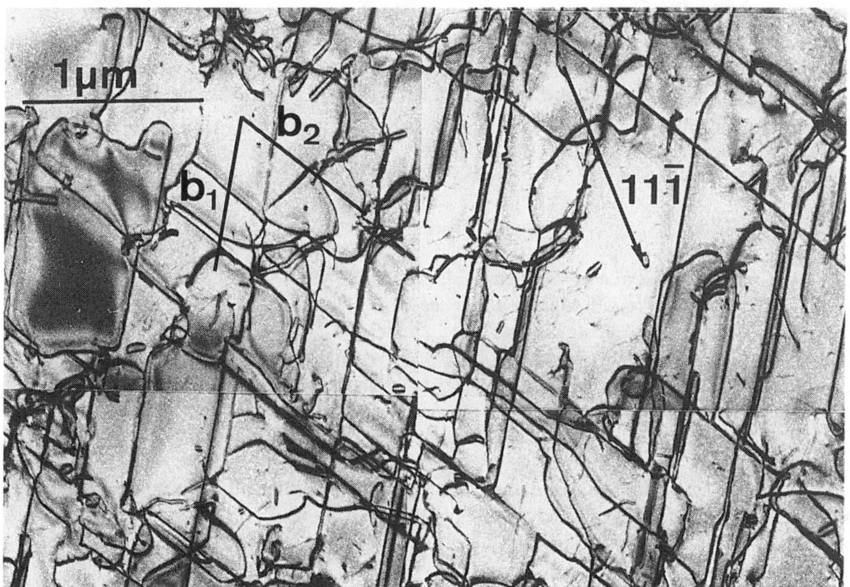

Figure 4: Ni_3Al, [001] compr. axis, 400°C, FN=[111]. Homogeneous dislocation structure, screws and screw dipoles of $\mathbf{b_1}$ and $\mathbf{b_2}$, the edge segments lie on (111).

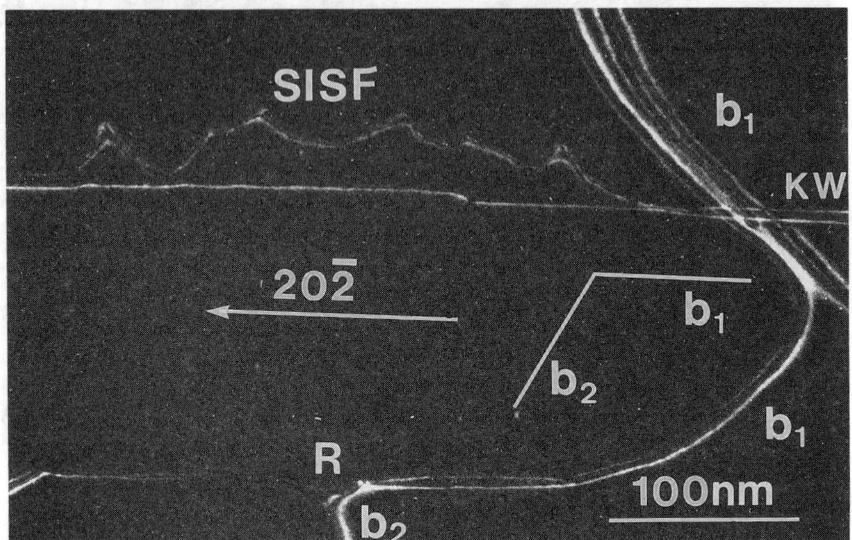

Figure 5: Ni_3Al, [001] compr. axis, 400°C, FN=[111], BD~[121]. Detail of Fig. 4; KW lock, SISF converted KW lock and a dislocation reaction R.

elsewhere. When the stress is sufficiently high the dislocation can bow out between obstacles, move on and leave the locked segments behind, which are stored.

Finally, it seems safe to assume that the strongly locked configurations were not formed as an artefact during unloading and specimen preparation since it needs high stresses to form these segments. Therefore an Orowan type source model is proposed to explain the observed dislocation structures.

Financial assistance from the 'Austrian FWF ' is acknowledged.

REFERENCES

[1] M. Khantha, J. Cserti and V. Vitek, Scripta Met. **27**, 481 (1992).
[2] P. B. Hirsch, Progress in Materials Science **36**, 63 (1992).
[3] M. J. Mills and D. C. Chrzan, Acta metall. mater. **40**, No. 11, 3051 (1992)
[4] P. Veyssiere, in High Temperature Ordered Intermetallic Alloys III, edited by C. C. Koch, C. T. Liu, N. S. Stoloff and A. I. Taub, (Mater. Res. Soc. Proc. **133**, Pittsburgh, PA, 1989) pp. 175-188.
[5] H. P. Karnthaler, M. J. Mills and N. Baluc, Proc. IVth APEM, Bangkok, 1988, Vol. 1, p. 241
[6] D. M. Dimiduk, Ph. D. Dissertation, Carnegie Mellon University (1989)
[7] K. J. Hemker, M. J. Mills and W. D. Nix, Acta metall. mater. **39**, No. 8, 1901 (1991)
[8] V. Gerold and H. P. Karnthaler, Acta metall. **37**, No. 8, 2177 (1989)
[9] P. Veyssiere, J. Douin and P. Beauchamp, Phil. Mag. A, **51**, 469 (1985)
[10] M. J. Mills, N. Baluc and H. P. Karnthaler, in High Temperature Ordered Intermetallic Alloys III, edited by C. C. Koch, C. T. Liu, N. S. Stoloff and A. I. Taub, (Mater. Res. Soc. Proc. **133**, Pittsburgh, PA, 1989) pp. 203-208.

ON THE INTERACTIONS BETWEEN LATTICE DISLOCATIONS AND GRAIN BOUNDARIES IN ORDERED COMPOUNDS

J. Th. M. DE HOSSON*, B.J. PESTMAN*, V. VITEK**
*Department of Applied Physics, University of Groningen, Zernike Complex, Nijenborgh 4, 9747 AG Groningen, The Netherlands.
**Department of Materials Science & Engineering, University of Pennsylvania, Philadelphia, PA 19104, U.S.A.

ABSTRACT

The interaction of 1/2<1 1 0> screw dislocations with symmetric [1 1 0] tilt boundaries was investigated by atomistic simulations using many-body potentials representing ordered compounds. The calculations were performed with and without an applied shear stress. The observations were: absorption into the grain boundary, attraction of a lattice Shockley partial dislocation towards the grain boundary and transmission through the grain boundary under the influence of a shear stress.

INTRODUCTION

Polycrystalline Ni_3Al, with a $L1_2$ structure, is one of the ordered compounds that is a very promising candidate for various applications, such as turbine blades used by the aircraft industry. If useful structural materials based on these intermetallic ordered compounds are ever to be further developed, it is crucial to scrutinize the reasons for their brittle behaviour. A possible reason could be found in low cohesion of the grain boundaries. If the energy of a grain boundary is not much lower than the sum of the energies of the two surfaces that are created by fracturing along the grain boundaries, inter-granular fracture can occur by simple de-cohesion of the grain boundary.

This paper, however, deals with a different approach by considering explicitly the plasticity near grain boundaries. Consequently, the ultimate objective of the work is to establish a link between inter-granular brittleness and the interaction between dislocations and grain boundaries.

Indeed, there exists experimental evidence that in Ni_3Al the dislocation mobility in the vicinity of grain boundaries may be strongly enhanced when ductilization takes place [1][2] and that plastic flow precedes inter-granular fracture [3]. Considering these experiments, it might be reasoned that the passage of gliding dislocations arriving from the lattice might be hindered by grain boundaries.

COMPUTER MODELLING EXPERIMENTS

In the modelling study, Finnis-Sinclair potentials representing Ni_3Al [4] were used for the description of interatomic forces. For the simulations, the following procedure was used. First, the grain boundary was relaxed, using a standard gradient method; details are described elsewhere [5].
Secondly, a computational block for the relaxation of the dislocation near the grain boundary was constructed. The computational block of the relaxed grain boundary was extended, according to the periodicity of the CSL, to form a block of more than 40 b x 40 b (b is the magnitude of the Burgers vector) perpendicular to the dislocation line. Next, the displacement field of a 1/2[1 1 0] dislocation was imposed with its elastic center initially positioned at such a distance from the grain boundary that there was no strong effect of the grain boundary on the relaxation of the dislocation core. Along the dislocation line, periodic boundary conditions were applied. The anisotropic elastic solution (as if there was only one grain present) was used for the boundary conditions perpendicular to the dislocation line. The displacement field of a 1/2[1 1 0] superpartial was imposed with its elastic centre near the boundary plane, connected by a ribbon of Anti Phase Boundary (APB) to another superpartial at elastic equilibrium distance, according to the APB energy. The initial position of the core was always chosen such, that dissociation would occur on the glide plane [6][7]. The dislocation-grain boundary relaxation was carried out in the usual way for dislocation relaxation [8].

After relaxation of the dislocation core, a homogeneous shear strain was imposed on the computational block, corresponding to a shear stress, as prescribed by anisotropic elasticity theory (as if only the grain initially containing the dislocation was present). The shear stress was applied in the direction of the Burgers vector, such that the dislocation would move towards the grain boundary plane. The simulations started with imposing a shear strain corresponding to a small stress. Larger stresses were built up by repeating this process.

RESULTS

Cu_3Au

In the kinematical simulations in perfect lattice in Cu_3Au, the movement of the superpartial screw dislocations started at 200 MPa. It has to be emphasized that in the following figures the symbols indicating the atom positions are drawn as if there is no dislocation present. The results for the kinematical simulations are depicted using the differential displacement method [9]. This method indicates the relative displacement of each atom with respect to its neighbours in a certain crystallographic direction (usually the direction of the Burgers vector). If the absolute value of the relative displacement exceeds half of the periodicity of the lattice in that direction (here, 1/2[1 1 0] was used), an integer number times the period is added or subtracted. The position of the APB is indicated by a line. The relative displacements are indicated by arrows drawn between the atoms.

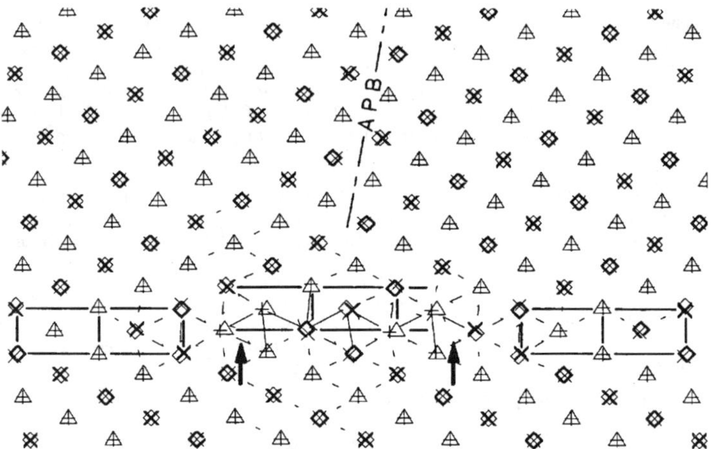

Fig. 1 *Example of the configurations of the Σ=11 boundary in Cu_3Au, showing absorption in the boundary plane. The two thick arrows indicate the position of the grain boundary dislocations. The different symbols indicate different heights. The B-atoms are indicated by thicker lines. Following order of the symbols:* ◇, +, x, Δ.

In the Σ=11 boundary, a similar configuration as in Cu [10] was found, showing absorption in the boundary plane and dissociation into two grain boundary dislocations of the disordered (or fcc) Σ=11 boundary, the 1/22[4 7 $\bar{1}$] and the 1/22[7 4 1]. See fig. 1. As the Burgers vectors of these grain boundary dislocations do not belong to the DSC lattice of the ordered boundary, the part of the boundary between the two grain boundary dislocations again has an ordering configuration which is different from the original boundary. The separation of the grain boundary dislocations was smaller than in Cu.

The Σ=27 boundary showed different types of interactions. A kinematical simulation has been performed with the $(1\,\bar{1}\,1)$ glide plane of the superpartial dislocation ending in the middle of structural unit no. 1 (fig. 2).

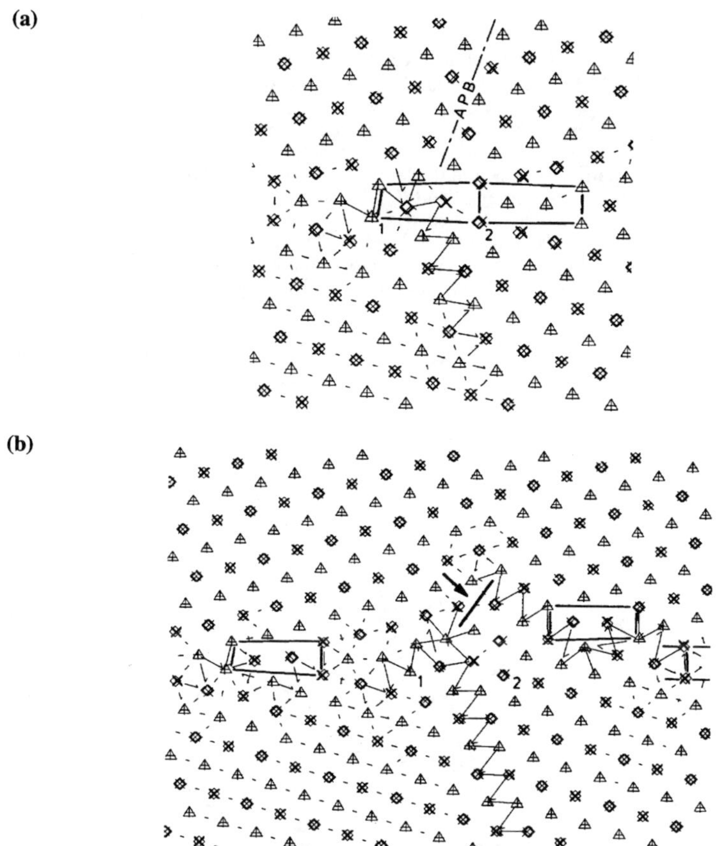

Fig. 2 *Interaction between the screw dislocation and the Σ=27 boundary in Cu_3Au.*
(a) *First stage (around 500 to 1000 MPa). A Shockley partial dislocation has formed in the lower grain, connected to another Shockley partial in the boundary plane. This partial has spread out its displacement field in the boundary plane.*
(b) *Second stage (1250 MPa). The numbers 1 and 2 indicate the same positions as in* (a). *The trailing superpartial dislocation has cross slipped to the right and has been absorbed in the boundary plane, creating a step in the boundary plane. The arrow indicates the APB connecting the two superpartial dislocations.*

At a shear stress around 500 MPa, a new Shockley partial dislocation had formed on the symmetric slip plane in the lower grain connected by a ribbon of a Complex Stacking Fault (CSF) to another Shockley partial dislocation which was still in the grain boundary. On increasing the stress, this trailing Shockley gradually spread out its core in the grain boundary plane (fig. 2 (a)). Upon further increase of the stress level, the Shockley partial dislocation in the lower grain moved further into the lower grain, with the other Shockley still in the grain boundary and at a stress level of 1250 MPa, the trailing superpartial dislocation approached the leading one, cross-slipped to the right of structural unit no. 2, and dissociated into two grain boundary dislocations, in the same way as the case of absorption in the static simulations. See fig. 2 (b).

Ni_3Al

In the kinematical simulations in perfect lattice, the movement of the superpartial screw dislocations started at 350 MPa. The Σ=11 boundary again showed the same configuration as was found in Cu and Cu_3Au (see fig. 1 for the configuration of Cu_3Au). The separation between the grain boundary dislocations was smaller than the separations in Cu and Cu_3Au. In the kinematical simulations, at first, one Shockley partial dislocation was attracted to the boundary while the other Shockley partial remained in the lattice at 3 a_0 from the boundary plane. Thus, a relatively large area of Complex Stacking Fault (CSF) was created. When the stress level was increased further, this configuration remained the same, with only a slight decrease in the separation of the Shockley partials, until a stress level of 950 MPa was reached. At this level, the superpartial was absorbed in the boundary and dissociated into two grain boundary dislocations belonging to the DSC lattice of the disordered boundary, analogous to the mechanism in Cu and Cu_3Au.

For the study of the interaction with the Σ=27 boundary in the kinematical simulations, two different initial configurations were chosen. In the first configuration, the $(1\,\bar{1}\,1)$ slip plane of the superpartial ended in the middle of structural unit no. 1. See fig. 3.

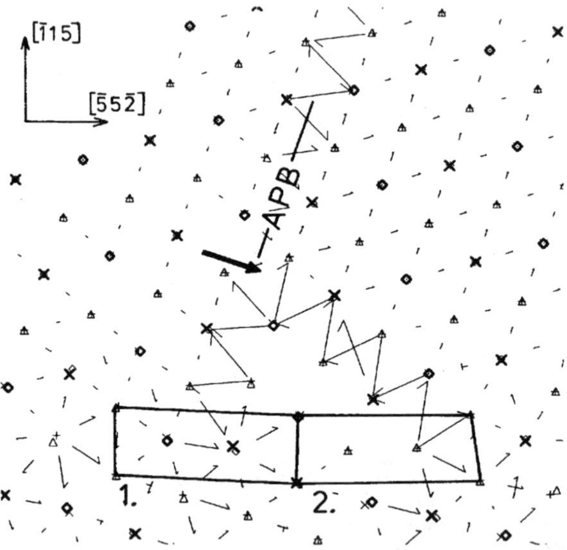

Fig. 3 *Interaction between the screw dislocation and the Σ=27 boundary in the kinematical simulations in Ni_3Al.*
First stage of the interaction (1700 MPa). Both the leading and the trailing superpartial dislocations are visible. The arrow indicates the position of the stair rod dislocation.

When the leading 1/2 [1 1 0] superpartial reached the boundary, it was halted, with one Shockley merged in the boundary plane and one Shockley in the lattice, very close to the boundary plane. When the stress level increased, the Shockley partial in the lattice gradually spread its core onto the $(\bar{1}\,1\,1)$ plane. At a stress level of 1700 MPa, the Shockley partial (originally 1/6[1 2 1]) had dissociated into a 1/6[$\bar{1}$ 1 0] stair rod dislocation (this type of dislocation is observed frequently as the connection between two stacking faults on different {1 1 1} planes), located approximately at the original position of the Shockley, and a new Shockley partial, 1/6[2 1 1], which had merged into the right part of structural unit no. 2. In this way, a new, second region of Complex Stacking Fault (CSF) had formed on the ($\bar{1}$ 1 1) plane, connecting the stair rod dislocation and the newly formed Shockley partial. See fig. 3 (a). At a stress level of 1950 MPa, the trailing 1/2 [1 1 0] superpartial approached the configuration and a reaction between the stair rod dislocation and the leading Shockley of this superpartial took place, in which the 1/6 [1 2 $\bar{1}$] Shockley partial was created. This 1/6 [1 2 $\bar{1}$] Shockley partial cross-slipped away along the ($\bar{1}$ 1 1) plane, thus creating an APB on the ($\bar{1}$ 1 1) plane. Finally, at a stress level of 2100 MPa, a CSF was formed on a $(1\ \bar{1}\ \bar{1})_{II}$ plane (where $_{II}$ indicates the coordinate system of the lower grain) in the other grain.

In the second configuration, the slip plane of the superpartial ended in the left part of structural unit no. 2. The superpartial cross-slipped onto the ($\bar{1}$ 1 1) plane before it reached the boundary and merged in the right part of the structural unit no. 2. At an applied shear stress of 1550 MPa, the superpartial dislocation was transmitted into the other grain, onto a $(1\ \bar{1}\ \bar{1})_{II}$ plane.(See fig. 4).

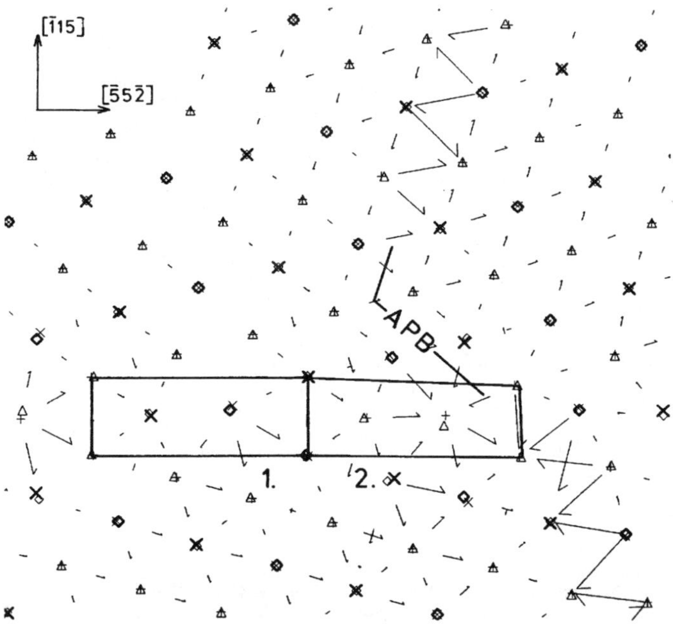

Fig. 4 *Transmission of the leading* 1/2[1 1 0] *superpartial dislocation through the* Σ=27 *boundary in* Ni_3Al. *In the upper grain, the trailing superpartial is visible.*

DISCUSSION AND CONCLUSIONS

The applied shear stresses necessary to start the movement of the screw dislocations in a perfect lattice are 0.004 μ for Cu_3Au and 0.004 μ for Ni_3Al (μ is the shear modulus). Analogous to Cu, these values are high in comparison to experimental values of the friction stress. The friction stress is equal to the yield stress for a single crystal, resolved on a {1 1 1} plane in a <1 1 0> direction. The experimental values at low temperatures are 0.0004 μ [11] for Cu_3Au and 0.0005 μ for Ni_3Al [12].

The absorption in the Σ=11 boundary compares well to the static simulations in Cu_3Au and Ni_3Al and to the static and kinematical simulations for Cu, in which the tendency of absorption and dissociation into DSC dislocations has been observed. Because of the large reduction in elastic energy, the tendency to absorption is presumably so strong, that transmission onto the $(1\,\bar{1}\,\bar{1})_{II}$ plane does not occur, in spite of the small deviation angle with respect to the original slip plane.

The kinematical simulation of the interaction with the Σ=27 boundary in Cu_3Au shows similarities to the interaction in Cu: the leading superpartial dislocation can enter the grain boundary relatively easily. However, transmission does not occur as easily as in Cu, as one Shockley partial dislocation seems to be trapped in the boundary plane. This trapped dislocation blocks the way for the trailing 1/2[1 1 0] superpartial dislocation and this trailing superpartial dislocation cross-slips to the right part of structural unit no. 2 and is absorbed in the boundary plane, similarly to the static simulations.

When we compare the results of the Σ=27 boundary between Cu and Ni_3Al, there are large differences between the interaction of the screw dislocation with the middle part of the structural unit in Cu and the interaction of the superpartial with the middle part of structural unit no. 1 in Ni_3Al. In Cu, transmission is observed at a stress level of less than 3 times the friction stress, while in Ni_3Al the leading superpartial dislocation only enters the boundary at a very high stress level. At 6 times the friction stress a complicated reaction takes place and transmission occurs at a different location in the grain boundary. When we compare the interaction of the screw dislocation gliding towards the left part of the structural unit in Cu with the interaction of the superpartial gliding towards the left part of structural unit no. 2 in Ni_3Al, we observe the same mechanism of cross slip towards the right part of the structural unit.

The results of the kinematical simulations indicate that the resistance of grain boundaries against passage of dislocations increases with increasing ordering tendency. The superlattice dislocation will first reduce the width of the APB in the lattice separating the two superpartials in response to an applied shear stress.

Finally, one should be careful in generalizing our results obtained for all $L1_2$ ordered systems. Indeed, as the mechanical properties of more $L1_2$ materials have been investigated experimentally, it has been found that several of them are intrinsically brittle because of intergranular failure as Ni_3Al. However, it has to be emphasized that in intermetallics, as in Al_3Ti, cleavage fracture is also found to be the failure mode, rather than intergranular failure. In cleavage fracture it is feasible that a mechanism for crack tip plasticity exists that does not require macroscopic dislocation motion at all [13].

References

[1] B. J. Pestman and J.Th.M. De Hosson, Acta Met et Mat, 40, 2511 (1992)
[2] I. Baker, E.M. Schulson and J.A. Horton, *Acta Metall. et Mat. 35*(1987) 1533.
[3] S. Hanada, T. Ogura, S. Watanabe, O. Izumi and T. Masumoto, *Acta Metall. et Mat. 34* (1986) 13.
[4] V. Vitek, G.J. Ackland and J. Cserti, *Alloy Phase Stability and Design, editors G.M. Stocks, D.P. Pope, A.F. Giamei, MRS 186* (1991) 237.
[5] A.P. Sutton and V. Vitek, Philos. *Trans. R. Soc. Ser. A 309* (1983) 37.
[6] D. Farkas and E.J. Savino, *Scripta Metall. et Mat. et Materialia 22* (1988) 557.
[7] M. Yamaguchi, V. Paidar, D.P. Pope and V. Vitek, *Phil. Mag. A 45* (1982) 867.
[8] Z.S. Basinski, M.S. Duesbery and R. Taylor, *Phil. Mag. 21* (1970) 1201.
[9] V. Vitek, R.C. Perrin and D.K.Bowen, *Phil. Mag. 21* (1970) 1049.
[10] B.J. Pestman, J.Th.M. De Hosson, V. Vitek and F.W. Schapink, *Phil. Mag. A, 64* (1991) 951.
[11] D.P. Pope, Philos. Mag., 27, 541 (1973).
[12] E.M. Schulson, T.P. Weihs, D.V. Viens, I. Baker, Acta Metall. et Mat., 33, 1587 (1985)
[13] M.L. Jokl, V. Vitek, C.J. McMahon, P. Burgers, *Acta Metall et Mat. 37* (1989) 87.

Structure and Energetics of Vacancies, Antisites and Divacancy Complexes in the Ni-Al system

Zhao Yang Xie and Diana Farkas
Department of Materials Science and Engineering,
Virginia Polytechnic Institute and State University,
Blacksburg, VA 24061

ABSTRACT

We studied the energetics and structure of simple point defects and defect clusters in the Ni-Al system. Atomistic computer simulation was used for this purpose with embedded atom interatomic potentials. In both, Ni_3Al and NiAl the energetics are such that excess Ni is accommodated by anitisite defects more easily than excess Al. Structural vacancies are therefore more likely with stoichiometry deviations in the Al-rich side than in the Ni-rich side. The interaction of vacancies of different types to form divacancies was studied. Several configurations were found with a very similar energy to that of the two monovacancies far apart. Some configurations show attraction and others show repulsion. The vacancy interaction distance is limited to a few interatomic distances.

The relaxation around these defects was studied in detail and the results can be interpreted in terms of the larger size of the Al atom.

Introduction

There is great interest in understanding the effects of compositional stoichiometry deviations and increasing disorder on the behavior of intermetallic compounds in the Ni-Al system. These effects are very important in mechanical behavior and alloy response to ternary additions. For example, it has been shown that B additions can only ductilize Ni-rich Ni_3Al [1]. There is also great interest in understanding the changes in mechanical behavior of B2 NiAl with deviations from stoichiometry [2]. Point defects and point defect interaction , and in particular structural vacancies are very important in this respect.

The purpose of the present work is to obtain energies and relaxation behavior corresponding to various point defects that may play a role in accommodating stoichiometry deviations and disorder. Computer simulation results are presented for vacancies and antisite defects in Ni_3Al and NiAl. The interaction of vacancies of different types to form divacancies was also studied.

The interatomic potential used in the present calculation was derived by Voter and Chen [3] within the spirit of the embedded atom model [4], and has proved successful in a variety of applications. They reproduce the experimental lattice constant, cohesive energy, and bulk modulus, as well as the elastic constants, the vacancy formation energy, diatomic bond length and bond energy, stacking-fault and antiphase boundary energy in Ni_3Al. The potentials also give reasonable values for the cohesive energy, lattice parameter and elastic constants of NiAl.

The computer simulation program for the calculation is based on the DEVIL code developed by Norgett, Perrin, and Savino [5]. The method does not include the entropy contributions to free energy and therefore gives vacancy structures at zero temperature.

Table I: Energies of vacancies and antisites in Ni_3Al

Defect	Energy (eV) in the Ni sublattice		Energy (eV) in the Al sublattice
Vacancy	1.64		1.87
Antisite	0.46		-0.14
Disorder		0.32	

Results

Vacancies and antisites in Ni_3Al

Two types of vacancies and two types of antisite defects were simulated. The two types of each case correspond to the defect being located in the Al sublattice or in one of the three Ni sublattices.

Table I presents the energetics of the various vacancies and antisite defects in Ni_3Al. The vacancy formation energies were part of the fitting procedure in the development of the interatomic potentials and therefore reproduce experimental values. These values are included here for completeness. The antisite defect energies refer to the location of an atom of the opposite kind in the sublattice under consideration.

The energies were calculated taking into account the different cohesive energies that can be attributed to each of the constituent atoms. The different cohesive energy attributable to Al was calculated simply adding the contributions to the cohesive energies of an Al atom arising from the interaction with all other atoms in the block up to the range of the interatomic potential. A similar process was carried out for Ni. These calculations yielded 4.5 eV/atom for Ni and 4.8 eV/atom for Al. The cohesive energy of Ni_3Al is the sum of 0.25 times the cohesive energy of the Al atoms and 0.75 times the cohesive energy of the Ni atoms, giving 4.59 eV/atom.

In calculating the antisite defect energy for the Al sublattice, an aluminum atom was removed from the simulation block and it was substituted by a Ni atom. The antisite defect energy was obtained as the difference in energy of the new block minus that of the old block (perfect lattice) corrected (subtracted) 0.3 ev, due to the different cohesive energies of the two types of atoms. The energy of antisite defects was calculated to be -0.14 eV for the substitution of Ni for Al and 0.46 eV for the substitution of Al for Ni. The negative value of -0.14 eV indicates that a Ni atom has actually lower energy in the Al sublattice than in the Ni sublattices. The alloy is ordered due to the very high energy involved in locating Al atoms in the Ni sublattices.

The disordering energy can be estimated as the sum of the two antisite energies, which is the energy of interchanging sublattice types with no interaction among the two antisites involved. These results suggest that an ordered Ni_3Al accommodates extra Ni in the Al sublattice more easily than it accommodates extra Al in the Ni sublattice. For the latter case it may be more favorable to accommodate the excess Al in the form of structural Ni vacancies.

The relaxation observed for these defects for the first and second nearest neighbors can be interpreted in terms of the larger size of the Al atom. Table II shows some relaxation data around the point defects. The large outward relaxation for the antisite that locates Al in a Ni site correlates with the high energy of the defect. The relaxation around the antisite that locates Ni in Al sites is inwards. These results could be interpreted as due to

Table II: Observed relaxation for the first two atomic layers

Defect	First layer relaxation (in % of the nearest neighbor distance)	Second layer relaxation
Ni vacancy	-2.0/-0.18	+0.57/+0.24
Al vacancy	-1.7	≈ 0
Al in Ni site	+2.57/+0.57	+0.3
Ni in Al site	-0.95	-0.25

Table III: Energies of vacancies and antisites in NiAl

Defect	Energy (eV) in the Ni sublattice	Energy (eV) in the Al sublattice
Vacancy	1.20	1.03
Antisite	1.98	-0.60
Disorder	1.38	

the larger size of the Al atom.

Vacancies and antisites in NiAl

Similar results were obtained for NiAl. The pattern of energy and relaxation is very similar in both compounds. We note that the nearest neighbor distance is almost the same in both cases, about 0.25 nm.

Table III presents the energetics of various vacancies and antisite defects in NiAl. The energies for the antisite defects were calculated taking into account the different contributions to the cohesive energy of the two types of atoms in NiAl, in the manner explained above. In NiAl one can also see that the disordering energy is basically due to the high energy involved in locating Al in a Ni site. The high energy of the Al in Ni site defect is again correlated with the larger outward displacement around the defect. size of the Al atom. Furthermore, in NiAl this defect also implies the loss of Al-Al bonds that occur in this structure at about the same distance as the first neighbors in pure Al (0.28 nm). Instead, new Al-Al bonds are created with shorter bond lengths, of about 0.25 nm.

The relaxation behavior again follows the same pattern, with a large outward displacement of the first layer around an Al atom located in a Ni site. The displacement is of the opposite sign and not as large for a Ni atom located in an Al site. The vacancies show inward relaxation of the first layer and outward relaxation of the second layer. The magnitudes are larger for an Al vacancy than for a Ni vacancy.

Table IV shows the relaxation data for the first two layers. The pattern of oscillatory relaxation as a function of the distance from the vacancy continues for the subsequent layers and will be reported in detail elsewhere [6].

Table IV: Observed relaxation for the first two atomic layers

Defect	First layer relaxation (in % of the nearest neighbor distance)	Second layer relaxation
Ni vacancy	-3.6	+2.5
Al vacancy	-7.6	+2.24
Al in Ni site	+2.96	-0.5
Ni in Al site	-1.04	+0.84

Figure 1: Formation energy of various divacancy configurations in Ni_3Al

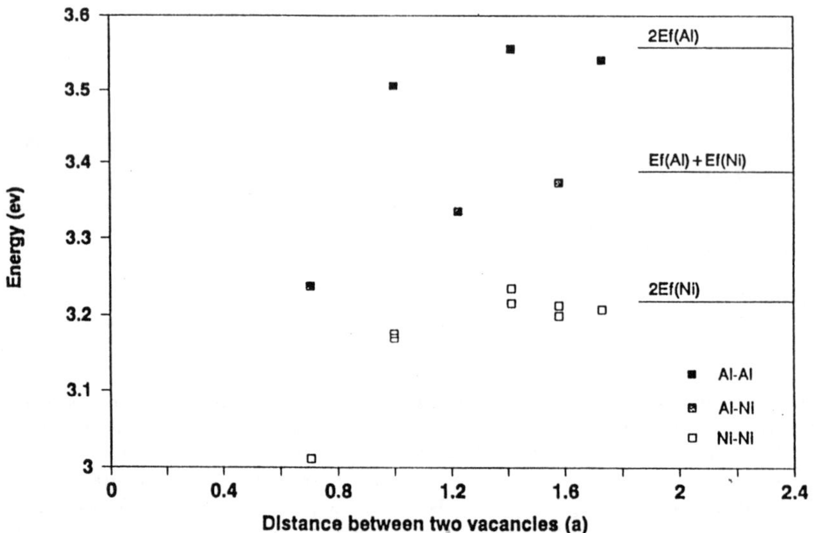

Di-vacancies in Ni_3Al

Several di-vacancy configurations were studied with inter-vacancy distances up to the sixth nearest neighbor. Note that some of the di-vacancies with the same inter-vacancy distance and the same atom types can still have different formation energy, because of the different environment or defect symmetry. As expected, the formation energy of di-vacancies is generally different from the sum of that of the two mono-vacancies. The difference is defined as the interaction energy, i.e. $\epsilon = E_{ab} - E_a - E_b$. If ϵ is less than zero, which means the energy of the pair is less than the sum of the individual vacancy energies, the two vacancies attract each other.

Table V: Observed relaxation for the first two atomic layers

Defect	First layer relaxation (in % of the nearest neighbor distance)	Second layer relaxation
Al(000)-Al(100)	-2.9	-1.6
Ni(000)-Ni($\frac{1}{2}0\frac{1}{2}$)	-5.5/1	+0.5

Figure 1 shows the formation energy of various divacancy configurations in Ni_3Al as a function of the separation distance. The most stable di-vacancy found is that in which two vacancies are at the first nearest-neighbor Ni lattice sites, and the formation energy is 3.01 eV (interaction energy of -0.2 eV). This interaction is larger than that of two vacancies at nearest neighbor sites that are of unlike chemical type.

The interactions at second neighbor positions (always of the same atom type) are already much lower. It is interesting to note that vacancy-vacancy interactions beyond fourth neighbor positions are extremely small. In this material this means interaction of vacancies about 0.5 nm apart.

Except for the case in which the geometric center of the di-vacancy is on a Ni atom (v-Ni-v), all the other two vacancies pairs tend to attract each other. As mentioned above, this interaction is very short range, since the interaction energy vanishes quickly with the increase of the separation distance between two vacancies.

The relaxation around di-vacancies is complex. The maximum inward displacement appears at $(0\bar{\frac{1}{2}}\bar{\frac{1}{2}})$., $(0\frac{1}{2}\frac{1}{2})$ (Al atoms) for Ni(000)-Ni($\frac{1}{2}0\frac{1}{2}$) di-vacancy. This relaxation is 5.54% of the distance from the middle point of the di-vacancy. For the Al(000)-Ni ($\frac{1}{2}0\frac{1}{2}$) di-vacancy the maximum displacement occurs at $(\frac{1}{2}\bar{\frac{1}{2}}0)$, $(\frac{1}{2}\frac{1}{2}0)$, $(0\frac{1}{2}\frac{1}{2})$, $(0\bar{\frac{1}{2}}\frac{1}{2})$ (Ni atoms) in the amount of 3.71%. The maximum outward displacement appears at (001),$(\bar{\frac{1}{2}}0\frac{1}{2})$ (Ni atoms) for the Ni(000)-Ni($\frac{1}{2}0\frac{1}{2}$) di-vacancy in the amount of 0.64%; at $(\frac{1}{2}1\frac{1}{2})$,$(\frac{1}{2}\bar{1}\frac{1}{2})$ (Ni atoms) for Al(000)-Ni ($\frac{1}{2}0\frac{1}{2}$). di-vacancy by the amount of 0.27%. For Al(000)-Al(100) di-vacancy every atom surrounding the vacancy relaxes inward; the maximum inward displacement appears at $(\frac{1}{2}0\frac{1}{2})$,$(\frac{1}{2}\frac{1}{2}0)$,$(\frac{1}{2}\bar{\frac{1}{2}}0)$,$(\frac{1}{2}0\bar{\frac{1}{2}})$ (Ni atoms) by the amount of 2.92%. Table V shows the values of relaxation observed for the first and second neighbor vacancy complexes. The relaxations are calculated for the first and second layer starting at the center of the complex. It is interesting to note that the Al-Al complex presents relaxation inwards for both layers whereas the Ni-Ni complex presents outwards relaxation for the second layer.

Discussion

The energetics and relaxation around simple point defects in the two compounds NiAl and Ni_3Al were calculated using atomistic computer simulation. The results obtained for the relaxation could be rationalized in terms of the effect of the larger size of the Al atom. Al atoms located in Ni sublattices produce large outwards relaxation displacements whereas Ni atoms located in Al sublattices produce inward displacements. The present results are consistent with the relaxation behavior observed for the APB in Ni_3Al. In the APB case, the largest relaxation displacements observed are those of the nearest neighbor Al-Al bonds created in the APB region [7].

The importance of the size effect in this system can be attributed to the fact that the lattice parameters of the compounds do not follow Vegard's law. Instead, the nearest neighbor distances in the two compounds considered here are about the same as those in pure Ni.

It was found that the basic contribution to the disordering energy in these compounds is that of the location of Al atoms in Ni sites. It is predicted by these simulations that excess Ni in these compounds can be accommodated by antisite defects whereas excess Al will probably generate structural Ni vacancies. This prediction is confirmed experimentally in the structure of the compound Ni_2Al_3. This structure can be rationalized as a basic B2 structure with one Ni vacancy every **3** unit cells. The resulting symmetry of the structure is orthorombic [8]. Evidence for structural vacancies in Ni_3Al has been reported from positron annihilation studies [9] and diffusion measurements [10]. Structural vacancies in NiAl have also been reported in the literature [11].

The extent of the relaxation around vacancies is of the order of a few percent and in principle it may have effects on properties, for example through some influence on the dislocation core structures.

The interaction between vacancies was found to be short range, mostly limited to a few interatomic distances. Strong effects only occur at distances smaller or of the order of the lattice parameter.

Acknowledgements

This work was supported by the Office of Naval Research, Division of Materials Science and the National Science Foundation, FAW Program.

REFERENCES

1. C. Liu, Scripta Met. **25**, 1231 (1991), Viewpoint set No 17.

2. I. Baker and P. Munroe, in *High Temperature Alumides and Intermetallics*, edited by S. H. Whang, C. T. Liu, D. P. Pope, and J. O. Stiegler, page 425, TMS, Warrendale, PA 15086, 1990.

3. A. F. Voter and S. P. Chen, MRS Symposia Proceedings **82**, 175 (1987).

4. M. S. Daw and M. Baskes, Physical Review B **29**, 6443 (1984).

5. M. J. Norgett, R. C. Perrin, and E. J. Savino, Journal de Physique **F2**, L73 (1972).

6. Z. Xie and D. Farkas, to be published.

7. D. Farkas et al., Philosophical Magazine A **60**, 433 (1989).

8. D. Farkas and V. Rangarajan, Acta Metallurgica **35**, 353 (1987).

9. A. DasGupta, L. C. Smedskjaer, D. G. Legnini, and R. W. Siegel, Materials Letters **3**, 457 (1985).

10. K. Hoshino, S. J. Rothman, and R. S. Averback, Acta Metallurgica **36**, 1271 (1988).

11. S. M. Kim., Acta Metallurgica **40**, 2793 (1992).

MOLECULAR STATICS SIMULATIONS OF THE MOTION OF A SINGLE KINK IN NiAl

T.A.PARTHASARATHY*, D.M.DIMIDUK# AND G.SAADA@
* UES, Dayton, OH-45432
# Wright Laboratory, WL/MLLM, WPAFB, OH-45433
@ LEM-CNRS, ONERA , France

ABSTRACT

Atomistic simulations of dislocation motion in intermetallics have so far been limited to straight dislocations. In this work, this limitation is relaxed by developing a technique to construct and study dislocation kinks. Using the EAM method, the intermetallic compound, NiAl, is studied using 0K simulations of single kinks on a mixed dislocation in the <001>{110} slip system. The threshold stress for the motion of the kink is calculated as 0.00125-0.00185μ (~160-230 MPa), which is fairly close to the CRSS measured at 77 K. These results, suggest that kink motion may be a contributory factor to the slip response of NiAl at low temperatures.

INTRODUCTION

The intermetallic NiAl is now recognized as a potential structural material for aerospace applications in the intermediate temperature regime, between 600 & 1000°C [1], although very little is understood about its mechanical behavior. The slip response of NiAl is of particular interest due to its unusual sensitivity to dilute ternary solutes [1,2], heat treatment, and prior thermo-mechanical history [1]. Attempts to develop an understanding of the slip response based on continuum elasticity theories have failed (for a review see[3]). Recent work by Parthasarathy et al. [4] suggests that atomistic simulations of dislocation motion may prove to be a useful technique in understanding the slip response of NiAl. Although, as observed in ref.[4], the atomistic simulations are not without limitations, it is expected that these limitations will be relaxed, moving gradually towards more realistic simulations. The present work addresses one of these limitations and how it may be overcome to yield new details about dislocation motion in this compound.

With only a few exceptions, all atomistic simulations of dislocations made to date have been on straight dislocations. However, in reality the motion of dislocations is known to be dominated by the motion of kinks [5]. Further, there are reasons to believe that the high sensitivity of ductility in NiAl single crystals to small amounts of solutes may be related to the interaction of the solutes with dislocation kinks. Thus simulation of the motion of straight dislocations may be an important first step but it cannot be considered complete in capturing all the relevant details. The only work on the simulation of kinks has been that of Duesbery [6,7]. Motivated by the fact that the calculated threshold stress for the motion of the screw dislocation was much larger than the observed flow stress of bcc metals, he studied the behavior of an isolated kink on the screw dislocations of bcc potassium and iron using pair-potentials [6]. He found that the stress to initiate motion of kinks can be much smaller than that required to move straight screw segments. However, this threshold stress for kink motion was too low to explain the experimentally observed flow stress at low temperatures, and thus Duesbery [7] simulated and calculated energies of kink pairs. From these calculations he concluded that glide in these bcc metals was dominated by kink pair nucleation rather than their motion. In the case of intermetallics, in particular those with the B2 structure, there is no experimental or theoretical data on dislocation kinks. In particular, it is not known whether kink motion plays a significant role in comparison to kink pair nucleation. Thus, simulation of a single kink and the threshold stress for its motion is considered the first step in the study of kinks in NiAl and is the focus of this study.

In this work, a procedure to generate a single kink along a dislocation is developed and the threshold stress to move the kink has been studied. The calculations are done at 0K and hence thermal effects are not captured. In ref.[4] it was reported that in the [001](110) slip system (the most favored at low temperatures), the mixed dislocation with the line direction of [$\bar{1}$11] had the highest threshold stress for motion compared with the edge and screw dislocations of the same slip system. The edge and screw dislocations moved at 0.0017 (~220 MPa) and 0.0025μ (~320 MPa) while the mixed dislocation moved at 0.009μ (~1.1 GPa). Experimental works show that the mixed dislocation is the least mobile (is present as long straight segments in the sub-structure) consistent with the calculations, but the experimental flow stress for this slip system is between 100 and 400 MPa, significantly lower than the threshold stress calculated for the mixed dislocation. For this reason, a mixed dislocation of the [001] Burgers vector on the (110) plane was selected for kink studies.

SIMULATION METHODS

Potential and Energy Minimization

The potential used to simulate NiAl was that generated by Rao et al. [8]. This potential is a many body EAM potential fit specifically to the properties of NiAl. The simulation method used for a straight dislocation has been described in ref. [4]. Briefly, the simulation cell is a cylinder of a finite radius, with the dislocation lying along the axis of the cylinder with periodic boundary conditions along this direction. In the plane perpendicular to the dislocation line the boundary atoms are fixed at the positions determined by the anisotropic elasticity solution to the displacement field of the dislocation. The boundary region has a thickness equal to twice the cut-off radius of the potential. The atoms within this boundary are relaxed using an energy minimization code, "MDYN", obtained from the Sandia Labs. The differential displacement method of Vitek [9] is used to study the location and structure of the dislocation core. To study the effect of stress, a homogeneous shear strain is applied (on the slip plane along the direction of the Burgers vector) to the relaxed dislocated crystal, and the boundary regions are fixed while the atoms inside are relaxed. Using the above method as a basis, several different approaches were explored to create and study kinks.

Procedure for Locating the Kink

In order to study the structure and motion of kinks, it was necessary to devise a method to study the dislocation morphology in addition to the core structure. This was done by determining the location of the center of the core for each atomic plane perpendicular to the dislocation line direction. The location of the maximum in the differential displacement was taken to be the center of the core. Thus the dislocation core center was traced along the dislocation line. The presence of a single kink will gradually shift the center by a distance equal to the lattice periodicity.

To construct and simulate a single kink, several approaches were tried. Starting with the approach of Duesbery [6], the effects of boundary conditions were analyzed. Following this the development of a new method, which appears to yield the most reliable results, is introduced.

SIMULATIONS AND RESULTS

In simulating a single kink in bcc iron, Duesbery [6] stuck together slabs of relaxed straight dislocations with the appropriate shift at the location of the kink. In this procedure, some of the atoms near the location of the kink may have displacements far removed from their equilibrium position, but upon relaxation the kink could be stabilized and studied [6]. The atoms bounding the cell in the x-y plane (plane perpendicular to dislocation line) as well as along the z-direction (along the dislocation line) were fixed while the remaining atoms were relaxed. In simulating kink motion, Duesbery found that upon removal of stress, the kink would always return to the center (initial position) indicating that the boundary condition exerted a force on the kink. Duesbery proceeded with a rather lengthy procedure (iteratively moving the boundary interface with the kink) to get around this problem before determining the stress to initiate motion. Although low stresses were found, the main problem with this method is that the boundary conditions corresponding to two dislocation segments with an offset at the center, tend to stabilize the kink at the center and exert a force on the kink during subsequent motion studies.

In the present work, a different approach was taken, in order to avoid the possible boundary effects discussed above. The new approach attempts to create a kink during the

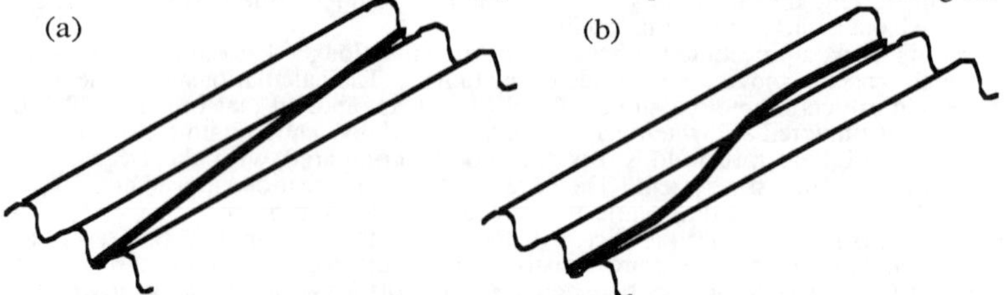

Fig.1: An illustration of the approach used in the present work to create a single kink. An skewed dislocation , (a) , on relaxation is forced down the Peierls hill , (b), to form a kink..

energy minimization procedure. In these simulations, the line direction of the dislocation was initially set along a direction slightly away from the regular lattice direction, as illustrated in Fig.1a. Upon relaxation, the dislocation is forced down the Peierls hill reflected by the lattice periodicity. Due to the skewness of the original dislocation, segments on either side of the center were expected to "fall down" adjacent sides of the Peierls hill resulting in a kink at the center, Fig.1b.

In the first approach, the atoms at the boundaries along all the three directions were frozen at the locations dictated by the anisotropic elasticity calculations corresponding to a dislocation of Burgers vector [001] on the (1$\bar{1}$0) plane, with the dislocation line along a direction that had a slope of 0.05 (~2.86°) with respect to the [111] direction. Thus the length of the dislocation was 20 times the periodicity along [112], which is the direction of glide of the dislocation on the (110) plane. The radius of the cylinder was 3.0 nm with a total of more than 33000 atoms in the cell. The energy minimization procedure resulted in the formation of a kink at the center of the simulation cell, but the fixed boundaries were found to exert a torque (as interpreted from the morphology of the dislocation) at the ends of the dislocation that opposes the formation of the kink.

In the second approach, three dislocation segments were introduced. The middle dislocation was the same as in the first approach, the second and third were short segments lying within the fixed boundary regions. The segments lying within the boundaries were placed along the [111] direction such that they do not create the torque that resulted from the first simulation. This procedure did result in a kink with no apparent end effects from the boundaries along the z-direction. The width of the relaxed kink was nearly one third of the length of the cell. However when a shear stress was applied to the kink, it was found that a force opposing the kink motion existed. The kink would move for stresses of magnitude 0.002μ and above, but when the stress was relaxed, the kink would move back to the original position. This, as pointed out by Duesbery [6], indicates that the kink is a potential well at the center of the cylinder and a restraining force on the kink exists. The fixed boundaries along the dislocation line were thought to be responsible for the presence of the potential well.

In order to avoid the possible effect of the boundaries, in the third approach the boundaries along the dislocation line were made periodic. Periodicity along the dislocation line allows translational freedom which permits the kink to lie within the deepest valley in the Peierls potential. On the other hand, periodicity introduces periodic image kinks that might interact with the kink under consideration, (Fig.2). In addition to having periodic boundaries

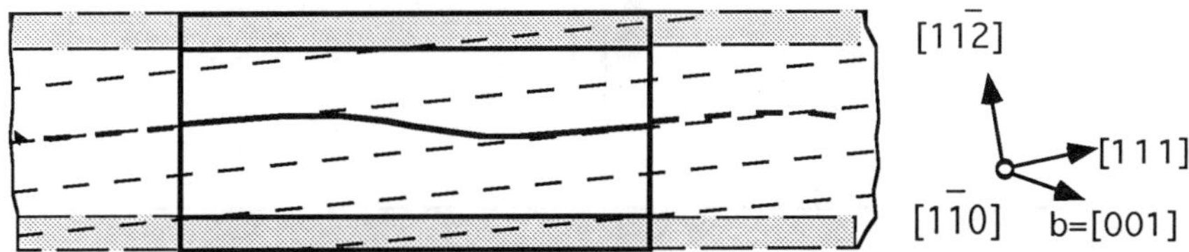

Fig.2: The use of periodic boundary conditions removes the effects of the fixed boundaries (in the z-direction.) on the motion of the kink. The kink moves under stress, and upon relaxing the stress the kink does not return to the original position,.

along z, the radius of the cylinder was increased to 4.0 nm. To obtain periodicity, the line direction of the starting dislocation configuration was taken to be along the [10,10,11] direction with a periodicity of ~5.16 nm. With three periodic units in the simulation cell a step height equal to the periodicity along [112] was obtained. The boundaries in the x-y plane were kept frozen as before at the anisotropic elasticity solution for the straight dislocation. Relaxation resulted in the formation an isolated kink as in the second approach. The location of the center of the core (obtained by the procedure described earlier) as a function of distance along the direction [111] is shown in Fig.3a. Within the kinked region, the center of the core could not be determined with sufficient precision to obtain the true structure of the kink. But the center of the kink and its width could be determined with reasonable accuracy (Fig.3a). Note that the distances along the directions are not on the same scale. The kink is found to have a width about 6 times the height, thus subtending an angle of ~44° with the Burgers vector. There were no kinks with a height less than the full periodic length along [112]. Attempts to simulate such kinks with a height of 1/3[112] did not result in a kinked dislocation. Thus the periodicity of the deep Peierls valley is ~0.7 nm, however local minima at distances of ~(0.7/3) nm are possible, as suggested by the (110) gamma surface of the EAM potential, shown in ref.[4].

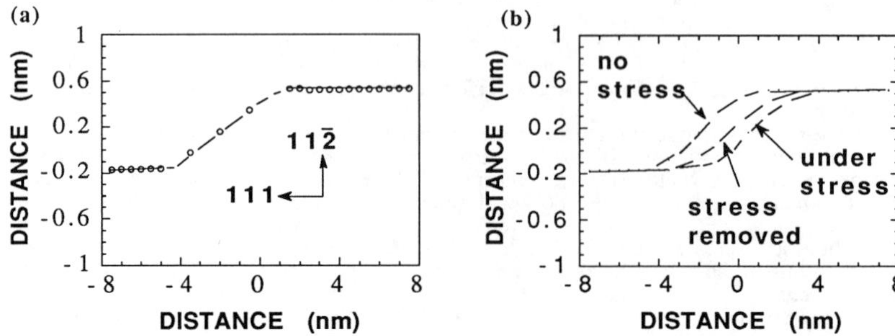

Fig.3 (a) The relaxed morphology of the dislocation containing a kink showing that the kink width is nearly 5 times the kink height. In (b) the effect of stress on the position of the kink and the effect of stress removal on the position of the kink are shown. The kink relaxes but does not return to the initial position, indicating that the boundary effects are insignificant.

Upon application of a stress of 0.0025μ, the kink moved a distance of ~2 nm (Fig.3b). Removal of the stress resulted in the kink moving back about half the distance to the initial position (Fig.3b). The fact that the kink did not return to the original position indicates that the boundary is not playing a significant role in determining the position of the kink. When a

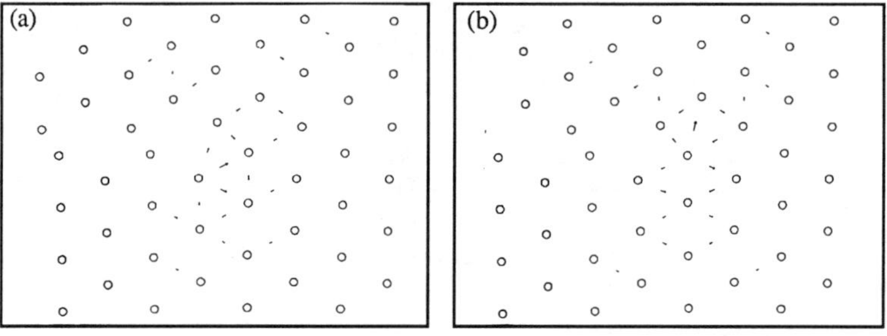

Fig.4: The core structures of the dislocation at the kink (a) and away from the kink (b). The core at the kink appears to be more constricted relative to the core away from the kink, but significant non-planarity is still present.

stress was applied once again, the threshold stress for renewed kink motion was the same as that for the starting kink. The stress required to initiate the motion of the kink was found to be between 0.00125 and 0.00175μ (~160-225 MPa).

Fig.4 compares the core structures of the dislocation at the kink (Fig.4a) and away from the kink (Fig.4b). The cores are depicted using the differential displacement method of Vitek, and both views are perpendicular to the (111) plane. At the kink the dislocation core appears to be somewhat constricted, but significant non-planarity is still present.

A kink of the opposite sign for the same dislocation was found to behave similary to the one discussed above, but its width was larger, ~10 times the kink height (Fig.5). Thus the kink subtends ~60° to the Burgers vector. The kink moved at a stress of 0.00185μ (~240 MPa), but did not move under a stress of 0.00125μ (160 MPa). Thus the threshold stress is nearly the same as that for the other kink. Elasticity calculations [10] reveal that the <001>{110} dislocation has the lowest energy factor near the 45° orientaion, favoring the mixed orientation in preference to the edge or screw. Thus the width of the kink in Fig.3 may have been influenced (so as to avoid elastic instabilities.) at least partly by the elastic energy variation with orientation. Note however that the two kinks are nearly the same in character, explaining the similarity in the threshold stress for their motion.

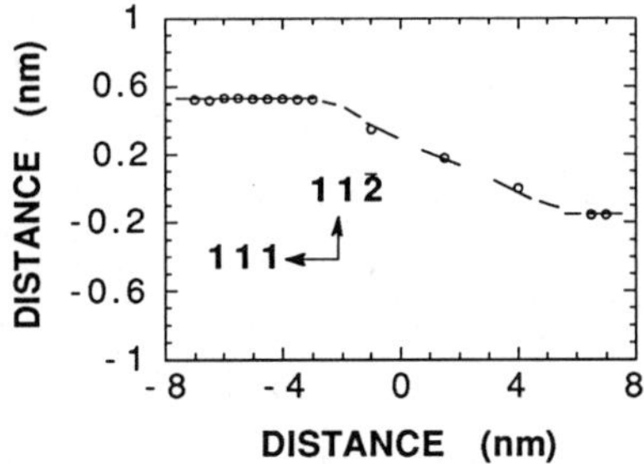

Fig.5: The relaxed kink of the sign opposite to that shown in Fig.3 The width is larger than that in Fig.3a.

DISCUSSION

The simulation results are discussed below under two sections; one deals with the effects of boundary conditions on the simulation results, and the other compares the present results with past theoretical and experimental work.

Boundary Effects:

The results show that the boundary conditions can have significant effects on the structure and motion of kinks in NiAl. The approach used in the present work eliminates the effect of fixed boundaries that tend to stabilize the location of the kink and make it difficult to study the motion of the kink. Although fixed boundaries along the z-direction can be used to isolate the kink from image or periodic kinks, they do not permit translational freedom that would allow the kinked dislocation to lie along the deepest valley of the Peierls potential. With periodic boundaries, it is possible to relax the kinked dislocation to its minimum energy configuration in the Peierls potential. The present study also finds that the periodic boundaries do not play a significant role in determining the location of the kink permitting a calculation of the threshold stress for motion of a periodic array of kinks. The periodic boundaries do introduce the possibility of interaction between kinks that may influence the structure of the kink (but not the motion). However, the interaction force between kinks falls off as ($1/L^2$), where L is the distance between the kinks [5]. Using the expression ($\mu b^2h^2/8\pi L^2$) for the interaction force between kinks [5], the interaction leads to an equivalent stress (=Force/bh) of ~5 MPa, which is much smaller than the threshold stress for motion.

Given the above limitations of the present work, the following section discusses some useful inferences that can be made.

Threshold Stress for Motion

The flow stress-temperature curve observed in bcc metals can be rationalized on the basis of the activation energy for nucleation of kink pairs [11]. This is due to the fact that the activation energy for motion of individual kinks is significantly lower than that for nucleation of kink pairs. In intermetallics with the B2 structure, the relative magnitudes of the activation energies for nucleation and motion are not known. Thus one could start by trying to rationalize the flow behavior using the threshold stress for motion for a single kink.

For the case of NiAl, the [001](110) slip system is the operative system at the lowest test temperatures. TEM studies have shown that the mixed dislocation with line direction along [111] is the least mobile. Elasticity calculations also show that the mixed character is energetically the most stable; and screw orientation is actually unstable [10]. Thus one may postulate that motion of the mixed segment may govern the flow stress of NiAl at low temperatures. Previous calculations of the threshold stress for motion of straight dislocations in NiAl [4] found the mixed dislocation to have the highest threshold stress, but the stresses were too high (~1 GPa) to explain the experimental CRSS of 100-400 MPa [12]. In this work, the threshold stress for kink motion is found to be 160-240 MPa, which is one fifth that needed to move the straight dislocation. This is close to the experimental CRSS at very low temperatures. Further, using a simple estimate of σb^2h (upper bound), for the activation energy for kink motion, one obtains 0.025-0.038 eV, which is ~kT at 300-440 K. This range compares favorably with the temperature (~300-500 K) above which the flow stress becomes athermal in NiAl [12]. Thus, assuming that kinks on the edge or screw dislocations are more mobile, it is quite possible that in the case of NiAl, the motion of individual kinks along the mixed dislocation play a role in determining the flow stress in the "soft" orientation at low temperatures. This implies that the flow of NiAl at lower temperatures may be limited by kink availability, which would explain the limited ductility in NiAl. Much of the ductility (~1%) may be provided by the more mobile edge and screw segments, leaving behind kinked mixed dislocations which move at significantly lower rates determined by the kink properties. In the presence of dilute concentrations of solutes, kink pair nucleation might be enhanced thereby improving mixed-segment mobilities. At higher solute concentrations, solute atoms might hinder kink motion and thus retard mixed segment motion. The observed enhancement in ductility with little change in flow stress [13] is also consistent with this explanation, since the flow stress is controlled by the stress for kink motion. However, these suggestions are preliminary and will have to be re-examined when more experimental and theoretical results are available.

It must be pointed out that the threshold stress for kink motion calculated in this study is significantly higher than that calculated by Duesbery [6] for bcc metals potassium and iron. This difference may be surprising but there is no reason to expect the ordered compound to behave exactly like a bcc metal (e.g., the preferred slip system of [111](110) is not observed in NiAl, except in the "hard" orientation). A high threshold stress for kink motion in NiAl implies a high secondary Peierls potential (periodic along [111]). Such a high secondary potential is expected in strongly bonded ionic or covalent materials [5], but the case of intermetallics is not considered in the literature. Internal friction studies that explore kink nucleation/motion and dislocation velocity measurements as a function of stress and temperature may be necessary to validate the calculations in this work.

SUMMARY

The structure and motion of an individual kink along the mixed dislocation of the [001](110) slip system of NiAl has been simulated using molecular statics (0K) calculations. The approach used in the present work reduces the boundary effects associated with simulations of kinks. The threshold stress for motion of the kink is found to be reasonably close to the critical resolved shear stress measured for this slip system at 77 K, suggesting that the motion of isolated kinks may be considered a contributing factor to the flow stress of NiAl at low temperatures.

ACKNOWLEDGMENTS

It is a pleasure to acknowledge Dr. M.S.Duesbery of Fairfax Materials Research, and Drs. P.M.Hazzledine, S.I.Rao and C.Woodward of UES, Inc. for several useful discussions. We would like to thank Dr.M.S.Daw for sharing the energy minimization code, "MDYN" and Dr.M.H.Yoo for sharing the code for anisotropic elasticity solution of dislocations. This work was sponsored by the U.S.Air Force under the contract #F33615-91-C-5663.

REFERENCES

1) R.Darolia, Jl of Metals, **43**, 44 (1991) - and Private Communications.
2) R.D.Field, D.F.Lahrman and R.Darolia, Acta Metall., **39**, 2960, (1991).
3) M.H.Yoo, T.Takasugi, S.Hanada and O.Izumi, Materials Transactions, JIM, **31**, 435 (1990).
4) T.A.Parthasarathy, S.I.Rao and D.M.Dimiduk, "Molecular Statics Simulations of core structures and motion of dislocations in NiAl", Phil. Mag., in press.
5) J.P.Hirth and J.Lothe, "Theory of Dislocations", John-Wiley & Sons, pp242-265, (1982).
T.Suzuki, S.Takeuchi and H.Yoshinagawa, "Dislocation Dynamics and Plasticity", Springer Series in Materials Science 12, Springer-Verlag, pp63-72 (1989).
6) M.S.Duesbery, Acta Metall, v**31**, No.10, pp1747-1758 (1983)
7) M.S.Duesbery, Acta Metall, v**31**, No.10, pp1759-1770 (1983)
8) S.I.Rao, C.Woodward and T.A.Parthasarathy, Mat. Res. Soc. Symp. Proc., *High Temperature Ordered Intermetallic Alloys IV,* Edited by L.A.Johnson, D.P.Pope and J.O.Stiegler, **213**, 125 (1991).
9) V.Vitek, R.C.Perrin and D.K.Bowen, Phil. Mag., **21**, 1049, (1970).
V.Vitek, Crystal Lattice Defects, **5**, 1, (1974).
10) J.Douin, LEM-CNRS, ONERA, unpublished work.
11) M.S.Duesbery, Z.S.Basinski, "The Flow Stress of Potassium", Acta Metall.-in press.
12) D.B.Miracle, "The Physical and Mechanical Properties of NiAl", Acta Metall., in press.
13) R.Darolia, D.Lahrman, R.Field, Scripta Metall. et Mater., v**26**, pp1007-1012 (1992).

DISSOCIATION PROCESSES IN THE ORTHORHOMBIC O PHASE

JOËL DOUIN*, SHIGEHISA NAKA** AND MARC THOMAS**

* LEM, CNRS/ONERA, 29, avenue de la division Leclerc, BP 72, 92332 Châtillon Cedex, France

** OM, ONERA, 29, avenue de la division Leclerc, BP 72, 92332 Châtillon Cedex, France.

ABSTRACT

In the orthorhombic phase of Ti_2AlNb-type, the occurrence of different dissociation modes of dislocations, glissile in the (001) basal plane is reported. [100] and 1/2[110] dislocations usually dissociate in order to form antiphase boundaries. Detailed weak-beam observations show that [100] dislocations, as well as [010] dislocations formed by interaction of 1/2<110> dislocations, can dissociate into three-fold configurations containing a pair of stacking faults.

INTRODUCTION

Recently, Nb-rich titanium aluminides near the Ti_2AlNb composition have received an increasing attention since they were found to exhibit an extremely high temperature strength associated with a good room temperature ductility [1]. It is presumed that these promising mechanical properties are related to the intermetallic ordered orthorhombic phase designated as O [2], since a high amount of this phase is present in Nb-rich alloys.

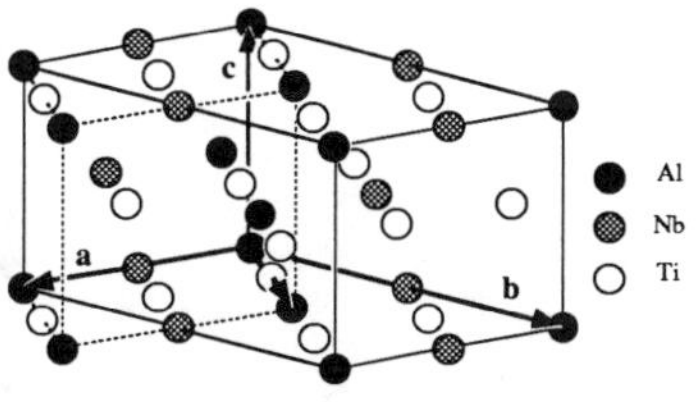

Figure 1 : The unit cell of the O phase. The shortest perfect translations have been outlined (cell parameters : a = 5.9 Å, b = 9.6 Å, c = 4.5 Å [2]).

Since then, a number of people have identified the O phase in a large range of compositions [3-9]. One of the best known of these alloys is the so-called Super α_2 alloy which, depending upon thermomechanical and thermal treatments, may exhibit a microstructure dominated by the O phase. For instance, Super α_2 alloy is transformed into the O phase almost completely after extrusion [10] and the O matrix has the same composition as the bulk material, here Ti-25Al-10Nb-3V-1Mo at.%.

As a preliminary investigation aimed at elucidating the deformation mechanisms of this phase, our observations were focussed on the dissociation mode of dislocations gliding in the basal plane during high temperature extrusion. Since the O phase has an ordered unit cell with large parameters (Fig.1), superdislocations are expected to dissociate in order to reduce their energy.

MATERIAL AND METHODS

Dislocations formed in the O phase during extrusion at 1050°C of a Super α_2 alloy have been studied using weak-beam transmission electron microscopy. After extrusion at a strain rate of 0.37m/s (ratio 16:1), the samples were aged 1 h at 800°C. This treatment promotes large grain formation which helps to identify the dislocation networks which originate from the deformation. The foils have been electropolished using a solution of 5% $HClO_4$, 35% 2-butoxyethanol and 60% methanol.

RESULTS

General deformation microstructure

According to previous investigations [7, 11], the deformation microstructure of the O phase is essentially characterized by the interaction of screw superdislocations with [100] and 1/2[110] (and equivalent 1/2[$\bar{1}$10]) Burgers vectors in the (001) basal plane (Figs. 2, 4-5). In spite of a much larger Burgers vector than the previous ones, [010] dislocations are also observed, resulting from the interaction of 1/2[110] and 1/2[$\bar{1}$10] dislocations (see Fig.5).

All these dislocations are dissociated in the (001) plane where they glide. [100] and 1/2[110] dislocations are commonly split into two identical superpartials separated by an anti-phase boundary (APB) which is out of contrast under fundamental reflections. However, [100] as well as [010] superdislocations can also dissociate into three superpartials separated by stacking faults characterized by a typical fringe contrast.

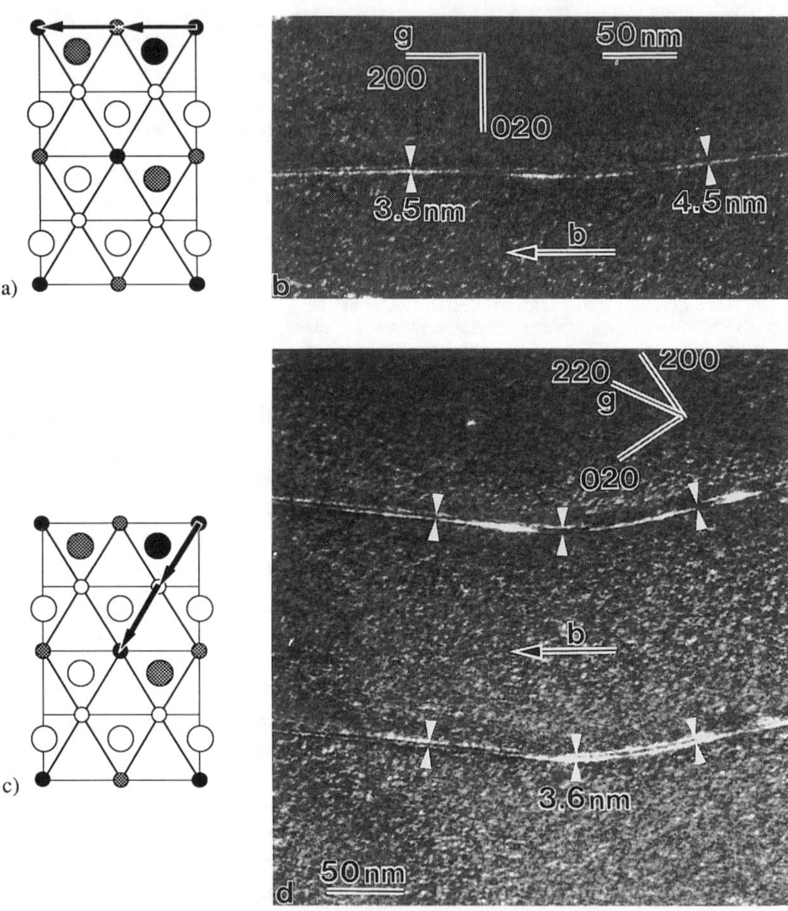

Figure 2 : Dissociations resulting in APB formation in the basal plane.

a : Scheme of the dissociation of a [100] superdislocation leading to APB formation (reaction (1));

b : [100] dislocation dissociated according to reaction (1). Normal to the plane : **N** = [001], diffraction vector **g** = [200]* ;

c : Scheme of the dissociation of a 1/2[110] superdislocation leading to APB formation (reaction (2)) ;

d : 1/2[110] dislocations dissociated according to reaction (2). **N** = [001], **g** = [220]* ; Indicated are the normals to the crystallographic planes and the direction of the Burgers vectors **b** in the O phase.

Dissociation leading to APB formation

[100] and 1/2[110] superdislocations usually dissociate in the (001) plane according to:

$$[100] \rightarrow 1/2[100] + 1/2[100], \quad (1)$$

$$1/2[110] \rightarrow 1/4[110] + 1/4[110]. \quad (2)$$

Figure 2 reveals that in this case the crystallographically unequivalent [100] and 1/2[110] Burgers vectors are dissociated over approximately the same extent, indicating that, since the Burgers vectors have about the same length, 1/2[100] and 1/4[110] fault vectors lead to very close APBs energywise.

Assuming isotropic elasticity (elastic constants are unknown), that the shear modulus is about the same as the Young's modulus $\mu = 0.8\ 10^5$ MPa determined from strain/stress curves [12], and a Poisson's coefficient $\nu = 1/3$, we found that a rough estimate of the energy of the APB on (001), which corresponds to a dissociation distance of 3.5 ± 1 nm in the screw orientation, is 350 ± 100 mJ/m^2.

Dissociation leading to intrinsic/extrinsic faults pair formation

Although APB formation is the most common dissociation mode in this alloy, [100] and [010] dislocations may also dissociate into three fold configurations involving a pair of stacking faults (Figs. 4 and 5). We have considered all the possible dissociation modes involving [100] and [010] and, for the sake of completeness, 1/2<110> Burgers vectors, and we have found using diffraction contrast analysis that the observed configurations, respectively in Figs 4 and 5, originate from the following dissociations (Fig. 3) :

$$[100] \rightarrow 1/12[310] + 1/2[100] + 1/12[3\bar{1}\bar{0}], \quad (3)$$

$$[010] \rightarrow 1/12[310] + 1/2[010] + 1/12[\bar{3}50] \quad (4)$$

As indicated in the captions of Figs. 4 and 5, it is worth emphasizing that contrast analysis of such dissociated dislocations is sometimes difficult to interpret, for example when g.R = ± 1/6 (Fig. 4 a and b and Fig. 5 a and d), and that image simulations are necessary for discriminating between the different dissociation alternatives [14].

The 1/12[310] fault vector yields a change in the stacking sequence which, by analogy with stacking sequence of the {111} planes in the hexagonal lattice, can be regarded as an intrinsic stacking fault of the O superlattice. This fault also induces a change in the nearest atom bonding. Accordingly, $1/12[3\bar{1}0]$ and $1/12[\bar{3}50]$ fault vectors can be associated with extrinsic stacking faults.

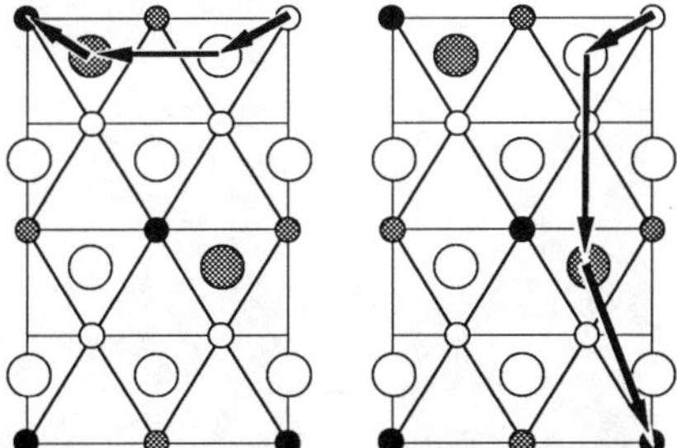

Figure 3 : Observed dissociation schemes of [100] and [010] superdislocations into three partials in the basal plane of the O phase.

Both from the observed dissociation distances and from considerations of Burgers vectors lengths, it is reasonable to believe that the energies of the intrinsic and of the extrinsic stacking faults are approximately the same. Thus, still assuming isotropic elasticity, it is possible to analytically calculate the total energy of the three fold dissociated configuration [13]. It is found ([14]) that at equilibrium the three fold configuration is symmetrical and that the energy of the planar defects γ is related to the total dissociation distance d in reaction (3) by :

$$\gamma = \frac{\mu\, a_{iso}^2}{8\,\pi\, d}\left(1 + \frac{2 - 3\nu}{12\,(1 - \nu)}\right)$$

leading to a stacking fault energy of about 14 ± 5 mJ/m^2 (here a_{iso} = 0.42 nm).

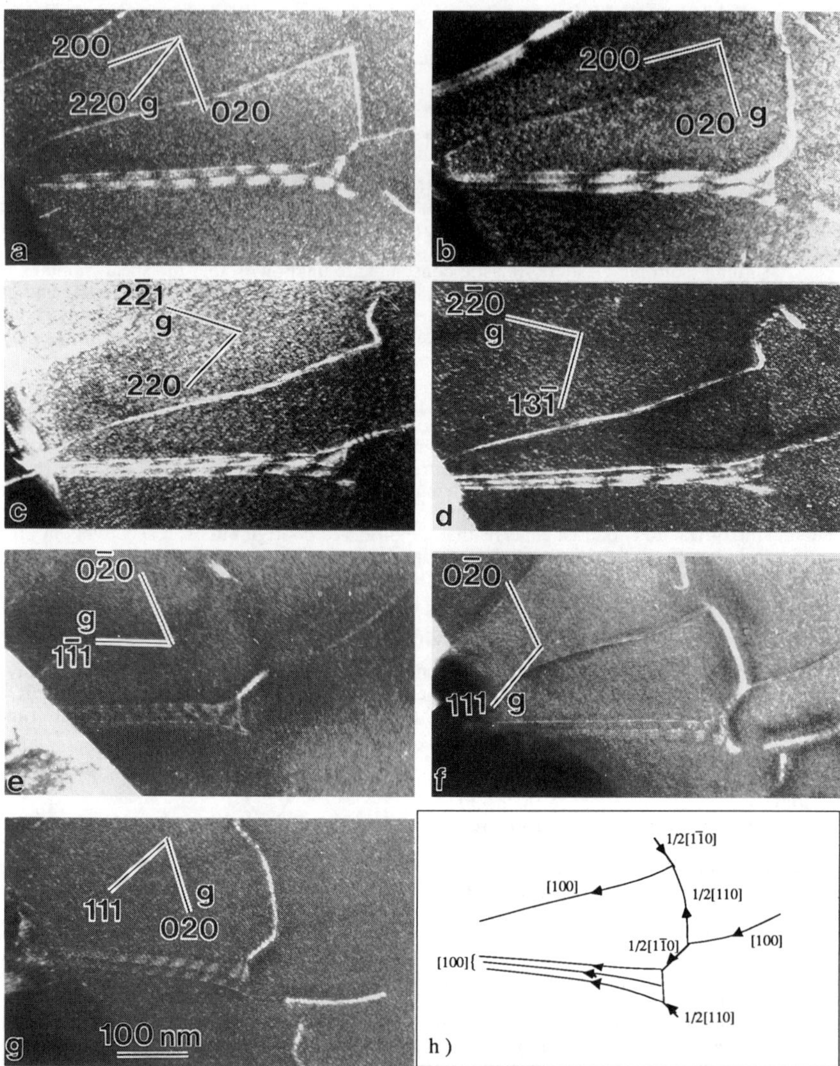

Figure 4 : Dissociation of a [100] dislocation according to reaction (3). a-b : Normal to the plane : **N** = [001] ; c : **N** = [114] ; d : **N** = $[\bar{1}14]$; e-g : **N** = $[\bar{1}01]$; h : scheme of the various Burgers vectors. Indicated are the normals to the cristallographic planes as well as the reflections used for the obtention of the weak-beam image.

Notice that extinction rules are highly dependent on the excitation conditions. In (g), the 1/2[100] partial is out of contrast for the g-4g condition, but visible in (b) for the g-2g condition. In (h), since g = [111]* is a superlattice reflection, a small deviation from the Bragg angle has been used in order to ensure visibility of the dissociated [100] superdislocation. In that case, 1/2[110] superdislocations appear under a strong contrast (g.b = 1), while $1/2[1\bar{1}0]$ dislocations, which should be out of contrast, are still visible with a faint contrast. Also, when $\mathbf{g}.\mathbf{R_1}$ = 1/3 and $\mathbf{g}.\mathbf{R_2}$ = 2/3, the expected shift between fringes is visible (a), while a more complicated case arises when $\mathbf{g}.\mathbf{R_1}$ = 1/6 and $\mathbf{g}.\mathbf{R_2}$ = -1/6 (b).

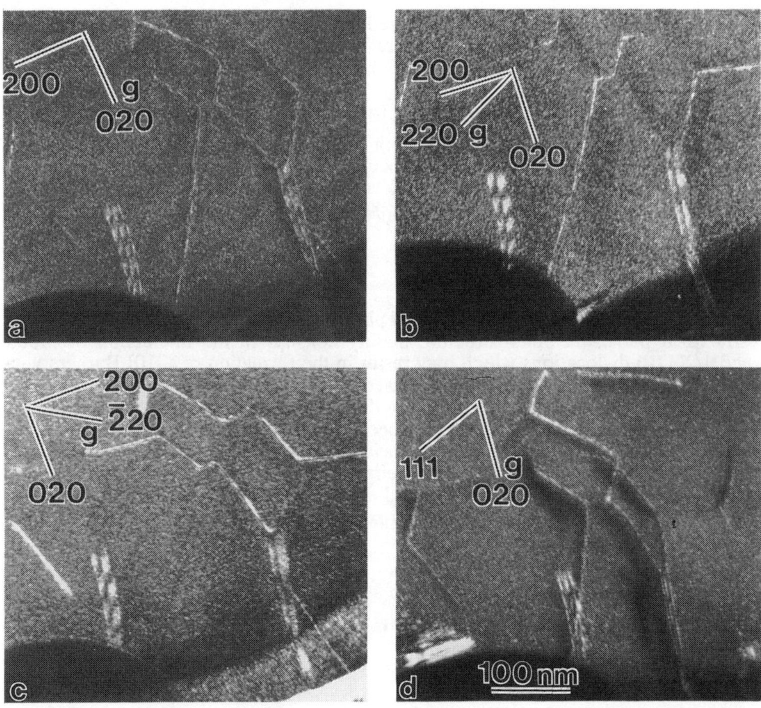

Figure 5 : Dissociation of a [010] dislocation, resulting from the interaction of 1/2<110> dislocations, according to reaction (4). a-c : Normals to the plane : **N** = [001] ; d : **N** = [$\bar{1}$01] ; e : schematic of the various Burgers vectors.

Notice that for **g** = [020]*, fringes corresponding to **g.R** = ± 1/6 are clearly visible under the g-4g conditions (a) and almost out of contrast under g-2.5g conditions (d)

DISCUSSION

Comparison of the total energy for each of the two possible dissociation modes of a [100], a [010] or a 1/2<110] dislocation shows that the formation of APB (two fold configuration) is energetically more favorable than the formation of stacking faults (three fold configuration) [14]. On the other hand, [010] dislocations are not really expected to be often observed as they should decompose according to the reaction:

$$[010] \rightarrow 1/2[110] + 1/2[\overline{11}0] \qquad (6)$$

since, from the b^2 criterion, the self energy simply reduces from b^2 to 1/2 ($a^2 + b^2$).

However, during deformation energy considerations may be outperformed by dynamical effects. Since no occurrence of dissociation of 1/2[110] dislocation into the following reaction :

$$1/2[110] \rightarrow 1/12[310] + 1/4[110] + 1/6[010] \qquad (7)$$

has been observed, 1/2<110] dislocations appear not to be disturbed by grain boundaries that they can cross easily. By contrast, three fold dissociated dislocations are always observed to interact with a grain boundary (see Figs. 4 and 5), leading us to believe that interfaces may assist in the dissociation process of [100] and [010] dislocations, or else act as obstacles against their propagation.

SUMMARY

Dissociation modes of dislocations gliding in the O phase have been analyzed.

- In the (001) basal plane, the microstructure mainly consists in the interaction of [100], 1/2[110] and 1/2[$1\bar{1}0$] dislocations which may result in the formation of [010] Burgers vector, never reported until now in other Ti_2AlNb-based materials.
- [100], 1/2[110] and 1/2[$1\bar{1}0$] dislocations usually dissociate in order to form APBs and the energy of the APBs on the (001) basal plane has been estimated to be 350 ± 100 mJ/m^2.
- [100] and [010] superdislocations can also dissociate into three fold configurations. These dissociation modes, which, to the authors' knowledge, are reported for the first time, are energetically unfavorable, but nevertheless observed due to the interactions of the dislocations with the grain boundaries. The intrinsic and extrinsic stacking fault energy in the O phase has been estimated to be 14 ± 5 mJ/m^2.

ACKNOWLEDGEMENTS

This work was supported by the French Ministry of Defense (contract DRET n° 89.34.001).

REFERENCES

1. R.G. Rowe, in 2nd ASM Paris Conf. on Synthesis, Processing and Modelling of Advanced Materials edited by F.H. Froes and T. Khan (Trans Tech Publications, 1993), **77**, p.61.
2. D. Banerjee, A.K. Gogia, T.K. Nandi and V.A. Joshi, Acta Met., **36**, 871 (1988).
3. M.J. Kaufman, T.F. Broderick, C.H. Ward, J.K. Kim, R.G. Rowe and F.H. Froes, in Proc. 6th World Conference on Titanium edited by P. Lacombe, R.Tricot and G.Béranger (Société Française de Métallurgie, 1988), p. 985.
4. D. Banerjee, T.K. Nandy, A.K. Gogia and K. Muraleedharan, Proc. 6th Int. Conf. on Titanium, eds P. Lacombe, R.Tricot and G.Béranger (Société Française de Métallurgie, 1988), p. 1091.
5. R.G. Rowe, E.L. Hall, G.K. Scarr, E.F. Koch and M.F. Garbauskas, ASM/TMS-AIME Symposium presentation, Indianapolis (1989).
6. L.A. Bendersky, W.J. Boettinger and A. Roytburd, Acta Met., **39**, 1959 (1991).
7. D. Banerjee, R.G. Rowe and E.L. Hall in High-Temperature Ordered Intermetallic Alloys IV edited by L.A. Johnson, D.P. Pope and J.O. Stiegler (Mater. Res. Soc. Proc. **213**, Pittsburgh, PA, 1991) pp. 285-290.
8. J.A. Peters and C.Bassi, Scripta Met., **24**, 915 (1990).
9. L.M. Hsiung and H.N.G. Wadley, Scripta Met., **27**, 605 (1992).
10. M. Thomas, S. Naka, M. Marty, W.G. Smarshy and T. Khan, Proc. 7th World Conference on Titanium, San Diego (1992).
11. D.A. Koss, D. Banerjee, D.A. Lukasak and A.K. Gogia in High Temperature Aluminides & Intermetallics edited by S.H. Whang, C.T. Liu, D.P. Pope and J.O. Stiegler (TMS Publication, 1990), p. 175.
12. R.G. Rowe, D.G. Konitzer, A.P. Woodfield and J.C. Chesnutt in High-Temperature Ordered Intermetallic Alloys IV edited by L.A. Johnson, D.P. Pope and J.O. Stiegler (Mater. Res. Soc. Proc. **213**, Pittsburgh, PA, 1991) pp. 703-708.
13. J. Bonneville and J. Douin, Journal de Physique I, in the press (1992).
14. J. Douin, to be published.

IN SITU OBSERVATIONS OF PRISMATIC GLIDE IN Ti_3Al

MARC LEGROS, ALAIN COURET AND DANIEL CAILLARD
CEMES-LOE/CNRS, 29, rue Jeanne Marvig, B.P. 4347, 31055 Toulouse Cedex, France

ABSTRACT

In situ straining experiments are performed in Ti_3Al, in order to study the glide of 1/3 $<11\bar{2}0>$ superdislocations in prismatic planes. Two different dislocation behaviours are observed, corresponding to two different antiphase boundaries (APBs) in two parallel prismatic planes.

INTRODUCTION

Ti_3Al with the DO_{19} structure deforms by the glide of dislocations on prismatic, basal and pyramidal planes. Rather little is known about the micro-mechanisms which control dislocation glide and thus the mechanical properties. Deformation properties of pure HCP (hexagonal close-packed) metals have been studied extensively, e.g. in Mg, Be and Ti, but the effect of ordering from HCP on mechanical properties is largely unknown and can be revealed by studying Ti_3Al. An in situ study of the plasticity of Ti_3Al polycrystals has thus been conducted, and the first results on the prismatic glide are reported in this article.

The critical resolved shear stress (CRSS) for prismatic glide has been measured in single crystals by Minonishi [1], and Minonishi, Otsuka, and Tanaka [2]. The CRSS is about 70 MPa at room temperature, and decreases with increasing temperature. Stress-strain curves are very similar in form to those of F.C.C metals in stage I deformation.

Several microscopic observations (Minonishi et al [2], Thomas, Vassel and Veyssière [3], Court, Lofvander, Loretto and Fraser [4]) have shown superdislocations dissociated into two superpartials with the same Burgers vector $1/6<11\bar{2}0>$, separated by an antiphase plane boundary (APB) ribbon in the prismatic plane. The dissociation width ranges between 6 nm and 10 nm. The substructure is dominated either by a high density of edge dislocations [1] or by a high density of rectilinear screw dipoles [2,4].

According to Umakoshi and Yamaguchi [5], on the prismatic plane, two distinct APBs (I and II) can be defined according to the position of the corresponding cutting plane. Their energies are expected to be different, and estimates for Mg_3Cd give $\gamma_{II}/\gamma_I = 5$. More detailed calculations have been performed for Ti_3Al by Cserti, Khantha, Vitek and Pope [6] giving $\gamma_{II}/\gamma_I = 9$ with $\gamma_{II} = 101\ mJ^{-2}$ and $\gamma_I = 11.2\ mJ^{-2}$. These authors have also simulated the core configuration of screw superpartials bounding type I APBs. A core extended in the prismatic plane with a secondary spreading in the basal plane has been obtained, leading to a high Peierls frictional stress. It is similar to that obtained by Legrand [7] and Vitek and Igarashi [8] for pure titanium.

High frictional forces along the screw orientation may also arise from covalent bonding, according to Court et als [4]. Both types of frictional forces are consistant with the observations of numerous rectilinear screw dislocations. They however cannot explain the absence of screw dislocations which has been reported in other cases. The exact mechanism controlling the glide of dislocations in prismatic planes thus remains to be determined.

EXPERIMENTAL TECHNIQUE

In situ deformation experiments were conducted in a JEOL 200CX transmission electron microscope operating at 200kV, using a JEOL room temperature straining device and a high temperature straining device described in [9]. Stoichiometric Ti_3Al polycrystals with large grains were supplied by ONERA. They were thinned down by electrochemical polishing, using the A_3 Struers electrolyte.

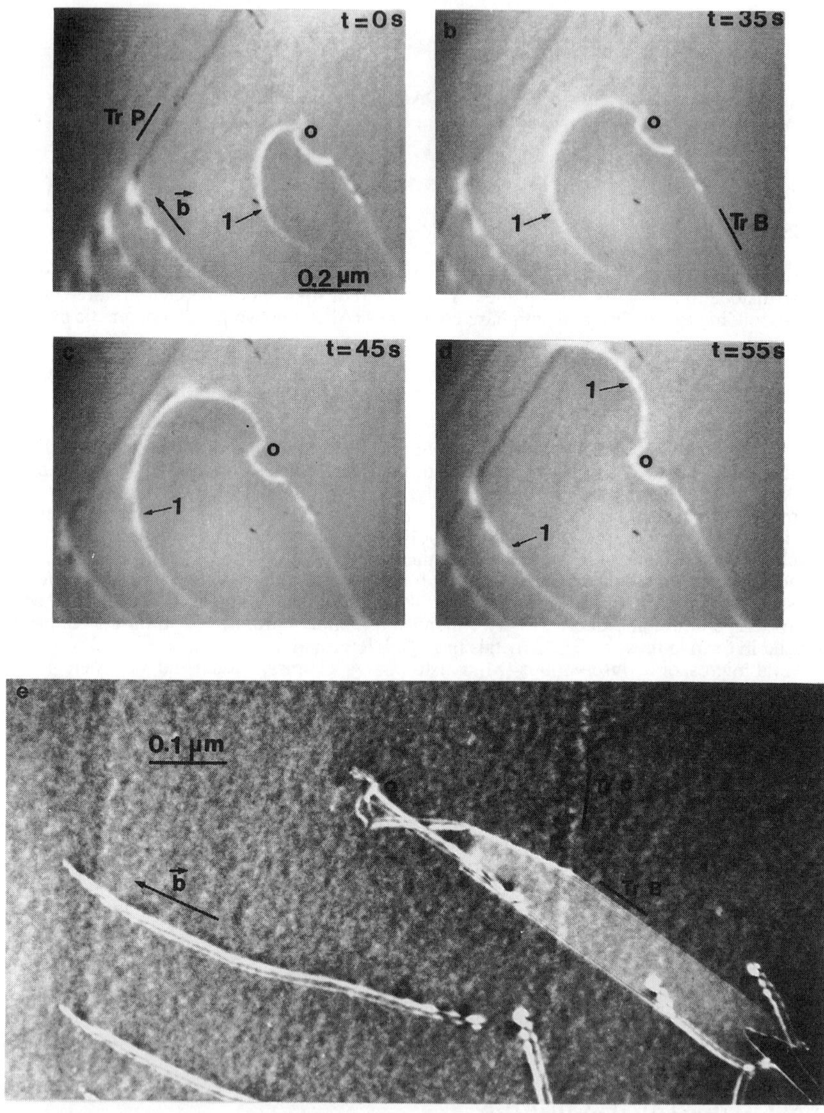

Fig. 1 : Steady movement of curved superdislocations with a small dissociation width. T=300 K
Burgers vector : b = 1/3[$11\bar{2}0$] ; glide plane : ($\bar{1}100$) ; g = [$20\bar{2}\bar{1}$]
a - d : Single-ended dislocation source;
e : Dislocation emitted (Weak-beam).

The images were recorded on videotapes, and further analysed frame by frame.
Local values of the local resolved shear stress have been estimated from the radii R of curved dislocations using the classical formulae : $\sigma = \tau/bR$
The line tension τ was approximated by :

$$\tau = \frac{\mu b^2}{2\pi} \mathrm{Log} \frac{R^2}{bd}$$ where d is the dissociation width.

In the present case, since the exact value of the shear modulus in the prismatic plane is not known, the shear modulus for the polycrystal, μ_{300K} = 51,9 GPa, has been used. Under such conditions, only orders of magnitude of σ can be obtained.

RESULTS

The results of in situ deformation experiments at room temperature are described first.
The most remarkable feature of prismatic slip is that two clearly different behaviours are observed in the glide of superdislocations with 1/3 <1120> Burgers vectors.

i) The first behaviour is the movement of a high density of dislocation loops with a rounded shape. Under weak-beam conditions, they appear to be dissociated into two superpartials, with a separation of about 9 nm along the screw orientation (Fig. 1e). These dislocations move over large distances and account for a large fraction of the total strain. They multiply easily from sources with one anchoring point, as shown in fig. 1. The dislocation segments that are emitted glide very rapidly to the top and pile up to the bottom. The local stress can be estimated from the radius of curvature of the source (R = 0,240 μm) and has been found around 100 MPa.

ii) The second dislocation behaviour that has been observed in the same sample is shown in fig. 2. Dislocation loops exhibit very rectilinear and widely split screw segments, with a separation between 20 and 80 nm. Both superpartials appear to be equally rectilinear under weak-beam conditions. Rectilinear screw dislocations exhibit very jerky movements consisting of a series of very fast jumps over short distances (~ a few 100 nm) between waiting times (Fig. 2). The shear stress necessary to move these dislocations at room temperature has been estimated to be of the order of 140-190 MPa from the radii of curved non screw segments ranging between 0,10 and 0,15 μm. It is worth noting that the large dissociation width of rectilinear screws remains unchanged upon unloading inside the microscope.
The two distinct behaviours coexist in the same dislocation groups and even in some cases along the same dislocation loops (Fig. 2). Screw dislocations are however seldomly observed to change their behaviour in both circumstances.
When the temperature is increased to 573 K and 773 K, the results are qualitatively the same. There is no clear variation in the respective densities of both types of dislocations. Fig. 3 shows widely dissociated rectilinear screw dislocations at 573 K. A jump has been observed on the leading superpartial only. It is worth noting that unlocking is initiated in this case along a rectilinear screw segment which is not intersecting the surfaces, indicating that unlocking is not initiated at the surfaces. Fig. 4 shows another locked screw dislocation widely dissociated at 773 K.

DISCUSSION

It is clear from the present observations and from the stress measurements that only rectilinear and widely dissociated screw dislocations experience substantial frictional forces.
The dissociation width measured on curved screw segments (9 nm) is consistent with that measured post-mortem. However, the larger dissociation width measured on rectilinear screw dislocations has never been reported to the authors knowledge. Such a wide dissociation might be an effect of the applied stress, for instance when only the trailing superpartial is subjected to a high frictional force. This possibility can however be ruled out because several observations have indicated that both superpartials experience a high frictional force (both superpartials are very rectilinear, and uncorrelated movements have been observed) and because the dissociation width remains unchanged upon unloading.

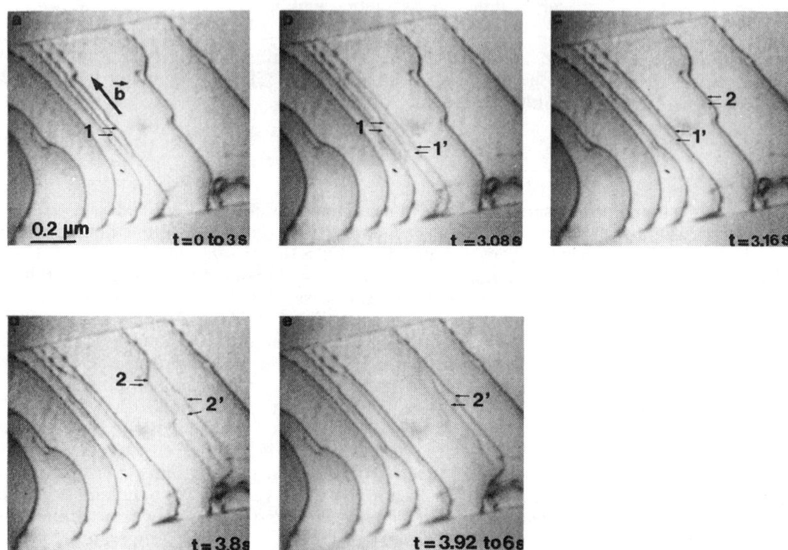

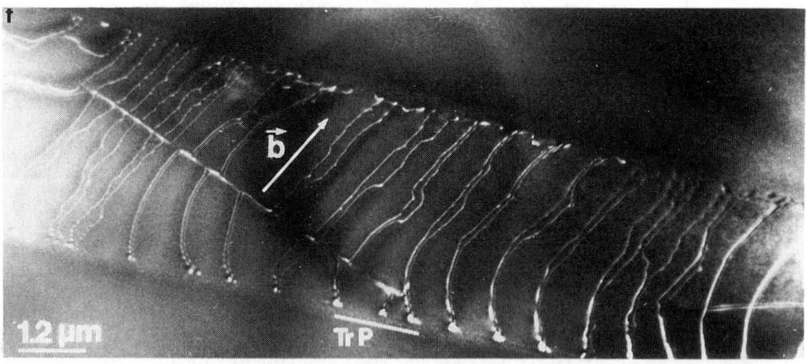

Fig. 2 : Jerky movement of rectilinear screw superdislocations with a large dissociation width. T=300 K
Burgers vector : b = 1/3[$11\bar{2}0$] ; glide plane : ($\bar{1}100$) ; g = [$\bar{2}021$]
a - e : Movement of two superdislocations (1 and 2);
f : Same area under weak-beam conditions.

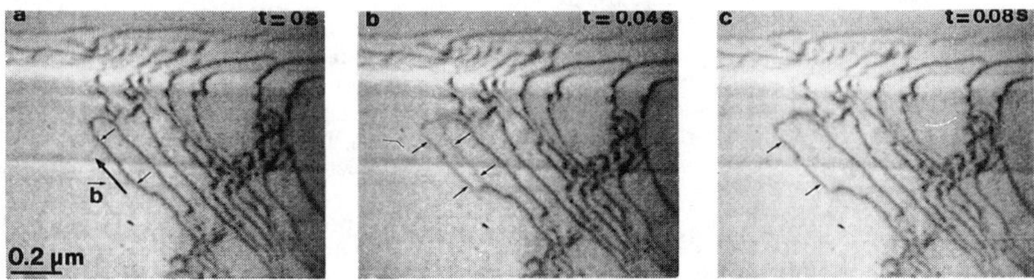

Fig. 3 : T=573K. Rectilinear screw superdislocation with a large dissociation width. Movement of the leading superpartial.Same conditions as in Fig. 2

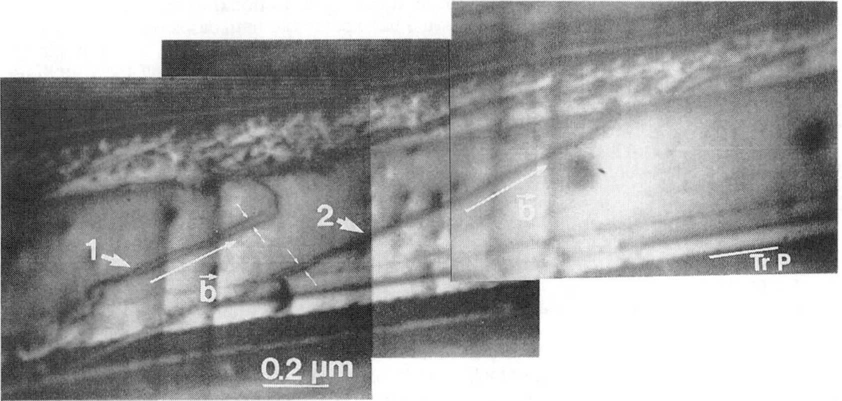

Fig. 4 : T=773K. Rectilinear screw superdislocation with a large dissociation width. Same conditions as in Fig. 2 ,
Burgers vector : b = 1/3<11$\bar{2}$0> ; glide plane : (1$\bar{1}$00) ; g = [20$\bar{2}\bar{1}$].

It is thus proposed that the two observed situations correspond to two different values of the APB energy γ. Since the dissociation width varies inversely with γ, the ratio of the APB energies is therefore about 5. This is consistent with the calculations of Umakoshi and Yamagushi[5] and Czerti, Khantha, Vitek and Pope[6], who ghave $\gamma_{II}/\gamma_I = 5$ and 9, respectively. The absolute values of γ_I and γ_{II}, which can only be very roughly estimated, owing to the error on μ, are $\gamma_I = 15$ mJm^{-2} and $\gamma_{II} = 70$ mJm^{-2}, again in a good agreement with the calculations of Czerti et al, which gave $\gamma_I = 11{,}2$ mJm^{-2} and $\gamma_{II} = 101$mJm^{-2}.
Under such conditions, the frictional force acting on the superpartials bounding the low energy APB clearly corresponds to the non-planar core configuration simulated by Czerti et al.[6]. Conversely, the absence of any frictional force on superpartials bounding the high energy APB indicates that the corresponding core structure may be planar. This configuration has however not been simulated.
Since the decrease of the dislocation line energy due to the dissociation is given by :

$$\Delta E = \frac{\mu b^2}{2\pi}\left(\text{Log}\,\frac{d}{b} - 1\right)$$

where d is the dissociation width, dislocations dissociated with the lowest APB energy (type I) have the lowest energy. Dislocations are however able to multiply and glide over large distances

in type-II planes, which shows that they are very stable in their configuration of high energy. Under such conditions, the CRSS may depend only on the mobility of these high energy dislocations which do not experience any measurable frictional force.
Dislocations are transformed into their low energy configuration only in piled-up groups, i.e. at high values of the local strain and the internal stress. This transformation is expected to occur through a cross-slip mechanism on screw dislocation segments. It is however not clear why this cross-slip process is not easier at high temperatures, as expected from a thermally activated process.
This behaviour is difficult to compare with the results of post-mortem observations. The observation of edge dislocations [1] may correspond to the activation of mobile dislocation loops in type II planes. The observation of dipoles of narrowly dissociated rectilinear screw dislocations [2,4] is however not explained at present.
The movement of widely dissociated rectilinear screw dislocations in type-I planes is controlled by frictional forces acting on both superpartials. This movement is however very jerky, as already observed in pure titanium for instance [10]. Such a behaviour is inconsistent with either a Peierls mechanism due to a non planar dislocation core or a frictional force originating from covalent bonding, since steady movements would be expected in both cases (see [11] in the case of Peierls forces, and [12] in the case of covalent forces in semiconductors). On the contrary, jerky movements of rectilinear screw dislocations can easily be interpreted by a locking-unlocking mechanism which is an extension of the Peierls mechanism [11]. Such a mechanism has already been proposed to explain the dislocation behaviour and the mechanical properties of pure Ti [10].

CONCLUSIONS

These first observations of dislocation glide in prismatic planes of Ti_3Al polycrystals have led to the following results :
- dislocations behave differently according to whether they lie in type-I or type-II parallel prismatic planes.
- the APB energy on type-I planes is about 5 times lower than that on type-II planes, in agreement with theoretical estimates.
- screw dislocations experience a strong frictional force only in type-I planes. This frictional force can be described by a locking-unlocking mechanism similar to that occurring in pure Ti, and is not controlled by frictional forces originating from covalent bonding.
- the CRSS of prismatic glide is proposed to be governed by the glide of dislocations on type-II planes only.

REFERENCES

1. Y. Minonishi, Phil. Mag. A **63**, 1085 (1991).
2. Y. Minonishi, M. Otsuka, and K. Tanaka, in Proceeding of International Symposium on Intermetallic Compounds - Structure and Mechanical Properties - Ed. by O. Izumi (Sendai, 1991, the Japan Institute of Metals), pp. 543-547.
3. M. Thomas, A. Vassel, and P. Veyssière, Scripta Met. **21**, 501 (1987).
4. S.A. Court, J.P. Lofvander, M.H. Loretto and H.L. Fraser, Phil. Mag. A **61**, 109 (1990).
5. Y. Umakoshi, and M. Yamaguchi, Phys. Stat. Sol. (a) **68**, 457 (1981).
6. J. Cserti, M. Khanta, V. Vitek, and D.P. Pope, Mat. Sci. and Eng. A **152**, 95 (1992).
7. B. Legrand, Phil. Mag. A **52**, 83 (1985).
8. V. Vitek and M. Igarashi, Phil. Mag. A **63**, 1059 (1991).
9. A. Couret and D. Caillard, Acta Met. **36**, 2515 (1988);
10. S. Farenc, D. Caillard and A. Couret, to appear in Acta Met.
11. D. Caillard, A. Couret, N. Clément, S. Farenc, and G. Molénat, in Strength of Metals and Alloys, Ed. by D.G. Brandon, R. Chaim and A. Rosen, (Freund Publish. House Ltd., London, 1991) vol. 1, pp. 139-154.
12. D. Caillard, N. Clément, A. Couret, Y. Androussi, A. Lefebvre and G. Vanderschaeve, in Proc. Microsc. Semicond. Mat. Conf. (Inst. Phys. Conf. Ser. n° 100, sect. 5, IOP Publishing Ltd 1989),pp. 403-408.

CHARACTERIZING DEFECT STRUCTURES IN ORDERED ALLOYS USING ELECTRON MICROSCOPY AND COMPUTED IMAGE SIMULATIONS

K.J. Hemker, B. Viguier, R. Schäublin and M.J. Mills†

Institut Genie Atomique, EPFL, CH-1015 Lausanne, Switzerland

†Sandia National Laboratories, Livermore, CA, 94551-0969

Abstract

The resolution offered by weak-beam transmission electron microscopy allows for the direct observation of partial dislocations in a large number of intermetallic alloys. In many instances, the comparison of these observations with computer simulated images leads to results that are much more quantitative and descriptive than are possible with ordinary analyses. Examples from several parallel studies are presented: i) comparisons of experimental and simulated images that have been used to characterize extrinsic stacking faults in TiAl are shown, ii) the need to correct the experimentally observed antiphase boundary dissociations (d^{APB}) in Ni_3Al in order to account for the image shifts that occur during microscopy are highlighted, and iii) dissociation distances of less than 2 nm, which were observed with (2g·5g) diffraction conditions, were verified and quantified by comparisons with computer simulated images. Antiphase boundary energies (γ^{APB}) and complex stacking fault energies (γ^{CSF}) were calculated from the corrected observations, and it was found that the APB energies did not depend on alloy content, but that the CSF energy of binary Ni_3Al is less than it is for a boron containing alloy.

1. INTRODUCTION

The dislocation mobility, and thus the unusual mechanical properties and inherent brittleness, of many ordered intermetallic alloys can be related to their dislocation core structure. The Burgers vectors of the dislocations that control deformation in ordered alloys are generally much larger than they are in ordinary metals and alloys, and the strain energy associated with these "super" dislocations leads to their dissociation into smaller partial dislocations that are connected by crystallographic faults. Dislocation motion is restricted when these dissociations results in a non-planar configuration, and there is general agreement in the literature that many of the interesting mechanical properties exhibited by intermetallic alloys can best be understood in terms of their dislocation mobility and thus their dislocation core geometry [1,2].

Transmission electron microscopy (TEM) is an especially powerful tool for characterizing dislocation structures in ordered alloys. The dissociations in ordered alloys are often too small to be seen with conventional electron microscopy, but the resolution offered by weak-beam microscopy allows for the direct observation of partial dislocations in a large number of these alloys. Weak-beam tilting experiments are commonly used to identify the partial dislocations and stacking faults that exist in intermetallic alloys, and weak-beam measurements of the separation distance between partial dislocations are considered to be the most reliable way of determining fault energies in these alloys. However, the close proximity of these partial dislocations can result in an overlapping of their strain fields and the formation of rather complicated TEM images. When this occurs, the quantitative nature of the TEM observations can be greatly improved by comparing the experimental images with computer simulated images.

The development of a new simulation program for conventional and weak-beam microscopy [3], provides us with the opportunity to do quantitative electron microscopy. We have used computer simulations to: i) characterize stacking faults associated with low temperature deformation in TiAl, ii) demonstrate the effect of image shifts on weak-beam measurements of d^{APB} in Ni_3Al, and iii) identify supplementary peaks and determine d^{CSF} from (2g·5g) weak-beam images of 4-fold dissociations in two different Ni_3Al alloys. The results of these projects will be presented and the benefits that were gained by comparing the experimental images with the computer simulated images will be highlighted in the remainder of this paper.

2. COMPUTER SIMULATIONS

Head *et al.* [4] developed a two beam program to simulate the images of one or two straight and parallel dislocations. Schäublin and Stadelmann [3] have extended the Head program to a many beam program that calculates the images of up to four dislocations and three planar faults. Schäublin's many beam program includes anisotropic elasticity by using the "ancalc" subroutine of Head *et al.* [4] to calculate the displacement fields around the dislocations. It also uses the column approximation to construct the images. Using this approximation simplifies the integration of the Schrödinger equations, but it also places a physical limit on the resolution of the simulations. Comparisons of images simulated with and without the column approximation [5] have shown that the column approximation is valid to at least 1 nm for diffraction conditions similar to those used in this paper. The success of Schäublin's many beam program in earlier studies [3,6,7] precedes its use in this paper.

3. COMPARISONS WITH EXPERIMENTAL OBSERVATIONS

Extrinsic stacking faults in TiAl. The existence of long straight extrinsic stacking faults that are bounded by dislocation dipoles have been observed in TiAl [8-10]. We have observed extrinsic faults in $Ti_{46}Al_{53}Nb_1$ specimens that were deformed at 80 K, and we have characterized them with Schäublin's program. The characterization of these faults becomes much easier when then experimental images are compared with computer simulated images, because the general rules that were developed to identify the intrinsic/extrinsic nature of stacking faults were derived for much wider faults where the contrast is much simpler. Comparing the contrast of the experimental images with a series of calculated images not only gives a much clearer indication of the nature of the faults, it also allows for the determination of the magnitude of the Burgers vector of the partial dislocations that surround them. The many beam program is especially useful for this type of analysis, because the partial dislocations that bound the narrow faults in TiAl and Ni_3Al can best be imaged with weak-beam conditions and this program is able to calculate weak-beam images.

Fig. 1 is a comparison of the experimental and computer simulated images for an extrinsic stacking fault that is associated with a (**b**=a<101]) super dislocation. The good correlation between these images has been taken as a strong confirmation of the validity of the simulations, and comparing the experimental images to different series of stacking fault images has allowed us to identify, unambiguously, the nature of this stacking fault and the Burgers vector of the surrounding dislocation. The results of this study are in good agreement with those reported earlier [8,9], namely that $^{a}/_{2}$<112] superdislocations lead to the formation of extrinsic stacking faults that are surrounded by $^{a}/_{6}$<112] partial dislocations while a<101] superdislocations lead to the formation of extrinsic stacking faults that can be surrounded by either $^{a}/_{6}$<112] or $^{a}/_{3}$<211] partial dislocations.

APB Energies. The fault energy associated with the dissociation of a dislocation into two partial dislocations can be calculated by balancing the repulsion force of the partial dislocations with the energy required to create the fault. Observations of d^{APB} and calculations of γ^{APB} have been reported for a number of Ni_3Al alloys [6,11-16]. The determination of precise values of γ^{APB} require: i) that the force balance be calculated using anisotropic elasticity, and ii) that the measured values of d^{APB} be corrected to account for the image shift that occurs when the strain fields of the partial dislocations overlap . Stroh [17] calculated the

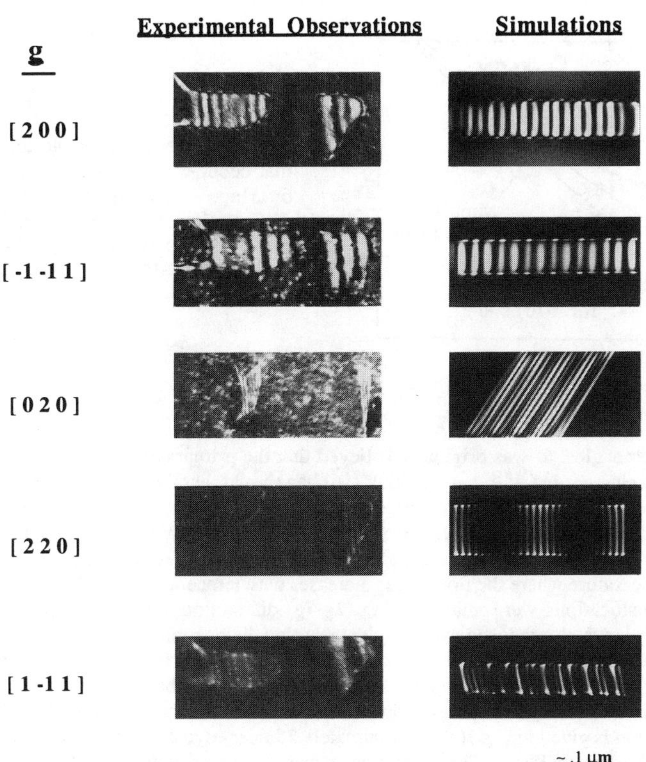

Fig.1: A comparison of experimental and simulated stacking fault images in $Ti_{46}Al_{53}Nb_1$. The weak-beam images have $\mathbf{R}_f$=-$^a/_3$[111] and a bounding dislocation with $\mathbf{b}_{partial}$ = $^a/_3$[-211]. The image simulations use 6 beams (-**g** -> +4**g**) and: $\mathbf{N}_{fault}$ =[111]; $\mathbf{N}_{foil}$=[012]; t $_{foil}$ =200 nm; **B** and **g** variable; c_{11}=190, c_{12}=105, c_{13}=90, c_{33}=185, c_{44}=120, c_{66}=50 GPa [18]; and d^{ESF}=50 nm.

repulsive forces between dislocations using anisotropic elasticity and the Stroh method has been used for all of the energy calculations in this paper.

Cockayne *et al.* [19] had derived a correction for the relative image shift that occurs between two partial dislocations in an isotropically elastic material, but the many beam simulation program now allows us to observe and correct for the image shifts within the framework of anisotropic elasticity. In order to determine the relative importance of these corrections, both the Cockayne correction and the image simulations have been performed for the case of two superpartial edge dislocations that were imaged with **g** parallel to **b** and $|s_g|$=0.28 nm^{-1}. The results of this comparison, which are very similar to those obtained in an independent study [20], are given in Fig. 2. Notice that both corrections indicate that the observed values of "d" are greater than the actual distance between the dislocations and that the correction is greater for smaller separations. The simulations also yield different corrections for the (111) and the (010) planes, a point that is not observable with the isotropic correction. Baluc *et al.* [6] used the different correction methods to determine values for γ^{APB}, and their results indicate that the corrections in Ni_3(Al,Ta) are approximately 20%.

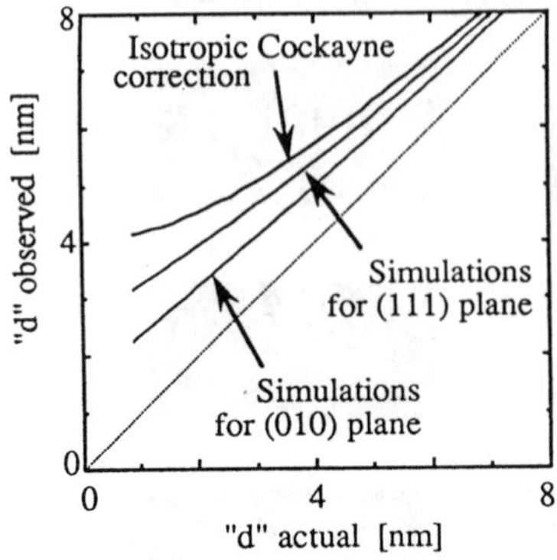

Fig.2: Corrections for the image shifts that occur when dislocation strain fields overlap. Calculated for an APB dissociation in Ni_3Al (s_g=0.28 nm^{-1}; **g**=$\mathbf{b}_{sp}$=$^a/_2$[-101]; ξ=[1-21]; c_{11}=223, c_{12}=148, c_{44}=125 GPa or ν=0.399 [21]).

CSF Energies. It was originally believed that the primary effect of alloying additions was related to changes in γ^{APB}, but Dimiduk [16] has shown that the effect of alloying on γ^{APB} is minimal. It is now believed that the role of the alloying elements is related to changes in the CSF energy. In an attempt to compare γ^{CSF} of a binary alloy with one containing a small amount of boron, two alloys ($Ni_{76}Al_{24}$ and $Ni_{75.5}Al_{23.8}B_{0.7}$) were deformed to 10% plastic strain at a temperature where the flow stress increases with temperature (485 K).

The dislocations were imaged using (2**g**·5**g**) diffraction conditions (**g**=**b**=[-202] and s_{2g}=+0.37 nm^{-1}) which produced images with very fine dislocation detail but also resulted in a very low image intensity. A large number of dislocations were imaged with exposure times between 20-70 seconds, but the majority of these were rejected because of image drift or the absence of the weakest peak. An experimental observation of the four fold dissociation of a super dislocation is given in Fig.3(a); approximately 25 near edge dislocations like this one were selected for further analysis. The same image was simulated with Schäublin's many beam program, and the calculated image is given in Fig. 3(b). The similarity between the calculated image and the experimental observations allows for direct comparisons between the images, and the simulations have been used to correct for image shifts.

Close inspection of both the experimental and the simulated images indicate that there are more that four image peaks in the (2**g**·5**g**) image. Humphreys *et al.* [22] have reported the existence of supplementary peaks in weak-beam images, especially images formed with a second-order reflection (2**g**). A series of simulations, where the input value for d^{CSF} was varied from 0.5 - 4.0 nm, has been used to identify the supplementary peaks. It was observed that the distance between the brighter pair of peaks did not change with d^{CSF}, but that the distance between the weaker pair of peaks was related to d^{CSF}. This observation underscores a very important consideration for the measurement of dissociation distances with (2**g**·5**g**) weak-beam microscopy. Namely, that d^{CSF} can only be measured by using the weaker pair of intensity peaks in the experimental image.

The distance between the peaks in the experimental image (x_1-x_3) were measured as a function of dislocation character and the representative distances for pure edge dislocations are given in Table I. These experimental values have been corrected by matching them with simulated values (r_1-r_3) and a comparison is also given in Table I. It is important to note that the simulation only uses two input values (d^{APB} and d^{CSF}) to predict all three of the peak separations (r_1-r_3). Thus, the simulations are able to predict not only the presence of supplementary peaks but their position as well. Values for γ^{APB} and γ^{CSF} have been calculated using both the corrected and the uncorrected distances. Comparisons of the two alloys indicate

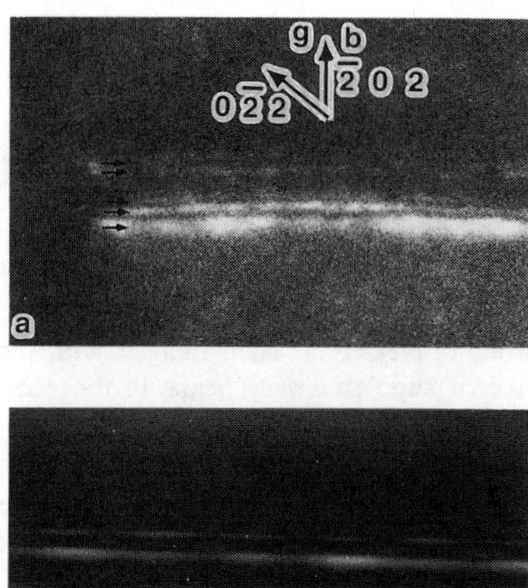

Fig.3: (**a**) Experimental and (**b**) simulated images of the 4-fold dissociation of an edge super dislocation in Ni_3Al. These (2g·5g) images were formed with $|s_{2g}|$ =+0.37 nm $^{-1}$. The simulations use 7 beams (**-g** -> +**5g**) and the following input parameters: **B**=$\mathbf{N}_{foil}$ = [212]; t_{foil} =50 nm; $\mathbf{N}_{fault}$ =[111]; the c_{ij}'s in Fig.2; d^{APB}=5.0 and d^{CSF}=2.0 nm.

that γ^{APB} is not dependent on alloy composition but that γ^{CSF} is higher in the boron containing alloy, see Table I. The boron alloy exhibits a higher flow strength than the binary alloy, and the observation of a higher γ^{CSF} in the boron alloy suggests that the cross-slip process, and thus the locking of dislocations, is assisted by alloying additions that will result in higher γ^{CSF}'s.

Table I

(Four fold dissociations of edge superdislocations in two Ni_3Al alloys)

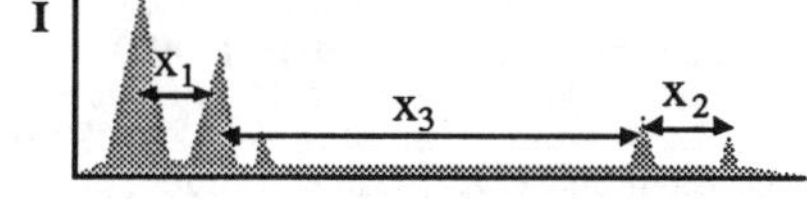

	$Ni_{76}Al_{24}$	$Ni_{75.5}Al_{23.8}B_{0.7}$
<u>Separations [nm]</u>		
Experimental (x_1,x_2,x_3)	2.5±.3 2.1±.2 6.0±.7	2.5±.3 1.5±.15 6.3±.4
Simulation output (r_1,r_2,r_3)	2.5 2.1 6.0	2.5 1.5 6.3
Simulation input (d^{CSF}, d^{APB})	1.8 4.8	0.9 5.8
<u>Energies [mJ/m²]</u>		
Uncorrected $(\gamma^{APB}, \gamma^{CSF})$	147 ±15 177 ±20	153 ±10 228 ±25
Corrected $(\gamma^{APB}, \gamma^{CSF})$	180 ±20 206 ±30	173 ±15 335 ±60

4. SUMMARY AND CONCLUSIONS

The studies presented in this paper suggest that the quantitative nature of weak-beam electron microscopy is greatly enhanced when the experimental images are supplemented with computer simulations. The major conclusions of these comparisons are that:

1. Both the intrinsic/extrinsic nature of extended stacking faults in TiAl and the Burgers vectors of the dislocations that surround these faults can be determined with simulations.
2. Precise measurement of d^{APB} in Ni_3Al require that the experimental observations be corrected for image shifts. These corrections are anisotropic and can best be done through comparisons with simulated images.
3. The 4-fold dissociation of super dislocations in Ni_3Al has been imaged with (2g·5g) weak-beam conditions. But , the presence of supplementary peaks in these images require that d^{CSF} be measured on the weaker pair of image peaks.
4. Measured dissociation distances, for a binary and a boron containing alloy of Ni_3Al, have been corrected with simulations and the fault energies have been calculated using anisotropic elasticity. The results indicate that γ^{APB} does not change with alloy content but that γ^{CSF} is higher for the boron containing alloy.

5. ACKNOWLEDGEMENTS

The authors would like to thank P. Spätig and T. Hessler for assistance with the experimental measurements. Funding was provided by Fonds National Suisse under contract number 20-31 302 91.

6. REFERENCES

1. P. Veyssière, *Rev. phys. appliquée*, **23**, 673 (1988).
2. M.S. Duesbery, *Dislocations in Solids*, ed. by F.R.N.Nabarro, Elsevier Science Publ.,Amsterdam, 67, (1989).
3. R. Schäublin and P. Stadelmann, *to be published in Mat. Sci and Eng. A,* (1992).
4. A.K. Head, P. Humble, L.M. Clarebrough, A.J. Morton, and C.T. Forwood, *Computed Electron Micrographs and Defect Identification*, ed. by North-Holland Publishing Company, Amsterdam, (1973).
5. A. Howie and C.H. Sworn, *Phil. Mag.*, **22**, 861, (1970).
6. N. Baluc, R. Schäublin and K.J. Hemker, *Phil. Mag. Letters*, **64**, 327, (1991).
7. K.J. Hemker and M.J. Mills, *submitted to Phil Mag.*, (1992).
8. D. Schechtman, M.J. Blackburn and H.A. Lipsitt, *Metall. Trans.*, **5**, 1373, (1974).
9. G. Hug, A. Loiseau and A Lasalmonie, *Phil. Mag.* **54**, 47, (1986).
10. Y.G. Zhang, Q. Xu, C.Q. Chen and H.X. Li, *Scripta Met.*, **26**, 865, (1991).
11. P. Veyssière, J. Douin, and P. Beauchamp, *Phil. Mag A*, **51**, 469, (1985).
12. J. Douin, P. Veyssière and P. Beauchamp, *Phil Mag. A*, **54**, 375, (1986).
13. A. Korner, *Phil Mag.* **58**, 507, (1988).
14. N. Baluc, H.P. Karnthaler and M.J. Mills, *Phil. Mag A*, **64**, 137, (1991).
15. C. Neveu, Thèse de Doctorat de l'Universite de Paris-Sud Centre d'Orsay, (1992).
16. D. Dimiduk, A.W. Thompson and J.C. Williams, *Accepted in Phil Mag. A*, (1992).
17. A.N. Stroh, *Phil. Mag.*, **3**, 625, (1958).
18. C.L. Fu and M.H. Yoo, *Phil. Mag. Letters*, **62**, 159, (1990).
19. D.J.H. Cockayne, I.L.F. Ray and M.J. Whelan, *Phil. Mag.*, **20**, 1265, (1969).
20. J. Oliver, Thèse de Doctorat de l'Universite de Paris-Sud Centre d'Orsay, (1992).
21. M.H. Yoo, *Acta. Metall.*, **35**, 1559, (1987).
22. C.J. Humphreys, R.A. Drummond and A. Hart-Davis, *Phil. Mag.*, **35**, 1543, (1977).

EFFECT OF PLANAR FAULT ENERGIES ON DISLOCATION CORE STRUCTURES AND MOBILITIES IN $L1_0$ COMPOUNDS

J.P. SIMMONS*, S.I. RAO**, D.M. DIMIDUK***

*NRC Research Associate, Wright Laboratory, WL/MLLM, Wright-Patterson AFB, OH 45433
**Universal Energy Systems, Inc., Dayton, OH 45432
***Wright Laboratory, WL/MLLM, Wright-Patterson AFB, OH 45433

ABSTRACT

A set of three Embedded Atom Method potentials are presented that produce a stable $L1_0$ structure. All three were nominally fitted to experimental parameters of the γ-TiAl phase; variations in the fit were allowed in order to give potentials with differing values of planar fault energies. These potentials were evaluated and found to produce stable APB(111), APB(100), CSF(111), and SISF(111) faults. Dislocation core structures were computed for 1/2<110>-type dislocations in both edge and screw orientation with all three potentials. The screw orientation of the highest fault energy potential was found to have a non-planar configuration, being spread about equally on two {111} close packed planes. All other cores were found to be planar and spread on a single {111} plane. No significant edge components of strain developed for the screw orientation, suggesting that these dislocations would not appear to be spread in Atomic Resolution TEM. Preliminary evaluations showed that dislocation mobilities were low, requiring stresses on the order of $10^{-3}\,\mu$ for motion.

INTRODUCTION

γ-Titanium aluminide alloys have attracted much interest in recent years as candidates for new generation high temperature alloys because their high temperature strength retention, superior oxidation resistance, and low density.[1,2,3] However, these alloys suffer from low strength overall and poor ductility and toughness at room temperatures. Many experimental efforts have been directed towards modifying alloy parameters in order to increase ductility.[4]

Experimental results to date indicate that the dislocations important for deformation processes are of type 1/2<110>, <101>, and 1/2<112>.[5] The first of these is an 'ordinary' dislocation of the disordered parent FCC structure, while the other two are superdislocations of the ordered structure. Experiments have shown that the alloys exhibiting higher ductility have increased activity of the 1/2<110> dislocations.[5] Microstructures often consist of these type dislocation in screw orientations, indicating that the screw orientation has a lower mobility than the other orientations.[5]

Atomistic modeling studies by Vitek[6,7] and Duesbery[8] and theoretical work by Greenberg,[9,10] indicate that dislocation mobility in most alloys and intermetallic compounds is controlled by the atomic structure of the dislocation cores. Atomistic simulations have been performed by pair potentials that were fit to give stable crystal structures and faults. More recently, the Embedded Atom Method (EAM), originally developed by Daw and Baskes,[11] has been successfully applied to B2 and $L1_2$ structures by Pasianot, *et al.*[12] and to Ni_3Al by Foiles, *et al.*,[13] Voter and Chen,[14] and Parthasarathy, *et al.*[15] Parthasarathy, *et al.*[16] used EAM to simulate dislocation core structures in NiAl.

Previously, an EAM potential set was developed by Rao, *et al.*[17] that produces a stable $L1_0$ structure by fitting the bulk properties of γ-TiAl. In this work, we develop two additional $L1_0$ potentials that give higher planar fault energies. Together, these three potentials form the basis of a parametric study of dislocation behavior in $L1_0$ structures, similar to the approach taken by Yamaguchi, *et al.*[18,19] towards understanding dislocation properties in $L1_2$ structures. In this phase of the work, we evaluate the three potentials in terms of fault stability and compute some properties of 'ordinary' 1/2<110> dislocations; further efforts to model the <101> and 1/2<112> dislocations are underway in our group.

METHODS OF COMPUTATION

Potentials

Rao, *et al.*[17] reported a set of interatomic potentials that described Al-Al, Ti-Ti and Ti-Al interactions for which the $L1_0$ structure was stable. The Al-Al interaction was taken from the work of Voter and Chen.[14] Potentials describing the Ti-Ti and Ti-Al interactions were developed by fitting selected experimental data of pure Ti and γ-TiAl, respectively. Here, we used the same approach to develop two additional Ti-Al potentials that gave higher values of planar fault energies. For details of the methodology, the reader is referred to the papers by Rao, *et al.*[17] and Voter and Chen.[14] Table I shows the calculated values of selected data, along with the values from the literature used for the fit. In this table, the symbols *NTA01, NTA02,* and *NTA03* represent the potentials with the lowest, intermediate, and the highest fault energies, respectively.

Table I. Table of parameters realized for each of the three potentials. Lattice parameters given in nm, elastic constants given in GPa, coesive and vacancy energies given in eV/atom, and ΔE_{B2} givven in eV/atom.

	Literature	*NTA01*	*NTA02*	*NTA03*
a_0	0.3997 [a]	0.4048	0.4033	0.4052
c/a	1.017 [a]	0.9802	0.9912	0.9814
C_{11}	190.0 [c]	188.4	201.8	233.7
C_{33}	185.0 [c]	216.3	237.0	326.1
C_{44}	120 [c]	71.77	83.26	150.0
C_{66}	50.0 [c]	50.14	54.00	60.0
C_{12}	105 [c]	90.99	94.59	69.3
C_{13}	90.0 [c]	112.3	123.6	158.5
E_{coh}	−4.51 [b]	−4.79	−4.87	−4.43
E_{vac}	—	1.80	1.88	1.87
ΔE_{B2}	—	+0.04	+0.12	+0.16

(a) Pearson, (1987), (b) Hultgren, et al., (1963), (c) Fu and Yoo, (1990)

For all cases, we used a Morse potential to describe the two-body atomic interactions. All functions were modified so that their values and derivatives approached zero smoothly at a cutoff distance, R_{cut}. A hydrogenic 4s distribution was assumed for the electronic contribution of each of the atoms. The embedding function was implicitly defined for the pure metals by requiring that Rose's equation of state be obeyed.[20]

We used 3 adjustable constants in the Morse potential, 1 to describe how rapidly the electron density falls off with distance, 3 embedding function constants, and R_{cut}, making a total of 8 adjustable constants. These were fit to select data from the Ti-Al system by a weighted least squares technique, giving a greater weight to the more important empirical data. The lattice constants, (a and c), the elastic constants (C_{11}, C_{12}, C_{13}, C_{33}, C_{44}, and C_{66}), the vacancy formation energy (E_V), and the four fault energies (APB^{111}, APB^{100}, $SISF^{111}$, and CSF^{111}) were used for this purpose. Since it was our aim to vary the fault energies, these quantities were given a large weighting in the least squares; weights were given to the other data according to our assessment of the quality of the data and its relative import on dislocation properties. Table I summarizes the quality of fit that we achieved for all three potentials.

Computations and Simulations

γ-surfaces were computed for both the (100) and (111) planes. Crystals were constructed that were at least two cut-off distances square by at least 6.0 nm in length. The crystal was then divided along the center of the long axis and then shifted by a fixed displacement. The internal energy was minimized by using a conjugate gradient technique. Only the positions of the atoms normal to the fault plane were allowed to vary; those parallel to the fault plane were not modified in the computation. Similar computations that we have performed that allowed this tangential motion of the atoms indicated that our approximation here amounted to an error of approximately 1-2% in evaluating the energy.

For the (100) γ-surfaces, periodic boundary conditions were applied to the [010]- and [001]-directions, respectively. The bounding (100)-surfaces were allowed to relax as free surfaces. Crystals for this computation were 4 [010] repeat distances by 4 [001] repeat distances by 16 [100] repeat distances. For the (111) γ-surfaces, periodic boundary conditions were applied to the $[\bar{1}\bar{1}2]$- and $[\bar{1}10]$-directions, respectively and the (111)-surface was allowed to relax as a free surface. Crystals that were 2 $[\bar{1}\bar{1}2]$ periodic units by 3 $[\bar{1}10]$ periodic units by 10 [111] periodic units were used for this computation.

For dislocation computations, cylindrical crystals were constructed with the cylindrical axis aligned with the dislocation line direction. The results of anisotropic elasticity were used in order to assign initial locations of the atoms in the simulation. This served also to fix the boundary values for the simulation, since the positions of the atoms within two cutoff distances of the cylindrical surface were not allowed to vary. Periodic boundary conditions were applied along the cylindrical axis. The atoms within the cylinder were then allowed to relax by the method mentioned above. The lengths of the crystals along the cylindrical axes were taken to be at least two cutoff distances.

Typically, the dimensions of the cylinders were 10.0 nm diameter × approximately 1.5 nm in length. For the screw dislocations, this length was 2 $[\bar{1}10]$ periodic units and for the edge dislocations, this length was 2 $[1\bar{1}2]$ periods. Atoms within 1.25 nm of the cylindrical surface were fixed. For the screw dislocations, the crystals consisted of 5532 atoms and for the edge dislocation simulations, these crystals consisted of 9524 atoms.

A preliminary study of stress effects on dislocations was performed on the potential *NTA01* screw dislocation. After screw dislocation was allowed to relax fully, a strain was put on the crystal that would produce a stress that was exactly parallel to the glide plane. The crystal was then allowed to relax and the final configuration was examined to determine whether or not the dislocation moved. A binary search method was employed in order to determine the minimum stress required to move the dislocation from its initial location.

RESULTS

Potentials

Quality of Fit. Table I gives the experimental and calculated values of the fitting parameters for each of the three potentials. A wide range of fault energies was achieved, along with some variation in the elastic constants. Somewhat better fits were achieved with the cohesive energies, vacancy formation energies, and the lattice parameters. In all cases, the B2 structure was metastable with respect to the $L1_0$ structure.

The fault energies could be varied by as much as a factor of approximately 5, in the case of SISF, but more typically by a factor of 2.5-3.5. This was achieved at the expense of the quality of fit for the empirical data. The largest variation this fit was in the elastic parameter, C_{44}, which varied by a factor of 2 when going from potential *NTA01* to *NTA03*. The variation of the other elastic parameters was on the order of 20-40%. In all cases, the *c/a* ratio is less than unity; the experimentally determined value was greater than unity. The cohesive energies were fit to within 2-8%. Variation of the lattice parameters was only of the order of 1%. No attempt was made to restrain the magnitude of the energy difference between B2 and $L1_0$, other than what was necessary to insure that the $L1_0$ structure was thermodynamically favored.

γ-surfaces. γ-surfaces were developed for the (100) and (111) fault planes and are given in Figures 1 and 2, respectively. It was found that all four faults, APB^{111}, APB^{100}, $SISF^{111}$, and CSF^{111} corresponded to minima and that their energies increased monotonically from potential *NTA01* to *NTA03*. Figure 1 shows that the APB position of the {100} γ-surfaces is a minimum and occurs where the fault vector **f** = 1/2[101]. Figure 2 shows that the APB, SISF, and the CSF positions are minima for all three {111} γ-surfaces, and occur where the fault vector is approximately 1/2$[\bar{1}01]$, 1/3$[\bar{1}\bar{1}2]$, and 1/6$[\bar{2}11]$, respectively. Potential *NTA03* developed shallow minima at the 'atom-on-atom' positions for both (111) and (100) γ-surfaces. We will discuss this point further below.

Dislocation Properties

Core Relaxation Structures

Figures 3, 4, and 5 show differential displacement maps[6] of the dislocation core structures. In these figures, the magnitude of the difference in displacement between neighboring atoms is depicted by the length of the arrows. Each arrow points to the atom that has been displaced by the greater amount. Since the lattice has a periodicity of the lattice translation vector τ, a differential displacement between two lattice points may be represented by the displacement ± any number of multiples of τ. In all of the differential displacement maps given here, the differential displacements between two lattice points is adjusted by adding or subtracting multiples of τ until the differential displacement is between $-\tau/2$ and $+\tau/2$. For clarity, differential displacements that amount to less than 2% of the final distance between atoms have not been plotted.

1/2<110>$(\bar{1}11)$ Screw Dislocations. Figure 3 shows the screw component of the displacements for the spontaneously relaxed screw dislocation cores generated by the three potentials. There is a monotonic decrease in core spreading, going from potential *NTA01* to *NTA03*. The cores generated by potentials *NTA01* and *NTA02* are planar, with the plane of spreading being the $(\bar{1}11)$

crystallographic plane. That generated by potential *NTA03* is evenly spread the on $(\bar{1}11)$ and $(1\bar{1}1)$ planes. The width of the spreading of the dislocation generated by potential *NTA01*, as measured by the full width at half maximum method, is approximately 1.2 nm. For potential *NTA02*, the spreading is more localized and measures about 0.5 nm. No planar spreading is observed in the core structure generated by potential *NTA03*.

The edge components of these three dislocations are shown in Figure 4. For potential *NTA01*, there is a significant amount of edge displacement. This displacement is planar and lies on the $(\bar{1}11)$ plane. For the other two potentials, this edge displacement is considerably smaller, but still lies in the {111}-type planes.

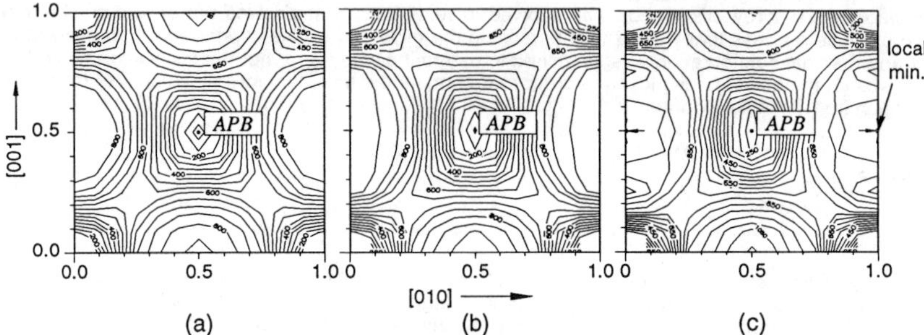

Figure 1. (100) γ-surfaces computed from (a) potential NTA01, (b) potential NTA02, and (c) potential NTA03.

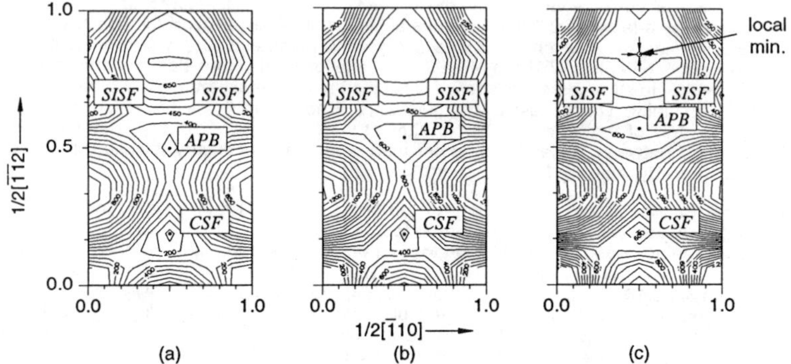

Figure 2. (111) γ-surfaces computed from (a) potential NTA01, (b) potential NTA02, and (c) potential NTA03.

1/2 <110>$(\bar{1}11)$ Edge Dislocations. Figure 5 shows a differential displacement map of the edge components of the spontaneously relaxed cores of edge dislocations generated by each of the potentials. Here, the arrow between any two atoms points to the atom that was displaced to the right by a greater amount. In contrast to the screw dislocation cases, all three cores have spread somewhat along the $(\bar{1}11)$ slip plane. The spreading of the dislocation generated by potential *NTA01* is 2.0 nm, that corresponding to potential *NTA02* is 1.45 nm, and that corresponding to potential *NTA03* is 1.0 nm.

Stress Effects

A number of simulations were employed with stresses varying from $10^{-2}\mu$ to $10^{-3}\mu$ on the potential *NTA01* screw dislocation. For a value of $1.25\times10^{-3}\mu$, the dislocation remained in its initial circuit of atoms, while for a value of $2.5\times10^{-3}\mu$, it moved by one circuit, indicating that the barrier to motion was overcome. Takeuchi[21] pointed out that as a dislocation moves the stress relaxes. By comparing the distances the dislocation moved for the higher applied stresses, we estimated that approximately $0.8\times10^{-3}\mu$ of stress is relaxed per circuit traversed by the dislocation and that the barrier to motion that we determined was above this value.

DISCUSSION

Fault Energies and Core Structures

Fault Energies. Inspection of the γ-surfaces given in Figures 1 and 2 shows that, for all three potentials, the APB^{111}, APB^{100}, CSF^{111}, and $SISF^{111}$ are stable faults in all cases. Table II gives the realized fault energies calculated according to each of the potentials, along with values reported by various experimental and theoretical workers.[22,23,24] While our potentials do not reproduce the properties of γ-TiAl, the range of values of fault energies spanned by the three potentials encompasses the behaviors reported in Table II.

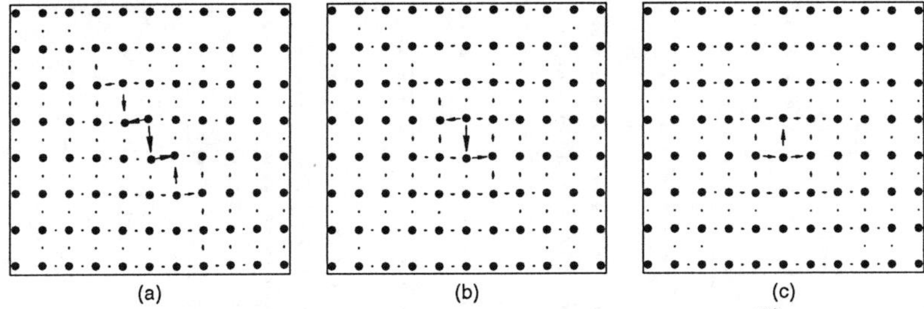

Figure 3. Differential displacement map of the screw components of screw dislocations computed from potentials (a) NTA01, (b) NTA02, and (c) NTA03. Arrows point to atoms displaced upward.

For the $(\bar{1}11)$ faults, the fault vectors generally vary from the 'ideal' locations by a translation parallel to $[11\bar{2}]$-direction. This was originally pointed out by Yamaguchi, *et al.*[25] to be due to the lack sufficient symmetry to guarantee the position of the minima to be in the exact locations as would be predicted by hard sphere models.

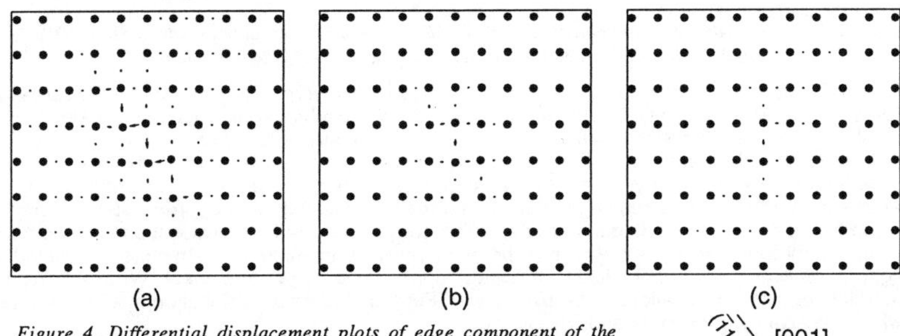

Figure 4. Differential displacement plots of edge component of the dislocations shown in Figures 3a, b, and c, respctively. Arrows point towards atoms displaced in the $[\bar{1}1\bar{2}]$-direction.

Core Structures. The core spreadings followed a predictable trend, the highest energy potentials producing dislocations with smallest spreadings. These results compared reasonably well with the predictions from elastic theory and differed with predictions of earlier work by Yamaguchi, *et al.*[25]

The two lower fault energy screw dislocations had a planar core, being spread on the $(\bar{1}11)$ plane. The higher fault energy core was evenly spread on the $(\bar{1}11)$ and $(1\bar{1}1)$ planes, owing to its high fault energy. The value reported above for the spreading of the *NTA01* dislocation, 1.2 nm, compares with the value of 0.873 nm predicted from anisotropic continuum elastic theory[26] reasonably well, since continuum theory is no longer valid on the scale of dimensions of the lattice parameter. Continuum theory predicts that the other two screw dislocations should not dissociate, a discrepancy that can also be attributed to the scale of dimension involved here.

The edge dislocations also showed a monotonic decrease in core spreading as the fault energies increased. All were found to be dissociated on the $(\bar{1}11)$ plane. The spreadings of these cores were 2.0 nm, 1.45 nm, and 1.0 nm for potentials *NTA01*, *NTA02*, *NTA03*, respectively. The values predicted from elastic theory are 1.64 nm, 0.66 nm, and no spreading, respectively. These results also compare favorably, differing by the same lattice parameter-scale dimension.

In an earlier pair potentials study of dislocations in $L1_0$ structures, Yamaguchi, *et al.*,[25] found that multilayer faults were possible on the {110} planes and suggested that spreading should occur on these planes. We did not observe any spreading of this type in any of our simulations: they were either planar or spread equally on two <111>-type planes.

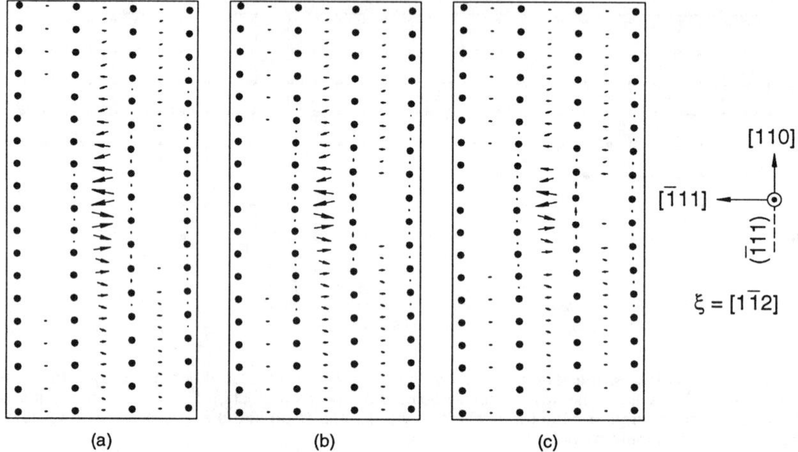

Figure 5. Differential displacement plot of edge dislocations computed according to potentials (a) NTA01, (b) NTA02, (c) NTA03. Arrows point to atoms displaced in the [110]-direction.

Stress Effects. Structurally, the dislocations produced here were rather ordinary: all dissociated structures having essentially planar structures. However, the results of our stress effects simulations on potential *NTA01* screw dislocations show that the minimum stress to move a dislocation is on the order of $10^{-3}\mu$, which is somewhat higher than we had expected for an FCC derivative structure. Earlier results by Parthasarathy, *et al.*[15] in Ni_3Al gave an upper limit on this stress of $5\times10^{-4}\mu$ for 1/2<110> superpartials. The spreading of the superpartials in this study was approximately 2.0 nm, as compared to the 1.2 nm spreading in our case. The core structure of this dislocation showed some edge component to the spreading, whereas the other screw dislocation core structures predicted for the higher fault energy potentials did not have any appreciable edge component. We would expect the mobilities of these dislocations to be even lower. Further studies on dislocation motion are under way which may clarify our understanding.

Table II. Table of fault energies realized by the three potentials in this work compared with calculated and experimental values from the literature. Values reported in mJ/m^2.

	APB (100)	APB (111)	CSF (111)	SISF (111)
NTA01	66	306	119	71
NTA02	91	550	320	220
NTA03	203	770	580	250
LKKR[a]	350	670	280	110
FLAPW[b]	430	510	—	90
WBTEM[c]	210	250	—	140

(a) Woodward, et al., (1992), (b) Fu and Yoo, (1990), (c) Hug, et al., (1988)

Limitations. The potentials used in this work were generated by fitting available data for the γ-TiAl phase, so that the results would approximate the behavior of this phase. Deviations from TiAl do exist, however. In particular, it is a general feature of EAM potentials that the Cauchy pressures must be positive. It has been suggested from a quantum mechanical computation of Fu and Yoo[23] that $C_{12} - C_{66} > 0$ and $C_{13} - C_{44} < 0$. It should also be remarked that if the stiffness constants for *NTA01* given in Table I are inverted, the compliance constant S_{12} has the value $+0.316\times10^{-3}$ GPa^{-1}. In this case, a tensile stress in the $\mathbf{a_1}$ direction causes a dilatation in the $\mathbf{a_2}$ direction. This does not violate the stability criteria of the lattice,[27] which require that $|S_{12}| < S_{11}$ (= 6.40×10^{-3} GPa^{-1}), but it is a rather unusual property. Finally, the small depressions in the γ-surfaces shown in Figures 1(c) and 2(c) were unexpected. We believe the reason for this behavior is that potential *NTA03* has a much stronger Al-Ti interaction than the other two potentials, so that stretching in the $\mathbf{a_1}$ direction allows the (002) planes to come closer together, which in turn creates the dilatation in the $\mathbf{a_2}$ direction.

In view of this discussion, as well as the values of the fault energies predicted from electronic structure computations listed in Table II, we believe that potential *NTA02* represents the best fit to Ti-Al.

Comparison with Experimental Results. The results of the simulations for potential *NTA02* presented here are in accords with experimental observations made of dislocation core structures in γ-TiAl. Work by Hug, *et al.*[24,28,29] shows that any spreading of 1/2<110> dislocations is small, being beyond the limit of resolution of the weak beam technique (~1nm). Hug concluded that the dislocations had, at best, a core structure that was extended in the {111} planes, which is what is observed here for all potentials. Earlier work, Schechtman, *et al.*[30] found no evidence for dissociation of these dislocations.

Recent work by Hemker, *et al.*[31] utilizing Atomic Resolution TEM (ARTEM) indicates no spreading of the dislocation core is observable for the 1/2[110] 'normal' dislocations in the 60° mixed orientation. We find for the screw dislocations simulated with potential *NTA02*, that there is some spreading of the core, but the edge components of this spreading are very small, as shown in Figure 4. Based on this, we would predict that spreading of screw dislocations may occur, but would not be observable in ARTEM. Simulations of other orientations, including the 60° orientation are under way.

SUMMARY AND CONCLUSIONS

We have constructed two additional EAM potentials by fitting selected experimental data from the γ-TiAl phase which produce a stable $L1_0$ structure. When added to the potential that was published earlier by Rao, *et al.*,[17] this forms a set of three potentials with low, intermediate, and high fault energies which can be used to indicate trends in dislocation behavior as the planar fault energies are varied. γ-surfaces were constructed that showed that all four planar faults, APB(111), APB(100), CSF(111), and SISF(111), were stable. Core structures corresponding to the 1/2<110> 'ordinary' dislocations in both screw and edge orientation were simulated with these three potential sets. The effect of stress on the screw orientation of one of these was determined. The following conclusions may be drawn:

(1) Dislocations were not observed to spread on {110} planes, as was suggested by Yamaguchi, *et al.*[25]

(2) Although the core structures observed here were not out of the ordinary, the mobility of the low fault energy screw dislocation was lower than expected. This result is preliminary, but we expect the mobilities for the other two potentials to be even lower.

(3) Spreading was observed for the screw dislocation simulated according to the potential that most closely matches the γ-TiAl phase, but only a negligible amount edge component of this spreading was found. For this reason, this spreading should not be observable in Atomic Resolution TEM for the screw oriented 1/2<110> dislocation.

ACKNOWLEDGMENTS

This work was performed while J.P. Simmons held a National Research Council/Air Force Office of Scientific Research (NRC/AFOSR) research associateship and S.I. Rao held an NRC/AFOSR senior research associateship. We would like to thank Dr. M.S. Daw, of Sandia National Laboratories, for providing us with a copy of the "MDYN" code, a modification of which was used in the EAM computations. We would also like to thank Dr. M. Yoo, of Oak Ridge National Laboratories, for making available to us code for solving anisotropic elasticity problems for straight dislocations. Finally, we would like to acknowledge Drs. T.A. Parthasarathy and C. Woodward for the helpful discussions during the course of this work.

REFERENCES

1. Y.W. Kim, Mater. Res. Soc. Symp. Proc. **213**, 777, (1991).
2. Y.W. Kim and D.M. Dimiduk, J. of Metals, **43**(8), 40, (1991).
3. Y.W. Kim, J. of Metals, **41**(7), 24, (1989).
4. R.L. Fleischer, D.M. Dimiduk, and H.A. Lipsitt, Annual Reviews in materials Science **19**, 231, (1989).
5. S.A. Court, V.K. Vasudevan, and H.L. Fraser, Phil. Mag., **61**, 141, (1990).
6. V. Vitek, Crystal Lattice Defects **5**, 1, (1974).
7. V. Vitek, in Dislocations and Properties of Real Materials, Proceedings of the Conference to Celebrate the Fiftieth Anniversary of the Concept of Dislocations in Crystals, (The Institute of Metals, London, 1985), p. 30.
8. M.S. Duesbery, in Dislocations in Solids, vol. 8, edited by F.R.N. Nabarro (North-Holland, NY, 1989), p. 67.
9. B.A. Grinberg, O.V. Antonova, V.N. Indenbaum, L.E. Karkina, A.B. Notkin, and M.V. Ponomarev, Dislocations in TiAl-Part I, Institute of Metals Physics, Ural Division of the USSR Academy of Science, 3, (1989).
10. B. A. Grinberg, M.A. Ivanov, Yu. N. Gornostyrev, and L.I. Yakovenko, Phys. Met. Metall., **39**, 117, (1978).
11. M.S. Daw and M.I. Baskes, Phys. Rev. Lett. **50**(17), 1285, (1983).
12. R. Pasianot, D. Farkas, and E.J. Savino, J. de Physique, special issue, 'Mechanisms of Deformation and Strength of Advanced Materials,' (1990).
13. S.M. Foiles and M.S. Daw, J. Mater. Res. **2**, 5, (1987).
14. A.F. Voter and S.P. Chen, Mater, Res. Soc. Symp. Proc. **82**, 175, (1990).
15. T.A. Parthasarathy, D.M. Dimiduk, C. Woodward, and D. Diller, Mater. Res. Soc. Symp. Proc. **213**, 337, (1991).
16. T.A. Parthasarathy, S.I. Rao, D.M. Dimiduk, accepted for publication in Phil. Mag.
17. S.I. Rao, C. Woodward, and T.A. Parthasarathy, Mater. Res. Soc. Symp. Proc. **213**, 125, (1991).
18. M. Yamaguchi, V. Vitek, and D.P. Pope, Phil. Mag. A **43**(4), 1027, (1981).
19. M. Yamaguchi, V. Paidar, D.P. Pope, and V. Vitek, Phil. Mag. A **45**(5), 867, (1982).
20. J.H. Rose, J.R. Smith, F. Guinea, and J. Ferrante, Phys. Rev. B **29**, 2963, (1984).
21. S. Takeuchi, Phil. Mag. A, **41**(4), 541, (1980).
22. C. Woodward, J. McLauren, and S.I. Rao, J. Mater. Res. **7**, 1735, (1992).
23. C.L. Fu and M. Yoo, Mater. Res. Soc. Symp. Proc., **186**, 265, (1990).
24. G. Hug, Ph.D. Dissertation (1988), University of Orsay (France).
25. M. Yamaguchi, Y. Umakoshi, and T. Yamane, Dislocations in Solids, 77, (1985).
26. J.P. Hirth and J. Lothe, Theory of Dislocations, (McGraw-Hill, HY, 1968), p. 298.
27. J. F. Nye, Physical Properties of Crystals, (Oxford University Press, London, 1957), p. 142.
28. G. Hug, A. Loiseau, and A. Lasalmonie, Phil. Mag. **A54**, 47, (1986).
29. G. Hug, A. Loiseau, and P. Veyssière, Philos. Mag. **A57**, 499, (1988).
30. D. Schechtman, M.J. Blackburn, and H.A. Lipsitt, Metall. Trans. **5**, 1373, (1974).
31. K.J. Hemker, V. Viguier, and M.J. Mills, in press.

MAGNETIC MOMENT DETERMINATION OF SITE SELECTION OF ADDITIVES (V, Cr, Mn) IN γ-TiAl ALLOYS

V. Suresh Babu, P.K. Khowash*, M.S. Seehra, Bernard R. Cooper, and D.L. Price**
Dept. of Physics, West Virginia University, Morgantown, WV 26506-6315
*Present address, AT&T Bell Labs., Middletown, NJ 07748
**Dept. of Physics, Memphis State University, Memphis, TN 07748

ABSTRACT

We discuss our experiments and coordinated computational modeling that show the value of magnetic moment measurements for helping to identify the site selection of an additive to an intermetallic compound. Magnetic susceptibility measurements (5K-300K) were used to determine additive (V, Cr, Mn) moments; and the site distribution was determined experimentally by measuring the relative variations of the (001) and (110) x-ray diffraction superlattice lines and comparing to the expected behavior for occupation of either site, or a distribution between sites. This gives us a predicted relationship of measured additive moment to site distribution which was then further validated by comparison to our ab initio calculation of the moments for Ti site occupation (as expected at low additive concentrations from total energy considerations). The moments (calculated using the cluster discrete variational (LCAO) method with isolated impurities) of 1.21, 2.36 and 2.52 μ_B for V, Cr and Mn, respectively compare favorably to the lowest-concentration experimental values of 1.01 and 2.3 μ_B for V and Mn, respectively; while the disagreement for Cr with low-concentration experimental moment of 0.55 μ_B may be due to additive clustering effects not included in the single-additive-atom cluster calculations or to multivalent behavior of Cr. The experimental decrease of V and Cr moment with concentration indicates a shift from Ti site occupancy toward increasing Al site occupancy with increase in V and Cr concentration.

INTRODUCTION

TiAl-based alloys are promising for high-temperature aerospace applications because of their high elastic modulii, low densities and high melting point. However, the pure material is brittle at room temperature; and it is of great interest to examine the mechanism by which small additions of other transition metal elements can improve the ductility. An important aspect of this mechanism is the site selectivity of the additive. In this paper we discuss our coordinated experimental[1,2] and computational modeling[3] study that demonstrates the use of magnetic moment measurements as an efficient and accurate means to identify the site selection of the additives. We have examined the series of additive Mn, Cr, V in γ-TiAl since the role of these additives is of practical interest[4,5].

EXPERIMENT

Experimental Procedures

Button ingots of $Al_{50}Ti_{50-x}Mn_x$ (x=0,0.77,1.85,3.30), $(Ti_{50}Al_{50})_{100-x}Cr_x$ (x=0.17, 0.41,0.88,1.14,1.91), and $(Ti_{50}Al_{50})_{100-x}V_x$ (x≠0.12,0.33,0.71,1.21) were prepared by arc melting in argon atmosphere. Thin slices from the ingots were used for x-ray diffraction measurements[1], and small pieces of these slices were used for magnetic measurements[2]. All x-ray diffraction patterns were taken with a fully automated Rigaku/DMax diffractometer using Cu-K_α radiation. For measurements of the magnetization and magnetic susceptibility χ, a SQUID magnetometer was used and data were taken from 5K to 300K using a field of 2*k*Oe. (The magnetic susceptibilities were found to be field independent in the range from 0 to 2*k*Oe.)

Determination of site occupancy using x-ray diffraction

The procedure for determining the additive site occupancy is based on comparing experimental and calculated changes in the intensities of the superlattice peaks, such as (001) and (110), as a function of additive concentrations[1]. The structure factor (F) used in the calculations has been taken for three cases: 100% Ti occupancy, 100% Al occupancy, and random occupancy. Then we compare the relative intensity $I_R\ (hkl)$

$$I_R(hkl) = \frac{I_{SL}(hkl)/I(111) \quad \text{for } x > 0}{I_{SL}(hkl)/I(111) \quad \text{for } x = 0} \tag{1}$$

for the (hkl) superlattice intensity ($I_{SL}(hkl)$) to experiment for each of the three possibilities as shown in Fig. 1 for the (001) superlattice peak. The intensity $I \propto |F|^2$, and the calculated F is different for the three cases[1].

Determination of additive magnetic moment

The data for the magnetic susceptibility χ versus temperature were analyzed using the relationship

$$\chi = \chi_P + C(x)/[T - \Theta(x)] \tag{2}$$

where χ_P represents all temperature independent contributions such as the Pauli and diamagnetic susceptibilities and the second term is the Curie-Weiss contribution from the dopant with concentration x. Note that $C(x) = N_o\mu^2x/3k_B$ where N_o is the Avogadro's number and μ is the magnetic moment. χ_P was evaluated by plotting the experimental data of χ vs 1/T and determining the limit for 1/T going to zero. C(x) and Θ(x) were determined by plotting $(\chi - \chi_P)^{-1}$ vs temperature T and using a least-squares linear fit to determine the slope C(x) and intercept Θ(x). The magnetic moments versus additive concentration x are plotted in Fig 2.

CALCULATION OF ADDITIVE MOMENT

In earlier calculations[3] using all-electron full-potential total energy calculations for a linear combination of muffin-tin orbitals (LMTO) method, large supercells were used to identify the site occupied by the additive. To do this, the total energy with the different transition metals substituted first at the Ti site and then at the Al site were calculated. It was concluded that all the additives discussed here (V, Cr, Mn) prefer to substitute at the Ti site in γ-TiAl, but with widely differing energies. Thus at present in calculating the additive moments in cluster calculations we have placed the additive atoms subsitutionally at Ti sites.

We have calculated the charge distributions, magnetic behavior and density of states in pure and Mn, Cr, and V-added TiAl (with the additive at a Ti site) using the cluster discrete variational combination of atomic orbitals method[6,7]. In these calculations a thirteen-atom cluster, comprised of a central Ti atom, eight nearest-

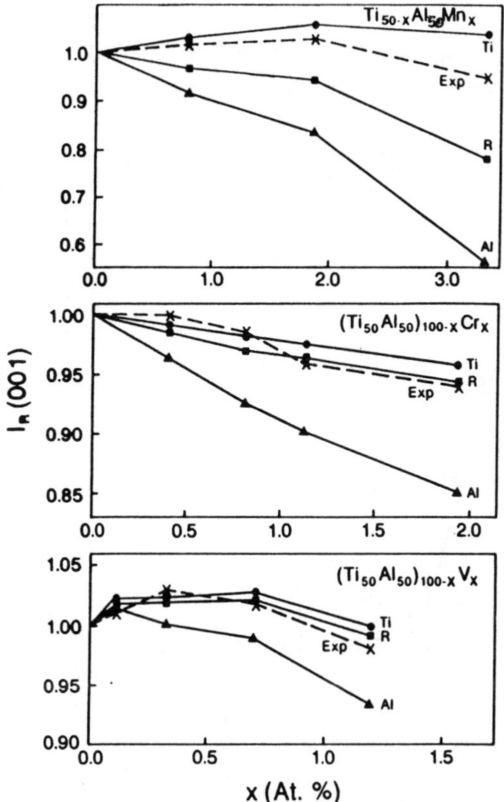

Fig. 1. The experimental and calculated variations of the relative intensities of the (001) superlattice peak as a function of the concentration x of V, Cr and Mn.

neighbor Al atoms and four titanium atoms in the next-nearest-neighbor shell is extracted from the solid. A microcrystal consisting of 250-300 atoms is generated around the cluster to mimic the overall solid environment in which the cluster is placed. The one-electron Schrodenger equation is solved self-consistently with the exchange potential as given in the Slater $X\alpha$ method. Each atom in the microcrystal contributes a spherically symmetrized coulomb potential and also contributes to the exchange interaction. The sum of these potentials results in a non-spherical cluster potential. The microcrystal (250-300 atoms) extends out to a radius of about 20 a.u., thereby making the peripheral atoms in the cluster (13 atoms as described above) sense a realistic potential as in the bulk. Within this potential, the cluster eigenfunctions are expanded as a linear combination of atomic orbitals (LCAO). The basis sets for Ti and the additives are Ar core, 3d, 4s, 4p and for Al are Ne core, 3s, 3p.

To investigate the additive-induced change in charge distributions, magnetic

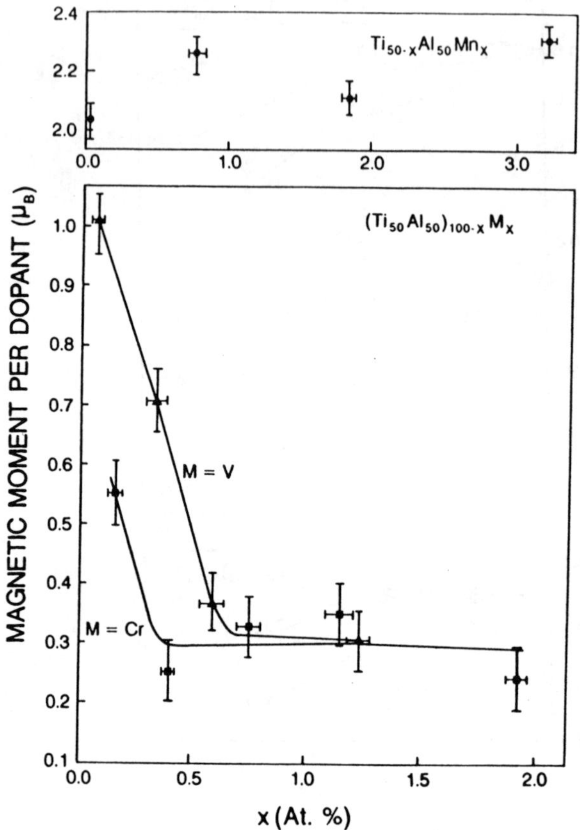

Fig. 2. Variations of the magnetic moments with the concentration x of the dopants V, Cr and Mn. The solid lines are drawn through the points as guides.

behavior and density of states, we have allowed the possibility of two different types of Ti potential---one for the central Ti replaced by the additive, and the other for the second-nearest-neighbor Ti atoms. The centrally located Ti atom is then conveniently replaced by an additive atom. As shown in Table 1, after a charge redistribution within its own valence states and with the Al neighbors, the 3d states of the central Ti before substitution settle down with 2.11 electrons and almost no magnetic moment (0.001 μ_B). As shown in Table 1 substituting V, Cr and Mn for the central Ti in the cluster gives rise to charge redistributions within the 13-atom cluster, and results in net moments of 1.21 μ_B, 2.36 μ_B and 2.52 μ_B, respectively. The density of states (DOS) for the additive-substituted clusters have 3d states more narrowly distributed in energy and with shifts between the up and down-spin DOS giving rise to the calculated moments.

Table I. Charge (electrons, e^-) and spin (μ_B) distribution for γ-TiAℓ and for different transition metals substituted at the Ti site in γ-TiAℓ. (Calculated for 13-atom cluster embedded in microcrystal.).

Central Site	(a) Ti		(b) V		(c) Cr		(d) Mn	
	charge (e^-)	spin (μ_B)	charge	spin	charge	spin	charge	spin
3d	2.11	0.001	3.24	0.846	4.23	2.260	5.24	2.417
4s	1.44	0.001	1.42	0.272	1.43	0.040	1.44	0.044
4p	0.27	0.000	0.25	0.093	0.21	0.054	0.25	0.060
Net	0.19	0.002	0.09	1.211	0.06	2.358	0.07	2.522
Nearest-neighbor Aℓ (8 such atoms in cluster)								
3s	1.94	0.000	1.93	0.002	1.94	0.002	1.94	0.001
3p	1.22	0.001	1.21	0.057	1.18	0.034	1.19	0.015
Net	-0.16	0.001	-0.14	0.059	-0.12	0.036	-0.13	0.016
Nearest-neighbor Ti (4 such atoms in cluster)								
3d	2.05	0.001	2.06	0.085	2.06	0.151	2.07	0.132
4s	1.59	0.000	1.58	0.006	1.61	0.011	1.59	0.013
4p	0.09	0.000	0.09	0.009	0.10	0.009	0.11	0.007
Net	0.27	0.001	0.27	0.099	0.22	0.171	0.236	0.152

RESULTS AND DISCUSSION

The concentration dependence of the (001) superlattice peak intensity as shown in Fig. 1, as well as that for the (110) peak, indicates that, at least up to 3 at %, Mn substitutes for Ti. However, while for initial doping Cr and V prefer the Ti site, for higher concentrations Ti and Al sites seem to be occupied with nearly equal preference. This is consistent with the magnetic moment behavior shown in Fig. 2. For Mn, the moment per Mn atom is nearly constant with concentration at about 2.3 μ_B atom, while for V and Cr, the moment decreases with concentration being 0.55 μ_B/Cr and 1.01 μ_B/V at the lowest concentrations. We are led to conclude that the dopants carry larger moments when occupying the Ti site and practically no moment when occupying the Al site. For Mn and V the calculated moments of 2.52 μ_B and 1.21 μ_B from our cluster calculations for isolated impurities compare favorably to the experimental values of 2.3 μ_B and 1.01 μ_B, respectively. However, for Cr there is a substantial discrepancy between the calculated value of 2.36 μ_B and the experimental value of 0.55 μ_B. This discrepancy could arise from Cr clustering effects, which lower the moment, being present experimentally even at the lowest concentrations or from the complications of multivalent electronic behavior for Cr not being accurately captured by the one-electron-potential theory used.

In conclusion, the results presented here have shown that for initial dopings of γ-TiAl, Mn, V and Cr occupy the Ti sites; and that the dopants carry localized moment when occupying the Ti sites and negligible moment when occupying the Al sites. Therefore, in this system, the observed magnetic moment provides a measure of the site occupancy of the dopant. As further confirmation, calculations should be done to verify that Al site occupancy is expected to yield negligible magnetic moments.

REFERENCES

1. V. Suresh Babu and M.S. Seehra, J. Mater. Res. 6, 339 (1991).

2. J. Coletti, V. Suresh Babu, A.S. Pavlovic and M.S. Seehra, Phys. Rev. B 42, 10,754 (1990).

3. P.K. Khowash, D.L. Price and B.R. Cooper, Mat. Res. Soc. Symp. Proc. 213, 31 (1991).

4. T. Hanamura, R. Uemori and M. Tanino, J. Mater. Res. 3, 356 (1988).

5. E.L. Hall and S.C. Huang, J. Mater. Res. 4, 595 (1989).

6. D.E. Ellis and G.S. Painter, Phys. Rev. B 2, 2887 (1970).

7. P.K. Khowash and D.E. Ellis, Phys. Rev. B 36, 3394 (1987), Phys. Rev. B 39, 1908 (1989).

THE ROLE OF LOCAL DISORDER NEAR GRAIN BOUNDARY IN DUCTILIZATION OF Ni_3Al

JIAN SUN AND DONGLIANG LIN(T.L. Lin)
Department of Materials Science and Engineering, Shanghai JiaoTong University, Shanghai 200030, P.R.China

ABSTRACT

The local chemical order near grain boundaries (GBs) in Ni_3Al was verified by measuring the local degree of order and chemistry with microdiffraction and EDS microanalysis. The composition of alloys are (a) Ni-24 at%Al with and without 700ppm wt boron;(b) Ni-25 at%Al with 700ppm wt boron;(c) Ni-26 at%Al with 700ppm wt boron. GBs studied in the present work are high angle boundaries. The results show that there is a local disordered region about 5-10nm wide on the side of the GB in non-stoichiometric Ni_3Al with or without boron, and no disordered region in stoichiometric Ni_3Al. EDS microanalysis results show that the composition of GB almost keeps the matrix concentration of Ni in stoichiometric Ni_3Al with boron, and is Ni-rich in hypostoichiometric and Al-rich in hyperstoichiometric Ni_3Al with and without boron. The local disordered region existing in the vicinity of GBs is related to Ni enrichment or depletion near GBs in hypostoichiometric or hyperstoichiometric Ni_3Al alloys. Boron seems not to be the main factor to control the local disorder and composition at GBs in Ni_3Al alloys. It can be concluded that the local disorder at GBs is not a decisive factor for ductilizing GBs in Ni_3Al alloys.

INTRODUCTION

The Ni_3Al compound has received considerable attention during the past decade owing to its potential high temperature applications. However, grain boundaries (GBs) in Ni_3Al are intrinsically brittle. Izumi et al.[1] first discovered that microalloying with a trace of boron increases the ductility of polycrystalline Ni_3Al. Liu et al.[2] have systematically investigated the beneficial effect of boron on ductility and fracture behaviour of Ni_3Al alloys and have found that boron tends to segregate to GBs. The concentration of boron at GBs decreases with increasing bulk aluminium concentration from 24 at% to 25 at%, with a corresponding reduction of ductility from 50% to 6%. Two hypotheses have been suggested to explain the ductilizing effect of boron: (1) Boron enhanced cohesive strength of GB, and (2) boron facilitated slip transfer across GB. It was suggested that disordering in the GB regions might lead to enhance ductility by an increased number of permissible dislocation reactions with these brittle interfaces. Several experimental studies have attempted to check these hypotheses. Early results from high resolution energy dispersive x-ray spectroscopy and atom probe analysis[3,4] showed that the boundary regions were Ni-rich in boron doped Ni_3Al, whereas in boron free Ni_3Al the boundaries had the bulk composition, suggesting that the excess Ni at the boundary made it effectively partially disordered. However George et al.[5] and Krzanoski[6], using AES and STEM respectively, showed that GB in both boron doped and boron free Ni_3Al had a similar Ni enrichment. Their results indicated no strong co-segregation between boron and nickel. Similar uncertainty exists in the determination of the local structure at the GB region. Baker et al. have observed a 20 nm wide second phase along GB in hypostoichiometric Ni_3Al with boron. Zhu et al.[7] reported no disordered phase at GB, and a decrease in order in the vicinity of a small angle tilt GB in hypostoichiometric Ni_3Al with boron.

Mackenzie and Sass[8], using the HRTEM technique, showed a 1.5nm disordered region at high angle boundaries in a hypostoichiometric Ni_3Al containing 300 wt. ppm boron. Mills' observation suggested that the disordering phenomenon may not be a general one[9]. Recently, Kung et al. also founded a disordered region about 1.5-2.0 nm thick at large angle general boundaries, but no such disordered region at twist and low Σ values boundaries[10]. Horton et al.[11] observed the disordered precipitates on grain boundaries in Ni_3Al alloys with Al content from 21.5 at% to 22.5 at%, but no evidence of any second phases at higher aluminum levels. Since similar high ductilities were found for alloys with and without grain boundary phases, they suggested that the ductility improvements in boron-doped Ni_3Al is not related to the disordered phases on grain boundaries.

However, the experiments mentioned above were not involved in stoichiometric or hyperstoichiometric Ni_3Al alloys. In order to fully understand the role of boron in ductilizing Ni_3Al, we need a comparison between GBs in hypostoichiometric Ni_3Al with boron or without boron, and stoichiometric and hyperstoichiometric Ni_3Al with boron, regarding the local chemistry and structure at the GBs. In this work, the local degree of order and chemistry was measured by comparing the intensity of fundamental spots with the intensity of superlattice spots in microdiffraction and EDS microanalysis. The mechanisms of the ductilizing effect of boron are then discussed with these results.

EXPERIMENTAL

The alloys used in this study were prepared by directional solidification, which results in a columnar grain structure with a moderate [001] texture. All of them were annealed at 1423 K for 50 hours, followed by furnace cooling. The compositions of the alloys are (a) Ni-24 at% Al with or without 700ppm wt boron; (b) Ni-25 at% Al with 700ppm wt boron; (c) Ni-26 at% Al with 700ppm wt boron. Pieces of 3mm in diameter were cut off normal to the growth direction, and mechanically ground to a thickness of 0.05mm, then electropolished using a 10% sulfuric-methanol solution at 243 K and 25 V. Specimens were ion milled for 2 hours immediately prior to examination in the TEM. This foil orientation is suitable for determining the degree of chemical disorder. The GB microanalysis were conducted using a Philips CM-12 Analytical Electron Microscope under nano-probe mode. A probe with diameter of 3.3nm was aligned on the specimen to scan the vicinity of a GB to conduct a series of microdiffraction and EDS microanalysis step by step. The typical counting time was 300 seconds.

RESULTS AND DISSCUSSION

A general examination of the TEM specimens revealed only a single phase structure. GBs studied in the present work were random high angle boundaries, as shown in Fig. 1.

Fig. 1 The image of random high angle grain boundary.

Figure 2 shows the [001] microdiffraction patterns of the sample, where the weaker diffraction spots around the transmission spot were the superlattice ones and the stronger diffraction spots are fundamental ones. The change of the ratio between the intensity of the superlattice and fundamental spots can be considered as a criterion for the change of the long range order in the vicinity of GBs. Figures 2a-2c show that there exists a region of partial disorder within 5-10 nm near GBs in hypostoichiometric Ni_3Al with or without 700ppm boron, and hyperstoichiometric Ni_3Al with 700ppm boron. The degree of order almost approaches zero at the region very close to the GB, then gradually increases as a function of distance from the boundary. The degree of order in stoichiometric Ni_3Al with 700ppm boron stays higher in the interior of grains and at the GB, as shown in Fig.2d. These results suggest that there is a region of partial disorder about 5-10nm wide on the side of the GB in non-stoichiometric Ni_3Al with or without boron, except for the stoichiometric Ni_3Al alloy.

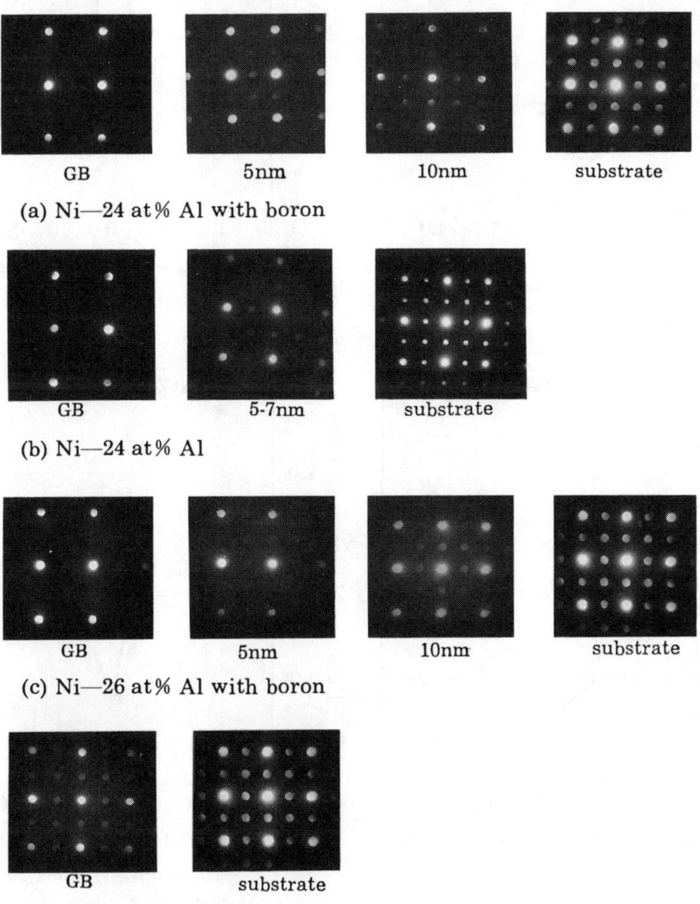

Fig. 2 The [001] microdiffraction patterns show a region of partial disorder near grain boundary in Ni_3Al alloys.

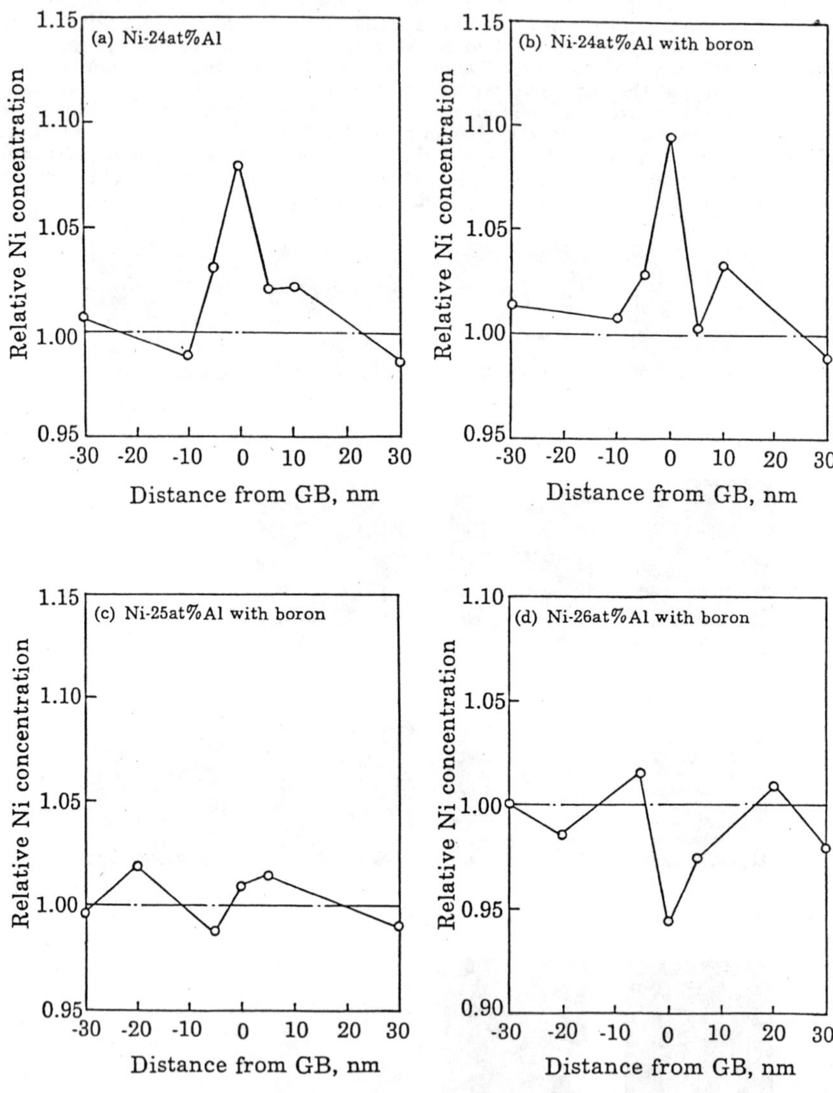

Fig. 3 EDS—microanalysis results of grain boundary and adjacent Ni_3Al alloys.

In order to understand the reason why a region of partial disorder exists near grain boundaries in non-stoichiometric Ni_3Al alloys, the compositions of grain boundaries in these four alloys were analysed. The EDS microanalysis results are shown in Figure 3. In each case, the ratio of Ni concentration at a GB and in the matrix measure the enrichment or depletion of Ni in GBs. Figures. 3a and 3b show the GBs to be Ni-rich in hypostoichiometric Ni_3Al with or without boron. There is no tendency for co-segregation between boron and nickel. The composition of GB almost keeps the matrix concentration of Ni in stoichiometric Ni_3Al with boron, and is depleted in Ni in hyperstoichiometric Ni_3Al with boron, as shown in Figs. 3c and 3d.

The microdiffraction and EDS microanalysis of GBs in four alloys imply that the local disordered region existing in the vicinity of grain boundaries is related to Ni enrichment or depletion in grain boundaries in hypostoichiometric or hyperstoichiometric Ni_3Al alloys. Boron seems not to be the main factor to control the local disorder and composition at GBs in Ni_3Al alloys. Monte Carlo calculations[12,13] using Embedded Atom Method potentials predict that the grain boundary acts as a sink for antisite defects, and the grain boundary composition exhibits nickel or aluminum enrichment compared to the bulk composition in hypostoichiometric or hyperstoichiometric Ni_3Al alloys, which results in a region of partial disorder near grain boundaries. Our results seem to be in agreement with these calculations, but the region of partial disorder is broader than that predicted theoretically. On the other hand, boron is effective in ductilizing Ni_3Al only with hypostoichiometric compositions but is ineffective in stoichiometric and hyperstoichiometric compositions. From the results of this study, it can be concluded that local disorder at grain boundaries is not a decisive factor for ductilizing grain boundaries in Ni_3Al alloys and avoiding intergranular fracture. The results presented here support the conclusion that boron segregation at grain boundaries plays the key role in the ductilization Ni_3Al alloys, which may lead to the increase of GB cohesive strength and/or more effectiveness for dislocation absorption and emission from the GBs by the changes of GB structure itself.

CONCLUSIONS

There is a local disordered region about 5-10 nm thick on the side of the GBs in non-stoichiometric Ni_3Al with and without boron, and no disordered region in stoichiometric Ni_3Al with boron. The local disordered region existing in the vicinity of GBs is related to Ni enrichment or deletion near GBs in hypostoichiometric or hyperstoichiometric Ni_3Al alloys. Boron seems not to be the main factor to control the local disorder and composition at GBs in Ni_3Al alloys. The local disorder near GBs does not play an important role in ductilization of Ni_3Al alloys.

ACKNOWLEDGEMENTS

This research was partially supported by the National Key Laboratory of Metal Matrix Composite Materials.

REFERENCES

1. K. Aoki and O. Izumi, Nippon Kinzoku Gakkaishi, **43**, 1190 (1979).
2. C.T. Liu, C.L. White and J.A. Horton, Acta Metall. **33**, 213 (1985).
3. I. Baker, E.M. Schulson, and J.R. Michael, Phil. Mag. B, **57**, 379 (1988).
4. D.D. Sieloff, S.S. Brenner, and M.G. Burke, in High Temperature Ordered Intermetallic Alloys edited by N.S. Stoloff, C,C, Koch, C.T. Liu and O. Izumi (Mat. Res. Soc. Symp. **81**, Pittsburgh, PA, 1987) pp. 87.
5. E.P. George, C.T. Liu and R.A. Padgett, Scripta Metall. **23**, 979 (1989).
6. Jame E. Krzanowski, Scripta Metall. **22**, 1219 (1988).
7. Jing Zhu, Z.Y. Cheng and D.X. Zou, Scripta Metall. **24**, 439 (1990).
8. R.A.D. Mackenzie and S.L. Sass, Scripta Metall. **22**, 1807 (1988).
9. M.J. Mills, Scripta Metall. **23**, 2061 (1989).
10. H. Kung, D.R. Rasmussen, and S.L. Sass, Scrita Metall. et Mater. **25**, 1277 (1991).
11. J.A. Horton, C.T. Liu, S.J. Pennycook, in High Temperature Ordered Intermetallic Alloys edited by L.A. Johnson, D.P. Pope and J.O. Stiegler (Mat. Res. Soc. Symp. **213**, Pittsburgh, PA, 1991) pp. 417.
12. S.M. Foils, in High Temperature Ordered Intermetallic Alloys edited by N.S. Stoloff, C,C, Koch, C.T. Liu and O. Izumi (Mat. Res. Soc. Symp. **81**, Pittsburgh, PA, 1987) pp. 51.
13. M.J. Mills, S.H. Goods and S.M. Foils, in High Temperature Ordered Intermetallic Alloys edited by L.A. Johnson, D.P. Pope and J.O. Stiegler (Mat. Res. Soc. Symp. **213**, Pittsburgh, PA, 1991) pp. 423.

CHARACTERIZATION OF DOPED NiAl BY ATOM PROBE FIELD ION MICROSCOPY

RAMAN JAYARAM AND M.K. MILLER,
Metals and Ceramics Division, Oak Ridge National Laboratory, Oak Ridge, TN 37831-6376.

ABSTRACT

The atom probe field ion microscope (APFIM) has been used to characterize grain boundaries and matrix in NiAl doped with either boron, carbon or beryllium. Boron was observed to segregate to grain boundaries whereas carbon and beryllium did not. Atom probe analyses of the matrix revealed that the matrix was severely depleted of the solute in the boron- and carbon-doped alloys. Field ion imaging and matrix analyses also revealed ultrafine MB_2 - and MC-type precipitates ranging in size between 2 and 20 nm in diameter in the boron- and carbon-doped alloys. These precipitates occurred in significant number densities. Atom probe analyses of beryllium-doped NiAl did not reveal ultrafine precipitates and was consistent with the fact that almost all the beryllium was in solid solution. The enormous increase in yield stress in the boron- and carbon-doped alloys is predominantly due to a precipitation hardening effect. The small increase in yield stress in beryllium-doped NiAl is due to a mild substitutional solid solution hardening effect.

INTRODUCTION

The ordered intermetallics such as Ni_3Al and NiAl have been studied extensively on account of their interesting mechanical properties at elevated temperatures. Typically, these materials exhibit an increase in plastic strength with increase in temperature. This behavior is of great significance for technological applications, such as turbine blades, where high creep strength at high temperatures is a critical requirement. NiAl is particularly attractive from that standpoint since it has a lower density and higher melting point than Ni_3Al. A serious drawback of these ordered intermetallics is their brittleness at room temperature. This has been successfully overcome in the case of Ni_3Al where the addition of a microalloying element such as boron in excess of 200 ppm results in a significant improvement in room temperature ductility [1,2]. This improvement in ductility has been attributed to boron segregation at grain boundaries which results in the transition from an intergranular to transgranular fracture mode. Similar efforts to ductilize NiAl have not been successful [3].

In the present work, the results of recent atom probe analysis of microalloyed NiAl are summarized. The atom probe field ion microscope (APFIM) combines atomic resolution imaging with single atom detection capability making it a powerful tool for the investigation of grain boundary and matrix chemistry [4]. Alloys of NiAl doped with either boron, carbon or beryllium have been analyzed in the atom probe and the results correlated with tensile property and fracture mode reported elsewhere [3].

EXPERIMENTAL

The alloys used in this investigation were prepared by arc melting high purity elemental materials and drop casting into cylindrical copper chill molds. The purity levels of aluminum and nickel were 99.99 wt. % and 99.97 wt. %, respectively. After homogenizing in vacuum for 24 h at 1100°C, the ingots were canned in mild steel and extruded at 900°C to obtain alloys with nominal bulk compositions of: (1) NiAl + 0.04 at. % B, (2) NiAl + 0.12 at. % B, (3) NiAl + 0.1 at. % C and (4) NiAl + 0.24 at. % Be [3]. The 0.04 at. % B alloy was given a heat treatment of 0.5 h at 800°C. The remaining alloy samples were given two separate heat treatments: (1) 1 h at 500°C and (2) 1 h at 1100°C. These temperatures correspond to the partially recrystallized and fully recrystallized microstructures, respectively with the average grain size in the fully recrystallized condition being 30 ± 5 μm [3]. The samples were then water quenched to room temperature. Field ion specimens were prepared from failed tensile specimens using standard techniques [4]. Since the apex region of the field ion microscope (FIM) specimen measures approximately 100 nm in diameter, the chances of randomly producing a specimen with a grain boundary in the apex region are very small in a material of such large grain size. Therefore, the specimens were preexamined in the transmission electron microscope (TEM) and subsequently either back-polished or ion-milled in a GATAN 645 precision ion milling system (PIMS) in order to position a grain boundary in the analyzable volume of the specimen. The specimens were then imaged and analyzed in the Oak Ridge National Laboratory energy-compensated atom probe with a specimen temperature of 50 K and pulse fraction in excess of 15 % [5].

RESULTS AND DISCUSSION

A field ion micrograph of a grain boundary in a specimen of the boron-doped alloy is shown in Fig. 1. As previously reported, grain boundary images in this material are characterized by substantial decoration as indicated by the brightly-imaging spots [6,7,8]. Atom probe analyses revealed these brightly-imaging spots were boron atoms. The boron coverage ranged from 8 to 22 % of a monolayer in the 0.04 at. % B alloy. These measurements indicated that in the alloy containing 0.04 at. % boron, approximately 5 - 6 % of the total amount of boron in the bulk segregated to grain boundaries. Field ion images of grain boundaries in carbon-doped NiAl were marked by an absence of decoration as shown in Fig. 2. This observation was consistent with atom probe analyses [9] and Auger measurements [3] which did not detect any significant carbon segregation to the grain boundaries. No decoration was observed in field ion images of grain boundaries in beryllium-doped NiAl [12]. Atom probe [12] and Auger analyses [10] of grain boundaries in beryllium-doped NiAl also failed to detect any significant segregation [10].

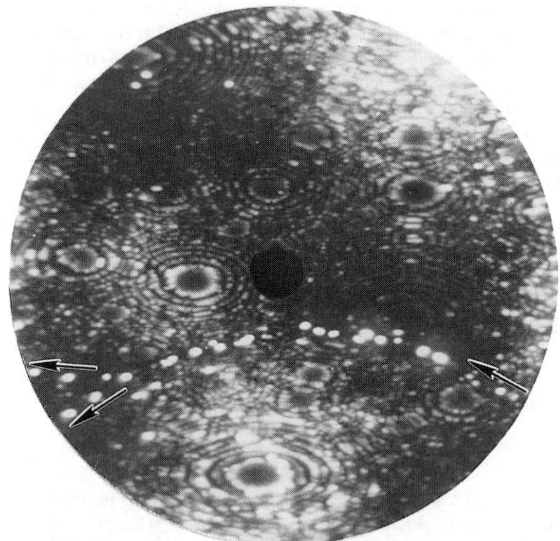

Fig. 1. Field ion micrograph of a NiAl + 0.04 at. % B specimen showing brightly-imaging boron atoms along a grain boundary (arrowed).

Atom probe measurements of the matrix revealed that in the boron and carbon-doped alloys the matrix was severely depleted of the solute. The boron concentration that remained in solid solution in the 0.04 at. % alloy was 0.026 ± 0.003 at. % B [6]. In the 0.12 at. % B alloy, the boron concentrations were 0.003 ± 0.0007 at. % and 0.026 ± 0.003 at. % B in the samples that were

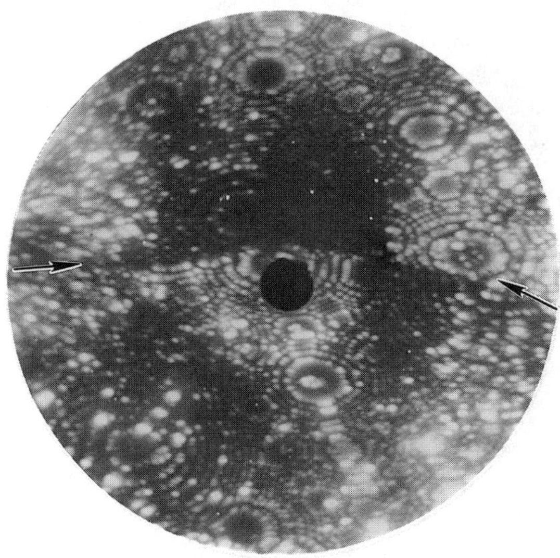

Fig. 2. Field ion micrograph of carbon-doped NiAl specimen showing the absence of decoration along a grain boundary (arrowed).

annealed for 1 h at 500° C and 1 h at 1100° C, respectively [7]. Similar trends were observed in the carbon-doped alloy where the carbon concentrations were 0.003 ± 0.001 at. % and 0.006 ± 0.002 at. % C in the materials that were annealed for 1 h at 500° C and 1 h at 1100° C, respectively [8]. In contrast, in the beryllium-doped alloy almost all of the beryllium remained in solid solution. The beryllium concentration was between 0.16 ± 0.04 at. % and 0.24 ± 0.03 at. % Be in the alloy that was annealed for 1 h at 500° C and 0.16 ± 0.02 at. % Be in the alloy that was annealed for 1 h at 1100° C [11]. The variations in beryllium level are presumably due to local inhomogeneities in the beryllium distribution.

The low solubility levels of boron and carbon in NiAl at the annealing temperatures used in this study and the fact that only about 5 % of the boron segregates to grain boundaries suggested that precipitate phases must have formed in the matrix. This was confirmed by extensive field ion imaging of the matrix which revealed ultrafine brightly-imaging precipitates, an example of which is shown in Fig. 3 [6,7]. The size of these particles typically ranged from 2 to 20 nm in diameter. Atom probe analysis of the larger precipitates showed the composition to be consistent with a MB_2-type stoichiometry with the metallic component consisting predominantly of titanium with smaller proportions of vanadium, chromium and tungsten. The amount of boron in precipitates is determined by subtracting the sum of the amount of boron in solid solution and at grain boundaries from the bulk value. The number densities of the precipitates for various particle sizes can then be estimated using this result and the assumption that the MB_2 precipitates have a crystal structure and stoichiometry identical to that of TiB_2. Thus, the two extreme estimated values for the number densities are 1 x 10^{20} m^{-3} and 1 x 10^{23} m^{-3} based on precipitate diameters of 20 nm and 2 nm, respectively. However, since both atom probe and TEM measurements revealed that there is a distribution in the particle size, the actual number density of precipitates is between these two extreme values. A typical size distribution is shown in a transmission electron micrograph in Fig. 4

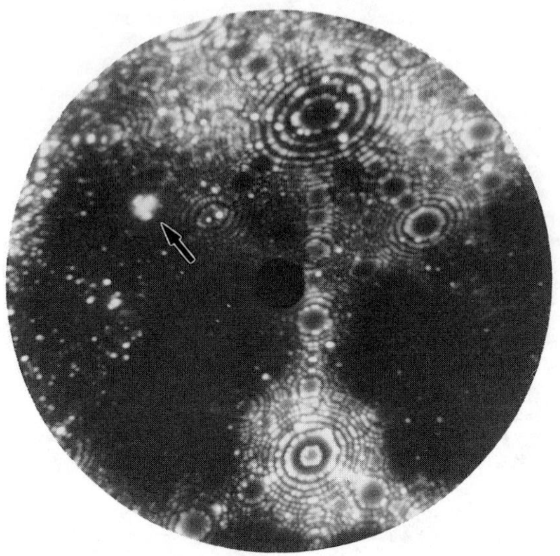

Fig. 3. Field ion micrograph of a boron-doped NiAl specimen showing a brightly-imaging MB_2-type precipitate (arrowed) in the matrix.

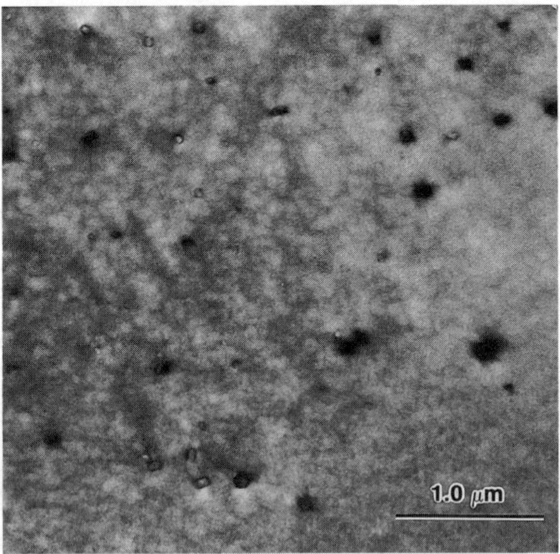

Fig. 4. Bright-field transmission electron micrograph of a NiAl + 0.04 at. % B specimen showing MB_2-type precipitates.

of a 0.04 at. % B specimen where the number density was estimated at 1 x 10^{22} m^{-3}. Brightly-imaging ultrafine precipitates have also been observed in FIM images of carbon-doped NiAl specimens and their compositions were consistent with a MC-type stoichiometry [9]. Similar to the boron-doped material, the metallic component consisted of predominantly titanium along with vanadium, chromium and tungsten in smaller proportions. The precipitate sizes and number densities were also similar to those encountered in the boron-doped material [9]. Atom probe analysis of the matrix in beryllium-doped NiAl did not indicate the presence of ultrafine precipitates containing beryllium. TEM examination, however, revealed the presence of coarse, micron-sized MC precipitates and the number density was qualitatively estimated to be significantly lower than those in the boron- and carbon-doped NiAl. The metallic component predominantly consisted of zirconium with smaller amounts of titanium and tungsten [11]. In all three alloys, the precipitates presumably originated from trace metallic impurities present in nickel. Titanium and chromium at concentration levels ranging from a few tens upto a hundred ppm are often detected in typical bulk analyses of commercial grade nickel at the level of purity used in this study. Based on atom probe analyses, it is estimated that total trace impurity levels of approximately 300 - 500 appm are sufficient to generate precipitates in number densities encountered here [9]. This estimate appears to be consistent with the fact that the total trace impurity concentration in the nickel used to fabricate these alloys was approximately 300 - 400 appm. The extremely low solubility of boron and carbon in the NiAl matrix causes the excess solute to react with trace impurities during thermal aging to form precipitates.

The tensile properties of microalloyed NiAl are significantly different from those of boron-doped Ni_3Al [3]. The yield stress increases from 154 MPa for the undoped NiAl to 329 MPa for the boron-doped NiAl and to 336 MPa for the carbon-doped NiAl [3]. The enormous increase in yield stress is accompanied by an absence of tensile elongation with fracture occurring prior to macroscopic yielding. This result is, at first, somewhat unexpected for the boron-doped alloy since scanning electron microscope (SEM) fractographs have revealed that boron does indeed strengthen grain boundaries by forcing the transition from an intergranular to transgranular fracture mode [3]. SEM fractographs of carbon-doped NiAl and beryllium-doped NiAl revealed the persistence of the intergranular fracture mode and were consistent with atom probe and Auger results which did not indicate segregation to the boundaries. The tensile behavior of beryllium-doped NiAl is quite different from that of the other two alloys in that the yield stress increase is only 24 MPa and is accompanied by a tensile elongation of 3.0 % [3].

It is clear that the observed increase in yield stress in these alloys must come from a combination of solid solution and precipitation hardening. Using the solute concentration as measured by atom probe analyses, the contributions to yield stress increase from solid solution hardening in boron- and carbon-doped NiAl have been made based on a model that assumes that boron and carbon occupy octahedral interstitial sites in the NiAl lattice and behave similar to carbon in iron [6]. According to this estimate, 0.026 at. % B and 0.003 at. % B in solution contribute 50 MPa and 17 MPa, respectively to the yield stress increase. The corresponding estimates for carbon are 17 MPa and 24 MPa for 0.003 at. % C and 0.006 at. % C, respectively [9]. The contribution to the yield stress increase arising from precipitation hardening has been estimated based on a simple Orowan mechanism [6,7]. In both boron- and carbon-doped NiAl, this contribution ranges between 280 MPa to 50 MPa for particle diameters lying between 2 and 20 nm, respectively. Hence, it is clear that a precipitation hardening mechanism can easily account for the tensile behavior of the boron- and carbon-doped NiAl alloys. It has been previously shown by APFIM that beryllium resides almost exclusively at aluminum sites in the NiAl lattice [11]. The small increase in yield stress in the beryllium-doped NiAl, therefore, comes solely from a relatively weak substitutional solid solution hardening mechanism.

CONCLUSIONS

APFIM measurements have shown that boron segregates to grain boundaries in NiAl whereas carbon and beryllium do not. Segregation of boron is accompanied by transition to a transgranular fracture mode but the intergranular fracture mode persists in carbon- and beryllium-doped NiAl. The matrix is substantially depleted of boron and carbon in the boron- and carbon-doped alloys, respectively. The majority of the excess solute, given by the difference between the bulk and the measured concentration, is in the form of highly stable MB_2- and MC-type refractory phases. These ultrafine precipitates occur in number densities significant enough to make substantial contributions to the yield stress increase. The beneficial effect of boron at the grain boundaries is, therefore, masked by the tremendous strengthening of the grains due to precipitation hardening. In the carbon-doped NiAl, the combination of precipitation hardening and the absence of grain boundary segregation results in a brittle material. Since the precipitates originate from trace impurities in NiAl, these results have important implications for the technological applications of NiAl. Beryllium additions to NiAl do not strengthen grain boundaries and contribute to a mild substitutional solid solution hardening.

ACKNOWLEDGEMENTS

The authors would like to thank Dr. C.T. Liu for supplying the alloys used in this investigation and K.F. Russell for her technical assistance. This research was sponsored by the Division of Materials Sciences, U.S. Department of Energy, under contract DE-AC05-84OR21400 with Martin Marietta Energy Systems, Inc. and through the Postgraduate Research Program administered by Oak Ridge Institute for Science and Education.

REFERENCES

1. K. Aoki and O. Izumi, J. Jpn. Inst. Met., **43**, 1190, (1979).
2. C.T. Liu, C.L. White and J.A. Horton, Acta Metall., **33**, 213, (1985).
3. E.P. George and C.T. Liu, J. Mater. Res., **5**, 754, (1990).
4. M.K. Miller and G.D.W. Smith, Atom Probe Microanalysis: Principles and Applications to Materials Problems, (Materials Research Society Publishers, Pittsburgh, PA, 1989).
5. M.K. Miller, J. de Physique, **47-C2**, 493, (1986).
6. R. Jayaram and M.K. Miller, Surface Science, **266**, 310, (1992).
7. R. Jayaram and M.K. Miller in Structure and Properties of Interfaces in Materials, edited by W.A.T. Clark, U. Dahmen and C.L. Briant (Mater. Res. Soc. Proc. **238**, Pittsburgh, PA 1991) pp. 445.
8. M.K. Miller, Raman Jayaram and P.P. Camus, Scripta Metall., **26**, 679, (1992).
9. R. Jayaram and M.K. Miller, Applied Surface Science, **66**, (1993), in press.
10. E.P. George and C.T. Liu, unpublished results.
11. R. Jayaram and M.K. Miller, Applied Surface Science, **66**, (1993), in press.
12. R. Jayaram and M.K. Miller, Acta Metall., submitted.

ATOMIC STRUCTURE OF THE NIAL $\Sigma=5$ (310) INTERFACE

RICHARD W. FONDA AND DAVID E. LUZZI
Laboratory for Research on the Structure of Matter and Department of Materials Science and Engineering, University of Pennsylvania, Philadelphia, PA 19104-6272

ABSTRACT

The structure of a highly oriented NiAl $\Sigma=5$ [001] (310) grain boundary has been imaged by high resolution electron microscopy and analyzed by multislice image calculations. This grain boundary exhibits an asymmetry produced by a rigid body displacement of approximately ½ $d_{1\bar{3}0}$ along the grain boundary, but does not show any measurable expansion of the grain boundary. The match between the simulated and experimental images of this grain boundary is improved by changing the stoichiometry of the boundary, which can be accomplished by adding antisite defects and vacancies. The final model structure incorporates 50% constitutional vacancies on the nickel sites adjacent to the boundary plane and a replacement of the aluminum sites adjacent to the boundary plane with 50% nickel antisite defects and 50% vacancies.

INTRODUCTION

There has been much recent interest in the development of intermetallic compounds for high temperature applications due to their high temperature strength and resistance to corrosion and oxidation. One alloy under current development for these high temperature applications is the B2-ordered NiAl phase, which has a lower density, higher thermal conductivity, and higher melting point than currently used nickel-base superalloys.[1] However, as in most intermetallic compounds, the grain boundaries in NiAl are intrinsically brittle at low temperatures.[2] Further development of this alloy therefore requires an understanding of the atomic grain boundary structure; this structure can best be determined by high resolution electron microscopy and multislice image simulation.

Although many theoretical and experimental studies have been conducted on Ni_3Al grain boundaries, there have been few studies on the structure of NiAl grain boundaries. The effect of non-stoichiometry within the bulk, however, has been well characterized in a classic study by Bradley and Taylor.[3] This study showed that while nickel-rich compositions are produced by nickel antisite defects on the aluminum sublattice, aluminum-rich compositions are generated by constitutional vacancies on the nickel sublattice. The grain boundary energies of many structures for the stoichiometric NiAl [001] (310) grain boundary (the orientation examined in the current study) have been calculated by Paten and Farkas.[4] Using the embedded atom method, Chen *et al.*[5] determined that the effect of non-stoichiometry in this alloy is to produce vacancies of the minority element at grain boundary sites. This is consistent with recent field ion microscopy results on nickel-rich alloys, which show a sharp drop in the Al:Ni ratio within a few Angstroms of the boundary.[6-7]

In this paper we present the preliminary results of an HREM examination of the $\Sigma=5$ [001] (310) grain boundary in NiAl. This grain boundary was prepared by diffusion bonding two single crystals with a 36.87° rotation about the common [001] axis. The subgrains produced during diffusion bonding provide many regions of well-aligned $\Sigma=5$ [001] (310) grain boundary which are suitable for high resolution imaging. A high resolution image of this boundary was compared to simulated images generated by the multislice algorithm in order to address the atomic structure of the grain boundary. This analysis considered the effects of grain boundary expansion, rigid body displacements along the boundary, grain boundary stoichiometry, and point defects at the boundary on the simulated image. It was found that although this analysis was based on a single experimental image, the number of physically possible grain boundary structures was severely limited by the observed rigid body displacements. Further research, involving the comparison of simulated images to a defocal series to confirm the validity of the proposed grain boundary structure followed by atomistic simulations to determine the relaxed configuration of this structure, will be conducted upon the structure which best matches the experimental image.

EXPERIMENTAL

The $\Sigma=5$ bicrystal was prepared from a single crystal of NiAl which was provided by Dr. David Pope of the University of Pennsylvania. After orienting the crystal using Laue diffraction, two slices 1.5 mm thick and 12 mm in diameter were cut along (310) on a Charmilles EDM. More accurate orientation of the slices was accomplished by Laue diffraction and mechanical polishing, which also removed any surface contamination produced during spark cutting. One slice was rotated about [310] by 180° to generate the orientation relationship [001] ‖ [00$\bar{1}$] with (310) ‖ (310). The two slices were diffusion bonded at 280 p.s.i. uniaxial pressure, 1000 °C, and 10^{-5} torr for 3 h in a graphite furnace. The resultant bicrystal was sectioned perpendicular to the common [001] axis, cut into 3 mm disks, and mechanically thinned to 125 μm. These samples were dimpled with 10% perchloric acid in ethanol at 44 V, followed by electropolishing in the same solution at 44 V (320 mA) and −25 °C. Initial examination of the foils was on a Philips 400T TEM, followed by high resolution imaging on a JEOL 4000EX.

The simulated images were generated using the NUMIS multislice simulation program on an Apollo workstation, with the output configured for the Krig Picture Station image analysis program. Image simulations of the matrix, and later the grain boundary, used a spherical aberration coefficient of 1.0 mm, a focal spread of 100 Å FWHM, and a beam convergence of 2.0 mrad. Matrix images were calculated for thicknesses ranging from 14 to 86 Å and objective defocus settings from -300 to -1004 Å. The initial sampling of 3.7 $Å^{-1}$, however, proved inadequate to resolve many features within the images, so a sampling of 7.4 $Å^{-1}$ was used for simulations of the matrix. The grain boundary simulations were able to reproduce these features within the matrix with samplings of 4.7 $Å^{-1}$ along the grain boundary and 8.5 $Å^{-1}$ normal to the boundary. For both the experimental and simulated results, images of moderate contrast and intensity were used in order to prevent saturation or underexposure of the very bright or dark regions.

RESULTS AND DISCUSSION

A high resolution image of the highly oriented grain boundary is shown in Figure 1a. In this [001] orientation, the B2 crystal structure consists of columns of pure Ni and pure Al aligned with the electron beam direction. The thickness and defocus were determined to be approximately 58 Å and -700 Å, respectively. Under these conditions, the matrix displays a strong superlattice of white atom contrast, with the nickel sites brighter than the aluminum sites. Planes of this superlattice contrast extend up to the grain boundary, indicating that the chemical ordering of this alloy is continuous up to the boundary. The grain boundary structural unit of this image was determined by inspection and is schematically shown in Figure 1b. It is diamond-shaped and has a length equivalent to the periodicity of the grain boundary contrast. This structural unit, which is 10 $d_{1\bar{3}0}$ (9.1 Å) long and approximately 4 d_{310} (3.6 Å) wide, displays an asymmetry produced by a ½ $d_{1\bar{3}0}$ (0.46 Å) rigid body translation of the top grain toward the right. The sites within the grain boundary structural unit are labeled in order to clarify discussion of this boundary.

The individual grain boundary lattice sites within the grain boundary region of the experimental image are discrete, not blurred together. This clarity appears to be at least partially due to the lower average intensity surrounding the grain boundary. Along the grain boundary, periodic points of bright contrast are commonly observed. These bright points are usually separated by 10 $d_{1\bar{3}0}$ plane spacings along the grain boundary, but are occasionally separated by only 5 $d_{1\bar{3}0}$ plane spacings, corresponding to the full and half period of the grain boundary structural unit. As the grain boundary penetrates into thicker regions of the foil some blurring of lattice sites is evident, producing diamond-like outlines along the grain boundary. The structure of the grain boundary will be analyzed by comparing these and other characteristic features with the corresponding features in simulated images.

The grain boundary structures used in the multislice image simulations contain the rigid body displacements measured from Figure 1a, which show a displacement along the grain boundary of approximately ½ $d_{1\bar{3}0}$. A grain boundary expansion of 0.18 d_{310} (0.16 Å), which is within the experimental measurements, was used in order to limit the number of interatomic separations which are closer than those within the matrix. The lateral displacement of the two grains removes the mirror symmetry of the boundary (see Figure 2) and causes the grain boundary lattice sites of the upper and

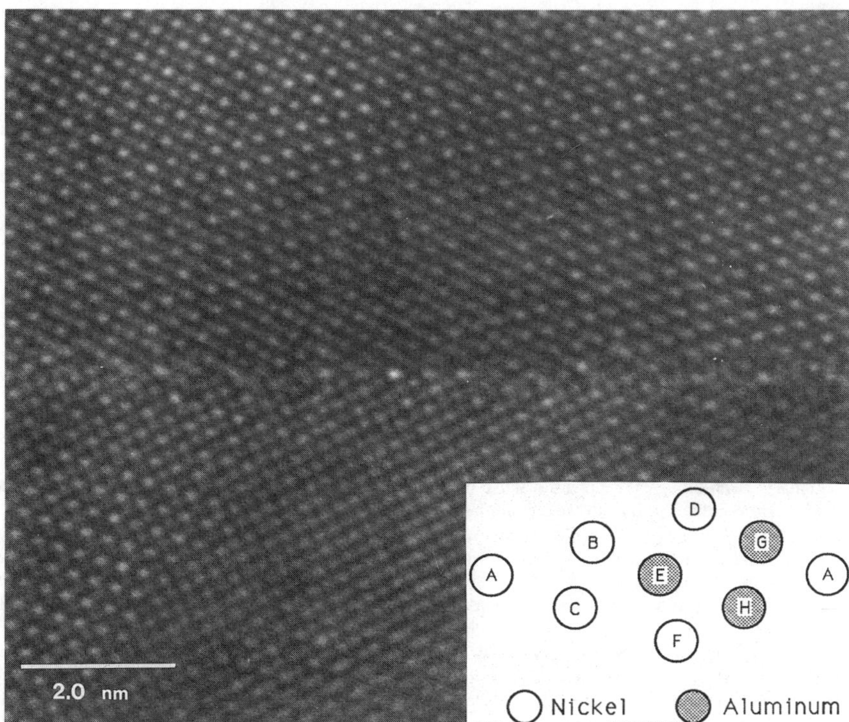

Figure 1. (a) Typical region of the highly oriented $\Sigma=5$ [001] (310) grain boundary. Thickness $\approx$ 58 Å and defocus $\approx$ -700 Å. (b) Schematic of the structural unit of the grain boundary with labels of the grain boundary sites.

lower grains to become distinct. The structure of the grain boundary will therefore be different for grain boundary atoms located at lattice sites of the upper grain, at lattice sites of the lower grain, or at some intermediate position. While the distinction between upper and lower grains applies to the observed shift of the upper grain towards the right, it is possible for this shift to be in either direction. However, these two displacements are not related by either a fundamental or superlattice dislocation and are therefore unlikely to be observed along the same boundary segment.

Associating the grain boundary atoms with the upper grain retains the matrix interatomic spacings around these atoms. However, the separations between sites B and E and between A and G are significantly reduced when the grain boundary atoms are located at lattice sites of the lower grain or between the lattice sites of the two grains. These separations are up to 9% shorter than those within the matrix. In addition, these structures also produce a bright streak along the grain boundary in simulated images, a feature which is not experimentally observed.

The separation between atoms adjacent to the grain boundary plane (e. g. between B and C) is fixed by the rigid body displacements and is therefore equivalent in all of these structures. In these structures, the nickel sites B and C and the aluminum sites G and H are almost 30% closer to each other than sites within the matrix. The near-symmetry of the boundary causes atoms on either side of the grain boundary plane to be much closer to each other than they would be within the matrix. Although this closeness can be relieved with a rigid body displacement along the common [001] direction (parallel to the beam direction), it also decreases the interatomic distances surrounding the grain boundary atoms at A and E and does not improve the match between simulated and experimental images.

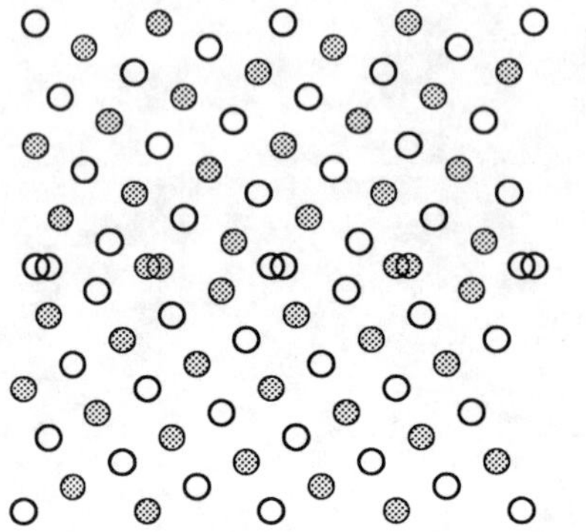

Figure 2. Schematic of grain boundary structure in which the grain boundary sites of both the upper and lower lattices are shown. Note the different environments of the grain boundary atoms in the two cases.

The structure with the grain boundary atoms at lattice sites of the upper grain was used in further refinements. This structure displays a very even distribution of atoms across the grain boundary, even though it retains the steric problems between B and C and between G and H. This smooth atomic distribution is reflected in the contrast of the simulated image (Figure 3). This grain boundary structure is very similar to a structure examined by Paten and Farkas.[4] The rigid body displacement parallel to the boundary which is exhibited by that structure is identical to the displacement shown in Figure 3. The structure calculated by Paten and Farkas was shown to have only a slightly (3.6%) higher energy than the lowest energy structure, which was a symmetrical boundary with no rigid body displacement along the boundary. However, this calculation also predicts a large (0.8 d_{310}) grain boundary expansion, which is not experimentally observed. This large predicted grain boundary expansion is due to the very strong interactions between sites B and C, and between G and H. Even though these sites have an interatomic separation of over 2 Å in the structure used in the image simulation, this separation is still smaller than the 2.88 Å matrix separation between similar atoms.

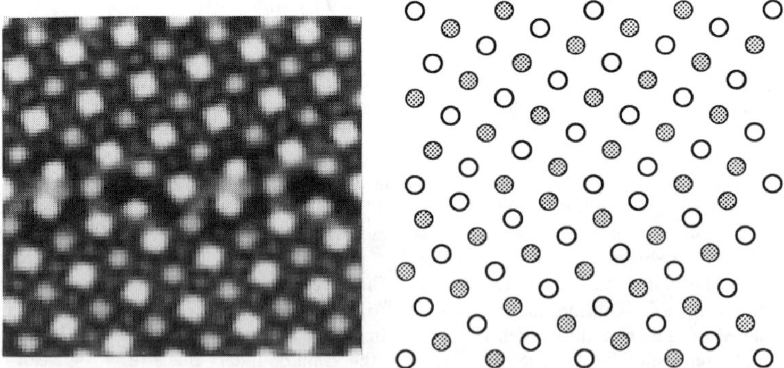

Figure 3. Simulated image of the $\Sigma=5$ grain boundary with the experimentally determined 0.5 $d_{1\bar{3}0}$ rigid body translation and a 0.18 d_{310} grain boundary expansion. The grain boundary atoms are located at lattice sites of the upper grain.

Although this simulated image is very similar to the experimental image in general form, there are differences in the brightness of some atom sites at and near the boundary. In the experimental image, the grain boundary aluminum sites at E typically have a brightness comparable to the nickel sites within the bulk, whereas the simulated image shows practically no increase in brightness at this position. Conversely, nickel sites B and C exhibit a very high intensity in the simulated image, and a streak between them produced by that brightness, which are not displayed in the experimental image. Although the intensity of any one site is related to the imaging conditions and the potential of that column of atoms in a complex manner, it was found that the intensity of these sites increases with increasing average atomic number of these columns. In NiAl, therefore, increasing the nickel content of an aluminum column will increase the intensity of that site while increasing the aluminum or vacancy content of a nickel column will decrease the intensity.

In order to improve the match between the experimental and simulated images, the intensity of the aluminum site E must increase while the intensity of nickel sites B and C must decrease. Attempts to increase the brightness of the aluminum grain boundary site E by adding nickel antisite defects to that atomic column have proved unsuccessful. While this substitution increases the intensity at that site, the increased potential produces a diffuse brightness surrounding that defect which is not experimentally observed. In order to decrease the intensity of nickel sites B and C, and likely decrease the amount of streaking between them, the average atomic number of these atomic columns must be reduced. This can be accomplished by replacing the nickel atoms at B and C with either aluminum antisite defects or nickel vacancies. However, Bradley and Taylor[3] have determined that the only defects on the nickel sublattice are constitutional vacancies; aluminum antisite defects are not observed on the nickel sublattice. In addition, placing the larger aluminum atoms at these already crowded nickel sites is physically unrealistic.

The simulated image of the grain boundary structure in which the nickel sites have a 50% probability of containing a vacancy is very similar to the experimental image. The brightness of sites B and C is significantly reduced, which removes the streaking between these sites. This image also displays an increase in the intensity of the grain boundary aluminum sites (E) relative to the initial stoichiometric structure, as is experimentally observed. Although there is also a slight decrease in the intensity surrounding the grain boundary which is not present in the experimental image, the match between the experimental and simulated images is quite good.

In this structure, the removal of one nickel atom from the two sites leaves the pair of nickel sites occupied by a nickel atom and a nickel site vacancy, removing the steric crowding of those sites. An obvious extension of this reasoning is to decrease the occupancy of the aluminum sites at G and H as well, removing the steric crowding of these sites and increasing the intensity of the grain boundary sites even further. However, even though the large atomic size of aluminum should cause this structure to be favored, the reduced clarity of the grain boundary region and lower intensity of the grain boundary plane this produces in the simulated image are not experimentally observed. Images were also simulated for other structures which would relieve the compression of these sites. The best match is produced for the grain boundary structure in which the Al sites G and H are one-half occupied by Ni antisite defects and one-half occupied by vacancies (shown in Figure 4). In this structure, the sites B, C, G, and H are identically occupied by nickel atoms and vacancies. The incorporation of nickel antisite defects adjacent to the boundary causes the grain boundary composition to be nickel-rich. The image produced for this structure is nearly identical to the image of the structure containing vacancies at B and C. This similarity is at least partially due to the similarity in the average atomic number of sites G and H. The only observable difference between these simulated images is the increased brightness of the grain boundary region of Figure 4, which more closely matches the experimental contrast.

Simulated images from the structure shown in Figure 4 also match the experimental image at other thicknesses. For the 58 Å thickness discussed above, the matrix displays a moderately strong nickel superlattice contrast and a correspondingly bright nickel grain boundary position at A. The periodic bright contrast of grain boundary sites which is exhibited at this and most other thicknesses has a periodicity equal to the length of the grain boundary structural unit. As the thickness increases, the intensity of the nickel sites decrease concomitant with an increase in the aluminum site brightness. At a thickness of 69 Å the two sublattices exhibit a nearly equivalent intensity within the matrix, which is reflected in the balanced intensity of the grain boundary sites. At this thickness the bright grain boundary sites are separated by one-half the length of the grain boundary structural unit, as is occasionally observed in the experimental image. This one-half period brightness is confined to a thickness range of approximately 5 Å, which is consistent with the uncommon nature of this feature.

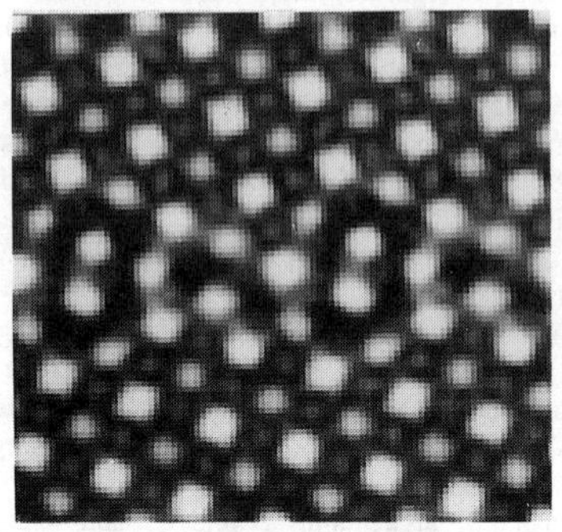

Figure 4. Simulated image of the $\Sigma=5$ NiAl grain boundary which best matches the experimental image. In this structure, all sites adjacent to the grain boundary plane (sites B, C, G, and H) are one-half occupied by Ni antisite defects and one-half occupied by vacancies.

Although this analysis was based on a single experimental image, there are few realistic grain boundary structures which conform to this image. The experimental image accurately displays the rigid body displacements of the two grains, which define much of the grain boundary structure. The ½ $d_{1\bar{3}0}$ displacement along the grain boundary requires the grain boundary atoms to occupy lattice sites of the upper grain in order to maintain the equilibrium interatomic separations. The lack of significant grain boundary expansion produces steric crowding between sites B and C and between sites G and H which can only be decreased by introducing vacancies into these sites. This would lead to a predicted structure in which sites B and C contain only one nickel atom between them and sites G and H similarly share an aluminum atom. It may therefore be more appropriate to view these pairs of sites as solitary atomic sites with two low energy positions for the atoms. The relative energies of these positions will be determined later with atomistic modeling. This description of the grain boundary, which is based solely on the rigid body displacements, is consistent with the image simulations in all respects except the occupancies of sites G and H. Image simulations indicate that the atoms which partially occupy these sites are nickel rather than aluminum. The simulated image of the resultant structure is very similar to the experimental image in all respects, and is consistent at different thicknesses as well.

ACKNOWLEDGEMENTS

The authors would like to acknowledge support of this project by the National Science Foundation under grants 91-11775 and 91-20668 and support for central facilities through the Penn-MRL. They would also like to thank Ms. D. Ricketts-Foot for the maintenance of the microscopes.

REFERENCES

1. R. Darolia, *JOM*, **43**, 44 (1991).
2. E. P. George and C. T. Liu, *J. Mater. Res.*, **5**, 754 (1990).
3. A. J. Bradley and A. Taylor, *Proc. Roy. Soc.*, **A159**, 56 (1937).
4. G. Petton and D. Farkas, *Scripta Met.*, **25**, 55 (1991).
5. S. P. Chen, A. F. Voter, A. M. Boring, R. C. Albers, and P. J. Hay, *Mat. Res. Soc. Symp. Proc.*, **133**, 149 (1989).
6. P. P. Camus, I. Baker, J A. Horton, and M. K. Miller, *J. de Physique*, **49**, C6-329 (1988).
7. M. K. Miller, R. Jayaram, and P. P. Camus, *Scripta Metall.*, **26**, 679 (1992).

MICROSTRUCTURES IN $L1_2$ TITANIUM TRIALUMINIDES CONTAINING IRON

Z.L. Wu, D.P.Pope and V.Vitek
Department of Materials Science and Engineering,
University of Pennsylvania, Philadelphia, PA 19104-6202.

ABSTRACT

The microstructures of the $L1_2$ titanium trialuminides at low temperatures were studied using a number of single crystals with various Al-Ti-Fe compositions, all of which lie in the nominal single phase $L1_2$ field at 1200°C. Five different second phases were found to be in equilibrium with the $L1_2$ matrix, namely, $(Al,Ti)_3Fe$, Al_3Ti, Al_2FeTi, Ti_2NAl and Al_2Ti+Fe. Small volume fractions of the first two phases are often seen in compounds containing relatively low Ti contents. The Al_2FeTi, a so-called T phase, was observed at relatively high Fe contents. The Ti_2NAl does not seems to be sensitive to the Al-Ti-Fe composition, it exists to some extent in all the alloys used in this study. Ti_2NAl and the interface with the $L1_2$ matrix are found to be brittle and provide the sites for crack initiation. The Al_2Ti+Fe phase has been observed in many compounds containing high Ti contents (>25 at.%), and has a large hardening effect. Like binary Al_2Ti, the phase possesses a tetragonal structure of the Ga_2Hf-type, and forms plates on the cube planes of the $L1_2$ matrix. The crystallographic relation between the Al_2Ti and the $L1_2$ matrix was determined to be $(100)_p//(100)_m$.and $(010)_p//(010)_m$. Porosity is also commonly seen in these alloys, and has a very destructive impact on ductility.

INTRODUCTION

A single phase $L1_2$ field has been identified in many Al-Ti-X (X=Cr, Mn, Fe, Ni, etc.) ternary systems [1-5]. In the case of the Al-Ti-Fe system, a substantial single phase $L1_2$ field was found at 1200°C with a composition range of 7 at.%; however, the phase field is considerably smaller at 800°C [1]. Such a shrinkage may continue when the temperature is further decreased, but no phase diagrams have been prepared for temperatures below 800°C. The low temperature microstructures in the alloys which are single phase at high temperatures are not well understood, nor is the effect of the microstructure on the deformation behavior of the alloys.

The purpose of this study is to investigate the low temperature microstructures in a number of Fe-modified Al_3Ti-based alloys. Single crystalline specimens were used for the study because the precipitates in single crystalline specimens were commonly found to be bigger and more easily recognized than those in polycrystals. Quantitative energy dispersive x-ray analysis in a SEM can, therefore, be utilized to determine the compositions of these phases. The compositions of the alloys used for the study all lie in the single phase $L1_2$ field at 1200°C, as determined by Mazdiyasni et al [1]. The effects of the microstructures on the mechanical properties were also studied.

EXPERIMENTAL

A number of single crystalline specimens of various compositions were prepared in the same procedure as described in [6]. X-ray powder diffraction using a Rigaku D/MAX II diffractometer was employed to identify the crystal structure of the alloys. The compositions of the phases of interest were determined using quantitative energy dispersive x-ray analysis in a scanning electron microscope (SEM). The habit planes of the Al_2Ti precipitates were determined using the two-surface trace analysis technique and a Phillips 400 transmission electron microscope (TEM). The TEM specimens were prepared using twin jet polishing at -40°C in an electrolytic polishing solution of one part of HNO_3 and 2 parts of methonol.

RESULTS AND DISCUSSION

Fig.1 shows second phase $(Al,Ti)_3Fe$ and Al_3Ti often seen in compounds containing relatively low Ti contents, such as $Al_{67}Fe_8Ti_{25}$, which lies on the boundary of the nominal single phase $L1_2$ field. The $(Al,Ti)_3Fe$ precipitate seen in Fig.1(a) has a composition $Al_{66.7}Ti_8Fe_{25.3}$ which is liquid at 1200°C according to [1]. Fig.1(b) shows several Al_3Ti precipitates of composition $Al_{75.6}Ti_{24.0}Fe_{0.4}$, which form clusters scattered throughout the matrix. Although the precipitates seem to be oriented along certain directions in the matrix, there appears to be no orientation relation between the DO_{22} phase and the $L1_2$ matrix. Computer simulated x-ray

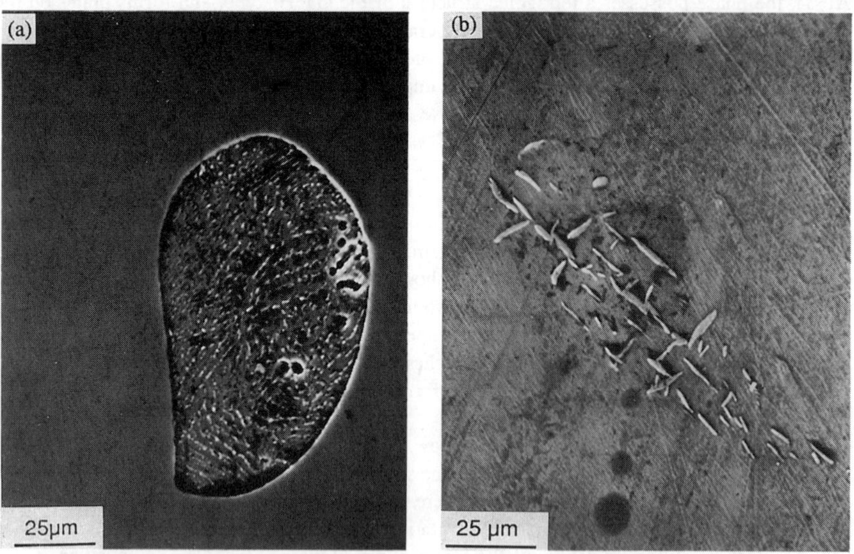

Figure 1. (a) $(Al,Ti)_3Fe$ and (b) Al_3Ti precipitates are seen in single crystalline $Al_{67}Fe_8Ti_{25}$ specimens.

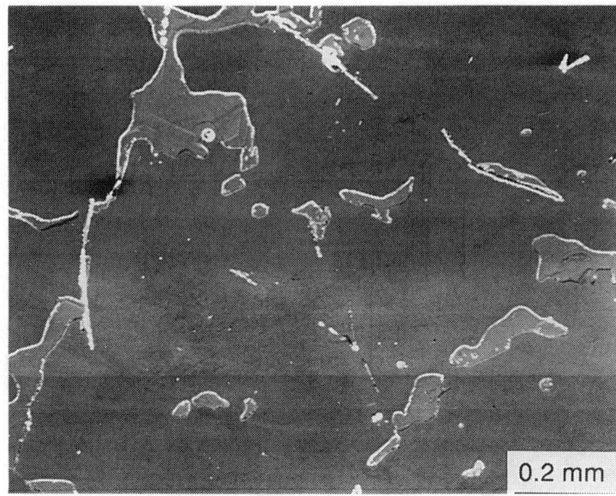

Figure 2 Al_2FeTi precipitates in a single crystalline $Al_{62.5}Fe_{9.5}Ti_{28}$ specimen.

diffraction spectra of the DO_{22} Al_3Ti and the $L1_2$ $Al_{267}Fe_8Ti_{25}$ phases have shown that the difference in peak positions is minimum between the {112} plane of the DO_{22} (the Miller indices refer to the tetragonal unit cell of the DO_{22} structure, not to a cubic unit cell, as is commonly done in the literature) and the {111} plane of the $L1_2$. However, even for these two planes, the lattice mismatch is still about 4.5%, which eliminates the possibility of coherent growth of one phase on the other.

The Al_2FeTi (actual composition is $Al_{50.3}Ti_{25.9}Fe_{23.8}$), the so-called T phase, was observed in compounds containing relatively high Fe contents. The precipitates shown in Fig.2 orient randomly in the $L1_2$ matrix of an alloy of composition $Al_{62.5}Fe_{9.5}Ti_{28}$. A similar phase, identified as Al_2NiTi, has been seen in a Ni-modified Al_3Ti alloy [7].

The Ti_2NAl phase was previously identified to be $Ti_{66.7}Al_{33.3}$, using a conventional SEM [8,9]. The formation of the $Ti_{66.7}Al_{33.3}$ in the Al-rich matrix was, however, difficult to understand, since it is not consistent with the overall composition of the compounds and was hard to understand based on considerations of the solidification sequence. In addition, the phase is not reported in the high temperature isotherms of the system [1,2]. It was later found that the phase actually contains N, which was detectable using the SEM but with the beryllium window in front of the x-ray detector removed. An x-ray powder diffraction spectrum and a computer simulation of the spectrum revealed that the phase has a hexagonal structure of the Cr_2CAl-type, with lattice parameters a=2.995Å and c=13.61Å. The Ti_2NAl constitutes a small volume fraction

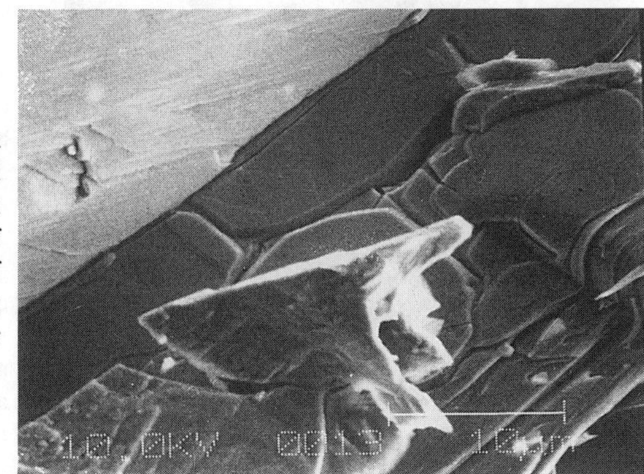

Figure 3. A fracture surface of a $Al_{67}Fe_8Ti_{25}$ specimen deformed in an Auger microscope. The dark area is Ti_2NAl where a number of cracks were clearly seen. Some of the cracks probably formed during processing and then grew under the applied stress. A gap is observed at the interface between the Ti_2NAl and the matrix.

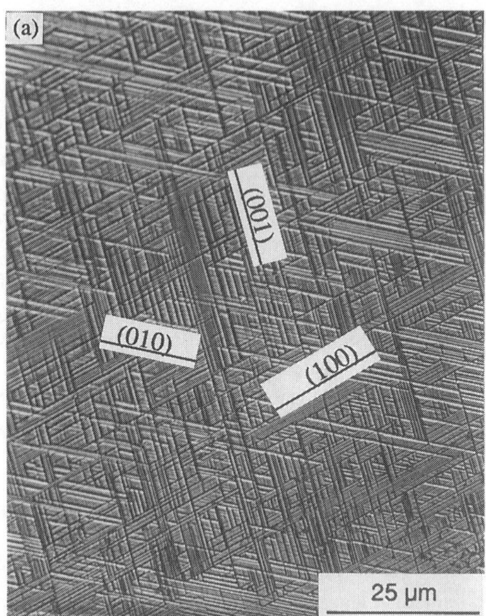

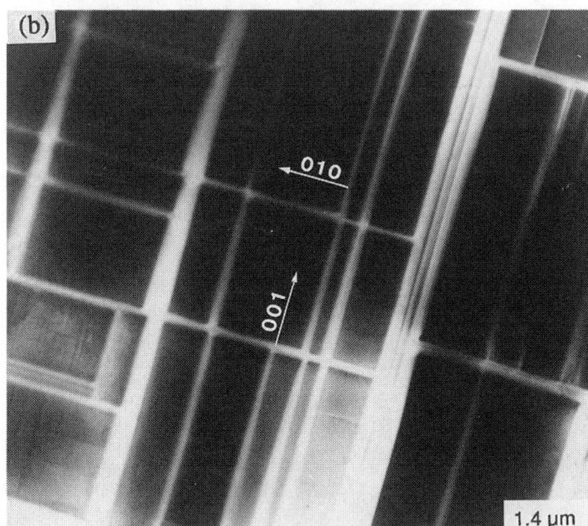

Figure 4. Al_3Ti platelets on the cube planes of the $L1_2$ matrix, as seen (a) on the surface of a single crystalline $Al_{64.8}Fe_{6.8}Ti_{28.4}$ specimen, and (b) in an $Al_{66.8}Fe_{5.8}Ti_{27.4}$ TEM thin foil.

in all the alloys used for the study, but has a large embrittling effect. Cracks are often initiated within the phase or at the interface between the phase and the matrix. Fig.3 shows a fracture surface of a $Al_{67}Fe_8Ti_{25}$ specimen deformed in an Auger microscope. The dark area is Ti_2NAl. A number of cracks are clearly seen in the Ti_2NAl. Some of the cracks do not seem to have resulted from the mechanical deformation, but probably formed during previous processing and then grew under the applied stress. Moreover, a gap exists at the interface between the phase and the matrix, indicating a weak bond between the two phases. All these factors contribute to the embrittlement.

Of the five second phases, the Al_2Ti+Fe is probably the most important as far as the structural and mechanical properties are concerned. Based on two-surface trace analysis, the phase forms plates on the cube planes of the $L1_2$ matrix (Fig.4(a)). The volume fraction of the phase depends primarily on Ti-content: the higher the Ti content in the overall composition, the larger the volume fraction of the phase. The phase has a complex tetragonal structure of the Ga_2Hf-type, the same as that of binary Al_2Ti. With small additions of Fe, the a-lattice parameter of Al_2Ti decreases from 3.976Å to 3.95Å, making coherent growth of the precipitates on the cube planes of the matrix possible, since the mismatch between the a-parameter and that of the matrix is then only about 0.3%. Fig.4(b) shows a TEM micrograph of the Al_2Ti precipitates. The crystallographic relation between the second phase and the $L1_2$ matrix was determined to be $(100)_p//(100)_m$.and $(010)_p//(010)_m$, based on the electron diffraction patterns obtained in the TEM. Because of its complex tetragonal structure, the phase acts as a barrier to the motion of dislocations on the primary octahedral slip planes, the major deformation mode in the $L1_2$ matrix, and results in a strong hardening effect. More detailed discussions can bc found in [10].

In addition to the second phases mentioned above, porosity is also commonly observed. Some of the pores apparently form during interdendritic solidification, as shown in Fig.5(a), while others are probably Kirkendall pores which form in isolated $(Al,Ti)_3Fe$ particles (see Fig.5(b)). As has been pointed out by others [11,12] the Kirkendall mechanism might be responsible for the increased void volume fraction seen after a homogenization treatment. The porosity has a profound impact on the brittleness of the alloys, providing crack initiation sites leading to premature failure, as can be seen in Fig.6.

CONCLUSIONS

Five different second phases, namely, $(Al,Ti)_3Fe$, Al_3Ti, Al_2FeTi, Ti_2NAl and Al_2Ti, have been identified at low temperatures in a number of $L1_2$ Fe-modified Al_3Ti-based alloys, the compositions of which all lie in the nominal single phase $L1_2$ field at 1200°C. Of the five phases, Ti_2NAl and Al_2Ti are probably the most important ones. The former phase exists in virtually all the alloys used for the study and is very deleterious to ductility. The latter forms plate-like precipitates on the cube planes of the $L1_2$ matrix in alloys containing relatively high Ti contents, and has a large hardening effect. In addition to the second phases, porosity also exists and cracks are often seen to initiate at the pores, leading to premature failure.

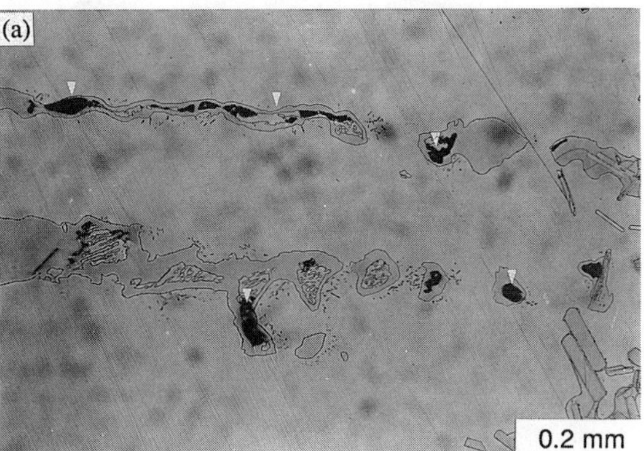

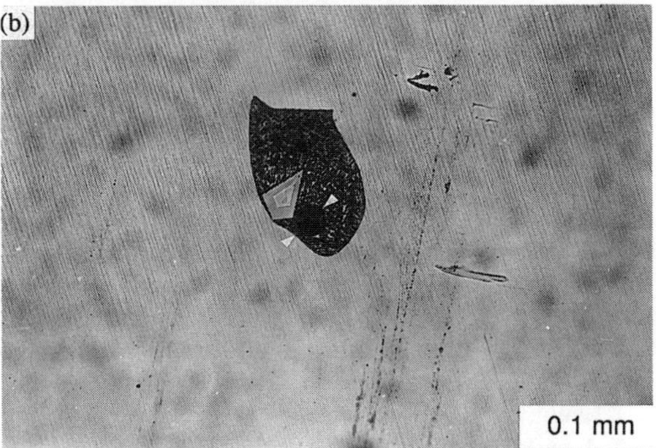

Figure 5. (a) Residual porosity in second phase regions of a $Al_{65.2}Fe_{8.2}Ti_{26.6}$ specimen. (b) Void formed by the Kirkendall mechanism in a $Al_{67}Fe_8Ti_{25}$ specimen.

ACKNOWLEDGEMENT

The work was supported by the AFOSR under Grant AFOSR-89-0062. Research facilities were provided by the LRSM supported by the NSF MRL program under Grant DMR88-19885. The authors would like to thank K.S. Kumar for valuable discussions, and W.J. Romanow and R. Hsiao for technical assistance.

Figure 6. Cracks initiated at pores.

REFERENCES

1. S. Mazdiyasni, D.B. Miracle, D.M. Dimiduk, M.G. Mendiratta, and P.R. Subramanian, Scrip. Metall., **223**, 327, 1989.
2. V. Ya. Markiv, V. V. Burnashova, and V. P. Ryabov, Akad. Nauk Ukr. SSR, Metallofizika, **4**, 103, 1973.
3. A. Raman, and K. Schubert, Z. Metallkd., **56**, 99, 1965.
4. H. Mabuchi, K. Hirukawa, and Y. Makayama, Scrip. Metall., **23**, 1761, 1989.
5. S. Zhang, J.P. Nic, W.W. Milligan, and D.E. Mikkola, Scrip. Metall., **24**, 57, 1990.
6. Z.L. Wu, D.P. Pope, and V. Vitek, submitted to this symposium.
7. C.D. Turner, W.O. Power, and J.A. West, Acta Metall., **37**, 2635, 1989.
8. Z.L. Wu, D.P. Pope, and V. Vitek, High Temperature Ordered Intermetallic Alloys, IV, (ed. by Johnson, L., Pope, D.P. and Stiegler, J.O.), MRS Symp., **213**, MRS Pittsburgh, 487, 1991.
9. Z.L. Wu, D.P. Pope, and V. Vitek, Scrip. Metall., **24**, 2187, 1990.
10. Z.L. Wu, Ph.D thesis, University of Pennsylvania, 1992.
11. E.P. George, J.A. Horton, W.D.Porter and J.H. Schneibel, J. Mater. Res., **5**, 1639, 1990.
12. D.D. Mysko, J.B. Lumsden, W.O. Powers and J.A. Wert, Scrip. Metall., **23**, 1827, 1989.

MORPHOLOGY, DEFORMATION, AND DEFECT STRUCTURES OF $TiCr_2$ IN Ti-Cr ALLOYS

Katherine C. Chen, Samuel M. Allen, and James D. Livingston, Department of Materials Science and Engineering, Massachusetts Institute of Technology, Cambridge, MA 02139

ABSTRACT

The morphologies and defect structures of $TiCr_2$ in several Ti-Cr alloys have been examined by optical metallography, x-ray diffraction, and transmission electron microscopy (TEM), in order to explore the room-temperature deformability of the Laves phase $TiCr_2$. The morphology of the Laves phase was found to be dependent upon alloy composition and annealing temperature. Samples deformed by compression have also been studied using TEM. Comparisons of microstructures before and after deformation suggest an increase in twin, stacking fault, and dislocation density within the Laves phase, indicating some but not extensive room-temperature deformability.

INTRODUCTION

In prior studies, several alloys containing the $TiCr_2$ Laves phase have shown promising mechanical properties and oxidation resistance at elevated temperatures [1,2]. Although Laves phases have the well-deserved reputation for low-temperature brittleness, recent studies have demonstrated room-temperature deformability of Laves phases in two-phase V-Hf-Nb [3] and Fe-Zr alloys [4]. HfV_2 deforms by twinning and through bands of concentrated shear, while $ZrFe_2$ experiences a stress-induced phase transformation between the C36 and C15 crystal structures. The Laves phase will most likely have to be in a two-phase alloy to be used as high-temperature structural material [5].

The current study considers Ti-Cr alloys strengthened by precipitation hardening of $TiCr_2$. The two-phase alloy consists of the hard and strong Laves phase $TiCr_2$ reinforcing the more ductile β-Ti matrix. In addition to studying the microstructural dependencies of the mechanical behavior of the two-phase alloys, the question of whether the Laves phase deforms was also addressed. Compressed samples are studied to identify deformation mechanisms, and to offer insight into improving Laves phase ductility.

EXPERIMENTAL PROCEDURES

Alloy compositions of Ti-40 at pct Cr and Ti-30 at pct Cr were prepared by arc-casting. For heat treatments, samples were encapsulated in vacuum with tantalum getter and back-filled with argon. A range of times and temperatures were employed in the β-Ti(Cr) + α-$TiCr_2$ phase field, followed by air cooling. The Ti-Cr binary phase diagram is shown in Figure 1 [6].

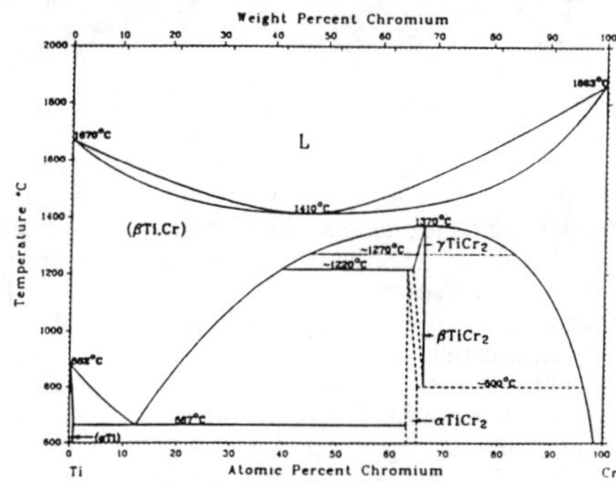

Fig.1 The titanium-chromium phase diagram [6].

Cubes of 5 mm were spark-cut for compression tests. The compression was done at room-temperature on an Instron machine at a crosshead speed of 0.0025 cm/min. Loading was stopped before complete fracture of the test cubes. TEM samples were made by a combination of mechanically grinding, dimpling, jet polishing (using a Blackburn and Williams electrolyte [7] or Spurling etch [8]), and ion milling. A JEOL 200CX electron microscope operating at 200kV was used for TEM. An etching solution of 25% HF, 25% HNO_3 and 50% glycerin was used for optical metallography.

RESULTS AND DISCUSSION

Microstructure

The as-cast alloys were single phase bcc β-Ti, with no evidence of any $TiCr_2$ from x-ray or electron diffraction. Optical microscopy revealed significant subgrain structure. The Ti-40Cr alloy contained etch pits that outlined the subgrains. These etch pits may represent dislocations that later serve as potent nucleation sites for precipitation, as the Ti-30Cr alloy showed very fine precipitates located along the subgrain boundaries.

The β-Ti phase itself is metastable and forms an hcp omega phase (ω) during the quench from anneals [9,10]. TEM revealed a high density of the omega phase particles aligned along <111> directions (Figure 2). Electron diffraction produced the characteristic diffuse streaking from the linear displacement defects. The omega phase hardens and embrittles the β-Ti phase, but is also unavoidable during the quench.

Fig.2 The omega phase (ω) in the metastable β-titanium.

Various heat treatments were performed on the Ti-Cr alloys in order to precipitate the $TiCr_2$ intermetallic and to establish microstructural control of the alloys. The strengthening effects of a second phase are related to the size, shape, number and distribution of the particles, and the orientation relation and interfacial coherency between the phases. The two alloy compositions produced strikingly different microstructures.

Figure 3(a) shows an optical micrograph of the Ti-40Cr alloy annealed at 1000°C for 24 hours. The Laves phase $TiCr_2$ forms a bimodal distribution of equiaxed precipitates averaging in sizes of 1 μm and 10 μm and constitutes about 38% of the specimen volume. This bimodal distribution persisted despite other heat treatments that involved solutionizing and water-quenching steps. The bimodal distribution also coarsened during a long anneal (168 hours), and thus it is not likely that the very small precipitates were formed during the cooling period. Several of the larger, blocky precipitates were comprised of mid-sized precipitates grown together into a massive aggregate. The large precipitates also contain twins, clearly seen in Figure 3(a).

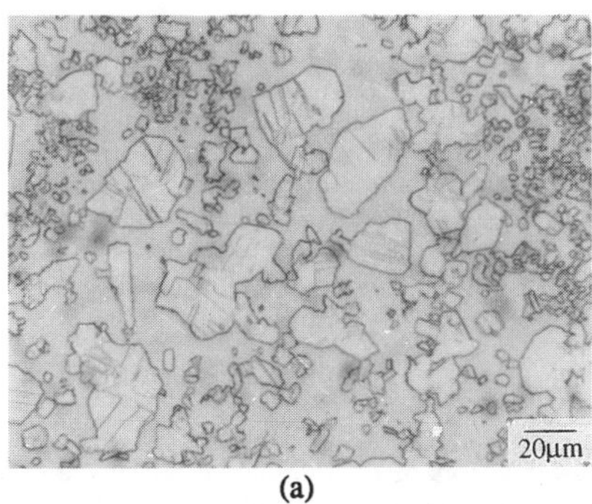

(a)

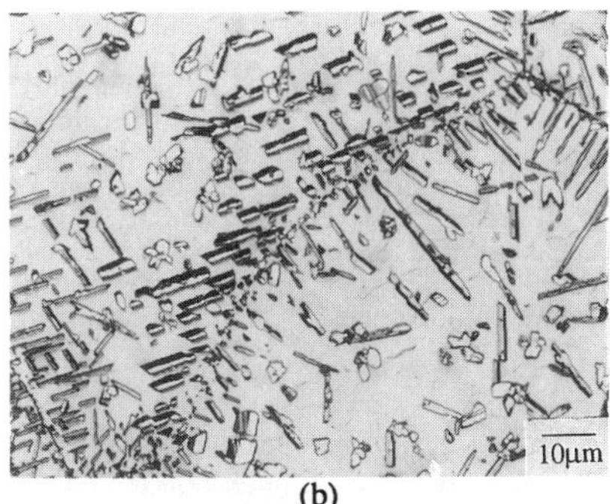

(b)

Fig.3 Optical microstructure of (a) Ti-40Cr alloy annealed at 1000°C and (b) Ti-30Cr alloy annealed at 950°C. Notice the twinning found in the equiaxed precipitates in the Ti-40Cr alloy and in the lath-shaped precipitates in the Ti-30Cr alloy.

Figure 3(b) is the Ti-30Cr alloy annealed at 950°C for 24 hours and shows a markedly different morphology. Here, the Laves phase has a lath-like shape with an aspect ratio of about 10 to 1 and has a 17% volume fraction. Some lath particles also appear to be twinned lengthwise and to have a preferred orientation within each grain of the alloy. The long straight edges and the orientation of the laths suggest greater coherency between the matrix and intermetallic than the equiaxed precipitates. There is also significant precipitation along the grain boundaries.

Although the two different compositions have distinctively different microstructures, the morphology of the Laves phase did have some similarities among the various anneals at different times and temperatures. For instance, towards the edges of the annealed Ti-30Cr sample (which was exposed to possible impurities and a faster quench), the precipitates were more equiaxed. And, in the Ti-40Cr alloy, some of the small precipitates in the lower temperature anneals had large aspect ratios.

TEM revealed dense dislocation tangles in the matrix adjacent to the Laves phases, as seen in Figure 4. These dislocations are thought to form during the quench after the heat treatments since most dislocations should be annealed out during the high temperature anneals. Dislocations may be generated upon cooling due to the mismatch of thermal expansion coefficients of the matrix and the precipitate. Another possibility may be a temperature-dependent phase transition. The high temperature phase of $TiCr_2$ is hexagonal (C14) and the low temperature phase is cubic (C15). Allen [11] has proposed dislocation models for the shear transformation of the crystal structures. Many investigators have identified this hexagonal phase although the exact temperature and composition boundaries are not clearly established [6]. The dislocations may have been produced by a martensitic transformation of the hexagonal to cubic phase during the quench.

Fig.4 Dislocations in the β-Ti matrix adjacent to the $TiCr_2$ Laves phase in the annealed Ti-30Cr alloy.

The Laves phase particles in both alloys were determined to have the cubic structure C15 (α-$TiCr_2$) by x-ray and electron diffraction, in agreement with the equilibrium phase diagram in Figure 1. The large precipitates of the Ti-40Cr alloy were chosen to examine the deformability of the Laves phase by TEM comparisons of the deformed and undeformed samples since the small precipitates were very heavily faulted. The annealing twins seen optically in the large equiaxed particles appear as wide twin bands in TEM. The twins conform to the {111} <112> twinning system found in fcc. Twins and faults on different variants of the {111} planes could be seen simultaneously. The comparison will be discussed in the following section.

Deformation

Results of the compression tests are shown in the plots of nominal stress vs. crosshead displacement (Figure 5). Strengthening of the titanium alloy due to precipitation of the Laves phase is evident. The Ti-40Cr alloy could be compressed to a strain of only about 6% before a load drop, indicating severe cracking through the sample. On the other hand, the Ti-30Cr alloy could be compressed to much larger strains, and loading was stopped at 26% strain.

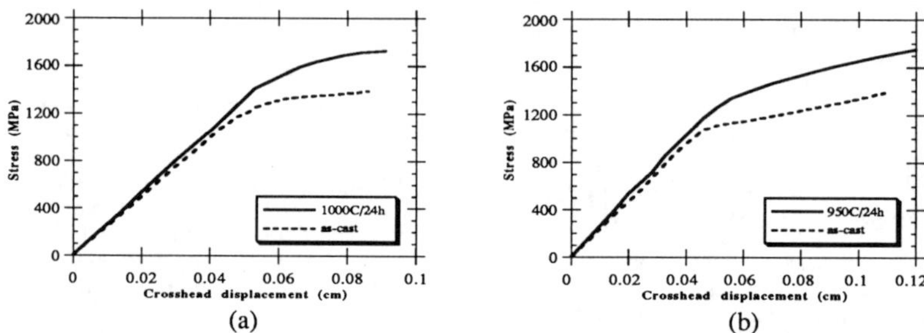

Fig 5 Yielding behavior of Ti-Cr alloys compressed at room temperature for (a) Ti-40Cr and (b) Ti-30Cr alloys. Precipitation of the $TiCr_2$ Laves phase strengthens the alloys.

Optical examination revealed that the Laves phase had cracked many times in both alloys with the cracks roughly parallel to the direction of compression. The lath shaped particles oriented with their length normal to the compression axis were cracked several times (reminiscent of fiber-reinforced composites behavior), while those parallel to the compression axis were not as likely to be cracked. The titanium matrix apparently serves to dull the crack as no further crack propagation in the matrix was found by TEM. Debonding of the precipitate and the matrix was sometimes seen, and may have been another mechanism to relieve some of the stress concentration at the interface and the strain energy caused by the deformation. No change in the crystal structure was found, in contrast to the behavior of $ZrFe_2$.

TEM analysis of the compressed sample showed heavy deformation in the matrix and some signs of deformation in the Laves phase. Although some faulting was present in the undeformed condition, the deformed samples appear to have more complex faulting and twinning. Figure 6 depicts the interaction of different defects. The change in direction or orientation of faults across twin boundaries suggest some type of interaction or sequence of defect formation. Possibly the faults were reoriented with the deformation that produced the twins, or the fault had to change directions when traversing through the twin.

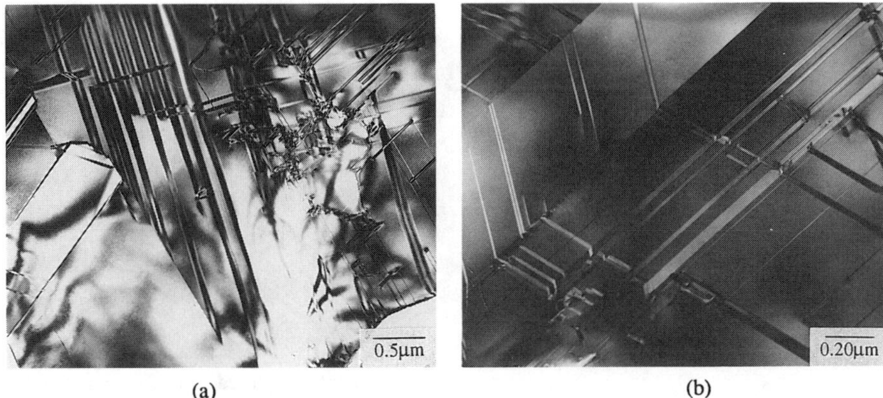

Fig.6 TEM images of compressed Laves phases found in the annealed (a) Ti-40Cr alloy and (b) Ti-30Cr alloy. Twins conform to the {111}<112> system found in fcc material.

The abundance of twins indicates that twinning may be an important deformation mechanism, as has been found in other studies [3]. The interface of the precipitate is often jagged or protruding where there is a twin band. Dislocation pile-ups in the matrix near a twin boundary suggest that these twins come from the deformation rather than from annealing (Figure 7).

Occasionally one could find areas of concentrated dislocation structure in the Laves phase (Figure 8). Livingston [3] has termed these planar defects as "shear bands", which form subgrain boundaries and mark areas of intense shear. Faults can be seen to have been sheared or displaced across the shear bands.

The matrix contained significant amounts of dislocations concentrated in slip bands, tangles, and networks. Because the Laves precipitates are dispersed throughout the alloy, distribution of the compressive stresses becomes quite complex and nonuniform. This may explain the lack of correlation between the dislocation slip bands in the matrix and the faults in the Laves phase. Also, no apparent correlation of the direction of faults and twins among the Laves phase particles could be found.

Although some deformation of the Laves phase has appeared to have occurred, there was no extensive deformation at room temperature in $TiCr_2$. Greater deformation may be imparted to the Laves phase by deforming at higher temperatures, which may allow thermally activated deformation mechanisms, or by alloying. These further steps will be pursued to better understand the deformability of Laves phases.

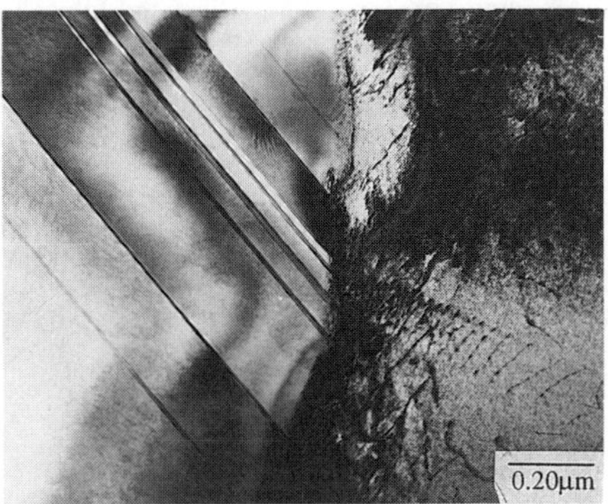

Fig.7 Dislocation pile-ups in the matrix, near a twin boundary in the $TiCr_2$ Laves phase.

SUMMARY

Ti-Cr alloys can be strengthened by the precipitation of $TiCr_2$ Laves phases. The Ti-40Cr alloy typically contained a bimodal distribution of equiaxed precipitates and could only be compressed to small strains. In contrast, the Ti-30Cr alloy had lath-like precipitates and could be deformed to relatively much larger strains. The $TiCr_2$ Laves phases were seen to have deformed by forming twins, faults, and shear bands. Higher temperature deformation or alloying may be needed to realize more Laves phase ductility.

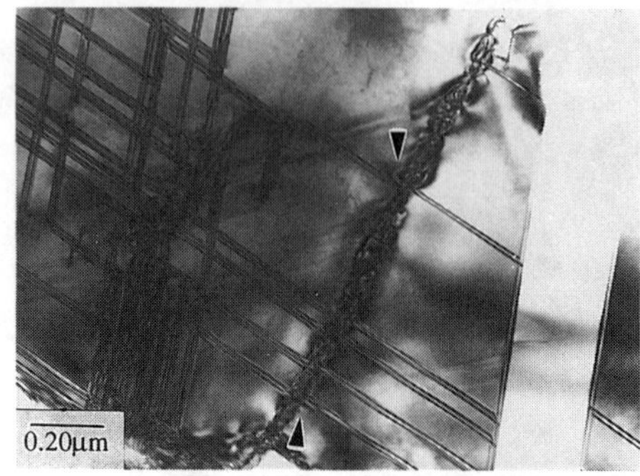

Fig.8 Faults displaced about 20 nm across the shear band found in the deformed $TiCr_2$.

ACKNOWLEDGEMENTS

We would like to acknowledge Y. Liu for his assistance and helpful discussions. Support for K. Chen comes from the DARPA-NDSEG fellowship program. This research has been supported by the Basic Energy Sciences Division of the Department of Energy, grant #DE-FG02-90ER45426.

REFERENCES

1. G. Sauthoff, Z. Metallkd. **80**, 337 (1989).
2. R.L. Fleischer and R.J. Zabala, Metall. Trans. **21A**, 2149 (1990).
3. J.D. Livingston and E.L. Hall, J. Mater. Res., **5**, 5 (1990).
4. Yaping Liu, Samuel M. Allen, and James D. Livingston, Metall. Trans. **23A** (1992).
5. J.D. Livingston, Phys. Stat. Sol. (a) **131**, 415 (1992).
6. J.L. Murray, Bull. Alloy Phase Diagr. **2**(2), 174 (1981); Binary Alloy Phase Diagrams, 2nd edition, Thaddeus B. Massalski, ed., (ASM International), pp. 1345-1348.
7. M.J. Blackburn and J.C. Williams, Trans. TMS-AIME, **239**, 287 (1967).
8. R.A. Spurling, Metall. Trans. **6A**, 1660 (1975).
9. S.L. Sass, J. Less-Common Met. **28**, 157 (1972).
10. J.C. Williams, D. DeFontaine and N.E. Paton, Metall. Trans. **4**, 2701 (1973).
11. C.W. Allen, Mat. Res. Soc. Symp. Proc. Vol 39, 141 (1985).

DEFORMATION STRUCTURES IN ORIENTED NiAl AND CoAl SINGLE CRYSTALS DEFORMED AT ELEVATED TEMPERATURE

Y. Zhang, S.C. Tonn and M.A. Crimp
Department of Materials Science and Mechanics, Michigan State University, East Lansing, MI 48824-1226

ABSTRACT

The deformation characteristics of the B2 intermetallic alloys NiAl and CoAl have been examined using single crystals deformed at 673 K in the <110> orientation. Slip trace analysis and transmission electron microscopy was used to characterize the deformation process. While many aspects of the deformation are similar, some distinct differences were observed. All of the alloys were found to deform by <001> slip. NiAl was found to deform by slip on a combination of {110} and {100} slip planes with the predominant slip plane being a function of alloy stoichiometry. In contrast, CoAl was found to deform by slip only on {100} planes.

INTRODUCTION

In the search for new high temperature structural materials, the B2 aluminides have received considerable attention due to their high melting temperatures, moderate densities and potential for excellent oxidation and corrosion resistance. Of these B2 alloys, NiAl has received the most attention (for review see [1]). FeAl has also received a considerable amount of research attention. However, CoAl has been the focus of very few investigations. Presumably, this is due to the reputation CoAl has for extreme brittleness.

In many respects CoAl and NiAl are very similar. Both of these materials form the B2 structure via a congruent melting point and have a wide range of solubility. The resulting structures are almost identical in terms of melting temperature (1932 K for NiAl and 1921 K for CoAl) and lattice parameter (0.288 nm and 0.286 nm respectively [2,3]). Consequently, the densities are also similar, 5.92 g/cc for NiAl versus 6.08 g/cc for CoAl [2,3] for the stoichiometric alloys.

Despite these similarities, significant differences are noted between the mechanical behavior of NiAl and CoAl. These differences were first noted by Westbrook [4] who measured the hardness of single crystals of B2 aluminides. While NiAl and CoAl both display increases in hardness with deviation in alloy stoichiometry from 50 at% Al, Westbrook found that CoAl is significantly harder than NiAl at temperatures ranging from room temperature to 1073 K. Additionally, the hardness of CoAl was found to increase at a greater rate with deviations from perfect stoichiometry. The higher hardness of CoAl is reflected in yield strength measurements recently performed by Fleisher [5] on a wide range of polycrystalline Co-Al alloys tested at temperatures up to 1223 K. When compared with similar tests for NiAl polycrystals [6,7], it is clear that the yield strength of CoAl is much greater than NiAl.

In addition to being harder and stronger than NiAl at low to moderate temperatures, CoAl also displays much greater creep strengths. Working with single crystals, Hocking, Strutt and Dodd [8] found that, depending on stress level, creep strain rates at 1323 K were at least 1.5 orders of magnitude greater in stoichiometric NiAl than stoichiometric CoAl. This is despite the fact that the diffusion rates for these alloys are very similar at this temperature [8]. In polycrystalline materials, Whittenberger [9] also found CoAl to be much more creep resistant between 1100 K and 1400 K. However, it was found that the magnitude of the difference in creep strain rate became smaller at very high temperature [10]. These observations suggest that the enhanced creep strength of CoAl may be attributed to differences in dislocation slip behavior (and associated dislocation creep) and not the diffusional creep rate. These creep study results, along with the observed differences in hardness, indicate that there may be significant differences in the stresses necessary to initiate dislocation slip and plastic flow in these materials.

The nature of slip in B2 NiAl alloys has been studied extensively [6,11-19]. At moderate to high temperatures NiAl displays <100>{001} and <100>{011} slip, depending on orientation. If NiAl is stressed in "hard" orientations which inhibit the motion of <100> dislocations, <111> slip may be initiated at room temperature [16,20]. Additionally, <011> dislocations have been reported in NiAl [21-22], although it is not clear whether these dislocations are involved in the plastic flow or are the result of interactions of <100> dislocations.

As opposed to the large number of slip studies in NiAl, observations of dislocations in CoAl have been limited [8,23-25]. Yaney, Pelton and Nix [24] found that both <100> and <111> dislocations were present in polycrystalline CoAl extruded at 1505 K. Both the <100> and <111> dislocations were observed with long, straight segments indicating that both types were contributing to the plastic deformation [24]. Interestingly, considerable numbers of <111> edge dislocations were observed on {110} planes. This is in contrast to most bcc and ordered bcc metals, where <111> dislocations are typically observed as screw dislocations. Drelles [26] examined the deformation structures in stoichiometric CoAl single crystal deformed at room temperature. Except in orientations near <111> where some <001> slip was observed, the structure was dominated by <111> Burgers vector dislocations. No examples of <110> dislocations have been observed in CoAl.

In order to understand the pronounced differences in strength and plasticity between CoAl and NiAl, the objective of the current study is to characterize the dislocation substructure in deformed single crystal CoAl and to contrast it with that observed in deformed NiAl. This is part of a larger program devoted to the study of the dislocation core structure of B2 alloys, with the goal of understanding the differences in the mechanical properties of these materials.

EXPERIMENTAL PROCEDURE

Single crystal NiAl and CoAl, grown by the Bridgeman method, was obtained from the Naval Air Development Center and GE Aircraft Engines. The nominal composition of these alloys were Ni and Co with aluminum additions of 48, 50 and 52 atomic percent. The crystals were oriented to the [011] orientation and sectioned into 10 X 3.5 X 3.5 mm compression samples. To facilitate slip trace analysis, the specimens were mechanically polished through 0.1 μm diamond paste. Compression testing was performed at 673 K in a vacuum of $5x10^{-5}$ pa. The deformation was carried out at a nominal strain rate of $1x10^{-4}$ s^{-1} to approximately 5% plastic strain.

Standard slip trace analysis was performed on the deformed samples to determine the active slip planes. Sections were then taken parallel to the slip plane for TEM thin foil preparation. Thin foils were prepared by thinning in a twin jet electropolisher. NiAl samples were prepared using a solution of two parts methanol to one part HNO_3 at -30°C under a potential of 12 volts. The CoAl foils were electropolished at -20°C in a solution of 10% perchloric acid in methanol at 10 volts.

The deformation structures were characterized using standard diffraction contrast techniques including **g•b**=0 analysis. The dislocation line directions, slip planes and Burger's vectors were determined.

RESULTS AND DISCUSSION

The [011] orientation was chosen for the compression axis in this study in an effort to enhance <001>{100} slip for the purposes of dislocation core examination. Unfortunately, this orientation may lead to slip on a number of different planes as two <001>{100} slip systems and/or up to four <001>{110} slip systems may be activated. The slip trace analysis of the deformed NiAl single crystals revealed the presence of both duplex {100} and {101} slip planes (Fig. 1). In the stoichiometric Ni-50Al, the {110} planes appeared to be dominant, whereas in the off-stoichiometric Ni-48Al and Ni-52Al, the {001} planes appeared to be the most active. In contrast, only duplex {001} slip was observed in all of the CoAl alloys as shown in Figure 2. In these alloys, the primary effect of off-stoichiometric deviations in chemistry was to decrease the overall amount of observed slip. This was particularly true for the Co-48Al where it was difficult to

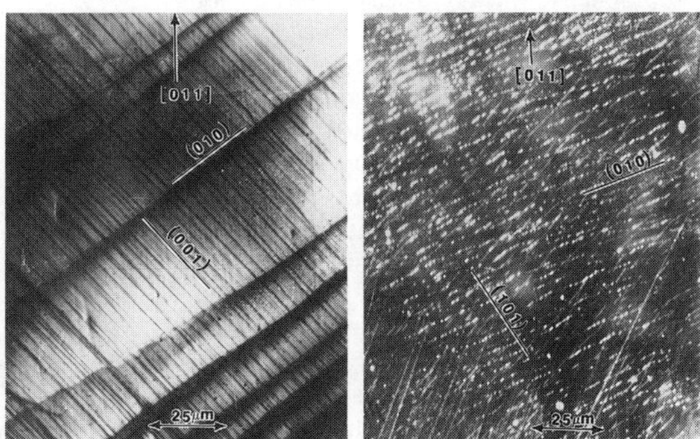

Figure 1. Examples of slip lines on two perpendicular faces of deformed stoichiometric <110> NiAl single crystal.

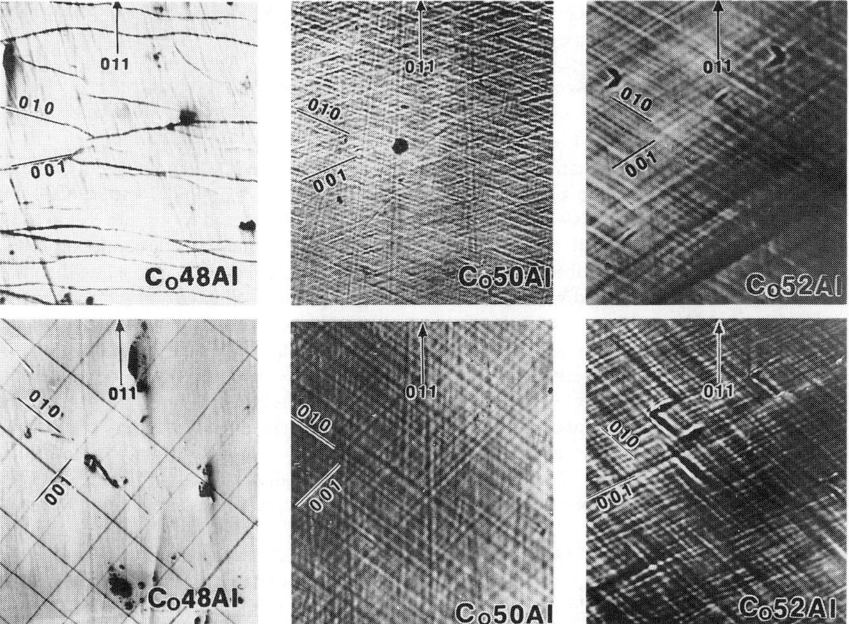

Figure 2. Examples of slip lines on two perpendicular faces of <110> single crystal CoAl with various alloy stoichiometries.

deform the materials beyond 2% plastic strain. This behavior corresponds to the increases in hardness observed with increased deviation in alloy stoichiometry [4]. In all of the Co-Al compression samples, a significant kinking was observed.

TEM examination of sections parallel to the {001} slip planes revealed a number of different dislocation configurations in the stoichiometric Ni-50Al as illustrated in Figure 3. Two types of long straight dislocations lying within the plane of the foil with line directions of [100] and [001] were observed. Additionally, heavy dense bands of dislocations inclined to the foil, with an overall band direction in the foil of [001] were observed. Diffraction contrast **g•b**=0 analysis revealed the straight dislocations lying in the plane of the foil were [001] and [010] dislocations with either pure edge or pure screw orientations. These dislocations would have moved primarily on the {100} slip planes. The inclined dislocations in the thick deformation bands were found to also have [010] Burgers vectors. These dislocations which were predominantly mixed in character were presumably slipping on the (110) plane.

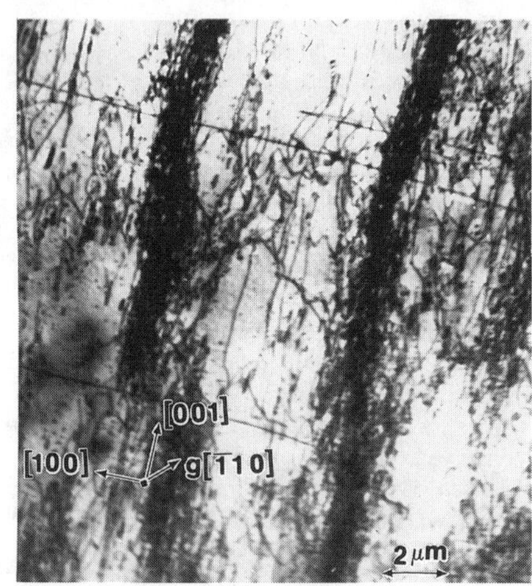

Figure 3. Bright field micrograph showing deformation structure of Ni-50Al.

Electron microscopy observation for the Ni-rich Ni-48Al and the Ni-poor Ni-52Al alloys revealed similar types and configurations of dislocations. However, consistent with the observed shift in slip plane, these off-stoichiometric alloys showed an increased density of <001> dislocations slipping on the {100} slip planes.

The deformed CoAl samples were sectioned parallel to the [010] slip plane in the same manner as the NiAl materials. The resulting observations for the stoichiometric Co-50Al summarized in Figure 4. Dislocations, many in the form of elongated loops, are found to lie within the plane of the foil and are characterized by a general line direction of [100]. Additionally, significant numbers of dislocations appear normal to the foil with a line direction of [010]. Contrast analysis reveals the dislocations lying in the foil have Burger's vectors of both [001] and [010]. These dislocations would be edge dislocations slipping on the (010) and (001) slip planes respectively. A small number of screw dislocations with Burger's vector of [001] have also been observed in the plane of the foil. The dislocations lying normal to the plane of the foil are a mixture of [010] screw dislocations, corresponding with slip on the (001) plane, and a small number of [001] edge dislocations, presumably slipping or climbing on the (100) plane. For the conditions of [011] deformation at 673 K, only these <001> dislocations were observed. However, it should be noted that <001>, <110>, and <111> dislocations have recently been observed in CoAl deformed at higher temperatures using a <123> orientation [27].

Analysis of the Co-52Al (fig. 5) revealed the same operative dislocations as in the stoichiometric alloy. Again, only [100] and [001] screw and edge dislocations were present in the deformed structure. Unfortunately, due to the extreme brittleness of the Co-48Al it has not been possible to successfully prepare a thin foil. This difficulty is consistent with the increased hardness and limited slip lines observed in this alloy.

The results presented here indicate that the deformation of both NiAl and CoAl alloys

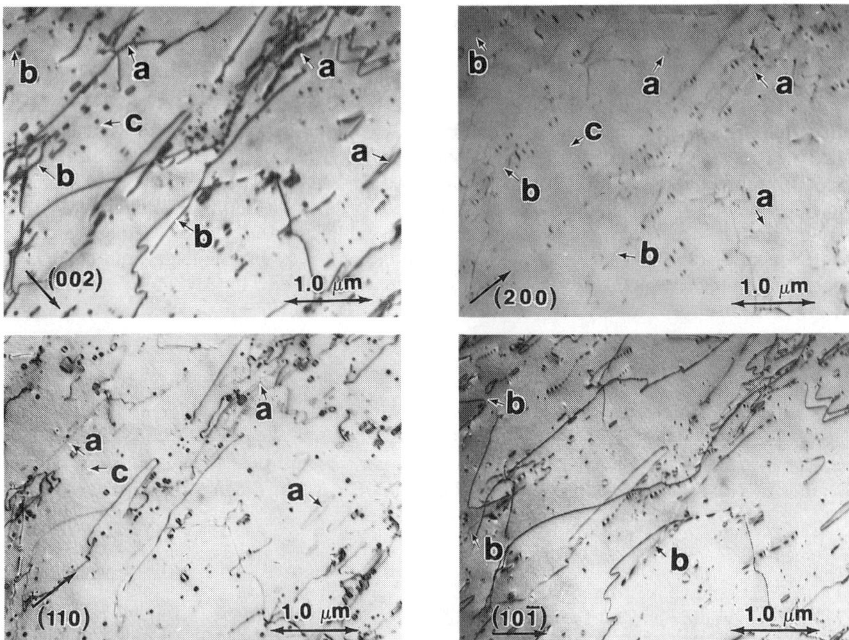

Figure 4. Bright field micrographs illustrating the deformation structure observed in stoichiometric Co-50Al.

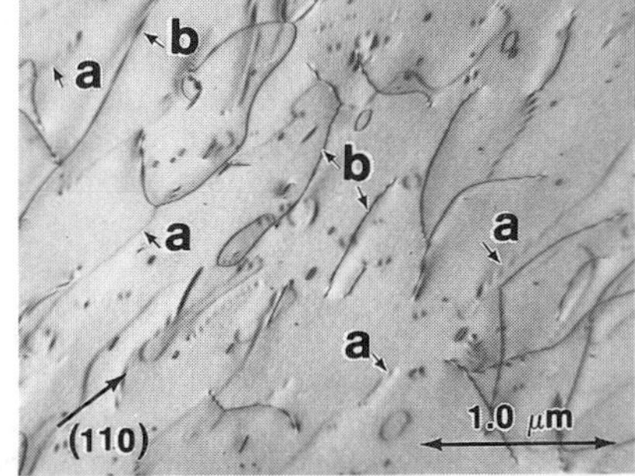

Figure 5. Bright field micrograph showing the dislocation structure in deformed Co-52Al.

deformed at 673 K occurs predominantly by the slip of <001> dislocations. With the exception of stoichiometric Ni-50Al, the dominant slip plane is {100} for these <110> oriented single crystals. Thus it appears that both the CoAl and NiAl alloys are deforming by essentially the same slip mechanisms. In order to rationalize the much greater strength and hardness of CoAl, it must be assumed that the <001> dislocations are fundamentally much harder to move. That is, they have a greater critical resolved shear stress. This occurs although NiAl and CoAl have the same crystal structure and almost identical lattice parameters, and many of their physical properties, such as melting temperature are similar. It is likely that the differences in dislocation slip behavior result from subtle differences in the dislocation core structures in these materials. Recently [28], we have shown changes in the core structure of <001> dislocations as a function of stoichiometry for B2 NiAl alloys. These differences have been related to changes in strength and hardness with changes in stoichiometry [28]. Currently, the structure of the <001> dislocation cores in CoAl are being examined to determine what role this structure plays in controlling the mechanical properties of this class of alloys.

CONCLUSIONS

The deformation behavior of B2 CoAl and NiAl single crystals has been examined as a function of alloy stoichiometry at 673 K using a <110> crystal orientation. Deformation in NiAl was characterized by <001> slip on both {100} and {110} slip planes. A shift in slip plane from predominantly {110} for stoichiometric Ni-50Al to predominantly {100} for off stoichiometric Ni-48Al and Ni-52Al was observed. Although the CoAl alloys were more brittle than the NiAl alloys, the CoAl materials also exhibited <001> slip. The observed slip plane for all of the CoAl alloys was {100} regardless of alloy stoichiometry. In general, a decrease in slip, corresponding to increased strength, was observed for deviations in chemistry form stoichiometry.

ACKNOWLEDGEMENTS

This work was supported by the Office of Naval Research, through grant #N00014-90-J-1910, Dr. G. Yoder contract monitor. Single Crystals of NiAl and CoAl were provided by R. Mahapatra at the Naval Air Development Center, Warminster, PA. Additional Single Crystals of NiAl were provided by Dr. R. Darolia, GE Aircraft Engines, Cincimmati, OH.

REFERENCES

1) R. Darolia, J. of Mat. Sci., **43**, 44 (1991).
2) A.J. Bradley and A. Taylor, Proc. Roy. Soc., **A159**, 56 (1937).
3) A.J. Bradley and G.D. Seager, J. Inst. Metals, **64**, 81 (1939).
4) J.H. Westbrook, J. Electro. Chem. Soc., **103**, 54 (1956).
5) R.L. Fleisher, GE Research and Development Center Alloys Properties Lab., 92CRD181 (1992); J. Mat. Res., in press.
6) A. Ball and R.E. Smallman, Acta Met., **14**, 1349 (1966).
7) R.T. Pascoe and C.W.A. Newey, Met. Sci. J., **2**, 138 (1968).
8) L.A. Hocking, P.R. Strutt, and R.A. Dodd, J. of Inst. Met., **99**, 98 (1971).
9) J.D. Whittenberger, Mat. Sci. and Eng., **73**, 87 (1985); ibid **85**, 91 (1987).
10) J.D. Whittenberger, private communication.
11) R.J. Wasilewski, S.R. Butler and J.E. Hanlon, Trans. Metal. Soc. AIME, **237**, 1357 (1967).
12) N.J. Zaluzec and H.L. Fraser, Script Met. Mat., **8**, 1049 (1974).
13) C.W. Marshal and J.O. Brittain, Met. Trans., **7A**, 1013 (1976).
14) A. Lasalmonie, J. Mat. Sci., **17**, 2419 (1982).
15) J. Baker and E.M. Schulson, Met. Trans., **15A**, 1129 (1984).
16) P.R. Munroe and I. Baker, Scripta Met. Mat., **23**, 495 (1989).
17) M.G. Mendiratand and C.C. Law, J. Mat. Sci., **22**, 607 (1987).
18) C.R. Feng and K. Sadananda, Scripta Met. Mat., **24**, 2107 (1990).
19) P.R. Munroe and I. Baker, Acta Met. Mat., **39**, 1011 (1991).
20) N.H. Lorretto and R.J. Wasilewski, Phil. Mag., **23**, 1311 (1971).
21) D.B. Miracle, Acta Met., **39**, 1457 (1991).
22) M. Dollar, S. Dymek, S.J. Hwang, and P. Nash, Scripta Met. Mat., **26**, 29 (1992).
23) W.D. Nix COSAM Prog. Overview, NASA TM-53006, 183 (1982).
24) D.L. Yaney, A.R. Pelton and W.D. Nix, J. Mat. Sci., **21**, 2083 (1986).
25) D.L. Yaney and W.D. Nix, J. May. Sci., **23**, 3088 (1988).
26) C. J. Drelles, M.S. Thesis, Mich. Tech. Univ., (1985).
27) Y. Zhang and M.A. Crimp, in prep.
28) S.C. Tonn, Y. Zhang and M.A. Crimp, submitted to Mat. Sci. and Eng.

TEM OBSERVATIONS OF DISLOCATION STRUCTURES IN SINGLE CRYSTALS OF NiAl AFTER HIGH TEMPERATURE CREEP

Uwe Glatzel*, Keith R. Forbes** and William D. Nix**
*Inst. Metallforschung, Technische Universität Berlin, 1000 Berlin 12, Germany
**Dept. Mat. Sci. Eng., Stanford University, Stanford, CA 94305-2205, USA

ABSTRACT

The evolution of dislocation substructures formed in single crystals of NiAl by tension creep testing at temperatures between 1123 K and 1473 K has been studied. Significant differences in the dislocation structure are evident in samples tested in different orientations.

For samples tested in soft orientations, where glide of $\langle 001 \rangle$ dislocations is easy, some interacting $\langle 001 \rangle$ dislocations can be found. In hard oriented samples, the density of dislocation networks is high and low angle subgrain boundaries are observed. The subgrain size in hard oriented crystals is strongly stress dependent.

In hard oriented crystals, $\langle 001 \rangle$ dislocations have no resolved shear stress for glide and must move by the slower process of climb; or by glide of $\langle 110 \rangle$ dislocations. The $\langle 001 \rangle$ dislocations are captured in networks, leading to a high total dislocation density in hard oriented samples compared to soft oriented crystals, whereas the dislocation density within the volume of subgrains is constant for all orientations.

INTRODUCTION

The compound NiAl offers significant promise for high temperature applications, but has limited plasticity at lower temperatures. Only three independent slip systems operate at lower temperature thus limiting ductility [1].

The ductility of polycrystalline as well as single crystal NiAl increases significantly at high temperatures. The transition from brittle to ductile behavior can be explained by activating either glide of $\vec{b} = \langle 110 \rangle$ or $\langle 111 \rangle$ or by climb of $\langle 001 \rangle$ dislocations. Evidence for $\langle 110 \rangle$ glide was found in [001] oriented samples tested at temperatures above $673K$ [2,3].

Tension creep tests carried out on single crystals with a [001] stress axis have zero resolved shear stress for $\langle 001 \rangle$ glide. Again, deformation must occur by either glide of $\langle 110 \rangle$ or climb of $\langle 001 \rangle$ dislocations. Samples with different orientations tested at the same temperature should result in significant differences in dislocation configurations.

Anisotropic elastic calculations of line energies [4] show that dislocation configurations in samples tested at 1273 K are very close to their energetic minimum, indicating that dislocations at this temperature are free enough to take up their minimum energy configuration.

EXPERIMENTAL

Creep Experiments

Constant stress creep tests have been carried out in the temperature range of 1123 K and 1473 K in vacuum. Single crystals of stoichiometric NiAl with stress axes parallel to [001], near [110], [223] and [111] crystal axes have been deformed up to 20% strain. The test parameters are described in detail in [5]. The creep response in these different orientations can be grouped as follows:

- Hard oriented samples (stress axis parallel to [001]): the steady state strain rate is about 100 times smaller than for other orientations. [001] samples show sigmoidal or Haasen-creep behavior [6]. The strain rate increases rapidly at the beginning of the test, up to an inflection point. Thereafter the strain rate decreases to a steady state value.

- Soft oriented samples (here [110], [223] and [111] stress axis): show a much higher steady state strain rate and no or very weak sigmoidal type behavior. Especially for higher stress levels the strain rate shows no inflection point.

For all orientations, the activation energy for steady state creep is close to the activation energy of self diffusion .

Transmission Electron Microscopy (TEM) Observations

After halting the creep test samples, were cooled under load. TEM foil preparation and determination of Burgers and line vectors is described in [4]. Undeformed NiAl shows a density of $10^{11}m^{-2}$ of typically straight dislocations.

Samples tested at $1473K$

Figure 1 shows dislocation networks in a sample tested in hard orientation with a stress σ of 10 MPa. The networks connect subgrains which are slightly tilted compared to each other. The subgrain size was determined to be $50 \pm 20 \mu m$. The subgrain size distribution is very broad and the statistical error in determining average sizes is large, depending on how many subgrains are visible in a TEM foil.

By imaging the surface of a TEM foil with a Scanning Electron Microscope (SEM), subgrain boundaries in samples with small subgrain sizes can be made visible as thin lines, see Fig. 2.

Samples tested in soft orientations show significantly higher subgrain sizes. Networks, similar to those in hard orientation, are observed, see Fig. 3. The right, horizontal network has a {111} habit plane and the other two networks, a {110} plane.

For all orientations, the networks consist primarily of ⟨001⟩ dislocations.

Figure 4 shows a dislocation configuration within the volume of a subgrain in asample oriented near [110]. All dislocations, except the marked one, have a ⟨001⟩ Burgers vector and an angle between line and Burgers vector of 40–60°, which is very close to their absolute minimum energy configuration, as calculated by anisotropic elastic theory [4]. These calculations show that line energies of straight dislocations vary in a complex manner with the direction of the line vector. The marked dislocation in Fig. 4 is a reaction of two ⟨001⟩ dislocations leading to a Burgers vector $\vec{b}=[1\bar{1}0]$.

All test parameters and observed subgrain sizes are listed in Table I. The dislocation density within the volume of the subgrains is constant for all orientations at $2 \cdot 10^{11}m^{-2}$.

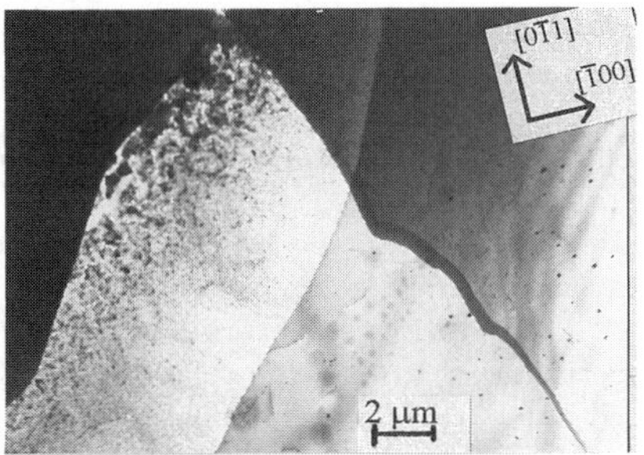

Figure 1: TEM overview of a hard oriented sample. T= $1473K$, σ = 10MPa, ϵ_{final} = 15%. Stress axis is [100].

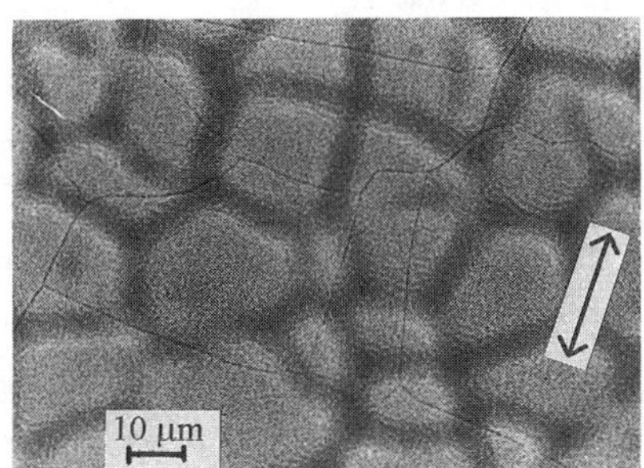

Figure 2: Lower magn. SEM picture. T=$1473K$, σ = 15MPa, ϵ_{final} = 17%. Very fine lines show etched dislocation networks. [100] stress axis is indicated.

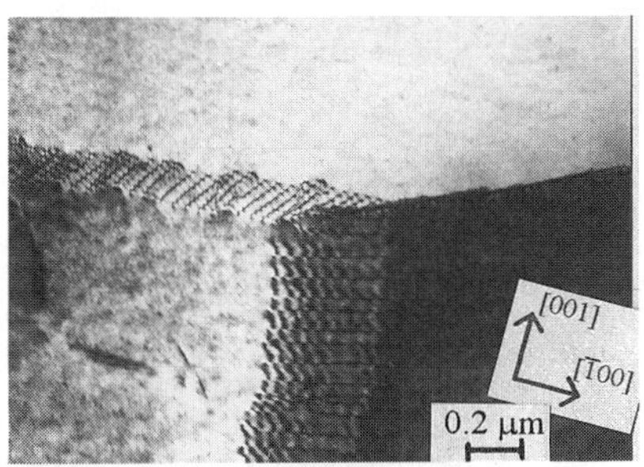

Figure 3: Network reactions in a soft oriented sample. T=$1473K$, σ = 10MPa, ϵ_{final} = 15%. Stress axis is near [101].

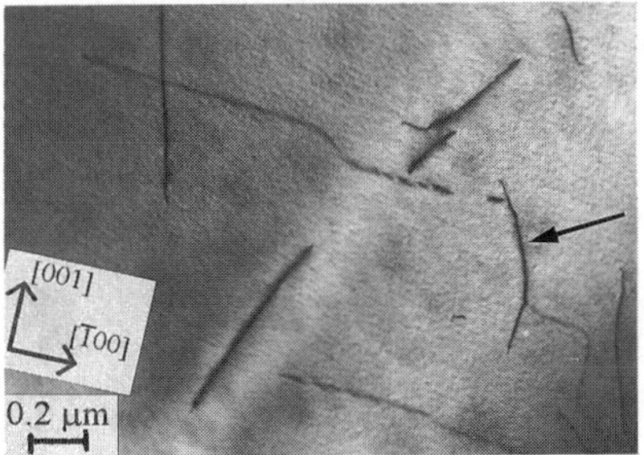

Figure 4: Dislocation configuration in the sample of Fig. 3. The marked dislocation has $\vec{b}$=⟨110⟩. Stress axis is near [101].

Table I: Test Parameters and Observed Subgrain Size at T=1473 K

Stress Axis	Temp. [K]	σ [MPa]	ϵ_{final} [%]	$\dot{\epsilon}_{\text{steady state}}$ [$1/s$]	Subgrain Size [μm]
[001]	1473	6	10	$3 \cdot 10^{-7}$	120 ± 40
[001]	1473	10	15	$2 \cdot 10^{-6}$	50 ± 20
[001]	1473	15	17	$1 \cdot 10^{-5}$	20 ± 10
[223]	1473	4	4	$2 \cdot 10^{-6}$	> 500
[223]	1473	8	23	$3 \cdot 10^{-5}$	400 ± 200
[110]	1473	4	5	$5 \cdot 10^{-7}$	> 500
[110]	1473	10	15	$3 \cdot 10^{-5}$	100 ± 40
[111]	1473	6	15	$2 \cdot 10^{-5}$	> 500

Samples tested at 1273K

Dislocation configurations in samples tested at $1273K$ are qualitatively similar to observations for $1473K$. Small subgrains are visible in hard oriented crystals and for increasing stress the subgrain sizes decrease. Table II lists the test parameters and subgrain sizes. Figure 5 shows three subgrains which are rotated compared to each other around a $\langle 100 \rangle$ axis which is perpendicular to the external stress axis.

The relative rotations can be determined by the change of the Kikuchi lines of a selected area diffraction pattern by moving from one subgrain to the other. The tilting angles betwen subgrains can be as large as 8°. The lower left network has a $\{110\}$ habit plane, the other two a $\{111\}$ plane. Dislocation configurations in a sample tested with 20 MPa in a hard orientation are discussed in [4]. Dislocations are observed with $\vec{b} = a\langle 111 \rangle$. Splitting into pairs with $\vec{b} = a/2\langle 111 \rangle$ could not be resolved, using weak beam imaging techniques.

If the creep test is halted during inflection, dislocation networks have already formed and subgrain sizes are comparable to sizes in samples with higher strains.

Soft orientationed creep samples show macroscopic twisting. Analysing the orientation of stress axis as a function of plastic deformation leads to movement of the stress axis towards a [001] direction, indicating $\langle 001 \rangle$ glide. Soft orientations have large subgrain sizes, but the density of dislocations within subgrains is about the same for all orientations, $3 \cdot 10^{11} m^{-2}$.

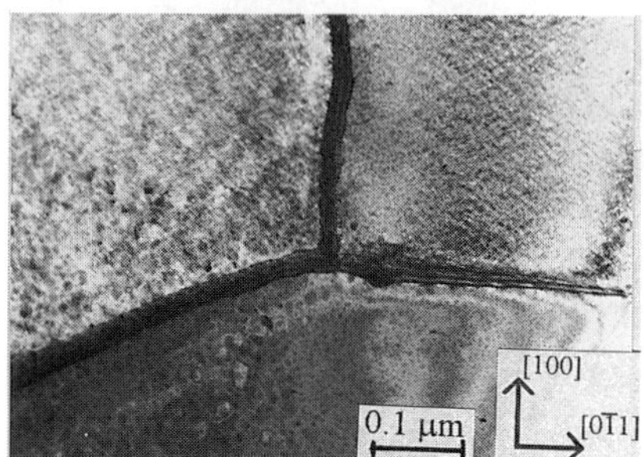

Figure 5: Three subgrains in a hard oriented sample deformed at $1273K$ with σ = 40MPa, $\epsilon_{\mathrm{final}}$ = 18%. Stress axis is [100].

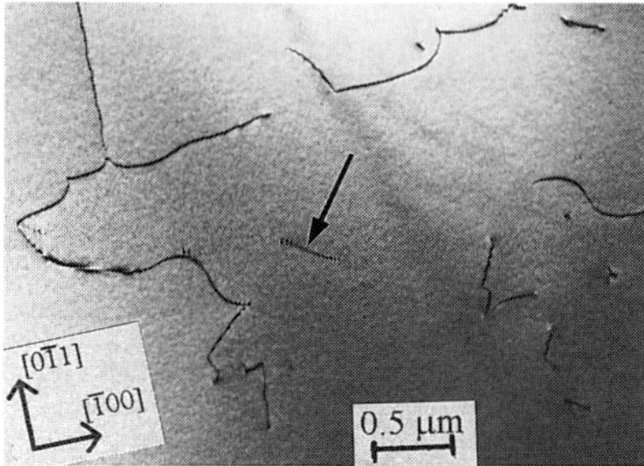

Figure 6: Dislocations in a hard oriented sample. T=$1123K$, σ = 40MPa, $\epsilon_{\mathrm{final}}$ = 9%. One dislocation with $\vec{b} = [110]$ is marked. Stress axis is [100].

Samples tested at 1123K

For all orientations, the stress exponent at $1123K$ is higher (≈ 12) than for samples tested at higher temperatures (≈ 4). It was not possible to prepare a well defined TEM foil for soft oriented samples, because soft orientations necked very quickly. Figure 6 shows a dislocation configuration in a hard oriented sample loaded with 40 MPa. The dislocations are very curved compared to samples deformed at higher temperatures. All dislocations, except one are $\langle 001 \rangle$ dislocations. This was the only isolated $\langle 110 \rangle$ dislocation observed. Table III lists test parameters and observed subgrain sizes.

The high curvature of dislocations explains the strong stress dependence at this temperature. Dislocations are pinned frequently and it is more difficult for them to overcome obstacles by climb. The dislocation density within the volume of subgrains is $10^{12} m^{-2}$.

Table II: Test Parameters and Observed Subgrain Size at T=1273 K

Stress Axis	Temp. $[K]$	σ [MPa]	ϵ_{final} [%]	$\dot{\epsilon}_{\text{steady state}}$ $[1/s]$	Subgrain Size $[\mu m]$
[001]	1273	20	9	$3 \cdot 10^{-7}$	120 ± 40
[001]	1273	40	19	$7 \cdot 10^{-6}$	20 ± 10
[223]	1273	10	7	$1 \cdot 10^{-6}$	200 ± 80
[223]	1273	15	8	$9 \cdot 10^{-6}$	> 500
[223]	1273	20	22	$2 \cdot 10^{-5}$	> 500

Table III: Test Parameters and Observed Subgrain Size at T=1123 K

Stress Axis	Temp. $[K]$	σ [MPa]	ϵ_{final} [%]	$\dot{\epsilon}_{\text{steady state}}$ $[1/s]$	Subgrain Size $[\mu m]$
[001]	1123	40	9	$3 \cdot 10^{-7}$	120 ± 40
[001]	1123	50	15	$7 \cdot 10^{-6}$	20 ± 10
[223]	1123	all samples necked			

Summary of TEM observations

For all orientations and at all temperatures $\langle 001 \rangle$ and $\langle 110 \rangle$ dislocations and dislocation networks have been observed. Dislocations with $\vec{b}=\langle 110 \rangle$ were only rarely observed in hard oriented samples. As explained in [5] there is other strong evidence for $\langle 110 \rangle$ glide to occur.

Temperature Influence

The density of dislocations within the volume of subgrains decreases with increasing temperature. At 1473K and 1273K dislocations are close to their absolute energetic minimum configurations. At 1123K dislocations are pinned frequently and bowed out between the pinning points.

Orientation Influence

The density of regular networks is much higher in hard oriented samples. Dislocation nodes, connecting $\langle 001 \rangle$ and $\langle 110 \rangle$ dislocations, are observed for all orientations but are more frequent in hard oriented crystals.

DISCUSSION

Solely from TEM observations, it cannot be clearly stated if $\langle 110 \rangle$ glide or $\langle 001 \rangle$ climb is the deformation carrying process in hard oriented crystals. Only a small number of $\langle 110 \rangle$ dislocations with non-zero Schmid factors are observed. As discussed elsewhere, $\langle 110 \rangle$ glide can be intended from cross-sectional shape changes [5]. The $\langle 110 \rangle$ gliding dislocations decompose in the early stages of creep testing leaving behind regular networks consisting of $\langle 001 \rangle$ dislocations. The structure formed by dislocation networks hardens the material.

The few dislocation interactions observed in soft oriented samples indicate that easy glide of $\langle 001 \rangle$ dislocations occurs, with a low probability for dislocation interactions. This is confirmed by the macroscopic twisting of samples and by analysing the change in stress axis with increasing plastic deformation.

Dislocation configurations at higher temperature ($1273K$ and $1473K$) indicate, that dislocations are free to move by glide or climb, orienting themselves in minimum energy configurations, as predicted by anisotropic elastic theory. At lower temperatures ($1123K$) dislocations are pinned very often, thus climb of dislocations over obstacles is hindered; this may explain the higher stress exponent at this temperature.

Acknowledgements

The authors gratefully acknowledge the support of Dr. Alan Rosenstein and the AFOSR under Grant No. 89- 0201. One of the authors (U. Glatzel) thanks the Alexander von Humboldt Foundation for the support with a scholarship in the Feodor-Lynen-Programm.

REFERENCES

1. A. Ball and R. E. Smallman, Acta Met. **14**, 1349 and 1517 (1966).

2. R. D. Field, D. F. Lahrman and R. Darolia, Acta Met. Mat. **39**, 2951 (1991).

3. J. T. Kim and R. Gibala in High Temperature Ordered Intermetallic Alloys IV (Mater. Res. Soc. Proc. **213**), 261 (1991).

4 U. Glatzel, K. R. Forbes and W. D. Nix, Phil. Mag., accepted for publication.

5. K. R. Forbes, W. D. Nix, U. Glatzel and R. Darolia in High Temperature Ordered Intermetallic Alloys V (Mat. Res. Soc. Proc. **288**), these proceedings.

6. E. Peissker, P. Haasen and H. Alexander, Phil. Mag. **7**, 1279 (1962).

THE EFFECT OF SCANDIUM ON THE MICROSTRUCTURE OF Ti_3Al

PAUL R.MUNROE
School of Materials Science and Engineering, University of New South Wales, P.O.Box 1, Kensington, NSW 2033, Australia.

ABSTRACT

Using Pettifor Maps as a guide to alloying, scandium additions were made to Ti_3Al to try and transform the crystal structure of this intermetallic alloy from ordered hexagonal to ordered face centred cubic. Scandium was found to have a solid solubility limit of about 6.4at.% in Ti_3Al and preferentially occupy the titanium sublattice. According to the A_3B Pettifor Map, these are criteria which should effect a change in crystal structure, however no such transformation was observed. Scandium additions in excess of the solid solubility limit resulted in the formation of Sc_2Al at grain boundaries and triple points.

INTRODUCTION

The ordered intermetallic compound Ti_3Al has a low density, good high temperature strength and good oxidation resistance at elevated temperatures and is envisaged as a candidate material for aerospace applications. However, its commercial exploitation has been inhibited by its poor room temperature ductility and fracture toughness. Research aimed at improving the ductility and toughness of this material has focused upon the addition of strong β-stabilizers, such as niobium and molybdenum, to Ti_3Al. Such alloys, when subject to appropriate thermomechanical treatment, may exhibit two-phase, ordered hexagonal plus body-centred cubic (α_2+β), microstructures [1-5]. The ductile β phase inhibits crack propagation and thus improves ductility [6].

A more novel approach to improving the ductility of Ti_3Al is to change the crystal structure, from an ordered hexagonal (or DO_{19}) structure to the higher symmetry ordered face-centred cubic (or $L1_2$) structure, through selective alloying. Liu and his co-workers have shown that the DO_{19}-structured compound Co_3V can be transformed to a $L1_2$ structure by alloying with iron, with significant improvements in ductility [7,8]. Furthermore, the tetragonal structured compound Al_3Ti can be transformed to a $L1_2$ structure through additions of transition metals and other elements [9-14], although it is worth noting that no significant increases in ductility have been achieved through this transformation [15]. It should be noted that transformations to more symmetrical crystal structures do not alone guarantee ductility.

One theoretical basis for structural modification through selective alloying is the Pettifor Map [16]. These maps are semi-empirical representations of the crystal structures of alloy phases based upon a single parameter, the Mendeleev number, of the constituent elements. It has been demonstrated elsewhere that ternary additions with low Mendeleev numbers, which substitute on to the titanium sublattice of Ti_3Al and thus lower the average Mendeleev number of this sublattice, should effect a change in the crystal structure of Ti_3Al from DO_{19} to $L1_2$ [17,18]. Scandium was identified as an element, with a lower Mendeleev number than titanium, which could fulfil this criterion. However, for a change in crystal structure to occur scandium must also exhibit some solid solubility in Ti_3Al and preferentially occupy the titanium sublattice. Liu et al briefly investigated the effect of scandium on the crystal structure of Ti_3Al, through examination of an alloy of composition $Ti_{67}Al_{25}Sc_8$ [17]. They noted that scandium had some solubility (6.3at.%) in Ti_3Al, but no change in crystal structure was observed. However, the lattice site occupancy of scandium in Ti_3Al was not determined. In this study, a range of Ti_3Al+Sc alloys were prepared, the solubility limit of scandium in Ti_3Al, its sublattice occupancy, and effect on crystal structure and microstructure were determined. Preliminary mechanical property evaluations were also performed.

EXPERIMENTAL

Five alloys were studied, their compositions are given in Table 1. Ten gram melts of each alloy were prepared from high purity starting materials by arc melting under an argon atmosphere. Each alloy was remelted four times to minimise heterogeneities. Section of the as-cast alloys were removed and homogenised by annealing at 1000°C for three hours in flowing argon followed by furnace cooling to room temperature.

Table 1. Compositions of Alloys Studied (in atomic %)

Designation	Titanium	Aluminium	Scandium
Ti_3Al	75.0	25.0	-
LSc	72.75	25.0	2.25
MSc	69.75	25.0	5.25
HSc	67.5	25.0	7.5
ScAl	75.0	19.0	6.0

Optical metallography and transmission electron microscopy (TEM) was performed on each alloy. Details of specimen preparation techniques have been given elsewhere [18]. Thin foils were examined in a JEOL 2000FX, operating at 200kV, furnished with an energy dispersive x-ray (EDS) detector. The sublattice occupancy of the scandium atoms in Ti_3Al was determined by a TEM technique called Atom Location by Channelling Enhanced Microanalysis or ALCHEMI [19,20], experimental conditions have been given elsewhere [18,21].

Preliminary mechanical property assessment was carried out using a Vickers microhardness indentor, using a load of 300g. The values given are an average of 10 tests.

RESULTS AND DISCUSSION

Optical microscopy revealed that the microstructure of Ti_3Al in both the as-cast and homogenised states consisted of a single phase of large (≈500μm diameter), non-equiaxed grains typical of an as-cast structure. By comparison, LSc, MSc and HSc all exhibited a basketweave structure within large (≈200μm diameter) prior β grains, for example Figure 1a. As the scandium content increased the basketweave microstructure was observed to become finer. Following the homogenisation anneal the basketweave structure was observed to have coarsened, and in both LSc and MSc a number of fine, equiaxed grains were observed to have nucleated, particularly on the prior β grain boundaries (Figure 1b). In contrast, the alloy designated ScAl exhibited a single phase microstructure of fine (≈50μm diameter), equiaxed grains in both the as-cast state and following the homogenisation anneal.

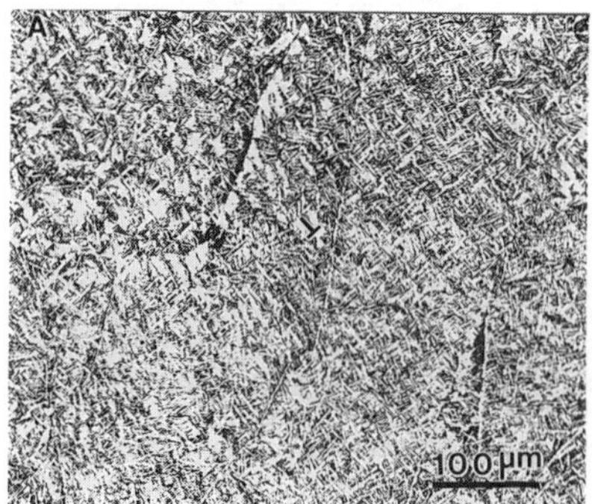

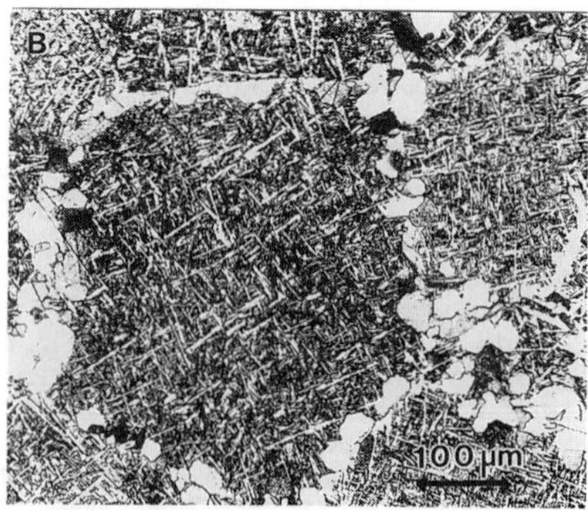

Figure 1. Optical micrographs of a) MSc as-cast b) MSc homogenised

Transmission electron microscopy of Ti_3Al in the as-cast state revealed a single phase microstructure, with a DO_{19} crystal structure, containing a low density of residual dislocations. Within the α_2 grains thermal antiphase domains, ≈100nm in diameter, were observed (Figure 2). Following homogenisation the overall structure of this alloy was largely unchanged, however, no antiphase domain contrast was observed, suggesting that the domains had coarsened and coalesced during annealing.

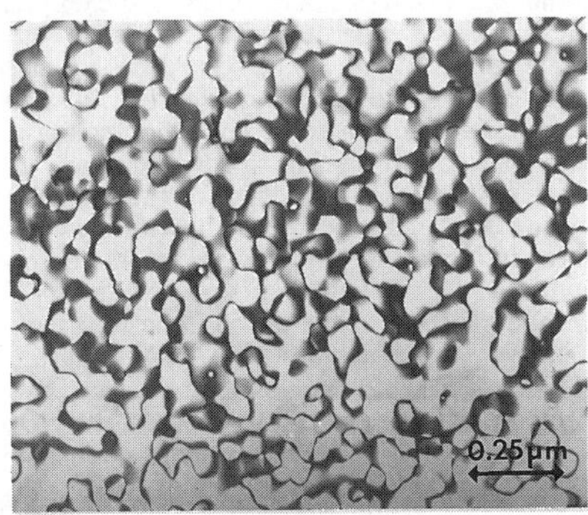

Figure 2. Dark field TEM image of thermal anti-phase domains in Ti_3Al in the as-cast state. $\mathbf{g} = (1\bar{2}10)_{\alpha 2}$

TEM studies of LSc in the as-cast state showed that it contained coarse laths, these were shown by electron diffraction to be DO_{19}-structured. These laths contained thermal antiphase domains (≈50nm in diameter). However, for both MSc and HSc more complex microstructures were observed in the as-cast state (Figure 3a). Both alloys contained laths, with a low density of dislocations, which electron diffraction studies confirmed to possess an ordered hexagonal crystal structure, together with regions containing finer, highly dislocated plates, which electron diffraction studies suggested were α_2' martensite. Anti-phase domains were again observed in the α_2 laths. EDS analysis showed that in HSc the scandium concentration of the α_2' martensite was slightly higher than that of the α_2 laths. In some regions of both alloys where the growth of the α_2 laths was more extensive, a scandium-rich phase, which yielded diffraction patterns consistent with an ordered b.c.c. structure, was observed at the lath boundaries (Figure 3b). It would appear that increasing scandium additions to Ti_3Al increasingly lowers the β-transus, and so the equilibrium β to α_2 transformation becomes increasingly inhibited during the relatively rapid cooling which follows arc melting.

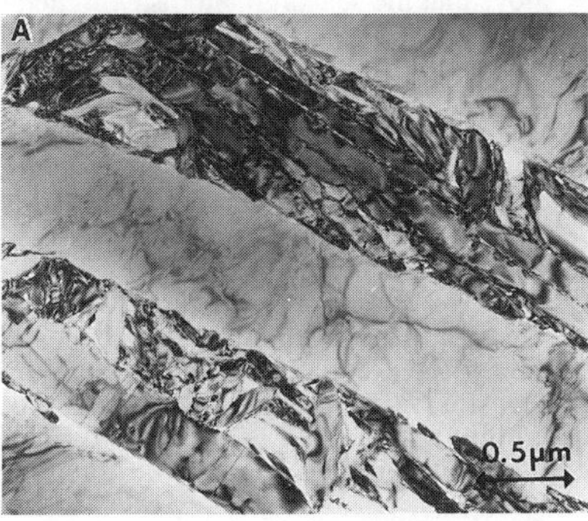

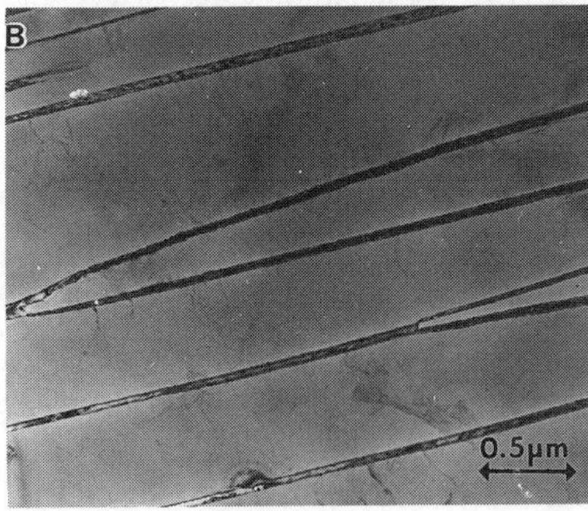

Figure 3. Bright field TEM micrographs of HSc in the as-cast state. a) shows laths of α_2 together with laths of α_2' martensite b) shows more extensive α_2 regions, the phase at the lath boundaries is believed to be β.

Following homogenisation, the microstructure of both LSc and MSc consisted of large, equiaxed grains and large laths, the crystal structure of both the equiaxed grains and the laths were shown by electron diffraction to be ordered hexagonal. No APB contrast was observed in the homogenised alloys.

However, HSc exhibited a two phase microstructure following homogenisation (Figure 4). The principal phase consisted of laths of α_2, again shown by electron diffraction to be ordered hexagonal. EDS analysis performed on the α_2 plates showed that the scandium concentration was approximately 6.4at.%±0.2%, this is consistent with the solubility limit of scandium in Ti_3Al determined by Liu et al [17]. A second phase was observed at lath boundaries and triple points, for example labelled X in Figure 4. EDS analysis of this phase indicated that this phase was essentially titanium-free, with a composition of approximately 67at.%Sc and 33at.%Al. Electron diffraction studies were performed on this phase, the patterns obtained were consistent with a hexagonal crystal structure with the lattice parameters a = 0.489nm and c = 0.617nm. Clearly, this is consistent with the compound Sc_2Al which has a hexagonal crystal structure and similar lattice parameters to those observed here [22]. A more detailed description of the analysis of this phase will be published elsewhere [23]. In contrast, electron microprobe analysis of the second phase observed in $Ti_{67}Al_{25}Sc_8$ by Liu et al suggested a composition consistent with a Ti(Sc,Al) phase [17]. It is possible that the accuracy of the composition of the fine second phase observed by Liu et al was limited the spatial resolution of the microprobe. It is pertinent to note that the melting point of Sc_2Al is only 1195°C [24]. It is clear that that this phase would be undesirable in a high temperature alloy.

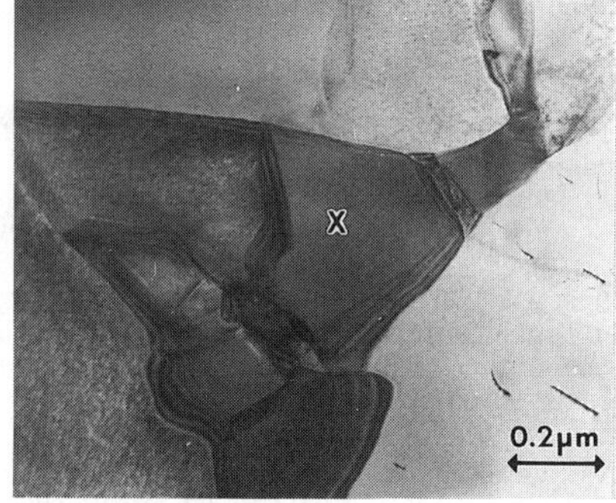

Figure 4. Bright field TEM micrograph of HSc homogenised. The region marked X is Sc_2Al.

The site occupancy of the scandium atoms in LSc, MSc and HSc in the homogenised state was deduced by ALCHEMI. The experimental observations have been described elsewhere in detail [18]. In brief, it was observed that in all three alloys that the scandium atoms exhibited a strong preference for the titanium sublattice.

TEM examination of as-cast ScAl revealed a dispersion of fine (≈5nm diameter) α_2 particles within an α matrix (Figure 5). Following homogenisation the α_2 phase had coarsened and become slightly ellipsoidal (Figure 6a). The major axis of elongation of the α_2 particles was parallel to the [0001] direction. It has been suggested that these particles become elongated in this direction because the lattice misfit between the lattice parameters of α and α_2 is smaller in the c direction [25]. Regions of α were observed between the α_2 particles. Furthermore, fringe contrast was observed around the α_2 particles (Figure 6b). Similar contrast has been observed in hypostoichiometric Ti_3Al, where it was suggested that this effect is associated with segregation at the α/α_2interfaces [26]. ALCHEMI studies suggested that the scandium atoms exhibited a strong preference for the titanium sublattice. Since this alloy contains only 19at.% aluminium, this suggests that stoichiometry is maintained either by the presence of titanium antisite defects on the aluminium sublattice or by constitutional vacancies on the aluminium sublattice. Clearly, replacing aluminium atoms with scandium atoms reduces the degree of order in this alloy.

This study has shown that scandium atoms have a solid solubility limit of ≈6.4at.% in Ti_3Al and also preferentially occupy the titanium sublattice. However, no changes in crystal structure were observed. If it is assumed that all the scandium atoms are located on the titanium sublattice, then according to Pettifor Map predictions both MSc and HSc should exhibit $L1_2$ crystal structures. The validity of these maps for theoretical predictions of structure was recently re-evaluated by Pettifor who concluded, from quantum mechanical calculations, that the transformation of Ti_3Al to a $L1_2$ structure was energetically unfeasible [27]. Other models which have been used to rationalise the effect of selective alloying on crystal structure, such as atomic radii [28,29] or electron atom ratio [30,31] have also been shown to be unapplicable to this alloying system [18]. However, it has been demonstrated that the effect of scandium on the crystal structure of Ti_3Al can be rationalised using the Engel-Brewer theories of electron concentration [18,32]. However, these theories also suggested that ternary additions, such as tin or indium, would effect a change in crystal structure from DO_{19} to $L1_2$. However, preliminary results suggest the the additions of tin and indium do not affect the crystal structure of Ti_3Al [23]. It is clear that the current models used to predict the effect of composition on crystal structure are inapplicable to this alloy system.

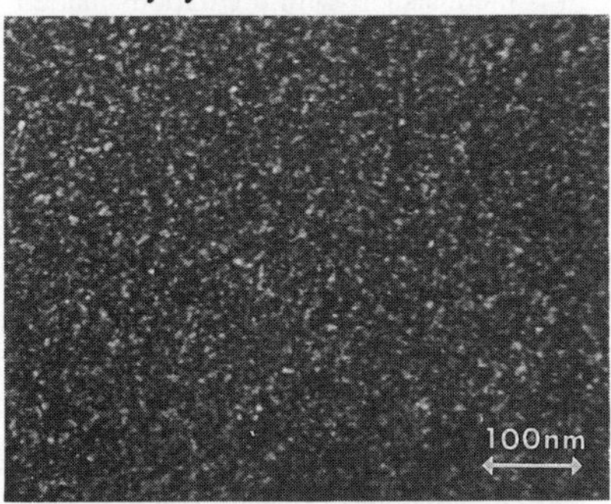

Figure 5. Dark field TEM micrograph of ScAl in the as-cast state, shows α_2 particles in an α matrix, $\mathbf{g} = (11\bar{2}0)_{\alpha 2}$.

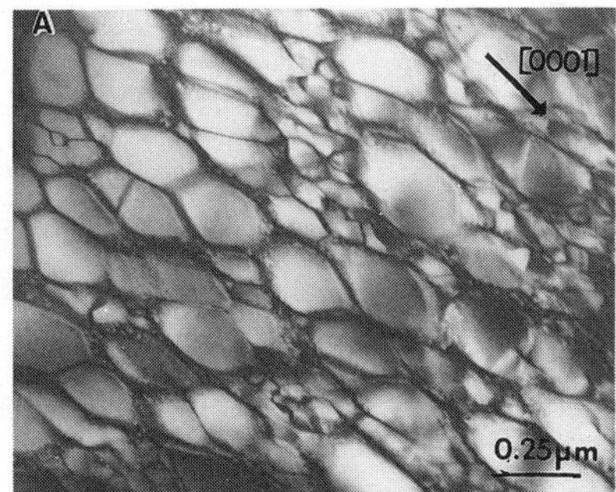

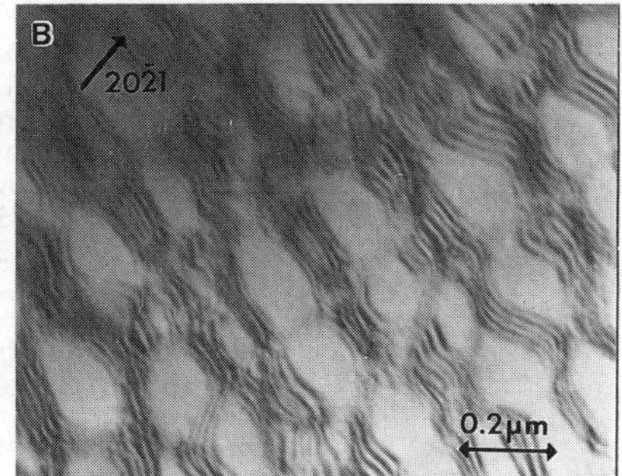

Figure 6. TEM micrographs of ScAl following homogenisation, a) Dark field image shows ellipsoidal α_2 particles in an α matrix, $\mathbf{g} = (11\bar{2}0)_{\alpha 2}$, beam direction near $[1\bar{1}00]$, b) Bright field image shows fringe contrast around α_2 particles, $\mathbf{g} = (20\bar{2}1)_{\alpha 2}$.

Vickers microhardness data from all five alloy in both the as-cast state and after homogenisation are shown in Table 2.

Table 2. Vickers Microhardness of Alloys Studied (Load 300g)

Designation	As-cast	Homogenised
Ti_3Al	352±8	250±7
LSc	449±6	293±6
MSc	422±5	340±7
HSc	515±7	357±6
ScAl	402±6	380±4

It can be seen that scandium increases the hardness of the Ti_3Al. This can be attributed to the effects of solid solution strengthening, and refinement of the microstructure. The presence of martensite in MSc and HSc in the as-cast state also contributes to the hardness. In contrast to the scandium-free alloy, the alloys containing scandium did not exhibit cracking when higher loads were applied.

CONCLUSIONS

Although scandium has significant solid solubility in Ti_3Al and preferentially occupies the titanium sublattice, no transformation in crystal structure from DO_{19} to $L1_2$ was observed, which is inconsistent with predictions made from the A_3B Pettifor map. The solid solubility limit of scandium in Ti_3Al is approximately 6.4at.%; scandium concentrations in excess of this value resulted in the formation of Sc_2Al.

ACKNOWLEDGEMENTS

This work is funded by the Australian Research Council. The author would like to thank Kym Farrell for assistance with preparation of the alloys.

REFERENCES

1. H.A.Lipsitt, D.Schectman and R.E.Schafrik, Metall. Trans. A **11A**, 1369 (1980).
2. M.J.Blackburn and M.P.Smith, U.S.Patent No. 4 292 077 (29 September 1981).
3. R.Strycor, J.C.Williams and W.A.Soffa, Metall. Trans. A **19A**, 225 (1988).
4. D.Banerjee, A.K.Gogia, T.K.Nandi and V.A.Joshi, Acta Metall. **36**, 874 (1988).
5. S.Djanaethany, C.Servant and R.Penelle, J. Mater. Res. **6**, 969 (1991).
6. K.S.Chan, Metall. Trans. A **23A**, 183 (1992).
7. C.T.Liu and H.Inouye, Metall. Trans. A **10A**, 1515 (1979).
8. C.T.Liu, Int. Metall. Rev. **29**, 168 (1984).
9. J. Tarnacki and Y-W. Kim, Scripta Metall. **22**, 329 (1988).
10. K.S. Kumar and J.R. Pickens, Scripta Metall. **22**, 1015 (1988).
11. S.C. Huang, E.L. Hall and M.F.X. Gigliotti, J. Mater. Res. **3**, 1 (1988).
12. C.D.Turner, W.O.Powers and J.A Wert, Acta Metall. **37**, 2635 (1989).
13. I.S.Virk and R.A.Varin, Scripta Metall. **25**, 85 (1991).
14. R.Lerf and D.G.Morris, Acta Metall. **39**, 2419 (1991).
15. D.G.Morris, J. Mater. Res. **7**, 303 (1992).
16. D.G. Pettifor, Mater. Sci. Tech. **4**, 675 (1988).
17. C.T.Liu, J.A.Horton and D.G.Pettifor in High Temperature Ordered Intermetallic Alloys III, edited by C.T.Liu, A.I.Taub, N.S.Stoloff and C.C.Koch (Mater. Res. Soc. Proc. **133**, Pittsburgh, PA, 1989) pp. 37-42.
18. P.R.Munroe, Scripta Metall., in press (1992).
19. J.C.H. Spence and J. Taftø, J. Microscopy **130**, 147 (1983).
20. J. Taftø and J.C.H. Spence, Science **218**, 49 (1982).
21. D.G. Konitzer, I.P. Jones and H.L. Fraser, Scripta Metall. **20**, 265 (1985).
22. S.Eymond and E.Parthe, J. Less-Common Metals **19**, 441 (1969).
23. P.R.Munroe, to be submitted to Materials Forum (1992).
24. T.B.Massalski, Binary Alloy Phase Diagrams, vol. 1 (ASM Metals Park, 1986) p.162.
25. J.C. Williams, Titanium Science and Technology, edited by R.I.Jaffee and H.M.Burte, (Plenum Press, New York, NY 1973) pp. 1433-1492.
26. M.J.Blackburn, Trans. AIME **239**, 1200 (1967).
27. D.G.Pettifor, Mater. Sci. Tech. **8**, 345 (1992).
28. J.H.N.van Vucht and K.H.J.Buschow, J. Less-Common Metals **10**, 98 (1965).
29. J.H.N.van Vucht , J. Less-Common Metals **11**, 308 (1965).
30. A.K.Sinha, TMS-AIME **245**, 237 (1969).
31. A.K.Sinha, TMS-AIME **245**, 911 (1969).
32. N.Durlu and O.T.Inal, J. Mater. Sci. **27**, 3225 (1992).

STRUCTURE AND MICROSTRUCTURE OF Al_2Ti INTERMETALLIC ALLOY

JOHN E. BENCI*, JOHN C. MA*, AND THOMAS P. FEIST**
*Department of Materials Science and Engineering, Wayne State University, Detroit, MI 48202.
**Central Research and Development, E.I. DuPont de Nemours & Company, Inc., Wilmington, DE 19880.

ABSTRACT

Alloys based on intermetallic compounds in the Ti-Al binary system have been extensively investigated as candidate light weight structural materials for elevated temperature applications. However, one compound in this system, Al_2Ti, has been virtually ignored. Al_2Ti, like Al_3Ti, is expected to possess lower density and better oxidation resistance compared to Ti_3Al and TiAl based alloys. Since the structure and stability of Al_2Ti are poorly understood, this work was undertaken to determine the compound's equilibrium crystal structure and the microstructures in the as-cast, melt-spun, and annealed conditions. Powder X-ray diffraction shows that annealed melt-spun Al_2Ti has the tetragonal Ga_2Hf crystal structure containing 24 atoms per unit cell with a = 3.9705 Å, c = 24.322 Å and a calculated density of 3.530 g/cm^3. Optical microscopy and scanning electron microscopy with energy dispersive spectroscopy were used to investigate the material's microstructures. Cast Al_2Ti has a small volume fraction of TiAl second phase particles. Rapidly solidified and annealed Al_2Ti is single phase. Powder processing (melt spinning and hot isostatic pressing) Al_2Ti yields material with a finer microstructure, a higher hardness, and increased resistance to room temperature cracking than as-cast material. Additionally, microhardness results indicate that Al_2Ti has an increased resistance to cracking at room temperature than identically processed Al_3Ti with a similar hardness.

INTRODUCTION

Low densities, high moduli, and high melting temperatures make intermetallic compounds in the Ti-Al binary system attractive for high temperature structural applications. The majority of research and development work has concentrated on Ti_3Al-based alloys,[1,2] with a substantial amount of attention also focused on TiAl[3,4] and Al_3Ti alloys.[1,5-7] By contrast Al_2Ti is a virtually unexplored intermetallic compound in this system. The higher aluminum content of Al_2Ti would be expected to impart oxidation resistance superior to, and result in a lower density than, Ti_3Al and TiAl. Al_2Ti is also likely to possess a higher fracture toughness than Al_3Ti.

The Al_2Ti phase was first identified by Potzschke and Schubert[8] who determined it has a crystal structure isotypic with Ga_2Hf. The binary Ti - Al phase diagram published by Murray[9] included an Al_2Ti phase with the Ga_2Hf structure, fig. 1. However, the Al_2Ti portion of the phase diagram was bounded by dashed lines, indicating uncertainty associated with its composition and stability. According to Murray's diagram, upon heating, Al_2Ti first decomposes into two phases, γ-TiAl and δ-Ti_2Al_5/Ti_9Al_{23}, at approximately 1240°C, and the first liquid phase starts to form at a temperature greater than 1380°C. Loiseau and Vannuffel[10] investigated the structures present in the Ti - Al system between 24 and 42 at. % Ti. They assigned the orthorhombic Ga_2Zr cystal structure to Al_2Ti below 650 and above 1250°C and the tetragonal Ga_2Hf crystal structure at intermediate temperatures. Schuster and Ipser[11] also concluded that two polymorphs of Al_2Ti exist, however they determined the Ga_2Hf structure is stable to room temperature, with lattice parameters a = 3.967Å and c = 24.2968Å, while the Ga_2Zr structure is stable only at temperatures greater than 1216°C. Mabuchi, et al.[12] produced Al_3Ti coatings on TiAl samples and found that an intermediate Al_2Ti layer formed during annealing. They determined the crystal structure to be that of Ga_2Hf with lattice parameters a = 3.971Å and c = 24.32 Å.

The incongruent melting indicated by all of the proposed phase diagrams suggests that conventional melting and casting of 67 at. % Al - 33% Ti would yield a multiphase nonequilibrium microstructure. Rapid solidification processing can be utilized in order to achieve chemical homogeneity. Rapid solidification by melt spinning has been successfully

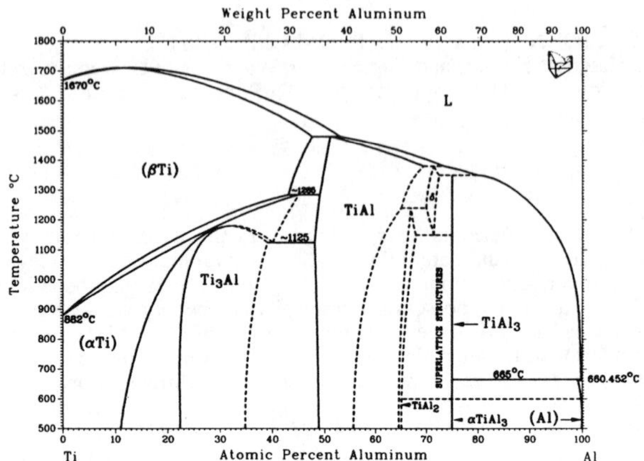

Fig. 1 Ti-Al binary phase diagram showing the presence of an Al_2Ti phase between TiAl and Al_3Ti.[9]

employed[5] to produce nearly single phase $Al_{5.5}Cu_{0.5}Ti_2$, whereas casting resulted in a three phase microstructure. In the current study, Al_2Ti has been prepared by melt spinning to ensure maximum homogeneity.

EXPERIMENTAL

High purity elemental aluminum (99.99%) and titanium (99.7%) were arc-melted in a water cooled copper hearth in an argon atmosphere to produce a cast ingot with a mass of about 300 grams measuring approximately 75 mm in diameter and 20 mm in height. The ingot was remelted several times to ensure chemical homogeneity. A photograph of a cast ingot is shown in fig. 2. Several melts were rapidly solidified by melt spinning, which resulted in Al_2Ti ribbon. Most of the ribbon was comminuted into powder using a hammer mill. Figure 3 is a view of the melt-spun ribbon, which has a needle-like shape, and the comminuted powder. Al_2Ti powder was consolidated by hot isostatic pressing at either 1000 or 1100°C for 4 hours at 25 ksi gas pressure. The result was a dense rod of Al_2Ti. A portion of the Al_2Ti powder was annealed at 1100°C for 68 hrs in oxygen-gettered high-purity argon.

X-ray diffraction was performed on the annealed powder to determine the equilibrium crystal structure and lattice parameters of Al_2Ti. X-ray analysis was performed using a Rigaku DMAX-B automated powder diffractometer equipped with a $\theta/2\theta$ goniometer and a graphite monochromator. A 0.02° step size, a scan rate of 0.5°/min and a Si internal standard were used. Lattice parameters were obtained by least squares refinement of X-ray diffraction data using the program FINAX,[13] which was also used for diffraction pattern calculations.

Material from each step of the processing, i.e., cast ingot, melt-spun ribbon, unannealed and annealed powder, and consolidated powder, was encapsulated in phenolic metallographic mounts. The mounted samples were ground and polished and etched using Kroll's etchant.

Optical and scanning electron microscopy were used to characterize the microstructure of Al_2Ti samples subject to different processing conditions. A SEM equipped with a Kevex energy dispersive spectrometer was used to perform semiquantitative compositional analysis of the cast and powder samples.

Microhardness testing of each type of sample was performed under various loads, from 25 g to 1000 g, and the indents were examined to determine the relationship between the processing and the resistance to microcracking.

RESULTS AND DISCUSSION

X-ray diffraction was performed on annealed Al_2Ti powder in order to determine its equilibrium crystal structure and lattice parameters. Table I shows the observed and calculated peak positions and intensities for the 14 most intense diffraction peaks. The observed pattern closely matches the calculated pattern for Al_2Ti with the Ga_2Hf crystal structure. The average difference between the observed and calculated 2θ values for these 14 peaks is less than 0.02°, the experimental step size. The agreement between the observed and calculated peak intensities is also very good. Figure 4 shows an idealized version of the unit cell for the body-centered tetragonal, Ga_2Hf structure of Al_2Ti. The lattice parameters determined from this analysis are a = 3.9705 Å and c = 24.322 Å, giving a theoretical density of 3.530 g/cm^3. These lattice parameters correlate well with those determined by Schuster and Ipser[11] and Mabuchi.[12] The 3.530 g/cm^3 density of Al_2Ti is lower than the 3.9 g/cm^3 and 4.2 g/cm^3 values for TiAl and Ti_3Al,[4] respectively, and only slightly higher than the 3.35 g/cm^3 density of Al_3Ti.[5] When these compounds are alloyed with elements such as Nb, Mo, V, Cr, etc., to improve the oxidation resistance or fracture toughness, their densities increase, e.g., the density of Cu-modified $L1_2$ Al_3Ti (Al_5CuTi_2) is 4.00 g/cm^3.[5]

The average grain size of cast Al_2Ti is seen to be about 1 mm from the photograph of a cast ingot in fig. 2. Optical and SEM micrographs of the cast microstructure are shown in figs. 5 and 6. Cracks and a large amount of porosity are present in the cast ingot due to nonuniform cooling rates. One large crack can be seen running across the ingot shown in fig. 2. A small volume fraction of a second phase can also be seen in figs. 5 and 6. The approximate compositions of the ingot and of the second phase particles was determined using semiquantitative (± 2 at. %) EDS. The overall composition is 65.6 at. % Al and 34.4 % Ti, while the composition of the second phase is 51.2 at. % Al and 48.8 at. % Ti, essentially TiAl. As expected, the as-cast microstructure is not single phase.

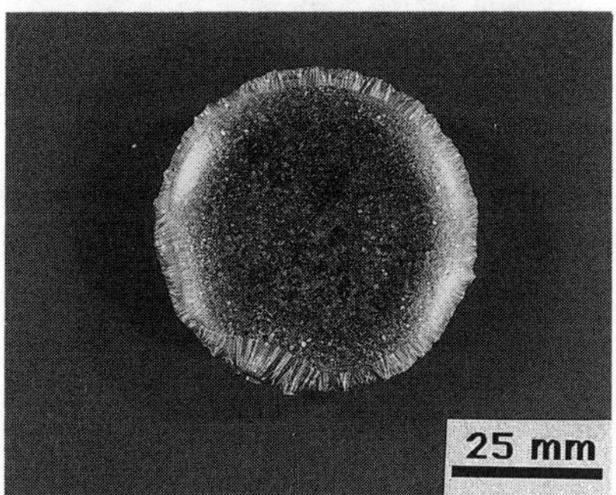

Fig. 2 Photogaph of an Al_2Ti cast ingot.

Fig. 3 Al_2Ti melt spun ribbon and comminuted powder.

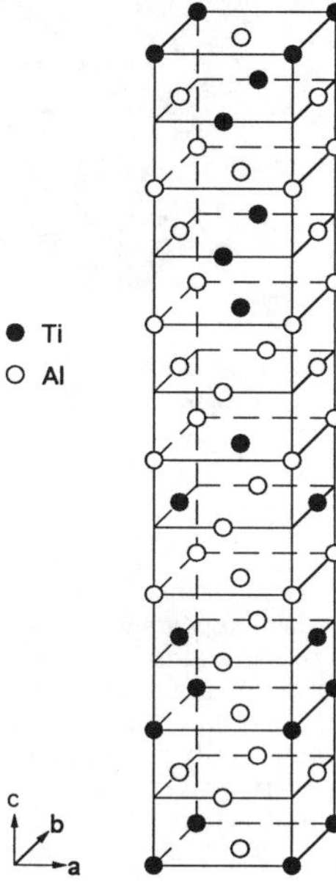

Fig. 4 Idealized version of the Al_2Ti unit cell with the Ga_2Hf crystal structure.

Table I
Observed and Calculated Diffraction Peak Positions and Normalized Intensities for Annealed Al_2Ti with the Ga_2Hf Crystal Structure.

	h k l	Obs. 2θ (degrees)	Calc. 2θ (degrees)	Obs. I/I_{max} (%)	Calc. I/I_{max} (%)
1	1 0 1	22.64	22.67	5.7	8.8
2	1 1 6	38.96	38.96	100.0	100.0
3	0 0 12	44.64	44.67	12.7	13.5
4	2 0 0	45.62	45.66	34.8	32.1
5	2 0 12	65.78	65.79	13.4	14.1
6	2 2 0	66.54	66.56	9.9	8.6
7	1 1 18	78.50	78.49	6.9	5.6
8	3 1 6	79.90	79.92	25.1	17.0
9	2 2 12	83.74	83.75	9.2	6.1
10	4 0 0	101.80	101.79	6.4	2.4
11	3 1 18	113.78	113.75	8.7	5.0
12	3 3 6	115.30	115.29	5.1	4.1
13	4 0 12	119.56	119.55	5.9	3.4
14	4 2 0	120.38	120.36	6.3	4.3

Body-centered tetragonal, space group = I 4_1/amd, a = 3.9705 Å, c = 24.322 Å.
(Cu K_α radiation, λ = 1.54056Å)

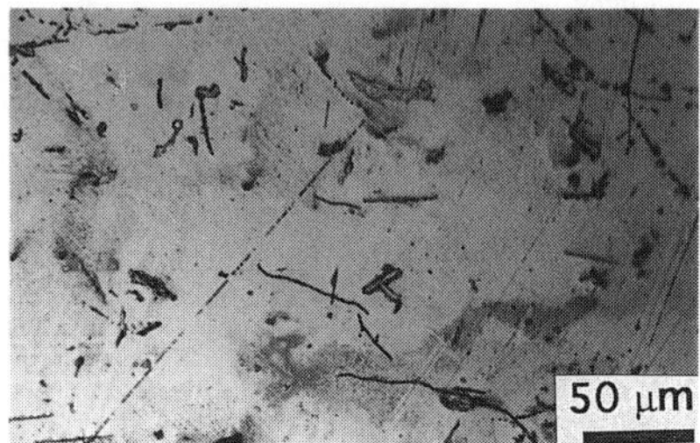

Fig. 5 Optical micrograph of the Al_2Ti cast microstructure showing the TiAl second phase particles.

Fig. 6 SEM micrograph of cast Al_2Ti.

Optical and SEM micrographs of mounted, polished and etched Al_2Ti melt-spun ribbon are presented in figs. 7 and 8. Several distinct morphologies were observed depending on the distance from and orientation with repect to the wheel surface during solidification. The region nearest the wheel appears featureless due to the rapid solidification rates. In the central portion of the ribbon, the grains have a columnar morphology. Near the surface of the ribbon away from the wheel, the grains are relatively equiaxed. The piece of ribbon shown in fig. 8 has this structure repeated because a layer of ribbon was solidified over a previously solidified ribbon which was temporarily stuck to the wheel. Equiaxed grains with an average grain size of about 5 to 20 µm were also found on the cross section of the ribbon which is perpendicular to the solidification direction. The melt-spun ribbons do not show evidence of any second phase. EDS compositional analysis of the ribbon yielded an average compostion of 65.5 at. % Al and 34.5 % Ti.

Comminuting Al_2Ti ribbon yielded irregularly-shaped powder with an average particle size of about 50 µm. The majority of the particles fall into the size range of 10 to 90 µm. The microstructure of the comminuted powder is identical to that of the ribbon. The grain size of annealed powder is approximately 12 - 16 µm.

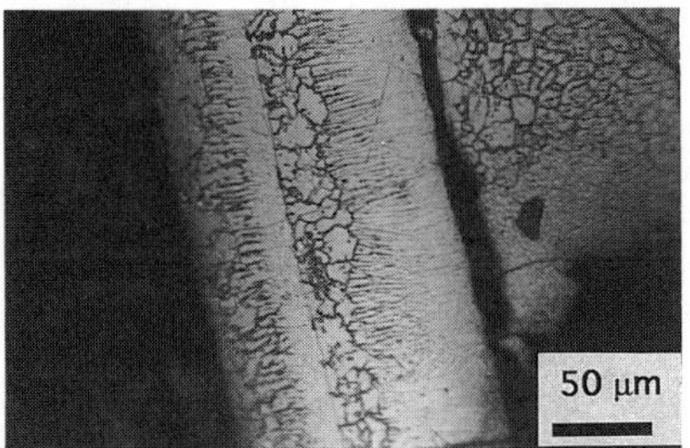

Fig. 7 Optical micrograph of the Al_2Ti melt-spun microstructure showing a featureless region, and columnar and equiaxed grains.

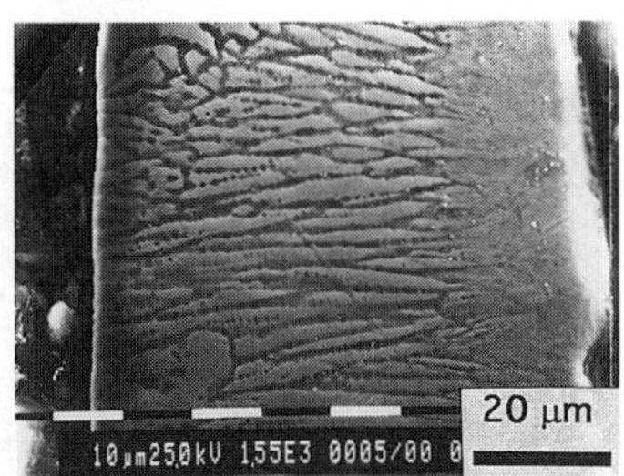

Fig. 8 SEM micrograph of melt-spun Al_2Ti.

The microhardness results are summarized in fig. 9. For a 200 g testing load, hot isostatically pressed material has the highest hardness, 484 - 507 VHN, cast Al_2Ti has an intermediate hardness of 397 VHN, while melt-spun ribbon has the lowest hardness - 349 VHN. The load at which cracking initiated is also plotted in fig. 9. Cracking was detected around the 300 g microhardness indents in melt-spun ribbon, at 500 g cracking was initiated in cast Al_2Ti, while there was no cracking detected around microhardness indents in the hot isostatically pressed samples at loads up to the maximum achievable, 1000 g. Microhardness indents from a 1000 g applied load in identically processed (melt spun and hot isostatically pressed) Al_2Ti and Al_3Ti are shown in fig. 10. The hardness values are 456 and 484 VHN for Al_2Ti and Al_3Ti, respectively. Cracking is visible around the indent in Al_3Ti, but no cracking is seen in the Al_2Ti sample.

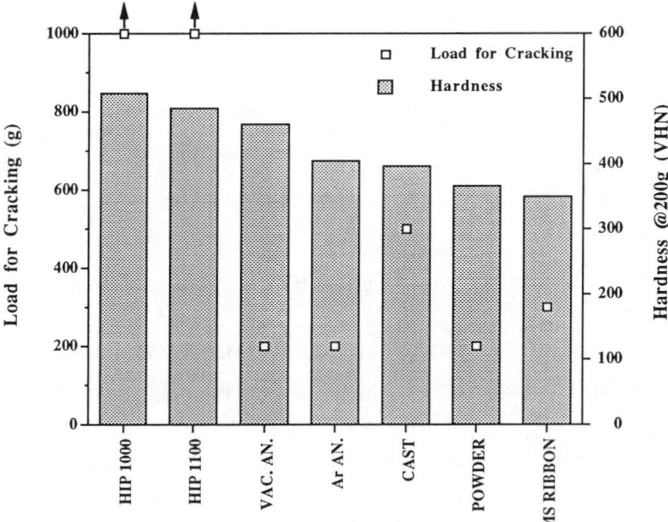

Fig. 9 Plot of microhardness values for Al_2Ti at a 200 g applied load for powder hot isostatically pressed at 1000°C (HIP 1000) and 1100°C (HIP 1100), vacuum annealed (Vac. An.) and argon annealed (Ar An.) powder, the cast ingot (Cast), and melt-spun powder (Powder) and ribbon (MS Ribbon). The resistance to cracking as a function of processing is also plotted as the minimum load at which cracking was observed for each material condition.

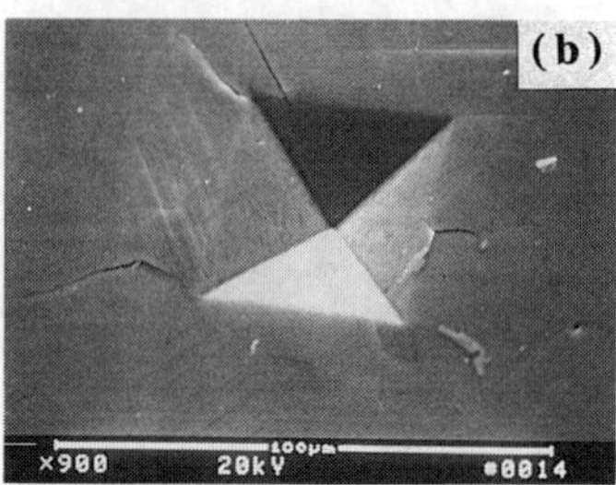

Fig 10 1000 g microhardness indents in identically processed (melt-spun and hot isostatically pressed) a.) Al_2Ti and b.) Al_3Ti showing cracking around the indent in Al_3Ti but none around the indent in Al_2Ti.

CONCLUSIONS

The phase Al_2Ti does exist, and melt-spun and annealed Al_2Ti has the Ga_2Hf crystal structure with a = 3.9705 Å, c = 24.322 Å, and a theoretical density of 3.530 g/cm^3. Producing Al_2Ti by casting yields a two-phase microstructure with TiAl as the small volume fraction second phase. However, rapidly solidified and annealed Al_2Ti is single phase. Near 100% dense Al_2Ti can be produced by hot isostatically pressing powder, yielding measured densities of up to 3.517 g/cm^3 (99.6% ρ_{Th}). Powder processing (melt spinning and hot isostatically pressing) Al_2Ti results in material with a finer grain size, a higher hardness, and an increased resistance to cracking than cast material. Al_2Ti has an increased resistance to cracking at room temperature than identically processed Al_3Ti with similar hardness.

ACKNOWLEDGEMENTS

The authors would like to acknowledge the support of the Office of Naval Technology through a postdoctoral fellowship (JEB), W. Frazier and J. Waldman at the Naval Air Development Center for the use of processing facilities, the NSF MRL program - Grant DMR91-20668 for facilities at the LRSM - Univ. of PA, and Wayne State University's Institute for Manufacturing Research and Thomas Rumble Fellowship program (JCM).

REFERENCES

1. M. Yamaguchi and Y. Umakoshi, Progress Mater. Sci., 34, 1-148 (1990).
2. A.K. Gogia, D. Banerjee and T.K. Nandy, Metall. Trans. A, 21A, 609-625 (1990).
3. E.L. Hall and S.-C. Huang, Acta Metall. Mater., 38, 539-549 (1990).
4. J.C. Beddoes, W. Wallace, and M.C. de Malherbe, Mater. & Manufac. Processes, 7, 527-559 (1992).
5. W.E. Frazier, J.E. Benci and J.W. Zanter, in Low Density, High Temperature Powder Metallurgy Alloys, eds. W.E. Frazier, M.J. Koczak and P.W. Lee, The Metallurgical Society, Warrendale, PA, 49-69 (1991).
6. E.P. George, D.P. Pope, C.L. Fu and J.H. Schneibel, ISIJ International, 31, 1062-1074 (1991).
7. Z.L. Wu, D.P. Pope and V. Vitek, Scripta Metall. Mater., 24, 2187-2190 (1990).
8. M. Potzchke and K. Schubert, Z. Metallkde., 53, 548-561 (1962).
9. J.L. Murray, in Binary Alloy Phase Diagrams, ed. T.B. Massalski, American Society for Metals, Metals Park, OH, 173-176 (1986).
10. A. Loiseau and C. Vannuffel, Phys.Stat. Sol. A, 107, 655-671 (1988).
11. J.C. Schuster and H. Ipser, Z. Metallkde., 81, 389-396 (1990).
12. H. Mabuchi, T. Asai, and Y. Nakayama, Scripta Metall., 23, 685-689 (1989).
13. E.R. Hovestreydt, J. Appl. Cryst., 16, 651-653 (1983).

DISLOCATION INTERACTIONS IN TWO-PHASE TiAl ALLOYS

M.A. MORRIS
Institute of Structural Metallurgy, University of Neuchâtel
Av. Bellevaux 51, 2000 Neuchâtel, Switzerland,

ABSTRACT

Dislocation configurations produced by room and high temperature deformation of two alloys with compositions Ti-48Al and Ti-47.5Al-2.5Cr have been compared. At room temperature the major differences between the two alloys is the increased twinning activity and that of 1/2 <112] superdislocations in the chromium containing alloy while the binary alloy contains a larger contribution from <101] superdislocations. At 400°C, the temperature at which a peak in flow stress is observed in both alloys, an increase in twinning activity and emission of 1/2<110] dislocations on (001) planes is observed from twin intersections in both materials. The contribution of superdislocations to the deformation process is limited and their effect on the flow stress increase appears irrelevant. Instead the anomalous strengthening effect has been related to the cross-slip of single dislocations from cube planes, on which they are emitted, to octahedral planes where they can also glide.

INTRODUCTION

It is now well established that γ-TiAl alloys exhibit very different ductilities and flow stress dependencies with temperature according to their compositions and microstructures [1-4]. At the present time is appears that alloys with slightly hypo-stoichiometric compositions, small additions of selected transition metals and with mixed lamellar/globular microstructure have the best combination of properties in terms of ductility and strength.

Some studies of the dislocation configuration of γ-TiAl in two phase ($\gamma+\alpha_2$) alloys have shown that superdislocations of the type <101] do not contribute to the deformation process at room temperature [5,6]. Instead glide of 1/2 <110] and 1/2 <112] dislocations and twinning of the type {111}<11$\bar{2}$] is observed. These features were only observed in single phase TiAl alloys after deformation at temperatures above 600°C [7,8]. This implies that the dislocation configurations responsible for ductility depend on initial microstructure and this is a function of composition and initial heat treatment.

The effect that alloying elements (such as chromium) have on dislocation configurations has not been systematically studied and this represents a major requirement in order to understand how such elements influence plasticity of the material. In the present study the deformed microstructure produced by compression at room temperature and at 400°C from alloys with two compositions, namely Ti-48Al and Ti-47.5Al-2.5Cr, have been compared. Dislocation configurations and dissociations have been analysed and their different activities have been related to the anomalous strengthening effect observed with increasing temperature in both alloys. The role played by the α_2 phase within the lamellar structure has been discussed in terms of the mircrocrack initiation occurring at the α_2/γ interface.

EXPERIMENTAL TECHNIQUES

The nominal compositions of the alloys used for this study were Ti-48Al and Ti-47.5Al-2.5Cr (at%). They were prepared by the Osprey deposition technique and were supplied by AluSuisse after HIP at 1100°C. X-ray diffraction analysis was performed from both alloys in order to measure the lattice parameters and to compare the effect of the chromium addition on the tetragonality of the L1o lattice and on the misfit strain between the α_2 and the γ phases. A Philips EPD 1880 diffractometer was used.

Microstructural analysis of the annealed and deformed materials was performed by transmission electron microscopy (TEM) using a Philips CM12 microscope equipped with energy-dispersive spectroscopy facilities (EDS). Thin foil preparation for TEM analysis was carried out using a solution of 5% sulphuric acid in methanol. Dislocation analysis was made from projected images

obtained by tilting the specimens to different known orientations (zone axes) using a kikuchi map corresponding to the L1o structure. From each zone axis different diffraction vectors were chosen to obtain invisibility under some contrast conditions. Weak beam images were taken using the g: 3g conditions for different diffraction vectors. Scanning electron microscopy (SEM) studies were made using a Cambridge 360 Stereoscan microscope also equipped with EDS facilities. Quantitative analysis of the volume fractions of the different phases present was made using an image analyser Cambridge Quantimet 720.
Bend and compression tests were carried out on rectangular bars and cylindrical specimens respectively. The compression tests were performed as a function of testing temperature between 20 and 700°C in an argon atmosphere.

RESULTS

The two alloys studied had a homogeneous distribution of equiaxed grains with a fine lamellar structure of α_2 (Ti_3Al) phase distributed within the γ (L1o) matrix. The volume fraction of α_2 laths (seen white in Figure 1) was 22% in the binary alloy and 15% in the chromium containing material. The measured grain sizes were 40 and 80 μm respectively. Also a fine distribution of 2% volume fraction of aluminium oxide particles was present in the two alloys. Surface observation made by SEM after room temperature deformation (by hardness indentations, bend and compression tests) have provided information about the role played by the α_2 laths in the deformation process and during crack initiation and propagation [9]. In all cases crack initiation was observed at the α_2/γ interface while its propagation was observed both in transverse directions across several lamellae, as shown in Figure 1a, or longitudinally along α_2/γ interfaces (1b). Crack propagation was rather slow during the compression tests with little change being observed between 1 and 5% strain. The first microcracks formed during compression were observed after 1% strain in the binary alloy but only after 2% in the chromium containing material. Neither of the alloys exhibited much ductility during the bend tests with fracture occurring after 0.1 and 0.2 % strain in the binary and chromium containing alloy respectively.
The presence of dislocation networks has been detected at the α_2/γ interfaces of the annealed materials [9] due to the misfit strain between the crystallographic structures of the two phases. Also the misfit parameter between the two phases calculated from X-ray diffraction data has been confirmed to be smaller for the chromium containing alloy [9]. This indicates that in the latter, a lower density of misfit dislocations will be required at the α_2/γ interface to accommodate the lower misfit strain. Therefore a better ductility of the chromium containing alloy can be understood if more strain accommodation is possible by accumulation of extra dislocations at the α_2/γ interface before crack initiation occurs.
During compression the values of flow stress have been measured as a function of testing temperature for both alloys and these are shown in Figure 2. We note that the chromium containing alloy has higher strength at all temperatures and also that a peak in flow stress is observed at 400°C in both alloys.
Detailed analysis of the deformed structures has been carried out in both alloys. Since the α_2 phase does not deform, only the deformation mechanisms taking place within the γ phase will

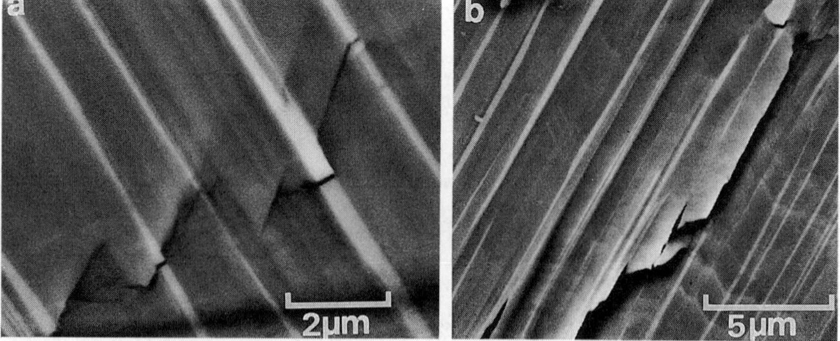

Figure 1.- Examples of crack propagation observed in both alloys.a) Crack traversing several lamellae.b) Longitudinal propagation along the α_2/γ interface.

be discussed, these being responsible for the stress concentrations leading to crack initiation at the α_2/γ interfaces. The deformation mechanisms occurring within the different γ matrix zones have been shown to depend on the rotation relationships between these grains. These have been analysed between different sets of adjacent γ regions of either lamellar or globular types. Either rotations of 90°C around <010> axes or of 120° around <111> axes, exist. The compatibility of deformation across several of these ordered domains determines the shear direction producing the deformation process. In some cases this produces twinning of the type <112]{111} while in adjacent domains the equivalent <112] {111} system only produces slip of superdislocations [9]. Similar active mechanisms have been analysed in both alloys after compression tests performed at room temperature, namely twinning activity and slip of single and superdislocations. The twinning process occurs in both alloys by the passage of Shockley partials 1/6<112] on {111} planes.

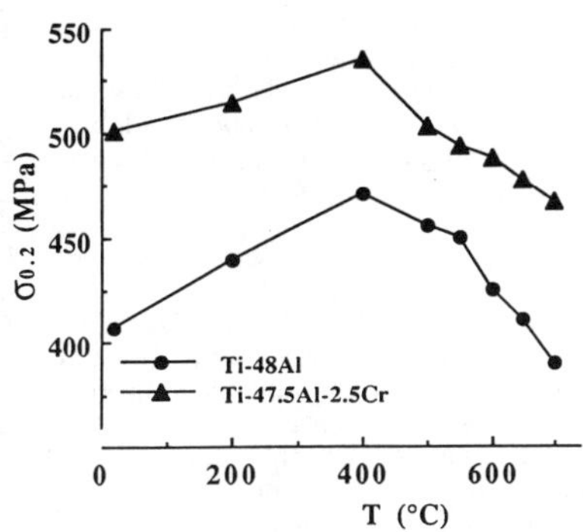

Figure 2.- Plots of flow stress as a function of testing temperature for both alloys

The twin density is however more important in the chromium containing alloy where several twinning systems can be present within a grain and where cross-twinning occurs. At fault intersections, reactions between Shockley partials occur and emission of 1/2<112] superdislocations has been observed.
Also both alloys contain single 1/2<110] dislocations, some gliding on {111} planes but others are observed having edge character and contained on (001) planes. Figure 3 shows an example of a grain from the deformed chromium containing alloy where we see the twin systems analysed and some of the bowed $1/2[1\bar{1}0]$ edge segments gliding on (001) planes. The twin systems seen here are produced by the passage of Shockley partials $1/6[1\bar{1}2]$ on the plane $(\bar{1}11)$ (horizontal twin) and by $1/6[\bar{1}\bar{1}2]$ on plane (111) (vertical twin).
The major difference between the two alloys studied is the presence of different types of superdislocations after room temperature deformation. The binary alloy contains segments of the type <101] while the chromium containing alloy deforms by glide of 1/2<112] superdislocations that trail faulted dipoles. Examples of the latter are shown in Figure 4. The analysis of these loops has confirmed that their Burgers vector is parallel to $1/2[1\bar{1}2]$ and that they glide on plane $(\bar{1}11)$. The faulted dipoles seen are also contained on plane $(\bar{1}11)$ and defined by the two directions [101] and [110]. The two partials bounding the faulted dipole have also been indentified as $1/3[1\bar{1}2]$ and $1/6[1\bar{1}2]$ as evidenced by their invisibility with the diffraction condition $g[\bar{1}11]$. The dissociation of this type of segment is consistent with:
$1/2[1\bar{1}2] = 1/3[1\bar{1}2]$ + SEFS $+1/6[1\bar{1}2]$. By contrast, the superdislocations of the type <101] observed in the binary alloy have been confirmed to dissociate according to two possible reactions :

(1) $<\bar{1}01] = 1/2 <\bar{1}01]$ + APB + $1/2 <\bar{1}01]$

(2) $<\bar{1}01] = 1/2 <\bar{1}01]$ + APB + $1/6 <[\bar{2}\bar{1}1]$ + SISF + $1/6<\bar{1}12]$

Figure 5 shows examples of these two types of dissociations for segments of Burgers vector $<\bar{1}01]$. Here segment 1 has a line direction at 60° from screw orientation. The dissociated segments are invisible with g [020] and $g\,[1\bar{1}1]$ and visible with $g\,[\bar{1}11]$, [002] and [200]; their dissociation is consistent with that shown in reaction (1). The width measured from the APB dissociation is $d_{APB\{111\}}$ = 3.5 nm. The dislocation labelled 2 has Burgers vector parallel to [011] and screw orientation. In that case we see that the presence of a stacking fault makes part of the dissociation and the line is split such that both an APB and a SISF are seen. The plane of the fault has been determined by trace analysis as the $(1\bar{1}1)$ which confirms the visibility of the partials analysed and the dissociation to be : [011] = 1/2[011]+APB+ $1/6[1\bar{2}1]$+ SISF+$1/6[\bar{1}12]$.
The different dissociations exhibited by segments of screw or at 60° orientation indicate that the core of screw segments can spread in different planes, thus, becoming immobile at room temperature. The deformed microstructures observed after compression at 400°C (the temperature

at which the peak in flow stress is observed) are mainly characterised by an increased twinning activity and also by the presence of large densities of single 1/2 <110] dislocations. This is shown in Figure 6. Also large number of debris are present produced by pinching off small loops of dipoles produced by annihilation of 1/2<110] segments. All these features were observed even in specimens deformed to 0.5% strain indicating that twinning and 1/2<110] dislocations are active from the moment at which plastic deformation is initiated.
One very interesting aspect of the analysis has shown that emission of 1/2<110] bowed segments occurs at the intersections between different active twin systems.

Figure 3.- Twin activity and $1/2[1\bar{1}0]$ edge dislocations observed after room temperature deformation of the ternary alloy.

In some cases, as that seen in Figure 7a, the bowed $1/2[\bar{1}10]$ segment has edge character and glides on plane (001). In other cases, as that shown in Figure 7b, the emitted segments glide on {111} planes. It is interesting to note that the loops expand by bowing out the edge segment while the screw parts become longer and very straight. In the case shown in Figure 7a, the straight segments seen correspond to the screw parts of such loops where the edge segments have been removed during foil preparation. In many cases several twin systems have been observed to impinge at α_2/γ interfaces and produce crack initiation [9]. This indicates that the twin intersections lead to stress concentrations responsible for emission of ordinary dislocations in the γ phase or for crack initiation at the α_2 phase.

DISCUSSION AND CONCLUSIONS

Comparison of the dislocation configurations analysed after deformation at room temperature and at 400°C shows that for both alloys the characteristic features are very similar. Also, the mechanisms of deformation responsible for the higher value of flow stress measured at 400°C are related to the increased activity of 1/2<110] dislocations in both alloys.
The presence of dislocation segments 1/2<110] on (001) planes have also been detected by Greenberg et al. [8] after room temperature deformation of single phase alloys. These authors

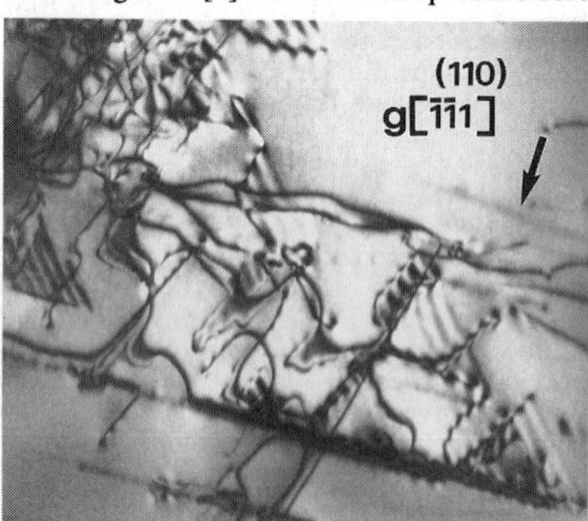

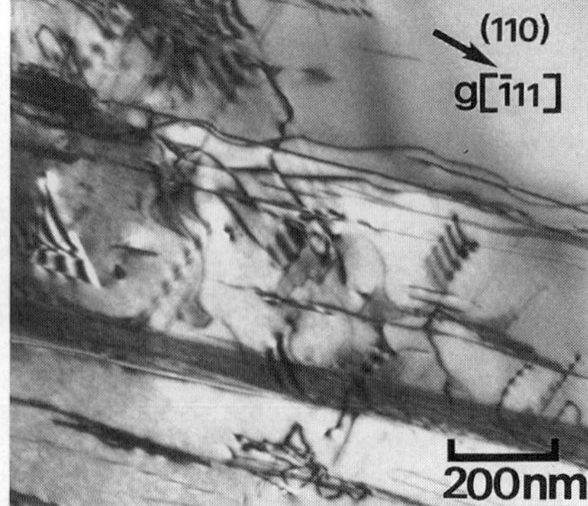

Figure 4.- Examples of 1/2<112] dislocations trailing faulted dipoles analysed from the chromium containing alloy.

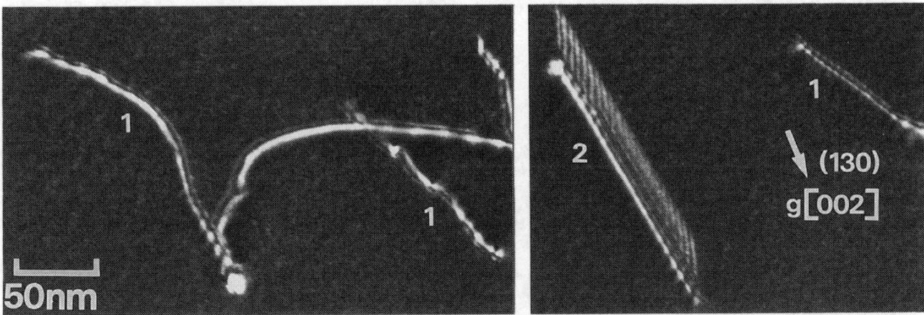

Figure 5.- Different examples of dissociations observed from <011] dislocations (see text)

noted the appearance of pile-ups of edge single dislocations on cube planes when twins of different systems got intersected. Although these segments have also been observed after room temperature deformation in connexion with twin intersections in our alloys (see Figure 3), their density is rather low compared to that observed after deformation at 400°C. Indeed at this temperature the twin density is also much higher and the possible intersections between different twin systems and even with lamellar boundaries are increased. It is interesting to note that the stress concentrations produced by such twin activity lead to crack initiation at the α_2/γ interfaces where the emission of 1/2<110] dislocations is not possible.

Although the binary alloy also presented twin activity during room temperature deformation, the twin density was lower than in the chromium containing alloy and the latter exhibited a higher flow stress. Of course at room temperature also different superdislocations have been observed in the two alloys and it is difficult to attribute the different strengths to a specific controlling mechanism. Indeed blocking processes can be produced by the type of dissociations observed from <101] segments in the binary alloy (see Figure 5) as well as by the faulted dipoles trailed by the 1/2<112] segments seen in the chromium containing alloy (see Fig 4). At 400°C, however, the contribution of superdislocations to the deformed microstructures is rather low in both alloys. The absence or the immobility of superdislocations at the peak temperature has also been confirmed by other authors in aluminium rich alloys [7,8,10].

According to Hug et al. [4] in Al-rich alloys, the mechanism responsible for the anomalous strengthening effect with increasing temperature is the cross-slip of dissociated 1/2<011] segments onto cube planes where they become immobile. In our alloys, the dislocation configurations analysed at room temperature are different in terms of the active superdislocations

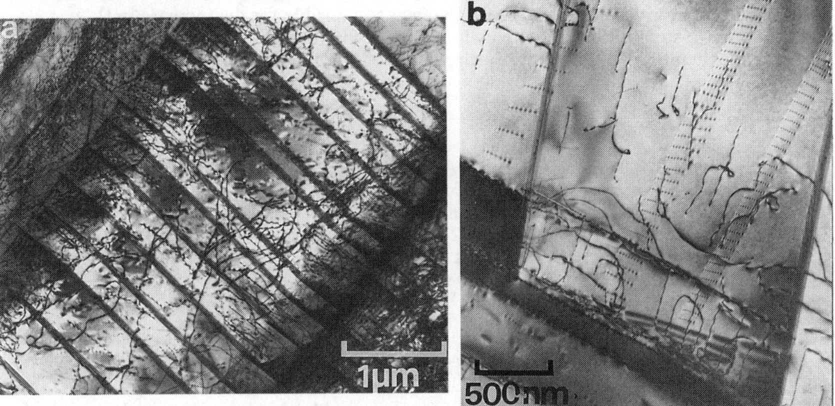

Figure 6.- Large density of twins, single dislocations and debris observed after deformation at 400°C. a) Binary alloy. b) Chromium containing alloy.

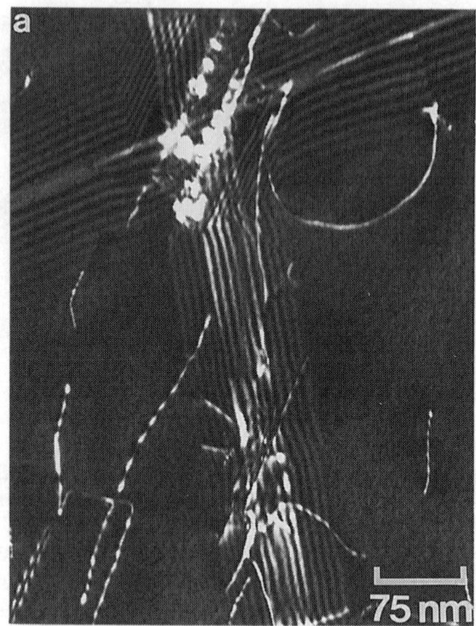

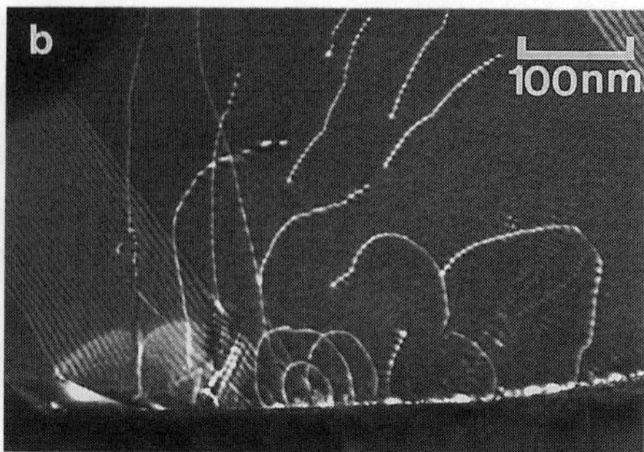

Figure 7.-Examples of single dislocations being emited from twin intersections after deformation at 400°C (see text for details).

present but in both alloys the peak in flow stress is observed at 400°C. At this temperature, the presence of superdislocations is rare and the large density of twins, of 1/2<110] dislocations and of debris is equally important in both alloys. Therefore the mechanism responsible for anomalous strengthening cannot be related to the only feature rarely observed in the deformed microstructure of both alloys. We believe that the anomalous strengthening effect observed at 400°C must be related to the large density of 1/2<110] dislocations that become active on (001) planes. Since some of these single dislocations have been analysed gliding on {111} planes in the matrix, cross-slip of 1/2<110] segments from cube to octahedral planes might be considered the mechanism responsible for the flow stress anomaly. However, our initial studies from the annealed microstructures have shown that single 1/2<110] dislocations on {111} planes are present due to the thermal stresses produced during cooling the alloys [9]. Therefore it is not possible to separate the dislocations gliding on {111} planes through the action of the stress at the peak temperature from those already present on {111} planes in the annealed materials. However the large number of debris present indicate that cross-slip mechanisms are indeed active producing dislocations annihilation and leading to debris formation. Further work in this area is needed in order to isolate the mechanism responsible for the increase in flow stress with temperature in such microstructures where cross-twinning and glide of single dislocations on cube planes continously occur. Measurements of activation energy at the peak temperature should help elucidate the present results.

REFERENCES

1. T. Kawabata, T. Kanai and O. Izumi. Acta Metall., 33, 1355, (1985)
2. S.C. Huang, Scripta Metall., 22, 1885, (1987)
3. S.C. Huang and E.L. Hall, Metall. Trans. A, 22, 427, (1991)
4. Y.W. Kim, Acta Metall. Mat., 40, 1121, (1992)
5. V. K. Vasudevan, M.A. Stucke, S.A. Court and H.L. Fraser, Phil. Mag. Letters, 59, 299, (1989)
6. M. Inui, A. Nakamura, M.H. Oh and M. Yamaguchi, Phil. Mag. 66, 557, (1992)
7. G. Hug, A. Loiseau and P. Veyssière, Phil. Mag. A, 57, 499, (1988)
8. B.A. Greenberg, O.V. Antonova, V.N. Indenbaum, L.A. Karkina, A.B. Notkin, M.V. Ponomarev, Acta Metall. Mat., 40, 815, (1992)
9. M.A. Morris, Phil. Mag. in press
10. G. Hug and P. Veyssière in Electron Microscopy in Plasticity and Fracture Research of Materials, edited by U. Messeschmidt, F. Appel and V. Schmidt, Akademic-Verlag (Berlin), 451, (1990)

PART IV

Deformation and Fracture

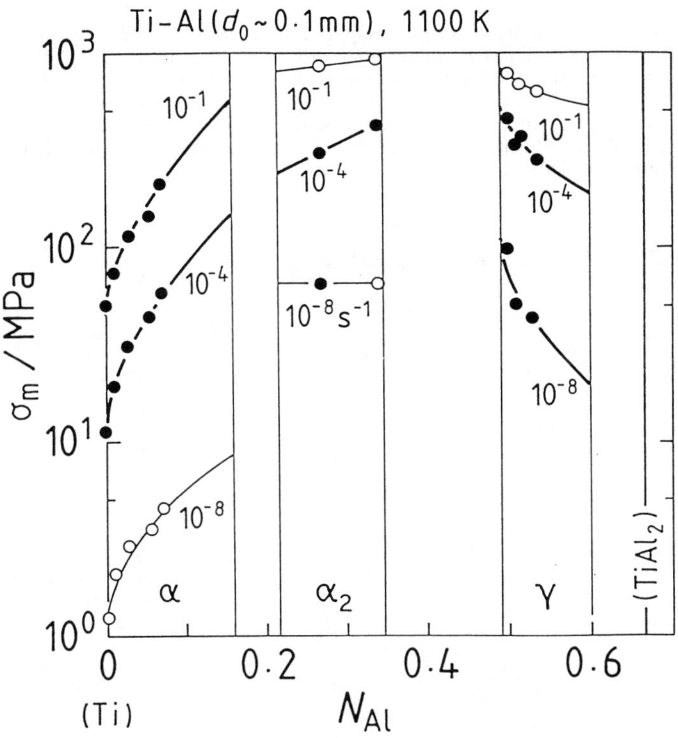

The figure illustrates the flow stresses required to produce several strain rates in approximately 100 μm grain size, single phase alloys at 1100 K in the Ti-Al system. The strength of the Ti terminal solid solution increases with an increasing Al content while the deformation resistance of TiAl phase demonstrates the opposite behavior with strength decreasing as the Al level increases. Although some solid solution strengthening of the Ti_3Al phase occurs at the faster strain rates, it appears that for the slowest rate strength of this phase is independent of composition.

Figure courtesy of K. Maruyama and H. Oikawa.

KEAR-WILSDORF LOCKS AND MECHANICAL PROPERTIES OF $L1_2$ ALLOYS

GEORGES SAADA and PATRICK VEYSSIÈRE
LEM*, CNRS-ONERA, BP 72, 92322 Châtillon Cedex, France.

ABSTRACT

There is no generally accepted model of the atypical mechanical properties of $L1_2$ alloys such as the flow stress peak. Since its introduction in the early sixties, the so-called Kear-Wilsdorf (KW) lock has played a central role in the analysis of the mechanical properties of this category of alloys. We analyze the mechanical stability of a split configuration, intermediate between the superdislocation fully extended in the primary slip plane and the KW lock. The influence of composition on mechanical properties is stressed, based on experimental determinations of the flow stress dependencies upon load orientation when composition is varied.

INTRODUCTION

More than 30 years of research has been devoted to the analysis of the atypical mechanical behavior of ordered intermetallic alloys and in particular to the flow stress peak in $L1_2$ alloys. There is much to learn from the available theoretical analyses of the mechanisms that control the positive temperature dependence of the flow stress (TDFS); however, the detailed origins of this atypical mechanical property and of related phenomena remain to be elucidated.

On an experimental standpoint, it makes little doubt that every property that could shed light on the positive TDFS is now documented - quite extensively in many instances - but our understanding of the overall problem still suffers from uncertainties in the determination of several crucial parameters. In the domain of positive TDFS, deformation occurs on the octahedral planes and it is agreed that

(i) the Schmid law, including the tension-compression asymmetry, is significantly violated,
(ii) the flow stress is very little sensitive to the strain rate,
(iii) there exists a microstrain level below which the TDFS is normal,
(iv) the rate of strain hardening is anomalously large,
(v) the flow stress is partly reversible,
(vi) the microstructure is dominated by APB-coupled superdislocations which are elongated in the screw direction where they are dissociated in the cube plane, a configuration known as the Kear-Wilsdorf (KW) lock.

This paper is not aimed at presenting a new model of the positive TDFS. Rather, we intend to reexamine some aspects of the phenomenon whose complexity may not have been fully appreciated in the past, such as

- the importance of composition, based on measurements of the flow stress dependence upon orientation which point to the role of lattice friction on dislocation behavior.
- properties of KW obstacles including the driving forces for transformation and KW stability,
- the framework of a tentative model based on the fact that strain is essentially provided by non-screw segments and the consequences of this on deformation at a constant strain rate, flow stress reversibility, creep and on relaxation experiments.

SOME CONSIDERATIONS ON COMPOSITION-RELATED EFFECTS

Because of technical limitations in growing stoichiometric single crystals, experiments have been carried out in a variety of $L1_2$ - mostly Ni_3Al-based - alloys, that show varied and

* UMR 104 : Unité Mixte de Recherche. Centre National de la Recherche Scientifique (CNRS) - Office National d'Etude et de Recherche Aérospatiale (ONERA).

sometimes pronounced deviations from stoichiometry, including significant amounts of ternary additions. This is at the origin of noticeable changes in mechanical behavior which are easy to identify but quite rarely pointed to attention. It is for instance interesting to note that the good consistency between the so-called PPV model[1] and the flow stress asymmetry under tension-compression - from which the PPV model was considered to be fully validated,[2,3] - has been recently questioned and the sources of the observed deviations between theory and experiments then explained in terms of impurity effects[4].

Shown in figure 1 is an example of a composition-related mechanical effect of a Ni_3Al-based alloy deformed in compression, taken from the contributions of Bontemps[5] and Heredia[6] in the case of Hf additions and of two orientations of the load. It can be seen that the stress levels are considerably affected by ternary additions and that the orientation effect can be significantly hindered :

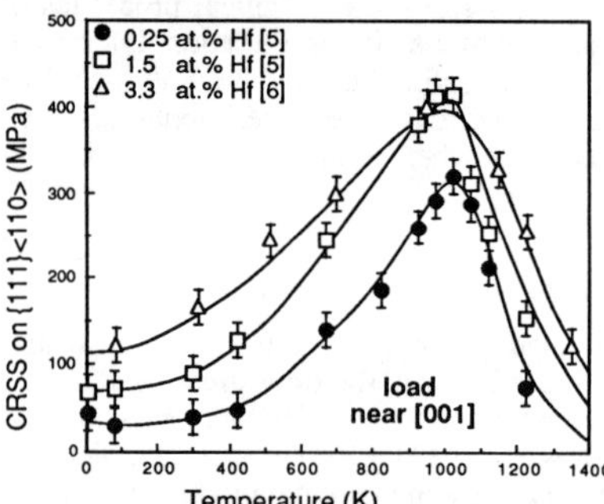

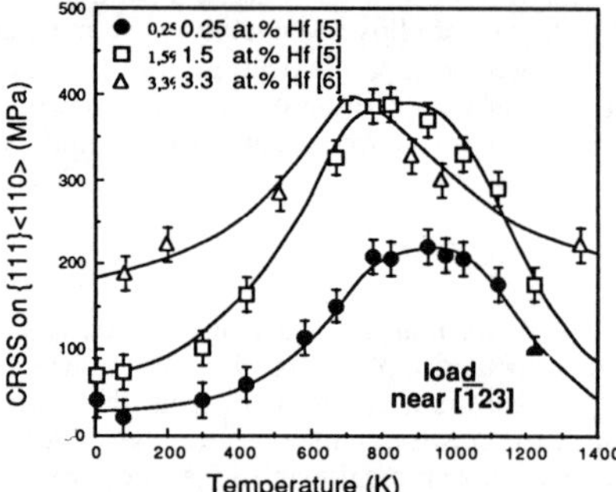

Figure 1. Comparison of the dependence of the flow stress at 0.2% of permanent strain on temperature for three concentrations of hafnium in ternary Ni_3Al-based single crystals and two orientation of the applied compressive load.

- the lower the Hf concentration the lesser the flow stress and the lesser the rate at which it increases (at least up to the peak temperature),
- at the flow stress peak, the violation of the Schmid law is the most pronounced in the case of reduced ternary additions : the ratio of resolved shear stresses ($\tau^P_{001}/\tau^P_{123}$) is changed from 1.5 to 1.06 when the concentration in Hf is varied from 0.25% to 1.5% and above.
- in the ascending part of the curve, the flow stress sensitivity upon Hf concentration is the greatest between 0.25% and 1.5% of Hf; the effect of ternary additions being approximately saturated for more than 1.5 at.% Hf, between 600K and the peak for both orientations.

PROPERTIES OF KEAR-WILSDORF BARRIERS

In the following, the subscripts o and c refer to quantities defined in the octahedral and in the cube plane, respectively. We take the ratio of antiphase boundaries in these planes as $z = \gamma_o/\gamma_c$. λ_o and λ_c are the equilibrium distances of a screw superdislocation when it is ascribed to be dissociated in the appropriate plane and we introduce a convenient elasticity parameter $a = (A + 2)/A\sqrt{3}$ where A is the Zener coefficient. To the authors' knowledge, a is of the order of 1 in most $L1_2$ alloys[7]. We consider the incomplete KW barrier shown in figure 2, where S_o is a 1/2[101] superpartial gliding in the octahedral plane and where the sign conventions for the forces applied to the superpartials are considered. In the absence of an

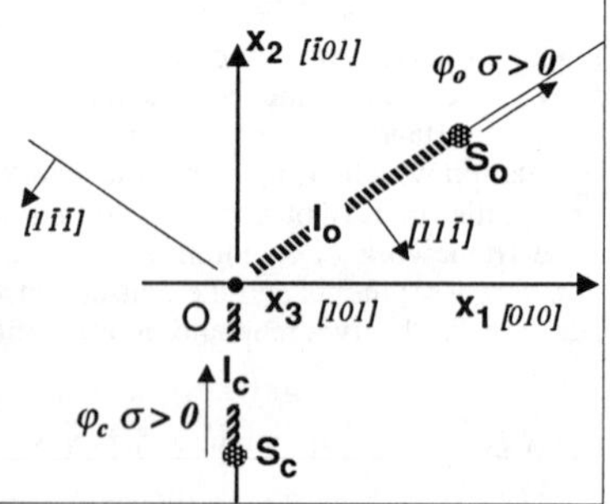

Figure 2. Schematic end-on representation of a dissociated screw superdislocation whose APB straddles the octahedral and the cube planes over l_o and l_c, respectively.

external stress, the forces in the primary octahedral and in the cube cross-slip plane, applied to the companion superpartials, S_o and S_c, write

$$F_o = E\left[\frac{a\,l_o + l_c}{\phi} - \frac{1}{\lambda_o}\right] \tag{1-a}$$

$$F_c = E\left[\frac{1}{\lambda_c} - \frac{l_o + 3^{1/2}\,l_c}{\phi}\right] \tag{1-b}$$

where 2E is the prelogarithmic term in the expression of the self-energy of an undissociated screw superdislocation and $\phi = a\,l_o^2 + 3^{1/2}\,l_c^2 + 2\,l_o\,l_c$. When dissociation equilibrium is restricted to occur in one given plane, the above forces reduce to

$$l_c = \lambda_c,\quad F_o^* = \gamma_c\,(1/\sqrt{3} - z) \tag{2-a}$$

$$l_o = \lambda_o,\quad F_c^* = \gamma_c\,(z/a - 1) \tag{2-b}$$

which coincide with expressions previously derived[8]. Obviously, the driving force for the nucleation of a KW barrier (F_c*) is not the same as for its destruction (F_o*). Three ranges of values of the parameter z can thus be defined[7]

- $z < 1\sqrt{3}$, the only stable configuration occurs in the octahedral plane (state O),
- $z > a$ (≈ 1), the KW barrier (state C) is the only stable state,
- $1\sqrt{3} < z < a$, the screw superdislocation is bistable and it can be shown that in the absence of lattice friction, an incomplete configuration such as that shown in figure 2 cannot be formed at equilibrium. Rather, the screw segment can be dissociated in either one of the O and C states, both of these being separated by a configuration of higher energy which, for the transition from one state to the other to occur, has to be overcome.

The expression of the height of the energy barrier that controls the C $\rightarrow$O transition has been derived elsewhere[7]. It shows that, given a and z slightly lesser and slightly larger than unity, respectively, once KW locks are fully formed, their destruction cannot be activated thermally. This is the case of most documented $L1_2$ alloys that exhibit a flow stress peak.

On the other hand, expression (2-b) implies that, in the absence of an external load, there exists a driving force for the nucleation of a KW configuration, a property first pointed out by Yoo[9]. This driving force remains significant up to an APB extension in the cube plane of the order of $\lambda_c/3$ (eqn. (1-b)) and this yields incomplete barriers which become stabilized as $F_c\{l_o, l_c\}$ becomes of the order of the friction force in the cube plane. In the course of the transition O $\rightarrow$ C, the leading partial is submitted to the force $F_o\{l_o, l_c\} + \varphi_o\sigma b$, where φ_i is the Schmid factor on plane i along the [101] direction. This partial may thus cross-slip back to the primary octahedral plane (to form a OCO or a more complex manifold configuration). Since the probability of cross-slip should be largely controlled by the core structure of the superpartials, the above analysis can describe but in a simplified manner the so-called APB-jump mechanism which has been observed to accompany deformation in $L1_2$ alloys[10].

It should be realized that the driving force F_c* which governs the dynamics of nucleation of KW barriers, and which eventually dictates the rate of TDFS, is proportional to the quantity (z -a) (eqn. 2-b). Therefore, estimates of this driving force are founded on measurements of APB energies which are currently subjected to uncertainties[11] which, in turn, may yield considerable misinterpretations in this particular case[7]. Indeed, since the anisotropy parameter a lies in the vicinity of unity - a more exact value can be taken at 0.93 - it can be shown that (z - a) varies between almost zero (0.07) and 0.87, which represents an experimental uncertainty on F_c* of one order of magnitude.

The above development remains unchanged upon application of an uniaxial stress, σ, provided the quantity $Z = (\gamma_o - \varphi_o\sigma b)/(\gamma_c - \varphi_c\sigma b)$ is substituted for z in the equations describing the dihedral configuration. The driving force for the nucleation of a KW configuration is obviously the largest when $N = \varphi_c/\varphi_o$ is the largest. On the other hand, the condition implying the athermal destruction of a KW lock by application of a stress writes $Z < 1/\sqrt{3}$. The condition $Z = 1/\sqrt{3}$ is equivalent to the relation

$$\tau_o\{z, N\} = \frac{\gamma_c}{b}\,\frac{z\,3^{1/2} - 1}{N + 3^{1/2}} = \frac{\sigma_o\{z, 0\}}{1 + N\,3^{-1/2}} \tag{3}$$

where $\sigma_o\{z, N\}$ is the level of shear stress resolved in the octahedral plane above which the KW barrier is automatically destroyed in the absence of any further obstacles.

Equation (3) is qualitatively consistent with the experimental evidence that the smaller N, the larger the flow stress at the peak. As pointed out in the introduction, a quantitative comparison between measurements and predictions based on eqn. (3) is rendered difficult by compositional effects and it is only in rather "pure" alloys[5,6,12] that the agreement can be regarded as satisfactory[7].

DISCUSSION

As made clear from a steady rate of theoretical contributions to this field, it is difficult to offer a comprehensive and consistent theory of the positive TDFS that would account for every anomaly-related experimental property identified in $L1_2$ alloys so far. This is mostly since, whatever detailed and stimulating these models are, they rely on physical quantities which are not all measurable or else whose theoretical estimates are not all reasonable. This can be illustrated with the following selected examples :

♣ in the pioneering model of Takeuchi and Kuramoto[13] and in its PPV[1] derivative — which includes the so-called tension-compression asymmetry — the predicted strain rate sensivity of the flow stress is at least one order of magnitude larger than in experiments. Though, as pointed out in the introduction, the agreement between the PPV model and selected experiments could be regarded as encouraging, a closer inspection of this model together with more precise comparisons between its implications and experimental data, such as the tension-compression asymmetry in "pure" alloys and the temperature dependence of the activation volume, showed such strong inconsistencies that the model was revisited twice in the last two years[4,14,15]. In fact, such revisions are largely facilitated by the flexibility provided by the many unknown parameters.

♦ since it is established experimentally that the spreading of the core of a 1/2<110> superpartial is less than 1 nm, it must be made very clear that no sensible quantitative theory can be worked out based on cross-slip processes calculated in the frame of linear elasticity[16]. Moreover, the temperature range over which a core spreading of less than 1 nm should control the cross slip process of a superpartial, is unexpectedly wide in Ni_3Al (about 600°C).

♥ in the upgrade of the PPV model recently released by Vitek and Sodani[14], the mean distance between obstacles reaches several kilometers at 1000K (taken here as the order of magnitude of the peak temperature). It has been since pointed to the authors' attention[17] that the activation enthalpy provided in refernce[14] was too high; nevertheless, with the suggested correction for the activation enthalpy of $4.8\ 10^{-20}$ J, the mean distance between obstacles at room temperature is decreased to about 10 µm, which is still too high in order to be rate limiting. On the other hand, in the model of Hirsch[8], the dislocation free flight velocity would be one fourth of sound velocity, which again is far too large.

♠ in an attempt at reconciling the PPV model with experiments, the conditions by which screw segments may unpin have been re-examined and a modified point of view, which implies two distinct values of the activation volumes in the domain of positive TDFS has been proposed[15]. Meanwhile, the experimental measurements of the apparent activation volume[18] have been corrected and what is believed to represent the true activation volume has been found to exhibit about the same level on either side of the discontinuity at 410°C[19].

It should be noted that in the models in which the screw superdislocations assume a finite average mobility against local obstacles, such as thermally activated cross-slip pinning events[1,13] or edge dipoles[8], the screws are only momentarily sessile. In our opinion, the original approach that is being developed by Mills and Chzran[20] offers an alternative description of the dynamical unpinning processes which carries a promising potential of development. This approach which does not claim to constitute a complete theory, makes use of a dynamical simulation of the evolution of octahedral dislocation sources, under the assumption that screw dislocation motion is hindered by individual pinning events which form at screw and near screw segments according to a specific probability rule. To be more specific, these simulations show that the pinning decreases the number of mobile dislocations, that the dynamical sequence may eventually lead to a complete exhaustion of mobile dislocations and that the screw dislocations have a finite lifetime after which they can be regarded as permanently arrested.

It is clear that this problem of a positive TDSF should still be regarded as a widely open field and this is why we limit the following part of this discussion to a few comments.

In a previous work[21], we have explored the implications on the flow stress of kink mobility in relation with the temperature dependent curvature of KW barriers in the cube plane and we have pointed out the limitations of this process. We continue here our investigation of a physical frame, based on microstructural observations, whereby the origin of TDFS could be reasonably investigated; we try to validate the consistency of the following set of remarks

(i) since the threshold stress for KW destruction (equation (3), transition C $\rightarrow$ O) is attained in the vicinity of the peak[8,21], since the corresponding activation energy is prohibitive and since KW motion in the cube plane is limited by lattice friction (though the bending of KW segments on cube planes increases gradually with temperature, it remains significantly limited), a KW lock can be considered as intrinsically sessile. However, some local motion cannot be excluded and this may occur by means of several processes such as local annihilation by cross-slip, APB jump of incomplete KW barriers[22] and KW transport by motion of kinks[23]. In short, KW segments are characterized by relatively modest velocities and KW segments, as a whole, carry but insignificant plastic strain.

(ii) consequently, deformation in Ni_3Al is essentially ensured by motion of mixed segments which, themselves, behave normally.

(iii) KW formation (O $\rightarrow$ C) is stress assisted as well as thermally activated.

(iv) as the density of KW is increased, dislocation multiplication is gradually hindered, that is, the role of KW segments during deformation is to control the density of mixed dislocations, which are glissile, but not to affect the intrinsic mobility of these.

(v) the flow stress but not the strain rate (as this would be in normal alloys), is controlled by the rate of formation of KWs.

It is important to bring to attention that the observed strain hardening rate, which is of the order of $\mu/20$ to $\mu/50$ in Ni_3Al in the domain of stress anomaly, is much too large in order to be explained by any documented hardening mechanism involving dislocation interactions. It seems reasonable to interpret such atypical hardening rates as the consequence of the rapid saturation of the density of mobile dislocations to rather moderate levels. In this sense, the flow stress would be quite similar to a preyield regime.

The above hypotheses, and in particular the permanent locking of KW segments, should reconcile the apparent opposition between the fact that, under creep, extensive cube slip — and more precisely slip on the cube cross-slip plane — may occur at temperatures well below the temperature where the flow stress peaks, under constant strain rate (CSR). It is indeed at this temperature and above, that one starts to detect significant cube slip in the latter samples, almost regardless of load orientation[24].

CONCLUSION

Because of our imperfect knowledge of the parameters that contribute to the phenomenon, it is still difficult to offer a consistent theory of the positive temperature dependence of the flow stress and to account for the many anomaly-related properties. In our opinion the major difficulties that should be solved are the following

- we need an estimate of the mobility of dislocations in the cube plane and/or some measurements of lattice friction forces and of their stress and temperature dependencies.
- the situation is obscured by the absence of reliable determinations of dislocation densities as a function of strain at various temperatures, and in particular of the fraction of KW segments in the deformation microstructure.
- the stability of KW segments remains to be studied theoretically as a function of their length.
- the effects of a departure from the stoichiometric composition as well as that of ternary additions are known qualitatively at best but it is clear that comparisons with experiments should be conducted on data obtained on rather "pure" alloys.
- the uncertainty on the experimental determination of $z = \gamma_{001}/\gamma_{111}$ is large. We have shown that this parameter is instrumental in the estimate of the driving force for KW nucleation; un-

fortunately, it is quite unlikely that we can measure z with an improved precision in the near future.
- it is fair to state that the theoretical analysis of the stability of a twofold screw superdislocation should be complemented by a detailed and a comprehensive model of the collective behaviour of dislocations in $L1_2$ alloys.

We have shown that KW locks are essentially undestructible under an applied stress, except however at those levels of critical shear stress that are reached at the peak temperature and this is what limits the extent of the anomaly. Hence, we are naturally led to propose a point of view based upon the idea that once transformed under a KW barrier, a screw superdislocation cannot be restored in its glissile configuration and thus remains definitively sessile. Accordingly, it is the density of KW locks that would govern the dynamics of glide in complete contrast with the foundation of many of the models proposed so far.

REFERENCES

[1] V. Paidar, D.P. Pope and V. Vitek, Acta metall. 32 (1984) 435.
[2] V. Vitek, in Dislocations and the Properties of Real Materials, edited by The Institute of Metals (Arrowsmith, Bristol, 1985), p. 30.
[3] D.P. Pope, in High Temperature Aluminides & Intermetallics, edited by S.H. Whang, C.T. Liu, D.P. Pope and J.O. Stiegler (TMS Publication, 1990), p. 51.
[4] M. Kantha, J. Cserti and V. Vitek, Scripta Met. Mater., 27 (1992) 487.
[5] C. Bontemps-Neveu,, PhD Thesis, University of Paris Sud (France) (1991).
[6] F.E. Heredia, PhD Thesis, University of Pennsylvania (1990).
[7] G. Saada, and P. Veyssière, Phil. Mag. A, 66 (1992) in press.
[8] P.B. Hirsch, J. Phys. III 1 (1991) 989 ; Phil. Mag. A 65 (1992) 569 ; in Ordered Intermetallics - Physical Metallurgy and Mechanical Behaviour, edited by C.T. Liu, R.W. Cahn and G. Sauthoff (Kluwer Academic Publishers, The Nederlands : NATO ASI series, 1992), p. 47 ; Progr. Mater. Sci., 36 (1992) 63.
[9] Yoo, M.H., Scripta Met., 20 (1986) 915.
[10] A. Couret and D. Caillard, Phil. Mag. Letters, (1992) in press.
[11] P. Veyssière, in Ordered Intermetallics - Physical Metallurgy and Mechanical Behaviour, edited by C.T. Liu, R.W. Cahn and G. Sauthoff (Kluwer Academic Publishers, The Nederlands: NATO ASI series, 1992), p. 165.
[12] Y.Q. Sun, PhD Thesis, University of Oxford (1990).
[13] S. Takeuchi and E. Kuramoto, Acta Met., 21 (1973) 415.
[14] V. Vitek and Y. Sodani, Scripta Met mater., 25 (1991) 939.
[15] M. Kantha, J. Cserti and V. Vitek, Scripta Met. Mater., 27 (1992) 481.
[16] G. Saada and J. Douin, Phil. Mag. Lett., 64 (1991) 67.
[17] V. Vitek, private comunication (1992).
[18] J. Bonneville, N. Baluc and J.L. Martin, in Intermetallic Compounds -Structure and Mechanical Properties, ed. O. Izumi (Sendai, Japan: The Japan Institute of Metals, 1991), p. 323. J. Bonneville, and J.L. Martin, High-Temperature Ordered Intermetallic Alloys IV, MRS Proceedings, 213, ed. L.A. Johnson, D.P. Pope and J.O. Stiegler (Pittsburg, PA: Materials Research Society, 1991), p. 629.
[19] N. Baluc, J. Bonneville, J.L. Martin and P. Späetig, Mater. Sci. & Engg. A, (1993) in press.
[20] M.J. Mills and D.C. Chzran, Acta Met., 40 (1992) 3051.
[21] G. Saada, and P. Veyssière, Mater. Sci. & Engg. A, (1993) in press.
[22] C. Bontemps, C. and P. Veyssière, Phil. Mag. A., 61 (1990) 259.
[23] G. Saada and P. Veyssière, Phil. Mag. Letters., 64 (1991) 365.
[24] K. Hemker, M.J. Mills and W.D. Nix, Acat Met., 39 (1992) 1901.

THERMALLY ACTIVATED UNPINNING OF SCREW DISLOCATIONS IN THE ANOMALOUS REGIME IN $L1_2$ COMPOUNDS

M. Khantha, J. Cserti and V. Vitek
Department of Materials Science and Engineering
University of Pennsylvania
Philadelphia, PA 19104-6272.

ABSTRACT

We present a model for the anomalous increase of the yield stress exhibited by many $L1_2$ compounds. It is based on two thermally activated processes that describe respectively the pinning and unpinning of $[\bar{1}01]$ screw dislocations in the $(1\bar{1}1)$ plane. The model explains all the important characteristic features observed in the anomalous regime. We discuss the applications of the model to Ni_3Ga and $Ni_3(Al,Ta)$.

INTRODUCTION

The anomalous increase of the yield stress with increase in temperature is well known in several $L1_2$ compounds and has been studied extensively. It is now recognized that the factors causing this behavior are also responsible for the many characteristic features in the anomalous regime such as (i) Strong orientation dependence of the yield stress and tension/compression (T/C) asymmetry [1-3]; (ii) A very low strain-rate sensitivity in the anomalous regime extending up to the peak temperature (T_p) [1, 4]; and (iii) A large discontinuity in the activation volume (v_a) at a temperature (T_c) well below T_p [4, 5]. While it has been known that a thermally activated mechanism obstructing the motion of screw dislocations in the (111) plane is the primary cause for the anomalous behavior, different approaches have been proposed to describe the unpinning or release of screw dislocations from the 'obstacles' that impede its motion. Many of the earlier models [1, 6] were based on a thermally activated process that 'pins' the screw dislocations but assumed an athermal process for the release. Such an approach could explain the feature (i) listed above as it is related to the details of the pinning mechanism but could not explain features (ii) and (iii) as they are related specifically to the thermally activated aspects of the release mechanism [7]. Both the pinning and unpinning mechanisms are assumed to be thermally activated in some recent models [8, 9] and agreement between theory and experiment has been demonstrated at certain selected temperatures and stresses. The starting point for all the models is common and is based on the thermally activated pinning mechanism proposed in the PPV model [6]. However, the models differ in the final dislocation configuration in the pinned state which then leads to differences in the unpinning mechanism as well. Recently, we have proposed a model [10, 11] which is also based on two thermally activated mechanisms and includes a simple unpinning process based on the concept of major breakaway that accounts for all the macroscopic characteristic features associated with the yield stress anomaly.

There are three main parts in our approach. The first describes a thermally activated process that results in the *local* pinning of screw dislocations in the (111) plane at random positions such that their movement in the glide plane is impeded. This part is qualitatively identical to the PPV model but includes some modifications in the saddle-point configuration in order to explain differences in the T/C asymmetry between binary and ternary compounds. Due to these changes, the activation enthalpy is somewhat different from that derived in the PPV model and has been described in [11]. It is based on the following description: Dislocation core transformations from a glissile to a sessile state are nucleated at random positions on mobile screw dislocations and result in the formation of a small cross-slipped segments on the (010) plane. The rest of the dislocation between the segments is bowed in the presence of stress.

Figure 1. The pinned state of the dislocation

The resulting configuration is shown schematically in Figure 1 in which the bow-out angle between two pinned segments is larger than the equilibrium bow-out angle determined by a single cross-slipped segment. This however does not imply that the cross-slipped segment shrinks and disappears [12] since the path for unpinning a pinned segment is generally different and not necessarily related to the path which led to the nucleation of the segment.

In the second part, we consider the *major breakaway* (i.e., release from many pinning points and the subsequent return to the initial glissile state) of the dislocation via a thermally activated process by enumerating the various possible reaction paths for the release via different saddle-point configurations. The activation enthalpy for the release is obtained by choosing that path which gives the maximum rate of release while also enabling a major breakaway. We find that at low stresses/temperatures, when the mean separation between pinning points is large, the release from a single pin is rate controlling for major breakaway. However, as the stress (in effect, the temperature) increases, major breakaway becomes less likely after the release from a single pin due to re-pinning (or backward jumps) being more probable than further unpinning from the remaining pins. The release is then controlled by more difficult, but alternate, reaction paths that involve *multi-pin activation* such as the simultaneous activation at two or more pins. The change in the reaction path causes a discontinuity in the activation volume. As the temperature increases, eventually one obtains the condition that describes the transition from a dislocation that is pinned at discrete points to a dislocation that is pinned continuously all along its length. We identify the temperature at which such a transformation ensues to be the peak temperature.

The third part of the model prescribes the steady state of a system which is governed by two thermally activated mechanisms with very different waiting times. The thermally activated pinning and release mechanisms in steady state determine the resolved shear stress (RSS) on the (111) plane as a function of temperature for a given strain rate. We have tested the model in the entire anomalous regime by calculating the yield stress and the activation volume for different orientations in binary Ni_3Ga, Ni_3Al and ternary $Ni_3(Al,Ta)$. The results compare well with observations including the prediction of the temperatures T_c and T_p. In addition, we predict at least one more discontinuity in the activation volume at a temperature, T_c', between T_c and T_p. All the other important characteristics such as the orientation dependences, T/C asymmetry and the very low strain-rate sensitivity are also explained by this model [10].

THE UNPINNING MECHANISM

The concept of major breakaway and alternate reaction paths was first discussed in the context of internal friction phenomena [13-15] and we adopt a similar approach here. The basic features which distinguish major breakaway as opposed to the release from a single pin can be understood by considering a simple example of the thermally activated release of a dislocation with Burgers vector b and effective line tension τ pinned at two equidistant points separated by a distance L. In the presence of an applied stress σ, the dislocation bows out away from the pins, and let y(x) be the displacement from the stress-free position in which the dislocation lies parallel to the x-axis. The equilibrium configurations of the pinned dislocation can be calculated by minimizing the energy within the line-tension model assuming a suitable dislocation-pin interaction energy, U(y).

There are six non-equivalent equilibrium configurations [14] (see Figure 2) corresponding to the three displacement states possible at each pin, namely, pinned, saddle-point and unpinned states represented by 1, 2 and 3, respectively. A particular configuration is labelled as (ij) (i, j = 1, 2, 3) which denotes that the displacements at the two pins are in states i and j respectively. The configurations (11), (31) and (33) correspond to energy minima, (21) and (32) to saddle-points and (22) to a energy maximum.

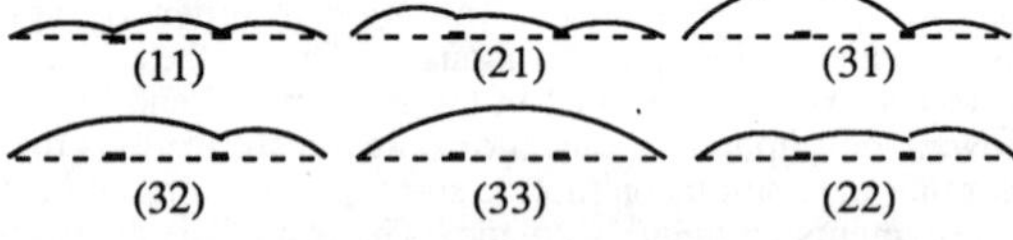

Fig. 2. The equilibrium configurations of a dislocation pinned at two points

The saddle-point state (21) involves activation at one pin while the saddle-point (32) involves simultaneous activation at both pins. Major breakaway corresponds to thermally activated transitions from state (11) to state (33) and this can occur via two distinct reaction paths: (i) (11)-(21)-(31)-(32)-(33) or, (ii) (11)-(32)-(33).

The (21) saddle-point is rate-controlling for transitions via the first path (since the subsequent passage through the state (32) is easier) while the (32) saddle-point controls the activation energy for the second path. We now examine which of these two alternatives provides the dominant contribution to the major breakaway by considering the activation enthalpy for transitions in the forward and backward directions (denoted by H_u and H_b respectively) for each path. The activation enthalpies are functions of the stress and the separation L. At low temperatures, when pinning is infrequent and the value of L is large, major breakaway occurs only via the first path because the activation enthalpy for the second path is actually negative and represents an unphysical path because the simultaneous activation of two pins separated by a large distance is unlikely to be achieved. The path via (21) is, therefore, rate controlling provided the activation enthalpy for forward transitions leading to unpinning, $H_u(1)$, is smaller than the enthalpy $H_b(1)$, for the backward movement from (31) to (11) via (21) which leads to re-pinning.

As the value of L decreases (which is the case for the $L1_2$ alloys as the temperature increases), a situation when $H_u(1) \geq H_b(1)$ will ensue at a certain stress. When this happens, the release of the dislocation via the (21) path is no longer likely since the probability of re-pinning is larger than the probability of unpinning. However, it is precisely in this regime that major breakaway via the second path becomes feasible and in fact represents the dominant contribution for the following two reasons: (i) the enthalpy for forward jumps via the second path, $H_u(2)$, is equal to $H_u(1)$, which is the corresponding enthalpy for the first path and (ii) the enthalpy for backward jumps via the second path, $H_b(2)$, is much larger than $H_u(2)$, making re-pinning unfavorable for this path. Thus, while considering major breakaway, a transition from one dominant reaction path to another occurs when the release via the first path becomes unlikely due to re-pinning, i.e., when $H_u(1) \approx H_b(1)$.

This simple picture of release can now be generalized to the case of N pinning points [15]. Consider the configuration in which the dislocation has simultaneously broken away from (m-1) adjacent pinning points and is in the saddle-point state at the mth pin (m =1,2,..N) [15]. (The dislocation is assumed to be pinned with zero displacements beyond the mth pin). The displacements at the pins and the activation enthalpy can be expressed as functions of two dimensionless variables α and β given by $\alpha = (\sigma L^2/2\tau)$ and $\beta = (U_0 L/\tau b^2)$. Here U_0 represents the maximum value of U(y) and α and β represent the combined effect of the stress σ and the separation L. Since the pinning points act as obstacles, the resisting force, $(-\partial U/\partial y)$, exerted by them on the dislocation during unpinning, must be positive everywhere. A simple form for U(y) that exhibits such a behavior is the linear 'triangle-force' proposed in [13]. The interaction is of the type $U(Y)/U_0 = Y^2/pq$ for $o \leq Y \leq p$, $U(Y)/U_0 = 1 - n(q - Y)^2$ for $p \leq Y \leq q$ and $U(Y)/U_0 = 1$ for $q \leq Y$; here p, q and n are constants and Y (= y/b) represents the normalized displacement at the pin. We use this form of interaction in all our calculations with the same parameter values as in [13] (n=1/8, p=1/3 and q=3). The qualitative features of the unpinning process however, remain the same for all forms of U(y) which rise monotonically from zero at y=0 to a certain maximum [15]. Let V_p, $V_{s.p.}$ and V_u represent the energies of the pinned (111...11), saddle-point (333...32) and the unpinned states (333...33) respectively. Using the triangle-force interaction, they can be expressed as

$$\frac{V_p}{U_0} = -\frac{\alpha^2}{\beta}\left[\frac{m+1}{6} + \frac{m^2}{m\beta+1}\right] \; ; \; \frac{V_{s.p.}}{U_0} = (m-1) + \left(1 - nq^2\right) - \frac{\alpha^2}{6\beta}\left(m^3+1\right) + \frac{[2\beta nq - (m+1)\alpha]^2}{2\beta\left(2\beta n - \left(\frac{m+1}{m}\right)\right)}$$

$$\frac{V_u}{U_0} = m - \frac{\alpha^2}{\beta}\left[m^2 + \left(\frac{m+1}{6}\right)\right] \tag{1}$$

The activation enthalpies for forward and backward jumps are given by $(V_{s.p.} - V_p)$ and $(V_{s.p.} - V_u)$ respectively. For large values of L, or more generally α, the controlling event for the unpinning of a dislocation is the release from a single pin (m = 1) while subsequent releases from adjacent pins require no further activation. When α decreases such that the backward jumps become as likely as the forward jumps, the major breakaway via this path becomes unlikely. The unpinning then proceeds by the next easiest path which is the simultaneous activation over two pins (m = 2) leading to major breakaway. However, when the backward jumps for m = 2 become easier than the forward jumps, the controlling path becomes simultaneous unpinning from three or more pins. A discontinuity in the activation volume $v_a = -\partial H_u/\partial\sigma$ occurs whenever there is a change in the dominant reaction path; v_a is usually much bigger for simultaneous activation at (m+1) sites than for m sites.

THE STEADY STATE

Let H_p represent the activation enthalpy for the pinning mechanism that results in the random nucleation of pinning points (or, segments of length ℓ_0) separated by a distance, ℓ, on average, along a segment of length L_s of a screw dislocation. We assume that after major breakaway, the dislocation glides freely on the (111) plane with a velocity V, travelling a distance *d* until enough pinning points have been nucleated on the unpinned segment that it again becomes sessile. The pinning process is more dominant than the unpinning process (although both are thermally activated) and is in fact the cause for the anomalous behavior. The rates of the two processes are therefore very different and hence a steady state cannot be achieved by simply equating their rates. If this is done, only the dominant process survives at long times while the weaker unpinning mechanism dies out completely in the steady state. In the present case, this would imply a complete transformation of screw dislocations into the sessile form. In order to attain a steady state in which both pinned and unpinned segments are present, a gain-loss rate equation must couple the dominant (pinning) and weak (unpinning) processes. The rate equations of predator-prey models, logistic equation and auto-catalytic reactions are examples in which strong and weak processes are coupled in a non-linear manner [16]. We adopt a similar but simpler approach here and use the characteristic time scale for the intermediate free-flight state between the pinned and unpinned configurations to write a linear coupled equation describing the rate of growth of pinned (or unpinned) segments at any time. In steady state, we approximate the average distance travelled in free flight *d* to the mean separation between pinning points, ℓ. We then obtain the steady-state condition (for details, see [10])

$$d \approx \ell = \sqrt[3]{(\ell_0 bV / \nu_0)} \exp(H_p / 3kT) \quad . \qquad (2)$$

where ν_0 is the Debye frequency, *k* is the Boltzmann constant and T is the temperature.

SCREW DISLOCATION MOTION AND THE STRAIN RATE IN THE ANOMALOUS REGIME

The rate of thermally assisted unpinning of screw dislocations moving on the (111) plane determines the strain rate which can be written as

$$\dot{\varepsilon} = \dot{\varepsilon}_0 \exp(-H_u / kT) \qquad (3)$$

with $\dot{\varepsilon}_0 = \rho bA(\nu_0 b/\ell)N$, ρ is the density of mobile dislocations, A the area swept by the unpinned dislocation and N ($\approx 1/\ell$) the number of nucleation sites for unpinning per unit length of the dislocation. Since A is proportional to ℓ^2 (the proportionality constant may be large due to the major breakaway), $\dot{\varepsilon}_0$ can be treated as a constant. The RSS on the (111) plane can be written as $\sigma_{pb} = \sigma^0_{pb} + \sigma^T_{pb}$, where the superscripts 0 and T distinguish the athermal and thermal components of the stress, respectively. (In the following we identify σ^0_{pb} with the low-temperature RSS.) The activation enthalpy H_u is calculated from equation (1) with L replaced by ℓ and σ replaced by σ^T_{pb} in α and β. For a given strain rate, the temperature dependence of σ^T_{pb} can be obtained from Equation 3 by finding the roots of the function $\Psi = \left[kT\ln(\dot{\varepsilon}_0 / \dot{\varepsilon}) - H_u\right]$ for the rate-controlling path chosen according to the criterion discussed above. For a given m, $H_u < H_b$ if $\alpha \geq \alpha_c \approx \sqrt{6\beta / (m+1)(m+2)}$ and thus this inequality determines when the simultaneous unpinning from m pins is the dominant process. The activation path for major breakaway changes from simultaneous activation at m pins to that at m+1 pins when α - α_c changes sign from positive to negative. When the reaction path via (332) (i.e., m = 3) becomes unlikely, difficult paths via other saddle-point configurations such as (323), (213), etc., have been considered [14]. Since the major breakaway is unlikely to occur in this regime due to the high probability for pinning at high temperatures we approximate $d \approx \ell/3$ and calculate ℓ directly from the rate-equation [10].

According to equation (2) the separation of the pinning points decreases with increase in temperature and eventually their separation becomes comparable with their size, ℓ_0, so that the dislocation is completely transformed into the sessile form all along its length. We identify the temperature at which this occurs as the peak temperature T_p. In the present context, this happens when all the configurations (saddle points and intermediate minima) become unstable except the states (11...1), (22...2) and (33...3). For true point-like pins, $\beta n < 1$ corresponds to a completely pinned dislocation. However, in the present case, the pinning points are actually segments of length 2b - 3b and thus when $\beta n \approx 3$-4 the notion of discrete pins is no longer meaningful. Hence, we identify T_p by the condition that βn is in the range 3-4.

APPLICATION TO Ni_3Ga AND $Ni_3(Al,Ta)$

The activation enthalpy H_p and the length ℓ_0 for the pinning process have been described in [11]. We choose the Burgers vector of the screw dislocation to be 2.52A in both the compounds. The separation between the Shockley partials which controls the constriction energy, W_c, has been determined using anisotropic elasticity. The values of APB energies on the (010) and (111) planes (in units of mJm^{-2}) have been taken as 200 and 237 in Ni_3Al containing 1 at.% Ta [17] and 20 and 100, respectively, in binary Ni_3Ga (an estimate); the CSF energy on the (111) planes has been estimated as 290, and 220, respectively. The energy gain, ΔE, in equation (2) of [11], has not been evaluated explicitly but we use an order of magnitude estimate, 0.01 (in units of $Gb^2/2\pi$) in $Ni_3(Al,Ta)$ and 0.015 in Ni_3Ga. The factor, β_c, which takes into account the effect of impurities, introduced in [11], is zero in the binary compound and in $Ni_3(Al,Ta)$ it is estimated to be -0.09 on the basis of the orientation dependence determined in [18]. The values of U_0 have been taken as 2.8eV in both Ni_3Ga and $Ni_3(Al,Ta)$, and $\dot{\varepsilon}_0=10^{12}s^{-1}$. For the free flight velocity, V, we use Leibfried's expression $V = 10\sigma_{pb}^T b^4 v_0/3kT$ [19]. The results are not sensitive to the actual values of U_0, $\dot{\varepsilon}_0$, ΔE and β_c but are more sensitive to the value of the CSF energy which is to be expected since small changes in the constriction energy and hence, H_p, produce large changes in ℓ.

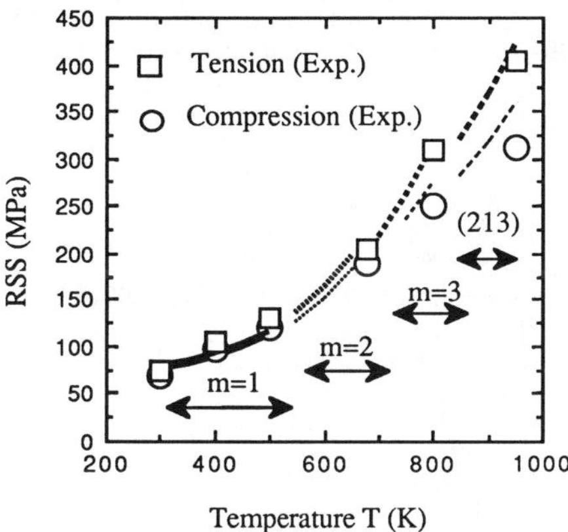

Figure 3. The calculated RSS vs T in Ni_3Ga.

Figure 3 shows the calculated temperature dependence of the RSS in the binary Ni_3Ga along with the experimental values [20] at $\dot{\varepsilon} = 1.3\times10^{-3}\ s^{-1}$ for an orientation in which the ratio of the RSS on $(111)[1\bar{2}1]$ to the RSS on $(111)[\bar{1}01]$ is 0.204. Following the experimental data σ_{pb}^0 has been chosen as 70 MPa. For this orientation the RSS in tension is higher than in compression which is borne out by the calculations. There are three distinct temperature regimes with different rate controlling paths. The m = 1 path is controlling from 300K to ~ 500K. Between 500K and 550K, when $\ell \approx 100b$, the m = 2 path starts to dominate and the activation volume exhibits a jump from $\approx 240b^3$ to $1000b^3$. This path gives the dominant contribution until 700 K. Between 700K and 750K, when $\ell \approx 45b$, there is another transition to the m = 3 path with a smaller jump in the activation volume.

Beyond 850K, paths such as (332) are no longer probable as H_b is less than H_u. Several alternate paths, (323), (213), etc., become possible [14] and it is difficult to decide unambiguously which one dominates. The value of the RSS is roughly the same for all the alternate paths but their activation volumes are slightly different. We show here the results obtained using the (213) saddle-point configuration. T_p determined from the condition that βn is in the range 3-4, is $\approx$ 950K agrees well with observations [20].

Figure 4 shows the calculated RSS in compression in ternary $Ni_3(Al,Ta)$ for the $[\bar{1}\ 2\ 3]$ orientation at $\dot{\varepsilon} = 1.3\times10^{-5}\ s^{-1}$ (σ_{pb}^0 = 40 MPa) along with the experimental values [5]. The corresponding variation of the activation volume with temperature is shown in Figure 5 along with the apparent values measured in stress-relaxation experiments [5]. The first jump discontinuity at Tc ~ 470K agrees well with observations. A second discontinuity is predicted to occur between 550 ~ 650K resulting in a smaller jump magnitude. The orientation dependence, tension/compression asymmetry, the prediction of the peak temperature and the low strain-rate sensitivity are reproduced well by the model calculations [10].

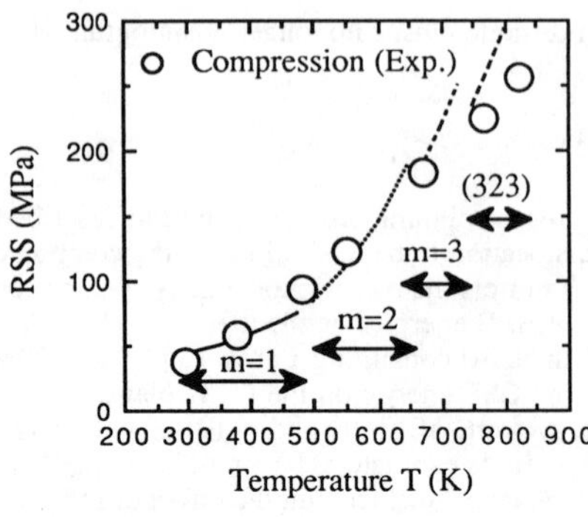

Fig. 4. The calculated RSS in compression vs T for the [$\bar{1}$ 2 3] orientation in $Ni_3(Al,Ta)$.

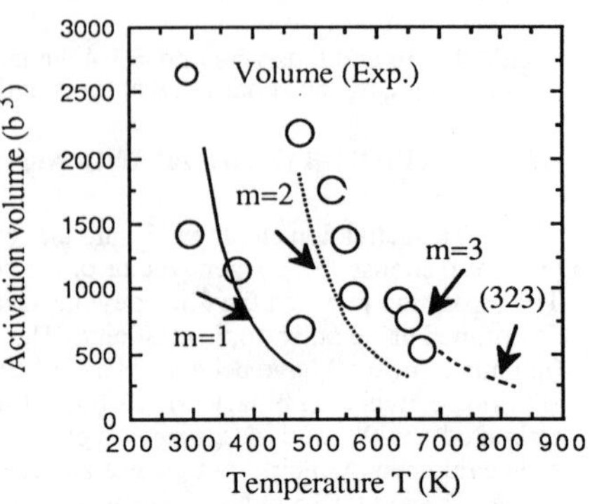

Fig. 5. The activation volume vs T in $Ni_3(Al,Ta)$ for different dominant rate-controlling paths.

ACKNOWLEDGEMENTS

We would like to thank Prof. Sir P. B. Hirsch, Prof. D. P. Pope, Drs. F. Heredia, Y. Sodani and G. Tichy for many useful discussions. This work was supported by the U.S. Air Force Office of Scientific Research (AFOSR-89-0062) and the National Science Foundation (DMR-88-22858).

REFERENCES

1. S. Takeuchi and E. Kuramoto, Acta Metall. **21**, 415 (1973).
2. D. P. Pope and S. S. Ezz, Int. Met. Rev. **25**, 233 (1984).
3. F. E. Heredia and D. P. Pope, Acta Metall. Mater. **39**, 2027 (1991).
4. P. H. Thornton, R. G. Davies and T. L. Johnston, Metall. Trans. A **1**, 207 (1970).
5. J. Bonneville, N. Baluc and J. L. Martin, Int. Symp. on Intermetallic Compounds - Structure and Mechanical Properties - (JIMIS-6), edited by O. Izumi (Sendai, Japan, The Japan Institute of Metals), Vol. p. 323-330, (1991).
6. V. Paidar, D. P. Pope and V. Vitek, Acta Metall. **32**, 435 (1984).
7. P. B. Hirsch, Journal de Physique III **1**, 989 (1991).
8. V. Vitek and Y. Sodani, Scripta. Metall. Mater. **25**, 939 (1991).
9. P. B. Hirsch, Philos. Mag. A **65**, 569-612 (1992).
10. M. Khantha, J. Cserti and V. Vitek, Scripta. Metall. Mater. **27**, 481-486 (1992).
11. M. Khantha, J. Cserti and V. Vitek, Scripta. Metall. Mater. **27**, 487-492 (1992).
12. P. B. Hirsch, to appear in High-Temperature Ordered Intermetallic Alloys V, edited by (Pittsburgh, Materials Research Society), (1993).
13. L. J. Teutonico, A. V. Granato and K. Lücke, J. Appl. Phys. **35**, 220 (1964).
14. L. J. Teutonico, K. Lücke, F. W. Heuser and A. V. Granato, Journal of the Acoustical Society of America **45**, 1401 (1969).
15. D. G. Blair, T. S. Hutchinson and D. H. Rogers, J. Appl. Phys. **40**, 97 (1969).
16. R. Haberman, Mathematical Models, (Prentice Hall Inc., Englewood Cliffs, 1977).
17. N. Baluc, H. P. Karnthaler and M. J. Mills, Philos. Mag. A **64**, 137-150 (1991).
18. F. H. Heredia, Ph.D. Thesis, University of Pennsylvania, (1990).
19. G. Leibfried, Z. Phys. **127**, 344 (1950).
20. S. S. Ezz, D. P. Pope and V. Vitek, Acta Metall. **35**, 1879 (1987).

A MECHANISTIC STUDY OF THE MICROALLOYING EFFECT IN NiAl BASE ALLOYS

R.D. Field, D.F. Lahrman, R. Darolia

Engineering Materials Technology Laboratories, GE Aircraft Engines,
1 Neumann Way, Cincinnati, OH 45215.

ABSTRACT

Alloys based on the B2 compound NiAl have significant potential for applications in hot sections of aircraft engines due to their low density, high melting point, and high thermal conductivity. A major disadvantage of this class of materials is low ductility at ambient temperatures. Recently, it was discovered that small levels of certain elements (eg. Fe, Ga, Mo) result in dramatic improvements in room temperature ductility. In this paper, results are presented from a mechanistic investigation of the "microalloying" effect. Tensile and compression testing as a function of temperature and orientation has been performed on both the binary compound and microalloyed material. Data on ductile to brittle transition temperatures, critical resolved shear stress values as a function of temperature on the different slip systems, and dislocation structures from TEM analysis of the tested specimens are presented. These data are discussed in terms of possible mechanisms for the microalloying effect in NiAl alloys.

INTRODUCTION

A major limitation to the application of NiAl based alloys in turbine engines is low room temperature ductility. For single crystal NiAl, no RT plastic elongation is seen in <100> oriented tensile specimens, whereas for <110> and <111> oriented NiAl (so called "soft" orientations), plastic elongation up to 2% can be obtained [1]. The plastic behavior is highly sensitive to composition and impurity content. The stoichiometric dependence of strength and ductility has been explained in terms of vacancies and anti-site defects within the structure [2]; however, the effects of impurities are not well understood.

Alloying studies have shown beneficial effects of microalloying on the room temperature tensile ductility of "soft" oriented NiAl single crystals [3]. In room temperature tensile tests, up to 6% plastic elongation to failure has been obtained in <110> oriented NiAl alloys containing 0.1 and 0.25 at% Fe. Significant improvements in ductility have also been made by alloying with Ga or Mo. For the Fe additions, the ductility improvements are sometimes accompanied by a reduction in yield strength. While ductility enhancements are obtained at microalloying levels, the beneficial effects are either reduced or disappear at higher levels. Low temperature heat treatments have also been found to increase RT ductility in NiAl. This was first observed in polycrystalline specimens [4], and more recent work has confirmed a ductility enhancement in "soft" oriented single crystal specimens [5]. The increased ductility is accompanied by a decrease in yield strength. In this paper, some early results are presented from a study to determine the mechanisms for these ductility enhancements in NiAl alloys.

EXPERIMENTAL PROCEDURE

Two alloys were studied: stoichiometric binary NiAl (D5) and an alloy containing 0.1at% Fe (D183). All of the specimens received a standard 1316°C/50 hour heat treatment. Blanks from some of the D5 specimens were given a subsequent 800°C/24 hour exposure, found to increase ductility and decrease yield stress in NiAl. These specimens will be referred to as D5HT.

Single crystal slabs of the alloys were grown in argon by a Bridgman method and subsequently homogenized in an argon atmosphere at 1316°C for 50 hours. Button head tensile specimens were produced with gage diameters of 2.5mm and gage lengths of 19.1mm. The tensile specimens were electropolished to remove residual grinding strains. For compression testing, 6.4mm diameter specimens were used, with a length of 15.9mm, resulting in a length to diameter ratio of 2.5/1. Both the tensile and compression tests were run at a strain rate of 8.3×10^{-5}/s. The tensile tests were run to failure, while the compression tests were discontinued after approximately 2% plastic strain.

RESULTS

Results from the tests are shown in Figures 1-2. The yield points for the compression tests were often not well defined, particularly for the low temperature tests. This may have resulted from slight misalignments of the platens and specimen, causing distortions of the stress-strain curve in the early stages of the test. As a result, reliable measurements of the 0.02 and 0.2% yield stresses were not always obtainable. However, the flow stresses, defined in this case as the point at which the stress-strain curve enters linear work hardening, were very reproducible. Flow stress values measured from tests with well-behaved yielding were comparable to those measured in tests conducted under the same conditions (ie. temperature and orientation) which exhibited poorly defined yielding. In the former, the flow stress, as defined above, was generally equal to the 0.2% yield stress. Therefore, the flow stress is used in lieu of an offset stress for the compression tests. For the tensile tests, 0.2% yield stresses are used. As can be seen in the plots, the compression flow stress and tensile yield stress values are essentially equivalent at temperatures for which both quantities were measured.

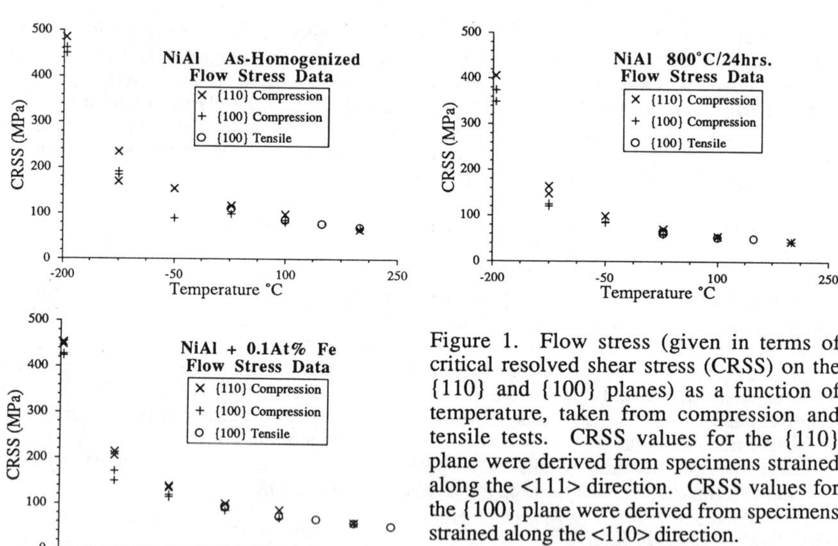

Figure 1. Flow stress (given in terms of critical resolved shear stress (CRSS) on the {110} and {100} planes) as a function of temperature, taken from compression and tensile tests. CRSS values for the {110} plane were derived from specimens strained along the <111> direction. CRSS values for the {100} plane were derived from specimens strained along the <110> direction.

The data in Figure 1 are given in terms of the critical resolved shear stress (CRSS) values for the {110} and {100} planes. These were measured in <111> and <110> oriented specimens, respectively. As these are both "soft" orientations for NiAl, the slip direction is <100> in both cases. The behavior of the CRSS as a function of temperature and orientation was similar in all of the alloys/heat treatments investigated. Inspection of the data reveals that the CRSS falls rapidly between –196 and -50°C for both orientations. The rate at which the CRSS falls with temperature decreases with increasing temperature. Above 0°C, the decline in CRSS with temperature is approximately linear. The CRSS for the {110} plane is generally slightly higher than that for {100}.

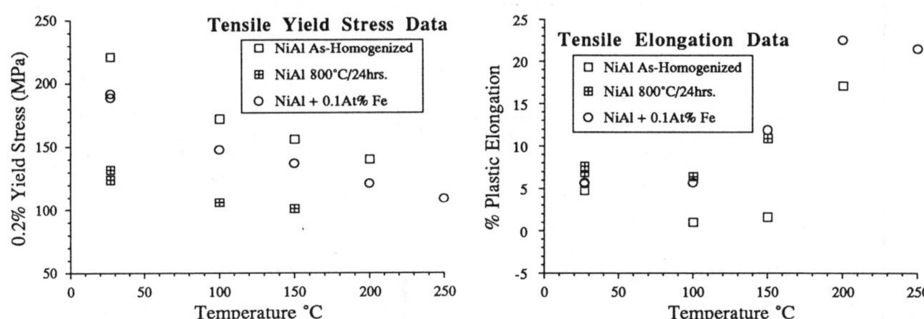

Figure 2. 0.2% yield stress and plastic elongation as a function of temperature for <110> oriented tensile specimens.

Yield stress and plastic elongation values from the tensile tests are presented in Figure 2. The room temperature elongation for the D5 specimen was unusually high for this alloy/heat treatment. However, at 100 and 150°C, the D5HT and microalloyed specimens displayed markedly higher plasticity than the D5 specimens. The results also indicate that the ductile to brittle transition temperature (DBTT) for D5HT and D183 is lower than that of D5. Serrated yielding was observed in two different temperature ranges for the D5 specimens: in the -125°C tests and in the 100 to 250°C range (ie. around the DBTT). This was particularly evident in the compression specimens, but was also observed during the tensile tests. The effect disappeared at -125°C for the D5HT specimens, and was significantly reduced for the D183 specimens; however, the serrated yielding persisted for all of the alloys/heat treatments at the higher temperatures. Examples of compression stress-strain curves for D5 and D183 are shown in Figure 3.

TEM dislocation studies were performed on compression specimens tested to approximately 2% plastic strain at –196 and 100°C in both <111> and <110> orientations. The –196°C specimens represent the simplest cases in which thermal activation is minimized. At 100°C, the D5HT and D183 specimens displayed high levels of plastic elongation in tensile tests, while D5 was still below its DBTT. Burgers vector determinations were made using the invisibility criterion. Slip plane identifications were achieved by determining the planes on which bowed dislocations were lying, by means of trace analyses on two or more segments of each dislocation. Trace analyses of straight dislocations were also used, as well as the elongation directions of dipole loops. The dipole loops were assumed to have been deposited from superjogs on moving screw dislocations, and thus elongated along directions perpendicular to their Burgers vectors in the slip plane [6-8].

The dislocation structures in these specimens were similar for the same orientations and temperature, regardless of alloy and heat treatment. Most of the dislocations present in the specimens were debris, rather than actively slipping dislocations. As expected, <100> Burgers vectors were observed in all of the specimens. All of the data indicate that the {110} slip planes are active in <111> oriented specimens and the {100} slip planes are active in <110> oriented specimens, at both temperatures studied. This is consistent with the resolved shear stresses on these planes for the two orientations: the {110} plane has the higher Schmid factor for the <111> orientation (0.47 vs. 0.33), while the {100} plane has the higher Schmid factor for the <110> orientation (0.5 vs. 0.35).

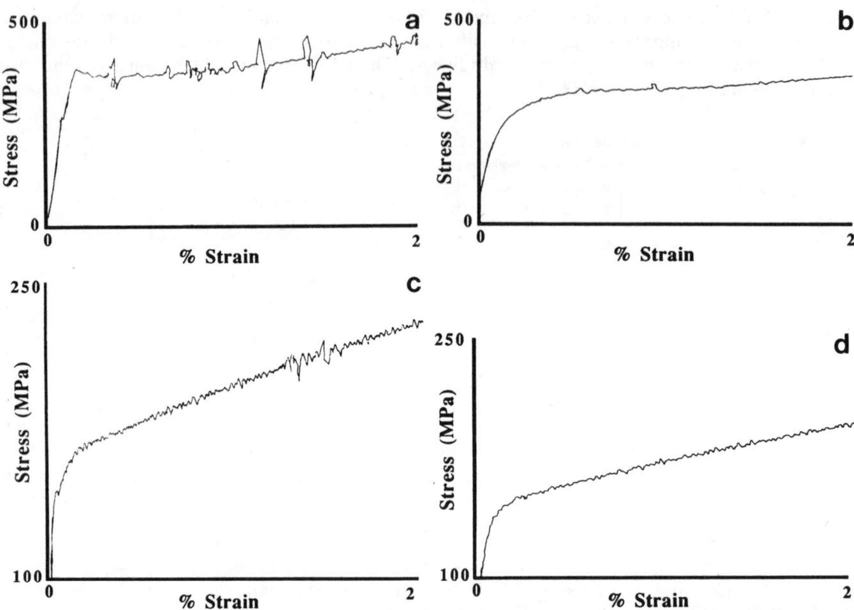

Figure 3. Stress-strain curves showing serrated yielding in compression tests. a) NiAl at –125°C, b) NiAl + 0.1 at% Fe.at –125°C, c) NiAl at 100°C, d) NiAl + 0.1 at% Fe.at 100°C

At –196°C, the <111> oriented specimens exhibited distinct slip bands of dislocations possessing the same Burger vector. This orientation and temperature yielded the highest percentage of dislocations actually lying on slip planes, and thus the most easily interpreted dislocation structures. It is interesting to note that the work hardening rates for <111> oriented specimens tested at –196°C were lower than those at any other temperature and orientation. This is consistent with the lower degree of dislocation interaction observed in these specimens. At 100°C, a cellular structure developed, with tight tangles surrounding areas of lower dislocation density. This may represent a further development of the slip band structure, forming as individual slip bands intersect and dislocation density increases. Many of the dislocations outside the tangles were found to lie on or close to {110} slip planes. The <110> oriented specimens tested at –196°C had a very homogeneous dislocation distribution, consisting mostly of debris, with no evidence of slip bands or dislocation tangles. At 100°C, a cellular structure developed, similar to that for the <111> oriented specimens tested at the same temperature. As at –196°C, bowed dislocations seldom lay entirely on slip planes. Most of them appeared to have high densities of superjogs.

DISCUSSION

The temperature dependence of the critical resolved shear stresses (CRSS) for the {100} and {110} slip planes does not appear to be affected by the low temperature heat treatment or microalloying, other than an overall decrease in strength over the standard heat treated binary material. Other aspects of the stress-strain curves (eg. yield behavior and work hardening rates) also are unaffected. Lastly, no differences in dislocation structures were noted for the different alloys/heat treatments.

The CRSS data and the dislocation structures for all of the alloys/heat treatments studied indicate that the {100} and {110} slip planes are about equally active in NiAl within the temperature range from –196 to 200°C. Slip on {100} is slightly easier, particularly at the lower temperatures. This is contrary to the historical view that {110} is the preferred slip plane in NiAl [9-11], but consistent with previous results obtained by us for stoichiometric NiAl [6-8]. Molecular dynamic calculations of dislocation motion in NiAl also indicate that slip should be extremely difficult on {100} planes [12,13]. It was thought that {110}, which is the closest packed plane in the B2 structure, might become more active relative to {100} at lower temperatures, where thermal activation plays a lesser role. If anything, the opposite effect is demonstrated by the data. Clearly, slip in this material is still not well understood.

Serrated yielding was observed in two temperature ranges for stoichiometric NiAl given the standard heat treatment: in the -125°C tests and in the 100 to 250°C range (ie. around the DBTT). This type of behavior has been associated with point defect/dislocation interactions, specifically the strain aging effect [14,15], and is believed to be caused by pinning of the dislocation by interstitial impurities. Serrated yielding occurs when the dislocation velocity is comparable to the diffusion rate of the pinning species. Work by Margevicius, et al [16] on polycrystalline specimens and Hack, et al [17] on single crystals have suggested that such an effect might be occurring in the 100 to 400°C range.

Besides serrated yielding and the effects of low temperature heat treatments on ductility and strength, other phenomena can provide evidence of strain aging effects. Some of these have been observed in NiAl. For example, strain rate sensitivity has been found to be very low in the strain aging regime. In earlier work [1], we observed that the strain rate sensitivity of NiAl near the DBTT is quite low. Also, a "plateau" in the temperature dependence of the yield strength is often associated with the strain aging effect. The flow stress drops monotonically around the DBTT in NiAl, although at a slower rate than at lower temperatures. Sharp yield points followed by regions of zero work hardening, also associated with strain aging effects, were not observed in the single crystal specimens of this study, although yield point phenomena have been observed in polycrystalline specimens [16].

A major difference in the behavior of the microalloyed and 800°C heat treated specimens as compared to the standard heat treated NiAl specimens is the decrease or elimination of the low temperature (-125°C) serrated yielding effects. It is difficult to understand what species could be diffusing rapidly enough to cause such an effect in this temperature range. Whatever the controlling species is, it must involve only very short range diffusion. Such a short diffusion path would be characteristic of a Snoek type mechanism, in which interstitials jump to adjacent sites in order to rotate their strain fields under the influence of a dislocation strain field. Such a hypothesis is purely speculative, however, until the effect can be more fully characterized. Further mechanical testing is planned, including an investigation of the effects of strain rate on serrated yielding. Internal friction studies are also being conducted by Prof. Granato's group at the University of Illinois as part of this investigation. It is hoped that this work will provide a better understanding of dislocation/point defect interactions in NiAl.

Faceted voids were observed during TEM studies of the 800°C heat treated specimens. The presence of these voids suggests that vacancies are annealing out at this temperature. Vacancies can affect the strength and ductility of B2 alloys. The effect is particularly potent in FeAl [18], but NiAl is also known to be affected by thermal and constitutive vacancies. It is very likely that the strength reduction observed in the D5HT specimens is due to a decrease in vacancy concentration. Preliminary work by us has indicated that vacancy concentrations may be reduced by microalloying [19]. However, more detailed work in this area is needed.

CONCLUSIONS

The effects of heat treatment and microalloying on ductility and yield strength in NiAl alloys have been studied. Both 800°C annealing and microalloying result in decreased yield strength, increased RT ductility and a lowering of the DBTT. Neither heat treatment nor microalloying affect the overall slip behavior (ie. slip planes or directions) or stress-strain behavior (eg. yield behavior and work hardening rates) in NiAl. The ratio of the CRSS values for <100> slip on {100} versus {110} planes was also unaffected by alloying or heat treatment. This ratio remains relatively constant as a function of test temperature. Serrated yielding was observed in two temperature ranges for the stoichiometric alloy given only a homogenization anneal: -125°C and 100-250°C, suggesting a dynamic strain aging phenomenon. This effect was reduced by both the 800°C heat treatment and Fe microalloying additions. Further study is being carried out to more fully characterize these effects in NiAl alloys.

ACKNOWLEDGEMENTS

The authors are grateful to Jim Nickley, Gary McCabe, Mike Agnello, Bill Davis, and Dr. Hang Yan for their assistance in specimen preparation. Thanks also go to Joe Wilson (Thread Rite Tool) for specimen machining and Jim Rossi and Gerry Boice (Westmoreland Mechanical Testing and Research) for testing. This work was supported by the Air Force Office of Scientific Research under contract #F49620-91-C-0077 with Dr. A. Rosenstein as the Program Manager.

REFERENCES

1. D. F. Lahrman, R. D. Field and R. Darolia, High Temperature Ordered Intermetallic Alloys IV, Materials Research Society Symposium Proceedings **213**, 603 (1991).
2. R.T. Pascoe and C.W.A. Newey, Met. Sci. J. **2**, 138 (1968).
3. R. Darolia, D. Lahrman and R. Field, Scripta Metall. Mater. **26**, 1007 (1992).
4. W. Wang, R.A. Dodd and P.R. Strutt, Met. Trans. **3**, 2049 (1972).
5. R. D. Field, D. F. Lahrman, R. Darolia, and A.J. Freeman, "Alloy Modeling and Experimental Correlation for Ductility Enhancement in Near Stoichiometric Single Crystal Nickel Aluminide", Final Report, AFOSR contract #F49620-88-C-0052, July, 1991.
6. R. D. Field, D. F. Lahrman and R. Darolia, High Temperature Ordered Intermetallic Alloys IV, Materials Research Society Symposium Proceedings **213**, 255 (1991).
7. R.D. Field, D.F. Lahrman and R. Darolia, Acta. Met. **39**, p 2951 (1991).
8. R.D. Field, D.F. Lahrman and R. Darolia, Acta. Met. **39,** p 2961 (1991).
9. R.J. Wasilewski, S.R. Butler and J.E. Hanlon, Trans. AIME **239**, 1357 (1967).
10. R.T. Pascoe and C.W.A. Newey, Met. Sci. J. **2**, 138 (1968).
11. R.T. Pascoe and C.W.A. Newey, Phys. Stat. Sol. **29**, 357 (1968).
12. D. Farkas, R. Pasianot, E.J. Savino, and D.B. Miracle, High Temperature Ordered Intermetallic Alloys IV, Materials Research Society Symposium Proceedings **213**, 223 (1991).
13. T.A. Parthasarathy, S.I. Rao and D.M. Dimiduk, presented at TMS Fall Meeting, Cincinnati, OH, Oct. 1991.
14. J.P. Hirth and J. Lothe, Theory of Dislocations, McGraw-Hill (1968).
15. R.E. Reed-Hill, Physical Metallurgy Principles, 2nd Edition., D. Van Nostrand Co. (1973).
16. R.W. Margevicius, J.J. Lewandowski and I. Locci, Scripta Metall. Mater. **26**, 1733 (1992).
17. J.E. Hack, J.M. Brzeski and R. Darolia, Scripta Metall. Mater. **27**, 1259 (1992).
18. M.A. Crimp and K. Vedula, Phil. Mag. A **63**, 559 (1991).
19. R.D. Field, D.F. Lahrman, R. Darolia, and A. Granato, "A Mechanistic Study of Microalloying Effects in NiAl", 1st Annual Report, AFOSR contract #F49620-91-C-0077, August, 1992.

THE CRITICAL STRESS FOR PLASTIC DEFORMATION IN $Ni_3(Al,Hf)$ SINGLE CRYSTALS.

P. Spätig, J. Bonneville and J-L. Martin

Ecole Polytechnique Fédérale de Lausanne, Institut de Génie Atomique, Département de Physique, 1015 Lausanne (Switzerland).

ABSTRACT

A new method of relaxation series is proposed which allows for the measurement of the apparent activation volumes and of the hardening correction terms. This method is successfully applied to the measurement of the effective activation volumes along the stress strain curves. A transition stress between a microplastic and a macroplastic domain can be unambiguously defined on the plots of the effective volume as a function of strain. The corresponding transition stress measured at three temperatures in the range of the flow stress anomaly (293K, 423K, 573K), agrees reasonably well with the values of $\tau_{0.2\%}$. The latter stress was usually considered as the critical stress for plastic deformation, in former studies.

1. INTRODUCTION

The flow strength anomaly of Ni_3Al compounds has been studied extensively and a number of models have been proposed to explain this peculiar behaviour [1]. It has generally been accepted that : i) below the 2.10^{-3} yield stress ($\tau_{0.2\%}$, *i.e.* the stress necessary to give a plastic strain γ_p of 2.10^{-3}), deformation corresponds to the rapid motion of edge dislocations that lead to a normal temperature dependence while ii) above $\tau_{0.2\%}$, flow is associated with the more difficult propagation of screw dislocations that yields the anomalous temperature dependence [2]. The models attempt to describe the temperature, sense and orientation dependences of $\tau_{0.2\%}$.

The validity of using the $\tau_{0.2\%}$ stress as the critical stress for screw dislocation motion in Ni_3Al compounds has been recently questioned [3, 4]. The main objections, that were addressed, are based on the following experimental observations : i) the 10^{-4} yield stress (*i.e.* $\tau_{0.01\%}$) exhibits already a clear anomalous behaviour with temperature [5], ii) extensive octahedral glide occurs during primary creep at stresses below $\tau_{0.2\%}$, iii) the stress-strain curves do not exhibit any distinct yield point.

Previous studies on the strain rate sensitivity of stress [6, 7] have indicated that, at least during stress relaxation experiments performed along the stress strain curve of $Ni_3(Al,Ta)$ single crystals, a normal thermal activation of dislocation motion takes place. In the framework of the thermal activation theory, this allows us to determine an activation volume that characterizes the strain rate sensitivity of the stress. It is expected that this parameter will undergo an abrupt change when there is a change in the microscopic mechanism that controls deformation. It has been shown, for instance, that the strain dependence of the activation volume clearly indicates the transition stress between the micro- and macro-plastic domains in various type of materials [8, 9, 10].

The present paper is devoted to the characterization of a critical stress between the micro- and macro- plastic domains. For this, activation volume measurements were performed as a function of strain by using a new technique of stress relaxation series [11]. This technique allows for the measurement of an effective activation volume that takes into account strain hardening [12]. Three temperatures have been investigated in the anomaly domain of the flow stress. These measurements are reported in this paper in order to provide quantitative results for the understanding of the unusual properties of intermetallic compounds.

2. EXPERIMENTAL PROCEDURES

Rectangular specimens with a $[\bar{1}23]$ compression axis were cut by spark erosion from single crystal rods of Ni_3(Al, 3% at. Hf), supplied by Professor Vitek at the University of

Pennsylvania. The specimens were mechanically polished with successively finer grades of diamond paste and subsequently electro-chemically polished. The final sizes of the specimens were around 2.9mmx2.9mmx6.5mm. The compression tests were performed on a Schenck RMC 100 machine. Complete descriptions of the deformation set up and data processing have already been given elsewhere [13].

The stress relaxation technique was used to investigate the strain rate sensitivity of stress. A stress relaxation experiment consists, during a deformation test performed at constant strain-rate, in stopping the crosshead at a given stage of the flow stress. The load applied on the specimen is allowed to relax and recorded as a function of time so that, for each time, the stress and the stress rate (equivalent to the plastic strain rate) are known. This yields a complete knowledge of the strain rate sensitivity of stress that, when flow is thermally activated, can be analyzed in terms of an activation volume [14]. When a single relaxation test is performed an apparent activation volume (V_a) is measured:

$$V_a = -\frac{kT}{\Delta \tau_a} \log\left(\frac{t}{c} + 1\right)$$

where c is an integration constant and where the other parameters have their usual meanings. To obtain the true activation volume (V_{eff}), V_a must be corrected by a correction term (V_h) that accounts for strain hardening and machine stiffness [12] :

$$V_{eff} = V_a - V_h \quad (1)$$

Kubin [15] has shown that V_h can be measured if the relaxation test was followed by a series of stress relaxations performed at a similar stress drop. This technique has been successfully applied, for instance, to study the activation volume of cross-slip in Cu [16]. But, this procedure leads to long time experiments that are very sensitive to the thermal stability of the deformation device. A new procedure has been used that also consists in performing a series of stress relaxations but at constant duration (figure 1). Assuming that the applied stress (τ_a) can be decomposed as:

$$\tau_a = \tau^* + \tau_\mu \quad (2)$$

where τ^* and τ_μ are the thermal (or effective) and the athermal stress components, respectively, then each relaxation starts at the same applied stress (see figure 1) but with a lower τ^*, if strain hardening occurs (*i.e.* a larger τ_μ) during the preceding relaxation. This hardening effect can be

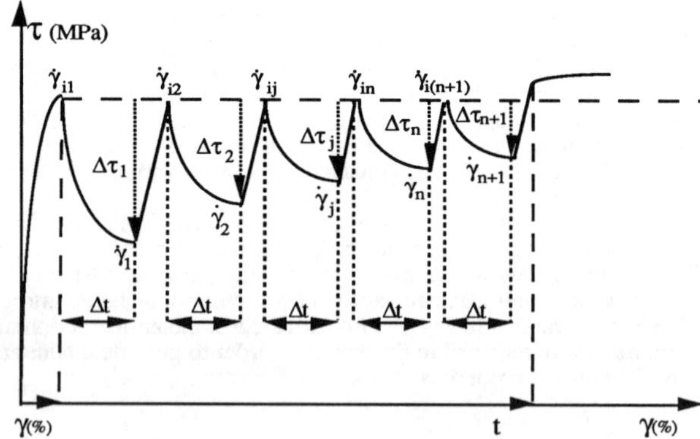

Fig. 1 : Schematic representation of a relaxation series performed at constant duration (Δt).

measured and related to V_h. In the framework of the thermal activation theory, we have shown that for n>1 (n is the number of relaxations) and with using the notation of figure 1, V_h can be deduced from such a repeated relaxation test by the relation [11]:

$$(n-1)\, V_h = \frac{kT}{\overline{\Delta\tau}} \log\left(\frac{\exp(-\beta\,\Delta\tau_n)-1}{\exp(-\beta\,\Delta\tau_1)-1}\right) \quad (3)$$

with

$$\beta = \frac{V_a}{kT} \qquad\qquad \overline{\Delta\tau} = \frac{1}{n-1}\sum_{j}^{n-1} \Delta\tau_j$$

where V_a is determined with the first relaxation. Therefore, a plot of the right hand side of expression (3) versus (n-1) must be linear with a slope equal to V_h.

3. EXPERIMENTAL RESULTS

Three temperatures (293K, 423K, 573K) have been investigated in the range of the flow stress anomaly. At least three samples have been deformed at each temperature and a good reproductibility has been observed between the different tests. Since we have also observed essentially the same features at all the tested temperatures, we shall first present in more details the results obtained at room temperature and then, more briefly, those corresponding to the other temperatures.

A) Results at room temperature.

A representative stress-strain curve is given in figure 2. As emphasized in [3], the stress strain curve exhibits a gradual transition from the micro to the macro-plastic domains without any evidence of a critical stress for macroplastic flow.

The repeated relaxation tests, performed along the stress-strain curves, were in fair agreement with the predictions of the thermal activation theory of dislocation motion. (1) During the relaxations, the stress was observed to decrease as a logarithmic function of time [14]. (2) The plots of the successive stress drops of the relaxation series, as a function of the relaxation number, were found to be linear, according to equation (3).

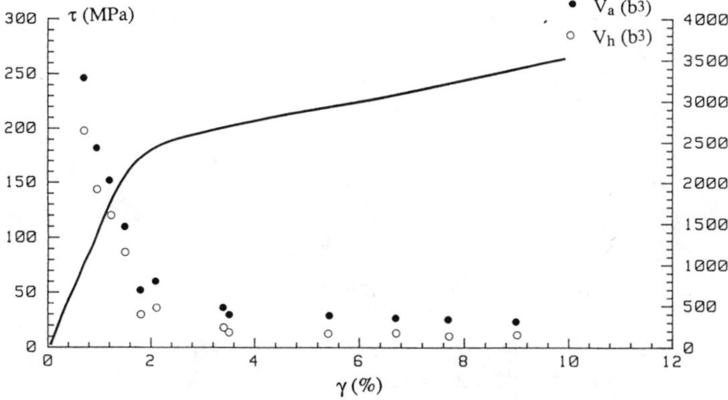

Fig. 2 : Representative stress strain curve and the corresponding apparent activation volume (V_a) and correction term (V_h) obtained at room temperature.

The measured V_a and V_h values are reported as a function of strain in figure 2. Both the V_a and V_h values start at a rather high level but decrease rapidly in the early part of the stress strain curve up to a transition strain $\gamma_p(V_a^{tr}) \simeq 1.2\%$, beyond which they become much less strain dependent. The corresponding V_{eff} values, that are directly deduced from the V_a and V_h values with relation (1), are reported in figure 3. The strain dependence of V_{eff} also evidences two distinct domains, but with a transition strain that is clearly shifted towards lower strains ($0.1\% < \gamma_p(V_{eff}^{tr}) < 0.5\%$) as compared to $\gamma_p(V_a^{tr})$. On the stress strain curve, this strain interval corresponds to a transition stress $\tau_{cr} = 170 \pm 15$ MPa that has to be compared with the mean value of $\tau_{0.2\%} = 173 \pm 2$ MPa measured on the stress strain curves of the three tests.

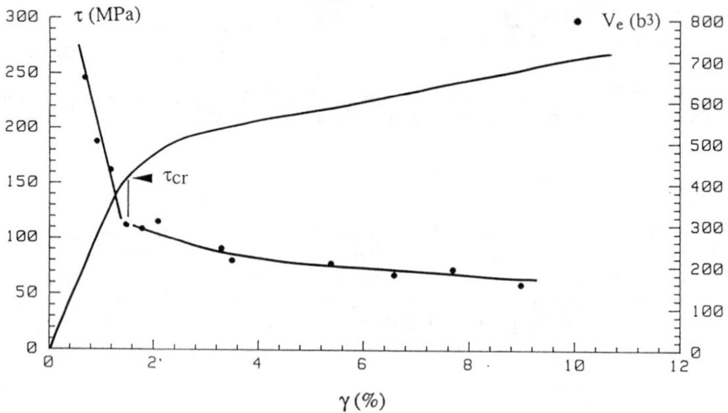

Fig. 3 : Representative stress strain curve and effective volumes (V_{eff}) at 293K.

B) Results at the other temperatures.

The above observations, that were made at room temperature, hold for the two other temperatures, 423K and 573K. The two domains of V_{eff} strain dependence are also clearly observed. The corresponding transition stresses (τ_{cr}) that are defined as previously are reported as a function of the $\tau_{0.2\%}$ stresses in figure 4. In this figure, the dotted line corresponds

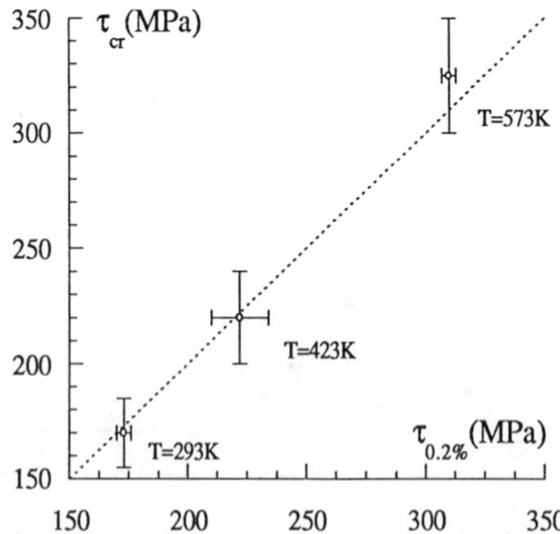

Figure 4. Transition stresses τ_{cr} as a function of $\tau_{0.2\%}$ at three different temperatures.

to $\tau_{cr} = \tau_{0.2\%}$. A reasonable correlation appears between the conventional $\tau_{0.2\%}$ stress and the more reliable τ_{cr} stress that is deduced from the strain dependence of the activation volume (V_{eff}).

4. DISCUSSION

Before discussing the transition stress (τ_{cr}) results presented above, it is necessary to establish that the activation volume measurements are both meaningful and reliable. During a relaxation test, V_h is related to V_{eff} by [11]:

$$V_h = V_{eff}\frac{K}{M} \quad (4)$$

where $K = \delta\tau_\mu / \delta\gamma_p$ and M is the elastic modulus of the sample-machine assembly. By using relation (2) and the equation of the relaxation ($\delta\gamma_p = - \delta\tau_a / M$), expression (4) can be rewritten as:

$$V_h = V_{eff}\frac{\delta\tau_\mu}{\delta\tau_a}$$

The high values of the correction term V_h as compared to V_{eff}, especially at the beginning of the stress strain curve, indicate that a very strong athermal strain hardening takes place during the stress relaxation tests. Therefore, the apparent activation volumes do not reflect only the direct effect of stress on the strain rate, but also its effect upon the internal stress and structure (see relations (1) and (5)). This explains the difference that has been found between the two transition strains $\gamma_p(V_a^{tr})$ and $\gamma_p(V_{eff}^{tr})$.

The V_h / V_{eff} ratio decreases a function of strain similarly to the strain hardening ($\theta = d\tau_a/d\gamma$) that is measured on the stress strain curve. But, it can be seen that, for instance at room temperature, even in the macro-plastic domain, the V_h and the V_{eff} values which are reported in figure 2 and 3, respectively, are approximately equal. This yields an athermal hardening coefficient $K \simeq M$ that is always much larger than θ (the maximum value of θ is M, *i.e.* the elastic slope). It is believed that the observed discrepancy between the two strain hardening coefficients does not invalidate the V_{eff} measurements since the same thermally activated process should control the plastic deformation during both the stress relaxation and the deformation at constant strain rate. Therefore, a larger strain hardening occurs during the stress relaxation tests (decreasing applied stress) than under constant strain rate conditions (increasing applied stress). But, the repeated stress relaxation technique measures correctly the correction term (V_h), corresponding to the strain hardening occurring during the relaxation, that must be deduced from the apparent activation volume (V_a) in order to obtain the true microscopic activation volume (V_{eff}).

This description is consistent with the strain rate jump experiments of Thornton *et al.* [5] showing that, for a given range of temperatures in the anomaly domain, a strain rate decrease was initially accompanied by a decrease of the applied stress (*i.e.* a positive strain rate sensitivity of stress) followed by a transient domain of strain rehardening.

The strain dependence of V_{eff} suggests that two deformation domains exist along the stress strain curve of $Ni_3(Al,Hf)$ single crystals. It must be noted that a plot of V_{eff} as a function of stress also exhibits the two domains. The strong similarity of the present results with previous works [8, 9, 10] indicates that the transition stress (τ_{cr}) may be considered as the critical stress for macro-plasticity. We may assume that the first domain (below τ_{cr}) corresponds to micro-plasticity that is associated with the motion of a small density of very mobile dislocations whereas in the second domain (above τ_{cr}) flow corresponds to macro-plasticity that is related to a large scale dislocation multiplication. Consequently, the good agreement that has been obtained between the stresses τ_{cr} and $\tau_{0.2\%}$ indicates that, in the case of $Ni_3(Al,Hf)$ single crystals, $\tau_{0.2\%}$ can be considered as representative of the critical stress for macrocopic deformation.

5. CONCLUSION

A new method of stress relaxation series has been used to measure the correction term V_h that accounts for strain hardening during a stress relaxation test. This technique allowed us to measure, at three different temperatures, the effective activation volume (V_{eff}) along the stress strain curves of $Ni_3(Al,Hf)$ [$\bar{1}23$] oriented single crystals. In summary:

i) The strain dependence of V_{eff} along a stress strain curve exhibits two domains that were identified as a micro- and a macro-plastic domain.

ii) A reasonable agreement is found between the critical stress τ_{cr}, as defined at the transition between the two domains, and the conventional stress $\tau_{0.2\%}$, that was usually considered as the critical stress for plastic flow.

iii) The strain hardening that takes place during a relaxation experiment (unloading conditions) is considerably larger that the strain hardening produced during a constant strain rate test.

ACKNOWLEDGMENTS

The authors acknowledge the financial support of Fonds National Suisse (contract n° 20-31302.91).

REFERENCES

1. D. M. Dimiduk, *J. de Phys. III* **1**, 1025 (1991).
2. R. A. Mulford and D. P. Pope, *Acta Metall.* **21**, 1375 (1973).
3. K. J. Hemker, M. J. Mills and W. D. Nix, *J. Mat. Res.* **7**, 2059 (1992).
4. K. J. Hemker, M. J. Mills, K. R. Forbes, D. D. Sternbergh and W. D. Nix, *Proc. 9th ICSMA*, Ed. D. G. Brandon, R. Chaim and A. Rosen, Freund Publishing, **1**, 271 (1991).
5. P. H. Thornton, R. G. Davies and T. L. Johnson, *Metall. Trans.* **1**, 207 (1970).
6. J. Stoiber, J. Bonneville and J-L. Martin, *Proc. 8th ICSMA,* Ed. P. O. Kettunen, T. K. Lepistö and M.E. Lehtonen, Pergamon Press, **1**, 457 (1988).
7. J. Bonneville and J-L. Martin, *Mat. Res. Soc. Symp. Proc.* **213**, 629 (1991).
8. B. Escaig, *J. de Phys.* **35**, C7-151 (1974).
9. J-L. Farvacque, J. Crampon, J-C. Doukhan and B. Ecsaig, *Phys. Stat. Sol.* **14**, 623 (1972).
10. J. D. Meakin, *Can. J. of Phys.* **45**, 1121 (1967).
11 P. Spätig, J. Bonneville and J-L. Martin, *submitted to Mat. Scie. Eng.* (1992).
12. P. Groh and R. Conte, *Acta. Metall.* **19**, 895 (1971).
13. J. Bonneville, N. Baluc and J-L. Martin, *Proc. 6th JIMIS on Intermetallic Coumpounds,* Ed. by O. Izumi, Sendai, Japan, 323 (1991).
14. F. Guiu and P. L. Pratt, *Phys. Stat. Sol.* **6**, 111 (1964).
15. L. P. Kubin, *Phil. Mag.* **30**, 705 (1974).
16. J. Bonneville, J-L. Martin and B. Escaig, *Acta Metall.* **36**, 1989 (1988).

POSSIBILITIES OF SLIP MODIFICATION IN B2 NiAl

Diana Farkas and Zhao Yang Xie
Department of Materials Science and Engineering,
Virginia Polytechnic Institute and State University,
Blacksburg, VA 24061

ABSTRACT

We discuss the possibilities of stabilizing ⟨111⟩ slip in NiAl based on recent results of the atomistic structure of dislocation cores in this material. In previous work it was found that the ⟨111⟩ complete dislocation is stable with respect to partials of 1/2 ⟨111⟩ Burgers vector. However, split configurations were found with a very similar energy. In particular, split configurations that are nearly planar were found, indicating that if this type of dislocation could be stabilized they would possibly result in ductile alloys.

We discuss the possible effects of local disorder and deviations from stoichiometry on core structure concentrating on the objective of stabilizing ⟨111⟩slip. Atomistic simulations performed to study these effects show that local compositional changes and disorder may have significant effects on the core structure. These effects are most important in the regions of the core where the point defects are localized and it is unlikely that such local changes could be used to successfully stabilize ⟨111⟩ slip. Furthermore, it is not clear whether local changes in ordering and compositional states can follow the dislocation as it moves during the deformation process. We therefore propose that the alloying efforts should be directed towards changing the APB energy of the NiAl- base alloy to a lower value.

INTRODUCTION

A large research effort has been devoted in recent years to the study of different possible methods to improve ductility in ordered intermetallic alloys, and in particular NiAl. These methods include second phase formation[1] and utilizing the role of surface and interface dislocations [2]. We recall that the basic reason for the lack of ductility in NiAl is usually believed to be the lack of sufficient slip systems, which is a consequence of the fact that NiAl deforms through ⟨100⟩ slip. Other compounds, such as CuZn with the same structure that deform through ⟨111⟩ slip do exhibit ductile behavior. It follows that in order to improve ductility one could investigate the possibilities of stabilizing ⟨111⟩ slip in this material, for example, through alloying.

It has been shown that when the crystal orientation is such that the applied stress on the existing ⟨100⟩ dislocations is zero, the material indeed deforms through ⟨111⟩ slip [3]. As a matter of fact, the occurrence of ⟨111⟩ slip has also been proposed as an explanation for the limited ductility of NiAl observed at room temperature [4]. The instability of ⟨111⟩ slip is certainly connected to the fact that the ordering energy is high and to the fact that the ⟨111⟩ dislocations apparently do not split into partials [5].

In agreement with experimental observations, computer simulation also shows that the stable form for screw ⟨111⟩ dislocations is for these dislocations to remain as complete ⟨111⟩. In more ductile B2 materials the stable form is that of two superpartials of 1/2⟨111⟩ burguers vector with an APB in between. The energetics and stability of the dissociated plane and other possible planes for core spreading. In previous work [6] we have computed these γ surfaces in different planes including (110) and (112) with particular interest in the vicinity of the APB. The important features of the γ surfaces in these planes are the minima found in the APB region, which are deviated from the $1/2\langle111\rangle$ position in both cases. For the $\{110\}\gamma$ surface the $1/2\langle111\rangle$ fault is an unstable maximum. Two minima occur on both sides of the $1/2a[1\bar{1}1]$ fault, deviated in the $[1\bar{1}0]$ direction. The γ surface in the (112) plane shows that the minimum corresponding to the APB is shifted in the $[\bar{1}\bar{1}1]$ direction from the perfect $1/2a[\bar{1}\bar{1}1]$ position. In our previous work, we studied metastable dissociations of the complete $\langle111\rangle$ dislocation into $1/2\langle111\rangle$ partials and the alternative splitting suggested by the γ surface calculations to have a stable fault between the two partials. Even using this alternative splitting, the complete $\langle111\rangle$ dislocation was still stable with respect to the dissociation into partials.

In the present work, we analyze the possibilities of stabilizing $\langle111\rangle$ slip in NiAl based on the results of the atomistic structure of dislocation cores in this material. We present results for the dislocation core structure as affected by the interaction with point defects in the center of the core region. The implications of these results for the possibilities of stabilizing $\langle111\rangle$ slip are discussed. We particularly discuss the role of the APB energy and the separation between partial dislocations for the dissociated dislocation.

EFFECTS OF STOICHIOMETRY DEVIATIONS AND LOCAL DISORDER ON CORE STRUCTURE AND MOBILITY

We computed core structures of $\langle100\rangle$ and$\langle111\rangle$ dislocations as affected by the interaction with various point defects present in the material. The point defects considered are, in the first place those that result from stoichiometry deviations. These are vacancies in Ni sites, which accommodate excess Al and Ni in Al sites, which accommodate extra Ni [4]. We also considered the effect of disorder, which will occur increasingly with temperature. Disorder is represented by a pair of antisite defects (a sublattice exchange). The effects of Al vacancies and Al in Ni sites were studied as well.

Due to the periodic boundary conditions used in the simulations along the dislocation line, we added the point defects as a row of defects parallel to the dislocation line. In order to avoid strong interaction among the defects we considered two unit cells in the direction of the dislocation line. The point defects were located in the center of the dislocation core. The configuration included one unit cell containing a point defect and one without a point defect in the periodic direction. In order to asses the extent of the influence that these point defects may have on the core structure of both $\langle100\rangle$ and $\langle111\rangle$ dislocations we studied the variations of the cores of a pure edge $\langle100\rangle$ dislocation with a cube slip plane and a pure screw undissociated $\langle111\rangle$ dislocation.

Figure 1 shows the effect of the various point defects on the core structure of the edge $\langle100\rangle$ dislocation. The calcualtion method and interatomic potentials used are similar to those used in our previous work. The core structures are represented using the invariants of the strain tensor, as described in our previous work [6, 7, 8] The most important effects occur for excess Al. The effects are confined to the region of very high deformation level. This is probably due do the fact that the point defects were located in that region.

From these results, it is clear that the basic core structure remains largely unaffected. This is probably also due to the fact that the shape of the core is greatly influenced by elasticity itself and therefore the effect of point defects on core structure are of lesser importance. We note that this does not mean that there ar no significant effects from the

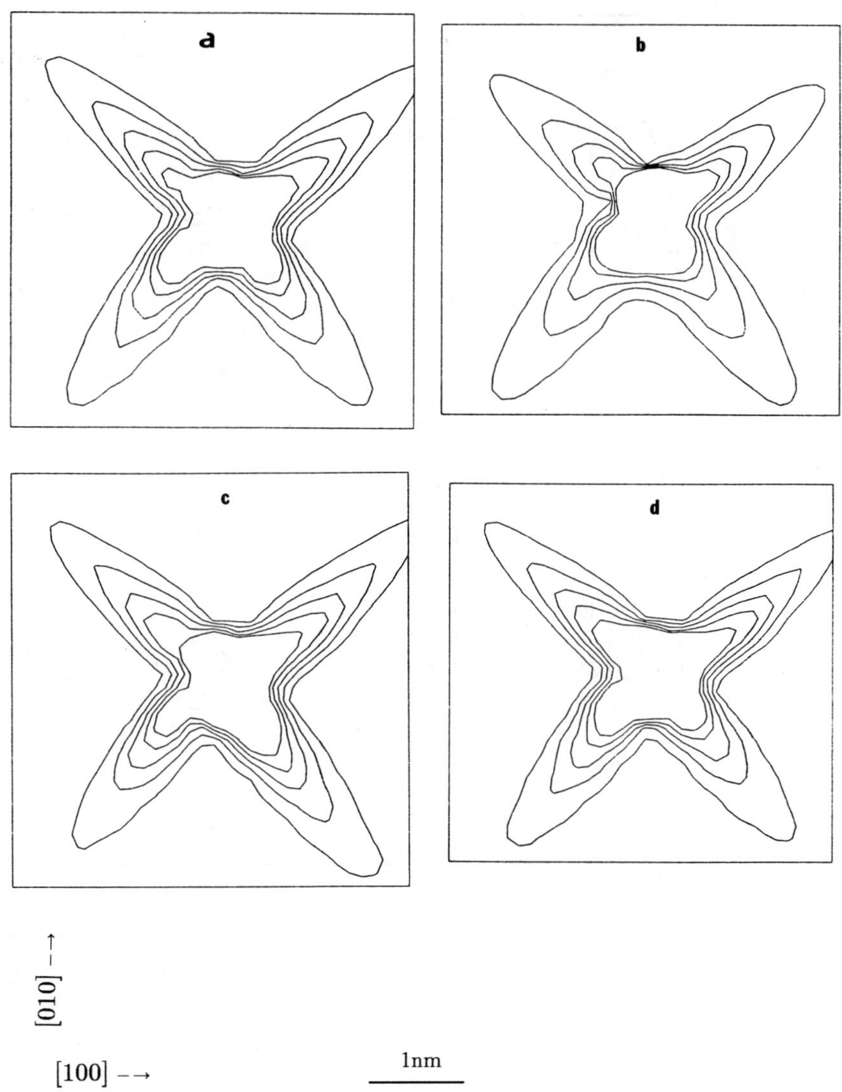

Figure 1: Core structure of a pure edge [100] dislocation along [001] with point defects. (a) Stoichiometric perfect alloy. (b)Core with a row of Ni vacancies. (c)Core with a row of Ni in Al sites. (d)Core with a row of antisite defects.

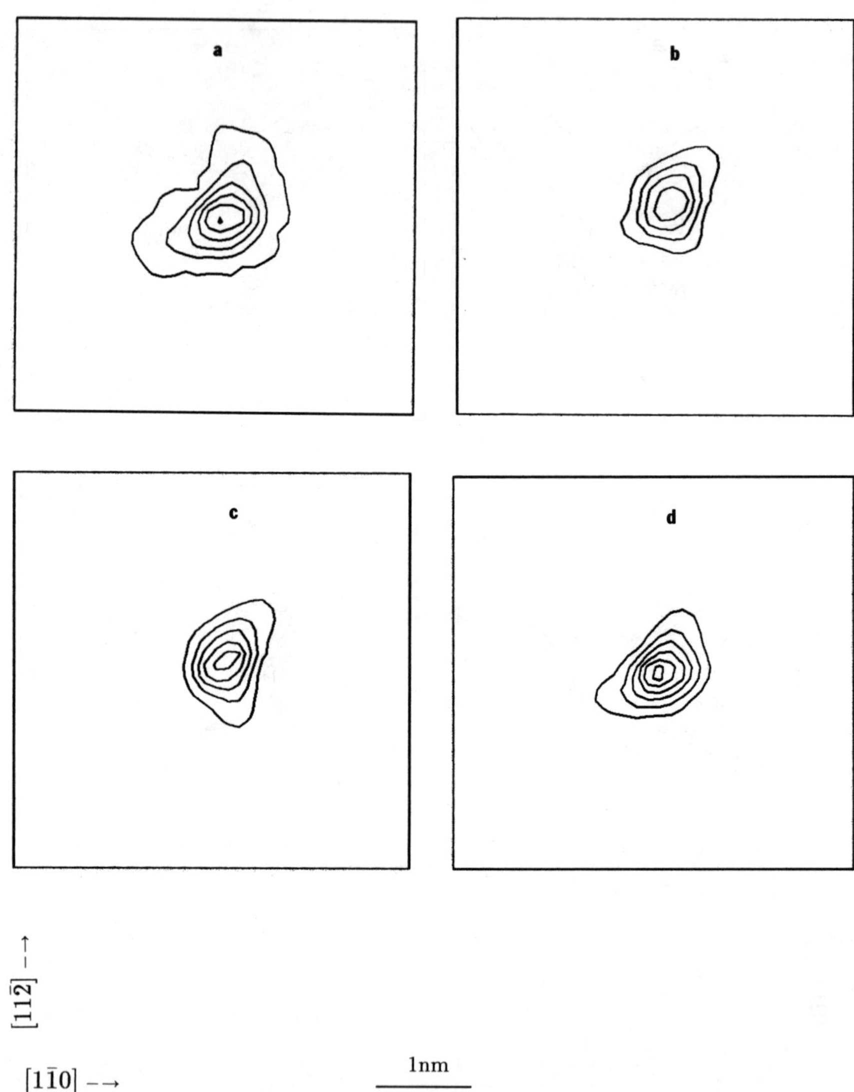

Figure 2: Core structure of a pure screw [111] dislocation along [111] with point defects. (a) Stoichiometric perfect alloy. (b)Core with a row of Ni vacancies. (c)Core with a row of Ni in Al sites. (d)Core with a row of antisite defects.

screw ⟨111⟩ dislocation.

In this case the effects are more important than for the ⟨100⟩ slip. In particular, the shape of the core changes. This is probably related to the fact that for ⟨111⟩ fault vectors, the gamma surface calculation shows that faults in various planes containing the ⟨111⟩ direction have very similar energies [6]. The shape of the dislocation core continues to be highly non-planar and very difficult to move. No spontaneous splitting of this dislocation occurred in the simulations.

We also studied the effects of Al located on Ni sublattices and of Al vacancies on the core structure of both dislocations. Although these point defects are higher energy defects, the results are very similar to those obtained for Ni in Al sites and Ni vacancies.

DISCUSSION

The possibilities of stabilizing ⟨111⟩ slip in NiAl through core changes due to point defects are predicted to be slim on the basis of atomistic simulations of dislocation core structure. The effects of point defects on the cores structure are very localized to the immediate vicinity of the point defect. In most cases these effects do not alter the basic shape and characteristic of the dislocation core. The observed effects may explain the very high rate of increase of the yield strain with stoichiometry deviations [4]. This rate was reported to be much higher than what could be explained on the basis of solution hardening theories. It seems possible that core effects contribute to the hardening process in off-stoichiometric alloys. In an experimental observation of dislocation cores of edge ⟨100⟩ dislocations [9] the core was found to be more extended in the off-stoichiometric alloys than in the stoichiometric material. The basic structure of the core which is extended in two {110} type planes, remained the same and in agreement with our simulations.

In analyzing several compounds with the B2 structure it is clear that ⟨100⟩ slip is preferred for compounds with higher values of the ordering energy whereas ⟨111⟩ slip is preferred for compounds with relatively lower values of the ordering energy. In compounds with intermediate values of the ordering energy both slip systems are observed in different temperature regimes. This fact is probably related to the fact that the complete ⟨111⟩ dislocation is very non-planar and difficult to move, as shown in our simulations. At lower values of the APB energy the dislocations split into partials that leave the APB in between and these partials are easier to move. In our simulations of metastable core structures, 1/2 ⟨111⟩ dislocations transformed to a more planar core structure under applied stress. It follows that in order to stabilize the ⟨111⟩ slip system one probably needs to have the partials stable with respect to the complete ⟨111⟩ dislocation. This is, in the light of the present simulations, probably difficult to accomplish through changes in the core structure produced by stoichiometry deviations. This is in agreement with the known fact that the ⟨100⟩ slip system is still stable for non-stoichiometric alloys.

The present results suggest that alloying is more likely to introduce qualitative changes in mechanical behavior through changes in the planar fault energies, (the APB in this case) than through local changes in core structure. Furthermore, the question always remains as to the possibility of the atmosphere of point defects following the dislocation line as it moves in the deformation process.

It seems that in order to stabilize ⟨111⟩ slip it is necessary to decrease the value of the APB energy. This can be attempted through alloying with a third element that would substitute for Al. In our previous studies of planar faults in the Ni-Al system it was found that the primary relaxation observed in the APB is the movement from the Al-Al first decreasing the APB energy would be alloying with a third element that would substitute for Al and be a smaller atom.

Acknowledgements

This work was supported by the Office of Naval Research, Division of Materials Science and the National Science Foundation, FAW Program.

REFERENCES

1. T. Khan, P. Caron, and S. Naka, in *High Temperature Alumides and Intermetallics*, edited by S. H. Whang, C. T. Liu, D. P. Pope, and J. O. Stiegler, page 219, TMS, Warrendale, PA 15086, 1990.

2. R. Noebe, R. Bowman, J. Kim, M. Larsen, and R. Gibala, in *High Temperature Aluminides and Intermetallics*, edited by S. Whang, C. Liu, D. Pope, and J. Stiegler, page 271, The Minerals, Metals and Materials Society, 1990.

3. M. Loretto and R. Wasilewski, Phil. Mag. **23**, 1311 (1971).

4. K. Vedula and P. S. Khadkikar, in *High Temperature Alumides and Intermetallics*, edited by S. H. Whang, C. T. Liu, D. P. Pope, and J. O. Stiegler, page 197, TMS, Warrendale, PA 15086, 1990.

5. I. Baker and P. Munroe, in *High Temperature Alumides and Intermetallics*, edited by S. H. Whang, C. T. Liu, D. P. Pope, and J. O. Stiegler, page 425, TMS, Warrendale, PA 15086, 1990.

6. D. Farkas and C. Vaihle, Journal of Materials Research , to be published.

7. D. Farkas, R. Pasianot, D. B. Miracle, and E. J. Savino, Proceedings of the Materials Research Society Symposia **213**, 2177 (1991).

8. R. Pasianot, Z. Xie, D. Farkas, and E. Savino, Scripta Materialia , to appear.

9. M. Crimp, Materials Science and Engineering , to be published.

PLASTIC ANISOTROPY OF Ti_3Al SINGLE CRYSTALS

Y. UMAKOSHI, T.NAKANO, K.SUMIMOTO AND Y.MAEDA
Department of Materials Science and Engineering, Faculty of Engineering, Osaka University, 2–1, Yamada–Oka, Suita, Osaka 565, Japan

ABSTRACT

The orientation and temperature dependence of the operative slip systems and critical resolved shear stress (CRSS) were investigated in Ti_3Al single crystals containing 24.4at%Al and 33.0at%Al. Prism $\{10\bar{1}0\}<1\bar{2}10>$ slip occurs preferentially in comparison with basal $(0001)<1\bar{2}10>$ and pyramidal $\{11\bar{2}1\}<11\bar{2}\bar{6}>$ slip. The CRSS for prism and basal slip decreases with increasing temperature, while that for pyramidal slip exhibits positive temperature dependence having an anomalous peak around 500°C. The temperature dependence of the CRSS for these slip systems does not strongly depend on Al content. The ductility of Ti_3Al single crystals is influenced by operative slip systems. Coarse slip bands on the basal plane often act as a trigger for crack nucleation. The compositional deviation to the Al–rich side is harmful for ductility of Ti_3Al alloys.

INTRODUCTION

Ti_3Al is a very attractive material with potential application for high–temperature structural use because of its high strength at high temperatures and low density. It has the hexagonal–based $D0_{19}$ structure (α_2 phase) at low temperatures and exhibits strong plastic anisotropy. The prism $\{10\bar{1}0\}<1\bar{2}10>$, basal $(0001)<1\bar{2}10>$ and pyramidal $\{11\bar{2}1\}<11\bar{2}\bar{6}>$ slip systems can be activated in ordered Ti_3Al with a near stoichiometric composition depending on crystal orientation [1–3]. The critical resolved shear stress (CRSS) for pyramidal slip is much higher than that for prism slip and activation of pyramidal slip is limited only for an orientation very close to [0001]. Basal slip is also seldom observed due to a high CRSS [1–3]. Such a finite number of available activated slip systems curtails the material's use in structural components.

It is well known that TiAl alloys exhibit a superior strength–to–weight ratio and excellent oxidation resistance. In particular, a good balance of higher strength and better ductility is found in two phase TiAl alloys composed of a γ matrix and a small amount of α_2 in a lamellar structure [4–9]. Most recently, Umakoshi et al. found that the anisotropy of deformation mode of the α_2 phase strongly affects the plastic behavior of lamellar TiAl alloys and they suggested that the control of plastic anisotropy of the α_2 phase holds the key to improvement of ductility of these alloys [10, 11].

In this paper, the temperature and orientation dependence of the activated slip systems and mechanical properties of Ti_3Al were investigated using single crystals of a stoichiometric and an Al–rich off–stoichiometric Ti_3Al having a composition similar to that which is in equilibrium with the γ phase.

EXPERIMENTAL PROCEDURE

Master ingots of Ti_3Al containing 24.4at%Al and 33.0at%Al were prepared by melting high purity Ti and Al in a plasma arc furnace. Single crystals with dimensions of 10mm in diameter and 100mm long were grown by the floating zone method at a rate of 5mm/h. The rods of the crystals were homogenized at 1000°C for 72h, then cooled in a furnace to 500°C and subsequently annealed at 500°C for 168h to obtain the long–range ordered $D0_{19}$ structure. Compression specimens, approximately 2mmx2mm in cross–section and 5mm long were cut by spark machining. After electrolytic polishing, compression tests were performed on an Instron–type testing machine at temperatures between 20 and 950°C in a vacuum. Slip traces were observed with an optical microscope using Nomarski interference contrast and operative slip systems were determined. Cold rolling tests were carried out at room temperature to evaluate the ductility of Ti_3Al crystals using an oriented sheet approximately 2mm in thickness, 3mm in width and 7mm long.

RESULTS AND DISCUSSION

Figure 1 shows four selected compression axes which have favorable stress components for the pyramidal, basal and prism slips. The Schmid factors for these orientations on the various slip systems of Ti–33.0at%Al are given in Table 1.

Figure 2 shows the slip traces of Ti–33.0at%Al deformed at room temperature. The orientation dependence of the observed slip geometry in Ti–24.4at%Al is similar to that in Ti–33.0at%Al [3]. For the specimen with orientation A, pyramidal $\{11\bar{2}1\}<11\bar{2}\bar{6}>$–type slips were activated at temperatures between 20 and 900°C but they were observed only at limited orientations very close to [0001]. Basal $(0001)<1\bar{2}10>$ slip was primarily observed at orientation B and prism slip appeared with further plastic flow. The activation of the basal slip is limited because of its high Peierls stress. Specimens with orientations C and D deformed by prismatic $\{10\bar{1}0\}<1\bar{2}10>$ slip, the dominant deformation mode in Ti_3Al crystals. The orientation dependence of the operative slip systems is similar at all test temperatures.

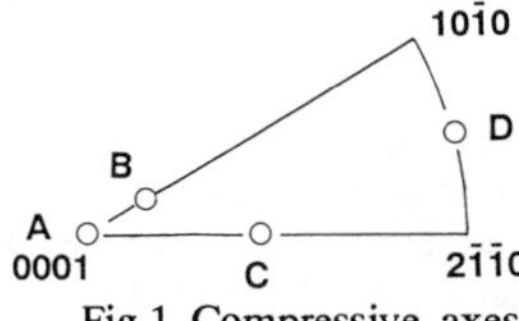

Fig.1 Compressive axes

Compressive Axis	Slip system		
	$\{10\bar{1}0\}<1\bar{2}10>$	$(0001)<1\bar{2}10>$	$\{11\bar{2}1\}<11\bar{2}\bar{6}>$
A	0	0	0.45
B	0.027	0.22	0.49
C	0.22	0.50	0.28
D	0.50	0	0.42

Table 1 Schmid factors for the various slip systems

Fig.2 Slip traces on Ti–33.0at%Al crystals deformed at room temperature.
(a) orientation A, (b) orienatation B, (c) orientation C, (d) orientation D

The difference in temperature dependence of the CRSS for the three slip systems is very striking. Figure 3 shows the CRSS for {10$\bar{1}$0}<1$\bar{2}$10>–slip of Ti–24.4at%Al and Ti–33.0at%Al single crystals as a function of temperature. The CRSS for prism slip of both crystals decreases monotonically with increasing temperature although compositional deviation to the Al–rich side strengthens Ti_3Al crystals. The CRSS for prism slip of both alloys obeys Schmid's law at all temperatures [3]. On the prism plane a <1$\bar{2}$10>–type dislocation is thought to dissociate into two superpartials bound by an APB which is always stable. The core configuration of the superpartials controls the plastic behavior of the prism slip.

According to our recent results [3, 12], additions of manganese or vanadium which are thought to decrease the stacking fault energy in Ti_3Al provide a significant violation of Schmid's law for prism slip. The CRSS increases with an increase in the stress component for (0001)<1$\bar{2}$10>–slip relative to {10$\bar{1}$0}<1$\bar{2}$10>–slip. This suggests that a part of the core of <1$\bar{2}$10>–type superpartials may extend on the basal plane forming a multi–layer fault, and this extended fault may act as an obstacle to the motion of the entire dislocation on the prism plane. However, in binary Ti_3Al alloys <1$\bar{2}$10> dislocation can easily move on the prism plane under a low applied stress and it may form a planar core with glissile configuration on the prism plane.

The CRSS for (0001)<1$\bar{2}$10>–slip is much higher than that for prism slip as shown in Fig.4. According to recent results of atomistic calculations of core structures in the $D0_{19}$ lattice [13], the 1/3<1$\bar{2}$10> superpartial on the basal plane forms two alternative sessile and glissile core configurations. One of them is non–planary with some parts of the core spread onto the prism plane and becoming sessile, while the second is a planar core spread on the basal plane. The high CRSS for basal slip suggests the existence of the non–planar core configuration in Ti_3Al alloys. The coarse and inhomogeneously distributed slip bands observed on the basal plane may be due to this non–planar core. The CRSS for the basal slip maintains a high value up to high temperatures where the CRSS decreases rapidly and the ductility is remarkably improved. Compositional deviations to the Al–rich side from stoichiometry shifts the temperature of the sharp drop in CRSS to higher temperatures, although there is no significant difference in the CRSS of

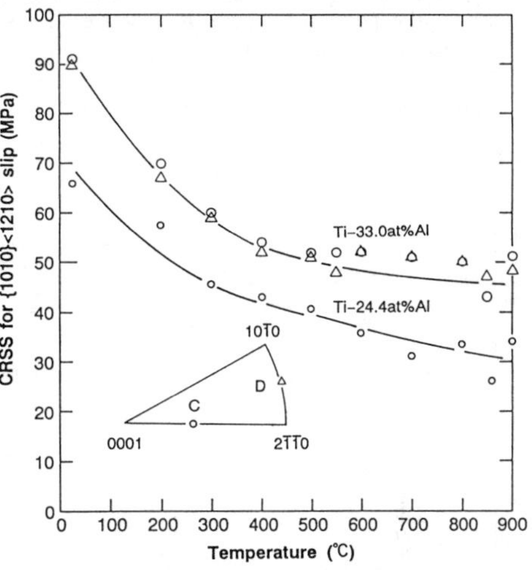

Fig.3 CRSS for prism slip of Ti_3Al as a function of temperature.

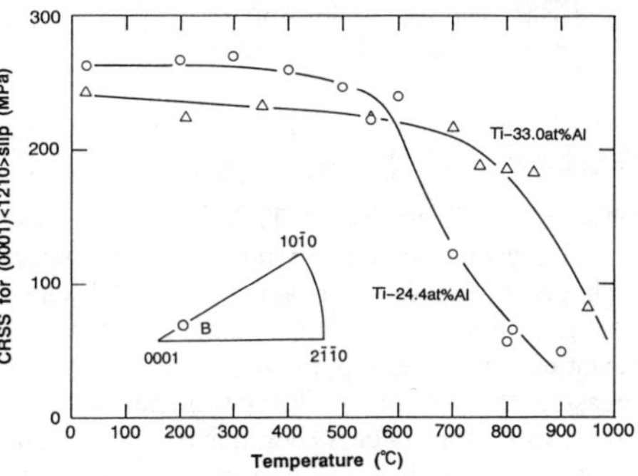

Fig.4 CRSS for basal slip of Ti_3Al as a function of temperature.

Ti–24.4at%Al and Ti–33.0at%Al crystals below the transition.

Figure 5 shows the temperature dependence of the CRSS for pyramidal {11$\bar{2}$1}<11$\bar{2}\bar{6}$> slip. The CRSS increases rapidly with increasing temperature, reaching an anomalous peak around 500°C. No significant difference can be found in the shape of the CRSS–temperature curves, or the anomalous peak temperature, although increasing Al content strengthens the crystals. In the $D0_{19}$ structure a 1/3<11$\bar{2}\bar{6}$> superlattice dislocation can be dissociated into four superpartials whose climb dissociation could be responsible for the anomalous strengthening [1, 3]. The anomalous strengthening is very sensitive to the addition of alloying elements. In the ternary Ti_3Al–V alloy the extent of positive temperature dependence of the CRSS is decreased [3]. The addition of vanadium, which is thought to decrease the stacking fault energy, may affect the core structure and the motion of <11$\bar{2}\bar{6}$>–type dislocations by changing the profile of the γ–surface and the restoring force. A similar behavior in stoichiometric and Al–rich off–stoichiometric Ti_3Al suggests that small changes in the degree of long–range order can affect the anomalous strengthening mechanism based on climb dissociation of <11$\bar{2}\bar{6}$>–type superpartials. An increase in the anomalous peak is necessary for structural use at high temperatures. Since the increase of about 100°C in the peak temperature was obtained in ternary Ti_3Al–Nb and Ti_3Al–Mn single crystals, Nb and Mn must be effective alloying elements to improve the strength of Ti_3Al at high temperatures [12, 14].

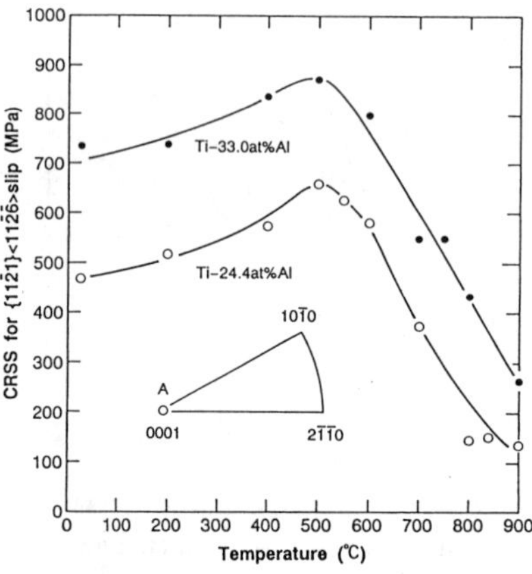

Fig.5 CRSS for pyramidal slip of Ti_3Al as a function of temperature.

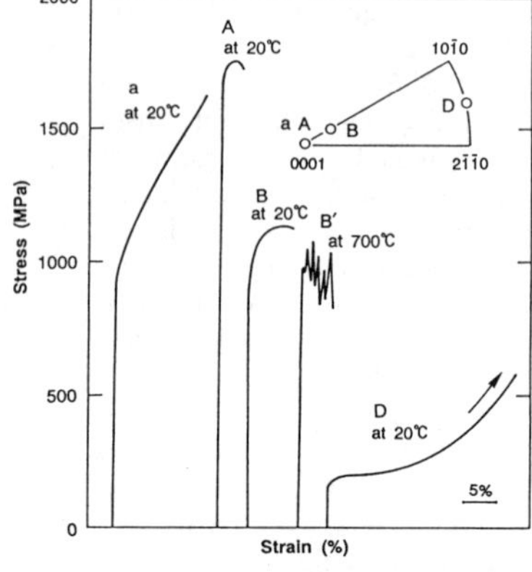

Fig.6 Compressive stress–strain curves of Ti–24.4at%Al(orientation a) and Ti–33.0at%Al (orientation A, B, B' and D).

The plastic behavior of Ti_3Al single crystals depends strongly on aluminium content and the crystallographic orientation of the sample. As shown in Fig.6, specimen D, in which prism slip is dominant, can be substantially deformed without cracking. At this orientation both compositions exhibit the same large deformability in compression. At orientation A, where pyramidal slip is operative, less plastic flow is obtained, and increasing the Al content from stoichiometry to the Al–rich side decreases the ductility. When basal slip occurs dominantly at orientation B and B', coarse slip bands are observed. Serrated flow appears in the stress–strain curves at high temperatures, particularly for Ti–33.0at%Al. At temperatures where the CRSS maintains a constant value (see Fig.4), the serrations are large and increase with increasing temperature. Also the magnitude of the serrations increase with increasing Al content. The motion of dislocations on the basal plane may be interrupted by solute atmospheres composed of segregating atomic species. The high stress concentrations at coarse slip bands often initiate cracks, especially in Ti–33.0at%Al, resulting in low plastic flow after yielding. At higher temperatures where the CRSS decreases rapidly, the ductility is remarkably improved, accompanied by fine and homogeneously distributed slip traces. The ductile–brittle transition temperature increases with deviation from stoichiometry.

On the prism plane <1$\bar{2}$10>–type dislocation can move easily and adequate plastic strain can be obtained in compression. When slip occurs only on prism planes, cold rolling may be possible. Figure 7 shows the variation of micro–Vickers hardness with reduction in thickness after cold rolling. The hardness was measured on the rolled surface. The rolling direction was chosen to be parallel to one of the [$\bar{1}2\bar{1}0$] Burgers vectors of the dislocations on the prism plane. The axis perpendicular to the rolled surface was [$10\bar{1}0$]. Slip occurs predominantly on the ($1\bar{1}00$) and ($01\bar{1}0$) planes and the density of the slip traces increases with increasing reduction in thickness accompanied by work–hardening. Interaction between both slip systems increases the work–hardening rate and finally the specimen breaks. Ti–24.4at%Al shows better deformability than Ti–33.0at%Al and it can be cold rolled up to 46% thickness reduction without annealing. The stoichiometric composition is the most ductile, even when deformation occurs by prism slip.

Annealing between cold rolling intervals is required to obtain thin sheets of Ti_3Al. After annealing at 900°C for 10000s, the hardness of rolled specimens is restored to the initial value prior to rolling. Further rolling is possible since only sub–boundaries initiated at slip bands are formed. New grains, due to recrystallization, did not appear.

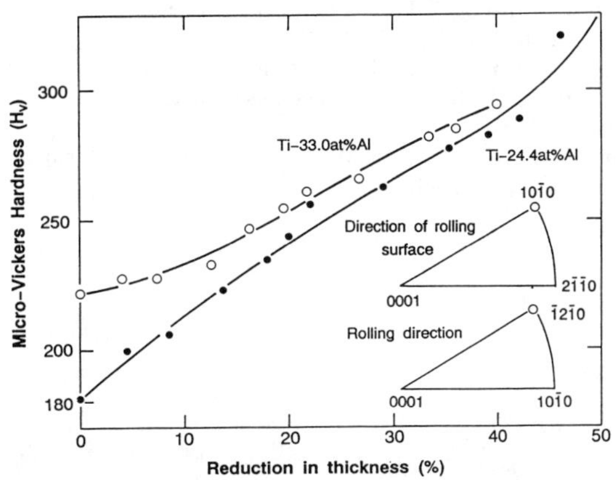

Fig.7 Variation of micro–Vickers hardness of cold rolled Ti_3Al with reduction in thickness.

In two phase TiAl alloys containing the α_2 phase, cracks often nucleate in the α_2 phase and/or at the interface between the γ and α_2 phases. Ti–24.4at%Al is more ductile than Ti–33.0at%Al. Since the α_2 phase in TiAl alloys is Al–rich, the addition of alloying elements which change the solibility limit of the (α_2 + γ) two phase region to near stoichiometric composition may improve the ductility of TiAl alloys.

ACKNOWLEDGEMENTS

Y. Umakoshi would like to thank the IKETANI SCIENCE AND TECHNOLOGY FOUNDATION for its research grant. This work was partly supported by a Grant–in Aid for scientific research and development from the Ministry of Education, Science and Culture of Japan.

REFERENCES

1. Y. Minonishi and M.H. Yoo, Phil. Mag. Lett. **61**, 203(1990).
2. Y. Minonishi, Phil. Mag. **63A**, 1085(1991).
3. Y. Umakoshi, T. Nakano, T. Takenaka, K. Sumimoto and T. Yamane, Acta Met. et Mater. in press.
4. J.B. McAndrew and H.D. Kessler, J. Met. **8**, 1348(1956).
5. H.A. Lipsitt, D. Shechtman and R.E. Schafrik, Met. Trans. **6A**, 1991(1975).
6. T. Fujiwara, A. Nakamura, M. Hosomi, S.R. Nishitani, Y. Shirai and M. Yamaguchi, Phil. Mag. **61A**, 591(1990).
7. S.R. Nishitani, M.H. Oh, A. Nakamura, T. Fujiwara and M. Yamaguchi, J. Mater. Res. **5**, 484(1990).
8. H. Inui, A. Nakamura, M.H. Oh and M. Yamaguchi, Proc. of Inter. Sympo. on Intermetallic Compounds–Structure and Mechanical Properties– (edited by O. Izumi), p.495 (1991), Japan Inst. of Metals, Sendai.
9. Y. Umakoshi, T. Nakano and T. Yamane, Mater. Sci. and Engn. **152A**, 81(1992).
10. Y. Umakoshi and T. Nakano, ISIJ Inter. **32**, 1339(1992).
11. Y. Umakoshi and T. Nakano, Acta Met. et Mater. in press.
12. T. Nakano, K. Sumimoto and Y. Umakoshi, Proc. 1992 Tokyo Meeting Japan Inst. of Metals, Sendai, the Japan Inst. of Metals, p.343(1992).
13. V. Vitek, M. Khantha, J. Cserti and Y. Sodani, Proc. of Inter. Sympo. on Intermetallic Compounds–Structure and Mechanical Properties– (edited by O. Izumi), p.3 (1991), Japan Inst. of Metals, Sendai.
14. T. Nakano and Y. Umakoshi, Proc. 1992 Toyama Meeting Japan Inst. of Metals, Toyama, the Japan Inst. of Metals, p.288(1992).

FLOW BEHAVIOR OF THE $L1_2$ $(Al,Fe)_3Ti$ SINGLE CRYSTALS

Z.L. Wu, D.P.Pope and V.Vitek
Department of Materials Science and Engineering,
University of Pennsylvania, Philadelphia, PA 19104-6202.

ABSTRACT

The compressive flow behavior of single crystalline $L1_2$ $Al_{67}Fe_8Ti_{25}$ was investigated as a function of temperature and orientation at temperatures from 77K to about 1250K, using specimens with compressive axes orientated near [001], $[\bar{1}13]$, [011], $[\bar{1}22]$ and $[\bar{1}11]$. The operating slip systems seen in these specimens after 0.4% plastic deformation are predominantly of the octahedral type at all temperatures, even in near-$[\bar{1}22]$ and $[\bar{1}11]$ specimens in which the Schmid factors for the primary cube slip system are larger than that for the primary octahedral slip system. The yield stress increases rapidly with decreasing temperature at low temperatures, while it decreases gradually from room temperature to higher temperatures. The critical resolved shear stress (CRSS) on the $[\bar{1}01](111)$ slip system does not seem to be orientation-dependent over a wide range of temperatures, except at temperatures from 1050K to 1250K where the CRSS exhibits a mild orientation-dependence. Fracture tests at room temperature were also conducted. No special orientation-dependence of the ductility was observed.

INTRODUCTION

The flow behavior of various $L1_2$ titanium trialuminides has been studied by many groups, however, the results often differ. For instance, Kumar and Pickens observed an anomalous flow stress peak at around 700K in polycrystalline $Al_{22}Fe_3Ti_8$ [1]. Similar peaks were also reported in $Al_{67}Mn_8Ti_{25}$ [2], $Al_{67}Cr_8Ti_{25}$ [3], and $Al_5Cu_3Ti_2$ [4]. Wu et al, on the other hand, did not observe an anomalous peak in Fe-modified alloys [5-7]. Instead, they observed a flow behavior which is similar to that of $L1_2$ Pt_3Al-type alloys [8], and the flow stress vs temperature curve at intermediate temperatures is plateau-like. The Pt_3Al-type flow behavior has also been recently seen in $Al_{67}Cr_8Ti_{25}$ and $Al_{66}Mn_9Ti_{25}$ [9], results that, however, differ from the ones reported earlier in [2, 3], even though the compositions of the alloys used in these studies were about the same.

The structure of [110] superdislocations in deformed alloys has also been investigated by several groups using weak beam TEM. Interestingly, some of the studies showed that the superdislocations are dissociated into antiphase boundary (APB)-associated superpartials, and lie on octahedral slip planes at low and intermediate temperatures [10-12], and mainly on cube slip planes at high temperatures [13]. Such a transition in slip plane from low temperatures to high temperatures has been seen in $L1_2$ Ni_3Al-type alloys and is believed to be resulted from an anisotropy of the APB-energy. According to the so-called PPV model [14], the anisotropy of the APB-energy on the two different slip planes provides a driving force for the cross-slip pinning of <110>{111} superdislocations onto the cube planes (with the aid of thermal activation). The cross-slip pinning in turn, forms obstacles to dislocation motion and leads to the strong anomalous peak seen in this type of alloy [15]. The reported anomaly in Al_3Ti-based alloys might also be explained in this way. However, in comparison with the strong anomaly seen in

Ni_3Al (the peak yield stress can be five times as high as the low temperature yield stress [16]), the reported anomaly in $L1_2$ Al_3Ti is substantially weaker (the peak yield stresses are commonly no more than 5 to 20% higher). The big difference seems to imply that the mechanisms which control the flow behaviors of the two different $L1_2$ alloys are not the same.

Due to the limited size of the single $L1_2$ phase field at low temperatures [17], the deformation behavior of the $L1_2$ Al_3Ti-based alloys is often altered by the influence of various second phases. The purpose of this study is to investigate the inherent deformation behavior of the $L1_2$ Al_3Ti-based alloys, using single crystalline $L1_2$ $Al_{67}Fe_8Ti_{25}$, a nearly single phase alloy. The flow behavior and operating slip systems are characterized over a wide range temperatures.

EXPERIMENTAL PROCEDURES

High aluminum buttons, iron flakes and titanium rods were used to cast an ingot of nominal composition $Al_{67}Ti_{25}Fe_8$, weighting about 250g, in an induction furnace under an argon atmosphere. A part of the ingot was pulverized and used as a starting material for crystal growth in a Bridgman furnace. The as-grown crystal was homogenized at 1100°C for 60 h in an argon atmosphere followed by slow cooling at a rate about 65°C/h. The growth axis of the crystal was found to be only a few degrees away from the [011] orientation. Compression specimens of dimension 3x3x5 mm, with compressive axes near [001], [$\bar{1}$13], [011], [122] and [$\bar{1}$11] directions, were prepared from the crystal. The phase distribution in the crystal was identified using quantitative energy dispersive x-ray analysis in the SEM. Compression tests were performed at a nominal strain rate of about $1.7x10^{-4}$ s^{-1} on an Instron Universal Testing Machine under a vacuum of $5x10^{-5}$ torr, except for the tests at 77K and 297K which were conducted in liquid nitrogen and air, respectively. Operating slip systems activated were determined using the two-surface slip trace analysis technique.

RESULTS AND DISCUSSION

MICROSTRUCTURES

The crystal structure of the alloy was examined using the x-ray powder diffraction technique, and confirmed to be of $L1_2$ type as was expected. Small quantities of three types of second phases were observed in the specimens: Al_3Ti, $(Al,Ti)_3Fe$ and Ti_2NAl. The existence of the first two is not surprising, since they have already been identified in the high temperature isotherms of the Al-Ti-Fe ternary system [18]. The formation of the third phase seems to be related to N existing in the Ti starting material, and was previously observed in other Fe-modified alloys [5,6]. In addition to these second phases, a certain amount of porosity is also present in these specimens. Some of the porosity apparently formed during interdendritic solidification, while others are probably Kirkendall pores. A more detailed discussion on these defects is given in [17].

OPERATING SLIP SYSTEMS

Fig. 1(a-d) show the slip traces observed from the two orthogonal surfaces of the specimens of four different orientations deformed to 0.4% plastic strain at 77K, 300K and 825K. The slip

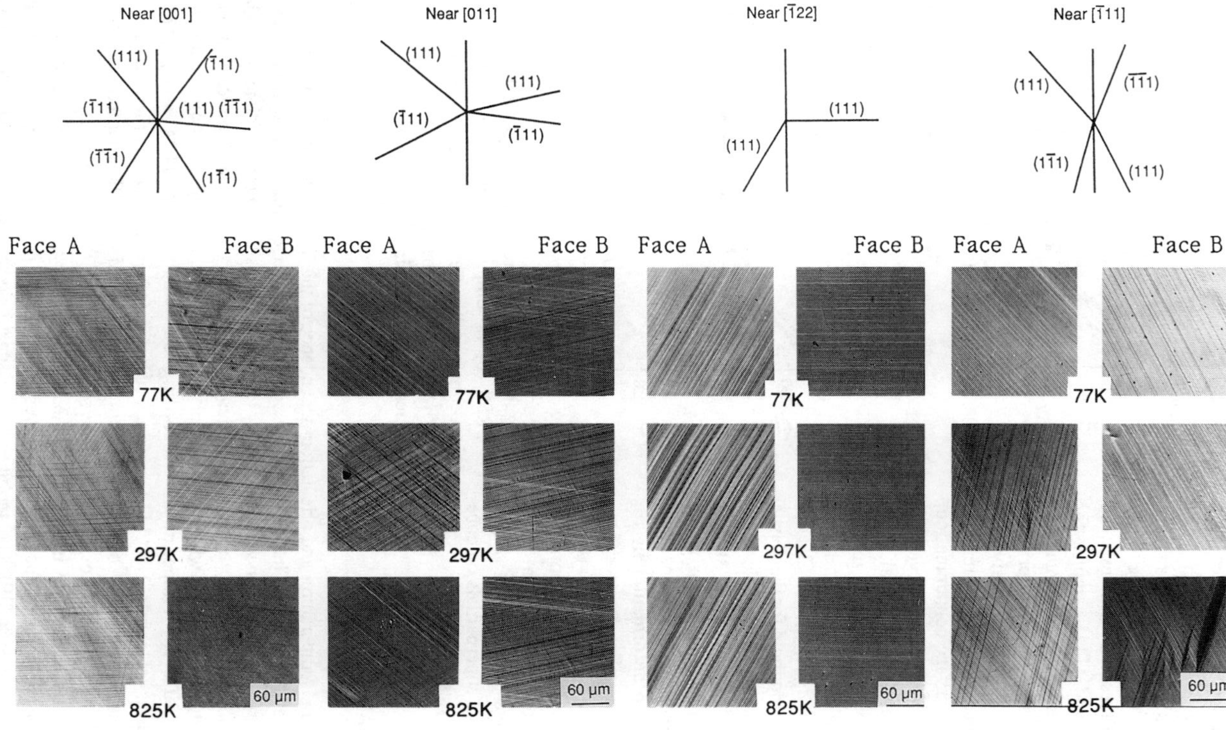

Figure 1. Slip traces seen on the two orthogonal surfaces of specimens with the orientations as indicated. The operating slip systems are predominantly octahedral type at all temperatures.

bands seen are fine, straight and uniformly distributed across the surfaces of the specimens, indicating a homogeneous deformation. The operating slip systems are predominantly octahedral type in all specimens at all test temperatures, even in near-[I12] and [I11] samples for which the Schmid factors are higher for primary cube slip than for primary octahedral slip. Slip trace analysis at higher deformation temperatures was difficult due to surface contamination, however, the slip bands observed are still predominantly octahedral type. These observations are totally consistent with those made on single crystalline $Al_{67}Cr_8Ti_{25}$ [18]. Clearly, cube slip is not important for the plastic deformation of this nearly single phase $L1_2$ alloy. Why some TEM studies have shown that the majority of dislocations seen in specimens after high temperature deformation lie on cube planes is not well understood. One possibility is that the dislocations seen are immobile, since computer simulations of dislocation core structures have shown that APB-coupled <110>/2 superpartials on cube planes are always sessile [19].

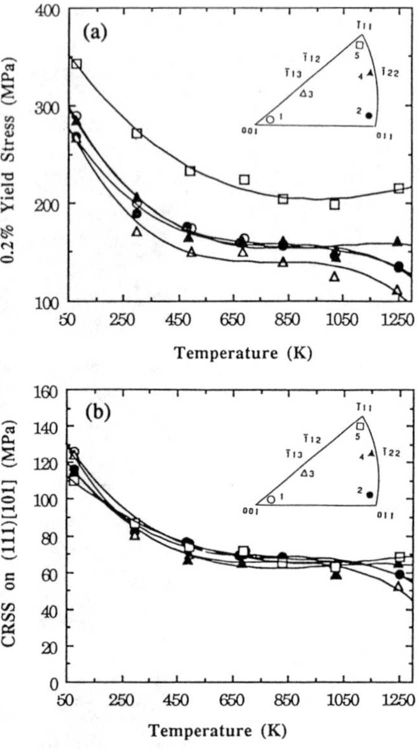

Figure 2. Yield stress (a) and corresponding CRSS (b) vs temperature curves measured on specimens with the five different orientations shown. For specimens 4 and 5 the CRSS increase slightly at high temperatures.

FLOW BEHAVIOR

Fig.2(a) shows the temperature- and orientation-dependence of the flow behavior at temperatures between 77K and 1250K. The yield stress is seen to increase rapidly with decreasing temperature at low temperatures, while is not very temperature-sensitive at intermediate and high temperatures. The two distinct responses of the flow stress to temperature implies that the <110>{111} superdislocations which carry the plastic deformation may possess different core structures at different temperatures. According our weak beam TEM study [18], superdislocations of the <110>{111} type are dissociated into SISF-coupled <112>/3 superpartials at temperatures below room temperature, and into APB-coupled <110>/2 superpartials at high temperatures. Because of the sessile nature of the <112>/3 partials [19], the motion the of superdislocation is difficult but can be thermally aided, leading to a strongly negatively temperature dependence of the flow stress. On the other hand, the core structure of the <110>/2 partials is glissile [19]. The mobility of the superdislocations dissociated into APB-

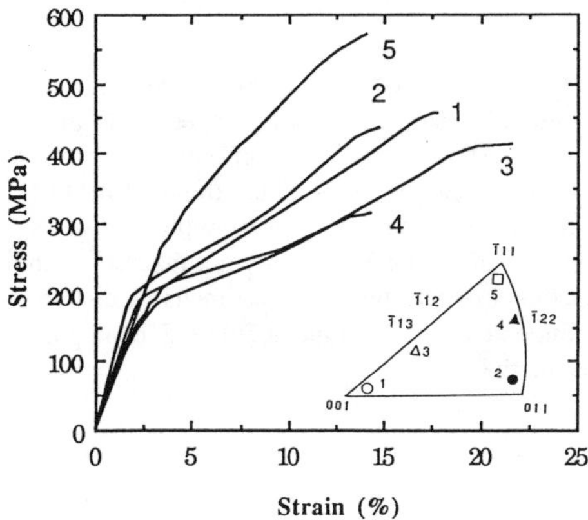

Figure 3. The deformation curves measured at room temperature on specimens of five different orientations.

coupled <110>/2 superpartials should not be temperature-sensitive, and therefore, an athermal flow behavior at intermediate and high temperatures is expected. In comparison with the flow behavior previously seen in other Fe-modified Al_3Ti-based alloys containing high Ti contents [5-7], the flow stress observed here is much lower and does not exhibit a sharp decrease at high temperatures. The differences result from the different amount of Ti contents in these alloys. In the alloys containing high Ti contents Al_2Ti precipitates exist which form a dense array on the cube planes of the $L1_2$ matrix and cause a strong hardening effect [18,20]. At high temperatures, however, these precipitates dissolve and the flow stress decreases rapidly.

Fig.2(b) shows the CRSS on $[\bar{1}01](111)$ primary slip system as a function of temperature. The difference in CRSS for different orientations is small at temperatures below 1050K, and is likely due to activation of multiple slip rather than a violation of Schmid's law. At temperatures higher than 1050K the CRSS shows a mild orientation-dependence. For specimens oriented at [001], [011] and $[\bar{1}13]$, the CRSS continually decreases, while for near-$[\bar{1}22]$ and $[\bar{1}11]$ specimens the CRSS shows a small increase. Since the Schmid factors for the specimens of the latter two orientations are higher for the primary $[1\bar{1}0](001)$ cube slip system than for the octahedral systems, the mild increase in flow stress may therefore be explained in terms of cross-slip pinning mechanism, similar to the one which successfully explains the anomalous flow stress increase in $L1_2$ Ni_3Al. The difference, however, is that in Ni_3Al such a cross-slip pinning starts at a much lower temperature and causes a strong anomalous peak between 600°C and 800°C, while in $L1_2$ Al_3Ti the cross-slip pinning starts at a high temperature, approaching the melting temperature, and causes only a small increase. From these results it is clear that the flow behavior of single phase $L1_2$ Al_3Ti-based alloys resemble that of the Pt_3Al-type alloy. The anomalous peak reported by a number of investigators does not appear to be inherent to this type of alloy, but could be caused by some other reason, for example, a dynamic strain aging effect [18,21].

Fig. 3 shows some of stress-strain curves tested at ambient temperature for specimens of five different orientations. No particular orientation-dependence of the ductility was observed. It has been seen that second phase $(Al,Ti)_3Fe$, and especially porosity, are very destructive. Cracks are easily initiated at these macroscopic defects and result in premature failure. Therefore, the deformation curves presented here are expected to show a lower ductility than for a defect-free alloy of the same composition.

CONCLUSIONS

The operating slip systems in the $Al_{67}Fe_8Ti_{25}$, an alloy of nearly single phase $L1_2$, are predominantly octahedral type at all temperatures in single crystalline specimens of all orientations. The flow behavior of the alloy is similar to that of Pt_3Al. The different behaviors of the flow stress in the low and high temperature regions imply that the <110>{111} superdislocations possess different dislocation core structures at these temperatures. The anomalous peak in flow stress reported by a number of groups is probably not inherent to the structure, but is probably caused by other reasons. Fracture tests on specimens of different orientations did not reveal any particular orientation dependence of the ductility. Second phase $(Al,Ti)_3Fe$, and especially porosity, are very destructive.

ACKNOWLEDGEMENTS

The work was supported by the AFOSR under Grant AFOSR-89-0062. Research facilities were provided by the LRSM supported by the NSF MRL program under Grant DMR88-19885.

REFERENCES

1. K.S. Kumar and J.R. Pickens, Scrip. Metall., **22**,1015, 1988.
2. S. Zhang, J.P. Nic, W.W. Milligan, and D.E. Mikkola, Scrip. Metall., **24**, 57, 1990.
3. H. Mabuchi, K.Hirukawa and Y. Makayama, Scrip. Metall., **23**, 1761, 1989.
4. M.B. Winnicha, and R.A. Varin, Scrip. Metall., **24**, 611, 1990.
5. Z.L. Wu, D.P. Pope, and V. Vitek, High Temperature Ordered Intermetallic Alloys, IV, (ed. by L. Johnson, D.P. Pope, and J.O. Stiegler), MRS Symp. **213**, MRS Pittsburgh, 487, 1991.
6. Z.L. Wu, D.P. Pope, and V. Vitek, Scrip. Metall., **24**, 2187, 1990.
7. Z.L. Wu, D.P. Pope, and V. Vitek, Scrip. Metall., **24**, 2191, 1990.
8. D.M. Wee, D.P. Pope, and V. Vitek, Acta Metall., **32**, No.6, 829, 1984.
9. K.S. Kumar, S.A. Brown, and J.D. Whittenberger, High Temperature Ordered Intermetallic Alloys, IV, (ed. by L. Johnson, D.P. Pope, and J.O. Stiegler), MRS Symp., **213**, MRS Pittsburgh, 481, 1991.
10. E.P. George, J.A. Horton, W.D. Porter, and J.H. Schneibel, J. Mater. Res., **5**, 1639, 1990.
11. D.G. Morris, Scrip. Metall. **25**, 713, 1991.
12. K.S. Kumar, S.A. Brown, Philos. Mag.A, **65**, No.1, 91, 1992.
13. H.Inui, D. E. Luzzi, W. D. Porter, D. P. Pope, V. Vitek, and M. Yamaguchi, Phil. Mag. A, **65**, No. 1, 245, 1992.
14. V. Paidar, D.P. Pope, and V. Vitek, Acta Metall., **32**, 435, 1984.
15. R.A. Mulford, D.P. Pope, Acta Metall., **21**, 1375, 1973.
16. D.M. Wee and T. Suzuki, Trans. Japan Inst. Metals, **20**, 634, 1979.
17. Z.L. Wu, D.P. Pope, and V. Vitek, submitted to this symposium.
18. Z.L. Wu, Ph.D thesis, University of Pennsylvania, 1992.
19. V. Vitek, M. Khantha, J. Cserti and Y. Sodani, Proc. JIMIS-6, 1991.
20. L. Potez, A.Loiseau, S. Naka and G. Lapasset, J. Mater. Res., **7**, No. 4, 876, 1992.
21. L. Potez, G. Lapasset, and L. P. Kubin, Scrip. Metall., **26**, 841, 1992.

DEFORMATION MECHANISMS IN $Be_{12}X$ COMPOUNDS

S. SONDHI*, R. G. HOAGLAND*, J. P. HIRTH*
J. L. BRIMHALL**, L. A. CHARLOT** and S. M. BRUEMMER**
*Pacific Northwest Laboratory, Richland, WA 99352
**Department of Mechanical & Materials Engineering,
Washington State University, Pullman, WA 99164-2920.

ABSTRACT

Dislocation structures have been examined, and active slip systems identified, in $Be_{12}Nb$ after compressive deformation at 20, 800, 900,1000 and 1200°C. A large number of slip systems are active at 1200°C, but these decrease significantly at temperatures below 1000°C. Dislocation structures at low temperatures are limited to 1/2<101]{101) partial dislocations either paired or creating isolated planar faults. Significant ductility is not observed until 1200°C when a second type of partial dislocation, 1/2<100]{011) is present. Dislocations observed in the body-centered tetragonal $Be_{12}X$ compounds (where X can be Nb, Ta, Mo, V, Fe etc.) have been modelled atomistically using molecular dynamics. Simulations corroborate the stability of these dislocation systems and indicate that the stacking faults associated with these partial dislocations have very low fault energy.

INTRODUCTION

Refractory metal beryllides, such as $Be_{12}X$, exhibit a unique combination of high melting point and high strength to weight ratio [1,2], which makes them a potential candidate for aerospace applications. Unfortunately, these compounds have a complex crystal structure which leads to poor low-temperature toughness. $Be_{12}X$ compounds have a body-centered tetragonal (bct) structure with 26 atoms per unit cell. Elements such as Nb, Mo, Ta, Ti, Cr, and V all form isostructural $Be_{12}X$ compounds with little change in lattice dimensions. The shortest lattice vectors for slip in bct $Be_{12}X$ are <001] and 1/2<111] with lengths of 0.43 and 0.56 nm, respectively. On the basis of self-energy considerations (proportional to b^2), these Burgers vectors are far more probable than the next shortest vectors, <100] at 0.74nm and 1/2<101] at 0.85 nm. Close-packed planes on which slip might be expected are {110), {101), {100) and {121).

Direct measurements of slip characteristics in $Be_{12}X$ compounds have been limited. Bruemmer et al. [3], investigated dislocation and fault structures in polycrystalline $Be_{12}Nb$ after compressive deformation at 1200°C. A high density of 1/2<101] and 1/2<100] dislocations were observed bounding planar faults on {101) planes. Partial dislocations and their faults were postulated to relate to a local phase transformation.

Lewis [1] reported slip traces near hardness indents in $Be_{12}Nb$ and proposed that low temperature slip may be possible on the basis of the zonal dislocation concepts of Kronberg [4]. More recent evaluations [5] of $Be_{12}Nb$ mechanical properties have shown no indications of macroscopic plasticity at temperatures below 800°C. Ductile-brittle transition temperatures of 770, 850 and 1100°C were documented from compression, bend and hot hardness tests, respectively. Thus, recent experiments have not revealed any macroscopic plastic deformation at low temperatures, contrary to the isolated observation of Lewis.

In the present work, dislocation structures in polycrystalline $Be_{12}Nb$ were examined and analyzed after compressive deformation at 20, 800, 900 and 1000°C. Comparisons are made to structures already documented at 1200°C. The 1/2<111], 1/2<101] and 1/2<100] dislocations have also been modelled atomistically using molecular dynamics. These calculations are useful in determining the equilibrium configurations of these dislocations. The results of these atomic scale calculations are presented here along with the dislocation structures in these compounds, obtained from the compression tests.

EXPERIMENTAL PROCEDURE

Samples for compressive loading were machined as 4x4x4-mm cubes or as 3-mm diameter cylinders, 3-mm in height. Deformation to plastic strain levels from 0.5 to 3.0% was produced in air using an Instron testing machine. The cross-head rate was controlled to give a strain rate of 1e-04 s^{-1}. Plastic deformation was not possible in specimens deformed at temperatures below 800°C. Therefore loads were applied equal to ~90% of the fracture stress (~2500 MPa) in compression. Thin slices were cut from the deformed samples, mechanically to a thickness of ~100 μm, and 3-mm disks were ultrasonically cut for TEM examination. Electron-transparent thin films were prepared by jet polishing in a solution of ethanol (93 vol%) and perchloric acid (7 vol%) at -40°C, at a potential adjusted to give a current of 100 mA.

Specimens were examined using a Phillips EM400T electron microscope at an operating voltage of 120 kV. The Burgers vectors (b) of observed dislocations imaged at various diffracting conditions (g) were determined by standard g . b analysis methods. In addition, contrast changes for deviations from the Bragg condition were documented for several diffracting vectors.

COMPUTATIONAL PROCEDURE

The computational procedure for determining the atomic interaction potentials and generating the equilibrium configurations of dislocations has been described in detail elsewhere [6], and only a very brief overview is given here. A schematic of the model used is shown in Fig. 1. It comprises of a cylindrical array of atoms divided in three concentric regions. The dislocation is contained in the innermost region, Region I. The atoms in Region I are free to move classically according to Newton's laws, while the atoms in Region II and Region III are held fixed in positions determined by the linear elastic displacement field of the dislocation. Presence of region II is necessary to be able to calculate the energies of boundary atoms in region I and region III is the infinite linear elastic continuum in which regions I and II are contained. A set of embedded atom method (EAM) potentials describes the interaction between atoms. Periodic boundary conditions were imposed along the axis of the cylindrical array which makes the length of the dislocation infinite.

A three dimensional array of atoms which are positioned according to the linear elastic field of the dislocation is produced. Although the linear elastic positions are also the equilibrium positions for atoms which are far from the dislocation, the same is not true for atoms in and near the core. The purpose of the calculation is to find the equilibrium arrangement of atoms in the core region. This is accomplished by moving the atoms in Region I by molecular dynamics, their motion governed by the EAM atomic interaction potentials. An analysis of the motion of the atoms during the calculation provides information about the behavior of the dislocation.

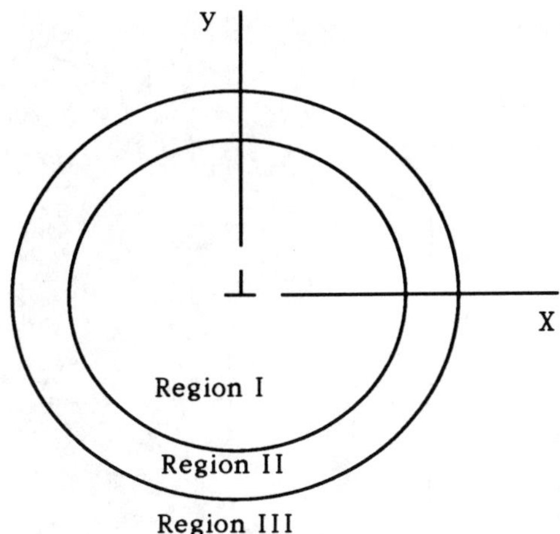

Figure 1. The Schematic of the model used. Atoms in Region I are free to move, whereas atoms in the outer regions are held fixed in positions determined by linear elasticity.

RESULTS AND DISCUSSIONS

The characterization of dislocation structures produced at 1200°C indicates a wide range of operating slip systems. Typical dislocation structures in $Be_{12}Nb$ after deformation at 1200°C are shown in Fig. 2(a). The observed dislocations are primarily partials bounding planar faults. 1/2<101]{101) and 1/2<100]{011) were found to be predominant slip systems and are also shown in Fig. 2(b) and 2(c). Slip systems, on which deformation has been documented at high temperatures, are listed in Table 1. It can be seen that more than enough slip systems are available in $Be_{12}Nb$ to satisfy the von Mises criterion for polycrystalline deformation at high temperature.

At 1000°C or lower temperatures, only 1/2<101]{101) partial dislocations were found to be present. These dislocations appear to be emitted from various interfaces and create long planar faults in the grain interiors similar to Fig. 2(c). In this case all faults were found to be tied to boundaries suggesting that the temperature is insufficient for the emission of the trailing partial. Furthermore, for compressive loading at room temperature, the individual bounding partials were found to be split into 1/4<101]{101) superpartials as shown in Fig. 3.

A number of slip systems, observed in the $Be_{12}X$ compounds in the TEM analysis have been modelled atomistically using molecular dynamics type calculations and some specific results which pertain to the predominant slip systems (as observed in the TEM studies) are discussed here.

The results of the calculation of the 1/2<111]{011) perfect dislocation in $Be_{12}X$ are presented in Figure 4. This is an energetically favorable dislocation from b^2 considerations. However, the results show that this dislocation splits into two edge superpartial dislocations

TABLE 1: Slip systems observed in TEM studies in $Be_{12}Nb$ after deformation at 1200°C.

Perfect dislocations:
$1/2\langle 111](101)$
$1/2\langle 111](121)$
$\langle 001](100)$
Partial dislocations:
$1/2\langle 101](101)$
$1/2\langle 101](121)$
$1/2\langle 101](010)$
$1/2\langle 100](011)$
$1/2\langle 100](001)$

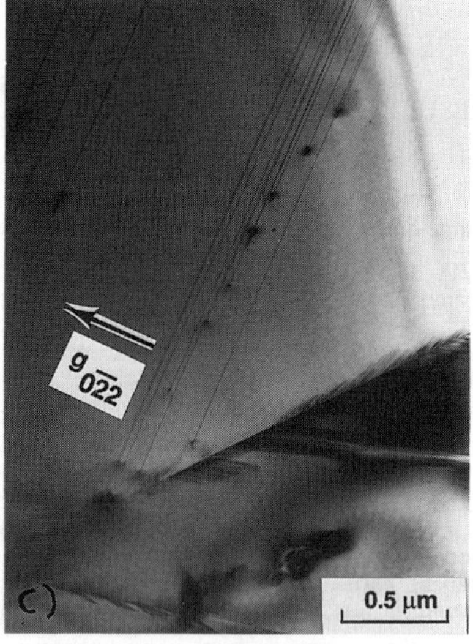

Figure 2: Dislocation structures in $Be_{12}Nb$ after high-temperature deformation at 1200°C. (a) typical dislocation structures in $Be_{12}Nb$, (b) extended (101) faults and short range (100) faults, and (c) long (101) faults with bounding partials. The (101) and the (100) faults are present in $Be_{12}Nb$ in high density.

Figure 3: 1/4<101] partial dislocation pairs observed in $Be_{12}Nb$ after deformation at 20°C.

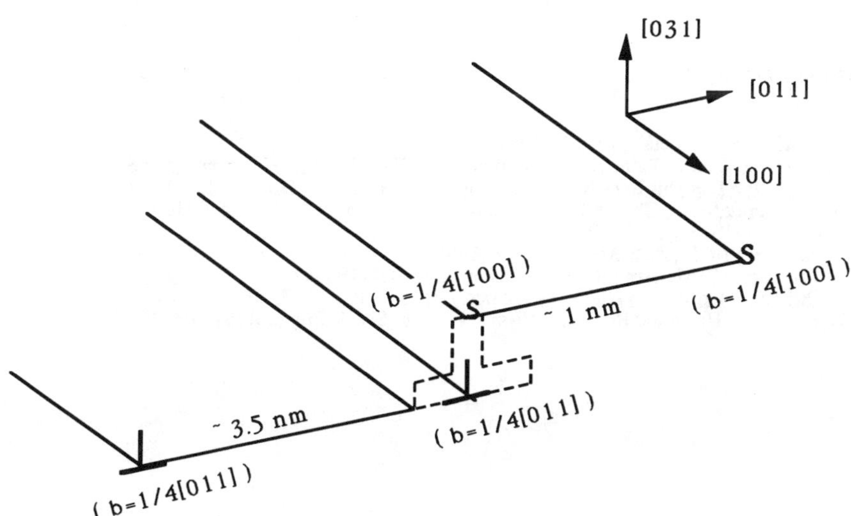

Figure 4: In our model, the 1/2[111] dislocation in $Be_{12}X$ was found to split into four partial dislocations. The final arrangement of these partials is shown here.

having Burgers vectors 1/4<011], and two screw superpartial dislocations having Burgers vectors 1/4<100]. Also all the partial dislocations do not lie on the same plane which makes the cross-slip of this dislocation difficult. This indicates that the 1/2<111] dislocation will not be a primary slip system in $Be_{12}X$ and, indeed, they are rarely observed in the TEM. The calculations of the <001] dislocation are still underway.

The 1/2<101]{101) and the 1/2<100]{011) partials, which were found to be present in high density in $Be_{12}X$, were also modelled atomistically. The calculation of the 1/2<101] dislocation suggested that this dislocation splits into two 1/4<101] superpartials in agreement with the TEM observations (Figure 4). The 1/2<100] dislocation did not show any tendency of splitting during the modelling calculation. This result is also consistent with the TEM observations which find the 1/2<100] partials to be stable i.e., the 1/4<100] superpartials are not observed.

The stacking fault energy for the faults associated with the 1/2<011] and the 1/2<100] partials was also calculated in this model by total energy calculations and was found to be zero within the errors of the calculation. This result predicts the presence of wide stacking faults separating these partials in agreement with the TEM observations.

ACKNOWLEDGEMENTS

This work is supported by the Defense Advanced Research Projects Agency through the Office of Naval Research and under U. S. Department of Energy contract DE-AC06-76RL0 1830 with the Pacific Northwest Laboratory, which is operated by Battelle Memorial Institute.

REFERENCES

1) Lewis J. R., J. Metals, **13** 829 (1961).
2) Stonehouse A. J., Paine R. M. and Beaver W. W., Mechanical Properties of Some Transition Element Beryllides, edited by J. H. Westbrook, (John Wiley, 1960) p. 297.
3) Bruemmer S. M., Charlot L. A., Brimhall J. L., Henager Jr. C. H. and Hirth J. P., Phil. Mag. A, **65** 1083 (1992).
4) Kronberg M. L., Acta Metall., **9** 327 (1961).
5) Bruemmer S. M., Arey B. W., Jacobson R. E. and Henager Jr. C. H., High-Temperature Ordered Intermetallic Alloys IV (Mater. Res. Soc. Proc. **213**, 1991) p. 475.
6) Sondhi S., Hoagland R. G. and Hirth J. P., Mat. Sci. & Engg., **A152** 103 (1992).

STRESS ANISOTROPY OF STOICHIOMETRIC NiAl SINGLE CRYSTALS

T. Takasugi, J. Kishino and S. Hanada
Institute for Materials Research, Tohoku University, Katahira 2-1-1, Aoba-ku, Sendai 980, Japan.

ABSTRACT

The yield stress properties and the associated slip systems of stoichiometric NiAl single crystals were investigated in terms of crystal orientation, temperature and the deformation mode. The CRSS was, in a wide range of experimental conditions, higher in the sequence of {110}<100>, {100}<100> and {hk0}<100> slips. In all the crystal orientations studied, the CRSS in compression was higher than the CRSS in tension particularly at low temperatures. The tension-compression asymmetry on the CRSS was understood qualitatively as being due to the effect of the normal stress on the core structure of a <001> dislocation.

INTRODUCTION

In the B2-type NiAl which possesses a high ordering energy and deforms by <100> slip, it is generally believed that the <100> dislocation is not dissociated into any (super-) partials. However, some indications showing the strength anomaly, which has been generally postulated to be due to the dissociated dislocations, can be found. For example, the variation of the flow stress with temperature in the NiAl single crystals was not so monotonous at intermediate temperatures and also was very much dependent on the crystal orientation. Also, the NiAl single crystals exhibited significant elongation at ambient temperature by the addition of a small amount of Fe, Mo and Ga [1]. In addition, it was shown in the NiAl polycrystals that an asymmetry in the yield stress was shown with the compressive yield stress being higher than the tensile stress [2]. These results may be associated with the characteristic of the <100> dislocation. In this study, the yield stresses of fairly stoichiometric NiAl single crystals were measured in terms of crystal orientation, temperature and the deformation mode whether tension or compression. Also, the activated slip systems were analyzed in the single crystals with various orientations. Based on these results, it is proposed that the yield stress anomaly observed in this study can be associated with the core effect of the <100> dislocation.

EXPERIMENTAL PROCEDURES

The just-stoichiometric NiAl single crystals used in this study were prepared by the Bridgman technique in the induction furnace using an alumina crucible. The analyzed chemical composition indicates that the NiAl single crystals were fairly stoichiometric (the deviation from stoichiometry was less than 0.4 at%). Both the tensile and the compressive specimens were taken from the same single crystal rod. The tensile specimens with 1x2x13 mm gage dimensions and the compressive specimens with dimensions of approximately 2x2x5 mm were prepared in desired orientations (i.e. [001], $[\bar{1}11]$, [011] and $[\bar{1}23]$) using a wire-slitting and an electron discharge machine. Here, it is however noted that the compressive specimens with orientations [001] and [011] were deviated from their exact positions in a standard stereographic unit triangle. The faces of the tensile and compressive specimens were abraded on SiC paper and then were electrolytically polished.

Mechanical tests of tension and compression were conducted on an Instron-type testing machine at temperatures from 77K to 1073 K. The tests at 77 K were carried out in a liquid nitrogen and tests at other temperatures including a room temperature were carried out in a vacuum. The initial strain rates adopted in this study were 7.6×10^{-5}/s and 1.5×10^{-4}/s for the tensile tests, and 3.3×10^{-4}/s for the compressive tests, and therefore were conducted at almost identical strain rates. The slip trace analysis was done particularly for the single crystals with soft orientations deformed up to approximately 2~3% plastic strains, using the Nomarski-type interference optical microscopy(OM). Also, the activated dislocations were investigated by transmission electron microscopy (TEM) to determine their Burgers vectors.

EXPERIMENTAL RESULTS

The yield stresses

Figures 1(a)-(d) represent the variations of the yield stress (defined at 0.2% plastic strain) with temperature and the deformation mode for the single crystals with orientations [001], $[\bar{1}11]$, [011] and $[\bar{1}23]$, respectively. Here, a large number of data reported in the previous works [3-8] were included for comparison.

First, for the single crystals with orientation [001](i.e. a hard orientation), both the yield stresses by compression and tension steadily decreased with increasing temperature. Kinking deformation took place at a wide range of temperatures up to 773 K on the macroscopically elastic range of the compressive specimens (or on the plastic range promptly after the yielding). In these cases, data points drawn for compression in Fig. 1(a) correspond to the onset values of the flow stress drop due to the kinking. On the other hand, the kink deformations were not clearly observed on the tensile specimens partly because they failed at relatively small strains. Here, the most striking result is that the tension-compression anisotropy can be recognized on the yield stress. The yield stresses evaluated by compression were much higher than those evaluated by tension. This difference was quite remarkable at low temperatures and tended to decrease with increasing temperature. Also, Fig. 1(a) indicates that the values of the yield (or kink) stresses observed in the previous works [4-8] were higher than the values observed in the present study not only for the compression data but also for the tension data.

The other single crystals with orientations $[\bar{1}11]$, [011] and $[\bar{1}23]$ (i.e. soft orientations) did not exhibit the kink deformations. The yield stress levels of the single crystals with these soft orientations were much lower than those of the single crystals with a hard orientation. This difference is primarily due to the difference in the Schmid factor on the $(\bar{1}10)$[001] slip system which is the preferable slip system activated in these orientations of NiAl single crystal, as actually identified by OM and TEM observations in this study. Also in the single crystals with these orientations, the yield stresses decreased rapidly in a low temperature region and then steadily in a high temperature region with increasing temperature. The tension-compression anisotropy of the yield stresses was again observed for these soft orientations. For the single crystals with orientation $[\bar{1}11]$, the tension-compression anisotropy was very significant at low temperatures and tended to disappear at higher temperatures (~900 K). For the single crystals with orientation [011], the tension-compression anisotropy was not large but consistently observed up to temperatures of about 1000 K. For the single crystals with orientation $[\bar{1}23]$, the tension-compression anisotropy was observed in a limited region of an ambient temperature. It was again shown in these figures that the values of the yield stresses observed in the previous work [3-8] were higher than the values observed in the present study for both the compression and tension data.

Activated slip systems

Activated slip plane was determined in a wide range of experimental conditions. In the single crystals with a hard orientation [001], it was difficult to determine the activated slip planes although the kinks were well identified. In the single crystals with a soft orientation $[\bar{1}11]$, {110} slip planes were largely observed at a wide range of temperatures and in the compression and the tension tests. In the single crystals with orientation [011], which was deformed at a room temperature and by the compression test, {100} slip planes were identified. Here, it must be noted that many previous observations reported {110} slip planes in this crystal orientation [9] but Wasilewski et al.[4] observed {100} slip planes. In the single crystals with orientation $[\bar{1}23]$ which were deformed at room temperature in compression, the activated slip planes were identified to be (hk0). This result is quite in contrast to the previous observation for which {110} slip planes were observed in single crystals with orientation $[\bar{1}22]$[10] which is close to the orientation of $[\bar{1}23]$. It must be noted here that the slip lines observed in the crystal orientation of $[\bar{1}23]$ were quite wavy and occasionally cross-slipped. This result means that the slip planes activated in the single crystals with orientation $[\bar{1}23]$ are not principal slip planes, but are formed from two orthogonal slip planes of {110}.

Among the three soft orientations observed in this study, the slip planes activated in the single crystals with orientations $[\bar{1}11]$ and [011] were primarily independent of test temperature and the deformation mode while those activated in the single crystals with orientation $[\bar{1}23]$ were dependent on test temperature and the deformation mode. The result in

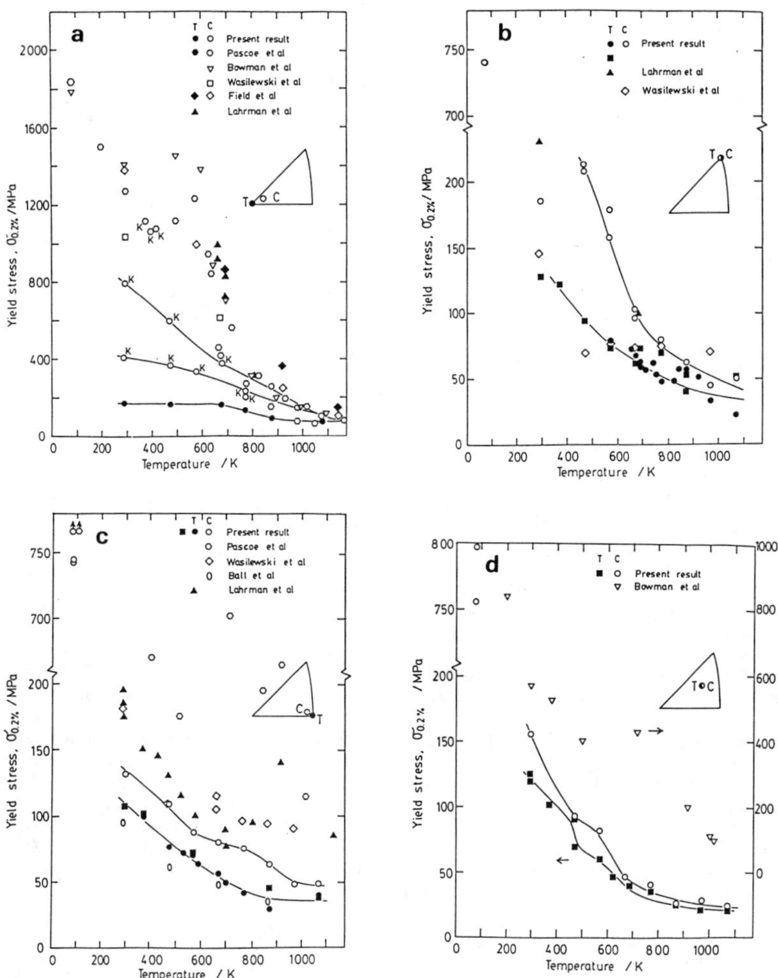

Fig. 1 Variations of the yield stress with temperature and the deformation mode in the single crystals with orientations (a) [001], (b) [$\bar{1}$11], (c) [011] and (d) [$\bar{1}$23] respectively, together with data reported in the previous works [3-8]. Data points obtained in this study were connected by the solid lines. Note in Fig. 1(a) that the marks denoted by K mean the kink deformation and the compression data were drawn as the upper and the lower bounds.

the latter orientation showed that the activated slip plane changed toward their principal slip, i.e. ($\bar{1}$10) plane with decreasing temperature and also its slope was much larger in tension mode than in compression mode. Thus, the tension-compression anisotropy can be seen also on the activated slip plane.

The Burgers vectors of activated dislocations were determined using the conventional **g.b** criterion by TEM analysis and were identified to be the <001>-type dislocations except for the single crystals with orientation [001]. Detailed analysis was not performed on the single

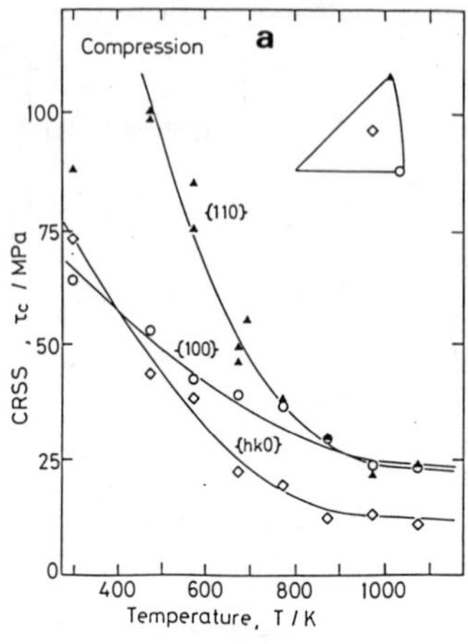

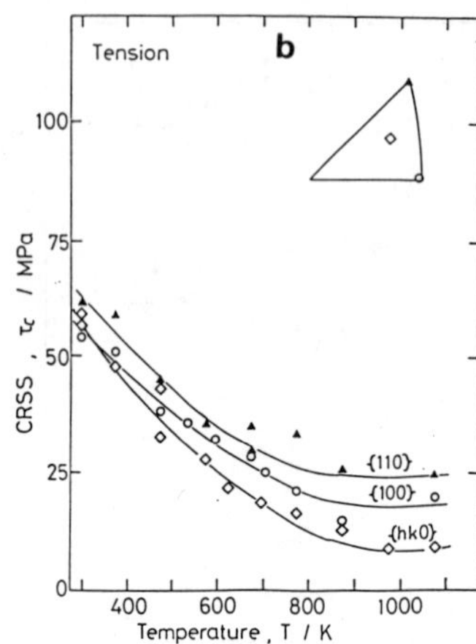

Fig. 2 The CRSS vs. temperature plots in the single crystals which were deformed by (a) compression and (b) tension, and had crystal orientations [$\bar{1}$11], [011] and [$\bar{1}$23] (therefore, slip systems <100>{110}, <100>{100} and <100>{hk0}), respectively.

crystals with orientation [001] because of the introduction of kink deformation. Therefore, whether the single crystals with orientation [001] involve the <111>-type dislocations particularly at low temperatures can not be concluded in this study.

Figures 2(a) and (b) represent, based on the determined slip systems, the CRSS vs. temperature plots for the two deformation modes of compression and tension in the single crystals with orientations [$\bar{1}$11], [011] and [$\bar{1}$23], respectively. The levels of the CRSS measured in compression were higher in the sequence, crystal orientations of [$\bar{1}$11], [$\bar{1}$23] and [011] (i.e. slip systems of <100>{110}, <100>{hk0} and <100>{100}) at low temperatures, and in the sequence, crystal orientations of [$\bar{1}$11], [011] and [$\bar{1}$23] (i.e. slip systems of <100>{110}, <100>{100} and <100>{hk0}) at intermediate and high temperatures. On the other hand, the levels of the CRSS measured in tension were higher in the sequence, crystal orientations of [$\bar{1}$11], [011] and [$\bar{1}$23](i.e. slip systems of <100>{110}, <100>{100} and <100>{hk0}) for all test temperatures. Thus, the present result indicates that the <100> slip on {hk0} is more preferable to on {110} or {100} planes regardless of compression or tension, therefore being primarily consistent with Field et al's recent observation [11] but inconsistent with Wasilewski et al's observation [4].

DISCUSSION

Figure 1 certainly showed that the yield stress levels observed in previous work were mostly higher particularly at low temperatures than those observed in this study. Some reasons responsible for this discrepancy should be considered from the point of view of the experimental and material conditions. It appears that stoichiometry and impurity strongly affect the stress properties of the NiAl single crystals through interacting with the activated dislocations. Non-stoichiometry leads to the rapid increase of the flow stress at low temperatures on both sides of the non-stoichiometric composition. Also, impurity atoms such as gaseous elements and metal elements, which may occupy the interstitial sites and interact with non-stoichiometric defects, must play an important role in the strength of NiAl single crystals and can contribute to defect hardening at low temperatures. The NiAl single crystals prepared in this study were stoichiometric (50.1 and 50.4 mol%Al) and did not contain high contents of the impurity atoms (i.e. O;0.004~5mass%, N;0.0003~6mass%, H; 0.0003~4 mass%). However, a number of previous studies dealt with NiAl single crystals, which nominally consist of stoichiomertric compositions, mostly have not specified their chemical

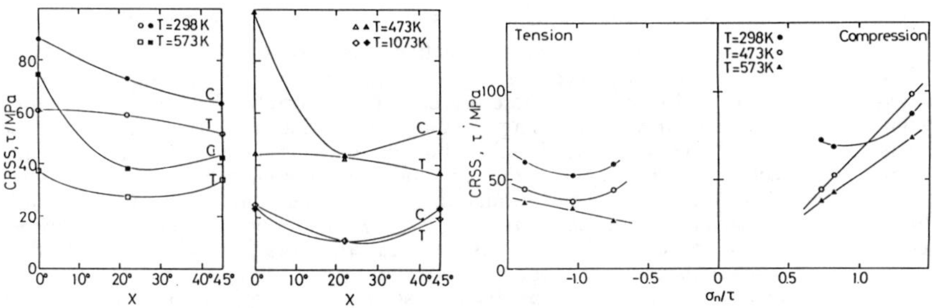

Fig. 3 The CRSS vs. χ curves calculated in the single crystals with soft orientations which were deformed at various temperatures.

Fig. 4 The CRSS for activated slip planes as a function of the stress ratio σ_n/τ.

compositions [12-17]. Therefore, it is very likely that the NiAl single crystals observed in the previous works may be deviated from the stoichiometric composition and also contained certain amounts of impurity elements, resulting in the higher stress. Also, thermal vacancies can be considered to produce such a discrepancy of the yield stresses. The presence of supersaturated thermal vacancies has been reported to enhance the hardness [18] and the yield strength [19]. The furnace cooling rate adopted in this study was very low (an initial cooling rate of about 0.37 K/s and then a slower cooling rate at lower temperatures) and therefore is postulated to produce a low concentration of vacancies.

In order to understand the stress anomaly in the B2-type intermetallic compounds, the plots in the CRSS vs. χ are very useful. Here, χ is the angle between the primary (110) slip plane and the maximum resolved shear stress plane (MRSSP) along a [001] great circle in a standard stereographic projection. Figure 3 illustrates the CRSS vs. χ curves calculated for the single crystals which were deformed by the <100>-type dislocations at various temperatures. A general trend is that the CRSS was higher in crystal orientations with small χ values and tended to decrease with increasing χ values in both deformation modes of compression and tension. As temperature increases, a minimum in the CRSS occurred in crystal orientations with an intermediate χ value. The tension-compression anisotropy on the CRSS was significant in the two limiting orientations, i.e. χ=0° and χ=45° and also became less significant high temperatures. Thus, the stress anomaly disclosed in the present NiAl single crystals is different from the stress anomaly observed in the B2 intermetallic compounds or b.c.c. metals which deform by the 1/2<111> dislocations (or connected by the anti-phase boundary (APB)). It is well known that the stress anomaly in such materials is associated with twinning-anti twinning asymmetry in slip on {112} planes. Thus, it is considered that the tension-compression anisotropy observed in the NiAl single crystals is associated with the non-glide stresses or the normal stresses on the activated slip plane by affecting the dislocation core of the <100>-type dislocations.

The yield stress, σ_n, normal to the observed slip plane can be derived by the expression of $\sigma_n = \tau\cos\theta\sec\lambda$ where τ is the resolved shear stress on the observed slip plane, θ is the angle between the stress axis and the observed slip plane normal and λ is the angle between the stress axis and the slip direction. The stress ratio σ_n/τ is, therefore, a function of crystal orientation. The effect of the normal stress on the CRSS at various temperatures is represented in Fig. 4, where the positive area of horizontal axis corresponds to the deformation mode of compression while the negative area corresponds to the deformation mode of tension. The general trend is that with increasing the compressive stress the CRSS increased while with increasing the tensile stress the CRSS decreased and then increased after showing the minimum. Thus, the compression result appears to give steeper slopes and higher levels of the CRSS. However, data points were not able to be connected by a set of straight lines through both of the positive and negative areas in σ_n/τ.

The normal stress effect observed in the NiAl single crystal is associated with an undissociated <100> dislocation and therefore can not be correlated with twinning-anti-twinning asymmetry. A likely mechanism producing the CRSS asymmetry in the NiAl single crystals is that the Burgers vector of the <100>-type dislocation is relatively large and therefore is expected to have a complex core structure, i.e. large inelastic strain field, which may be affected sensitively by the deformation mode, temperature, glide- (or non-glide stresses) and their levels. Under the normal stress, the lattice constriction (or expansion) occurs and leads to more sessile (or glissile) core configurations of the <100>-type dislocation, resulting in the increased (or decreased) stress.

The single crystals with orientation [001] also showed large tension-compression anisotropy of the yield stress. The <111>-type dislocations are typically activated in single crystals with this orientation [20]. The Burgers vector of the <111>-type dislocation is larger than that of the <001>-type dislocation and therefore is similarly expected to have a complex core structure, by which the stress anomaly may occur. Certainly, more work, for instance, high resolution TEM observations or core structure calculations are needed to understand the stress anomaly in NiAl.

CONCLUSION

The yield stress and the associated slip systems of fairly stoichiometric NiAl single crystals were investigated in terms of crystal orientation, temperature and the deformation mode. The following results were obtained.

(1) The critical resolved shear stress (CRSS) was strong function of crystal orientation, temperature and the deformation mode. The CRSS was primarily higher in the sequence of {110}<100>, {100}<100> and {hk0}<100> slips, regardless of tension or compression. The tension-compression asymmetry was observed for all crystal orientations and was significant at low temperatures. The CRSS in compression was higher than the CRSS in tension.

(2) The results can be understood qualitatively as being due to the effect of the normal stress on the core structures of the <100>-type dislocation (or the <111>-type dislocation).

REFERENCES

1. R. Darolia, J. of Metals, **43**(3), 44(1991).
2. P. Nagpal, I. Baker, F. Liu and P. R. Munroe in High Temperature Ordered Intermetallic Alloys IV, edited by L. Johnson et al., (MRS Symp. Proc. **213**, Boston, MA, 1991) p.533.
3. A. Ball and R. E. Smallman, Acta Metall., **14,** 1517(1966).
4. R. J. Wasilewski, S. R. Butler and J. E. Hanlon, Trans. Met. Soc. AIME, **239**, 1357(1967).
5. R. T. Pascoe and C. W. A. Newey, Metal Science Journal, **2,** 138(1968).
6. R. R. Bowman, R. D. Noebe and R. Darolia, HITEMP Review - 1989, p.47-1, (1989) NASA CP-10039.
7. R. D. Field, D. F. Lahrman and R. Darolia, Acta Metall. Mater., **39**, 2951(1991).
8. D. F. Lahrman, R. D. Field and R. Darolia in High Temperature Ordered Intermetallic Alloys IV, edited by L. Johnson et al., (MRS Symp. Proc. **213**, Boston, MA, 1991) p. 603.
9. R. D. Noebe, A. Misra and R. Gibala, ISIJ -International, **31**, 1172(1991).
10. R. T. Pascoe and C. W. A. Newey, Phys. Stat. Solidi, **29**, 357(1968).
11. R. D. Field, D. F. Lahrman and R. Darolia in High Temperature Ordered Intermetallic Alloys IV, edited by L. Johnson et al., (MRS Symp. Proc. **213**, Boston, MA, 1991) p. 255.
12. A. Ball and R. E. Smallman, Acta Metall., **14,** 1517(1966).
13. A. Ball and R. E. Smallman, Acta Metall., **14,** 1349(1966).
14. R. J. Wasilewski, S. R. Butler and J. E. Hanlon, Trans. Met. Soc. AIME, **239**, 1357 (1967).
15. R. T. Pascoe and C. W. A. Newey, Metal Science Journal, **2,** 138(1968).
16. R. T. Pascoe and C. W. A. Newey, Phys. Stat. Solidi, **29**, 357(1968).
17. M. H. Loretto and R. J. Wasilewski, Phil. Mag., **23**, 1311(1971).
18. P. Nagpal and I. Baker, Metall. Trans., A, **21A**, 2281(1990).
19. R. R. Bowman, R. D. Noebe, S. V. Raj and I. E. Locci, Metall. Trans. A, in press.
20. M. H. Loretto and R. J. Wasilewski, Phil. Mag., **23**, 1311(1971).

GLIDE OF PERFECT DISLOCATIONS IN TiAl

SYLVIE FARENC AND ALAIN COURET
CEMES-LOE/CNRS, BP 4347, 31055 Toulouse Cedex, France.

ABSTRACT

In situ deformation experiments have been performed in TiAl in order to study the glide of perfect dislocations. The deformation is accomodated by the jerky movement of rectilinear screw dislocations moving slower than the edge segments. This behaviour is interpreted in terms of frictional forces which originate in a non-planar structure of the dislocation core.

INTRODUCTION

TiAl is an $L1_0$ ordered intermetallic alloy which exhibits a yield stress anomaly with temperature at various orientations of the deformation axis [1], similar to the well known anomaly of some $L1_2$ alloys. It deforms either by twinning or by the glide of perfect dislocations [2,3] (simple a/2<110] or super a<101]). The parentheses used with Miller indices in this paper follow the convention introduced in [4]. Results concerning twinning will not be included here and can be found elsewhere [5,6]. In this paper, the glide of perfect dislocations in the range of the yield stress anomaly (20°C - 600°C) is studied mainly by *in situ* deformation experiments.

Except in [7], simple dislocations have been observed by *post-mortem* observations in TEM [8-14]. Whereas Whang and co-workers [13,14] have stressed that simple dislocations lie parallel to the screw orientation, Greenberg and co-workers [10-12] have shown that they also lie parallel to the $<32\bar{1}]$ direction. It has been not possible to observe the dissociation or spreading of simple dislocations under weak-beam conditions or by the lattice imaging technique [8,15]. The superdislocation core structure has been analysed in details by Hug [4,8] under weak-beam conditions. They are dissociated into three partial dislocations according the following relation :

$$a[011] \rightarrow a/6[\bar{1}12] + (SISF) + a/6[121] + (APB) + a/2[011] .$$

At low temperature the APB lies on the octahedral glide plane whereas at higher temperature, it lies on the cube cross-slip plane. These results have been confirmed recently by Li and Whang [16]. It is now generally accepted that the motion of perfect dislocations is controlled by frictional forces. The frictional force has been given different interpretations by various authors, arising from covalent bonding [7,10,11,12] or from non-planar dislocation core structures [4,8,16].

EXPERIMENTAL PROCEDURE

In situ deformation experiments have been performed between 20°C and 600°C in a JEOL 200 CX electron microscope operating at 200 kV. Microsamples were cut from a $Ti_{46}Al_{54}$ polycrystal, provided by Gilles Hug (LEM-ONERA, France).

EXPERIMENTAL RESULTS

Specimen orientations

In figure 1 is shown two unit standard triangles adapted to the $L1_0$ structure in which are shown the directions of the tensile axes of all the specimens in which the glide of either simple dislocations (Fig. 1,a), or superdislocations (Fig. 1,b) was detected. Various symbols are used to represent the different deformation temperatures. The indexation of the active glide systems during *in situ* straining has been checked *post-mortem*. In experiment, the dislocation motions are sometimes difficult to observe, because of difficulties caused either by the use of polycrystalline samples or by the nature of dislocation movement.

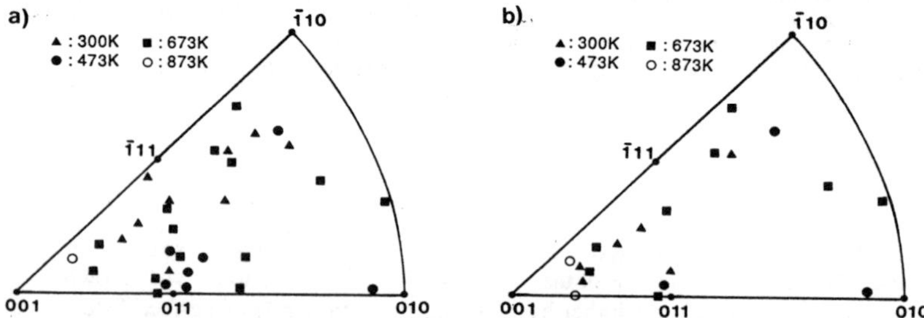

Fig. 1 : Unit standard triangles in which are shown the directions of the tensile axes of the specimens in which the glide of either simple dislocations (a) or superdislocations (b) was detected.

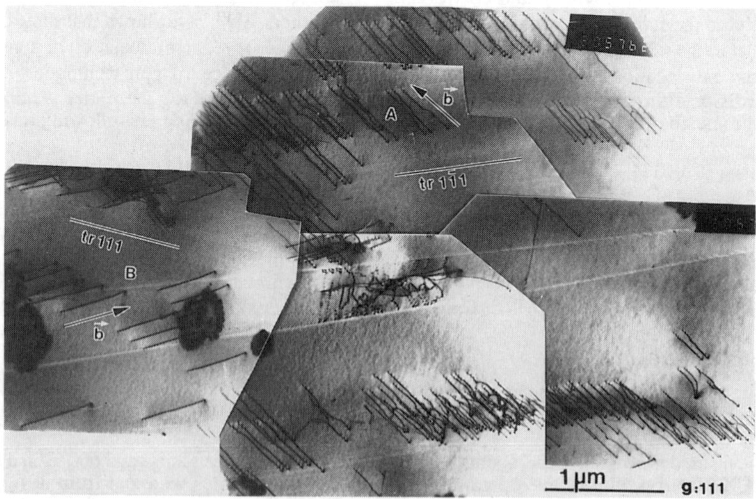

Fig. 2 : Substructure of deformation, A : simple dislocations, B : superdislocations.

Substructure of deformation

Figure 2 shows the substructure of deformation for a specimen in which the glide of simple dislocations and superdislocations was activated. The dislocations of these two families are rectilinear and parallel to their respective screw direction.

Dynamic observations

i) Simple dislocations

The movement of simple dislocations at room temperature is described in figure 3 which shows a dislocation moving in the primary octahedral glide plane. The dislocation which is rectilinear along the screw direction glides by series of jumps between locking positions. Positions marked 1, 2, 3 and 4 are locking positions and in pictures b, d and e the dislocation jumps between these positions. In pictures b and d, for instance, the dislocation jumps between positions 1 and 2, and 2 and 3, respectively. In each case, the initial and final positions of the jumps are visible because of the remanence of the fluorescent screen.

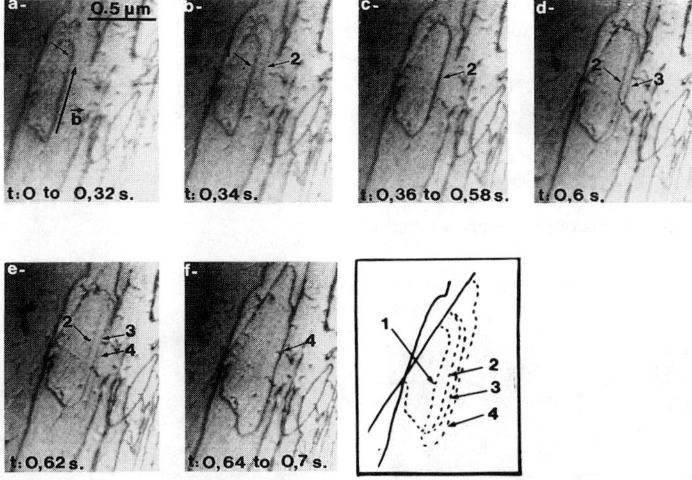

Fig. 3 : Jerky movements of screw simple dislocations at 20°C.

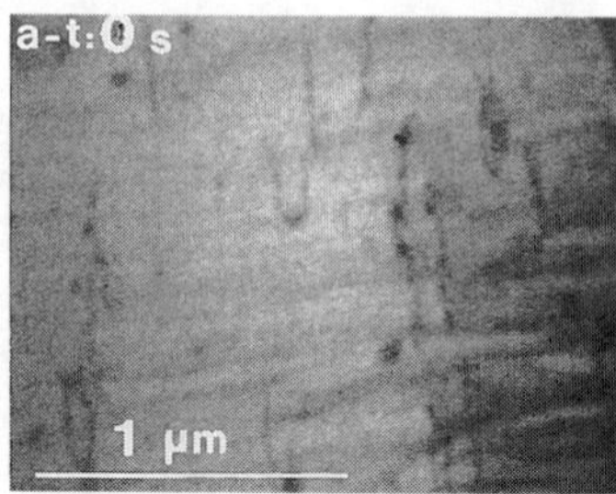

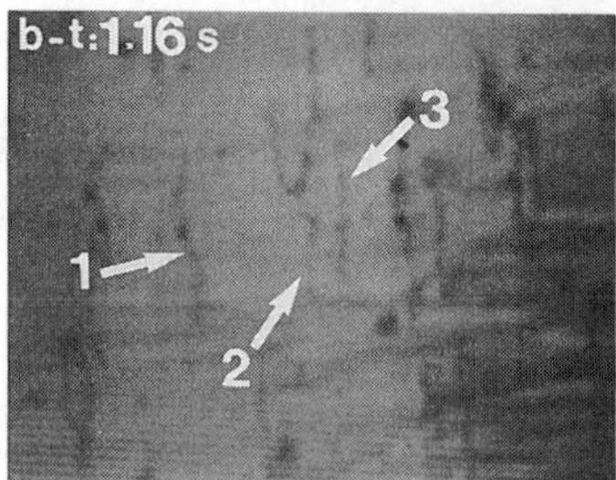

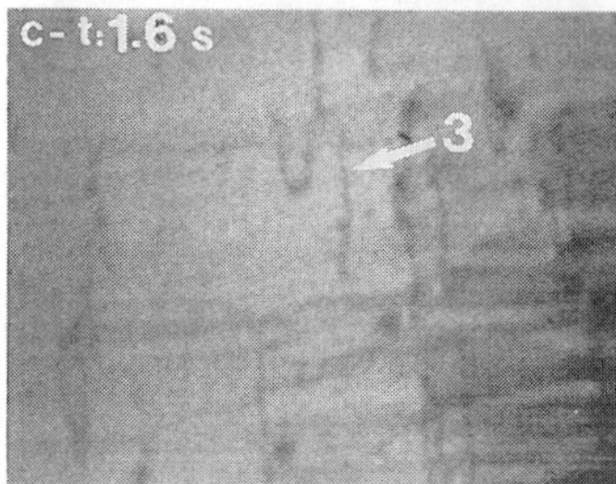

Fig. 4 : Movements of simple dislocations at 200°C.

At 200°C, similar jumps have been observed sometimes. The deformation is however more often owing to a quick and collective movement of groups of dislocations. Multiplication occurs very quickly. The dislocations jump over long distances and once locking takes place, subsequent unlocking seems to be very difficult. This behaviour is partly illustrated in Fig. 4. Very quickly, three dislocations (noted 1,2,3) glide rapidly and then lock (picture b). Between pictures b and c, two of them (1 and 2) jump over a distance larger than the field of view.

At 400°C, the movement is more discontinuous and heterogeneous than at 200°C. To date, it has not been possible to observe the movement of individual dislocations. When the deformation occurs large groups of dislocations glide rapidly and then lock . If more strain is applied to the specimen, dislocations which have already moved will remain inactive, but new groups will glide rapidly and then lock. This behaviour is described in figure 5 which shows a group of dislocations moving rapidly and stopping.

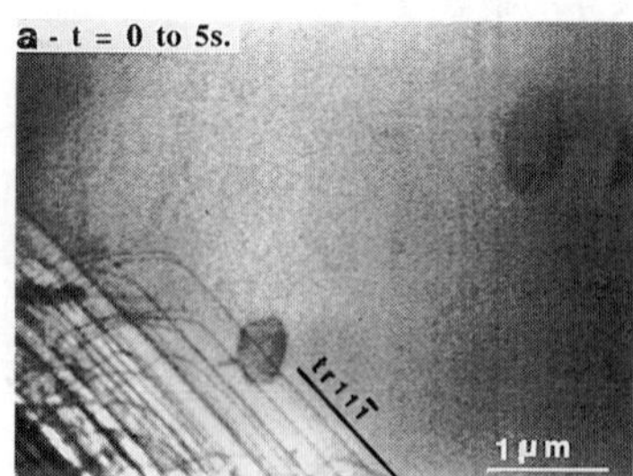

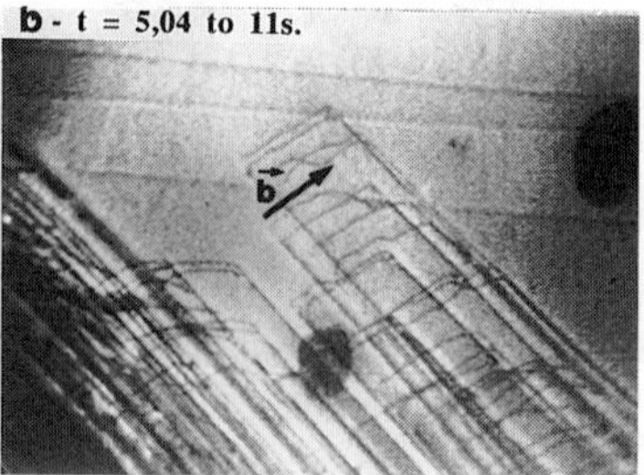

Fig. 5 : Rapid movement and stopping of a group of simple dislocations at 400°C.

ii) Superdislocation

The movement of superdislocations appears very discontinuous and heterogeneous like simple dislocations at high temperature. Only once, it has been possible to observe a few individual movements of superdislocations. These movements are similar to that presented here in the case of simple dislocations at room temperature (Fig. 3). This example of superdislocation movement has been published elsewhere [3].

Core structure of superdislocations

The core structure of superdislocations has been analysed under weak-beam conditions in specimens previously deformed *in situ*. Whatever is the deformation temperature, the dissociation into two superpartial dislocations bordering an APB has been observed (Fig. 6). No further dissociation of the superpartial dislocations has been detected. Details of this investigation can be found in [2]. At low temperatures (20°C - 400°C), the APB lies in the octahedral glide plane, whereas at 600°C it lies partly in the octahedral glide plane and partly in the cube cross-slip plane.

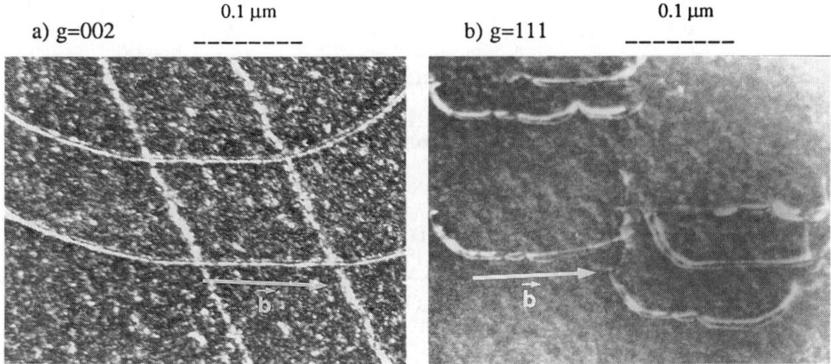

Fig. 6 : Weak beam pictures of the cores of superdislocations, a: 400°C, b : 600°C.

DISCUSSION

Fig. 1 shows the specimen orientations at which simple dislocations and superdislocations have been activated. It has been found that simple dislocations contribute significantly to the deformation for many specimen orientations at all the temperatures investigated. From the comparison of this figure with Schmid factor maps, it has been possible to determine a critical Schmid factor value for the activation of simple dislocations and superdislocations ; these values are 0.2 and 0.4, respectively.

The present observations have shown that the deformation is governed by the motion of rectilinear screw dislocations which glide slower than the edge segments. This motion is therefore controlled by a frictional force acting only on screw segments. The jerky feature of this movement indicates that the covalent bonding is probably not at the origin of this frictional force. Rather, when the deformation is controlled by frictional force originating from covalent bonds, e. g. in [17], the dislocation motion is steady and not jerky. It is thus noted that the observation of movement of simple dislocations at room temperature is inconsistent with the covalent frictional force models [7,10-12].

It is therefore proposed that the frictional force acting on perfect dislocations results from non planar configurations of screw segments. The jerky movement is interpreted in terms of a double transition between the sessile and glissile configurations of the dislocation in the framework of the locking-unlocking mechanism [18]. For simple dislocations, by analogy with the dislocations in titanium [19], the sessile configuration is supposed to result from a spreading in several planes. This view is supported by the observations of cross-slip of simple dislocations in the cube plane [2,3]. For superdislocations, the sessile configurations are those which have been found in figure 6. At high temperature, the sessile character of dislocations results from the cross-slip of the APB in the cube plane in a similar way than the Kear-Wilsdorf lock in the case of $L1_2$ alloys. At low temperatures, since the APB lies in the octaedral glide plane a further spreading of the superpartial dislocation out of the glide plane is thought to be at the origin of the sessile character of the dislocation.

CONCLUSION

It has been shown in this work that the deformation of TiAl is accomodated by the glide of simple dislocations or superdislocations at favorable stress orientations. Their movement is jerky and more or less discontinuous and heterogeneous depending on the circomstances. It has been interpreted in terms of frictional forces governed by a non-planar structure of the dislocation core.

Acknowledgements : The authors are gratefull to Gilles Hug (LEM-ONERA, France) for discussions and for providing TiAL crystals.

REFERENCES

1. T. Kawataba, T. Kanai and O. Izumi, Acta. Metall. **33**, 1355 (1985).
2. S. Farenc, Thèse de Doctorat, University Paul Sabatier, Toulouse-France, (1992).
3. S. Farenc, D. Caillard and A. Couret, Proc. of 6th JIMIS on Intermetallic Compounds Structure and Mechanical Properties, 791 (1991).
4. G. Hug, A. Loiseau and P. Veyssière, Phil. Mag. , **57**, 499 (1988).
5. S. Farenc, A. Coujou and A. Couret, Phil. Mag., in press (1993).
6. S. Farenc, A. Coujou and A. Couret, Mater. Sci and Eng. A., in press (1993).
7. S. A. Court, V. Vasudevan and H. L. Fraser, Phil. Mag. A, **61**, 141 (1990).
8. G. Hug, Thèse de Doctorat, University of Paris Sud-France, (1988).
9. H.A. Lipsitt, D. Shechtman and R.E. Schafrik, Metal. Trans. A, **6**, 1991 (1975).
10. B.A. Greenberg, O.V. Antonova, V.N. Indenbaum, L.I. Karkina, A.B. Notkin, M.V. Ponomarev and Smirnov L.V., Acta Met. Mater., **39**, 233 (1991).
11. B.A. Greenberg, O.V. Antonova, V.N. Indenbaum, L.I. Karkina, A.B. Notkin, M.V. Ponomarev and Smirnov L.V., Acta Met. Mater., **39**, 243 (1991).
12. B.A. Greenberg, O.V. Antonova, L.I. Karkina, A.B. Notkin, and M.V. Ponomarev, Proc. of 6th JIMIS on Intermetallic Compounds Structure and Mechanical Properties, 355 (1991).
13. S.H. Whang and Y.D. Hahn, Scripta Met., **24**, 1679 (1990).
14. S.H. Whang, Y.D. Hahn, Z.X. Li.and Z.C. Li, Proc. of 6th JIMIS on Intermetallic Compounds Structure and Mechanical Properties, 763 (1991).
15. K.J. Hemker, B. Viguier and M.J. Mills, Accepted in Mater. Sci and Eng. A.(1993).
16. Z.X. Li and S.H. Whang, Mat. Sci. and Eng. A, **152**, 182 (1992).
17. D. Caillard, N. Clément, A. Couret, Y. Androussi, A. Lefebvre and G. Vanderschaeve, Proc. of the Microsc. Semicond. Mater. Conf., Oxford, 361 (1987).
18. A. Couret and D. Caillard,Proc. Int. Conf. on Dislocation Mechanisms and the Strength of Advanced Materials, Aussois, France, J. Phys. (III), 885 (1991).
19. S. Farenc, D. Caillard and A. Couret, Acta Met. Mater., in press, (1993).

THE INFLUENCE OF PRESTRAIN ON THE FLOW STRESS ANOMALY IN TiAl SINGLE CRYSTALS

M. A. Stucke*, D. M. Dimiduk*, and P. M. Hazzledine**
*Wright Laboratories, WL/MLLM, Wright-Patterson AFB, OH 45433-6533
**UES, Inc., Materials Research Division, 4401 Dayton Xenia Rd., Dayton, OH 45432

ABSTRACT

Single crystal specimens of $Ti_{44}Al_{56}$, grown by the optical float zone method, were compression tested at room and elevated temperatures. The compression axis was close to [010] but the slip lines show essentially single slip. The material exhibits the anomalous yield behavior and a high work hardening rate. The work hardening rate at 1% strain is ≈3.5 GPa and is higher at 773K than at 296K. Specimens tested first at 773K and then at 296K retained their high temperature yield stress at the lower temperature, unlike $L1_2$ Ni_3Al, but the work hardening rate was thermally reversible. The experiments suggest that the dislocations which first move at the yield point are 'permanently' locked at high temperature but that the dislocation interactions responsible for work hardening are characteristic of the deformation temperature.

INTRODUCTION

Gamma titanium-aluminide alloys have been the subject of interest for the last several years because of their high strength to density ratio [1]. Much of the past work on these alloys has included processing, heat treatment and microstructural studies [2,3]. The deformation behavior in binary gamma alloys has been studied extensively by several authors [4-6]. However, all of this work is on polycrystalline materials or polysynthetically twinned materials. Hug et al. reported an extensive study of dislocation behavior in polycrystalline Ti-56Al at room and elevated temperatures [7,8]. In 1985, Kawabata et al. tested single crystal Ti-56Al and found a yield stress anomaly, similar to that in Ni_3Al alloys [9]. Dislocation behavior and mechanical properties have been observed in single crystals from 4-1273K [9,10]. Whang and co-workers [11-14] found that the yield stress anomaly also occurs in single crystal Ti-55Al with ternary additions of V and Nb. In the deformed crystals, both perfect 1/2<110] and <101] super dislocations are observed to be active. After deformation at high temperatures but below the peak in the yield stress, both types of dislocations are predominantly screws and the super dislocations have non-planar cores as in Ni_3Al. However, the locks in $L1_0$ materials are geometrically different from the Kear-Wilsdorf locks observed in Ni_3Al and it is not obvious that they behave in exactly the same way [15]. One empirical approach to gain an understanding of these alloys, is to perform Cottrell-Stokes type experiments in which a sample is tested first at a high temperature and then retested at a low temperature to determine if the dislocation processes are reversible [16]. In Ni_3Al, several studies have shown that the yield stress anomaly is essentially thermally reversible [17-21]. The objective of this paper is to describe prestraining experiments in single crystal Ti-56Al to determine whether the yield stress and work hardening rates are thermally reversible in TiAl.

EXPERIMENTAL PROCEDURE

The material used in this investigation is a single crystal binary Ti-56Al (atomic %) alloy, grown by the optical float zone method [22], with a final diameter of 7-8 mm. The aluminum content was found to be approximately 56.4 atomic percent aluminum by microprobe experiments. The specimens were heat treated in a vacuum furnace back filled with argon at 1573K (1300°C) for 24 hours, then 1273K (1000°C) for 100 hours. This heat treatment was used to adequately precipitate the inevitable nitrides occurring in standard purity single phase materials [23]. One end of the crystal was polished and oriented toward a <010] zone using the Laue back reflection x-ray diffraction technique. Each specimen was between 4 and 9 degrees from the major zone. Specimens were then wire EDM machined into small compression specimens, approximately 2.5 x 2.5 x 5.0 mm. These were carefully ground on 600 grit silicon carbide paper and electropolished prior to testing.

All of the compression specimens were tested on an Instron mechanical testing machine to approximately 1.5-2.0% total strain at a strain rate of $1x10^{-4}$/sec. No more than 2.0% strain was used so that it was possible to determine yield stress and work hardening rate values and also analyze the dislocation structure in the single crystals. Two specimens were separately tested at room temperature and 773K (500°C). Then, two more specimens were tested at 773K to about 1.3% strain, cooled to room temperature and tested again at room temperature to 2.0% total strain. One specimen was successfully tested at 200, 400 and 500°C. The surface of the samples were analyzed for slip lines. Work-hardening rates were measured at a standard plastic strain of 1%.

RESULTS AND DISCUSSION

Figure 1 is an optical photograph of one side of a single crystal compression specimen after testing at room temperature near the <010] zone. Fine slip lines along the {111} plane are visible. Most of the specimen showed single slip behavior although slip on other {111} planes was also visible. As mentioned in the experimental section, the tests were conducted at 296K and 773K. Then, two specimens were prestrained at 773K, cooled to room temperature and tested to 2.2% total strain. This type of experiment, first performed by Cottrell and Stokes to demonstrate the effects of substructure on flow stress in metal alloys [16], gives an indication of whether the dislocation locking processes are recoverable. For example, in Ni_3Al, in this type of experiment the yield stress has been shown to be essentially reversible, except for a small amount of work hardening [17-21]. Figure 2 is a schematic of the typical room temperature and two-step compression tests. At room temperature, the specimen yields at about 165 MPa. At 773K, the specimen yields at a higher stress (240 MPa). The high temperature test was stopped at about 290 MPa, the specimen was unloaded, allowed to cool to room temperature, and again tested. The specimen yielded the second time at about 300 MPa.

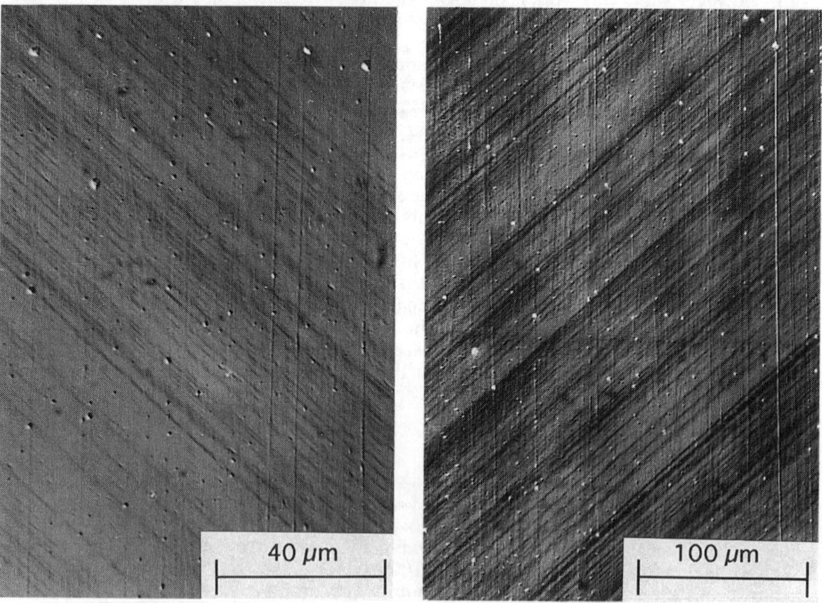

Figure 1. Optical photographs of one side of a Ti-56Al compression specimen. {111} slip lines are visible on the surface of the specimen.

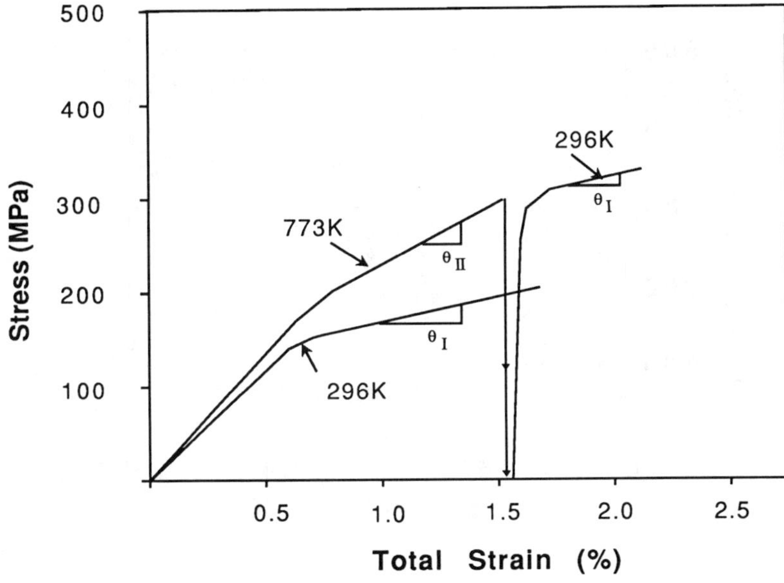

Figure 2. Schematic of the stress-strain curve for the single room temperature test and the two temperature compression test.

Figures 3 and 4 are the plots of the 0.2% yield stress and the critical resolved shear stress versus temperature, respectively. The critical resolved shear stress was calculated with a Schmid factor of 0.440, assuming ordinary dislocations are operating although a <01$\bar{1}$]{$\bar{1}$11} system has a comparable Schmid factor of 0.426. The dashed line represents the data from Kawabata et al. [9,10]. The room and elevated temperature tests show that these data agree with Kawabata's data except one test at 773K which shows a slightly higher critical resolved shear stress. The reversibility tests, shown as solid circles and diamonds in Figures 3 and 4, indicate that this material is not thermally reversible; the yield stress and the critical resolved shear stress remain high after prestraining at high temperature and retesting at room temperature. This data is in contrast to previously reported tests on Ni_3Al single crystal alloys [17-21]. The data would suggest that the substructure formation is sensitive to temperature and that deformation processes are "permanent" in TiAl single crystals. A test that shows this material to be structure sensitive is the data shown as an "x". The sample was tested at three different temperatures: 473K, 673K and 873K in succession. The specimen was deformed at 473K to less than 0.5% total strain, additionally deformed by 0.5% at 673K and finally deformed 0.5% at 873K. This data suggests that at each increasing temperature, the substructure attained through work hardening at the previous temperature remains in the sample and the specimen yields at much higher yield stresses than would be measured if three unyielded samples were tested at three temperatures. Clearly, some component of the stress at a given temperature is due to the substructure induced by prior straining at a lower temperature.

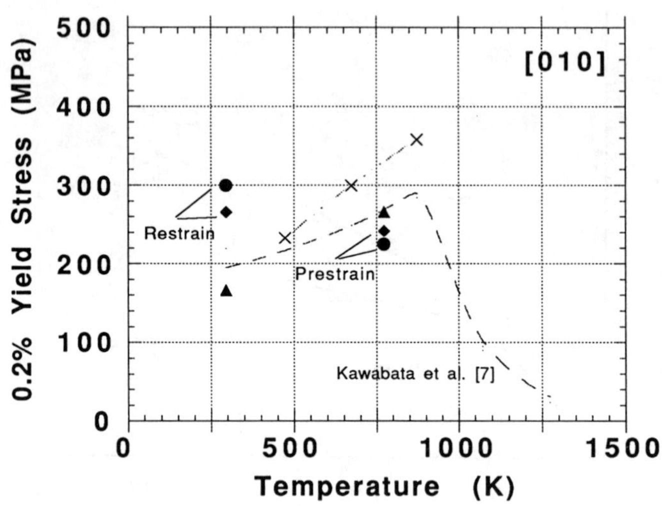

Figure 3. Plot of yield stress versus temperature for Ti-56Al single crystal.

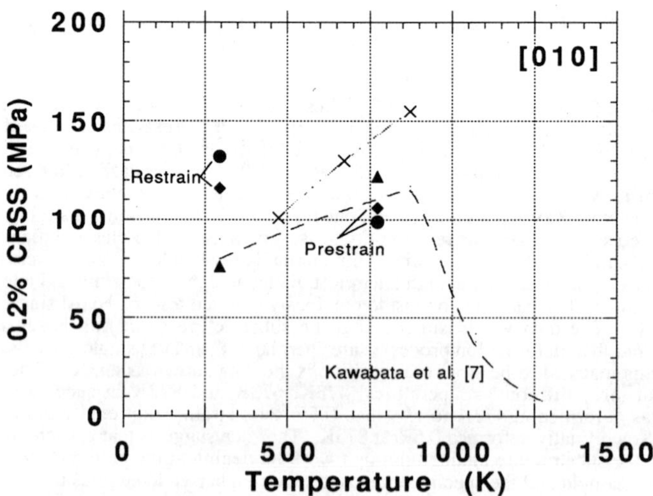

Figure 4. Plot of critical resolved shear stress versus temperature for Ti-56Al single crystal.

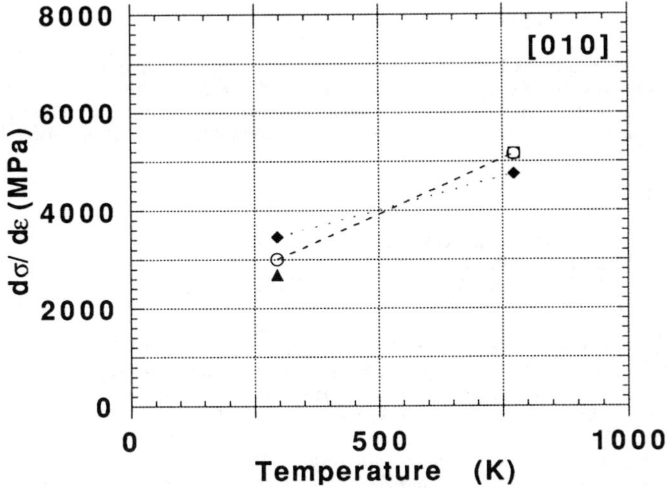

Figure 5. Plot of work hardening rate dσ/dε versus temperature for single crystal Ti-56Al. (σ - tensile stress; ε - tensile strain)

Figure 5 shows a plot of work hardening rate versus temperature. The data show that the work hardening rate is low at room temperature and high at elevated temperature. For the prestraining experiments, the high temperature work hardening rate is not retained upon restraining at room temperature. Thus, phenomenological results show that the work hardening rates for these crystals are thermally reversible, but the yield stress and the critical resolved shear stress are strong functions of the retained substructure.

In Ti-56Al, the temperature dependence of the yield stress is anomalous, as in many $L1_2$ alloys e.g. Ni_3Al. Greenberg et al. has suggested that in both materials a Kear-Wilsdorf type lock is forming [24-26], although additional types of locked dislocations have been proposed for TiAl [26]. For the case of Kear-Wilsdorf locking, deformation begins on the {111} planes and segments become locked on cube planes. In Ni_3Al, such substructure is essentially instantaneously recoverable since both the CRSS and work hardening rate return to their room temperature values with no load drop upon restraining [17-21]. In single crystal TiAl, however, the critical resolved shear stress remains high, suggesting that whatever locks are formed, they cannot be unlocked by the same processes as in Ni_3Al single crystals. It is speculated that the locked dislocations are less easily broken than in Ni_3Al or that interstitials may precipitate onto the dislocations at high temperature thereby locking them at lower temperature. Such "indestructible" barriers result in the CRSS remaining high. Two possibilities are that it is caused by either precipitation of interstitials on the dislocations or by a particularly stable non-planar dissociation of the dislocations. Electron microscopy is required to determine what causes this permanent locking.

CONCLUSIONS

A single crystal Ti-56Al alloy was compression tested at room and elevated temperatures. A yield stress anomaly was found for <010] orientation in agreement with previous results. However, the Cottrell-Stokes type of prestraining experiments, in which the sample is deformed at high temperature and then deformed again at room temperature, show that the yield stress anomaly is not reversible. The dislocations may have become 'permanently' locked by high temperature deformation in TiAl. Further, the successive straining of a single crystal at increasing temperatures, produces an additivity of the strain hardening increments. Such experiments provide phenomenological evidence to suggest that the dislocation locking mechanism(s) governing flow in Ti-56Al are permanent on the time scales ordinarily used in flow stress experiments. Extensive transmission electron microscopy is being performed on these samples to determine the nature of the dislocation motion and locking.

ACKNOWLEDGEMENTS

The authors wish to thank Prof. D. Pope, University of Pennsylvania, for the use of the optical float zone system to grow the single crystals and Mr. J. Barlowe for careful wire EDM machining of the single crystal samples. One of the authors, P. Hazzledine, acknowledges support under U.S Air Force Materials Directorate contract #F33615-91-C-5663.

REFERENCES

1. Y.-W. Kim and D.M. Dimiduk, JOM **43**, 40 (1991).
2. Y.-W. Kim, Acta Metall. **40**, 1121 (1992).
3. E.L. Hall and S.C. Huang, J. Mater. Res. **4**, 595 (1989).
4. H.A. Lipsitt, D. Schectman and R.O. Schrafik, Metall .Trans. **6A**, 1991 (1975).
5. M. Yamaguchi and Y. Umakoshi, Prog. Mater. Sci. **34**, 1 (1990).
6. S.A. Court, V.K. Vasudevan and H.L. Fraser, Philos. Mag. **A61**, 141 (1990).
7. G. Hug, A. Loiseau and A. Lasalmonie, Philos. Mag. **A54**, 47 (1986).
8. G. Hug, A. Loiseau and P. Veyssiere, Philos. Mag. **A57**, 499 (1988).
9. T. Kawabata, T. Kanai and O. Izumi, Acta Metall. **33**, 1355 (1985).
10. T. Kawabata, T. Abumiya, T. Kanai and O. Izumi, Acta. Metall. **38**, 1381 (1990).
11. Y.D. Hahn, Z.X. Li and S.H. Whang, MRS Symp. Proc., **213**, 291 (1991).
12. S.H. Whang, Y.D. Hahn, Z.X. Li and Z.C. Li, in Proc. of Int. Sym. on Intermetallics, edited by O. Izumi (The Japanese Institute of Metals, Japan, 1991), p. 763.
13. Z.X. Li and S.H. Whang, Mat. Sci. and Eng., **A152**, 182 (1992).
14. S.H. Whang, in Ordered Intermetallics - Physical Metallurgy and Mechanical Behaviour, edited by C.T. Liu, R.W. Cahn and G. Sauthoff (Kluwer Academic Publishers, The Netherlands, 1992), **213**, p.279.
15. B.A Greenberg, Phys. Stat. Sol. **42**, 459 (1970).
16. R.J. Stokes and A.H. Cottrell, Acta Metall. **2**, 341 (1954).
17. R.G. Davies and N.S. Stoloff, Trans AIME **233**, 714 (1965).
18. M.H. Yoo and C.T. Liu, J. Mater. Res. **3**, 845 (1988).
19. W.E. Dowling and R. Gibala, MRS Symposium **133**, 209 (1989).
20. D.M. Dimiduk, PhD Thesis, Carnegie-Mellon University (1989).
21. K.J. Hemker, M.J. Mills and W.D. Nix, J. Mater. Res. **7**, 2059 (1992).
22. C.S. Chang and D.P. Pope, MRS Symposium, **213**, 745 (1991).
23. G. Hug, PhD Thesis, University of Paris (1988).
24. B.A Greenberg, Phys. Stat. Sol. **42**, 459 (1970).
25. B.A. Greenberg and Y. N. Gornostirev, Scripta Metall. **22**, 853 (1988).
26. B.A. Greenberg et al., Acta Metall. **39**, 233 (1991).

DISLOCATIONS AND SLIP SYSTEMS IN V_3Si

L.S. SMITH, M. AINDOW AND M.H. LORETTO
IRC in Materials for High Performance Applications, and School of Metallurgy and Materials, The University of Birmingham, Edgbaston, Birmingham, B15 2TT, UK.

ABSTRACT

A transmission electron microscopy study of the defect structure in Czochralski single crystals of V_3Si is presented. As grown crystals contain occasional V_5Si_3 precipitates and a low density of edge dislocations with **b** = a<100>. Following deformation to 4% strain at 1600°C in compression much higher densities of precipitates and screw dislocations are present. Images of the dislocations are doubled under certain conditions and it is suggested that this corresponds to a dissociation of the form; a<100> ⇒ a/2<100> + CSF + a/2<100>, as observed previously in Nb_3Al. The significance of this dissociation for deformation mechanisms in V_3Si is discussed.

INTRODUCTION

A great many intermetallic compounds have been considered recently as potential high temperature structural materials. The highest operating temperatures are probably offered by the "exotics" whose structures involve simple lattices with complex bases rather than those based on ordering of different species onto the sites of a simple structure. One example of such a structure is the A15 phase which is formed in certain binary systems at A_3B compositions, e.g. Nb_3Al, Cr_3Si and V_3Si. As with most exotics, A15 compounds are rather brittle but the reasons for this are not well understood.

In V_3Si, for example, there have been many attempts to identify the nature of defects and deformation mechanisms but no clear picture has emerged. Etch pit studies in deformed single crystals have been used to suggest that deformation occurs by slip on the systems <100>{001} [1] and <100>{012} [2]. Transmission electron microscopy (TEM) studies of grown-in dislocations appear to confirm that they have **b** = a<100> [3] but the observation of double images [4] has caused some controversy. These features have been explained as **g**.**b** = 2 contrast [5], dipoles [6] and even pairs of dispirations on common {012} planes [7], but none of these explanations can account fully for the observed contrast. A TEM analysis of dislocation structures produced during creep deformation showed many dislocations with **b** = a<100>, often arranged as arrays in subgrain boundaries [8]. The consensus would appear to be that slip must occur on the system <100>{001} since the lattice is simple cubic, but this has not been proved conclusively. In this paper we describe a TEM study of the dislocation microstructure in V_3Si single crystals as grown and after compression at 1600°C. The implications of these observations for deformation mechanisms in V_3Si and isostructural compounds are discussed

EXPERIMENTAL PROCEDURE

Single crystals of V_3Si were prepared by the Czochralski method in a tri-arc furnace with a 30g charge and using a water cooled seed rod. Compression specimens 3mm in diameter and 10mm in length were cut from the crystals using electro-discharge machining (EDM). The specimens were heated to 1600°C under an background pressure of Argon in an Instron test rig and deformed to 2% total plastic strain at 5×10^{-5} s^{-1}. Discs for TEM specimens were cut from both deformed and undeformed portions of the crystal using EDM and ground mechanically to a thickness of 125μm. Final thinning was achieved using the twin jet electropolishing technique with a solution containing 1 part hydrofluoric acid to three parts orthophosphoric acid in a mixed ethanol/ethandiol base at -5°C and 0.1A. The foils were examined in a Philips CM20 TEM operating at an accelerating voltage of 200kV.

RESULTS

Undeformed Specimens

Samples produced from as-grown crystals appear almost featureless in the TEM and give rise to selected area diffraction patterns (SADPs) corresponding to the A15 phase. A small number of precipitates is observed; these have a spheroidal morphology and SADPs obtained from them indicate that they consist of V_5Si_3 with the hexagonal $D8_8$ structure. A very low density of dislocations is observed ($<5x10^5cm^{-2}$) and two such features are shown in Fig. 1. Figs.1(a-e) are weak beam dark field (WBDF) images obtained with diffraction vectors, **g**, given by **g** = 200, $\bar{2}00$, 210, 021 and $0\bar{2}0$, respectively. Both defects are out of contrast when imaged either with **g** = 021 (Fig.1(d)), or with **g** = $0\bar{2}0$ (Fig.1(e)); and this is consistent with the dislocations having their Burgers vectors, **b**, parallel to [100]. In Figures 1(a) - (c) the images appear to split into two lines ≈7nm apart. It is unlikely that the doubling arises as a result of **g.b** = 2 contrast since similar effects are also observed with **g**=$1\bar{2}0$ and other reflections for which the value of **g.b** is half that in the images shown here. Furthermore, no change of line spacing is observed on reversing the sign of **g** (e.g. Figs. 1(a) and (b)) and thus the features are not closely spaced dipoles. It can be deduced from these observations that these are pairs of dislocations whose Burgers vectors are parallel and of the same sense and thus they must be partial dislocations bounding faults. Stereographic analysis indicates that the line direction, **u**, of the dislocations is given by **u** = [010], thus the dislocations have almost pure edge character. A similar analysis indicates that the direction which connects the ends of the two long segments as they terminate at the foil surface corresponds to ≈[110] and thus the partial dislocations probably lie on a common (001) plane. Analysis of many such

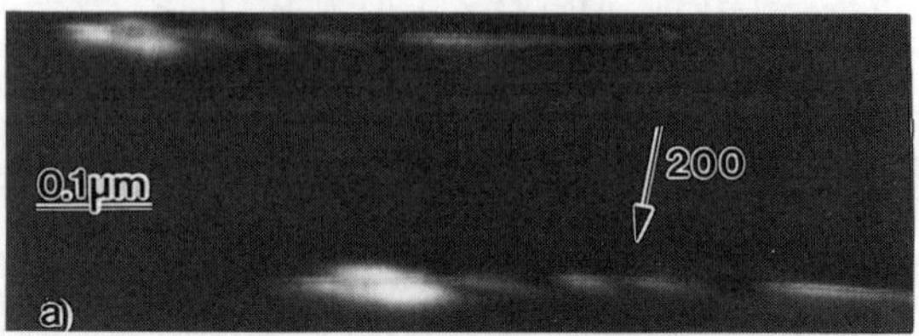

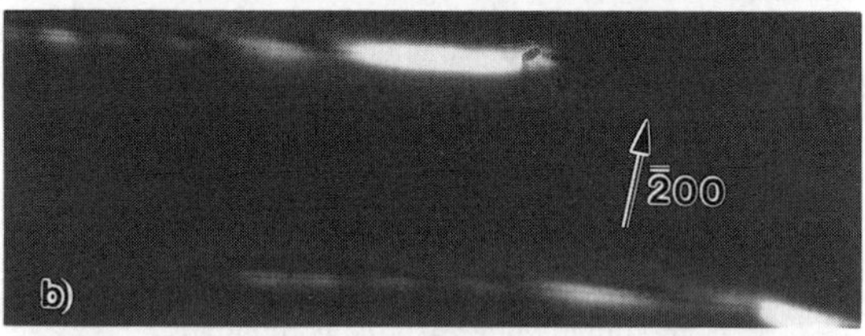

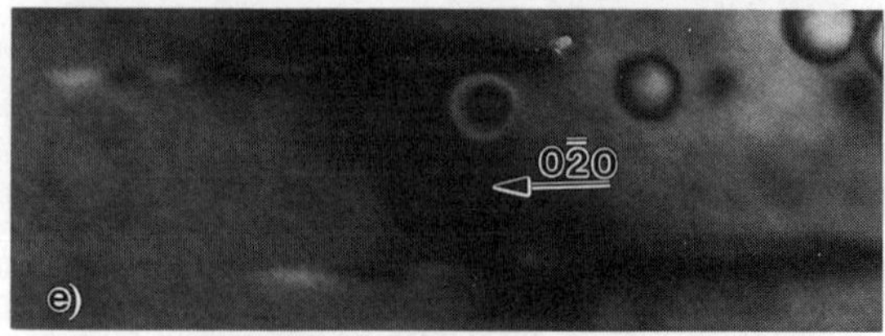

Figure 1. WBDF TEM micrographs from two edge dislocations in undeformed V_3Si.

isolated features shows that, in the vast majority of cases, the partial dislocations are almost pure edge with **b** parallel to <100> lying on {001} planes.

It can be seen that, in Fig. 1(e) and, in particular, Fig. 1(d), the dislocation feature exhibits significant residual contrast. In the latter image, residual **g**.(**bxu**) contrast is to be expected, but for Fig. 1(e) **g**.(**bxu**) = 0 and yet some residual contrast is still observed. This contrast could arise because of; elastic anisotropy, segregation of interstitial impurities to the fault plane, or the presence of an additional supplementary displacement associated with the ribbon of fault due to elastic relaxations across the fault plane. We can discount the first of these since {001} are elastic symmetry planes in the A15 structure but we cannot distinguish between the other alternatives at this stage.

Deformed Specimens

Samples produced from the deformed crystals show much higher densities of V_5Si_3 precipitates than do the as-grown samples. The dislocation content is also much higher but the majority of them are present as arrays in subgrain boundaries. Estimates of the dislocation density have an upper bound of $\approx 10^9 cm^{-2}$ and a lower bound of $\approx 5x10^7 cm^{-2}$ including and excluding those contained in boundaries respectively. Both the isolated dislocations and those in subgrain boundaries show double images in WBDF images, as in the undeformed specimens, but in the present case the isolated dislocations are predominantly screw in character.

Fig. 2 is a bright field image obtained using **g** = 210 from a region containing a typical subgrain boundary. The dislocations are present in two regular orthogonal arrays with a spacing of ≈80nm. Analysis of this boundary shows that the dislocations in the vertical array have **b** // **u** // [100] whilst the orthogonal set have **b** // **u** // [010] and the boundary plane is thus (001). Diffraction pattern analysis showed that the rotation across the boundary was a pure twist of 0.3° which is consistent with the spacing of the moiré fringes lying parallel to **g**. Applying the Frank model for low angle grain boundaries to these values confirms that the Burgers vectors of the dislocations must have a magnitude equal to the lattice parameter, i.e. **b** = a<100> in each case as expected. All of the subgrain boundaries examined in this material exhibited regular structures, with most being either symmetrical tilt boundaries consisting of edge dislocations, or pure twist boundaries with orthogonal arrays of screw dislocations.

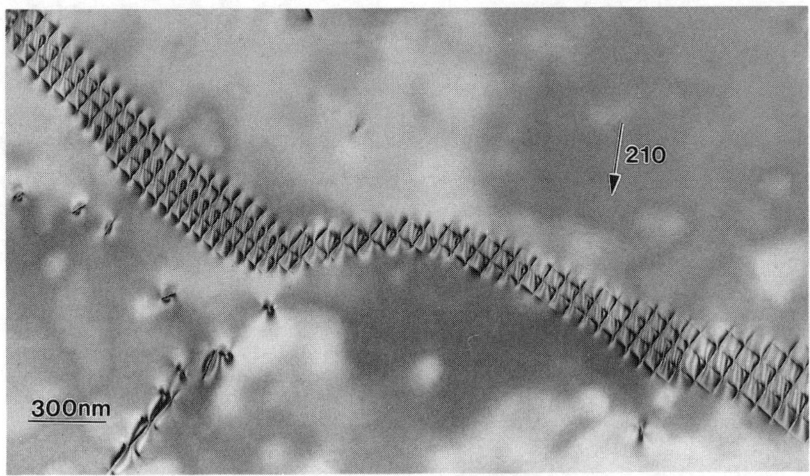

Figure 2. Bright field TEM micrograph obtained from a subgrain boundary in deformed V_3Si.

Figs.3(a-e) are WBDF images obtained from a region containing a subgrain boundary and two long isolated dislocations using **g** = 200, $\bar{2}00$, 210, 020 and 021, respectively. The two long dislocations have **b** // [100] since they are out of contrast when imaged either with **g** = 020 (Fig.3(d)), or with **g** = 021 (Fig.3(e)). In Figures 3(a) - (c) they appear to split into two lines $\approx$4nm apart and here again we deduce that the doubling arises from a dissociation since the spacing does not change on reversing **g** (Figs 3(a) and(b)) and thus they cannot be dipoles. Stereographic analysis indicates that **u** = [100], i.e. the isolated dislocations are pure screw, but it was not possible to obtain accurate measurements of the dissociation plane since the partial dislocations were not separated sufficiently at their ends. In Fig 3(a) it can be seen that the partial dislocations constrict at various points along their length. Analysis of many such features shows that, in the vast majority of cases, the partial dislocations are almost pure screw with **b** // <100>. Here again we suggest that residual contrast in Figs. 3(d) and (e) must arise from segregation or an additional supplementary displacement across the fault plane since **g**.(**bxu**) = 0 in both cases. We note that, in this case, the rotation at the subgrain boundary is a mixture of tilt and twist components and the boundary is comprised of dislocations with mixed character having **b** // [100].

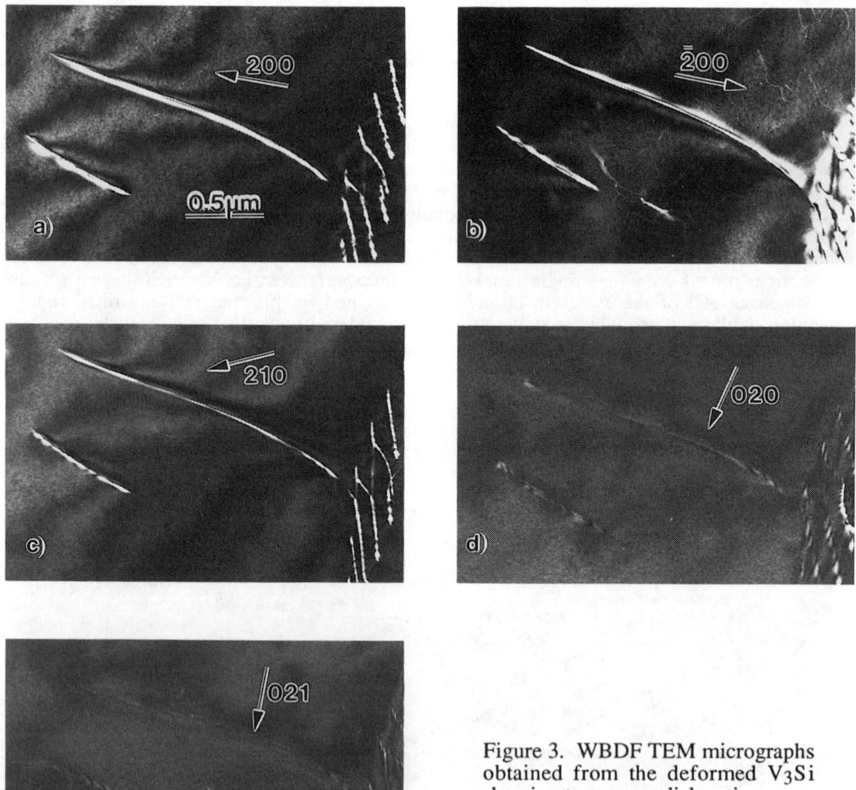

Figure 3. WBDF TEM micrographs obtained from the deformed V_3Si showing two screw dislocations near a subgrain boundary

DISCUSSION

The density of dislocations in the deformed samples is much higher than that in the as grown crystal and therefore it seems likely that dislocation glide is the dominant deformation mechanism. The presence of higher densities of precipitates and well ordered subgrain boundaries does, however, suggest that diffusion and significant recovery has taken place and thus it is difficult to deduce the slip systems directly. Instead we consider the origins and possible consequences of the observed dissociations.

If the observed dislocation features do indeed originate by the dissociation of perfect dislocations with **b** = a<100>, then the simplest reaction consistent with the experimental data from both the undeformed and deformed specimens is;

$$a\langle 100\rangle \Rightarrow a/2\langle 100\rangle + a/2\langle 100\rangle$$

which gives partial dislocations bounding a complex stacking fault (Stacking fault plus APB) and satisfies the Frank rule. If the dislocations are in edge orientation, as observed in the undeformed specimens, then a perfect dislocation can undergo a shear dissociation of the above form only on the plane containing **b** and **u**. This is consistent with the dissociation plane {001} observed experimentally. If, however, the dislocations are in screw orientations they can undergo a shear dissociation on any planes which contain the Burgers vector, i.e. on {hkl} where **b**.[hkl] = 0. Clearly the plane on which this occurs will be that which corresponds to the lowest fault energy. Since the A15 structure has the space group $Pm\bar{3}n$, the atomic arrangements across the fault produced on {001} by the shear dissociation of edge and screw dislocations with the same **b** will be different. This may account for the differences in partial dislocation separation observed in Figs 1 and 3, however we note that pairs of line defects in screw orientations have been observed on common {012} planes in V_3Si [4]. Furthermore, a similar dissociation has been observed on {012} for screw dislocations in Nb_3Al [9]; in this study constrictions were also observed as the dissociation plane changed from one {012} to another. We conclude that the screw dislocations in the deformed specimens probably dissociate on {012} planes which are the second most closely packed planes in the A15 structure.

The diffraction contrast experiments outlined in the previous section demonstrate that, at least in our specimens, double images at dislocation features are not simply **g**.**b**=2 contrast or closely spaced dipoles as suggested previously [5,6]. Reynaud and Lamine [7] suggested that a perfect dislocation can dissociate into two perfect dispirations thus:

$$a\langle 100\rangle \Rightarrow (4_2^+ / a/2\langle 111\rangle) + (4_2^- / a/2\langle 1\bar{1}\bar{1}\rangle)$$

A simple analysis of the elastic strain fields [10] shows that this defect would have an energy three orders of magnitude higher than that for the partial dislocations proposed here. It is, therefore, the present authors' contention that a dissociation into two partial dislocations is a more satisfactory explanation for the double line contrast exhibited by dislocations in V_3Si than those proposed previously.

The inherent brittleness of compounds with the A15 structure is usually ascribed to the operation of the system <100>{001} such that the Von Mises criterion is not fulfilled since only three independent slip systems are available. If the dislocations are dissociated similarly during plastic deformation, then the slip system which operates will be <100>{012}. It was suggested previously that this gives many distinct combinations of five slip systems and that Von Mises criterion is fulfilled [9]; a subsequent analysis by Shyue [11] shows that this is not the case and that here again, for any combination of systems, only three are independent. Furthermore, all simple models of the Peierls barriers to dislocation motion indicate that these will be greater for <100>{012} slip than for <100>{001}slip. Use of a hard sphere model suggests that the maximum dilation perpendicular to the slip plane as a dislocation with **b** = a<100> passes is greater on {012} than on {001}. If we consider the number of bonds broken as a dislocation glides, the preferred slip plane will depend on the relative strength of V-V and V-Si bonds; more V-V bonds are broken for glide on {001} whereas more V-Si bonds are broken for glide on {012}. It has been proposed that V-Si bonds are stronger at elevated temperatures favouring slip on <100>{001} over that on other systems [6]. The significance of the dissociation proposed here is that, if the screw dislocations responsible for glide in V_3Si are dissociated on {012} when at rest, then there may be an additional resistance to cross slip from {012} to {001} planes since there will be an associated change in fault energy. Such an effect could help to explain the brittleness of V_3Si even when in the form of single crystals where Von Mises criterion is not appropriate.

Clearly, for deformation at 1600°C the deformation mechanisms which operate will correspond to creep. Nghiep *et al.* [8] have concluded that, under such conditions the rate controlling processes were dislocation glide and dynamic recovery by dislocation climb. In order to determine the details of slip system activity and test the above hypotheses, we are now performing similar experiments with deformation at lower temperatures.

CONCLUSIONS

TEM studies of Czochralski V_3Si single crystals have shown that the dislocations present are pure edge whereas deformation in compression introduces much higher densities of screw dislocations. Both the edge and screw dislocations have Burgers vectors **b** = a<100> and are dissociated into two partial dislocations with parallel Burgers vectors **b** = a/2<100> bounding a complex stacking fault on {001} and {012} respectively. It has been suggested that a dissociation on {012} will inhibit <100>{001} slip since the fault energy may change on cross-slip from {012} to {001}.

ACKNOWLEDGEMENTS

This work has been carried out with the support of SERC. The authors would like to thank Dr. Y.J. Bi for preparation of the single crystals and Dr. J. Shyue and Prof. H.L. Fraser for helpful discussions. Financial support to attend this meeting was provided by the Institute of Physics EMAG group (LSS + MA), C.R. Barber Trust Fund (LSS) and the Institute of Materials Andrew Carnegie Grant Scheme (LSS).

REFERENCES

[1] H.J. Levinstein, E.S. Greiner and H. Mason Jr., J. Appl. Phys., **37**,164 (1966).
[2] R.N. Wright and K.A. Bok, Met. Trans. A, **19**, 1125 (1988).
[3] U. Essman, H. Haag and G. Zerweck, Comm. on Phys., **2**,127 (1977).
[4] A.B. Lamine and F. Reynaud, Phil. Mag. A, **38**, 359 (1978).
[5] S. Nakahara, S. Mahajan, J.H. Wernick and G.Y. Chin, J.Appl. Phys., **50**,3552 (1979).
[6] U. Kramer, Phil. Mag. A, **47**, 721 (1983).
[7] F. Reynaud and A.B. Lamine, Acta Met., **29**, 1485 (1981).
[8] D.M. Nghiep, P. Paufler, U. Kramer and K. Kleinstuck, J. Mat. Sci., **15**, 1140 (1980).
[9] M. Aindow, J. Shyue, T.A. Gaspar and H.L. Fraser, Phil. Mag. Lett., **64**, 59 (1991).
[10] L.S. Smith, M. Aindow and M.H. Loretto, in preparation.
[11] J. Shyue, Ph.D. Thesis, The Ohio State University (1992).

THE ROLE OF INTERFACES IN THE ROOM TEMPERATURE MECHANICAL BEHAVIOR OF DIRECTIONALLY SOLIDIFIED $\beta+\gamma'$- $Ni_{70}Al_{30}$ AND $\beta+(\gamma+\gamma')$-$Ni_{50}Fe_{30}Al_{20}$ ALLOYS

A. Misra*, R.D. Noebe** and R. Gibala*
* Department of Materials Science and Engineering, The University of Michigan, Ann Arbor, MI 48109-2136
** NASA Lewis Research Center, Cleveland, OH 44135.

ABSTRACT

We have shown previously that directionally solidified $\beta+\gamma'$-$Ni_{70}Al_{30}$ and $\beta+(\gamma+\gamma')$-$Ni_{50}Fe_{30}Al_{20}$ alloys with quasi-lamellar microstructures exhibit up to 10% tensile ductility at 300 K. The ductility enhancement was partly attributed to a slip transfer mechanism which is favored by the Kurdjumov-Sachs orientation relationship and strong interphase interface that shows no debonding. The slip systems in the β phase were of the type {110}<001> even though the loading axis was parallel to the <001> growth direction in β. In the present investigation, an electron microscopy study of the β / γ' and $\beta / (\gamma+\gamma')$ interfaces has been conducted to understand the interfacial structure and its relation to room temperature mechanical properties. Furthermore, the stress concentration at the interface is treated analytically to understand how {110}<001> slip occurs in the β phase, even though the loading axis is along <001> .

INTRODUCTION

It is well established that brittle materials such as intermetallics and ceramics can be substantially toughened by ductile phase reinforcements [1]. For very brittle matrix materials like ceramics, the ductile phase toughening obtained is much greater for weak interfaces where decohesion can occur than for strongly bonded, highly constrained, interfaces [1]. However, for some intermetallics like β-NiAl where room temperature ductility is limited by dislocation mobility and density, the highly constrained ductile reinforcement may cause ductility enhancement by efficient dislocation generation into the matrix [2]. Earlier work on two directionally solidified model materials, $\beta+\gamma'$-$Ni_{70}Al_{30}$ and $\beta+(\gamma+\gamma')$-$Ni_{50}Fe_{30}Al_{20}$, has shown that up to 10% tensile ductility at 300 K may be achieved in these ductile-phase reinforced composites [2,3]. This ductility enhancement was partly attributed to a slip transfer mechanism which is favored by the Kurdjumov-Sachs orientation relationship (slip planes in the two phases are nearly parallel) and the strong interphase interface.

In the present investigation, an electron microscopy study of the interfaces in these directionally-solidified alloys has been conducted to understand the interfacial structure and its relation to the dislocation generation and slip transfer processes. Furthermore, the elastic and plastic constraints at the interfaces were modelled to explain the mechanism for {110}<001> slip in the β phase, even though the loading axis was <001>. Transmission electron microscope (TEM) investigations were performed on JEOL 2000FX and JEOL 4000EX microscopes.

INTERFACIAL STRUCTURE - MECHANICAL BEHAVIOR RELATIONSHIPS

Misfit Accommodation

The β / γ' interfaces in the $Ni_{70}Al_{30}$ alloy and the $\beta / (\gamma+\gamma')$ interfaces in the $Ni_{50}Fe_{30}Al_{20}$ alloy exhibit the Kurdjumov-Sachs orientation relationship :

$$(111)\,\gamma \,/\!/\, (110)\,\beta \,/\!/\, \text{interface plane and } [0\bar{1}1]\,\gamma \,/\!/\, [1\bar{1}1]\,\beta.$$

Van der Merwe *et al.* [4] have shown that for $\{111\}_{fcc}$ // $\{110\}_{bcc}$ interfaces with either Kurdjumov-Sachs or Nishiyama-Wasserman orientation relationships, misfit accommodation by structural ledges is energetically preferable to planar misfitting interfaces. In general, the planar interfaces were found to be stable only above a critical misfit strain. Fig. 1 shows a HREM (high resolution electron microscope) micrograph of the β / γ' interface viewed along the $[001]\beta$ // $[\bar{1}01]\gamma'$ direction. Approximate positions of the monoatomic structural ledges (spacing ~ 1.2 nm) are marked by arrows. No extra half plane is seen. This is consistent with the original geometric model of Rigsbee *et al.* [5] for fcc/bcc interfaces. Note that the interface plane // $(111)\,\gamma$ // $(110)\,\beta$. Fig. 2 shows an atomic model of the β / γ' interface viewed along the $[001]\,\beta$ // $[\bar{1}01]\gamma'$ direction. Without any structural ledges, an extra $(\bar{1}10)\beta$ plane is needed every 13 $(1\bar{1}1)\gamma'$ planes. However, with the introduction of monoatomic structural ledges, the misfit in the plane of the drawing is accommodated. Rigsbee *et al.* [5] have predicted that a single array of misfit dislocations will also exist in the interface plane , normal to the structural ledge direction, to accommodate misfit in the third dimension. These misfit dislocations are not visible in the orientation shown in Fig. 1. No HREM work could be performed on the strongly ferromagnetic $Ni_{50}Fe_{30}Al_{20}$ alloy but the same misfit accommodation mode is expected.

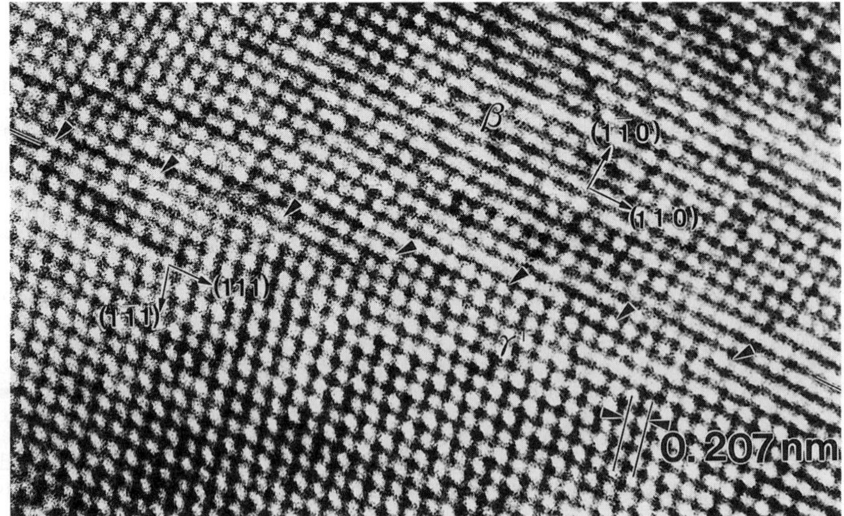

Fig. 1 HREM micrograph of the β / γ' interface viewed along the $[001]\,\beta$ // $[\bar{1}01]\gamma'$ direction. Monoatomic structural ledges at the interface are marked by arrows. No extra half plane is seen.

These low energy, semi-coherent β / γ' and β / (γ+γ') interfaces are expected to be very strong and indeed, no decohesion at the interface was observed after tension and compression tests [2]. Thus, the strong interfaces in these composites provide a case of high constraint which is extremely important for the dislocation nucleation and slip transfer processes discussed later in this paper. Van der Merwe *et al.* [4] have suggested that the elastic stresses in the vicinity of the structural ledges, coupled with the externally applied stress, may initiate early plastic behavior. Depending on the material and misfit of the two crystals, plastic behavior may result from the presence of steps alone [6]. However, microscopy study of the deformed specimens has not provided any clear evidence of dislocation nucleation from <u>structural</u> ledges in our materials.

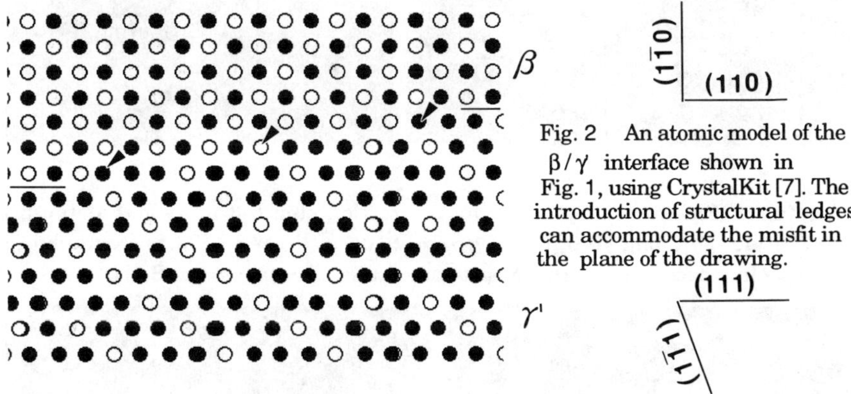

Fig. 2 An atomic model of the β / γ' interface shown in Fig. 1, using CrystalKit [7]. The introduction of structural ledges can accommodate the misfit in the plane of the drawing.

<u>Interface Facets (Direction Steps)</u>

While the structural ledges are present for misfit compensation, other larger steps may also be present at the interface to compensate for interface curvature or to allow the lamellae to adjust spacing while preserving the low energy habit plane between the two phases [8] . Fig 3(a) is a HREM micrograph that shows a direction step (~ 3 nm wide) at the β / γ' interface. Thus, even though the interface in the lamellar microstructure may show smooth curvature when viewed by either optical microscope, or scanning electron microscope or low magnification TEM , the interface maintains its low energy habit plane on the atomic level. The accommodation of a complex curvature may involve two or more non-parallel direction steps as shown in Fig. 3 (b), which is a weak beam (WB) TEM micrograph of a curved β / γ' interface tilted to reveal the structure.

The direction steps (interface facets) provide excellent sources for dislocation nucleation at the interfaces in these composites, similar to what has been postulated for steps at surface film-substrate interfaces. Fig. 4 shows an example of dislocation nucleation in the β phase from the direction steps at the interface in the $Ni_{50}Fe_{30}Al_{20}$ alloy. The slip systems for these dislocations were of the {100}<010> type, as determined by TEM trace analysis. Similar nucleation of mobile <001> dislocations from the direction steps was observed in the $Ni_{70}Al_{30}$ alloy also. The nucleation of mobile dislocations from the interfacial steps plays a significant role in the reduction of flow stress and ductility enhancement observed in these composites [2,3]. Analytical treatment of the dislocation nucleation at steps will be presented in a future paper.

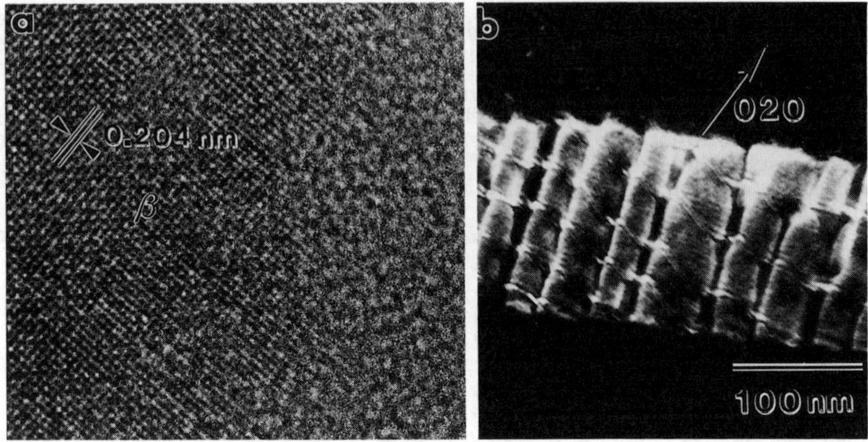

Fig. 3 (a) HREM micrograph showing a direction step (~3 nm wide) at the β/γ interface viewed along [100]β , ~10° from [010]γ . (b) WB TEM micrograph of a curved β/γ interface showing two sets of non parallel direction steps. **B**= [101]

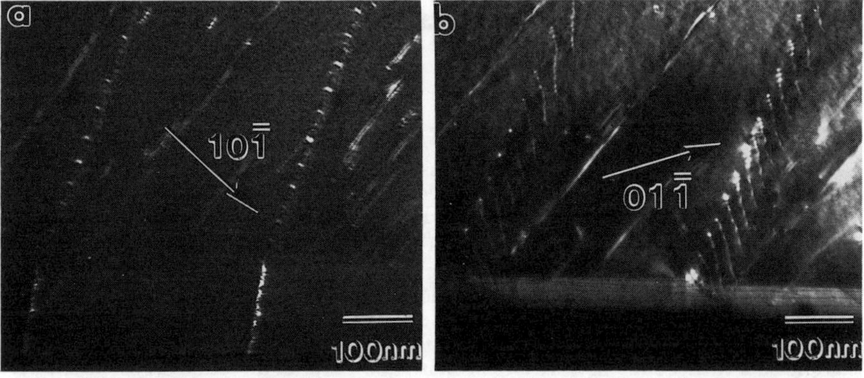

Fig. 4 (a) WB TEM micrograph showing dislocation nucleation from direction steps into the β phase in the $Ni_{50}Fe_{30}Al_{20}$ alloy. The angle between the slip plane normal and beam direction is close to 90°. **B**=[131]. (b) WB TEM micrograph of the same area as in (a) , tilted to reveal the individual dislocations. **B**=[111].

SLIP TRANSFER

The observations of slip traces and dislocation substructures in the deformed alloys have shown that the enhanced ductility of these alloys may be partly attributed to easy slip transfer from the ductile reinforcing phase to the brittle matrix [2,3]. This slip transfer is favored by the Kurdjumov-Sachs orientation relationship since the slip planes in the two phases are nearly parallel. The slip systems in the β phase as a result of slip transfer were of the {101}<010> type, even though the loading axis was parallel to [001]β. Thus, the resolved shear stress in the β phase comes from the elastic and plastic incompatibility stresses at the interface. A stress model, similar to the one used by Miracle [10] to explain the generation of <110> dislocations at grain boundaries in NiAl bicrystals, is formulated to determine whether the incompatibility stresses can preferentially nucleate <010> dislocations in β-NiAl otherwise oriented for <111> slip. The crystallography used in the calculations is shown schematically in Fig. 5.

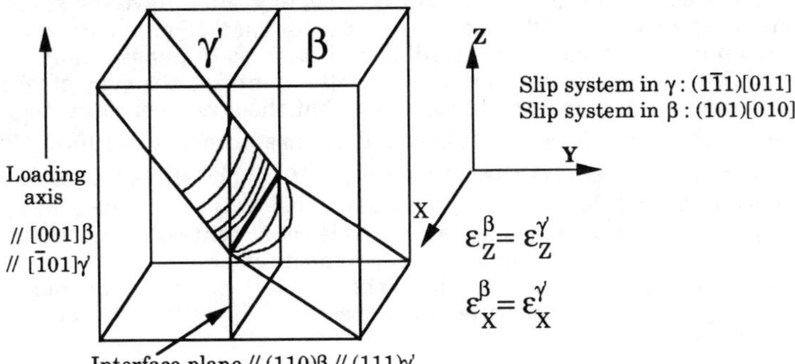

Fig. 5 Schematic illustration of slip transfer across β / γ and β / (γ+γ') interfaces.

Elastic Incompatibility

When the composite is loaded uniaxially to the yield stress of γ' , the strain tensors in decoupled β and γ' are calculated using the generalized form of Hooke's law. These are then transformed to the boundary coordinates and the difference between the two strain tensors gives the elastic incompatibility strain tensor. The elastic incompatibility stress tensor in the β phase is then calculated using the stiffness constants for NiAl and then resolved onto the possible slip systems i.e. {110}<111>, {110}<001> and {100}<010>. It was found that the resolved shear stress due to elastic incompatibility alone does not exceed the critical resolved shear stress (CRSS) of any slip system in the β phase. The CRSS data was obtained from [10].

Plastic Incompatibility

The resolved shear stress, τ^{β}, on any slip system in β due to a dislocation pile-up in the γ' phase may be calculated following the approach of Livingston and Chalmers [11] :

$$\tau^{\beta} = m_{ij} \,.\, N \,.\, \tau^{\gamma'} \tag{1}$$

where N is the number of dislocations in the pile-up , $\tau^{\gamma'}$ is the shear stress on the active slip system in γ' , $m_{ij} = (e_\gamma \,.\, e_\beta)(g_\gamma \,.\, g_\beta) + (e_\gamma \,.\, g_\beta)(g_\gamma \,.\, e_\beta)$; e=slip plane normal and g= slip direction. At the composite yield stress, the normal stress in the γ' phase is calculated assuming that γ' is plastic and β is elastic. The resolved shear stress on the active slip system in the γ' is calculated by Schmid's law. For each possible slip system in the β phase, the value of N required to make τ^{β} equal to the quantity

(CRSS - $\tau^{\beta}_{elastic\ incompatibility}$ - $\tau^{\beta}_{applied\ stress}$) is then calculated using equation (1).

For the specific γ' slip system shown in Fig. 5, the calculations predict that both (011)[100] and (101)[010] systems need only 2 dislocations in the pile-up to be activated. The m_{ij} factor for the former is 0.49 while that of the latter is 0.41 and so according to the pure shear approach of Livingston amd Chalmers [11] , (011)[100] should be the active slip system. But the (101)[010] slip system was observed. The failure of this approach to correctly predict the slip system activated in grain 2 by a pile-up in grain 1 when all the possible slip systems in grain 2 have the same CRSS was shown by Shen *et al.* [12]. However, in the case of slip transfer across interphase boundaries where all the possible slip systems in phase 2 may not have the same CRSS, the pure shear approach correctly identifies the type of slip to be activated. In other words, the calculations show that the stress concentration at the interface will activate {110}<001> slip in β in preference to {100}<010> and {110}<111> slip systems. Out of the 12 possible {110}<111> slip systems in β, the one with the highest m_{ij} factor needs at least 7 dislocations in the pile-up to be activated. The number of dislocations calculated is consistent with that observed by TEM [3] and also with Chou's formulation [13] of the number of dislocations in a pile-up at an interphase boundary. The details of the calculations and refinements in the model to correctly predict the experimentally observed slip system will be presented in a separate paper.

SUMMARY

The misfit accommodation at the β / γ' and β / (γ+γ') interfaces occurs by the structural ledge mode. The direction steps compensate for interface curvature to maintain the low energy habit plane and act as sources of mobile <001> dislocations in the β phase. A model has been developed to show that the elastic and plastic incompatibility stresses at the β/ γ' and β / (γ+γ') interfaces can nucleate <001> dislocations in the β phase otherwise oriented for hard <111> slip.

REFERENCES

1. M.F. Ashby, F.J. Blunt and M. Bannister, Acta Met., **39**, 2575 (1991).
2. R.D. Noebe, A. Misra and R. Gibala, ISIJ International, **31**, 1172 (1991).
3. A. Misra, R.D. Noebe and R. Gibala, in Intermetallic Matrix Composites II, ed. by D.B. Miracle, D.L.Anton and J.A. Graves (Mater. Res. Soc. Proc. **273** , Pittsburgh, PA 1992) pp. 205-210.
4. J.H. van der Merwe, G.J. Shiflet and P.M. Stoop, Met. Trans., **22A**, 1165 (1991).
5. J. M. Rigsbee and H.I. Aaronson, Acta Met., **27**, 351 (1979).
6. G. J. Shiflet, in Structure and Properties of Interfaces in Materials, ed. by W.A.T. Clark, U. Dahmen and C.L.Briant (Mater.Res.Soc.Proc. **238**, Pittsburgh, PA 1992) pp. 53-63.
7. CryatalKit, Software by Roar Kilaas, Total Resolution, Berkeley,CA.
8. D.S. Zhou and G.J. Shiflet, Met. Trans., **22A**, 1349 (1991).
9. R.D. Noebe and R. Gibala, Scripta Met., **20**, 1635 (1986).
10. D.B. Miracle, Acta Met., **39**, 1457 (1991).
11. J.D. Livingston and B. Chalmers, Acta Met., **5**, 322 (1957).
12. Z. Shen, R.H. Wagoner and W. A. T. Clark, Scripta Met., **20**, 921 (1986).
13. Y. T. Chou, Acta Met., **13**, 779 (1965).

This research is funded by NSF Grant No. DMR- 9102414.

SLIP TRANSFER FROM γ' TO β PHASE IN A Ni-Al-Ti ALLOY

RUI YANG, JOHN A. LEAKE and ROBERT W. CAHN
Cambridge University, Materials Science Department, Cambridge CB2 3QZ, United Kingdom.

ABSTRACT

Two-phase alloys comprising γ'-$Ni_3(Al,Ti)$ and β-Ni(Al,Ti) deform in a non-homogeneous manner at room temperature. Plastic deformation initiates in the softer γ' phase, resulting in internal stresses which build up at the γ'/β phase boundaries. These stresses can be relieved by a number of mechanisms including cracking, void formation, and slip propagating across the boundary into the harder β phase. We have investigated the interaction between mobile dislocations and the γ'/β interface by TEM. We found that compatible deformation occurs at boundaries where a favourable orientation relationship exists between the adjoining phases. In particular, the $\gamma'[0\bar{1}1](111)$ dislocations can be injected into the β phase to become $\beta[100](011)$ dislocations, leaving residual dislocations at the interfaces. The effect of phase distribution on the operation of this mechanism is analysed, and discussed in relation to the enhancement of ductility of the β phase.

INTRODUCTION

During the plastic deformation of two-phase materials, internal stresses gradually build up in the neighbourhood of the phase boundaries where decohension would occur unless the stresses can be effectively relieved. In the case of composites where the harder phase is plastically undeformable, the relaxation is realised by means of dislocation activity only in the softer phase; the various mechanisms were analysed by Brown and Stobbs [1]. Metallic alloys differ from composites in that both phases are plastically deformable and therefore dislocation activities in the harder phase can be expected during plastic relaxation. In the case of two-phase alloys comprising γ'-$Ni_3(Al,Ti)$ and β-Ni(Al,Ti), we may call γ' the softer and β the harder phase, since polycrystalline NiAl [2] has much higher yield strength than polycrystalline Ni_3Al [3] at room temperature.

Figure 1 illustrates the microstructure of many γ'+β two-phase alloys with appropriate compositions. The β phase solidifies dendritically, the remnant liquid becoming the *primary* γ' after crystallisation. A homogenisation anneal changes the β dendrites into round β grains, and a further anneal at a lower temperature results in the precipitation of lath-shaped γ' particles in the β (we call these particles *secondary* γ'). An important feature is that the primary γ' phase is not completely wrapped by the β phase, in contrast with the γ' precipitates which are embedded in the β matrix. The primary γ' can respond plastically to an applied load without the β phase doing so in the very early stage. Later, glide dislocations in the primary γ' gradually move into the phase boundaries, and may generate glide dislocations in the β phase if the stress and geometrical conditions are favourable. We regard this process as slip transfer from the primary γ' to the β phase. The γ' precipitates, on the other hand, are "shielded" by the more rigid β phase, and cannot deform unless the β phase does. Such γ' inclusions cause stress concentration, and the precipitate/matrix interface can also provide dislocation sources for the β. The mechanism, however, is different from the slip transfer process as defined above, and will not be discussed here.

Slip transfer across high-angle grain boundaries has been studied by several authors using TEM (for example [4-6]; recently reviewed in [7] by Forwood and Clarebrough). Such events relieve the incompatibility stress introduced at grain boundaries because of elastic anisotropy of the crystals [8]. Detailed studies of slip transfer across phase boundaries are rarely reported, but there is good evidence that such a mechanism may be responsible for the ductility enhancement observed in a directionally solidified multiphase Ni-Fe-Al alloy [9]. In this paper, we report results of a TEM investigation of slip transfer from the γ' to the β phase in a Ni-Al-Ti alloy.

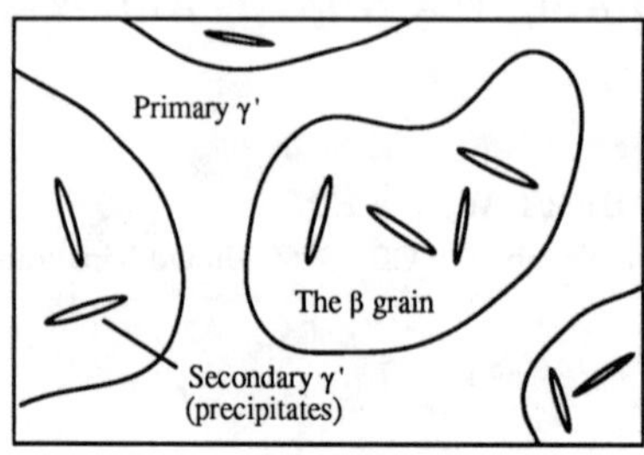

FIG.1 Schematic diagram showing the microstructure of γ'+β two-phase Ni-Al-Ti alloys.

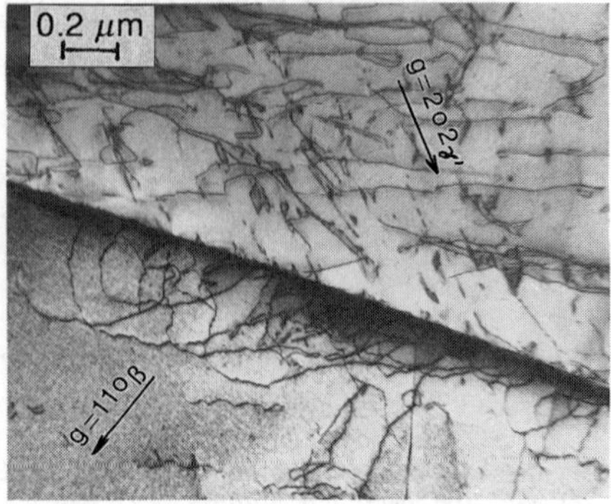

FIG.2 Double two-beam bright-field micrograph showing dislocation configuration near a γ'/β boundary in a sample deformed 2%. The two diffraction vectors are indicated.

EXPERIMENTAL

The alloy has a nominal composition of $Ni_{63}Al_{28}Ti_9$ (at.%) which sits on the γ'-β edge of the γ'+β+(β'-Ni_2AlTi) three-phase triangle in the 900°C isotherm [10]. The melting was done in an arc furnace under argon protection. After a homogenisation treatment of 4 h at 1050°C and a further anneal at 900°C for 90 h, the alloy has a microstructure as shown in Figure 1.

Rectangular blocks were cut from the alloy and compressed at room temperature to plastic strains up to 5% prior to the preparation of thin-foil specimens for electron microscope observation. The electropolishing was done in a solution of 8% perchloric acid in ethanol at about -50°C. The samples were examined on a Philips 400T microscope operating at 120 kV, using a ±45° double-tilting holder.

There are several procedures for determining the orientation relationship between two crystals using Kikuchi patterns (e.g., [11, 12]); we use Ball's [11] procedure. The dislocation Burgers vectors were identified using the $\boldsymbol{g}\cdot\boldsymbol{b}=0$ invisibility criterion [13]. Slip planes and boundary geometry were determined by trace analysis.

RESULTS

Figure 2 shows a boundary between primary γ' (top) and β (bottom) phase in a sample strained plastically to 2%. The γ' phase has been moderately deformed, as evidenced by the glide dislocations seen in the micrograph. In the harder β phase, plastic deformation was being

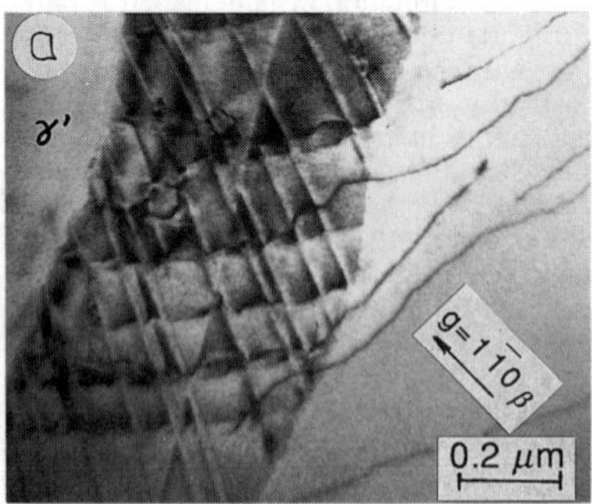

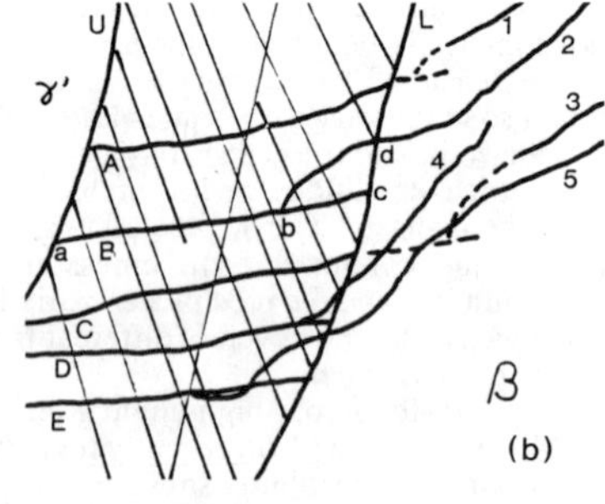

FIG.3 (a) Bright-field micrograph ($\boldsymbol{g}$= $1\bar{1}0_\beta$, as indicated) illustrating dislocation reactions at a γ'/β boundary in a deformed sample. (b) Schematic of (a); the intersection of the phase boundary with the top and bottom surfaces of the foil is marked U and L, respectively.

Table I.
Crystallographic Data of Slip Transfer from γ' to β Phase.

Case	Orientation Matrix ${}_{\beta}R_{\gamma'}$			Dislocation Burgers Vector			Slip Plane		
				$b_{\gamma'}$	b_{β}	Angle	in γ'	in β	Angle
1	-0.04263	0.97025	-0.22542	$[01\bar{1}]$	$[\bar{1}00]$	2.6°	(111)	(011)	3.4°
	-0.69979	0.13115	0.70205						
	0.71265	0.18941	0.67530						
2	0.84940	-0.21607	0.48135	$[10\bar{1}]$	[100]	14.5°	$(1\bar{1}1)$	$(0\bar{1}1)$	10.8°
	0.09150	0.95829	0.26934						
	-0.51971	-0.18396	0.83385						
3	-0.71048	-0.30517	-0.63377	$[1\bar{1}0]$	$[00\bar{1}]$	7.2°	$(11\bar{1})$	$(\bar{1}\bar{1}0)$	11.8°
	-0.62062	-0.15303	0.76861						
	-0.33166	0.93972	-0.07993						

initiated by the dislocation half loops generated from the phase boundary which relieve the incompatibility stresses there. When the dislocation half loops are not entirely contained in the thin foil, it is necessary to examine the dislocation reactions at the interface in order to establish in which direction the slip was propagating. An example is shown in Figure 3, where five extrinsic boundary dislocations originally coming from the γ' phase are labelled A through E. Each of these dissociates at the interface, emitting a mobile dislocation into β (numbered 1 to 5), and leaving a residual dislocation at the interface. In the second pair, for example, segment *ab* is the incoming γ' dislocation, segment *bd* the outgoing β dislocation, and segment *bc* the residual dislocation whose Burgers vector compensates for the Burgers vector difference of the incoming and outgoing dislocations. In the first and third pairs, the junction of the incoming and outgoing dislocations is outside the foil, and is represented by dotted lines in Figure 3(b). Other line defects seen in Figure 3 are due to the intrinsic structure of the phase boundary and are not of interest here.

In order to understand under what geometrical conditions slip can transfer from the γ' to β phases, we have examined three cases in detail and list the results in Table I. The slip planes in β, however, cannot be determined since the outgoing dislocations do not have a well-defined line direction and usually do not form a pile-up. We assume the slip plane in β to be of {110} type, and have listed in Table I that specific one which has the smallest misorientation angle with the corresponding slip plane in γ'. The results suggest that slip transfer from γ' to β can

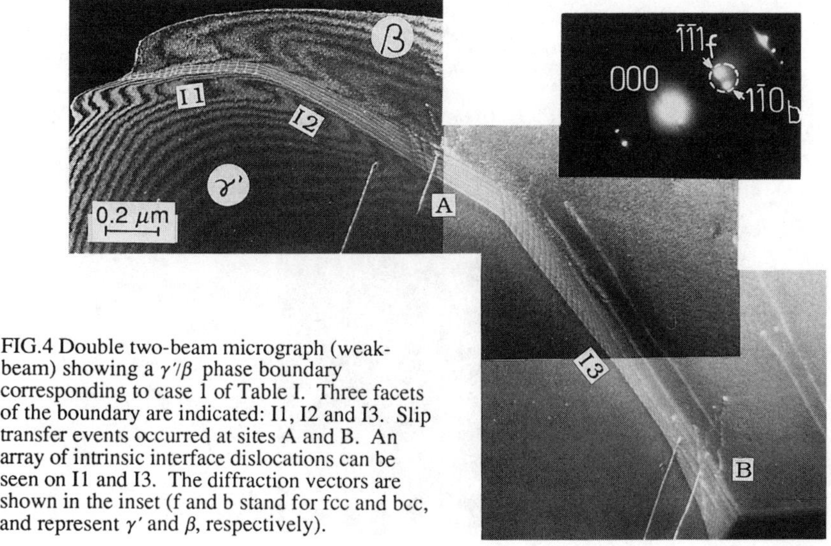

FIG.4 Double two-beam micrograph (weak-beam) showing a γ'/β phase boundary corresponding to case 1 of Table I. Three facets of the boundary are indicated: I1, I2 and I3. Slip transfer events occurred at sites A and B. An array of intrinsic interface dislocations can be seen on I1 and I3. The diffraction vectors are shown in the inset (f and b stand for fcc and bcc, and represent γ' and β, respectively).

operate even when the Burgers vectors of incoming and outgoing dislocations are misoriented about 15°. In the following, we illustrate the mechanism of the slip transfer process using case 1 as an example.

Figure 4 shows three facets of the γ'/β boundary investigated in case 1. The weak-beam micrograph was recorded using an objective aperture which enclosed both $(\bar{1}\bar{1}1)_{\gamma'}$ and $(1\bar{1}0)_{\beta}$ reflections (they are misoriented 14.5°, as can be measured from the diffraction pattern shown in the inset, or calculated using the rotation matrix given in Table I). The normals to facets I1, I2 and I3, as well as the line directions of the intrinsic boundary dislocations seen on I1 and I3, have all been determined, but we were unable to interpret these using any model because the lattice constants of the two phases are not known precisely.

At site A (Figure 5), seven slip transfer events have taken place, generating dislocations 1 to 7 in β, and leaving residual dislocations r_1 through r_7. The outgoing dislocation 7 was generated by the incoming dislocation G, so that r_7 runs only half way through the boundary. The middle part of r_6 is out of contrast in Figure 5(b) (but visible in other micrographs) owing to its strain field overlapping with dislocation G under this specific diffraction condition.

At site B (Figure 6), two dislocation transfer events occurred on the same slip plane, and the second event is due to the incoming dislocation H. Segment *ef* is the residual dislocation left

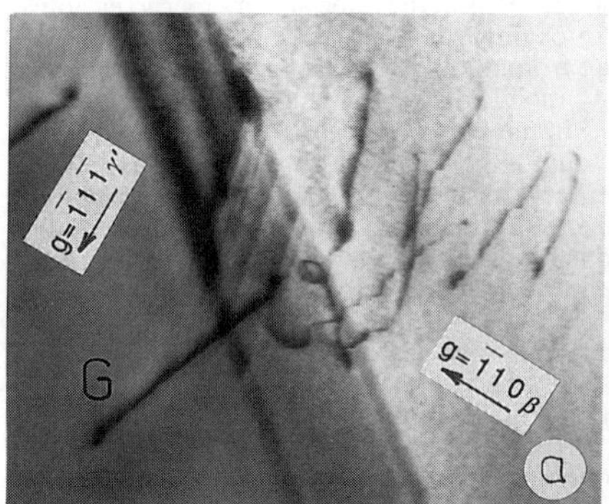

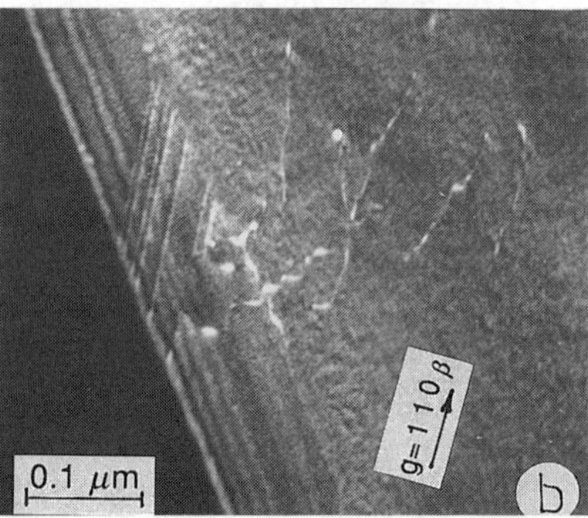

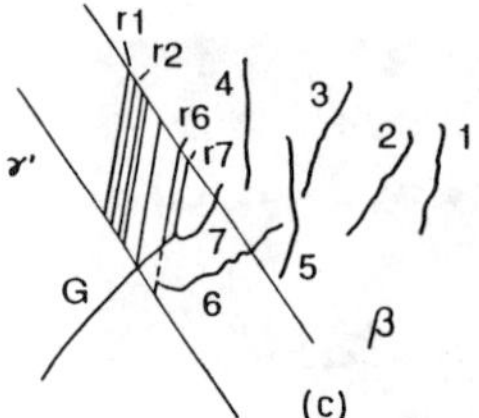

FIG.5 (*Above*) Slip transfer events at site A of Figure 4. (a) Double two-beam bright-field micrograph; (b) weak-beam micrograph; (c) schematic. The scale bar is the same for Figures 5 through 7.

FIG.6 (*Below*) Slip transfer events at site B of Figure 4. (a) Bright-field micrograph, diffraction conditions being the same as in Figure 5(a); (b) weak-beam micrograph; (c) schematic. (Dislocation I in (a) did not participate in the reaction.)

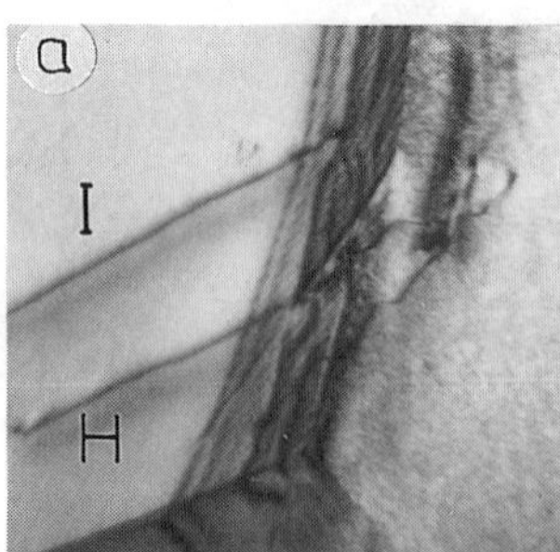

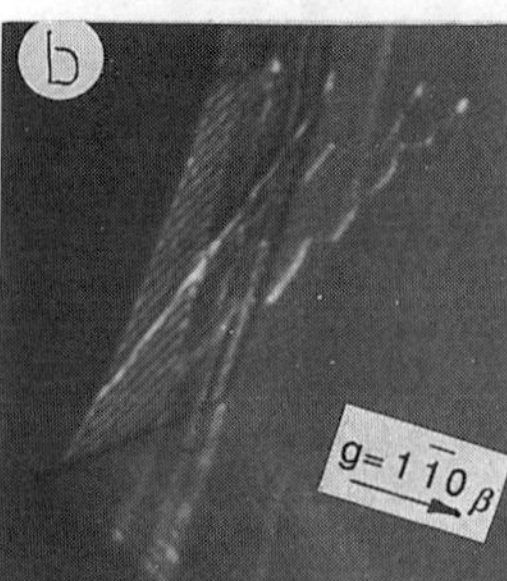

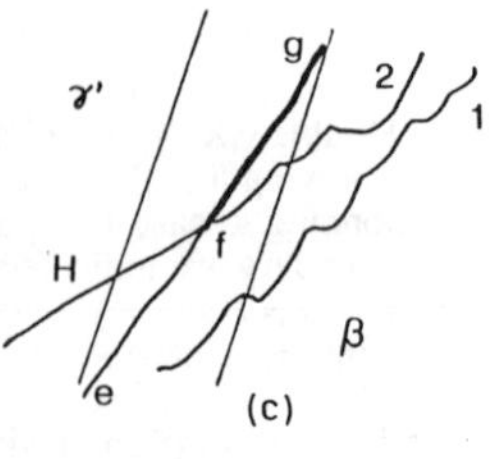

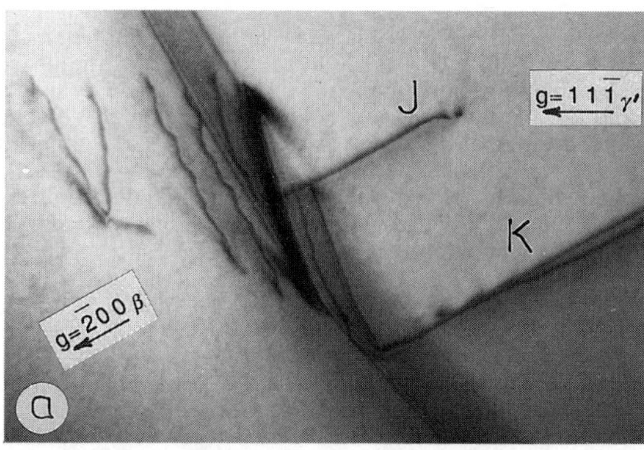

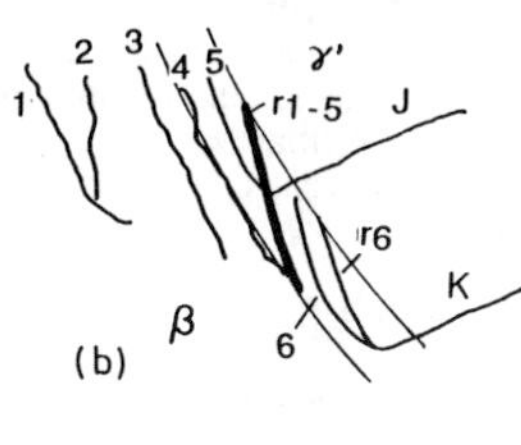

FIG.7 Slip transfer from γ' to β, at another boundary facet in case 1 of Table I. (a) Double two-beam bright-field micrograph; (b) schematic.

after the first transfer event, r_1, while segment *fg* is the sum of r_1 and r_2. Note that in both micrographs of Figure 6, dislocation segment *fg* appears darker than *ef*, indicating that they have different Burgers vectors.

A final example of slip transfer from γ' to β, also taken from case 1 but not seen in Figure 4, is illustrated in Figure 7. Here the incoming dislocation K generates the outgoing dislocation 6, leaving residual dislocation r_6. At a different place, the γ' dislocation J was generating the β dislocation 5. The dark, wide contrast is due to residual dislocations left by repeated transfer events there.

DISCUSSION

In all the cases examined, we found <110> type screw dislocations in γ' and no dissociation into 1/2<110> superpartials could be resolved. Dislocations in β invariably have <100> Burgers vectors; the dislocation lines tend to be "sharply bent" (see Figures 5 and 6), which is typical of <100> dislocations in NiAl [14]. Since the magnitude of γ'<110> Burgers vector (~0.50 nm) is nearly twice that of β<100> (which is ~0.29 nm), the residual dislocation resulted from an individual transfer event has a fairly large Burgers vector, which appears to explain the good visibility of residual dislocations in our micrographs. This helps us to discern the dislocation reactions at the γ'/β phase boundaries, and to identify in which direction the slip was propagating.

From a purely geometrical point of view, slip transfer is easiest when slip planes and dislocation Burgers vectors on either side of the boundary are parallel. In our alloy, this means that $(111)_{\gamma'} \| (011)_\beta$ and $[01\bar{1}]_{\gamma'} \| [\bar{1}00]_\beta$, i.e., the two phases obey a Nishiyama-Wassermann orientation relationship. For a Kurdjumov-Sachs orientation relationship, the slip planes are still parallel, but the Burgers vectors are misoriented 5.26°. For other non-special orientation relations, both the slip planes and the Burgers vectors are misoriented, and the two misorientation angles can be taken as characteristic parameters when judging whether or not the geometry is favourable for slip transfer. Thus, in Table I, slip transfer is relatively easy in case 1 since both misorientation angles are small; it is more difficult in case 2 because both angles are fairly large.

When repeated transfer events occur on the same slip plane, a subsequent event will always be more difficult than the preceding one, since the new incoming dislocation has to overcome the repulsion from the residual dislocation left by preceding events. This repulsive force increases with the total residual Burgers vector which is the sum of those resulted from all preceding transfer events [7]. Such a mechanism of slip transfer from primary γ' to β therefore cannot operate indefinitely, and will be important only in initiating the plastic deformation of the β phase. Nevertheless, it can supply an appreciable quantity of mobile dislocations to β which multiply afterwards. This provides a basis for the argument of ductility enhancement of β phase by a softer second phase.

Since slip transfer events have been observed in boundaries where the characteristic misorientation angles are as high as 14.5° (Table I), a fairly large proportion of the phase boundaries in the material would be capable of generating mobile dislocations for β by such a mechanism, even when the orientation of the two phases is completely random. We have however observed that for many boundaries, a pair of close-packed planes in the two phases tend to have a small misorientation angle rather than be oriented completely at random. This perhaps has resulted from a tendency to epitaxial growth when the remnant liquid crystallises into the primary γ' following the dendritic solidification of the β phase. Such an epitaxial tendency creates a more favourable geometrical condition for the mechanism of slip transfer to operate.

CONCLUSION

The mechanism of slip transfer from the softer γ' phase to the harder β phase can operate when the γ' phase is not completely enclosed by the β phase. The process is driven by the incompatibility stresses which have built up at the phase boundaries as a result of prior plastic deformation of the soft phase, and is aided by a favourable orientation relationship between the two phases. We have shown that the $\gamma'[0\bar{1}1](111)$ dislocations can generate $\beta[100](011)$ dislocations when the misorientation angles of Burgers vectors and slip planes are smaller than about 15°, which implies that a significant proportion of phase boundaries in our alloy is vulnerable to the slip transfer process.

ACKNOWLEDGEMENTS

We are grateful to Professor C.J. Humphreys for the provision of laboratory facilities and to Rolls-Royce plc for supporting the project. R. Yang expresses his gratitude to St. John's College, Cambridge, for offering stipend and discounted accommodation.

REFERENCES

1. L.M. Brown and W.M. Stobbs, Phil. Mag. **23**, 1185; 1201 (1971).
2. R.T. Pascoe and C.W.A. Newey, Metal Sci. J. **2**, 138 (1968).
3. P.A. Flinn, Trans. AIME **218**, 145 (1960).
4. C.T. Forwood and L.M. Clarebrough, Phil. Mag. A **44**, 31 (1981).
5. Z. Shen, R.H. Wagoner and W.A.T. Clark, Scripta metall. **20**, 921 (1986).
6. T.C. Lee, I.M. Robertson and H.K. Birnbaum, Ultramicroscopy **29**, 212 (1989).
7. C.T. Forwood and L.M. Clarebrough, Electron Microscopy of Interfaces in Metals and Alloys (Adam Higher, Bristol, 1991), p.336.
8. J.P. Hirth, Metall. Trans. **3**, 3047 (1972).
9. M. Larsen, A. Misra, S. Hartfield-Wunsch, R. Noebe and R. Gibala, in Intermetallic Matrix Composites, edited by D.L. Anton, P.L. Martin, D.B. Miracle and R. McMeeking (Mater. Res. Soc. Symp. Proc. **194**, Pittsburgh, PA, 1990), pp.191-198.
10. R. Yang, N. Saunders, J.A. Leake and R.W. Cahn, Acta metall. mater. **40**, 1553 (1992).
11. C.J. Ball, Phil. Mag. A 44, 1307 (1981).
12. P. Heilmann, W.A.T. Clark and D.A. Rigney, Ultramicroscopy **9**, 365 (1982).
13. P.B. Hirsch, A. Howie, R.B. Nicholson, D.W. Pashley and M.J. Whelan, Electron Microscopy of Thin Crystals (Butterworths, London, 1965).
14. M.H. Loretto and R.J. Wasilewski, Phil. Mag. **23**, 1311 (1971).

LAMELLAR INTERFACES AND THEIR CONTRIBUTION TO PLASTIC FLOW ANISOTROPY IN TiAl-BASED ALLOYS.

Bimal K. Kad, AMES-0411, University of California-San Diego, LaJolla, CA 92093.
Peter M. Hazzledine, UES Inc. 4401 Dayton-Xenia Road, Dayton, OH 45432. and
Hamish L. Fraser, 2041 College Road, Ohio State University, Columbus, OH 43210

ABSTRACT

Lamellar γ/α_2 TiAl deforms much more easily in the plane of the lamellae (soft mode) than across the lamellae (hard mode). This plastic anisotropy is caused by the soft mode being softer and the hard mode being harder than in monolithic TiAl. The suggested explanations are that the γ phase is softened by interstitial gettering, that an extra soft mode exists because superdislocations can channel along interfaces and that the hard mode is dominated by the Hall-Petch effect.

INTRODUCTION

Two phase γ-TiAl ($L1_0$) + α_2-Ti_3Al ($D0_{19}$) alloys with fine scale ($\approx$1-2μm) lamellar microstructures exhibit a lower soft mode yield stress (1), increased strain to failure (2) and improved fracture toughness over their monolithic constituents. The beneficial effect of a small volume fraction (5-15%) of Ti_3Al and the subsequent lamellar morphology has been the subject of considerable debate in recent years (3). While the exact reason for this behavior is not well understood, two phase microstructures are characterized to have an abundance of 1/2<110] perfect dislocations in the TiAl phase (4).

The lamellar structure (containing a variety of γ/γ_{twin}, $\gamma/\gamma120^o$ (pseudo twin) and the γ/α_2 interfaces) confers plastic anisotropy such that shear deformation parallel to the lamellae (soft mode) is considerably easier than deformation across (hard mode) the lamellae (1,2). Lamellar structures show a soft mode yield stress of $\leq$100 MPa compared to the equivalent soft yield stress of $\approx$125 MPa and $\approx$250 MPa in single crystals of TiAl (5) and Ti_3Al (6) respectively. The simple rule of mixtures to predict composite properties is invalidated in this case. Possible softening effects may arise from 1) selective gettering of the TiAl phase (7) by Ti_3Al, 2) additional deformation modes available in the lamellar form or 3) a contribution from the additional variable: the γ/γ_{twin}, $\gamma/\gamma120^o$ and γ/α_2 lamellar interfaces. In hard mode, the yield stress may be as high as 500MPa. There are several reasons why this should be higher than the yield stress of monolithic material including i) the difficulty of operating Frank-Read sources in the narrow lamellae, ii) the high Hall-Petch stress resulting from the difficulty of propagating deformation from lamella to lamella, and iii) the presence of internal stresses generated by the mismatches at lamellar interfaces.

In this paper, we elucidate the contributions made by the various lamellar interfaces to the observed plastic anisotropies in TiAl, considering only dislocation (not twinning) deformation. We then evaluate two mechanisms which are specific to the particular microstructure of TiAl, one which softens the soft mode and one which hardens the hard mode: the first involves an extra slip mode which occurs in the lamellar interfaces themselves and the second involves the difficulty of passing a perfect dislocation from one lamella into another where it is a superpartial.

EXPERIMENTAL PROCEDURE

Master ingots of Ti-50at%Al alloy prepared from individual elements of 99.99% or better purity were levitation zone melted in a static helium atmosphere, yielding large crystals. All examinations were performed on a single zone melted bar, in the as-processed as well as deformed condition. Thin foil specimens were prepared using a twin-jet electropolishing technique in a polishing solution of 7 vol.% sulphuric acid in methanol and were examined in Phillips transmission electron microscopes operating at 100kV and 300kV.

EXPERIMENTAL RESULTS

Figure 1 shows a mixed mode deformation microstructure. The compression axis (C.A.) is $\approx$65° from [111], the lamellar normal. The dislocations are characterized as soft mode perfect dislocations 1/2[1$\bar{1}$0] (A), soft mode superdislocation [10$\bar{1}$] (B) and hard mode dislocation 1/2[110] (C) with maximum applicable schmid factors of 0.21, 0.39 and 0.38 respectively. It is apparent that soft mode perfect dislocations are activated at significantly lower schmid factors.

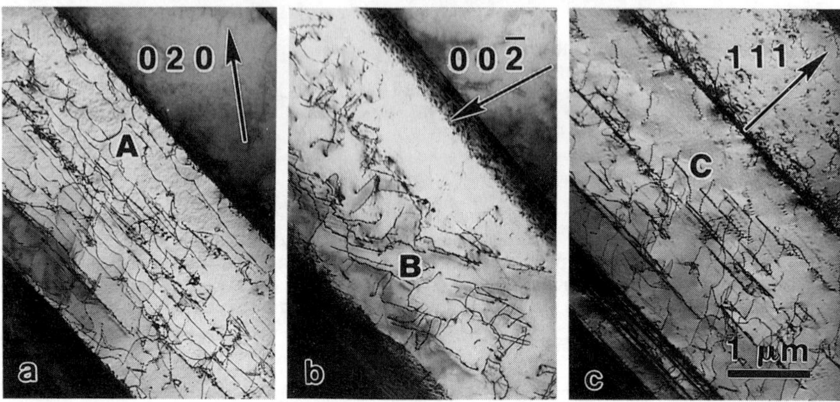

Figure 1. Mixed mode deformation for C.A.≈65° from [111], the lamella normal. The dislocations are A: soft mode 1/2[1$\bar{1}$0] (fig.1a), B: soft mode superdislocation [10$\bar{1}$] (fig.1b), and C: hard mode perfect dislocation 1/2[110] (fig.1c). The soft mode dislocations (i.e all dislocations with Burgers vectors contained in the lamella plane) are out of contrast for **g**=111, (fig.1c), the lamella normal, and this distinguishes soft mode from the hard mode.

Significant microstructural differences are observed for variations of the C.A. from [111], the lamellar normal, especially when the schmid factor for the soft mode perfect dislocation is high. Fig. 2 shows a sample where C.A. is ≈30° from [111]. The maximum applicable schmid factors for 1/2[1$\bar{1}$0](soft mode), [10$\bar{1}$](soft mode), [110](hard mode) and [101](hard mode) dislocations are 0.40, 0.36, 0.45, and 0.41 respectively. However, the deformation microstructure is completely dominated by soft mode 1/2[1$\bar{1}$0] perfect dislocations. Figure 3 shows a hard mode

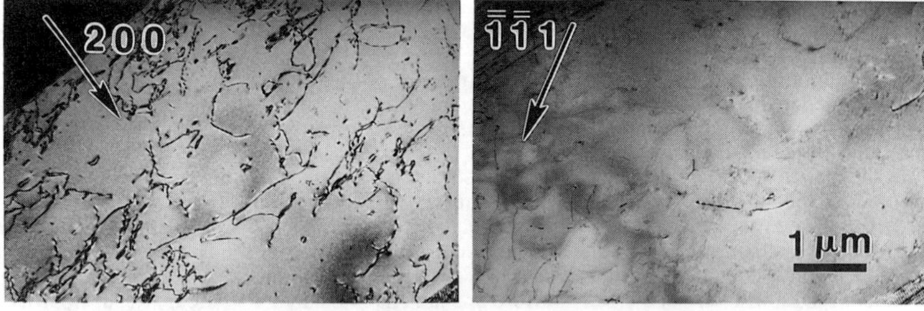

Figure 2. Predominant soft mode deformation for C.A.≈30° from [111]. Schmid factors for soft mode 1/2[1$\bar{1}$0] and hard mode 1/2[110] dislocations are 0.40 and 0.45 respectively.

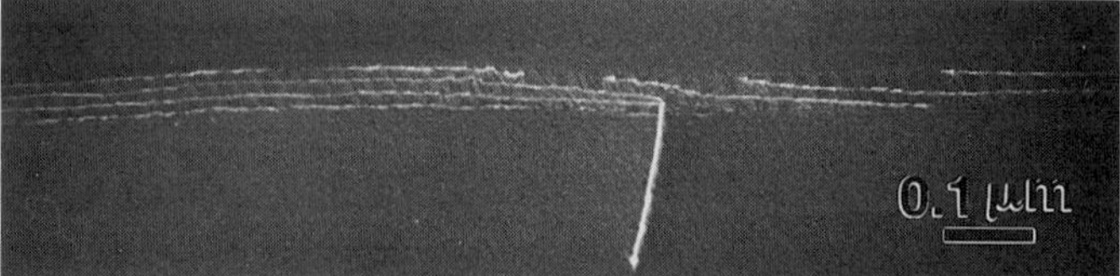

Figure 3. A hard mode superdislocation dissociating into equal components at the 120° interface.

superdislocation intersecting a 120° interface where it dissociates into component partials with wide separations. These observations are discussed in the following sections.

CHARACTERISTICS OF γ/γ and γ/α_2 INTERFACES

γ/γ interface In the lamellar form of TiAl (3) the γ grains are flat parallel plates with {111} normals. Adjacent lamellae are rotated about [111] with respect to their neighbours in multiples of ≈60° (8) forming three kinds of boundaries: 60°, 120° and 180°. Across these boundaries, the close packed directions $<\bar{1}01>$ match exactly when the rotation is 180° (and a twin is formed) but for all other rotations, because of the tetragonality, the matching is only approximate. For 60° and 180° boundaries the hard mode slip planes on either side of the the boundary meet at an angle of ≈39° whereas in the 120° boundaries, the slip planes are almost continuous across the boundary. At 60° and 120° boundaries, since there are lattice parameter mismatches, internal stresses are generated in the lamellae (9) which are partly relaxed by interfacial dislocations (10). The interfaces present a range of obstacles to dislocations attempting to cross them:

1) The slip plane may 'bend' through a large angle (≈39°)
2) The Burgers vector may be required to double: if a perfect 1/2<110] dislocation in one lamella crosses to a lamella in which it is one half of a superdislocation, the leading dislocation must drag an APB fault and a second dislocation is required.
3) The Burgers vector may be required to change by a small amount ≈ 1/2(c/a-1) ≈1%.
4) The slip plane may be nearly continuous, but in fact tilted through an angle ≈1/2(c/a-1)≈0.5°
5) The crossing dislocations may interact elastically with the interfacial dislocations

All of these effects contribute to the difficulty of transmitting hard mode dislocations through the lamellar interfaces.

γ/α_2 interface At the γ/α_2 interface close matching is achieved between the two phases in any one of 6 orientations, each rotated by a multiple of 60° with respect to the others. Because of the hexad axis in α_2, however all the interfaces are physically the same. Since the atomic spacings in α_2 are larger than in γ, there is a biaxial tension in γ and a biaxial compression in α_2. In addition, the tetragonality of γ induces shear strains of opposite signs in the two phases (9). These internal stresses are partly relaxed by interfacial dislocations which have some edge and some screw character. The γ/α_2 interfaces act as barriers to slip in all the same ways as the γ/γ interfaces but in this case the gross bending of the slip planes is always present: the smallest angle between corresponding slip planes, a {111} in γ and $\{10\bar{1}0\}$ in α_2 is ≈19°

CONTRIBUTIONS TO THE YIELD STRESS

The shear stress required to move dislocations in lamellar TiAl has four main components:

i) The lattice resistance of the material or Peierls stress τ_p. Four different values of τ_p are expected for perfect dislocations and superdislocations in γ and for superdislocations in α_2 on the basal and the prism planes.
ii) The stress to operate dislocation sources, the Frank Read stress, τ_F
iii) The stress required to transmit dislocations through interfaces, the Hall Petch stress τ_H,
iv) The internal stress generated by lattice mismatches in the lamellar interfaces, τ_I.

These stresses have very different effects on soft and hard-mode deformation and account for some of the plastic anisotropy.

HARD MODE DEFORMATION

The Peierls stress has two values in γ, one for perfect dislocations and one for super dislocations; in α_2 there is a third value for slip on the prism plane. The Frank Read stress, τ_F, has a minimum value set by the requirement to operate a source in a lamella of width h. The order of magnitude is Gb/h, where G=shear modulus and b= burgers vector, (perfect or super). The internal stresses do act on the hard mode dislocations and the order of magnitude is the the same as τ_F , but it could have either sign, depending on the slip system, $\tau_I = \pm Gb/h$ (10). The Hall Petch stress is high, since the lamellar boundaries are close together for hard mode dislocations, and difficult to calculate because of the many kinds of obstacles presented by lamellar interfaces. A sample calculation may be made by considering a pile up of perfect screw dislocations at a 120° interface. The leading dislocation is passing into a lamella in which it is a superpartial dragging an antiphase boundary (APB) with energy/unit area=γ_{APB}. The concentrated stress from the pile up of n dislocations forces the leading screw to create an APB;

$$nb\tau_H = \gamma_{APB} \qquad ...(1)$$

the length of the pile up is of the order h, so

$$h = nGb/\pi\tau_H \qquad ...(2)$$

Combining equations (1) and (2), the Hall-Petch stress is of the usual form;

$$\tau_H = (G\gamma_{APB}/\pi h)^{1/2} \qquad ...(3)$$

with a slope ≈ 0.10 MPa√m for γ_{APB}= 0.5Jm^{-2}. This is lower than the experimental values for the Hall-Petch slope in lamellar TiAl, which are in the range of 0.2 to 0.5 MPa√m (11). But theoretical estimates can be brought up to the experimental range by considering other aspects of the barrier. Both the elastic interaction with the interface dislocations and the 'tilted plane' effect contribute Hall-Petch slopes of order $(G/2)(b(c/a-1))^{1/2}$ ≈0.05 MPa√m and the need to create a 'difference dislocation' in the interface contributes a slope of about $Gb^{1/2}(c/a-1)/\pi$ ≈0.005 MPa√m. Assuming that these effects are additive, the theoretical estimate for the slope is ≈0.2MPa√m. Further effects which should also be included in a full calculation are i) the difficulty of cross-slipping at the interface when the slip plane bends, and ii) the difficulty of crossing an γ/α_2 boundary. It is evident from equation 3 that τ_H is large compared with either the Frank-Read stress or the internal stress in lamellae of width ≈1μm, since the ratio

$$\tau_H/\tau_F = (h\gamma_{APB}/\pi Gb^2)^{1/2} \approx 10 \qquad ...(4)$$

we conclude that the hard mode deformation is hardened principally by the Hall-Petch effect.

SOFT MODE DEFORMATION

The dislocations glide in the plane of the lamellae. In α_2 there are three equivalent superdislocation <11$\bar{2}$0> Burgers vectors in the basal plane. In γ there are also three Burgers vectors, two superdislocations of type <101] and one perfect dislocation of type <110]. For these, three different values of τ_p are expected, τ_I is exactly zero in both phases and τ_F and τ_H are small because the barriers (domains as opposed to lamellar interfaces) are far apart.

The Peierls stress is therefore the dominant stress in soft mode deformation and this agrees with the observation that the yield stress is independent of lamellar thickness (12). It would be expected that slip would be concentrated in the softer phase, namely γ and that it would be carried by the most mobile dislocations, the 1/2[1$\bar{1}$0] dislocations, figure 2. These dislocations are known to move at low stresses (in monolithic TiAl) when the interstitial content is low (13) as it probably is in γ plates precipitated from the parent α_2 phase (7). But there is only one Burgers vector available in the (111) lamella, 1/2[1$\bar{1}$0] and so, for some lamellae, (or for some domains within a lamella) the superdislocations [$\bar{1}$01] and [0$\bar{1}$1] have the highest schmid factors. Such lamellae could either not contribute to slip (1,2) or else they may be ductilized by channeling their superdislocations into their lamellar interfaces, where they become mobile. Either way the yield stress would be insensitive to rotations of the stress axis about [111], as observed (1) despite the variation in dislocation activity as shown in figure 1 and figure 2. Superdislocations may move more easily if they are channelled into γ/γ 120° interfaces because the APB energy is lowered there enabling the superpartials to uncouple and glide or climb essentially as perfect dislocations.

CHANNELED DISLOCATION MOTION

For a dislocation that lies in the interface plane between two lamellae, the irregular bonding in some cases changes the values of the energies of faults between partial dislocations and this may uncouple pairs of superpartials.

γ/γ twin interfaces The interface is perfectly coherent and the structure of the interface and the neighboring layers in the matrix (or the twin) are identical to the simple L1o stacking as shown in fig. 4a. The interface confers no additional advantages to dislocations with burgers vectors contained in the interface plane.

γ/γ 120° interface The interface plane is best represented by two layers of (111) planes with one layer each in respective adjacent slabs, figure 4b. In an L1o lattice an Al atom has an environment of 4 Al-Al and 8 Al-Ti bonds. This environment remains unchanged for a perfect twin. For the 120° interface, equal numbers of two different atomic environments exist both for

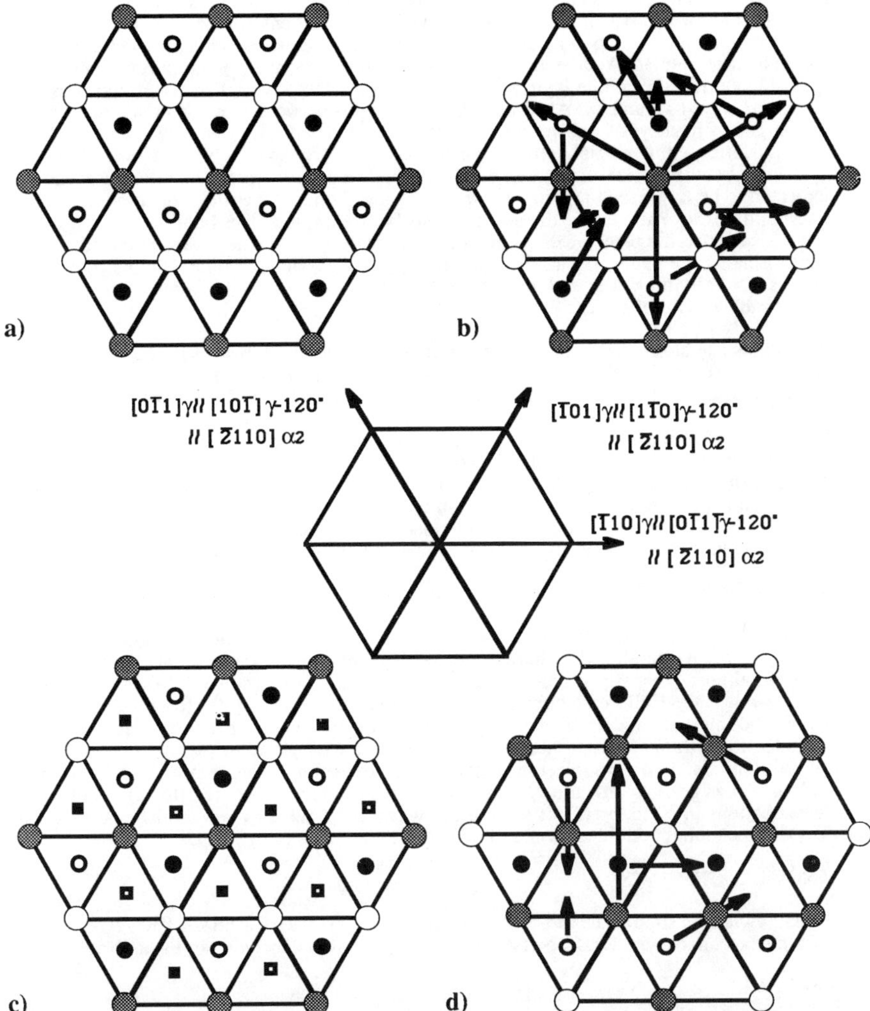

Figure 4. In all figures the layers are: A (large circles), B (small circles) and C (squares); Black atoms are Ti and white atoms are Al. a) Two layers AB of the L1o (γ) structure representative of that at the perfect twin interface. b) Two layers of (111) plane of atoms containing the 120° interface between them. The arrows represent slip vectors that preserve the chemical environment. c) Three layers (ABC) of (111) plane of atoms. The 120° interface is between layer A (large circles) and layer B (small circles). The C layer (squares) is in the same lath as layer B. The C layer of atoms has an identical average chemical environment across the interface as does layer B. d) Composite layer of the γ/α_2 interface with a layer of (0001) plane of atoms in Ti_3Al (large circles) and a (111) plane of atoms in TiAl (small circle) across the γ/α_2 interface. In the Ti_3Al layer black is Ti and white is Al. The arrows represent slip vectors that preserve the chemical environment. The horizontal arrow is an APB vector in the α_2 phase

the Al and Ti atoms; i) with 4 Al-Al and 8 Al-Ti nearest neighbor (NN) bonds and ii) with 5 Al-Al and 7 Al-Ti N:N bonds such that the average bond environment in the vicinity of the interface is 4 1/2 Al-Al and 7 1/2 Al-Ti bonds. This change in chemical environment occurs because of the 120° rotation across the two layer interface and the average environments for Al, Ti atoms across the interface are 3/2 Al-Al and 3/2 Al-Ti bonds. In fact the average atomic environment for any possible atomic site in one layer with respect to the other layer of the interface is exactly equivalent as 3/2 Al-Al and 3/2 Al-Ti neighbours. It is this particular characteristic that confers additional glide advantages for dislocations contained in this two layer interface. In Figure 4b, all lattice translations and partial lattice translations (see arrows) involve no net change in bond/chemical environment. Thus fault energies for all types of superdislocation dissociations are expected to be low, assuming that it is the number of bonds rather than the directions of the bonds which is the main factor determining the fault energies. The bonding across a 120° interface is such that if a plane of atoms immediately adjacent to the interface is removed, the NN bonding remains the same, fig 4c. For this reason hard mode superdislocations may climb dissociate in the interface and consequently climb more easily than they could as superdislocations. This effect is evident in figure 3.

γ/α_2 interface The atomic structure of the two layer interface is shown in figure 4d. The large circles are the Ti_3Al layer, black as Ti and white as Al. The small circles are the TiAl layer, black as Ti and white as Al. Two different environments; i) 2Ti + 1Al and ii) 3Ti atoms exist in equal numbers for an Al atom (in TiAl) across the interface, and only one environment i.e. 2Ti + 1Al for a Ti atom (in TiAl) across the interface. Hence translations which preserve the chemical/bond environment for the Ti atoms need only be considered. It is evident that the interface confers no special advantages for $L1_0$ slip vectors in the interface plane. However, the $\langle 11\bar{2}0\rangle$ superdislocations in the α_2 can dissociate on the basal plane with low APB fault energy. This may promote basal slip activity in lamellar interfaces in contrast to monolithic Ti_3Al where basal slip is severely limited (6) and is invariably accompanied by shear cracking.

SUMMARY

The plastic anisotropy of lamellar TiAl is caused by both the hardening of hard mode slip and the softening of soft mode slip. The hardening mechanisms are extremely complicated, being different for almost every slip system. It seems, however, that the dominant effect in lamellae of $\approx 1\mu m$ thickness is the Hall-Petch effect. In finer lamellae, the contributions from the Frank-Read stress and the internal stresses (which may not be of the same sign) become larger since they vary with lamellar thickness as h^{-1}. The explanation for the observed softening of the soft mode is not so clear but two possible mechanisms are advanced i) the removal of interstitials (principally oxygen) from the γ phase by the α_2 phase, and ii) the channeling of superdislocations into the interfaces themselves where they move as perfect dislocations. This process may be thought of either as grain boundary sliding or as coarse slip.

ACKNOWLEDGEMENTS

PMH acknowledges support from USAF contract no. F33615-91-C-5663. BKK acknowledges stimulating exchanges with Dr. Robert Asaro, Dr. Richard Beanland and Dr. Pavan Nagpal.

REFERENCES

1. T. Fujiwara, A. Nakamura, M. Hososmi, S.R. Nishitani, Y. Shirai and M.Yamaguchi, Phil. Mag. A, **61**, No. 4, 591, (1990)
2. H. Inui, M.H. Oh, A. Nakamura and M. Yamaguchi, Acta. Met., **40**, (11), 3095, (1992)
3. Y. W. Kim, J. Metals, **41**, No. 7, (1989)
4. E.L. Hall and S.C.Huang, J. Mat. Res., **4**, 595, (1989).
5. R.D. Reviere, B.F. Oliver and D.D. Bruns, Mat. & Manufacturing Proc., **4**, (1), 103, (1989)
6. Y. Minonishi, Phil. Mag., **63**, (5), 1085, (1991)
7. V.K. Vasudevan, M.A. Stucke, S.A.Court and H.L. Fraser, Phil. Mag. Lett. **59**, 299, (1989)
8. Y.S. Yang and S.K. Wu, Scripta Met., **24**, 1801, (1990), Phil. Mag. **A65** , 15, (1992).
9. P.M. Hazzledine, B. Kad, D.M. Dimiduk and H.L. Fraser, MRS Conf. Proc. **273**, 81, (1992)
10. B.K. Kad and P.M. Hazzledine, Phil. Mag. Lett. **66**, 133, (1992)
11. Y. Umakoshi, T. Nakano, and T. Yamane, Mat. Sci. Eng., **A152**, 81, (1992).
12. Y. Umakoshi, T. Nakano, and T. Yamane, Scripta Met. **25** (7), 1525, (1991).
13. B.K. Kad. S.Swaminathan and H.L. Fraser. unpublished results.

ROOM TEMPERATURE FRACTURE OF FeCo

L. ZHAO*, I. BAKER* AND E. P. GEORGE**
*Thayer School of Engineering, Dartmouth College, Hanover, NH 03755
**Metals and Ceramics Division, Oak Ridge National Laboratory, Oak Ridge, TN 37831

ABSTRACT

FeCo is a B2 intermetallic compound which undergoes an order-disorder transformation. In this paper, the effects of changes in both constitutional and thermal disorder on the room temperature fracture of FeCo are presented. Tensile tests were performed on three compositions of FeCo, $Fe_{30}Co_{70}$, $Fe_{50}Co_{50}$ and $Fe_{70}Co_{30}$. The resulting fracture surfaces were examined by SEM and the grain boundary chemistry was investigated by Auger electron spectroscopy, AES. Ordered $Fe_{50}Co_{50}$ and $Fe_{70}Co_{30}$ were very brittle, fracture occurring before yielding, with intergranular fracture occurring in most grains. In contrast, ordered $Fe_{30}Co_{70}$ showed about 18% elongation and exhibited a dimple-type fracture. It was also found that disordering improved the ductility of each composition but had little influence on the fracture mode. AES showed that a low level of sulfur segregation was present at the grain boundaries, suggesting that sulfur segregation alone was not responsible for the brittle behavior of the ordered alloys.

INTRODUCTION

For an ordered polycrystal it has been suggested [1] that five independent slip systems are not a sufficient criterion for ductility. In addition, a (partially) disordered grain boundary region may be needed. This suggestion implies that both thermal and constitutional disorder could improve ductility for an ordered alloy. This paper presents a study designed to test this hypothesis.

FeCo is an ideal B2 intermetallic compound with which to study the effects of constitutional and thermal disorder. It undergoes an order-disorder transformation over a considerable range, from 23 at.% Co to 74 at.% Co. It is well known that stoichiometric FeCo has less ductility in the ordered state than in the disordered state. Accompanying this decrease in ductility is a change from transgranular cleavage to intergranular fracture [2,3]. Since grain boundary fracture in bcc metals is usually caused by grain boundary segregation, the entirely intergranular fracture of the ordered $Fe_{50}Co_{50}$ raises the possibility that segregation of impurities to the grain boundaries occurs during the relatively slow cooling required to induce long range order, in turn weakening the boundaries and causing intergranular fracture [4,5]. However, Glezer and Maleyeva [3] argued that impurities are not responsible for grain boundary fracture and that the order of grain boundary itself is important feature.

EXPERIMENTAL

Dumbbell-shaped specimens for tensile tests were made from hot-extruded ingots of $Fe_{30}Co_{70}$, $Fe_{50}Co_{50}$ and $Fe_{70}Co_{30}$ annealed at 860°C for 5 hrs in order to produce fully-recrystallized microstructures with a grain size of about 25-35 μm. Details of the extrusions are given elsewhere [6]. The specimens were further heat-treated to obtain either the ordered or the disordered states. The disordered state was obtained by annealing samples at 800°C for 2 hrs followed by a quench into iced saline solution. The ordered samples were obtained by annealing at 800°C for 2 hrs, then furnace-cooling to 500°C, where they were kept for 10 hrs before furnace-cooling to room temperature. Electron diffraction and differential scanning calorimetry were used to confirm that the specimens were indeed ordered after the ordering heat treatment [7].

Tensile tests were performed on an Instron Testing Machine with a crosshead velocity of $1.7x10^{-3}$ mm sec^{-1} resulting in an initial strain rate of $6.0x10^{-4}$ sec^{-1}. The resulting fracture surfaces were examined using a Zeiss DSM 962 SEM. The grain boundary chemistry was investigated by Auger electron spectroscopy, AES, of in-situ fractured specimens that were given either the ordering or disordering heat treatment.

RESULTS

Fig.1 shows the yield strengths of the tensile samples of the three alloys. However, the tensile samples of ordered $Fe_{50}Co_{50}$ and ordered $Fe_{70}Co_{30}$ fractured before yielding, so the strengths shown are fracture strengths. Yield strengths obtained from the tensile tests were in good agreement with those obtained from compression tests [7]: disordering decreases the yield strength of cobalt-rich alloy $Fe_{30}Co_{70}$ and the disordered stoichiometric alloy has a higher yield strength than the two disordered off-stoichiometric alloys. The fracture strengths of ordered $Fe_{50}Co_{50}$ and ordered $Fe_{70}Co_{30}$ in tension were about 70 % and 75 % of their yield strengths obtained from compression tests [7].

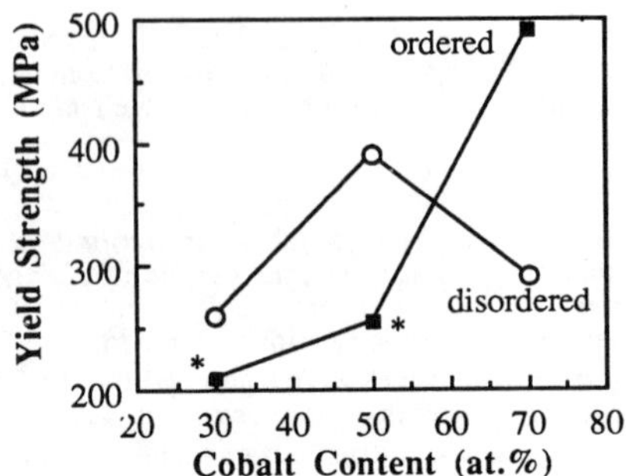

Fig.1 Yield strengths of FeCo.
* Indicates fracture strength

The elongations to fracture of the tensile samples are shown in Fig.2. As noted above, no elongation was obtained for ordered $Fe_{50}Co_{50}$. However, on disordering 4 % elongation could be obtained. $Fe_{30}Co_{70}$ and $Fe_{70}Co_{30}$ also showed more elongation in the disordered state than in the ordered state although the effect of disordering on their yield strengths was different. Two interesting features were noted. First, as noted earlier, ordered $Fe_{70}Co_{30}$ showed no elongation whilst 8 % elongation was obtained in the disordered state. Second, $Fe_{30}Co_{70}$ is ductile; 17 % elongation could be obtained even in the ordered state and this increased to 21 % in the disordered state.

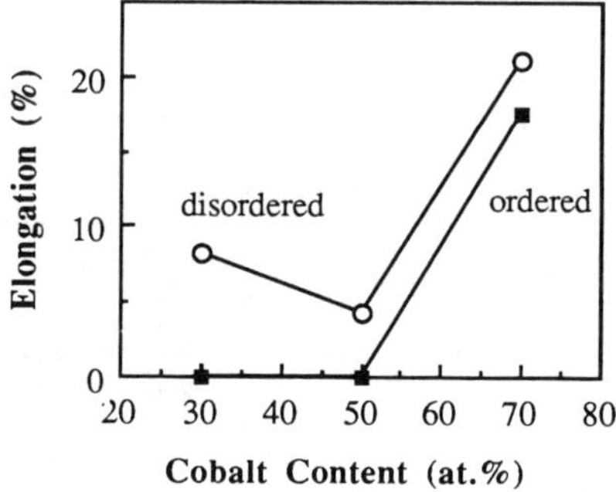

Fig.2 Elongations of FeCo tensile samples.

$Fe_{50}Co_{50}$ exhibited a brittle fracture mode in both the ordered and disordered states. Fracture was mostly intergranular with a small amount of transgranular cleavage, see Fig.3. Obtaining high quality channeling patterns can be used as a qualitative measure of the degree of deformation on the fracture surfaces. In the ordered state, channeling patterns were obtained from not only intergranular facets but also the few transgranular facets, suggesting that there was little or no plastic deformation accompanying crack propagation. As expected, in the disordered state, since about 4% elongation was produced in the tensile tests, channeling patterns could not be obtained. Grain boundary particles were observed in both the ordered and disordered states. EDS analysis showed that these particles consisted of Si and Al [8].

Intergranular fracture also dominated in both the ordered and disordered states of $Fe_{70}Co_{30}$, although a few transgranular facets were also present, see Fig.4. Compared to $Fe_{50}Co_{50}$, there were fewer particles in this alloy. Again, channeling patterns could be obtained only from the ordered samples, which failed before yield.

Both ordered and disordered $Fe_{30}Co_{70}$ failed in a cup-and-cone, dimple, ductile-rupture mode accompanied by necking, see Fig.5. Small particles could be observed at the bottom of many dimples.

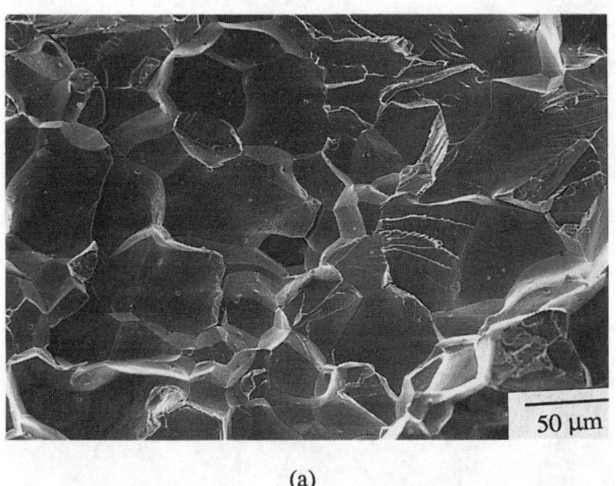

(a)

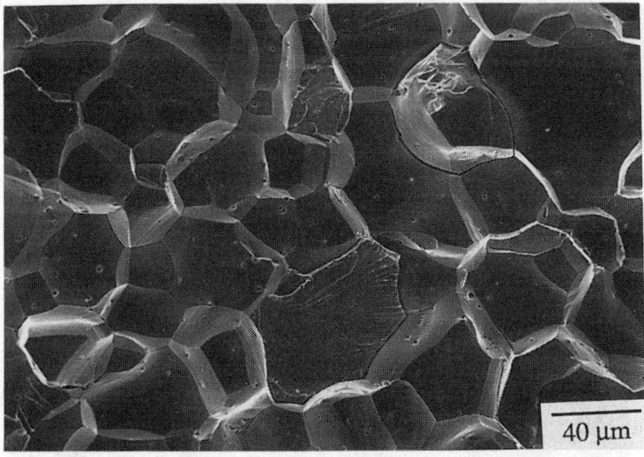

(b)

Fig.3 SEM micrographs of fractured surfaces of ordered (a) and disordered (b) $Fe_{50}Co_{50}$.

(a)

(b)

Fig.4 SEM micrographs of fractured surfaces of ordered (a) and disordered (b) $Fe_{70}Co_{30}$.

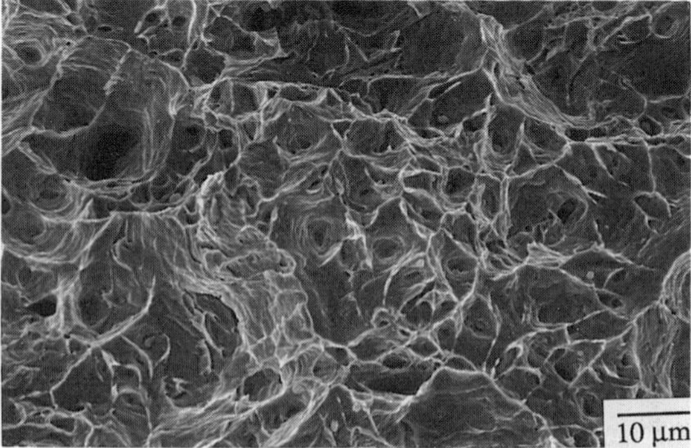

Fig.5 SEM micrograph of fractured surfaces of ordered $Fe_{30}Co_{70}$.

AES was used to investigate the grain boundary chemistry of the alloys. As shown in Table 1, about 3.6±3.9 at.% sulfur segregation and 11.4±2.3 at.% carbon were observed average on grain boundaries of the ordered stoichiometric alloy. For the disordered stoichiometric alloy, a lower level of sulfur segregation (1.8±0.7 at.%) and 6.2±0.7 at.% nitrogen were present on grain boundaries whilst no carbon was observed. Using the peak height ratios from transgranular facets to determine the electron sensitivity factor, the grain boundary composition was determined to be the same in the ordered and disordered states and is slightly Co enriched with respect to the bulk. Table 1 also shows that 5.8±2.8 at.% sulfur segregation and 3.7±1.9 at.% nitrogen were present on grain boundaries of disordered $Fe_{70}Co_{30}$. Grain boundary chemistry analysis on the cobalt-rich alloy $Fe_{30}Co_{70}$ and the ordered iron-rich alloy $Fe_{70}Co_{30}$ is currently under way.

Table 1 AES results on grain boundaries of FeCo.

alloy	state	sulfur (at.%)	nitrogen (at.%)	carbon (at.%)	grain boundary chemistry
$Fe_{50}Co_{50}$	ordered	3.6±3.9	not detected	11.4±2.3	Fe-52 at.% Co
	disordered	1.8±0.7	6.2±0.7	not detected	Fe-52.1 at.% Co
$Fe_{70}Co_{30}$	disordered	5.8±2.8	3.7±1.9	not detected	Fe-28.5 at.% Co

DISCUSSION

All FeCo alloys have more ductility in the disordered state than in the ordered state. The stoichiometric alloy has less ductility than the off-stoichiometric alloys in both states. These results show that both thermal and constitutional disorder improve ductility for FeCo. Thus, they are consistent with the hypothesis proposed by Baker and Schulson [1] that a (partially) disordered grain boundary region may be needed for ductility of an ordered alloy. It is noted that cobalt has more effect than iron in improving the ductility of FeCo alloys.

Munroe and Baker [9] studied the dislocation structure in FeAl which is also a B2 intermetallic compound. Based on their observation of the presence of <001> dislocation pairs, in addition to APB-coupled <111> dislocations which are normally observed in FeAl at room temperature, they proposed that the <001> dislocations formed through the interaction of the <111> dislocations. The <001> edge dislocation pair may act as a crack nucleus. For FeCo, no observations of <001> dislocations have been reported. However, in the disordered states, single dislocations were present. Slip mode observations [7] showed that these dislocations could cross-slip easily, therefore, the possibility of crack nucleation by dislocation interactions is small. Besides, in the disordered state, grain boundaries have less resistance to slip transmission, that is the Hall-Petch parameter which can be regarded as the resistance for dislocations to slip transmittal across grain boundaries, is smaller in the disordered state [9]. Therefore, disorder always improves the ductility of FeCo alloys.

Sulfur segregation was found in both the ordered and disordered states of the stoichiometric alloy. Surprisingly, higher sulfur segregation was found in the disordered $Fe_{70}Co_{30}$ than in the ordered $Fe_{50}Co_{50}$, since disordered $Fe_{70}Co_{30}$ has more ductility than ordered $Fe_{50}Co_{50}$. Sulfur contamination is also expected in $Fe_{30}Co_{70}$ since it was produced from the same Fe and Co sources. However, this alloy is ductile even in the ordered state. All these suggest that sulfur segregation alone was not responsible for the brittle behavior of the ordered alloys. Glezer et al.[10] also studied fracture surfaces of ordered FeCo and ordered FeCo-2% V. Intergranular fracture was observed in ordered $Fe_{50}Co_{50}$ whilst transgranular fracture was present in ordered FeCo-2% V, which has more ductility than the ordered $Fe_{50}Co_{50}$. Furthermore, it was found that FeCo-2% V had a higher impurity concentration on the grain boundaries than FeCo. Therefore, they suggested that ductility of FeCo-2%V was improved by grain boundary disordering rather than a decrease in the impurity segregation to the grain boundaries. However, the suggestion that segregants are not responsible for the brittle intergranular fracture of FeCo should be tempered by the realization that segregants may be more deleterious in the stoichiometric alloy than in off-stoichiometric alloys.

Previous studies on fracture modes of ordered and disordered $Fe_{50}Co_{50}$ showed that entirely intergranular fracture was observed in the ordered state whilst entirely transgranular fracture occurred in the disordered state [2,3]. In this research little difference was observed between the ordered state and disordered state of each alloy. The reason for this may be related to the particles in the alloys. EDS analysis [8] showed that particles changed the matrix chemical composition around them. Therefore, it is reasonable to assume that the particles could produce local constitutional disorder in the ordered $Fe_{50}Co_{50}$ in the small area around the particles. The local constitutional disorder might be high enough on some grain boundaries. Thus, transgranular fracture occurred in these grains. In the disordered state the particles did not help to produce the disorder. In contrast, they could weaken the grain boundaries and make intergranular fracture occur.

SUMMARY

Three FeCo alloys, $Fe_{30}Co_{70}$, $Fe_{50}Co_{50}$ and $Fe_{70}Co_{30}$, were used to study how constitutional and thermal disorder affect the room temperature fracture of FeCo. It was found that the cobalt-rich alloy $Fe_{30}Co_{70}$ exhibited a cup-and-cone fracture in both the ordered and disordered states whilst brittle fracture was observed in the stoichiometric alloy $Fe_{50}Co_{50}$ and in the iron-rich alloy, $Fe_{70}Co_{30}$, in both their ordered states as well as their disordered states: intergranular fracture was present in most grains whilst transgranular fracture was present in a few grains. Sulfur and nitrogen were present on the grain boundaries. The disordered alloys had more ductility than the ordered alloys and that the stoichiometric alloy had less ductility that the off-stoichiometric alloys. These results are consistent with the hypothesis proposed by Baker and Schulson [1] that a (partially) disordered grain boundary region is needed for the ductility of an ordered alloy. Cobalt is more effective than iron in improving the ductility of FeCo.

ACKNOWLEDGMENTS

This research was supported by the U.S. Department of Energy, Office of Basic Energy Science, Division of Materials Sciences through grant DE-FG02-87ER45311 with Dartmouth College and through contract DE-AC05-84OR21400 with Martin Marietta Energy Systems, and by the SHaRE program of the Oak Ridge Associated Universities through grant DE-AC05-75OR00033.

REFERENCES

1. I. Baker and E.M. Schulson, Scripta Metall. **23**, 345 (1989).
2. M.J. Marcinkowski and J. Larsen, Metall. Trans. **1**, 1034 (1970).
3. A.M. Glezer and L.V. Maleyeva, Phys. Met. Metall. **66**, 174 (1988).
4. J.H. Westbrook and D.L. Wood, J. Inst. Metals **91**, 174 (1962).
5. J.H. Westbrook, *Met. Rev.* **9**, 415 (1964).
6. L. Zhao and I. Baker, submitted to J. Mat. Sci..
7. L. Zhao and I. Baker, submitted to Acta Metall..
8. L. Zhao, M.S. Thesis, Dartmouth College, 1992.
9. P.R. Munroe and I. Baker, Acta Metall. **39**, 1011 (1991).
10. A.M. Glezer and I.V. Maleyeva, Phy.Met.Metall. **68**, 65 (1989).

STUDY OF FRACTURE IN Nb–Al ALLOYS AND PURE METALS BY COMPUTER MOLECULAR DYNAMIC SIMULATION

DONGHYUN KIM, P. C. CLAPP AND J. A. RIFKIN
Center for Materials Simulation, Institute of Materials Science, University of Connecticut Storrs, CT 06268

ABSTRACT

In molecular dynamic studies of 15,000 atom arrays of A15 Nb_3Al, BCC Nb and FCC Al containing crack under external stress in Mode I loading, it has been verified that fracture behavior can be predicted (starting in the vicinity of the crack) in terms of competition between dislocation nucleation and Griffith crack propagation. BCC Nb and FCC Al appears to be ductile and A15 Nb_3Al appears to be brittle, in agreement with theoretical predictions. The elastic solution used for predictions was proposed by Rice[1] for the anisotropic material.

The interatomic interactions used in the simulation were Embedded Atom Method (EAM) potentials developed by Voter and Chen[2] for the Al–Al and by Rifkin[3] for the Nb–Nb and Nb–Al.

INTRODUCTION

1) Background and motivation

Computer Molecular Dynamics (CMD) simulation is an atomistic simulation that is based on interatomic potentials and volume forces developed semi–empirically via Embedded Atom Method and tested by comparing simulation predictions to other data not used in constructing the potentials. The EAM is a simple procedure for computing the electronic contribution to cohesion in transition metals. The EAM is based on the Hohenberg–Kohn theorem[4] which states that the energy contribution of an atom in an array of interacting atoms is a function of the local electron density, due to all other atoms. Unlike the band structure approach, the EAM does not require a periodic array of atoms. Consequently it can be used for disordered alloys, surfaces, cracks, dislocation cores, grain boundaries, stacking faults, and liquid–solid phase interfaces. By observing microscopic mechanisms like crack propagation, and dislocation emission during CMD simulation, we can get important input for predicting fracture.

Nb_3Al was selected for study, because it is one of the candidate materials for high temperature applications in aerospace jet engines, but is limited by unacceptable levels of fracture toughness at ambient temperature. Thus we were motivated to examine the atomistic mechanisms of fracture in Nb_3Al at various temperatures; an examination which is uniquely possible with the CMD simulation.

2) Main focus of investigation

Rice[1] recently proposed a new model which can calculate the level of applied stress intensity factors(K_{IC}) required for the dislocation nucleation at the crack tip of anisotropic material based on the Peierls concept[5]. K_{IC} is shown by them to be proportional to $(\gamma_{us})^{1/2}$, where γ_{us}, the unstable stacking energy, is a new solid state parameter identified by the analysis. It is the maximum energy needed in the block–like sliding along a slip plane.

Based on the above information, the main concern of this study is to predict the brittle versus ductile response in terms of the competition between dislocation nucleation and Griffith ideally brittle cleavage at a crack tip.

SIMULATION CONDITION

1) Initial configuration

Molecular dynamics simulations were done for 3 dimensional arrays representing an A15 Nb3Al alloy using the Embedded Atomic Potentials. The Nb3Al arrays were limited to 6000–16000 atoms due to the speed and memory limits of the IBM 3081 and Stardent 3040 computers that were used.

Periodic boundary conditions were used in the Z direction, so the crack was effectively of infinite length. The outer 2 planes of atoms in the X and Y direction were fixed except for the surface where the crack was introduced. To simulate a uniaxial tension on the array (region I), a "force" was applied in the Y direction by displacing the outer fixed atoms (region II) at each surface according to the anisotropic elastic solution for a stressed solid(Fig.1).

Each molecular dynamic time step was taken as 2.5E–15 seconds, or 0.083 times the average atomic vibration period, which was estimated to be 3E–14 seconds.

2) Boundary condition

Molecular Dynamics simulations can not be performed without the establishment of proper boundary conditions for the simulated atomic array. One must employ a boundary condition to impose the stress so that the atomic region behaves as though it was embedded in a much larger macroscopic material. Three different boundary conditions were used in our CMD simulations, they are:

1. Free surface boundary
2. Periodic boundary
3. Modified fixed boundary(MFB)

In the free surface boundary, the array is considered as a large, free molecule. This boundary condition was used only for calculating the surface energy of the array. A periodic boundary or modified fixed boundary should be used to simulate macroscopic material. The periodic condition means that atoms on one side of the array interact with those on the other. This means if an atom leaves the array during the simulation, it is assumed to reappear in the array from the opposite side. Therefore, the density and the number of atoms of the array are kept constant in a CMD simulation.

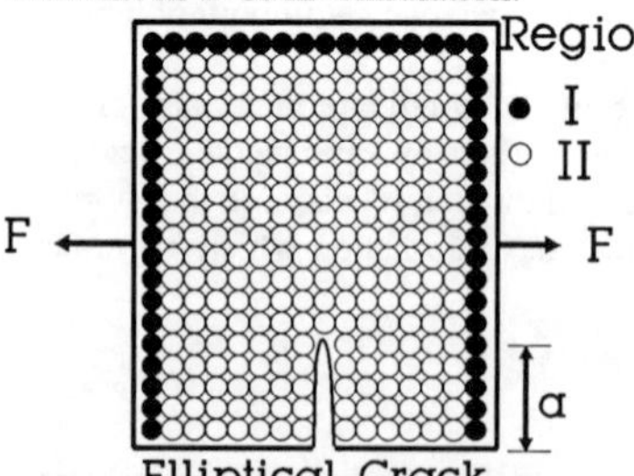

Figure 1. Modified fixed boundary condition with elliptical crack

In the modified fixed boundary condition, the array is divided into 2 regions(Fig.1). If the crack size and stress intensity factor K have been decided, the stress field around the crack tip and the displacement of individual atoms can be calculated by the continuum elastic solution. The starting configuration for the CMD simulations will begin with this displacement of the individual atoms. The atoms in region I are then fixed for a stabilizing period which is usually 2000 steps (about 125 atomic vibrations) and the region II atoms move according to Newton's equation of motion. By monitoring this stabilizing period, the stress intensity factor at which the crack closes or opens is decided. After the stabilizing period, the region I fixed boundary atoms are displaced according to a stress intensity factor increase of the anisotropic elastic solution, and keeping the region II atoms moving according to Newton's equation. The main advantage of this condition is that it is a good way of calculating K_{IC}^{MD} (cleavage) and relating these numbers to continuum elasticity parameters.

INTERATOMIC POTENTIAL

1) Atomic potential for Nb_3Al

The EAM potentials were used to simulate two pure metals and binary alloy systems: Nb, Al, Nb_3Al. Voter and Chen[2]'s Al potential was used for Nb_3Al and Al. The Nb–Nb and Nb–Al cross potentials were determined as follows:
1) The Nb potential is slightly modified from R.A.Johnson's[6,7] potential. This potential has a core repulsion term with a cutoff at the Nb bcc first neighbor distance. This term was translated inward so that the cutoff was positioned at the Nb–Nb A15 first neighbor distance(from 2.858 A° to 2.591 A°).
2) The procedure for the Nb–Al cross potential was different. There were no single crystal elastic constant data available for Nb_3Al. The data used for fitting the Nb–Al cross potential were the A15 Nb_3Al structure and its lattice constant of 5.183 A°. This was done by insuring; (1) that the A15 structure was stable, (2) that two other possible structures with the same stoichiometry(DO_{22} and $L1_2$) had higher energies when their lattice constants were varied to give the lowest energy and (3) that the A15 structure had a lower energy than Nb_3Al segregated into pure Nb and pure Al.

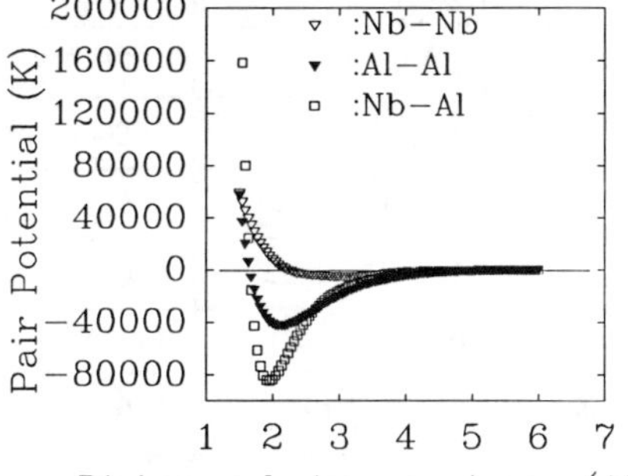

Figure 2. Interatomic potential curve for Nb, Al and Nb–Al

For an alloy model, the embedding function $F(\rho)$ and an atomic electron density function $\rho(r)$ must be specified for each atomic species, and a two–body potential $\phi(r)$ specified for each possible combination of atomic species. Since the electron density at any location is taken as a linear superposition of atomic electron densities and the embedding energy is assumed to be independent of the source of the electron density, these two functions can be directly taken from monatomic models. Johnson's way was used to construct the Nb–Al cross potential:

$$\phi^{ab}(r)=\frac{1}{2}[\frac{f^b(r)}{f^a(r)}\phi^{aa}(r)+\frac{f^a(r)}{f^b(r)}\phi^{bb}(r)] \quad (1)$$

where f^a and f^b are functions required for EAM alloy calculation. For a binary alloy with a– and b–type atoms, ϕ^{aa} and ϕ^{bb} are potentials given by the monatomic models and ϕ^{ab} is a cross potential. The only free parameter in this scheme for the fitting A15 is the relative weight of the Nb and Al electron densities. When this parameter was varied to give a stable A15 structure at 5.183 A°, it was found that condition(3) mentioned above was not satisfied. This was corrected by modifying the Nb pair potential as previously described and by removing the imposed cutoffs from the Nb and the Al electron density functions for the purpose of determining the cross potential. The original electron density functions were retained for actual energy and force calculations.

PREDICTION OF SLIP SYSTEMS

The energy required for dislocation nucleation is proportional to γ_{us}(the relaxed unstable stacking energy) according to Rice's theory. The unstable stacking energy and the surface energy in Nb, Al and Nb_3Al were determined by the CMD in order to predict dislocation nucleation by this method. Then the dislocation nucleation criteria was evaluated using elastic constants, surface energies and unstable stacking energies obtained from 0K simulations (Table 1).

Table 1. Ductility ratios of pure metals determined from CMD and based on Rice's theory

Material	TYPE	orientation	Slip system	Ductility Ratio ($\gamma_s/(\gamma_{us}f\beta)$)
BCC Nb	A	$\langle 010\rangle\langle 001\rangle\langle 100\rangle$	$(0\bar{1}1)\langle 011\rangle$	0.395
	B	$\langle 010\rangle\langle 001\rangle\langle 100\rangle$	$1/2(0\bar{1}\bar{1})\langle 11\bar{1}\rangle$	0.933
	C	$\langle 110\rangle\langle 001\rangle\langle 1\bar{1}0\rangle$	$1/2(0\bar{1}\bar{1})\langle 11\bar{1}\rangle$	0.514
	D	$\langle 011\rangle\langle 0\bar{1}1\rangle\langle 100\rangle$	$(011)\langle 0\bar{1}1\rangle$	0.388
	E	$\langle 011\rangle\langle 0\bar{1}1\rangle\langle 100\rangle$	$1/2(011)\langle 1\bar{1}1\rangle$	0.916
	F	$\langle 1\bar{1}0\rangle\langle 111\rangle\langle 11\bar{2}\rangle$	$(1\bar{1}0)\langle 110\rangle$	0.376
	G	$\langle 1\bar{1}0\rangle\langle 111\rangle\langle 11\bar{2}\rangle$	$1/2(\bar{1}\bar{1}0)\langle 111\rangle$	1.480
FCC Al	H	$\langle 110\rangle\langle 001\rangle\langle 1\bar{1}0\rangle$	$1/6(\bar{1}\bar{1}\bar{1})\langle 112\rangle$	2.679
	I	$\langle 111\rangle\langle 1\bar{1}0\rangle\langle 11\bar{2}\rangle$	$1/6(111)\langle 1\bar{2}1\rangle$	2.604
A15 Nb_3Al	S	$\langle 010\rangle\langle 001\rangle\langle 100\rangle$	$(010)\langle 001\rangle$	0.19
	T	$\langle 011\rangle\langle \bar{0}11\rangle\langle 100\rangle$	$(010)\langle 001\rangle$	0.16

According to Rice's theory, dislocation nucleation will happen before the Griffith cleavage condition is reached when the value of the "ductility ratio" $\gamma_s/(\gamma_{us}\cdot f\cdot \beta)$ is greater than 1. It is assumed that the slip system with the highest value of this ductility ratio will prevail if several possible slip system exists.

1) Analysis of Nb fracture

Four possible slip systems, $(010)\langle 100\rangle$, $(1\bar{1}0)\langle 110\rangle$, $(1\bar{1}0)\langle 001\rangle$ and $(1\bar{1}0)\langle 111\rangle$, were chosen, and the ductility ratios were calculated by the CMD simulation. The ductility ratios were calculated in several different crack orientations. Only one slip system (type G) shows a ductility ratio greater than 1, predicting that a dislocation nucleates first before the Griffith crack propagates at 0K. $1/2(1\bar{1}0)\langle 11\bar{1}\rangle$ slip systems were observed in the CMD simulations of the $\langle 100\rangle\langle 010\rangle\langle 001\rangle$ orientation. Two other slip systems have ductility ratios close to 1, type B and E. $1/2(1\bar{1}0)\langle 11\bar{1}\rangle$ slip systems were observed in the CMD simulations of the $\langle 1\bar{1}0\rangle\langle 110\rangle\langle 001\rangle$ orientation and the $\langle 1\bar{1}0\rangle\langle 111\rangle\langle 11\bar{2}\rangle$ orientation. All these slip systems were observed to operate in the CMD simulation above 10K. Type B and E slip system which are borderline cases, can be understood on the basis that the thermal energy plays a favorable role in the dislocation nucleation.

2) **Analysis of Al fracture**

The Al CMD simulations show very high ductility values compared with Nb ones. Two slip systems are possible dislocation generators according to the ductility ratio at 0K, type H and type I. Becquart[8] reported Al CMD simulations showing very ductile behavior. In her simulations no crack propagation was observed due to crack tip blunting by copious plastic deformation. The (111)<110> slip system nucleating two 1/6 (111)<112> type Shockley partial dislocations was observed. This slip system is the same slip system which is predicted by the ductility ratio.

3) **Analysis of Nb_3Al fracture**

It shows the unstable stacking energy plot determined by the CMD simulation. The slip distance refers to the block-like sliding movement along the slip direction and the relaxed distance means the movement along the relaxing direction which is normal to the slip direction. The Nb_3Al stacking energy plot shows flat surface after 5.5A° relaxed distance due to the cut-off distance of the Al-Al pair potential(Fig.3).

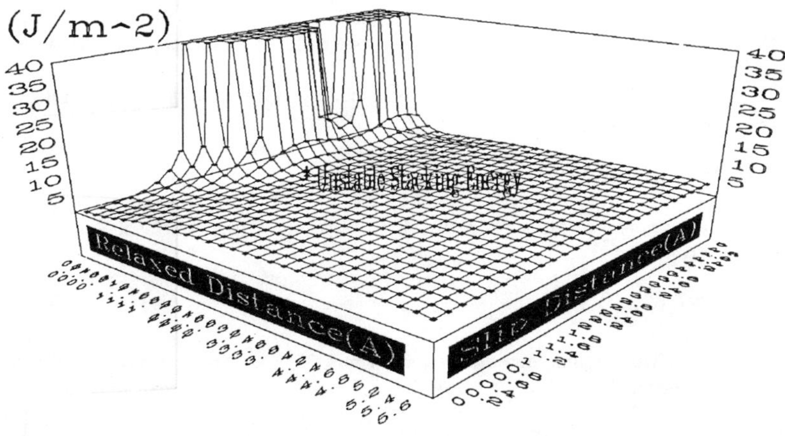

Figure 3. Unstable stacking energy plot of Nb_3Al (111)<1$\bar{1}$0>

It shows a minimum energy path for relaxations of 1-1.7 A° (Fig.4). This relaxation is expected to cause difficulties in the shear due to the large energy needed for this movement. Type T and Type S slip systems are expected to have the highest ductility ratios according to the γ_s/γ_{us} ratio. Both ductility ratios turn out to be below 0.2, which is a low value compared with the highest ductility ratio of NiAl. In CMD simulations of Nb_3Al, we observed very brittle behavior and did not observe any dislocation nucleations.

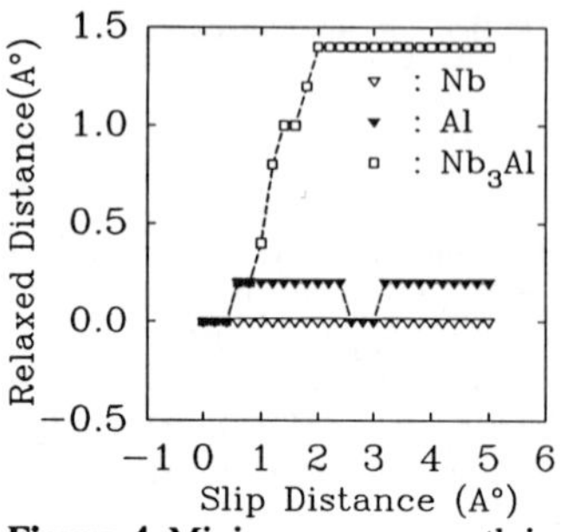

Figure 4. Minimum energy path in 3 D stacking energy plot

Very brittle behavior was observed up to 1500K, which is close to the melting point(1500–1600K) determined by the CMD simulations. However shear bond breaking was only observed above 300K in our CMD simulations. The same shear bond breaking can not be observed at low temperatures. This means that thermal energy is also a very important factor in the competition between slip induced planar defects and crack propagation.

CONCLUSIONS

In molecular dynamic studies of A15 Nb_3Al, BCC Nb and FCC Al containing a crack under external stress in Mode I loading, following results emerge:
1) It has been verified that the fracture behavior of these systems can be successfully predicted in terms of the competition between dislocation nucleation and crack propagation using Rice's model and material parameters determined by CMD. For instance, the 1/2 (011)<111> slip system was observed in the Nb CMD simulations in agreement with the theoretical predictions and experimental results. Also the CMD simulations showed that BCC Nb and FCC Al appear to be ductile and A15 Nb_3Al appears to be brittle, in agreement with the theoretical predictions and available experimental data.
2) The Brittle behavior of Nb_3Al alloy was explained by the relation between the interatomic potential and the structure.

ACKNOWLEDGEMENT

We very much appreciate the support of this work by the Office of Naval Research under grant # N00014–90–J–1223

REFERENCE

1 (a) Rice,J.R.,"Dislocation Nucleation from a Crack Tip: An Analysis based on the Peierls Concept",J.Mech.Phys.Solids., 40, 239–271 (1992).
(b) Rice,J.R., Beltz,G.E. and Sun Yuemin,"Peierls Framework for Analysis for Dislocation Nucleation from a Crack Tip" in Critical Problems in the Fracture of Solids (1992).
2 Voter, A.F., and Chen, S.P., in Accurate Interatomic Potentials for Ni, Al and Ni_3Al, Mater. Res.Soc.Proc., 82, 175–180 (1987).
3 Rifkin,J.A.,Becquart,C.S.,Kim,D. and Clapp,P.C.,"Dislocation generation and Crack propag ation in Metals examined in Modecular Dynamics Simulation", Mater.Res.Soc.Proc., 278,173(1992).
4 Hohenberg, H., and Kohn, W., Phys. Rev., 136, B864 (1964).
5 Peierls,R.E.,"The Size of a Dislocation",Proc.Physics.Soc., 52, 34–37 (1940).
6 Johnson, R.A., Physical Review B, 39, 554–559 (1989).
7 Johnson, R.A., and Oh, D.J., J. Mater. Res., 4, 1195–1201 (1989).
8 Becquart, C.S., Clapp, P.C., Kim, D.,"Brittle to ductile behavior of High Temperature Intermetallics studied in Computer Simulation", TMS Fall Meeting (1991).

DEFORMATION AND FRACTURE CHARACTERISTICS OF A GAMMA TITANIUM ALUMINIDE AT HIGH TEMPERATURES

V. SEETHARAMAN*, S. L. SEMIATIN#, C. M. LOMBARD# AND N. D. FREY‡
*UES, Inc., 4401 Dayton-Xenia Road, Dayton, OH 45432.
#Wright Laboratory, Wright-Patterson AFB, OH 45433.
‡Battelle Memorial Institute, 505 King Avenue, Columbus, OH 43201.

ABSTRACT

The hot workability of a γ titanium aluminide alloy was investigated using hot tension tests in the temperature range 900-1350°C and for deformation rates varying from 0.0001 to 5 s^{-1}. One series of tests was performed on cast + HIP'ped material and was designed to characterize fracture behavior during primary ingot break-down. Another series of tests was conducted on extruded and heat treated material in order to investigate the fracture processes occurring during secondary processing. The deformation and fracture characteristics were analyzed to assess the influence of microstructural features as well as the relative volume fractions of the γ and α/α_2 phases.

INTRODUCTION

Deformation processing of γ titanium aluminides constitutes a crucial step in the successful development and application of these materials. High strength and resistance to plastic deformation exhibited by γ titanium aluminides at high homologous temperatures make them very attractive for high temperature service, but drastically reduce their hot workability. In general, the hot working regime in which the material can be processed with adequate control over both microstructural evolution, and nucleation and propagation of cracks, is quite narrow. This is especially true for cast materials with relatively coarse grain size, which are prone to cracking and fracture under the influence of tensile stresses. Even though primary hot working processes such as forging or extrusion used for the break-down of the ingot structure involve nominally compressive states of loading, secondary tensile stresses can be generated due to geometrical, frictional or thermal effects. Examples of failures induced by tensile stresses include free surface bulging and fracture in open die forging, edge cracking during rolling and nose fracture during canned extrusion [1,2].

Tensile deformation and fracture behavior of γ or near γ titanium aluminides has been extensively investigated. The majority of these studies has been devoted to the evaluation of the alloys under potential service conditions, i.e., up to ~ 800°C. Notable exceptions are the studies by Lipsitt et al. [3], Huang and Hall [4] and by Krishnamurthy and Kim [5]. While these studies extended the test temperatures up to 1000°C, the strain rates used were typically in the range of 10^{-4} - 10^{-3} s^{-1}. The only published study on the tensile behavior of a near γ alloy over a wide range of temperatures and strain rates relevant to hot working was performed by Nobuki et al. [6,7]. These researchers evaluated the temperature and strain rate dependence of flow stress and fracture strain of a two-phase alloy containing fine, equiaxed grains of γ and α_2 phases obtained by isothermal forging. The objective of the present work was to investigate the tensile deformation and fracture behavior of a typical multi-component γ alloy in two different conditions: (a) cast and hot isostatically pressed and (b) extruded and heat treated. The temperature and strain rate dependence of the peak stress and the tensile ductility have been analyzed and interpreted in terms of the phase stability in this alloy.

MATERIAL AND PROCEDURES

A γ titanium aluminide alloy with a nominal composition of Ti-49.5Al-2.5Nb-1.1Mn (at.%) was obtained in the form of cast and hot isostatically pressed (HIP'ped) ingots, ~ 70 mm in diameter. The microstructure of these ingots consisted primarily of equiaxed γ grains, with a mean grain size of ~ 125 µm, and 10-15% of lamellar γ/α_2 colonies [Fig. 1(a)]. Billets 60 mm in diameter and 125 mm long were machined and encapsulated in evacuated and sealed cans of type 304 stainless steel. These preforms were extruded at 1170°C using streamlined dies and then cooled slowly. The extruded bars were 40 mm wide and 20 mm thick, corresponding to an extrusion ratio of 6:1. These bars were heat treated at 1050°C for one hour, followed by furnace cooling. The microstructure of the extruded and heat treated material was quite uniform and contained relatively fine and equiaxed γ grains [mean grain size = 35 µm, Fig. 1 (b)].

Cast + HIP'ped ingots and extruded + heat treated bars were machined and crush ground to yield tension test specimens with a gage diameter of 5 mm and gage length of 20 mm. Uniaxial tension tests were conducted at nominal strain rates ranging from 10^{-4} to 5.0 s^{-1} and at temperatures varying from 850°C to 1377°C. The specimens were coated with a protective layer of glass, induction heated and equilibrated at test temperature for 10 minutes before testing commenced. After testing, the specimens were cooled rapidly by forced convection.

Figure 1. Microstructures of (a) Cast+HIP'ped and (b) Extruded and heat treated specimens.

RESULTS AND DISCUSSION

Engineering stress-strain curves obtained at a nominal strain rate of 0.001 s^{-1} and at different temperatures for cast + HIP'ped and extruded + heat treated conditions are shown in Figs. 2(a) and 2(b), respectively. All the curves (except at 950°C) exhibit sharp maxima at low strain levels (0.01-0.05) regardless of the test temperature or the initial microstructure. This suggests that uniform elongation in these tests is quite low and that the specimens undergo necking and the resultant strain localization from very early stages of plastic deformation. The severity of necking and the associated extent of non-uniform elongation are dependent on the strain rate sensitivity characteristic of the initial microstructure as well as the test conditions. Earlier work by Seetharaman and Lombard [8] on the compressive deformation of the cast material showed that the values of strain rate sensitivity, m varied from 0.19 to 0.38 for the temperature range 1000-1400°C. A similar trend for the variation of m with test temperature was also obtained in the case of extruded and heat treated material [9]. The pronounced increase in non-uniform elongation with the test temperature for both cast + HIP'ped and extruded + heat treated structures is consistent with the above-mentioned temperature dependence of strain rate sensitivity.

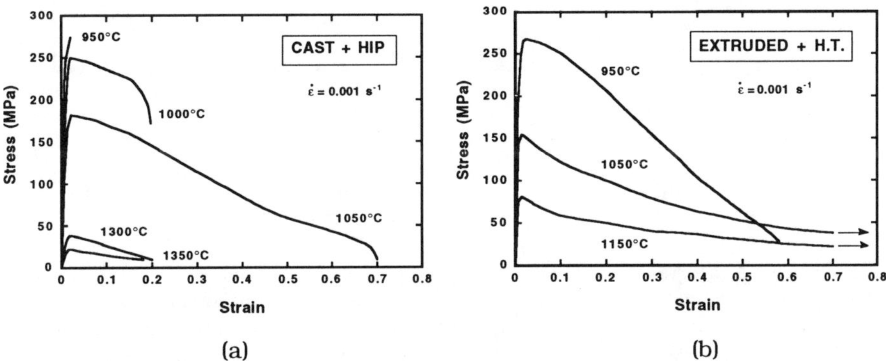

Figure 2. Engineering stress-strain curves obtained at a nominal strain rate of 0.001 s^{-1}: (a) Cast+HIP'ped and (b) Extruded and heat treated.

Peak stress values decrease strongly with an increase in the test temperature for the cast and the wrought microstructures. For example, the peak stress of the cast material at a strain rate of 0.001 s^{-1} decreases from 250 MPa at 1000°C to 23 MPa at 1350°C. Similarly, the extruded and heat treated material showed peak stress values of 193 and 21 MPa at 1000 and 1350°C, respectively. The difference in the peak stress values between cast and wrought microstructures, particularly in the temperature range 1000-1050°C can be attributed primarily to the fine size of the γ grains and the resultant enhancement in the kinetics of dynamic recrystallization in the wrought material. The absence of lamellar colonies in the wrought microstructure may also have contributed to a reduction in the flow stress.

Comparison of stress-strain curves obtained at high strain rates shows that (i) the dependence of peak stress on temperature and initial microstructure is similar to that observed at the low strain rate and (ii) strain rate exerts a strong influence on the peak stress and total elongation values. Figure 3 shows plots of log (strain rate) vs. T^{-1} corresponding to constant value of peak stress, σ_p in the range $75 \leq \sigma_p \leq 200$ MPa. These plots are linear and parallel to each other in the strain rate range, 10^{-4} to 10^{-2} s^{-1}. At higher strain rates, the plots become distinctly non-linear and slope upward. From the mean slope of the linear segments, an activation energy value, Q = 411 kJ/mol was obtained. This value is somewhat higher than 327 kJ/mol obtained in previous work [8] on this material. However, Semiatin et al. [10] have reported Q = 627 kJ/mol for an alloy containing a mixture of γ and α/α_2 phases.

When severe localized deformation occurs during tension testing, reduction in area in the fractured region of the specimen provides a good measure of the hot ductility [1,11]. Figure 4 shows the variation in reduction in area with temperature and strain rate for cast and wrought microstructures. At 0.001 s^{-1}, this parameter

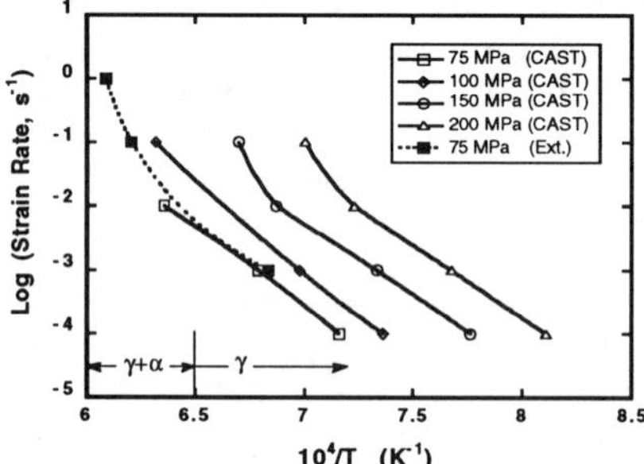

Figure 3. Log $\dot{\varepsilon}$ vs 1/T plots at constant values of σ_p.

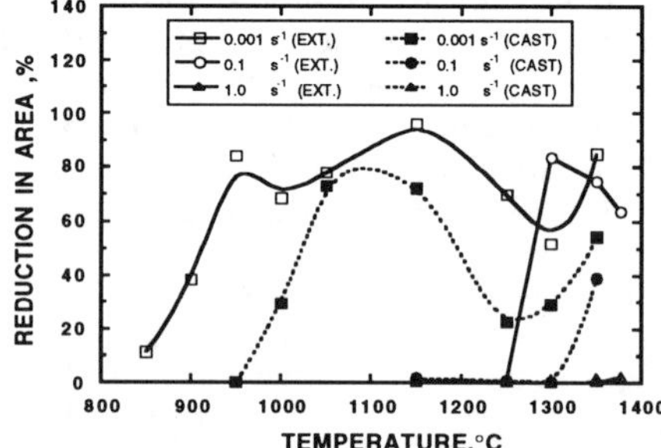

Figure 4. Variations in reduction in area with temperature and strain rate.

for the cast material increases with temperature, reaches a maximum at approximately 1100°C, decreases and appears to attain a minimum value at 1250°C and thereafter increases steadily up to 1350°C. In contrast, the behavior at 0.1 and 1.0 s^{-1} is characterized by nearly zero ductility and a sharp increase above this temperature. The extruded and heat treated material also exhibits a similar dependence of reduction in area on temperature and strain rate; however, the ductility values for the wrought material are significantly higher than those for the cast material. Factors responsible for the increased ductility of the wrought material include (i) fine, equiaxed morphology of the γ grains (35 μm) and (ii) microduplex distribution of equiaxed crystals of γ and α/α_2 phases resulting from the transformation of lamellar grains during hot extrusion and the subsequent heat treatment. The wrought microstructure prevents the nucleation and growth of wedge cracks and enables the formation of a uniform and finely divided array of voids at large strain levels. The drop in ductility noted above ~ 1100°C can be attributed to the rapid grain growth kinetics of the γ phase and to a decrease in the amount of the α phase. This trend continues up to the γ transus, 1270°C, for this alloy. Above 1270°C, the volume fraction of the α phase increases gradually with temperature, leading to a net decrease in the grain growth kinetics and hence a reversal in the temperature dependence of ductility.

The ductile-brittle transition temperature, T_{DB} and its dependence on strain rate are illustrated in Fig. 5. Data for $\dot{\varepsilon}= 10^{-4}\ s^{-1}$ for the wrought material were obtained from the work of Lipsitt et al. [3]. Both plots appear to be linear within the experimental scatter. An Arrhenius type of analysis of these data yields activation energy values of 275 and 189 kJ/mol for cast and wrought materials, respectively. Ductile-brittle transition at hot working strain rates is controlled by dynamic recrystallization of the gamma grains. Thus, it is appropriate to note that the Q value of 275 kJ/mol for the cast material compares reasonably well with Q = 327 kJ/mol for the dynamical recrystallization process obtained from compression tests [8]. The low value of Q for the wrought material suggests a large contribution to deformation by grain boundaries. The strongest influence of grain size on the D-B transition is observed in the diffusional flow regime, i.e., at low strain rates and low temperatures. Under these conditions, $\dot{\varepsilon}\alpha(d)^{-p}$, where d is the grain size. The value of p decreases from 2-3 in the diffusional flow regime to 0 in the power-law creep regime. The extensive amount of damage accumulation in the form of elongated voids, together with the microstructural development that occur near the fracture zone of a specimen tested at 1050°C and $10^{-3}\ s^{-1}$ are illustrated in Fig. 6. A large fraction of dynamically recrystallized γ grains is discernible. The size, morphology and distribution of these voids depend critically on the temperature and strain rate. At high strain rates, the voids link to form wedge-type cracks along interphase boundaries [2].

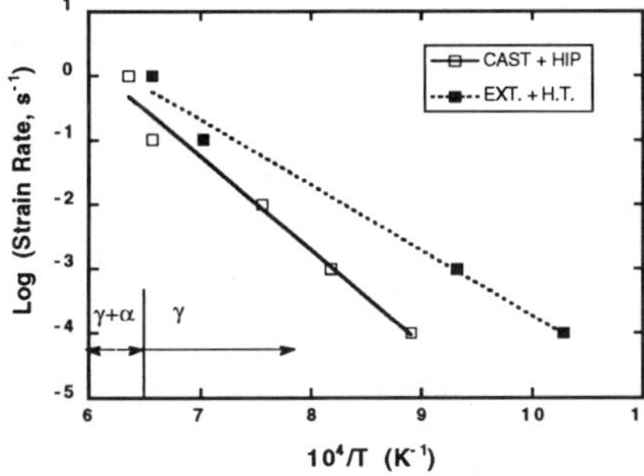

Figure 5. Log $\dot{\varepsilon}$ vs $1/T_{DB}$ plots based on the ductile-brittle transition behavior.

Figure 6. Microstructure of the cast + HIP'ped specimen tested at 1050°C and $10^{-3}\ s^{-1}$.

SUMMARY

Hot tension tests were performed on a γ titanium aluminide alloy in both the cast and wrought conditions. Peak stress values for cast as well as wrought microstructures decreased systematically with an increase in temperature and a decrease in the strain rate. The values for the cast material are slightly higher than those associated with the extruded and heat treated material, particularly at temperatures below 1200°C. The temperature dependence of peak stress follows an Arrhenius behavior with an activation energy of 441 kJ/mol. The tensile ductilities of both microstructures determined at a strain rate of 0.001 s^{-1} exhibit a complex temperature dependence involving ductility maxima at 1100-1150°C, and minima at ~ 1250°C. In contrast, deformation at high strain rates resulted in brittle fracture up to ~1250°C and ductile fracture at higher temperatures. Higher ductility values associated with the wrought alloy could be explained in terms of the enhanced kinetics of dynamic recrystallization in a previously hot worked microstructure.

ACKNOWLEDGEMENTS

This work was performed as part of the in-house research activities of the Processing Science Group, Materials Laboratory, Wright-Patterson AFB, OH. One of the authors (V.S.) was supported by Air Force Contract F33615-92-C-5900. The authors appreciate the cooperation extended by Messrs. J. T. Morgan and D. R. Barker during the course of the work. They also acknowledge the assistance of Ms. J. F. Hickman in preparation of the manuscript.

REFERENCES

1. S. L. Semiatin, in Workability Testing Techniques, edited by G. E. Dieter (Amer. Soc. Met., Metals Park, OH, 1984) p. 197.
2. V. Seetharaman, R. L. Goetz, and S. L. Semiatin, in High Temperature Ordered Intermetallic Alloys IV, edited by J. O. Stiegler, L. A. Johnson and D. P. Pope (Mater. Res. Soc. Proc., **213**, Pittsburgh, PA, 1991), p. 895.
3. H. A. Lipsitt, D. Shectman, and R. E. Schafrik, Metall. Trans. A, **6**A, 1991 (1975).
4. S. C. Huang and E. L. Hall, Metall. Trans. A, **22**A, 427 (1991).
5. S. Krishnamurthy and Y-W. Kim, in Microstructure/Property Relationships in Titanium Aluminides and Alloys, edited by Y-W. Kim and R. R. Boyer (The Minerals, Metals and Materials Society, Warrendale, PA, 1991), p. 149.
6. M. Nobuki, K. Hashimoto, and T. Tsujimoto, Bull. Japan Inst. Met., **30,** 49 (1991).
7. M. Yamaguchi, Mater. Sci. Tech., **8**, 199 (1992).
8. V. Seetharaman and C. M. Lombard, in Microstructure/Property Relationships in Titanium Aluminides and Alloys, edited by Y-W. Kim and R. R. Boyer (The Minerals, Metals and Material Society, Warrendale, PA, 1991), p. 237.
9. C. M. Lombard, V. Seetharaman, and S. L. Semiatin, Materials Laboratory, Wright-Patterson Air Force Base, Ohio, unpublished research (1992).
10. S. L. Semiatin, N. Frey, S. M. El-Soudani, and J. D. Bryant, Metall. Trans. A, **23**A, 1719 (1992).
11. W. Roberts, in Deformation, Processing and Structure, edited by G. Krauss (Amer. Soc. Met., Metals Park, OH, 1984) p. 109.

MOLECULAR DYNAMICS SIMULATION OF FRACTURE IN RUAL

C.S. BECQUART, P.C. CLAPP and J.A. RIFKIN
Center for Materials Simulations, Institute of Materials Science, University of Connecticut, Storrs, CT 06269-3136

ABSTRACT

High temperature intermetallic compounds show promise for aerospace technology. Unfortunately many of them have the property of being brittle at low temperatures and have therefore to be modified to be used. However among the binary high temperature materials, stoichiometric RuAl has shown some significant room temperature toughness [1].

Computer simulations at the atomistic level have shown to be of significant help in the understanding of many phenomena (phase transformations, etc). In this work, Molecular Dynamics and potentials derived by the Embedded Atom Method (EAM) have been used to study the behavior of a microcrack embedded in a three-dimensional lattice of RuAl. Values of the critical toughness factor (value at which the crack starts propagating) have been established for different orientations of the crack and for different temperatures. Dislocation emission and propagation have been observed in some cases. Parameters such as the *unstable stacking energy* (which characterizes the resistance to slip) and the *surface energy* (which characterizes the resistance to cleavage) have also been calculated for this material. A recent theory developped by Rice [2] will be used to interpret the behavior of the crack upon orientation.

Introduction

Since the advent of the Griffith criterion [1] which predicts the condition for stability of a crack in a purely brittle material, much research has been done to extend this criterion to ductile and semibrittle materials. A pionneering work by Rice and Thomson [2] in 1974 established the important contribution of dislocation emission and propagation at the crack tip to the understanding of the behavior of a microcrack under load. Since then many models have been proposed to predict the brittle versus ductile behavior. One of the most recent one was proposed recently by Rice [2]. Molecular Dynamics simulations give access to all the parameters needed in this theory and this paper will present the results of simulations performed on crack-embedded arrays of RuAl (a prospective candidate for high-temperature use) and compare them with Rice's predictions.

Conditions of the simulations

The simulations were performed on arrays of approximately 8,000 particles with several different orientations. In this array an initial crack was created and a tensile load normal to the crack applied. Orientations are given as a the direction of crack propagation / crack plane pair. For example [001](1-10) means that the crack would normally propagate in the [001] direction (if crack propagation occurs at all) and (1-10) is the crack plane normal. In these two directions there are free surfaces at the array boundary. In the remaining direction there is a repeating boundary. The dimensions of the arrays were on the order of 70 Å by 70 Å by 20 Å, where the shorter direction is the one with the repeating boundary.

The interatomic potentials (which describes the forces between atoms) are from Voter and Chen [3] for the Al-Al interactions and from Rifkin [4] for the Ru-Ru and Ru-Al interactions. All potentials were computed using the Embedded Atom Method (EAM) from Baskes and Daw [5]. These potentials were empirically fit to the cohesive energies, vacancy energies, lattice constants and elastic constants for the simulated systems.

Atomic motion is obtained from the calculated forces via numerical integration, which is performed in discrete intervals called a time step. In our simulations each time step represents 3.10^{-15} seconds simulated time. Throughout this paper simulation time is given in units of time steps.

A crack was formed by displacing the atoms of the otherwise perfect array according to the continuum mechanics equations for anisotropic elasticity due to Sih et al. [6]. The following equations give the displacements of the atoms in a crack-embedded lattice under pure tensile load (applied along the Y axis) and for plane strain conditions along the Z direction:

$$u_x = K_I\sqrt{2r}\,Re\left\{\frac{1}{s_1 - s_2}\left[s_1 p_2(\cos\theta + s_2\sin\theta)^{\frac{1}{2}} - s_2 p_1(\cos\theta + s_1\sin\theta)^{\frac{1}{2}}\right]\right\}$$

$$u_y = K_I\sqrt{2r}\,Re\left\{\frac{1}{s_1 - s_2}\left[s_1 q_2(\cos\theta + s_2\sin\theta)^{\frac{1}{2}} - s_2 q_1(\cos\theta + s_1\sin\theta)^{\frac{1}{2}}\right]\right\}$$

where r and θ are the polar coordinates of the atom for which the displacement is calculated with the origin at the crack tip (figure 1). K_I is the stress intensity factor :

$$K_I \equiv \sigma\sqrt{a\pi} \tag{2}$$

where σ is the applied load and a is the length of the crack. The variables s_1, s_2, p_1, p_2, q_1,and q_2, depend upon the compliances of the materials [6].

After the atoms have been positioned to form a crack of given length (25Å in our case) under a given value of σ, the lattice is let to relax for 2,000 steps. During these 2,000 steps, the atoms move according to the Molecular Dynamics algorithm (that is, according to the classical equations of motion) for some chosen temperature. To prevent the crack from closing, the outer-most two layers of atoms on three of the four free surfaces (not including the surface with the crack opening) are constrained to remain in their initial positions (given by equation 1). These constrained atoms are referred to as the *frozen layers*.

During the relaxation period the crack will either close, stabilize or propagate. When a value of the applied tensile load σ has been found at which the crack remains stable, the stress intensity factor is slowly increased by displacing the atoms of the "frozen layers" by 1% of their original displacements (given by equation 1) every 20 steps. With this scheme, the stress intensity factor is doubled in 2,000 computational steps. Under the increasing load, the crack may propagates (brittle behavior), be blunted by dislocation emission (ductile behavior) or show both behaviors.. The value of the stress intensity factor at the time the crack propagates (assuming it does propagate) is taken to be the critical value.

The molecular dynamics simulations were run typically for between 4,000 and 5,000 time steps for four orientations:[110](1-10), [1-10](111), [100](010) and [110](001). Depending upon the orientation of the crack plane and of the tensile load, different behaviors were observed in our simulations. Some cases showed only crack propagation and were considered purely brittle. Others showed only dislocation emission and crack tip blunting and were considered purely ductile. Still others showed a combination varying degrees of both crack propagation and dislocation emission. These are an intermediate case. In all instances of dislocation emission the relevant slip system was determined. For those cases which showed crack propagation a value of the critical stress intensity factor was determined.

Results of the simulations

Out of the four orientations studied ([110](1-10), [1-10](111), [100](010) and [110](001)), the first two exhibit purely brittle behavior under mode I loading. The crack simply propagates on the original crack plane at some value of the stress intensity factor K_I, and the lattice behaves in a very brittle way (Fig.1). Very important surface reconstruction (rearrangement of the atoms of the external layers) is visible (Fig.1). On the free surfaces containing the crack opening, one {111} plane remains at its original position, while the next two planes get closer to each other (very similar to a ω-type motion). On the free surfaces of the crack, the Al atoms move away from each other, while the Ru atoms get closer, resulting in a rippled layer : the phenomenon of surface reconstruction in RuAl is discussed in more details in another paper published by this group [7]

For the third orientation [110](001), the crack does not propagate and dislocation emission (on {110} planes with <001> type Burgers vector) blunts the crack tip (figure 2). However, before dislocation emission, one can definitely see one broken bond on each side of the crack just below the tip as if the crack tended to propagate at a 90 degree angle. This may be due to surface reconstruction similar to the surface reconstruction observed on the free surface containing the crack opening of the array (Ru atoms are getting closer to each other, while the Al move away from each other) or to the fact that the {110} planes have lower surface energies than the {100} planes. This will play an important role in concentrating the strain around the crack tip.

Orientation [100](010) shows a combination of events: the crack veers a little onto {110} planes (planes of lowest surface energy for our array) which are not the original crack plane; soon afterwards {100}<010> dislocations are emitted in those cases where there is a little thermal energy available (T > 100K). At higher temperature (600K) the crack starts its propagation on {100} for one or two bonds, but sometimes it tries to propagate on {110} before emitting {100}<010> dislocations (Fig.3).

The values of stress intensity factor at which the crack started to propagate is summarized in Table I for different temperatures.

Table I : Variation of the critical stress intensity factor with temperature for RuAl

Crack orientation	K_{Ic}^{G} $(Mpa\sqrt{m})$	K_{Ic} $(Mpa\sqrt{m})$ from simulations at different temperatures				
		1K	100K	300K	600K	1000K
[100](010)	**1.26**	2.2-2.4			<2.1	<2.1
[110](1-10)	**1.10**	1.3	1.7	1.7	1.7	1.7
[1-10](111)	**1.26**	2.0	1.6	1.6	1.6	1.6
[110](001)	**1.31**	**NO CRACK PROPAGATION**				

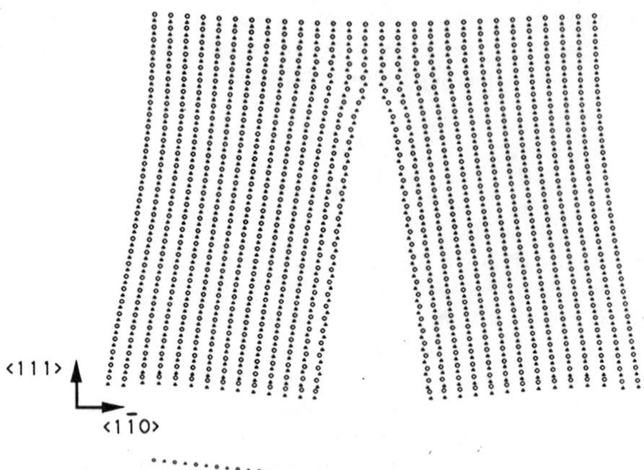

Figure 1
The crack propagates: RuAl in [1-10](111) orientation. Step 4000. Triangles are Al, circles Ru.

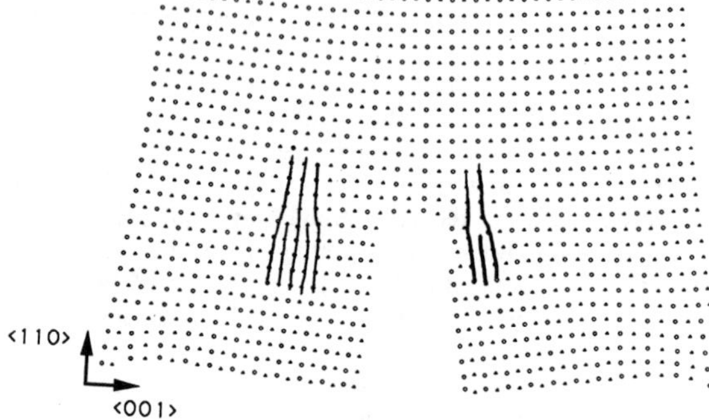

Figure 2
Blunting of the crack tip: RuAl in [110](001) orientation. Emission of dislocations at right angles to crack plane has blunted the crack tip. Step 3700. Triangles are Al, circles Ru.

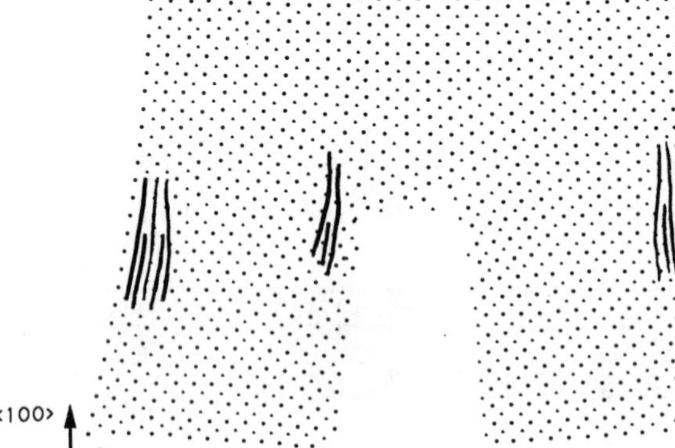

Figure 3
Crack propagation combined with blunting of the crack tip: RuAl in [100](010) orientation. The crack has tried to propagate on different {110} planes, dislocations have also been emitted from the crack tip Step 4300. Triangles are Al, circles Ru.

K_{Ic}^{G} is the theoretical critical stress intensity factor according to Griffith criterion. It is calculated assuming that the energy supplied by the propagating crack is only used to create the two new free surfaces of the crack, and ignores the additional energy required for dislocation emission or blunting the crack tip.

From this table one can see that the crack starts propagating at values of the stress intensity factor higher (120% to 170% higher) than the theoretical value determined with the Griffith criterion K_{Ic}^{G}. Depending upon the orientation, thermal energy helps crack propagation (orientations [1-10](111) and [100](001)) or hinders it (orientation [110](1-10)) and it is not yet clear why both behaviors were observed.

Comparison with Fracture Theory

A theory on the competition between brittle and ductile behavior of a crack has been recently published by J.R. Rice[2]. This theory assumes a periodic relationship between the applied shear stress and the atomic displacement. Because of this relation, when a small load is applied a small amount of slip can take place at the crack tip which can lead to a fully formed dislocation. The crack will be blunted by dislocation emission if the resistance to dislocation emission is overcome before the resistance to cleavage. The resistance to dislocation nucleation is taken to be proportional to a new solid state parameter called *the unstable stacking energy* while the resistance to cleavage is *the surface energy*.

The values for the *unstable stacking energy* were calculated for RuAl according to Rice's scheme (Table II). The *unstable stacking energy* is the energy barrier encountered in a block-like sliding of one half of a lattice relative to the other half along a slip plane. The *surface energy* (Table III) is calculated by taking the difference in energy between an array with periodic boundaries on and the same array with free surfaces parallel to the plane which energy is needed.

Table II: Unstable stacking energies

Slip orientation	Unrelaxed unstable stacking energy (mJ/m^2)	Relaxed unstable stacking energy (mJ/m^2)
{100}<010>	3070	2498
{100}<011>	7224	3981
{110}<001>	2268	1882
{110}<1-10>	5014	2190
{110}<1-11>	1202	1130

Table III: Surface energies

Surface orientation	Surface energy (mJ/m^2)
{100}	3203
{110}	2446
{111}	3029
{210}	2998

Using these values Rice made predictions on the behavior of the crack for each orientation. RuAl was predicted to be brittle for all the orientations studied. The source of the discrepancy between prediction and simulation is not yet fully understood. It is believed [8] that local deformations induced by phenomena such as surface reconstruction, veering of the crack onto planes of lower energies, etc, change the distribution of the strain pattern ahead of the crack (Fig.4). This new configuration of the strain differs thus from the one assumed by J.R Rice in his model. Furthermore, more importance should be accorded to the coupling between shear and tension and more investigations are being done by Rice and his group on this particular problem [9].

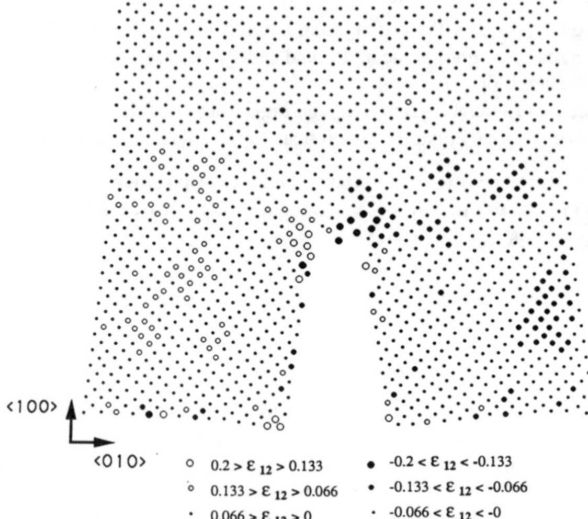

Figure 4
Shear strain distribution around the crack tip: high values of the shear strain are concentrated where the crack tried to veer and where dislocations will be emitted.

Conclusion

Molecular Dynamics simulations of a microcrack embedded in a single crystal of RuAl and under tensile stress were done. Depending upon the orientation of the crack plane and of the crack line, different behaviors were observed : brittleness, ductility and combination of both. A theory derived by Rice [2] was used to try to interpret the various behaviors. It appears that consideration of local deformations at the crack tip (due to surface reconstruction or crack veering), as well as the coupling between shear and tension, must be made to characterize more completely the area of the crack tip.

Acknowledgment

This work was supported by the Office of Naval Research Materials Division Grant N00014-90-J1223.

REFERENCES

1 R.L. Fleischer, R.D. Field and C.L. Briant, Met. Trans. A, **22**, 403 (1991)
2 J.R. Rice, Jour. Mech. Phys. Solids, **40**, 239 (1991)
3 A.F. Voter and S.P. Chen, Mat. Res. Soc. Symp. Proc. **82**, 175 (1987)
4 J.A. Rifkin, C.S. Becquart, D. Kim and P.C. Clapp, Mat. Res. Soc. Symp. Proc. **278**, 173 (1992)
5 M.S. Daw and M.I. Baskes, Phys. Rev. B, **29**, 6443 (1984)
6 G.C. Sih and H. Liebowitz, *Fracture: An Advanced Treatise*, *Vol.2*, H. Liebowitz editor, Academic Press New york and London (1968)
7 C.S. Becquart, J.A. Rifkin and P.C. Clapp, to be published.
8 C. S. Becquart, Ph.D Thesis, University of Connecticut, (1993)
9 J.R. Rice, private communication (1992)

{111} ZIG-ZAG TRANSGRANULAR CRACKING IN THE $L1_2$ INTERMETALLIC COMPOUND Ni_3Ge

Jianxin Fang and Erland M. Schulson
Thayer School of Engineering, Dartmouth College, Hanover, NH 03755

ABSTRACT

Slip and fracture behavior in Ni_3Ge with and without 0.06 at.% Boron has been examined by *in-situ* straining in a TEM. Crack propagate in a zig-zag, transgranular manner and the crack planes are coincident with the slip planes. Slip is activated on two major systems and a transition occurs ahead of dislocation pile-ups which occur on the primary system. Due to the thinning of material from the motion of dislocations, the crack follows alternately on these two slip planes, giving rise to the zig-zag transgranular path. Calculations show that the stress field due to the pile-up drives the slip transition.

I. INTRODUCTION

Direct observations by transmission electron microscopy have been extensively made of the formation of zig-zag cracks, which are typical rupture modes in various ductile disordered metals and alloys [1-5]. It was found that the edges of the cracks are alternately parallel to <110> and <112>, and that the zig-zag cracks are formed by intense slip along <110> and mechanical twinning parallel to <112> [2, 4, 6, 7].

Recently, zig-zag cracks have been observed in the TEM by *in-situ* straining the $L1_2$ intermetallic compound Ni_3Al [8] and by *ex-situ* straining of thin films of the $L1_2$ alloy Ni_3Ge at atmospheric conditions [9]. In contrast to the fracture morphology in the disordered materials, the traces of the crack surfaces are along <110> in these $L1_2$ alloys. It has been suggested [8, 9] that the zig-zag pattern arises from the facts that the crack propagation follows slip bands within which the foils thin due to the slip of dislocations, and that the slip switches from one system to another. However, no attempt to identify the controlling factor for the slip switching has been made yet.

In the present work, direct observations were made through TEM *in-situ* straining of the slip and fracture behavior in polycrystalline thin films of the $L1_2$ intermetallic compound Ni_3Ge with and without 0.06 at.% boron. The purpose of the work was to determine the factor controlling the slip system transition and, hence, to highlight the mechanism for the formation of the zig-zag transgranular cracks. In this paper, we briefly report the main results obtained. The complete analysis can be found elsewhere [10].

II. EXPERIMENTAL

Miniature tensile samples for the *in-situ* deformation experiments were cut from extruded rods of Ni-23.5 at.% Ge with 0.06 at.% boron and without boron. The processing to produce the rods and the analyzed chemical compositions were described earlier [11]. The geometry of the samples was designed to ensure that deformation and fracture occur within the thin area [8]. The specimens were annealed at 1223 K for 2 hours in an atmosphere of flowing dried and de-oxygenated argon to produce a grain size about 15 μm. The samples were electropolished in a modified specimen holder to produce a center hole surrounded by a uniformly thin area. The electrolytes and polishing conditions were given earlier [11]. The tensile deformation and fracture were carried out *in-situ* in a JEOL 2000FX microscope operated at an accelerating voltage of 200 kV.

III. RESULTS

3.1 Crack propagation and slip transition

Although a modest beneficial effect of boron on the ductility and fracture mode was observed in polycrystalline Ni_3Ge, the general behavior of a crack in a thin region was similar for both alloys. Thus, while most of the observations to be presented were made on the doped alloy, the behavior is considered to be characteristic of thin foils of both alloys.

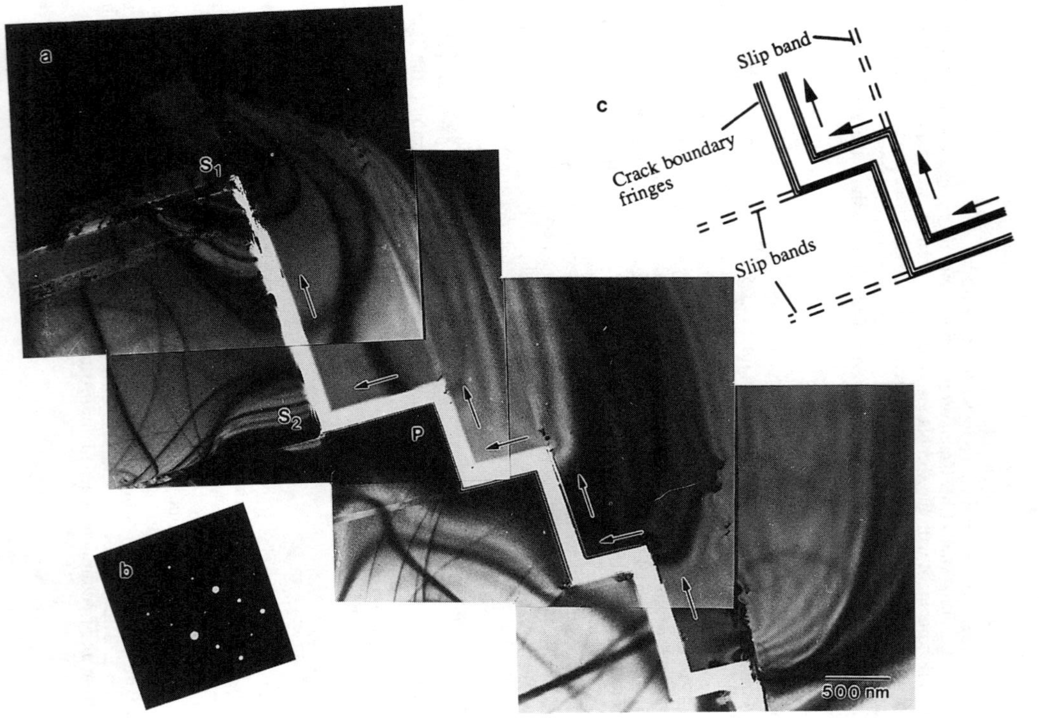

Fig. 1. (a) Zig-zag transgranular crack within the interior of a grain in boron-doped Ni_3Ge. Arrows indicate the direction of the crack propagation. (b) Diffraction pattern showing that foil normal is [001]. (c) Sketch of a portion of the crack showing that slip bands extend from the corners on the crack path in the direction of crack propagation and that no slip bands transmit through the crack in the reverse direction of crack propagation.

Fig. 1 (a) shows a transgranular crack within the boron-doped foil. Note its zig-zag shape. This image was taken with the beam direction close to [001] by slight tilting because the foil normal was [001] as shown by diffraction pattern in Fig. 1 (b). The arrows indicate the direction of the crack propagation. In front of the crack tip at site S_1, several slip bands were observed: one parallel to and the others normal to the crack propagation direction at that site. Note that the slip bands became more electron transparent owing to the thinning of the foil through the dislocation motion within the band. As discussed below, these bands indicate that the crack propagation was preceded by localized plastic deformation. That the cracks and the slip bands were parallel to each other implies that mainly two slip systems were active and that the crack followed the slip planes. Unfortunately, the sequence of slip and crack propagation could not be observed step by step, owing to the speed with which the process occurred.

The scheme for explaining the zig-zag pattern is that slip first occurred locally on one system (primary system), switched locally onto another slip system (secondary system), and then switched back at some other site onto the primary system. Owing to the thinning of the material within the bands, the crack then advanced along these bands, forming a jog. Repetition of this process produced the zig-zag pattern. In this scheme, extensions of the parent slip band beyond the corners of the crack would be expected to occur only in the direction of the crack propagation, as observed (Fig. 1 (a)) and schematically shown in Fig. 1 (c).

Upon continued straining, the crack shown in Fig. 1 (a) bifurcated at site S_1; also an extension formed at site S_2. In addition, slip was observed to switch from one slip band to another, as shown in Fig. 2, indicating a transition from one slip system to another. This slip transition further supports the idea that prior to zig-zag cracking, slip occurred in a zig-zag manner. Blurry though it is, Fig. 2 also suggests that the slip transition occurred near a dislocation pile-up within the parent slip bands.

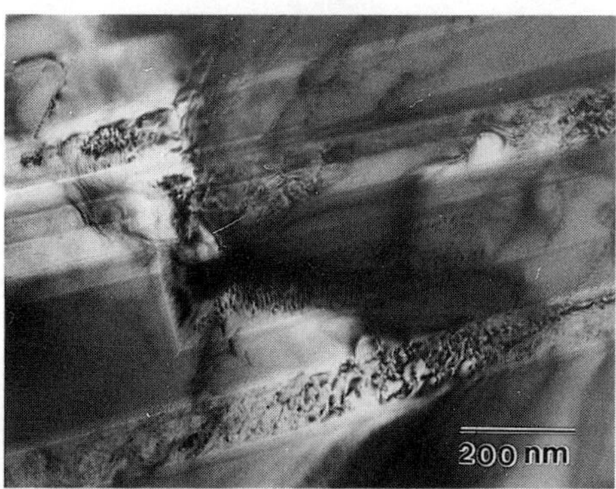

Fig. 2. Electron micrograph shows slip transition from one band to another and occurring near the tip of dislocation pile-up within the parent slip band.

3.2 Crack plane and slip systems

Fig. 3 shows a higher magnification dark-field image of the crack boundary at (P) in Fig. 1 (a). The boundary fringes clearly show that the crack surfaces were inclined to [001]; i.e., the crack did not propagate on {001} planes. The crack plane was determined by calculating the angle, α, between the normals to the crack surface and to the foil, which is given by:

$$\alpha=\tan^{-1}\left(\frac{t}{w_{pro}}\right) \qquad (1)$$

where t is the foil thickness and w_{pro} the projected crack width. The localized foil thickness, t, was determined to be 814±16 Å using the convergent beam technique [12] and the projected crack width, w_{pro}, was measured to be about 560 Å. Substituting these values into Eq. (1) yields $\alpha = 55 \pm 1$ degrees, which is in good accord with the angle between the [001] orientation and the normal to [111]. It is thus concluded that thin foils of Ni_3Ge cleave on {111} planes.

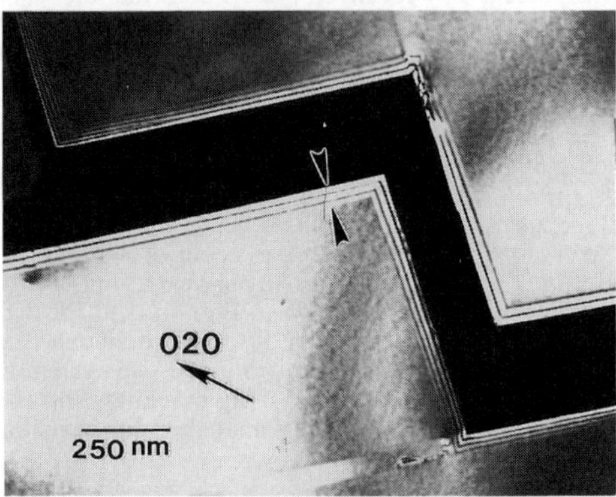

Fig. 3. Higher magnification dark image at (P) in Fig. 1 (a). The crack surface is {111} and coincident with operative slip plane.

The fact that slip bands extended from the corners of the crack to the grain boundaries, which impeded dislocations, allowed us to trace the dislocations and then to determine the slip systems involved. Fig. 4 (a) and (b) show two sets of dislocations, A and B, within slip bands which extended from the upper and from the lower crack surfaces of Fig. 1 (a), respectively. Contrast and trace analysis revealed that both dislocations A and B are screw in character with the same Burgers' vector $[\bar{1}01]$. Trace analysis further showed that the slip planes for

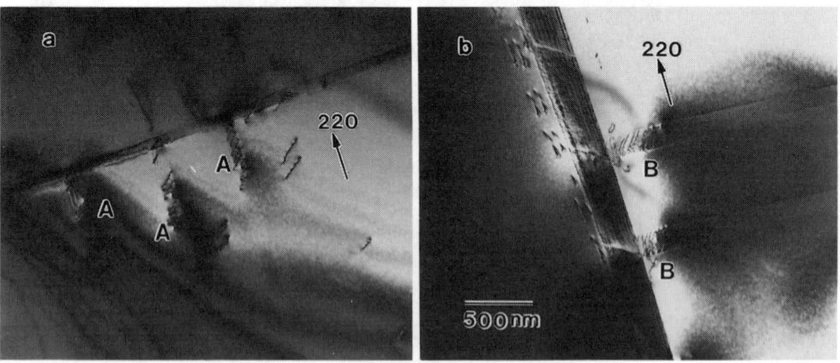

Fig. 4 (a) and (b) showing two sets of dislocations A and B within slip bands extended from the upper and from the lower crack surfaces of Fig. 1 (a), respectively.

dislocations A and B were (111) and $(1\bar{1}1)$, respectively, which were coincident with the crack surface, as expected. Therefore, the two active slip systems in the grain are $(111)[\bar{1}01]$ and $(1\bar{1}1)[\bar{1}01]$. In other words, the zig-zag slip which preceded the zig-zag cracking was the result of repetitive cross slip.

IV. DISCUSSION

Before considering the origin of the {111} zig-zag cracking, it is noted again that Inoue et al. [9] discovered the same feature following the deformation of thin foils of Ni_3Ge at atmospheric pressure. It appears, therefore, that the environment is not a factor.

It has been established from the present investigation that the zig-zag cracking occurs by the slip transition which precedes cracking and by the attendant thinning of the foil within the slip bands. However, the question is the origin of the shear stress which drives the slip transition.

Since the observations (Fig. 2) suggest that dislocations pile up within the grain and that the slip transition occurs ahead of the pile-up, the most likely stress field which drives cross slip is one arising from a dislocation pile-up. In the present case, system A $((111)[\bar{1}01])$ on which the A dislocations glide is assumed to be the primary slip system and then B $((1\bar{1}1)[\bar{1}01])$ is the observed secondary slip system. The question is whether the attendant shear stress is higher on the observed cross slip system B than on any of the other ten possible {111}<110> systems. The resolved shear stress fields on secondary slip systems resulting from a pile-up of screw dislocations on the primary slip plane have been calculated and plotted (albeit on different scales for different systems) by Mitchell [13] and by Basinski and Mitchell [14] for fcc crystals. We confirmed their calculations, but plotted stress contours on one scale for easier comparison [10]. The calculations show that the resolved shear stress on system B (Only the stress field on system B is shown in Fig. 5 and stress fields on the other ten secondary systems can be found in ref. [10]) is indeed greater than on any other slip system.

Therefore, the process of {111} zig-zag cleavage cracking may be summarized as follows. Systems A $(111)[\bar{1}01]$ and B $(1\bar{1}1)[\bar{1}01]$ are activated alternately due to the stress concentration of dislocation pile-ups. Since the gliding of dislocations on the alternate slip planes within the grain reduces the thickness of the specimen, the crack then follows these planes, producing the zig-zag path.

V. CONCLUSIONS

TEM *in-situ* straining experiments were carried out on polycrystalline Ni_3Ge with and without 0.06 at.% boron to investigate the mechanism for the formation of a transgranular crack. The main results are summarized as follows:

(1) Crack propagation occurred in a zig-zag, transgranular manner and was preceded by a localized plastic deformation. The crack planes were {111} and were coincident with the slip planes.

(2) Slip was activated alternately on two major systems and slip transition occurred via cross slip ahead of dislocation pile-ups which occurred on the primary slip system. Due to the thinning of material from the slip of dislocations, the crack propagated alternately on these two slip planes, leading to the {111} zig-zag transgranular path. Calculations of the stress fields showed that the choice of the slip system was dictated by the stress concentration due to the pile-up.

VI. ACKNOWLEDGEMENTS

The authors wish to thank Dr. C. Briant of the GE company for preparing the materials, Prof. I. Baker for valuable discussions and Dr. C. P. Daghlian for the help during experiment. The use of the Dartmouth College Electron Microscope Facility is gratefully acknowledged. This work was supported by a grant from U. S. Dept. of Energy, Contract No. DE-FG02-86ER45260.

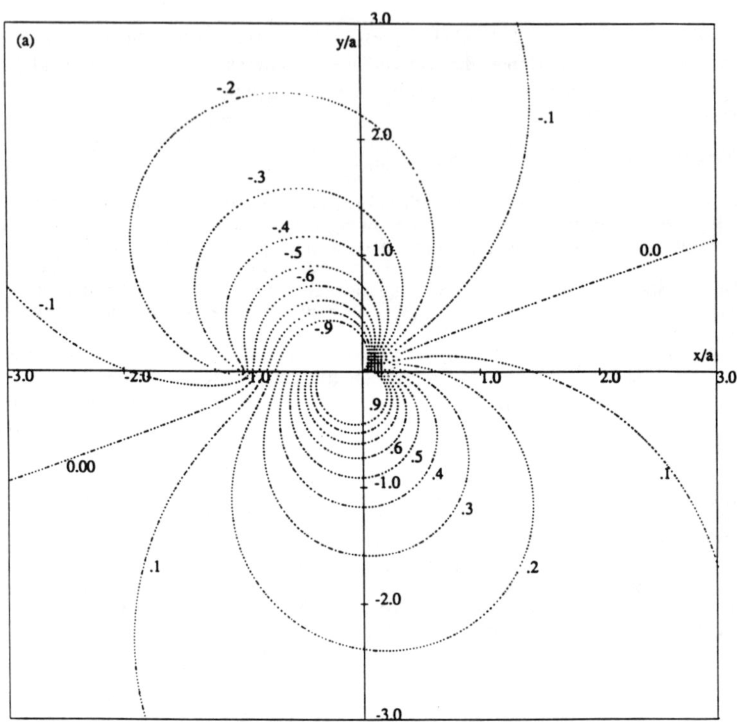

Fig. 5 The stress-fields on the observed secondary system B due to a screw dislocation pile-up on the primary system A between x/a =-1 and x/a = 0, in units of τ_a.

VII. REFERENCES

1. R. L. Lyles, Jr. and H. G. F. Wilsdorf, Acta Metall. 23, 269 (1975).
2. H. G. F. Wilsdorf, Acta Metall. 30, 1247 (1982).
3. J. A. Horton and S. M. Ohr, J. Mater. Sci. 17, 3140 1982.
4. Robertson and Birnbaum, Acta Metall. 34, 353 (1986).
5. S. M. Ohr, J. A. Horton, S.-J. Chang, in G. C. Sih and J. W. Provan (eds.), *Defects, Fracture and Fatigue*, Martinus Nijhoff, The Hague, p3 (1983).
6. H. G. F. Wilsdorf, Scripta Metall. 17, 1209 (1983).
7. K. Jagannadham and H. G. F. Wilsdorf, Mater. Sci. and Eng. 81, 273 (1986).
8. I. Baker, E. M. Schulson and J. A. Horton, Acta Metall. 35, 1533 (1987).
9. H. R. P. Inoue, C. M. Wayman and T. Saburi, Mat. Res. Soc. Symp. Proc. 213, p. 521 (1991).
10. J. Fang and E. M. Schulson, Phil. Mag., in press, (1992).
11. J. Fang and E. M. Schulson, Mater. Sci. and Eng. A152, 138 (1992).
12. P. M. Kelly, A. Jostsons, R. G. Blake and J. G. Napier, Phys. Stat. Sol. (a) 31, 771 (1975).
13. T. E. Mitchell, Phil. Mag. 10, 301 (1964).
14. Z. S. Basinski and T. E. Mitchell, Phil. Mag. 13, 103 (1966).

DEFORMATION BEHAVIOR OF POLYCRYSTALLINE NIAL CYCLICLY DEFORMED NEAR THE BRITTLE-TO-DUCTILE TRANSITION TEMPERATURE

Cheryl L. Cullers and Stephen D. Antolovich*, Georgia Institute of Technology, Atlanta, GA; *presently at Washington State University, Pullman, WA
Ronald D. Noebe, NASA Lewis Research Center, Cleveland, OH.

ABSTRACT

Low cycle fatigue (LCF) behavior of polycrystalline NiAl was investigated near the monotonic brittle-to-ductile transition temperature (BDTT) at plastic strain ranges of 0.5 and 1.0%. Between 600 and 700 K, NiAl exhibited rapid hardening for the first few cycles followed by a stress plateau and a subsequent return to hardening. Slip traces were observed on the gage surfaces of most LCF specimens using scanning electron microscopy (SEM). The fatigue properties in this intermediate temperature range (600 to 700 K) were found to be a logical transition between previously reported ambient and elevated temperature properties. Transmission electron microscopy (TEM) confirmed that the dislocations had typical <100> Burgers vectors. A cellular dislocation structure began developing before saturation was achieved. This structure transformed at longer lives to elongated cells and eventually to veins of dislocation tangles. The resulting dislocation morphology did not change from 600 to 700 K, but the dislocation density decreased noticeably.

INTRODUCTION

Stoichiometric NiAl possesses excellent oxidation resistance, low density, and high melting temperature. Despite these advantages, binary NiAl exhibits poor high-temperature strength and low ambient-temperature ductility and fracture toughness. However, NiAl experiences a sharp transformation from brittle to ductile behavior. This change occurs at a low temperature compared to other intermetallics and is an important phenomenon in the behavior of NiAl. Above about 600 K, fracture strength and tensile ductility increase markedly and the percentage of transgranular fracture increases [1]. While agreement on the mechanism responsible for this change in behavior has not been reached, several studies [1-5] have suggested that thermally activated dislocation climb provides the additional independent deformation mechanisms necessary to accommodate deformation near the grain boundaries.

Fatigue properties of NiAl alloys have been investigated only to a limited extent. Bain, et al. [6] have examined the cyclic behavior of <001> single crystal Ni-49.9 Al-0.1 Mo at room temperature and 1033 K. Preliminary results by Smith, et al. [7] of room temperature fatigue properties for <121> oriented NiAl single crystals have also been reported. Lerch and Noebe have performed LCF tests on polycrystalline NiAl at both room temperature [8] and 1000 K [9]. Some general fatigue behavior trends can be drawn from the previous NiAl fatigue studies. Both single and polycrystalline NiAl cyclicly hardened and fractured with little or no stable crack growth when tested at room temperature . In fact, the polycrystalline samples cyclicly hardened to stresses ~60% greater than the monotonic tensile fracture stress. At temperatures near 1000 K, NiAl exhibited hardening for a few cycles then slight cyclical softening until failure. The increase in fatigue life for polycrystalline NiAl was sizable between 300 and 1000 K; lives increased about two orders of magnitude for a plastic strain range of 0.2% [8,9].

When a Coffin/Manson model is employed [10], the abrupt change in monotonic properties near the BDTT predicts an equally precipitous transition in fatigue life. The purpose of this study was to determine the fatigue behavior very near the monotonic BDTT and examine the dislocation structures which developed.

EXPERIMENTAL PROCEDURES

A Ni-49.5Al billet was prepared by hot extrusion of vacuum atomized, prealloyed powders. Optical metallography revealed that the extruded material had a recrystallized, equiaxed grain structure with an average linear intercept grain size of 39 μm. Sample blanks were electro-discharge machined parallel to the extrusion axis and ground into cylindrical, button-head tension and fatigue specimens. Monotonic tension specimens were tested in air between room temperature and 700 K in a screw-driven test frame at an initial strain rate of 10^{-4} sec^{-1}. Load versus time data were recorded and converted to true stress and true strain values by assuming constant volume conditions during plastic deformation. Fully reversed low cycle fatigue tests were conducted on a computer controlled, servo-hydraulic test frame at a total strain rate of 10^{-4} sec^{-1} between plastic strain ranges of 0.5 and 1.0%. Temperatures for the fatigue tests were chosen based on the monotonic test results and achieved using a 2 KW induction furnace. Scanning electron microscopy and transmission electron microscopy techniques were used to characterize specimens from interrupted and failed tests. Further details of the material composition, sample dimensions, and testing procedures can be found elsewhere [11].

RESULTS AND DISCUSSION

Monotonic true stress-strain curves for the NiAl alloy used in this study are shown in Figure 1. As the test temperature was increased, the yield stress dropped slightly, the fracture stress and tensile elongation increased markedly and the fracture morphology shifted from predominantly intergranular fracture to transgranular cleavage. The changes in the tensile properties observed for this heat of material closely resembled those commonly found in nominally stoichiometric extrusions of NiAl [1,4,5]. The tensile BDTT for this extrusion, defined here as the lowest temperature for which 5% plastic strain was achieved, was approximately 650 K. The monotonic tensile ductility increased by an order of magnitude from 600 to 700 K. From these monotonic results, fatigue test temperatures of 600, 675, and 700 K were chosen such that deformation would be examined slightly below and above the BDTT.

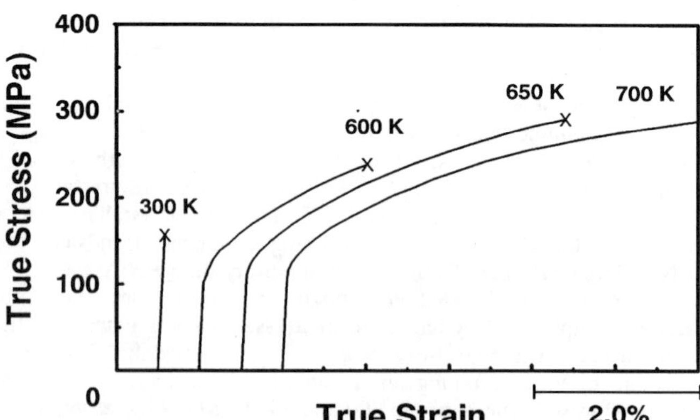

Figure 1 - True stress versus strain for monotonic tensile tests, $\dot{e}_t = 10^{-4}$ s^{-1}. Test at 700 K continued hardening to 373 MPa and 18% tensile elongation.

Cyclic hardening curves, Figure 2(a-b), demonstrated that the fatigue response of NiAl changed gradually within the temperature range of 600 to 700 K at plastic strain ranges of 0.5 and 1.0%. The horizontal axis in Figure 2(a-b) is the number of cycles multiplied by the imposed plastic strain range. The samples cyclicly hardened for the first few (up to 15) cycles followed by a stress plateau. Secondary hardening later in the fatigue lives occurred for all test conditions except at 600 K and $\Delta\epsilon_p$ = 1.0%, Figure 2(b). Cyclic softening was not observed in any of the test conditions and tension/compression stress asymmetry was minimal. At higher test temperatures for a given strain range, the rate of initial hardening and the stress at the plateau decreased. A similar shift in fatigue response occurred when the temperature was held constant and plastic strain range was decreased from 1.0 to 0.5%. The only significant decrease in fatigue life occurred at 600 K and $\Delta\epsilon_p$ = 1.0% where lives were approximately ten cycles. However, ten cycles under these conditions represents significant deformation at a temperature for which the monotonic tensile ductility was slightly less than two percent.

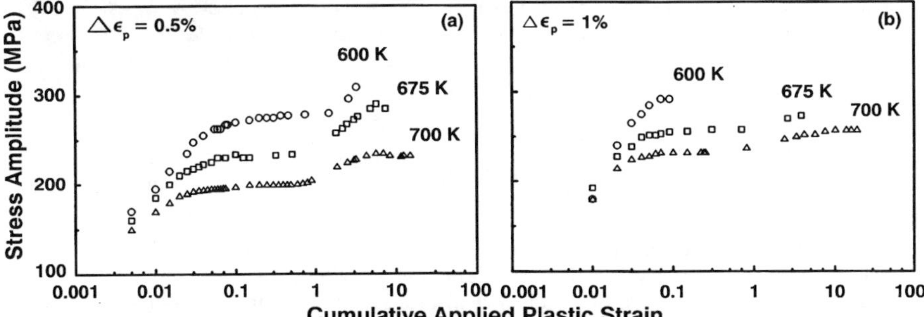

Figure 2 Stress amplitude versus cumulative plastic strain (cycles x strain range) as a function of temperature for (a) $\Delta\epsilon_p$ = 0.5% and (b) $\Delta\epsilon_p$ = 1.0%.

SEM inspection of fatigue specimens revealed that all fatigue failures at 600 K initiated at internal inclusions. Fatigue specimens tested at 675 and 700 K failed either at internal inclusions or fracture was initiated near the specimen surface. The inclusions appeared to be small silica particles which may have been introduced during production of the NiAl powders. Samples deformed in fatigue were more sensitive to these pre-existing flaws than the tensile tests based on the percentage of flaw initiations. SEM examination of the fracture surfaces revealed the morphology to be mixed intergranular fracture and transgranular cleavage. Transgranular cleavage facets were distinctly more prevalent very near the initiation site. There were, however, no clear demarcations separating regions of stable crack growth from those of fast fracture. Analysis of the bulk fracture surface, remote from the initiation site, revealed that the percentage of intergranular fracture decreased from 47.0% (±6.2) at 600 K to 32.5% (±3.6) at 700 K, following the same trends found in monotonic tensile fracture surfaces.

Slip traces and secondary cracks were observed by SEM on the gage surface of LCF specimens. Scattered slip traces could be seen as early as the onset of the stress plateau seen in Figure 2. The density and height of the traces increased with additional deformation (i.e. greater number of cycles or larger strain range) or lower test temperatures. Figure 3 is a SEM micrograph which shows an example of distinct slip extrusions on a specimen tested to failure at 700 K and $\Delta\epsilon_p$ = 0.5%. In many cases, slip traces were obvious initiation sites for either intergranular cracks at the slip trace/grain boundary intersections, region A shown in Figure 3, or transgranular cracks within slip bands, region B.

The fatigue properties of NiAl between 600 and 700 K resembled those of the room temperature tests [8,9]. Significant cyclic hardening occurred in this intermediate temperature range similar to the behavior at room temperature and unlike that found near 1000 K [9]. Likewise, the amount of stable crack growth was very small between 600 and 700 K and at room temperature. The fatigue behavior changed gradually with increasing temperature. The specimens tested at 600 K hardened to stresses approximately 20% greater than the tensile fracture stress analogous to hardening seen in polycrystalline NiAl at room temperature. However, the fatigue samples at 675 and 700 K hardened to stress levels less than their monotonic fracture stresses (almost 40% less at 700 K). Lives increased substantially between 600 and 700 K, but increased considerably less than Coffin/Manson estimates based on tensile ductility. Apparently the ductility of the material does not control the fatigue life and discernment of deformation mechanisms is imperative.

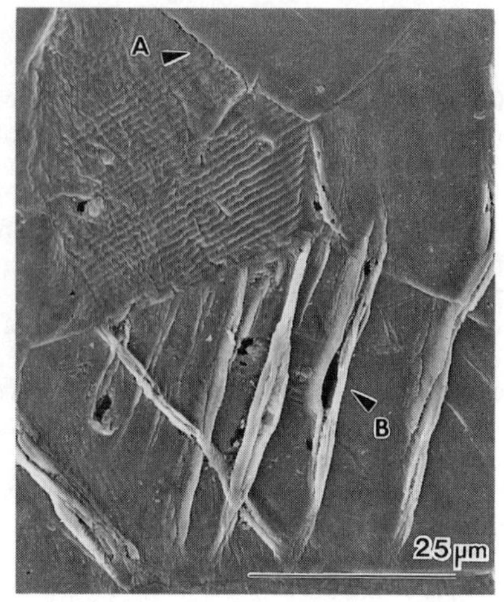

Figure 3- SEM micrograph of the gage of specimen deformed to failure at 700 K and 0.5%. Cracks are pointed out (A) intergranularly and (B) at a slip extrusion.

Limited Burgers vector determination was performed on individual dislocations from specimens interrupted early in fatigue deformation at 600 and 675 K and $\Delta\epsilon_p$ = 0.5%. Fields of view including at least 10 distinct dislocations were analyzed and only <100> Burgers vectors were identified. TEM observations of fatigued material revealed several dislocation configurations. Equiaxed cellular arrays of dislocations developed at the beginning of fatigue life and elongated cells, distinct veins, and regions of uniform tangles were seen later in sample life.

Figure 4 shows a cyclic hardening curve for a specimen tested to failure at 600 K and $\Delta\epsilon_p$ = 0.5% along with micrographs representing the prominent dislocation morphologies at increasing fractions of life. All TEM micrographs in this paper were taken under 2 beam conditions with **g** = <110>. After as little as one fully reversed cycle at 600 K, a cellular structure began to emerge. Bowman et al. [5] and Nagpal and Baker [12] have reported similar cellular dislocation structures from monotonic tests in near-stoichiometric NiAl plastically deformed 1 to 2%. After five cycles at 600 K, cells with walls of dense dislocation tangles were observed, Figure 4(a). Throughout the stress plateau, a mixed morphology of equiaxed and elongated dislocation cells appeared. Seen in Figure 4(b) is a region of elongated cells from a specimen tested for 15 cycles. At the end of the specimen life, Figure 4(c), regions of long, well aligned dislocation veins dominated the microstructure with regions of uniform dislocation tangles also present. These veins sometimes permeated an entire grain. The orientation of the veins frequently corresponded to a <100> line direction. The previously described trend in changing dislocation morphology was the same at all three temperatures. However, Figure 5 shows an obvious decrease in dislocation density, in both areas of dislocation veins and uniform tangles, as test temperature increased for samples deformed to failure.

The dislocation morphologies described above, in conjunction with the shape of the cyclic hardening curve, suggest the following evolution of fatigue deformation. During the period of initial hardening, a uniform cell structure developed. Throughout the stress plateau, the cells became elongated until a well defined structure of parallel veins formed.

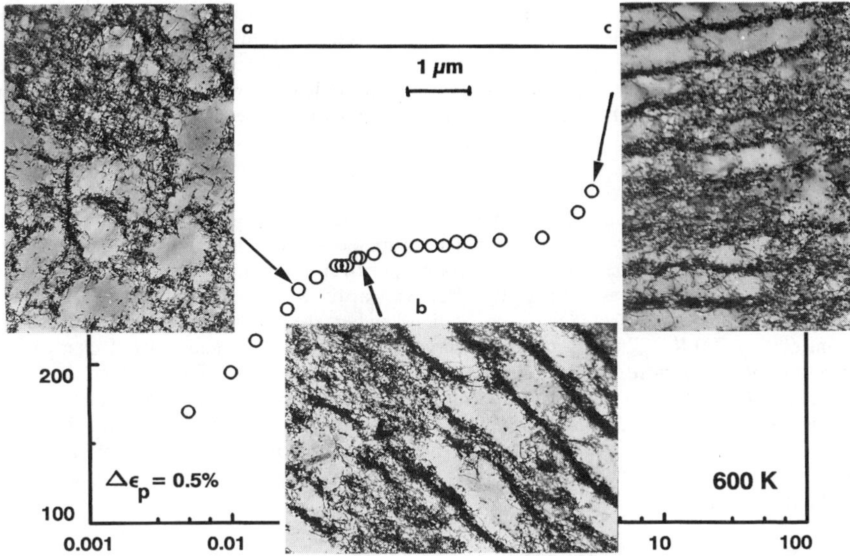

Figure 4 Cyclic hardening curve at 600 K, $\Delta\epsilon_p = 0.5\%$ with micrographs for specimens tested to (a) 5, (b) 15 and (c) 781 cycles.

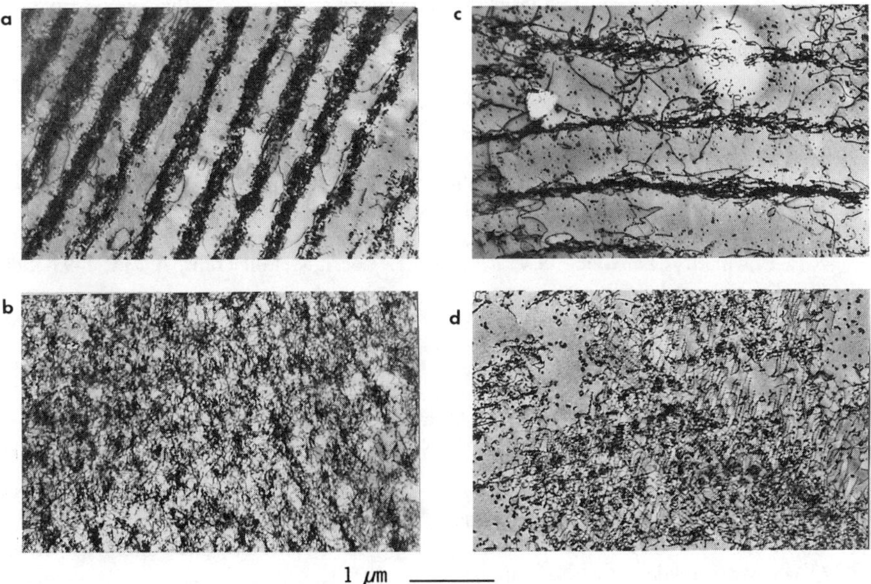

Figure 5 Different dislocation morphologies in specimens tested to failure at $\Delta\epsilon_p = 0.5\%$, (a,b) 600 K and (c,d) 700 K showing decreasing density.

The hardening near the end of life and the increasing areas of uniform dislocation tangles indicated that the veins must be a metastable dislocation structure that eventually broke down with increasing cycles. The lower density of dislocations in specimens tested at higher temperatures was consistent with the amount of surface damage observed. The increasing dislocation mobility within this temperature range could lead to increasing dynamic recovery and, more importantly, decreasing dislocation densities for a given imposed strain.

SUMMARY

The fatigue properties of polycrystalline NiAl at temperatures near the monotonic BDTT were comparable to the properties determined at room temperature. Even though there was a drastic change in monotonic properties over a narrow temperature range, there were only gradual changes in the fatigue behavior at the same temperatures. Fatigue lives improved and over-all strengths dropped slightly when the test temperature was increased from 600 to 700 K. The dislocation structures developed in fatigue did not change as a function of temperature within this range, but the dislocation density decreased with increasing temperature. The dislocation structure began as cellular arrays similar to those found after monotonic deformation. These cells became well-aligned veins of dislocation tangles during the stress plateau region of the cyclic hardening curves. A breakdown in the vein structure late in life caused an additional region of hardening followed by fracture due to the increasing stress levels.

ACKNOWLEDGEMENTS

The authors appreciate the support of M. V. Nathal and the NASA Lewis Research Center under Grant NCC3-116. Thanks is also extended to R. R. Bowman, B. A. Lerch, and J. D. Whittenberger for providing insightful discussions and reviewing this manuscript.

REFERENCES

1. R.D. Noebe, C.L. Cullers and R.R. Bowman, J. Mater. Res. **7**, 605-612 (1992).
2. A. Ball and R.E. Smallman, Acta Metall. **14**, 1349-1355 (1966).
3. H.L. Fraser, M.H. Loretto and R.E. Smallman, Phil. Mag. **28**, 667-677 (1973).
4. R.D. Noebe, R.R. Bowman, C.L. Cullers and S.V. Raj, in *High Temperature Ordered Intermetallic Alloys IV*, edited by L.A. Johnson, D.P. Pope and J.O. Stiegler (Mater. Res. Soc. Proc. **213**, Pittsburgh, PA, 1991), pp. 589-596.
5. R.R. Bowman, R.D. Noebe, S.V. Raj and I.E. Locci, Metall. Trans. A **23A**, 1493-1508 (1992).
6. K.R. Bain, R.D. Field and D.F. Lahrman, presented at the 1991 TMS Fall Meeting, Cincinnati, OH 1991 (unpublished).
7. T.R. Smith, C.G. Kallingal, K. Rajan and N.S. Stoloff, Scripta Metall. Mater. **27**, 1389-1393 (1992).
8. R.D. Noebe and B.A. Lerch, Scripta Metall. Mater., **27**, 1161-1166, 1992.
9. B.A. Lerch and R.D. Noebe, in *HITEMP Review 1992*, (NASA CP-10104 1992) pp. 47-1 to 47-15.
10. J.B. Conway and L.H. Sjodahl, *Analysis and Representation of Fatigue Data*, (ASM International, Materials Park, OH, 1991), pp. 13-14.
11. C.L. Cullers and S.D. Antolovich, in *Superalloys 1992*, edited by S.D Antolovich, et. al. (TMS, Warrendale, PA, 1992), pp. 351-359.
12. P. Nagpal and I. Baker, J Mater. Sci. Let. **11**, 1209-2110 (1992).

THE EFFECT OF DISLOCATION DISSOCIATION ON CRACK TIP PLASTICITY IN $L1_2$ AND B2 INTERMETALLIC ALLOYS

MICHAEL F. BARTHOLOMEUSZ, WEIGANG MENG AND JOHN A. WERT
Department of Materials Science and Engineering
University of Virginia, Charlottesville, VA 22903

ABSTRACT

Low toughness at ambient temperatures is an inherent problem in many ordered intermetallic alloys that otherwise have attractive properties for high temperature applications. One potential explanation of low toughness is the existence of an energy barrier for crack tip dislocation emission. A model describing the energy associated with emission of a dissociated superlattice dislocation from a crack tip in ordered intermetallic alloys has been formulated. Application of the model to a wide variety of intermetallic alloys with the $L1_2$ and B2 crystal structures has revealed a correlation between the observed macroscopic fracture mode and the calculated range of slip system orientations for which emission of a dissociated superlattice dislocation at ambient temperatures is possible. Additionally, the model has been used to predict the effects of lowering the stacking fault energy, increasing the thermal energy available for thermally activated dislocation emission, and changing the active slip system. Preliminary fractographic analyses have been conducted with a chromium-modified Al_3Ti alloy to verify the model predictions.

1 INTRODUCTION AND MODEL

Cleavage fracture in materials possessing a high critical resolved shear stress (CRSS) is relatively easy to understand: the high resistance to slip inhibits crack tip plasticity, resulting in cleavage fracture. However, many intermetallic alloys that fail by cleavage fracture possess low to moderate values of CRSS, requiring an alternate explanation for cleavage fracture. Rice and Thomson (RT) [1] proposed a model for cleavage fracture independent of CRSS in which the Griffith criterion is invoked to determine the critical remote stress required for cleavage fracture. The RT model proposes that cleavage fracture will not occur if it is energetically favorable for an atomistically sharp crack under mode I loading to emit a dislocation at the critical stress. However, if an energy barrier for dislocation emission exists at the crack tip, dislocation emission is inhibited resulting in cleavage fracture.

Bartholomeusz and Wert (BW) [2, 3] extended the RT analysis to treat emission of dissociated superlattice dislocations from crack tips in $L1_2$ and B2 intermetallic alloys. The extended model can be used to examine the effects of temperature and variations of material properties on the emission of superlattice and dissociated superlattice dislocations from crack tips. For a detailed description of the mathematics of the model and a discussion of the material parameters selected, see Refs.[2, 3]. The purpose of the present paper is to describe model results, and place them in the context of available experimental results.

In the case of a dissociated superlattice dislocation, the total energy of the dislocation as a function of distance ahead of the crack tip is dependent on five material and two geometrical parameters. The material parameters are: shear modulus, G; Burgers vector, b; true fracture surface energy, γ; bulk modulus, K; and stacking fault energy, Γ. Two geometrical parameters are also required: ϕ describes the angle between the slip plane and the crack plane, and ψ_s describes the superlattice Burgers vector orientation relative to the crack. For fixed values of these parameters, the energy barrier for superlattice dislocation emission, U_{act}, is determined numerically. If U_{act} is equal to zero, there is no energy barrier to superlattice dislocation emission, plastic crack blunting occurs and cleavage fracture is averted. For a given material, it is possible to determine U_{act} as a function of ϕ and ψ_s, the parameters describing the slip system orientation relative to the crack tip. Variation of ϕ and ψ_s accounts for the variety of grain orientations encountered along a crack front.

2 RESULTS

The model results for each material in this study can be represented using a plot such as the one shown in Figure 1. The vertical axis of the plot represents the activation energy for dislocation emission. The two horizontal axes represent slip system orientations relative to the crack tip. The plateau at 30 eV is a suitable cut-off for the purpose of representing the numerical results. The plateau at 1 eV (shaded region) represents the range of slip system orientations for which spontaneous or thermally activated dislocation emission can occur at room temperature (293K). We assume that at room temperature, there is enough thermal energy available to the dislocation half loop to overcome an energy barrier of 1 eV. This corresponds approximately to the total thermal energy of the atoms in a dislocation core 10b in length.

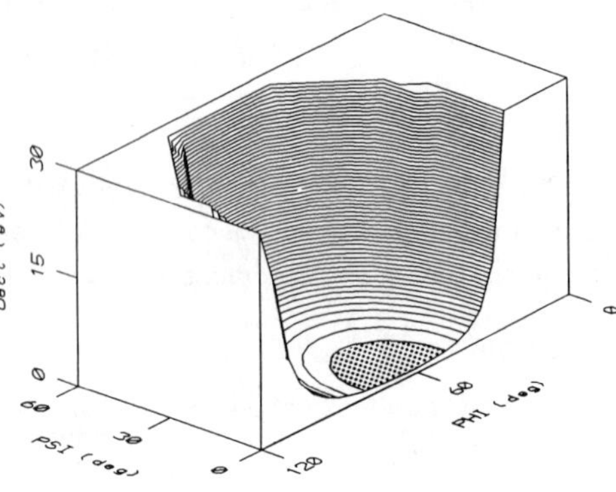

Fig. 1. Activation energy for emission of a dissociated superlattice dislocation as a function of slip system orientation for the B2 intermetallic alloy Fe-40Al.

3 DISCUSSION

3.1 Correlation between spontaneous emission of superlattice dislocations and fracture mode

The parameter $\varsigma(293)$, first introduced by BW, describes the fractional angular ranges of ϕ and ψ_s corresponding to spontaneous or thermally activated emission of dislocations at room temperature for a given material. *Fractional* implies the solid angle range for which emission is favorable, normalized by the solid angle range of all possible orientations of a superlattice dislocation lying on a slip plane that contains the crack front. This normalization factor encompasses a ϕ range from -180° to +180° and a ψ_s range from -90° to +90°.

Table I shows the calculated values of $\varsigma(293)$ for the $L1_2$ and B2 intermetallic alloys studied. The observed dislocation dissociation mode and experimentally-determined single crystal fracture mode are listed for each alloy. Comparing the values of $\varsigma(293)$ and the fracture modes listed in Table I reveals a strong correlation. Materials having a value of $\varsigma(293)$ greater than 0.08 are resistant to cleavage fracture. This suggests that spontaneous or thermally activated emission of dislocations must occur over a range of slip system orientations greater than some critical range for a material to resist cleavage fracture. We do not attach a physical significance to the value of $\varsigma(293) = 0.08$.

3.2 Effect of changing material properties on fracture mode

3.2.1 Effect of changing the Burgers vector

Reducing the magnitude of the Burgers vector renders dislocation emission from a crack tip more energetically favorable. This can be accomplished either by dissociation of superlattice dislocations (as discussed in section 3.2.2) or by a change in slip vector. Analysis of the effect of changing the slip vector on dislocation emission is applicable to B2 intermetallic alloys, since these alloys may deform by either <100> or <111> slip [4]. For a given B2 intermetallic alloy, the model results indicate that

Table I. Correlation Between ς(293) and Fracture Mode for the $L1_2$ and B2 Intermetallic Alloys Considered in the Present Investigation*.

Material and dislocation dissociation mode		ς(293)	Fracture mode	Crystal structure
Cu_3Au	APB	0.262	Ductile	$L1_2$
Ni_3Al	APB	0.196	Ductile	$L1_2$
Ni_3Al	SISF	0.090	Ductile	$L1_2$
Ni_3Fe	APB	0.156	Ductile	$L1_2$
Zr_3Al	SISF	0.085	Ductile	$L1_2$
Al_3Sc	APB	0.069	Brittle	$L1_2$
$Al_{67}Cr_8Ti_{25}$		0.049	Brittle	$L1_2$
$Al_{67}Ni_8Ti_{25}$	None	0.043	Brittle	$L1_2$
$Al_{67}Mn_8Ti_{25}$	APB	0.042	Brittle	$L1_2$
$Al_{66}Fe_6Ti_{23}V_5$	SISF	0.042	Brittle	$L1_2$
AuCd		0.168		B2
CuZn	APB{112}	0.161	Ductile	B2
CuZn	APB{011}	0.154	Ductile	B2
AuZn		0.137		B2
Fe-37Al	APB	0.037	Brittle	B2
Fe-40Al	APB	0.029	Brittle	B2
Fe-40Al		0	Brittle	B2
NiAl<100>		0.025	Brittle	B2
NiAl<111>		0		B2
AgCd		0.020		B2
AgZn		0.005		B2
CoTi		0.004	Brittle	B2
AgMg		0	Brittle	B2
CoAl		0	Brittle	B2

* For data on $Al_{67}Cr_8Ti_{25}$ and $Al_{66}Fe_6Ti_{23}V_5$ see Ref.[5]. For the sources of the material parameters and fracture modes of all the other alloys listed see Refs. [2, 3].

emission of <100> dislocations is more favorable than emission of <111> superlattice dislocations.

A considerable research undertaking has been devoted to promotion of <111> slip in NiAl [6, 7]. The objective of this research has been to overcome the lack of compatibility associated with <100> slip, which was presumed to restrict ductility in NiAl. Results of the present investigation indicate that changing the active slip system from <100>{011} to <111>{01$\bar{1}$} in NiAl significantly reduces the value of ς(293), as shown in Table I. Thus, model results suggest that changing the slip vector in NiAl cannot achieve the desired result of improving ductility. Although compatibility requirements may be satisfied for <111> slip, spontaneous or thermally activated emission of <11$\bar{1}$>{01$\bar{1}$} superlattice dislocations from a crack tip in NiAl does not occur for any slip system orientations at room temperature.

3.2.2 Effect of changing the stacking fault energy

Figure 2 is a plot of ς(293) as a function of fault energy for APB and SISF dissociated superlattice dislocations in $Al_{67}Ni_8Ti_{25}$. This plot suggests that emission of APB coupled superlattice partial dislocations is more energetically favorable than emission of undissociated superlattice dislocations. Lowering the APB energy of $Al_{67}Ni_8Ti_{25}$ below approximately 330 mJ/m^2 would increase ς(293) into the range associated with ductile fracture in other $L1_2$ and B2 intermetallic alloys at room temperature. This implies that lowering the APB energy is a possible means of inducing resistance to cleavage failure in intermetallic alloys at ambient temperatures. In contrast to this, Figure 2 demonstrates that lowering the SISF energy does not have a significant effect on $\varsigma(293)$. The different results for APB and SISF dissociated superlattice dislocations stem from the difference in Burgers vector orientations of the two superlattice partial dislocations in the case of SISF-dissociation.

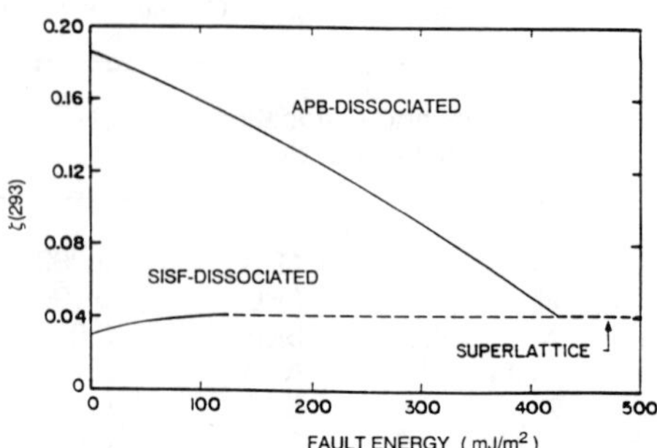

Fig. 2. Effect of {111} APB and SISF energy on $\varsigma(293)$ in $Al_{67}Ni_8Ti_{25}$. For values of SISF energy greater than about 150 mJ/m^2, it is appropriate to use the value of $\varsigma(293)$ associated with emission of an undissociated superlattice dislocation due to the narrow equilibrium separation between the two partial superlattice dislocations.

Although direct evidence for the effect of APB fault energy on fracture mode has not yet been reported, some insight into the correlation between these two factors can be gained by comparing $Al_{67}Ni_8Ti_{25}$ and $Al_{67}Mn_8Ti_{25}$. These alloys have very similar material parameters, except that $Al_{67}Ni_8Ti_{25}$ deforms by glide of undissociated <110>{1$\bar{1}$1} superlattice dislocations while $Al_{67}Mn_8Ti_{25}$ deforms by glide of APB dissociated superlattice dislocations [8]. From Table I it can be seen that both $Al_{67}Ni_8Ti_{25}$ and $Al_{67}Mn_8Ti_{25}$ have approximately the same value of $\varsigma(293)$ for superlattice dislocations. Calculations show that APB dissociation in the case of $Al_{67}Mn_8Ti_{25}$ raises its value of $\varsigma(293)$, making it more resistant to cleavage fracture. In support of this, recent results by Mikkola et al. [9] demonstrate that $Al_{67}Mn_8Ti_{25}$ has a significantly higher fracture stress than $Al_{67}Ni_8Ti_{25}$.

3.2.3 Effect of changing the shear and bulk moduli

Pugh [10] proposed that the ratio of the bulk modulus to the shear modulus (K/G) could be used to describe the relationship between the fracture strength and the ease of plastic flow in a material. Resistance to cleavage fracture correlates with increasing values of K/G. The model described in the present paper is consistent with Pugh's criterion. $\varsigma(293)$ decreases with increasing values of G and/or decreasing values of K. In the Pugh criterion, however, the fracture mode is equally sensitive to changes in K and G. This dependency is not reflected by the results of our model.

The relative importance of variations of K and G can be illustrated in the case of $Al_{67}Ni_8Ti_{25}$ and $Al_{67}Cr_8Ti_{25}$. $Al_{67}Cr_8Ti_{25}$ has values of shear and bulk modulus that are 7 GPa and 41 GPA lower than those of $Al_{67}Ni_8Ti_{25}$, respectively [5]. Application of the Pugh criterion as mentioned by Mikkola et al. [5], indicates that $Al_{67}Ni_8Ti_{25}$ is less susceptible to cleavage fracture than $Al_{67}Cr_8Ti_{25}$ (K/G = 1.4 and 1.0 respectively). However, BW model calculations show that the lower value of shear modulus for $Al_{67}Cr_8Ti_{25}$ offsets its lower bulk modulus and predicts that $Al_{67}Cr_8Ti_{25}$ should be more resistant to cleavage fracture than $Al_{67}Ni_8Ti_{25}$. This prediction is consistent with the previously mentioned report that $Al_{67}Cr_8Ti_{25}$ exhibits a higher fracture stress than $Al_{67}Ni_8Ti_{25}$ [9]. Thus, the dislocation emission model indicates that crack tip plasticity is more sensitive to changes in G than in K.

3.3 Thermal activation effects

In defining $\varsigma(293)$, it was proposed that 1 eV of thermal energy associated with the core atoms of a dislocation at room temperature enables the dislocation to overcome energy barriers of 1 eV or less [2]. At higher temperatures, more thermal energy is available and the value of $\varsigma(T)$ increases. The thermal activation of dislocations over a crack tip energy barrier is a possible explanation for the change in fracture mode from cleavage to a ductile or intergranular (non-cleavage) mode, as observed in many ordered intermetallic alloys with increasing temperature.

The gradient of the surface represented by contour lines in Figure 1 is related to the potential for significant thermal activation effects. Decreasing steepness of the sides of the activation energy well indicates a greater sensitivity to thermal activation. Within the scope of the present model and the alloy systems studied, three factors found to affect significantly thermally activated dislocation emission are: superlattice dislocation dissociation, shear modulus, and slip system. Examples of the effects of dislocation dissociation and shear modulus on thermally activated dislocation emission are presented below.

Figure 3(a) depicts $\varsigma(T)$ as a function of available thermal energy for three $L1_2$ modified Al_3Ti intermetallic alloys. $\varsigma(T)$ increases with increasing availability of thermal energy. In the case of $Al_{66}Fe_6Ti_{23}V_5$ and $Al_{67}Cr_8Ti_{25}$, $\varsigma(T)$ attains values near 0.08 that are characteristic of $\varsigma(293)$ for the $L1_2$ and B2 intermetallic alloys that do not exhibit cleavage fracture at room temperature. This suggests that a transition in fracture mode from cleavage to non-cleavage might occur in $Al_{66}Fe_6Ti_{23}V_5$ near 1100K and in $Al_{67}Cr_8Ti_{25}$ near 800K [2]. For $Al_{67}Ni_8Ti_{25}$, the temperature required to increase $\varsigma(T)$ to 0.08 exceeds the melting temperature of the material, suggesting that no fracture mode transition would be observed in this case. Comparison of $Al_{66}Fe_6Ti_{23}V_5$ and $Al_{67}Ni_8Ti_{25}$ shows that both alloys have similar material parameters but $Al_{66}Fe_6Ti_{23}V_5$ has SISF dissociated superlattice dislocations while $Al_{67}Ni_8Ti_{25}$ has undissociated superlattice dislocations [2, 8]. Model calculations show that APB or SISF dissociation of superlattice dislocations reduce the gradient of the activation energy well, compared with that found for undissociated superlattice dislocations. This suggests that superlattice dislocation dissociation renders crack tip emission more sensitive to thermal activation and thus increases the resistance to cleavage fracture at elevated temperatures. The greater susceptibility of $Al_{67}Cr_8Ti_{25}$ to thermally activated superlattice dislocation emission compared to that of $Al_{67}Ni_8Ti_{25}$ is due to its lower value of shear modulus [5]. If future experimental observations confirm the presence of dissociated superlattice dislocations in $Al_{67}Cr_8Ti_{25}$, this would simply reduce the calculated fracture mode transition temperature.

Figure 3(b) shows the result of fractographic measurements done on the $Al_{67}Cr_8Ti_{25}$ alloy as a function of temperature. There is a clear transition in fracture mode at 800K from transgranular cleavage to intergranular fracture. A correlation between model and experimental results suggests that with increasing

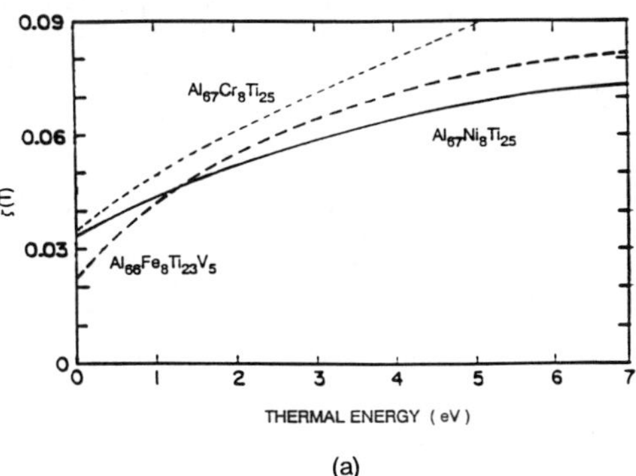

(a)

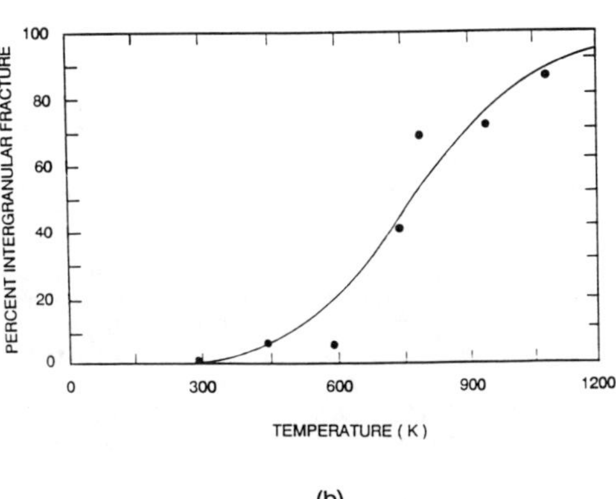

(b)

Fig. 3. (a) Effect of thermal energy on $\varsigma(T)$ in several $L1_2$ intermetallic alloys. (b) Percent intergranular fracture as a function of temperature in $Al_{67}Cr_8Ti_{25}$

temperature the $Al_{67}Cr_8Ti_{25}$ matrix becomes resistant to transgranular cleavage and subsequent failure occurs at the grain boundaries. A future paper will describe detailed analyses of the fracture mode transition in $Al_{67}Cr_8Ti_{25}$ [11].

4 CONCLUSIONS

The results presented in this paper can be summarized by a few broad observations that transcend the details of individual intermetallic alloys.
(1) A correlation exists between the angular range of slip system orientations for which spontaneous or thermally activated emission of dislocations is predicted and the macroscopic fracture mode at room temperature.
(2) Decreasing the magnitude of the Burgers vector by inducing a change in the slip vector is predicted to enhance a material's ability to exhibit plastic crack blunting.
(3) For low to moderate values of APB energy, the emission of APB-dissociated superlattice dislocations is always more energetically favorable than the emission of undissociated superlattice dislocations.
(4) Lowering the APB energy for $L1_2$ and B2 intermetallic alloys can generate sufficient spontaneous or thermally activated emission of dissociated superlattice dislocations to avoid cleavage fracture at room temperature. In contrast, lowering the SISF energy for $L1_2$ intermetallic alloys does not have a significant effect on spontaneous or thermally activated superlattice dislocation emission, and is not expected to affect the fracture mode.
(5) The model predicts that crack tip plasticity is more sensitive to variations in the shear modulus than equivalent variations in the bulk modulus. Reducing the shear modulus increases resistance to cleavage fracture.
(6) The thermally activated emission of undissociated and dissociated superlattice dislocations from a crack tip with increasing temperature is a possible explanation for a transition in fracture mode from cleavage to non-cleavage, as observed in many $L1_2$ and B2 intermetallic alloys.

5 ACKNOWLEDGEMENTS

This work was sponsored by AFOSR under contract number AFOSR-90-0143; Dr. A. Rosenstein was the contract monitor.

6 REFERENCES

1. J.R. Rice and R. Thomson, Phil. Mag. **29**, 73 (1974).
2. M.F. Bartholomeusz and J.A. Wert, Acta Metall. et Mater., **40**, 673 (1992).
3. M.F. Bartholomeusz and J.A. Wert, J. Mater. Res., **7**, 919 (1992).
4. I. Baker and P.R. Munroe, in *High Temperature Aluminides & Intermetallics*, S.H.Whang, C.T. Liu, D.P. Pope and J.O. Stiegler (eds), MRS, Warrendale, 1990, p. 425.
5. D.E. Mikkola, J.P. Nic, S. Zhang and W.W. Milligan, ISIJ International, **31**, 1076 (1991).
6. D.B. Miracle, S. Russell and C.C. Law, in *High Temperature Ordered Intermetallic Alloys III*, C.T. Liu, A.I. Taub, N.S. Stoloff and C.C. Koch (eds), MRS, Pittsburgh, 1989, p. 225.
7. R. Darolia, D.F. Lahrman, R.D. Field and A.J. Freeman in *High Temperature Ordered Intermetallic Alloys III*, C.T. Liu, A.I. Taub, N.S. Stoloff and C.C. Koch (eds), MRS, Pittsburgh, 1989, p. 113.
8. E.P. George, D.P. Pope, C.L. Fu and J.H. Schneibel, ISIJ International, **31**, 1063 (1991).
9. J.P. Nic, S. Zhang and D.E. Mikkola, in *High Temperature Ordered Intermetallic Alloys IV*, L.A. Johnson, D.P. Pope and J.O. Stiegler (eds), MRS, Pittsburgh, 1991, p. 697.
10. S.F. Pugh, Phil. Mag., **45**, 823 (1954).
11. W. Meng, M. Vaudin, M.F. Bartholomeusz and J.A. Wert, "Experimental assessment of a fracture mechanism model for an $Al_{67}Cr_8Ti_{27}$ intermetallic alloy", in preparation.

LOADING RATE EFFECTS AND FRACTURE IN A TiAl ALLOY

MADAN G. MENDIRATTA, YOUNG-WON KIM* AND DENNIS M. DIMIDUK**
*UES, Inc., 4401 Dayton-Xenia Road, Dayton, OH 45432
**Materials Directorate, WL/MLLM, Wright-Patterson AFB, OH 45433

ABSTRACT

The room-temperature fracture toughness and tensile properties were evaluated as a function of loading rate, on a fully lamellar and nearly lamellar γ titanium aluminide alloy. The fully lamellar microstructure exhibited increasing toughness with increasing loading rate, however, all other mechanical property data indicated insensitivity to the loading rate. The fracture processes were dominated by translamellar, Mode I separation, and exhibited a process zone consisting of parallel slip traces/twins leading to formation of microcracks.

INTRODUCTION

Numerous studies [1,2] have shown that different thermomechanical processes yield different classes of microstructures in gamma TiAl-based alloys, and these have a strong influence on the balance of mechanical properties. One of the properties which is strongly influenced by the microstructure is the fracture toughness and associated fracture micro-mechanisms and resistance-curve behavior [3,4]. The observed degree of toughening in the two phase (γ + α_2) alloys has been interpreted within a mechanics analyses of crack tip plastic blunting, large scale crack bridging involving shear ligament formation, and crack deflection [3,4] and crack bridging involving a twinned process zone in the wake of an advancing crack [5]. These analyses, however, are based upon the continuum mechanics approach where in local crystallography and its effect on fracture processes are not taken into account. In addition, these studies do not consider the dislocation/twinning plastic processes which occur prior to and may be responsible for crack nucleation. Recent research on polysythetically twinned (PST) γ crystals [6,7] has shown a strong influence of crystallographic orientation with respect to the applied stress direction on plastic and fracture processes of the fully lamellar microstructure. Such intrinsically high anisotropy in the flow of single grains is expected to dominate the fracture properties, just as it does in some hcp metals.

In previous studies, the reported toughness values of the γTiAl alloys have been measured within the range of loading rates permitted by the ASTM Standard. The objective of the present study is to perform a systematic determination of toughness as a function of loading rate encompassing ASTM range as well as significantly higher rates, and along with this, to understand the fracture micro-mechanisms of a polycrystalline alloy taking into account the strong plastic and fracture anisotropy exhibited by PST crystals.

EXPERIMENTAL PROCEDURE

A two-phase γTiAl alloy with a nominal composition of Ti-48Al-1.6Cr-0.9V-2.3Nb (all compositions in atomic percent) was cast using induction skull melting. The casting was hot isostatically pressed at 1200°C for three hours under an argon pressure of 270 MPa and then isothermally forged at 1180°C, in a two step process, to a reduction in height of 70 percent in each step. The forged plates were subjected to the following two different heat treatments in air atmosphere: i) 1380°C/2 h → furnace cool (30°C/min. cooling rate) to 900°C/6 h → air cool. ii) 1290°C/2 h → furnace cool (30°C/min. cooling rate) to 900°C/6 h → air cool. A thin oxide layer formed during the heat treatments, however, the specimens for mechanical property tests were machined by removing ~ 1.9 mm from the outer surface. The chemical analysis of the machined specimens revealed the composition to be Ti-48.65Al-1.6Cr-0.96V-2.77Nb with small amount of metallic impurities (< 1000 ppmw). Interstitial contents were found to be: 500 ppmw O, 150 ppmw N, and 240 ppmw C.

The fracture toughness tests were carried out on fatigue-precracked compact tension specimens (thickness direction parallel to the forging direction) using ASTM standards for normal loading rates. Similar test procedures were also employed for up to two orders of magnitude faster loading rates. Tension tests were carried out on threaded tensile specimens (with axis of the specimens perpendicular to the forging direction) at loading rates which were approximately similar to those for fracture toughness tests. All mechanical property tests were conducted in the load control mode at room temperature and in laboratory air. Extensive fractography was carried out using scanning electron microscopy (SEM).

RESULTS AND DISCUSSION

The two different heat treatments produced to different microstructures: fully lamellar (FL), Fig. 1(a) and nearly lamellar (NL), Fig. 1(b); average sizes of the lamellar grains were determined to be ~ 1400 μm and 600 μm, respectively. For the two microstructures, Fig. 2 shows the variation of fracture toughness, K_Q, with loading rates. In all the fracture toughness tests, the maximum load, P_{max}, was between 20 to 40% higher than the load, P_S, which represents intersection of 95% secant line with the load-deflection curve. These results imply occurrence of significant plastic deformation and stable crack growth before fast fracture. Therefore, the fracture toughness -- which was calculated using the loads P_S -- is given as K_Q rather than K_{IC}.

For the FL microstructure the toughness increased significantly with increasing loading rates, however, for the NL microstructure the toughness did not change except for the highest loading rate. The fractographic observations (to be described below) did not reveal any significant differences in the crack initiation and propagation modes as a function of loading rate for either of the two microstructures. At present, therefore, we do not have an explanation for the K_Q dependence of the loading rate as exhibited by Fig. 2.

Figures 3 and 4 show variation of tensile properties with strain rate for the FL and NL microstructures, respectively. Except for the highest strain rate used, the tensile properties were insensitive to the strain rate; at the highest strain rate

Figure 1. Light micrographs showing grain structure of a) nearly lamellar microstructure and b) fully lamellar microstructure.

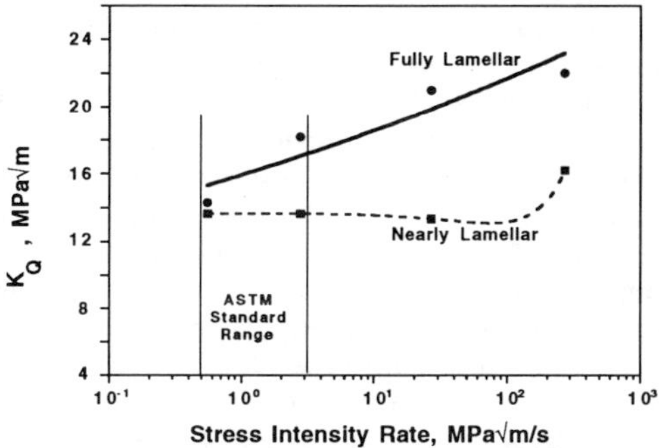

Figure 2. Variation of fracture toughness, K_Q, with loading rate for fully lamellar and nearly lamellar microstructures.

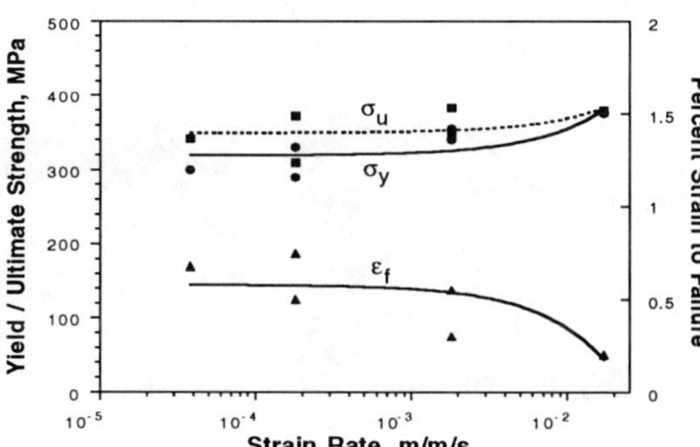

Figure 3. Dependence of tensile properties, i.e., ultimate strength (σ_u), yield strength (σ_y), and total elongation to fracture (ε_f) on strain rate for the fully lamellar microstructure.

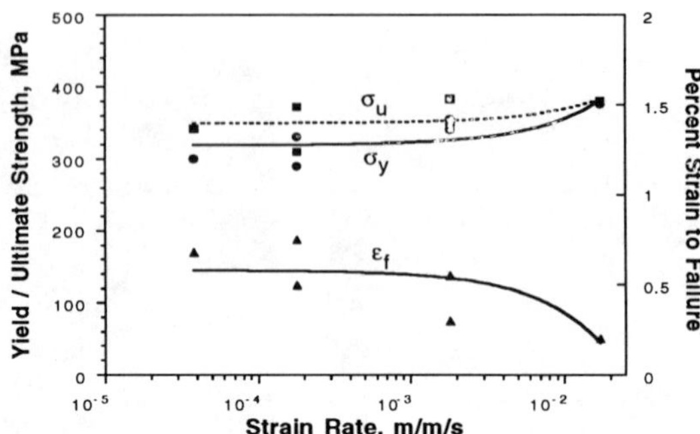

Figure 4. Dependence of tensile properties on strain rate for the nearly lamellar microstructure.

the yield and ultimate strengths increased slightly accompanied by a small decrease in elongation to fracture.

Figures 5(a) and 5(b) are the low magnification SEM fractographs showing typical fracture appearance for the FL microstructure for two loading rates. The crack propagation mode was predominantly translamellar (~ 85% of the fracture surfaces) with a low fraction of interspersed "flat faceted" regions signifying interlamellar fracture. SEM channeling patterns taken from the "flat" appearing fracture regions showed them to be {111} γ planes [8]. Published results on the tensile behavior of the PST TiAl crystals [7] indicate that for a wide range of lamellar interface plane orientations with respect to the applied stress direction (i.e., 30-60° angles), both the yield and fracture strengths are quite low and that extensive tensile plasticity occurs (a maximum of ~ 20% for 37.5° angles) prior to failure. Further, the fracture occurs by separation of the {111} lamellar interface planes. These results might suggest that the incidence of interlamellar fracture should be higher (~ 33%) than that observed in the present study (~ 15%). However, in polycrystalline samples the constraint imposed by the unfavorably oriented neighboring grains very likely leads to high local stresses apparently favoring translamellar crack extension. Another possibility is that the present forged alloy may have a specific texture which might suppress the frequency of interlamellar separation.

$\dot{K}$ = 0.55 MPa $\sqrt{m}$/s $\dot{K}$ = 26.7 MPa $\sqrt{m}$/s

Figure 5. Effect of loading rate on the fracture appearance for the fully lamellar microstructure.

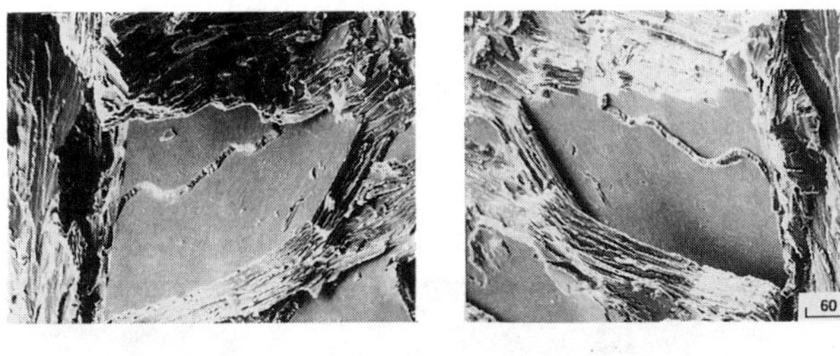

(a) (b)

Figure 6. Matching fracture surfaces showing occurrence of Mode I failure.

The low yield strength and the occurrence of extensive plasticity exhibited by some orientations of the PST crystals might suggest that the flat interlamellar fracture may occur prior to a shear separation. In the present study extensive observations of the matching fracture surfaces of two halves of broken specimens did not present any evidence of a shear failure mode. Figures 6(a) and 6(b) are one example of matching surfaces; no relative displacements of the fine fracture features could be detected indicating that the final failure must have occurred in Mode I opening. Even though PST crystals might possibly fracture in a shear mode in some orientations, again the constraints inherent in the polycrystalline materials should suppress shear-type failures.

In the fracture toughness test specimens surface observations revealed the presence of a process zone ~ 150 μm wide parallel to the main propagating crack. The features of the process zone are shown in the SEM micrographs of Figs. 7(a) and 7(b). A previous study [5] based upon light microscopic observations, concluded that the process zone consisted mainly of microtwins whose formation dissipated energy and consequently provided toughening. The features in Figs. 7(a) and 7(b) are parallel thin traces which may be either twins or slip lines. The important observation is that numerous microcracks open up aligned with the traces. This indicates that slip/twinning precedes microcracking and that the energy dissipative process(es) consists not only twinning, but should also include microcrack formation. It is suspected that the microcracks open up on {111} crystallographic planes, however, a more thorough characterization is the subject of continuing research.

SUMMARY

1. For fully lamellar microstructure, K_Q increases with increasing loading rate. For the nearly lamellar microstructure, however, the K_Q values are relatively insensitive to loading rate. For both microstructures, the tensile properties (i.e., σ_y, σ_u, and ε_f) are insensitive to loading rate.
2. The overall fracture process is dominated by a high proportion of translamellar separation as opposed to interlamellar separation.

3. Under the global tensile loading, the final failure occurs mostly in a Mode I opening manner rather than possibly by a shear separation. This is independent of the plastic flow and fracture anisotropy of the lamellar orientation.
4. Slip lines, twins and microcracks all seem to be aligned parallel to each other in a process zone.

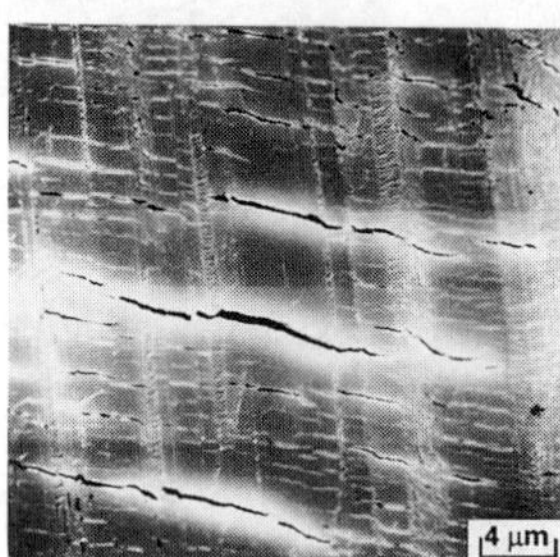

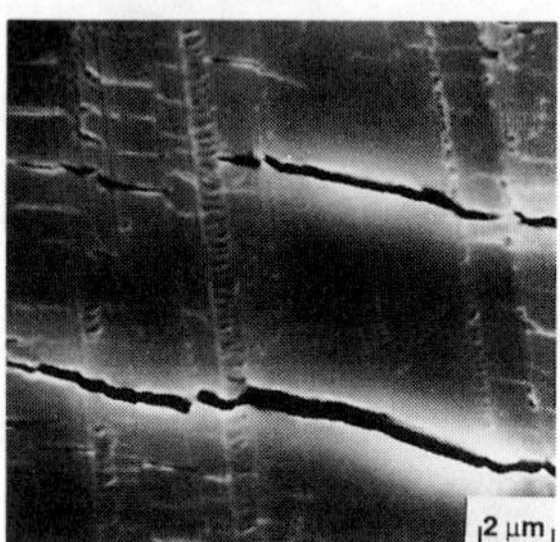

Figure 7. The process zone in the vicinity of main crack contains traces of numerous parallel slip lines/twins which open to form microcracks.

REFERENCES

1. Y-W. Kim, JOM, **41**(7), 24-30 (1989).
2. Y-W. Kim and D. M. Dimiduk, JOM, **43**(8), 40-47 (1991).
3. K. S. Chan, JOM, **44**(5) , 30-38, (1992).
4. W. O. Soboyejo, S. J. Midea, D. S. Schwartz, and J. J. Parzukowski, in Microstructure /Property Relationship in Titanium Aluminides and Alloys, edited by Y-W. Kim and R. R. Boyer (TMS, Warrendale, PA, 1991), pp. 197-212.
5. E. E. Dève, A. G. Evans, and D. S. Shih, Acta Metall. Mater., **40**(6), 1259-1265 (1992).
6. Y. Umakoshi, T. Nakano, and T. Yamane, Scripta Metall. et Mater., **25**(7), 1525-1528 (1991).
7. H. Inui, A. Nakamura, and M. Yamaguchi, in High Temperature Ordered Intermetallic Alloys IV, edited by L. A. Johnson, (Mater. Res. Soc. Proc. **213**, Pittsburgh, PA, 1990), 569-574.
8. J. W. Woodhouse and M. G. Mendiratta, UES, Inc., Dayton, OH, unpublished research.

CRACK PROPAGATION IN NiAl AND FeAl

J. H. SCHNEIBEL, M. G. JENKINS, AND P. J. MAZIASZ
Metals and Ceramics Division, Oak Ridge National Laboratory, Oak Ridge, TN 37831-6114

ABSTRACT

The crack-propagation behavior and fracture toughness at room temperature of extruded and heat-treated NiAl and FeAl were examined by testing chevron-notched, three-point flexural specimens at constant crosshead speeds. In Ni-50 at. % Al, sudden load drops occurred repeatedly, indicating run-arrest crack propagation. The fracture resistance was not found to depend on the crosshead speed. Iron additions of up to 1 at. % and boron additions of 0.01 at. % did generally not improve the fracture toughness. By contrast, crack propagation in Fe-40 at. % Al occurred in a stable manner. In agreement with the environmental sensitivity of this intermetallic alloy, fracture resistance did depend on the crack-propagation velocity, indicative of the kinetic nature of this process. While the crack-growth resistance of iron aluminides was reduced by changing the aluminum content from 40 to 45 at. %, it was increased significantly by small additions of boron.

INTRODUCTION

The B2 intermetallics NiAl and, to a lesser extent, FeAl, suffer from low ductilities and fracture toughnesses at room temperature which limit their processing and hinder practical use. The physical and mechanical properties of NiAl have recently been reviewed by Noebe et al. [1]. Depending on crystallographic orientation, single crystal toughnesses varying between 4 and 9 MPa $m^{1/2}$ have been measured [2]. Recent results obtained by Hack et al. [3] indicate that heat treatment involving fast cooling from 673 down to 293 K increases the fracture toughnesses substantially. This effect is presumably due to the presence of mobile dislocations (the fast cool-down did not allow all dislocations to be locked by the segregation of interstitials). Pronounced effects on the yield stress of NiAl due to mobile dislocations introduced by prestraining have also been found by Margevicius et al. [4,5,6]. Fracture toughnesses for polycrystalline NiAl prepared via casting (which was in some cases followed by extrusion) vary between about 4 and 7 MPa $m^{1/2}$ [7,8].

Rigney and Lewandowski [9] noted limited stable, as well as intermittent, crack propagation in powder-processed NiAl. Their finding is not directly applicable to cast and extruded materials, which exhibit much lower yield stresses (only about 25% of Rigney and Lewandowski's values). The question therefore arises whether crack propagation in cast and extruded NiAl occurs in a stable or intermittent manner. Also, since related intermetallic alloys, namely Fe_3Al and FeAl, exhibit moisture-induced embrittlement due to oxidation of aluminum by water vapor and the associated release of atomic hydrogen [10,11,12], one might expect a similar situation in NiAl. Similar to the behavior observed in an Fe_3Al intermetallic [13], one might expect a dependence of the crack-growth velocity on the applied stress-intensity factor.

Whereas stoichiometric FeAl is very brittle, hypo-stoichiometric Fe-40Al [= Fe-40 at. % Al] exhibits a room-temperature ductility in (humid) air of 1 to 2%, which is increased by adding small amounts of boron and by heat treating followed by slow cooling [14]. Since moisture-induced embrittlement is a major cause of the low ductilities of iron aluminides, and since this type of embrittlement is a kinetic process involving several steps before embrittling hydrogen arrives at crack tips, it is expected that ductilities and fracture toughnesses depend on strain or crack-propagation rates, respectively. Thus, at high rates, a lesser degree of moisture-induced embrittlement is expected. Nagpal and Baker [15] have shown this effect by carrying out tensile tests with Fe-45Al at widely different strain rates. As the strain rate was increased from 10^{-2} to 1 s^{-1}, the ductility increased from about

2 to 9%. Corresponding results have been obtained for Fe_3Al-based materials tested either in tension or in cyclic fatigue [13,16].

The purpose of the present work is to measure the fracture toughnesses of B2 iron aluminides with different stoichiometries, as well as with and without boron, and to examine the dependence of the stress-intensity factor on the crack-propagation velocity for the case of monotonic crack propagation.

EXPERIMENTAL PROCEDURE

Several NiAl and FeAl intermetallics were arc-cast from elemental constituents with typical purities of 99.95%. The castings, with diameters and lengths of 25 and 120 mm, respectively, were extruded in mild steel cans at 1173 K with an area reduction of 9:1. Chevron-notched specimens with cross sections of 5 mm x 5 mm and lengths of 45 mm were electro discharge-machined and ground. The geometry of the chevron notches is shown in Fig. 1. The notched specimens were annealed for 1 h at 1273 K in a vacuum (10^{-4} Pa) and quickly cooled by removing them from the hot zone without breaking the vacuum. Three-point bend testing with a span of 40 mm was carried out at room temperature in laboratory air in an Instron 4501 testing machine. The bend fixture was equipped with two capacitance transducers. Its compliance was determined and used to correct the measured load-displacement curves in order to obtain load-load point displacement (LPD) curves. The load-LPD curves were analyzed using the method of Jenkins et al. [17]. Jenkins et al. defined a dimensionless compliance C_d as:

$$C_d = E' \, B \, LPD / P, \tag{1}$$

where E' is the plane strain Young's modulus, B the sample width in a direction perpendicular to the loading direction, and P the applied load. For the present specimen dimensions and a span of 40 mm, the dimensionless compliance may be represented by the following expression, which fits the theoretical results of Jenkins et al.:

$$C_d = m_1 + m_2 \times \alpha + m_3 \times \alpha^2 + m_4 \times \alpha^4 + m_5/(1-\alpha) + m_6/(1-\alpha)^2 , \tag{2}$$

where $m_1 = 485.6776$, $m_2 = -484.35697$, $m_3 = 587.3787$, $m_4 = 857.759$, $m_5 = -154.7605$, and $m_6 = 57.30218$. The normalized crack length α defined in Fig. 1 is given by that value, for which the theoretical and the experimentally determined compliances match. The stress-intensity factor is given by:

$$K_I = \left(0.5 \times \frac{dC_d}{d\alpha} \times \frac{1-\alpha_0}{\alpha-\alpha_0}\right)^{1/2} \times \frac{P}{BW^{1/2}}, \tag{3}$$

where $\alpha_0 = 0.35$ in our case. Furthermore, the instantaneous work of fracture (IWOF) may be obtained by evaluating the load versus LPD curve up to a point corresponding to the desired crack length, subtracting the energy contribution due to the elastic compliance of the

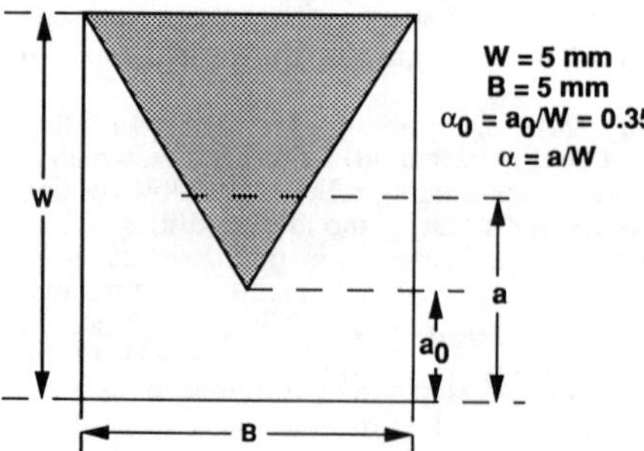

Fig.1. Geometry of the chevron notch in the center of a bend specimen

specimen, and dividing by two, since two surfaces are created during fracture. Average fracture toughnesses were determined from the IWOF at $\alpha = 0.9$ and the relationship:

$$K_I = (2\ \mathrm{IWOF}\ E')^{1/2}. \quad (4)$$

E'-values of 200 and 180 GPa were chosen for NiAl and FeAl, respectively. Average crack-propagation velocities were determined from plots of the crack length as a function of time.

RESULTS AND DISCUSSION

Figure 2 shows a typical load-LPD curve for NiAl. Initially, the load increases steeply, indicating little or no crack growth. Eventually, a first distinct load drop occurs. Owing to the stability of the chevron-notch configuration, the crack arrests, the load increases again, another load drop occurs, and so on. Further analysis of the raw data in Fig. 2 is shown in Figs. 3(*a*) and (*b*). These figures show the stress-intensity factor, as well as the IWOF, as a function of the normalized crack length. In agreement with Fig. 2, Fig. 3(*a*) shows successive increases in the stress-intensity factor (full points) and crack jumps (solid lines). Figures 3(*a*) and (*b*) both indicate a rise in the crack-growth resistance as the crack advances. Since the IWOF plot in Fig. 3(*b*) involves the integral of the load-LPD curve, it is not as sensitive to the run-arrest crack propagation as the stress-intensity factor in Fig. 3(a) and is therefore a preferred measure of the average stress-intensity factor required to propagate the crack. The average stress-intensity factors (crack-growth resistances) measured in this work were all calculated from the IWOF at a normalized crack length $\alpha = 0.9$.

The fracture surface of a Ni-50Al specimen is shown in Fig. 4. Fracture involved a mixture of transgranular and intergranular separation, with the former being more predominant. Optical microscopy showed the grain size in the center of the specimens to be somewhat larger than that near the edges. However, the sharp load drops observed during testing could not unambiguously be correlated with microstructural features such as grain size or fracture-surface features.

Fracture toughnesses measured for NiAl are shown in Fig. 5 as a function of iron content and crosshead speed. Two conclusions can be drawn from this plot. First, within the experimental scatter, the stress-intensity factor is not sensitive to the imposed crosshead speed. In view of the run-arrest crack propagation, this is not entirely surprising, since the rate with which the cracks run following arrest is not directly related to the imposed crosshead speed. Second, the fracture toughness tends to decrease with increasing iron content. In <110>-oriented NiAl single crystals, small iron additions lead to substantial ductility increases [18]. Therefore, fracture-toughness increases might be expected due to additions of iron. Since NiAl exhibits only three easily activated independent slip systems, and since our experiments were conducted with polycrystals, iron additions are not capable of enhancing the fracture toughness of NiAl. One exception is a specimen containing 0.26 at. % Fe, which exhibited a fracture toughness of 15.7 MPa $m^{1/2}$. The reasons for its high toughness are not clear at the present time and require further investigation.

One heat with the nominal composition Ni-50Al-0.01B was also investigated. Its fracture behavior was found to be comparable to that of the binary, boron-free NiAl.

We will now turn to the fracture behavior of iron aluminides. Figure 6 shows that the grains of Fe-40Al are generally much larger than those of NiAl, for identical heat treatments. However, agglomerates of fine grains, indicating a bimodal size distribution, are also found. Fracture is almost completely intergranular. Detailed inspection at higher magnifications shows that the grain facets are not completely smooth. Rather, the possibility exists that the fracture path deviates locally from the grain boundaries, leaving small depressions or raised areas behind (Fig. 7).

The load-LPD curves for FeAl generally extended to greater loads and displacements than those for NiAl. Figure 8(*a*) shows a typical load-LPD curve for an Fe-40Al specimen. Figure 8(*b*) shows an evaluation of stress-intensity factors for a range of crack propagation velocities and three different compositions. In particular, at the greater crack-propagation velocities, the measured stress-intensity factors are in the same range as the fracture toughnesses which Chang et al. [2] determined for single-crystal Fe-40Al. Depending on orientation, these toughnesses varied between 33 and 56 MPa $m^{1/2}$. In the present work, the

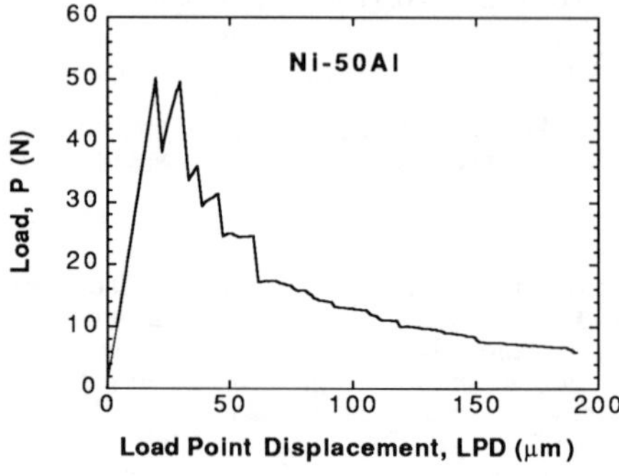

Fig. 2. Load-LPD curve for Ni-50Al; crosshead speed = 10 μm/s.

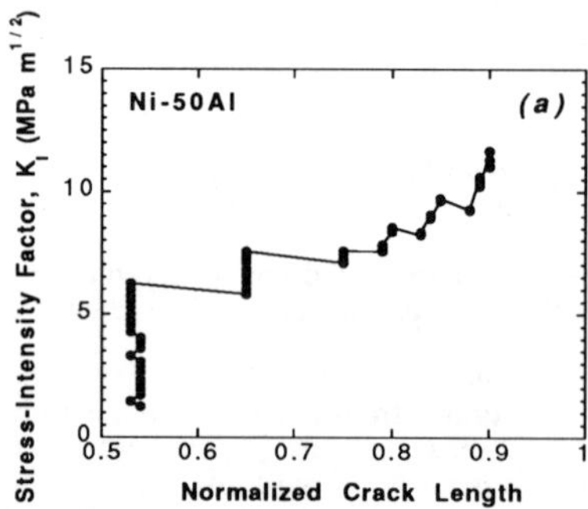

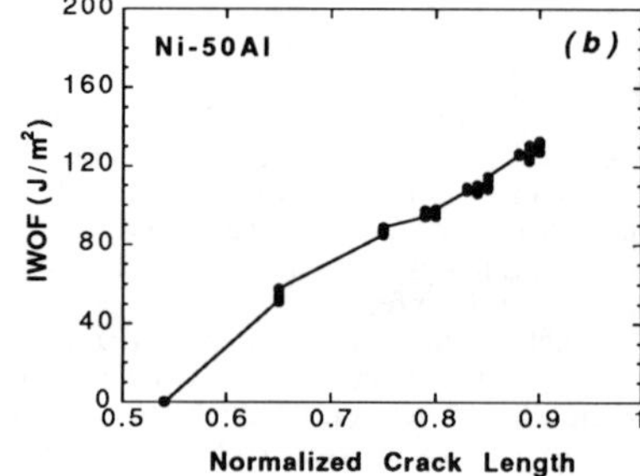

Fig. 3. Evaluation of load-LPD curve in Fig. 2 to obtain (*a*) stress-intensity factor and (*b*) instantaneous work of fracture.

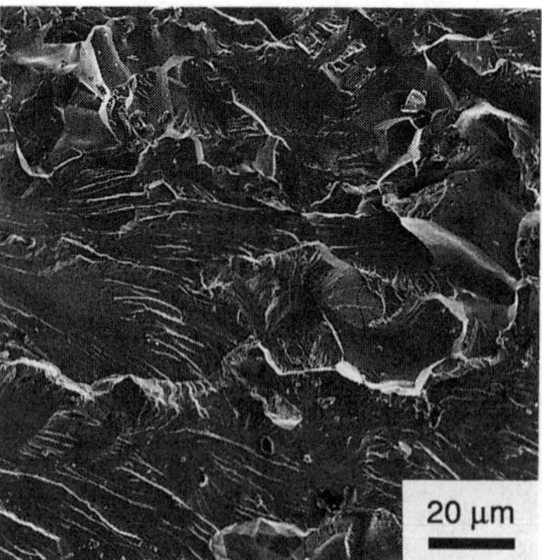

Fig. 4. Fracture surface of a Ni-50Al specimen fractured at a crosshead speed of 1 μm/s

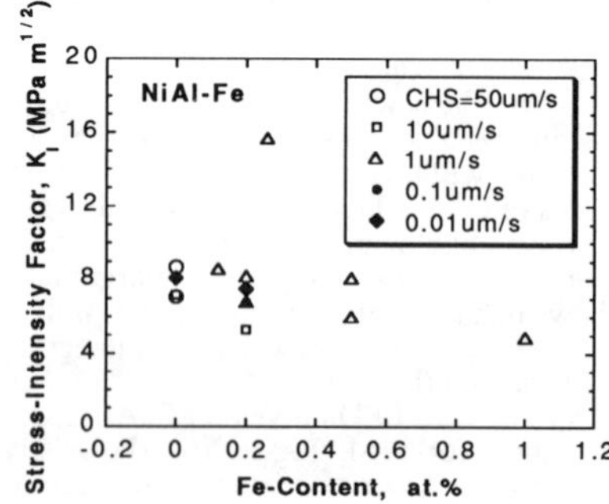

Fig. 5. Fracture toughness of NiAl as a function of iron content and crosshead speed.

stress-intensity factors for polycrystalline Fe-40 Al and Fe-45Al decrease by more than a factor of two as the crack velocity is reduced by three orders of magnitude. The dependence of the ductility of Fe-45Al on the strain rate [15] follows a qualitatively similar pattern. These results are consistent with the moisture-induced embrittlement observed by Liu and co-workers in FeAl intermetallics [10,11]. As the crack velocity is reduced, more time is available for water vapor to migrate to, and to react with, newly created surfaces. Similarly, more time is available for the generated atomic hydrogen to diffuse into the material.

The stress-intensity factors required for crack propagation in Fe-45Al are much less than those for Fe-40Al. However, they do show a qualitatively similar dependence on the crack-growth velocity. Boron, which is known to be beneficial for improving the ductility of iron aluminides and Ni_3Al [11,14,19,20], improves the fracture toughness of Fe-40Al, as expected. The limited data obtained suggests that boron increases the threshold value of the stress-intensity factor reached at very low strain rates. Also, within the range of crack-growth velocities examined, the variation in crack-growth resistance is smaller than that found for crack propagation in undoped Fe-40Al and Fe-45Al.

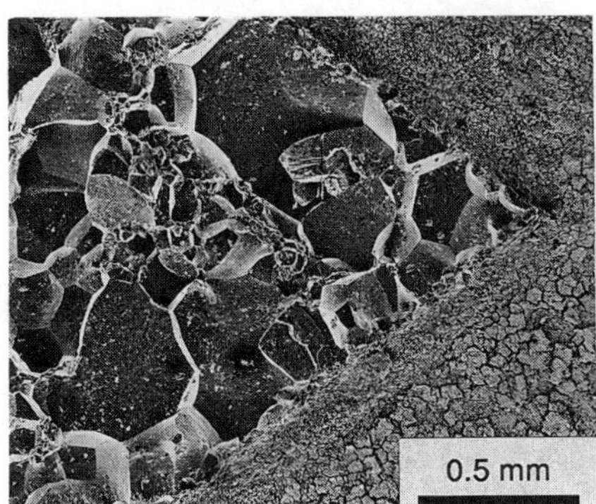

Fig. 6. Fracture surface of an Fe-40Al specimen fractured at a crosshead speed of 1 μm/s. The crack propagation started at the tip of the chevron on the right-hand side of the micrograph.

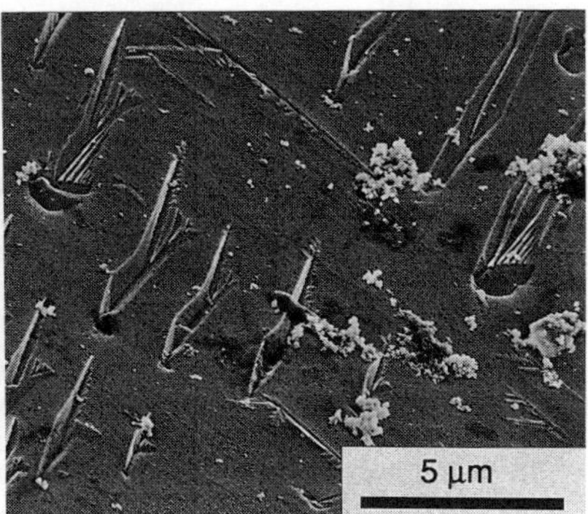

Fig. 7. Grain facets of Fe-40Al fractured at a crosshead speed of 0.1 μm/s.

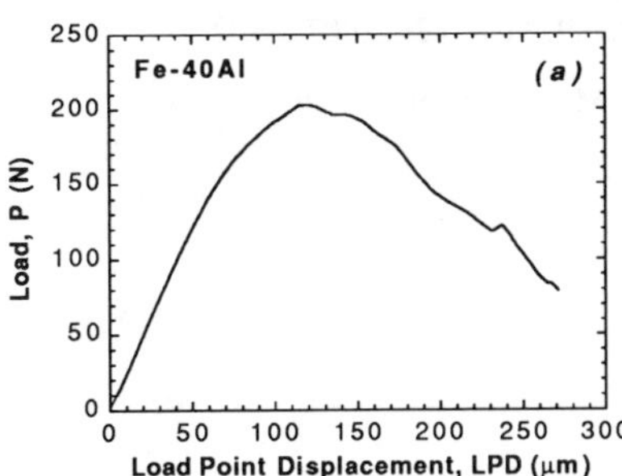

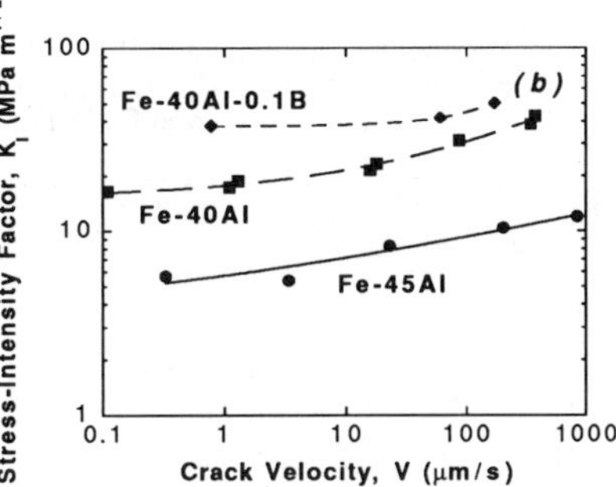

Fig. 8. Crack-growth behavior of Fe-Al: (*a*) load-LPD curve for Fe-40Al and (*b*) stress-intensity factor as a function of crack-propagation velocity for Fe-40Al, Fe-40Al-0.1B, and Fe-45Al.

CONCLUSIONS

The crack-propagation behavior and crack-growth resistance in B2 NiAl and FeAl intermetallics have been found to be quite different. Cracks in NiAl show distinct run-arrest behavior, and the fracture toughness of NiAl is insensitive to the average rate of crack propagation, to iron additions, and to small boron additions. By contrast, the crack-growth resistance of FeAl intermetallics decreases as the crack-propagation velocity decreases, and small boron additions increase the crack-growth resistance distinctly.

ACKNOWLEDGEMENTS

This work was sponsored by the Division of Materials Sciences, United States Department of Energy, under contract DE-AC05-84OR21400 with Martin Marietta Energy Systems, Inc. The review of this manuscript by C. G. McKamey and E. P. George is appreciated. The authors thank B. E. Mercer for manuscript preparation and K. Spence for editing.

REFERENCES

1. R. D. Noebe, R. R. Bowman, and M. V. Nathal, accepted for publication in Int. Mater. Rev.
2. K.-M. Chang, R. Darolia, and H. A. Lipsitt, Acta Metall. Mater. **40**, 2727 (1992).
3. J. E. Hack, J. M. Brzeski, and R. Darolia, Scr. Metall. Mater. **27**, 1259 (1992).
4. R. W. Margevicius and J. J. Lewandowski, Scr. Metall. Mater. **25**, 2017 (1991).
5. R. W. Margevicius, J. J. Lewandowski, and I. Locci, Scr. Metall. Mater. **26**, 1733 (1992).
6. R. W. Margevicius, J. J. Lewandowski, I. Locci, and G. M. Michael, this Symposium.
7. S. M. Russel, C. C. Lawn, M. J. Blackburn, P. C. Clapp, and D. M. Pease, Lightweight Disk Alloy Development, PWA-FR-19577-8, Pratt & Whitney, 1989.
8. S. Reuss and H. Vehoff, Scr. Metall. Mater. **24**, 1021 (1990).
9. J. D. Rigney and J. J. Lewandowski, Mater. Sci. Eng. **A149**, 143 (1992).
10. C. T. Liu, E. H. Lee, and C. G. McKamey, Scr. Metall. **23**, 875 (1989).
11. C. T. Liu and E. P. George, Scr. Metall. Mater. **24**, 1285 (1990).
12. D. J. Gaydosh and M. V. Nathal, Scr. Metall. Mater. **24**, 1281 (1990).
13. A. Castagna and N. S. Stoloff, Scr. Metall. Mater. **26**, 673 (1992).
14. M. A. Crimp, K. M. Vedula, and D. J. Gaydosh, Mater. Res. Soc. Proc. **81**, 499 (1987).
15. P. Nagpal and I. Baker, Scr. Metall. **25**, 2577 (1991).
16. Aidang Shan and Donglian Lin, Scr. Metall. Mater. **27**, 95 (1992).
17. M. G. Jenkins, A. S. Kobayashi, K. W. White, and R. C. Bradt, Int. J. Fract. **34**, 281 (1987).
18. R. Darolia, D. Lahrman, and R. Field, Scr. Metall. Mater. **26**, 1007 (1992).
19. C. T. Liu, Scr. Metall. Mater. **27**, 25 (1992).
20. E. P. George, C. T. Liu, and D. P. Pope, Scr. Metall. Mater. **27**, 365 (1992).

EFFECTS OF PRESSURE ON THE FLOW AND FRACTURE OF POLYCRYSTALLINE NiAl

R. W. MARGEVICIUS, J. J. LEWANDOWSKI, I. E. LOCCI, AND G. M. MICHAL,
Department of Materials Science and Engineering, Case Western Reserve University, Cleveland, Ohio 44106

ABSTRACT

The effects of testing NiAl in tension under a superimposed pressure are described. The flow stress decreases when subjected to pressure due to the generation of mobile dislocations. These dislocations can become pinned when subjected to aging at moderate temperatures and times. The ductility increases substantially when tested under a superimposed hydrostatic pressure. A new method for performing fracture toughness tests under superimposed pressure is also described.

INTRODUCTION

Nickel aluminide, NiAl, is a candidate for high temperature structural applications because of its low density, high melting temperature, and good oxidation resistance. However, inadequate low temperature ductility and toughness may limit its use in structural applications. Several attempts have been designed to look at this low ductility including examination of slip systems[1-3], grain size effects[4,5], grain boundary fracture[6], and alloying additions[7,8]. The purpose of this study was to carefully modify the stress state to examine how ductility changes with altering the applied stress. These experiments were done by uniaxially loading tension and compression samples under a superimposed hydrostatic pressure.

Recently published work has given the details of how hydrostatic pressure influences the flow properties of NiAl[9-12]. Both pressurization (i.e., the application of a hydrostatic pressure to a sample, its removal and subsequent testing at atmospheric pressure) and axial loading under an applied pressure decreased the flow stress by up to 40% below that of samples not subjected to pressure. Detailed TEM analyses indicated that dislocations, presumably mobile, were injected into the material. They were seen to be generated at second phase particles (viz., inclusions), as well as grain boundaries.[10,11]

The mechanism of dislocation generation at second phases is well known. These dislocations are generated because of the shear stresses which arise from the differences in the bulk moduli between the matrix and the inclusion when the system is subjected to pressure. Another manifestation of this mechanism also became apparent in the grain boundary regions. The dislocations generated at grain boundaries were done so in the absence of any second phases. Energy dispersive spectroscopy showed that slight variations in composition existed from region to region in the cast material. (Similar tests conducted on a powder metallurgy material, which showed no decrease in the yield stress and no increase in dislocation density, revealed no such compositional variations.)[11] These results suggested that processing of the cast material (i.e., extrusion and annealing) did not completely homogenize the initially compositionally segregated material. Work is currently in progress to determine the mechanism of dislocation generation arising from compositional variations[13].

The decrease in the flow stress has potentially important consequences when considering the fracture events of a material which undergoes a brittle-to-ductile transition. For example, pressurization of Cr was also observed to generate mobile dislocations at second phase particles[14]. The Cr that was pressurized failed at 60% elongation compared to the sample that was unpressurized where the fracture strain was less than 1%. The authors suggested that the injection of mobile dislocations at room temperature effectively lowered the brittle-to-ductile-transition temperature below room temperature, thereby causing the large increases in ductility. Similar effects have been seen for Mo[15].

The suggestion that the generated dislocations observed in the TEM are mobile is supported by both theoretical and experimental considerations. Theoretical work[16] examined the upper yield point in b.c.c. metals and showed that as the mobile dislocation density increased, the upper yield point decreased. Experimental work performed on samples of Armco iron showed similar decreases in flow stress[17]. Additional experiments were done on the pressurized iron in which the samples were aged at 150 ^{o}C for 2 hours[18]. After aging, the upper yield point returned to the level of the unpressurized samples. Repeated pressurization to the previous levels did not change the upper yield point, and only after pressurization to a value 30% higher than the initial pressurization was the yield point eliminated. These experiments indicated that initial pressurization generated mobile dislocations at dislocation sources which were most easily activated. Aging pinned those dislocations generated after the initial pressurization, and the upper yield point returned to its previous value.

EXPERIMENTAL

After the decrease in the upper yield point in NiAl was observed, it was decided to conduct a similar aging experiment. The details of the pressurization as well as the subsequent testing and analyses have been given elsewhere[8-10]. In additional to those samples only pressurized, some compression samples were pressurized and then aged at 200 ^{o}C for 2 hours. The results of the aging experiment are shown in Figure 1. Strain for all of the samples was measured as the arithmetic mean of two strain gages mounted 180° apart on cylindrical compression samples. The flow stress at 0.1% offset for the unpressurized sample (Compression + 0.1 MPa) was 150 MPa. The 0.1% offset flow stress for the pressurized sample (1400 MPa/Compression + 0.1 MPa) decreased to 110 MPa. The yield stress for the sample pressurized to 1400 MPa, aged at 200 ^{o}C for 2 hours and tested at ambient conditions rose to 135 MPa but did not recover to the initial level of 150 MPa. Work reported elsewhere[19] reveals that aging for 2 hours at 400 °C is sufficient to return their flow properties to their unpressurized values. This indicates that re-

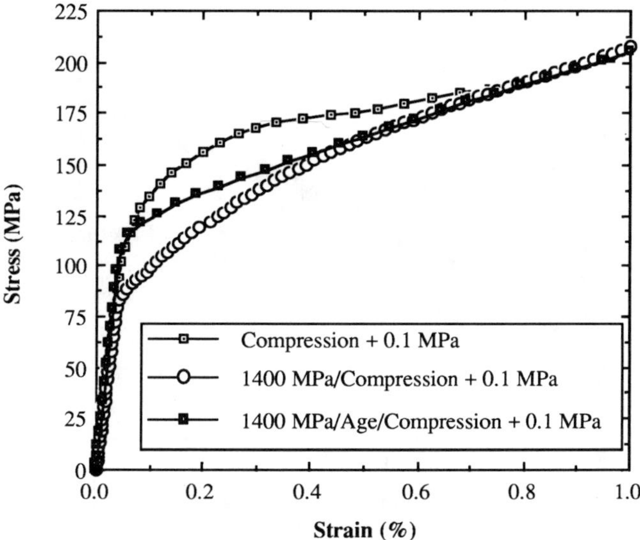

Figure 1. Compressive stress-strain curves for annealed NiAl samples that were both pressurized and aged at 200 ^{o}C for two hours and tested at 0.1 MPa.

pinning of mobile dislocations is possible in NiAl as in Fe and Cr, although the mechanisms responsible in NiAl are under investigation. Consequently, a pressure dependence of the brittle-to-ductile-transition temperature might also exist in NiAl.

Our work to this point on polycrystalline NiAl has not shown this to be true.[9] Figure 2 shows the results for the tension samples pressurized to1400 MPa. While the flow stress is seen to decrease greatly for the pressurized sample, there are no large increases in tensile ductility. It is apparent that a weak fracture path still exists in the polycrystalline NiAl pressurized to different levels despite the presence of mobile dislocations. Additional tests are in progress in order to modify the fracture path. Specimens tested under a superimposed pressure exhibited large increases in tensile ductility (e.g., > 10%). A change in the fracture surface from transgranular fracture to intergranular fracture accompanied the large increases in ductility. Details of the aspects of fracture will be presented elsewhere[20].

WORK IN PROGRESS

The change in the fracture response exhibited while testing under pressure led to a study of fracture toughness under pressure. A compression fixture which had been used in the high pressure vessel was modified to hold a three-point-bend sample. The sample geometry is shown in Figure 3. The span S is fixed at 20 mm, making the depth B and height W (according to

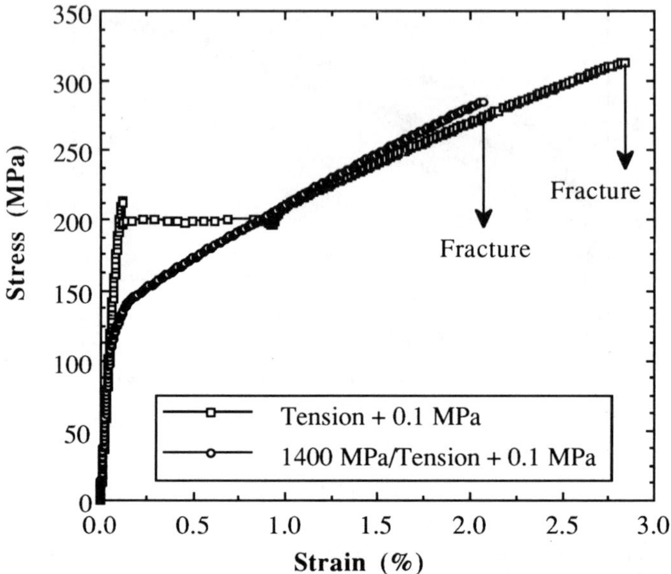

Figure 2. Tensile stress-strain curves for annealed NiAl samples that were unpressurized and pressurized to 1400 MPa and tested at 0.1 MPa.

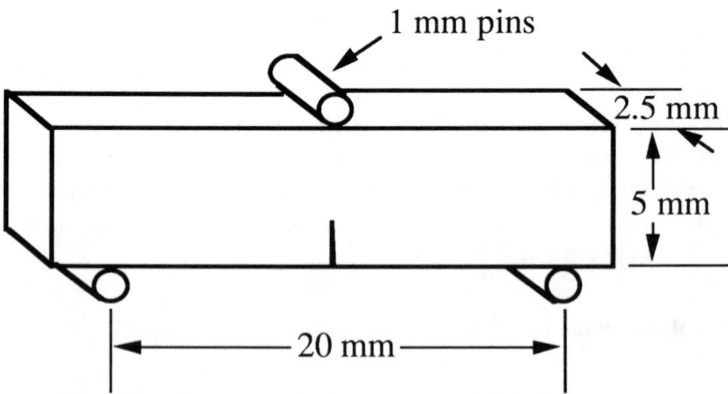

Figure 3. Sample geometry for toughness specimens to be tested under a superimposed pressure.

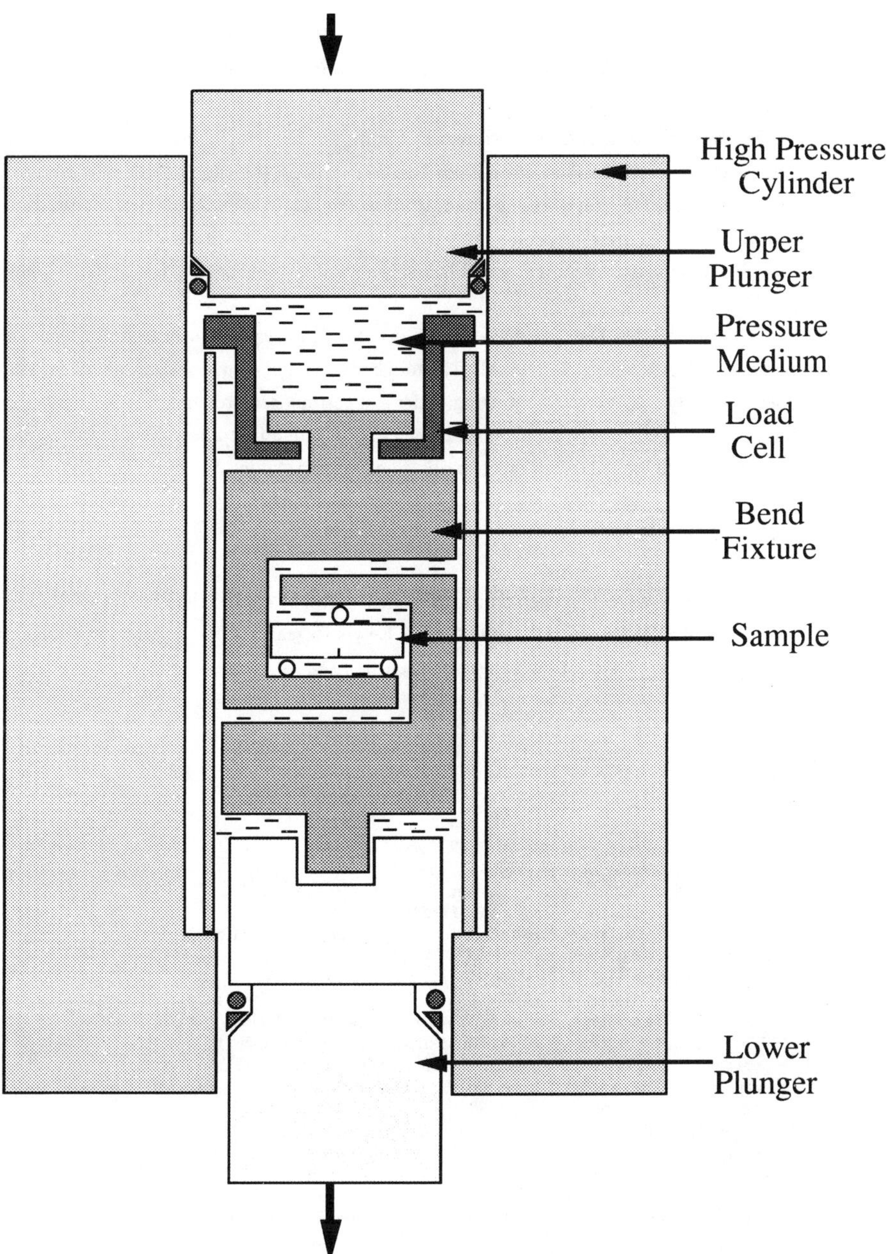

Figure 4. Schematic of the three-point-bend fixture for use in the high pressure chamber.

ASTM E399[21]) fixed at 2.5 mm and 5 mm, respectively. The maximum load that the fixture can accommodate is 250 N. With these conditions the maximum toughness that can be measured for a materials is approximately 12 MPa(m)$^{0.5}$.

A schematic of the apparatus is shown in Figure 4. The single-edged notched fracture toughness specimen is loaded in three-point bending through the cage. (The cage fixture is necessary to transfer a compressive load from a tensile load as the internal, pressure-compensated load cell measures only tensile loads.) Load point displacement is measured externally through the use of a linear variable displacement transducer mounted onto the bottom plunger.

Preliminary work has been conducted on an off-stoichiometry, single crystal NiAl having a composition of 62 at% Ni and 38 at% Al. The toughnesses attained under atmospheric pressure were approximately 2 MPa(m)^$^{0.5}$. Testing under a superimposed pressure changed neither the toughness nor the fracture appearance for the level of pressure utilized here. Additional work is continuing in order to look at stoichiometric single and polycrystalline NiAl.

ACKNOWLEDGMENTS

Support for this work has been from NSF-DMR- PYI-89-58326 (RWM and JJL), ONR-N00014-91-J-1370 (RWM and JJL), and NASA Lewis Research Center. Supply of material was also obtained from NASA Lewis Research Center.

REFERENCES

[1] A. Ball and R. E. Smallman, *Acta Met.* **14**, 1349 (1966).
[2] A. Ball and R. E. Smallman, *Acta Met.* **14**, 1517 (1966).
[3] R. J. Wasilewski, S. R. Butler, and J. E. Hanlon, *Trans. Met. Soc. AIME* **239**, 1357 (1967).
[4] E. M. Schulson and D. R. Barker, *Scripta Met.* **17**, 519 (1983).
[5] I. Baker, P. Nagpal, F. Liu, and P. R. Monroe, *Acta Met. et Mat.* **39**, 1637 (1991).
[6] K. H. Hahn and K. Vedula, Scripta Met. 23, 7 (1989).
[7] S. V. Raj, R. D. Noebe, and R. R. Bowman, *Scripta Met.* **23**, 2049 (1989).
[8] R. D. Field, D. F. Lahrman, and R. Darolia, *Acta Met. et Mat.* **39**, 2961 (1991).
[9] R. W. Margevicius and J. J Lewandowski, *Scripta Met. et Mat.* **25**, 2017 (1991).
[10] R. W. Margevicius, I. Locci, and J. J Lewandowski, *Scripta Met. et Mat.* **26**, 1733 (1992).
[11] R. W. Margevicius and J. J Lewandowski, *Acta Met. et Mat.*, **41**, 485 (1993).
[12] J.J. Lewandowski, G. M. Michal, I. Locci, and J. D. Rigney, *Proc. Mat Res. Symp.* (edited by G. M Stocks and D. Pope) MRS Pittsburgh, Pa, 341 (1990)
[13] R. W. Margevicius and J. J Lewandowski, unpublished research (1993).
[14] F. P. Bullen, F. Henderson, H. L. Wain, and M. S. Patterson, *Phil. Mag.* **9**, 803 (1964).
[15] H. G. Mellor and A. S. Wronski, *Ocean Eng.* **1** 233 (1969).
[16] G. T. Hahn, *Acta Met.* **10**, 727 (1962).
[17] F. P. Bullen, F. Henderson, M. M. Hutchison, and H. L. Wain, *Phil. Mag.* **9**, 285 (1964).
[18] F. M. C. Besag and F. P. Bullen, *Phil. Mag.* **10**, 41 (1964).
[19] R. W. Margevicius and J. J Lewandowski, *Scripta Met. et Mat.* submitted (1992)
[20] R. W. Margevicius and J. J Lewandowski, in preparation (1993).
[21] "Standard Test Method for Plane Strain Fracture Toughness of Metallic Materials," E399, *Annual Book of ASTM Standards*, Vol. **0301**, ASTM, Philadelphia, PA, 788 (1984).

DEFORMATION OF A HF-V-NB C15 LAVES PHASE

Fuming Chu and David P. Pope
Department of Materials Science and Engineering, University of Pennsylvania, Philadelphia, PA 19104, U.S.A.

ABSTRACT

We have performed compression tests on $Hf_{14}V_{64}Nb_{22}$ at both room temperature and elevated temperatures with different strain rates to study the deformation modes of the ternary C15 Laves phase based on HfV_2. At room temperature the alloy exhibits fair ductility and is insensitive to strain rate. TEM observation revealed that the major deformation modes at room temperature are {111}<11$\overline{2}$> mechanical twinning. Compression tests up to 1000°C, combined with SEM and TEM examinations, revealed that the mechanical properties can be divided into three temperature regions: at low temperatures the alloy is fairly ductile due to the formation of deformation twins, in the intermediate temperature region the alloy exhibits a ductility minimum, and at high temperatures (>$0.65T_m$) the alloy shows large plastic deformations via 1/2<110> dislocation slip. A new deformation twinning mechanism is proposed based on synchroshear mechanism.

INTRODUCTION

Due to growing interest in intermetallic compounds as potential high temperature structural materials, there have been many investigations carried out on various compounds, but mostly on structures that are ordered forms of fcc, bcc or hcp; however, most intermetallics have more complex structures and Laves phases constitute the single largest group. Laves phases have either the cubic C15($MgCu_2$), hexagonal C14($MgZn_2$) or mixed C36($MgNi_2$) structure. Generally speaking, they have high melting temperatures and fairly low specific gravities (1), and therefore could be attractive candidates for high temperature structural applications if the deformability could be increased at low temperatures. Inoue et al (2-3) have shown that C15 Hf-V-Nb alloy can be deformed by large amounts at room temperature and subsequent work by Livingston and Hall (4) and Chu and Pope (5) has shown that twinning is the main deformation mode. We report here the compression test results obtained at different temperatures and strain rates, TEM and SEM observations of deformation modes for different temperature regions and a new mechanism for C15 mechanical twinning based on the synchroshear mechanism.

EXPERIMENTAL PROCEDURES

Based on the Hf-V-Nb phase diagram (6), a C15 alloy of composition $Hf_{14}V_{64}Nb_{22}$ was chosen for this study. The arc-melted buttons were made using elemental Hf, V, and Nb with nominal purities 99.99%, 99.6% and 99.7%, respectively. The buttons were turned over and remelted five times to ensure homogeneity. The buttons were homogenized at 1200°C for 48 hours and then argon quenched.

Compression specimens of dimension 3x3x5 mm were cut using EDM and mechanically polished. Before mechanical testing and SEM observation polished sections of these samples were examined in an optical microscope after etching with a reagent consisting 30 ml HNO_3, 30 ml lactic acid, 2 ml HF and a few drops of water. Compression tests were performed in an Instron machine at different strain rates ranging from 1.69×10^{-4} s^{-1} to 4.23×10^{-3} s^{-1}. A vacuum of about 10^{-6} Torr was used to prevent oxidations during the high temperature compression tests.

Specimens were prepared for transmission electron microscopy from the mechanically deformed samples by EDM cutting, mechanically polishing to a thickness of about 70 ~ 90 µm, mechanically dimpling to a thickness of about 20 ~ 30 µm, and ion milling using a liquid nitrogen cold stage on a Gatan ion milling machine. The samples were viewed in a Phillip 400T TEM operating at 120kV.

RESULTS AND DISCUSSION

The single phase C15 alloy prepared in the course of determining the phase diagram showed substantially different degrees of brittleness as indicated by sensitivity to thermal shock and handling stress. Those alloys in the region of V/Hf>2 and with substantial ternary Nb additions are less brittle than those for which V/Hf<2. With this qualitative data in mind we performed mechanical tests on two phase C15 alloys consisting of a matrix of the C15 Laves phase with V/Hf>2 surrounding the V-rich bcc solid solution. A microstructure of a $Hf_{14}V_{64}Nb_{22}$ alloy is shown in Fig. 1.

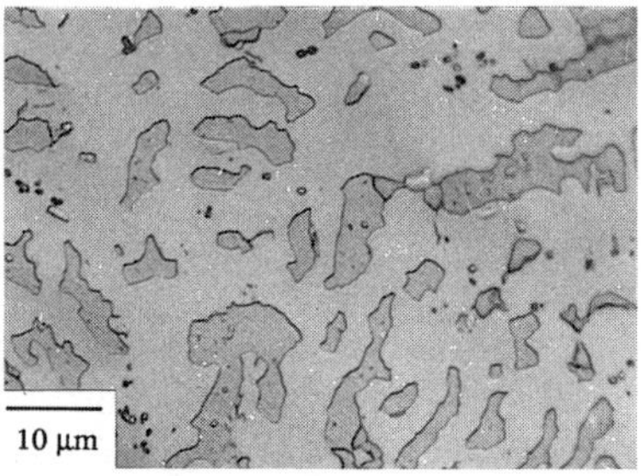

Fig. 1. The microstructure of a $Hf_{14}V_{64}Nb_{22}$ alloy. The matrix is C15, the second phase is bcc solution.

The strain rate dependences of σ_Y and ε_{pl} are shown in Fig. 2. The results show that the mechanical behavior of the C15 alloy is strain rate insensitive.

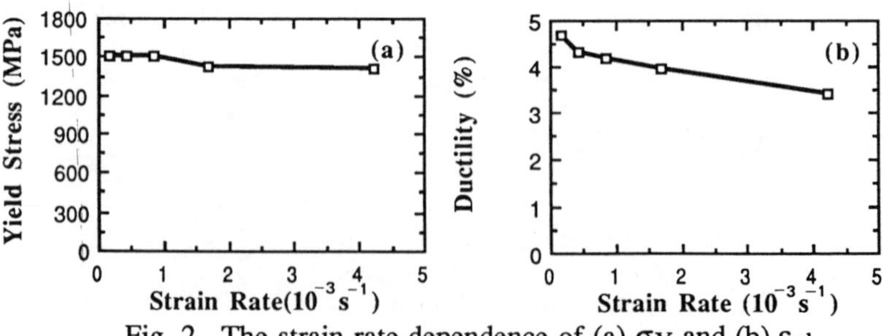

Fig. 2. The strain rate dependence of (a) σ_Y and (b) ε_{pl}.

The stress-strain curves obtained on $Hf_{14}V_{64}Nb_{22}$ at a strain rate 8.4×10^{-4} s^{-1}, and at different temperatures are shown in Fig. 3. Based on Fig. 3 the plots of σ_Y vs T, UTS vs T, σ_F vs T, and ε_{pl} vs T are obtained as shown in Fig. 4. It can be seen that the mechanical properties of the C15 alloy as a function of temperature can be divided into three distinct regions. In the low temperature region, from room temperature to 300°C, the alloy is fairly ductile, in the medium temperature region, from 300°C to 750°C, the alloy exhibits a ductility minimum and at high temperatures (>$0.65T_m$) the alloy shows large plastic deformations

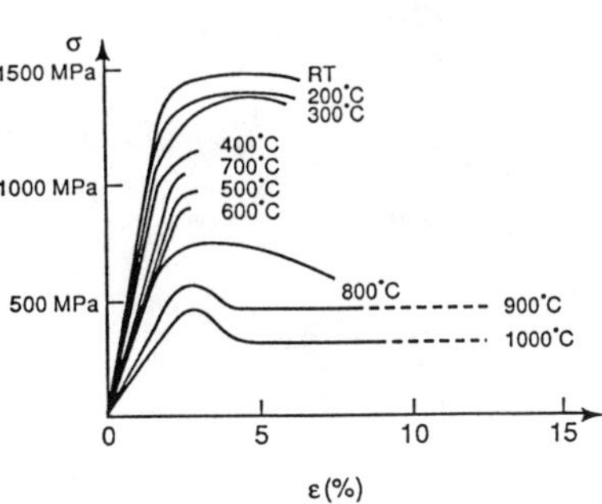

Fig. 3. Compression stress-strain curves of a $Hf_{14}V_{64}Nb_{22}$ alloy at various temperatures.

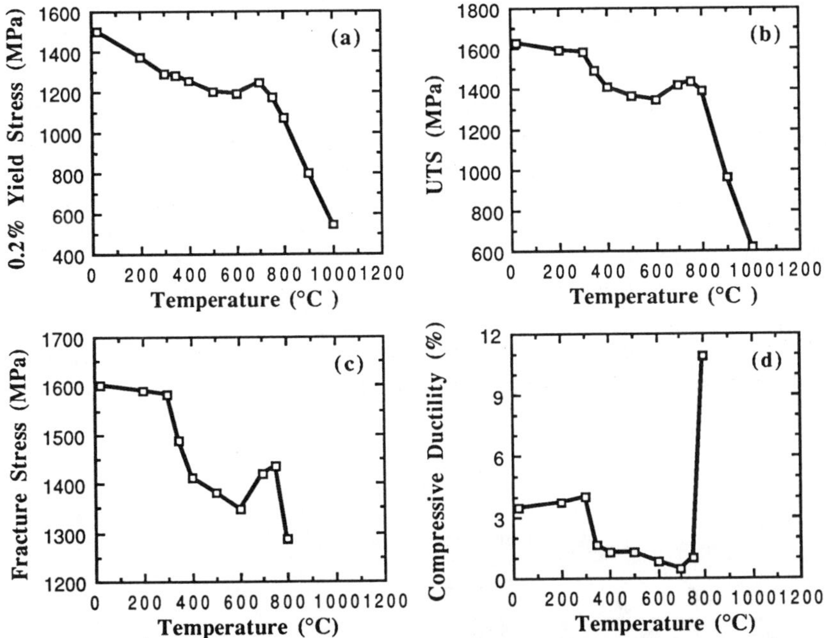

Fig. 4. Plots of (a) σ_Y, (b) UTS, (c) σ_F and (d) ε_{pl} vs T for $Hf_{14}V_{64}Nb_{22}$.

TEM examination of the sample tested at the room temperature showed that the major deformation mode in the ternary C15 Laves phase is mechanical twinning, as shown in Fig. 5(a). The electron diffraction pattern using a <110> zone axis, as demonstrated in Fig. 5(b), unambiguously identifies it as a $\{111\}/\langle 11\bar{2} \rangle$ twin, the one commonly seen in fcc materials.

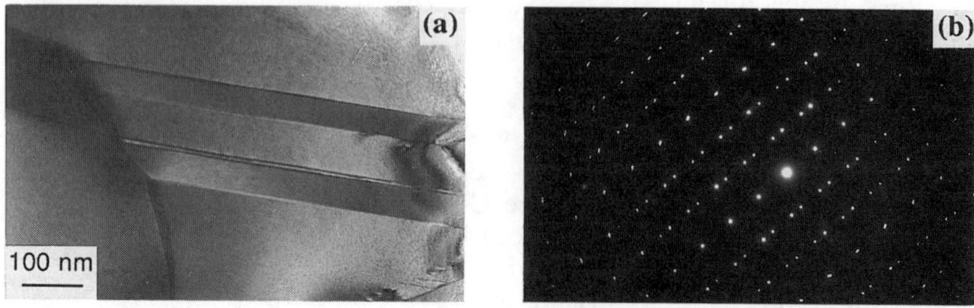

Fig. 5. (a) Microstructure of a plastically deformed C15 alloy. The zone axis is [110]; twin interface is $(1\bar{1}1)$. (b) [110] diffraction pattern of $\{111\}<11\bar{2}>$ twin in this system.

Twinning elements have been developed for various superlattices (7). For these twins, long shear vectors and/or shuffles are commonly used in the description, but this does not mean that such long shears and complex shuffles actually occur. For example, in C15 intermetallic compounds the shuffling of a large number of atoms in this topologically close packed structure at low temperatures is probably physically unreasonable. The synchroshear of selected layers has been proposed to describe the kinematic process of the formation of $\{111\}<11\bar{2}>$ twins in the C15 Laves phase (4-5). We suppose that in the stacking sequence along [111] direction, $\cdots\gamma C \gamma b \alpha A \alpha c \beta B \beta a \cdots$, the $\alpha A \alpha$ type sandwich is a rigid group because of the longer shear vector and the directional bonding along [111] between α - α atoms. But the $\alpha c \beta$ type sandwich is more deformable as long as synchroshear is operative within the sandwich. In this process, the smaller atoms shear by $1/6[\bar{2}11]$, the larger atoms move synchronously but with a shear displacement of $1/6[11\bar{2}]$, as shown in Fig. 6. The net results is that the upper block moves by $1/6[\bar{1}2\bar{1}]$ on the (111) plane relative to the lower block, thereby producing the following stacking sequence, which obviously produces a $\{111\}<11\bar{2}>$ twin:

$$\cdots \gamma C \gamma b \alpha A \alpha c \beta B \beta a \cdots$$

$$\cdots \gamma C \gamma b \alpha A \alpha b \gamma C \gamma \cdots$$

Matrix →|← Twin

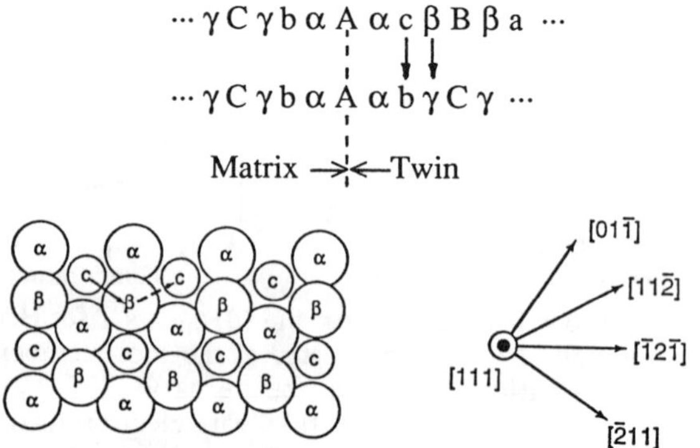

Fig. 6. Synchroshear within the $\alpha c \beta$ type sandwich.

The reason for increased twinning with Nb additions can be partially understood based on the combination study of the experimental and theoretical x-ray diffraction pattern of the ternary C15 Laves phase(5) and the fact that Nb substitutes both for Hf and V. There is a relatively high density of Hf atoms but a low density of V atoms in the $\alpha c\beta$–type sandwiches. Thus when Nb atoms substitute for Hf and V, free volume is produced in the $\alpha c\beta$-type sandwiches since more Nb atoms are on Hf sites than on V sites in these layers (remember $R_{Hf}>R_{Nb}>R_V$ and since V/Hf>2 more Nb atoms go to Hf sites than to V sites). Consequently, the additional free volume in the $\alpha c\beta$-type sandwiches leads to easier synchroshear and, therefore, easier twinning.

TEM observations of the samples tested in the intermediate temperature region show that deformation twins in the ternary C15 Laves phase are substantially decreased at these temperatures. SEM photographs of well preserved fracture surfaces taken from different temperature regimes are shown in Fig. 7, showing that the alloy in the intermediate temperature region is more brittle.

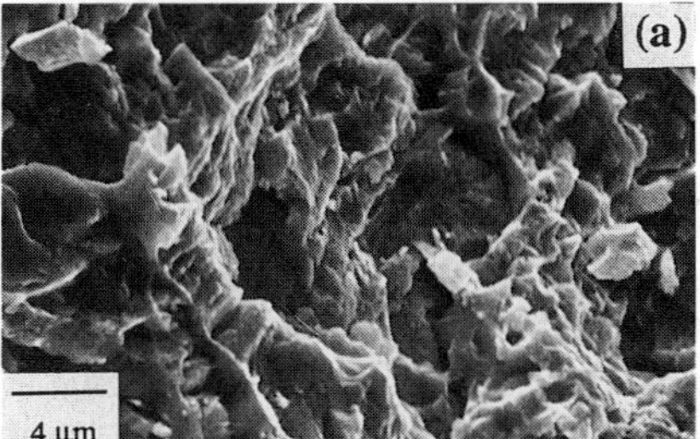

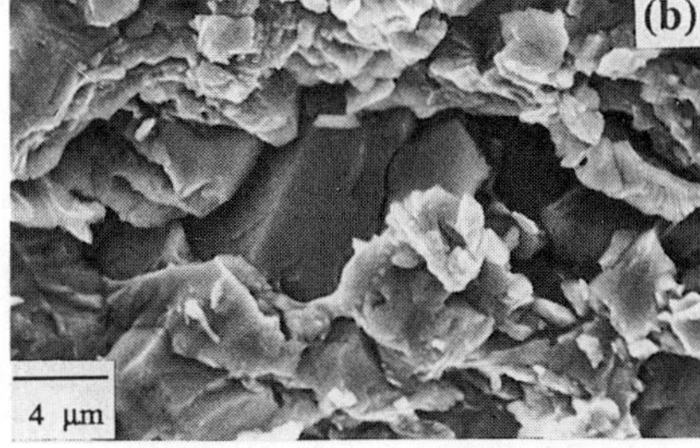

Fig. 7. Fracture surfaces of a sample deformed at (a) 300°C and (b) 400°C.

Three possibilities for the ductility drop in the intermediate temperature region were considered: oxidation, phase transformation and thermal properties. The high vacuum($\sim 10^{-6}$ Torr) has been used to prevent oxidation during the high temperature compression tests so oxidation was ruled out. High temperature x-ray diffraction experiments showed no phase transformation between 300°C and 750°C. These same experiments suggest that the lack of twinning in this temperature range is probably the result of atomic interference across the (111) plane due to the low thermal expansion coefficient and relatively large atomic thermal vibrations which inhibit synchroshear (8).

TEM observations of the samples plastically deformed above 800°C showed that the major deformation mode in the C15 phase at high temperatures($>0.65T_m$) is **b**=1/2<110> dislocation slip, as shown in Fig. 8. The slip plane has not been determined. These results are consistent with high temperature deformation studies of other cubic Laves phases (9-10).

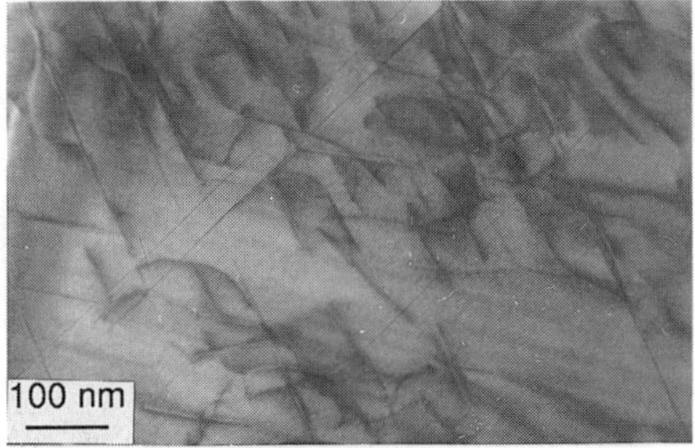

Fig. 8. Dislocations in the C15 alloy.

CONCLUSIONS

1. Twinning is the major deformation mode of plastic deformation at low temperatures in C15 Hf-V-Nb alloys based on HfV_2. The twinning occurs on {111} planes by a shear in the <112> directions. This twinning mode can be accomplished via the synchroshear mechanism resulting in shear on selected {111} planes, i.e. the shear is inhomogeneous. This synchroshear involves only short displacements, 1/6<112>, and requires no atomic shuffles. The substitution of Nb into HfV_2 appears to facilitate deformation twinning by increasing the free volume in the planes undergoing synchroshear.
2. In the intermediate temperature region(300°C - 750°C), the C15 alloy exhibits a ductility drop. This is not the result of a phase transformation but is probably the result of atomic interference across the (111) plane due to the low thermal expansion coefficient and relatively large atomic thermal vibrations which inhibit synchroshear.
3. Above 800°(>$0.65T_m$), the $Hf_{14}V_{64}Nb_{22}$ alloy exhibits a good ductility. The major deformation mode in this temperature regime is **b**=1/2<110> dislocation slip.

ACKNOWLEDGEMENT

This research is supported by the Office of Naval Research (Grant no. N00014-91-J-1165PO2). Research facilities were supported by National Science Foundation MRL program (Grant no. DMR91-20668) through the LRSM at the University of Pennsylvania. Technical help from Dr. S. Chang is greatly appreciated.

REFERENCE

1. R. Fleischer, J. Mater. Sci. **22**, 2281 (1987).
2. K. Inoue and T. Tackikawa, IEEE Trans. Magn. MAG-**13**, 840 (1977).
3. K. Inoue, T. Kuroka and K. Tackikawa, IEEE Trans. Magn. MAG-**15**, 635(1979).
4. J. D. Livingston and E. L. Hall, J. Mater. Res. **5**, No.1, 5 (1990).
5. F. Chu and D. Pope, submitted to Mat. Sci. Eng.
6. F. Chu and D. Pope, Scripta Metall. et Mater., **26**, 399 (1992).
7. M. Khantha, V. Vitek and D. Pope, unpublished research.
8. F. Chu and D. Pope, to be published in Scripta Metall. et Mater.
9. J. B. Moran, Trans. Met. Soc. AIME, **233**, 1473 (1965).
10. J. D. Livingston, E. L. Hall and E. F. Koch in High Temperature Ordered Intermetallic Alloys III, edited by C. T. Liu, A. I. Taub, N. S. Stoloff and C. C. Koch (Mat. Res. Soc. Sym. Proc., **133**,1989) pp. 243-248.

DEFORMATION MECHANISMS IN THE INTERMETALLIC COMPOUND $MoSi_2$

D.J. EVANS[1,2], S.A. COURT[3], P.M. HAZZLEDINE[4], H.L. FRASER[1]
1 : Department of Materials Science and Engineering, The Ohio State University, 2041 College Road Columbus,OH, 43210
2 : Wright Laboratory, Materials Directorate, Wright-Patterson AFB, Ohio, 45433
3 : Alcan Banbury Laboratories, Banbury, United Kingdom
4 : UES, Inc., 4401 Dayton - Xenia Road, Dayton, Ohio 45432

ABSTRACT

The nature of dislocations which have been activated during plastic deformation of $MoSi_2$ at 1400°C in compression have been identified using transmission electron microscopy. Particular attention has been paid to the possible dissociation of dislocations, and it has been confirmed that dislocations with Burgers vectors lying parallel to <111> are dissociated. The dissociation is represented by:

$$1/2\langle 111\rangle \rightarrow 1/4\langle 111\rangle + \text{SISF} + 1/4\langle 111\rangle,$$

where SISF stands for superlattice intrinsic stacking fault. The SISF energy has been estimated from the separation of the partial dislocations to be $\approx$ 255 mJ.m^{-2}. A simple explanation to account for the occurrence of dissociation for particular dislocations is presented.

INTRODUCTION

$MoSi_2$, an intermetallic with the $C11_b$-type ordered structure (i.e. structure derived by stacking three b.c.c. lattices and compressing along the c-axis) has been the subject of study in recent years with a view to its use as a high temperature material in aerospace applications. Its high melting point, approximately 2020°C, and excellent oxidation resistance are among its most attractive properties. However, like most ordered compounds, it also lacks ductility or plasticity except at very high temperatures, above 1000°C. It is of interest to determine the factors that influence this brittle behavior and to that end, the deformation mechanisms, in this compound slip systems and nature of dislocations, have been studied.

The deformation behavior of single crystals of $MoSi_2$ has been studied recently by Umakoshi, Sakagami, Yamane, and Hirano (1). Single crystals compressed at a nominal strain rate of 1.4 x $10^{-4}s^{-1}$ along a number of orientations between 900°C and 1200°C were found to slip predominately by the <331]{110) and <331]{103) systems; these dislocations were found to be dissociated narrowly, which is attributed to a high value of the anti-phase boundary (APB) energy. Above 1200°C, ordinary dislocations with **b** parallel to <100] and <110] directions were observed and were suggested to have some mobility through climb; these dislocations were not observed to be dissociated. Stacking faults, presumably formed through the dissociation of dislocations with **b**=1/2<111> on {110) planes, were also observed in $MoSi_2$ single crystals compressed above 1200°C.

Recently, Unal, Petrovic, Carter, and Mitchell (2) have studied the slip behavior of polycrystalline $MoSi_2$ (reinforced by SiC whiskers) that had been deformed at high temperature (1200°C and above). Their analysis has shown that dislocations with **b** parallel to <100>, <110> and <111> glide on a variety of crystallographic planes. In contrast to the results of Umakoshi *et al.* (1), no dissociation of the dislocations with **b**=1/2<111> was detected. The absence of detectable dissociation of dislocations by Unal, et al. (2) is a surprising result, especially in view of the Japanese work (1), and the aim of the present study has been an understanding of this difference in results.

EXPERIMENTAL PROCEDURE

Specimens of polycrystalline $MoSi_2$ were prepared by hot pressing 99.9% pure powder at 1700°C under 4 ksi pressure in grafoil-lined graphite dies. The resulting material was approximately 97% dense. Coupons of 5 mm x 5 mm x 10 mm were spark machined from the hot-pressed pucks and were compressed at 1400°C under vacuum to a strain of approximately 25% at a nominal strain rate of $1x10^{-3}s^{-1}$. Foils were cut from the deformed coupons and were thinned by mechanical polishing and ion milling until perforation. The thin foils were examined in both a Phillips CM12 transmission electron microscope operating at 120 kV and a JEOL 200CX transmission electron microscope operating at 200 kV.

RESULTS AND DISCUSSION

In samples compressed to ≈25% strain at 1400°C, the deformation microstructure of the $MoSi_2$ was found to consist of matrix dislocations with **b**=<100] and networks made up of dislocations with **b**=<100], **b**=<110] and **b**=<111>, and has been discussed elsewhere (3). Interest here is focused on the nature of dislocations with Burgers vectors parallel to <111>. Imaging of these defects using weak-beam dark-field conditions reveals that they are in fact dissociated on {110) planes, as shown in Figures 1 and 2. It is interesting to note that segments such as those marked A_2 in the figure do not exhibit dissociation; in fact, these segments are dissociated but the plane of dissociation is another {110), inclined to that upon which the segments A_1 and A_3 are dissociated. The contrast exhibited by these defects when imaged with a variety of diffraction vectors, **g**, as well as ±**g**, is consistent with each partial dislocation having parallel Burgers vectors, and it is reasonable to assume that the following dissociation has occurred:

$$1/2\langle 111\rangle \rightarrow 1/4\langle 111\rangle + 1/4\langle 111\rangle$$

The nature of the fault separating these defects is discussed below.

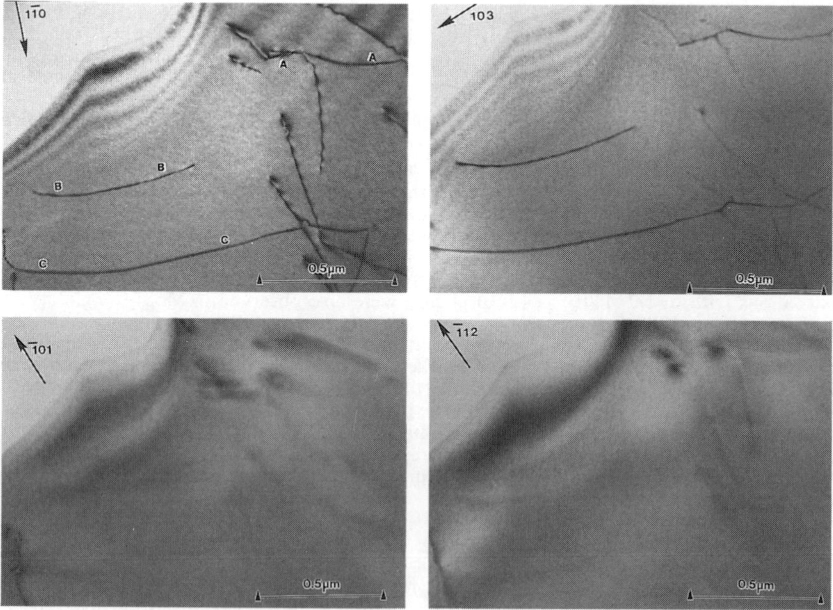

Figure 1 TEM bright field micrographs showing $MoSi_2$ deformed at 1400°C. The dislocation segments labeled A, B, and C all have **b** parallel to 1/2<111>. All other segments have **b** parallel to <100]. (**g** vectors are as indicated).

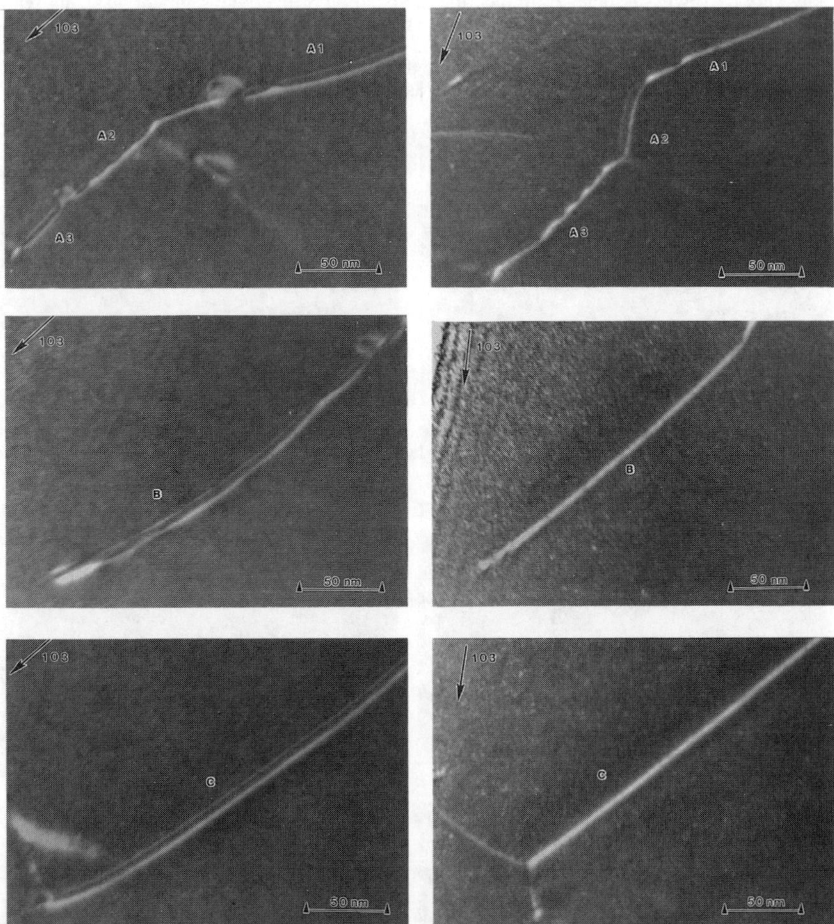

Figure 2 TEM weak beam dark field micrographs showing $MoSi_2$ deformed at 1400°C. The following are the character, slip system and dissociation plane for each segment. **A1** : 15° from screw, 1/4<111>{110), {110), **A2** : 40° from screw, 1/4<111>{112), {1$\bar{1}$0), **A3** : same as A1, **B** : 7° from screw, 1/4<111>{112), {110), **C** : same as B. For the figures in which A1, A3, B and C appear clearly dissociated, the incident beam direction is close to [110]. For the figures in which A2 appears clearly dissociated while B and C do not, the incident beam direction is near [$\bar{1}$10]. (**g** vectors are as indicated).

Figure 3 shows three segments of dislocations with **b**=1/2<111>; detailed diffraction contrast experiments and crystallographic analyses have revealed that segments marked B and C in the figure lie close to the screw orientation whereas segment A lies nearer to edge. It is readily apparent from Figure 3 that the screw segments are dissociated appreciably whereas the edge is only very narrowly dissociated; in fact, the dissociation of the edge segment was only realized when extreme tilting of the foil was effected. The fact that most segments observed were close to screw orientation implies that edge segments were considerably more mobile than the screw dislocations. It is also interesting to speculate that the narrow dissociation of edge segments may account for the apparent difference in results, as discussed above.

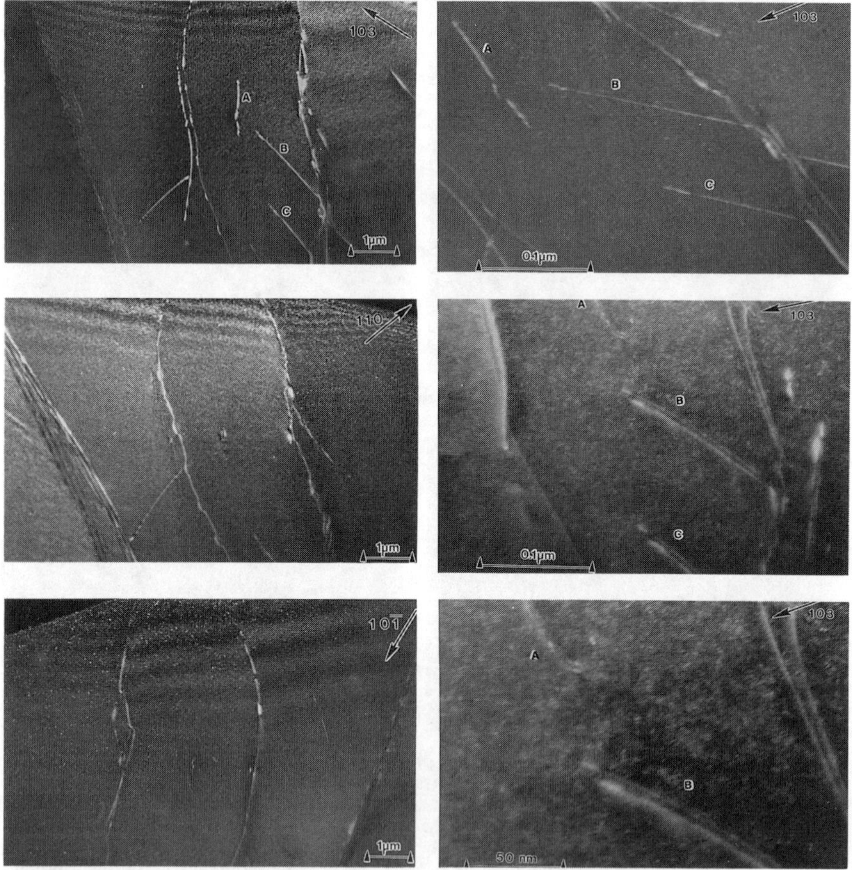

Figure 3 TEM weak beam dark field micrographs showing $MoSi_2$ deformed at 1400C. The following are the character, slip system and dissociation plane for each segment. **A** : 40° from screw, 1/4<111>{110),{110) **B** : 6° from screw, 1/4<111>{110) **C** : same as B. In the images in which the dislocations do not appear dissociated (upper right hand corner), the incident beam direction is close to [010]. In the images in which the segments are clearly dissociated (middle and lower right), the incident beam direction is near [110], (**g** vectors are as indicated).

In the present work, it has not been possible in the case of dislocations with Burgers vectors other than those parallel to <111> to resolve images which would be consistent with dissociation. It is interesting that some defects are dissociated but others appear to be constricted. It is instructive to examine the possible dissociations of dislocations on a given {110); various possibilities are shown in Figure 4. In the figure, two adjacent plane sections taken parallel to {110) are shown with Mo atoms being depicted by large circles and Si atoms by small diamonds; open and closed symbols indicate the higher and lower planes, respectively. Using this figure, possible dissociation modes for dislocations with **b** given by **b**=1/2<111>, **b**=1/2<331], **b**=<110] and **b**=<001] may be deduced. The various dissociation schemes involve the production of either anti-phase boundaries (APBs), complex stacking faults (CSFs, i.e. when a given Mo atom is sheared to a position such that it sits between both a Mo atom and a Si atom in the adjacent {110) layer) or SISFs (i.e. when a given Mo atom is sheared to a position such that it sits between two Si atoms in the adjacent {1$\bar{1}$0) layer). It can be deduced that, on this slip plane, the following dissociations may be possible:

<110] $\rightarrow$ 1/12<331] + CSF + 1/12<331] + APB + 1/12<33$\bar{1}$] + CSF + 1/12<33$\bar{1}$]

1/2<111> $\rightarrow$ 1/12<331] + CSF + 1/6[001] + SISF + 1/6<001] + CSF + 1/12<331]

1/2<331] $\rightarrow$ 1/12<331] + CSF + 1/12<331] + APB + 1/12<331] + SISF +
1/12<331] + APB + 1/12<331] + CSF + 1/12<331].

Since dissociation of dislocations with **b**=<110] has not been observed, it is reasonable to assume that the energies of APBs and CSFs are very high. In the case of the other two dislocations, while significant separations between the partials bounding the APBs and CSFs are not expected, it is possible that the SISF has a sufficiently low energy to account for the observed dissociation. Thus, the dissociations may be represented by:

1/2<111> $\rightarrow$ 1/4<111] + SISF +1/4<111]

1/2<331] $\rightarrow$ 1/4<331] + SISF + 1/4<331],

and the dissociation of dislocations with **b**=1/2<331] is influenced by the SISF energy rather than the APB energy, as suggested previously (1). An approximate value of the energy of SISFs (γ_{SISF}) on {110) has been determined from the separation of the images of the partial dislocations. This separation has been determined to be 65 Å, and taking the shear modulus to be 208 GPa (4), the γ_{SISF} is found to have a value given by γ_{SISF}=255 mJ.m^{-2}.

SUMMARY / CONCLUSIONS

The main results of the present study may be summarized as follows.

1. Dislocations with **b**=1/2<111> are dissociated into two partial dislocations, each with **b**=1/4<111>, which bound a ribbon of superlattice intrinsic stacking fault.

2. Screw segments of dislocations with **b**=1/2<111> are more widely dissociated than edge segments. The separation of screw segments of partial dislocations each with **b**=1/4<111> on {110) implies a γ_{SISF}= 255 mJ.m^{-2}.

3. A hard sphere model has been used to predict the occurrence of dissociation of dislocations in the compound $MoSi_2$. Dissociation of dislocations is predicted to occur for those defects in which the dissociation involves a ribbon of SISF. The observations of the dissociation of dislocations with **b** = 1/2<111> and **b** = 1/2<331> are consistent with this model.

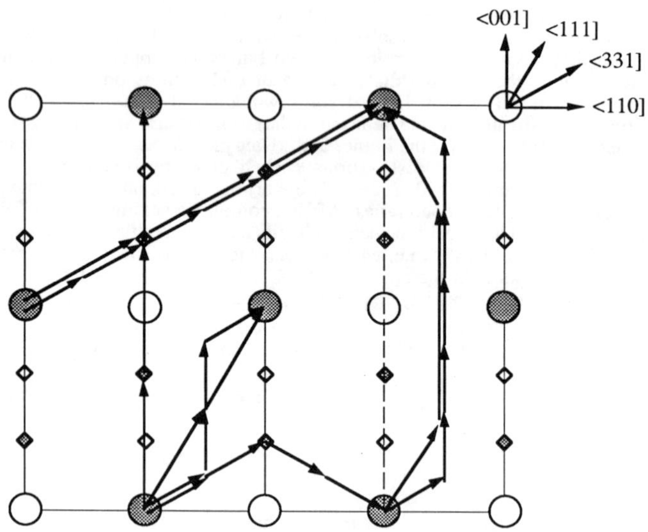

Figure 4 Stacking and dissociation scheme for {110) planes

ACKNOWLEDGEMENTS

This work has been supported in part by the U.S. Office of Naval Research, with Dr. George Yoder as Program Manager, and Wright Laboratory (Wright-Patterson Air Force Base) with Dr. Dennis Dimiduk as Program Manager.

REFERENCES

1. Y. Umakoshi,, T. Sakagami,, T. Hirano, T. Yamane, Acta Metall. Mater., **38**, 909, (1990).
2. O. Unal, J. Petrovic, D.H. Carter, T.E. Mitchell, J. Am. Ceram. Soc., **73**, 1752, (1990).
3. D. J. Evans, S. A. Court, H. L. Fraser, Phil. Mag. Letts, 1992, in press.
4. M. Alouani, R. C. Albers, M. Methfessel, Physical Review B, **43**, 6500, (1991).

DEFORMATION MECHANISMS AND MECHANICAL PROPERTIES OF B2 COMPOUNDS IN Ti-MODIFIED Nb-Al ALLOYS

Jein Shyue, Duen-Huei Hou, Steve Johnson, Mark Aindow*, and Hamish Fraser
Department of Materials Science and Engineering
The Ohio State University, 2041 College Road, Columbus, OH 43210

*Now at School of Metallurgy and Materials, and the IRC in Materials for High Performance Applications, University of Birmingham, Elms Road, Edgbaston, Birmingham B15 2TT, UK.

ABSTRACT

The deformation mechanisms have been determined of two compounds based on Nb_3Al containing additions of Ti (Nb-15Al-10Ti and Nb-15Al-40Ti, in at.%) such that they exhibited the B2 crystal structure. The compounds are strong over a range of temperatures, at least up to 900°C, and are inherently ductile. The compounds deform by activation of one or more of the following slip systems, namely <111>{110}, <111>{112} and <111>{123}. The dislocations are present in the form of dissociated superpartial pairs, each with Burgers vector, **b**, given by **b**=1/2<111>, which bound a ribbon of antiphase boundary(APB). From the separation of the partial dislocations, the APB energy decreases with increasing amounts of Ti.

INTRODUCTION

There has been considerable attention paid to the general area of intermetallic compounds for advanced structural applications. In the main, attention has been focused on the aluminides of Ti and Ni. However, in response to the demand made by advanced propulsion systems, it has been necessary to study other intermetallic compounds with higher melting points and therefore potentially improved mechanical properties. Nb_3Al is one such compound, and this has been the subject of the present study.

Nb_3Al exhibits the A15 crystal structure and is very brittle at ambient temperatures [1]. The origin of this brittle behavior has been shown to involve the limited number of slip systems operating in the compound, namely <001>{210} [2]. In previous work, it has been found that in melt-spun ribbons of binary compositions, which had been heat-treated subsequently, grains of Nb_3Al which had the B2 crystal structure were formed [3]. Similar observations have been reported in the literature [4,5]. It has been shown that the stability of this B2 structure can be enhanced by the addition of Ti [6,7]. The present study is aimed at establishing whether the Ti modified Nb_3Al is indeed inherently ductile, and at determining the deformation mechanisms which operate in these materials.

EXPERIMENTAL PROCEDURE

Two alloys were prepared by transferred-arc plasma melting in an Ar atmosphere on a water-cooled hearth. The nominal compositions of the alloys are Nb-15Al-10Ti and Nb-15Al-40Ti, both in atomic %, and are designated alloy I and II, respectively. Compression and tensile samples were prepared by electro-discharge machining, and tests were carried out at a strain rate of approximately $10^{-4}s^{-1}$. Thin foils for transmission electron microscopy (TEM) were prepared by twin jet electropolishing using 7.5 vol.% H_2SO_4 in methanol at -40°C and 90mA. The specimens were studied in a JEOL 200CX and also a Philips CM12 TEM, operating at accelerating voltages of 200kV and 120kV, respectively

RESULTS AND DISCUSSION

In this study, elementary mechanical properties have been assessed and appropriate deformation mechanisms have been determined. These two aspects are considered in turn.

Mechanical Properties

The first part of this study has involved the determination of elementary mechanical properties, namely yield strengths and ductilities. Yield strengths were determined by performing compression tests over a range of temperatures (room temperature to 950°C) on materials in the as-cast condition or following heat-treatment. Samples of alloy I were treated at 1100°C for 50 hours and those of alloy II at 800°C for 20 hours. The microstructure of the heat treated samples consists of A15 precipitates in a B2 matrix for alloy I [8,9], and very fine precipitates which may be the Ti_2NbAl ortho-phase in a B2 matrix for alloy II [9,10]. The results of these tests are shown in graphical form in Figure 1. The compounds are very strong over a wide range of temperatures; for example, alloy I is comparable in strength with IN 718 at all temperatures [11], and this difference is even greater when density-corrected yield strengths are considered.

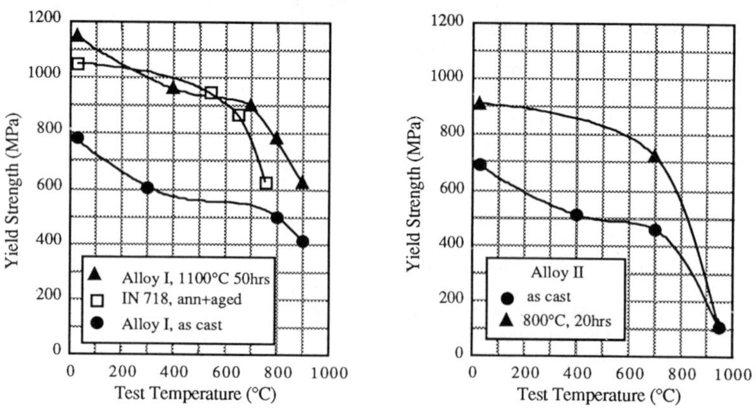

Figure 1. Compressive yield stress for the Nb-15Al-10Ti alloy (alloy I), Nb-15Al-40Ti alloy (alloy II), and IN 718 alloy at different temperatures.

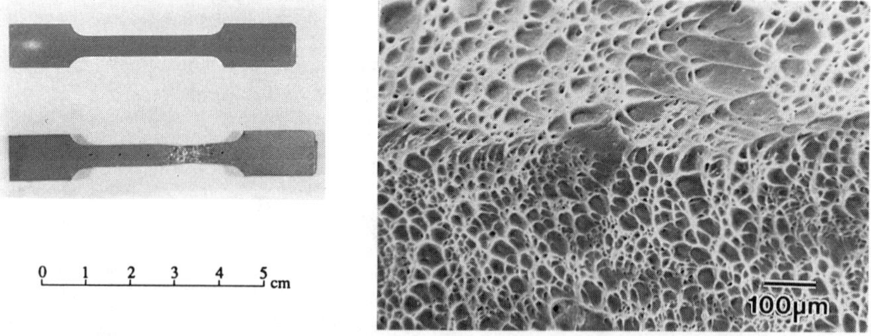

Figure 2. (a) Tensile test specimens of Nb-15Al-40Ti alloy (alloy II) tested at room temperature showing 20% elongation to failure with 1 inch gage length. (b) SEM fractograph taken from the fracture surface of (a).

Tension tests have been performed so that room temperature ductility can be assessed. In the case of alloy II, cold-rolling to a reduction of 40% followed by annealing at 1100°C for 1 hour resulted in an average grain size of ≈ 200µm. Samples of this material were tested in tension,

Figure 2 (a), and elongation to failure of ≈20% with 1 inch gauge length were recorded. The inherent ductility of these samples may be noted from the fractograph shown in Figure 2 (b); clearly, failure occurred in a completely ductile fashion.

Deformation Mechanisms

The deformation mechanisms which operate in samples of both alloy I and II with the B2 structure have been determined. Samples taken from the as-cast ingots were deformed in compression to ≈5% strain at room temperature. The dislocation microstructures are typified by that shown in Figure 3 for alloy I; diffraction contrast experiments have been performed to determine the nature of these defects. The dislocations marked A in Figure 3 (a) are out of contrast in Figure 3 (d and e) with **g**=$1\bar{1}0$ and **g**=$\bar{1}0\bar{1}$, respectively, yielding Burgers vector, **b**, given by **b**=$[\bar{1}\bar{1}1]$. The dislocations marked B in Figure 3 (a) are out of contrast in Figure 3 (b and c) with **g**=$10\bar{1}$ and **g**=110, respectively, yielding **b**=$[1\bar{1}1]$. The slip planes for the dislocations marked A have been identified by determining the line direction of the dislocations using crystallographic analysis and combining this information with the Burgers vector yields slip planes parallel to (101).

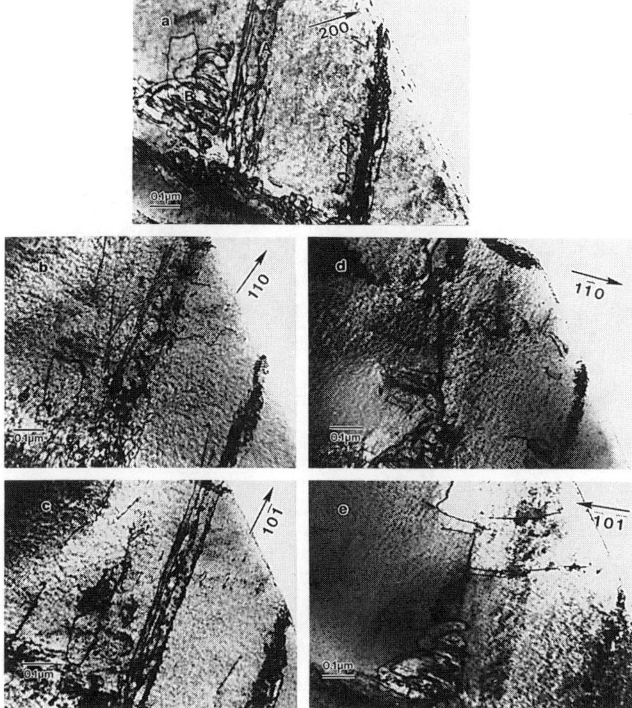

Figure 3. TEM micrographs taken from a thin foil of alloy I with B2 structure and compressed to ≈5% strain at room temperature; the dislocations marked A and B in (a) have Burgers vectors **b** parallel to $[\bar{1}\bar{1}1]$ and $[1\bar{1}1]$ respectively. (b) and (c) show the invisible conditions for dislocations marked with B while dislocations A are out of contrast under diffraction conditions (d) and (e).

Similar experiments have also been performed on alloy II and a typical result is shown in Figure 4. The dislocations in Figure 4 (a) are found to have **b** parallel to [$\bar{1}\bar{1}$1] by the two invisibilities given by **g**=1$\bar{1}$0 and **g**=011 (Figure 4 (c and d)). Furthermore, the dislocations glide on two different slip planes, which are marked A and B in Figure 4 (a) and are identified as (112) and (213), respectively. Examination of many such foils confirmed that glide of dislocations with **b** parallel to <111> on {110}, {112} and {123} is common in both B2 compounds. Also, it has been found that the B2 phase in heat-treated samples of alloy I deforms in this way. This result, namely that the slip systems are <111>{110}, <111>{112} and <111>{123} is completely consistent with the observation that the compound is inherently ductile since many combinations of five independent shears are available and Von Mises criterion is fulfilled.

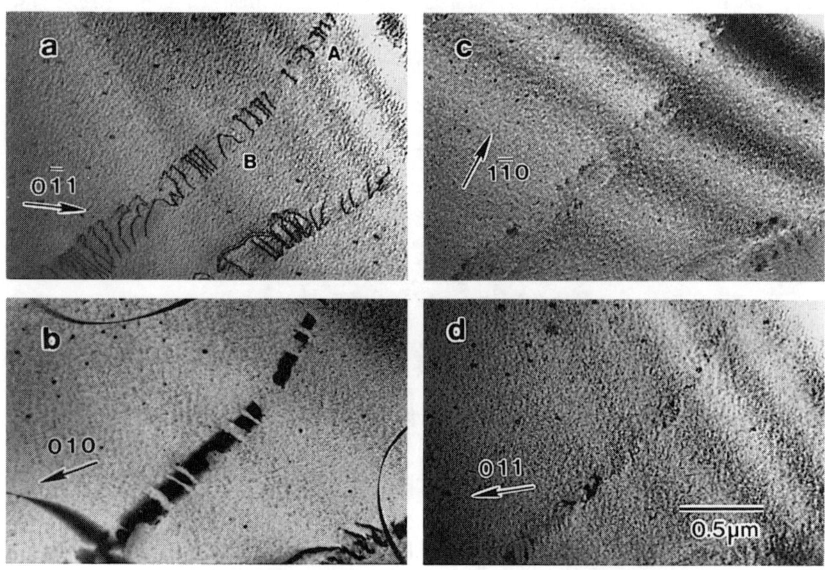

Figure 4. TEM micrographs taken from a thin foil of alloy II with B2 structure compressed to ≈5% strain at room temperature; the dislocations in (a) are with Burgers vectors **b** parallel to [$\bar{1}\bar{1}$1] and glide on two slip planes marked as A and B which are identified as (112) and (213), respectively. (b) is imaged with (010) super-lattice reflection to show the APB bounded by pairs of superpartials. (c) and (d) show the invisible conditions.

It is important to establish whether the dislocations with **b**=<111> are in fact dissociated into pairs of superpartials each with **b**=1/2<111>. This may be determined by making observations such as in Figure 5 involving alloy I. Pairs of dislocations are seen in Figure 5 (a) and (b); the Burgers vectors of these defects are established by the absence of contrast in Figure 5 (c and d), revealing that **b** is parallel to [1$\bar{1}$1], and upon reversing the sign of **g** used to form the images (Figure 5 (a and b)), the spacing and contrast between the dislocations sets changes but the spacing between the individual pairs does not. These contrast effects are consistent with the individual pairs being superpartials with **b**=1/2[1$\bar{1}$1], and so the dislocation pairs are in fact dissociated dipoles. The separation between the superpartial pairs is approximately 100 Å. This is a very large separation and implies a relatively low value of the antiphase boundary (APB) energy, which is consistent with the composition of the sample, and the possible site occupancies, described elsewhere [6].

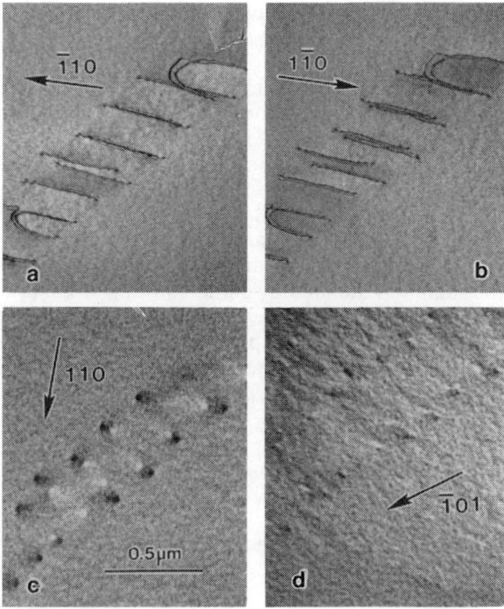

Figure 5. TEM micrographs showing the presence of dissociated dipoles in a specimen of alloy I which had been heat treated at 1500°C for 20 hours and deformed at room temperature.

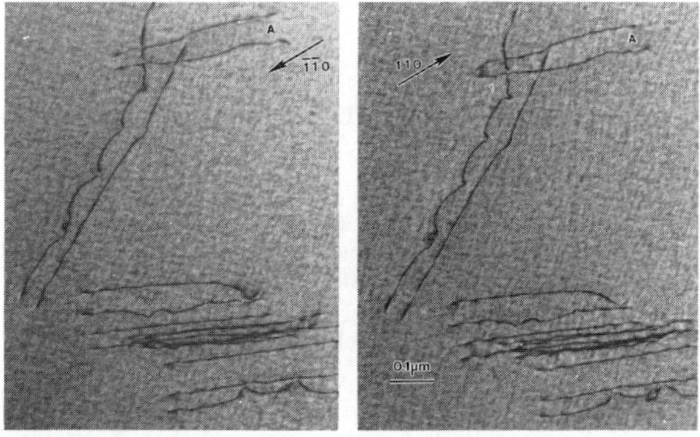

Figure 6. TEM micrographs taken from alloy I compressed to ≈5% at 800°C showing the dynamical dissociation of $[11\bar{1}]$ screw dislocations *in-situ*. The dislocations marked A are dipoles.

If the superdislocations with **b**=<111> are indeed dissociated then it might be expected that the ribbons of APB bounded by the partials might be imaged in dark-field with a superlattice reflection. Such imaging has been attempted using samples of alloy I but without success. This lack of contrast of the APB ribbons may be explained on the basis of the structure factors of the superlattice reflections which are expected to be fairly small in value because of the relatively high concentrations of Nb on each sublattice. However, such small values of the structure factors of reflections in samples of alloy II are not expected, and indeed when dislocations are imaged in this material in dark-field using a superlattice reflection, the APB ribbons bounded by the superpartial dislocations may be observed, Figure 4 (b). The spacing of these superpartial pairs is considerably larger than the corresponding separations in alloy I, and this may imply the APB energy is lower than that in alloy I.

In-situ TEM observations have been made of dislocation motion in thin foils which were taken from heat treated samples deformed at room temperature. The shear stress required to cause dislocation motion was induced by focusing the electron beam on the foil; presumably the stresses arise from heating and subsequent buckling of the foils. These observations show that the dislocations do indeed exist as coupled pairs, supporting the conclusion reached above that the individual Burgers vectors are given by **b**=1/2<111>. More interesting was the observation of the motion of these coupled dislocations. It appears that the leading superpartial dislocation often glides ahead of the trailing superpartial so that the dynamical dissociation of the coupled pair can result in a larger separation than that dictated by static equilibrium. An example of this effect in a sample that had been deformed at 800°C is given in Figure 6. The leading superpartial is bowed out at several points along its length, and the separation of the partials varies quite markedly. It is tempting to speculate that the pinning points are APB domain boundaries. Presumably, it is the relatively low value of the ordering energy that permits the dynamical dissociation to vary in this way.

SUMMARY

In summary, it appears that the B2 phase found in samples of Nb-15Al-10Ti and Nb-15Al-40Ti is inherently ductile. These compounds deform by the activation of the slip systems <111>{101}, <111>{112} and <111>{123}. These slip systems provide sufficient numbers of independent shears to provide for an arbitrary change of shape, and so are consistent with the inherent ductility of the compounds. The dislocations with **b**=<111> are in fact present as dissociated superdislocations, each superpartial having **b**=1/2<111>. The energy of the APB is much lower in the compound with 40% Ti.

Acknowledgments

This work has been supported by the Office of Naval Research, Dr. George Yoder as Program Manager.

References

1. K. Yasuda.T. Fujiwara, H. Kodoma and M. Suwa, in High Temperature Ordered Intermetallic Alloys IV, edited by L.A. Johnson, D.P. Pope and J.O. Stiegler (*Mater. Res. Soc. Proc.*, **213**, Pittsburgh, PA, 1991) pp.865
2. M.Aindow, J. Shyue, T. A. Gaspar and H.L. Fraser, *Phil. Mag. Lett.*, **64**, 59 (1991)
3. H. Kohmoto, J. Shyue, M. Aindow and H.L. Fraser, submitted to *Scripta Metall.* (1992)
4. V.M. Pan, V.I Latysheva, Ye. N. Khusid, V.N. Minakov and Ye. V. Turtsevich, *Physics of Metals and Metallography*, **40,** 52(1975)
5. T.N. Marieb, A.D. Kaiser, S.R. Nutt, D.L. Anton and D.M. Shah, in High Temperature Ordered Intermetallic Alloys IV, edited by L.A. Johnson, D.P. Pope and J.O. Stiegler (Mater. Res. Soc. Proc., **213**, Pittsburgh, PA, 1991) pp.329
6. J. Shyue, D.-H. Hou, M. Aindow and H.L. Fraser, submitted to *Phil. Mag. Lett.* (1992)
7. E.S.K. Menon, P.R. Subramanian and D.M. Dimiduk, *Scripta Met.all* , **27**, 265 (1992)
8. J. Shyue, Ph. D. Thesis, The Ohio State Univ. (1992)
9. J. Shyue, D.-H. Hou, S.C. Johnson, M. Aindow and H.L. Fraser, these proceedings.
10. L.A. Bendersky, W.J. Boettinger and A. Roytburd, *Acta Metall.*, **39**, 1959 (1991)
11. Metals Handbook (Desk Edition), Eds. H.E. Boyer and T.L. Gall, ASM, 16.15(1985)

EFFECTS OF STRAIN RATE AND PRESTRAINING ON TENSILE BEHAVIOR OF DUPLEX GAMMA TITANIUM ALUMINIDES

D. S. SHIH, D. S. SCHWARTZ, and J. E. O'NEAL
McDonnell Douglas Aerospace, MC 111 1041, P. O. Box 516, St. Louis, MO 63166-0516

ABSTRACT

The effects of strain rate and prestraining on tensile behavior of two-phase ($\gamma+\alpha_2$) titanium aluminides at 20 and 730°C have been investigated. At 20°C the elongation remains at about 1.4% as the strain rate increases from $5x10^{-5}$ to $5x10^{-2}$ s^{-1} and it drops to nearly zero at $5x10^{-1}$ s^{-1}. At 730°C (*i.e.* above DBTT) the plastic strain is about 13% when tested at $5x10^{-5}$ s^{-1}, while it reduces significantly to less than 3% at $5x10^{-4}$ and $5x10^{-2}$ s^{-1}. Again, the elongation is about zero at the highest strain rate tested, $5x10^{-1}$ s^{-1}. Regardless of the strain rate, fracture by an intergranular mode of the primary equiaxed γ appears to increasingly dominate as temperature changes from 20 to 730°C. Introduction of prior plastic deformation by prestraining beyond yielding at 945°C obviously increases the 20°C yield stress, however, with little influence on ductility. Transmission electron microscopy reveals that a number of dislocation loops are produced during prestraining. These loops are generally immobile resulting in the observed increase of flow stress and unchanged ductility.

INTRODUCTION

For the past 5-6 years, research and development efforts on two-phase ($\gamma+\alpha_2$) titanium aluminides have made a significant progress in advancing the technology. The progress is mostly reflected on alloy development, microstructure-property relationship, and deformation behavior [1-4]. While the mechanical behavior of two-phase ($\gamma+\alpha_2$) titanium aluminides is believed to be very sensitive to strain rate, limited work has been conducted and the understanding is far from being complete.

It is well established that a ductile-brittle transition temperature (DBTT) exists in tensile behavior of two-phase ($\gamma+\alpha_2$) titanium aluminides; the DBTT typically ranges between 550 between 750°C depending strongly on alloy chemistry and microstructure [1]. For a duplex microstructure, the DBTT is generally about 650°C [2]. Furthermore among other factors, strain rate is considered to have an impact on DBTT. An increased strain rate tends to increase the friction stress, which should raise the DBTT. For steels the DBTT increases by about 15°C for each order magnitude of increase in strain rate [5]. Therefore, one purpose of this study is to investigate the effects of strain rate on tensile behavior at room temperature and at 730°C (which is above the DBTT under a normal strain rate of $1\text{-}5x10^{-4}$ s^{-1}) of a binary near γ-TiAl alloy having a duplex microstructure.

Introduction of prior plastic deformation at 945°C could possibly result in two beneficial effects in γ-TiAl based aluminides. Firstly, it may increase room temperature ductility by activating normally immobile slip systems, *e.g.* on non-{111) planes. Secondly, prestraining may promote strain hardening by increasing specific fracture energy, thereby lowering DBTT. The first effect will be examined in the present study, while the second effect will be addressed in a separate publication.

EXPERIMENTAL PROCEDURES

The starting Ti-48 at.% Al ingots were made using a vacuum induction skull melting method at Duriron, Inc. Each ingot weighed about 16 kg and had dimensions of 70 mm in diameter and 900 mm in length. The materials also contained about 0.062 wt% oxygen and 0.016 wt% carbon. Prior to hot-working ingots were hot-isostatic-pressed at 1175°C/105

MPa/3 h. The ingots were then extruded at 1340°C (in the α+γ phase field), at a reduction ratio of 10:1 (*i.e.* a true strain of ≈230%), and a strain rate of about $1x10^{-1}$ s^{-1}.

The α-transus of this alloy composition is approximately 1380°C. Therefore in order to produce a duplex microstructure, all the hot-worked materials in this study were recrystallized at 1300°C for 2 hours in an inert atmosphere, followed by furnace-cooling, and then aged at 900°C for 24 hours prior to furnace-cooling. Tensile tests on round-gauge specimens were carried out at 20 and 730°C in air at four different strain rates: $5x10^{-5}$, $5x10^{-4}$, $5x10^{-2}$, and $5x10^{-1}$ s^{-1}. Additionally, some specimens were prestrained to introduce some initial plastic deformation prior to 20°C tensile tests at three strain rates: $5x10^{-5}$, $5x10^{-4}$, and $5x10^{-2}$ s^{-1}. Tests at $5x10^{-1}$ s^{-1} are continuing. The prestraining was performed beyond yielding to about 3.4% plastic strain at a strain rate of $5x10^{-4}$ s^{-1} at 945°C in a flowing argon environment. Fracture and deformation morphologies were characterized using SEM and TEM.

RESULTS

Microstructure

Figure 1 shows the optical microstructure of the Ti-48Al alloy in the recrystallized (1300°C/2h/FC + 900°C/24h/FC) condition. Since both of the extrusion and recrystallization temperatures are within the γ+α phase field, the structure is of a duplex type that consists of primary equiaxed γ grains (≈ 25 μm) and lamellar grains (≈ 40 μm) coantaining of γ and α_2 platelets.

Figure 1. The duplex structure of the recrystallized Ti-48Al.

Strain Rate Effects at 20°C and 730°C

The relationships between yield stress, elongation, and strain rate at 20°C and 730°C are shown in Figures 2 and 3, respectively. At 20°C the yield stress slightly increases with an increasing strain rate up to $5x10^{-2}$ s^{-1} and the plastic strain basically remains at about 1.4%. Nevertheless at the strain rate of $5x10^{-4}$ s^{-1}, the yield stress raises to 532 MPa concomitant with a sharp decline in elongation to about 0.05%. Similar relationships exist for the 730°C tensile behavior: increased strain rates raise yield stress and lower elongation with a large decrease at $5x10^{-1}$ s^{-1}. Additionally, there is a significant decrease in elongation (from 13.2% to 2.8%) when the strain rate is increased from $5x10^{-5}$ s^{-1} to $5x10^{-4}$ s^{-1}.

Post-fracture examination of the fracture surfaces shows that fracture essentially proceeds at 20°C by a mixed mode of transgranular fracture of γ grains and ligament shearing / interface separation of lamellar grains, Figure 4. On the other hand, the primary γ constituent fractures predominantly in an intergranular fashion at 730°C, especially at high strain rate, Figure 5. It is also observed that there is a large amount of localized plasticity in the intergranularly-failed grains, Figure 6. Qualitatively differences in fracture morphology due to strain rate variations are insignificant at either 20 or 730°C, despite of the large decrease in elongation at $5x10^{-1}$ s^{-1} strain rate.

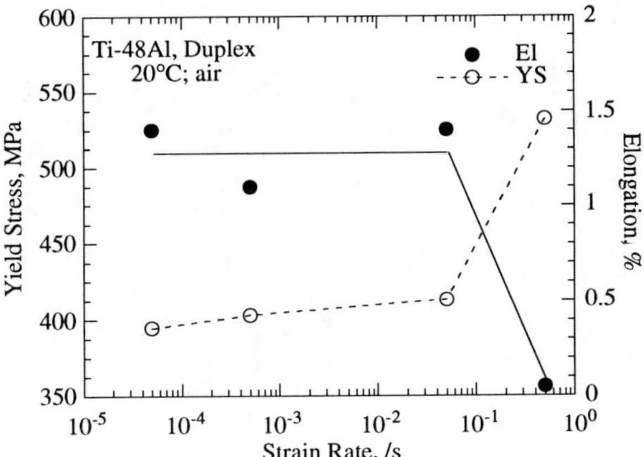

Figure 2. The yield stress and elongation as a function of strain rate at 20°C.

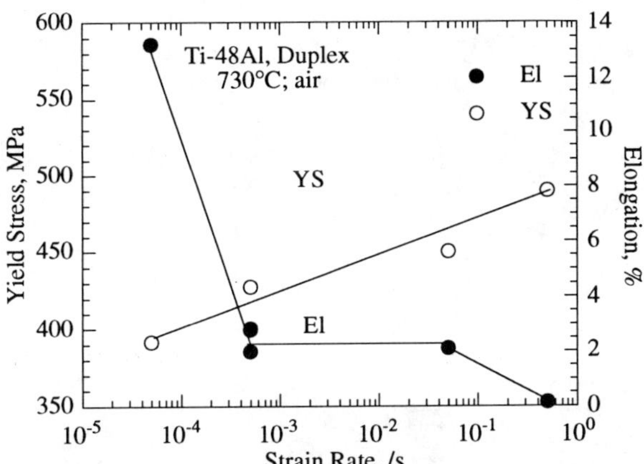

Figure 3. The yield stress and elongation as a function of strain rate at 730°C.

The strain rate sensitivity, m, can be obtained form the power relationship between yield stress, σ_{YS}, and strain rate, $\dot{\varepsilon}$, according to Eq. (1):

$$\sigma_{YS} = C\,(\dot{\varepsilon})^{m} \qquad (1)$$

where C is a constant. The m values are 0.027 and 0.023 for the 20°C and 730°C behavior, respectively.

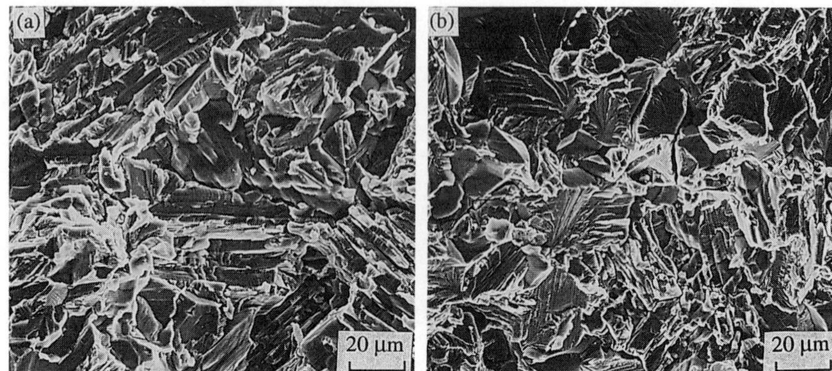

Figure 4. The fracture morphology at 20°C for (a) $5x10^{-5}$ s^{-1} and (b) $5x10^{-1}$ s^{-1} strain rates.

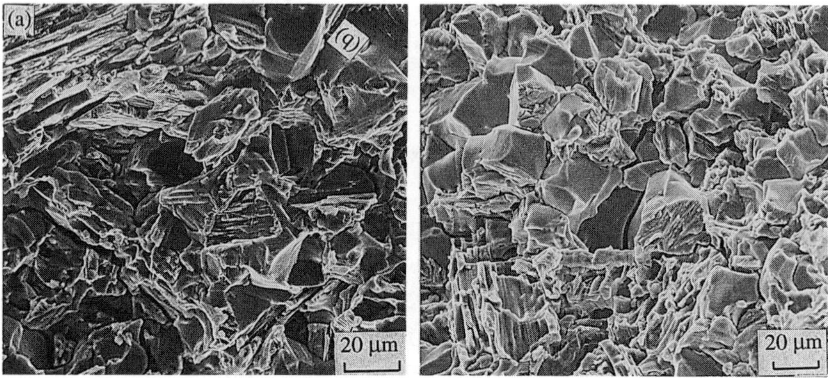

Figure 5. The fracture morphology at 730°C for (a) $5x10^{-5}$ s^{-1} and (b) $5x10^{-1}$ s^{-1} strain rates.

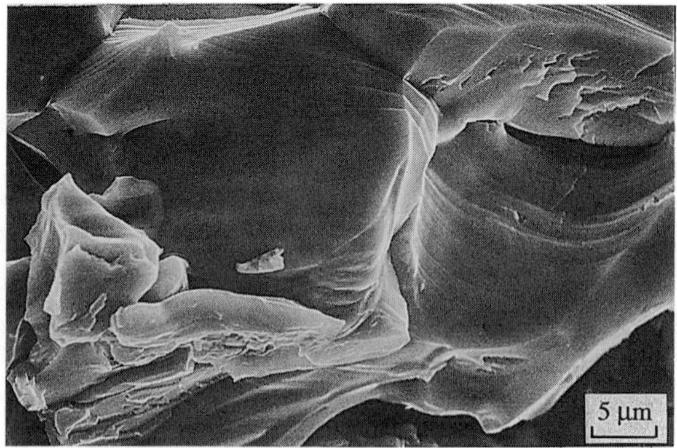

Figure 6. An SEM micrographs showing extensive localizedplasticity in the intergranularly-failed regions at 730°C and $5x10^{-2}$ s^{-1}.

Prestraining Effects at 20°C

Prior plastic deformation at 945°C in a flowing argon atmosphere results in an enhanced yield stress; greater than 15% increase compared to that for the un-prestrained materials at corresponding strain rates, Figure 7. The elongation of the prestrained materials indicates small improvement except at the strain rate of 5×10^{-5} s^{-1} where a reduced plastic strain and a greatly increased (by 33%) yield stress are noted. The fracture morphology is similar to that for the un-prestrained materials.

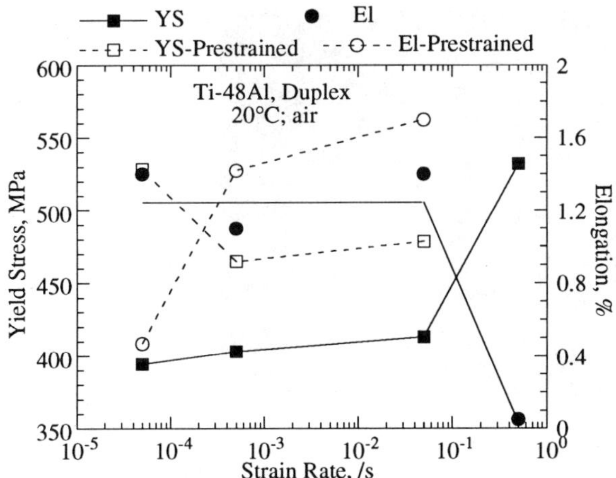

Figure 7. A comparison of the yield stress and elongation as a function of strain rate at 20°C as affected by 945°C prestraining.

DISCUSSION

The results clearly show that strain rate has a harmful effect on tensile ductility of two-phase ($\gamma+\alpha_2$) titanium aluminides Ti-48Al, especially at strain rate higher than 5×10^{-1} s^{-1} both at 20°C and 730°C. Strain rate also raises the yield stress gradually up to 5×10^{-1} s^{-1} where a large increase occurs. While the difference in fracture morphology is negligible on the scanning electron microscopy scale, the harmful effect may be explained by a critical crack size argument. The critical crack size for cleavage fracture would decrease with an increasing strain rate, thus during high strain rate deformation, there may be numerous critical-sized cracks available to nucleate a truly brittle failure.

At 730°C which is normally above the DBTT of the Ti-48Al duplex microstructure, strain rate plays an even more important role. As pointed out by the Davidenkov diagram [6], a higher strain rate, causing to raise flow stress, will increase DBTT. However, a quantitative relationship between strain rate and DBTT enhancement has not been determined.

It is also noted that the fracture mode of the primary equiaxed γ grains changes from a transgranular cleavage at 20°C to an intergranular mode at 730°C. Moreover, a large amount of plastic deformation is often associated with the intergranular fracture. The localized plastic deformation within the γ grains may cause dislocation pile-ups against grain boundaries, which will either further weaken the grain boundaries or nucleate crack along boundaries resulting in an intergranular fracture.

The strain rate sensitivity values at 20 and 730°C are about the same, and very small (0.027 and 0.023). This may suggest that even at 730°C dislocation activity is not as important as critical crack size in determining the mechanical behavior at various strain rates. At higher temperatures (>1000°C), Shih and Scarr [7] have shown that the strain rate sensitivity increases to 0.16-0.28 and that a great deal of dynamic recrystallization occurs.

There are four possible effects of introducing prior plastic deformation at 945°C in γ-TiAl: 1) nucleate normally immobile slip systems to increase low-temperature ductility, 2) increase specific fracture energy by causing strain-hardening, thereby lower DBTT, 3) increase flow stress, therefore raise DBTT, and 4) raise triaxial stress state, therefore increase DBTT. The first effect will be discussed here, while the rest will be addressed in another article. No additional operating slip systems are observed at 20°C due to the 945°C prestraining. However, a large number of dislocation loops were produced during the elevated temperature prestraining, Figure 8. The loops are expected to be rather immobile, especially at low temperatures. Therefore, the results of increased flow stress along with unchanged elongation are obtained at 20°C.

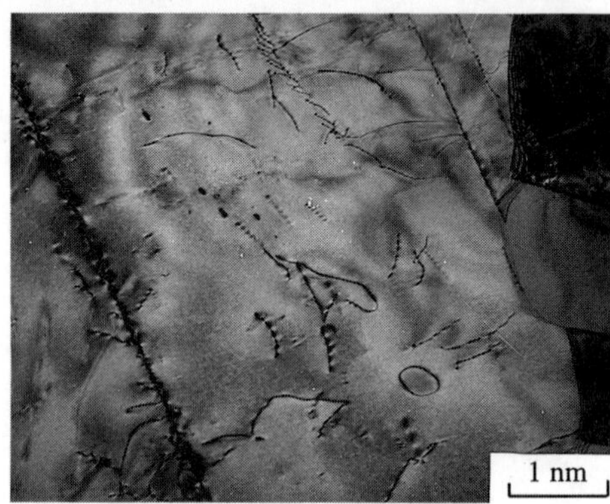

Figure 8. A TEM micrograph showing the dislocation loops in the primary γ grains produced during the 945°C prestraining.

CONCLUDING REMARKS

1. At 20°C yield stress increases, but elongation remains the same with an increasing strain rate up to $5x10^{-2}$ s^{-1}. At 730°C an increased strain rate raises yield stress, but degrades elongation. At the highest strain rate tested, $5x10^{-1}$ s^{-1}, ductility effectively drops to zero and yield stress rises sharply.
2. No additional slip systems are activated at 20°C due to the 945°C prestraining beyond yielding. No improvement on 20°C ductility by prestraining is realized. However, the yield stress shows a marked enhancement.
3. Regardless of strain rate, fracture of the primary γ grains proceeds transgranularly at 20°C and intergranularly at 730°C.

ACKNOWLEDGMENTS

This research is performed under the McDonnell Douglas Independent Research and Development program. We appreciate the assistance that R.J. Lederich, J.D. Keyes, and J.J. Evans have provided.

REFERENCES

1. S-C. Huang and D.S. Shih, in Microstructure/Property Relationships in Titanium Aluminides and Alloys, edited by Y-W. Kim and R.R. Boyer, TMS, p. 105 (1991).
2. D.S. Shih, S-C. Huang, G.K. Scarr, H. Jang and J.C. Chesnutt, *ibid*, p. 135.
3. Y-W. Kim and D.M. Dimiduk, JOM, **43**, 40 (1991).
4. H. Inui, A. Nakamura, M.H. Oh and M. Yamaguchi, Phil. Mag., **66**, 557 (1992).
5. F.A. McClintock and A.S. Argon, editors, Mechanical Behavior of Materials, Addison-Wesley Pub Co., p. 560 (1966).
6. N.N. Davidenkov, Diannicheskaya Ispytania Metallov, Moscow (1936).
7. D.S. Shih and G.K. Scarr, High-Temperature Ordered Intermetallic Alloys IV, edited by L.A. Johnson, D.P. Pope, J.O. Stiegler, MRS Symp. Proc., **213**, 727 (1991).

THE BEHAVIOR OF INTERMETALLIC COMPOUNDS AT LARGE PLASTIC STRAINS

GEORGE T. GRAY III* AND J. DAVID EMBURY**
* Los Alamos National Laboratory, Los Alamos, New Mexico 87545
**McMaster University, Materials Science Department, Hamilton, Canada.

ABSTRACT

Much effort has been devoted to the study of ordered materials at modest plastic strains and the problem of premature failure. However by utilizing stress states other than simple tension it is possible to study the deformation of intermetallic compounds up to large plastic strains and to consider the behavior of these materials in the regime where stresses approach the theoretical stress. The current work outlines studies of the work hardening rate of a number of titanium and nickel-based intermetallic compounds deformed in compression. Attention is given to the structural basis of the sustained work hardening. The large strain plasticity of these materials is summarized in a series of diagrams. Fracture in these materials in compression occurs via catastrophic shear at stresses of the order of E/80 (where E is the elastic modulus).

Introduction

Much effort has been devoted to the elucidation of detailed dislocation configurations in a variety of ordered compounds and to the temperature dependence of the flow stress. However to date little attention has been paid to the behavior of these materials at large plastic strains. In large part this reflects the limited ductility of many of these compounds in tension. In the current work we have considered the plasticity of ordered intermetallics in terms of their response to compressive deformation. The objective of this work can be summarized by considering the deformation history in the following sequence of events.

a) How do intermetallics differ from pure metals in terms of their ability to sustain work hardening?

b) What stress levels can be attained in intermetallic compounds after large strains and what configurations of lattice defects are required to attain these stress levels?

c) At large plastic strains what are the essential competitive processes between continuing plasticity and fracture?

Clearly in the limitations of a conference paper only a basic outline of these aspects can be covered. This paper contains a summary of a broad study of intermetallics which includes the following materials, Ni_3Al, Ti-48Al-1V, Ti-24Al-11Nb, Ti-48Al-2Cr-2Nb, and Ti-24.5Al-10.5Nb-1.5Mo.

Experimental Procedure

An experimental program was conducted to examine the deformation and work-hardening response of a number of Ti-aluminides and Ni_3Al deformed to large strains in compression. The essential microstructural features and initial yield stresses of these materials are summarized in Table I. The quasi-static stress-strain response of the materials was measured using an Instron machine for strain rates in the range 10^{-3} to 0.1 s^{-1}. The dynamic constitutive response, strain rates of 1000 to 8000 s^{-1}, was measured with a Split-Hopkinson pressure bar. Testing to large true strains was accomplished using repeated reloading of samples with intermediate relubrication and remachining. Specifics of the mechanical testing samples and procedures have been detailed previously[1-3].

Results and Discussion

The stress-strain responses of the intermetallics studied are shown in Figures 1 and 2. All the intermetallics exhibited extended work-hardening to high stress levels of order 1.5 to 2 GPa.

The stress levels achieved at compressive strains > 0.25, normalized with respect to the elastic modulus[**E**], are E/105 for Ni_3Al and E/68 for Ti-24Al-11Nb deformed quasi-statically and E/80 for Ti-48Al-2Cr-2Nb deformed dynamically. The rates of sustained hardening are seen to be surprisingly similar in all the intermetallic compounds. Normalizing the quasi-static hardening rate θ with the Taylor Factor, **M**, for a random polycrystal, $[\theta/(3.07)^2]$, yields a strain hardening rate of ~μ/175 for Ti-48Al-1V and Ti-48Al-2Cr-2Nb, μ/130 for Ni_3Al, and μ/165 for Ti-24.5Al-10.5Nb-1.5Mo, where μ is the shear modulus.. All of these hardening rates are consistent with Stage-II hardening in FCC metals and also that typically exhibited by α+β-Ti-alloys[1-3]. Defect storage at large strains in Ni_3Al, Ti-24Al-11Nb, and Ti-48Al-2Cr-2Nb was terminated by catastrophic shear failure.

Table I.

Material	Composition (atomic %)	Microstructure	Yield Stress (MPa)
Ti-48Al-1V	Ti + 48 Al, 1.07 V 0.2 C, 0.07 O_2	equiaxed γ (50μm) + lamellar α_2 / γ	650
Ni_3Al	75.9 Ni, 24.1 Al 0.095 B	equiaxed (40 μm)	240
Ti-24Al-11Nb	Ti + 24.6 Al, 10.5 Nb 0.016 O_2	elongated α_2 + transformed β	600
Ti-24.5Al-10.5Nb-1.5Mo	Ti + 23.8Al, 10.6 Nb 1.43 Mo, 0.19 O_2	equiaxed α_2 + fine secondary α_2 & β	900
Ti-48Al-2Cr-2Nb	Ti + 47.8 Al, 2 Cr, 1.7 Nb 0.2 O_2, 0.1 C	primary γ (~20 μm)+ α_2 / γ lamellae (~30-40 μm)	350

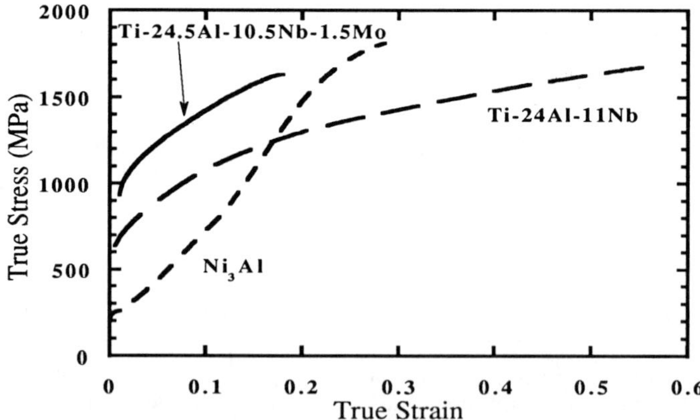

Figure 1: Compressive stress-strain response of Ni_3Al, Ti-24Al-11Nb, and Ti-24.5Al-10.5Nb-1.5Mo at a strain rate of 0.001 s^{-1} at 298K.

Examination of the defect storage in Ti-48Al-1V deformed quasi-statically to a true strain of 0.25 revealed a high density of deformation twins in the equiaxed primary γ and the γ laths and a high density of dislocations in both the γ and α_2 phases. Many of the twins are seen to be curved when viewed edge-on in a <101> zone axis. The observation of a large number of bent twins and the presence of large densities of dislocations between the twin lamellae indicate that both slip and twinning contribute to the deformation and that the twinning may serve to subdivide the plastic regions and hence contribute to the overall strength level.

In discussing the behavior of intermetallics at large plastic strains it is important to emphasize that the current work represents only an initial view. Much detailed TEM work remains to be done before any detailed mechanistic base for the mechanical property observations can be proposed. However, three salient features emerge from the current work. Firstly, most of the intermetallic systems in this study are able to attain strength levels of order 1.5 - 2 GPa if fracture processes are suppressed by the use of a compressive deformation mode. Second, the rate of work hardening is sustained over a large strain range due to the suppression of dynamic recovery. Finally, at very high stresses the intermetallics studied fail by catastrophic shear rather than by a damage accumulation mode involving grain boundary or cleavage failure.

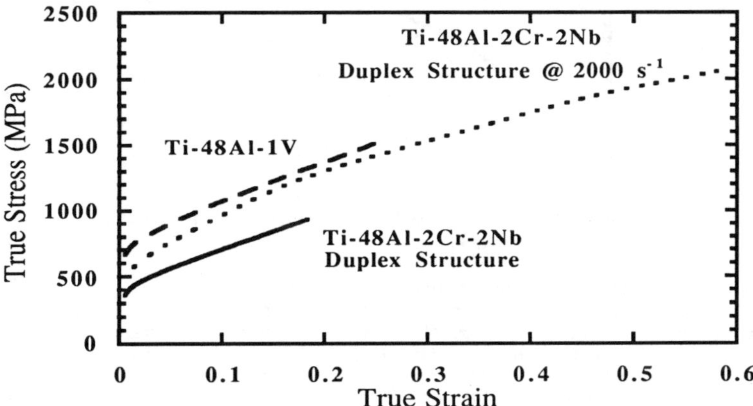

Figure 2: Compressive stress-strain response of Ti-48Al-1V and Ti-48Al-2Cr-2Nb.

When discussing the work hardening behavior it is useful to refer to the formalism developed by Mecking and Estrin[4] in which the current hardening rate θ can be related to the initial hardening rateθ_{II} (which for polycrystals at low $\mathbf{T / T_m}$ can expressed as $\mathbf{M^2}\ \theta_{II}$ where θ_{II} is the stage II single crystal hardening rate, nominally = μ / 200 for FCC metals, and **M** is the Taylor factor = 3.0).

The expression used in this relationship is of the form:

$$\theta = \theta_{II}\,(1 - \sigma / \sigma_{sat}) \qquad [1]$$

where σ is the current flow stress and σ_{sat} a hypothetical saturation stress at which the rate of dynamic recovery is equal to the rate of hardening. This formalism can be expressed in $\sigma\theta$ versus σ plots for the intermetallics in this study as shown in Figures 3-5. Comparison of these plots reveals that the work-hardening response differs between the Ni_3Al, and α_2 and γ -based intermetallics. Ni_3Al, Ti-48Al-1V, and Ti-48Al-2Cr-2Nb each exhibit increasing $\theta\sigma$ vs. σ plots indicating sustained defect storage, i e., the material behaves as though Stage II hardening continues to high stresses. Ti-24Al-11Nb and Ti-24.5Al-10.5Nb-1.5Mo on the contrary show gradually decreasing hardening rates although both alloys reach moderately high absolute flow stresses. This is particularly pronounced in the Ti-24Al-11Nb alloy which sustained continued hardening, although at a decreasing rate, to a true strain of 0.55. The hardening response of the Ni_3Al and the Ti-48Al-2Cr-2Nb deformed dynamically display hardening responses suggesting both Stage-II and Stage-III behavior. Initially they show increasing hardening rates followed by sustained hardening but at a decreasing rate. In both instances the absolute flow stress levels reached are approaching 2 GPa. At these stress levels plastic flow is terminated by catastrophic shear rather than by any distributed damage mechanism such as grain boundary cracking, void formation, or cleavage cracks.

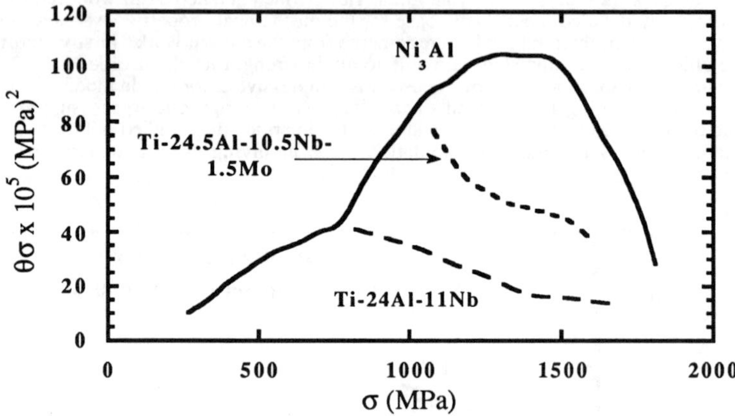

Figure 3: A diagram of σθ versus σ based on compression data for Ni3Al, Ti-24Al-11Nb, and Ti-24.5Al-10.5Nb-1.5Mo.

While it is difficult to compare the behavior of ordered intermetallics at large strains with pure metals several insightful observations are evident from such a comparison. In a simplistic sense this can be considered in terms of a very high value of the saturation stress, σ_{sat}. In Ni3Al sustained hardening at quasi-static rates under ambient conditions is observed to stresses approaching 2 GPa. In contrast pure polycrystalline nickel exhibits a saturation stress, i e, asymptotic saturation of the flow stress, at a stress level of approximately 450 MPa. Given the nearly identical elastic and shear modulii of Ni and Ni3Al the approximately 4x increase in the "saturation stress" between the two materials clearly demonstrates that drastically different dynamic recovery mechanisms are controlling the transition from Stage II to III in these materials.

Clearly the details of dislocation storage in ordered intermetallics are very different from disordered FCC pure copper or nickel however it is instructive to compare the form of the θσ versus σ plots. In the intermetallics the stress levels after large plastic strains are much higher and the period of nearly linear hardening is extended. This may reflect the difficulty of concomitant dislocation annihilation due to the dislocation configuration in ordered materials. In dynamic recovery some form of short range dislocation motion either by cross-slip or the mechanical interaction of edge segments must result in removal of dislocation line length. The difficulty of this process is reflected in the magnitude of σ_{sat} in equation 1.

In addition in ordered structures the basic mechanism of dislocation accumulation may depend more on the core configuration of the dislocation and dislocation trapping via the associated anti-phase domains, such as Kear-Wilsdorf locks in Ni3Al, than by the interaction of dislocations in the cell wall structure as in the case of nickel or copper. Thus the dynamic recovery events appear to be very different in ordered structures and the local stresses needed to promote annihilation may be very high (i e., significant fraction of the theoretical strength). However, in the absence of a detailed mechanism for dynamic recovery in the ordered intermetallics it is sufficient to state that the rate of dynamic recovery is very low. In intermetallics this results in extension of the strain hardening and the attainment of stress levels after large plastic strains which are order E / 50.

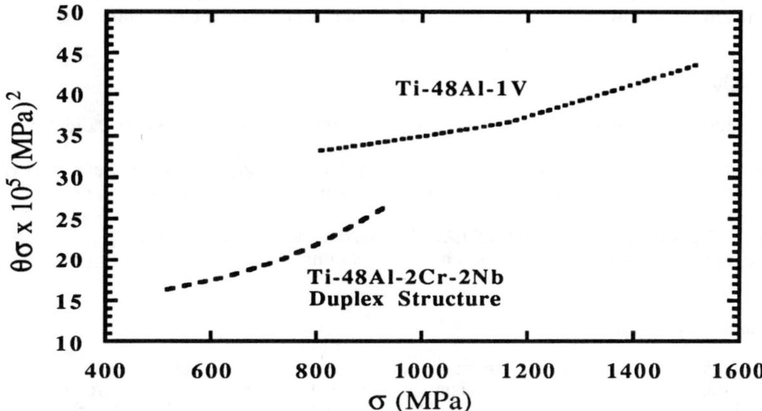

Figure 4: A diagram of σθ versus σ based on compression data for Ti-48Al-1V and Ti-48Al-2Cr-2Nb

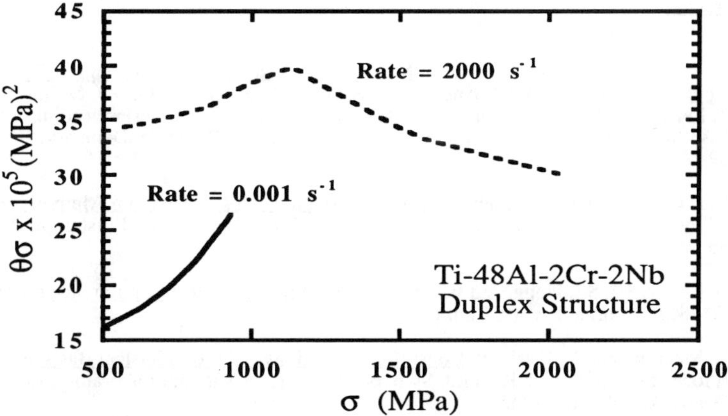

Figure 5: A diagram of σθ versus σ based on compression data for Ti-48Al-2Cr-2Nb deformed at low and high strain rate.

An exception to this behavior is the data for the two α_2 -based intermetallics which have a large volume fraction of β-phase. In this case the rate of hardening decreases over a wide range of strain but at a slow rate. Thus it would be of value to measure the large strain properties of single-phase intermetallics to examine how the process of dynamic recovery is related to crystal structure and the detailed dislocation configuration.

The scenario outlined above gives only a very preliminary view and a number of other factors may play an important role. For example, in Figure 5 it is clear that the hardening rate in Ti-48Al-2Cr-2Nb is also dependent on the applied strain rate. This may reflect the increased rate of twin production which is observed at high strain rate and/or low temperature in this intermetallic. These twins may assist the creation of sessile dislocations from the existing structure or simply reduce the mean-free-path and hence raise the hardening rate.

Conclusions

The current study of the large strain plasticity of intermetallic compounds indicates:

a) Intermetallics deformed in compression exhibit prolonged work hardening to very high stress levels, E/80, due to the suppression of dynamic recovery processes.

b) At stress levels of the order of E/80 localized shear becomes a limiting process in ordered compounds deformed in compression to large strains.

Acknowledgments

This work was supported under the auspices of the United States Department of Energy. The authors acknowledge the assistance of M.F. Lopez and W. Wright for conducting the mechanical tests. One of the authors (JDE) is grateful for the award of a Mattias fellowship at Los Alamos National Laboratory where this paper was prepared. The authors gratefully acknowledge GE Aircraft Engines for providing the Ti-24.5Al-10.5Nb-1.5Mo and Ti-48Al-2Nb-2Cr used in this study.

References

1. J.D. Embury and G.T. Gray III, in Modeling of Plastic Deformation and Its Engineering Applications, edited by S.I. Anderson, J.B. Bilde-Sorensen, N. Hansen, D. Juul Jensen, T. Leffers, H. Lilholt, T. Lorentzen, O.B. Pedersen, and B. Ralph (Proceedings of 13th Riso Int. Sym. on Matls. Sci., Riso National Laboratory, Roskilde, Denmark, 1992) pp. 39-56.

2. G.T. Gray III, in Microstructure / Property Relationships in Titanium Aluminides and Alloys, edited by Young-Won (Y-W.) Kim and R.R. Boyer (TMS, Pittsburgh, PA, 1991), pp. 263-274.

3. G.T. Gray III, S.I. Hong, and B.J. Marquardt, in Seventh World Conference on Titanium, (TMS, Pittsburgh, PA, 1993) in press.

4. H. Mecking and Y. Estrin, in Constitutive Relations and their Physical Basis, edited by (Proceedings of the 8th Riso Int. Sym. on Matls., Riso National Laboratory, Roskilde, Denmark, 1987) pp. 123.

SYNCHROSHEAR OF LAVES PHASES

P.M.HAZZLEDINE[1], K.S.KUMAR[2], D.B.MIRACLE[3] and A.G.JACKSON[1]

1. UES Inc, 4401 Dayton-Xenia Road, Dayton, OH 45432
2. Martin Marietta Laboratories, 1450 South Rolling Road, Baltimore, MD 21227
3. Wright Laboratory, Materials Directorate, Wright-Patterson AFB, OH 45433.

ABSTRACT

The three Laves phases consist of alternating single layers and triple layers of atoms. Shear of the structure within the triple layers may be achieved by moving synchrodislocations. The dislocation with the smallest Burgers vector is the synchroshockley a/6<112> which has a core split over two planes. If a synchroshockley sweeps every triple layer of the cubic C15 it is twinned, if it sweeps every other triple layer, C15 is transformed into hexagonal C14. If the synchroshockley sweeps two triple layers, leaves out two, sweeps two etc. C15 is transformed into C36. Synchroshockleys travelling in pairs in any of the structures form dissociated perfect dislocations capable of giving slip.

INTRODUCTION

The Laves phases form the most numerous group of intermetallic compounds. Some of them have promisingly high melting temperatures and contain elements which could provide oxidation resistance. Ductility has been observed at low temperatures [1] and two-phase alloys containing Laves phase have some attractive mechanical properties [2]. Often Laves phases are hexagonal at high temperatures and transform to a cubic structure at low temperatures [3]. Electron microscopy shows crystals which are heavily faulted [3] or twinned [1] on the close packed planes. The relationship between the three crystal structures [4] and dislocation models for shear transformations [5] are both generally described by means of shears on these planes. However, it is important to realise that ordinary shear cannot occur in Laves phases and that synchroshear is necessary instead. In this paper we use the concept of synchroshear [6] and the established notation [3] to give a unified description of the crystal structures, phase transformations, twinning and slip in Laves phases.

LAYERED STRUCTURE OF LAVES PHASES

The three Laves phases C14, C15 and C36 (whose archetypes are $MgZn_2$, $MgCu_2$ and $MgNi_2$ respectively) have ideal chemical compositions S_2L containing small atoms S and large atoms L with an ideal radius ratio of $\sqrt{3/2}$. The crystal structures are layered and have just two structural units, a single layer of S and a triple layer in two variants. The layers may be thought of in the following way (Fig.1): Form a close packed layer of S atoms (Fig.1A) and remove from that one quarter of the atoms. The remaining S atoms form a single layer, s, and the crystallography is defined by a hexagonal lattice with lattice points A on the holes in the s layer (Fig.1B).

The triple layers are made by keeping the quarter of S atoms on the same two-dimensional lattice and filling the interstices with L atoms. Because the L atoms are larger than S atoms they do not quite form a flat plane; half of the L atoms are slightly above the s plane (L^+) and half are slightly below (L^-). If, for example, the S atoms of the triple layer occupy B sites (Fig.1C) then L^- atoms could occupy C sites and L^+ atoms could occupy A sites; this is a triple layer t. Alternatively L^+ atoms could occupy C sites and L^- atoms could occupy A sites; this is a triple layer t' (Fig. 1D). Thus two crystallographically different, but structurally identical, triple layers can be made.

The Laves phases each consist of alternating single and triple layers stacked normal to the plane of Fig. 1 and this implies also that alternate layers of the crystal are composed of S and L atoms, see Fig.2B. The rules governing the possible stacking sequences derive from the necessity to keep the structures close packed: Starting from a single layer with its holes on A sites, the next triple layer must have its L- atoms on A sites fitting into the holes. Likewise the

L^+ atoms of the triple layer fit into the holes of the next single layer up and so on. This form of packing permits only a restricted range of stacking sequences.

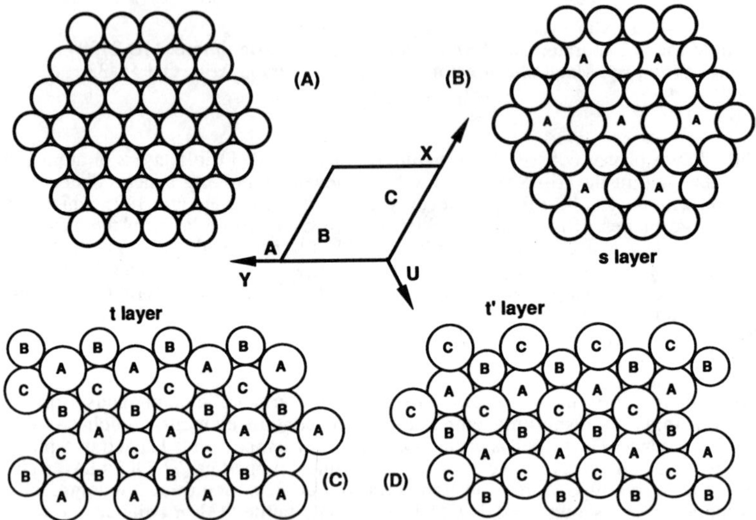

Figure 1. (A) A close packed plane of S atoms. (B) Single 'A' layer of S atoms. (C) Triple layer t with L^+ atoms (on A sites) above and L^- atoms (on C sites) below S. (D) Triple layer t' with L^+ and L^- reversed from (C).

CRYSTALLOGRAPHY OF THE UNIT CELL

It is convenient to use a hexagonal unit cell for all three Laves structures, to illustrate their similarities, despite the fact that C15 is cubic. The unit cell is shown in Fig.2A with both hexagonal (projected on (0001)) and cubic (projected on (111)) indices and with x,y,u axes parallel to close packed rows of S atoms. (It is worth noting that these are <u>not</u> close packed rows in the triple layers).

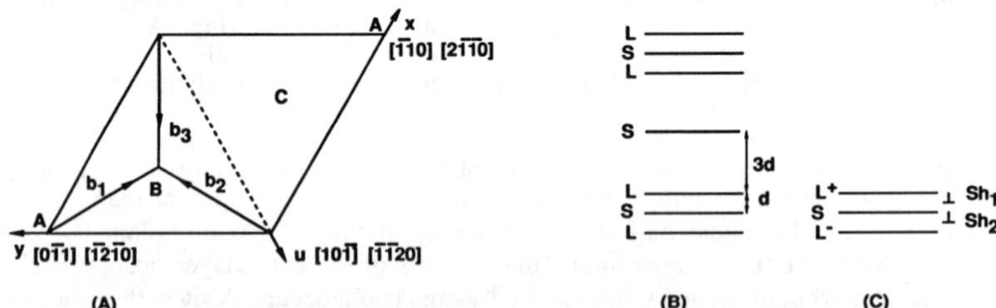

Figure 2. (A) Hexagonal unit cell projected on (0001) for C14 and C36 and on (111) for C15. <u>b_i</u> are Shockley burgers vectors $a/6\langle 112\rangle$ or $a/3\langle 1\bar{1}00\rangle$
(B) Layer structures showing alternation of single layers with triple layers and also of large L and small S atoms
(C) Glide of a synchroshockley partial dislocation through a triple layer. The dislocation consists of two ordinary Shockleys on two separate slip planes.

Within the unit cell, the projected atomic positions are as follows: The triple layers have atoms on sites A,B,C where the upper case letters are used in a generic sense to label the layers, so an 'a' layer of S atoms and an 'α' layer of L atoms have atoms at the A sites etc.The single layers are labelled by the fact that they have holes on sites A, B and C; they have atoms midway between the holes. Collecting this together, and using a hexagonal coordinate system, with the unit cell shown in Fig.2, the atomic positions are:

Single layers
A layer: S atoms at (1/2,0) , (0,1/2) , (1/2,1/2)
B layer: S atoms at (1/3,1/6) , (5/6,1/6) , (5/6,2/3)
C layer: S atoms at (1/6,1/3) , (1/6,5/6) , (2/3,5/6)

Triple layers
a layer of S atoms or α layer of L atoms at (0,0)
b layer of S atoms or β layer of L atoms at (1/3, 2/3)
c layer of S atoms or γ layer of L atoms at (2/3, 1/3)

The difference between triple layers t and t' is that in t the sequence of atoms B ->C ->A ->B is S->L^- ->L^+ ->S whereas in t' the sequence is S->L^+ ->L^- ->S, see Fig.1(C,D respectively).

The layers defined in this way are stacked so that triple layers alternate with single layers and small atoms alternate with large atoms (Fig.2B). The larger spacing between planes is three times larger than the smaller spacing.

STACKING SEQUENCES

Starting with a single layer of S atoms at the bottom of the unit cell having its holes on A sites, the L^- atoms of the next triple layer must necessarily be on the A sites so it is labelled α. The triple layer must use all three letters so the S atoms could occupy either b sites or c sites. If they occupy b, then L^+ must be on γ sites (forming a t' triple layer) and if they occupy c then L^+ must be on β sites. Above L^+ on γ the sequence must be Cγ and above L^+ on β the sequence must be Bβ. The two sequences so far are AαbγCγ and AαcβBβ. At this level, there are again two choices above γ and above β. As every quadruple (single plus triple) layer is added, the choice of sequences doubles as each triple layer may be t or t'.

A useful abbreviation is to call an A layer with t above it a quadruple X layer and an A layer with t' above it, a quadruple X' layer. Similarly B and C layers with t triple layers above them are termed Y and Z quadruple layers and if they have t' above them, they are termed Y' and Z'. An X quadruple layer has a Y or a Y' quadruple layer above it and an X' layer has a Z or a Z' quadruple layer above it. With this notation the possible crystal structures can be generated starting from an X layer at the bottom:

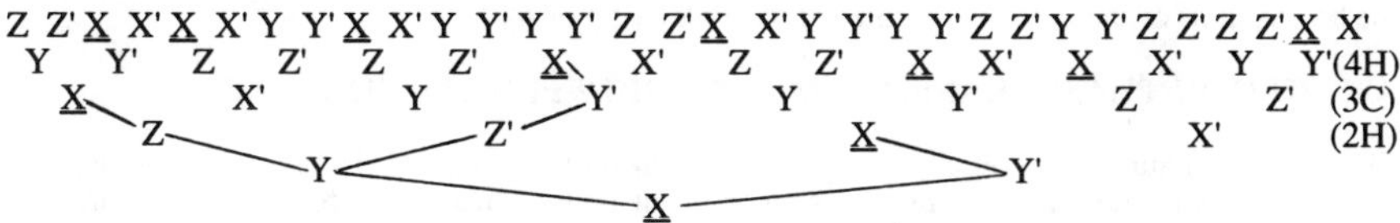

Moving up from X, every time another X is reached, a crystal structure has been generated. So the simple structures are:

XY'X... which is C14, a 2H structure
XYZX... which is C15, a 3C structure
XYZ'Y'X.... which is C36, a 4H structure
Clearly, 5H, 6H etc. structures are also possible, see [7]

The full stacking sequences implied by this notation are

C14 AαcβBβcαA....

C15 AαcβBβaγCγbαA....

C36 AαcβBβaγCγaβBβcαA....

One point to note is that, because the quadruple layer Y' is not identical to Y the stacking sequence of C14 is not quite analagous to the stacking sequence of hcp, whereas there is a close analogy between C15 and fcc.

SYNCHROSHEAR

Within a triple layer, the planes of the L^-, S and L^+ atoms are very close, much less than an atom diameter. Consequently shear within a triple layer could only occur with difficulty. Refering to Fig. 1C, and considering the upper L^+ atoms in the t layer to be fixed, the S atoms can move only in three directions through the gaps between the L^+ atoms. These directions are $\underline{AB}$ or, in Fig.2A, they are the 'Shockley' vectors $\underline{b}_1$, $\underline{b}_2$,$\underline{b}_3$ of type 1/6<112>.But this movement can only occur if the L^- atoms move in synchronism, also through the gaps between the L^+ atoms; their displacements are along the vectors $-\underline{b}_1$, $-\underline{b}_2$, $-\underline{b}_3$. In this way the L^- and S atoms swap positions and the t layer is transformed into a t' layer.

From a dislocation standpoint, these movements correspond to the propagation of a Shockley dislocation Sh_1 with burgers vector $\underline{b}_1$, $\underline{b}_2$ or $\underline{b}_3$ between the two upper layers and a second Shockley Sh_2 between the two lower layers (Fig.2C). Now, the first Shockley shifts the S layer and the L^- layer by, say, $\underline{b}_1$ but the L^- layer is required to move in, say, the $-\underline{b}_3$ direction. Consequently the Burgers vector of the second Shockley is $-\underline{b}_1 - \underline{b}_3 = \underline{b}_2$. The two dislocations must propagate together and so they are in fact an ordinary Shockley dislocation with $\underline{b} = \underline{b}1 + \underline{b}_2 = -\underline{b}_3$ whose core is split over two planes. Propagation of this split Shockley has two effects (1) it transforms a t layer into a t' layer, and (2) it shifts the entire crystal below the S layer by $-\underline{b}_3$ relative to the L^+ layer and the upper part of the crystal. In the electron microscope, therefore, the contrast is that of an ordinary Shockley dislocation trailing a stacking fault. Only by analysing the fault or using lattice resolution microscopy is there any chance of distinguishing between an ordinary Shockley and a synchroshockley.

The Burgers vector of the first Shockley may be $\underline{b}_1$, $\underline{b}_2$ or $\underline{b}_3$ and the Burgers vector of the second Shockley may also be $\underline{b}_1$, $\underline{b}_2$ or $\underline{b}_3$. All six combinations of unlike dislocations convert a t triple layer into a t' triple layer and, macroscopically, they may displace the crystal through $-\underline{b}_1$, $-\underline{b}_2$ or $-\underline{b}_3$.

Reference to Fig.1D shows that the same arguments may be applied to the shear of a t' triple layer except that the Burgers vectors of both Shockleys Sh_1 and Sh_2 are reversed in sign. The Burgers vectors may be $-\underline{b}_1$, $-\underline{b}_2$ or $-\underline{b}_3$. Synchroshear of a t' layer converts it into a t layer and the bottom half of the crystal is displaced through $\underline{b}_1$,$\underline{b}_2$ or $\underline{b}_3$.

ALLOTROPIC PHASE TRANSFORMATIONS BETWEEN C14, C15 AND C36

The cubic structure C15 contains t triple layers but no t' layers. The hexagonal structures do contain t' layers but at different levels in the unit cells. Since synchroshear is required to convert t layers into t' layers and vice versa, it is clear that transformations between these phases must involve synchroshear. Since there are several synchroshockleys whose propagation through triple layers can convert t to t', a number of schemes for the transformations are possible, and the simplest, involving the propagation of one type of synchroshockley across successive planes, is illustrated in Fig.3. The transformation C14 / C15 can occur if a synchroshockley sweeps alternate triple layers and the C36 / C15 transformation occurs by a synchroshockley sweeping two successive triple layers, then missing two triple layers and so on. Both these transformations require a shear of $1/\sqrt{8}$ as in the fcc to hcp transformation. It follows that the C14 / C36 transformation can occur with no macroscopic shear by a synchroshockley dipole sweeping across two adjacent triple layers, and this is also shown in Fig.3.

If the same synchroshockley propagates on all the active triple layers macroscopic shear appears in the hexagonal to cubic transformations but, since there is a choice of three synchroshockley burgers vectors which convert t to t', if these were to propagate in turn the macroscopic shear could be zero.

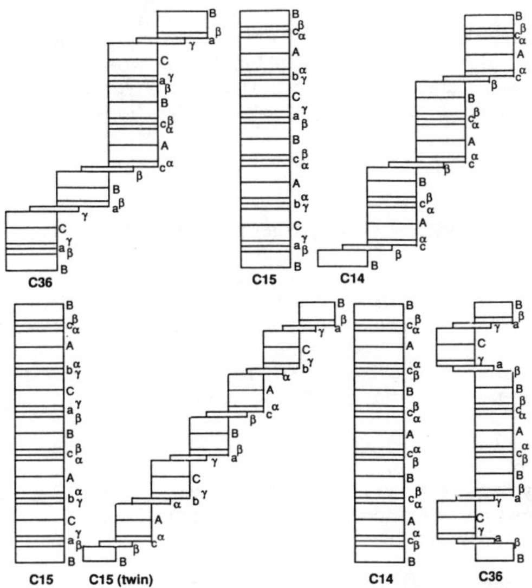

Figure 3.

Phase transformations between C14, C15 and C36 showing single layers A,B,C and triple layers of type $\alpha c \beta$. Synchroshear is indicated by the steps. C15 can be twinned by synchroshear on every triple layer and the cubic to hexagonal transformations require synchroshear on alternate triple planes.

PLASTIC MECHANISMS: GLIDE AND TWINNING OF C15

As shown above, the leading synchroshockley propagating into a t triple layer converts the t layer into a t' layer and it may displace the lower half of the crystal by $-\underline{b}_1$, $-\underline{b}_2$ or $-\underline{b}_3$. If this is followed by a second synchroshockley in the same triple layer, the second dislocation is propagating into a t' layer. This second synchroshockley converts the t' layer into a t layer and it may displace the lower half of the crystal by $\underline{b}_1$, $\underline{b}_2$ or $\underline{b}_3$. After the passage of both synchroshockleys, the t layer is intact and the crystal has been displaced by any one of the combinations $(\underline{b}_1 - \underline{b}_2)$, $(\underline{b}_3 - \underline{b}_1)$ etc. and these are the regular glide burgers vectors 1/2<110> in C15 and $1/3\langle 11\bar{2}0\rangle$ in C14 or C36. Glide can take place in all three structures by glide dislocations dissociating into synchroshockley pairs separated by t' if they are gliding in t and by t if they are gliding in t'.

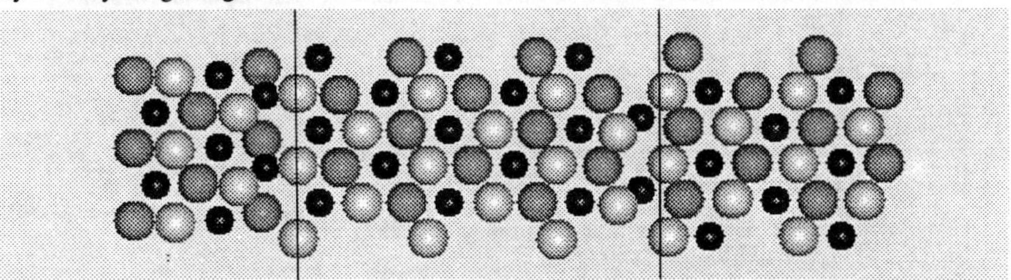

Figure 4. Atomic positions in a triple layer t near a glide dislocation dissociated into two synchroshockleys. The faulted region between the two partials is a triple layer t'.

An example of a dissociated dislocation in C15 is shown in Fig.4. Here the atomic positions are worked out using isotropic elasticity. Since C15 contains only t triple layers, the faulted region between the two synchroshockleys is a t' layer, as may be verified from the sequence of S->L^+ ->L^- ->S... atoms.

As an alternative to glide, the cubic Laves phase C15 may be twinned by the propagation of synchroshockleys through each triple layer of the structure. Both the shear vector a/6<112> and the plane spacing $a/\sqrt{3}$ are the same as in fcc metals so the twin shear is $1/\sqrt{2}$. As in fcc metals, twinning may occur either by the same synchroshockley sweeping each triple layer (which creates a macroscopic shear) or by sets of three synchroshockleys with zero net Burgers vector (which create no macroscopic shear). An illustration of twinning by propagation of the same synchrohockley on successive triple layers is shown in Fig.3. It is worth noting that synchroshear produces one crystal whose stacking sequence is an exact mirror image of the other crystal whereas the same macroscopic shear achieved by ordinary Shockley dislocations on every fourth plane would not do this.

DISCUSSION AND CONCLUSIONS

The three common Laves phases, and other structures with longer periods perpendicular to the close packed planes [7], are all related to each other by synchroshear operations. The vehicle for the transformation of one structure into another is a synchroshockley dislocation which may formally be thought of as a pair of ordinary Shockley dislocations with different Burgers vectors gliding on slip planes only 0.5Å apart. The same synchroshockley dislocations are capable of twinning the cubic C15 Laves phase. Pairs of synchroshockley dislocations form perfect dislocations and therefore provide a mechanism for slip in all the Laves phases. The geometry of these dislocations is analagous to the glide set dislocations in silicon [8]

If the same synchroshockley sweeps all the active planes during a phase transformation or during twinning, macroscopic shears of $1/\sqrt{8}$ and $1/\sqrt{2}$ are generated. But, on the close packed plane, there are three synchroshockley burgers vectors and if all three were to operate equally, no shear would be generated by either transformation or twinning. Unequal operation of the three dislocations could lead to any shear between zero and the maximum.

The descriptions of transformations, twinning and glide are in terms of what is geometrically possible. However the rates at which the processes can occur are in each case determined by the mobility of the synchroshockley dislocations. Inside the core of these dislocations three sets of atoms are moving in three directions within a crystal structure which is nearly close packed. The mechanism by which the dislocations move, whether by kink propagation or by short range diffusion, would be a good subject for molecular dynamics simulation. It is no surprise that the Laves phases are generally brittle at low temperatures or that the phase transformations are sluggish. It can be concluded that all three processes, phase transformations, glide and twinning are possible only at temperatures above which synchroshockleys become mobile.

ACKNOWLEDGEMENTS

PMH and AGJ wish to acknowledge support from the USAF contract F 33615-91-C-5663
We would like to thank Fuming Chu and D.P.Pope for sending their paper on twinning of C15 before publication.

REFERENCES

1. J.D.Livingston and E.L.Hall, J.Mater. Res. 5, 5 (1990)
2. M.Takeyama and C.T.Liu, Mater. Sci. Engnr., A132, 61 (1991)
3. C.W.Allen, P.Delavignette, and S.Amelinckx, Phys. Stat. Sol., a9, 237 (1972)
4. C.S.Barrett and T.B.Massalski, 'Structure of Metals' Pergamon Press, Oxford, p.256 (1980)
5. C.W.Allen and K.C.Liao, Phys. Stat. Sol, a74, 673 (1982)
6. M.L.Kronberg, Acta Metall., 5, 507 (1957)
7. Y.Komura and Y.Kitano, Acta Cryst.,B33, 2496 (1977)
8. J.P.Hirth and J.Lothe, 'Theory of dislocations' Wiley, New York, p.376 (1982)

SURFACE FILM SOFTENING EFFECTS IN $MoSi_2$

C. M. CZARNIK, R. GIBALA, M.A. NASTASI*, R. B. SCHWARZ*, S. R. SRINIVASAN*, J. J. PETROVIC*
The University of Michigan, Department of Materials Science and Engineering, Ann Arbor, MI 48109-2136 and *Los Alamos National Laboratory, Los Alamos, NM 87545

ABSTRACT

Surface film softening, which is associated with the generation of mobile dislocations at a film-substrate interface and their motion into the substrate, has been demonstrated previously for bcc metals and B2 ordered alloy substrates. $MoSi_2$ exhibits features of limited dislocation mobility and plasticity similar to these materials and would also be expected to exhibit film-enhanced plasticity, but at much higher temperatures. In this investigation, 150 nm thick films of SiO_2, $MoSi_2$ and ZrO_2 were deposited on $MoSi_2$ substrates by electron beam deposition at ~50°C in an attempt to observe film softening. In initial experiments, hot hardness testing was utilized primarily to maintain deformation conditions characteristic of dislocation glide. ZrO_2-coated $MoSi_2$ demonstrates a large film-induced softening in hardness tests carried out at 0.5 - 1 kgf loads over the temperature range 25 - 1300°C.

INTRODUCTION

$MoSi_2$ has properties of high stiffness, high thermal conductivity, low density, excellent oxidation resistance and potential high strength at elevated temperatures which make it attractive for investigation as a structural material. However, a major drawback to $MoSi_2$ has been its lack of ductility at temperatures to about 1100°C. The brittle-to-ductile transition temperature (BDTT) is generally reported to be in the range of 900-1400 °C [1], and can be attributed to the onset of dislocation climb and/or creep. There have been attempts to reduce the BDTT of this material or to enhance its plasticity, including second-phase toughening, grain size manipulation, and solid solution alloying [2].

In this paper, we examine surface film softening as a possible method to increase plasticity in polycrystalline and single crystal $MoSi_2$. This phenomenon has been demonstrated in many bcc metals and B2 ordered alloys at relatively low homologous temperatures $T < 0.2\ T_m$, where T_m is the absolute melting temperature. At these temperatures, dislocation glide mechanisms operate exclusively [3]. The interface between the surface film and substrate acts as an operative dislocation source, in effect pumping dislocations into the material under applied stress. Deformation of the film-substrate composite introduces geometrically necessary dislocations in the substrate. If these dislocations are also mobile, enhanced plasticity may result. Enhanced plasticity has been demonstrated for bcc metals and B2 ordered alloys in tension, compression and cyclic deformation experiments [4,5]. In the case of NiAl, for which additional plastic strains on the order of 10% have been reported, an Al_2O_3 surface oxide film has the effect of generating <001> dislocations at the interface which then move into the substrate material. In the case of $MoSi_2$, the relative types and mobilities of dislocations are less well-known than for bcc metals or B2 ordered alloys [6,7], but the interface mechanics for properly selected surface films should result in sufficient compatibility stresses to generate dislocations by this process. However, the high BDTT of $MoSi_2$ places the requisite tension or compression test temperatures well above room temperature, perhaps into the dislocation climb or creep regime [2].

EXPERIMENTAL

Materials

Polycrystalline $MoSi_2$ came from two sources: a commercial grade $MoSi_2$ powder from CERAC, Inc., and a laboratory-produced $MoSi_2$ powder which was mechanically alloyed from high purity elemental powders [8]. All powders were hot pressed uniaxially in grafoil-lined graphite dies in a high purity Ar atmosphere at 1600°C for 1.5 hours at pressures approaching 30 MPa. The equiaxed grain size of the commercial powders was 25 μm and the mechanical alloyed material was 12 μm. Both materials were approximately 95% of theoretical density. Samples were mechanically polished to a 0.5 μm finish.

The single crystalline $MoSi_2$ was supplied by Mr. James Garrett at McMaster University. It was grown at a rate of 240 mm/hr from high purity Si and Mo arc-melted in an Ar atmosphere to form a single button which was used for crystal growth. The Czochralski method of crystal growing with a tri-arc furnace was employed. The crystal was nominally oriented in the [001] direction, as confirmed by Laue x-ray patterns. Samples were cut from this crystal to be aligned as closely as possible with the [001] direction. Samples were polished to a 0.5 μm finish and electropolished to remove the mechanically deformed layer.

Coatings

All coatings were deposited by an electron beam evaporation system under high vacuum conditions. SiO_2 and ZrO_2 were deposited from their respective commercial powders, while control coatings of $MoSi_2$ were co-deposited from high-purity elemental powders. The coating parameters from elemental sources were adjusted during deposition to ensure that the Mo:Si ratio was maintained at 1:2, which was confirmed by Rutherford Back Scattering (RBS). All three coatings were deposited in an amorphous condition, to thicknesses of about 150 nm. The composition of the coatings was checked by RBS techniques to determine the composition of the oxide coatings and the final thickness of deposited films. All films were of the stoichiometric composition within experimental error.

Surface films of SiO_2 and ZrO_2 were chosen because of factors outlined previously by Noebe and Gibala [3]. (1) Both films form adherently on $MoSi_2$. (2) Their oxidation resistance is very good up to at least the maximum test temperature of 1300°C. (3) Neither film reacts chemically with the $MoSi_2$ substrate. SiO_2 was also chosen because it is the natural oxide layer which is formed on $MoSi_2$, although the thickness of the natural occurring layer of SiO_2 is much smaller than the tens to hundreds of nanometers of thickness which are typically necessary for observation of film softening [3-5]. Thus, additional SiO_2 was deposited to achieve these thicknesses. The ZrO_2 was additionally chosen based on the observations of Petrovic and co-workers [9] who have demonstrated that ZrO_2 particles increase the toughness of $MoSi_2$ and generate substantial densities of dislocations in $MoSi_2$. They also confirmed the absence of any reaction at the ZrO_2-$MoSi_2$ interface. The deposition of $MoSi_2$ as a film onto the same substrate was used as a check to ensure that there were no artifacts resulting from the coating process which affected subsequent test results.

Mechanical Testing

A Nikon QM1 hot hardness tester was employed to measure the microhardness of coated materials as a function of temperature. Indentations at loads 0.5-1 kgf were made at temperatures ranging from 25-1300°C. In most cases, five measurements were taken at each temperature on both coated and uncoated surfaces. Half of each polycrystalline and single crystalline sample was coated to allow for hardness measurements of coated and uncoated materials on the same specimen at the same time. This procedure also had the effect of preserving the orientation between indenter edges and directions in the single crystal for both coated and uncoated specimens [10]. The single crystal tests were performed on [001] surfaces of $MoSi_2$ to study effects in this hard orientation [6].

Hardness testing was employed in part for convenience and efficient use of $MoSi_2$ specimens, but mainly for maintaining deformation conditions characteristic of dislocation glide. Hardness testing utilizes deformation under a large, primarily hydrostatic stress state at large effective stresses and high strain rates. Although the stress state beneath the indenter is complex and difficult to characterize exactly, it can be shown that the major mode of deformation for most indentation experiments involves dislocation glide at various temperatures and loads [11].

RESULTS AND DISCUSSION

Figure 1 gives some of the results obtained for the temperature dependence of the hardness of $MoSi_2$-coated polycrystalline $MoSi_2$ substrates and compares these results to ones for the uncoated material. There is no significant effect of the $MoSi_2$ coating on the hardness of $MoSi_2$. Although one might expect subtle effects on the hardness due to structural differences and modulus and plasticity mismatch, no effects are observed. These results also suggest that there are no artifacts associated with the coating processes utilized that might produce unexpected effects.

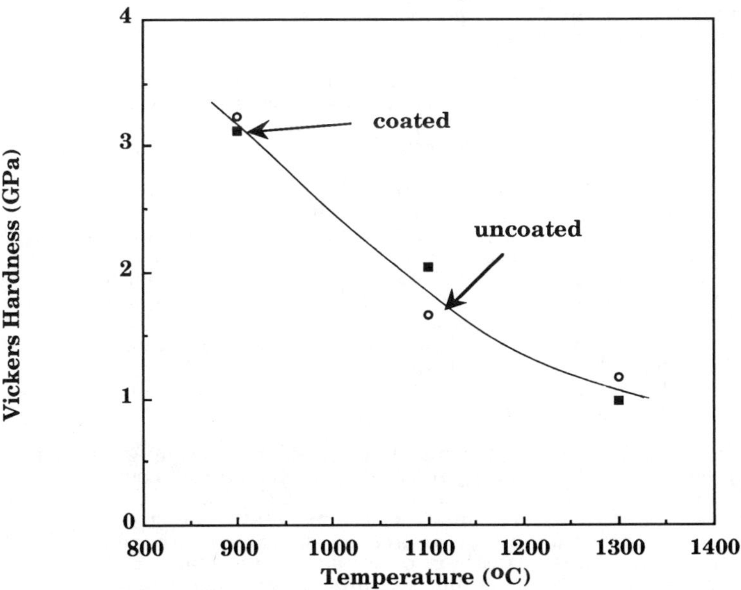

Figure 1. The effect of 0.15 μm thick $MoSi_2$ films on the Vickers hardness of a polycrystalline $MoSi_2$ substrate.

The temperature dependence of the hardness of the two ZrO_2-coated polycrystalline materials is given in Figure 2. The hardnesses of both ZrO_2-coated materials are consistently smaller than those of the corresponding uncoated materials at all temperatures. Note that the hardness of the uncoated high purity mechanically alloyed $MoSi_2$ is less than that of the uncoated commercially pure $MoSi_2$. Also, the higher purity mechanically alloyed material exhibits a proportionately greater film-induced hardness decrease at all temperatures, as expected from previous investigations on film softening of bcc metals and B2 ordered alloys [3].

A film-induced decrease in hardness has also been observed in SiO_2-coated polycrystalline $MoSi_2$, but the magnitude of the effect is smaller than for ZrO_2-coated material. These results will be presented in a subsequent paper.

The decreased hardness of the coated specimens is consistent with previous experiments on film softening [3-5], and can be attributed to similar dislocation generation mechanisms associated with the mechanical interaction between film and substrate at the interface. The hardness difference cannot be attributed to the film alone, because even an infinitely soft film with zero hardness would result in an additional indentation depth equal to the coating thickness of only 0.15 μm. This added depth translates to an additional 0.8 μm increase in diagonal length according to the regular Vickers indenter geometry, which can be compared with an indentation diagonal length of 30-40 μm for average room temperature indentations. The reduction in hardness due to this effect would be on the order of only 60 MPa for low temperature indentations, and even smaller for higher temperature indentations. This does not nearly account for the large decreases measured. Thus, there must be an additional interaction between the film and substrate, such as film-induced plasticity, which is allowing larger indentations of the coated material compared to that of the uncoated material.

The ZrO_2 film induced decrease in hardness presented in Figure 2 is also observed in single crystals of $MoSi_2$. Figure 3 shows the effect of a 0.15 μm ZrO_2 coating on the hardness of an [001]-oriented single crystal over the temperature range of 25-1300°C. The hardness of the film-coated hard-orientation single crystal is substantially less than that of the same uncoated material, i.e. the uncoated portion of the same specimen held in the same orientation relative to the position of the hardness indenter. The decreased hardnesses in Figure 3 are proportionately smaller than those observed for the polycrystalline materials given in Figure 2. This result is possibly surprising, given the anticipated larger mean free path of mobile dislocations and higher purity expected for the single crystals. On the other hand, it should be noted that the hardness indentation depths of 10 ± 5 μm are of a comparable size to the typical grain diameter of the polycrystalline materials, so that the dislocation mean free paths may not be so different in the two sets of materials. Also, the orientation dependence of the hardness of single crystal $MoSi_2$ [10] may alter the magnitude of the observed film-induced softening. We intend to investigate these effects in future experiments.

In summary, we have demonstrated that there is a significant effect of a 0.15 μm deposited ZrO_2 film on $MoSi_2$ substrates which results in lower hardness values over the range of temperatures from 25 to 1300°C. This effect is too large to be attributed to the film hardness itself, as the indentation depth far exceeds the film thickness. The reduced hardness is probably associated with enhanced dislocation generation at the film-substrate interface as the material is deformed under the hardness indenter. Thus, the film softening effect which has been demonstrated previously in bcc metals and B2 ordered alloys also seems to be effective in $MoSi_2$. Future experiments will be required to determine whether the film-induced reduction in hardness is also manifested in monotonic tension, compression, or bend tests at elevated temperatures in the vicinity of the BDTT, e.g. $T \geq 1000$°C.

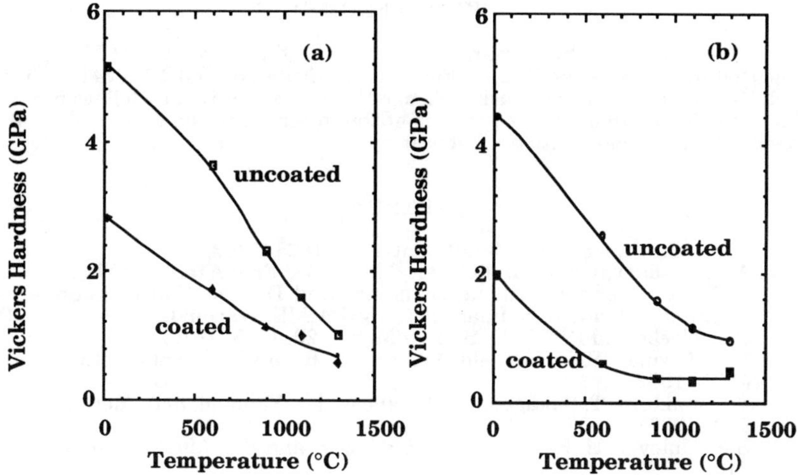

Figure 2. The temperature dependence of the Vickers hardness of uncoated and ZrO_2-coated polycrystalline $MoSi_2$. (a) Commercially pure hot-pressed $MoSi_2$. (b) High Purity mechanically alloyed hot-pressed $MoSi_2$.

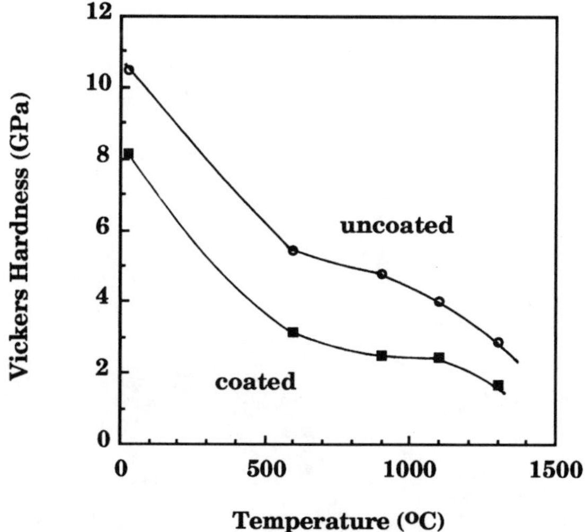

Figure 3. The temperature dependence of the Vickers hardness of uncoated and ZrO_2-coated [001] oriented single crystalline $MoSi_2$.

ACKNOWLEDGMENTS

The portions of this research carried out at the University of Michigan were supported by the National Science Foundation, Grant No. DMR 9102414. Two of us (C.M. Czarnik and R. Gibala) acknowledge Department of Energy fellowship support which enabled us to do major portions of this research in the Center for Materials Science at Los Alamos National Laboratory.

REFERENCES

1. R.M. Aikin, Jr. Scripta Metall. Mater. 26, 1025 (1992).
2. A.K. Vasudevan and J.J. Petrovic, Mater. Sci. Eng. A155, 1 (1992).
3. R.D. Noebe and R. Gibala, Structure and Deformation of Boundaries, K. Subramanian and M.S. Iman, eds., TMS-AIME, Warrendale, PA, p. 89, 1986.
4. R.D. Noebe and R. Gibala, Scripta Metall. 20, 1635 (1986).
5. K.J. Bowman, S.E. Hartfield-Wünsch and R. Gibala, Scripta Metall. Mater. 26, 1529 (1992).
6. Y. Umakoshi, T. Sakagami, T. Hirano and T. Yamane, Acta Metall. Mater. 38, 909 (1990).
7. S.A. Maloy, A.H. Heuer, J.J. Lewandowski and T.E. Mitchell, Acta Metall. 40, 3159 (1992).
8. R.B. Schwarz, S.R. Srinivasan, J.J. Petrovic and C.J. Maggiore, Mater. Sci. Eng. A155, 75 (1992).
9. J.J. Petrovic, A.K. Bhattacharya, R.E. Honnell, T.E. Mitchell, R.K. Wade and K.J. McClellan, Mater. Sci. Eng. A155, 259 (1992).
10. P.H. Boldt, J.D. Embury and G.C. Weatherly, Mater. Sci. Eng. A155, 251 (1992).
11. W.B. Li, J.L. Henshall, R.M. Hooper and K.E. Easterling, Acta Metall. Mater. 39, 3099 (1991).

PART V

Mechanical Properties

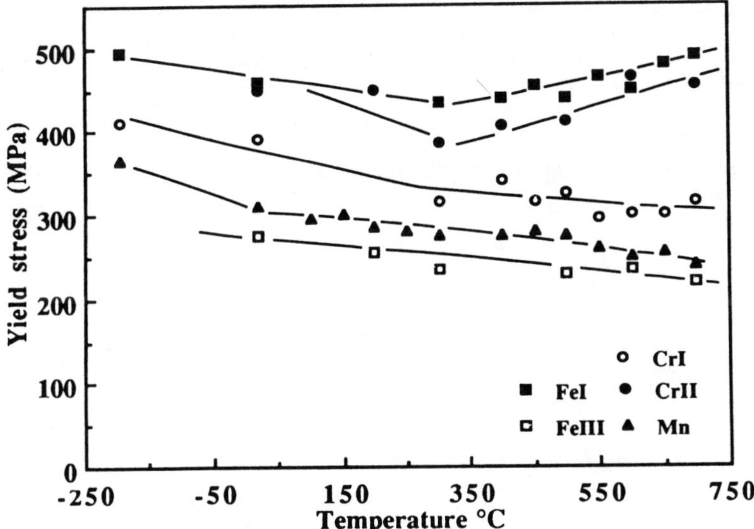

The figure shows the temperature dependence of the compressive yield strength for several Al_3Ti-based materials which have been alloyed with ~8% of Cr, Fe or Mn to promote the formation of the $L1_2$ cubic crystal structure. Irrespective of the third element alloying addition, strength anomalies will occur if the Ti content is above 28% (CrII and FeI), while the usual decreasing strength with increasing temperature behavior takes place if the Ti level is below ~27% (CrI, FeIII and Mn).

Figure courtesy of D.G. Morris, S. Gunther and R. Lerf.

TEMPERATURE DEPENDENCE OF DEFORMATION BEHAVIOUR AND DISLOCATION STRUCTURE OF $Al_{67}Mn_8Ti_{25}$ $L1_2$ INTERMETALLIC ALLOY

XIAOHUA WU, SHIPU CHEN, YONGHUA RONG, XIAOFU CHEN AND GENGXIANG HU
Department of Materials Science, Shanghai Jiao Tong University, Shanghai 200030, China

ABSTRACT

The Al_3Ti-based $L1_2$ alloy $Al_{67}Mn_8Ti_{25}$ shows different temperature dependence of the yield stress and ductility from that of the model $L1_2$ alloy Ni_3Al. The dislocation structures and dissociation modes of the alloy deformed at room temperature, 673 and 873K were investigated by TEM. It is determined that at room-temperature and 673K, a⟨110⟩ dislocation dissociated into two a/3⟨112⟩ superpartials with SISF between them on {111} plane, while at 873K, the dissociation is a/2⟨110⟩ pairs on {111} planes separated by APB. The temperature dependence of yield stress and ductility may be related to the dislocation structures at various temperatures.

INTRODUCTION

Recently, the $L1_2$ ordered Al_3Ti-based alloys have attracted much attention because of their low density, relatively good strength, excellent oxidation resistance and potential ductility. The transformation of DO_{22} Al_3Ti into $L1_2$ structure by alloying has developed considerable ductility in compression, but these $L1_2$ alloys still remain brittle in tension at ambient temperatures [1–2].

The temperature dependence of yield strength of the $L1_2$ Al_3Ti alloys has been reported by several investigators. Unlike some $L1_2$ compounds such as Ni_3Al, which exhibit a strong positive temperature dependence of strength, these alloys only show a weak positive temperature dependence [3–4] or a plateau [5–6] in the anomalous regime. The dislocation structure of the deformed $L1_2$ Al_3Ti alloys studied by transmission electron microscopy (TEM) so far is controversial. Earlier observations on room temperature deformed $Al_{67}Ni_8Ti_{25}$ [7–8] or Al_5Ti_2Fe [9] samples suggested that the a⟨110⟩ dislocations moving on {111} planes are undissociated. Further work on the Fe-modified $L1_2$ alloy have shown that at room temperature the a⟨110⟩ dislocations tend to dissociate into a/2⟨110⟩ partials on {111} planes separated by an antiphase boundary (APB) [10–12], while others cited that the dissociation is of the a/3⟨112⟩-type with the superlattice intrinsic stacking faults (SISF) between two partials on {111} planes [13,14]. TEM studies on the dislocations in Al_3Ti-based $L1_2$ alloy deformed at elevated temperatures are very limited. Lerf and Morris [9] reported that majority of the dislocations in the Fe-modified alloy after 500℃ deformation are dissociated as pairs of a/2⟨110⟩ on {111} planes with APB between, and the result of Inui et al [14] showed also the APB-coupled a/2⟨110⟩ superpartials at 600℃ but on {001} planes.

In the present study, the temperature dependence of yield stress and ductility of a Mn-modified $L1_2$ alloy was examined by compression tests conducted at temperatures ranging from R.T. to 1023K. The dislocation structures of the samples deformed at selected temperatures were investigated by TEM in order to elucidate the deformation mechanisms of this alloy in various temperature regimes.

EXPRIMENTAL

The Al_3Ti-based ternary alloy with nominal composition $Al_{67}Ti_{25}Mn_8$ (at%) was prepared by nonconsumable arc melting in argon on a water cooled copper hearth. The button ingots were homogenized at 1373K for 60h.

Compression specimens with dimensions 4mm × 4mm × 7mm were cut from the homogenized ingots. Compression tests were carried out at room temperature, 423, 523, 673, 773, 873 and 1023K with a constant crosshead speed of 0.1mm/min, i.e., a nominal strain rate of 2×10^{-4}/s. Crystal structure of the alloy was determined by X-ray diffraction. Thin foil samples for TEM studies were prepared from compression specimens deformed to ~2% plastic strain at room temperature, 673K and 873K respectively.

RESULTS AND DISCUSSION

Temperature dependence of yield stress and ductility

The cast and homogenized material, which was shown by X-ray diffration to be a single $L1_2$ phase alloy, was compression tested from ambient to 1023K. The 0.2% offset compressive yield stress and the compressive ductility (plastic strain at fracture) were plotted as a function of test temperature, as shown in Fig.1. The yield stress declines slowly from 293 to 523K, with a very faint peak at about 670K, and then remains almost constant up to 1023K. This result agrees very well with that reported by Zhang et al [4], and is comparable with the result of Brown and Kumar [6], but in our case the yield stress data are at a lower level. It is interesting to note that the strong positive temperature dependence observed in some $L1_2$ intermetallic compounds such as Ni_3Al does not happen in this alloy. Although a small peak has been plotted, the increment up to the peak is only ~15MPa and appears only in a narrow temperature range (from 673 to 773K), so it is reasonable to consider the temperature dependence a plateau. The compressive ductility as a function of temperature is also shown in Fig.1. The plastic strain increases slightly from room temperature to about 773K, then changes abruptly to a rapid increase stage, i.e. from 22.4% at 773K to 32.5% at 873K, and much higher ductility is obtained at 1023K so that the specimens were compressed to 40% plastic strain without cracking. It should be noted that there is no apparent change of yield stress at the corresponding temperatures, the above phenomenon seems to be related to the change of dislocation behaviours.

The work-hardening rate at each temperature was measured as the average slope of the stress-strain curve over the plastic strain range up to 5%. The temperature dependence of work-hardening rate is plotted in Fig.2, it shows a similar relationship to that of yield stress, and is different from that of Ni_3Al, which increases rapidly with temperature. This implies that a different deformation mechanism may take place in the present $L1_2$ alloy.

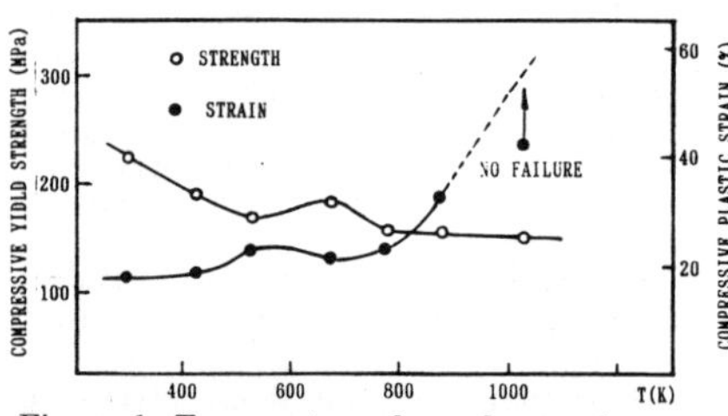

Figure 1. Temperature dependence of compressive yield stress and compressive ductility for $Al_{67}Mn_8Ti_{25}$

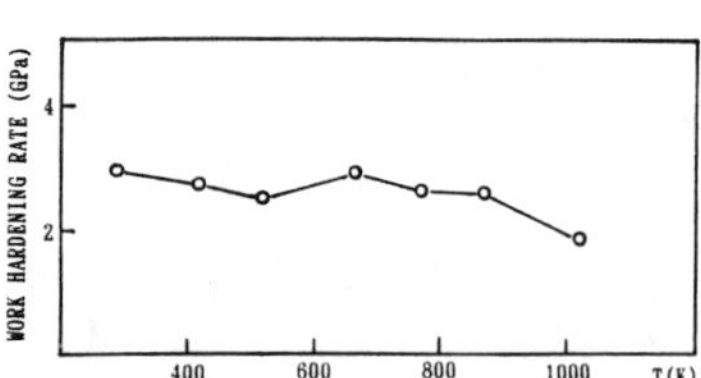

Figure 2. Work hardening rate as a function of compression testing temperature

Dislocation structure at various temperatures

The samples after deformation to ~2% plastic strain at room temperature, 673K and 873K were studied by TEM using conventional and weak beam techniques. Fig.3 shows the general features of dislocation structure in these samples. In contrast to the long, straight dislocations commonly observed in the room temperature deformed alloy (Fig.3(a)), the dislocations in 673K deformed sample are curved , and more dipoles are seen (Fig.3(b)), while the 873K deformed sample shows random arrangement of short, curved dislocation segments and dipoles (Fig.3(c)).

Detailed analysis were done to determine the slip system and dissociation mode of the superdislocations in the above mentioned samples. The dislocations in the sample deformed at room temperature consisting of paired partials , labeled A and B in Fig.4, were chosen for more careful analysis. Trace analysis through a series of dislocation images taken with different beam directions determined that both A and B are lying on ($\bar{1}11$) plane. The Burgers vectors of the dislocations were examined by diffraction contrast analysis using various reflections. Table I gives the **g·b** values for the partials under different reflections and among them typical micrographs are shown in Fig.4. It is found that the superdislocations formed during compressive deformation at room temperature are dissociated into two a/3⟨112⟩ superpartials on {111} planes, with a SISF between them. For the dislocations A :
$a[1\bar{1}0] \rightarrow a/3[\underline{21}1] + SISF + a/3[1\underline{2}\bar{1}]$, and for B:
$a[011] \rightarrow a/3[\bar{1}\bar{2}1] + SISF + a/3[1\bar{1}2]$.

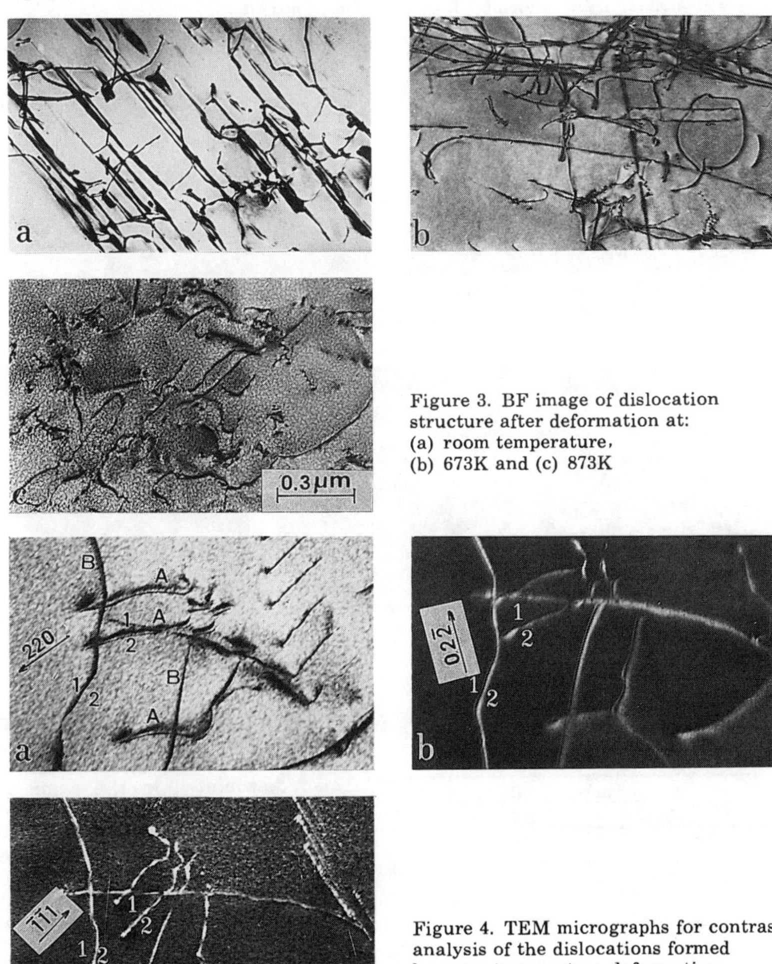

Figure 3. BF image of dislocation structure after deformation at:
(a) room temperature,
(b) 673K and (c) 873K

Figure 4. TEM micrographs for contrast analysis of the dislocations formed by room temperature deformation

Table I. **g·b** values for the dislocations formed by room temperature deformation under different reflections

g	A				B				Fig.
	1 (**b**=1/3[211])		2 (**b**=1/3[$12\bar{1}$])		1 (**b**=1/3[$\bar{1}\bar{2}1$])		2 (**b**=1/3[$1\bar{1}2$])		
	obser.	**g·b**	obser.	**g·b**	obser.	**g·b**	obser.	**g·b**	
220	V.	2	V.	2	V.	−2	I.V.	0	4(a)
$02\bar{2}$	I.V.	0	V.	2	V.	−2	V.	−2	4(b)
$31\bar{1}$	V.	2	V.	2	V.	−2	I.V.	0	
$3\bar{1}1$	V.	2	I.V.	0	I.V.	0	V.	2	
$1\bar{3}1$	I.V.	0	V.	−2	V.	2	V.	2	
$\bar{1}\bar{1}1$	R.C.	−2/3	V.	−4/3	R.C.	4/3	V.	2/3	4(c)

V. : visible, I.V. : invisible, R.C. : residual contrast

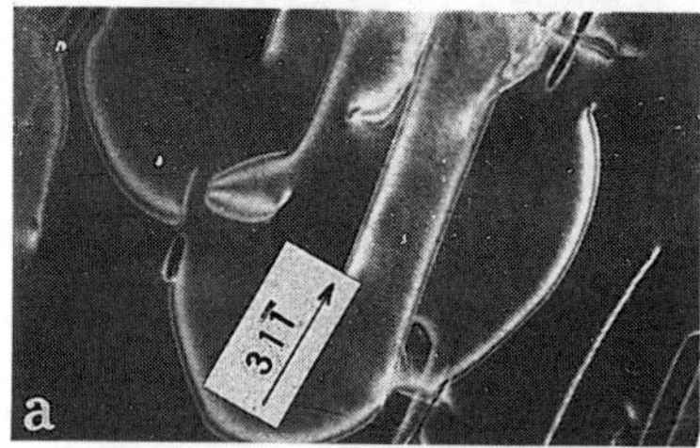

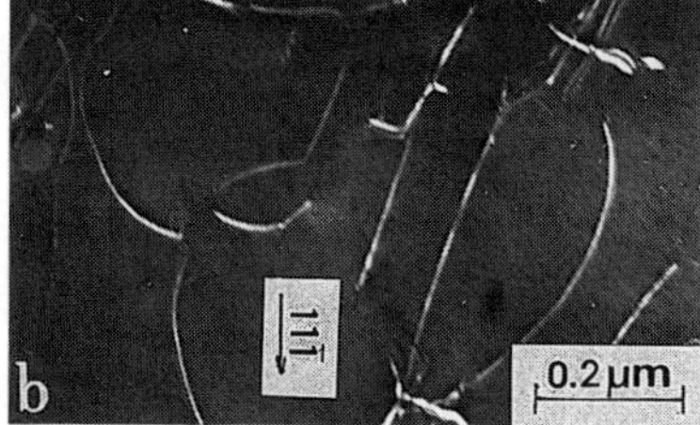

Figure 5. TEM micrographs for contrast analysis of the dislocations formed during 673K deformation

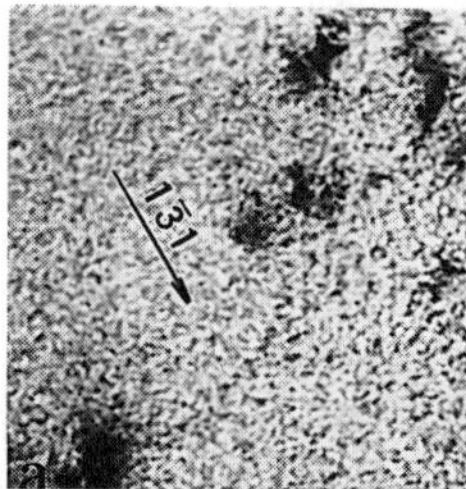

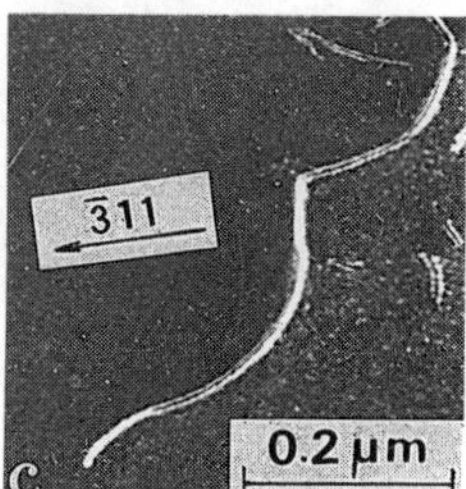

Figure 6. TEM micrographs for contrast analysis of the dislocations formed at 873K deformation

Table II. **g·b** values for the dislocations in 873K deformed sample

g	**b**=$\mathbf{b}_1$=$\mathbf{b}_2$=1/2[$\bar{1}01$]		Fig.
	obser.	**g·b**	
020	I.V.	0	
$1\bar{3}1$	I.V.	0	6(a)
$11\bar{1}$	V.	−1	
$\bar{2}20$	V.	1	
$\bar{2}02$	V.	2	6(b)
$\bar{3}11$	V.	2	6(c)

V. : visible, I.V. : invisible

Analysis of the dislocations in samples deformed at 673K indicated that superdislocations are also dissociated in the same way as those formed during R.T. deformation, some examples of the TEM micrographs for this analysis are shown in Fig.5.

The dislocations in the sample deformed at 873K are apparently dissociated, but the contrast analysis shows that both partials are in contrast or out of contrast simultaneously for all the reflections used. This suggests that the dissociation is of the a/2⟨110⟩ type. Table II lists the **g·b** values and the weak-beam images used for the analysis are shown in Fig.6. The Burgers vector of the superdislocation shown in the micrograph was determined by the "invisibility criterion" as a[$\bar{1}$01], and the dissociation is a[$\bar{1}$01]→a/2[$\bar{1}$01]+APB +a/2[$\bar{1}$01]. The dissociation plane was determined by the widest separation of the dissociated pair through tilting of the samples.The result shows that the APB is on (111) plane (see Fig.6(b)), and the measured separation is ~17nm. The APB plane is checked further by observing the change of dislocation configuration when viewed in different orientations. The possible dissociation planes for the [$\bar{1}$01] superdislocation are (111), (1$\bar{1}$1) of {111} and (010) of {001}. Since the curved dislocations did not appear straight when the sample was tilted to a beam direction [001] or [011], the dislocation plane must not be (010) or (1$\bar{1}$1) respectively.

The TEM studies of dislocation structures in deformed $Al_{67}Mn_8Ti_{25}$ indicated that at ambient and intermediate temperatures, e.g. 673K, the a⟨110⟩ superdislocations dissociate into a/3⟨112⟩ superpartials bounding SISF on {111} planes, and at higher temperatures, e.g. 873K, they dissociate into a/2⟨110⟩ bounding APB on {111} planes. It has been shown by atomistic modeling [15] that the dislocation core structure of a/3⟨112⟩ superpartials is nonplanar and is, therefore, sessile. Their motion requires a thermally activated process, so the yield stress will decrease rapidly with increasing temperature. But in the present case, the yield stress declines slowly from room temperature to 523K, then shows a plateau with increasing temperature. This phenomenon may be related to the strong tendency of dipole formation with increasing temperature in this alloy (Fig.3). Thus, a high stress is required to break these dipoles and a plateau temperature dependence of $L1_2$ Al_3Ti alloy is resulted.

Another interesting feature of the alloy studied is the rapid increase of compressive ductility with temperature above 773K, while the yield stress is almost constant (Fig.1). This behavior can be explained by the dissociation mode of superdislocations at the higher temperature. It has been generally accepted that the symmetry of the $L1_2$ structure assures the stability of a/2⟨110⟩ APB on {001} but not on {111}, and, therefore, cross slip of a/2⟨110⟩ screw dislocations from {111} to {001} leads to the anomalous temperature dependence of yield stress in some alloys such as Ni_3Al. In the present case, the a/2⟨110⟩ superpartials are on {111} planes, no {001} cross slip was examined by the weak-beam electron microscopy. It is also shown by atomistic modeling [15] that when a/2⟨110⟩ superpartial with APB is on the (111) plane, one of the three core configurations of a/2⟨110⟩ is glissile with the core spreading in the plane of APB. Thus a⟨110⟩ superdislocations can move on the {111} plane and contribute to the plastic deformation of $Al_{67}Mn_8Ti_{25}$ at higher temperatures, the ductility of this alloy is greatly improved.

CONCLUSIONS

1. Compression tests at room and elevated temperatures on $Al_{67}Mn_8Ti_{25}$ $L1_2$ alloy showed that: The yield stress declines slowly from 293 to 523K, with a very faint peak at ~670K, then is almost constant up to 1023K. Compressive ductility increases slightly from rooom temperature to ~773K, then changes abruptly to a rapid increase stage. The work-hardening rate shows a similar relationship with temperature as does yield stress.

2. TEM investigation revealed that: at room temperature and 673K, the a⟨110⟩ superdislocations dissociate into two a/3⟨112⟩ superpartials with an SISF between them on {111} planes, while at 873K, the dissociation changes into a/2⟨110⟩ partials separated by APB on {111} planes. The temperature dependence of mechanical properties can be explained by the dislocation structures and dissociation modes at different temperatures.

ACKNOWLEDGMENT

This research was supported by the National Advanced Materials Committee of China.

REFERENCES

1. K.S. Kumar and S.A. Brown, Phil. Mag. A **65**, 91 (1992)
2. Xiaofu Chen, Xiaohua Wu, Shipu Chen and Gengxiang Hu, Scripta Metall et Mater. **26**, 1775 (1992)
3. K.S. Kumar and J.R. Pickens, Scripta Metall **22**, 1015 (1988)
4. S. Zhang, J.P. Nic, W.W. Milligan and D.E. Mikkola, Scripta Metall **24**, 1441 (1990)
5. Z.L. Wu, D.P. Pope and V. Vitek in High-Temperature Ordered Intermetallic Alloys, edited by L. Johnson, D.P. Pope and J.O.Stiegler (Mater. Res. Soc. Symp. Proc. **213**, Pittsburgh, PA, 1991) pp.487
6. S.A. Brown, K.S. Kumar, Scripta Metall **24**, 2001 (1990)
7. V.K.Vasudevan R.Wheeler and H.L.Fraser in High-Temperature Ordered Intermetallic Alloys, edited by C.T. Liu, A.I. Taub, N.S. Stoloff and C.C. Koch (Mater. Res. Soc. Symp. Proc. **133**, Pittsburgh, PA, 1989) pp.705
8. C.D. Tuner, W.O. Powers and J.A. Wert, Acta Metall **37**, 2635 (1989)
9. R. Lerf and D.G. Morris, Acta Metall **39**, 2419 (1991)
10. E.P. George, J.A. Horton, W.D. Porter and J.H. Schneibel, J. Mater. Res. **5**, 1639 (1990)
11. H.R.P.Inoue, C.V.Cooper, L.H.Favrow, Y Hamada, and C.M.Wayman in High-Temperature Ordered Intermetallic Alloys, edited by L. Johnson, D.P. Pope and J.O.Stiegler (Mater. Res. Soc. Symp. Proc. **213**, Pittsburgh, PA, 1991) pp.493
12. D.G. Morris, J. Mater. Res. **7**, 303 (1992)
13. Hu Gengxiang, Chen Shipu, Wu Xiaohua and Chen Xiaofu, J. Mater. Res. **6**, 957 (1991)
14. H. Inui, D.E. Luzzi, W.D. Porters, D.P. Pope, V. Vitek and M. Yamaguchi, Phil. Mag. A. **65**, 1245 (1992)
15. V. Vitek, Y. Sodani and J. Cserti in High-Temperature Ordered Intermetallic Alloys, edited by L. Johnson, D.P. Pope and J.O.Stiegler (Mater. Res. Soc. Symp. Proc. **213**, Pittsburgh, PA, 1991) pp.195

CREEP BEHAVIOR OF A DIRECTIONALLY SOLIDIFIED Ni_3Al ALLOY

Yun Zhang, Dongliang Lin(T.L.Lin)
Department of Materials Science & Engineering, Shanghai Jiao Tong University, Shanghai 200030, P.R.China

ABSTRACT

The creep behavior of a directionally solidified multicomponent Ni_3Al alloy was investigated in a temperature range from 923 to 1173 K. The dislocation structure during secondary stage creep has been examined by transmission electron microscopy. At lower temperatures from 923 to 1023K under a stress of 500 MPa, there exist a number of dense three–dimensional dislocation networks in the Ni_3Al creep specimen,while at a higher temperature of 1173 K under a stress of 200 MPa the dislocation structure degenerates to regular two–dimensional dislocation networks.Climb of dislocations occurs in the overall test temperture range. The stress dependence and temperature dependence of creep rates for the Ni_3Al alloy were also determined.It was found that power law creep is obeyed with the stress exponent equal to 4.7 and the activation energy equal to 326.6 kJ/mol.The climb of dislocations was suggested to be the rate controlling factor for the secondary stage creep rate.

INTRODUCTION

The intermetallic Ni_3Al alloy is considered for high temperature structural application,due to an anomalous increase of yield stress with temperature.Therefore, the study of creep behavior of Ni_3Al alloy becomes important to its future high temperature service. Although the short–time mechanical properties of Ni_3Al alloys are well documented, the understanding of their long–time creep properties is far less complete. Up to now there have been only a few investigations in which secondary stage creep in Ni_3Al alloy has been reported[1–4]. Therefore, extensive work is needed to understand the creep mechanisms operating in Ni_3Al alloy. In the present invetigation, the secondary stage creep characteristics of a directionally solidified (DS) nickel–rich off–stoichiometric Ni_3Al alloy containing Zr, Hf, Ti, Nb, Ta, W, Mo, Cr, Mg, and B were determined in the temperature range from 923 to 1173K by analyzing the stress dependence of creep rates, the activation energy, and the dislocation structure during secondary stage creep.

EXPERIMENTAL

The chemical composition of the Ni_3Al alloy used for testing is as follows (at%): Zr 0.20, Hf 0.80, Ti 0.40, Nb 0.20, Ta 0.60, W 0.20, Mo 0.60, Cr 8.53, Al 16.55, Mg 0.02, B 0.15, Ni balance. The elemental additions were within their solubility limits in Ni_3Al, as transmission electron microscopy(TEM) results showed that the alloyed Ni_3Al maintained the single phase state. Using the starting materials, which were electrolytic nickel (99.9wt% purity) and aluminum (99.9wt% purity) and other high purity materials as available, the alloys were prepared by induction melting and directional solidifying in argon gas.It should be noted that the boron was added by an intermediate alloy Ni–B with a boron content of 18.57wt%. The resulting DS Ni_3Al alloy plates (170mm×70mm×15mm) were homogenized in air at 1473K, 5h/furnace cool, and 1373K,48h/furnace cool. The width of the columnar grains was approximately 500–1000μm.

The round bar creep specimens (5 mm diameter ×25 mm gage length)were machined such that the applied stress axis was parallel to the axis of the columnar structure. The constant load creep tests were carried out over the temperature range 923–1173K and the initial stress range 200–550 MPa.

When the creep specimens were deformed into secondary stage, the tests were interupted with rapid unloading and cooling. Thin foils were sliced normal to the specimen axis by spark machine and mechanically thined to ~0.1 mm.TEM foils were electrochemically polished with a twin jet thinner using a solution of 5% perchloric acid and 95% ethyl alcohol at 243K. Dislocation structures were then studied in a H800 instrument at 200 kV.

RESULTS

Dislocation Structure During Secondary Stage Creep

Figure 1 shows the dislocation structures for the DS Ni_3Al alloys during secondary stage of creep at temperatures of 923K, 973K and 1023K under a stress of 500MPa. The Burgers vectors were of type 1/2<110>.Figure 1 shows that creep-induced dense three-dimensional dislocation networks are distributed in the creep specimen at these temperatures. The dislocation networks are formed mainly by the 1/2<110> dislocation pairs shown in Fig.2 by an arrow. Between the dislocation walls, there exist a number of edge dislocation dipoles and rows of dislocation loops resulting from the collapse of the dipoles, as shown in Fig.3. Occasionally, it is seen that straight 1/2<110> screw dislocation pairs,

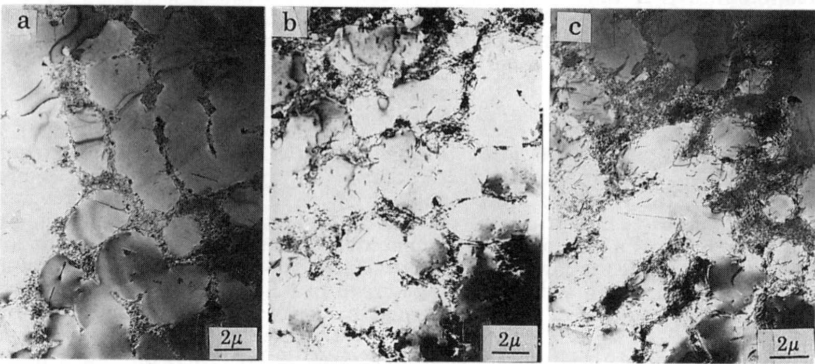

Fig.1 Dislocation substructures in the secondary stage of creep under 500 MPa at (a) 923 K(ε=0.32%),(b) 973 K(ε=0.19%), and (c) 1023 K(ε=0.48%).

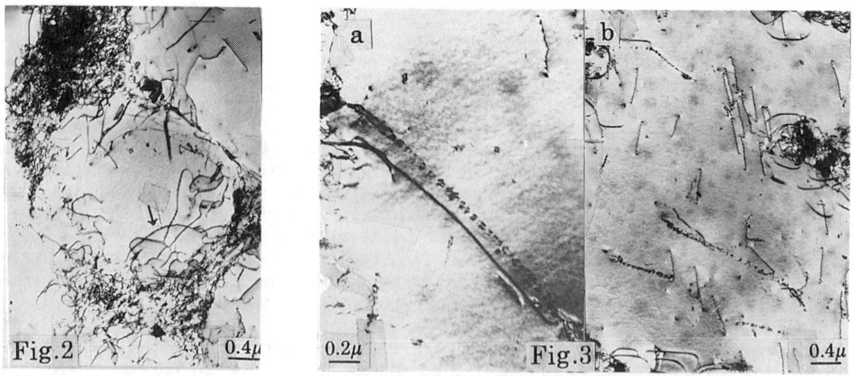

Fig.2 Dislocation substructures with dislocation pairs after 0.48% strain in the secondary stage of creep at 1023 K, 500MPa.

Fig.3 Dislocation structure between dislocation walls in the secondary stage of creep under 500 MPa at (a) 973 K(ε=0.19%), and (b) 1023 K(ε=0.48%).

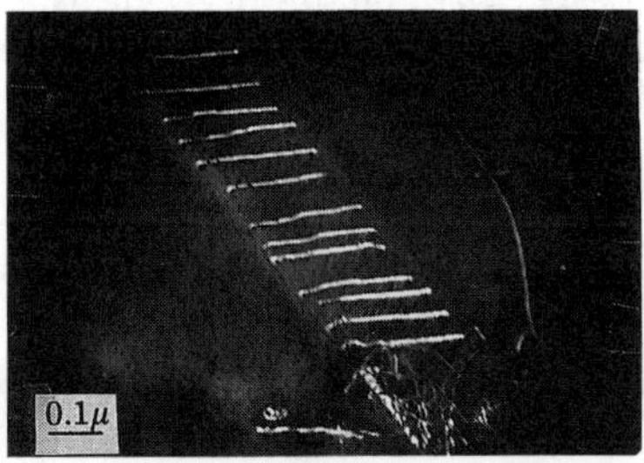

Fig.4 Weak beam image of screw dislocations piled up at a dislocation wall after 0.19% strain in the secondary stage of creep at 973 K, 500MPa.

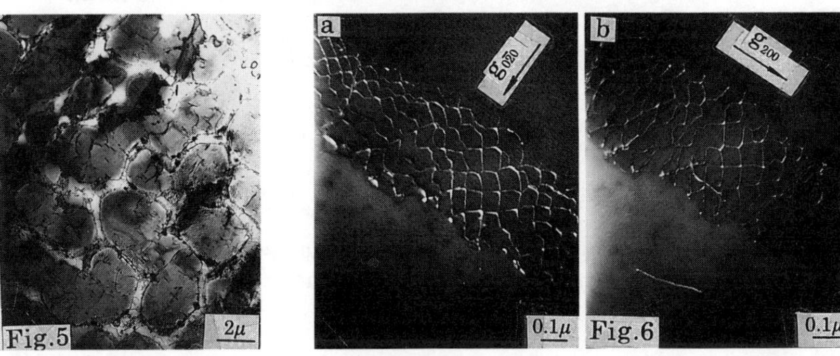

Fig.5 Dislocation substructures after 1.28% strain in the secondary stage of creep at 1173K, 200MPa.
Fig.6 Weak beam images of the dislocation substructure,composed of two rows of edge dislocations, after 1.28% strain in the secondary stage of creep at 1173K, 200MPa.

slipped along {111} plane, are piled up at the dislocation wall (fig.4).The dislocation networks can act as obstacles to the dislocation slip.

The creep – induced dislocation at 1173 K under a stress of 200 MPa mainly consists of loose and regular two – dimensional dislocation networks, shown in Fig.5. Figure 6 shows that such regular dislocation networks are formed by a slip process in which a row of 1/2 <101> edge dislocations meets another row of 1/2 <011> edge dislocations. Besides, it is found that many curved dislocation lines are randomly distributed between the dislocation walls (fig.5). Therefore,the movement of the dislocations can primarily proceed by climbing of dislocations along the walls and climbing of dislocations of the networks untill they meet other dislocations of the opposite sign climbing in the opposite direction and eventually annihilate. The dislocation density is kept constant by the glide of dislocations between the dislocation walls.

Secondary Stage Creep Rates

When climb of edge dislocation is the controlling process for creep, the secondary stage creep rate should obey Dorn equation[5]

$$\dot{\varepsilon}= \frac{ADbG}{kT} \left(\frac{\sigma}{G}\right)^{n} \tag{1}$$

Substitution from $D = D_0 \exp(-Q/kT)$ into eqn.(1) yields

$$\dot{\varepsilon} = \frac{K}{T} \left(\frac{\sigma}{G}\right)^n \exp\left(-\frac{Q}{kT}\right) \tag{2}$$

with the stress exponent n in the range 4–5 and where $K = (AD_0bG)/k$, $\dot{\varepsilon}$ is the sedondary stage creep rate, σ is the applied stress, T is the absolute temperature, D is the volume self–diffusion coefficient, D_0 is the frequency factor, Q is the activation energy of self–diffusion, b is Burgers vector,G is the shear modulus,k is Boltzmann's constant and A is a structural constant. From above equations, the relation between the secondary stage creep rates and the applied stress and the temperature can be analyzed.

A. Stress dependence of secondary stage creep rates

Figure 7 shows a set of typical creep curves for the DS Ni_3Al alloy at 1023K under stresses of 450 MPa, 500MPa, 525MPa and 550MPa. In fig. 8, according to eqn.(2) $\ln\dot{\varepsilon}$ was plotted with $\ln\sigma$ at 1023K, at which a well–defined linear

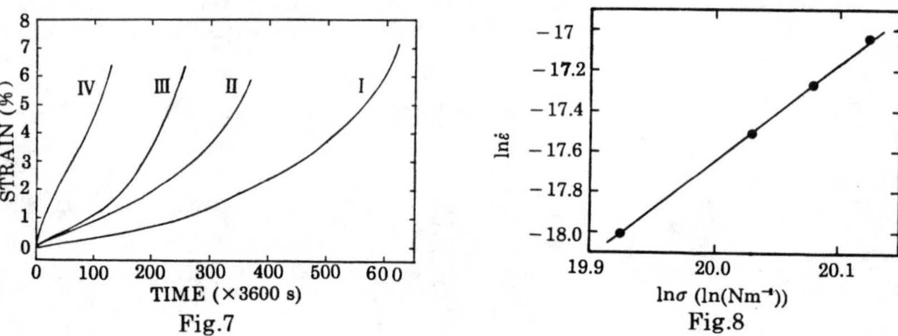

Fig.7 Typical creep curves of the DS Ni_3Al alloy at 1023 K under I –450MPa, II –500MPa, III – 525MPa and IV – 550MPa.

Fig.8 Stress dependence of the secondary stage creep rates for the DS Ni_3Al alloy at 1023 K.

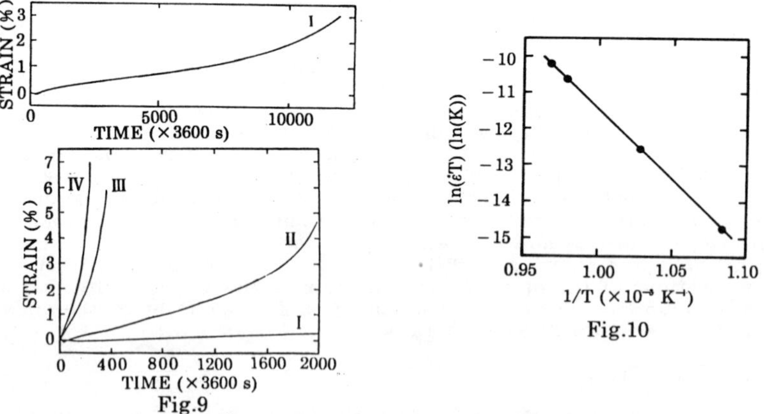

Fig.9 Typical creep curves of the DS Ni_3Al alloy under 500MPa at I –923 K, II –973 K, III – 1023 K and IV – 1033 K.

Fig.10 Temperature dependence of the secondary stage creep rates for the DS Ni_3Al alloy under 500 MPa.

relation was obtained. The stress exponent was determined to be 4.7 from the slope of the curve in fig.9.

B. Temperature dependence of secondary stage creep rates

Figure 9 shows a set of typical creep curves for the DS Ni_3Al alloy under a stress of 500 MPa at temperatures of 923 K, 973 K, 1023 K and 1033 K. According to eqn.(2), an Arrhenius type curve for the DS Ni_3Al alloy was plotted for $\ln(\dot{\varepsilon}T)$ with reciprocal test temperature,1/T, as shown in fig.10, at which a well – defined linear relation was also obtained with a slope of − 39.28K, and a intercept of 27.81. From the above slope value, the activation energy of volume self – diffusion was determined to be 326.6kJ/mol. Besides, from the above intercept value and the experimental value of n, the constant, K, in eqn.(2) was determined to be $10^{-28.808} \cdot G^{4.7}(Nm^{-2})^{-4.7}$.

DISCUSSION

Disordered metal alloys are generally classified in two classes[6]: class 1—alloys with stress exponent 3 resulting from viscous dislocation glide without substructure formation, and class 2—pure metals and alloys with stress exponent 4 – 5 resulting from dislocation climb with substructure formation. Flinn[1], and Nicholls and Rawlings[2] measured the creep properties of the same polycrystalline Ni_3Al – Fe alloy. At the temperature range 953 – 1203K, it was found that $n \approx 3$. The rate – controlling creep mechanism was considered to be the collapse of the edge dipoles, which occurs by dislocation climb and the removal/production of APBs by diffusion, and consequently, a microcreep theory proposed by Weertman[7] would be applicable.

In the present study, when the DS Ni_3Al alloy is deformed into the secondary stage of creep at 923K, 973K and 1023K under a stress of 500MPa, there exist a number of dense three – dimensional dislocation networks (fig.1). Between the dislocation walls, 1/2<110> edge dislocation dipoles and rows of loops as a result of the collapse of the dipoles can be seen at the same time (fig.3). These reveal that the two mechanisms of dislocation movements, i.e. the climb of edge dislocations, and the collapse of edge dipoles, should be taken into account. However, it is known that the edge dipoles are formed as a result of the slip of 1/2<110> screw dislocation pairs along {111} planes[8]. When the dislocation networks are formed during secondary stage creep, the role of glide of dislocations between the dislocation walls is only to keep the dislocation density constant. Hence, the climb of dislocations along the walls and in the networks occurs and controls the strain rate during the secondary stage creep, while the motion of the dislocation dipoles will be limited by the above dislocation climbing process.

At 1173K under the stress of 200MPa, the dislocation networks in creep specimen are found to be loose and regular (fig.5) as a result of more rapid climb of dislocations along the walls and in the networks, and annihilation with opposite – sign dislocations climbing in the opposite direction. Therefore, the climb becomes the predominant process to control the creep rate for the DS Ni_3Al alloy in the overall temperature range from 923 to 1173K.

Furthermore, the stress dependence analysis at the temperature of 1023K shows that the experimentally determined stress exponent n (4.7) is within the range 4 – 5 predicted by the dislocation climbing mechanism, while the stress exponent of 3, which characterizes the viscous dislocation glide mechanism, is not observed in the present study. Besides, the temperature dependence analysis under a stress of 500MPa shows that the experimental activation energy for creep (326.6kJ/mol) is very close to the range 272 – 301kJ/mol for the activation energy of volume difusion for Ni_3Al alloys[9].The higher value of the activation energy for the multicomponent Ni_3Al alloy may be attributed to the elemental additions increasing the difficulty in diffusion. Further work is needed to clarify the mechanisms of diffusion and creep.

Finally,with respect to the stress exponent value, the activation energy, and the dislocation structure during creep, the DS Ni_3Al alloy would belong in a material of class 2, in which the secondary stage creep rates are controlled by the dislocation climbing at the temperature of 1023K. From the dislocation structure results, the same mechanism would be operative in the overall temperature range from 923 to 1173K.

CONCLUSIONS

(1) Dislocation climbing occurs during secondary stage creep for the DS Ni_3Al alloy in the temperature range from 923 to 1023K under the stress of 500MPa, and at the temperature of 1173K under the stress of 200MPa.
(2) At the temperature range from 923 to 1173K, the secondary stage creep rates for the DS Ni_3Al alloy are governed by dislocation climbing, and the activation energy for creep is 326.6kJ/mol.

ACKNOWLEDGEMENTS

This research was sponsored by the National Advanced Materials Committee of China.

REFERENCES

1. P.Flinn,Trans,AIME,218,145(1960)
2. J.R.Nicholls and R.D.Rawlings, J. Mater. Res., 1,68(1986).
3. J.H.Schneibel, G.F.Peterson and C.T.Liu, J. Mater. Res., 1, 68(1986).
4. J.H.Schneible and W.D. Porter, J. Mater. Res., 3, 403(1988).
5. B. Ilschner, Hochtemperaturplastizität Springer Verlag Berlin,(1973).
6. O.D.Sherby and P.M.Buke, Progr.Mat.Sci., 13, 325(1968).
7. J.Weertman, J. Appl. Phys., 28, 1185(1957).
8. A.E.Staton – Bevan and R.D. Rawlings, Phil.Mag.,32,787(1975).
9. T.C.Chou and Y.T.Chou, in High Temperature Ordered Intermetallic Alloys,edited by C.C.Koch, C.T.Liu and N.S.Stoloff (Mater. Res. Soc.Proc.,39,Pittsburgh, PA, 1985),pp461 – 474.

DUCTILITY ENHANCEMENT VIA MICROSTRUCTURAL CONTROL IN Cr MODIFIED Ni_3Al SINGLE CRYSTAL

S.E. HSU, Y.P. WU, T.S. LEE AND S.C. YANG
Materials Research and Development Center, Chung Shan Institute of Science and Technology, Lung-Tan, Taiwan, R.O.C.

ABSTRACT

An attempt to improve the ductility in Cr modified Ni_3Al alloys is made by means of microstructural control. In order to select an alloy composition such that it locates in the γ' single phase region at 1000 °C and in the $\gamma + \gamma'$ two-phase region at 1200 °C of the Ni-Al-Cr ternary phase diagram, we first calculated γ and γ' phase boundaries, using the Cluster Variation Method. Single crystal specimens with desired composition of Ni-19.5Al-7.5Cr at% were grown and homogenized at the above two temperatures for 72 hours and the mechanical properties were tested at room temperature. It is found that the elongation of specimen homogenized at 1200 °C is almost twice as large as that of one homogenized at 1000 °C. The theoretically predicted microstructures and the γ precipitates are also confirmed by optical, SEM micrographs and EDS analysis. It is concluded that a small amount of disordered γ phase precipitates in the γ' matrix can drastically enhance the tensile ductility of Ni_3Al ($L1_2$)-base alloys

INTRODUCTION

Aluminide intermetallic alloys are currently of considerable interest as a basis for a new generation of high temperature structural materials. In particular, alloys based on Ni_3Al are especially attractive because of their positive temperature dependence of yield strength and excellent high temperature oxidation resistance. However, most ordered intermetallics are, in nature, brittle, which provides one reason why they have had limited applications as structural materials yet. In this regard, a large amount of research studies have been directing at the ductility improvement in ordered intermetallics.

Based on our past experience in the study of a number of different ternary additions to Ni_3Al, we have empirically found that, in addition to Ni-rich alloys, specimens with good ductility were always obtained if there are small amount of disordered γ phase precipitates in the γ' matrix [1]. However, since the alloy compositions were not fixed, it is difficult to determine whether the ductilization is due to different chemistry or microstructure. The purpose of this work is to test the idea that a new method to ductilize an $L1_2$ type ordered intermetallics can be accomplished by means of a proper microstructural control (i.e. control the thermal treatment).

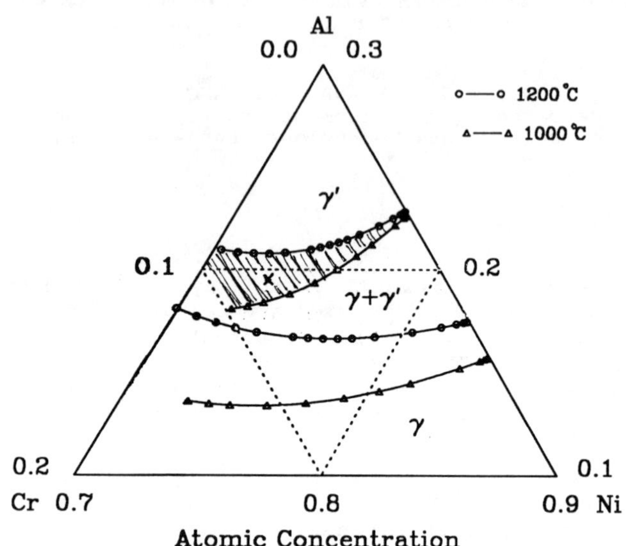

Figure 1: Calculated isothermal section of γ/γ' phase boundaries in the Ni-Al-Cr system at 1000 °C (triangle symbol) and 1200 °C (circle symbol). The composition of Ni-19.5Al-7.5Cr alloy is labelled "x" within the shaded area.

ALLOY DESIGN

As can be seen from the binary Ni-Al phase diagram that the solubility of the γ' phase almost does not change within a reasonable homogenization temperature range and in the light of the fact that ternary additions can have drastic effect on γ' solubility lobe [2,3], a ternary system is introduced in this study. Since additions of Cr in Ni_3Al were found to have beneficial effects on tensile strength and the creep resistance [1,4], the ternary element used in this study is, therefore, chromium.

In order to find a suitable alloy composition such that it can locate in the γ' single phase region at one temperature and in the γ + γ' two phase region at another temperature, we begin with a reliable thermodynamic calculation based on the tetrahedron approximation of Cluster Variation Method (CVM) [5,6]. Figure 1 shows the calculated Ni-Al-Cr partial ternary phase diagram at two temperatures of 1000 °C and 1200 °C. The shaded area in Fig. 1 is of particular interest, since an alloy with its composition inside this area will be monolithic γ' phase if homogenized at 1000 °C, whereas it will be γ + γ' two phases if homogenized at 1200 °C. An alloy with its composition of Ni-19.5Al-7.5Cr in atomic percent is, accordingly, selected in this study.

EXPERIMENTAL

The Ni-19.5Al-7.5Cr alloy was prepared by vacuum induction melting. Button-head type single crystal specimens were grown through a modified Bridgeman technique. The orientation of the single crystal specimens was determined by Laue back-reflection method and only specimens with their normal directions within 10° from a <100> direction were used in this study. Tensile specimens were machined to a gauge length of 30 mm and a diameter of 6.25 mm. The tensile test was performed on an Instron testing machine with a constant cross-head speed of 1 mm/min in air.

Since the accuracy of the alloy composition is very critical, it was determined by inductively coupled plasma spectroscopy before and after single crystal growth. Chemical analysis of the specimens shows a slight variation in compositions after single crystal growth but it can still meet the theoretical requirements (i.e. within the shaded area in Fig. 1). The real composition is 73.2Ni-19.4Al-7.4Cr. The test bar specimens were homogenized at 1200 and 1000 °C, respectively, in an argon atmosphere for 72 hours and were followed by a rapid argon gas quench treatment from the furnace temperature. Both the γ and γ' phases were identified from the SEM microstructures and energy dispersive X-ray spectroscopy (EDS).

RESULTS AND DISCUSSION

The microstructures of the homogenized Ni-19.5Al-7.5Cr alloy are illustrated in Fig. 2a and 2b for 72 hours at 1000 °C and 1200 °C, respectively. As expected, the microstructure of the alloy homogenized at 1000 °C is γ' single phase only and at 1200 °C is $\gamma + \gamma'$ two phases. It is interesting to note that the precipitation of γ phase gives a microstructure which is very similar to an eutectic microstructure. The region A in Fig. 2b is monolithic γ' phase, while the region B is a two phase mixture of γ and γ'. In order to further distinguish the γ and γ' phase, the region B of Fig. 2b is magnified as shown in Figure 3, where the dark cuboidal particles are γ' phase and the white thin film between cuboidal particles is γ phase. The EDS analysis shows that the γ precipitates are Cr-rich and Al-deficit, which is in agreement with our theoretical calculations as can be deduced from the tie-line in Fig. 1. More accurate composition measurement, using the EPMA technique, which will allow us not only to qualitatively determine the γ and γ' phases but also to quantitatively compare the tie line composition data between CVM calculations and the experimental measurements, are currently underway.

The mechanical properties of the alloys homogenized at different temperatures are listed in Table I. As can be seen from Table I, without sacrifice in tensile strength, the two phase specimen has almost twice the elongation as the single phase one, which provides a new way to effectively enhance the ductility in $L1_2$ intermetallics. The tensile specimens after the test are shown in Figure 4.

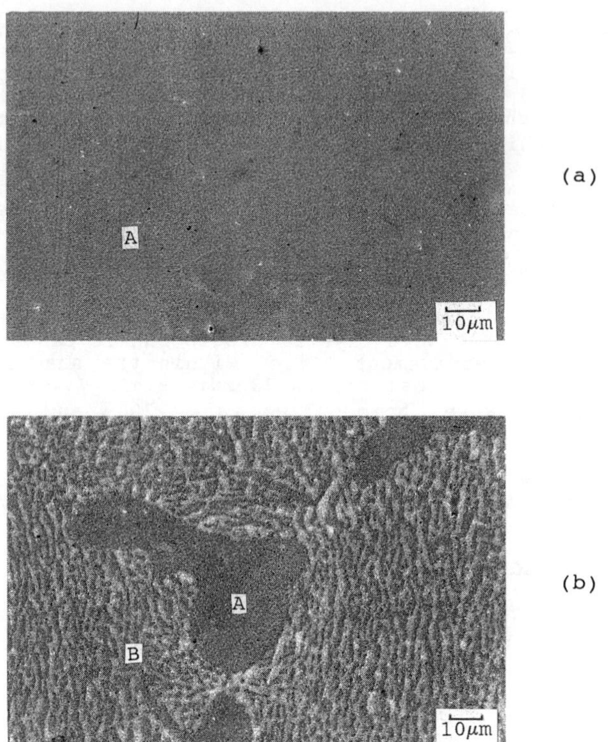

Figure 2: SEM micrograph of Ni-19.5Al-7.5Cr alloy for 72 hour homogenized at (a) 1000 and (b) 1200 °C. Region A consists of γ' and region B is a two phase mixture of γ and γ'.

Table I Effect of microstructure on room temperature tensile properties of the Ni-19.5Al-7.5Cr alloy

	homogenized at 1000 °C (γ' phase only)	homogenized at 1200 °C (with γ precipitates in γ')
Yield Strength	219.5 MPa	292.9 MPa
UTS	494.9 MPa	416.5 MPa
Elongation	21.0 %	39.1 %

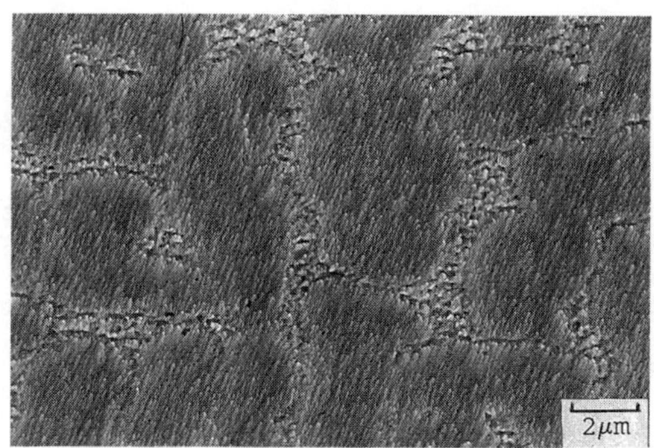

Figure 3: Magnified SEM micrograph of region B in Fig. 2. This micrograph reveals cuboidal γ' particles surrounded by thin layer of γ phase.

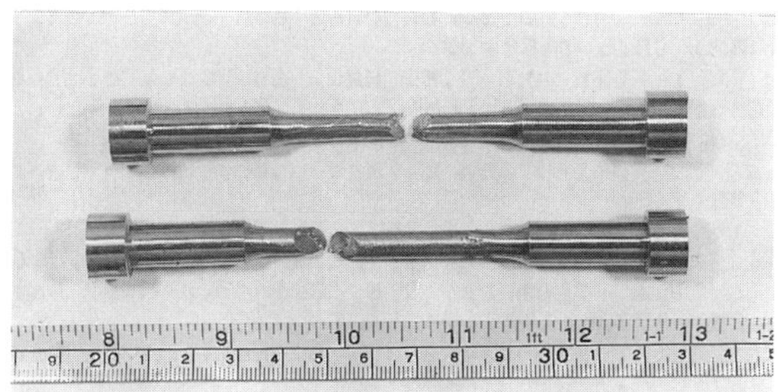

Figure 4: Tensile specimens of Ni-19.5Al-7.5Cr alloy, top one homogenized at 1000 °C and bottom one at 1200 °C.

The ductilization mechanism is not clear yet, however, as can be seen from Fig. 3 that, since the $\gamma + \gamma'$ eutectic structure is about micron scale, the thin layer of the ductile γ phase surrounding the cuboidal γ' particles must have a significant effect on the ductility enhancement of $L1_2$ intermetallics. It should be mentioned that, recently, Cahn et al [7,8] have found that the disordered phase tends to form from APBs in Ni-rich alloys and argued that this disordered APBs could enhance the ductility of Ni_3Al alloy. It will be an interesting subject to study whether the nucleation sites of the disordered γ phase observed in our work is exactly developing from the APBs as suggested by Cahn et al. More study should be done in this respect.

CONCLUSIONS

We have found and confirmed that, by means of the precipitation of a small amount of disordered γ phase, it is able to increase the ductility of $L1_2$ based Ni_3Al ordered intermetallic alloys. We have also found that the Cluster Variation Method can be served as a useful tool in the microstructure design of ductile $L1_2$ intermetallics, since it provides accurate thermodynamic information between γ and γ' phase equilibrium.

ACKNOWLEDGEMENTS

The authors would like to thank R.R. Jeng, T.S. Liu, L.C. Wang, C.Y. Wu and S.H. Pong for their assistance in the preparation and testing of the material.

REFERENCES

1. S.E. Hsu, C.H. Tong, T.S. Lee and T.S. Liu, in High-Temperature Ordered Intermetallic Alloys III, MRS symposia, ed. by C.T. Liu, A.I. Taub, N.S. Stoloff and C.C. Koch, Vol 133, pp. 275-280 (1989).
2. Y.P. Wu, N.C. Tso, J.M. Sanchez and J.K. Tien, Acta Metall. **37**, 2835 (1989).
3. Y.P. Wu, Y.L. Lin and S.E. Hsu, in Proc. of The 1992 Annual Conf. of The Chinese Soc. for Mat. Sci., pp. 98-99 (1992).
4. A. Inoue, H. Tomiku and T. Masumoto, Met. Trans. A **14**, 1367 (1983).
5. R. Kikuchi, Acta Metall. **25**, 195 (1977).
6. C. Sigili and J.M. Sanchez, Acta Metall. **33**, 1097 (1985).
7. R.W. Cahn, P.A. Siemers, J.E. Gieger and P. Bardhan, Acta Metall. **35**, 2737 (1987).
8. R.W. Cahn, P.A. Siemers and E.L. Hall, Acta Metall. **35**, 2753 (1987).

STRENGTH AND DUCTILITY OF Fe_3Al WITH ADDITION OF Cr

D.G. MORRIS, M.M. DADRAS and M.A. MORRIS
Institute of Structural Metallurgy, Avenue de Bellevaux 51, University of Neuchâtel, NEUCHATEL 2000, SWITZERLAND

ABSTRACT

The development of Fe_3Al aluminides has been restricted in the past by poor ductility at ambient temperatures, and it is only recently that possible solutions to this problem have been found. It was shown a few years ago (1) that the addition of 2-6% Cr to a Fe_3Al base alloy led to good ductility and this improvement was explained by a reduction of the APB energy, increasing the separation of the superpartial dislocations and thereby allowing easier dissociation of these dislocations, easier cross slip and a reduced tendency to stress and strain concentrations. However, at a later stage, an alternative explanation was proposed (2,3) based on examinations of both FeAl and Fe_3Al alloys under different environments, and the ductility change was explained in terms of chemical attack at the tip of a crack leading to local hydrogen embrittlement.

The present study re-examines the behaviour of a Fe_3Al alloy both with and without the addition of Cr. Strength, work hardening behaviour and failure ductility are examined under conditions where environmental effects should not be important, and the mechanical behaviour is interpreted in terms of significant variations in the type of order, the ordered domain structure and the resulting dislocation structures. It is seen that the addition of Cr can lead to a better ordered material and the differences in ordered state between the two materials can significantly affect dislocation behaviour and mechanical properties.

EXPERIMENTAL TECHNIQUES

Strips of two alloys of composition Fe-28%Al and Fe-28%Al-4%Cr were supplied for study by C.T. Liu and C.G. McKamey of ORNL. These strips were produced by arc-melting, drop casting and hot rolling to a thickness of about 0.7mm. Recrystallization treatments were given to remove the worked structure remaining after rolling and to give a range of grain sizes for the study of mechanical properties, and tensile samples were cut from the sheets by spark machining. Before mechanical testing the samples were electropolished to remove any surface oxide. Based on earlier studies (2,3) it is known that such polishing largely removes any environmental protection due to iron or chromium oxides formed on the surface during heat treatment such that mechanical properties are largely determined by intrinsic dislocation behaviour. Mechanical testing was carried out, at room temperature, under an inert atmosphere of argon in order to further ensure the absence of any different environmental effects between the two materials.

For the ternary, Cr containing alloy, recrystallization was carried out at temperatures in the range 750-850°C for various times giving a range of grain sizes from about 90μm to 230μm. The binary alloy contained a small amount of TiB_2 which had been added to avoid grain growth and it was difficult to obtain a large range of grain sizes. Recrystallization temperatures of 1100-1300°C were used to give a grain size that ranged from about 45μm to about 100μm. Following the recrystallization heat treatments, all samples were given standard ordering treatments based on those recommended in previous work (1). All alloys treated at temperatures other than 850°C were given a short anneal of 15mins at this temperature to ensure an equilibrium degree of B2 order for this temperature; the samples were cooled to room temperature and then given a second ordering treatment of 1 week at 500°C to ensure that the DO_3 order was achieved. These treatments were carried out with the temperature controlled to a precision of about ±1°C. Following tensile testing the state of order and dislocation microstructures were examined by TEM, making use of EDS analysis, electron microdiffraction, high resolution imaging and x-ray diffraction as necessary . In addition fracture surface morphologies were examined by SEM to deduce the mode of rupture.

RESULTS

Ordered State

The Fe-Al-Cr alloy was found to have a high degree of order after the 850/500°C annealing treatments and showed large domain sizes with fine domain wall images in the TEM. The degree of order was determined by x-ray diffraction from the relative intensity of 111 and 444 peaks (to deduce the degree of DO_3 order) and from the relative intensity of 200 and 400 peaks (to deduce the degree of B2 order). When calculating the degree of order it was assumed that 25%Al lies on the Al sites in the Fe_3Al lattice and the excess Al as well as the Cr lie on Fe sites. The degree of order was deduced in this way to be 0.95, the size of B2 domains was very large (estimated to be about the same as the grain size), and the size of the DO_3 domains was about 1μm. It is thus clear that ordering has proceeded well at both 850°C and 500°C for this material.

The binary alloy showed an ordered microstructure that varied from case to case. To a large extent the variation in ordered structure could be correlated with the grain size, or more specifically with the temperature used for recrystallization, as illustrated in Fig. 1. In this figure it is seen that an increase in recrystallization temperature from 1100°C to 1250°C (Fig. 1(a) and Fig. 1(c)) leads generally to a decrease in domain size and to a thickening of the domain walls. It may also be seen that there were variations of the ordered state for the same recrystallization and ordering conditions (Fig. 1(a) and Fig. 1(b)). Analysis of chemical composition by EDS indicated that these materials all had the same composition, as did the domain interiors and the domain walls, and it thus appears that ordering is highly sensitive to minor fluctuations in furnace temperature or alloy composition. In the case where the material was ordered with fine domain walls (Fig. 1(a)) the B2 domain size was 2-3 times larger than that of the DO_3 domains, whilst for the other cases the B2 and DO_3 domain sizes were identical, as confirmed by dark field imaging using superlattice reflections corresponding to each of these lattices. For the cases where thick domain walls were seen (Fig 1(b) and (c)) examination of the walls by microdiffraction confirmed that the walls were in fact disordered, as illustrated in Fig. 2, where it may be seen that both B2 and DO_3 superlattice reflections found when examining the domain interior are missing when examining the domain wall. For those cases of the Fe-Al-Cr alloy where the domain walls were very thin (eg Fig. 1(a)) the domain walls show no sign of containing a layer of disorder. Imaging such domain walls using a superlattice reflection of the B2 lattice makes many of these walls invisible, as would be expected from a shear-produced fault not containing

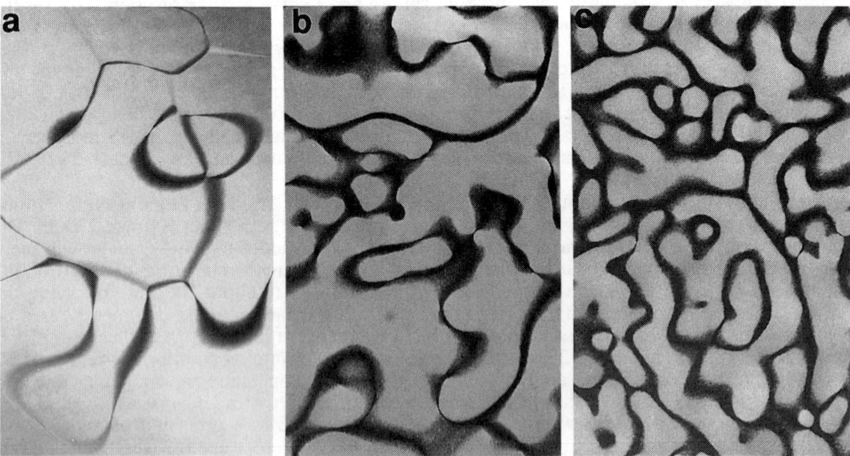

FIG. 1: Dark field electron micrographs (111**g** vector) showing domain boundaries and disordered domain walls in Fe-Al material: (a) recrystallized for 2h at 1100°C giving a degree of order of 0.6; (b) 2h at 1100°C, order 0.87; (c) 4h at 1250°C, order 0.87.

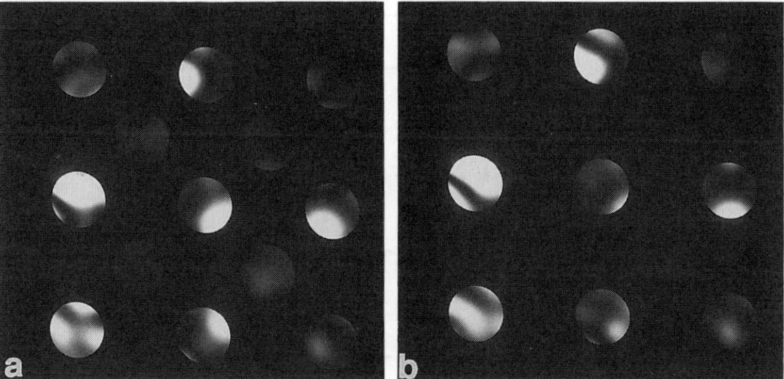

FIG. 2: Converging beam electron diffraction patterns showing {001} diffraction pattern of (a) domain interior, and (b) domain wall of Fe_3Al material showing thick, disordered domain walls. Sample shown in similar to that of Fig. 1(c).

any disorder. In addition, imaging such boundaries by high resolution electron microscopy, Fig. 3, confirms that the ordered structure composed of planes of different atomic species remains ordered up to the fault plane, within the resolution possible by this mode of imaging. The degree of order of the binary Fe_3Al materials as deduced by x-ray analysis varied from sample to sample in a manner that seemed to be unrelated to the grain size, recrystallization temperature or the domain size. This variation in degree of order (both B2 and DO_3) from about 0.6 to 0.87 is apparently explained by the slight, uncontrolled fluctuations in composition and temperature.

Mechanical Behaviour

The mechanical behaviour of samples depended sensitively on the degree of order as well as the domain size and the grain size. Fig. 4 illustrates, for the binary Fe_3Al alloy, that the yield stress can be related to these structural parameters and most particularly to the degree of order. The high values of yield stress correlate well with a low ($\approx$0.6) degree of order, low values of yield stress

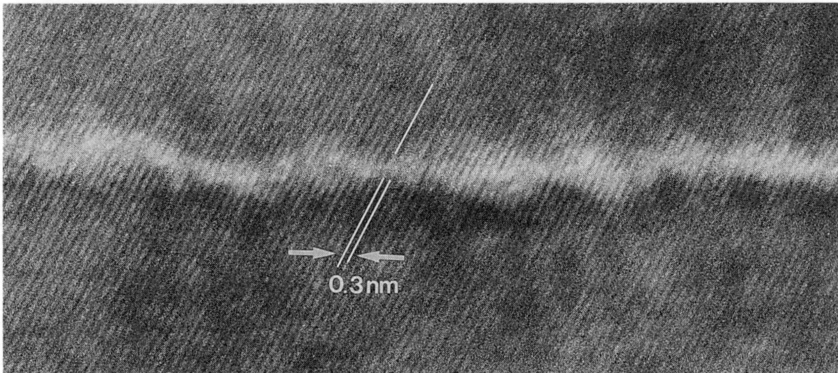

FIG. 3: High resolution electron micrograph showing ordered atomic planes (200) remaining continuous up to the domain boundary which crosses the middle of the field. The sample shown is binary Fe_3Al, shown in Fig. 1(a).

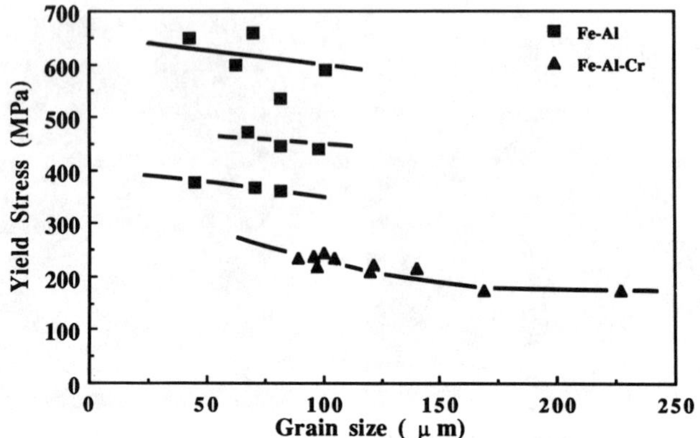

FIG. 4: Yield stresses of binary Fe-Al and ternary Fe-Al-Cr samples as a function of the grain size. The ternary alloy samples were all well ordered with a large domain size. The binary alloy samples show different behaviour according to the order parameter: the upper line corresponds to samples with an order parameter of ≈0.6; the middle line to those with an order parameter of ≈0.75; and the lower line to those with an order parameter of ≈0.87.

with a high degree of order (≈0.87), with intermediate order values leading to intermediate stresses. At the same time it is seen that grain size plays a minor role over the range of grain sizes encountered here (45-100μm for the binary alloy and 90-230μm for the Fe-Al-Cr alloy). For the binary alloy the increase in grain size is accompanied by a decrease in the domain size, typically from 300nm at the small grain size (45μm) to 125nm at the large grain size (100μm): this decrease in domain size is accompanied by an increase in domain wall thickness from 0-10nm to about 30nm. The effect of domain size or wall thickness on yield stress is not large and cannot be distinguished from that brought about by grain size change.

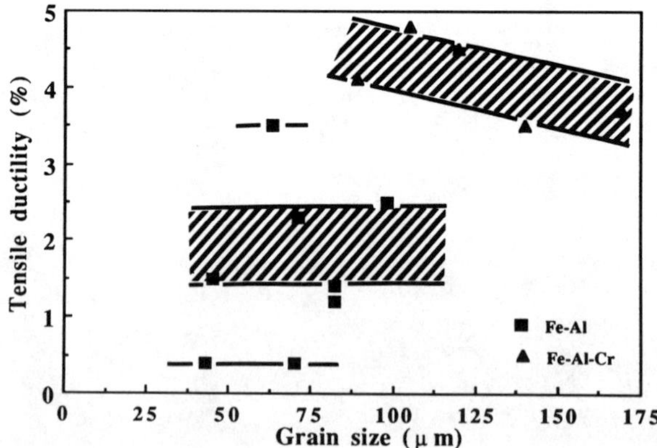

FIG. 5: Ductility of Fe-Al and Fe-Al-Cr as a function of the grain size. The ternary alloy shows good ductility which decreases continuously as the grain size increases. The ductility of the binary Fe-Al samples depends on the degree of order. Samples of low order (0.6) have low ductility; samples of high order (≈0.87) have medium ductility; samples of medium order (≈0.75) have high ductility.

The same microstructural parameters affect the tensile ductility as shown in Figure 5. While the grain size plays a noticeable role in determining ductility of the ternary alloy, which always shows reasonable ductility, the range of grain sizes covered for the binary alloy is too small for a clear influence to be distinguished. For the binary alloy, the effect of order parameter on ductility is evident. The lowest ductility was found on samples with low order (order parameter $\approx$0.6); samples with highest order ($\approx$0.87) led to medium-level ductilities; while samples with medium order parameters ($\approx$0.75) gave the highest levels of ductility. The reasons for high ductility are understood in terms of yield stress, work hardening behaviour and failure stress, as will be described later.

Dislocation Structures

After straining by about 1% the Fe-Al-Cr alloy shows superdislocations composed of four partial dislocations each of Burgers vector $1/4\langle 111\rangle$, as illustrated in Fig. 6(a). Such dislocations are typical of material ordered to the DO_3 structure (4) and the spacing of the closer partials found is 6-16nm for screw-edge dislocations while the spacing of the wider pairs is 28-34nm. Using suitable values for elastic constants (4) the APB energies can be estimated as 80-90mJ/m^2 for the B2 fault and 60mJ/m^2 for the DO_3 fault. These values are in fact close to those determined by Crawford and Ray (4) on well-ordered Fe-28%Al, and much higher than the low values reported earlier by McKamey et al (1). It thus seems unlikely that the Cr has a major effect on the APB energies of well-ordered material. After straining to larger strain, for example 4%, these superdislocations uncouple into pairs of 2-fold dislocations trailing DO_3 APB faults.

The dislocations found in the binary Fe-Al samples vary according to the degree of order and the domain structure. For the samples of low order (Fig. 6(b)) and with large, shear-like domain walls, the dislocations are seen in arrays of single dislocations with virtually no tendency to coupling as loose pairs. For such materials it is clear that the dislocations move with the destruction of short or long range order and thus tend to remain in such planar arrays. It is only as the degree of order increases to higher values that the dislocations become loosely coupled as pairs. For all the materials containing thick, disordered domain walls, the dislocations seen are single dislocations which tend to remain in the disordered regions after deformation, as seen in Fig. 6(c). The dislocations seen here show a certain resemblence to those found in conventional superalloys where the superdislocations present in the ordered regions dissociate in the disordered matrix regions.

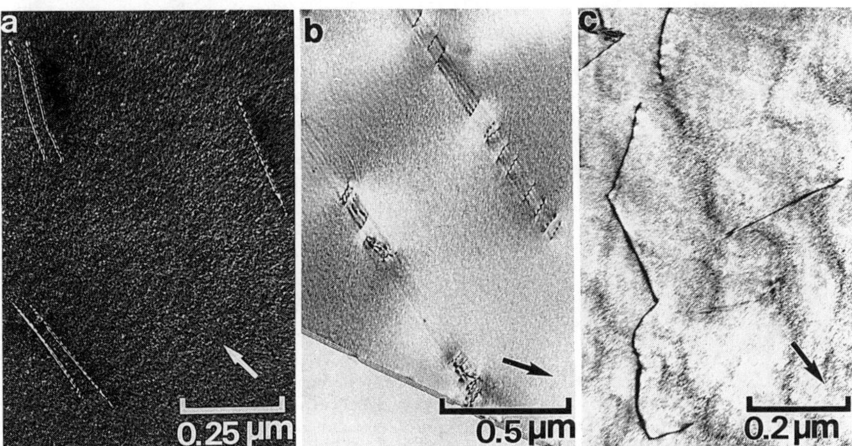

FIG. 6: Dislocations in Fe-Al-Cr and Fe-Al after slight room temperature deformation: (a) Fe-Al-Cr deformed to 1% strain; weak beam image, zone axis near (112), **g** vector $2\bar{2}0$; (b) Fe-Al ordered to 0.6 with large, fine domain walls, strained 1%; zone axis (001), **g** vector 220; (c) Fe-Al ordered to 0.75 with small domains and disordered walls, strained 1%; zone axis (110), **g** vector $2\bar{2}0$.

DISCUSSION

The mechanical behaviour of the binary and ternary alloy can be explained by the difference between yield and failure stresses and by the work hardening behaviour. The ternary alloy shows a low yield stress with a fairly low work hardening rate which means that a significant strain can be accumulated before failure. Yield occurs in the well ordered lattice by the easy movement of superdislocations, which also leads to a low work hardening rate since dislocation debris builds up slowly. Samples of the binary alloy show a yield stress which is medium or high, depending on how much disorder is created by dislocation passage. The work hardening rate of such materials is initially very high, but falls parabolically with strain to very low values: this can be understood since the dislocations are readily blocked in domain walls creating obstacles to other dislocations, but these obstacles are not as stable at high stresses and strains as those composed of superdislocations. It is for an intermediate range of conditions that a low yield stress and medium work hardening rate allows significant strain before failure. The failure stress was in fact approximately the same for all materials, irrespective of whether these were of the binary or ternary alloy, and independent of grain size, occurring largely by cleavage. It is interesting to note that deformation and failure behaviour is determined by the state of order and the domain boundaries, and not directly by the addition of Cr to the binary Fe_3Al alloy. The quantitative analysis of stresses and work hardening rates will be reported elsewhere (5).

The unexpected discovery of the present work is the varying partial order for the binary alloy since an alloy with 28%Al should be single phase (ordered or disordered) at all temperatures (6). The microstructures seen are similar to those found in hypo-stoichiometric Fe_3Al alloys (7). For a significant amount of disorder to arise on annealing at 500°C the average composition should be 24-25%Al, and the ordered and disordered phases should have different compositions, by about 3% (6). While EDS in a TEM cannot determine the precise absolute composition, the difference between the confirmed nominal composition (28%Al) and that necessary to observe disorder (24-25%Al) seems excessive: in addition, comparative measurements of the compositions of domain centres and domain walls on nearby areas of thin foils without changing beam or sample configuration (which should give a precise comparison) failed to find any difference.

The unusual disorder observed may be explained by the dissolution of a small amount of TiB_2 during high temperature recrystallization. A small amount of substitutional Ti and interstitial B may change the phase diagram, as already pointed out in the case of Ti (8). Depending on the recrystallization temperature a different amount of dissolution will occur leading to changes in the subsequent ordering, as seen (Fig. 1). This deduction implies that the ordered structure of these materials, and therefore their mechanical behaviour, depends rather sensitively on the amount of certain alloying additions and comparisons between alloys should be conducted with caution.

As a final comment, it should be remembered that annealing an ordered material containing APB's at temperatures below but close to the disordering temperature may be responsible for the creation of a thin disordered film along the boundaries (9,10). The amount of this disorder may depend on the composition of the alloy, how close it is to stoichiometry, and perhaps on additional alloying elements. Such disorder has been seen along domain boundaries (9,10) but may equally be expected to form along grain boundaries. This disorder may greatly affect mechanical behaviour, both through destabilising and dissociating a superdislocation, as observed in the present study, as well as affecting the tendency to crack formation at grain boundaries, as has been postulated to explain the ductility of Ni_3Al with B additions (11,12).

REFERENCES

1. C.G. McKamey, J.A. Horton and C.T. Liu, J. Mater. Res. **4**, 1156 (1989).
2. C.G. McKamey and C.T. Liu, Scripta Metall. et Mater. **24**, 2119 (1990).
3. C.T. Liu, E.H. Lee and C.G. McKamey, Scripta Metall. **23**, 875 (1989).
4. R.C. Crawford and I.L.F. Ray, Phil. Mag. **35**, 549 (1977).
5. D.G. Morris, M.M. Dadras and M.A. Morris, Acta Metall. et Mater., in press.
6. P.R. Swann, W.R. Duff and R.M. Fisher, Met. Trans. **3**, 409 (1972).
7. J.W. Park and I.G. Moon, Mater. Sci. and Eng. **A152**, 341 (1992).
8. M.G. Mendiratta, S.K. Ehlers and M.V. Nathal, Met. Trans. **20A**, 1701 (1989).
9. D.G. Morris, Phys. Stat. Sol. **(a) 32**, 145 (1975).
10. C. Ricolleau and A. Loiseau, Proc. Conf. "iib 92", Thessalonica, Greece, 21June(1992).
11. K. Aoki and O. Izumi, Nippon Kinzoku Gakkaishi, **43**, 1190 (1979).
12. C.T. Liu, C.L. White and J.A. Horton, Acta Metall. **33**, 213 (1985).

THE EFFECT OF CARBON AND THERMAL EXPOSURE ON THE TENSILE BEHAVIOR OF Ti-48Al-1V (at%)

WILLIAM T. DONLON AND W.E. DOWLING, Jr.
Ford Motor Company, Research Laboratory, S-2065, P.O. Box 2053, Dearborn, MI 48121

ABSTRACT

Room temperature tensile properties of Ti-47.5Al-1V-0.2C (at%) and Ti-48.2Al-1V-.06C (at%) alloys were measured after thermal exposure at 775°C (in air and vacuum) to evaluate the influence of carbide precipitation and environment. Both alloys possess ≈2% tensile ductility in the unexposed condition. Thermal exposure of fully machined tensile samples consistently reduces the ductility of the 0.2 carbon alloy to ≈0.5%. Exposure of the low carbon (0.06) alloy in vacuum results in no ductility loss, while exposure in air reduces the ductility to 1.3%. Yield strengths are unaffected by thermal exposure and are 450 MPa and 300 MPa for the 0.2 and the low carbon alloys, respectively. The pre-exposure ductility is recovered for tests in which thermally exposed machined tensile samples had ≈10μm of their surfaces removed by polishing. Thermal exposure prior to machining results in no change in tensile behavior. SEM and TEM examination of thermally exposed surfaces show that below the oxide surface layer, a layer close in composition to Ti_2Al, having a simple cubic (a_o=6.85Å) structure is present. Beneath this layer, Ti_3AlC carbides are observed in the 0.2 carbon alloy. The density of these carbides is observed to decrease away from the surface. No carbides were observed in the low carbon alloy. Although carbon significantly enhances the yield strength of this alloy it also makes it much more susceptible to embrittlement from thermal exposure.

INTRODUCTION

Recent research on alloy development and microstructural optimization of two-phase ($\gamma + \alpha_2$) Ti-Al alloys has produced material with consistently greater than 2% room temperature tensile ductility[1,2]. Maintaining these properties after repeated thermal exposures has not yet been demonstrated. Precipitation of carbides[3,4], and/or the formation of various surface phases[5,6] during thermal exposure, can alter mechanical properties. The purpose of this study is to evaluate the effect of thermal exposure on microstructure and the tensile properties of a Ti-48Al-1V (at%) alloy as a function of carbon content.

EXPERIMENTAL PROCEDURE

Castings (0.2m dia x 0.9m) with nominal compositions of Ti-48Al-1V but with different carbon contents were acquired from TIMET after hot isostatic pressing. The composition of these castings as determined by TIMET are shown in Table I.

TABLE I Material Chemistry (at%)

	Aluminum	Vanadium	Carbon	Oxygen
0.2 C Alloy	47.5	1.1	0.21	0.13
Low C Alloy	48.2	1.1	0.06	0.15

The 0.2 C ingot was sectioned into 80 mm tall mults and the low C ingot was sectioned into 150 mm tall mults. Both alloys were isothermally forged at 1150°C to thicknesses of 19 mm and 44 mm, respectively. Different multiple step heat treatments and cooling rates were utilized to produce microstructures with a range of α_2 and ($\gamma + \alpha_2$) lath contents for both castings. Stabilization heat treatments for the castings were selected to stay above the carbide solvus[4].

Quantitative microscopy was performed on backscattered electron (BSE) images from a scanning electron microscope (SEM) with a Kontron image analyzer. Heat treatments were chosen to produce similar microstructures for both alloys to isolate the effect of carbon on mechanical properties. Grain size measurements were made using the linear intercept method. BSE images of the microstructures used in this study are shown in FIG 1. The corresponding heat treatments, grain intercept sizes, α_2 and lath contents are listed in TABLE II. It is noted that the lower α_2 content in the 0.06 carbon alloy is the result of the higher aluminum content (the α-Ti formed during the heat treatment is more difficult to retain as α_2 in subsequent heat treatment steps).

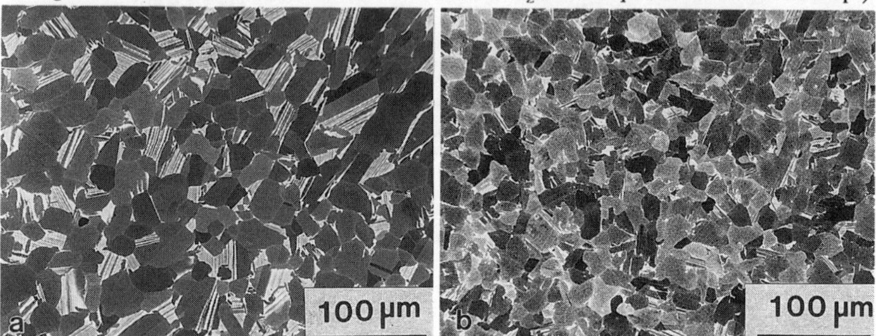

FIG 1. BSE images of (a) the 0.2 C alloy, and (b) the low C alloy (TABLE I) used in this study. Heat treatments used to produce these microstructures are shown in TABLE II. Cooling rates were 1°C/s and 2°C/s for the 0.2 C and the low carbon alloys, respectively.

TABLE II. Microstructural Parameters

Alloy	Heat Treatment	Intercept Size	α_2 Vol	Lath Vol
0.2 C	1300°C/24h + 1275°C/16h + 1000°C/8h	35 μm	15 %	20 %
Low C	1335°C/8h + 1350°C/8h + 925°C/24h	20 μm	4 %	8 %

To isolate surface effects, fully machined tensile samples and/or blanks were thermally exposed in lab air, dry air or vacuum (10^{-6}Pa) between 725°C - 925°C. Tensile samples were machined to a concentricity of better than .012mm and tested on a MTS servohydraulic system, ($\dot{\epsilon}=8{\cdot}10^{-3}$/s in stroke control mode) with fixed hydraulic collet grips and a precision alignment fixture[7].

Coupons (5 × 1.9 × 0.2 cm) from the forged pancake, were heat treated and thermally exposed for SEM and transmission electron microscopy (TEM) examination. Cross section metallographic samples were prepared from the coupons by cutting 0.5 cm × 1.9 cm strips from each coupon and cementing surfaces of interest together. These cross sectional specimens were examined in SEM. TEM of the surface layers was performed by back thinning one side of the coupons using either ion milling or electropolishing as the final step. Subsequent ion milling was performed on the oxidized surface to prior to examination to remove surface oxides.

RESULTS

Precipitation of Ti_3AlC[3,4] (Perovskite) has been observed in the 0.2 C alloy following thermal exposure in the temperature range of 625°C to 925°C. At temperatures below 800°C the precipitates occur in the matrix with precipitate free zones near grain boundaries and dislocations (FIG 2a), above approximately 800°C, the carbides only occur at dislocations and grain boundaries (FIG's 2b & 2c). Both morphologies of precipitates have been unambiguously identified by electron diffraction as Ti_3AlC having the Perovskite structure, and are referred to as homogeneously and heterogeneously nucleated Ti_3AlC, respectively. The range of temperatures and times over which these precipitates form was determined by TEM observations as part of a separate investigation of precipitation.

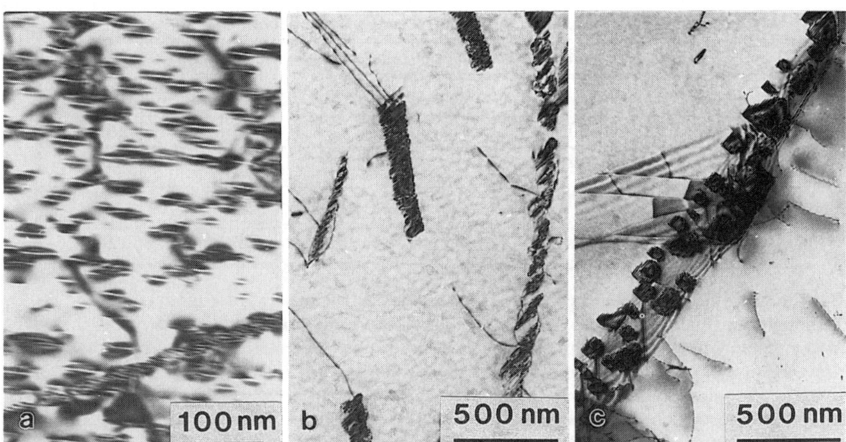

FIG 2. Different morphologies of the Perovskite Ti_3AlC precipitate observed by TEM in the 0.2C alloy. The sample shown in (a) was aged at 725°C for 100h, the sample shown in (b) was aged at 825°C for 100h, while the sample shown in (c) was aged at 875°C for 100h.

Fully machined tensile samples from both alloys were thermally exposed and tested at room temperature. Table III summarizes the mechanical properties (yield stress, ultimate tensile stress, total plastic elongation) obtained from these specimens. The marked ductility decrease of the 0.2 C alloy for all exposure conditions is in sharp contrast to the results obtained for the low C alloy.

TABLE III. Tensile data from thermally exposed tensile samples.

Alloy	Exposure / Number Samples	σ_{ys} (MPa)	σ_{uts} (MPa)	ϵ_p (%)
0.2 C	None / (6)	440	530	1.7
0.2 C	775°C/50h (air) / (2)	425	450	0.4
0.2 C	775°C/100h (vac) / (1)	445	475	0.6
0.2 C	775°C/100h (dry air) / (1)	450	453	0.3
0.2 C	815°C/4h (air) / (2)	465	480	0.35
Low C	None / (5)	303	402	2.20
Low C	775°C/100h (air) / (2)	299	343	1.3
Low C	775°C/20h (dry air) / (1)	302	367	1.9
Low C	775/100h (dry air) / (1)	310	358	1.3
Low C	775/100h (vac) / (2)	286	371	2.2

In a study described elsewhere[8], additional tensile samples from the 0.2 carbon alloy with a different microstructure, were thermally exposed in air for 4h at 815°C and two were tensile tested. The ductility decreased to 0.4% from 1.1% in the unexposed material. The remaining exposed samples had their gauge sections abraded with 600 grit SiC paper in order to reduce the gauge section diameter by $\approx$ 10μm. The tensile ductility after this surface removal returned to greater than 1% implying that embrittlement is a near surface phenomena.

Samples of the 0.2 carbon alloy were thermally exposed (725°C $\leq$ T $\leq$ 925°C) after stabilization but prior to machining. No embrittlement was seen in these samples. Results of these tensile tests are summarized in TABLE IV. Identical results were obtained for a microstructure having volume fractions of 16% α_2 and 40% lath.

TABLE IV. Tensile data from exposed then machined samples.

Alloy	Exposure / Number of samples	σ_{ys} (MPa)	σ_{uts} (MPa)	ϵ_p (%)
0.2 C	None / (2)	440	530	1.7
0.2 C	725°C/100h (air) / (1)	460	565	2.15
0.2 C	775°C/100h (air) / (1)	456	540	1.75
0.2 C	825°C/100h (air) / (1)	440	550	2.24
0.2 C	925°C/100h (air) / (1)	457	533	1.80

Fracture surfaces of unexposed and exposed tensile samples were examined in the SEM. Although the general fracture characteristics of all the tensile samples were identical, the edges of the fracture surfaces from the thermally exposed tensile samples (both alloys) showed the presence of a 0.5 - 1 μm surface film (FIG 3a, 3c). This feature was obscured by the thick oxide layer for samples exposed in air. Numerous surface cracks were also visible on the surfaces tensile samples which were thermally exposed (FIG 3b, 3d). These features were absent on fractured unexposed samples (FIG 3e and 3f, respectively).

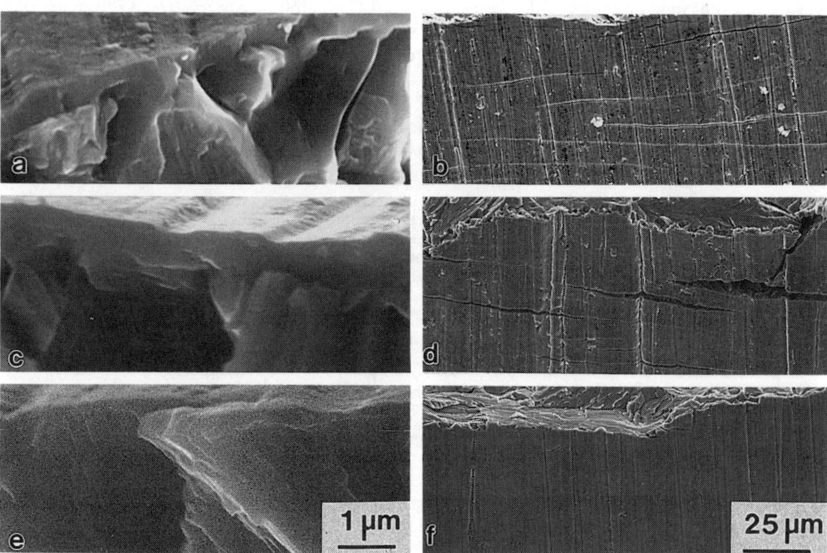

FIG 3. Secondary electron images of fractured tensile samples. (a) and (b) 0.2 C alloy exposed to 775°C for 100h in vacuum, (c) and (d) 0.06 C alloy exposed to 775°C for 100h in vacuum, while (e) and (f) correspond to a 0.2 C unexposed sample.

Polished cross sections of the surface were imaged with back scattered electrons in the SEM. No surface film is observed for the unexposed sample (FIG 4a). Thermal exposure to 775°C at 100h yields a 0.3 μm and a 0.8 μm thick layer, beneath the surface oxide, for exposures in vacuum (FIG 4b) and air (FIG 4c) respectively, for both alloys.

Back thinned specimens of the surface film from the 0.2 carbon alloy were examined in the TEM to identify the surface film. Electron diffraction and EDX spectroscopy showed that α-Al_2O_3 and TiO_2 (Rutile) were present on the surfaces of samples heat treated in air and vacuum. Removal of these oxides by subsequent ion milling of both sides of the specimen for short times showed that a phase (FIG. 5) having a composition close to Ti_2Al as determined by EDX spectroscopy was present. Analysis of electron diffraction patterns of this phase identified it as simple cubic with a lattice parameter of 6.85Å. This phase is what is observed as the bright surface layer shown in Figures 4b and 4c. Further ion milling of the sample revealed the microstructure of the underlying Ti-Al material.

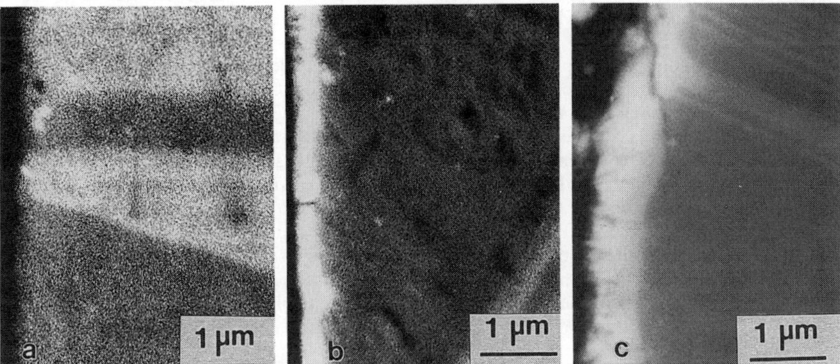

FIG 4. BSE images of the surface layers produced by thermal exposure to 775°C for 100h. (a), (b) and (c) correspond to no exposure, exposure in vacuum, and exposure in air, respectively.

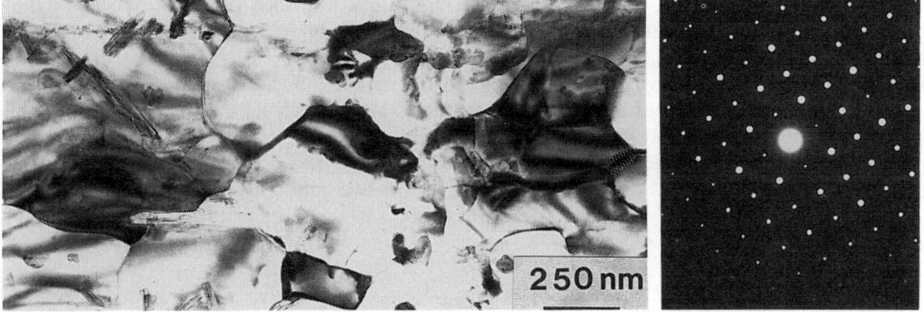

FIG 5. Back-thinned TEM image and diffraction pattern of cubic Ti_2Al surface layer formed by exposure to 775°C for 100h in vacuum.

DISCUSSION

Thermal exposure at 775°C of fully machined Ti-48Al-1V-0.2C tensile samples markedly decreases the room temperature ductility. Thermal exposure of a similar 0.06 carbon alloy results in a lower ductility loss. Cursory examination of these results would suggest that precipitation of Ti_3AlC is the cause for the ductility loss. However, no embrittlement is observed for samples which have had the thermally exposed surfaces removed by grinding or machining. Examination of the surfaces after thermal exposure shows that a layer of high atomic number density is created beneath the oxide layer. This layer was identified for the 0.2 carbon alloy, by TEM/EDX, as simple cubic (a=6.85Å) Ti_2Al. Extensive cracking is observed greater than the depth of the surface layer itself. Since these features are observed on both alloys, the presence of this surface phase and ensuing surface cracking cannot be solely responsible for the decreased ductilities. The formation of this surface layer during thermal exposure is not only the result of substrate aluminum depletion due to the formation of Al_2O_3[6,9] or sublimation of aluminum in vacuum. If either of these scenarios occurred the surface phase would be Ti_3Al (DO_{19}) rather than the cubic phase observed. The formation of the surface phase was probably the result of aluminum depletion plus oxygen diffusion inward to stabilize the cubic microstructure. The different thickness of this surface layer resulting from thermal exposure in air versus vacuum suggests that oxygen plays a role in the formation of this phase. Electron microprobe analysis and scanning Auger microanalysis of a 10μm thick surface layer formed at 925°C showed that the phase was enriched with 2 to 5 wt% oxygen. In both conventional titanium[5] and α_2 titanium aluminide[6] alloys, oxygen enriched layers from thermal exposure produce large reductions in room temperature tensile ductility with no change in tensile yield strength.

The apparent different susceptibility of each alloy to embrittlement from thermal exposure is probably caused by the significantly higher strength of the 0.2 C alloy and also possibly through a reduction in toughness with the additional carbon. Assuming that the Ti_2Al layer is isostrain with the substrate and fractures at the same strain level in both alloys (the thicknesses are similar for the same exposure condition), the stress at a given plastic strain level is much higher in the 0.2 C alloy (yield strength 450 MPa) than the 0.06 C alloy (yield strength (300 MPa). Therefore the driving force to propagate the crack in the high carbon alloy is much greater and the critical flaw size for fracture is reduced at a higher stress. If the addition of carbon also reduces the toughness of the alloy the ductility differences from cracking of the Ti_2Al surface layer would be further accentuated. The mechanism of ductility reduction is the same for each alloy, but appears as a much greater effect in the higher strength carbon containing material.

SUMMARY

Thermal exposure of fully machined tensile samples of Ti-48Al-1V-0.2C (at%) at 775°C results in a 3 to 4 fold decrease in room temperature ductility. Identical thermal exposures to a similar alloy containing 0.06 carbon, results in less ductility loss. Surface removal experiments, and thermal exposure to samples prior to machining indicate that embrittlement is caused by changes occurring near the surface. A cubic phase having a Ti:Al ratio of 2:1 is observed to form between the surface oxides and the $(\gamma+\alpha_2)$ TiAl in the 0.2 carbon alloy. A similar surface phase is observed on exposed surfaces in the 0.06 carbon alloy. Since this layer is observed on both alloys, we believe that it is not solely responsible for the magnitude of the ductility loss. The marked differences observed in the effect thermal exposure has on the 0.2 and the 0.06 carbon alloys suggests that the enhanced ductility loss is due to the strengthening effect of carbon and also possibly to a reduction in toughness. Removal of the surface layer restores the lost ductility.

ACKNOWLEDGMENTS

The authors thank J.E. Allison, W.B. Copple and F.A. Alberts for their contributions to this paper.

REFERENCES

1. Y.-W. Kim, Journal of Metals, **41**, 24,(1989)
2. D.S. Shih, S.C. Huang, G.K. Scarr, H. Jang and J.C. Chestnutt, in Microstructure/Property Relationships in Titanium Aluminides & Alloys, edited by W-Y Kim and R.R. Boyer, TMS, **135** (1991).
3. S. Chen, P.A. Beaven and R. Wagner, Scripta Metall., **26**, 1205,(1992).
4. W.T. Donlon, and W.E. Dowling (unpublished results).
5. K.C. Antony, Journal of Materials, **1**, 456, (1966).
6. G.H. Meier and F.S. Pettit, Mats. Sci. and Eng., **A153**, 548, (1992).
7. W.E. Dowling, Jr. and J.E. Allison, in Proc. 4th Int. Conf. on Fatigue and Fatigue Thresholds, Part III, edited by H. Kitagawa and T. Tanaka, 1923, (1990).
8. W.E. Dowling, Jr. and W.T. Donlon, Scripta Metall., **27**, 1663, (1992).
9. K. Maki, M. Shioda, M. Sayashi, T. Shimizu and S. Isobe, Mats. Sci. and Eng., **A153**, 591, (1992).

A PRELIMINARY STUDY ON THE BEHAVIOR OF NOTCHED $Ni_3Al(B)$: FRACTURE MODE TRANSITION

YIXIN XU AND ERLAND M. SCHULSON
Thayer School of Engineering, Dartmouth College, Hanover, NH 03755

ABSTRACT

Tensile tests have been carried out on notched $Ni_3Al(B)$ specimens with different thickness/width ratios . The result indicate a transition in the fracture mode upon increasing the specimen thickness, from ductile transgranular to intergranular fracture. Experiments at high strain rates show that the transition cannot be explained in terms of the strain rate at the root of the notch. Rather, it is explained in terms of an increasing degree of triaxiality of the stress state.

I. INTRODUCTION

As a potential structural material, $Ni_3Al(B)$ may suffer from the tendency to fracture intergranularly in the presence of a notch [1,2]. Notch-induced intergranular fracture has also been seen in another potentially useful, strongly ordered $L1_2$ alloy, Zr_3Al, where a notch strength ratio (the ratio of maximum stress of notched specimen to the ultimate tensile stress of unnotched specimen), NSR, of 0.72 was reported [3]. For $Ni_3Al(B)$, NSR=0.56 [1]. In comparison austenitic stainless steel does not show this effect. The question concerns the origin of the phenomenon.

One factor may be strain rate at the notch tip. Since the radius of the notch is small, plastic deformation will be concentrated in a narrow region, resulting in a high and possibly embrittling strain rate locally. Another factor is the stress state behind a loaded notch [4]. Under uniaxial loading, a triaxial stress state will develop if the specimen is thick enough. Correspondingly, the material will deform plastically at higher applied stresses. As a result, brittle fracture could occur before extensive plastic strain is imparted, particularly if the fracture criterion involves a critical tensile stress. A third factor is the environment. Many intermetallic compounds fracture prematurely even in non-aggressive environments like air and water [5]. A notch could exacerbate the situation. Another factor is the work hardening rate. Strongly ordered $L1_2$ intermetallic compounds generally have high work hardening rates [6-10], which, when normalized with respect to the shear modulus, are about 3 to 4 times higher than those of elemental metals and random f.c.c. solid solutions. A high work hardening rate will probably reduce the plastic strain which can be imparted locally, should brittle fracture be triggered by a critical tensile stress.

This paper focuses on $Ni_3Al(B)$ and on the first two possible factors; namely, strain rate and stress state. The strain rate is varied in the usual manner. The stress state is varied through the use of notches.

II. EXPERIMENTAL

The material used in this study was the same as that used earlier [11]; namely, Ni - 24at.% Al - 0.3at.% B, hot-extruded from powder. All notched specimens were machined and/or ground from the same as-extruded rod and were subsequently annealed at 700 °C for 0.5 hour. Their grain size was 9.5 μm. Smooth-bar specimens were tested at different strain rates. They were cut from another rod of the same composition and were annealed at 1000 °C for 1 hour. Their grain size was 18 μm.

Strip tensile specimens with two 90° V-shape notches were used, Fig. 1. The thickness, t. and width. w. were varied. The surfaces of smooth-bar specimens. used as references. were electro-chemically polished, while those of the notched specimens were as-machined.

For the strain rate series cylindrical specimens were used, with a diameter of 2.9 mm and a gauge length of 16.5 mm. Their surfaces also were electro-chemically polished.

Tensile tests were carried out using an Instron machine, at room temperature. The displacement rate was 8 x 10^{-4} mm/sec.. Load-displacement curves were recorded on strip charts.

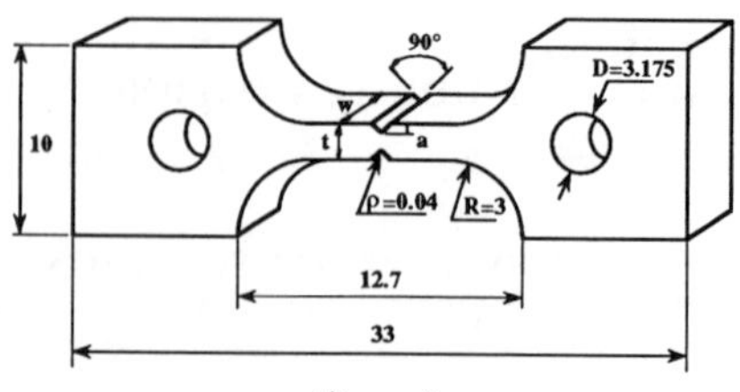

a = t/4, units: mm

Fig. 1 Geometry of specimen used for stress state testing

Fracture modes were examined using a ZEISS 962 scanning electron microscope (SEM) operated at 15 kv.

III. RESULTS

The load-displacement curves for all notched specimens were similar to those of smooth bars. Upon increasing strain, specimens went through yielding, work hardening, necking and finally fracturing. The reduction in area of specimens b and d (Table 1) was about 20% and 37%, respectively. Owing to a lower constraint, the percent reduction in the dimension parallel to the notches was about 2 to 3 times larger than that in the perpendicular direction. For specimen b in the direction parallel to the notch, the related reduction in dimension was larger in the center than along the notches. These results clearly show the constraint of the notches. The reduction in area was not able to be measured on specimens f and g owing to shearing failure along the notches (more below).

Table 1: Fracture mode and NSR of Ni_3Al(B)

	a	b	c	d	e (= c)	f	g
specimen	t=2mm w=2mm	t=2mm w=2mm	t=2mm w=0.2mm	t=2mm w=0.2mm	t=0.2mm w=2mm	t=0.2mm w=2mm	t=0.3mm w=0.3mm
fracture mode	transgranular ductile	intergranular	transgranular ductile	transgranular ductile	transgranular ductile	transgranular ductile	transgranular ductile
fracture strength (MPa)	1370	1060	1240 ± 20	900 ± 5	1240 ± 20	1520 ± 60	1370
NSR	0.77		0.73		1.22		—

The NSR's and the fracture modes are shown in Table 1 and Fig. 2. It can be seen that the fracture mode for thin (small t and/or w), notched specimens was essentially transgranular, ductile, dimple fracture. Near the center of these specimens was some cleavage-like facets, presumably of lower fracture energy. Thick, notched specimens fractured differently. The edge of the specimens (near the root of the notches) showed a rim of transgranular ductile fracture (Fig. 2, (b), right side), while the remainder of the cross-section showed only a small amount (≈ 15%) of ductile fracture and was dominated by intergranular fracture. The thickness of the edge, ductile rim was 50~100 μm, in confirmation of the observation by Khadkikar et al. [2]. Other than for specimen d (Table 1), the NSR's showed the trend the fracture modes suggested. NSR>1 corresponded to transgranular ductile fracture while NSR<1 corresponded to

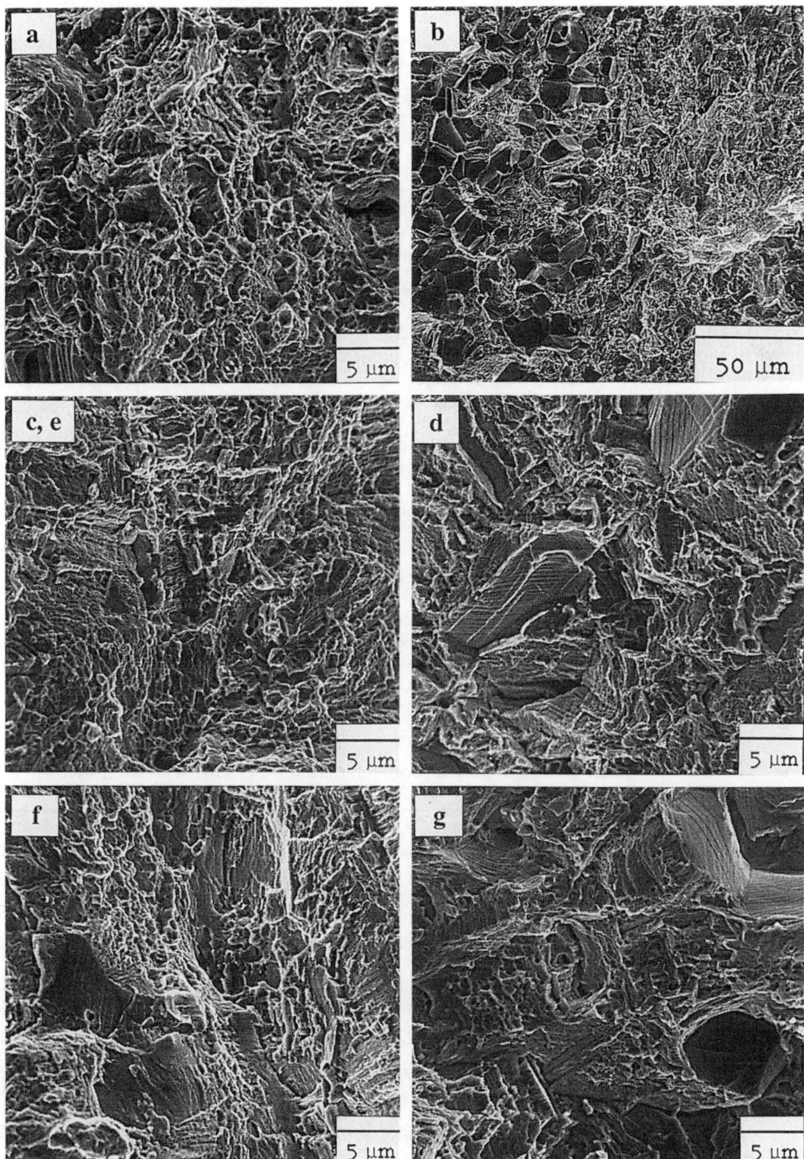

Fig. 2 Fracture modes of $Ni_3Al(B)$; (a) to (g) correspond to the letter in Table 1

intergranular fracture. For specimen d, although the fracture mode was transgranular, the NSR was only 0.73. This lack of correspondence suggests that the NSR is not a good parameter to use in describing the notch effect.

The NSR for specimen b (Table 1) was 0.77, which was higher than 0.56 reported by Stoloff et al. [1]. This difference probably reflects the difference in the specimen geometry, which resulted in a different constraint.

IV. DISCUSSION

4.1 Strain rate effect ?

The strain rate at the notch could be estimated approximately by taking the ratio of the gauge length to the notch root diameter; that is, $\dot{\varepsilon}_{notch} \approx \dot{\varepsilon}_{nominal} \cdot (L/2\rho)$. For the present specimen dimensions and the nominal strain rate of $10^{-4}\ s^{-1}$, the strain rate at the notch root was about $10^{-2}\ s^{-1}$. Table 2 shows the tensile properties and fracture modes of $Ni_3Al(B)$ at strain rates of 10^{-4} and $7.7\ s^{-1}$. These results show no significant difference in properties. The fracture mode was 100% ductile transgranular under both conditions, with a typical cup-and-cone appearance. The transition, therefore, can not be attributed to an effect of strain rate.

The notch root radius in the present work was the same for all specimens, implying that the effective strain rate at the notch root was essentially the same. If a high strain rate was the cause of the change in the fracture mode, then all specimens should have fractured intergranularly. That they did not is further proof that the transition in fracture mode is not related to strain rate.

4.2 Stress state effect ?

In the notched specimen (b, Table 1), owing to the constraint of the notch material behind the notch was stressed triaxially; i.e. the stresses along the thickness and width were both greater than zero. On the other hand, in the thin specimens (d, f, g, Table 1), when t and/or w was reduced by a factor of ~10, the notch constrained the material mainly in one direction. In the other direction, the small t and/or w did not allow a high tensile stress to build up. Therefore, the

Table 2: Tensile properties and fracture mode of $Ni_3Al(B)$ tested at two different strain rates at room temperature (grain size = 18 μm)

smooth bar strain rate (s^{-1})	10^{-4}	7.7 *
behavior	ductile ($\varepsilon_f = 33\%$)**	ductile ($\varepsilon_f = 36\%$)**
yield stress (MPa)	390	430
fracture stress (MPa)	1480	1540
fracture mode	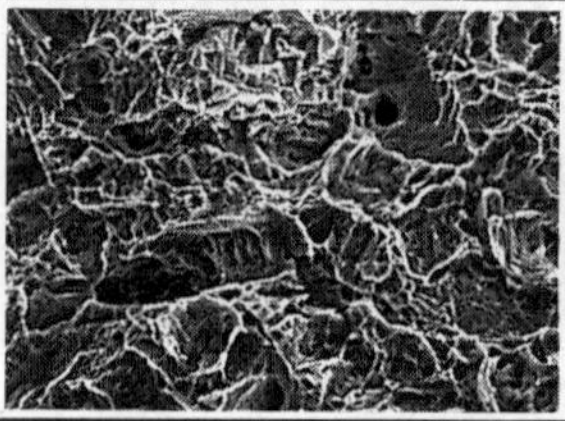	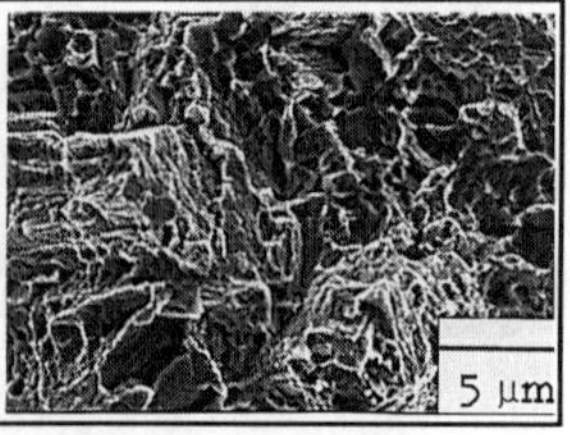 5 μm

* The higher strain rate of $7.7\ s^{-1}$ was obtained using a MTS model 810-14 testing system equipped with an accumulator and computer data acquisition system

** ε_f is elongation to fracture

stress state behind the notches in the thin specimens was closer to biaxial. That premature (intergranular) fracture did not occur under the biaxial stress state is consistent with Liu and White's [12] being able to deeply draw a thin-walled cup of $Ni_3Al(B)$ at room temperature, for during the forming process a biaxial tensile state of stress existed within the material.

A triaxial stress state suppresses yielding, while for plastically isotropic material a biaxial stress state does not. Therefore, assuming that brittle fracture occurs once a critical tensile stress is reached, thick specimens, owing to the high degree of triaxility, will fail when one of the principal stresses reaches the fracture stress and with less plastic strain. Thin specimens, on the other hand, can be plastically deformed easily, and the plastic deformation can relax the stress. Therefore, thin specimens will fail when the plastic strain reaches a critical value. This could explain why specimen d (Table 1) failed at a stress not as high as that of specimen b, but at a higher strain. In fact, the fracture strain imparted to specimen d was similar to that of a smooth tensile bar.

Further insight into the stress state within the notched specimens was obtained from the macroscopic appearance of the fracture surface. In the case of the thick specimens, the dominant fracture mode was intergranular. For the thin specimens, the failure was dominated by ductile transgranular fracture and shearing occurred on a plane inclined ~ 45° to the maximum and minimum stress directions. This analysis is summarized in Table 3, where the multiaxial character of the stress state is deduced from the fracture surface.

Table 3: Stress state behind a loaded notch

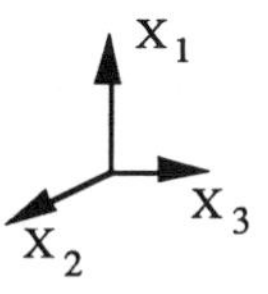 X_1, X_2, X_3	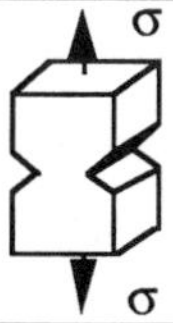 σ	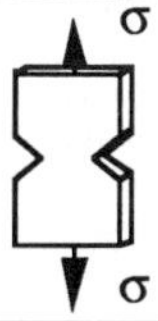 σ	 σ	 σ
σ_{11}	> 0	> 0	> 0	> 0
σ_{22}	> 0	→ 0	> 0	→ 0
σ_{33}	> 0	> 0	→ 0	→ 0
stress state	triaxial	biaxial	biaxial	—
macroscopic failure appearance		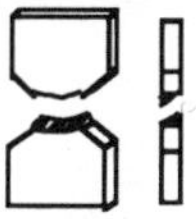	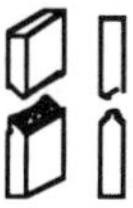	

The present results suggest that there is a critical thickness which controls the transition of the fracture mode, Fig. 3. This critical thickness is probably of the order of twice the plastic zone size under plane stress: i.e.,

$$2 \times 1/2\pi \times (K_{IC} / \sigma_y)^2$$

If the fracture toughness is taken to be $K_{IC} = 30$ MPa.m$^{1/2}$ [13] and the yield strength to be $\sigma_y = 450$ MPa [11], then the critical thickness is estimated to be approximately 1.4 mm.

4.3 Environment effect?

The environmental embrittlement of some intermetallic compounds has been explained by the reaction of atomic hydrogen with active elements, such as Si and Al at the crack tip. The reaction may not be able to take place if the strain rate is high enough [14]. As mentioned above, a strain rate as high as 7.7 s^{-1} does not effect the fracture strength nor the fracture mode of $Ni_3Al(B)$.

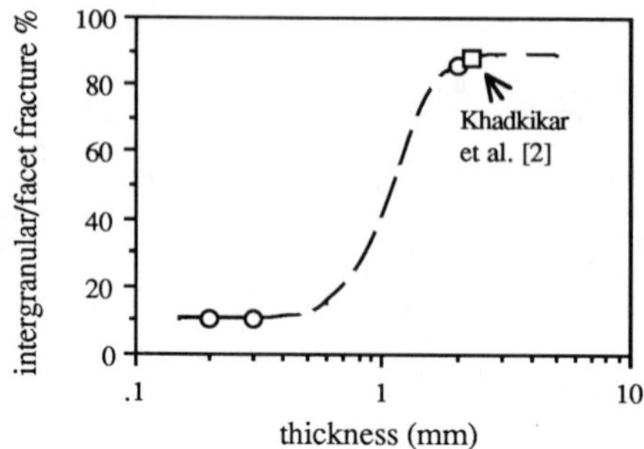

Fig. 3 Transition in fracture mode vs specimen thickness

Liu and coworkers' [15] recent work on $Ni_3Al(B)$ concluded that at least under the uniaxial stress state, this particular material does not show environmental embrittlement. Whether a triaxial stress state will sensitise the material is not clear. An experiment to test this point is in progress.

V. CONCLUSION

It is concluded that the triaxial stress state behind a loaded notch accounts for the transition in fracture mode from ductile transgranular fracture to intergranular fracture upon increasing the thickness of $Ni_3Al(B)$. Neither a biaxial stress state nor a strain rate as high as 7.7 s^{-1} affects the fracture mode. Thin specimens can not support a through-thickness tensile stress and, therefore, are not susceptible to intergranular fracture.

VI. ACKNOWLEDGMENTS

The authors acknowledge Mr. Gary Kuehn for performing the high strain rate experiment. This work was supported by the Office of Basic Energy Sciences, U.S. Department of Energy, contract no. DE-FG02-86ER45260.

VII. REFERENCES

1. N. S. Stoloff, G. E. Fuchs, A. K. Kuruvilla and S. J. Choe, Proc MRS, **81**, 247 (1987).
2. P. S. Khadkikar, J. J. Lewandowski and K. Vedula, Metall. Trans. **A, 20A**, 1247 (1989).
3. E. M. Schulson and J. A. Roy, J. Nuclear Mats. **71**, 124 (1977).
4. J. F. Knott, Fundamentals of Fracture Mechanics, Halsted Press, John Wiley & Sons Inc., New York (1973).
5. C. T. Liu, Proc. Symp. Intermetallic Compounds, JIMIS-**6**, 7003 (1991).
6. E. M. Schulson and J. A. Roy, Acta metall. **26**, 15 (1978).
7. T. P. weihs, V. Zinoview, D. V. Viens and E. M. Schulson, Acta metall., **35**,1109 (1987).
8. R. Lerf and D. G. Morris, Acta metall. et mater. **39**, 2419 (1991).
9. J. H. Schneibel and E. P. George, Scripta metall. **24**, 1069 (1990).
10. S. J. Zhang, P. Nic, W. W. Milligan and D. E. Mikkola, Scripta metall. **24**, 1099 (1990).
11. E. M. Schulson, Y. Xu, R. P. Munroe, S. Guha and I. Baker, Acta metall. et mater., **39**, 2971 (1991).
12. C. T. Liu and C. L. White, Proc MRS, **39**, 365 (1985).
13. J. D. Rigney and J. J. Lewandowski, Mat. Sci. and Eng. **A149**, 143 (1992).
14. P. Nagpal and I. Baker, Scripta metall. **25**, 2577 (1991).
15. C. T. Liu, Proc. Symp. High-Temperature Ordered Intermetallic Alloys, MRS (1992). In press.

MECHANICAL BEHAVIOR OF MONOCRYSTALLINE NiAl USING A MINIATURIZED DISK-BEND TEST

Ha K. DeMarco and Alan J. Ardell

Department of Materials Science and Engineering, University of California, Los Angeles, California 90024-1595

ABSTRACT

Miniaturized disk-bend tests were conducted on stoichiometric monocrystalline NiAl alloys of two different nominal purities. Disks 3 mm in diameter and ~250 mm thick were prepared with faces oriented parallel to either (100) or (110) and tested in biaxial bending. The specimens exhibited some ductility, even in the "hard" (100) orientation, prior to catastrophic failure. The yield strength of the specimens was higher in the (100) orientation than in the (110) orientation, as expected, but the specimens in (110) orientation were considerably more ductile. The higher purity alloy was considerably more ductile in both orientations. The estimated CRSS of the samples (110) in orientation is ~80 to 85 MPa, which is somewhat lower than reported values for deformation on the ⟨001⟩{010} slip system. For the most part, the fracture surfaces are similar in both alloys, with cleavage being the dominant mode of fracture. There is no visual evidence on the fracture surfaces that can account for the differences in ductility of the two alloys tested.

INTRODUCTION

The ordered intermetallic compound NiAl has stimulated considerable interest in the aerospace industry as a high-temperature structural material due to its low density and good oxidation resistance. However, the fabrication of NiAl into useful components is hampered by its lack of room-temperature ductility. It has recently been shown that the ductility of monocrystalline NiAl is enhanced by minor alloying additions (~ 0.2 at. %) of Fe, Mo or Ga [1]. Micro-alloying of polycrystalline NiAl, however, does not produce the same enhancement of ductility [2]. This suggests that microalloying affects cleavage, but not the factors affecting crack propagation along grain boundaries. The mechanism responsible for the enhanced ductility of microalloyed single crystals has not been identified.

In the present investigation a miniaturized disk-bend test (MDBT) is being used to study the changes in ductility of monocrystalline NiAl brought about by micro-alloying additions of Fe and other elements. The idea is to exploit the small size of the typical MDBT specimen (3 mm in diameter and about 250 μm in thickness) to prepare alloys of various compositions using the same starting material. The proposed method of alloy preparation involves deposition of controlled amounts of the desired alloying element onto disks of specific orientation, followed by annealing in vacuum at high temperatures (~1300 °C) and slow cooling to avoid complications associated with excess vacancy concentrations. Since mechanical testing using the MDBT apparatus has never been performed on single crystals of any material, let alone NiAl, it is essential to establish baseline data with which the results of future experiments can be compared. The results of our initial experiments are reported in this paper.

EXPERIMENTAL METHODS

Stoichiometric single crystals of NiAl were obtained from two sources, General Electric Aircraft Engines (GE), Cincinnati, OH and the NASA Lewis Research Center, which provided material prepared at the University of Tennessee (UT). Although the precise composition and the purity of the GE alloy have not yet been analyzed, the major impurity is typically 600 to 800 ppm Si and it is nominally stoichiometric. The UT material is a stoichiometric high-purity alloy with <10 ppm of interstitial impurities, which is prepared by a containerless processing method [3].

Disk specimens of both alloys were prepared in either (100) or (110) orientation for MDBT testing. Slices ~500 μm thick were cut from the bulk crystals using an electric spark discharge machine and ground using emery paper to a thickness of ~350 μm. The disks were then cut from the slices using either an ultrasonic disk cutter or an abrasive slurry disk cutter. They were further ground and finally polished to a mirror finish using 0.05 μm alumina. The final thicknesses of the disks tested were generally between 230 and 280 μm. The orientations of the surfaces of the individual disk specimens were not checked, but the original slices were all cut to within 1° of either (100) or (110).

The MDBT apparatus is described in detail elsewhere [4-6]. The tests were conducted in the ring-on-ring mode using a table model Instron testing machine at a cross-head displacement rate of 8.3 μm/s.

RESULTS AND DISCUSSION

Typical curves of load, P, vs. displacement, w, for both alloys in the (100) and (110) orientations are shown in Fig. 1. The loads are compensated for the small differences in the thicknesses, t, of the specimens by actually plotting P/t^2 vs. w (this is justified by the fact that during the elastic deformation of isotropic thin plates in the MDBT the stress is proportional to P/t^2 [4,5]). The curves for the specimens oriented (100), Fig. 1a, are similar in form, with a linear elastic region at low loads, followed by curvature indicating plastic deformation up to the point of catastrophic fracture (indicated by ×). The UT material is somewhat more ductile in this orientation, and fractures at a larger value of P/t^2.

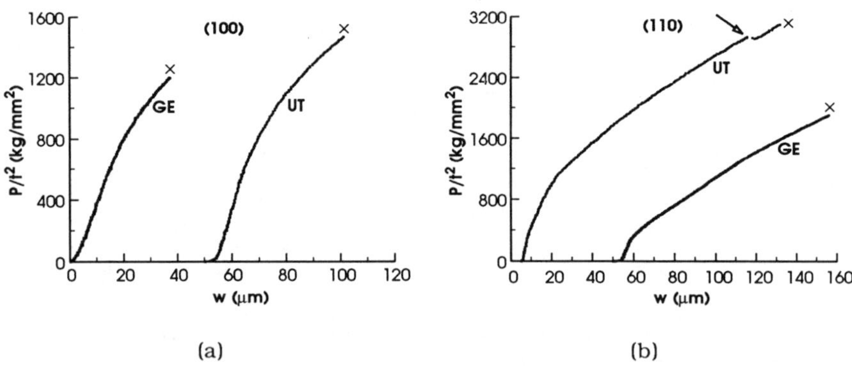

FIG. 1. Curves of thickness-compensated load, P/t^2, vs. displacement, w, for monocrystalline NiAl MDBT specimens oriented (a) (100) and (b) (110). Catastrophic fracture is indicated by × on all the curves. The arrow in (b) indicates a load-drop.

The behavior of the two materials is very different in the (110) orientation (Fig. 1b). Both alloys exhibit considerably more plasticity than in the (100) orientation, and the ductility of the UT alloy far exceeds that of the GE alloy. Furthermore, the UT alloy fractures catastrophically at a much larger value of P/t^2 (also indicated by the ×) than the GE alloy. A comparison of the curves in Figs. 1a and 1b indicates that the UT alloy fractures at a much larger load in the (110) orientation than in the (100) orientation (note that the scale in Fig. 1b is double that in Fig. 1a). The ultimate strength of the GE alloy is also greater in the (110) orientation, but the difference is not as striking as it is for the UT alloy.

Interestingly, all the (110)-oriented specimens of the UT alloy experienced a small load drop prior to catastrophic failure, as indicated by the arrow in Fig. 1b. In tests run on other nominally brittle materials, such load drops are always associated with crack initiation [4,5]. Because of the relatively small difference between the displacements at the load drop and onset of catastrophic failure, we have not yet succeeded in unloading a specimen immediately after the initial load drop to confirm that it is associated with the initiation of a crack. Assuming that our interpretation is correct, however, it appears that small cracks in the UT alloy are capable of being arrested prior to ultimate failure. This is consistent with the greater ductility of the alloy.

A comparison of the yielding behavior of the GE alloy in the two orientations is shown in Fig. 2. Yielding in the MDBT occurs at the departure from linearity in the load-displacement curve [7]; these points are indicated by the open arrows in Fig. 2. The filled arrows indicate the end of the region of the so-called low-load nonlinearity, which is a ubiquitous feature of our miniaturized disk-bend tests [3,5]. The yield stress is calculated from the difference between the loads at these two points of the load-displacement curves. For tests conducted in the ROR mode [5] under freely supported boundary conditions, the yield stress is given by the formula

$$\sigma_y = \frac{3P_y}{2\pi t^2}\left\{(1+\nu)\ln\frac{a}{b} + \frac{(1-\nu)}{2}\frac{a^2}{R^2}\left[1-\frac{b^2}{a^2}\right]\right\}, \qquad (1)$$

where P_y is the load at yielding, a is the radius of the loading ring, b is the radius of the lower supporting die, R is the radius of the disk (3 mm) and ν is Poisson's ratio (ν = 0.3135 for polycrystalline NiAl at room temperature [8]).

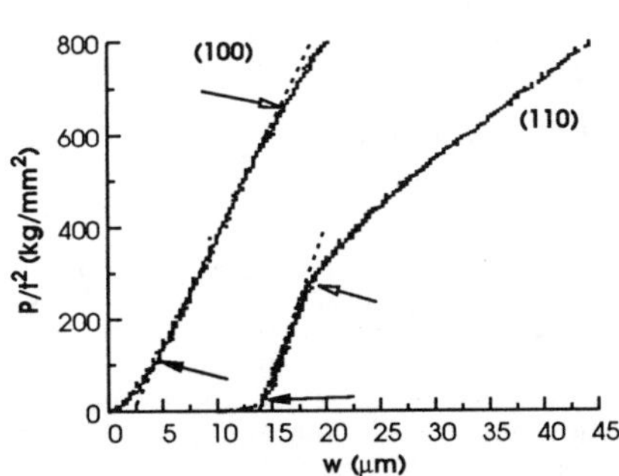

FIG. 2. Curves of thickness-compensated load, P/t^2, vs. displacement, w, for the GE alloy in (100) and (110) orientations. The open arrows indicate the onset of yielding and the filled arrows indicate the end of the low-load nonlinearity.

Equation (1) is valid for elastically isotropic specimens that yield at the tensile surface in pure bending. The shapes of the deformed specimens in previous work on polycrystals [4,6] and bicrystals of Ni_3Al [8] are consistent with this assumption, but verification is necessary for purely monocrystalline NiAl specimens deformed in this investigation. We have used equation (1) to calculate σ_y despite these qualifications because we are unaware of a corresponding formula for elastically anisotropic materials deformed in biaxial bending, and have just begun to evaluate the shapes of the yielded, but unfractured samples. The yield stresses calculated for monocrystalline NiAl are therefore approximate at best, although for disks in a specific orientation a comparison of the relative strengths of the two alloys is undoubtedly meaningful. The values of σ_y are summarized in Table 1. It is evident that σ_y for the GE alloy in (100) orientation is considerably larger than that of the UT alloy. The same is not true of the samples in (110) orientation; the yield stresses of the two alloys are nearly identical in this case. Both alloys yield at much higher stresses in the (100) than in the (110) orientations.

The orientation dependence of yielding in the MDBT is consistent with previously published results. Values ranging from 980 to 1400 MPa have been reported for the compressive yield stress (0.2 % offset) of monocrystalline NiAl tested in the "hard" [100] orientation [9-11]. Accurate values of the tensile yield stress at room temperature for the [100], i. e. "hard" orientation, have not been reported because the plastic strains rarely, if ever, reach the necessary 0.2 % level in regular tensile specimens [12,13]. Our results are therefore quite unusual in that the yield stresses measured in the MDBT are well defined; these are perhaps the first values reported for the tensile yield stresses of NiAl monocrystals tested in the hard orientation. The values of σ_y obtained in our tests are much lower than the previously reported compressive yield stresses, but this can be attributed to the sensitivity of the technique at very small plastic strains.

Table 1. Yield Stresses, σ_y (MPa), measured in this investigation.

Alloy	Specimen Orientation	σ_y
GE	(100)	346.7 ± 32.5
UT	(100)	298.3 ± 14.6
GE	(110)	162.4 ± 10.2
UT	(110)	168.5 ± 8.8

The values of σ_y obtained in the MDBT for specimens in the (110) orientation (Table 1) compare quite favorably with the results of previous experiments on specimens of other lots of GE alloys tested either in uniaxial tension or compression. Field et al. [14] reported 0.2 % compressive yield strengths of 217 ± 0.7 MPa, and Lahrman et al. [12] reported 0.2 % tensile yield stresses ranging from ~175 to 200 MPa, both sets of data having been obtained from specimens deformed at a strain rate of $8 \times 10^{-5}\ s^{-1}$. These exceed the values of σ_y in Table 1 (~165 MPa), but the discrepancy here could be due to the assumption of elastic isotropy implicit in the use of equation (1) to calculate σ_y.

The critical resolved shear stresses (CRSS) can be estimated for specific slip systems from the values of σ_y in Table 1, assuming that yielding occurs by pure bending. For example, for the MDBT specimens in (100) orientation the resolved shear stress on the most common slip systems $\langle 001\rangle\{110\}$ and $\langle 001\rangle\{100\}$ are zero, and the Schmid factors on the most highly stressed slip systems of the type $\langle 111\rangle\{1\bar{1}0\}$, $\langle 111\rangle\{11\bar{2}\}$, and $\langle 111\rangle\{12\bar{3}\}$, are 0.4082, 0.2357 and 0.4629, respectively. However, since the stress for plastic flow by the motion of dislocations with $\langle 111\rangle$ Burgers vectors exceeds the values of σ_y in Table 1 [15], we expect that some other mode of yielding of the specimens is operative. A process analogous to membrane stretching [7], but involving shear on the {001} or {011} planes parallel to the axis of the applied stress might be operative.

For the specimens in (110) orientation, the Schmid factors on the $\langle 001\rangle\{110\}$ and $\langle 001\rangle\{010\}$ slip systems are 0.3536 and 0.5, respectively. These suggest values of the CRSS of about 81.2 (GE) and 84.3 (UT) MPa on the more highly stressed slip system. Field et al. [14] have reported a value of 109 ± 0.4 MPa for the CRSS on the $\langle 001\rangle\{010\}$ slip system in GE material, which is 20 to 25 % larger than our measurements.

There is no significant difference between the CRSS of the two alloys. It is surprising that both alloys are equally strong in this orientation, because the greater purity of the UT alloy should make it the weaker of the two, regardless of orientation. However, recent work [15] has shown that the mechanical behavior of NiAl monocrystals is sensitive not only to impurity content but also to heat-treatment after crystal growth. Our GE alloy was heat-treated in the same manner as the alloys used in the experiments of Field et al. [14], so differences in the CRSS, for example, must be attributed to some other factor. The UT alloy has not yet been heat-treated. We are currently exploring the effects of these factors on the mechanical behavior of MDBT specimens.

Only limited fractography has been performed. The fracture surfaces of (100) samples are shown in Fig. 3. It is evident that the appearances of the fracture surfaces of the two alloys are not very different. The principal mode of fracture in both alloys is cleavage, as evidenced by the smooth fracture planes containing river patterns. Further work is necessary before we can identify the reasons for the significant differences between the ductilities of the GE and UT alloys.

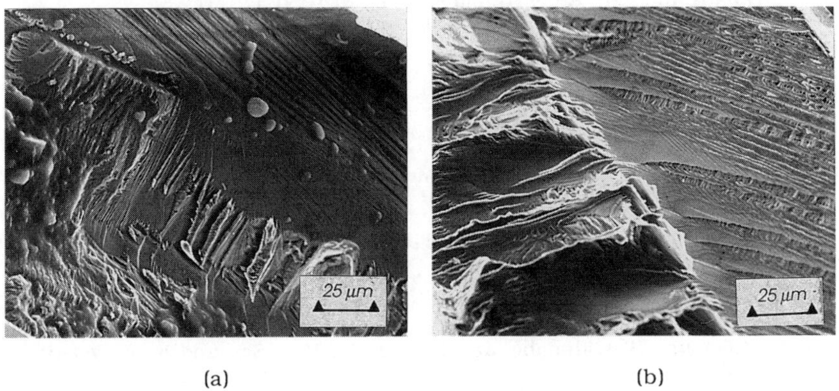

FIG. 3. Fractographs taken in the scanning electron microscope of specimens oriented (100) of the (a) GE and (b) UT alloys.

SUMMARY

Our initial experiments on the mechanical behavior of small specimens of monocrystalline NiAl deformed in the MDBT have produced several interesting results. Perhaps the most unexpected is that both specimens tested in (100) orientation exhibit enough plasticity to define a yield stress in biaxial tension. In the (100) orientation the GE alloy is stronger, but less ductile than the UT alloy. In the (110) orientation both alloys yield at about the same stress, but the UT alloy is significantly more ductile. Both alloys have higher yield stresses in the (100) orientation than in the (110) orientation.

Assuming that the operative slip system in the specimens oriented (110) is $\langle 001\rangle\{010\}$, the CRSS is between 80 and 85 MPa for both alloys. This value is about 20 % smaller than those reported in the literature. The reason for the equal strength of both alloys in the (110) orientation, but not in the (100) orientation, is not known.

The principal mode of fracture of both alloys in the (100) orientation is cleavage. The features characteristic of the fracture surfaces are similar, featuring flat crystallographic facets and river patterns.

ACKNOWLEDGEMENTS

We are grateful to the NASA Lewis Research Center for financial support of this research under grant No. NAG 3-1325. The GE alloy was provided by Dr. R. Darolia and the UT alloy was provided by Mr. R. D. Noebe, whose assistance throughout the course of this research is appreciated.

REFERENCES

1. R. Darolia, D. Lahrman, and R. Field, Scripta Metall. Mater. **26**, 1007 (1992).
2. R. D. Noebe and M. K. Behehani, Scripta Metall. Mater. (submitted 1992).
3. R. E. Reviere, B. F. Oliver and D. D. Burns, Mater. and Manufac. Proc. **4**, 103 (1989).
4. H. Li, F. C. Chen, and A. J. Ardell, Met. Trans. **22A**, 206 (1991).
5. J. Zhang and A. J. Ardell, J. Mater. Res., **6**, 1950 (1991).
6. D. L. Meyers, F. C. Chen, J. Zhang and A. J. Ardell, J. Test. and Eval., to be published (1993).
7. O. K. Harling, M. Lee, D.-S. Sohn, G. Kohse and C. W. Lau, in The Use of Small-Scale Specimens for Testing Irradiated Material, edited by W. R. Corwin and G. E. Lucas (ASTM STP 888, American Society for Testing and Materials, Philadelphia, PA, 1986), pp. 50-65.
8. D. E. Meyers and A. J. Ardell, Acta Metall. Mater. (submitted 1992).
9. R. T. Pascoe and C. W. A. Newey, Met. Sci. J., **2**, 138 (1968).
10. R. R. Bowman, R. D. Noebe, and R. Darolia, HITEMP Review–1989, NASA CP-10039, 47-1 (1989).
11. R. J. Wasilewski, S. R. Butler, and J. E. Hanlon, Trans. Met. Soc. AIME, **239**, 1357 (1967).
12. D. F. Lahrman, R. D. Field, and R. Darolia, in High-Temperature Ordered Intermetallic Alloys IV, edited by L. A. Johnson, D. P. Pope and J. O. Stiegler, (Mat. Res. Soc. Symp. Proc. **213**, Pittsburgh, PA, 1991) pp. 603-607.
13. T. Takasugi, S. Watanabe, and S. Hanada, Mat. Sci. and Engr., **A149**, 183 (1992).
14. R. D. Field, D. F. Lahrman and R. Darolia in High-Temperature Ordered Intermetallic Alloys IV, edited by L. A. Johnson, D. P. Pope and J. O. Stiegler, (Mat. Res. Soc. Symp. Proc. **213**, Pittsburgh, PA, 1991) pp. 255-260.
15. J. E. Hack, J. M. Brzeski and R. Darolia, Scripta Metall. Mater., **27**, 1259 (1992).
16. R. D. Noebe, R. R. Bowman and M. V. Nathal, Int. Mater. Rev. (accepted for publication, 1992).

OBSERVATION OF A NEW CREEP REGIME IN POLYCRYSTALLINE Ni-50(at.%)Al INTERMETALLIC ALLOY

S. V. Raj and Serene C. Farmer
NASA Lewis Research Center, MS 49-1, Cleveland, OH 44135.

ABSTRACT

Constant load creep tests were conducted on fine-grained (~ 23 μm) polycrystalline Ni-50(at.%) Al in the temperature range 1000 - 1400 K. Power-law creep with an average stress exponent, n, of 6.6 and an average activation energy, Q_c, of about 300 kJ mol^{-1} was observed above 25 MPa, while $n \approx 2$ and $Q_c \approx 95$ kJ mol^{-1} for $\sigma < 25$ MPa. Primary creep was observed in both regions thereby signifying dislocation activity during the initial period of the test. Preliminary experiments with coarse-grained Ni-50Al suggested that the mechanism in the n = 2 region is dependent on grain size. Transmission electron microscopy observations of the deformed specimens revealed dislocation tangles, dipoles, loops and networks in the power-law creep regime. The Burgers vector was determined to be <100> with the dislocations lying on the {100} and {110} planes. Although well-defined subgrains were not always observed, there was a greater tendency towards subgrain formation at stresses above 25 MPa. The deformation microstructures were inhomogeneous in the n = 2 creep regime and many grains did not reveal any dislocation activity. The observed characteristics of the low stress region suggest the dominance of an accommodated grain boundary sliding mechanism.

INTRODUCTION

Stoichiometric Ni-50(at.%)Al is being developed as a potential aircraft engine material to replace superalloys. Certain components, such as turbine blades and vanes, are subjected to high stresses and high temperatures during use so that good creep properties are an important design requirement. As a result, an understanding of the mechanisms governing creep of NiAl is essential in order to design alloys with superior creep properties. Earlier investigations on the creep behavior of NiAl have been largely restricted to the power-law creep regime [1-5] and no data exist at lower stresses, where other processes, such as diffusion creep, may dominate. This paper reports the observation of a new mechanism in Ni-50Al at low stresses.

EXPERIMENTAL

Pre-alloyed vacuum atomized Ni-50Al powder was procured from Homogeneous Metals, Inc., Clayville, NY. The powders were sealed under vacuum in mild steel cans and extruded at 1400 K using a ratio of 16:1. The average linear intercept equiaxed grain size, d, was 22.9 ± 0.7μm at the 95% confidence interval. Compression specimens of nominal dimensions 6.4 mm in diameter and 12.7 mm long were machined from the extruded rods with the compression axis parallel to the extrusion direction. The specimens were tested under constant load between 1000 and 1400 K with the initial stresses varying between 3.0 and 52.0 MPa. Additional data were obtained from stress increase tests. Microstructural observations were conducted on selected tested specimens using optical and transmission electron microscopy (TEM).

RESULTS AND DISCUSSION

Two distinct creep regimes were observed at high and low stresses where the transition stress was about 9 to 25 MPa depending on the absolute

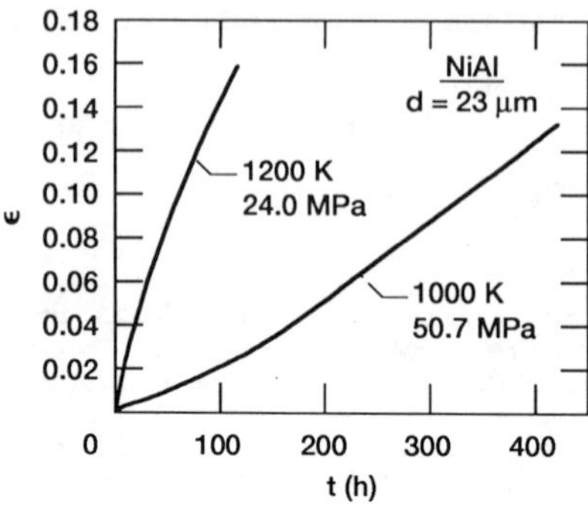

Figure 1. Plot of compressive true strain against time showing typical creep curves in the high stress region.

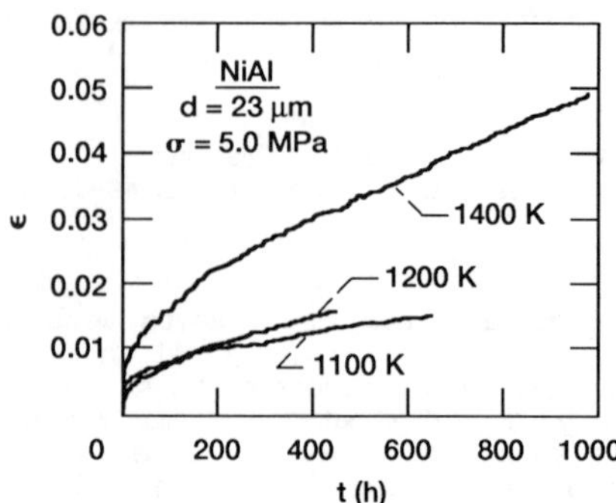

Figure 2. Plot of compressive true strain against time showing typical creep curves in the low stress region.

temperature, T. The nature of the primary creep curve was dependent on stress and temperature in the high stress regime. Inverse and sigmoidal primary creep was observed at 1000 and 1100 K, respectively, while normal primary creep occurred at 1200 and 1300 K. However, the primary creep region was always followed by a steady-state region in all instances. Figure 1 shows typical creep curves at 1000 and 1200 K under initial stresses of 50.7 and 24.0 MPa, respectively. Observations of inverse and sigmoidal creep in solid solution alloys [6] and LiF single crystals [7] have been attributed to the inhibition of dislocation nucleation or mobility in the material. Since dislocation nucleation is an athermal process, the transition from inverse to sigmoidal to normal primary creep in traversing from 1000 to 1200 K can be attributed to an increase in dislocation mobility with increasing temperature. Normal primary creep was observed between 1100 and 1400 K at low stresses, where serrations were observed in the creep curves (Fig. 2). It is unclear at present whether these serrations are due to dynamic recrystallization and grain growth similar to those observed in other materials [8] or due to variations in room temperature during the duration of the experiment.

Figure 3 shows the variation of the true steady-state creep rate, $\dot{\epsilon}$, against true stress, σ, between 1000 and 1300 K. The present data are in excellent agreement with those reported by Whittenberger [4] and Bowman _et al._ [9] for powder extruded NiAl alloys. As shown in Fig. 3, the data fall into two creep regimes: a high stress creep region with the stress exponent, $n \approx 5.8$-7.4 and a low stress region with $n \approx 1.6$-2.1. There appears to be no previous report of a similar distinct transition in creep behavior in binary NiAl although Jung _et al._ [5] reported a change in mechanism from $n \approx 4$ to $n \approx 1$ in a coarse-grained Ni-10Fe-50Al ternary alloy below 80 MPa at 1023 K. In comparison, the transition stresses for binary NiAl, while being temperature dependent, occur below 25 MPa (Fig. 3).

The values of n and the true activation energy for creep, Q_c, determined by multiple regression analysis using the data in Fig. 3, are tabulated in Table I for the two regimes. These values have been corrected for the temperature dependence of the Young's modulus, E, and the pre-exponential factor in accordance with the Bird-Mukherjee-Dorn (BMD) equation [10]:

$$\dot{\epsilon} = A\ (D_o Eb/kT)(\sigma/E)^n \exp(-Q_c/RT) \qquad (1)$$

where D_o is the frequency factor, b is the Burgers vector, k is Boltzmann's constant, R is the universal gas constant and A is a dimensionless constant.

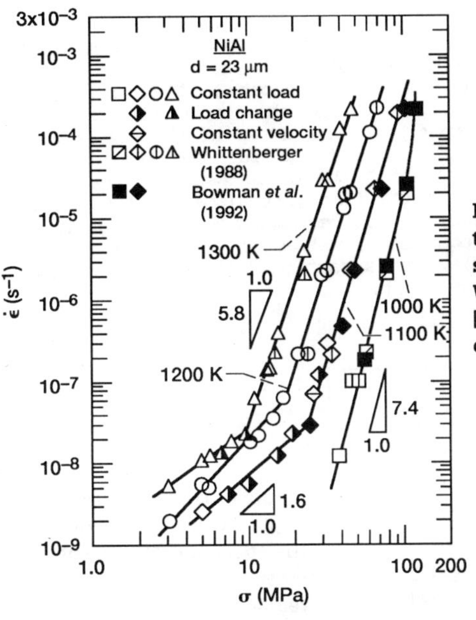

Figure 3. Variation of the compressive true strain rate with applied true stress for NiAl. The data reported by Whittenberger [4] and Bowman *et al.* [9] are also included in the figure for comparison.

The values of Young's modulus for Ni-50Al used in this study are based on the experimental data of Harmouche and Wolfenden [11]. The values of Q_c obtained by multiple regression are about 280 kJ mol^{-1} in the high stress power-law creep region and about 100 kJ mol^{-1} in the low stress creep regime (Table 1). Similar values of n and Q_c are obtained using plots of $\log(\dot{\epsilon})$ versus $\log(\sigma)$ and $\ln(\dot{\epsilon})$ versus 1/T, respectively (Table 1). The values of Q_c shown in Table I for the high stress creep regime compares well with other estimates reported in the literature [12] and with the activation energy for self diffusion of Ni, Q_{Ni}, in Ni-50Al, which is about 305 ± 10 kJ mol^{-1} [13]. No comparable values corresponding to $Q_c \approx 95$ kJ mol^{-1} appear to have been reported in the creep or diffusion literature on NiAl. Since activation energies for creep and diffusion are similar for a large number of materials [6], a value of 95 kJ mol^{-1} probably represents a measure of the activation energy for grain boundary diffusion in NiAl.

In order to verify if the mechanism in the n = 2 regime is dependent on grain size, a specimen was crept for about 1000 h at 1400 K under an initial stress of about 5.0 MPa in order to permit grain growth to 117.8 ± 6.1 μm (Fig. 2). The temperature was then reduced to 1200 K and the steady-state creep rates were determined from stress increase tests. Figure 4 compares these creep rates with the regression lines obtained in the uninterrupted

Table I: Magnitudes of n and Q_c in the two creep regimes.

Technique	High stress region		Low stress region	
	n	Q_c (kJ mol^{-1})	n	Q_c (kJ mol^{-1})
1. Graphical	5.8-7.4	295-310	1.6-2.1	90-100
2. Multiple regression[1]	6.2 ± 0.4	289.2 ± 18.8	1.8 ± 0.2	95.9 ± 22.2

(1) The values of error represent the 95% confidence intervals.

tests on the fine-grained specimens. It is clear from Fig. 4 that the creep rates for the coarse and fine-grained specimens are in good agreement in the power-law creep regime with n ≈ 5.9. However, while the fine-grained material deviates from power-law creep below a stress of about 16 MPa, the coarse-grained specimen continues to exhibit this creep behavior until a stress of about 9 MPa. A change in mechanism from power-law creep to the n ≈ 2 regime is expected to occur below this value of stress for the coarse-grained material. Figure 4 clearly demonstrates that the creep rate for the coarse-grained specimen is lower than that for the fine-grained material by about a factor of 15 for $\sigma \approx 5.0$ MPa. An estimate of the magnitude of the grain size exponent, p, using the ratios of logarithmic values of the creep rates and grain sizes suggests that $\dot{\epsilon} \propto (1/d)^2$. It is cautioned that this is only an approximate value of the grain size exponent.

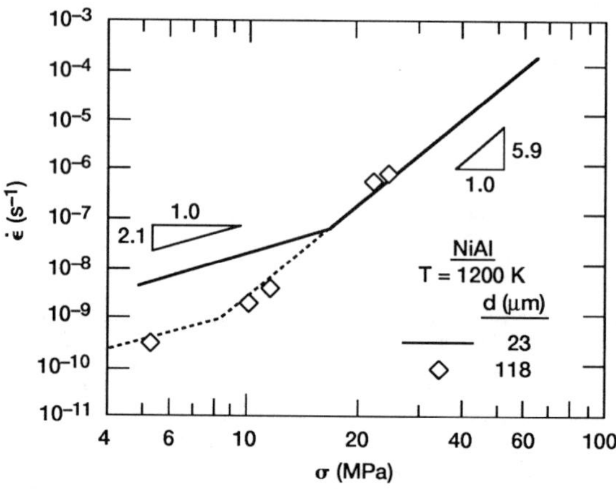

Figure 4. True creep rate vs. applied true stress showing the effect of grain size on the creep rate in the low stress region.

Transmission electron microscopy of crept specimens showed a mixture of subboundaries, dislocation tangles, and random dislocations at stresses corresponding to the power-law creep region with n ≈ 5.8–7.4 (Fig. 5(a)). Most of the dislocations had a <100> Burgers vector with the glide planes being either {100} or {110}. Hexagonal dislocation networks were sometimes observed and found to consist of three <100> type dislocations. As expected, smaller subgrains were observed in specimens subjected to higher stresses. A variety of substructures were observed in the n = 2 region and the microstructure was heterogeneous. Ill-formed cells, dipoles, loops, dislocation tangles and hexagonal dislocation networks were observed at stresses corresponding to the upper part of the n = 2 region. For example, Fig. 5(b) shows a mixture of hexagonal networks, dislocation tangles and loops in a specimen deformed to a true strain of 2.7% at 1200 K under an initial stress of about 9.8 MPa. In contrast, no substructure was evident in most of the grains in a specimen deformed to a true strain of 3.1% at 1200 K under an initial stress of 5.4 MPa corresponding to the lower part of the n = 2 region, although isolated bands of dislocation loops were seen in a few grains (Fig. 5(c)). Similar observations were also made at 1300 K. On close examination, a number of these dislocation loops appeared to have been punched out at Al_2O_3 particles strung along the extrusion direction.

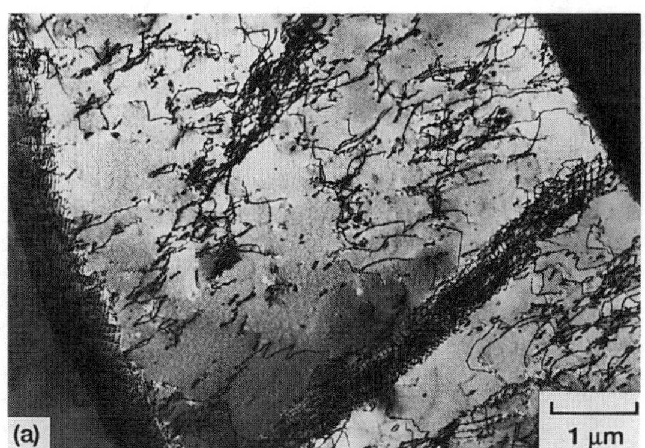

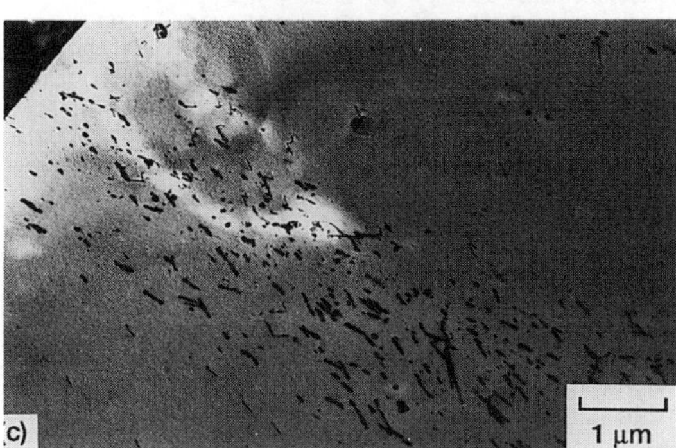

Figure 5. Transmission electron micrographs of NiAl specimens after deformation in the (a) power-law creep and (b) and (c) low stress regimes. (a) T = 1200 K, σ = 41.3 MPa, ϵ = 17.3 %; (b) T = 1200 K, σ = 9.8 MPa, ϵ = 2.7 %; (c) T = 1200 K, σ = 5.4 MPa, ϵ = 3.1 %.

The values of $n \approx 2$, $Q_c \approx 95$ kJ mol^{-1} and a probable value of $p \approx 2$ are in agreement with several accommodated grain boundary sliding models proposed in the superplastic literature [14]. The close connection between creep and superplasticity implies that one of these mechanisms may be applicable to the present results.

CONCLUSIONS

1. Power-law creep with $n \approx 6.6$ and $Q_c \approx 295$ kJ mol^{-1} is observed above 25 MPa, where the shape of the primary creep region appears to be influenced by changes in dislocation mobility with temperature in this regime.
2. A new grain size dependent creep regime with $n \approx 2$, $Q_c \approx 95$ kJ mol^{-1} and a probable value of $p \approx 2$ is observed below 25 MPa. A long primary creep region indicates that it is a dislocation controlled process. It is suggested that grain boundary sliding accommodated by lattice slip is dominant in this region.

REFERENCES

1. R. R. Vandervoort, A. K. Mukherjee and J. E. Dorn, Trans. ASM 59, 930 (1966).
2. W. J. Yang and R. A. Dodd, Met. Sci. J. 7, 41 (1973).
3. J. D. Whittenberger, J. Mater. Sci. 22, 394 (1987).
4. J. D. Whittenberger, J. Mater. Sci. 23, 235 (1988).
5. I. Jung, M. Rudy and G. Sauthoff, in High-Temperature Ordered Intermetallic Alloys II, edited by N. S. Stoloff, C. C. Koch, C. T. Liu and O. Izumui (Mater. Res. Soc. Proc. 81, Pittsburgh, PA 1987) p. 263.
6. W. D. Nix and B. Ilschner, in Strength of Metals and Alloys, Vol. 3, edited by P. Haasen, V. Gerold and G. Kostorz (Pergamon Press, Oxford, 1980), p. 1503.
7. W. A. Coghlan, R. A. Menezes and W. D. Nix, Phil. Mag. 23, 1515 (1971).
8. H. J. Frost and M. F. Ashby, in Deformation-Mechanism Maps: The Plasticity and Creep of Metals and Ceramics (Pergamon Press, Oxford, 1982), p. 14.
9. R. R. Bowman, R. D. Noebe, S. V. Raj and I. E. Locci, Metall. Trans. 23A, 1493 (1992).
10. J. E. Bird, A. K. Mukherjee and J. E. Dorn, in Quantitative Relations Between Properties and Microstructure, edited by D. G. Brandon and A. Rosen (University Press, Jerusalem, 1969), p. 255.
11. M. R. Harmouche and A. Wolfenden, J. Testing. Eval. 15, 101 (1987).
12. M. V. Nathal, Ordered Intermetallics - Physical and Mechanical Behaviour edited by C. T. Liu, R. W. Cahn and G. Sauthoff (Kluwer Academic Publishers, Dordrecht, The Netherlands, 1992), p. 541.
13. G. F. Hancock and B. R. McDonnell, Phys. Stat. Sol. 4(a), 143 (1971).
14. O. D. Sherby and J. Wadsworth, Prog. Mater. Sci. 33, 169 (1989).

COMPARATIVE CREEP PROPERTIES OF SINGLE-PHASE INTERMETALLICS OF THE Ti-Al SYSTEM

Kouichi MARUYAMA and Hiroshi OIKAWA
Department of Materials Science, Tohoku University, Sendai, 980, Japan

ABSTRACT

Creep characteristics of polycrystalline single-phase intermetallics, α_2-Ti_3Al and γ-TiAl, at 1100 K have been summarized to evaluate their relative strength at elevated temperature at three strain-rate levels, 10^{-1}, 10^{-4} and $10^{-8}s^{-1}$, and are compared with those of α Ti-Al terminal solid solution alloys. The strength of these intermetallics is obviously higher than that of α, random solid solutions, especially at low strain-rates. However, the difference in high-temperature long-term strength among intermetallic alloys is not significant at any strain-rate level.

INTRODUCTION

Alloys based on the intermetallic phases of the Ti-Al system have been studied by many investigators as promising high-temperature materials. One of the essential properties in materials for high-temperature use is their creep feature. However, systematic studies of creep in intermetallics of the Ti-Al system are still limited. Although the evaluation of the long-term strength of matrices will give a sound basis for alloy design, the relative strength among basic phases in the Ti-Al system has not been investigated systematically.

In this study, the high-temperature strength of three phases in the Ti-Al system has been compared. Most experimental data used here have been obtained on single-phase polycrystalline materials in the authors' laboratory during recent years.

The strength of these single-phase materials gives a guaranteed lower-bound elevated-temperature strength of intermetallic alloys based on these matrices.

In this report, the minimum creep rates $\dot{\varepsilon}_m$ are taken as the measure of creep strength and the effects of material conditions, such as the composition N (the mole fraction of solute) and the initial grain size d_0, and external conditions, such as stress σ and temperature T, on the minimum creep rate are analyzed according to the following equation,

$$\dot{\varepsilon}_m \propto (1+X)^m \, d_0^{\,p} \, \sigma^n \exp(-Q/RT). \qquad (1)$$

Here, X is the off-stoichiometric factor which is related with N by $X=(N/N_0-1)$ (N_0: the stoichiometric composition), R and T have the usual meanings. Parameters, m, p, n and Q, are numerical constants which are determined experimentally.

CREEP CHARACTERISTICS OF THE γ-PHASE

A fair amount of data has been reported on the creep behavior of the γ-TiAl, for example, see ref.1. Typical examples of the $\dot{\varepsilon}_m$-σ relationship in three single-phase polycrystalline γ-phase materials is shown in Fig.1 [2], in which three regions can be recognized clearly in every material.

Creep in each region has unique characteristics. The values of parameters in eq.(1) for creep of single-phase γ-alloys are cited in Table I. These values are based mostly on ref.1. Creep characteristics in the low stress region are not yet determined firmly, but tentative values, based on preliminary experiments [3,4], are cited with parentheses in the table. Deformation characteristics in the range of higher strain-rates can be estimated from data of conventional tests up to about $10^0 s^{-1}$ [5,6].

CREEP CHARACTERISTICS OF THE α_2-PHASE

Only limited data have been reported on creep of the α_2-phase of the simple binary system. Typical examples of the $\dot{\varepsilon}_m$-σ relationship are shown in Fig.2 for two materials, Ti-27mol%Al and Ti-34mol%Al [7,8].

The data points may be divided into three groups in each material as in the case of the γ-phase, though the parameters in eq.(1) have not been firmly determined yet. It is worth noting that the relative strength of these two materials depends greatly on the stress level; no significant difference can be seen in the low stress region, whereas the strength of the aluminum-poor alloy is obviously weaker than that of aluminum-rich alloy in the high stress region.

Tentative values of the parameters in eq.(1) are derived from preliminary experimental results and shown in Table I.

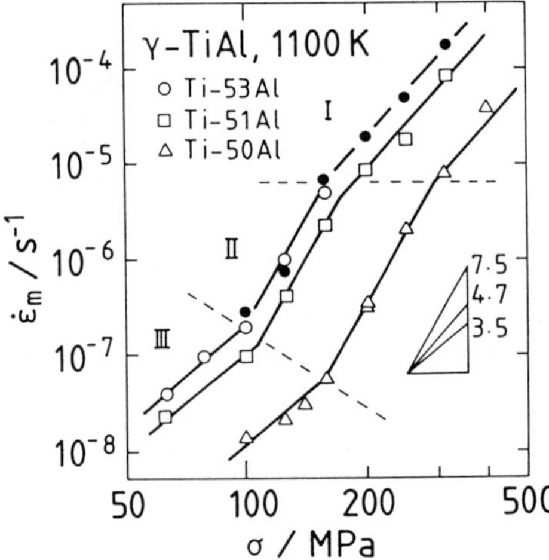

Fig.1
Minimum creep-rate as a function of stress in polycrystalline single-phase TiAl intermetallics at 1100 K [2].

Table I
Minimum Creep-Rate Characteristics in Single-Phase γ and α_2

	Stress level	low	intermediate	high
γ	n	3.5	7.5	4.7
	Q (kJ/mol)	(350)	600	360
	m	(~10)	~70	10~20
	p	(~0)	−2~−2.5	~0
α_2	n	(4)	(9)	(5)
	Q (kJ/mol)			
	m	(0)	(−50)	(−70)
	p			

Preliminary results are shown in parentheses

HIGH-TEMPERATURE DEFORMATION CHARACTERISTICS OF THE α-PHASE

Conventional tensile tests at elevated temperatures have been made extensively on the Ti-Al terminal solid solutions [9,10,11]. A plateau (σ_s) is observed in most stress-strain (σ-ε) curves. That is, the steady-state deformation appears under these conditions. The σ_s-$\dot{\varepsilon}$ (or σ_{peak}-$\dot{\varepsilon}$) relation obtained in constant-rate straining ($\dot{\varepsilon}$) tests is equivalent to the σ-$\dot{\varepsilon}_s$ (or σ-$\dot{\varepsilon}_{min}$) relation obtained in constant stress (σ) tests and both data can be complied as a single σ-$\dot{\varepsilon}$ relationship.

Typical examples of the $\dot{\varepsilon}$-σ_s relationship are shown in Fig.3 on alloys containing 1.0~6.9mol%Al [9]. Two characteristic regions are clearly recognized in high-temperature deformation and the transition condition between these two regions has been well established [9].

General characteristics of creep in solid solution alloys are well known, and mostly based on data obtained in Al-Mg [12] and Fe-Mo [13] alloys. Three regions are observed in the $\dot{\varepsilon}_s$-σ relationship in these alloys as seen in Fig.4. The (quasi) steady-state deformation characteristics were analyzed by the Dorn-type equation,

$$\dot{\varepsilon} \propto N^m (\sigma/G)^n \exp(-Q/RT). \qquad (2)$$

The parameters in each alloy system are listed in Table II.

The characteristics observed in Ti-Al solid solution alloys shown in Fig.3 correspond to those observed in the high and intermediate stress regions in Al-Mg and Fe-Mo alloys. Although no direct experimental data are reported, the existence of another region in the low stress range is also expected in Ti-Al alloys. Deformation behavior under high strain-rates or high stresses can be represented by an exponential function of stress, rather than a power function.

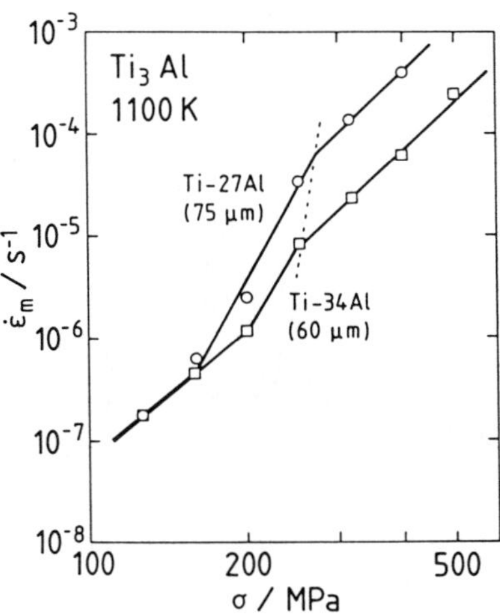

Fig.2
Minimum creep-rate as a function of stress in polycrystalline single-phase Ti_3Al intermetallics at 1100 K [7,8].

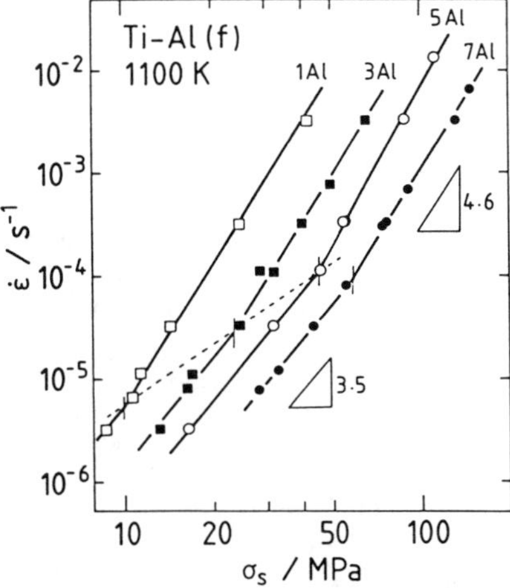

Fig.3
Relation between the true strain-rate and the steady-state (or maximum) stress in fine-grained α Ti-Al alloys at 1100 K [9].

High-temperature deformation of pure titanium in the α-phase range has been reported by several investigators [14,15] and the steady-state deformation has been represented by the following equation [15], that is,

$$\dot{\varepsilon}_m \propto (\sigma/G)^{4.2} \exp(-169 \text{ kJ·mol}^{-1}/RT). \tag{3}$$

STRENGTH AT 1100 K

The composition and temperature ranges of experimental data available in the three phases of the Ti-Al binary system are indicated in the phase diagram, see Fig.5. The strength of each phase cited in the preceding sections can be estimated as a function of the composition and we will compare the strength under specific conditions among materials.

Table II
Steady-State Deformation Characteristics in Terminal Solid Solution Alloys.

	Stress level	low	intermediate	high
α Ti-Al	*n*		~3.5	~4.5
	Q (kJ/mol)		~290	~270
	m		–1.3	–2.3
Al-Mg	*n*	4.5	3.3	4.3
	Q (kJ/mol)	174	136	137
	m	–0.9	–1.0	–1.9
α Fe-Mo	*n*	4.4	3.4	4.6
	Q (kJ/mol)	(>330)	284	(~280)
	m	–1.0	–1.0	–1.8

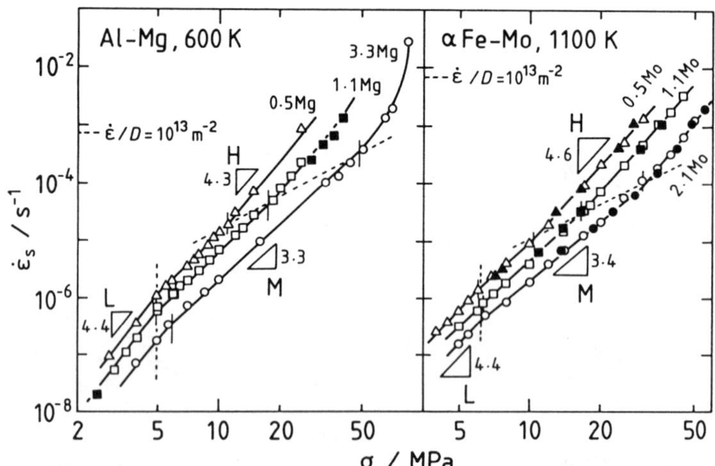

Fig.4 Stress dependence of steady-state creep rates in solid solution alloys.
a) Al-Mg alloys at 600 K [12], b) Fe-Mo alloys at 1100 K [13].

The (quasi) steady-state stress or the maximum stress is estimated for constant strain-rate deformation. The strain rate selected as examples are 10^{-1}, 10^{-4} and $10^{-8}s^{-1}$, which are representative conditions of slow forging, conventional mechanical tests, and creep, respectively. The grain size is known to be an important materials factor under some conditions, but here one condition, $d_0 \approx 100\mu m$, is selected.

The stresses necessary to keep settled strain-rates are shown in Fig.6 as a function of the concentration of aluminum at 1100 K. In high strain-rate deformation, solid solution hardening in the α phase is significant and the strength of concentrated terminal alloys is comparable to that of intermetallics. With decreasing strain-rate, however, the strength of terminal alloys decreases significantly, whereas that of intermetallics decreases only moderately. The difference in strength between terminal solid solutions and two intermetallics becomes larger as strain-rate decreases.

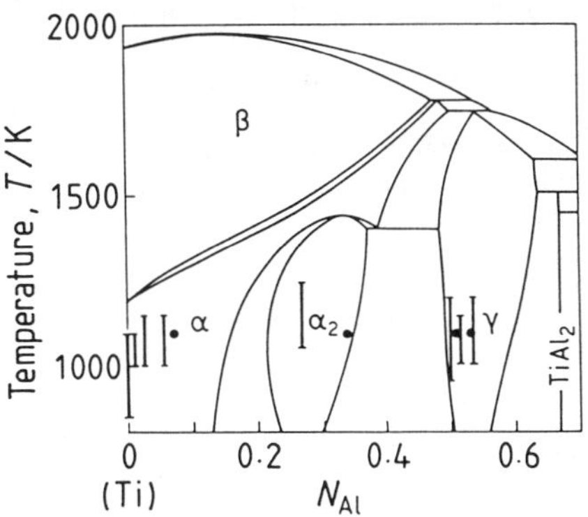

Fig.5
Ti-Al binary phase diagram [16] and the conditions of available data on elevated-temperature deformation.

The solid solution strengthening observed, or estimated, in the alloys containing 7mol%Al, or more, is obviously larger than that expected from the Cottrell-atmosphere drag strengthening [9]. This significant solution strengthening at the higher solute concentration range seems to be a characteristic feature of the hcp αTi-Al alloys, since no such significant strengthening has been observed in other concentrated alloys of β-Ti solid solutions, for examples, 1~7mol%Al alloys [17] and 5~30mol%V alloys [18].

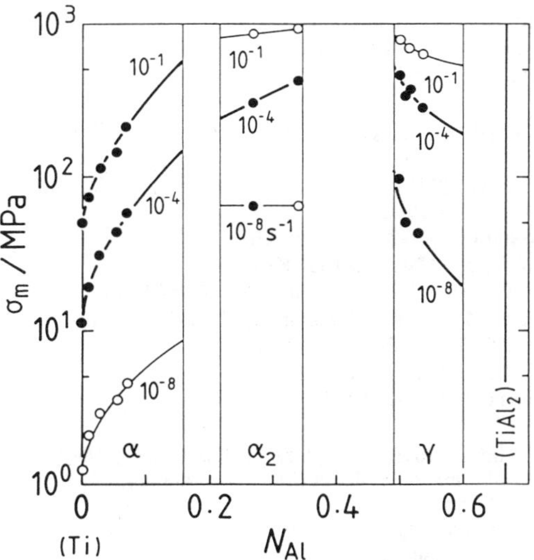

Fig.6
Stress necessary to deform in given strain-rates as a function of composition in polycrystalline alloys (d_0~0.1mm) in the Ti-Al system at 1100 K.

The strength of the α_2-phase and the γ-phase is not significantly different. In high strain-rate deformation, the strength increases in the α_2-phase as the composition deviates from the stoichiometry, whereas it decreases in the γ-phase. The strength of both phases is located roughly in the same range. With decreasing strain-rate, the effect of composition becomes insignificant in the α_2-phase, whereas the effect of composition is always significant in the γ-phase.

CONCLUDING REMARKS

Under creep conditions, the strength of the two ordered phases appearing in the Ti-side of the Ti-Al system is roughly the same. At the phase-boundary composition the strength of the γ-phase is obviously higher than that of the α_2-phase. However, this priority of the γ-phase disappears soon with a slight increase in aluminum concentration, because of the significant influence of the composition on the strength in the γ-phase. Therefore, no special priority in high-temperature long-term strength can be found between single-phase polycrystalline materials based on either of these intermetallics.

One of the challenging problems for developing practical materials based on intermetallics in the Ti-Al system is how much materials scientists/engineers can improve the strength by metallurgical means.

Acknowledgments

The work was supported in part by a Grant-in-Aid of the Ministry of Education (Grant No.04239105) and by Light Metals Educational Foundation, Incorp.

References

1. H. Oikawa, Mater. Sci. Eng. **A152/153**, 427-432 (1992).
2. Y. Ishikawa, K. Maruyama and H. Oikawa, Mater. Trans. JIM **33**, (1992) (in press).
3. M. Koo, K. Mashino, T. Matsuo and M. Kikuchi, Tainetsu Kinzoku Zairyo **33**, 59-67 (1992).
4. Y. Ishikawa, K. Maruyama and H. Oikawa, unpublished work (1992).
5. M. Nobuki, K. Hashimoto, T. Tsujimoto and Y. Asai, Nippon Kinzoku Gakkai-shi **50**, 840-844 (1986).
6. S. Mitao, Y. Kohsaka and C. Ouchi, in THERMEC-88, edited by I. Tamura (ISIJ, Tokyo, 1988), pp.620-627.
7. M. Ohtsuka and H. Oikawa, in Proc. 7th World Conf. on Titanium, edited by I.L. Caplan and F.H. Froes (TMS, Warrendale, 1992), (in press).
8. T. Fukuda, M. Ohtsuka and H. Oikawa: unpublished work (1992).
9. H. Oikawa and T. Oomori, Mater. Sci. Eng. **A104**, 125-130 (1988).
10. T. Oomori, T. Yoneyama and H. Oikawa, Trans. Jpn. Inst. Met. **29**, 399-405 (1988).
11. H. Oikawa, K. Sasaki, A. Sagara and K. Maruyama, J. Mater. Sci. Lett. **7**, 987-988 (1988).
12. H. Sato and H. Oikawa, Scr. Metall. **22**, 87-92 (1988).
13. S. Nanba and H. Oikawa, Mater. Sci. Eng. **A101**, 31-37 (1988).
14. M. Doner and H. Conrad, Metall. Trans. **4**, 2809-2817 (1973).
15. M.X. Cui and H. Oikawa, Nippon Kinzoku Gakkai-shi **49**, 195-202 (1985).
16. J.C. Mishurda and J.H. Perepezko, in Microstructure/Property Relationships in Titanium Aluminides and Alloys, edited by Y.W. Kim and R.R. Boyer (TMS, Warrendale, 1991), pp.3-30.
17. M. Seki and H. Oikawa, Mater. Trans. JIM **30**, 1027-32 (1989).
18. H. Oikawa, Y. Ishikawa and K. Masada, unpublished work (1989).

CREEP BEHAVIOR OF THE PSEUDO-BINARY Ni_3Si AND Co_3Ti

Souji Hasegawa*, Seiya Wada*, Takayuki Takasugi** and Osamu Izumi***
*Sumitomo Metal Industry Ltd,Minatomachi 2-12-1, Joetsu, Japan.
**Institute for Materials Research, Tohoku University, Katahira 2-1-1, Aobaku, Sendai, Japan.
***Professor Emeritus, Tohoku University, Sendai, Japan.

ABSTRACT

Creep behavior was investigated along the pseudo-binary line between $L1_2$-type Ni_3Si and Co_3Ti system with constant Si content. Addition of Co to Ni_3Si and addition of Ni to Co_3Ti enhanced creep strength due to precipitation hardening, respectively. Maximum creep strength on the pseudo-binary line was found at the composition which Co/(Ni+Co)=0.3 with 11 at.% Si content. The peak of creep strength on the pseudo-binary line was shifted to higher Co/(Ni+Co) value with decreasing Si content. Variation of creep strength was discussed based on metallographic observations.

Also, the creep strength of these pseudo-binary alloys was improved by further addition of other elements, such as Hf and Ta.

INTRODUCTION

The Ni_3Si alloy shows an increased strength with temperature and excellent corrosion and oxidation resistances at ambient and elevated temperatures, respectively. Recently, the ductilization of this alloy has been achieved by the addition of Ti [1]. The peak temperature in yield stress vs. temperature curve increased with increasing Ti concentration. Also, it was shown that the $Ni_3(Si,Ti)$ alloy exhibited high creep resistance[2].

On the other hand, the Co_3Ti alloy shows sufficient ductility at a wide range of temperatures and the positive temperature dependence of yield stress, accompanied with a high peak temperature in the yield stress vs. temperature curve[3].

It is well known that diffusion coefficients for ordered intermetallics are smaller than those for constituent elements. Therefore, ordered intermetallics are expected to have high creep resistance. It has been suggested that the volume fraction of $L1_2$ phase should be around 65 vol.% to have the maximum precipitation hardening for Ni_3Al-base superalloys. Nevertheless, data on creep behavior of Ni_3Si are very few as compared with Ni_3Al and superalloys.

The stress exponent n in the relationship, $\dot{\varepsilon} \propto \sigma^n$ for $L1_2$-type Co_3Ti has been determined to 3.0[4]. The stress exponent for $Ni_3(Si,Ti,Hf)$ was 3.0[2]. We've reported that precipitation of a small amount of metallic Ni phase in $Ni_3(Si,Ti)$ were effective to enhance its creep resistance and the alloy had superior creep resistance to conventional Ni-base superalloy, Inconel 718[2].

The present work aims to find the possibility of developing new heat-resistant alloys, in which the advantages of both the nickel-based alloys and cobalt-based alloys are combined. The phase stability and the effect of composition on creep strength of $(Ni,Co)_3(Si,Ti)$ are discussed.

EXPERIMENTAL

Starting materials were nickel of 99.9 mass % purity, silicon of 99.999 mass %, titanium of 99.9 mass %, cobalt of 99.8 mass %. The compositions tested in this work were selected along the pseudo-binary lines between $Ni_3(Si,Ti)$ and $Co_3(Ti,Si)$ systems with constant Si content of 4 and 11 at.%.

Alloy buttons were prepared by non-consumable arc melting in an argon atmosphere. In order to get the uniform alloy composition, remelting was repeated several times, and then homogenized in a vacuum of about 1.3×10^{-3} Pa at 1323 K for 1 day. Tensile specimens with 1 x 2 x 14 mm were prepared by electro-erosion technique. Specimens were then annealed at 1173 K for 1 day in a vacuum.

Metallographic observation and X-ray diffraction measurement were performed to characterize the microstructures of these alloys. Creep tests were carried out using a lever-type testing machine at temperatures from 1073 K to 1173 K in either air or argon atmosphere. The lattice parameters were determined precisely by the Debye-Scherrer method using full-annealed specimens.

RESULTS AND DISCUSSION

Figure 1 illustrates the pseudo-binary lines between Ni_3Si and Co_3Ti which were creep-tested in this work. This figure also shows the results for micrographic observations. In this figure, the open circles and full circles mean the samples which showed the microstructures of mono-phase and two-phase (or multiple phase), respectively. There were $L1_2$ mono-phase regions around Co_3Ti composition which extended toward Ni_3Ti[5] and around Ni_3Si composition which extended toward Ni_3Ti[6]. In the case of pseudo-binary Ni_3Al and Co_3Ti alloy system, $L1_2$-type continuous solid solution has been found [7]. However, $L1_2$ continuous solid solution between Ni_3Si and Co_3Ti could not be found under this experimental condition. This results may be associated with lattice parameter differences between two pseudo-binary alloy systems. The lattice parameter difference between Ni_3Si(3.609Å) and Co_3Ti(3.504Å) is larger than that between Ni_3Al(3.570Å) and Co_3Ti. This difference may enhance the tendency of phase decomposition.

Figure 2 illustrates creep curves of pseudo-binary alloys with constant 11 at.% Si content between Ni_3Si and Co_3Ti which were creep-tested under constant stress in an argon atmosphere at 1073 K. All these alloys showed the normal-type primary creep in which strain rate of these alloys decreased with time to the minimum.

Figure 3 shows the minimum creep rates of these alloys along the pseudo-binary lines with 4 and 11 at.% Si contents. Maximum creep strength on the pseudo-binary line was found at the composition which Co/(Ni+Co)=0.3 with 11 at.% Si content. The peak of creep strength on the pseudo-binary line was shifted to higher pseudo-binary ratio, Co/(Ni+Co) value with decreasing Si content. Figure 4 shows the variation of stress exponents with Co/(Ni+Co) value with 11 at.% Si content. The maximum of stress exponent was located at the pseudo-binary ratio nearly equal to 0.3, and much larger than 5. This result implies that precipitation hardening is effective in the alloys with this composition.

Figure 5 shows the microstructure of the alloy with pseudo-binary ratio, 0.3 at constant Si content of 11 at.% which was observed by scanning electron microscopy(SEM). Matrix phase contained homogeneously distributed fine precipitates on along crystallographic directions. It was largely observed in these

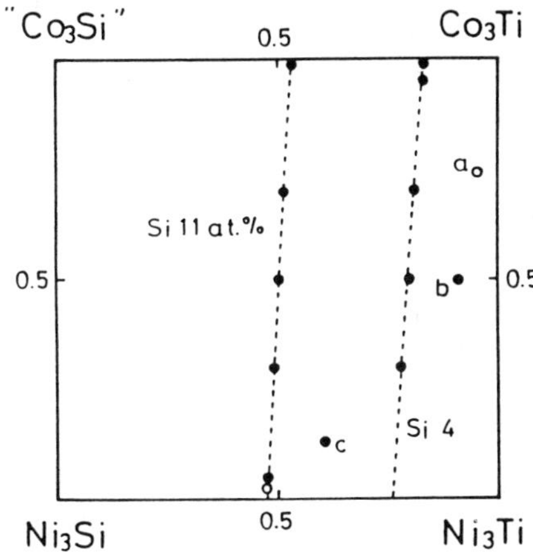

Fig.1 The quaternary system Ni_3Si-Ni_3Ti-Co_3Ti-Co_3Si, showing the stable compositional range for the $L1_2$ solid solution. 'Co_3Si' is the hypothetical phase.

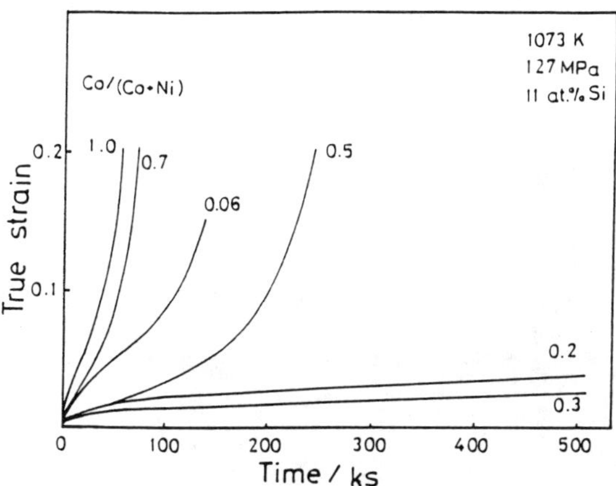

Fig.2 Creep curves of alloys along the pseudo-binary line with constant Si content of 11 at.%.

pseudo-binary alloys that dynamic recrystallization occurred within very short time and played an important role in creep behavior of $Ni_3(Si,Ti)$. Thus, these second phases reduce the stress concentration due to grain boundaries and retard the dynamic recrystallization introduced by the stress concentration at grain boundaries.

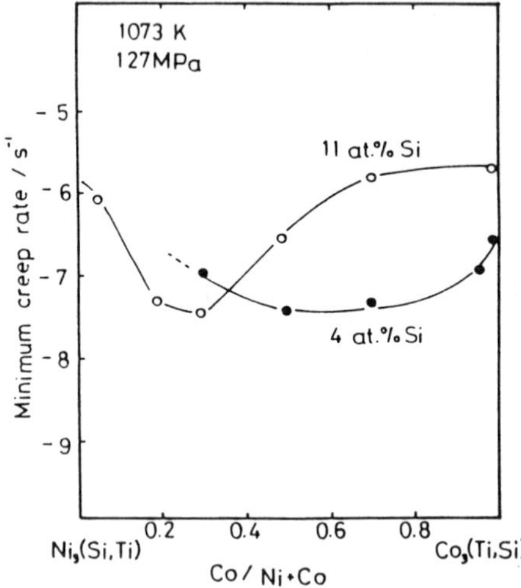

Fig.3 Minimum creep rates of alloys along the pseudo-binary lines between $Ni_3(Si,Ti)$ and $Co_3(Ti,Si)$.

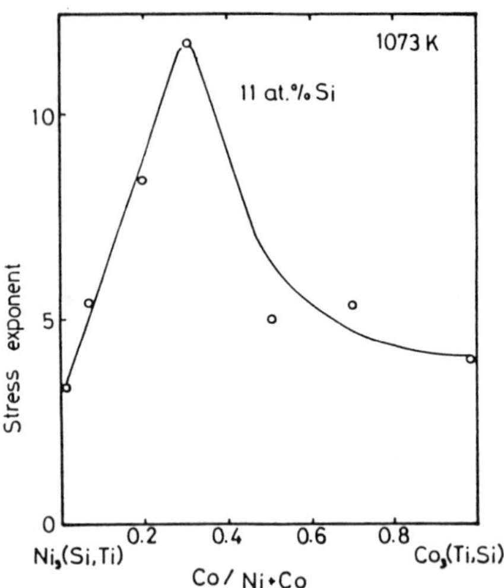

Fig.4 Stress exponents of alloys along the pseudo-binary line with constant Si content of 11 at.%.

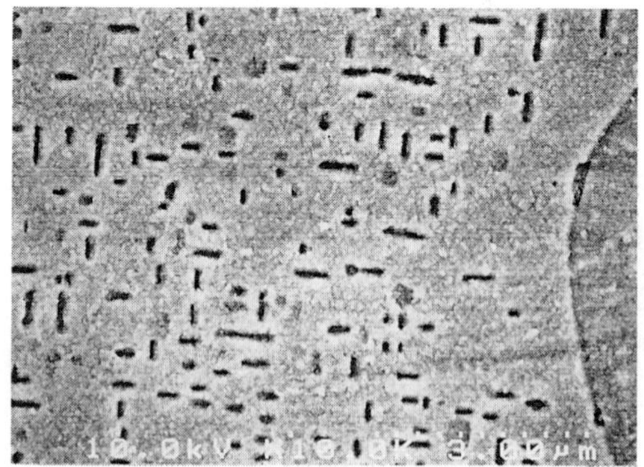

Fig.5 Microstructure obtained by SEM observation for the alloy which showed the maximum creep strength.

Figure 6 shows the Larson-Miller diagram for the alloys tested in this work. The broken line corresponds to $Ni_3(Si,Ti,Hf,Ta)$ alloy with a small amount of metallic Ni phase. Pseudo-binary alloy tested in this work exhibited superior creep resistance to that alloy and In 718, under this experimental condition.

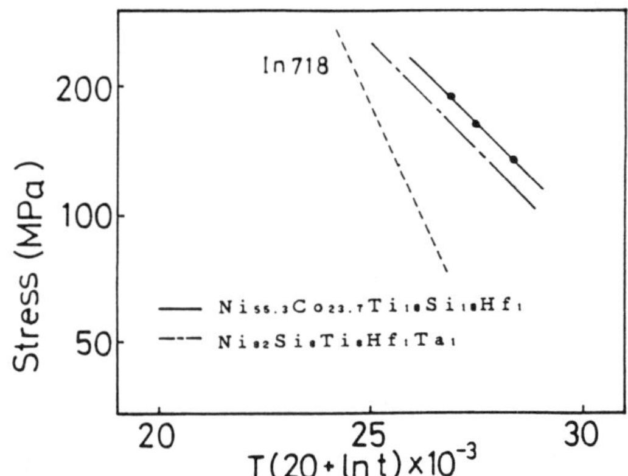

Fig.6 The Larson-Miller diagram for the pseudo-binary alloy used in this work.

CONCLUSION

Results are summarized as follows;

1. Maximum creep strength on the pseudo-binary line between Ni_3Si and Co_3Ti was found at the composition which Co/(Ni+Co)=0.3 with 11 at.% Si content.
2. At the above composition, matrix phase contained fine precipitates aligned crystallographically.
3. The peak of creep strength on the pseudo-binary line between Ni_3Si and Co_3Ti was shifted to high Co/(Ni+Co) value with decreasing Si content.
4. Further addition of Ta (or Hf) to these pseudo-binary alloys improved their creep resistances.
5. The solubility limit of Co to $Ni_3(Si,Ti)$ (11 at.% Si) is smaller than 3 at.%.

REFERENCES

1.T.Takasugi, M.Nagashima and O.Izumi, Acta Metall. **38**,747(1990).
2.S.Hasegawa, S.Wada, T.Takasugi and O.Izumi, Proc. Intermetallic compounds - Structure and Mechanical Properties, Japan Inst. of Metals 427(1991).
3.T.Takasugi and O.Izumi, Acta Metall. **33**,39(1985).
4.S.Hasegawa, S.Wada, T.Takasugi and O.Izumi, to be published.
5.Y.Liu, T.Takasugi and O.Izumi, Metal Trans. **17A**,1433(1986).
6.T.Takasugi, D.Shindo, O.Izumi and M.Hirabayashi, Acta Metall.**38** 1417(1990).
7.S.Miura, H.Kuriyama, Y.Mishima and T.Suzuki, J.Japan Inst. Metals, **56**,943(1992).

LOW-CYCLE FATIGUE BEHAVIOR OF POLYCRYSTALLINE NiAl AT ROOM TEMPERATURE

KEITH M. EDWARDS and R. GIBALA
Department of Materials Science and Engineering
The University of Michigan, Ann Arbor, MI 48109

ABSTRACT

The room temperature cyclic deformation of a cast and extruded NiAl alloy has been investigated. Low-cycle fatigue tests were performed under plastic strain control at strain ranges from 0.0002 to 0.0016. Cyclic hardening behavior has been analyzed and compared to the monotonic tensile and compressive behavior of the same material. Electroless nickel films applied to NiAl samples have little effect on monotonic deformation behavior, but do have a significant effect on cyclic deformation behavior, particularly cyclic stress asymmetry. Fractographic analysis has suggested that cycling at low plastic strain ranges may promote stable microcracking, while step testing indicates that cycling at lower plastic strain ranges can improve the cyclic life and/or stress levels achieved at the higher plastic strain ranges. Low plastic strain range cyclic prestrain can also significantly increase the monotonic tensile yield and ultimate strength of the material while most of the tensile ductility is retained.

INTRODUCTION

Nickel aluminide (β-NiAl) with the B_2 (CsCl) crystal structure is an intermetallic compound which is currently a candidate for high temperature structural applications. Unfortunately, its lack of ductility and toughness at ambient temperatures and its lack of sufficient strength and creep resistance at elevated temperatures has thus far limited it to primarily non-structural applications. However, NiAl does possess many properties which would make it superior to currently used materials if its low-temperature ductility and high-temperature strength could be improved. Additionally, as a candidate for structural applications, an understanding of the fatigue behavior of this material at ambient, as well as elevated temperatures, is necessary. The purpose of this study is to characterize the room temperature cyclic deformation behavior of polycrystalline NiAl and relate this to monotonic deformation behavior. As a part of the characterization, the effect of thin nickel surface films on deformation behavior has also been investigated. Surface films have been shown to enhance the room temperature ductility of several bcc metals by inducing mobile dislocation generation at the film/substrate interface [1].

EXPERIMENTAL

Three identical NiAl castings 150 mm long and 15 mm in diameter were canned in 72 mm diameter mild steel billets and extruded at 1150°C at a 6:1 reduction in area. The pre-extrusion composition analysis showed that the material was stoichiometric within the 0.5 wt.% accuracy of measurement and contained low levels of oxygen (0.0347 wt.%). Metallographic examination revealed that the NiAl was essentially equiaxed and had an average grain size of approximately 100 μm. A texture analysis indicated a <111> texture of 2 to 3 times random along the extrusion axis. Fatigue/tension specimens and compression specimens were centerless ground from sections of the extrusion. The gauge diameter of all specimens was 3 mm, and for fatigue/tension specimens, grip sections were 8 mm in diameter. The last 0.05 mm of the gauge surface was removed by electropolishing [2]. For all experiments investigating surface film effects, ~170 nm electroless nickel films were applied, following guidelines described in the Metals Handbook [2, 3].

All fatigue experiments were performed on an MTS servohydraulic machine equipped with a 100 kN load cell and hydraulic collet grips. Strains were measured with an 8 mm gauge length extensometer (± 6% accuracy) attached to the NiAl samples with four dental elastics. Tests were started in compression and run in plastic strain control with an imposed strain rate of approximately 1.4 x 10^{-4} /s. All tension and compression tests were performed on a screw-driven Instron Model 1137 instrument using a 10,000 lb. load cell. The crosshead speed was maintained at a constant 0.0021 mm/s which produced initial strain rates of 3.5x10^{-4} /s and 2.1 x 10^{-4} /s for the 6 mm long compression samples and the 10 mm gauge length tension samples, respectively. After failure, the gauge and fracture surfaces of all samples were examined with Hitachi S-520 and/or S-800 scanning electron microscopes.

RESULTS AND DISCUSSION

In general, the ductility and fracture of polycrystalline NiAl at room temperature is governed by its limited slip systems and hence, the von Mises criterion. Also, factors such as initial dislocation density and efficiency of dislocation generation as well as the mobility of these dislocations play an important role in determining the extent of deformation in this material. This holds true for both monotonic and cyclic deformation, although each deformation mode poses somewhat different conditions for dislocation generation and damage accumulation processes.

At room temperature, the as-extruded material had a 0.2% offset yield stress of ~210 MPa in both monotonic tension and compression. This value is relatively low compared to some previously studied NiAl alloys but is expected for a material that has low impurity content and a large grain size. Monotonic tensile plasticity was ~0.8% and fracture occurred at ~265 MPa. In monotonic compression, samples hardened to ~800 MPa and plastic strains to failure were >15%. However, a large amount of cracking occurred in compression samples so that true compressive ductility could not be determined.

Figure 1 depicts several stress-plastic strain hysteresis loops generated during fatigue cycling at a plastic strain range of 0.0016. Although the loops generated at other plastic strain ranges were quantitatively different, these loops qualitatively represent the change in stress-strain behavior of all of the samples subjected to fatigue cycling. On the first loop, the elastic modulus was determined to be 217 ± 5 GPa which was subsequently used to calculate plastic strains. Yielding occurred at ~180 MPa and showed no discontinuous behavior. Note that the cyclic yield stress is less than the typically reported 0.2% offset monotonic yield stress and that all of the plastic strain ranges used in this study were less than 0.2% (0.0020). Peak stresses at low accumulated strain were less than 200 MPa, but increased quickly as cycling was continued. This hardening was accompanied by a change in hysteresis loop shape which showed that deviation from linear elastic behavior was occurring consistently earlier as accumulated strain increased. However, no significant modulus change occurred during cycling as evidenced by the nearly constant (vertical) slope in the hysteresis loops immediately following strain reversals.

In Figure 2, the hardening behavior of the same sample described in Figure 1 is given in more detail. As was true of all fatigue samples, the peak stress levels at low accumulated strains showed a small positive asymmetry ($\sigma_T > \sigma_C$). A recent study by Noebe and Lerch [4] on the low-cycle fatigue behavior of NiAl indicated no such peak stress asymmetry. We believe that the initial peak stress asymmetry seen in our material may be due to the specific composition or processing techniques used in this study and that the observed asymmetry is not intrinsic to NiAl. Nevertheless, this small positive asymmetry decreased as cycling was continued and eventually became negative, reaching values on the order of -50 MPa for uncoated samples at the larger plastic strain ranges. The asymmetries were generally somewhat larger at higher plastic strain ranges than at the lower plastic strain ranges. This type of asymmetry was also seen by Noebe and Lerch [4] and was interpreted to be the result of cracking. In their fatigue study, no observations of cracking were reported and fatigue fracture initiation occurred at preexisting defects. Also, fracture surfaces were predominantly intergranular in both fatigue testing and monotonic tensile testing. In our study, however, post-fracture examination of samples which had been cycled at the 0.0016 plastic strain range showed evidence of cracking on side surfaces both near and away from the final fracture surface. Also, while fracture propagation in unprestrained monotonic tension samples was primarily transgranular, samples which had been cyclically deformed exhibited significantly greater amounts of intergranular fracture propagation.

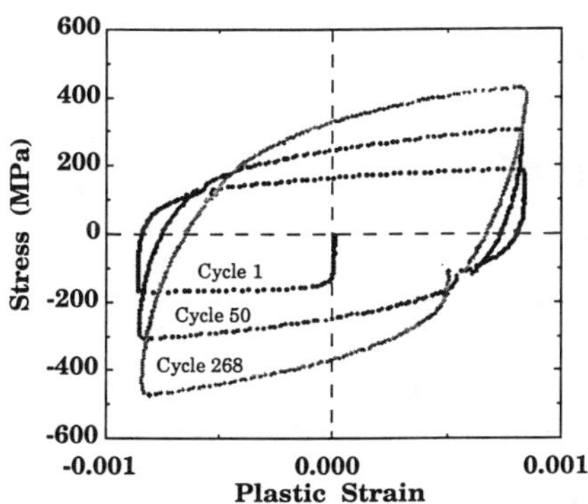

Figure 1. Several stress-plastic strain hysteresis loops generated during room temperature cyclic deformation testing of NiAl at a plastic strain range of 0.0016.

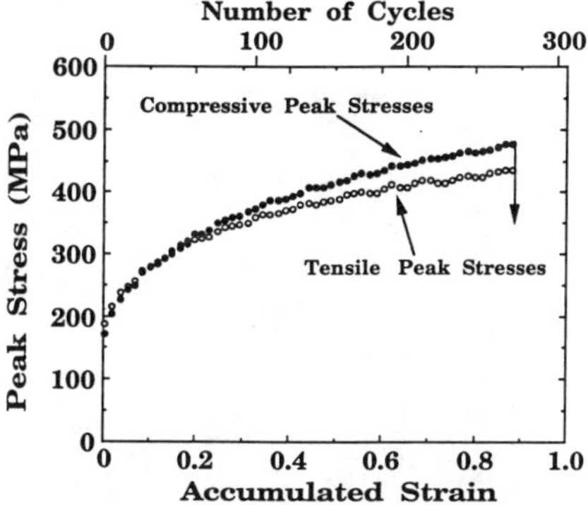

Figure 2. Absolute tensile and compressive peak stresses as a function of accumulated strain for a NiAl sample cycled at a plastic strain range of 0.0016 at room temperature.

As shown in Figure 3, the fracture surface of a sample deformed to failure in monotonic tension indicated an intergranular initiation site away from the sample surface, but primarily transgranular quasi-cleavage fracture propagation. In contrast, Figure 4 shows the fracture surface of a sample which failed after being deformed 273 cycles at a plastic strain range of 0.0016. In this case, fracture initiated at the surface and propagated primarily intergranularly. For additional comparison, Figure 5 shows the fracture surface of a sample which was fatigue prestrained 1000 cycles at the 0.0002 plastic strain range and 1000 at 0.0004 and then deformed to failure in monotonic tension. Similar to the unprestrained sample fractured during monotonic tension testing, fracture initiation was intergranular. However, similar to the sample fractured during cyclic deformation testing, fracture propagation was primarily intergranular. Overall, these observations suggest that stable microcracking may be initiated by low-cycle fatigue deformation, although a more thorough study of the deformation preceding failure is necessary. We should note also that although fatigue fracture initiation often occurred at the sample surface, the initiation sometimes occurred well away from the surface.

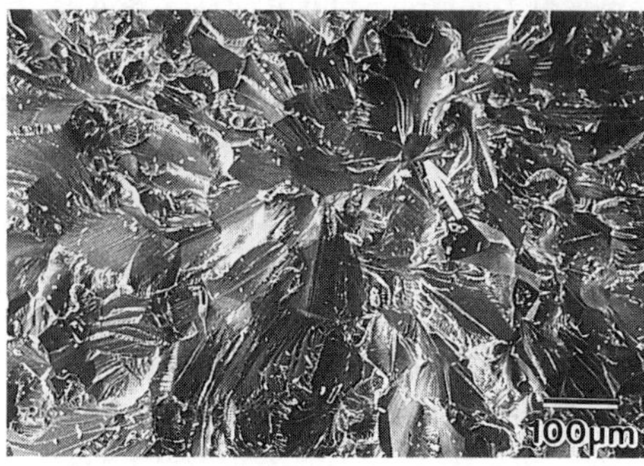

Figure 3. SEM micrograph of the fracture surface of an as-extruded NiAl sample deformed to failure in monotonic tension. Fracture initiation (indicated by the arrow) was intergranular, but fracture propagation was primarily transgranular.

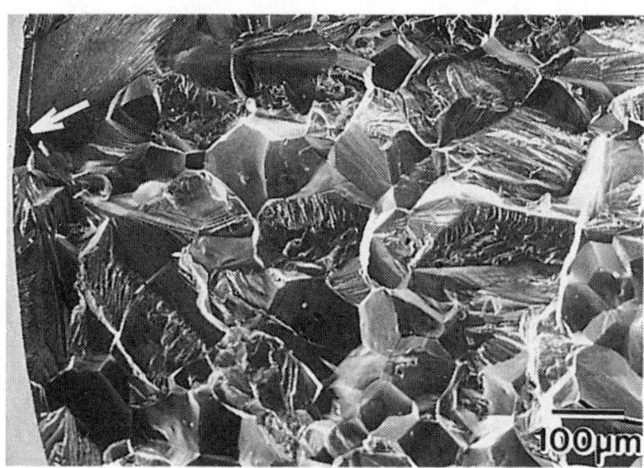

Figure 4. SEM micrograph of the fracture surface of an NiAl sample which had been cyclically deformed 273 cycles at 0.0016 to failure. Fracture initiation (indicated by the arrow) was at the sample surface and fracture propagation was primarily intergranular.

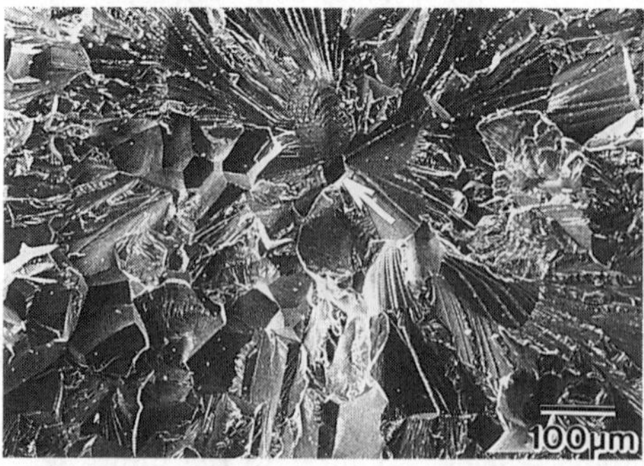

Figure 5. SEM micrograph of the fracture surface of an NiAl sample which had been cyclically deformed 1000 cycles at the 0.0002 plastic strain range and 1000 cycles at 0.0004 and then deformed in monotonic tension to failure. Fracture initiation (indicated by the arrow) was intergranular and fracture propagation was primarily intergranular.

In Figure 6, the monotonic stress-plastic strain behavior of the fatigue prestrained tensile sample described in Figure 5 is compared to the stress-plastic strain behavior of the unprestrained monotonic tension sample described in Figure 3. Since the fatigue prestrained sample was in a strain-hardened condition, its yield stress was approximately double that of the unprestrained sample. Its initial hardening rate was greater and it achieved an ultimate fracture stress that was approximately double that of the unprestrained sample. The fatigue prestrained sample had a somewhat lower ductility than the unprestrained sample, failing after ~0.5% plastic strain. Thus, low plastic strain range fatigue cycling may be an effective way of introducing dislocations and hardening the material while retaining compatibility at grain boundaries and hence, a measurable ductility. Also, if we take a rough estimate of toughness as the area under the stress-strain curve, the fatigue prestrained sample appears to have a higher toughness than the sample which had not been fatigue prestrained.

Also shown in Figure 6 are the average peak stresses attained by fatigue samples as a function of plastic strain range. The maximum stresses reached by this material were roughly 500 MPa, achieved only by samples which had been cyclically prestrained at the 0.0002 or 0.0004 plastic strain ranges and eventually cycled at 0.0016. Samples cycled only at the lower plastic strain ranges could not harden to these stresses because a saturated state was approached well below 500 MPa. Samples cycled only at the higher plastic strain ranges did not attain 500 MPa peak stresses because failure occurred before this stress level could be reached. Therefore, if samples are deformed at only the higher plastic strain ranges (or in monotonic tension), sufficient damage occurs to initiate a critical sized crack at low stress levels. However, if samples are hardened by step testing through the less damaging low plastic strain ranges, they are more likely to have substantial cyclic lives at the higher plastic strain ranges and can attain significantly higher stress levels.

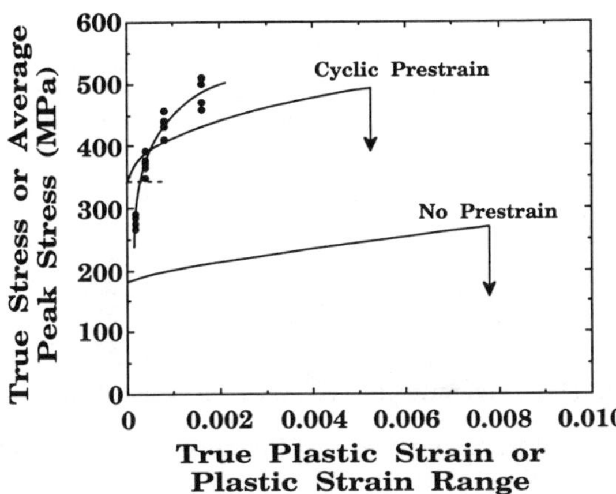

Figure 6. Room temperature monotonic tensile response for an as-extruded NiAl sample and an NiAl sample which had been cyclically prestrained 1000 cycles at the 0.0002 plastic strain range and 1000 cycles at 0.0004. Also shown are the average peak stresses attained by NiAl fatigue samples as a function of plastic strain range.

In general, NiAl samples coated with thin nickel films behaved similarly to uncoated samples in their reaction to cyclic deformation testing and cyclic prestrain (step testing) experiments. At low accumulated strains, peak stresses were below 200 MPa and a small positive peak stress asymmetry existed. With increasing accumulated strain, samples hardened rapidly and attained average peak stresses which were not significantly different than uncoated specimens for a given plastic strain range. However, the negative asymmetry which developed during cycling reached values on the order of -100 MPa, or roughly double that developed in uncoated specimens. Furthermore, NiAl samples coated with thin nickel films generally had lower fatigue lives than uncoated specimens which were subjected to identical deformation conditions. Lastly, it should be noted that the nickel films had no significant effect on monotonic deformation behavior or on the fracture behavior of samples subjected to either monotonic or cyclic deformation. These results are different than ones obtained on higher strength Ni-Fe-Al and Fe-Al ordered alloys [5]

SUMMARY AND CONCLUSIONS

1. The plastic strain controlled low-cycle fatigue behavior of a cast and extruded NiAl alloy has been evaluated at room temperature for several plastic strain ranges from 0.0002 to 0.0016.

2. Cyclic prestraining NiAl samples at the lower plastic strain ranges was successful in strain hardening the material and changed damage accumulation characteristics so that cyclic lives were enhanced and/or higher peak stresses were attainable.

3. Cyclic prestraining approximately doubles the monotonic yield and ultimate tensile strengths of NiAl while a significant amount ductility is retained.

4. Nickel coated samples tended to develop larger negative peak stress asymmetries and generally had shorter fatigue lives than uncoated samples which were subjected to identical deformation conditions.

ACKNOWLEDGMENTS

The support of this research by the National Science Foundation under Grant No. DMR-9102414 is gratefully acknowledged. The authors would also like to thank Dr. Vinod Sikka and Oak Ridge National Laboratory for extruding the materials used in this study.

REFERENCES

1. R.D. Noebe and R. Gibala in Structure and Deformation of Boundaries, edited by K.N. Subramanian and M. A. Imam (The Metallurgical Society of AIME, 1985), p. 89.

2. K. M. Edwards, M.S. thesis, The University of Michigan, Ann Arbor, MI, 1992.

3. W.D. Fields, R.N. Duncan, and J.R. Zickgraf, Metals Handbook, 9th ed., **5**, 219 (1982).

4. R.D. Noebe and B.A. Lerch, Scripta Metall. Mater. **27**, 1161 (1992).

5. K.J. Bowman, S.E. Hartfield-Wünsch and R. Gibala, Scripta Metall. Mater. **26**, 1529 (1992).

STRAIN-RATE EFFECTS ON THE ROOM TEMPERATURE TENSILE PROPERTIES OF A TiAl ALLOY

Y-W. Kim* and D.M. Dimiduk**,
*UES, Materials Research Division, 4401 Dayton-Xenia Rd., Dayton, OH 45432
**Wright-Laboratory, WL/MLLM, WPAFB, OH 45433

ABSTRACT

The room temperature tensile properties and fracture behavior of a Ti-47Al-1.5Cr-1V-2.3Nb gamma TiAl alloy, have been investigated for two duplex and a fully-lamellar microstructural conditions. These microstructures were controlled through forging and subsequent recrystallization/solution treatments followed by furnace cooling and aging treatment at 900°C. Tensile testing was conducted on either as-machined or electropolished round-bar specimens over a range of strain rates from 10^{-5} to 10^{-1} sec^{-1} in laboratory air. Experiments showed that both tensile ductility and strength levels vary not only with microstructure and surface condition, but also with applied strain rate in an unexpected way. The complex relationships between these variables are under continuing investigation. The results to date, which were analyzed using microstructural observations made on the deformed and fractured specimens, indicate that the deformation and fracture modes remain essentially the same within the strain rate range used. The possible involvement of environmental effects in the fracture process is suggested.

INTRODUCTION

Gamma titanium-aluminide alloys of engineering importance consist of $L1_0$ gamma TiAl as a matrix phase and small amounts of DO_{19} Ti_3Al. Collectively, prior studies show, at least qualitatively, that the mechanical properties of gamma alloys are controlled by the distribution of these phases including morphologies, relative amounts, and sizes [1,2]. Depending upon the distribution and morphology of the phases, the microstructures can be classified into two types–standard and engineered [1-4]. Duplex and fully-lamellar(FL) microstructures are two typical standard types and have been characterized quite extensively for mechanical properties [2,5,6]. From these studies, some characteristic deficiencies were found [2,7], including the inverse relationship between ductility and toughness, in that duplex microstructures exhibit reasonable ductility but poor toughness while the opposite is true for FL structures. The engineered microstructures, which are essentially modified lamellar structures, have been developed to improve upon these deficiencies with some success [3,4], but their optimization is far from being complete. Clearly, the effects of micro structure on the mechanical properties, especially RT tensile properties, are far greater than we have known and expected. In connection with the microstructural effects, other factors such as strain rate, specimen surface conditions and test environment, appear to significantly influence tensile properties. Growing evidence indicates that tensile testing in laboratory air reduces the RT ductility of gamma TiAl in the duplex[8] and single-phase gamma[9] microstructure conditions, as well as in unidirectional FL PST crystals[10]. This paper reports the preliminary results of the variation in RT tensile properties and associated deformation and fracture behavior that were observed upon testing selected microstructures in laboratory air over a range of strain rates. As is shown later, the strain-rate effect appears to be microstructurally sensitive, even within the small subset of the numerous microstructural conditions available.

EXPERIMENTAL

The starting material was an induction-skull melted/cast and isothermally forged two-phase alloy having the composition, Ti-47.0-1.6Cr-0.9V-2.3Nb (in at%), with

550ppm O and 90ppm (N+C+H) by weight. The cast ingot was HIP'ed at 1200°C under 172MPa pressure for 3h. A billet was sectioned from the ingot and hot deformed, uniaxially, into a pancake form with a total height reduction of 91%, using a two-step forging process at 1150°C. The alpha-transus temperature (Tα) was 1362°C as determined by DTA analysis and metallographic observations. Three microstructures were produced using a standard heat-treatment process consisting of annealing and aging. Two duplex microstructures, "A" and "B", were obtained by two different annealing treatments, 1275°C/4h and 1290°C/3h, respectively, followed by furnace cooling (FC) and aging at 900°C for 4h. A FL microstructure, FL "A", was obtained by annealing at 1380°C for 3h followed by FC and aging. Round-bar tensile specimens having a 1.4cm long and 3.3 mm diameter gage section were machined from the heat-treated materials. Both duplex "A" and FL "A" specimens were electro-polished and duplex "B" specimens were tested in as-machined condition. Tensile testing was conducted at room temperature (RT) in laboratory air over a range of strain rates from $5x10^{-5}$ to $3x10^{-1}s^{-1}$. Microstructual changes after tensile deformation and fracture were observed using light microscopy, both under bright-field and polarized-light conditions. Fracture surfaces were investigated using both conventional and high-resolution SEM.

RESULTS and DISCUSSION

Examples of the duplex microstructures, taken from areas away from the fracture surfaces after tensile testing, are shown in Fig. 1. The deformed structures observed within grains did not cause any appreciable changes in the grain size and morphology. The grain sizes of duplex A and B range from 20-100 μm and 10-60 μm, respectively, with an equiaxed morphology. Fig. 1 also demonstrate the tensile deformation provides some uniform elongation to the duplex specimens prior to failure. The measured tensile properties are plotted against strain rate for both the duplex A and B structures in Fig. 2. Clearly, the maximum tensile fracture strains for both microstructures occur at a strain rate around $2x10^{-2}s^{-1}$. At lower or higher strain rates,

Fig. 1. Optical microstructures of the longitudinal metallographic sections below the fracture surfaces of (a) duplex "A" and (b) duplex "B" tensile specimens tested at RT and at a strain rate of $1.4x10^{-3}s^{-1}$

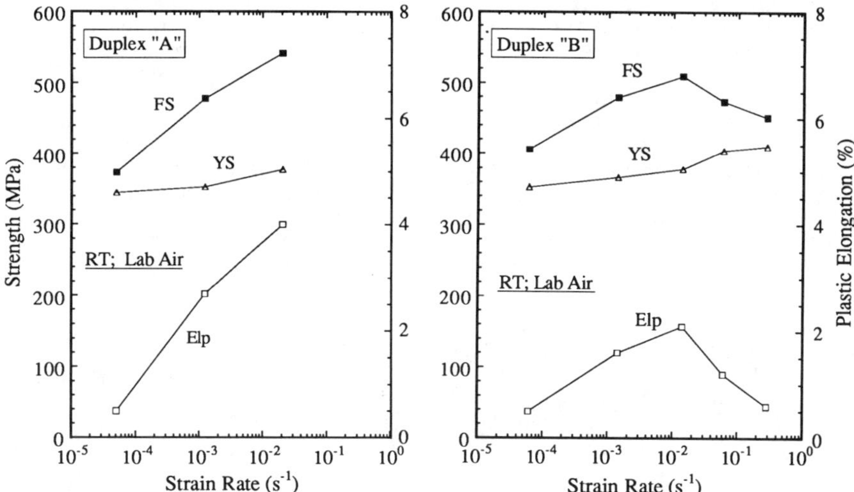

Fig. 2. Tensile properties – FS (fracture stress), YS (yield stress) and Elp (plastic strain to failure) – as functions of strain rate for duplex "A" (a) and duplex "B" (b) microstructure specimens at RT and in laboratory air.

the ductility decreases considerably, especially at lower rates as shown for the duplex A microstructure. Most of tensile properties reported in the literature have been obtained at strain rates around $1x10^{-3}s^{-1}$. For both microstructures, the yield stress showed only small and gradual increases with increasing strain rate, but the fracture stress followed the trend of variation in the tensile elongation with strain rate. This tensile behavior indicates that for a given microstructure, the stress-strain curves at different strain rates are essentially identical with equivalent strain-hardening rates, but fracture occurs at different points. Nevertheless, there is a small strain-rate sensitivity of flow stress, m, defined by $\sigma_T=K'(\varepsilon_T)^m$ [11] where T stands for "true". The m values were estimated to be 0.015 and 0.02, respectively, for duplex the A and B structures. Another feature from Fig. 2 is an apparent effect of specimen surface condition on tensile elongation. The electro-polished duplex A specimens exhibit larger elongations at intermediate strain rates (10^{-3}–$10^{-2}s^{-1}$) than the as-machined duplex B specimens which have finer grains than duplex A. Since, for a given class of microstructures finer grains yield higher elongations than coarser grains, in general, the higher elongations of the duplex A than of the duplex B are likely due to the electro-polishing. However, the strain rate effect is clearly evident regardless of the specimen surface condition.

Figure 3 shows both SEM fracture surface fractographs (a and b) and polarized-light optical micrographs, taken from the areas just below the fracture surfaces on the metallographic sections (c and d) of the duplex "A" specimens, after testing at strain rates of $5x10^{-5}$ and $2x10^{-2}s^{-1}$. Both specimens show rather brittle fracture, consisting mostly of cleavage-like and some intergranular cracking, and extensive amounts of plastic deformation activity. Some detailed features of the fracture surfaces in Fig. 3 are shown at higher magnifications in Fig. 4. Both strain rates result in significant crystal plasticity, most likely consisting of twinning and slip, which cause the formation of deformation bands imparting a layered structure to the duplex grains. At the low strain rate, multiple variants of fine and dense cross-slip/twin activity are observed across the layered structure, which may be the cause of the dominant fracture mode, trans-layer fracture (Fig. 4a). At the higher rate, the fracture modes are similar, however, with increased inter-layer fracture events or fracture parallel to primary deformation bands (Fig. 4b). Material having duplex B structure revealed similar deformation and fracture modes, showing only slight differences between different strain rates at very high magnifications.

Figure 5a shows the fully-lamellar microstructure, FL "A", taken below the fracture surface of a tested specimen. No deformation activity is apparent at this magnification. The grain size ranges from 800 to 1500μm with little interruption of the lamellae. The tensile properties of FL "A" specimens are plotted against strain rate in Fig. 5b. Generally, the trends are similar to those of the duplex microstructures shown in Fig. 2; however, the variation in tensile ductility with strain rate was much less, while the strain-rate sensitivity of flow stress, m=0.03, was slightly higher than for

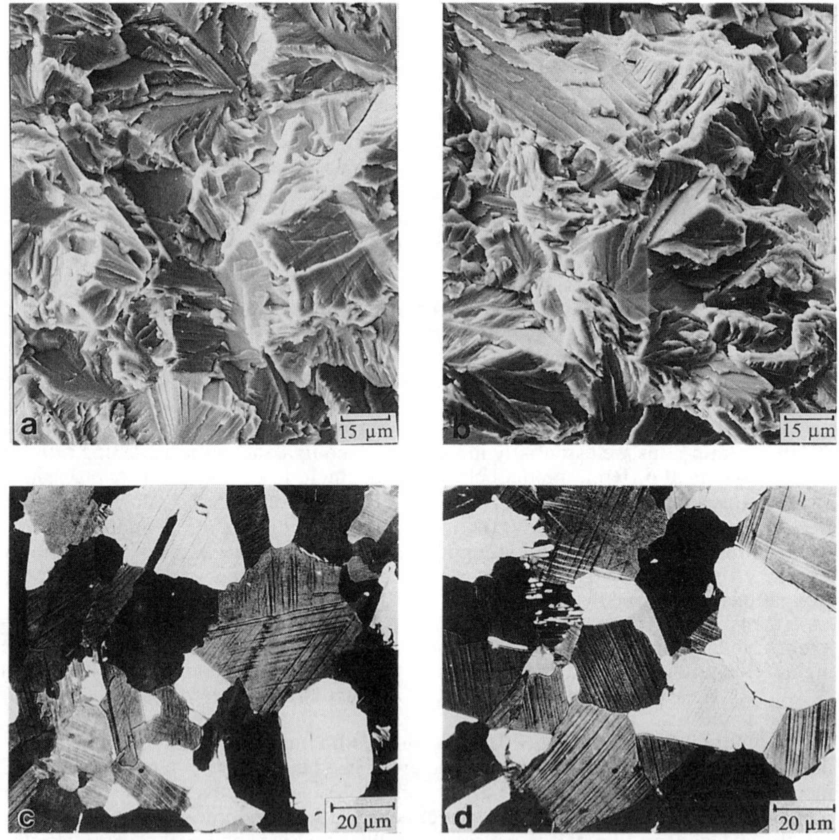

Fig. 3. SEM fracture surfaces (a, b) and optical microstructures of the longitudinal metallographic sections just below the fracture surfaces (c, d) of the duplex "A" tensile specimens tested at RT and at two different strain rates, resulting in different plastic elongations to fracture (Elp): (a, c) $5x10^{-5}s^{-1}$ and Elp=0.5% and (b, d) $2x10^{-2}s^{-1}$ and Elp=4.0%.

the duplex microstructures. Limited amounts of deformation activity were observed below the fracture surfaces under both light microscopy and SEM, occurring in the form of fine translamellar slip/twins confined within a narrow zone adjacent to the fracture surface. On the electro-polished surfaces, however, deformation bands or slip steps with 10-20μm spacing were abundant, occurring parallel to the laths.

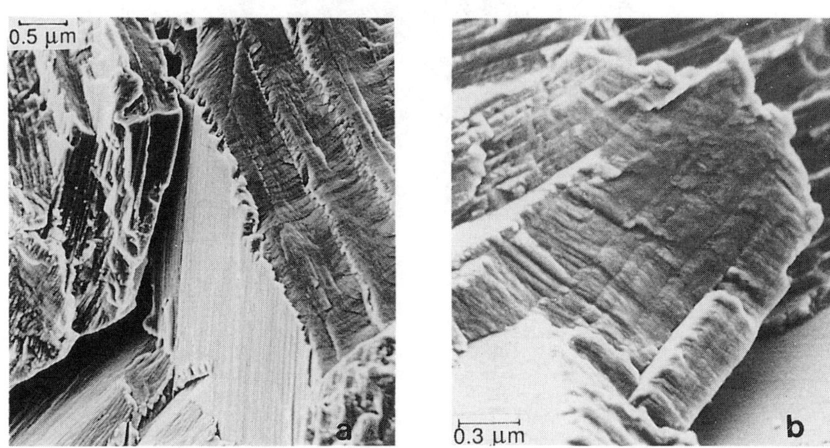

Fig. 4. Details of the Fig. 3 fracture surfaces at high magnifications: (a) an area in Fig. 3a and (b) an area of Fig. 3b, showing deformation-induced layered structures and both inter-layer and trans-layer fractures

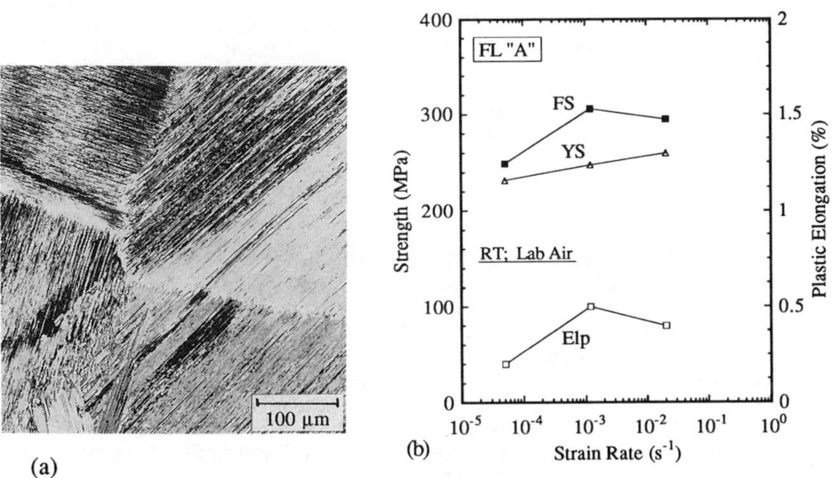

Fig. 5. (a) A fully-lamellar microstructure taken from the longitudinal metallographic section near the fracture surface of a FL "A" tensile specimen fractured at Elp=0.3% and (b) RT tensile properties as a function of strain rate for FL "A" tensile specimens.

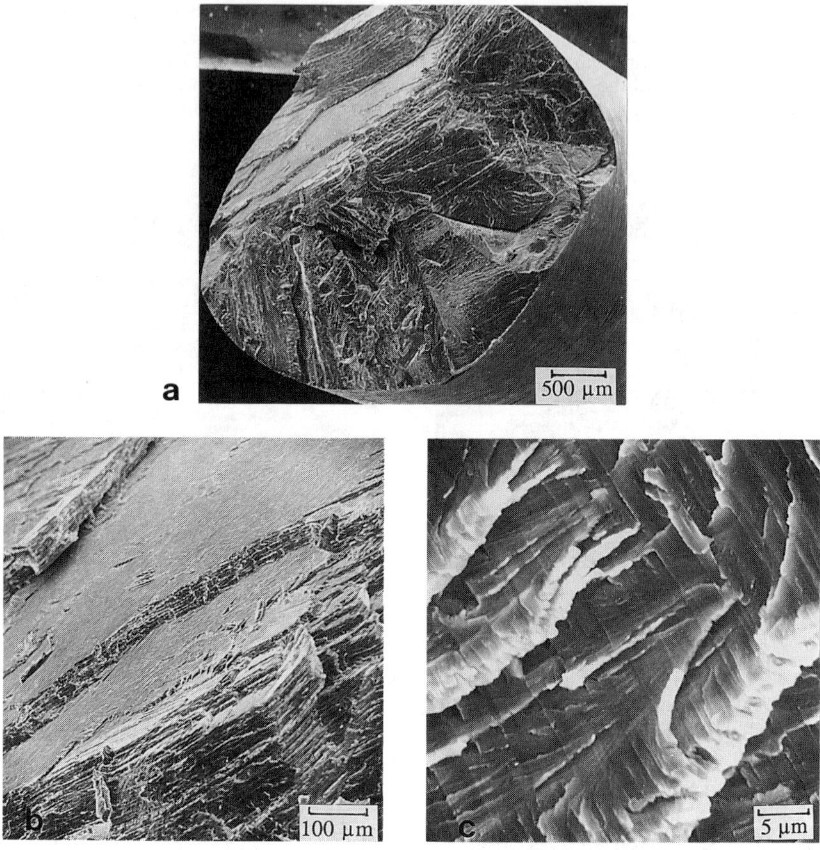

Fig. 6. SEM fractographs of a FL "A" tensile specimen tested at $1.4 \times 10^{-3} s^{-1}$ and fractured at Elp=0.36%, showing: (a) the overall surface at a low magnification, (b) both interlamellar and translamellar fractures, and (c) details of an edge-on translamellar fracture exhibiting interlamellar slip or slip/twin bands.

The fracture surfaces of the FL A tensile specimens exhibit interlamellar fracture, translamellar cleavage-like fracture and mixed-mode fracture, as shown for the case of the 0.36% plastic elongation at a strain rate of $1.4 \times 10^{-3} s^{-1}$ in Fig. 6a. Interlamellar fracture surfaces usually show little evidence of deformation (Fig. 6b), while translamellar fracture surfaces, especially when formed by the crack whose edge is perpendicular to the lamellar edges, usually reveal the interlamellar slip/twin activity (Fig. 6c). This indicates that in polycrystalline materials, conditions are such that more than one slip system may usually be activated to be involved in the fracture process. The fracture surfaces at all other strain rates showed similar fracture features, which is also consistent with other recent observations [12].

Another important observation made in this study is that remarkable similarities exist in the deformation and fracture features occurring in the duplex (Fig. 4) and the fully-lamellar (Fig. 6) microstructures. This is not too surprising if one considers that in the early stage of plastic deformation single-phase gamma grains in the duplex structures develop twins and localized slip bands (see Fig. 3c), which create internally layered structures having a similar morphology to the lamellar structures.

SUMMARY

The effects of strain rate ($5x10^{-5}$ - $3x10^{-1}s^{-1}$) on the RT tensile properties of a gamma alloy were investigated in laboratory air for three selected microstructures. The strain-rate dependence of the tensile ductility was observed, with the maximum elongation for each microstructure occurring at strain rate of $2x10^{-2}s^{-1}$. However, the ductility varies with specific microstructure and specimen surface condition. Electropolished duplex specimens showed the most pronounced strain rate dependence of the ductility, while the fully-lamellar specimens revealed the least. No apparent differences in deformation and fracture behavior were observed between different strain rates for a given microstructure. The strain-rate sensitivity of flow stress(m) was very low, ranging from 0.01 to 0.03. One may interpret the decreases in ductility with decreasing strain rate, observed for the duplex microstructures, as evidence for some extrinsic effect of the test environment, though more complete studies are required.

ACKNOWLEDGMENTS

The authors appreciate useful discussions and comments from Dr. M.G. Mendiratta and the assistance of S.E. Boone, J.E. Henry and J. Williams for metallography and fractography. One of the authors (Y-W. Kim) acknowledges the support from the U.S. Air Force Wright Laboratory, Materials Directorate under contract No. F33615-91-C-5663.

REFERENCES

1. Y-W. Kim and D.M. Dimiduk, JOM, 43 (August, 1991), pp. 40-47.
2. Y-W. Kim, Acta Metall., 40(6) (1992), pp. 1121-1134.
3. U.S. Patent, filed (1992), Y-W. Kim and D.M. Dimiduk.
4. Y-W. research in progress (1992).
5. D.S. Shih and S.C. Huang, in "Microstructure/Property Relationships in Titanium Aluminides and Alloys"' ed., Y-W. Kim and R.R. Boyer, A TMS Publication, TMS, Warrendale, 1990, pp.135-148.
6. W.E. Dowling, et al., in Ref.5, pp. 123-133.
7. K.S. Chan and Y-W. Kim, Metall. Trans. A, 23A(June, 1992), pp. 1663-1677.
8. C.T. Liu and Y-W. Kim, Scripta Metall., 27 (1992), pp. 599-603.
9. T. Takasugi and S. Hanamura, J. Mater. Res., 7(10) (1992), pp. 2739-2746.
10. M. Oh, H. Inui, M. Misaki and M. Yamaguchi, Acta Metall., Submitted (1992).
11. Mechanical Behavior of Materials, T.H. Courtney, McGraw-Hill, N.Y., 1990.
12. M.G. Mendiratta, Y-W. Kim and D.M. Dimiduk, This Proceedings.

THE EFFECT OF CRYSTALLOGRAPHIC ORIENTATION ON THE MECHANICAL PROPERTIES OF A SINGLE CRYSTAL NiAl+Fe Alloy

D.F. Lahrman, R.D. Field, and R. Darolia
Engineering Materials Technology Laboratories, GE Aircraft Engines,
1 Neumann Way, Cincinnati, OH 45215.

ABSTRACT

In this study, the room temperature tensile properties of a single crystal NiAl alloy were investigated as a function of orientation. Fifteen crystallographic orientations were tested, including the <001>, <110> and <111>. The tensile properties measured include yield strength, plastic strain to failure, and ultimate tensile strength. Room temperature ductility as high as 1.4% was measured as close as 10° from the <001> orientation.

INTRODUCTION

The development of advanced gas turbines depends upon achieving higher operational temperatures and thrust to weight ratios. In order to meet these performance requirements, intermetallic compounds such NiAl are being evaluated because of their lower densities compared to nickel-based superalloys. The predicted payoff for an intermetallic alloy such as NiAl with a density of 5.8 g/cc compared to 8.3 g/cc for state of the art superalloys, is a rotor weight savings of up to 50%.

Stoichiometric NiAl has been extensively evaluated by several investigators to determine its operative slip systems and deformation properties [1-5], and to identify the reason for its lack of low temperature ductility. These investigators have measured the mechanical properties of polycrystalline and single crystal NiAl as a function of temperature and stoichiometry in tension and compression [6-9]. Other investigators [10-15] have alloyed NiAl to improve such properties as low temperature ductility and elevated temperature rupture strength. Darolia et al [13-15] have successfully increased the ductility of <110> oriented NiAl single crystals from 1% to ~6% by microalloying with Mo, Ga or Fe, although these additions have not increased the ductility in the <001> orientation. The mechanisms for these effects are currently being investigated [16]. Cr has been added to NiAl with the hope of lowering the anti-phase boundary energy and thus providing additional slip systems [17,18,19]. None of these alloying additions have been shown to increase the number of slip systems.

If single crystal NiAl alloys are to be used as structural materials such as turbine airfoils, the mechanical properties as a function of crystallographic orientation need to be characterized. In this study, the room temperature tensile properties of a NiAl+Fe alloy were measured for several orientations to determine strength, ductility and the applicability of Schmid's Law. The alloy selected for this study was Ni-49.75Al-0.25Fe (atomic percent), which has ~6% plastic strain to failure in the <110> orientation in a room temperature tensile test.

EXPERIMENTAL PROCEDURE

Single crystal slabs, 25mm x 32mm x 100mm, of Ni-49.75Al-0.25Fe were grown in argon by a

Bridgman method. The slabs were homogenized in a flowing argon atmosphere at 1316°C for 50 hours. They were then oriented using the back reflection Laue technique, and two specimens were EDM wire cut from the slabs for each of the desired crystallographic orientations. Button head type tensile specimens with a gage diameter of 2.54mm and gage length of 19.05mm were produced using low stress grinding techniques. The tensile specimens were electropolished in a solution of 10% perchloric acid and 90% methanol at –30°C to remove the residual grinding strains on the surface of the gage section. Tensile tests were conducted at room temperature at a strain rate of $8.3x10^{-5}$/s. Both specimens for each orientation were tested and the average for these tests is presented in the figures and table.

The orientations were selected on the basis of two parameters: 1) Schmid factor for the active slip systems and 2) distance from the <001>. In Figure 1, iso-Schmid lines are drawn within a standard stereographic triangle. For NiAl, there are two dominant slip systems, <100>/{110} and <100>/{100}, expected to operate in specimens oriented near the <111> and the <110>, respectively. For this study, the critical resolved shear stress (CRSS) values for the {100} and {110} slip planes were assumed to be equal. Iso-Schmid factor lines for both slip systems were plotted in the stereographic triangle and a line was drawn where the values for both systems are the same. Thus, this line represents the boundary between regions in which the {110} and {100} slip planes are expected to dominate at room temperature. In addition to providing mechanical property data (YS & %εp) as a function of orientation, this study was expected to provide information on applicability of Schmid's Law in NiAl based alloys. The iso-Schmid factor lines in the stereographic triangle in Figure 1 were used for selection of the 15 orientations. These orientations are identified in Figure 1 by the italic numbers.

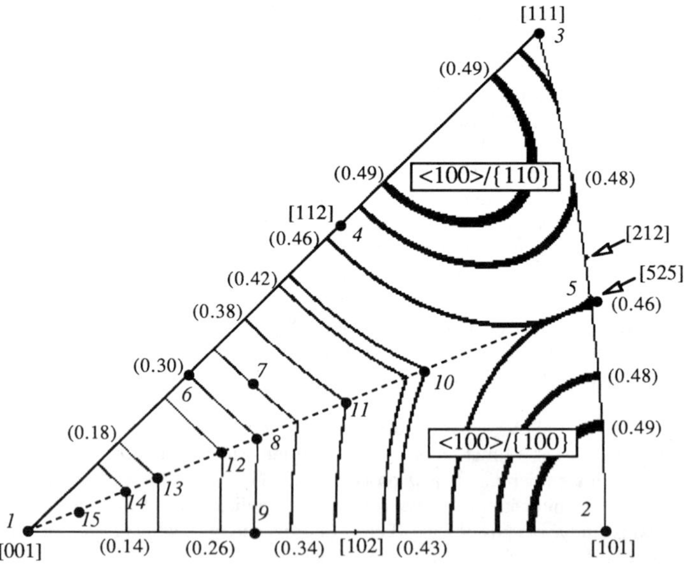

Figure 1 Stereographic triangle showing the iso-Schmid factor plot and the orientations tested. The CRSS for both slip systems were assumed to be equal. The slip system for each region within the stereographic triangle is shown in the enclosed box. The numbers in parentheses are the Schmid Factors values and the italic numbers are the specific orientations tested in this study.

RESULTS

The yield strength and tensile ductility results are shown in the stereographic triangle of Figure 2. As can be seen, the highest ductility was measured from the specimens with the soft <110> and <111> orientations, where the plastic elongations are 4.9% and 4.5%, respectively. Examination of the data shows that ductility is achieved over most of the triangle, except for those orientations within 9° of the hard <001> orientation, where the ductility falls to 1.4%. This is encouraging, because it was anticipated that ductility would not be obtained this close to the <001> orientation. Within the first 20° of the <001> orientation, the yield stress decreases quite rapidly, with a more gradual decrease at larger distances from the <001>.

Using the Schmid factors calculated for each of the slip systems and the yield stresses measured for each orientation as shown in Table I, the CRSS was determined for each orientation. The CRSS values were calculated for the slip plane with the higher Schmid value; ie {110} for the upper area of the triangle and {100} for the lower area. On the line separating these areas, the Schmid factors for both planes are equal. The CRSS values are also shown in the bottom of each box in Figure 2. The CRSS values can be grouped into three regions. The first is near the [111] orientation and contains orientations 3 and 4, the second is near the [101] orientation which contains orientations 2 and 9. The third region contains the remaining orientations most of which are near the [001]. In the first two regions, the CRSS values are fairly consistent: 95.0 MPa and 83.5 MPa for the {110} and {100} planes, respectively. In the third region, there is considerable scatter in the the CRSS values. This may be caused by slight deviations of the actual orientations of the tensile specimens relative to the nominal orientations. As seen in Figure 1, the Schmid Factor changes rapidly near the [001], so that the calculated CRSS value is extremely sensitive to small changes in orientation.

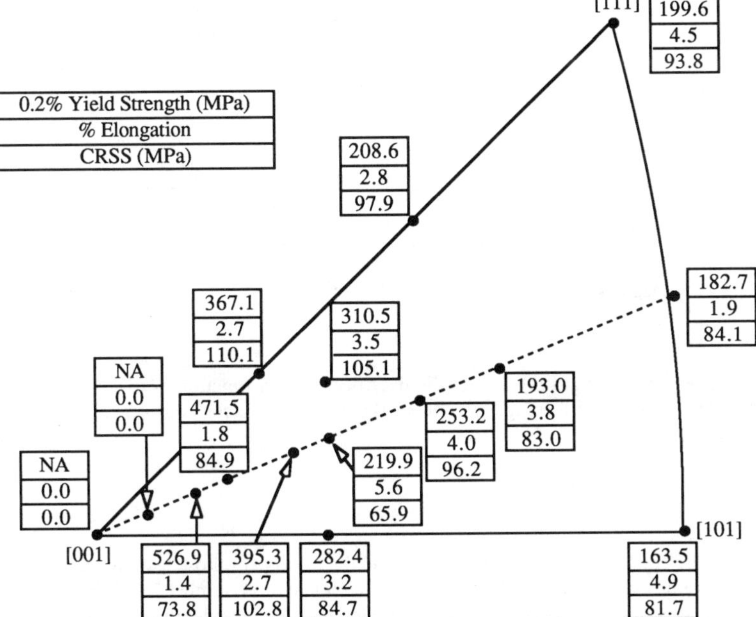

Figure 2 The yield strength, ductility and resolved shear stress plotted for each orientation. The value in each box is the average of two tests.

The data from Figure 2 are presented in another format in Figure 3 to show the tensile properties as a function of angular distance from the <001> orientation. Since the <001> oriented tensile specimens failed prior to yielding, the yield stress shown in Figure 3a is the compressive yield strength (1428.6 MPa) of the alloy. As can be seen, the yield strength drops from 1206.6 to 517.1 MPa within the first 10° of the <001> orientation. Between 10° and 35° from the <001>, the yield stress drops from 517.1 to 206.9 MPa. Beyond 35°, yield strength decreases less rapidly, consistent with the gradual increase of the Schmid factor, varying from 206.9 to 137.9 MPa. The two orientations tested near the <001> orientation show that there is limited tensile ductility in this area, as shown in Figure 3b. The specimens tested at 5° from the <001> (orientation # 15 in Table I) did not show any tensile ductility while the specimens tested at 10° from the <001> (orientation # 14 in Table I) had an average ductility of 1.4%. Beyond 10°, the tensile ductility increases very quickly with increasing distance from the <001> orientation. For the two plots along the <001>-<110> and <001>-<111> lines, the tensile ductility continuously increases, reaching a maximum for the <110> and <111> orientations. The tensile ductility of the specimens tested along the <001>-<255> boundary line between the two competing slip systems first increases with increasing distance from the <001> orientation, but at about 30° the tensile ductility begins to decrease to slightly less that 2.0% for the <255> oriented specimens.

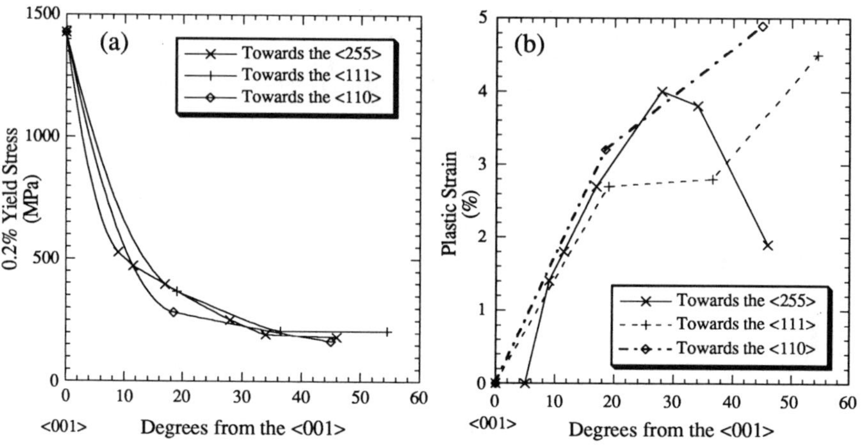

Figure 3 The yield strength (a) and tensile ductility (b) as a function of distance (in degrees) from the <001>. The yield strength for the <001> orientation is the compressive yield strength of the alloy.

DISCUSSION

Two tests were conducted for each orientation and the average of those tests are shown in Table I. The tensile properties for the <110> orientation (orientation number 2 in Table I) produced similar results to an earlier study [13] indicating that the room temperature tensile properties of the Ni-49.75Al-0.25Fe alloy are reproducible. The average 0.2% yield strength for the <110> specimens tested in this study is within 13 MPa of the results obtained in the earlier study (163.5

MPa for this study vs 176.2 MPa for the original study). The scatter observed in the yield strength and ductility is similar to that obtained for stoichiometric NiAl [20,21]. The yield strengths measured for each orientation are within a 15 MPa of each other except for orientations 6 an 14 where there is a 21 MPa difference.

The average CRSS for the points along the <001>-<111> line (3, 4 and 6), where the {110} slip plane is operative, is 100.6 MPa. For the points along the <001>-<110> line (2 and 9), where the {100} slip plane is operative, the average CRSS is 83.2 MPa. As for orientation number 7 between the <001>-<111> and <001>-<255> lines, the resolved shear stress (RSS) is 105.5 MPa for the {110} plane and 93.1 MPa for the {100} plane. Although the exact values are higher, the ratio of these RSS's is similar to the ratio of the CRSS's for the {110}/{100} slip planes as measured in the <111> and <110> orientations. Thus, it is likely that both planes are equally active for this orientation. There is considerable scatter in the CRSS values for the orientations along the <001>-<255> line. This may be due to errors in the orientations of the specimens. Minor deviations of the specimen orientation from the nominal orientation can significantly change the Schmid factor, particularly in the regions nearer the <001>. The actual orientation of each specimen needs to be determined to verify that the Schmid factor value used is correct. This work is in progress and, once completed, will allow a more reliable assessment of the applicability of Schmid's Law for the system.

The CRSS values near the [111] and [101] orientations are fairly consistent. The CRSS for the {100} slip plane is lower than that for the {110} slip plane, which is consistent with results obtained for stoichiometric NiAl [5, 22] in compression. The CRSS ratio, {100}/{110}, found in the previous study was 0.88, similar to the value of 0.83 obtained in this study, indicating that the alloying does not affect the ratio.

Table I Room Temperature Tensile Properties, Schmid Factor and Resolved Shear Stress for Each Orientation.

Orientation Number	0.2%YS MPa*	Plastic Strain %*	Schmid Factor	CRSS MPa**	Operative Slip Plane
1	N/A	0.0	0.00	0.0	
2	163.5	4.9	0.50	81.7	{100}
3	199.6	4.6	0.47	93.8	{110}
4	208.3	2.8	0.47	97.9	{110}
5	182.7	1.9	0.46	84.1	{110}/{100}
6	367.1	3.9	0.30	110.1	{110}
7	310.4	3.6	0.34	105.5	{110}/{100}
8	219.8	5.6	0.30	65.9	{110}/{100}
9	282.4	3.2	0.30	84.7	{100}
10	193.0	3.8	0.43	83.0	{110}/{100}
11	253.2	4.0	0.38	96.2	{110}/{100}
12	395.3	2.7	0.26	102.8	{110}/{100}
13	471.4	1.8	0.18	84.9	{110}/{100}
14	527.0	1.4	0.14	73.8	{110}/{100}
15	N/A	0.0	0.00	0.0	{110}/{100}

* average value of two test per orientation

** based upon the average yield strength values

N/A = not available, stress strain curve was linear to failure

CONCLUSIONS

For the Ni-49.75Al-0.25Fe alloy, relatively high room temperature tensile ductility was obtained for most of the orientations tested. The only region where tensile ductility was not obtained was for the point 5° from the <001> orientation. Tensile ductility as high as 1.4% was measured as close as 10° from the <001> orientation. Also, the ratio of the CRSS ({100}/{110}) appears to be independent of testing mode (tension vs compression) and alloying effects.

ACKNOWLEDGEMENTS

The authors would like to acknowledge Air Force Wright Laboratory and Naval Air Warfare Center Aircraft Division Trenton for funding this investigation through contract F33615-90-C-2006. We are also thankful to Jim Nickely, Jim Jackman, Gary McCabe, Mike Agnello and Bill Davis for their assistance in preparing the specimens and to Joe Wilson (Thread Rite Tool) for machining the specimens and Mike Huber (Mar-Test Inc.) for testing.

REFERENCES

1. A. Ball, and R.E. Smallman, *Acta Met.* **14**, 1349 (1966).
2. A. Ball, and R.E. Smallman, *Acta Met.* **14**, 1517 (1966).
3. H.L. Fraser, R.E. Smallman, and M.H. Loretto, *Phil. Mag*. 28, 651 (1973).
4. H.L. Fraser, R.E. Smallman, and M.H. Loretto, *Phil. Mag*. 28, 667 (1973).
5. R.D. Field, D.F. Lahrman and R. Darolia, *Mat. Res. Symp. Proc.* **213**, 255 (1991).
6. R.T. Pascoe and C.W.A. Newey, *Metal Science Journal* **2**, 138 (1968).
7. R.T. Pascoe and C.W.A. Newey, *Phys. Stat. Sol.* 29. 357 (1968).
8. R.J. Wasilewski, S.R. Butler, and J.E. Hanlon, *Trans. Met. Soc.AIME* **239**, 1357 (1967).
9. A.G. Rozner and R.J. Wasilewski, *Journal of the Institute of Metals* **94**, 169 (1966).
10. E.M. Grala, Mechanical Properties of Intermetallic Compounds, edited by J.H Westbrook, p.358, (1960).
11. G. Sauthoff, *Z. Metallkde*. **80**, 337 (1989).
12. I. Jung, M. Rudy and G. Sauthoff, *Mat. Res. Symp. Proc.* **81**, 263, (1987).
13. R. Darolia, D.F. Lahrman and R.D. Field, *Scripta Met.* **26**, 1007 (1992).
14. R. Darolia, *Journal of Metals,* **43**, 44 (1991).
15. R. Darolia, D.F. Lahrman, R.D. Field, J.R. Dobbs, K.M. Chang, E.H. Goldman and D.G. Konitzer, Ordered Intermetallics: Physical Properties and Mechanical Behavior, edited by C.T. Liu, R.W. Cahn and G. Sauthoff, NATO ASI Series, p. 669 (1992).
16. R. D. Field, D.F. Lahrman and R. Darolia, *Mat. Res. Symp. Proc.* Symposium L (1992).
17. R.D. Field, D.F. Lahrman and R. Darolia, *Acta Met. Mater.* **39**, 2969 (1991).
18. J.D. Cotton, The Influence of Cr on the Structure and Mechanical Properties of B2 Nickel Aluminide Alloys, PhD Thesis, University of Florida (1991).
19. D.B Miracle. S Russel, and CC Law *Mat. Res. Symp. Proc.* **133**, 225 (1991).
20. D.F. Lahrman, R. D. Field and R. Darolia, *Mat. Res. Symp. Proc.* **213**, 603, (1991).
21. D.F. Lahrman, R. D. Field and R. Darolia, submitted to *Scripta Met.* (1993).
22. R. Darolia, R.D. Field, D.F. Lahrman and A.J. Freeman, Final Contract Report, F49620-88-C-0052.

MICROSTRUCTURE AND MECHANICAL PROPERTIES OF CAST, HOMOGENIZED AND AGED NiAl SINGLE CRYSTAL CONTAINING Hf

I. E. LOCCI[*], R. DICKERSON[*], R. R. BOWMAN[*], J. D. WHITTENBERGER[*], M. V. NATHAL[*] AND R. DAROLIA[**]
[*]NASA-Lewis Research Center, Cleveland, OH 44135
[**]GE Aircraft Engines, Cincinnati, OH 45215.

ABSTRACT

Small additions of Hf to [001] oriented single crystal NiAl are shown to be effective in improving high temperature creep strength. The presence of Hf-rich second phases and/or solid solution strengthening are responsible for the improved behavior observed at high temperatures. In the as-cast condition, large Hf-rich interdendritic regions were found. Homogenization heat-treatments (1590 K for 50 hours) substantially reduced this interdendritic segregation. TEM observations of the as-homogenized microstructure revealed fine G-phase ($Ni_{16}Hf_6Si_7$) precipitates, with plate or cuboidal shapes. Varying the cooling rates after homogenization resulted in the refinement or complete suppression of the G-phase. Aging the homogenized material at 1300 K resulted in the formation of Heusler precipitates (β'-Ni_2AlHf), preferentially nucleating at the G-phase sites. These Heusler precipitates were more stable at this or higher temperatures and coarsened at the expense of the less stable G-phase. Post-test analyses of compression tested specimens, conducted at 1200, 1300 and 1400 K, revealed extensive changes in the distribution and size of the second phases. Deformation at 1300 K appears to occur by two distinct mechanisms: at high strain rates the stress exponent is $\cong$ 4 while at slower rates ($< 10^{-6}s^{-1}$) a much higher exponent ($\sim$ 12) was found. Testing at 1300 K of specimens over-aged at 1400 K reduced the creep resistance of the alloy which suggests a contribution by precipitation strengthening to the overall strength of the alloy.

INTRODUCTION

Binary NiAl alloys, regarded as potential candidates for high temperature applications[1-2], require improvement in room temperature ductility or toughness as well as high temperature creep resistance to compete with today's superalloys. Ternary additions, especially from group IVA and VA elements have been shown to provide significant strengthening at high temperatures.[1-6] Vedula et al.[7] suggested that Hf additions to polycrystalline NiAl would be effective; however subsequent testing[8] of a polycrystalline small grain sized Ni-47Al-0.8Hf (at.%) alloy failed to confirm their results. Because grain boundary sliding could contribute to the creep rate, it is possible that the observed weakness in the latter study could be circumvented through the use of single crystals.[1-3] Hence a study of the elevated temperature deformation properties of Hf modified NiAl single crystals was initiated to characterize the phases, their stability and their possible influence on the high temperature strength of NiAl alloys.

EXPERIMENTAL

Single crystal [001] oriented NiAl alloys containing 0.3 or 1.0 at.% Hf substituted for Al were prepared at General Electric Aircraft Engines (GEAE) by directional solidification using alumina-silicate shell molds. Samples parallel to the [001] direction were wire EDM (Electric Discharge Machined) from the ingot and ground into compression and tensile creep specimens. Cylindrical compression specimens were 3 or 5 mm in diameter and 6 or 10 mm in length, with the compression axis parallel to the [001] direction. The specimens were given a standard homogenization heat-treatment of 1590 K for 50 h followed by furnace cooling (10 K/min in Ar). In some cases, the specimens were air cooled or water quenched. Constant velocity compression and constant load compression tests were conducted in air. Constant velocity compression tests were conducted between 1200 and 1400 K at strain rates ranging from 3.3 x 10^{-7} to 1.3 x 10^{-4} s^{-1}. Constant load compression tests were performed to obtain information at lower strain rates.

RESULTS AND DISCUSSION

Microstructural Characterization of the NiAl+1 at.% Hf Single Crystal:

Homogenized Condition

Large interdendritic Hf-rich segregated regions were observed in the as-cast samples (Fig.1a). The homogenization heat-treatment reduced, but did not eliminate segregation (Fig. 1b), as a finer and more uniform distribution of precipitates was produced. Contrary to what was expected based on the phase diagram, the precipitates were not the Heusler (β'-Ni_2AlHf) phase, but instead were found to be G-phase ($Ni_{16}Hf_6Si_7$)[3]. As reported earlier[3], substantial Si from the shell mold dissolved in the molten metal alloy during the directional solidification process resulting in the formation of fine silicide precipitates (G-phase) during crystal growth similar to the one shown in Figure 2.

Effect Of Cooling Rates

Figure 2 shows transmission electron microscope (TEM) microstructures of homogenized samples cooled at different rates. Furnace cooling produced large G-phase plates (Fig. 2a) associated with denuded regions and 10 to 40 nm cuboidal G-phase precipitates away from the plates. Air cooling produced a fine and uniform distribution of cuboidal G-phase precipitates (10 nm) throughout the entire sample (Fig.2b). The large G-phase cuboidal precipitates and plates (Fig. 2a) could be redissolved by a rapid homogenization heat-treatment (20 min. at 1590 K) and reprecipitated as fine precipitates (like in Fig. 2b) if air cooled. A water quench from the homogenization temperature completely suppressed precipitation. These observations indicate that the G-phase precipitates dissolved at 1590 K and reprecipitated during cooling.

Aging Studies

The alloy containing 1 at.% Hf was aged at 1300 K for 1 to 500 h after homogenization and air cooling. Figure 3 shows the presence of large second phase particles that nucleated and coarsened as a function of time. Denuded regions are also associated with these particles. Microdiffraction, as well as regular selective aperture diffraction (SAD), indicated that these are β' precipitates. G-phase precipitates were still present at this aging temperature but their density decreased with increasing time. At intermediate times (Fig.3b) the β' precipitates coalesced into "worm-like" morphologies, while after 500 h a more equiaxed shape was achieved (Fig. 3c).

Creep Behavior of the NiAl+1 at.% Hf Single Crystal:

Constant velocity compression tests indicates that initial work hardening occurs until an equilibrium stress level is reached after approximately 1% strain. In the case of constant load compression and tension, steady state creep was achieved at less than 1 % strain. Figure 4 is a plot of the temperature dependent flow stress-strain rate data for the homogenized and furnace cooled NiAl+1.0 at.% Hf alloy tested in compression compared to binary NiAl and the single crystal superalloy NASAIR 100.[9] The true stress (σ) and true strain rates ($\dot{\varepsilon}$) are average values from nominally constant flow regimes. The creep strength of the 1 at.% Hf modified single crystal is vastly superior to binary NiAl and is approaching the behavior of the single crystal superalloy. Examination of the 1300 K data (Fig. 4) indicates that there is little difference in the flow stress-strain rate data for NiAl+1.0 at.% Hf under either constant velocity or constant load conditions. Additionally, no difference in the creep response at 1300 K was observed for samples that were homogenized and furnace or air cooled, even though their starting microstructures were different (Fig. 2). A temperature compensated power law relationship can describe the 1200-1400 K data for strain rates $\gtrsim 10^{-6}$ s^{-1}, where

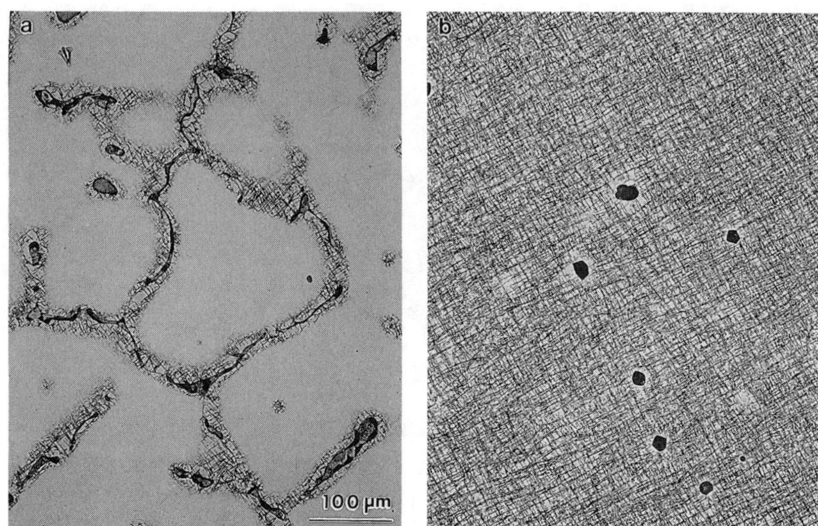

Figure 1. Optical microstructure of NiAl+1 at.% Hf single crystal in the (a) as-cast and (b) homogenized at 1590 K for 50 h followed by furnace cooling.

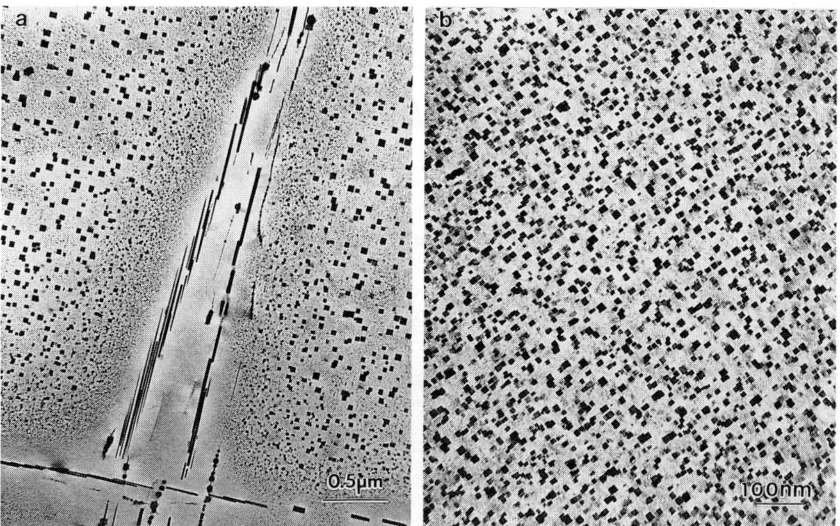

Figure 2. TEM bright field images of a NiAl+1 at.% Hf single crystal homogenized at 1590 K for 50 h followed by (a) furnace cooling and (b) air cooling.

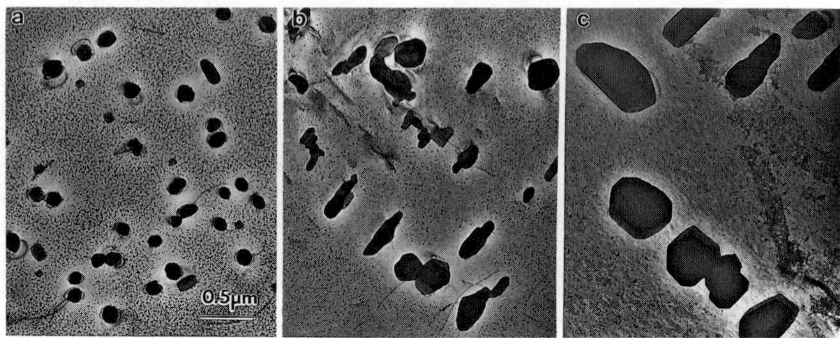

Figure 3. TEM bright field images of a NiAl+1 at.% Hf single crystal aged at 1300 K for (a) 1h (b) 100 h and (c) 500 h.

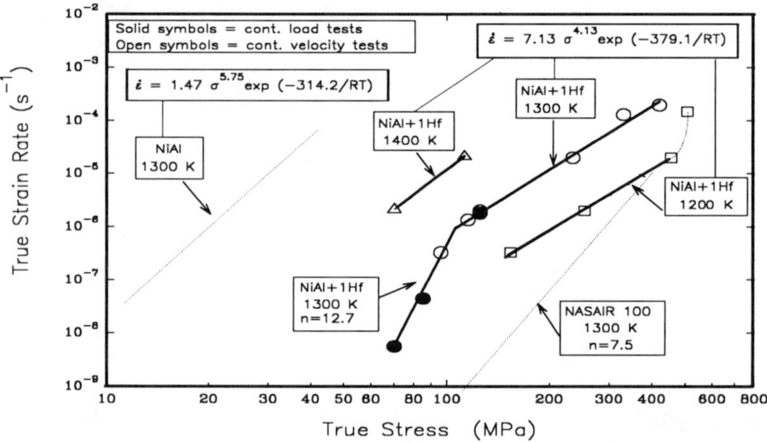

Figure 4. True compressive flow stress-strain rate behavior for [001] oriented NiAl+1 at.% Hf single crystal compared to binary NiAl and NASAIR 100.

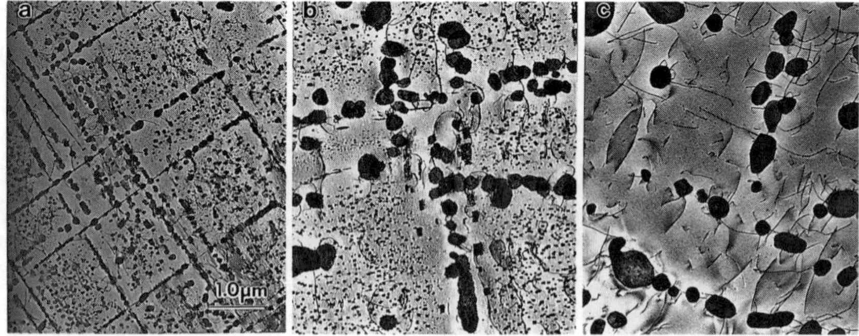

Figure 5. TEM bright field images of a NiAl + 1at.% Hf single crystal tested in compression at $\dot{\varepsilon} \cong 2x10^{-6}\ s^{-1}$ *(a)* σ*= 253.6 MPa, T= 1200 K (b)* σ*= 125 MPa, T=1300 K (c)* σ*= 70 MPa , T=1400 K.*

$$\dot{\varepsilon} = 7.13\ \sigma(MPa)^{4.13} exp(-379.1(KJ/mol)/RT)$$

with R being the universal gas constant and T being the absolute temperature and the coefficient of determination r^2 = 0.98. At strain rates exceeding $2x10^{-5}$ s^{-1} at 1200 K, a power law description is no longer adequate. At strain rates less than 10^{-6} at 1300 K a different deformation regime exists as indicated by a stress exponent of ~ 12. This higher stress exponent compared to that of NiAl (n ≅ 6)[10] may indicate that precipitation hardening is an effective strengthening mechanism at lower strain rates. Testing currently in progress has indicated that the behavior in tension agrees well with that observed for compression.

Microstructure of Creep Tested Specimens

Figure 5 shows TEM bright field images of specimens compression tested at a strain rate of ~ 2 $x10^{-6}$ s^{-1} at 1200, 1300 and 1400 K. All three samples experienced a total testing time of approximately 16 h and an accumulated strain of 5 %. β' precipitates were observed to have nucleated preferentially at the original G-phase plate sites during 1200 and 1300 K testing. Also, some β' nucleation had occurred on the finer cuboidal G-phase precipitates at 1200 K (Fig. 5a). The microstructure for the specimen tested at 1300 K showed large "worm-like" β' particles. Regions with higher concentrations of β' were devoid of G-phase. The 1400 K compression tested microstructure shows mostly coarser equiaxed β' particles and essentially no G-phase. The very few G-phase particles which were observed after 1400 K testing are believed to have formed during cooling. Evidence of both G-phase and β' precipitates interacting with dislocations also was observed. Higher dislocation densities were found in the specimens tested at faster strain rates or higher stresses. In summary, β' precipitates formed in specimens tested between 1200 and 1400 K. For slower strain rates and/or higher temperatures, coarsening of the β' and G-phase precipitates, as well as a decrease in the density of the G-phase, occurred.

Over-Aging Studies

Exposure to 1400 K, as illustrated in Fig 5c, dissolved most of the G-phase precipitates and produced blocky and worm-like β' precipitates. This indicates that the G-phase dissolved at 1400 K and was unable to reprecipitate during cooling since the Hf originally contained in the G-phase is probably tied-up in the coarser β' particles. Compression testing at 1300 K of specimens that were over-aged at 1400 K for 20 hours, revealed lower strengths than the homogenized 1.0 at.% Hf containing alloy (Fig.6). Post-test observations of the over-aged samples indicated that the microstructure consisted mostly of blocky β' precipitates and essentially no G-phase particles. Several possibilities exist for the decrease in creep resistance. The first option is that the G-phase precipitates contribute to the high temperature strengthening behavior of this alloy and their absence eliminates that contribution. The second possibility is that the β' phase has coarsened too much to be an effective precipitate strengthener. Thirdly, it is possible that during aging at 1400 K, some of the Hf that was in solid solution was used to form the coarser β'-particles and therefore, is not available for solid solution strengthening. Preliminary data from GEAE has indicated that alloys containing only β' precipitates and no G-phase have creep strengths higher than those depicted in figure 6. Creep data for the NiAl+0.3 at.% Hf [001] single crystal indicated that this alloy is weaker by two orders of magnitude compared to NiAl+1.0 at.% Hf; however heat-treatment of the 0.3 at.% Hf modified alloy at 1400 K did not change the creep response. This behavior of the 0.3 Hf alloy indicates that it is primarily strengthened by Hf in solid solution. To fully ascertain the factors influencing the strengthening of the NiAl+1.0 at. % Hf single crystals, further heat-treatment studies and microstructural characterization are planned.

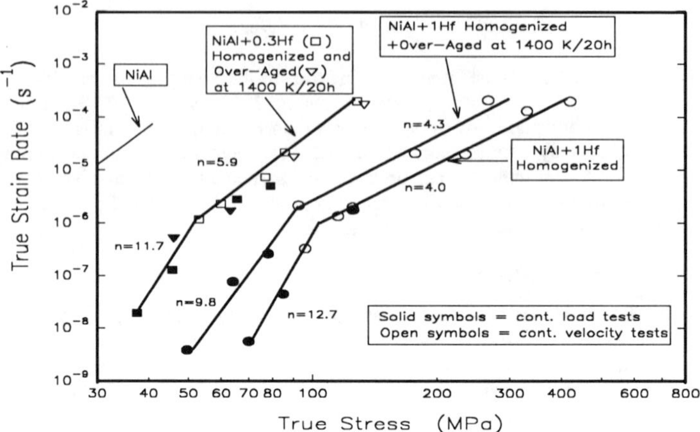

Figure 6. Compression creep strength of Hf modified NiAl single crystals tested at 1300 K.

Summary

Addition of 1 at.% Hf to single crystal NiAl has been shown to improve the high temperature creep resistance in compression. However, a change in deformation mechanisms at about $10^{-6}s^{-1}$ from a low stress exponent (~4) to a higher stress exponent process (~12) was observed. The Si introduced during processing reacted with Ni and Hf to form G-phase precipitates, whose size and distribution can be modified by cooling rates. Aging and creep testing at 1300 K revealed a gradual replacement of the G-phase by the β' precipitates. Over-aging at 1400 K completely dissolved the G-phase precipitates, coarsened the β' particles, and resulted in lower creep resistance. At this stage, it is difficult to fully assess the individual contributions of the solid solution and precipitation strengthening of this alloy.

References

1. R. Darolia, JOM, 43, № 3, 44 (1991).
2. R. Darolia et al., in *Ordered Intermetallics - Physical Metallurgy and Mechanical Behavior,* (C. T. Liu, R. W. Cahn, G. Sauthoff, eds.), NATO ASI Series, 213, Irsee, Germany, pp. 679-698 (1992).
3. I. E. Locci et al., in *High Temperature Ordered Intermetallics Alloys IV,* (L. A. Johnson, D. P. Pope, and J. O. Stiegler, eds.) Mat. Res. Soc. Symp. Proc., Vol. 213, Boston, Mass., pp. 1013-1018 (1991).
4. M. Takeyama, C.T. Liu , J. Mat. Res., 5, 1189 (1990)
5. P.R.Strutt and B. H. Kear, in *High Temperature ordered Intermetallic Alloys,* Mat. Res. Soc. Symp. Proc., Vol. 39, pp. 279-292 (1985).
6. R.S. Polvani, Wen-Shian Tzeng, and P.R.Strutt, Metall. Trans. A,7A, 33 (1976).
7. K. Vedula et al., in *High Temperature Ordered Intermetallic Alloys,* (C. C. Kock, C. T. Liu, and N. S. Stoloff, eds.) Mat. Res. Soc. Symp. Proc., Vol. 39, pp. 411-421 (1985).
8. J. D. Whittenberger et al., Mater. Lett., 11, 267 (1991).
9. M. V. Nathal, as in reference 2, pp. 541-563.
10. J. D. Whittenberger, J. Mater. Sci., 23, 235 (1988).

MICROSTRUCTURE AND ROOM-TEMPERATURE DUCTILITY OF UNIDIRECTIONALLY GROWN Ni_3Al

TOSHIYUKI HIRANO AND TOSHIO MAWARI
National Research Institute for Metals, 1-2-1, Sengen, Tsukuba, Ibaraki 305, Japan

ABSTRACT

The formation mechanism of the columnar-grained stoichiometric Ni_3Al by FZ-UDS (unidirectional solidification using a floating zone method) was investigated. The quenched solidification interface showed that Ni_3Al and β-NiAl formed simultaneously in the form of lamellar structure at the interface. The β-NiAl dissolved into the Ni_3Al matrix during cooling, resulting in the columnar-grained structure of single phase Ni_3Al at room temperature. The columnar-grained Ni_3Al had a large number of low angle boundaries and exhibited a large tensile ductility. The characteristics of the deformation behavior were also investigated. Slip lines easily transferred across grain boundaries without crack initiation. Deformed bands with many stacking faults were produced by deformation.

INTRODUCTION

Polycrystalline Ni_3Al is brittle at room temperature owing to intrinsic grain-boundary brittleness, and unless doped with a small amount of boron [1], it easily suffers from intergranular fracture, showing much limited ductility.

We recently found that unidirectional solidification (UDS) using a floating zone method (FZ) overcomes this problem without boron doping. We call this method FZ-UDS [2,3]. The stoichiometric Ni_3Al grown by FZ-UDS has a columnar-grained structure and shows about a 60% larger tensile elongation at room temperature. The fracture mode is a transgranular type. Of particular interest is that even Al-rich Ni_3Al can be ductilized [4], which is quite different from the boron-doping method [5]. These results indicate the FZ-UDS is a promising way to improve the ductility of intermetallic compounds. It is important to examine the formation mechanism of the columnar-grained structure since the ductilization by FZ-UDS has a close relation with this structure.

In this study we present the formation mechanism of the columnar-grained Ni_3Al in the FZ-UDS process and some features of the deformation behavior at room temperature.

EXPERIMENTAL

The feed rod preparation and the FZ-UDS procedure were the same as previously described [2]. Polycrystalline Ni_3Al rods (Ni-25.2 at%Al) were grown at a growth rate of 25 mm/h by FZ-UDS in a flowing argon atmosphere. When a steady-rate was attained the process was rapidly stopped by deenergizing the halogen lamps which were used as a heating source in the FZ furnace. The molten zone, the grown alloy, and the solidification interface were quenched. The grown rods were cut along the growth direction and the longitudinal sections were subjected to metallographic observations. The crystal structure was examined by X-ray diffraction (XRD) and transmission electron diffraction (TED). The grain orientation and grain-boundary character were examined by electron channeling pattern using scanning electron microscope.

Sheet tensile specimens with a gauge section of 0.7x4.8x10

mm were cut from the grown rods with the longest dimension parallel to the growth direction. Tensile test was carried out at room temperature with an initial strain rate of $1.7x10^{-3}$ s^{-1}. The tensile test was interrupted five times after yielding and the surface slip trace and the substructure were observed with optical microscope. Transmission electron microscopic (TEM) observation was carried out after 49.7% elongation. The specimens were thinned mechanically and finally electropolished using 20% perchloric acid-ethanol solution at 253K. A JEM 2010 microscope was used for the observation, which was operated at 200 kV.

RESULTS AND DISCUSSION

Formation mechanism of the columnar-grained Ni_3Al

Figure 1 shows the optical micrograph of longitudinal section quenched from the steady-state FZ-UDS condition. Although no forced cooling was utilized, the molten zone and the grown alloy were frozen by this procedure. The solidification interface is clearly visible. Figure 2 shows the magnified micrographs of the four indicated regions in Fig.1. The quenched molten zone

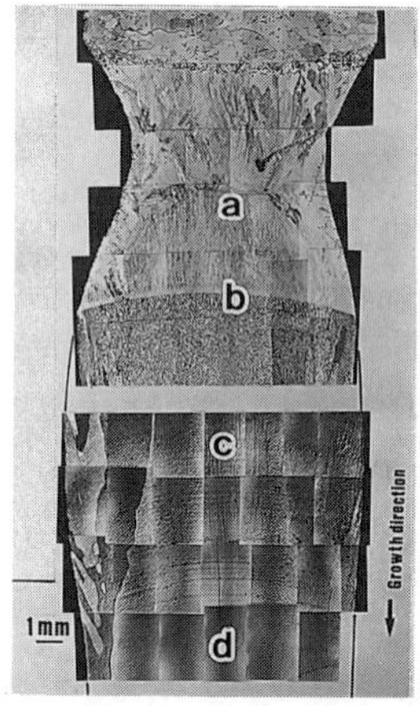

Fig.1 Optical micrograph of a longitudinal section of Ni_3Al quenched from the steady-state FZ-UDS.

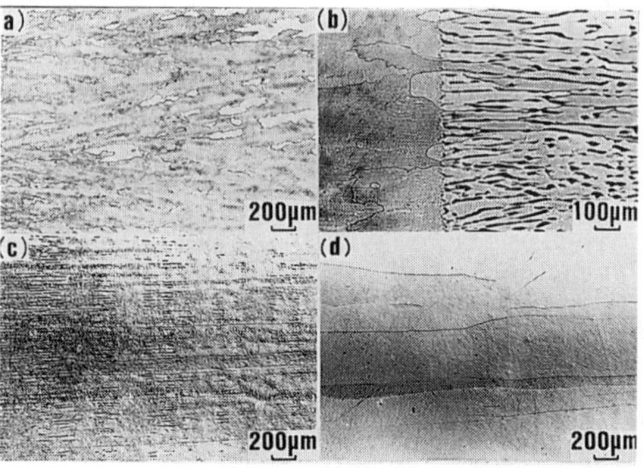

Fig.2 Magnified micrographs of the four indicated regions in Fig.1. (a) Molten zone, (b) solidification interface, and (c) and (d) grown alloy.

(Fig.2(a)) has an almost homogeneous structure which was identified as $L1_2$-type Ni_3Al from the XRD pattern. On the other hand, the grown alloy shows a lamellar structure at the solidification interface, as shown in Fig.2(b). The structure's matrix (bright phase in Fig.2(b)) was determined to be $L1_2$-type Ni_3Al from the TED pattern. The precipitates (dark phase in Fig.2(b)) contain numerous planar faults, as shown in Fig.3(a), which are similar to that observed in bulk quenched Ni-36.8 at%Al-1 at%Co [6] and in melt quenched Ni-33.8 at%Al using the hammer and anvil technique [7]. Figure 3(b) shows that the precipitates are $L1_0$-type β'-NiAl. The chemical composition ranges from 25.2 to 39.2 at%Al which corresponds to the composition of β-NiAl [8]. From the Ni-Al phase diagram [8], it is considered that β-NiAl phase formed at the interface during FZ-UDS and underwent martensitic transformation into the β'-NiAl

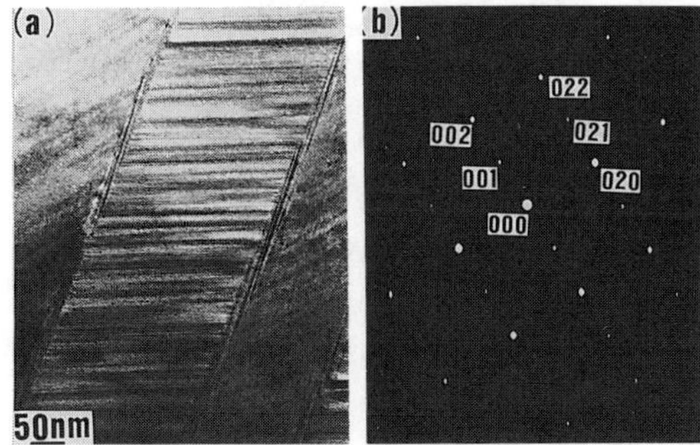

Fig.3 (a) Transmission electron micrograph and (b) selected-area TED pattern of the dark phase of the lamellar structure in Fig.2(b), showing $L1_0$ structure.

phase by the quenching. These results indicates that the quenching rate in this study was fast enough to preserve the solidification interface during FZ-UDS. Therefore, it is possible to infer the formation mechanism of the columnar-grained structure from the quenched interface.

Note the morphology of the solidification interface. The lamellar structure of Ni_3Al and β-NiAl is formed at the planar solidification interface. It indicates that both the phases grew simultaneously from the liquid similar to eutectic alloys, although peritectic reaction, liquid+ β-NiAl→Ni_3Al, occurs for stoichiometric Ni_3Al under equilibrium condition [8]. The precipitation of β-NiAl plays an important role in the formation process of the columnar-grained Ni_3Al. Note that the precipitated β'-NiAl occupies all the Ni_3Al grain boundaries (Fig.2(b)). It suggests that the precipitated β-NiAl suppresses the grain growth of Ni_3Al perpendicular to the growth direction, resulting in the columnar-grained structure of Ni_3Al along the growth direction. It is known that large single crystal of stoichiometric Ni_3Al is extremely difficult to grow from liquid. Probably the reason is because of the simultaneous growth of β-NiAl with Ni_3Al from liquid. Figures 2(c) and (d) show that the precipitated β-NiAl is not stable below the solidification temperature. It dissolved into the Ni_3Al matrix when cooling started after solidification (Fig.2(c)), and completely disappeared at room temperature (Fig.2(d)). The grain shape of the matrix did not change in this process and consequently the columnar-grained structure of single-phase Ni_3Al was produced at room temperature. Thus, the formation mechanism of the columnar-grained structure of stoichiometric Ni_3Al by FZ-UDS is concluded to be as follows: liquid→ Ni_3Al + β-NiAl (during solidification)→ Ni_3Al (during cooling).

The columnar-grained Ni_3Al grown by FZ-UDS contains a large amount of low angle boundaries and low sigma coincidence boundaries [9]. Figure 4, which is the microstructure of the tensile specimen used, shows the columnar grains and their misorientation. The misoriention ranges from 1.1 to 13.8 degree and the growth orientation is close to [15 9 1].

Deformation behavior

Some room-temperature deformation features of the columnar-grained Ni_3Al are described here. The tensile axis was parallel to the columnar grains, as shown in Fig.4. As previously

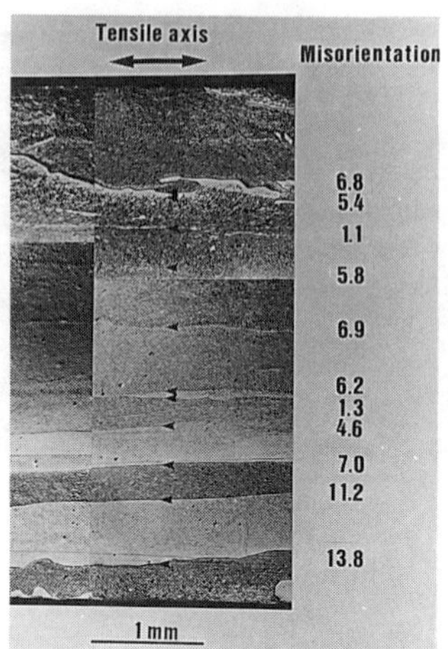

Fig.4 Optical micrograph of the tensile specimen used and the misorientation between the columnar grains (degree).

reported, the specimen showed a large elongation (larger than 60% elongation) after yielding [2-4]. Figure 5 shows the surface slip trace after 3.1 and 49.7% elongation. The primary slip is observed after yielding and the secondary slip after about 3% elongation. A significant feature is that the slip lines easily transferred across grain boundary. The slip lines bent near grain boundary after about 10% elongation. Nevertheless no cracks were initiated at grain boundary even after 49.7% elongation. The specimen surface was corrugated after about 3% elongation, as shown by the arrows in Fig.5(a). The surface corrugation developed with increasing elongation. It looks like martensite plate, as shown in Fig.5(b). Another feature is the substructure of the tensile-tested specimen. Figure 6(a) shows the optical micrograph of the polished and etched surface after 49.7% elongation. Heavily etched bands where a large amount of strain

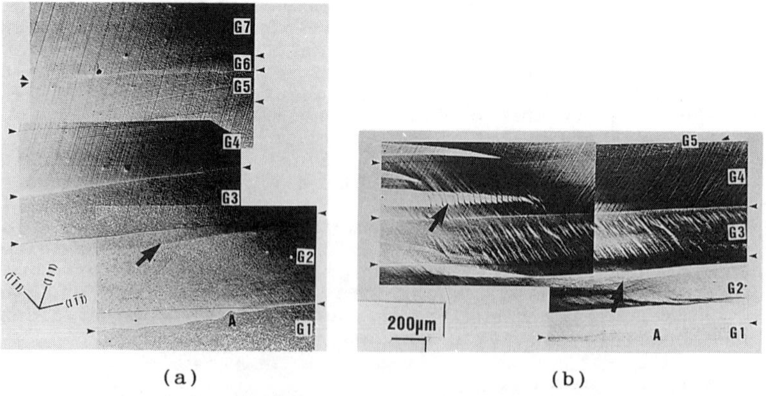

(a) (b)

Fig.5 Surface slip trace after (a) 3.1% elongation and (b) 49.7% elongation. As-grown grain boundary is shown by ► and ◄.

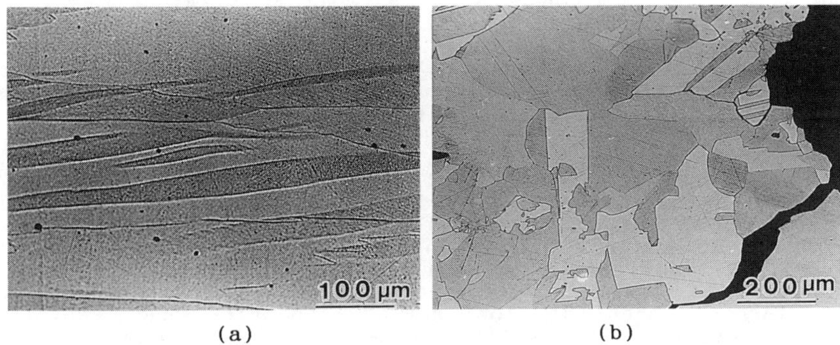

Fig.6 Optical micrographs of the substructure of the tensile-tested (a) Ni_3Al grown by FZ-UDS(49.7% elongation), comparing with (b) that of conventionally cast Ni_3Al.

may be accumulated are observed. As shown in Fig.6(b), such substructure is not observed for the conventionally cast Ni_3Al which exhibits no ductility at room temperature. TEM observation revealed that many stacking faults are formed in these bands, as shown in Fig.7. The nature of the stacking faults is superlattice intrinsic type (SISF) which is similar to the previous reports [10-12]. Until now, however, there have been no previous reports about such band-like substructure with many SISFs in Ni_3Al. This substructure is a significant feature of the Ni_3Al grown by FZ-UDS as well as its large room-temperature ductility. It suggests a close relation between the formation of many SISFs and the large tensile ductility.

From these results the large room-temperature ductility of Ni_3Al grown by FZ-UDS can be qualitatively explained as follows. Since the Ni_3Al grown by FZ-UDS mainly consists of low angle boundary and low sigma coincidence boundaries, dislocation transfer across grain boundaries is essentially easy. Therefore, crack hardly initiates even after some deformation proceeds and

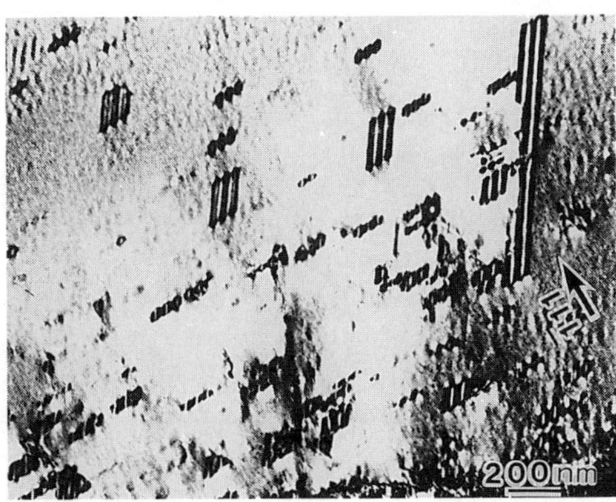

Fig.7 Transmission electron micrograph of the deformed specimen, showing formation of many SISFs.

the tensile ductility is increased. Nevertheless, as deformation proceeds considerable strain is accumulated at grain boundary because of the misorientation between the grains. Crack may initiate at grain boundary and intergranular fracture may occur unless the accumulated strain is released. It is considered that the accumulated strain is released through the formation of SISFs, resulting in the large tensile ductility. The detailed mechanism is now under investigation.

SUMMARY

The formation mechanism of the columnar-grained Ni_3Al by FZ-UDS was investigated. Ni_3Al and β-NiAl formed simultaneously in the form of lamellar structure at the solidification interface. Subsequently, the β-NiAl dissolved into the Ni_3Al matrix during cooling, resulting in the columnar-grained structure of single phase Ni_3Al at room temperature. The columnar-grained Ni_3Al had a large number of low angle boundaries and exhibited characteristic deformation behavior. Slip lines easily transferred across grain boundary without crack initiation. The band-like substructure with many SISFs was produced by deformation.

ACKNOWLEDGEMENT

We would like to thank to Dr.Tadao Watanabe, Tohoku University, for measuring the electron channeling pattern and helpful discussions.

REFERENCES

1. K.Aoki and O.Izumi, J. Jpn. Inst. Metals **43**,1190(1979).
2. T.Hirano, Acta metall. mater. **38**,2667(1990).
3. T.Hirano and T.Kainuma, ISIJ International **31**,1134(1991).
4. T.Hirano, S.S.Chung, Y.Mishima, and T.Suzuki, Mat. Res. Soc. Symp. Proc. **213**,635(1991).
5. C.T.Liu, C.L.White, and J.A.Horton, Acta metall. **33**,213(1985).
6. K.Enami, S.Nenno, and K.Shimizu, Trans.JIM **14**,161(1973).
7. S.Nourbakhshi and P.Chen, Acta metall. **37**,1573(1989).
8. T.B.Massalski, ed. Binary Alloy Phase Diagrams, (American Soc. for Metals, Metals Park, OH, 1986) pp.140.
9. T.Watanabe,in 6th Int. Conf. on Intergranular and Interphase Boundaries in Materials, edited by E.K.Polychroniadis, June 22-26, 1992, Thessaloniki, Greece, (Materials Science Forum) in press.
10. K.Enami and S.Nenno, J. Phys. Soc. Japan, **25**,1517(1968).
11. P.Veyssiere, J.Douin, and P.Beauchamp, Phil. Mag. A, **51**,469(1985).
12. B.J.Pestman and J.Th.M.de Hosson, Acta metall. mater. **40**,2511(1992).

MICROSTRUCTURE AND MECHANICAL PROPERTIES OF B2 (Co,Ni)Al BASED ALLOYS

YOSHISATO KIMURA, TOMOO SUZUKI* and YOSHINAO MISHIMA**
Graduate Student, Department of Materials Science and Engineering, Tokyo Institute of Technology, Nagatsuta, Midori-ku, Yokohama 227, Japan
*Department of Metallurgical Engineering, Tokyo Institute of Technology, O-okayama Tokyo 152, Japan
**Precision and Intelligence Laboratory, Tokyo Institute of Technology, Nagatsuta, Midori-ku, Yokohama 227, Japan

ABSTRACT

Two-phase alloys based on B2 type intermetallic compound (Co,Ni)Al with either Co primary solid solution or $L1_2$ type Ni_3Al, γ', phase are investigated for their variations in microstructure and mechanical properties. The aim of the present work is to improve ductility and to enhance fracture resistance in extremely brittle B2 CoAl or NiAl by introducing second phase. To start with, microstructure-mechanical property relationship is investigated in two-phase alloys consisting of B2 CoAl and Co primary solid solution in the Co-Al binary system. Subsequently, the similar strategy is extended into the Co-Ni-Al ternary system. Mechanical properties are measured by compression test conducted over the wide temperature range from 77 K to 1273 K at strain rates ranging from $1.4x10^{-3}$ to $1.4x10^{-4}s^{-1}$. Microstructure observations are carried out using both optical and transmission electron microscope.

INTRODUCTION

B2 type intermetallic compounds such as CoAl and NiAl have been given attention as candidates for high temperature structural materials because of their superior properties such as high melting point over 1900 K, excellent high temperature oxidation resistance, and relatively low densities. However, there remains two major problems which must be overcome for the practical application. One is lack of ductility at room temperature and the other is insufficiency in strength at high temperature over 1273 K. Although CoAl has higher strength than NiAl at elevated temperature, it is known to be extremely brittle, which is one of the reasons for the fact that there have been little work on the compound[1]. In the case of NiAl, there have been some attempts on improving ductility by introducing $L1_2$ Ni_3Al as a second phase[2,3] and ternary addition of Co[4]. The objective of the present work is to seek for the method to provide some ductility to intrinsically brittle CoAl by introducing and controlling the morphology of a second phase. From the Co-Al binary phase diagram[5], it can easily be seen that the second phases which could be introduced into CoAl are the intermetallic compound, δ , and the Co primary solid solution. It is then obvious that Co primary solid solution should be introduced as a second phase because of the higher symmetry of the crystal structure than of δ to ensure ability for plastic deformation. In the case of the ternary Co-Ni-Al system, the possible second phases are ternary Co primary solid solution at Co rich side and $L1_2$ intermetallic compound based on Ni_3Al, γ', at Ni rich side both with high crystallographic symmetry. In both binary and ternary alloys, the microstructure-mechanical property relationship is investigated for series of alloys. Then the methods of microstructural control to impart ductility are suggested. Note that in the following text we denote Co primary solid solution in the Co-Al binary system as (Co) and that in the Co-Ni-Al system as (Co,Ni) unless otherwise specified.

EXPERIMENTAL PROCEDURE

For the Co-Al binary system, alloys over a wide composition range from Co-20 at%Al to Co-60 at%Al were prepared by the arc-melting under an argon atmosphere. All the ingots were homogenized at 1273 K for 7 days and/or at 1423 K for 48 h. Phase boundaries for δ/B2 and B2/(Co) two-phase and B2 single phase regions were determined by optical microscopy and X-ray diffractometry. Ternary Co-Ni-Al alloys were also prepared by the arc-melting for compositions shown in Fig. 1[6], in which Al concentration is fixed at 30 at%. All the ternary alloys were solution treated in the B2 single phase field at 1573 K for an hour, air cooled, and subsequently aged for the precipitation of second phase at 1373 K for an hour.

Compression tests were carried out using specimens with a dimension of 3 x 3 x 6 mm^3 at nominal strain rates of 1.4×10^{-3} and/or $1.4 \times 10^{-4} s^{-1}$ over a temperature range from 77 K to 1273 K under an argon atmosphere.

Microstructures of the alloys were observed using both optical and transmission electron microscope(TEM). Discs for TEM observation were prepared by electropolishing in a 10% perchloric/acetic acid solution at room temperature.

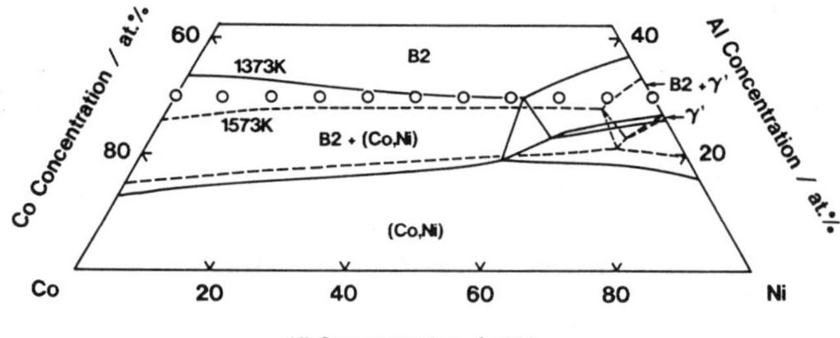

Fig. 1 A portion of isothermal section of the Co-Ni-Al ternary system at 1373 and 1573 K[6]. Open circles represent alloy compositions investigated in the present work.

RESULTS AND DISCUSSIONS

Characterization of Microstructures

(1) Co-Al Binary System

Microstructure observation has been conducted for the B2/(Co) two-phase alloys with 68, 72, 76 and 80 at%Co. Typical microstructures observed by optical microscopy are shown in Fig. 2. The (Co) precipitates predominantly at grain boundary of B2 phase in alloys with 68 and 72 at%Co. It can be clearly seen that (Co) precipitates also inside of the B2 grain as the Co content increases from 72 to 80 at%Co. Note that the microstructure of an 80 at%Co alloy differs from others showing a typical eutectic microstructure, which is understood from the Co-Al binary phase diagram[5].

For a 76 at%Co alloy, TEM observation was conducted and typical microstructure is shown in Fig. 3. It is clearly shown that blocky (Co) exists in B2 matrix, in which stacking faults are frequently observed indicating close phase stability between fcc and hcp crystal structure. To be noted is that very fine precipitation of the (Co) has also been confirmed in B2 matrix, for which the origin can be regarded to be the precipitation during cooling due to the change in solubility of (Co) into B2 phase as temperature decreases. The existance of precipitate free zones of the fine particles surrounding blocky (Co) would support this speculation.

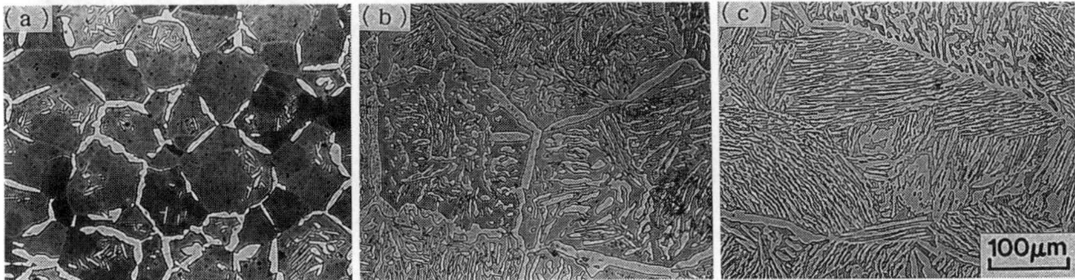

Fig. 2 Optical micrographs of binary Co-Al alloys being B2/(Co) two-phase, (a)72, (b)76 and (c)80 at%Co.

Fig. 3 A bright field micrograph of a binary Co-Al alloy with 76 at%Co.

(2) Co-Ni-Al Ternary System

By the heat treatment described in the experimental procedure, the second phases (Co,Ni) and/or $L1_2$, γ', are introduced as expected providing a homogeneous distribution in the B2 matrix. Varieties of optical microstructures are shown in Figs. 4(a) through (f). In alloys with Ni concentration from 0 to 35 at%, similar microstructures are observed where almost B2 single phase is observed in as-quenched condition and homogeneous precipitation of (Co,Ni) is found in the B2 matrix by the subsequent aging, as shown in (a) and (b) for a 35 at%Ni alloy. With increasing Ni to more than

42 at%Ni, quenching results in martensitic transformation from B2 to $L1_0$, for instance, as shown in (c) for a 42 at%Ni alloy. By the subsequent aging, as shown in (d), the second phase precipitates in a different morphology from what are seen in (b) having certain crystallographic variant related to the martensite phase. With further increase in Ni concentration toward the Ni-Al binary edge, γ' precipitation occurs in place of (Co,Ni) as a second phase in the B2 matrix. Examples of optical micrographs for such two-phase alloys are shown in (e) and (f) for a 63 at%Ni alloy. A homogeneous precipitation of γ' is observed mainly in the grain after the quenching and aging. It has been suggested that $L1_0$ martensite decomposed into γ' and B2 in the Ni-Al binary system[2,3], for which the confirmation has not been made in the present case.

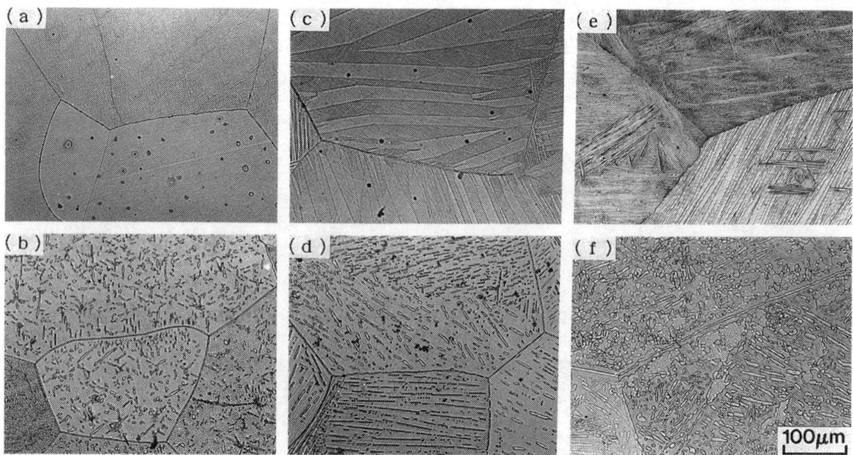

Fig. 4 Optical micrographs of ternary Co-Ni-Al alloys; (a) through (d) being B2/(Co,Ni) two-phase, and (e) and (f) B2/γ' two-phase. (a), (c) and (e) are as-quenched, and (b), (d) and (f) are after subsequent aging. Ni concentration is 35 at% in (a) and (b), 49 at% in (c) and (d), and 56 at% in (e) and (f).

Compression Tests

(1) Co-Al Binary System

Compression test specimens were prepared for the B2/(Co) two-phase alloys with 72 to 80 at%Co. For single phase B2 alloys and for a B2/(Co) two-phase alloys with 68 at%Co, preparation of test pieces by mechanical cutting was unsuccessful because of their extreme brittleness. With increasing of Co content from 72 to 80 at%Co, however, an improved ductility is observed and plastic deformation reaches 3% for an 80 at%Co alloy. The temperature dependence of compressive 0.2% flow stress for 72, 76 and 80 at%Co are shown in Fig. 5 over a temperature range from 77 K to 1273 K. All the alloys exhibit a similar trend in the temperature dependence of strength and a region is found at between about 400 to 800 K where the softening with increasing temperature is suppressed. Since such behavior shown in many B2 type intermetallics including CoAl and NiAl, it must be concluded that the temperature dependence of strength of the two-phase alloys shown in Fig. 5 comes from the nature of the matrix B2 phase. From these results together with the microstructural observation, more homogeneous distribution of the second phase would provide further ductility in the Co-Al binary alloys based on B2 phase.

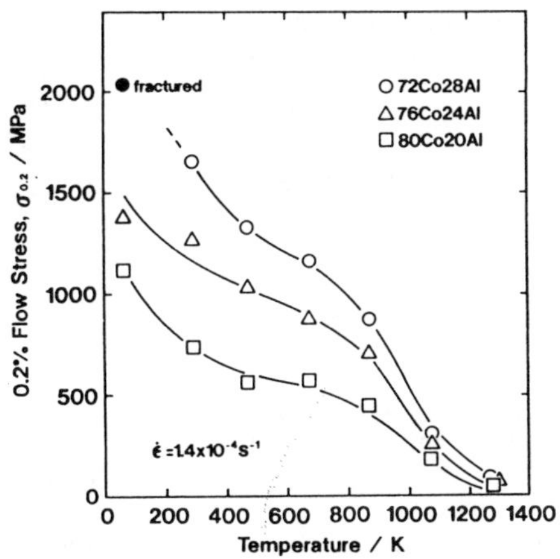

Fig. 5 Temperature dependence of compressive 0.2% flow stress in binary Co-Al alloys being B2/(Co) two-phase.

(2) Co-Ni-Al Ternary System

Fig. 6 shows temperature dependence of compressive 0.2% flow stress for three B2/(Co,Ni) two-phase alloys. Plastic strain of 2 to 3% is achieved at room temperature with high strength over 1200 MPa in all the alloys. The fracture strain increases with increasing temperature except in a temperature range between 873 K and 1073 K, where all the alloys exhibit minimal plastic deformation. At 873 K, the alloy with 21 at%Ni fractured before yielding which is indicated by a solid triangle in the figure. In 21 and 28 at%Ni alloys the similar temperature dependences are observed to those found in the binary two-phase alloys consisting of B2 and (Co) shown in Fig. 5. At below 673 K, 0.2% flow stress of an alloy with 35 at%Ni is found much lower than what is expected by the extraporation from higher temperature region. Judging from the stress-strain curve, this would be attributed to the occurence of stress induced martensite transformation of the B2 matrix to $L1_0$. The stress-strain curve exhibits two step yielding and the values shown by half-filled square in Fig. 6 are for the initial yielding.

For the alloys with γ' as the second phase in the B2 matrix, excellent ductility, over 10%, is achieved at all the test temperatures. Temperature dependence of compressive 0.2% flow stress in 56 and 63 at%Ni alloys are shown in Fig. 7 with the result previously reported by Pank et al[4] for a Co-61 at%Ni-30 at%Al alloy prepared by consolidation of pulverized melt-spun ribbon. The anomalous positive temperature dependence of strength found below 873 K obviously originated from the γ' phase in the alloy. By the present work, a significantly higher strength levels are achieved as compared to the previous work at intermediate temperatures and excellent ductility is obtained at all the test temperatures, however, further work is necessary to improve the high temperature strength in this alloy system.

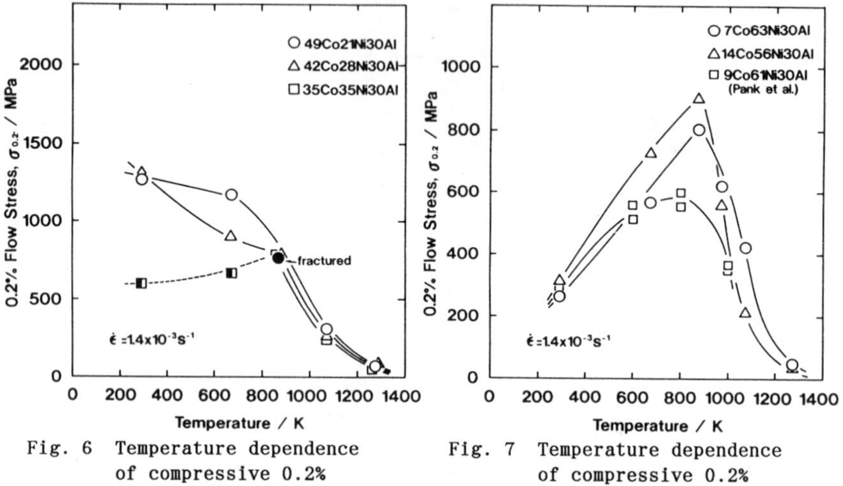

Fig. 6 Temperature dependence of compressive 0.2% flow stress in ternary Co-Ni-Al alloys being B2/(Co,Ni) two-phase.

Fig. 7 Temperature dependence of compressive 0.2% flow stress in ternary Co-Ni-Al alloys being B2/γ' two-phase.

CONCLUSIONS

Methods to improve ductility of extremely brittle B2 phase CoAl and (Co,Ni)Al based alloys are shown by introducing second phase through simple heat treatment of conventionally cast materials. Especially, B2/γ' two-phase alloys in the Co-Ni-Al ternary system exhibit excellent compressive ductility at all the temperatures ranging from room temperature to 1273 K. In B2/(Co) or B2/(Co,Ni) two phase alloys, some indication is obtained for the possibility of further improving ductility by achieving finer and more homogeneous distribution of the second phase.

REFERENCES

1. S.K.Mannan, K.S.Kumar and J.D.Whittenberger, Metal.Trans.A.21A, 2179(1990).
2. S.Ochiai, I.Yamada and Y.Kojima, J.JPN.Inst.Met.54, 301(1990).
3. S.Ochiai, T.Shirokura, Y.Doi and Y.Kojima, ISIJ Intnl.31, 1106(1991).
4. D.R.Pank, M.V.Nathal and D.A.Koss, J.Mater.Res.5, 942(1990).
5. A.J.McAlister, in Binary Alloy Phase Diagrams vol.1, 2nd ed. edited by T.B.Massalski (ASM Intnl.,Materials Park,1990), p.136.
6. M.Hubert-P. and H.Hubert, in Ternary Alloys vol.4, edited by G.Petzow and G.Effenberg (VCH,Weinheim,1991), p.234.

EFFECTS OF MICROSTRUCTURE ON THE ROOM-TEMPERATURE STRENGTH AND DUCTILITY OF THE ALLOY Ti-25Al-10Nb-3V-1Mo

FRANÇOIS-CHARLES DARY AND ANTHONY W. THOMPSON
Carnegie Mellon University, Materials Science and Engineering Dept., Pittsburgh, PA 15213

ABSTRACT

The role of the microstructure on the room-temperature tensile properties of the alloy Ti-25Al-10Nb-3V-1Mo (at.%) was investigated. A wide spectrum of microstructures was obtained by varying the cooling rate after a solutionizing treatment in the β phase field. The strength was found to increase monotonically with increasing cooling rate. It is proposed that strength is controlled by (α_2+β_R) boundary strengthening in β-solutionized microstructures. The ductility was observed to go through a maximum at a cooling rate of 1.5°C/s. Different deformation and failure mechanisms were identified, depending on the cooling rate regime, and correlated to the ductility trends. The β_R nature, α_2 lath size, α_2 lath arrangement and prior-β grain boundaries appeared to be the principal features governing the deformation and failure mechanisms.

INTRODUCTION

For many years, attention has been focused on developing Ti_3Al-based titanium aluminides for aerospace applications. Indeed, the low density of binary Ti_3Al alloys and their ability to retain excellent strength and creep properties at high temperature made them very attractive. However, the necessity to improve the poor ductility and toughness of these alloys called for additions of β-stabilizing elements, such as Nb, V or Mo, which in turn increased significantly the complexity of the microstructure. To date, although mechanical properties have been found to be very sensitive to microstructural changes, only limited research has been aimed at trying to understand the microstructure/mechanical property relationships for given alloy compositions. The disparity of microstructures used in most studies has precluded thorough understanding of the role of microstructure upon mechanical properties.

The focus of this work was to examine the dependence of room-temperature tensile properties upon microstructure in the titanium aluminide alloy, Ti-25Al-10Nb-3V-1Mo (at.%). Microstructural features were varied by means of heat treatments, and their effect on room temperature strength and ductility were investigated.

EXPERIMENTAL PROCEDURE

Details concerning the as-received material, heat treatments and method of tensile specimen preparation have been described in an earlier paper [1]. Specimens were solutionized in air in the β-phase field, at 1150°C, for one hour and cooled at different rates, in order to vary the microstructure. Seven cooling rates were chosen, i.e., 25, 7.5, 4, 1.5, 0.5, 0.1 and 0.01°C/s. Round tensile specimens, with a 2.5 mm gage diameter and 10 mm gage length, and compression specimens, with a 6x6 mm square cross area and 9 mm length, were used. Prior to testing, tensile specimens were electropolished and compression specimens were mechanically polished to 0.5 μm alumina. Room temperature tests were conducted at a cross-head speed of 0.5 mm/min., using a universal machine. Compression tests were interrupted after 3% plastic elongation. After testing, fractured tensile specimens and deformed compression specimens were examined, using scanning electron microscopy, to provide clues for a better understanding of deformation and failure mechanisms. The investigation included observations of fracture surfaces and longitudinal sections of tensile specimens, as well as, specimen surfaces of compression samples.

RESULTS

Microstructures

The seven microstructures are shown in Figures 1a through 1g. α_2 volume fractions were determined by X-ray powder diffraction using an adaptation of the Miller's method by Venkataraman, *et al.* [2]. For all cooling rates up to 7.5°C/s, the α_2 volume fraction was found to remain rather constant, ranging from 68 to 76%. However at 25°C/s, it was estimated to be below 10%. α_2 lath widths were also measured, using the linear intercept method. They are plotted in Figure 2 as a function of cooling rate.

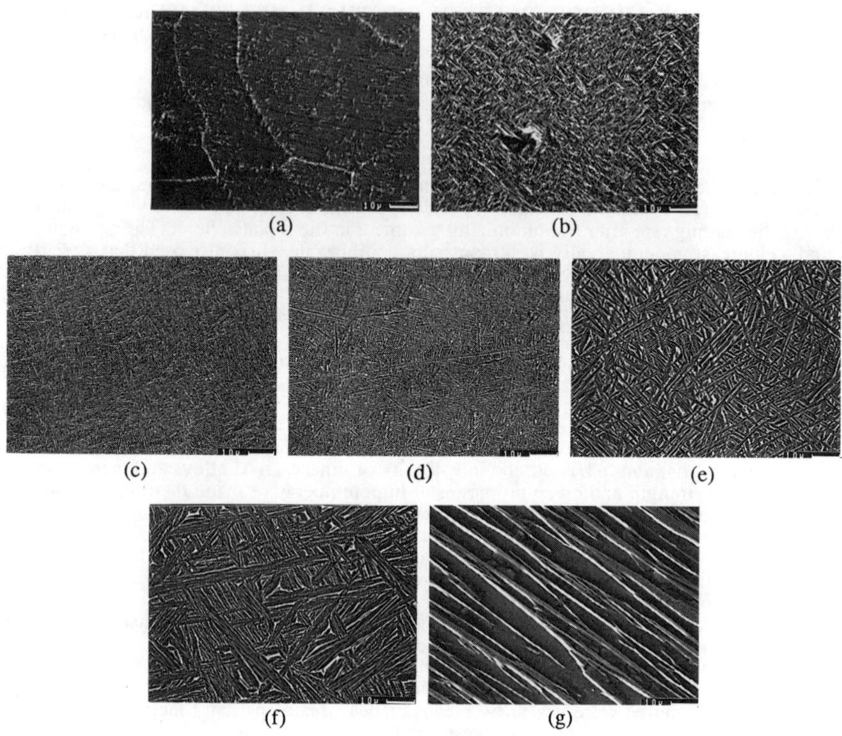

Figure 1 - Microstructures: (a) "1150,25" ; (b) "1150,7.5" ; (c) "1150,4" ; (d) "1150,1.5" ; (e) "1150,0.5" ; (f) "1150,0.1" ; (g) "1150,0.01".

The "1150,25" microstructure exhibited very limited, non-uniform, precipitation of fine α_2 laths within prior-β grains (Figure 1a). The α_2 phase precipitated essentially at prior-β grain boundaries and subgrain boundaries. The matrix was found to be made up of the ordered B2 phase and very fine precipitates. This phase mixture will herein be called β_R. AEM revealed that they were "ω-type" precipitates resulting from the decomposition of the highly unstable B2 phase (Figure 3a).

For all cooling rates below 25°C/s, microstructures exhibited extensive α_2 precipitation throughout prior-β grains. The α_2 lath width increased monotonically with decreasing cooling rates (Figure 2), but again, the α_2 volume fraction remained essentially constant.

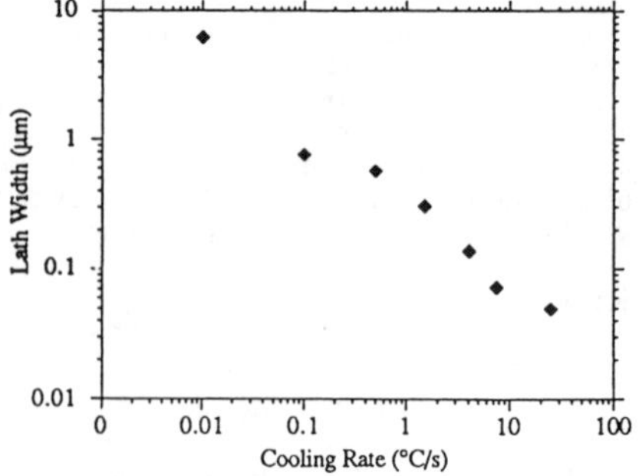

Figure 2 - Width of α_2 laths *vs.* cooling rate.

Between 7.5 and 0.5°C/s, α_2 laths were arranged in a basketweave manner (Figures 1b through 1e). In the "1150,7.5" microstructure, small α_2 deficient regions were observed revealing that the α_2 precipitation was still not uniform at this cooling rate (Figure 1b). Moreover at 7.5°C/s, the matrix was still composed of the same β_R phase mixture observed at 25°C/s. However, below 7.5°C/s,the β matrix was solely made up of the ordered B2 phase. Precipitation of a grain-boundary α_2 film was observed in all microstructures, along with, in many cases, aligned α_2 laths precipitating from prior-β grain boundaries, towards the inside of the grains (Figure 3b). The extent of the alignments and size of α_2 laths within them, increased with decreasing cooling rates. Finally, Figure 3c shows an α_2 lath arrangement, herein called "cross-type", a unique feature frequently observed in basketweave-type microstructures. TEM observations revealed that the interfaces between α_2 laths at the cross intersection were of the α_2/α_2 type, and that the matrix, otherwise continuous throughout the microstructure, was therefore discontinuous in these areas.

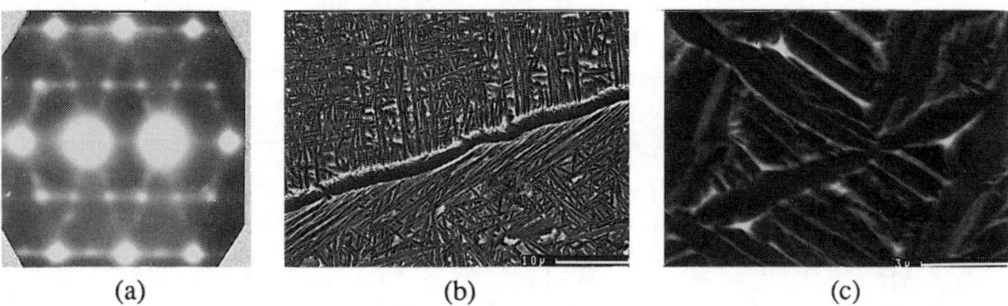

(a) (b) (c)

Figure 3 - Remarkable microstructural features: (a) $[\bar{1}13]_{B2}$ zone axis in "1150,25", note the diffuse intensity maxima at 1/2[301], 1/2[110] positions and streaking in <301>, <110> directions characteristic of "ω-type" precipitation ; (b) Prior-β grain boundary in "1150,1.5", note the continuous α_2 film and aligned α_2 laths ; (c) "Cross-type" arrangement of α_2 laths in "1150,0.5", note the absence of $B2/\beta_R$ at the intersection.

α_2 laths in the "1150,0.1" and "1150,0.01" microstructures were now arranged in packets or colonies (Figures 1f and 1g). α_2 laths were separated by thick regions of ordered B2 phase. In the "1150,0.1" microstructure, small α_2 packets were observed in the middle of prior-β grains and larger α_2 colonies were present at prior-β grain boundaries (Figure 1f). The "1150,0.01" microstructure exhibited very thick α_2 laths arranged in colonies, each prior-β grain containing only a few large colonies (Figure 1g).

Mechanical behavior

Room temperature tensile properties are summarized in Figure 4 as a function of cooling rate. The strength, shown in Figures 4a and 4b, increased monotonically with increasing cooling rates, up to 7.5°C/s. From 7.5 to 25°C/s, the fracture stress dropped dramatically and no yield stress was recorded at 25°C/s. Figures 4c and 4d show ductility, i.e., elongation and reduction of area, *vs*. cooling rate. The ductility exhibited a steep narrow peak at intermediate cooling rates, with a maximum at about 1.5°C/s. The three microstructures making up the peak exhibited the basketweave arrangement, whereas the two colony-type microstructures and the fast cooled "1150,25" and "1150,7.5" microstructures displayed limited plastic elongation.

Observations of Fractured Tensile Specimens and Deformed Compression Specimens

The surface of "1150,25" compression samples exhibited well defined, widely separated, coarse slip bands, intersecting each other (Figure 5a). These slip bands were observed extending throughout prior-β grains. Slip transmission at prior-β grain boundaries occurred only occasionally. A mixed intergranular and transgranular fracture was observed on tensile specimens (Figure 5b). Transgranular regions were made up of cleavage-like facets. The fracture profile was very straight but contained many steps (Figure 5c). Below the fracture surface, the few cracks observed were along prior-β grain boundaries or along B2 slip bands (Figure 5d). By contrast, the surface of "1150,7.5" specimens exhibited mixed features, that is long slip bands linked to the B2 phase and surface offsets linked to the α_2 laths (Figure 5e). The B2 slip band spacing was now substantially smaller. The fracture surface also had a mixed intergranular/transgranular character and the fracture profile appeared virtually identical to that of "1150,25" samples.

In the "1150,4", "1150,1.5" and "1150,0.5" basketweave-type microstructures deformation concentrated at or near prior-β grain boundaries (Figure 6a). Very intense slip bands were observed crossing the aligned α_2 laths present at prior-β grain boundaries (Figure 6b). Extensive α_2/B2 surface offsets, along with occasional α_2 shear instabilities, were observed

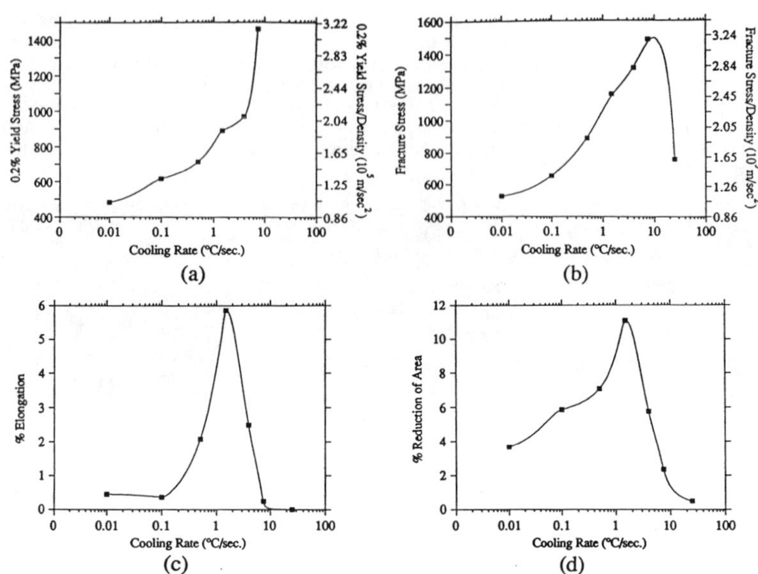

Figure 4 - Room-temperature tensile properties *vs.* cooling rate: (a) Yield stress ; (b) Fracture stress ; (c) % elongation ; (d) % reduction of area. Note the specific strength values.

within prior-β grains. A mixed intergranular and transgranular fracture was observed except for the "1150,0.5" microstructure which failed transgranularly (Figure 6c). The transgranular regions were not faceted, but the α_2 lath basketweave arrangement could easily be observed, indicating that the fracture scale was that of individual α_2 laths, which failed by cleavage-like

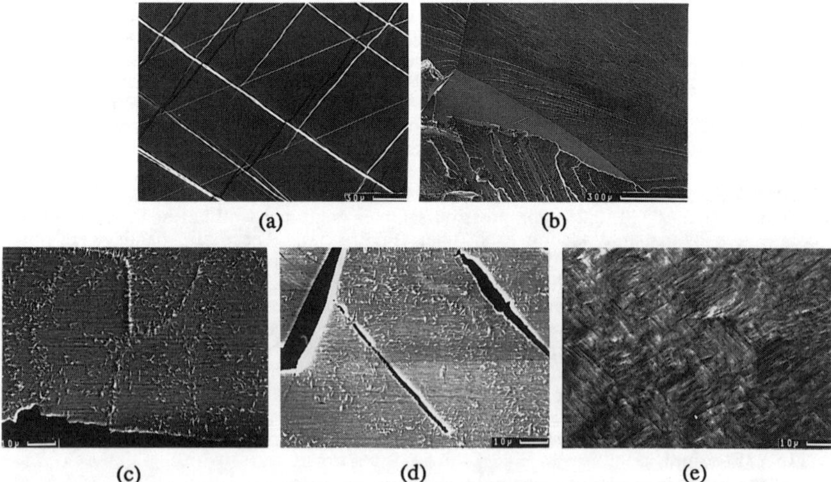

Figure 5 - High cooling rates: (a) Compression specimen surface in "1150,25" ; (b) Fracture surface in "1150,25" ; (c) and (d) Longitudinal section in "1150,25" ; (e) Compression specimen surface in "1150,7.5".

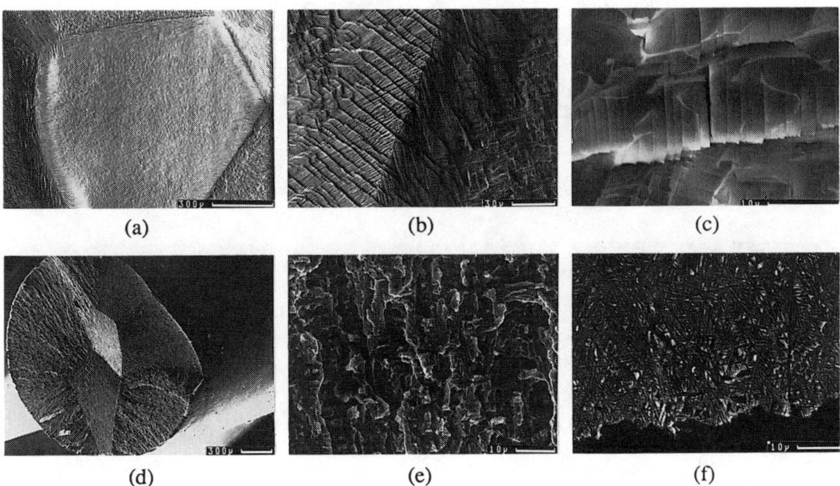

Figure 6 - Intermediate cooling rates: (a) through (c) Compression specimen surface ; (d) and (e) Fracture surface in "1150,1.5" ; (f) Longitudinal section in "1150,1.5".

fracture or by decohesion of the α_2/B2 interface (Figure 6d). Large tear ridges were observed in the B2 phase, bounding the α_2 laths. A relatively high microcrack concentration was observed, within prior-β grains, close to the fracture surface (Figure 6e). Microcracks were observed initiating at the α_2/α_2 interface in α_2 "cross-type" arrangements (Figure 6e) or within α_2 laths, along α_2 slip bands or at α_2/B2 interfaces (Figure 6f). Bridging of the B2 phase and crack-tip blunting by the B2 phase were also observed. The fracture path was very tortuous, following α_2/B2 interfaces or crossing α_2 laths almost perpendicularly to their long axis (Figure 6e).

At low cooling rates, i.e., 0.1 and 0.01°C/s, surface slip was observed concentrated in coarse α_2 slip bands extending throughout entire colonies (Figure 7a). Slip transmission between α_2 laths, through B2, was frequently observed. Only colonies oriented favorably for slip exhibited slip band networks. Slip transmission across colony boundaries was rarely observed. Cracks were observed running along colony boundaries (Figure 7b). Failure occurred by transgranular fracture. The fracture was faceted and contained, for the most part, macrofacets and microfacets, which corresponded to α_2 colonies and cleaved aligned individual α_2 laths, respectively (Figure 7c). The fracture path, secondary cracks and the few microcracks were always similarly oriented with respect to the orientation of the α_2 laths, i.e., almost perpendicular to α_2 long axis (Figure 7d and 7e). The microcrack concentration below the fracture surface was small. Extensive crack-tip blunting by the B2 phase and bridging of the B2 phase were observed (Figure 7d).

DISCUSSION

Strength

In Figure 8, an attempt was made to fit the yield stress with a Hall-Petch relation. Although this graph is not an attempt to prove the validity of the Hall-Petch relation for these microstructures, we believe that it clearly shows that boundary strengthening of the two phase mixture ($\alpha_2+\beta_R$/B2), and thus α_2 lath thickness, control the strength of β-solutionized microstructures. Thus, as cooling rates increase, α_2 lath thickness decreases, boundary strengthening increases, and material's strength increases.

In addition, the α_2 lath arrangement did not seem to influence the strength behavior of these alloys and we believe that its contribution to strengthening mechanisms is insignificant.

The singular behavior observed with the "1150,25" microstructure (Figures 4a and 4b) is believed to be due to the fact that, at a cross-head speed of 0.5 mm/s, the samples fractured in Mode I brittle manner, well before reaching macroscopic yielding.

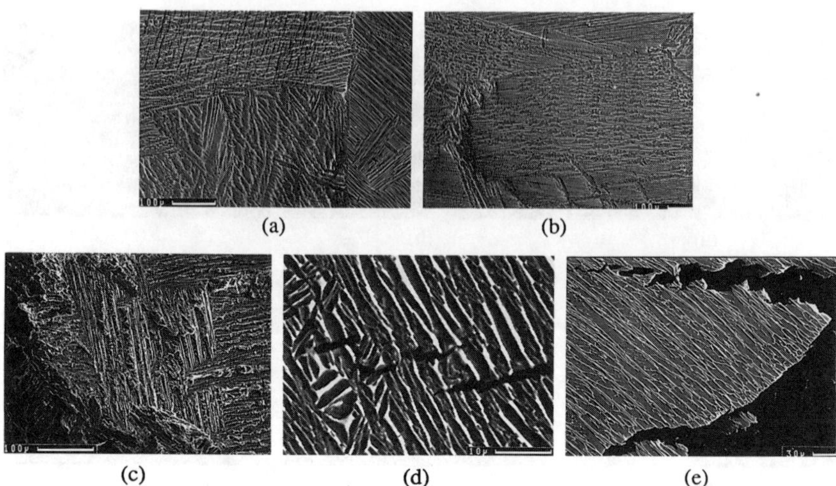

Figure 7 - Low cooling rates: (a) and (b) Compression specimen surface in "1150,0.01" ; (c) Fracture surface (in "1150,0.01") ; (d) and (e) Longitudinal section in "1150,0.1" and "1150,0.01", respectively.

Deformation and Failure Mechanisms

At all cooling rates fracture is brittle and the failure mechanism can be decomposed in three stages : slip band formation, crack initiation and crack propagation.

At high cooling rates, i.e., 25 and 7.5°C/s, slip concentrates in long, coarse B2 slip bands, extending throughout prior-β grains and interactions between α_2 laths and B2 slip bands are virtually non existent at 25°C/s and limited at 7.5°C/s. Thus, deformation appears to be controlled by the B2 phase. In addition, slip band formation leads to important dislocation pile-ups at slip band intersections or prior-β grain boundaries, rapidly followed by crack initiation. The limited amount of sub-fracture surface microcracks indicates that crack propagation, along prior-β grain boundaries or along B2 slip bands, occurs rather quickly and thus that failure is fully nucleation controlled. We believe that at 25°C/s the shear instability onset at the microscopic level occurs at very low macroscopic stresses resulting in the mode I failure observed. The small ductility recorded at 7.5°C/s is thought to be indirectly due to the extensive precipitation of α_2 laths, which affects the β_R matrix by retarding the onset of shear instability and narrowing the B2 slip band spacing, thus decreasing the dislocation density in each slip bands. Finally, based on TEM observations, Banerjee *et al.* have proposed that the heterogeneous slip observed in the β_R matrix might result from the presence of "ω-type" precipitates [3]. Our findings, that B2 deforms homogeneously at lower cooling rates where no "ω-type" precipitation is observed, seem to support their findings.

At intermediate and low cooling rates, deformation appears to be controlled by the α_2 phase. At intermediate cooling rates, i.e., from 4 to 0.5°C/s, the deformation of α_2 laths within prior-β grain concentrates essentially in the α_2 laths oriented favorably for slip, as evidenced by α_2/B2 surface offsets. Shear instability in α_2 laths results in planar, parallel slip bands, oriented almost perpendicular to the α_2 lath long axis and stopped at α_2/B2 or α_2/α_2 interfaces. Crack nucleation at interfaces depends then on

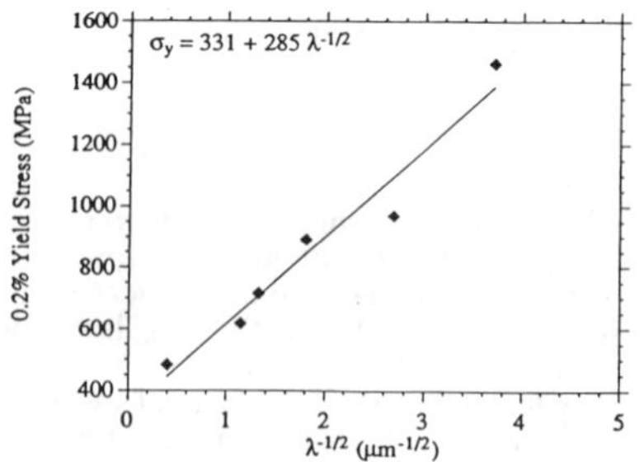

Figure 8 - Yield stress *vs*. (α_2 lath width)$^{-1/2}$.

the thicknesses of both α_2 laths and surrounding B2 film: the α_2 lath thickness because it determines the α_2 slip band length, and the B2 phase thickness because it determines how much strain the B2 phase can sustain ahead of α_2 slip bands. The observation of bridging of the B2 phase, crack-tip blunting by the B2 phase and the fact that most microcracks remain small shows that once cracks are generated, the B2 phase effectively toughens the microstructure. However, in basketweave-type microstructures, the largest surface offsets are located at or near prior-β grain boundaries. Indeed, inside prior-β grains, α_2 laths precipitate with an orientation relationship with respect to their parent β phase. It results in a "texture" between prior-β grains. The plastic deformation, controlled by α_2 laths, is therefore anisotropic at the scale of prior-β grains and important strain accumulations develop at or near prior-β grain boundaries, aggravated, when aligned α_2 laths are present at prior-β grain boundaries, by the formation of coarse slip bands throughout the alignments. Easy crack initiation results in the grain-boundary B2 phase. Crack propagation however differs between basketweave-like microstructures. Thus in the "1150,0.5" microstructure, the coarse α_2 laths present within prior-β grains generate microcracks with lengths equal or greater than that of grain boundary microcracks, leading to intergranular failure, whereas in the "1150,4" and "1150,1.5" microstructures grain boundaries exhibit the largest microcracks, which results in intergranular/transgranular failure. The increase in ductility from 4°C/s to 1.5°C/s is proposed to be due to the increase in α_2 lath thickness which decreases the yield stress, allowing for extended plastic deformation before strain concentrations ahead of dislocation pile-ups induce intergranular failure. The decrease in ductility from 1.5°C/s to 0.5°C/s is also believed to be imputable to α_2 lath thickening, which results in large microcracks within prior-β grains and leads to premature transgranular failure.

At slow cooling rates, i.e., at 0.1 and 0.01°C/s, slip takes place only in colonies favorably oriented and concentrates in widely separated slip bands. α_2 slip easily propagates throughout colonies due to the alignment of α_2 laths, and thus to the constancy in crystallographic orientations, and to the ease of transmission through the B2 phase. The efficacy of B2 as a toughening phase thus appears to be very limited in colony-type microstructures. The very large colony boundary dislocation pile-ups, resulting from the lack of slip transmission across colony boundaries, are expected to lead to early crack initiation at colony boundaries at low plastic strains. In addition, due to the virtual absence of microcracks below the fracture surface it is believed that crack propagation, along colony boundaries or slip bands, is very rapid and that failure is entirely nucleation controlled. Colonies, thus, behave as units during plastic deformation and failure. The low ductility observed with these microstructures probably results from the toughening action of the thick B2 films located at colony boundaries.

CONCLUSION

The microstructural features of β-solutionized microstructures were varied by changing the cooling rate. Their role on the room temperature strengthening, deformation and fracture mechanisms were then investigated. The following conclusions were drawn:

1. Strength is controlled by α_2 lath width through $(\alpha_2+\beta_R)$ boundary strengthening. The α_2 lath arrangement does not appear to contribute significantly to the strengthening mechanisms.
2. The deformation and failure mechanisms differ depending on the cooling rate regime:
   A. At high cooling rates, deformation is controlled by the B2 phase. The heterogeneous character of B2 slip, probably due to the presence of "ω-type" precipitates in the matrix, results in low ductilities.
   B. At intermediate and low cooling rates, deformation is controlled by the α_2 phase. The deformation and failure mechanisms are, however, very sensitive to α_2 lath arrangement. Therefore, on one hand the colony-type α_2 arrangement observed at low cooling rates leads to low ductilities, because colonies behave as units during deformation and failure, but on the other hand, the basketweave-type α_2 arrangement present at intermediate cooling rates yields the highest ductilities, because α_2 laths now act individually during deformation and failure, and the B2 phase efficiently behaves as the ductile constituent .
   C. At intermediate cooling rates, the α_2 lath thickness, which affects the onset of yielding and indirectly the size of intragranular microcracks, is the dominant factor influencing tensile elongation. Indeed, the amount of plastic deformation prior to failure can be described as resulting from a trade off between yield stress, size of intragranular microcracks and strength of prior-β grain boundaries. A maximum in ductility results at 1.5°C/s.

REFERENCES

1. F.-C. Dary and A.W. Thompson, *7th World Conf. on Titanium*, 1992, in Press.
2. G. Venkataraman, K.R. Teal and F.H. Froes, in *Sixth World Conference on Titanium*, edited by P. Lacombe, R. Tricot and G. Béranger (Les Éditions de Physique, Les Ulis, France, **2**, 1989) pp. 967-972.
3. D. Banerjee, A.K. Gogia and T.K. Nandy, *Metall. Trans. A*, **21A**, 627-639 (1990).

METALLOGRAPHIC STRUCTURE AND MECHANICAL PROPERTIES IN VANADIUM MODIFIED TITANIUM TRIALUMINIDES

TOHRU TAKAHASHI*, KATUYUKI ENDO*, a), SUSUMU KAIZU**, b) AND TADASHI HASEGAWA*
* Department of Mechanical Systems Engineering, Faculty of Technology, Tokyo University of Agriculture and Technology, Koganei, Tokyo 184, JAPAN
** Graduate School, Tokyo University of Agriculture and Technology, Koganei, Tokyo 184, JAPAN
a) now with SEIKO EPSON Corp., Okaya, JAPAN
b) now with NKK Corp., Kawasaki, JAPAN

ABSTRACT

Two types of aluminum-titanium-vanadium ternary intermetallic compounds have been prepared by arc melting under argon atmosphere. Their compositions were nominally $Al_{65}Ti_{25}V_{10}$ and $Al_{57}Ti_{22}V_{21}$; the numbers represent the molar fractions in mol%. These two alloys stand within the gamma-phase field where the $L1_0$ face-centered tetragonal structure is produced. Metallographic structure and mechanical properties have been investigated. It is suggested that vanadium addition to titanium aluminides and titanium trialuminides can enhance their strength.

INTRODUCTION

Nowadays, titanium aluminide intermetallics are widely accepted as potential light-weight heat-resisting materials for aerospace applications in the near future[1,2]. The pioneering work to investigate the mechanical properties of titanium aluminides at elevated temperature first appeared in 1950s [3]. More and more experimental information is becoming available[4-11]. One of the present authors has recently carried out a systematic study about the creep properties of polycrystalline materials of TiAl single phase in compressive creep tests [5-8]. Basic understanding, however, about the well-known brittleness in an ambient temperature range is not satisfactory enough. The fundamental mechanisms of high-temperature deformation and of creep resistance have not been made clear so far.

In order to alleviate the room-temperature brittleness and to improve the mechanical strength of titanium aluminides, many experimental efforts have been made, in which small amounts of third alloying elements are added. It was reported that the third-element alloying with vanadium brought about effective improvement in room temperature ductility of TiAl [4]. If further improvement is achieved by micro- and macroalloying, it can be expected that the merits of TiAl intermetallics as heat-resisting materials are fully realized.

As far as the present authors know, the wide composition range of the gamma-phase field in the aluminum-titanium-vanadium ternary intermetallic alloys has not been fully exploited so far. Concerning the aluminum-titanium-vanadium ternary alloy system, several phase diagrams are available[12-14]. On the other hand, there have been not much data published about the mechanical properties of these ternary intermetallic alloys. Aluminum-titanium-vanadium ternary alloys were mainly developed near the titanium corner as the beta-phase and/or alpha-beta dual-phase titanium alloys for jet engine materials. In contrast with these titanium alloys, the gamma intermetallic phase has not attracted much attention. According to the isothermal sections of Hashimoto's phase diagrams at 1073K and 1273K [14], the gamma-phase field can be seen to expand toward chemical compositions containing more aluminum when increasing the vanadium content up to a maximum solid solubility of about 20mol%. It can be expected that increasing aluminum content leads to reduced density, to improved oxidation resistance, and, possibly, to enhanced strength.

In the present study, metallographic structures and mechanical properties were investigated on two types of $L1_0$ ternary aluminum-titanium-vanadium intermetallic alloys: $Al_{65}Ti_{25}V_{10}$ and $Al_{57}Ti_{22}V_{21}$ (numbers represent molar fractions in mol%).

EXPERIMENTAL PROCEDURE

The materials used in the present study were prepared as button ingots by arc melting, and the mass of the ingot was around 0.06~0.07 kg. Some other intermetallics such as titanium aluminides and titanium trialuminides were also used as references. $L1_2$ modified ternary titanium trialuminides containing chromium ($Al_{66}Ti_{25}Cr_9$) or manganese ($Al_{66}Ti_{25}Mn_9$) or iron ($Al_{65}Ti_{25}Fe_{10}$) were also utilized; the subscripts represent molar fractions in mol%. All these materials were prepared in the same way as described above.

The following experiments were carried out;

(1) Metallographic observations on optical microscope
(2) X–ray diffraction analysis
(3) Sound velocity measurements
(4) Compressive tests at room temperature
(5) Compressive tests at elevated temperatures
(6) Compressive creep tests in vacuum
(7) Metallographic observation of the hot–deformed structure

RESULTS AND DISCUSSION

Metallographic Structure

Fig. 1 (a) and (b) show optical micrographs of the as–cast structure in the $Al_{65}Ti_{25}V_{10}$ and in the $Al_{57}Ti_{22}V_{21}$ ternary intermetallics; the observation was made under polarized light conditions. The macro–structure in the $Al_{65}Ti_{25}V_{10}$ material is largely made of coarse columnar grains, but this material contained plate–like structures oriented in a few directions within the columnar grains. Structures in the $Al_{57}Ti_{22}V_{21}$ material contained bundles of thin plates, which are also oriented in a few directions. The thickness of these plates were in the range of a few micrometers. Detailed analysis has not yet been carried out, but these plates might be related to some kind of twins, which probably formed during cooling from the melt.

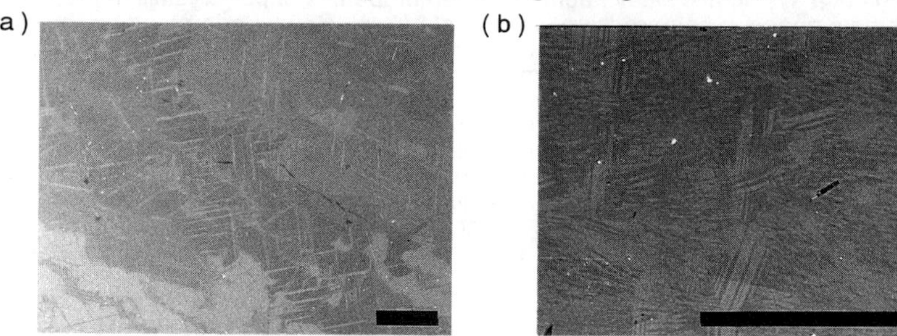

Fig. 1 Optical micrographs of the metallographic structures in (a) $Al_{65}Ti_{25}V_{10}$ and in (b) $Al_{57}Ti_{22}V_{21}$. Scales indicate 50μm.

X–Ray Diffraction Analysis

The crystal structures of the $Al_{65}Ti_{25}V_{10}$ and the $Al_{57}Ti_{22}V_{21}$ materials were investigated with the ordinary X–ray diffraction analysis. The characteristic X–ray of copper (Cu Kα) and a nickel filter were used.

Almost all the prominent diffraction peaks were successfully indexed assuming the $L1_0$ tetragonal structure for both the $Al_{65}Ti_{25}V_{10}$ and the $Al_{57}Ti_{22}V_{21}$ materials. The lattice parameters, *a* and *c*, were calculated by a numerical calculation program which was based on the Cohen's method (least square method). The diffraction pattern of the $Al_{57}Ti_{22}V_{21}$ material is shown in Fig. 2.

In the $Al_{65}Ti_{25}V_{10}$ material, *a* and *c* were calculated as 0.3916nm and 0.4081nm, respectively. This *c* value is almost the same as that in the stoichiometric binary titanium aluminide. On the other hand, the *a* value is about 2% smaller than that in the stoichiometric titanium aluminide. Therefore, the axial ratio *c* / *a* is 1.042 which is about 2% greater than that in the stoichiometric titanium aluminide.

In the $Al_{57}Ti_{22}V_{21}$ material the *a* value is not very different from that in the other material; *a* = 0.3913nm. The *c* value is about 1% smaller than that in the $Al_{65}Ti_{25}V_{10}$; *c* = 0.4051nm. Therefore, the axial ratio becomes *c* / *a* = 1.035, which is about 1% smaller than that in the other material. The axial ratio is still greater than that in the stoichiometric binary titanium aluminide.

The chemical compositions of the $Al_{65}Ti_{25}V_{10}$ and the $Al_{57}Ti_{22}V_{21}$ materials are converted to mass% as $Al_{50.7}Ti_{34.6}V_{14.7}$ and as $Al_{42.0}Ti_{28.8}V_{29.2}$. These composition are plotted in the isothermal diagram which was reported by Hashimoto et al. [14]. They are both near the periphery of the gamma single–phase field.

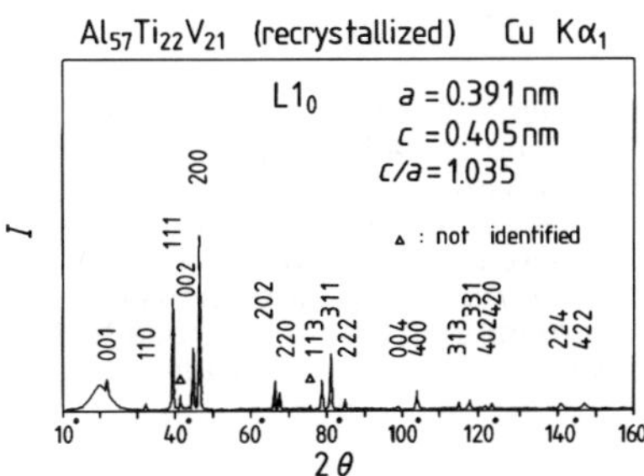

Fig. 2 X–ray diffraction pattern of $Al_{57}Ti_{22}V_{21}$.

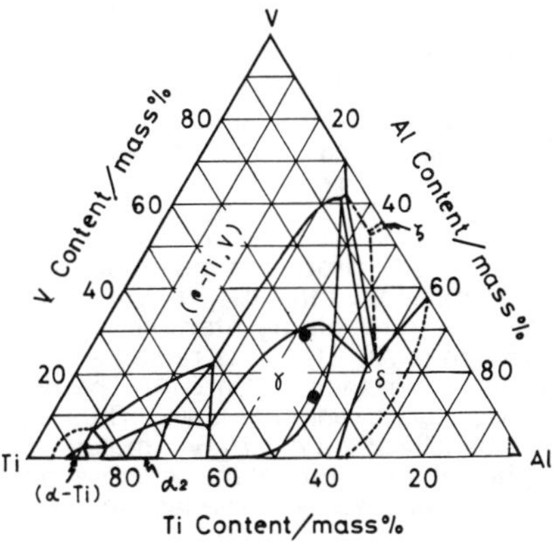

Fig. 3 Al–Ti–V ternary alloy phase diagram at 1273K.

Sound Velocity Measurements

Sound velocity measurements of the longitudinal and the transverse elastic waves were conducted using the pulse–echo method. The velocity of longitudinal waves was around 7500m/s, and that of transverse waves was around 4500m/s. From these values and the density values the elastic moduli were calculated, assuming the formulae for an elastically isotropic media. The results are shown in Table 1 on the next page.

Compression Tests at Room Temperature

The strength and deformability were evaluated in compression tests at room temperature. The stress–strain curves are shown in Fig. 4. The curves in this figure are true stress – true strain curves.

The yield strength values are given by the 0.2% proof stress and are shown in Table 1. The deformability values read from Fig. 4 are also shown in Table 1.

The $Al_{65}Ti_{25}V_{10}$ material is poor in deformability, however, its deformability is improved as compared with Al_3Ti.

The $Al_{57}Ti_{22}V_{21}$ material shows a superior value of yield stress, and at the

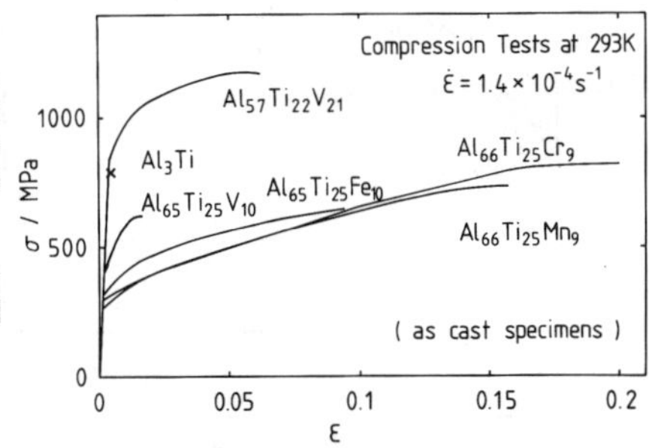

Fig. 4 Stress–strain curves at room temperature.

Table 1 Basic Data of Titanium Aluminide Intermetallics

	TiAl	Al_3Ti	$Al_{66}Ti_{25}Mn_9$	$Al_{65}Ti_{25}V_{10}$	$Al_{57}Ti_{22}V_{21}$
Density ρ (10^3kg/m^3)	3.80	3.35	3.70	3.67	3.92
Crystal Structure	$L1_0$	$D0_{22}$	$L1_2$	$L1_0$	$L1_0$
Lattice Parameters					
a (nm)	0.400	0.385	0.396	0.3916	0.3913
c (nm)	0.408	0.859	====	0.4081	0.4051
c/a	1.02	2.23	====	1.042	1.035
Elastic Moduli					
E (GPa)	170~180	~200	170~180	~180	180~190
G (GPa)	70~75	75~85	70~75	75~85	~75
Yield Stress					
σ_y (MPa)	200~300	600~800	~300	300~400	~800
Deformability (Compression) at Room Temp.	20~40%	negligible	~20%	3~5%	~10%

same time, it produces a significant amount of plastic deformation. The plastic deformability of the $Al_{57}Ti_{22}V_{21}$ material is somewhat smaller than that in the stoichiometric binary titanium aluminide intermetallics, however, the superior strength of $Al_{57}Ti_{22}V_{21}$ is remarkable. The yield strength of $Al_{57}Ti_{22}V_{21}$ is about 3 times greater than that of TiAl. The deformability is greater than that of $Al_{65}Ti_{25}V_{10}$ whose yield strength is lower. Thus the superiority of $Al_{57}Ti_{22}V_{21}$ is demonstrated. The density of the $Al_{57}Ti_{22}V_{21}$ material is only 3% greater than that of TiAl, due to its higher aluminum content, which almost cancels the increase in average atomic weight produced by the vanadium addition.

Compressive Tests at Elevated Temperatures

Compressive tests were carried out at elevated temperature, and the high temperature deformation behavior has been investigated. The tests were conducted in air. Flow curves of the $Al_{65}Ti_{25}V_{10}$ and the $Al_{57}Ti_{22}V_{21}$ materials are displayed in Fig. 5 (a) and (b), respectively. The test temperatures were 293, 900, 1100 and 1300K. In both the $Al_{65}Ti_{25}V_{10}$ and the $Al_{57}Ti_{22}V_{21}$ materials large deformability could be attained at temperature higher than about 1100K, and at such high temperatures both materials showed flow curves of dynamic recovery type. This could be related to the structure changes probably due to dynamic recrystallization.

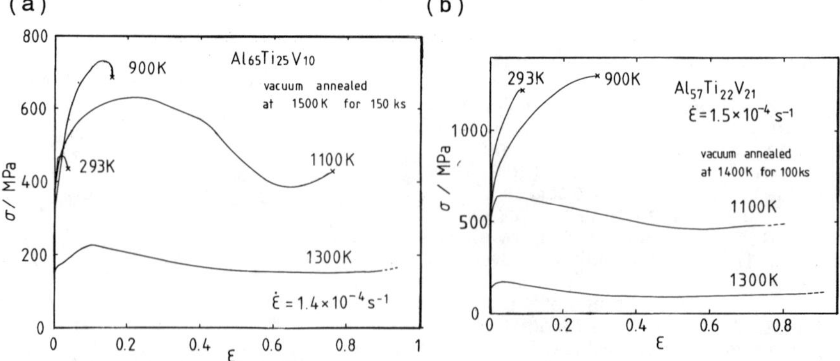

Fig. 5 Flow curves at various temperatures in (a) $Al_{65}Ti_{25}V_{10}$ and (b) $Al_{57}Ti_{22}V_{21}$, respectively.

The temperature dependence of the yield strength in the $Al_{65}Ti_{25}V_{10}$ and the $Al_{57}Ti_{22}V_{21}$ materials are shown in Fig. 6 together with the corresponding data for Al_3Ti. The deformability in Al_3Ti materials is negligibly small so that the yield stress data are uncertain. They would rather correspond to the fracture stress than the yield stress. In the lower temperature range $Al_{57}Ti_{22}V_{21}$ materials show excellent strength which is greater than that in Al_3Ti. The yield strength of the Al_3Ti and the $Al_{57}Ti_{22}V_{21}$ materials are seen to fall off remarkably at temperatures higher than about 1000K. At the highest temperature used in this study (1300K), the $Al_{65}Ti_{25}V_{10}$ material becomes stronger than the $Al_{57}Ti_{22}V_{21}$ material.

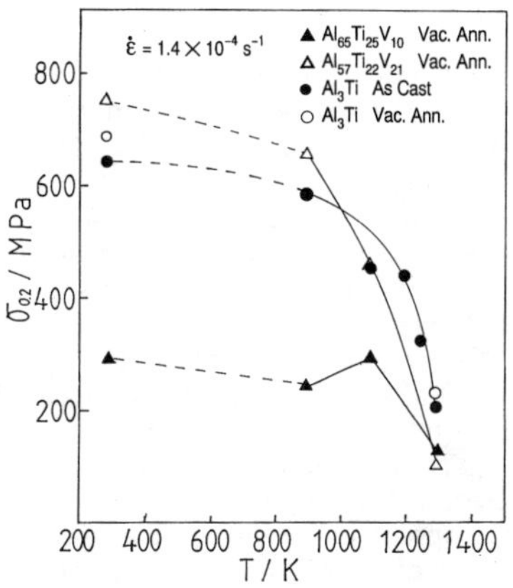

Fig. 6 Temperature dependence of yield stress

Compressive Creep Behavior

Compressive creep tests have been performed in vacuum of 10^{-3} Pa. The test temperatures ranged from 1100K to 1250K, and the applied stress ranged from 80MPa to 251MPa. The load was intermittently adjusted so that the true stress was kept constant.

Curves in Fig. 7 are indicating the creep rates as a function of strain. The creep curve comprises a small normal primary, a minimum creep rate region and an accelerating portion. The stress dependence of the minimum creep rates is shown in Fig. 8. $Al_{65}Ti_{25}V_{10}$ and $Al_{57}Ti_{22}V_{21}$ materials both indicate stress exponents of about 5, suggesting dislocation mechanisms.

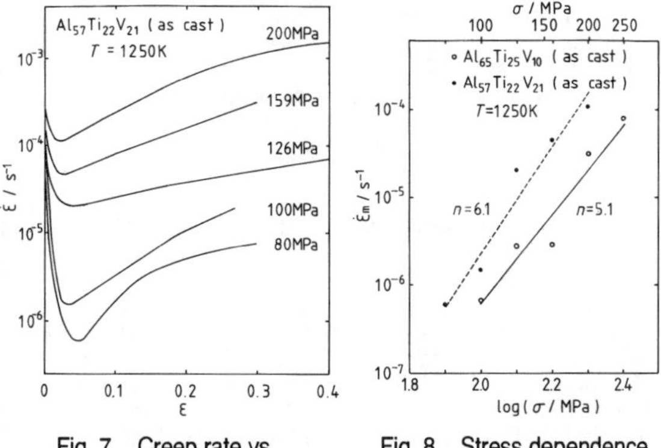

Fig. 7 Creep rate vs. creep strain.

Fig. 8 Stress dependence of creep rates.

The creep resistance is enhanced in $Al_{65}Ti_{25}V_{10}$ as compared to $Al_{57}Ti_{22}V_{21}$. This is seen from the minimum creep rates, which coincide with the yield stress data at the highest temperature. It is not clear if such a merit can be realized, as its deformability is considerably poorer.

Hot-Deformed Structure

The flow curves in Fig. 5 and the creep rate acceleration in Fig. 7 both indicate the dynamic recovery behavior. Fig. 9 shows the metallographic structure as observed in the $Al_{65}Ti_{25}V_{10}$ material after hot deformation at 1300K : the amount of strain was about 0.7 (50% compression). The coarse columnar grains in the as-cast condition are almost entirely changed into fine equiaxed grains, indicating dynamic recrystallization.

Fig. 9 Deformed structure in $Al_{65}Ti_{25}V_{10}$. Scale indicates 100μm.

SUMMARY

Metallographic observations and mechanical tests have been carried out on two types of Al–Ti–V ternary intermetallic compounds. The results are as follows ;

① $Al_{65}Ti_{25}V_{10}$ and $Al_{57}Ti_{22}V_{21}$ ternary intermetallics showed $L1_0$ tetragonal structure.

② The as-cast materials of $Al_{65}Ti_{25}V_{10}$ and $Al_{57}Ti_{22}V_{21}$ contained many plate-like structures which were possibly twin bands.

③ Elastic moduli determined by sound velocity measurements were comparable to those in steels.

④ The yield strength of $Al_{65}Ti_{25}V_{10}$ and $Al_{57}Ti_{22}V_{21}$ at room temperature were ~400MPa and ~800MPa, respectively.

⑤ Deformability under compression at room temperature were ~5% and ~10% in $Al_{65}Ti_{25}V_{10}$ and $Al_{57}Ti_{22}V_{21}$, respectively.

⑥ Dynamic recrystallization occurred during high temperature deformation.

⑦ Compressive creep properties of the $Al_{65}Ti_{25}V_{10}$ material were superior to those of the $Al_{57}Ti_{22}V_{21}$ material, in contrast to the room temperature strength.

ACKNOWLEDGMENT

Materials supply and financial support from Nippon Steel Corp. are gratefully acknowledged. Sound velocity measurements were performed with the kind help of Professor Nishiwaki in the Department of Mechanical Systems Engineering, Tokyo University of Agriculture and Technology.

REFERENCE

1. H.A. Lipsitt, in High–Temperature Ordered Intermetallic Alloys, edited by C.C. Koch, C.T. Liu, and N.S. Stoloff (Mater. Res. Soc. Proc. 39, Pittsburgh, PA, 1985) pp. 351–364.
2. Y.–W. Kim, J. of Metals 41 (7), 24–30, (1989).
3. J.B. McAndrew and H.D. Kessler, Trans. AIME 206, 1348, (1956).
4. M.J. Blackburn and M.P. Smith, U.S. Patent No. 4 294 615, (13 Oct. 1981).
5. T. Takahashi, H. Nagai and H. Oikawa, Mater. Trans., JIM 30, 1044 (1989); Mater. Sci. Eng. A114, 13 (1989); Mater. Sci. Eng. A128, 195 (1990).
6. H. Nagai, T. Takahashi and H. Oikawa, J. Mater. Sci. 25, 629 (1990).
7. T. Takahashi and H. Oikawa, in High–Temperature Ordered Intermetallic Alloys III, edited by C.T. Liu, A.I. Taub, N.S. Stoloff, and C.C. Koch (Mater. Res. Soc. Proc. 133, Pittsburgh, PA 1989) pp. 699–704.
8. T. Takahashi and H. Oikawa, in Creep and Fracture of Engineering Materials and Structures, edited by B. Wilshire (The Institute of Metals, London, UK 1990) pp. 237–253.
9. P.L. Martin, M.G. Mendiratta, and H.A. Lipsitt, Metall.Trans. 14A, 2170 (1983).
10. M. Nobuki, K. Hashimoto, T. Tsujimoto and Y. Asai, Journal of the Japan Institute of Metals, 50, 840 (1986).
11. S. Mitao, Y. Kohsaka and C. Ouchi, in THERMEC–88, edited by I. Tamura (Iron and Steel Institute of Japan, Tokyo, 1988), p. 620.
12. P. Villars and L.D. Calvert (eds.), in Pearson's Handbook of Crystallographic Data for Intermetallic Phases (second edition), ASM International, (1991).
13. A. Raman, Zeitschrift fuer Metallkunde, 57, 535, (1966).
14. K. Hashimoto, H. Doi and T. Tsujimoto, Journal of the Japan Institute of Metals, 49, 410 (1985).

MICROSTRUCTURAL AND MECHANICAL EVALUATION OF NEAR γ-TiAl ALLOYS CONTAINING MOLYBDENUM AND RHENIUM

D. B. SNOW*, D. L. ANTON*, M. Y. NAZMY‡ and M. STAUBLI‡
*United Technologies Research Center, 411 Silver Lane, East Hartford, CT 06108 USA
‡ABB Power Generation Ltd., Baden CH-5401, Switzerland

ABSTRACT

Near γ-TiAl alloys which incorporated Cr, and Mo or Re additions [Ti-48Al-2Cr-1.5Mo-0.1Y, Ti-48Al-2Cr-1.5Re-0.1Y, and Ti-46Al-2Cr-1.5Re-0.1Y(at.%)], were prepared by arc casting and hot isostatic pressing; then heat treated at 1200 and 900°C to produce a duplex microstructure. All three alloys contained a bimodal size distribution of B2 phase, which was enriched in Cr+Mo or Cr+Re. Very little α_2 was present in the 48Al alloys; whereas the 46Al, 1.5Re alloy contained areas of transformed lamellar α_2/γ, within which short lengths of B2 periodically interrupted the α_2 plates. The composition of both the equiaxed and lamellar γ in each alloy was 49±0.5Al, 1.6±0.4Cr and ≈1.2Mo or <1Re(at%). Similarly, the composition of the B2 in the three alloys was experimentally equivalent: 33±1Al, 8.5±2Cr and 6.5±1.5Mo or Re(at%). In the Re containing alloys, the α_2 composition was ≈33Al, 1.7±0.3Cr, and 0.5±0.3Re. Room temperature tensile ductilities were low, ranging from 0.5 to 1%; while ultimate strengths approached 500MPa [$7\times10^{-4}s^{-1}$ strain rate]. Testing at 700°C produced little change in alloy strength, while ductility increased to as much as 2.5%.

INTRODUCTION

High temperature tensile and creep strength enhancements have been made in near-γ TiAl alloys through the introduction of transition metals to achieve solid solution strengthening [1-3]. The most likely mechanism for the increase in ductility which has been observed in these alloys is either the influence of alloying on tetragonality [4], or on the $\alpha + \gamma$ phase field boundaries [5]. Alloy additions such as V and Cr have been found to substantially increase room temperature ductility [3,6,7]. Mn additions have also been found to improve strength and increase ductility in these materials [4]. Consequently, it appeared possible that the addition of Re, another group VIIA element, might have similar beneficial effects. This study was conducted to better define the effects of Re additions, in comparison to Mo, on the microstructure and mechanical properties of a near γ, TiAl + 2Cr alloy.

EXPERIMENTAL PROCEDURE

The base alloy compositions of this investigation were Ti-48Al-2Cr-1.5Mo-0.1Y and Ti-48Al-2Cr-1.5Re-0.1Y (at%). The 0.1Y was added to scavenge oxygen from the matrix. These alloys contained unexpectedly low volume fractions of α_2; consequently, an alloy of lower aluminum concentration, Ti-46Al-2Cr-1.5Re-0.1Y, was prepared in the anticipation that it would contain a significant quantity of lamellar α_2 after heat treatment. High purity titanium (400 wtppm O by neutron activation analysis) was used for all alloys; all other alloy constituents were of commercial purity. Ingots of ≈5.7x5.7x1.3cm dimensions were prepared by nonconsumable electrode arc melting, remelted once, and solidified on a water cooled copper hearth under continuously purified argon. Porosity closure was achieved by hot isostatic pressing (HIP) at 1260°C and 138 MPa (20 ksi) for 4 hours. The HIP'ed ingots were wrapped in Ti foil and heat treated in an atmosphere of Ti-gettered helium at 1200°C for 72 hours, then at 900°C for 4 hours. The post-heat treatment oxygen content of the alloys was measured by neutron activation analysis. Tensile tests were conducted on specimens ground to a 10 mm gage length and 3 mm diameter at ambient temperature, and at 700°C in air at a strain rate of $7\times10^{-4}s^{-1}$. The average alloy and specific phase compositions were determined by wavelength-dispersive X-ray spectroscopy (electron microprobe; unetched surfaces; pure element standards); and by energy-dispersive X-ray spectroscopy of thin foils, using a TiAl standard of known composition and conditions of constant geometry, beam current, and foil thickness [8]. The alloy microstructure was characterized primarily by light microscopy and analytical transmission electron microscopy. The volume fraction of the B2 phase was determined from backscattered electron images using a Noran TN8502 image analyzer with standard analytical software.

RESULTS AND DISCUSSION

Microstructure

Since all three alloys exhibited a duplex microstructure (Figs. 1-3), it is reasonable to assume that 1260°C was below the α transus in each case , as suggested by phase transformation data for Ti-Al-2Cr [9]. In the case of the 46Al, 1.5Re alloy, the equiaxed γ grains were noticeably smaller, both in size and quantity. The presence of a population of larger (5-20µm) B2 grains in all three alloys, often in contact with the primary, equiaxed γ, suggests that β may also have

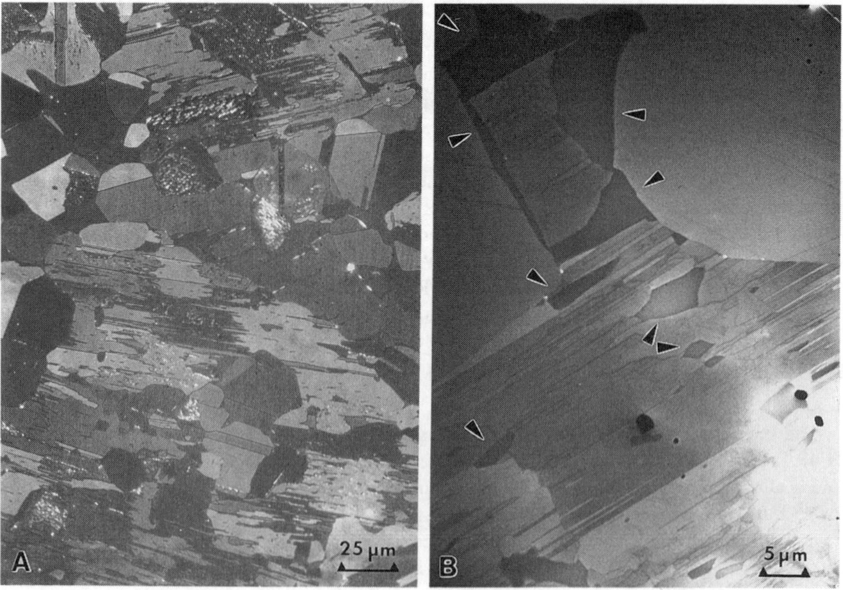

Fig. 1. Ti-48Al-2Cr-1.5Mo. (A) Polarized light micrograph. (B) Bright field STEM image, with regions of B2 marked by arrows. Small, black areas are Y_2O_3

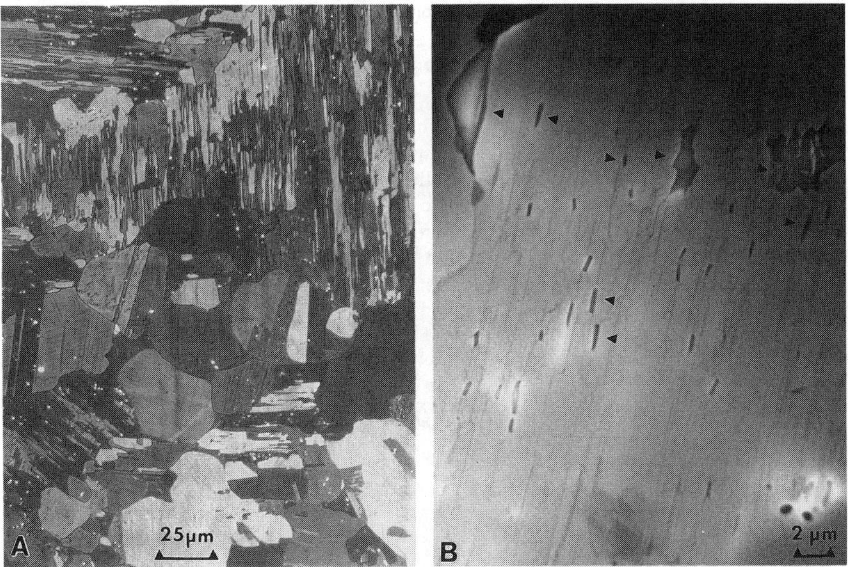

Fig. 2. Ti-48Al-2Cr-1.5Re. (A) Polarized light micrograph. (B) Bright field STEM image, with two sizes of B2 (arrows) in a lamellar γ matrix. Very little α_2 is present.

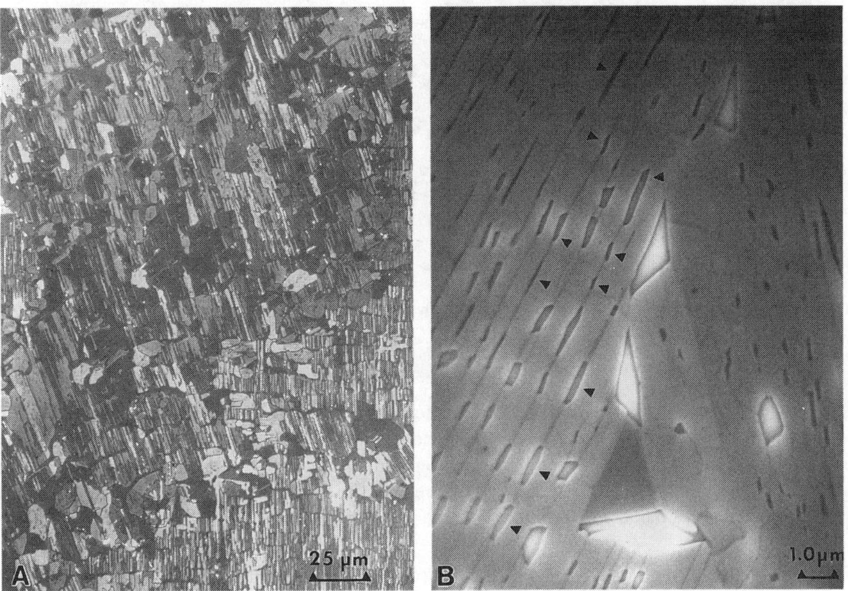

Fig. 3. Ti-46Al-2Cr-1.5Re. (A) Polarized light micrograph. (B) Bright field STEM image. Large and small, rectangular B2 particles, the latter (arrows) interconnected by lamellar α_2.

been stable at 1260°C. Upon annealing at 900°C, the α in each alloy transformed almost completely to either twinned γ (1.5Mo alloy), twinned γ with some α_2 laths (48Al, 1.5Re alloy), or an α_2/γ lamellar microstructure (46Al, 1.5Re alloy). In each alloy, the β transformed to an ordered, cubic B2 structure.

No lamellar α_2 was observed in thin foils of the Mo containing alloy; very little could be detected in the 48Al, 1.5Re alloy. A smaller population ($\leq$1μm) of elongated B2 grains was present at γ/γ interfaces throughout the lamellar γ regions in all three alloys. It exhibited a consistent orientation approximately parallel to the γ/γ and α_2/γ interfaces. This suggested that it was a solid state transformation product, presumably having formed at 900°C. The 46Al, 1.5Re alloy contained extensive areas of α_2/γ lamellae, within which the α_2 laths were periodically terminated by elongated regions of B2 (Figs. 3B, 4).

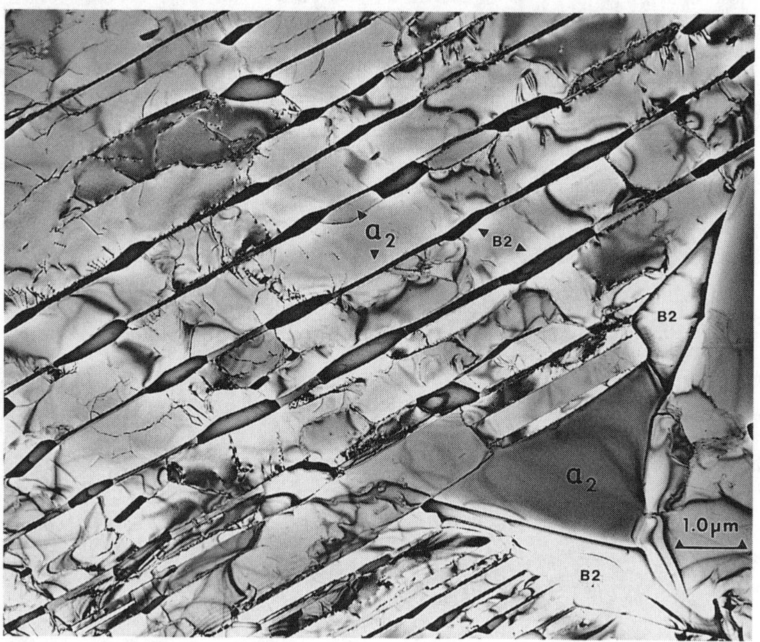

Fig. 4. Ti-46Al-2Cr-1.5Re. Bright field CTEM image, area of Fig. 3B. The specimen was tilted so that the direction of view is near <110>γ; thus the α_2/γ interfaces are viewed approximately edge-on.

The identification of the B2 in both the larger and smaller size populations as ordered cubic rather than disordered β was based on the presence of 100 spots in <011> and <001> selected area diffraction patterns. Also, both the large and small B2 grains had approximately the same composition (Table I). Confirmation of phase identity and distribution within the areas of STEM images was made by simultaneous digital X-ray mapping.

Composition

The Re containing alloys were found to contain approximately 1200 wtppm {O}, while the oxygen content of the Mo containing alloy was found to be 446 wtppm. Although these concentrations were determined by neutron activation analysis, thus measuring the oxygen incorporated in the Y_2O_3 as well as the alloy, 1200wtppm was perceived to be unusually high. Subsequent analyses showed that the Re originally contained 23,500 wtppm {O}. This was assumed to be the source of the high oxygen level in the Re containing alloys.

The overall composition of the alloys and the specific composition of the individual phases were determined by both WDXS (electron microprobe) and EDXS (STEM) (Table I). This data showed that the Cr and Mo or Re had dissolved to a limited, and approximately equivalent, extent in the γ and α_2; and had segregated to (and undoubtedly helped to stabilize) the B2 phase. Both the refractory element partitioning and the ≈10% volume fraction of B2 seemed unusual, in view of the $2(\alpha_2):1(\gamma)$ Cr partitioning and low β phase volume fraction observed in Ti-48Al-2Cr-2Nb (at%), albeit after heat treatment at 1400°C, above its α transus [10], which probably achieved a more complete homogenization [11] than in the present investigation.

Table I. Alloy and Phase Compositions

Nominal at%:	48Al-2Cr-1.5Mo	48Al-2Cr-1.5Re	46Al-2Cr-1.5Re
Volume % B2:	10.7	7.8	14.4
Overall at% Al:	47.8 ± 0.8*	47.8 ± 0.8*	46.5 ± 0.8*
Gamma; at%			
Al:	49.5 ± 0.8*/49.7 ± 1.5‡	49.1 ± 0.8*/49.3 ± 0.8‡	48.7 ± 0.8*/48.2 ± 1.5‡
Cr:	1.8 ± 0.3*/1.6 ± 0.5‡	1.5 ± 0.3*/1.4 ± 0.5‡	1.4 ± 0.3*/1.2 ± 0.5‡
Mo:	1.2 ± 0.3*/1.0 ± 0.5‡	-----	-----
Re:	-----	0.9 ± 0.3*/0.7 ± 0.5‡	0.7 ± 0.3*/0.5 ± 0.5‡
B2; at% Al:	33.4 ± 0.8*/34.0 ± 1.5‡	32.6 ± 0.8*/34.5 ± 1.5‡	33.6 ± 0.8*/34.7 ± 1.5‡
Cr:	8.8 ± 0.3*/8.2 ± 0.5‡	10.7 ± 0.5*/8.2 ± 1‡	6.6 ± 0.5*/7.6 ± 1‡
Mo:	6.2 ± 0.5*/6.3 ± 1‡	-----	-----
Re:	-----	6.5 ± 0.5*/8.0 ± 1‡	6.5 ± 0.5*/6.7 ± 1‡
α_2; at% Al:		33 ±3‡	32 ± 3‡
Cr:	(not observed)	1.7 ± 0.3‡	1.7 ± 0.3‡
Mo:		-----	-----
Re:		0.5 ± 0.3‡	0.4 ± 0.3‡

*Wavelength Dispersive X-Ray Spectroscopy (pure element standards)
‡Energy Dispersive X-Ray Spectroscopy (thin foil, TiAl standard)

Tensile Properties

Specimens from each of the three alloys were tested at ambient temperature and at 700°C. The results of these tests are given in Table II. The 1.5% Mo alloy displayed the greatest ductility (> 1%), while both Re alloys displayed < 0.5% strain to failure. Room temperature strengths were comparable, with the 46Al, 1.5Re alloy being the strongest. At 700°C, both of the Re bearing alloys displayed improved strength compared to the Mo alloy; but again their ductilities were significantly lower. For the 1.5Re alloys, this was most probably due to the abnormally high oxygen. In the case of the 1.5Mo alloy , the explanation for a tensile ductility significantly less than 2% it is not apparent.

Table II. Tensile Properties [$7\times10^{-4}s^{-1}$strain rate]

Alloy (nominal at%)	Test Temp. (°C)	Yield Strength (MPa)	Ultimate Strength (MPa)	Strain to Failure (%)
48Al-2Cr-1.5Mo	25	414	465	1.10
	700	355	461	2.25
48Al-2Cr-1.5Re	25	----	449	0.38
	700	390	512	1.57
46Al-2Cr-1.5Re	25	445	481	0.50
	700	384	485	0.98

SUMMARY

1. Heat treatment at 1260°C for 4 hours; 1200°C for 72 hours; and 900°C for 4 hours resulted in a duplex microstructure in all three alloys.
2. The Ti-48Al-2Cr-1.5Mo alloy contained no observable lamellar α_2; the Ti-48Al-2Cr-1.5Mo alloy contained very little. All alloys contained ≈10vol% B2 phase.
3. The tensile properties for all alloys at 25°C and a strain rate of $7\times10^{-4}s^{-1}$ were ≤ 500MPa UTS and 0.5-1% ductility. At 700°C, the ductility increased to ≤2.5% elongation, while strength did not change significantly.

REFERENCES

1. T. Kawatabe, T. Tamura, and O. Izumi, *High Temperature Ordered Intermetallic Alloys III*, edited by C. T. Liu *et al.* (Mater. Res. Soc. Proc. **133**, Pittsburgh, PA, 1989), pp. 329-334.
2. S-C. Huang and D. S. Shih, op. cit. 1, pp. 105-111.
3. Y-W. Kim, *High Temperature Ordered Intermetallic Alloys IV,* edited by L. A Johnson, D. P. Pope, and J. O. Stiegler (Mater. Res. Soc. Proc. **213**, Pittsburgh, PA, 1991), pp. 777-794.
4. T. Tsujimoto and K. Hashimoto, *High Temperature Ordered Intermetallic Alloys III*, edited by C. T. Liu *et al.* (Mater. Res. Soc. Proc. **133**, Pittsburgh, PA, 1989), pp. 391-396.
5. Y-W. Kim and F. H. Froes, *High Temperature Aluminides and Intermetallics*, edited by S. H. Whang *et al.* (TMS, Warrendale, PA, 1990), pp. 465-492.
6. M. Morinaga *et al*, Acta metall. mater. **38**, 25 (1990).
7. S-C. Huang and E. L. Hall, Acta metall. mater. **39**, 1053 (1991).
8. C. G. Rhodes, *Proceedings, 38th Annual Meeting EMSA*, edited by G. W. Bailey (Claitor's Publishing Div., Baton Rouge, 1980), pp. 142-143.
9. S-C. Huang and E. L. Hall, *Alloy Phase Stability and Design*, edited by G. M. Stocks, D. P. Pope, and A. F. Giamei (Mater. Res. Soc. Proc. **186**, Pittsburgh, PA, 1991), pp. 381-386.
10. D. S. Shih *et al*, *Microstructure/Property Relationships in Titanium Aluminides and Alloys*, edited by Y-W. Kim and R. Boyer (TMS, Warrendale, PA, 1991), pp. 135-148.
11. S. L. Semiatin et al, *High Temperature Ordered Intermetallic Alloys IV,* edited by L. A Johnson, D. P. Pope, and J. O. Stiegler (Mater. Res. Soc. Proc. **213**, Pittsburgh, PA, 1991), pp. 883-888.

MECHANICAL BEHAVIOR OF $L1_2$ SINGLE CRYSTAL $Al_{66}Ti_{25}Mn_9$

S. A. BROWN*, D. P. POPE** AND K. S. KUMAR*
*Martin Marietta Laboratories, Baltimore, MD 21227
**Dept. of Materials Science and Engineering, University of Pennsylvania, Philadelphia, PA 19104

ABSTRACT

Single crystal $Al_{66}Ti_{25}Mn_9$ has been produced and tested in compression as a function of temperature and orientation. Yield strengths for orientations near [001] and $[\bar{1}11]$ continuously decrease with increasing temperature in a manner similar to other single crystal $L1_2$ trialuminides, and also to polycrystals of these materials. Slip was determined to occur on the {111} octahedral planes using two-surface analysis. Critical resolved shear stress (CRSS) variation with temperature, calculated on {111} planes, overlap closely for both orientations. Dislocation analysis confirmed the Burgers vectors to be of the type a<110> at both 298K and 1073K. A limited number of uniaxial tension tests were conducted near [001] at 1073K; the HIPed specimens contained a small amount of residual porosity and second phases which resulted in elastic failure even at this high temperature.

INTRODUCTION

The mechanical response of ternary $L1_2$ trialuminides derived by macroalloying binary Al_3Ti (DO_{22} structure) with any of several transition elements has generated interest in their deformation mechanisms and fracture behavior [1-9]. Production of completely single-phase microstructures, however, has been difficult in these systems; frequently second phases such as Al_2Ti are present [3, 4]. Polycrystals of these compounds have been tested in compression, bending and in tension as a function of temperature; low temperature tensile ductility has been confirmed only in the Cr- and Mn-alloyed variants thus far [5-8]. In addition, isothermally forged $Al_{66}Ti_{25}Mn_9$ tested in uniaxial tension revealed a ductility minimum at ~773K [5]. Reasons for the lack of ambient ductility and the occurrence of the intermediate temperature ductility minimum are not yet clear. The potential role of grain boundaries in impeding slip transfer from one grain to another at low temperatures and the possibility of these boundaries being embrittled at the intermediate temperatures, leading to a ductility minimum, prompted the present study.

In this study, single crystals of $Al_{66}Ti_{25}Mn_9$ were characterized in terms of their microstructure and mechanical properties with the intentions of identifying the role of grain boundaries, or lack thereof, on ambient temperature ductility as well as on the intermediate temperature ductility minimum. To date, compressive yield strength variation with temperature and orientation has been determined and operative slip systems identified. Persistence of a small amount of porosity in the HIPed single crystal has limited tensile testing to elevated temperatures (1073K) and even then, failure occurred elastically.

EXPERIMENTAL PROCEDURE

Cylinders (19 mm diameter) were machined from a 3-kg casting of nominal composition $Al_{66}Ti_{25}Mn_9$ and used for single crystal growth via a modified Bridgeman technique. Crystals (for tension tests only) were HIPed (1473K; 104 MPa; 4 hours) in an attempt to eliminate residual porosity. Crystal orientations were determined by Laue backscattered X-ray diffraction operating at 40kV.

Compression tests were conducted as a function of temperature from 77K to 1073K on electrodischarge machined and mechanically polished 3 x 3 x 6.5 mm specimens at an initial strain rate of 1.3 x 10^{-4} s^{-1} for compressive axis orientations near [001] and [$\bar{1}$11]. Slip bands were observed optically, and two-surface slip analyses were performed on the deformed specimens. Critical resolved shear stresses were calculated for the [$\bar{1}$01](111) slip system using the appropriate Schmid factors. Transmission electron microscopy (TEM) was employed to determine Burgers vectors of dislocations present in compression specimens deformed at 298K and 1073K; thin foils for TEM were prepared using 20% nitric acid in methanol solution at ~240K (12V, 60 mA). Additionally, a few tension specimens machined with wedge-shaped grip sections were tested at 1073K with specially-made Inconel grips.

RESULTS AND DISCUSSION

1. Single Crystal Microstructure

Several crystals were grown whose measured compositions were nominally in the single phase field described by the 1473K Al-Ti-Mn ternary isotherm [9]; however, a small amount of an interdendritic phase was present and could not be dissolved upon homogenization (Fig. 1a). Based on energy-dispersive X-ray analysis, this interdendritic phase was found to contain substantially more Mn than the $L1_2$ phase, with an approximate composition, 57 at.% Al, 39 at.% Mn and 3 at.%Ti. Compression tests in this study were performed on specimens obtained from a crystal of this type (crystal 1). This interdendritic phase, in small amounts, is not expected to significantly influence compressive yield strength measurements, whereas, in tension, this brittle phase forming a semi-continuous network would undoubtedly lead to premature failure. In an attempt to eliminate this interdendritic second phase, a small amount of Al-65 wt.% Ti master alloy was added to one of these crystals and remelted. The "new" crystal (crystal 2) thus obtained was devoid of the previously observed interdendritic phase. A thin-foil suitable for TEM examination was made to ensure that the microstructure was indeed truly single phase; however, fine Al_2Ti precipitates were observed (Fig. 1b), which are not frequently resolvable at the optical microscopy level. Differential thermal analysis (DTA) was performed at a rate of 10K per minute on crystal 1 and it did not reveal any sharp peaks indicative of precipitate dissolution upon heating, and only a small exothermic peak at ~1280K upon cooling. This is supportive of the fact that this crystal (crystal 1) did not contain the Al_2Ti phase; its

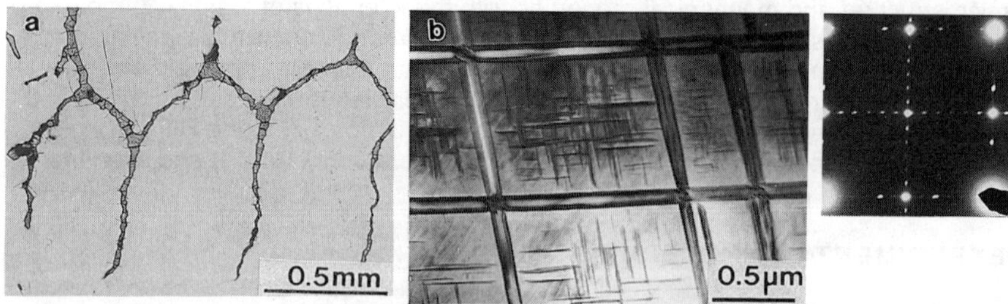

Fig. 1. Optical micrographs of (a) the Mn-enriched interdendritic phase, and (b) the Al_2Ti phase. $[100]_m$ zone axis diffraction pattern confirms the presence of Al_2Ti.

absence was subsequently confirmed by TEM. Thus, crystal 1 used in the compression studies contained the interdendritic phase but no Al_2Ti, whereas, crystal 2 contained Al_2Ti but no interdendritic phase. In an attempt to obtain a completely single-phase material, a first generation crystal (crystal 1) was re-melted, this time with a small amount of the Al-65 wt.% Ti master alloy and some pure Al. The resulting crystal (crystal 3) was examined optically. In contrast to crystal 1, a significantly reduced amount of the interdendritic phase was noted, no longer in the form of a semi-continuous network, but rather present as isolated precipitates. Crystal 3 was used for tensile studies. Thin foils of this crystal have not been examined to date in the TEM for verifying the absence of Al_2Ti.

2. Compression Testing

The compressive properties of single crystal $Al_{66}Ti_{25}Mn_9$ (crystal 1) as a function of temperature were examined in two orientations using compression axes near [001] and [$\bar{1}$11]. Yield stresses measured at 0.2% offset continuously decrease with increasing temperature (Fig. 2). This yield strength-temperature profile is similar to the behavior exhibited by other $L1_2$ trialuminides in both single crystal and polycrystal forms. Two-surface slip trace analysis at various temperatures showed the slip planes

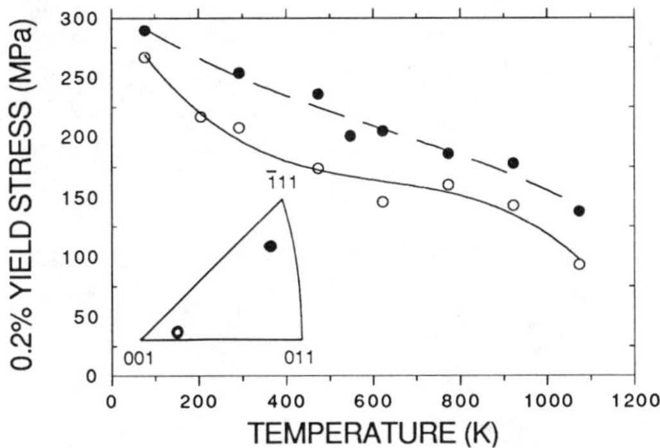

Fig. 2. Compressive yield strength variation with temperature for single crystal $Al_{66}Ti_{25}Mn_9$ intermetallic.

to be of the {111} type for all temperatures (Fig. 3). Schmid factors were obtained for the various possible slip systems for both orientations (near [001] and [$\bar{1}$11]), Table I. Critical resolved shear stresses for [$\bar{1}$01](111) slip were calculated, and the variation in CRSS with test temperature is given in Fig. 4.

Dislocation analysis was performed on samples deformed at 298K and 1073K. Burgers vectors were identified to be of the a<110> type. The **g • b** analysis for the 1073K specimen is shown in Fig. 5. Several dipoles were observed after deformation at both these temperatures.

3. Tension Testing

Preliminary tensile tests were conducted at 1073K on the assumption that microstructural flaws that were present (e.g. isolated interdendritic phases, unhealed pores) would be more readily tolerated at this temperature, thereby permitting

Table I. Schmid Factors for Compression Tests

Orientation	Slip System	Schmid Factor
Near [001]	[$\bar{1}$01](111)	0.485
	[101]($\bar{1}$11)	0.467
	[0$\bar{1}$1]($\bar{1}$11)	0.390
	[0$\bar{1}$1](111)	0.379
Near [$\bar{1}$11]	[$\bar{1}$01](111)	0.382
	[$\bar{1}$10](111)	0.334
	[011]($\bar{1}\bar{1}$1)	0.270

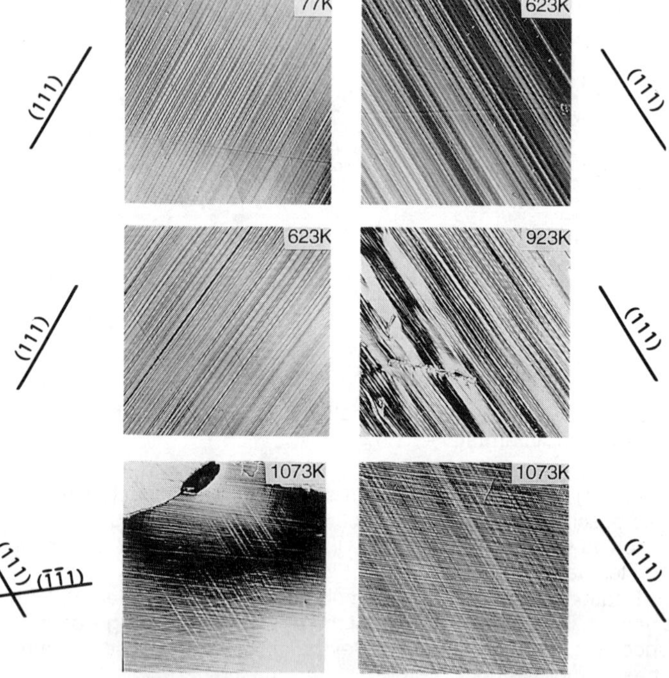

Fig. 3. Slip traces for compression specimens tested at various temperatures indicating {111}-type slip behavior.

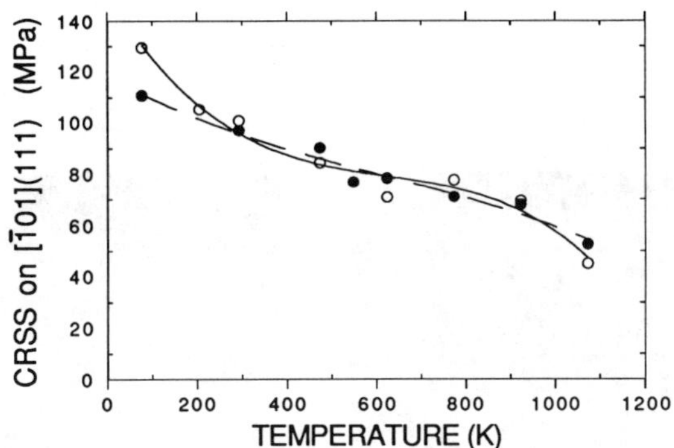

Fig. 4. Critical resolved shear stress variation as a function of temperature for orientations near [001] and [$\bar{1}$11] (same symbols as Fig. 2).

measurable ductility. The tensile axis of these specimens was within a few degrees of the [001] compression specimens' axis and thus, the Schmid factors for the two cases were similar. The best tensile test produced a fracture stress of 91 MPa with no measurable macroscopic plasticity. This tensile fracture stress is similar to the compressive yield stress for the material at 1073K in the same orientation, suggesting that the tensile specimen was close to yielding; however, slip lines were not observed on the polished faces of this specimen. Improved specimen quality is clearly mandatory, even for successful high temperature testing, reflecting the extreme flaw sensitivity of these materials.

Conclusions

Single crystal $Al_{66}Ti_{25}Mn_9$ shows decreasing compressive yield stress with temperature in both [001] and [$\bar{1}$11] orientations. CRSS variation with temperature is similar to that of other trialuminide crystals. Two-surface trace analysis together with **g • b** analysis showed the <110>{111} slip systems to be operative at all temperatures examined. Stringent requirements on crystal quality (e.g. flaws, second phase, etc.) make it extremely difficult to conduct reliable tensile tests, even at temperatures as high as 1073K.

Acknowledgement

This research is supported by AFOSR contract #F49620-91-C-0099 with Dr. Alan Rosenstein as program monitor. The authors gratefully acknowledge W. Romanow, R. Hsiao and the NSF MRL program through the LRSM at the University of Pennsylvania for the crystal growth facilities. Also, we acknowledge Drs. D. Dimiduk and D. B. Miracle from the Wright Materials Laboratory of the Wright Patterson Air Force Base for assistance with HIPing these crystals.

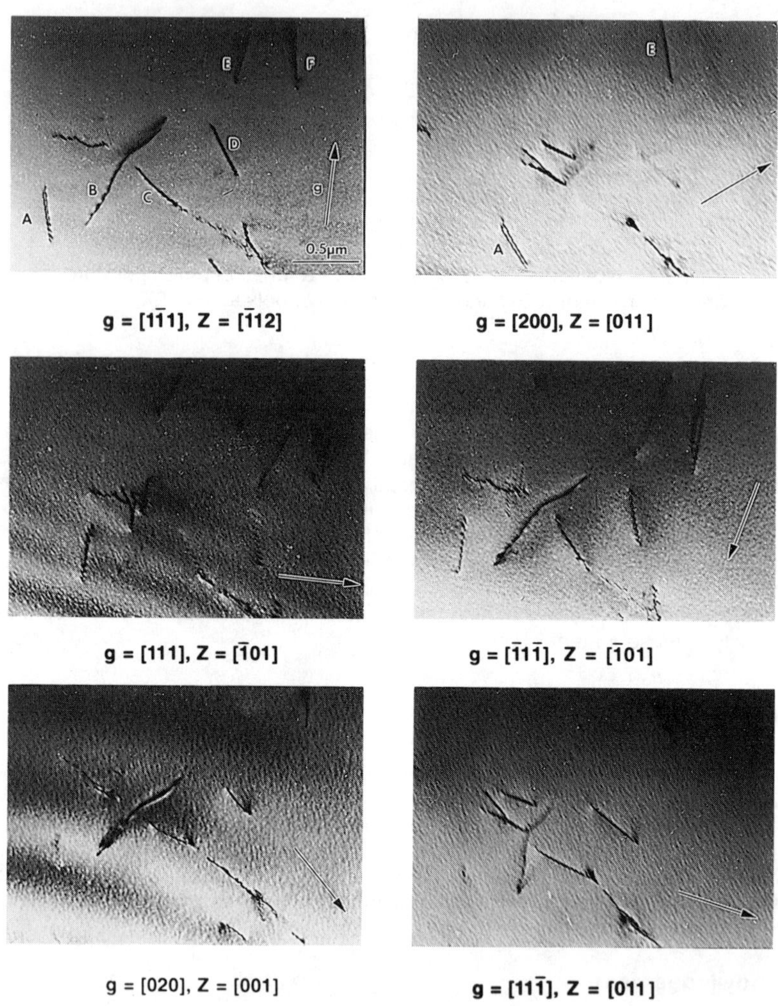

Fig. 5. Dislocation analysis for the 1073K specimen (compressive axis near [001]) using various two-beam imaging conditions. Dislocations B, C, D and F have **b** = [011] and A, E are of the **b** = [101] type.

References

1. E. P. George, D. P. Pope, C. L. Fu and J. H. Schneibel, ISIJ International **31**, 1063-1075 (1991).
2. K. S. Kumar and S. A. Brown, Scripta Metall. et Mater. **26**, 197-202 (1992).
3. L. Potez, A. Loiseau, S. Naka and G. Lapasset, J. Mater. Res. **7**, 876-882 (1992).
4. Z. Wu, Ph. D. Thesis, University of Pennsylvania (1992).
5. K. S. Kumar and S. A. Brown, Philos. Mag. A **65**, 91-109 (1992).
6. K. S. Kumar and S. A. Brown, Acta Metall. et Mater. **40**, 1923-1932 (1992).
7. J. H. Schneibel, H. A. Horton and W. D. Porter, Mater. Sci. and Eng. 1992, Special issue: Proceedings of the International Conference on High-Temperature Aluminides and Intermetallics held in San Diego, CA, Sept. 1991 - in press.
8. S. Zhang, J. P. Nic and D. E. Mikkola, Scripta Metall. et Mater. **24**, 57-62 (1990).
9. S. Zhang and D. E. Mikkola, Scripta Metall. et Mater. **26**, 1315-1320 (1992).

TRANSFORMATIONS, MICROSTRUCTURE AND PROPERTIES OF Nb-(10-16) At.% Al ALLOYS

Sen-Shan Yang* and Vijay K. Vasudevan*
*Department of Materials Science & Engineering, University of Cincinnati, Cincinnati,OH 45221

ABSTRACT

Transformations, microstructures and mechanical properties of Nb-(10-16) at.% Al alloys are reported. Cast buttons of the alloys were solutionized in the high temperature single phase Nb-Al region and cooled to retain supersaturated solid solutions (SS). These SSs are observed to show B2 ordering, with both the degree of ordering and hardness increasing with increase in Al. The precipitation of Nb_3Al from the supersaturated SS, which was studied as a function of temperature (1000-1600°C) and time (to 100h) by hardness measurements and microscopy, is sluggish, proceeds by initial heterogeneous nucleation at grain boundaries followed by growth into the grains in a discontinuous fashion, follows C-curve kinetics, and leads to significant hardening. The size, spacing and amount of the Nb_3Al precipitates depend strongly on Al content, temperature and time. The compressive yield strength of the SS increases with %Al and high values of 900 and 500 MPa at 25 and 900°C, respectively, are observed. The presence of Nb_3Al leads to a very significant increase in yield strength, with values being as high as 1550 and 1100 MPa at 25 and 900°C, respectively. Also, a strong dependence of strength on the size and spacing of the Nb_3Al precipitates is observed. These results are presented and discussed.

INTRODUCTION

Alloys based on the refractory intermetallic compound Nb_3Al have recently become of interest for applications at very high temperatures (1000-1600°C) in the new generation of aircraft propulsion systems [1]. This compound has an A15 structure, a density of 7.29 g/cm^3 and forms by a peritectic reaction from the liquid below 1960°C [2]. Moreover, it exists in a two-phase equilibrium with a bcc Nb-Al solid solution over a wide composition range between ~9 and 18 at.% Al [2]. Recent results have shown that monolithic Nb_3Al is quite strong and displays superior creep resistance above 1000°C, although it is very brittle at low temperatures [1]. With regard to the latter, it has been shown recently that improved toughness at ambient temperatures can be achieved by the presence of the Nb solid solution as a second phase [3]. The system offers significant opportunity for manipulating microstructure by controlling phase transformations in the solid state. However, apart from some recent work [3-7], limited studies of transformations and microstructure development have been reported. In this paper some results of a study of phase transformations, microstructures and mechanical properties in Nb-(10-16) at.% Al alloys are presented.

EXPERIMENTAL

Buttons of Nb-10, 13 and 16%Al (at.%) alloys were prepared by arc-melting high purity Nb and Al, with extra Al to compensate for evaporation losses. The actual compositions were determined to be 10.07, 13.12, and 15.99%Al, respectively, with less than 200 ppm weight each of O_2, N_2 and C. The cast buttons were solutionized in vacuum for 0.5h in the high temperature Nb-Al single phase field (1700°C for 10, 13%Al alloys; 1800°C for 16%Al alloy) and furnace cooled to obtain supersaturated SSs. These were aged in vacuum between 1000–1600°C for times from 0.25 to 100h. The microhardness of samples prepared from the heat treated alloys was measured. Microstructures were observed by optical (OM), scanning (SEM) and transmission electron microscopy (TEM), the latter in a Philips CM20 TEM operated at 200 kV. Both as-solutionized and selected aged samples (3x3x8mm) were deformed in compression to a permanent strain of ~5% between room temperature (RT) and 900°C under a strain rate of 10^{-4} s^{-1} and the 0.2% yield stress determined.

RESULTS

Effect of Temperature and Time on Microhardness and Microstructural Changes

Solution Treated Samples. OM observations of the solution treated 10, 13 and 16 at.% Al alloys indicated that a homogeneous, supersaturated SS was achieved in each case. A representative example is shown in Figure 1 for the 16%Al alloy. Also, a significant increase in microhardness is observed with an increase in Al content in the SSs. For example, compared with a hardness of 100 kg/mm^2 for pure Nb [2], addition of 10%Al increases the hardness to 259 kg/mm^2; the values increase further to 337 and 354 kg/mm^2 in the 13 and 16%Al alloys, respectively. TEM examination indicated that the as-solutionized samples had a single phase bcc structure. An example is shown in Figure 2 for the 16%Al alloy. Apart from the strong bcc reflections, another prominent feature visible in the [001] (Figure 2b) and [011] (Figure 2c) diffraction patterns is the presence of extra reflections at the {100} positions, suggesting the occurrence of B2 type ordering. A mottled contrast, presumably due to B2 domains, is also visible in Figure 2a. The B2 reflections were also observed in the 13 and 10%Al SS, except that their intensities were lower, i.e., the degree of B2 ordering decreases with decrease in Al.

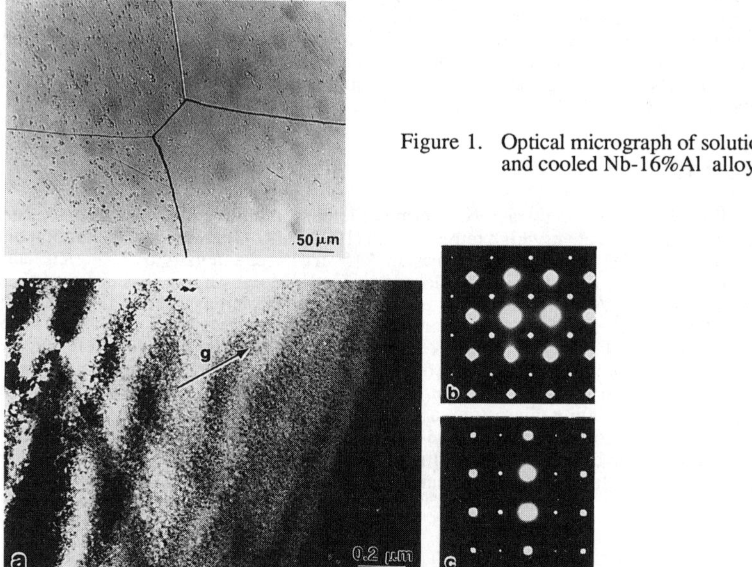

Figure 1. Optical micrograph of solution treated and cooled Nb-16%Al alloy.

Figure 2. TEM micrographs of solution treated and cooled Nb-16%Al alloy. (a) Bright field micrograph, (b), (c) are respectively [001] and [011] diffraction patterns.

Aged Samples. Samples of the solutionized 10%Al alloy were aged at 800, 1000 and 1200°C. At 800°C, no change in hardness from the as-solutionized value was observed even upto 100h, and correspondingly no Nb_3Al precipitates were observed. Similar behavior was noted on aging at 1000°C to 24h. At 100h, the hardness had increased to 284 kg/mm^2 and this change could be associated with the precipitation of Nb_3Al. At 1200°C, no appreciable increase in hardness was observed to 24h, although after 100h the hardness increased to 310 kg/mm^2 and this change could again be associated with the formation of plate shaped Nb_3Al precipitates at the grain boundaries.

Samples of the solutionized 13%Al alloy were aged at 1000, 1400 and 1600°C. At 1000°C, no change in either microstructure or hardness was observed to 24h. At 100h, OM showed the presence of plate-shaped Nb_3Al precipitates emanating from the grain boundaries and occupying a small region around them. The hardness of the transformed regions had increased to 428 kg/mm^2. The kinetics was faster at 1400°C. Although no changes in either hardness or microstructure was observed at 0.25h, at 10h the hardness had increased to 402 kg/mm^2. This change could be associated with plates of Nb_3Al at and near grain boundaries. At 24h, the hardness did not change appreciably, but the amount of Nb_3Al had increased and its distribution had become more homogeneous (Figure 3(a, b)). In Figure 3b, the dark and light regions correspond to Nb_3Al and Nb-Al SS, respectively. At 1600°C, very few Nb_3Al precipitates were observed and the hardness was unchanged, because of the close proximity of this temperature to the solvus [2].

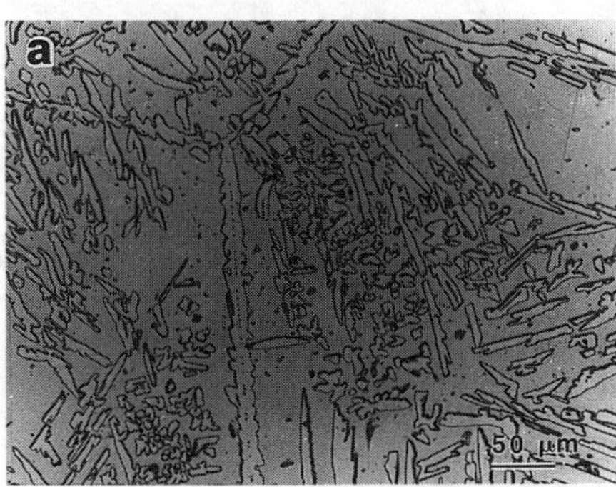

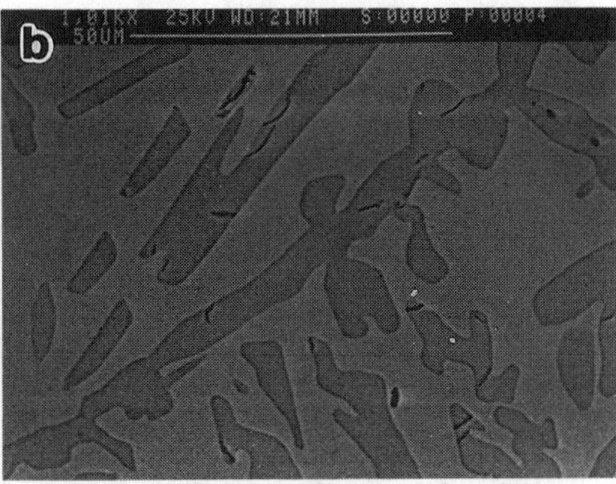

Figure 3. Nb-13%Al alloy solution treated and aged at 1400°C for 24h. (a) Optical and (b) BSE micrographs.

Samples of the 16%Al alloy were also aged at 1000, 1400 and 1600°C. Compared with the 10 and 13%Al alloys, both the kinetics and amount of precipitation of Nb_3Al were much higher in this alloy. At 1000°C, fine-scale, plate-shaped Nb_3Al precipitates were observed at and around grain boundaries within 24h (Figure 4(a,b)). At 100h, Nb_3Al precipitates had formed homogeneously throughout the grains (Figure 4(c,d) and the hardness had increased appreciably to 550 kg/mm^2. The kinetics of precipitation was faster at 1400°C than at either 1000 or 1600°C. Within 15 mins, considerable precipitation of Nb_3Al had occurred at and around grain boundaries, whereas at 1600°C, the Nb_3Al precipitates were present only at grain boundaries and at 1000°C none were observed. After 10h at 1400°C, precipitation of Nb_3Al was practically complete, and the only change on further aging to 24h was thickening of the plates (Figure 5a); the hardnesses for the 10 and 24h aging were high, viz, 548 and 555 kg/mm^2. The microstructure following aging at 1600°C for 24h is shown in Figure 5b. One notable feature is the rapid thickening of the precipitates.

Mechanical Properties of Solution Treated and Representative Aged Materials

The RT yield strengths of the as-solutionized alloys are quite high and increase with an increase in Al. Compared with the value of 200 MPa for pure Nb, the values for the 10, 13 and 16%Al alloys were 700, 800 and 900 MPa, respectively. Also, a modest drop in strength between RT and 350°C was seen in all cases. Beyond 350°C, the rate of decrease was not as marked, with the relative difference between the values for different Al contents at any given temperature being about the same and values as high as 500 MPa being attained at 900°C [8]. Representative micrographs of samples deformed at RT and 350°C are shown in Figures 6a–b. One notable feature is the presence of twins following RT deformation and their absence at higher temperatures. The occurrence of twinning on {112} planes was established by TEM [8]. Similar behavior was noted in the lower Al alloys, except that in the 10%Al alloy twins were observed only at large strains. Slip by **b**=1/2<111> dislocations was also observed [8].

To evaluate the effect of Nb_3Al precipitates on strength, samples of the 16%Al alloy aged at 1000°C, 100h and 1400°C, 24h were tested in compression. The results, including those for the as-solutionized samples, are presented in Figure 7. The volume fraction of Nb_3Al precipitates was 62% in both cases, with the main difference being in the average size/width of the precipitates (0.5 vs 3.3 µm for the 1000 and 1400°C aging). As can be seen in Figure 7, the presence of Nb_3Al precipitates leads to a dramatic increase in strength at all test temperatures compared with the supersaturated SS. For instance, RT strength increases from 900 MPa in the as-solutionized condition to about 1260 and 1540 MPa for the 1400°C,24h and 1000°C,100h aged samples, respectively. Although strength decreases at high temperatures in these aged materials, the absolute values are quite high—at 900°C these were ~980 and 1090 MPa. Also, the samples aged at 1000°C are much stronger than those aged at 1400°C at all test temperatures. SEM imaging of deformed two-phase samples showed the presence of cracks confined within the Nb_3Al plates, indicating that the SS regions around arrest their propagation and hence could have a potential toughening effect.

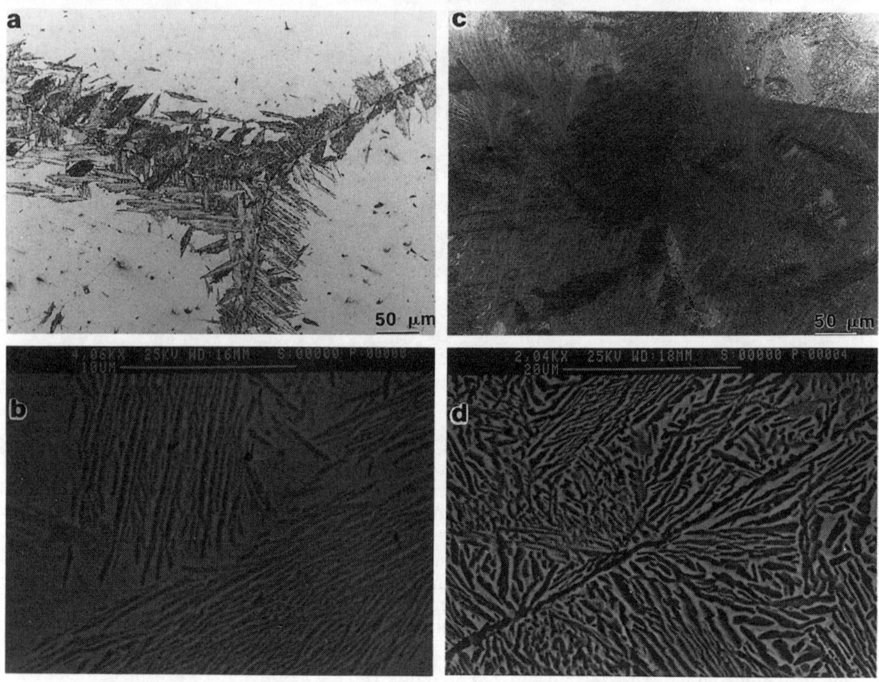

Figure 4. Nb-16%Al alloy solution treated and aged at 1000°C. (a) Optical, (b) BSE micrographs, 24h aging; (c) optical, (d) BSE micrographs, 100h aging.

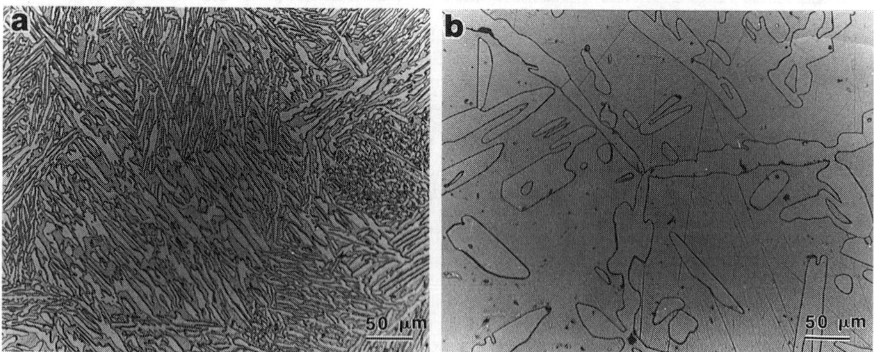

Figure 5. Optical micrographs of solution treated and aged Nb-16%Al alloy. (a) 1400°C, 24h. (b) 1600°C, 24h.

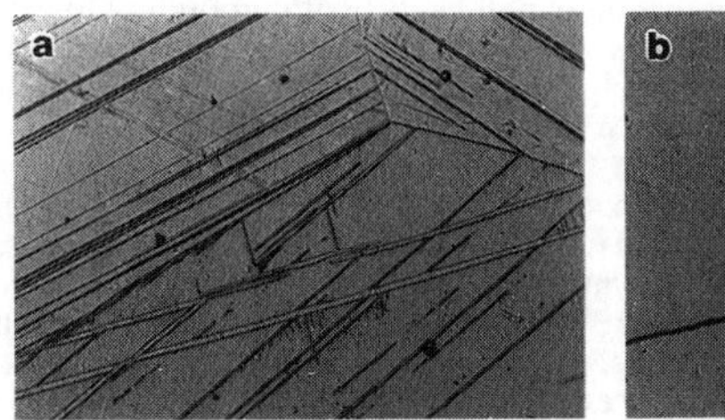

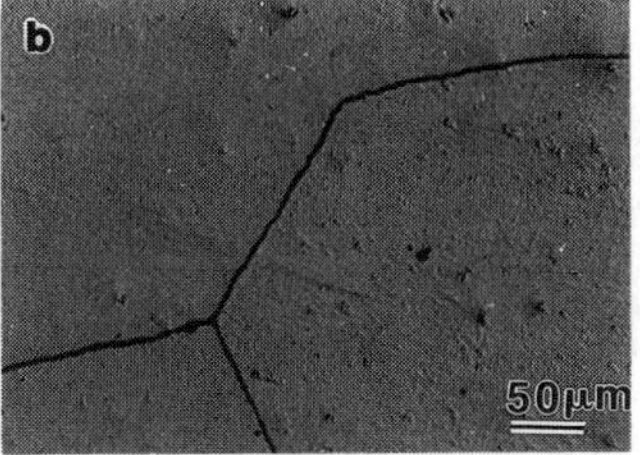

Figure 6. Optical micrographs of Nb-16%Al SS alloy deformed at (a) RT and (b) 350°C.

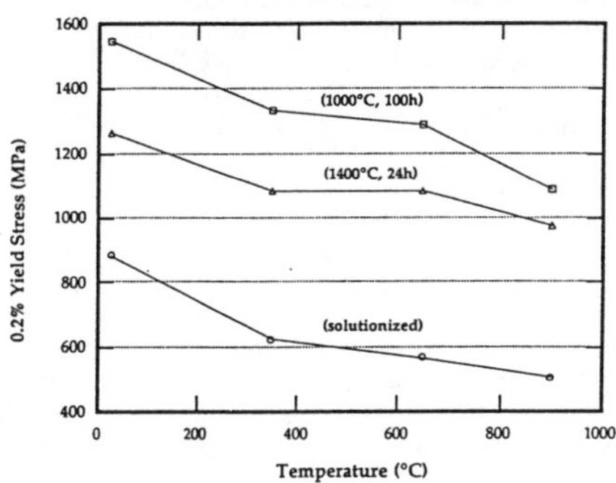

Figure 7. 0.2% yield strength vs temperature for solutionized and aged Nb-16%Al alloy.

DISCUSSION

The results presented above have shown that it is possible to successfully produce homogeneous, supersaturated SSs of Nb-Al alloys by simple solutionization in the high temperature Nb-rich single phase field followed by even slow cooling. Both the high strengths of the SSs and the increase with increase in Al are thought to be due to SS strengthening, with differences in elastic moduli between the Nb and Al atoms and electronic effects rather than atom size difference providing the major contribution [4]. Another important point is that supersaturated SSs show B2 ordering, with the degree of ordering increasing with increase in Al. This type of ordering was not detected in earlier studies on cast binary alloys [2,3], but has been reported recently in rapidly solidified binary alloys [4,5], cast binary Nb-Al and ternary Nb-Al-Ti alloys [6,7]. The present results suggest that a second order bcc to B2 transformation occurs on cooling from the single phase Nb-Al SS region; the exact nature of the B2 phase and associated bcc-B2 phase boundaries require investigation. It is likely that apart from the SS effects discussed above, the presence of B2 ordering also contributes to the strengthening, although the large increase in strength at the 10%Al level (compared with pure Nb), where the B2 ordering is quite weak, suggests that this contribution is not large. Although detailed results are reported elsewhere [8], deformation at room temperature of the SSs is observed to proceed by twinning on {112} bcc planes and slip of 1/2<111> dislocations. The extent of twinning increases, and ductility decreases, with increase in Al content [8]. The temperature dependence of strength of the SSs is similar to other bcc/B2 alloys in that there is modest decrease between 25 and 350°C and a more gradual decrease at higher temperatures.

The present results have also shown that the supersaturated SSs can be aged to cause precipitation of Nb_3Al. The nucleation of Nb_3Al first occurs uniformly at the grain boundaries, and then new nuclei form at locations away within grains and grow to consume the matrix. The final microstructure appears as alternate plates/lamellae of Nb_3Al with the intervening regions composed of the Nb SS with a lower Al content, near that predicted by the phase diagram. The fact that general intragrain precipitation is sluggish and does not occur in the early stages implies that there is significant free energy barrier to homogeneous nucleation. This is most likely associated with the complex A15 structure of Nb_3Al compared with the simple bcc structure of the Nb-Al SS and the large difference in lattice parameters between the two phases (5.19Å in former vs 3.28Å in latter).

Consequently, it is reasonable to expect that there will be difficulty in finding low index crystallographic matching between planes and directions in the two phases and that the nuclei will be incoherent. The oriented appearance of Nb_3Al precipitates may be associated with growth along certain preferred planes/directions to minimize surface energy. The predominance of initial precipitation at grain boundaries can also be understood on the above basis, since it is well known that grain surfaces can lower the critical free energy of nucleation and thus provide a substrate for nucleation.

Apart from these features, the precipitation of Nb_3Al is observed to be a strong function of Al content and temperature. Both the amount and rate of precipitation at a given temperature increase with an increase in Al content, primarily because of the associated increase in the degree of supersaturation and hence driving force. Also, the precipitation appears to show C-curve kinetics, as expected for nucleation and growth processes because of the competing temperature dependence of driving force and diffusion. For example, the nose of the TTT curve for the 16%Al alloy appears to be around 1400°C, well below the equilibrium solvus temperature of ~1700°C [2]. A detailed discussion of the mechanisms and kinetics of Nb_3Al precipitation is given elsewhere [8].

As to the mechanical properties of the aged alloys, the results show that the strength of two-phase structures is substantially higher than those of the supersaturated SSs. The increase in strength in these can be attributed to the high hardness and strength of Nb_3Al [1-3], and the higher strength of the higher Al alloys can be associated with a larger volume fraction of Nb_3Al. A notable point is that samples of the 16%Al alloy aged at 1000°C for 100h showed significantly higher strengths at all test temperatures used than those aged at 1400°C for 24h. Although the volume fraction of Nb_3Al in these two samples is about the same (~62%), there is appreciable difference in size (0.5 vs 3.3 μm) and spacing. This behavior is difficult to explain from a simple rule-of-mixtures standpoint. Thus, the difference in strength of these two microstructures must be associated with the difference in size or, more importantly, spacing of the Nb_3Al precipitates. It is quite likely that in the two-phase microstructure, at room temperature and even at 900°C, the brittle Nb_3Al phase will not undergo any plastic deformation, so that most of the deformation occurs in the Nb solid solution. Thus, when the spacing of the Nb_3Al precipitates is small, the slip lengths for dislocations moving in the softer SS in between the precipitates will be small. This may result in pile-ups at the interfaces and increased work hardening, which could lead to higher strengths. When the spacing of the Nb_3Al precipitates is larger, the dislocations in the SS will be able to glide more easily over longer distances and accordingly strength is lower. The decrease in strength at higher temperatures may be associated with: 1) decrease in the elastic modulus of Nb_3Al [1], and 2) decrease in the strength of the Nb-Al SS regions due to thermally activated dislocation motion.

CONCLUSIONS

In summary, the results of the present study have brought to light a number of interesting features of transformations and properties of Nb-(10-16)%Al. The results indicate that supersaturated SSs of Nb-Al alloys can be produced by simple heat treatment. These SSs show B2 type ordering and high strength. Both the degree of order and hardness/strength increase with an increase in Al content. Aging of the supersaturated SSs leads to substantial increase in hardness due to the formation of plate-shaped precipitates of Nb_3Al. The precipitation depends strongly on Al content and aging temperature, initially commences at grain boundaries and appears to show C-curve kinetics. The presence of Nb_3Al precipitates also leads to enormous increase in strength at ambient and high temperatures, with the degree of strengthening being a strong function of volume fraction, size, and more importantly, spacing of the precipitates. Finally, the results of the transformation characteristics in this system suggest that significant potential exists for thermo-mechanical processing and manipulation of microstructure for improved properties.

REFERENCES

1. D. L. Anton and D. M. Shah, MRS Symp. Proc., 133, 361 (1989).
2. C. E. Lundin and A. S. Yamamoto, Trans. TMS-AIME, 236, 863 (1966).
3. D. L. Anton and D. M. Shah, MRS Symp. Proc., 194, 45 (1990).
4. T. N. Marieb et al., MRS Symp. Proc., 213, 329 (1991).
5. M. Aindow, J. Shyue, T. A. Gasper and H. L. Fraser, Phil. Mag. Lett., 64, 59 (1991),
6. J. Shyue and H. L. Fraser, unpublished research (1992).
7. E.S.K. Menon, P.R. Subramanian and D.M. Dimiduk, Scripta Met. Mater., 27, 265 (1992).
8. S. S. Yang and V. K. Vasudevan, J. Mater. Res., to be submitted (1992).

MECHANICAL PROPERTIES AND DISLOCATION STRUCTURES IN TiAl ALLOYS WITH VARYING ALUMINUM CONTENTS

S. Sriram*, Vijay K. Vasudevan* and Dennis M. Dimiduk**.
*Dept. of Materials Science and Engineering, University of Cincinnati, Cincinnati, OH, 45221
**Wright Laboratory, Materials Directorate, Wright-Patterson AFB, Dayton, OH, 45433.

ABSTRACT

Binary, coarse-grained polycrystalline Ti-48, 50 and 52 Al (in at.%) alloys, containing low (~250 wt.ppm) levels of interstitials (O+N) have been deformed at various temperatures in compression and four-point bending. The 0.2% proof strength-temperature profiles are observed to comprise of three distinct regimes. Differences in the material response between bending and compression modes of deformation have been observed. Importantly, in both cases a flow stress anomaly is observed in the 50 and 52 Al alloys. Dislocation fine structures of samples deformed in compression, have been observed in the TEM. Analyses suggests the possible influence of Al contents on the line directions of dislocations, superdislocation dissociation modes, SISF dissociation distances and hence planar fault energies. Distinct differences in the deformation structures at RT and at the flow stress peak temperature (800°C) have been observed. These results are presented and discussed in relation with the observed mechanical properties.

INTRODUCTION

There have been several studies [1-5] aimed at understanding the fundamental deformation behavior of TiAl as a function of temperature. The initial observation of a flow stress anomaly in single crystals of Ti-56Al [5] has been followed by observations of the same in polycrystalline binary alloys [6-8]. The presence of the flow anomaly in the latter has been suggested to be dependent on the grain size [6,7], the heat treatment temperature and Al content [8]. On the basis of dislocation fine structure investigations [3,4] it has been suggested that the flow anomaly is due to thermally activated cross slip of the <011] superdislocations, similar to the Kear Wilsdorf (K-W) locking mechanism which occurs in the $L1_2$ alloy Ni_3Al [9]. However, from a fundamental structural point of view, there is still a lack of understanding of the deformation characteristics of these materials as a function of (i) Al content, (ii) interstitial content and (iii) grain size, over a range of deformation temperatures. Hence a systematic study was undertaken to understand the deformation behavior of polycrystalline TiAl as a function of (i) Al content and (ii) interstitial content. Alloys containing 48, 50 and 52 Al, each with low (~250ppm) and nominal (~1000ppm) interstitial (O+N) levels have been deformed at temperatures from RT-1000°C in four-point bending and in compression, and the fine structure of dislocations in deformed samples been observed. In this paper we report some results on low interstitial content alloys.

EXPERIMENTAL

Binary Ti-48, 50 and 52 Al alloy buttons were prepared by vacuum arc-melting of high purity electrolytic Ti sponge and Al ingots. The cast buttons were given a 1300°C, 25h+1150°C, 50h+1050°C, 2h heat treatment in vacuum to homogenize the alloys and to obtain a coarse grain size (48Al-150 μm; 50Al-300 μm; 52Al-500 μm). The analyzed Al contents in the alloys were very close to the aimed values and the O+N contents were ~250 wt. ppm in each. Compression (5x5x12.5 mm) and four-point bend (3x5x25 mm) samples were deformed in air (to a plastic strain of ~1-1.5%) and in vacuum (to failure), respectively, at temperatures from room (RT) to 1000°C under a strain rate of 1.7×10^{-4} s^{-1}. Thin foils from deformed compression samples were observed in a Philips CM20 TEM operated at 200kV by weak beam dark field (WBDF) imaging.

RESULTS

The results of the 0.2% proof stress vs temperature for the alloys tested in compression are shown in Figure 1, whereas those of the bend proof stress and maximum outer fiber strain vs temperature are shown in Figures 2a and 2b, respectively. Several important observations can be made. Firstly, in both compression and bending, the proof stress values at all temperatures are the highest for the 48 Al alloy followed by 52 Al and finally the 50 Al alloy. Secondly, for both 50 and 52 Al alloys, the compressive strength (Figure 1) appears to decrease from RT to 300°C. Thirdly, between 300 and 600°C the compressive proof stress remains roughly the same in 50 Al, whereas in the 52 Al alloy an increase is seen. In contrast with the behavior in compression, the bend proof stress (Figure 2a) does not show the characteristic minimum between RT and intermediate temperatures. Further, between RT and 600°C, the strength either remains the same (in 50 Al) or increases marginally (in 52 Al). Lastly, in both the 50 and 52 Al alloys, an increase in proof stress occurs between 600 and 800°C in both compression and bending, followed by a drop above 800°C. The increase in strength between 600 and 800°C is most pronounced for the 50Al alloy. It should be emphasized that maximum strength in compression and bending occurs at 800°C in both 50 and 52 Al alloys. In the 48Al alloy, the strength remains high and relatively constant up to 800°C with no indication of a maximum at 800°C as in the other two alloys; beyond 800°C, the strength drops appreciably. The high strength of the 48Al alloy is most likely associated with microstructure, viz, the finer grain size and presence of lamellar grains.

The bend ductility vs temperature plot, Figure 2b, shows that the ductilities for the 48 Al alloy are highest followed by 50 Al and lastly 52 Al alloy. In all the alloys, the ductility values either remained relatively constant or increased slightly between RT and 800°C, followed by a significant increase at temperatures above 800°C, especially in the 48 and 50 Al alloys. In the 52 Al alloy, even at 1000°C the ductility is only ~1.0%.

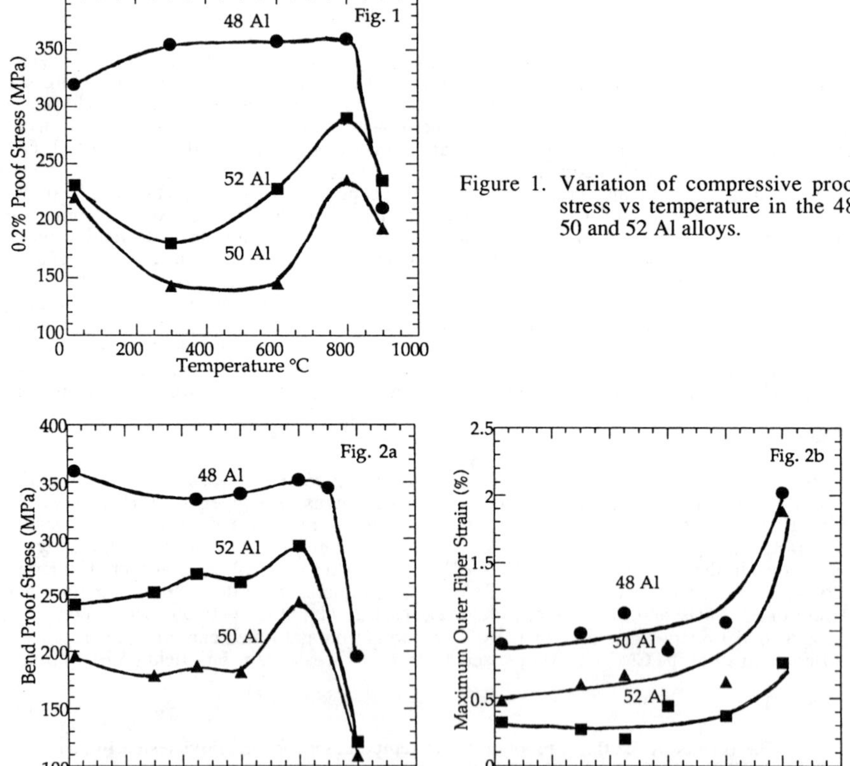

Figure 1. Variation of compressive proof stress vs temperature in the 48, 50 and 52 Al alloys.

Figure 2. Variation of (a) bend proof stress and (b) bend ductility vs temperature in the 48, 50 and 52 Al alloys.

TEM examination showed that following deformation at RT all dislocations, namely, 1/2<110], <011] and 1/2<112] are present in the 50 and 52Al alloys. Only 1/2<110] dislocations and twinning were observed in the 48Al alloy. Twinning was also observed in the Ti-50Al alloy at all tempertures, but only at high temperatures in the 52 Al alloy. The 1/2<110] dislocations appeared predominantly as long, relatively straight screw segments [10]. The characteristic line directions of the various superdislocation segments are shown in Table I. Figures 3a and 3b, corresponding to <011] superdislocations in the 50 and 52 Al alloys, respectively, clearly show a two-fold dissociation of the superdislocation. It is generally observed that the screw portions are relatively straight (segments A), and the curved portions are non-screw oriented (segments B and C). Figures 3c and 3d show the 1/2<112] superdislocation dissociated into a triplet. Following deformation at 800°C, the 1/2<112] dislocations are not present. The structure comprises of 1/2<110] and <011] dislocations only. Figures 4a and 4b show <011] superdislocations in 50 and 52 Al alloys, respectively, again showing the characteristic two-fold dissociation. In general, the 1/2<110] dislocations appeared more curved at this temperature.

DISCUSSION

Mechanical Properties

Referring to Figures 1 and 2a, it is perhaps convenient to distinguish three temperature regimes in the proof stress- temperature curves: (I). RT-300°C, (II). 300-800°C and (III). above 800°C. The reason for the decrease in the compressive proof stress in regime I, in contrast to no such decrease in bending (which has a tensile force component) is not clear. One possibility is that subtle microstructural and grain orientation effects are responsible. Alternatively, it is perhaps indicative of the existence of a tension/compression asymmetry in the material. Chu and Thompson [11] have reported a large difference in RT tensile and compressive strength in a Ti-50Al alloy. The possibility of existence of a tension/compression asymmetry in TiAl has been suggested by Hazzledine and Sun [12] and related to the fully dissociated configuration of the <011] dislocations, which involve planar CSF's and SISF's. They have pointed out that the large energy difference between these planar faults would lead to cross slip of only one of the (CSF-coupled) partials, and hence the behavior of the superdislocation is expected to be very different when the glide direction is reversed. Whether a true tension/compression asymmetry exists in these materials can only be established positively by experimental studies in single crystals.

In both the test modes, the proof strength in regime II between 300 and 600°C reveals either a small increase (for 52Al) or remains more or less constant (for 50 Al), followed by the flow anomaly between 600 and 800°C. The higher Al alloy has a higher strength over the entire temperature range. There are several factors to be considered as being responsible for this. Firstly, it is possible that the higher "effective" interstitials present in 52 Al leads to greater solute hardening ('effective' here implying that the small amount of α_2 present in 50 Al depletes the matrix γ of interstitial elements and thereby decreases their influence). A second contribution comes from point defect hardening by excess Al atoms. A final possibility is that the higher Al content reduces the planar fault energies (experimentally verified here), thus necessitating a higher stress to recombine the partials before any cross slip event could occur and leading to a higher strength value. The existence of the flow stress anomaly in coarse-grained, single phase TiAl is in agreement with other earlier reports [6,7].

Dislocation Structures

Table I shows experimentally determined dislocation line directions. Upon deformation at RT, the 1/2<110], <011] and 1/2<112] dislocations have been observed, as mentioned earlier. Also shown in Table II are experimentally measured, but uncorrected, SISF dissociation distances. The results clearly indicate that two effects of Al content need to be considered: (a) on dislocation line directions and (b) on superdislocation dissociations and planar fault distances/energies and hence their concomitant influence on the flow properties

Regarding (a), comparing the present results (Table I) with earlier results of Hug et al. [3,4] on polycrystalline Ti-54 Al alloy and considering predictions of choice of line directions due to Court et al. [13] and Woodward et al. [14], an interesting trend emerges. With increasing Al contents, there is an increasing propensity for the dislocations to lie in directions along which Peierls forces may be significant.. This implies that as Al content increases, the directionality of interatomic bonding increases thereby dictating the dislocation line direction. For example, in Ti-

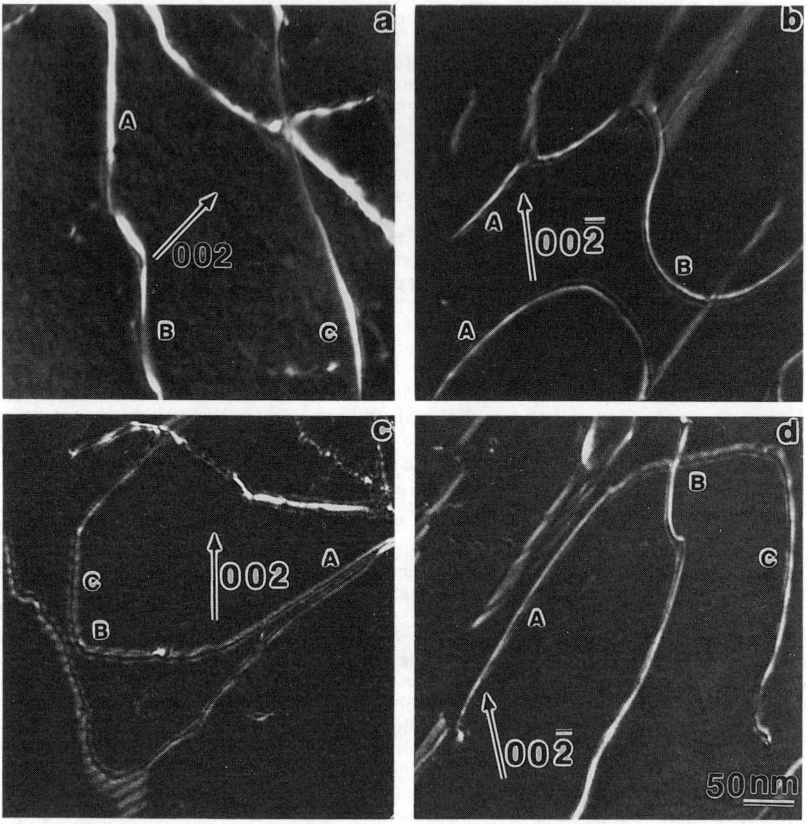

Figure 3. WBDF micrographs of dislocation structures observed at RT. (a), (b) <011] dislocations in 50 and 52Al, respt.; (c), (d) 1/2<112] dislocations in 50 and 52Al, respt.

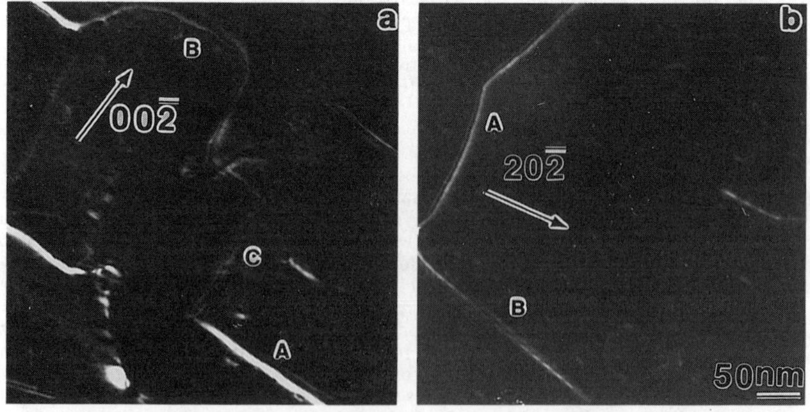

Figure 4. WBDF micrographs of <011] dislocations observed at 800°C. (a) 50Al and (b) 52Al.

50Al, <011] screw segments have been identified (direction along which the Peierls forces have no effect [13,14]), whereas Hug et al. [4] reported that they could identify only the ±60°segments off screw orientation (direction along which Peierls forces maybe operational). Further, the presence of screw segments in 50Al at RT would imply that these segments are favorably oriented to undergo cube cross-slip at elevated temperatures, thereby readily forming the equivalent of the K-W lock. Perhaps, this explains the pronounced increase in the strength in 50 Al between 600 and 800°C (Figure 1). However after deformation at 800°C, line directions in 50 Al are not in the screw orientation (Table I). The results suggest that apart from the cube-cross slip process line directions are also affected at elevated temperatures. Detailed TEM work is underway to address these issues and to obtain better statistics of the character of dislocations.

Regarding (b) it is seen that as the Al content increases, the SISF dissociation distance also increases (Table II). Comparing these values with those for Ti-54 Al [4] indicates that the fault energies decrease with an increase in Al content. These results are consistent with the calculations of Woodward et al. [14], who reported that both APB and SISF energies decrease (i.e., implying an increase in the fault width) with an increase in Al content, being highest at the 50Al composition. Furthermore, Woodward et al. [14] from their computations have shown that directional bonding can lead to an APB energy higher than the CSF energy. However, it has not been possible to confirm this in the present study, since the entire dissociation involving all the faults and component partial dislocations could not be resolved (discussed below). Lastly, kink-like features are also observed along the dislocation lines, as seen for example in the 50 Al alloy (Figure 3a). Details of these features and their role is the subject of ongoing research.

TABLE I. Experimentally Determined Dislocation Line Directions

Alloy	Deformation Temp., °C	Burgers Vector	Dislocation Line Directions*		
			Segment A	Segment B	Segment C
52 Al	RT	[011]	[011] / (Screw)	~$[21\bar{1}]$ / (Edge)	------
52 Al	RT	[112]	$[0\bar{1}\bar{1}]$ / (-30°)	$[1\bar{1}0]$ / (Edge)	~[211] /(73°)
50 Al	RT	[101]	[101] / (Screw)	[101] / (Screw)	[101] / (Screw)
50 Al	RT	$[11\bar{2}]$	~$[10\bar{1}]$ / (~30°)	$[1\bar{1}2]$ / (~-48°)	~$[10\bar{1}]$ / (~30°)
52 Al	800	$[\bar{1}01]$	$[\bar{1}01]$ / (Screw)	[112] / (~73°)	------
50 Al	800	$[\bar{1}01]$	[110] /(-60°)	[211] / (~-73°)	$[1\bar{1}2]$ / (30°)

*Segment characters are shown in brackets

TABLE II. Measured SISF Dissociation Distances

Alloy/ Deformation Temp.	<011] Dislocation (Screw) (nm)	1/2<112] Dislocation (Edge) (nm)
50 Al / RT	2.8	2.7
52Al / RT	3.3	3.2
52Al / 800°C	4.5	------

Dislocation Dissociations

<011] superdislocations. Contrast analysis of the <011] dislocation in both 50 and 52 Al, following deformation at RT and 800°C indicates a SISF- coupled two-fold dissociation. This is seen in Figures 3 (a,b) and 4(a,b). Analysis suggests that the dissociation is as follows :

$$<011] \rightarrow 1/6<\bar{1}12] + SISF + 1/6<121] + APB + 1/6<\bar{1}12] + CSF + 1/6<121] \quad (1)$$

Since the APB and the CSF energies are presumably higher compared with the higher Al alloys [14], only the outer 1/6<112] partial is visible, with the SISF separating the remaining dissociation. The observation of an increase in the SISF spacing following deformation at 800°C compared with that at room temperature suggests that the <011] superdislocation has undergone a non-planar sessile core transformation, producing a configuration similar to the K-W lock in $L1_2$ alloys. This is because after deformation at 800°C and cooling to room temperature, the superpartial dissociation distance is expected to reflect the "equilibrium" value at the temperature of observation, namely, room temperature. It appears therefore, that these superpartials are not relaxing to their equilibrium spacings, because a cross slip event/core transformation has occurred, so that they do not lie on the same slip plane. Perhaps, this process is also responsible for the pronounced increase in strength between 600 and 800°C in the 50 and 52Al alloys. Detailed TEM experiments are in progress to fully characterize this configuration.

1/2<112] superdislocations. These dislocations have been observed in both 50 Al (Figure 4a) and 52 Al (Figure 4b) alloys after deformation at RT, but they disappear following deformation at 800°C. Contrast analyses is consistent with the following dissociation scheme:

$$1/2\langle\bar{1}12] \rightarrow 1/6\langle\bar{1}12] + \text{SISF} + 1/6\langle\bar{1}12] + \text{SESF} + 1/6\langle\bar{1}12] \qquad (2)$$

It appears that the above dissociation path is favored over the

$$1/2\langle\bar{1}12] \rightarrow 1/6\langle\bar{1}12] + \text{SISF} + 1/6\langle 121] + \text{APB} + 1/2\langle\bar{1}01] \qquad (3)$$

dissociation path reported earlier by Hug et al. [3,4] because the lower Al content in the present alloys tends to increase the APB energy such that scheme (3) becomes unstable and an alternative, stable configuration would be according to (2). This configuration is the same as that reported in a Ti-54 Al alloy containing Mn by Hug et al. [15], where it has been proposed that the Mn addition results in a decrease in the stacking fault energy .

CONCLUSIONS

The mechanical properties of high purity TiAl alloys containing 48, 50 and 52 at.% Al respectively, have been determined in both bending and compression over a range of temperatures. It is found that the strength values can be described over three temperature regimes, particularly in the 50 and 52 Al alloys. Differences in the strength-temperature curves between the two modes of deformation has been observed. It is also seen that in the 50 and 52 Al alloys a flow stress anomaly exists. Dislocation line directions, SISF dissociation distances and the dissociation schemes have been shown to be sensitive to alloy compositions. Finally, detailed WB-TEM work is underway to establish the mechanisms responsible for the overall strength-temperature characteristics of these alloys as a function of Al and interstitial contents.

ACKNOWLEDGEMENTS

The authors would like to thank the Univ. Research Council of the Univ. of Cincinnati, the Dept. of Mat. Sci. and Engg., Univ.of Cincinnati and the Materials Directorate, Wright Laboratory, Wright-Patterson AFB, Dayton, for supporting this work.

REFERENCES

1. D. Shechtman, M. J. Blackburn and H. A. Lipsitt, Metall. Trans., 5A, 1373 (1974).
2. H. A. Lipsitt, D. Shechtman and R. E. Schafrik, Metall. Trans., 6A, 1991 (1975).
3. G. Hug, A. Loiseau and P. Veyssiere, Phil. Mag., 54, 47 (1986).
4. G. Hug, A. Loiseau and P. Veyssiere, Phil. Mag., 57, 499 (1988).
5. T. Kawabata, T. Kanai and O. Izumi, Acta Metall., 33, 1355 (1985).
6. V. K. Vasudevan, S. A. Court, P. Kurath and H. L. Fraser, Scripta Metall., 23, 467 (1989).
7. P. Prasad Rao and K. Tangri, Mat .Sci. Engg., A132, 49 (1991).
8. S. C. Huang and E. L. Hall, Scripta Met et Mat, 25, 1805 (1991).
9. D. P. Pope and S. S. Ezz, Int. Met. Rev., 29, 136 (1984).
10. S. Sriram, V. K. Vasudevan and D. M. Dimiduk, Mat. Res. Soc. Symp., 213, 375 (1991).
11. W. U. Yang Chu and A. W. Thompson, Scripta Met. 25, 641 (1991).
12. P. M. Hazzledine and Y. Q. Sun, Mat. Res. Soc. Symp., 213, 209 (1991).
13. S. A. Court, V. K. Vasudevan and H. L. Fraser, Phil. Mag., 61, 141 (1990).
14. C. Woodward, J. M. MacLaren and S. Rao, J.Mat. Res., 7 , 1735 (1992).
15. G. Hug and P. Veyssiere, in: Proc. Int. Symp. of Electron Microscopy in Plasticity and Fracture Research of Materials (1989).

LOW CYCLE FATIGUE BEHAVIOR OF POLYCRYSTALLINE NiAl AT 1000 K

R.D. Noebe and B.A. Lerch
NASA Lewis Research Center, M.S. 49-3, Cleveland, OH, 44135.

ABSTRACT

The low cycle fatigue response of polycrystalline NiAl above the brittle-to-ductile transition temperature (BDTT) was investigated. Samples of nominally stoichiometric NiAl were prepared from extruded ingots and from hot isostatically pressed (HIP'ed) prealloyed powders. The fatigue samples were cycled in a fully reversible fashion at plastic strain ranges between 0.06 and 1.0% in air. The HIP'ed NiAl material was also tested in vacuum at 1000 K. Both processing route and environment were found to have an effect on fatigue life. The lives of the powder samples were about a factor of three less than the cast and extruded material, which was attributed to the lower flow stress of the wrought NiAl. An environmental effect was noted for the HIP'ed material with a factor of 2-3 increase in fatigue life when samples were tested in vacuum compared to samples tested in air. In general, fatigue behavior for NiAl at high plastic strain ranges was typical of most metals, exhibiting a Coffin-Manson strain life behavior with a slope of -0.7. However, fatigue life of the HIP'ed powder material at low plastic strain ranges was controlled by intergranular cavitation and creep processes leading to a change in the slope of the fatigue life curve. Both materials exhibited cyclic softening over the majority of their fatigue lives with fatigue crack propagation occurring predominantly by intergranular mechanisms and final fracture by tensile overload occurring in a transgranular manner. Overall, the 1000 K fatigue life of NiAl was superior to conventional superalloys on a plastic strain range basis partly because of its high ductility and low flow stress but NiAl was not competitive on a stress range basis. The results indicate that binary NiAl has excellent plastic strain cycling capabilities and would be an acceptable material for the matrix phase of intermetallic matrix composites.

INTRODUCTION

In response to the need for high temperature structural materials, ordered intermetallic alloys and intermetallic matrix composites have been earnestly studied for the past decade. Of the many materials under consideration, NiAl is one of the few systems that has emerged as a promising candidate for further development as either a single crystal alloy or the matrix phase for intermetallic matrix composites. This is due to a number of property advantages including low density, high melting point, high thermal conductivity, excellent environmental resistance and the potential for significantly improving creep resistance through alloying [1,2].

While a reasonable understanding of the monotonic flow and fracture behavior of polycrystalline NiAl has been achieved over the years [2-4], cyclic properties have been neglected until recently. Room temperature fatigue tests have been performed on powder processed NiAl material and $NiAl/Al_2O_3$ composites under displacement control between set stress ranges and on cast and extruded NiAl under fully reversed strain control conditions [5,6]. In general, the room temperature fatigue life of polycrystalline NiAl was surprisingly good, exceeding that of most other B2 intermetallics. Cullers and Antolovich [7] have performed fully reversed low cycle fatigue tests on polycrystalline NiAl between 600 and 700 K at the rather large plastic strain ranges of 0.5 and 1%. These tests were performed very near the BDTT of the material and fatigue lives were limited to 1000 - 3000 cycles. The fatigue behavior of polycrystalline NiAl above the BDTT has not been reported, though preliminary testing of single crystal NiAl above and below the BDTT has been performed by Bain et al. [8] on <001> Ni-49.9Al-0.1Mo crystals.

While there is a lack of polycrystalline NiAl fatigue data above the BDTT, such information is necessary to evaluate whether fatigue is a limiting property for this intermetallic. Fatigue properties over a range of temperatures are also needed to provide insight into the behavior of the NiAl matrix phase in thermally cycled composites, where strains develop as a result of thermal expansion mismatch between the matrix and fiber. Identification of the mechanisms controlling fatigue life at both low and high temperatures will also guide second generation alloy development and processing efforts. Consequently, a study of the fully reversed fatigue behavior of polycrystalline NiAl tested between constant plastic strain limits has been performed at 1000 K and is summarized in the following paper.

MATERIALS AND PROCEDURES

Vacuum atomized -20/+325 mesh prealloyed powders of stoichiometric NiAl were obtained from Homogeneous Metals, Inc. The powder was packed in 304 stainless steel cans and HIP'ed at 1533 K and 241 MPa pressure for 5 hours. These HIP conditions are close to the minimum necessary to ensure full consolidation while minimizing time at temperature [9] and are similar to the final processing step for the fabrication of continuous fiber, NiAl-based composites produced by the powder cloth technique [10]. Final grain size of the HIP'ed powder material was 70 ± 14 μm. Wrought NiAl was produced by extruding vacuum induction melted ingots that were originally 38 mm in diameter and 95 mm in length. The ingots were placed in mild steel extrusion cans and extruded at 1200 K at an area reduction ratio of 12:1. The cast and extruded NiAl had a recrystallized, equiaxed grain structure and final grain size of 18 ± 3 μm. Both materials were stoichiometric in composition and contained less than 0.042 at.% total interstitials.

Cylindrical fatigue specimens with gage dimensions of 6.4 mm in diameter by 22 mm in length were tested in a closed loop, servohydraulic load frame, equipped with an environmental chamber capable of a vacuum of 1.3×10^{-4} Pa. Specimens were subjected to low cycle fatigue loading in a fully reversed fashion between constant plastic strain limits at a constant total strain rate of 10^{-3}/s. Plastic strain range was determined by measuring the total strain range and subtracting out the elastic contribution. Strain was measured using a 13 mm gage length, clip-on extensometer, which utilized alumina probes to allow testing at 1000 K. Temperature was maintained through induction heating using a three turn coil powered by a 5 kW RF generator and was measured and controlled by an infrared pyrometer. The temperature gradient was not more than 5 degrees at the test temperature of 1000 K. After failure of the specimens, fractography was performed by typical scanning electron (SEM) and optical microscopy techniques.

RESULTS AND DISCUSSION

At a strain rate of 1×10^{-3}/s, the HIP'ed NiAl powder material had a BDTT of 900 K and the cast and extruded alloy exhibited a BDTT near 650 K, with both materials displaying greater than 25% tensile elongation at 1000 K under monotonic loading conditions. Therefore, a test temperature of 1000 K represents a temperature that is above the BDTT for both materials. The 1000 K low cycle fatigue life versus plastic strain range for these two materials is plotted in Fig. 1. Both NiAl materials exhibited the following strain life relationship:

$$\Delta\epsilon_p = C\,N_f^{-n}$$

where $\Delta\epsilon_p$ is the plastic strain range, N_f is the number of cycles to failure and C and n are material constants. At a plastic strain range greater than about 0.3% for the HIP'ed NiAl material, n was calculated to be -0.7 and is in close agreement with a slope of -0.6, which is commonly used in the universal slopes equation for predicting fatigue life [11]. However, at lower plastic strain ranges the slope of the HIP'ed powder material became substantially more negative. The cast and extruded NiAl also exhibited a slope of -0.7 but the slope did not appear to change at lower plastic strain ranges, although more tests would be needed to confirm this behavior. The cast and extruded NiAl also had a fatigue life that was approximately three times greater than the HIP'ed powder material. Finally, when the powder material was tested in vacuum (1.3×10^{-4} Pa), the fatigue life improved by a factor of 2-3, equaling that of the cast and extruded NiAl at high plastic strain ranges. While the fatigue life was different, the strain-life behavior for samples tested in vacuum was assumed to be similar to those tested in air, therefore the life curve for samples tested in air was simply translated to the longer lives exhibited by the vacuum tests as shown in Fig. 1.

A downward trend in the fatigue life curve at low plastic strain ranges, although unusual, has been predicted to occur in materials where the life of the specimen depends on two competing damage mechanisms such as fatigue crack growth and grain boundary cavity growth

[12]. At high plastic strain ranges the material should fail by fatigue crack growth mechanisms. At lower strain ranges, cavity growth dominates and the life curve is expected to exhibit a downward trend. This type of behavior also has been observed experimentally in the case of a magnesium alloy [13]. In the magnesium alloy, grain boundary cavities reduced the effective cross-sectional area of the specimen, which increased the crack growth rate and led to premature failure at lower plastic strain ranges. A similar phenomenon has occurred in the HIP'ed NiAl material shown in Fig. 1, though it is possible that differences in microstructure made the cast and extruded NiAl less sensitive to this problem. In the wrought material, the grain boundaries were clean and free from any grain boundary particles and in fact the material underwent dynamic grain growth during testing reaching a final grain size of greater than 70 μm from an initial grain size of about 18 μm. In the HIP'ed material the grain size was stable throughout testing because the grain boundaries were pinned by prior particle boundaries. The presence of these oxide particles at the grain boundaries combined with the higher stress levels in the HIP'ed material (Fig. 2) would result in high local stress concentrations making cavity growth much more prevalent in the powder material.

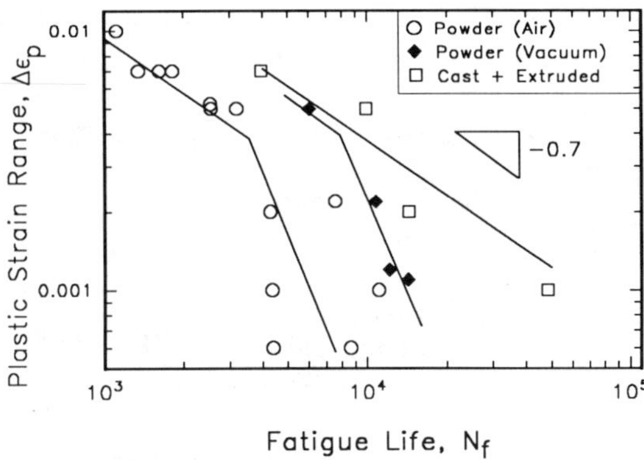

Fig. 1. Fatigue life of polycrystalline NiAl at 1000 K.

The amount of softening or hardening that occurred during fatigue testing at 1000 K was small, with the largest change in stress being only 35 MPa. This is considerably less than the 200-300 MPa increase in stress that was observed during room temperature fatigue testing of cast and extruded NiAl [6]. At 1000 K, NiAl samples tested in air reached a peak stress at less than 1 percent of their total life, Fig. 3. This is consistent with the observation that most of the fatigue life for air tested NiAl was spent in crack propagation instead of crack initiation. Fatigue tests interrupted at approximately 10% of the life of the sample displayed considerable intergranular cracking along the gage surface. These cracks were generally limited to 50-200 μm in length and only occurred along grain boundaries that were oriented perpendicular to the loading axis of the sample. On the other hand, samples tested in vacuum work hardened slightly for about the first half of their life before reaching a plateau or softening slightly. Also, a slight (<10%) stress asymmetry was noted in all samples with a slightly higher stress observed during compressive loading.

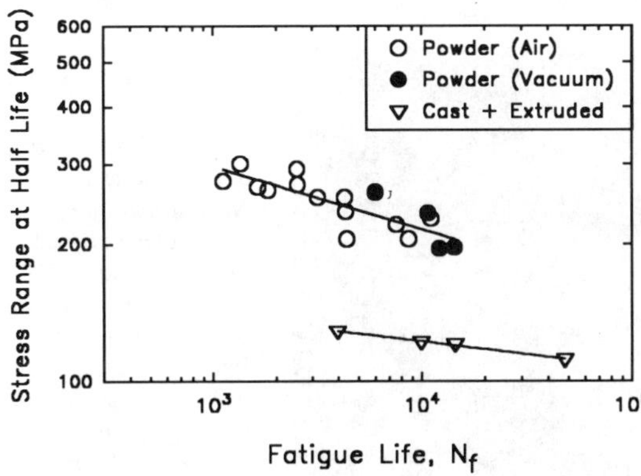

Fig. 2. Stress range (σ_{max} - σ_{min}) at half life as a function of fatigue life for NiAl at 1000 K.

Post-test examination of the fatigue samples revealed a great deal about the low cycle fatigue behavior of NiAl at 1000 K. It was observed that intergranular cracking was typical of slow, stable crack growth, while fast fracture, characteristic of final tensile overload failure, occurred by transgranular cleavage fracture. Roughly 20 - 50% of the fracture

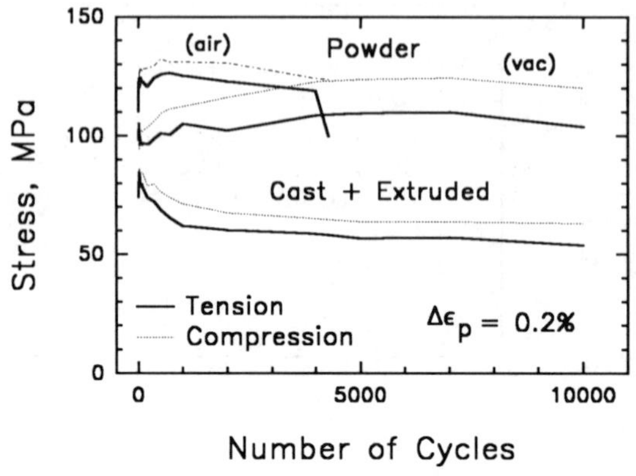

Fig. 3. Absolute cyclic hardening response for NiAl materials at 1000 K and a plastic strain range of 0.2%.

surface of the HIP'ed powder samples was due to stable fatigue crack growth, which could be easily identified by a dark ring on the sample surface as a result of oxidation. Metallographic sections taken from tested samples also revealed numerous, intergranular cracks along the gage length that originated at the sample surface. In addition, there was a high percentage of grains whose boundaries were decorated with tiny pores or voids, typical of diffusional creep damage. These pores were observed in all specimens, though they generally covered a greater area in samples deformed at lower plastic strain ranges. The stress-assisted diffusional growth of these pores contributed significantly to fracture of the material at the lower plastic strain ranges, resulting in the downward slope in the life curve shown in Fig. 1. Finally, intergranular cracking originating from the surface and propagating into the sample was commonly observed on all samples, even those tested in vacuum. However, a vacuum of 1.3×10^{-4} Pa would result in a lowered oxidation rate. Therefore, crack wedging by oxides and other accelerated environmental crack growth processes would be similarly reduced. Consequently, the samples tested in vacuum had a longer fatigue life even though they also succumbed to intergranular cavitation due to creep processes at low plastic strain ranges.

The cast and extruded NiAl also exhibited intergranular fatigue crack growth. But for this wrought material, fatigue crack growth existed over at least 80% of the specimen cross-section regardless of the applied plastic strain range. This was a much higher percentage of the fracture surface than was observed in the powder samples and was probably a result of the much lower flow stress (Figs. 2 & 3) and subsequently a lower crack driving force in the cast and extruded material. As in the HIP'ed powder material, numerous grain boundary cavities were observed in metallographic sections of failed fatigue samples and significant amounts of intergranular cracking could be observed along the entire length of the sample gage section as

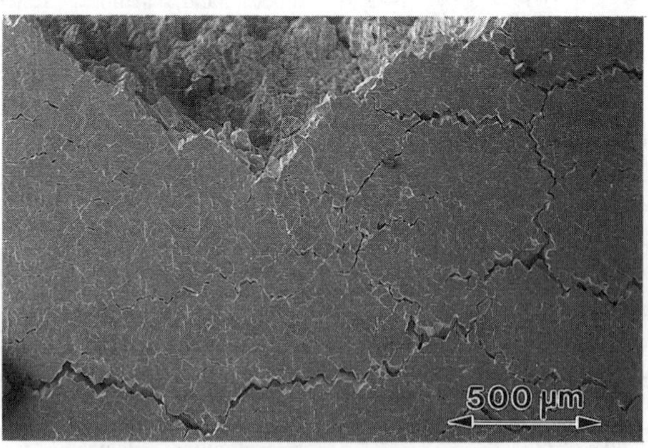

Fig. 4. Gage section of a cast and extruded NiAl fatigue sample near the fracture surface. Sample tested at 1000 K at a plastic strain range of 0.5%. Extensive intergranular cracking was observed along the entire gage length.

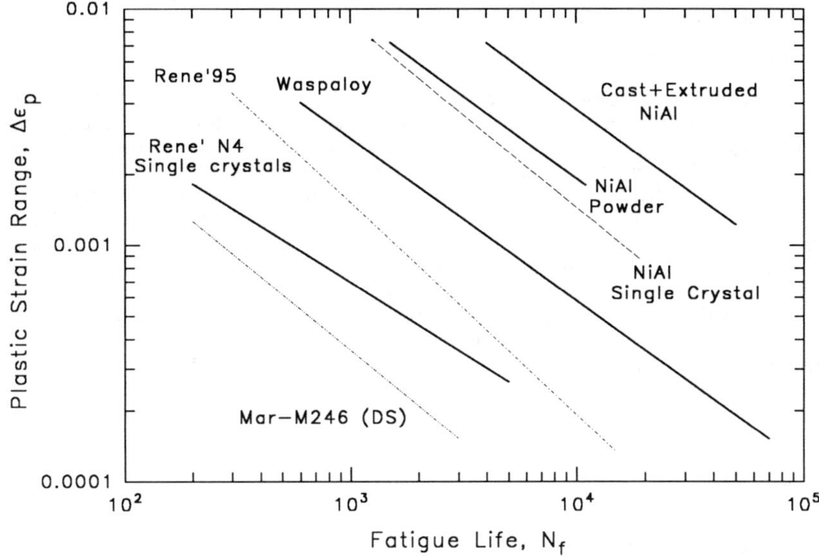

Fig. 5. Fatigue life at nominally 1000 K for NiAl compared to Ni-base superalloys [14-16] and single crystal NiAl [8] on a plastic strain range basis.

demonstrated in Fig. 4.

In Fig. 5, the fatigue life of NiAl has been compared to various turbine disk and blade alloys at a temperature of nominally 1000 K. The following nickel-base superalloys were used for comparison in Fig. 5: Waspaloy [14] Rene' 95 [14], single crystal Rene' N4 [15] and directionally solidified MAR-M246 [16]. Also included on this plot is the fatigue life for single crystal NiAl [8]. It is evident from Fig. 5 that all forms of NiAl have superior fatigue properties compared to conventional nickel-base superalloys on a plastic strain range basis. However, if these same alloys were compared on a stress range basis, NiAl would be inferior because of its low yield stress. The superalloys would fall in a band between 600 and 1000 MPa on a plot such as Fig. 2. However, the excellent plastic strain, fatigue life properties of NiAl compared to superalloys cannot be explained solely by differences in yield strength or ductility and is under further investigation.

In an intermetallic matrix composite, the matrix needs to withstand total strains equal to or greater than what the reinforcing fibers can support. Recent work [17] has shown that Saphikon single crystal Al_2O_3 fibers within an intermetallic matrix fracture at an elastic tensile strain of about 0.17%. In a fully reversed fatigue test, this implies the Saphikon fibers can withstand a total strain range of up to 0.34%. Based on the total strain range in these plastic strain controlled tests, this would result in average matrix lives of approximately 10^4 cycles for NiAl. Therefore, NiAl would not appear to be a life limiting constituent in this particular composite system.

It is worth noting one final observation concerning Fig. 5. Single crystal NiAl has about the same fatigue life as polycrystalline NiAl. While the stress levels during elevated temperature low cycle fatigue testing were also comparable for the [001] oriented single crystal and polycrystalline material, the fracture mechanisms were completely dissimilar. For the single crystals, cracks initiated and propagated basically parallel to the loading direction until they crosslinked and caused failure of the specimens [8]. This is in contrast to the intergranular crack growth that was observed in the polycrystalline material and shown in Fig. 4. Therefore, the correlation in fatigue lives demonstrated in Fig. 5 is either a coincidence or indicates that fatigue crack initiation occurs early in the life of both materials and that intergranular and transgranular fatigue crack growth rates are similar at elevated temperatures. Future work will hopefully clarify this observation.

SUMMARY & CONCLUSIONS

Low cycle fatigue tests were performed at 1000 K on polycrystalline NiAl fabricated by two techniques: HIP consolidation of prealloyed powders and extrusion of vacuum induction melted ingots. The powder NiAl material exhibits a typical Coffin-Manson strain life behavior at plastic strain ranges greater than about 0.3% but at lower plastic strain ranges the slope of the fatigue life curve changes due to grain boundary cavity growth processes. The fatigue life of the cast and extruded material was about a factor of three greater than that for the HIP'ed powder material. All specimens exhibited surface initiation of fatigue cracks and subsequent intergranular fatigue crack growth. A factor of almost three increase in life was observed for the HIP'ed powder material when run in vacuum compared to testing in air indicating an environmentally assisted fatigue damage mechanism. In air, cyclic hardening was observed over only the first few cycles and then the materials softened/and or reached a plateau in stress until fracture, while samples tested in vacuum hardened over the first half of their life. On a plastic strain range basis the fatigue behavior of NiAl is superior to that of superalloys; and would be a comparatively good candidate material for a composite matrix where strengthening would be achieved through the fiber phase.

ACKNOWLEDGEMENTS

The authors would like to thank Ken Bain of General Electric Aircraft Engines for supplying us with a copy of his TMS talk and A.D. Tenteris-Noebe, J.D. Whittenberger and T. Gabb for reviewing this manuscript. The authors are also grateful to G. Halford for many stimulating and insightful conversations concerning this work.

REFERENCES

1. R. Darolia, *J. Metals*, vol. 43, no. 3, 44-49, 1991.
2. R.D. Noebe, R.R. Bowman and M.V. Nathal, "Review of the Physical and Mechanical Properties and Potential Applications of the B2 Compound NiAl," NASA TM-105598,1992.
3. A. Ball and R.E. Smallman, *Acta Metall.*, vol. 14, 1517-1526, 1966.
4. R.R. Bowman, R.D. Noebe, I.E. Locci and S.V. Raj, *Metall. Trans. A*, vol. 23A, 1493-1508, 1992.
5. R.R. Bowman, in Intermetallic Matrix Composites II, eds. D.B. Miracle, D.L. Anton and J.A. Graves, MRS Symposia Proc. Vol. 273, 145-155, 1992.
6. R.D. Noebe and B.A. Lerch, *Scripta Metall. Mater.*, vol. 27, 1161-1166, 1992.
7. C.L. Cullers and S.D. Antolovich, in Superalloys 1992, eds. S.D. Antolovich et al., The Metallurgical Society, Inc., Warrendale, PA., 351-359, 1992.
8. K.R. Bain, R.D. Field and D.F. Lahrman, "Fatigue Behavior of NiAl Single Crystals," Paper presented at the Fall TMS Meeting, Cincinnati, OH., 1991.
9. R.N. Wright, J.R. Knibloe and R.D. Noebe, *Maters. Sci. Eng.*, vol. A141, 79-83, 1991.
10. J.W. Pickens, R.D. Noebe, G.K. Watson, P.K. Brindley and S.L. Draper, "Fabrication of Intermetallic Matrix Composites by the Powder Cloth Process, " NASA TM-102060, 1989.
11. S.S. Manson, *Exp. Mechanics*, July, 1-34, 1965.
12. W. Beere and G. Roberts, *Acta Metall.*, vol. 30, 571-580, 1982.
13. J. Wareing, B. Tomkins and G. Summer, in Fatigue at Elevated Temperature, ASTM STP 520, eds. A.E. Carden, ASTM, Philadelphia, PA., 123-137, 1973.
14. Characterization of Low Cycle High Temperature Fatigue by the Strainrange Partitioning Method. AGARD CP-243, AGARD, Paris, France , 1978. (Available NTIS, AD-A059900).
15. T.P. Gabb, J. Gayda and R.V. Miner, *Metall. Trans A*, vol. 17A, 497-505, 1986.
16. E.S. Huron and S.D. Antolovich, in Structures and Deformation of Boundaries, eds. K.N. Subramanian and M.A. Imam, TMS-AIME, Warrendale, PA., 185-203, 1986.
17. S.L. Draper, D.J. Gaydosh and A. Chulya, in HITEMP Review 1991, NASA CP-10082, pp. 42-1 to 42-14, 1991.

TRANSIENT DEFORMATION OF SINGLE CRYSTAL NiAl AT HIGH TEMPERATURES

Keith R. Forbes and William D Nix
Department of Materials Science and Engineering, Stanford University, Stanford, CA 94305

Abstract

The deformation transients associated with changes in stress and strain rate have been studied as a means of determining the controlling deformation mechanisms in NiAl. Stress change experiments in tension, and strain rate change tests in compression have been performed on single crystals of NiAl at temperatures between 850 and 1200°C. The orientation dependence of these transients was studied by testing in the hard [001] orientation and a soft [223] orientation. Strain rate change experiments suggest increased contribution from structure-controlled mechanisms in hard oriented crystals. Stress change experiments in samples tested along the hard orientation produce transients that are characteristic of the evolution of a stable dislocation substructure. In soft oriented crystals, however, stress change transients suggest that deformation is not significantly affected by the formation of a dislocation substructure. These results are consistent with observations of dislocation substructure formation and strain hardening in hard oriented crystals.

Introduction

The high temperature strength of single crystal NiAl has been found to be strongly orientation dependent[1, 2, 3]. Soft orientations deform by the glide of **b**=<100> dislocations. In crystals deformed along the hard [001] orientation, easy **b**=<100> glide is prohibited and deformation occurs by **b**=<110> glide[4]. The core structure of **b**=<100> dislocations has been observed to be more compact than that of **b**=<110> dislocations; thus **b**=<100> should have a higher mobility in NiAl[5]. The low mobility of **b**=<110> dislocations can account, at least in part, for the increased strength of hard oriented crystals.

The creep curves of soft oriented crystals are characterized by immediate steady state creep with no significant primary creep transient. Strain hardening is minimal in both tension creep and constant strain rate compression. However, creep of hard oriented crystals is characterized by significant strain hardening. Dislocation networks, forming subgrains and tangles, are observed in NiAl single crystals deformed along the hard orientation but are not common in soft oriented crystals[6]. This suggests that dislocation substructure additionally strengthens hard oriented crystals.

Deformation in soft oriented crystals is limited by the mobility **b**=<100> dislocations. However, deformation in the hard orientation is controlled by the extensive dislocation substructure that is formed. This investigation focuses on the difference in the controlling deformation mechanism in a soft [223] orientation and the hard [001] orientation on the basis of insight provided from deformation transients.

The preparation of the tension and compresion samples used in these tests is explained elsewhere[4]. In compression tests, samples were deformed between two flat alumina platens, onto which extensometery was attached in order to measure the strain in the sample without the complications of subtracting machine displacements. In tension tests, the strain in the sample was measured with extensometry attached to pins passing through the grip region of the sample.

Studying Deformation Mechanisms with Deformation Transients

The deformation state equation can be written as:

$$\dot{\varepsilon}_{pl} = f\left(\sigma, T, \rho_m, \hat{\tau}, \ldots\right).$$

The plastic strain rate, $\dot{\varepsilon}_{pl}$, in a material is determined by the externally applied stress, σ and temperature, T, as well as the internal dislocation structure described here by the density of mobile dislocations, ρ_m, and the strength of the immobile dislocation substructure, $\hat{\tau}$. During a deformation experiment, the stress and temperature (creep test) or strain rate and temperature (constant strain rate test) are controlled and the resulting strain rate or stress is monitored. During deformation, the internal state of the material evolves, modifying the density of mobile dislocations and the strength of the dislocation substructure. The internal state of the material, however, can not be measured directly during the test.

During a transient test, a rapid change in one of the external state variables is made and the resulting change in the others is monitored. In a strain rate decrease experiment, the strain rate is rapidly dropped to a low value while the stress is measured. In a stress change experiment, the stress is varied and the strain rate monitored. The manner in which the measured variables respond implies qualitative changes in the internal state of the sample.

In materials where deformation is limited by the mobility of dislocations, such as bcc metals or solid solution alloys, the strength of the dislocation substructure will be small and will vary only slightly throughout the test. The deformation transient is thus characterized by changes in the density of mobile dislocations. If a material develops a dislocation substructure that limits deformation, deformation transients will be characterized by the processes that modify this substructure. Pure fcc metals typically show structure-controlled deformation.

A mobility-controlled material will exhibit deformation transients that are distinct from those of a structure-controlled material. Transient tests, however must be conducted rapidly, especially at high temperatures, in order to distinguish between deformation mechanisms. In this investigation we utilize both strain rate decrease and stress change experiments.

Strain Rate Decrease

Strain rate change experiments have been shown to be useful in characterizing deformation mechanisms[7]. The response of NiAl samples was studied during a strain rate decrease from 10^{-4} to 10^{-5} s^{-1} in compression. In soft [223] oriented samples, steady state deformation was reached before the strain rate decrease was initiated. Steady state could be reached in the hard [001] oriented samples only during 1200°C tests and at large strains. Thus in hard oriented crystals, strain rate decreases at various strains were compared. No significant variations in the results were observed with varying strain in the sample.

In a material in which dislocation structure controls the deformation, higher strain rates form a finer network of immobile dislocations. The dislocation substructure is thus harder at the higher strain rate than at lower rates. If the strain rate is rapidly dropped, this harder structure will persist until the dislocation structure can recover. Thus the strain rate initially drops quickly with a small change in the stress. As the structure recovers, the stress decreases to its steady state value. The response of a structure-controlled material is illustrated in Fig. 1a.

The response of a mobility-controlled material is distinct from that of structure-controlled materials. In a mobility-controlled material, many dislocations are moving rapidly at high strain rates. When the strain rate is decreased, these dislocations initially maintain their high mobilities so that the stress drops rapidly. In time, the density of mobile dislocations decreases and the stress increases to the new steady state value (Fig. 1b).

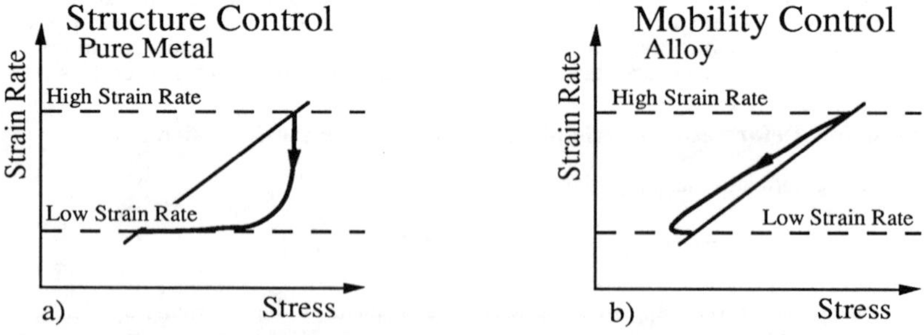

Figure 1. Characteristic a) structure-controlled and b) mobility-controlled responses during strain rate decrease experiments.

The strain rate decrease of soft [223] and hard [001] oriented NiAl single crystals, shown in Fig. 2, do not match either of the idealized curves in Fig. 1. However, the transients of the hard oriented crystals show a greater tendency toward structure-controlled behavior than the soft oriented crystals. This is evident in comparing the strain rate decreases of the crystals at 1200°C. In this case, the soft orientation remains close to the steady state line resembling the mobility-controlled response while the hard orientation shows clear structure control. At lower temperatures, there is evidence for increased stucture-control in both the hard and soft orientations. These results suggest that between 850°C and 1200°C, structure-controlled deformation is more prevalent in hard oriented crystals.

The strain rate decrease experiment is hindered by the compliance in the testing machine, which prevents instantaneous changes in strain rate in the sample. The relaxation of the machine hides some of the sample response. Thus, we attempted to produce more rapid and dramatic changes in the material with stress change experiments.

Stress Removal

The strain rate vs. strain curve for a hard [001] oriented NiAl crystal in tension creep is shown in Fig. 3a. The strain rate initially increases as mobile dislocations multiply, but then begins to decrease as the dislocation substructure hardens the material. In Fig. 3b, the sample was allowed to deform to about 2% strain and then the load was removed for about 20 hours. Upon reloading, the sample was found to have softened, rising to a higher strain rate than that before the stress removal. The strain rate then decreases as the dislocation structure forms and hardens the sample.

The response shown in Fig. 3b is clear indication that dislocation substructure limits deformation in hard oriented crystals. At 1200°C, the dislocation substructure can be annealed out of the sample, which results in a softer material. If deformation were mobility-controlled, then annealing would not soften the material, but would rather harden it, as mobile dislocations recover into less mobile configurations. Thus, the strain hardening seen in hard oriented crystals is a result of increasing dislocation substructure and not decreasing mobile dislocation denisity.

The 21 hour anneal time is not large compared to the time of creep tests at this temperature. Thus, the processes by which the dislocation substructure recovers should also contribute significantly to creep deformation at 1200°C. At 1000°C, the recovery of the dislocation substructure is not significant as shown by the stress removal experiments in Fig. 4.

Stress removal tests were also performed on soft [223] crystals, which show no strain hardening during creep. There was no significant softening or hardening of soft oriented crystals resulting from a 20 hour anneal.

Stress Change

Stress change experiments were accomplished on NiAl single crystals in tension creep. A steady state strain rate was established before the stress was changed.

In a structure-controlled material, the dislocation substructure is coarse at low stresses and thus the material is soft during a stress increase. Initially the sample should rise to a high strain rate before the dislocation structure hardens and the strain rate decreases to the new steady state (Fig. 5a). Similarly, during a load decrease, the material initially seems hard and the strain rate drops rapidly before rising to steady state as the dislocation structure softens. A structure-controlled material will overshoot the steady state strain rate during a stress increase and undershoot it during a stress decrease.

In a mobility-controlled material, the mobile dislocation density is low at low stresses. When the stress is increased, the mobile density increases slowly so that the strain rate rises gradually to steady state (Fig. 5b). In a stress decrease, the strain rate approaches steady state gradually as the mobile dislocation density decreases.

Hard oriented NiAl crystals show structure-controlled behavior at both 1000°C and 1200°C (Fig 6). In a stress increase, the strain rate consistently rises above the steady state value before the material hardens. The strain rate also undershoots the steady state value, in a stress decrease, until the dislocation structure softens. In contrast, soft oriented NiAl crystals show mobility-controlled behavior with a smooth transition between high and low stresses.

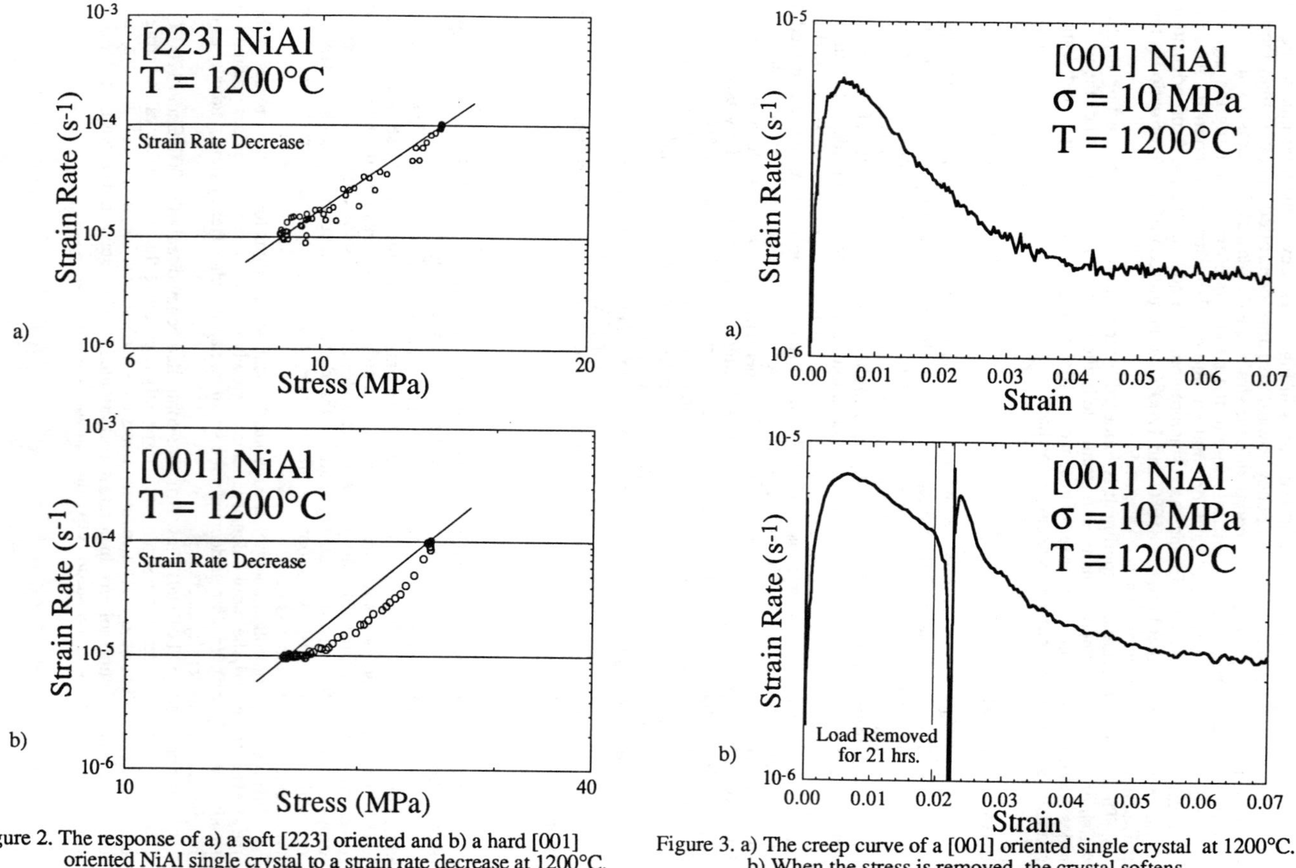

Figure 2. The response of a) a soft [223] oriented and b) a hard [001] oriented NiAl single crystal to a strain rate decrease at 1200°C.

Figure 3. a) The creep curve of a [001] oriented single crystal at 1200°C. b) When the stress is removed, the crystal softens.

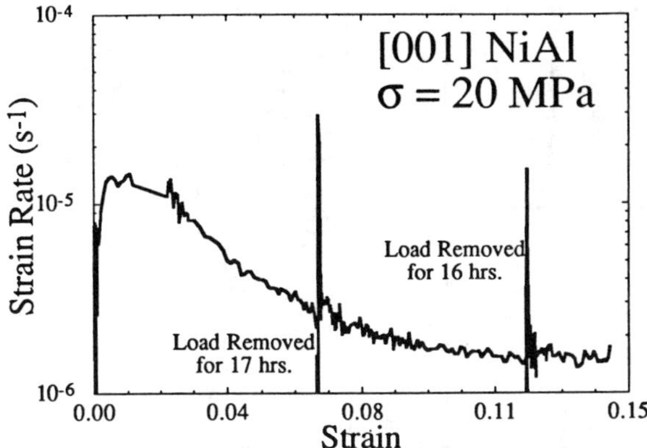

Figure 4. The creep curve of a [001] oriented single crystal at 1000°C during which the stress was removed and reapplied.

Conclusions

Deformation transients are useful in describing the controlling deformation mechanisms in soft and hard oriented NiAl single crystals. However, at high temperatures, it is difficult to produce to the transients necessary to distinguish between deformation mechanisms. Strain rate change tests can provide some insight but stress change tests were found to provide a more rapid and dramatic stimulus to the sample and thus provided clearer indications of the mechanisms involved.

The deformation transients of soft [223] oriented NiAl are characteristic of mobility-controlled deformation. Dislocation substructure is not significant in soft orientations at these temperatures and stresses.

The deformation transients of hard [001] oriented NiAl are characteristic of structure controlled deformation. The dislocation substructure that has been observed in hard oriented NiAl contributes significantly to the strength of these crystals. The dislocation substructure can be annealed out at high temperatures suggesting that the climb of dislocations contributes to deformation in hard oriented crystals at 1200°C.

Acknowledgments

The authors gratefully acknowledge the support of the Air Force Office of Scientific Research under AFOSR Grant No. F49620-92-J-0009. The support of Dr. Alan Rosenstein of AFOSR is very much appreciated. The authors thank Dr. Ram Darolia of the Engineering Materials Technology Laboratories of GE Aircraft Engines for his assistance in providing NiAl crystals for this study. The authors also thank Dr. Uwe Glatzel for his helpful discussions about these tests.

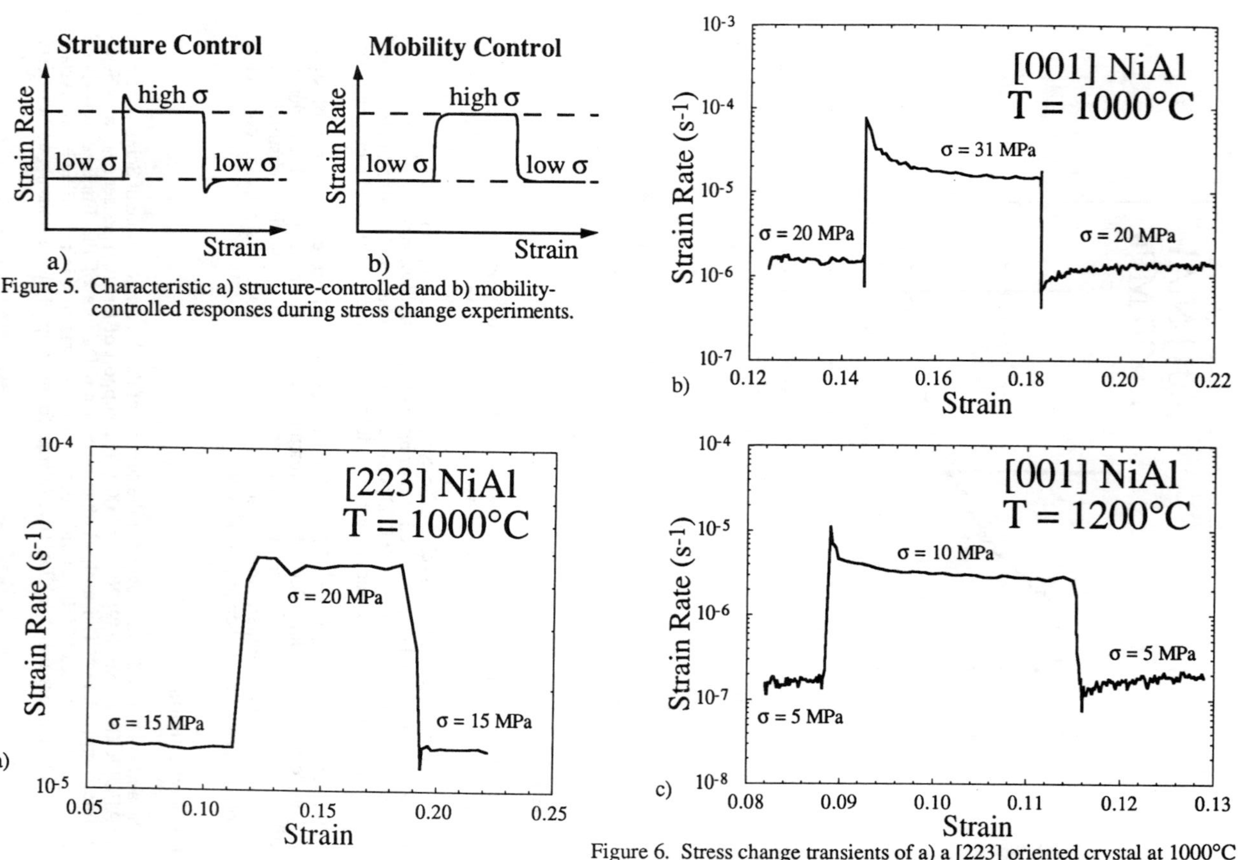

Figure 5. Characteristic a) structure-controlled and b) mobility-controlled responses during stress change experiments.

Figure 6. Stress change transients of a) a [223] oriented crystal at 1000°C and [001] oriented crystals at b) 1000°C and c) 1200°C.

References

1. A. Ball and R. E. Smallman, Acta metall. **14**, 1349 (1966).
2. A. Ball and R. E. Smallman, Acta metall. **14**, 1517 (1966).
3. J. Bevk, R. A. Dodd and P. R. Strutt, Met. Trans **4**, 159 (1973).
4. K. R. Forbes, U. Glatzel, R. Darolia and W. D. Nix, in these proceedings.
5. M. J. Mills and D. B. Miracle, submitted to Acta metall. mater.
6. U. Glatzel, K. R. Forbes and W. D. Nix, in these proceedings.
7. D. L. Yaney, J. C. Gibeling and W. D. Nix, Acta metall. **35**, 1391 (1987).

ELEVATED AND ROOM TEMPERATURE MECHANICAL BEHAVIOR OF BORON DOPED NiAl ALLOYS

YI TAN[+], TETSUMORI SHINODA[++], YOSHINAO MISHIMA, AND TOMOO SUZUKI[*]
Precision and Intelligence Laboratory, Tokyo Institute of Technology, Nagatsuta, Midori-ku, Yokohama 227, Japan.
*Department of Metallurgical Engineering, Tokyo Institute of Technology, O-okayama, Meguro-ku, Tokyo 152, Japan.
+On leave from Department of Materials Engineering, Dailian University of Technology, Dailian, China.
++On leave from Hitachi Research Laboratory, Hitachi Ltd., Saiwai-cho, Hitachi-shi, Ibaraki 317, Japan

ABSTRACT

The effects of boron addition on defect hardening at room temperature and on high temperature creep properties are investigated in the B2 NiAl intermetallic compound. It is found that boron addition is effective in increasing the room temperature hardness on the Ni-rich side but has no effect on the Al-rich side of stoichiometry. These observations are attributed to interstitial dissolution of boron in Ni-rich NiAl and due to a lack of solubility, and consequently an enrichment at grain boundaries on the Al-rich side. The similar effect is found for high temperature creep resistance of NiAl by boron addition, where it is increased at Ni-rich side but is unaffected at Al-rich side of offstoiciometry.

INTRODUCTION

An intermetallic compound NiAl is of Berthollide type having a high melting temperature and a wide compositional range for the B2 ordered structure. It is consequently expected to be a candidate as a heat resisting material by itself. However, efforts in designing this compound for this purpose has not been extensive mainly due to such drawbacks as poor room temperature ductility and insufficient high temperature strength.

It has been shown that the defect structure in NiAl as the composition deviates from stoichiometry differs between the Ni-rich and the Al-rich side, where it consists of anti-structure defects in the former and vacancies on Ni-site in the latter[1]. Accordingly, various properties of the compound show characteristic change as the composition deviates from stoichiometry. For example, it has been shown by Vedula and Khadkikar[2] that the room temperature strength normalized by shear modulus is twice as high on the Al-rich side than on the Ni-rich side when compared at identical deviations from stoichiometry. Although there have been several studies on the high temperature creep properties of NiAl[3-7], it should be recognized that the effect of deviation from stoichiometry, which is very important in practice, has not been systematically investigated.

The low temperature brittle nature of NiAl is characterized by grain boundary fracture[8,9]. As one of the efforts to improve room temperature ductility of the compound, boron addition has been attempted [10] because the method was successful in eliminating the grain boundary brittleness in Ni_3Al. However, there has been no work to correlate the effect of boron addition on the mechanical properties of off-stoichiometric NiAl.

The objective of the present investigation is to examine the effect of boron addition on the both room temperature strength and the high temperature creep properties for a range of compositions deviating from stoichiometry.

EXPERIMENTAL PROCEDURE

The alloys were prepared either by vacuum induction or by arc-melting techniques to produce a 5 kg and 25g ingots, respectively. Wet chemical analysis was performed for all the alloys prepared by induction furnace but the nominal compositions are accepted for those by arc-melting because the weight loss after the melting was minimal. The alloys were homogenized in vacuum at 1373 K for 24 h. After homogenization, grain size of the alloys are found to be about 300 to 580 μm. Room temperature hardness was measured using a micro-Vickers hardness tester with a 300 g load. Lattice parameter of the alloys were determined by X-ray powder diffractometry using (310) peak by CuK$_\alpha$ radiation.

High temperature compressive creep test was carried out using a cantilever type testing machine at temperatures ranging from 1023 to 1223 K under an initial applied stress ranging from 8 to 300 MPa. Test specimen has a dimension of approximately 3 x 3 x 6 mm^3.

RESULTS AND DISCUSSION

Effect of Boron Addition on the Defect Hardening

The effect of deviation from stoichiometry on the room temperature hardness of NiAl with or without 0.20 at% boron is shown in Fig. 1, in which together shown are the data from literature[11-13]. The deviation from stoichiometry is expressed both by Al concentration and a deviation parameter being defined as, η = 1/2 - x/100, where x is concentration of aluminum in at%. It is clearly seen that hardeness increases as the composition deviates from stoichiometry and that hardening is more extensive on the Al-rich side. It is evident that the data are in very good agreement from the different investigators in the case of boron-free NiAl. It is also shown that the boron addition is very effective in increasing the hardness on the Ni-rich side but not on the Al-rich side.

In Figure 2 is shown the variation in lattice parameter of NiAl with or without boron, in which the available data from the literature are also shown[1,14-17]. The agreement among the data by different investigators is again well for boron free NiAl except the fact that there is a slight misfit in the composition to give the maximum lattice parameter, being 50.5 at% by Bradley and Taylor[1] and 51 at% by Guseva and Markov[16] and by Cooper[17]. It is clearly seen that the lattice parameter increases by boron addition only on the Ni-rich side of stoichiometry. The results shown in Figs. 1 and 2 are consistent, by which the boron addition is confirmed to provide solid solution hardening on the Ni-rich side of stoichiometry.

From the available ternary Ni-B-Al phase diagram at 1073 K[18], it can be seen that Al-rich NiAl equilibrates with pure boron, while Ni-rich NiAl with some ternary borides as shown in Fig. 3. In the presnt investigation, optical microscopy was carried out on alloys with different compositions.

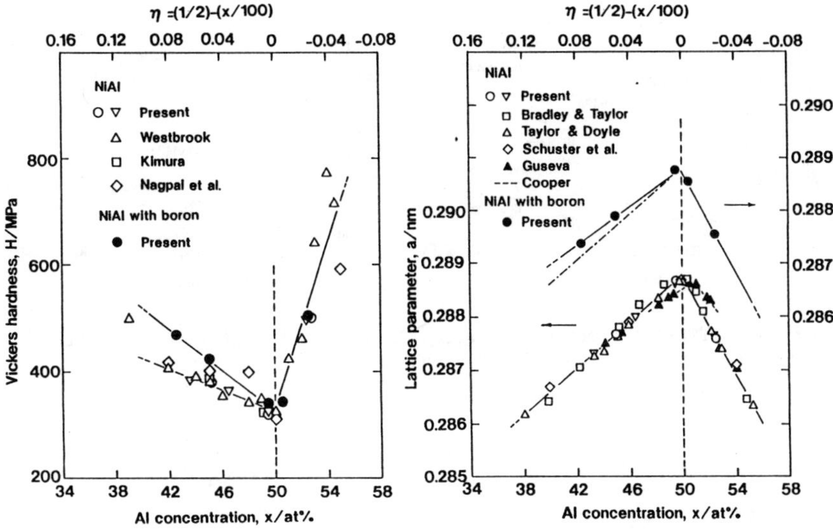

Fig. 1 Effect of deviation from stoichiometry on the room temperature hardness of NiAl.

Fig. 2 Effect of deviation from stoichiometry on the lattice parameter of NiAl.

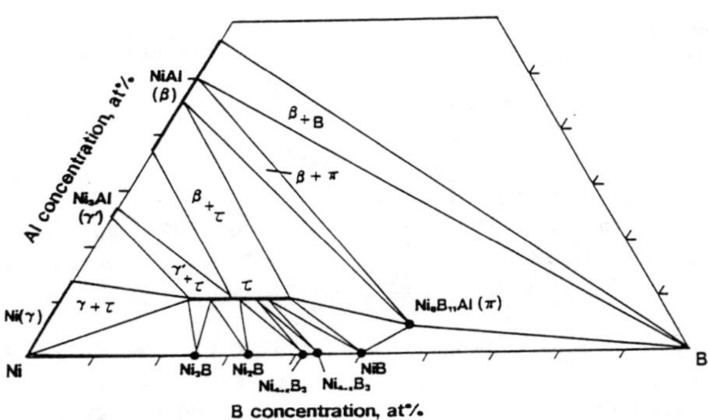

Fig. 3 The Ni-B-Al ternary phase diagram at 1073 K[18].

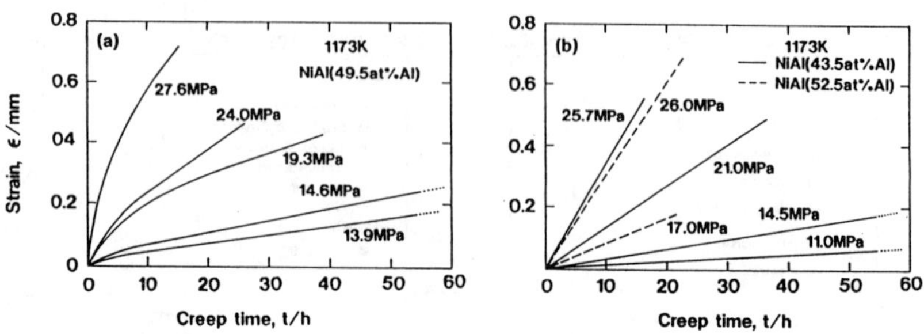

Fig. 4 Typical creep curves for a) near stoichiometric(49.5 at%Al) and (b) off stoichiometric(43.5 and 52.5 at%Al) NiAl at 1173 K.

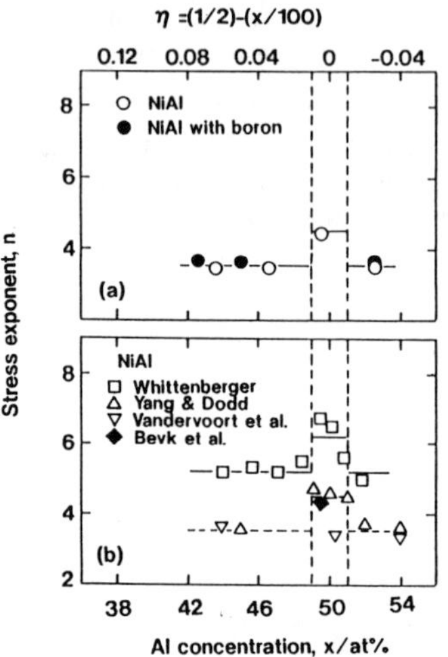

Fig. 5 Effect of deviation from stoichiometry on the stress exponent, n; (a) the present results to see the effect of boron, and (b) data from literature in NiAl.(c) shows the data in CoAl for comparison

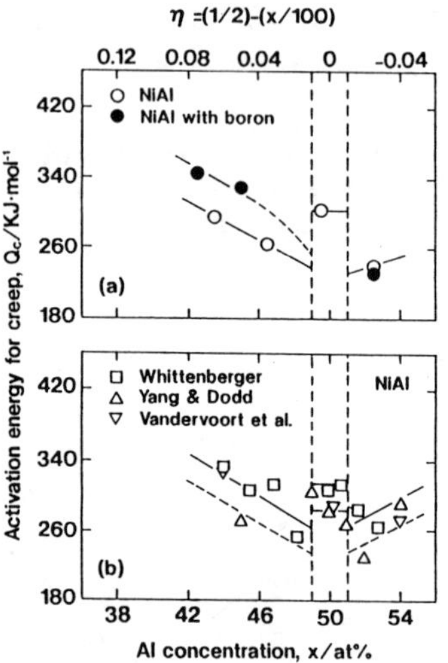

Fig. 6 Effect of deviation from stoichiometry on the activation energy for creep, Q_c, in NiAl; (a) the present results to see the effect of boron, and (b) data from literature

It was found that on the Ni-rich side, there is an increased tendency for second phase precipitation both at the grain boudaries and in the matrix. No particles were observed on the Al-rich side. It is suggested that on the Ni-rich side boron atoms dissolve interstitially in the matrix to NiAl up to the solubility limit beyond which they precipitate as borides. On the Al-rich side, they can not dissolve into solution and therefore they would probably be present at the grain boundaries. Additional results will be published elsewhere to confirm this explanation[19,20].

High Temperature Creep Behavior

Figures 4(a) and (b) show typical creep curves for near stoichiometric (49.5 at%Al) and off-stoichiometric(43.5 at% and 52.5 at%Al) boron-free NiAl at 1173 K, respectively. It can be seen that near stoichiometric NiAl exhibits class II type creep behavior, whereas off-stoichiometric alloys class I type. This is in good accordance with the result reported by Yaney and Nix[21]. Then by evaluating the steady state creep rate from Fig. 4, it is found that the applied stress and the steady state creep rate obey a power-law type relation in all the alloys examined, from which stress exponent, n, and activation energy for creep, Q_c, are estimated. The composition dependences of n and Q_c are shown in Figs. 5 and 6, in which the effect of boron is shown in (a) and a compilation of data from previous investigations in (b)[3-5, 7] of both figures. From Fig. 5, the observed differences in creep behaviors between near and off-stoichiometric composition are confirmed by comparison with previous results for NiAl. Note that from the present work boron addition is found to have no effect on n at least at off-stoichiometric compositions. Activation energy for creep is found to increase as the composition deviates further from stoichiometry but seems to be constant at near stoichiometry as shown in Fig. 6. The effect of boron in this case is similar to that on the solid solution hardening described earlier so that there seems to be no effect on the Al-rich side but has an effect to increase Q_c on the at Ni-rich side.

Finally, the creep resistance, defined as the creep stress evaluated at a steady-state creep rate of $1 \times 10^{-7}\ s^{-1}$, is shown as a function of deviation from stoichiometry in Fig. 7 in order to illustrate the effect of boron on the creep behavior of NiAl. It is clearly seen that the boron addition enhances the creep resistance of NiAl on the Ni-rich side although the degree of the enhancement decreases with increasing temperature.

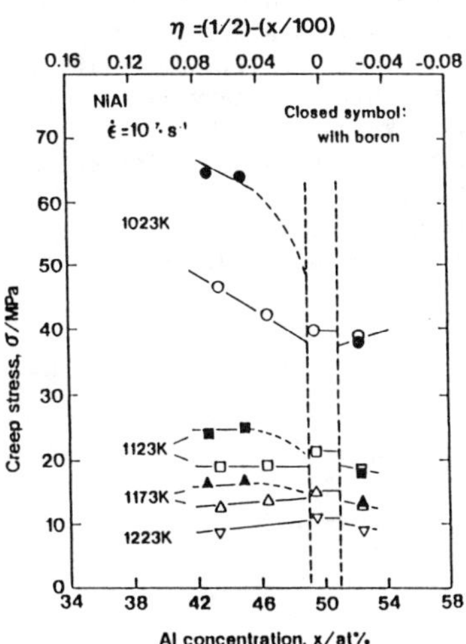

Fig. 7 Effect of boron addition and deviation from stoichiometry on the creep resistance of NiAl

CONCLUSIONS

It is found that the solid solution hardening by addition of 0.20 at% boron at room temperature is only found on the Ni-rich side of stoichiometry but not on the Al-rich side. The similar observation is made on the high temperature creep resistance of NiAl, where the effect of boron addition is to enhance it only on the Ni-rich side due to an increase in the activation energy for power-law creep.

REFERENCES

1 A.J.Bradley and A.Taylor: Proc.Roy.Soc.London, A159(1937), 56.
2 K.Vedula and P.S.Khadkikar: High Temperature Aluminides and Intermetallics, Ed. by S.H.Huang, C.T.Liu, D.P.Pope and J.O.Stiegler, TMS, Ohio(1990), 197.
3 J.D.Whittenberger: J.Mater.Sci., 22(1987), 394.
4 W.J.Yang and R.A.Dodd: Metal Sci.J, 7(1973), 41.
5 R.R.Vandervoort, A.K.Mukherjee and J.E.Dorn: Trans.ASM, 59(1966), 930.
6 D.L.Yaney and W.D.Nix: J.Mater.Sci., 23(1988), 3088.
7 J.Bevk, R.A.Dodd and P.R.Strutt: Metall.Trans., 4(1973), 159.
8 T.Ogura, S.Hanada, T.Masumoto and O.Izumi: Metall.Trans.A, 16A(1985), 441.
9 C.T.Liu, C.L.White and J.A.Horton: Acta Metall., 33(1985), 213.
10 E.P.George and C.T.Liu: J.Mater.Res., 5(1990), 754.
11 J.H.Westbrook: J.Electr.Soc., 103(1956), 54.
12 Y.Kimura: private communication, Tokyo Institute of Technology, 1992.
13 P.Nagpal and I.Baker: Metall.Trans.A, 21A(19900, 2281.
14 A.Taylor and N.J.Doyle: J.Appl.Crst., 5(1972), 201.
15 J.C.Schuster and H.Nowotny: Monatsh.Chem., 113(1982), 163.
16 L.N.Guseva and E.S.Markov: Dokl.Akad.Nauk.SSSR, 77(1951), 615.
17 M.J.Cooper: Phil.Mag., 8(1963), 805.
18 E.E.Schmid: Ternary Alloys, Vol.3, Ed. by G.Petzow and G.Effenberg, VCH, Weinheim, (1990), 201.
19 Y.Tan, T.Shinoda, Y.Mishima and T.Suzuki: J.Japan Inst.Metals, 32(1993), in press.
20 idem, ibid, in press.
21 D.L.Yaney and W.D.Nix: J.Mater.Sci., 23(1988), 3088.

CREEP AND STRESS RUPTURE BEHAVIOR OF Ti-24Al-11Nb

WEGO WANG* AND LI-TUSNG CHIEN**
*U.S. Army Research Laboratory, Materials Directorate, Watertown, MA 02172-0001, USA
**Du Pont Taiwan Limited, Taiwan, ROC.

ABSTRACT

This work concerns the temperature and stress dependencies of the steady-state creep rate of a Ti-24Al-11Nb alloy in the temperature range between 593 and 760°C (1,100 and 1,400°F) and stress levels 137.9 to 275.8 MPa (20 to 40 ksi). Tests were conducted on both as-extruded and heat-treated specimens. A transition occurs in the creep curves at 704°C (1,300°F) and 206.9 MPa (30 ksi), indicating a creep mechanism change. Generally speaking, the fracture surface of as-extruded specimens shows a ductile transgranular fracture mechanism; a more ductile fracture mode is observed for heat-treated specimens.

INTRODUCTION

The relatively low density and superior high-temperature strength of titanium aluminide alloys have made them attractive candidates for aircraft engine materials. However, data on the high-temperature creep and stress rupture properties of the Ti-24Al-11Nb alloy are still sparse, and little systematic analysis has been reported on the high-temperature deformation characteristics of this alloy. The major objective of the current study is to investigate the fundamental creep/stress rupture characteristics of this alloy and to establish the stress and temperature dependencies of the steady-state creep rate.

MATERIAL

The material for this study was supplied by Nuclear Metals, Inc. Raw material powders atomized by the Plasma Rotating Electrode Process (PREP) were first compacted in two low carbon steel canisters and then extruded with an area reduction ratio of 13.76. The extrusion die had a diameter of 1.40 cm (0.55 inch) and the maximum force applied during this extrusion was 270 tons. The two extruded rods were 93.7 and 96.3 cm (36.9 and 37.9 inches) long, respectively, and had the low carbon steel canning material as an intimate cladding.

Slices of materials were machined from the extruded rods for heat treatment and hardness tests. The heat treatment was conducted at 1,000°C (1,832°F) for 4 hours in nitrogen, followed by a furnace cool to room temperature. The microstructures of specimens prior to and after heat treatment are shown in Figs. 1 and 2, respectively. A mixed microstructure of primary α_2 phase and α_2+β matrix is observed in Fig. 1. The largest primary α_2 size is about 10 μm in the as-extruded sample. The α_2+β matrix shows a moderately deformed microstructure resulting from the prior extrusion process. The α_2+β

matrix that has been annealed and transformed to the coarse equiaxed α_2 phase as a result of the heat treatment is shown in Fig. 2. A fine Widmanstatten structure is also observed within the α_2 grains. The average size of α_2 increases to about 15 μm after the sample had been heat treated. The microstructural difference is also reflected by their respective hardness values. The Rockwell C (HRC) hardness value for the heat-treated specimen is 25.1 and the as-extruded specimen shows a higher value of 35.3 HRC.

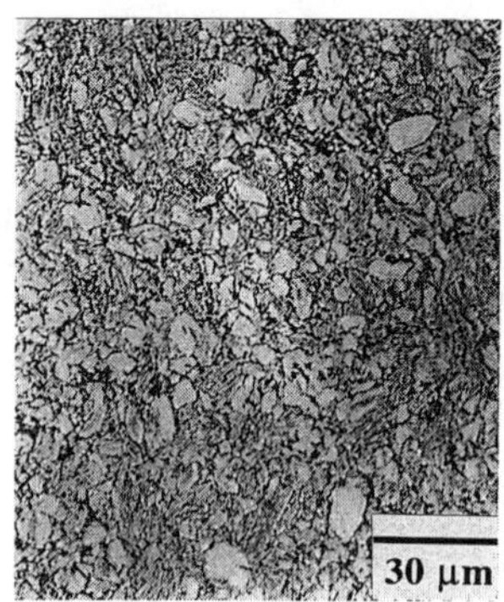

Fig. 1 Microstructure of as-extruded sample.

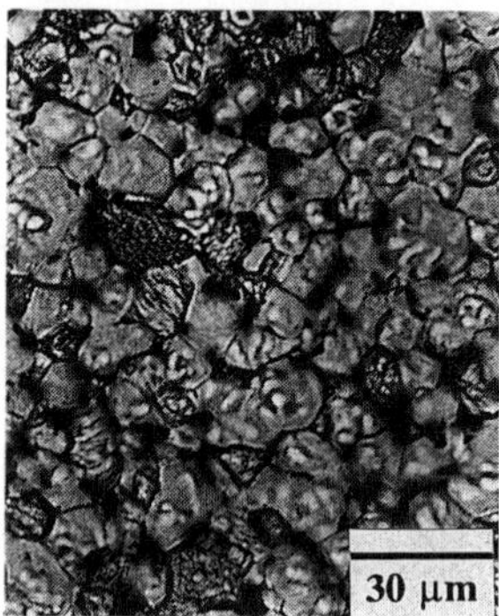

Fig. 2 Microstructure of heat-treated sample.

CREEP/STRESS RUPTURE

The temperature and stress dependencies of minimum (steady-state) creep rate can be expressed as: $\dot{\varepsilon}_s = A\sigma^n \exp(-Q/RT)$, where $\dot{\varepsilon}_s$ = minimum creep rate, A = material parameter, σ = applied stress, n = stress exponent for creep, Q = activation energy for creep, R = universal gas constant, T = absolute temperature. Fig. 3 shows the variations of steady-state creep rates for heat-treated specimens tested at 760°C (1,400°F) while under different stresses of 137.9, 206.9 and 275.8 MPa (20, 30 and 40 ksi). At 137.9 MPa (20 ksi), three distinct stages are observed during the creep. The steady-state stage is established after about 2 hours. The minimum creep rate is 3.56×10^{-3}/hr. The creep rate shows a rapid increase at about 23 hours indicating that tertiary creep is becoming predominant. At 206.9 MPa (30 ksi), a higher initial elastic strain is observed; the minimum creep rate in the steady-state stage has increased to 1.32×10^{-2}/hr. The creep rate rapidly increases at about 6 hours with a strain of 20.8%. At 275.8 MPa (40 ksi), the steady-state creep is completely eliminated and the specimen fails after a short period of tertiary creep.

The minimum creep rates are plotted vs the reciprocal of absolute temperatures in Fig. 4. It shows the temperature dependence of minimum creep rate for stresses of 137.9, 206.9, 275.8 MPa (20, 30 and 40 ksi) at temperatures of 593, 649, 704 and 760°C (1,100, 1,200, 1,300 and 1,400°F), respectively. The curves corresponding to each stress are divided into two regimes represented by two straight lines having different slopes. The slope for the high temperature regime is steeper than that of the low temperature regime.

Above a temperature of 704°C (1,300°F), the activation energies for creep, Q, at 137.9, 206.9 and 275.8 MPa (20, 30 and 40 ksi) were experimentally measured as 321, 328 and 407 kJ/mol (76.8, 78.4 and 97.2 kcal/mol), respectively. Below 704°C (1,300°F), the Q values are 181, 189 and 218 kJ/mol (43.3, 45.1 and 52.1 kcal/mol), respectively. The activation energy above 704°C (1,300°F) is about 1.8 times that below 704°C (1,300°F). This difference results from a change of the rate-controlling creep mechanism.

The minimum creep rates are plotted in Fig. 5 against the ratios of stress/Young's modulus for three different test temperatures, 704, 649 and 593°C (1,300, 1,200 and 1,100°F). A room temperature Young's modulus of $1x10^5$ MPa ($14.5x10^3$ ksi) was used. The linear relationship between the logarithm value of the minimum creep rate and the stress/Young's modulus ratio can be mathematically described in a power-law formula as: $\varepsilon_s = K(\sigma/E)^n$, where K = material parameter and E = Young's modulus.

A discontinuity of slope is observed at 206.9 MPa (30 ksi) for all three test temperatures. This discontinuity of slope reflects the metallurgical instability resulting from a microstructure change or a shift of the predominant creep mechanism. The numerical value of the stress component, n, was calculated by measuring the slope in Fig. 5. For a stress of 206.9 MPa (30 ksi) or higher, the average n value is 5.6 for temperatures ranging from 593 to 704°C (1,100 to 1,300°F). For a stress lower than 206.9 MPa (30 ksi), the average value is 5.1. One recent study on the steady-state creep of the Ti-24Al-11Nb alloy observed that the creep behavior was a function of microstructure, temperature and stress [1]: the stress exponent, n, varied from 1 to 5 and higher. Another study reported that the activation energy and the stress exponent for dislocation-climb controlled creep of a two-phase Ti-24Al-11Nb alloy, from 650 to 760°C (1,200 to 1,400°F) with a constant stress from 69 to 172 MPa (10 to 25 ksi) [2], were about 260 kJ/mol and 4.3, respectively.

FRACTOGRAPHY

The fracture mechanisms and microstructural characteristics of the fracture surface are a function of processing, heat treatment and testing conditions. The fracture surface of an as-extruded specimen tested at 649°C/275.8MPa (1,200°F/40ksi) shows a ductile dimple transgranular fracture mechanism. Second phase particles and inclusions can be observed on the fracture surface. These particles and inclusions are usually relatively small, in the range of about 2 μm as shown in Fig. 6. Due to their small size, they are harmless in terms of brittle crack initiation. Most of them remain uncracked after the specimen fails. Nevertheless, these particles help to initiate microvoids which subsequently coalescence and grow into macro dimples because of the weak bonding force between them and the matrix.

A more ductile fracture mode is observed for heat-treated specimen tested at 704°C/241.3MPa (1,300°F/35ksi), deeper and larger dimples are found throughout the fracture surface, as shown in Fig. 7. The average dimple size is about 15 μm compared with 7 μm for the as-extruded specimen tested at 649°C/275.8MPa (1,200°F/40ksi). This more ductile dimple fracture surface is consistent with the higher ductility of 17.02% for the heat-treated specimen compared with 7.14% for the as-extruded specimen. However, the second phase particles and inclusions found in the heat-treated specimen are about the same size as in the as-extruded one. The larger dimple size is a consequence of the increase of ductility resulting from the heat treatment.

Compared with the fractograph for an as-extruded specimen tested at 649°C/275.8MPa (1,200°F/40ksi), the fracture surface of a heat-treated specimen tested at 649°C/172.4MPa (1,200°F/25ksi) shows deeper and larger dimples due to a more ductile matrix. However, brittle cleavage fracture can also be observed in the heat-treated specimen which has an elongation of 15.42%, as shown in Fig. 8. A much more ductile mode prevails for a similar specimen tested at 760°C/137.9MPa (1,400°F/20ksi), for which the elongation at fracture is 26.38%.

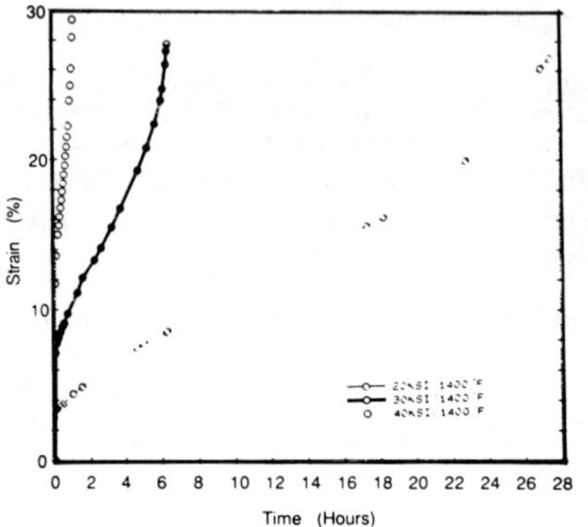

Fig. 3 Creep curves at 760°C (1,400°F).

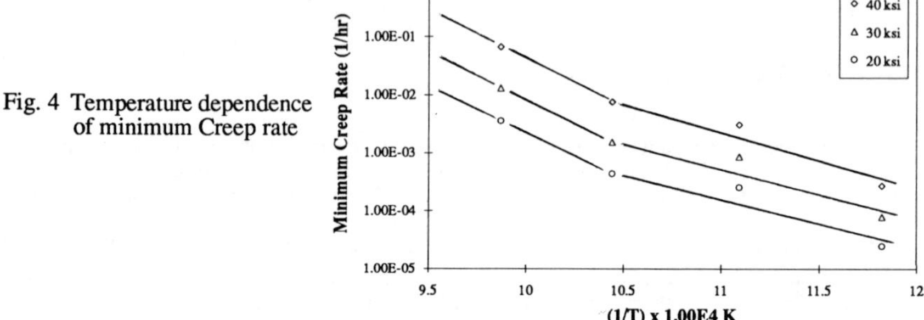

Fig. 4 Temperature dependence of minimum Creep rate

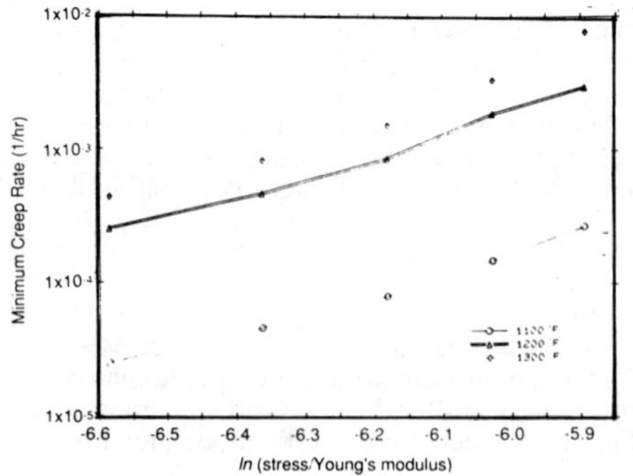

Fig. 5 Stress dependence of minimum creep rate.

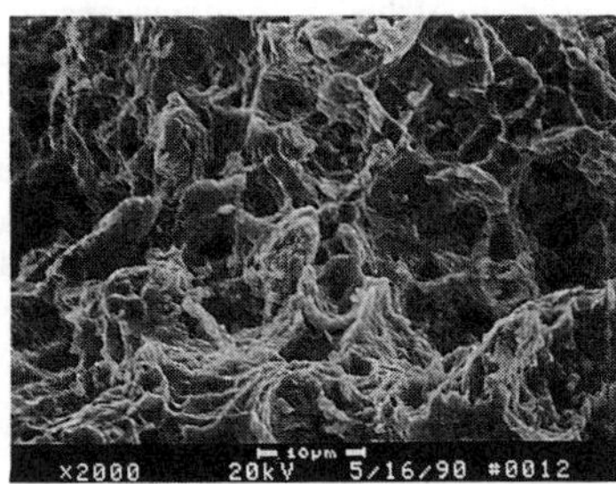

Fig. 6 SEM Fractograph of an as-extruded specimen tested at 649°C/137.9MPa showing ductile dimple fracture.

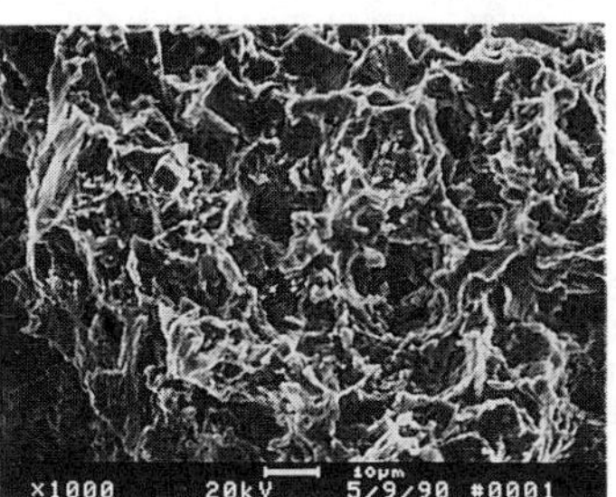

Fig. 7 A ductile fracture surface of a heat-treated specimen tested at 704°C/241.3MPa.

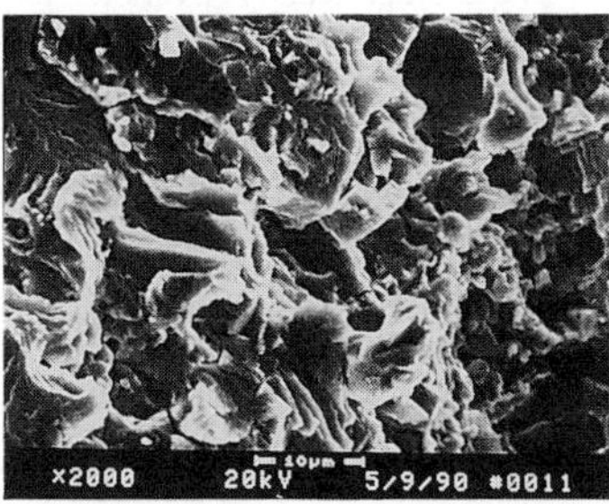

Fig. 8 Fractograph of a heat-treated specimen tested at 649°C/172.4MPa showing a mixed ductile and brittle cleavage fracture.

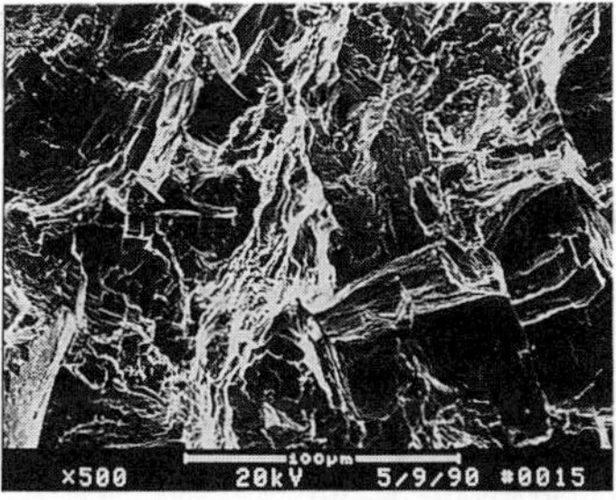

Fig. 9 A mixed fracture mode observed in a heat-treated specimen tested at 704°C/275.8MPa.

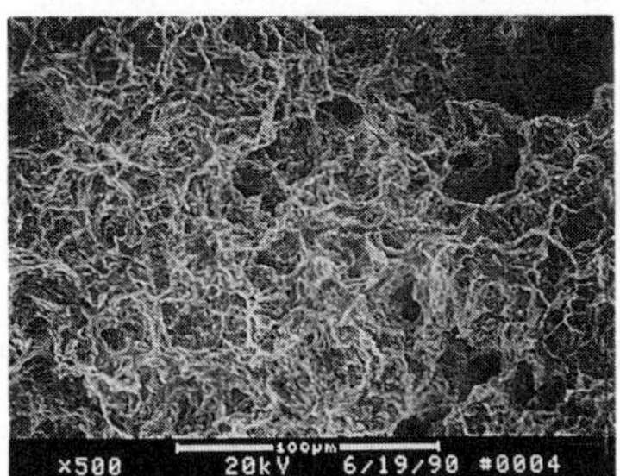

Fig. 10 Fractograph of a heat-treated specimen tested at 760°C/137.9MPa showing a wide spectrum of ductile dimple size.

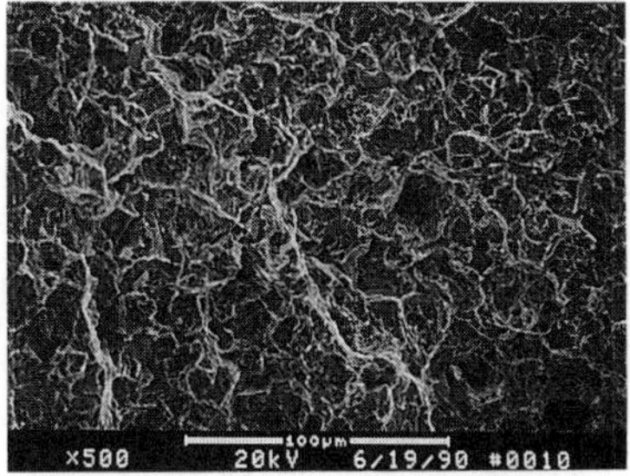

Fig. 11 Scattered brittle cleavages in a heat-treated specimen tested at 760°C/275.8MPa.

Fig. 9 shows a mixed ductile dimple and brittle cleavage transgranular fracture mode for a heat-treated specimen tested at 704°C/275.8MPa (1,300°F/40ksi). The dimple are formed by the coalescence of microvoids due to the weak matrix strength. The brittle cleavage is a result of shearing caused by the high stress applied to the specimen. This specimen failed at 11.45 hours and had an elongation of 11.74% at fracture. Figs. 10 and 11 show the fracture surfaces of heat-treated specimens tested at 760°C/137.9MPa (1,400°F/20ksi) and 760°C/275.8MPa (1,400°F/40ksi), respectively. The fracture surface of a specimen tested at 137.9 MPa (20 ksi) is mainly composed of ductile dimples, ranging from about 36 to 15 μm. This specimen failed at 27.45 hours with an elongation of 26.3%. The specimen tested at 275.8 MPa (40 ksi) displays more brittle features and less ductile dimples with an average dimple size of about 14 μm. It failed at only 1.08 hours and with an elongation of 15.37%. A higher test stress makes the material more brittle.

CONCLUSIONS

The steady-state creep behavior of the Ti-24Al-11Nb alloy can be described by the power-law equation in the stress range of 137.9 to 275.8 MPa (20 to 40 ksi) and the temperature range of 593 to 760°C (1,100 to 1,400°F). The average creep activation energy is approximately 352 kJ/mol (84.1 kcal/mol) for temperatures above 704°C (1,300°F) and decreases to 196 kJ/mol (46.8 kcal/mol) below 704°C (1,300°F). The average stress exponent value, n, is 5.6 which decreases to 5.1 below the stress 206.9 MPa (30 ksi).

The fracture surface of the as-extruded specimen shows that usually a ductile dimple transgranular fracture mechanism is predominant. Under a similar testing condition, the fracture surface of a heat-treated specimen shows deeper and larger dimples due to the increase of ductility . Small second phase particles and inclusions are observed on the fracture surface. Most of them remain uncracked during the creep/stress rupture test. Comparing the fracture surfaces of specimens tested at the same temperature for different stresses, it is found that the specimen subject to a higher stress shows more cleavage features and less dimples than the specimen under a lower stress. Thus a higher applied stress during the creep/stress-rupture test results in lower ductility.

ACKNOWLEDGMENTS

The authors wish to express their appreciation to Nuclear Metals, Inc. for providing the experimental material and Northeastern University for conducting the experiments.

REFERENCES

1. R.S. Mishra and D. Banerjee, Mat. Sci. Eng., **A130**, 151 (1990).
2. R.W. Hayes, Scripta Metall., **23**, 1931 (1989).

NOTCH EFFECT ON STRESS RUPTURE BEHAVIOR OF Ti-24Al-11Nb AT 650 °C

Jin Liang * and R. M. Pelloux *
*Department of Materials Science and Engineering, Massachusetts Institute of Technology Cambridge, MA 02139, USA.

ABSTRACT

Notch stress rupture behavior of a titanium aluminide Ti_3Al alloy, Ti-24Al-11Nb, was investigated. Three axisymetric test bar geometries were used: smooth bar (k_t = 1), circular notch (U-notch, k_t = 1.6) and British standard notch (V-notch, k_t = 4.2). Tests were performed at 650 °C, with a real-time DC potential drop (DCPD) automated data acquisition system continuously monitoring local creep deformation and damage accumulation in the notched specimens. Two microstructures were investigated, an equiaxed grain structure produced by an α_2+β heat treatment and a transformed β microstructure produced by a β heat treatment. It was found that the effects of notches on the stress rupture lives are dependent on microstructure, notch geometry, and applied stress level. For the α_2+β treated Ti-24Al-11Nb alloy, a U-notch has a notch strengthening effect in the high stress or short life region and a notch weakening effect in the low stress, long life region; a V-notch always has a notch weakening effect. The β treated Ti-24Al-11Nb alloy shows notch weakening effects both for U-notch and V-notch, with V-notch and high stress levels being the most deleterious. The theoretical DCPD response corresponding to a FEM-calculated creep deformation at a given time t was estimated and compared with the real time recorded DCPD changes. This combined DCPD-FEM analysis method provided quantitative information about the relative contributions of creep deformation and damage mechanisms (cavitation, micro- and macro-cracking) to the experimental DCPD curve in a notch stress rupture test.

INTRODUCTION

The newly developed intermetallic alloys, based on titanium aluminide Ti_3Al, have attractive properties which make them excellent candidates for application in gas turbine engines and airframe structures in advanced aircraft.[1, 2, 3] The primary advantages of Ti_3Al alloys are low density, high strength to weight ratio at high temperatures, good creep rupture strength, high moduli and an oxidation resistance which is better than that of the conventional titanium alloys.[4] The use of Ti_3Al base alloys at temperatures as high as 750 °C, where they could replace nickel base superalloys, would decrease the engine weight, increase fuel efficiency and decrease the engine operating stresses, thus achieving high durability and maximizing engine performance. However, in all the intermetallic alloys as well as in Ti_3Al alloys, the problems of limited ductility, low fracture toughness and low impact resistance are still of great concern. The successful use of the new Ti_3Al alloys for advanced engine components requires a good understanding of long time deformation and fracture behavior, and also of the notch stress rupture behavior, since these alloys with limited ductility are expected to be creep notch sensitive.[5]

In this study of the notch stress rupture behavior, a Ti_3Al alloy, Ti-24Al-11Nb, was investigated at elevated temperatures. Smooth bar and notched bars (both U-Notch and V-Notch) were creep tested at 650 °C in air. The damage processes due to creep deformation, microcracking or cavitation, macro creep crack growth, and microstructural instabilities under applied loads during long time exposure to high temperature were studied with a DC potential drop (DCPD) technique in notched specimens. Theoretical modelling with finite elements both for creep deformation history, stress redistribution and for the corresponding DCPD responses were carried out to account for the relative contribution to the overall damage by each damage process during the creep rupture tests of notched specimens.

MATERIALS AND EXPERIMENTAL PROCEDURES

Two heat treatments were given to Ti-24Al-11Nb; the first was an α_2+β heat treatment; the second heat treatment was the β treatment. The details of thermomechanical processing and heat treatments were given in Reference 6. The α_2+β heat treatment of the Ti-24Al-11Nb alloy resulted in a fine equiaxed grain structure. Microstructure for the β treated Ti-24Al-11Nb alloy has a prior β grain size of around 0.35 mm. The microstructure consists of fine acicular α_2 phase in the matrix β phase (basket-weave structure).[6]

The creep test specimen geometries are: a) a blunt notch (U-notch) with a semicircular notch (k_t = 1.6), which was selected to study the deformation and rupture behavior under a triaxial state of stress; (b) a sharp notch (V-notch) specimen with a British standard notch geometry (k_t = 4.2).[6] Each notch specimen has two identical notches 0.5" apart, so that when the specimen breaks at one notch, the unfailed notch may be sectioned to assess the state of creep damage. All tests were carried out on a 20,000 lb capacity Lever Arm Creep Tester with a 20:1 lever ratio. The temperature was maintained at 650 °C ± 3 °C.

Both U-notch and V-notch specimens were monitored in real time with the DCPD technique to investigate the processes of creep deformation, cavitation, microcracks and macrocracks. A effort was made to assess the relative damage due to creep deformation and to microcavities or microcracks, which contributed to the total recorded DCPD data. A complete description of this automated DCPD data acquisition system is given in References 6, 7 and 8.

EXPERIMENTAL RESULTS AND DISCUSSIONS

Stress Rupture Behavior

The stress rupture results of smooth, U-notched and V-notched specimens of the α_2+β and β alloys are shown in Fig. 1. For the α_2+β alloy, it is clear that notch effect on the rupture life at 650 °C is relatively small; the material showed a notch strengthening effect for U-notches at high stresses or in the short life region. However, V-notch always decreased the stress rupture life for the α_2+β alloy in the applied stress regime used in this study. For the β alloy, both U-notch and V-notch decreased the stress rupture life, with V-notch having a remarkable detrimental effect, indicating that the β alloy is more sensitive to geometric discontinuity. Comparing the stress rupture behavior of the α_2+β and β alloys at 650 °C for both smooth and notched specimens clearly shows that the β alloy has a stress rupture capability superior to the α_2+β alloy in both smooth and notch stress rupture tests.

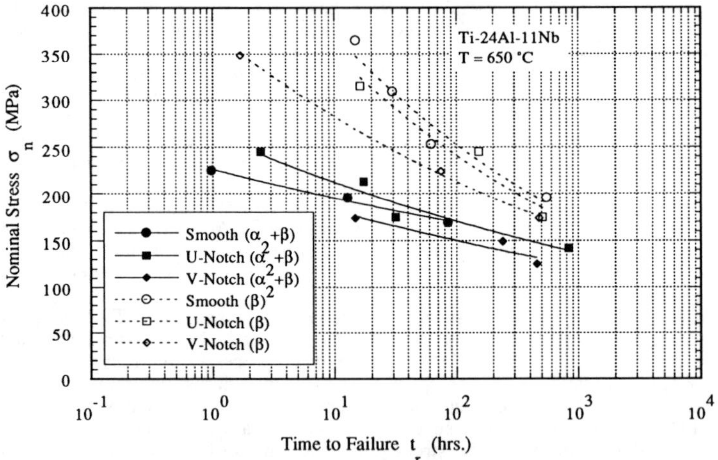

Fig.1 Stress rupture lives of smooth, U-notch and V-notch specimens for α_2+β treated and β treated Ti-24Al-11Nb at 650 °C.

DCPD Results of Notched Specimens

Fig. 2 shows the DC potential versus time curves recorded for the U-notch and V-notch specimens of the α_2+β and β alloys at 650 °C. It can be seen from these figures that the potential curves are similar to typical uniaxial creep curves. The initial transient stages resemble the primary creep stage of a uniaxial creep test, where the initial DCPD rates are high and then decrease continuously to more or less constant rates, depending on the applied stress and the alloy.

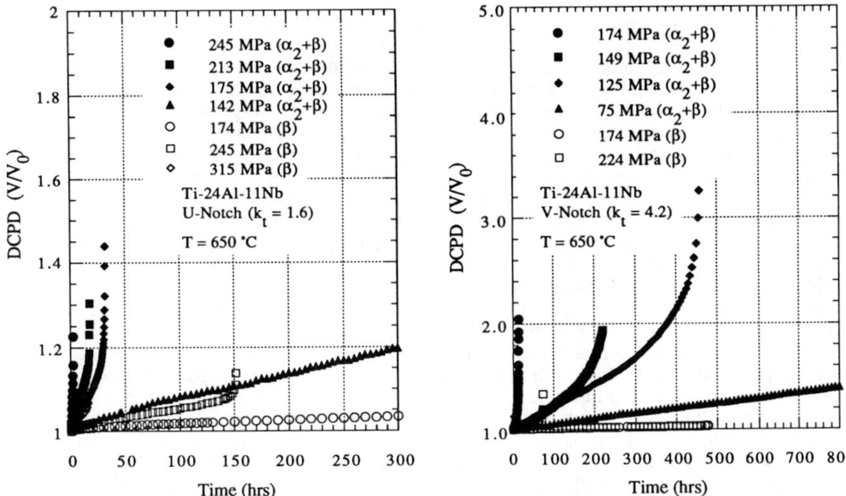

Fig. 2. DCPD (V/V_0) vs time results for U-notch (left) and V-notch (right) specimens at 650 °C for α_2+β and β treated Ti-24Al-11Nb alloys.

For the α_2+β alloy, it can be seen that following a very short primary stage, a constant DCPD rate stage (secondary stage) was very well established in all the tests. This stage usually lasted up to 0.6-0.8 time fraction to failure, and was followed by an acceleration of the potential rates in the last stage (tertiary stage). The maximum attainable DC potentials measured from the failed notches are dependent on the applied stress level, increasing with decrease in the applied stress (longer rupture life). This indicates that either creep rupture ductility was enhanced during the long time creep test or local damage (mainly macro-crack) can extend to a very large degree in the notch regions for the low stress tests. The test under σ = 142 MPa lasted 833 hours and developed a V/V_0 of about 1.8 at failure in the failed notch (80 % increase in the potential during the creep test).

The β alloy showed a similar three stage DCPD behavior, as shown in Fig. 2. However, the secondary stage lasted longer, compared with the α_2+β alloy. It is clear that the β alloy has a higher resistance to creep damage accumulation during the crack initiation stage. The tertiary stage is relatively short, indicating that the β alloy is more sensitive and less tolerant to local damage. This is further confirmed by the final DC potential values at the failed notch, which are much smaller than those for the α_2+β alloy tests.

The DCPD results for the V-notch stress rupture tests of the α_2+β alloy are given in Fig. 2 (right). The DCPD versus time curves also have three stages, with the secondary stage decreasing with increasing applied stress, indicating that cracks might be initiated in the early stages in the high stress tests. Compared to the U-notch creep rupture tests, the V-notch tests have longer tertiary stages. The attainable maximum DC potentials are also increased with a decrease in the applied stress, and they are all much higher than the U-notch tests (300% increase in the potential for the V-notch test at σ = 125 MPa as shown in Fig. 2 right). Such large DCPD increases are apparently due to creep crack growth. Thus, in the case of V-notch stress rupture tests, the transition of the DCPD rates from an approximate constant rate to an accelerated rate was associated with initiation of macro-cracks at the notch root. The creep crack growth rates after the transition points were also evaluated from the V-notch stress rupture test data in Reference 6.

The V-notch DCPD results for the β alloy (see Fig. 2 right) showed a less extensive change in the recorded DCPD, compared with the α_2+β alloy. The maximum attainable DC potential change (V/Vo) at fracture point, which represents the critical damage (either the rupture ductility or the critical crack length), is relatively small for the V-notch specimens of the β alloy, compared with the α_2+β alloy. This clearly indicates that the β alloy has a much lower fracture toughness or defect tolerance than the α_2+β alloy.

Damage Assessment in Notch Stress Rupture Tests

Notch stress rupture tests have been used to qualitatively assess the suitability of materials for designs that will contain deliberate or accidental stress raisers.[6, 9, 10] The notch stress rupture behavior of a material is the reflection of the combined effects of many factors, such as the local plastic and creep deformation, damage accumulation under triaxial stress state, the capability for the alloy to transfer local load from the highly damaged zone to the neighbor material, the resistance to local fracture and the resistance to creep crack growth. All these factors are related not only to the materials intrinsic properties but also to the specific notch geometry, specimen or component size, and the level of applied load. Thus a variety of notch stress rupture behaviors could be demonstrated in a same material. In spite of these complex metallurgical and mechanical interactions in the notch stress rupture tests, it is useful to divide the stress rupture life into two stages, that is, life to crack initiation and life for crack growth to the critical size. Two situations exist. First, crack initiation in the vicinity of the notch root takes place very early and the rupture life is mainly due to creep crack growth. Secondly, crack initiation can take most of the life if the damage evolution in the notch effected zone and in the remaining section are slow and if the material has very poor resistance to creep crack growth and a low fracture toughness. It is clear that both the crack initiation and crack growth lives are also determined by notch geometry, the remaining ligament, load levels and intrinsic properties.

Preceding the events of crack initiation and crack growth, there is a stress redistribution process due to creep deformation. The local damage accumulation is largely determined by this process. Since the DCPD change during a notch stress rupture test can come from elasto-plastic deformation on loading, creep deformation, damage processes (cavities and void), cracking as well as microstructural changes, the recorded DCPD curves can not be elucidated unless the contribution from each single process is estimated quantitatively. In the early stage of the test, the real-time measured DCPD change is mainly due to creep deformation. When damage has developed throughout the whole or part of the notch section, the DCPD will come from both creep deformation and damage accumulation. Finally, when a macro-creep crack initiates and grows, the signal will come mainly from the crack, with a small contribution from the deformation and from global damage process. The instrumented DCPD reliably records the whole stress rupture processes in the notch region. A correct explanation of the DCPD curves recorded for all the notch stress rupture tests can give an insight into the processes of creep deformation, damage accumulation and creep crack growth in the notch regions.

Assuming that the volume and resistivity of the specimen remain unchanged during creep across a portion of the straight section of a smooth bar, the DCPD equivalent axial strain can be given by the following relation: [6]

$$\varepsilon = \frac{\Delta L}{L_0} = \left[\left(\frac{V}{V_0}\right)^{\frac{1}{2}} - 1\right] x\ 100\ \ (\%) \tag{1}$$

where V_0 and V are the voltages across the gage section at the beginning of the test and at time t. Tests on unloaded smooth specimens and comparison of LVDT and DCPD measured creep curves confirmed that the above assumptions and Equation (1) are valid for Ti-24Al-11Nb under the test conditions used in this study.[6] For notched specimens, the relation between creep deformation and DCPD can not be estimated in such a simple way due to nonuniformity both for deformation and DC potential in the notch region. Thus, in order to assess the damage processes in the notch stress rupture tests of Ti_3Al alloys, a combined DCPD-FEM modelling was developed.[6] First, an elasto-viscoplastic finite element analysis was used to calculate creep deformation and stress redistribution for both U-notch and V-notch specimens by using a power law creep constitutive equation, [9, 10, 11] given as the following:

$$\dot{\varepsilon}_{ij}^{c} = \frac{3}{2} A\, \sigma_e^{n-1}\, \dot{\sigma}_{ij} \tag{2}$$

which simplifies to $\dot{\varepsilon} = A\,\sigma^n$ (Norton's creep law) in uniaxial tension; where $\dot{\sigma}_{ij}$ is the deviatoric stress tensor, σ_e is the von Mises equivalent stress. The theoretical DCPD corresponding to the calculated deformation history was solved through a finite element method, which solves the following equation in a cylindrical coordinate system:[6, 12, 13]

$$\frac{1}{r}\frac{\partial}{\partial r}\left(k_r \frac{\partial \phi}{\partial r}\right) + \frac{\partial}{\partial z}\left(k_z \frac{\partial \phi}{\partial z}\right) = 0 \tag{3}$$

where ϕ is potential, k_r and k_z are the electrical conductivities along r and z directions. Therefore, a theoretical DCPD (V/V_0) versus time curve is derived, which is the contribution of creep deformation in a notch stress rupture test. Fig. 3 (left) compares the calculated DCPD curves with the real-time measured curves for U-notch specimens of the $\alpha_2+\beta$ alloy. A good agreement is obvious for the two lower stress cases in the early stage of the tests. Taking into consideration the complexity of the notch stress rupture processes and the calculations themselves, as well as the simplified constitutive relation utilized, the estimated DCPD accounts fairly well for the main source of the recorded real-time DCPD change during the early stage of the tests. It is clear that the deviation of the measured DCPD curves from the calculated one indicates that other creep damages took place. In this case, macro-creep crack initiation and propagation might be the main source of the acceleration of DCPD in the U-notch specimens of the $\alpha_2+\beta$ alloy. For the V-notch stress rupture tests, it is expected that a creep crack could be initiated very early during the test, thus the main source of the DCPD change comes from crack extension. The calculated DCPD curves corresponding to the FEM creep deformation are given in Fig. 3 (right) for V-notch specimens of the $\alpha_2+\beta$ alloy. It is clear that the FEM calculated and measured DCPD curves showed only a very short period of consistence. It is in agreement with the fact that the V-notch stress rupture lives were largely determined by the creep crack growth.

Metallographic examinations were also carried out longitudinally through unfailed or failed notches, as shown in Fig. 4. For the $\alpha_2+\beta$ alloy, multiple site creep crack initiation was found both for U-notch and V-notch specimens (Fig. 4a). Cavitation along the $\alpha_2+\beta$ interfaces also can be found away from the notch tip for U-notch specimens. Crack growth is not necessarily along the intergranular path. For the β alloy, creep cracks were found to be formed by the coalescence of cavities (Fig. 4b) in the subsurface of notch roots both for U-notch and V-notch specimens. A thorough analysis in Reference 6 showed that the cracking modes and damage accumulation observed near the notch roots are well related to the stress redistribution history, which is dependent on the alloys' creep constitutive relation (mainly the stress exponent n in a Norton's power law) and the stress rupture ductility. It is found that for the blunt notch (U-notch), creep deformation and damages could be concentrated near the notch tip or near the specimens centers or in between.[6] A high n or a large primary creep facilitates transfer of the high stress from the notch tip towards the center and develops a rather smooth stress distribution across the notch throat and may lead to a notch strengthening effect. For the V-notch, both creep deformation and stress are highly concentrated in the notch tip region, leading to early crack initiation and to notch weakening.

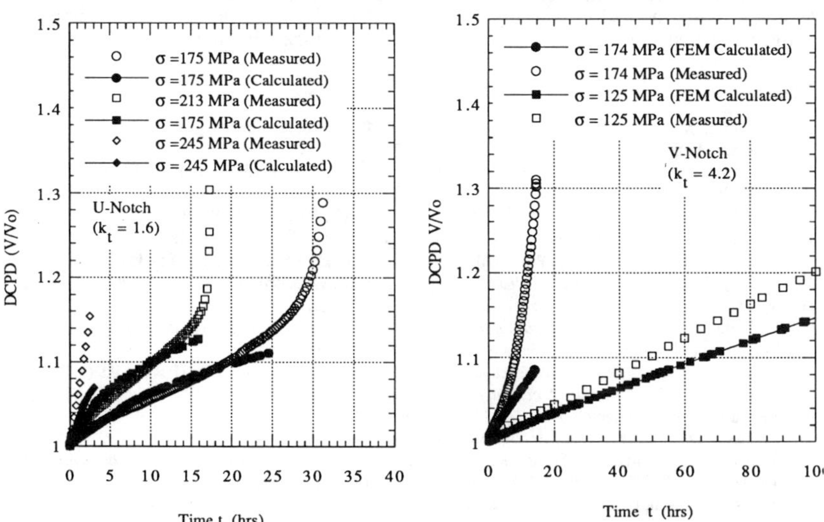

Fig. 3 Comparison of calculated DCPD curves with the real-time measured curves for U-notch specimens of the $\alpha_2+\beta$ alloy.

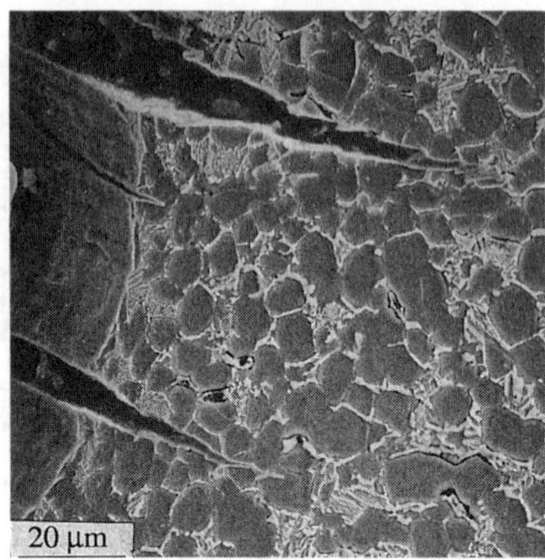

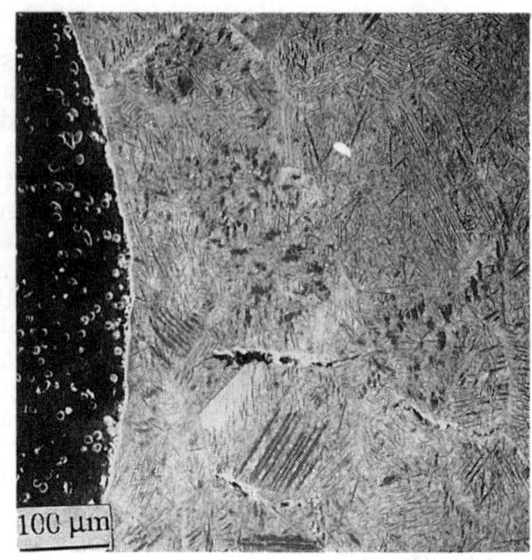

(a) α_2+β (U-notch, 142 MPa/833 hrs) (b) β (U-notch, 315 MPa/16.2hrs)

Fig. 4. Metallographs of creep cracks at 650 °C in Ti-24Al-11Nb alloy.

CONCLUSIONS

1). Effects of notches on the stress rupture lives of Ti_3Al alloys are dependent on microstructure, notch geometry and applied stress level. The β treated Ti-24Al-11Nb has a superior stress rupture capability to the α_2+β alloy, but it is also more sensitive and less tolerant to local damage. For the α_2+β treated Ti-24Al-11Nb alloy, a U-notch has a notch strengthening effect in the high stress or short life region and it shows a notch weakening effect in the low stress or long life region; a V-notch always has a notch weakening effect. The β treated Ti-24Al-11Nb alloy shows notch weakening effects both for U-notch and V-notch, with V-notch being the most deleterious at high stress levels.

2). The combined DCPD-FEM analysis method provided quantitative information about the contribution of creep deformation and other damages (cavitation, micro- and macro-cracking) to the real-time recorded DCPD changes. Coupled with the experimental DCPD technique, this method can be used to verify the validity of the three dimensional creep deformation constitutive relations and the criterion of crack initiation in a notched specimen.

References:

1. Blackburn, M. J., Ruckle, D. L., Bevan, C. E., AFML-TR-78-18, Air Force Materials Laboratory, Wright-Patterson AFB, Ohio, (1978).
2. Blackburn, M. J., Smith, M. P., AFML-TR-82-4086, Air Force Materials Laboratory, Wright-Patterson AFB, Ohio, (1982).
3. Blackburn, M. J., Smith, M. P., WDDC-TR-89-4095, Wright-Patterson AFB, Ohio, (1989).
4. Lipsitt, H. A., *High-Temperature Ordered Intermetallic Alloys*, edited by C. C Koch, C. T. Liu, N. S. Stoloff, (Mat. Res. Soc. Symp. Proc. **Vol. 39**, MRS, Pittsburgh, Pennsylvania, 1985) 351.
5. Stoloff, N. S., *High-Temperature Ordered Intermetallic Alloys*, edited by C. C. Koch, C. T. Liu, N. S. Stoloff, (Mat. Res. Soc. Symp. Proc. **Vol. 39**, MRS, Pittsburgh, Pennsylvania, 1985) 3.
6. Jin Liang, *Creep Deformation and Rupture Behavior of Titanium Aluminide Ti_3Al Alloys*, Ph.D. Thesis, Department of Materials Science and Engineering, MIT (1992).
7. Vasatis, I. P., *The Creep Rupture Behavior of Notched Bars of IN-X750*, Ph.D. Thesis, MIT (1986).
8. Peltier, J. M. , *Creep Rupture Mechanisms in Notched Specimens of René 95*, Ph.D. Thesis, MIT (1987).
9. Hayhurst, D. R., Henderson, J. T., *Int. J. Mech. Sci.*, **19** (1977) 133.
10. Hayhurst, D. R., Leckie, I. A , Henderson, J. T., *Int. J. Mech. Sci.*, **19** (1977) 147.
11. Yoshida, M., Levaillant, C., Piques, R., and Pineau, A., *High Temperature Fracture Mechanisms and Mechanics*, EGF6, Edited by P. Bensussan and J. P. Mascarrell (1990) 3.
12. Hinton, E., Owen, D. R. J., *An Introduction to Finite Element Computations*, Pineridge Press Limited, Swansea, U. K. (1979).
13. Ritter, M. A., and Ritchie, R. O., *Fatig. Engin. Mater. Struct.*, **5** (1982) 91.

MECHANICAL TWINNING IN DUPLEX Ti-48Al-2Nb-2Cr AT 1038K

Zhe Jin and Thomas R. Bieler

Department of Materials Science and Mechanics
Michigan State University, East Lansing, Michigan 48824-1226

ABSTRACT

As a part of an ongoing study of creep deformation mechanisms, mechanical twinning behavior was investigated in the lamellar region of a creep specimen. A multi-stress jump creep test was performed and the microstructures before and after deformation were investigated using optical and transmission electron microscopy. The identification of mechanical twins in a lamellar microstructure is discussed. Extensive mechanical twinning was observed in lamellar regions, in addition to slip and subgrain formation. The occurrence of mechanical twinning depended on the lamellar orientation with respect to the tensile axis. The mechanical twins are analyzed and discussed in terms of possible crystallographic twinning systems. In this case, a maximum resolved shear stress criterion for mechanical twinning is proposed to account for the observed orientations.

INTRODUCTION

Gamma Titanium aluminide (TiAl) has long been considered as a potential candidate material in high temperature and high performance applications. In the past decade, much work has been done in order to understand its deformation mechanisms both at room temperature and high temperatures and to improve its ductility at low temperature. Most experiments have been conducted in short times, such as tensile strength measurements [1-3]. Much less has been done with long term measurements such as creep experiments. Since one goal of TiAl components will be to replace nickel and cobalt base superalloys in aircraft applications, understanding creep deformation at high temperature is necessary.

Some microstructure investigations in creep deformed TiAl [4-9] have shown that creep deformation occurred by a combination of dislocation slip, mechanical twinning, subgrain formation, grain boundary sliding, and recrystallization. It has been clear from the these previous studies that the contribution of these deformation mechanisms to creep deformation will be different for the different testing conditions. For example, creep tests in an investment cast Ti-48Al-2Nb-2Cr show power-law creep with stress exponent $n=3$ at low stresses and $n=7$ at higher stresses [4], but the rate limiting mechanisms of creep deformation have not yet been identified.

For an individual deformation behavior, there are some differing observations and interpretations. For instance, the mechanical twinning was observed in some specimens [5,8,9] but not in others [6]; pseudo-twinning was found in a creep deformed TiAl specimen [8]. The objective of this paper is to present some preliminary observation and analysis of mechanical twinning in creep deformed investment cast Ti-48Al-2Nb-2Cr.

EXPERIMENTAL PROCEDURES

The material used in this study was investment cast γ TiAl made by Vacuum Arc Remelting (VAR) and casting at Howmet Corp., Whitehall, MI. Rods 152mm in length and 16mm in diameter were HIPed at 1533 K, 172 MPa for 4 hours to eliminate solidification porosity. The composition of this material is shown in Table I. The material was then heat treated in inert Ar atmosphere at 1573 K for 20 hours and gas fan cooled with a cooling rate of 65 K/min to produce a duplex microstructure with $(\gamma+\alpha_2)$ lamellar grains plus equiaxed γ grains. The rods were machined into 62 mm long tensile specimens having a 25 mm gage length and diameter of 5 mm.

TABLE I Composition of the γ TiAl specimen

	Ti	Al	Nb	Cr	Fe	Cu	Si	O	N	H
(wt%)	bal	32.85	4.47	2.95	0.03	<0.01	0.02	555ppm	53ppm	23ppm
(at%)	bal	47.4	1.9	2.2						

The specimen was deformed in tension in an ATS Stress-Relaxation/Creep testing machine. Strain was measured with an external extensometer. A multi-stress drop test was performed at 1038 K in air, starting at 276 MPa and dropping the stress in increments to a final stress of 103 MPa at about 20% strain. The specimen was furnace cooled to room temperature at the final stress. The maximum temperature difference between the two ends of the specimen was within 3K.

Microstructural investigation on the specimen before and after deformation was carried out using optical and transmission electron microscopy. Specimens for optical microstructure were prepared by making longitudinal sections from the deformed specimen and the original bar. The specimens were etched using Kroll's reagent to view grain structures after the specimens were ground and polished. For TEM investigation, however, 0.7 mm thick slices were cut in both longitudinal and transverse directions from deformed and undeformed samples. 3 mm diameter disks were cut from the slices using an ultrasonic cutting machine and ground to about 0.1 mm thick. The disks were finally thinned using a double jets electropolishing system using a 10% sulfuric acid + methanol solution at 253K. The TEM investigation was performed on a HITACHI H800 transmission electron microscope with an accelerating voltage 200 kV.

RESULTS

Deformed and undeformed specimens were investigated. The initial microstructure after heat treatment was a duplex structure with $(\gamma+\alpha_2)$ lamellar colonies and equiaxed γ grains (Fig. 1(a)). After deformation, cross twinning configurations were observed within lamellar colonies as indicated by the arrows in Fig. 1(b). One possible form of mechanical twinning occurs parallel to the existing lamellar interfaces, which is denoted as "parallel twinning" hereafter. But it is difficult to distinguish parallel twins from original lamellae in their images since both have the same image characteristics.

In TEM images of deformed microstructures, mechanical twinning was more evident. In Fig. 2(a), both parallel and cross deformation twins were observed in an area where two differently oriented lamellar colonies intersected one another. The diffraction pattern in Fig. 2(b) shows a true twin relationship typical of the interface observed in most original lamellar laths. In Fig. 2(a), the vertically oriented lamellae were tilted about 5° away from the tensile axis, and the inclined lamellae were 42° from the tensile axis. This relationship was observed in several places in the specimen, such as the images reported in [8]. Mechanical twinning occurred in the inclined lamellar colony parallel to the lamellar interfaces and across the vertical lamellar colony by twinning shear in (111)[21$\bar{1}$] and (111)[1$\bar{2}$1] systems. Therefore, this is parallel twinning with respect to the inclined lamellae, but it is also cross twinning with respect to the vertical lamellar colony. An analysis of twinning systems is described in reference [8]. The mechanical twinning is also evident as a reduction of lamellar lath width. By comparison of lath widths of inclined lamellar colony in intensively twinned area "A" to those of the same lamellar colony in untwinned or less twinned area "B" in Fig. 2(a), it is obvious that the parallel twinning results in the reduction in lamellar width.

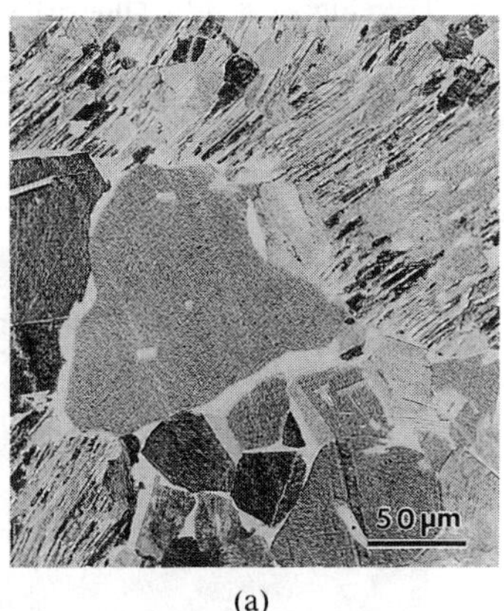

(a)

(b)

Figure 1 Optical micrographs of (a) initial microstructure and (b) after creep deformation

ANALYSIS

To clearly identify mechanical twins, one must consider two different lamellar structures. There are two types of lamellar structures, depending on phase transformation procedures. One is formed by an α_2-Ti_3Al plate growing into γ-TiAl phase. This lamellar structure has the orientation relationship $\{111]_{TiAl}$ // $(0001)_{Ti3Al}$ and $<110>_{TiAl}$ // $<1120]_{Ti3Al}$ between the γ and α_2. Since all $<1120]$ directions in Ti_3Al are equivalent to one another, there is only one α_2 crystal orientation and only the original γ crystal orientation in the grain [9]. The diffraction pattern of this type of lamellar structure does not show any twin relationship, so one can easily distinguish the mechanical twins from this type of lamellae.

The second type of lamellar structure is formed by the growth of γ laths or plates into the α or α_2 phase (in two phase $\alpha+\gamma$ or $\alpha_2+\gamma$ regions). The γ phase laths in this structure have four different orientation relationships [11]:

(i) When $[110]_1//[110]_2$ in (111) interface, where the subscripts "1" and "2" refer to two adjacent lamellae under consideration, either an antiphase boundary or no interface is formed between the two adjacent γ laths.

(ii) When $[110]_1//[101]_2$ or $[110]_1//[011]_2$ (these two relations are equivalent in $L1_o$

crystals), a 120° rotated antiphase boundary is formed where the c-axis of the two adjacent γ laths are in the same cube plane but perpendicular to each other. These two orientation relationships between the two adjacent γ laths are not twin-related. Therefore, identification of mechanical twins from these two lamellae is not difficult

(iii) When $[\bar{1}10]_1//[\bar{1}10]_2$, the two lamellar laths have a $(111)[11\bar{2}]$ true twin relationship.

(iv) When $[110]_1//[101]_2$ or $[110]_1//[011]_2$, which are equivalent to each other, a pseudo twin is formed. For these two orientation relationships between two lamellar laths, however, the diffraction patterns are not different than those from the mechanical twins. Therefore, identifying the mechanical twins from the twin-related lamellae by their diffraction patterns is not possible [1,12].

In this study, since the initial lamellae are twin-related, the mechanical twins can be identified by their morphologies, that is, by their cross twinning configurations, by their lath width reduction or by their twin end morphologies. Therefore, the diffraction pattern method is reliable for the lamellar structure formed by α_2 plates growing into γ phase. For the twin-related lamellar structure, however, the diffraction pattern method cannot be used, so mechanical twins must be identified by their geometric configurations [8,9].

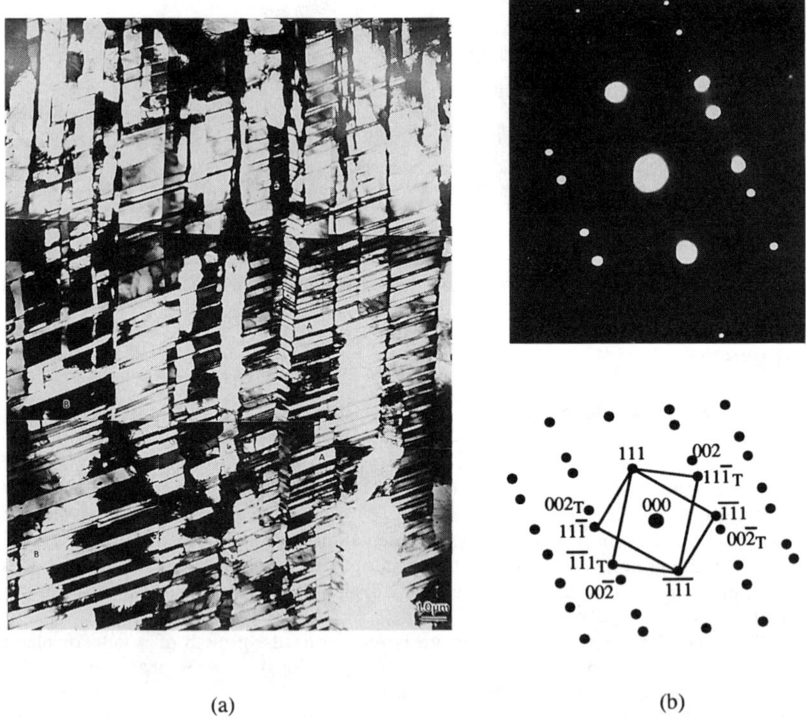

Figure 2 Twinning of two intersected lamellar colonies (a) and diffraction pattern of true twins.

DISCUSSION

Analyzing the atomic rearrangement after the operation of $(111)[2\overline{1}\overline{1}]$ and $(111)[1\overline{2}1]$ twinning systems, it was found that the resultant crystal structure was pseudo-twin related across the twin plane [8]. This result is contrary to a traditional energetic consideration since the pseudo-twinning is associated with a creation of an antiphase boundary element, in addition to the normal twinning shear deformation observed in disordered fcc crystals. Based on this consideration, this type twinning has been described as "forbidden" [13,14]. The observation of this "forbidden" twinning can not be interpreted on the basis of an energy criterion. Since the mechanical twinning is associated with a stress induced homogeneous shear on {111} twin planes, and the strain energy does not vary much between true- and pseudo-twinning, the shear strain energy dominates the energy to form the twin rather than anti-phase boundary energy. Therefore, a maximum resolved shear stress criterion for mechanical twinning in TiAl has been proposed [8].

The high density of mechanical twins (rather than a high dislocation density) was observed in the area of high local stress concentrations where the two lamellar colonies were intersected (Fig. 2(a)). This indicates that the stress concentration is relaxed by the mechanical twinning rather than the dislocation emission. The contribution of mechanical twinning to accommodate stress concentrations at triple grain points has been observed in specimens deformed into the steady state creep regime [9]. This observation of twin-toughening is contrary to the more conventional idea of twins generating micro-cracks [14]. In the case of creep deformation of TiAl, the stress resulting from mechanical twinning is relaxed either by another twinning process at ends of the deformation twins (the resultant twins are called "accommodation twins" in order to distinguish them from other twins) or by initiating dislocation slip. Accommodation twins become thinner in morphology as the stress relaxation proceeds by twinning. The accommodation twin size also depends on the local stress gradients; the sharper the local stress concentration, the thinner and the smaller the accommodation twins [9].

It has been observed that the mechanical twinning results in a finer lamellar structure during creep deformation [8,9,15]. Since the mechanical twinning follows the maximum resolved shear stress criterion, both true-twinning and pseudo-twinning occur depending on the crystal orientations. In the case of pseudo-twinning, since the pseudo-twin plane is in a higher energy configuration due to the antiphase boundary component (and therefore unstable), a very thin α_2 plate may precipitate along the pseudo-twin plane with an orientation of $(0001)_{\alpha\text{-}2}$ // $\{111\}_{twin\ plane}$ and $<11\overline{2}0]_{\alpha\text{-}2}$ // $<110>_{twin}$ to reduce the interface energy [9].

Mechanical twinning across lamellar interfaces or grain boundaries strongly depends on the boundary structures. Large angle grain boundaries are more effective barriers for twinning. Fig. 3 shows

Figure 3 Interactions of mechanical twinning with lamellar interfaces and grain boundaries.

that the extensive twinning has occurred across the lamellar interfaces, which are usually symmetrically tilted or twisted Σ3, Σ2 or Σ6 boundaries [8,16], but the twinning is effectively blocked by the lamellae-equiaxed γ grain boundary. The high density of dislocations instead of twins in the equiaxed γ grain illustrates this phenomenon. Therefore, mechanical twinning differs in lamellar structures and in equiaxed γ grains.

CONCLUSIONS

1. Mechanical twinning is an important deformation mechanism within the lamellar colonies during creep deformation in duplex γ-TiAl.
2. Mechanical twins are formed by either true twinning or pseudo twinning. The formation of mechanical twins follows the maximum resolved shear stress criterion.
3. Mechanical twinning more easily traverses the lamellar interfaces than the large angle grain boundaries.
4. Mechanical twins can be identified by diffraction patterns from the lamellar structures formed by γ --> $\gamma+\alpha_2$ reaction, but for the lamellar structures formed by α --> $\alpha+\gamma$ or α_2 --> $\alpha_2+\gamma$ reaction, the mechanical twins must be identified by their geometric configurations.

ACKNOWLEDGEMENT

This work was supported by a research contract with Howmet Corp., Whitehall, Michigan.

REFERENCES

1. D. Shechtman, M.J. Blackburn and H.A. Lipsitt, Metall. Trans., 5, 1373 (1974).
2. H.A. Lipsitt, D. Shechtman and R.E. Schafrik, Metall. Trans., 6A, 1991 (1975).
3. T. Fujiwara, et.al., Phil. Mag., 61A, 591 (1990).
4. D.A. Wheeler, B. London, and D.E. Larsen, Jr., Scripta Metall., 26, 939 (1992).
5. A. Loiseau and A. Lasalmonie, Mater. Sci. Eng., 67, 163 (1984).
6. J.S. Huang and Y-W. Kim, Scripta Metall., 25, 1901 (1991).
7. P.L. Martin, M.G. Mendiratta, and H.A. Lipsitt, Metall. Trans., 14A, 2170 (1983).
8. Z. Jin, and T.R. Bieler, Scripta Metall., 27, 1301 (1992).
9. Z. Jin, unpublished research.
10. Y-W. Kim, in Microstructure/Property Relationships in Titanium Aluminides and Alloys, ed. by Y-W. Kim, and Rodney R. Boyer, (TMS, Warrendale PA 1990), p. 91.
11. M. Yamaguchi, and Y. Umakoshi, Prog. Mater. Sci., 34, 1 (1990).
12. C.R. Feng, D.J. Michel, and C.R. Crowe, Scripta Metall., 23, 1135 (1989).
13. M. Yamaguchi, S.R. Nishitani, and Y. Shirai, in High Temperature Aluminides & Intermetallics, ed. by S.H. Whang, C.T. Liu, C.P. Pope, and J.O. Stiegler, (TMS, Warrendale PA 1989), p. 63.
14. M.V. Klassen-Neklyudova, Mechanical Twinning of Crystals, (Consultants Bureau, New York, 1964).
15. C.R. Feng, et. al., in Mater. Res. Soc. Proc., Vol. 194, MRS, 1990, p. 219.
16. D.S. Schwartz, and W.O. Soboyejo, in Microstructure/Property Relationships in Titanium Aluminides and Alloys, ed. by Y-W. Kim, and Rodney R. Boyer, (TMS, Warrendale PA 1990), p. 65.

BEND TESTING OF FORGED $Al_{66}Ti_{25}Cr_9$

K.S. KUMAR AND S.A. BROWN
Martin Marietta Laboratories,1450 South Rolling Road, Baltimore, MD 21227

ABSTRACT

Several recent reports on the mechanical behavior of polycrystalline $L1_2$ trialuminides, particularly those based on the ternary Al-Ti-Cr and Al-Ti-Mn systems, have indicated limited tensile ductility at low temperatures (≤623K) as measured in uniaxial tension or in three- and four-point bending. While plastic elongation in bending and in uniaxial tension was not observed for *forged* $Al_{66}Ti_{25}Cr_9$ at and below 473K, some room temperature ductility was reported in bending for this compound in the *cast and HIPed* as well as in the *extruded and annealed* conditions. This study attempts to understand this discrepancy by subjecting the forged compound to a variety of heat treatments that influence the residual dislocation content and the grain size, and then evaluating bend specimens at 473K. After identifying the heat treatment that provided the maximum ductility, a second set of bend specimens were heat treated and tested at 473K as a function of strain rate (crosshead speed). All bend tests were conducted in air. Bend specimens that were optimally heat treated were also tested at 473K at a slow strain rate in diffusion pump oil to isolate the possible role of environment on ductility. Notched bend specimens were independently tested in the as-forged condition as a function of temperature to obtain a measure of fracture toughness.

INTRODUCTION

The $L1_2$ trialuminides $Al_{66}Ti_{25}Cr_8$ and $Al_{66}Ti_{25}Mn_9$ have been evaluated in three- and four-point bending at room- and elevated temperatures [1-3]. In one of these studies [1], the specimens from the two compounds were in the as-forged condition and were evaluated in three-point bending. In the other two studies [2,3], four-point bend tests were conducted and specimens were either in the extruded and annealed condition [2] or, in the cast and HIPed condition [3]. Subsequently, these compounds have been characterized in the as-forged condition in uniaxial tension as a function of temperature [4,5]. From these studies [1,3,4], some room-temperature ductility was noted in $Al_{66}Ti_{25}Mn_9$ irrespective of processing route; in the Cr-containing counterpart, while limited bend ductility at room temperature was reported in two studies [2,3], plastic deformation was not recorded in the forged material at 298K or 473K [1,5]. No explanation has been provided to date for this discrepancy, in part due to the lack of in-depth microstructural characterization of the test specimens and in part due to the absence of information on the sensitivity of mechanical properties of these materials to strain rate, grain size, dislocation content, flaw size and population, and environment.

In this paper, some of these issues have been addressed by evaluating forged $Al_{66}Ti_{25}Cr_9$ in three-point bending after it had been subjected to a variety of heat treatments. Heat-treated bend specimens were also tested at various crosshead speeds to understand the role of strain rate, and in diffusion-pump oil to examine the role of environment. To obtain an initial appreciation for fracture toughness variation with temperature, notched bend specimens of the as-forged material were tested in the range 298K - 973K and the resulting fracture surfaces were characterized.

EXPERIMENTAL PROCEDURE

Details of the casting and forging procedures as well as the measured compositions of the forgings have been previously described [1,5]. A preliminary metallographic study of small specimens that were cut from an isothermally forged pancake of nominal composition $Al_{66}Ti_{25}Cr_9$, and heat treated according to the schedules indicated in Table I, provided a range of grain sizes. Subsequently, rectangular bend specimens (32 mm x 7.9 mm x 3.1 mm) cut from the forging were similarly heat treated. The first of these heat treatments (973K/1 day) was selected to provide some stress relief (if any) in the forging without affecting the grain size. The other heat-treatment schedules were intended to provide various degrees of grain growth. Two-step heat treatments, always including the 973K/1day step, were used to better understand observed results.

TABLE I: Heat-Treatment Schedules and Measured Grain Sizes for $Al_{66}Ti_{25}Cr_9$

Processing Schedule	Grain Size (μm)
As-Forged	20
As-Forged + 973K anneal/1 day	20
As-Forged + 973K anneal/1 day + 1323K/10 min.	23
As-Forged + 973K anneal/1 day + 1373K/10 min.	53
As-Forged + 973K anneal/1 day + 1423K/10 min.	70
As-Forged + 973K anneal/1 day + 1473K/10 min.	95
As-Forged + 973K anneal/1 day + 1500K/1 day	122

The microstructure of the heat-treated, bend specimens were examined to verify that the grain sizes were consistent with those provided in Table I. Three-point bend tests were conducted at a crosshead speed of 4.2×10^{-5} mm/s at 473K in air. Load-strain curves were generated in each instance using resistance strain guages that were glued to the polished tensile surface of these specimens. Specimens suitable for transmission electron microscopy (TEM) were obtained from the undeformed regions (outside of the three-point bend span) of the bend specimens and examined using a JEOL 100CX. From this study, an optimal heat-treatment schedule was identified and a second set of bend specimens were heat treated accordingly. These specimens were tested at 473K in air at crosshead speeds ranging from 4.2×10^{-5} mm/s to 4.2 mm/s. In an effort to understand the role of environment on tensile ductility, a bend specimen was tested at 473K in diffusion-pump oil. It was hoped that the oil would continuously fill the propagating crack thereby insulating it from environmental oxygen, nitrogen and water vapor, and at 473K, it was assumed that the oil was free of moisture. Finally, as-forged, notched bend specimens were tested as a function of temperature at a constant crosshead speed, fracture toughness was calculated and the associated fracture mode was examined.

RESULTS AND DISCUSSION

Representative optical micrographs illustrating the change in grain size following various heat treatments of the forged $Al_{66}Ti_{25}Cr_9$ bend specimens are shown in Figure 1a-c. The average grain sizes in the bend specimens are about what they should be for the different heat treatments as suggested by the preliminary experiments, Table I. Electron microscopy studies revealed the presence of subgrains

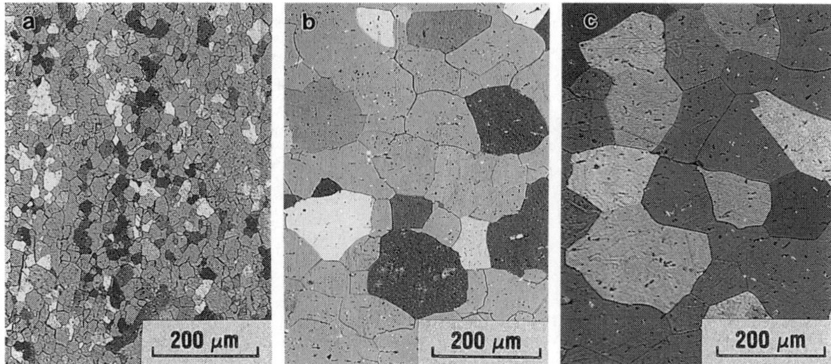

Figure 1. Effect of heat treatment on the grain size of the bend specimens: (a) forged + 973K/1 day, (b) forged + 973K/1 day + 1423K/10 min, and (c) forged + 973K/1 day + 1500K/1 day.

and matrix dislocations in the as-forged material (Figure 2a), that were not significantly affected by the 973K/1 day heat treatment (figure 2b). An additional high temperature exposure (i.e. forged + 973K/1 day + 1373K/10 min) was successful in decreasing the matrix dislocation content (Figure 2c). A heat-treatment schedule corresponding to forged + 973K/1 day + 1473K/10 min was successful in dramatically decreasing the

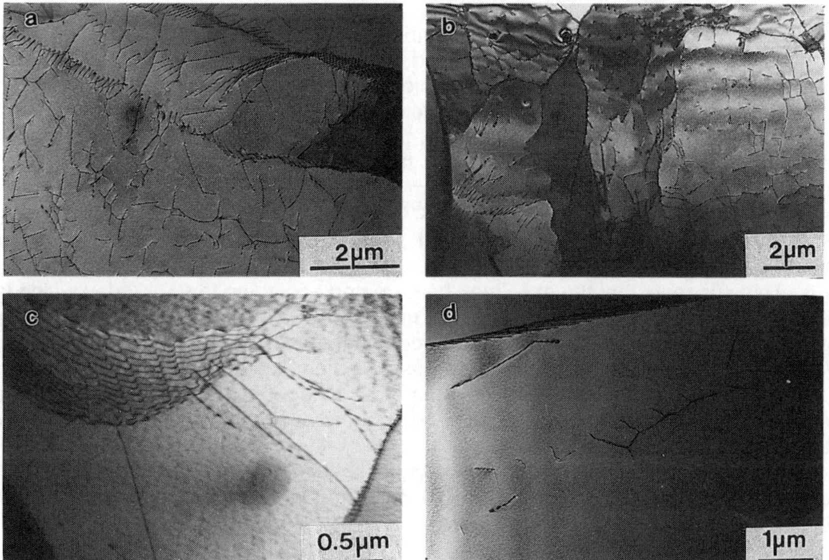

Figure 2. Transmission electron micrographs of (a) the as-forged and (b,c and d) the forged and heat treated specimens. [(b) 973K/1 day, (c) 973K/1 day + 1373K/10 min and (d) 973K/1 day + 1473K/10 min].

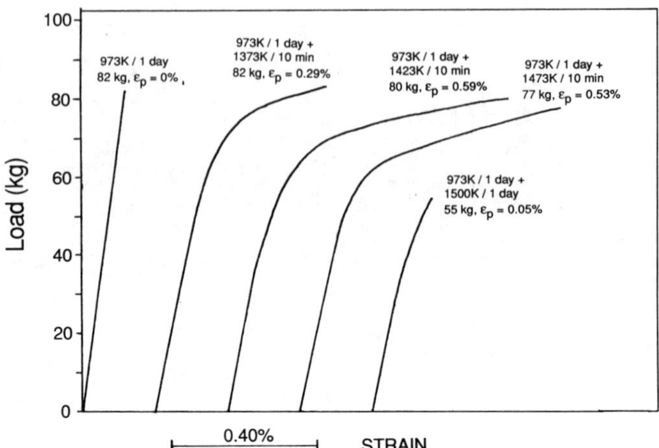

Figure 3. Load-strain curves from three-point bend tests at 473K in air of $Al_{66}Ti_{25}Cr_9$ specimens that had been subjected to various heat treatments. Crosshead speed = 4.2×10^{-5} mm/s.

matrix dislocation content and in eliminating a substantial number of subgrain boundaries (Figure 2d). Thus, while there is grain growth occurring on a more macroscopic scale, on a local level, dislocation content of the matrix is significantly affected by these heat treatments. Normally, these two factors work in opposite directions in affecting ductility.

Load-strain curves obtained from the 473K, three-point bend tests are shown in Figure 3 for the specimens subjected to different heat treatments. Specimen dimensions were maintained within a close tolerance so as to facilitate direct load comparisons within any one set of curves. Several observations can be made from the data in Figure 3. The specimen heat treated for 1 day at 973K failed elastically at a fracture load of 82 Kg. Additional high-temperature heat treatments provide measureable plasticity with a maximum measured at ~0.6% plastic strain for the specimen heat treated according to 973K/1 day + 1423K/10 min. The largest grain size material repeatedly failed without significant plasticity and at a much lower fracture load (55 Kg). Except for this last case, fracture load remains constant at ~80 Kg. This suggests that the specimen heat treated according to 973K/1day which failed at 82 Kg must have been very close to macroscopic yielding. The load at which the curve deviates from linearity gradually decreases with an increasing grain size (i.e. increasing heat-treatment temperature), as would be expected. It is interesting to note that the material fails at a fairly constant load independent of the amount of plastic strain, including none at all. This may be suggestive of a critical flaw size that can be tolerated at this temperature.

Load-strain curves for a set of bend specimens that had been heat treated according to 973K/1 day + 1473K/10 min and tested at 473K in air at various crosshead speeds are shown in Figure 4. Ductility, fracture load and load at which the curves deviate from linearity are all fairly constant over a large range of crosshead speeds (4.2×10^{-5} mm/s to 0.85 mm/s) before an abrupt loss in ductility is observed and is accompanied by elastic failure at significantly lower loads. Exact reasons for this behavior remains unknown at present. Neither the polished specimen surfaces prior to testing nor the fracture surfaces bore any signs that were indicative of the presence of abnormal flaws that could account for the elastic failure at such low loads.

In an attempt to understand if the environment posed an inherent limitation on the maximum achievable ductility at 473K, a bend specimen was immersed in diffusion pump oil (this is stable at 473K) and tested at a crosshead speed of 8.5 x 10^{-4} mm/s. The load-strain curve is compared to its counterpart tested in air (Figure 5). The

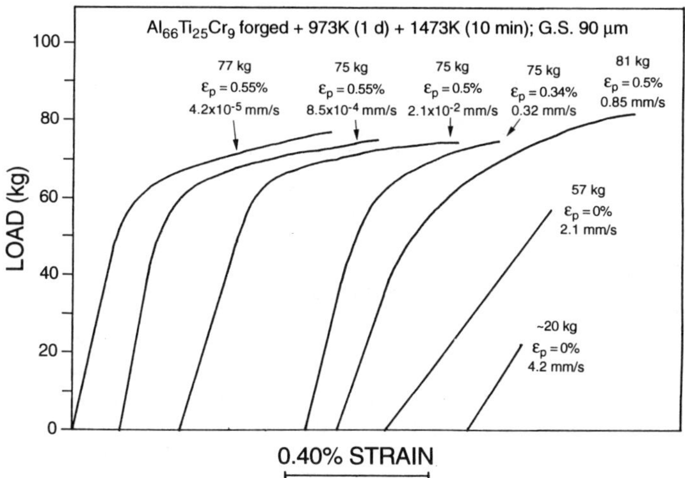

Figure 4. Load-strain curves at 473K at various crosshead speeds. Bend specimens were forged + heat treated (973K/1 day + 1473K/10 min); grain size ~90 μm.

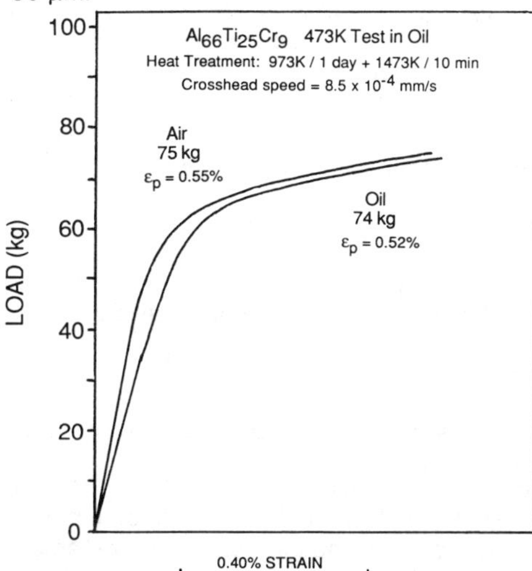

Figure 5. A comparison of the load-strain curves of heat-treated bend specimens tested at 473K in air and in diffusion pump oil using a similar crosshead speed.

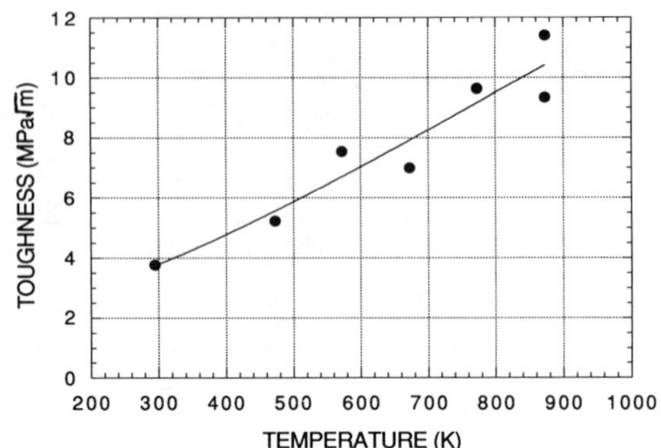

Figure 6. Effect of temperature on fracture toughness of as-forged $Al_{66}Ti_{25}Cr_9$.

two curves almost overlap, fracture in both cases occurring at about the same load and plastic strain; this suggests the likely absence of environmental embrittlement at 473K.

Fracture toughness of the as-forged material was measured using notched (notch-tip radius of ~75 μm), three-point bend specimens. The 298K value measured by this method is in good agreement with those previously obtained using a chevron-notched, short-rod specimen. Toughness increases, but only gradually with temperature, reflecting the poor crack propagation resistance of these materials even at high temperatures. Fracture mode transitions from transgranular cleavage at 298K to predominantly intergranular failure at 873K.

CONCLUSIONS

Extensive bend testing at 473K of forged and heat treated specimens reveals that ~0.5% plastic deformation is achievable, but only in the well-annealed condition. Residual dislocations from prior working are detrimental to plasticity. This ductility can be achieved over a large range of strain rates and does not appear to be adversely influenced by the environment (air). Fracture toughness is low (~4 MPa$\sqrt{m}$) at 298K and does not increase appreciably up to ~900K (~10 MPa$\sqrt{m}$).

Acknowledgments

This effort was supported by the NASA Lewis Research Center under contract # NAS3-26069 with Dr. J.D. Whittenberger as monitor.

REFERENCES

1. K.S. Kumar, S.A. Brown and J.D. Whittenberger in High Temperature Ordered Intermetallic Alloys IV, edited by L. Johnson, D.P. Pope and J.O. Stiegler, (Mater. Res. Soc. Proc. 213, Pittsburgh, PA, 1991) pp. 481-486.
2. J.H. Schneibel, H.A. Horton and W.D. Porter, Mater. Sci. and Eng., (1992) - in press. (Proceedings of the Int. Conf. on High-Temp. Aluminides and Intermetallics held in San Diego, CA in Sept. 1991).
3. S. Zhang, J.P. Nic and D.E. Mikkola, Scripta Metall. Mater., **24**, 57 (1990).
4. K.S. Kumar and S.A. Brown, Philos Mag. A, **65**, 91 (1992).
5. K.S. Kumar and S.A. Brown, Acta Metall. Mater., **40**, 1923 (1992).

TENSILE, CREEP PROPERTIES AND MICROSTRUCTURAL CORRELATIONS IN AN EXTRUDED Ti-48Al ALLOY

G. Babu Viswanathan and Vijay K.Vasudevan
Dept. of Materials Science and Engineering, University of Cincinnati, Cincinnati, OH 45221

ABSTRACT

In this study, new combinations of heat treatment and extrusion conditions was employed to produce fine-grained, nearly lamellar microstructures in a binary Ti-48 at.% Al alloy. The effect of subsequent heat treatment on microstructure was studied and room temperature tensile properties were determined for some representative microstructures with relatively high volume fractions of lamellar grains. High tensile strengths and ductilities (upto 2.3%) were obtained and microstructural factors thought to be responsible for these are discussed. Constant load tensile creep properties for the same microstructures were evaluated in the temperature range of 700 to 815°C for stresses from 103 to 241 MPa. This data was used to determine minimum creep rates, activation energies and stress exponents and their dependence on microstructure. A significant effect of microstructure, particularly volume fraction of lamellar grains on the creep rates was observed, and creep rates comparable to those reported in quarternary Ti-48Al alloys with similar microstructures were obtained. The various results suggest that it is possible to obtain an attractive and balanced combination of room temperature ductility and high temperature creep resistance in these materials by suitably varying thermomechanical processing and heat treatment conditions.

INTRODUCTION

Alloys based on the TiAl compound are candidate materials for high temperature aerospace applications because of their unique combination of properties. The most promising alloys, which are based on the Ti-48 Al composition (compositions in at.%) with ternary or quarternary additions, are characterized by the two-phase Ti_3Al+TiAl (α_2+γ) lamellar microstructure [1]. The properties of these alloys are quite sensitive to microstructure; duplex structures show higher tensile ductilities, for example up to 2.2% in binary Ti-48Al [2,3] and up to 4% in ternary or quarternary alloys[2,4-6], whereas fully lamellar structures have poor ductility [3,6]. On the other hand, the creep resistance of the fully lamellar structure is markedly superior than that of the duplex structure [6]. One factor that controls ductility of these materials is grain size, and generally finer the grain size higher is the ductility [4]. A major challenge with regard to the application of these materials is how to balance room temperature ductility and high temperature creep resistance. In this regard, fine-grained, fully lamellar microstructures have the potential to provide this balance. However, the heat treatment conditions generally used to produce the fully lamellar microstructure lead to a coarse grain size, so that alternative processing routes for producing fine-grained, fully lamellar microstructures must be explored. In this paper, results of such a study are presented and it is shown that it is possible to obtain an attractive balance of tensile and creep properties.

EXPERIMENTAL

A Ti-48Al alloy ingot was obtained from the Duriron Company in the cast + HIP (1177°C, 2h, 105 MPa) condition. The alloy composition was determined by chemical analysis to be: Al-47.86 at.%, O-0.049 wt.%, N-0.006 wt.%, C-0.013 wt.%, H-0.002 wt.% and Ti-bal. The α-transus was determined by DTA to be 1380°C. A cut section of the ingot was solutionized at 1400°C for 4h and furnace cooled. The piece was then canned, sealed in vacuum and extruded at a temperature near the α-transus. Samples cut from the extruded material were subjected to a number of heat treatments and the microstructures observed by optical and electron optical techniques. Cylindrical blanks were obtained from the extrusion by electro-discharge machining (EDM), given selected heat treatments and machined into tensile and creep test specimens. The specimen gage length and gage diameter were 16 mm and 3.5 mm, respectively. Tensile tests were carried out in an Instron under a strain rate of 10^{-4} s^{-1}. Constant load tensile creep tests were conducted in air at 700, 768 and 815°C under stresses ranging from 105 to 241 MPa. The sample elongation was monitored by an extensometer attached directly to the sample ends. The output from the extensometer was interfaced to a Mac-IIx computer with Labview 2 software for obtaining programmed, real-time, continuous output/display of strain versus time and data storage. To determine the stress exponents, stress increment tests were used.

RESULTS

The optical microstructure of the cast+HIP material consisted largely of primary γ grains with few lamellar $\alpha_2+\gamma$ grains, whereas those of samples subsequently solutionized at 1400°C and furnace cooled were large-grained fully lamellar. The as-extruded microstructure, Figures 1a and 1b, shows that extrusion at the high temperature has resulted in a microstructure consisting of a high volume fraction of fine-grained lamellar grains, together with some primary γ grains, i.e, is nearly fully lamellar, fine-grained microstructure. The grain size of lamellar grains ranges from about 50 to 100 μm, whereas those of the primary γ grains (minor constituent) is about 10 to 30 μm. The fact that a high volume fraction of lamellar grains with a fine grain size was obtained suggests that the α grains that have formed on reheating to near the transus dynamically recrystallize during extrusion. These α grains upon subsequent cooling transform to lamellar $\alpha+\gamma$, followed by ordering of α to α_2 below the eutectoid temperature (1125°C) to produce the fine-grained $\alpha_2+\gamma$ lamellar structure.

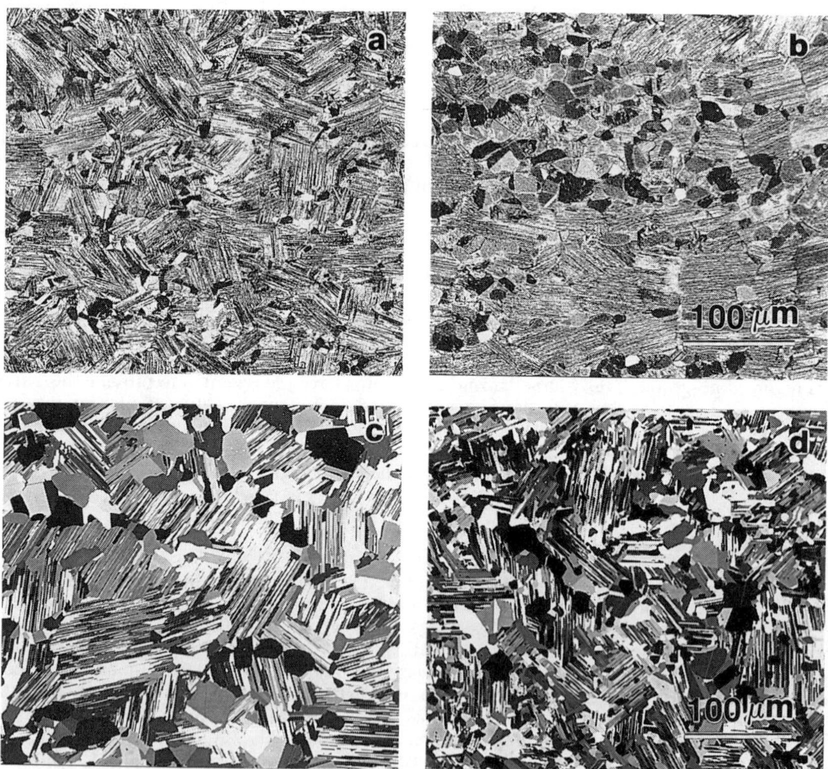

Figure 1. Optical microstructures of Ti-47.86Al alloy. (a) As-extruded transverse, (b) As-extruded longitudinal, (c) 1350°C, 0.5h + 900°C, 6h, FC, (d) 900°C, 24h, FC.

In order to assess the stability of these microstructures, extruded samples were given a variety of heat treatments both in the α+γ region as well as in the α_2+γ region. Representative optical micrographs of samples given two types of heat treatments are shown in Figures 1c and 1d. The first, Figure 1c, corresponds to a 1350°C, 0.5h + 900°C, 6h heat treatment, whereas the second, Figure 1d, corresponds to a single stabilization treatment at 900°C for 24h. The essential effect of these heat treatments is to alter grain size and volume fraction of the constituents. The micrograph in Figure 1d shows that additional formation of equiaxed γ grains has occurred during the 900°C, 24h hold, thus suggesting that the as-extruded microstructure is unstable. The 1350°C, 0.5h + 900°C, 6h treatment, Figure 1c, has resulted in some increase of both the volume fraction of primary γ grains and the lamellar grain size compared with the as-extruded microstructure. Details of the microstructure, namely, volume fraction and grain size of the constituents, and room temperature tensile properties for the two heat treatments are summarized in Table 1. As can be seen, the essential difference between the 900°C, 24h and 1350°C, 0.5h + 900°C, 6h heat treated microstructures is that the volume fraction of lamellar grains is lower and the lamellar grain size is finer in the former. Correspondingly, both ductility and strength are higher in the former. In either case, Table 1 shows that despite the relatively high volume fractions of lamellar grains, high tensile ductilities (>2.0%) are obtained.

TABLE 1

Heat Treatment, Microstructure and Room Temperature Tensile Properties

Heat Treatment	0.2%YS (MPa)	UTS (MPa)	$\%e_p$	Vol. % γ_p	Vol. % L	γ_p Grain Size (μm)	L Grain Size (μm)
1350°C, 0.5h + 900°C, 6h	369	468	2.1	35	65	20-30	50-200
900°C, 24h	437	537	2.3	45	55	10-30	50-150

Typical creep curves for the 900°C, 24h and the 1350°C, 0.5h + 900°C, 6h heat treated samples showing both the effect of stress at 768°C and of temperature at a stress of 207 MPa (30 ksi) are shown in Figures 2(a,b) and 3(a,b), respectively. The minimum creep rates are also indicated. As can be seen from Figures 2 and 3, both stress and temperature have a marked effect on the creep rate. For example, the minimum creep rate increases by a little over two orders of magnitude with either an increase in temperature from 700 to 815°C at 207 MPa or by an increase in stress from 105 to 241 MPa at both 768 and 815°C (not shown). The results of minimum creep rates versus reciprocal temperature are shown plotted in Figures 4a and 4b, and those of minimum creep rates versus stress are shown plotted in Figures 5a and 5b. It can be seen that in general the 1350°C, 0.5h heat treated material with the higher volume fraction of lamellar grains has lower creep rates. This points to the beneficial effect of the lamellar structure on creep resistance. Based on Figures 4a and 4b, the apparent activation energy, Q, for the two cases, namely, the 900°C, 24h and the 1350°C, 0.5h heat treated materials is obtained as ~280 and 350 kJ/mole, respectively. Also, the stress exponents, n, determined from Figures 5a and 5b, have values around 5 for both microstructures.

DISCUSSION

The results presented in the foregoing suggest that fine-grained, fully/nearly-fully lamellar microstructures can be produced successfully by controlling the thermo-mechanical processing conditions. These microstructures are, however, unstable and formation of additional grains of γ can occur on heat treatment in the α_2+γ region. Nevertheless, it is possible by subsequent heat treatment in the α+γ region at a suitable temperature to produce stable, relatively fine-grained, fully/nearly-fully lamellar microstructures with attractive and balanced room temperature ductility and creep properties. Although the general effects of microstructure on tensile properties are known, a basic understanding of the influence of various microstructural parameters on the deformation mechanisms is yet to emerge. This is partly because changes in one parameter, for instance, grain size, are also accompanied by changes in another, namely, volume fraction of

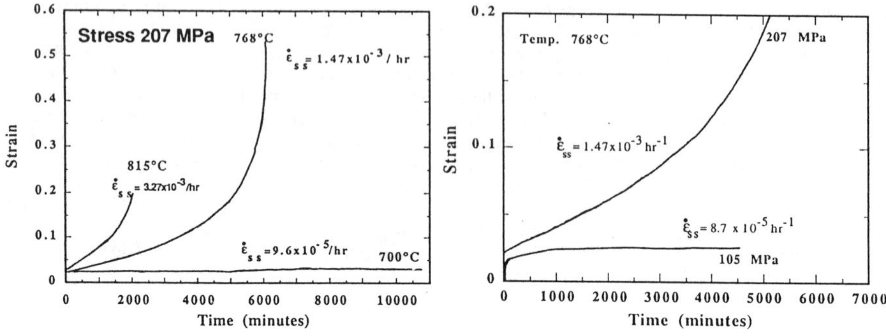

Figure 2. Typical creep curves for the Ti-48Al alloy extruded and heat treated at 900°C for 24h showing the effect of temperature and stress.

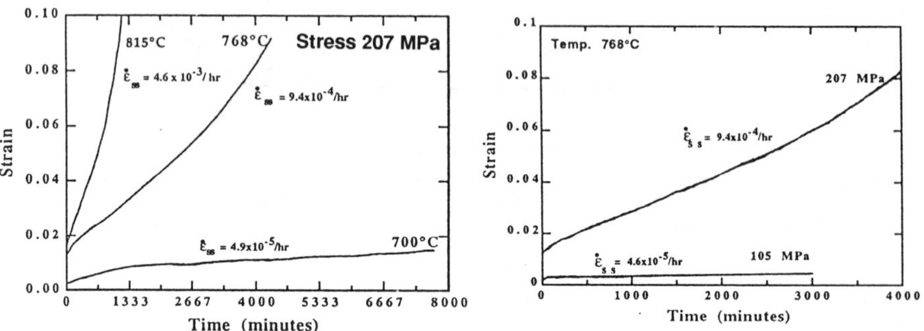

Figure 3. Typical creep curves for the Ti-48Al alloy extruded and given a 1350°C, 0.5h + 900°C, 24h heat treatment showing the effect of temperature and stress.

constituents. Within the scope of the present work, one factor that can be stated as important for high ductility and strength is fine grain size (Table 1). The numerous interfaces between $\gamma/\gamma/\alpha_2$ in the lamellar grains, which can serve as sources for slip activity, are also likely to be important in governing strength and ductility, as suggested in some recent reports [7,8]. Detailed TEM analysis of the microstructure of both heat treated and deformed samples is required to shed light on these issues, which is the subject of ongoing research.

With regard to the creep properties, the present results indicate that in duplex structures, creep resistance increases with an increase in volume fraction of lamellar grains. This holds true especially at lower test temperatures (700 and 768°C), whereas at higher temperatures (815°C) the minimum creep rates appear to approach each other (results not shown herein). The values of both Q and n obtained in the present study fall within the range reported in TiAl alloys by other investigators [6, 9,10]. Furthermore, the Q values are near 291 kJ/mole reported recently for self diffusion of Ti in TiAl [11]. Thus, based on the measured Q and n values, it appears reasonable to conclude that dislocation power law creep, i.e., climb of dislocations [12] most likely dominates

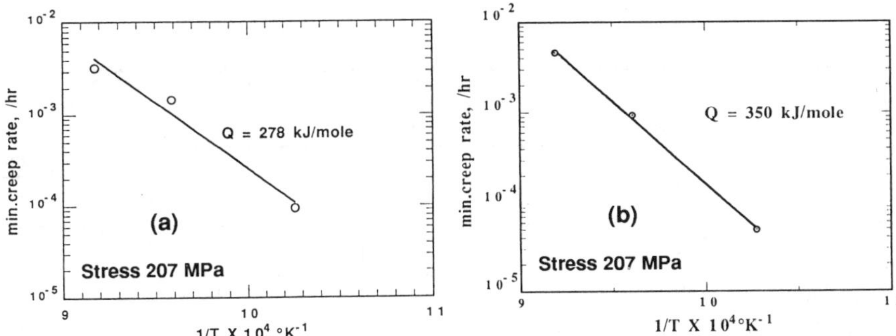

Figure 4. Minimum creep rates versus reciprocal temperature for the Ti-48Al alloy extruded and heat treated. (a) 900°C, 24h and (b) 1350°C, 0.5h + 900°C, 6h.

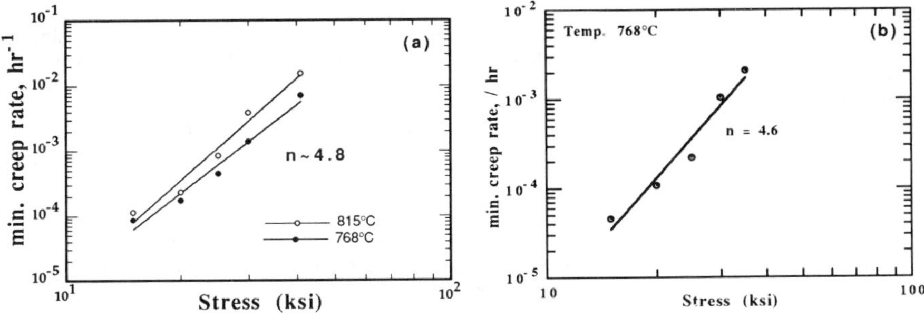

Figure 5. Minimum creep rates versus stress for the Ti-48Al alloy extruded and heat treated. (a) 900°C, 24h and (b) 1350°C, 0.5h + 900°C, 6h.

the creep process within the stress and temperature range used in the present study. TEM studies of the deformation structures are being carried out to obtain further confirmation. A comparison of tensile properties and minimum creep rates obtained in the present study with those from other investigations on TiAl alloys is shown in Table 2. It is significant to note that the creep rates under the same test conditions in the binary Ti-48Al investigated in the present study are comparable to those in the quarternary cast as well as extruded Ti-48Al-2Cr-2Nb alloy. The microstructures in the latter are also duplex, but a direct comparison is difficult because the volume fraction and grain size of the constituents is not known. These results, together with the tensile properties shown in Tables 1 and 2, indicate that it is possible to obtain a high room temperature ductility together with good creep properties by controlling microstructure, which was one of the goals of the present work. It is anticipated that the absolute property values can be further enhanced in ternary/quarternary alloys by using a similar approach. Finally, a systematic study of the creep behavior with microstructures covering a wide range from fully primary γ to fully lamellar, coupled with detailed TEM analysis of deformed samples, is in progress to obtain a basic understanding of both the mechanisms of creep and the influence of microstructure.

TABLE 2

Comparison of Room Temperature Tensile Properties and Creep rates for some TiAl alloys

Alloy	Heat Treatment	0.2%YS (MPa)	%e_p	Steady State Creep Rate/hr (760°C,105 MPa)	Ref. #
Extruded Ti-48Al	1350°C, 0.5h+900°C, 6h	369	2.1	4.6 x 10^{-5}	this study
	900°C, 24h	437	2.3	8.7 x 10^{-5}	
Cast Ti-48Al-2Cr-2Nb	1350, 0.5h + 1300°C, 10h (Duplex)	–	–	2.2 x 10^{-5}	13
Extruded Ti-48Al-2Cr-2Nb	1300°C, 2h, FC (Duplex)	480	3.1	3.3 x 10^{-5}	6
Extruded Ti-48Al-2Cr-2Nb	1400°C, 2h, FC (Lamellar)	455	0.4	4.0 x 10^{-5}	6
Extruded Ti-50Al	1150°C, 2h + 950°C, 8h (Duplex)	–	–	9.0 x 10^{-5}	9

CONCLUSIONS

In summary, the results of this study have shown that it is possible to obtain fine-grained, fullly lamellar microstructures in Ti-48Al based alloys by controlling pre-heat treatment and extrusion conditions. Using these procedures and subsequent heat treatments, it is possible to control microstructure such that attractive combinations of room temperature ductility and high temperature creep resistance are obtained. The results have also shown that microstructural parameters, particularly grain size and volume fraction of lamellar grains, have a marked influence on tensile and creep properties. Determination of operative deformation mechanisms and developing a basic understanding of microstructural effects is the subject of ongoing research.

ACKNOWLEDGEMENTS

Financial support for this research from the State of Ohio Edison Materials Technology Center is deeply appreciated. The authors are thankful to the Duriron Company for supplying the alloy ingot used in this study and to Drs. D. M. Dimiduk and Y. W. Kim for helpful discussions.

REFERENCES

1. Y. W. Kim, J. of Metals, 41, 24 (1989).
2. S. C. Huang and E. L. Hall, MRS Symp. Proc., 133, 373 (1989).
3. S. C. Huang and E. L. Hall, Metall. Trans, vol. 22A, 427 (1991).
4. Y. W. Kim, Acta Metall. Mater., 40, 1121 (1992).
5. T. Tsujimoto and K. Hashimoto, MRS Symp. Proc., 133. 391 (1989).
6. D. S. Shih, S. C. Huang, G. K. Scarr, H. Jang and J. C. Chesnutt: in: Microstructure/ Property Relations in Titanium Aloys and Titanium Aluminides, Y. W. Kim and R. R. Boyer (eds.), p. 135, TMS-AIME, Warrendale, PA (1991).
7. B. Kad and P. M. Hazzledine, these proceedings.
8. P. A. Beavan, these proceedings.
9. P. L. Martin, M. G. Mendiratta and H. A. Lipsitt, Metall. Trans., 14A, 2170 (1983).
10. R. W. Hayes and B. London, Acta Metall. et Mater., 40, 2167 (1992).
11. O. A. Ruano, J. Wadsworth and O. D. Sherby, Acta Metall., 36, 1117 (1988).
12. S. Kroll et al., Z. Metallk., 83, 8 (1992).
13. D. A. Wheeler, B. London and D. E. Larsen, Jr., Scripta Metall. et Mater., 26, 939 (1992).

MECHANICAL PROPERTIES OF Co ALLOYS BASED ON A $E2_1$ TYPE Co_3AlC INTERMETALLIC COMPOUND

HIDEKI HOSODA, MASARU TAKAHASHI, TOMOO SUZUKI* AND YOSHINAO MISHIMA**

Graduate Student, Department of Materials Science and Engineering, Tokyo Institute of Technology, Nagatsuta, Midori-ku, Yokohama 227, Japan.
*Department of Metallurgical Engineering, Tokyo Institute of Technology, O-okayama, Meguro-ku, Tokyo 152, Japan.
**Precision and Intelligence Laboratory, Tokyo Institute of Technology, Nagatsuta, Midori-ku, Yokohama 227, Japan.

ABSTRACT

Alloy design of a new Co-base heat resisting alloy is being pursued. Similar to Ni_3Al in the Ni-base superalloys, the introduction of $E2_1$ type intermetallic compound Co_3AlC as a strengthening phase in a Co solid solution matrix is being attempted. Three experimental alloys are prepared which consist of either two or three phases. Two phase alloys consist of the Co terminal solid solution and the Co_3AlC, while a three phase alloy contains additional B2 phase based on CoAl. Temperature dependence of strength is measured for all the three alloys to reveal that the alloys could exhibit comparable intermediate to high temperature strength with Ni-base superalloys and that, most astonishingly, are found to be quite ductile over all the test temperature including 77 K.

INTRODUCTION

It is well known that the excellent creep resistance as well as high temperature strength in commercial Ni-base superalloys is achieved by the dispersion of a $L1_2$ type intermetallic compound based on Ni_3Al taking up 60 % or more in volume in a Ni solid solution matrix. The dipersoid Ni_3Al may have been better known because it exhibits an anomalous positive temperature dependence of strength arising from its intrinsic ordered structure.

Commercial Co-base superalloys, although they show some superior properties to Ni-base superalloys such as resistance to thermal fatigue and hot corrosion[1,2], they have been limited in usage because of insufficient high temperature strength mainly due to lack of effective dispersion strengthener as Ni_3Al in Ni-base superalloys. There has been a series of work on the mechanical properties of $L1_2$ Co_3Ti[3], but practical importance has not been further investigated since the compound is not stable over 1200 K.

In the present work, the $E2_1$ type Co_3AlC intermetallic compound, whose mechanical properties have rarely been investigated, is chosen for the possible dispersoid in the Co-base alloys. The Co-Al-C ternary phase diagram is shown in Fig. 1 in which the phase field for Co_3AlC as well as its relation to the Co terminal solid solution at 1173 K can be found[4]. The crystal structure of $E2_1$ Co_3AlC is constructed by Co atoms at the face center site, Al atoms at the corner site, and C atoms at the body center site as shown in Fig. 2. It is therefore clear that Co_3AlC can be regarded basically to be $L1_2$ Co_3Al containing a certain amount of carbon atoms, analogous to the situation for the NaCl type transition metal monocarbides

(TMMC) such as TiC or TaC, which are fcc metals that contain carbon. Although ionic NaCl type compounds primarily deform by {110}<110>, {110}<100> and {100}<100> slip systems under the electro-neutrality condition, the deformation of the TMMC generally occurs by {111}<110> slip system, similar to fcc metals at even room temperature[5]. Therefore, it is highly likely that the slip system in the Co_3AlC is similar to that in Ni_3Al. There have been little available data for the deformation mechanism of Co_3AlC to the authors' knowledge, except the fact that it is known as the most stable phase with 0.57 carbon atoms per unit cell[4].

The above consideration would let us expect to be able to design Co-base alloys dispersion hardened by an intermetallic phase with a favorable ordered structure, for which a fundamental investigation has been carried out and the preliminary results are described herein.

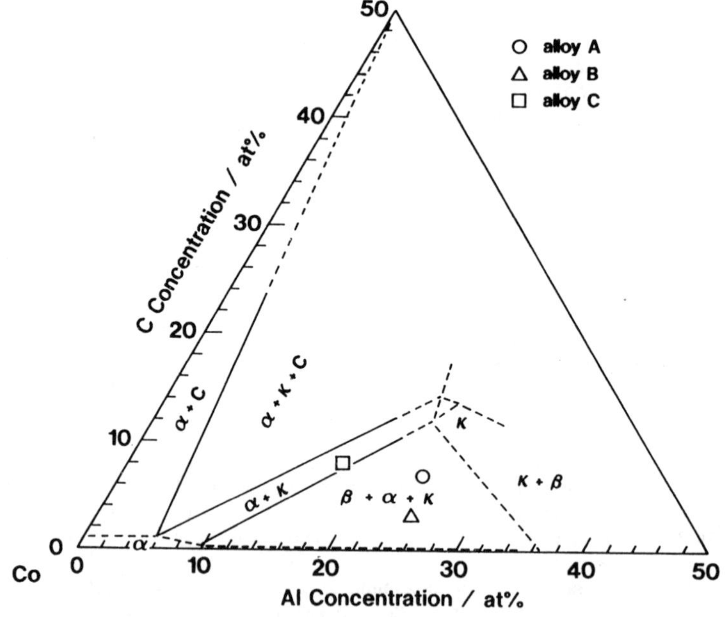

Fig.1 The Co-rich portion of the Co-Al-C ternary phase diagram at 1173 K. α:disorder fcc Co terminal solid solution, β: B2 type CoAl, κ:$E2_1$ type Co_3AlC and C: graphite.

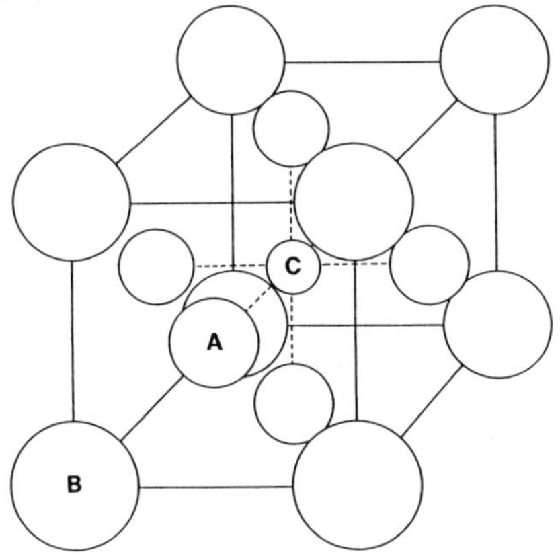

Fig.2 $E2_1$ (perovskite) type crystal structure.
A : the face center site (Co site),
B : the corner site (Al site),
C : the body center site (C site).

EXPERIMENTAL PROCEDURE

Three Co-Al-C ternary alloy ingots of approximately 200g each were prepared by a Tammann furnace under an argon atmosphere. They were cut into pieces of about 25g and were remelted by arc-melting under the same atmosphere. The alloys were homogenized for 6.05×10^6 s (1 week) at 1273 K in a vaccum and quenched in water. Compression specimens were machined to approximately 3 x 3 x 6 mm^3 from the annealed ingots. After electropolishing to remove a deformed surface layer, the compression tests were performed on an Instron-type testing machine at a strain rate of approximately 1.4×10^{-3} q^{-1}. Test temperatures ranged from 77 to 1273 K. The tests at 77K were carried out in liquid nitrogen and those at elevated temperature in an argon atmosphere. X-ray analysis was carried out to determine the phases present by using CuK_α radiation.

RESULTS AND DISCUSSIONS

Compositions and Phases Present in the Alloys

The results of the chemical analysis of the three alloys are shown in table 1. The compositions of the alloys, A, B, and C in table 1, are also shown on the isothermal section at 1173 K of the Co-Al-C ternary phase diagram shown in Fig.1 [4]. X-ray analysis revealed that three phases, $E2_1$, B2 and a Co terminal solid solution, coexist in alloy A and B, while two phases of $E2_1$ and the terminal solid solution of Co coexist in alloy C. The results are in good agreement with the phase diagram in Fig. 1.

Table 1 Chemical composition of the Co-Al-C alloys investigated.

alloy	%	Co	Al	C
A	wt%	84.59	13.65	1.76
	at%	68.75	24.23	7.02
B	wt%	85.65	13.55	0.80
	at%	71.86	24.84	3.30
C	wt%	89.14	8.98	1.88
	at%	75.55	16.62	7.83

Mechanical Properties and Microstructure

An example of a set of compressive stress-strain curves at various test temperatures are shown in Fig. 3 for alloy A. To be noted is that the end of each stress strain curve does not indicate the point of fracture and therefore the alloy exhibits at least 10 % compressive plastic strain at all the test temperatures. Similar stress-strain curves were obtained for the rest of the alloys, indicating that all the experimental alloys are ductile regardless of whether the structure is two or three phases.

Fig. 4 shows the temperature dependence of compressive 0.2% flow stress for all the alloys investigated at a strain rate of $1.4 \times 10^{-3}\ s^{-1}$. To be noted first is that a very high strength of over 1 GPa is obtained in all the alloys at low temperatures even though there is an appreciable ductility as shown for alloy A in Fig. 3. Then there was a tendency to show a positive temperature dependence of strength between 600 K and 1000 K, paticularly in alloy C. Since such anomalous behavior is best known in many $L1_2$ type compounds, including Ni_3Al, and since alloy C is the two phase alloy containing Co_3AlC by a considerable volume percent (Fig. 1), it is speculated that Co_3AlC would also exhibit the mechanical anomaly because of the resemblance its crystal structure with $L1_2$, as discussed in introduction. In alloys A and B, the tendency to exhibit less mechanical anomaly compared with alloy C could be explained by the smaller amount of the compound being replaced by the B2 compound based on CoAl. It should be noted that the high temperature 0.2% flow stress of alloys A and C is as high as 300 MPa at 1273 K, which is comparable to the commercial Ni-base superalloys[6].

Optical micrographs of the alloys after homogenization by heat treatment are shown in Fig. 5 for a three phase alloy A and in Fig. 6 for a two phase alloy C. The microstructure found in Fig. 5 of a three phase mixture seems to retain the dendritic structure formed during solidification. On the other hand, the two phase alloy in Fig. 6 exhibits a fine eutectic-like microstructure for which the formation process is unknown. Although the identification of the constituent phase has not yet been made, a bright phase would be $E2_1$ Co_3AlC, and a dark phase would be a Co terminal solid solution.

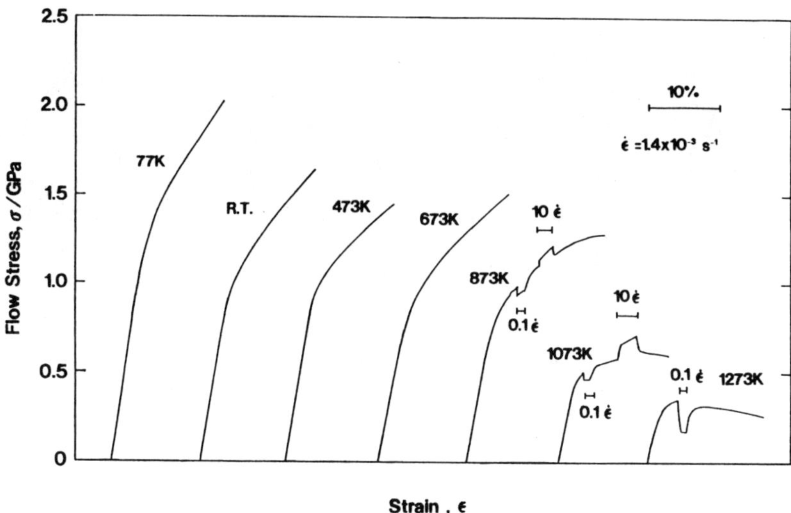

Fig.3 Compressive stress-strain curves of a three phase Co alloy containing $E2_1$, Co_3AlC and B2, CoAl phase (alloy A) at various test temperatures.

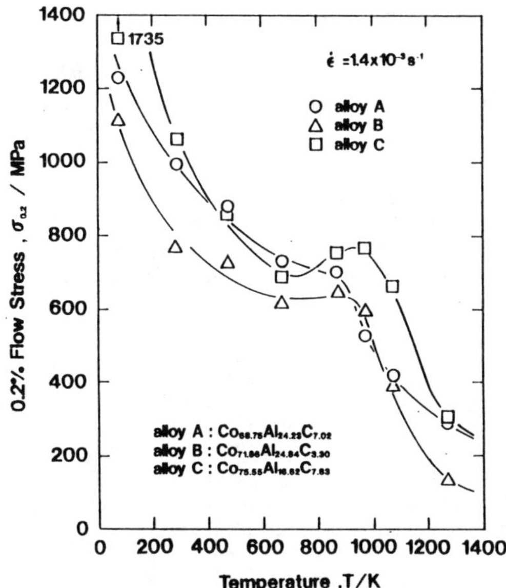

Fig.4 Temperature dependence of 0.2% flow stress in Co alloys containing $E2_1$, Co_3AlC.

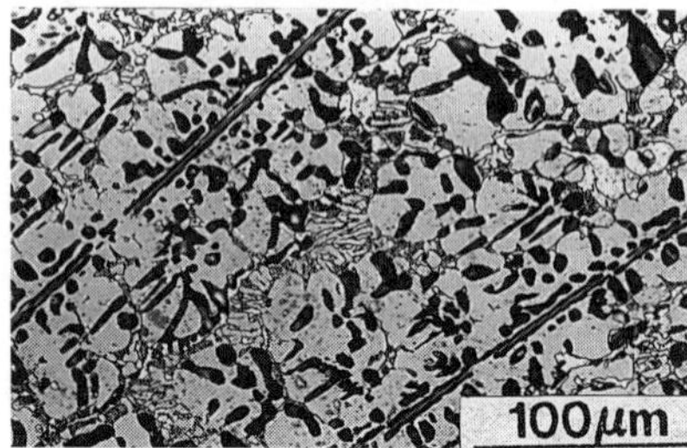

Fig.5 An optical micrograph for three phase alloy containing Co terminal solid solution, $E2_1$, Co_3AlC and B2, CoAl, (alloy A) after a homogenization heat treatment.

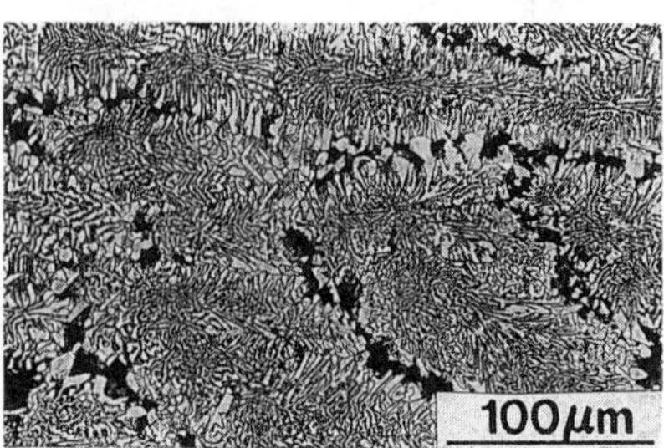

Fig.6 An optical micrograph for two phase alloy containing Co terminal solid solution and $E2_1$, Co_3AlC, (alloy C) after a homogenization heat treatment.

CONCLUSIONS

An investigation has been carried out to develop a new Co base alloy which is dispersion strengthened by a proper intermetallic compound for application as a high temperature heat resisting material. An intermetallic compound of $E2_1$, Co_3AlC was chosen for the purpose and alloys containing this phase were prepared to examine their mechanical properties. The followings are the conclusions drawn:

(1)Two or three phase alloys containing $E2_1$ type Co_3AlC exhibit good ductility at all the temperatures investigated.

(2)Two phase alloys consisting of the Co terminal solid solutions and the compound shows the positive temperature dependence of the 0.2% flow stress, where the peak temperature is approximately 1000K.

(3)0.2% flow stress at low temperatures ranges from 1200 MPa to 1700 MPa in all the alloys investigated containing the compound.

(4)$E2_1$ Co_3AlC intermetallic compound would be an effective dispersoid in Co base alloys to create a new Co base superalloys.

REFERENCES

[1] R.V.Miner,in Superalloys II, ed. by C.T.Sims, N.S.Stoloff and W.C.Hagel, (John Wiley & Sons, New York) p.263.

[2] F.S.Pettit and C.S.Giggins, ibid, p.327

[3] T.Takasugi and O.Izumi, Acta Metall.,33(1985)p.39.

[4] L.J.Huetter, H.H.Stadelmaier and A.C.Fracker, Metall.,14(1960)p.113.

[5] G.E.Hollox, Mater. Sci. Eng.,3(1968/1969)p.121.

[6] T.P.Gabb and R.B.Dreshfield, in Superalloys II, ed by C.T.Sims, N.S.Stoloff and W.C.Hagel,(John Wiley & Sons, New York) p.575.

Properties and Potential of High-Temperature Niobium Beryllides

S. M. Bruemmer, J. L. Brimhall, C. H. Henager, Jr. and J. P. Hirth*
Pacific Northwest Laboratory, Richland, WA 99352
*Washington State University, Pullman, WA 99164

Abstract

Recent research on the low- and high-temperature properties of two beryllium-niobium intermetallic compounds, $Be_{12}Nb$ and $Be_{17}Nb_2$, is reviewed and discussed. Strength (bend and compression), hardness and fracture toughness has been mapped as a function of test temperature up to 1200°C. Results for hot-isostatically-pressed $Be_{12}Nb$ and $Be_{17}Nb_2$ are highlighted illustrating the potential for reasonable strength at both low and high temperatures. Chemical compatibility between $Be_{12}Nb$ and several high-temperature materials such as SiC, Al_2O_3, $MoSi_2$ and various refractory metals is evaluated. Limitations for the structural use of the beryllides are identified and discussed including low-temperature toughness, intermediate-temperature embrittlement, high-temperature creep strength and composite compatibility.

Introduction

Beryllium-refractory metal intermetallic compounds, i.e. $Be_{12}X$, $Be_{13}X$ and $Be_{17}X_2$, exhibit excellent strength and oxidation resistance at high temperatures. Because of these properties and their very low densities, beryllides are promising candidate materials for aerospace applications. The combination of properties exceeds other intermetallics such as refractory metal silicides and aluminides in most cases. Unfortunately, similar to many other intermetallics, beryllides have complex crystal structures which lead to poor low-temperature toughness.

The $Be_{12}X$ compound is the simplest with a body-centered-tetragonal (bct) structure (I4/mmm) and 26 atoms per unit cell. Elements such as Nb, Mo, Ta, Ti, Cr, V and W all form isostructural $Be_{12}X$ compounds with little change in the lattice dimensions. A number of metallic elements also form isostructural $Be_{13}X$ (cubic, Fm3c) and $Be_{17}X_2$ (hexagonal-rhombahedral, R3m) compounds as summarized in Figure 1. Phase stability in beryllides (e.g., formation of $Be_{13}X$ versus $Be_{12}X$) depends on atom size of the metal atom (X) reacting with Be and, to a lesser extent, its bonding character.[1] Face-centered-cubic (fcc) $Be_{13}X$ is favored when the diameter of metal atom (d_X) is greater than ~0.3 nm, while $Be_{12}X$ is stable for smaller atom sizes, d_X = 0.25 to 0.29 nm. Among the $Be_{12}X$ compounds, larger (>0.28 nm) metal atoms (e.g., Nb, Ta and Ti) also form $Be_{17}X_2$, but smaller ones (e.g., Cr, Mn, V, Mo and W) do not.

Recent work by the authors on $Be_{12}Nb$ and $Be_{17}Nb_2$ intermetallics is reviewed with emphasis on: (1) phase stability, (2) mechanical properties, (3) deformation mechanisms and (4) composite compatibility. Specific properties of these compounds are discussed in relation to their potential for high-temperature structural applications.

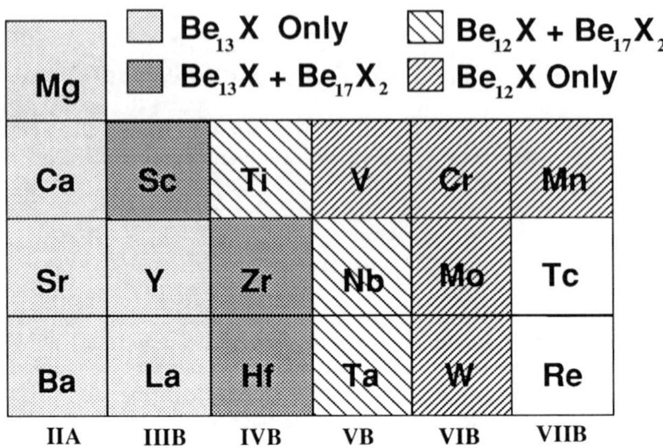

Figure 1. Refractory and transition metal atoms forming various beryllides.

Phase Relations and Stability

Because of published uncertainties in the phase diagram,[2,3] phase relations in Be-Nb alloys at high Be concentration have been evaluated by Brimhall et al.[4-6] With proper target design, sputter deposition could be used to make the alloys with variable compositions. In addition to the equilibrium phases, several metastable phases were found. Deposition on a cold substrate produces an amorphous phase for all compositions above ~5% Nb. Annealing the amorphous structure at 500-750°C or by direct deposition at 350-400°C produces metastable crystalline phases. A metastable phase ($Be_{12}Nb'$) with an apparent bcc structure forms which can be viewed as the bct $Be_{12}Nb$ phase with a very high fault density on the (001) planes. Because of the structural relationship between the $Be_{12}Nb$ and $Be_{17}Nb_2$ phases, $Be_{12}Nb'$ can serve as a precursor phase to the formation of either compound. Another metastable phase, also consistent with a bcc lattice ($a_o = 1.208$ nm), forms near the compositional range of $Be_{17}Nb_2$ during high temperature deposition. The structure does not correspond to any known Be-Nb phases. Single-phase $Be_{12}Nb$ appeared to exist over the stoichiometric range of 5.5 to 7.7% Nb. The apparent stoichiometric range probably results from Nb vacancies in the lattice. No evidence of the Be_5Nb phase was found and the $Be_{17}Nb_2$ phase was stable to room temperature. The results support the phase diagram originally published by Massalski[2] and suggest that $Be_{12}X$ may not be a true line compound.

Ternary alloying and the stability between $Be_{12}X$ and $Be_{13}X$ phases has also been examined.[7] In a sputter-deposited Be-Nb-Zr ternary alloy, the Nb-to-Zr composition ratio determined whether $Be_{12}Nb$ incorporated Zr or $Be_{13}Zr$ incorporated Nb. Upon annealing, a $Be_{17}X_2$ phase forms with $Be_{12}X$ or $Be_{13}X$, but $Be_{12}X$ and $Be_{13}X$ phases are not found together. This is probably due to the lack of a structural relationship between the $Be_{13}X$ and the $Be_{12}X$ phases as is found between the $Be_{12}X$ and the $Be_{17}X_2$. The formation of the phase depends strongly on the size of the X atom, with small atoms preferring the $Be_{12}X$ structure. The $Be_{12}X$ phase can accommodate Zr atoms until the average size of the Nb+Zr atoms reaches a critical value at which point the $Be_{13}X$ phase forms. Relative stabilities between the various stable and metastable crystal structures is believed to play an important role in the deformation behavior and mechanical properties of $Be_{12}X$ compounds as discussed in the following sections.

Mechanical Properties

Bend Strength and Fracture Toughness

Four-point-bend and chevron-notch-bend specimens have been used to determine strength and toughness of VHP and HIP beryllides.[8-10] HIP $Be_{12}Nb$ exhibited the best properties in bending with strengths of 210-250 MPa measured at 20°C to 650°C (Figure 2). Strength dropped sharply from 650°C to 900°C, and then increased to ~250 MPa at higher temperatures. Limited plasticity was observed below 1100°C. At low temperatures, a fine, granular fracture surface is observed consisting of intergranular (IG) facets and small cleavage regions across individual grains. The fracture surface became brittle intergranular and macroscopically flat at temperatures from 800°C to 1000°C. Temperature effects on fracture toughness closely matched the bend strength results as shown in Figure 2. Low-temperature K_{IC} ranged from 4 to 4.6 MPa$\sqrt{m}$ corresponding to the highest measured bend strengths. Toughness values decreased to less than 0.5 MPa$\sqrt{m}$ at intermediate temperatures before increasing above 900°C. This embrittlement occurs at temperatures where accelerated oxidation (pest) is seen in static tests, but cannot be directly linked to the fracture process.[11]

HIP $Be_{17}Nb_2$ exhibited only elastic behavior in bend tests at all temperatures up to 1100°C. Slight inelastic deflection was observed at 1100°C and measurable plastic deformation was seen at 1200°C. Bend strengths increased from 100 MPa at 20°C to a maximum of 740 MPa at 1100°C (Figure 2). High-temperature strengths for $Be_{17}Nb_2$ were much greater than for $Be_{12}Nb$. Measured K_{IC} for $Be_{17}Nb_2$ material was less than 2.5 MPa$\sqrt{m}$ at test temperatures below 1100°C, which indicates that the measured bend strengths are limited by the toughness of the material. $Be_{17}Nb_2$ failed by transgranular cleavage (flat, featureless appearance) with no obvious IG features.

Compression Strength

Compression tests have been performed on various beryllide heats using small cylindrical or cube specimens.[9,10] A maximum uniaxial compressive strength of 2750 MPa was measured for the HIP $Be_{12}Nb$ at 20°C. Strength remained quite high (2380 MPa) at 800°C and then decreased rapidly at temperatures greater than 1000°C (Figure 3). Compression strengths at temperatures from 20°C to 800°C were significantly improved over that for $Be_{17}Nb_2$. Plasticity was indicated from non-linearities in the load-displacement curves for $Be_{12}Nb$ at ~800°C, while $Be_{17}Nb_2$ revealed brittle behavior at temperatures up to 1000°C. Ambient temperature strength for $Be_{17}Nb_2$ was only 1200 MPa, less than one-half that for the HIP $Be_{12}Nb$. Strength increased with temperature, reaching a maximum value of 1550 MPa at 800°C (Figure 3). A very low strength of 400 MPa, below that for $Be_{12}Nb$, was measured at 1000°C. Compressive strength at 1200°C was identical to $Be_{12}Nb$ (~200 MPa), in contrast to the higher strength observed in the bend tests.

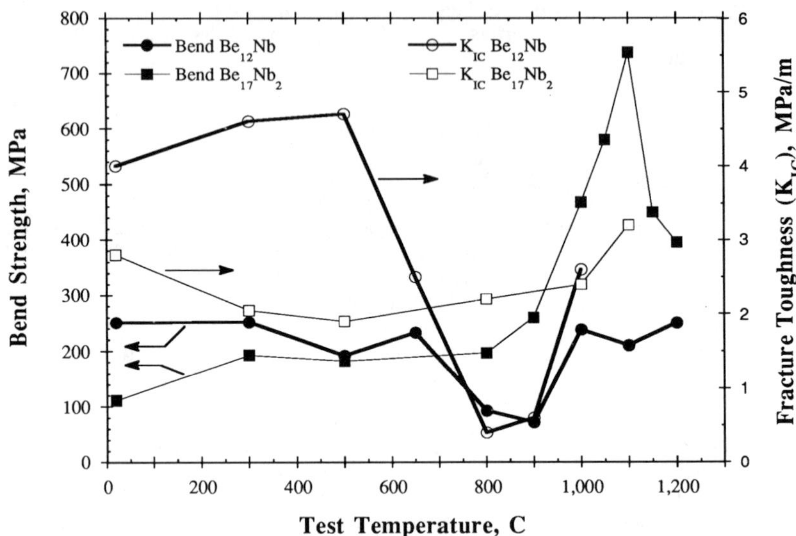

Figure 2. Bend Strength and Fracture Toughness for HIP $Be_{12}Nb$ and $Be_{17}Nb_2$ Compounds.

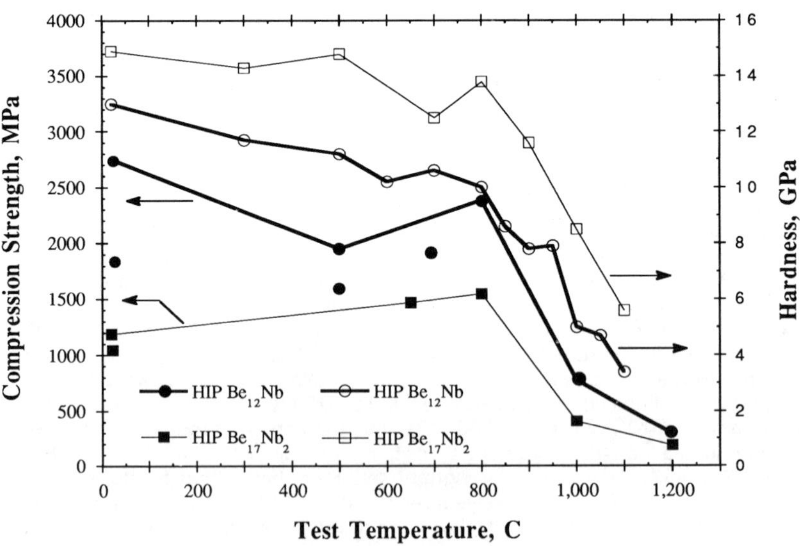

Figure 3. Compression Strength and Hot Hardness for HIP $Be_{12}Nb$ and $Be_{17}Nb_2$ Compounds.

High-Temperature Microhardness and Indentation Creep Tests

High-temperature microhardness tests have been performed on VHP and HIP $Be_{12}Nb$ and $Be_{17}Nb_2$ heats[10,12,13] and on several isostructural VHP $Be_{12}X$ materials.[12] HIP heats showed a higher hardness than VHP heats at temperatures below 900°C, due to their finer grain size and reduced porosity. The largest difference was observed at 20°C where the HIP materials show hardness values from 12.8 GPa to 14.0 GPa, while hardness for the VHP materials range from 8.8 GPa to 9.8 GPa. Hardness results for $Be_{17}Nb_2$ were consistently above that for $Be_{12}Nb$ (Figure 3) with a maximum difference of 3 GPa occurring at 800°C to 950°C. Test temperatures of about 1000°C were required before a significant decrease in hardness was observed, a slightly higher transition than for $Be_{12}Nb$ (~850°C). Differences in ductile-brittle transition behavior between $Be_{12}X$ compounds (Cr, Mn, Ti, Mo, V, Ta and Nb) were observed and related to $Be_{12}X$-$Be_{17}X_2$ phase stability. Elements such as V and Mo (which form $Be_{12}X$, but not $Be_{17}X_2$), were found to exhibit lower transition temperatures. Secondary slip systems were found to be activated in $Be_{12}V$ at temperatures ~300°C below that for $Be_{12}Nb$ even though both compounds have similar melting temperatures.[13]

Indentation creep tests were performed on $Be_{12}Nb$ and $Be_{17}Nb_2$ materials using the high-temperature microhardness tester and varying the indentation dwell time. This data was used to calculate creep parameters for this material.[10,12] Power-law creep exponents of 4.3 and 4.1 were determined for the HIP $Be_{12}Nb$ and VHP $Be_{17}Nb_2$, respectively. Activation energies for dislocation creep were about 290 kJ/mole for HIP $Be_{12}Nb$ and 230 kJ/mole for VHP $Be_{17}Nb_2$. $Be_{17}Nb_2$ required a slightly higher temperature (950°C) for the transition to power-law creep as compared to that for $Be_{12}Nb$. Although individual hardness values were consistently higher for $Be_{17}Nb_2$, the rate of change with temperature was quite similar to that for $Be_{12}Nb$. Specific properties for $Be_{12}Nb$ and $Be_{17}Nb_2$ are summarized in Table 1.

Table 1. Selected Properties for $Be_{12}Nb$ and $Be_{17}Nb_2$ Intermetallics

Property	$Be_{12}Nb$	$Be_{17}Nb_2$
Crystal Structure	tetragonal	rhomohedral
Melting Temperature, °C	1672	1800
Density, g/cm^3	2.9	3.2
Elastic Modulus, GPa	220	320
Maximum Strength at 23°C		
• Bending, MPa	250	110
• Compression, MPa	2750	1200
• Hardness, GPa	13.7	14.9
Maximum Strength at 1200°C		
• Bending, MPa	250	390
• Compression, MPa	295	190
• Hardness (1100°C), GPa	3.8	5.8
Ductile-Brittle Transition Temperature, °C		
• Bending	~1100	~1100
• Hardness	~850	~900
Activation Energy for Dislocation Creep, kJ/mole	290	230

Deformation Mechanisms in $Be_{12}Nb$

Transmission electron microscopy (TEM) observations of deformed material have revealed substantial dislocation activity in $Be_{12}Nb$ at high temperatures.[14,15] Few dislocations are observed in as-processed $Be_{12}Nb$,[16] but a significant density of partial and some perfect dislocations (< 111] type) are produced after deformation at 1200°C. Partial dislocations imaged in the deformed material were predominantly 1/2<101]{101), 1/2<101]{121) or 1/2<100]{011), and the majority of these were observed to be bounding planar faults in the material. For the grains examined, the 1/2<101] type partials are more common and correspond to extended faults in the as-received material. Because of the complex bct crystal structure, faults cannot be created by a simple shift of Nb atoms to Be positions and vice versa. The bct structure has an open site or "anti-site" position for Nb atoms at the {100) face-centered positions or their equivalents, the mid-points of the unit cell edges. Displacements required to shift Nb atoms to this position (1/2<101] or 1/2<100]) are consistent with the observed partial dislocations.

Dislocation structures in $Be_{12}Nb$ compressed at 800, 900 or 1000°C were somewhat different than documented at 1200°C.[14] The vast majority of dislocations observed at these temperatures were identified as leading partial dislocations bounding extended planar faults that were anchored at one end to grain boundaries. Very few trailing partials were observed at the lower temperatures, in contrast to the 1200°C deformation where both bounding partials were visible in the grain interiors. Slip was limited to fewer systems with 1/2<101] partials seen mainly on {101) planes. At 1000°C, 1/2<101{121) and a few 1/2<100]{011) partial dislocations are observed. Interestingly, deformation of $Be_{12}V$ revealed all three systems active at 800°C.[12]

The onset of ductility in $Be_{12}Nb$ (and $Be_{12}V$) is related to the activation of the 1/2<101] and 1/2<100] partial dislocations to produce stacking faults on {101), {121) and {011) planes. Mobility of these partials will depend on the stacking fault energy, since the motion of a single partial is opposed by a force related to this energy. Also, the Peierls stress decreases as a dislocation dissociates, so the mobility of the dissociated perfect dislocation should also increase as the stacking fault energy decreases. The high density of planar faults in deformed $Be_{12}Nb$ suggests that the fault energy is low at 1200°C. Fault energies would also seem to be low at 800°C, but the mobility of the partial dislocations appears to be reduced from that at 1200°C. The apparent immobility of the trailing partials at 800°C suggests an interesting asymmetry in mobilities from leading to trailing partials. This may be related to the formation of very stable stacking faults associated with the crystallography of the $Be_{12}X$ to $Be_{17}X_2$ phase transformation.

At low temperatures, the numerous extended dislocations should form faults of the Lomer-Cottrell type, and core locking associated with the Peierls barrier should be enhanced. Both of these effects are analogous to the situation for TiAl.[16] Even though dislocation motion may occur in the microstrain region, this locking could limit any macroscopic ductility. $Be_{12}Nb$ does show dislocation activity even during ambient temperature deformation. HIP specimens were compressed to ~90% of its fracture stress at 20°C, and examined by TEM. The only dislocations that were created during this deformation were 1/2<101]{101) partials again bounding planar faults. In contrast to what was observed at 800°C, many of the individual partials were further split into what appear to be 1/4<101] partials. This splitting is consistent with the "zonal" dislocation concept proposed by Kronberg[18] and is in agreement with atomistic modeling predictions[19] for dislocation stability in $Be_{12}Nb$.

Composite Compatibility

To gain some indication of the feasibility of using a beryllide in a composite, the chemical compatibility between $Be_{12}Nb$ and several high temperature materials has been examined.[20] After annealing at 1150°C for 24 hours, the interaction zone was considerable in all cases, ranging from 25 to 150 microns. Of the materials tested, $MoSi_2$ showed the least amount of reaction and SiC the greatest. All of the reactions studied primarily involved diffusion of Be from the $Be_{12}Nb$ into, and subsequent reaction with, the adjacent material. The Be depletion in the beryllide resulted in formation of higher Nb-containing beryllide compounds such as $Be_{17}Nb_2$. There was no evidence of diffusion for any of the refractory metals (Nb, Ta, Mo or W) into the beryllide; however, there was evidence of Si from SiC and from $MoSi_2$ diffusing into the beryllide. The reaction with Al_2O_3 produced a complex oxide which could be beneficial in the long run if the kinetics are sufficiently slow in this product oxide. It is important to note that excellent compatibility exists between $Be_{12}Nb$ and other beryllides ($Be_{17}Nb_2$ and $Be_{13}Zr$), BeO and Be.

Conclusions

Refractory metal beryllides, such as $Be_{12}Nb$ and $Be_{17}Nb_2$, exhibit many excellent characteristics for high-temperature structural applications. $Be_{17}Nb_2$ showed the best strength above 1000°C (factor of 3 better than $Be_{12}Nb$ in bending), but had very poor low-temperature toughness. The HIP $Be_{12}Nb$ was found to have the highest toughness (K_{IC} = 4.6 MPa$\sqrt{m}$) and a good combination of low- and high-temperature strength, but was embrittled at intermediate temperatures. Multiple slip systems were observed during high-temperature deformation of $Be_{12}Nb$. However, only 1/2<101]{101) partial dislocations appear to be mobile below 1000°C. Macro-plasticity requires activation of secondary systems, i.e. <101]{121) and <100]{011). Stacking fault energies may play a critical role in the deformation process and are believed to be related to phase stabilities. Bend, compression and indentation creep measurements suggest that reinforcement will probably be required for both high temperature creep strength and low-temperature toughness.

References

1. G.V. Raynor, *Proc. 4th Int. Conf. on Beryllium*, The Metals Society, 1977, p. 2/1.
2. T. B. Massalski, ed., *Binary Alloy Phase Diagrams*, Vol. 1, American Society for Metals, Metals Park, OH, 1986.
3. H. Okamoto and L. Tanner, ed., *Phase Diagrams of Binary Beryllium Alloys*, ASM International, Metals Park, OH, 1987.
4. J.L. Brimhall, L.A. Charlot and S.M. Bruemmer, Scripta Metall. et Mat., 25 (1991) 553.
5. J.L. Brimhall, L.A. Charlot and S.M. Bruemmer, J. Mater. Res., 7 (1992) 89.
6. J.L. Brimhall, L.A. Charlot and S.M. Bruemmer, *High-Temperature Ordered Intermetallic Alloys*, Materials Research Society, 1991, p. 175.

7. J.L. Brimhall, L.A. Charlot and S.M. Bruemmer, Mater. Sci. Eng., A152 (1992) 76.
8. C.H. Henager, Jr., R.E. Jacobson and S.M. Bruemmer, Mater. Sci. Eng., A152 (1992) 416.
9. S.M. Bruemmer, B.W. Arey and C.H. Henager, Jr., *Intermetallic Matrix Composites II ,* MRS Spring Meeting, Materials Research Society, 1992.
10. C.H. Henager, Jr., J.P. Hirth and S.M. Bruemmer, submitted to Mat. Sci. Eng., 1992.
11. S.M. Bruemmer and L.A. Charlot, submitted to Scripta Metall. et Mat., 1992.
12. R.E. Jacobson, *High Temperature Strength and Creep Resistance of* $Be_{12}Nb$ *and* $Be_{17}Nb_2$ *Intermetallics,* Masters Thesis, Washington State University, 1991.
13. S.M. Bruemmer, B.W. Arey, J.L. Brimhall and J.P. Hirth, submitted to J. Mater. Research, 1992.
14. S.M. Bruemmer, L.A. Charlot, J.L. Brimhall, C.H. Henager, Jr., and J.P. Hirth, Phil. Mag. A, 65 (1992) 1083.
15. S.M. Bruemmer, L.A. Charlot, C.H. Henager, Jr., and J.P. Hirth, submitted to Scripta Metall., 1992.
16. L.A. Charlot, J.L. Brimhall, L.E. Thomas, S.M. Bruemmer and J.P. Hirth, Scripta Metall., 25 (1991) 99.
17. B.A. Greenberg, O.V. Antonova, V.N. Indenbom, L.E. Karkina, A.B. Notkin, M.V. Panomarev and L.V. Smirnov, Acta Metall., 35 (1991) 233.
18. M.L. Kronberg, J. Nucl. Mater., 1 (1959) 85.
19. S. Sondhi, R.G. Hoagland, J.P. Hirth and S.M. Bruemmer, *High-Temperature Ordered Intermetallic Alloys*, Materials Research Society, submitted 1992.
20. J.L. Brimhall and S.M. Bruemmer, Scripta Metall, in press, 1992.

Acknowledgements

Contributions of R. Jacobson, S. Sondhi, L. A. Charlot and B. W. Arey, as well as critical discussions with R. G. Hoagland, are acknowledged. Work is supported by the Defense Advanced Research Projects Agency through the Office of Naval Research and under U.S. Department of Energy contract DE-AC06-76RLO 1830 with Pacific Northwest Laboratory, which is operated by Battelle Memorial Institute.

Fatigue and Fracture of Nb_3Al/Nb *In Situ* Composites Based on an Nb 18 at.% Al Alloy

D.L. Davidson, Southwest Research Institute, San Antonio, TX 78228
D.L. Anton, United Technologies Research Center, East Hartford, CT 06108

ABSTRACT

Near Nb_3Al alloys, having Nb_3Al with the ordered A15 crystalographic structure within regions of Nb solid solution, can be considered as *in-situ* reinforced composites. Nb-18 at.%Al ingots were arc melted and heat treated to provide several microstructures. Small compact tension specimens were cracked in compression- compression cyclic loading and subsequently grown in tension-tension cyclic loading. Fatigue crack growth rates were found to be similar to crack growth through a TiAl-based alloy with a lamellar microstructure. Threshold stress intensity factor was estimated as 5.3 MPa√m, and fracture toughness as about 10 MPa√m.

INTRODUCTION

The creep resistance of Nb_3Al (an A15 compound) is comparable to some nickel based superalloys at temperatures up to 1100°C,[1] making this material a candidate for gas turbine engines that would operate at temperatures well above the current maximum temperature of ≈ 1100°C. The physical metallurgy and slip characteristics of A15 compounds (Nb_3Al, V_3Si, Cr_3Si) and near-Nb_3Al compositions have been determined.[2] Elastic modulus, bend and yield strengthes, and elongation to fracture have recently been determined[3] and found to be within the range useful in turbine applications. Furthermore, Nb_3Al has been found to be more creep resistant than other intermetallic compounds melting above 1400°C.[4] These results appear sufficiently promising that studies of possible alloying additions have been started[3] and a study of the phase stability of Nb-Al-Ti alloys is in progress.[5] A large government sponsored initiative is currently underway in Japan to refine processing and study the properties of Nb_3Al alloys.[6] The intermetallic Nb_3Al is very brittle at ambient temperature. The present study was undertaken to evaluate the fatigue and fracture toughness characteristics of a ductile phase toughened Nb_3Al alloy at ambient temperature. Through the introduction of Nb in solid solution, which is thermodynamically stable in this case, it is anticipated that both fatigue and fracture resistance can be significantly enhanced.

DESCRIPTION OF MATERIALS AND EXPERIMENTAL PROCEDURES

Several ingots of Nb-Al alloys were arc melted and cast in argon. Nominal composition of these alloys was Nb-18 at.%Al. The ingots, approximately 110 mm square by 8 mm thick, were then aged, cut into specimen blanks and examined for cracks and compositional uniformity. Small CT specimens (20 mm square, by ≈ 5 mm thick) were removed from the remaining ingots by electric discharge machining. After polishing by typical metallographic preparation techniques, specimens were notched with a diamond saw to a depth (measured from the center line of the pin holes) of about 3 mm. Compression-compression (C-C) cyclic loading was used to initiate and grow a crack from the notch. The specimen was then cycled in tension-tension (T-T) for data collection. Unfortunately, several of the specimens broke during C-C loading because it was necessary to use unusually large loads while trying to induce fatigue crack initiation from the notch. An example of the specimen used, together with the type of fracture caused by an overload in C-C loading, is shown in Fig. 1.

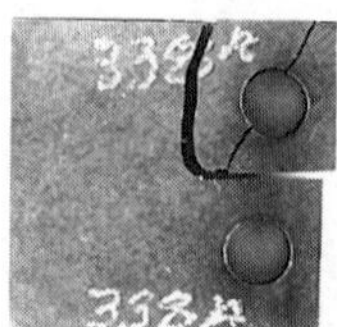

Fig. 1 Compact tension specimen used for fracture studies. Several of the specimens broke under compression-compression loading used to initiate cracks.

A total of 5 specimens were prepared, but successful tests with fatigue crack growth data were obtained from only 2 specimens, and from these, only a limited range in ΔK was covered. Information on the materials tested and a summary of the results obtained are listed in Table I.

Table I

Heat Treatments and Test Results

Material Designation	Heat treatment Temp.(°C)	Time (Hrs)	Hardness Rockwell C	Test Results
A	1000	4	37	some da/dN data pin hole failure: T-T
B3	1000 1400	4 100	47	(1) pin hole failure: C-C (2) some da/dN data
B4	1000 1200	4 100	57	(1) pin hole failure: C-C (2) $K_c = 5.5$ MPa$\sqrt{m}$

T-T = Tension-tension loading with R = 0.1
C-C = Compression-compression loading with R = 10

RESULTS

Microstructure

Optical micrographs of the materials tested are shown in Fig. 2. The two phases of materials B3 and B4 are fairly homogeneously distributed but for material A there appear to be isolated regions of fine precipitates within a more uniform structure. Porosity on this scale was not found. Dark lines seen in the micrographs are cracks. Ion etching was used to examine further the microstructure of material A. Shown in Fig. 3 are the microstructures before and after ion etching, as revealed by optical and back-scattered electron microscopy (at 30 keV). This examination indicated that the isolated regions had a very fine eutectic-like structure which formed prior to the grain boundaries (these regions crossed the grain boundaries). Thus, material A had a much coarser and inhomogeneous microstructure than either B3 or B4. A more detailed description of the microstructures and mechanisms of formation has been given previously.[7-9]

Interpretation of these micrographs, and further information about how these alloys were made, may be found in the results of Anton and Shah.[1] The composition chosen, according to the phase diagram, is on the edge of the Nb_3Al phase field. Study of this system has shown that

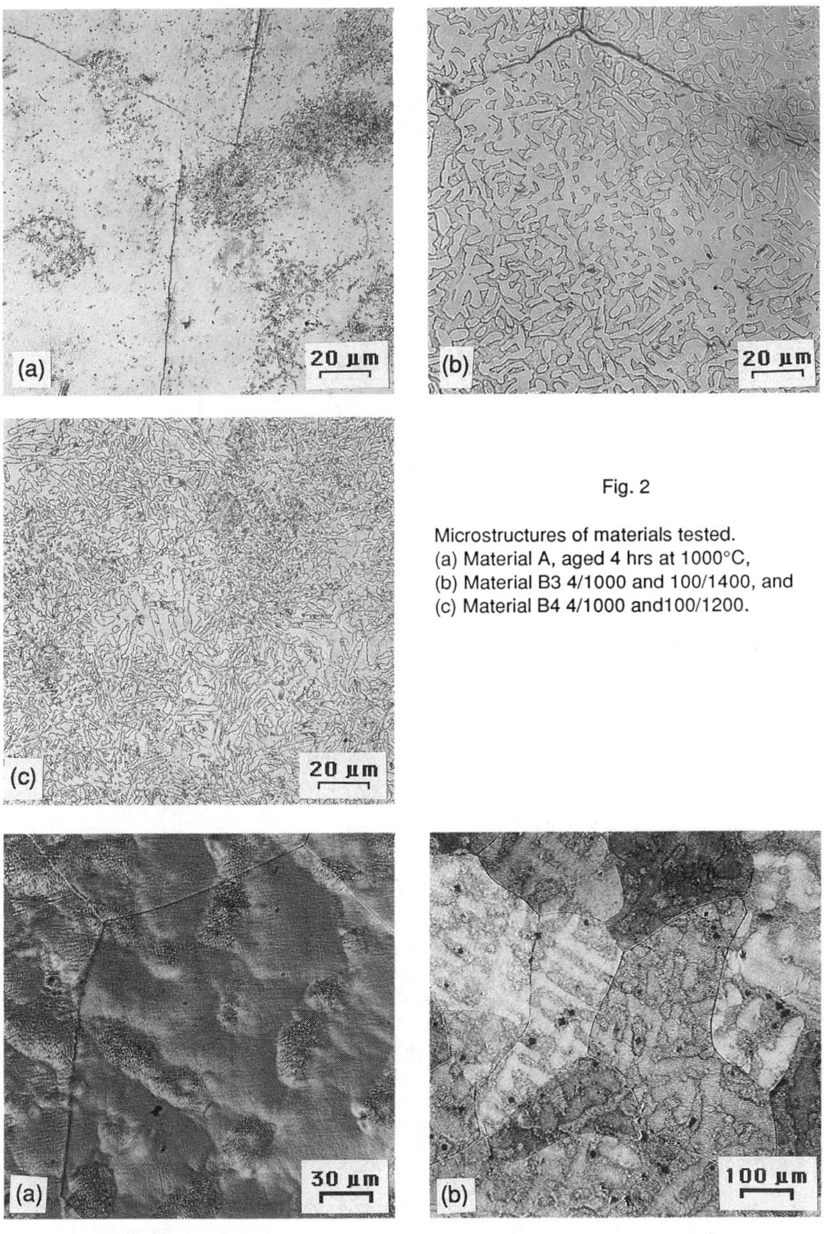

Fig. 2

Microstructures of materials tested.
(a) Material A, aged 4 hrs at 1000°C,
(b) Material B3 4/1000 and 100/1400, and
(c) Material B4 4/1000 and100/1200.

Fig. 3

Microstructure of Material A: (a) nomarski (optical) image after ion etching, and (b) backscattered electron image after ion etching.

quenching from elevated temperature produces a Nb solid solution stable to ambient temperature. Subsequent ageing results in the precipitation of Nb_3Al through a peritectic reaction. The nature of the precipitation is that of a massive phase transformation that "results in a highly uniform, fine distribution of filamentary Nb within a contiguous Nb_3Al matrix.[1]" Both materials B3 and B4 appear to be composed of greater than 50% Nb_3Al.

Fatigue Crack Growth

The fatigue crack growth results, obtained from several specimens, are compared to the fatigue crack growth data for a titanium aluminide alloy in **Fig. 4.** The line shown describes the trend in the TiAl-alloy data and fits the relation

$$da/dN = 4.2\times10^{-14}\ \Delta K^{5.4} \qquad (1)$$

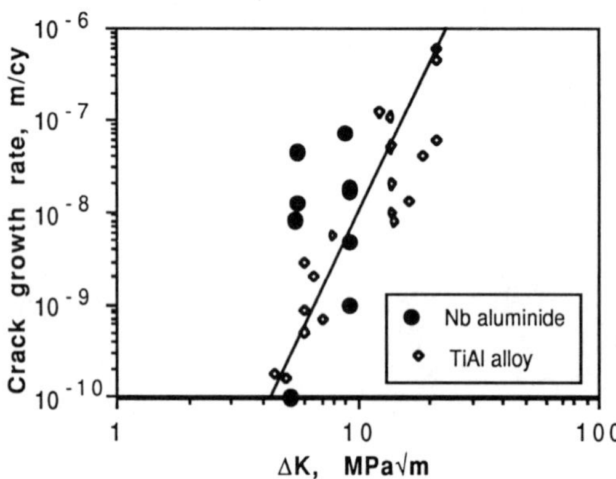

Fig. 4 Comparison of crack growth rates in the Nb_3Al-based alloys with those for a lamellar TiAl-based alloy.

Fatigue cracks grew very irratically and with great variation within the range of ΔK values for which data were obtained. Most of the data were taken from one specimen of material A, but some data are included for material B3. For material A, a crack was initiated from the notch in C-C loading and grown until it stopped. In subsequent T-T loading, the crack would not grow at ΔK less than 5.5 MPa$\sqrt{m}$; this value is assumed to be slightly greater than ΔK_{th}. The largest K_{max} sustained during these experiments was 10.3 MPa$\sqrt{m}$. Material B4 broke in T-T loading at K_{max} = 5.5 MPa$\sqrt{m}$ prior to any known fatigue crack growth. Several crack growth rate values were obtained from a specimen of material B3 at $\Delta K \approx 5.7$ MPa$\sqrt{m}$. If it is assumed that $da/dN \approx 10^{-10}$ m/cycle at ΔK_{th} = 5.3 MPa$\sqrt{m}$ (R = 0.1) and $da/dN \approx 10^{-7}$ m/cy at K_c = 10 (ΔK = 9) MPa$\sqrt{m}$, then the constants in eq (1) would be: s = 11 and $B = 1\times10^{-18}$ m/cy. These values are intermediate between those from metals ($3 < s < 5$) and ceramics. For the TiAl alloy s = 5.4, while for partially stabilized zirconia s = 25.[10]

Fractography was performed on material B3 within the zone of fatigue crack growth. The results are shown in Fig. 5. The fine parallel lines approximately perpendicular to the direction of crack growth in some parts of the microstructure could be interpreted as periodic crack arrest lines (striations). To examine this possibility in more detail, two stage acetate carbon replicas were made from this region, shadowed, and examined by TEM under much higher resolution conditions. Typical results of this examination are shown in Fig. 5(b). No striations could be found on the replicas. The fine lines seen formed mainly in the Nb solid solution portion of the microstructure, as determined from the shape of the region, rather than in fractured Nb_3Al.

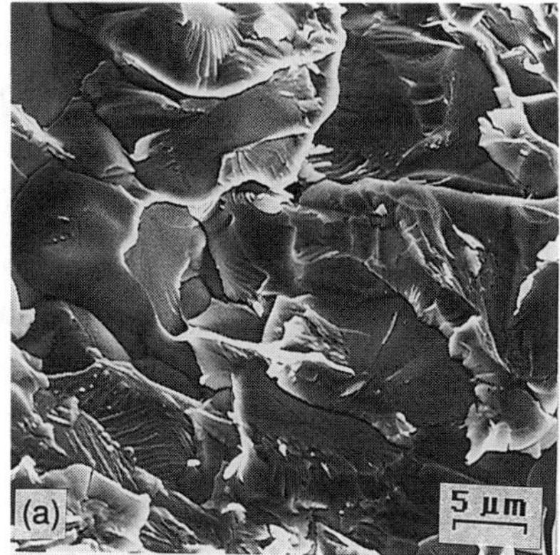

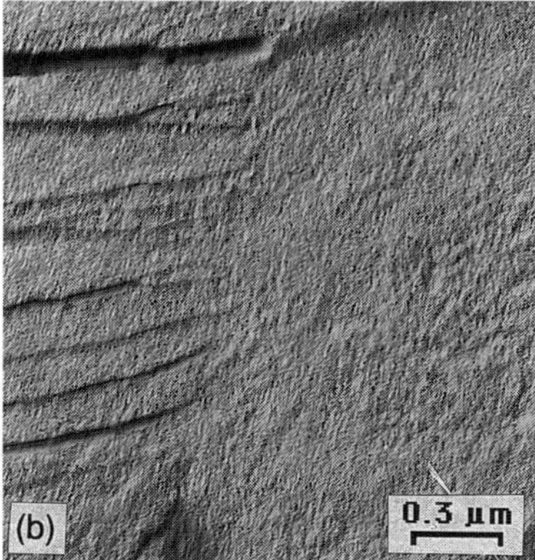

Fig. 5

Fractography of Material B3 in fatigue region: (a) overall secondary electron view showing small regions with linear regions, and (b) two stage acetate carbon replica.

FATIGUE CRACK TIP ANALYSIS

One of the notched specimens was cycled 565,000 cycles in C-C loading, R = 10, $\Delta K \approx 34$ MPa$\sqrt{m}$, attempting to initiate a fatigue crack, but no fatigue crack formed. Vickers hardness indents were made at the ends of the notch on both sides, but no visible cracks were formed by indentation. Cycling was continued in C-C at $\Delta K \approx 36$ MPa$\sqrt{m}$ for another 295,000 cycles whereupon it was discovered that a fatigue crack had initiated at the hardness indention on one side of the specimen within the last 30,000 cycles, giving an approximate crack growth rate of 2×10^{-9} m/cy. The crack had not linked to the notch, however. Loading was then switched to tension-tension, R = 0.1, starting at $\Delta K = 2$ MPa$\sqrt{m}$, then raised to 3, then 4 MPa$\sqrt{m}$ after no growth occurred in 10,000 cycles at each level. A photograph of the loaded fatigue crack tip was made at $K_{max} = 2$ MPa$\sqrt{m}$ (assuming a through crack). Displacements around each end of this crack were measured using stereoimaging, and strains were computed. A micrograph of the tip of the crack fartherest from the indention is shown in Fig. 6 with measured displacements (magnified 8 times) superimposed. The Mohrs circles of strain for the same area covered by the displacements are shown in the figure. The diameter of a Mohrs circle indicates the maximum shear strain; thus, values of shear strain may easily be determined using the strain scale in the lower right of the figure. Comparison of the features on the photograph and the Mohrs circles indicates that the maximum strains were in a region to the left of the crack tip where few precipitates are visible. The crack opening displacement, shown in the inset has the same form as fatigue cracks in conventional materials with COD = $C_o\sqrt{d}$ where d = distance behind the crack tip. However, the same relationship also pertains to elastic cracks, where the relationship for C_o is:

$$C_o = K(1/2\pi)^{1/2}[8(1-\upsilon)(1+\upsilon)/E] \qquad (2)$$

υ = Poisson's ratio (≈ 0.25), E = Tensile modulus ($\approx$ 250 GPa)[3], and K = Mode I stress intensity factor. Using $C_o = 1.7\times10^{-5}\sqrt{m}$ and the above values in eq. (2) gives K = 1.42 MPa$\sqrt{m}$ which is the same approximate value as estimated from stress and crack length. However, the strain determination indicated that the crack tip (effective) strain was $\approx$

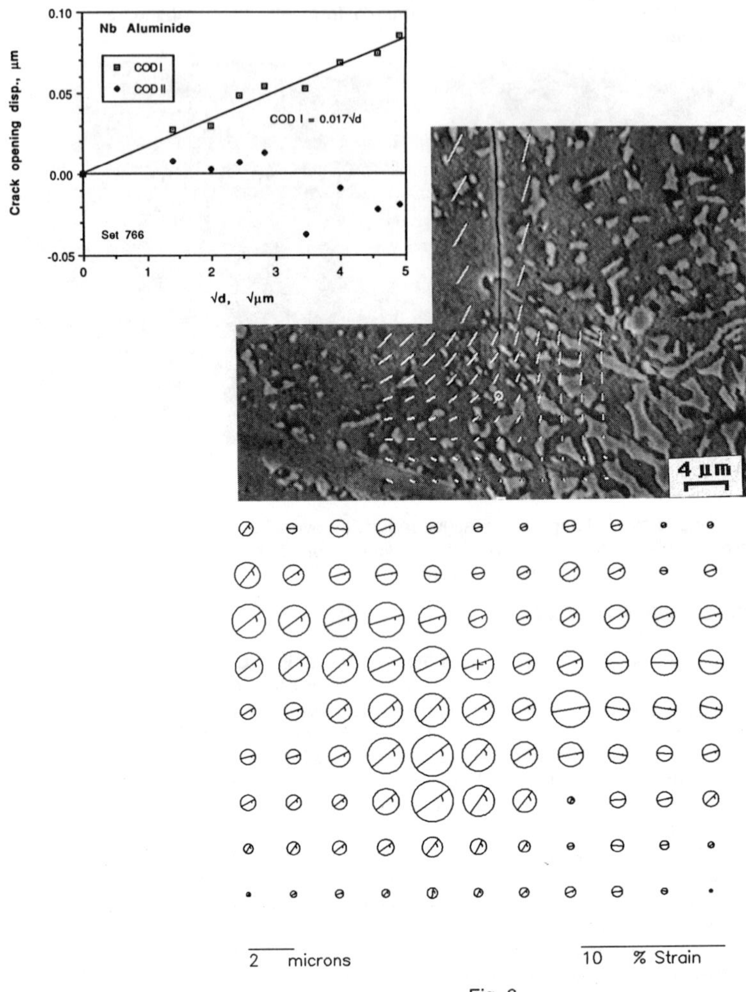

Fig. 6

Analysis of fatigue crack tip in material B3 at $K_{max} \approx 2$ MPa√m.
Displacements are magnified by a factor of 8. Strains are shown as Mohrs circles. Inset shows crack opening displacements.

0.015, which is well within the plastic region if $\sigma_y \approx 500$ MPa [3] (elastic strain $2\sigma_y/E \approx$ 0.004). The fact that this crack was not growing argues for validation of the low level of K calculated and may indicate that the yield stress is higher than has been estimated.

DISCUSSION

Anton and Shah[2,3] estimate the yield stress of these alloys to be approximately 500 MPa. Using the linear elastic fracture mechanics equation for crack tip "process zone size" (r_s)

$$r_s = (1/2\pi)\ (K_{max}/\sigma_y)^2 \quad (3)$$

where σ_y = yield stress and K_{max} = stress intensity factor at the threshold for fatigue crack growth (5.3 MPa√m), gives $r_s \approx 20$ µm. Some microstructural features in material A might be found of this approximate size scale, but the microstructure is much finer than this for materials B3 (for which ΔK_{th} was closely approximated) and material B4. Thus, it is not clear which microstructural feature is controlling fracture in these materials. Conversely, if the controlling microstructural feature is approximately 4µm, as estimated from Fig. 2, then the yield stress would be closer to 1100 MPa. Although the Rockwell C hardness values listed in Table I cannot be directly converted to strength for these alloys, the magnitudes of hardness are not incompatible with strengths of approximately 1000 MPa, based on the relationship for steels. Using the microstructural value of 4 µm and yield value of 1000 MPa in eq. (3) gives $\Delta K_{th} \approx 5$ MPa√m. It is concluded that the measured fatigue threshold is about what would be expected, at least for materials B3 and B4, and that the yield stress is greater than 500 MPa, perhaps as high as 1000 MPa.

SUMMARY

The log-log curve of fatigue crack growth versus ΔK had a slope of ≈ 11 for the Nb-18 at.%Al alloys tested, with $\Delta K_{th} \approx 5.3$ and $K_c \approx 10$ MPa√m. Using these values and microstructural dimensions, yield stress was estiamated as approaching 1000 MPa. Little effect of microstructure was found on these properties. Analysis indicated crack tip strains in excess of elastic values.

ACKNOWLEDGEMENTS

This collaborative research effort was funded at SwRI by the Air Force Office of Scientific Research, Dr. Alan Rosenstein, Contract Monitor, Contract F49620-89-C-0032, and at UTRC by WRDC Contract F33615-87-C-5214. The assistance of J.B. Campbell with fatigue crack growth and J.E. Spencer with stereoimaging measurements is acknowledged.

REFERENCES

1. D.L. Anton and D.M. Shah in **High Temperature Ordered Alloys III,** C.T. Liu, et al., eds., Mat. Res. Soc. v. 133, Pittsburgh, PA, 1986, pp. 361-371.
2. D.L. Anton and D.M. Shah "Ductile Phase Toughening of Brittle Intermetallics" in **Intermetallic Matrix Composites** D.L. Anton, P.L. Martin, et al., eds., Mat. Res. Soc. v. 194, Pittsburgh, PA, 1990, pp. 45-52.
3. D.M. Shah and D.L. Anton, Mater. Sci. and Eng. A153 402 (1992).
4. E.P. Barth, J.K. Tien, S. Uejo and S. Kambara, Mater. Sci. and Eng. A153, 398 (1992).
5. E.S.K. Menon, P.R. Subramanian and D.M. Dimiduk, Scripta metall. mater. 27, 265 (1992).
6. Symposium Proceedings of **High Performance Materials for Severe Environments,** K. Inabe and S. Shindo, eds., NEDO, Tokyo, Japan, Nov. 28, 1992.
7. T.N. Marieb, et al., in High Temperature Ordered Intermetallic Alloys IV, L. Johnson, et. al., eds., MRS, Pittsburgh, PA, 1991, pp.
8. C.E. Lundin, et al., Trans. Met. Soc. AIME, 236 863 (1966).
9. L. Cort, et al., J. Less Common Metals, 44 215 (1976).
10. D.L. Davidson, J.B. Campbell and J. Lankford, Acta metall.mater. 39,1319 (1991).

PART VI

Processing

Processing of Al_2Ti

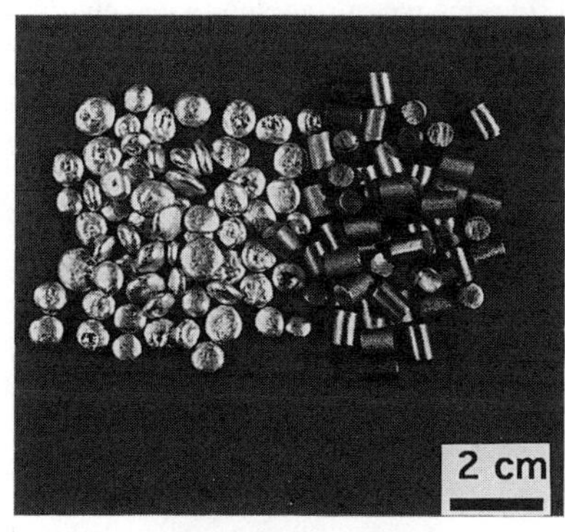

Al Buttons and Ti Cut Rod

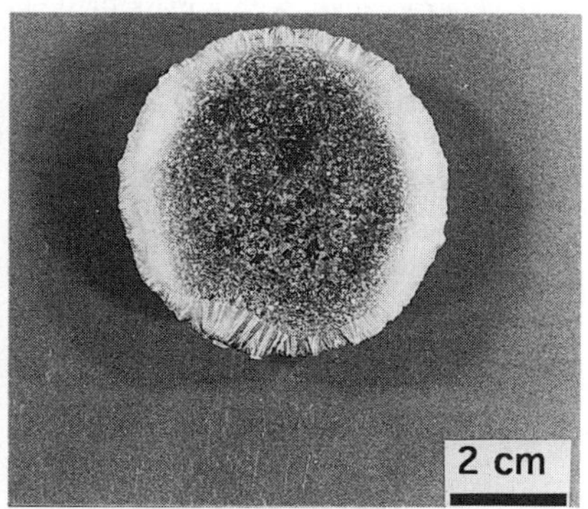

Cast Al_2Ti Ingot

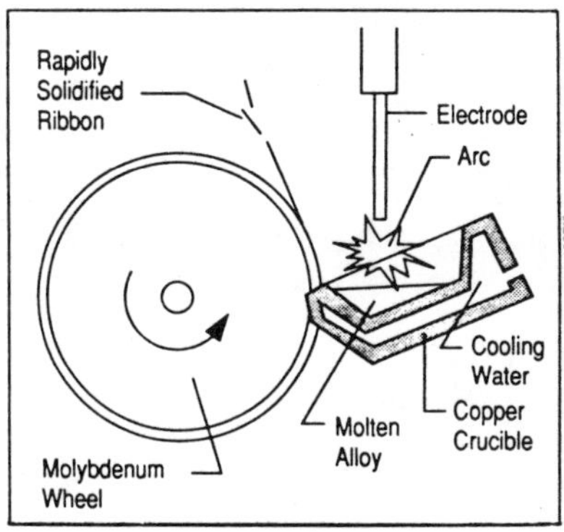

Melt Spinner

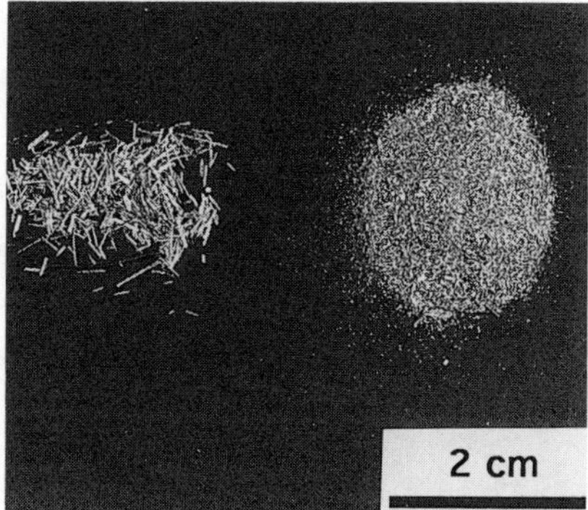

Al_2Ti Melt-spun Ribbon and Powder

Al_2Ti was prepared starting with high purity aluminium (99.99%) and titanium (99.7%) which were arc-melted on a water cooled copper hearth in an argon atmosphere to produce a cast ingot with a mass of about 300 grams measuring approximately 75 mm in diameter and 20 mm in height. The ingot was remelted several times to ensure chemical homogeneity. Several melts were rapidly solidified by melt spinning which yielded Al_2Ti ribbon with a needle-like shape. The ribbon was then comminuted into powder using a hammer mill which, in turn, was consolidated by hot isostatic pressing.

Figure courtesy of J.E. Benci, J.C. Ma and T.P. Feist.

STUDY OF INTERMETALLIC FeAl—ALUMINA WETTABILITY AFFECTED BY MICROALLOYING Y AND Nb

DONG XU, DENING WANG, WEIMIN ZHOU, HONG YANG
AND DONGLIANG LIN(T. L. LIN)
Department of Materials Science, Shanghai Jiao Tong University, Shanghai 200030, P. R. China

ABSTRACT

The effects of microalloying elements,Y(0.02at%, 0.04at% and 0.06at%) and Nb(0.2at%, 0.5at% and 0.8at%) on the wettability of intermetallic Fe-40Al with polycrystalline α-Al_2O_3 were investigated experimentally by means of sessile drop method. The addition of 0.02at%Y and 0.04at%Y showed no significant effects on the wettability. However, the addition of 0.06at%Y improved the wettability in temperature range from 1673K to 1823K. The addition of different amount of Nb were not beneficial to the FeAl-Al_2O_3 wettability under conditions investigated. The SEM study of the shear fractured surfaces showed that the additions of Y and Nb coursed Al_2O_3 particles more adherent to FeAl matrix, compared with the case of pure Fe-40Al. The degree of adherence varied with the amount of Y and Nb. These phenomena are discussed in terms of interfacial reactions and interfacial bonding.

INTRODUCTION

Intermetallic compound Fe-40Al has a good oxidation and corrosion resistance and low density [1]. However, its strength is unacceptably low at temperatures above 700K[2]. One way to solve this problem may beto develop Fe-40Al matrix composites [3].

Compatibility of a series of ceramics, including oxides, carbides, nitrides and borides, with FeAl compounds has been studied systematically by A. K. Misra in order to acquire proper systems of IMCs[4]. Alumina is found to be the most thermodynamically stable reinforcement. S. L. Draper et al.[2] investigated the compatibility of Fe-40Al with alumina by hot processing method, no chemical reaction at the interface was observed up to 1473K. This is in coincidence with the thermodynamical study. Hence, alumina is considered to be one of the best candidates of reinforcements for Fe-40Al compound.

A fundamental understanding of bonding at the interface between intermetallis and ceramic is required not only to predict the mechanical behavior of IMCs but also to select constituents for the IMCs so that their properties and processing can be optimized. A lot of woks has been contributed on the metal-ceramic interface reactions and interfacial strength[5, 6], but researches of intermetallic-ceramic interface phenomena, which is also of great importance, are far from sufficiency. This paper will devote to the preliminary work on interface phenomena between Fe-40Al and alumina by means of sessile drop method[7]. Special interests are focused on the wetting behaviors and interfacial bonding affected by active elements additions.

EXPERIMENTAL

Sessile drop method[7] was used to study the wetting behaviors of Fe-40Al alloys, microalloyed with yttrium (0.02 at%, 0.04 at% and 0.06 at% respectively) and niobium (0.2 at%, 0.5 at% and 0.8 at% respectively), with polycrystalline α-Al_2O_3. Fe-40Al and its alloys were prepared in an arc-melting furnace, followed by homogenization at 1273K for 48 hours. Afterwards, the alloys were cut into cylinder specimens with the size of diameter 3×3.5 mm which were polished and cleaned with acids, and also, the α-Al_2O_3 substrates with size of 12×12×4 mm were polished and cleaned by ultrasonic vibration in acetone for 30 minutes. Heating rate in sessile drop experiment was controlled within 40K per minute under vacuum kept better than 3×10^{-3} Pa. Specimens were degassed by holding at 1273K for 30 minutes prior to be heated to the wetting temperature. Wetting angles were measured at temperatures of 1673K, 1723K, 1773K and 1823K, with holding time of 30 minutes at 1673K, 1723K and 1773K, and of 5 minutes at 1823 K. Scanning electron microscope (SEM) was used to examine the fracture surfaces of interfaces between the Fe-40Al alloys and alumina after the sessile drop experiments.

RESULTS

The results of sessile drop experiments showed that the contact angles of intermetallic Fe-40Al with alumina were affected by alloying element additions. The values of contact angles of Fe-40Al-0.02Y, Fe-40Al-0.04Y and Fe-40Al-0.06Y alloys on alumina substrates were measured at all temperatures studied. The Fe-40Al-0.06Y had lower contact angles on alumina substrates in temperature range from 1673 K to 1823 K in comparison. It appears that when the amount of yttrium is more than 0.04 at% the wettability is enhanced significantly. The contact angles of yttrium alloyed Fe-40Al on alumina against holding time at 1723 K are shown in Figure 1. Niobium additions did not affect the wettability of Fe-40Al with alumina significantly under the conditions investigated, as shown in Figure 2. The curves of yttrium and niobium alloyed Fe-40Al at temperatures of 1673 K, 1773 K and 1823 K were similar to those in Figures 1 and 2.

The wetting process of the system of Fe-40Al and alumina reaches equilibrium within 30 minutes, as shown in Figures 1 and 2. The equlibrium contact angles of yttrium and niobium alloyed Fe-40Al on alumina at all temperatures studied are plotted in Figures 3 and 4 respectively. In figure 3, the 0.06 at % yttrium addition induced Fe-40Al contact angles a decrease of 10° or so at all temperatures studied.

The shear fracture surfaces of the interfaces of Fe-40Al alloys and alumina were examined by scanning electron microscope. The SEM results showed that

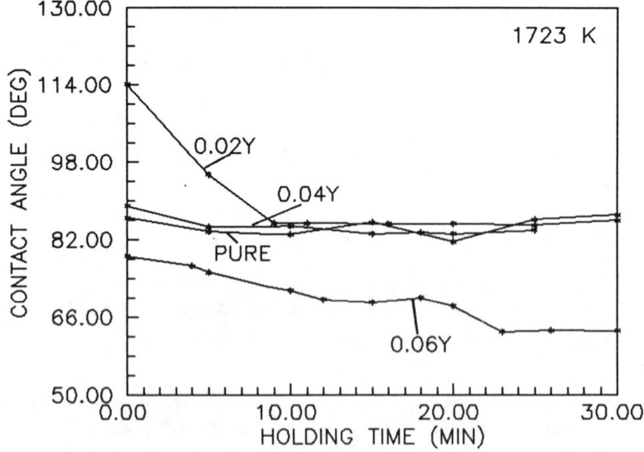

Figure 1 Contact angles vs holding time of Fe-40Al/alumina with Y

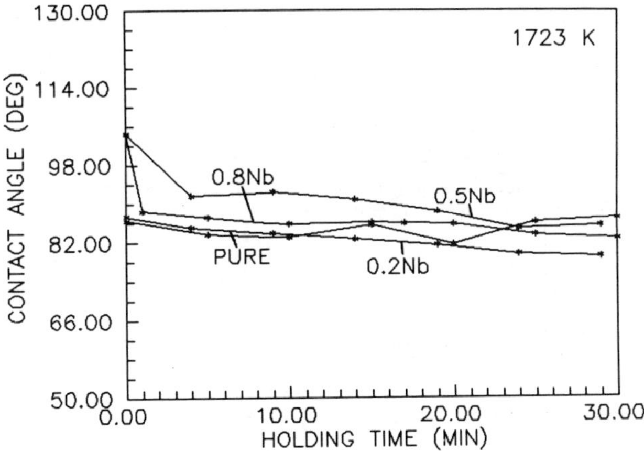

Figure 2 Contact angles vs holding time of Fe-40Al/alumina with Nb

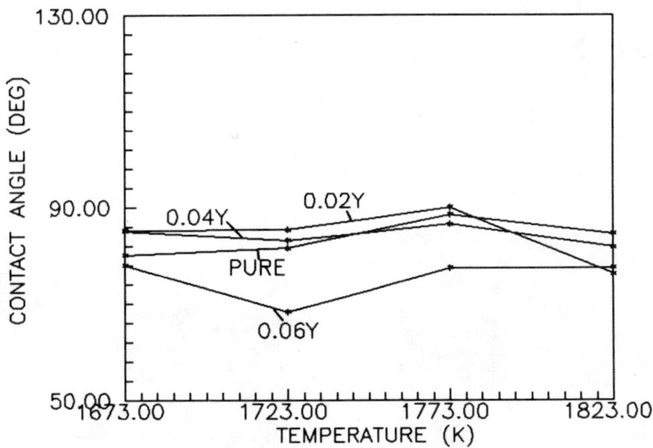

Figure 3 Contact angles at equilibrium of Fe-40Al/alumina with Y

some alumina particles were sticked to the drop bottom surfaces. The number of sticked particles increased with the increases of yttrium and niobium additions, as shown in Figures 5 and 6.

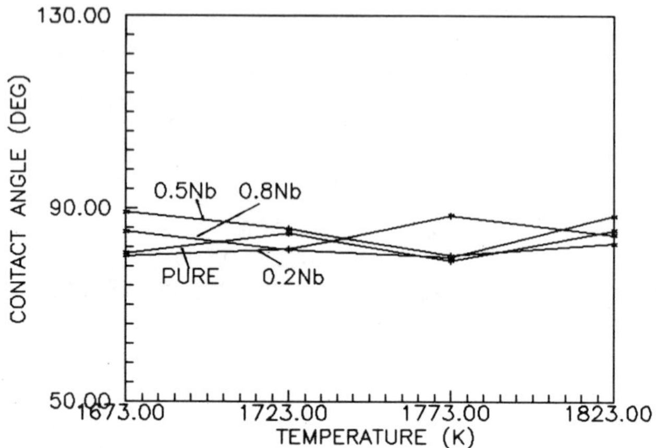

Figure 4 Contact angles at equilibrium of Fe-40Al/alumina with Nb

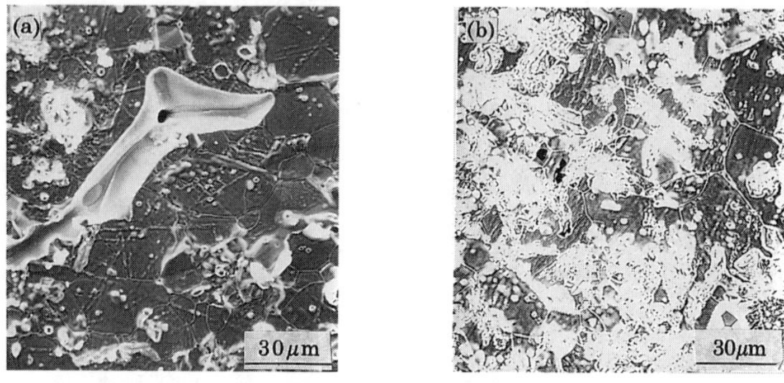

Figure 5 Shear fracture surfaces of interfaces of (a)Fe-40Al--0.02Y(b)Fe-40Al-0.06Y with alumina, 1772 K, ×500

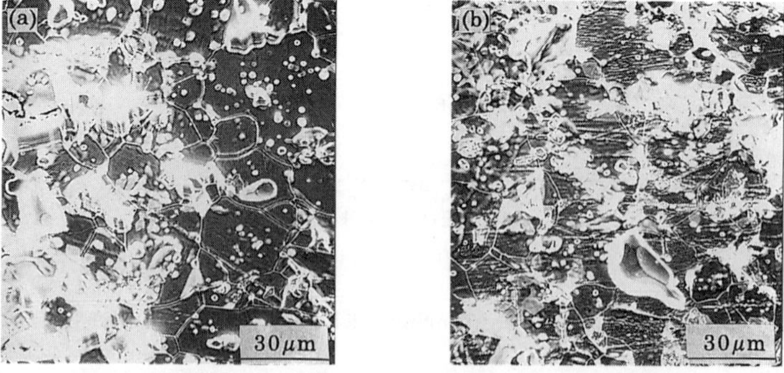

Figure 6 Shear fracture surfaces of interfaces of(a)Fe-40Al-0.5Nb(b)Fe-40Al-0.8 Nb with alumina, 1723 K, ×500

DISCUSSION

Wetting is the process of interactions between liquid metal and ceramic substrate at interface. The interface interactions are usually of either physical type, which induces no interlayers at the interface, or chemical type which induces interlayers at the interface. However, the interface interaction between Fe-40Al and alumina appears more complicated. The auther's previous work[8] revealed that liquid intermetallic Fe-40Al reacted with alumina above the melting point, but the reaction product was in the gas state, thus no interlayer was developed at the interface. This means that although the wetting behavior of liquid Fe-40Al on alumina should be preferably related to the interface reaction, the interface adhesion is more likely associated with atomic bondings of Fe-40Al and alumina at interface.

Some researchers[9] discussed that for metal-oxide systems, the main factor which affects the wettability would be oxygen affinity of metals, and better oxygen affinity corresponds to better wettability. Additions of yttrium and chronium were found to be effective to improve wettability of nickel alloy-alumina systems, and niobium was found to be beneficial to the improvement in wettability of steel and alumina systems.

In Table 1, the Gibbs free energy of oxide formation(ΔG_f°) of metals investigated in this paper are presented[10]. The Fe-40Al contact angles on alumina are around 80°(1823 K) in this paper, and those of Fe and Al are 121° (1823 K) and 47° (1528 K) respectively. This can probably be understood that aluminium decreases the contact angles of iron in Fe-40Al during wetting process, and the fairly good wettability of Fe-40Al with alumina is mainly due to aluminium.

TABLE 1 Gibbs free energy of oxide formation of sevsral metals[10]

Metals	Al	Fe	Y	Nb
$-\Delta G_f^\circ$ (kJ/ mol atom oxygen)	526.7	247.0	605.0	370.0

The improved wettability of yttrium-contained Fe-40Al with alumina is probably due to its low value of ΔG_f° compared with that of aluminium. Since ΔG_f° of niobium is higher than that of aluminium, the Nb additions showed no significant effect on wettability even though it is an oxide-forming metal. For steel, Nb is a good wetting agent as in ref[11].

It is known that microalloying of yttrium and niobium are effective for improving the interface bonding strength in nickel-alumina and steel-alumina systems[9, 11]. The bond strength is also considered to be associated with the oxygen affinity of the metals. SEM examinations showed that both yttrium and niobium additions affected the interface bonding behavior for Fe-40Al and alumina systems, as shown in Figures 5 and 6. No interlayers were found in cross-section of the interfaces [8], but the increase in the amount of the sticked alumina particles indicate that the interface bonding strength is improved by the additions of yttrium and niobium. Note that ΔG_f° of yttrium and niobium are both lower than that of iron, better interface bonding would be due to the better oxygen affinity of the alloying elements. Quantitative work on Fe-40Al interface cohesive strength is in progress in our labratory.

Wetting behavior and interface cohesive strength are the two key problems for intermetallic matrix composites, especially when manufactured by melt-processing. The results in this paper reveals that alloying with some oxide-forming metals may be an effective way to improve both of these two problems.

CONCLUSION

(1). Intermetallic Fe-40Al has a fairly good wettability with α-alumina.
(2). Addition of 0.06 at%Y is beneficial to the wettability of Fe-40Al with α-alumina, and additions of 0.04 at% or lower Y have no significant effect on the wettability.
(3). Additions of Nb show no significant effect on the wettability of Fe-40Al with α-alumina under conditions investigated.
(4). Both Y and Nb are effective addition elements to cause stronger cohesive strength between Fe-40Al and α-alumina.

REFERENCES

1. C. T. Liu, C. G. McKamey and E. H. Lee, Scripta Metall. Mater. **24**, 385(1990).
2. S. L. Draper, D. J. Gaydosh and M. V. Nathal, J. Mater. Res. **5**, 1976(1990).
3. D. E. Alman, and N. S. Stoloff in Intermetallic Matrix Composites, edited by D. L. Anton, P. L. Martin, D. B. Miracle, and R. McMeeking(Mater. Res. Soc. Symp. Proc. **194**, Pittsburgh, PA, 1990)p31.
4. A. K. Misra, Metall. Trans. **21**A, 441(1990).
5. M. A. Smith and D. P. Pope, Mater. Sci. Eng.A**145**, 79(1991).
6. M. Nicholas, R. R. D. Forgan and D. M. Poole, J. Mater. Sci. **3**, 9(1968).
7. A. Mortensen, Mater. Sci. Eng. A**135**, 1(1991).
8. D. Xu, D. N. Wang and T. L. Lin, to be published in Scripta Metall.
9. M. G. Nicholas, Mater. Sci. Forum **29**, 127(1988).
10. J. A. Dean, Lange's Handbook of Chemistry, 13th ed. (McGraw-Hill Book Company, 1985).
11. Y. H. Liu, Ph. D. Thesis, Harbin Polytech. Univ., Harbin, PRC, 1989.

DYNAMIC RECRYSTALLIZATION IN TIAL ALLOYS

D.S. LEE*, D.M. DIMIDUK*, S. KRISHNAMURTHY†
*Materials Directorate, WPAFB, OH 45433
†UES, Inc, Materials Research Division, Dayton, OH 45432

ABSTRACT

TiAl alloys with Cr, V, and Nb additions show promise as high temperature materials due to their high temperature strength and modulus. Dynamic recrystallization has been shown to be important for the processing and superplastic forming of these materials. Earlier studies have indicated that dynamic recrystallization may also occur during tensile straining at or near the proposed use temperature of these alloys. A systematic study has been conducted to determine the effects of various parameters - temperature, strain, strain rate, and microstructure, on the occurrence of dynamic recrystallization near expected use conditions. The conditions near the ductile to brittle transition temperature which bound the onset of dynamic recrystallization in tension and compression were investigated. The findings will be presented and correlations will be drawn between the contributions of dynamic recrystallization and high temperature to the observed ductility.

INTRODUCTION

Titanium aluminides are currently being studied for a variety of military and commercial applications. The alloys of most interest combine some combination of gamma and alpha-2 phases, in either a duplex or fully lamellar microstructure, in order to obtain a balance of properties. While recent advances in processing and alloy development have made these alloys contenders to applications for 1500°F, many basic questions remain which may ultimately affect their performance in use.

One of the most basic problems with this class of alloys is its lack of room temperature ductility. In fact, the tensile ductility does not rise appreciably above 2-4% until the material reaches its ductile to brittle transition temperature. This is typically around 700-800°C, depending upon the alloy composition and prior thermomechanical history. This phenomenon, although widely known, is not well understood. The mechanisms which might account for this change in behavior include increased dislocation activity [9], increased twinning, dynamic recrystallization, or some combination these mechanisms.

A study by Krishnamurthy et al. [1] of a near gamma alloy, showed that dynamic recrystallization could be found in tensile specimens near the necked region for specimens tested above 900°C. In specimens tested at 1000°C the dynamic recrystallization consumed the entire necked region in a duplex alloy. The fully lamellar alloy studied was also fully recrystallized in the necked region, except for lamellar grains, which were oriented in the direction of the tensile axis. The true strains calculated in the affected region were all above 1.0.

Given that gamma/alpha-2 alloys are expected to be used at or near their ductile to brittle transition temperature, it is of interest to determine whether dynamic recrystallization is occurring at these temperatures, and if so, under what conditions. Additionally, if dynamic recrystallization is occurring, whether this is expected to deleteriously affect long term stability and properties of the alloy under service loads.

APPROACH

The material used in this study was a near gamma alloy of nominal composition Ti-47.0Al-1.6Cr-0.9V-2.4Nb. The alloy was induction skull melted into a cylindrical ingot and then two-step, isothermally forged into a 9" diameter x 0.4" thick pancake. Wedges of the pancake were heat treated to two microstructural conditions and stabilized with a direct age treatment: Condition A- a nearly gamma structure (1280°C/3 hrs/direct cooled to 900°C/6 hrs/air cool) and Condition B- a nearly lamellar structure (1360°C/2hrs/direct cooled to + 900°C/6 hrs/air cooled). Wet chemistries were taken from the heat treated pancakes. Compression specimens 0.4" long x 0.25" diameter were then cut from the heat treated wedges. Specimens were tested at 800°C, at a strain rate of 5×10^{-2} in/in, and to various amounts of total strain (0.05 - .25). The specimens were then sectioned and analyzed using optical, back-scattered electron (BSE), transmission electron microscopy (TEM), and energy-dispersive spectroscopic analysis techniques.

RESULTS

The as forged microstructures indicated that the material had fully recrystallized during hot working. There was little or no evidence of banding, as has been seen in single step forging. Wet chemistries of the as heat treated alloys showed that the Al content was closer to 49 a/o and total interstitial levels less than 800 ppm. Representative pictures of the as heat treated microstructures can be found in Figures 1a and 1b. Condition A shows the nearly gamma structure with a grain size of 50-75μm. This alloy appears to be approximately 95% gamma from BSE. Condition B is a nearly lamellar structure with a lamellar colony size of about 700-1000μm, and small gamma particles decorating the grain boundaries. The lamellar structure contains very little alpha-2, approximately 5%, which is found as fine lathes in the lamellar regions and at triple points.

Figure 1 - Optical micrographs of as heat treated specimens: (a) Condition A: 1280°C/3hrs/DC + 900°C/6 hrs/AC and (b) Condition B: 1360°C/2hrs/DC + 900°C/6 hrs/AC.

The compression yield strength data from the tests indicated that these specimens behaved in a fashion typical for alloys of this class. Measured yield strengths averaged 49 ksi. Typical flow curves for each microstructure are found in Figure 2. None of the tests, regardless of microstructure, exhibited any flow softening.

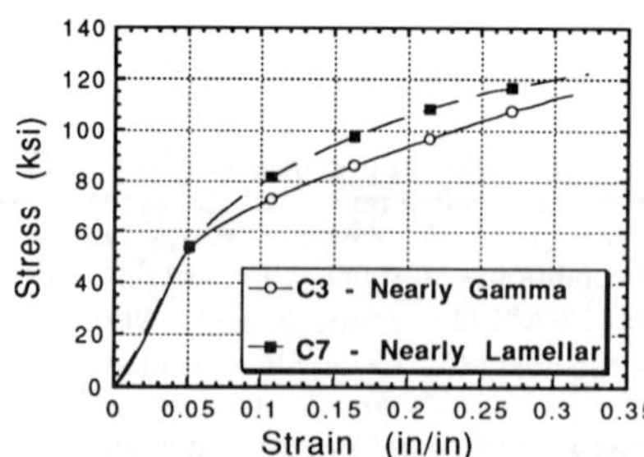

Figure 2: Example of flow stress diagram for both the nearly gamma and nearly lamellar microstructures tested at 800°C.

Optically, Condition A showed massive twinning in the as tested state. Twinning and cross-twinning increased with increasing total strain. At 25% total deformation (Figure 3a), the specimen exhibited bent twins, slip bands, and obviously deformed grains. Condition B showed twinning in the gamma grains, twinning across the lamellar grams, and twinning within the lamellae. As deformation increased to 25% total strain (Figure 3b), more evidence was found of bent lamellae and slip bands across the lamellar grains. No evidence was found, at the optical or BSE level, of dynamic recrystallization in either of the conditions.

Alloys A and B with 25% deformation were examined in the TEM. Condition A had a deformation structure consisting of massively twinned areas and dense dislocation tangles (Figure 4a). A few recrystallized grains were found at triple points and along grain boundaries (Figure 5a). These grains contained none, or very little, of the deformation activity found in the remainder of the sample. Condition B had a deformation structure consisting of massive amounts of twinning and dislocations within the lathes themselves (Figure 4b). Very few recrystallized grains were found in this sample and were only located at grain boundaries (Figure 5b). No recrystallized grains were found at twin boundaries.

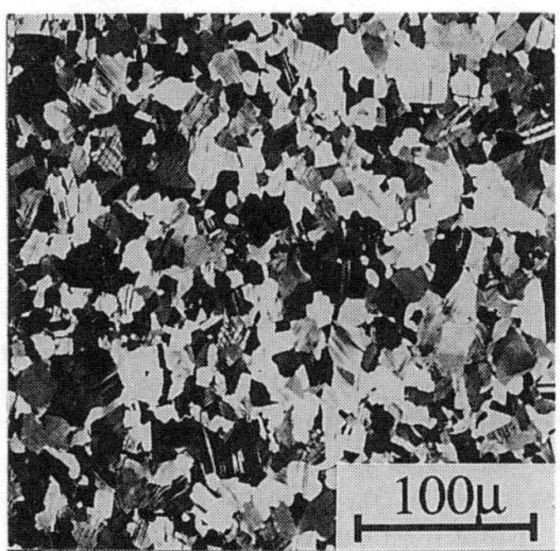

Figure 3 - Optical micrographs of as tested specimens deformed to 25% total strain (a) Condition A, (b) Condition B.

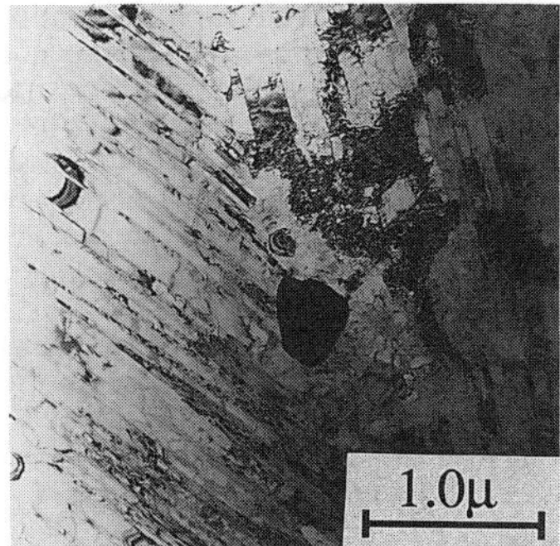

Figure 4- TEM bright field micrographs showing dense dislocation substructures in (a) Condition A and (b) Condition B. Both have been deformed to 25% strain

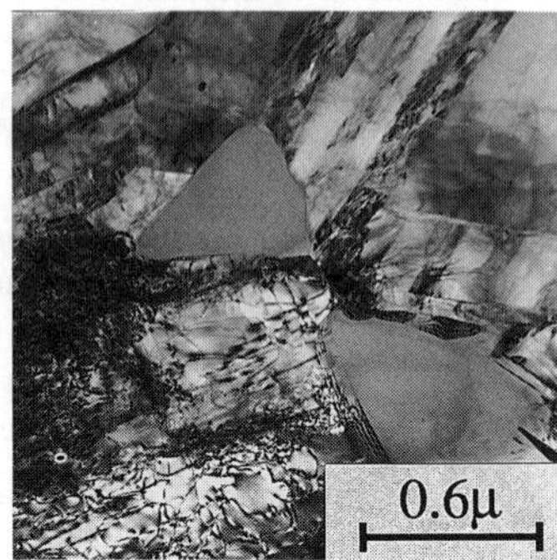

Figure 5 - TEM bright field micrographs showing undeformed, recrystallized grains in (a) a triple point in Condition A and (b) in a grain boundary in Condition B.

DISCUSSION

The recrystallized grains found in this study were observed only at areas of high localized strain, i.e. at grain boundaries and triple points. They were never found within the grains, despite the high amounts of deformation substructure which was in evidence. In all cases the measured flow curves are continuously rising, suggesting a lack of significant dynamic recrystallization. In addition, relatively small amounts (<<1%) of recrystallized grains were found at all. These facts lead us to believe that the dynamic recrystallization process is just beginning for these conditions. Recrystallized grains in the nearly gamma microstructure were found primarily at triple points while in the lamellar microstructure they were found only at grain boundaries. It is suggested that the constraints on the grain boundaries due to the lamellar plates in the nearly lamellar condition make this the preferred site for the onset of recrystallization, as opposed to at the triple points for the nearly gamma structure.

Numerous studies have found that the major deformation mechanisms in gamma/alpha-2 alloys are {111} twinning and slip of 1/2[110] dislocations [5,7,8]. At higher temperatures the 1/2[110] superdislocations and [101] slip systems are active [6]. Although the exact nature of the twins and dislocations in this study were not determined, the deformed substructures are morphologically consistent with the previous works. Dynamic recrystallization of this class of alloys has been reported upon earlier by many authors [2,3,4]. However, the emphasis in their studies was to investigate microstructural changes under processing conditions. The study by Krishnamurthy mapped dynamic recrystallization in the necked regions of tensile specimens tested at 900°C, but no evidence was shown for dynamic recrystallization at lower temperatures. The evidence obtained in the present work, for the beginnings of the dynamic recrystallization is therefore significant, at least for these conditions of temperature, strain, and strain rate. Despite the evidence for the onset of dynamic recrystallization at temperatures near the proposed use temperatures, it is felt that the stability or properties of the alloy will not be adversely affected during service conditions. Further, we expect that dynamic recrystallization plays little to no role on the mechanism of the ductile-brittle transition.

CONCLUSIONS

A study was conducted to determine if dynamic recrystallization was occurring at or near the use temperatures in a typical gamma/alpha-2 alloy. Compression testing at 800°C was conducted and specimens were analyzed by optical microscopy, SEM, and TEM. Dynamic recrystallization was found in both near gamma and near lamellar microstructures at triple points and grain boundaries respectively. The amount of recrystallization was small enough to be negligible with respect to the resultant properties.

ACKNOWLEDGMENTS

The authors thank G. Cornish for help testing assistance, F. Scheltens for microscopy discussions, S. Apt for foil preparation, and M. Scott and E. Fletcher for specimen preparation. S.K. acknowledges the support of the Air Force Materials Directorate under contract F33615-91-C-5663.

REFERENCES

1. S. Krishnamurthy and Y-W. Kim in Microstructure/Property Relationships in Titanium Aluminides and Alloys, edited by Y-W. Kim and R. R. Boyer (TMS proceedings, Warrendale, PA: , 1990) 149-163.

2. N. Fujitsuna, H. Ohyama, Y. Miyamoto, Y. Ashida, ISIJ International, **31** (10), 1147-1153 (1991).

3. R. M. Imayev, O. A. Kaibyshev, G. A. Salishchev, Acta metall. mater., **40**, (3), 581-587 (1992).

4. V. Seetharaman and C. M. Lombard in Microstructure/Property Relationships in Titanium Aluminides and Alloys, edited by Y-W. Kim and R. R. Boyer (TMS proceedings, Warrendale, PA: , 1990).

5. E.L. Hall, S-C. Huang, J. Mater. Res., **4**, (3), 595-602 (1989).

6. G. Hug, A. Loiseau, P. Vessiere, Phil. Mag. A, 57, (3), 499-523 (1988).

7. Shechtman, M.J. Blackburn, H.A. Lipsitt, Met. Trans., 5, 1373-1381 (1974).

8. C. R. Feng, D, J, Michel, C. R. Crowe, Scripta Met., **23**, 1135-1140 (1989).

9. 7. H.A. Lipsitt, D. Shechtman, R.E. Schafrik, Met. Trans. A, **6A,** 1991-1996 (1975).

EFFECT OF PROCESSING ON THE HIGH TEMPERATURE MECHANICAL PROPERTIES OF $MoSi_2$

S.N. Patankar and J.J. Lewandowski
Dept. of Materials Science and Engineering
Case Western Reserve University
Cleveland OH 44106

ABSTRACT

Polycrystalline $MoSi_2$ has been processed via spex milling and the notched fracture behavior has been determined at 800 - 1400 °C. Elemental Mo and Si powders were mechanically alloyed to produce $MoSi_2$ and were subsequently vacuum hot pressed to fully dense $MoSi_2$ compacts. Microstructures were analyzed and both hardness and notched fracture tests were conducted at high temperatures. The material toughness was constant at 4.5 MPa $m^{1/2}$ between 800 and 1000 °C, with significant plasticity exhibited at 1200 °C. The results will be compared to recent work on the effects of C on high temperature properties of $MoSi_2$.

INTRODUCTION

Molybdenum disilicide is currently being considered as a potential high temperature material because of its high melting point, good strength to density ratio, and its reasonable ductility at high temperatures as compared to other intermetallics and refractory metals. Molybdenum disilicide is often prepared by arc melting or P/M routes, followed by consolidation and sintering at elevated temperatures. In the recent past there has been renewed interest in the development of molybdenum disilicide and molybdenum disilicide based composites with the emergence of more advanced milling and synthesis techniques like spex milling and reaction synthesis [1-7]. Considerable discussion exists as to the brittle to ductile transition temperature, which has been reported as 1000 °C[8], although recent studies utilizing smooth four point bend tests have indicated that this transition temperature may be as high as 1300 to 1400 °C[9] after certain processing conditions. It is clear that additional studies are necessary to establish the effects of processing conditions on the brittle to ductile transition temperature in $MoSi_2$, while lowering the BDTT will improve the performance of $MoSi_2$ for use as a matrix material whose high temperature strength can be enhanced with suitable reinforcement. The current investigation was conducted to study the high temperature properties of $MoSi_2$ prepared by spex milling.

EXPERIMENTAL PROCEDURE

The materials used in the present study were 99.9% pure Molybdenum powder with size range of 3-7 micron and -325 mesh (< 44 micron) pure Silicon powder. A spex mill/model 8000 was used for mechanical alloying. Milling was carried out in a carbide-lined steel vial with argon atmosphere to minimize oxidation. The charge to ball ratio was 1:10 while the charge was comprised of 10 grams of powder mixture taken in a proper stoichiometry. The milling was conducted for 3 hours and 13 minutes, followed by x-ray analysis to determine the reaction products present. The milled powders were subsequently degassed at 800 °C for 6 hours and hot pressed at 1500 °C for 30 minutes using 51.7 MPa pressure. Hot micro-hardness experiments were performed on the as hot pressed $MoSi_2$ using a 1000 gram load over temperatures ranging from room temeprature to 1000 °C. Toughness was estimated via the indentation cracks produced while both the toughness and the flexural strength were determined from four point bend tests performed on the notched and smooth $MoSi_2$ bars. The specimen dimensions were 25 mm x 2mm x 5mm for the notched and 25 mm x 2mm x 2.5mm for the smooth specimens. Notching was accomplished with a diamond saw blade. The four point bend tests were performed at temperatures ranging from 25 to 1400 °C. The notched four point bend tests were conducted using a displacement rate of 50 μm per minute while the smooth bars were tested at 127 μm per minute.

RESULTS AND DISCUSSION

Figure 1 contains x-ray diffraction (XRD) patterns of Mo and Si powders milled for different time intervals, indicating that times in excess of 3 hours and 12 minutes of spex milling is sufficient to produce abrupt formation of $MoSi_2$. The details of such formation are discussed elsewhere [10]. For the conditions utilized presently, conversion to $MoSi_2$ reaches completion only after hot pressing, as shown in Fig. 2. The $MoSi_2$ obtained following hot pressing had a density of 6.21 gm/cc, representing 98.4 % theoretical density. The grain size of the as hot pressed material was in the range of 1-5 μm.

The room temperature hardness of $MoSi_2$ was 1300 Kg/mm^2, while Figure 3 shows the variation of hardness of $MoSi_2$ as a function of temperature. The hardness value decreases steadily from the room temperature value of 1300 Kg/mm^2 to 400 Kg/mm^2 at 1000 °C. Comparison to previous data [11] reveals that the hardness of the spex milled $MoSi_2$ at the temperatures ranging from 1000 to 1400 °C is higher than the hardness of $MoSi_2$ produced from commercially available Cerac powder, and is comparable to the hardness of $MoSi_2$ with 2% C. The toughness values computed using the expression derived by Evans and Charles [12] are plotted as a function of temperature for $MoSi_2$ and are shown in Fig. 4. The toughness slightly increases at 800 °C, with a significant rise at temperatures greater than 900 °C. Figure 5 shows the toughness (derived from four point bend tests of the notched specimens) as a function of temperature. In contrast to the Cerac powders and similar to the $MoSi_2$ + 2 wt% C material [11], the toughness of the spex milled $MoSi_2$ remains constant from 800 °C to 1000 °C. The toughness of the Cerac powder material decreases at higher temperatures, while those of the spex milled and C-containing materials increase. The K_{q1} values were computed using the deviation from linearity on the load-displacement curve, obtained from the four point bend test of notched specimens. The K_{q2} values were computed in a similar manner but using maximum load values from the load - displacement curves. The trend shown by the toughness values computed using the Evans and Charles expression [12] shown in Fig.4 is identical to the trend exhibited in Fig. 5.

Visual examination of the spex milled $MoSi_2$ specimens tested at temperatures above 1200 °C showed a significant amount of plastic deformation. The fracture surface of the $MoSi_2$ subjected to four point bend tests at 1000 and 1100 °C showed transgranular as well intergranular fracture. The brittle to ductile transition

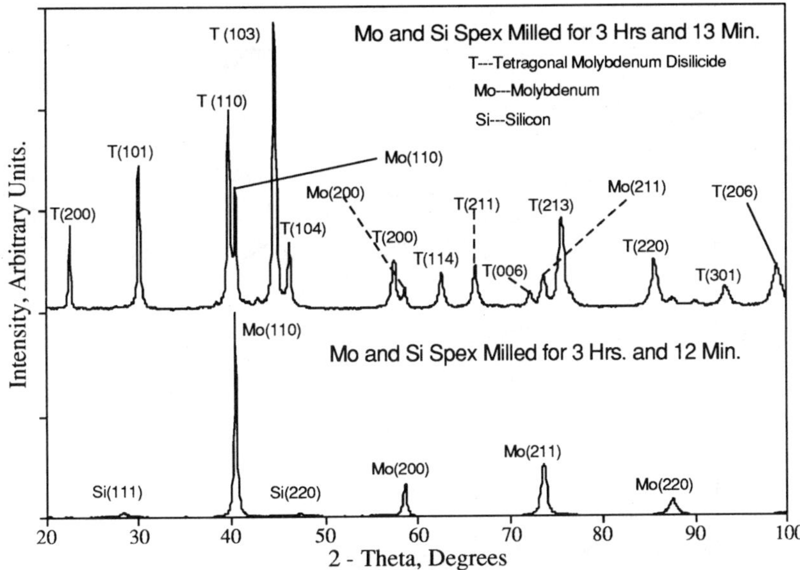

Fig. 1 XRD Pattern of Mo-Si Powder Milled for Different Time Intervals.

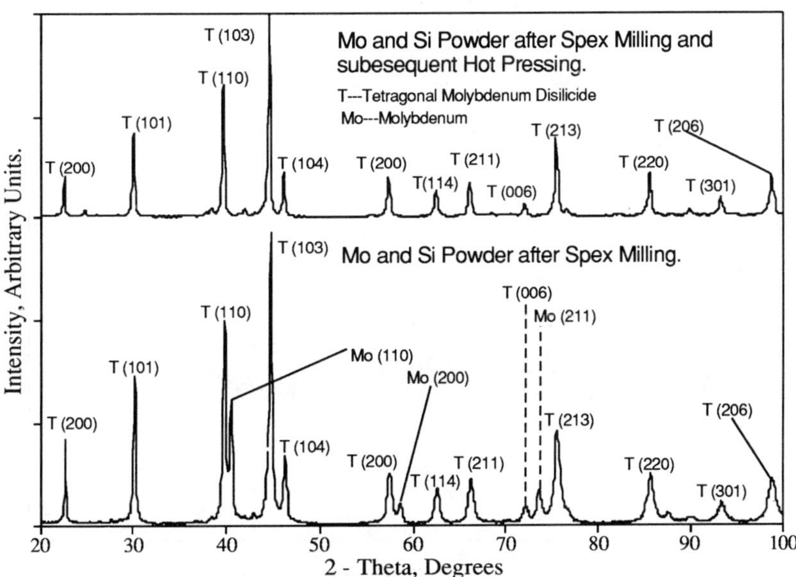

Fig. 2 XRD Pattern of Milled Powder and Hot Pressed Compact.

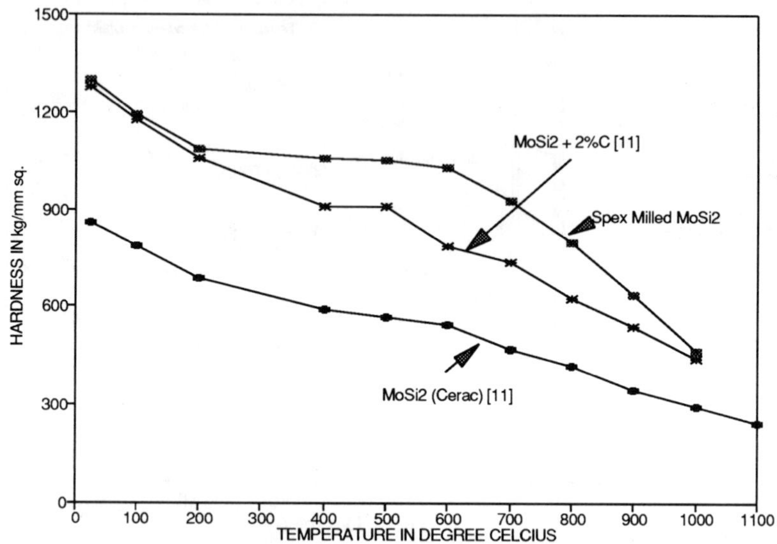

Fig. 3 Variation of Hardness with Temperature.

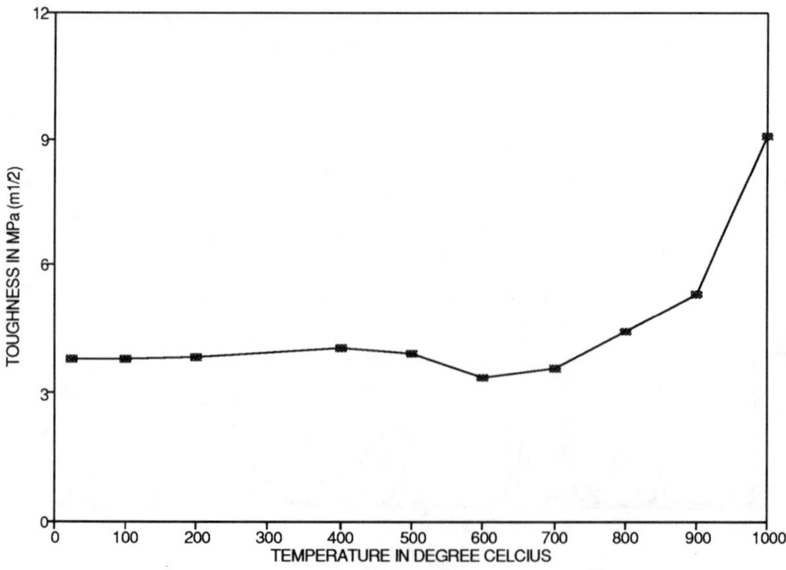

Fig. 4 Variation of Toughness (Derived from Hardness) with Temperature.

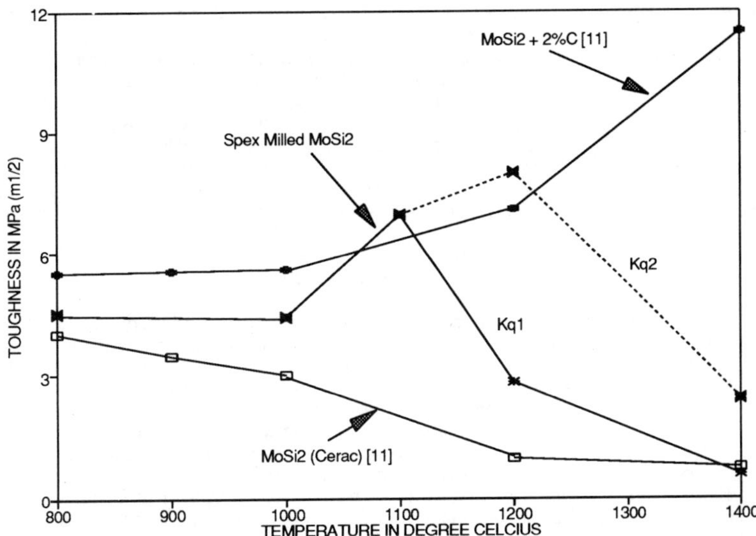

Fig. 5 Variation of Toughness (obtained from Notched Four Point Bend Test) as a function of Temperature.

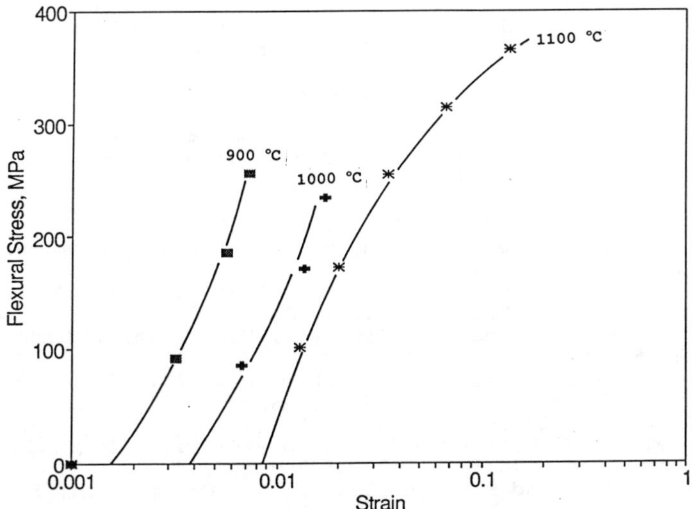

Fig. 6 Flexural Stress-Stress Curves obtained on Smooth Four Point Bend Bars.

temperature in the notched specimens was 1200 °C, significantly greater than that in the smooth specimens, as shown in Figure 6. Significant deformation was obtained in $MoSi_2$ tested at 1100 °C, while brittle behaviour was obtained at temperatures less than 1000 °C as shown in Figure 6. The flexural stress as well as amount of plastic deformation at 1100 °C in the present case is much higher than the corresponding flexural stress and deformation at 1200 °C reported by Aikin [9] in material with significantly larger grain size. The sources of the differences in transition temperatures may be related to the amount of SiO_2 present as well as differences in grain size. These topics are under discussion.

CONCLUSIONS

It can be concluded from our preliminary work on the processing of $MoSi_2$ that spex milling results in abrupt formation of $MoSi_2$ after a critical milling time. The conversion of remaining elemental Mo and Si to $MoSi_2$ reaches completion during the hot pressing of the milled powder. The $MoSi_2$ obtained is fine grained with a room temperature hardness of 1300 Kg/mm^2. The fracture toughness measurement from the bend tests of notched specimens showed a increase in toughness up to 8 MPa $m^{1/2}$ at 1200 °C from the room temperature value of around 4 MPa $m^{1/2}$. The brittle to ductile transition temperature obtained on smooth four point bend bars in the present study was 1100 °C, although the sources of this brittle to ductile transition are under investigation.

ACKNOWLEDGEMENTS

Support for the authors was provided by AFOSR - 89 - 0508 with Dr. Alan Rosenstein as Program manager.

REFERENCES

1) P.J. Meschter and D.S. Schwartz, JOM, **41**(1989), 52.
2) A. Calka,A.P. Radlinski,R.A. Shanks and A.P. Pogany, J. Mater. Sc. Letters, **10**(1991),734-737.
3) Z.A. Munir, High Temp. Sc., **27**(1990), 279-293.
4) M. Atzmon, Phy. Rev. Letters, **64**(1990), 487-490.
5) K.S. Kumar and S.K. Mannan, Mater. Res. Soc. Symp. Proc., **133**(1989), 415-421.
6) M. Atzmon, Mater. Sci. and Eng., **A134**(1991), 1326-1329.
7) S.C. Deevi, J. Mater. Sci., **26**(1991), 3343-3353.
8) A.K. Vasudevan and J.J. Petrovic, Mater. Sc. and Eng., **A155**(1992), 1-17.
9) R.M. Aikin, Scripta Metall. et Mater., **26**(1992), 1025-1030.
10) S.N. Patankar,S.Q. Xiao,J.J. Lewandowski, and A.H. Heuer, J. of Materials Research, (1992),Submitted.
11) S. Maloy, A.H. Heuer,J.J. Lewandowski and J. Petrovic, J. Am. Ceram. Soc., **74**(1991), 2704-2706.
12) A.G. Evans and E.A. Charles, J. Am. Ceram. Soc., **59**(1976), 371.

CHARACTERIZATION OF VACUUM PLASMA SPRAYED MECHANOFUSED NiAl

Z.J. Chen and H. Herman,
Dept. of Materials Science and Engineering, State University of New York, Stony Brook, NY 11794-2275, USA;
C.C. Huang and R. Cohen,
Micron Powder Systems, 10 Chatham Road, Summit, NJ 07901, USA.

ABSTRACT

Vacuum plasma spray is a melt spray process capable of producing coatings or near-net shaped deposits with ultra-fine grained structures. Mechanofusion is a novel technique for processing powdered materials. In this study, mixtures of elemental Ni and Al powders, either blended or mechanofused, were vacuum plasma sprayed, yielding a series of Ni-Al intermetallic phases. Mechanofusion resulted in significant effects on the phase distribution in subsequently as-sprayed deposits. After heat treatment at 1100 °C, single phase NiAl was obtained for both the mechanofused and blended samples. The microhardness of the NiAl is comparable to that produced by other methods. These preliminary experiments indicate that mechanofusion and vacuum plasma spraying can be employed to produce various compounds using mixed elemental powders.

INTRODUCTION

Intermetallic compounds are candidates for high temperature structural materials. Of particular interest is NiAl, which has a high melting temperature, low density, and excellent oxidation resistance. The wide single-phase region of NiAl allows deviations from stoichiometry and alloying for improvements of the mechanical properties without entering a two-phase field (1). Furthermore, NiAl is a promising matrix for composite materials.

There are various methods for producing Ni-aluminides, such as reactive sintering of mixed elemental powders (2), high-pressure shock loading (3), and mechanical alloying (4). Ni-aluminides produced by plasma spray have been reported by a number of researchers using different types of spray processing under various conditions (5-7).

In this paper is described an alternative means for producing intermetallic NiAl: vacuum plasma spray (VPS) of mechanofusion (MF) processed mixtures of elemental Ni and Al powders. In addition, simply blended powders have been VPS processed, and these deposits are compared with those formed from MF processed samples.

PROCESSES DESCRIPTION

1). Vacuum Plasma Spray operates in a reduced pressure (e.g., $6x10^3$ Pa) inert gas environment, Fig. 1(a) (8,9). A distinct attribute of the VPS process is that it combines melting, quenching and consolidation in one operation. The high solidification rate (10^7 K/s) of the process can result in ultra-fine grained microstructures (10). Moreover, the structure of sprayed deposits is homogeneous without the large scale macrosegregation resulted from conventional casting. While VPS has generally been used to produce protective coatings, it also has the advantages of being capable of fabricating thick, dense and oxide-free bulk forms and near-net shapes. It has been shown that VPS-aluminides (11) have properties which can be superior to such materials formed using other techniques. In this paper VPS processing of intermetallic feedstock formed by mechanofusion is described. Mechanofusion also leads to spheriodization of powders, which adds to the utility for forming powder for VPS, since spherically shaped powders are free flowing.

2). Mechanofusion is a novel powder processing technique (12), in which raw materials are mechanically intermingled with sufficient energy to create a powder with different properties. Two views of the mechanofusion device are displayed schematically in Fig. 1(b), showing the rotary shallow cylindrical chamber which contains the powder, a fixed semi-cylindrical "inner piece", and a fixed "scraper blade". The gap between the inner-piece and the chamber wall is adjustable. An investigation of mechanofused powders for plasma spray has been reported (14). In addition, MF can create a variety of composited powders for the production of plasma spray-formed composites (15). Experiments on the VPS-forming of intermetallic matrix composites are currently underway.

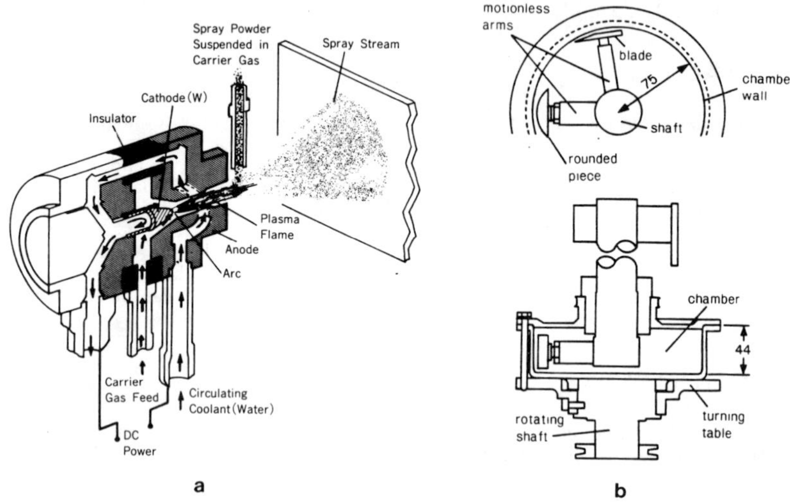

Fig. 1. Schematic illustration of (a) plasma spraying process (9), (b) mechanofusion (13).

EXPERIMENTAL PROCEDURE

1). Sample Preparation: The samples are coded by NAXXXX. NA refers to the Ni-Al compound. The first two digits represent the atomic percentage of Ni in the starting powders. The second two digits represent the nominal time (in minutes) of MF processing. The target composition was equiatomic single phase NiAl. Since more elemental Al powder than Ni powder is lost during MF and VPS, starting powders with more than 50 at% Al were chosen for compensation. Two groups of samples were prepared. In Group I, NA3200 (32 at.% Ni + Al) and NA4800 (48 at.% Ni + Al), the feedstock powders for VPS were simply blended. In Group II, NA3240 (32 at.% Ni + Al) and NA4840 (48 at.% Ni + Al), the feedstock powders for VPS were MF processed for 40 minutes at 700 RPM. The average particle sizes for the Ni and Al powders were 34 micrometers and 19 micrometers, respectively.

Vacuum plasma spray was performed with a Plasma Technik VPS unit using a PT-F4V plasma gun. The feedstock powder was spray-deposited onto mild steel substrates. It took about 2 minutes for VPS spraying to form the deposits of ~500 micrometers in thickness. All Group I and II deposits were also annealed at 1100 °C in flowing argon for 2, 4 or 12 hours and then furnace cooled.

2). Characterizations Technique: X-ray diffractometry with CuKα radiation was performed to identify the phases. Scanning electron microscopy (SEM) analysis was also carried out. Energy dispersive spectroscopy (EDS) was used to analyze elemental distributions. Vickers microhardness measurements (VHN) were conducted on polished cross-section of the deposits using a 500 gm load for 15 seconds.

RESULTS AND DISCUSSION

1). Phases Analysis and Microstructures: The results of X-ray diffraction and SEM/EDS studies are given in Table I. Before VPS spraying, no Ni-Al intermetallic compounds were observed in MF-processed powders. The apparent absence of these phases may be due to the low proportion of compound present.

Table I. Phase distributions from X-ray diffraction

Group	Sample	Before VPS	As-sprayed		2 hours		4 hours	12 hours
			Major	others	Major	others		
I	NA3200	Ni, Al	Ni_2Al_3	Ni, Ni_3Al NiAl?	NiAl		NiAl	NiAl
I	NA4800	Ni, Al	Ni_2Al_3	Ni, Ni_3Al NiAl	NiAl	Ni, Ni_3Al	NiAl	NiAl
II	NA3240	Ni, Al	NiAl	Ni, Ni_3Al Ni_2Al_3?	NiAl		NiAl	NiAl
II	NA4840	Ni, Al	NiAl	Ni_3Al, Ni	NiAl	Ni_3Al	NiAl	NiAl

Following the VPS process, the phase distributions of MF-processed deposits (NA3240 and NA4840) were significantly different from those of the blended deposits (NA3200 and NA4800). The strongest phase in the blended as-sprayed deposits was Ni_2Al_3, with incompletely reacted Ni, Ni_3Al and perhaps NiAl, which had peaks which overlap with Ni_2Al_3, Fig. 2(a). But, for MF-treated as-sprayed deposits, NiAl was the dominant phase, with some Ni_3Al and Ni present, while the Ni_2Al_3 was almost undetectable, Fig. 2(b).

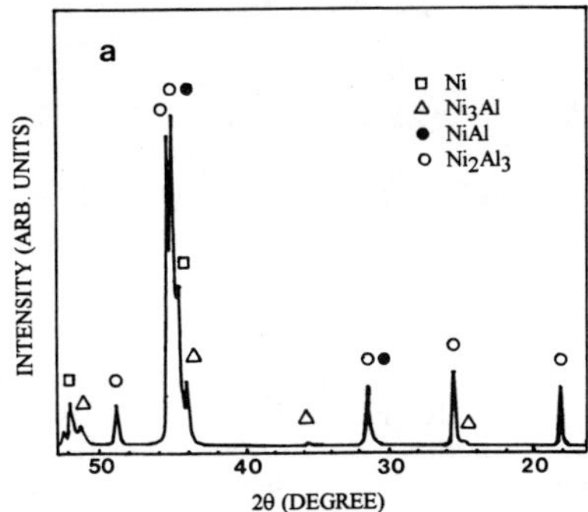

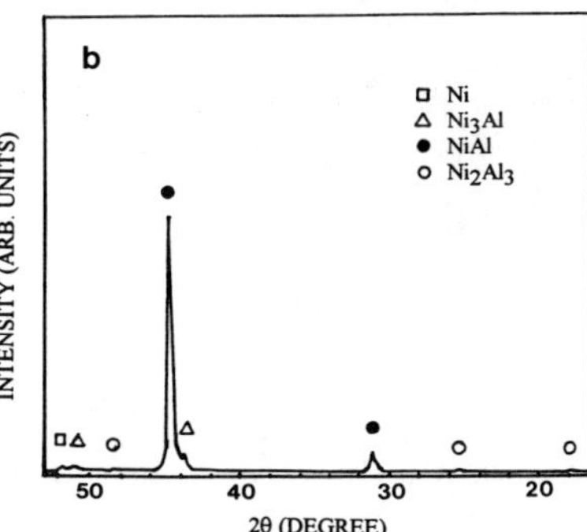

Fig. 2. (a) X-ray diffraction pattern of blended as-sprayed deposit NA3200. The strongest phase was Ni_2Al_3, with Ni, Ni_3Al and possibly NiAl. (b) X-ray diffraction pattern of MF-processed as-sprayed deposits NA3240. NiAl was the dominant phase, with Ni_3Al and Ni present. The Ni_2Al_3 was almost undetectable.

SEM micrographs disclosed the microstructure of as-sprayed deposits containing aluminides with some incompletely reacted Ni (bright spots), Fig. 3(a). After annealing at 1100 °C for 2 hours, X-ray diffraction analysis demonstrated that NiAl was the only phase present in the samples with 32 at.% Ni in the starting powders, Fig. 3(b). When annealed at 1100 °C for 4 hours or longer, all samples were transformed to the single phase NiAl.

Annealing also improved the deposit density. The as-sprayed or short-time annealed samples showed a porous structure, Figs. 3(a) and 4(a). It is suggested here that most of pores resulted from a Kirkendall-type effect due to the different Ni and Al diffusivities as well as from the volumetric change due to phase reactions, as suggested by Newbery et al. (7). Annealing at 1100 °C for 12 hours enhanced the deposit density, as shown in Fig. 4(b).

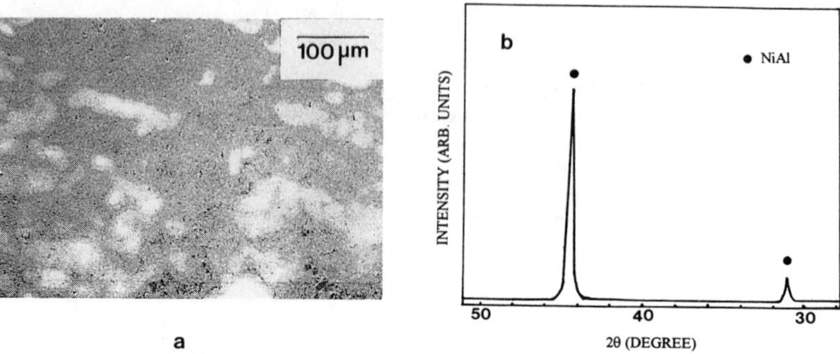

Fig. 3. (a) SEM secondary electron micrograph of MF-processed as-sprayed deposit NA3240. The bright spots were incompletely reacted Ni. (b) After annealing at 1100 °C for 2 hours, X-ray diffraction pattern showing that NiAl was the only phase present.

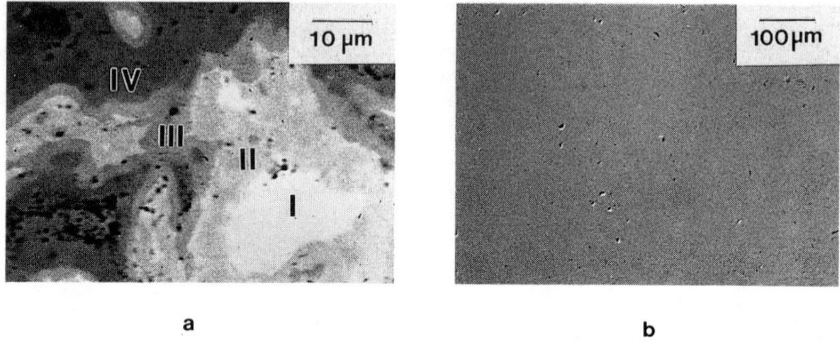

Fig. 4. (a) SEM backscattered electron micrograph of blended as-sprayed deposit NA4800 showing four regions of phase distribution with porosity. Region I is Ni, II Ni_3Al, III NiAl and IV Ni_2Al_3. (b) After annealing at 1100 °C for 12 hours, SEM secondary electron micrograph of blended NA4800 sample showing homogeneous microstructure and enhanced density.

2). Phase Formation: X-ray diffraction analysis of the over-sprayed powders, which were collected from the VPS chamber after spraying, revealed that elemental Ni and Al remained present with no apparent formation of Ni-aluminide compounds. This result indicates that the intermetallic phases were not formed in-flight from the nozzle to the substrate. It may be that the time-of-flight to the substrate was too short and droplets of elemental Ni and Al were not in intimate contact for interdiffusion to occur. So, intermetallic phases were formed after droplets were deposited on the substrate.

Fig. 4(a) SEM backscattered electron micrographs of the as-sprayed deposits of NA4800 also show four regions (I-IV) of the phase distribution. Based on SEM/EDS and X-ray diffraction studies, the phases in those regions can be identified. Fig. 4(a) reveals that incompletely reacted Ni particles (region I) are surrounded by Ni_3Al (region II), which is in turn surrounded by a thin layer of NiAl (region III) in an Ni_2Al_3 matrix (region IV).

On the basis of the above results, a mechanism for the formation of Ni-aluminides by VPS is proposed as follows. The plasma flame melts the Al and Ni powders. After the droplets of Ni and Al impact on the substrate or previously deposited layers, they come into intimate contact. Since an intense exothermic reaction begins at about 660 °C (5), the system temperature during VPS is high enough to initiate such reactions. The exothermic reaction of Ni-Al provides a thermodynamic driving force which favors the formation of the Ni_2Al_3, [ΔH(Ni_2Al_3)=–282.6 KJ/mol] instead of NiAl [ΔH(NiAl)=–118.5 KJ/mol] or Ni_3Al [ΔH(Ni_3Al=–150.8 KJ/mol] (16). In this study, $NiAl_3$ phase was not found. $NiAl_3$, with its low melting temperature of 850 °C, would disappear soon if it were formed, since the substrate temperature was about 900 °C during VPS spraying (10). Such high temperatures also promote interdiffusion between phases, resulting in a series of Ni-aluminides.

Although all samples eventually transformed to single phase NiAl after annealing, the MF process showed significant effects on phase formation. The Ni-aluminides present in MF-processed as-sprayed deposits were closer to the equilibrium NiAl phase. In fact, MF has been successfully used to prepare composited powders with uniform agglomeration from micrometer-sized Ni and Al. Subsequently, VPS processing yielded only single phase intermetallics in the as-sprayed deposit without annealing (15).

The manner in which MF influences the formation of intermetallic phases using VPS process is unclear. One possible benefit of MF may be the creation of a more homogeneous distribution of Ni and Al powders. This would maximize elemental particle contact and decrease Ni or Al powder segregation. Therefore, the chemical reaction of Ni and Al would be faster, and the diffusion reaction would be more complete, as compared with simply blended samples. The argument here may in some way relate to Atzmon's study of NiAl formation (4). He suggested that when elemental Ni and Al are well mixed, the formation of NiAl is enhanced and intermediate steps in Ni-aluminide formation can be bypassed.

3). Microhardness: Microhardness measurements of the samples are given in Table II. All as-sprayed deposits have higher Vicker Hardness Numbers (VHNs) than the annealed samples. This may result from the difference in the phases present, the fine grain size and/or residual stresses. After annealing, the NA4800 and NA4840 samples appear to have higher VHNs than do the NA3200 and NA3240 samples, as seen in Table II. This is probably due to the fact that during MF and VPS, there is more Al lost than Ni. Therefore, the chemical compositions of NA3200 and NA3240 deposits, through analysis of lattice parameters, are closer to the stoichiometric alloy, which has the lowest VHN (17). However, the hardness values of all samples are comparable to those produced by other methods (17,18).

Table II. Microhardness measurements (VHN)

	As-sprayed	2 hours	4 hours	12 hours
NA3200	568±101	347±32	283±23	310±21
NA4800	518±67	391±25	367±41	340±50
NA3240	479±62	334±17	363±15	329±14
NA4840	433±35	381±15	398±17	376±15

SUMMARY

1). A series of Ni-aluminides containing Ni_2Al_3 as the major phase were formed by VPS using simply blended Ni and Al. For the deposits that have been treated with mechanofusing, the desired NiAl became dominant. After annealing at 1100 °C for various times, single phase NiAl was obtained in all samples. The microhardness of the NiAl produced in this study is comparable to that produced by other methods.

2). It is believed that the Ni-aluminides are formed after the molten droplets of Ni and Al are deposited on the substrate. High substrate temperatures during vacuum plasma spray processing is a critical factor for intermetallic formation. Mechanofusion appears

to assist intermetallic phase formation by creating a homogeneous distribution of Ni and Al powders.
3). Vacuum plasma spray in conjunction with mechanofusion is a useful process for producing compounds using mixed elemental powders or for fabricating composites from mixtures of selected materials.

REFERENCES

1. R. Darolia, J. of Metals, 44-49, March (1991).
2. A. Bose, B. Moore, R.M. German and N.S. Stoloff, J. of Metals, 14-17, Sept. (1988).
3. Y. Horie, R.A. Graham and I.K. Simonsen. Materials Letters, Vol. 3, No. 9,10, 354-359, July (1985).
4. M. Atzmon, Mat. Sci. & Eng. A134, 1326-1329 (1991).
5. J.M. Houben, Jr and J.H. Zaat, Proc. 7th Int. Metal Spraying Conf., London, 77-88, 296-303 (1973).
6. S. Kozerski, 11th Int. Therm. Spraying Conf., Montreal, 845-852 (1986).
7. A. P. Newbery, B. Cantor, R. M. Jordan and A. R. E. Singer, Scripta Met., Vol. 27, 915-918 (1992).
8. H. Herman, MRS Bulletin, Vol. XIII, No. 12, 60-67 (1988).
9. H. Herman, KONA No. 9, 187-199 (1991).
10. S. Sampath, Ph. D., Thesis, State University of New York, Stony Brook, NY, (1989).
11. A.I. Taub, S.C. Huang and K.M. Chang, in High-Temp. Ordered Intermetallic Alloys, Materials Research Society Sym. Proc., Vol. 39, edited by C. C. Koch, C. T. Liu and N.S. Stoloff, Materials Research Society, Pittsburgh, PA, 221-228 (1985).
12. T. Yokoyama, K. Urayama, M. Naito, M. Kato and T. Yokoyama, KONA No.5, 59-68 (1987).
13. M. Alonso, M. Satoh and K. Miyanami, KONA, No. 7, 97-105 (1989). L. Hixon and J. Van Cleef, R & D Magazine, 60-63, April (1988).
14. H. Herman, Z.J. Chen, C.C. Huang and R. Chen, J. of Thermal Spray Technology, Vol. 1, No. 2, 129-135 (1992).
15. Z. J. Chen, H. Herman, R. Tawari, C. C. Huang and R. Cohen, Proc. Inter. Therm. Spraying Conf., Orlando, edited by C. C. Berndt, ASM International, Metals Park, OH, 355-361 (1992).
16. E.A. Brandes, editor, 'Smithells Metals Reference Book', 6th ed. Butterworths, Boston, 8-9 (1983).
17. J.H. Westbrook, J. Electrochem. Soc., Vol. 103, No. 1, 54-63, (1956).
18. S. Nourbakhsh and P. Chen, Acta Metall., Vol. 37, No. 6, 1573-1583 (1989).

REACTIVE SINTERING OF Ni_3Al INTERMETALLIC ALLOYS UNDER COMPRESSIVE STRESS*

K. H. WU* AND C. T. LIU***
**Florida International University, Dept. of Mechanical Engineering, University Park Campus, Miami, FL 33199
***Metals and Ceramic Division, Oak Ridge National Laboratory, Oak Ridge, TN 37831-6115

ABSTRACT

Synthesis of the Ni-25% at. Al intermetallic compound by the thermal explosion mode subject to the application of compressive stress was systematically investigated. The stress was applied prior to the reaction. It was proved that the compressive stress had a profound effect on the reaction, as well as the densification. At a relative low stress, a full reaction could be completed and the major phase of the final product is Ni_3Al. At higher stresses, a partial reaction was resulted and the final product was primary the intermediate phases, e.g. Ni and NiAl, etc. Three distinct reaction products were obtained in this study. When the stress was below approximately 5 MPa, a fully reacted product was obtained. Between 5 and 15 MPa, a fully and partially reacted zone coexisted. Above 15 MPa, only a partial reaction prevailed. In this experiment, a final product of 99.9% relative density was obtained at the high stress level.

INTRODUCTION

Exothermic reaction is an innovative way to produce low cost, near net-shaped products at a temperature much lower than the melting point of the compound [1-5]. German et al. [2,6-9] conducted substantial studies on the reaction sintering of Ni_3Al under stress-free conditions. The maximum relative density reported in their study was 97%. A great deal of effort has been exerted to improve the density of the final product through different means, such as explosive compacting [10]. However, the ultimate density reported rarely exceeds that of the German et al. study, unless other processes are also involved, e.g., Hiping [9,11]. It is the goal of many researchers to accomplish a fully dense product from this novel technique. Reaction sintering of intermetallic compounds with the presence of compressive stress has been investigated by a few researchers [12-14]. Rawers et al. [12] applied a pressure of 1.4 MPa to TiAl specimens. The density of the product achieved using this method did not exceed 95%. Nishimura and Liu [13,14] recently applied a 50 MPa compressive stress during the processing of Ni_3Al using alumina blocks smaller than the diameter of the specimen, and they were able to achieve 99.3% relative density. It appears that the compressive stress has a significant effect on the reaction process. The purpose of this study is to conduct a systematic investigation of the effect of compressive stress on the reaction sintering of Ni_3Al and to obtain very dense products.

THE EXPERIMENTAL PROCEDURE

The average diameter of the spherically shaped, atomized aluminum powder was approximately 12 μm, and the irregularly shaped, general-purpose nickel powder was 8 μm. The 25.4 mm diameter, 20-g green compacts were prepared from stoichiometric (75 at. % Ni and 25 at. % Al) elemental powders using a hydraulic press and a cylindrical die. At an applied stress of 220 MPa, a relative density of about 65% was achieved.

Tests were carried out in a vacuum on an Instron Model 1321 servo-hydraulic machine. The load was applied on the specimen through two pieces of 19.6 mm diameter, 12.7 mm-thick 99.5% alumina (see Fig. 1). The specimen was degassed in the vacuum chamber for at least two hours and than heated at a rate of 60 K/min until it reached a temperature of 620°C and held there for 30 min. The load was applied right at the moment when the temperature reached 620°C.

Throughout this study, the test conditions, particularly the boundary conditions, remained unchanged, except for the applied load, which varied from 0 to 100 MPa. Specimens removed from the vacuum chamber after testing were sliced and fractured, and examined under optical and scanning electron microscopes (SEM). Density measurements were performed after the

* Research sponsored by the Division of Materials Science, US Department of Energy under contract DE-AC05-840R21400 with Martin Marieeta Energy Systems, Inc.

specimen was heat-treated at 1100° C for five hours in a vacuum, using the principle of buoyancy by immersing the specimen in toluene. Only the portion of the material under compression (center part) was used as a sample for the density measurement. The relative density was calculated based on the theoretical value of 7.44 g/cm^3 [15].

RESULTS

A. Density

As the compressive stress varied from 0 to 100 MPa, three distinctly different final products were obtained from this experiment (see Table I).

Table I

Results of Compressive Reaction Sintering of Ni_3Al

Stress (MPa)	Degree of Reaction	Final Products: Major	Final Products: Minor	Relative Density	Grain Size
100	Partial				
50	Partial	$NiAl + Ni$	Ni_3Al	99.3	
25	Partial			97.9	
20	Full + Partial			96.3	
15	Full + Partial			-	13
10	Full + Partial			-	15
5	Full + Partial			-	16
3	Full			98.5	27
1	Full	Ni_3Al	NiAl, Ni	97.8	45
0	Full			97.0	50

At a stress level below 5 MPa, a fully reacted product was formed with Ni_3Al as the predominant phase. Between 5 and 15 MPa, fully reacted zones coexisted with the partially reacted zones. This was demonstrated by a specimen subjected to 5 MPa compressive stress, as shown in Fig 2. In this sample, a flame-shaped, fully reacted zone (Zone A) was accompanied by the surrounding partially reacted product (Zone B). Once the stress was above 15 MPa, the full reaction was completely suppressed and only a partial reaction prevailed. Intermediate phases predominated in these final products.

Here, a full reaction is defined as a reaction that directly leads to a nearly fully crystallized Ni_3Al phase, and a partial reaction refers to the cases where a Ni_3Al reaction did not complete and only intermediate phases were obtained, e.g., AlNi, and Ni, etc. These points will be further elaborated in the discussion section.

At a low stress level (0 to 5 MPa), increasing the compressive stress leads to better densification and a slight increase in the relative density (see Table 1). The minimum density of 97.0% was obtained at the stress-free condition, which agrees with the results obtained by other researchers [2,6,13,14]. Between 5 and 15 Mpa, no density measurements were carried out because there were two zones involved and the density could not be justified. At high stress levels (15 MPa and above), the density depends strongly on the applied stress. At 20 MPa, a theoretical density of 96.3% was obtained. As the compressive stress was increased to 50 MPa, a relative density of 99.3% was obtained. This agrees well with Nishimura and Liu's results [13,14]. At 100 MPa, a final product with a relative of theoretical density as high as 99.9% was achieved.

Table I clearly indicates that near full density can be achieved at very high stress levels, where only a partial reaction was obtained. Lowering the compressive stress could lead to poor densification, as demonstrated by the specimen subjected to 20 MPa stress, where the relative density was only 96.3% (which was worse than that in the stress-free case).

B. Microstructure

The specimen subjected to a stress of 3 MPa shows some porosity. SEM examination of the specimen bent-fractured at room temperature shows a fully crystallized structure, implying that a full reaction occurred during this process. Equiaxial grains prevailed with an average grain size of about 27 μm. This grain size is almost only half of that observed in the specimen under the stress-free condition. The density of the specimen is higher than that of the stress-free specimen.

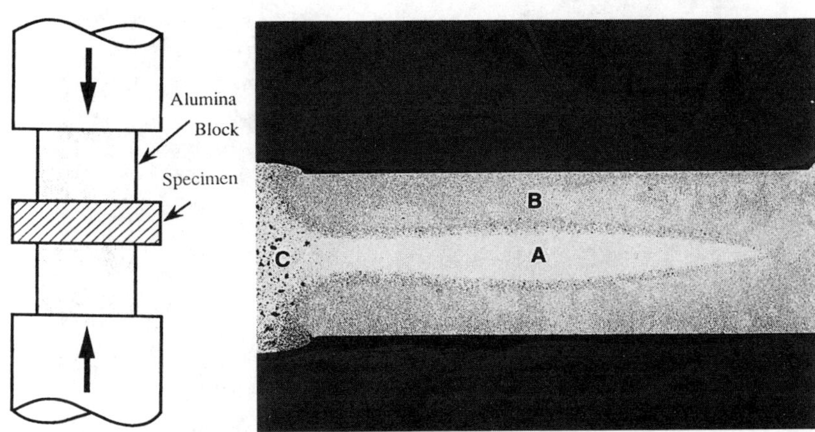

Fig. 1 Loading system

Fig. 2 Macroscopic view of the specimen subjected to 5MPa. Three zones were observed: (A) A compressed fully reacted zone, (B) a compressed, partially reacted zone, (C) a stress-free reacted zone (5X).

The specimen subjected to 5 MPa shows three distinct zones (see Fig. 2). Zone A is a near full density structure with very minor second phases present. Figure 3(a) shows the microstructure of this area after etching. Some second phases were noticed in the very dense matrix. No big pores could be detected. The average grain size in this region is 16 μm, which is even smaller than the one subjected to 3 MPa.

A full reaction did not take place in Zone B. In contrast, a partially reacted product with ample porosity characterizes the microstructure in this region, as evidenced in Fig. 3(b). A Ni-rich (89 ± 5% Ni) phase and NiAl (62.3 ±5% Ni) were the major phases detected in this region using electron probe microanalyses. A minor portion of Ni_3Al was present between the Ni and NiAl phases (see Table I.) Note that no aluminum-rich phases, such as Al_3Ni and Al_3Ni_2, were detected in this sample.

A fully reacted structure with big pores existed in Zone C, which was typical for stress-free specimens. This portion of the material was not subjected to any compressive stress because the alumina blocks were not big enough to cover this region. As the compressive load gradually increased, Zone A progressively became smaller and thinner.

DISCUSSION

A. Reaction

The reaction sintering of Ni_3Al can be considered as consisting of a complex reaction with several sequential steps. Based on the results of this investigation and those from other studies [1,13,14], it is proposed that a complete reaction sintering should undergo the following steps:

$3Al + Ni$	----->	Al_3Ni	(1)
$Al_3Ni + Ni$	----->	Al_3Ni_2	(2)
$Al_3Ni_2 + Ni$	----->	$3NiAl$	(3)
$NiAl + 2Ni$	----->	Ni_3Al	(4)

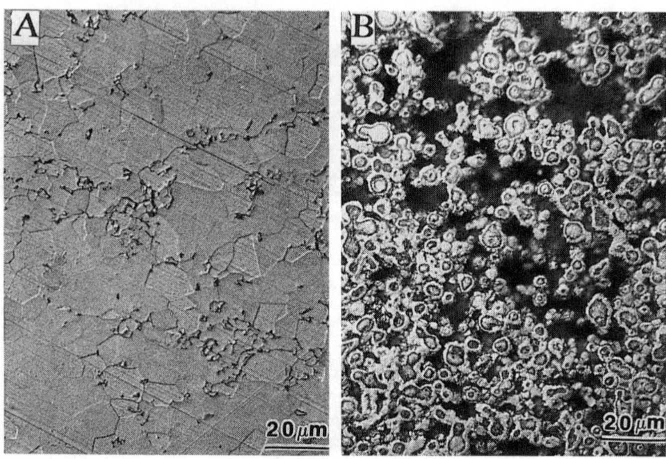

Fig. 3 The fully reacted zone (a) and the partially reacted zone (b) of the specimen subjected to 5 MPa.

Based on the electron probe microanalysis, the partially reacted sample showed that Ni and NiAl were the major phases. This indicates that Steps (1), (2) and (3) were all involved and completed in the partial reaction. Nishimura and Liu [13,14] proposed that Steps (1) and (2) were the preliminary stage of the reaction, and they were the steps that accumulated heat for the onset of Step (3). Their postulate was based on samples without the onset of self-propagation reaction, where only the products of Steps (1) and (2) were detected.

As mentioned before, a major Ni_3Al phase associated with minor Ni and NiAl phases characterized the fully reacted product. The result implies that in a full reaction, all of the four steps above have been completed, though 100% Ni_3Al has not been achieved. The full conversion to Ni_3Al can be accomplished by a proper annealing process. It thus becomes clear that a full reaction is featured with completion of Step (4), and a partial reaction is characterized by the completion of Step (3). Steps (1) and (2) serve as a launch pad for reactions (3) and (4). The key factor that links all four steps of reaction is the heat accumulation (or heat loss) within the sample. With sufficient heat accumulation, Steps (1) to (4) can be completed within a very short period of time and a full reaction is achieved. Slightly less heat accumulation leads to the completion of Steps (1) to (3) and consequently, a partial reaction. With lack of heat accumulation, only Steps (1) and (2) can be completed. Table 1 summarizes the final phases observed in these two type of reactions.

The specimens failing to initiate the self-propagation reaction generally displayed Al_3Ni and Al_3Ni_2 phases, and NiAl and Ni_3Al were essentially not present in the specimen [13,14]. In this systematic experiment, a self-sustained reaction took place in all the specimens tested, and Al_3Ni and Al_3Ni_2 were not detected. Instead, only Ni_3Al, NiAl were noticed, regardless of whether it was a partially reacted or fully reacted sample. This provided the evidence that after the onset of a self-sustained reaction, a specimen should have completed step (3). Therefore, step (3) is regarded as essential for a self-propagating reaction.

The uniqueness of the presence of compression stress in reaction sintering is the involvement of both stress level and heat loss. At a relatively low stress level, the accompanying heat loss is lower, which leads to a greater amount of heat accumulation, and, hence, the reaction completes to a greater extent. As a result, a greater amount of Ni_3Al was detected. Only a minimum amount of NiAl could be detected. With increasing compressive stress, the percentage of Ni_3Al in a specimen diminishes and the fraction of NiAl increases, as indicated by a microprobe study. As the stress is further increased, the full reaction is suppressed and a partial reaction prevails. In these cases, NiAl and Ni were generally observed. Ultimately, the self-propagation reaction could be suppressed at a relatively high stress level and only a preliminary reaction, steps (1) and (2), could occur. It is clear from this investigation that the degree of reaction is controlled by the level of the compressive stress and the associated heat transfer process. Heat accumulation turns out to be a key factor that controls the degree of the reaction.

B. The Effect of Compressive Stress

From Table I, it is obvious that increasing the compressive stress reduces the degree of reaction by enhancing the heat out-flow from the specimen to the alumina blocks, and therefore, the loading trends. According to German [9], more contact surface between the powder could enhance the inter-diffusion, as well as the speed and temperature of the reaction, which increases the density. In the compression cases studied here, more contact among the powders is provided because of the presence of the compressive stress, particularly at very high stress levels. However, it did not lead to a more thorough reaction. Partially reacted products were obtained at a relatively high stress level. It appears that heat was dissipated out through the contact elements leading to insufficient heat accumulation. Under these circumstances, since diffusion is a function of temperature, sufficient liquid could not be generated and a full reaction could not be propagated. The specimen subjected to a stress of 5 MPa provides the best evidence of this speculation, where enough heat accumulation was achieved at the center and a wholly crystallized structure was obtained. The outside layer, on the other hand, lost a great deal of heat because it was in direct contact with the alumina blocks. As a result, only a semi-reacted zone was developed. Higher compressive stress leads to more contact areas and more inter-diffusion between the powders, as well as more heat conduction and more heat loss. It is obvious that the effect of heat conduction outweighs the enhancement of interdiffusion and heat flow becomes the major factor in this kind of process.

C. Densification

The self-propagation stage is essential to the densification process of the product. At this self-propagating stage, a sufficient amount of liquid is formed [9], and the diffusion process takes place rapidly. If the heat loss is minimal, a full reaction can be obtained, and a better densification can be achieved. This is evident by comparing the specimen subjected to a stress of 3 MPa with one that was subjected to a stress of 20 MPa. The specimen with the 3 MPa stress achieved a 98.5% relative density with a fully-reacted final phase, whereas a partial reaction and a density of 96.2% was obtained in the specimen subjected to 20 MPa stress.

Compressive stress apparently plays an important role in the densification of the compound. The exertion of compressive stress helps to eliminate the big pores often observed in the stress-free specimen. This is supported by the increasing density at increased compressive stress both in the fully reacted (1 - 3 Mpa) and partially reacted (20 - 100 Mpa) products (see Table I). The densification of the final product cannot be justified solely from the level of the compressive stress. As indicated before, a heat loss always accompanies compressive stress, and a higher stress promotes a greater degree of heat loss. Greater heat loss retards the diffusion process and hinders the densification of the material. Hence, it is detrimental to the densification, as opposed to the compressive stress in itself which is beneficial. These two competing mechanisms can be manifested by comparing the specimens subjected to 3 MPa (98.5% density) and 20 MPa (96.3% density). Nevertheless, it is safe to say that in the same type of reaction (full or partial), increasing compressive stress enhances the density of the final product.

CONCLUSIONS

The following conclusions can be drawn from this study:

1. Three distinct types of reactions were observed in this experiment as the compressive stress varied from 0 to 100 Mpa. A full reaction was obtained as a consequence of low stress level (< 5 Mpa) and small amount of heat loss. An increase in the stress level leads to the coexistence of a fully-reacted zone and a partially-reacted zone. A further increase of the compressive stress (> 15 Mpa) completely suppressed the full reaction and only a partial

reaction prevailed in the specimen.

2. Application of compressive stress will always be accompanied by an enhanced heat out-flow from the specimen through the contacting elements, and hence, it will reduce the degree of reaction.

3. The reaction sintering of Ni_3Al involves three stages: preliminary, self-propagation, and conversion, as proposed by Nishimura and Liu [13,14]. Self-propagation is the key to densification of the product. A full reaction can only be accomplished only if all three stages are completed. Otherwise, only a partial reaction can be achieved.
4. Densification depends on both compressive stress and the reaction product. There are two possible ways to obtain high density products; one is a very high stress with a partially reacted product; and the other is a relative low stress associated with a fully reacted product.

ACKNOWLEDGMENT

The authors would like to extend their grateful thanks to C. Nishimura for his input and helpful discussion. E. H. Lee, C. A. Carmichael, H. Pierce, and T. J. Hensen are also thanked for their technical assistance.

REFERENCES

1. K. A. Philpot, Z. A. Munir, and J. B. Holt, "An Investigation of the Synthesis of Nickel Aluminides through Gasless Combustion," *J. of Mater. Sci.*, **22**, 159-169, 1986.
2. R. M. German, "Reactive Sintering and Reactive Hot Isostatic Pressing of High Temperature Intermetallic Compounds," *Adv. Powder Metall.*, **2**, 115-132, 1991.
3. V. M. Maslov, I. P. Borovinskaya, and A. G. Merzhanov, "Problem of the Mechanism of Gasless Combustion," *Combust. Explos. Shock Wave*, **12**, 631, 1976.
4. Z. A. Munir, "Synthesis of High Temperature Materials by Self-Propagating Combustion Methods," *Ceramic Bulletin*, **67**(2), 342-349, 1988.
5. J. E. Crider, "Self-Propagating High Temperature Synthesis - A Soviet Method for Producing Ceramic Materials," *Ceram. Eng. Sci. Proc.*, **3**, 519, 1982.
6. D. M. Sims, A. Bose and R. M. German, "Reactive Sintering of Nickel Aluminide," *Progress in Powder Metall.*, **43**, 575-596, 1987.
7. B. H. Rabin, A. Bose, and R. M. German, "Processing Effects on Densification in Reactive Sintering of Nickel-Aluminum Powder Mixtures," *Modern Development in Powder Metall.*, **20**, 511-529, 1988.
8. A. Bose, B. H. Rabin, and R. M. German, "Reactive Sintering Nickel-Aluminide to Near Full Density," *Powder Metall. International*, **20**(3), 25-30, 1988.
9. R. M. German, A. Bose, and N. S. Stoloff in "Powder Processing of High Temperature Aluminides," ed. C.T. Liu et al. (Mater. Res. Soc. Proc., **133**, High Temperature Ordered Intermetallic Alloys III), pp. 403-411, 1989.
10. V. E. Panin, A. I. Slosman, B. B. Ovechkin, M. P. Bondar, and N. A. Kostyukov, *Poroshkovaya Metallurgiya*, **7**, 271, 1985.
11. M. Concannon, E. S. Hodge, A. C. Nuce, and C. P. Turnel in High-Temperature Ordered Intermetallic Alloys IV, ed. L.A. Johnson et al., **213**, *Mater. Res. Soc. Proc.* **213**, Pittsburgh, PA, pp. 913, 1991.
12. J. C. Rawers and W. Wrzesinski, *Scripta. Metall.* **24**, 1985 (1990).
13. C. Nishimura and C. T. Liu, *Scripta Metall.* **26**, 381-85, 1992.
14. C. Nishimura and C. T. Liu, *Acta Metall.*, 1991.
15. P. Villers and L. D. Calvert, *Pearson's Handbook of Crystallographic Data for Intermetallic Phases*, (ASM, Metals Parks, Ohio), pp. 1038, 1985.

POWDER METALLURGY PROCESSING OF Ti-48Al-2Nb-2Cr(at%) ALLOYS

G.E. Fuchs
General Electric Company, P.O. Box 1072, Schenectady, NY 12301-1072

ABSTRACT

The effect of processing on the microstructures and properties of powder metallurgy processed Ti-48Al-2Nb-2Cr alloys was examined. Both gas atomization (GA) and plasma rotating electrode process (PREP) techniques were used to produce pre-alloyed powder of the desired composition. The powders were then consolidated by either HIPing or extrusion. The effects of HIP temperature (1090°-1300°C) and HIP pressure (103MPa and 172MPa) were examined during HIP consolidation of GA powders. In addition, some of the PREP and GA HIPed materials were subsequently hot worked by isothermal forging. The tensile properties of these materials were determined in air in the temperature range 25°-1000°C. These results were then compared with previous data for I/M materials. The inter-relationship of processing, microstructure and properties was examined.

INTRODUCTION

Due to their low density, high strength, and oxidation resistance, TiAl-based alloys are candidate materials for elevated temperature applications [1]. A significant amount of alloy development has been reported and reviewed [2-4]. However, the vast majority of this work has been performed on ingot metallurgy (I/M) processed materials. Only a limited amount of work has been reported on powder metallurgy (P/M) processing [5-12]. Much of the P/M work has emphasized the microstructure of as-solidified and heat treated powder particles [6,7,10,11]. The purpose of this study is to evaluate the microstructure and tensile properties of near-gamma (α_2/γ) alloys with similar nominal compositions processed by two powder metallurgy techniques.

EXPERIMENTAL PROCEDURES

Pre-alloyed powders with a nominal composition of Ti-48Al-2Nb-2Cr (at%) were prepared by both gas atomization (GA) and plasma rotating electrode process (PREP) techniques (Tables 1 and 2) and consolidated by hot isostatic pressing (HIP) or extrusion.

The -35 mesh (<500μm) PREP powder yield was canned and HIPed (1230°C/103MPa/4hrs) to full density (ID TA-16A). A portion of -35 mesh (<500μm) GA powder yield was canned and extruded (1290°C/16:1, ID TA-43A) to full density. Another portion of the GA powder yield was canned and HIPed (1230°C/103MPa/4hrs) to full density (ID TA-11A). The remaining GA powder yield was canned in five separate HIP cans and HIPed to full density at various temperatures at 172MPa for 4 hours. The HIP temperatures included 1090°C (ID TA-55A), 1140°C (ID TA-56A), 1230°C (ID TA-57) and 1300°C (ID TA-58A). Note that the TA-55A and TA-56A HIP temperatures are 25°C above and below the eutectoid (1115°-1120°C) temperature, respectively. Therefore, the TA-55A material was HIPed in the $\alpha_2+\gamma$ phase field; whereas, the TA-56A, TA-57A, TA-58A and TA-59 samples were consolidated in the $\alpha+\gamma$ phase field. The effect of HIP pressure was examined in samples TA-11A (1230°C/103MPa/4hrs) and TA-57A (1230°C/172MPa/4hrs). In addition, a combination HIP/Heat Treatment (HIP/HT) was examined. The HIP/HT (ID TA-59) included the 1300°C/172MPa/4hrs HIP cycle of TA-58A, but included a 900°C hold for 6 hours during furnace cooling (FC) to yield a duplex heat treatment thermal cycle (1300°C/4hrs/FC → 900°C/6hrs/FC).

Table 1
Composition (in At%) of Powder Metallurgy Ti-48Al-2Nb-2Cr Alloys

Atomization Technique	Composition (at%)							
	Ti	Al	Nb	Cr	O_2	H_2	N_2	C
GA	Bal	48.25	2.00	1.78	0.243	0.122	0.025	0.045
PREP	Bal	48.32	2.01	2.21	0.162	0.062	0.023	0.056

Table 2
Processing History and Grain Size of P/M Ti-48Al-2Nb-2Cr Alloys

Alloy ID	Processing	Grain Size (μm)
TA-11A	GA+HIP(1230°C/103MPa/4hrs)	43
TA-38A	GA+HIP(1230°C/103MPa/4hrs)+Forge(1177°C/0.1min^{-1}/75%)	34
TA-43A	GA+extrude(1290°C/16:1)	24
TA-55A	GA+HIP(1090°C/172MPa/4hrs)	25
TA-56A	GA+HIP(1140°C/172MPa/4hrs)	22
TA-57A	GA+HIP(1230°C/172MPa/4hrs)	18
TA-58A	GA+HIP(1300°C/172MPa/4hrs)	20
TA-59	GA+HIP/HT(1300°C/172MPa/4hrs→900°C/6hrs/FC)	17
TA-16A	PREP+HIP(1230°C/103MPa/4hrs)	40
TA-34A	PREP+HIP(1230°C/103MPa/4hrs)+Forge(1177°C/0.1min^{-1}/75%)	32

Note: All samples, except TA-59, heat treated as follow:
1300°C/2hrs/FC→900°C/6hrs/FC

a.)

b.)

Figure 1 - Optical photomicrographs showing the microstructures of the heat treated P/M Ti-48Al-2Nb-2Cr alloys. The TA-43A GA + extruded alloy exhibited an equiaxed duplex microstrucutre. The as-HIP PREP (TA-16A) alloy also exhibited an equiaxed duplex microstrucutre; however, the PREP material exhibited a coarser grain size.
a.) TA-43A, GA + Extruded
b.) TA-16A, PREP + HIP

Samples from both PREP (ID TA-34A) material and GA (ID TA-38A) HIP consolidated (1230°C/103MPa/4hrs) materials were hot worked by isothermal forged (1175°C/0.1 min^{-1}/75%).

All of the samples, except the HIP/HT (TA-59) materials, were given a heat treatment (1300°C/2hrs/FC → 900°C/6hrs/FC) intended to form a duplex microstructure. The microstructure of the samples was examined by optical metallography and scanning electron microscopy (SEM). Threaded tensile samples with a 3.81mm diameter x 12.70mm gage section were tested in air at 25°, 500°, 600°, 700°, 850° and 1000°C at an initial strain rate of 0.1/minute. The fracture morphologies were characterized by SEM examination.

RESULTS AND DISCUSSION

Microstructures

The heat treatment in the $\alpha + \gamma$ phase field, at 1300°C, was intended to produce a duplex microstructure consisting of a mixture of single phase γ grains and lamellar α_2/γ grains (Figure 1). The single phase γ grains were observed in all of the alloys. However, the α_2 was not always present in the form of lamellar α_2/γ grains. The materials consolidated at high temperatures, such as the TA-58A and TA-59 samples HIPed at 1300°C and TA-43A extruded at 1290°C, exhibited a microstructure with the α_2 present primarily in the form of small equiaxed grains. Only a limited amount of α_2 phase was present in lamellar α_2/γ grains. The remainder of the materials exhibited duplex microstructures with higher volume fractions of lamellar α_2/γ grains. In general, the volume fraction of α_2 present in the lamellar α_2/γ grains versus equiaxed α_2 grains appeared to decrease with increasing HIP temperature.

In addition to affecting the α_2 phase morphology, the consolidation temperature also affected the grain size of the as-consolidated and heat treated conditions. An increase in HIP temperature decreased the heat treated grain size (Table 2). An increase in HIP pressure from 103MPa to 172MPa at 1230°C also appeared to refine the microstructure. A similar trend was observed in HIP consolidated PREP Ti-48Al (at%) alloys [9]. The finest grain size was observed in the extruded materials (TA-43A). The grain size of the HIP + forged (TA-34A and TA-38A) materials were intermediate to the as-HIP and extruded materials. In addition to refining the microstructure, the forging process homogenized the microstructure. The as-HIP materials exhibited a wider distribution of grain sizes than the HIP + forged materials.

Tensile Properties

The effects of HIP temperature and pressure on the tensile properties of the GA alloy were quite apparent at test temperatures up to 500°C (Figures 2a and 2b). The ductility and the strength increased with increasing HIP temperature up to a maximum at approximately 1230°C. SEM evaluation of the room temperature fracture surfaces indicated that failure of the samples consolidated at low temperatures frequently occured along prior particle boundaries. Mixed transgranular/intergranular cleavage was observed in samples HIPed at temperatures above 1140°C. Further increases in HIP temperature appeared to decrease the ductility and strength at room temperature. Increasing the HIP pressure did not appear to have a significant effect on the room temperature tensile properties. The HIP/HT (TA-59) samples exhibited tensile properties similar to the 1230°C HIP condition and superior to the 1300°C HIP material (TA-58A). In the intermediate temperature range (600°-700°C), variations in the HIP cycle did not appear to have a significant effect on the tensile properties (Figures 2c and 2d). At high test temperatures (850°-1000°C), increasing the HIP temperature and pressure increased the strength and ductility of the materials (Figures 2e and 2f) and the HIP/HT (TA-59) samples exhibited properties similar to the HIP and heat treated material (TA-58A). At test temperatures in excess of 500°C, HIP parameters did not appear to affect the fracture mechanisms.

The GA (TA-11A) and PREP (TA-16A) materials consolidated with the 1230°C/103MPa/4hr HIP cycle exhibited similar tensile properties (Figure 3). At test

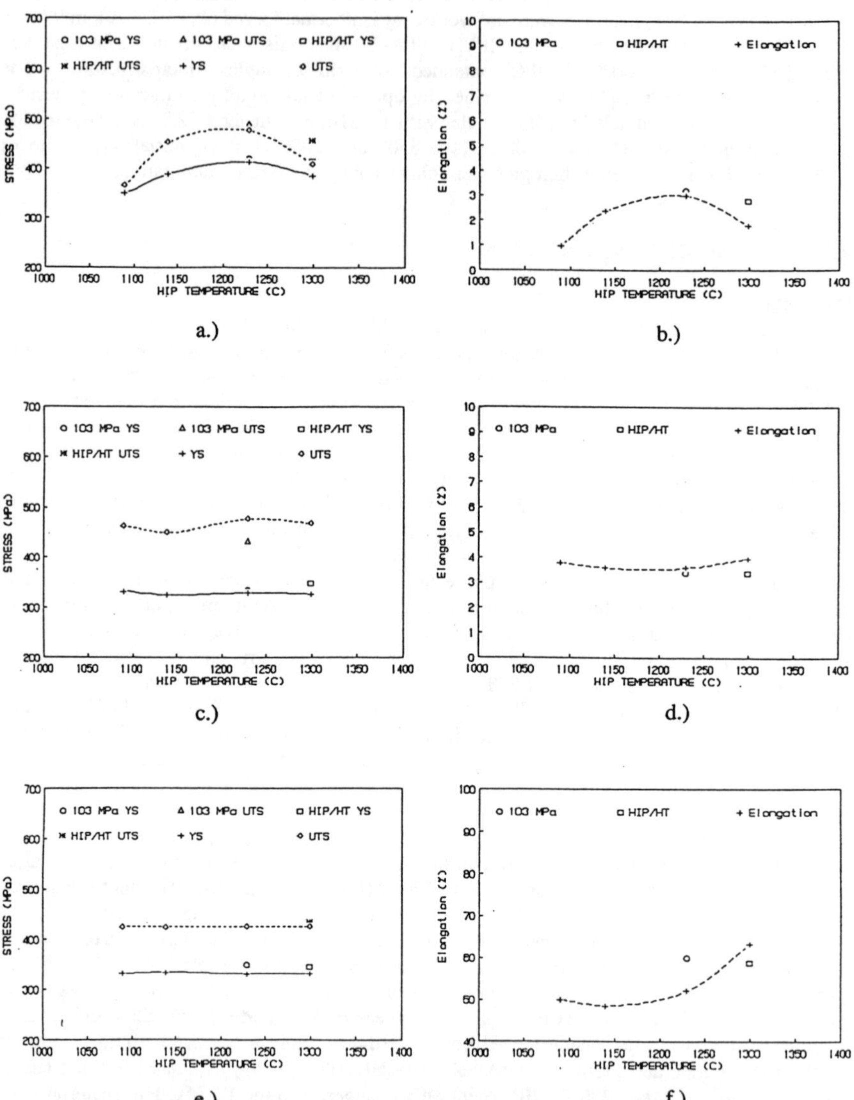

Figure 2 - The effect of HIP temperature and pressure on the tensile properties of GA P/M Ti-48Al-2Nb-2Cr alloys. The data lines represent samples HIPed at 172MPa and the data points represents samples HIPed at 103MPa and the HIP/HT samples.

a.)	25°C - Yield and Ultimate Tensile Stress	b.)	25°C - Elongation
c.)	600°C - Yield and Ultimate Tensile Stress	d.)	600°C - Elongation
e.)	850°C - Yield and Ultimate Tensile Stress	f.)	850°C - Elongation

temperatures less than 1000°C, the GA (TA-11A) samples exhibited higher strength and greater ductility. At 1000°C, the GA samples still exhibited higher strength, but greater ductility was observed in the PREP sample. Atomization technique did not appear to affect the fracture mechanisms at the test temperatures examined. The oxygen content of the GA alloy was significantly greater than the PREP material (Table 1). The high strength of the GA alloy may be due to interstitial hardening of the alloy. However, the higher oxygen content of the GA would be expected to result in reduced ductility [3,4].

Hot working the HIP microstructure by isothermal forging resulted in a slight increase in tensile properties at test temperatures less than 850°C (Figure 3). In comparison to the PREP

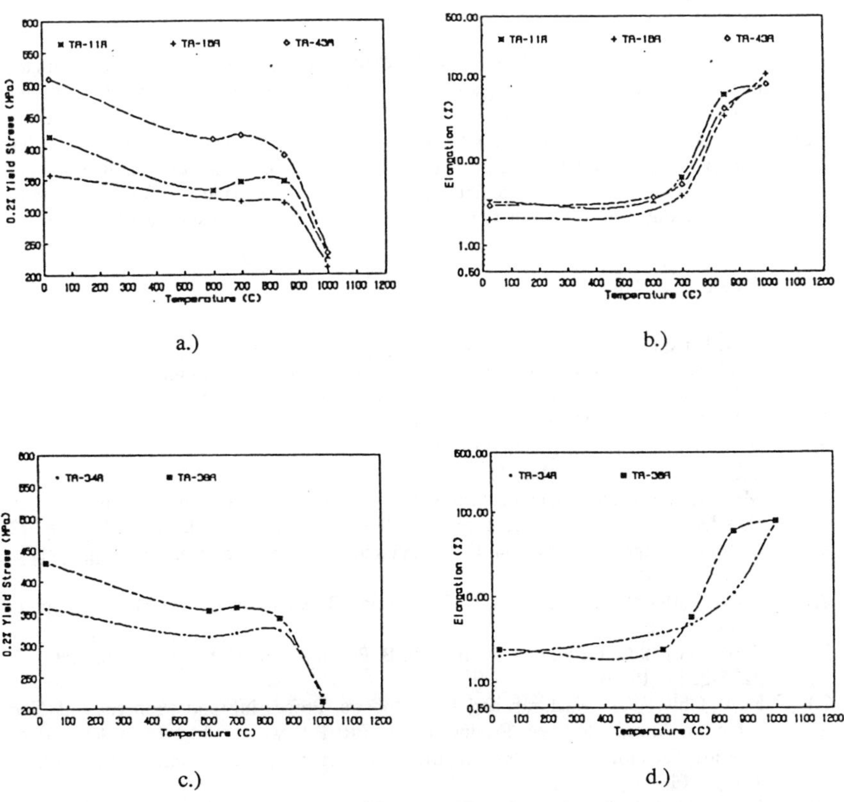

Figure 3 - Effect of temperature on the tensile properties of P/M Ti-48Al-2Nb-2Cr alloys. The highest strength was observed in the GA + Extruded (TA-43A). The HIP consolidated PREP (TA-16A) and GA (TA-11A) alloys exhibited similar properties at all test temperatures. Isothermal forging resulted in a slight increase in strength for both the PREP (TA-34A) and GA (TA-38A) alloys; however, no significant increase in ductility was observed.

a.) Yield Stress (TA-11A, TA-16A, TA-43A) b.) Elongation (TA-11A, TA-16A, TA-43A)

c.) Yield Stress (TA-34A, TA-38A) d.) Elongation (TA-34A, TA-38A)

HIP + forge samples (TA-34A), the GA alloy (TA-38A) exhibited a greater increase in strength after forging. At 850°C and 1000°C, the as-HIP samples exhibited slightly higher strength. In addition, the forging did not appear to have any significant effect on the ductility or the fracture mechanisms of the materials. Although the isothermal forging appeared to homogenize the microstructure, it does not appear to significantly alter the tensile properties in the temperature range examined.

The GA material consolidated by extrusion (TA-43A) exhibited the highest strength of any of the materials tested (Figure 3). In addition, the extruded material exhibited ductility and fracture mechanisms similar to the HIP consolidated materials at all test temperatures. The high strength and ductility may be due to the refinement of the microstructure during the extrusion process.

SUMMARY

Powder metallurgy processing of Ti-48Al-2Nb-2Cr alloys results in materials with tensile properties superior to ingot metallurgy materials [12]. The method of consolidation and the processing parameters appear to have a strong influence on the resulting properties. Hot working of HIP consolidated materials homogenized the microstructures, but does not have a significant effect on the tensile properties. The results of this study indicate that low temperature HIPing results in complete densification. Mechanical testing indicated that materials consolidated at low temperature exhibited only limited tensile ductility.

REFERENCES

1.) H.A. Lipsitt, Proc. Mat. Res. Soc., 33, pp. 351-364 (1984).

2.) Y.W. Kim, Microstructure/Property Relationships in Ti-Aluminides and Alloys, Eds. Y.W. Kim and R.R. Boyer, TMS, p. 91-103 (1991).

3.) Y.W. Kim, JOM, 41, pp. 24-30 (1989).

4.) Y.W. Kim, JOM, 43, pp. 40-47 (1991).

5.) C.F. Yolton and D. Eylon, "Effect of Processing History and Heat Treatment on the Microstructure and Mechanical Properties of Consolidated Gamma Titanium Aluminide Powder", Seventh World Conference on Titanium, TMS, San Diego, CA, 6/28-7/2/92.

6.) C. McCullough, J.J. Valencia, C.G. Levi and R. Mehrabian, Mat. Sci. Eng., A124, pp. 83-101 (1990).

7.) C. McCullough, J.J. Valencia, C.G. Levi and R. Mehrabian, Acta Met., 37, pp. 1321-1336 (1989).

8.) B.W. Choi, Y.G. Deng, C. McCullough, B. Paden and R. Mehrabian, Acta Met., 38, pp. 2225-2243 (1990).

9.) M.A. Ohls, W.T. Nachtrab and P.R. Roberts, Processing and Properties of Gamma Titanium Aluminide Sheet Produced from PREP Powders. In P/M in Aerospace and Defense Technologies Symposium, Metal Powder Industry Federation, March 4-6, 1991, Tampa, Fl.

10.) D.S. Shih, G.K. Scarr and J.C. Chesnutt, Mat. Res. Soc., 133, pp.. 167-172 (1989).

11.) G.E. Fuchs and S.Z. Hayden, Mat. Sci Eng., A152, pp 227-282 (1992).

12.) G.E. Fuchs, "The Effect of Processing on the Microstructure and Tensile Properties of a γ-TiAl Based Alloy", Seventh World Conference on Titanium, San Diego, CA, 6/28-7/2/92.

ON THE KINETICS OF Nb_5Si_3 COMPOUND FORMATION

Jan Kajuch, John W. Short, Changqi Liu and John J. Lewandowski
Dept. of Materials Science and Engineering
Case Western Reserve University, Cleveland, OH

ABSTRACT

The kinetics of intermetallic Nb_5Si_3 compound formation via the mechanical alloying process was investigated. Interrupted milling process, X-ray diffraction, SEM examination and TEM imaging and diffraction were utilized to characterize changes in the milled powders, while DTA analyses were used to determine the critical and onset temperatures of reaction as a function of milling time. On the basis of experimental results, a kinetic model was proposed for formation of Nb_5Si_3 via the interrupted MA process. It is suggested that precipitation of Nb_5Si_3 particles during cooling in the interrupted milling process is responsible for the exothermic reaction after resumption of milling.

INTRODUCTION

Mechanical alloying (MA) is a simple but effective process for the production of intermetallic compounds of high temperature refractory metals. Its main advantage as compared to the standard melting and casting process is in the capability to maintain exact composition (stoichiometry), low degree of contamination, and flexibility in producing monolithic and composite powders. MA is a non-equilibrium processing technique analogous to Rapid Solidification (RS). In contrast to the RS process, the MA process is entirely a solid state operation at or near room temperature. The MA process has been defined as a dry, high energy ball milling process that produces composite metal powders with extremely fine microstructures. Interdispersion of the powders occurs by the repeated cold welding and fracturing process of free powder particles, trapped between two colliding steel balls[1]. The force of the impact deforms the particles and creates atomically clean surfaces which weld together on contact. To prevent oxidation of these surfaces, the milling operation is carried out in an inert gas atmosphere. Refinement of the structure is approximately a logarithmic function of time and depends on the mechanical energy input into the milling process and the work hardening of the powders being processed[2]. The microstructural refinement continues into the steady-state period despite the fact that the hardness saturates and a constant agglomerate particle size distribution is achieved.

Several authors studied the formation of intermetallic compounds by the MA process. Atzmon determined the parameters affecting phase formation in the Al-Ni system[3]. While NiAl formed by an explosive, self-propagating reaction, Al_3Ni formed in a reaction with layer diffusion as a predominant factor. Kumar and his co-workers were the first to study the mechanism of MA in group V transition metal/silicon systems[4]. In order to study the progress of mechanical alloying, the ball mill was stopped periodically and cooled to room temperature in order to enable removal of small amounts of the powder for analysis. This "interrupted process" resulted in the formation of Nb_5Si_3 in 75 minutes, while milling for 73 minutes and cooling to room temperature produced elemental Nb and Si. Whittenberger in his analysis of the solid state processing of high temperature alloys and composites[5] believes that "enhanced diffusivity" plays a major role in the alloying process.

Schaffer and McCormick studied the mechanism of compound formation in the "interrupted process" in several systems with the conclusion that room temperature "enhanced diffusivity" facilitates the exothermic reaction which occurs almost instantaneously after milling is resumed[6].

This investigation concentrates on proposing a kinetic model of the formation of Nb_5Si_3 via interrupted milling utilizing Differential Thermal Analyses, X-ray diffraction and Scanning Electron and Transmission Electron Microscopy. The work represents a continuation of work reported elsewhere[7].

EXPERIMENTAL PROCEDURES

Elemental silicon (Aldrich Chemical Company) and niobium powders (Cabot Corporation) were obtained with a particle size of -325 mesh (less than 44 µm) and nominal purities exceeding 99% and 99.8% respectively. For MA, elemental powders with the proper Nb-Si ratio for the formation of Nb_5Si_3 (Nb-37.5 at% Si) were weighed and placed into a tungsten carbide vial while in an argon-gas-filled glove box. MA was carried out in a Spex model 8000 high intensity mixer-mill using 100 g of hardened 52100 steel balls (12 mm diameter) and 10 g of elemental powders for a 10:1 balls/powder weight ratio. The vial temperature was monitored with a portable digital thermometer with a contact thermocouple type J probe.

X-ray diffraction was performed on elemental powders and the mechanically alloyed powders for time intervals of 1 to 3.5 hrs. A Phillips X-ray Autodiffractometer operated in the continuous step scanning mode using a Cu Kα radiation source was used for the X-ray analyses. A JEOL 840A scanning electron microscope (SEM) and JEOL 200CX transmission electron microscope (TEM) were used for the microstructural examinations. SEM characterization of agglomerate size and microstructural refinement was performed on powders milled for various times as well as on reacted powders. Back scattered electron imaging was used for the studies of microstructural refinement.

A Netzsch STA 429/409 Differential Thermal Analyzer at NASA Lewis Research Laboratories and a modified DTA unit built at Case Western Reserve University were used to determined the critical and onset reaction temperatures on prealloyed powders as well as on reacted powders (Nb_5Si_3). Prealloyed powders (1 hr) aged at room temperature for times of 1 to 1000 hrs were held at -196°C when the DTA equipment was not immediately available. For the DTA tests at CWRU, powder samples (2 to 5 grams) were cold pressed into small discs and a center hole was drilled for the insertion of a K type thermocouple, while another K type thermocouple was placed in the alumina crucible as a reference. Argon gas was used to prevent powder oxidation during the analysis and post-analysis cooling. Some of these sintered samples were crushed and x-rayed to determine the phase evolution during DTA analysis.

For the TEM observations, Nb-Si milled powders were dispersed on a piece of carbon film which was supported on a copper grid (3mm in diameter). An additional layer of carbon film was deposited on top of the dispersed powders in order to avoid contamination of the microscope. Both bright field and dark field imaging techniques were employed to observe and identify the particles.

RESULTS AND DISCUSSION

In our previous work on the synthesis of Nb_5Si_3 by the "interrupted process" it was found that compound formation proceeded by self-propagating exothermic reaction upon resumption of the milling process[7]. Two major variables controlling compound formation were identified. A minimum milling time of 1 hour and a minimum cooling time of 2 hours was required before milling was resumed (Figure 1). The following observations were analyzed to determine the reasons for the critical milling time and hold time at room temperature.

Figure 2 shows the microstructural refinement in a Nb-Si agglomerate after MA for 1 hour. Particles within the agglomerate are not uniformly refined, with an average inter-particle spacing on the order of 1 µm although there are areas where refinement is on a much smaller scale. According to the theory of mechanical alloying, true alloying occurs when the microstructural refinement is no longer visible in an optical microscope, roughly a particle spacing of 0.5 µm. This leads us to conclude that there are small areas where

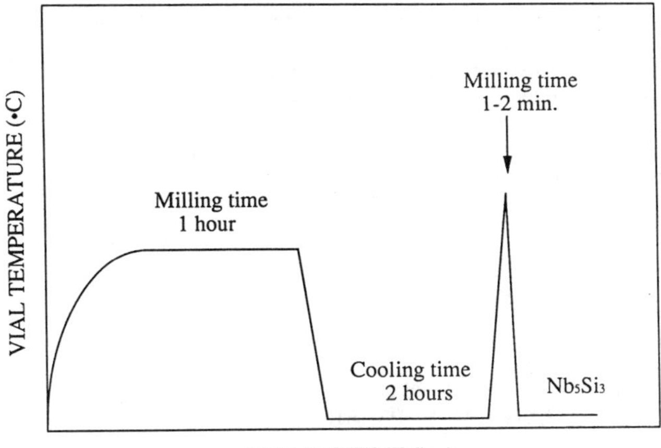

Figure 1. Schematic of interrupted milling process

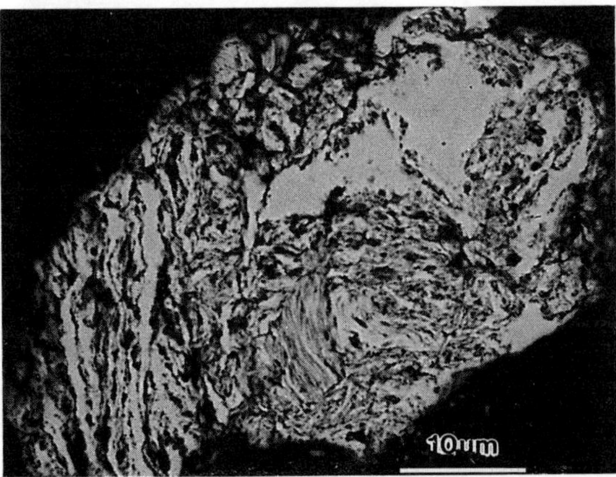

Figure 2. SEM micrograph of Nb-Si mixture milled for 1 hr

intensive MA energy input promotes Si dissolution in Nb, creating a supersaturated solid solution with respect to its equilibrium solubility at low processing temperatures (≈600-900 K). Upon cooling, Nb_5Si_3 particles precipitate from solid solution as expected from the equilibrium diagram. Upon resumption of milling, a self-propagating exothermic reaction takes place, due to a large heat of formation release upon growth of Nb_5Si_3 particles.

Differential Thermal Analyses (DTA) were conducted on powders milled for a total time of 1 hour immediately after the milling process was stopped as well as after room temperature "aging" for up to 1000 hours (Figure 3). Figure 4 shows the onset reaction temperature of Nb_5Si_3 vs. "aging" time at room temperature, with a total temperature differential of less than 9°C. This small reaction temperature drop suggests that although "enhanced diffusivity" occurred, it does not play a significant role in the compound formation in the interrupted process. In order for the reaction to take place as shown in Figure 5, enhanced diffusivity would have to decrease critical reaction temperature on the order of 50°C. Here, the critical reaction temperature is the temperature at which the heat of reaction is large enough to cause a positive increase in the temperature differential between the sample and the reference thermocouple, while the onset temperature is designated as the temperature at which the reaction is self-propagating.

TEM electron diffraction of Nb-Si powders milled for 1 hour (Figure 6) show particles of Nb+Si mixture as well as Nb_5Si_3 compound. No other metastable, amorphous, or equilibrium phases were observed.

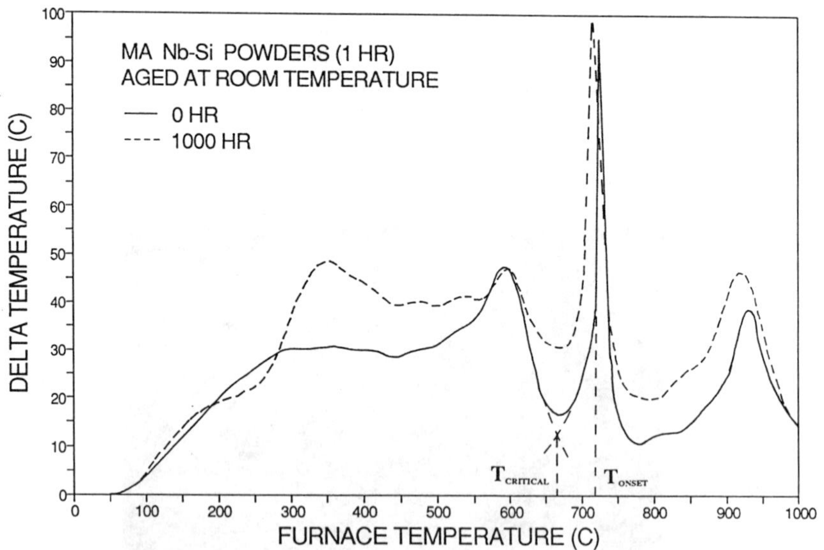

Figure 3. DTA trace of MA Nb-Si powders (1 hr) aged at RT

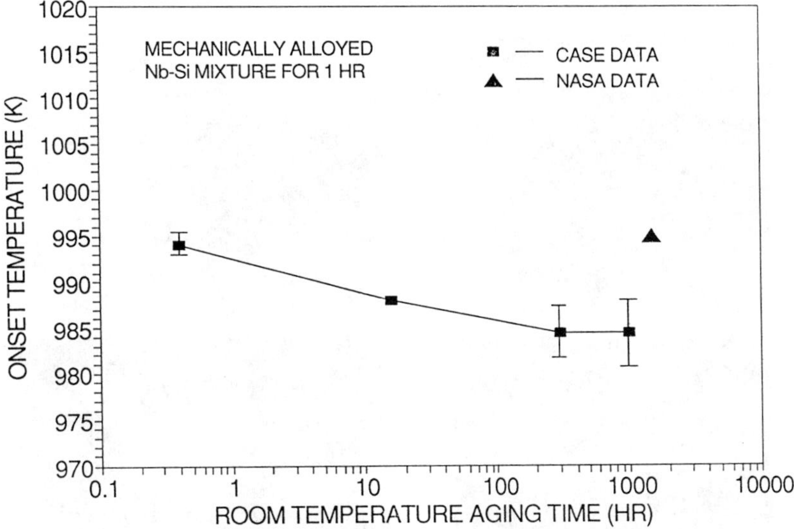

Figure 4. Reaction onset temperature vs. aging time at RT

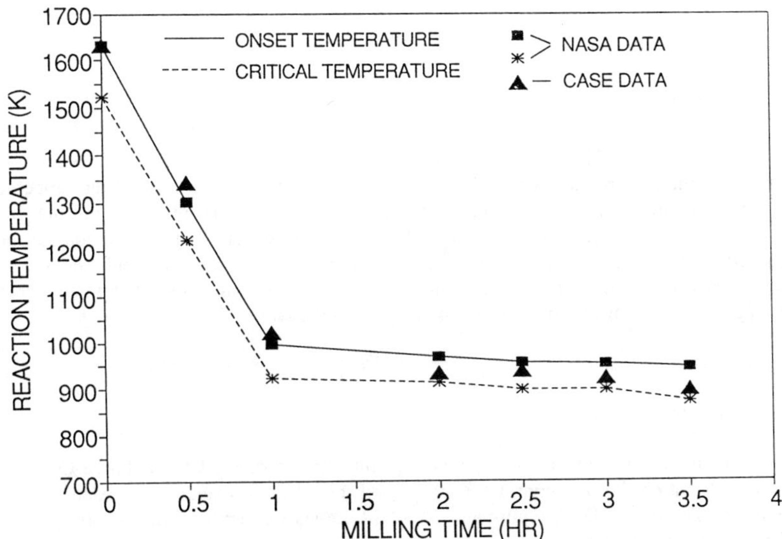

Figure 5. Critical and onset reaction temperature vs. milling time

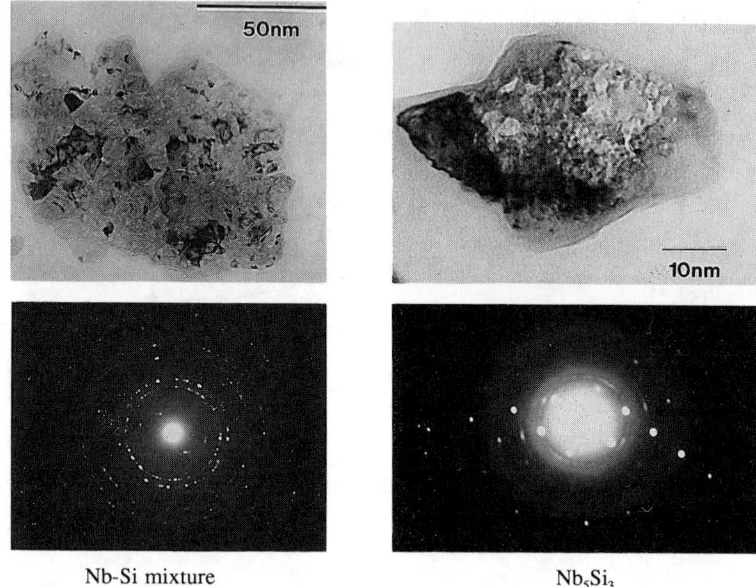

Nb-Si mixture Nb_5Si_3

Figure 6. TEM micrograph and diffraction of MA Nb-Si mixture (1 hr)

CONCLUSIONS

The initiation of Nb_5Si_3 compound reaction in the interrupted milling process occurred via the precipitation of Nb_5Si_3 particles upon cooling from the milling temperature. Two major parameters controlling the precipitation process are the minimum milling time of 1 hour and the minimum cooling time of 2 hours. The most plausible explanation for the precipitation process is the creation of a non-equilibrium supersaturated solid solution of Si in Nb. TEM observations of powders milled for 3 hours failed to show precipitates other than Nb_5Si_3, supporting the mechanism of Nb_5Si_3 precipitation.

ACKNOWLEDGEMENTS

The authors would like to acknowledge the support of the US Air Force Office of Scientific Research through contract 89-0508 and the program manager Dr. Alan Rosenstein. The use of DTA equipment at NASA Lewis Research Center and the assistance of Dr. Ivan Locci is also appreciated.

REFERENCES

[1]P.S. Gilman and J.S. Benjamin, *Annual Rev. Mater. Sci.*, 13, 279 (1983).
[2]J.S. Benjamin and T.E. Volin, *Metall. Trans.*, 5, 1929 (1974).
[3]M. Atzmon in A.H. Clauer and J.J. deBarbadillo (eds.), *Solid State Processing*, The Minerals, Metals & Materials Society, Warrendale, 173 (1990).
[4]K.S. Kumar and S.K. Mannan, *Mater. Res. Soc. Symp. Proc.*, 133, 415 (1989).
[5]J.D. Whittenberger in A.H. Clauer and J.J. deBarbadillo (eds.), *Solid State Powder Processing*, The Minerals, Metals & Materials Society, Warrendale, 137 (1990).
[6]G.B. Schaffer and P.G. McCormick, *Metall. Trans.A*, 22A, 3019 (1991).
[7]J. Kajuch, J.D. Rigney, and J.J. Lewandowski, *Materials Sci. and Eng.*, A155, 59 (1992).

DISPERSION STRENGTHENED INTERMETALLICS BY MECHANICAL ALLOYING: CREEP RESULTS AND DISLOCATION MECHANISMS

E. Arzt, E. Göhring and P. Grahle,
Max-Planck-Institut für Metallforschung and Institut für Metallkunde Stuttgart, Germany

Abstract

In order to increase the creep strength, small dispersoids were introduced into the intermetallic compound NiAl by mechanical alloying. Under favorable conditions, the resulting specific creep rates approach those of superalloys. To accompany these experimental effects, a new model has been developed which predicts tendencies for the success of dispersion strengthening in ordered alloys. It is shown that the dissociation distance of the partial dislocations, relative to the dispersoid particle size, has an important effect on the strengthening achievable.

Introduction

The creep strength of monolithic intermetallics such as aluminides is typically by far inferior to that of advanced superalloys [1]. Recent developments have sought to improve the high temperature strength of NiAl by preparing particulate composite materials containing second phase particles of more than 10 vol.% [2]. The probably first study on oxide dispersion strengthened (ODS) intermetallic alloys [3] showed the producibility, in principles of ODS-NiAl and ODS-FeAl based on a processing technique that is now called Mechanical Alloying (MA). The main advantage of ODS alloys lies in the retention of useful creep strength up to high homologous temperatures, where other strengthening mechanisms are no longer effective [4]. Model approaches of creep in dispersion strengthened, disordered alloys show that the choice of the dispersoid parameters has a strong influence on the efficacy of dispersion strengthening [4, 5]. However, no real theoretical understanding of creep in dispersion strengthened ordered alloys exists [6] and thus, the promises and limitations of dispersion strengthened intermetallics are unclear. The present paper adresses new experimental and theoretical aspects of this promising new class of high-temperature materials.

Experimental

Processing of dispersion strengthened and composite materials

ODS NiAl materials were prepared from prealloyed, gas-atomised Ni_3Al and Al-24 at.% Ni powders supplied by Homogeneous Metals, Inc.. Mechanical Alloying was carried out with 1 to 2 vol.% Y_2O_3 dispersoid powder for 50 hours in a high energy ball mill at the Metallgesellschaft AG Frankfurt. Milled powders were vacuum canned in stainless steel and hot extruded at 1200°C [7]. Further details can be found in table 1.

The preparation of NiAl-NiAlNb composite materials is described elsewhere in these proceedings [8]; it resulted in globular second phase particles, randomly distributed in the NiAl matrix, with a mean diameter of 18 µm. As an alternative to the cryomilling process [2], a NiAl alloy containing about 10 Vol.% AlN was prepared by a gas-metal absorption reaction: atomised NiAl powder was milled for 1h under argon to an average particle size of about 7 µm; the as milled powder was heat treated at 1200°C in a nitrogen atmosphere with a partial N_2-pressure of 300 hPa. X-ray diffraction analysis confirmed the formation of AlN after heat treatment. The nitrogen concentration was determined by WDS analysis (see table 1). The resulting AlN volume fraction can be estimated, assuming all nitrogen had reacted to AlN, to be maximum 10 vol.%. Consolidation was performed by HIP [8]. The formation of AlN took place at the NiAl-powder surfaces. In the bulk material, regions near grain boundaries show high amounts of fine AlN dispersoids (mean diameter about 50 nm) while grains are mostly free of AlN.

Creep tests

Cylindrical compression specimens were electrodischarge machined from the compacted materials. Constant strain rate compression tests ranging from 10^{-4} to 10^{-8} s^{-1} were conducted at 1200 to 1300 K in a universal testing machine to about 5% plastic strain. The recorded stress-strain data were converted to true values. In every case a steady state creep behaviour was observed. After the test, the specimens were cooled down under the current steady state stress to avoid any relaxation of dislocation structures.

TEM investigations

TEM investigations were carried out on sections perpendicular to the extrusion axis. Specimens were prepared by a combination of electrodischarge machining (250 μm thick), soft mechanical grinding (100 μm, Ø 3 mm) and dimpling to 20 μm, followed by ion milling.

Alloy	Composition	Consolidation	Heat treatment	Matrix grain size
IP 6000	Ni-50 at.%Al + 1 vol.% Y_2O_3	hot extrusion	-	1 μm
IP 6000 A	Ni-50 at.%Al + 1 vol.% Y_2O_3	hot extrusion	1400°C/10h	10 μm
IP 6002	Ni-50 at.%Al + 2 vol.% Y_2O_3	hot extrusion	-	< 1 μm
ODS-Fe40Al [9]	Fe 40 at.% Al + 2 vol.% Y_2O_3	HIP	-	15 μm
NiAl/NiAlNb	Ni-50 at.%Al + 10 wt.% Nb	HIP	1200°C/12h	36 μm
NiAl/AlN	Ni-50 at.% Al + 2 wt.% N	HIP	-	10 μm

Table 1: Chemical composition, processing parameters and grain sizes

Microstructure of ODS-Aluminides

The matrix grain size was found to lie in the micrometer range as is common after mechanical alloying (see IP6000 and IP6002 in table 1). The dispersoids, of mean size 40 nm, are homogeneously distributed with a mean free spacing of about 300 nm. Only in alloys with coarser grains, realized by heat treatment, we did find dislocation configurations as in Fig. 1. These configurations suggest an interaction between dislocations and the yttria-dispersoids in both alloys, as expected for ODS materials [10, 11].

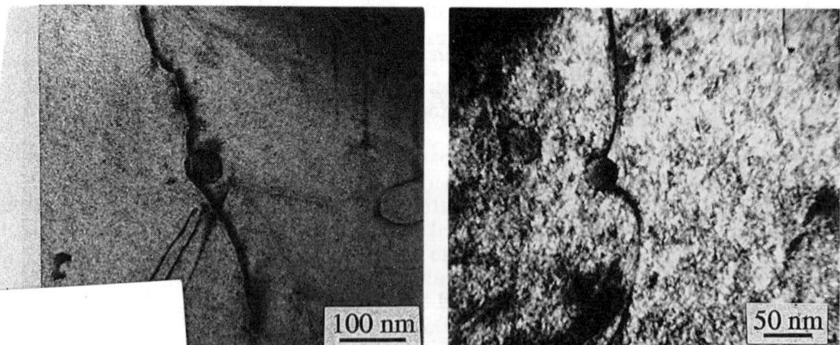

...raction between a dispersoid and a dislocation in ODS-Fe40Al (left) and ...d at moderate creep rates and 1200K and 1300K resp. Dispersoid diameter ... d≈40 nm. The dislocations are bowed out in their glide directions.

Creep strength of ODS - Aluminides and NiAl composite materials

A high stress exponent (n=27), typical for ODS materials, could be observed for IP6000 A at test temperatures of 1200 and 1300 K (see Fig. 2). The grain size of this material could be increased during a heat treatment (see table 1). Increasing the dispersoid volume fraction from 1 to 2 Vol.% resulted in a higher stress level at 1200 K and a higher stress exponent in the fine grained, as extruded material IP6002 (see Fig. 3). Attempts to coarsen grain sizes in IP6002 were not yet successful. The ODS-NiAl materials investigated show a clearly higher creep resistance at 1200K than ODS-Fe40Al. The NiAl/AlN composite material exhibits a higher creep resistance than NiAl/NiAlNb with comparable second phase volume fractions, mainly due

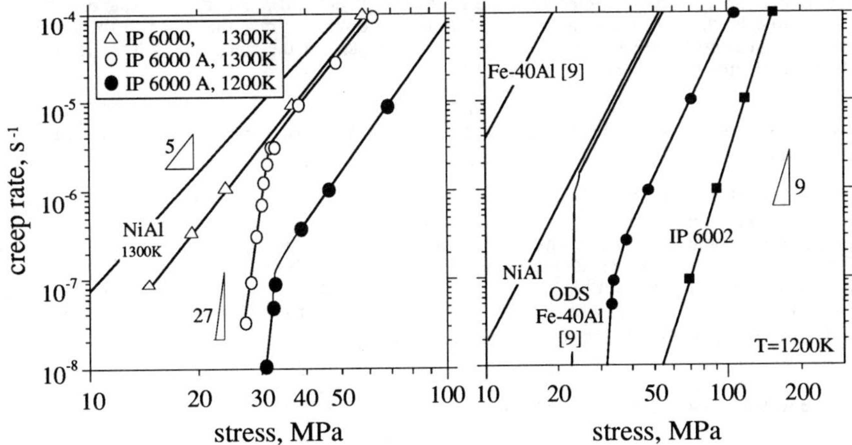

Fig. 2. Effect of oxide dispersion strengthening, grain size and test temperature on creep strength of NiAl

Fig. 3. Influence of dispersoid volume fraction on creep strength of NiAl in comparison to ODS-FeAl [9].

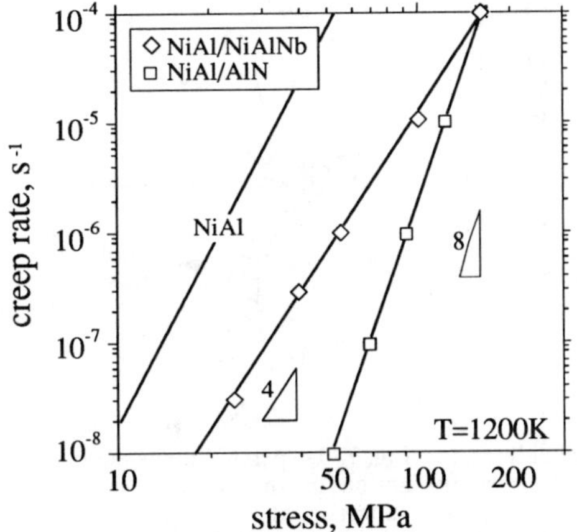

Fig. 4. Creep strength of two particulate composite materials, tested at 1200 K in air.

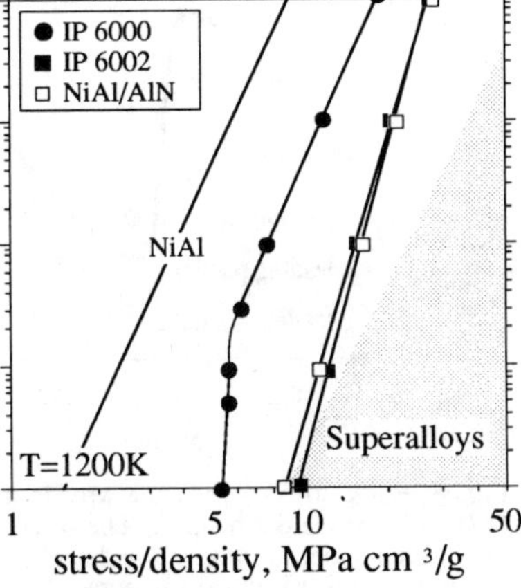

Fig. 5. Density compensated comparison of advanced NiAl materials with superalloys at 1200 K

to a higher stress exponent (Fig. 4). In a density compensated point of view, ODS NiAl materials reach the creep resistance of some superalloys at 1200 K, especially at low strain rates. Note that dispersion strengthened, but still fine grained IP6002 nearly shows the same creep behaviour at 1200 K as the NiAl/AlN composite material.

Modelling of Dispersion Strengthening in Ordered Allyos: First Results

The theory of dispersion strengthening at high temperatures has been extended considerably in recent years. The main result is that the effect of creep retardation can be explained if an attractive dispersoid-dislocation interaction exists (for details see [4, 5, 12]). A consequence of this effect is that a dislocation which climbs over a dispersoid meets two "obstacles" (fig. 6): one is due to the necessary increase in line length during climb, the other to the barrier to detachment from the dispersoid. An important parameter for the creep strength is the athermal detachment stress τ_d, which is given by [12]

$$\tau_d = \tau_o \sqrt{1-k^2}$$

where τ_o is the Orowan shear stress, and k a factor measuring the strength of the attractive interaction.

This model has recently been extended to ordered matrix materials (see [15]). The major modification is due to the dissociation of dislocations into superpartials which have now to be taken into account (fig. 7): because the partials experience different back stresses (as shown in fig. 6) and because they interact elastically, the detachment stress - and thus the creep strength - now depend on the dissociation spacing relative to the dispersoid size. The simple analysis in [15] shows that the detachment stress for the leading partial τ_d is given by the arithmetic mean of its own back stress $\tau(d)$ and that on the trailing partial $\tau(d\text{-}w)$:

$$\tau_d = \frac{\tau(d) + \tau(d-w)}{2}$$

where w is the spacing of the superpartials.

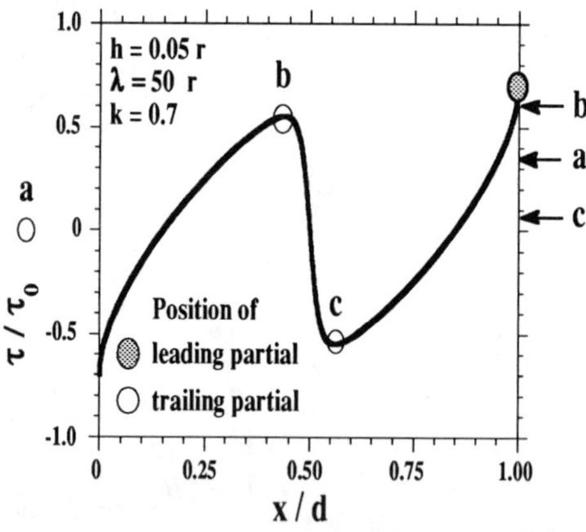

Fig. 6. Back stress - distance profile as calculated in [12] for a single dislocation. The effect of the spacing w between superpartials on the detachment stress of the leading partial: a) w > d (dispersoid diameter), b) w ≈ 0.55 d, c) w ≈ 0.45 d. The arrows indicate the resulting stress level for detachment in the three cases.

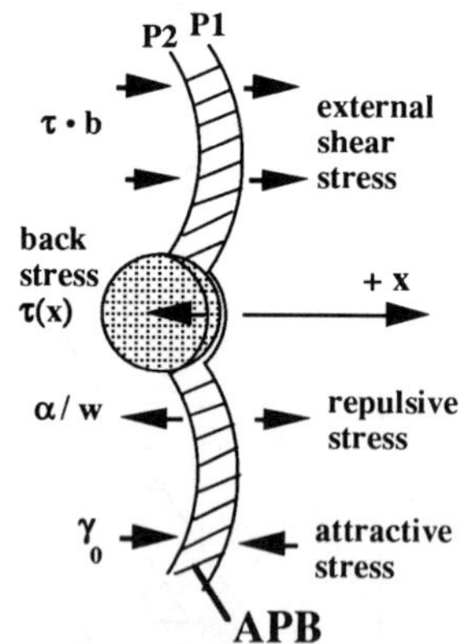

Fig. 7. Dispersoid-dislocation interaction in an ordered matrix. The superpartial spacing is fixed by mutual forces on the superpartials.

Fig. 6 shows in three special cases the change in the detachment stress of a leading partial as a function of the separation width of the superdislocations:

Case a) The detachment stress is reduced to half the value of a single dislocation if the partial spacing is greater than the dispersoid diameter (leading partial at x=d, trailing partial at "a").

Case b) The detachment stress of an extended superdislocation is relatively high if the trailing partial sits at the point of the climb barrier maximum (at "b").

Case c) The detachment stress decreases to even negative values, if the trailing partial resides in the valley of the stress profile ("c") The trailing partial facilitates the detachment process for the leading partial

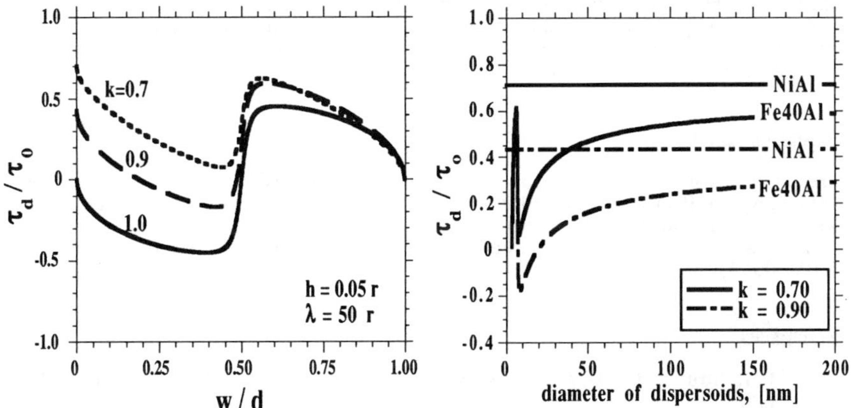

Fig. 8. The detachment stress for the leading partial under the influence of the trailing partial, as a function of the relation partial spacing to dispersoid diameter.

Fig. 9. Athermal detachment stress of the leading partial for NiAl with $\gamma \approx 810$ mJ/m^2 [13] and w=0 and Fe-40at%Al with $\gamma \approx 150$ mJ/m^2 and w $\approx$ 7 nm (Data extrapolated from [14]) for different values of relaxation parameters k.

The detachment stress of the leading partial is now a function of the relation partial spacing to dispersoid diameter (Fig. 8). An undissociated superdislocation has the maximum detachment stress (w/d = 0) whereas an extended superdislocation has two possibilities to achieve a high detachment stress: i) to approach the undissociated case, i.e. w/d $\approx$ 0 or ii) w/d $\approx$ 0.55. The second possibility dominates over the first one if the attractive interaction is small (k $\approx$ 1). For usual applications (0.7<k<1.0) the maximum variation in detachment stress depending on the relation w/d can be considerable. For further details the reader is refered to [15].

A first application of these results to intermetallics is shown in Fig. 9. For Fe-40at%Al and NiAl we can estimate the optimum dispersoid sizes to achieve an appreciable detachment stress. It is seen that the athermal detachment barrier in NiAl is practically independent of dispersoid size whereas in Fe-40at%Al the detachment barrier increases with dispersoid size. Therefore it is expected that NiAl would be more responsive to strengthening by small dispersoid particles than FeAl; this is in qualitative agreement with our experiments.

Conclusion

Dispersion strengthened aluminides can successfully be prepared by Mechanical Alloying. A high stress exponent, which is typical for the creep behaviour of dispersion strengthened materials, can be observed for NiAl and FeAl if the grain structure is coarsened after consolidation. An improvement in creep properties can be obtained even by moderate grain growth. A further increase in creep strength is attained by an optimised dispersoid volume fraction. Particulate composite materials also improve the creep properties of NiAl. A new processing technique leads to NiAl/AlN composite materials with controllable grain sizes and second phase volume fractions. Our results indicate that a finely dispersed AlN composite material is preferable to a coarser second phase NiAl/NiAlNb material, due to a higher stress exponent.

TEM investigations of ODS aluminides suggest that the mechanism of dispersion strengthening in ordered alloys is similar to that in disordered alloys but is affected by the dissociation of dislocations into superpartials. Our new model takes this effect into account. It is concluded that high creep strength of ordered dispersion-strengthened materials requires an optimum relation between the partial spacing, which is determined by the APB energy, and the dispersoid size. Alloy development anlong this line is currently in progress.

Acknowledgment

Special thanks are due to Dr. K. Zöltzer at Metallgesellschaft AG Frankfurt and Dr. H. Clemens at Metallwerke Plansee for support in the preparation and consolidation of powders. P. Grahle would like to acknowlege the assistance of B. Schietinger in preparing NiAl/AlN materials. The authors acknowledge financial support for this work by the BMFT (project number 030M3031A).

References

[1] P. R. Strutt and B. H. Kear, Proc. Mater. Soc. Symp., Boston Massachusetts, 39, 279 (1984)
[2] J. D. Whittenberger, E. Arzt and M . J. Luton, Scripta Metall. Mater. 26, 1925, (1992)
[3] A. U. Seybolt, Trans. ASM, 59, 860, (1966)
[4] E. Arzt, Res Mechanica, 31, 399, (1991)
[5] J. Rösler and E. Arzt, Acta Metall., 38, 671, (1990)
[6] C. C. Koch, Proc. Mater. Soc. Symp, Boston Massachusetts, 81, 369, (1987)
[7] K. Zöltzer, P. Grahle and E. Arzt, to be published
[8] H. Clemens, I. Rumberg, P. Schretter, P. Grahle, O. Lang, A. Wanner and E. Arzt, Proc. Mater. Soc. Symp., Boston Massachusetts (1992), this volume
[9] J. H. Schneibel, P. Grahle and J. Rösler, Mat. Sci. and Eng., A153, 684, (1992)
[10] V. C. Nardone and J. K. Tien, Scripta Metall., 17, (1983), 467
[11] J. H. Schröder and E. Arzt, Scripta Metall., 19, (1985), 1129
[12] E. Arzt and D. S. Wilkinson, Acta Metall., 34, (1986), 1893
[13] C. L. Fu and M. H. Yoo, Mat. Res. Soc. Symp. Proc.; 213, (1991), 667
[14] R. C. Crawford and I. L. F. Ray, Phil. Mag., 35, (1977), 549
[15] E. Arzt and E. Göhring, Scripta Metall. Mater. (1993) in press

FORMING OF γ-TiAl-ALLOYS

K.WURZWALLNER*, H.CLEMENS**, P.SCHRETTER**, A.BARTELS*** AND C.KOEPPE***

* BÖHLER Edelstahl GmbH, A-8605 Kapfenberg, Austria
** Metallwerk Plansee GmbH, A-6600 Reutte, Austria
*** Technische Universität Hamburg-Harburg, Werkstoffphysik und -technologie, D-2100 Hamburg 90, Germany

ABSTRACT

The forging and rolling behavior of binary Ti-48Al and ternary Ti-48Al-2Cr (at-%) on laboratory and industrial scale is investigated. Arc-melted and cast ingots are used for forging operations. In order to enable the determination of optimized forging and rolling conditions the plastic flow behavior was examined for low ($< 10^{-2}$ s^{-1}) and high (1 s^{-1}) deformation rates.

Forging under near-isothermal conditions was first performed on laboratory scale and upscaled as non-isothermal forging to pancakes of about 50 kg. Rolling tests on forged material are described. This paper presents results of microstructural examinations of forged ingots, rolled and annealed sheets and mechanical properties of sheet material under tensile test conditions.

INTRODUCTION

Intermetallics on the basis of γ-Titanium Aluminides play an important role in the international materials development sector. Some work concerning hot forming has already been published, but a lot of investigations were conducted on laboratory tests on a mm-scale. The present work describes basic research and some steps on the way to industrial-scale manufacturing Titanium Aluminides by melting, forging and rolling techniques in order to make them applicable at reasonable costs.

EXPERIMENTAL PROCEDURE

Ingots with a diameter of 190 mm and weigths up to 150 kg were produced by arc melting with non-consumable rotating electrode and casting into a skull. The analyzed compositions of some ingots are shown in Table I. The ingots show homogeneity of chemical composition over the entire ingot length and cross section. In the present work non-isothermal ingot-forging was carried out in the (α+γ) temperature region (1125-1360°C). Compression tests on primary forged material were conducted at temperatures between 1000 and 1250°C with various deformation ratios in a deformation dilatometer. Specimens with ϕ 4 mm and 8 mm in heigth were cut by spark erosion and upset with strain rates of 1 s^{-1}. During deformation the load-contraction curve is directly plotted and all other informations are obtained by computer. Flow stress data are presented.

Table I. Chemical analysis of some cast alloys in weight-% (the numbers in brackets represent atomic percentage)

Alloy	Ti	Al	Cr	C	Fe	N	O	H ppm
Ti48Al2Cr	62.60 (49.51)	34.27 (48.07)	2.79 (2.03)	0.012	0.10	0.019	0.051	5
Ti48Al2Cr	63.02 (49.96)	33.88 (47.69)	2.78 (2.03)	0.012	0.06	0.010	0.044	12
Ti48Al	65.54 (51.80)	34.09 (47.83)		0.014	0.06	0.026	0.055	7
Ti48Al	65.31 (51.53)	34.41 (48.19)		0.011	0.07	0.016	0.041	8

For subsequent rolling operations forged Ti-48Al-2Cr pancakes were cut into rectangular shape by means of spark erosion. Microcracks on the surface were removed by lapping. Rolling was conducted in the (α+γ) temperature region at low rolling speeds (<10m min^{-1}) using a modified pack rolling process.

The mechanical properties of the sheet material at room temperature were determined under tensile test conditions. The sheets were ground and lapped to avoid surface imperfections and to remove the surface near zone which exhibit an enhanced impurity content. The tensile specimens were cut by spark erosion. The surfaces of the specimens were electrolytically polished. The final depth of roughness turned out to be approximately 0.3 μm. Strain gauge length was about 19 mm and 10^{-4} s^{-1} was chosen as the strain rate.

RESULTS AND DISCUSSION

The present work describes near- and non-isothermal forging of cast Ti-48Al-2Cr ingots. Forging parameters are based on a lot of tests on specimens from ϕ 4mm up to ϕ 190 mm. The ingots have to be canned to minimize loss of heat and to protect them against oxidation. The upsetting operations start at specimen-temperatures higher than 1125°C, the eutectoid temperature. The lowest strain rate that can be achieved with our industrial equipment is about 1 s^{-1} at the beginning of the upsetting, which is very high compared to strain rates for isothermal deformation with about $10^{-3}s^{-1}$ [1]. Pancakes with diameters up to 570 mm have been forged [2].

Due to a low deformation ratio forging does not perfectly destroy the cast structure. Therefore lamellae which were orientated in forging direction are sliding along their boundaries or submit to buckling [Fig. 1(a)]. The material features a banded-type lamellar structure with equiaxed recrystallized γ-grains. As the ratio of reduction increases recrystallization starts at the buckling points of the lamellae packets and the lamellae begin to spheroidize. The banded structure consisting of coarse equiaxed γ-grains being embedded in a matrix of fine grained γ stabilized by the α_2-phase still exists [3]. Fig. 1(b) shows that the coarse grained stripes are orientated perpendicular to the forging direction. After homogenization the dendritic as-cast structure of the lamellar grains is broken up and the lamellae packages become randomly orientated. This microstructure proved to be optimum regarding to chemical homogeneity on a microscopic scale, deformation behavior and resulting microstructure [3]. In Fig. 1(c) the typical microstructure after homogenization, 85% forging ratio and an annealing treatment is illustrated. The coarse grained zones are remarkably refined but the banded microstructure can still be seen. Obviously there exists a necessity of multistep deformation to produce a more homogeneous microstructure which has been shown on laboratory scale [3].

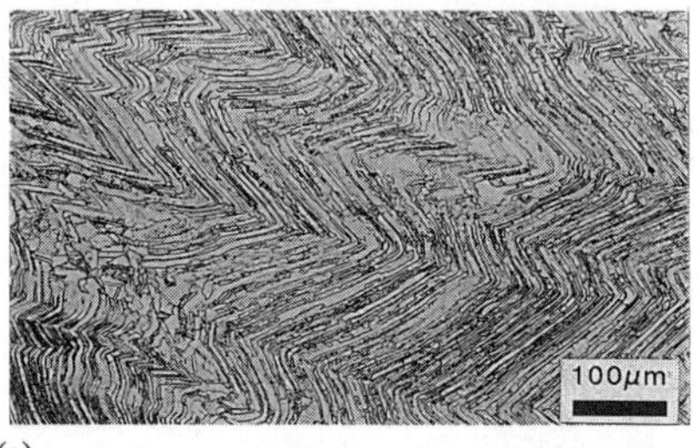

(a)

(b)

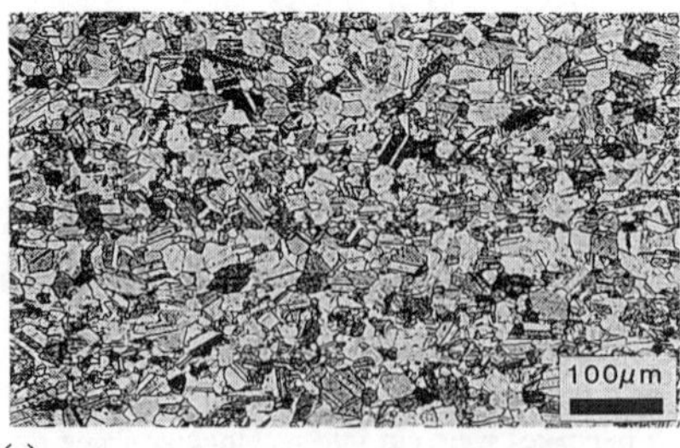

(c)

Fig. 1. Optical micrographs of forged Ti-48Al-2Cr
(a) after low deformation ratio (< 20 %)
(b) after high deformation ratio (> 50 %)
(c) homogenized before forging with a high deformation ratio (85 %)
Forging direction is vertical in all micrographs.

In order to realize deformation operations on industrial scale flow curves of preforged (85% as described above) Ti-48Al-2Cr were measured by means of a deformation dilatometer. The flow curves show a very strong dependence on temperature [Fig. 2]. The flow stress increases very rapidly with increasing strain rate and temperature. Remarkably there is no evidence of a distinct peak stress. Prior work [2, 4] shows there is a peak stress of around 400 MPa during upsetting as-cast material at 1200°C with deformation rates of about 1 s^{-1}, but after a certain amount of primary deformation, which means a changing of microstructure by recrystallization, the peak stress disappears.

To visualize the influence of strain rate, Fig. 3 shows maximum stress as a function of temperature. The four curves [5] with $\varepsilon = 1x10^{-4}$ s^{-1} up to $3x10^{-3}$ s^{-1} (ε = [ln (h_o/h_1)]/t) are the results of tests on multi-step thermomechanical treated material with a completely homogeneous microstructure consisting of equiaxed grains of about 10μm diameter. The σ - T - curve with ε = 1 s^{-1} from the present work was achieved on preforged material. Due to the phase transformations at 1125°C the curve for ε = 1 s^{-1} has to be devided into two parts. In the ($\alpha_2+\gamma$)-phase field all measured data fit together very well. The curves show practically identical courses only being shifted in the stress direction as a consequence of the different strain rates. In the ($\alpha+\gamma$)-phase field the σ_{max} vs. T curve shows a steeper decrease compared to the ($\alpha_2+\gamma$)-phase field. Up to now only few results are available for deformations in this temperature range.

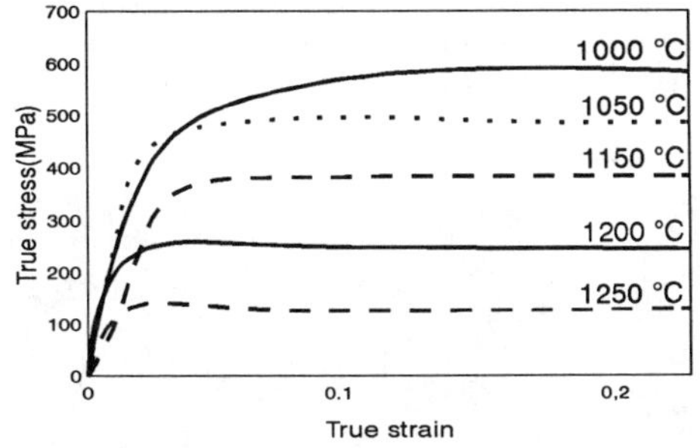

Fig. 2. Flow stress data of Ti-48Al-2Cr cast, homogenized, deformation ratio = 85%; compression tested in deformation dilatometer.

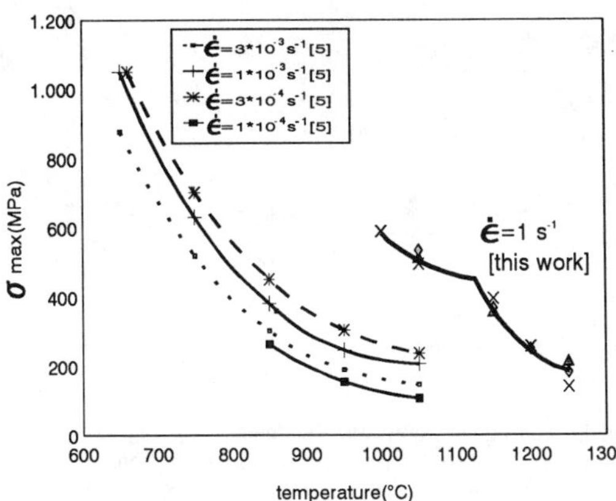

Fig. 3. Maximum stress versus temperature of Ti-48Al-2Cr for different strain rates [5]:as-cast and forged [this work]:same as Fig. 2.

The borderline conditions for successful rolling are by far more severe as compared to the upsetting process [1, 6]. In order to roll γ-TiAl alloy material which exceeds laboratory-size the following conditions must be fulfilled: (a) nearly isothermal conditions at high temperatures, (b) careful selection of rolling speed and reduction per pass in order to avoid overcritical strain rates which lead to cracking, and (c) prevention of oxidation. To meet these requirements as far as possible a modified pack rolling process was developed which enables rolling in the ($\alpha+\gamma$) temperature region at low rolling speeds (<10m min^{-1}) [7]. First experiments on an industrial rolling mill have proven the potential of this method. Up to now sheet material with dimensions of 420x150x2.2 mm^3 have been rolled. A Ti-48Al-2Cr-sheet with deformation ratio of more than 75% after removing from the capsule is shown in Fig. 4.

Fig. 4. Rolled Ti-48Al-2Cr sheet; deformation ratio > 75%; sheet thickness: 2,2mm

Fig. 5 illustrates the duplex microstructure of two sheets which have been rolled at different temperatures (sheet A: furnace temperature in the middle of the $(\alpha+\gamma)$-phase field; sheet B: furnace temperature below α-transus). Due to the higher rolling temperature in the case of B the amount of lamellar grains is considerably higher. The grain size in as-rolled material is 10-30μm. The sheets exhibit good homogeneity of microstructure over the entire length and cross section. However, some remainings of banded structure (see Fig. 1(c)) have been detected.

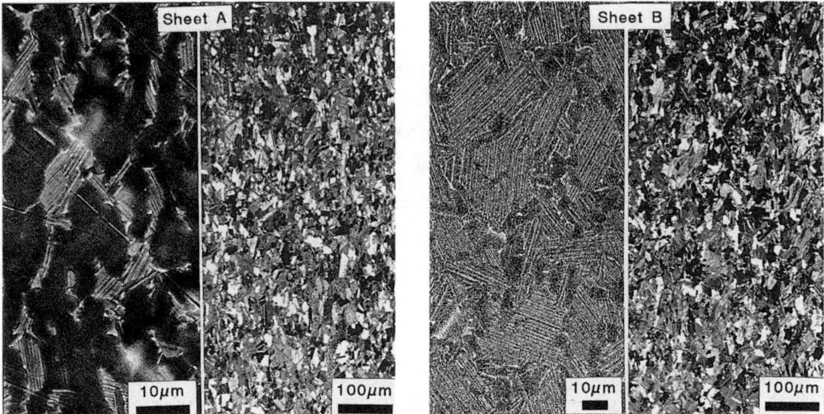

Fig. 5. Microstructure of as-rolled Ti-48Al-2Cr sheet material which was rolled under different conditions (see text). Left half: backscattered electron micrograph; right half: polarized light micrograph. Rolling direction is vertical in all micrographs.

The microstructure can be modified by subsequent annealing treatments [8]. Fig. 6 shows the phase distribution in sheet B which was annealed at 1200°C/1h under vacuum conditions. Due to this low temperature (only 75°C above the eutectoid temperature) the portion of α_2 is small and therefore most of the lamellar grains are dissolved.

The influence of short-time (1 hour) annealing treatments on microstructure and room temperature-hardness of as-rolled material was investigated in the temperature range of 900-1400°C (cooling rate approx. 100K min^{-1}). The results are summarized in Fig. 7 and show a minimum in hardness after annealing at 1200°C, which is a consequence of the decreased α_2-content (see Fig. 6). For higher annealing temperatures within the $(\alpha+\gamma)$-phase field the quantity of Ti_3Al is increased and lamellar grains appear again. No remarkable effect on grain growth was observed within the $(\alpha+\gamma)$-phase field. However, annealing at temperatures above the α-transus leads to a fully transformed microstructure and enormous grain growth was observed (grain size >100μm) which is attributed to the absence of the γ-phase at annealing temperature [9].

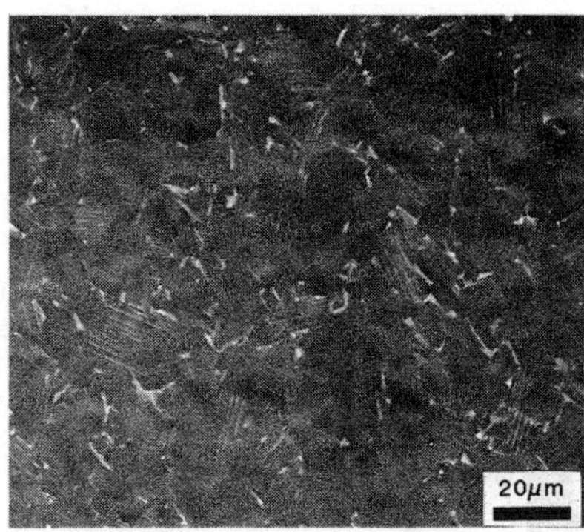

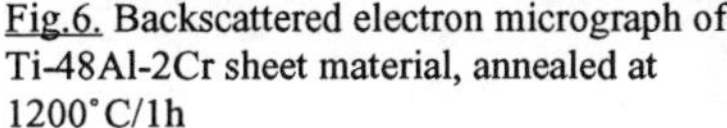
Fig.6. Backscattered electron micrograph of Ti-48Al-2Cr sheet material, annealed at 1200°C/1h

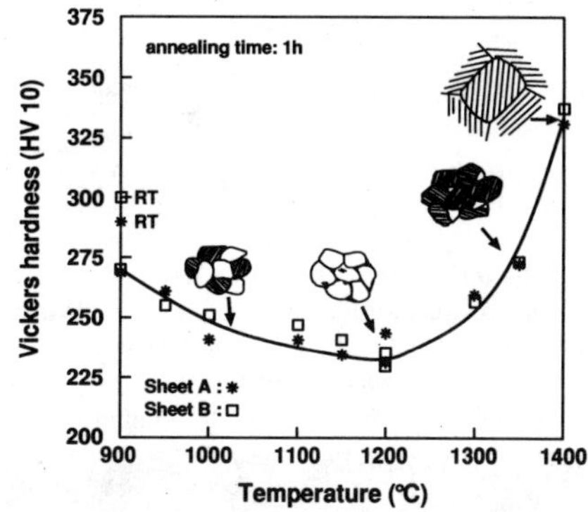

Fig.7. Dependence of room temperature-hardness and microstructure on annealing temperature; Annealing time: 1 hour

The mechanical properties of the sheet material at room temperature were determined under tensile conditions. Sheets A and B were tested in as-rolled condition and with a subsequent heat treatment (1200°C/2h + 1000°C/4h). Fig. 8 shows the typical difference in the stress vs. strain curve of the as-rolled or heat treated specimens (here shown for sheet B). In the as-rolled condition a continuous strain hardening was observed. With subsequent heat treatment the behavior of the strain hardening changes totally. After reaching the yield stress the stress remains approximately constant up to a true plastic strain of 0.008 followed by a constant strain hardening. Up to now the reasons for the different behavior of hardening are unknown.

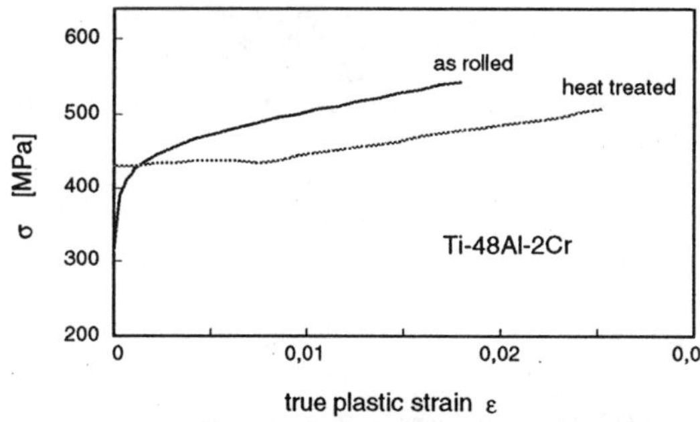

Fig.8. Typical difference of the stress vs. strain behavior during tensile test at room temperature under as-rolled and subsequently heat treated conditions (Ti-48Al-2Cr)

In Table II the value of yield stress (or 0.2% proof strength) and the highest measured values of fracture strain are listed. In the as-rolled condition of sheet A a higher strength due to a sligthly smaller grain size according to the lower furnace temperature was measured. The ductility was poorer than that of sheet B. After heat treatment sheet B exhibits the higher yield stress. In both series the heat treatment results in an improvement of ductility up to the plastic fracture strains of 0.025.

Fractography was carried out with SEM. In Fig.9 as a representative fracture surface the as-rolled specimen of Fig.8 is shown. The cleavage fracture type with a pure transgranular character is typical for the tested sheets. Surprisingly the mode of fracture did not change after heat treatment as found after forging with the same subsequent heat treatment where the fracture surfaces show partly intergranular character [10]. This change of fracture mode may be founded with a smaller grain size (typical 8 μm) in tested forged specimens.

Table II: Yield stress and maximum true plastic fracture strain of tensile tests at room temperature, subsequent heat treatment: 1200°C/2h + 1000°C/4h

Sheet	Condition	$\sigma_{0.2}$ [MPa]	ε_F
A	as rolled	462	≤ 0.006
B	as rolled	444	≤ 0.018
A	heat treated	416	≤ 0.025
B	heat treated	427	≤ 0.025

The fractography shows that flaws of microscopic size, inner micropores and inhomogeneities of the microstructure like remainings of banded structure with coarse Al-rich γ-grains lead to fracture initiation [3] and early failure of the specimen. Consequently the aim of development must be sheets with very homogeneous microstructure and homogeneous distribution of the α_2-phase without any flaws and pores exceeding the size of grains.

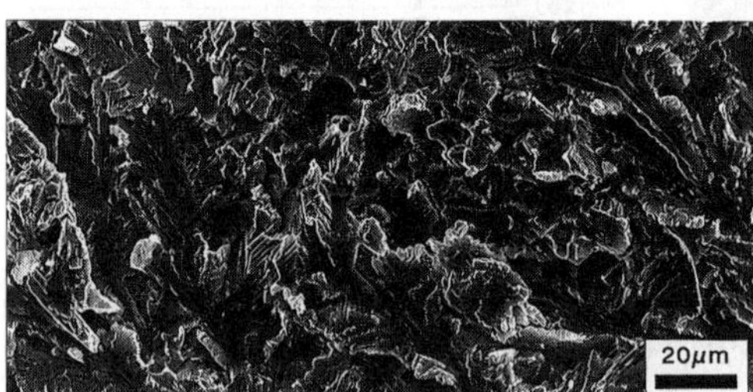

Fig.9. SEM fractography of as-rolled condition with ε_F=0.018. Only transgranular fracture is observed.

CONCLUSION

Cast γ-TiAl based alloys have successfully been deformed under non-isothermal conditions at high deformation rates by means of forging and rolling on an industrial scale. Primary deformation is the most critical step because of the appearance of a peak flow stress. Deformation ratio and temperature strongly influence the resulting microstructure and properties. Subsequent heat treatment allows the adjusting of microstructure and thus the mechanical properties over a wide range.

ACKNOWLEDGEMENTS

Part of this work was financially supported by the Bundesministerium für Forschung und Technologie (BMFT 03 M 30 29).

REFERENCES

1. S.L.Semiatin, N.Frey, S.M.El-Soudani, J.D.Bryant, Met. Trans. A, **23A** , 1719 (1992)
2. K.Wurzwallner, internal progress report Böhler Nr.2, 1991
3. C.Koeppe, J.Seeger, A.Bartels, H.Mecking, submitted to Metallurgical Transactions A
4. C.Koeppe, A.Bartels, internal progress report TUHH Nr.2, 1991
5. M.von Schwerin, thesis, Techn.Univ.Hamburg-Harburg, 1992
6. S.L.Semiatin, D.C.Vollmer, S.M.El-Soudani, C.Su, Scripta Metall. et Mater., **24**, 1409 (1990)
7. H.Clemens, P.Schretter, internal progress report Plansee Nr.2, 1991
8. for example see: S.-C.Huang, D.S.Shih in Microstructure/Property Relationships in Titanium Aluminides and Alloys, edited by Y.-W.Kim and R.R.Boyer (The Minerals, Metals & Materials Society, 1991) pp.105-122
9. H.Clemens, E.Tinzl, I.Rumberg, P.Schretter, to be published
10. C.Koeppe, A.Bartels, H.Mecking, to be published

TI-AL ALLOYS PREPARED BY BALL MILLING AND HOT ISOSTATIC PRESSING

M. OEHRING, T. KLASSEN AND R. BORMANN
GKSS-Research Center, Institute for Materials Research, Max-Planck-Str., D-2054 Geesthacht, Germany

ABSTRACT

For the powder metallurgical production of Ti-Al alloys, elemental powder blends and prealloyed powders were mechanically alloyed in a planetary ball mill. Thereby metastable solid solutions or amorphous phases are formed depending on the overall composition and the milling conditions. The crystalline phases exhibit an extremely small crystallite size of about 10 - 30 nm. This favors the consolidation of the powders which was performed by hot isostatic pressing in the temperature range between 500 °C and 800 °C. The results show that fully dense Ti-48 at.% Al alloys can be obtained only for a temperature of 800 °C. The crystallite sizes increase during compaction, but remain below about 0.15 µm. This allows the preparation of test specimens to investigate the mechanical properties of Ti-Al alloys with submicron crystallite sizes.

INTRODUCTION

The powder metallurgical production of alloy components can exhibit several advantages with respect to ingot metallurgy. For example, the coarse-grained microstructures and chemical inhomogeneities caused by solidification can be avoided. Since powders can be compacted in one step to complex parts with close tolerances, material consumption is reduced and machining can be saved. In addition, metal powders are required for advanced powder metallurgical techniques such as injection molding and plasma spraying.

Therefore, the powder metallurgical processing has been applied for the production of Nb-Al[1-3], Ti-Al[4-6] and Ti-Si[3,4,7] alloys. In particular, mechanical alloying of elemental powder blends and of intermetallic compounds have been employed in order to prepare homogeneous alloy powders with a submicron grain size. Due to the non-equilibrium processing, mechanically alloyed Nb-Al, Ti-Al, and Ti-Si powders consist of metastable solid solutions or amorphous phases, depending on the composition of the alloy. Details of the preparation and the mechanisms of phase formation have been described in previous publications[1-5,7,8]. In this article we will report on first consolidation experiments by hot isostatic pressing performed on mechanically alloyed Ti-Al powders.

EXPERIMENTAL

Powders of the pure elements Ti and Al were blended in the desired composition for mechanical alloying. The milling of the powder blends and of powders of intermetallic phases was performed with a planetary ball mill of type Fritsch pulverisette 5 using Cr steel vials and balls. The milling intensity (an arbitrary rotation speed scale) was adjusted to 5 or 7 corresponding to rotation speeds of 150 min^{-1} and 230 min^{-1}, respectively. To avoid atmospheric impurities all handling of the powders including the milling was carried out in glove boxes supplied with Ar via a gas purification system which maintains the oxygen and moisture content of the atmosphere below 1 ppm. After milling, the powders were sealed inside the glove boxes into Ti-tubes which subsequently were evacuated to 10^{-3} Pa, degassed at 350 °C and hot isostatically pressed at temperatures between 500 °C and 800 °C and a pressure of 200 MPa for 2 hours. For structural characterization a Siemens D5000 diffractometer was employed, whereas for transmission electron microscopy (TEM), specimens were prepared by jet polishing and characterized in a Philips 400 T microscope. The thermal phase stability of metastable structures obtained by ball milling was determined by a differential scanning calorimeter (DSC) modified to handle highly reactive powders.

RESULTS AND DISCUSSION

In the initial stages mechanical alloying of elemental powder blends leads to a lamellar microstructure consisting of Ti and Al layers. Upon further milling, this microstructure is refined (Fig. 1) and phase formation occurs at the interface between the Ti and Al layers[8]. Finally, a homogeneous phase is formed whose structure depends on the overall composition and the milling intensity.

The structural evolution during milling was followed by X-ray diffraction methods. As an example, the results obtained on a Ti-Al alloy with an overall composition of 50 at.% Al are displayed in Fig. 2. The X-ray diffraction patterns taken after selected milling times show the

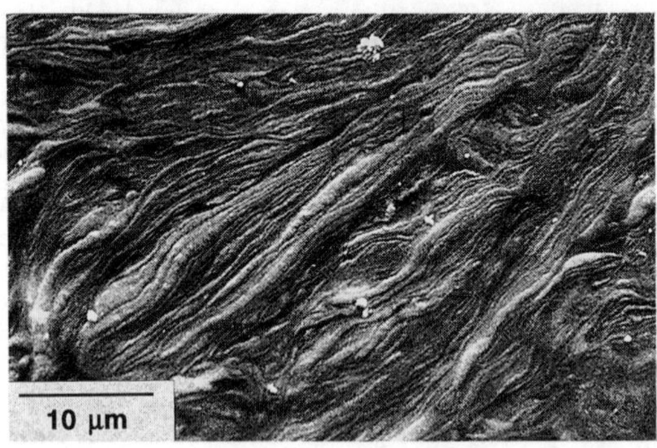

Fig. 1:
Cross-section of a Ti-50 at.% Al powder blend mechanically alloyed for 10 hours (scanning electron micrograph).

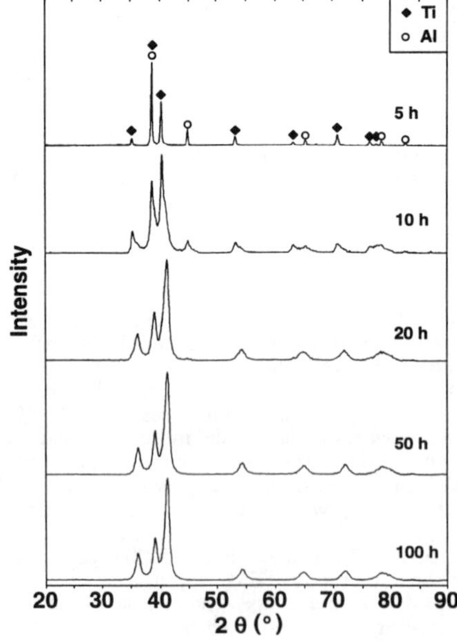

Fig. 2:
X-ray diffraction patterns of a Ti-50 at.% Al powder mechanically alloyed for different milling times.

formation of an hcp solid solution. The average crystallite size is about 15 nm, as derived from a Williamson-Hall analysis of the X-ray line broadening[9]. The mechanism of phase formation was investigated by TEM and is reported in ref. 8 in detail. As determined by DSC measurements the metastable phase transforms into the equilibrium compound γ-TiAl at a temperature of about 580 °C, whereby the crystallite size is slightly enhanced.

The results for other selected compositions are summarized in Table I. They demonstrate that metastable hcp solid solutions can be obtained up to 60 at.% Al, whereas fcc solid solutions are formed for Al-concentrations higher than 75 at.%[5]. Amorphous phases are only observed for low milling intensities and for Al concentrations of 50-65 at.%. Similar results are obtained for ball milling of intermetallic compounds indicating that the final structures are independent of the structure of the starting powder under our milling conditions[5].

Table I: Characterization of solid solutions formed by mechanical alloying of Ti and Al. The grain size is determined by the method of Williamson and Hall[9], whereas the thermal stability is obtained by DSC measurements with a heating rate of 20 K/min.

Composition	Structure	Grain size	Thermal stability
$Ti_{75}Al_{25}$	hcp	35 nm	460 °C
$Ti_{50}Al_{50}$	hcp	13 nm	580 °C
$Ti_{40}Al_{60}$	hcp	22 nm	560 °C
$Ti_{25}Al_{75}$	fcc	9 nm	410 °C

The formation of the metastable phases can be understood by considering the thermodynamics of the Ti-Al system which have been calculated by the CALPHAD method[5]. As the intermetallic compounds are energetically destabilized due to the reduction of their long-range order upon milling, solid solution phases exhibit the highest stability of all phases during milling[4]. As a consequence, they are favored in an interdiffusion process. This thermodynamic description of the phase formation is supported by the phase transformation of the intermetallic compounds observed during milling and by the homogeneity ranges of the hcp and fcc solid solutions which can be predicted by their Gibbs free energy curves.

Consolidation of Ti-Al powders

Compaction of Ti-Al powders was performed by hot isostatic pressing (HIP). As an example, the results obtained for a Ti-48 at.% Al alloy will be reported. After milling, the powder exhibits a disordered hcp structure with a crystallite size of about 15 nm. As the metastable structure is stable up to about 580 °C, temperatures of 500 °C and 600 °C for compaction were chosen. Fig. 3 represents the X-ray diffraction patterns after HIP. They show that for a temperature of 500 °C a large volume fraction of the metastable hcp structure is still maintained, whereas the material is totally transformed into the equilibrium compounds α_2-Ti_3Al and γ-TiAl for temperatures above 600 °C. As seen from the decrease in the line broadening of the X-ray diffraction pattern, the crystallite sizes are increased at higher processing temperatures. However, they remain below about 0.15 µm, as shown by the Williamson-Hall analysis[9] and TEM investigations (Fig. 4).

Cross-sections of the Ti-48 at.% Al powders compacted at different temperatures are illustrated in Fig. 5. They demonstrate that HIP at 500 °C and 600 °C is not sufficient to obtain a fully dense material. In particular for a temperature of 500 °C, a large fraction of pores is visible, indicating that the metastable hcp structure exhibits no significant advantages for the consolidation process, although it exhibits a higher crystal symmetry. This result is probably

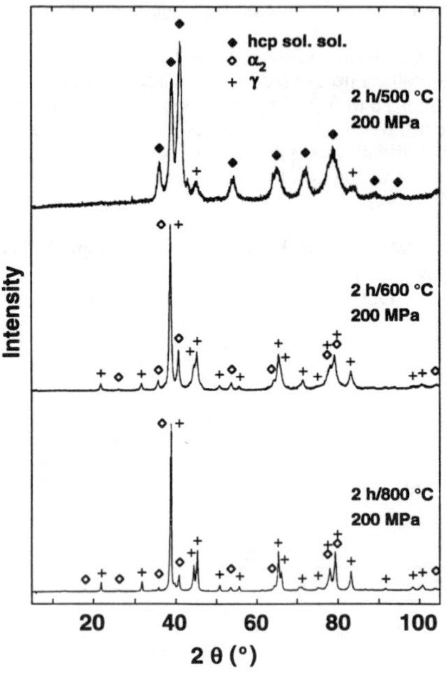

Fig. 3:
X-ray diffraction patterns of mechanically alloyed Ti-48 at.% powder compacts subjected to hot isotstatic pressing for 2h at temperatures between 500 °C and 800 °C.

Fig. 4:
Microstructure of Ti-48 at.% Al alloy hot isostatically pressed at 800 °C (TEM).

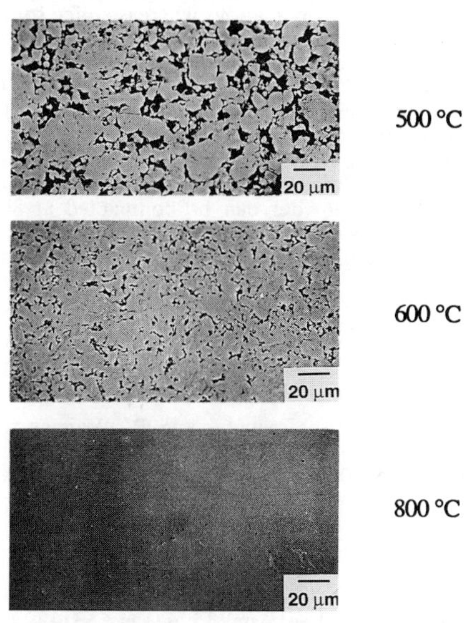

Fig. 5:
Cross-sections of mechanically alloyed Ti-48 at.% Al powder compacts subjected to hot isostatic pressing at temperatures between 500 °C and 800 °C, and a pressure of 200 MPa (SEM).

related to the small crystallite sizes of the hcp phase which deteriorates the deformation behavior at temperatures below the thermal stability limit of about 580 °C. On the other hand, for alloys processed at 800 °C, almost no pores in the cross-section could be detected by light microscopy or by scanning electron microscopy. We therefore conclude that these samples are almost fully dense with a porosity significantly below 1%. For rapidly solidified powders of similar composition almost full density is only achieved for HIPing at substantially higher temperatures above 1000 °C[10], whereas the consolidation of the mechanically alloyed powder is probably favored by the small crystallite size allowing for superplastic deformation at temperatures of about 800 °C[6]. From this material specimens were prepared (Fig. 6) in order to measure mechanical properties at different temperatures in detail. The results will be reported in a forthcoming paper.

Fig. 6:
Ti- 48 at.% Al sample encapsulated in a Ti-tube after hot isostatic pressing at 800 °C and specimen for a tensile test.

SUMMARY AND CONCLUSIONS

Ti-Al alloys were prepared by mechanical alloying and ball milling. Depending on the overall composition, an hcp solid solution is formed for Ti-rich alloys whereas an fcc solid solution is observed for Al-rich alloys. An amorphous phase is only obtained in the case of a low milling intensity and for Al-concentrations between 50 and 65 at.%. The phase formation observed is in agreement with the thermodynamics of the Ti-Al system as derived from the CALPHAD method. Unfortunately, the crystalline metastable structures exhibit a relatively low thermal stability and transform into the equilibrium compounds at temperatures between 400 and 600°C. This is below the temperature where a fully dense material can be obtained by hot isostatic pressing (for pressures of 200 MPa). However, due to the extremely small crystallite size of about 15 nm after the milling, the powder can be compacted at 800 °C. The consolidation is probably favored by superplastic deformation as the crystallite size is still below about 0.15 μm after densification. This allows the preparation of test specimens and the determination of the mechanical properties of Ti-Al alloys with submicron crystallite sizes, which is interesting from a basic point of view as well as for technical applications.

ACKNOWLEDGEMENTS

We thank P.A. Beaven, M. Dahms, F.P. Schimansky and R. Wagner for valuable discussions, F. Schmelzer and D. Potrykus for chemical analyses. The research has been supported by the BMFT under grant No. 03M0040 and the Deutsche Forschungsgemeinschaft (Leibniz-Programm).

REFERENCES

1. E. Hellstern, L. Schultz, R. Bormann and D. Lee, Appl. Phys. Lett. **53**, 1399 (1988)
2. M. Oehring and R. Bormann, J. Physique **C4**, 169 (1990)
3. M. Oehring and R. Bormann, Mater. Science and Engg. **A134,** 1330 (1991)
4. M. Oehring, Z.H. Yan, T. Klassen and R. Bormann, phys. stat. sol. (a) **131**, 671 (1992)
5. M. Oehring, T. Klassen and R. Bormann, submitted to J. Mater. Res. (1992)
6. M. Tokizane, K. Ameyama and H. Sugimoto, in *Solid State Powder Processing*, edited by A.H. Clauer and J.J. deBarbadillo (The Minerals, Metals and Materials Society, 1990), p. 67
7. Z.H. Yan, M. Oehring and R. Bormann, J. Appl. Phys. 72, 2478 (1992)
8. T. Klassen, M. Oehring and R. Bormann, submitted to J. Mater. Res.
9. G.K. Williamson and W.H. Hall, Acta Metall. 1, 22 (1953)
10. B.W. Choi, Y.G. Deng, C. McCullough, B. Paden, R. Mehrabian, Acta Metall. Mater. 38, 2225 (1990)

MICROSTRUCTURAL EVOLUTION DURING SUPERPLASTIC DEFORMATION OF TITANIUM ALUMINIDES

H.S. YANG, M.G. ZELIN AND A.K. MUKHERJEE

Department of Mechanical, Aeronautical and Materials Engineering
University of California, Davis, CA 95616, U.S.A.

ABSTRACT

Microstructural studies have been performed on two superplastically deformed Ti_3Al-based alloys, regular α_2 and super α_2. Dynamic grain growth, formation of phase stringers, and changes in the apparent volume fraction of the phases occurred in both alloys. The lath-like α_2 phase in the super α_2 alloy broke up during deformation. All these features are examined from the viewpoint of cooperative grain boundary sliding (CGBS). Cooperative grain boundary migration (CGBM), which is coupled with CGBS, is proposed to be the mechanism for strain induced grain growth and changes in the phase volume fraction.

INTRODUCTION

The current interest in superplasticity of titanium aluminides based on Ti_3Al stems from their potential for application in advanced aerospace industries due to their attractive high-temperature strength, low density, and good oxidation resistance [1, 2]. While the mechanical behavior of two Ti_3Al-based titanium aluminides, Ti-25Al-10Nb-3V-1Mo (super α_2 alloy) and Ti-24Al-11Nb (regular α_2 alloy) during superplastic deformation has been relatively well-understood [3-6], microstructural evolution due to superplastic strain has not been thoroughly studied [6].

Ti_3Al alloys usually have a duplex α_2 (ordered DO_{19} structure) + β microstructure within the temperature regime of potential engineering applications [7, 8]. Upon cooling after superplastic deformation (α_2+ β region), the β phase transforms to α_2 and the untransformed part of β phase in the case of intermediate to fast cooling rates will revert to the ordered structure, B2 [8]. Such basic knowledge of both microstructures and transformation kinetics is helpful in understanding the microstructural evolution of the Ti_3Al alloys during superplastic deformation.

Typical features of microstructural changes during superplastic deformation of common metallic materials are: (1) strain enhanced grain growth; (2) equiaxed grain structure even after very high superplastic strain; and (3) breaking-up of a lath-like phase [9, 10]. The focus of this paper is to address whether these microstructural features are also common for superplastic intermetallics such as Ti_3Al.

EXPERIMENTAL DETAILS

Two Ti_3Al alloys, regular α_2 (Ti-24Al-11Nb) and super α_2 (Ti-25Al-10Nb-3V-1Mo) were used in this study. The regular α_2 and super α_2 alloys were supplied by Rockwell International Science Center in the form of sheets 1.25 mm and 1 mm thick, respectively. The details of high temperature tensile testing and a special etching technique for microstructural studies of Ti_3Al are given in a previous paper [6]. The tested specimens generally exhibited significant necking. The local strains, $2\ln(t_o/t)$ (where t_o and t are the initial and final local thickness, respectively), along the gauge were measured and related to the corresponding micorstructures. Measurement of the area fraction of phases and the average interphase spacing as a function of strain were performed on scanning electron micrographs of deformed Ti_3Al by using a digitizer.

It is necessary to point out one experimental limitation of this investigation, i.e., the microstructural study of the deformed specimen was carried out after it had been cooled down to room temperature. Nonetheless, the comparative study of microstructures at different locations along the gauge section of a specimen can still provide clues about microstructural evolution at testing temperature.

RESULTS

Back scattered SEM (scanning electron microscopy) micrographs taken at different local strains are shown in Fig. 1 for a regular α_2 sample superplastically deformed under optimum conditions (1253 K and a strain rate of $1.5 \times 10^{-4} s^{-1}$) [6]. The corresponding quantitative result on the changes in the area fraction of α_2 phase as a function of local strain is shown in Fig. 2. As seen in Fig. 1, oxidation occurred during testing, which caused a very high area fraction of α_2 (darker) phase near the specimen surfaces. This layer with high α_2 area fraction was relatively thin and, hence, would not noticeably affect the value of area fraction measured on high magnification micrographs usually taken from the center region of specimen thickness. The area fraction of α_2 phase decreased continuously with superplastic strain. Initially ($\varepsilon = 0$), the area fraction of the α_2 phase, 75%, was noticeably higher than that of β phase and the β phase was dispersed in the continuous α_2 matrix. At a high strain level ($\varepsilon > \sim 3.0$), the β phase became the continuous phase and enveloped the α_2 particles. Stringers of α_2 - α_2 and β - β, oriented at ~45° to the tensile axis, were observed after a strain of ~1, as shown by the arrows in Fig. 1(b).

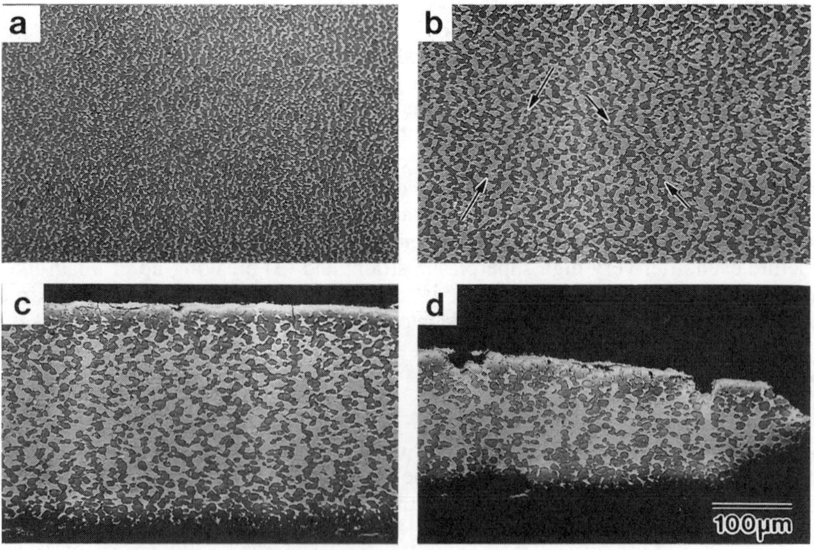

Fig. 1 SEM micrographs taken along the gauge of a regular α_2 specimen after superplastic deformation at 1253 K and $\dot{\varepsilon} = 1.5 \times 10^{-4} s^{-1}$. The local true strains are : (a) 0.0, (b) 2.0, (c) 3.0, and (d) 4.1.

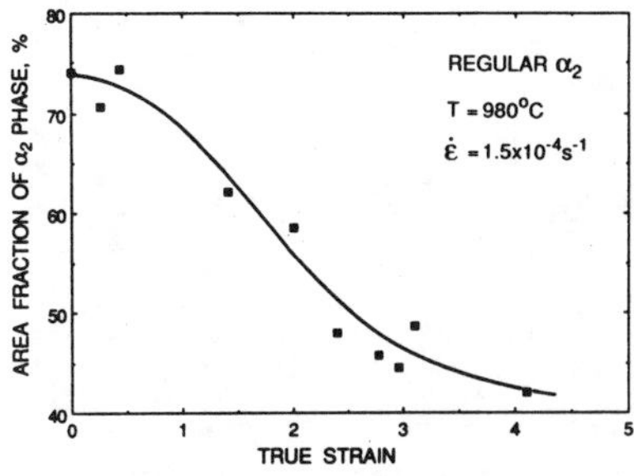

Fig. 2 Area fraction change of α_2 phase of regular α_2 alloy as a function of strain.

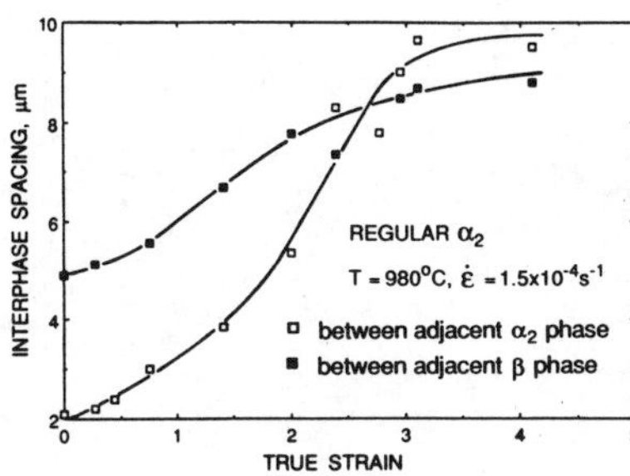

Fig. 3 The average interphase spacing of regular α_2 alloy as a function of strain.

(a) TEM

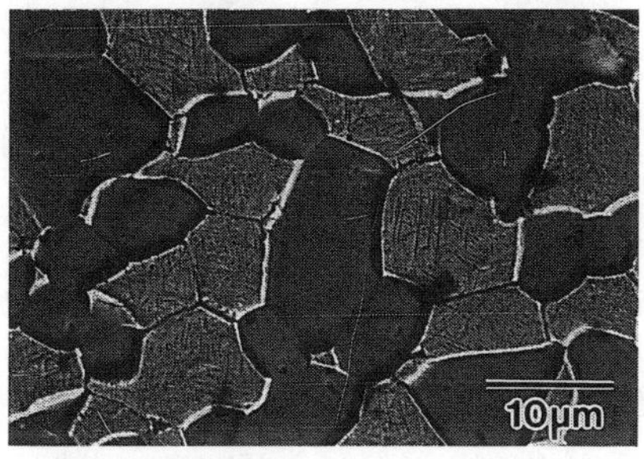

(b) SEM

Fig. 4 Fine Widmanstätten plates of α_2 surrounded by regions of untransformed β in regular α_2. (T = 1253 K and $\dot{\varepsilon} = 1.5 \times 10^{-4} s^{-1}$).

The interphase spacing between adjacent α_2 phases and β phases along the tensile axis is plotted in Fig. 3 as a function of strain. The spacing between both α_2 phases and β phases increased significantly with strain while the grain size of α_2 phase was almost constant and β grain size increased slightly with increasing strain.

Fig. 4(a) and 4(b) are SEM and TEM micrographs showing fine Widmanstätten plates of α_2 surrounded by regions of untransformed β of the regular α_2 sample. The importance of this observation in understanding whether there were indeed changes in the true phase volume fractions is addressed in the Discussion section.

The microstructural evolution of super α_2 with superplastic strain at 1253 K and a strain rate of $4 \times 10^{-4} s^{-1}$ is shown in Fig. 5. The lath-like α_2 grains broke up during the early stages of deformation whereas the fine spheroidal α_2 particles coarsened continuously with increasing superplastic strain. Stringers of α_2 grains (arrowed in Fig. 5(c)) were observed in super α_2, similar to those in regular α_2. In addition, the volume fraction of the α_2 phase in the super α_2 alloy also decreased with increasing strain. The size of β grains increased with strain. The average length of α_2 particles measured along the tensile axis is plotted in Fig. 6 against the strain. Three stages, I, II, and III, of the change in α_2 particle length can be identified, the physical origin of which will be discussed in the next section.

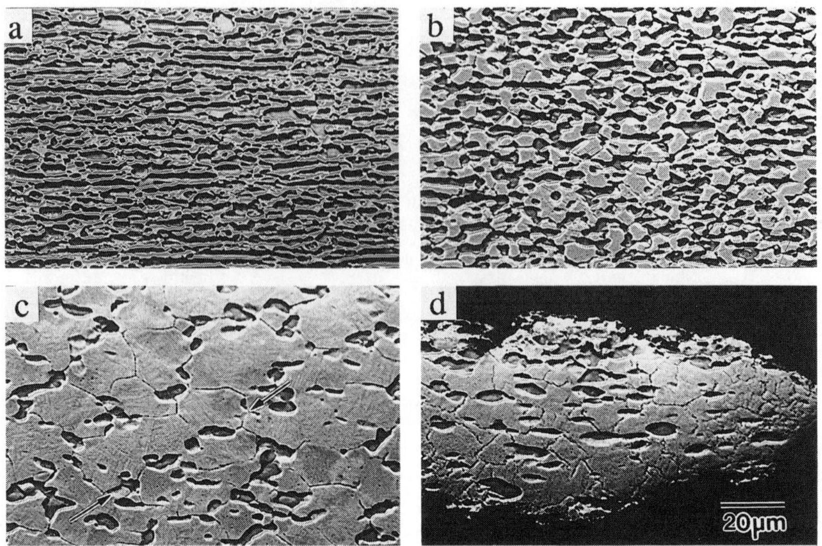

Fig. 5 The microstructural evolution of super α_2 with superplastic strain: (a) 0.0, (b) 1.2, (c) 2.9, and (d) 5.5. (T = 1253 K and $\dot{\varepsilon}$ = $4 \times 10^{-4} s^{-1}$).

DISCUSSION

It is generally accepted that grain boundary sliding (GBS) is the main mechanism of deformation for superplastic materials. Equiaxed grain microstructure, which was also observed in this work, is usually referred to the operation of GBS of individual grains. Recently, long range cooperation of GBS has been reported [11, 12]. This cooperative nature of GBS refers to the sliding of grain groups as a unit along grain boundary surfaces. On the basis of both qualitative microstructural observations and quantitative results, the breaking up of the initial lath-like α_2 in super α_2, the formation of phase stringers, and phase volume fraction changes in both alloys during superplastic deformation can be explained from the point of view of cooperative grain boundary sliding (CGBS).

The average length of α_2 particles is nearly constant at small strains (stage I in Fig. 6) because a certain minimal strain (~ 1) is required to completely break a α_2 particle. The tendency for some lath-like α_2 particles to break at very small strains ($\varepsilon < 0.5$) rules out the possibility of dynamic-recrystallization-induced breaking up of α_2 which requires significant strain. In stage II, the breaking up of α_2 particles is due to: (1) the decrease in the spacing of active CGBS surfaces; (2) continuous initiation of new CGBS surfaces with superplastic strain. The length of the broken α_2 reflects the spacing of active surfaces of CGBS. The microstructure becomes more and more equiaxed as the average α_2 particle length decreases with strain. However, in the final stage III, the process of breaking up diminishes and grain agglomeration/grain growth of α_2 causes the increase in the average length of α_2 particle.

It is noted that, after a certain amount of straining ($\varepsilon \approx 1$, required for the process of breaking-up), stringers of α_2 and β phases in super α_2 alloy have formed in the same fashion as the regular α_2 alloy which has an initially equiaxed grain structure. The process of stringer formation has been explained by the concept of CGBS in a previous paper [13]. Such microstructure can be considered as optimal for superplasticity because CGBS is easy to proceed along the interfaces between α_2 - α_2 and β - β phase stringers due to their favorite orientation and less accommodation required.

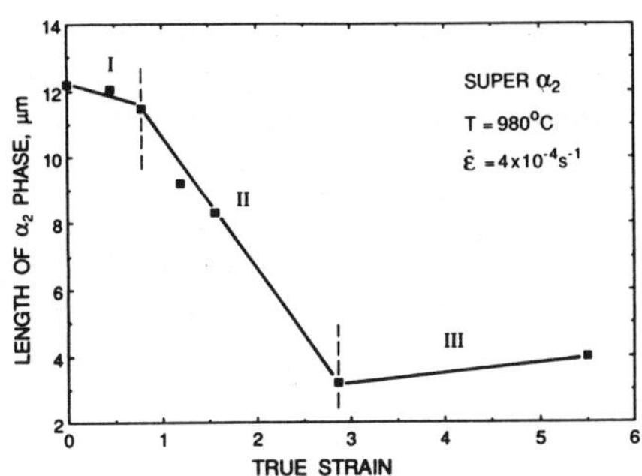

Fig. 6 Average particle length of α_2 of super alloy α_2 vs. strain.

From Fig. 1 and Fig. 5, it is seen that both the super α_2 and regular α_2 alloys show an increase in the area fraction of the β phase. The acicular-shaped Widmanstätten α_2 shown in Fig. 4 are not usually discernible under optical or scanning electron microscopy and not accounted for in the determination of phase area fractions Therefore, it seems that the area fraction change of phases is only an apparent effect. The observed change in the apparent area fraction of phases and grain growth can be explained from the viewpoint of cooperative grain boundary migration (CGBM), which is coupled with CGBS [14]. Movement of grain boundary dislocations, which causes grain boundary sliding, is, in general terms, accompanied by grain boundary migration [15]. Such grain boundary migration will result in grain growth in a single phase material [16]. In the case of cooperative grain boundary sliding along the stringer interfaces, the concurrent migration of stringer interface will cause the growth of one phase and gradual dissolution of the other phase, resulting in the "apparent" changes in phase volume fractions. The changes in the phase volume fractions are apparent because a nonequilibrium chemical composition is introduced into the newly grown β region in the process of dissolution of α_2 phase and precipitation of α_2 occurs during cooling from the testing temperature, Fig. 4.

The superplastic ductility of regular α_2 is higher than that of super α_2 [6]. Along with the possible causes such as different initial microstructures and compositions, the difference in ductility could be due to the apparent area fraction changes. For super α_2, the area fraction of β phase is initially over 50% and the β phase is the continuous phase. Therefore, the increase in the volume fraction of β phase in super α_2 will lower the number of α_2/β interphase boundaries and may therefore decrease the extent of grain boundary sliding. As a result, the superplastic elongation can be lowered. In contrast, the β area fraction is initially low (25%), and the "pinching off" and enveloping of α_2 grains by β grains as the β area fraction increases in the case of regular α_2 alloy may have a beneficial effect as the more diffusive and softer β phase easily accommodate the grain boundary sliding process.

CONCLUSIONS

Along with the typical features of microstructural evolution of superplastic materials (equiaxed grains, grain growth), formation of phase stringers, breaking up of lath-like α_2 phases, and changes of the apparent area fraction of phases during superplastic deformation have been observed in Ti_3Al alloys. The apparent α_2 volume fraction decreases with superplastic strain. Stringers of phases with an orientation of ~45° to the tensile axis are noticed after strain of ~1.0. All of these observations have been explained from the view point of cooperative grain boundary sliding coupled with cooperative grain boundary migration.

ACKNOWLEDGMENTS

This work is supported by the National Science Foundation under grant number DMR-90-1333. The authors would like to thank Dr. C. C. Bampton of Rockwell International Science Center for supplying the Ti_3Al materials.

REFERENCES

1. R.W. Cahn, *Metals, Materials and Processes,* **1**, 1 (1989).
2. H.A. Lipsitt, *Mater. Res. Soc. Symp. Proc.*, **39**, 351 (1985).
3. A. Dutta and D. Bannerjee, *Scripta Metal. Meter.*, **25**, 1223 (1991).
4. H. S. Yang, P. Jin, E. Dalder and A.K. Mukherjee, *Scripta Metal. Mater.*, **25**, 1223 (1991).
5. A.K. Ghosh and C-H Cheng, in *Superplasticity in Advanced Materials,* S. Hori, M. Tokizane and N. Furushiro, eds, The Japan Society for Research on Superplasticity, Osaka, Japan, 299 (1991).
6. H.S. Yang, P. Jin and A.K. Mukherjee, *Mater. Sci. Eng.*, **A153**, 457 (1992).
7. H.T. Kestner-Weykamp, C.H. Ward, T.F. Broderick and M.J. Kaufman, *Scripta Metall.*, **23**, 1697 (1989).
8. C. Bassi, J.A. Peters and J. Wittenauer, *JOM*, **41**(9), 18 (1989).
9. O.D. Sherby and J. Wadsworth,, *Prog. Mater. Sci.*, **33**, 169 (1989).
10. A.H. Chokshi, A.K. Mukherjee and T. G. Langdon, to be published in *Materials Science Reports*, Elsevier Science Publisher, (1992).
11. V.V. Astanin and O.A. Kaibyshev and S.N. Faizava, *Scripta Metall. Mater.*, **25**, 2663 (1991).
12. M.G. Zelin and M.V. Alexsandrova, in *Superplasticity in Advanced Materials*, S. Hori, M. Tokizane and N. Furushiro, eds, The Japan Society for Research on Superplasticity, Osaka, Japan, 63 (1991).
13. H.S. Yang, M. G. Zelin, R.Z. Valiev and A.K. Mukherjee, *Scripta Metall. Mater.*, **26**, 1707 (1992).
14. M.G. Zelin, R. James and A.K. Mukerjee, *J. Mater. Sci. Lttr*, in press.
15. R.C. Pond, D.A. Smith and P.W.J. Southerden, *Phil. Mag.*, **A37**, 27 (1978).
16. H. J. Frost and R. Raj, in *Mat. Res. Soc. Symp. Proc.*, **vol. 196**, Materials Research Society, 21 (1990).

MICROSTRUCTURE ENGINEERING FOR OPTIMIZING THE ROOM TEMPERATURE MECHANICAL PROPERTIES OF Fe_3Al-BASED ALUMINIDES

Z. Q. SUN, Y. D. HUANG, W. Y. YANG, AND G. L. CHEN
Department of Materials Science & Engineering, University of Science and Technology Beijing
Beijing 100083, P. R. China.

ABSTRACT

A new strategy of microstructure design for improvement of the mechanical properties of Fe_3Al based aluminides was developed. This new approach emphasizes microstructure control on the basis of ordered B2 phase structure matrix instead of the conventional $D0_3$ structure. This approach is characterized by the improvement of ductility without lose of strength. The observed data from various Fe_3Al based alloys illustrated that the flattened pancake-shaped grains with ordered B2 phase structure increased both yield strength and elongation of sheet specimens compared to the equiaxed grains. The three dimensional configuration of the flattened grains and their annealing temperature dependence as well as fractography were studied.

INTRODUCTION

Iron aluminides based on Fe_3Al have relatively low density,high specific strength,low cost, good oxidation resistance and excellent sulphidation resistance at high temperature in comparison with stainless steels. Based on these benefits iron aluminides have potential applications instead of stainless steels in many different fields, such as heat elements, automotive applications, and chemistry industry applications [1]. But the poor ductility at ambient temperature and a sharp drop in strength at 600°C greatly hinder its development for commercial use. From the 1940's to 80's, considerable work has been devoted to improvement of the room temperature ductility of iron aluminides, but no magnificent progress was made, the best elongation was about 4~6%[2,3]. Recently, C. T. Liu et al of ORNL indicated that iron aluminides were susceptible to environmental embrittlement at ambient temperature in the presence of water vapor[4,5]. This discovery is expected to help in alloy design of ductile iron aluminides as well as other aluminides for structural applications. By means of the addition of alloying elements such as chromium and the control of thermomechanical processing, the room temperature ductility of iron aluminides can be increased to 10% or more[6,7,8].

With a view toward possible development of Fe_3Al based alloys for corrosion resistant structural applications, systematic work on the relationship between chemical composition, processing, structure and properties of Fe_3Al based intermetallic alloys is in progress in our research group. The basic strategy is improving the ambient temperature ductility by softening first, and then increasing the strength at both ambient and elevated temperatures by optimizing the processing and chemical composition. In the present work, Fe-28Al-5Cr iron aluminide has been chosen as a softened alloy for study of the effects of complex alloying and the effects of processing. The best room temperature mechanical properties in air were yield strength of up to 520MPa~530MPa, and elongation of 19.0%.

EXPERIMENTAL PROCEDURES

The Fe_3Al based alloys used in this study were prepared by vacuum induction melting. High purity raw materials were used. The ingots weigh 5kg. The composition of the basic alloy is: Al=28at%, Cr=5at%, balance Fe. Several heats were doped with Zr, Mo, Nb, B. The heats were homogenized at 1000°C for 24h. All heats were hot rolled into plates with a thickness of 5mm at 850°C~1000°C. A special thermomechanical treatment was given. No cracks were found on the sheets.

The tensile specimens having a gage length of 15mm were cut from the sheets along the rolling direction. Two different heat treatments were performed for obtaining different phase structures. The first set (A), designated as B2 treatment, was annealed at various temperatures and then oil quenched. The second set (B), designated as $D0_3$ treatment, was annealed at 850°C plus a ordering heat treatment(500°C/5days). The specimens were tested in tension in air at a strain rate of 2mm/min. Fracture surfaces were examined by scanning electron microscope.

Specimens were etched in a solution of 50% CH_3COOH, 30%HNO_3, and 20%HCl or in a solution of 36.5%HCl. The phase structures of the specimens were investigated by means of X-ray diffractometer, and the SEM, TEM, optical microscope studies as well.

RESULTS AND DISCUSSIONS

Controlled Thermomechanical Treatment (B2 Treatment)

The most important fact is perhaps that controlled thermomechanical treatment is a dominant factor in achieving excellent mechanical properties, especially the ductility at room temperature. Table I and II show the mechanical properties of Fe_3Al based alloys with B2 and with $D0_3$ treatments, respectively. The data clearly illustrates the significant effects of the treatment. The thermomechanical B2 treatment can greatly improve both the room temperature ductility and strength. An ambient temperature tensile elongation of over 12.0% has been observed even for binary Fe-28at%Al alloy only by optimizing the thermomechanical processing(see Table I). Actually the beneficial effects of thermomechanical treatment keep importance for all Fe_3Al based alloys with different chemical composition.This can be easily confirmed by comparing the data in Table I with that showed in Table II. The excellent combinations of mechanical properties at ambient temperature can be achieved up to yield strength of 520MPa and elongation of 19.1%.

Table I Room temperature tensile properties and the recrystallization temperatures(T_S) of Fe_3Al-based alloys

Alloy	σ_S (MPa)	σ_b (MPa)	ε%	T_S(°C)	fracture mode
Fe-28Al	350	890	12.3	700	cleavage[a]
Fe-28Al-5Cr	345	777	16.8	750	cleavage
Fe-28Al-5CrB	330	865	18.1	700	cleavage
Fe-28Al-5CrZrB	520	1220	19.1	750	cleavage
Fe-28Al-5CrZrMoNbB	530	950	14.1	750	cleavage

Note: a----transgranular

Table II Tensile properties of $D0_3$ treatment Fe_3Al-based alloys

Alloy	σ_S (MPa)	σ_b (MPa)	ε%	fracture mode
Fe-28Al	339	428	2.0	cleavage[a]
Fe-28Al-5Cr	280	513	6.4	mixed[b]
Fe-28Al-5CrB	281	530	7.1	mixed
Fe-28Al-5CrZrB	288	448	4.0	cleavage
Fe-28Al-5CrZrMoNbB	329	546	5.0	cleavage

Note: a----transgranular b----intergranular and transgranular

An annealing treatment after thermomechanical processing was employed for optimizing the combination of the mechanical properties. Fig. 1 shows a typical example of the effect of annealing temperature on the mechanical properties. A critical annealing temperature for maximum elongation always existed for various alloys while the yield strength continuously decreases with increasing annealing temperature.

The crystallographic structure after thermomechanical treatment was analyzed by X-ray diffraction. The structure of oil quenched iron aluminide is B2 structure. Fig. 2 shows highly deformed grains, elongated in the direction of hot working. The effect of annealing treatment on the microstructure is dependent on the recrystallization temperature. The recrystallization starting temperatures for alloys with various chemical compositions were determined by optical microscopy. The data is shown in Table I. It can be seen that when the annealing temperature is

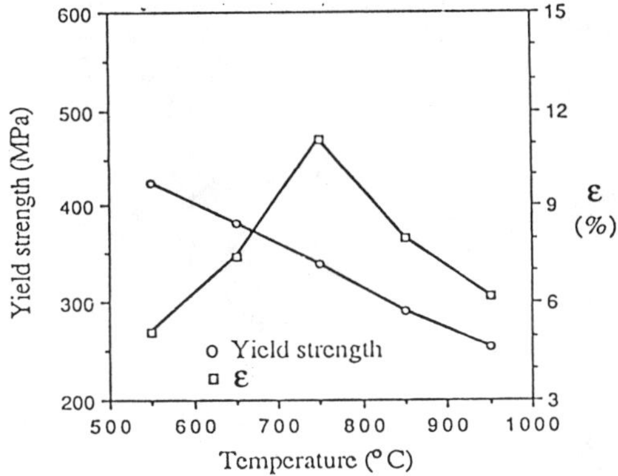

Fig. 1 Effect of annealing temperature on room temperature tensile properties of Fe-28Al-Cr iron aluminides.

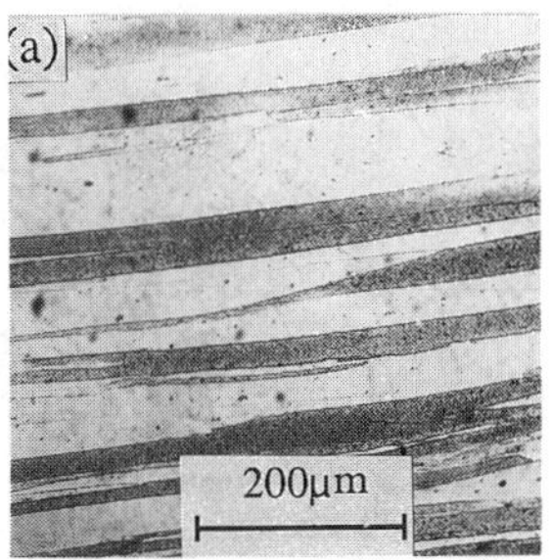

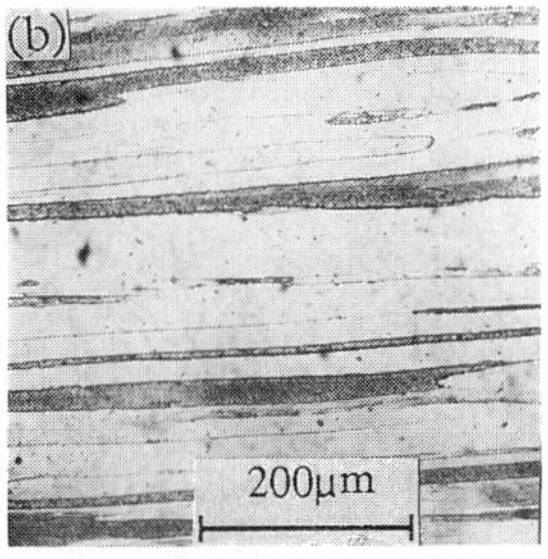

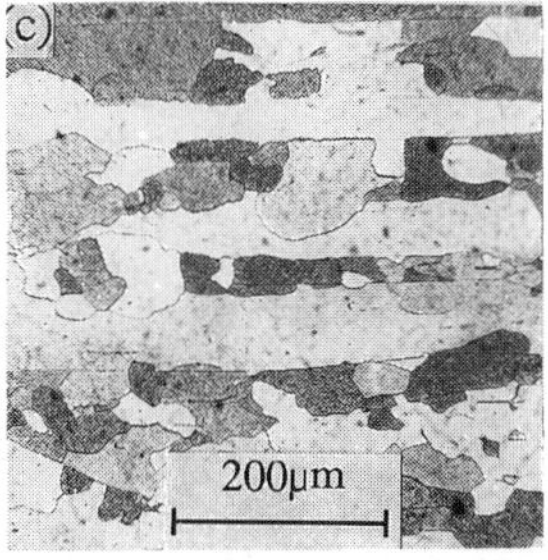

Fig. 2 Optical microstructure of the Fe-28Al-Cr iron aluminide annealed for 1h at (a): 550°C; (b): 750°C; (c): 850°C.

lower than recrystallization starting temperature, the deformed elongated grain morphology is maintained, but dislocation recovery always takes place. If the annealing temperature is high enough that recrystallization occurs, then ductility decreases as the grain becomes equiaxed.

Fig. 3 a, b, c show the typical examples of fracture modes of room temperature tensile specimens. All fractures for the specimens with the B2 treatment are transgranular or cleavage. It seems that the number of secondary cracks on the fracture surface increases with increasing elongation. It implies that the propagation of the main crack is difficult for the specimens with more strain. The fracture modes of the specimens with D03 treatment vary with the composition of alloys(see Table II). For the alloys containing Cr the fracture mode is mixed. It seems that grain boundary fracture occured easily in the specimens with the D03 treatment.

It is believed that the main effects of B2 thermomechanical treatment lie on the change in the morphology of the grains and in keeping the B2 structure. The thermomechanical processing deformed the grain into a pancake shape, resulting in parallel to the surface of the sheet specimen(Fig. 4). If the annealing temperature exceeds the recrystallization temperature, then grain becomes equiaxed and corresponding ductility decreases. However, we consider keeping B2 structure instead of D03 structure may improve the grain boundary structure, resulting in the improvement of the ductility. Table III clearly suggests the grain boundaries of the polycrystalline D03 structure are weak. In order to understand this phenomenon, further simulation works of modeling the detail of the grain boundary structure of B2 and D03 structure needs to be done.

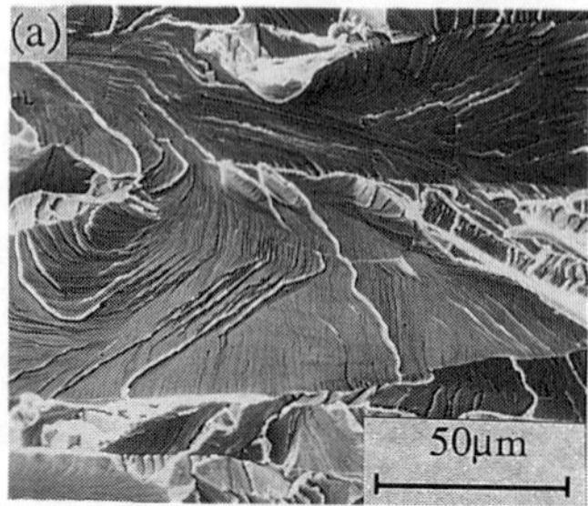

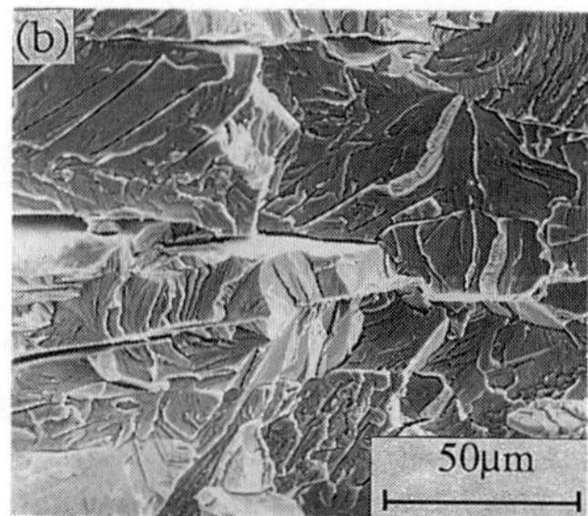

Fig. 3 SEM fractographs of the Fe-28Al-Cr iron aluminide annealed for 1h at (a): 550°C; (b): 750°C; (c): 850°C.

Table III Room temperature tensile properties for Fe-28Al-5Cr

	σ_s (MPa)	σ_b (MPa)	ε%	fracture mode
$D0_3$ heat treatment	280	513	6.4	mixed[a]
B_2 thermomechanical treatment	345	777	16.8	cleavage[b]
B_2 thermomechanical treatment + $D0_3$ heat treatment	320	639	9.7	mixed

Note: a----intergranular and transgranular b----transgranular

Environment Response

It is well known that C. T. Liu et al have demonstrated the unique effect of the environment on ductility for iron aluminides[4,5,9]. It is believed that the advantages of the thermomechanical treatment are closely related to the decrease of the environment sensitivity of the alloy. An interesting hydrogen charging test was performed for a Fe-28Al-5Cr alloy during B2 or $D0_3$ treatment to confirm the effects of environment sensitivity. The results are shown in Table IV. For the alloy with $D0_3$ structure, the ductility is almost completely lost in the hydrogen-charged condition. The corresponding tensile elongation is reduced from 6.4% to 1.6%. For the alloy with B2 treatment, the tensile elongation only partially decreases from 16.8% to 9.4%. A relatively high elongation of 9.4% could be retained under the hydrogen charging condition. The results undoubtedly prove the important effects of environment sensitivity and the relationship between microstructure of the alloy and the environment sensitivity. It is believed that both the morphology and the structure of the grain boundaries are particularly important concern for the environment sensitivity. The thermomechanical B2

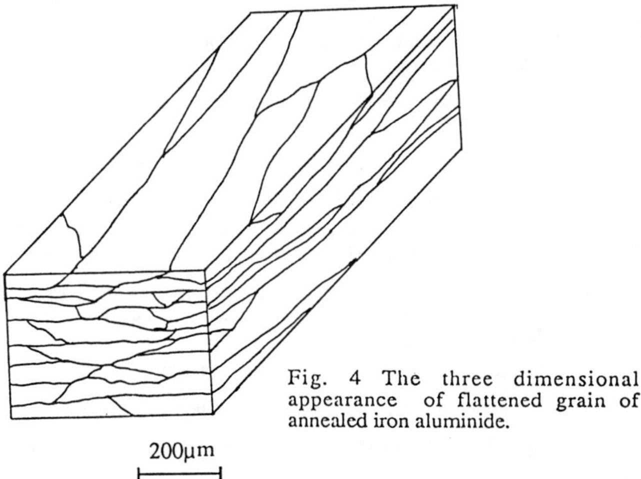

Fig. 4 The three dimensional appearance of flattened grain of annealed iron aluminide.

treatment changes the configuration of the grains to the pancake shape, resulting in shorter and discontinuous transverse grain boundaries(see Fig. 4), and the grain boundaries for B2 structure seem stronger than that for $D0_3$ structure. All these factors are favorable to weaken the environment effects, and this is why the thermomechanical B2 treatment reduces the environment sensitivity. A.Castagna and N.S. Stoloff also show similar resistance of B2 phase to environmental sensitivity compared to $D0_3$ phase in Fe-28Al-5CrNbC alloy[10]. However, if the annealing temperature is higher than the recrystallization starting temperature, new recrystallization grains form, resulting in the formation of the continuous transverse boundaries and equiaxed grains. Therefore the diffusion of atomic hydrogen towards interior material becomes easy, leading to the decrease in the ductility.

The data shown in Table IV also proved that the environment effect is perhaps not the only factor for controlling the ductility. It is believed that both the shorter and discontinuous transverse boundaries and the more stronger grain boundaries themselves provided favorable microstructural conditions for preventing the crack formation and propagation. This is proved by the fact that the number of the secondary cracks on the fracture surface increases as the ductility to fracture increases.

Table IV Effects of hydrogen charging on room temperature tensile properties

	σ_s (MPa)	σ_b (MPa)	ε%
$D0_3$ heat treatment	223	247	1.6
B_2 thermomechanical treatment	333	537	9.4
B_2 thermomechanical treatment without hydrogen charging	345	777	16.8

CONCLUSIONS

1. The thermomechanical B2 treatment can effectively improve the mechanical properties of Fe_3Al based alloys. Even for the binary Fe-28at%Al alloy, an ambient temperature tensile elongation of over 12% has been obtained only by optimizing the thermomechanical processing.

2. The main effects of B2 thermomechanical treatment lie in changes in configuration of the grain to a pancake shape and retaining the B2 structure. The grain boundary of B2 structure is stronger than that of $D0_3$ structure,the pancake grain morphology makes the transverse grain boundaries short and discontinuous.

3. The configuration and structure of the grain boundary are a particularly important concern for the environment sensitivity.

REFERENCES

1. C.G.McKamey, J.H.DeVan, P.F.Tortorelli and V.K.Sikka, J. Mater. Res.6, 1779(1991).
2. G.Culbertson and C.S.Kortovich, AFWAL-TR-4155, AIR FORCE WRIGHT AERONAUTICAL LABORATORIES, Wright Patlerson Air Force Base OH, March, 1986.
3. M.G.Mendiratta and H.A.Lipsitt, in High Temperature Ordered Intermetallic Alloys, edited by C.C.Koch, C.T.Liu and N.S.Stoloff(Materials Research Society, Pittsburgh, PA, 1985), Vol39, pp155-162.
4. C.T.Liu, E.H.Lee and C.G.McKamey, Scripta. Metall. 23, 875(1989).
5. C.T.Liu, C.G.McKamey and E.H.Lee, Scripta. Metall. 24, 385(1990).
6. C.G.McKamey, J.A.Horton and C.T.Liu, J. Mater. Res. 4, 1156(1989).
7. Z.Q.Sun, Y.D.Huang, W.Y.Yang and G.L.Chen, J of USTB 13,539(1991).
8. C.G.McKamey and C.T.Liu, Scripta. Metall. 24, 2119(1990).
9. D.J.Goydosh and M.V.Nathal, Scripta. Metall. 24, 1281(1990).
10. A .Castagna and N.S.Stoloff, Scripta. Metall. 26, 673(1992).

PROCESSING OF $FeAl/Al_2O_3$ IMCs USING ADVANCED ELECTRODEPOSITION METHODS

Brett Wilson, C.J. Suydam, M.J. Crimp and M.A. Crimp
Department of Materials Science and Mechanics, Michigan State University, East Lansing, MI 48824-1226

ABSTRACT

A theoretical examination of the $FeAl/Al_2O_3$ fiber intermetallic matrix composite (IMC) system was coupled with experimental results to determine optimum processing conditions for maximum FeAl adhesion to Al_2O_3 fibers and minimum attraction of FeAl and Al_2O_3 fibers to themselves. Optimizing the processing conditions leads to a more uniform green composite without matrix or reinforcement rich zones. The theoretical examination consisted of applying a model developed using traditional colloidal approaches to suspensions while accounting for the complexities of a multicomponent composite system. The model uses the suspension properties of the individual materials such as size and surface potential along with the processing conditions for the system as the basis for calculation. This model describes the interaction potentials between components of the suspension and the stability of and between various components of the system in terms of a stability ratio, W. The optimum processing conditions were found by determining the conditions under which the calculated values of W are ideal. Experimental results utilizing the model predictions have been examined and include verification of the FeAl particle adhesion to the Al_2O_3 fiber and preliminary consolidation studies.

INTRODUCTION

As evidenced by this fifth MRS meeting on high temperature intermetallic alloys, intermetallics continue to receive considerable attention as potential moderate to high temperature materials. The B2 aluminides have been the focus of much of this attention because, in addition to their attractive high melting temperatures, they offer relatively low densities and the promise of excellent oxidation and environmental properties. Unfortunately, like many intermetallics, this class of alloys is limited by poor low temperature toughness and ductility, and limited high temperature strength. As a result, a large fraction of the current research now being focused on these materials is not for monolithic applications, but instead as matrix materials for advanced composites. The objective of these composites is to use large volume fractions of high strength fiber as the load carrying component in high temperature creep applications while taking advantage of the desirable properties of the intermetallic matrix.

One of the hurdles in developing intermetallic matrix composites (IMCs) has been processing the reinforcement and matrix materials into a fully dense, well distributed product. A wide range of processes have been used to produce short fiber, whisker and particulant reinforced materials including hot pressing [1-4], hot extrusion [5], reactive sintering [1-3], powder injection [3] and XD™ [6]. Many of these methods have been used to successfully synthesize near fully dense, well dispersed composites. Production of continuous fiber reinforced composites has proven to be somewhat more difficult. Composites with continuous fiber reinforcements have been made using a number of methods including powder cloth [7,8], foil-fiber-foil [9] and thermal spray processing [8]. All of these processes can successfully produce highly dense composites but are limited to single filament, large diameter fibers (>100 μm). Unfortunately, the powder cloth method requires a binder which is a possible source of contamination, while thermal spray techniques can be expensive.

If the composite strengthening results from frictional sliding of fractured fibers, the toughness increases as the fiber diameter decreases [10,11]. Fine diameter fibers (<25 μm) are readily available in the form of multistrand fiber tows. Unfortunately, attempts to produce IMCs with these fibers have met with limited success. Liquid infiltration/pressure casting has been used to produce IMCs with fine, multistrand reinforcements [12,13]. This type of processing is limited by the need for the liquid intermetallic to wet the fiber. For example, liquid Ni_3Al does not wet Al_2O_3 reinforcements and requires alloying additions to improve the wetting [12]. Furthermore, pressure cast IMCs display an inhomogeneously dispersed reinforcement, often resulting in the individual fibers sintering together, leading to a brittle fracture path [12,13].

The purpose of the current work is to develop a process for producing IMCs with continuous, fine diameter reinforcements without the addition of binders. This will be achieved by adhering FeAl particles to the Al_2O_3 fiber surface in a powder suspension. The impregnated fiber tows will then be consolidated into a fully dense IMC. This fiber electrophoretic deposition (FED) process builds upon established colloid chemistry theory as well as knowledge of the intermetallic powder surface to yield a composite in which the reinforcements are well distributed within the matrix materials without the drawbacks associated with current processing methods. The FED process is being optimized for use with continuous ceramic fibers as the reinforcement material but has the advantage of being easily adapted to chopped fiber or whisker additions.

The forces acting upon approaching colloidal, micron-sized particles include: van der Waals attractive forces, steric forces caused by co-polymers which are adsorbed onto the particle surface, and electrostatic repulsive forces which are due to the presence of electrical charges on particle surfaces. Assuming conditions in which steric forces will not be present, the net force on the particle then becomes a balance between the attractive forces which cause particles to flocculate and the repulsive forces which cause particles to disperse. Therefore, in a composite $FeAl/Al_2O_3$ suspension, if the forces may be manipulated, a two-component system may have attractive forces existing between the matrix and the reinforcement materials while having repulsive forces preventing the clumping of either matrix or reinforcement material.

Dispersion of the powder/reinforcement materials consists of two elements: dispersion into a suspension and the stability of this suspension. While it is ideal to have a disperse, stable suspension in order to improve the final microstructure by removing agglomerates, new processing methods contain steps which call for alternating from stable suspensions to unstable, coagulated suspensions (containing large, loosely bound agglomerates). This prevents macrosegregation of the different system components.

Pioneering work in the area of colloidal surface forces was performed in the 1940's resulting in the well known DLVO theory [14,15]. This theory describes the total interaction energy between particles in a single component suspension. From this information, some insight into the stability behavior of the suspension can be inferred. However, with the increasing use of multi-component systems due to addition of reinforcements and of processing additives (e.g. sintering aids, stabilizers, composites, etc.), it is necessary to develop a theory for the prediction of the stability of multicomponent systems.

In the current study, the stability behavior of multicomponent IMC systems will be examined and prediction of the stability studied by inputing system and material data (such as pH, electrolyte concentration, particle size, Hamaker constant, and zeta potential) into a computer program which used a method that is an adaptation of a method originally developed by Healy, Hogg, and Furstenau [16]. Particles which spontaneously ionize in a medium result in surface charges yielding a surface potential. The surface potential may be measured by acoustophoresis and results in the zeta potential. A complete description of this program has been reported in detail elsewhere [17,18].

For development purposes, the current paper will describe the coating and infiltration of fine diameter multistrand Al_2O_3 fiber with B2 FeAl powder particles. Consolidation to a fully dense composite, which is currently in progress, will be described in a later work.

EXPERIMENTAL PROCEDURE

The starting material for the matrix is prealloyed FeAl powder manufactured by Alloy Metals Inc., Troy, MI, with a particle size of <177 μm. To be suitable for suspension, the powder was sieved and sedimented to remove the coarse particles. SEM photomicrographs (Figure 1) taken on a Hitachi S-2500C at an accelerating voltage of 15 KeV showed an average particle size of 2.3μm (±1.6 μm). Multistrand Al_2O_3 DuPont FP fiber bundles having an average filament diameter of 21 μm (±4 μm) and an average of 60-70 filaments per bundle, was used as the reinforcement material for these composites and is shown in Figure 2. The Al_2O_3 fibers were dipped into a suspension of FeAl matrix particles. The Al_2O_3 fibers were cut and layed up onto a tape for dipping. A laboratory scale-up FED processing system using continuous fibers is currently under development.

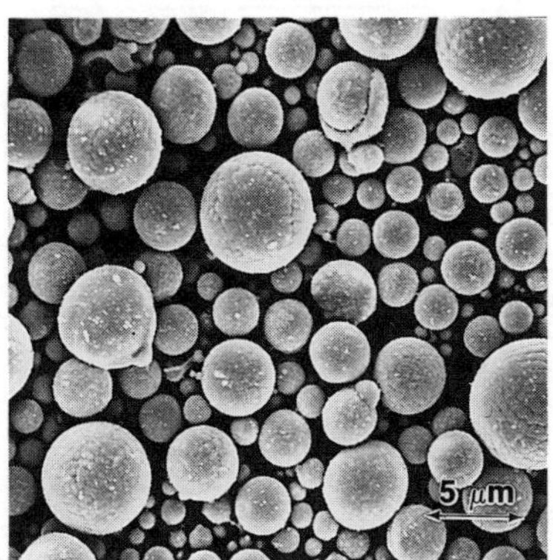

Figure 1. SEM micrograph of sedimented and sieved prealloyed FeAl powder.

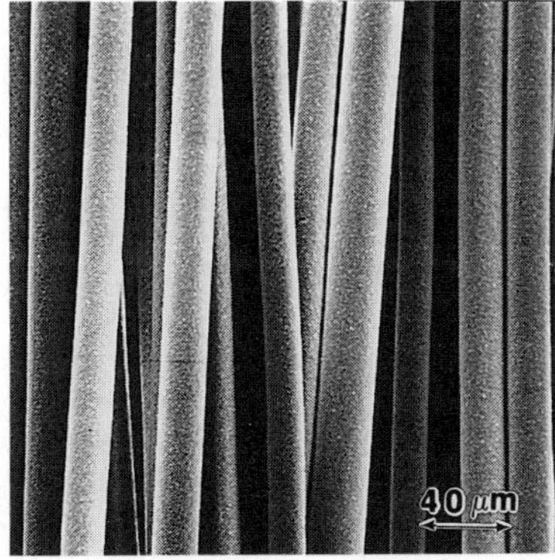

Figure 2. SEM micrograph showing the Al_2O_3 multistrand fiber tow.

While the FED process was developed for continuous fiber reinforcement, the experimental measurements necessary for the green processing of the composite were made using zeta potential data obtained from the ground Al_2O_3 fiber. The zeta potential was calculated from acoustophoretic measurements made on a Matec 8000 Electrokinetic Sonic Amplitude apparatus. Suspensions were made in 0.001 M KNO_3 where the pH was adjusted using either HNO_3 or KOH. The zeta potential of Al_2O_3 was measured using fibers which had been ground. The grinding was carried out in a ball mill using methanol and Al_2O_3 grinding media down to an average diameter of 1.06 μm.

RESULTS AND DISCUSSION

The ESA zeta potential data versus pH for both the FeAl and Al_2O_3 are shown in Figure 3. The zeta potential for the FeAl shows a relatively flat response of potential to changes in pH with small values of zeta potential. Additionally, no change in the zeta potential behavior was observed with repeated titrations. The scatter observed for the Al_2O_3 powder zeta potential curve was due to the elongated shape (aspect ratio of approximately 4) of the ground fibers used for ESA testing. From these preliminary results, the optimum fiber coating was believed to be in the pH range from 4.5 to 7.

The zeta potential results were input into the stability prediction program with the stability curves shown in Figure 4. From this plot several points may be inferred. Stability, meaning a stable dispersion of unflocculated particles, is approximated by logW values of >10. This value of W corresponds to an approximate interparticle potential of 20kT, which has often been used to indicate suspension stability [19]. Using this criterion, the system may be evaluated as follows. The Al_2O_3 fibers would be stable (unflocculated) with respect to each other for pHs less than approximately 5. The FeAl powder should coat (heterocoagulation) the Al_2O_3 fibers at all pH values in the range of 4 to 11. Finally, the FeAl powder is predicted to be unstable with respect to itself (homocoagulate) so that it would form FeAl flocs. However, this instability was not experimentally observed at any pH. Combining both the predictions from the ESA zeta potential data (Figure 3) with the values from the stability plots (Figure 4) leads to the optimized coating conditions. The optimum experimentally determined FED conditions involved dipping the fibers into a 17wt% FeAl/deionized water suspension at pH 5.3. A SEM photomicrograph illustrating the powder coating the fibers is shown in Figure 5. From these micrographs it can be seen that the Al_2O_3 fiber has been coated by the spheres of FeAl powder. Individual fibers were not only separated but relatively uniformly coated by the matrix particles. Additionally, the matrix FeAl particles did not appear to flocculate (homocoagulate) over the pH range tested and the

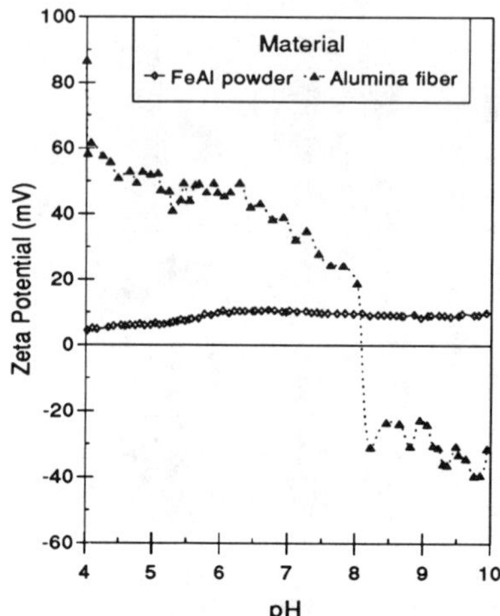

Figure 3. Zeta Potential vs. pH for FeAl and Al_2O_3 from ESA measurements at electrolyte concentration of 0.001M KNO_3.

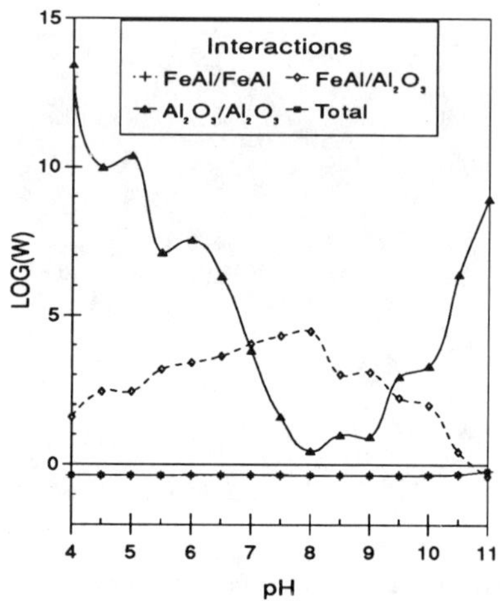

Figure 4. Predicted stability ratio vs. pH for FeAl/Al_2O_3 system where: Ce=0.001 M, r(FeAl)=1.7 μm, r(Al_2O_3)=15 μm.

fibers appear to be covered by roughly a monolayer of particles. Other suspensions were prepared in 0.5 pH increments from pH 4.5 to 7 but were not as successful at matrix/particle heterocoagulation.

Combining both the experimentally collected zeta potential results with the computer predictions for stability allows a processing scheme to be realized in which continuous fibers may be separated from each other and surrounded by matrix particles. In this manner, relatively uniform green composites are possible. Consolidation studies along with upgrading the processing to include continuous fibers are underway.

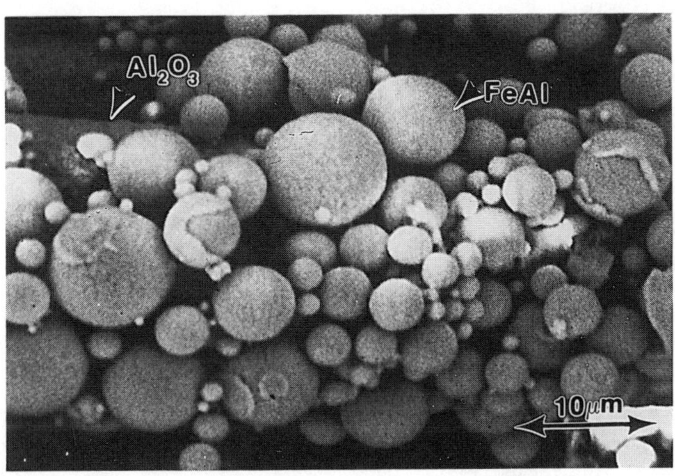

Figure 5. SEM micrograph showing the coating of the multistrand Al_2O_3 fiber with prealloyed FeAl intermetallic powder.

CONCLUSIONS

Optimum experimental conditions for FED coating of FeAl onto Al_2O_3 occurred between pH 5 and 6. This resulted in: [1] Al_2O_3 fibers which remained separated within the fiber bundle, [2] no homocoagulation between the FeAl particles thereby eliminating groups of matrix particles which would not remain adhered to the Al_2O_3 fiber and [3] uniform coating of the Al_2O_3 fiber by the matrix. These conditions were verified by the FED computer program which predicted stability of the Al_2O_3 fibers with respect to themselves and instability (heterocoagulation) of the FeAl/Al_2O_3 using zeta potential data.

ACKNOWLEDGEMENTS

The authors wish to acknowledge the support of the State of Michigan Research Excellence Fund. The work of J. Ambriz and A. Martin has also contributed to the results presented in this paper. The authors also wish to thank Dr. M. Nathal at NASA-Lewis Research Center, Cleveland, OH and Professor K. Vedula of Iowa State University, Ames, IA for supplying the FeAl powder, and Dr. Chambers and Dr. Weddell at the E. I. Dupont deNemours Co., Wilmington, DE for supplying the Al_2O_3 fiber.

REFERENCES

1) A.Bose, B. Moore, R.M. German and N.S. Stoloff, JOM, **9**, 14 (1988).
2) R.M. German and A. Bose, Mat. Sci. and Eng., **A107**, 107 (1989).
3) G.L. Povirk, J.A. Horton, C.G. McKamey, T.N. Tiegs and S.R. Nutt, J. Mat. Sci., **23**, 3945 (1988).
4) D.E. Alman and N.S. Stoloff, Int. J. of Powder Met., **27**, 29 (1991).
5) G.E. Fuchs and W.H. Kao, Inter. P/M Conf., Orlando, FL, June (1988).
6) K.S. Kumar, M.S. DiPietro, S.A. Brown and J.D. Whittenberger, NASA Technical Mem. 103724.
7) P.K. Brindley, Proc. Mat. Res. Soc., **81**, N.S. Stoloff, C.C. Koch, C.T. Liu and O. Izumi eds., 419 (1987).
8) G.K. Watson and J.W. Pickens, Proc. of 2nd. Ann. HiTemp Rev., Westlake, OH, NASA CP 10039, 50-1 (1989).
9) R.A. MacKay, P.K. Brindley and F.H. Froes, JOM, **43**, 23 (1991).
10) M.D. Thouless, O. Sbaizera, L.S. Sigl and A.G. Evans, J. Am. Ceram. Soc., **72**,4, 525 (1989).
11) A.G. Evans, Mat. Sci. and Eng., **A105/106**, 65 (1988).
12) S.Nourbakhsh, S.L. Liang and H. Margolin, Ad. Manuf. Proc.,**3**, 37 (1988).
13) S.Nourbakhsh, H. Margolin, O. Sahin and W.H. Rhee., Mat. Sci. and Eng. **A152**, 619 (1991).
14) B.V. Derjaguin and L.D. Landau, Acta Physichim, URSS, **14**, 633 (1941).
15) E.J.W. Verwey and J.Th.G. Overbeek, *Theory of the Stability of Lyophobic Colloids*, (Elsevier, Amsterdam, 1948).
16) T. Hogg, T.W. Healy, and D.W. Fuerstenau, Trans. Faraday Soc., **62**, 1638 (1966).
17) B.A. Wilson, M. S. Thesis, Michigan State University, 1992.
18) B.A. Wilson and M.J. Crimp, submitted for publication in Langmuir.
19) R.J. Hunter, *Foundations of Colloid Science* (Clarendon Press, Oxford, 1987), p. 444.

MICROSTRUCTURE AND TEXTURES IN ROLLED AND RECRYSTALLIZED Ni_3Al + B

JOACHIM BALL AND G. GOTTSTEIN
Institut für Metallkunde und Metallphysik, RWTH Aachen
Kopernikusstr. 14, 5100 Aachen, FRG

ABSTRACT

The deformation and recrystallization structure as well as the evolution of crystallographic textures of heavily deformed, partially and fully recrystallized Ni_3Al, ductilized with minor additions of boron, were investigated. The macroscopic orientation distribution was determined by X-ray pole-figure measurements and subsequent computation of the orientation distribution function (ODF). Convergent electron beam diffraction (CEBD) was employed to obtain information on the orientation topography and the orientation relationship of volume fractions adjacent to deformation inhomogeneities. Single grain orientation measurements by electron back scattering patterns (EBSP) were carried out to study the orientation distribution of new grains in the recrystallized microstructure. From this data the microtexture and the misorientation distribution function were calculated and the grain boundary character distribution was analysed.

The results are discussed with regard to the mechanisms which govern the evolution of the conspicuously inhomogeneous deformation microstructure and deformation textures. A model for the interpretation of the observed microstructure and texture evolution is proposed.

INTRODUCTION

The current study is concerned with the microstructural evolution and texture development of ductilized Ni_3Al during cold rolling at low homologous temperatures and during recrystallization. The investigation is aimed at a deeper understanding of the mechanisms which govern the evolution of the rather complex deformation microstructure and resulting weak deformation texture as well as the occurrence of recrystallization in ordered intermetallics. For this purpose systematic studies have been carried out employing optical microscopy, TEM investigations (CBED), texture determination by X-ray pole-figure measurements with ODF computation and individual grain orientation measurements by EBSP in a SEM.

EXPERIMENTAL PROCEDURE

A nickel aluminide with the composition $Ni_{76}Al_{24}$+0.24at% B (IC15 according to Oak Ridge National Laboratory terminology) was investigated. A detailed description of the specimen preparation and thermomechanical heat treatment is given elsewhere [1]. Macrotexture measurements for various rolling reductions at room temperature as well as for progressive stages of recrystallization were performed. Metallographic specimens were prepared from deformed, partially and fully recrystallized material. All optical and electron microscopy observations were made on the transverse plane of the rolling specimen. The local orientation distribution in deformed and partially recrystallized Ni_3Al was determined in the TEM by means of CBED. For microtexture measurements of recrystallized specimens EBSP in a SEM was employed.

RESULTS

During cold rolling the microstructural evolution of Ni_3Al is characterized by extensive formation of deformation inhomogeneities. Already at low strains ($\varepsilon \leq 30\%$) very thin microbands were formed. At intermediate strain a second, differently oriented set of microbands was generated and with increasing strain the two sets linked up to form a 3D-network of microbands (Fig.1a). The width of these microbands is approximately 0.05μm and marked shear offsets can be observed along their lines of intersection. With increasing rolling reduction shear bands develop (Fig.1b). On a microscopic level two different types of shear bands could be distinguished from their morphology as well as from their crystallographic orientation: brass-type and copper-type shear bands. The brass-type shear bands comprised most of the microstructure. They consisted of very small equiaxed crystallites, similar to shear bands found in brass [2]. Copper-type shear bands provided only a minor volume fraction. Their structure consisted of elongated and very thin subgrains comparable with shear bands found in copper [2]. The width of both types of shear bands varied between 0.4 to 0.8μm. TEM investigation revealed a random dislocation distribution at any strain. Cell formation was not observed even at high strains.
The main evolutionary stages of the complex deformation microstructure are schematically illustrated in Figs.2a-c.

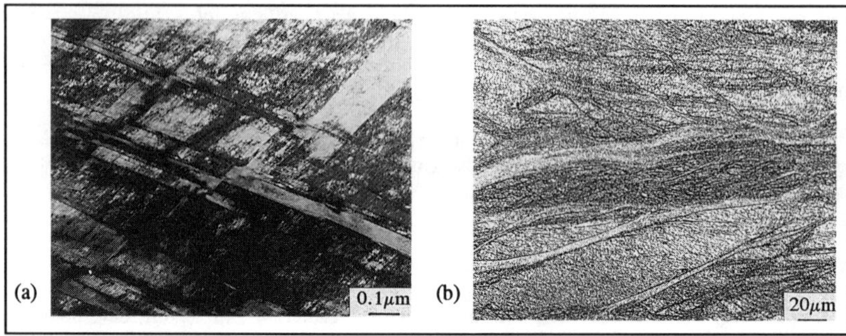

Fig.1 TEM bright field image of the longitudinal section of (a) 40% cold rolled Ni_3Al and (b) optical micrograph of 90% cold rolled coarse grained material.

A local orientation analysis within brass-type shear bands yielded orientations in the scatter of {112}<110>, {314}<485> and {638}<746>. The orientations in the matrix and in regions of dense microband clusters could be associated with the components {100}<011>, {111}<112>, {110}<100> and {112}<111>. The latter two, the Goss- and Copper orientations respectively, were also found as major macroscopic texture components.

The microstructure at incipient stages of recrystallization is schematically shown in Fig.2d. TEM investigations of partially recrystallized Ni_3Al, annealed for 30min. at 723K after 70% rolling reductions, revealed that nucleation started preferentially at shear bands, but also at microband intersections and microband/matrix interfaces. The highest nucleation rate was found within shear bands, where the recrystallized grains attained a grain size of 0.2μm. In matrix regions grains grew mainly parallel to microband interfaces with an average grain size of approximately 0.5μm, i.e. larger than in shear bands. Annealing twins were not uncommon in the recrystallized microstructure. Single grain orientation measurements on partially recrystallized material did neither reveal preferred crystallographic orientations nor orientations close to the macroscopic texture components.

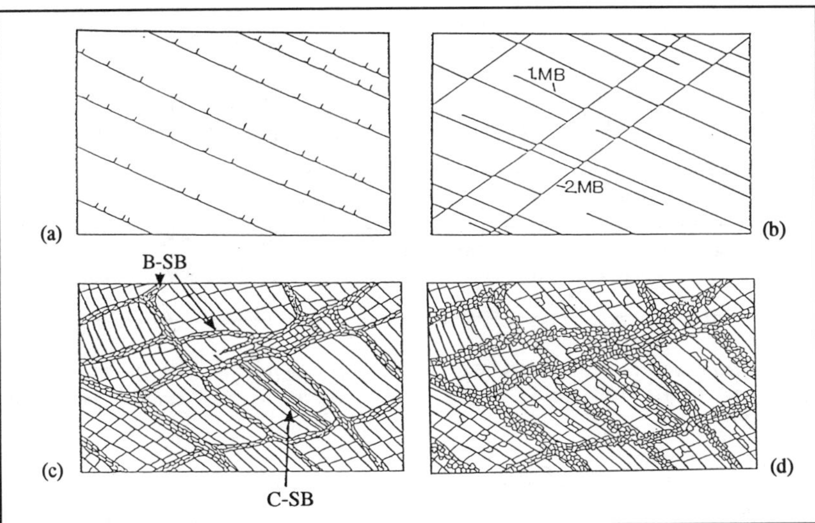

Fig.2 Schematic representation of the microstructure development in Ni_3Al during cold rolling and incipient stages of recrystallization
(a) Slip bands and non-cellular dislocation arrangement
(b) Ultrathin microbands of 1. and 2. generation with "shear-offsets" along their lines of intersections
(c) Microband clusters already formed at intermediate strain and brass-type and copper-type shear bands after large strains
(d) Locally different driving forces for nucleation cause an inhomogeneous recrystallization microstructure

ODFs of 92% rolled, partially and fully recrystallized material are given in Fig.3. The deformation texture of Ni_3Al is a weak copper-type texture consisting mainly of the Brass- (B:{011}<211>), Copper- (C:{112}<111>) and S-orientation ({123}<486>). Especially noteworthy is the characteristic anisotropic spread of the Brass component toward the Goss- and the S-orientation. The appearance of new orientations during incipient stages of recrystallization could already be recognized after annealing for 3min. at 743K. While the C- and S-orientations of the deformation texture quickly vanished, the remaining B-orientation simply showed an increased scatter about ND (normal direction) with progressing annealing time. However, a completely recrystallized state could only be established after annealing for 30min. at 1023K. The recrystallization texture was found to be very weak with a maximum intensity of about two times random. Quantitative ODF analysis revealed 90% of the volume to be randomly oriented, but the non-random components were found reproducibly. Three texture components could invariably be recognized after primary recrystallization with the ideal orientations of approximately {013}<100>, {102}<201>, both located in Euler space in the $\varphi_2=0°$ section, and {112}<294>, in the $\varphi_2=40°$ section.

The fully recrystallized microstructure was composed of alternating bands of fine and coarse grained regions depending on the underlying former deformation microstructure. On a fully recrystallized specimen (deformed 85% and annealed for 1h at 973K) individual grain orientation measurements were carried out. By associating each orientation with a Gauss-type scatter of 5° an ODF could be calculated (Fig.4). The microtexture calculated from the individual grain orientations of fine grained

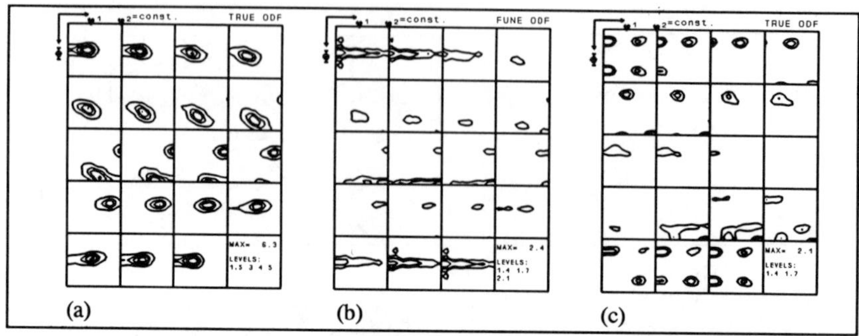

Fig.3 *Crystallographic texture (ODF) of Ni_3Al: (a) rolling texture (92% cold rolled), (b) texture of partially recrystallized material (3 min. at 743K) and (c) recrystallization texture (1h at 1023K).*

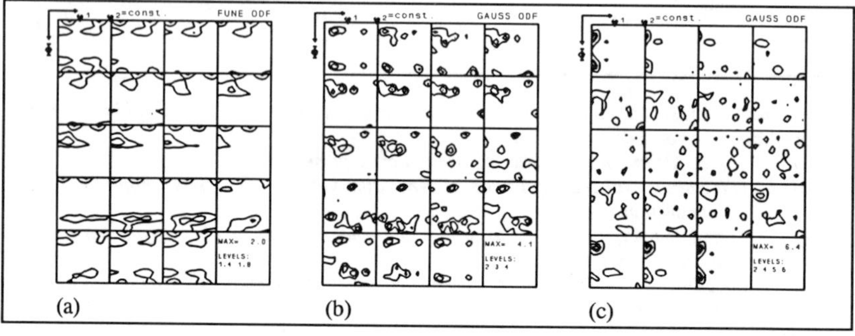

Fig.4 *Crystallographic texture in Ni_3Al: (a) macroscopic recrystallization texture (85% cold rolled and annealed for 1h at 973K), microtextures of (b) fine and (c) coarse grained recrystallized regions.*

regions revealed orientation maxima for about the same orientations as found in the macroscopic recrystallization texture (Fig.4b). The microtexture calculated from orientations inside coarse grained regions (Fig.4c) was found to be different from the macroscopic recrystallization texture (Fig.4a), such that the two macroscopic recrystallization texture components ({112}<294> and {102}<210>) were completely absent, and the third component was found rotated about RD.

For the same sample the misorientation distribution function (MODF) was calculated from the orientation relationship between next neighbour grains by associating each orientation with a Gauss type scatter of 2°. The MODF shows the highest maxima for small angle boundaries ($\Sigma 1$) and first order twin boundaries ($\Sigma 3$) (Fig.5).

The frequency of low Σ boundaries with progressing annealing time was determined on 85% cold rolled specimens, subsequently annealed for different annealing times at 973K. The Brandon criterion [3] was used for the maximum allowable deviation from the exact CSL orientation relationship. The low Σ grain boundaries were grouped into three categories, namely $\Sigma 1$, $\Sigma 3^n$ (first, second and third twins) and $5 \leq \Sigma \leq 25$ (except $\Sigma 9$) grain boundaries. The shaded bars in Fig.6 correspond to the statistical frequency of occurrence in the given orientation distribution. With increasing annealing time the frequency of small angle boundaries eventually decreases, $\Sigma 3^n$ boundaries remain virtually unaffected and other low Σ boundaries occur at statistical frequency.

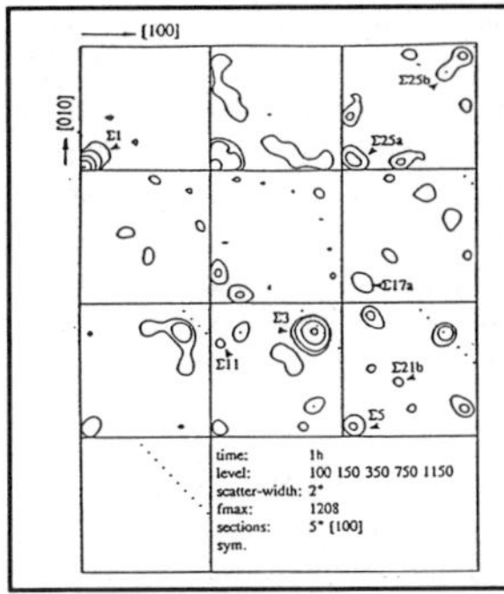

Fig.5 MODF of single grain orientations represented in sections through Rodrigues space.

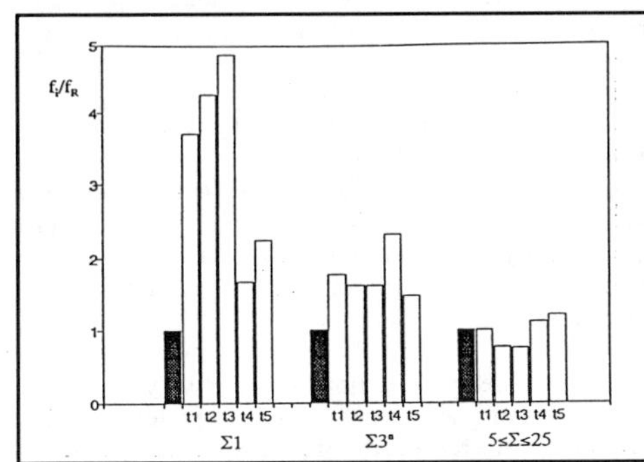

Fig.6 Frequency f_i normalized with the frequency of statistical occurrence f_R of special Σ boundaries (shaded bars) in recrystallized Ni_3Al, cold rolled 85% and subsequently annealed for t_1=1h, t_2=11h, t_3=35h, t_4=107h and t_5=175h at 973K.

DISCUSSION

The most remarkable difference between the deformation structure of Ni_3Al and pure metals or single phase alloys is the dislocation arrangement. Whereas even in metals with low stacking fault energy the dislocation mobility for non-planar slip is high enough to form a cellular structure, there is no indication of cell formation in Ni_3Al even at high strains. Instead, forced planar slip leads to the formation of ultrathin microbands, ten times narrower than in pure copper, which demonstrates the strong slip localisation in Ni_3Al even at moderate strains. From detailed X-ray diffraction studies [1] we conclude that the passage of many dislocations on a slip plane greatly enhances the cross slip mobility of dislocations owing to a local reduction of long range order. Avalanches of local cross slip supposedly lead to the observed extremely thin microbands [4]. Disordering of heavily active slip planes was also recently reported by other authors [5].

With increasing strain the dense spacial arrangement of microbands severely confines slip to the progressively smaller space between the microbands and promotes macroscopic flow instabilities by shear band formation. The copper-type shear bands, which occupy only a small volume fraction in rolled material, were found to comprise clusters of microbands and carry only little misorientation to the surrounding matrix. In contrast, Brass-type shear bands carry extremely large strains and according to the microcrystalline morphology with drastically different texture and the noncrystallographic trajectory of ± 30 to 35° to the rolling direction, they are caused by local mechanical instabilities. The inhomogeneous microstructure at large strains also affects the texture evolution such that a very large random texture component (60% background) and an extensive spread of the discrete texture components develop. The obvious absence of deformation twinning in Ni_3Al proves that brass-type shear bands also can occur in an untwinned matrix and thus, apparently involve all texture components. Therefore, shear band formation contributes to the formation of random orientations and with increasing degree of rolling the brass orientation is formed, which was found to notably gain intensity at large rolling reductions.

The almost random recrystallization texture of Ni_3Al is attributed to the high nucleation rate in deformation inhomogeneities in combination with a low grain boundary mobility, which renders growth selection during recrystallization less important. But, the recrystallization texture components, although weak, were found reproducibly. Model ODF-calculations [1] reveal that the two recrystallization components ({013}<100> and {012}<021>) are related to the brass orientation by a 27°<332> rotation. The third recrystallization component has a second order twin orientation relationship, 38.9°<110>, to the brass orientation. These results show that the brass rolling texture component, which most likely represents the main orientation within the shear bands, and annealing twin formation play an important role during recrystallization of Ni_3Al. First order annealing twinning during incipient recrystallization may even cause the remarkably enhanced spread of the brass orientation at short annealing times, since one primary twin of the brass orientation is also a complementary brass orientation.

Although the recrystallization texture is very weak and almost random, the MODF proves that the arrangement of grains is not random. A preference of certain low Σ boundaries, namely $\Sigma1$ and $\Sigma3^n$, was found in the recrystallized microstructure. Owing to the complex deformation microstructure and to the large random fraction of the recrystallization texture, it was not possible to correlate the occurrence of special boundaries with the preferred orientations in the macrotexture. Low angle boundaries ($\Sigma1$) could either be due to a strong local recovery or to similar nucleation events in closely spaced similar environments. Although noticeable recovery of the microstructure was not observed, we cannot entirely rule out that local recovery may have produced subgrains during annealing. The origin of $\Sigma3^n$ boundaries is evidently due to multiple twinning as also reported for metals and alloys with low stacking fault energy [6].

CONCLUSIONS

1. The microstructure of Ni_3Al after large strains is characterized by the development of a conspicuously inhomogeneous structure in terms of microband clusters as well as shear bands. It is proposed that forced planar slip promotes local disordering in heavily activated slip bands. This leads to extensive microband formation which eventually promotes shear band formation. No deformation twinning and dislocation cell formation was observed.
2. The rolling texture is a weak copper-type texture with strongly scattering components. The brass orientation becomes the strongest component after large rolling reductions.
3. Recrystallization proceeds with different kinetics locally. The highest nucleation rate occurs at shear bands. The high nucleation rate in combination with a low grain boundary mobility does not only produce a very small recrystallized grain size, but also leads to an almost random recrystallization texture.
4. The recrystallization texture components can be related to the brass orientation by special orientation relationships and 2nd order annealing twin formation. The preference of low Σ boundaries ($\Sigma1$ and $\Sigma3^n$) is attributed to preferred nucleation sites and to annealing twin formation.

LITERATURE

[1] J. Ball, Doctoral Dissertation, RWTH Aachen (1992)
[2] J. Hirsch, Doctoral Dissertation, RWTH Aachen (1984)
[3] D.G. Brandon, Acta Metall. **14**, 1479 (1966)
[4] P.J. Jackson, Scripta Metall. **17**, 199 (1983)
[5] J.A. Horton, I. Baker and M.H. Yoo, Phil.Mag.A **63**, 319 (1985)
[6] G. Gottstein, Acta Met. **32**, 1117 (1984)

"RESULTS ON POWDER INJECTION MOLDING OF Ni_3Al AND APPLICATION TO OTHER INTERMETALLIC COMPOSITIONS"

RENE M. COOPER,
Xform Incorporated, 100 North Mohawk Street, P.O. Box A9, Cohoes, N.Y. 12047

ABSTRACT

Net shape forming processes are under development to allow affordable production of intermetallic components. Powder injection molding (PIM) may be employed for the production of complex-shaped intermetallic geometries. Proper choice of powder parameters and processing conditions can lead to the formation of fully dense structures through pressure-less sintering. In this study, Ni_3Al with 0.04 wt.-% boron has been successfully injection molded and sintered to full density. A yield strength of 340 MPa, ultimate tensile strength (UTS) of 591 MPa, and 8 % elongation were attained for injection molded and sintered tensile bars. Powder characteristics and sintering behavior are given for the nickel aluminide employed in this study to highlight the powder attributes needed for injection molding. Molding parameters, debinding and sintering schedules, along with mechanical properties are presented to indicate the viability of PIM for intermetallics. This approach based on the understanding of key powder characteristics and use of the reactive synthesis powder process may be extended to the successful injection molding of other intermetallic systems.

INTRODUCTION

Recent work by German and Hens[1] highlight the powder attributes that are of concern for achieving defect-free, high density injection molded components: These attributes include particle size, particle shape, interparticle friction and packing density. Typical attributes for several powders including stainless steels, iron, copper, nickel, and nickel aluminide are presented and the attributes for an ideal powder for injection molding are presented in their analysis. This paper will highlight the injection molding approach based on focusing on powder characteristics and will report on resulting properties for Ni_3Al with 0.04 wt.-% boron. The powder attributes for nickel aluminide will be compared against those attributes of an ideal injection moldable powder as described by German and Hens.

Intermetallics are considered highly desirable and receive considerable attention and interest because of favorable properties such as increasing strength with increasing temperature, low density, excellent oxidation and corrosion resistance, and high stiffness which make them candidates for applications such as corrosion-resistant coatings, metal-bonds for diamonds and carbides, chemical filters, and automotive engine components. Some of the drawbacks to these materials include low ductility, relatively high costs for raw materials, and expensive; and as of yet, not fully developed processing routes. Alloying of the intermetallics has brought significant improvement to the ductility of these alloys, and processing advances have achieved enhancement in properties while reducing fabrication costs.

This paper explores the various methods that are available for the powder consolidation of nickel aluminides and then outlines a processing route for injection molding. The study described reports on the successful injection molding of nickel aluminide (Ni_3Al) and pressureless-sintering to full density. The development of an injection-moldable grade of powder is discussed and attributes such as particle size distribution, particle morphology, packing characteristics, and sintering behavior are provided. Processing steps for injection molding are shown, followed by the debinding and sintering schedules employed, and resulting mechanical properties. A comparison is then made showing the general agreement between the powder characteristics of the powder utilized in this study and the model by German and Hens.

Powder Consolidation Techniques for Nickel Aluminides

Intermetallics may be shaped and consolidated by traditional powder metallurgical processes such as hot pressing, hot isostatic pressing (HIP) and powder extrusion. The brittle nature of the intermetallics along with their ordered structure normally requires the application of stress and temperature to achieve full density. Emergence of new applications is pushing development of alternative consolidation approaches. One application for nickel aluminide powder is as a metal-bond for diamonds and carbides in cutting tools. Figure 1 shows a nickel aluminide matrix/diamond abrasive composite that was consolidated by hot isostatic pressing at 1150° C and 35 MPa for 15 minutes. The powder employed for this use is produced by Xform, Inc. and is the same grade of powder employed for injection molding[2].

Figure 1
Ni_3Al/Diamond Cutting Tool Composites

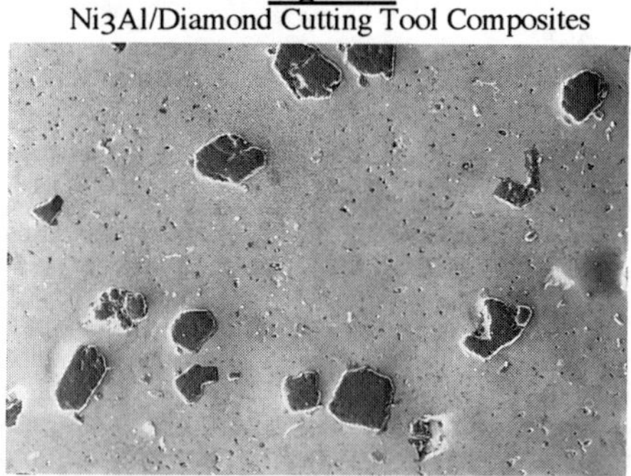

Ni_3Al INJECTION MOLDING STUDY

Powder Characteristics

PIM offers the advantage of a net-shape forming process which is especially important because of the difficulty in machining many of the intermetallic compositions. Because of this flexibility, PIM has been applied for fabricating intermetallic composites.[3] For intermetallic components to be processed by injection molding, it is critical to use powders possessing the needed particle characteristics of proper size, shape, chemistry and packing density. Small, high interparticle friction powders are needed for injection molding. The required powder size for injection molding is below 20 micrometers.

Advantages to spherical powders include high packing densities which minimizes the amount of binder needed for molding and reduces the amount of shrinkage during sintering. Ligamental powders give a lower packing density but exhibit better particle interlocking which translates into better shape retention during debinding.[1]

A few processes are in place for producing intermetallic powders. Atomized and PREP intermetallic powders are typically coarse which hinders shape retention during debinding. Additionally, coarser particles are also more difficult to sinter to full density. Atomization of some intermetallics is difficult because of the propensity for segregation and because of the high melting temperatures for some of the intermetallics. The *Xform Process* involving reactive synthesis may be employed for producing powders with a mean particle size under 20 micrometers and with a non-spherical structure which benefits shape retention. And because the process does not rely on melting, high melting point compounds such as NiAl and $MoSi_2$ may be produced.

Emphasis in this study has focused on achieving full density of powder injection molded components by pressure-less sintering. This necessitates that powder attributes such as particle size and shape be chosen to enhance particle packing and sintering.

<u>Experimental Work</u>

As previously mentioned, the Xform Process for producing an injection moldable grade of Ni_3Al by reactive sintering was employed in this study. The following approach was taken:

- Production of reactive sintered powder
- Characterization of powders
- Mixing of powder with binder
- Injection mold
- Debinding and sintering
- Assessment of properties

Ni_3Al with 0.04 wt.-% boron intermetallic powder is produced from mixed elemental powders utilizing the reactive sintering process. Powder was processed to a mean value of 16 μm. The particle size distribution employed in this study is included in Figure 2.

Figure 2 Particle Size Distribution for Powder Employed for PIM

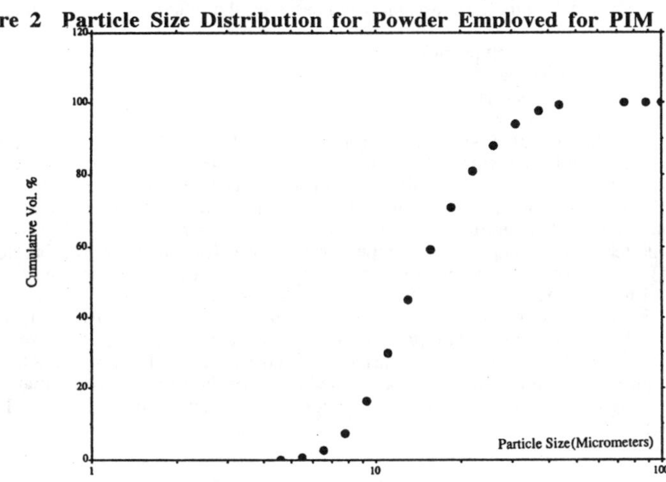

This powder has an apparent density of 43% and a tap density of 52%. Figure 3 is an SEM of the Ni_3Al powder and shows the morphology of the powder.

Figure 3
SEM of Ni_3Al Powder

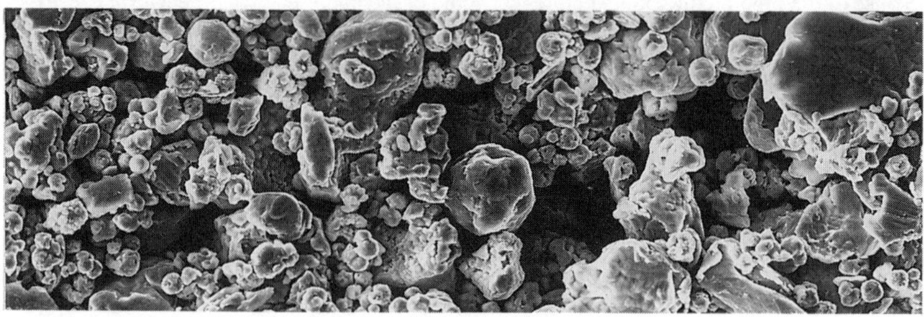

Prior to injection molding, the sinterability of the powder should be determined. Dilatometry on the Ni_3Al powder indicated maximum sinterability in the range of 1320° to 1350° C. Based on these results, loosely tapped samples were sintered in vacuum at 1340° C for one hour. The resulting density was 97% (7.2 g/cc) theoretical density and the Vickers Hardness with a 100gf load averaged 286 (28 Rockwell C). Once these characteristics were determined, injection molding was performed.

Binder for injection molding was prepared based on the following composition:

35% polypropylene
60% paraffin
5% stearic acid

The loading selected was 56% volume fraction powder. Powder and binder were mixed using a Haake Torque Rheometer. The mixture was granulated and fed into a Battenfeld injection molding machine. Both tensile bars and wrenches were fabricated. Wrenches were formed to demonstrate the high complexity nickel aluminide parts that can be formed by PIM. The injection temperature was 120° C and an injection pressure of 140 MPa was used. Solvent debinding in heptane at a temperature of 38° C for four hours was employed for removing the majority of the binder. Following solvent debinding, the components were sintered in hydrogen. Three different sintering temperatures were chosen, 1320, 1335 and 1350° C. Densities of 99% of theoretical were obtained for sintering at 1350° C for one hour. The maximum density obtained for sintering at 1320° and 1335° C was 92% of theoretical. The final sintered parts consisting of tensile bars and 9 mm wrenches are shown in Figure 4. Tensile testing was performed only for the samples sintered at 1335 and 1350° C. Figure 5 shows actual stress-strain curves for a sample sintered at at 1335° C and one sintered at 1350° C.

Figure 4
Injection Molded and
Sintered Wrenches and Tensile Bars

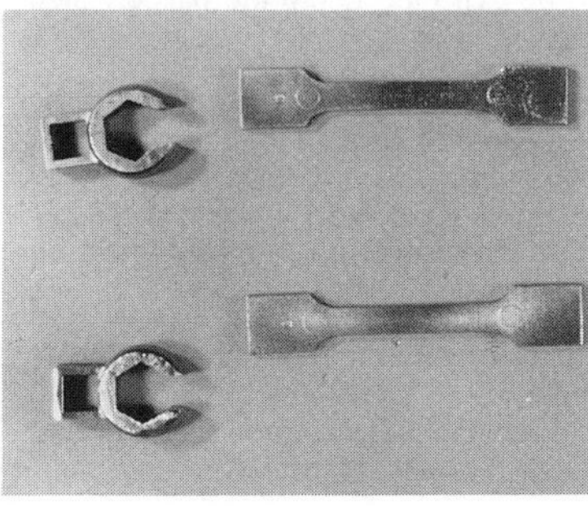

Figure 5
Tensile Test Results

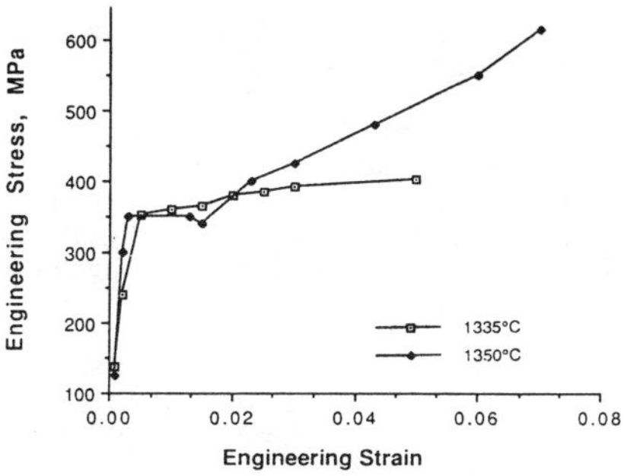

Table I.	
Mechanical Properties	
Sintering temperature 1335° C	Sintering temperature 1350° C
Density = 99%	Density = 92%
Avg. YS = 298 MPa	Avg. YS = 340 MPa
Avg. UTS = 343 MPa	Avg. UTS = 591 MPa
% Elongation = 5	% Elongation = 8

The measured properties compare favorably with those from nickel aluminide produced by traditional melting and casting techniques followed by further processing for property refinement. Yield strength reported for IC50 (wt. %- 86.2 Ni, 11.3 Al, 1.88 Hf, 0.6 Zr, 0.02 B) is 333 MPa and a UTS of 625 MPa.[4]

DISCUSSION

The powder developed for this study was successfully injection molded and sintered to 99% of theoretical density without the application of pressure. Two indicators based on particle size used by German and Hens for assessing powder attributes for injection molding is the D_{50} which is the medium particle size based on weight distribution and S_W which is the width of the particle size distribution. S_W is determined from the particle size at the 10% and 90% cumulative distribution points, denoted as D_{10} and D_{90}.
$S_W = 2.56/\log(D_{90}/D_{10})$

In the analysis performed by German and Hens, an optimal median particle size of 2 to 8 for injection molding was given with values of 8 to 16 being more typical. The median size for the Ni_3Al powder was 14 μm. This particle size allowed good sinterability and attainment of high as-sintered densities. Table II summarizes the Sw and D_{10} and D_{90} values for the Ni_3Al powder for D_{50} = 14 μm.

Table II		
Particle Size Distribution Width for D_{10} and D_{90} with D_{50} = 14 μm		
S_W	D_{10}, μm	D_{90}, μm
4.85	27	8

The width of the particle size distribution is represented by a S_W value of 4.85. Values between 3 and 6 for S_W are common[1]. Best moldability is achievable with very wide particle size distributions, a value of S_W near 2, or with a narrow distribution, S_W greater than 7. For this case, improved moldability would be expected if the the particle size distribution were narrowed.

A slightly non-spherical shape helps injection molding by reducing the viscosity during molding and by helping shape retention during debinding. It should be noted that the molding pressure employed for the Ni_3Al, 140 MPa, is high compared to other materials and may result from the ligamental morphology of the powder which necessitates a higher injection pressure. A less ligamental structure may lower the required injection pressures.

A tap density of 52% and an apparent density of 43% for the Ni_3Al powder compare quite favorably with ideal tap and apparent density values of at least 50% and 40 to 45%, respectively for a model injection moldable powder based on the German and Hens analysis.

These results may be extended to other intermetallic systems. The *Xform Process* has also been employed for the production of TiAl powder. Powder with a mean particle size of 13 μm has been produced. Based on the D_{10} (4 μm), D_{90} (22 μm) and D_{50} (14 μm) values, the powder has a S_W of 3.5. A particle size less than 20 μm and a particle width between 3 and 6, based on previous results, indicate that this powder is a candidate for injection molding.

SUMMARY AND CONCLUSIONS

Good final properties for Ni_3Al intermetallic has been attained by injection molding. The results from this study demonstrate the importance of particle size, shape, and particle size distribution and show good agreement with the model of an ideal injection moldable powder provided by German and Hens. It is believed that this approach employing the reactive synthesis powder fabrication route may be used to successfully injection mold other intermetallics such as TiAl.

ACKNOWLEDGEMENTS

In addition to the author, this work represents the contributions and efforts of Dr. David Alman, Dr. Randall German, Dr. Karl Hens, Mr. Kaz McCoy, Dr. Karl Shaw and Mr. Haorong Zhang.

REFERENCES

1. R. M. German, K. F. Hens, Identification of the Effects of Key Powder characteristics on Powder Injection Molding, In-press, Proceedings of 1992 World Powder Metallurgy Congress
2. K. G. Shaw, K. P. McCoy, J. A. Trogolo, Reduction of Free Nickel During the Reaction Synthesis of Nickel Aluminide Powder, In-press, Proceedings of 1992 World Powder Metallurgy Congress
3. R. M. German, R. G. Iacocca, Powder Metallurgy Processing of Intermetallics, Chapter from Intermetallics, In-press. Editors Drs. Norm Stoloff and Vinod Sikka (Van Nostrand and Co., publishers)
4. C. T. Liu et al.Initial Development of Nickel and Nickel-Iron Aluminides for Structural Uses, ORNL-6067, Martin Marietta Energy Systems, Aug 1984

ANALYSIS OF RESIDUAL PHASES IN NICKEL ALUMINIDE POWDERS PRODUCED BY REACTION SYNTHESIS

K. P. MCCOY* , K. G. SHAW** and J. A. TROGOLO†
*Xform, Inc. Cohoes, NY
**Xform, Inc. Cohoes, NY
†Rensselaer Polytechnic Institute

ABSTRACT

The use of x-ray diffraction has been used to determine phases present after reaction synthesis of Ni_3Al powder. The complex diffraction spectra produced by the powder prompted the development of a simulator. The simulator uses nonlinear regression to determine the weight percent of the phases present. The simulator also determines the broadening of each peak in the spectrum. The phases present in Ni_3Al powder produced by reaction synthesis has been determined with the simulator. The simulator has been used to monitor the progress of phase transformation during various thermal treatments of Ni_3Al powder. A thermal cycle of 1200°C for two hours has been shown to produce a phase-pure product. The activation energy for the interdiffusion of nickel and aluminum has been determined to be 260 ± 35 kJ/mole.

INTRODUCTION

It is the purpose of this paper to describe a method for determining the residual phases in powder produced by reaction synthesis. X-ray diffraction was chosen to determine the phases present after reaction synthesis. Ni_3Al powders produced by reaction synthesis have a multiphase composite microstructure[1]. The phases can have diffraction spectra with very close or overlapping peaks. Some phases present do not have any unique, non overlapping peaks. This makes the diffraction spectra of the powders very difficult to interpret. A simulator was developed to model experimental spectra. With this simulator the phases that are present and the weight percent of each phase can be determined. This method has been applied to monitor the progress of a post heat treatment of Ni_3Al powder. A series of post heat treatments were conducted, the time and temperature dependence of the resulting transformation were determined.

To produce intermetallic powders by reaction synthesis, it is necessary to control the reaction exotherm. Control of the exotherm can be achieved by heat extraction methods. The result of heat extraction is a multiphase, composite product that represents the reaction in a transient condition[2-9]. In many ways the composite powders produced through controlled reaction synthesis are beneficial. If a ductile material is used as a reactant, this component may be preserved in the composite powder structure resulting in a powder that is easily die-pressed. This is advantageous as very often intermetallic powders are too brittle or hard to effectively shape by die compaction. The concentration gradient present in a composite powder also may aid in sintering by providing a diffusive driving force. In thermal spray applications the lack of phase purity may present problems if a resident phase is lower melting or prone to oxidation. Consequently, it is desirable to determine and control the composition, and phase purity of a reactively sintered powder, tailoring it to the application.

THEORY

The first step in developing the simulator was to chose a model. Using Bragg's Law, the wavelength of the x-ray used, and D-spacings of the compounds, the diffraction angles of each peak can be calculated. Each peak can then be fitted with a gaussian distribution, and all the peaks added up to make a simulated spectrum. The model used for this simulator is then a summation of gaussians placed at appropriate angles. The intensity at any angle θ is given by Eq. 1.

$$y(\theta) = \sum_{j=1}^{J} W_j \sum_{k=1}^{K_j} H_k \exp\left[-\left(\frac{\theta - T_k}{B_k}\right)^2 \right] \tag{1}$$

The parameters H_k, T_k, B_k are; peak height, angle in two-theta, and broadening respectfully. K_j is the number of peaks in phase j. W_j represents the weight percent of phase j, where there are J phases present in the powder. The initial height of each peak, H_k, is determined from the intensity values given in the powder diffraction files. The height of each peak belonging to a particular phase is adjusted by a fractional weight, W_j, this weight can be interpreted as the physical weight percent of that phase present in the powder. The weight of each phase and the broadening of each peak can all be independently varied to match the experimental spectrum. The method used for reducing the error between the simulated and the experimental spectrum was the Levenberg-Marquardt Method for nonlinear regression[10]. This method is advantageous in that it varies smoothly between the inverse Hessian method and the steepest descent method. The latter method is used when the initial guess is far from minimum, and switches smoothly to the former as the minimum is approached. This means that this method will work for both reasonable and unreasonable guesses. Even though this is a robust regression technique, some care has to be taken in selecting the initial guesses. An initial guess with too many phases will lead to long regression times, or incorrect solutions. For accurate regression the initial guess should consist of only the most likely phases.

The method defines a χ^2 merit function

$$\chi^2 = \sum_{i=1}^{N} \left[\frac{y_i - y(x_i;\mathbf{a})}{\sigma_i}\right]^2 \tag{2}$$

where $y = y(x_i;\mathbf{a})$ is the model being fitted. The model depends nonlinearly on a set of M parameters $\mathbf{a}$. In this case the set of parameters $\mathbf{a} = \mathbf{H_k} \cup \mathbf{T_k} \cup \mathbf{B_k} \cup \mathbf{W_j}$. The number of parameters $\mathbf{a}$ is consequently three times the number of peaks in the simulation plus the number of phases simulated. Lets say a powder has two phases and each phase has three peaks. This means that the parameter set $\mathbf{a}$ has $3 \times 6 + 2$ or 20 elements. This method has to solve an $N \times N$ matrix, were N is the number parameters being minimized, attempting to minimize all parameters at once can lead to very large matrices. An advantage of the method is the ability to minimize only a subset of the parameters $\mathbf{a}$. This is very helpful in matching diffraction spectra, for example, minimization may be needed on the broadening B_k and not the peak heights H_k Eq. 1. The ability to minimize only a subset of parameters also helps to overcome memory and speed constraints. For spectra with many peaks, memory constraints can be overcome by minimizing only a few parameters at a time. The output of the simulator is the adjusted parameter list $\mathbf{a}$, consisting of: the weight percent of each phase, the height of each peak, the broadening of each peak, the position of each peak, and the background intensity. With this information it may be possible to determine the amount of ordering, and strain in a particular phase. Error bars and a probability of good fit are also available from the simulator.

EXPERIMENTAL

Novamet nickel 123 with a 8μm average particle size was mixed to a stoichiometry of Ni_3Al with Valimet H15, 15μm average. The powders were reacted using heat extraction methods which yielded the highest percentage of fine particles. The product was milled to 60μm average and the resulting powder was examined by x-ray diffraction. The x-ray spectrum was simulated to determine the weight percent of each phase. The x-ray scan was run from 20 to 100 degrees 2θ in steps of 0.05 degrees, Fig. 1. The simulator covered the same range of 2θ. In order to investigate the homogenization of the powder, samples of the original powder were heat treated at various temperatures, and times. The temperature dependence of the homogenization process

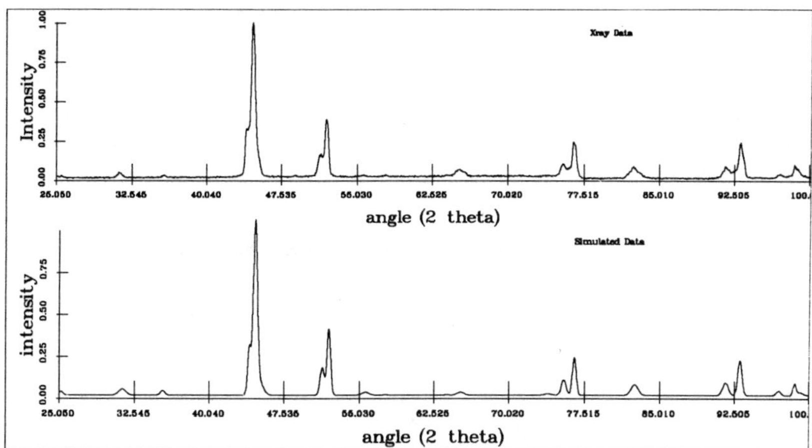

Figure 1: X-ray diffraction scan of powder produced by reaction synthesis with heat extraction methods. Simulator output 0.540 wt.% Ni .190 wt.% Ni_3Al .270 wt.% NiAl

was determined by heat treating samples at temperatures of 900, 950, 1000, 1050, 1100, 1200°C for 2 hours. The time dependence was determined by heat treatments of 0.5, 1, 2, 3, 4, and 5 hours at 1000°C . After the heat treatment the samples were remilled, and x-ray diffraction was performed on each of the samples. The samples were also investigated by electron microscopy to determine the morphology and the phase composition of the fabricated powders. Energy dispersive spectroscopy and back-scattered electron imaging were used to map the compositional variations. The compositional and morphological transformations could be monitored as functions of temperature and time.

ANALYSIS

The diffraction spectrum of the reacted powder, along with the simulated spectrum, can be seen in Fig. 1. There were no aluminum peaks found in the reacted powder, but strong nickel peaks were observed. Other peaks observed in the spectrum correspond to NiAl and Ni_3Al . Nearly all of the Ni_3Al peaks align with the nickel peaks. These peaks can be seen at approximately 44, 52, 76, 93, and 98 degrees. They appear in most cases to be closely spaced doublets, in which the Ni_3Al peaks are the left hand peaks. The Ni_3Al peaks at approximately 35 and 58 degrees are uncharacteristically small due to the fact that the Ni_3Al is still disordered. The rest of the peaks are due to NiAl. The diffraction spectrum of a samples heat treated at 950°C and 1000°C for 2 hours can be seen in Figs. 2 and 3 respectfully. It can be seen that the Ni_3Al peaks grow and that the peaks for nickel and NiAl decrease with increasing temperature. The simulator output shows that the nickel content has decreased to 37 wt% and that the Ni_3Al has increased to 44.5 wt%, for the sample annealed at 1000°C . As can be seen in Table I the amount of residual phases decreases with increasing time and temperature. Complete transformation of the powder to Ni_3Al was achieved with a heat treatment of 1200°C for two hours Fig. 4. This is below the typical sintering temperature for Ni_3Al . Products made by conventional press and sinter techniques will then be phase-pure Ni_3Al following sintering. Plots of the amount of residual nickel vs $time^{1/2}$ show linear dependence as would be expected from a diffusional type mechanism. The activation energy for this mechanism has been determined to be 260 ± 35kJ/mole[1]. This is consistent results of E.A. Brandes [11].

The microstructure of the reacted powder had two distinct forms. The first is one where

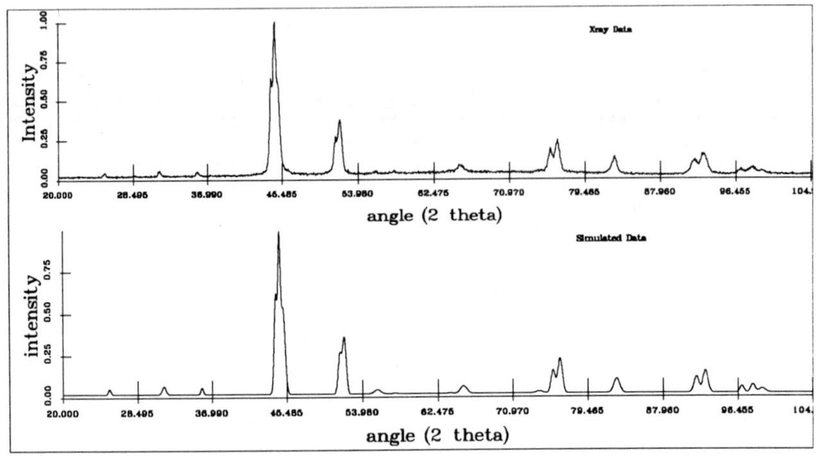

Figure 2: X-ray diffraction of powder after 950°C heat treatment. Simulator output 49.3 wt% Ni 26.0 wt% Ni_3Al 24.7 wt% NiAl.

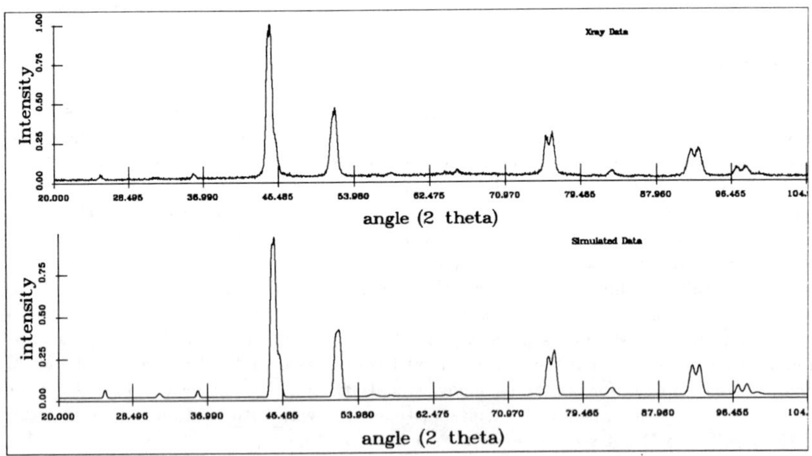

Figure 3: X-ray diffraction of powder after 1000°C heat treatment. Simulator output 37.0 wt% Ni 44.5 wt% Ni_3Al 18.5 wt% NiAl.

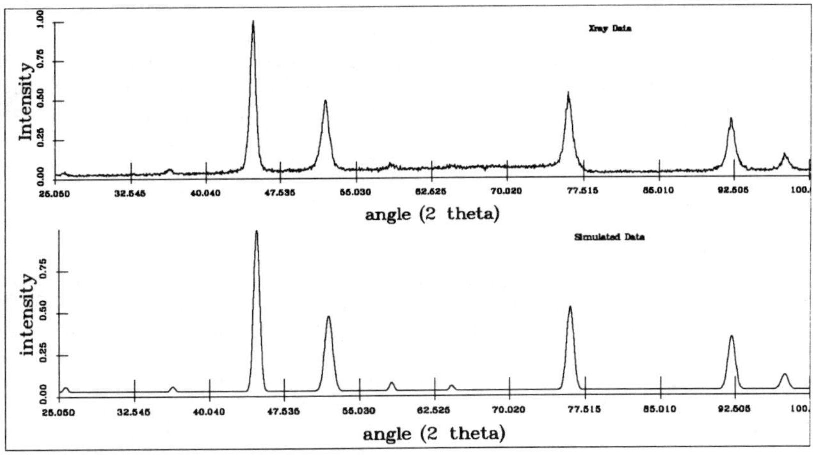

Figure 4: X-ray diffraction of powder after 1200°C heat treatment. Simulator output 0.0 wt% Ni 100 wt% Ni_3Al 0.0 wt% NiAl.

Table I: Simulator output vs time and temperature data

Diffusion Temp. (°C)	Diffusion Time (hr)	Ni	Ni_3Al	NiAl
As reacted	—	0.540	.190	.270
900	2	0.533	0.200	0.267
950	2	0.493	0.260	0.247
975	2	0.427	0.360	0.213
1000	2	0.370	0.445	0.185
1025	2	0.277	0.585	0.138
1050	2	0.200	0.700	0.100
1100	2	0.180	0.730	0.090
1200	2	0.000	1.00	0.000
1000	0.5	0.473	0.290	0.237
1000	1	0.413	0.380	0.207
1000	2	0.370	0.445	0.185
1000	3	0.347	0.480	0.173
1000	4	0.300	0.550	0.150
1000	5	0.280	0.580	0.140

the reaction temperature remained below the melting temperature of nickel forming a coated structure. The second is where the reaction temperature reached the melting point of nickel forming a large ligamental structure. Both of these have composite interior structures, consisting of: nickel, nickel with 10 wt% aluminum dissolved in it, Ni_3Al , NiAl, and trace amounts of Al_3Ni_2 [1].

CONCLUSIONS

The developed simulator has been used to determine the residual phases in Ni_3Al produced by reaction synthesis, and used to monitor the phase transformations occurring during thermal treatments of the product. The simulator has also been applied to reaction products of TiAl and Fe_3Al. The simulator is well suited for use where there are many phases present or were phases have similar diffraction spectra. The results of this investigation of Ni_3Al show that homogenization can take place at temperatures below the typical sintering temperature. Two important conclusions can be made from these results. First, compacts pressed from this powder will be homogenized well before the sintering cycle is finished. Secondly, the powder can be homogenized for application that need a phase pure powder. This shows that reaction synthesis can be used to produce powders for both press and sinter applications as well as thermal spray applications.

REFERENCES

1. K. G. Shaw, K. P. McCoy, and J. Trogolo, Reduction of Free Nickel During the Reaction Synthesis of Nickel Aluminide Powder, in *Proceesings of Powder Metallurgy World Congress*, Metal Powder Industries Federation, 1992.

2. R. M. German, A. Bose, and D. M. Sims, Production of Reactive Sintered Nickel Aluminide Material, U. S. Patent 4,762,5582, 1990.

3. B. H. Rabin, A. Bose, and R. M. German, *Modern Developments in Powder Metallurgy*, volume 20, Metal Powder Industries Federation, Princeton, NJ, 1988.

4. A. Bose, B. H. Rabin, and R. M. German, Powder Met. Inter. **20**, 255 (1988).

5. C. T. Liu, A. I. Taub, N. S. Stoloff, and C. C. Koch, editors, *High-Temperature Ordered Intermetallic Alloys III*, Materials Research Society, Pittsburgh, PA, 1989.

6. Z. A. Munir, *Sintering '87*, volume 1, Elsevier, London, UK, 1988.

7. R. P. Santadnrea, R. G. Behrens, , and M. A. King, Reaction Chemistry and Thermodynamics of the Ni-Al and Fe-Al Systems, in *Mat. Res. Soc. Symp. Proc.*, volume 81, page 467, Materials Research Society, 1987.

8. A. G. Strunina, T. M. Martemyanova, V. V. Garzykin, and V. I. Ermakov, Comb. Explos. Shock Wave **10**, 449 (1974).

9. J. C. Withers, S. Guha, and R. O. Loutfy, The development of plasma synthesis to produce pre-alloyed, ultrafine intermetallic aluminide powders for injection molding, Technical Report MTL TR 91-47, U. S. Army Materials Technology Laboratory, Watertown, MA, 1991.

10. W. H. Press, B. P. flannery, S. A. Teukolsky, and W. T. Vetterling, *Numerical Recipies in C*, Cambridge University Press, 6th edition, 1990.

11. E. A. Brandes, editor, *Smithells Metals Reference Book*, Butterworths, 6th edition, 1983.

SOLIDIFICATION MODELING OF NiAl SINGLE CRYSTAL CASTINGS

K.O. YU*, J.A. OTI* AND W.S. WALSTON**
*PCC Airfoils, Inc., 23555 Euclid Avenue, Cleveland, OH 44117
**GE Aircraft Engines, 1 Neumann Way, MD 85, Cincinnati, OH 45215

ABSTRACT

Solidification conditions of a 0.64 cm x 2.5 cm x 15.2 cm slab and a 1.3 cm ϕ x 15.2 cm bar were simulated for two NiAl alloys. Simulation results were correlated with experimental inspection results to establish a relationship between dendrite arm spacing and local solidification time. The results were also used to construct a defects map for NiAl single crystal castings. Results have been compared to a similar model for single crystal nickel-base superalloys.

INTRODUCTION

NiAl is one of the most promising intermetallic systems for replacing superalloys as a high temperature structural material in high performance jet engines. While considerable work is being done on developing and optimizing the properties of NiAl alloys, relatively less attention is being paid to the production-scale processing of these alloys. For single crystal components, the investment casting process appears to be the only viable approach. However, because of their higher melting temperatures, higher thermal conductivities, higher ductile-to-brittle-transition temperatures, and lower room temperature ductilities than those of nickel-based superalloys, modifications of the investment casting process from that producing superalloy single crystals are required to produce NiAl single crystal components.

Finite-element thermal modeling offers the capability of simulating casting heat transfer conditions and understanding the solidification sequence. As a result, it provides valuable information regarding the investment casting processing of NiAl single crystals. Efforts were undertaken at GE Aircraft Engines and at PCC Airfoils to simulate the solidification conditions of cylindrical bars and slabs, correlate dendrite arm spacing with local solidification time (DAS-LST) and construct a defects map for two selected NiAl alloys. The approach combined finite-element thermal analyses and experimental casting results. The purpose of this paper is to document results from these studies.

EXPERIMENTAL CASTINGS

Two NiAl alloys were cast in two clusters (one for each alloy) of variously shaped bars, slabs and generic airfoils at specified furnace temperatures and withdrawal speeds. Thermocouples were placed at selected locations in the mold to record the metal and shell temperatures as a function of time. After shake-out of the mold, all castings were inspected for grain defects. Several pieces of castings were cut to measure the primary and secondary dendrite arm spacings (PDAS and SDAS).

FINITE-ELEMENT MODELING

The modeling approaches are described in References 1-3. Two casting simulation models (one for 1.3 cm diameter bar and one for 0.64 cm x 2.5 cm x 15.2 cm slab) were built. Figure 1 shows the finite-element model for the slabs. Due to symmetry, it was only necessary to model one slab on the mold to simulate the entire cluster mold of slabs shown in Figure 2. These geometric models were built in PATRAN, and the finite-element solution was performed in ProCAST. These simulations were post processed in ProCAST and plotted in PATRAN. Model calculated results include cooling curves, isotherms, isochrons, temperature gradients (G), solidification rates (R) and combinations of G and R (e.g., G/R).

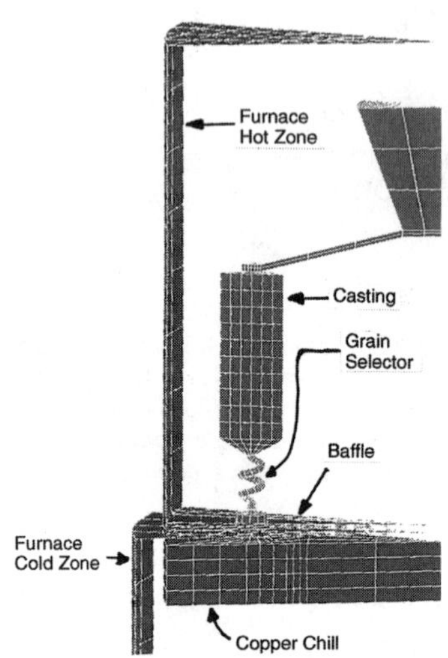

Figure 1. Finite-element model of slab.

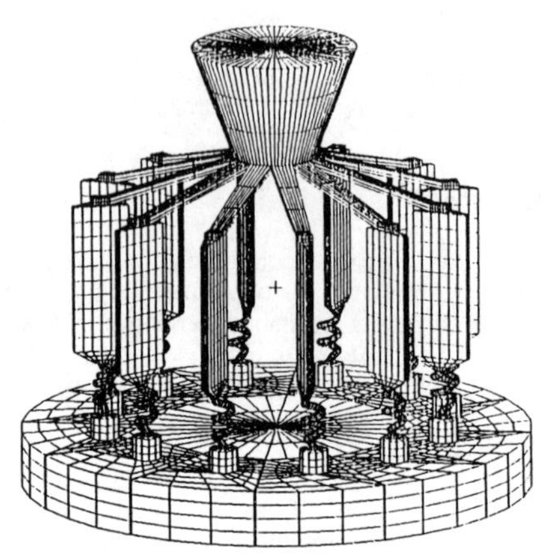

Figure 2. Cluster configuration of the slab model.

RESULTS AND DISCUSSION

Effects of Boundary Condition

The initial temperature conditions of the metal and mold were assigned based on the thermocouple readings of the experimental castings. The boundary conditions of the furnace wall temperature were also applied based on thermocouple data. The calculated isotherms of a slab of NiAl alloy A at various withdrawal times are shown in Figure 3. Based on this figure, the relationship between mushy zone and baffle locations is plotted in Figure 4. Also shown in this figure is the mushy zone-baffle location relationship based on theoretical boundary conditions. From the figure, it can be seen that for the real production boundary condition, during the entire withdrawal process, the mushy zone is above the baffle and stays inside the hot zone. Initially, the mushy zone size gradually increases with the increase of withdrawal distance. When withdrawal distance is greater than 4 cm, the mushy zone size is approximately a constant.

Theoretically, when a preheat temperature is specified, the entire graphite susceptor should maintain that specified temperature throughout the complete withdrawal cycle. In an actual furnace setting, however, the graphite susceptor temperature is not only non-uniformly distributed along the height of the susceptor but also changes with withdrawal time. The magnitude of the temperature differences between the actual furnace susceptor conditions and the theoretical conditions depends on the amount of furnace insulation and the baffle geometry. The better the insulating condition and the smaller the baffle-part gap size, the smaller the magnitude of the difference between the actual furnace susceptor condition and the above mentioned theoretical condition. From Figure 4 it can be seen that the mushy zone is located very close to the baffle for the theoretical condition, whereas it is located more than 1 inch above the baffle (most of the time) for the actual furnace condition. This is because the graphite susceptor temperatures of actual furnace conditions are usually lower than that of the theoretical condition and, hence, the casting cools faster. Figure 4 also shows that the mushy zone size of the theoretical condition is significantly smaller than that of the actual furnace condition, which in turn indicates that the theoretical condition has a significantly higher temperature gradient than the actual furnace condition.

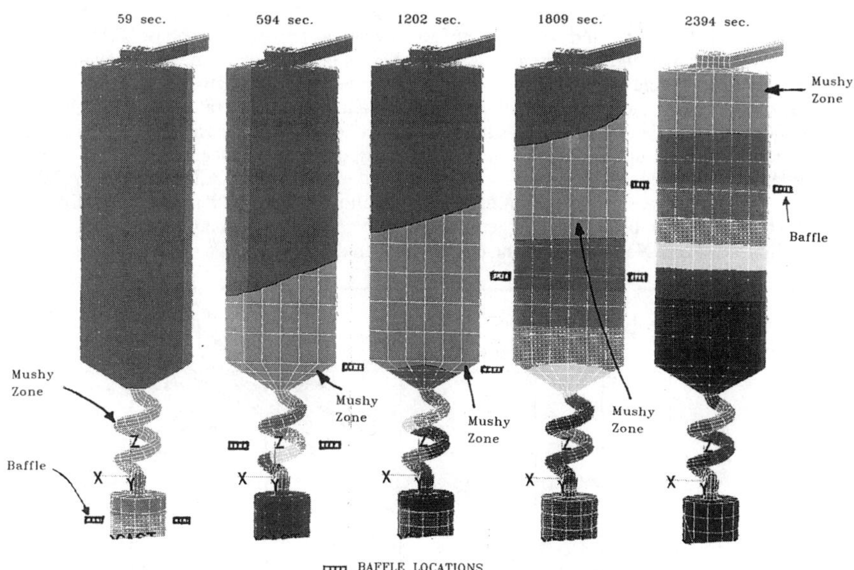

Figure 3. Calculated isotherms of a slab of NiAl alloy A at various withdrawal times.

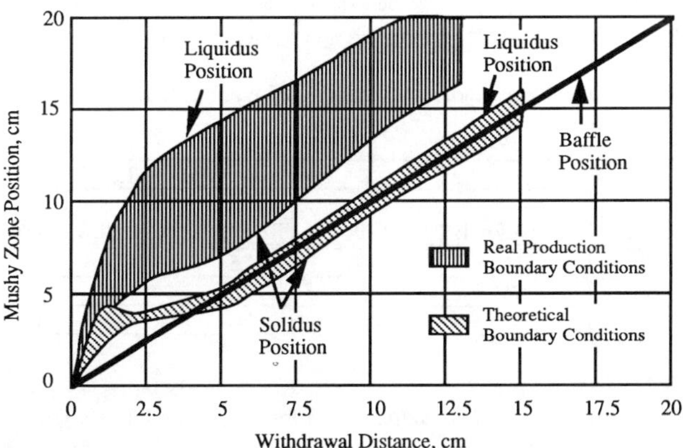

Figure 4. Solidification conditions of NiAl alloy A for two different boundary conditions.

Effect of Thermal Conductivity

NiAl alloys have thermal conductivities which are 3x-7x higher than single crystal superalloys, such as René N5. For the two particular alloys studied, alloy A had 4x higher conductivity than René N5 and alloy B had 3x higher conductivity. Figure 5 shows the solidification conditions of two simulations with an identical withdrawal cycle for NiAl alloy A and René N5. The preheat and pour temperatures for NiAl alloy A were based on theoretical boundary conditions, whereas René N5 castings were based on theoretical conditions for typical superalloy castings. From the figure it can be seen that the main difference of solidification conditions among these two runs is mushy zone location. The mushy zone of René N5 is completely below the baffle and out of the hot zone, whereas the mushy zone of NiAl alloy A is mostly above the baffle and stays in the hot zone. This is because when the withdrawal cycle is identical for two different alloys, the mushy zone locations are primary functions of alloy thermal conductivities. Lower thermal conductivity results in a lower cooling rate which then pulls the mushy zone out of the hot zone, whereas high thermal conductivity increases the cooling rate and pushes the mushy zone inside the hot zone. Another important difference among these two runs are temperature gradients. Table I shows the calculated temperature gradient of the NiAl alloys and René N5 slabs. The simulation condition of NiAl alloy B was identical to that of NiAl alloy A. From the table it can be seen that in general, the higher the thermal conductivity (NiAl alloy A > NiAl alloy B > René N5), the lower the temperature gradient.

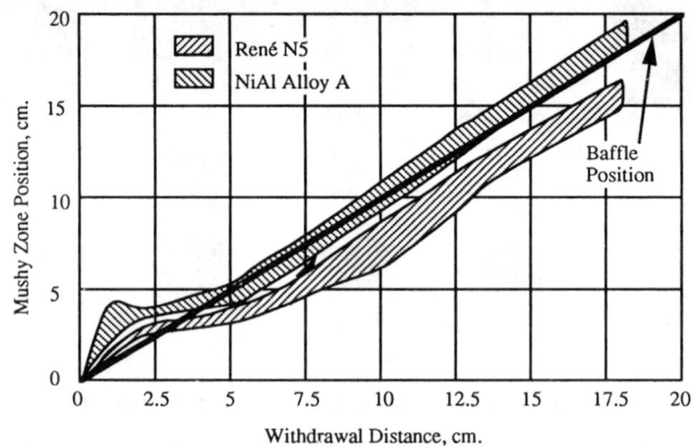

Figure 5. Mushy zone locations of NiAl alloy A and René N5 with identical withdrawal cycles.

Table I. Calculated temperature gradient, G, of NiAl alloys and René N5 slabs for an identical withdrawal cycle.

	G at T_{liq} (°C/cm)			G at T_{sol} (°C/cm)		
Slab Position	René N5	NiAl alloy B	NiAl alloy A	René N5	NiAl alloy B	NiAl alloy A
Top	60	51	38	26	25	17
Middle	60	53	43	28	28	19
Bottom	44	54	43	20	28	19

Dendrite Arm Spacing

Figure 6 shows the measured SDAS as a function of model calculated LST. For the purpose of comparison, DAS data of René N5 are also shown in this figure. It can be seen that the SDAS of the NiAl alloys are higher than those of René N5. This was also observed for the PDAS data. This is believed to be due to the differences in the solidification temperature between NiAl alloys and René N5. Dendrite coarsening is a diffusion process which strongly depends on the temperature. The solidification temperature of NiAl alloys is about 250°C higher than that of René N5, as shown in Table II. This results in significantly faster diffusion and dendrite coarsening rates, and, consequently, NiAl alloys have higher DAS values than those of René N5. Figure 6 also shows that SDAS of NiAl alloy A is higher than that of NiAl alloy B. Since the solidification temperature of NiAl alloy A is higher than that of NiAl alloy B, this result is consistent with the above mentioned theory.

Table II. Solidification temperatures of alloys studied.

Alloy	Liquidus	Solidus
René N5	1430°C	1357°C
NiAl alloy B	1662°C	1601°C
NiAl alloy A	1678°C	1617°C

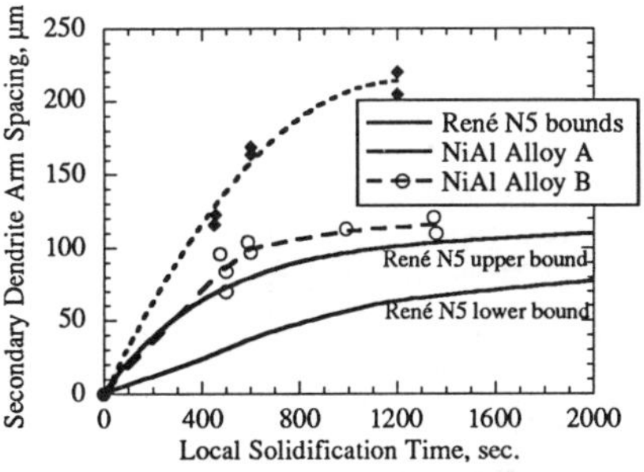

Figure 6. Measured secondary DAS as a function of calculated LST.

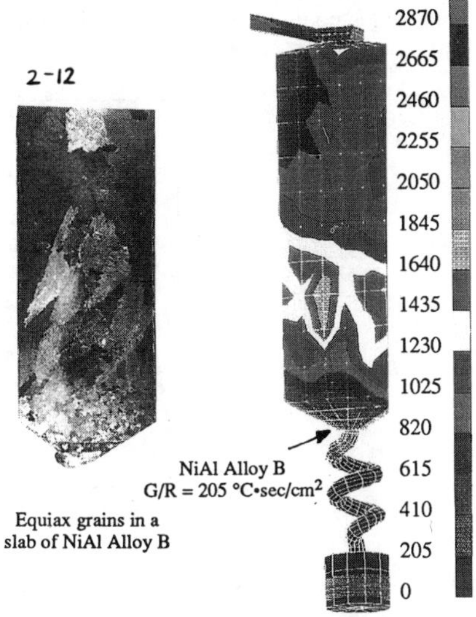

Figure 7. Equiaxed grains in a slab of NiAl alloy B correlate with areas of low G/R values.

Defects

Grain inspection results showed that the major grain defects were equiaxed grains (Figure 7). These equiaxed grains formed near the top of the grain selector. Equiaxed grain formation tendency[1-3] decreases with the increase of G/R at the liquidus temperature. When the G/R value is lower than a critical number, equiaxed grains form. Model results show that values of G/R at the liquidus temperature in the region of equiaxed grains formation are 205°C•sec/cm^2 and 515°C•sec/cm^2 for NiAl alloy B and NiAl alloy A, respectively. Thus, these values are selected as the equiaxed grains formation criteria for these two alloys. When G/R values are lower than these values, equiaxed grains will form.

Freckles are small chains of equiaxed grains oriented parallel to the <001> solidification direction[4] and can be predicted by the average cooling rate during solidification. Since no freckles were found in these castings, the lowest values of cooling rate (i.e., 0.033°C/sec for NiAl alloy B and 0.044°C/sec for NiAl alloy A) were selected as the criteria for freckle formation. When the average cooling rates are lower than these values, the likelihood of freckle formation increases.

Figure 8 shows the defects map for the NiAl alloys. For comparison purposes, data for René N5 are also shown. Note while lower G/R values are necessary to grow single crystals in NiAl alloys compared to superalloy René N5, it is much more difficult to obtain these G/R values for NiAl alloys because of their high thermal conductivities.

Macroshrink formation is related to casting bulk heat transfer conditions. In the casting, if there is an isolated region which is last to solidify, then macroshrink forms in that region. Equiaxed castings usually have a higher tendency to form macroshrink than directionally solidified and single crystal castings. The formation of macroshrink is predicted by the contour of the solidus isochron. An isochron is a contour plot which illustrates regions of constant time for castings to cool to a specified temperature . In a solidus isochron plot, an isolated spot with a higher time to reach solidus temperature means metal in that spot is the last to solidify and macroshrink is expected to form in that spot. There were no isolations found in such isochron plots which could create macroshrink. Thus, we do not expect a macroshrink problem for these castings; this was confirmed by the results of the actual castings.

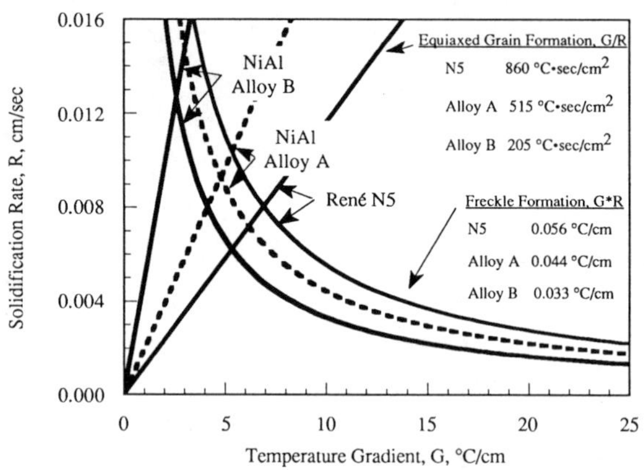

Figure 8. Defects map for NiAl alloys compared to René N5 data[3].

CONCLUSIONS

Finite-element modeling has been performed for simple shaped slabs and bars of two NiAl alloys. NiAl alloys have higher thermal conductivities and solidification temperatures than superalloys, such as René N5. This results in higher cooling rates, higher mushy zone positions, and larger dendrite arm spacings compared to superalloys. The defects (equiaxed grains and freckles) map for both NiAl alloys has been established. While lower G/R values are necessary to grow NiAl single crystals compared to superalloys, it is more difficult to obtain these values because of the effect of the high thermal conductivity.

REFERENCES

1. K.O. Yu, et al., AFS Transactions, 417-428 (1990).
2. K.O. Yu, J.A. Oti, M. Robinson and R.G. Carlson, in Proc. 7th Int. Symp. on Superalloys, edited by S.D. Antolovich, et al (TMS, Warrendale, PA, 1992), pp. 135-144.
3. "Manufacturing Technology for Advance Propulsion Materials, Phase IX - Advanced Turbine Airfoil Casting Technology, Task 2: Casting Simulation", AFWAL, No. F33615-85-C-5014, Final Report, August 1986 - May 1990.
4. A.F. Giamei and B.H. Kear, Metall. Trans., Vol. 1A, 2185-2191 (1970).

ACKNOWLEDGMENTS

This work was supported by the U.S. Air Force under contract F33615-90-C-5938, Tom Broderick, program manager. Discussions and guidance from Ed Goldman of GE Aircraft Engines is gratefully appreciated.

THE DEFORMATION AND DYNAMIC RECRYSTALLIZATION IN A HOT ISOSTATICALLY PRESSED Ti-48Al-2W POWDER ALLOY

L. ZHAO*, J. BEDDOES† and W. WALLACE*
* Structures and Materials Laboratory, Institute for Aerospace Research, National Research Council Canada, Ottawa, Canada
† Department of Mechanical and Aerospace Engineering, Carleton University, Ottawa, Canada

ABSTRACT

Titanium aluminide powder of composition Ti-48Al-2W (at%) was consolidated by hot isostatic pressing (HIP) at 200 MPa and 1050°C, 1100°C, 1150°C and 1250°C, respectively. The as-HIP'ed microstructure and substructure were characterized by optical microscopy and transmission electron microscopy (TEM). The densified powder alloy possesses an inhomogeneous microstructure consisting of fine-grained prior dendrites and coarse-grained interdendritic regions. The former contain the γ, α_2 and β_o phases. The HIP processing induced substantial deformation in the latter regions where dislocation segments and tangles, deformation twins and dislocation sub-boundaries were formed, indicative of plastic yielding and creep deformation. Both 1/2<110] unit dislocations and superdislocations are activated during HIP, but the former are the predominant dislocations in the sub-boundaries. With increasing HIP temperature, dynamic recrystallization occurs to a greater extent, resulting in a relatively uniform microstructure. The microstructure can be divided into hard and soft regions in which the densification is dominated by different mechanisms. Methods of optimizing the HIP cycle parameters to produce a more homogeneous microstructure are discussed.

INTRODUCTION

Recently, considerable work has been carried out on the consolidation of titanium aluminide powders by hot isostatic pressing (HIP) [1-4]. The densification of these powders has been successfully controlled in accordance with the processing models proposed by Ashby and his co-workers [5,6]. However, to achieve optimum mechanical properties, the microstructural changes occurring during HIP must be understood and controlled. Previous studies [4] have shown that a gas-atomized Ti-48Al-2W powder alloy can be fully densified by HIP, but under the conditions employed the material remained microstructurally and mechanically inhomogeneous. This illustrates that microstructural control is an important consideration in specifying HIP cycle parameters.

During HIP, plastic yielding, creep deformation and diffusion can contribute to the densification of powders [5,6]. Moreover, microstructural changes may occur, including grain growth, phase transformation and dynamic recrystallization [1,3,4]. For the TiAl+W powder alloy studied [4], however, it is not yet fully understood how these microstructural processes are associated with the densification. In the present study, the microstructure and deformation substructure of Ti-48Al-2W powder subjected to HIP are characterized and correlated to the HIP conditions. The observations reported may also provide useful information to improve the intelligent process control [2] for the HIP consolidation of intermetallic powders.

EXPERIMENTAL

A gas-atomized powder with composition of Ti-48Al-2W (at%) was vacuum encapsulated in 304 stainless-steel tubes and consolidated by HIP at 200 MPa and 1050°C, 1100°C, 1150°C and 1250°C, respectively. Figure 1 shows a typical HIP cycle in which the specimen was held at 200 MPa and 1100°C for 2 hours. The microstructures of the as-received powder and as-HIP'ed bulk material were examined in an optical microscope and the dendrite area fraction was determined using an image analyzer. TEM foils of the powder were prepared by nickel plating, mechanical thinning and electrical jet polishing using 7% sulphuric acid in methanol at 75 mA and -35°C [7]. The preparation of TEM specimens for the consolidated material included mechanical thinning followed by electrical jet polishing in 64% methanol+ 31% butanol+5% perchloric acid at 13~15 V and -30~-35°C. The TEM experiments were carried out with Philips EM201 and EM400T microscopes, the latter equipped with an EDAX 9100/60EDS system for energy dispersive analysis of x-ray (EDX).

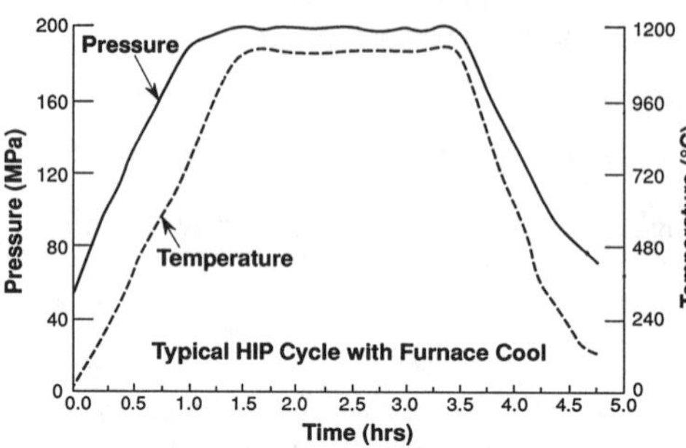

Fig.1 The HIP cycle applied to consolidation of the Ti-48Al-2W (at%) powder.

RESULTS AND DISCUSSION

Figure 2 shows that the microstructure of the as-received powder is characterized by a dendritic structure. The interdendritic regions (Fig.2(b)), containing higher Al and lower W [4], consist of γ-TiAl phase with a low density of dislocations and deformation twins. The W-enriched [4] dendrite (Fig.2(c)) comprises α_2-Ti_3Al phase with antiphase domain boundaries present.

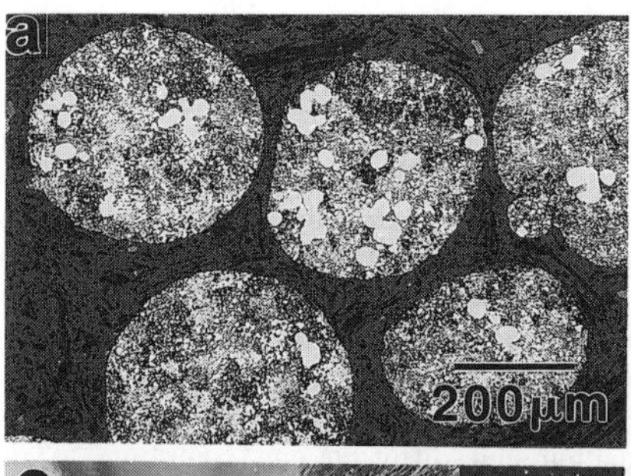

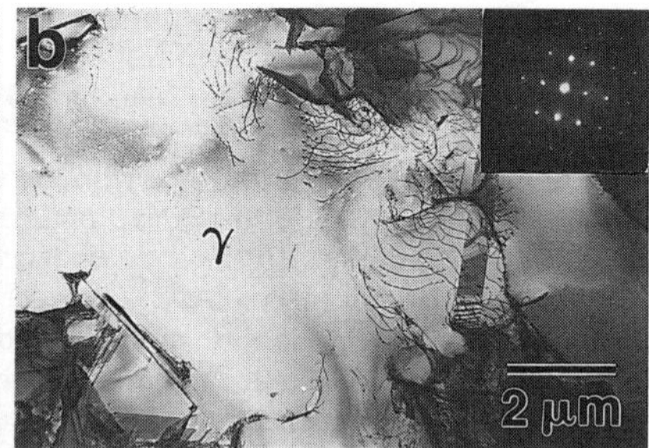

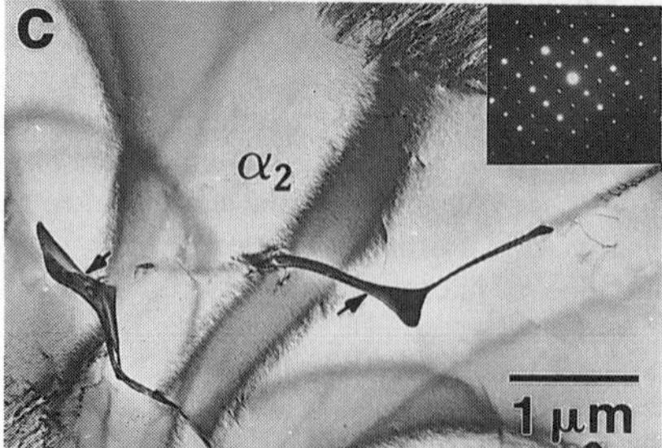

Fig.2 (a) The dendritic structure of the as-received powder; (b) γ-TiAl phase of the interdendrite with $[110]_\gamma$ diffraction pattern (inset); (c) α_2-Ti_3Al phase of the dendrite with $[11\bar{2}0]_{\alpha 2}$ diffraction pattern (inset). Arrows identify antiphase domain boundaries in α_2.

The powder material can be fully densified by HIP at 1100°C and higher temperatures. However, the as-HIP'ed microstructure is inhomogeneous, as seen in Fig.3, with clearly discernable prior dendritic and interdendritic regions. Although increasing HIP temperature improves the homogeneity (Fig.3(b)), the dendritic structure exists in specimens consolidated at all HIP temperatures.

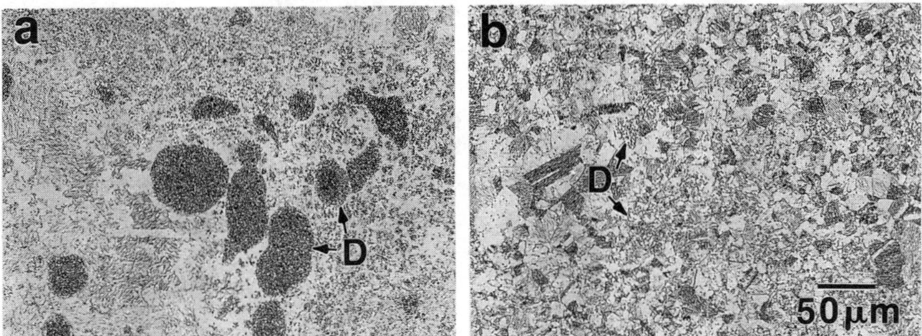

Fig.3 HIP consolidated microstructures at 200 MPa and (a) 1050°C and (b) 1250°C. The prior dendrites are indicated by "D".

The dendritic area fraction decreases with HIP temperature, as shown in Fig.4. Within the prior dendrites, the original W-enriched α_2 phase of the powder (Fig.2(c)) decomposes into three fine-grained phases during HIP. Figure 5 illustrates this fine-grained structure in which γ, α_2 and β_o phases are identified by electron diffraction and EDX analysis. It can be noted that superlattice diffraction spots appear in the $[110]_{\beta o}$ pattern, indicating that the β_o phase has an ordered bcc (B2) structure. Furthermore, consistent with the previous observations [4,8], β_o is enriched with W as shown by the EDX spectrum. The average grain size of the phases in the prior dendritic structure increases with HIP temperature (Fig.4). The decomposition of the dendrites during HIP can be summarized as: α_2(powder) $\rightarrow \gamma+\alpha_2+\beta_o$. TEM experiments also revealed that the consolidated prior dendrites had a low dislocation density. The dislocations present are predominantly in the γ phase, which is the softest phase [9].

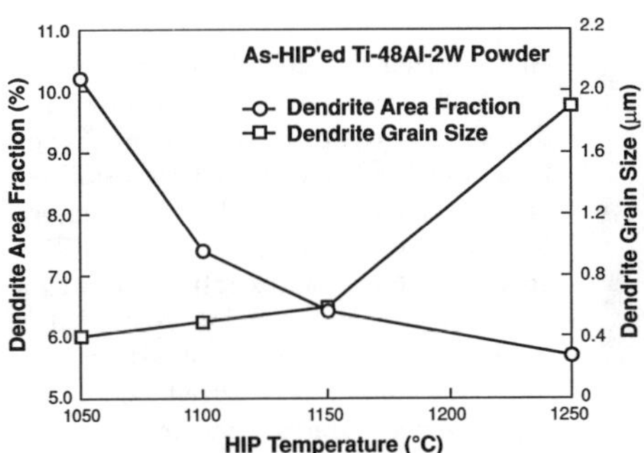

Fig.4 Variation of the prior dendritic area fraction and the average dendrite grain size with HIP temperatures.

Diffraction experiments showed that the interdendritic areas consist predominantly of γ phase. Figure 6 illustrates the interdendritic structure with both recrystallized and non-recrystallized regions distinguishable. In the latter region, irregularly shaped coarse grains contain a high density of dislocation segments and tangles, with deformation twins and dislocation sub-boundaries. Diffraction contrast experiments revealed that both 1/2<110] unit dislocations and super-dislocations exist in this region, the former being the primary constituents of the sub-boundaries. Figure 7 illustrates that the sub-boundaries are visible under **g**=$\bar{1}11$, but out of contrast with **g**=002, indicating that they are made up of 1/2<110] unit dislocations. Superdislocations exhibit strong contrast in Fig.7(b) and detailed analysis identified that both <101] and 1/2<112] superdislocations exist.

The recrystallized interdendritic regions resulting from different HIP temperatures are

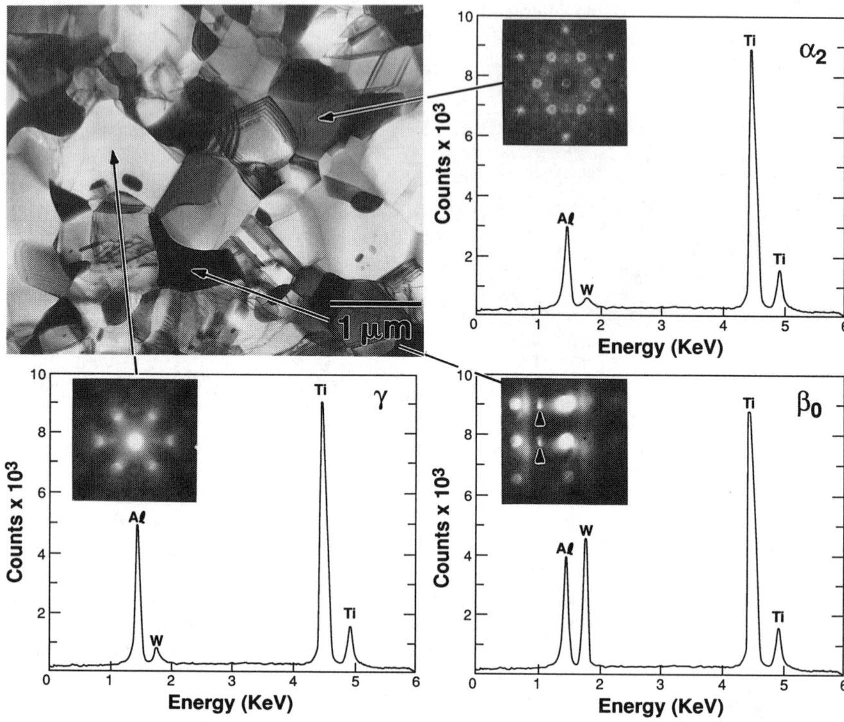

Fig.5 A prior dendrite containing γ, α_2 and β_o fine-grained phases (1050°C HIP). The diffraction patterns shown are $[101]_\gamma$, $[0001]_{\alpha 2}$ and $[110]_{\beta o}$, respectively. Arrows in $[110]_{\beta o}$ indicate superlattice diffraction spots.

Fig.6 An interdendritic region consisting of recrystallized, marked "R", and non-recrystallized regions.

characterized by refined equiaxed grains with a substantially lower dislocation density than the non-recrystallized regions (Fig.6). As seen in Fig.8, the average recrystallized grain size increases with HIP temperature. Moreover, recrystallization during HIP at all temperatures causes refinement of the coarse interdendritic grains. Note that in Fig.8, the interdendrite grain size plotted includes both recrystallized and non-recrystallized grains. In recrystallized regions, W-enriched particles precipitate more uniformly at the lower HIP temperatures. With increasing HIP temperature, the W-enriched particles coarsen and tend to form on grain boundaries or at triple points, thus retarding the coarsening of the recrystallized grains. The predominant substructure in this region

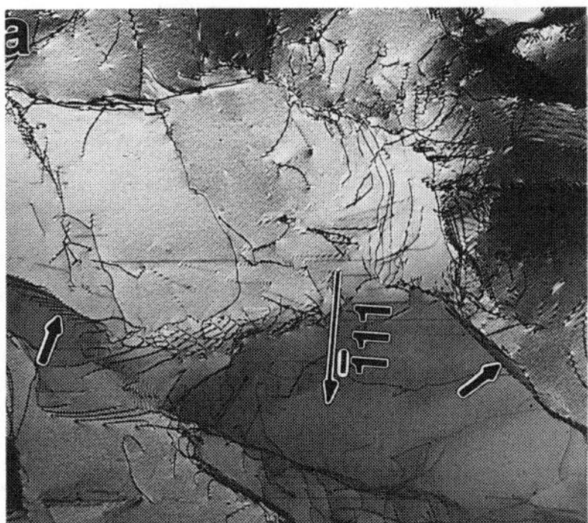

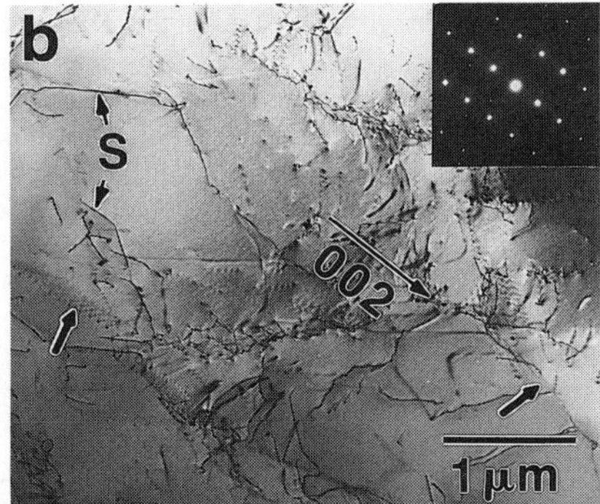

Fig.7 The dislocation substructure within the non-recrystallized region: (a) **g**=$\bar{1}11$ and (b) **g**=002 with the incident electron beam near $[110]_\gamma$ (inset). Arrows identify the same sub-boundaries in (a) and (b). Superdislocations are indicated by "S".

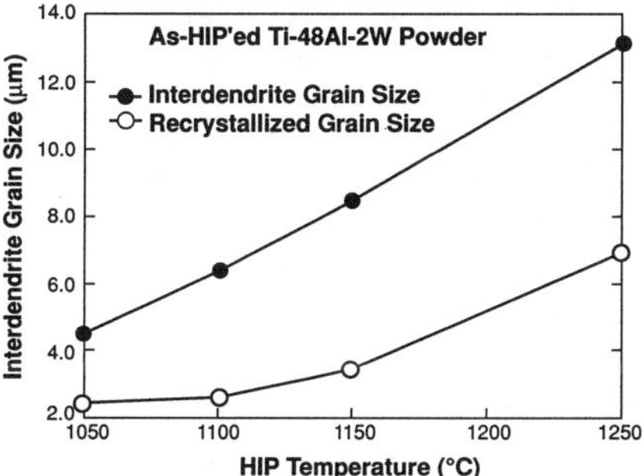

Fig.8 Variation of the average interdendrite grain size and recrystallized grain size with HIP temperatures.

are dislocation segments including 1/2<110] unit dislocations and superdislocations.

As previously reported [6], the densification of powders can be divided into two stages. During the *initial stage* where the relative density (D) is less than ~0.9, densification occurs by plastic yielding at contact points between powder particles, causing growth of the contact areas. During the *final stage* (0.9<D<1), densification is achieved by the shrinkage of remaining interparticle voids predominantly by time-dependent processes. For the TiAl+W powder studied, plastic yielding dominating the initial densification occurs in the softer interdendritic regions, producing a high density of dislocation segments and tangles, and deformation twins. Both 1/2<110] unit dislocations and superdislocations are activated to accommodate this deformation. After the powder compact density resulting from plastic yielding reaches its maximum value [6], a final densification associated with shrinkage of interparticle voids by time-dependent mechanisms occurs. These include dislocation (power-law) creep, Nabarro-Herring and Coble (diffusional) creep, and interparticle boundary diffusion. The sub-boundary structure seen in the interdendritic regions (Figs. 6 and 7) results from dislocation creep [10], in which 1/2<110] unit dislocations are predominant. Dynamic recrystallization occurs in the interdendritic regions, driven by the release of the free energy stored by the high density of defects. The recrystallized grains are softer and finer, compared to the non-recrystallized grains (Figs.6 and 8), and thus can accommodate further HIP deformation. The dendrites have a higher hardness because of the higher W content [4] and concentration of the hardest β_o phase [9]. Given this nature and the small fraction of the dendrites (Fig.4), their contribution to the overall densification is minimal.

To produce a homogeneous microstructure in the consolidated TiAl+W powder, the HIP cycle parameters should be designed to completely eliminate the original dendritic structure. This could be achieved by either complete recrystallization or development of a fully transformed microstructure at higher (>1250°C) HIP temperatures. However, to prevent extensive grain growth (Fig.8), the HIP time at such high temperatures may have to be relatively short. Consequently, to ensure microstructural stabilization, as well as full densification, may require subsequently holding the powder at a lower HIP temperature for a period of time. An alternative way to produce a homogeneous microstructure may be mechanical working of the powder through ball milling prior to HIP consolidation. The cold work introduced could break up the dendritic structure in the powder and provide a large amount of strain energy stored in the ball-milled powder to promote recrystallization during HIP. A research program is presently underway to explore mechanical cold working of powders followed by HIP processing.

CONCLUSIONS

1. The HIP consolidated microstructure is inhomogeneous in character, consisting of fine-grained prior dendrites and coarse-grained interdendritic regions.
2. The HIP'ed prior dendrites contain a fine-grained structure of γ, α_2 and β_o phases. The average grain size increases with HIP temperature.
3. The interdendritic regions consist of recrystallized and non-recrystallized regions, the latter contain high density of defects including dislocation segments and tangles, deformation twins and dislocation sub-boundaries. Both 1/2<110] unit dislocations and superdislocations are active during HIP, but the former are the primary constituents of the sub-boundaries.
4. Increasing HIP temperature promotes dynamic recrystallization, improving the homogeneity of the consolidated microstructure. A completely recrystallized or fully transformed microstructure with controlled grain size may be achieved by optimizing the HIP cycle parameters. Mechanical cold working of powders followed by HIP consolidation could be a potential process to produce a homogeneous fine microstructure.

ACKNOWLEDGEMENTS

The authors would like to thank Mr. R. Dainty, Mr. D. Chow and Mr. D. Morphy for providing technical assistance. One of the authors (L. Zhao) is also appreciative to Dr. G.J.C. Carpenter and Mr. M. Charest of the Metal Technology Laboratories, Energy, Mines and Resources Canada for granting access to their Philips EM400T microscope. The support of the Department of National Defence (Canada) is greatfully acknowledged. This work was completed as part of the project NRCC-IAR JHR-01.

REFERENCES

1. B.W. Choi, Y.G. Deng, C. McCullough, B. Paden and R. Mehrabian, Acta metall. mater. **38**, 2225 (1990).
2. R.J. Schaefer, The International Journal of Powder Metallurgy **28**, 161 (1992).
3. R.J. Schaefer and G.M. Janowski, Acta metall. mater. **40**, 1645 (1992).
4. J. Beddoes, W. Wallace and M.C.de Malherbe, The International Journal of Powder Metallurgy **28**, 313 (1992).
5. F.B. Swinkels, D.S. Wilkinson, E. Arzt and M.F. Ashby, Acta metall. **31**, 1829 (1983).
6. A.S. Helle, K.E. Easterling and M.F. Ashby, Acta metall. **33**, 2163 (1985).
7. C. McCullough, J.J. Valencia, C.G. Levi and R. Mehrabian, Mater. Sci. Eng. **A124**, 83 (1990).
8. P.L. Martin, M.G. Mendiratta and H.A. Lipsitt, Metall. Trans. **14A**, 2170 (1983).
9. Y. Zheng, L. Zhao and K. Tangri, Scripta metall. mater. **26**, 219 (1992).
10. M. Koo, T. Matsuo and M. Kikuchi, in Intermetallic Compounds - Structure and Mechanical Properties (JIMIS-6), edited by O. Izumi (The Japan Institute of Metals, Aoba Aramaki, Sendai 1991), pp.519-523.

MICROSTRUCTURE EVOLUTION AND TEXTURE DEVELOPMENT IN HOT-ROLLED SUPER α_2 TITANIUM ALUMINIDE

C. J. Yu, D. Zhao, J. J. Valencia, and P. K. Chaudhury
Concurrent Technologies Corporation, 1450 Scalp Avenue, Johnstown, PA 15904, U.S.A.

ABSTRACT

Textural and microstructural changes in super α_2 titanium aluminide (Ti-14Al-20Nb-3.2V-2Mo in wt%) subjected to hot-rolling process were investigated. X-ray pole figure measurements on sheet specimens were conducted to quantify the crystallographic texture development during the hot-rolling process. Results show strong texture development in both the α_2 and β phases. High temperature compression tests at various strain rates were conducted on specimens taken from hot-rolled plates, and the microstructural changes induced by the deformation were examined using transmission electron microscopy (TEM). Results indicated that deformation at low strain rates retains the hot rolled microstructure with elongated and strongly aligned α_2 phase. At higher strain rates, the characteristic α_2 phase appears to break down into equiaxed and/or recrystallized grains during deformation. Furthermore, results are correlated between the macroscopic orientations of texture components and the local orientations of α_2 phase.

INTRODUCTION

It has been widely recognized that the influence of texture, or preferred crystallographic orientation causes microstructural anisotropy which can lead to fracture during metal forming operations [1-4]. However, the degree and/or absence of anisotropy that dictates product performance depends on the manufacturing process and the material. Current interests in thermomechanical processing of super α_2 titanium aluminide require the knowledge of microstructural changes and textural development during high temperature deformation for quality assurance of the final product. The process parameters such as strain, strain rate and temperature, all of which affect the final material characteristics, should be properly controlled in order to obtain desired product.

Super α_2 titanium aluminide sheet in the hot-rolled condition consists of an α_2 (Ti_3Al) phase with an ordered hcp ($D0_{19}$) structure and a β phase with a bcc crystal structure [1,3]. Plastic deformation on these two phases develops different types of crystallographic texture [3,4]. However, for a given processing condition one type of texture may be dominant. The objective of this study is to determine the texture development during hot rolling and the effect of high temperature compressive deformation on microstructural changes.

EXPERIMENTAL PROCEDURE

The material used in this investigation was Ti-14Al-20Nb-3.2V-2Mo in both plate and sheet forms. The plate was forged and hot-rolled to a thickness of 20.32 mm, and were further reduced by straight rolling and finally cross rolled to thickness of 2.54 and 0.51 mm at 1283 - 1310 K. During the final rolling process, most of the reduction was made in the initial straight rolling steps prior to the cross rolling operation. This procedure had a direct effect in the microstructure and texture development of the final sheet material.

Texture examinations of the sheet materials were conducted using conventional X-ray diffraction techniques including pole figure analysis. X-ray diffraction pattern of a titanium aluminide powder standard was also obtained and compared with those of the rolled sheets.

Cylindrical compression samples of 12.7 mm diameter and 15.87 mm height were machined from the 20.32 mm thick plate. Axial compression tests were performed at 1227 K in the $(\alpha_2+\beta)$ field below the β transus (approximately 1273 K) [5]. Three strain rates, 0.1, 2 and 8 s^{-1} were selected to simulate the hot-rolling speeds. Immediately after testing, the deformed samples were water-quenched to retain the microstructure. The load and displacement were recorded during testing, and transformed into true stress-true strain curves.

Specimens for TEM examination were taken from the center plane perpendicular to the load direction of the compressed samples. Three millimeter diameter disks were cut from 100 - 150 μm thick samples. The disks were then electropolished using a solution of 10 vol% H_2SO_4 and methanol at 225 K to produce thin foils to be examined by TEM.

RESULTS AND DISCUSSION

Texture Development during Hot-Rolling

Figure 1 shows the X-ray diffraction pattern of a titanium aluminide powder standard having the same composition as the rolled material, and

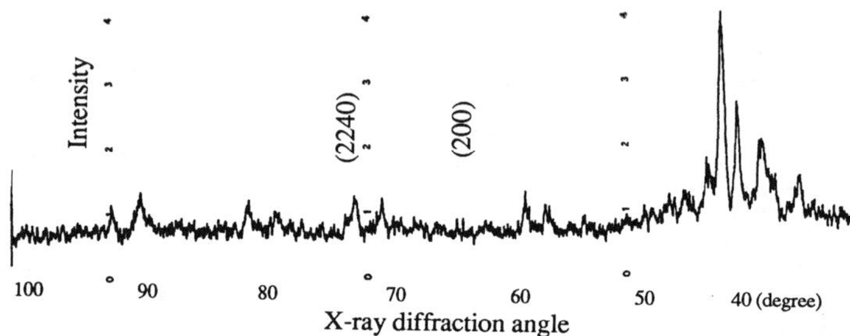

Figure 1. X-ray diffraction pattern of the titanium aluminide powder standard having the same composition as the rolled material.

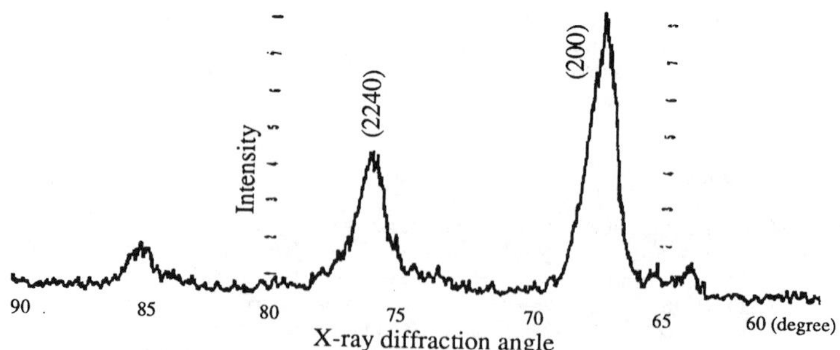

Figure 2. X-ray diffraction pattern of the rolled super α_2 titanium aluminide sheet.

Figure 2 shows the X-ray diffraction pattern of the 2.54 mm-thick titanium aluminide sheet. Comparison of these patterns indicates a substantial increase in the intensity level for the {2240} and {200} reflections. This indicates texture development of the α_2 phase and the β phase during rolling. The diffraction angles were then fixed on the α_2 {2240} and β {200} peaks to construct α_2 phase and β phase pole figures.

Figure 3 shows the X-ray {2240} and {200} pole figures obtained from 2.54 mm-thick titanium aluminide sheet. Two texture components $(11\bar{2}0)[0001]$ and $(11\bar{2}0)[\bar{1}100]$ for the α_2 phase and one texture component (100)[011] for the β phase were found. The detected maximum X-ray peak intensity for this thickness was found to be 12R (i.e. 12 times random intensity). Pole figure analysis on the sample obtained from the edge of the sheet showed same type of texture, but slightly rotated (4° to 6°) toward the transverse direction.

For the 0.51 mm thick-sheet material the {200} pole figure of the β phase presented a strong (100)[011] texture (Figure 4b) with an intensity of 24R. It was also observed that there was another (554)[225] bcc texture component with an intensity of 7R. This was expected since the (554)[225] component is near the $(111)[11\bar{2}]$ orientation and is commonly found in straight rolled bcc metals [6]. The presence of the (554)[225] texture component conforms the fact that the material was predominantly straight rolled, and the deformation induced by cross rolling was limited. However, the history appeared different for texture development in the α_2 phase. In this case, a maximum intensity of 39 R was observed for the $(11\bar{2}0)[\bar{1}100]$ component, while an intensity less than 2R was detected for the

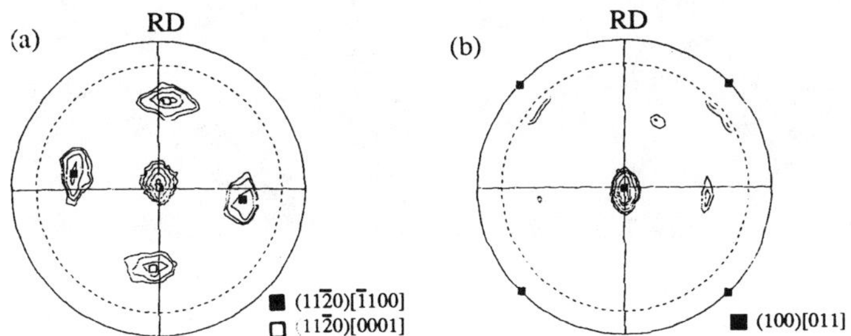

Figure 3. X-ray pole figures obtained from 2.54 mm thick super α_2 titanium aluminide sheet, (a) {2240} pole figure, (b) {200} pole figure.

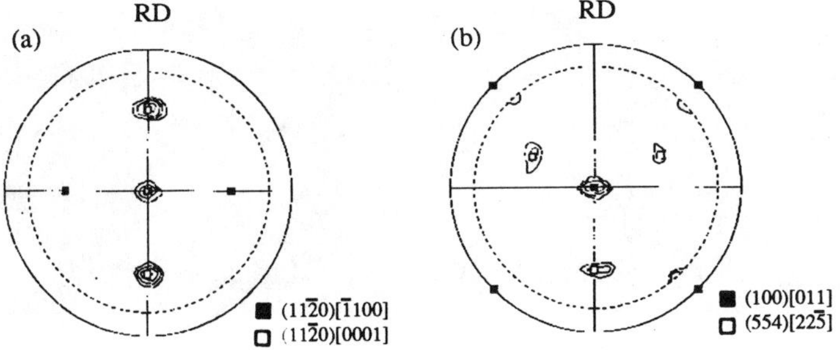

Figure 4. X-ray pole figures obtained from 0.51 mm thick super α_2 titanium aluminide sheet, (a) {2240} pole figure, (b) {200} pole figure.

$(11\bar{2}0)[0001]$ component (Figure 4a). It appears that α_2 phase is more prone to textural changes than β phase during the final rolling operation that includes straight rolling followed by cross rolling.

Microstructural Changes during High Temperature Compressive Deformation

The TEM micrograph of the 0.51 mm-thick β annealed titanium aluminide sheet is shown in Figure 5. This consists of primary α_2 globules, retained β, and $(\alpha_2+\beta)$ widmanstätten type phases indicating a moderate cooling rate from the annealing temperature [7]. A close examination of the β phase (Figure 6) shows high density of dislocations which appears as dislocation network.

Scanning electron microscopy (SEM) of the as-received plate (20.32 mm-thick) material (Figure 7) shows elongated α_2 grains, retained β, and $(\alpha_2+\beta)$ widmanstätten colonies. Figure 8 to Figure 10 show TEM microstructures for the compression samples deformed at 1227 K at strain rates of 0.1 s^{-1} to 8 s^{-1}. Since the compression testing temperature is slightly below the β transus and the rolling temperature, the widmanstätten structure transforms to $\alpha_2+\beta$ constituents during deformation. The water quenching of the compressed samples retains the microstructure at deformed temperature for

Figure 5. TEM micrographs of the 0.51 mm thick titanium aluminide sheet. Microstructure consists of primary α_2 globules, retained β, and $(\alpha_2+\beta)$ widmanstätten type phases.

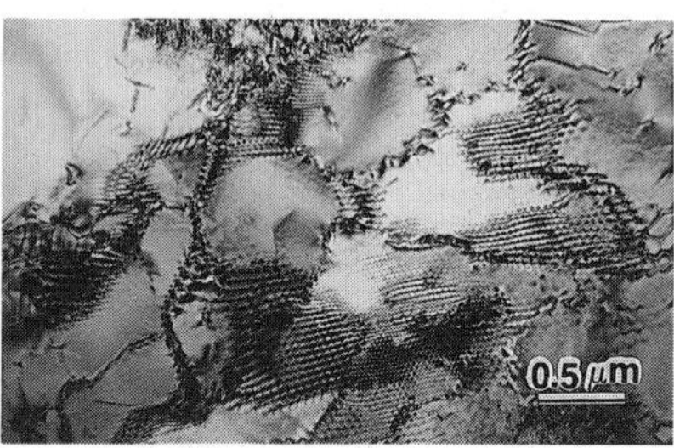

Figure 6. TEM micrographs of the 0.51 mm thick titanium aluminide sheet showing dislocation networks in the retained β phase.

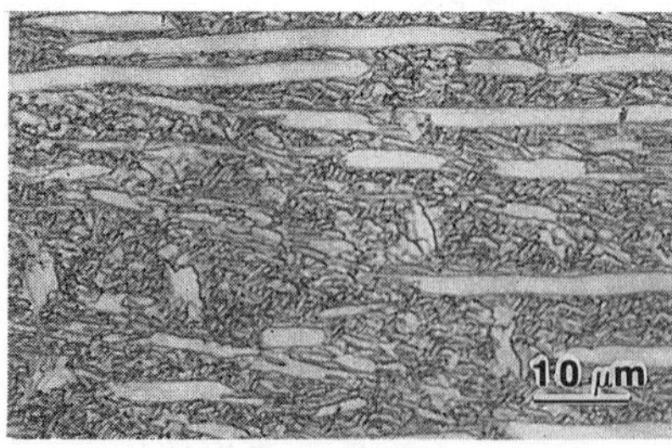

Figure 7. SEM micrograph of the as received titanium aluminide plate (20.32 mm thick).

Figure 8. TEM micrograph of the specimen deformed at strain rate of 0.1 s^{-1}.

all strain rates. In Figure 8, at the lowest strain rate (0.1 s^{-1}), the strong alignment of the α_2 phase, that resulted from rolling was still evident (Figure 7). There was some breakdown of α_2 grains into small well-aligned grains as indicated by the arrows in Figure 8 to mainly accommodate the deformation. At a higher strain rate, 2 s^{-1}, there is a reduced amount of alignment of the α_2 phase (Figure 9), possibly due to increased amount of breakdown of elongated α_2 grains. The microstructure also shows α_2 and β equiaxed granular regions, where the β grains are heavily deformed. However, at highest strain rate of 8 s^{-1}, a totally equiaxed grain structure resulted from the thermomechanical treatment, as seen in Figure 10. The equiaxed microstructure consists of heavily deformed α_2 and β grains (as evident from the dark striations inside β grain). The presence of this microstructure is possibly due to the combined effect of shearing and recrystallization.

Figure 9. TEM micrograph of the specimen deformed at strain rate of 2 s^{-1}.

Figure 10. TEM micrograph of the specimen deformed at strain rate of 8 s^{-1}.

TEM diffraction patterns in Figure 11 represents the locus of the orientations of α_2 phase observed ([$\bar{2}110$] and [$\bar{1}100$]) in the compression sample tested at 0.1 s-1 and 1227 K). It is evident that there is a strong alignment of the α_2 phase in the preferred orientation of [$\bar{3}210$] near [$\bar{1}100$] and [$\bar{2}110$]. This orientation also appears to be consistent throughout the whole TEM foil examined. These results strongly suggest

Figure 11. Local orientation analysis of the α_2 precipitate alignment at (a) [$\bar{2}110$] orientation, and (b) [$\bar{1}100$] orientation.

that the type of texture developed during the final hot rolling processes and compression testing is similar.

SUMMARY

Textural and microstructural changes in hot-rolled super α_2 titanium aluminide (Ti-14Al-20Nb-3.2V-2Mo) were investigated. X-ray pole figure measurements show the development of strong crystallographic texture components particularly in the α_2 phase. High temperature compression tests were conducted at three different strain rate conditions (0.1, 2 and 8 s^{-1}) to simulate the hot rolling process. TEM microstructural examination indicated that the specimen deformed at 0.1 s^{-1} still showed a strong alignment of the α_2 phase in the microstructure. The specimen deformed at the strain rate of 8 s^{-1} showed a microstructure consisting of sheared and recrystallized α_2 and β phases.

ACKNOWLEDGMENTS

This work was conducted by the National Center for Excellence in Metalworking Technology, operated by Concurrent Technologies Corporation (formerly Metalworking Technology, Inc.), under contract to the U.S. Navy as part of the U.S. Navy Manufacturing Technology Program. The authors would like to thank the RMI Titanium company for providing the material, Dr. B.K. Kad of University of California at San Diego for the TEM work, Dr. H. Hu of University of Pittsburgh for the X-ray work, and Mr. G. Holt of CTC for conducting the mechanical testing.

REFERENCES

1. C. Bassi, J.A. Peters and J. Wittenauer, JOM, September, 19 (1989).
2. K.S. Chan and D.A. Koss, Met. Trans. A 14A, 1333 (1983).
3. D.B. Knorr and N.S. Stoloff, J. Mater. Sci. and Eng. A123, 81 (1990).
4. K. Morii, C. Hartig, H. Mecking, Y. Nakayama and G. Luetjering, in Eighth International Conference on Textures of Materials, edited by J. S. Kallend and G. Gottstein (Springer-Verlag, Berlin, 1988), p. 991.
5. R. Strychor, J.C. Williams, and W.A. Soffa, Met. Trans. A 19A, 225 (1988)
6. H. Hu, Texture, 1, 233 (1974).
7. C. Bassi, J.A. Peters and J. Wittenauer, JOM September, 18 (1989)

PART VII

Environment Resistance

Stress Dependence of the Steady-State Creep Rate Between 1000 and 1300 K

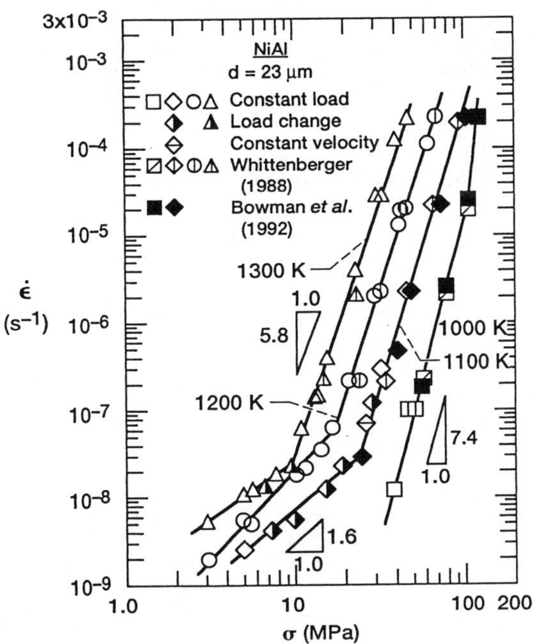

The figure illustrates the true strain rate-flow stress behavior for a 23 μm grain size NiAl tested in compression between 1000 and 1300 K. Two distinct deformation regimes are visible: at higher stresses and strain rates the stress exponent is about 6 and the activation energy is ~290 kJ/mol while at lower stresses and strain rates the stress exponent is about 2 with an activation energy of ~95 kJ/mol. Since the activation energy for lattice diffusion in NiAl is on the order of 300 kJ/mol, the lower activation energy value is believed to be associated with a grain boundary diffusion process. While the estimate of 95 kJ/mol is from an indirect experiment, it represents the first known determination of the activation energy for grain boundary diffusion in NiAl.

Figure courtesy of S.V. Raj and S.C. Farmer.

ENVIRONMENTAL EFFECTS IN B2 FeAl ALLOYS

O. KLEIN, P. NAGPAL and I. BAKER
Thayer School of Engineering, Dartmouth College, Hanover, NH 03755

ABSTRACT

The low room-temperature ductility of iron-rich B2-structured FeAl has been attributed to atomic hydrogen arising from a reaction between water vapor and aluminum atoms at crack tips [C.T. Liu and C.G. McKamey, in "High Temperature Aluminides and Intermetallics", eds. S.H. Whang et al (TMS, Warrendale, PA, 1990), p. 133]. Since hydrogen diffusion is time-dependent, the ductility might be expected to improve with increasing strain rate. Such behavior has been confirmed by tensile tests on Fe-45Al and Fe-40Al-5Cr in air at strain rates ranging from 1 x 10^{-6} s^{-1} to 1 s^{-1}. A brittle-to-ductile transition was observed at a strain rate of ~10^{-2} s^{-1}. The material with Cr appeared to be slightly less susceptible to environmental embrittlement, as reflected by the shift of the brittle-to-ductile transition to a lower strain rate. Fracture toughness tests on Fe-45Al in air demonstrated a similar strain rate effect.

INTRODUCTION

Recent work has shown that the room-temperature ductility of polycrystalline B2-structured FeAl is influenced by intrinsic factors such as Fe:Al ratio as well as extrinsic factors such as addition of ternary elements, test environment, and heat treatment [1-4]. With respect to Fe:Al ratio, the ductility has been shown to increase with increasing iron-rich deviation from stoichiometry, and is greatest near Fe-35Al (compositions are given in atomic percent throughout) [2,3]. Recently, it was demonstrated that Cr additions lowered the hardness of B2 Fe-40Al [5]. However, the effect of Cr on ductility of Fe-40Al was not investigated. An environmental effect was first demonstrated for Fe-36.5Al, which exhibited a higher elongation (~5%) when tested in vacuum than when tested in air (~2%) [6]. Similarly, Fe-40Al showed higher ductility (~5%) in vacuum than in air (~1%) [7,8]. It has also been shown that the cooling rate after annealing can have a dramatic effect on ductility. Air-cooled samples of FeAl, with aluminum contents from 34 to 50 at.%, were demonstrated to be more brittle than similar samples which were cooled more slowly [9,10]. The lower ductility presumably arose from hardening by quenched-in vacancies, and it was shown that a low-temperature anneal (400 °C) could remove this hardening effect [11]. It was demonstrated that after this low-temperature anneal, Fe-45Al could exhibit ~3% elongation when tested in air [2], whereas without this anneal failure before yield occurs in tension [3]. Furthermore, it appears that the cooling-rate effect is additive to the effect of the test environment. For example, Fe-40Al when furnace-cooled and tested in vacuum exhibited ~9% elongation [7].

It has been suggested that the brittle behavior of iron-rich FeAl is due to atomic hydrogen which arises from a reaction between aluminum atoms and water vapor [1]. The hydrogen atoms drive into the metal and thereby enhance crack propagation. Since hydrogen diffusion is time dependent, a strain-rate dependence of the ductility in air might be expected. The present results on B2-structured Fe-45Al with and without Cr confirmed this expectation.

EXPERIMENTAL

Single- and double-extruded rods of Fe-45Al and Fe-40Al-5Cr were produced from cast ingots. Details of the extrusion conditions are given elsewhere [12]. All extrusions, as received, were fully recrystallized. Dumbbell-shaped cylindrical tensile specimens (gauge length ~18 mm; gauge diameter ~3 mm) were made from the double-extruded rod by grinding, and annealed at 660 °C for 2 days, followed by an anneal at 400 °C for ~120 h to remove retained vacancies. The specimens were electropolished in a 10% perchloric acid-methanol mixture at – 25 °C prior to testing. Tensile tests were performed at room temperature at various strain rates from 1 x 10^{-6} s^{-1} to 1 s^{-1}, using an MTS 810 hydraulic testing machine interfaced with an IBM data acquisition system. The resulting fracture surfaces were examined with a Zeiss 962 scanning electron microscope.

Bend specimens for fracture toughness testing (B (height) ≈ 5 mm, W (width) ≈ 7 mm, L (length) ≈ 35 mm) were prepared from single-extruded rod by grinding, followed by mechanical polishing with SiC paper and 5 μm alumina powder. Most bend specimens were given the 400 °C vacancy-relieving anneal, while some were left unannealed. All samples were notched using a diamond saw, which produced a notch-root radius of ~100 μm. Some samples were Chevron-notched to a depth of ~0.3 W followed by fatigue precracking to give a sharp, microscopic crack. However, this invariably resulted in asymmetrical precracks. Other samples were given only a straight-across notch with the diamond saw to depths of 0.45 W - 0.55 W. Three-point bend fracture toughness tests conforming to ASTM E399 were conducted on an Instron Hydraulic testing machine interfaced with a MacIntosh computer. The fracture toughness, K_Q, was calculated using the equation

$$K_Q = \frac{P_Q S}{B W^{3/2}} f\left(\frac{a}{W}\right)$$

where P_Q is the peak load, a is the initial crack length, f(a/W) is a geometrical factor related to crack depth ratio, and S = 4W is the separation between three-point bend supports. The time (t) to peak load was recorded and used to calculate the average rate of increase in stress intensity factor, $\dot{K}_Q = K_Q/t$, at the notch tip before failure [13]. Crosshead velocities of 10^{-5} mm/s and 100 mm/s were used, resulting in $\dot{K}_Q \approx 0.1$ MPa·m$^{1/2}$s^{-1} and $\dot{K}_Q \approx 6 \times 10^3$ MPa·m$^{1/2}$s^{-1}, respectively.

Metallographic samples of Fe-45Al were etched with Nital, while the Fe-40Al-5Cr was etched with Aqua-Regia. Using the line-intercept method, the grain size was estimated to be approximately 28 μm for Fe-45Al used in tensile tests and 5 μm for Fe-40Al-5Cr. The Fe-45Al samples used in fracture toughness testing originated from the single-extruded rod and had grains of ~80 μm diameter. In the Fe-40Al-5Cr, second-phase particles of about 200 - 400 nm diameter were observed throughout the matrix and on grain boundaries. EDS analysis revealed greater Cr concentration in the particles than in the matrix. Reference to the ternary Fe-Al-Cr phase diagram [14] suggests that the particles were Cr_2Al.

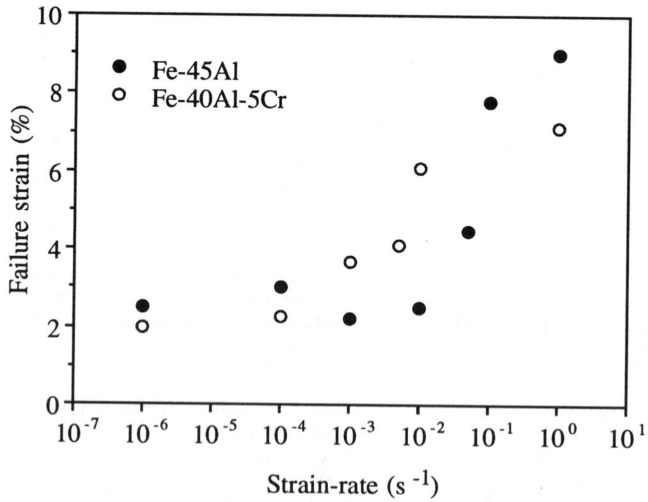

FIGURE 1. Graph of strain to failure versus strain rate.

RESULTS

Figure 1 shows the strain to failure versus strain rate for Fe-45Al and Fe-40Al-5Cr. For Fe-45Al, the elongation at ~3% was shown to be independent of strain rate up to 1 x 10^{-2} s^{-1}, but increased at faster strain rates, reaching a value of ~9% at a strain rate of 1 s^{-1}. For Fe-40Al-5Cr, the elongation at ~2% was independent of strain rate up to 1 x 10^{-4} s^{-1}. For strain rates greater than 1 x 10^{-4} s^{-1}, the ductility increased with strain rate, and the highest elongation (~7%) was obtained at a strain rate of 1 s^{-1}. Compared with Fe-45Al, the brittle-to-ductile transition appeared to move to a lower strain rate, from ~5 x 10^{-2} s^{-1} to ~5 x 10^{-3} s^{-1} for Fe-40Al-5Cr, suggesting that Fe-40Al-5Cr is less susceptible to environmental embrittlement. Note, however, that the maximum elongation at the highest strain rate (1 s^{-1}) was less than that for Fe-45Al.

The yield stress was independent of strain rate in both materials, at 560 MPa with Cr and 300 MPa without Cr. The fracture strength of Fe-40Al-5Cr increased from 690 MPa at 1 x 10^{-6} s^{-1} to 1020 MPa at 1 s^{-1}, compared with a peak value of 760 MPa for Fe-45Al at 1 s^{-1}. In both materials, the fracture mode was insensitive to strain rate, being mainly intergranular with a little transgranular cleavage, as shown in Figure 2. The Cr_2Al particles were observed on all Fe-40Al-5Cr fracture surfaces.

The fracture toughness tests on Fe-45Al mirrored the environmental sensitivity observed in the tensile tests (see Table I). A typical graph of load vs. time for a high strain rate test is shown in Figure 3. The tests on fatigue-precracked samples resulted in K_Q values of 12 - 16 MPa·$m^{1/2}$ at $\dot{K}_Q$ = 0.1 MPa·$m^{1/2}s^{-1}$, and increased to K_Q values between 24 and 27 MPa·$m^{1/2}$ at $\dot{K}_Q$ = 6 x10^3 MPa·$m^{1/2}s^{-1}$. The results for samples with macroscopic straight-across notches showed a similar trend, where K_Q increased from ~23 to ~30 MPa·$m^{1/2}$ over the same change in $\dot{K}_Q$. For specimens not given the low-temperature anneal, K_Q increased from ~21 to ~28 MPa·$m^{1/2}$.

DISCUSSION

The above results show an environmental effect governing the mechanical behavior of iron-rich B2 FeAl in air. Comparing the behavior of the Fe-40Al-5Cr with that of Fe-45Al in tensile tests, it appears that Fe-45Al is intrinsically more ductile. On the other hand, the lower strain rate required for a brittle-to-ductile transition with the addition of Cr signifies greater environmental resistance. In both cases, strain rate has no influence on the yield stress, indicating that the observed effects are not due to changes in dislocation mobility. It is difficult, however, to ascribe the observed behavior to the Cr additions alone, since the grain size, at 5 µm, is smaller in the material with Cr than in the Cr-free material (28 µm). In separate work, compression tests on Fe-40Al-5Cr with a 28 µm grain size resulted in a yield strength of 380 MPa, compared to 300 MPa observed in the Fe-45Al with a similar grain size [15]. At first sight this is perhaps a surprising result, since instead of the expected lattice softening [4], it suggests that Cr increases the grain size strengthening effect in FeAl. But it is worth noting that although the intrinsic ductility of Fe-40Al-5Cr is less than that of Fe-45Al, the fracture strength is much higher, i.e. 1020 MPa for Fe-40Al-5Cr compared with 760 MPa for Fe-45Al at 1 s^{-1}.

In both materials, the improvement in ductility with increasing strain rate does not result in a change of fracture mode, which remains mainly intergranular. This result is also similar to that in Fe-40Al, where the fracture mode remains intergranular when an increase in ductility occurs on testing in either oxygen or vacuum compared with testing in air [2]. These results also reinforce earlier conclusions [1-3] that grain boundaries are intrinsically weaker in iron aluminides with higher Al content, and that these weak boundaries are the reason for the low ductility and brittle fracture behavior of Fe-Al with higher Al contents.

The results from the fracture toughness tests provide further evidence for an environmental effect. All fracture toughness values are presented as K_Q values either because of the blunt notches or asymmetrical precracks used in tests. Also, owing to the difficulty of determining exact precrack lengths, a/W was taken to equal 0.5 in the precracked samples (see Table I).

(a)

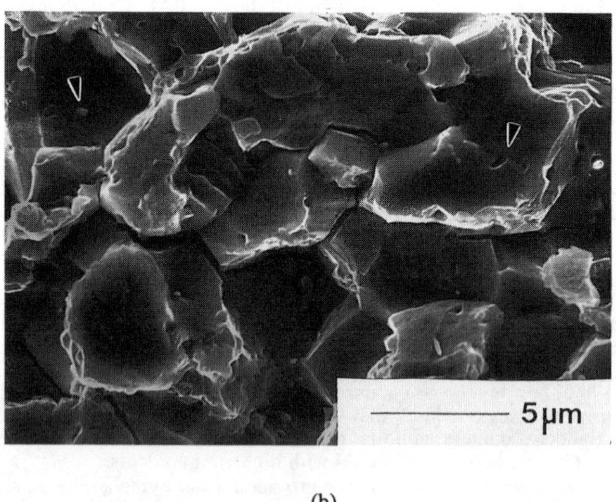

(b)

FIGURE 2. Fracture surface of tensile-tested (a) Fe-45Al and (b) Fe-40Al-5Cr. In (b), second-phase particles appear as both light specks and dark craters where particles once sat, as indicated by arrows.

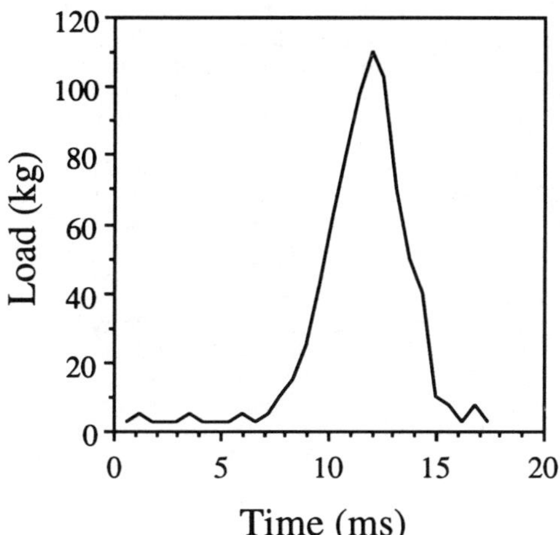

FIGURE 3. Load vs. time plot for fracture toughness test at crosshead velocity of 100 $mm{\cdot}s^{-1}$.

TABLE I

Crosshead Velocity*	P_Q (kg)	a/W	f(a/W)	K_Q ($MPa{\cdot}m^{1/2}$)
Precrack + 400 °C anneal				
slow	42.0	0.50	2.66	11.6
slow	59.1	0.50	2.66	16.1
fast	92.5	0.50	2.66	25.1
fast	100.2	0.50	2.66	27.2
fast	89.3	0.50	2.66	24.3
fast	94.3	0.50	2.66	25.6
Straight-across notch + 400 °C anneal				
slow	95.5	0.46	2.34	22.9
fast	107.6	0.51	2.77	30.8
fast	85.9	0.58	3.48	30.7
fast	110.7	0.51	2.73	30.4
Straight-across notch, no anneal				
slow	81.8	0.47	2.45	20.5
fast	103.2	0.52	2.82	29.7
fast	82.0	0.54	2.98	25.4
fast	84.5	0.58	3.44	29.7

* The slow crosshead velocity is 10^{-5} $mm{\cdot}s^{-1}$, corresponding to a rate of increase of stress intensity $\dot{K}_Q \approx 0.1$ $MPa{\cdot}m^{1/2}s^{-1}$, and the fast crosshead velocity is 100 $mm{\cdot}s^{-1}$, corresponding to $\dot{K}_Q \approx 6 \times 10^3$ $MPa{\cdot}m^{1/2}s^{-1}$.

Taking the averages of 2 - 4 tests, the K_Q values for precracked samples approximately doubled from ~14 to ~26 $MPa{\cdot}m^{1/2}$, while those for straight-across notched samples grew from ~23 to ~30 $MPa{\cdot}m^{1/2}$, as $\dot{K}_Q$ increased from 0.1 to 6 $x10^3$ $MPa{\cdot}m^{1/2}s^{-1}$. The somewhat higher values of K_Q in the latter case reflect the dullness of the root notches compared to sharp precracks. Nevertheless, the results from the three tests performed on bluntly-notched samples at high strain rate were quite similar. The earlier observation [11] that a 400 °C vacancy-relieving anneal improved the ductility of Fe-45Al is only weakly reinforced in the fracture toughness results (see Table I).

While Fe-45Al fractures mostly intergranularly, some transgranular cleavage areas do occur on fracture surfaces [2]. A possible nucleation mechanism for transgranular fracture on {100} in FeAl has been proposed by Munroe and Baker [16] based on their observations of pairs of <100> dislocations which appear to have formed by the reaction of gliding a/2<111> anti-phase boundary (APB) coupled partial dislocations during room-temperature deformation. Hydrogen from water vapor might enhance the formation of <100> dislocations in FeAl and hence promote cleavage cracking at lower strains. The hydrogen could do this by segregating to the dislocation cores (both the a<100>'s and the a/2<111>'s) and changing their mobility. The hydrogen might also be expected to segregate to the APB, coupling the a/2<111> partials, with the effect of promoting <100> dislocation formation. Thus, the mechanism based on the production of sessile <100> dislocations may lead to crack nucleation, whilst subsequent crack propagation is intergranular.

CONCLUSIONS

The tensile ductility and fracture toughness of Fe-45Al and Fe-40Al-5Cr were found to increase with increasing strain rate. Specifically, the strain to failure increased from ~2% to ~9% for Fe-45Al and from ~3% to ~7% for Fe-40Al-5Cr as the strain rate increased from $1 \times 10^{-6}\ s^{-1}$ to $1\ s^{-1}$. The fracture toughness also increased over a similar strain rate range, from ~23 to ~30 $MPa{\cdot}m^{1/2}$ for bluntly-notched samples and from ~14 to ~26 $MPa{\cdot}m^{1/2}$ for sharp-precracked samples.

ACKNOWLEDGEMENTS

The authors gratefully acknowledge Dr. J. D. Whittenberger of the NASA-Lewis Research center for providing and extruding ingots. We would like to thank Gary Kuehn and Jim McNamara for their help with mechanical testing. The use of the Dartmouth Electron Microscopy Facility is also acknowledged. This work was supported by the U.S. Department of Energy, Office of Basic Energy Sciences, Division of Materials Sciences, Grant No. DE-FG02-87ER45311.

REFERENCES

1. C.T. Liu and C.G. McKamey, in High Temperature Aluminides and Intermetallics, edited by S.H. Whang et al (TMS, Warrendale, PA, 1990), p. 133.
2. P. Nagpal and I. Baker, Mater. Char. **27**, 167 (1991).
3. C.T. Liu and E.P. George, Mater. Res. Soc. Proc. **213**, 527 (1991).
4. C.G. McKamey, J.A. Horton and C.T. Liu, J. Mater. Res. **4**, 1156 (1989).
5. P.R. Munroe and I. Baker, Scripta Metall. **24**, 2273 (1990).
6. C.T. Liu, E.H. Lee and C.G. McKamey, Scripta Metall. **23**, 875 (1989).
7. D.J. Gaydosh and M.V. Nathal, Scripta Metall. **24**, 1281 (1990).
8. C.T. Liu and E.P. George, Scripta Metall. **24**, 1285 (1990).
9. B. Schmidt, P. Nagpal and I. Baker, Mater. Res. Soc. Proc. **133**, 755 (1989).
10. M.A. Crimp, K.M. Vedula, and D.J. Gaydosh, Mater. Res. Soc. Proc. **81**, 499 (1987).
11. P. Nagpal and I Baker, Metall. Trans. **21A**, 2281 (1990).
12. P.R. Munroe and I. Baker, J. Mat. Sci. **24**, 4246 (1990).
13. J.D. Rigney et al, Mater. Res. Soc. Proc. **213**, 1001 (1991).
14. V.G. Rivlin and G.V. Raynor, Int. Met. Rev. **4**, 139 (1980).
15. O. Klein and I. Baker, unpublished work.
16. P.R. Munroe and I. Baker, Acta Metall. **39**, 1011 (1991).

ENVIRONMENTAL EMBRITTLEMENT OF BINARY AND Zr-DOPED Ni_3Al

E. P. GEORGE,[1] C. T. LIU,[1] and D. P. Pope[2]
[1]*Metals and Ceramics Division, Oak Ridge National Laboratory, Oak Ridge, TN 37831-6093*
[2]*Dept.of Materials Science and Engineering, Univ. of Pennsylvania, Philadelphia, PA 19104-6272*

ABSTRACT

This paper summarizes results of our recent work on moisture-induced environmental embrittlement in Ni_3Al. We took the unconventional approach of starting with single crystals of B-free Ni_3Al, which were cold rolled and recrystallized to produce crack-free polycrystalline material. Our results show that the intrinsic ductility (~16%) of Ni_3Al (23.4 at.% Al) is considerably higher than previously thought; however, it is severely embrittled by moisture in air (ductility dropping from a high of ~16% when tested in oxygen to a low of ~3% in air). Since B-doped Ni_3Al does not show such embrittlement (i.e., ductilities high in both air and oxygen), we conclude that a significant part of the beneficial effect of B must be related to suppression of this environmental effect. However, B must also improve grain boundary (GB) cohesion in Ni_3Al, since our B-free alloy fractures intergranularly whereas B-doped alloys in general fracture transgranularly. Addition of a small amount (0.26 at.%) of Zr to Ni_3Al significantly improves its ductility: to 11-13% in air, and 48-51% in oxygen. The ductilities observed in oxygen are comparable to the highest ever ductility observed in B-doped Ni_3Al, indicating that the GBs in this Zr-doped alloy are not intrinsically brittle (rather, environmental embrittlement is the main reason for its brittleness). Zr dramatically increases the resistance of Ni_3Al to GB fracture—perhaps by increasing GB cohesion. However, Auger analysis shows little or no Zr segregation on the GBs of Ni_3Al, making it unclear how it might actually affect GB cohesion. Zr does not significantly increase the resistance of Ni_3Al to environmental embrittlement; nor does it suppress intergranular fracture. In both these respects Zr behaves differently than B.

INTRODUCTION

Polycrystalline Ni_3Al fractures intergranularly with limited tensile ductility [e.g., 1-4], in contrast to single crystals which undergo extensive plastic deformation before fracture [e.g., 5-7]. This behavior persists even in high-purity Ni_3Al in which the GBs are quite clean and free of potentially harmful impurities [4,8,9]. It has been commonly believed, therefore, that the GBs in Ni_3Al are "intrinsically" brittle—unlike GBs in conventional metals which rarely constitute a serious source of weakness unless embrittled by (extrinsic) impurities.

We were prompted to revisit the subject of brittle fracture in Ni_3Al because of mounting recent evidence that the moisture in ordinary ambient air can severely embrittle many ordered intermetallics [10-24] [in the absence of such (extrinsic) embrittlement some of these intermetallics can actually exhibit extensive ductilities]. The proposed mechanism involves the reduction of H_2O in air by reactive elements in the intermetallics, resulting in the generation of atomic H, which then embrittles the crack-tip region. Our goal here is to summarize results of our recent studies [25,26] aimed at investigating whether such an environmental effect contributes significantly to the room-temperature brittleness of Ni_3Al.

EXPERIMENTAL

Two different Ni_3Al single crystals were used as starting materials in our studies, a <100> oriented plate of Ni-23.4Al [26], and a <110> oriented cylinder of Ni-22.65Al-0.26Zr [25]. Both crystals were first homogenized for 24 h at 1200°C. The Zr-doped crystal was then forged (to flatten the cylinder), and subsequently cold rolled, for a total reduction of ~60%. The binary crystal was rolled directly (total reduction in thickness ~40%). All the deformation processing operations were carried out at room temperature. Tensile specimens having a gauge section of $7.9 \times 3.2 \times 0.6$ mm were electro-discharge machined (EDM) from the rolled plates at 0° and 45° to the rolling direction, and ~0.08 mm was then removed from all the EDM surfaces by grinding. The specimens were annealed at 1000°C for 1 h + 800°C for 5 h, in a vacuum better than ~7×10^{-4} Pa. After this heat treatment, the gauge sections of the specimens were polished with SiC paper (finishing with 2/0 grit) to remove any oxide that may have formed during annealing.

Room temperature tensile tests were performed on a screw-driven Instron machine equipped with a vacuum chamber. The tests were conducted in 3 different environments (air, distilled water, and oxygen) at a constant cross-head speed of 4.2×10^{-2} mm·s^{-1}, which corresponds to an engineering strain rate of 5.3×10^{-3} s^{-1}. The water tests were conducted with the specimen gauge sections completely immersed in distilled water. For the oxygen tests, the test chamber was first evacuated to a pressure of ~4×10^{-4} Pa, and then oxygen from a gas bottle was leaked into the test chamber through a Varian leak valve until the oxygen pressure (monitored with a Granville Philips convectron gauge) reached 6.7×10^{4} Pa.

Grain-boundary chemistry was analyzed by Auger electron spectroscopy in a PHI 590 scanning Auger microprobe operated at a beam voltage of 5 kV and beam currents of ~8 nA. Fracture surfaces were examined in an ISI 40 scanning electron microscope operated at 15 kV.

RESULTS AND DISCUSSION

Table 1 summarizes our results on binary Ni_3Al [26]. For specimens oriented with their tensile axes at 0° to the rolling direction, the measured ductilities (average of 2 tests) are 3.1% in air and 15.8% in (dry) oxygen. Similarly, for the 45° specimens, the ductilities are 4.8% in air and 12.6% in oxygen. Two important conclusions can be drawn from these results: (*a*) the intrinsic ductility (~16%) of B-free Ni_3Al is not as low as was once thought, and (*b*) Ni_3Al is indeed severely embrittled by moisture in air. Earlier work [27] on conventionally produced polycrystalline Ni_3Al (Ni-23.5Al) had also demonstrated an environmental effect, but ductility in oxygen was considerably less in that case (7-8%). A possible reason for this difference is that specimens produced from single crystals contain fewer defects (such as microcracks) than those produced conventionally. In other words, ductility is limited if there are other brittle paths for premature fracture such as preexisting microcracks—consistent with the view that fracture occurs at the "weakest link." (Since polycrystalline Ni_3Al is brittle at both low *and* high temperatures, it is extremely difficult by conventional means to process a cast ingot of Ni_3Al down to sheet or plate stock without introducing cracks.)

TABLE 1. Effect of test environment and specimen orientation on the room temperature tensile properties of Ni–23.4Al tested at a strain rate of 5.3×10^{-3} s^{-1}.

Specimen Orientation	Test Environment	Elongation to Fracture (%)	Yield Strength (MPa)	Ultimate Tensile Strength (MPa)
0°	Air	3.1	308	392
	Oxygen	15.8	336	681
45°	Air	4.8	327	401
	Oxygen	12.6	345	642

Figs. 1(a) and 1(b) show the fracture surfaces of the 45° specimens tested in air and oxygen, respectively. Fracture was predominantly intergranular independent of test environment, and the exposed GBs were smooth and showed only occasional traces of slip. GBs are the weak links in Ni_3Al when moisture is allowed to embrittle them [Fig. 1(a)]. Interestingly, they continue to be weak links even after environmental embrittlement is suppressed (by testing in oxygen) [Fig. 1(b)]. Despite this, however, it is clear that the GBs in Ni_3Al are not as brittle as was once thought (provided that the extrinsic sources of embrittlement are scrupulously eliminated).

Fig. 1. Room temperature fracture surfaces of 45° specimens of Ni–23.4Al tensile tested in (a) air, and (b) oxygen. (The fracture surfaces of the 0° specimens looked similar.)

It is instructive to reexamine the role of B in Ni_3Al in light of our present results. Liu [28] has recently shown that when a sufficient amount of B is added to Ni_3Al (24 at.% Al) it essentially eliminates environmental embrittlement (ductilities of 39 and 43% in air and oxygen, respectively). Similar results were obtained by Mashashi et al. [29] when they tested as-cast Ni-24Al containing 500 wppm B (ductilities of 22 and 28% in air and vacuum, respectively). However, when less than the optimum level of B is added, Ni_3Al does exhibit environmental embrittlement, and its ductility depends on the testing environment (10 and 26% in air and vacuum, respectively) [30]. Similarly, if atomic hydrogen is forcibly introduced into the GBs of Ni_3Al (say, by cathodic charging), ductility drops precipitously and fracture becomes intergranular even in alloys containing high levels of boron [31,32]. These results, taken together with our results on B-free Ni_3Al, suggest that a significant part of the so-called B effect is related to the suppression of environmental embrittlement—perhaps by blocking H penetration along the GBs. However, B also has a dual role of strengthening the GBs in Ni_3Al {based on the observation that fracture in B-doped Ni_3Al is almost entirely transgranular [4], whereas B-free Ni_3Al fractures predominantly intergranularly (Fig. 1)}.

An even more dramatic environmental effect is seen in Zr-doped Ni_3Al [25]. As summarized in Table 2, specimens oriented with their tensile axes at 0° to the rolling direction have ductilities of 8.7% in water, 13.2% in air, and 50.6% in oxygen. Similarly, the 45° specimens have ductilities of 6.3% in water, 10.7% in air, and 47.8% in oxygen. The ductilities observed in dry oxygen are comparable to the highest ever ductilities observed in B-doped Ni_3Al, indicating that the GBs in this (B-free) alloy are not intrinsically brittle. However, the ductilities decrease dramatically in moisture-containing environments, indicating a substantial environmental effect.

Despite the relatively high ductilities listed in Table 2, fracture in the Zr-doped alloy remains predominantly intergranular regardless of test environment [25]. Given that, however, there are subtle differences in the fracture behavior as a function of test environment. Thus, for example, as shown in Fig. 2, there is somewhat more transgranular fracture in the oxygen-tested sample than in the air-tested one; it also shows more evidence of slip on the exposed GBs.

Eventhough the GBs appear to be the weak links in both the binary and Zr-doped alloys (because both fracture predominantly intergranularly), Zr has clearly increased the resistance of Ni_3Al to GB fracture—perhaps by increasing GB cohesion. Unfortunately, our Auger analysis shows little or no Zr segregation on the GBs of Ni_3Al. So it is unclear how it might affect GB cohesion without actually being present in detectable amounts on the GBs. Interestingly, Zr does not significantly increase the resistance of Ni_3Al to environmental embrittlement; nor does it suppress intergranular fracture. In both these respects Zr behaves differently than B. Clearly, additional work is needed to unravel the detailed mechanism by which Zr improves the ductility of Ni_3Al.

TABLE 2. Effect of test environment and specimen orientation on the room temperature tensile properties of Ni–22.65Al–0.26Zr tested at a strain rate of 5.3×10^{-3} s^{-1}.

Specimen Orientation	Test Environment	Elongation to Fracture (%)	Yield Strength (MPa)	Ultimate Tensile Strength (MPa)
0°	Water	8.7	322	528
	Air	13.2	324	661
	Oxygen	50.6	326	1451
45°	Water	6.3	331	473
	Air	10.7	341	603
	Oxygen	47.8	327	1438

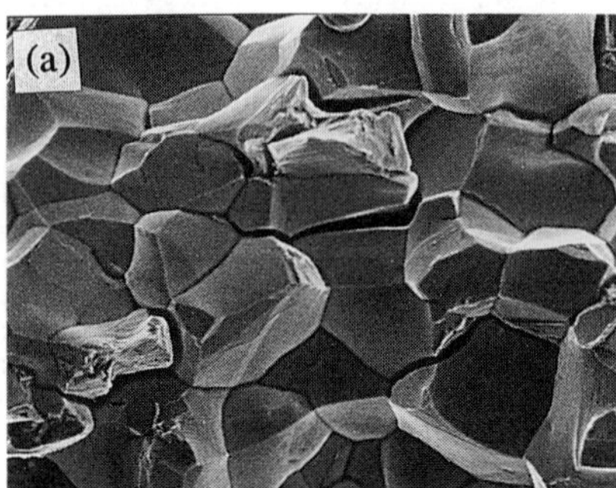

Fig. 2. Room temperature fracture surfaces of 0° specimens of Ni–22.65Al–0.26Zr tensile tested in (a) air, and (b) oxygen. (The fracture surfaces of the 45° specimens looked similar.)

CONCLUSIONS

1. Polycrystalline, B-free Ni_3Al (23.4 at.% Al), produced by cold rolling and recrystallizing a single crystal, has room-temperature tensile ductilities of 3-5% in air and 13-16% in oxygen. Thus, the "intrinsic" ductility of Ni_3Al is considerably higher than previously thought. To obtain such high ductilities, however, extrinsic sources of embrittlement (impurities, embrittling environments, microcracks, etc.) have to be carefully eliminated.
2. Ni_3Al is severely embrittled by the moisture present in ordinary ambient air (ductility decreasing from a high of ~16% in oxygen to a low of ~3% in air).
3. Fracture is predominantly intergranular in both air and oxygen indicating that, while moisture can further embrittle the GBs in Ni_3Al, they remain the weak links even in the absence of environmental embrittlement. However, they are not as intrinsically brittle as once thought.
4. A significant portion of the so-called B effect in Ni_3Al is related to suppression of moisture-induced environmental embrittlement. In addition, B also enhances GB cohesion and suppresses GB fracture.

5. The addition of a small amount (0.26 at.%) of Zr has a dramatic effect on the ductility of Ni_3Al (22.65 at.% Al): ductilities increase to 11-13% and 48-51% in air and oxygen, respectively.
6. Although the detailed mechanism by which Zr improves ductility is still unclear, some distinguishing features have so far been identified: Zr does not segregate strongly to the GBs in Ni_3Al; nor does it suppress intergranular fracture or environmental embrittlement. In these respects it is quite different from B. It does appear to increase the resistance of Ni_3Al to GB fracture—probably by enhancing GB cohesion (like B)—although how it does that without segregating to the GBs remains a puzzle.

ACKNOWLEDGMENTS

We thank E. H. Lee and D. H. Pierce for technical assistance, and J. H. Schneibel and C. G. McKamey for manuscript review. This research was sponsored by the Division of Materials Sciences, U.S. Department of Energy under contract DE-AC05-84OR21400 with Martin Marietta Energy Systems, Inc., and by the LRSM at the University of Pennsylvania supported by the National Science Foundation under Grant No. DMR/MRL88-19885.

REFERENCES

1. R. G. Davies and N. S. Stoloff, Trans. TMS-AIME, **233**, 714 (1965).
2. P. H. Thornton, R. G. Davies, and T. L. Johnston, Metall. Trans. **1**, 207 (1970).
3. K. Aoki and O. Izumi, *Nippon Kinzoku Gakkaishi*, **41**, 170 (1977).
4. C. T. Liu, C. L. White, and J. A. Horton, Acta Metall. **33**, 213 (1985).
5. S. M. Copley and B. H. Kear, Trans. TMS-AIME, **239**, 977 (1967).
6. K. Aoki and O. Izumi, Trans. Jpn. Inst. Met. **19**, 203 (1978).
7. F. E. Heredia and D. P. Pope, Acta Metall. **39**, 2017 (1991).
8. T. Takasugi, E. P. George, D. P. Pope, and O. Izumi, Scripta Metall. **19**, 551 (1985).
9. T. Ogura, S. Hanada, T. Masumoto, and O. Izumi, Metall. Trans. **16A**, 441 (1985).
10. C. T. Liu, E. H. Lee, and C. G. McKamey, Scripta. Metall. **23**, 875 (1989).
11. C. T. Liu and E. P. George, Scripta Metall. **24**, 1285 (1990).
12. D. J. Gaydosh and M. V. Nathal, Scripta Metall. **24**, 1281 (1990).
13. R. J. Lynch, L. A. Heldt, and W. W. Milligan, Scripta. Metall. **25**, 2147 (1991).
14. C. T. Liu, C. G. McKamey, and E. H. Lee, Scripta. Metall. **24**, 385 (1990).
15. C. G. McKamey and C. T. Liu, Scripta. Metall. **24**, 2119 (1990).
16. C. T. Liu and W. C. Oliver, Scripta. Metall. **25**, 1933 (1991).
17. T. Takasugi, H. Suenaga, and O. Izumi, J. Mater. Sci. **26**, 1179 (1991).
18. T. Takasugi, in *High-Temperature Ordered Intermetallic Alloys IV*, eds. L. A. Johnson, D. P. Pope, and J. O. Stiegler (Materials Research Society, Pittsburgh, PA, 1991), Vol. 213, p. 403.
19. N. Masahashi, T. Takasugi, and O. Izumi, Metall. Trans. **19A**, 353 (1988).
20. T. Takasugi and O. Izumi, Acta Metall. **34**, 607 (1986).
21. T. Takasugi and O. Izumi, Scripta Metall. **19**, 903 (1985).
22. C. Nishimura and C. T. Liu, Scripta Metall. **25**, 791 (1991).
23. C. Nishimura and C. T. Liu, Acta Metall. et Mater. **40**, 723 (1992).
24. C. T. Liu, in *Proc. Int. Symp. on Intermetallic Compounds – Structure and Mechanical Properties (JIMIS-6),* ed. O. Izumi (The Japan Inst. Met., Sendai, Japan, 1991) p. 703.
25. E. P. George, C. T. Liu, and D. P. Pope, Scripta Metall. **27**, 365 (1992).
26. E. P. George, C. T. Liu, and D. P. Pope, Scripta Metall., to be published.
27. C. T. Liu, Scripta Metall. **27**, 25 (1992).
28. C. T. Liu, unpublished research, ORNL (1992).
29. N. Masahashi, T. Takasugi, and O. Izumi, Acta Metall. **36**, 1823 (1988).
30. X. J. Wan, J. H. Zhu, and K. L. Jing, Scripta Metall. **26**, 473 (1992).
31. A. K. Kuruvilla, S. Ashok, and N. S. Stoloff, in *Proc. Third Intl. Congress on Hydrogen in Metals* (Pergamon, Paris, 1982) Vol. 2, p. 629.
32. A. K. Kuruvilla and N. S. Stoloff, Scripta Metall. **19**, 83 (1985).

HYDROGEN EFFECTS ON FRACTURE IN α_2 TITANIUM ALUMINIDES

ANTHONY W. THOMPSON
Dept. of Materials Science and Engineering, Carnegie Mellon University, Pittsburgh, PA 15213

ABSTRACT

The effects of hydrogen on titanium aluminide alloys based on Ti_3Al or α_2 are now beginning to be understood. It has been established that large amounts of hydrogen are readily absorbed into these alloys, and on cooling to room temperature, virtually all this hydrogen is precipitated as a hydride phase or phases. These hydrides in turn affect mechanical properties much as in other hydride-forming materials. Extensive data now exist on these effects in the alloys based on α_2. Fracture in particular is now being studied in detail, including work on micromechanisms of hydrogen fracture, which can be quantitatively compared to experiment. Needs for additional work are identified.

INTRODUCTION

In recent years, considerable understanding has emerged of the metallurgy and mechanical properties of titanium aluminide alloys based on the Ti_3Al compound [1-5], which is called α_2 and has the DO_{19} crystal structure. Additional interest has been attached to the effects of hydrogen on these alloys, in part from concern for environmental effects on the material in service conditions. As the present paper summarizes, it has become established that hydrogen affects the properties of titanium aluminide alloys, in ways which are broadly similar to effects in other titanium alloys.

For titanium alloys, the hydride phases have been reviewed in some detail [6-8]. The primary effects occur in the HCP α phase,which has quite limited hydrogen solubility and readily precipitates hydrides. Three hydride phases are known, a δ phase with FCC structure, and two FCT phases, called γ and ε, which have, respectively, c/a ratios greater than and less than unity [8]. (The FCT nomenclature is used to emphasize the similarity of γ and ε to the δ phase.) The δ is usually regarded as the dominant hydride, and is reported [8,9] to exhibit H/Ti ratios from 1.5 to 1.99.

Evidence to date indicates that hydrogen interactions with titanium aluminide alloys have a number of similarities to effects in titanium alloys. Hydride precipitation in a titanium aluminide was apparently first observed in binary Ti_3Al [10], while the first hydride observations in alloys were for the α_2 phase [11,12]. As has been reviewed [13-16], hydride formation in α_2 alloys has now been widely observed, although a number of discrepancies and questions remain on the published record. Occasional confusion about amounts of hydrogen absorption has been clarified by the determination [17] of the equilibrium solvus for hydrogen in the α_2 alloy Ti-24 Al-11 Nb (atomic pct), called Ti-24-11. This solvus is shown in Fig. 1.

As has been reviewed, there has now been established a fairly clear pattern of behavior for α_2 alloys containing hydrogen and precipitated hydrides [13-16,18-20]. Most studies have concentrated on tensile properties, but there have also been efforts to study fundamentals of fracture behavior. Those fracture studies, and recent results extending them, are the subject of the present paper.

HYDROGEN AND HYDRIDES IN α_2 ALLOYS

Hydrogen in titanium aluminide alloys not only is capable of precipitating as a hydride phase, but also affects the stability of the microstructure. Study of stabilization of the beta phase in the α_2 aluminide alloy Ti-24-11 has been reported [21]; the primary effect appeared to be kinetic rather than thermodynamic. This conclusion is based on the observation that even the absorption of 3000 wppm

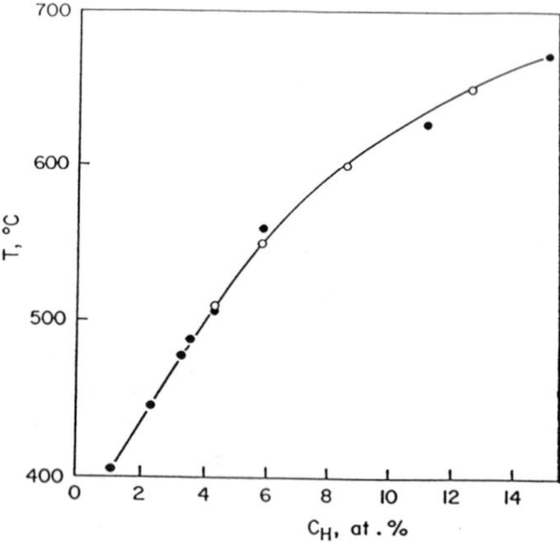

Fig 1. Experimentally measured solvus for hydrogen in Ti-24-11 by Chu, *et al.* [17].

hydrogen decreased the beta transus temperature by only about 50 K, while very considerable refinement of microstructure occurred in the presence of this much hydrogen [21], evidently through kinetic postponement of transformation to lower temperatures. This effect of hydrogen on β stability in Ti-24-11 is rather less than what is observed in titanium alloys [8].

Hydrogen also appears to have a second effect on α_2 alloys which exhibit substantial content of β phase (which then often exhibits the ordered B2 structure) at low temperatures. It appears that hydrogen absorption assists formation of the orthorhombic or O phase [22-25], which may reflect hydrogen assistance of O phase formation kinetics. This O phase then appears to remain present after hydrogen has been removed by vacuum heat treatment. Whenever the O phase forms, it can form in both β and α_2 phases, but appears [25] to occur primarily at the expense of β. It is also possible [25] for the O phase to revert to the stable α_2 phase.

The literature contains a number of disagreements about the structure and composition of the hydride phase(s) in the α_2 aluminide alloys [15,16]. The FCC δ phase hydride is often observed, just as in titanium alloys; but when hydrogen content is lower, a distorted (enlarged) α_2 phase, with lattice parameters corresponding to about a 5 pct volume increase, is observed [24-26]. This phase is apparently unknown in titanium alloys. Although H/Ti ratios have not usually been determined for hydrides in aluminides, the likelihood of participation by other elements in the hydrides is becoming clarified by recent work. Rudman, *et al.* [10] reported that the hydride appeared to form as TiH_x, without any aluminum participation in the hydride, and confirmatory evidence has now been reported by Legzdina, *et al.* [27] for TiAl alloys. Thus it appears that Al does not participate in the hydride [15,16] of titanium aluminide alloys.

HYDROGEN EFFECTS ON FRACTURE

Mechanical properties of the α_2 titanium aluminide alloys have been reviewed in some detail [2-5,19,21,28]. Studies of the effects of hydrogen are now extensive enough that fracture and other properties have also been reviewed [13-16,19,20,25,29]. The effect of microstructure in Ti-24-11 was summarized by Chan [19] as in Fig. 2. Mechanical properties, beyond the ductility shown in Fig. 2, have been determined in some detail for Ti-24-11 [20] and are shown in Fig. 3 to illustrate what is expected to be the typical dependence of these properties on hydrogen content.

As fracture behavior receives increased attention, it may be appropriate to point out the valuable characteristics of notched specimen testing in evaluating fracture processes. The advantages of notched specimens may be summarized as follows. First, the spatial distribution of stresses and strains is well

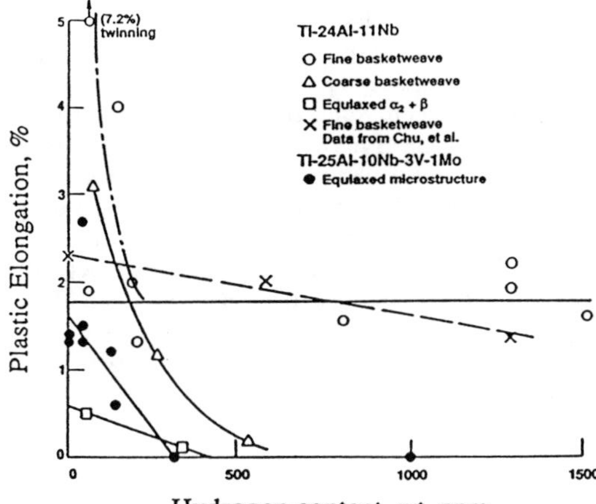

Fig. 2. Microstructure dependence of hydrogen effects on fracture, from Chan [19]. The data of Chu, *et al.* are in ref. 20.

known [30-34], permitting easy analysis of local conditions near the notch. Second, variation in stress state, e.g. the ratio of normal to shear stress, is readily accessible through variation in notch geometry [31,32]. Third, since peak stresses and strains occur at quite different locations, separated by a distance of approximately a notch root radius, it is relatively easy to determine the critical parameter, e.g. for crack initiation, from location beneath the notch [32,34,35]. Figure 4 is a depiction of the notch root slip line field, with the peak strain location and approximate location of peak stress indicated. Finally,

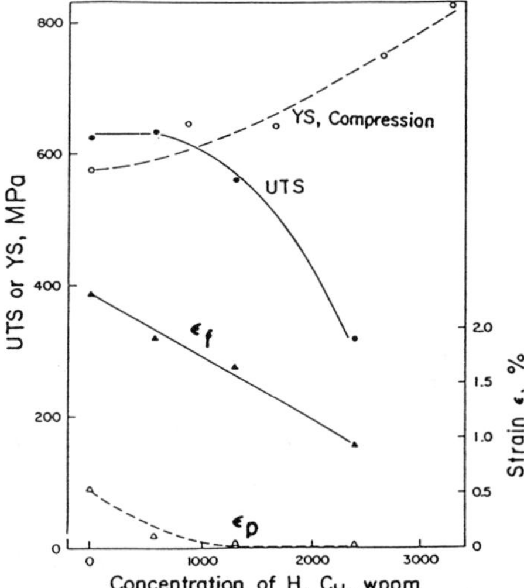

Fig. 3. Basic mechanical properties of Ti-24-11 with hydrogen. Compressive yield strength (YS), failure stress (either ultimate tensile strength, UTS, or fracture stress), and ductilities (uniform true strain ε_p and true fracture strain ε_f) shown as a function of hydrogen concentration in wt. ppm. Each plotted point averages two to five tests. From ref. 20.

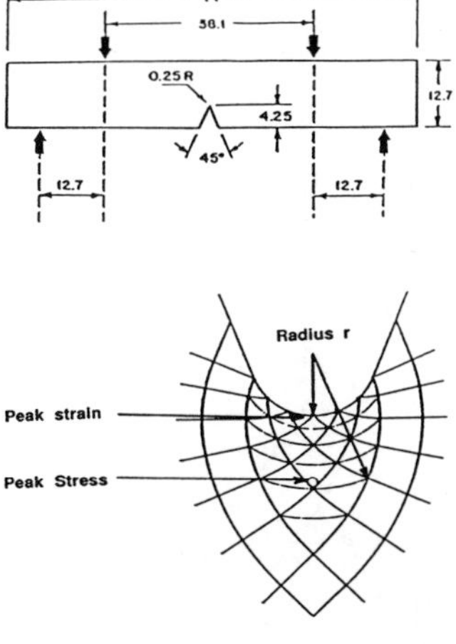

Figure 4. Characteristics of notched specimens which facilitate fracture studies. At top is shown a typical 4-point bend bar, with dimensions in millimeters [34,35]. At bottom is the slip line field characteristic of local yielding at the notch root, with indications of notch root radius r, and approximate locations of peak stress and strain [30,31].

because the notch root radius is relatively large compared to microstructural dimensions, e.g. 250 μm, the scale is convenient for study of microstructural effects [32,34,35].

Notched specimens can provide a variety of fundamental fracture information [31,32] and can be used to supplement the more conventional (and sometimes less productive of insight) parameters such as fracture toughness and ductility [36-38]. An example of the use to which this type of specimen can be put is the result found for Ti-24-11 [20], that a temperature-independent, true local fracture stress σ_F was observed in cleavage-like fracture at room temperature. The "transgranular" cleavage model of Smith [39] was used to calculate the effective surface energy [20]; the magnitude was about 27 J/m^2, and the fractographic observations appeared to be generally consistent with the micromechanism selected [20], although some question is permissible [16,20,28,40] as to whether the failure process on a local scale is in fact classical, crystallographic cleavage. The initiation process corresponding to this description is schematically indicated in Fig. 5, based on concepts developed elsewhere [20,28]. The

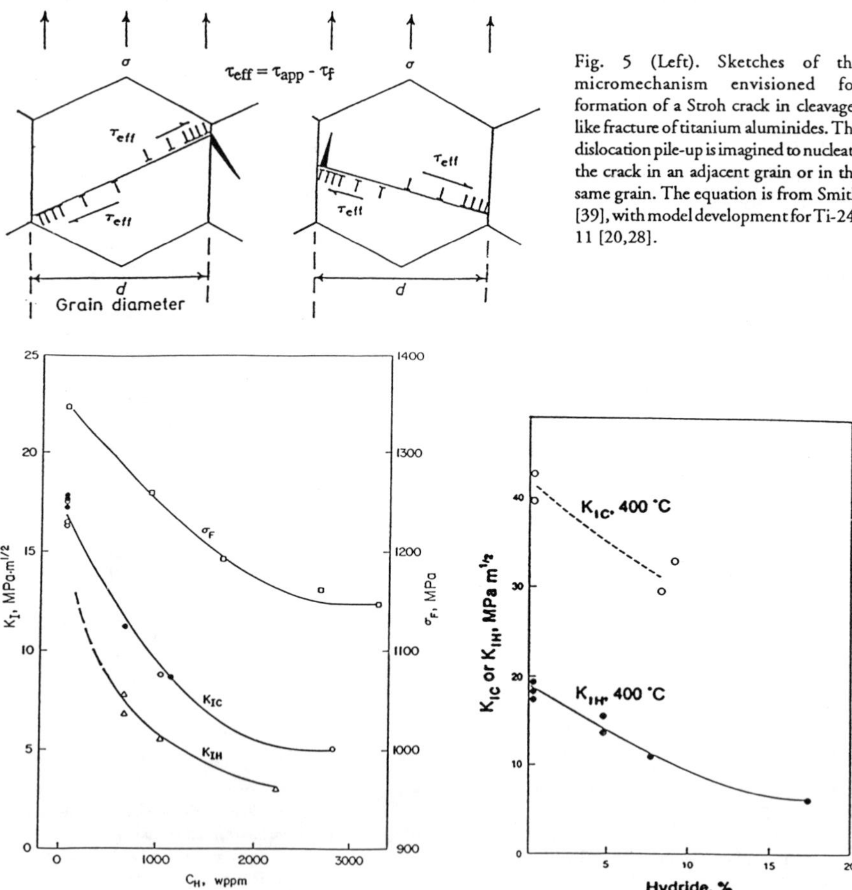

Fig. 5 (Left). Sketches of the micromechanism envisioned for formation of a Stroh crack in cleavage-like fracture of titanium aluminides. The dislocation pile-up is imagined to nucleate the crack in an adjacent grain or in the same grain. The equation is from Smith [39], with model development for Ti-24-11 [20,28].

Fig. 6. Dependence of local fracture stress σ_F, plane strain fracture toughness K_{Ic}, and threshold stress intensity for hydrogen cracking K_{IH}, as a function of hydrogen content in Ti-24-11.

Fig. 7. Variation of fracture toughness and hydrogen cracking threshold as a function of hydrogen content, expresssed as vol. pct hydride, for tests at 400°C [29].

role, if any, of hydrides in this fracture process remains to be determined. It may resemble that familiar in other hydride-forming materials [7,38,41], or may differ in significant ways [16,20,29]. Study of these effects is complicated by the microstructure of Ti-24-11 and Ti-25-10-3-1, the most widely studied α_2 aluminide alloys, along with the newer alloys of similar compositions [25,42]. Extension of the same study to orthorhombic and related alloys, which show considerable promise [43], will also be a challenge.

Extension of ambient temperature properties to intermediate temperatures which may relate to aerospace service conditions [2,44,45] are of interest for α_2 aluminides. Fig. 6 presents some of the fracture results obtained as a function of hydrogen content at room temperature, comparing the threshold stress intensity, K_{IH}, as a function of hydrogen content to the K_{Ic} and σ_F results previously published [20,25]. Fig. 7 present additional fracture results [29] for Ti-24-11, for tests at 400°C. Most of these data are for a single microstructure of this alloy, so considerable work remains to be done, both to expand the range of microstructures which are characterized, and to generalize the findings. Yet there is an encouraging indication that understanding to date is consistent with other materials of similar behavior, including the extensively-studied titanium alloys. More thorough knowledge of the effects of hydrogen on α_2 titanium aluminide alloys should continue to develop with added study.

ACKNOWLEDGEMENTS

I am grateful to NASA, the Alcoa Center for Engineering Materials at Carnegie Mellon, to DARPA, and to the Air Force Office of Scientific Research, for support of work on titanium aluminides.

REFERENCES

1. R. Strychor, J.C. Williams and W.A. Soffa, *Metall. Trans. A*, 19A, 225-234 (1988).
2. R. L. Fleischer, D.M. Dimiduk and H.A. Lipsitt, *Ann. Rev. Mater. Sci.*, 19, 231-263 (1989).
3. D.A. Koss, D. Banerjee, D.A. Lukasak and A.K. Gogia, in *High Temperature Aluminides and Intermetallics*, edited by S.H. Whang, C.T. Liu, D.P Pope and J.O. Stiegler (TMS-AIME, Warrendale, PA, 1990), pp. 175-196.
4. J. M. Larsen, K. A. Williams, S. J. Balsone and M. A. Stucke, in *High Temperature Aluminides and Intermetallics*, edited by S.H. Whang, C.T. Liu, D.P Pope and J.O. Stiegler (TMS-AIME, Warrendale, PA, 1990), pp. 521-556.
5. D.M. Dimiduk, D.B. Miracle, Y.-W. Kim and M.G. Mendiratta, *ISIJ Internat. Jnl.*,31, 1223-1234 (1991).
6. W.M. Mueller, in *Metal Hydrides*, edited by W.M. Mueller, J.P. Blackledge and G.G. Libowitz (Academic Press, NY, 1968), pp. 337-383.
7. N.E. Paton and J.C. Williams, in *Hydrogen in Metals*, edited by I.M. Bernstein and A.W. Thompson (Amer. Soc. Metals, Metals Park, OH, 1974), pp. 409-431.
8. A. San-Martin and F.D. Manchester, in *Phase Diagrams of Binary Titanium Alloys*, edited by J.L. Murray (ASM Int., Metals Park, OH, 1987), pp. 123-135.
9. P.E. Irving and C.J. Beevers, *J. Mater. Sci.*, 7, 23-30 (1972).
10. P.S. Rudman, J.J. Reilly and R.H. Wiswall, *J. Less-Common Metals*, 58, 231-240 (1978).
11. A.W. Thompson, W.-Y. Chu and J.C. Williams, in *Summary Proceedings of the 2nd Workshop on Hydrogen-Materials Interaction*, edited by H.G. Nelson, NASP Joint Program Office Workshop Pub. 1004 (NASA-Ames, Moffett Field, CA, 1988), pp. 133-135.
12. D.E. Matejczyk and R.P. Jewett, in *Summary Proceedings of the 2nd Workshop on Hydrogen-Materials Interaction*, edited by H.G. Nelson, NASP Joint Program Office Workshop Pub. 1004 (NASA-Ames, Moffett Field, CA, 1988), pp. 137-139.
13. N.S. Stoloff, M. Shea and A. Castagna, in *Environmental Effects on Advanced Materials*, edited by R.H. Jones and R.E. Ricker (TMS-AIME, Warrendale, PA, 1991), pp. 3-19.

14. A.W. Thompson, in *Environmental Effects on Advanced Materials*, edited by R.H. Jones and R.E. Ricker (TMS-AIME, Warrendale, PA, 1991), pp. 21-33.
15. D. Eliezer, F.H. Froes and C. Suryanarayana, *J. Metals*, 43, No. 3, 59-62 (1991).
16. A.W. Thompson, in *High Temperature Aluminides and Intermetallics*, edited by S.H. Whang, D.P. Pope, and C.T. Liu (Elsevier, New York, 1992), pp. 578-583.
17. W.-Y. Chu, A.W. Thompson and J.C. Williams, *Acta Metall. Mater.*, 40, 455-462 (1992)
18. B.S. Majumdar, H.J. Cialone and J.H. Holbrook, in *Summary Proceedings of the 3rd Workshop on Hydrogen-Materials Interaction*, edited by H.G. Nelson, NASP Joint Program Office Workshop Pub. 1007 (NASA-Ames, Moffett Field, CA, 1990), pp. 73-87.
19. K.S. Chan, *Metall. Trans. A*, 23A, 497-507 (1992).
20. W.-Y. Chu and A.W. Thompson, *Metall. Trans. A*, 23A, 1299-1312 (1992).
21. W.-Y. Chu and A.W. Thompson, *Metall. Trans. A*, 22A, 71-81 (1991).
22. M. Saqib, L.S. Apgar, D. Eylon and I. Weiss, "Effects of Hydrogen on Microstructure and Phase Stability of α_2 Titanium Aluminides," in *Proc. 7th Int. Conf. on Titanium*, edited by F.H. Froes (TMS-AIME, Warrendale, PA), in press.
23. D.A. Hardwick (personal communication), 1991.
24. X. Pierron, D.P. Allen and A.W. Thompson (unpublished research), 1991.
25. A.W. Thompson, in *Summary Proceedings of the 5th Workshop on Hydrogen-Materials Interaction*, edited by H.G. Nelson, (NASA-Ames, Moffett Field, CA), in press.
26. I.M. Robertson and H.K. Birnbaum (personal communication), 1992.
27. D. Legzdina, I.M. Robertson and H.K. Birnbaum, *Scripta Metall. Mater.*, 26, 1737-1741 (1992).
28. A.W. Thompson, "Micromechanisms of Fracture in Titanium Aluminides," in *Proc. 7th Int. Conf. on Titanium*, edited by F.H. Froes (TMS-AIME, Warrendale, PA), in press.
29. W.-Y. Chu and A.W. Thompson, in *High Performance Composites for the 1990's*, edited by S.K. Das, C.P. Ballard and F. Marikar (TMS-AIME, Warrendale, PA, 1991), pp. 143-157.
30. J.R. Griffiths and D.R.J. Owen, *J. Mech. Phys. Solids*, 19, 419-431 (1971).
31. J.F. Knott, *Fundamentals of Fracture Mechanics*, Ch. 7 and 8 (Butterworths, London, 1979), pp. 176-233.
32. A.W. Thompson, *Mater. Sci. and Technol.*, 1, 711-718 (1985).
33. D.J. Alexander, J.J. Lewandowski, W.J. Sisak and A.W. Thompson, *J. Mech. Phys. Solids*, 34, 433-454 (1986).
34. J.J. Lewandowski and A.W. Thompson, *Acta Metall.*, 35, 1453-1462 (1987).
35. I.-G. Park and A.W. Thompson, *Metall. Trans. A*, 22A, 1615-1626 (1991).
36. N.E. Paton and R.A. Spurling, *Metall. Trans A*, 7A, 1769-1774 (1976).
37. R.O. Ritchie and A.W. Thompson, *Metall. Trans. A*, 16A 232-248 (1985).
38. H.K. Birnbaum, in *Hydrogen Effects on Material Behavior*, edited by N.R. Moody and A.W. Thompson (TMS-AIME, Warrendale, PA, 1990), pp. 639-658.
39. E. Smith, *Acta Metall.*, 14, 985-989 and 991-996 (1966).
40. A.W. Thompson and W.-Y. Chu, in *Microstructure/Property Relationships in Titanium Alloys and Titanium Aluminides*, edited by Y.-W. Kim and R.R. Boyer (TMS-AIME, Warrendale, PA, 1991), pp. 165-177.
41. D.S. Shih, I.M. Robertson and H.K. Birnbaum, *Acta Metall.*, 36, 111-124 (1988).
42. A.W. Thompson and W.-Y. Chu, in *Summary Proceedings of the 4th Workshop on Hydrogen-Materials Interaction*, edited by H.G. Nelson, NASP Joint Program Office (NASA-Ames, Moffett Field, CA), in press.
43. W.G. Rowe, in *Microstructure/Property Relationships in Titanium Alloys and Titanium Aluminides*, edited by Y.-W. Kim and R.R. Boyer (TMS-AIME, Warrendale, PA, 1991), pp. 387-398.
44. T.M. Ronald, in *Summary Proceedings of the 2nd Workshop on Hydrogen-Materials Interaction*, edited by H.G. Nelson, (NASA-Ames, Moffett Field, CA, 1988), pp. 3-17.
45. H.G. Nelson, *SAMPE Quarterly*, 20, No. 1, 20-23 (1988).

OXIDATION AND PROTECTION OF Ti3AL-BASED INTERMETALLIC ALLOYS

DOUGLAS W. McKEE
General Electric Corporate Research and Development, P.O. Box 8, Schenectady, New York 12301.

ABSTRACT

Titanium aluminides containing 20-30 atom percent Al can suffer from severe subsurface embrittlement (alpha-case) when exposed to air at elevated temperatures for extended periods. For example,the alloys Ti-24Al-12.5Nb-1.5Mo and Ti-24Al-8Nb-2Mo-2Ta (based on the Ti_3Al alpha-2 intermetallic compound) embrittle to a depth of about 80 microns during 1000 hours exposure to air at 815°C (1500°F), resulting in major reductions in tensile properties. Plasma-sprayed coatings of the MCrAlY and MCr types deposited over a thin diffusion barrier of chromium or tungsten, have been found effective in protecting these alloys against oxidation and embrittlement at this temperatures.

INTRODUCTION

Alloys based on the alpha-2 intermetallic compound Ti_3Al are of increasing interest for aircraft and aerospace applications because of their low density and strength at elevated temperatures [1]. Although the binary alloy is brittle, improved ductility can be achieved by adding small amounts of other elements, such as niobium, tantalum or molybdenum [2]. These materials are also being investigated as possible matrices for fiber-reinforced metal matrix composites [3]. However, titanium alloys in general are susceptible to oxidation in air above 600°C and the alpha-2 family of alloys are particularly vulnerable to sub-surface embrittlement ("alpha-case") during elevated temperature exposure as a result of dissolution of oxygen in the metal lattice [4]. This hardening effect can cause serious degradation of mechanical properties [5]. It is likely, therefore, that protective coatings will be required for applications of alpha-2 alloys that involve exposure to air at temperatures exceeding 600°C. In an earlier study [6], plasma-spray coatings of the MCrAlY- and MCr- types were found effective in protecting (alpha+beta) titanium aluminide alloys against oxidation in air for long periods of time at temperatures as high as 850°C. This paper summarizes the results of an investigation of the behavior of these coatings on two typical alpha-2 type alloys.

Experimental

The alpha-2 alloys used as substrates for the coating studies had the nominal compositions (atom percent): Ti-24Al-12.5Nb-1.5Mo and Ti-24Al-8Nb-2Mo-2Ta.

Cast-and-forged ingots of these alloys were prepared and given appropriate heat and annealing treatments. Pins 1.25 in. long and 0.125 in. diameter were cut from these ingots by electrical discharge machining. The pins were centerless ground, the ends rounded and the surfaces given a buffing treatment before the coating procedure. Previous work [7] had shown that the diffusion of coating elements into the alloy substrates during high temperature exposure could be minimized by interposing a thin barrier layer of tungsten or chromium between the coating and the substrate. This layer (1-2 microns thick) was applied to the pins by RF sputtering before the low pressure plasma spray procedure.

Two coating compositions were evaluated in this work. Powders (-325 mesh) were obtained from Alloy Metals Inc. (Troy, MI) with the nominal compositions Fe-24Cr-8Al-0.5Y (AMI 970, wt.%) and Co-30Cr (AMI 326, wt.%) and applied to the Cr-coated pins by the low-pressure dc arc plasma-spray process to give a coating thickness of approximately 5 mils. On oxidation these coatings form protective surface films of alumina and chromium, respectively (6).

Oxidation experiments on uncoated and coated alloy pins were carried out in microprocessor-controlled test rigs using furnaces which were rapidly raised and lowered around the suspended samples. During the heating cycle, laboratory air was passed over the samples at a rate of 300 ml/min. The heating cycle used was 45 minutes at the test temperature (815°C in the work reported in this paper), followed by cooling in 2-3 minutes to room temperature for a period of 15 minutes. On raising the furnace again the samples attained the test temperature within 5 minutes. Periodically the samples were weighed on an analytical balance to +0.02 mg and the thermal cycling was continued for periods of 1000 hours. Some oxidation experiments were also carried out by heating the specimens at constant temperature in air.

Following the oxidation exposures, the pin specimens were sectioned and the condition of the coatings examined metallographically. Hardness traverses from the coating/alloy interface into the substrate were made as described previously [8] to determine the extent of oxygen embrittlement ("alpha-case") as a function of depth.

Results and Discussion

Fig. 1 shows weight changes for one uncoated and two coated alloy pins during thermal cycling exposure in air at 815°C. The uncoated specimen gave a small but linear weight increase for the 1000 hour duration of the test, whereas pins coated with FeCrAlY and CoCr, over a 1-2 micron W or Cr barrier layer showed much smaller weight gains. Two separate microhardness traverses beneath the surface of an uncoated alloy specimen after this exposure are shown in Fig. 2. A zone of embrittled "alpha-case" clearly has formed in the alloy to a depth of about 60 microns. This hardening effect is sufficient to cause serious reductions in tensile properties.

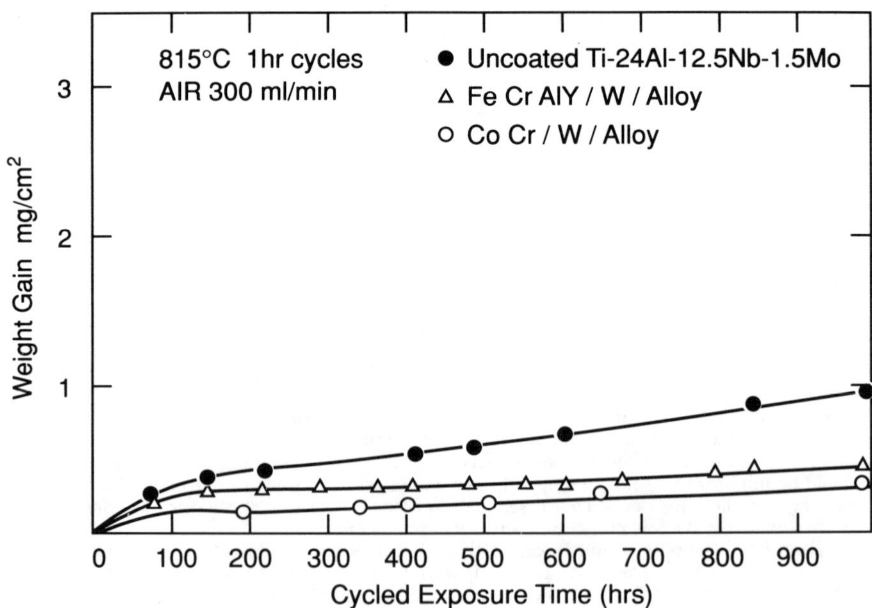

Fig. 1. Weight gain vs. time data for coated and uncoated alpha-2 alloy pins during rapid thermal cycling in air to 815°C.

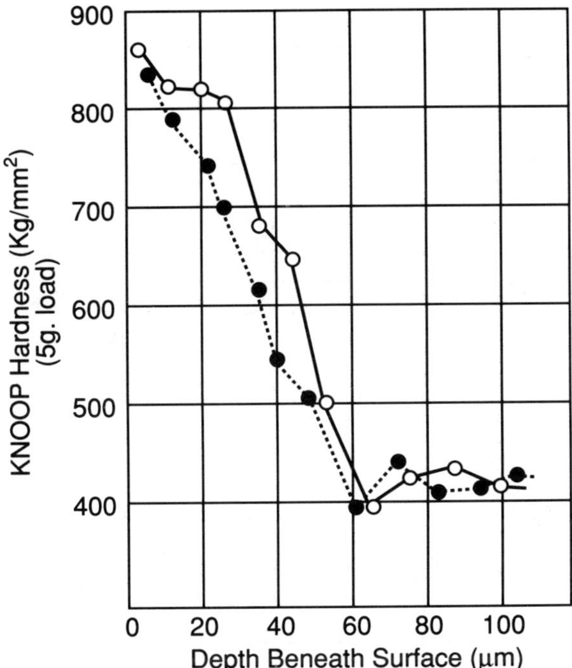

Fig. 2. Microhardness profiles beneath surface of uncoated Ti-24Al-12.5Nb-1.5Mo alloy after 1000 hrs rapid thermal cycling in air to 815°C.

Microhardness profiles beneath the alloy surface for the samples coated with FeCrAlY/W and CoCr/W after the same oxidation exposure are shown in Figs. 3 and 4. No significant "alpha-case" embrittlement was detected beneath the coatings with these specimens. The specimens used to produce the data shown in Figs. 2 and 3 were prepared from two different ingots of the alpha-2 alloy. This may explain the slightly different basal hardness values in the two figures. Fig. 5 shows a section through the FeCrAlY/W-coated pin after the 1008 hr/815°C exposure. The thin layer of tungsten between the coating and the alloy substrate is visible. The constant length of the hardness indentations beneath the coating is evident. In the absence of the thin chromium or tungsten interlayer, diffusion of the elements iron and cobalt from the coating into the substrate occurred to a depth of 15-20 microns during exposure for 1000 hours at 815°C which resulted in increases in hardness of the substrate to this depth. The application of the thin sputtered barrier layers before the plasma-spray coatings were applied effectively eliminated this problem.

Similar behavior was observed using the Ti-24Al-8Nb-2Mo-2Ta alpha-2 alloy. In this case also, heating in air for 1000 hours at 815°C resulted in the formation of an embrittled "alpha-case" zone to a depth of 80 microns as a result of dissolution of oxygen in the alloy lattice. Hardness profiles as functions of depth beneath the coating/alloy interface for a CoCr/Cr-coated pin specimen of this alloy after this oxidation exposure are shown in Fig. 6. No embrittled zone could be detected in this case and the duplex coating is therefore an effective barrier to the ingress of oxygen as well as for preventing diffusion of coating elements into the substrates.

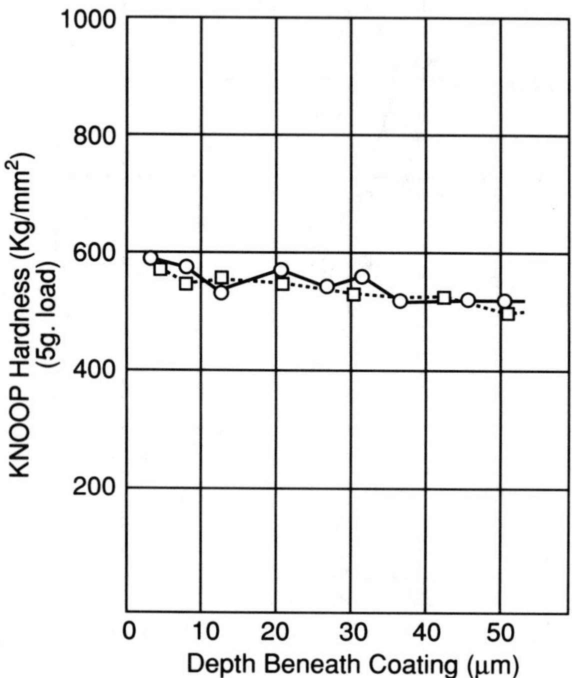

Fig.3. Microhardness profiles beneath coating/alloy interface in a FeCrAlY/W-coated Ti-24Al-12.5Nb-1.5Mo alloy pin after 1008 hrs rapid thermal cycling in air to 815°C.

Conclusions

The problem of "alpha-case" embrittlement of alpha-2 and orthorhombic titanium aluminide alloys during exposure at elevated temperatures in air can be alleviated by coating the alloy surfaces, first with a thin sputtered layer of chromium or tungsten, and then with a plasma-sprayed coating of the MCrAlY or MCr types. Coatings of the type FeCrAlY/W or CoCr/W applied to the alloy Ti-24Al-12Nb-2Mo, have been found to be adherent and protective for periods of at least 1000 hours during rapid thermal cycling in air to 815°C, with no observed embrittlement of the substrate alloy. Similar results have been obtained using the orthorhombic alloy Ti-22Al-27Nb as substrate.

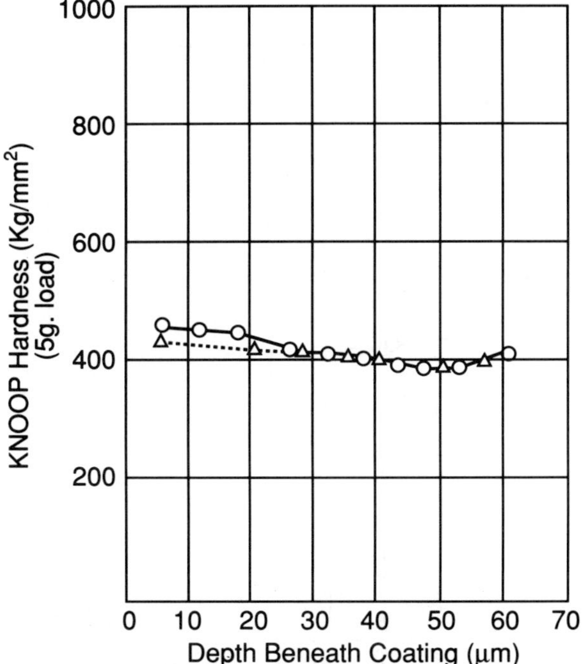

Fig. 4. Microhardness profiles beneath coating/alloy interface in a CoCr/W-coated Ti-24Al-12.5Nb-1.5Mo alloy pin after 1008 hrs rapid thermal cycling in air to 815°C.

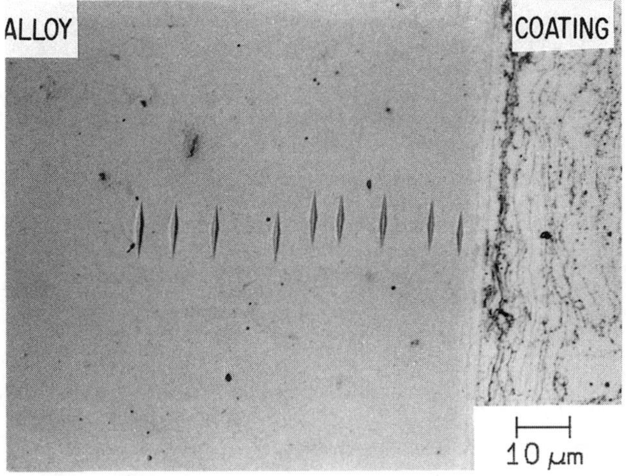

Fig. 5. Section through FeCrAlY/W coating on a Ti-24Al-12.5Nb-1.5Mo alloy pin after 1008 hrs rapid thermal cycling in air to 815°C.

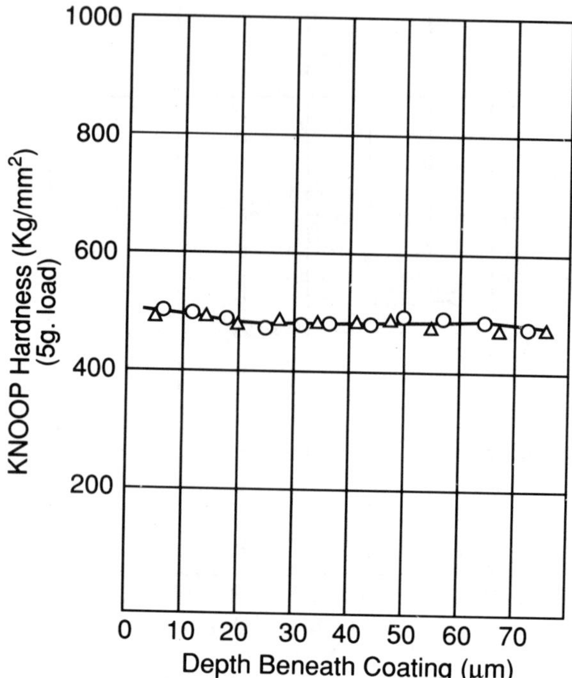

Fig. 6. Microhardness profiles beneath CoCr/Cr coating on Ti-24Al-8Nb- 2Mo-2Ta pin after 1000 hrs rapid thermal cycling in air to 815°C.

References

1. H.A. Lipsett, MRS Symposium Proc., 39, 351 (1985).
2. M.F.X. Gigliotti and B.J. Marquardt, U.S.Patent 4,788,035, 1988.
3. J.W. Pickens and G.K. Watson, paper at 14th. Conf.,on Metal Matrix, Carbon, and Ceramic Composites, Cocoa Beach, FL, Jan. 1990.
4. Z. Liu and G. Welsch, Met. Trans. 19A, 1121 (1988).
5. S. Balsone in "Oxidation of High Temperature Intermetallics" (T. Grobstein and J. Doychak, Eds.) p.219, TMS, 1988.
6. D.W. McKee and K.L. Luthra, Surf. Coatings Tech., to be published.
7. D.W. McKee, paper at AeroMat '92 Conference, Anaheim, CA, May 1992.
8. A. Barbuto and D.W. McKee, paper at IMS Conf., Monterey, CA, July 1991.

HIGH-TEMPERATURE OXIDATION BEHAVIOR OF Nb_3Al BASED ALLOYS

TETSUO FUJIWARA, KEN YASUDA AND HIDEYO KODAMA
Hitachi Research Laboratory, Hitachi, Ltd., 3-1-1 Saiwaicho Hitachi-shi 317 Japan

ABSTRACT

Oxidation behavior of Nb_3Al has been studied in the temperature range from room temperature to 1500°C. It was revealed in this study that the oxide scale that formed at elevated temperatures was not protective against the propagation of the oxidation front. The products of oxidized Nb_3Al alloys melted below 1500 °C, even though the melting point of Nb_3Al is 1960°C.

We expected that the addition of third elements would improve oxidation resistance because atomic sizes smaller than Nb and Al might reduce the solubility and diffusivity of oxygen in the alloy. The elements V, Mn, Fe, Mo, Ta, W and Re were added as the third element for the examination of oxidation resistance. The oxidation resistance tests were carried out by measuring weight gain during temperature elevation and during isothermal heating in air. It was concluded that the addition of Re or Mo improved the oxidation resistance for Nb_3Al alloys.

INTRODUCTION

Intermetallic compound Nb_3Al is a candidate for high-temperature structural applications because of its high melting point, 1960°C. This compound has attractive high-temperature strength even at 1500°C [1]. However, oxidation resistance is a matter of concern. The oxidation properties of Nb_3Al have been presented previously as one material among many Nb based intermetallic compounds [2]. The results indicated that Nb_3Al did not show oxidation resistance at 1200°C in air. However, high-temperature oxidation behavior and oxidation resistance, especially above 1200°C, of this material is almost unknown. Furthermore, the effect of third elements on oxidation resistance of Nb_3Al based alloys has not been investigated by previous researchers.

For oxidation protection, materials must form a slow-growing, dense, and well-adherent oxide layer, such as Al_2O_3. According to Wagner's theory [3], formability of external Al_2O_3 scale is increasing with decreasing of the solubility and diffusivity of oxygen in the alloy. We expected that the addition of third elements might improve oxidation resistance because the smaller atomic size than Nb and Al might reduce the solubility and diffusivity of oxygen in the alloy due to decreasing the lattice size of A15-Nb_3Al.

In this work, oxidation behavior at the temperature range from room temperature to 1500°C of binary Nb_3Al based alloys was studied. Moreover, effects of some third elements, which have smaller atomic size than Nb and Al, on oxidation resistance were investigated.

EXPERIMENTAL PROCEDURE

The materials used in this study were prepared as 30 g button ingots by arc melting under argon atmosphere. The nominal alloy compositions in this study were the following: Nb-20, 25 at.%Al (binary alloys) and Nb-25 at.% Al-5 at.% X (ternary alloy: X= V, Mn, Fe, Mo, Ta, W and Re). The ingots were annealed for 10 hr at 1600°C under vacuum. The samples were cut from the homogenized ingot into 2 mm cubes by spark cutting. Samples were polished with #1000 emery paper just before oxidation test.

Oxidation tests were carried out by the thermobalance apparatus with flowing laboratory air, 100 cc/min flow rate. The weight gains by high-temperature oxidation were measured by thermogravimetric measurement during temperature elevation from room temperature to 1500°C in flowing air. Heating rate was 20 °C/min. Oxidation tests by isothermal annealing at 900°C in air were also carried out. The cross-section and surface morphology of the oxide layers were observed by scanning electron microscopy. Formed phases by oxidation were identified by X-ray diffraction (XRD) analysis. Lattice constants of ternary alloys were also determined by XRD analysis. Effect of doping element on oxidation resistance was evaluated by comparing weight gain of ternary alloys with that of binary alloys at 900°C and 1200°C during temperature elevation.

RESULTS AND DISCUSSIONS

Oxidation behavior of Nb_3Al based binary alloy

Figure 1 shows the oxidation curve of Nb_3Al. The specimens oxidized at 900°C, 1200°C and 1500°C are indicated as (a), (b), and (c) in Fig. 1. The corresponding micrographs are shown in Fig. 2. The weight of Nb_3Al specimens gradually increased from about 800°C to about 1400°C. The oxidation rapidly proceeded above about 1400°C. Nb_3Al was completely oxidized and melted before the temperature reached 1500°C.

Micrographs of cross-sections of Nb_3Al oxidized at 900°C and 1200°C during temperature elevation are shown in Fig. 3 (a) and (b), respectively. The oxide scale can be divided into two layers; outer and inner layer. With increasing temperature, thickness of the inner layer increased, even though that of the outer layer was almost constant. Nb_2O_5, NbO and $AlNbO_4$ were identified by X-ray diffraction analysis for the oxide layer. Fig. 4 shows micrographs of the oxidized Nb_3Al surface after heating up to 900°C and 1200°C. The outer layer has a porous surface at both oxidation temperatures. It implies that the outer layer is not protective against oxygen penetration. Therefore, the oxidation front penetrates rapidly inward as shown in Fig. 3 .

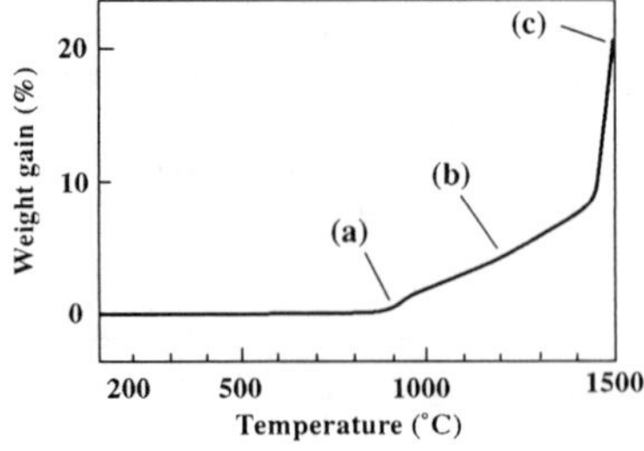

Fig.1 Oxidation test curve of Nb_3Al during temperature elevation.
(a), (b) and (c) in figure indicates the observation in Fig.2.

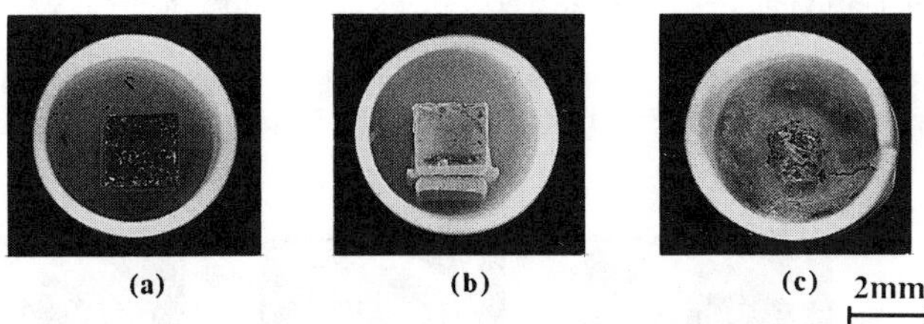

Fig.2 Photographs of Nb_3Al after oxidation test up to (a) 900°C, (b) 1200°C, and (c) 1500°C during temperature elevation, corresponding with the indications on Fig.1 respectively. Each sample was put in a cylindrical alumina holder.

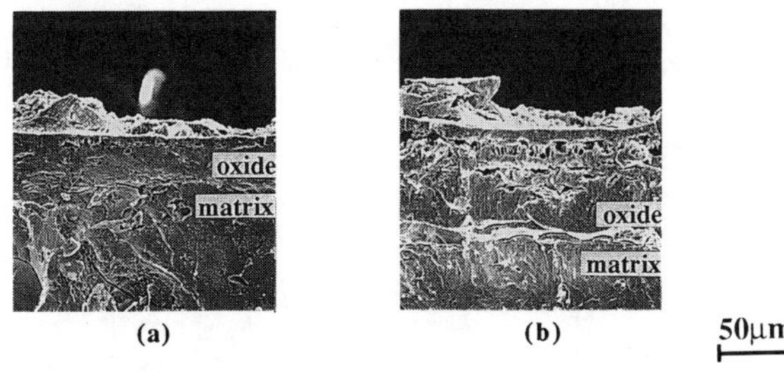

Fig.3 Cross-sectional micrographs of oxidized Nb_3Al after heating up to (a) 900°C and (b) 1200°C.

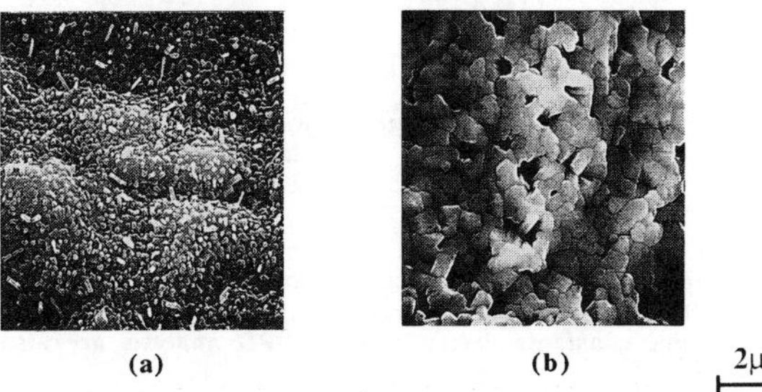

Fig.4 Micrographs of oxidized Nb_3Al surface after heating up to (a) 900°C and (b) 1200°C.

Upon reaching 1500 °C, Nb_3Al was melted as shown in Fig. 2 (c). The XRD result indicated that the main products during oxidation were Nb_2O_5 and Al_2O_3-0.9Nb_2O_5 . The melting point of Nb_2O_5 is 1510°C [4] and the eutectic temperature between Nb_2O_5 and Al_2O_3 is 1400°C [3]. Therefore, the oxidized Nb_3Al based alloys have melted below 1500°C, which is about 500° lower than the melting point of Nb_3Al, 1960°C.

Fig. 5 shows the photographs of Nb_3Al after isothermal heating for 1 hour and 3 hours at 900 °C in air. A thick layer had spalled and disintegrated into flake-like pieces after annealing for 1 hr at 900°C as shown in Fig. 5 (b). The sample was completely oxidized and disintegrated into fine flake-like pieces after annealing for 3 hours at 900°C as shown in Fig. 5 (c).

It can be concluded that a coating or an alloying of a third element which improves oxidation resistance is necessary for the use of binary Nb_3Al intermetallic compound.

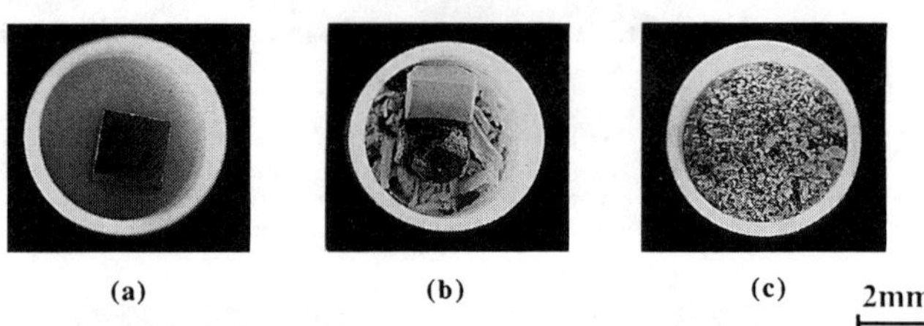

Fig.5 Photographs of Nb_3Al after isothermal annealing at 900°C in air. (a) before test, (b) after annealing for 1 hour and (c) 3 hours.

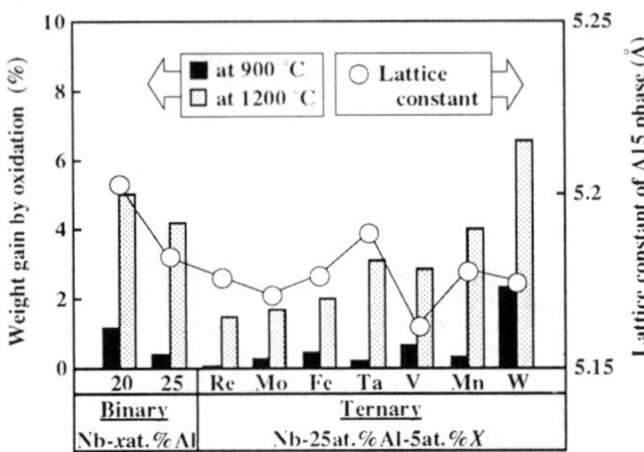

Fig.6 Effects of the third elements on oxidation resistance and A15 lattice constants of Nb_3Al based alloy. Weight gain by oxidation up to 900°C and 1200°C during temperature elevation are shown. A15 lattice constants determined by XRD analysis are also shown.

Effects of third elements on oxidation resistance

Enhancement of oxidation resistance via third element additions in Nb_3Al intermetallic compounds were examined.

The weight gain by oxidation of binary and ternary Nb_3Al based alloys up to 900°C and 1200°C during temperature elevation are shown in Fig. 6. The reduction of A15 lattice constant by addition of ternary elements with smaller atomic radii are also presented in Fig. 6 as open circles. The oxidation resistance of most ternary alloys except Nb-25Al-5W was better than that of Nb-20Al and Nb-25Al binary alloys. The weight gain at 1200°C of Nb-25Al-5Re or Nb-25Al-5Mo was less than half of the magnitude of the Nb-25Al binary alloy. Therefore Re and Mo are effective elements for oxidation resistance. The lattice constants of ternary alloys except for Nb-25Al-5Ta were smaller than that of Nb-25Al and Nb-20Al. Although Nb-25Al-5V has the smallest lattice size in this study, the alloy does not exhibit the best oxidation resistance. The relationship between the lattice constants of A15 phase and the oxidation resistance is still under investigation.

In the case of isothermal oxidation tests at 900°C, addition of Re or Mo, which have smaller atomic size, suppressed the propagation of the disintegration. Fig. 7 (a) and (b) show that these ternary alloys still exhibited sound morphology after isothermal annealing for 1 hour at 900 °C in air, even though binary Nb_3Al was disintegrated into small pieces as shown in Fig. 5 (b). Effect of other third elements (e.g., Fe, Ta, and W) are also illustrated in Fig. 7. These elements were not as effective on disintegration resistance as Re and Mo. Fig. 8 shows the oxidation curves of Nb-25Al, Nb-25Al-5Re and Nb-25Al-5Mo during isothermal oxidation at 900°C in air. Weight gain by oxidation of Nb-25Al-5Re and Nb-25Al-5Mo were quite lower than that of binary Nb-25Al. Therefore addition of Re or Mo improves the oxidation resistance of Nb_3Al alloy by suppressing disintegration.

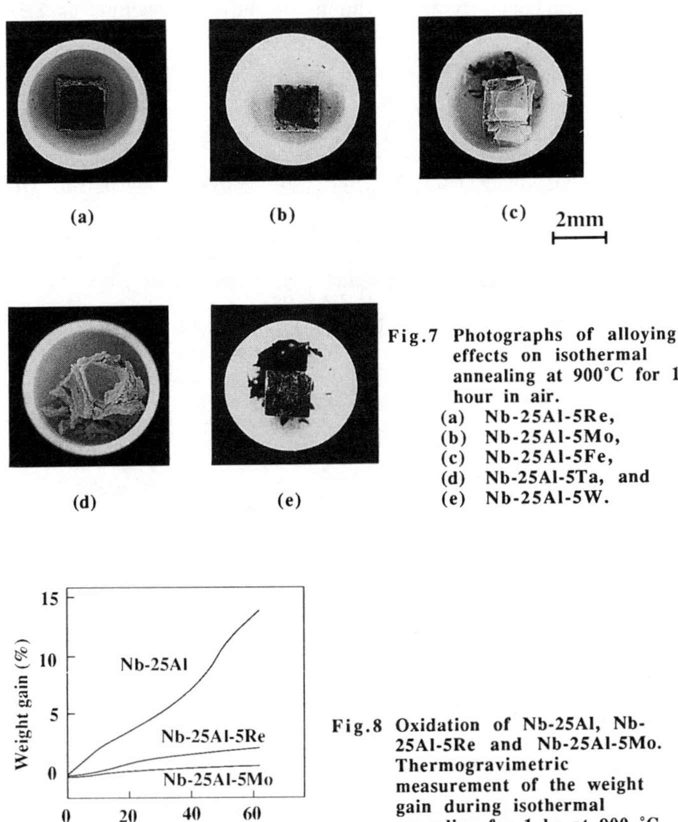

Fig.7 Photographs of alloying effects on isothermal annealing at 900°C for 1 hour in air.
(a) Nb-25Al-5Re,
(b) Nb-25Al-5Mo,
(c) Nb-25Al-5Fe,
(d) Nb-25Al-5Ta, and
(e) Nb-25Al-5W.

Fig.8 Oxidation of Nb-25Al, Nb-25Al-5Re and Nb-25Al-5Mo. Thermogravimetric measurement of the weight gain during isothermal annealing for 1 hr at 900 °C.

However, the protective oxide scale of Al_2O_3 did not form in any ternary alloy. Nb_2O_5 and $AlNbO_5$ were identified in oxidized Nb-25Al-5Re and Nb-25Al-5Mo by XRD analysis. Moreover, the weight gain by oxidation of Nb-25Al-5Re and Nb-25Al-5Mo up to 1500°C totaled over 30 %. The morphologies after oxidation tests were similar to that of Nb-25Al binary alloys as mentioned earlier (Fig. 2(c)). Therefore, it would be difficult to use Nb_3Al based alloys over 1400 °C in air without coating, even though alloying with Re or Mo.

CONCLUSIONS

Nb_3Al based binary intermetallic alloys oxidized significantly above 1400°C. The oxide scale that formed at elevated temperatures was not protective against the propagation of the oxidation front. The products of oxidized Nb_3Al alloys melted below 1500°C, even though the melting point of Nb_3Al is 1960°C.

Addition of smaller size elements, especially Re or Mo, was effective in increasing oxidation resistance for Nb_3Al alloy. This fact is confirmed by lower weight gain by oxidation during temperature elevation and during isothermal annealing at 900°C. However, protective oxide scales were not formed.

REFERENCE

1. T. Fujiwara, K. Yasuda, H. Kodama and M. Suwa, in Proceedings of International Symposium on Intermetallic Compounds (JIMIS-6) edited by O. Izumi (JIM, Sendai, Japan, 1991), p. 633.
2. R. C. Svedberg, in Properties of High Temperature Alloys, edited by Z. A. Forroulis and F . S. Pettit (Electrochem. Soc., Pennington, NJ, 1976), p. 331.
3. G. H. Meier, in Oxidation of High-Temperature Intermetallics, edited by T. Grobstein and J. Doychak (TMS, Warrendale, PA, 1988), p. 16.
4. T. B. Massalski, in Binary Alloy Phase Diagrams, ASM, 139 (1986).

NUCLEATION AND GROWTH OF $MoSi_2$ PEST DURING LOW TEMPERATURE OXIDATION

T. C. CHOU* AND T. G. NIEH**
*Lockheed Research and Development Division, O/93-60, B/204, Palo Alto, CA 94304
**Lawrence Livermore National Laboratory, L-350, P. O. Box 808, Livermore, CA 94550

ABSTRACT

The phenomenon of "$MoSi_2$ pest" has been studied in detail by low temperature oxidation of various polycrystalline $MoSi_2$ monoliths and composites. Single crystalline samples were studied to address the nucleation and growth of pest reaction. Pest of $MoSi_2$ was observed to occur at temperatures between 375 and 500°C. Samples after pest reactions yielded to powdery reaction products of MoO_3 whiskers, SiO_2 clusters, and residual $MoSi_2$ particles. Although all the samples showed complete disintegration after sufficient oxidation durations were rendered, the kinetics of pest disintegration varied among different samples. By performing interrupted oxidation tests, pest reaction was found to consist of nucleation and growth stages, with the former stage acting as the rate-limiting step. Furthermore, the nucleation of pest reaction was found to be closely related to the formation of MoO_3, which was preferentially formed at such defect sites as intergranular boundaries and cracks. Mechanisms leading to the pest disintegration of $MoSi_2$ are discussed in light of the nucleation and growth of pest reaction.

INTRODUCTION

Molybdenum disilicide ($MoSi_2$), possessing a high melting point (2020°C), excellent oxidation and corrosion resistant properties at high temperatures [1-6], good thermal conductivity, and reasonable strength and density, is being considered as a promising material for structural applications above 1200°C. Although $MoSi_2$ exhibits superior environmental protective capabilities, it suffers from low temperature brittleness and decreased strength at high temperatures. Specifically, the ductile-to-brittle transition temperature (DBTT) of $MoSi_2$ is about 1000°C or higher [7,8], and the yield stress (250~350 MPa) decreases drastically at temperatures above 1200°C. To improve the low temperature toughness and high temperature strength of $MoSi_2$, a composite route has been widely adopted. The most notable reinforcements for $MoSi_2$ being explored are ductile refractory metal Nb [9-11] and SiC ceramic whiskers [7]. Despite these extensive efforts, a fundamental understanding is much needed for the environmental compatibility of $MoSi_2$.

The high temperature oxidation properties of $MoSi_2$ have been well-studied during the past 30 years. The excellent oxidation resistance of $MoSi_2$ at high temperatures is attributed to the formation of a self-passivating, glassy silica (SiO_2) layer. Low temperature oxidation of $MoSi_2$, on the other hand, has not been adequately addressed; its mechanism is still poorly understood [12]. A well-known, yet unresolved, phenomenon occurring during low temperature oxidation of $MoSi_2$ is pest disintegration, which has been reported to occur at temperatures around 400 - 600°C [1-2]. (Pest-disintegrated $MoSi_2$ is commonly featured as a lump of powdery products.) Although $MoSi_2$ pest was first discovered by Fitzer [1] over three decades ago and the subject has been fairly studied ever since [4,6,12-17], the fundamental understanding of its origin is still much needed. In this paper we summarize some of our recent findings concerning $MoSi_2$ pest.

EXPERIMENTAL PROCEDURES

Polycrystalline, monolithic bulks, thin films and composites, and single crystals were used for the present study. These samples were produced by a variety of methods. Detailed descriptions of preparation methods, and resulting sample densities and microstructures are summarized in Table I. It is particularly noted that the arc-melted samples were very brittle; numerous cracks were visually observed.

The samples were cut into various dimensions; polycrystalline samples were sliced into rectangular-shaped coupons, and rod-shaped, single crystalline samples were cut into thin discs (8mm in diameter and 2 mm in thickness). The cut-surfaces of all the samples (except thin films) were polished with 600-grit emery papers, ultrasonically cleaned in methanol and then dried in air. The samples were then placed in Al_2O_3 trays and loaded into a box furnace, which was stabilized at a pre-determined temperature. The temperature control of the furnace was accurate within ± 3°C. Oxidation tests were conducted at temperatures ranging from 350 to 675°C in air. (The relative humidity of air is about 55%, which corresponds to a water vapor pressure of 1740 Pa.) To address the kinetics of $MoSi_2$ pest, interrupted oxidation tests were performed at 500°C by periodically withdrawing the samples from the furnace after oxidizing for certain periods of time, and the samples were examined by various methods.

The morphological characteristics of oxidized samples were examined using optical metallography (OM) and scanning electron microscopy (SEM) with energy dispersive spectroscopy (EDS) and wavelength dispersive spectroscopy (WDS) x-ray microanalyses. The composition and structure of oxidation products were analyzed by Auger electron spectroscopy (AES), x-ray diffraction (XRD), and transmission electron microscopy (TEM, JEOL 2000FX) with EDS analysis.

Table I. A summary of $MoSi_2$ samples used for pest oxidation study

sample	preparation method	density (%)	microstructure
polycrystalline $MoSi_2$ monolith	hot press*	90	$MoSi_2$; minor SiO_2 & Mo_5Si_3
polycrystalline $MoSi_2$ monolith	arc melt & vacuum anneal	97	$MoSi_2$; minor Mo_5Si_3
$MoSi_2$ single crystal	floating zone**	~100+	$MoSi_2$
$MoSi_2$-AlN (80-20 wt.%)	hot press & vacuum anneal	92	$MoSi_2$; AlN
$MoSi_2$-Al_2O_3 (80-20 wt.%)	hot press & vacuum anneal	94	$MoSi_2$; Al_2O_3
$MoSi_2$ thin film@	sputter deposition	NA	amorphous; minor hcp-$MoSi_2$

*purchased from Cerac Inc. +based on visual estimation @on Si(111) substrates
**produced by Dr. T. Hirano at National Research Institute for Metals, Japan

RESULTS AND DISCUSSION

Hot-pressed Monolithic Polycrystals

Experimental results revealed that samples subjected to oxidation between 375 and 500°C were susceptible to pest disintegration (within the time period of 100 h). All the pested samples showed invariably powdery characteristics. On the other hand, samples were intact after oxidation either at and below 350°C or at and above 600°C, although severe cracking was observed in samples after oxidation at 550°C. Figure 1 presents a series of photographs illustrating the morphological characteristics of hot-pressed $MoSi_2$ after oxidation tests under various conditions. It is interesting to note in Fig.1(c) that, although the sample still exhibit a rectangular morphology, it has already disintegrated into powders. On most of the surfaces of unpested samples, AES analysis indicated the formation of an oxide layer containing compositions O > Si >Mo (denoted as Si-Mo-O hereafter). Such an oxide layer also formed on pested samples, but only at an early stage.

It appears that pest reaction is most vigorous at 500°C; most samples completely disintegrate and collapse into lumps of powders within a reasonably short period of time (in the range of 10 to 30 h). In contrast, samples oxidized at lower temperatures either showed partial disintegration (with the same period of oxidation as samples oxidized at 500°C) or retained their morphological shape even after complete disintegration. In the following, for the sake of simplicity, we will focus on the results obtained at 500°C.

The morphological characteristics of the powdery reaction products from a pest-disintegrated (at 500°C) $MoSi_2$ are presented in Fig.2. The powdery products contain mainly whiskers and clusters of irregular particles (denoted as clusters hereafter). In some cases, residual $MoSi_2$ particles are also observed. EDS and WDS microanalyses revealed that the whiskers contained primarily Mo and O, while the clusters contained Si and O. Further analysis by AES indicated that the whiskers were MoO_3 and the clusters were SiO_2. The presence of MoO_3 whiskers and residual $MoSi_2$ particles was further confirmed by XRD and TEM. The observation of MoO_3 and SiO_2 as primary reaction products is consistent with the fact that the reaction of $2MoSi_2(s) + 7O_2(g) \rightarrow 2MoO_3(s) + 4SiO_2(s)$ occurring at low temperatures is thermodynamically favored over the reaction of $5MoSi_2(s) + 7O_2(s) \rightarrow Mo_5Si_3(s) + 7SiO_2(s)$ occurring at high temperatures [18,19].

Interrupted oxidation tests revealed that pest reaction took place, initially, at some intergranular boundaries [see Fig.3(a)], which was characterized by the preferential formation of MoO_3 whiskers therein. The reaction, then, extended to local areas, which was manifested by significant volume expansion [see Fig.3(b)]. Numerous microcracks were generated in the neighborhood as a result of the volume expansion. Eventually, it propagated throughout the whole sample by accompanying with overall volume expansion of the sample. It was found that the amount of sample volume expansion incurred during pest oxidation could serve as an indicator for the measurements of pest-reaction kinetics. A more elaborate study on this subject will be reported elsewhere [20]. Close examination of an unpested part of a partially disintegrated sample revealed that pest reaction products (i.e., MoO_3 whiskers and SiO_2 clusters) had already formed [see Fig.3(c)] in the sample interiors with intergranular boundaries being the preferred formation sites. The occurrence of internal pest reactions is attributable to concurrent internal oxidation resulting from the short-circuit diffusion of O_2 through intergranular boundaries and pores. Grain boundary diffusion usually takes place more effectively at low temperature regimes than volume diffusion because of a lower activation energy associated with the process.

Fig.1 Optical micrographs showing the morphological characteristics of hot-pressed $MoSi_2$ after oxidation at various temperatures for different periods of time.

Fig.2 An SEM micrograph showing the morphological features of the reaction products from a sample after oxidation at 500°C for 21 h.

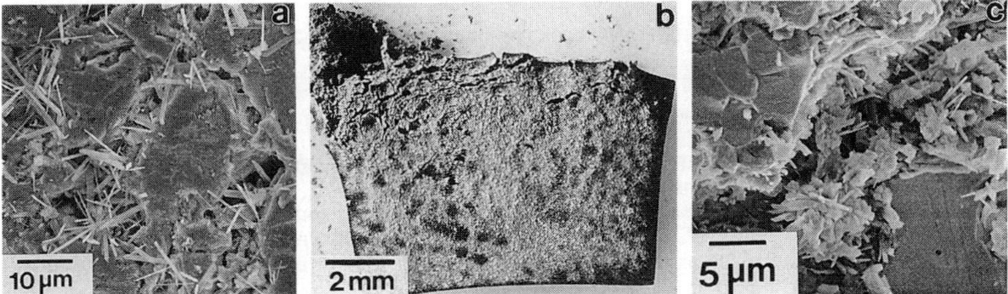

Fig.3 SEM micrographs showing (a) the preferential formation of MoO_3 whiskers at intergranular boundaries, (b) the local volume expansion and disintegration of a slightly-pested sample after oxidation at 500°C for 11 h, and (c) the internal microstructure of an unpested region of a partially disintegrated sample after oxidation at 500°C for 21 h.

It is further pointed out that, during the interrupted oxidation tests, most samples did not exhibit noticeable changes in surface morphology for a rather long period of time until localized volume expansion became noticeable. It is at those local areas in which samples start to disintegrate. Once localize disintegration takes place, it requires only a short period of time for the entire sample to collapse catastrophically. All these observations indicate that pest reaction is characterized by a lengthy nucleation period followed by a relatively fast growth period.

Arc-melted, Monolithic Polycrystals

Arc-melted samples also exhibited pest disintegration after oxidation at 500°C for various periods of time. Further examination of the disintegrated, powdery samples revealed that the reaction products consisted of MoO_3 whiskers, SiO_2 clusters, and residual $MoSi_2$ crystals. Although most of the $MoSi_2$ crystals were clusters of a few grains, single-grain crystals were occasionally found. All these $MoSi_2$ crystals were surrounded by surfaces either with a clean, thin oxide layer or decorated with distinctive, pest reaction products. The composition of the thin oxide layer is similar to Si-Mo-O as mentioned before. The presence of only a thin oxide layer (compared to a significantly thicker oxide layer which was formed on the sample surface) on the surfaces of those crystals suggests that those interfaces were produced by intergranular fracture or cleavage at a later stage of pest disintegration.

Close examination of the oxide-decorating surfaces revealed the presence of numerous blisters with MoO_3 whiskers growing thereon [see Fig.4(a)]. These blisters were essentially Si-Mo-O oxide, according to AES analysis, and most of them were burst-open [see Fig.4(b)], suggesting the existence of internal stresses which disrupted the surface oxide layer formed during oxidation. The sources of the internal stresses will be discussed later. In some areas MoO_3 whiskers were noted to form preferentially at certain sites, see Fig.4(c). These regions were believed enriched of defects which facilitated the nucleation and growth of MoO_3. It is particularly noted in Fig.4(c) that areas in the vicinity of the MoO_3-forming regions show darker contrast than the remote areas. EDS microanalysis revealed that the darker areas were depleted of Mo, compared to the light areas. This suggests that the formation of MoO_3 whiskers is attributed to the fast diffusion of Mo, which reacts with O_2 and forms MoO_3. Since few SiO_2 clusters are present in the neighborhood, the formation kinetics of SiO_2 may be relatively slower than that of MoO_3 (the diffusion kinetics of Si is believed to be less important because of the rich source of Si in $MoSi_2$). Based upon the above results, the occurrence of pest reaction is suggested to be closely related to the formation of MoO_3.

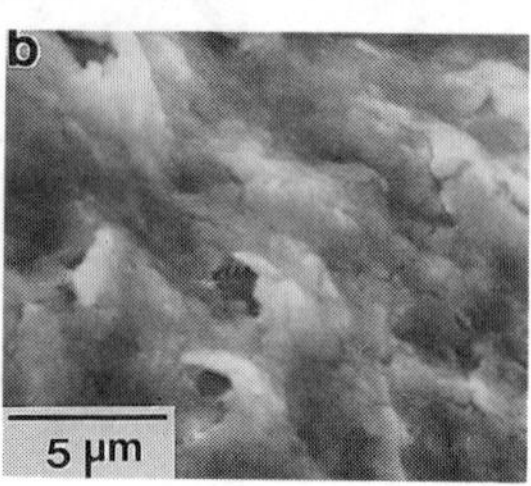

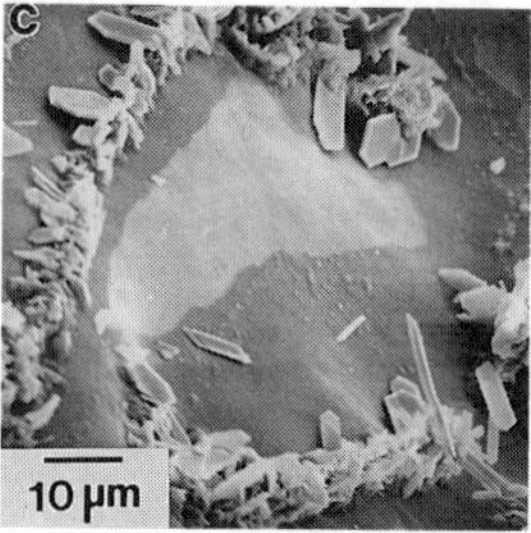

Fig.4 SEM micrographs showing various microstructures of the oxide-decorating surfaces.

Single Crystals

Pest reactions in single crystalline $MoSi_2$ took place with much slower kinetics than in polycrystalline counterparts. Samples oxidized at 500°C for a few hrs resulted in no noticeable change in the surface morphology, although AES analysis indicated the formation of a thin Si-Mo-O surface oxide. The same surfaces after longer periods (a few tens hrs) of oxidation started to show signs of pest reactions; local (heterogeneous) formation of burst-open blisters were observed [see Fig.5(a)]. Occasionally, extensive pest reactions were noted [see Figs.5(b) and 5(c)] to have taken place at microcrack sites, confirming that the nucleation and growth of pest reaction is indeed highly favored at defect sites.

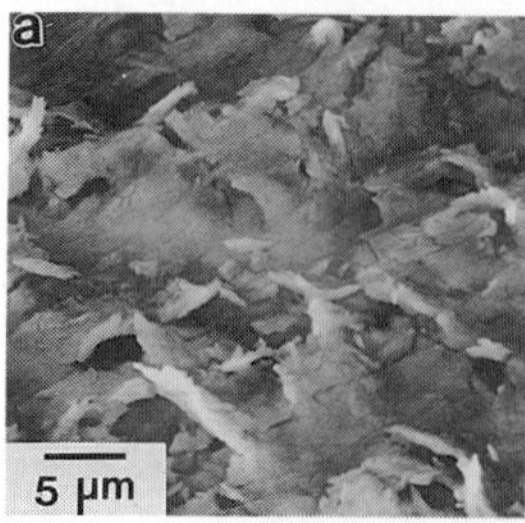

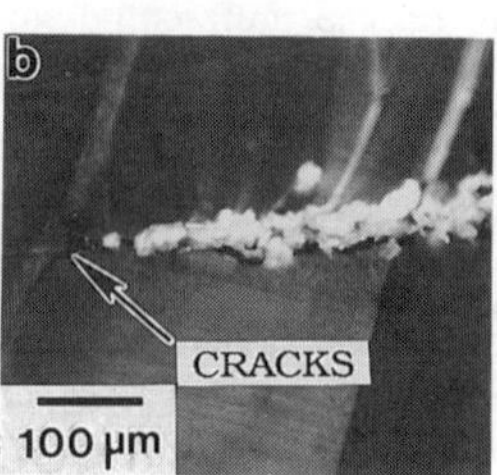

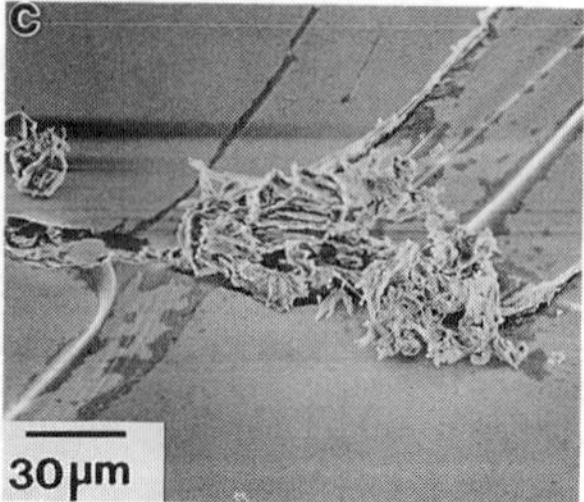

Fig.5 SEM (a & c) and optical (b) photographs showing various surface morphologies of the single crystalline $MoSi_2$ after oxidation at 500°C for (a) 15 h and (b & c) 22 h.

Further oxidation of the single crystalline samples resulted in the evolution of MoO_3 whiskers and SiO_2 clusters from those areas that were covered with blisters; blisters were no longer present. This suggests that the blisters may represent a transient state which characterizes the very early stage of pest reactions. Complete disintegration of the single crystals occurred only after prolonged (typically around 1000 h) oxidation. The disintegrated samples were found to contain mostly residual $MoSi_2$ crystals with fair amounts of MoO_3 whiskers and SiO_2 clusters.

$MoSi_2$ thin film

The $MoSi_2$ films also exhibited pest disintegration after oxidation at 500°C for various periods of time. Figures 6(a) and 6(b) present SEM micrographs showing a disintegrated $MoSi_2$ film and its microstructural characteristics. Initially, the film broke into pieces after a short period (5 min) of oxidation. Upon further oxidation, pest reaction products of MoO_3 and SiO_2 started to evolve from the sample. The sample eventually turned into powdery stuff with a yellowish color. In Fig.6(b) some fibers with less-regular surfaces are also noted. AES analysis indicated that those fibrous structures also contained various amounts of Si, in addition of Mo and O, suggesting that they may be in a transient state before evolving into MoO_3 whiskers and SiO_2 clusters.

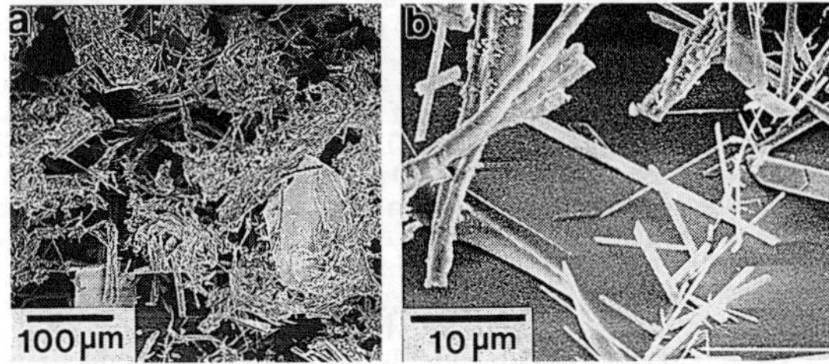

Fig.6 (a) Low and (b) high magnification SEM micrographs of a $MoSi_2$ film after oxidation at 500°C for 14 h.

$MoSi_2$-Al_2O_3 and $MoSi_2$-AlN composites

The composite samples also exhibited pest disintegration after various periods of oxidation at 500°C. However, the time required for the complete disintegration of the $MoSi_2$-Al_2O_3 was considerably longer (140 h) than that (20 h) for the $MoSi_2$-AlN, suggesting that the foreign additives do have an effect on the kinetics of pest disintegration. XRD analysis indicated that the pest-disintegrated composites contained mainly MoO_3, SiO_2, residual $MoSi_2$, and Al_2O_3 (Corundum) or AlN. No chemical reactions appear to have taken place at 500°C. It is postulated that the kinetics of pest disintegration may be further modified if chemical reactions between $MoSi_2$ and the foreign additives should occur.

The present experimental results indicate that pest reaction is likely to nucleate in the form of blisters. The fact that most blisters were burst open leads us to believe that the internal stresses developed during pest oxidation must be significant. The internal stresses may arise from three sources: thermal mismatch between the surface oxide (under tension) and the $MoSi_2$ substrate (under compression); molar volume expansion by oxidation; the vapor pressures of MoO_3. According to our calculations, mismatch between the thermal expansion coefficients of $MoSi_2$ (5×10^{-6} F^{-1}) and fused silica glass (0.5×10^{-6} F^{-1}) is about 90%, the amount of volume expansion (compared to $MoSi_2$) by the formation of MoO_3 and SiO_2 is about 250% [19], and the vapor pressure of MoO_3 at 500°C is about 10^{-2} Pa [18].

Since disintegrated samples contained loosely-bound, MoO_3 and SiO_2 reaction products, and residual $MoSi_2$ crystals, pest disintegration is likely to involve volume diffusion and concurrent, heterogeneous, interface diffusion of O_2 along grain boundaries or cracks. Schematic drawings showing the effects of volume and interface diffusion of O_2 on pest disintegration of a polycrystalline $MoSi_2$ are presented in Fig.7. The volume diffusion of O_2 gives rise to the massive oxidation of the sample (a & b) and forms loosely - bound, nonadherent MoO_3 and SiO_2 (c & d). On the other hand, the interface diffusion of O_2

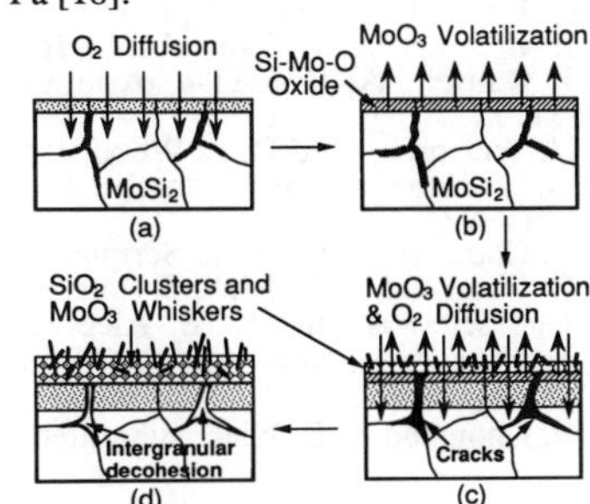

Fig.7 Schema showing pest disintegration of a polycrystalline $MoSi_2$.

causes preferential oxidation of intergranular or crack interfaces and the formation of protruding MoO_3 whiskers thereon (b & c), which in turn result in intergranular decohesion or fracture (caused by internal tensile stresses from the formation of protruding MoO_3 and volume expansion by oxide formation) of the sample (d) and form residual $MoSi_2$ crystals. The interplay of these two kinetic processes is expected to give rise to the pest disintegration of $MoSi_2$.

SUMMARY

Pest disintegration occurred at temperatures between 375 and 500°C in both poly- and single- crystalline $MoSi_2$ monoliths and composites. Pested samples failed catastrophically, and powdery reaction products of MoO_3 whiskers, SiO_2 clusters, and $MoSi_2$ residues were formed. The occurrence of pest disintegration was little affected by sample density as well as foreign additives, although the kinetics of disintegration were modified to various extents. Pest reactions, characterized by a lengthy nucleation period followed by a fast growth period, were found to take place preferentially at heterogeneous, defect sites, such as cracks and intergranular boundaries. The pest reactions nucleated in the form of blisters as a result of: thermal mismatch between the Si-Mo-O surface oxide and $MoSi_2$ substrate; volume expansion by oxides formation; the vapor pressures of MoO_3. These blisters evolved into MoO_3 whiskers and SiO_2 clusters upon further oxidation. The physical disintegration of the $MoSi_2$ can be ascribed to the formation of loosely-bound MoO_3 and SiO_2 by massive oxidation and the generation of internal tensile stresses by local oxidation on cracks and grain boundary interfaces. The internal tensile stresses, arising from substantial volume expansion by oxides formation and preferential growth of protruding MoO_3 whiskers at interfaces, result in the reproduction of more internal interfaces and cracks. The synergistic effects of interface creation and preferred pest nucleation thereon can lead to catastrophic failure of the samples. Above all, the present study indicates that the kinetics of pest disintegration can be extensively suppressed by minimizing the presence of grain boundaries and microstructural defects.

ACKNOWLEDGEMENT

The authors are grateful to C. G. McKamey (Oak Ridge National Laboratory) for providing the single crystal. This work was supported by the Lockheed Independent Research Program.

REFERENCES

1. Von E. Fitzer, "Molybdenum Disilicide as High-Temperature Material," Plansee Proc., 2nd Seminar, Reutte/Tyrol, 1955, pp. 56-79 (published in 1956).
2. J. Schilichting, High Temperature-High Pressure, **10**, 241 (1978).
3. A. W. Searcy, J. Am. Ceram. Soc. **40**, 431 (1957).
4. J. B. Berkowitz-Mattuck and R. R. Dils, J. Electrochem. Soc. **112**, 583 (1965).
5. C. D. Wirkus and D. R. Wilder, J. Am. Ceram. Soc. **49**, 173 (1966).
6. G. V. Samsonov, "Silicides and Their Uses in Engineering," (Akad. Nauk. Ukr. SSR, Kiev, 1959).
7. D. H. Carter, "SiC Whisker-Reinforced $MoSi_2$," LA-11411-T (Los Alamos National Laboratory, New Mexico, 1988).
8. R. M. Aikin, Jr., Scripta Met. et Mat. 26, 1025 (1992).
9. E. Fitzer and W. Remmele, 5th International Conference on Composite Materials (Metallurgical Society Inc., Warrendale, PA, 1985) pp. 515-530.
10. E. Fitzer and F. K. Schmidt, High Temperature-High Pressure, **3**, 445 (1971).
11. T. C. Lu, A. G. Evans, R. J. Hecht, and R. Merabian, Acta Met. **39**, 1853 (1991).
12. P. J. Meschter, Met. Trans. **A23**, 1763 (1992).
13. J. J. Rausch, ARF 2981-4, Armour Research Foundation, August 31, 1961, NSA 15-31171.
14. J. H. Westbrook and D. L. Wood, J. Nucl. Mater. **12**, 208 (1964).
15. C. G. McKamey, P. F. Tortorelli, J. H. DeVan, and C. A. Carmichael, J. Mater. Res. **7**, 2747 (1992).
16. T. C. Chou and T. G. Nieh, Scripta Met. et Mat. **26**, 1637 (1992).
17. T. C. Chou and T. G. Nieh, Scripta Met. et Mat. **27**, 19 (1992).
18. O. Kubaschewski and C. B. Alcock, "Metallurgical Thermochemistry," 5th edition, (Pergamon Press, New York, 1979), p. 366.
19. T. C. Chou and T. G. Nieh, J. Mater. Res. (in press).
20. T. C. Chou and T. G. Nieh, submitted to J. Mater. Res.

ACCEPTABLE ALUMINUM ADDITIONS FOR MINIMAL ENVIRONMENTAL EFFECT IN IRON-ALUMINUM ALLOYS*

VINOD K. SIKKA,† SRINATH VISWANATHAN,† AND SANJAY VYAS‡
†Oak Ridge National Laboratory, P.O. Box 2008, Oak Ridge, TN 37831-6083
‡Carnegie Mellon University, 5000 Forbes Avenue, Pittsburghs, PA 15213

ABSTRACT

A systematic study of iron-aluminum alloys has shown that Fe-16 at. % Al alloys are not very sensitive to environmental embrittlement. The Fe-22 and -28 at. % Al alloys are sensitive to environmental embrittlement, and the effect can be reduced by the addition of chromium and through the control of grain size by additions of zirconium and carbon. The Fe-16 at. % Al binary, and alloys based on it, yielded over 20% room-temperature (RT) elongation even after high-temperature annealing treatments at 1100°C. The best values for the Fe-22 and -28 at. % Al-base alloys after similar annealing treatments were 5 and 10%, respectively. A multi-component alloy, FAP, based on Fe-16 at. % Al was designed, which gave an RT ductility of over 25%.

INTRODUCTION

The Fe_3Al-based alloys are low-cost materials with excellent resistance to high-temperature oxidation and sulfidation. They suffer from low ductility at RT and poor high-temperature strength.[1] The low RT ductility has been associated with environmental embrittlement,[1-4] which results from the reaction of aluminum in the alloy with moisture in the air. The nascent hydrogen that is generated diffuses into the metal and causes loss in ductility. The hydrogen ingress can be reduced by a simple barrier such as oil film between the metal surface and the moisture. When hydrogen is generated (e.g., on a bare surface), its embrittling effect can be reduced[1,4,5] by producing an elongated microstructure with minimum transverse boundaries for diffusion paths. However, neither the surface barrier nor the elongated microstructure is an acceptable commercial solutions to the environmental effect. The present paper will describe a systematic study of environmental effects in binary iron-aluminum alloys over a range of Fe-16 to -28 at. % Al. (All percentages in the remaining text are in atomic percent.) This study established an acceptable aluminum limit for minimal environmental effects and developed an alloy composition with RT ductility values exceeding 25%.

EXPERIMENTAL DETAILS

This study is based on binary iron-aluminum alloys selected from three different regions of the phase diagram: disordered α (Fe-16% Al), disordered α plus ordered $D0_3$ (Fe-22 % Al), and ordered $D0_3$ (Fe-28% Al). The binary alloys were also systematically modified by additions of Cr; Cr and C; and Cr, Zr, and C. A final iron-aluminum alloy containing Cr, Zr, C, and Mo was investigated for structural applications. The nominal composition of the alloys investigated is shown in Table I.

Alloys of 500 g were prepared by nonconsumable arc melting in a copper crucible and casting into a 12.7 × 25.4 × 127-mm copper mold. Each ingot was cut into two pieces and processed into 0.76-mm-thick sheet by forging 50% at 1000°C, hot rolling 60% at 800°C, and warm rolling 70% at 650°C.

*Research sponsored by the U.S. Department of Energy, Fossil Energy Advanced Research and Technology Development Materials Program [DOE/FE AA 15 10 10 0, Work Breakdown Structure Element ORNL-2(F)] under contract DE-AC05-84OR21400 with Martin Marietta Energy Systems, Inc.

The warm-rolled sheets were given a stress-relief treatment of 1 h at 700°C followed by oil quenching. Tensile specimens of 12.7-mm gage length, with pin holes for loading, were punched from the sheets. The punched specimens were given a final heat treatment of 1 h at 700°C followed by either quenching in oil or cooling in air. A few specimens were also given other heat treatments, which will be described in the results and discussion section. The oil-quenched and air-cooled specimens were tested in air. The air-cooled samples were also tested in vacuum. All of the tests were at a strain rate of $3.3 \times 10^{-3}/s^{-1}$. All tests were run in duplicate, and average values are used for plots.

Table I. Chemical analysis of various alloys

	Atomic percent					
Alloy	Iron	Aluminum	Chromium	Carbon	Zirconium	Molybdenum
FAL-B8	84	16	--	--	--	--
FAL-B12	78	22	--	--	--	--
FAL-B16	72	28	--	--	--	--
FAL-T8	79	16	5	--	--	--
FAL-T12	73	22	5	--	--	--
FAL-T16	67	28	5	--	--	--
FAL-TC8	78.80	16	5	0.2	--	--
FAL-TC12	72.80	22	5	0.2	--	--
FAL-TC16	66.80	28	5	0.2	--	--
FAL-TCZ8	78.78	16	5	0.11	0.11	--
FAL-TCZ12	72.78	22	5	0.11	0.11	--
FAL-TCZ16	66.78	28	5	0.11	0.11	--
FAP	77.21	16.1	5.4	0.11	0.11	1.07

RESULTS AND DISCUSSION

Binary Alloys

The total elongation data at RT for the three binary alloys, tested in air and vacuum, are compared in Figure 1, which shows that the Fe-16% Al alloy possesses the highest ductility and is independent of the test environment, air, or vacuum. The oil-quenched sample, when tested in air, showed a slightly lower value. The ductility values of Fe-22 and -28 % Al alloys are significantly lower than Fe-16% Al alloys, irrespective of cooling media or test environment. Both Fe-22 and -28% Al alloys showed higher ductility values when tested in vacuum. The oil-quenched samples of these alloys also gave higher values than the air-cooled samples when tested in air. The improvement in ductility, when tested in vacuum, is a direct consequence of the absence of environmentally damaging species (moisture).[2,3] The oil film in the oil-quenched sample eliminates the direct contact of the specimen surface with the moisture in the air. In fact, the ductility improvement for specimens tested in vacuum after air cooling and in air after oil quenching are the same. Thus, the comparison of ductility data for specimens tested in air after oil quenching and air cooling is used for investigating the environmental effects in the remaining part of the paper.

The results in Figure 1 can be interpreted to conclude that the Fe-16% Al alloy is free from major environmental effects[6] and that the Fe-28% Al alloy shows the most sensitivity. The intermediate alloy with Fe-22% Al shows the intermediate environmental sensitivity.

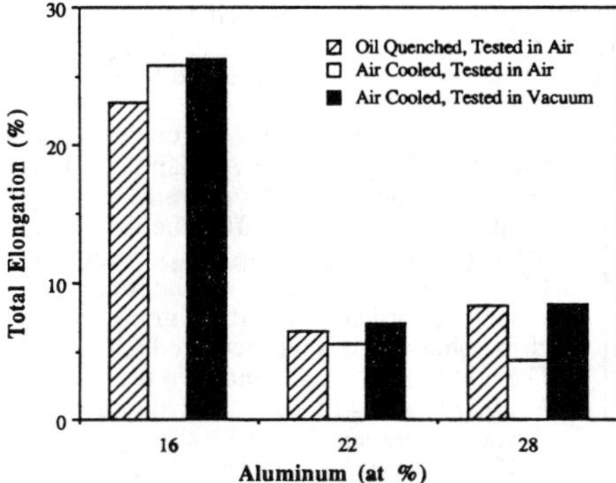

Figure 1. Effect of cooling medium from annealing temperature, test environment, and aluminum content on RT ductility of iron-aluminum binary alloys.

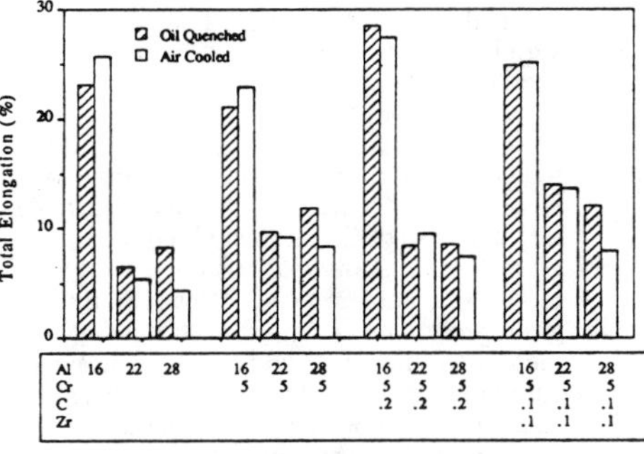

Figure 2. Effect of cooling medium from annealing temperature and alloying addition on RT ductility of Fe-16, -22, and -28 at. % Al alloys. The alloying additions were Cr; Cr and C; and Cr, Zr, and C.

Alloying Element Effects

A systematic addition of Cr; Cr and C; and Cr, Zr, and C were made to the three binary iron-aluminum alloys discussed above. Exact alloying amounts are shown in Table I. All of the material for alloying effects was prepared and processed identically and tested at RT after cooling in oil and in air from the heat-treating temperature. Tensile elongation data tested in air for the alloys are compared with the binary compositions in Figure 2.

The Fe-16% Al alloy showed over 20% elongation for all alloying additions and in both oil-quenched and air-cooled conditions. This confirms again that neither the binary Fe-16% Al nor alloys based on this composition are very sensitive to environmental embrittlement. The chromium addition to the Fe-22% Al alloy increased its ductility in both oil-quenched and air-cooled conditions. Even with increased ductility from the chromium addition, some indication of embrittlement exists. This is further confirmed by tests conducted in vacuum on an air-cooled sample that yielded 4% higher elongation than either the oil-quenched or air-cooled samples tested in air. The addition of carbon and chromium to the Fe-22% Al alloy also increased the ductility but less than the chromium addition only. The carbon addition ties up part of the chromium as carbides and, thus, lowers the amount of chromium available in reducing the embrittlement. The zirconium addition to the alloy containing Cr was the most effective in maximizing the ductility of the Fe-22% Al alloy in both the oil-quenched and air-cooled conditions. This observation implies that the combination of Cr, Zr, and C has the potential of enhancing the ductility and eliminating environmental embrittlement in the Fe-22% Al alloy.

The effects of Cr and Cr plus C additions to the Fe-28% Al are similar to those observed for the Fe-22% Al alloy. For the combined addition of Cr, Zr, and C, ductility values in the oil-quenched condition were greatly enhanced, but the air-cooled values were sufficiently lower to still indicate an embrittlement effect.

Heat Treatment Effects

The results presented in Figures 1 and 2 are for an annealing temperature of 700°C. Such a treatment produces a highly elongated and unrecrystallized microstructure which is considered to be resistant to environmental effects.[4,5] In order to further understand the environmental resistance of the alloys investigated in this study, they were also tested after various heat treatments. The heat treatments were at temperatures of 1000, 1100, 1200, and 1300°C for 1 h followed by oil quenching. Tensile ductility for the binary alloys containing Fe-16, -22, and -28% Al, and the same alloys containing Cr, Zr, and C, are compared in Figure 3. The grain-size measurements for the same alloys after different heat treatments are compared in Figure 4.

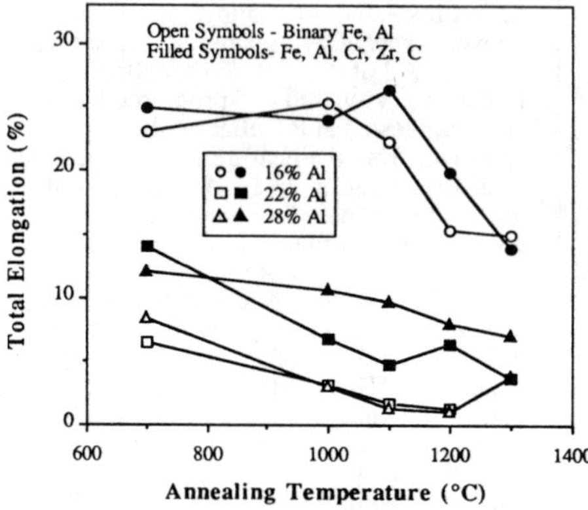

Figure 3. Effect of annealing temperature on RT ductilities of binary Fe-16, -22, and -28 at. % Al alloys and the same alloys containing the same additions of Cr, Zr, and C. All of the specimens were quenched in oil from annealing temperature and tested in air.

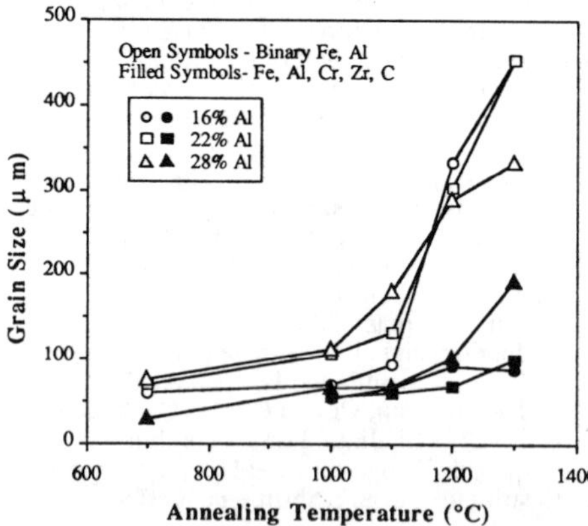

Figure 4. Effect of annealing temperature on grain size of binary Fe-16, -22, and -28 at. % Al alloys and the same alloys containing the same additions of Cr, Zr, and C. All of the specimens were quenched in oil from annealing temperature and tested in air.

Data in Figure 3 show that the binary alloy containing Fe-16% Al maintains a high ductility value of ≥ 20% up to an annealing temperature of 1100°C. The ductility values reach 15% after annealing temperatures of 1200 and 1300°C. These are the same annealing temperatures where grain growth of the binary alloys has occurred (Figure 4). The addition of Cr, Zr, and C to the Fe-16% Al alloy resulted in ≥ 20% ductility up to an annealing temperature of 1200°C. A value of 15% was reached after an annealing temperature of 1300°C. The higher ductility value of the Fe-16% Al alloy containing Cr, Zr, and C is a result of its finer grain size, which remains stable up to an annealing temperature of 1200°C. From these observations, it can be concluded that not only is the Fe-16% Al alloy containing Cr, Zr, and C free from the environmental embrittlement effect, but it also maintains a high ductility value of 20% even after an annealing temperature of 1200°C.

Compared to the Fe-16% Al alloy, the grain size of the binary Fe-22 and -28% Al alloys increased more rapidly (Figure 4). The addition of Cr, Zr, and C reduced the grain size of all three alloys containing Fe-16, -22, and -28% Al. However, the grain growth again occurred somewhat rapidly for the Fe-22 and -28% Al alloys as compared to the Fe-16% Al alloy. The ductility of the binary alloys containing Fe-22 and -28% Al dropped sharply with an increase in grain size. In the grain-refined alloys containing Fe-22 and -28% Al, the ductility values were significantly increased (Figure 3). These values dropped as the grain growth occurred for the higher annealing temperatures (1100 to 1300°C). It is interesting to note that the Fe-16% Al alloys were free of environmental embrittlement and showed the highest ductility value of 15% even in the grain-coarsened condition. The ductility values of Fe-22 and -28% Al alloys containing Cr, Zr, and C in the grain-coarsened condition were 5 and 10%, respectively.

Structural Alloy

After establishing that the Fe-16% Al alloy is free of environmental effects, it was used as a basis to design an engineering alloy. This alloy is designated as FAP in Table I. The chromium is used in the alloy for aqueous corrosion resistance.[7] The Zr and C additions are made to form ZrC which controls the grain size and can provide precipitation strengthening. The molybdenum additions are made to obtain pitting resistance in chloride solutions.[7] It can also provide solid-solution hardening. A 7.5-kg heat of FAP was prepared by air-induction melting and cast into a 75-mm-diam ingot. This ingot was processed into a 0.76-mm-thick sheet by using the same processing steps that were used for the 500-g heats of the base compositions in Table I. Based on a study of the effect of annealing temperature on RT properties, 800°C for 1 h followed by air cooling was used as the final heat treatment for the FAP alloy. Tensile properties of the FAP alloy as a function of test temperature are shown in Figures 5 and 6. It can be observed from these figures that the final alloy based on Fe-16% Al has an RT elongation of over 25%, which remains nearly constant up to 500°C and increases sharply at higher test temperatures. The strength values are considered acceptable for many applications up to 600°C.

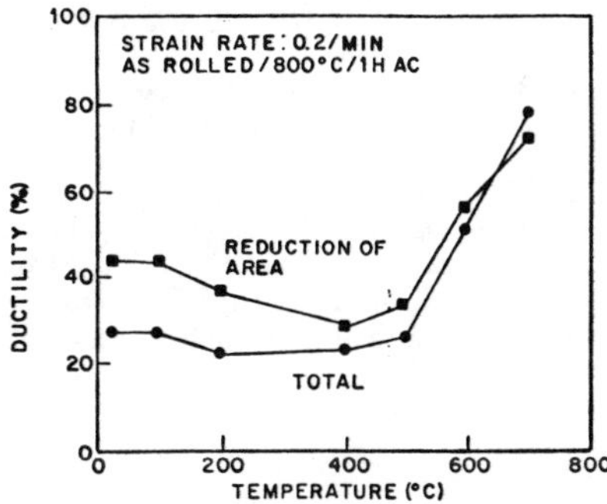

Figure 5. Effect of test temperature on total elongation and reduction of area values of a multicomponent alloy (FAP) based on Fe-16 at. % Al.

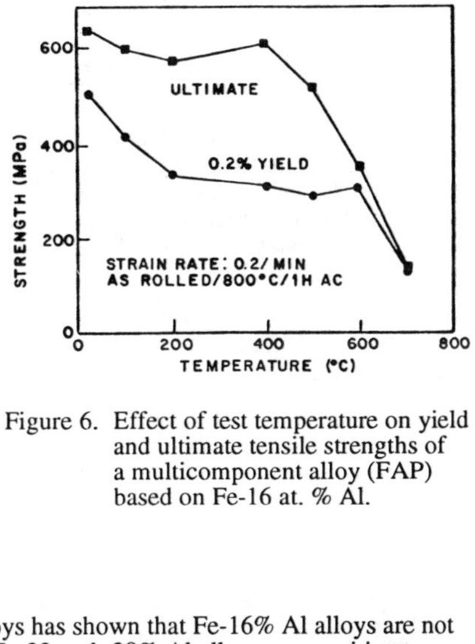

Figure 6. Effect of test temperature on yield and ultimate tensile strengths of a multicomponent alloy (FAP) based on Fe-16 at. % Al.

CONCLUSIONS

A systematic study of iron-aluminum alloys has shown that Fe-16% Al alloys are not sensitive to environmental embrittlement. The Fe-22 and -28% Al alloys are sensitive to environmental embrittlement, and the effect can be reduced by the addition of chromium and through the control of grain size by additions of zirconium and carbon. The Fe-16% Al binary, and alloys based on it, yielded over 20% RT elongation even after high-temperature annealing treatments at 1100°C. The best values for the Fe-22 and -28% Al-base alloys for similar annealing treatments were 5 and 10%, respectively. A multicomponent alloy, FAP, based on Fe-16% Al was designed and gave an RT ductility of over 25%.

The most likely reasons for Fe-16% Al alloys to be free of environmental embrittlement are: (1) disordered solid-solution alloys versus ordered $D0_3$ structure of the Fe-28% Al alloys and (2) significantly lower aluminum activity (0.16 vs 0.28 atomic fraction) in Fe-28% Al alloys.

ACKNOWLEDGMENTS

The authors thank J. D. Vought and E. C. Hatfield for processing the alloys into sheet, C. R. Howell for punching and heat treating the specimens, R. H. Baldwin for tensile testing, H. W. Hayden for paper review, K. Spence for editing, and M. L. Atchley for preparation of the manuscript.

REFERENCES

1. C. G. McKamey, J. H. DeVan, P. F. Tortorelli, and V. K. Sikka, "A Review of Recent Developments on Fe_3Al-Based Alloys," *J. Mater. Res.* **6**(8), 1779-1805, 1991.
2. C. T. Liu, C. G. McKamey, and E. H. Lee, "Environmental Effects on Room-Temperature Ductility and Fracture in Fe_3Al," *Scr. Metall.* **24**(2), 385, 1990.
3. C. T. Liu and C. G. McKamey, "Environmental Embrittlement – A Major Cause for Low Ductility of Ordered Intermetallics," pp. 133-51 in *High Temperature Aluminides and Intermetallics*, ed. S. H. Whang, C. T. Liu, D. P. Pope, and J. O. Stiegler, TMS-AIME (The Metallurgical Society of AIME), 1990.
4. V. K. Sikka, "Ductility Enhancement of Iron-Aluminide Alloys," *SAMPE Quarterly* **22**(4), 2-10, 1991.
5. P. G. Sanders, V. K. Sikka, C. R. Howell, and R. H. Baldwin, "A Processing Method to Reduce Environmental Embrittlement in Fe_3Al-Based Alloys," *Scr. Metall.* **25**, 2365-2369, 1991.
6. S. Vyas, S. Viswanathan, and V. K. Sikka, "Effect of Aluminum Content on Environmental Embrittlement in Binary Iron-Aluminum Alloys," *Scr. Metall.* **27**, 185-90, 1992.
7. R. A. Buchanan and J. G. Kim, "Chloride-Induced Localized Corrosion of Fe_3Al-Type Iron Aluminides: Beneficial Effects of Cr and Mo Additions," pp. 945-56 in *High-Temperature Ordered Intermetallic Alloys IV*, ed. L. A. Johnson, D. P. Pope, and J. O. Stiegler, Materials Research Society, 1990.

ENVIRONMENTAL EFFECTS ON MECHANICAL DISORDERING OF Ni_3Al-BASE ALLOY

S. GIALANELLA*, M. GUELLA*, F. MARINO°, M.D. BARO'°°, J. MALAGELADA°°, S. SURINACH°°, R.W. CAHN**
*Dipto di Ingegneria dei Materiali, Università di Trento, 38050, Mesiano, Trento, Italy
°Dipto di Scienza dei Materiali e Ingegneria Chimica, Pol. di Torino, 10129, Torino, Italy
°°Dept. de Fisica, Universitat Aut. de Barcelona, 08193, Bellaterra, Barcelona, Spain
**Dept. of Mat. Sci. and Metallurgy, Univ. of Cambridge, CB23QZ, Cambridge, U.K.

ABSTRACT

We present some results regarding the influence of grinding atmosphere on the phenomena occurring during mechanical milling of a Ni_3Al-base alloy. We have disordered this permanently ordered alloy by ball-milling in a vibratory mill, using alternatively argon and hydrogen atmospheres. We observed a slower reduction of the long-range order parameter during experiments carried out under a hydrogen atmosphere, regardless of the used grinding media. We discuss our findings in terms of the effects, such as in grain boundary cohesive strength or localize ductility enhancement, that hydrogen may induce in such alloy. The microstructural parameters of the powders at different stages of disordering have been investigated using calorimetric tests, x-ray diffraction analysis and microscopic observations.

INTRODUCTION

The main driving force for the study and development of advanced ordered alloys and intermetallics is the potential that this class of materials offers for high temperature applications.

The extreme service conditions faced in aerospace engines, electrical power plants turbines, etc., require good mechanical strength, surface stability and creep resistance at elevated temperatures and under environmentally aggressive atmospheres.

Apart from the obvious consequences that such environments may have on the oxidation [1] and corrosion [2] resistance of intermetallics, they have also been studied in view of their effects on the mechanical properties of these materials. For some of them testing atmospheres can cause dramatic changes in the observed results [3]. So, for example, according to the cited study, at room temperature oxygen reduces dislocation mobility in γ-TiAl and, consequently, ductility. Analogous effect was observed in Ni_3Al compounds deformed at high temperatures: oxygen grain boundary diffusion leads to the formation of brittle regions which facilitate intergranular crack growth.

Hydrogen too has been shown to produce embrittlement in a number of intermetallics, through different mechanisms: by interacting with dislocations and therefore hindering their motion under applied stresses by reducing grain boundary strength and increasing stress concentration. Hydrogen may be cathodically charged into the alloys or adsorbed from the gaseous products of the reaction between water vapour and the alloy. The latter phenomenon is particularly active in the case of Al alloys and leads to the contemporary formation of aluminum oxide. Apart from these detrimental effects, hydrogen has to be cited in connection with its exploitation in the so called thermochemical processing techniques of titanium alloys [4].

In this study we are mainly concerned with the effect that an inert (Ar) and an reactive (H_2) atmosphere may have on the disordering kinetics, during ball-milling, of boron doped Ni_3Al powders.

Large strains induced into the material by the grinding media and the continuous production of clean, oxide-free metallic surfaces render the working atmosphere particularly important in this sort of experiments. An in depth investigation on hydrogen occupation sites was carried out on Ni-Zr mechanical alloyed amorphous powders [5]. Dislocations introduced into the heavily strained material under a hydrogen atmosphere create hydrogen free volumes which promote further inward diffusion of this gas. In this respect mechanical milling can be considered an effective medium for hydrogen charging, as demonstrated with accurate calorimetric analysis in the experimental study on ball-milling of the TiCu intermetallic phase [6]. This process can be included in the wider category of dynamic hydrogen trapping, occurring in metallic materials under intense mechanical deformations [7]. Milling in hydrogen atmosphere was employed in an investigation aiming at the production of nanograined fcc metals, even though no specific attention was paid to the role that this gas might have had on the observed microstructures [8].

EXPERIMENTAL METHODS

Ni_3Al-base gas atomized powders (Ni-23.3 at % Al+1000 ppm B, kindly supplied by J.R. Knibloe and R.N. Wright, INEL-Idaho) were milled in a Fritsch Pulverisette 0 vibratory mill. The vial lid was modified in order to allow vacuum and gas connections. All experiments were conducted under static Ar atmosphere or flux and, alternatively, in H_2 flux. In both cases a degassing, using a rotary pump, was carried out to eliminate water vapor and other adsorbed gases. We used both agate and hardened Cr stainless steels grinding media. Agate gave significant contamination of the ground powders, whereas no detectable contamination from the grinding media was observed with Cr steel, as evaluated from the lack of powder weight increase after milling and EDXS analysis. XRD and DSC analysis were used to follow the evolution of microstructural characteristics and stored energy due to milling. Full experimental details regarding XRD and DSC methods can be found elsewhere [9]. The morphology of the milled powders was examined with SEM observations. Metallographic sections of the powders, mounted in resin, were prepared and observed after a selective etching with a 1%HF-1%HNO_3 water solution.

EXPERIMENTAL RESULTS AND DISCUSSION

In the first reported study on mechanical disordering of Ni_3Al ordered alloys [10] Jang and Koch explored several nonstoichiometric compositions and found that in the $Ni_{76}Al_{24}$ compound the disordered state was attained earlier than in stoichiometric and Al-rich alloys, under identical milling conditions. Incidentally also the occurrence of amorphous regions was detected earlier in such alloy. These results were interpreted in terms of the higher ductility of the $Ni_{76}Al_{24}$ alloy and its higher capability to withstand plastic deformations before fracturing.

The raw material used in the present study is a fairly ductile powder, due to the slight off-stoichiometry and to the presence of boron. Figure 1a shows the different disordering behaviour of powders milled under Ar and H_2 atmospheres using agate (silicon oxide) vial and ball. The relative long-range order parameter, S_r, as measured from the ratio of the integrated intensities of the (110) and (220) diffraction peaks, divided by the same ratio for an unmilled powder, has been evaluated for increasing milling times. It is evident that the H_2 atmosphere has a retarding effect on the disordering kinetics, compared to argon. To rule out any possible effect of the contamination coming from the grinding media we repeated the experiment using Cr steel media, which gave qualitatively similar results, as regards disordering (figure 1b). No significant differences between the two atmospheres were detected except in the evolution of average crystallite size and lattice strain, as measured from XRD peak broadening, shown by figures 2a and 2b for the Cr steel milled samples.

Our findings seem to agree with the cited results by Jang and Koch. We expect in fact the H_2-milled powders to be embrittled by the atmosphere and therefore to be less prone to disordering than those milled in a inert Ar atmosphere. Indeed this is the case: regardless of the grinding media, the series of specimens prepared under hydrogen flux were still partially ordered for times long enough to completely destroy long-range order in the specimens milled under an argon atmosphere. According to the reported results [11] the main reason for hydrogen embrittlement in boron doped Ni_3Al is the stress concentration at grain boundaries which become preferential routes for crack propagation. Several models have been proposed to explain hydrogen embrittlement in terms of grain boundary weakening. With reference to non-hydride forming systems, as Ni_3Al-base alloys are, there are basically two ways in which this phenomenon can occur: either for a decrease in the cohesive strength of the atomic bonds, or through a localized enhancement of dislocation mobility, near the crack tip, so that

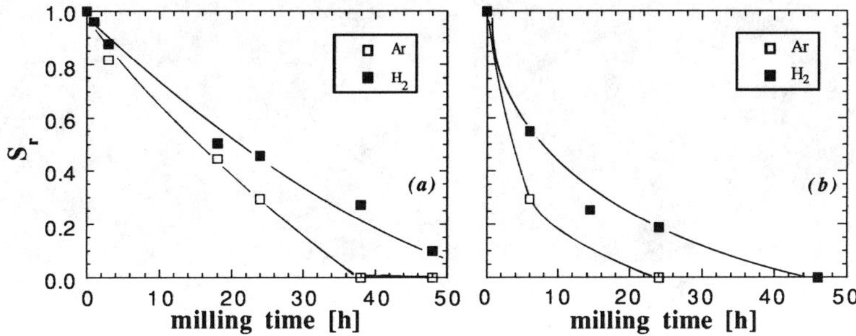

Figure 1: Evolution of the relative long-range order parameter, S_r, vs milling time for Ni_3Al powders milled with Agate (a) and Cr steel media (b) in Ar and H_2 atmosphere.

slip becomes easier in the nearby regions [12]. In the light of this model we can argue that, as grain boundaries become preferential sites for slip, lattice defect concentration, due to mechanical attrition, increases more slowly inside the grains when hydrogen is used rather than argon. For this reason, as disorder is directly related to point defects, namely vacancies and anti-sites concentration [13], we observe a slower disordering kinetics in hydrogen.

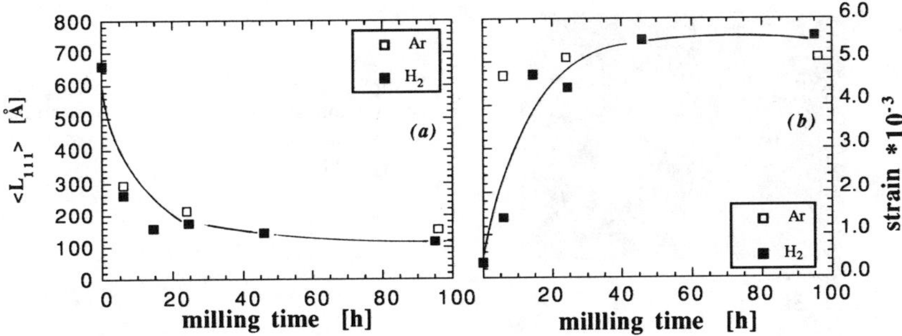

Figure 2: Variation of the average crystallite size (a) and lattice strain (b) for Ni_3Al powder milled with Cr-steel media.

Optical and electron microscopy observations revealed the sequence of changes in the morphology induced by the intense deformation in the originally spherical grains of powder. They are initially flattened (figure 3a); subsequently fractures are visible on the edges of highly deformed and work-hardened grains (figure 3b). In view of the particular loading conditions realized by the milling device, fatigue is certainly one of the active phenomena, and may initiate and favour the propagation of fracture. A typical fatigue induced failure morphology, characterized by delamination, is shown by figure 3c. In the long run it is possible to observe a diffuse grain fragmentation with grain size much smaller than the original one, as visible in figure 3d.

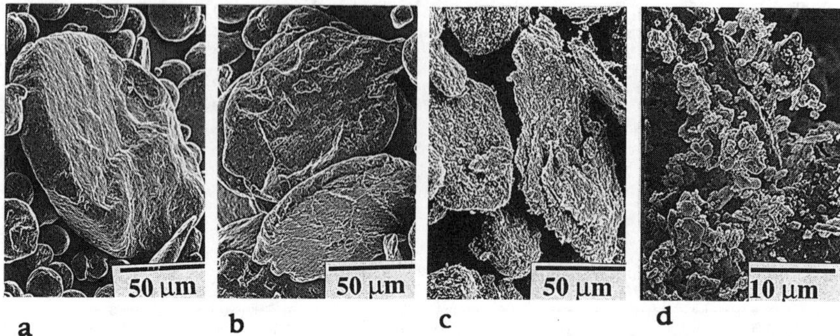

Figure 3: SEM micrographs showing the evolution of Ni_3Al powder milled for 3(a), 7(b), 18(c) and 24(d) hours under Ar atmosphere, using agate media.

The relevant role played by fatigue in these experiments is also proved by figure 4, where a typical shear band structure is visible in a heavily deformed particle. This deformation mode usually occurs in condition of high frequency complex loading [14], when deformation dynamics is quite insensitive to the slip systems and grain boundary structure of the material. A shear band-based mechanism has been recently proposed to explain the deformation behaviour of milled metallic powders[15].

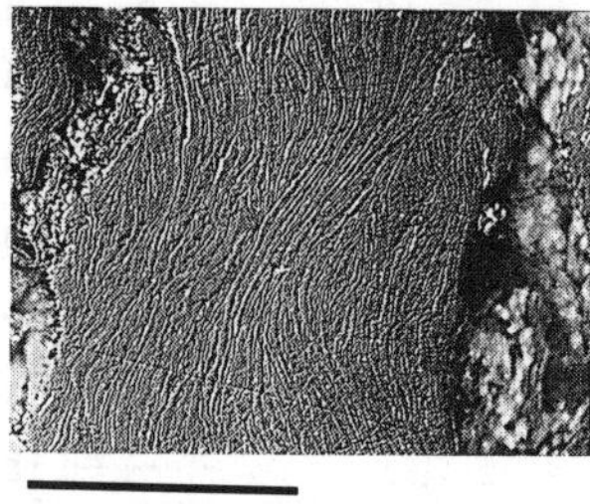

Figure 4: Optical micrograph showing the shear band structure in a Ni_3Al particle milled for 24h in H_2 atmosphere using Cr steel media.

Calorimetric tests, carried out to investigate the reordering kinetics, revealed the occurrence of the complex exothermic transformations already reported for Ni_3Al [10,16] milled powders and vapour deposited thin films [17] of the same alloy, see figure 5a. No significant differences were detected in the thermograms obtained with the Ar- and H_2-milled specimens. In particular no (endothermic) signals which could be ascribed to gas emission or desorption were detected in the analyzed specimens. The only remarkable difference regards the ultimate values of the enthalpy released in the reordering processes. They are in fact higher for the powders milled in hydrogen flux. Figure 5b shows the evolution of the total enthalpy of the process occurring upon calorimetric heat treatments for powders milled with agate media (same specimens as figure 1a). The value of enthalpy after 48h of milling for the H_2 series is higher than a slightly ordered specimen. These results add a further proof to the proposed explanation for the different disordering behaviour of the H_2-milled powders with respect to those milled in Ar. As pointed out in the first study on the mechanical disordering of Ni_3Al [10], the total enthalpy released cannot be ascribed to reordering only, but also to the large grain boundary contribution, on account of the small average grain size attained through the grinding action.

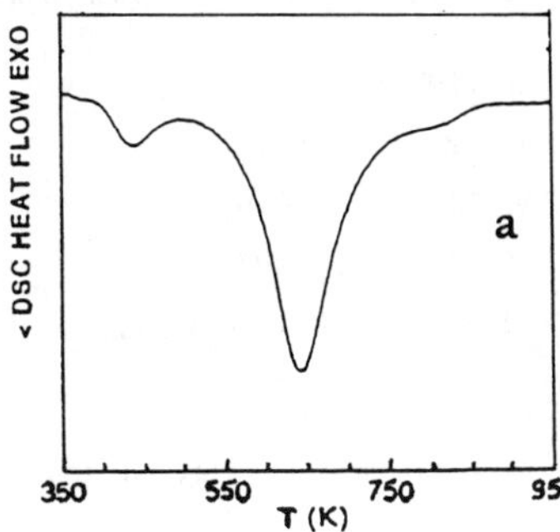

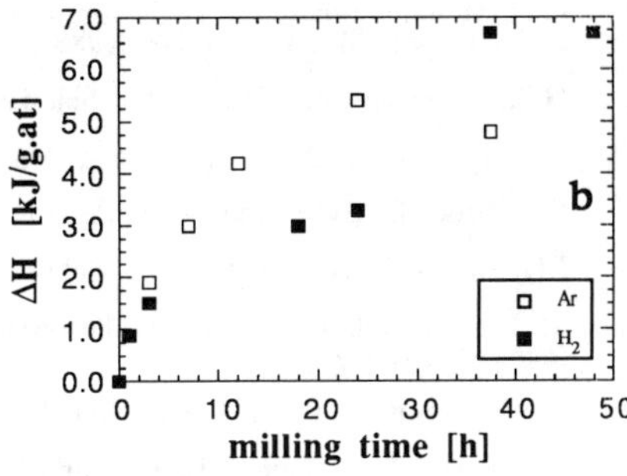

Figure 5: a) Typical DSC thermogram for a completely disordered Ni_3Al milled alloy. b) Enthalpy released in the exothermic process as a function of the milling time for the samples milled under Ar and H_2 atmosphere using agate media.

In the light of this consideration we can argue that in the powders milled under a hydrogen atmosphere the ultimate value of the average grain size should be smaller than the final grain size achieved in the Ar-milled powders. This would be in agreement with the reported weakening action of hydrogen on grain boundary cohesive strength of Ni_3Al base alloys [11]. By assuming that a consistent part of the mechanical energy is stored in the material as grain boundary surface energy, it is also possible to explain the slower disordering kinetics observed in hydrogen. As said, grain boundary sliding and fracturing, occurring to a larger extent in the hydrogen experiments, does not contribute to disordering. Indeed the basic mechanism for mechanical disordering seems to be the creation of anti-site point defects, assisted by dislocation movements, occurring within each grain under the effect of the applied loads. Accurate microscopic investigations aiming at the confirmation of this idea are underway.

CONCLUSIONS

The effect of different atmospheres on the response of a Ni_3Al base alloy to heavy mechanical deformations, during ball-milling, have been studied. The main difference between the powders milled under an argon and hydrogen atmosphere is the slower disordering observed in the powders milled under this latter atmosphere. The proposed explanation to this effect is based upon the hydrogen induced weakening action of Ni_3Al grain boundaries. Therefore a higher contribution of grain boundary sliding and fracturing diverts part of the mechanical energy from the disordering process of the alloy.

ACKNOWLEDGMENTS

We thank the European Community for the financial support provided to some of the research reported here and for making possible a fruitful multinational scientific collaboration on the exciting topic of "Ordering Kinetics in Alloys".
We'd like to thank M. Baldessari and S. Setti for their help with XRD analysis.

REFERENCES

1. G.H. Meier, Oxidation of High-Temperature Intermetallics, edited by T. Grobstein and J. Doychak (TMS, Warrendale, 1988).
2. G.H. Meier and F.S. Pettit, Mat. Sci. Tech., **8**, 331 (1992).
3. C.A. Hippsley and M. Strangwood, Mat. Sci. Tech., **8**, 350 (1992)
4. F.H. Froes D. Eylon and C. Suryanarayana, JOM, **42**(3), 26 (1990)
5. J.H. Harris, W.A. Curtin and L. Schultz, J. Mat. Res., **3**, 872 (1992)
6. M. Baricco etal, in Ordering and Disordering of Alloys" edited by A.R. Yavari, (Elsevier , 1992)
7. H. Hashimoto and R.M. Latanision, Acta Met., **36**, 1837 (1988)
8. J. Eckert, J.C. Holzer, C.E. Krill and W.L. Johnson, Mat. Sci. For., **88-90**, 505 (1992)
9. M.D. Barò, J.Malgelada, S. Suriñach, M.T. Mora, S. Gialanella and R.W. Cahn, Acta Met. Mat. in press (1992)
10. J.S.C. Jang and C.C. Koch, J. Mat. Res., **5**, 498 (1990)
11. A.K. Kuruvilla and N.S. Stoloff, Scripta Met., **19**, 83 (1985)
12. I.M. Robertson and H.K. Birnbaum, Acta Met., **34**, 353 (1986)
13. Bakker and L.M. Di, Mat. Sci. For., **88-90**, 27 (1992)
14. A. Korbel and P. Martin, Acta Met., **36**, 2575 (1988)
15. S. Li, K. Wang, L. Sun and Z. Wang, Scripta Met. Mat., **27**, 437 (1992)
16. J. Malagelada, S. Suriñach, M.D. Barò, S. Gialanella, R.W. Cahn, Mat. Sci. Forum, **88-90**, 497 (1992)
17. S.R. Harris, D.H. Pearson, C.M. Garland, B. Fultz, J. Mat. Res., **6**, 2019 (1991)

MEASUREMENT OF THE CRITICAL LEVEL OF MOISTURE FOR INITIATION OF WATER-VAPOR-INDUCED ENVIRONMENTAL EMBRITTLEMENT IN AN Fe_3Al-BASED ALLOY

C. G. McKAMEY AND E. H. LEE
Metals and Ceramics Division, Oak Ridge National Laboratory, Oak Ridge, TN 37831-6114

ABSTRACT

Iron aluminides based on Fe_3Al are of interest as structural materials because of their excellent corrosion resistance in many environments. However, studies have shown that one of the major causes of low room temperature tensile ductility, which so far has limited the use of these alloys, is an environmental reaction involving aluminum in the presence of water vapor. During this reaction, atomic hydrogen is released, moves into the sample ahead of the crack tip during stressing, and causes failure before the true ultimate tensile strength of the material is reached. This reaction is reduced significantly by testing in oxygen or vacuum. In the present study, an Fe_3Al-based alloy was tensile tested as a function of the level of water vapor in the test environment, from a vacuum of 10^{-4} Pa to a water vapor partial pressure of 1330 Pa. The results show that a water vapor level of as low as 133 Pa (1 torr) can result in significant embrittlement. The fracture mode remains transgranular cleavage, but the scale of the cleavage facets changes with the water vapor level.

INTRODUCTION

Iron aluminides based on Fe_3Al afford excellent corrosion properties at relatively low cost, making them candidates for use as structural material in corrosive environments [1,2]. However, these alloys have not yet found widespread use because they exhibit poor ductility at ambient temperatures, accompanied by brittle fracture [3]. Recently, efforts have been devoted to understanding the reason for their brittle behavior and to improving their ductility through control of grain structure, alloy additions, and material processing [3-5].

Recent studies of FeAl and Fe_3Al in various tensile testing environments have indicated that both alloy systems are more ductile at room temperature when tested in vacuum or dry oxygen [6-8]. Ductilities of 12-18% were attained in both iron aluminide systems in an oxygen pressure of 6.7×10^4 Pa, while only 2-4% ductility was achieved in normal laboratory air. The low ductility in air tests was attributed to environmental embrittlement involving generation of atomic hydrogen at the crack tip which is then transported into the specimen during stressing producing brittle cleavage failure. This atomic hydrogen is produced by the reaction of aluminum atoms at the crack tips with water molecules in the air. More recent studies have indicated that this type of environmental embrittlement also contributes to the low ductility of many other intermetallic systems [9-17].

The purpose of the present study is to determine the partial pressure of water vapor in the tensile test atmosphere necessary to produce severe embrittlement at room temperature.

EXPERIMENTAL PROCEDURES

A 500-g ingot of an Fe_3Al alloy (nominal composition Fe-28Al-5Cr-0.1Zr-0.5B, at.% [3-5,18]) was prepared by arc melting and drop casting, using commercially pure starting materials. The ingot was clad in stainless steel and hot rolled at 1000-850°C, then warm-rolled bare at 650-600°C to 0.75 mm sheet. The sheet was heat treated for 1 h at 700°C to relieve stress produced by the fabrication process. Tensile specimens with a gage section of 12.7 × 3.2 × 0.75 mm were then punched from the sheet, parallel to the rolling direction, and given a heat treatment of 1 h at 750°C.

Tensile tests were performed on an Instron testing machine equipped with a vacuum system. For control of the test environment, the system was first pumped to a vacuum of approximately 1×10^{-4} Pa and then water vapor was leaked into the system through a Varian leak valve. All specimens were tested at room temperature at a strain rate of 3.3×10^{-3} s^{-1}. Fracture surfaces were examined using a scanning electron microscope (SEM) operated at 25 kV.

RESULTS AND DISCUSSION

Figure 1 shows the microstructure of the alloy after a heat treatment of 1 h at 750°C. The specimens were almost completely recrystallized, with grain sizes ranging from 5 to 70 μm, and texturing produced by hot rolling was evident.

Table I lists the results of the room temperature tensile tests conducted in this study. Representative test curves for the range of partial pressures studied are also shown in Fig. 2. For comparison, tests conducted in air are also included. The yield strength of all tests conducted in the controlled environment ranged from 364 to 401 MPa; slightly higher scatter was observed for those tests conducted in ordinary laboratory air. Such repeatable yield strengths are expected in the presence of environmental embrittling agents, which manifest themselves primarily during the fracture process. The fracture strength scaled with the elongation. However, none of the tests reached an actual ultimate tensile point, as evidenced by continued work-hardening to fracture in all of the curves in Fig. 2. The measured elongations ranged from 24.2% in a vacuum of 10^{-4} Pa to 3-6% in tests conducted in a water vapor partial pressure of 665 or 1330 Pa (5 or 10 torr). The embrittling effect of the water vapor was apparent even at 1.3 Pa (10 mtorr) and appeared to saturate at 665 Pa; no difference was noted in the data for partial pressures of 665 and 1330 Pa. A plot of ductility

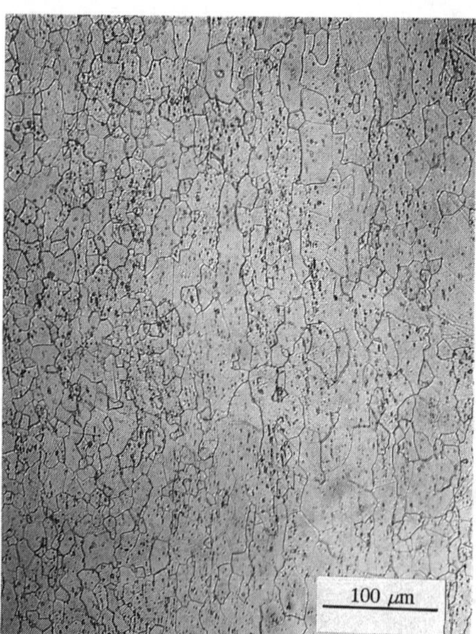

Fig. 1. Optical microstructure of as-heat-treated Fe-28Al-5Cr-0.1Zr-0.5B (at.%).

Table I. Effect of water vapor level on tensile properties of Fe_3Al[a]

Water vapor level [Pa (torr)]	Yield strength (MPa)	Fracture strength (MPa)	Elongation (%)
Vacuum, 10^{-4} (10^{-6})	389	1051	24.2
	372	1005	22.2
1.3 (0.01)	381	944	19.6
	398	948	18.4
	391	914	17.6
133 (1)	390	714	9.0
665 (5)	364	521	3.1
	381	540	3.9
1330 (10)	396	566	4.4
	401	617	5.8
Lab air, 30% RH	407	718	8.9
	336	743	11.1
	352	801	12.7

[a]Composition of test alloy was Fe-28Al-5Cr-0.1Zr-0.5B (at.%); heat treatment before testing was 1 h at 750°C, followed by air cooling.

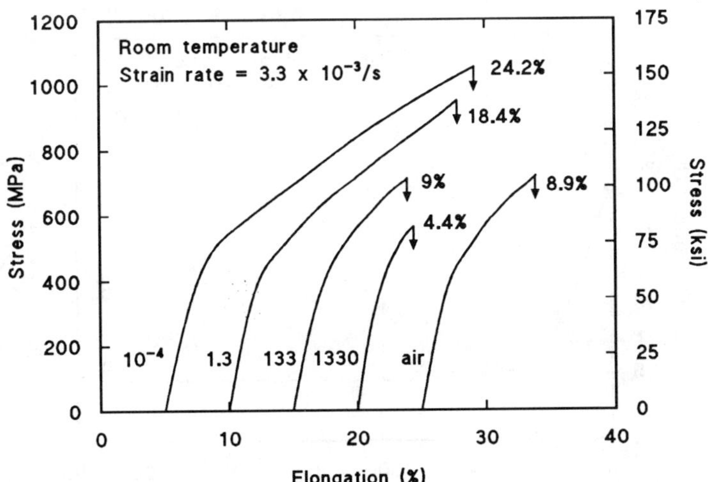

Fig. 2. Tensile curves showing the effect of water vapor on room temperature properties. The controlled test environment ranged from a vacuum of 10^{-4} Pa to a water vapor partial pressure of 1330 Pa.

as a function of the amount of water vapor (Fig. 3) indicated that 133 Pa (1 torr) of water vapor resulted in an embrittlement similar to laboratory air tests in which the relative humidity was approximately 30%. Similar results were also observed for an FeAl-based alloy [6].

Observation of the fracture surfaces using SEM techniques (Fig. 4) indicated that fracture in each case was predominantly by transgranular cleavage, although some small fraction of intergranular failure was noted in all specimens. The presence of a predominantly cleavage fracture in the tests conducted in the higher levels of water vapor suggested that the alloys contained mostly B2 order rather than $D0_3$; the presence of $D0_3$ order in hydrogen-embrittling test conditions has been shown to result in grain boundary failure, rather than transgranular cleavage [8]. The specimens tested in vacuum exhibited a much finer cleavage, bordering on a quasi-cleavage description of the fracture [Fig. 4(a,b)]. With increased levels of water vapor in the test environment, the scale of the cleavage facets became larger [Fig. 4(c,d)]. This change in cleavage facet size with test environment was also noted by Castagna and Stoloff for B2 ordered iron aluminides [8].

CONCLUSIONS

An Fe_3Al-based iron aluminide was tensile tested in various partial pressures of water vapor to determine the minimum amount of water vapor necessary in the environment to cause room temperature embrittlement. Tests were conducted from a vacuum environment of 10^{-4} Pa to a water vapor partial pressure of 1330 Pa. It was

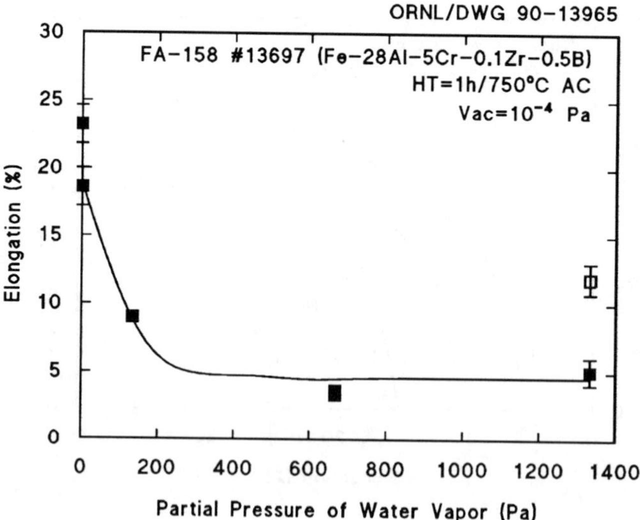

Fig. 3. Tensile elongation as a function of the partial pressure of water vapor in controlled environment tests of Fe_3Al. The open symbol represents data taken in ordinary laboratory air of approximately 30% relative humidity.

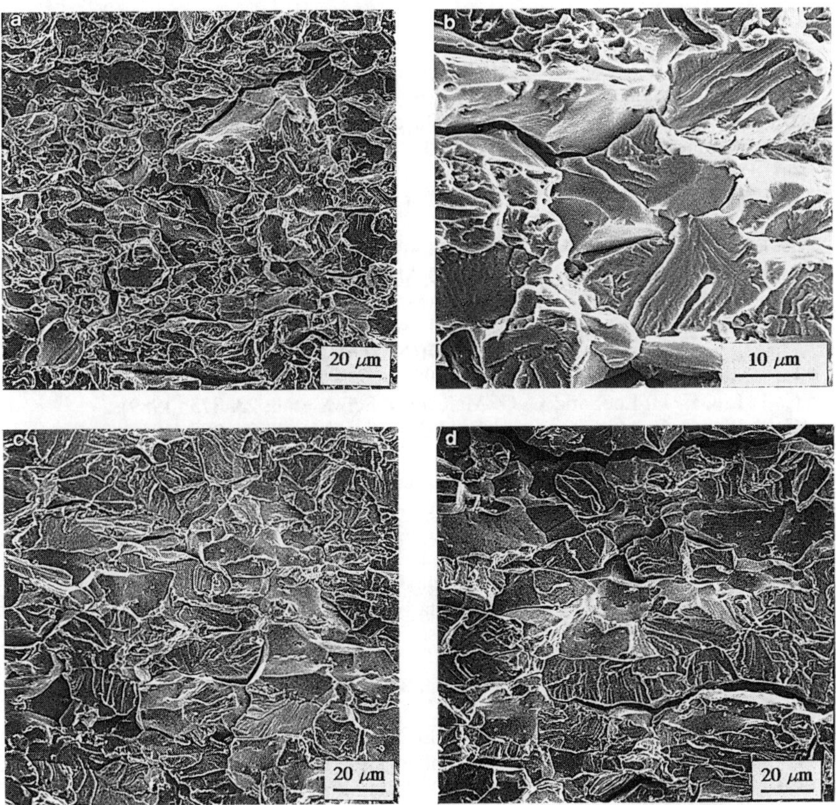

Fig. 4. SEM fractographs of Fe_3Al tensile tested at room temperature in (a,b) a vacuum of 10^{-4} Pa, or a partial pressure of water vapor of (c) 133 Pa or (d) 1330 Pa.

found that even 133 Pa (1 torr) of water vapor caused significant embrittlement, reducing ductility to 9% compared to 24% measured in vacuum. The fracture mode remained predominantly transgranular cleavage in all levels of water vapor. However, the scale of the cleavage facets tended to become larger with increasing water vapor.

ACKNOWLEDGEMENTS

The authors would like to thank J. A. Horton, Jr. and P. J. Maziasz for reviewing the manuscript. Research sponsored by the U.S. Department of Energy, Fossil Energy AR&TD Materials Program under contract DE-AC05-84OR21400 with Martin Marietta Energy Systems, Inc.

REFERENCES

1. J. H. DeVan, in Oxidation of High-Temperature Intermetallics, ed. by T. Grobstein and J. Doychak (TMS, Warrendale, PA, 1989), pp. 107-15.
2. J. L. Smialek, J. Doychak, and D. J. Gaydosh, in Oxidation of High-Temperature Intermetallics, ed. by T. Grobstein and J. Doychak, (TMS, Warrendale, PA, 1989), pp. 83-95.
3. C. G. McKamey, J. H. DeVan, P. F. Tortorelli, and V. K. Sikka, J. Mater. Res. **6**, 1779 (1991).
4. V. K. Sikka, C. G. McKamey, C. R. Howell, and R. H. Baldwin, Fabrication and Mechanical Properties of Fe_3Al-Based Aluminides, ORNL/TM-11465 (Oak Ridge National Laboratory, Oak Ridge, TN, March 1990).
5. V. K. Sikka, C. G. McKamey, C. R. Howell, and R. H. Baldwin, Properties of Large Heats of Fe_3Al-Based Alloys, ORNL/TM-11796 (Oak Ridge National Laboratory, Oak Ridge, TN, March 1991).
6. C. T. Liu, E. H. Lee, and C. G. McKamey, Scr. Metall. **23**, 875 (1989).
7. C. T. Liu, C. G. McKamey, and E. H. Lee, Scr. Metall. **24**, 385 (1990).
8. A. Castagna and N. S. Stoloff, Scripta Metall. **26**, 273 (1992).
9. C. G. McKamey and C. T. Liu, to be published in Environmental Effects on Advanced Materials, proceedings of ADVMAT/91 (NACE, 1992).
10. C. T. Liu and E. P. George, Scr. Metall. **24**, 1285 (1990).
11. C. Nishimura and C. T. Liu, Scr. Metall. **25**, 791 (1991).
12. T. Takasugi, H. Suenaga, and O. Izumi, J. Mater. Sci. **26**, 1179 (1991).
13. C. Nishimura and C. T. Liu, Acta Metall. **40**, 723 (1992).
14. A. K. Kuruvilla and N. S. Stoloff, Scr. Metall. **19**, 83 (1985).
15. N. S. Stoloff, J. Metals, 40, **23** (1988).
16. T. Takasugi and O. Izumi, Acta Metall. **34**, 607 (1986).
17. N. Masahashi, T. Takasugi, and O. Izumi, Metall. Trans. **19A**, 353 (1988).
18. C. G. McKamey, C. T. Liu, S. A. David, J. A. Horton, D. H. Pierce, and J. J. Campbell, Development of Iron Aluminides for Coal Conversion Systems, ORNL/TM-10793 (Oak Ridge National Laboratory, Oak Ridge, TN, July 1988).

HYDROGEN EMBRITTLEMENT OF Ni_3Al AT VARIOUS HYDROGEN CONTENTS AND STRAIN RATES

HUAXIN LI AND T. K. CHAKI
State University of New York, Department of Mechanical and Aerospace Engineering, Buffalo, NY 14260

ABSTRACT

Hydrogen embrittlement has been studied in continuous cast sheet of a Ni_3Al alloy ($Ni_{77.83}Al_{21.73}Zr_{0.34}B_{0.1}$), known as IC-50, after introducing various amounts of hydrogen cathodically. The elongation and UTS decreased with the increasing content of hydrogen. When tensile-tested at a strain rate of 5.8×10^{-5} s^{-1}, the elongation decreased from 32.7% for no charging to 1.9% for 330 min of charging with 50 mA cm^{-2} current. The yield stress, however, did not change. When tested at a higher strain rate of 5.8×10^{-3} s^{-1}, the embrittlement was less, but the yield stress increased with the hydrogen content. With increasing hydrogen content the fracture mode changed from dimpled to intergranular and cleavage modes.

INTRODUCTION

Nickel aluminide, Ni_3Al, an $L1_2$ compound, has attractive mechanical properties. The yield strength of single crystals of Ni_3Al increases with temperature [1], but polycrystalline material is severely brittle and fractures intergranularly [2]. In 1979 Aoki and Izumi [3] showed that alloying with a small amount of boron (a few hundred p.p.m.) increases room temperature ductility in Ni-rich Ni_3Al dramatically. By Auger electron microscopic study, it has been shown [4] that B segregates at the grain boundaries of Ni_3Al. Liu et al. [4] argued that B atoms at the grain boundaries increase the cohesive strength of the boundaries. Chaki [5,6] has proposed that interstitial B atoms distort the directionality of Ni-Al bonds in Ni_3Al lattice and, as a result, the atoms can relax easily and fill up the cavities at the grain boundaries. This makes movement of dislocations easier within the grains as well as across the grain boundaries.

However, boron-doped Ni_3Al again becomes brittle upon cathodic charging of hydrogen. Kuruvilla and Stoloff [7] reported a 70% decrease in tensile ductility in cathodically charged Ni-24 at.% Al-0.2% B specimens and fracture was intergranular. Bond et al. [8] claimed from their transmission electron microscope (TEM) observations that the dislocation pile-ups at the grain boundaries of Ni_3Al were similar with and without hydrogen, and argued that hydrogen embrittles the grain boundaries. Wan et al. [9] elongated cathodically charged Ni-23 at.% Al-120 wt. ppm B specimens inside a scanning electron microscope (SEM) and observed that the slip bands had difficulties in crossing the grain boundaries and that the cracks developed along boundaries. Here we present the results of hydrogen embrittlement in a ductile Ni_3Al alloy at various hydrogen contents and strain rates, and make a detailed fractographic analysis. The results will shed light on the mechanism of hydrogen embrittlement in Ni_3Al.

EXPERIMENTAL PROCEDURE

Ni_3Al alloy used in this study contained by weight 11.30% Al, 0.60% Zr, 0.02% B and balance Ni (by atom 21.73% Al, 0.34% Zr, 0.1% B and the balance Ni). The alloy, designated as IC-50, was obtained from Oak Ridge Natioanl Laboraory, Tennessee. Thin sheets (about 1 mm thick) of IC-50 were prepared by continuous sheet casting. Tensile specimens with the gauge section of dimension 25.4 $mm \times 4.4\ mm \times 1\ mm$ were prepared by cutting from the sheet by SiC blade and then grinding. The specimens were annealed at 1000°C for 2 h in argon. Fig. 1 shows the photograph of the specimens. The specimens were charged with hydrogen cathodically in 1N H_2SO_4 solution kept at room temperature. 0.05 g l^{-1} of $NaAsO_3$ was added to the solution to poison hydrogen recombination. Just before charging, the surfaces of the specimen were ground to 600 grit with the help of SiC abrasive paper and cleaned in methanol. The thickness of the sheet after grinding was about 0.8 mm. The charging current density was 50 mA cm^{-2} throughout this study. To reduce the loss of hydrogen the gauge section of the specimen was wrapped with Scotch tape (made by 3M Company) immediately after charging. The specimen was then pulled in tension at a constant displacement rate in ambient temperature and atmosphere. The strain was measured with an extensometer. The content of diffusible hydrogen in the charged specimen was measured by the glycerol volumetric technique [10]. The glycerol bath was maintained at a constant temperature of 40°C and hydrogen gas was collected for 48 h inside a glass tube having an inner diameter of 4 mm.

RESULTS AND DISCUSSION

The diffusible hydrogen content in Ni_3Al specimens initially increased with the charging time, reaching a plateau (Fig. 2) at (15.5 ± 0.1) ml/100g upon 330 min of charging. Fig. 3 shows typical engineering stress-strain curves at room temperature after various amounts of charging. Elongation decreased with the increasing hydrogen content. Fig. 4 compares elongations, yield stresses and ultimate tensile stresses (UTS) of charged specimens tensile-tested at two initial strain rates, namely, 5.8×10^{-5} and $5.8 \times 10^{-3}\ s^{-1}$. At the slow strain rate the yield stress and the flow stress of the pre-charged specimens remained the same as for the corresponding uncharged specimen. However, the yield stress of the charged specimens at the high strain rate increased upon charging. Upon 120 min of charging, the 0.2% yield stress at the strain rate of $5.8 \times 10^{-3}\ s^{-1}$ was (525 ± 1) MPa (7% increase), compared to (498 ± 2) MPa for the uncharged specimen. This shows that the mobility of dislcoations is affected by hydrogen. At both strain rates and under all charging conditions, the specimens failed only after yielding.

Hydrogen embrittement was less at the higher strain rate. For 120 min of charging, elongations at the strain rates of 5.8×10^{-3} and $5.8 \times 10^{-5}\ s^{-1}$ were $(11.1 \pm 0.8)\%$ and $(5.4 \pm 0.2)\%$, respectively. At the high strain rate the elongations for 50 and 120 min of charging were almost identical (Fig. 4), even though the diffusible hydrogen contents were different. At the high strain rate there is less time for hydrogen to diffuse to high stress regions at crack tips, and consequently, hydrogen embrittlement is decreased. Moreover, at the high strain rate multiple cracks (as shown in Fig. 1) were generated, because there was not enough diffusion of hydrogen to assist growth of a single crack. At the high strain rate the work hardening rate of the charged specimen was less (curve e in Fig. 3) than that of the uncharged specimens, due to development of a large number of multiple cracks.

The fracture surface of charged Ni_3Al specimens contained regions of various fracture modes, as depicted in the schematic of Fig. 5. At the initial part of the crack growth with lower stress intensity factor (K_I), the fracture surface was predominantly intergranular with occasional ductile tearing. The middle portion of the fracture surface, corresponding to moderate K_I, had quasi-cleavage failure mode comprising of transgranular cleavage mixed with ductile tearing. The final part of the fracture surface, corresponding to high values of K_I, had a dimpled failure mode. Typical appearances of intergranular, quasi-cleavage and dimple modes in the fracture surface of a Ni_3Al specimen, pre-charged for 50 min and pulled in tension at the strain rate of $5.8 \times 10^{-5}\ s^{-1}$, are shown in Fig. 6(a), (b) and (c), respectively. Table 1 summarizes the proportion (measured from scanning electron micrographs of the fracture surface) of three modes of fracture under various charging conditions. With increasing hydrogen content the proportion of intergranular and quasi-cleavage failures increased. In Table 1 we have tabulated the values of the true fracture stress, which

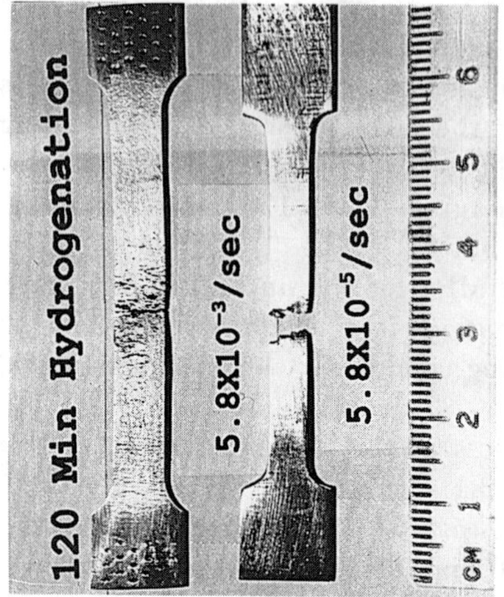

Fig. 1. Photographs of specimens for hydrogen charging. Multiple cracks are seen at the strain rate of 5.8X $10^{-3}\,s^{-1}$ upon 120 min of charging.

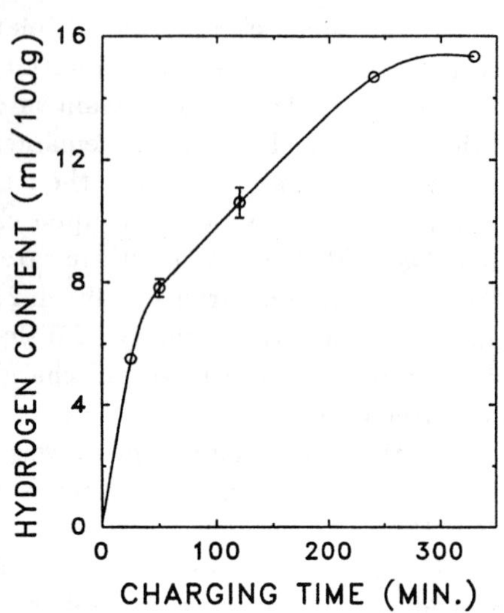

Fig. 2. Diffusible hydrogen content against the charging time at a current density of 50 mA cm^{-2}.

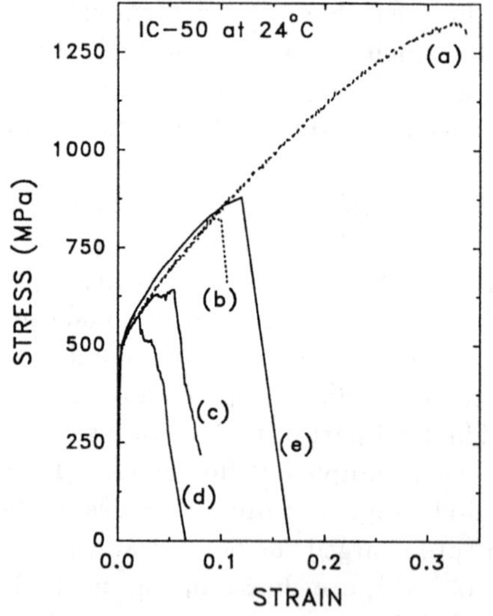

Fig. 3. Engineering stress-strain curves. Curves a, b, c and d were obtained at the strain rate of $5.8 \times 10^{-5}\,s^{-1}$ with no charging, 50, 120 and 330 min of charging, respecively. Curve e was obtained at $5.8 \times 10^{-3}\,s^{-1}$ upon 120 min of charging.

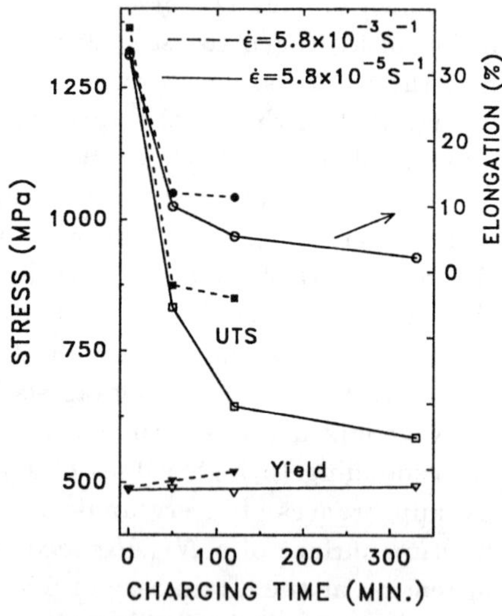

Fig. 4. Elongation, UTS and 0.2% yield stress against charging time. The dash and solid curves correspond to strain rates of 5.8×10^{-3} and $5.8 \times 10^{-5}\,s^{-1}$, respectively.

different cases at the strain rate of $5.8 \times 10^{-5}\ s^{-1}$, we notice a trend that, as the fracture stress increased, the proportion of the dimple fracture increased. In the cases of fracture with high K_I values, the large plastic zone in front of the crack tip caused dimples. When tested at the high strain rate, the charged specimens failed by joining of small cracks which could produce only low K_I values and, as a result, the fracture surface showed very little dimple.

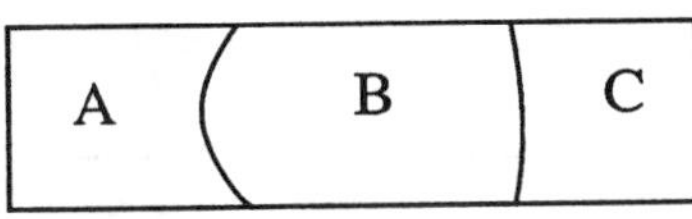

CRACK GROWTH →

Fig. 5. Schematic diagram of fracture surface, showing various modes of fracture in specimens charged with hydrogen. (A) Intergranular. (B) Quasi-cleavage. (C) Dimple.

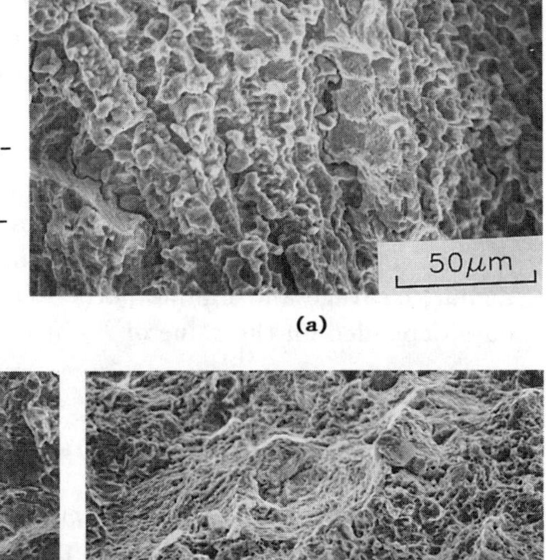

(a)

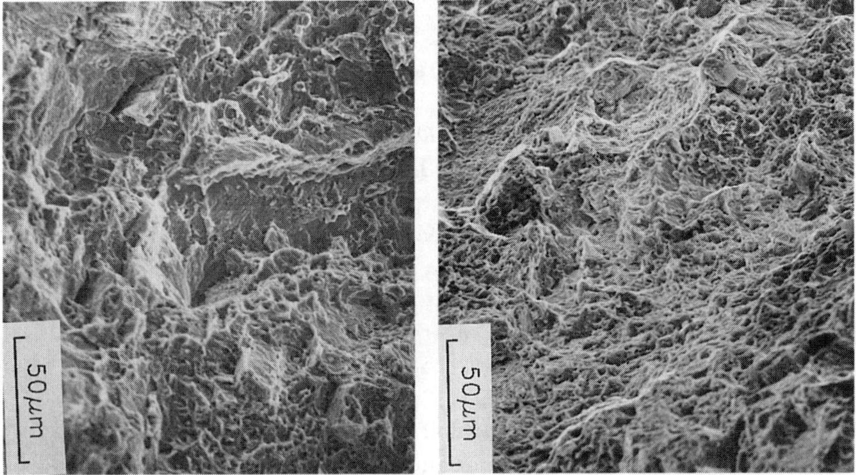

(b) (c)

Fig. 6a, b, c. Fracture surface of a specimen charged for 50 min and pulled in tension at the strain rate of $5.8 \times 10^{-5}\ s^{-1}$. (a) Inter-granular, (b) Quasi-cleavage and (c) Dimple.

Table 1. Proportion of Fracture Modes for Various Charging Conditions.

Strain rate (s^{-1})	Charging time (min)	Intergranular (%)	Quasi-cleavage (%)	Dimple (%)	True fracture stress (MPa)
	0	0	0	100	1753
$5.8X10^{-5}$	50	20	53	27	971
	120	25	57	18	680
	330	38	47	15	594
	0	0	0	100	1831
$5.8X10^{-3}$	50	50	40	10	980
	120	55	40	5	944

CONCLUSION

Cathodic charging of hydrogen caused serious embrittlement of a ductile Ni_3Al alloy. Embrittlement was less at faster strain rates. At the strain rate of 5.8×10^{-3} s^{-1}, the yield stress increased by 7% upon charging for 120 min. The fracture surface of the charged specimen contained intergranular, cleavage and dimple parts. The proportion of various fracture modes depended on the value of K_I at the crack tip.

REFERENCES

[1] S. M. Copley and B. H. Kear, Trans. The Metall. Soc. - AIME, **239**, 977 (1967).

[2] C. T. Liu, in *High-Temperature Ordered Intermetallic Alloys II*, edited by N. S. Stoloff, C. C. Koch, C. T. Liu, and O. Izumi (Materials Research Society, Pittsburgh, 1987), p. 355.

[3] K. Aoki and O. Izumi, Nippon Kinzoku Gakkaishi, **43**, 1190 (1979).

[4] C. T. Liu, C. L. White, and J. A. Horton, Acta Metall., **33**, 213 (1985)

[5] T. K. Chaki, Philos. Mag. Lett., **61**, 5 (1990); **63**, 123 (1991).

[6] T. K. Chaki, in *High-Temperature Ordered Intermetallic Alloys IV*, edited by L. A. Johnson, D. P. Pope, and J. O. Stiegler (Materials Research Society, Pittsburgh, 1991), p. 769.

[7] A. K. Kuruvilla and N. S. Stoloff, Scripta Metall., **19**, 83 (1985).

[8] G. M. Bond, I. M. Robertson, and H. K. Birnbaum, Acta Metall., **37**, 1407 (1989).

[9] X. J. Wan, J. H. Zhu, and K. L. Jing, Scripta Metall. Mater., **26**, 479 (1992).

[10] R. C. Shutt and D. A. Fink, Welding Journal, **64**, 19 (1985).

EFFECTS OF ELEVATED TEMPERATURE HYDROGEN EXPOSURE ON THE MICROSTRUCTURE OF α_2- AND γ-BASED TITANIUM ALUMINIDE ALLOYS

D. S. SCHWARTZ*, R. J. LEDERICH*, W. B. YELON**, Y.-Y. TANG**, AND S. M. L. SASTRY***
*McDonnell Douglas Aerospace, PO Box 516, m/c 111-1041, St. Louis, MO 63166-0516
**University of Missouri-Columbia, Research Reactor Facility, Columbia, MO 65211
***Washington University, St. Louis, MO 63130

ABSTRACT

Ti-25Al, Ti-25Al-10Nb-2V, and Ti-25Al-10Nb-3V-1Mo (at. %) α_2-based alloys, and Ti-48Al and Ti-52Al γ-based alloys were exposed to gaseous hydrogen at elevated temperatures. A novel ternary hydride was observed in Ti-25Al and Ti-25Al-10Nb-3V-1Mo, identified as Ti_3AlH. A highly faulted ternary hydride was seen in two phase $\alpha_2 + \gamma$ Ti-48Al which did not have the crystal structure or chemistry of any known Ti- or Ti-Al-hydride. Very fine, oriented, needle-shaped hydrides were observed in single-phase γ Ti-52Al.

INTRODUCTION

Titanium and its alloys are known to interact strongly with gaseous hydrogen at elevated temperatures. A variety of Ti-hydrides have been identified in Ti-based alloys [1-5], and the β-phase stabilization effect is well-known [6,7]. The effects of hydrogen on Ti-aluminide alloys are significantly different and less understood. A negative effect on mechanical properties due to hydrogen exposure has been reported [8-10] for both α_2 (Ti_3Al) and γ (TiAl) based Ti-aluminides. The presence of γ-Ti-hydrides (C1 structure) has been reported in hydrogen charged α_2 [11-13] and γ [10,14], as well as the formation of a new hexagonal hydride in γ [15]. Recent work at McDonnell Douglas has uncovered novel ternary hydrides in hydrogen-charged α_2 and γ alloys, and microstructural details about these new hydrides will be discussed.

EXPERIMENTAL DETAILS

Ti-25Al, Ti-25Al-10Nb-2V, and Ti-25Al-10Nb-3V-1Mo (at. %) α_2-based alloys were exposed to gaseous hydrogen at 600°C. Two different levels of hydrogen exposure were used: 0.1MPa (1 atm) pure H_2, and a mixture of 0.1MPa He + 0.5% H_2. A high pressure, 10.5MPa pure H_2 environment was used for Ti-48Al (a two-phase α_2+γ alloy) and Ti-52Al (single-phase γ). Hydrogen contents were measured by a combustometric technique at Analytical Associates, Inc.

Microstructural examination of the charged specimens was performed by optical microscopy, TEM and large-angle neutron diffraction at the University of Missouri Research Reactor. Large-angle neutron diffraction is a powerful technique for examining hydrides, because unlike electrons and X-rays, neutrons are strongly scattered by hydrogen atoms and therefore provide diffraction information about hydrogen atom sites. No unusual specimen preparation is required for neutron diffraction, and the specimens were directly examined in their as-charged state. TEM examination was done in a JEOL 2000FX at 200kV. Specimens were prepared for TEM by dimpling and ion milling, rather than electropolishing, to avoid any hydrogen contamination due to electrolytic charging.

RESULTS AND DISCUSSION

Ti-25Al

Prior to hydrogen charging, the binary Ti-25Al alloy consisted of equiaxed 5μm-10μm diameter α_2 (Ti_3Al) grains. The Ti-25Al alloy was charged at 600°C/16h in 0.1 MPa He + 0.5% H_2, and also at 600°C/18h in pure H_2. The effects of these different levels of hydrogen exposure were quite different. Exposure to the lower hydrogen environment resulted in a hydrogen concentration of 480 wppm, and had virtually no effect on the microstructure. Neutron diffraction indicated only a small expansion (0.1%) of the α_2 unit cell, and no other changes were observed by TEM or optical microscopy. On the other hand, specimens charged in the pure H_2 environment absorbed 4800 wppm hydrogen, and reacted strongly. Neutron diffraction revealed that 64 vol. % of the material transformed to a new, recently reported [16] hydride phase (Figure 1). This phase has the ordered $E2_1$ structure, with ideal stoichiometry Ti_3AlH. TEM examination showed that the Ti_3AlH phase was densely faulted, and frequently had a faceted morphology. The faults and facets were found to lie parallel to (111) Ti_3AlH planes. The untransformed α_2 phase showed a 0.7% unit cell expansion in the highly charged specimens.

Figure 1. Ti_3AlH phase in hydrogen charged Ti-25Al alloy, showing faulted and faceted structure.

Ti-25Al-10Nb-3V-1Mo

The Ti-25Al-10Nb-3V-1Mo specimens given an initial homogenization anneal of 650°C/24h, producing a microstructure consisting of transformed α_2 Widmanstätten platelets and equiaxed α_2 grains, absorbed 2600 wppm hydrogen when charged at 600°C/18h in 0.1 MPa pure H_2. Neutron diffraction showed no peaks except those due to α_2 in the charged material, and only a

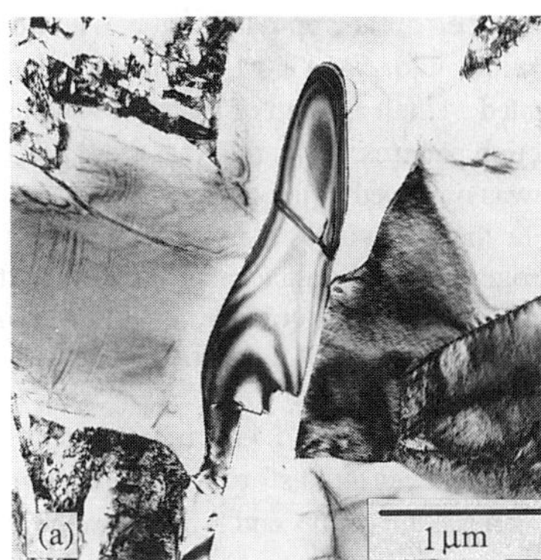

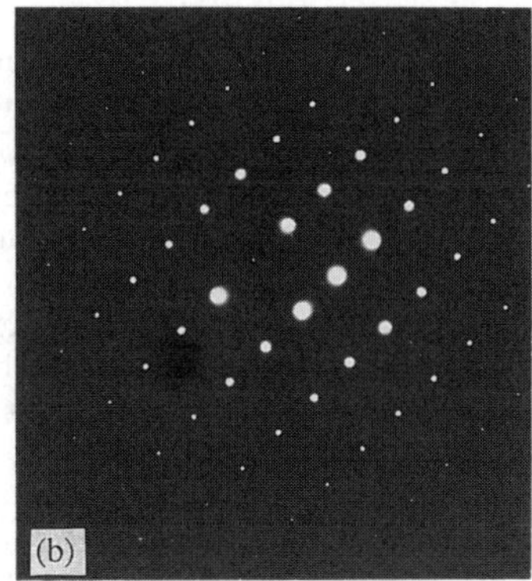

Figure 2. (a) Ti_3AlH phase in Ti-25Al-10Nb-3V-1Mo alloy, (b) <110> diffraction electron diffraction pattern, confirming identity of phase.

0.1% α_2 unit cell volume increase was detected. Closer inspection by TEM showed the presence of ~2 vol. % of a second phase (Figure 2a). Analysis of several sets of electron diffraction patterns identified this phase as the Ti_3AlH phase (the <110> diffraction pattern is shown in Figure 2b). Faults on {111} planes were observed in the Ti_3AlH grains, identical to the faults seen in hydrides in Ti-25Al, but somewhat less densely distributed. EDX showed the presence of Nb in the hydride, and indicated that the metal components of the hydride were present in the ratio (Ti:Al:Nb) 65:25:10. This is consistent with a hydride stoichiometry of $(Ti,Nb)_3AlH$. The Ti-25Al-10Nb-3V-1Mo alloy was also annealed at 982°C/0.5h to produce an equiaxed $\alpha_2 + \beta$ microstructure prior to hydrogen exposure at 600°C/18h in 0.1 MPa pure H_2. Although material with this microstructure absorbed 2600 wppm hydrogen when charged, no hydrides were observed.

Figure 3. Faults in hydrogen charged Ti-25Al-10Nb-2V alloy on $\{1\bar{1}00\}$ planes.

Ti-25Al-10Nb-2V

The Ti-25Al-10Nb-2V specimens with a pre-exposure microstructure consisting of Widmanstätten α_2 + equiaxed β were charged at 600°C/16h in 0.1MPa He + 0.5% H_2, and also at 600°C/16h in 0.1 MPa H_2. No hydrides were observed in the specimens exposed to the low hydrogen environment, which absorbed 610 wppm hydrogen. Neutron diffraction data indicated the specimens were mostly α_2 with a small amount of β phase. Peak broadening was observed in the neutron diffraction peaks for α_2, indicating strain or faulting in the α_2 lattice. The specimens exposed to the high hydrogen environment absorbed 5000 wppm hydrogen. Neutron diffraction showed that the specimen was primarily α_2, which underwent a 0.1% unit cell volume expansion due to the hydrogen. TEM examination revealed a small amount of β phase, which displayed faint superlattice reflections in electron diffraction patterns, indicating that the β had the ordered B2 structure. A high density of stacking faults was observed in the α_2 phase, lying parallel to $\{1\bar{1}00\}$ planes (Figure 3). These faults may be the early stages of hydride formation on $\{1\bar{1}00\}$ planes, or possibly the early stages of a hydrogen-induced transformation of α_2 to the orthorhombic phase. Similar faults have been reported in Ti-24Al-11Nb containing 4400 wppm hydrogen [12], where they were interpreted as hydrides. The presence of the Ti_3AlH phase in some Ti-Al-Nb alloys, but not in others may be explained by differences in the pre-hydrogen exposure microstructure. The Ti-25Al-10Nb-3V-1Mo annealed at 650°C/24h started as an all α_2 material, and the relatively low solubility of hydrogen in α_2 induces a transformation of α_2 to Ti_3AlH. On the other hand, the alloys initially containing a mixture of α_2 and β phases were able to dissolve enough hydrogen in the β phase to keep the α_2 phase from exceeding the hydrogen solubility limit and transforming to Ti_3AlH. The material used in reference [12] was also α_2 + β initially, so this may also explain why no Ti_3AlH was observed in their specimens.

Ti-48Al

Alloys containing γ as a dominant phase have a much lower hydrogen solubility than Ti- or Ti_3Al-based alloys. As a result, γ has been considered as a candidate for high pressure hydrogen applications. With this in mind, Ti-48Al specimens were charged at 800°C in 10.5 MPa pure H_2, for 10 and 94 hours. The material had an initial microstructure consisting of a mixture of lamellar α_2 + γ colonies and equiaxed γ grains. Both charging conditions produced a similar microstructural effect, although the specimen charged for 10h absorbed 1000 wppm H, and the 94h specimen absorbed 2600 wppm H. After exposure, no α_2 phase was observed, and a large volume fraction of the microstructure transformed to a new ternary hydride phase (Figure 4). Roughly 25 vol. % of the 94h specimen transformed to the new phase, based on TEM observations. The hydride phase had an elongated lamellar morphology, on average parallel to {111} planes in the γ phase. Locally, the hydride deviated significantly from this habit plane, and showed no consistent orientation relationship with the γ lamellae. This lamellar morphology and {111} habit plane indicates that the hydride forms by transformation from the α_2 phase, leaving the γ unaffected. Prior to hydrogen charging, the α_2 and γ phases in Ti-48Al have a lamellar morphology with the well-known orientation relationship: $(0001)\alpha_2 \parallel \{111\}\gamma$. When the α_2 lamellae transform to the

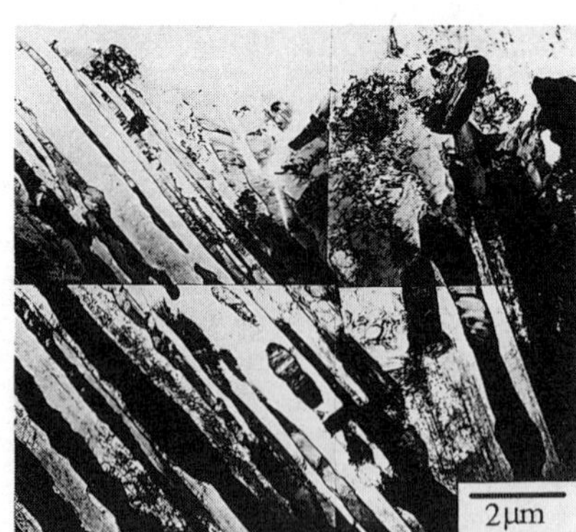

Figure 4. Typical microstructure of highly charged Ti-48Al alloy.

hydride, the orientation of the γ lamellae is preserved, as well as the general lamellar morphology of the transformed α_2 lamellae, as observed. The hydride contained a high density of parallel faults (Figure 5), with a fairly uniform spacing of 4.0nm. The d-spacing of crystal planes parallel to the fault plane in the hydrides was determined to be 0.7nm, so the faulting occurs every 5 to 6 repeats of the stacking sequence in this direction. Fault variants on other planes were not seen, suggesting that the fault plane is a unique crystallographic plane . The crystal structure of the ternary hydride has not yet been determined, although it was confirmed by electron diffraction that the hydride was neither cubic, tetragonal, nor hexagonal, which rules out all of the known Ti- or Ti-Al-hydrides. EDX indicated that a considerable redistribution of Al took place in the charged specimens. The γ phase in the charged specimens had a Ti:Al ratio of 0.72, which is near the Al-rich limit of the γ phase field. The hydride had a Ti:Al ratio of 1.38, much lower than Ti_3AlH, and also lower than any of the Ti:Al ratios measured by Legzdina et al. [17] for deuterides in deuterium-charged Ti-52Al-2Ta.

Figure 5. Densely faulted microstructure of hydride in highly charged Ti-48Al alloy.

Figure 6. Needle shaped hydrides in hydrogen charged single-phase γ alloys.

Ti-52Al

The Ti-52Al specimens had a single phase starting microstructure, composed of large, equiaxed γ grains. Charging at 800°C/18h in 10.5 MPa H_2 produced a fine, uniformly distributed array of 10-50nm long needle shaped hydrides throughout the material (Figure 6). TEM diffraction analysis showed that the needles were aligned with the <001] direction (i.e. parallel to the c-axis of the tetragonal unit cell) in the γ grains. Precipitate free zones were observed around dislocations and adjacent to grain boundaries. It has been suggested that these hydrides have a FCC crystal structure [14], but this has not yet been confirmed because their small size and low volume fraction makes identification very difficult. The fact that single phase γ does not form large scale hydrides after hydrogen exposure is further evidence that the hydride which forms in Ti-48Al results from transformation of the α_2 phase.

CONCLUSIONS

Hydrogen exposure produced the following microstructural effects in α_2 and γ alloys:

1. A ternary hydride (Ti_3AlH), detailed in [16], was observed in Ti-25Al and Ti-25Al-10Nb-3V-1Mo charged at 600°C in pure H_2.
2. Ti_3AlH was not seen in Ti-25Al-10Nb-2V that was $\alpha_2 + \beta$ initially. Evidence suggests that material must be fully α_2 for the Ti_3AlH phase to form, because the solubility of hydrogen in β-containing alloys is high enough to prevent large-scale hydride formation.
3. A high density of faults on $\{1\bar{1}00\}$ α_2 planes was observed in Ti-25Al-10Nb-2V.
4. A new ternary hydride was observed in γ with Ti:Al ratio of 1.38, produced by transformation of the α_2 phase. The ternary hydride was highly faulted, had a minimum planar spacing of 0.7nm, and was determined not to be any of the known ternary or Ti-H hydrides.

ACKNOWLEDGMENTS

This research was performed for NASA contract NAS2-13182. The authors would like to thank the program manager, H. G. Nelson, for his kind support.

REFERENCES

1. S. S. Sidhu, L. Heaton, and D. D. Zaubensi, Acta Crystall. 9 (56) 607.
2. H. L. Yakel, Jr., Acta Crystall. 11 (58) 46.
3. N. E. Paton, B. S. Hickman, and D. H. Leslie, Metall. Trans. 2 (71) 2791.
4. O. T. Woo, G. C. Weatherly, C. E. Coleman, and R. W. Gilbert, Acta Metall. 33 (85) 1897.
5. D. S. Shih, and H. K. Birnbaum, Scripta Metall. 20 (86) 1261.
6. A. D. McQuillan, Proc. Roy. Soc. London A204 (50) 309.
7. G. A. Lenning, C. M. Craighead, and R. I. Jaffee, Trans. AIME 200 (54) 367.
8. W.-Y. Chu and A. W. Thompson, Metall. Trans. A 23A (92) 1299.
9. S. M. L. Sastry, W. O. Soboyejo, and R. J. Lederich, Summary Proc. 3d Workshop on Hydrogen-Materials Interaction, NASP Joint Prog. Office Workshop Pub. 1007, (H. G. Nelson, ed., Moffet Field, CA, 1990) 191.
10. W.-Y. Chu and A. W. Thompson, Scripta Metall. Mater. 25 (91) 2133.
11. E. Manor and D. Eliezer, Scripta Metall. 23 (89) 1313.
12. D. S. Shih, G. K. Scarr, and G. E. Wasielewski, Scripta Metall. 23 (89) 973.
13. W.-Y. Chu, A. W. Thompson, and J. C. Williams, Acta Metall. 40 (92) 455.
14. D. Legzdina, I. M. Robertson, and H. K. Birnbaum, J. Mater. Res. 6 (91) 1230.
15. D. E. Matejczyk and C. G. Rhodes, Scripta Metall. Mater. 24 (90) 1369.
16. D. S. Schwartz, W. B. Yelon, R. R. Berliner, R. J. Lederich, and S. M. L. Sastry, Acta Metall. Mater. 39 (91) 2799.
17. D. Legzdina, I. M. Robertson, and H. K. Birnbaum, Scripta Metall. Mater. 26 (92) 1737.

ENVIRONMENTAL EFFECTS ON THE ROOM TEMPERATURE DUCTILITY OF POLYSYNTHETICALLY TWINNED (PST) CRYSTALS OF BINARY AND SOME TERNARY TiAl COMPOUNDS

M. H. OH, H. INUI, M. MISAKI, M. KOBAYASHI AND M. YAMAGUCHI

Department of Metal Science and Technology, Kyoto University, Sakyo-ku, Kyoto 606, Japan.

ABSTRACT

Environmental effects on the room temperature ductility of polysynthetically twinned (PST) crystals of binary and some ternary TiAl compounds have been investigated through tensile tests conducted in four different atmospheres. The tensile ductility of TiAl PST crystals is sensitive to test environment. It is higher when tested in vacuum or in dry air than in air or in hydrogen gas. The environmental loss in ductility of PST crystals decreases with increasing strain rate. The environmental embrittlement of binary TiAl PST crystals can be interpreted in terms of hydrogen embrittlement. The ternary TiAl PST crystals containing Cr, Mo or Mn exhibit higher ductility than the binary TiAl PST crystals when tested in air, while the ternary ones show lower ductility than the binary ones in vacuum. Additions of alloying elements such as Cr, Mo and Mn seem to be effective in reducing the environmental loss in ductility of TiAl PST crystals.

INTRODUCTION

Ti-rich TiAl compounds with the two-phase TiAl/Ti_3Al lamellar structure have been of considerable interest in the last few years as a new class of high-temperature structural materials [1-3]. One of the major limitations for the practical use of the compounds is the poor ductility at ambient temperature. Resent researches on environmental effects on the ductility of intermetallic compounds such as Ni_3Al and FeAl have shown that environmental embrittlement can be a possible reason for their limited ductility at ambient temperature [4]. There is some evidence to indicate that TiAl-based compounds are susceptible to environmental embrittlement [5,6]. However, the two-phase TiAl compounds exhibit only a limited ductility in polycrystalline form and the difference in the tensile elongation between in air and in vacuum is rather small to characterize their environmental embrittlement behavior [7-9]. With the use of so-called polysynthetically twinned (PST) crystals of TiAl in which only a single grain with the TiAl/ Ti_3Al lamellar structure is contained, we have made a systematic study on the deformation behavior as a function of the angle (ϕ) between the lamellar boundaries and loading axis [10-12] and demonstrated that a tensile elongation as large as 20% can be obtained in air at room temperature for ϕ=31° where the easy type of deformation (shear deformation parallel to the lamellar boundaries) occurs [12]. This is far larger than any other values of room temperature ductility ever reported on two-phase TiAl compounds. We have thus decided to investigate environmental effects on the ductility of two-phase TiAl compounds using PST crystals to characterize the environmental embrittlement behavior of the two-phase TiAl compounds.

In this paper, we report the results of deformation experiments of binary and some ternary TiAl PST crystals in tension at room temperature in four different atmospheres in order to clarify whether the two-phase TiAl compounds are susceptible to environmental embrittlement.

EXPERIMENTAL

Compositions of the master ingots used in the present study were Ti-49.3at.%Al and Ti-48.4at.%Al-0.6at.%X (X=Cr, Mo and Mn), respectively. Binary and ternary TiAl PST crystals, 10 mm in diameter and 100 mm long, were grown from the master ingots using an ASGAL FZ-SS35W optical floating zone furnace. Oriented tensile specimens with ϕ=31°, approximately 2 mm x 0.5 mm in cross section and 5 mm in gauge length, were cut from as grown PST crystals. Tensile tests were carried out at room temperature in the strain rate range of $2.0\times10^{-4}\,s^{-1}$ to $1.0\times10^{-1}\,s^{-1}$, using an Instron-type testing machine equipped with a vacuum chamber. Four different atmospheres were employed; air, vacuum, dry air (an artificial mixture of 79% nitrogen and 21% oxygen gas) and dry hydrogen gas. Deformation structures were

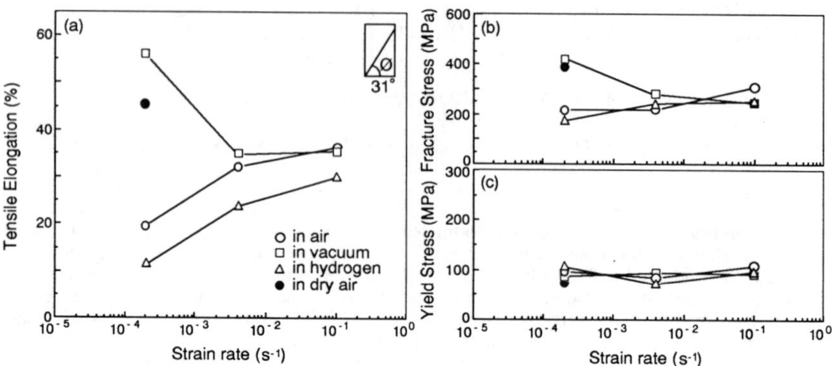

Fig. 1 Strain rate dependence of (a) tensile elongation, (b) fracture stress and (c) yield stress of binary TiAl PST crystals.

examined by optical microscopy and transmission electron microscopy (TEM). Fracture surfaces were examined by scanning electron microscopy (SEM).

RESULTS

Binary TiAl PST crystals

Figure 1 shows the strain rate dependence of the tensile elongation, fracture stress and yield stress for binary TiAl PST crystals in four different atmosphere. As seen in Fig. 1, the yield stress is almost insensitive to test environment and strain rate but the tensile ductility and the fracture stress are sensitive to both test environment and strain rate. For example, at the lowest strain rate of $2.0\times10^{-4}s^{-1}$, the value of tensile elongation observed in vacuum is as large as 56% which is far larger than those in air (20%) and in hydrogen gas (11%). However, at the highest strain rate of $1.0\times10^{-1}\ s^{-1}$, the tensile elongation in vacuum decreases to 35%, which is comparable to those in air (36%) and in hydrogen gas (30%). When tested in vacuum, the tensile elongation and fracture stress decrease with increasing strain rate. This is normally observed in many metals. In contrast, the tensile elongation and fracture stress increase with increasing strain rate in air and in hydrogen gas. In other words, the environmental loss in ductility of PST crystals of TiAl decreases with increasing strain rate, indicating that a diffusion process contributes to the embrittlement. The environmental loss in ductility is larger when tested in hydrogen gas than in air for all the strain rates used. In dry air at the lowest strain rate of $2.0\times10^{-4}s^{-1}$, the tensile elongation is 46%, which is somewhat smaller than that in vacuum but much larger than those observed in air and in hydrogen gas. This indicates that the environmental embrittlement in air may be due to moisture-induced hydrogen. Thus, the environmental embrittlement in TiAl may be interpreted in terms of hydrogen embrittlement.

When tested at the lowest strain rate at which the environmental loss in ductility is largest, two distinct types of fracture mode were observed depending on test environment. Figures 2(a)-(c) show side views of fracture surfaces of specimens tested in air, in hydrogen gas and in vacuum, respectively. The values of tensile elongation of specimens shown in Figs. 2(a)-(c) were 20%, 11% and 56%, respectively. In air and in hydrogen gas, fracture occurs in a cleavage-like mode with a macroscopic habit plane parallel to the lamellar boundaries, whereas fracture occurs across the lamellar boundaries when tested in vacuum. The fracture mode observed in specimens tested in dry air was essentially similar to that observed in specimens tested in vacuum. It should be noted in Fig. 2(c) that fracture occurs in a brittle manner without showing any local contraction even after deformed to 56% in vacuum.

Top views of fracture surfaces of specimens of Figs. 2(a)-(c) are presented in Figs. 2(d) -(f), respectively. Specimens tested in air (a) and in hydrogen gas (b) show a flat fracture

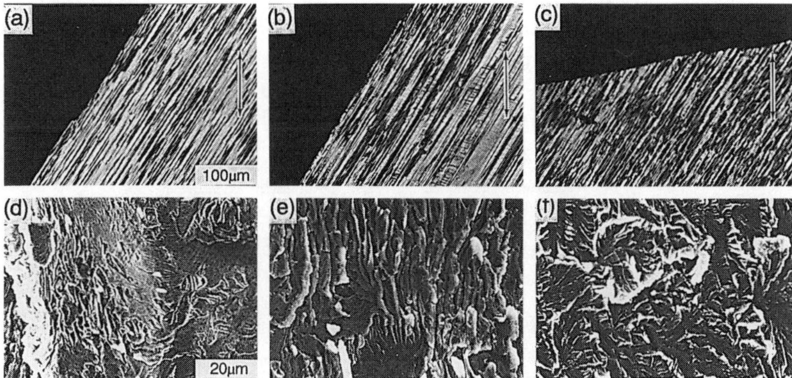

Fig. 2 Side (a-c) and top (d-f) views of fracture surfaces of binary TiAl PST crystals tested in air (a,d), in hydrogen gas (b,e) and in vacuum (c,f), respectively, at the lowest strain rate of $2.0 \times 10^{-4}\ s^{-1}$. The tensile axis is shown by arrows in the figures.

surface composed of terraces of various sizes with the terrace plane parallel to $(111)_{TiAl}$, whereas those tested in vacuum (c) show a relatively rough and tortuous fracture surface. The environmental loss in ductility decreases with increasing strain rate as shown in Fig. 1, and the fracture behavior becomes accordingly less dependent on test environment with increasing strain rate. Thus, at higher strain rates, cracks propagate not parallel to the lamellar boundaries but across the lamellar boundaries regardless of test environment.

Thus, the environmental loss in ductility of binary TiAl PST crystals which deform by the easy type of deformation is clearly related to the change in fracture mode. When the environmental loss in ductility is large, PST crystals of TiAl fail by a cleavage-like mode with a habit plane parallel to the lamellar boundaries. Otherwise, fracture occurs across the lamellar boundaries.

Ternary TiAl PST crystals

Figures 3(a)-(c) show stress-strain curves obtained in air and in vacuum at the lowest strain rate of $2.0 \times 10^{-4} s^{-1}$ for binary and ternary TiAl PST crystals containing Cr and Mo, respectively. The stress-strain curves shown in Fig. 3 are those for specimens with the largest elongation for each testing condition. As in the case of binary PST crystals, the tensile elongation of ternary PST crystals is sensitive to test environment; it is higher when tested in vacuum than in air. When tested in air, the values of tensile elongation obtained for ternary PST crystals doped with Cr and Mo are respectively 35% and 28%, which are larger than that for binary PST crystals (20%). This result is in agreement with the recent reports on the beneficial effects of ternary alloying additions on the room temperature ductility of polycrystalline two-phase TiAl compounds [1-2]. In contrast, when tested in vacuum, the values of tensile elongation obtained for ternary PST crystals (52% and 38% for Cr- and Mo-doped, respectively) are lower than that for binary PST crystals (56%). The same trend in tensile ductility was observed also for Mn-doped ternary PST crystals.

Figures 4(a)-(d) show side views of fracture surfaces of Cr- (a,b) and Mo-doped (c,d) ternary PST crystals tested at the lowest strain rate of $2.0 \times 10^{-4} s^{-1}$ in air (a,c) and in vacuum (b,d), respectively. As seen in the figures, fracture occurs across the lamellar boundaries regardless of test environment for ternary PST crystals doped with Cr and Mo. In contrast, fracture occurs in a cleavage-like mode with a habit plane parallel to the lamellar boundaries for binary PST crystals tested in air (Fig 2(a)). Thus, the increased tensile ductility of ternary PST crystals in air is related to the change in fracture mode. TEM examinations of deformation structures made on binary and ternary PST crystals tested in air and in vacuum have revealed no significant difference, i.e., the major deformation modes are deformation twinning of the

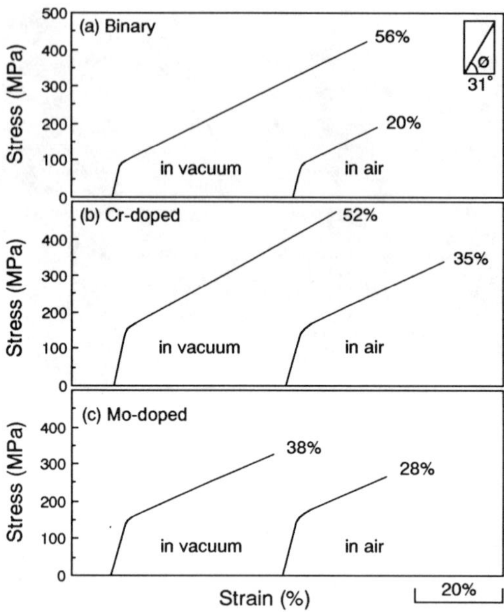

Fig. 3 Stress-strain curves obtained for binary and ternary TiAl PST crystals tested in air and in vacuum at the lowest strain rate of 2.0×10^{-4} s^{-1}.

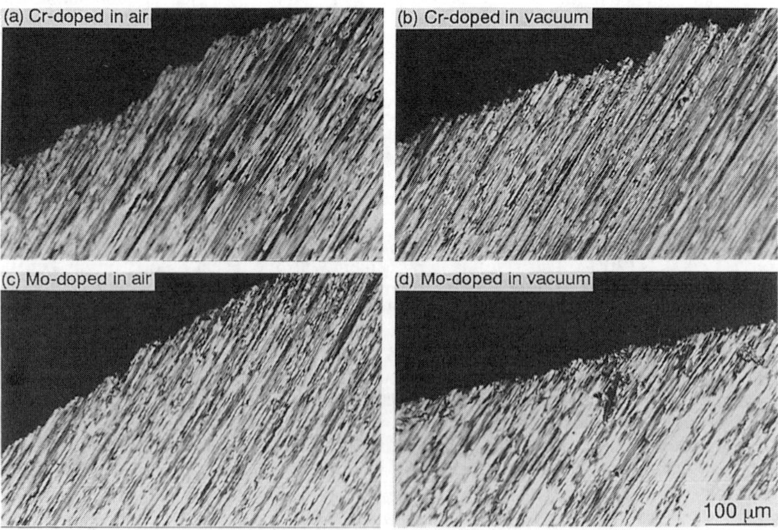

Fig. 4 Side views of fracture surfaces of Cr- (a,b) and Mo-doped (c,d) ternary TiAl PST crystals tested in air (a,c) and in vacuum (b,d) at the lowest strain rate of 2.0×10^{-4} s^{-1}.

{111}<11$\bar{2}$]-type and slip along <1$\bar{1}$0] for all the cases. Thus, the increase in tensile ductility of ternary PST crystals in air seems to be largely due to the reduction in the environmental loss in ductility.

DISCUSSION

The tensile ductility of binary TiAl PST crystals is higher when tested in vacuum or in dry air than in air or in hydrogen gas. The environmental loss in ductility decreases with increasing strain rate. Such features are essentially similar to those observed in many intermetallic compounds which are susceptible to environmental embrittlement [4]. Thus, the environmental embrittlement of two-phase TiAl compounds may be interpreted in terms of hydrogen embrittlement as in many intermetallic compounds. Hydride formation was not observed in any of deformed PST crystals by TEM observations [13]. Thus, in contrast to the Ti_3Al-based alloys [14], the environmental embrittlement of PST crystals of TiAl may not be attributed to hydride formation. Many investigations of hydrogen embrittlement in intermetallics have postulated that a decohesion mechanism, in which the lowering of bond strength occurs in the presence of hydrogen gas, is responsible for the embrittlement [4]. As shown in Fig. 2, binary PST crystals fail by a cleavage-like mode with a habit plane parallel to the lamellar boundaries when the environmental loss in ductility is large, that is when tested in air and in hydrogen gas at lower strain rates, otherwise fracture occurs across the lamellar boundaries. This indicates that the embrittlement of binary PST crystals can be also explained by a decohesion model in which the bond strength along the lamellar boundaries is lowered in the presence of hydrogen gas.

Due to the presence of six different types of ordered domains in the TiAl phase, three different types of lamellar domain boundaries can exist in the TiAl lamellae [15,16]; (I) true-twin type (energy : γ_T), (II) pseudo-twin type (energy : γ_P) and (III) 120°-rotational order-fault type (energy : γ_R). Using the values of $\gamma_{APB} / \gamma_{SISF}$ obtained theoretically [17,18] and experimentally [19], the ratio of energies of these three boundaries ($\gamma_T : \gamma_P : \gamma_R$) has been estimated to be in the range of 1:3:2 to 1:7:6 on the basis of a hard-sphere model [16], where γ_{APB} and γ_{SISF} are the energy of the antiphase boundary (APB) on {111} with a fault vector of 1/2<110] and that of the superlattice intrinsic stacking fault on {111}, respectively. In accord with the estimate, true-twin boundaries are most frequently observed. However, the other two types of boundaries are also often observed. This suggests that the lamellar domain boundaries of pseudo-twin and 120°-rotational types may have much smaller energies than those estimated on the basis of a hard-sphere model [16] most probably due to the relaxation of atoms in the close vicinity to the fault plane. Otherwise, the lattice mismatches would generate huge elastic stresses at the lamellar domain boundaries of these two types [20] and they may not exist in the TiAl phase. Thus, the lamellar domain boundaries of the pseudo-twin and the 120°-rotational types would involve such relaxations of atoms and may have potential to accept hydrogen atoms along them, offering the preferential path for hydrogen penetration. Then, the bond strength of these boundaries will be reduced in the presence of hydrogen, resulting in a cleavage-like fracture along them, that is, parallel to the lamellar boundaries. The environmental embrittlement in TiAl PST crystals can thus be understood in terms of the high susceptibility of the pseudo-twin and 120°-rotational lamellar domain boundaries to hydrogen attack.

These lamellar domain boundaries should have potential to accept not only hydrogen atoms but also ternary alloying atoms. In fact, as shown in Fig.4, PST crystals of TiAl vary in the room-temperature fracture mode from cleavage-like parallel to the lamellar boundaries to cracking in zigzag across the lamellar boundaries by adding ternary alloying elements such as Cr, Mo and Mn, and their environmental embrittlement is reduced as a result. This may be because ternary alloying atoms segregate preferentially to the pseudo-twin and 120°-rotational lamellar domain boundaries and reduce their vulnerability to hydrogen attack.

CONCLUSIONS

(1) TiAl PST crystals are susceptible to environmental embrittlement. The tensile ductility of TiAl PST crystals is higher when tested in vacuum or in dry air than in air or in hydrogen gas. Thus, the embrittlement can be interpreted in terms of hydrogen embrittlement.
(2) The environmental loss in ductility decreases with increasing strain rate. When the

environmental loss is large, that is when tested in air or in hydrogen gas at lower strain rates, binary TiAl PST crystals fail in a cleavage-like mode with a habit plane parallel to the lamellar boundaries, otherwise fracture occurs across the lamellar boundaries.
(3) The ternary TiAl PST crystals containing Cr, Mo and Mn exhibit higher ductility than the binary TiAl PST crystals when tested in air, while the ternary ones show lower ductility than the binary ones in vacuum. The ternary TiAl PST crystals fail by cracking across the lamellar boundaries regardless of test environment. Thus, the increase in tensile ductility of the ternary PST crystals can be interpreted largely in terms of the reduction in the environmental loss in ductility of TiAl PST crystals.

ACKNOWLEDGEMENTS

This work was supported by Grant-in-Aid for Scientific Research on the priority Area "Intermetallic compounds as New High Temperature Structural Materials" from the Ministry of Education, Science and Culture, Japan and in part by the research grant from R & D Institute of Metals and Composites for Future Industries.

REFERENCES

[1] Y.W. Kim and D.W. Dimiduk, JOM 43 (8), 40 (1991).
[2] S.C. Huang and D.S. Shih in Microstructure/Property Relationships in Titanium Aluminides and Alloys, edited by Y.W. Kim and R.R. Boyer (TMS, Warrendale, Pa, 1991) pp. 105-122.
[3] M. Yamaguchi and H. Inui in Ordered Intermetallics-Physical Metallurgy and Mechanical Behavior, edited by C.T. Liu, R.W. Cahn and J.O. Sauthoff (Kluwer Academic Publishers, Dordrecht, 1992) pp. 217-235.
[4] for example, C.T. Liu and C.G. McKamey in High Temperature Aluminides and Intermetallics, edited by S.H. Whang, C.T. Liu, D.P. Pope and J.O. Stiegler (TMS, Warrendale, Pa, 1990) pp. 133-151.
[5] C.T. Liu and Y.W. Kim, Scripta metall. mater. 27, 599 (1992).
[6] T. Takasugi, S. Hanada and M. Yoshida, J. Mater. Res. 7, 2739 (1992).
[7] R.D. Kane and E.A. Chakachery in Environmental Effects on Advanced Materials, edited by R.H. Jones and R.E. Ricker (TMS, Warrendale, Pa, 1991) pp. 35-43.
[8] D.A. Meyn in Microstructure/Property Relationships in Titanium Aluminides and Alloys, edited by Y.W. Kim and R.R. Boyer (TMS, Warrendale, Pa, 1991) pp. 275-281.
[9] W. Y. Chu and A. W. Thomson, Scripta metall. mater. 25, 2133 (1991).
[10] T. Fujiwara, A. Nakamura, M. Hosomi, S.R. Nishitani, Y. Shirai and M. Yamaguchi, Phil. Mag. A 61, 591 (1990).
[11] H. Inui, A. Nakamura, M.H. Oh and M. Yamaguchi, Phil. Mag. A 66, 557 (1992).
[12] H. Inui, M.H. Oh, A. Nakamura and M. Yamaguchi, Acta metall. mater.40, 3095 (1992).
[13] M.H. Oh, H. Inui, M. Misaki and M. Yamaguchi, Acta metall. mater., in press (1992).
[14] W.Y. Chu, A.W. Thomson and J.C. Williams in Hydrogen Effects on Material Behavior, edited by N.R. Moody and A.W. Thomson (TMS, Warrendale, Pa, 1990) pp. 543-554.
[15] H. Inui, A. Nakamura, M.H. Oh and M. Yamaguchi, Ultramicroscopy 39, 298 (1991).
[16] H. Inui, M.H. Oh, A. Nakamura and M. Yamaguchi, Phil. Mag. A 66, 539 (1992).
[17] C.L. Fu and M.H. Yoo, Phil. Mag. Lett., 62, 159 (1990).
[18] C. Woodward, J.M. MacLaren and S. Lao in High-Temperature Ordered Intermetallic Alloys IV, edited by L.A. Johnson, D.E. Pope and J.O. Stiegler (Mater. Res. Soc. Proc. 213, Pittsburgh, Pa, 1991) pp. 715-720.
[19] G. Hug, A. Loiseau and P. Veyssiere, Phil. Mag. A 57, 499 (1988).
[20] P.M. Hazzledine, B.K. Kad, H.L. Fraser and D.M. Dimiduk, presented at the 1992 MRS Spring Meeting, San Francisco, CA, 1992, in press.

A STUDY OF THE PROCESSING AND COMPOSITIONAL VARIABLES ON THE MICROSTRUCTURAL DEVELOPMENT AND OXIDATION CHARACTERISTICS OF RAPIDLY SOLIDIFIED NiAlCo ALLOYS

M.L. STEWART AND P.B. ASWATH
University of Texas at Arlington, Department of Mechanical and Aerospace Engineering, Materials Science and Engineering Program, Arlington, Tx.

ABSTRACT

Oxidation tests were performed isothermally at 1000° C on chill block melt spun metallic ribbons with alloy compositions consisting of Ni(46-50 at%), Al(31-33 at%), Co(17-20 at%) and Hf(0-1 at%). The oxidation rates were calculated and the microstructure was examined. Results indicated the difference in microstructural and oxidation characteristics could largely be attributed to the differences in the superheat of the melt. There was no indication that hafnium retarded grain growth. The oxidation behavior of two of the ribbons exhibited a logarithmic rate law.

INTRODUCTION

Ordered intermetallic nickel aluminides are currently receiving a great deal of attention for high temperature structural applications due to several outstanding properties including an increase in strength with increasing temperature, good oxidation resistance and a density about 10% less than nickel based superalloys.[1,2,3,4] However, the inability to process these materials due to their low ductility and low fracture toughness had, until recently, discouraged their development for structural applications.[3,4]

Processing ordered intermetallic nickel aluminides by rapid solidification techniques has the potential to improve ductility by increasing homogeneity, producing very fine grain sizes and creating novel microstructures.[5,6] However, the characteristics of rapid solidified materials which lead to the enhanced properties can be lost during high temperature consolidation. The primary objective of the research contained herein is to study the microstructural changes and oxidation behavior of six rapidly solidified intermetallic nickel aluminide alloys exposed to air at 1000°C for 4 hours and 68 hours.

EXPERIMENTAL PROCEDURE

Five slightly different NiAlCoHf alloy compositions were rapidly solidified using the chill block melt spinning process (CBMS) in order to produce metallic ribbons, by the NASA Lewis Research Center in Cleveland Ohio. Alloy compositions are listed in Table I and amount of melt super heat in Table II. After CBMS processing, each of the six types of metallic ribbons were examined for variations in the thickness, width, and length dimensions as well as the differences porosity and surface texture. An SEM analysis was made of the ribbons in the as-received state in two orientations: (a) horizontally mounted so as to expose the edge of the ribbon, through the thickness, parallel to the wheel direction and (b) vertically mounted and then broken, perpendicular to the wheel direction in order to observe fracture surfaces.

Individual and multiple samples were prepared for oxidation testing. Individual ribbon samples were approximately 2mm in width and 3mm in length. Each multi-piece sample contained 15 pieces (approximately 1mm x 2mm) of the same material. All oxidation test

TABLE I: METALLIC RIBBONS COMPOSITIONS (atomic percent)				
ALLOY	Ni	Al	Co	Hf
0076A	46	33	20	1
0106A	46	33	20	1
0112A	50	31	19	0
0114A	50	31	18	1
0108B	50	32	18	0
0110B	50	32	17	1

TABLE II: EJECTION TEMPERATURE AND SUPER HEAT FOR EACH ALLOY		
Alloy	Ejection Temp (°F)	Super Heat (°F)
0076A	2660	not determined
0106A	2825	165
0112A	2775	200
0114A	2750	225
0108B	2800	260
0110B	2825	230

samples were inspected for porosity and the physical characteristics; including thickness, weight, surface area and perimeter, were determined.

Three series of oxidation tests were performed using a Thermal Gravimetric Analyzer (TGA) at 1000°C: (a) single samples were oxidized for 234 minutes in a mixed argon/air (98% Ar, 2% air) atmosphere, (b) single samples of ribbon 0114 were oxidized for various times from 10 to 300 minutes in a mixed argon/air atmosphere (98% Ar, 2% air), (c) single and multiple samples were oxidized in a compressed air (70% N_2, 30% O_2) atmosphere for 68 hours.

There were two phases in the analysis of the oxidized samples. The first involved the calculation of the oxidation constant and the second was a detailed SEM analysis of the oxidized samples. The data collected by the TGA during the oxidation testing consisted of measuring the amount of weight gained per unit time. This data was plotted versus time and log time in order to determine if there was a parabolic or logarithmic oxidation constant relationship.

The SEM analysis of the oxidized samples was divided into two phases: (a) an analysis of an edge view of the ribbon through the thickness and (b) an analysis of fractured surfaces of the ribbons which were broken by bending. This analysis provided information on the penetration of the oxide into the parent material and the changes in the original microstructure after oxidation testing.

RESULTS AND DISCUSSION

Based on the results of this research, it was determined the six ribbon types (varying in composition and processing parameters) could be divided into two groups. The remaining portion of this paper will concentrate on the differences between these two groups with group (A) representing ribbons 0076, 0106, 0112 and 0114 and group (B) representing ribbons 0108 and 0110.

Two limiting characteristics of these ribbons are their short lengths, averaging 2 to 3 inches, (compared to research by Hemker and Glasgow at NASA Lewis which produced ribbons as long as 15 meters[7]) and the lack of ductility. The combination of these characteristics indicates the optimum processing parameters for these materials have yet to be determined.

Polished and etched cross section samples and fracture samples from each of the alloys were studied and photographed using a scanning electron microscope (SEM). Figure 1 is a typical fracture specimen for the ribbons in group (A). The two zone microstructure of this ribbon is typical for a number of rapidly solidified materials. Hebsur and Locci found a similar structure in melt spun ribbons of niobium aluminidies with the relative size of each of the zones being determined by differences in processing parameters.[8]

The main characteristics of the ribbons in group (A) are: (1) the microstructure is divided into two zones consisting of columnar grains on the chill side of the ribbons and equiaxed grains extending from a transition region to the free side of the ribbons and (2) a small amount of a very fine precipitate is present in the equiaxed section of the ribbons which contain hafnium. Hafnium, which forms very stable MC carbides, is added to pin the grain boundaries thereby reducing grain growth during exposure to high temperatures.

The microstructure of the ribbons in group (B) are very different from group (A). In these materials the columnar and equiaxed regions have been replaced with relatively large featureless equiaxed grains as shown in a typical fracture sample, Figure 2. The large equiaxed grains are easily identifiable due to the predominantly intergranular fracture. In some sections, the equiaxed grains are replaced by a dendritic growth approximately midway through the ribbon thickness. Based on the detail on the free surface, the dendritic growth is not continuous but runs in a random pattern.There is no second phase or precipitate visible in the grains or in the grain boundaries.

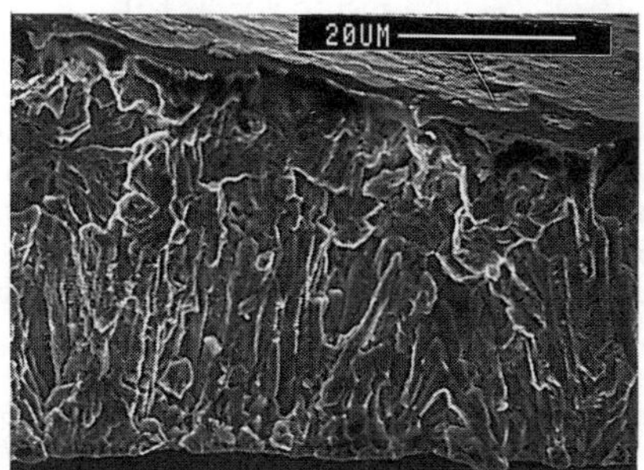

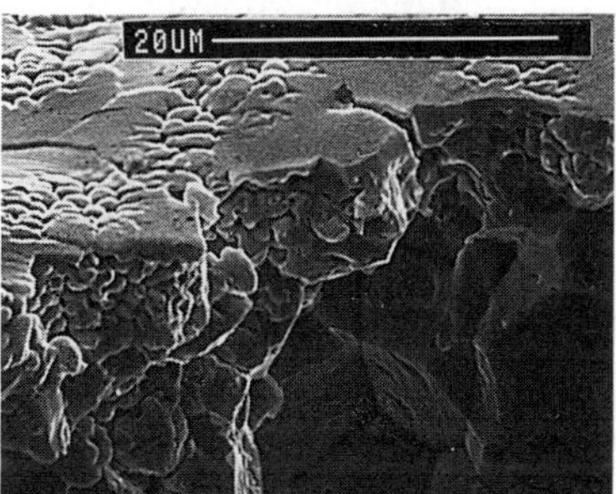

Figure 1: Group (A) Ribbon   Figure 2: Group (B) Ribbon

As-received fracture samples, chill side at the bottom.

Variations in the processing parameters resulted in the differences in the microstructure between the ribbons in group (A) and group (B). The most influential processing parameter was the amount of superheat experienced before the material was ejected from the crucible onto the chilled wheel. Group (B) ribbons had the highest combined ejection temperatures and superheat. These two properties increased the amount of fluidity of these materials leading to the formation of a stable planar solidification front and a larger microsegregation free region.

Microstructure Analysis of Oxidation Test Samples

Figure 3 (a-d) is a series of photographs which detail the changes in the microstructure of fractured samples of ribbons in group (A) at various time during oxidation testing. In Figure 3a, after only 11 minutes of oxidation time, the microstructure quickly becomes muddled with less definition between the columnar and the equiaxed grain regions. On the bottom surface of the ribbon, the small zone of chill grains are more obvious and clearly line the chilled edge of the ribbon. In the next two photographs, Figures 3b (179 minutes) and 3c (309 minutes), there is a progression of grain growth from a single row of grains on the chill side and free side of the ribbon. These grains continue to grow until they have consumed approximately half of the ribbon thickness, 1/4 on top and 1/4 on the bottom.

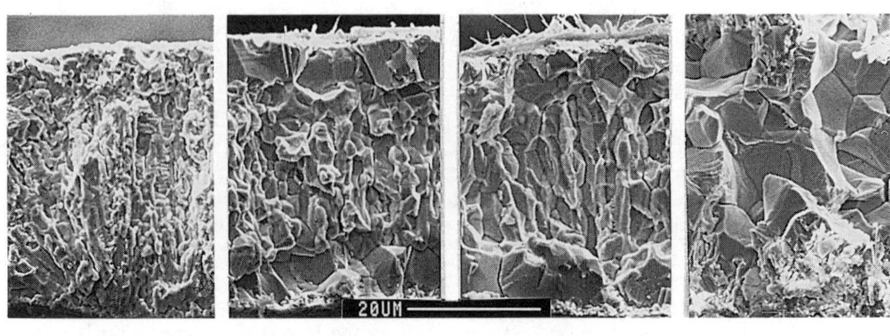

(3a): 11 minutes (3b): 179 minutes (3c): 309 minutes (3d): 68 hours

Figure 3 (a-d): Group (A) Ribbons, oxidized fracture samples, chill side at bottom
Oxidizing Atmosphere: Figure 3(a-c): 2% air Figure 3d: 30% 0_2

During this grain growth, the columnar and equiaxed grain regions shrink in size but retain their features through 179 minutes of testing. By 234 minutes, the center of the ribbon changes from the dendritic and equiaxed grain regions to a region containing only small equiaxed grains. These grains begin to grow and continue to grow until there is no distinction between the grains originally formed on the upper and lower edges of the ribbon and the grains formed later in the interior of the ribbon. Figure 3d (68 hours) show an intergranular fracture mode which resembles the as-received material for the ribbons in group (B) and the penetration of the oxide into the ribbon interior. Figures 3(b) and 3(c) show the progression of oxidation with whisker development on the ribbon surface. These whiskers are the result of outward diffusion of aluminum and can be found on NiAl and Ni_3Al alloys.[1] The precipitates that were visible in the as-processed ribbons containing hafnium, were visible in 2 of the ribbons after 4 hours of oxidation testing but were not detectable in any of the ribbons after 68 hours.

The ribbons in group (B) were oxidized for 4 hours and 68 hours only. After the 4 hour oxidation test the microstructure contained: (1) an extensive amount of grain growth with some grains as large as the thickness of the ribbon, (2) no second phase present and (3) a large percentage of microporosity. After 68 hours of oxidation testing the microstructure had the same general appearance as the ribbons in group (A) except the group (B) ribbons had: (1) much larger grains which in many areas cover the entire thickness of the ribbon and (2) a large decrease in the amount of oxide penetration into the ribbons.

Calculation of the Oxidation Rate Constants

To calculate the oxidation rate constant, the weight gain/surface area data points were plotted versus log time and the slope of the best fit line was determined using linear regression analysis. The individual tests for each material were combined and a minimum, maximum and median value of the oxidation rate constants were calculated.

Normally the oxide scale growth rate can be described by one of three rate equations which depend in part on the thickness of the oxide and the Pelling-Bedworth ratio[9]. For the high temperature oxidation of nickel based alloys, the parabolic rate equation usually describes the growth rate of the oxide scale, however, the alloys used in this study exhibit primarily a logarithmic oxidation rate.

There are limited instances where the logarithmic rate equation applies at high temperatures. In one instance, when an oxide scale is highly adherent, few physical defects appear in the oxide layer; the oxide has a low concentration of point defects and only slow ionic transport takes place. In this particular instance, the corrosion rate is gradually reduced with increasing oxide thickness and the diffusion controlled kinetics can be either logarithmic or parabolic.[10] A different explanation for the observance of the logarithmic law at high temperatures involves the compressive narrowing of grain boundaries which slows down cation diffusion. As the oxide film grows, compressional stresses block the porosity associated with the grain boundaries in the oxide which decreases ion diffusion. Any decrease in dynamic diffusion would result in a logarithmic rate of oxidation.[11]

A comparison between the weight gain/surface area for the 4 hour (2% air) and 68 hour (30% O_2) oxidation test samples is given in Figure 4. This figure illustrates the differences in the magnitude of the amount of weight gain/surface area between the ribbons in group (A) and group (B).

As the oxidation time and the partial pressure of oxygen increased so did the difference in the weight gain/surface area between the ribbons in group (A) and group (B). Reasons for difference between group (A) and group (B) are given below:

(1) The type of oxide formation found on the 4 hour samples. The oxide formation in group (A) consisted of a semi-continuous oxide that covered most of the surface of these ribbons with secondary single crystal oxides which penetrated through this oxide scale. In group (B), there was a continuous oxide scale with a slightly different morphology than in group (A). The ribbons in group (B) also had a secondary form of oxide on the surface, but the morphology is very irregular, not crystalline as in group (A).

(2) The ribbons in group (A) had a greater oxidation rate constant than group (B) in the 4 hour oxidation tests.

(3) After the 4 hour tests, grain boundary oxide penetration can be seen in all of the ribbons in group (A) and is not present in group (B).

(4) After 68 hours of testing all of the ribbons experienced deep oxide penetration along the grain boundaries. The main difference between the ribbons in group (A) and group (B) is the change in the fracture pattern along the outer edges of ribbons in group (A) due to oxygen diffusion into the base metal.

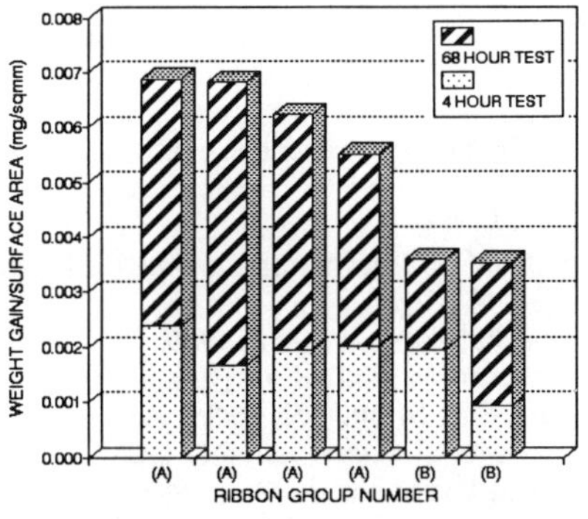

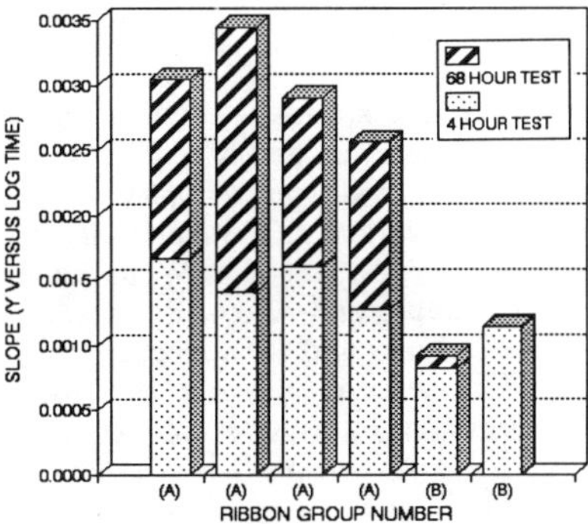

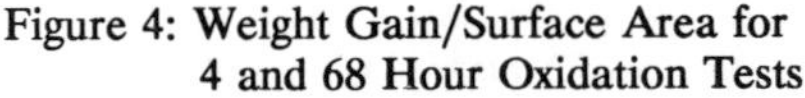

Figure 4: Weight Gain/Surface Area for 4 and 68 Hour Oxidation Tests

Figure 5: 4 and 68 Hour Logarithmic Oxidation Rate Constants

In Figure 5, a comparison is given between the 4 hour and 68 hour oxidation rate constants. The oxidation rates for group (B) are almost identical. This indicates that the oxidation rate can be described as logarithmic for these particular materials and is independent of the increase of oxygen partial pressure. The 68 hour oxidation rates for ribbons in group (A) have increased substantially from the rates calculated for the 4 hour samples. This indicates that the oxidation does not follow a logarithmic rate equation for these conditions. Further evaluation concluded that the oxidation data did not follow a parabolic rate equation. Therefore, based on the calculations of the logarithmic rate constant, the lack of linearity associated with the parabolic graph, the presence of several different types of oxidation scales on the surface of the 68 hour samples and the large amount of oxide penetration into the material, the equation which would describe the oxidation rate for ribbons in group (A) most likely incorporates both logarithmic and parabolic rate equations and is dependent on the oxygen partial pressure.

CONCLUSIONS

(1) The differences in microstructure and oxidation characteristics of the metallic ribbons used in this research can be attributed to differences in processing parameters and to a lesser extent to the differences in composition.
(2) There was no indication that the addition of hafnium retarded the grain growth at 1000°C in any of the ribbons.
(3) The 4 hour and 68 hour logarithmic oxidation rates for the ribbons in group (B) are nearly identical indicating that the rate of oxide growth can be described using a logarithmic rate equation for these two materials and is independent of oxygen partial pressure.
(4) The ribbons in group (B) had an improved overall quality, better oxidation resistance and a higher relative ductility than the ribbons in group (A). The improved properties associated with these ribbons is directly linked to the combination of a high ejection temperature and a large amount of melt superheat which encourages the formation of a more homogeneous microstructure. However, ribbons 0108 and 0110 have two problems. First, they experienced extreme grain growth in only 4 hours at 1000°C with grains growing as large as the ribbon thickness. Second, the presence of dissolved gasses form voids at the grain boundaries during solidification enhancing intergranular fracture.

REFERENCES

[1] D.R. Pank, Master thesis, Penn. State, 1988.
[2] M.A. Meyers and K.K. Chawla, Mechanical Metallurgy, (Prentice-Hall, Inc., New Jersey, 1984), pp. 283, 538.
[3] C.T. Liu and J.O. Steigler, Science, **226**, 636 (1984).
[4] R.A. Varin and M.B. Winnicka, presented at the 5th International Symp. on Plasticity of Metals and Alloys, Prague, 1990.
[5] I.E. Locci and R.D. Noebe, NASA Technical Paper No. 101450, 1989.
[6] I.E. Locci, R.D. Noebe, J.A. Moser, D.S. Lee and M. Nathal, in High-Temperature Ordered Intermetallic Alloys III, (Mater. Res. Soc. Symp., **133**, 1988), 639.
[7] K.J. Hemker and T.K. Glasgow, Journal of Material Science **22**, 798 (1987).
[8] M.G. Hebsur and I.E. Locci, NASA Technical Paper No. 100264, 1988.
[9] H.H. Uhlig and R.W. Revie, Corrosion and Corrosion Control, (John Wiley and Sons, New York, 1985), p. 187.
[10] J.L. Carrasco, P. Adeva and M. Aballe, Oxidation of Metals, **33**, 1 (1990).
[11] V.L. Cabrera and M.B. Maple, Oxidation of Metals, **32**, 207 (1989).

THE EFFECT OF ION-IMPLANTED ELEMENTS ON THE θ TO α PHASE TRANSFORMATION OF Al_2O_3 SCALES GROWN ON β-NiAl

B. A. PINT*, A. JAIN**, AND L. W. HOBBS*
*H. H. Uhlig Corrosion Laboratory, Massachusetts Institute of Technology, Cambridge, MA 02139
**ITMMEC, Indian Institute of Technology, Delhi 110016, India

ABSTRACT

During oxidation at 1000°C in dry, flowing O_2, ß-NiAl first forms an oxide scale of primarily θ-Al_2O_3. After 1-4 hr of continued exposure this metastable oxide transforms into the stable α-Al_2O_3 structure. An implant of 2 x 10^{16}Y/cm^2 at 70kV was found to stabilize the first-forming θ-Al_2O_3 scale up to 100hr at 1000°C. β-NiAl was also implanted with Al and Cr to study the effect of the implantation damage and of another more noble element on the oxidation behavior. At 1200°C, only a short term effect of the Y-implant was observed.

INTRODUCTION

As reported previously[1-3], ion-implanted Y retards the phase transformation of metastable θ-Al_2O_3 to the stable α-Al_2O_3 in the scales grown on β-NiAl at 1000°C. Ion implantation is a commonly-used technique to study the effect of certain so-called "reactive" elements (REs) on the formation of protective Cr_2O_3 and Al_2O_3 scales[4]. Using ion implantation, a wide variety of elements can be added uniformly and in controlled, measurable quantities.

Initial studies on Y implantation in β-NiAl had demonstrated that a fluence of 2 x 10^{16}Y/cm^2 implanted at 70kV was effective in reducing the parabolic oxidation rate and improving scale adhesion at 1200°C[5-6]. Recent work also showed a reduction in the oxidation rate, but only a short term improvement in oxide adhesion at 1200°C[1,3]. In general, ion implantation is not as effective for alumina-formers as has been demonstrated for chromia-formers[4,7].

At higher temperatures (≥1200°C), the ineffectiveness is most likely due to the shallow implant being unable to have a long-term effect on the high oxidation rate of α-Al_2O_3[7]. However, at lower temperatures where the oxidation rate is much slower, Y implantation does not produce the RE effect because of its effect on the θ to α-Al_2O_3 phase transformation. The set of experiments presented here was designed to determine if the implantation process itself affected the phase transformation or if another element, Cr, would also affect the phase transformation[8-9].

EXPERIMENTAL PROCEDURE

Ni-50.2at%Al(31.7wt%Al) ingots were obtained from NASA Lewis Research Center, Cleveland, OH. Ion implantation was performed at M.I.T. (Y) and at the Naval Research Laboratory, Washington, D.C.(Y, Al, Cr) using identical parameters: 70keV and a fluence of 2 x $10^{16}cm^{-2}$ (estimated Y peak concentration of 4.2at% at 23nm[3]). For direct comparison, the Al and Cr implants were only on one side of the coupon.

Coupons (≈1.5cm x 1.5cm x 0.1cm) were polished through 0.3μm alumina powder prior to implantation and ultrasonically cleaned in acetone and alcohol prior to oxidation. Isothermal weight gain was measured with a Cahn model 1000 microbalance. Isothermal experiments were performed at 1000°C and 1200°C in a dry, flowing atmosphere of pure O_2.

Prior to oxidation, the specimens were characterized using Rutherford backscattering spectroscopy (RBS), X-ray photoelectron spectroscopy (XPS) and secondary ion mass spectroscopy (SIMS). After oxidation, specimens were characterized using glancing angle (0.5° incident angle) X-ray diffraction (GAXRD) and scanning electron microscopy with X-ray energy dispersive spectroscopy (SEM/XEDS).

RESULTS & DISCUSSION

The most useful technique for characterizing the oxidation product was GAXRD. Using a 0.5° incident angle, the majority of diffraction information comes from the sample surface where the oxide scales in this study varied in thickness from 0.1-3μm. Table I lists the percentage of θ-Al_2O_3 in the scales after oxidation for 1 and 50hr at 1000°C. The percentage was obtained by a simple peak height comparison of θ-Al_2O_3 (2θ = 32.8° with Cu Kα) and α-Al_2O_3 (57.5°). Without an implant, a wide range of behavior was observed after a 1hr exposure at 1000°C. However, after 50hr the undoped scales were always fully α-Al_2O_3. The implantation of Y stabilized the first-forming, metastable θ-Al_2O_3 much more effectively than the implanted Cr or Al. After a 100hr exposure at 1000°C, there was still 82% θ-Al_2O_3 present, and even after 2hr at 1200°C the scale was 19% θ-Al_2O_3.

Table I Phase composition of the scales grown on various alloys at 1000°C after 1 and 50hr exposures.

Base Alloy:	Ni-50.2at%Al				Ni-48Al	Ni-55Al
Implant:	None	Y	Cr	Al	None	None
After 1hr:	25%*	99%	70%	57%	30%	78%
After 50hr:	0%	95%	3%	0%	0%	0%

* average of 4 samples

Cr had a slight effect on the phase transformation, with a small amount of θ being present after 50hr. However, the effect of Cr was substantially less than that of Y. Surprisingly, the implantation of Al had a significant effect on the initial phase transformation. This may indicate a potential effect of the implantation process itself on the phase transformation, occurring after 1hr. However, additional experiments on different Al compositions within the β phase field revealed that increased Al contents initially formed a higher fraction of θ-Al_2O_3. Thus, enriching the surface with Al, rather than the implantation process, may be responsible for the initial change in the phase transformation. In general, the other implanted species had very little effect on the $\theta-\alpha$ phase transformation, compared to the efficacy of Y.

SIMS sputter-depth profile analysis of the samples prior to oxidation revealed only a slight increase in the surface oxide after implantation. Both XPS and SIMS revealed a ≈100Å oxide layer on the surface after implantation. Its presence could potentially affect the phase transformation, but as judged by the Al implantation it is not a significant effect. The chemical effect of Y appears to be more significant than any effect of the implantation process itself.

The effect of Y on the scale phase composition was also observed in the oxidation kinetics, the θ phase growing more quickly than α-Al_2O_3[10]. Figure 1 shows the continuous sample weight gains plotted against the square root of time. For unimplanted Ni-50.2Al, the weight gain appears to follow two separate parabolic rates. Based on the GAXRD data, these would be initially (1-4hr) for the growth of θ and subsequently for the growth of α. Y-implanted NiAl appears to fall between these two rates, and at longer times approaches the second rate asymptotically as the growing scale begins to transform to α.

SEM/XEDS analysis of the scale surface morphology qualitatively confirmed the GAXRD results. For instance, the scale on Cr-implanted NiAl after 50hr did show signs that a small fraction of θ-Al_2O_3 was retained as a result of the Cr addition. Figure 2 shows an area across a substrate grain boundary where one grain has a ridge-type structure indicative of α while the other has oxide blades typical of θ. Previous studies of single crystal β-NiAl have shown an epitaxial effect on the growth of metastable alumina scales[11]. Apparently, there is also a slight effect of substrate orientation on the phase transformation.

The stabilization of θ-Al_2O_3 has also been observed locally on a ZrO_2-dispersed FeCrAl alloy. At 1000°C, FeCrAl alloys typically give no indication of θ formation[3]. However, in anomalous areas where a large ZrO_2 particle is found at the alloy surface, a "nest" of θ blades is retained around the particle after oxidation for 100hr at 1000°C, Figure 3. This localized structure is presumably a result similar to the effect of Y, where Zr from the particle stabilizes the θ structure. Similar structures have been observed in other studies of RE-doped FeCrAl alloys[12]

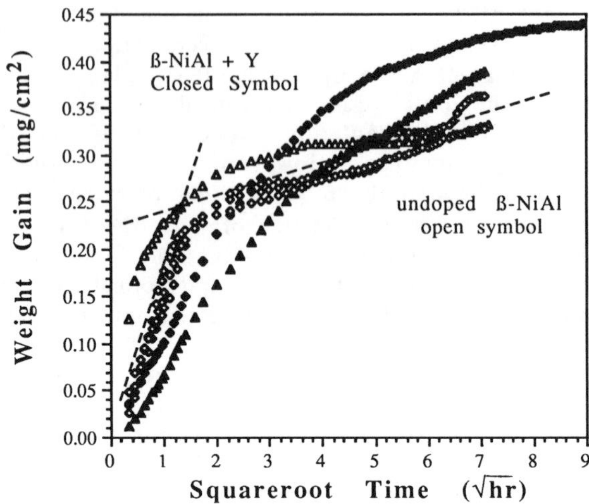

Figure 1. Parabolic plot of weight gain versus square root of time for unimplanted and Y-implanted NiAl at 1000°C. The two dashed lines denote the two parabolic rate periods for undoped NiAl.

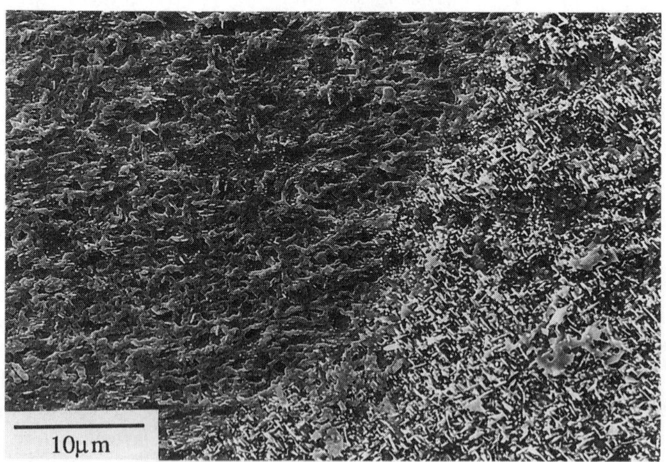

Figure 2. SEM secondary electron image of Cr-implanted NiAl after oxidation at 1000°C for 50hr. Adjacent substrate grains have oxide ridges typical of α-Al_2O_3 and oxide blades typical of θ.

Wynnyckyj and Morris[13] have proposed a diffusionless martensitic-type transformation to explain the rapid kinetics of the $\theta-\alpha$ transformation. Y and Zr are found to segregate to α-Al_2O_3 grain boundaries[14], presumably because these larger cations cannot easily fit within the α lattice structure[15]. There are indications from the catalysis literature that larger ions are more easily incorporated into the θ lattice, a more defective, cubic close-packed structure[16-18]. Located within the lattice, large ions such as Y, Zr, La and Ce appear to inhibit the transformation to the α structure. The retention of high-surface area metastable aluminas is desirable for catalyst carriers; however, because of the faster growth rate of θ, it is undesirable for oxidation resistance.

The effect of Y on the phase transformation has also hampered efforts to study the growth mechanism of scales on alumina-formers using an ^{18}O tracer. Earlier studies identified an effect of Y on the phase transformation[19]. However, it was necessary to separate the phase transformation effect from the growth mechanism effect, in order to demonstrate that an RE addition inhibits the Al transport in α-Al_2O_3 resulting in growth predominantly by O inward diffusion. Undoped α-Al_2O_3 grows by the combined transport of Al and O, while the more open θ structure grows by an outward growth mechanism[14].

CONCLUSIONS

1. Yttrium retards the θ to α-Al_2O_3 phase transformation when ion implanted into β-NiAl.

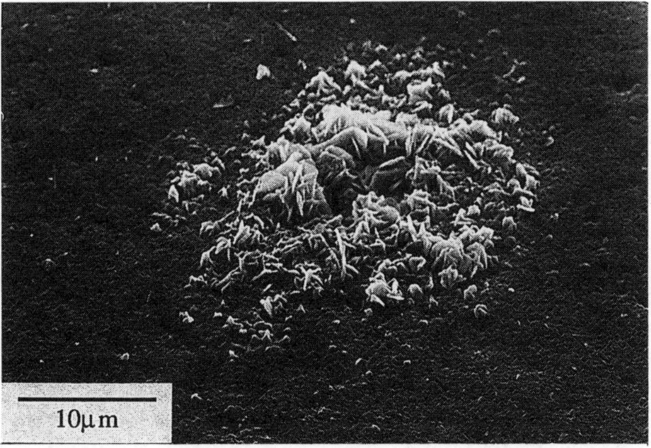

Figure 3. SEM secondary electron image of the scale on a ZrO_2-dispersed Fe-20wt%Cr-4Al alloy (Kanthal APM) after oxidation at 1000°C for 100hr. The central particle was identified by XEDS as Zr-rich.

2. The effect of Y is a chemical effect and is not a result of the implantation process.
3. Implanted Cr and Al have a noticeable, but lesser, effect on the $\theta-\alpha$ phase transformation.

ACKNOWLEDGMENTS

The authors are very grateful to Dr. J. Doychak at the NASA Lewis Research Center for providing the undoped NiAl and Zr-doped NiAl; and to Dr. C. M. Cotell at the Naval Research Laboratory (NRL) for arranging the ion implantation work at NRL. Among our colleagues at MIT, thanks goes to Mr. J. Adario for assistance with the GAXRD measurements, and to Mr. G. Arndt with the oxidation equipment. The Electric Power Research Institute provided financial support under contract RP242644.

REFERENCES

1. B. A. Pint, A. Jain and L. W. Hobbs, in *High-Temperature Ordered Intermetallic Alloys IV*, edited by L. A. Johnson et al., MRS Symp. Proc. **213**, 981 (1991).
2. B. A. Pint and L. W. Hobbs, Electrochem. Soc. Extended Abstracts **91-2**, 915 (1991).
3. B. A. Pint, Ph.D. Thesis, Massachusetts Institute of Technology, Cambridge, MA, 1992.
4. M. J. Bennett, in *Environmental Degredation of Ion and Laser Beam Treated Surfaces*, edited by G. S. Was and K. S. Grabowski (TMS, Materials Park, OH, 1989), p.261.
5. J. Jedlinski and S. Mrowec, Mat. Sci. and Eng. **87**, 281 (1987).
6. S. Mrowec, A. Gil and J. Jedlinski, Werk. Korr. **38**, 563 (1987).
7. B. A. Pint and L. W. Hobbs, submitted to the Spring 1993 meeting of the Electrochemical Society.
8. G. C. Bye and G. T. Simpken, J. Amer. Cer. Soc. **57**, 367 (1974).
9. W. C. Hagel, Corrosion **21**, 316 (1965).
10. G. C. Rybicki and J. L. Smialek, Oxid. Met. **31**, 275 (1989).
11. J. Doychak, J. L. Smialek and T. E. Mitchell, Met. Trans. **20A**, 499 (1989).
12. J. Jedlinski, G. Borchardt and S. Mrowec, Werk. Korr. **41**, 701 (1990).
13. J. R. Wynnyckyj and C. G. Morris, Met. Trans. **16B**, 345 (1985).
14. B. A. Pint, J. R. Martin and L. W. Hobbs, submitted to Oxid. Met..
15. J. D. Cawley and J. W. Halloran, J. Am. Cer. Soc. **69**, C195 (1986).
16. H. Schaper, E. B. M. Doesburg and L. L. Van Reijen, App. Catal. **7**, 211 (1983).
17. P. Burtin, J. P. Brunelle and M. Soustelle, App. Catal. **34**, 225 (1987).
18. M. Ozawa, M. Kimura and A. Isogai, J. Mat. Sci. Letter **9**, 709 (1990).
19. P. A. van Manen, E. W. A. Young, D. Schalkoord, C. J. van der Wekken and J. H. W. de Wit, Surf. Inter. Anal. **12**, 391 (1987).

THE SURFACE INTERACTION OF OXYGEN WITH A GAMMA TiAl

T. N. TAYLOR AND M. T. PAFFETT
Chemical and Laser Sciences Division, Los Alamos National Laboratory
Los Alamos, NM 87545

ABSTRACT

The composition and bonding of incipient and atmospheric oxides grown on a Ti - 47.0 Al - 2.2 Nb alloy were examined using Auger, x-ray, and low-energy ion scattering spectroscopies. The depth distribution of various components was determined with secondary-ion mass spectroscopy (SIMS). An oxide grown on the material in air at room temperature showed advanced Al and Ti oxide bonding states, which had very different stabilities during heating in vacuum up to 600°C. The Ti-oxygen bonding state was markedly reduced while the Al oxide was somewhat enhanced during this process. Further controlled studies at 600°C, involving oxygen adsorption at pressures below ~ 10^{-6} Torr, showed an initial Al cation surface enrichment until the detectable Al 2p peak was 90% oxidized. Further oxygen exposure gradually produced a fully oxidized Ti valence state with a pronounced enrichment of the Ti cation at the vacuum-solid interface. The results are discussed in terms of the kinetic and thermodynamic factors of the rate-limiting steps in the oxidation process.

INTRODUCTION

The thrust of the present research is to employ surface-science probes to measure the chemical properties of a γ-TiAl during its interaction with oxygen. This approach is important for understanding the fundamental processes that are involved in bulk hydrogen [1] and oxygen [2] environmental embrittlement of these light-weight structural materials because it is the surface oxide coating that is first encountered by the gas-phase species.

A number of atmospheric oxidation experiments have shown how the reactivity and oxide composition for a binary titanium aluminide depend on the Ti-Al bulk stoichiometry [2,3]. It has been shown that, as the Al content is increased, a point will eventually be reached where the outward diffusion of Al limits Ti oxide formation and produces a continuous alumina scale, which acts as both an adsorption and diffusion barrier to oxygen and hydrogen. Dopants, such as Nb, are typically added to titanium-aluminum alloys because they enhance this effect, while also inhibiting the net oxide growth [2,4]. A Nb oxide complex is believed to bond with the immiscible oxides of Ti and Al to block oxygen fast diffusion paths. Ti-oxide surface enrichment is one of the major concerns relating to the use of TiAl-based materials at elevated temperatures in a hydrogen environment. The hydrogen reduction of this oxide layer to produce chemically active Ti sites is one potential pathway for bulk hydride embrittlement. By examining the controlled adsorption of oxygen on a well characterized titanium aluminide surface, information is obtained on how the Ti and Al cations compete for the oxygen and how the surface region is reconfigured during the early stages of oxide formation. Understanding these chemical processes as a function of bulk and surface composition is quite important to tailoring the growth of a preferred surface coating.

The current study includes measurements to determine the chemical identity and concentration of the surface species during oxidation of a Ti - 47.0 Al - 2.2 Nb alloy. By doing so, the work extends our understanding of environmental embrittlement by examining the atomic-level interfacial chemistry that is the initial step in the observed bulk mechanical failure.

EXPERIMENTAL APPROACH AND TECHNIQUES

A number of surface-sensitive probes were used to measure the atomic-level composition and bonding on the γ-TiAl as a function of controlled surface treatment in an ultra-high vacuum

(UHV) chamber. The apparatus was equipped with a hemispherical energy analyzer that acquired the energy distribution of either electrons or positive ions emitted or scattered by the sample. A vacuum transfer mechanism permitted rapid motion between atmospheric and vacuum conditions.

The standard techniques of Auger electron spectroscopy (AES) and x-ray photoelectron spectroscopy (XPS) were often utilized. The XPS data were taken using a Mg-K_α source ($h\nu$ = 1253.6 eV) and the photoemitted electrons were detected in a pulse-counting mode. Spectrometer calibration was performed according to ASTM standards. Low-energy ion scattering spectroscopy (ISS) measurements were made using incident He ions and detecting their scattered kinetic energy distribution with the hemispherical analyzer. These spectra are sensitive to the atomic composition in the topmost surface layer. Information on the depth distribution of the atomic species in the surface region was obtained with secondary-ion mass spectroscopy (SIMS). The technique, which uses a mass analyzer in place of the energy analyzer to detect the atomic and molecular ion species sputtered from the sample, was employed in a dynamic mode (rapid material removal of material at 5 - 10 Å per minute) using Xe ions. These measurements were performed in a second apparatus.

CLEAN SURFACE PROPERTIES

A sample disk (1.5 mm thick x 9.5 mm dia.) was spark-cut from a parent material having a composition of Ti = 50.48, Al = 46.95, and Nb = 2.24 at. %. The faces were subsequently polished and then mildly etched in a Kroll's solution (HF, HNO_3, H_2O) at room temperature. Two Ta wires were spotwelded to opposite edges of the disk and this spider-web assembly was mounted on a holder that permitted ohmic heating. The sample temperature was monitored by a chromel-alumel thermocouple that had also been spotwelded to the edge of the disk.

Following vacuum insertion the sample was bombarded with argon ions and annealed to give a surface with only C, O, and Ar as detectable impurity species. We estimate the concentration of the C and O to be < 10% of a monolayer. These tenacious O and C baseline impurities are possibly the result of segregants in the exposed grain boundaries of the polycrystalline material. The XPS data from the clean surface gave Ti $2p_{3/2}$ (453.85 eV BE) and Al 2p (72.05 eV BE) peaks that were located at binding energies comparable to those from elemental Ti (453.80 eV BE) and Al (72.65 eV BE) [5]. The $3d_{3/2,5/2}$ energy levels from Nb were readily identifiable.

Significant changes were observed in the Ti:Al surface ratio during sample cleaning treatments. In particular, the ISS showed that the surface concentration of the Ti was appreciably larger after sputtering at room temperature than it was after subsequent heating to 500°C in UHV. The decrease in surface Al content is consistent with preferential sputtering that results from the relatively easier removal of the less massive Al compared to Ti. When the ISS intensities were adjusted by the ratio of differential scattering cross-sections for Ti compared to Al, the annealed surface was found to be within 10% of the bulk stoichiometry.

THERMAL DECOMPOSITION OF AN ATMOSPHERIC OXIDE

As shown in Fig. 1, the Ti 2p and Al 2s peaks undergo a pronounced change in shape following room temperature atmospheric oxidation (10 minutes, 585 Torr, 40% humidity). We have displayed the Al 2s transition in Fig. 1 as, for that series of measurements, there was some minor interference with the Al 2p lineshape from a Ta component in the analysis area. The Ti region shows a dominant peak at 459.2 eV BE, which is the $2p_{3/2}$ value typically assigned to the 4+ valence state for elemental oxidation to TiO_2 [6]. The Al 2s transition after the oxidation exhibits a doublet peak shape resulting from the oxide film and the underlying unoxidized metallic species.

The changes produced in the Ti 2p and Al 2s spectra after progressively higher temperature annealing of the atmospheric oxide in 10^{-8} Torr vacuum are shown in Fig. 1. For annealing up to 600°C the Al 2s peak undergoes a relative change in doublet intensities that gives added weight to the oxide component at the higher binding energy. The Ti 2p spectrum, on the other hand, returns to a lineshape that is similar to that for the clean surface. However, the lineshape after annealing to 600°C is somewhat broader and shifted so that the $2p_{3/2}$ peak is found at 454.6 eV BE, at the lower range of values normally associated with Ti^{2+} bonding [6]. Thus,

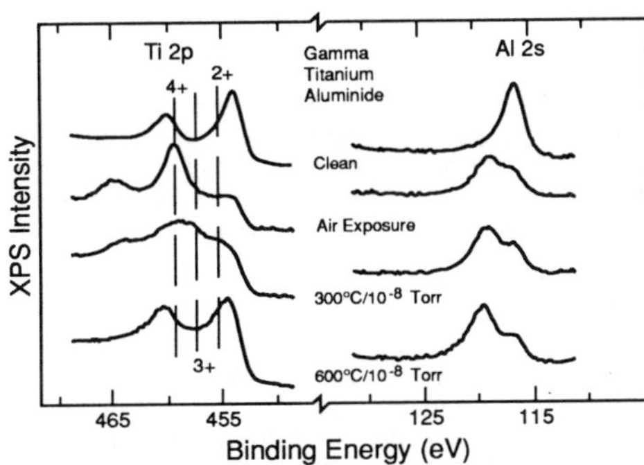

Fig. 1. Compilation of the Ti 2p and Al 2s spectra for a clean surface and after atmospheric exposure followed by anneal in UHV. The vertical lines show Ti $2p_{3/2}$ peak positions for the indicated valence

the Ti-oxygen bonding state is reduced by the thermal treatment, in contrast to the behavior seen for the Al bonding to oxygen.

Titanium has four primary valence states (metallic, 2+, 3+, 4+), which to first order give 2p lineshapes like that of the clean surface. Higher valence states are shifted to progressively larger binding energies and they exhibit different $2p_{1/2}$ - $2p_{3/2}$ separations. In order to obtain a first-order estimate of the mixture of oxidation states that produce a given Ti 2p lineshape, we have used a simulation routine that includes a pair of Gaussian peaks at the doublet energies to approximate the lineshape for each valence state. Four of these sets have then been combined to fit the actual data. The $2p_{3/2}$ binding energies used for the 2+ to 4+ oxidation states were respectively 455.2, 457.4, and 459.2 eV, assignments that agree with values listed in the literature for the oxidation of pure Ti [6]. Table I lists the fractional valence state content for the three oxidized conditions of Fig. 1 as obtained from the relative amount of spectral content in the simulations.

TABLE I

Fractional Ti Valence State Content for a 25°C Atmospheric Oxide Grown on γ-TiAl as a Function of Heating in UHV.

	25°C	300°C	600°C
Ti^{2+}	0.18	0.26	0.39
Ti^{3+}	0.19	0.33	0.19
Ti^{4+}	0.47	0.21	0.10

The Ti-oxide reduction and Al-oxide stability are in agreement with the behavior expected from the heats of formation for the various oxides. For oxidation of elemental materials the heats of formation for TiO, Ti_2O_3, TiO_2, and Al_2O_3 are 495.0, 717.1, 889.5, and 791.2 KJ/mole per metal atom, respectively. With an increase in temperature the kinetically stabilized atmospheric oxide layer is undermined as the Ti oxides begin to react with the underlying metal. That is, when the Ti oxidation states are reduced by oxygen bonding to previously metallic Ti or Al, the net free energy of the involved cation bonding is lowered. For the same treatment the Al oxide is stable because its single valence state has a higher heat of formation than the Ti suboxides. The reduction process most likely involves the diffusion of metallic Ti and Al from the metal into the oxide film, as oxygen migration would require breaking of the strong oxide bond. Because no oxygen is lost after heating to 300°C and there is only a 10% decrease up to 600°C, which could be due to surface reconfiguration, only minor reduction effects are possible from desorption of oxygen into the vacuum. The final anneal also produces a 20% decrease in the Ti:Al ratio. Compared to Ti, the Al has a much stronger tendency to scavange oxygen in a reaction sequence with the Ti suboxides. This chemical driving force is most likely responsible for the surface enrichment of the Al component. The formation of Ti and Al oxides effectively depletes the surface region of Nb, as might be expected from the relatively low heat of formation for NbO. The vacuum anneal reduces the Nb oxide, and this is seen by return of the Nb 3d spectra to a more metallic lineshape (not shown).

Lastly, we believe that the Al spectra following oxidation are principally due to formation of an alumina (like) bond. There may be water-related bonding, involving hydroxyls, following the initial atmospheric exposure. In the absence of a distinct experimental signature for this type of bonding, we have only invoked the oxide. Another possible oxide entity may involve some mixed bonding between Ti and Al, which could perhaps exhibit a binding energy similar to pure Al oxide.

CONTROLLED OXIDATION

A set of controlled oxygen exposures up to 1000 Langmuirs (L) was taken on the γ-TiAl at 25, 300, and 600°C. One Langmuir is approximately equal to the incidence of a monolayer of oxygen on the surface. The exposures were performed at pressures no higher than ~ 10^{-6} Torr. The 25°C data gave some insight into the initial stages of growth for the previously analyzed atmospheric oxide. For these 25°C exposure conditions the oxygen surface signal increased rapidly and attained a nearly saturated value after 25L. The AES signal for the highly surface-sensitive metallic Al (68 eV) peak was rapidly attenuated with the concurrent appearance of the Al-oxide peak at 57 eV. Figure 2 shows the result of a 1000L exposure at 25°C as measured with ISS. The starting condition for the clean surface was a nearly 1:1 atomic ratio of Ti:Al. Note that the adsorption of oxygen produces a significant O peak and greatly lowers the cation ISS intensity, presumably due to shadowing from its placement at the vacuum-solid interface. When the oxidized sample is heated to 600°C in UHV the ISS shows an enrichment of the Al cation species, in analogy to the behavior observed for the heated atmospheric oxide during the Ti reduction process.

The initially acquired adsorption data for 1000L exposure demonstrated a more complete oxidation of the Ti and Al as the temperature was increased from 300 to 600°C. The Ti^{4+} and Al^{3+} states dominated the XPS spectra at 600°C. Most remarkably, after the 1000L exposure at 600°C, the surface became heavily enriched in the Ti cation (see bottom ISS spectrum in Fig. 2). Given the Al-enrichment behavior seen in the ISS after the 25°C exposure plus heating, an exposure region must exist for oxygen adsorption at 600°C that produces an Al cation surface enhancement prior to the eventual Ti enrichment seen at 1000L.

Such an effect was indeed observed at 600°C when the surface was subsequently exposed in increments terminating at 1000L. When the Ti:Al ratio obtained from the integrated Ti 2p and Al 2p peaks in XPS is plotted as a function of exposure at 600°C, the surface is found to be enriched in the Al cation until approximately 350L (Region I), as seen in Fig. 3. At this exposure the Al is more than 90% oxidized within the detected sampling depth (< 100Å). The

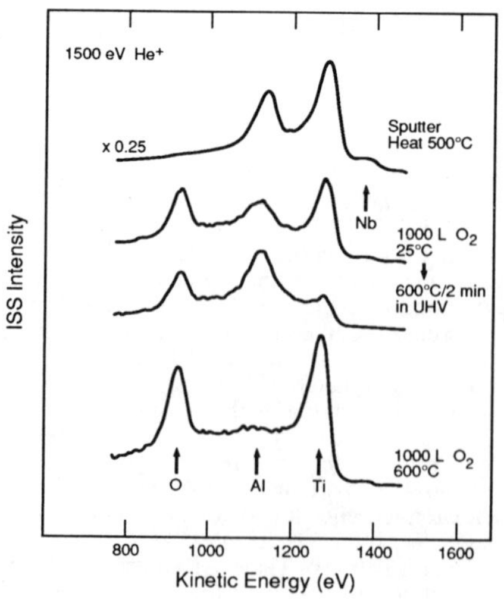

Fig. 2. ISS spectra taken from a clean surface (Ti/Al ~ 1:1) and after various oxygen/anneal treatments.

Ti enrichment process (Region II) primarily involves the complete oxidation of the Ti to the 4+ state as it diffuses through the developing ceramic layer. An analogous compositional behavior as a function of exposure is found in the ISS analysis of the topmost surface layer.

Our model for the two-region adsorption process at 600°C is consistent with the thermodynamic and kinetic arguments offered for the thermal decomposition of the atmospheric oxide. Because of its relatively large heat of formation, the Al is initially enriched at the surface. Its single valence state serves as the lowest free energy trap for this portion of the oxidation (< 350L) because mostly suboxides of Ti are formed. However, beyond a certain

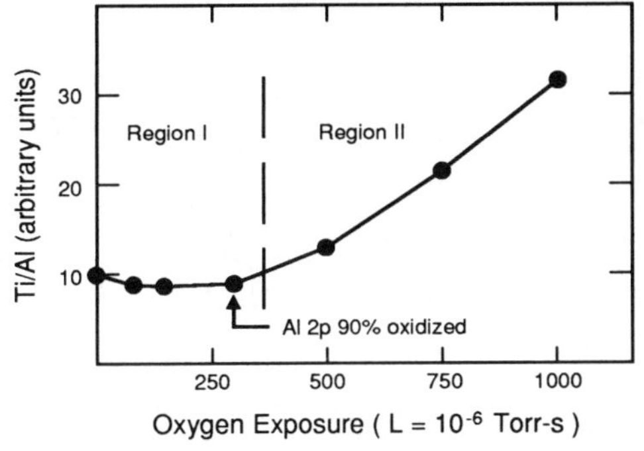

Fig. 3. The Ti 2p - Al 2p integrated XPS intensity ratio as a function of oxygen exposure at 600°C.

point in the exposure the availablility of the metallic Al is greatly reduced due to the diffusion distance required from the unoxidized Al below the surface. In any oxidation process the Ti would eventually become fully oxidized. However, this step is facilitated on the titanium aluminide by the thermally induced migration of the Ti through the ceramic layer.

More information was obtained on the 600°C oxides using SIMS to determine the depth distribution of the Ti, Al, and Nb cation species for 150 and 1000L oxygen exposures. Figure 4 gives a clear indication of the reversal of the Ti and Al surface concentrations as the oxidation proceeds from the initial Al-rich (150L) to the final Ti-rich (1000L) conditions. Comparison with the depth profiling of standards gives a measure of the oxide thicknesses. From this we estimate that the 1000L oxide is breached after removal of 100 - 150 Å of material at ~ 2500 seconds when the Al and Ti signals no longer appreciably change. Correspondingly, the Al-rich oxide is some five times thinner. The behavior of the Nb is somewhat different for the 150 and 1000L conditions. At 150L it is depleted at the surface and gradually increases with depth into the material. After the 1000L exposure the Nb is also depleted at the surface. However, the SIMS depth-profile shows an increase in signal intensity above that for the bulk material before the oxide is breached.

The temporal dependences of the Ti and Al signals for the SIMS data are much smoother than would be found for planar growth of a uniform oxide layer involving one dominant cation entity. The enrichment signatures are quite likely due to a variable mixture of the two cations as a function of depth into the surface. The data were taken in flowing oxygen at a pressure of 2.5×10^{-7} Torr. This mitigated a drastic change in the neutralization rate as the oxide was

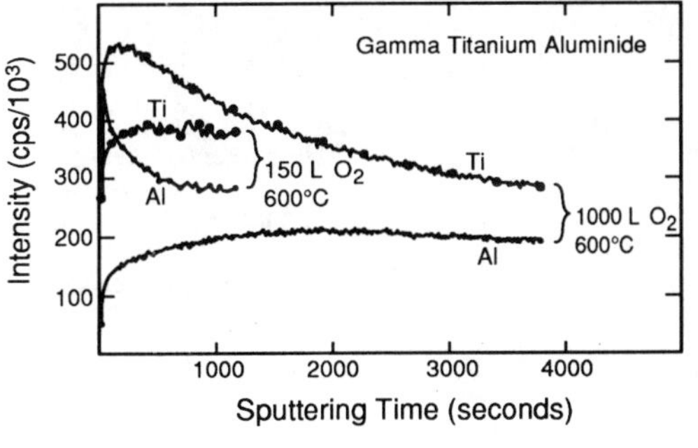

Fig. 4. Positive ion SIMS spectra showing the depth distribution of Ti (M/Z = 48), Al M/Z = 27) after 150 and 1000L of O_2 exposure at 600°C.

removed. Without the flowing oxygen all the signals were found to decrease in the region where the oxide was breached. The SIMS data were taken in a separate apparatus from where the oxides were grown. This required transfer in air for ~ 15 minutes. Prior to the transfer the as-prepared surface was exposed to air at room temperature for 10 minutes and then returned to the growth chamber for XPS analysis. For both the 150 and 1000L conditions the surface became more heavily oxidized under atmospheric conditions. However, in both cases the integrated XPS intensity ratio of Ti:Al changed < 10 %. We therefore believe that the depth distributions of the cations were only marginally different from the as-grown condition.

DISCUSSION

The present surface measurements on the two-phase ($\gamma + \alpha_2$) alloy provide a rather complete picture of the chemical states formed during its interaction with oxygen. The data are explicable from analysis of the available valence states and their heats of formation in the context of kinetic factors that vary as a function of temperature. The valencies of the Ti oxide are found in varying amounts as a function of exposure conditions, with the maximally valent Ti^{4+} oxide gradually increasing as the surface reaches a saturated oxide thickness. The dominant Al thermodynamic factor in the bonding of oxygen to the alloy mixture is very apparent in the thermal destabilization of an oxide grown at low temperature.

High-temperature oxidation is complex due to the strong activation of diffusion controlled processes. The enrichment of Al at elevated temperatures in the initial stages of adsorption occurs because the Al migration to the surface competes strongly with the growth of the Ti suboxides and allows the thermodynamic factor or heat of formation to become dominant. This was clearly in evidence for the measurements at 600°C for exposures below 350L. At this point due to the presence of thermally activated cation diffusion there is a striking change in the oxidation process, which leads to a dramatic enrichment of the Ti cation species. The work to date has underscored some important ideas concerning the role played by the Nb additive in altering the oxidation process. Measurements with SIMS have shown that the Nb cation is enriched in the oxide coating, even though it is largely depleted at the gas-solid interface. This is consistent with the notion that the oxidation is altered by a Nb oxide complex that acts to glue together the insoluble Ti and Al oxides and minimize diffusion pathways to the surface during the oxidation.

ACKNOWLEDGMENTS

Thanks are given to M. R. Shanabarger (UCSB) for supplying the titanium aluminide sample used in this study. Appreciation is also extended to R. D. Reiswig (Los Alamos) for sample polishing and preparation. This research was supported by NASA-Ames under contract # W-7405-ENG-36.

REFERENCES

1. H. G. Nelson, Proc. 2nd Int. SAMPE Metals and Metals Processing Conf., Dayton, OH, August 1988, p. 310.
2. G. Welsch and A. I. Kahveci, Oxidation of High-Temperature Intermetallics, T. Grobstein and J. Doychak (eds.); The Minerals, Metals, and Materials Society, Warrendale, PA, 1989, p. 207.
3. G. H. Meier, D. Appalonia, R. A. Perkins, and K. T. Chiang, Oxidation of High-Temperature Intermetallics, T. Grobstein and J. Doychak (eds.); The Minerals, Metals, and Materials Society, Warrendale, PA, 1989, p.157.
4. M. Khobaib and F. W. Vahldiek, Proc. 2nd Int. SAMPE Metals and Metals Processing Conf., Dayton, OH, August 1988, p. 262.
5. C. D. Wagner, W. M. Riggs, L. E. Davis, J. F. Molder, and G. E. Muilenberg, Handbook of X-Ray Photoelectron Spectroscopy, Physical Electronics Division, Perkin-Elmer Corp., Eden Prairie, MN, 1979.
6. A. F. Carley. P. R. Chalker, J. C. Riviere, and M. W. Roberts, J. Chem. Soc., Faraday Trans. 1, **83**, 351 (1987).

EFFECT OF TERNARY ADDITIONS ON THE MICROSTRUCTURAL STABILITY AND OXIDATION CHARACTERISTICS OF Ti-48 Al

S. A. Kekare, D. K. Shelton & P. B. Aswath,
Department of Mechanical & Aerospace Engineering and Materials Science & Engineering Program, University of Texas at Arlington

ABSTRACT

Oxidation behavior of binary Ti-48Al and ternary alloys with 1.5 at.% Cr, 1.4 at.%Mn, 0.2 at.% W, 2.2 at% V, 20 vol.% TiNb particles were examined in an air atmosphere at 704°C, 815°C and 982°C. Results indicate that the addition of Cr is detrimental at all temperatures. The Cr and V containing alloys exhibited a linear oxidation behavior at 815°C while the Mn containing alloy showed a linear behavior at 982°C. At 982 °C the alloys with W and V exhibit the best oxidation behavior. The mechanism of oxidation has been explained with the help of a simple vacancy model for oxygen diffusion.

INTRODUCTION

Titanium aluminide intermetallics have been a topic of special interest for the past few years in the stream of aerospace materials. These alloys are prominent candidate materials for applications in the hypersonic structures and advanced turbo-jet engines. The high specific strength even at high operating temperatures has made these alloys especially lucrative for these applications. However components like these are expected to face extreme oxidizing environments in combination with high temperatures. This necessitates a careful investigation of the oxidation response of these alloys.

Oxidation of any metal proceeds as a combination of two processes, oxygen dissolution and oxide scale formation. The total oxidation rate is thus controlled by the mass transport characteristics of the oxide film. In most common metallic materials, the oxidation proceeds in a parabolic manner, although in a few isolated cases one comes across a linear oxidation behavior. Wagner's [1] analysis relates the rate of oxidation to the diffusivities of the metal & oxygen ions in the oxide film and the partial pressures of oxygen at the oxide-gas & oxide-metal interfaces. In the case of multilayered oxides, this analysis has been modified to accommodate the partial pressures at oxide-oxide interfaces. Since the oxide formed are ionic crystals, their defect structure monitors the diffusion behavior of the metallic and oxygen ions. Through a detailed analysis of marker experiments on titania crystals, Kofstad [2] and coworkers determined that oxygen ion vacancies are highly mobile in comparison with titanium ions. This result indicates that doubly ionized oxygen ion vacancies are the dominant defects in this oxide. Thus diffusivity of oxygen ion vacancy governs the rate of oxidation. Sankaran et.al. [3] provided an analytical model explaining the importance of the concentration of the doubly ionized oxygen ion vacancies and their dependence on the valence of the associated cation in the oxide film.

MECHANISM OF OXIDATION IN γ TiAl ALLOYS

Static oxidation of binary two phase γ-TiAl alloys can occur leading to the formation of oxide of the kind, TiO, Ti_2O_3, TiO_2 and Al_2O_3.

From studies of oxidation of pure titanium it is known that the main product is TiO_2 and the other oxide states are metastable and decompose into TiO_2. In binary alloys it has been shown by various authors [4-6] that the main products of the oxidation process are TiO_2 and Al_2O_3.The key to improving the oxidation properties of these alloys is to form a protective Al_2O_3 oxide scale on the surface. However, the selective oxidation of aluminum would leave behind a titanium enriched layer which promotes the formation TiO_2 . Increasing the Al content up to the TiAl/$TiAl_3$ boundary is not feasible due to the poor mechanical properties. The alternative to this problem is the addition of a ternary alloying element which can potentially decrease the activity of Ti and increase the activity of Al and at the same time tie up the oxygen vacancies in the oxide layer.

It is generally well accepted that TiO_2 is a n-type semiconductor, [2]. Kofstad [2], concluded based on studies of nonstoichiometric TiO_2 that both interstitial cations and oxygen vacancies are present. He also determined that the oxygen ions exist in the doubly ionized form and the Ti ions are present in the trivalent and quatravalent forms, Ti^{3+} and Ti^{4+} forms. It has also been shown [7,8], that the primary mobile ion is one of oxygen which has a higher diffusivity compared to the titanium ions. The formation of doubly ionized oxygen vacancies in the TiO_2 lattice has been described by Sankaran et. al., [3] and Kofstad, [8] as

$$O_O \Leftrightarrow V_{\ddot{O}} + \frac{1}{2}O_2 + 2e' \qquad [1]$$
$$K = P_{O_2}^{1/2}.[V_{\ddot{O}}].n^2$$

where O_O is an oxygen ion occupying a normal site in the oxide, $V_{\ddot{O}}$ is the doubly ionized oxygen ion vacancy from which both electrons are freed. e' is an electron, n the concentration of electrons and K the equilibrium constant. The electrical neutrality requirements of the crystal require that positive and negative charges be balanced, i.e. $2V_{\ddot{O}}$ = n. Hence the concentration of vacancies can be written as [3,8],

$$[V_{\ddot{O}}] = \frac{K^{1/3}}{4^{1/3}}P_{O_2}^{-1/6} \qquad [2]$$

as the value of K is a constant at a given temperature the oxygen partial pressure depends on the oxygen ion vacancy concentration in a pure TiO_2 crystal.

The addition of ternary elements can control the oxygen pressure dependence of the vacancies by either annihilation of vacancies or generation of vacancies. As an example the addition of tungsten can potentially form WO_3 with a higher valent cation which can dissolve in TiO_2 and the process would result in the annihilation of vacancies by the reaction

$$WO_3 + V_{\ddot{O}} \Leftrightarrow W_{\ddot{T}i} + 3O_O \qquad [3]$$

$W_{\ddot{T}i}$ is the W^{6+} ion occupying a normal Ti^{4+} site, which has +2 charge relative to the lattice. The electroneutrality condition is satisfied by

$$2[V_{\ddot{O}}] + [W_{\ddot{T}i}] = n \qquad [4]$$

when $[W_{\ddot{T}i}] >> [V_{\ddot{O}}]$, $[W_{\ddot{T}i}] = n$, the concentration of vacancies is given by

$$[V_{\ddot{O}}] = \frac{K}{[W_{\ddot{T}i}]^2}P_{O_2}^{-1/2} \qquad [5]$$

However, if the concentration of vacancies is much larger than the level of dopant added or if the dopant does not dissolve in the oxide ($[V_{\ddot{O}}] >> [W_{\ddot{T}i}]$)then $2[V_{\ddot{O}}] = n$ then the vacancy concentration is given by eqn. [2]. The beneficial aspects of adding an element of higher valence is only fruitful if the element dissolves in the oxide scale of TiO_2 and ties up the vacancies, thus inhibiting the diffusion of oxygen ions. This type of an analysis can be carried out for the various ternary elements that have been added and it can be seen that from a theoretical point of view addition of V which forms V_2O_5 is beneficial and so is Nb which forms Nb_2O_5. The addition of Mn which forms MnO_2 should not lead to any differences in the oxidation behavior.

In a similar fashion it can be shown that the presence of a lower valent cation would lead to the creation of vacancies. For example, if Cr_2O_3 is dissolved into TiO_2 it will result in the generation of vacancies by a reaction of the kind,

$$Cr_2O_3 \Leftrightarrow 2Cr'_{Ti} + 3O_O + V_{\ddot{O}} \qquad [6]$$

The presence of these addtional vacancies in the TiO_2 oxide scale can increase the rate of diffusion of oxygen leading to an increase in oxidation rate. In a similar fashion it can be argued that if Al_2O_3 is dissolved in TiO_2 it will lead to an excess of vacancies and an increased oxidation rate.

EXPERIMENTAL PROCEDURE

Extruded bars with the various compositions were provided by McDonnell Douglas Research Labs, samples were sectioned perpendicular to the extrusion direction. The final dimensions of the specimen were noted down to an accuracy of 0.1 mm. Average specimen dimensions were of the order of 5 mm square with 1 mm thickness.

Oxidation Tests: The oxidation was carried out in a Perkin-Elmer™ integrated thermal analysis system. Continuous weight gain data was acquired and concomitant analysis was performed with the help of a 7000-series computer in conjunction with the thermal analysis system. A flow of compressed air was introduced into the furnace at a controlled flow rate when the test temperature was reached and maintained constant throughout the duration of the test. The alloy was tested at three different temperatures, 704°, 815° and 982° C.

Analysis : Cross section of the mounted oxidized specimens were taken at approximately half the width to reveal the different oxide layers and the oxide metal interface. The specimen were then examined using scanning electron microscope and electron probe micro analyzer to get elemental distribution in the oxide scale and base metal.

Kinetics: Data that showed a parabolic increase in weight gain with time was reduced to a linear data of the square of the weight gain versus time. Slope of the linear plot provided the value of the rate constant K_p for oxidation at that temperature. In case of a linear oxidation curve, the rate constant value was directly obtained from the oxidation curve. When the alloy showed a parabolic oxidation at all three temperatures, the activation energy for oxidation was obtained from the slope of an Arrhenius type plot of $\ln(K_p)$ versus inverse of absolute temperature. The activation energy for the alloys showing linear behavior could not be calculated due to the difference in the units of the rate constants for the linear and parabolic oxidation.

RESULTS AND DISCUSSION

Table I provides a list of the composition of the alloys tested. The results of this project will be discussed from the perspective of the role of ternary addition on the oxidation behavior at the three temperatures, 704°C, 815°C and 982°C. Table II lists the mechanism of oxidation in each of the alloys at the various temperatures.

Alloy composition
Ti-48 Al {600 ppm Oxygen}
Ti-48 Al {400 ppm Oxygen}
Ti-48 Al + 1.5 Cr (at. %)
Ti-48 Al + 2.2 V (at. %)
Ti-48 Al + 0.2 W (at. %)
Ti-48 Al + 1.4 Mn (at. %)
Ti-48 Al + 20 vol. % TiNb

Table I. Alloy composition

Alloy Stoichiometry	704°C	815°C	982°C
Ti-48 Al (600 ppm Oxygen)	Parabolic	Parabolic	Parabolic
Ti-48 Al (400 ppm Oxygen)	Parabolic	Parabolic	Parabolic
Ti-48 Al + 1.5 at.% Cr	Parabolic	Linear	Parabolic
Ti-48 Al + 2.2 at.% V	Parabolic	Linear	Parabolic
Ti-48 Al + 0.2 at. % W	Parabolic	Parabolic	Parabolic
Ti-48 Al + 1.4 at. % Mn	Parabolic	Parabolic	Linear
Ti-48 Al + 20 vol.% TiNb	Parabolic	Parabolic	Parabolic

Table II. Mechanisms of Oxidation

Oxidation at 704°C: All the alloys exhibit a parabolic oxidation behavior as evidenced in Fig. 1. It is immediately apparent that the ternary alloy with 1.4 at% Mn exhibits the best oxidation behavior and the alloy with 1.5 at% Cr and the alloy with 20 vol.% TiNb is the poorest (the alloys with Cr and Mn have a fully transformed $\alpha_2+\gamma$ microstructure). This is also apparent from the quality of the oxide scales formed on the surface. The alloy with Mn exhibited very small patches of oxidation which was very limited in extent while the Cr containing alloy showed a continuous oxide scale. It is important to note that at 704°C the worst case scenario exhibited a total weight gain of only 0.53 mg/cm^2 on exposure for 100 hours. There was very little influence of the presence of 400 or 600 ppm of predissolved oxygen in the base Ti-48Al (at%) alloy on the oxidation kinetics at 704°C (both these alloys have a mixture of single phase γ and two phase $\alpha_2+\gamma$). The alloy with V exhibited very good oxidation resistance. As the oxide layers are very thin in all the samples, cross sectioning was not attempted.

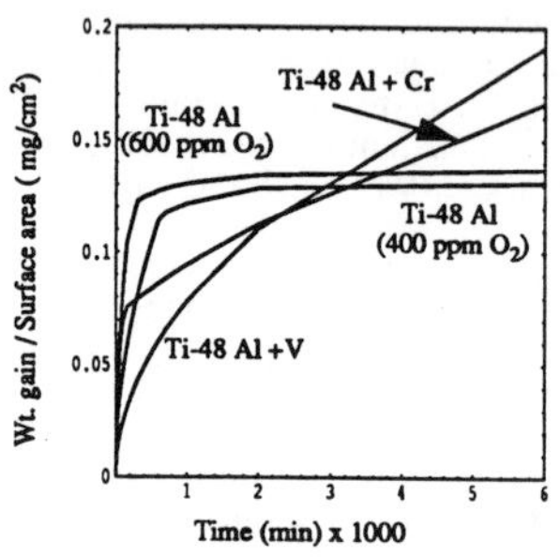

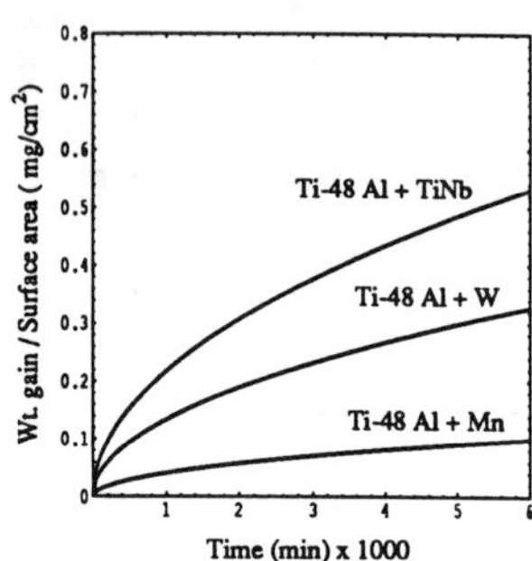

Fig.1. Weight gain per unit area at 704°C as a function of time for all the alloys.

Oxidation at 815°C: All the alloys with the exception of the alloys with Cr and V exhibit a parabolic oxidation behavior while the alloys with Cr and V exhibit a linear behavior, Fig. 2. The best oxidation response was exhibited by the base Ti-48Al alloys and the alloys with W and Mn which showed the lowest weight gain of under 2 mg/cm^2 while both the V and Cr containing alloys had a weight gain of the order of 7-8 mg/cm^2 after 100 hours of exposure. The alloy with 20 vol.% TiNb particles exhibited a total weight gain of about 3 mg/cm^2.

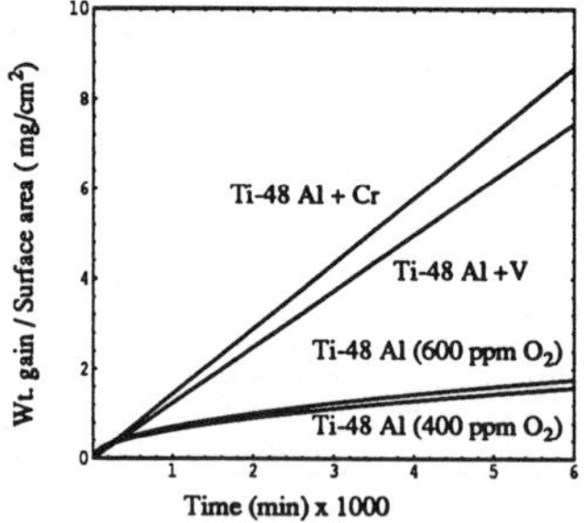

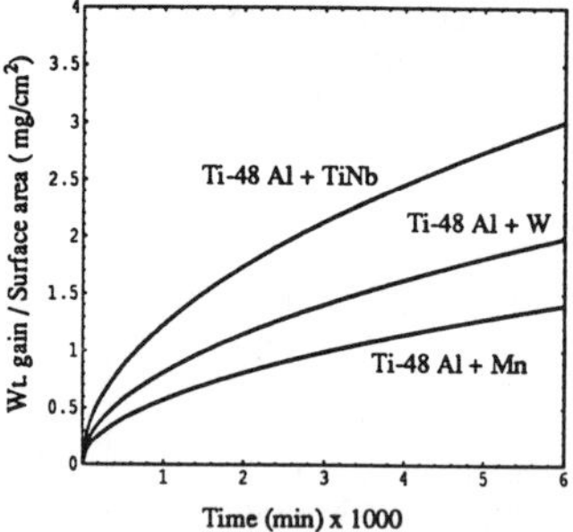

Fig.2. Weight gain per unit area at 815°C as a function of time for all the alloys.

Oxidation at 982°C: All the alloys exhibited a parabolic type oxidation characteristic except the alloy with Mn which exhibited a linear oxidation rate. This is shown in Fig. 3. Due to paucity of space all the characteristic micrographs cannot be exhibited. Fig. 4 is a sequence of photographs showing the cross section of the oxidation product in the Ti-48Al-0.2W alloy. Also included is a Al map and Ti map of the microstructure delineating the regions of partitioning of the two elements in the oxide as well as the microstructure. In all the alloys tested the outermost oxide was pure TiO_2 while the layer immediately below was a single phase Al_2O_3. Below the two outermost layers was a thick intermediate layer made up of a mixture of aluminum and titanium oxide which was relatively porous compared to the outer

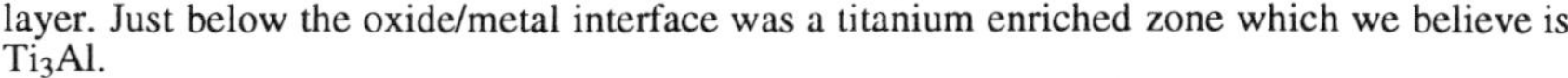

layer. Just below the oxide/metal interface was a titanium enriched zone which we believe is Ti_3Al.

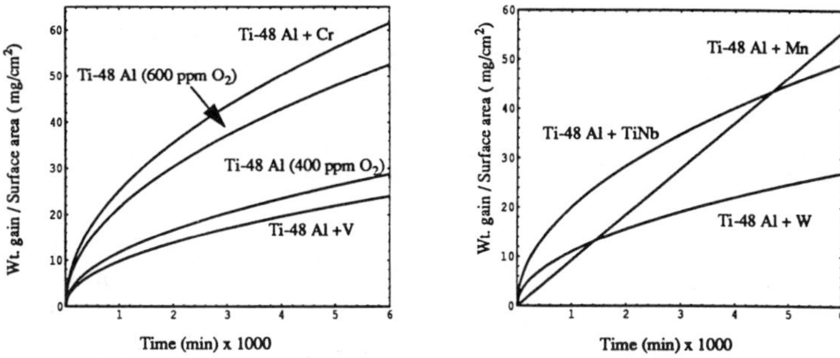

Fig.3. Weight gain per unit area at 982°C as a function of time for all the alloys.

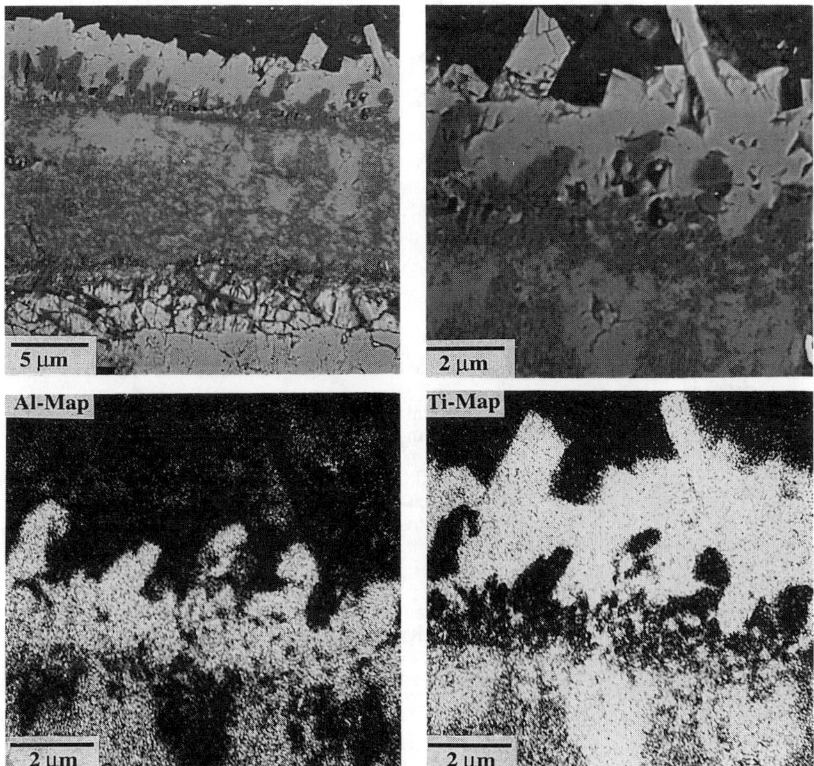

Fig.4. Oxidation cross section of the Ti-48Al-0.2 W alloy oxidized at 982°C for 100 hours. From the left upper corner are, low magnification back scattered image, high magnification image of same region. Al X-ray dot map and Ti X-ray dot map.

The total weight gain after 100 hours was larger in the Ti-48Al alloy with 600 ppm oxygen in comparison to the Ti-48Al alloy with 400 ppm oxygen. This can be attributed to a larger proportion of porosity in the outside oxide layers and delamination of the outer layers. Amongst the alloys oxidized at 982°C the alloys with Cr and Mn showed the largest weight gain values. Both these alloys exhibited a large proportion of porosity and cracking. Cr is a trivalent element and favors the formation of excess vacancies in the oxide leading to an accelerated oxidation kinetics. The oxidation behavior of the Ti-48Al-20Vol.% TiNb particles exhibited an oxidation behavior very similar to the base Ti-48Al alloys. It was however observed that the TiNb particles exhibited very little oxidation even when they were completely engulfed in oxide. The alloy with W exhibited the lowest weight gain at 982°C, followed by the alloy with V.

Activation Energy for Oxidation: As three of the alloys exhibited linear behavior (Cr and V containing alloys at 815°C and Mn containing alloy at 982°C) the activation energies could not be calculated for these alloys. However, the activation energies for the other four alloys which exhibited a parabolic behavior at all temperatures could be calculated.

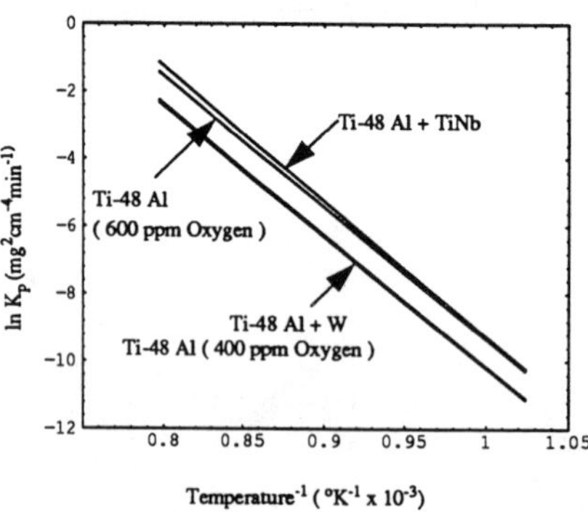

Fig. 5. Oxidation rate constants as a function of temperature.

Fig. 5, is a plot of the rate constants as a function of temperatures from which activation energies were calculated.

The activation energy of Ti-48Al (400 ppm. oxygen) is 324.3 kJ/mol and is 324.4 kJ/mol for Ti-48Al (600 ppm. oxygen). The Ti-48Al-0.2W has a activation energy of 324.6 kJ/mol and Ti-48Al- 20 Vol. % TiNb has an activation energy of 315 kJ/mol. It is clear that the activation energies do not significantly vary with ternary alloy additions.

CONCLUSIONS

[1] At 704°C the ternary alloy with Mn and the binary alloys exhibit the best oxidation resistance.
[2] The addition of 1.5 at% Cr is detrimental to the oxidation behavior at all temperatures.
[3] All alloys except the alloys with Cr and V exhibit a parabolic oxidation behavior at 815°C.
[4] At 982°C the alloys with W and V exhibit the best oxidation behavior indicating that they may be tying up the doubly ionized oxygen vacancies.
[5] Overall there is not much of an improvement of oxidation behavior of Ti-48Al with ternary additions.

REFERENCES

[1] K. Wagner, Atom Movements, American Soc. for Materials, 153-173, (1951).
[2] P. Kofstad, P.B. Anderson and O.J. Krudtaa., J. Less-Common Met., **3** (2), 89-97, (1961).
[3] S.N. Sankaran, R.L. Clark, J. Unnam, and K.E. Wiedemann., NASA Technical Paper 3012, (1990).
[4] Y. Umakoshi, M. Yamagushi, T. Sakagami and T. Yamani, J. Mater. Sci., **24**, 1599, (1989).
[5] E.U. Lee and J. Waldman., Scripta Metall., **22**, 1389, (1988).
[6] R.A. Perkins, K.T. Chiang and G.H. Meir, Scripta Metall., **21**, 1505, (1987).
[7] K. Hauffe., Oxidation of Metals., (Plenum Press 1965), pp. 217-225.
[8] P. Kofstad, Nonstoichiometry, Diffusion and Electrical Conductivity in Binary Metal Oxides. (John Wiley and Sons Inc. 1972.), pp. 15-46.

COATING OF SIC ON TITANIUM ALUMINIDE

SHYI-KAAN WU* AND RAY Y. LIN**

* Institute of Materials Engineering, National Taiwan University, Taipei, Taiwan 106, ROC
** Department of Materials Science and Engineering, M.L. #12, University of Cincinnati, Cincinnati, Ohio 45221-0012, USA

ABSTRACT

Coating of silicon carbide on the surface of structural materials is being used in industry to improve the oxidation resistance and wear resistance properties. In this study, silicon carbide coatings were applied on titanium aluminide by both chemical vapor deposition and sputter deposition techniques. Chemical vapor deposition of SiC was accomplished by reacting methyl trichlorosilane (MTS) and hydrogen in the presence of argon on the substrate surface at 1100 C under 1 atm pressure. Sputter deposition of SiC was completed in a Cooke sputter deposition system using a SiC target at 10 mTorr argon total pressure and 100 watt power. Under the experimental conditions of this study, the deposition rate of CVD SiC is about 10 μm per hour whereas that of sputtered SiC is about 1 μm per hour. To minimize the thermal residual stress between the aluminide substrate and SiC coating as a result of the coefficient of thermal expansion mismatch and temperature variation, a thin layer of TiC was coated prior to the SiC coating. For sputter deposited SiC, it was found that when the coating was done at room temperature, the adhesion of coating to the aluminide substrate was poor. The coating layer spalled off when the coated specimen was thermal cycled to 900 C. By coating the carbides at an elevated temperature, e.g. 560 C, strong adhesion between carbide coating and the substrate was obtained. The coating survived many thermal cyclings. The degree of surface roughness ranging from #80 to #800 grit surface polish did not affect the adhesion of coating.

1. INTRODUCTION

Titanium aluminides is a new class of material receiving significant attention lately for structural applications at elevated temperatures**[1]** especially in the aerospace industry. This is primarily due to the high temperature strength retention of this class materials. However, it has been reported that the oxidation resistance of the two commonly used titanium aluminides, i.e. Ti_3Al and TiAl, is far from being acceptable for use in air at the anticipated application temperature, namely, up to 1000 C.**[2,3]** To increase the oxidation resistance of titanium aluminides, alloying elements which may form protective oxides have been added in aluminides. Niobium addition is one of the examples in the anticipation of forming niobium oxides or other complex oxides. The other approach of improving the oxidation resistance of aluminides is the application of oxidation resisting coatings. Silicon carbide, with its excellent oxidation resisting property, has the potential of being used as a protective coating for titanium aluminides. The objective of the present study is to investigate silicon carbide coatings on TiAl applying the chemical vapor deposition and the sputter deposition techniques.

2. EXPERIMENTAL PROCEDURES

A. Chemical Vapor Deposition (CVD)

The CVD system used in this study is illustrated in Figure 1. For CVD coating of SiC, specimens were polished to 600 grit, cleaned ultrasonically in acetone and, then, placed in the CVD furnace. Before deposition, the furnace and the specimen were heated up to 1100 C in argon. As the temperature of the deposition reached 1100 C, the reaction gas was introduced. All deposition was carried out at 50 ml/min Ar and 50 ml/min H_2. MTS was introduced into the system by directing a small portion of the argon gas into the MTS container and bubbling through MTS liquid. By controlling the amount of the bubbling Ar, the amount of MTS carried out is fixed since the equilibrium partial pressure of MTS is well documented.**[4]**

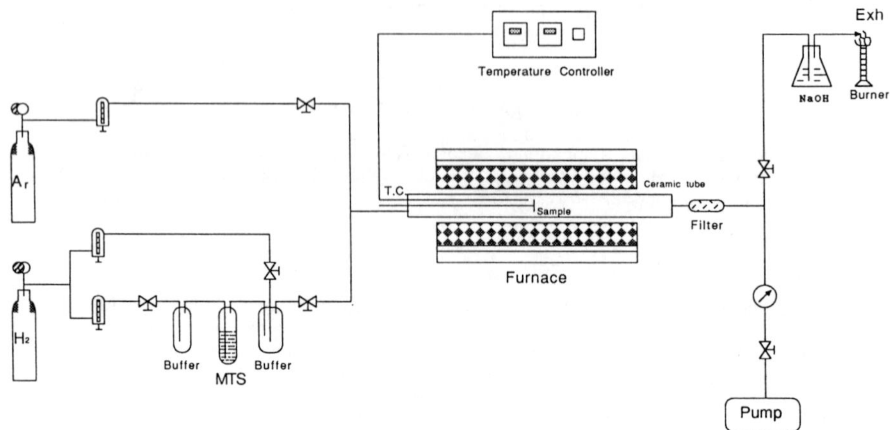

Figure 1. Chemical vapor deposition system.

At the end of deposition, the specimen was moved to a lower temperature zone in the reaction chamber and the MTS and H_2 streams were shut off immediately to prevent further reaction under uncontrolled conditions. Cooled specimens were removed from the reaction chamber and the amount of deposition was determined from the weight gain of the specimen. The structure of the deposit was characterized with a Philips x-ray diffractometer. The coating morphology was examined with a scanning electron microscope(SEM).

B. Sputter deposition

Sputter deposition is done in a Cooke sputter deposition station. Figure 2 shows a schematic diagram of the sputter deposition station used in this study. It consists of a gas

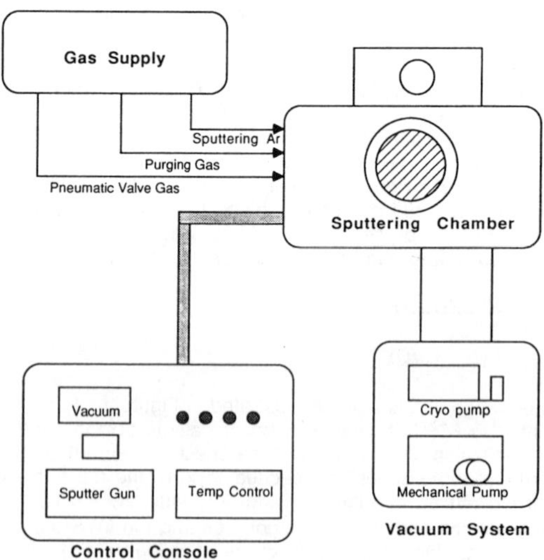

Figure 2. A schematic drawing of the sputter deposition system.

supply system, a vacuum system, a control console and the sputtering chamber. Specimens were polished to 80, 240, 400 and 800 grits, respectively, and cleaned ultrasonically in acetone before deposition. During deposition, the sputtering chamber was, first, evacuated to 10^{-7} torr for at least 3 hours to ensure minimal moisture contamination from the chamber surface. Sputter deposition was done at various argon pressure, time, power, temperature and target-substrate distance. After deposition, the chamber was back-filled with nitrogen and, then, the specimen was removed from the substrate holder. The amount of the coating was determined in-situ with a vibrating crystal thickness monitor installed near the specimen in the deposition chamber. The coating structure was determined with the Philips x-ray diffractometer. The coating morphology was characterized with the SEM.

3. EXPERIMENTAL RESULTS AND DISCUSSION

A. CVD SiC Coatings

Table I shows the summary of the CVD study. Both experimental conditions and observations are presented. It includes only results from the 1100 C experiments since they belong to the most systematically investigated group of specimens. As shown in Table I, reaction products of most experiments are β-SiC. However, when the MTS content is low,

Table I. Experimental Conditions and Observations of CVD SiC

Sample No.	Gas Composition* MTS (ml/min)	Time (hr)	X-Ray Analysis	Coating Amt (mg/cm^2)
2	10.0	2	C + SiC	--
3	6.7	4	β-SiC	7.5
5	6.7	2	SiO2	--
7	3.3	2	Si	--
8	8.3	6	β-SiC	22.3
10	6.7	5	β-SiC	8.4
11	6.7	5	β-SiC	10.4
13	8.3	5	β-SiC	14.0
14	5.0	5.25	β-SiC	10.2
15	5.0	5.75	β-SiC	14.9
16	6.7	4	β-SiC	10.9
17	6.7	2	β-SiC	2.5

* Flow rates of both Ar and H_2 were maintained to be 50 ml/min.

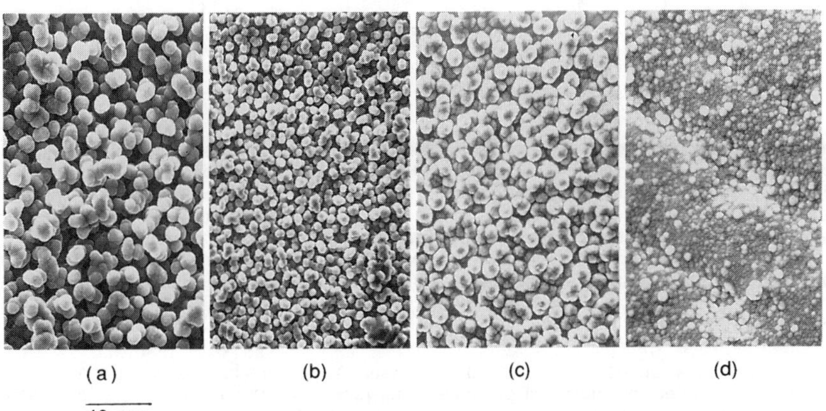

(a) (b) (c) (d)

40 μm

Figure 3. Surface morphology of CVD SiC coatings.

pure Si was form. When the MTS content is high, the reaction product contained free carbon as well. Experiment 5 showed SiO_2 as the reaction product. During this experiment, minor leakage was found in the reaction chamber. The experimental setup was very carefully assembled since that observation.

Figure 3-a, b, c and d show the surface morphology of CVD SiC products from Experiments 8, 14, 16 and 17, respectively. It appeared that both the time and the MTS concentration affected the surface morphology. Figure 3-c and d represent the effect of deposition time on the size of the crystalline on the coating surface. Under identical experimental conditions, the longer it deposits , the larger the grain size. On the other hand, comparison between Figure 3-a and b shows that the higher the MTS concentration, the larger the grain size of the SiC coating. A closer examination of the coating morphology shows that the coating layer is not fully dense. Pores exist in the coating. More studies are needed to establish the proper coating conditions for fully dense coatings in the experimental setup of this study.

From Table I, it can be seen that the deposition rate is about 2 mg/cm^2 per hour. Assuming a fully dense SiC coating with a density of 3.17 g/cm^3, the rate of coating is equivalent to about 10 μm/hour. From limited data shown in Table I, it appeared that the deposition rate increased with the increasing MTS content in the reaction gas.

To determine the rate of SiC deposition at long deposition time, a series of CVD experiments were carried out at 1100 C with the MTS concentration controlled to 6.7 ml/min for 2, 4, 8, and 16 hours. Figure 4 shows the amount of deposition as a function of total deposition time. The original point and data at 2 and 4 hours form a straight line with a slope of 25 mg/hour. Since all specimens have a surface area of 12.0 cm^2, the deposition rate is 2.0 $mg/cm^2 \cdot h$ for deposition up to 4 hours. Data beyond 4 hours form another straight line with a slope equal to 15 mg/h, which is equivalent to a deposition rate of 1.25 $mg/cm^2 \cdot h$. The linear

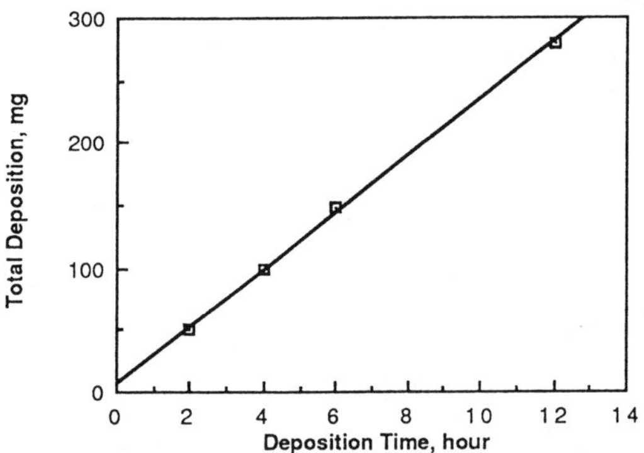

Figure 4. CVD SiC as a function of time at 1100 C.

deposition dependence on time suggests that the deposition reaction has reached a steady state condition after 4 hours under the experimental conditions of this study. Comparing with the amount of the MTS entering the CVD reactor, 657 mg/h, the deposition efficiency is only about 2.3%. This is because the CVD reactor in this study is a hot wall device. The total hot surface area in the reactor which can deposit SiC and consume MTS is far more than the specimen surface area. In a production line CVD facility in industry, the specimen to vessel wall surface area ratio will be increased significantly and the production yield can, thus, be improved.

B. Sputter Deposition

For sputter deposited SiC coating, systematic experiments have been carried to determine the effect of sputtering pressure, time, power, substrate temperature, and target-substrate distance on the rate of deposition. Tables II.1-II.5 show the result of such experiments. The sputtering pressure appears to have the most pronounced effect on the deposition rate. From 5 to 60 mTorrs, the lower the pressure, the higher the deposition rate. This observation agrees well with those reported in the literature. The deposition time and substrate temperature both show little effect on the rate of deposition.

Table II.1 Sputter coating of SiC at various pressures

Argon Pressure (mTorr)	Thickness (Å)	Rate(Å/min)
5	3000	50
15	2500	42
30	300	5
60	<150	-

Table II.2 Sputter coating of SiC at varying deposition Time

Deposition Time (min)	Thickness (Å)	Rate(Å/min)
120	5000	42
60	2500	42
30	1500	50

Table II.3 Sputter coating of SiC with different powers

Sputtering Power (W)	Thickness (Å)	Rate(Å/min)
100	3500	58
70	3000	50
40	750	13

Table II.4 Sputter coating of SiC at different substrate temperatures

Substrate Temp. (C)	Thickness (Å)	Rate(Å/min)
850	3000	50
550	2000	33
45	2500	42

Table II.5 Sputter coating of SiC at varying Target-Substrate Distances

Distance (cm)	Thickness (Å)	Rate(Å/min)
3.6	4000	67
5.1	3000	50
10.2	2500	42

Power, on the other hand, does affect the deposition rate. The higher the deposition power, the faster the deposition. The target-substrate distance also affects the deposition rate significantly in that shorter distances give higher deposition rates.

For a set of specimens deposited at a substrate temperature of 700 C, the as deposited surface of samples appears rather smooth. This is illustrated in the SEM surface morphology of the deposited film in Figure 5a-d for samples which were polished to 80, 240, 400 and 800 grits prior to deposition. It appears also that the coating adhere to the substrate very well. The surface roughness of the as polished samples seems to have no effect on the deposition. Since the coating is only about 1 μm, the polishing lines are clearly observable in the photos of Figure 5. The x-ray diffraction pattern of the film shows that silicon carbide peaks can be

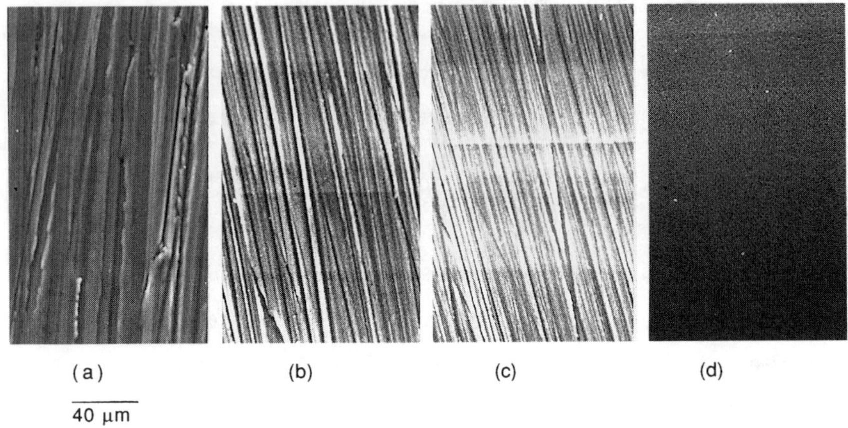

Figure 5. SEM surface morphology of sputter deposited SiC.

identified although the film thickness is only about 1 μm. To ensure proper adhesion of the SiC coatings on the specimen, a layer of TiC has been coated on TiAl prior to the coating of SiC. It has been observed that, without the TiC coating, the SiC coating on TiAl spalled off easily after a single thermal cycling of up to 900 C.

4. SUMMARY

Ceramic coatings play an increasingly important role in applications where high temperature corrosion resistance, oxidation resistance, and wear resistance are of concerns.**[5]** CVD and sputter deposition were successfully used to coat titanium aluminide coupons. It has been observed that the rate of CVD coating is much faster than that of sputter deposition. On the other hand, the coating surface morphology on the sputter deposited film is superior to those of CVD. The effect of deposition parameters on the deposition has been investigated for sputter deposition. These parameters include deposition pressure, power, time, temperature and target-substrate distance. Information obtained in this study would be valuable to industry for providing oxidation resisting SiC coatings on titanium aluminides.

Acknowledgement

The work was supported by a grant from Taiwan's National Science Council. The assistance from graduate students at both National Taiwan University and University of Cincinnati is greatly appreciated.

References

1. J.M. Larsen, K.A. Williams, S.J. Balsone and M.A. Stucke, in "High Temperature Aluminides and Intermetallics," ed. S.H. Whang, C.T. Liu, D.P. Pope and J.O. Stiegler, TMS, Warrendale, PA, pp. 521-556, 1990.
2. T.M.F. Ronald, Adv. Mat. & Proc., vol. 89(5), pp. 29-37, 1989.
3. F.A. Smidt, Adv. Mat. & Proc., vol. 91(1), pp. 61-62, 1990.
4. R.C. Weast, "Handbook of Chemistry and Physics, 55th Ed.," CRC Press, Cleveland, OH, 1974, p. D-171.
5. G. Geiger, Ceramic Bulletin, vol. 71(10), pp. 1470-1481, 1992.

EFFECT OF PROCESSING ON OXIDATION OF Ti_5Si_3

ANDREW J. THOM, YOUNGMAN KIM, AND MUFIT AKINC
Ames Laboratory and Department of Materials Science and Engineering, Iowa State University, Ames, IA 50011

ABSTRACT

The mechanical properties and oxidation resistance of HIPed Ti_5Si_3 have been measured. HIPing submicron size powder compacts produces a crack-free, fine-grained microstructure with significantly higher hardness and toughness than a coarse-grained microstructure which contains microcracks within larger grains. Oxidation resistance is influenced by the grain size. Coarse-grained material has much lower mass gain than fine-grained material in an oxidizing atmosphere and exhibits parabolic oxidation kinetics. The oxidation resistance of fine-grained material was measured between 700°C and 1000°C in air. Mass gain at 120 hours was measured to be 0.07 mg/cm^2 at 700°C. At 900°C cracking of the scale leads to linear oxidation kinetics and significantly higher mass gain.

INTRODUCTION

Silicides have received considerable attention as potential high temperature structural materials. Of the less commonly investigated silicides[1], Mo_5Si_3, Nb_5Si_3, Cr_5Si_3, Zr_5Si_3, Ti_5Si_3, V_5Si_3, and Ta_5Si_3 are particularly interesting as potential high temperature materials. Among this group of materials, Ti_5Si_3 is attractive due to its low density, high melting point, potential oxidation resistance, and interesting interstitial chemistry. The previous work indicates that the monolithic material suffers from microcracking [2-7] which degrades the mechanical properties. Additionally, the oxidative behavior of Ti_5Si_3 has received only limited attention[2,7,8].

In this paper, the effect of processing upon the mechanical and oxidative behavior of Ti_5Si_3 is studied. In particular, the influence of grain size on the mechanical and oxidative behavior is discussed.

EXPERIMENTAL

Ti_5Si_3 was synthesized in 125g batches. A stoichiometric mixture of titanium and silicon was arc-melted under an argon atmosphere using a non-consumable electrode. The buttons were milled to -325 mesh (<43 μm) powder in a WC-lined mill. X-ray diffraction patterns indicated the powders were single phase Ti_5Si_3 with the Mn_5Si_3-type structure. Powders were then milled for up to an additional six hours. Submicron size powders were pressed into 15mm x 15mm x 60mm bars, coated with boron nitride, and HIPed in a borosilicate encapsulating glass at 206 MPa at 1250°C for 10 hours. A slow cooling rate of 1-2°C/min was used to minimize bulk thermal stresses and reduce cracking. Room temperature hardness and fracture toughness were measured by Vickers indentation technique.

Specimens for thermal oxidation studies were prepared by cutting 10mm x 5mm x 2mm coupons from the HIPed bar. These coupons were polished through 0.05μm Al_2O_3 abrasive. Samples were suspended from a sapphire hangdown wire in a TGA and heated at 20°C/min in a flowing stream of synthetic air (mixture of 79%N_2-21%O_2, free of H_2O). Continuous isothermal oxidation was measured for up to 120 hours.

RESULTS AND DISCUSSION

The grain size of the HIPed monoliths varied significantly with initial powder size. A starting powder of -325 mesh produced a coarse-grained microstructure of 10-25μm which exhibited transgranular microcracks. A starting powder which was milled for an additional six hours produced a fine-grained microstructure of 1-2μm and was free of microcracks. The room temperature hardness and fracture toughness increased as the grain size decreased. The fine-grained material had hardness and toughness values up to 17.1±0.7 GPa and 2.9±0.2 MPa•m½, respectively, for an indentation load of 500 g. The corresponding values for the coarse-grained material were 9.1±0.5 GPa and 1.6±0.2 MPa•m½ . The lower values of hardness and fracture toughness exhibited by the large-grained material are attributed to the transgranular microcracking. This microcracking is believed to arise from the large thermal expansion anisotropy of the material [9]. The coefficient of thermal expansion (CTE) as measured by high temperature x-ray diffraction is: α_a = 8.7±0.2 ppm/K and α_c = 20.4±0.4 ppm/K. The large difference between α_a and α_c (thermal expansion anisotropy) results in large thermal stresses during processing which may exceed the strength of the material when the grain size exceeds a critical value[10]. Residual stresses which develop during cooling from the processing temperature are believed to cause transgranular microcracking in large-grained microstructures.

Oxidation resistance measurements indicated that the coarse-grained material showed good oxidation resistance at 1000°C. The mass gain after 72 hours was 1.5 mg/cm^2 and was parabolic with a rate constant of 0.027 mg^2/cm^4/hr. Fine-grained material showed significantly lower oxidation resistance, experiencing nearly an order of magnitude increase in mass gain at 1000°C after 72 hours. The fine-grained material showed a complex scale development. An outer 10 μm thick layer of TiO_2 formed. Below this layer was a 10 μm layer composed of TiO_2 particles mixed with a discontinuous silicon rich, titanium depleted phase, presumably SiO_2. The lower oxidation resistance of fine-grained material is likely related to the increased grain boundary area. A continuous, protective SiO_2 layer was not observed to form.

The oxidative behavior of the fine-grained material was further investigated from 700°C-1000°C in air. For each temperature, two flow rates were used, 30 and 150 ml/min. If linear oxidation kinetics occurred, a dependence of the linear rate on gas flow would indicate that the supply of oxygen at the surface is the rate determining step[11]. Figure 1 shows the isothermal oxidation at 700°C. The mass gain is quite small, less than 0.07 mg/cm^2 after 120 hours. Figure 2 shows the isothermal oxidation at 800, 900, and 1000°C. Mass gain increases as temperature increases. The complex nature of the scale development is evident from the changing oxidation kinetics outlined in Table I. Two equations are listed when the data are intermediate between two rate equations. There is similar behavior in the oxidation kinetics between flow rates at each temperature except at 800°C which indicates reproducibility of the data.

All samples show an initial transient oxidation. This initial rate changes as the scale grows inward, and a second rate-controlling mechanism dominates. At 700°C and 800°C, the steady state oxidation rate is between parabolic and cubic. The low level of mass gain at 700°C approaches the sensitivity of the microbalance and prevents an accurate

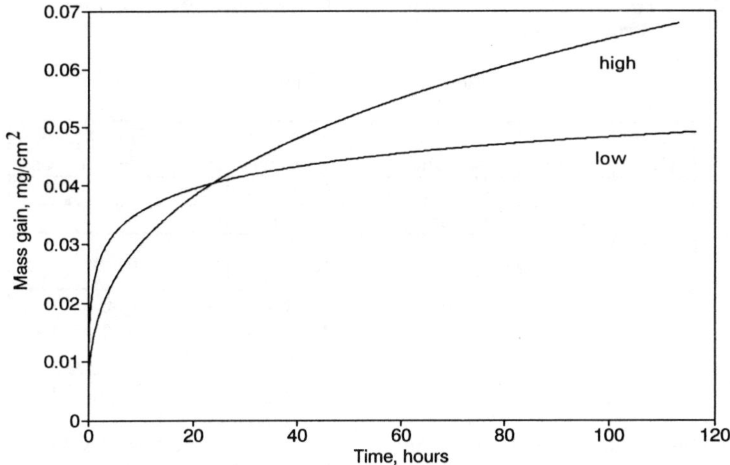

Figure 1: Isothermal oxidation of fine-grained Ti_5Si_3 at 700°C in air at low and high flow rate (30 and 150 ml/min).

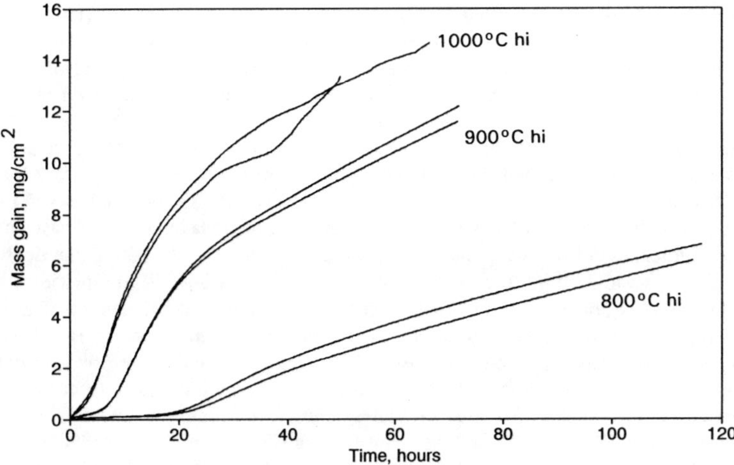

Figure 2: Isothermal oxidation of fine-grained Ti_5Si_3 in air from 800°C-1000°C at low and high flow (30 and 150 ml/min). Curves for high flow rate at each temperature are labelled.

determination of the steady state reaction kinetics. At 800°C and low flow, transient paralinear oxidation occurs up to 15 hours and changes to cubic oxidation at 22 hours. At 800°C and high flow, the transient oxidation is parabolic and changes to a higher parabolic rate after 22 hours. Since these are not linear mechanisms, flow rate is not expected to cause this different behavior.

Table I: Oxidation Kinetics*

Temp (°C)	Flow Rate	Time Scale/ Rate Equation	Time Scale/ Rate Equation	Time Scale/ Rate Equation
700	low	0-3 hrs/Log	3-120 hrs/PB,C	
	high	0-3 hrs/Log	3-120 hrs/PB,C	
800	low	0-15 hrs/PL	22-70 hrs/C	
	high	0-15 hrs/PB	22-70 hrs/PB	
900	low	0-3 hrs/PL,PB	10-30 hrs/Log	>30 hrs/L
	high	0-3 hrs/PL,PB	10-30 hrs/Log	>30 hrs/L
1000	low	0-3 hrs/L	7-35 hrs/Log	>40 hrs/L
	high	0-3 hrs/L	7-35 hrs/Log	>40 hrs/L

* Abbreviations used for the rate equations fit to the data are as follows: Log=direct logarithmic, C=cubic, PB=parabolic, PL=paralinear, and L=linear.

The linear rate constants at 900°C for long times (t≥30 hours) and 1000°C for short times (t=0-3 hours) for low and high flow rate are nearly identical, indicating that the rate determining step is not associated with an adsorption of oxygen. The increased rate for 1000°C for low flow and (t>40 hours) may be due to formation of stress cracks in the scale which exposed large amounts of new surface. The linear rate at (t>40 hours) is slightly less than the initial rate for (t=0-3 hours), indicating this to be a plausible event. For both 900 and 1000°C, the linear rates are associated with the cracking of the scales, causing short circuit paths to the base material and allowing for an increased rate of mass gain.

No scale formation was observed for samples exposed to air at 700°C. The external surface was slightly tarnished, indicating the presense of a thin film. The scale formation from 800 to 1000°C is shown in Figure 3. A dense outer 15 μm thick layer forms at 800°C. Dendrites about 12 μm wide grow from this outer layer into the base material. The total thickness of the layer and dendrites is about 150 μm. At 900°C the dense outer layer has grown to 35 μm. The dendrites have coarsened to about 50 μm in width, and the total thickness is about 275 μm. At 900°C the mass gain is about twice that at 800°C. The increased thickness at 900°C accounts for most of the mass gain. By 1000°C the dense outer layer has separated into a two phase region as seen in the previous oxidation study of the fine-grained material. The dendrites have grown in width to consume all of the base material so that no base material extends to the dense outer scale. The total thickness is about 300 μm. The mass gain has increased by 15% from 900°C, and this increase is reflected in the increase in total thickness to 300μm and the coarsening of the dendrites.

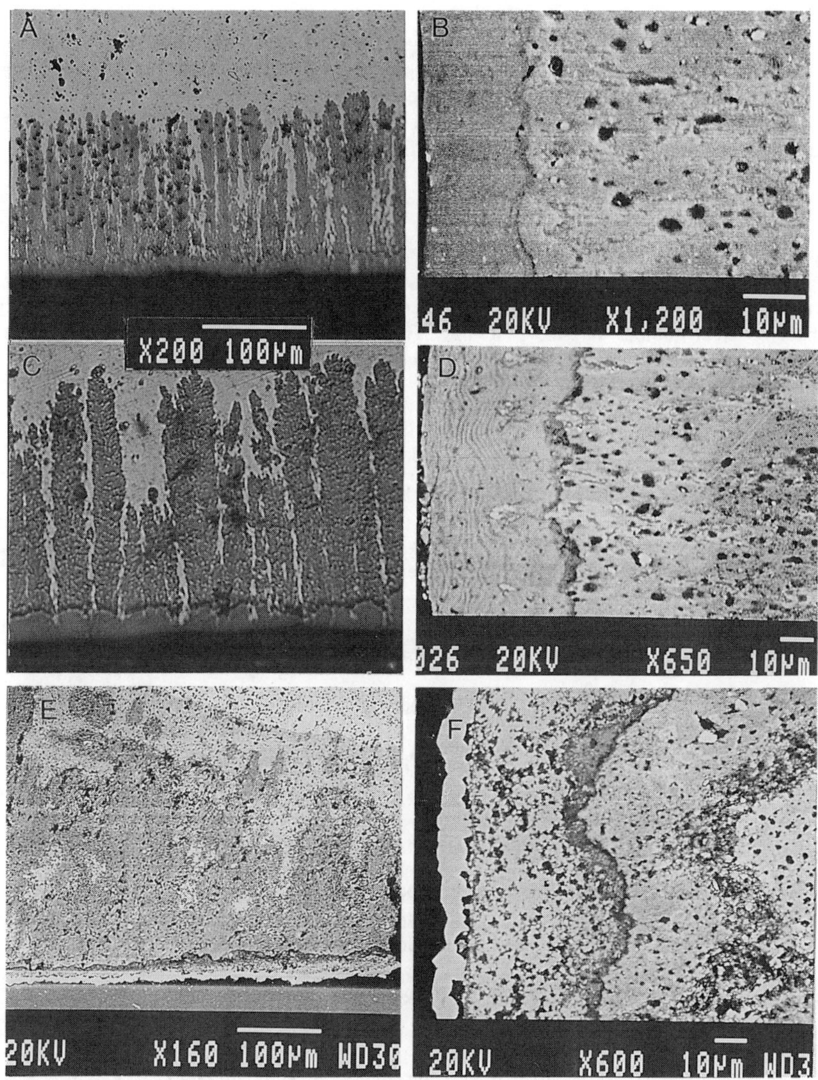

Figure 3: Microstructures of fine-grained Ti_5Si_3 oxidized in air. (a),(b) 800°C for 115 hours. (c),(d) 900°C for 72 hours. (e),(f) 1000°C for 60 hours.

CONCLUSION

Dense, microcrack-free Ti_5Si_3 was produced by HIPing powder milled to submicron size. Hardness and fracture toughness improves as grain size decreases. Oxidation resistance appears to decrease as grain size decreases, presumably due to increased grain boundary diffusion. For fine grained material, oxidation resistance at 700°C is excellent. Above 700°C oxygen uptake increases significantly as temperature increases. By 900°C substantial cracking occurs to prevent the formation of a protective scale and linear oxidation kinetics dominate.

ACKNOWLEDGMENTS

Ames Laboratory is operated for the U.S. Department of Energy by Iowa State University under contract number W-7405-ENG-82. This research was supported by the Office of Basic Energy Science, Materials Science Division.

REFERENCES

1. D.M. Shah, D. Berczik, D.L. Anton, and R. Hecht, Mater. Sci. Eng. **A155**, 45 (1992).

2. D.L. Anton and D.M. Shah, US Air Force Rep., contract no. WRDC-TR-90-4122.OH,USA. (1991)

3. W. Smarsly, R. Rosenkranz, G. Frommeyer, Mater. Sci. Eng., **A152**, 288 (1992).

4. G. Frommeyer, R. Rosenkranz, and C. Ludecke, Z. Metallkde., **81**, 307 (1990).

5. C. T. Liu, E.H. Lee, and T.J. Henson, Report, ORNL-6435; Order No. DE88007860, 1988.

6. S. Reuss and H. Vehoff, Scripta Met., **24**, 1021 (1990).

7. Y. Kim, A.J. Thom, and M. Akinc, in Processing and Fabrication of Advanced Materials for High Temperature Applications-II, edited by T.S. Srivatsan and V.A. Lavi (The Min., Met., and Mater. Soc. Proc. Chicago, IL, 1992) In Press.

8. R.M. Paine,A.J. Stonehouse, and W.W. Beaver, WADC Tech. Rep. 59-29-Part I (1960).

9. A.J. Thom and M. Akinc, Iowa State Unversity, C.R. Hubbard and O.B. Cavin, Oak Ridge National Laboratory, Unpublished data.

10. J.A. Kuszyk and R.C. Bradt, J. Am. Ceram. Soc., **56**, 420 (1973).

11. P. Kofstad, High-Temperature Oxidation of Metals, John Wiley and Sons, Inc., New York, 1966, p.13.

INFLUENCE OF ENVIRONMENT ON CRACK GROWTH RESISTANCE OF AN Fe_3Al,Cr ALLOY

A. CASTAGNA, P.J. MAZIASZ, AND N.S. STOLOFF

ABSTRACT

The effect of hydrogen on the tensile and fatigue behavior of an Fe_3Al, Cr type intermetallic alloy is examined. Hydrogen due to moisture in the air is found to be a major cause of embrittlement. The rates and mechanisms of the observed embrittlement appear to be temperature dependant. In addition, the alloy was found to have no notch sensitivity.

INTRODUCTION

Iron aluminides are currently being studied for use in applications where low cost, relatively low weight, and excellent corrosion resistance are desired. The main disadvantage of the alloys is low room temperature ductility in air due to moisture which causes hydrogen embrittlement [1-3]. This investigation examines changes in tensile and fatigue behavior of an Fe_3Al,Cr type intermetallic compound in environments of various hydrogen content and at various temperatures. The alloy was studied in two ordered states, the DO_3 and the B2. The DO_3 is a highly ordered superlattice, while the B2 is a superlattice of imperfect order.

EXPERIMENTAL MATERIALS AND PROCEDURE

Material

The material used in this study, designated FA129, was supplied by Oak Ridge National Labs and was composed of 28.6 at.%Al, 4.8%Cr, 0.21%C, 0.5%Nb, balance Fe. Material fabrication and annealing conditions to obtain B2 and DO_3 order are described in [4].

Notch Sensitivity

Notch sensitivity experiments were carried out on electro-discharge machined cylindrical specimens with a 22.6mm long by 4.6mm diameter gauge section. Unnotched specimens and specimens with a 1.0mm deep circumferential notch of 0.089mm radius were tested. This notch geometry resulted in a stress concentration factor of 3.9. The unnotched specimens were mechanically polished to a 0.3μm finish.

Tests were performed at a strain rate of 3.3×10^{-4} sec^{-1} on an Instron servo-hydraulic machine fitted with a vacuum chamber. Two environmental conditions were studied in each of the two ordered states: dry oxygen and air. In the oxygen tests, the vacuum chamber was first evacuated to $< 1.3 \times 10^{-4}$ Pa before admitting constantly flowing oxygen at 1.01×10^5 Pa. All tests were conducted at room temperature.

Fatigue Crack Growth

Room temperature fatigue testing was performed on compact tension specimens measuring 31.6mm x 30.5mm x 5.1mm thick with a machined notch of 60° angle and 0.127mm radius cut by EDM techniques. Specimen faces were polished to a 0.3μm finish.

Fatigue crack growth tests were performed on an Instron servo-hydraulic testing apparatus and crack length was measured via the d.c. potential drop method. Experimental details are as described in [4], except for a load shedding routine which was incorporated into the procedure to obtain more reliable threshold data. Five environmental conditions were previously studied in each of the two ordered states: dry oxygen, air, vacuum, hydrogen gas, and hydrogen charged. Fatigue tests in air at temperatures of 150°C, 300°C, and 450°C have been performed on DO_3 ordered FA129, and at 150°C on B2 ordered FA129. The compact tension specimens for these tests measured 25.4mm x 24.4mm x 5.1mm thick.

Microscopy

Fracture surfaces were examined under an AMR-1000 SEM. In addition, the fatigue zones of the compact tension specimens were examined under a 300 kV Phillips TEM.

TEM foils were prepared from the fatigued regions of the compact tension specimens parallel to and just under the fracture surface. The foils were mechanically ground to 150 μm, and electro-thinned until perforated. Electro-thinning conditions were as follows: an electrolyte of 6% perchloric acid in methanol at 100 mA and -20°C for 30 seconds.

RESULTS

Notch Sensitivity

Tensile results are listed in Table I. Comparison of unnotched data shows the expected embrittling effect of hydrogen on ε_f and RA in both ordered states. In the B2 state ε_f doubles when tested in oxygen, and in the DO_3 state there is a five-fold increase. As expected, ductility in the DO_3 state is lower than that in the less ordered B2 material. However, neither condition displays notch sensitivity. In fact, the DO_3 material shows notch strengthening, i.e. the notch sensitivity ratio is greater than one.

B2 samples showed mixed mode failure whether notched or unnotched in air or in oxygen, e.g. see Fig. 1. Notched and unnotched DO_3 tensile samples (air and oxygen) showed predominantly cleavage fracture, as seen in Fig. 2. However, mixed mode regions near the notched surfaces were observed.

TABLE I - Notch Sensitivity Data for FA129

Condition	Environment	UTS (MPa)	YS (MPa)	ε_f (%)	RA (%)	NSR
B2	Air	645	407	5.9	4.4	-
B2 notched	Air	641	512*	-	1.4	0.993
DO_3	Air	507	411	1.5	3.3	-
DO_3 notched	Air	623	596*	-	2.8	> 1
B2	O_2	851	432	11.6	11.8	-
B2 notched	O_2	840	628*	-	2.8	0.987
DO_3	O_2	824	513	7.3	10.4	-
DO_3 notched	O_2	801	457*	-	2.8	0.972

* - denotes stress taken at yield point

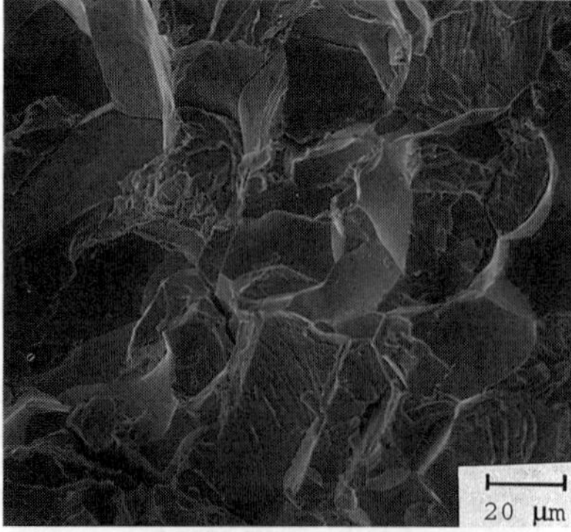

Figure 1. Unnotched B2 Ordered FA129 Tensile Tested in Air at 25°C.

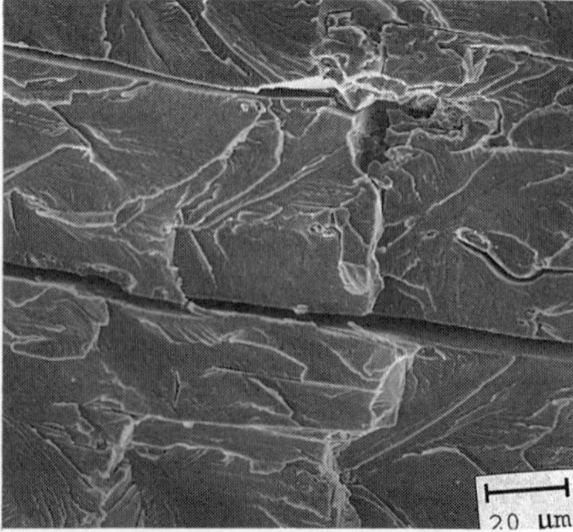

Figure 2. Unnotched DO_3 Ordered FA129 Tensile Tested in Air at 25°C.

Fatigue Crack Growth

Figs. 3 and 4 show the da/dN curves generated for DO_3 and B2 material, respectively, at 25°C. Figs. 5 and 6 show the da/dN curves of the DO_3 and B2 material, respectively, at elevated temperature. Table II lists the fracture mode for each test condition and summarizes the threshold and critical stress intensities, as well as the constant m of the Paris equation,

$$da/dN = C(\Delta K)^m .$$

The fracture surfaces of the two ordered states demonstrate different fracture modes. The B2 ordered specimens fail entirely transgranularly, with occasional fatigue striations. Note the large decrease in growth rate in air between 25°C and 150°C. In 150°C air, as well as 25°C oxygen, there is a dimpled appearance suggesting microvoid coalescence. The 150°C specimen also displayed wide, irregularly spaced striations with a cleavage character as seen in Fig. 7.

The DO_3 ordered specimens demonstrate a shift in fracture mode from transgranular in oxygen to mixed mode to predominantly intergranular with more aggressive environments. A typical intergranular fracture surface for DO_3 ordered material is shown in Fig. 8.

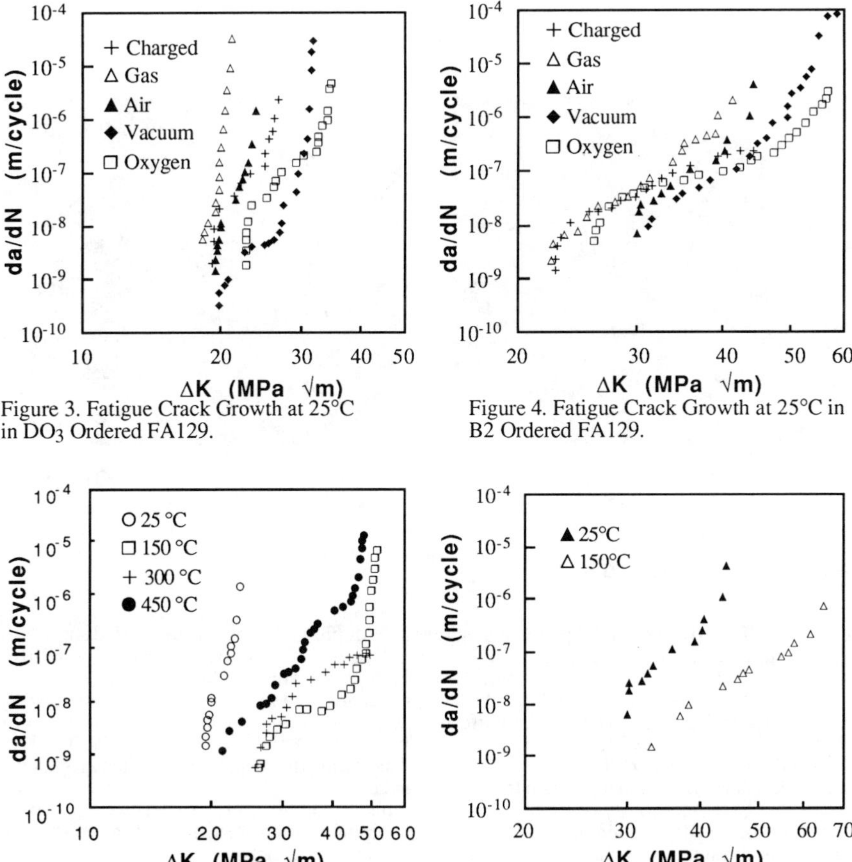

Figure 3. Fatigue Crack Growth at 25°C in DO_3 Ordered FA129.

Figure 4. Fatigue Crack Growth at 25°C in B2 Ordered FA129.

Figure 5. Fatigue Crack Growth at Elevated Temperature in DO_3 Ordered FA129.

Figure 6. Fatigue Crack Growth at Elevated Temperature in B2 Ordered FA129.

TABLE II - Fatigue Crack Growth Data for FA129

Condition	Temperature (°C)	Fracture Surface	m	ΔK_{TH} (MPa√m)	ΔK_C (MPa√m)
B2 Oxygen	25	Dimpled, few striations	2.9	26	61
B2 Vacuum	25	TG, many striation	5.1	31	59
B2 Air	25	TG, many striation	9.4	30	44
B2 Charged	25	TG, few striations	6.8	23	44
B2 Gas	25	TG, few striations	8.6	22	41
DO_3 Oxygen	25	Dimpled + Cleavage	8.8	23	35
DO_3 Vacuum	25	TG, few striations	5.2	20	32
DO_3 Air	25	Mixed, few striations	22.8	19	24
DO_3 Charged	25	Mixed, no striations	9.1	19	27
DO_3 Gas	25	Mixed, no striations	37.6	18	21
B2 Air	150	Dimpled, few striations	5.3	32	65
DO_3 Air	150	-	0.8	26	52
DO_3 Air	300	-	3.1	25	49
DO_3 Air	450	-	8.7	21	48

Figure 7. Fracture Surface of B2 Ordered FA129 Fatigued in Air at 150°C.

Figure 8. Fracture Surface of DO_3 Ordered FA129 Fatigued in Air at 25°C.

Transmission Electron Microscopy

The B2 ordered specimens showed a duplex microstructure: predominantly unrecrystallized with a cellular subgrain structure (see Fig. 9). However, some recrystallized regions were observed with a lower dislocation density, more paired dislocation, and APB's.

The DO_3 specimens exhibited few or no subgrains, a much lower dislocation density in general than the B2 material, and a uniform network of APB's. The dislocations which were observed were also more likely to be paired than in the B2 material, as seen in Fig. 10. The DO_3 grain boundaries tended to be free of dislocations while the B2 material often had dense dislocation networks emanating from the grain boundaries.

Heterogeneously distributed coarse precipitates (up to 5 µm diameter) were observed, which are believed to be niobium carbides. The precipitates often acted as dislocation sources, possibly as a result of mismatched thermal contraction during quenching.

At elevated temperatures large arrays of dislocations were also observed (Fig. 11), sometimes forming subgrains or cells (Fig. 12)

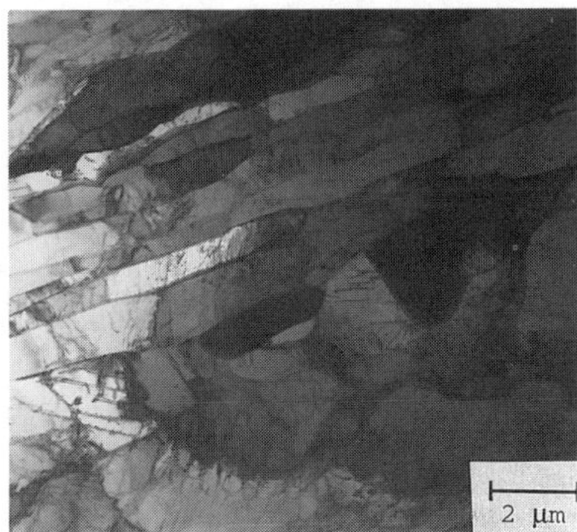

Figure 9. Subgrains in B2 ordered FA129 Fatigued in Air at 25°C.

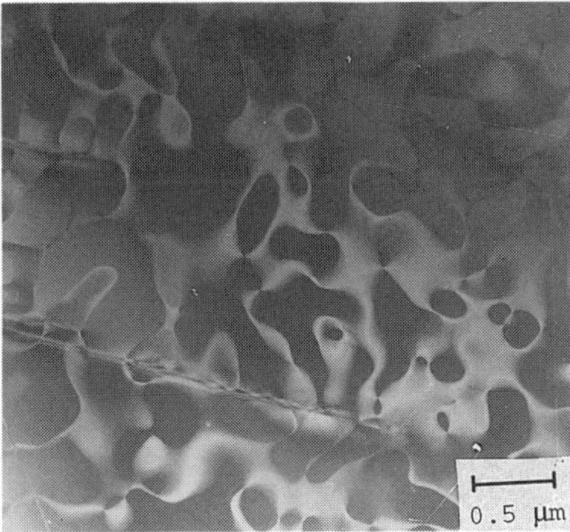

Figure 10. APB's in DO_3 Ordered FA129 Fatigued in H_2 Gas at 25°C.

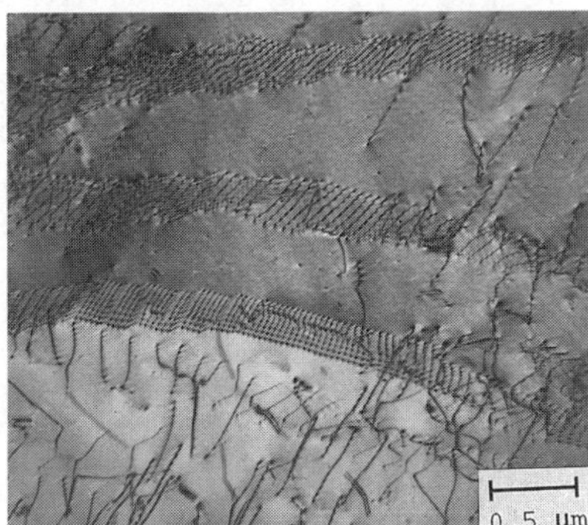

Figure 11. Dislocation Array in DO_3 Ordered FA129 Fatigued in Air at 450°C.

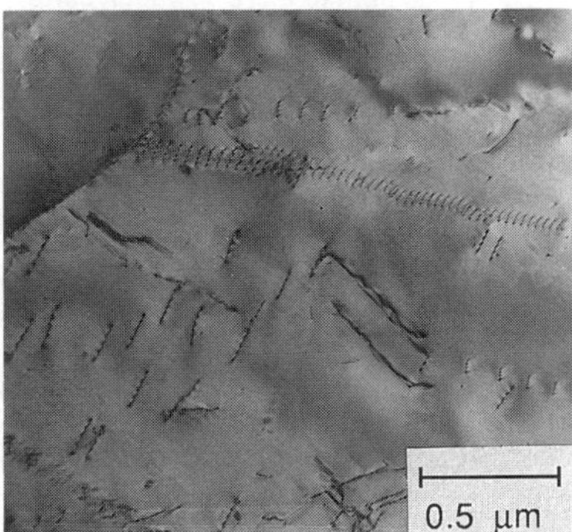

Figure 12. Dislocation Array in DO_3 Ordered FA129 Fatigued in Air at 450°C.

DISCUSSION

The NSR (see Table I) was found to be approximately 1, indicating no notch sensitivity for both the B2 and DO_3 conditions. This is in contrast to previous data on binary Fe-22at.%Al which showed notch sensitivity ratios between 0.60 and 0.75 [3]. The lack of notch sensitivity may be a result of the chromium addition in FA129 [5,6].

The increased tensile ductilities measured in oxygen as compared to air (see Table 1) demonstrate the embrittling effect of moisture in laboratory air, which is consistent with previous findings that hydrogen embrittles Fe_3Al. In a recent study of Fe_3Al alloys [2] a decrease in ductility from 19.9% in vacuum to 3.3% in air was reported. The ΔK_C measured in oxygen is slightly higher than that measured in vacuum, which may be explained by oxidation of aluminum without the release of hydrogen. The shift to a dimpled fracture surface for both ordered states in oxygen is also indicative of enhanced protection from hydrogen.

In the B2 ordered material at 150°C in air a decrease in crack growth rate and an increase in fracture toughness is seen, which is consistent with an increase in intrinsic toughness of Fe_3Al as temperature increases. The fracture surface appearance was similar to that of the B2 ordered FA129 fatigued in room temperature oxygen, another environment which yields high fracture toughness. A similar increase in fracture toughness is seen between DO_3 ordered material tested in air at 25°C and 150°C. However, as the temperature was increased further there was an unexpected result: the fracture toughness decreased slightly and growth rates and stage II slopes increased dramatically.

The increase in toughness at 150°C may be due in part to a decrease in the embrittling effect of hydrogen at high temperatures, as well as the increased intrinsic toughness. The mechanism of environmental embrittlement can be considered to be composed of several stages: transport of the embrittling species to the crack tip, a chemical reaction and absorption at the freshly exposed fracture surface, diffusion to the fracture zone, and H_2 - metal interactions [7]. A change in the rate of any of these processes can affect the degree of embrittlement.

At room temperature the mechanism of embrittlement appears to be decohesion of slip planes or weakening of grain boundaries, as evidenced by the cleavage fracture seen in B2 ordered FA129 and the intergranular fracture seen in DO_3 ordered FA129 fatigued in air. This suggests rapid H_2 transport along grain boundaries, or dislocation-assisted diffusion along slip planes. When tested in oxygen, both revert to a dimpled, transgranular appearance. At 150°C, hydrogen diffusion in the matrix becomes a significant mode of transport, and build-up of hydrogen at grain boundaries or slip planes no longer occurs, resulting in a more uniform hydrogen distribution and no excessively weakened regions.

The increased crack growth in higher temperatures in DO_3 ordered material may be explained by a further increase in the amount of hydrogen reaching the matrix, either by increased diffusion rates, or increased reaction rates at the newly exposed fracture surface.

TEM analysis indicates that environment also affects the dislocation density. B2 ordered material revealed a distinct increase in dislocation density as the fatigue environment changed from air to vacuum to oxygen, which corresponds to the increasing ductility measured in each of these environments. Specimens tested in vacuum and in oxygen displayed very dense dislocation networks in the cellular regions, and more paired dislocations in the recrystallized regions.

The DO_3 specimens fatigued in hydrogen gas and in air showed little increase in dislocation density over the control (undeformed) material, and dislocation densities were much lower than in B2 ordered material tested in similar environments. This decreased dislocation density is consistent with the theory of grain boundaries acting as the main transport path in DO_3 ordered FA129, while dislocation-assisted transport on slip planes occurs in the higher dislocation density B2 ordered FA129.

CONCLUSIONS

Fe_3Al,Cr undergoes an increase in crack propagation rate and a loss of fracture toughness in hydrogen and moisture-bearing environments. Oxygen proved to be a very effective inhibitor of embrittlement, due to the oxidation of aluminum at the crack tip. TEM and SEM analyses of microstructure and fracture surfaces were consistent with the change in fracture toughness with order and environment. Elevated temperature affects the degree of embrittlement in a complex manner, possibly changing the rates of several of the processes involved. Notched tensile tests indicate no notch sensitivity of this alloy.

ACKNOWLEDGEMENTS

The authors are grateful to the Department of Energy, Fossil Energy Program for providing financial support under Martin Marietta Energy Subsystems, Subcontract No. 19X-SF521C.

REFERENCES

1. C.T. Liu, H.E. Lee and C.G. McKamey, Scripta Metall. **23**, 875 (1989).
2. M. Shea, A. Castagna and N.S. Stoloff in High Temperature Ordered Intermetallic Alloys IV, edited by L.A. Johnson, D.P. Pope, and J.O. Stiegler (Mater. Res. Soc. Proc. **213**, Pittsburgh, PA, 1991) pp 609-616.
3. G.E. Fuchs and N.S. Stoloff, Acta Metall. **36**, 1311(1988).
4. A. Castagna and N.S. Stoloff, Scripta Met. **26**, 673 (1991)
5. C.G. McKamey and C.T. Liu, Scripta Metall. **24**, 2119 (1990).
6. C.G. McKamey, J.A. Horton and C.T. Liu, Scripta Metall. **22**, 1679 (1990).
7. R.P. Wei and M. Gao in Hydrogen Effects on Material Behavior, edited by N.R. Moody and A.W. Thompson (The Min., Met., and Mater. Soc., 1990) pp 789-815

PART VIII

Multiphase Materials, Composites and Joining

POWDER STRUCTURES FORMATION PROCESS BY MECHANOFUSION

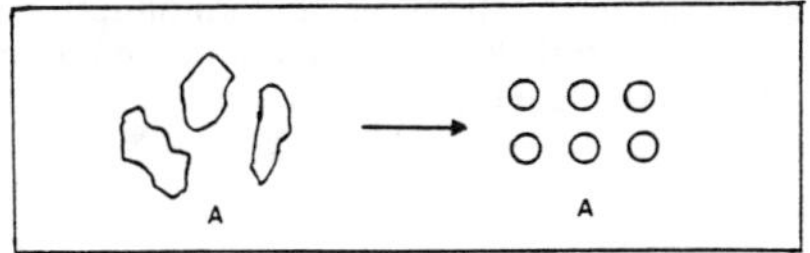

(a) Morphology Change.

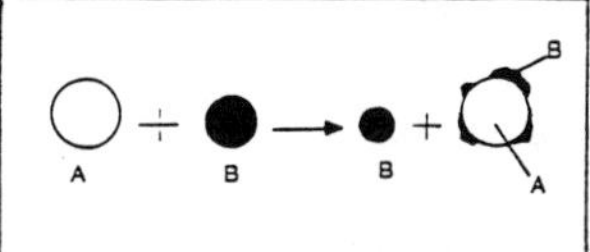

(b) Coated Structure.

(c) Lamellar Structure.

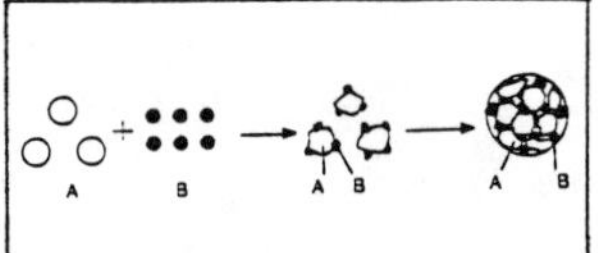

(d) Agglomerate Structure.

(e) Shell Structure.

Mechanofusion is a novel technology for processing powdered materials. During operation, the powders are intensively mixed and subjected to compression, attrition and frictional shearing. Therefore, mechanical-thermal energies are generated, resulting in a variety of distinct effects on powdered materials. (a) Particle size reduction and spheriodization from large, irregular powder by mechanofusion. The spherical powder will have better flowability which is an important factor in many processes. (b)-(e) A range of homogeneously mixed composite powders can be created by mechanofusion with no need for binders, thereby eliminating contamination. The individual particles are well-bonded, and therefore the different components in the composited particles will be in intimate contact during subsequent processes.

Figure courtesy of Z.J. Chen, H. Herman, C.C. Huang and R. Cohen.

MICROSTRUCTURES OF XD™ $MoSi_2$+SiC_p COMPOSITES

C.R. Feng and D.J. Michel
Materials Science and Technology Division, Naval Research Laboratory, Washington, DC 20375

ABSTRACT

The microstructure of $MoSi_2$ composite materials reinforced with 30vol% SiC particulate were investigated by TEM. These materials were initially produced by the XD™ process followed by either HIPing or forging.

The microstructural study revealed no significant differences between the HIPed and forged materials. In both materials, the size of the SiC particulate was found to vary from submicron to 30μm and twins and stacking faults were observed in the SiC. No orientation relationships between the SiC and $MoSi_2$ were found and dislocations emitted from the SiC/$MoSi_2$ interfaces were noticed.

The distribution of dislocations in the $MoSi_2$ matrix was non-uniform. The morphology of the dislocation structures may be categorized into three groups: 1) individual; 2) array; and 3) networks. [100]/[010] prismatic loops and (110) twins also were observed in some grains.

INTRODUCTION

Because of the high melting point (2020°C) and the outstanding high temperature oxidation resistance, molybdenum disilicide ($MoSi_2$, $C11_b$ structure) has been considered as a promising high temperature (1600°C) structural material [1]. However, for $MoSi_2$, the toughness at low temperatures (below the ductile to brittle transition temperature, ~1000°C) is poor and the strength and the creep resistance at high temperatures (above 1250°C), are modest [2].

To improve the mechanical properties, $MoSi_2$ reinforced with SiC, TiC or TiB_2 whiskers/particles have been studied by several research groups [3-8]. This paper is to report the results of a transmission electron microscopy (TEM) study of the microstructures of XD™ processed $MoSi_2$ reinforced with SiC particulate. This work is part of an extensive investigation of the microstructures, the mechanical properties and the oxidation behavior of XD™ $MoSi_2$ reinforced with various discontinuous particles, such as SiC, TiB_2, ZrB_2 and HfB_2.

MATERIALS

The materials used in this investigation were $MoSi_2$ (base alloy) and $MoSi_2$ reinforced with 30vol% SiC (XD™ alloy). The SiC was a B3 structure (β-SiC) single crystal and in the form of particulate. The XD™ alloy was initially produced by Martin Marietta Laboratories through the exothermal dispersion (XD™) process [9], followed by either hot isostatic pressing (HIPing) or forging and the base material was produced by HIPing 325 mesh $MoSi_2$ [7].

TEM foils were prepared by electrodischarge machining followed by hand grinding and dimpling. The foils were thinned to perforation by ion milling at 5kV. All foils were examined in a JEOL 200CX TEM operated at 200kV.

RESULTS AND DISCUSSION

The grain sizes of $MoSi_2$ in the base alloy and the XD^{TM} alloy were 15μm and 10μm, respectively. Although it was not observed in this study, amorphous SiO_2 was found in the base alloy [7].

The size of SiC particulates in the XD^{TM} alloy was found to vary from submicron to 30μm, Fig. 1a. Although the shape of the SiC was approximately hexagonal, no specific facet planes could be determined and these planes were usually curled. Twins, stacking faults and dislocations were observed in the SiC. No orientation relationships between SiC and $MoSi_2$ were found. Dislocations with 1/2<111> and <100]/<010] Burgers vectors emitted from the SiC/$MoSi_2$ interfaces were observed, Fig. 1b.

The distribution of dislocations in $MoSi_2$ grains was not uniform. However, the dislocation densities in the XD^{TM} alloy were higher than those in the base material. The dislocation structures were similar in either base alloy or XD^{TM} alloy. The dislocation structures may be categorized into three groups: 1) individual dislocations; 2) dislocation arrays; and 3) dislocation networks, Figs. 2a-2c. Those individual dislocations and the dislocations arrays were <100] and <010] dislocations. Two types of dislocation reactions were involved in the dislocation networks:

$$[100] + [0\bar{1}0] \rightarrow [1\bar{1}0] \qquad (1); \text{ and}$$

$$1/2[\bar{1}1\bar{1}] + 1/2[\bar{1}\bar{1}1] \rightarrow [\bar{1}00], \text{ and} \qquad (2)$$
$$1/2[111] + 1/2[\bar{1}1\bar{1}] \rightarrow [010].$$

Complex dislocation structures which involved both types of dislocation reactions were also observed, Fig. 1b. These observations are in agreement with previous findings [10].

Two additional features observed in the as-forged XD^{TM} alloy were prismatic dislocation loops and twins, Figs. 3 and 4, respectively. The Burgers vectors of the dislocation loops were [100] and [010], and trace analysis indicated that these loops were lying on (011) and (011) planes and (101) and (101) planes, respectively. Identical dislocation loops were observed in a indented single crystal $MoSi_2$ [11]. However, in this study, the source that generated these prismatic loops was unknown.

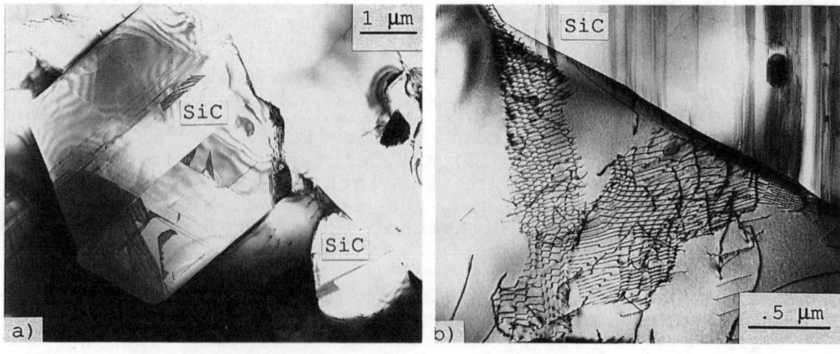

Fig. 1. a) SiC particulate; and b) dislocations emitted from the SiC/$MoSi_2$ interfaces.

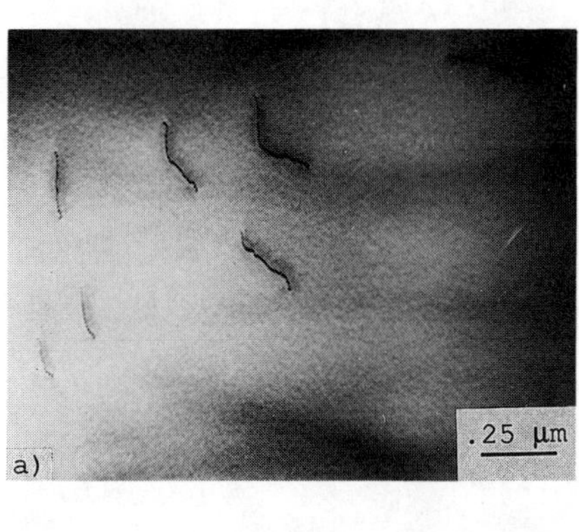

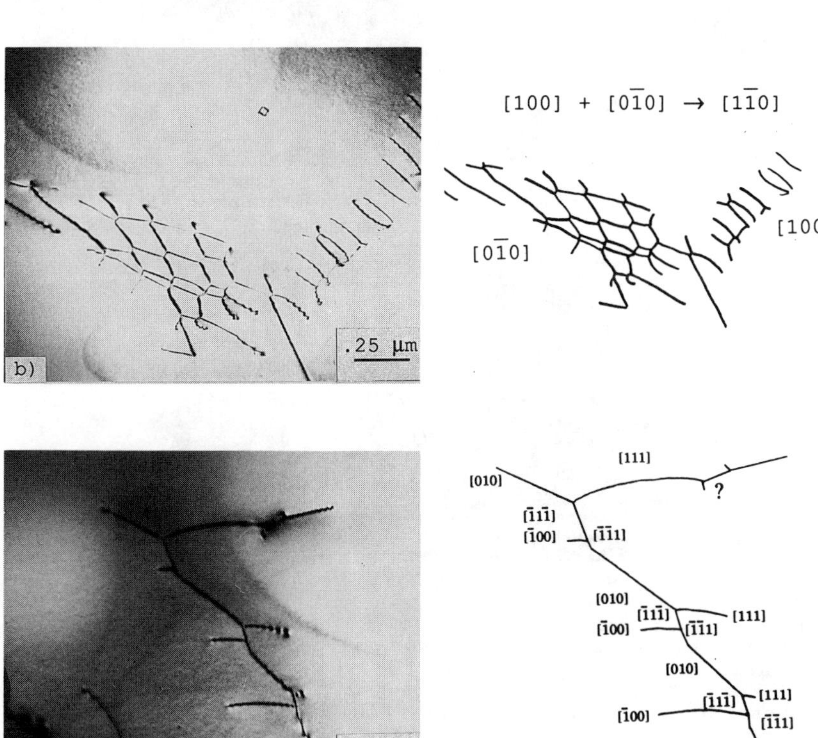

Fig. 2. Dislocations in the $MoSi_2$ grains: a) individual dislocations; b) dislocations arrays and network; and c) dislocations network .

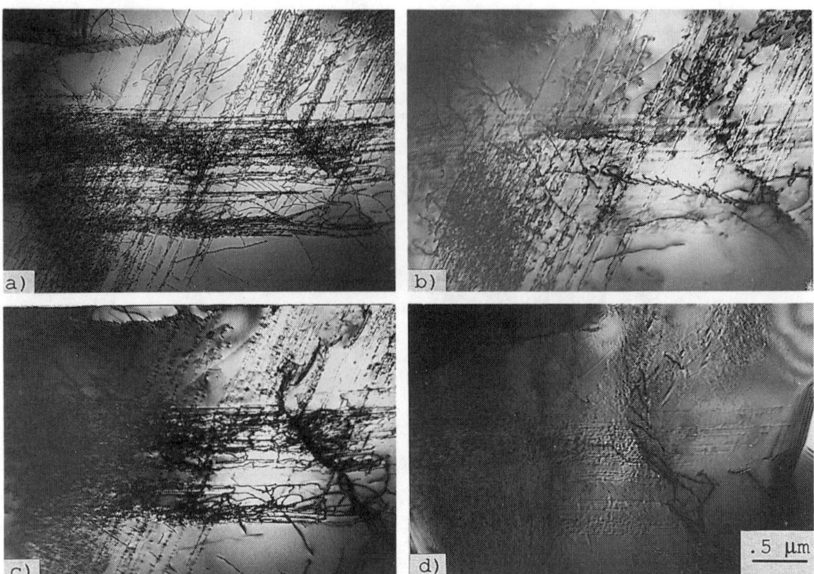

Fig. 3. Prismatic dislocation loops in the $MoSi_2$ grain: a) g[1$\bar{1}$0] and B[111]; b) g[10$\bar{1}$] and B[111]; c) g[01$\bar{1}$] and B[111]; and d) g[002] and B[120].

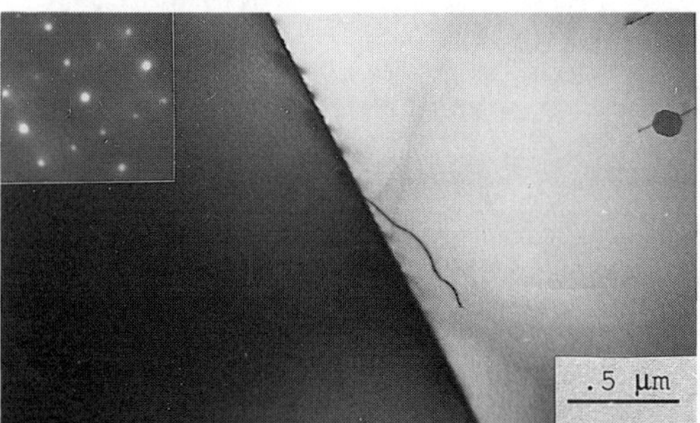

Fig. 4. Twin in the $MoSi_2$ grain. B[110]

The twin boundaries were parallel to the (110) plane which was shown to be a 60° twist boundary. The corresponding orientations of the parent and the twin regions are shown in Fig. 5, and the select area diffraction patterns (SADPs) of some of these orientations are shown in Figs. 6. The twinning elements are:

$$K_1 = (112),\ \eta_1 = [11\bar{1}];\ K_2 = (110),\ \text{and}\ \eta_2 = [001].$$

Because the magnitude of the required <111> shear to producing a twin is large, the shear is unlikely to operate as a deformation twinning mode [12]. However, <111> dislocations were observed at the twin boundary, Fig. 7.

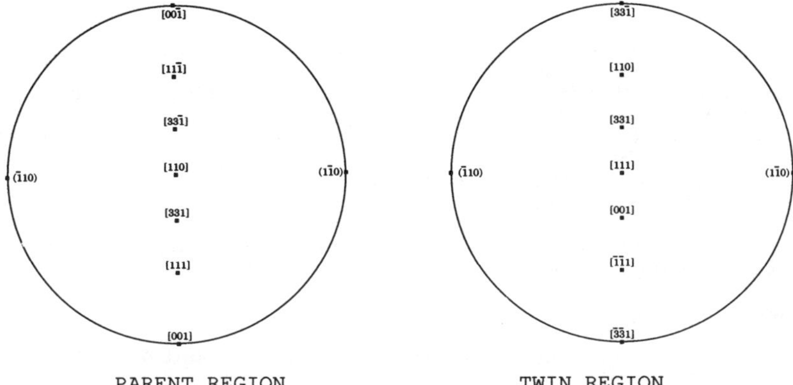

Fig. 5. The corresponding orientations of the parent and the twin regions.

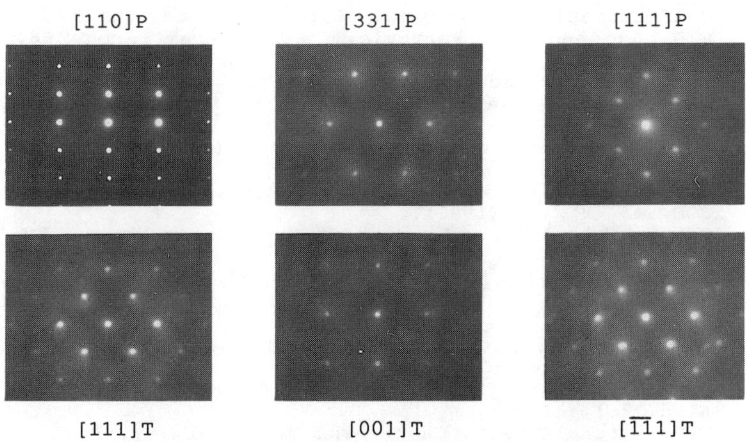

Fig. 6. SADPs of parent and twin regions.

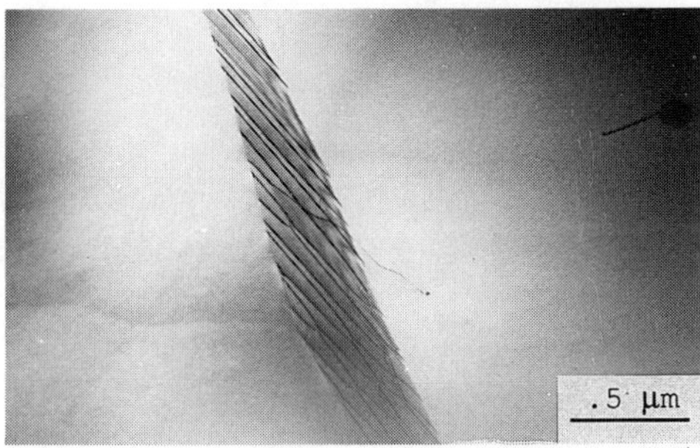

Fig 7. <111> dislocations at twin boundary. $g[2\bar{1}\bar{1}]$ and B[120].

SUMMARY

The microstructures of HIPed or forged XD™ $MoSi_2$+30vol%SiCp were investigated by TEM.

1. Individual dislocations, dislocations arrays and dislocations network were observed in $MoSi_2$ grains.
2. <100], <010] and <111> dislocations were emitted from the $MoSi_2$/SiC interfaces.
3. Two types of dislocations reactions were involved in the dislocations networks; these are <100] + <010] → <110] and 1/2<111> + 1/2 <111> → <100]/<010]
4. Prismatic dislocation loops and twins were observed in the forged material on {101} and {011} planes.

ACKNOWLEDGMENT

This work was supported in part by the Defense Advanced Research Projects Agency and in part by the Office of Naval Research at the Naval Research Laboratory.

REFERENCES

1. J. Schlichting, High Temp.- High Pressures, **10**, 241 (1978).
2. F.D. Gac and J.J. Petrovic, J. Am. Ceram. Soc., **68**, C-200 (1985).
3. P.J. Meschter and D.S. Schwartz, JOM, **41** (11), 52, (1989).
4. J.-M. Yang, W. Kai and S.M. Jeng, Scripta Metall., 23, 1953 (1989).
5. C.B. Lim, T. Yano and T. Iseki, J. Mater. Sci., **24**, 4144 (1989).
6. E.W. Lee, J. Cook, A Khan, R. Mahapatra and J. Waldman, JOM, **43** (3), 54 (1991).
7. R.M. Aikin, Jr., J. Ceram. Eng. Sci. Proc., **12**, 1643 (1991).
8. R.M. Aikin, Jr., Mat. Sci. Eng., **A155**, 121 (1992).
9. E.W. Lee, J. Cook, A. Khan, R. Mahapatra and J. Waldman, JOM, **42**, No. 3, 54 (1991).
10. O. Unal, J.J. Petrovic, D.H. Carter and T.E. Mitchell, J. Am. Ceram. Soc., **73**, 1752 (1990).
11. P.H. Boldt, J.D. Embury and G.C. Weatherly, Mat. Sci. Eng., **A155**, 251, (1992).
12. T.E. Mitchell, R.G. Castro and M.M. Chadwick, Phil. Mag. A, **65**, 1339, (1992).

DEBOND CRACK PROPAGATION AND BRANCHING UNDER VARYING SHEAR STRESSES

C. M. KENNEFICK, R. M. DICKERSON AND P. O. DICKERSON
NASA Lewis Research Center, 21000 Brookpark Road, Cleveland, OH 44135

ABSTRACT

Silicon carbide fibers in a Ti24Al+11Nb matrix were debonded under varying combinations of shear and compression in off axis loading in a pushout test. Evidence of debond crack branching was recorded for etched specimens using scanning electron microscopy and the interface microstructure was examined by transmission electron microscopy. The regions of high shear stress and multiple weak interfaces promoting crack branching were analyzed in terms of an energy balance for brittle fracture.

INTRODUCTION

The deflection of cracks at fiber-matrix interfaces in both intermetallic and ceramic matrix composites with subsequent fiber pullout has been shown to be an important fracture toughening mechanism. Interfacial chemical reaction layers from composite fabrication and also the use of fiber coatings to tailor the bond strength or morphology at the interfaces opens the possibility of not one but multiple debonding cracks. Such multiple cracking may increase crack deflection, but it may at the same time increase the number of asperities at a fiber-matrix interface, thereby increasing the stress required for frictional pullout. This paper examines experimental evidence for crack branching in an intermetallic matrix composite and analyzes the energy release rate for crack branching for a main crack under a combination of compressive and shear loading.

Theories on the cause of crack branching have focussed on critical crack tip velocities and stress intensity factors, and on the interaction of elastic waves with running cracks. Studies have so far been mainly on homogenous materials and have included a review of fast fracture [1], the interaction of stress waves with cracks [2, 3], and analysis and experimental results of crack branching in polycarbonate [4], in glass [5, 6], and at inclusions inside a matrix [7].

In this study, crack branching during debonding along a fiber-matrix interface in a pushout test is examined. The energy expended in overcoming friction both from residual compressive stresses on the fibers and from dragging along fiber coating fragments during debonding are compared with the overall strain energy release to clarify which frictional mechanism may retard branching.

EXPERIMENTAL

A three-ply composite of silicon carbide SSC-6 fibers in a Ti24Al+11Nb matrix, made by a wire arc spray process [8], was successively sliced to produce pushout specimens whose fibers made an orientation angle Θ from 0 to 40 degrees with respect to the loading axis in a pushout test (Figure 1). The fibers comprised 44 volume percent of the composite, had a diameter of 142 μm, and had an average interedge spacing of 30.3 $\pm$ 0.5 μm within each row. Each specimen was 401 $\pm$ 3 to 468 $\pm$ 8μm thick and the groove in the sample holder in Figure 1 was an average of 366 $\pm$ 6 μm wide.

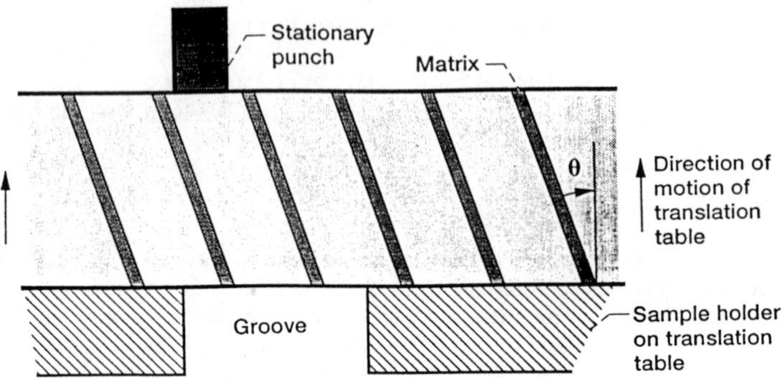

1. Off axis loading of the fibers in a thin specimen pushout test.

Details of the pushout apparatus [9] and the testing procedure [10] are described elsewhere. What is important is that for the specimen and groove dimensions and debonding loads used, the tensile bending stresses on the lower sample edge are negligible, particularly for orientation angles near 30 degrees. The debonding crack therefore initiated mode II fracture at the fiber-matrix boundary and was also under a compressive stress in the direction of the fiber radius arising both from composite fabrication [8] and from the applied load. The effect of the fiber orientation was to place a part of the fiber-matrix interface more directly under the load as the orientation angle θ increased [10], which resulted in an increasing maximum interfacial shear stress as θ increased. The oriented fibers were thus debonded under varying combinations of shear and compression.

Evidence of the fiber debonding crack path and fiber damage was examined by etching away the matrix of the pushout specimens with a cold bath of 10 percent HF, 5 percent HNO_3, and 85 percent water. For the specimens with fiber orientation angles of 4 and 30 degrees that were used for examination of crack branching, at least ten pushed out fibers on each specimen were available for study. Scanning electron microscopy was then used to record fiber morphology of both debonded and undebonded fibers. The structure of the fiber-matrix interface that arose from the wire arc spray process to produce the composite originally was examined by transmission electron microscopy.

RESULTS AND DISCUSSION

Instead of one debonding crack, two or three cracks propagated down the interface in regions of high shear stress. For an orientation angle of 30 degrees, the region of maximum shear stress was on the lower side of the debonded fiber in Figure 2 [10]. The spalling seen in Figure 2 is that of the two carbon rich layers on the outside of the SiC SCS-6 fiber. Undebonded fibers far away from tested fibers in specimens with fiber tilts of both 4 and 30 degrees showed no or very little damage. Damage adjacent to the region of maximum shear stress gave evidence that crack interaction with elastic stress waves from the main debonding crack causes cracks to propagate in adjacent fibers. For fibers with an orientation angle of 4 degrees, surface damage was often at the top of the debonded fiber and was uniformly distributed around its circumference, corresponding again to the region of maximum shear stress [10].

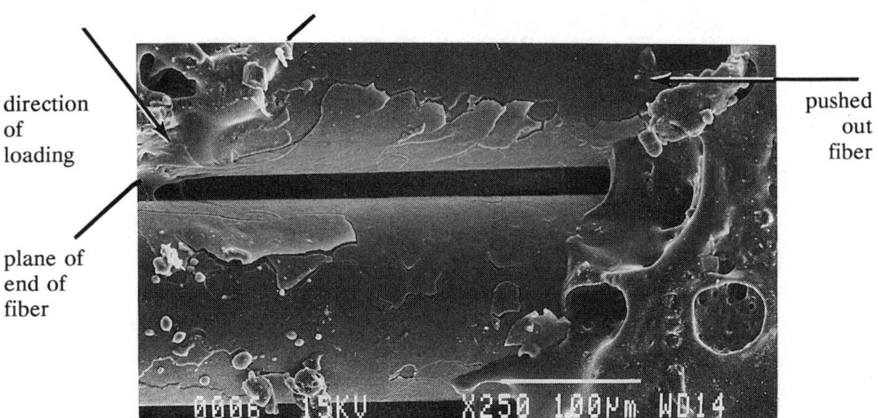

2. SEM micrograph of etched fibers with an orientation angle of 30 degrees showing damage both on the side of maximum shear stress on the pushed out fiber and on the fiber adjacent to this side.

The ease with which multiple debonding occurs may be partially attributed to a weak interfacial bond at the titanium carbide-carbon interface (Figure 3). Scanning electron micrographs have shown that the main debonding crack and subsequent sliding occurs at the outer carbon layer-titanium carbide boundary. For this composite batch made by the wire arc spray process, this same weak interface also appears over large areas between the two carbon layers (Figure 3). The microstructure and composition of the reaction layers is similar to that found for the same composite made by a powder cloth technique [11], except two thin carbon rich layers remain intact and TiC forms between the two layers.

The kinetic energy KE provided by the elastic waves emitted from the main debonding crack that may be used to propagate additional cracks may be calculated from a difference between the total strain energy in the system and the energy needed to open up a crack of length C [13]. Neglecting the small bending stresses and assuming the geometrical area providing strain energy from a crack of length C is divided equally between fiber and matrix material, the total energy U of the system may be written [13-15]

$$U = \frac{-\sigma_a A_f}{2E_f} - \frac{\sigma_a^2 A_m}{2E_m} - \frac{1}{2}\pi C^2 \tau^2 \left[\frac{1}{E_m} + \frac{1}{E_f}\right] + 2\gamma C + KE + U_f \qquad \textbf{(1)}$$

In the above expression σ_a is the average applied compressive stress in the pushout test, τ the average interfacial shear stress, A_f, A_m, E_f, and E_m the area under the punch of the pushout test and Young's modulus of the fiber and matrix material respectively, γ the energy needed to expose a unit area of crack surface, and U_f the work done by the shear stress in overcoming friction to propagate the crack.

The resistance to propagating the crack in shear mode can come from two sources. One is the frictional resistance due to the radial compressive stress at the interface from both composite fabrication and the applied contact load. Since the matrix in this case has a higher thermal expansion coefficient than the fiber, a radial compressive stress σ_p develops at the interface upon cooling from the consolidation temperature. If G_f is the shear modulus of the fiber, μ the interfacial frictional coefficient, and δ the approximate height of crack opening, then the work done per unit crack length by the shear stress in overcoming the friction is given by $\mu(\tau \tan \Theta + \sigma_p)(\tau/G_f)\delta C$, where τ is the average interfacial

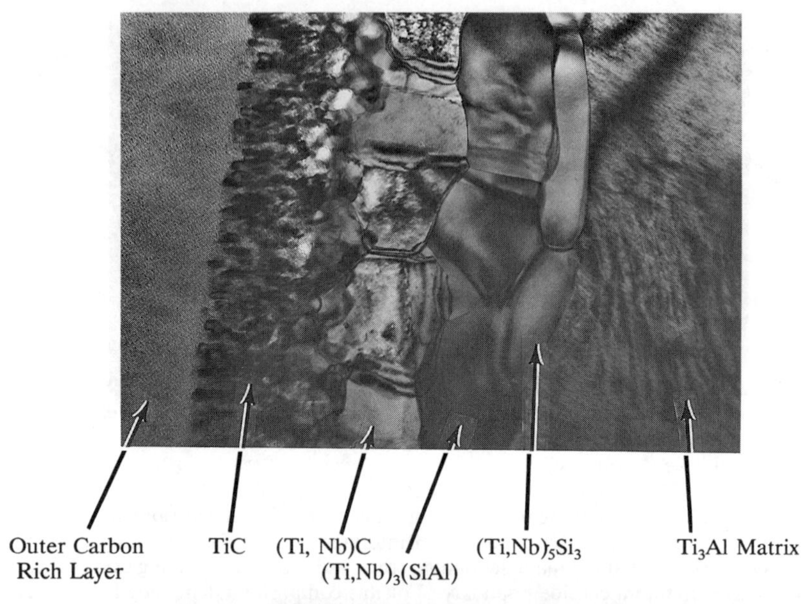

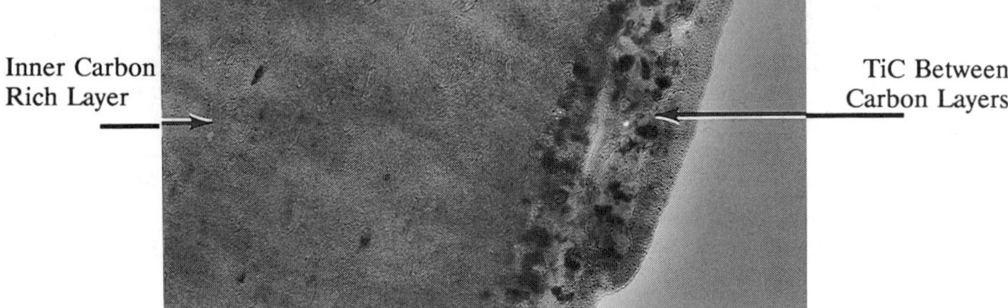

3. TEM micrographs showing the reaction layers between the outer carbon rich layer and the matrix *(upper)* and the TiC layer found in most areas between the two carbon coatings *(lower)*.

debonding shear stress and $\tau \tan \Theta$ is the average normal compressive component of applied stress at the interface.

The other additional source of work done by the shear stress is from friction due to asperities. Fragmentation of the fiber coating can occur from composite fabrication itself, or from the intersection of the debonding crack with radial cracks (Figure 4). If w is the width of a gap between two coating fragments (Figure 4), then the additional work done by the shear stress per unit crack length is τ wC.

If the total energy U is considered constant during crack propagation and the boundary condition that KE is zero for the originally stationary crack of length C_o is used, and if the relation [15]

$$2\gamma = \pi C_o \tau^2 \left[\frac{1}{E_m} + \frac{1}{E_f}\right] \tag{2}$$

holds for the stationary crack, then U may be solved for in equation (1) [15] and the kinetic energy per unit crack length KE written as

$$KE = \tau^2 \left\{\left[\frac{1}{E_m} + \frac{1}{E_f}\right]\left[\frac{\pi C^2}{2} + C_o^2\left(\frac{\pi}{2}-1\right) - CC_o\right] - \frac{\mu C\delta}{G_f}\left[\tan\Theta + \frac{\sigma_p}{\tau}\right] - \frac{Cw}{\tau}\right\} \tag{3}$$

The first six terms in the outer brackets in the equation above represent the strain energy available from a crack of length C and show that the original strain energy in the sample due to the fixed ends of the pushout specimen in equation (1) does not affect the kinetic energy emitted. The last three negative terms represent the energy expended to slow crack branching.

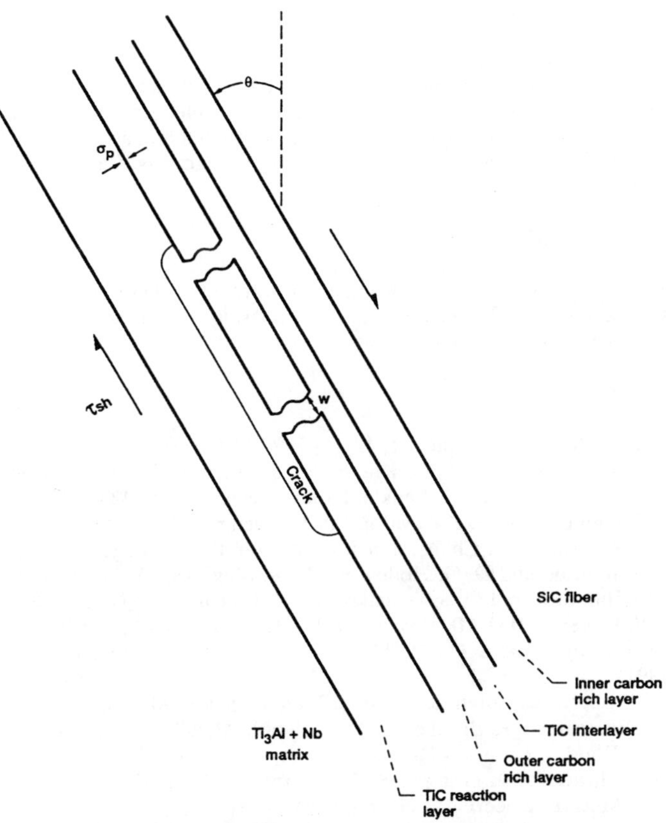

4. Schematic of the debonding and radial cracks at the interface.

The energy expended to overcome Coulombic friction in equation (3) is low compared with the available strain energy for crack propagation or the energy required to drag along a coating fragment. An original crack length C_o of 10 μm, a crack length C of 20μm, a debonding shear stress τ of 75 MPa, and E_m and E_f 110 GPa and 400 GPa respectively as in the SiC - Ti24Al+24Nb composite gives a strain energy term of 3.2×10^{-5} J/m. The term for Coulombic friction, $\tau^2(\mu C\delta/G_f)[\tan \Theta + \sigma_p/\tau]$, in contrast, is 3.6×10^{-7} J/m for a friction coefficient μ near 0.2, a crack opening δ of 1 μm, a fiber shear modulus G_f of 160 GPa, an orientation angle Θ of 30 degrees, and a residual compressive stress of 150 MPa. The Coulombic friction term may increase if μ is a bit higher, but the increase can be offset by a decrease in tan Θ for lower orientation angles. These differences in μ and Θ, however, will not increase the term by a factor of 100 to bring it near the strain energy term. The work done to drag along a coating fragment, $Cw\tau$, is 1.5×10^{-3} J/m and is large enough to overcome the strain energy term, thus using up energy available for crack branching. The energy balance resulting in equation 3 is in agreement with the experimental observation that crack branching ceases in areas of the fiber surface having low applied shear stresses and having radial cracks resulting in coating fragments in the wake of the debonding crack.

CONCLUSIONS

Crack branching can occur on a debonding fiber and adjacent fibers under mode II conditions. A high local shear stress intensity factor coupled with multiple weak interfaces promotes the multiple cracking in the presence of residual and applied radial compressive stresses on the fibers. Frictional drag from coating fragments is a possible cause for the cessation of the branching.

ACKNOWLEDGEMENTS

One of the authors (C. M. Kennefick) gratefully acknowledges the support of a National Research Council Postdoctoral Associateship for this study. The typing of the manuscript by M. Wolf is also appreciated.

REFERENCES

1. L. R. F. Rose, Int. J. Fracture, **12**, [6] 799-813 (1976).
2. K. Ravi-Chandar and W. G. Krauss, Int. J. Fracture, **26**, 189-200 (1984).
3. H. P. Rossmanith and A. Shukla, Exper. Mech., **21**, 415-422 (1981).
4. M. Ramulu and A. S. Kobayashi, Int. J. Fracture, **27**, 187-201 (1985).
5. S. R. Anthony, J. P. Chubb, and J. Congleton, Philos. Mag., **22**, 1201-1216 (1970).
6. J. W. Johnson and D. G. Holloway, Philos. Mag., **19**, 731-743 (1966).
7. P. S. Thoecaris and C. B. Demakos, Int. J. Fracture, **32**, 71-92 (1986).
8. J. W. Pickens, HITEMP Review 1991, **NASA CP-100082**, 40-1 to 40-18 (1991).
9. J. I. Eldridge, **NASA TM 105341**, NASA Lewis Research Center, Cleveland, Ohio (1991).
10. C. M. Kennefick, submitted to Acta Metallurgica et Materialia, September 1992.
11. S. F. Baumann, P.K. Brindley, and S. D. Smith, Metall. Trans., **21A**, 1559-1569 (1990).
12. C. G. Rhodes, Materials Research Society Symposium Proceedings, **273**, D. B. Miracle, J. A. Graves, and D. L. Anton, eds., 1992.
13. N. F. Mott, Engineering, **165**, 16 (1948).
14. J. P. Berry, J. Mech. Phys. Solids, **8**, 194-206 and 207-216 (1960).
15. B. R. Lawn and J. R. Wilshaw, Fracture of Brittle Solids, Cambridge University Press, 1975.

GAMMA-TITANIUM ALUMINIDE REINFORCED WITH Al_2O_3 AND TiB_2 FIBERS*

DARRELL DIXON** AND JOSEPH W. NEWKIRK†
**The Bionetics Corporation, Arnold Air Force Base, TN 37389-3400
†University of Missouri-Rolla, Metallurgical Engineering Department, Rolla, MO 65401

ABSTRACT

A two-phase Ti-47Al-2Ta composite reinforced with a titanium diboride (TiB_2) and two alumina fibers was processed by hot isotatic pressing and heat treated to simulate service conditions.Diffusion couples were heat treated between 593°C and 982°C for up to 500 h and analyzed using a scanning electron microscope and energy dispersive x-ray spectroscopy.

The fibers used were 3M® alumina, Sapphikon® sapphire, and Textron Specialty Materials TiB_2. Observations reveal reaction layers, cracking fibers due to thermal stress, and chemical interactions between ternary addition of Ta with the fibers. Alumina is thermodynamically stable in TiAl, and its coefficient of thermal expansion closely matches TiAl. Observations revealing alumina as a good potential reinforcement in TiAl are discussed. Thermodynamic information supports TiB_2 stability in two-phase TiAl. A three-fold increase of Ta at the fiber/matrix interface of TiB_2 is evidence of a lack of homogeneity due to Ta mobility. This indicates a possible loss of oxidation resistance provided by the Ta addition.

INTRODUCTION

Gamma-titanium aluminide is one of several materials possessing a combination of high-temperature strength and low weight. Because of these properties, it has excellent potential as a high-temperature material to replace current aerospace materials; however, it has low room-temperature ductility which must be managed by engineering solutions.

Two solutions to reduce the low-ductility problem include ternary alloy additions and composite reinforcements. Ternary additions include V, Cr, or Mn.[1] Increased ductility or other properties, such as oxidation resistance, dependent upon well-distributed alloying additions may be lost if the microstructure or alloying element distribution does not remain homogeneous. In this investigation, tantalum was added primarily for oxidation resistance (personal communication with C. Sabinash, McDonnell-Douglas Corporation, 1991). Besides chemical means to improve ductility, composite fiber reinforcements have been used to toughen materials, but interfacial integrity is important. Fiber dissolution, brittle reaction layers, and interfacial void formation are among the many causes of poor composite performance.

This paper will examine the performance and reaction layers of three fibers and the interaction of tantalum with three fibers. These fibers are 3M® alumina, Sapphikon® sapphire, and Textron Specialty Materials titanium diboride (TiB_2) fibers, impregnated into a two-phase Ti-47Al-2Ta (atomic percent) matrix. Diffusion couples were heat treated and analyzed by scanning electron microscope (SEM) and energy dispersive x-ray spectroscopy (EDS) techniques.

*Work and analysis for this research were done by personnel of the University of Missouri-Rolla, Metallurgical Engineering Department. Darrell Dixon was then employed by Bionetics Corporation, mission support subcontractor at the Arnold Engineering Development Center (AEDC), Air Force Materiel Command (AFMC). Further reproduction is authorized to satisfy needs of the U. S. Government.

EXPERIMENTAL PROCEDURE

The diffusion couples heat treated and investigated are detailed in Table I. Some of the systems were not evaluated under all heat treatment conditions, due to the limited availability of material and uncertainty in the fiber lengths and position in the samples. The alumina fibers included a large diameter fiber of 250 µm (Sapphikon sapphire) and a small diameter fiber of 10 to 15 µm (3M alumina). Textron Specialty Materials manufactured the multilayered 80-µm-diam titanium diboride fiber. The diffusion couples were made using a 6.35-mm-diam Ti-3Al-2.5V (weight percent) tube, and densification was achieved by hot isostatic pressing (HIP). Following the HIP operation, the specimens were sectioned, wrapped in tantalum foil, and encapsulated in silica tubes with the assistance of the McDonnell-Douglas Corporation. Nine sets of the specimens were heat treated at 593, 815, and 982°C for 100, 250 and 500 h. Finally, the specimens were mounted and metallographically prepared for SEM analysis in similar fashion to samples in a previous investigation by Newkirk and Dixon.[2] The compositional information was obtained by conducting line scans across the composite interface. Line scans were obtained in a manual point-by-point method. The line scan began 15 µm out in the matrix and ended where the fiber was identified to have its original composition. Two lines scans were obtained for each sample and averaged. The error bars were computed using the square root of the sum of three squared numbers. These numbers are the 2-σ errors reported by the semi-quantitative analysis for the two compositions of each point and the standard deviation of the average of those compositions. The analysis was performed on a JEOL T330A scanning electron microscope with a Kevex Super Quantum Energy Dispersive X-ray (EDS) system.

Table I. Diffusion Couples Prepared for Investigation

Fiber Name	Temp., °C	100 hr	250 hr	500 hr
3M alumina	593	n/a	Y	Y
	815	Y	Y	Y
	982	N	N	Y
Sapphikon	593	n/a	Y	Y
	815	Y	Y	Y
	982	Y	Y	Y
TiB_2	593	n/a	N	N
	815	Y	Y	Y
	982	Y	Y	Y

*All three fiber have an un-heat treated as-HIPped sample
Y: A good sample was available for investigation
n/a: No sample was prepared for the investigation
N: No fibers were present in the heat treated sample

RESULTS

The TiB_2 and two alumina fibers investigated were chosen based on the results of a previous study.[3] In this study, reaction products, interfacial separation, cracked fibers, and other physical aberrations were noted. Minor oxidation of the heat treated samples was noted, but it is believed the results were not biased by the oxidation. Untreated and heat treated results are reported separately.

Untreated Diffusion Couples

The Sapphikon fiber is a single crystal sapphire about 250 µm in diameter as seen in Figure 1. The interface between the fiber and matrix has a noticeable roughness. These concave indentations are not present in the original fiber surface and are believed to be caused by the HIP operation. No cracking or fiber/matrix separation was observed due to thermal or mechanical stresses of the HIP operation.

The 3M alumina fiber is a polycrystalline fiber about 10 to 15 µm in diameter with some fibers being oval with a long diameter of up to 25 µm as seen in Figure 2. These fiber are very flexible and are present in bundles. No cracks or interfacial voids were present due to the HIP operation.

The TiB_2 fiber is a multiple-layer fiber approximately 80 µm in diameter. Observations using the SEM found more than 90 percent of the fibers cracked due to thermal stress as seen in Figure 3. A characteristic crack is a radial crack from the center to the edge of the fiber, up to 2 µm wide at the edge. The as-HIPped compositional profiles for the titanium diboride fiber reveal a sulfur-rich core with 20 percent aluminum being present. Outer layers are 100 percent titanium except for the fiber/matrix interface containing minor levels of yttrium and sulfur. The as-HIPped sample has a yttrium peak at the fiber/matrix interface which is not detected anywhere except at the interface, and it is believed the yttrium is present from the manufacturing process.

Heat Treated Diffusion Couples

After 500 hr at 982°C, the sapphire fiber is dimensionally stable as it is at all other heat treatment conditions. There are no observable reaction layers or interfacial voids, and there appears to be good adhesion between the fiber and matrix. Compositional profiles for any time and temperature are nearly identical. An expected change in composition is seen at the fiber/matrix interface without any evidence of other phases forming between the fiber and matrix. An unexpected presence of tantalum was detected in the outer regions of 250 hr at 593°C and 250 h at 982°C samples. The tantalum level varies from 0 to 3 percent up to 4 µm into the fiber.

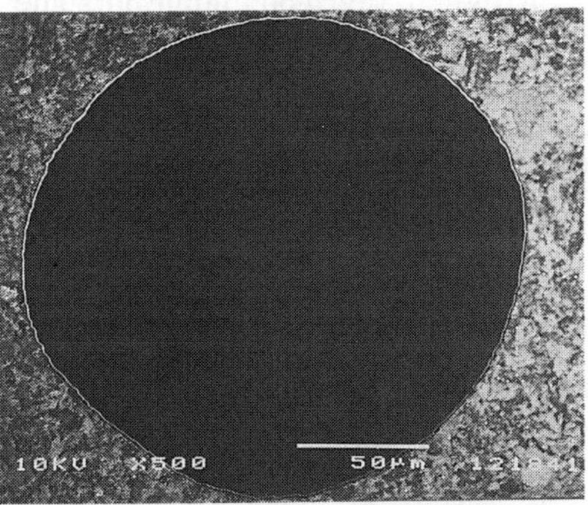

Fig. 1. No reaction zones or interfacial voids present in a Sapphikon fiber after heat treatment at 982°C for 500 h.

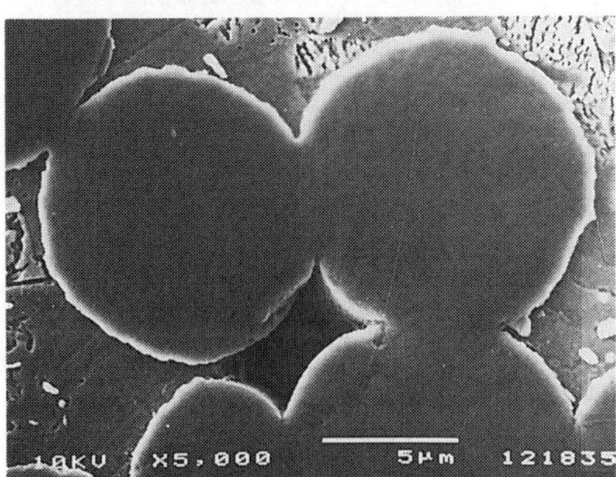

Fig. 2. Bundle of 3M alumina fibers after heat treatment at 815°C for 500 h.

Fig. 3. A TiB_2 fiber after exposure at 982°C for 500 h.

The 3M alumina fiber is dimensionally and compositionally stable up to 815°C for times up to 500 h; however, results at 982°C are inconclusive. Figure 4 shows partly dissolved fibers. These fibers in the 500-h 982°C sample are located between 50 and 150 μm from the Ti-3Al-2.5V (weight percent) canister. Compositional results reveal the regions adjacent to the partly dissolved fibers are 30 percent higher in aluminum than the canister. As-HIPped fibers compared to the dissolved fibers show 50 to 100 percent of an individual fiber dissolving leaving a void. Also, reduced presence of a Widsmanstatten structure around the fiber is observed in relation to other unreacted fibers.

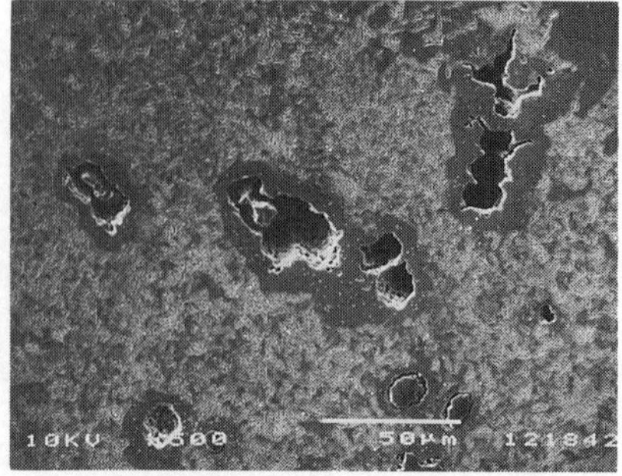

Fig. 4. Dissolved 3M alumina fibers after exposure to 982°C for 500 h.

Tantalum was detected in the 3M alumina samples at levels up to 2 percent at 4 μm into the fiber. The as-HIPped and 250 h at 815°C samples had tantalum in the exterior regions of the fibers.

Although thermodynamic data support TiB_2 being stable in two-phase TiAl, a discontinuous reaction zone around the fiber is present in the 982°C samples.[3] The presence of a reaction zone depends upon several variables such as exact composition and structure of the TiB_2, and a characterization study is required to determine these properties. The reaction zone grows up to 10 μm in thickness. In addition, boride needles 1 to 2 μm in length have been observed at 815°C after 100 h. These needles grow up to 8 μm in other samples with longer exposure times at higher temperatures.

Prior to heat treatment, the fiber/matrix interface has 2 to 10 percent sulfur. After heat treatment, sulfur is present with the reaction zone containing 20 percent sulfur and 6 percent tantalum. Yttrium is always present in modest amounts at the interface as seen in Figure 5. The presence of yttrium at the fiber/matrix interface is believed to be from the production process. No interactions are observed to involve the yttrium. Tantalum is found in the reaction zone, but is not found in interfacial regions free of a reaction product.

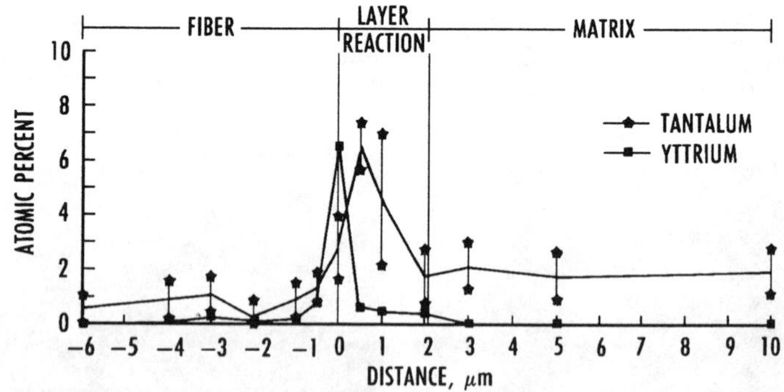

Fig. 5. TiB_2 fiber exposed at 982°C for 500 h showing a buildup of Ta in the reaction layer.

DISCUSSION

Several important issues were raised during the investigation. These include the diffusion of tantalum into the alumina fibers, the stability of alumina in two-phase TiAl, the buildup of tantalum at the TiB_2 fiber/matrix interface, and boride needles in the reaction zone of the TiB_2 fiber.

The Sapphikon fiber is stable at all temperatures for all times. However, minor amounts of tantalum are present in the outer 4 μm of the fiber. The poly-crystalline 3M alumina has tantalum present in the as-HIPped specimen in contrast to the sapphire where tantalum is not present, except in heat treated samples. Grain boundaries or other surface discontinuities are possible diffusion paths for tantalum into 3M alumina. At this time, it is unknown if tantalum is detrimental to the 3M fiber.

The location of alumina-based fibers in the composite is important. When the diffusion couples were made, the fibers lay against the side of the Ti-3Al-2.5V canister which is a titanium-rich region compared to the gamma-TiAl matrix. After heat treatment at 982°C, the fibers dissolved. Work by Misra suggests that alumina fibers may not be stable in the presence of a titanium-rich matrix because of the ability of titanium to dissolve large amounts of oxygen.[4] Misra concluded that alumina is stable with gamma-titanium aluminide alloys, but the aluminum concentration in the region where the fibers are located must be greater than 49 percent aluminum.

The buildup of tantalum was reported earlier in a paper by Newkirk and Dixon. A six-fold increase in the tantalum concentration at the TiB_2 coating/ matrix interface was reported. Evidence of tantalum mobility is observed again. According to Saqib, et al., TiB_2 is very stable in two-phase Ti-43Al.[3] After 1,000 h at 1,200°C, only minor dimensional changes were observed. Compositional investigations by EDX and electron energy loss spectroscopy reveal distinct compositional changes at interfaces and no detectable diffusion of boron across interfaces. Also, thermodynamic calculations for various reactions among titanium, aluminum, boron, and their compounds support the stability of TiB_2 in titanium aluminide. However, the presence of tantalum and two-phase TiAl compositional difference alters the thermodynamics of the system.

The presence of tantalum at the center of the TiB_2 fiber is further evidence of tantalum mobility as seen in Figure 6. In other specimens, a compound of sulfur and tantalum forms in the 100-h, 815°C specimen where 8-percent Ta is detected. The fiber has a porous appearance which may explain the diffusion of tantalum through the fiber; however, no tantalum was detected in the titanium-rich regions of the fiber between the center and fiber exterior.

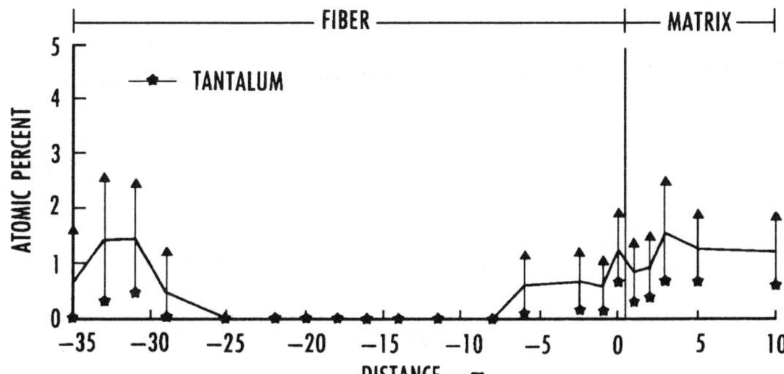

Fig. 6. TiB_2 fiber exposed at 815°C for 100 h showing Ta in the center of the fiber.

SUMMARY

Compositional profiles were obtained for diffusion couples of gamma-titanium aluminide reinforced with sapphire, alpha-alumina, and titanium diboride fibers. The chemical compatibility of the fibers and the matrix was observed using scanning electron microscopy and energy dispersive X-ray techniques. As-HIPped samples and samples heat treated at 593, 815, and 982°C for up to 500 h were observed. The results are summarized as follows:

1. The Sapphikon sapphire fiber is very stable at all temperatures for all times.
2. The 3M alumina fiber is stable up to 815°C for 500 h, but at 982°C the results are inconclusive.
3. Tantalum is observed to diffuse short distances into alumina fibers. The consequence of this is not known at this time.
4. Evidence of tantalum mobility is observed in the TiB_2 fiber.
5. Boride needles 8 μm in length are present in the reaction zone of the TiB_2 fiber.
6. Tantalum is present in the reaction zone of the TiB_2 fiber but is not present at interfacial regions free of a reaction product.

ACKNOWLEDGEMENTS

The authors acknowledge McDonnell-Douglas Corporation for their financial support. We are grateful to Catherine Sabinash of McDonnell-Douglas Corporation for her involvement in providing the material and sample encapsulation assistance. Additional support was received from the Missouri Research Assistance Act.

REFERENCES

1. Kim, Y.-W., *Journal of Metals*, pp. 24-29, (July 1989).
2. Newkirk, J. and Dixon, D. J., *Material Science and Engineering*, A153, pp. 662-667 (1992).
3. Saqib, M., Weiss, I., Mehrotra, G. M., Clevenger, E., Jackson, A. J., and Lipsitt, H. A., *Metallurgical Transactions A*, Vol. 22A, pp. 1721-1728 (August, 1991).
4. Misra, A. K.,*Metallurgical Transactions A*, Vol. 22A, p. 715 (March, 1991).

FRACTURE DURING HIGH TEMPERATURE DEFORMATION OF A Ti-44Al-3V-7.5v/o TiB_2 XD™ COMPOSITE

D. ZHAO, J. J. VALENCIA, K. G. ANAND, AND S. J. WOLFF
Concurrent Technologies Corporation, 1450 Scalp Avenue, Johnstown, PA 15904, U.S.A.

ABSTRACT

High temperature compression workability tests have been performed on a Ti-44a%Al-3a%V-7.5v%TiB_2 XD™ composite over a range of temperatures (1273 to 1473 K), strain rates (10^{-3} to 10 s^{-1}), and interrupted strains. Three types of specimen configurations were used in the tests. Fracture caused by secondary tensile stress at the surface of the specimens was investigated in terms of crack initiation and propagation. The fractured specimens were analyzed using both optical and SEM microscopy. Cracks started at the equator of the specimens and propagated into the specimens through their longitudinal axis. The fracture mode varied with temperature and strain rate. At the lowest temperature (1273 K) and the lower strain rate (0.1 s^{-1}), the fracture was transgranular with a crack arrester type. At the lowest and intermediate temperatures (1273 and 1373 K) and the highest strain rate (10 s^{-1}), intergranular cracks were observed. At the highest temperature (1473 K) and the highest strain rate (10 s^{-1}), transgranular failures with delamination and crack divider types were observed.

INTRODUCTION

Titanium aluminide matrix composites have the potential for aerospace high temperature structural applications because of their high modulus and high strength to weight ratio [1-3]. However, a major drawback of the composites is their low ductility and fracture toughness at low temperature. The fracture behavior of these materials have attracted some attention [4-6]. Larsen [5] conducted tensile tests on a Ti-47.5Al-2Nb-2Mn XD composite with 1v% TiB_2 reinforcement particles. TiB_2 particles were not observed on the fracture surfaces, and the fracture was mainly translamellar. Pao and coworkers [6] studied Ti-46Al and Ti-47Al, both with 7.5v% of TiB_2 particle reinforcement. They concluded that the presence of TiB_2 particles decreased the composite fracture toughness because of the brittleness of the particles and the ease of crack initiation at the particles. Fatigue cracks tend to follow weak twin-matrix interfaces, resulting in step-like features on the fracture surfaces.

Although the fracture behavior of XD composites has not been studied extensively. Sufficient information on the tensile fracture of (α_2+γ) titanium aluminides [7-12] is available to make reasonable correlations between the two materials. At room temperature, materials with an equiaxed γ microstructure tend to fracture by grain boundary decohesion and cleavage, while materials with a lamellar microstructure fracture by interface delamination and cracking across the lamellae [12]. Three fracture modes with different crack orientation and propagation have been summarized by Chan and Kim for transgranular fracture of lamellar grains [12]. These are: crack divider, crack delamination, and crack arrester orientations (Figure 1). Krishnamurthy and Kim [11] studied the fracture temperature dependence of a titanium aluminide with mixed equiaxed γ grains and lamellar microstructure. At temperatures below 973 K, the fracture occurred at grain boundaries and interlamellar regions, while intergranular fracture was predominant between 973-1073 K due to the primary γ boundary decohesion. Dynamic recrystallization occurred at 1173 K and 1273 K. However, heat treatment studies showed that XD composites with 7-15v% TiB_2 particles had more sluggish microstructural response than monolithic near-gamma alloys [13].

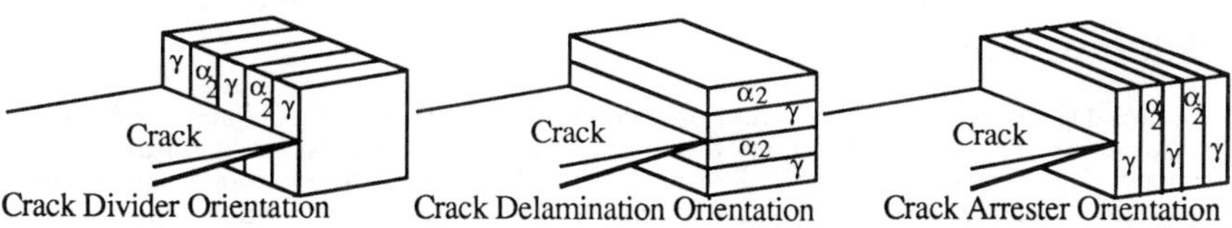

Figure 1.Fracture modes observed in Ti-47Al-2.6Nb-2(Cr+V) with a lamellar microstructure [12].

™ XD is a trade mark of Martin Marietta Corporation, Baltimore, MD.

One of the main interests in the studies of XD composites is to understand their mechanical behavior during secondary processing, such as forging and rolling. Previous investigations identified the optimum processing condition for high temperature deformation and the workability of a dual phase ($\alpha_2+\gamma$) Ti-44Al-3V-7.5v/oTiB_2 XD alloy [14-15]. Since most metalforming processes are carried out under compression, the high temperature fracture behavior of this material was studied on compression specimens, where the fracture was caused by secondary tensile stress. The emphasis of this work is to study the effect of temperature, strain rate, and TiB_2 particle reinforcement on the crack initiation and propagation in the XD composite.

EXPERIMENTAL PROCEDURE

The material used in this work was a Ti-44Al-3V-7.5v/oTiB_2 XD composite. The as-cast material was HIP'ed to reduce porosity and homogenize the ingot microstructure. Figure 2 shows a backscattered electron image (BEI) microstructure of the as-HIP'ed composite. The microstructure consists of a sub-transus structure [8] containing an ($\alpha_2+\gamma$) lamellar matrix with a uniform distribution of TiB_2 particulates (black) and, dispersed pockets of undissolved γ (gray) at the ($\alpha_2+\gamma$) lamellar grain boundaries. The lamellar grain size is approximately 30 μm, and the size of TiB_2 particulates is in the range of 1 to 15 μm.

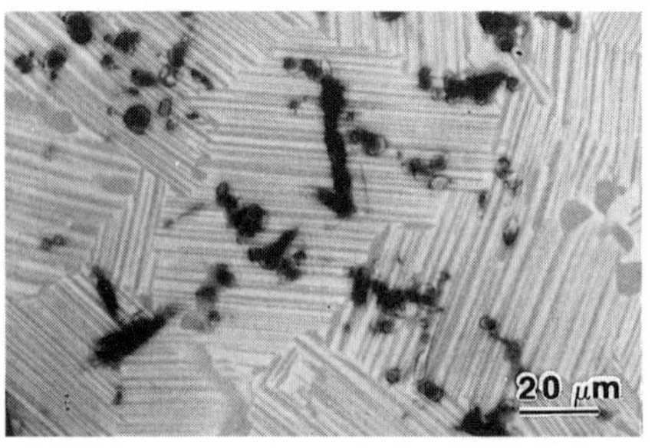

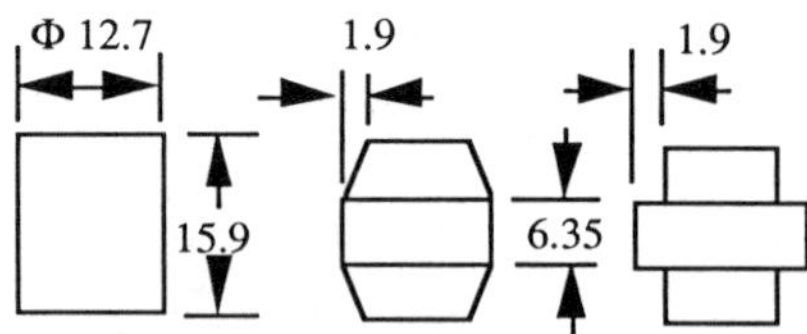

all dimensions are in milimeters

(a) cylindrical (b) tapered (c) flanged

Figure 2. SEM micrograph of the as-HIP'ed Ti-44 Al-3 V-7.5v/o TiB_2 XD composite.

Figure 3. Configurations of specimens used in compression tests.

Compression tests at temperatures of 1273, 1373, and 1473 K, and strain rates of 0.001, 0.1, and 10 s^{-1} were conducted on cylindrical, tapered and flanged specimens with dimensions shown in Figure 3. These test conditions were determined based on current metalworking practice [4,16]. The testing was interrupted at different strains (up to 6 strains for each test condition and specimen configuration) to identify crack propagation path. Testing was performed in vacuum and boron nitride was used as lubricant. Prior to deformation the specimens were held for 10 minutes in the vacuum chamber at the test temperature. Load and stroke data from the test was acquired by a microcomputer and later converted into true stress-true strain curves. Immediately after the test, the specimen was quenched with helium gas in an attempt to retain the microstructure. Fracture analysis of the specimens was performed using both optical and scanning electron microscopy.

RESULTS AND DISCUSSION

Figures 4 shows the typical fractures observed in cylindrical, tapered and flanged specimens. The cracks always start at the equator and propagate through the longitudinal axis of the specimen. The crack paths always look macrographically straight. Once the crack was initiated, it propagated quickly through the whole specimen. It took up to six interrupted tests to get a small crack which just initiated and propagated not far into the specimen. Flanged specimens were easier to fracture because of the geometric stress concentration. Fracture occurred in specimens tested at 1273, 1373, and 1473 K and a strain rate of 10 s^{-1}, and at 1273 K and a strain rate of 0.1 s^{-1} tested to a high enough strain.

Figures 5a and 5b show SEM fracture surfaces and optical microstructures for a specimen fractured at the lowest temperature and the lower strain rate condition (1273 K, 0.1 s^{-1}). The fracture surface at this test condition has a flat appearance (Figure 5a). The fracture was transgranular as shown by the crack path in Figure 5b. From this optical micrograph one can also observe that the fracture path is relatively straight and the fracture is mainly of the crack arrester type (Figure 1c)[12]. In contrast, a specimen fractured at the highest strain rate (10 s^{-1}) and the same temperature showed mainly intergranular fracture as evidenced by the fracture surface (Figures 6a, 6b) and the secondary cracks extensively appeared at lamellae grain boundaries (Figure 6c). In addition, the TiB_2 particle seems to serve as the sites for crack initiation at the grain boundaries, as shown in Figure 6c, and particle cracking was more evident as compared with the lower strain rate (0.1 s^{-1}). There was small portion of translamellar fracture, which exhibited a combination of the crack arrester and crack delamination modes suggested by Chan and Kim [12] (Figure 1).

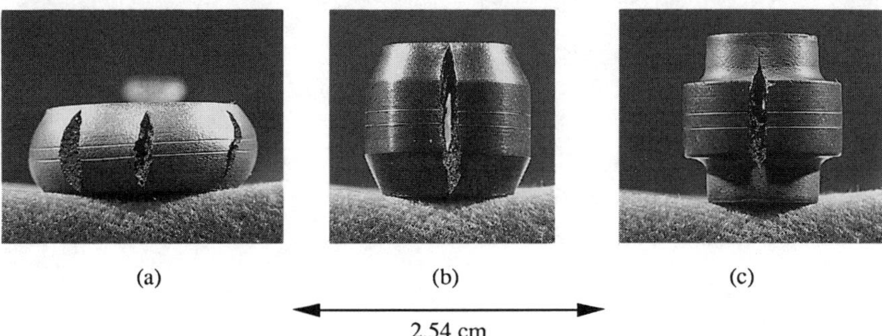

(a) (b) (c)

2.54 cm

Figure 4. Typical fracture appearance in compression specimens; (a) cylindrical, (b) tapered, and (c) flanged.

Intergranular fracture has been mainly observed for equiaxed near gamma alloys at room temperature, but very seldom in lamellar microstructures [11,12]. The XD composite with lamellar matrix showed transgranular fracture at the lowest temperature 1273 K and the lower strain rate 0.1 s^{-1}, which is in agreement with Chan and Kim's finding in the room temperature tensile testing at low strain rate (10^{-3}) [12]. It appears that lamellae interfaces were still strong to resist delamination at the higher temperatures. However, it did not seem to be the case at the highest strain rate of 10 s^{-1} and the same temperature. Similar observations were made from a specimen tested at the same strain rate (10 s^{-1}), but a higher temperature (1373 K, which possibly, is very close to the eutectoid transformation for a similar material ~1373 K [18]) (Figure 7). In both cases (1273 and 1373 K, 10 s^{-1}), the fracture surface morphology and fracture path was more tortuous because of the intergranular fracture. The TiB_2 particles seem to play an important role in crack initiation and propagation which is in agreement with the findings of Pao and co-workers [6]. TiB_2 particle fracture indicates load transfer from the matrix to the reinforcement particles. However, as the temperature increases to 1373 K, fewer fractured particles were observed, indicating plastic energy dissipation by crack propagation through the matrix.

Figure 8 shows the fracture for a specimen tested at 1473 K (above the eutectoid temperature) and the strain rate of 10 s^{-1}. The presence of translamellar cracks and a coarse ($\alpha_2 + \gamma$) lamellar structure suggests that a complete ($\alpha_2 + \gamma$) decomposition into disordered α did not occur. This is expected since the specimens were kept at temperature only for a short time which was not sufficient for the eutectoid decomposition to occur. Although both fractured transgranularly, some difference can be observed by comparing Figure 8 and Figure 5 (1273 K and 0.1 s^{-1}). At the high temperature, crack divider and delamination mechanisms were observed (Figure 8b) instead of crack arrester type shown in Figure 5. The delamination appear to be exacerbated by the weakening of the α_2/γ lath interface due to an incipient decomposition of the ($\alpha_2 + \gamma$) lamellae, and possibly due to the formation of some β at the α_2/γ lath interface [19]. These may induced lath interfacial stresses because of the phase transformation volume changes at α_2/γ lath interface. Throughout testing, dynamic recrystallization was not observed due to the sluggishness of the phase transformation kinetics of XD near gamma composites [13].

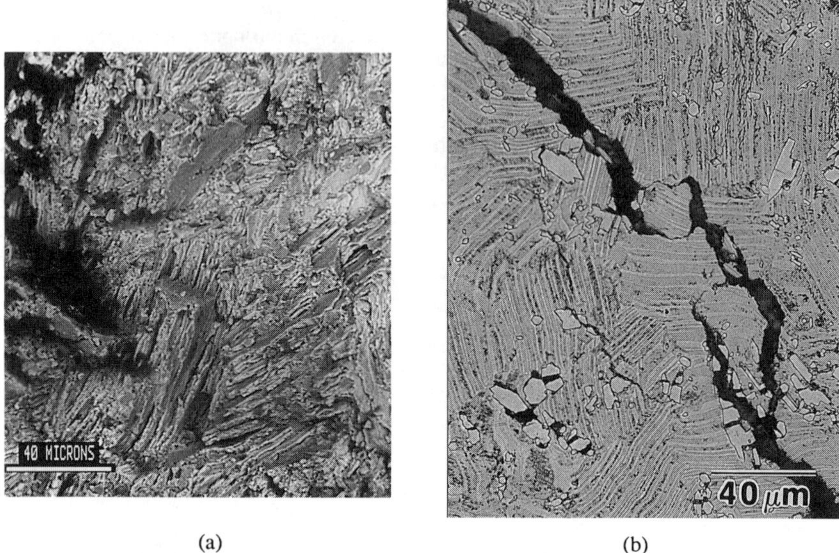

(a) (b)

Figure 5. (a) SEM fracture surface, and (b) optical view of the transverse section of the crack path of a specimen tested at 1273 K and at a strain rate of 0.1 s^{-1}.

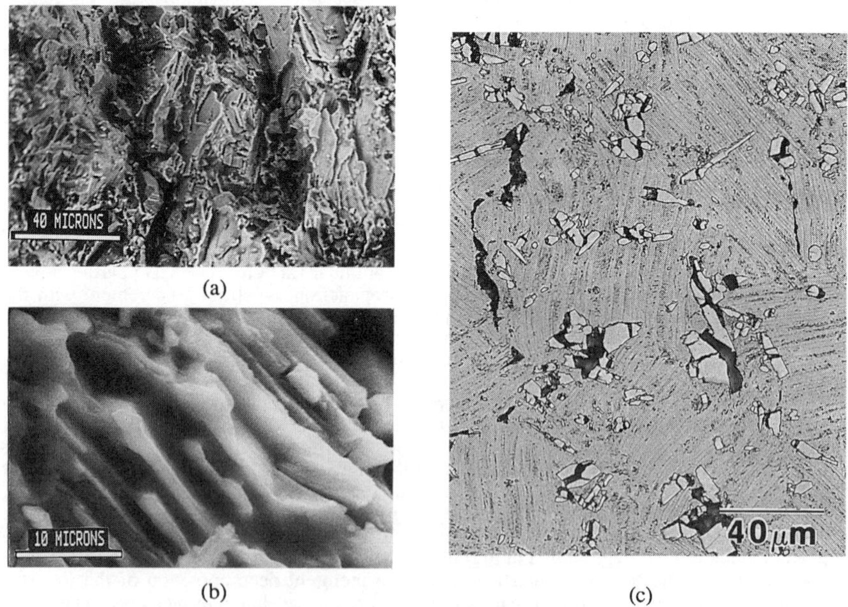

(a) (b) (c)

Figure 6. (a) and (b) SEM fracture surface, (c) optical view of the transverse section of the crack path of a specimen tested at 1273 K and at a strain rate of 10 s^{-1}.

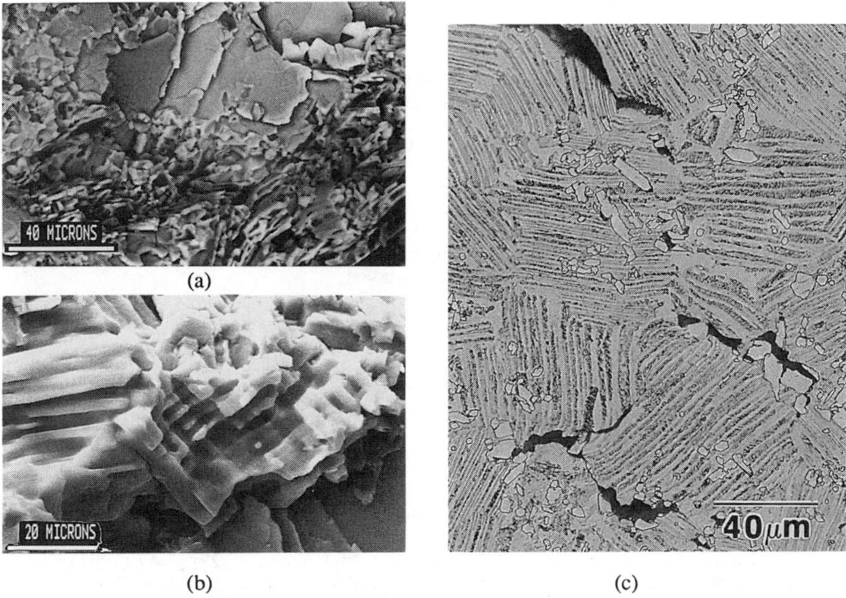

(a)

(b) (c)

Figure 7. (a) and (b) SEM fracture surface, (c) optical view of the transverse section of the crack path of a specimen tested at 1373 K and at strain rate of 10 s^{-1}.

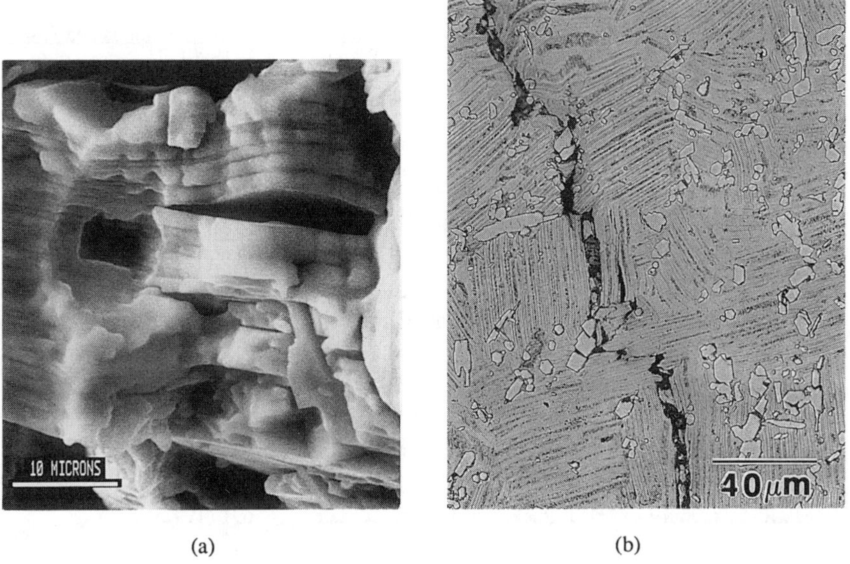

(a) (b)

Figure 8. Fractographs of the specimen tested at 1473 K and 10 s^{-1}, (a) high magnification of SEM fracture surface and (b) optical micrograph of the section through the crack.

SUMMARY

Investigation on the fracture behavior caused by surface cracking during high temperature compression was carried out on a dual phase (α_2+γ) Ti-44Al-3V-7.5v% TiB_2 XD composite. Cracks started at the equator of the specimens and propagated into the specimens through the specimen longitudinal axis. The fracture mode varied with temperature and strain rate. At the lowest temperature, 1273 K, and the lower strain rate, 0.1 s^{-1}, the fracture was transgranular with crack arrester type of crack path. At the lowest and intermediate temperatures, 1273-1373 K, but the highest strain rate, 10 s^{-1}, cracks occurred intergranularly. The TiB_2 particles seem to serve as crack initiation sites. At the highest temperature, 1473 K, and the highest strain rate, 10 s^{-1}, the fracture became transgranular again, but with crack divider and delamination types of crack paths. No dynamic recrystallization was observed.

ACKNOWLEDGEMENT

This work was conducted by the National Center for Excellence in Metalworking Technology, operated by *Concurrent Technologies Corporation* (Formerly Metalworking Technology, Inc.) under contract to the U.S. Navy as part of the U. S. Navy Manufacturing Technology Program.

REFERENCES

1. M. Saqib, I. Weiss, G.M. Mehrotra, E. Clevenger, A.G. Jackson, and H.A. Lipsitt, Met.. Trans. A, 22A, 1721 (1991).
2. S.L. Kampe, J.D. Bryant, and L. Christodoulou, Met. Trans. A, 22A, 447 (1991).
3. D.E. Larsen, M.L. Adams, S.L. Kampe, Scripta Met. et Mater., 24, 851 (1990).
4. J.D. Bryant, M.L. Adams, A.R.H. Barrett, J.A. Clarke, G.B. Gaskin, L. Christodoulou, and J. Brupbacher, NASP Contractor Report 1095, Martin Marietta Laboratories, (1990).
5. D. E. Larsen, in Microstructure/Property Relationships in Titanium Aluminides and Alloys, edited by Y-W. Kim and R. R. Boyer, (TMS, Warrendale, PA, 1990), p. 345.
6. P. S. Pao, A. Pattnaik, S. J. Gill, D. J. Michel, C. R. Feng, and C. R. Crowe, Scripta Met. et Mater., 24, 1895 (1990).
7. Y-W. Kim, in Microstructure/Property Relationships in Titanium Aluminides and Alloys, edited by Y-W. Kim and R. R. Boyer, (TMS, Warrendale, PA, 1990), 91.
8. S-C. Huang and D. S. Shih, ibid., p. 105.
9. W. E. Dowling, Jr., B. D. Worth, J. E. Allison, and J. W. Jones, S-C. Huang and D. S. Shih, ibid., p. 123.
10. D. S. Shih, S-C. Huang, G. K. Scarr, and J. C. Chesnut, S-C. Huang and D. S. Shih, ibid., p. 135.
11. S. Krishamurthy and Y-W. Kim, S-C. Huang and D. S. Shih, ibid., p. 149.
12. K. S. Chan and Y-W. Kim, S-C. Huang and D. S. Shih, ibid., p. 179.
13. L. Christodoulou, P. A. Parish, and C. R. Crowe, in High Temperature, High Performance Composites, edited by S. D. Lemkey et al (Mater. Res. Soc. Proc. 120, Pittsburgh, PA, 1988), p. 29.
14. K. G. Anand, D. Zhao, J. J. Valencia, and S. J. Wolff, in Application of Mechanics and Material Models to Design and Processing III, edited by E.S. Russell, (TMS, Warrendale, PA, 1992).
15. D. Zhao, K. G. Anand, J. J. Valencia, and S. J. Wolff, in Intermetallic Matrix Composites II, edited by D. B. Miracle, J. A. Graves, and D. L. Anton, (MRS, Pittsburgh, PA, 1992).
16. D. Popoola et al, in Interfaces in Metal-Ceramic Composites, ed. R. Y. Lin et al, (TMS, Warrendale, PA, 1989) p. 465.
17. M. Yamaguchi and Y. Umakoshi, Progress in Materials Science, 34, 1 (1990).
18. P. K. Chaudhury, M. Long and H. J. Rack, Materials Science and Engineering, A152, 37 (1992).
19. P. K. Chaudhury and H. J. Rack, Scripta Met. et Mater., 26, 691 (1992).

HIP SYNTHESIS OF TWO-PHASE MO-TI-SI ALLOYS

D. S. SCHWARTZ, R. J. LEDERICH, AND D. A. DEUSER
McDonnell Douglas Aerospace, m/c 111-1041, PO Box 516, St. Louis, MO 63166-0516

ABSTRACT

The possibility of producing novel microstructures in $MoSi_2$ materials was examined by HIP reacting $MoSi_2$ and Ti powders at 1400°C/207MPa to produce a two-phase $MoSi_2$-based alloy. The HIP reaction produced $MoSi_2$ + hexagonal $(Mo_{0.42},Ti_{0.58})_5Si_3$. The $(Mo_{0.42},Ti_{0.58})_5Si_3$ was composed of 0.5µm diam., equiaxed grains, and the average $MoSi_2$ grain size was reduced by the reaction. Preliminary compression testing showed no loss of strength compared to single-phase $MoSi_2$, and two-phase specimens were compressible to 24% strain without any cracking. The $(Mo_{0.42},Ti_{0.58})_5Si_3$ phase appeared to be plastically undeformable even at 1400°C.

INTRODUCTION

The intermetallic alloy $MoSi_2$ is a promising candidate for a number of aerospace applications in the 1200°C-1300°C temperature range. $MoSi_2$ retains significant strength (200MPa-300MPa) in this temperature regime, and has excellent oxidation resistance [1,2]. Developmental efforts have focused on increasing the low fracture toughness of this material by reinforcing it with Nb particles, wires and foils, and SiC particles and whiskers [see 3, 4 for good technical and scientific overviews]. $MoSi_2$ has been synthesized by arc melting, reaction synthesis, plasma spraying, mechanical alloying, and powder processing [5]. Although rapid advances are being made in the technology of $MoSi_2$ synthesis, the most successful and practical processing route is powder processing by isostatic and non-isostatic hot pressing, due to the maturity of powder processing technology. Large quantities of >97% dense, acceptably clean monolithic and composited $MoSi_2$ can be inexpensively produced by hot isostatic pressing (HIPing) and/or hot pressing.

One failing of the HIP and hot press methods for $MoSi_2$ synthesis is the relative lack of microstructural manipulation possible using these techniques. Pressure consolidation of $MoSi_2$ powder results in a predominantly single phase, equiaxed microstructure with an average grain size on the order of 50% of the original powder size. As a result, the only microstructural parameter that can be manipulated is the grain diameter, through control of the initial powder size. It is desirable to be capable of controlling a range of microstructural parameters, in order to explore the possibility of manipulating these parameters to control the mechanical behavior of $MoSi_2$. One means to achieve this control is through the introduction of a second phase into the microstructure, which will allow manipulation of grain morphology and size, $MoSi_2$/second phase volume ratio, and second phase chemistry. A simple, inexpensive approach being explored at McDonnell Douglas is blending $MoSi_2$ powder with a suitable ternary elemental powder and reacting the powders during HIPing. Our preliminary results and observations will be the subject of this paper. Ti was chosen as the ternary element for the reaction HIPing because it is highly reactive with $MoSi_2$ and can also produce a reduction in density, a goal that pervades all aerospace materials development.

EXPERIMENTAL DETAILS

All metal powders used in this work were purchased from Cerac, Inc. The $MoSi_2$ powder was 99.5% pure, -325 mesh (i.e. 44μm maximum diam.). The Ti powder was 99.5% purity vacuum deposition grade, -150, +325 mesh (i.e. particle diams. between 40μm and 100μm). The powder mixture was 90% $MoSi_2$ + 10% Ti by volume. All powder handling was done in high purity dry Ar, and powders were mechanically blended for 16 hours prior to HIPing. Blended powder was packed into Nb tubes under Ar, vacuum degassed for 16 hours, and electron-beam sealed. The packed tubes were then HIPed in an ASEA HIP apparatus at 1400°C for 4.25 and 8.25 hours under 207 MPa pressure. Cubes approximately 10mm on a side were cut from the consolidated material and compressed at 1300°C and 1400°C to ~24% total plastic strain in a vacuum of 10^{-5} torr. A strain rate of $2 \cdot 10^{-4} s^{-1}$ was used, and the compression tests were interrupted before fracture occurred. The microstructure of the specimens was examined by SEM (AMR 1400) and TEM (JEOL 2000FX). EDX was used in both SEM and TEM to obtain localized chemical information from the specimens. TEM electron diffraction was used to determine the crystal structure of phases present in the specimens. Specimens were prepared for TEM from both as-HIPed and compression tested specimens by dimple grinding, followed by ion milling to electron transparency.

Figure 1. Overall microstructure of as-HIPed $MoSi_2$.

RESULTS AND DISCUSSION

Microstructure

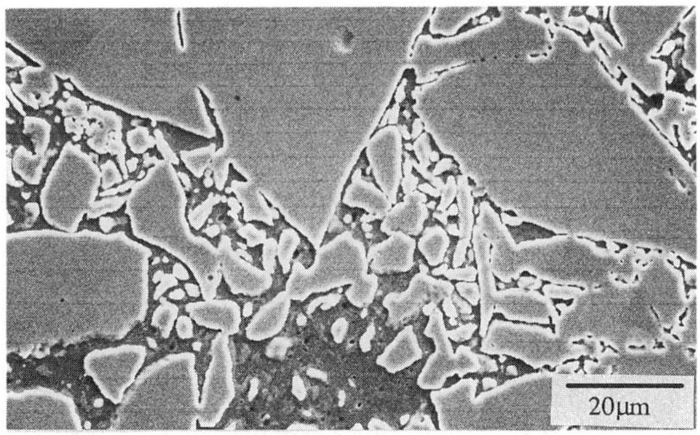

Figure 2. Typical distribution of phases in as-HIPed $MoSi_2$.

The quality of the $MoSi_2$ + Ti HIP reaction product was good, with 2-3% porosity. The general microstructure is shown in the optical micrograph in Figure 1, for a specimen HIPed at 1400°C/207MPa/8.25hr. The material is clearly two-phase, and the typical distribution of the phases can be seen in the SEM micrograph in Figure 2. The $MoSi_2$ phase tends to be outlined in white, and appears raised in Figure 2. Roughly 30% of the material transformed to a Mo-Ti-Si second phase. No differences were seen in specimens HIPed for 4.25 hours compared to those HIPed for 8.25 hours, indicating that equilibrium is reached within 4.25 hours at 1400°C. The micrographs in this paper are all from specimens HIPed for 8.25 hours.

A pronounced grain refinement effect was observed in the HIP reacted specimens. Figure 3 is a secondary electron image acquired in the TEM scanning mode. The difference in atomic number between Ti and Mo results in easily discernable contrast differences between the $MoSi_2$ and the Mo-Ti-Si phase, $MoSi_2$ being the lighter phase. It can be seen that many of the $MoSi_2$

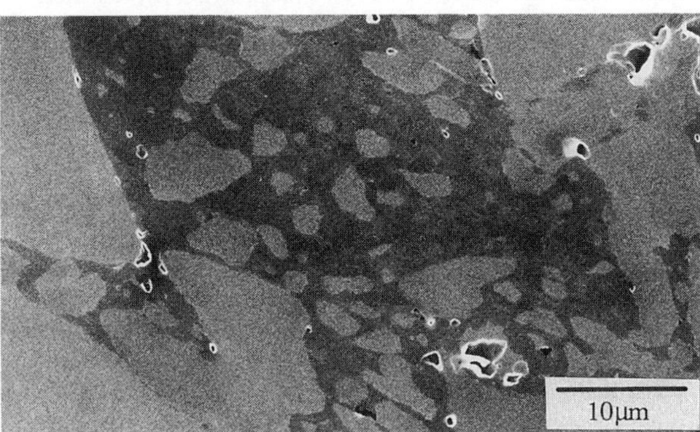

Figure 3. Secondary electron image of as-HIPed $MoSi_2$ + Ti ($MoSi_2$ is the light phase).

Figure 4. TEM micrograph of a typical second phase region in as-HIPed $MoSi_2$ + Ti.

grains in contact with the Mo-Ti-Si phase are on the order of 1μm diam, 10 times smaller than the average grain size seen in single phase $MoSi_2$ materials made from the same powder. Closer examination by TEM revealed that the second phase had an even greater degree of grain refinement. The TEM micrograph in Figure 4 shows the structure of a typical second phase region. The second phase regions were typically composed of clusters of equiaxed, ~0.5μm diam. grains.

Chemistry and Phase Identification

EDX spectroscopy was used to determine the distribution of Mo, Ti, and Si in the reacted specimens. No pockets of unreacted Ti were detected because the HIP reaction was done at 8.251% of the melting point of Ti, where Ti diffusivity is quite high. No Ti was detected in $MoSi_2$ grains, and the second phase chemistry was consistently measured to be 26 at. % Mo, 36 at. % Ti, and 38 at. % Si, within an error of 3 at. %. Electron diffraction unambiguously demonstrated that the second phase had the hexagonal $D8_8$ crystal structure, associated with a large number of 5-3 metal silicides. The measured lattice parameters of $a_0 = 0.744$nm and $c_0 = 0.507$nm are well within error of the parameters reported in [6] for $(Mo,Ti)_5Si_3$. Although a second hexagonal C40 ternary Mo-Ti-Si silicide exists [7], this phase was never observed in our specimens. In terms of the 5-3 stoichiometry, the second phase is $(Mo_{0.42},Ti_{0.58})_5Si_3$ with an ideal density of ~5.9 g/cm^3, based on the measured lattice parameters. Using this density and the measured Ti concentration in the second phase, it can be calculated that 24% of the material should have transformed to $(Mo_{0.42},Ti_{0.58})_5Si_3$, in good agreement with the observations. Unfortunately, $(Mo_{0.42},Ti_{0.58})_5Si_3$ is only 7% less dense than $MoSi_2$, so the calculated density of the two-phase mixture is 6.2g/cm^3, a very modest density reduction from 6.3g/cm^3 for pure $MoSi_2$.

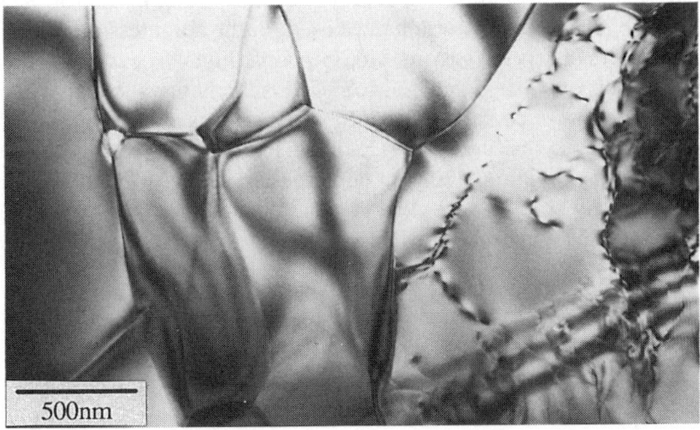

Figure 5. TEM micrograph of material compressed at 1300°C ($MoSi_2$ grain on right).

Mechanical Behavior

The mechanical testing specimens showed no signs of cracking after compression to 24% strain at 1300°C and 1400°C. The yield stresses at the two temperatures were 186MPa and 120MPa respectively, showing no loss of strength compared to single phase $MoSi_2$. TEM examination of compressed specimens revealed that deformation took place only in the $MoSi_2$ grains (Figure 5). The $(Mo_{0.42},Ti_{0.58})_5Si_3$ phase was always observed to be dislocation free, even when completely surrounded by highly deformed $MoSi_2$ grains. The lack of plastic deformation observed in the $(Mo_{0.42},Ti_{0.58})_5Si_3$ grains may indicate that these second phase

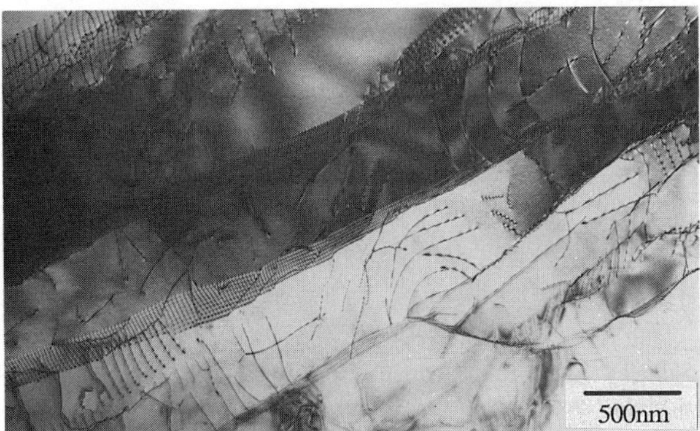

Figure 6. TEM micrograph showing subgrain boundaries in $MoSi_2$ grain.

regions are able to accomodate strain through grain boundary sliding, a deformation mode that may be assisted by the relatively slow strain rates used for the compression testing. Low-angle subgrain boundaries were observed in many $MoSi_2$ grains (Figure 6), at a higher density than normally observed in deformed single phase $MoSi_2$. It is likely that the hard $(Mo_{0.42},Ti_{0.58})_5Si_3$ grains in contact with $MoSi_2$ grains act as nucleation sites for many of these subgrain boundaries, so the fine grain size of the second phase particles serves to refine the larger $MoSi_2$ grains by dividing them into numerous subgrains. The ultimate result of the refined grain structure may be increased yield strength due to Hall-Petch strengthening, but more mechanical testing is required before this can be said with certainty.

CONCLUSIONS

Preliminary work has shown that high quality two phase material can be inexpensively produced by HIP reacting $MoSi_2$ + Ti powders. This process shows promise for producing novel, adjustable microstructures, and appears to be particularly promising for producing very fine grained material. In summary:

1. A second phase is formed with hexagonal $D8_8$ crystal structure and composition $(Mo_{0.42},Ti_{0.58})_5Si_3$,
2. 10 vol. % Ti powder in $MoSi_2$ reacted to produce ~30 vol. % $(Mo_{0.42},Ti_{0.58})_5Si_3$,
3. The second phase average grain diam. was ~0.5µm, and significant but inhomogeneous $MoSi_2$ grain refinement occurred,
4. Density reduction was not significant,
5. The $(Mo_{0.42},Ti_{0.58})_5Si_3$ phase showed no sign of plastic deformation in specimens compressed at 1300°C or 1400°C.

ACKNOWLEDGMENTS

This work was performed under AFOSR contract number F49620-90-C-0030. The authors would like to thank the contract monitor, Dr. A. Rosenstein, for his support.

REFERENCES

1. P. J. Meschter and D. S. Schwartz, JOM 41, no. 11 (89) 52-55.
2. E. Fitzer, J. Schlichting, and F. K. Schmidt, High Temp.-High Pres. 2 (70) 553-570.
3. J. Schlichting, High Temp.-High Pres. 10 (78) 241-269.
4. A. K. Vasudevan, J. J. Petrovic, Mat. Sci. Eng. A155 (92) 1-17.
5. e.g. see pp. 75-115 of Mat. Sci. Eng. A155 (92).
6. P. Villars, L. D. Calvert, Pearson's Handbook of Crystallographic Data for Intermetallic Phases, (American Society for Metals, Metals Park, OH, 1985) 2772.
7. V. N. Svechnikov, Yu. A. Kocherzhinsky, L. M. Yupko, Dokl. Akad. Nauk, Ukrain SSR, 6A (72) 566.

THE FRACTURE TOUGHNESS OF A BRITTLE NICKEL SILICIDE CONTAINING THE DUCTILE PHASES Ni(Si) AND Ni_3Si

ZHAOHUI LI AND ERLAND M. SCHULSON
Thayer School of Engineering, Dartmouth College, Hanover, NH 03755

ABSTRACT

The fracture toughness at room temperature of a brittle nickel silicide containing the ductile phases Ni(Si) and Ni_3Si has been investigated. The microstructure was comprised of a brittle nickel silicide matrix, ductile Ni(Si) particles distributed within the matrix, and Ni_3Si rims around the particles. It was obtained by annealing hot-extruded Ni-23 at.%Si (with and without 0.19 at.% boron) at 1100°C and then air cooling. The fracture toughness values were in the range of 13 MPa $m^{1/2}$ to 22 MPa $m^{1/2}$. Systematic effects of annealing time and of boron were not apparent. It would appear that the incorporation of the ductile particles has a beneficial effect on the fracture toughness of an otherwise brittle silicide.

I. INTRODUCTION

The incorporation of ductile inclusions into brittle matrix is an effective method for toughening materials [1-6]. In keeping with this procedure this paper shows that the incorporation of ductile particles within a nickel-silicide matrix imparts significant toughness to an otherwise brittle, but highly corrosion resistant alloy. The particles are a Ni(Si) solid solution which contain a sub-structure of Ni_3Si precipitates, and the matrix is a complex silicide denoted β_2 on the Ni-Si phase diagram. That the matrix is brittle is apparent from the fact that it cracks under thermally induced stresses upon quenching.

II. EXPERIMENTAL

The alloys used in this study were the same materials as used in an earlier study [7]. They contained Ni-23 at.% Si (with and without 0.19 at.% boron) and had been extruded from powders into rod (26 mm dia.). From the as-received rods specimens were machined for three-point, notched bend tests. The dimensions were S = 45.7 mm, W = 11.4 mm and B = 5.8 mm; the ratio of these dimensions is in compliance with ASTM specification E-399. The specimens were annealed in dried and de-oxygenated argon at 1100°C for 45 minutes to 48 hours to obtain different Ni(Si) particle sizes within the matrix, and then air cooled. (While quenching following the high-temperature anneal would have been desirable to suppress the peritectoid phase transformation (see below), such rapid cooling cracked the material.) Subsequently, the specimens were annealed at 300°C in dried and de-oxygenated argon for 1 h to reduce any internal stresses. The specimens were notched at the center of the bottom surface (S x B) using either a diamond saw or the single-edge-precracked-beam method (given by Nose and Fujii) [8]. The notches were 5.8mm deep (for both methods) and of root radii 0.11mm (diamond saw) or 0.0005mm (single-edge-precracked-beam), measured from optical photographs of the notch roots at 200x magnification. Three-point bend tests were performed at room temperature using a floor model MTS machine. The loading rate was around 1.0 MPa$(m)^{1/2}$/s. A clip displacement gage was mounted across the notches to measure the crack opening displacement (COD).

The microstructure was analyzed through quantitative metallography [9]. Marble's reagent revealed the phase boundaries. The deformation features and fracture surfaces were examined by scanning electron microscopy and by optical microscopy. The surfaces of some of the specimens were polished prior to testing to assist the analysis.

III. RESULTS AND DISCUSSION

3.1 Microstructure

Fig.1 shows optical micrographs. In keeping with the Ni-Si phase diagram [10] and TEM analysis [11], the microstructure consists of Ni(Si) particles (dark phase), Ni_3Si rims* (light phase) around the Ni(Si) particles, a complex silicide matrix** (grey), and Ni_3Si precipitates within the Ni(Si) particles. The precipitates can not be resolved optically, but are seen in transmission electron micrographs [11]; they account for the mottled appearance of the etched Ni(Si) particles. Whether boride particles are present in the doped alloy is not known. Table I lists the volume fractions, V_f, interface to interface distances, λ, apparent diameters, $L_{Ni(Si)}$, of the Ni(Si) particles, the thickness, L_{Ni_3Si}, of the Ni_3Si rims and the surface area to volume ratio, S/V, of the particles, as determined from at least six different photographic areas. With increasing time the Ni(Si) phase coarsened from small, micron-sized particles with a partially network-like morphology to larger, separate particles. Boron increased the Ni(Si) coarsening rate, as noted earlier [11], but decreased slightly its final volume fraction, implying that boron shifts the Ni(Si) + β_2 area on the phase diagram [10] to the Ni-rich end. The smaller the Ni(Si) particles, the larger is the surface to volume ratio and so the greater is the volume fraction of the peritectoid Ni_3Si(rim) and the smaller is the rim surface-to-surface distance. The volume fraction of Ni(Si) is lower than the equilibrium fraction (at 1100°C) calculated from the lever rule, owing to the reaction of Ni(Si) with β_2 during the air cooling to produce the Ni_3Si rim.

3.2 Fracture Toughness

The load vs COD curves from the bending tests (for both sharp and blunt notches) were essentially straight lines. Table II gives the fracture toughness or, K_Q, for both alloys for each heat treatment. The calculation of K_Q followed the standard procedure described in ASTM specification E-399. The K_Q measured is actually K_{Ic}, because the three-point bending specimens were thick enough to satisfy the condition: B, a > 2.5 $(K_Q/\sigma_y)^2$. For instance, K_Q of one of the specimens of Ni-Si-1100°C-0.75h was 19.7 MPa$(m)^{1/2}$, and the yield strength was 1290 MPa (obtained from separate measurements of the compression stress vs strain curve at 0.2% offset) [12], giving $2.5(K_Q/\sigma_y)^2 = 0.6$ mm; this value is much smaller than both the thickness and the crack length of 5.8 mm. Every result was analyzed through this procedure. The root radius of the crack tip (0.11 mm or 0.0005 mm) had little, if any, effect. Similarly, annealing and the addition of boron had no systematic effect. It appears, therefore, that the fracture toughness of the material is 17.5 ± 3.0 MPa.$m^{1/2}$.

---

* Formed via a peritectoid transformation [see reference 11]

** The matrix is assumed to be β_2 phase. This is based on the observation [11] that the eutectoid transformation $\beta_2 \rightarrow \beta_1+\gamma$ occurs upon annealing below 825°C. In the present experiments the air cooling suppressed the transformation, for no evidence of it was seen from optical microscopy.

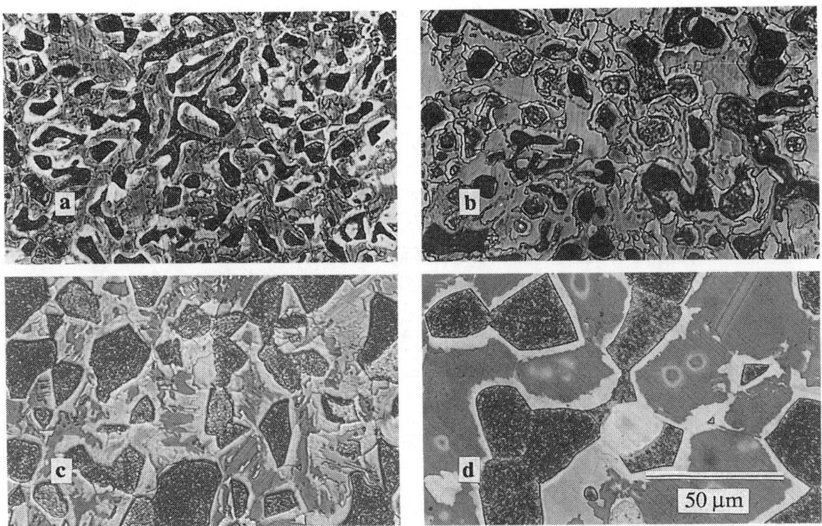

Fig. 1 - Optical micrographs of Ni-23Si with(b,d) and without 0.19at.% boron(a,c) after 0.75hrs(a,b), 28hrs(c), and 48hrs(d) at 1100°C, followed by air cooling; etched with Marbles reagent.

Table I. Microstructural Parameters of Four Selected Specimens

	Ni-Si-1100°C-0.75h	Ni-Si-1100°C-28h	Ni-Si+B-1100°C-0.75h	Ni-Si+B-1100°C-48h
$(V_f)_{Ni(Si)}$	0.22	0.33	0.25	0.28
$(V_f)_{Ni_3Si}$	0.31	0.22	0.23	0.1
$\lambda_{Ni(Si)}$	8.5 μm	20.2 μm	12.0μm	21.2 μm
λ_{Ni_3Si}	5.2 μm	13.5 μm	9.1 μm	15.4 μm
$L_{Ni(Si)}$	1.6 μm	8.2 μm	2.4 μm	21.2 μm
L_{Ni_3Si}	1.7 μm	3.4 μm	1.5 μm	2.9 μm
$S_{Ni(Si)}/V_{Ni(Si)}$	2472 mm^{-1}	487 mm^{-1}	1643 mm^{-1}	189 mm^{-1}

Table II. Fracture Toughness as a Function of Alloy and Heat Treatment

Alloy		Heat Treatment	K_Q (MPa.m$^{1/2}$) at Room Temperature	
			diamond saw notch[a]	pre-cracked notch[b]
Ni-23 at.%Si	1	1100°C-0.75h-air cool, then 300°C-1h	19.7,22.6	
	2	1100°C-28h-air cool, then 300°C-1h	16.3, 13.2	15.2
Ni-23at.%Si+B	3	1100°C-0.75h-air cool, then 300°C-1h	18.4, 17.7,22.0	18.6,17.1
	4	1100°C-48h-air cool, then 300°C-1h	13.0	16.4

a - Specimens were notched by a diamond saw with root radius of 110 μm.
b - Specimens were precracked by a single-edge-precracked-beam method with root radius of 0.5 μm [8].

This fracture toughness is relatively high for a silicide intermetallic. For Nb_5Si_3, for instance, K_{Ic} is much smaller; i.e. around 1-2 MPa.m$^{1/2}$ [reference 2]. Presumably, the fracture toughness of the particle-free, β_2 matrix, although not measured, is also small, given its propensity for cleavage (see below).

The relatively high fracture toughness of the present materials can be explained in terms of the microstructure. When cracks propagated during bending, they took four paths: through the matrix; through the matrix and along the matrix/rim interface; through the matrix, through the rim then along the rim/particle interface; and as the previous path, but through the particle instead of along the particle/rim boundary. Fig. 2 illustrates these paths. The particles, it appears, first bridged the cracks and then fractured in a ductile mode. The last point was evident from SEM fractography, Fig. 3, where on every specimen ductile regions (A) were seen within a cleaved matrix.The size of the ductile zones is similar to the size of the Ni(Si) particles, implying that these regions are ruptured particles. The SEM fractography also revealed crystallographic rupture zones (B) between the particles and the matrix (β_2, Fig. 3), suggesting that the Ni_3Si phase fractured on certain crystal planes. This is not surprising because the rim, as shown elsewhere [11], is a single crystal coherently attached to the particle.

It would appear, therefore, that the relatively high toughness of the present alloys originates primarily in crack bridging by the particles and in the subsequent ductile rupture of the particles.

Why the variations in the microstructure had little systematic effect on the fracture toughness is curious. Two theories(1,13), for instance, suggest that the fracture toughness enhancement (ΔK_c) is proportional to the square root of the product of the particle strength, σ_o, the volume fraction (V_f) of the particles and size (L) of the particles; i.e.

$$\Delta K_c \propto (\sigma_o V_f L)^{1/2}$$

where L is the apparent diameter of Ni(Si) phase or of the combined Ni(Si)+Ni_3Si phases. Assuming that the fracture toughness of the matrix is very small, then the value measured is essentially the same as the increment. Fig. 4 plots the fracture toughness of the alloy versus part of this function $(V_f L)^{1/2}$ using both measures of L. No effect is apparent. The origin of this discrepancy is not clear. Possibly, it can be explained in terms of an attendant reduction in the strength of the composite particles, for as the product $(V_f L)^{1/2}$ increased the volume fraction of the harder Ni_3Si phase decreased (The hardness [12] of the rim with and without boron is greater than that of the Ni(Si) core: Ni_3Si = 379±72, Ni_3Si(B) = 449±88; Ni(Si) = 249±30, and Ni(Si,B) = 241±18). Correspondingly, σ_o should have decreased [14].

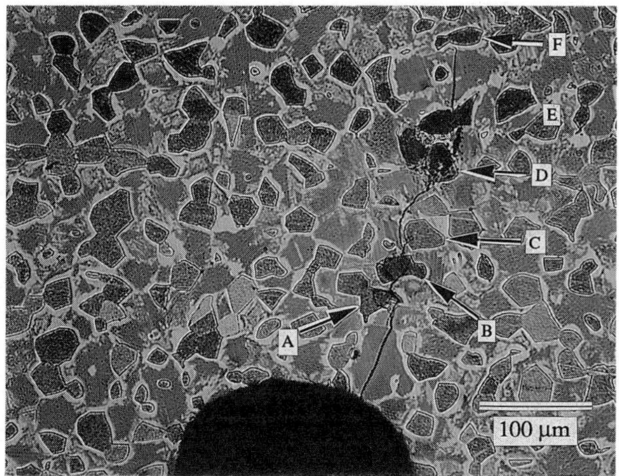

Fig. 2 - Optical micrograph showing the crack path at the tip of a notch observed in Ni-Si+B-1100°C-48h.The crack entered the matrix from the root of the notch, passed through a rim and through particle A, and back into the matrix; it appears to be bridged by particles B, D and E; along its path, one branch stopped within particle C while the main branch went along the matrix/rim interface (at C).

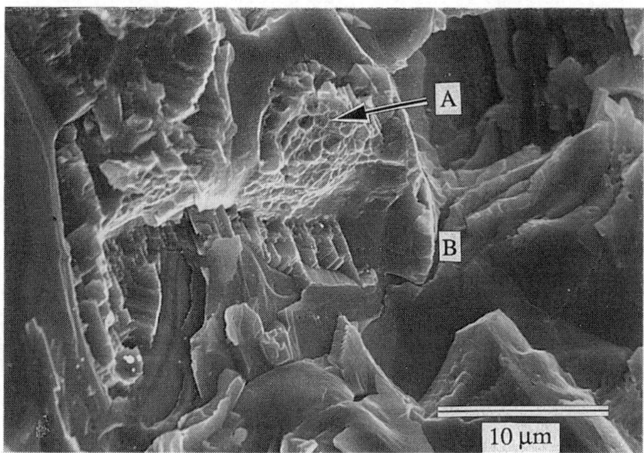

Fig. 3 - Scanning electron micrograph showing the fracture mode of the three phases observed in bending sample. Ni-Si-1100°C-28hrs.

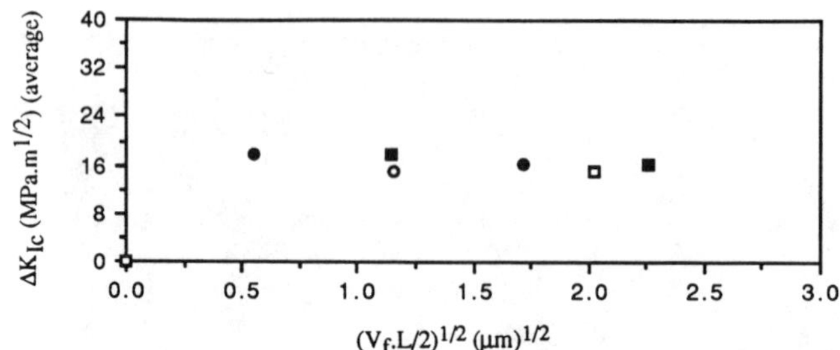

Fig. 4 - Fracture toughness increment as a function of $(V_f.L/2)^{1/2}$.
The open (o)/open (□), and closed (●)/closed (■) points were boron free and boron doped alloys, respectively.
The open (o)/closed (●), and open (□)/closed (■) points were obtained using $L=L_{Ni(Si)}$ and $L=L_{Ni(Si)}+L_{Ni_3Si}$, respectively.

IV. CONCLUSION

1. The incorporation of ductile particles of Ni(Si) is an effective method for toughening the complex β_2 nickel silicide.

2. The improvement in the toughness is not sensitive to the size and volume fraction of the ductile phase, at least for the Ni-23Si alloy with and without 0.19at.% boron.

ACKNOWLEDGEMENTS

The authors would like to thank Gary A. Kuehn for help and for the use of the Thayer School Ice Lab MTS facility. The use of the Dartmouth College Electron Microscope facility is gratefully acknowledged. This work was supported by the U.S. Department of Energy, Office of Basic Energy Science, grant number is DE-FG02-86ER 45260.

REFERENCES

1. M. F. Ashby, F. J. Blunt, and M. Bannister. Acta Metall., 1989, Vol. 7, pp 1847-57.
2. M.G. Mendiratta, J.J. Lewandowski, and D.M. Dimiduk, Metall. Trans A, 1991, Vol. 22A, July, pp 1573-83.
3. L.S. Sigl, P.A. Mataga, B.J. Dalgleish, R.M. McMeeking, and A.G. Evans, Acta Metall., 1988, Vol. 36(4),pp. 945-53.
4. L.S. Sigl and H.E. Exner, Metall. Trans. A, 1987, Vol. 18A, pp. 1299-1308.
5. V.D. Krstic, Phil. Mag., 1983, Vol. 48 (5), pp. 695-708.
6. C.K. Elliott, G.R. Odette, G.E. Lucas, and J.W. Sheckherd: High Temperature, High Performance Composites, MRS Proc., Lemkey, ed., Materials Research Society, Pittsburgh, PA, 1988, Vol. 120, pp. 95-101.
7. E. M. Schulson, L. J. Briggs and I. Baker, Acta. Metall. Mater. vol. 38, No. 2, pp 207 - 213,1990.
8. T. Nose and T. Fujii, J. Am. Ceram. Soc., Vol. 71, No. 5, 328-33 (1988).
9. Quantitative Microscopy, R. T. DeHoff and F. N. Rhines, eds., McGraw - Hill, Inc., New York, NY, 1968.
10. Binary Alloy Phase Diagrams, Ed-T.B. Massalski, ASM, (1986) p1754.
11. I. Baker, J. Yuan and E. M. Schulson. Metall Transactions A (in press).
12. Zhaohui Li, M.S. Thesis, Thayer School of Engineering, Dartmouth College. (1992).
13. K.S. Ravichandran, Scripta Metall., Vol. 26, pp. 1389-1393 (1992).
14. D. Tabor: Hardness of Metals, Clarenden Press, London, 1951.

INFLUENCE OF COARSE SECOND PHASE ADDITIONS ON MECHANICAL PROPERTIES OF NiAl

H.CLEMENS*, I.RUMBERG*, P.SCHRETTER*, P.GRAHLE**, O.LANG**, A.WANNER**, AND E.ARZT**
* Metallwerk Plansee GmbH, A-6600 Reutte, Austria.
** Max-Planck-Institut für Metallforschung, Institut für Werkstoffwissenschaft, Seestraße 92, D-7000 Stuttgart 1, Germany.

ABSTRACT

NiAl/Nb and NiAl/Cr composite materials were prepared by a powder metallurgical approach. The content of the second phases varied between 5 and 10 weigth percent. The thermodynamic stability of Nb and Cr particles in the NiAl matrix was studied at 1473K (1200°C) by means of SEM, XRD and microhardness measurements. In the case of Nb the formation of a Laves phase (NbNiAl) was observed whereas the Cr particles remained stable but were hardened to a remarkable extent due to in-diffusion of Ni and Al. This paper presents the results of three-point bending tests performed on as-HIPed and annealed material at room temperature and elevated temperatures. The yielding behavior at room temperature and the creep behavior at 1200K (927°C) were investigated by compression tests.

INTRODUCTION

Intermetallic aluminides, especially NiAl, are attractive candidate materials for high temperature applications. The main advantages of NiAl are the low density, excellent thermal conductivity and oxidation resistance. In order to improve mechanical properties the effects of Cr and Nb as second phase additions were investigated.

EXPERIMENTAL PROCEDURE

NiAl/Cr and NiAl/Nb composite material was prepared by blending gas-atomized NiAl powder with Cr or Nb powder. The content of the second phases varied between 5 and 10 weight percent (in volume percent: 4.1 - 8.3 for Cr; 3.4 - 7 for Nb). In the case of Cr, powders with two different particle fractions have been used. Typical impurity concentrations and average particle sizes of the powders are given in Table I.

Table I: Chemical analysis and average particle size of powders used.

	NiAl	Cr < 63 μm	Cr > 63 μm	Nb
O (ppm)	185	220	200	1000
N (ppm)	10	< 5	< 5	67
H (ppm)	3	3	3	40
average particle size (μm)	73	40	85	17

The blended powders were filled in steel cans (inner diameter: 45 mm; length: 100 mm) which were evacuated, sealed and compacted by hot isostatic pressing (HIP). After the HIP process (3 hours at 1100°C under a pressure of 2000 bar) fully densified material was obtained. The microstructure of NiAl-10 wt%Cr and NiAl-10wt%Nb is shown in Fig.1a and b.

The mechanical properties of the NiAl matrix and the composite materials as well the alteration of these properties due to annealing at 1473K (1200°C) were measured by means of three-point bending tests and compression tests. Three-point bending tests were performed at room temperature and temperatures between 773K (500°C) and 1373K (1100°C). From these

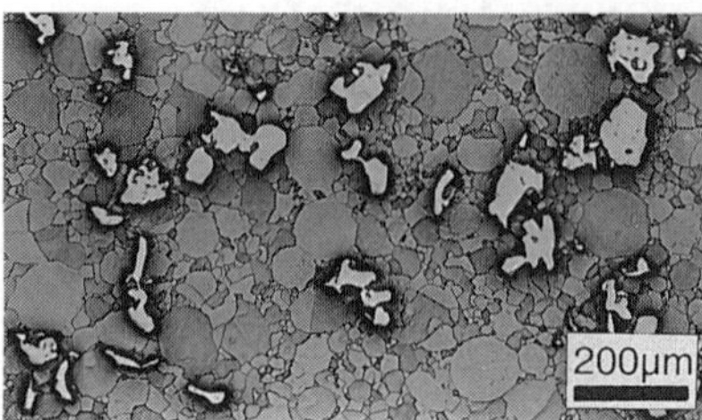

Fig.1a: NiAl-10wt%Cr after HIPing.

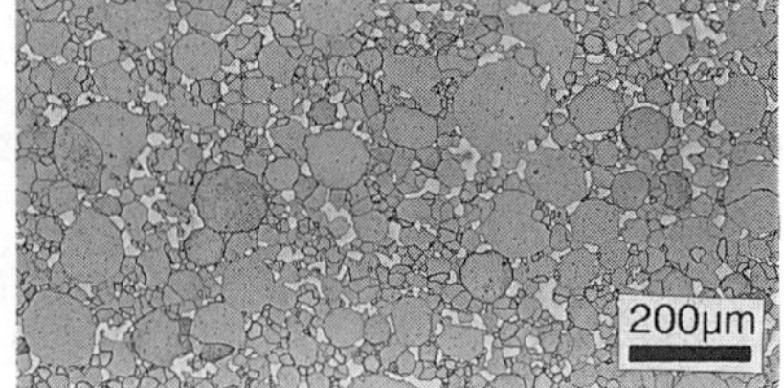

Fig.1b: NiAl-10wt%Nb after HIPing.

experiments the ductile-to-brittle transition temperatures (DBTT) were determined. The yielding behavior at room temperature (i.e. below DBTT) and the creep behavior at 1200K (i.e. above DBTT) were investigated by strain-rate controlled compression tests on cylindrically shaped samples.

RESULTS AND DISCUSSION

Phase stability

SEM examination of as-HIPed NiAl/Nb material indicated a reaction between NiAl and Nb which leads to the formation of additional phases. Fig.2 shows a SEM micrograph of a Nb particle which is covered by two phases. From XRD measurements the phase in contact with the NiAl matrix was identified as NbNiAl (Laves type, C14 structure), whereas the inner phase was referred to Nb_2NiAl [1,2]. Due to the formation of the NbNiAl phase the hardness in the reaction zone is increased to a threefold value as compared to the NiAl matrix.

In the case of as-HIPed NiAl/Cr material no indication about the presence of additional phases was found by XRD and SEM. However, due to diffusion of Ni and Al an increase in hardness was measured in the Cr particles near the NiAl/Cr interface.

In order to investigate the thermal stability of Nb and Cr within the NiAl matrix at elevated temperatures and longer durations, samples were annealed under vacuum conditions

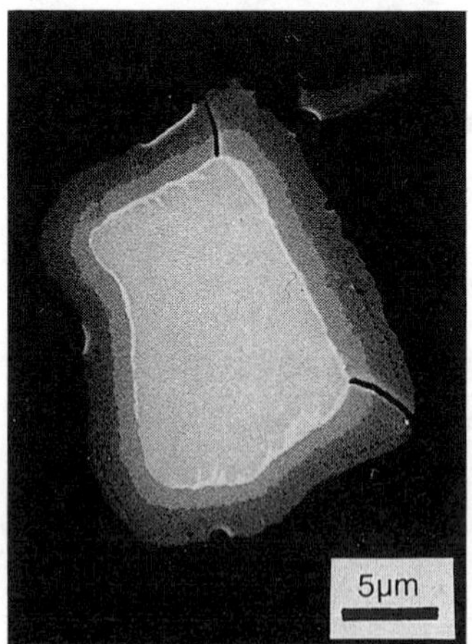

Fig.2: Backscatter electron micrograph of Nb particle after HIP process.

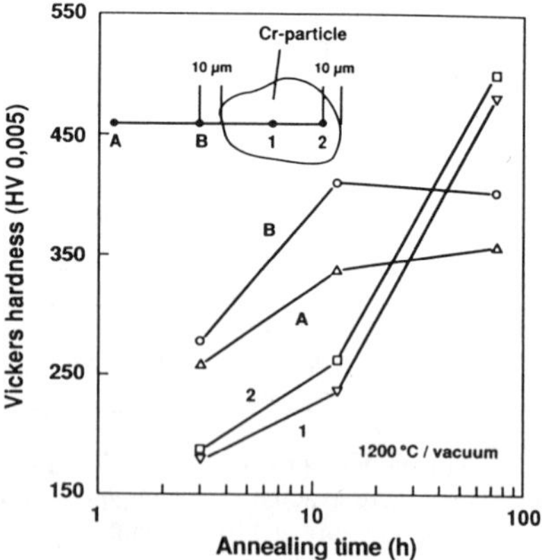

Fig.3: Room temperature hardness of Cr particles and NiAl matrix as a function of annealing time (annealing temperature: 1200°C)

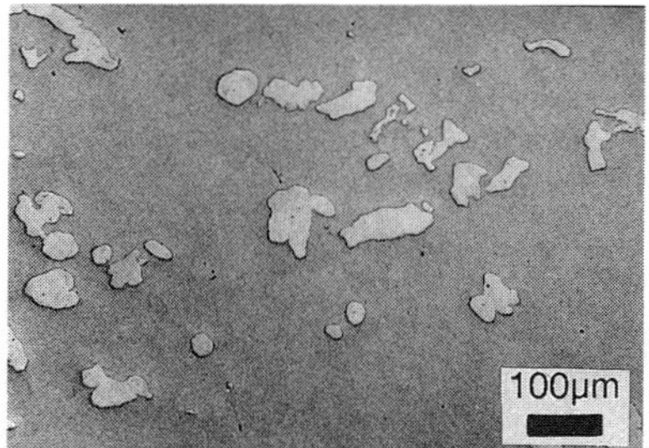

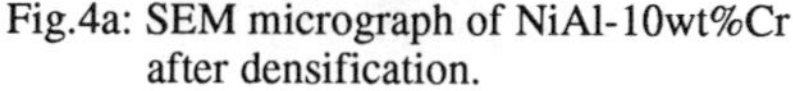

Fig.4a: SEM micrograph of NiAl-10wt%Cr after densification.

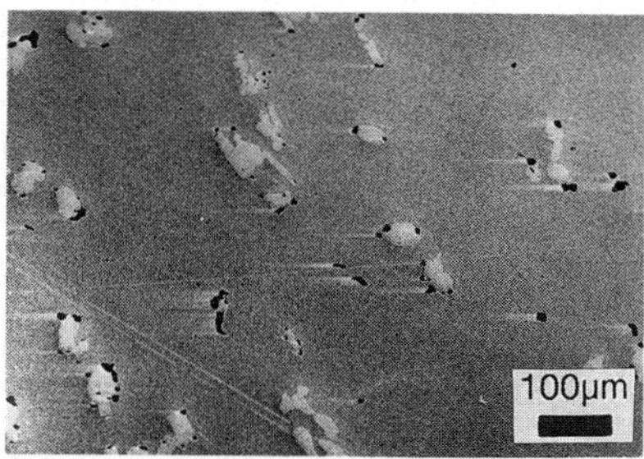

Fig.4b: SEM micrograph of NiAl-10wt%Cr after annealing at 1200°C for 70 hours.

at 1200°C for 10 and 70 hours. In the case of NiAl/Nb full transformation of the Nb particles to NbNiAl was observed after an annealing time of 70 hours. Additionally, due to the solubility of Nb in NiAl strengthening of the matrix by solid solution was found.

Whereas Nb exhibited a strong reaction with NiAl at elevated temperatures the Cr particles remained stable but were hardened to a remarkable extent due to further diffusion of Ni and Al. The dependence of room temperature hardness on annealing time for the Cr particles and the NiAl matrix is illustrated in Fig.3. After the HIP process the hardness of Cr is lower than the hardness of the NiAl matrix. After approximately 50 hours annealing time at 1200°C the hardness of the Cr particles surpasses that of the matrix. A further annealing effect is the appearance of pores at the NiAl/Cr interface which is attributed to the Kirkendall mechanism [2]. The size and the density of the pores is dependent on temperature and annealing time. Figs.4a and b show the microstructure of NiAl-10wt%Cr after HIPing (no porosity) and after a 70 hour heat treatment at 1200°C.

Mechanical Properties

The mechanical properties of the monolithic and the composite NiAl materials and the change of these properties due to annealing at 1200°C were investigated by three-point bending tests on plain beams and by compression tests on cylindrically shaped samples. The testing temperatures ranged from room temperature to 1100°C. The three-point bending tests were performed in vacuum whereas the compressive tests were performed in ambient atmosphere.

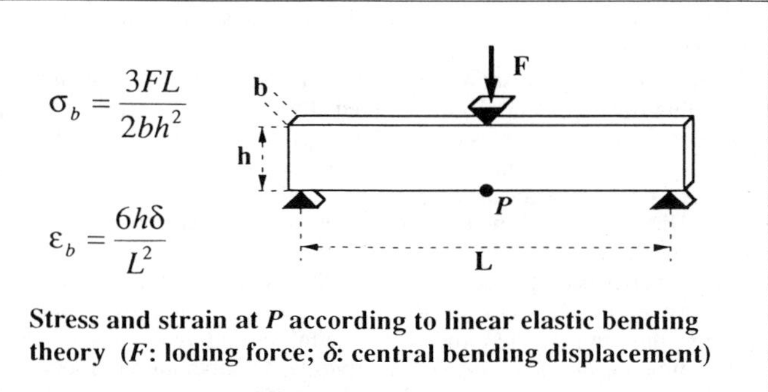

Fig.5: Evaluation of three-point bending tests (see text).

Bending samples of size 5x5x40 mm³ were prepared via spark erosion. In order to avoid premature fracture caused by surface flaws the samples were polished on the tensile side until roughness values less than 0.1µm were obtained. The span of the three-point bending fixture was 32 mm. The experiments were performed with an Instron type testing machine, the crosshead speed was chosen 0.1 mm/min throughout all tests. During each experiment the central loading point displacement and the load were monitored continously. Tests were performed at room temperature and at temperatures between 500°C and 1100°C (temperature variation in steps of 50° and 100°). As long as the material behavior is entirely linear-elastic the load displacement curves can easily be transformed to stress-strain curves according to the linear elastic bending theory (see equations in Fig.5). As these equations are no longer valid in the elastic-plastic regime, stresses and strains obtained from non-linear sections of the load-displacement curves are merely effective quantities. These quantities, further referred to with a "b" index, are useful to compare different materials tested in the same way. However, it must always be kept in mind that they are no true stresses and strains, respectively.

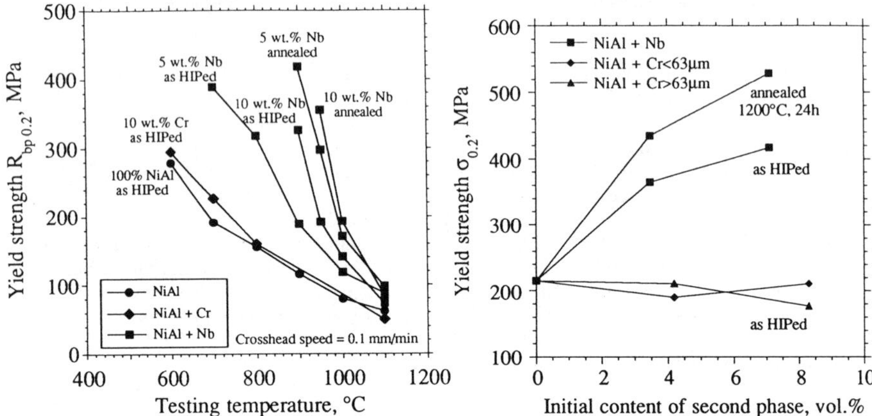

Fig.6: Bending yield strength $R_{bp0.2}$ as a function of testing temperature.

Fig.7: Room temperature compressive yield stress as a function of initial second phase content.

In this study the material behavior is regarded *brittle* if the plastic bending strain-to-failure is less than 0.2% and *ductile* if this limit is exceeded. In the case of *brittle* fracture the materials are characterized by the fracture stress R_{bF}, in the case of *ductile* behavior by their bending yield strength $R_{bp0.2}$ (wich is equal to the bending stress σ_b that corresponds to the plastic bending strain ε_{bp}=0.2%).

All materials investigated showed a *brittle* behavior both at room temperature and at 500°C. The bending fracture values R_{bF} measured at room temperature are listed in Table II. The room temperature strength seems to be raised slightly by Nb in the as-HIPed material and lowered during the subsequent anneal. However, this trend cannot be expressed quantitatively because of the statistical scatter of the data. All materials containing Cr exhibit lower strength values than pure NiAl but there is no clear influence from volume content, Cr particle size or material condition.

At more elevated testing temperatures both the monolithic and the composite materials turn *ductile* . However, the *ductile-to-brittle* transition temperature (DBTT) and the yield stresses just above this temperature turned out to be influenced by the type of second phase and by the annealing time (Table II and Fig.6). The transition temperature of monolithic NiAl is about 500°C. While the presence of Cr particles does not change this temperature considerably, it is raised by Nb. Both the amount of the DBTT shift and the yield stress just above the DBTT increase systematically with Nb content and annealing time.

Table II: Results of three-point-bending tests.

Material	Condition	DBTT (°C)	R_{bF} (MPa)* at room temperature
NiAl	as-HIPed	500-550	459
NiAl-5wt%Nb	as-HIPed	750-800	496
NiAl-5wt%Nb	annealed**	900-950	466
NiAl-10wt%Nb	as-HIPed	900-950	470
NiAl-10wt%Nb	annealed**	950-1000	297
NiAl-5wt%Cr (Cr>63μm)	as-HIPed	500-600	409
NiAl-5wt%Cr (Cr>63μm)	annealed**	600-700	350
NiAl-10wt%Cr (Cr>63μm)	as-HIPed	500-600	411
NiAl-10wt%Cr (Cr>63μm)	annealed**	600-700	404
NiAl-5wt%Cr (Cr<63μm)	as-HIPed	500-600	370
NiAl-5wt%Cr (Cr<63μm)	annealed**	600-700	405
NiAl-10wt%Cr (Cr<63μm)	as-HIPed	500-600	378
NiAl-10wt%Cr (Cr<63μm)	annealed**	600-700	452

*: Mean value of two tested samples
**: Annealing treatment: 1200°C/24h/vacuum

The yielding behavior at room temperature (i.e. below DBTT) and the creep behavior at 1200K (i.e. above DBTT) were investigated by strain-rate controlled compression tests on cylindrically shaped samples of diameter 9 mm and length 18 mm. Experimental details are described in Ref.[3].

Room temperature compression tests were performed at a constant strain rate of $1x10^{-4}$ s^{-1}. The yield strength was defined to be the compressive stress corresponding to 0.2 % plastic strain. In Fig.7 the results are shown as a function of the initial second phase volume fraction. Again it becomes obvious that the presence of Nb has a much greater influence than that of Cr. However, due to pore formation in the case of Cr no tests on annealed NiAl/Cr material were performed. Moreover, as shown in Fig.8 there is a clear correlation between the room temperature yield strength and the brittle-to-ductile transition temperature.

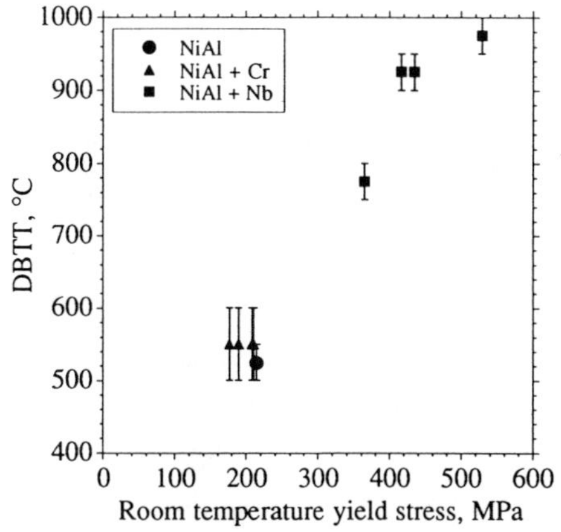

Fig.8: Correlation between compressive yield stress at room temperature and the DBTT measured in bending tests at elevated temperatures.

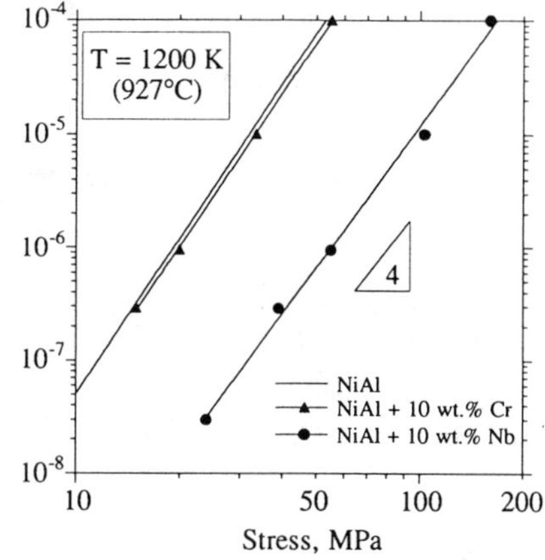

Fig.9: Creep behaviour at 1200K (927°C).

The creep behavior at 1200K (927°C) was investigated by measuring the steady state stress as a function of the applied strain rate, which was varied between $2x10^{-8}s^{-1}$ and $1x10^{-4}s^{-1}$. The results obtained for pure NiAl (as HIPed), NiAl-10wt% Cr (as-HIPed) and NiAl-10wt% Nb (annealed) are shown in Fig.9. Due to the increase of the DBTT in the case of NiAl/Nb the creep strength is considerably higher than in the case of monolithic NiAl and NiAl/Cr composite material. This improvement of creep strength is partly caused by the formation of Laves phase and by solid solution strengthening [4]. However, up to now the quantitative contributions of these two mechanisms are not fully understood.

CONCLUSION

NiAl/Nb and NiAl/Cr composite materials were prepared by a powder metallurgical approach. The thermodynamic stabilty of the second phases at 1200°C was studied. Whereas Nb reacts with NiAl to a ternary Laves phase, Cr shows no reaction with NiAl but is hardened due to in-diffusion of Ni and Al. A further annealing effect is the appearance of pores at the NiAl/Cr interface. The mechanical properties (i.e. DBTT, yield strength, creep behaviour) of the monolitihic and composite NiAl materials and the change of these properties due to annealing at 1200°C were investigated by means of three-point bending and compression tests. While Cr particles do not change the DBTT considerably, it is raised in the case of Nb. In addition, a clear correlation between the room temperature yield strength and the DBTT was found. Creep tests at 1200K (927°C) revealed that the NiAl/Nb composites are remarkably stronger than the NiAl/Cr composites and the pure NiAl material investigated.

Future Work

In bending tests performed below the ductile-to-brittle transition temperature both the pure and the composite NiAl materials investigated in this study show only little ductility. The lack of low temperature ductility is regarded to be an obstacle for their use in mechanical applications, even though the problem of brittleness does not exist in the potential service temperature range. In order to examine the brittle behaviour of this class of materials in detail controlled crack growth experiments and tensile tests at room temperature are currently being performed. By optimising the composition and the distribution of the second phase particles it should be possible to raise the fracture thoughness and achieve increased ductility which results from multiple microcracking even under tensile loading conditions.

ACKNOWLEDGEMENTS

Part of this work was financially supported by the Bundesministerium für Forschung und Technologie der Bundesrepublik Deutschland (BMFT 03 M 30 31).

REFERENCES

1. M.Sherman and K.Vendula, J.Mat.Sci. **21**, 1974 (1986).
2. H.Clemens and H.Bildstein, Z.Metallkd. **83**, 6 (1992).
3. E.Arzt, E.Göhring, and P.Grahle, Submitted to Proc. MRS Fall Meeting, Boston, Mass., Nov/Dec. 1992.
4. G.Sauthoff, Z.Metallkd. **81**, 855 (1990).

FRACTURE BEHAVIOR OF VANADIUM/VANADIUM SILICIDE IN-SITU COMPOSITES

M. J. Strum and G. A. Henshall
Lawrence Livermore National Laboratory, P.O. Box 808, Livermore, CA 94550.

ABSTRACT

The fracture behavior of V-V_3Si in-situ composite alloys has been evaluated for intermetallic phase fractions of 30, 50, and 70 volume percent. In addition, the mechanical properties of the ductile constituent in these composites have been evaluated from two alloys containing silicon at its solubility limit in vanadium at the eutectic temperature and at 1400°C. The fracture toughness was found to increase monotonically with increasing volume fraction of the ductile phase. Fractography has shown evidence of crack bridging by the vanadium solid solution phase and increased ductility within the composite relative to bulk measurements. The toughness results are compared with predictions based on an existing model of ductile-phase toughening using our measurements of component properties.

INTRODUCTION

Refractory metal intermetallic compounds could satisfy needs for high strength and creep resistance at extremely high temperatures if they can be toughened by the dispersion of a ductile second phase [1]. The ductile phase may increase toughness by "bridging" the crack faces [2], thereby inhibiting crack opening, or by blunting the crack tip. In this investigation, "in-situ" methods are being explored as a means of synthesizing V-V_3Si composites, in which the two components of the composite form by phase separation during solidification. Using this system, the intermetallic phase is V_3Si (A15 structure) and the ductile phase is a solid solution of Si in V, V(Si), which appears to precipitate out small particles of V_3Si upon heat treatment. Both phases have densities significantly lower than Ni-based superalloys, which is important in aerospace structures, and vanadium has excellent low temperature ductility, perhaps making it a good toughening agent.

EXPERIMENTAL PROCEDURES

The four alloy compositions produced in the series-1 castings were formulated to contain 30, 50, 70, and 100 percent ductile phase using an assessed V-Si phase diagram [3]. The compositions of the series-1 alloys are listed in Table I. The alloys were arc melted in an argon atmosphere using procedures which attempted to minimize interstitial levels. The melting stock consisted of crushed boules of high-purity silicon and high-purity vanadium sheet sheared into small pieces and acid cleaned in an HNO_3/HF solution. Due to the relatively high hydrogen levels measured in the as-cast material, all test specimens were vacuum degassed at 800°C for 1 h. The level of measured hydrogen decreased substantially after degassing with only marginal increases in oxygen and nitrogen levels. To minimize hydrogen absorption from requisite acid cleaning and to attempt further minimization of interstitial levels from the melt stock, as-received, high-purity vanadium chips were substituted as melting stock in the series-2 castings. The two additional compositions produced were a eutectic V-V_3Si (V-7.3%Si) and a 100% vanadium solid solution alloy (V-2.7%Si). All compositions are reported in weight percentages.

Tensile and fracture toughness specimens were fabricated by electrical discharge machining. The uniaxial tensile properties of the vanadium solid solution alloys were measured using specimens with 7.1 mm gage lengths. The fracture toughness was measured for all of the alloys using single-edge-notched bend (SENB) specimens loaded in three-point bending. The SENB specimens were 3.5x7x30 mm in dimension and tested at a span of 28 mm and a displacement rate of 7.6 mm/s. With the exception of the V-7.3%Si alloy which contained a straight-notch, all SENB specimens contained chevron starter notches. All specimens were precracked by crack extension under continuous loading and crack resistance data (for R-curves) were determined using unloading cycles after each crack growth increment. The crack lengths were determined using surface and COD compliance measurements.

Table I. Alloy chemical compositions.

Series	Condition	Nominal % V_3Si	Si (wt. %)	V (wt. %)	O (ppm)	N (ppm)	H (ppm)
1	As Cast	10	4.21	95.7	370	670	20
1	Degassed	10	4.21	95.7	470	810	9
1	As Cast	50	7.66	92.2	420	860	160
1	Degassed	50	7.66	92.2	410	870	4
1	As Cast	30	5.42	94.5	260	370	52
1	As Cast	70	10.0	89.9	240	480	20
2	As Cast	0	2.7	97.2	290	370	30
2	As Cast	50	7.3	92.6	260	280	72

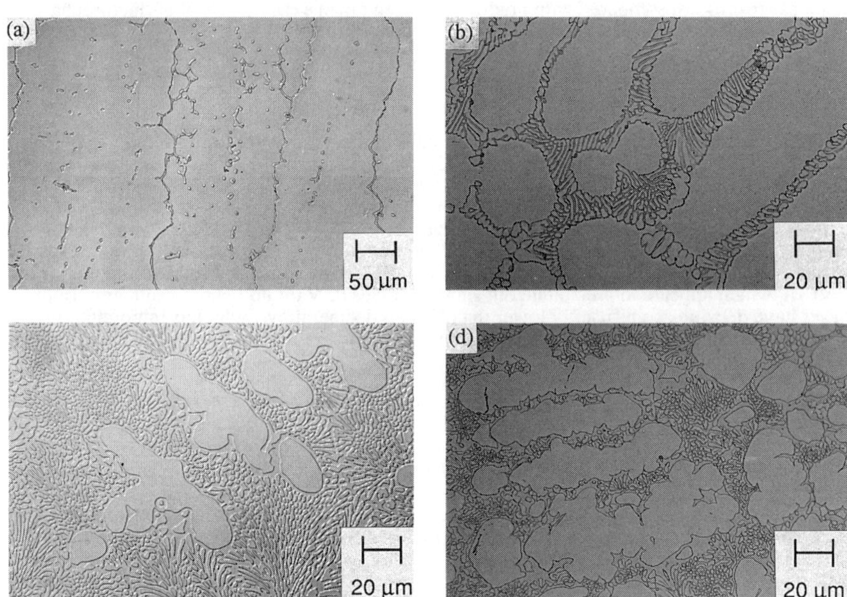

Figure 1. As-cast optical microstructures of series-1 alloys: a) V-4%Si, b) V-5.4%Si, c) V-7.6%Si, and d) V-10%Si.

RESULTS

Microstructures

The as-cast microstructures of the series-1 alloys are shown in the optical micrographs of Figure 1. The V-4%Si alloy matches the reported composition of maximum solid solubility of Si in V at the eutectic temperature [3]. This alloy is not single-phase, however, but contains a significant amount of intermetallic V_3Si which is semicontinuous at interdendritic and grain boundary locations. As shown in Table I, the volume fraction of V_3Si was estimated at 10 percent. The nominal 70% ductile-phase alloy, V-5.4%Si, contains eutectic interdendritic networks within a V(Si) matrix. The predicted eutectic composition alloy [3], V-7.6%Si, contains a matrix of eutectic V(Si) and V_3Si with a minor constituent of dendritic V_3Si islands estimated at 5 volume percent. The nominal 30% ductile-phase alloy, V-10%Si, contains a large fraction of dendritic islands of V_3Si and interdendritic eutectic. In all of the as-cast microstructures, the V_3Si intermetallic phase is the discontinuous phase. The morphology of the eutectic V_3Si is best described as ribbons with length to width ratios of up to 10 and width to thickness ratios of approximately 1. Transmission electron microscopy verified the component phases as bcc vanadium and V_3Si which is an A15 phase. A typical region of the V-7.6%Si alloy in bright-field is shown in Figure 2a. The multiple orientations of the V_3Si phase reveal the ribbon morphology. The presence of a low density of dislocations within the V_3Si phase and high density within the vanadium solid solution is illustrated in Figure 2b.

Figure 2. TEM micrographs of V-7.6%Si showing; a) a bright-field image of the V-V_3Si eutectic, and b) the presence of dislocations in both phases.

In a second series of castings, a V-2.7%Si alloy was selected to better represent the characteristics of the vanadium solid-solution constituent within the composite microstructures. This composition matches the terminal solubility limit of Si in V at 1400°C as predicted in the V-Si phase diagram [3]. In addition, a V-7.3%Si alloy was selected as a more probable eutectic composition. The V-2.7%Si alloy consists of a single-phase vanadium solid solution of large grain size (approximately 400 micron diameter) with a low-angle substructure revealed by numerous etch pits. The V-7.3%Si alloy produced a near-eutectic microstructure with less than 1 volume percent of pro-eutectic intermetallic islands. Chang *et al* [4] also observed that the eutectic composition is leaner in Si than reported by the phase diagram.

Component Properties

Although a 100% V_3Si intermetallic alloy was not prepared, the strength and toughness properties of proeutectic intermetallic islands were probed by microhardness techniques. As expected, the microhardness of the intermetallic is quite high, averaging 750 HV (200 gm), and produced cracks emanating from the corners of the pyramidal indentor. The fracture toughness of the V_3Si phase was estimated by assuming the formation of a penny-shaped crack beneath the indentor [5]. While the cracks were invariably blunted or bridged by the eutectic regions, an upper-bound estimate of the fracture toughness was made using the maximum observed crack lengths. The V_3Si toughness was calculated as 1.3 MPa· $m^{1/2}$.

The microhardness of the ductile component in series-1 and series-2 alloys was measured from the vanadium solid solution regions of the V-2.7%Si and V-4%Si alloys, with average values of 207 HV and 320 HV, respectively. The microhardness of a series-2 alloy of 100% vanadium was measured to be 107 HV, showing that the extent of solid solution strengthening from silicon additions is substantial. The uniaxial tensile behavior of the as-cast V-4%Si alloy was brittle with fracture occurring prior to bulk yielding at a maximum stress of 550 MPa. In the V-2.7%Si alloy, a yield strength of 430 MPa, fracture strength of 435 MPa, and maximum uniform elongation of 2.2% were measured in one sample while fracture occurred prior to yielding in another. The fracture toughness of the as-cast V-4%Si alloy was measured as 21.5 MPa· $m^{1/2}$. The fracture morphology was predominantly intergranular with grain diameters on the fracture surface between 0.4 and 2.2 mm. By examining metallographically polished and etched fracture profiles, preferential cracking along V_3Si networks was revealed. The fracture toughness measurements of the V-2.7%Si alloy ranged between 26 and 37 MPa· $m^{1/2}$.

Composite Fracture Behavior

The composite toughness of the $V-V_3Si$ specimens increases monotonically with increasing nominal percentage of ductile-phase constituent, as shown in Figure 3a. The crack propagation behavior was similar in all of the series-1 alloys. Pop-in type crack extensions occurred which were large enough, except for the V-5.4%Si alloy, to preclude R-curve analysis. The fracture surfaces contained large cleavage facets separated by micro-roughened regions containing changes in fracture path which match the coarseness of the eutectic microstructure.

The series-2 V-7.3%Si specimens showed mixed behavior with respect to crack propagation. In three specimens, the crack propagated through most of the specimen upon first initiation of a crack from the chevron starter notch. Replacing the chevron starter notch with a straight but deeper notch (a/w of 0.5) resulted in short pop-in crack extensions. The R-curve behavior of this sample is plotted in Figure 3b along with that of the series-1 V-5.4%Si alloy for crack extensions past the end of the chevron starter notch. Although the toughness values clearly increase with crack extension for the V-5.4%Si alloy, the toughness variations in the V-7.3%Si alloy are relatively minor. Ductile-phase plasticity was greatest in the series-2 V-7.3%Si alloy and, as shown in Figure 4, the fracture surface contains a mixture of micro-ductile and cleavage-dominated regions approximately equal in area fraction. In the micro-ductile areas, brittle V_3Si facets are separated by bridging ligaments of V(Si). These ligaments have clearly necked prior to fracture but interface debonding was not observed on the fracture surfaces and appears to be quite limited in extent.

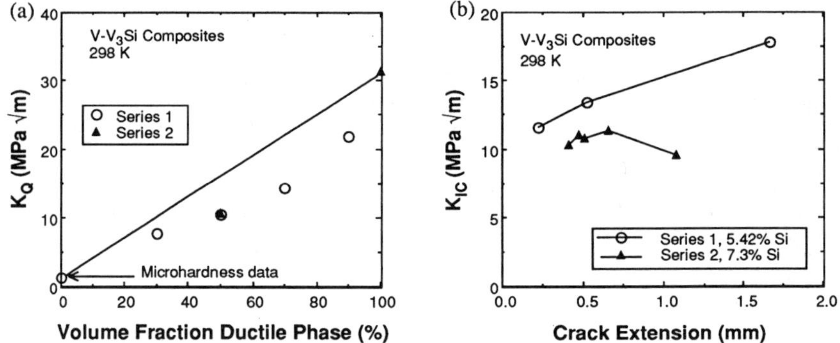

Figure 3: (a) the average fracture toughness of $V-V_3Si$ composites as a function of ductile phase fraction, and (b) fracture toughness vs crack extension for V-5.4%Si and V-7.3%Si alloys.

DISCUSSION

In all of the V-Si alloys prepared, including both hypo- and hyper-eutectic compositions with up to 70% volume fraction of V_3Si, the vanadium solid solution is the continuous phase in the as-cast microstructures. From the viewpoint of obtaining high creep resistance in addition to ductile-phase toughening, a continuous intermetallic phase is clearly desirable. Realization of these microstructures may require higher intermetallic phase fractions and/or thermomechanical processing to intersperse the solid solution phase within the intermetallic islands.

In measurements of component properties, the ductility of the vanadium solid solution alloys was unexpectedly low. This can be attributed to strengthening by the silicon additions and interstitial contamination which raise the ductile-brittle transition temperature (DBTT). For example, increases in nitrogen from 30 to 600 ppm raise the hardness of vanadium from 100 HV to 200 HV while raising the DBTT above -55°C [6]. The puzzle is to explain the enhanced ductility of the V(Si) phase within the composite relative to bulk measurements in either the V-2.7%Si or V-4%Si alloys. A likely explanation is that the total elongation in the bulk tensile

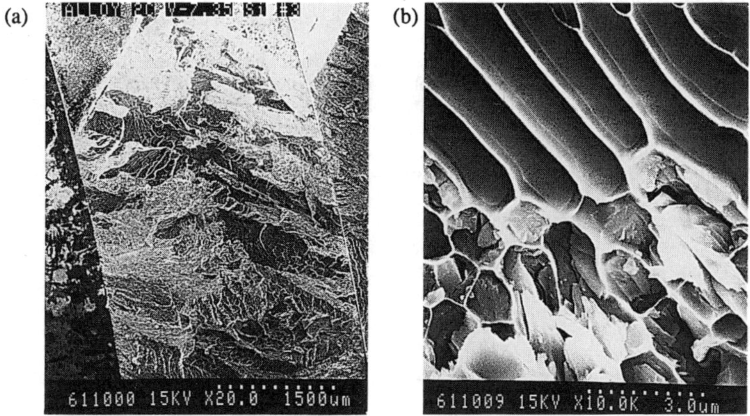

Figure 4. Fracture surfaces of the V-7.3%Si alloy toughness specimen.

specimens is limited by grain boundary and intermetallic network properties, and not by the intrinsic ductility. Intergranular fracture and fracture at intermetallic networks were both observed. Conversely, in the composite, the strains within the ductile ligaments are localized at points of intermetallic fracture. This is more likely to test the intrinsic properties of the V(Si) phase, even though the constraint remains high due to limited interface debonding.

To quantitatively evaluate the effectiveness of the V(Si) phase in bridging cracks, it is useful to compare the data with models for the increase in fracture toughness due to crack bridging. The most prevalent theory is that of Ashby *et al.* [7], or a modification thereof presented by Ravichandran [8]. The equation derived by Ashby *et al.* is:

$$K_{IC} = K_m + E\left[Cv_f \frac{\sigma_o}{E} a_o\right]^{1/2} \quad (1)$$

where K_m is the fracture toughness of the matrix and E, v_f, σ_o and a_o are the Young's modulus, volume fraction, yield strength and radius of the ductile particles. The constraint parameter, C, is given by

$$C = \frac{1}{\sigma_o a_o} \int \sigma(u)\,\mathrm{d}u \quad (2)$$

where σ is the stress and u is the displacement of the ductile particle. For the conditions of concern in the following discussion, the equation given by Ravichandran [8] produces results nearly identical to those of equations (1) and (2).

Since the series-2 eutectic alloy demonstrated some bridging behavior (c.f. Fig. 4), we consider here only the eutectic composition. For this alloy, $v_f = 0.5$, $a_o \approx 1.5$ μm and $\sigma_o = 430$ MPa. The elastic modulus for the ductile phase is approximately 130 GPa. The value of C, unfortunately, is not known. The experiments of Ashby *et al.* [7] on Pb wires encapsulated in glass suggest that C lies between 1.6 and 6, with lower values corresponding to higher constraint of the ductile phase. Using these values, equation (1) predicts that K_{IC} lies between 9.5 and 17.2 MPa$\cdot$ m$^{1/2}$. The experimental value is 10.5 MPa$\cdot$ m$^{1/2}$, suggesting that the constraint on the ductile phase may be high in the V-V_3Si composites. This implies that there is relatively little debonding of the V_3Si-V(Si) interface during fracture, which is consistent with our observations (c.f. Fig. 4). It is noteworthy that a value of $C = 2$ gives a predicted fracture toughness consistent with the data. This is well below the value of 4 used by Ravichandran [8] to model a wide range of ductile-phase toughened materials.

The model predictions must be interpreted with caution. First, the model corresponds to the case in which the brittle phase is continuous, with an interspersed ductile phase. As shown in Figs. 1-2, the brittle phase is discontinuous in the V-V_3Si composites. In addition, other possible mechanisms of toughening, including large scale bridging [9], may influence the measured K_{IC} but are not included in the model of Ashby *et al.* The model is useful, however, in suggesting the need to coarsen the microstructure. All other parameters remaining constant (and using $C = 2$), an increase in a_o from 1.5 to 10 μm would increase the toughness from 10.5 to 25 MPa· $m^{1/2}$. Promoting debonding and increasing σ_o also would improve toughness, but are probably more difficult to achieve.

CONCLUSIONS

A study of the fracture behavior of V-V_3Si in-situ composites has led to the following findings:

1. The fracture toughness of the as-cast V(Si)-V_3Si composites increases monotonically with increasing ductile-phase fraction.
2. The ductility of the V(Si) phase is much larger in the composite than in bulk measurements, despite the added constraint imposed by the intermetallic.
3. In the eutectic alloys, the extent of plastic stretching by the ductile-phase ligaments increases with reduced O and N interstitial levels.
4. The intermetallic phase is discontinuous in the as-cast eutectic microstructures, which is expected to result in poor creep resistance unless methods are developed to produce continuous intermetallic networks.
5. It appears that microstructural coarsening could significantly benefit the fracture toughness of the V(Si)-V_3Si composites.

ACKNOWLEDGEMENTS

The authors would like to thank D. Hiromoto, R. Kershaw, E. Sedillo, and M. Wall for their contributions to this investigation. This work was performed under the auspices of the U. S. DOE for the Lawrence Livermore National Laboratory under contract W-7405-Eng-48.

REFERENCES

1. M. G. Mendiratta, J. J. Lewandowski, and D. M. Dimiduk, *Metall. Trans.* **22A**, 1573-1583 (1991).
2. L. S. Sigl, P. A. Mataga, B. J. Dalgleish, R. M. McMeeking, and A. G. Evans, *Acta Metall.* **36**, 945-953 (1988).
3. J. F. Smith, Ed., *Phase Diagrams of Binary Vanadium Alloys*, ASM International, Metals Park, OH (1989).
4. K.-M. Chang, B.P. Bewlay, J.A. Sutliff and M.R. Jackson, *JOM* 44, 59-63 (1992).
5. A.G. Evans, "Fracture Toughness: The Role of Indentation Techniques", *Fracture Mechanics Applied to Brittle Materials, ASTM STP 678*, S. W. Freiman, Ed., American Society for Testing and Materials, pp. 112-135 (1979).
6. M.J. Strum and L.M. Wagner, "Interstitial Embrittlement in Vanadium Laser Welds," *Proc. Intl. Conf. on Lasers*, F. J. Duerte and D. G. Herns, eds. STS Press, McLean, VA, pp. 474-478 (1992).
7. M. F. Ashby, F. J. Blunt and M. Bannister, *Acta Metall.* **37**, 1847-1857 (1989).
8. K. S. Ravichandran, *Scr. Metall. Mater.* **26**, 1389-1393 (1992).
9. K. S. Ravichandran, *Acta Metall. Mater.* **40**, 1009-1022 (1992).

DUCTILE-TO-BRITTLE TRANSITION IN $MoSi_2$

S. R. SRINIVASAN, R. B. SCHWARZ, and J. D. EMBURY
Center for Materials Science, Mail Stop K-765
Los Alamos National Laboratory, Los Alamos, NM 87545, USA

ABSTRACT

We have studied the mechanical behavior of two fine grained $MoSi_2$ alloys containing 0.61 at% (0.19 wt%) and 0.29 at% (0.09 wt%) oxygen, respectively. By preparing these alloys in almost identical fashion, their only difference was their oxygen content. The mechanical behavior was studied by four-point flexure tests in unnotched specimens between 800°C and 1400°C. We interpret the mechanical behavior data in terms of Davidenkov diagrams which describe the dependence of the *apparent* ductile-to-brittle transition temperature on the SiO_2 content in the alloys.

INTRODUCTION

$MoSi_2$ and $MoSi_2$-based composites have potential for high-temperature structural applications due to their excellent oxidation resistance and high melting temperature (2030°C). In addition, $MoSi_2$ has a density of 6.24 g cm^{-3}, lower than that of Ni-based superalloys. Two main problems of $MoSi_2$ have slowed this development: a) low-temperature brittleness and b) low strength at high temperatures ($T > 1200$°C) [1,2]. The literature reports the ductile-to-brittle transition temperature (DBTT) of $MoSi_2$ ranging from 900 to 1300°C [1,3,4,5]. In considering this data, attention must be given to the experimental method and the stress state used to determine the transition temperature.

The *intrinsic* DBTT can be clearly defined for a single crystal deformed in simple shear, as the temperature at which the stress necessary to nucleate and glide dislocations becomes equal to the cleavage stress. For polycrystalline specimens, however, the comparison of plasticity and fracture is obscured by phenomena such as (a) a lack of a sufficient number of glide systems to accommodate polycrystalline plasticity (von Mises criterion), (b) dependence of flow and fracture stresses on grain size, (c) diffusional grain boundary sliding, and (d) the effect of intergranular phases on grain boundary sliding. In addition, tests are usually done in conditions other than pure shear (e.g., superimposed hydrostatic stress), which may further obfuscate the DBTT determination. Thus, the criterion for defining the DBTT from tests in polycrystals becomes highly subjective. In spite of these complications, various authors have defined the DBTT in polycrystalline $MoSi_2$ from the observation of plastic deformation during four-point bend tests.

Results from compression tests on single crystal $MoSi_2$ by Kimura et al. [6] show only 0.3 to 0.6% fracture strain at 1300°C depending on the orientation of the specimen. The strain to fracture is about 2% at 1400°C. Such a small fracture strain at 1300°C, even in compression, suggests that ductility is very much limited and supports the contention in ref. 5 that the DBTT is closer to 1300°C. However, similar experiments by Umakoshi et al. [7] show the presence of $\{110\}\langle 331\rangle$ and $\{013\}\langle 331\rangle$ dislocations in $MoSi_2$ deformed at 900°C. These authors argue that these dislocations provide at least the five independent slip systems necessary to satisfy the von Mises criterion for polycrystalline plasticity. Further, they report that single crystals of $MoSi_2$ tested in compression show about 0.3% ductility at 1000°C and between 2 and 2.6% ductility around 1350°C. These observations support the earlier value of the DBTT, around 900°C. However, it can be argued that the compressive stress state in these tests helps suppress fracture and causes an apparent decrease in the transition temperature. Thus, the presently available results on single crystal $MoSi_2$ do not conclusively establish the value of the DBTT.

Mitchell et al. [8] studied the dislocations present in polycrystalline $MoSi_2$ deformed in bending between 1200°C and 1500°C and found that they were of the type $\{110\}\langle 111\rangle/2$ and $\{0kl\}\langle 100\rangle$. They did not find $\langle 331\rangle/2$ dislocations in these specimens. Dislocations were also observed under hardness indents introduced in the temperature range from ambient temperature to 1000°C.

In a recent paper, Aikin [5] puts the DBTT between 1300°C and 1400°C. He deduced this value from the observed onset of ductility in a polycrystalline sample with 0.19 at% (0.061 wt%) oxygen and from a comparison of the mechanical properties measured in flexure and compression in two samples having different oxygen contents: 2.06 at% (0.65 wt%) and 0.19 at% (0.061 wt%) oxygen, respectively. However, in addition to the difference in oxygen content, these two samples also differed in their Mo/Si atom ratio (by 11%) and in their carbon content (by a factor of 34!). Aikin concluded that the early reports of the DBTT ($\approx$ 1000°C) are incorrect and that the discrepancy is due to the presence of a continuous glassy silica (SiO_2) phase at the grain boundaries of the $MoSi_2$ alloys which show the lower transition temperature. As this SiO_2 phase softens with the increase in the test temperature, the mechanical test would show a low apparent transition temperature.

EXPERIMENTAL

We prepared $MoSi_2$ powders by mechanical alloying (MA) starting from a mixture of 99.9% pure Mo and 99.9999% pure Si powders. MA is a high-energy ball-milling process where the intimate atomic-level mixing of the starting materials is accomplished by intense mechanical working. To minimize the oxygen content in the alloys, all powder processing was done under a high-purity argon atmosphere. The MA powders were then consolidated by hot pressing in graphite dies in a flowing argon atmosphere at 1500°C and a pressure of 12 MPa. Two alloys were prepared this way which, in agreement with our earlier notation [9], we designate as MA1 and MA5. Alloy MA1 was prepared by loading the graphite die in air. It had 0.61 at% (0.19 wt%) oxygen and 0.18 at% carbon. Alloy MA5 was prepared by loading the MA powder into the die under high-purity argon. It had 0.29 at% (0.09 wt%) oxygen and the same carbon content. Except for the way the die was loaded, all other synthesis steps were identical. The alloys had similar grain sizes of less than 5 μm. We have previously presented details of microstructural characterization and oxygen analysis, using nuclear (d,p) reactions, of these alloys [9]. Four-point flexure tests were performed at temperatures ranging from 800°C to 1400°C in air using a SiC fixture (10 and 20 mm loading points separations) and a screw-type testing machine at a crosshead speed of 0.5 mm min^{-1}. Yield strengths were determined from the load at a displacement of 0.05 mm offset from the linear portion of the loading curve. For the samples that fractured at smaller displacements, the fracture load was used for this calculation. These values are expected to be an underestimate of the yield strength. For the more ductile samples the tests were stopped at load-point displacements of between 1.2 and 1.6 mm.

RESULTS AND DISCUSSION

Figure 1 shows load-displacement curves at 1200°C for the MA1 and MA5 alloys. The origin for the second curve has been displaced horizontally for clarity. The figure clearly shows that the plastic strain is much higher for the MA1 alloy. As the grain sizes and composition of these two alloys were quite similar, the difference in strain must be attributed to the presence of glassy SiO_2 at or near the grain boundaries.

The outer fiber strains at fracture, E_{max}, was calculated using the formula in ref. 10:

$$E_{max} = \frac{h}{\left[\frac{1}{4\,y_L^2}(La - a^2)^2 + a^2\right]^{1/2} + h}$$

where h is the thickness of the beam, y_L is the deflection at the load points, and L and a are the outer and inner spans, respectively. Figure 2 shows the temperature dependence of the outer-fiber strain to fracture. For all test temperatures, the MA5 specimens showed smaller plastic strain at fracture compared to the MA1 samples. Further, the MA1 samples did not fracture at 1300°C and 1400°C, and only one of the three MA1 samples tested at 1200°C fractured at a displacement of 0.58 mm (2.7 % strain). The MA1 samples that did not fracture were deformed to a displacement of 1.5 mm (approx. 7% strain) before the test was stopped. In contrast, all the MA5 samples fractured at all test temperatures. In summary, the data in Figs. 1 and 2 clearly shows that the SiO_2 content has a profound influence on the high-temperature plasticity of $MoSi_2$.

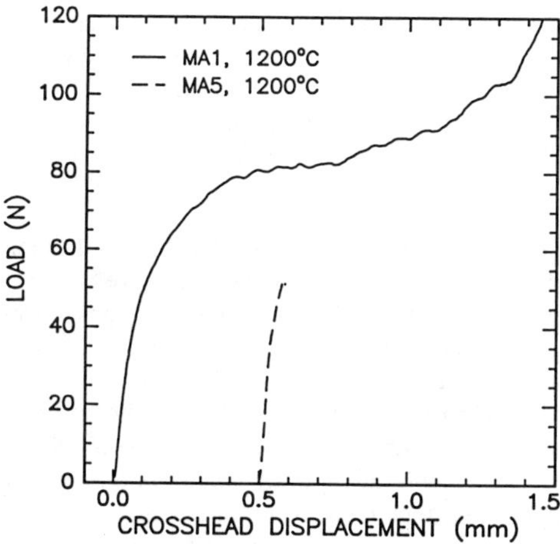

Figure 1. Load versus displacement curves for MA1 and MA5 tested in four-point flexure at 1200°C.

Figure 3 shows the temperature dependence of the yield stress. We have joined the data in Figs. 2 and 3 by curves but, because at present our data for the MA5 sample only covers the T > 1200 °C regime, and we only have one data point for 1200°C, the mechanical behavior expected below 1200°C has been shown by an horizontal dashed line in Fig. 3. The rationale for these curves will be explained below.

We have previously shown by TEM analysis that in slowly cooled $MoSi_2$ specimens the amorphous SiO_2 is present as discrete particles at the grain boundaries in both alloys [9]. The SEM micrographs in Figures 4 and 5 show the fracture surfaces from the MA1 and MA5 specimens deformed in four-point flexure at 1200°C and 1300°C, respectively (i.e., fractured at approximately the same stress). The fracture is intergranular in both alloys. However, there is difference in the finer detail. The surfaces of the grains in the MA1 alloy (Fig. 4) show a dimpled appearance whereas the grains in the MA5 alloy (Fig. 5) show a smooth appearance. Further, the MA5 alloy reveals discrete SiO_2 particles, between 500 and 700 nm in size, on the surfaces of the grains, whereas such particles are not as clear in the MA1 alloy. The micrographs suggest that the glassy SiO_2 in the MA1 alloy forms an almost *continuous* layer at the grain boundaries whereas in the MA5 alloy, it forms *discrete* pockets.

Aikin [5] reported lack of plasticity in polycrystalline $MoSi_2$ alloys containing 0.19 at% (0.061 wt%) oxygen tested in four-point flexure at 1000°C, 1200°C, and 1300°C. Further, he noticed a mixed mode of fracture at these temperatures. The MA5 alloy in our study has a similar oxygen content to Aikin's alloy but showed no transgranular failure when tested in flexure at 1200°C. This difference in fracture behavior is most likely due to the coarser grain size in Aikin's sample, over 20 μm with some grains as large as 100 μm, compared to our alloys with grain sizes less than 5 μm. In agreement with this, the yield strengths in our alloys are almost an order magnitude lower than that in Aikin's alloys.

In considering the mechanical behavior of polycrystalline $MoSi_2$, care must be taken to distinguish between an intrinsically brittle behavior and the competition which arises between fracture and extensive plasticity, the later influenced by factors such as grain size and the presence of SiO_2 inclusions at the grain boundaries. Although mechanical tests in polycrystalline specimens cannot provide the *intrinsic* DBTT of $MoSi_2$, the data can be rationalized to provide an *apparent* DBTT and trends in this transition, as we explain next.

The plastic strain to failure in Fig. 2 is based on the observed outer fiber strains and clearly shows that above 1200°C the material containing more SiO_2 at the boundaries is more ductile. This is because at the test temperature (and imposed strain rate) the SiO_2 can relax at a stress lower than that required to initiate fracture. In Figure 3 we rationalize the yield stress behavior by constructing diagrams of the type originally proposed by Davidenkov [11] to explain the brittle-to-ductile transition in steels. The horizontal lines are the brittle strength of the polycrystalline specimens at low temperatures in the absence of general yielding, which are assumed to be temperature independent. The temperature-dependent solid curves represent the general yielding which involves the deformation of the SiO_2 films at the boundaries. It is clear that the higher the SiO_2 content, the lower the yield stress. It is apparent that the brittle strength is also strongly affected by the change in the SiO_2 content. However, further low

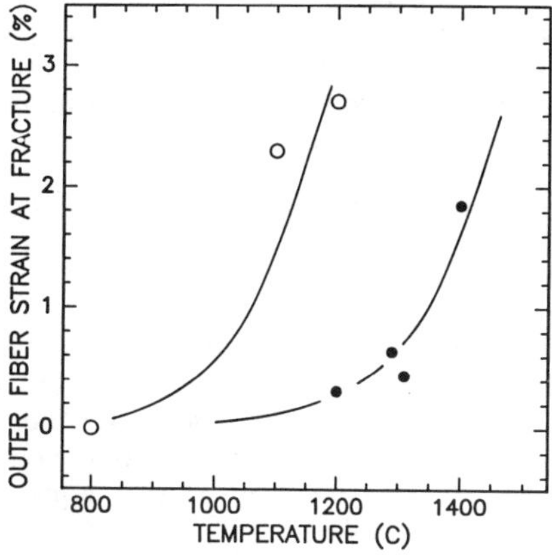

Figure 2. Temperature dependence of the plastic strain to failure in the MA1 (open symbols) and MA5 alloys (solid symbols).

LOG_{10} YIELD STRENGTH (MPa)
2.5
2.0
1.5
1.0
800
1000
1200
1400
TEMPERATURE (C)

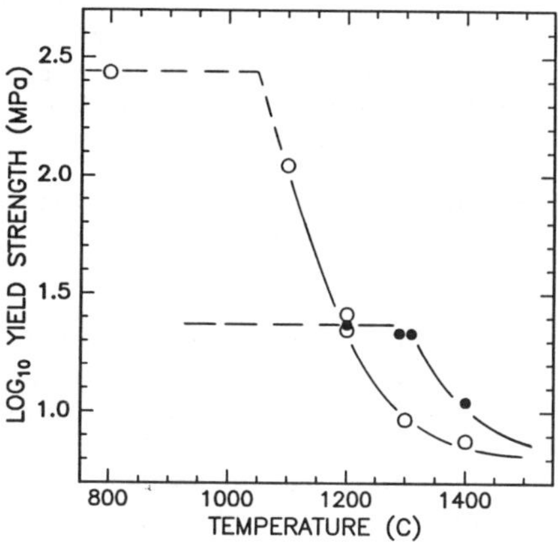

Figure 3. Temperature dependence of the yield stress in the MA1 (open symbols) and MA5 alloys (solid symbols).

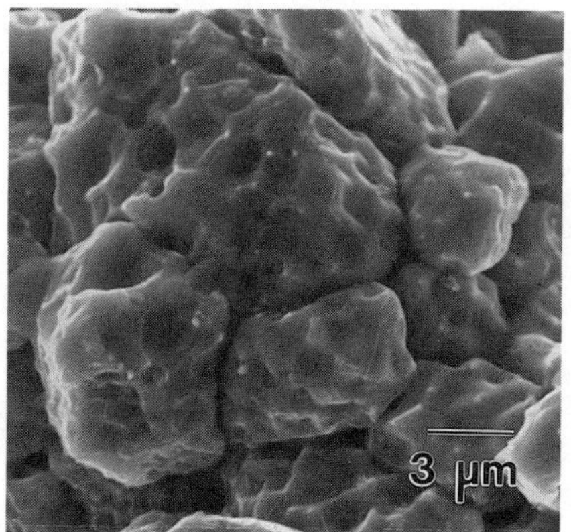

Figure 4. SEM micrograph of the fracture surface of the MA1 alloy tested in flexure at 1200°C.

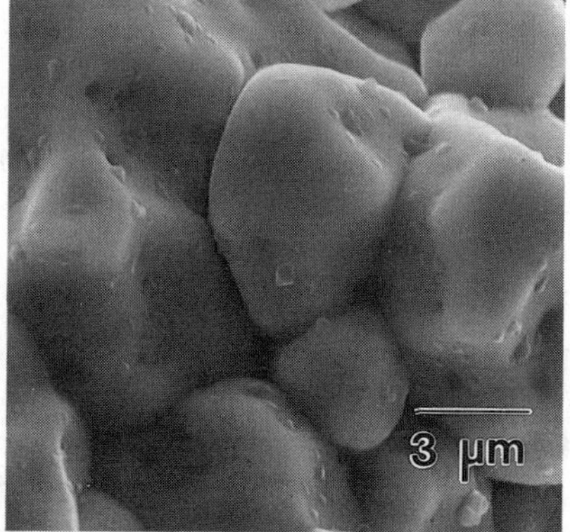

Figure 5. SEM micrograph of the fracture surface of the MA5 alloy tested in flexure at 1300°C.

temperature measurements on the MA5 alloy are needed to confirm this result. The intersection of the dashed and solid curves define *apparent* DBTTs. The salient feature of this plot is that an increase in oxygen content (by approximately a factor of two) results in a decrease in the DBTT of about 250°C. It should be remembered, however, that the yield strength and the *apparent* DBTT.

temperature values resulting from these diagrams are only meaningful in the context of the present fine grain sizes, test mode, imposed deformation rate and, of course, SiO_2 contents.

The data in Figs. 2 and 3 provide a rationalization for the important role of the intergranular SiO_2 in $MoSi_2$ and are in accord with conclusions of Aikin. It must be emphasized that they are an initial rationalization. It is certainly of value to extend this approach by constructing Davidenkov diagrams which encompass the influence of grain size and stress state in addition to oxygen content in order to provide a more comprehensive view of the ductile-to-brittle transition in $MoSi_2$.

ACKNOWLEDGEMENTS

Work supported by the US Department of Energy, Basic Energy Sciences. We thank Mr. R. Castro for useful discussions and Mr. Dick Hoover for hot pressing the $MoSi_2$ powders and help with the four-point bend tests.

REFERENCES

1. W. A. Maxwell, National Advisory Committee for Aeronautics, Reports RM E9G01 (1949).
2. J. Schlichting, *High Temp. - High Press.* 10, 241 (1978).
3. E. Fitzer, O. Rubisch, J. Schlichting, and I. Sewdas, *Sci. Ceram.* 6, XVIII (1973).
4. D. H. Carter, J. J. Petrovic, R. E. Honnell,and W. S. Gibbs, *Ceram. Eng. Sci. Proc.* 10, 1121 (1989).
5. R. M. Aikin, Jr., *Scripta Metall. Mater.*, 26, 1025 (1992).
6. K. Kimura, M. Nakamura, and T. Hirano, J. Mater. Sci. 25, 2487 (1990).
7. Y. Umakoshi, T. Sakagami, T. Hirano, and T. Yamane, *Acta Metall.* 38, 909 (1990).
8. T. E. Mitchell, R. G. Castro, J. J. Petrovic, S. A. Maloy, O. Unal, and M. M. Chadwick, *Mater. Sci. and Eng.* A155, 241 (1992).
9. R. B. Schwarz, S. R. Srinivasan, J. J. Petrovic, and C. J. Maggiore, *Mater. Sci. Eng,* A155, 75 (1992).
10. G. W. Hollenberg, G. R. Terwilliger, and R. S. Gordon, *J. Am. Ceram. Soc.* 54, 196 (1971).
11. N. N. Davidenkov, *Dinamicheskaya Ispytania Metallov*, Moscow, (1936). See also F. A. McClintock and A. S. Argon, editors, Mechanical Behavior of Materials (Addison Wesley, Reading, Massachussets, 1966), p. 564.

INTERFACE REACTION IN SiC REINFORCED Al_3Ti-BASED INTERMETALLIC MATRIX COMPOSITES

JIAN WANG, QIAN WANG, SHIPU CHEN, JIAN SUN AND GENGXIANG HU
Department of Materials Science, Shanghai Jiao Tong University,
Shanghai 200030, China

ABSTRACT

The Al_3Ti-based $L1_2$ intermetallic alloy $Al_{66}Fe_9Ti_{25}$ has shown low density, high oxidation resistance, good strength and appreciable compression ductility at room temperature. To explore the approach of combined toughening and strengthening of this alloy, composites are being studied. The interface reactions between $Al_{66}Fe_9Ti_{25}$ matrix and SiC reinforcement material during different fabrication processes were investigated using SEM, EDS, EPA and X-ray diffraction analyses. The extent of reaction was determined and the reaction products were identified. It was found that SiC is chemically incompatible with the $Al_{66}Fe_9Ti_{25}$ matrix, strong interfacial reactions occurred during fabrication and the formation of ternary compounds leads to the deterioration of the matrix alloy.

INTRODUCTION

Although the DO_{22} intermetallic compound Al_3Ti has attractive characteristics such as low density, excellent oxidation resistance and relatively good strength, it is extremely brittle at ambient temperature[1]. A generally accepted approach to improve the ductility of the compound is to change its DO_{22} tetragonal structure to the cubic $L1_2$ structure by alloying with Fe, Mn, Cr, Cu etc [2–4]. The $L1_2$ Al_3Ti-based intermetallic alloys have shown appreciable compression ductility at room temperature, while they are still brittle in tension [5,6]. Therefore, further investigations are being carried on to explore the approach of combined toughening and strengthening for these alloys. In addition to research on monolithic $L1_2$ Al_3Ti alloys [6], composites of Al_3Ti-based alloys are also under investigation in our laboratory. It is known that the interfacial reactions between reinforcement and matrix during fabrication, or at elevated service temperatures, play an important role in the mechanical properties of metal matrix composites. The purpose of the present paper is to address the interfacial reactions in the $SiC/Al_{66}Fe_9Ti_{25}$ composites and the effects of consolidation process on the reactions, so as to provide information required for the development of Al_3Ti-based intermetallic matrix composites.

EXPERIMENTAL

The matrix alloy is a $L1_2$ structure Al_3Ti-based compound with nominal composition $Al_{66}Fe_9Ti_{25}$(at.%), which was prepared by arc-melting in argon on a water-cooled copper hearth. The cast buttons were used as master material for fabricating the composites.

The composites were fabricated by different techniques for comparison:

(1) Pieces of $Al_{66}Fe_9Ti_{25}$ alloy mixed with tiny SiC crystallites ($\sim$1mm) were arc-remelted in argon on the water-cooled copper hearth. The reason for using SiC crystallites instead of particulates is that a larger area of interface can be obtained for the convenience of inspection.

(2) $Al_{66}Fe_9Ti_{25}$ powders (-320 mesh) produced by ball-milling were mechanically blended into a mixture with 15 vol.% SiC particles ($\sim$14 μm in average size). After compacted by cold-pressing and heated up to 1023K for 1 hour in vacuum, the powder metallurgy (PM) compacts were consolidated by: (a) remelting in induction furnace and cast, or (b) HIPing at 1473K under 70 MPa for 2 hours.

The consolidated composites were examined by X-ray diffraction, SEM, EDS and EPA analyses.

RESULTS AND DISCUSSION

Microstructures

The interface reaction zone in the arc melted sample is shown in Fig.1(a); a distinct layer of reaction products can be observed in the interface region which is approximately 2μm thick with some columnar products stretched into the matrix alloy. Fig.1(b) shows the microstructure in the induction-remelted PM compact; the reaction between the SiC and matrix is so strong that the reaction zone is no longer a distinct layer but spreads out into the matrix forming a wide reaction region. Much more strong reactions occurred in the HIP'ed material, the microstructure shown in Fig.1(c) revealed that the reaction products spread all over the compact and no distinct reaction region can be determined.

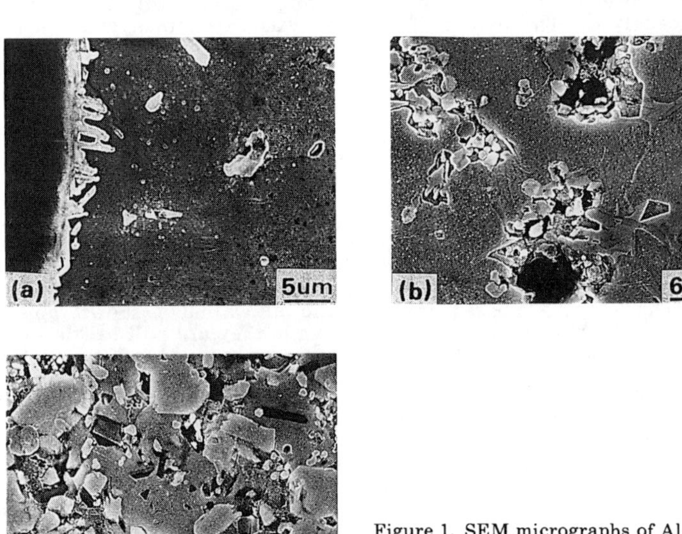

Figure 1. SEM micrographs of $Al_3Ti(L1_2)$/SiC interface after fabrication
(a) arc-melted compact
(b) induction remelted compact
(c) HIP'ed PM compact.

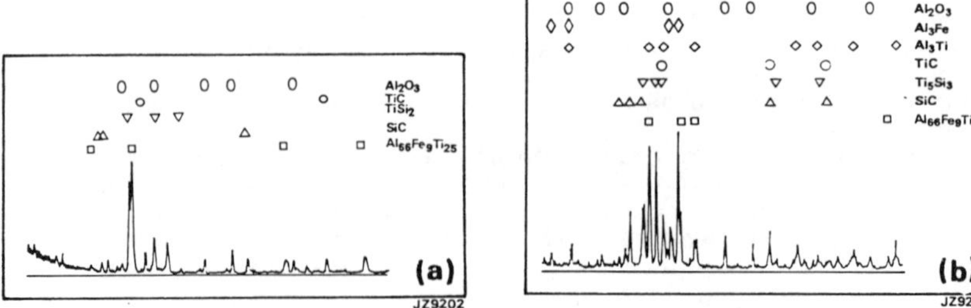

Figure 2. X-ray diffraction spectra of (a)induction remelted and (b)HIP'ed PM sample

X-ray diffraction spectra

The reaction products in the arc-melted sample could not be detected by X-ray diffraction because of the low volume present. In the composite fabricated by induction remelting of the PM compacts, the interfacial reaction products were identified as $TiSi_2$ and TiC (Fig.2a). Moreover, Al_2O_3 was also identified from the X-ray spectrum, which is not a product of the interfacial reaction but was formed by oxidation during fabrication. The diffraction spectrum of the HIP'ed sample is rather complicated (Fig.2b). Ti_5Si_3, TiC, Al_3Ti and Al_3Fe, as well as Al_2O_3 were identified.

Microanalyses

To determine the compositional variations in the interfacial region and identify the microstructural species of the reaction products, the elemental distributions were measured by EDS and shown by characteristic X-ray mapping and line-scanning results. The arc-melted composite was examined first since a distinct reaction layer was observed. The compositional analyses were performed by SEM/EDS on the SiC/matrix interface region. EDS data was not calibrated against standards and carbon could not be measured. The results were plotted versus distance to obtain a set of curves in Fig.3. The corresponding microstructures are shown in Fig.4, together with X-ray mapping and line scanning results examined by EPA which are consistent with the EDS results. It can be seen that the interfacial reaction has led to the redistribution of all the elements involved. The peaks of Ti and Si concentrations observed in the interfacial layer indicated that titanium silicide and titanium carbide may have formed. Since the cooling in the water cooled copper hearth after arc melting was very fast, it is believed that the interface layer is likely to be formed by the melt and SiC, i.e., the reaction between Ti atom and SiC. Based on the thermodynamical data obtained from Ref.[7], the Gibbs free energies for the reactions occurred, e.g., at 1623K (the melting point of Al_3Ti) have been calculated as follows:

$$3Ti(l) + 2SiC \rightarrow TiSi_2 + 2TiC \qquad \Delta G = -104.46\text{kcal}$$

$$8Ti(l) + 3SiC \rightarrow Ti_5Si_3 + 3TiC \qquad \Delta G = -282.64\text{kcal}$$

Although Ti_5Si_3 seems to be more stable than $TiSi_2$ in this case in the viewpoint of thermodynamical condition, the existence of $TiSi_2$ had been confirmed by the X-ray diffraction result (Fig.2a) and also by the EPA analysis (Fig.4) for the sample of remelted PM compact. The reason of such inconsistancy may attribute to the kinetic effect, for instance, the insufficient supply of Ti at the interface.

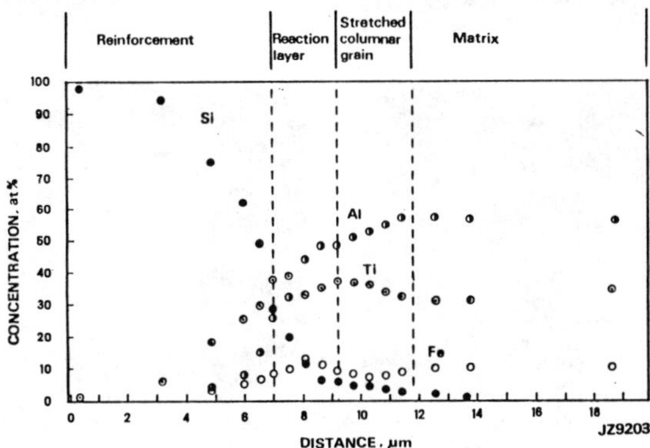

Figure 3. Interface analysis by EDS (carbon analysis not available)

The silicide in HIP'ed composite was identified as Ti_5Si_3, which can be explained from thermodynamical consideration. In this material fabricated by HIP'ing at 1473K, the reactions products should occur in the solid state and, therefore, the calculated results are:

$$8Al_3Ti+3SiC \rightarrow Ti_5Si_3+3TiC+24Al(s) \qquad \Delta G=-45.15kcal$$

$$3Al_3Ti+2SiC \rightarrow TiSi_2+2TiC+9Al(s) \qquad \Delta G=-15.80kcal$$

that is, Ti_5Si_3 is more stable than $TiSi_2$. The Al_3Ti and Al_3Fe are not the direct products of SiC/matrix reaction. Their formation was due to the depletion of Ti in the matrix. The Al-Ti-Fe phase diagram [8] shows that when Ti content becomes lower, the matrix composition will move from the $L1_2$ single phase region towards the $L1_2-Al_3Ti-Al_3Fe$ ternary phase region. The extent of reaction is rather strong in HIP'ed material due to the long holding time at 1473 K (2 hours). EPA analysis of the microstructure after HIP'ing is shown in Fig.6.

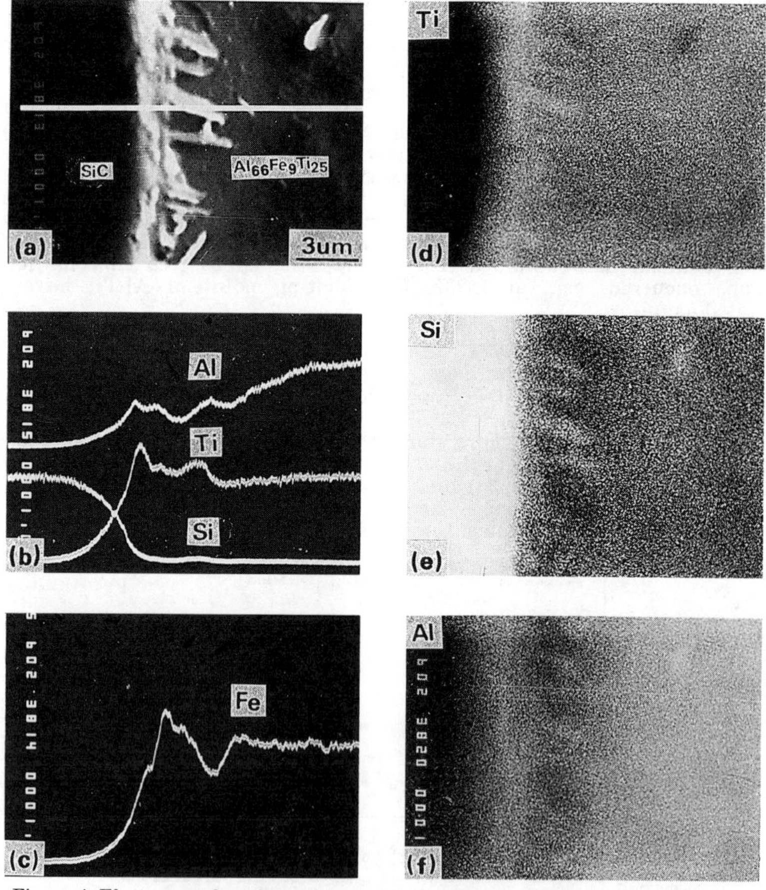

Figure 4. Electron probe microanlyses of the interface region (a)S.E.image, (b) and (c)line scanning along the path shown in (a), and (d), (e) and (f) are X-ray mapping for Ti, Si and Al respectively.

CONCLUSIONS

(1) Strong interfacial reactions occurred during fabrication, and the formation of ternary reaction products seriously deteriorated the matrix alloy. SiC has been found to be chemically incompatible with $Al_{66}Fe_9Ti_{25}$ alloy.

(2) The interfacial reaction products are $TiSi_2$ and TiC in arc-melted or induction-remelted PM compact,while in HIP'ed PM compact the products are Ti_5Si_3, TiC, Al_3Ti and Al_3Fe.

ACKNOWLEDGMENT

This research was supported by the National Science Foundation of China.

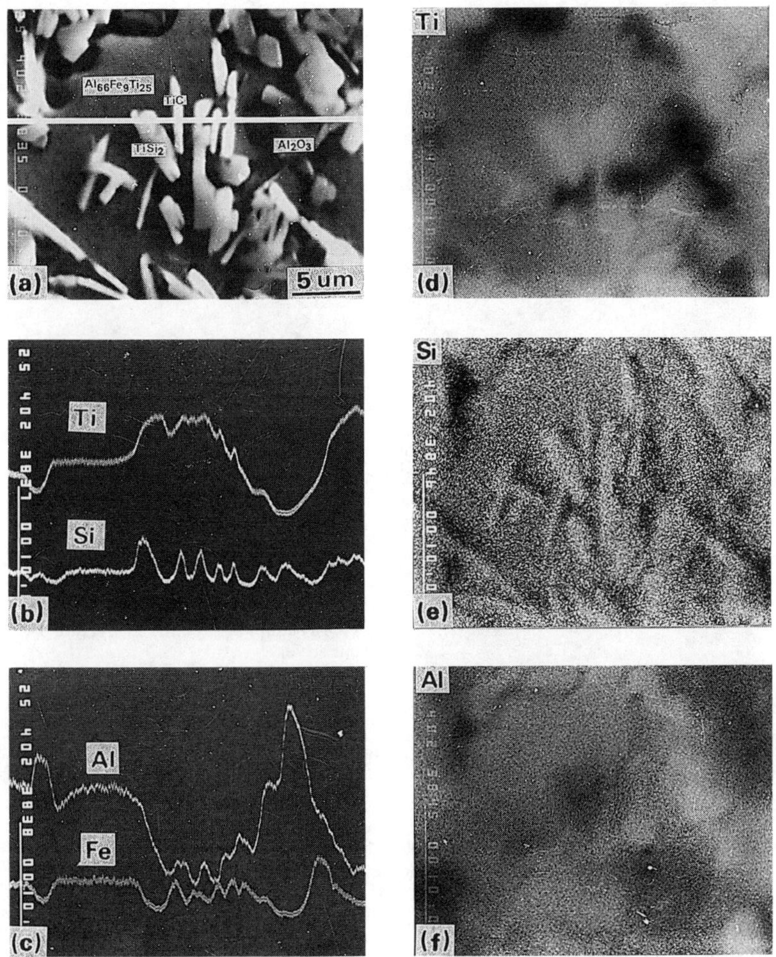

Figure 5. Electron probe microanalyses of the interface region for the induction remelted PM compacts: (a) S.E. image; (b) and (c) line scan along the path shown in (a); and (d), (e) and (f) are X-ray mapping for Ti, Si and Al respectively.

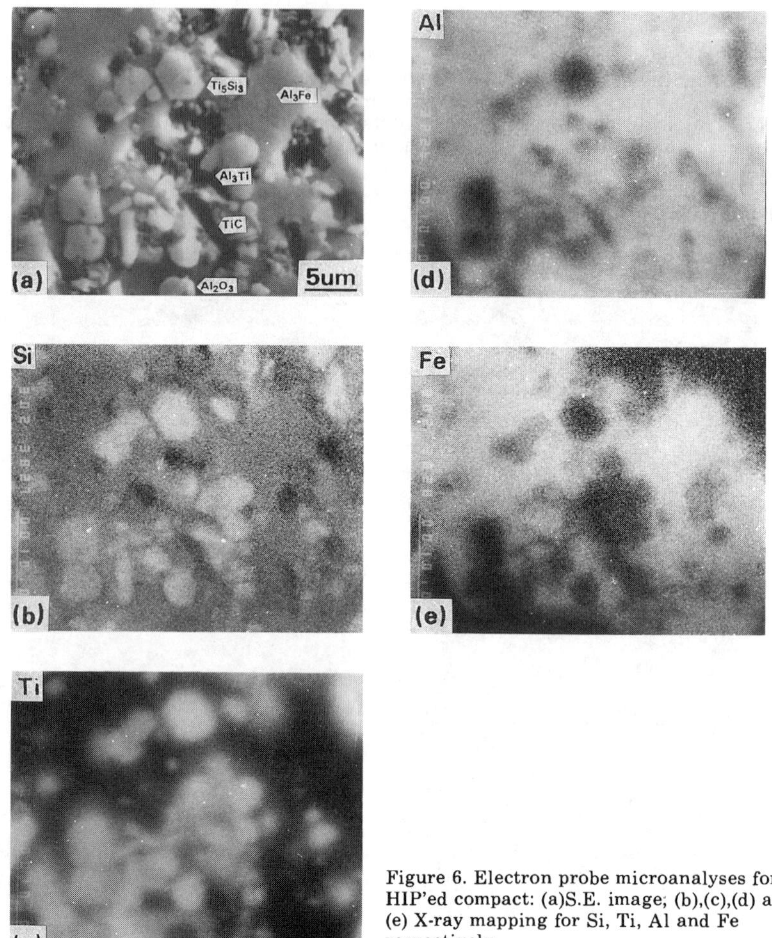

Figure 6. Electron probe microanalyses for the HIP'ed compact: (a)S.E. image; (b),(c),(d) and (e) X-ray mapping for Si, Ti, Al and Fe respectively.

REFERENCES

1. M. Yamaguchi, Y. Umakoshi, and T. Yamane in High-Temperture Ordered Intermetallic Alloys II, edited by N.S.Stoloff, C.C.Koch, C.T.Liu, and O. Izumi (Mater. Res. Soc. Symp. Proc. **81**, Pittsburgh, PA, 1987) p.275.
2. K.S. Kumar and J.R. Pickens, Scripta Metall. **22**, 1015 (1988)
3. J. Tarnacki and Y.W. Kim, Scripta Metall. **22**, 329 (1988)
4. S. Zhang, J.P. Nic, and D.E. Mikkola, Scripta Metall. **24**, 57 (1990)
5. K.S. Kumar and S.A. Brown, Phil. Mag. A **65**, 91 (1992)
6. Xiaofu Chen, Xiaohua Wu, Shipu Chen and Gengxiang Hu, Scripta Metall. et Mater. **26**, 1775 (1992)
7. I. Barin, O. Knacke, O. Kubascheweti, in Thermochemical Properties of Inorganic Substances (Springer-Verlag, Berlin, 1973, supplement 1977)
8. S. Mazdiyasni, D.B. Miracle, D.M. Dimiduk, M.G. Mendiratta and P.R. Subramanian, Scripta Metall. **23** 327 (1989)

MICROSTRUCTURE/PROPERTY RELATIONS IN AlN AND Al_2O_3 PARTICULATE STRENGTHENED NiAl.

M. G. Hebsur, Sverdrup Technology Inc., Brookpark, OH 44142
J. D. Whittenberger, NASA-Lewis Research Center, Cleveland, OH 44135,
R. M. Dickerson and B. J. M. Aikin, Case Western Reserve University, Cleveland OH 44106

ABSTRACT

The primary objective of this investigation was to confirm previous results on AlN particulate strengthened NiAl (J. Mater. Res 5, 271(1990)) produced by cryomilling and to determine if cryomilling modified to produce NiAl containing fine Al_2O_3 particles would also yield a material with equally attractive strength levels. Compression tests conducted on NiAl-AlN at a strain rate of 1.2×10^{-4} s^{-1} between 300-1300 K and as a function of strain rate at 1300 K indicated that its strength was comparable to previously produced material. While the NiAl-Al_2O_3 material was stronger than unreinforced NiAl at lower temperatures, at 1300 K it exhibited strength levels commensurate with fine grain sized NiAl. At all test temperatures and strain rates NiAl-AlN exhibited higher strengths compared to either NiAl-Al_2O_3 or unreinforced NiAl.

INTRODUCTION

Recent studies [1-2] have shown that cryomilling of NiAl powder can lead to remarkable improvements in elevated temperature creep strength due to inhomogeneously distributed fine AlN particles surrounding NiAl grains. The results [3] of 1473 K cyclic oxidation tests have also shown that AlN particulate reinforced NiAl has excellent cyclic oxidation resistance. Thus, this material appears to have excellent potential for development as an elevated temperature structural alloy. The primary objective of this investigation was to validate the previous results on creep properties of NiAl-AlN material and determine if NiAl containing fine Al_2O_3 particles also possessed attractive strength levels.

EXPERIMENTAL

Two lots of approximately 500 gm of prealloyed NiAl powder were cryomilled under two conditions in a modified Union Process Model 01-HD Research 1400 cc attritor mill. In both cases the attritor outer vessel was cooled by a continuous flow of liquid nitrogen. In the first case liquid nitrogen was introduced in the milling vessel to form AlN and in the second case the atmosphere was a mixture of nitrogen and air to promote the formation of Al_2O_3. At the completion of each milling cycle, the milling vessel was transferred to a glove box for warming to room temperature under dry argon. The powders were subsequently consolidated by extrusion at 1473 K and 12:1 reduction ratio after being vacuum canned in mild steel. From the extruded material, cylindrical compression specimens of 3 mm dia x 6 mm long, and bend bars of 1.25 mm thick x 2.0 mm wide x 50 mm long having their longitudinal axis parallel to the extrusion direction were machined by electrodischarge machining followed by grinding.

Compression and four point bend tests were conducted on an universal test machine between 300 and 1300 K in air at a constant crosshead speed of 0.5 mm/s. Compression tests were also conducted in air between 1200 to 1400 K under constant crosshead speeds ranging from 2.12×10^{-4} to 8.47×10^{-7} mm/s to approximately 8 % strain in order to determine the elevated temperature properties.

RESULTS AND DISCUSSIONS

Since the starting NiAl powder contained only small amounts (about 30 to 40 ppm) of nitrogen and oxygen, the substantial increases in nitrogen and oxygen contents (Table I) in the cryomilled materials can be attributed to the processing. Table I also reveals a wide difference in nitrogen and oxygen contents from milling in liquid nitrogen versus air. X ray diffraction (XRD) scans of both lots of cryomilled powders did not detect any ceramic phase. However, after extrusion distinct peaks of AlN (Fig. 1a) or Al_2O_3 (Fig. 1b) were evident in the XRD spectra. Assuming that all nitrogen formed AlN and all oxygen formed Al_2O_3, the volume fraction of each of these phases and the matrix level of Al were calculated and are included in Table I.

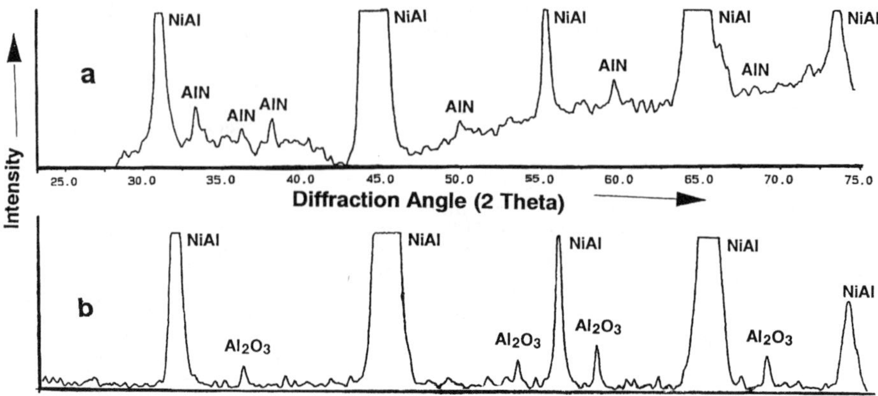

Figure 1. XRD intensity as a function of diffraction angle (2θ) for extruded materials after being milled in (a) liquid nitrogen and (b) air.

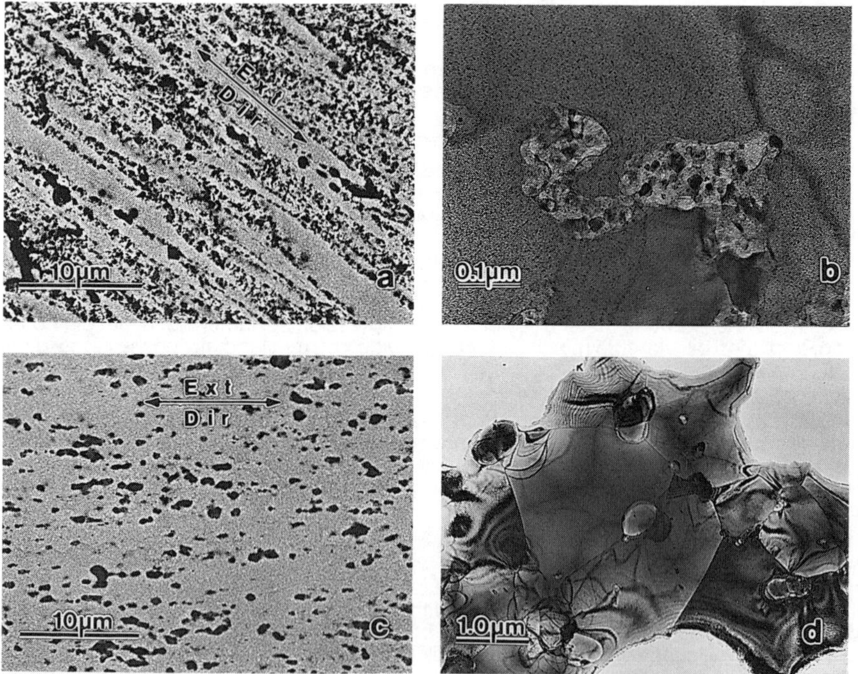

Figure 2. Typical microstructures of extruded materials after cryomilling: (a) BSE micrograph of NiAl-AlN; (b) TEM micrograph illustrating AlN in the mantle between NiAl grains; (c) BSE micrograph of NiAl-Al_2O_3; (d) TEM micrograph illustrating the size and distribution of Al_2O_3 particles.

Table I: Chemical composition of cryomilled NiAl powders

ID	As Cryomilled (wt. %)			Matrix. Comp Al, at.%	2nd Phase vol. %	
	Al	N	O		AlN	Al_2O_3
NiAl-AlN	30.90	1.60	0.49	47.1	8.3	1.1
NiAl-Al_2O_3	30.80	0.10	1.36	48.0	0.5	4.0

Figure 2 illustrates the typical microstructure found in the longitudinal section of as-extruded NiAl-AlN where elongated fine NiAl grains are surrounded by a dark mantle. Transmission electron microscopy (TEM) (Fig. 2b) of the extruded material indicated that the mantle contained a high density of extremely small (5-100 nm) AlN particles. This structure is identical to the Exxon processed material [1-2]. The extruded microstructure of NiAl-Al_2O_3 (Fig. 2c) on the contrary does not show a continuous mantle surrounding each grain. Although the grain size of about 5μm is similar to NiAl-AlN (Fig. 2a), the Al_2O_3 particles (Fig. 2d) are much larger than the AlN particles (Fig. 2b). Furthermore, the Al_2O_3 particles are more or less uniformly distributed throughout the microstructure and appear to be slightly elongated in the extruded direction.

MECHANICAL PROPERTIES

The temperature dependence of the compressive yield stress values for NiAl-AlN and NiAl-Al_2O_3 is shown in Fig. 3a. For comparison purposes, properties for binary Ni-50.6Al (at.%) with a 10μm grain size [5] and Exxon processed NiAl-AlN [1] are also included in this figure. The yield strength decreased with increasing temperature for all of these NiAl-base materials; however, both particulate strengthened alloys exhibited much higher strength than the binary NiAl at all temperatures even though the strength in these alloys started decreasing sharply at about 800 K. The NiAl-AlN produced in this investigation has almost identical strength compared to that of the Exxon produced material and both are clearly superior to either NiAl-Al_2O_3 or Ni-50.6Al.

The results of bend tests performed on NiAl-AlN between 300 and 1300 K are presented in Fig. 3(b). Between 300 and 573 K, the NiAl-AlN specimens failed in a brittle manner without exhibiting any deviation from linearity in the load-displacement data. Between 573 K and 773 K, specimens exhibited limited bend ductility. At and above 773 K, significant plasticity was observed, and the specimens deformed to the limits of the bend fixture. Comparison of bend yield strength data for NiAl-AlN to that for Ni-50.6Al [6] indicates that at lower temperatures (< 900 K), the NiAl-AlN is five times stronger than Ni-50.6Al; at higher temperatures NiAl-AlN becomes weak, but is still nearly three times stronger than Ni-50.6Al. Lastly, comparison of the compressive yield (Fig. 3a) with bend yield (Fig. 3b) for NiAl-AlN reveals similar strength levels and temperature dependence.

Figure 4a illustrates the true stress-strain rate behavior for the NiAl-AlN, NiAl-Al_2O_3, and NiAl as determined by constant velocity testing. The flow stresses, σ, and strain rates, $\dot{\epsilon}$, were fitted to the standard power law relationship,

$$\dot{\epsilon} = A\sigma^n \tag{1}$$

by linear regression techniques where A is a constant and n is the stress exponent. Results of the regression analysis along with the coefficients of determinations R_d^2 are given in Table II. Clearly, NiAl-AlN is much stronger than NiAl and is more creep resistant than NiAl-Al_2O_3. Fig. 4a also indicates that the NiAl-AlN produced in this investigation has almost identical behavior as that of the best Exxon processed NiAl-AlN [2]. However, the NiAl-Al_2O_3 material, only exhibited strength levels commensurate with fine grain sized NiAl [7]. Thus, it appears that Al_2O_3 particulate of the size and volume fraction produced by cryomilling is not an effective strengthener at elevated temperatures. Fig. 4(b) illustrates the 1200-1400 K stress-strain rate data for NiAl-AlN. The flow stress of NiAl-AlN is sensitive to temperature

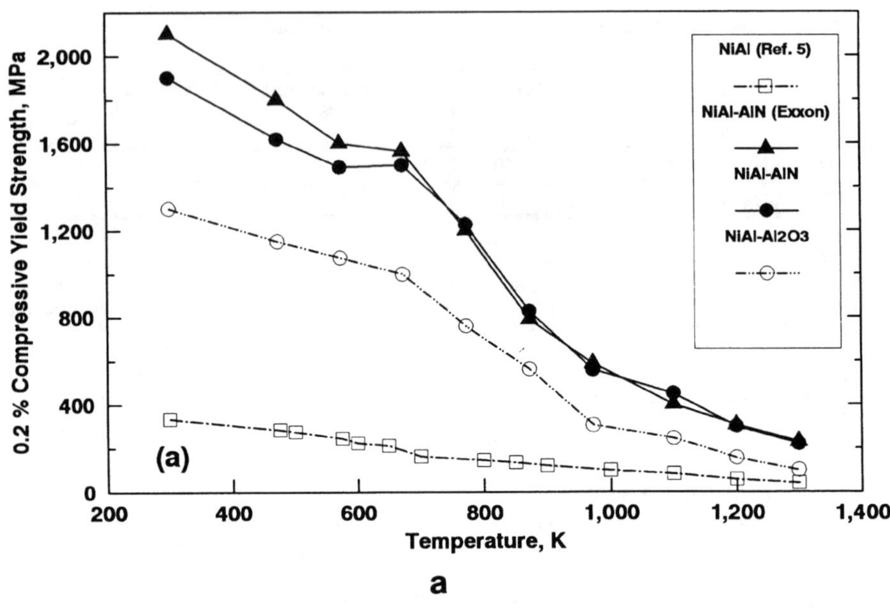

a

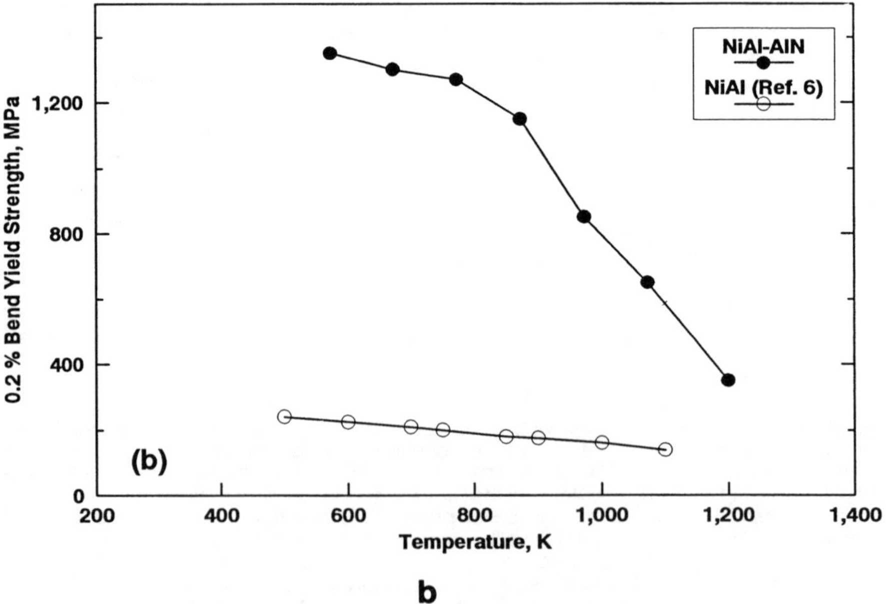

b

Figure 3. The temperature dependence of the (a) compressive yield strengths and (b) bend yield strengths for NiAl-AlN, NiAl-Al_2O_3 and Ni-50.6Al.

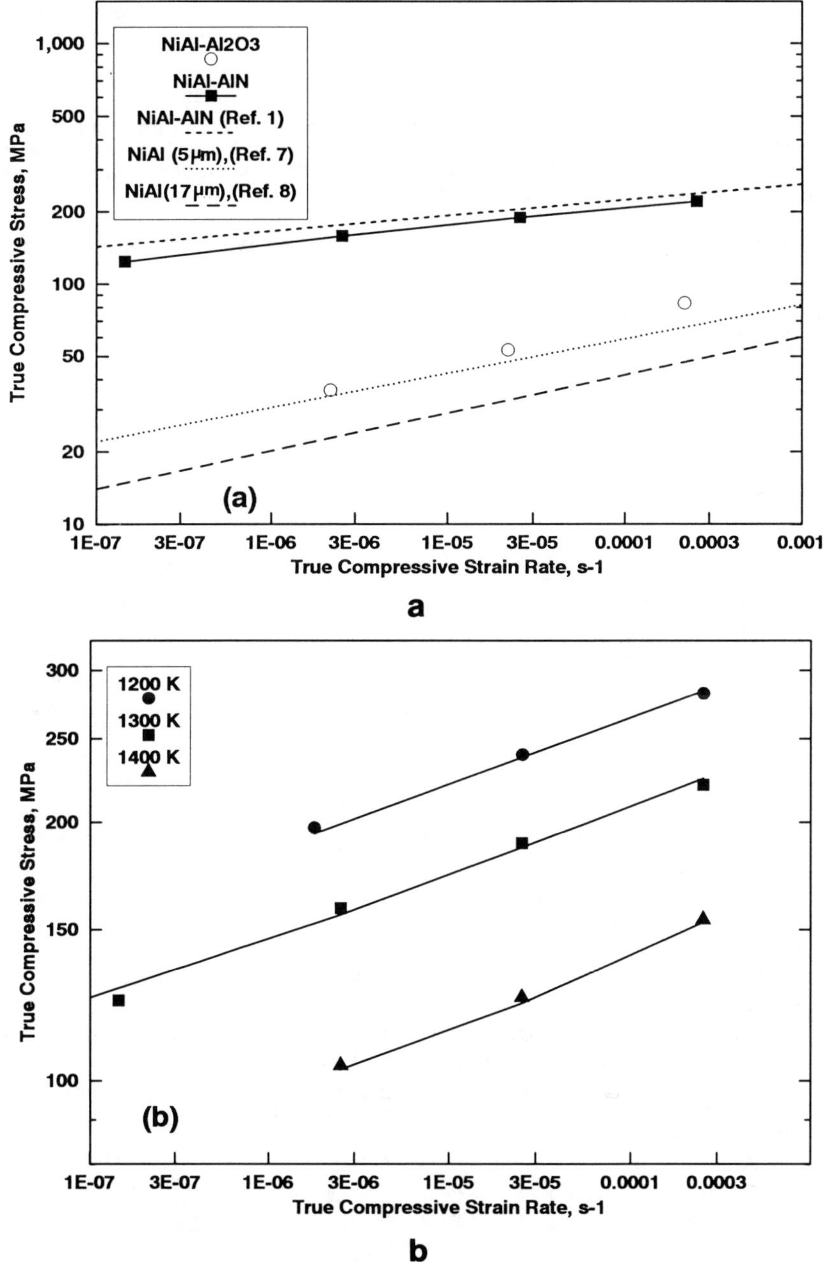

Figure 4. True flow stress-strain rate properties for (a) NiAl-AlN, NiAl-Al_2O_3 and Ni-50.6Al at 1300 K and (b) NiAl-AlN between 1200 and 1400 K.

and decreases nearly by a factor of 2 for every 100 K increase in test temperature. The multiple regression fit of σ, $\dot{\epsilon}$, and temperature (T) data to the temperature compensated power law equation

$$\dot{\epsilon} = B\sigma^n \exp(-Q/RT) \qquad (2)$$

where B is a constant and Q, the activation energy and R is the universal gas constant, is shown in Fig. 4(b). The regression parameters are given in Table II (b). While the stress exponents for the present and Exxon versions of NiAl-AlN are in good agreement (Table IIb), there is difference in the activation energies.

Table II: Power law and temperature compensated power law fits of creep Data

(a) *Power law fits, 1300 K data*

vol. %, 2nd Phase	A, s^{-1}	n	R_d^2
8 AlN	1.91×10^{-36}	13.27	0.971
10 AlN (Exxon)	5.1×10^{-34}	12.5	0.975
4 Al_2O_3	3.91×10^{-16}	5.01	0.98

(b) *Temperature compensated power law fits*

Vol. % 2nd Phase	Temp. Range	B s^{-1}	n	Q kJ/mol	R_d^2
8 AlN	1200-1400	6.5×10^{-11}	11.75	517.7	0.97
10 AlN (Exxon)	1200-1400	7.3×10^{-7}	12.0	644.5	0.95

SUMMARY OF RESULTS

NiAl matrices containing fine particles of predominantly either AlN or Al_2O_3 were produced by cryomilling followed by hot extrusion. While AlN particles were in a mantle surrounding particle free NiAl grains, the Al_2O_3 particles were much larger and more uniformly distributed. The NiAl-AlN possessed superior strength and elevated temperature plastic flow resistance as compared to either NiAl-Al_2O_3 or Ni-50.6Al. Thus, the attractive strength levels of the AlN containing NiAl are concluded to be the direct result of the AlN particles and not some other feature of cryomilling.

ACKNOWLEDGEMENT

This work was supported under the EPM program at NASA-Lewis Research Center.

REFERENCES

1. J. D. Whittenberger, E. Artz and M. J. Lutton, J. Mater. Res., 5 271 (1990)
2. J. D. Whittenberger, E. Artz and M. J. Lutton, J. Mater. Res., 5 2819 (1990)
3. C. E. Lowell, C. A. Barrett and J. D. Whittenberger, in Intermetallic Composites, (eds. D. L. Anton, R. McMeeking, D. Miracle and P. Martin) Mat. Res. Soc. Symp. Proc. 194, Pittsburgh, PA, 1990) pp. 355-60
4. J. D. Whittenberger, Mater. Sci. Eng. 57, 77 (1983)
5. R. R. Bowman, R. D. Noebe, S. V. Raj and I. E. Locci, Met. Trans. 23A 1493 (1992)
6. R. R. Bowman and R. D. Noebe, HITEMP Review 1990, NASA-CP 10051, pp.40-1/14
7. J. D. Whittenberger, J. Mat. Sci. 23, 235 (1988)
8. J. D. Whittenberger, J. Mat. Sci. 22, 394 (1987)

THE ROLE OF DISPERSOIDS IN MECHANICALLY ALLOYED NiAl

S. DYMEK, M. DOLLAR, S.J. HWANG AND P. NASH
Department of Metallurgical and Materials Engineering,
Illinois Institute of Technology, Chicago, Il 60616

ABSTRACT

Mechanical alloying followed by hot extrusion has been used to produce fully dense, crack free, very fine grained NiAl-based alloys containing a bimodal distribution of aluminum oxide dispersoids. The unique microstructure provides the materials with high strength and good compressive ductility at ambient and elevated temperatures. The emphasis of the paper is on the importance of the dispersion phase in controlling grain size, texture, deformation mechanisms and ultimately mechanical properties of the mechanically alloyed NiAl-based materials.

INTRODUCTION

Intermetallic compounds, such as nickel, iron and titanium aluminides, have recently emerged as a new class of potential structural materials for high temperature applications. Among others, the NiAl compound is a possible high temperature structural material, either in monolithic form or as a matrix phase in a composite, because of its low density, high melting temperature, good thermal conductivity and excellent oxidation resistance [1]. However, before this material can be of practical use a number of problems must be overcome, including lack of ductility at room temperature and poor strength at high temperatures [2].

Our approach has been to use mechanical alloying followed by hot extrusion to produce several very fine grained materials containing oxide dispersoids, to address both the ambient temperature brittleness and high temperature strength problems. In this paper, the results of our studies on microstructure, texture, deformation mechanisms and temperature-dependent mechanical properties of a selected, mechanically alloyed (MA), near-stoichiometric NiAl-based material are presented and contrasted with the analogous observations of its cast counterpart. The comparison is aimed at emphasizing the importance of the unique microstructure developed during mechanical alloying and hot extrusion.

In the course of our studies, we have become increasingly convinced of the importance of the dispersoid phase in controlling grain size, texture, deformation mechanisms and ultimately mechanical properties of the MA NiAl materials. Thus, the discussion focusses on the roles of dispersoids in the MA materials. The present paper is likely the first attempt to comprehend the significance of dispersion phase in MA NiAl-based materials.

EXPERIMENTAL DETAILS AND RESULTS

An NiAl-based alloy, with the chemical composition shown in Table I, was produced for the present study. The alloy was obtained from elemental Ni and Al powders, mechanically alloyed in an attritor mill, sieved, degassed and encapsulated under vacuum in a stainless steel can, and hot extruded at 1400 K at a ratio of 16:1. A cast NiAl ingot, with the chemical composition given in Table I, hot extruded under the same conditions, was also investigated for comparison with the MA NiAl.

Table I. Chemical composition of MA and cast NiAl (at. %). The content of Ni and Al in cast NiAl sums up to 100% since the analysis was conducted separately for metallic and interstitial elements.

Alloy	Ni	Al	Ti	Mo	C	H	N	O
MA	47.0	47.4	1.37	0.56	0.13	0.42	1.38	1.72
Cast	50.1	49.9	-	-	0.05	0.12	0.003	.007

The optical microscopy observations showed that the hot extruded MA material is fully dense and free from cracks, and revealed a fairly homogenous distribution of dispersoids throughout the matrix with average particle size of about 100 nm (Fig.1). The particles were identified by X-ray diffraction and energy dispersive spectrometry as aluminum oxides α-Al_2O_3. Transmission electron microscopy (TEM) studies of the hot-extruded MA NiAl revealed fine, equiaxed grains with an average size of about 0.5μm (Fig.2). The majority of grain boundaries were found to be low angle ones. The other important feature of the microstructure of the MA material is a bimodal distribution of aluminum oxides. There are two types of oxides: coarse, with diameter of about 100nm (revealed also by optical microscopy), preferentially distributed on grain boundaries and much smaller, with a mean size of about 10nm, dispersed uniformly throughout the grains. The optical and transmission electron microscopy observations showed that the cast NiAl is a single phase material with an average grain size of about 30 μm. The equiaxed grains observed on both longitudinal and transverse sections indicate that the material was fully recrystallized during hot extrusion.

The extrusion texture was determined on transverse sections by the Schulz back reflection method using Copper K_α radiation. The <110> and <200> pole figures were analyzed to give orientation distribution functions and inverse pole figures using a modified Williams-Imhof-Matthies-Vinel algorithm [3]. The analysis indicates that the MA alloys have a strong <110> fiber texture parallel to the extrusion axis (Fig.3) while the hot extruded cast material has a <111> fiber texture.

Mechanical properties of the as-extruded MA and cast NiAl were examined by compression tests in air from room temperature to 1100K and at a strain rate of $8.5 \times 10^{-4}s^{-1}$. The compression test specimens were electro-discharge machined cylinders 10mm long (parallel to the extrusion direction) by 5mm diameter. Selected results are shown in Table II.

Table II. Yield stress and strain to failure of MA and cast NiAl.

Alloy	Deformation Temperature (K)	Yield Stress (MPA)	Strain to Failure (%)
MA	300	1275	>11.5*
MA	800	950	6.8
MA	1100	234	>13.7*
cast	300	303	2.8

*Room temperature compression test of the MA material was stopped when the load on the specimen reached the limit of the Instron cell. The 1100K test was stopped as well since the compressive ductilities at high temperatures typically exceed 30%.

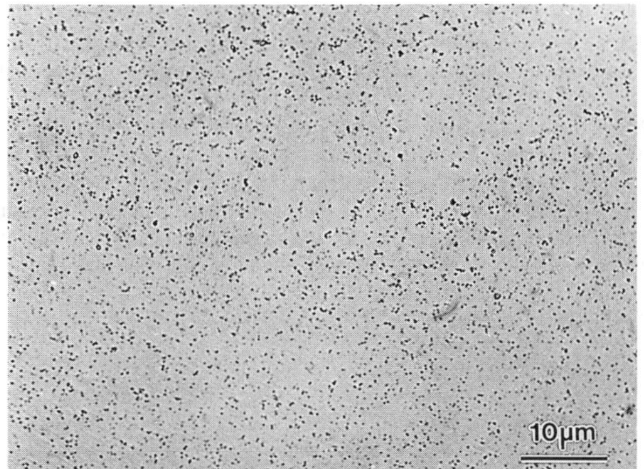

Fig. 1. Optical microstructure of the hot extruded MA NiAl.

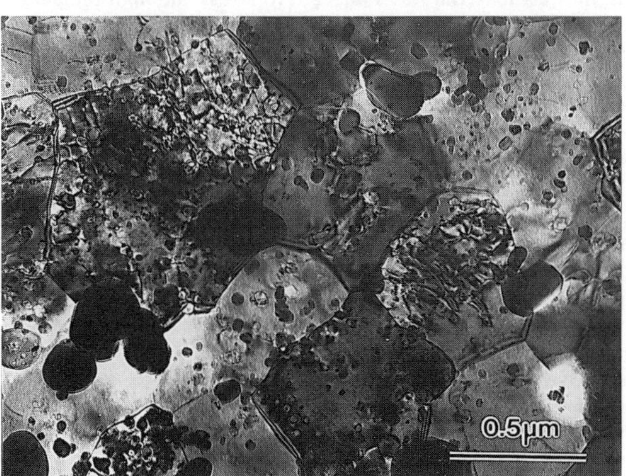

Fig. 2.Typical TEM microstructure of the hot extruded MA NiAl.

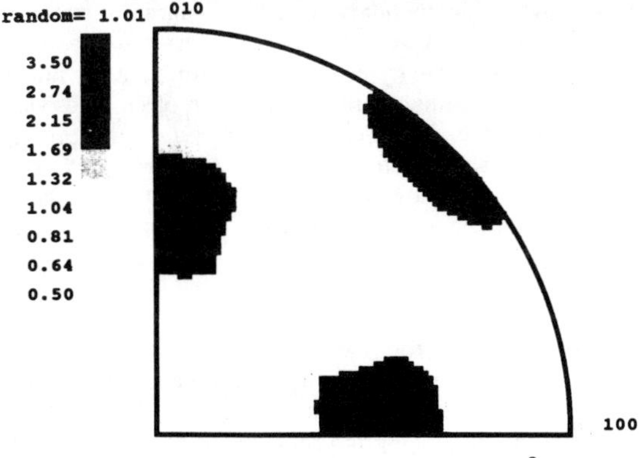

Fig. 3. The inverse pole figure of the hot extruded MA NiAl.

Strain rate change tests were also conducted. Strain rate changes $\dot{\epsilon}_2/\dot{\epsilon}_1=10$ were used to measure strain rate sensitivity, $\beta= \delta\sigma/\delta\ln\dot{\epsilon}$ [4,5]. A negative strain rate sensitivity was observed at room temperature and at 500K while at 800K β was found to be strongly positive.

Comprehensive TEM studies of the MA material, deformed to 2% at 300, 800 and 1100K, were carried out but only selected results are discussed here. At 300 and 800K, a high dislocation density, exceeding that in the as-extruded material by at least one order of magnitude, was observed. Specimens tested at 1100K exhibit much lower dislocation density than those tested at room temperature and at 800K after the same amount of deformation. No classical Orowan or cutting mechanisms were observed. Instead, the dislocations were frequently pinned on the oxides. This mechanism was found to be particularly pronounced in specimens deformed at 800K, where nearly all dislocations are pinned on the dispersoids.

DISCUSSION

Microstructure and texture

During hot extrusion, the cast material was subjected to stress and high temperature simultaneously and dynamic recrystallization took place. In fact, the microstructure of the as-extruded cast NiAl consists of recrystallized and equiaxed grains. There is a remarkable contrast between the microstructure of the coarse grained, single phase cast material and its fine grained MA counterpart. A unique feature of the MA material is the presence and a bimodal distribution of the aluminum oxides, with coarser (d=100nm), residing predominantly at grain boundaries and smaller (d=10nm), dispersed uniformly throughout the grains (Fig.2).

The texture developed during hot extrusion in cast and MA NiAl is also distinct. The present analysis of the pole figures and the orientation distribution functions revealed a strong <111> fiber texture along the extrusion direction in the as-extruded cast NiAl. This is a common recrystallization texture, observed in the cast NiAl by others as well [6,7]. On the other hand, the as-extruded MA NiAl showed a strong <110> fiber texture. This in turn is a common texture produced by cold drawing or extrusion in b.c.c. materials [8,9]. Since the texture of MA powders is essentially random and featureless, the <110> texture in MA NiAl is a direct consequence of the deformation during extrusion.

The <110> deformation texture is postulated to be retained during hot extrusion of the MA NiAl powder due to the presence of dispersoids, above all coarser ones. It is well established that the (discontinuous) recrystallization process involves the nucleation of strain-free regions with at least one high-angle boundary capable of migration and of a size bigger than a critical one [10]. The mean distance between the coarser oxides is in the present alloy much less than a critical nucleus size (typically about 1 μm) and the nucleation is prevented altogether. In such a case discontinuous recrystallization often gives way to a subgrain coarsening reaction, so called continuous recrystallization [9]. In the present MA alloys, continuous recrystallization takes place as evidenced by TEM observations, revealing very fine equiaxed grains typically separated by low angle boundaries. The grain growth is inhibited because low angle grain boundaries are known to have low mobility [11,12] and also are pinned by the coarser oxide particles.

Strength

The MA NiAl is much stronger than the cast NiAl at ambient and elevated temperatures (Table II). The room temperature yield strength of the MA and cast material was measured to be 1275 and 303 MPa, respectively. Several factors may be postulated to contribute to the high strength of the MA material: grain refinement, texture, the presence of dispersoids as well as interstitial and substitutional solute atoms.

Using Hall-Petch plots for NiAl published by Nagpal [13] and Barker [14], Hwang estimated that the increase in yield stress due to grain refinement in the present case does not exceed 10 MPa, thus it is insignificant [15]. Specimens cut parallel and perpendicular to the extrusion direction exhibit essentially the same yield stress, thus the <110> texture observed in the longitudinal specimens, does not contribute significantly to strengthening. Hwang showed also that the concentration of substitutional solute atoms does not significantly influence the stress levels in MA NiAl-based materials [15]. Thus, the interstitial atoms and dispersoids, above all the finer ones, are believed to significantly strengthen the MA material. However, because of the extremely complex chemistry and microstructure of the MA material it is not possible to quantify the strengthening contributions of dispersoids and interstitial atoms.

The strength of MA NiAl at the intermediate temperature of 800K ($\approx 0.4T_M$) is only about 10% lower than at room temperature, indicating that thermally activated climb does not yet occur. At 1100K ($\approx 0.55T_M$), thermally activated processes facilitate the dislocation motion, resulting in significantly decreased yield stress and work hardening and improved ductility.

Ductility

At room temperature, the MA material is not only a few times stronger than its cast counterpart, but it does not fail even after deformation of 12.5%, whereas the cast material fails after only about 3% strain. As discussed elsewhere [16], the <110> texture enables the activation of <100>{110} and <110>{110} slip systems. The occurrence of <100> and <110> slip dislocations satisfies the von Mises criterion for general plasticity and is postulated to contribute to notable compressive ductility of the MA materials. Other factors likely affecting the compressive ductility are the fine grain size and the predominant occurrence of low angle grain boundaries.

At 800K, the MA NiAl exhibited considerably lower compressive ductility than that measured in specimens deformed at lower and higher temperatures. The ductility trough was in fact observed in all MA NiAl-based materials processed in our laboratory, but not in cast NiAl. Typically, a loss of ductility at elevated temperatures is a result of dynamic strain aging (DSA). It is generally accepted that a negative strain rate sensitivity, β, is indicative of DSA [4,5]. However, in the present study, a negative strain rate sensitivity was found at room temperature and 500K while at 800K β was strongly positive. Thus, DSA resulting from the interactions between moving dislocations and diffusing interstitial atoms is not expected to play a significant role in the present MA materials at 800K. The drop in compressive ductility at 800K is postulated to be associated with pronounced dislocation - dispersoid interactions revealed by TEM studies. At 800K, virtually all dislocations were pinned on oxide particles, in a clear contrast to dislocation structures observed in specimens deformed at 293 and 1100K.

Similar dislocation structures were frequently observed in other dispersion strengthened materials deformed at elevated temperatures; pinning of dislocations on oxides, borides and carbides was reported [17-20]. Weak beam studies of dislocation configurations in the vicinity of particles revealed that the dislocations were pinned at the departure side of the particles [18]. The presence of such dislocation configurations implies that there is an attractive force between dislocations and particles. However, the analyses of dislocation - particle elastic interactions predict a repulsive force when the shear modulus of particles exceeds that of the matrix [21] which is clearly the case in dispersion strengthened systems. To explain the discrepancy, Srolovitz et al. [21,22] as well as Arzt and Wilkinson [23] proposed models predicting that the interaction of dislocations with particles can change from repulsive to attractive as a result of high temperature relaxation processes.

CONCLUSIONS

A unique feature of the fine grained MA NiAl is the presence of a bimodal distribution of aluminum oxide dispersoids. The coarse oxides, residing at grain boundaries, prevent grain growth, affect the progress of recrystallization and contribute to the preservation of the <110> deformation texture. By preserving the <110> texture, they play an indirect role in providing the material with a notable compressive ductility. The fine oxides, dispersed throughout the grains, contribute significantly to the material's high strength. Their attractive interactions with dislocations cause the drop in compressive ductility at 800K.

ACKNOWLEDGEMENTS

This research was supported by Air Force Office of Scientific Research, under the technical direction of Dr. Alan H. Rosenstein (grant No. 90-0152B). The authors wish to thank Dr. Rosenstein, Dr. J.S. Kallend (IIT), Dr. J. D. Whittenberger (NASA Lewis), and Dr. D. B. Miracle (AFWAL/MLLM) for helpful discussions, and Dr. Whittenberger and Captain Dr. Chuck Ward of AFWAL/MLLM for extruding the material. We wish to thank GE Aircraft Engines for supplying the cast ingot.

REFERENCES

1. R.Darolia, J. of Metals, **43**, #3, 44 (1991).
2. D.M.Dimiduk, D.B.Miracle and C.H.Ward, Mat.Sci.Techn., **8**, 367 (1992).
3. J.S. Kallend, U.F. Kocks, A.D. Rollet and H.-R. Wenk, Mat. Sci. Eng., **A132**, 1 (1991).
4. A. Van Den Beukel, phys. stat. sol., **30**, 197 (1975).
5. P. Wycliffe, U.F. Kocks and J.D. Embury, Scripta Met., **14**, 1349 (1980).
6. P.S. Khadkakir, G.M. Michal and K. Vedula, Metall. Trans., **21A**, 279 (1990).
7. S.V. Raj, R.D. Noebe and R. Bowman, Scripta Met., **23**, 2049 (1989).
8. M. Hatherly and W.B. Hutchinson, *An Introduction to Textures in Metals*, Institution of Metallurgists, London 1979.
9. R.W.K. Honeycombe, *The Plastic Deformation of Metals*, p. 327, Edward Arnold (Publ.), American Society for Metals (1984).
10. E. Hornbogen and U. Köster, *Recrystallization of Metallic Materials,* Rieder-Verlag GMBH, Stuttgart 1978.
11. B. Ralph, Mat. Sci. Techn., **6**, 1139 (1990).
12. G.S. Grest, D.J. Srolovitz and M.P. Anderson, Acta Met., **33**, 509 (1985).
13. P. Nagpal and I. Baker, Scripta Met., **24**, 2381 (1990).
14. D. Barker, M.S. Thesis, Dartmouth College, 1992.
15. S.H. Hwang, PhD. Thesis, Illinois Institute of Techn., Chicago, 1992.
16. S. Dymek, M. Dollar, S.J. Hwang and P. Nash, Mat. Sci. Eng., **A152**, 160 (1992).
17. V.C. Nardone, D.E. Matejczyk and J.K. Tien, Acta Met., **32**, 1509 (1984).
18. J.H. Schröder and E. Arzt, Scripta Met., **19**, 1129 (1985).
19. C.M. Sellars and R.A. Petkovic-Luton, Mat. Sci. Eng., **46**, 75 (1980).
20. S.C. Jha, R. Ray and D.J. Gaydosh, Scripta Met., **23**, 805 (1989).
21. D.J. Srolovitz, M.J. Luton, R. Petkovic-Luton, D.M. Barnett and W.D. Nix, Acta Met., **32**, 1079 (1984).
22. D. Srolovitz, R. Petkovic-Luton and M.J. Luton, Scripta Met., **16**, 1401 (1982).
23. E. Arzt and D.S. Wilkinson, Acta Met., **34**, 1893 (1986).

DEFECT STRUCTURES AND PLANAR FAULTS IN PLASMA SPRAY DEPOSITED $MoSi_2$

Bimal K. Kad, Kenneth S. Vecchio and Robert J. Asaro
Dept. of Applied Mechanics and Engineering Sciences, MC-0411
University of California-SanDiego, LaJolla, CA 92093-0411

ABSTRACT

Plasma-sprayed microstructures of $MoSi_2$ have been studied by electron microscopy. The as-deposited microstructures are metastable and inhomogenous. Two new dissociations of [001] and 1/2[33$\bar{1}$] dislocations have been observed for the first time. The 1/2[33$\bar{1}$] is dissociated on the (1$\bar{1}$0) plane bounding a superlattice intrinsic stacking fault (SISF). The reaction is given as: 1/2[33$\bar{1}$] = 1/4[33$\bar{1}$] + SISF + [33$\bar{1}$]. The [001] is also dissociated on (1$\bar{1}$0) plane into two symmetrical components, with the reaction being given as [001] = 1/2[001] + 1/2[001].

INTRODUCTION

Intermetallic matrix composites are attractive materials for high temperature structural applications. One of the most promising intermetallic in the silicides is $MoSi_2$ which combines excellent oxidation resistance, a high melting point (2020°C), reasonable strength and density and good thermal conductivity making it a promising material for structural applications above 1200°C. $MoSi_2$ essentially behaves like a ceramic, possessing little ductility below the brittle to ductile transformation temperature (BDTT) and exhibits plastic behavior above the BDTT (1,2).

The brittle behavior of $MoSi_2$ poses two problems: i) limits use as a potential structural material and ii) severe cracking problems in conventional processing/casting routes. While the former affects it end usage, the latter hinders the evaluation of its intrinsic properties in a rapidly evolving research effort. Current research initiatives are the incorporation of a wide variety of ductile and/or hard dispersoids in a $MoSi_2$ matrix to increase fracture toughness. Towards this end, plasma co-spraying is a promising near-net shape processing technology combining melting, blending and consolidation into a single operation. A wide variety of microstructures can thus be created in tetragonal ($C11_b$) $MoSi_2$ based alloys (3). Controlled compositions/microstructures of $MoSi_2$, $MoSi_2$+Mo_5Si_3 and $MoSi_2$ + ductile/brittle reinforced particulates are being successfully fabricated (4). Of particular interest is the incorporation of fine scale (≈100nm) dispersoids with lattice parameters, elastic modulii (E) and coefficient of thermal expansion (CTE) different than $MoSi_2$. These property mismatches create a myriad of matrix precipitate interactions that can be tailored to enhance matrix plasticity. Due to the somewhat rapid nature of processing, non-equilibrium effects and planar faults reminiscent of $C11_b$ (tetragonal) => C40 (hexagonal) transformations are frozen in these microstructures; these transformation related faults are suggested to enhance ductility (5,6).

In this paper we present our preliminary observations of the microstructural modifications obtained by plasma spraying and their possible contributions to enhancing plasticity and fracture toughness. The character of several quenched-in faults has been identified and is correlated with possible transformation mechanisms and/or predicted lattice dissociations.

EXPERIMENTAL PROCEDURE

The spray-deposited samples were prepared at the low pressure plasma spraying (LPPS) facility at Los Alamos National Laboratory. MoSi2 powders obtained from Metco, with a size distribution of -200 +325 mesh were internally injected into a SG-100 Plasma-dyne spray torch during the plasma spraying operation. The specific details of the deposition and the optimization parameters for the spray process are published elsewhere (4). The as-deposited samples were vacuum encapsulated in quartz tubing and homogenized at 1200°C for 100 hours. TEM disks were sliced and slurry drilled from the as-deposited and homogenized specimens. Thin foils were obtained by ion milling the samples in a GATAN duomill at operating conditions of 6 kV and 1mA gun current. Samples were examined in a Phillips CM-30 electron microscope operating at 300kV.

EXPERIMENTAL RESULTS

As-deposited microstructures

The as-spray-deposited microstructure was inhomogeneous and large deviations in morphology were obtained from region to region. The most common of microstructural features is a slight deviation of the chemistry to Mo-rich side (presumably because of the evaporation loss of silicon in the plasma) resulting in a decomposed eutectic $Mo_5Si_3 + MoSi_2$ mixture as shown in figure 1. Figure 1(a) shows a side view of the eutectic morphology and figure 1(b) shows a top view where clean precipitate free grains of $MoSi_2$ are pinned by trails of Mo_5Si_3 at the grain junctions. Two additional features observed frequently are i) large grains of $MoSi_2$ with a large density of randomly distributed dispersoids (either Mo_5Si_3 or SiO_2) in the grains, figure 2(a) and ii) a significant density of twins and stacking faults in the $MoSi_2$ grains, figure 2(b). In a following section we shall concentrate on the characteristics of the features shown in figure 2 in relation to their possible contribution to plasticity enhancement.

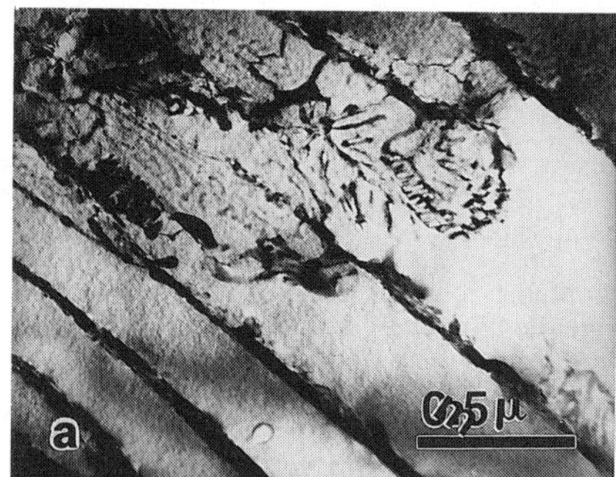

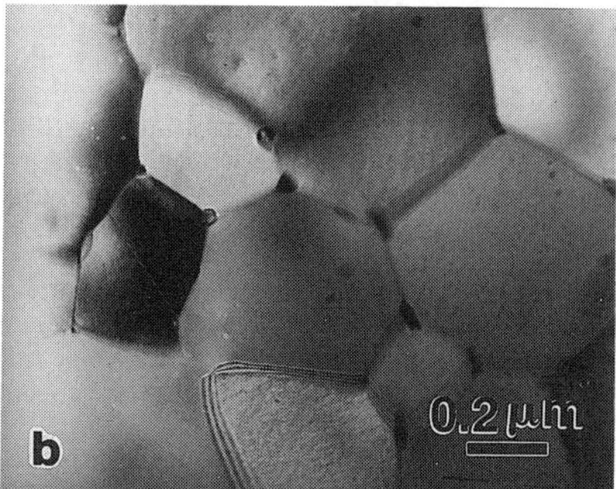

Figure 1. As-deposited microstructures in plasma-sprayed $MoSi_2$ showing the decomposed eutectic $Mo_5Si_3 + MoSi_2$. a) side view of the eutectic morphology, b) top view where precipitate free grains of $MoSi_2$ are pinned by trails if Mo_5Si_3 particles at grain junctions.

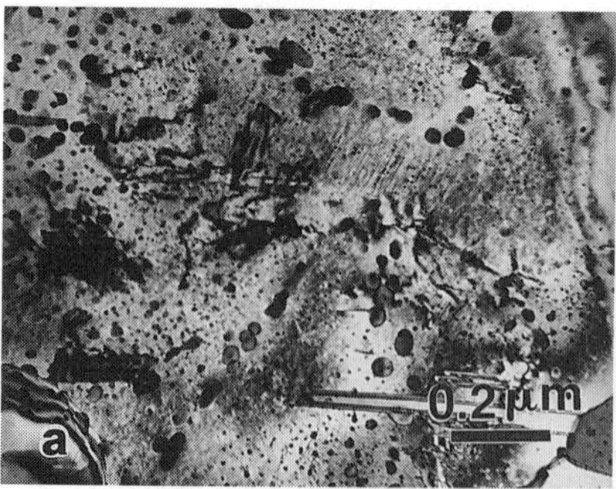

Figure 2. Two frequently observed features in the as-deposited microstructre. a) large grains of $MoSi_2$ with randomly distributed dispersoids Mo_5Si_3 or SiO_2 in the grains. b) Significant density of twins and stacking faults in the $MoSi_2$ grains.

Dispersoid-matrix interactions

Figure 3 shows a series of prismatic dislocation loops punched out into the $MoSi_2$ matrix possibly as a result of $MoSi_2$/dispersoid thermal mismatches. The loops were identified as pure edge with Burgers vector **b**=[100] lying close to (100) plane. The loops are seen to interact with several matrix dislocations (see arrow) and may be obstacles to their propagation. In addition reactions of the matrix dislocations with the prismatic loops are also observed (see arrow) which may be energetically favorable and provide similar pinning points for Orowan looping of the matrix dislocations.

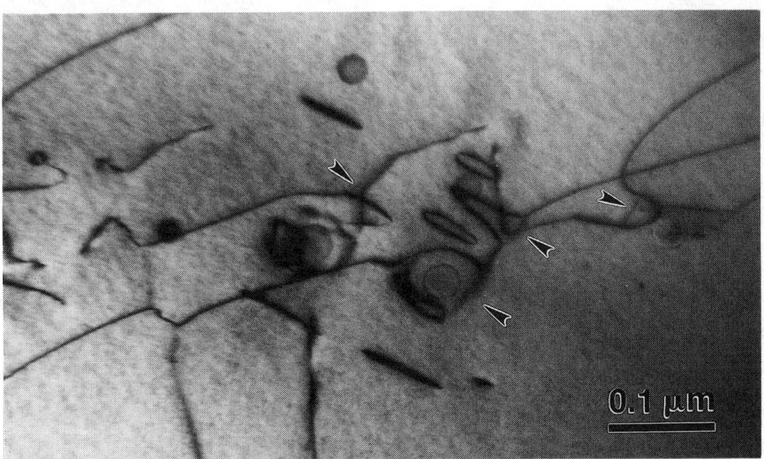

Figure 3. Prismatic loop punch out in the $MoSi_2$ matrix as a result of dispersoid/$MoSi_2$ matrix thermal mismatches. The arrow shows the entanglement and the subsequent reaction of the dislocation loop with matrix dislocations.

Stacking Faults in $MoSi_2$

The most significant and unique feature of the as-deposited specimens is the observation of a wide variety of symmetrical faults lying on $\{1\bar{1}0)$ planes. Figure 4 shows a few selected micrographs of the experimentally observed contrast for [001] and $[33\bar{1}]$ faults. The fault labeled A exhibits a $\mathbf{g}\cdot R_F$ = integer and $\mathbf{g}\cdot b = 0$ type contrast for $\mathbf{g} = 1\bar{1}0$, figure 4(b), and $\mathbf{g} = 200$, figure 4c, yielding a fault vector = 1/2[001]. Fault B exhibits a $\mathbf{g}\cdot R_F$ = integer and $\mathbf{g}\cdot b = 0$ type of contrast for $\mathbf{g} = 103$, figure 4a, and $\mathbf{g} = 1\bar{1}0$, figure 4b, yielding a fault vector = $1/4[33\bar{1}]$. Table 1 lists all the contrast observations (visible (V) and invisible (I)) for the fault (F) and the bounding partials (P) along with the absolute values of the expected contrast for faulted pairs of different feasible magnitudes.

Table 1. Comparison of experimentally obtained fault (F) and Partial (P) contrast with the absolute values of the expected contrast for faulted pairs of varying magnitudes. The experimentally observed contrast* is listed as visible (V) and invisible (I).

$g\Downarrow . R_F \Rightarrow$	$[33\bar{1}]$* F/P	$1/4[33\bar{1}]$	$1/6[33\bar{1}]$	$1/12[33\bar{1}]$	[001]* F/P	1/2[001]	1/3[001]	1/6[001]
$(1\bar{1}0)$	I / I	0	0	0	I / I	0	0	0
$(00\bar{6})$	V / V	3/2	1	1/2	I / V	3	2	1
(103)	I / I	0	0	0	V / V	3/2	1	1/2
(200)	V / V	3/2	1	1/2	I / I	0	0	0
$(\bar{1}03)$	V / V	3/2	1	1/2	V / V	3/2	1	1/2
$(\bar{1}\bar{1}0)$	V / V	3/2	1	1/2	I / I	0	0	0

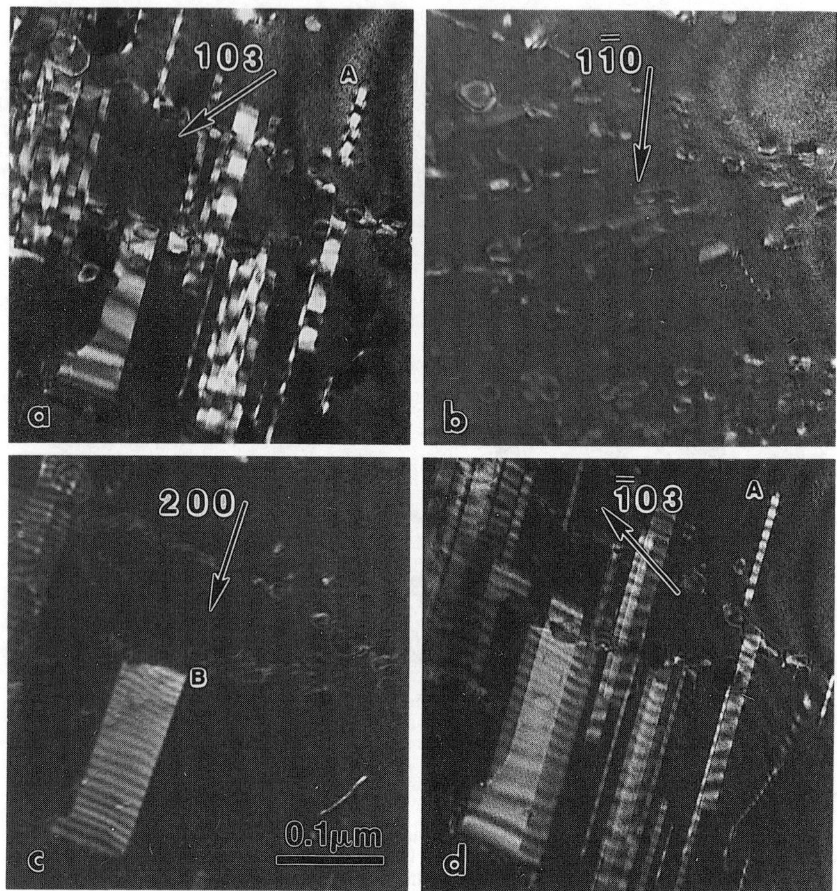

Figure 4. Weak-Beam Transmission Electron Micrographs of dissociated dislocations in as-deposited $MoSi_2$. a) g=103, B≈[010] b) g=1$\bar{1}$0, B≈[110] c) g=200, B≈[010] d) g=$\bar{1}$03, B≈[010]. The Fault labeled A is dissociated as [001](1$\bar{1}$0) = 1/2[001](1$\bar{1}$0) + 1/2[001](1$\bar{1}$0) and fault labeled B is dissociated as 1/2[33$\bar{1}$](1$\bar{1}$0) = 1/4[33$\bar{1}$](1$\bar{1}$0) + 1/4[33$\bar{1}$](1$\bar{1}$0).

A 1/6[33$\bar{1}$] fault is ruled out because of the strong fault contrast observed for **g** = $\bar{1}$03, figure 4(d) and **g** = 00$\bar{6}$ (see table 1). A 1/3[001] fault is ruled out because of the strong fault contrast observed for **g** = 103, figure 4(a) and **g** = $\bar{1}$03, Figure 4(d). The experimentally observed contrast is consistent with fault vectors of 1/4[33$\bar{1}$] and 1/2[001] bounded by equivalent partial dislocations on either side with Burgers vectors given as b = 1/4[33$\bar{1}$] and b = 1/2[001] respectively. The lattice dislocations 1/2[33$\bar{1}$] and [001] are dissociated as has been predicted in the literature (6,9,12). The experimental results obtained here are compared to the theoretically predicted dissociations in $MoSi_2$ in a later section. The dissociations are not faulted dipoles as has been determined by ±**g** analysis. Both faults have a line direction of [110], such that the [001] dissociated fault is pure edge and the 1/2[33$\bar{1}$] dissociated fault is mixed in character. It is interesting to note that the dissociations observed for 1/2[33$\bar{1}$] dislocations are consistently wider than those observed for [001] dislocation; this will be discussed in the next section.

DISCUSSION

Dispersoid-matrix dislocation generation

The prismatic dislocations at the precipitate matrix interface are present presumably because of thermal mismatches. Such dislocations, while increasing the total 'as-processed' dislocation density of $MoSi_2$ matrix, are not expected to contribute to the glide deformation. On the contrary they may cause tangling with the normal glide dislocations and hinder their propagation. Since no part of the edge loop lies on any 'suggested' slip plane of $MoSi_2$, they may only contribute to deformation at high temperatures by diffusion assisted climb processes.

Stacking-Faults in $MoSi_2$

The <331] and [001] faults have not been observed in the homogenized material or in any hot pressed samples prepared by us. It is likely that these defects are metastable and presumably anneal out during a homogenization treatment and/or a hot pressing operation. Since each of these dissociations occur on the {110) planes, it is tempting to suggest that their presence may be linked to the hexagonal (C40) to tetragonal (C11b) phase transformation (The transformation occurs with (0001) basal plane of the high temperature C40 phase parallel to the {110) plane of the low temp $C11_b$ phase). This is plausible since plasma processed samples go through the liquid state. However, there is much recent controversy about whether or not the high temperature phase indeed exists. Much of the recent support for the occurrence of the high temperature phase is circumstantial, in the form of observations of rotational twin variants (7) and in rapidly synthesized microstructures (8). The C11b <=> C40 transformations can be described by the operation of 1/4<111> partials on {110) planes in ($C11_b$) $MoSi_2$. The hexagonal C40 structure has an ABCABC layer stacking in (0001) planes whereas the tetragonal $C11_b$ has an ABAB stacking in {110) planes. The (0001) basal plane in C40 structure is compositionally identical to the {110) planes in $MoSi_2$. The C layer in the C40 structure can be translated to the A or B layer positions in $C11_b$ structure by ±1/4[111] and ±1/4[$\bar{1}\bar{1}$1] translations respectively. For the dislocation assisted growth of ($C11_b$) $MoSi_2$ both ±1/4[111] and ±1/4[$\bar{1}\bar{1}$1] poles are expected to operate (9). Hence combinations of different overlapping stacking faults may have fault vectors of 1/2<001> and 1/2<331> magnitude as observed here.

Alternatively these observations can be rationalized in terms of the dissociation schemes published in the literature (6, 11). Several different dissociation schemes are of interest for the 1/2<331> dislocation. The most commonly predicted (6) is:

1/2<331] => 1/6<331] + APB + 1/6<331] + APB + 1/6<331] [1]

However, our contrast observations are not consistent with such a dissociation. An alternate dissociation scheme for 1/2<331] dislocations is (11):

1/2<331] => 1/12<331] + CSF + 1/12<331] + APB + 1/12<331]
+ SISF +
1/12<331] + APB + 1/12<331] + CSF + 1/12<331] [2]

It is expected that E_{APB} and $E_{CSF} > E_{SISF}$, hence a simple observation may be

1/2<331] => 1/4<331] + SISF + 1/4<331] [3]

Our experimental observations of a two fold dissociation for 1/2[33$\bar{1}$] dislocation are consistent with the dissociation scheme of reaction [3]. This particular dissociation of 1/2<331] has been observed for the first time, and presumably occurs in an effort to reduce its elastic energy. It is suggested that the energy reduction obtained for this particular dissociation is more favorable than a dissociation of 1/2<331] => 1/2<111] + <110] recently observed in deformed single crystal specimens (10).

Similarly for the [001] dislocation the possible dissociation schemes are (6, 9):

[001] => 1/3[001] + APB + 1/3[001] + APB + 1/3[001] [4]

$$[001] \Rightarrow 1/12[3\bar{3}1] + CSF + 1/6[001] + SISF + 1/6[001] + CSF + 1/6[001] + CSF + 1/6[001] + SISF + 1/6[001] + CSF + 1/12[\bar{3}31] \quad [5]$$

For E_{APB} and $E_{CSF} > E_{SISF}$, Equation [5] may reduce to

$$[001] \Rightarrow 1/4[1\bar{1}1] + SISF + 1/2[001] + SISF + 1/4[\bar{1}11] \quad [6]$$

However, the experimental observations cannot be reconciled to either one of these dissociation schemes. The contrast is consistent with a fault vector of 1/2[001]. Theoretical computations reveal that the E_{APB} for [001] and [33$\bar{1}$] dissociation on (1$\bar{1}$0) planes are identical (11), hence comparing the dissociation distances of [001] and [33$\bar{1}$] faults, it is likely that the nature of dissociations are different. It is interesting to note that the 1/2[001] faults are located in close proximity to the trail of Mo_5Si_3 particles as shown in figure 4. Mo_5Si_3 forms either by eutectic decomposition, figure 1, or because of silicon evaporation losses during processing of single phase $MoSi_2$. Two consistent arguments suggest that 1/2[001] fault may be related to the $MoSi_2$ => Mo_5Si_3 transformation: i) the fault vector displacement creates first nearest neighbors of Mo-Mo atoms, a situation desirable for the formation of the Mo-rich Mo_5Si_3 compound and ii) the stacking fault creates a pure edge misfit of 1/2[001](1$\bar{1}$0) as well as creating a dilatational misfit perpendicular in the [1$\bar{1}$0] direction, both of which are required for the matching of Mo_5Si_3 to the $MoSi_2$ matrix. This is the subject of further study and will be reported at a later date.

Prior <111> stacking fault observations in $MoSi_2$, resulting from compressive deformation at 1173K (5,6), led to the suggestion that such faults may be responsible for low (≈1173K) BDTT. It is plausible that the occurrence of quenched-in 1/2<001> and 1/2<331> faults may have some effect on the toughness behavior of as-processed $MoSi_2$; this is the subject of further study.

SUMMARY

The microstructures of plasma sprayed MoSi2 have been examined. While high density deposits are obtained, the microstructures are essentially inhomogenous and metastable such that a homogenization treatment is necessary. [001] and 1/2[33$\bar{1}$] dissociations on (1$\bar{1}$0) have been observed for the first time. The 1/2[33$\bar{1}$] dissociation is consistent with theoretical predictions made earlier, while the [001] dissociation could not be reconciled to any previously suggested dissociation. These faults may be related to the C40 <=> $C11_b$ or $MoSi_2$ => Mo_5Si_3 transformations.

ACKNOWLEDGEMENTS

This study was conducted under AFOSR contract no: 91-0427 with Dr. Alan H. Rosenstein acting as program monitor. The authors thank Dr. G.T. Gray for kindly providing the samples used in this study. The authors thank Dr. Peter M. Hazzledine, Dr Satish Rao and Stuart Maloy for communicating their unpublished results. BKK acknowledges many helpful discussions with Dr. Peter M. Hazzledine.

REFERENCES

1. P.J. Mescheter and D.S. Schwartz, J. Met., 11 (1989) 52
2. R.M. Aikin, Jr., Mat. Sci. & Eng., **A155** (1992) 121
3. R. Tiwari, S. Sampath and H. Herman, Mater. Res. Soc. Symp. proc., **213** (1991) 807
4. R.G. Castro, R.W. Smith, A.D. Rollett and P. Stanek, Mat. Sci. & Eng., **A155** (1992) 101
5 K. Kimura, M. Nakamura and T. Hirano, J. Mat. Sci, **25** (1990) 2487
6. Y. Umakoshi, T. Sakagami, T. Hirano and T. Yamane, Acta. Met. **38** No.6 (1990) 909
7. T.E. Mitchell, R.G. Castro and M.M. Chadwick, Phil. Mag., **A65**, (1992) 1339
8. K.S. Vecchio, L. S. Yu and M.A. Meyers, Acta. Met., *in review*
9. Peter M. Hazzledine, *private communication*
10. S. A. Maloy, T.E. Mitchell, J.J. Lewandowski and A.H. Heuer, Phil. Mag. Lett. *in review*
11. Satish I. Rao, Dennis M. Dimiduk and Madan G. Mendiratta, Phil. Mag., *in review*

Creep Deformation Studies in Directionally Solidified $MoSi_2$-Mo_5Si_3 Eutectics

D. P. Mason and D. C. Van Aken, Department of Materials Science and Engineering, The University of Michigan, Ann Arbor, MI 48109-2125

ABSTRACT

The high temperature deformation behavior of directionally solidified (DS) $MoSi_2$-Mo_5Si_3 eutectics was studied and compared to powder processed $MoSi_2$-Mo_5Si_3 composites having the same volume fraction of Mo_5Si_3. Decremental step strain rate tests were performed in the temperature range of 1100-1300°C and at strain rates between 10^{-4} to 10^{-6}/s. A considerable increase in the flow stress was observed for the directionally solidified material. At 1200°C and a strain rate of 10^{-6}/s the flow stress of the DS eutectic was 255 MPa as compared to 20 MPa for the powder processed composite. The high temperature strength of the DS eutectic was unaffected by changes in the scale of the lamellar microstructure and these results were modeled using a constitutive relation for power law creep. A stress exponent of 4.5 and an activation energy of 300 kJ/mol was determined for the DS eutectic. Evidence of dislocation glide and climb was observed in the $MoSi_2$ lamellae whereas the dislocation density was small in the Mo_5Si_3 phase. The improved creep strength of the eutectic is believed to be a result of both the fibrous morphology and a stronger interface structure as compared to the powder processed composite.

INTRODUCTION

Molybdenum disilicide ($MoSi_2$) has recently received considerable attention as a possible matrix for high temperature structural materials [1,2]. The two main factors limiting its use are poor high temperature creep strength and low temperature ductility. Many researchers are applying composite engineering to address these issues and the viability of many different reinforcements, reinforcement morphologies, and processing paths are being assessed yet, only recently have structure-property relationships been addressed [2,3,4,5]. The authors have shown that the room temperature hardness of $MoSi_2$ and $MoSi_2$-Mo_5Si_3 eutectics is dependent upon the microcrystalline scale of the $MoSi_2$ and that the hardness follows a Hall-Petch relationship [6]. This scale effect was also observed at elevated temperatures, but the difference in hardness decreased continuously with increasing temperature. In the present study, the relationship between microstructural scale and mechanical behavior is further examined for directionally solidified $MoSi_2$-Mo_5Si_3 eutectics.

EXPERIMENTAL PROCEDURE

The materials used in this study were all of eutectic composition, i.e. 44.5 volume percent Mo_5Si_3 dispersed in a $MoSi_2$ matrix. Directional solidification (DS) was accomplished by a Czochralski method using a tri-arc furnace and a pull rate of 39 mm/hr. Two DS eutectic rods were produced by this method and one alloy contained 0.35 atomic percent erbium. A composite was also produced by hot pressing (HP) elemental molybdenum and Johnson Matthey $MoSi_2$ powders at 1625°C for 2 hours under a pressure of 23 MPa and two billets were produced in this manner. In an effort to reduce the amount of SiO_2 in the hot pressed material one sample was ball milled with erbium powder (50 μm in diameter) giving a final composition of 0.35 atomic percent erbium.

Room temperature fracture toughness of the DS eutectics was measured by a Vickers pyramid hardness indentation technique [7,8]. The uncertainty in the fracture toughness was minimized by only considering indentations that were "well behaved", i.e. cracks extending from the corners of the indent and crack lengths that were greater than 2-3 times the indent diagonal. The modulus of the eutectic was estimated to be 360 GPa by a rule of mixtures (isostrain) using Young's moduli of 440 GPa and 260 GPa for $MoSi_2$ and Mo_5Si_3, respectively [9]. An Instron model 4507 equipped with a Centorr high temperature furnace was used to conduct decremental step strain-rate compression tests in the temperature range 1100°C-1300°C, strain rates between $1x10^{-4}$/s and $1x10^{-6}$/s, and samples with dimensions of approximately 3 x 3 x 6 mm^3. Macroscopic cracks were evident in all of the DS eutectics prior to compression testing. In a few cases the samples buckled during the test as a result of cracks extending along the sample length. Results for the buckled samples were excluded from this presentation.

Electron microscopy studies were performed using the facilities of the Electron Microbeam Analysis Laboratory at the University of Michigan. Thin foils for transmission electron microscopy were prepared by mechanical dimpling and room temperature ion-milling.

RESULTS

A blocky microstructure consisting of discrete grains of $MoSi_2$ and Mo_5Si_3 was observed for the hot pressed material (see Fig. 1). The grain diameters of the constituent phases averaged 7.0 µm and 15 µm for the $MoSi_2$ and Mo_5Si_3, respectively. SiO_2 was evident at the grain corners in both billets and small $Er_2Mo_3Si_4$ particles ranging in size from 1 to 15 µm were evident in the microstructure of the erbium containing material. Porosity was on the order of 4% for each billet. A cellular microstructure was observed in the DS materials and the script size varied continuously along the length of the rods from an initial 2.5 µm lamellar spacing to as large as 11.4 µm in the erbium containing material (see Fig. 2). These growth perturbations were evident along the length of the DS rods and $Er_2Mo_3Si_4$ was present at the interfaces between the eutectic constituents (see Fig. 2b). Growth perturbations are believed to have resulted from variations in the pull rate during solidification. The volume fraction of $Er_2Mo_3Si_4$ also increased as the DS bar was pulled from the melt.

Decremental step strain-rate test results at 1200 and 1300°C are shown in figure 3. No apparent change in the creep strength was observed for materials containing erbium. The DS materials were significantly stronger than the hot pressed powders as shown by the following comparison; at 1200°C and a strain rate of 10^{-6}/s the flow stress of the DS eutectic was 255 MPa as compared to 20 MPa for the powder processed composite. Stress exponents for the hot pressed material varied between 2.5 and 3.5 whereas the DS eutectics

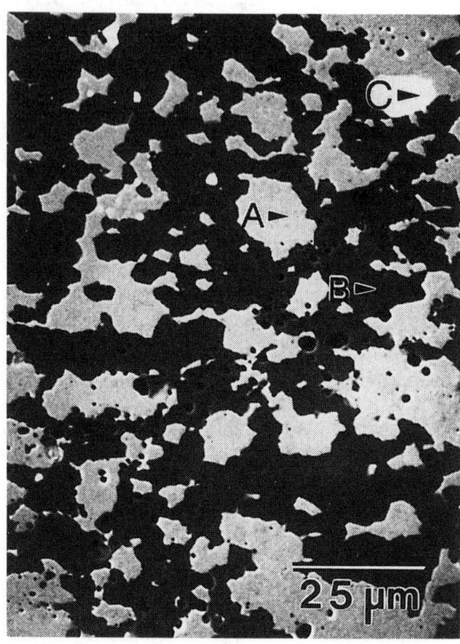

Fig. 1 Backscattered electron micrograph showing the reinforcement morphology in the HP + 0.35 at% Er alloy. (phase identification A-Mo_5Si_3, B-$MoSi_2$ and C-$Er_2Mo_3Si_4$)

ranged between 4.5 and 4.9. At 1300°C the stress exponent for the hot pressed material decreased from 3.5 to 2.5 at a strain rate of $1x10^{-5}$/s as the strain rate was decreased to $5x10^{-6}$/s. The stress exponent for the DS eutectic was constant for each temperature tested. Creep strength was not affected by variations in the lamellar spacing or erbium content since similar results were obtained from both DS eutectic rods and compression samples obtained from both ends of the erbium treated DS rod. An activation energy of 300 kJ/mol for the creep process was determined for the DS eutectic by assuming a power law relationship of the form

$$\dot{\varepsilon} = A \sigma^n \exp(-Q/RT) \qquad (1)$$

where $\dot{\varepsilon}$ is the strain rate, A is a rate constant, σ is the flow stress, n is the stress exponent, Q is the creep activation energy, R is the universal gas constant, and T is the absolute temperature. A fixed stress exponent of 4.5 was used in the regression analysis while Q and A were varied to obtain the best straight line fit (see Fig. 4).

Dislocation structures were examined in the DS material tested at 1300°C. Dislocation glide and climb is evident in the $MoSi_2$ phase

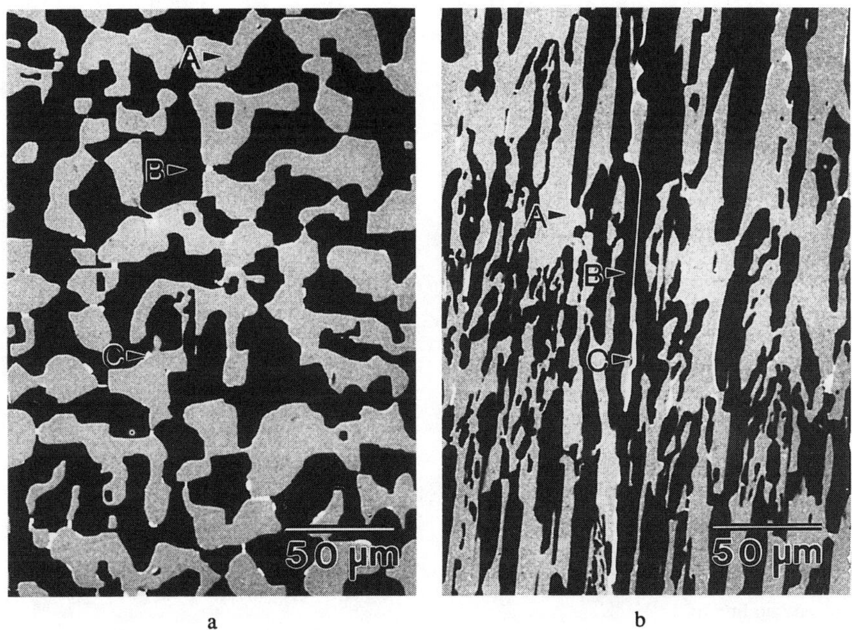

Fig. 2 Backscattered electron micrograph showing a) transverse and b) longitudinal sections of the DS + 0.35 at% Er alloy. Phase identification A-Mo_5Si_3, B-$MoSi_2$ and C-$Er_2Mo_3Si_4$.

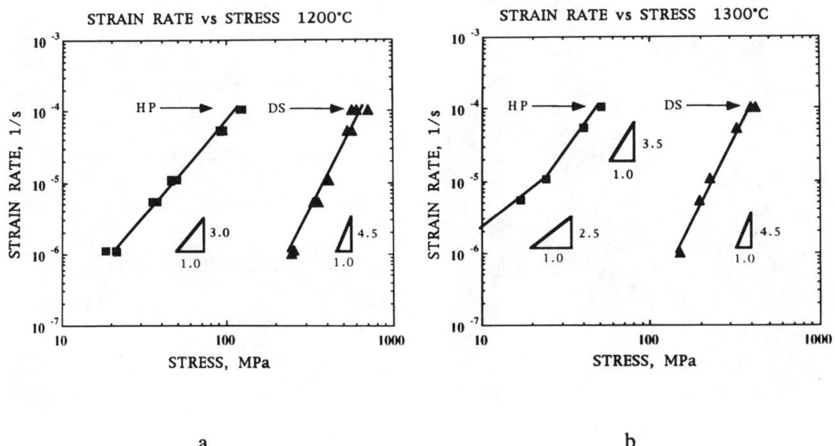

Fig. 3 Strain rate vs stress plots for a) 1200°C and b) 1300°C. The DS are the strongest followed by the HP at both temperatures. Note the change in slope for the HP alloys which contain SiO_2.

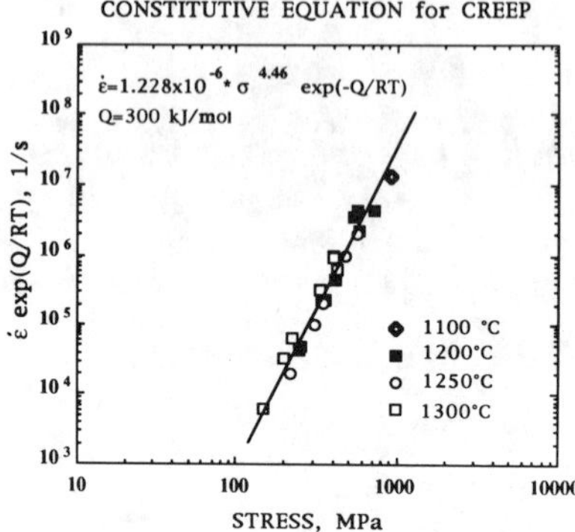

Fig. 4 Regression analysis of the DS eutectic creep data. The stress exponent was held constant at 4.5 while the activation energy, Q, and the rate constant A where varied to obtain the best straight line fit.

Table I
A Comparison of Creep Data for $MoSi_2$ Based Composites and DS $MoSi_2$-Mo_5Si_3 Eutectics

Material	σ @ 1200°C $\dot{\varepsilon}=1\times10^{-6}$	Q (kJ/mol)	n
Sadananda et al $MoSi_2$ + 20% SiC	220 MPa	596	3.2
Sadananda et al $MoSi_2$ + WSi_2	40 MPa	540	2.4
Suzuki et al $MoSi_2$ + 30% SiC	210 MPa	265-465	3.5
Bose et al $MoSi_2$ + 18% SiC	180 MPa	306	3.2
DS eutectics	255 MPa	300	4.46

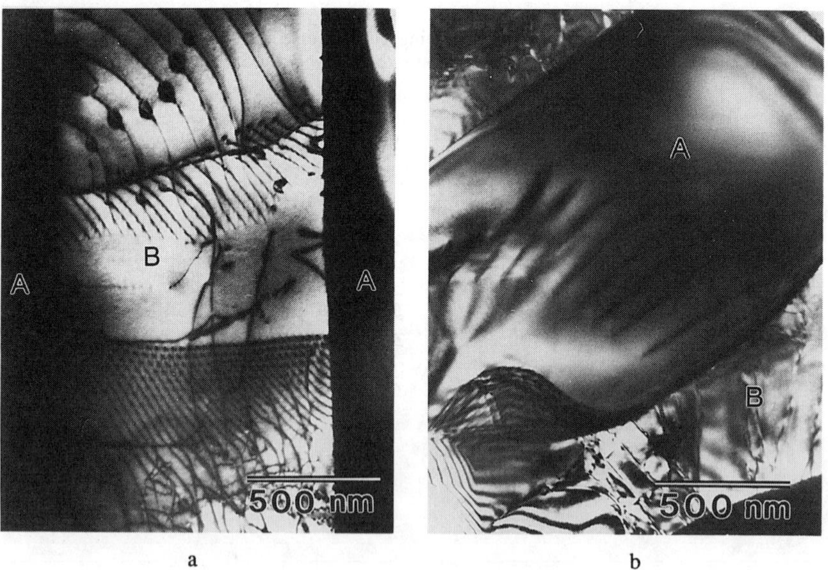

Fig. 5 TEM micrograph of a) the typical dislocation structures in the $MoSi_2$ phase and b) low dislocation density in the Mo_5Si_3 phase of DS alloys deformed at 1300 °C. Note the presence of dislocation network formation and bowing in the $MoSi_2$. Phase identification: A-Mo_5Si_3 and B-$MoSi_2$.

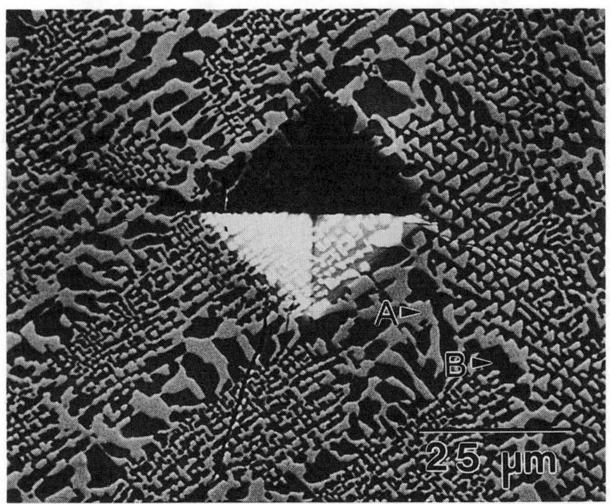

Fig. 6 Backscattered electron micrograph of a typical crack pattern obtained for fracture toughness measurements. (load=1kg) Phase identification: A-Mo_5Si_3 and B-$MoSi_2$.

whereas very little dislocation activity was observed in the Mo_5Si_3 (see Fig. 5). Interface cracking between the lamellae was not observed.

Room temperature fracture toughness values showed a slight dependence on the scale of the lamellar structure and varied between 2.0 MPa√m at a spacing of 6.56 μm to 3.0 MPa√m at 0.93 μm which approached that reported for monolithic $MoSi_2$ (2.85 MPa√m) [8]. A typical indentation with resultant crack paths is shown in Figure 6. It should also be noted that there was evidence of pre-existing microcracks in the $Er_2Mo_3Si_4$ phase and well behaved crack patterns were not obtained in most of the Er treated materials.

DISCUSSION

It is well documented that high concentrations of SiO_2 is detrimental to the high temperature creep resistance of $MoSi_2$ based materials [3,10,11]. In this study, the addition of erbium powders during the ball milling process was ineffective in reducing the SiO_2 concentration in the final hot pressed billet. Creep strengths of our hot pressed materials were typical of those reported for $MoSi_2$ materials containing SiO_2 [3]. Much higher strengths were observed for the DS eutectics and a summary of published creep strengths, stress exponents, and activation energies for $MoSi_2$ composites are shown in table I [12,13,14]. This comparison shows that the DS eutectics are stronger than the $MoSi_2$/SiC/20w composites. These eutectics also exhibit a stronger stress dependency and this may be a result of the fiber morphology and possible strengthening from a crystallographic texture.

In this study the creep strength of $MoSi_2$-Mo_5Si_3 eutectics was found to be independent of microstructural scale, i.e. $MoSi_2$ spacing. The creep mechanism, based upon the measured stress exponent and observed dislocation structures in the $MoSi_2$, appears to be dislocation climb and glide. Sadananda et al [12] have reported that the diffusion of Mo atoms in $MoSi_2$ should be the rate limiting factor for dislocation climb and have estimated an activation energy between 350 and 540 kJ/mol for this process. Experimental values of 596 kJ/mol and 465 kJ/mol reported by Sadananda et al [12] and Suzuki et al [13] in compression fall at the high end of this range. However, Bose et al [14] and Weiderhorn et al [4] have obtained activation energies of 306 kJ/mol and 312 kJ/mol in compression and 347 kJ/mol and 557 kJ/mol in tension, respectively. The present result of 300 kJ/mol in compression is in the range of published results for $MoSi_2$.

Deformation of the Mo_5Si_3 phase appears to be controlled by diffusional flow as evidenced by the low dislocation activity shown in Fig. 5. This observation is consistent with the work of Anton and Shaw [15] who have reported a stress exponent of 1.9 for Mo_5Si_3. A stress exponent greater than one usually indicates some contribution from dislocations. Anton and Shaw [15] also demonstrated that Mo_5Si_3 is more creep resistant than $MoSi_2$ at 1200°C. In the present study however, the deformation of the Mo_5Si_3 in the eutectic does not appear to be rate limiting despite the high volume fraction. Strengthening in the eutectic does appear to be a result of a strong interface structure and the fibrous morphology of the Mo_5Si_3 since both of these microstructural characteristics are absent in the powder processed material. The difference in the interface structure was previously shown by Kung et al [16].

A marginal improvement in the fracture toughness was observed for finer lamellar spacings. Although, increased hardness [6] and fracture toughness may be achieved at room temperature through microstructural refinement, the same principal does not apply to the high temperature creep strength.

CONCLUSIONS

The creep strength of $MoSi_2$- Mo_5Si_3 materials was found to be independent of microstructural scale but dependent upon the interface structure and morphology of the Mo_5Si_3 reinforcement. There was a slight dependence of room temperature fracture toughness on the scale of the $MoSi_2$ phase but, no advantage over monolithic $MoSi_2$ was realized.

ACKNOWLEDGMENTS

This work was supported in part by the Air Force Office of Scientific Research under the AFOSR-URI grant No. DoD-G-AFOSR-90-0141. The program manager was Dr. Alan H. Rosenstein. Also, the authors gratefully acknowledge Mr. Jim Garrett of McMaster University for providing the two Czochralski directionally solidified samples.

REFERENCES

1. J. J. Petrovic and A.K. Vasudevan, Intermetallic Matrix Composites II, eds. D.B. Miracle, D.L. Anton and J.A. Graves (Mat. Res. Soc. Proc.vol. 273, Pittsburg, PA 1992) p 229.
2. D.E. Alman, K.G. Shaw, N.S. Stoloff and K. Rajan, Mat. Sci. and Eng. **A155**, 95 (1992).
3. R. Gibala, A.K. Ghosh, D.C. Van Aken, D.J. Srolovitz, A. Basu, H. Chang, D.P. Mason and W. Yang, Mat. Sci. and Eng. **A155**, 147 (1992).
4. S.M. Wiederhorn, R.J. Gettings, D.E. Roberts and C. Ostertag, Mat. Sci. and Eng. **A155**, 147 (1992).
5. R.M. Aiken, Scripta Metall. **26**, 1025 (1992).
6. D. P. Mason and D. C. Van Aken, Scripta Metall., in press.
7. G.R. Anstis, P. Chantikul, B.R. Lawn and D.B. Marshall, J. Am. Ceram. Soc. **64**, 9 (1981).
8. A.K. Bhattacharya and J.J. Petrovic, J. Am. Ceram. Soc. **74**, 10 (1992).
9. S.R. Srinivasan and R.B. Schwartz, J. Mater. Res. **7**, 1610 (1992).
10. D. A. Hardwick, P. L. Martin and R. J. Moores,Scripta Metall. **27**, 391 (1992).
11. J. D. Cotton, Y. S. Kim and M. J. Kaufman, Mat. Sci. and Eng. **A144,** 287 (1991).
12. K. Sadananda, C. R. Feng, H. Jones and J. J. Petrovic, Mat. Sci. and Eng. **A155,** 227 (1992).
13. M. Suzuki, S. R. Nutt and R. M. Aiken, Jr, Intermetallic Matrix Composites II, eds. D. B. Miracle, D. L. Anton and J. A. Graves (Mat. Res. Soc. Proc.vol. 273, Pittsburg, PA 1992) p 267.
14. S. Bose, Mat. Sci. and Eng. **A155**, 217 (1992).
15. D. L. Anton and D. M. Shaw, High Temperature Ordered Intermetallic Alloys IV, eds. L. A. Johnson, D.P. Pope and J.O. Stiegler, (Mat. Res. Soc. Proc.Vol. 213, Pittsburg, PA 1991) p.733.
16. H. Kung, D.P. Mason, A. Basu, H. Chang, D.C. Van Aken, A.K. Ghosh and R. Gibala, Intermetallic Matrix Composites II, eds. D. B. Miracle, D. L. Anton and J. A. Graves (Mat. Res. Soc. Proc. vol. 273, Pittsburg, PA 1992) p 241.

PHYSICAL AND MECHANICAL PROPERTIES OF $MoSi_2$-$Er_2Mo_3Si_4$ COMPOSITES

D. Keith Patrick and David C. Van Aken, Department of Materials Science and Engineering, The University of Michigan, Ann Arbor, MI 48109-2136

ABSTRACT

A study has been conducted to determine the feasibility of using $Er_2Mo_3Si_4$ as a reinforcement phase to improve the high temperature strength of $MoSi_2$. The melting temperature of $Er_2Mo_3Si_4$ was determined to be 1930 ± 20°C whereas the eutectic of $Er_2Mo_3Si_4$-39 vol% $MoSi_2$ melted at 1790 ± 10°C. Elevated temperature microhardness tests show that $Er_2Mo_3Si_4$ has significantly higher hardness than $MoSi_2$ above 1000°C, e.g. approximately 5.8 GPa versus 1.5 GPa at 1300°C, respectively. A $MoSi_2$/$Er_2Mo_3Si_4$/20p composite was produced by ball milling and hot pressing arc-melted $MoSi_2$-20 vol% $Er_2Mo_3Si_4$ materials. At 1300°C the $MoSi_2$/$Er_2Mo_3Si_4$/20p composite and a directionally solidified $Er_2Mo_3Si_4$-$MoSi_2$ eutectic exhibited hardnesses of 2.4 GPa and 4 GPa, respectively. Preliminary results from compressive decremental step strain rate tests at 1300°C indicate that the creep strength of the $MoSi_2$/$Er_2Mo_3Si_4$/20p composite is comparable to that of a $MoSi_2$/SiC/20w composite. The creep stress exponent was determined to be 3.3 at 1200°C and 3.7 at 1300°C.

INTRODUCTION

The high temperature creep strength of $MoSi_2$ is adversely affected by the presence of amorphous SiO_2 [1]. This oxide may be present in the initial starting material or may develop as part of the processing of the $MoSi_2$ powders, e.g. ball milling. It has been shown that the SiO_2 may be eliminated by adding carbon prior to hot pressing [2], or by carefully processing elemental powders in an inert environment [3]. It has also been suggested that the improved properties of $MoSi_2$/SiC composites may in part be a result of the reduction of SiO_2 by free carbon associated with the SiC reinforcement [4]. Addition of rare earth metals during ball milling has not been as successful as the addition of carbon in reducing the SiO_2 [5].

The purpose of this study was to investigate some of the physical properties of $Er_2Mo_3Si_4$, explore processing paths to develop $MoSi_2$/$Er_2Mo_3Si_4$ particulate composites, and determine the high temperature creep properties of these composites. It has been reported that $Er_2Mo_3Si_4$ formed naturally in a $MoSi_2$-Mo_5Si_3 eutectic as a result of adding erbium (Er) to deoxidize the eutectic liquid [6]. However, information available about $Er_2Mo_3Si_4$ seems to be limited to crystallographic or x-ray diffraction data [6,7]. Many basic physical and mechanical properties of $Er_2Mo_3Si_4$ have not as yet been reported.

EXPERIMENTAL

Samples of $MoSi_2$ and $Er_2Mo_3Si_4$ were prepared by arc-melting high purity elemental Mo bar stock (99.9%), Si single crystal stock (99.999%), and Er chips (99.9%) in an argon atmosphere. These buttons were then broken up, mixed in the desired proportions, and remelted to produce the $MoSi_2$-20 vol% $Er_2Mo_3Si_4$ alloy and the $Er_2Mo_3Si_4$-39 vol% $MoSi_2$ eutectic used in this study. The eutectic arc-melted buttons were then directionally solidified by the Czochralski method at a pull-rate of 45 mm/hr using a tri-arc furnace at McMaster University. A $MoSi_2$/$Er_2Mo_3Si_4$/20p composite was produced by first ball milling the $MoSi_2$-20 vol% $Er_2Mo_3Si_4$ buttons in a freon slurry for 24 hours to produce a fine powder (less than 100 μm in diameter). This powder was then hot pressed at 1565°C for 4 hours under a pressure of 30 MPa in a flowing argon atmosphere. The reported grain diameters and constituent volume fractions were determined by the mean linear intercept method and by computer aided image analysis, respectively.

A density of 7.85 g/cm^3 was measured for the arc-melted $Er_2Mo_3Si_4$ using Archimedes principle in accordance with ASTM Standard D-3800-79. The calculated theoretical density is 8.25 g/cm^3. The hot pressed $MoSi_2/Er_2Mo_3Si_4/20p$ contained less than 1% porosity as measured by the linear intercept method. Melting temperatures for the eutectic and the $Er_2Mo_3Si_4$ phase were determined by differential thermal analysis performed at NASA-Lewis Research Center and by direct visual observation in a Centorr model CG-2.5X2-3X3-W-A-D6A3-A-22 furnace at the University of Michigan.

Room temperature Vickers microhardness tests were performed on each material using a Zwick microhardness tester with a load of 0.5 kg. Hot hardness tests (23°C to 1300°C) were performed using a Nikon QM-2 hot hardness test machine under a 0.5 kg load. The hardness of the eutectic was determined by indentations on a transverse face relative to the growth direction. Compressive decremental step strain rate tests were performed at 1200°C and 1300°C on the $MoSi_2/Er_2Mo_3Si_4/20p$ composite at constant crosshead speeds corresponding to strain rates of $5x10^{-5}$, $1x10^{-5}$, $5x10^{-6}$, and $1x10^{-6}$ sec^{-1}. Specimen dimensions were 3 x 3 x 6 mm^3.

The isothermal oxidation behavior of $Er_2Mo_3Si_4$ was examined by exposing arc-melted samples to air at 1400°C for 24 hours. The oxide layer was examined by optical and electron microscopy. Standardless x-ray energy-dispersive spectroscopy (EDS) was done using a KEVEX microanalyzer attached to a Hitachi S-520 SEM, while wavelength-dispersive spectroscopy (WDS) was performed on a CAMECA MBX electron microprobe at the University of Michigan Electron Microbeam Analysis Laboratory.

RESULTS AND DISCUSSION

Arc-melted samples of $MoSi_2$-10 vol% $Er_2Mo_3Si_4$ show a cellular solidification with continuous films of $Er_2Mo_3Si_4$ at the $MoSi_2$ cell boundaries (see Fig. 1a). At 20 volume percent $Er_2Mo_3Si_4$ these continuous boundary films were replaced by a eutectic structure consisting primarily of $Er_2Mo_3Si_4$ and $MoSi_2$ with small particles of Mo_5Si_3 dispersed randomly within the eutectic (see Fig. 1b and 1c). The presence of a small amount of Mo_5Si_3 within the eutectic (see Fig. 1c) may indicate a possible ternary eutectic reaction between $Er_2Mo_3Si_4$, $MoSi_2$, and Mo_5Si_3. A eutectic composition of 38.6 vol% (32.1 wt%) $MoSi_2$ was determined by volume fraction measurements. This composition was then directionally solidified (see Fig. 1d). A eutectic temperature of 1790 ± 10°C was determined both visually and by differential thermal analysis. The melting temperature of $Er_2Mo_3Si_4$ was determined visually to be 1930 ± 20°C.

The hot pressed $MoSi_2/Er_2Mo_3Si_4/20p$ composite is shown in Figure 2. The cellular nature of the as-solidified structure permitted easy fragmentation of the $MoSi_2$ and a powder size equivalent to the intercellular spacing was produced. An average $MoSi_2$ grain size of 9.3 ± 0.8 μm was measured in the hot pressed material (see Fig. 2b). Oxides are evident in the microstructure and a comparison of optical and backscattered images shows that the larger oxides contain erbium (see Fig. 2c and 2d). At present it is uncertain whether these oxides are crystalline or amorphous in nature. Most of the $Er_2Mo_3Si_4$ and oxide particles are on $MoSi_2$ grain boundaries.

The oxidation behavior of arc-melted $Er_2Mo_3Si_4$ samples was examined to determine if a catastrophic reaction would take place. A cross-sectional view of the surface reaction in a sample exposed to air at 1400°C for 24 hours is shown in Figure 3. An 80 μm thick reaction layer was formed consisting of an outer layer of oxide scale and a subsurface multiphase intermetallic layer. The surface of the sample was dark gray in color with pink regions indicating the presence of large amounts of Er_2O_3. The backscattered electron image (see Fig. 3b) shows several phases present in the reaction layer. The outermost layer (A) is an $Er_2O_3 + SiO_2$ mixture with a dark phase directly beneath the $Er_2O_3 + SiO_2$ layer identified as $MoSi_2$ (B). Two other phases, (C) and (D), are present beneath the $MoSi_2$ layer. The composition of these phases is currently under investigation. Some spalling of the outermost $Er_2O_3 + SiO_2$ oxide layer was observed, but the underlying reaction layer was adherent.

At room temperature the hardness of $Er_2Mo_3Si_4$ is slightly higher than that reported for $MoSi_2$ [1,8]; however, the $Er_2Mo_3Si_4$ intermetallic retains its hardness above 1000°C (see Fig. 4). A hardness of about 10.2 GPa was measured on the directionally solidified eutectic at

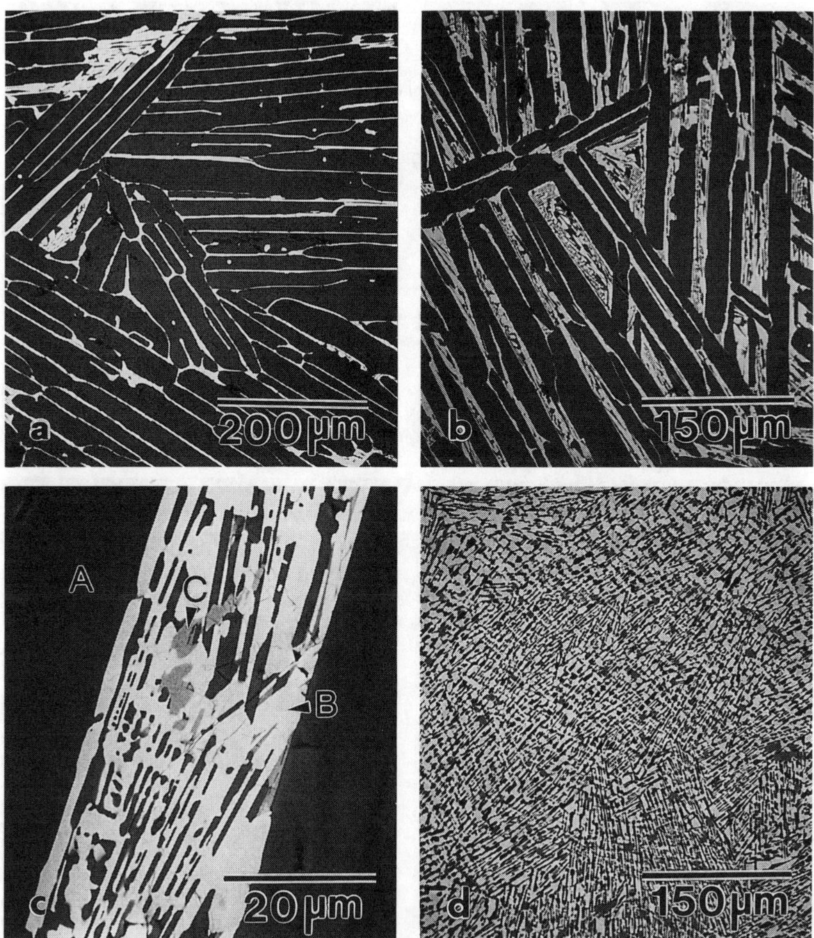

Fig. 1 Backscattered electron images of the arc-melted $MoSi_2$-$Er_2Mo_3Si_4$ materials. Micrographs show (a) the cellular solidification structure observed in the arc-melted $MoSi_2$-10 vol% $Er_2Mo_3Si_4$ sample; (b) the solidification structure of the $MoSi_2$-20 vol% $Er_2Mo_3Si_4$ material prior to ball milling; (c) the eutectic structure observed in the 20 vol% material; (d) a transverse section of the directionally solidified ingot, i.e. growth direction is normal to the micrograph. Phases are identified as A- $MoSi_2$, B- $Er_2Mo_3Si_4$, and C- Mo_5Si_3.

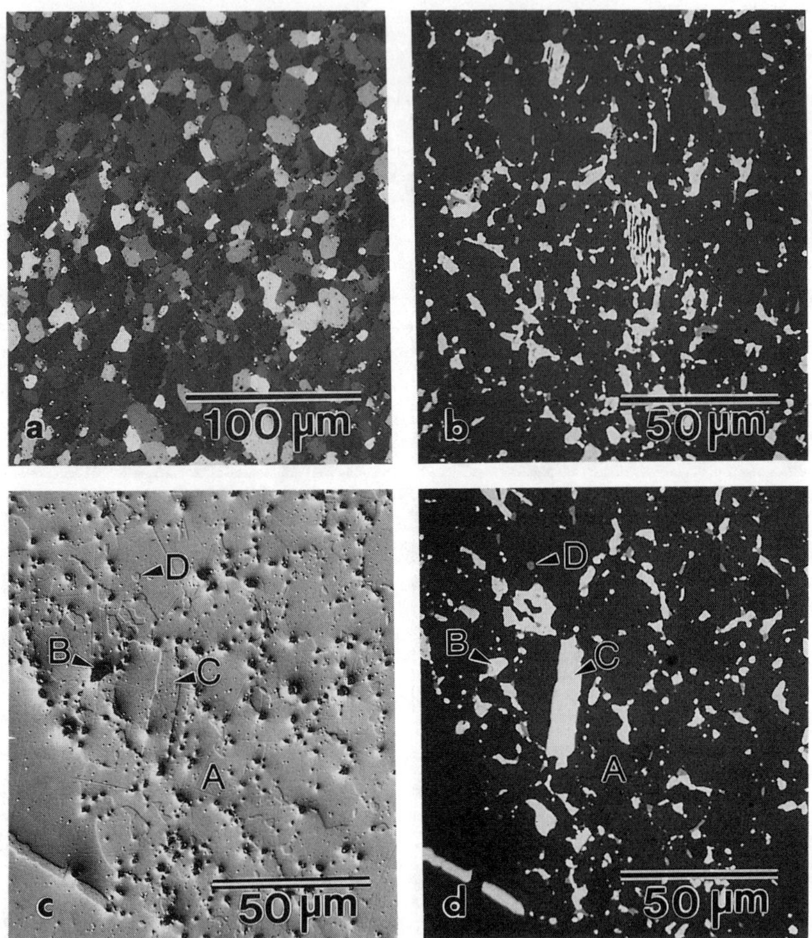

Fig. 2 Microstructures of the $MoSi_2/Er_2Mo_3Si_4$/20p composite. (a) Optical image using cross-polarized light showing the $MoSi_2$ grain structure. (b) Backscattered electron image showing the dispersion of erbium-rich particles. (c) Optical image using Nomarski phase contrast showing the oxide constituents. (d) A backscattered electron image of the region shown in (c). A comparison of both (c) and (d) shows that the oxide constituent is erbium rich. Phases are identified as A- $MoSi_2$, B- Er_2O_3, C- $Er_2Mo_3Si_4$, and D- Mo_5Si_3.

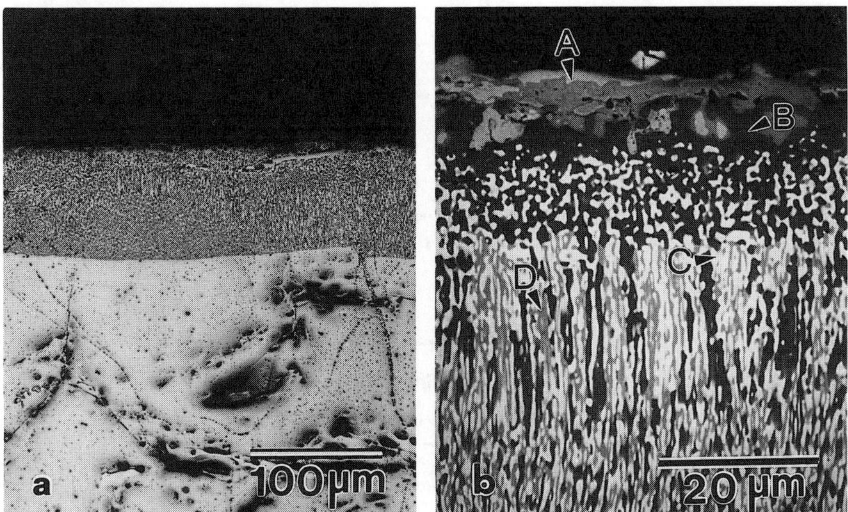

Fig. 3 (a) An optical micrograph of the 80μm reaction layer formed on a $Er_2Mo_3Si_4$ sample oxidized in air at 1400°C for 24 hours. (b) A backscattered electron image of the reaction layer. The reaction layer consists of 4 phases identified by WDS as A- an Er_2O_3 + SiO_2 surface layer and B- $MoSi_2$. The compositions of the phases identified at C and D are currently under investigation.

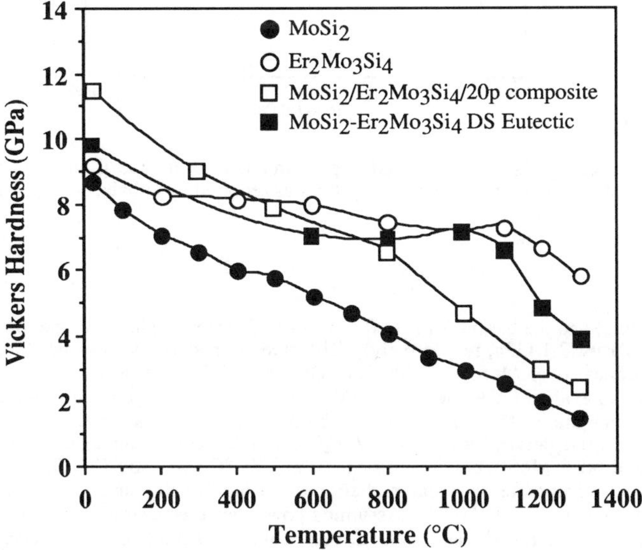

Fig. 4 Hot hardness of the $Er_2Mo_3Si_4$, $MoSi_2$/$Er_2Mo_3Si_4$/20p composite, the eutectic, and the reported values for $MoSi_2$ [1,8].

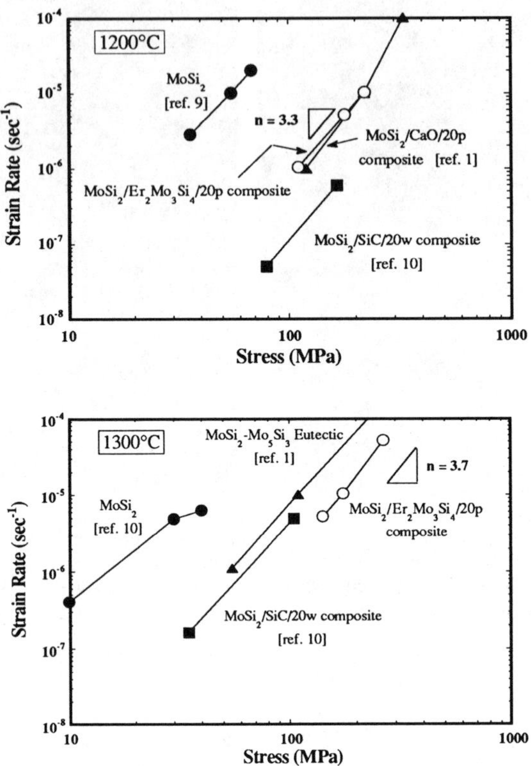

Fig. 5 Results of compressive decremental step strain rate tests at 1200°C and 1300°C as compared with $MoSi_2$ and other composite systems as reported by Gibala, et al. [1], Bose [9], and Sadananda, et al. [10].

room temperature, and the hardness at 1000°C was considerably higher than that of monolithic $MoSi_2$ (7.4 GPa vs. 3.1 GPa, respectively). The eutectic material also showed a substantial increase in hardness over $MoSi_2$ at 1300°C (4 GPa vs. 1.5 GPa, respectively). An improved hardness was also evident for the $MoSi_2/Er_2Mo_3Si_4/20p$ composite both at room temperature and elevated temperature. The composite had a hardness of 2.4 GPa at 1300°C.

Stress-strain rate curves for the $MoSi_2/Er_2Mo_3Si_4/20p$ composite are compared in Fig. 5 with $MoSi_2$ and other $MoSi_2$ composite systems at 1200°C and 1300°C [1,9,10]. The creep strength of the composite is greater than monolithic $MoSi_2$ and is comparable to a $MoSi_2/SiC/20w$ composite at 1300°C. Assuming power law creep, stress exponents of 3.3 and 3.7 were determined at 1200°C and 1300°C, respectively. An attempt was made to test the directionally solidified eutectic material at 1300°C, however, the sample fractured shortly after application of the second strain rate. A maximum flow stress of 625 MPa was recorded under the initial strain rate of $5x10^{-5}$ sec^{-1}.

The stress exponents determined from decremental step strain rate tests indicate a dislocation climb and glide mechanism is controlling the creep process which is consistent with previous investigations [1,9,10]. The mechanism of the improved strength is still under investigation, but it is thought to be related to either the pinning of the $MoSi_2$ grain boundaries or the formation of crystalline Er_2O_3-SiO_2 oxides. The strengthening attributes of $Er_2Mo_3Si_4$ is exemplified by the surprisingly high strength of the eutectic composition. At 1300°C, the 625 MPa flow stress exhibited by the eutectic is greater than any other $MoSi_2$ composite and approaches the strength of monolithic SiC.

SUMMARY

The intermetallic $Er_2Mo_3Si_4$ is thermodynamically compatible with $MoSi_2$, exhibits a higher hardness than $MoSi_2$ above 1000°C, and has approximately the same density as Mo_5Si_3. The low measured density of the $Er_2Mo_3Si_4$ is believed to be related to enclosed porosity in the arc-melted material. A fine dispersion of $Er_2Mo_3Si_4$ in $MoSi_2$ was successfully created by ball milling arc-melted buttons of $MoSi_2$-20 vol% $Er_2Mo_3Si_4$.

It has been shown that $Er_2Mo_3Si_4$ increases the high temperature hardness and creep strength of $MoSi_2$. The creep strength of a $MoSi_2/Er_2Mo_3Si_4$/20p composite at 1300°C is similar to that of a $MoSi_2$/SiC/20w composite. A composition of 38.6 vol% (32.1 wt%) $MoSi_2$ and a melting temperature of 1790 ± 10°C was determined for the eutectic.

ACKNOWLEDGMENTS

The authors would like to acknowledge Mr. Jim Garrett of McMaster University for processing of the directionally solidified samples, Mr. Ron Noebe at NASA - Lewis Research Center for providing the DTA measurements and Professor A.K. Ghosh for permitting the use of his equipment for measuring the creep strengths. Also, the assistance of Mr. Carl Henderson of the University of Michigan Electron Microbeam Analysis Laboratory is appreciated. This research was funded by the National Science Foundation, Grant No. MSM 86-57581.

REFERENCES

1. R. Gibala, A.K. Ghosh, D.C. Van Aken, D.J. Srolovitz, A. Basu, H. Chang, D.P. Mason, and W. Yang, Mat. Sci. and Engr. **A155**, 147 (1992).
2. S.A. Maloy, J.J. Lewandowski, A.H. Heuer, and J.J. Petrovic, Mat. Sci. and Engr. **A155**, 159 (1992).
3. D.A. Hardwick, P.L. Martin, and R.J. Moores, Scripta Metall. **27**, 391 (1992).
4. S. Jayashankar and M.J. Kaufman, Scripta Metall. **26**, 1245 (1992).
5. D.P. Mason and D.C. Van Aken, this proceedings.
6. D.P. Mason, D.C. Van Aken, and J.F.Mansfield, Intermetallic Matrix Composites II, ed. by D.B. Miracle, D.L. Anton, and J.A. Graves (Mat. Res. Soc. Proc. vol. 273, Pittsburg, PA, 1992), pp. 289-294.
7. O.I. Bodak, Yu. K. Gorelenko, V.I. Yarovets, and R.v. Skolozdra, Izvestiya AkademII Nauk SSSR, Neorgan Icheskie Materialy, **20** (5), 741 (1984).
8. S. Maloy, A.H. Heuer, J.J Lewandowski, and J.J. Petrovic, J. Amer. Cer. Soc. **74**, 2074 (1991).
9. S. Bose, Mat. Sci. and Engr. **A155**, 217 (1992).
10. K. Sadananda, C.R. Feng, H. Jones, and J. Petrovic, Mat. Sci. and Engr. **A155**, 227 (1992).

CHARACTERIZATION OF THE PLASTICITY ENHANCEMENT OF $MoSi_2$ OBTAINED AT ELEVATED TEMPERATURES BY THE ADDITION OF TiC

H. CHANG AND R. GIBALA
Department of Materials Science and Engineering
University of Michigan, Ann Arbor, MI 48109

ABSTRACT

Decremental step strain rate tests in compression were performed on $MoSi_2$ and $MoSi_2$-10 vol % TiC at 1150°C and 1200°C. The composite exhibits a lower stress exponent than $MoSi_2$ between 10^{-4} s^{-1} and 10^{-5} s^{-1}. TEM observations reveal the deformation of the carbide phase, as well as the efficient generation of dislocations into the matrix from the carbide-matrix interfaces. The substructures of both the matrix of the composite and monolithic $MoSi_2$ appear to consist of mixed dislocations of predominantly edge character which occur on the same slip systems.

INTRODUCTION

The brittle-to-ductile transition temperature (BDTT) of $MoSi_2$ has been reported by various authors to be between 900°C and 1300°C [1]. Within this temperature range, $MoSi_2$ exhibits limited deformation capability. The plasticity of $MoSi_2$ can be enhanced in this region of limited plasticity by the addition of a more deformable phase such as TiC. TiC deforms on fcc slip systems above its BDTT of 600°C to 800°C, and can be deformed to large strains in compression at 900°C and above [2]. Experimental observations [3] and thermodynamic calculations [4] have shown that $MoSi_2$ and TiC are chemically stable.

Previous compression test results have shown that $MoSi_2$-TiC composites can be deformed plastically at lower temperatures than $MoSi_2$ can before brittle fracture occurs [5]. The composites also attain zero strain hardening at much lower strains than in $MoSi_2$. These differences in plasticity and in strain hardening behavior between the composites and $MoSi_2$ are attributed to efficient generation of dislocations into the matrix from sources at the $MoSi_2$-TiC interfaces.

EXPERIMENTAL PROCEDURE

Two materials, monolithic $MoSi_2$ and a composite of $MoSi_2$ and 10 vol % TiC in particulate form, were mechanically tested in the current investigation. These materials were hot pressed and subsequently HIPed from commercially available powders. The $MoSi_2$ is nominally 5 to 6 micron powder from CERAC, Inc.; the TiC is 2.5 to 4 micron powder from Johnson Matthey, Inc., with a nominal stoichiometry of $TiC_{0.95}$. Hot pressing was performed in argon at 1700°C for one hour at 29.4 MPa. HIPing conditions were at 1800°C for 1.5 hours at 200 MPa. The densities of both materials after HIPing are approximately 98%. The microstructures of these materials have been described previously [5].

Decremental step strain rate tests in compression were performed on the two materials at 1150°C and 1200°C. These are constant crosshead speed tests with nominal strain rates of 10^{-4} s^{-1}, 5 x 10^{-5} s^{-1}, and 10^{-5} s^{-1}. Two tests were performed on each material at each temperature. The total plastic strain in these tests ranged from 8% to 13% for the composite, and 13% to 15% for the monolithic material. Testing was performed in an argon atmosphere furnace in an Instron Model 4507.

Conventional transmission electron microscopy (TEM) was performed using a JEOL 2000FX microscope operating at 200 kV. Foils were mechanically thinned by hand grinding, dimpling and ion milling. The dislocation substructures examined in this investigation are from foils of compression tests performed at constant crosshead speed at an initial strain rate 10^{-4} s^{-1}. These tests were performed on the two materials described above and on an unHIPed $MoSi_2$-15 vol % TiC composite [5].

RESULTS AND DISCUSSION

The flow curves of the two materials at 1150°C are shown in Figure 1. The composite shows a greater strain rate sensitivity in comparison to $MoSi_2$, as evidenced by the greater stress drop required to attain steady state flow after each change in strain rate. This difference in flow behavior is reflected in the stress exponents n, as shown in Figures 2 and 3. For the two trials performed on each material at each temperature, the maximum variation in flow stress between trials was approximately 40 MPa. $MoSi_2$ exhibits n values of approximately 5 in this temperature and strain rate range, while the composite exhibits exponent values of approximately 4. The possibility that these exponent values are the result of differences in creep mechanism between the two materials is currently under investigation. The strengths of the composite are greater than monolithic $MoSi_2$ in these tests and are in agreement with previous compression test results from 1050°C to 1200°C on HIPed material [5].

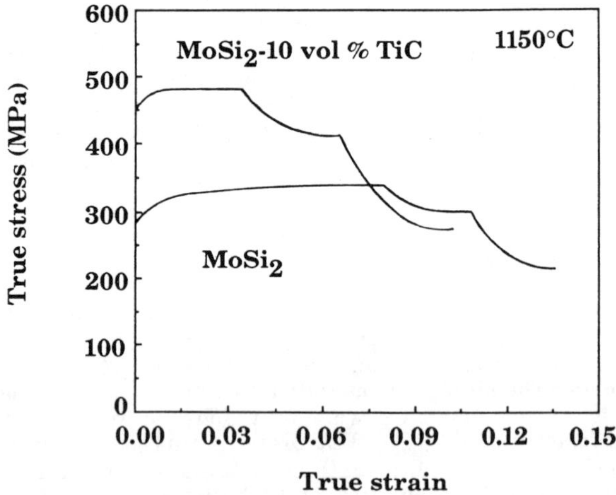

Figure 1. Flow curves of $MoSi_2$ and of $MoSi_2$-10 vol% TiC obtained from decremental step strain rate tests in compression at 1150°C at nominal strain rates of 10^{-4} s^{-1}, 5 x 10^{-5} s^{-1} and 10^{-5} s^{-1}.

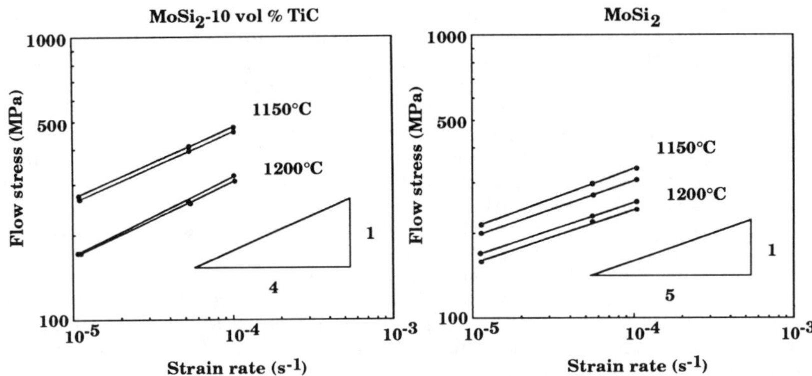

Figures 2 and 3. Stress exponents of $MoSi_2$-10 vol% TiC and of $MoSi_2$, respectively, from decremental step strain rate tests in compression at 1150°C and 1200°C.

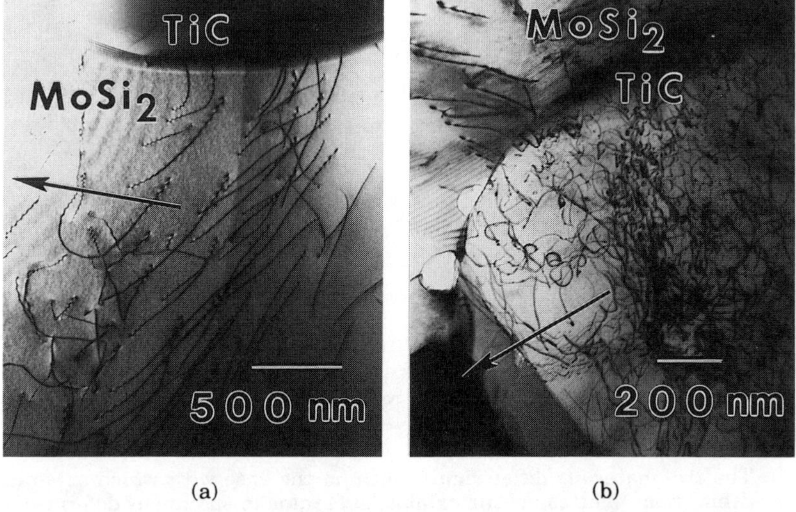

Figure 4. Substructures in (a) $MoSi_2$-10 vol% TiC deformed at 1150°C to 1.5% true strain, **B** = [331] **g** = $1\bar{1}0$, and in (b) $MoSi_2$-15 vol % TiC deformed at 1050°C to 5% true strain, **B** = [110] **g** = 002.

TEM observations reveal the generation of dislocations into the matrix from the interfaces of $MoSi_2$ and TiC (Figures 4a and 4b). The presence of the TiC introduces dislocations into the $MoSi_2$ matrix by generation processes not present in monolithic $MoSi_2$. This additional density may be a critical factor in the origin of the difference in stress exponents. The undeformed composite, by comparison, shows either no dislocations, or an extremely low density, in the carbide particles, and a very low density of dislocations in the matrix. In Figure 4a, there is no evidence of dislocation generation in the TiC particle, whereas in Figure 4b a high dislocation density in the TiC is readily observed. Figure 4b is representative of the observed deformation in the TiC particles. Within a given foil, observations reveal carbide particles both with and without a dislocation substructure. The majority of the deformed carbide particles are of larger size, 1 to 6 microns. Most of the submicron size TiC particles observed do not exhibit evidence of dislocations. The conditions leading to deformation of any given particle thus may depend upon the particle size, as well as the strain, strain rate, temperature and other variables.

In the absence of debonding or cracking at the interface, and neither has been observed to date, the stresses generated at the interface due to the deformation of the carbide, or simply the presence of the particle, must be accommodated by one or both phases forming the interface. Matrix dislocations are generated from the interfaces at both undeformed and deformed carbide particles. This suggests that multiple mechanisms exist for the generation of dislocations from the interface.

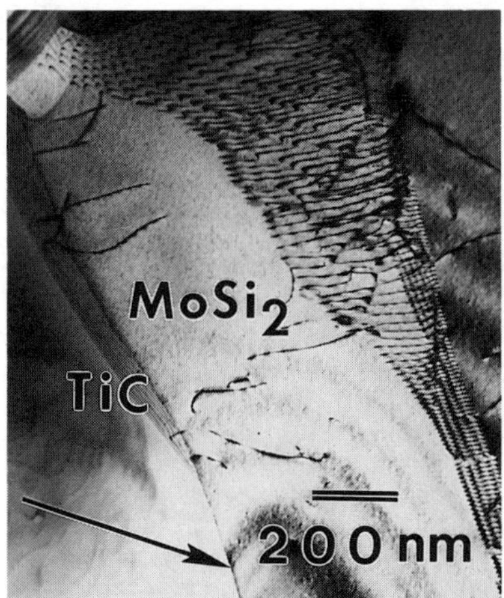

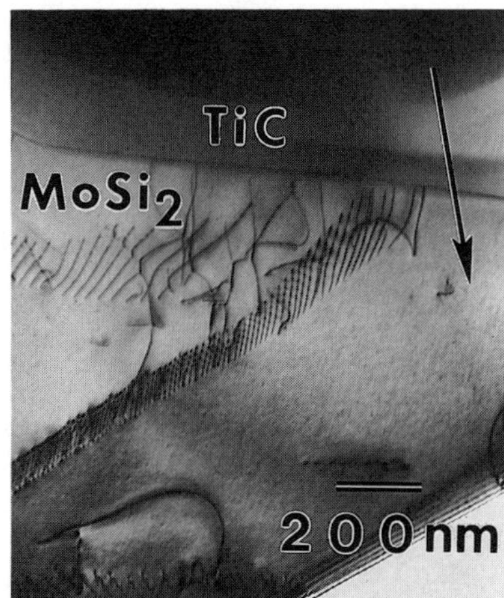

Figure 5. Subboundary formation in $MoSi_2$-10 vol % TiC at 1150°C and 5% true strain. **B** = [331] **g** = $1\bar{1}0$.

The two materials differ significantly in the ease with which subgrains form within them. The composite exhibits subgrains in specimens deformed 5% at 1050°C, 1100°C, 1150°C, and 1200°C (Figure 5). Subgrain boundaries are homogeneously distributed in the matrix, occurring both in the vicinity of the TiC particles and away from them. The substructure of the composite has also been investigated after 0.5% strain and 1.5% strain. Preliminary observations to date have not revealed the presence of subgrains for either of these two deformation conditions.

At 1100°C and 5% strain, $MoSi_2$ shows no evidence of subgrain formation. The substructure consists of a homogeneously distributed dislocation density throughout the material. Subgrains are observed, however, at 1200°C and 5% strain, and at 1100°C and 16% strain. This suggests that the formation of subgrains is dependent upon both the amount of deformation and the temperature. Qualitatively, the subboundaries, when they occur, occur in lower concentration in $MoSi_2$ compared to the composite. However, a quantitative measure of the density of subboundaries and the average size of subgrains in the two materials has not been performed.

The attainment of zero strain hardening occurs after much smaller amounts of deformation in the composite compared to $MoSi_2$, as shown in Figure 1. In previous tests [5], the prevalence of subboundaries in the composites at 5% strain but not at 1.5% strain corresponds macroscopically to the zero hardening region of the flow curve for 5% strain, and the hardening region at 1.5% strain. In $MoSi_2$ subboundaries occur in smaller concentration, or are absent for certain temperatures and strains. Macroscopically, hardening occurs to greater strains. Thus, the prevalence of subboundaries in the substructure is to a first approximation representative of the attainment of a steady state flow condition resulting from dynamic recovery processes.

Table I. Observed slip systems and relative occurrences

$MoSi_2$-15 vol % TiC at 1050°C $MoSi_2$-10 vol % TiC at 1150°C		$MoSi_2$ at 1100°C and 1200°C	
<100> {001}	8	<100> {001}	7
<100> {101}	11	<100> {101}	6
<100> {103}	27	<100> {103}	9
1/2 <111> {101}	1	<100> {100}	3

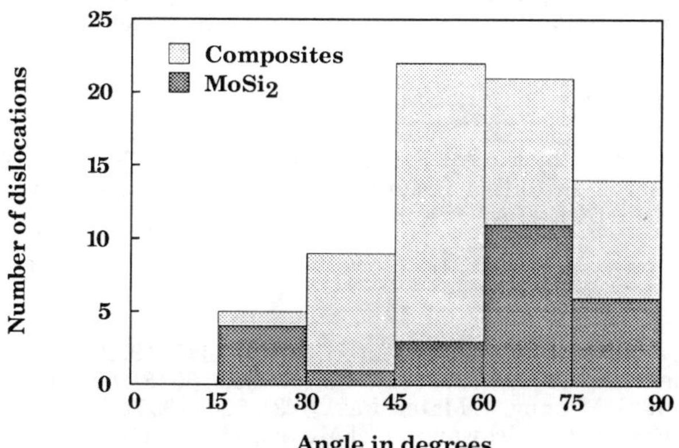

Figure 6. Numbers of dislocations versus angle between the Burgers vector and the dislocation line direction in $MoSi_2$-10 vol % TiC at 1150°C, in $MoSi_2$-15 vol% TiC at 1050°C and in $MoSi_2$ at 1100°C and 1200°C. (90° = edge dislocations).

Table I summarizes the slip systems and the numbers of observations of each system, as observed in the two materials from Burgers vector and trace analyses. These slip systems in polycrystalline $MoSi_2$ composites and $MoSi_2$ have been previously observed [6-8]. The two materials exhibit many of the same slip systems in the temperature range studied. The majority of the dislocations are of the <100> type. Examinations of 46 <100> dislocations in $MoSi_2$-15 vol % TiC at 1050°C and in $MoSi_2$-10 vol % TiC at 1150°C and of 25 <100> dislocations in $MoSi_2$ at 1100°C and 1200°C have been performed to date to determine their slip systems and their edge/screw character. The numbers of dislocations as a function of angle between the Burgers vector and the dislocation line direction are plotted in Figure 6. The dislocations are either of mixed type or of edge type. No screw dislocations were observed. The distributions in both materials are weighted towards mixed dislocations with a predominantly edge character.

The matrix of the composite deforms apparently with the same slip systems and types of dislocations as occur in monolithic $MoSi_2$. This suggests that the additional density of dislocations present in the composites is responsible for the observed differences in strain hardening behavior, the ease of subgrain formation, and possibly for the difference in stress exponent.

SUMMARY

Decremental step strain rate tests in compression on $MoSi_2$ and $MoSi_2$-10 vol % TiC composite at 1150°C and 1200°C between 10^{-4} s^{-1} and 10^{-5} s^{-1} have been performed. $MoSi_2$ exhibits stress exponents of approximately 5, and the composite exponents of approximately 4. This difference in stress exponent is possibly the result of a difference in creep mechanism. TEM observations reveal that many of the larger TiC particles exhibit a high density of dislocations due to deformation. The generation of dislocations into the matrix occurs from the matrix-carbide interfaces of both undeformed and deformed carbide particles. This suggests that multiple mechanisms of dislocation generation occur. The matrix of the composite deforms with the same slip systems and mixed dislocations of predominantly edge character as occur in monolithic $MoSi_2$.

ACKNOWLEDGEMENT

This research is funded by the Air Force Office of Scientific Research under the University Research Initiative Program, Dr. Alan H. Rosenstein, Program Director. Grant No. DOD-G-AFOSR-90-0141.

REFERENCES

1. R. M. Aikin, Jr., Scripta Metall. et Mater., 26 (7), 1025 (1992).
2. D. B. Miracle and H. A. Lipsitt, J. Am. Ceram. Soc., 66 (8), 592 (1983).
3. J-M. Yang and S. M. Jeng, J. Mater. Res., 6 (3), 505 (1991).
4. P. J. Meschter and D. S. Schwartz, J. of Metals, 41 (11), 52 (1989).
5. H. Chang, H. Kung and R. Gibala in Intermetallic Matrix Composites II, edited by D. B. Miracle, D. L. Anton and J. A. Graves (Mater. Res. Soc. Proc. 273, Pittsburgh, PA, 1992) pp. 253-258.
6. O. Unal, J. J. Petrovic, D. H. Carter and T. E. Mitchell, J. Amer. Ceram. Soc., 73 (6), 1752 (1990).
7. K. Sadananda, C. R. Feng, H. Jones and J. J. Petrovic, J. Mater. Sci. & Eng. A, A155 (1-2), 227 (1992).
8. S. A. Maloy, A. H. Heuer, J. J. Lewandowski and T. E. Mitchell, Acta Metall. Mater., 40 (11), 3159 (1992).

CHANGES IN CREEP DEFORMATION MECHANISMS IN AlN PARTICULATE STRENGTHENED NiAl

T.R. BIELER, J.D. WHITTENBERGER* and M.J. LUTON**
Department of Materials Science and Mechanics, Michigan State University, East Lansing, MI 48824
*NASA-Lewis Research Center, Cleveland OH 44135
** EXXON Research and Engineering, Annandale, NJ 08801

ABSTRACT

AlN particulate strengthened NiAl containing about 10vol% AlN was investigated in stress change experiments above, below, and through a transition stress near 80 MPa. All creep transients were normal, and incubation periods were observed in all stress drops near or below the transition stress. Stress change experiments indicated that the microstructure is very stable, and provided similar values for steady state strain rate as the monotonic tests. Analysis of the stress change data reconfirmed the existence of several deformation mechanisms in this material.

INTRODUCTION

Compression Creep of the NiAl-AlN particle strengthened material made by cryomilling has shown properties that are among the best obtained for NiAl based alloys [1-3]. The microstructure is unusual since the particles are heterogeneously dispersed throughout the matrix in mantle regions surrounding particle-free cores of NiAl. Though the power law creep behavior of monotonically deformed specimens has been characterized on the basis of empirical power law creep, little has been established about the rate limiting deformation mechanisms responsible for the creep deformation. In particular, this material exhibits a change in the stress exponent from about 5 to about 11, at a stress between 60 and 80 MPa. The change in the stress exponent is insensitive to temperature between 1000 and 1400 K, so it may be related to physical microstructure rather than thermally activated processes. Stress change experiments were done on one lot of samples to investigate the deformation mechanisms and compare the results of this test procedure with those determined by monotonic tests.

EXPERIMENTAL PROCEDURES

The NiAl-AlN composite was prepared by ball milling a slurry of NiAl and Y_2O_3 powders in a liquid nitrogen medium (cryomilling) at Exxon. The powder charge contained 99.5 wt% Ni-51Al (-325 mesh, from Homogeneous Metals, Inc.) and 0.5 wt% Y_2O_3 (<10μm powder from Research chemicals). Following 16 hours of milling, the powder charge was warmed to room temperature under dry argon and kept in a plastic container under an argon cover gas. Wet chemical analysis showed that no change in Ni/Al ratio occurred, and the oxygen increased from 135 ppm to 0.58wt%, and nitrogen increased from 4 ppm to 1.98wt% [1]. Approximately 350 g of the cryomilled powder was poured into a 75 mm diameter mild steel can with 6.4 mm thick walls which was subsequently sealed under vacuum. The can was heated to 1505 K in a furnace, and transferred within about 7 seconds to the press for extrusion. The first attempt to extrude was at 1505 K at a 12:1 reduction ratio; however only partial compaction under a pressure of 1310 MPa was achieved. After remachining to a 75mm diameter, the billet was successfully extruded at 1505 K with a 6:1 reduction ratio.

A cursory TEM investigation indicates that the ceramic particles are less than 50 nm in diameter making a net-like mantle about 300 nm thick around the original particle boundaries [1-3]. This processing history suggests that the nitrogen transforms into AlN particles at or near the surface of the cryomilled particles prior to extrusion. Oxygen and nitrogen react to form about 1.3v% alumina and 10.3 v% AlN. The microstructures are comparable to the composites extruded at higher ratios, but a lower grain aspect ratio is apparent in Figure 1 [1,2]. The strain is not homogeneous across the rod; the grain aspect ratio is near 10:1 in the center, and some grains have ratios as small as 5:1 near the interface with the steel can. During 16:1 extrusion of the same lot of powder, a 25:1 grain aspect ratio has been observed [1].

Specimens nominally 5mm in diameter and 10mm long were cut out of the extruded rod by electro-discharge machining, with the compression axis parallel to the extrusion axis. Such samples were deformed in compression creep machines through SiC push rods with extensometry built in to measure strains. A layer of

Ta foil was used as a compliant layer to minimize stress concentrations at the ends. The specimen was heated to temperature in about 3 hours, and kept at temperature for about an hour to permit thermal gradients in the extensometry to stabilize. The tests were conducted under constant load conditions well into the steady state regime prior to initiating stress change experiments which were accomplished by manually adding or removing weights.

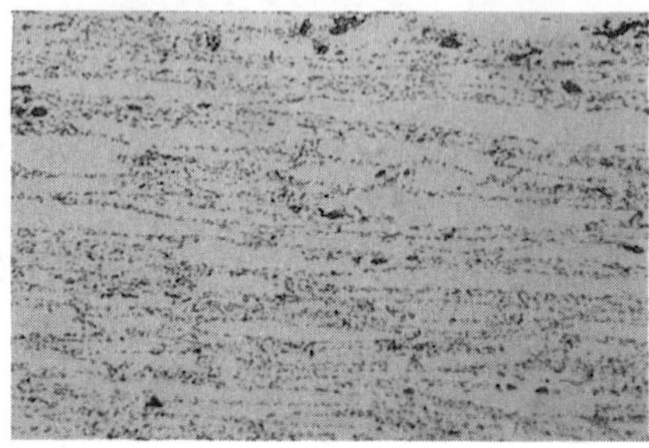

(a) Near the center 10μm

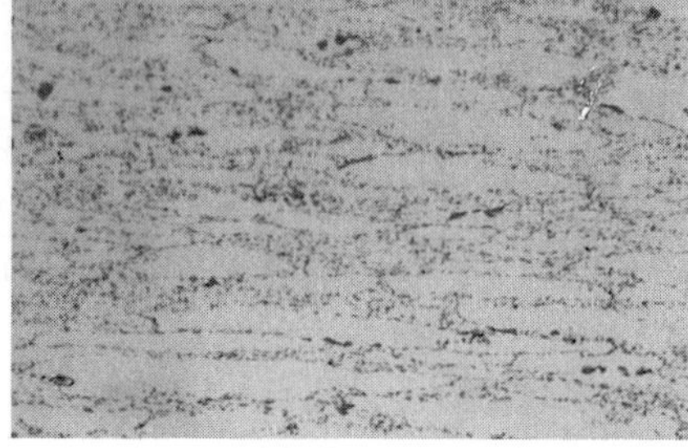

(b) Near the outer edge 10μm

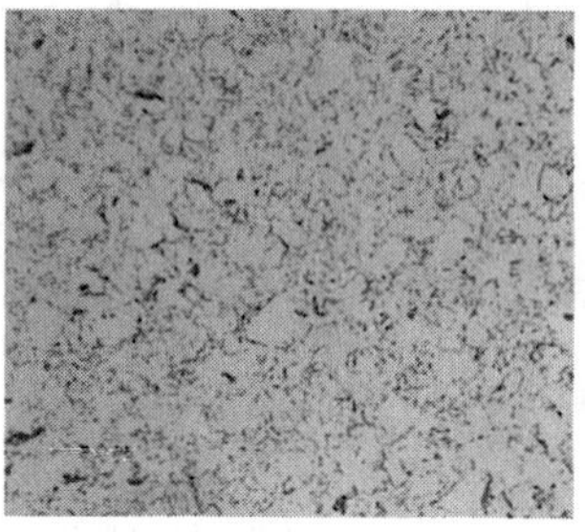

(c) Extrusion axis normal 10 μm

Figure 1 Light optical microstructure of an AlN particulate reinforced NiAl extruded at a 6:1 ratio, extrusion axis is horizontal in (a) and (b) and normal to page in (c)

RESULTS AND DISCUSSION

The results from a typical stress change experiment are presented in Fig. 2, where part (a) illustrates the strain-time behavior for a 1400 K experiment at an initial stress of 37 MPa followed by several load cycles (37 to 47 to 37 MPa). The corresponding strain rate - strain data from this experiment are shown in Fig. 2(b), indicating that large deviations in instantaneous deformation rate follow the load changes, but the subsequent steady state deformation rates are consistent with results from monotonically deformed specimens. The stress change transients exhibit "normal" behavior, e.g. decreasing deformation rate to the steady state value following a stress increase. During several experiments at 1350 K power failures occurred while the specimen was under stress, and this provided an additional factor in the study, where the sample was cooled to ambient temperature and then reheated to 1350 K under load. An example of this behavior in terms of strain rate - strain is given in Fig. 3(a), where the excursion to room temperature occurred at about 1.5% strain. The strain rate prior to and following the temperature excursion are almost identical once the influence of the transient deformation during cooling and heating was overcome.

Several important features can be drawn from the stress change (or temperature cycling) experiments (Figs. 2,3). First, stress increases or decreases between two levels always returned the deformation rate of the previous cycle (Fig. 2(b)). Likewise the deformation rate returned to nearly the same rate after a temperature excursion to room temperature (Fig. 3(a)). These observations suggest that any microstructural changes during creep are reversible. Furthermore this conclusion is supported by the recognition that the superposition of the stress change flow stress - strain rate results with those from monotonic loading tests (Fig. 4(a) yielded data which are indistinguishable on the basis of test method. Secondly, incubation periods occurred following stress decreases, and several examples are shown in Fig. 3(b-d). For stress values above the 80 MPa transition stress, the incubation times were very short (Fig. 3(b)). On the other hand, stress decreases conducted below the transition stress led to long interruptions before deformation recommenced (Figs 3(c,d)). In addition, reverse anelastic flow was seen in experiments where the stress dropped to levels below the transition stress value (Fig. 3(c)). These latter incubation time and anelastic strain results reinforce the concept that deformation of AlN-NiAl involves several different mechanisms.

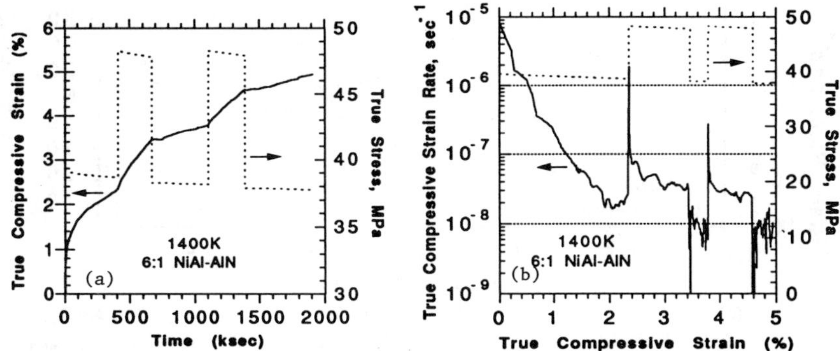

Figure 2 Effect of stress cycling on an AlN particulate reinforced NiAl at 1400 K. Imposed stress profile and the resultant creep curve (a), and imposed stress profile and the resultant strain rate - strain behavior (b).

Burton [4,5] has described how changes in the dimensions of the dislocation network can account for incubation times and normal stress transients. Burton's model is based on a steady state strain rate law given by

$$\dot{\varepsilon}_s kT/D_l Gb = (\alpha/(1+\alpha))^3(\sigma/G)^3$$

where α is the ratio of the glide mobility to the climb mobility, σ is the applied stress, G is shear modulus, b is Burgers vector, $\dot{\varepsilon}_s$ is the steady state strain rate, D_l is lattice diffusivity, and kT has its usual meaning. "Normal" transients are predicted when $\alpha>2$, or when glide is rapid compared to climb. The incubation time in this model results from the suppression of dislocation glide until a dislocation network grows to a new size defined by Gb/σ, where σ is the new stress.

Figure 5 shows that following a stress decrease, the incubation time increases with decreasing stress or temperature. The stress reductions are often normalized by the prior stress as shown in Fig. 6, where the incubation time increases with the magnitude of the stress drop in a temperature dependent manner. These observations are similar to those predicted in Burton's model. However, the incubation data above the transition stress do not follow the same power law (linear on plot) trend with respect to the normalized stress change. Stress changes above the transition stress did not exhibit anelastic reverse flow; whereas stress decreases to levels below the transition stress always exhibited measurable reverse anelastic flow (e.g. Fig. 3(c)). These features, in addition to the change in stress exponent, indicate that the rate limiting deformation mechanism is different above and below the transition stress.

Though Burton's model can describe transient phenomena, the steady state strain rate is much lower than Burton's model would predict. Dispersion strengthened alloys often exhibit superior creep resistance and higher stress exponents due to threshold stresses that must be overcome for a particular deformation mechanism to operate. Since the AlN particulate reinforced NiAl has similar kinds of dispersions as are found in other dispersion strengthened alloys, two strategies commonly used to identify particle induced threshold stresses were used to determine whether they are applicable to NiAl-AlN: The first method used was the use of an effective stress instead of the applied stress σ_a in existing models for monolithic materials. For example, $\sigma = \sigma_a - \sigma_p$ was substituted into Burton's model, and σ_p was used as a fitting parameter to minimize the difference between the data and the model. The second method requires assumption of a plausible stress exponent n, and the threshold stress

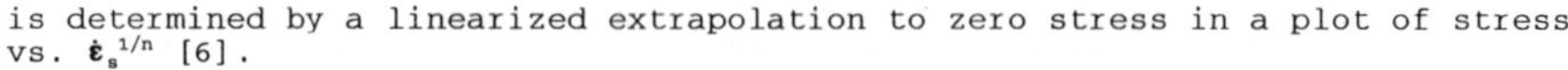
is determined by a linearized extrapolation to zero stress in a plot of stress vs. $\dot{\varepsilon}_s^{1/n}$ [6].

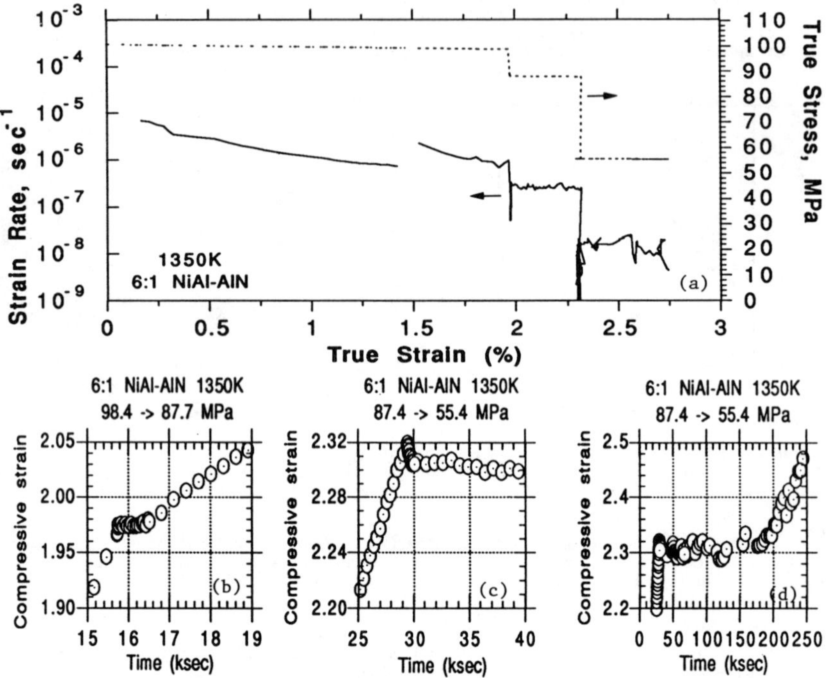

Figure 3 Effect of temperature and stress cycling on an AlN particulate reinforced NiAl deformed at 1350 K. (a) Imposed stress profile and the resultant strain rate - strain behavior; the break in the curve at about 1.5% strain was due to an excursion to room temperature, (b) creep curve during a stress drop experiment above the transition stress, and (c,d) creep curve during a stress drop experiment through the transition stress.

In evaluating the threshold stresses, G(T) was determined from the data of Harmouche and Wolfenden [7], and the lattice diffusivity values used were $D_o = 10^{-4}$ and an activation energy of -280kJ/mol, based upon an approximate fit of many data [8]. The best fit occurred when two values of σ_p, one for the data above the transition stress, and the other for the data below were used, for both methods. Similar threshold stresses were obtained in both cases for stresses above the transition stress, as shown in Table I. For the extrapolation method, n=3, representing the stress exponent in Burton's model, and n=4.5, representing climb controlled creep deformation were used.

TABLE I
Computed Threshold Stresses (MPa)

Temp	Burton	n=3	n=4.5
High stress regime			
1200 K	90	162	143
1300 K	86	96	80
1350 K	81	69	59
1400 K	68	68	52
Low stress regime			
1350 K	63	13	
1400 K	42	17	5

The threshold stresses obtained for the high stress regime (Table I) are similar to the observed transition stress, where the deformation mechanism changes. Since incubation times exhibit evidence for stress dependent dislocation networks, a more complete picture of deformation that includes both dislocation networks and the effects of particles in inhibiting dislocation glide needs to be developed.

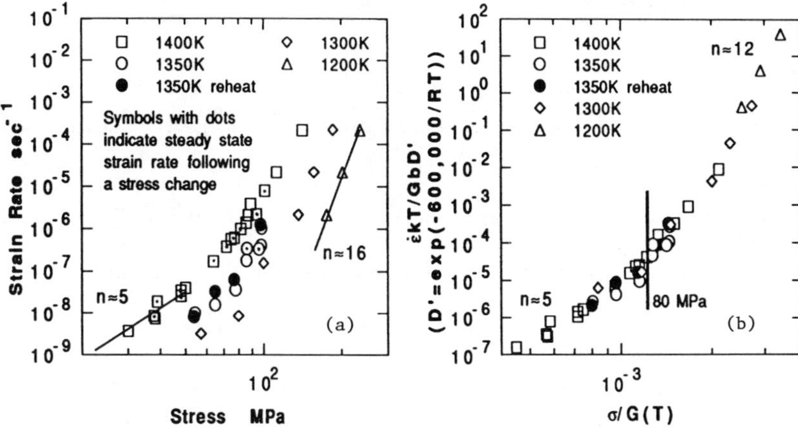

Figure 4 (a) The effect of temperature and stress on the steady state strain rate of an AlN particulate reinforced NiAl, and (b) the dependence of temperature compensated creep rate on normalized stress. Steady state creep rates following all stress change transients are included, denoted by symbols with centered dots.

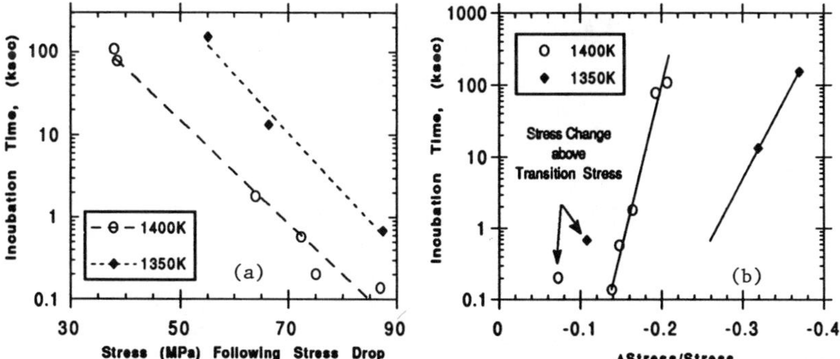

Figure 5 Incubation time following a stress decrease as functions of temperature and stress.

Figure 6 Incubation time as a function of temperature and normalized stress differential.

Lastly, creep results are often normalized with respect to diffusivity and modulus in order to coalesce the data onto a master plot. This has been accomplished in fig. 4(b) using the modulus data of Harmouche and Wolfenden [7]; however a 600 kJ/mol activation energy was needed to superimpose the data. Such a value is more than twice the activation energy of diffusion in NiAl [8] and is similar to activation energies for diffusion in carbides [9]. This suggests that diffusion within the AlN, perhaps through particle drag mechanisms, is the thermal rate controlling step for deformation of AlN reinforced NiAl. The superposition of the creep rate data in Fig. 4(b) also clearly identifies the two deformation regimes, where below about 80 MPa the stress exponent is about 5 while above 80 MPa, a much higher stress exponent of about 12 is observed.

CONCLUSIONS

Stress and temperature change experiments have been undertaken on an inhomogeneously AlN reinforced NiAl. All transient stress testing yielded the behavior consistent with two deformation regimes above and below the transition stress of about 80 MPa. Furthermore the strain rate - flow stress data determined by this method agreed well with those obtained from monotonic creep testing.

ACKNOWLEDGEMENTS

T.R. Bieler acknowledges support from the ASEE Summer Faculty Fellowship Program administered by Case Western Reserve University for NASA-Lewis Research Center.

REFERENCES

1. J.D. Whittenberger, E. Arzt, and M.J. Luton, J. Mater. Res., 5, 271, (1990).

2. J.D. Whittenberger, E. Arzt, and M.J. Luton, J. Mater. Res., 5, 2819, (1990).

3. J.D. Whittenberger, E. Arzt, and M.J. Luton, Scripta Metall. et Mater., 26, 1925, (1992).

4. B. Burton, Materials Science and Technology, 5, 1005, (1989).

5. B. Burton, Metal Science, 9, 297, (1975).

6. F.A. Mohamed, J. Mater. Sci. 18, 582, (1983).

7. M.R. Harmouche and A. Wolfenden, ASTM J. Test and Eval., 15, 101, (1987).

8. G.F. Hancock and B.R. McDonnell, Phys. Stat. Sol., A4, 143, (1971).

9. A.M. Brown and M.F. Ashby, Acta Metall., 28, 1085, (1980).

CREEP DEFORMATION OF TITANIUM ALUMINIDES WITH TiB_2

C. R. Feng and K. Sadananda
Materials Science and Technology Division, Naval Research Laboratory, Washington, DC 20375

ABSTRACT

Creep deformation of as-forged and heat-treated titanium aluminides, Ti-48Al with 5 and 10% volume fraction of TiB_2, has been studied under tension in the temperature range of 649°-871°C to determine the rate-controlling mechanism. Microstructure of these alloys correspond to a mixture of equiaxed grains of γ and lamellar regions containing eutectoid $\gamma+\alpha_2$.

Data indicate that with increase in load, stress exponent increases from one to seven. The activation energy for creep varies from 340kJ/mol to 455kJ/mol in the temperature range investigated. Creep data of these alloys were analyzed using several theoretical models.

INTRODUCTION

Tensile creep behavior of gamma-based titanium aluminides and composites was studied by several groups [1-11]. Most of these studies were limited to a narrow range of temperature or stress. Reported stress exponents vary from 3 to 5 [1-6 and 8-11], with an activation energy of 320kJ/mol [2-3, 5-7 and 9]. Occasionally, higher values of activation energy and stress exponent were also reported [6, 8 and 9]. Although there has been no direct evidence, power-law creep involving dislocation climb has been considered as the rate-controlling mechanism [3].

In this study, the creep tests were conducted in a broader load range. Systematic data evaluation was made and data indicated that as the load was increased, the stress exponent increased from 1 to 7. The stress dependence of creep rates was found to be better represented by a *sinh* function rather than by a power-law function. The implication of this in terms of creep mechanisms are discussed.

EXPERIMENTAL PROCEDURE

The materials used in this investigation were produced by a proprietary 'XD' process by Martin Marietta Laboratory and were forged. The nominal compositions were Ti-48at%Al+5vol%TiB_2 (Ti-48Al+5TiB_2) and Ti-48at%Al+10vol%TiB_2 (Ti-48Al+10TiB_2). Specimens were tested in as-forged condition. Some specimens of Ti-48Al+5TiB_2 were heat-treated at 1250°C/16h+900°C/8h. The microstructures of all materials contain about 85vol% of equiaxed γ-TiAl grains and 15vol% of grains with γ-TiAl/α_2-Ti_3Al lamellar structure. The average grain sizes in as-forged Ti-48Al+5TiB_2 and Ti-48Al+10TiB_2 were 20μm and 14μm, respectively. However, some large grains (~100μm) with lamellar structure were occasionally observed. The size of the TiB_2 particulate ranged from 1μm to 3μm.

The round button-head tensile specimens with gage length 25.4mm and gauge diameter of 3.8mm were machined from specimen blanks. The creep deformation behavior was studied in the temperature range of 649° - 871°C using a constant load lever arm creep-frame. The displacement as a function of time was measured using a LVDT with connecting rods fixed to the specimen shoulders. The specimens were thermally stabilized in the creep frame for one hour before the initial load was applied. The applied load was increased in increments and at each load the test was continued till steady state was reached. All tests were conducted in ambient air. After each test, the unbroken specimens were cooled under load to room temperature.

RESULTS AND DISCUSSION

Creep rates in as-forged Ti-48Al+5TiB_2 as a function of stress and temperature are presented in Fig. 1. Three regimes can be identified from this plot: a) a low stress regime (<80MPa) with stress exponent, n, equals to 1, b) an intermediate stress regime (80-300MPa) with n=4, and c) a high stress regime (>300MPa) with n=7. Since all curves are continuous rather than with discrete breaks, the division into three regimes is somewhat arbitrary. Results were similar for as-forged Ti-48Al+10TiB_2 and forged and heat-treated Ti-48Al+5TiB_2 and are shown in Figs. 2a and 2b, respectively.

The Arrhenius plot giving the activation energy, Q_C, for various stresses is shown in Fig. 3. At low stress, Q_C = 350kJ/mol. This value is in agreement with previous findings [2-3, 5-7 and 9], and also it is very close to the activation energy of Ti self-diffusion in TiAl, 291kJ/mol, [12]. At higher stresses the energy increases to 470kJ/mol.

Since the creep rates vary continuously with stress, all data were fitted in functional form using a *sinh* function

$$\dot{\varepsilon} = A \sinh(B\sigma) \exp(-Q_C/RT) \qquad (1)$$

where A and B are material constants. Using the activation energy values determined in Fig. 3 as guide lines and changing Q_C incrementally, values of A and B were determined by least square analysis. The best fit values of Q_C, A and B are those that give the largest value for correlation coefficient. However, two values of Q_C, 340kJ/mol at low stresses and 455kJ/mol at high stresses, were found to give the best correlation in as-forged Ti-48Al+5TiB_2, Fig. 4. Results were similar for as-forged Ti-48Al+10TiB_2. Selecting one Q_C value (455kJ/mol), we have shown the creep data of both alloys in a single plot, Fig. 5. For both alloys, the least square fit values of B are identical while A values fall in the same range. The functional form is represented by dotted line for each case. For forged and heat-treated Ti-48Al+5TiB_2, Fig 6, data fit better with Q_C=340kJ/mol. But this may be a reflection of lack of sufficient data for this condition. Thus the analysis indicates that higher activation energy is more representative of the materials behavior for higher stresses and lower temperatures.

Figure 5 indicates that contrary to the normal expectation, creep rates in the alloy increased rather than decreased with the increase in volume fraction of TiB_2. This clearly shows that unlike in the dispersion strengthened material, TiB_2 is not acting as reinforcement to enhance creep strength of the alloy. The particulates are contributing to a weakening effect, and this can be attributed to refinement of grain size [2].

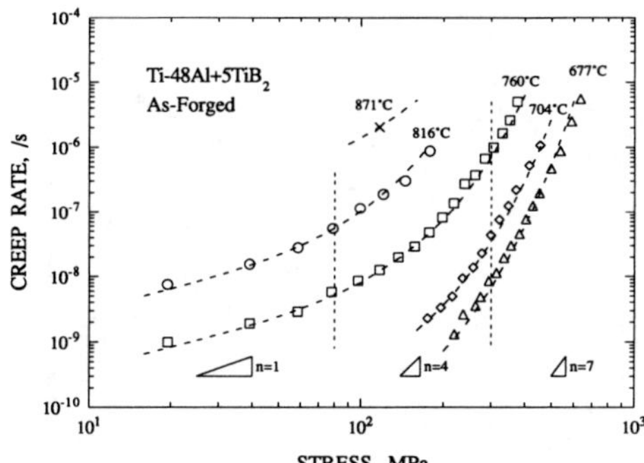

Fig. 1. Creep rate as a function of stress and temperature for as-forged Ti-48Al+5TiB_2.

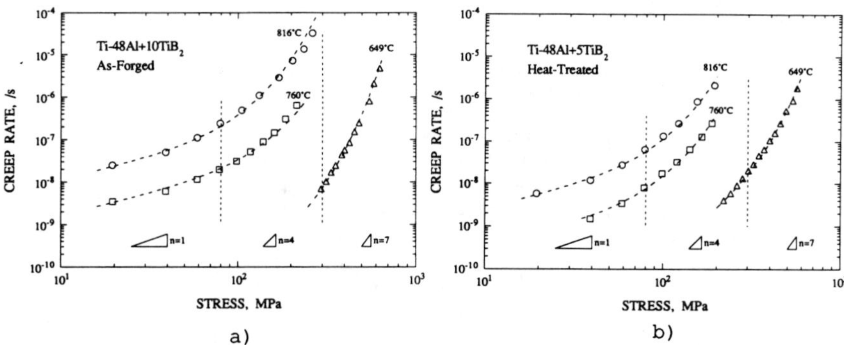

Fig. 2. Creep rate as a function of stress and temperature for: a) as-forged Ti-48Al+10TiB_2; and b) forged and heat-treated Ti-48Al+5TiB_2.

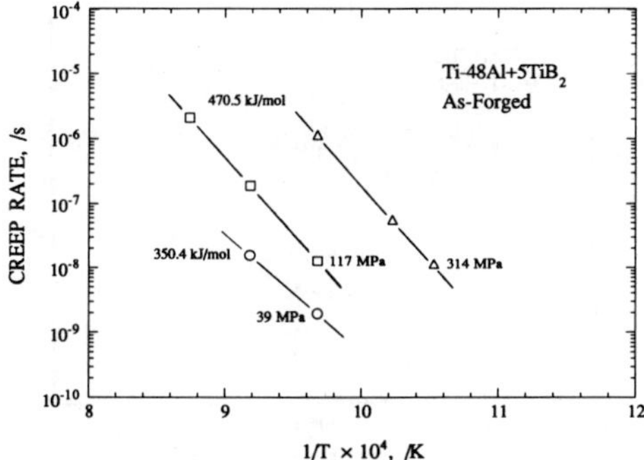

Fig. 3. Creep rate as a function of inverse temperature and stress for as-forged Ti-48Al+5TiB_2.

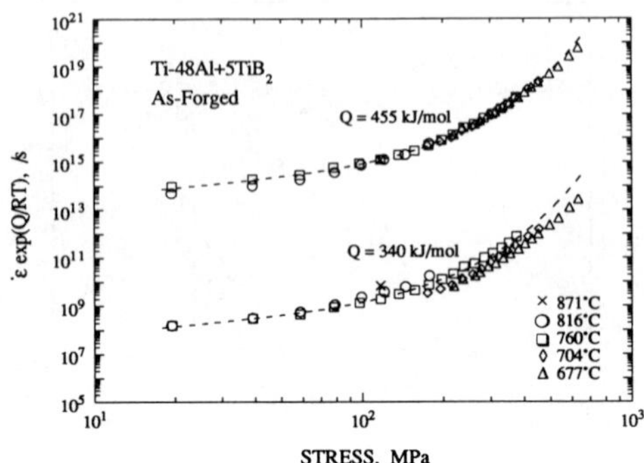

Fig. 4. Temperature compensated creep rate as a function of stress for as-forged Ti-48Al+5TiB_2.

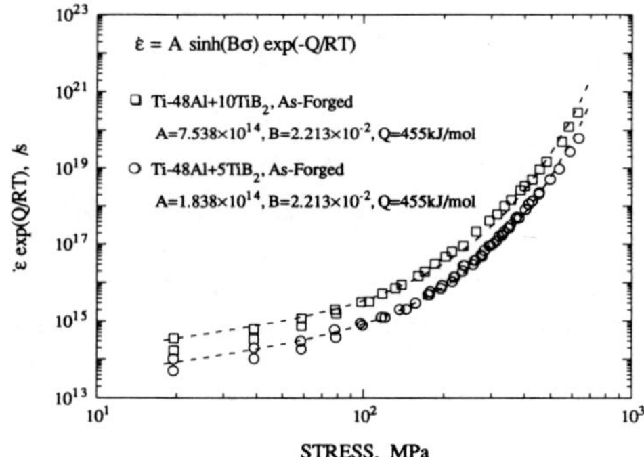

Fig. 5. Temperature compensated creep rate as a function of stress for as-forged Ti-48Al+5TiB_2 and Ti-48Al+10TiB_2.

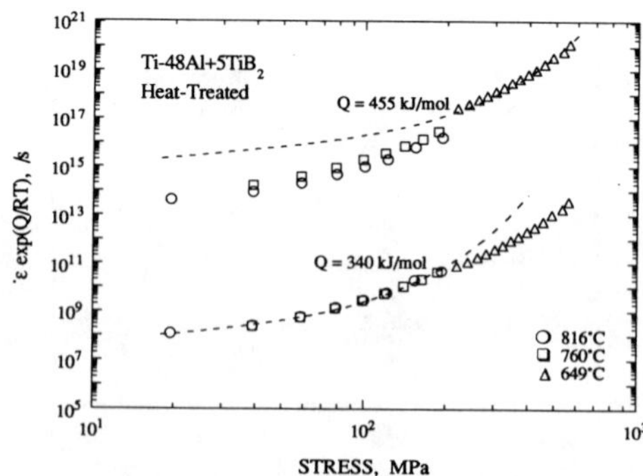

Fig. 6. Temperature compensated creep rate as a function of stress for forged and heat-treated Ti-48Al+5TiB_2.

The stress dependence of creep rates in the form of *sinh* function implies that there is no unique mechanism that is operating throughout the stress range. At low stresses with n=1 and Q_C=340kJ/mol the creep behavior is characteristic of Newtonian viscous regime with a bulk diffusion controlled process. With n=4 and Q_C=340kJ/mol at the intermediate stresses regime, the creep behavior is characteristic of dislocation glide with a climb process controlling the creep rate. The high value of Q_C, 455kJ/mol, at intermediate stresses and at high stresses regimes in the as-forged materials may relate to twinning. Jin and Bieler [13] report that mechanical twinning is one of the creep mechanisms in TiAl. Crossing twinning was found to occur when two differently oriented lamellar colonies met in such a way that one lamellar colony with a large Schmid factor twinned across the other one, or when the lamellar interfaces tended to be perpendicular or parallel to tensile axis [13]. It is not clear whether twinning is a thermally activated process and can be represented by an Arrhenius type of equation. On the other hand, the smooth functional relation in Figs. 1 and 2 implies there is an overlap of several mechanisms operating in parallel at each stress level.

Li [14] had proposed a creep model consisting of a number of dislocation-dislocation intersections giving rise to a network of two sets of dislocations. One set of dislocations are mobile and slip in the plane of the net, except at dislocation nodes. These nodes are formed by the interaction of the two sets of dislocations. The other set of dislocations in the network with Burgers vectors out of the plane of the net are immobile. Because of these non planar Burgers vectors, the resultant vectors at the nodes are also out of plane. Therefore, the nodes are sessile in the plane of the net. While creep occurs by motion of glissile dislocation segments in the net, the creep rates are controlled by the drag forces exerted by the nodes. This model predicts the stress dependence of creep rates in the form of *sinh* function, and the value of Q_C may be higher than the value of activation energy for bulk diffusion.

Recent studies of the dislocation structures and the dislocation reactions at the Ti_3Al/TiAl and TiAl/twin-related TiAl boundaries [15-17] indicate that for TiAl the creep model proposed by Li may be applicable. A detailed study is currently in progress.

SUMMARY

Tensile creep properties of TiAl with TiB_2 dispersions were investigated. TiB_2 does not act as reinforcement to enhance creep strength of the alloy. In fact, the particulates contribute to weakening by refining grain size. Data indicate that as the load increases, the stress exponent increases from 1 to 7. The stress dependence of creep rates is better represented in the form of *sinh* function than by a power-law function. The activation energy of creep varies from 340kJ/mol to 455kJ/mol, which may correspond to bulk diffusion and twinning, respectively. The *sinh* functional relation normally implies that there may not be a unique mechanism that is operating throughout the stress range. As an alternative, the *sinh* functional dependence can be described by a model based on the concept that the rate-controlling step is the non-conservative motion of short dislocation segments.

ACKNOWLEDGMENT

This research is supported by the Office of Naval Research.

REFERENCES

1. K. Maruyama, T. Takahashi and H. Oikawa, Mat. Sci. Eng., **A153**, 433 (1992).
2. S.L. Kampe, J.D. Bryant and L. Christodolou, Metall. Trans., **22A**, 447 (1991).
3. P.L. Martin, M.G. Mendiratta and H.A. Lipsitt, Metall. Trans., **14A**, 2170 (1983).
4. P.L. Martin and H.A. Lipsitt, in Proc. of the 4th Int. Conf. on Creep and Fracture of Engineering Materials and structures, edited by B.W. Wilshire and R.W. Evans (Institute of Metals, London, 1990), p. 265.
5. R.W. Hayes and B. London, Acta Metall. Mater., **40**, 2167 (1992).
6. D.A. Wheeler, B. London and D.E. Larsen, Jr., Scripta Metall., **26**, 939 (1992).
7. D.S. Shih, S-C. Huang, G.K. Scarr H. Jang and J.C. Chesnutt, in Microstructure/Property Relationships in Titanium Aluminides and Alloys, edited by Y-W. Kim and R. R. Boyer (TMS, Warrendale, PA, 1991), p. 353.
8. J.S. Huang and Y-W. Kim, Scripta Metall., **25**, 1901 (1991).
9. P.L. Martin, D.H. Carter, R.M. Aikin, Sr., R.M. Aikin, Jr. and L. Christodoulou, in Proc. of the 4th Int. Conf. on Creep and Fracture of Engineering Materials and structures, edited by B.W. Wilshire and R.W. Evans (Institute of Metals, London, 1990), p. 265.
10. S.L. Kampe, L. Christodoulou and J.A. Clarke, TMS Symposium Creep and Fatigue of Metal Matrix Composites, New Orleans, 1991.
11. C.R. Feng, H.H. Smith, D.J. Michel and C.R. Crowe, in Microstructure/Property Relationships in Titanium Aluminides and Alloys, edited by Y-W. Kim and R. R. Boyer (TMS, Warrendale, PA, 1991), p. 353.
12. S. Kroll, H. Mehrer, N. Stolwijk, C. Herzig, R. Rosenkranz and G. Frommeyer, Z. Metallkd. **83**, 591 (1992).
13. Z. Jin and T.R. Bieler, to be presented at Inter. Sym. on Structural Intermetallics, Champion, PA (1993).
14. J.C.M. Li, Trans. AIME, **227**, 1474 (1963).
15. L. Zhao and K. Tangri, Acta Metall. Mater., **39**, 2209 (1991).
16. L. Zhao and K. Tangri, Phil. Mag. A, **64**, 361 (1991).
17. L. Zhao and K. Tangri, Phil. Mag. A, **65**, 1065 (1992).

HIGH TEMPERATURE MECHANICAL BEHAVIOR OF SiCf/Ti–Al COMPOSITES

H.Y. CHOU*, S.C. YANG*, K.L. WANG*, C.I. CHEN* and S.E. HSU*
* Materials R&D Center, CSIST, P.O. Box 90008–8, Lungtan, Taiwan, R.O.C.

ABSTRACT

Titanium aluminide was reinforced by AVCO SCS–6 continuous SiC fibers in unidirectional or $0^o/+45^o/-45^o$ direction to make SiCf/Ti–Al composites through hot–pressing of Ti and Al powder mixture. The toughness of the composites is greatly improved at room temperature. The typical tensile elongation is over 0.8%. The room temperature strength of the composite can be sustained up to about 700°C. The processes developed in this study have two merits: (1) Ti and Al powders instead of Ti–Al powder can be directly used as raw materials, and (2) the mechanical properties of composites can be tailored in a similar fashion as those of conventional composites.

Introduction

Titanium aluminide has recently gained much attention due to its relatively high melting point and some other attributes, such as high specific strength and good corrosion resistance [1,2]. However, the biggest drawback of the Ti–Al intermetallic compound is its brittle nature. Therefore, finding ways to improve the ductility is an important effort of alloy development. One of the approaches to toughen the Ti–Al intermetallic compound, in both ductility and strength, is to incorporate the SiC–continuous fibers into the Ti–Al matrix to form SiCf/Ti–Al composites [3,4].

In this work, an innovative process was developed for the fabrication of unidirectional and $0^o/\pm45^o/0^o$ plied SiCf/Ti–Al composites by hot–pressing Ti and Al powder mixture through a spontaneous reaction [5] to form the titanium aluminide matrix as stated in our previous paper [6]. The objectives of this paper are: (1) to report the results of mechanical test both at room and at elevated temperatures, and (2) to investigate the formation of intermetallic compound matrix during hot–press process.

Experiments

Fibers used in this study were AVCO SCS–6 SiC fibers of a 143μm diameter with a double coated surface layer of 4μm thickness. The fibers were wound over a rotating drum and coated with an organic binder to form fiber tape [7]. The raw materials used to form the matrix of composite were titanium powder (3–5μm, >99.9%) and aluminum powder (–100 mesh, >99.9%) with a stoichiometric composition. An organic binder was added into the powder mixture to form a metal powder tape. The fiber tape (A) and metal powder tape (B) were stacked in an ABAB sequence with fiber reinforcement in unidirectional or $0^o/+45^o/-45^o$ direction. The stacks were put into a stainless steel case. After removing the binder by heating the unconsolidated laminates to 500°C for 4 hours, the case was degassed and vacuum–sealed. The case with laminates inside was heated to 950°C with a heating rate of 15°C/min., then hot–pressed at a pressure of 10000 psi.

Tensile and bending specimens were prepared from consolidated composites along the unidirectional and 90° transverse direction for $0^o/+45^o/-45^o$ reinforcement and were tested from room temperature to 820°C. The microstructure of the composite and the morphology of fracture surface were investigated by SEM. The phase analysis of matrix during heating process was conducted by X–ray diffraction.

Results and Discussion

A. Improvement in toughness and strength

Table I depicts the results of flexural and tensile tests of continuous SiC fiber reinforced Ti–Al composites and pure titanium aluminide at room temperature. From the comparison of flexural strength between unidirectional SiCf/Ti–Al composites and pure titanium aluminide, we know that the titanium aluminide reinforced by AVCO SCS–6 continuous SiC fibers has an apparent improvement in strength. The strength of unidirectional SiCf/Ti–Al composite is stronger than that of 0°/+45°/–45° SiCf/Ti–Al composite in the 0° direction, and the latter is slightly stronger than that in 90° direction.

Fig.1 shows the load–deflection curves of SiCf/Ti–Al composites and pure titanium aluminide in flexural and tensile tests at room temperature. It is evident that the toughness of titanium aluminide can be greatly improved by continuous SiC fiber reinforcement. As shown in Table I, the tensile elongation of the 0°/+45°/–45° SiCf/Ti–Al composite in the 90° direction is better than that in the 0° direction, and the latter is better than that of unidirectional reinforcement. Fig.2 shows the typical SEM morphology of a fracture surface of a SiCf/Ti–Al composite and of pure titanium aluminide. Pure titanium aluminide shows typical brittle fracture surface behavior as can be seen from an even fracture surface and a linear load–deflection curve. On the contrary, the fracture surface of SiCf/Ti–Al composites, whether of unidirectional or of 0°/+45°/–45° reinforcement, clearly demonstrates a fiber pull–out phenomenon which is irrefutable evidence of the improvement in toughness.

Table. I The flexural and tensile strength of SiCf/Ti–Al composites and pure titanium aluminide at room temperature.

	unidirectional SiCf/Ti–Al	0°/+45°/–45° SiCf/Ti–Al		TiAl
		0°	90°	
flexural strength (MPa)	630	400	240	450
tensile strength (MPa)	200	175	126	—
tensile elongation (%)	0.8	1.2	1.4	—

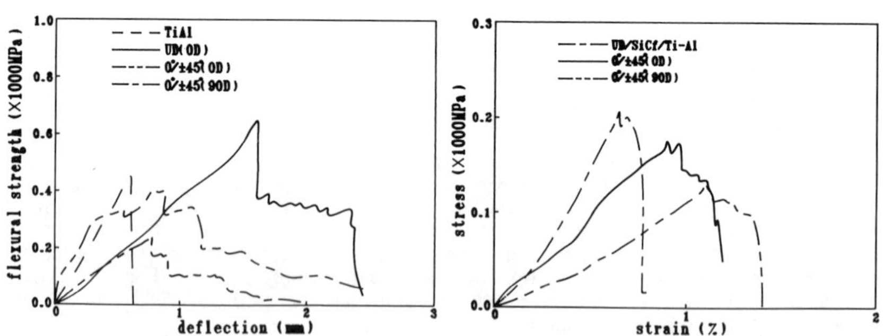

Fig.1. The load–deflection curves of SiCf/Ti–Al composites and pure titanium aluminide from flexural and tensile tests at room temperature.

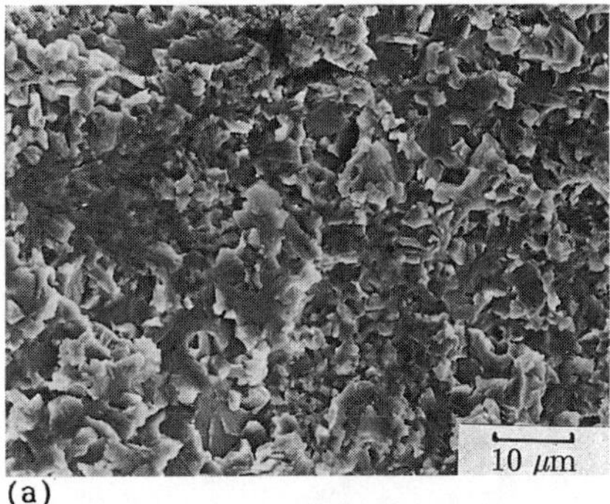

(a)

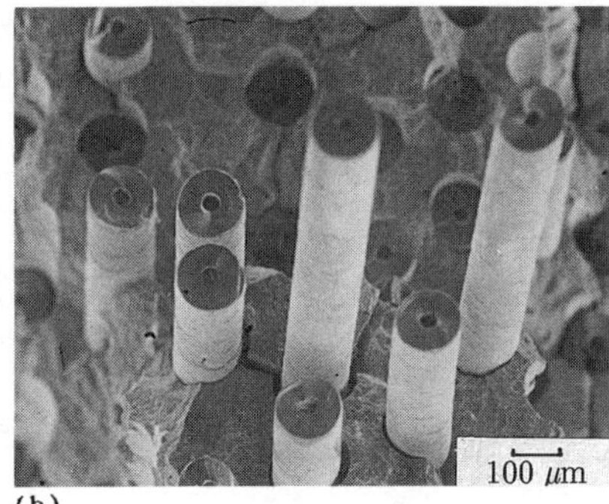

(b)

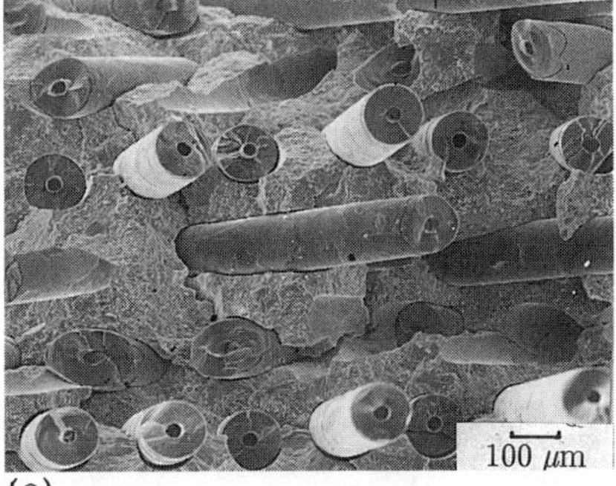

(c)

Fig.2. The SEM morphology of the fracture surfaces of pure titanium aluminide and unidirectional and 0°/+45°/–45° SiCf/Ti–Al composites after flexural tests.
a) pure titanium aluminide
b) unidirectional SiCf/Ti–Al composite
c) 0°/+45°/–45° SiCf/Ti–Al composite

B. High temperature mechanical behavior

Fig.3 shows the high temperature mechanical behavior of SiCf/Ti–Al composites and pure titanium aluminide. The tensile and flexural strengths increase gradually as temperature increases up to about 700°C, and then decrease when the temperature is over 700°C. Figure 3 also shows similar behavior for the unreinforced TiAl compound. Therefore, the room temperature strength of the composites for both unidirectional and 0°/+45°/–45° reinforcement remains up to about 700°C. The tensile elongation increases gradually when the temperature is over 650°C. An explanation for the increase is that the matrix exhibits brittle–ductile transition behavior around this temperature, which is evident as shown in Fig.4.

C. Matrix phase formation mechanism

The matrix of the SiCf/Ti–Al composite is a nearly fully dense titanium aluminide intermetallic compound which contains major TiAl and minor Ti3Al phases as shown in Fig.5. A DSC result of the heating simulation of the titanium and aluminum powder mixture is shown in Fig.6. An endothermic peak which appears at about 660°C is thought to correspond to the melting of Al powder, while the exothermic peak that appears around 710°C is evidence of the spontaneous reaction of Ti and Al to form titanium aluminide. In manufacturing SiCf/Ti–Al composites, a sample was taken when the raw materials were heated only up to 775°C and then cooled down to room temperature. The phase analysis result is shown in Fig.7. It contains four phases: Ti, TiAl, Ti3Al and TiAl3. The TiAl3 and Ti phases disappear when the temperature increases to 950°C and is held for over 2 hours. Therefore, the formation of the titanium aluminide matrix phase from Ti and Al powder mixtures is first through a spontaneous reaction synthesis (SRS) process to form various titanium aluminide phases, followed by the solid diffusion and densification processes.

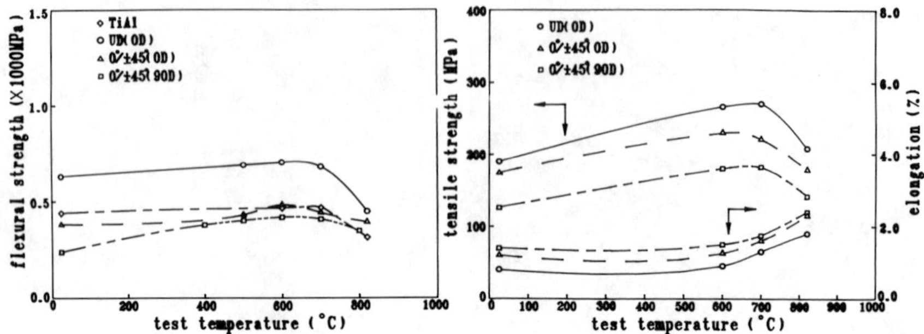

Fig.3. The high temperature mechanical behavior of SiCf/Ti–Al composite and pure titanium aluminide.

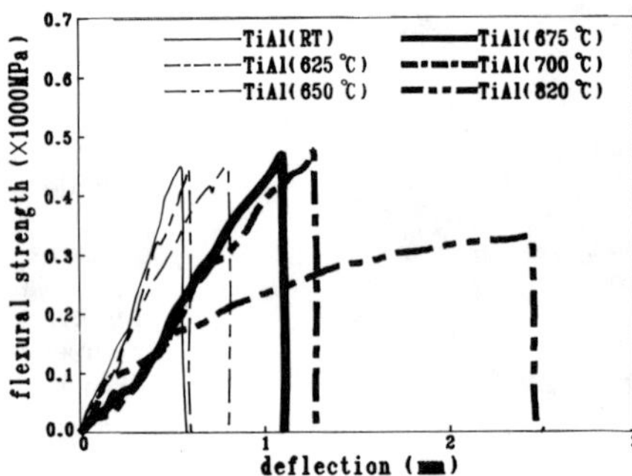

Fig.4. The load–deflection curves of titanium aluminide in flexural test at elevated temperature.

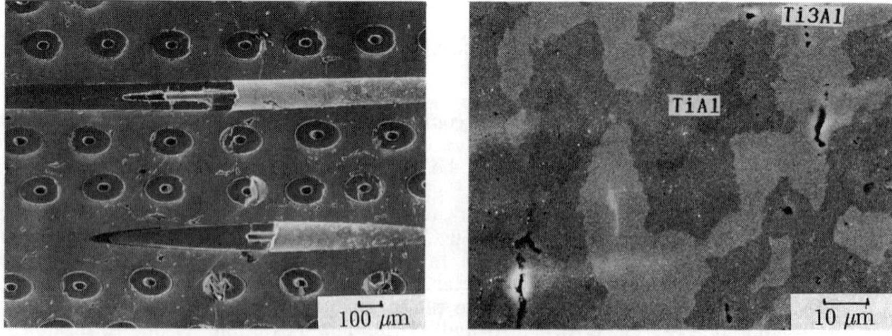

Fig.5. The SEM micrograph of 0°/+45°/–45° SiCf/Ti–Al composite showing the microstructure of matrix phase.

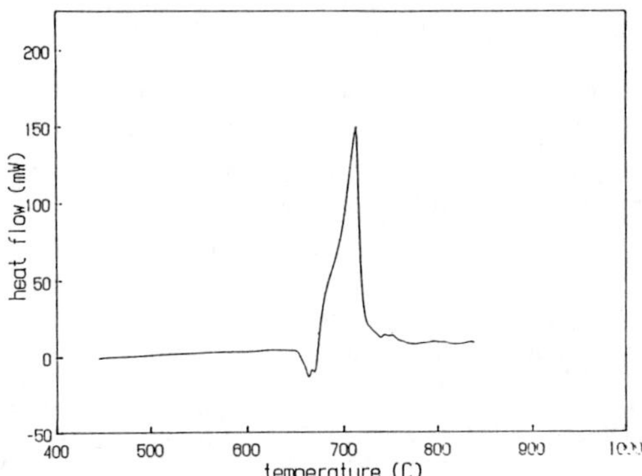

Fig.6. A DSC result of heating simulation of a Ti and Al powder mixture with a heating rate of 15°C/min.

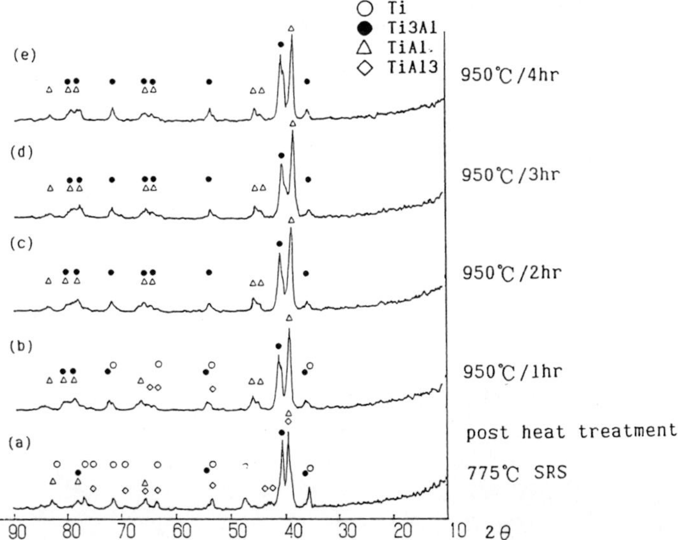

Fig.7. The results of X–ray phase analysis of samples which were taken when the raw materials were heated up to 775°C and then cooled down to room temperature (a), followed post heat treatment at 950°C for 1 to 4 hours (b), (c), (d) and (e).

Conclusions

1. The SiCf/Ti–Al composites developed in this study, whether of unidirectional or of 0°/+45°/–45° reinforcement, have considerable toughness at room temperature. The typical tensile elongation is over 0.8%.

2. Titanium aluminide reinforced by AVCO SCS–6 continuous SiC fibers also shows improvement in strength. The room temperature strength of the composite for both unidirectional and $0^o/+45^o/-45^o$ reinforcement can be sustained up to about 700^oC.
3. The processes developed in this study have the following two merits:
   (1) Ti and Al powders can be directly used as the raw materials, instead of Ti–Al powder.
   (2) The mechanical properties of the composites can be tailored in a fashion similar to that of conventional composites.

References

1. S.M.L. Sastry and H.A. Lipsitt, in Titanium 80' edited by H.Kimura and O.Izumi (Warrendale, PA: The metallurgical Society, pp1231–1243 (1980).
2. S.M. Barinov, Z.I. Konilova, Yu.L. Krasulin, E.M. Lazarev, T.T. Nartova, and L.V. Sapozhnikova, Poroshkovaya Metallurgiya, No.12(300), 61 (1987).
3. C.M. Mard–Close and P.G. Patridge, J. Mater. Sci., 25 4315 (1990).
4. P.K. Brindley in High–Temperature Ordered Intermetallic Alloys ": edited by N.S. Stoloff, C.C. Koch, C.T. Liu,, and O. Izumi (Material Research Society symposia proceedings, Vol.81, pp419–424 (1987)).
5. J.C. Rawers, W.R. Wrzesinski, E.K. Roub, and R.R. Brown, Mater. Sci. & Tech., Vol.6, No.2, 187 (1990).
6. S.E. Hsu, H.D. Wu, C.M. Li, H.Y. Chou, and K.L. Wang in Proceeding of International Symposium on Intermetallic Compounds edited by Osamu Izumi, Tohoku university, JIM, pp979–983 (1991).
7. P. Martineau, R. Pailler, M. Lahaye, and R. Naslain, J. Mater. Sci., 19, 2749, (1984).

WELDING OF A TWO-PHASE Ni_3Al ALLOY

HUAXIN LI AND T. K. CHAKI
State University of New York, Department of Mechanical and Aerospace Engineering, Buffalo, NY 14260

ABSTRACT

Autogeneous gas tungsten arc welding was performed on plates (about 3 mm thick) cut from cast ingot of a Ni_3Al alloy ($Ni_{73.80}Al_{15.82}Cr_{8.33}Mo_{1.68}Zr_{0.52}B_{0.05}$), known as IC-396M. The alloy was susceptible to cracking in the heat-affected zone (HAZ). The cracks in HAZ were liquation cracks and occurred along dendritic boundaries and around eutectic cells. The fusion zone (FZ) contained many fine eutectic cells (about 5-10 μm in diameter) and there were microcracks (1-5 μm in size) in the cells. Two types of eutectic structures were observed in the base metal, HAZ and FZ of IC-396M. One had web-like structure and the other had lamellar structure, the later having high enrichment with Zr. The role of Zr enrichment in dendritic boundaries and eutectic cells in causing liquation cracks will be discussed.

INTRODUCTION

Polycrystalline nickel aluminide (Ni_3Al), alloyed with a small amount of boron (a few hundred p.p.m.) has large room temperature ductilitity [1]. Besides boron, zirconium is added [2] to Ni_3Al in small proportion (less than 1 at. %) to provide solid solution strengthening. However, ductile Ni_3Al compound doped with B only or with B and Zr looses its ductility at intermediate temperatures of 400-800°C [3,4], when they are subjected to tensile stresses in air. The embrittlement, known as "dynamic embrittlement" (it occurs within the short time span of the test), can be overcome by addition of moderate amounts of chromium [2,5]. A series of advanced Ni_3Al alloys containing B, Zr, Cr and Mo have been developed [2,5] at Oak Ridge National Laboratory, TN.

Weldability is a key issue in the development of a commercial alloy, because welding is the primary means of joining parts. David and coworkers [6,7] performed autogeneous gas tungsten arc (GTA) and electron beam (EB) welding on several Ni_3Al alloys, some of them containing iron as an alloying element. The specimens were made by arc melting followed by repeated rolling and annealing. Their study showed that ductile Ni_3Al is highly susceptible to hot cracking in the HAZ with cracks occuring along the grain boundaries. The alloys can be successfully welded only by EB welding operated within a narrow range of welding speeds and beam focus conditions. In an earlier paper Li and Chaki [8] have reported observation of a large number of liquation cracks in the HAZ of a single-phase Ni_3Al alloy containing 0.34% Zr and 0.10% B by atom. In this paper, we report the nature of weld cracking in cast ingots of a two-phase (γ and γ') Ni_3Al alloy containing B, Zr, Cr and Mo.

EXPERIMENTAL PROCEDURE

Ni_3Al alloy used in this study contained (at.%) 15.62% Al, 8.33% Cr, 1.68% Mo, 0.52% Zr, 0.05% B and balance Ni. The alloy is called IC-396M. Cast ingots (250 mm x 120 mm x 18 mm) were obtained from Armco Inc., OH. Welding coupons of approximate size of 50 mm x 18 mm x 3 mm were cut by a SiC cutting wheel, and ground to remove oxide films and sharp edges. No further thermal or mechanical treatment was applied. Welding was done by a gas tungsten arc welder (Miller, DIALAR HF-P, constant current AC/DC) in alternating current mode. A tungsten rod of diameter 2.4 mm with r.m.s. current of 50 A was used. The flow rate of argon was 20 m^3/h and the welding speed was 12 cm/min. The welding was autogeneous, i.e., no filler metal was used. The weldment areas of a few specimens were ground by emery paper and polished with alumina slurry. The specimens were etched with an etchant containing 40 ml HCl, 30 ml HNO_3, 20 ml glacial acetic acid and 10 ml glycerol. The microstructure of the specimens was examined by optical microscope and scanning electron microscope (SEM). Chemical composition was analyzed by energy-dispersive x-ray spectroscopy (EDS) and by microprobe using wavelength-dispersive x-ray spectroscopy. Hardness in weldments was measured by Vicker's micro-indentation with 500 g load.

RESULTS AND DISCUSSION

As-cast IC-396 consisted of γ' (ordered Ni_3Al) and γ (solid solution of Ni) phases. The volume fraction of γ phase was estimated by Liu et al. [2] to be about 15%. The microstructure of the base metal contained typical cast dendrites with many small eutectic cells in interdendritic regions. Fig. 1(a) and (b) show typical eutectic structures in the cast material. There were two types (A and B) of eutectic structures. The eutectic cells of type A had web-like structure, while type B had lamellar structure [shown by arrows in Fig. 1(a) and (b)]. B-type cells were enriched with Zr (about 7 at.%) and possibly produced by an eutectic reaction of Ni and Ni_5Zr with an eutectoid temperature of 1175°C [9].

Type-A cells had chemical compositions similar to that of the nominal composition and were probably formed by eutectic reaction of γ and γ' [9]. There were micro-cracks [Fig. 1(a)] around the eutectic cells in the as-cast material and they were probably formed by late-stage solidification of Zr-rich regions which had lower melting point [9].

Weldability of two-phase Ni_3Al alloy (IC-396) is better than the single phase alloy [8]. Long, macroscopic cracks were not observed in the FZ. The FZ did not contain shrinkage pores or crater cracks at the end of the weld pass. The FZ contained eutectic cells of type A and B similar to those present in the base metal. At an intermediate magnification (1000X), no visible defect was found in the cells [Fig. 2(a)]. However, SEM investigation at high magnifications (5000X or more) revealed many microcracks in the eutectic cells, as shown in Fig. 2(b). Cuboidal ZrO_2 particles were occasionally found in the FZ [Fig. 2(a)].

There were many small cracks in the HAZ of IC-396M, as shown in Fig. 3(a). These cracks were found in the HAZ predominantly along the main dendritic boundaries [Fig. 3(a)]. On the average, 4 ± 1 cracks (50 μm or longer in length) per 3 cm length of weld were found. Maguire et al. [10] claimed to produce completely crack-free HAZ in electron-beam weld of a Ni_3Al alloy containing 8 at.% Cr, 0.5% Hf (instead of Zr) and 0.10% B.

The cracks in the HAZ resulted from the formation of a liquid film at dendritic boundaries, grain boundaries and eutectic cells during welding and the inability of this film to accomodate thermally induced stresses during weld cooling. This kind of cracking in the HAZ is known as liquation cracking and is common [11] in welding of Ni-base alloys. The reason for incipient liquation is usually segregation of minor alloying elements at interfaces. In a Ni_3Al containing 0.6 at.% Zr, Chuang et al. [12] observed, with the help of Auger electron microscopy, Zr segregation at the grain boundaries. In present study of welding of IC-396M we observed, with the help of microprobe analysis, Zr enrichment in the regions of HAZ where incipient melting had taken place. Inside the surface cracks made by liquation in the HAZ, Zr-rich eutectics with lamellar structures were observed [Fig. 3(b)]. These observations suggest that alloying element Zr might be responsible for liquation cracking in the HAZ of IC-396M.

Vicker's hardness values in the FZ, HAZ and the base metal of IC-396M were (319$\pm$12), (316$\pm$14) and (283$\pm$6) Kg mm^{-2}, respectively. Due to rapid cooling, the volume fractions of γ' in the FZ and HAZ were smaller than that in the base metal, and, as a consequence, the hardness in the FZ and HAZ was higher [13].

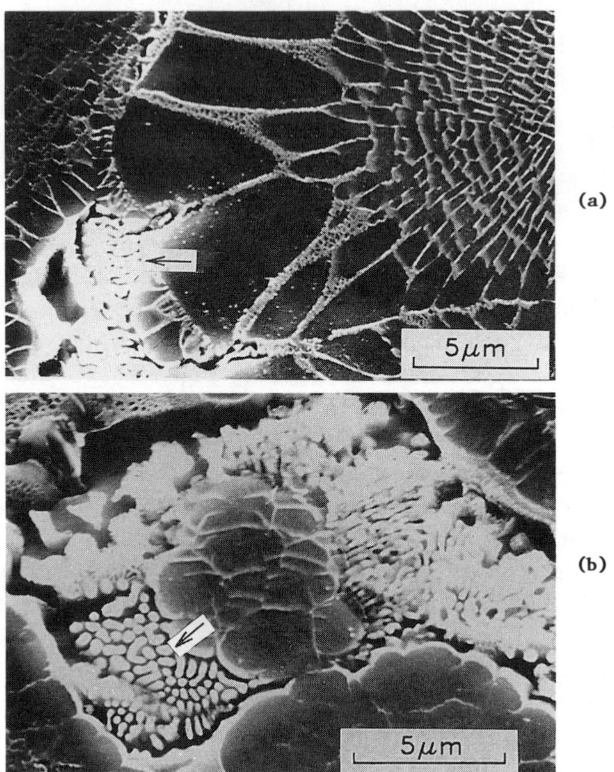

Fig. 1a, b. Scanning electron micrographs showing (a) web-like, type A and (b) lamellar, type B eutectic structures in base metal. The arrows indicate the eutectics of type B.

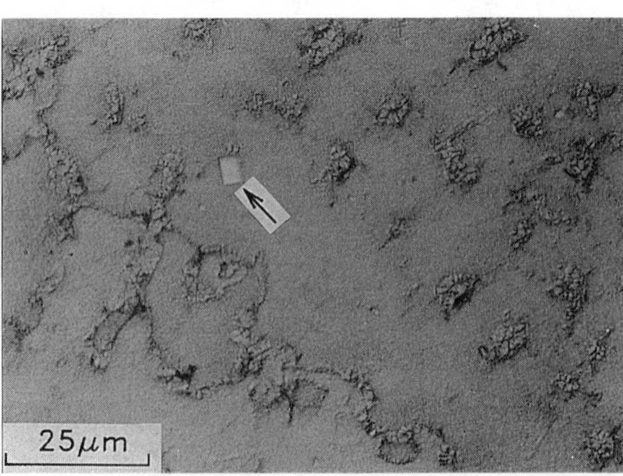

Fig. 2a
(Caption overleaf)

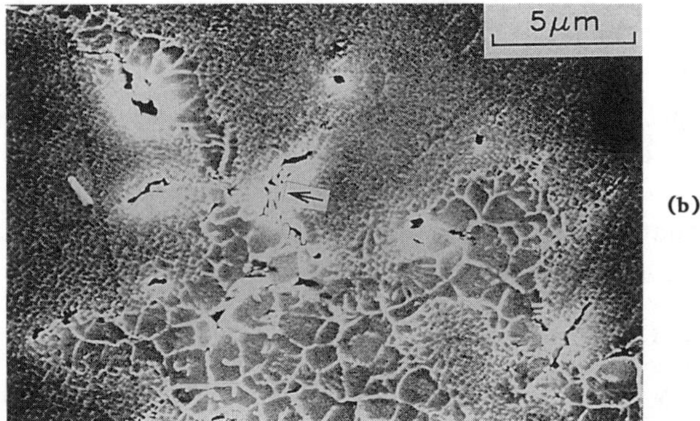

(b)

Fig. 2a, b. Eutectic structures in the fusion zone of IC-396M. (a) Optical micrograph with a ZrO_2 particle shown by an arrow. (b) Scanning electron micrograph with B-type eutectic shown by an arrow.

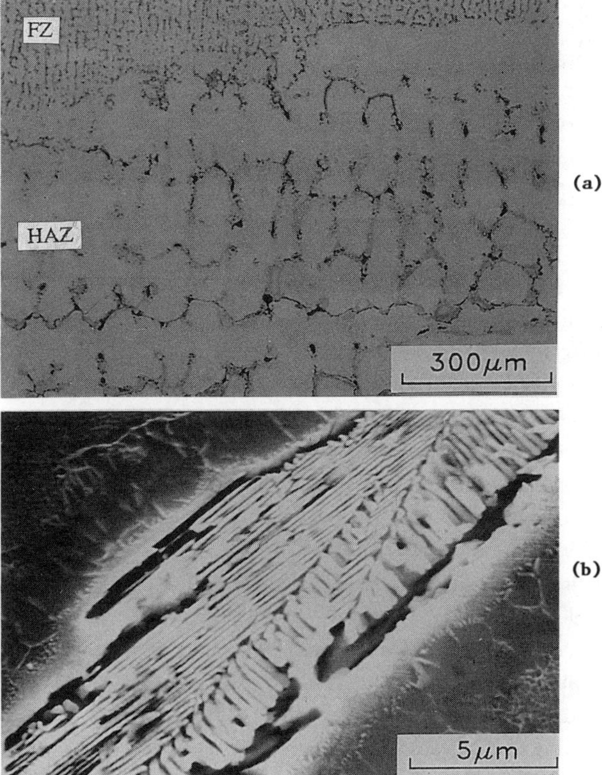

(a)

(b)

Fig. 3a, b. Microcracks in HAZ. (a) Optical micrograph showing cracks along dendritic boundaries. (b) Scanning electron micrograph showing lamellar eutectic structure (B-type) in the liquation crack.

CONCLUSION

Upon addition of Cr and Mo, the weldments of two-phase Ni_3Al alloys can be prepared without macroscopic cracks, but micro-cracks existed in the FZ and HAZ. The cracks in the FZ occurred in the eutectic cells and those in the HAZ were liquation cracks occurring primarily along the dendritic boundaries. Segregation and enrichment of Zr are thought to be responsible for the microcracks in the weldments of IC-396M.

ACKNOWLEDGEMENT

We would like to thank Mr. Philip Erfort of Armco Inc., Ohio for providing ingots of IC-396M.

REFERENCES

[1] K. Aoki and O. Izumi, Nippon Kinzoku Gakkaishi, **43**, 1190 (1979).
[2] C. T. Liu, V. K. Sikka, J. A. Horton, and E. H. Lee, in *Alloy Develop ment and Mechanical Properties of Nickel Aluminide (Ni_3Al) Alloys*, Report no. 6483, Oak Ridge National Laboratory, TN, 1988.
[3] C. T. Liu and C. L. White, Acta Metall., **35**, 643 (1987).
[4] A.I. Taub, K.M. Chang, and C.T. Liu, Scripta Metall., **20**, 1613 (1986).
[5] C. T. Liu and V. K. Sikka, J. of Metals, **38**, 19 (1986).
[6] S. A. David, W. A. Jemian, C. T. Liu, and J. A. Horton, Welding Journal, **64**, 22-s (1985).
[7] M. L. Santella and S. A. David, Welding Journal, **65**, 124-s (1986).
[8] H. Li and T. K. Chaki, in *High-Temperature Ordered Intermetallic Alloys IV*, edited by L. A. Johnson, D. P. Pope, and J. O. Stiegler (Materials Research Society, Pittsburgh, 1991), p. 919.
[9] T. B. Massalski, in *Binary Alloy Phase Diagrams* (ASM International, Materials Park, Ohio, 2nd ed., 1990).
[10] M. C. Maguire, G. R. Edwards, S. A. David, Welding Journal, **71**, 231-s (1992).
[11] B. RadhaKrishnan and R. G. Thompson, Scripta Metall. Mat., **24**, 537 (1990).
[12] T. H. Chuang, Y. C. Pan, and S. E. Hsu, Metall. Trans. A, **22A**, 1801 (1991).
[13] P. Beardmore, R. G. Davies, and T. L. Johnston, Trans. TMS-AIME, **245**, 1537 (1969).

DIFFUSION BRAZING NiAl WITH SELF-GENERATED FILLER METAL

Thomas J. Moore* and Joseph M. Kalinowski**
* NASA - Lewis Research Center, MS 49-3, Cleveland, OH 44135
** Sverdrup Technology, 2001 Aerospace Pkwy, MS 49-3, Brook Park, OH 44142

ABSTRACT

Conventional fusion welding procedures are not likely to be useful for joining NiAl because of its poor low temperature ductility. Thus, the potential use of this material could be limited. This study was designed to demonstrate the feasibility of using a diffusion brazing process to fabricate NiAl components. A unique feature of this process is the development of self-generated filler metal through elevated temperature vacuum exposure. Butt joints were made in extruded bar and joint quality was evaluated using metallographic techniques. Sound joints, which were produced with complete grain growth across the braze interface, were indistinguishable from the base metal. Thus, the feasibility of diffusion brazing NiAl was established.

INTRODUCTION

NiAl is of interest for advanced turbine engine applications because of its physical properties and oxidation resistance [1]; however, in order to fully utilize NiAl in structural applications, suitable welding procedures must be developed. NiAl is very difficult to join by fusion welding processes because of its poor ductility below the ductile-to-brittle transition temperature of $\sim$775 K [2]. Preliminary studies at NASA-Lewis have shown that cracking is a major problem in the electron beam welding of NiAl. For example, a preheat on the order of 1000 K may be required to avoid cracking in 2 mm thick sheet. Since this order of preheat would be difficult for industrial applications, diffusion brazing was thought to be a more viable process. The objective was to produce sound braze joints in NiAl without the addition of any foreign material.

Diffusion brazing NiAl was accomplished in a unique two-step process. In the first step, a glaze of self-generated filler metal was formed on the faying surfaces simply by heating the NiAl in vacuum. In the second step, the braze was made by bringing the faying surfaces together, holding them in place under slight pressure, and reheating to melt the self-generated filler metal and produce coalescence. Some brazed joints were hot isostatically pressed (HIP'ed) to further assure joint soundness. Joint quality was evaluated by metallography.

MATERIALS AND PROCEDURE

The NiAl material was in the form of 10 mm diameter bar resulting from the extrusion of cast billets. The composition in atomic percent was as follows: Ni - 49.6 Al - 0.20 C - 0.08 O, with N being $<$0.015 and S $<$0.004. In the as-extruded condition the grain size was about 63 μm. Square butt joint specimens were prepared using an automatic cutoff wheel. The faying surfaces of the braze specimens were then wet sanded on 240 grit paper. After mechanical preparation, the specimens were cleaned ultrasonically in ethanol and placed in a vacuum furnace for heat treatment to produce the self-generated filler metal.

Self-generated filler metal can be produced by the evaporation of Al from exposed NiAl surfaces in vacuum. At elevated temperature, the vapor pressure of Al is much higher than that of Ni [3]; hence, when NiAl is heated in vacuum, Al should be preferentially vaporized from the surface, forming a thin Ni-rich layer, or glaze. This lower melting glaze can subsequently be used as the filler metal for the brazing operation. To this end, two groups of specimens were prepared by heating in a vacuum of 6.7 X 10^{-3} Pa, or better. The first group was held at 1803 K for 180 s, while the second batch was exposed for 60 s at 1843 K.

After cooling to room temperature, the glazed surfaces were then placed in contact and held in place with a small dead weight producing a slight pressure at the joint. Specimens which had been glazed at 1803 K were diffusion brazed by heating in vacuum to 1803 K and holding for 900 s with a stress of 0.032 MPa at the joint. The samples which had been glazed at 1843 K were vacuum diffusion brazed at 1843 K under a stress of 0.011 MPa for either 60 or 120 seconds. One lot of specimens which was diffusion brazed at 1843 K for 120 s was hot isostatically pressed (HIP) at 1613 K and 138 MPa for 4 h. Following the brazing or brazing plus HIP cycles, welded samples were heat treated at 1803 K for 6 h in argon to promote chemical homogenization.

Weld soundness was evaluated by examining cross sections of the joints using light microscopy. White light was used to detect anticipated color changes in the Ni-rich braze regions.

RESULTS

The surfaces glazed with self-generated filler metal were bright and shiny in appearance and, as shown in Fig. 1, individual grains were visible. Diffusion brazed joints produced at 1803 K were sound with some grain growth across the braze interface (Fig. 2a), as well as grain growth throughout the base metal, where the grain size was an order of magnitude greater than the 63 μm grain size of the as-extruded bar. The only evidence that molten filler metal was present is the cast structure at the braze metal fillet shown in Fig. 2b. Examination under white light revealed that a copper color band, about 200 μm wide, was present at the joint. This color change is indicative of a Ni-rich aluminide composition [4,5]. In the postheated condition, the 1803 K brazed joint showed complete grain growth across the braze interface (Fig. 3) and elimination of the cast fillet microstructure shown in Fid. 2b. A diffuse pink-colored band was present at the joint, and the overall grain size was increased from ~600 μm (Fig. 2a) to ~3 mm (Fig.3). The transition from a copper color band in the as-brazed condition to the diffuse pink-colored band is an indication that partial chemical homogenization occurred during postheating.

Diffusion brazing at 1843 K for 60 s under 0.011 MPa resulted in the formation of shrinkage voids in the cast filler metal (Fig. 4). These voids are evidence that the filler metal had been molten.The copper-colored band at the joint was about 500 μm wide, compared to 200 μm after processing at 1803 K. Diffusion brazing at 1843 K for 120 s followed by HIP'ing at 1613 K and 138 MPa for 4 h yielded sound joints with some grain growth across the interface. The 6 h at 1803 K postheating of such HIP'ed joints produced complete grain growth across the braze interface and the largest grain size (~5 mm; Fig. 5) for any joint. Even after a total of three heat treatments at homologous temperatures exceeding 0.85, a diffuse pinkish color band remained at the joint.

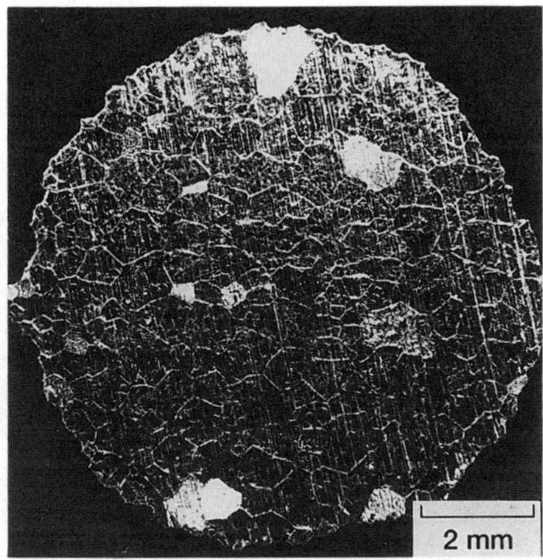

Figure 1.—Glazed faying surface produced by heating NiAl bar at 1803 K for 180 s in vacuum.

DISCUSSION

Diffusion brazing of NiAl with self-generated filler metal produced by exposing faying surfaces to vacuum at high temperature has proven to be a feasible process. It has the potential of producing the ideal joint; that is, a joint which can not be detected metallographically and which matches base metal properties. This finding is important because NiAl does not appear to be weldable using conventional fusion welding procedures.

While the current process tends to produce very large grains during diffusion brazing, this effect would not be of concern in the joining of single crystals. For applications where a smaller grain size needs to be maintained, lower temperatures and shorter processing times might be used. In a one-step brazing cycle, the glazed surfaces would be brought into contact as soon as they become Al-deficient. This would be accomplished in a single furnace run.

Since changes in composition affect the color of NiAl, it can be seen that complete chemical homogenization was not achieved across the joints. Procedural changes to reduce diffusion distances, such as glazing only one of the faying surfaces, could alleviate this situation. Lastly, hot isostatic pressing of brazed joints is an option which could be used to further assure joint soundness.

Considerably more work is required in order to fully characterize the diffusion brazing process for NiAl material. As part of the evaluation, scanning electron microscopy and chemical analysis are in progress. Mechanical testing of butt joints is also under way.

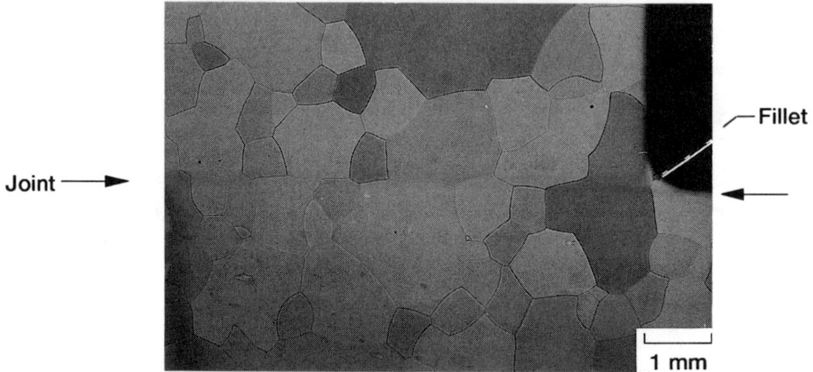

(a) Sound joint with some grain growth across the interface.

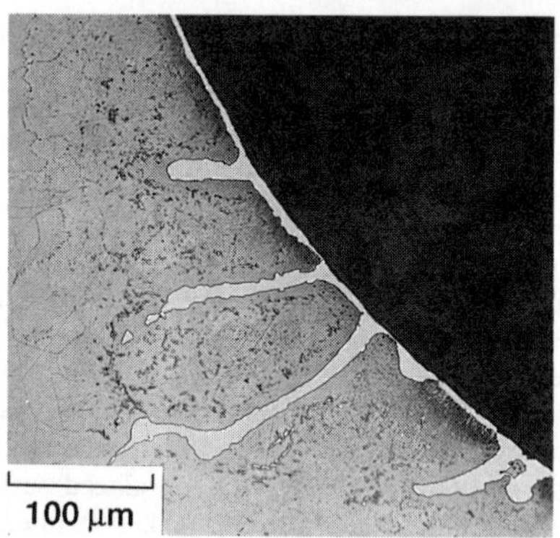

(b) Cast filler metal at fillet shown in figure 2a.

Figure 2.—NiAl joint diffusion brazed at 1803 K and 0.032 MPa for 900 s in vacuum. As-brazed condition. Etched in molybdic acid.

Figure 3.—NiAl joint diffusion brazed at 1803 K and 0.032 MPa for 900 s in vacuum, then postheated at 1803 K for 6 h in argon. Etched in molybdic acid.

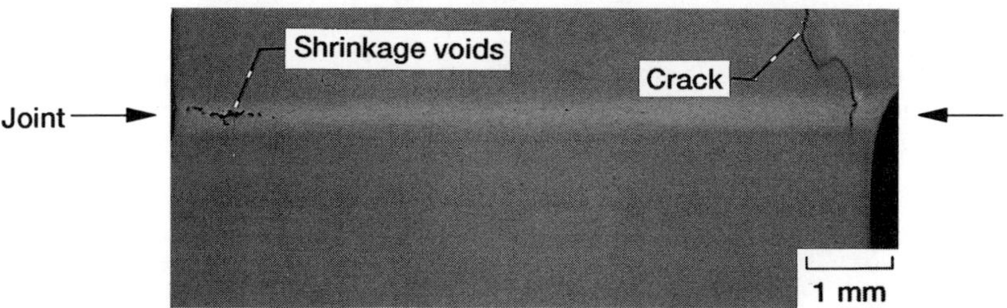

Figure 4.—NiAl joint diffusion brazed at 1843 K and 0.011 MPa for 60 s in vacuum. The crack is believed to have occurred during sample preparation. As-brazed condition. Unetched.

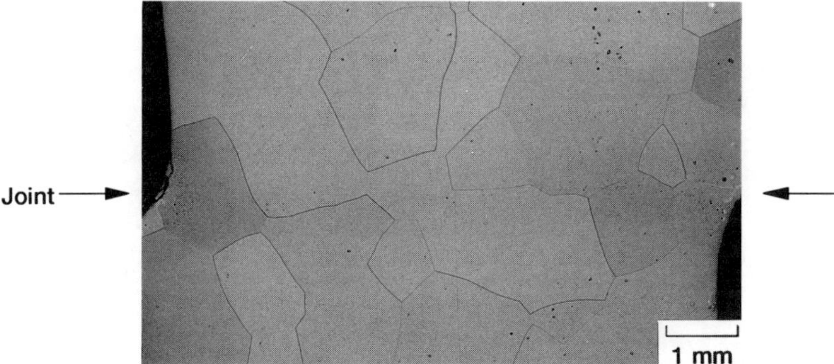

Figure 5.—NiAl joint diffusion brazed at 1843 K and 0.011 MPa for 120 s in vacuum, HIP'ed (1613 K, 138 MPa, 4 hrs), and postheated at 1803 K for 6 hrs in argon. Etched in molybdic acid.

CONCLUSIONS

A study of the feasibility of joining NiAl to itself has shown that diffusion brazing with self-generated filler metal is a practical and viable process.

REFERENCES

1. R. Darolia, JOM 43, 44-49, March(1991).
2. D. L. Anton, D. M. Shah, D. N. Duhl, & A.F. Giamei, JOM 41, 12-17, September(1989).
3. H. E. Pattee, High Temperature Brazing, Welding Research Council Bulletin No. 187, (1973).
4. J. J. Rechtein, et. al., J. Applied Physics 38, 8(1967).
5. H. Jacobi, and R. Stahl, J. Phys. Chem. Solids 34, 1737-1748, (1973). Pergamon Press.

CORRELATION OF STRAIN HARDENING AND CREEP BEHAVIOUR OF γ–TiAl

A. BARTELS, J. SEEGER AND H. MECKING
Devision of Physics and Technology of Materials, Technical University of Hamburg-Harburg, D-2100 Hamburg 90, Germany

ABSTRACT

Compression tests on Ti-48Al-2Cr were performed in the temperature range of 650 - 1000°C. The stress vs. strain curves show an initial increase with a decreasing rate due to dynamic recovery. In the second stage, correlated with dynamic recrystallization (DRX), the stress passes through a maximum and decreases to a steady state value. The characteristic steady state stress associated with dynamic recovery exhibits a lower activation energy (269 kJ/mol) than the steady state stress during DRX (400 kJ/mol). Constant-load creep experiments show a corresponding behaviour. After primary creep the creep rate passes through a minimum. The final steady state creep conditions are dominated by DRX as well.

INTRODUCTION

The investigations are performed on γ-TiAl with an Al-content of 48at% and 2at% Cr-addition to achieve good ductility [1]. Consequently the TiAl contains a small amount of the α_2–Ti_3Al phase and the as-cast material shows a coarse lamellar microstructure in combination with a strong casting texture. These give rise to very anisotropic and inhomogeneous deformation behaviour and to low values of the fracture strain. Therefore, a thermo-mechanical treatment is indispensable to develop a homogeneous fine grained microstructure, since this is correlated with acceptable properties of formability and ductility. These properties desirable for our application, i.e. rolling and near net shape forging, unfortunately are accompanied with poor creep properties especially during the technically relevant early stages of creep [2]. Formability and creep resistance will be topics of this paper.

EXPERIMENTAL

The Ti-48Al-2Cr (at%) alloy was produced at Böhler Corporation, Kapfenberg (Austria) by a vacuum skull melting technique and subsequently HIP'ed [3]. The ingots had a diameter of about 190mm and showed a strong casting texture. Main parts of the microstructure consist of dendrites with a lamellar two phase substructure of alternating laths of the γ-TiAl and α_2–Ti_3Al phases (figure 1). The lamellae show a preferred alignment perpendicular to the direction of maximum heat flow during solidification. Therefore, near the cylindrical surface of the ingot the lamellae are oriented parallel to the surface [4]. The lamellar dendrites reveal a higher concentration of Al in the outer zones and are surrounded by single phase γ-TiAl as a consequence of the peritectic solidification [5]. For compression tests cylindrical specimen with 4-6 mm in diameter and 6-8 mm in height were prepared by spark erosion

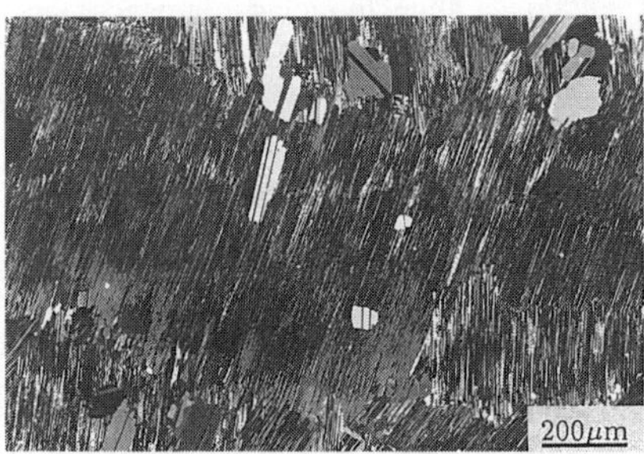

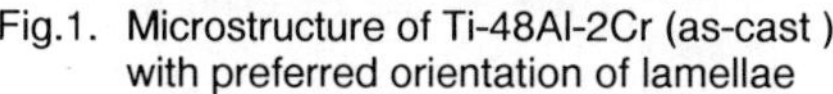

Fig.1. Microstructure of Ti-48Al-2Cr (as-cast) with preferred orientation of lamellae

Fig.2. Two step upset forged Ti-48Al-2Cr with an equiaxed microstructure

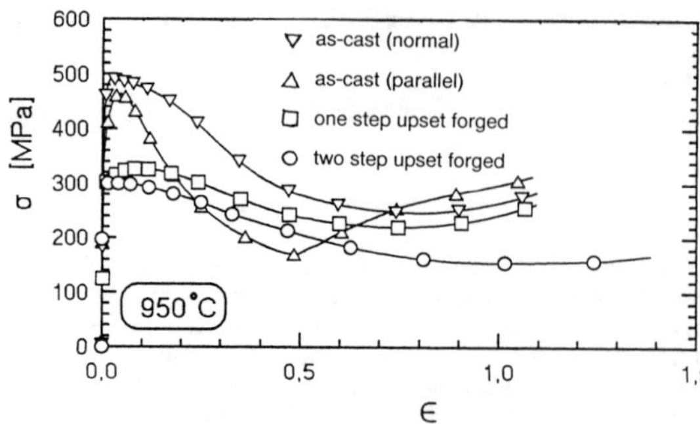

Fig.3. Stress vs. strain in compression of as-cast Ti-48Al-2Cr (fig.1) with two different orientations of lamellae compared with upset forged material. The lowest curve corresponds to the microstructure of figure 2.

with two preferred orientations of the lamellae - parallel and perpendicular to the axis of the cylinders, i.e. the compression direction.

Most of the compression tests were performed with upset forged material. The thermomechanical treatment was as follows: homogenisation (1400°C 1.5 h), upset forging (about 70% at 1000°C), annealing (1250°C 1.5 h), a second upset forging step (same direction about 60% at 1000°C) and final heat treatment (1200°C 1.5 h, 1000°C 4h, slow cooling) [5]. The resulting microstructure consists of fine equiaxed grains (grain size about 8-10µm) and is nearly homogeneous. A small amount of α_2–Ti_3Al is located preferably at the triple lines of the grain boundaries and stabilizes the microstructure. This upset forging and annealing sequence was developed for achieving optimum values of ductility (more than 4% fracture strain) and optimum formability in subsequent hot rolling and near net shape forging [3].

The compression tests were performed with constant strain rate in a closed loop computer-controlled testing machine. The measured stresses were corrected with regard to the friction forces during compression [6]. The creep experiments were conducted with constant load.

COMPRESSION TESTS: RESULTS AND DISCUSSION

Figure 3 demonstrates the strong influence of the microstructure on the stress versus strain behaviour during compression tests with a strain rate of $3x10^{-3}$ 1/s at 950°C. The specimen with orientation of the lamellae normal to the compression direction exhibits the highest stress levels and the stress decrease beyond the maximum is much smaller than in the specimen with the lamellae parallel to the compression direction. In the latter case the stress drops by a factor of more than two and the material becomes softer than the specimens of the upset forged materials which show much lower maximum stresses followed by only a slight decrease of the stress. The difference between the two curves of forged materials is caused by differences in the microstructure. The specimen with a homogeneous grain size distribution (as to be seen in figure 2) show the lowest stresses. Inhomogeneities like zones of coarse grained single phase γ-TiAl [7] lead to elevated stresses .

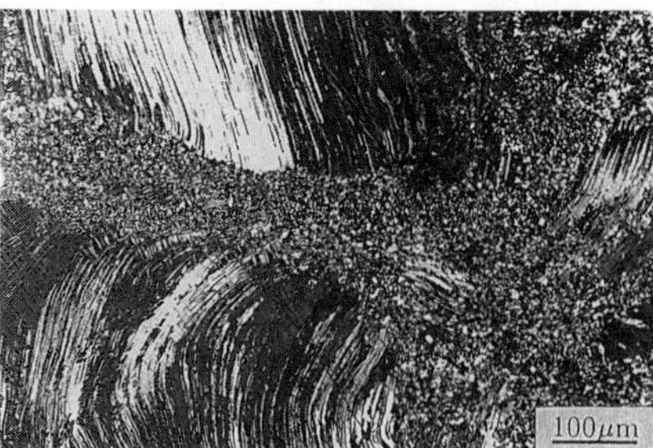

Fig.4. Microstructure after deformation. The fine grained zone between the lamellar grains is generated by dynamical recrystallization

To deform the lamellar grains (more than 100µm in diameter) high external stresses are necessary [5]. Between grains with slightly differently oriented lamellae zones of strong interaction develop during straining such as shown in figure 4 where along the grain boundaries dynamic recrystallization (DRX) has been initiated in a specimen with the lamellae parallel to the direction of compression. The DRX zones are very fine grained. They are assumed to be ductile and therefore to be able to compensate the misfit strains set up by anisotropically deforming lamellar grains. Consequently the stresses decrease when further deformation takes place preferentially in these dynamically recrystallized zones whereby these expand continuously at the expense of the lamellar areas. Under these circumstances, the deformation is very inhomogeneous. Although lamellar areas decrease with strain, even at a strain of 1.6 great parts of the microstructure are still lamellar.

The hot formability of upset forged material with an equiaxed microstructure (about 10μm) is superior to lamellar material [5]. The deformation behaviour in compression tests was systematically investigated [8]. For instance figure 5 shows the influence of the strain rate between $3x10^{-5}$ and $3x10^{-3}$ 1/s at 850°C and figure 6 the effect of temperature between 650°C and 1000°C at a strain rate of 10^{-3} 1/s. Characteristic stress values of these curves are the maximum stress σ_p indexed by p for peak) and the minimum stress σ_{DRX} which is assumed to represent the quasi steady state due to DRX (the slight increase at very large strains is attributed to imperfect correction of the friction effects). One can evaluate the influence of dynamic recovery by extrapolating the initial part of the stress-strain curve to a steady state level σ_s which corresponds to equilibrium between athermal hardening and thermally activated recovery. A linear extrapolation of $d\sigma/d\varepsilon$ to zero in a $d\sigma/d\varepsilon$ vs. σ-plot leads to this saturation stress σ_s [9].

Figures 7-9 show the dependence of these different stress values on strain rate in double logarithmic $\dot{\varepsilon}$ vs. σ-plots for different temperatures. Only

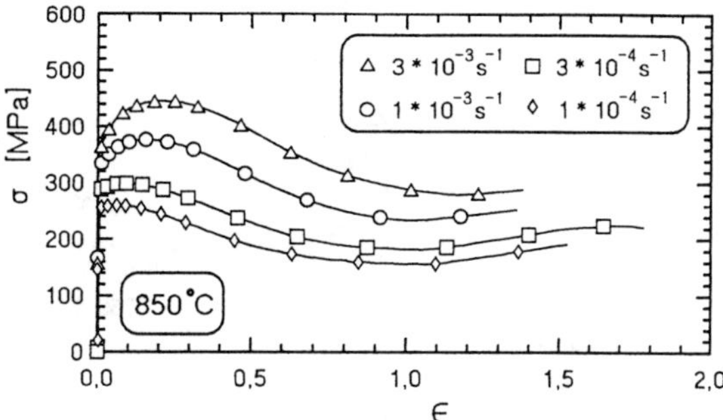

Fig.5. Stress vs.strain in compression with upset forged Ti-48Al-2Cr at 850°C with different strain rates

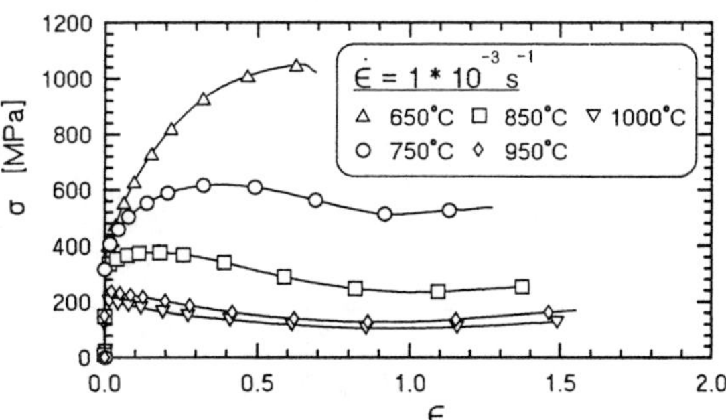

Fig.6 Stress vs.strain in compression with upset forged Ti-48Al-2Cr between 650°C and 1000°C at a strain rate of 10^{-3} s^{-1}

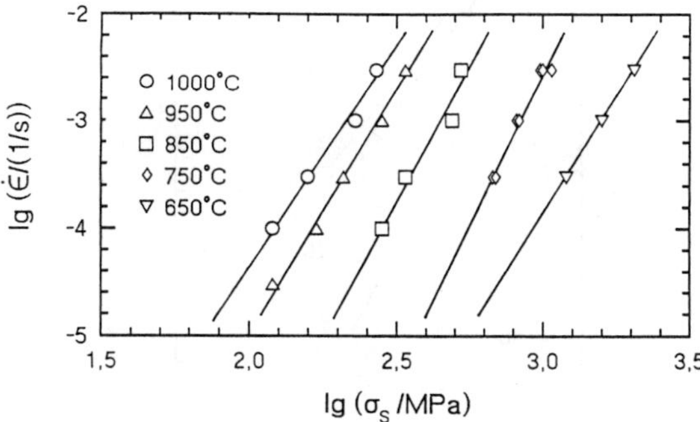

Fig.7. Double logarithmic plot of the strain rate vs. the aturation stress associated with the dynamical recovery

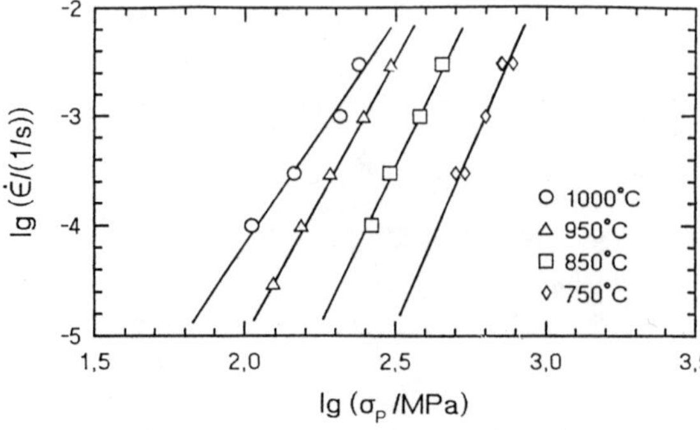

Fig.8. Double logarithmic plot of the strain rate vs. the peak stress

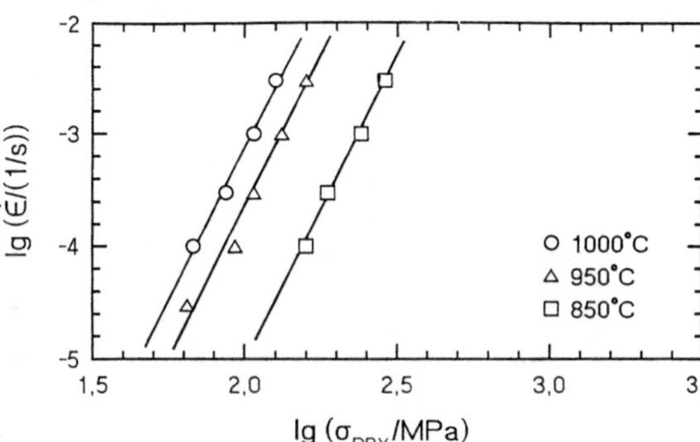

Fig.9. Double logarithmic plot of the strain rate vs. the steady state stress during dynamical recrystallization

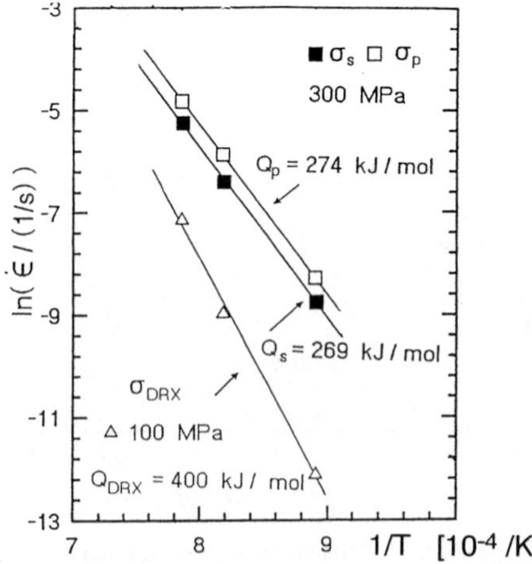

Fig.10. Arrhenius plot of the strain rates corresponding to the different stresses of figures 7-9

small differences between the peak stress σ_p and the saturation stress σ_s, i.e. the steady state stress of dynamic recovery, can be recognized. These results suggest that the maximum is mainly determined by the hardening and recovery processes but only to a minor extend by DRX. The slope n_s in the double logarithmic $\dot{\varepsilon}$ vs. σ_s-plots decreases from 5.9 at 750°C to 4.0 at 1000°C. These values are typical for dynamic recovery [10]. The values of n_p show slightly higher differences for the different temperatures. Only n_{DRX} (=5.2-5.5) seems to be independent of temperature but could only be determined in a reduced temperature range from 850° to 1000°C. In the thermal activation analysis a temperature dependence of the n-values leads to a stress dependence of the activation energies, which has to be taken into consideration by comparing results. The Arrhenius plots (figure 10) show the activation energies evaluated at 300 MPa for σ_s and σ_p and at 100 MPa for σ_{DRX}. The value for dynamic recovery is found to be Q_s=269 kJ/mol and can be compared with that of thermal diffusion. Ti-tracer diffusion data exhibit an activation energy of 291 kJ/mol [11] (measured in single-phase γ-Ti-54Al). The peak stress σ_p leads to a higher value, but the difference is small and does not exceed the limit of uncertainty. The activation energy evaluated for DRX, however, is much higher. At σ_{DRX} = 100 MPa the activation energy is 400 kJ/mol. Such a high activation energy combined with n-values around 5 is typical for DRX in conventional metals [10].

CREEP: RESULTS AND DISCUSSION

The creep behaviour of the upset forged Ti-48Al-2Cr was studied under constant load conditions which are more relevant for technical applications than constant stresses. Under constant compression load the decrease of stress is correlated with the widening of the specimen ($\sigma=\sigma_0\exp(-|\varepsilon|)$). Under steady state conditions, i.e. if constant stress would cause a constant creep rate $|\dot{\varepsilon}|$, the n value can be extracted directly from the creep curves in plots of $\dot{\varepsilon}$ vs. ε. Since the creep rate will then decrease with $\ln|\dot{\varepsilon}| = -n|\varepsilon|$ (quasi stationary), buckling and friction effects are neglected in this case.

Figure 11 compares the creep curves ($\dot{\varepsilon}$ vs. ε) of upset forged (upper curve) and as-cast material. The course of the creep rate corresponds very well with that of the stress-strain curves with constant strain rate. In the case of lamellar microstructure the strain decreases rapidly to a minimum. The expected decrease of the creep rate $-n|\varepsilon|$ due to the stress reduction by the cross section increase is not observed since it is overcompensated by the softening caused by the DRX. At $|\varepsilon|=0.4$ the constant decrease of the creep rate indicates stationary conditions and from the slope n=5.9 can be evaluated. The upset forged material behaves much softer in that it exhibits much higher creep rates than the as-cast material corresponding to the lower stress levels of the compresion tests of figure 3. The final stationary decrease corresponds to n=4.3. At 850°C the corresponding value is n=3.9.

An estimation of the actvation energy for these creep ranges resulted in Q=460 kJ/mol which is about 15% higher than the activation energy obtained from σ_{DRX} in the stress-strain tests (figure 10) and it can serve again as an indication of the presence of DRX [10,13]. There is no doubt that also in the upset forged material during creep deformation DRX took place, causing the deterioration of creep. A direct proof is the occurrence of a strong grain refinement during creep which can be easily seen from a comparison of the initial microstructure in figure 2 with that after a creep test at 850°C (in figure 12). Also in lamellar specimen the effect of DRX during creep is observed by the appearance of new grains between the different colonies of equally aligned lamellae. Even during creep under tension conditions the fine grained zones caused by DRX are observed [14].

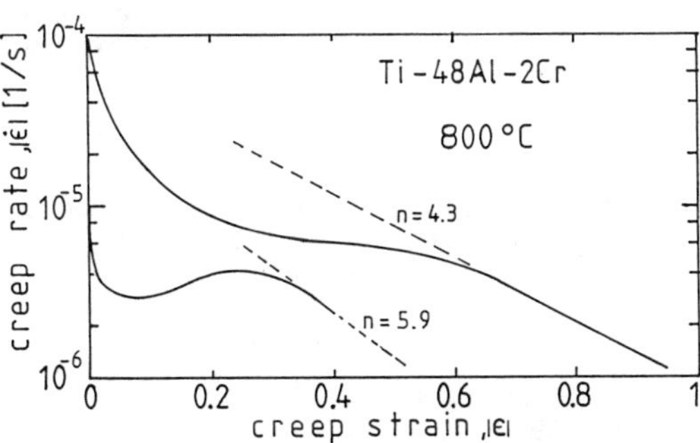

Fig.11. Creep rates vs. creep strain in compression for as-cast (lower curve) and for upset forged Ti-48Al-2Cr (upper curve) under constant load conditions (initial load or stress 330 MPa)

Fig.12 Microstructure after a creep strain of 1.16 (69%) at 850°C. A grain refinement takes place during dynamical recrystallization (Compare with the initial microstructure in figure 2)

CONCLUSIONS

Stress strain and creep tests show corresponding results for the change with strain of the stress and creep rate, respectively. Hardening effects become more significant if the temperature is lowered and the strain rate is increased, be it by an increase of the cross head speed or the applied load. Under both conditions an evaluation of the effect of dynamic recovery is possible and exhibits an activation energy of about 270 kJ/mol while the n-values range between 5.9 at 750°C and 4.0 at 1000°C. The final deformation state is controlled by DRX. Even at the lowest strain rates in our investigations, i.e. in creep tests, DRX finally became the predominant process.

The parameters of a thermal activation analysis are to be evaluated with caution. The maximum stress in stress strain curves and the minimum creep rate are influenced in a complex manner by a superposition of recovery and recrystallization effects, but, at least in our case they seem to reflect mainly the temperature and rate dependence of dynamic recovery. Under stress strain and under creep conditions identical steady state conditions are achieved by DRX processes. Stress exponents of n = 4 - 5 and activation energies of about 400 - 460 kJ/mol are typical for DRX in Ti-48Al-2Cr .

The lamellar microstructure is the more creep resistant one especially for small strain levels. Even in these microstructures after a strain of some percent of DRX is observed and causes increasing creep rates in combination with microstructural instabilities.

ACKNOWLEDGEMENT

The authors are grateful to M. v. Schwerin and M. Koeppe for providing unpublished results and to R. Behn for experimental assistance. Part of this work was financially supported by the Bundesminister für Forschung und Technologie (BMFT 03 M 3029).

REFERENCES

1. S. C. Huang and E. L. Hall, Metall. Trans. **A22,** 2619 (1991)
2. Y.-W. Kim and D. M. Dimiduk, JOM **43**, 40 (1991)
3. K. Wurzwallner, H. Clemens, P. Schretter, A. Bartels and C. Koeppe, in *High Temperature Ordered Intermetallic Alloys V,* (Mater. Res. Soc. Proc.) this proceedings
4. J. Seeger, A. Bartels and H. Mecking, Scripta Metall. Mater. **25**, 2523 (1991)
5. C. Koeppe, J. Seeger, A. Bartels and H. Mecking, submitted to Metall. Trans. A (1992)
6. G. E. Dieter, *Mechanical Metallurgy,* (MacGraw-Hill International Editions, New York, 1986)
7. A. Bartels and H. Mecking, in *Ordered Intermetallics - Physical Metallurgy and Mechanical Behaviour,* ed. by C. T. Liu, R. W. Cahn, G. Sauthoff, NATO ASI Series E, Vol. **213** (Kluwer Academics Publisher,Dordrecht,1992) pp. 663-678
8. J. Seeger, Doktor thesis Technical University of Hamburg-Harburg, 1993
9. H. Mecking, B. Nicklas, N.Zarubova and U. F. Kocks, Acta metall. **34**, 527 (1986)
10. H. Mecking and G. Gottstein, in *Recrystallization of Metallic Materials,* ed. F. Haessner (Dr. Rieder-Verlag GmbH, Stuttgart, 1978) pp.195-222
11. Kroll, H. Mehrer, N. Stowijk, C. Herzig, R. Rosenkranz and G. Frommeyer, Z. Metallkd. **83**, 591(1992)
12. H. Oikawa, Mater. Sci. Eng. **A153**, 427 (1992)
13. H. J. McQueen and J. J. Jonas, in *Treaties of Materials Science and Technology,* ed. R. Arsenault, (Academic Press, New York, 1975) pp.393-493
14. M. Es-Souni, A. Bartels and R. Wagner, to be published

HIGH TEMPERATURE FRACTURE TOUGHNESS OF Ni_3Al ALLOY IC-396M

DAVID J. ALEXANDER
Metals and Ceramics Division, Oak Ridge National Laboratory, P.O. Box 2008, Oak Ridge, TN 37831-6151.

ABSTRACT

The fracture toughness of a nickel aluminide alloy IC-396M has been measured from room temperature to 800°C. The material was tested in the as-cast condition. Specimens were oriented for crack growth either parallel or perpendicular to the dendritic structure (C-R and R-C orientations, respectively). Both orientations had acceptable levels of toughness from room temperature to 600°C. At 650°C the toughness decreased markedly with some recovery at 800°C. Examination of the fracture surfaces revealed that the fracture mode changed from a ductile tearing mechanism at temperatures up to 600°C to interdendritic fracture at 650°C and higher temperatures. At 650°C the crack grew along the dendrite axis for the C-R specimens, but for the R-C specimens the crack grew out-of-plane in attempting to follow the dendritic structure. It is believed that the presence of the sharp crack in the fracture toughness specimens allowed oxygen to penetrate to the crack tip, where straining exposed metal surfaces that were not protected by an oxide layer. The result was embrittlement of the interdendritic regions and low energy fracture, analogous to intergranular fracture in wrought materials.

INTRODUCTION

Nickel aluminide intermetallic alloys offer the potential for good mechanical properties and corrosion resistance at intermediate to high temperatures. One of the alloys developed at Oak Ridge National Laboratory for use as a casting alloy is designated IC-396M. The nominal composition of this alloy is 8Al-8Cr-3Mo-0.8Zr-0.005B (wt %). This work continues a previous investigation of the tensile and fracture properties of this alloy [1].

EXPERIMENTAL PROCEDURE

The IC-396M material was vacuum induction melted and cast into a 75-mm-diam ingot, followed by electroslag remelting into a 100-mm-diam ingot. All processing was performed by Haynes International. Slices approximately 15 mm thick were taken from the ingot, and compact specimens 12.7 mm thick were machined, with the crack plane either in the radial or circumferential directions (C-R and R-C orientations, respectively). These specimens were precracked at room temperature with a final maximum stress intensity of approximately 25 MPa$\sqrt{m}$, and then side grooved 10% of their thickness on each side.

The fracture toughness was determined from the J-integral-resistance (J-R) curves, which were obtained using the unloading compliance technique to monitor crack extension during the test, in general accordance with ASTM Standards E 813-89,

Standard Test Method for J_{lc}, a Measure of Fracture Toughness, and E 1152-87, Standard Test Method for Determining J-R Curves. A rod-in-tube device was used to transfer the load-line displacement of the specimen outside the furnace where it was measured with a conventional clip gage. At temperatures up to 650°C hardened steel razor blades spot-welded to the specimens were used to locate the rod-in-tube attachment. At 800°C knife edges fabricated from MAR-M 246 nickel-base superalloy were used, as the steel blades were expected to be degraded by very high temperatures. A thermocouple was spot-welded to the specimen to monitor temperature throughout the test. A split box furnace was used to control the temperature, which was maintained within ±3°C of the target temperature during the test. Laboratory air was used for all tests. After completion of the tests the specimens were cooled to room temperature, and broken open. Temperatures of 350°C or higher resulted in some heat tinting of the fracture surface that made crack extension during the test visible on the fracture surface. The specimens tested at room temperature were heat tinted by heating them on a hot plate until a noticeable color change was obtained. The fatigue precrack and the final crack length were measured with an optical measuring microscope.

A section of the ingot was polished and macroetched to reveal the solidification structure. The fracture surfaces were cut from several of the compact specimens, and the section immediately beneath the fracture surface was metallographically prepared and etched to determine if the test temperatures affected the microstructure.

RESULTS

The cross section of the ingot is shown in Fig. 1. The electroslag remelting process resulted in the growth of large grains in from the mold wall. As a result, the C-R and R-C specimens had crack growth either parallel or perpendicular,respectively, to the axis of the grains. Examination at higher magnification shows the dendritic structure within the grains (Fig. 2). The cuboidal gamma prime is visible within the dendrites. Also present are cells of gamma and gamma prime, and occasional patches

Fig. 1. Macroetched section through ingot, showing coarse columnar grains.

of an apparent third phase. Some microporosity is present between the dendrites, but no large-scale macroporosity was observed. No obvious differences were observed between the as-cast material and that exposed to high temperatures during testing.

The fracture toughness results are shown in Fig. 3 [$K_J = (JE)^{1/2}$ where E is Young's modulus]. The results show that the toughness of the as-cast material is approximately constant up to 600°C but drops significantly at 650°C, with some recovery at 800°C. There is no consistent difference between the toughness of the R-C and the C-R specimens, within the scatter of the data.

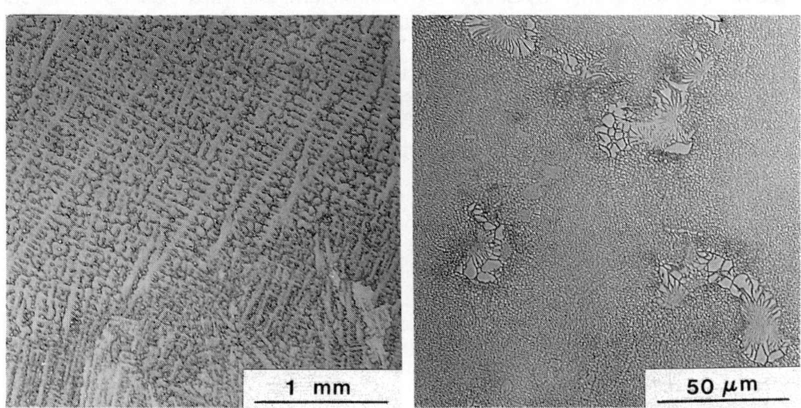

Fig. 2. Left: low magnification view of dendritic structure. Right: higher magnification view, showing cuboidal gamma prime, and gamma/gamma prime cells.

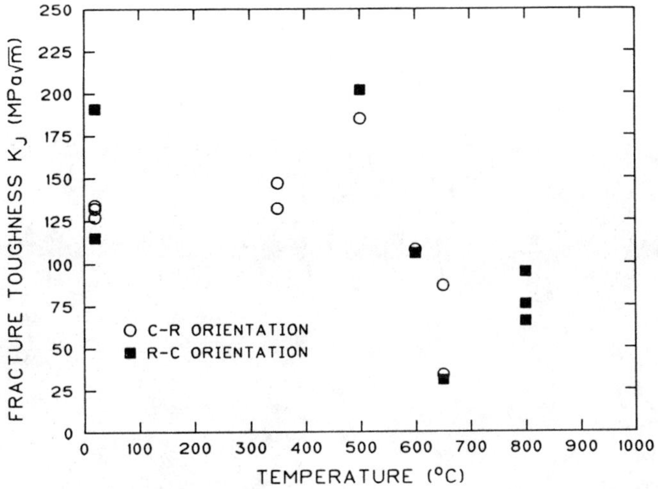

Fig. 3. Fracture toughness for cast IC-396M, showing a significant decrease at 650°C with some recovery at 800°C.

DISCUSSION

The fracture toughness of this material is approximately constant up to 600°C, drops off significantly at 650°C, and recovers at 800°C. Tensile tests of IC-396M show a similar trend, with a minimum in ductility followed by some recovery [2], but the minimum is shifted to slightly higher temperatures (about 750°C). Over the same range of temperatures the yield strength remains high (Fig. 4). Similar decreases in ductility at high temperature have been observed in other nickel aluminide alloys [3,4] and attributed to oxygen embrittlement of the grain boundaries. Chromium has been added to alleviate this problem. The present alloy does contain chromium, and retains useful tensile ductility at high temperatures. However, in the fracture toughness tests the specimens contain a sharp precrack. It has been suggested [1,5] that although the chromium will prevent embrittlement for smooth tensile specimens, the presence of the sharp crack will rupture the protective oxide coating at the crack tip as soon as strain is applied, exposing unprotected metal that can interact with oxygen to create embrittlement. A similar mechanism is believed to exist for these cast alloys.

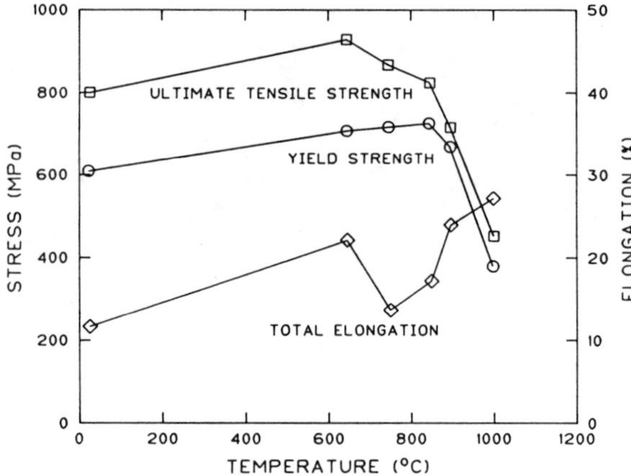

Fig. 4. Tensile data for IC-396M showing a minimum in ductility at 750°C.

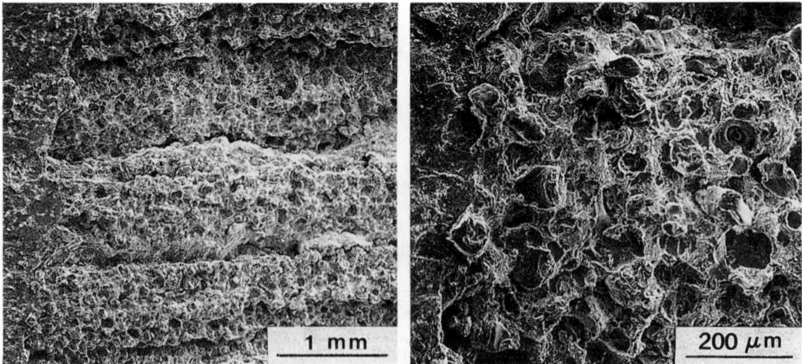

Fig. 5. The fracture surface of a C-R specimen tested at 650°C, showing the interdendritic crack path. Crack growth is from left to right.

Examination of the fracture surfaces shows that the fracture mode is ductile tearing at the lower temperatures. At 500°C there is some evidence of interdendritic fracture. At 650°C the fracture mode is entirely interdendritic. The coarse scale of the dendrites is readily apparent on the fracture surface (Fig. 5), particularly for the C-R specimens, with crack growth parallel to the axis of the dendrites. This change in fracture mode from transgranular to interdendritic is responsible for the drop in toughness at 650°C, consistent with the oxygen embrittlement mechanism.

The J-R curves for four of the R-C specimens are shown in Fig. 6. Crack growth will be perpendicular to the dendritic structure for these specimens. At room temperature the curve shows the typical initial rise associated with crack tip blunting. At 600°C the toughness is lower, but there is still some initial increase in J before large amounts of crack extension have occurred. At 650°C, the crack extension is immediate and large. There is little initial resistance to crack growth, as reflected by the low J values. With additional crack growth, the J-R curve does begin to rise, but this is actually an artifact of the crack's selection of a path. The crack grows out-of-plane, and tries to turn to follow the axis of the dendrites. The crack deflection results in the apparent rise in toughness. This change of direction further emphasizes the embrittlement of the interdendritic regions at this temperature. At 800°C the J-R curve again shows an initial rise in J before crack extension, and is very similar to the 600°C data. The crack also shows less tendency for deflection.

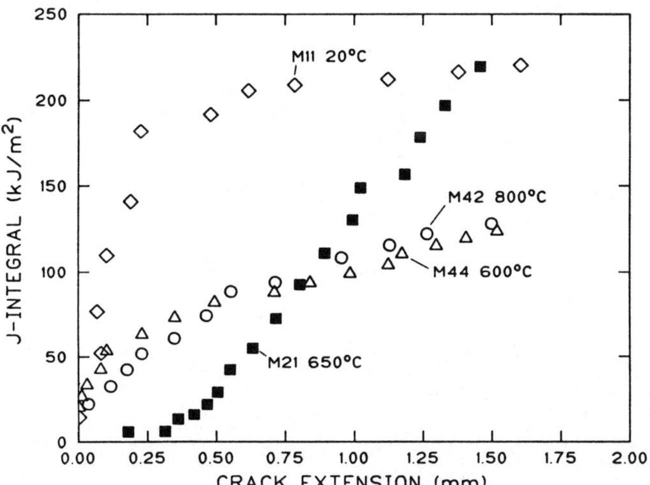

Fig. 6. J-R curves for four R-C specimens. At 650°C crack extension begins immediately. The apparent increase in toughness with extension is due to crack deflection as the crack tries to turn and follow the dendritic structure.

CONCLUSIONS

The fracture toughness of a cast nickel aluminide alloy IC-396M has been measured from room temperature to 800°C. The alloy has good toughness up to 600°C, but the toughness drops significantly at 650°C, with some recovery at 800°C. This decrease in toughness is accompanied by a change in fracture mode, from transgranular tearing at lower temperatures, to interdendritic failure at 650°C. It is believed that the sharp precrack in the specimens fractures the protective oxide layer and permits the access of oxygen to the metal, resulting in oxygen embrittlement and low fracture toughness at 650°C.

ACKNOWLEDGMENTS

This research was sponsored by the U.S. Department of Energy, Assistant Secretary for Conservation and Renewable Energy, Office of Industrial Technologies, Advanced Industrial Concepts (AIC) Materials Program, under contract DE-AC05-84OR21400 with Martin Marietta Energy Systems, Inc. The fracture toughness tests were conducted by R. L. Swain. We appreciate helpful discussions with V. K. Sikka and S. Viswanathan. The manuscript was reviewed by B. G. Gieseke, S. K. Iskander, and P. S. Sklad and was prepared by J. L. Bishop.

REFERENCES

1. D. J. Alexander and V. K. Sikka, Mat. Sci. Eng. **A152**, 114 (1992).

2. V. K. Sikka, Nickel-Aluminide Data Package, Oak Ridge National Laboratory, Oak Ridge, TN, 1992.

3. C. T. Liu, V. K. Sikka, J. A. Horton, and E. H. Lee, Development and Mechanical Properties of Nickel Aluminide (Ni_3Al) Alloys, ORNL-6483, Oak Ridge National Laboratory, Oak Ridge, TN, 1988.

4. C. T. Liu, in High Temperature Ordered Intermetallics, edited by N. S. Stoloff et al. (Mater. Res. Proc. **81**, Pittsburgh, PA, 1987) p. 355.

5. D. J. Alexander, Scripta Metall. Mater. **24**, 845 (1990).

MICRO- AND CRYSTAL-STRUCTURAL OBSERVATIONS OF Al-RICH Nb AND AL-RICH Ti ALLOYS

K. Mizuuchi*, Y. Okanda*, M. Kitamura**. C.V. Cooper***, A.F. Giamei*** and H.R.P. Inoue****
*Osaka Municipal Technical Research Institute, Osaka, 536, Japan
**Electrical and Electronic Engineering, Utsunomiya Univ., Utsunomiya 321, Japan
***United Technologies Research Center, East Hartford, CT 06108
****Materials Science and Engineering, Univ. of Washington, Seattle, WA 98195.

ABSTRACT

Microstructural observations, phase identification, and hardness measurements of several ternary and quaternary Al_3Nb- and Al_3Ti-based alloys containing Ni or Fe have been carried out after homogenization at 1100°C. An $L1_2$ phase is formed in an $(Al,Fe)_3Ti$-based alloy when a small amount of Nb is added. Conversely, in Al-Nb-Ni alloys whose compositions are based on Al_3Nb, no $L1_2$ phase has been detected. In this case, alloys are either three-phase(B2, DO_{22} and Laves) or two-phase(B2 and Laves). The absence of the $L1_2$ phase formation in Al_3Nb-based alloys containing Ni is attributed to limited solubility of Ni in the Al_3Nb compound. Rapidly solidified and slowly cooled specimens have also been analyzed.

INTRODUCTION

The trialuminide Al_3Nb is considered to be a promising high temperature structural material, in contrast with other trialuminides such as Al_3Ti, because of its high melting point (1600°C) and low density ($4.91 g/cm^3$), and is expected to be used at temperatures above 1000°C. The binary Al_3Nb compound has a tetragonal DO_{22} structure and is brittle at room temperature. Recently the DO_{22} to $L1_2$ structural modification has proved to increase room temperature compressive strain (14%) substantially especially in the case of Al_3Ti containing 7.5 at.%Fe [1]. Such structural modifications can be accomplished by adding transition elements to Al_3Ti. However, there is an intrinsic problem in the case of Al_3Ti; that is, these modified compounds all have rather low melting points, suggesting that they cannot be used at high temperatures. Hence a similar approach has been attempted to modify the crystal structure of Al_3Nb to $L1_2$. The phase stability of Al_3Ti and Al_3Nb has been investigated on the basis of an extended Hückel tight-binding method, resulting in the prediction that $L1_2$ is stable compared with DO_{22} in Al_3Nb when more than 12.5at%Ni is added, provided that Ni substitutes for Al [2]. Ternary and quaternary Al-rich Nb and Al-rich Ti alloys have been prepared by arc melting and rapid solidification in an attempt to investigate the phase stability of Al_3Nb and Al_3Ti-based compounds. Micro- and crystal-structural observations along with compositional analyses have been carried out by scanning electron microscopy, X-ray diffraction and EPMA techniques.

EXPERIMENTAL

Several ternary and quaternary Al-Nb and Al-Ti alloys containing Ni and Fe were prepared from high purity Al, Ti, Fe, and Nb metals by arc melting in an argon atmosphere (0.1 MPa). These alloys were produced in a copper crucible in the shape of buttons about 20 mm in diameter and 5 mm high. Alloy buttons were all arc melted twice and homogenized at 1100°C for 10 days in vacuum of 0.1 Pa. Table 1 shows nominal alloy compositions of the alloys prepared in this study. In the case of an Al-25at%Nb-25at%Ni alloy, a rapid solidification process (RSP) was also used [3], where a

Table 1 Nominal compositions of alloys produced and phases present after homogenization at 1100°C

Alloy	Compositions, at.%					Phases present
	Al	Nb	Ni	Ti	Fe	
A	67.5	25.0	7.5			DO_{22}-Al_3Nb + Laves-$Nb(Ni,Al)_2$ + B2-AlNi
B	62.5	25.0	12.5			DO_{22}-Al_3Nb + Laves-$Nb(Ni,Al)_2$ + B2-AlNi
C	50.0	25.0	25.0			Laves-$Nb(Ni,Al)_2$ + B2-AlNi
D	62.5	16.75	12.5	8.25		DO_{22}-$Al_3(Nb,Ti)$ + $L2_1$-Ni_2NbAl + unidentified
E	62.5	8.25	12.5	16.75		DO_{22}-$Al_3(Nb,Ti)$ + $L2_1$-Ni_2NbAl + unidentified
F	67.5	25.0			7.5	DO_{22}-Al_3Nb+ Al_3Fe(?)
G	67.5	16.75		8.25	7.5	DO_{22}-$Al_3(Nb,Ti)$+ Al_3Fe(?)
H	67.5	8.25		16.75	7.5	DO_{22}-$Al_3(Nb,Ti)$+ $L1_2$-$(Al,Fe)_3Ti$+ Al_3Fe(?)

portion of an arc-cast button of the alloy (about 5 g) was remelted in a silica tube with a small bore (1 mm in diameter) in the bottom end using an induction furnace and ejected onto the surface of a rotating stainless steel roll (300 mm in dia. and 30 mm thick) by applying argon gas pressure. The velocity of the rotating stainless roll was 40 m/s. After ejection, flakes were immediately collected, cleaned with methanol and dried in vacuum. Another portion of this arc-cast button was heated to 1100°C, held at temperature for 1 h, and slowly cooled to 400°C at a cooling rate of 6.25°C/h, followed by furnace cooling. Homogenized, RSPed, and slowly cooled samples were examined by SEM, x-ray diffraction, and EPMA techniques. A Shimazu EPMA-8705 microscope was used. Hardness measurements were carried out using a Vickers microhardness tester at a load of 25 g.

RESULTS AND DISCUSSION

Microstructure and EPMA analysis

According to a recent phase stability study by electronic band structure calculations on an extended Hückel tight-binding method, it has been predicted that the $L1_2$ structure is stable in ternary Al_3Nb compounds when nickel is added at a level of more than 12.5 at%; the model assumes that Ni substitutes for Al to form a single phase [2]. Hence, Alloys B and C in Table 1 are predicted to exhibit a single phase with the $L1_2$

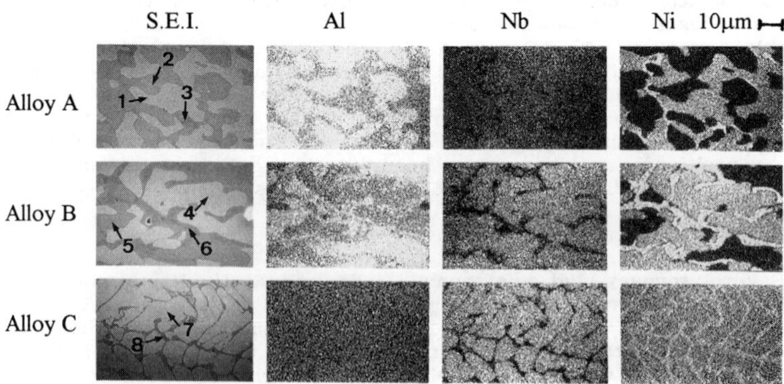

Figure 1 Scanning electron micrographs and x-ray dot mappings of Al, Nb, and Ni taken from Alloy A, B, and C after homogenization heat treatment.

structure. After homogenization treatment at 1100°C for 10 days, these alloys have unexpectedly exhibited multiphases as shown in Fig. 1, suggesting that no $L1_2$ phase exists. Multiphase structures have been observed in all the alloys produced in this study. As will be shown later, these alloys do not contain an $L1_2$ phase except for Alloy H, which is an essentially Al_3Ti compound containing 7.5 at.%Fe. In the case of Al_3Ti containing 7.5 at.%Fe, the alloy becomes single $L1_2$ phase after homogenization at 1100°C [4,5]. As shown in Table 1, Alloy H contains a small amount of Nb; hence, it shows an $L1_2$ phase but as a minor phase. In the case of Alloy G which contains higher Nb than Ti, no $L1_2$ phase has been detected. In the present paper, detailed observations of Alloys A, B, and C are presented, and those of alloys containing Fe and/or Ti (Alloy D to H) will be reported elsewhere.

Figure 1 shows scanning electron micrographs and x-ray dot mappings of Al, Nb, and Ni elements taken from Alloys A, B, and C after homogenization treatment following arc melting. As seen, Alloys A and B consist of three phases which appear as white, gray, and black areas. The volume fraction of white areas is large and that of gray areas is small when Ni content is large, and gray areas disappear in Alloy C showing a two-phase alloy. X-ray dot mappings of Al, Nb, and Ni in Fig. 1 show that white areas contain high content of Nb and gray areas contain high content of Al. To see more quantitative results for elemental partitioning of each phase, semi-quantitative analyses were performed by EPMA, which will help to explain the apparent inconsistencies between experiment and prediction based on an electronic tight-binding theory.

EPMA analyses of eight regions as marked in Fig. 1 have been tabulated in Table 2. As shown, white region 1 in Alloy A, 4 in B, and 7 in C show compositions of near $Nb(Al,Ni)_2$, gray region 2 in Alloy A, and 5 in B show Al_3Nb with less than 1 at.%Ni; and black region 3 in Alloy A, 6 in B, and 8 in C show nickel-poor NiAl with less than 4 at.%Nb. These results agree with x-ray analyses shown in Fig. 3. Only a small amount (less than 1 at.%) of Ni is found to dissolve in Al_3Nb, which dissatisfies the assumption used to predict the presence of an $L1_2(Al,Ni)_3Nb$ phase.

Table 2 EPMA results obtained from regions in Alloy A, B, and C in Figure 1 and in Alloy CI in Figure 2.

Alloy	Region	Elements (at.%)		
		Al	Nb	Ni
Alloy	1	53.3	32.9	13.8
Alloy A	2	74.6	24.8	0.6
Alloy	3	57.6	3.9	38.5
Alloy	4	52.6	32.9	14.5
Alloy B	5	74.9	24.2	0.9
Alloy	6	54.6	0.1	45.3
Alloy C	7	48.5	32.4	19.1
Alloy	8	51.2	0.1	48.7
Alloy CI	9	47.9	32.1	20.0

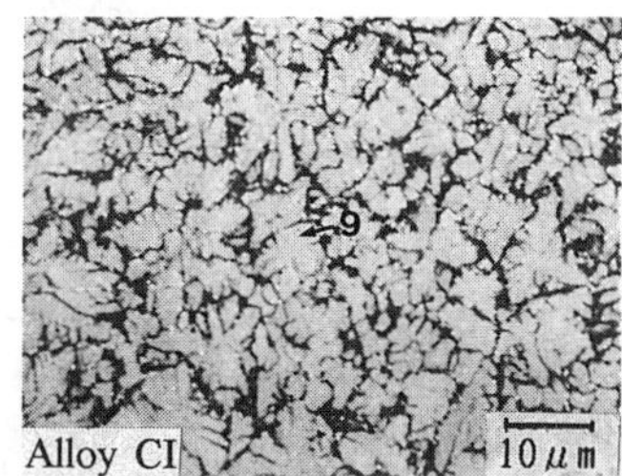

Figure 2 Scanning electron micrograph of as-rapidly-solidified flake (Alloy CI).

Since there is no stable $L1_2$ phase in the above-three alloys at 1100°C, the synthesis of a metastable, Ni-supersaturated Al_3Nb-based alloy was attempted by utilizing a rapid solidification process, where a portion of an arc-cast and homogenized button of Alloy C was used. Figure 2 is a scanning electron micrograph obtained from a flake of the RSPed alloy, denoted as Alloy CI. The flake is found to be two-phase. EPMA analyses have shown that there are no areas having compositions close to those of trialuminides containing Nb and Ni. An EPMA result obtained in region 9 in Fig. 2 is also tabulated in Table 2. The microstructure of Alloy C was also examined after slow cooling from 1100°C. It has been shown that the specimen is two-phase, displaying a microstructure which is essentially the same as those of homogenized specimens of this alloy.

X-ray diffraction analysis

The alloys produced in this study were all examined by x-ray diffraction to identify phases present in different thermal conditions. X-ray diffraction results obtained from homogenized specimens are tabulated in Table 1. Examples of diffraction patterns taken from Alloys A, B, C, and CI are shown in Fig. 3. All the phases present in these alloys can be identified as DO_{22}-Al_3Nb, $MgZn_2$-type Laves-$Nb(Ni,Al)_2$, and B2-AlNi structures. Lattice parameters measured from Fig. 3 are tabulated in Table 3. From this table, Table 2, and Fig. 1, gray, black, and white areas are correspond to phases with DO_{22}, B2, and Laves structures, respectively. These results are in good agreement with data reported by Benjamin et al [6]. From data of the Laves structure shown in Table 3, the change in axial ratio (c/a) has been determined, and is depicted as a function of Ni content in Fig. 4. This figure also contains hardness data obtained from Alloys A, B, and C after homogenization heat treatment and from Alloy CI.

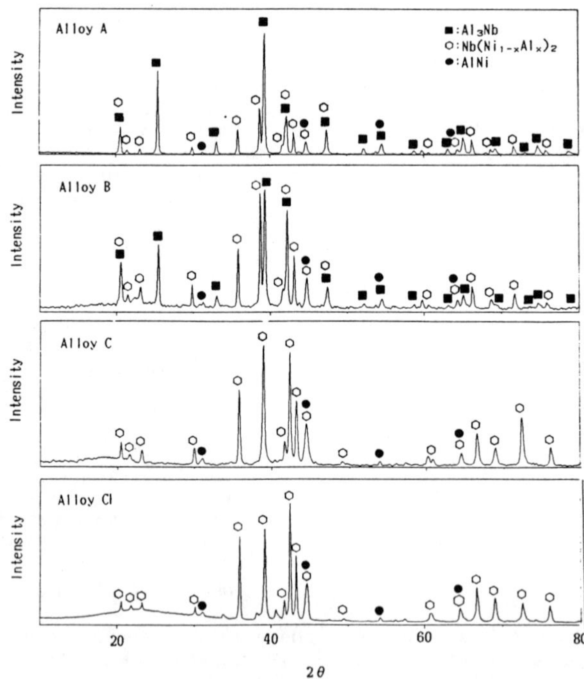

Figure 3 X-ray diffraction profiles taken from Alloy A, B, C, and CI.

Table 3 Lattice parameters measured of Alloy A, B, C, and CI

Alloy	Lattice parameters (A)				
	Al_3Nb		$Nb(Ni,Al)_2$		AlNi
	a	c	a	c	a
A	3.842	8.604	4.989	8.297	2.888
B	3.832	8.597	4.984	8.252	2.883
C	—	—	5.008	8.214	2.882
CI	—	—	4.989	8.117	2.882

Table 4 Vickers microhardness of phases in Alloy A, B, C, and CI.

Alloy	Vickers hardness (Hv)		
	Al_3Nb	$Nb(Ni,Al)_2$	AlNi
A	724	769	—
B	724	958	—
C	—	1152	270
CI	—	1310	

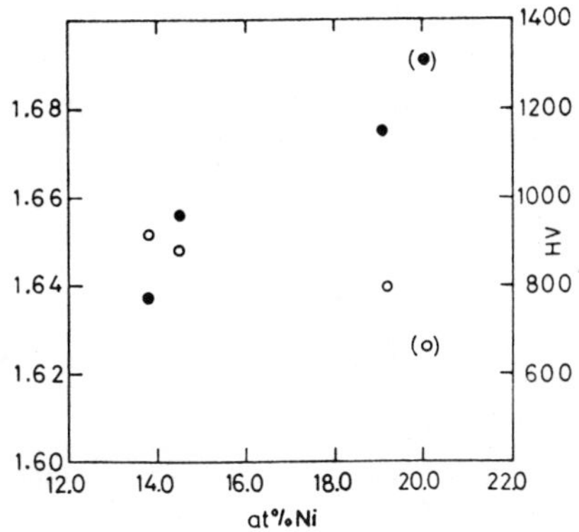

Figure 4 Change in axial ratio (c/a) and Vickers microhardness (Hv) of $Nb(Ni_{1-x}Al_x)_2$ as a function of Ni content. Solid circles for Hv; open circles for c/a; circles with parentheses for RSPed Alloy CI.

Hardness

Vickers microhardness measurements were performed on relatively large areas of phases in Alloys A, B, and C. In the case of Alloy CI, as shown in Fig. 2, it is difficult to differentiate hardness values of Laves from B2 phases because of the size of the indenter used. Hence, hardness values obtained in this case are considered to be an average value from these two phases. Hardness values obtained Alloys A, B, C, and CI are listed in Table 4. As shown, the Laves phase is the hardest, having hardness values from 769 to 1152kg/mm^2, the DO_{22} phase has 724kg/mm^2; and B2 phase is the softest and has 270kg/mm^2. The hardness value of Alloy CI is 1310kg/mm^2, which is higher than that for the Laves phase of Alloy C. Hardness values of the Laves phase are plotted in Fig. 4 as a function of Ni content. As indicated in this figure, the Laves phase becomes harder with increasing Ni content; conversely, the axial ratio becomes smaller with increasing Ni content. The increase in hardness of the Laves phase with the axial ratio of the hexagonal structure may result from the change in slip systems of this phase. It is widely accepted that a means to improve fracture toughness of materials is to incorporate hard inclusions as fine precipitates. Recently, some B2-type NiAl-based materials have shown some ductility at room temperature [7]; hence, these materials may possibly be strengthened by fine precipitates of this hard Laves phase.

SUMMARY

Microstructural observations, phase identification, and hardness measurements of several ternary and quaternary Al_3Nb and Al_3Ti-based alloys containing Ni or Fe have been carried out after homogenization at 1100°C, rapid solidification, and slow cooling. An $L1_2$ phase is formed in an $(Al,Fe)_3Ti$-based alloy when a small amount of Nb is added; conversely, in Al-Nb-Ni alloys whose compositions are based on Al_3Nb, no $L1_2$ phase has been detected, and alloys have been determined to be either three-phase(B2, DO_{22} and $MgZn_2$-type Laves) or two-phase(B2 and Laves). The absence of $L1_2$

phase formation in Al_3Nb-based alloys containing Ni is attributed to limited solubility of Ni in the Al_3Nb compound. Hardness measurements of the three phases have revealed that the Laves phase is the hardest and the B2 is the softest. The Laves phase becomes harder and the axial ratio (c/a) becomes smaller with increasing Ni content. This may be ascribable to the change in slip systems in the hexagonal structure of the Laves phase. This hard Laves phase can be used to strengthen soft B2-type NiAl-based materials.

ACKNOWLEDGEMENTS

Two authors (K.M. and Y.O.) would like to acknowledge partial support from Osaka Municipal Technical Research Institute, Osaka, Japan for conducting this research. Three authors (C.V.C., A.F.G., and H.R.P.I.) are also pleased to acknowledge the support of the Air Force Office of Scientific Research through contact F49620-89-C-0047, Dr. A.R. Rosenstein, technical monitor.

REFERENCES

[1] H.R.P. Inoue, C.V. Cooper, L.H. Favrow, Y. Hamada, and C.M. Wayman, Mat. Rec. Soc. Symp. Proc., 213, 493 (1991).

[2] H.R.P. Inoue, M. Kitamura, C.M. Wayman, and H. Chen, Philos. Mag. Letters, 63, 345 (1991).

[3] K. Mizuuchi, Y. Okanda, and I. Ohnaka, J. Japan Inst. Metals, 55, 874 (1991).

[4] A. Seibold, Zeitshrift fur Metallkunde, 72, 712 (1981).

[5] H.R. Pak, C.M. Wayman, L.H. Favrow, C.V. Cooper, and J.S.L. Pak, Mat. Res. Soc. Symp. Proc., 186, 357 (1991).

[6] J.S. Benjamin, B.C. Giessen, and N.J. Grant, Met. Trans. AIME, 236, 224 (1966).

[7] T. Ishida, R. Kainuma, N. Ueno, and Nisizawa, Metall. Trans., 22A, 441 (1991).

EVIDENCE OF INHERENT DUCTILITY IN SINGLE CRYSTAL NiAl

J. E. HACK*, J. M. BRZESKI*, R. DAROLIA** and R. D. FIELD**
*Department of Mechanical Engineering, Yale University, New Haven, CT 06520
**GE Aircraft Engines, Cincinnati, OH 45215

ABSTRACT

The ductility and fracture toughness of single crystal NiAl have been studied as functions of thermal treatments at moderate and high temperatures. The data indicate that fast cooling through the temperature range 400°C - 20°C results in a material with a tensile elongation of 7% and a fracture toughness in the range of 13 -17 $MPam^{1/2}$. It is concluded that prior reports of brittle behavior in single crystal NiAl may be a result of strain-age embrittlement, similar to that observed in mild steels. The data strongly suggest that ductility and toughness in NiAl are more strongly dependent upon mobile dislocation density rather than on the inherent mobility of dislocations in the ordered lattice. Similar behavior may also be possible in other intermetallic compounds.

INTRODUCTION

The relatively low densities of intermetallic compounds, combined with their ability to retain strength and stiffness at elevated temperatures, have made these materials attractive for aerospace applications for over three decades[1,2]. Most of the compounds which possess excellent high temperature properties have not achieved their potential in structural applications because of prohibitively low room temperature ductility and fracture toughness values. It has been believed that the high melting temperatures and high ordering energies found for these compounds are indicative of an inherently poor dislocation mobility at ambient temperatures.

However, several aspects of the mechanical behavior of intermetallic compounds suggest that, while perhaps limited, the room temperature dislocation mobility in these materials is sufficient to provide significant ductility. As one example, recent studies have shown that several ordered intermetallics, primarily aluminides, possess significant room temperature ductility if protected from moisture in the environment[3-5]. In addition, some compounds which appear to exhibit inherently low room temperature ductility show tensile behavior reminiscent of the brittle body-centered-cubic refractory metals. Such phenomena as yield drops and regions of Luder's deformation have been observed in NiAl[6]. These phenomena are generally attributed to difficulty in the generation of fresh dislocations rather than a low inherent dislocation mobility[7,8].

When dislocation generation is difficult, ductility can suffer if the initial mobile dislocation density cannot provide sufficient plastic deformation for a given loading state[9-11]. The situation is exacerbated at the tip of a loaded crack in a fracture toughness test where the steep gradients in stress severely restrict the volume of material which experiences stresses above the level necessary to cause dislocation motion[12]. The population of mobile dislocations can be further reduced by strain aging processes which pin potentially mobile dislocations by decorating their core regions with solute atoms[13]. This report is intended to summarize recent results concerning the influence of mobile dislocation density on the ambient temperature mechanical behavior of single crystals of the compound NiAl. Particular attention will be paid to the effects of annealing temperature and cooling rate on subsequent fracture toughness values.

EXPERIMENTAL PROCEDURE

Details of the experimental procedures used in this work are given in previous papers[14,15] but will be repeated here in part for the sake of clarity. All tests were performed on samples cut from single crystal bars of nominally stoichiometric NiAl. After growth, all bars were given a homogenization anneal for 48 hours at 1317°C in an argon environment. Cooling from the anneal consisted of either a furnace cool under vacuum (FC) or a relatively rapid cool under flowing argon (AC). Tension and compression samples with a <110> axis were machined from material in the AC condition. Electropolishing was used to remove residual stresses induced by the grinding. Cylindrical compression samples were also machined with a <110> axis. In addition, double-cantilever beam fracture toughness samples were cut from the bars by EDM with a (110) crack plane and $[1\bar{1}0]$ crack direction as shown in Figure 1. The notch root radius of the specimens was approximately 0.4mm.

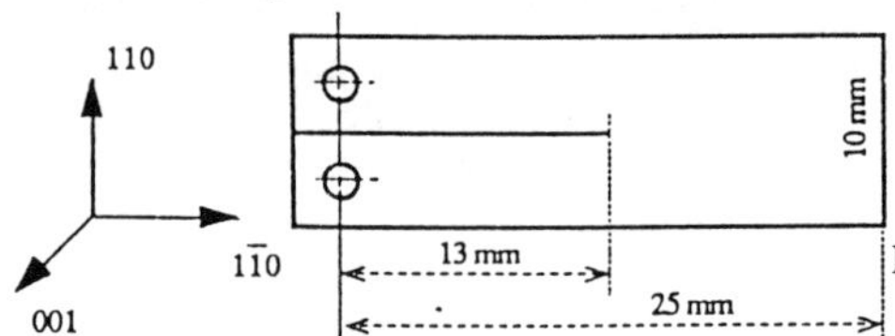

Figure 1 - Double Cantilever Beam Specimen Geometry

The broad faces of all the beams were ground and polished on SiC papers down to a 30 μm finish and subsequently electropolished. The fracture toughness samples were heat treated at various temperatures in either an argon or laboratory air environment. Chemical analyses of potentially important substitutional and interstitial impurities were performed by an independent testing laboratory.

RESULTS

Chemical Analysis

Table I gives the chemical compositions of four of the single crystal bars used in this study. These values are believed to be representative of all material tested in this study. The data show a reasonable consistency from bar to bar for all of the major constituents and impurity elements. The small concentration of Si results from an interaction between the mold and the molten metal. The primary interstitial impurities are C and O.

TABLE I - Representative Chemical Compositions of Single Crystals Used in This Study

Crystal	Cool Rate	wt% Ni	wt% Al	wt% Si	wppm C	wppm O	wppm N	wppm H
1	FC	66.7	bal	0.13	100	51	15	0.9
2	FC	66.7	bal	0.16	100	64	14	0.9
3	AC	67.0	bal	0.12	130	59	11	1.4
4	AC	67.1	bal	0.14	60	44	11	1.4
5	AC	66.8	bal	0.12	40	74	12	3.2

Fracture Testing

The results of fracture toughness tests as a function of thermal treatment are given in Table II. It is clear from the data that material in the AC condition yielded fracture

TABLE II - Fracture Toughness as a Function of Thermal Treatment

Crystal	Specimen	Cool Rate	Anneal	K_Q ($MPam^{1/2}$)
2	C	FC	none	4.0
3	I	AC	none	13.3
3	K	AC	none	13.2
3	L	AC	none	12.5
3	N	AC	none	11.0
5	A	AC	none	10.0
5	B	AC	none	9.5
5	C	AC	1300 C - 3 hrs fast cool	15.6
5	L	AC	1300 C - 3 hrs fast cool; 200 C - 1.5 hrs slow cool	2.8
6	A	FC	none	2.4
6	E	FC	400 C - 12 hrs air cool	16.7
6	H	FC	400 C - 12 hrs slow cool	5.8

toughness values 2.5 - 3 times higher than the 3 - 4 $MPam^{1/2}$ found for material which had been slowly cooled in vacuum. Thermal exposure at 1300°C for three hours followed by rapid cooling under rapidly flowing argon raised the toughness to a level of 15.6 $MPam^{1/2}$. A subsequent anneal at 200°C for 1.5 hours in air followed by furnace cooling reduced the fracture toughness of this material to 2.8 $MPam^{1/2}$. Conversely, a 12 hour anneal at 400°C in air followed by air cooling raised the fracture toughness samples in the FC condition from 2.4 $MPam^{1/2}$ to 16.7 $MPam^{1/2}$. Furnace cooling after the 400°C exposure showed only a minor increase in fracture toughness over the FC condition (5.8 $MPam^{1/2}$).

Fractography of the fracture toughness specimens showed that samples which exhibited toughness levels less than or equal to 4 $MPam^{1/2}$ exhibited a region approximately 300 μm wide near the notch tip which cleaved along a high index plane. Final separation under unstable fracture occurred on the {011} plane perpendicular to the stress axis. In samples where the fracture toughness exceeded 4 $MPam^{1/2}$, the region over which the high index fracture occurred grew to several mm. In these cases, final fracture was accomplished by the shearing off of the loading arms of the sample. Although more complete descriptions of the fractography are presented elsewhere[15,16], Figures 2 and 3 illustrate an important feature of the fracture process. In Figure 2a, typical {110} cleavage

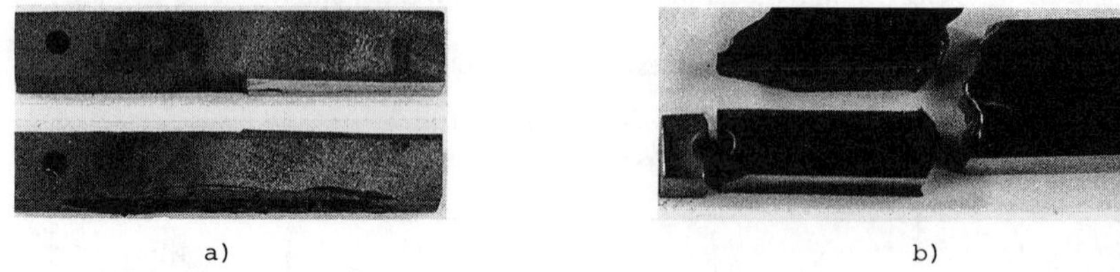

a) b)

Figure 2a & 2b - Failed Double Cantilever Beam Speciments Slow Cooled From The Homogenization Anneal: a) Sample 6A (FC); b) Sample 6E (FC plus 400°C and air cool). Magnification 2x.

is seen for sample 6A in the FC condition. The seven-fold increase in fracture toughness induced by a rapid cool from a 12 hour exposure at 400°C is accompanied by a complete change in failure mode, with {110} cleavage being completely suppressed (sample 6E in Figure 2b). Similarly, Figure 3a shows the remnants of sample 5C after rapid cooling from a three hour anneal at 1300°C. Again, the fracture toughness of 15.6 $MPam^{1/2}$ is associated with a complete suppression of {110} cleavage. Examination of sample 5L (Figure 3b), which was treated under the same conditions as sample 5C followed by an anneal at 200°C, revealed that final fracture once again occurred along {110} planes.

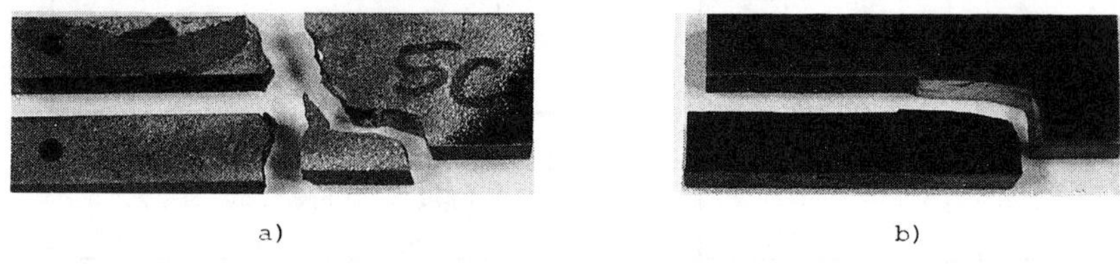

a) b)

Figure 3a & 3b - Failed Double Cantilever Beam Specimens Cooled Under Flowing Argon: a) Sample 5C (AC plus 1300°C and fast cool); b) Sample 5L (same as 5C plus 200°C). Magnification 2x.

Tensile and Compressive Behavior

Room temperature tensile ductilities reported for NiAl single crystals in the FC condition with a <110> axis have typically been reported as 1 - 2 %[17]. The corresponding value of the 0.2 % offset yield strength is approximately 200 MPa. Tensile tests on AC material during this investigation resulted in average values of the tensile ductility and yield strength of 7 - 8 % and 190 MPa, respectively. Therefore, the large increases in fracture toughness associated with the AC condition were accompanied by a four-fold increase in tensile ductility.

Compressive load-strain behavior of material with a <110> axis at room temperature and 200°C are presented in Figures 4a and 4b, respectively. It can be seen from the curves that smooth yielding takes place under ambient conditions, but pronounced serrations appear in the plastic portion of the curve at 200°C. Similar serrations were visible in tests run at 100°C[15].

The serrations always began with the onset of plastic deformation. This is typical of solute drag interactions between dislocations and interstitial solutes[13,18].

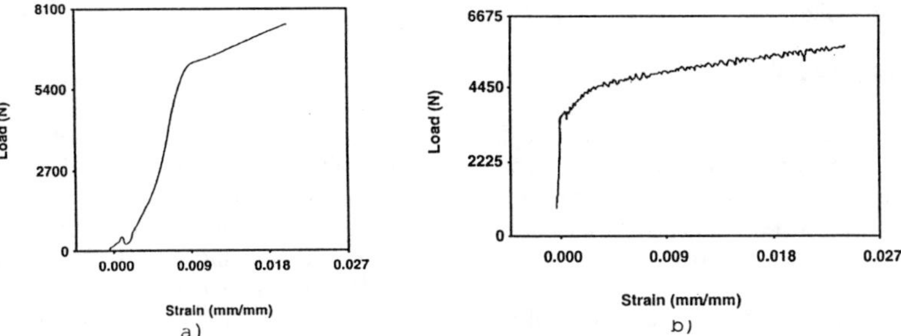

Figure 4a & 4b - Load-Strain Curves from Compression Tests: a) Room Temperature; b) 200°C.

DISCUSSION

It is clear from the data presented in Table II, Figures 2 and 3 and the results of the tensile tests on material in the AC condition that anneals at relatively low temperatures (200 - 400°C) have a profound impact on the tensile ductility and fracture toughness of single crystal NiAl. The most significant aspect of the thermal treatments used in this study seems to be the rate at which the material is cooled below about 400°C. Relatively fast cooling through this range leads to reasonable levels of ductility and fracture toughness while slow cooling or dwell times at 200°C result in severe embrittlement.

This behavior bears a remarkable resemblance to strain age and blue embrittlement observed in mild steels[18]. There, the strong interaction between interstitial C atoms and dislocation cores, coupled with the relatively high mobility of C in bcc iron at low to moderate temperatures, leads to the diffusion of C to dislocations at moderate temperatures. The C segregated to the dislocation cores tends to pin the dislocation, thus reducing the mobile dislocation density. The onset of serrated yielding observed at moderate temperatures (Figure 4) is also consistent with the occurrence of static strain aging in slowly cooled samples. Strain aging is also consistent with independent observations of recovery of the yield drop and flow stress in polycrystalline NiAl which has been pre-deformed by the application of a hydrostatic pressure by a subsequent anneal at 200°C[6,19] and the rapid rise in tensile ductility of single crystal NiAl in the <110> orientation when the temperature is raised above 200°C[17]. Although it may be inferred that strain aging is responsible for the brittle behavior induced in single crystal NiAl by slow cooling through low temperatures, more work must be done to identify the mobile species responsible. Likely suspects include interstitial C or O and vacancies induced by slight deviations from stoichiometry.

It should be noted that even though strain aging appears to be the phenomenon responsible for earlier reports of brittle behavior in single crystal NiAl, it is merely a symptom of a larger problem; e.g. a sensitivity to the density of mobile dislocations which arises from a difficulty in dislocation generation. Dislocation mobility appears to be sufficient for usable levels of ductility and toughness as long as a reasonable mobile dislocation density is maintained. In binary NiAl, the inherent levels of ductility and toughness can be significantly higher than those generally reported if the cooling rate can be controlled or the interstitial content can be reduced or rendered innocuous by the addition of gettering agents. Although other intermetallic compounds which are believed to be brittle may or may not be susceptible to strain aging, it is reasonable to expect that many of them could show a similar sensitivity to mobile dislocation density. Thus, it may prove fruitful to examine the effects of low temperature thermal and warm working treatments on room temperature properties in other ordered intermetallic compounds.

SUMMARY

The ductility and fracture toughness of single crystal NiAl have been studied as functions of thermal treatments. The data indicate that the inherent tensile ductility is on the order of 7 -8 % and the fracture toughness is approximately 15 - 16 $MPam^{1/2}$ for the <110> orientation. It is concluded that prior reports of brittle behavior for similar material may be a result of strain-age embrittlement. This would suggest that the material behavior is limited more by a lack of mobile dislocations rather than an inherent lack of dislocation mobility. The results of this study may have implications for other intermetallic compounds as well.

ACKNOWLEDGEMENTS

Two of the authors (JEH and JMB) would like to acknowledge the support of the Office of Naval Research under Contract N00014-89-J-1843 monitored by Dr. G. R. Yoder. The other two authors (RD and RDF) appreciate the support of the Air Force Office of Scientific Research under Contract F49620-91-C-0077 monitored by Dr. A. H. Rosenstein.

REFERENCES

1. H. A. Lipsitt, in **High Temperature Ordered Intermetallic Alloys**, C. C. Koch et al, (eds.), Materials Research Society, Pittsburgh, PA, 351 (1985).
2. **Scientific American, 255**, 4 (Oct. 1986).
3. C. T. Liu, **Scripta Metall., 27**, 25 (1992).
4. C. T. Liu, E. Lee and C. G. McKamey, **Scripta Metall., 23**, 875 (1989).
5. S. Miura and C. T. Liu, **Scripta Metall. et Mater., 26**, 1753 (1992).
6. R. W. Margevicius, J. J. Lewandowski and I. Locci, **Scripta Metall. et Mater., 26**, 1733 (1992).
7. W. G. Johnston and J. J. Gilman, **J. Appl. Phys., 30**, 129 (1959).
8. G. T. Hahn, **Acta Metall., 10**, 727 (1962).
9. A. Ball, F. P. Bullen, F. Henderson and H. L. Wain, in **Fracture 1969**, P. L. Pratt, ed., Chapman and Hall, London, 327 (1969).
10. G. Agte and J. Vacek, **Tungsten and Molybdenum**, NASA TT F-135, NASA, Washington, D. C. (1963).
11. **Metals Handbook, 9th Ed., 3**, ASM, Metals Park, OH 328 (1980).
12. M. F. Ashby and J. D. Embury, **Scripta Metall., 19**, 951 (1985).
13. A. H. Cottrell, **Dislocations and Plastic Flow in Crystals**, Oxford University Press, Fair Lawn, NJ, (1953).
14. J. E. Hack, J. M. Brzeski and R. Darolia, **Scripta Metall., 27**, 1259 (1992).
15. J. M. Brzeski, J. E. Hack, R. Darolia and R. D. Field, **Mater. Sci. Engrg.**, in press.
16. R. Darolia, K.-M. Chang and J. E. Hack, **J. Intermetallics**, in press.
17. R. Darolia, **JOM, 43**, 48 (March 1991).
18. J. P. Hirth and J. Lothe, **Theory of Dislocations**, McGraw-Hill, New York, NY, 613 (1968).
19. R. W. Margevicius and J. J. Lewandowski, **unpublished research**, Case Western Reserve University, Cleveland, OH (1992).

Indices

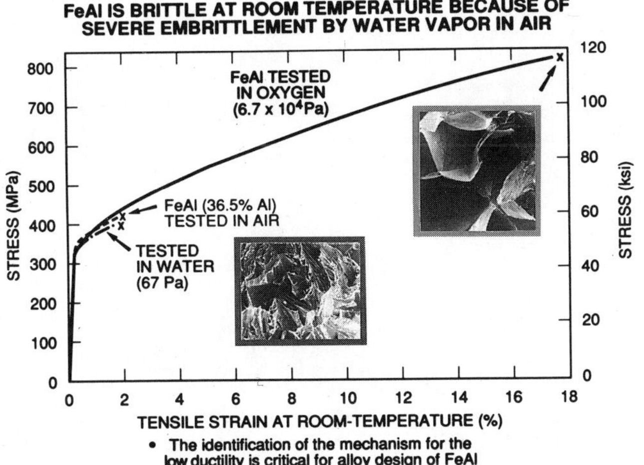

Figure courtesy of C.T. Liu.

Author Index

Subject Index